T0132633

COMPUTING HANDBOOK

THIRD EDITION

Information Systems and
Information Technology

COMPUTING HANDBOOK

THIRD EDITION

Information Systems and

Information Technology

EDITED BY

Heikki Topi
Bentley University
Waltham, Massachusetts, USA

EDITOR-IN-CHIEF

Allen Tucker
Bowdoin College
Brunswick, Maine, USA

CRC Press
Taylor & Francis Group
Boca Raton   London   New York

CRC Press is an imprint of the
Taylor & Francis Group, an **informa** business

A CHAPMAN & HALL BOOK

CRC Press
Taylor & Francis Group
6000 Broken Sound Parkway NW, Suite 300
Boca Raton, FL 33487-2742

© 2014 by Taylor & Francis Group, LLC
CRC Press is an imprint of Taylor & Francis Group, an Informa business

No claim to original U.S. Government works

Printed on acid-free paper
Version Date: 20130321

International Standard Book Number-13: 978-1-4398-9854-3 (Hardback)

This book contains information obtained from authentic and highly regarded sources. Reasonable efforts have been made to publish reliable data and information, but the author and publisher cannot assume responsibility for the validity of all materials or the consequences of their use. The authors and publishers have attempted to trace the copyright holders of all material reproduced in this publication and apologize to copyright holders if permission to publish in this form has not been obtained. If any copyright material has not been acknowledged please write and let us know so we may rectify in any future reprint.

Except as permitted under U.S. Copyright Law, no part of this book may be reprinted, reproduced, transmitted, or utilized in any form by any electronic, mechanical, or other means, now known or hereafter invented, including photocopying, microfilming, and recording, or in any information storage or retrieval system, without written permission from the publishers.

For permission to photocopy or use material electronically from this work, please access www.copyright.com (http://www.copyright.com/) or contact the Copyright Clearance Center, Inc. (CCC), 222 Rosewood Drive, Danvers, MA 01923, 978-750-8400. CCC is a not-for-profit organization that provides licenses and registration for a variety of users. For organizations that have been granted a photocopy license by the CCC, a separate system of payment has been arranged.

Trademark Notice: Product or corporate names may be trademarks or registered trademarks, and are used only for identification and explanation without intent to infringe.

Library of Congress Cataloging-in-Publication Data

Information systems and information technology / editors, Heikki Topi, Allen Tucker. -- Third
edition.
  pages cm. -- (Computing handbook ; volume 2)
  Includes bibliographical references and index.
  ISBN 978-1-4398-9854-3 (hardcover : alk. paper)
  1. Information technology--Handbooks, manuals, etc. I. Topi, Heikki. II. Tucker, Allen B.

T58.5.I5274 2013
658.4'038011--dc23   2013003954

**Visit the Taylor & Francis Web site at
http://www.taylorandfrancis.com**

**and the CRC Press Web site at
http://www.crcpress.com**

Contents

PART I Disciplinary Foundations and Global Impact

PART II Technical Foundations of Data and Database Management

PART III Data, Information, and Knowledge Management

PART IV   Analysis, Design, and Development of Organizational Systems

PART V   Human–Computer Interaction and User Experience

PART VI Using Information Systems and Technology to Support Individual and Group Tasks

PART VII Managing and Securing the IT Infrastructure and Systems

PART VIII Managing Organizational Information Systems and Technology Capabilities

PART IX Information Systems and the Domain of Business Intertwined

Preface to the *Computing Handbook Set*

The purpose of the *Computing Handbook Set* is to provide a single, comprehensive reference for specialists in computer science, information systems, information technology, software engineering, and other fields who wish to broaden or deepen their understanding in a particular subfield of the computing discipline. Our goal is to provide up-to-date information on a wide range of topics in a form that is accessible to students, faculty, and professionals.

The discipline of computing has developed rapidly since CRC Press published the second edition of the *Computer Science Handbook* in 2004 (Tucker, 2004). Indeed, it has developed so much that this third edition requires repartitioning and expanding the topic coverage into a two-volume set.

The need for two volumes recognizes not only the dramatic growth of computing as a discipline but also the relatively new delineation of computing as a family of five separate disciplines, as described by their professional societies—The Association for Computing Machinery (ACM), The IEEE Computer Society (IEEE-CS), and The Association for Information Systems (AIS) (Shackelford et al., 2005).

These separate disciplines are known today as computer engineering, computer science, information systems, information technology, and software engineering. These names more or less fully encompass the variety of undergraduate and graduate degree programs that have evolved around the world. The document "Computing curricula 2005: The overview report" describes computing this way (Shackelford et al., 2005, p. 9):

> In a general way, we can define computing to mean any goal-oriented activity requiring, benefiting from, or creating computers. Thus, computing includes designing and building hardware and software systems for a wide range of purposes; processing, structuring, and managing various kinds of information; doing scientific studies using computers; making computer systems behave intelligently; creating and using communications and entertainment media; finding and gathering information relevant to any particular purpose, and so on.

To add much flesh to the bones of this very broad definition, this handbook set describes in some depth what goes on in research laboratories, educational institutions, and public and private organizations to advance the effective development and utilization of computers and computing in today's world. The two volumes in this set cover four of the five disciplines in the following way:*

1. Volume I: Computer Science and Software Engineering
2. Volume II: Information Systems and Information Technology

---

* This handbook set does not currently cover *computer engineering*, though we hope such coverage can be developed and added to the set in the near future.

This set is not designed to be an easy read, as would be gained by browsing a collection of encyclopedia entries on computing and its various subtopics. On the contrary, it provides deep insights into the subject matter through research-level survey articles. Readers who will benefit most from these articles may be undergraduate or graduate students in computing or a related discipline, researchers in one area of computing aiming to expand their knowledge of another area, or other professionals interested in understanding the principles and practices that drive computing education, research, and development in the twenty-first century.

This set is designed as a professional reference that serves the interests of readers who wish to explore the subject matter by moving directly to a particular part and chapter of the appropriate volume. The chapters are organized with minimal interdependence, so that they can be read in any order. To facilitate rapid inquiry, each volume also contains a table of contents and a subject index, thus providing access to specific topics at various levels of detail.

Preface to Volume II: Information Systems and Information Technology

We are delighted to introduce this new volume of the *Computing Handbook Set* on information systems (IS) and information technology (IT). Both of these disciplines focus on computing in a context (such as an organization) that uses computing resources to achieve its goals and to transform how it reaches these goals. The older and more established discipline IS has existed since the 1960s. Its focus from the beginning has been organizational computing in a business context, with a specific emphasis on how changes in business processes and information management transform the way business is conducted. IT emerged as an independent academic discipline during the last decade, primarily through its contribution to undergraduate education. IT also focuses on the organizational context, but it primarily addresses questions related to the technology infrastructure.

This handbook set has numerous elements that discuss the nature and identity of the IS and IT disciplines (such as the Preface to Volume I, and see Chapters 1 through 5 in this volume). We will not attempt to replicate that conversation here; it is, however, safe to say that neither discipline has even internally converged into a fully unified understanding of its own identity. Throughout its almost 50 year history, the IS discipline has continuously debated its identity (see Chapter 1 in this volume). The same process has now begun for IT. We hope that this volume will contribute to the disciplinary processes that seek for clarity regarding the identities of both IS and IT.

When referring to organizational units and capabilities, the terms "information systems" and "information technology" are, in practice, used interchangeably. Even the research literature is often incapable of separating them in a consistent way. We would have liked to follow a consistent approach where IT refers primarily to the technology infrastructure and IS to the application of the technology in an organizational context, but doing so would have imposed overly restrictive constraints, given the current practice in the field. Therefore, we ask for your patience particularly in the context of management of IS capabilities—the use of these terms is not always systematic between (or even within) the chapters. Sometimes, we resort to the IS/IT moniker when we want to specifically acknowledge a consistent practice of using both terms interchangeably.

This volume of the *Computing Handbook Set* has nine parts, which collectively demonstrate the richness and breadth of the IS and IT disciplines and their close linkages to the practice of using, managing, and developing IT-based solutions to advance the goals of their environments. As the entire handbook set, this volume is targeted to both academics and practitioners. On the one hand, the chapters provide academic readers with introductions to the current status and future directions of academic research in the topic areas of the chapters. On the other hand, the chapters give advanced practitioners in-depth perspectives on the contributions of academic research to the practice of IS and IT development, use, and management. In this preface, we will first explain the background for the structure of the volume and then provide brief introductions to the core content of each of the parts of the volume. We conclude the preface by acknowledging and thanking the people who have been instrumental in making this effort possible.

Structure of the Volume

This volume includes nine parts: (I) Disciplinary Foundations and Global Impact; (II) Technical Foundations of Data and Database Management; (III) Data, Information, and Knowledge Management; (IV) Analysis, Design, and Development of Organizational Systems; (V) Human–Computer Interaction and User Experience; (VI) Using Information Systems and Technology to Support Individual and Group Tasks; (VII) Managing and Securing the IT Infrastructure and Systems; (VIII) Managing Organizational Information Systems and Technology Capabilities; and (IX) Information Systems and the Domain of Business Intertwined.

The selection of chapters and topic categories for this volume is based on an editorial decision process that integrates several perspectives. As with Volume I of this handbook set, this volume pays close attention to the undergraduate model curricula developed by ACM (in collaboration with AIS in the case of IS) for these disciplines. The latest IS model curriculum IS 2010 (Topi et al., 2010) identifies the following core topics: foundations of IS (Part I); data and information management (Parts II and III); enterprise architecture, IS project management, IT infrastructure (Part VII); systems analysis and design (Part IV); and IS strategy, management, and acquisition (Parts VIII and IX). Five of them (with slightly different titles) are directly implemented as parts of this volume, and the remaining two (enterprise architecture and IS project management) have their own chapters. The IT curriculum recommendation IT 2008 (Lunt et al., 2008) has a strong emphasis on human–computer interaction (HCI), information assurance and security, information management, system administration and management, and system integration and architecture. HCI has its own part in this volume (Part V), and data and information management two others (Parts II and III); the other core IT topics are covered in Part VII.

The curriculum recommendations were not, however, the only source of guidance for the structure of this volume. Several recent articles explore the identity of the IS discipline (e.g., Sidorova et al., 2008; Taylor et al., 2010); of these, particularly Sidorova et al. provides a useful high-level categorization that supports the structure that had emerged from the curriculum reports. The high-level categories in Sidorova et al. are IT and organizations (covered primarily in Part IX, but also in Part VIII), IS development (primarily Parts IV and V, but also Parts II and III), IT and individuals (Part VI), IT and markets (also Part IX), and IT and groups (included in Part VI). The final structure was formed iteratively throughout the development process. Our hope is that it serves as helpful conceptual framework for structuring the rich set of materials included in this handbook.

Introduction to the Parts of the Volume

A summary of the major contents of each of the nine parts of this volume is given next. Readers should feel free to engage these parts in any order they prefer.

Disciplinary Foundations and Global Impact

Part I, Disciplinary Foundations and Global Impact, explores the identity, characteristics, and boundaries of the IS and IT disciplines. In addition, it positions IS and IT in the societal context and discusses the opportunities IT offers as an enabler of global development and social transformation.

Chapter 1 discusses the core topics and development of the discipline of IS based on an analysis of most widely cited articles throughout the history of the discipline. Chapter 2 covers the history and development of the emerging IT discipline. Chapters 3 and 5 provide alternative perspectives on the identity, nature, and core theories of the IS discipline: Chapter 3 emphasizes the importance of the IT artifact and the connection to practice of IS development and management, whereas Chapter 5 focuses on the connection between IS and the social context in which they are used by introducing sociotechnical approaches to IS. Chapter 4 is one of the earliest published integrated reviews of principles, methods, and theories for the discipline of IT. Chapters 6 and 7 both explore the broader societal impact of information and communication technologies in the global context: the former discusses the role these technologies have in the global development processes, whereas the latter specifically focuses on digital divide through a case example.

Data and Information Management

One of the key focus areas of both the IS and IT disciplines is data and information management, an area that computer science also considers its own. In designing the *Computing Handbook Set*, we decided to locate the Technical Foundations of Data and Database Management (Part II) in this volume to keep it together with the chapters on Data, Information, and Knowledge Management (Part III). Both parts focus on issues related to the acquisition, storage, and retrieval of data, information, and knowledge. The perspective of chapters in Part II emphasizes on database design and implementation while Part III covers topics related to modeling, analyzing, and managing information.

Part II starts with a discussion on foundational data models (relational, semantic, object-oriented, and object-relational) in Chapter 8. It continues with topics related to the tuning of various aspects of database design to achieve high performance (Chapter 9). Chapter 10 covers fundamental access methods (B-tree with variations and extensions, and hashing), spatial access methods (specifically, R-tree), and advanced access methods topics. Chapters 11 and 12 address two additional essential technical data management topics: query optimization and concurrency control and recovery, respectively. In Chapter 13, the focus moves to issues that are particularly important for distributed and parallel database systems. Chapter 14 reviews the core computing questions related to the modeling, search, and storage of multimedia (image, audio, and video) data.

In addition to the technical data management issues discussed earlier, the broader topic of organizational data, information, and knowledge management includes a rich variety of essential practical and theoretical questions, which are covered in Part III. The part starts with a review of conceptual modeling, particularly focusing on the approaches that are built on the philosophical field of ontology (Chapter 15). Chapter 16 explores issues that are instrumental in ensuring organizational data and information quality. The following two chapters provide two different perspectives on the utilization of both structured and unstructured data: Chapter 17 includes a comprehensive overview of the area of knowledge management, covering both theoretical and practical issues, whereas Chapter 18 focuses on the management of digital content in digital libraries. The discussion continues with issues related to the effective use of large amounts of organizational data, in Chapter 19 in the context of knowledge discovery and data mining and in Chapter 20 with an emphasis on unstructured, very large and highly dynamic data sets under the umbrella of Big Data. The part concludes with a managerial perspective introduced in Chapter 21 that covers the high-level governance of organizational data and information.

Analysis and Design of Organizational Systems

Analyzing the requirements for and designing organizational IS has been a core element of the IS discipline since its early days in the 1960s. Part IV—Analysis, Design, and Development of Organizational Systems— provides comprehensive coverage extending from the role of IT as an enabler of organizational change to methods and approaches of systems analysis and design, as well as through topics such as enterprise architecture, business process management and analysis, and the impact of culture on IS design and use.

One of the topics that has been vigorously debated within the academic IS community is the role and importance of design for the discipline. Building on earlier seminal works in IS literature (such as Hevner et al., 2004), Chapter 22 makes a strong case advocating the importance of design and design science for the discipline, providing guidance for conducting this type of academic work, and linking it to practice. The next two chapters explore the role of enterprise systems and enterprise architecture as enablers of organizational transformation, separating the impact of large core systems from systems that improvise at the edges beyond the core. Chapter 23 has a more general focus on IT as an enabler of new opportunities, whereas Chapter 24 specifically presents a review of issues related to enterprise systems. Enterprise architecture has become an increasingly important topic within the IS field (as demonstrated by its inclusion in the undergraduate model curriculum in IS 2010), and, thus, it is natural for this volume to include a chapter on this topic (Chapter 25). The part continues with two chapters that explore the two key aspects of any systems specification effort: business process management and analysis (Chapter 26) and information requirements determination (Chapter 27). Chapter 28 provides an overview of the development of systems analysis and design approaches over the lifetime of the IS discipline, and Chapter 29 explores the systems development process with an emphasis on the individual developer's perspective. Chapter 30 is a high-level overview of the development and management of complex, large-scale information infrastructures. The part concludes with a review of a more specialized topic in Chapter 31, which studies the impact of culture on IS design, particularly in the context of e-business.

Human–Computer Interaction and User Experience

The shared interests of IS and IT are again in focus in Part V—Human–Computer Interaction and User Experience. The interaction between human users and IT systems is an area that IS, IT, and computer science (CS) all explore from their own perspectives that are connected but distinct. The chapters in this part represent mostly viewpoints that come from HCI and usability practice and from research communities connected to the disciplines of computer science, human factors, cognitive psychology, and industrial design.

Chapter 32 starts the part with a comprehensive overview of usability engineering, an important core area within HCI. The chapter provides a framework and offers process guidance for designing highly usable systems. Chapter 33 covers the role of detailed task analysis in the process of designing usable functionality. The following chapters focus on specific issues and contexts within the broad area of HCI: the design of multimedia applications (Chapter 34), the role of international usability standards in supporting the development of highly usable systems (Chapter 35), and the design of web applications (Chapter 36). The part ends with Chapter 37, which summarizes, analyzes, and supports the recent calls to transform the discipline so that its focus will be increasingly on a broader user experience (UX) concept with a full range of psychological, social, cultural, physiological, and emotional elements.

Using Information Systems and Technology to Support Individual and Group Tasks

Part VI—Using Information Systems and Technology to Support Individual and Group Tasks—also explores questions regarding the relationship between individual users (or groups of users) and IT solutions. This part represents perspectives and approaches most often associated with the IS discipline. Its chapters explore questions regarding the adoption and acceptance of technology by individuals, factors that influence the effectiveness of individuals' use of technology, and the development of individual users' computing capabilities. Another important broad topic area covered in this part is the use of IT to enable collaborative work through general communication capabilities and specifically designed task support.

Chapter 38 starts the part with a review and assessment of the research stream that is by far the most widely cited in IS (see Chapter 1): technology acceptance and adoption at the individual level. The chapter also articulates new opportunities for future research contributions in this area. The part continues with two chapters that consider the nature and development of individual capabilities in the use of IT. Chapter 39 explores the concept of computer self-efficacy (based on social cognitive theory), its impact on IT use and learning, and its continued importance even in the era of digital natives. Chapter 40 reviews issues related to the development of individual computing capabilities and provides an integrative framework to support future research and practical applications in this area.

Chapter 41 analyzes the literature on trust, a concept that has strong impact on the acceptance and continued use of IT solutions. One of the task categories most commonly supported by IT is decision making, and Chapter 42 gives a general overview of cognitive aspects of decision making and a summary of IT-based decision aids that are used to support it. The remaining chapters in this part emphasize collaborative processes. Chapter 43 reviews literature on computer supported collaborative work, a field that explores how computing technologies influence human collaboration at various levels. Within that same context, two specific forms of joint action at the team level—problem solving and decision making—are the focus of Chapter 44, which also reviews an influential long-term stream of IS research on group support systems. The remaining two chapters in this part explore organizational adoption of IT-based communication technologies, such as e-mail, text messaging, instant messaging, blogging, wikis, and other social media. Chapter 45 provides a broad, comprehensive overview on communication and collaboration technology adoption, whereas Chapter 46 takes a more focused view on internal use of social media.

Managing and Securing the IT Infrastructure and Systems

Part VII—Managing and Securing the IT Infrastructure and Systems—brings us to another broad topic area in which the disciplines of IS, IT, and CS share interests. The chapters explore issues related to organizing and managing IT infrastructure solutions so that both individual and organizational users can expect to have reliable and secure computing capabilities at their disposal without interruptions. The perspectives of the chapter vary from highly detailed technical to broadly managerial, but all of them share the focus on system continuity, reliability, security, and privacy.

Chapter 47 explores the role of different types of virtualization as key underlying capabilities of contemporary IT architectures. Building on the same conceptual ideas but at a higher level of abstraction is Chapter 48 on cloud computing and its implications on IT infrastructure development and acquisition. Computing has become pervasive and highly mobile, and this has a strong impact on the management of IT and its impact on enterprise characteristics, as described and discussed in Chapter 49 on enterprise mobility. Because of the pervasiveness and continuous need to have IT solutions available, it has become increasingly important to provide these technical capabilities in a sustainable way, as argued in Chapter 50. Chapter 51 reviews the characteristics and capabilities of organizational IT that make it possible to ensure business continuity in a computing-dependent environment.

The second half of Part VII focuses on security, starting from an overview of the technical foundations of IS security (Chapter 52) and continuing with an analysis of special requirements for securing organizational database resources (Chapter 53) and a policy-level review of information security from the behavioral and managerial perspective (Chapter 54). Chapter 55 explores the interaction between user privacy and system security with a focus on understanding the factors that affect privacy in digital business. Part VII concludes with Chapter 56 on digital forensics, a cross-disciplinary area that studies the identification, extraction, and analysis of digital artifacts for investigative purposes.

Managing Organizational Information Systems and Technology Capabilities

The emphasis shifts from systems and technology to organizations and management in Part VIII—Managing Organizational Information Systems and Technology Capabilities. The chapters in this part are all related to the management, organization, acquisition, development, and evaluation of people and IT as resources that make it possible for organizations to achieve their goals. Many of the chapters explore the processes that bring people and technology together as effectively as possible to serve the goals that the organization is striving to attain.

Chapter 57 reviews the current state-of-the-art understanding of how the IT function should be organized and configured in the broader context of an organization. The chief information office's (CIO) role as a bridge between the broader organizational context and the IT function is continuously evolving, and Chapter 58 makes the case for the contemporary CIO's role as a business integrator and innovator. Within the internal IT organization, various IT management frameworks (for IT governance, IT service management, information security management, and IT project management) have become increasingly important and widely used tools. Chapter 59 provides a broad review of these frameworks, with a particular focus on COBIT (governance), ITIL (service management), and ISO/IEC 27000 series (security management).

Chapter 60 builds on a long-term research stream on IT outsourcing and offshoring and presents an overview of scholarly work related to sourcing of IT services. IT project management has been identified as one of the most important capabilities for IS/IT professionals; Chapter 61 reviews core research and practice issues related to this topic area. The next two chapters are both related to human resource management in IS/IT: Chapter 62 focuses on development and retention of computing professionals in organizations, and Chapter 63 reviews what is known about the evaluation of the performance of IS/IT professionals. The part concludes with Chapter 64 on IS audit practices in the context of financial IS.

Information Systems and the Domain of Business Intertwined

This volume concludes with Part IX—Information Systems and the Domain of Business Intertwined—in which the chapters directly focus on the planning, use, organization, and selection of IS solutions in a way that optimally supports the organizational goals and leads to the discovery of new organizational opportunities. The chapters in this part are specifically focused on the application domain of business.

The part starts with Chapter 65 on strategic alignment, emphasizing a topic that IS executives have consistently found to be among the most important for them, that is, how to ensure that organizational IS and technology resources are fully aligned with the overall strategic goals of the organization. Chapter 66 explores IS strategizing and its development over the history of organizational use of IS, proposing a five-component framework for IS strategizing: IS strategy context, IS planning process, intended strategy, IS strategy impact, and emergent strategy. Chapter 67 reviews the literature on the relationship between IT and organizational characteristics, such as extent of centralization, size, and emergence of new organizational forms. It summarizes the research findings and suggests that there are practical implications to be considered in organizational IT planning and design.

Recent advances in computing and communication technologies have had a wide-ranging impact on a variety of organizational practices and capabilities. One of the areas affected strongly is innovation management. Chapter 68 demonstrates how open innovation processes enabled by IT-based infrastructure capabilities are becoming mainstream and radically change how organizations innovate. Chapter 69 explores another phenomenon that has been strongly affected by ubiquitous communication capabilities between organizations: interorganizational information systems (IOS). The chapter shows how IOS have transformed both industries and the way organizations conduct business, but it also identifies the areas where IOS implementations have been more rare and less successful.

The remaining three chapters of Part IX focus on organizational outcomes of IS. Chapter 70 builds directly on one of the most widely cited streams of IS research and presents a review of the current status of research on IS success. The chapter demonstrates that there are still significant opportunities for moving this rich research tradition forward and that both academics and practitioners can benefit from a deeper understanding of the definition of IS success and the factors affecting its achievement. Chapters 71 and 72 come from an important area within the IS discipline that uses the methods and conceptual models of economics to understand the economic impact of IT investments. Chapter 71 reviews and summarizes the literature on the business value of IS investments, pointing out the need to evaluate IS human resources and technology investments synergistically. Chapter 72 concludes the part and the book with a parallel but different view on the impact of IT on firm value, with a specific focus on the evolution of our collective understanding of this phenomenon, exploration of key paradox concepts (productivity and profitability), and an analysis of future directions.

References

Hevner, A. R., S. T. March, J. Park, and S. Ram. 2004. Design science in information systems research. *MIS Quarterly* 28(1): 75–105.

Joint Task Force on Computing Curricula. 2012 *Computer Science Curricula 2013*. Association for Computing Machinery (ACM) and the IEEE-Computer Society (IEEE-CS), December 2013. http://www.acm.org/education/curricula-recommendations

Lunt, B. M., J. J. Ekstrom, S. Gorka, G. Hislop, R. Kamali, E. Lawson, R. LeBlanc, J. Miller, and H. Reichgelt. 2008. *IT 2008: Curriculum Guidelines for Undergraduate Degree Programs in Information Technology*. ACM and IEEE Computer Society.

Shackelford, R., J. Cross, G. Davies, J. Impagliazzo, R. Kamali, R. LeBlanc, B. Lunt, A. McGettrick, R. Sloan, and H. Topi. 2005. *Computing Curricula 2005: The Overview Report*. http://www.acm.org/education/curricula-recommendations

Sidorova, A., N. Evangelopoulos, J. S. Valacich, and T. Ramakrishnan. 2008. Uncovering the intellectual core of the information systems discipline. *MIS Quarterly* 32(3): 467–482.

Taylor, H., S. Dillon, and M. Van Wingen. 2010. Focus and diversity in information systems research: Meeting the dual demands of a healthy applied discipline. *MIS Quarterly* 34(4): 647–667.

Topi, H., J. S. Valacich, R. T. Wright, K. Kaiser, J. F. Nunamaker, J. Janice, C. Sipior, and G. Jan de Vreede. 2010. IS 2010: Curriculum guidelines for undergraduate degree programs in information systems. *Communications of the Association for Information Systems* 26(18). http://aisel.aisnet.org/cais/vol26/iss1/18

Tucker, A. (ed.). 2004. *Computer Science Handbook*, 2nd edn. Chapman & Hall/CRC Press, in cooperation with the Association for Computing Machinery (ACM), Boca Raton, FL.

Acknowledgments

This volume would not have been possible without the tireless and dedicated work of the authors. We feel fortunate and are very thankful for the willingness of both established leading experts and influential younger researchers to join this effort and contribute their expertise to the process. Many of the authors are recognized international leaders within their areas of study. Overall, the chapters incorporate and bring to the readers an unparalleled level of expertise and deep insights.

We would also like to thank the editorial staff at Taylor & Francis/CRC Press, particularly the persistent and gentle leadership of Randi Cohen, computer science acquisitions editor, whose role has extended beyond any reasonable call of duty. Bob Stern, as the original computer science editor who helped create this handbook in its original and second editions, also deserves recognition for his continued support over the years.

Heikki gratefully acknowledges the support of Bentley University—without his sabbatical, Heikki's work on this project would not have been possible. Most of Heikki's work took place while he was visiting at Aalto University School of Business in Helsinki, Finland; he thanks the Information Systems Science Group at Aalto for their collegiality and support. Overall, Heikki acknowledges the advice and help from a number of colleagues throughout the project. Most importantly, he extends his sincere thanks to his wife, Anne-Louise, and daughters, Leila and Saara, who patiently accepted and lovingly supported the long hours and periods of very intensive work that a project like this requires!

Allen acknowledges Bowdoin College for providing the time and support for him to develop the first, second, and third editions of this handbook set over the last 15 years. He also thanks the many colleagues who have contributed to this handbook and/or provided valuable advice on its organization and content throughout that period. Finally, he thanks his wife, Meg, and children, Jenny and Brian, for their constant love and support over a lifetime of teaching and learning.

We hope that this volume will be actively and frequently used both by academics who want to explore the current status and new opportunities within specific research areas and by advanced practitioners who want to understand how they can benefit from the latest thinking within the disciplines of information systems and information technology. We also warmly welcome your feedback.

Heikki Topi
Bentley University
Waltham, Massachusetts

Allen Tucker
Bowdoin College
Brunswick, Maine

Editors

Dr. Heikki Topi is professor of computer information systems at Bentley University in Waltham, Massachusetts. He earned his MSc in economics and business administration (management information systems) from Helsinki School of Economics in 1991 and his PhD in management information systems from Indiana University Graduate School of Business in 1995. Prior to his current role, he served at Bentley as associate dean of business for graduate and executive programs and as chair of the Department of Computer Information Systems. His other academic appointments include associate dean (MBA Program) at Helsinki School of Economics and Business Administration and visiting assistant professor at Indiana University. His teaching interests cover a range of topics, including advanced systems analysis and design, systems modeling, data management, and IT infrastructure. His research focuses on human factors and usability in the context of enterprise systems, information search and data management, and the effects of time availability on human–computer interaction.

Dr. Topi's research has been published in journals such as *European Journal of Information Systems*, *JASIST*, *Information Processing & Management*, *International Journal of Human–Computer Studies*, *Journal of Database Management*, *Small Group Research*, and others. He is the coauthor of a leading data management textbook, *Modern Database Management*, with Jeffrey A. Hoffer and V. Ramesh, and has served as the coeditor of Auerbach's *IS Management Handbook*, with Carol V. Brown. He has contributed to national computing curriculum development and evaluation efforts in leadership roles (including IS 2002, CC2005 Overview Report, and as a task force cochair of IS 2010, the latest IS curriculum revision) since the early 2000s. He has been a member of ACM's Education Board since spring 2006 and has represented first AIS and then ACM on CSAB's board since 2005.

Allen B. Tucker is the Anne T. and Robert M. Bass Professor Emeritus in the Department of Computer Science at Bowdoin College, Brunswick, Maine, where he served on the faculty from 1988 to 2008. Prior to that, he held similar positions at Colgate and Georgetown Universities. Overall, he served for 18 years as a department chair and two years as an associate dean of the faculty. While at Colgate, he held the John D. and Catherine T. MacArthur Chair in Computer Science.

Professor Tucker earned a BA in mathematics from Wesleyan University in 1963 and an MS and PhD in computer science from Northwestern University in 1970. He is the author or coauthor of several books and articles in the areas of programming languages, natural language processing, and software engineering. He has given many talks, panel discussions, and workshop presentations in these areas and has served as a reviewer for various journals, NSF programs, and curriculum projects. He has also served as a consultant to colleges, universities, and other institutions in the areas of computer science curriculum, software development, programming languages, and natural language processing applications. Since retiring from his full-time academic position, Professor Tucker continues to write, teach, and develop open source software for nonprofit organizations.

A fellow of the ACM, Professor Tucker coauthored the 1986 Liberal Arts Model Curriculum in Computer Science and cochaired the ACM/IEEE-CS Joint Curriculum Task Force that developed

Computing Curricula 1991. For these and related efforts, he received the ACM's 1991 Outstanding Contribution Award, shared the IEEE's 1991 Meritorious Service Award, and received the ACM SIGCSE's 2001 Award for Outstanding Contributions to Computer Science Education. In 2001, he was a Fulbright Lecturer at the Ternopil Academy of National Economy in Ukraine, and in 2005, he was an Erskine Lecturer at the University of Canterbury in New Zealand. Professor Tucker has been a member of the ACM, the NSF CISE Advisory Committee, the IEEE Computer Society, Computer Professionals for Social Responsibility, the Liberal Arts Computer Science (LACS) Consortium, and the Humanitarian Free and Open Source Software (HFOSS) Project.

Contributors

Sherif Hanie El Meligy Abdelhamid
Department of
    Computer Science
Virginia Tech
Blacksburg, Virginia

Walid El Abed
Global Data Excellence
Geneva, Switzerland

Silvia Abrahão
Department of Information
    Systems and Computation
Universitat Politècnica de
    València
Valencia, Spain

Miguel I. Aguirre-Urreta
DePaul University
Chicago, Illinois

Monika Akbar
Department of
    Computer Science
Virginia Tech
Blacksburg, Virginia

Gabriel Alatorre
Storage Services Research
IBM Research—Almaden
San Jose, California

Micheal Axelsen
UQ Business School
The University of Queensland
Brisbane, Queensland,
    Australia

Nijaz Bajgoric
School of Economics and Business
University of Sarajevo
Sarajevo, Bosnia and
    Herzegovina

Hillol Bala
Operations and Decision
    Technologies
Indiana University
Bloomington, Indiana

Philippe Bonnet
Department of Software and
    Systems
IT University of Copenhagen
Copenhagen, Denmark

Robert Briggs
Department of Management
    Information Systems
San Diego State University
San Diego, California

Stephen Brobst
Teradata Corporation
San Diego, California

Jan vom Brocke
Institute of Information Systems
University of Liechtenstein
Vaduz, Liechtenstein

Carol V. Brown
Howe School of Technology
    Management
Stevens Institute of Technology
Hoboken, New Jersey

Glenn J. Browne
Information Systems and
    Quantitative Sciences
Rawls College of Business
Texas Tech University
Lubbock, Texas

Nicolas Bruno
Microsoft Corporation
Redmond, Washington

Andrew Burton-Jones
UQ Business School
The University of Queensland
Brisbane, Queensland,
    Australia

Keith A. Butler
University of Washington
Seattle, Washington

John M. Carroll
College of Information
    Sciences and Technology
Center for Human–Computer
    Interaction
The Pennsylvania State University
University Park, Pennsylvania

Ann Corrao
Delivery Technology and
    Engineering
IBM Global Technology
    Services
Research Triangle Park,
    North Carolina

Jedidiah Crandall
Department of Computer Science
University of New Mexico
Albuquerque, New Mexico

Edward Curry
Digital Enterprise Research
    Institute
National University of Ireland,
    Galway
Galway, Ireland

Dianne Cyr
Beedie School of Business
Simon Fraser University
Surrey, British Colombia, Canada

Fred D. Davis
Department of Information
    Systems
Walton College of Business
University of Arkansas
Fayetteville, Arkansas

**Sabrina De Capitani di
Vimercati**
Department of Computer Science
Università degli Studi di Milano
Crema, Italy

William DeLone
Department of Information
    Technology
American University
Washington, District of Columbia

Paul Devadoss
Lancaster University
    Management School
Lancaster University
Lancashire, United Kingdom

Brian Donnellan
Innovation Value Institute
National University of Ireland,
    Maynooth
Maynooth, Ireland

Chitra Dorai
IBM T.J. Watson Research Center
Newark, New Jersey

Joseph J. Ekstrom
Information Technology
Brigham Young University
Provo, Utah

Noha Ibrahim Elsherbiny
Department of
    Computer Science
Virginia Tech
Blacksburg, Virginia

**Mohamed Magdy Gharib
Farag**
Department of Computer Science
Virginia Tech
Blacksburg, Virginia

Adrian Fernandez
Department of Information
    Systems and Computation
Universitat Politècnica de
    València
Valencia, Spain

Thomas W. Ferratt
Department of MIS, OM and
    Decision Sciences
University of Dayton
Dayton, Ohio

Sara Foresti
Department of
    Computer Science
Università degli Studi di Milano
Crema, Italy

Edward A. Fox
Department of Computer Science
Virginia Tech
Blacksburg, Virginia

Michael J. Franklin
University of California at
    Berkeley
Berkeley, California

Bill Franks
Teradata Corporation
Atlanta, Georgia

Robert D. Galliers
Department of Information and
    Process Management
Bentley University
Waltham, Massachusetts

Joey F. George
Supply Chain and Information
    Systems
Iowa State University
Ames, Iowa

Ester S. Gonzalez
Department of Information
    Systems and Decision
    Sciences
California State University
    at Fullerton
Fullerton, California

Anastasios Gounaris
Department of Informatics
Aristotle University of
    Thessaloniki
Thessaloniki, Greece

Art Gowan
Department of Information
    Technology
Georgia Southern University
Statesboro, Georgia

Peter Green
UQ Business School
The University of Queensland
Brisbane, Queensland,
    Australia

William M. Gribbons
Human Factors
Bentley University
Waltham, Massachusetts

Jonathan Grudin
Natural Interaction Group
Microsoft Research
Redmond, Washington

Saurabh Gupta
Department of Management
University of North Florida
Jacksonville, Florida

Mikko Hallanoro
Department of Management
Turku School of Economics
University of Turku
Turku, Finland

Ole Hanseth
Department of Informatics
University of Oslo
Oslo, Norway

Markus Helfert
School of Computing
Dublin City University
Dublin, Ireland

Alan R. Hevner
Information Systems and
    Decision Sciences
College of Business
University of South Florida
Tampa, Florida

Ellen D. Hoadley
Loyola University Maryland
Baltimore, Maryland

Michael E. Houle
National Institute of
    Informatics
Tokyo, Japan

Roland Hübscher
Human Factors
Bentley University
Waltham, Massachusetts

Sharon E. Hunt
Graziadio School of Business
    and Management
Pepperdine University
Malibu, California

Juhani Iivari
Department of Information
    Processing Sciences
University of Oulu
Oulu, Finland

Emilio Insfran
Department of Information
    Systems and Computation
Universitat Politècnica de
    València
Valencia, Spain

Sushil Jajodia
Center for Secure Information
    Systems
George Mason University
Fairfax, Virginia

Mohammad Hossein Jarrahi
School of Information Studies
Syracuse University
Syracuse, New York

Sirkka L. Jarvenpaa
McCombs School of Business
University of Texas at Austin
Austin, Texas

and

School of Science
Aalto University
Espoo, Finland

Matthew L. Jensen
Division of Management
    Information Systems
University of Oklahoma
Norman, Oklahoma

Fang Jin
Department of Computer Science
Virginia Tech
Blacksburg, Virginia

Sherif Kamel
Department of Management
School of Business
The American University
    in Cairo
New Cairo, Egypt

Anna Karpovsky
Department of Information and
    Process Management
Bentley University
Waltham, Massachusetts

David Kieras
Department of Electrical
    Engineering and Computer
    Science
University of Michigan
Ann Arbor, Michigan

Laurie J. Kirsch
Joseph M. Katz Graduate
    School of Business
University of Pittsburgh
Pittsburgh, Pennsylvania

Hope Koch
Department of Information
    Systems
Baylor University
Waco, Texas

Rajiv Kohli
Mason School of Business
College of William & Mary
Williamsburg, Virginia

Henry F. Korth
Lehigh University
Bethlehem, Pennsylvania

Mary Lacity
College of Business
University of Missouri–St. Louis
St. Louis, Missouri

Kiljae Lee
University of Kansas
Lawrence, Kansas

Yang W. Lee
Supply Chain and
    Information Management
    Group
D'Amore-McKim School of
    Business
Northeastern University
Boston, Massachusetts

Jonathan P. Leidig
School of Computing and
    Information Systems
Grand Valley State University
Allendale, Michigan

Xin Li
College of Business
University of South Florida
–St. Petersburg
St. Petersburg, Florida

Ying Li
Service Operations
    Management and Analytics
IBM T.J. Watson Research Center
Yorktown Heights, New Jersey

Juho Lindman
Hanken School of Economics
Helsinki, Finland

Henry C. Lucas, Jr.
Department of Decision,
    Operations and Information
    Technologies
Robert H. Smith School of
    Business
University of Maryland
College Park, Maryland

Jerry Luftman
Global Institute for IT
    Management LLC
Fort Lee, New Jersey

Barry M. Lunt
Information Technology
Brigham Young University
Provo, Utah

Stuart E. Madnick
Information Technology Group
Sloan School of Management
    and Engineering Systems
    Division
School of Engineering
Massachusetts Institute of
    Technology
Cambridge, Massachusetts

Yannis Manolopoulos
Department of Informatics
Aristotle University of
    Thessaloniki
Thessaloniki, Greece

Marco Marabelli
Department of Information and
    Process Management
Bentley University
Waltham, Massachusetts

George M. Marakas
Department of Decision
    Sciences and Information
    Systems
Florida International University
Miami, Florida

M. Lynne Markus
Department of Information and
    Process Management
Bentley University
Waltham, Massachusetts

Ephraim McLean
Department of Computer
    Information Systems
Georgia State University
Atlanta, Georgia

Martin Meyer
Dublin City University
Dublin, Ireland

Sunil Mithas
Department of Decision,
    Operations and Information
    Technologies
Robert H. Smith School of
    Business
University of Maryland
College Park, Maryland

John G. Mooney
Graziadio School of Business
    and Management
Pepperdine University
Malibu, California

Michael G. Morris
McIntire School of Commerce
University of Virginia
Charlottesville, Virginia

Leigh Mutchler
Department of Accounting
    and Information
    Management
University of Tennessee
Knoxville, Tennessee

Sai Tulasi Neppali
Department of
    Computer Science
Virginia Tech
Blacksburg, Virginia

Susan Newell
Department of Information and
    Process Management
Bentley University
Waltham, Massachusetts

and

Department of Information
    Management
University of Warwick
Warwick, United Kingdom

Fred Niederman
Department of Operations and
    Information Technology
    Management
Cook School of Business
Saint Louis University
St. Louis, Missouri

Paul Brillant Feuto Njonko
Research Centre Lucien
    Tesnière
Natural Language Processing
University of Franche-Comté
Besançon, France

Onook Oh
The Center for Collaboration
    Science
University of Nebraska
Omaha, Nebraska

Daniela Oliveira
Department of
    Computer Science
Bowdoin College
Brunswick, Maine

Jim Olson
Delivery Technology and
  Engineering
IBM Global Technology Services
Southbury, Connecticut

Vincent Oria
Department of Computer Science
New Jersey Institute of
  Technology
Newark, New Jersey

M. Tamer Özsu
Cheriton School of Computer
  Science
University of Waterloo
Waterloo, Ontario, Canada

Apostolos N. Papadopoulos
Department of Informatics
Aristotle University of
  Thessaloniki
Thessaloniki, Greece

Joe Peppard
European School of
  Management and
  Technology
Berlin, Germany

Steven E. Poltrock
Bellevue, Washington

V. Ramesh
Operations and Decision
  Technologies
Indiana University
Bloomington, Indiana

Han Reichgelt
School of Computing and
  Software Engineering
Southern Polytechnic State
  University
Marietta, Georgia

Roni Reiter-Palmon
The Center for Collaboration
  Science
University of Nebraska at Omaha
Omaha, Nebraska

Gail Ridley
School of Accounting and
  Corporate Governance
University of Tasmania
Tasmania, Australia

Matti Rossi
Department of Information and
  Service Economy
School of Business
Aalto University
Helsinki, Finland

Mary Beth Rosson
The Pennsylvania State
  University
University Park, Pennsylvania

Vassil Roussev
Department of Computer
  Science
University of New Orleans
New Orleans, Louisiana

Pierangela Samarati
Department of Computer
  Science
Università degli Studi di
  Milano
Crema, Italy

Steve Sawyer
School of Information Studies
Syracuse University
Syracuse, New York

Darshana Sedera
School of Information
  Systems
The Queensland University
  of Technology
Brisbane, Queensland,
  Australia

Teresa Shaft
Division of Management
  Information Systems
University of Oklahoma
Norman, Oklahoma

Dennis Shasha
Department of Computer
  Science
Courant Institute of
  Mathematical Sciences
New York University
New York, New York

Avi Silberschatz
Yale University
New Haven, Connecticut

Aameek Singh
Storage Services Research
Services Innovation
  Laboratory
IBM Research—Almaden
San Jose, California

Sandra A. Slaughter
Scheller College of Business
Georgia Institute of
  Technology
Atlanta, Georgia

Yang Song
Storage Services Research
IBM Research—Almaden
San Jose, California

Christian Sonnenberg
Institute of Information Systems
University of Liechtenstein
Vaduz, Liechtenstein

Carsten Sørensen
Department of Management
The London School of Economics
  and Political Science
London, United Kingdom

Charles Steinfield
Department of
  Telecommunication,
  Information Studies, and
  Media
Michigan State University
East Lansing, Michigan

Tom Stewart
System Concepts
London, United Kingdom

S. Sudarshan
Indian Institute of Technology
  Bombay
Mumbai, India

Alistair Sutcliffe
Manchester Business School
University of Manchester
Manchester, United Kingdom

Erkki Sutinen
School of Computing
University of Eastern Finland
Joensuu, Finland

Heikki Topi
Department of Computer
  Information Systems
Bentley University
Waltham, Massachusetts

Eileen M. Trauth
College of Information Sciences
  and Technology
The Pennsylvania State University
University Park, Pennsylvania

Kostas Tsichlas
Department of Informatics
Aristotle University of
  Thessaloniki
Thessaloniki, Greece

Jennifer Tucker
National Organic Program
U.S. Department of Agriculture
Washington, District of Columbia

Patrick Valduriez
Le Laboratoire d'Informatique,
  de Robotique et de
  Microélectronique de
  Montpellier
Institut National de Recherche
  en Informatique et en
  Automatique
Montpellier, France

Viswanath Venkatesh
Walton College of Business
University of Arkansas
Fayetteville, Arkansas

Gert-Jan de Vreede
The Center for Collaboration
  Science
University of Nebraska
Omaha, Nebraska

Triparna de Vreede
The Center for Collaboration
  Science
University of Nebraska
Omaha, Nebraska

Richard Y. Wang
Information Quality Program
Massachusetts Institute of
  Technology
Cambridge, Massachusetts

Merrill Warkentin
Department of Management
  and Information Systems
Mississippi State University
Starkville, Mississippi

Ron Weber
Faculty of Information
  Technology
Monash University
Melbourne, Victoria,
  Australia

Benjamin Wigert
The Center for Collaboration
  Science
University of Nebraska
Omaha, Nebraska

Leslie Willcocks
Department of Management
London School of Economics
  and Political Science
London, United Kingdom

Michael L. Williams
Graziadio School of Business
  and Management
Pepperdine University
Malibu, California

David W. Wilson
Department of Management
  Information Systems
Eller College of Management
University of Arizona
Tucson, Arizona

Till J. Winkler
Department of IT
  Management
Copenhagen Business School
Copenhagen, Denmark

Ryan T. Wright
Department of Operations
  and Information
  Management
University of
  Massachusetts–Amherst
Amherst, Massachusetts

Jennifer Jie Xu
Computer Information
  Systems
Bentley University
Waltham, Massachusetts

Emre Yetgin
Division of Management
  Information Systems
University of Oklahoma
Norman, Oklahoma

Hongwei Zhu
Department of Operations and
  Information Systems
Manning School of Business
University of Massachusetts–
  Lowell
Lowell, Massachusetts

I

Disciplinary Foundations and Global Impact

1

Evolving Discipline of Information Systems

Heikki Topi
Bentley University

1.1 Introduction

This chapter gives its readers an overview of the Information Systems (IS) discipline and discusses how the discipline has evolved during in its relatively brief history. To achieve this goal, it uses two approaches: first, an analysis of the most highly cited articles within the discipline and second, an introduction to broad, self-reflective scholarly conversations that have taken place within the discipline.

This chapter does not, however, intend to present a comprehensive history of the field. There are many reasons underlying this choice. One of the most important ones is the recent *Journal of Association for Information Systems* special issue on the history of the IS field published in 2012 (Hirschheim et al., 2012). Particularly an article by Hirschheim and Klein (2012) provides a very thorough and comprehensive historical analysis of the IS field. It would be difficult, if not impossible, to add anything substantive to its 50 pages written over a 12 year period (Hirschheim et al., 2012, p. iv) in a short book chapter. Despite its length and depth, the Hirschheim and Klein (2012) article is highly readable, and it is an excellent starting point for anybody who wants to understand where the IS field came from, where it currently is, and where its future could be. A more condensed version of this history is available in Section 1.2.2 of Hirschheim and Klein (2003).

The analytical approaches chosen for this chapter have two goals: First, the chapter wants to provide its readers with an overview of the topics that have been most influential in the course of the history of the IS field. Second, the chapter introduces a contextual structure that helps in positioning the chapters in this volume in the broader context of the field. Both the analysis of the highly cited articles and the review of the broad conversations provide insights regarding the topics that the field itself and those using its results have considered most important.

1.2 Analysis of Most Highly Cited Scholarly Work in Information Systems

The first approach to describing the discipline of IS is based on a comprehensive attempt to identify about 100 most highly cited articles published by IS scholars at any time during the field's history. These scholarly works are then used to discover the patterns they reveal about the core topics and areas of interest within the field. There is no straightforward way to identify the pool of all scholarly articles within the IS field, and thus, a number of mechanisms were used to add articles to this pool. The first was to approach the problem through highly cited scholars and assume that their most highly cited articles would be among the target group. Consequently, one of the starting points for forming the article pool was Hu and Chen's (2011, Table 1.1) list of 85 highly productive IS researchers with high H index values. The other approach to forming the pool was to identify key IS journals and choose the most highly cited articles for each of them. The journals included the premier journals chosen by the IS senior scholars (in alphabetical order: *European Journal of Information Systems, Information Systems Journal, Information Systems Research, Journal of AIS, Journal of Information Technology, Journal of MIS, Journal of Strategic Information Systems,* and *MIS Quarterly*). In addition to these eight, *Decision Sciences, Decision Support Systems,* and *Management Science* were covered. Moreover, the citations for key articles of well-known authors not included in the first group were also evaluated. It is possible that this approach left out published works that should have been included, but it is likely that these would not have had a material impact on the story that the data tell.

TABLE 1.1   Classification Categories

	Category	Sum of Citation Index Values	Total Number of Articles	Average Citation Index Value
1.	User acceptance and use of IT	15,918	23	692
2.	Impact of IT on organizations	4,107	11	373
3.	Methodology	3,407	8	426
4.	Media and organizational communication	3,292	6	549
5.	Knowledge management	3,265	7	466
6.	IS success	2,628	4	657
7.	IT and business performance	2,619	7	374
8.	Trust	2,552	7	365
9.	Electronic marketplaces	1,962	4	491
10.	Adoption	1,881	3	627
11	Group support systems	1,843	4	461
12.	User satisfaction	1,375	3	458
13.	Design science	1,027	2	514
14.	IT implementation	1,019	3	340
15.	Self-efficacy	946	2	473
16.	Managerial IT use	761	2	381
17.	Artificial intelligence	701	1	701
18.	Coordination	647	1	647
19.	Enterprise systems	595	1	595
20.	IS continuance	365	1	365
21.	Discipline identity	356	1	356
22.	Data quality	316	1	316
23.	IS effectiveness	256	1	256
24.	Outsourcing	224	1	224
25.	Electronic commerce	173	1	173

The citation counts were retrieved from two sources: Google Scholar (using Publish or Perish; see Harzing, 2007) and Social Science Citation Index (SSCI). The citation counts were retrieved during a single day in June 2012. The value used in this analysis for ranking purposes was based on the formula (Google Scholar count/4 + SSCI count)/2. Obviously, this is not an attempt to determine a real citation count; instead, it provides a citation index that gives more weight to the SSCI count but still takes into account the significantly higher nominal values of Google Scholar (see Meho and Yang, 2007 for a justification to use both Google Scholar and SSCI). The citation counts have not been adjusted based on the publication year; this analysis will naturally leave out the most recent influential articles. This is not a problem, given the purpose of this chapter, which intends to identify the themes and topics that the field of IS has focused on and contributed to over its history.

Using an arbitrary cutoff point of 1000 Google Scholar citations, the analysis identified a list of the 105 articles, which is in Appendix 1.A. In addition to the year, author, and journal information, the table also presents a categorization for each of the papers. The categories used in this analysis are at a lower level of abstraction than, for example, those derived in Sidorova et al. (2008), because the purpose of the analysis was to illustrate the scholarship topics that the most highly cited IS articles during the past 30 years represent. The author of this chapter categorized the articles based on their title and abstract; if necessary, the full article was used to determine the proper category. The categorization and category development process was iterative: during the first round, each article was assigned to an initial category, which was potentially modified for clarity and consistency during subsequent rounds.

In this chapter, the focus is on the content of the most highly cited articles (instead of, for example, the authors or the institutions). Again, the purpose is to demonstrate through the most actively used scholarly contributions what topics the IS field, those external fields benefiting from it, and the practitioner community value most highly based on the use of the scholarship. Next, the key categories discovered in the analysis will be described.

1.2.1 Description of Identified Categories

Table 1.1 includes a list of the categories that emerged from the classification process; in addition to the category name, the table includes the sum of citation index values for the articles in the category, the number of articles in each category, and the average of citation index values in each category. The categories are sorted by the sum of the citation index values.

The rest of the section will discuss the key characteristics and the most important research themes featured in each of the categories.

1.2.1.1 User Acceptance and Use of IT

By far, the most actively cited category is *user acceptance and use of information technology (IT)*. The seminal works in this category are Davis (1985, 1989) and Davis et al. (1989), which introduced the Technology Acceptance Model (TAM). These articles were then followed up by a number of influential efforts to evaluate and extend TAM, such as Taylor and Todd (1995b), Gefen and Straub (1997), Venkatesh and Davis (2000), and Venkatesh et al. (2003). Applications of TAM to various contexts have also had a strong impact and been cited widely (e.g., Koufaris, 2002; Gefen et al., 2003; Pavlou, 2003). The field as a whole has spent a significant amount of time and effort discussing the contributions and the future of TAM (including a special issue in the *Journal of AIS*; Hirschheim, 2007). The core premise of TAM is simple: it posits that the probability of a user's behavioral intention to use a specific IT is based on two primary factors: perceived usefulness and perceived ease of use, the former directly and the latter both directly and mediated by perceived usefulness. The research building on the seminal articles over a period of 25 years has expanded the scope significantly and made the theory much more complex. TAM and related research specifically focus on individual users.

TAM has been so dominant that it has become the default conceptual model for any research on antecedents of IT adoption and use. For example, when Agarwal and Karahanna (2000) introduced an instrument for measuring the level of user's cognitive absorption with software (in their case, a web browser) and studied its impact on use, it appears that the choice to use TAM constructs to model the mediating variables between cognitive absorption and intention to use was, in practice, automatic. Thompson et al. (1991) proposed an alternative model for understanding use based on Triandis (1980). Even at that time, just a couple of years after the seminal TAM papers were published, the authors already positioned the reporting of their results as a comparison with TAM.

In addition to technology acceptance research discussed earlier that has focused on the antecedents of intention to use, IS researchers have also studied individual use of IT and its antecedents and consequences from other perspectives (often directly linked to TAM). Goodhue and Thompson (1995) introduced a model entitled "Technology-to-Performance Chain" and as part of it the task-technology fit model. Their article integrates two insights into one model: first, to have an impact on individual performance, a specific technology artifact must be used, and second, it has to have a good fit with the task in the context of which it is used. The study found strong empirical evidence to support these propositions that formed an important foundation for the stream of research that was based on it.

The dominance of TAM and the broader acceptance and use of IT category in IS research are also demonstrated by the citation metrics—articles related to user acceptance and use have a higher total citation index value than the following four categories together. These articles count for about 30% of the total of the citation index values in the top 105 articles. Some researchers (such as Benbasat and Barki, 2007) have questioned whether or not the strong focus on TAM has led to nonoptimal resource allocation patterns for the discipline.

This volume includes a comprehensive analysis of the current status and future opportunities of the individual technology acceptance research by Venkatesh et al. (Chapter 38). This chapter and the JAIS 2007 special issue (Hirschheim, 2007) are excellent starting points for a reader who wants to build a strong foundation for understanding this essential stream of research.

1.2.1.2 Impact of IT on Organizations

The second most widely cited category is labeled *Impact of IT on organizations,* and it is dominated by Wanda Orlikowski's work. Most of the articles in this category provide perspectives on and guidance for studying, understanding, and making sense of the role of IT in organizations (e.g., Orlikowski and Baroudi, 1991; Orlikowski and Gash, 1994). Many of them follow a structurational perspective, which, based on the structuration theory by Giddens, explores how structures emerge from the interactions between the users and the technology (Orlikowski and Robey, 1991; Orlikowski, 2000). For this perspective, it is particularly important to understand how the use of technology contributes to and is associated with innovation, learning, improvisation, and emergent practices in organizations. One of the key insights revealed through this stream of research is the essential role of unintended and unplanned uses of technology and the strong impact these uses have on organizations. Technology and its users interact within organizations in a myriad of highly complex ways, and the users' reactions to technology and technology-based solutions are typically impossible to fully anticipate. In this volume, Sawyer and Jarrahi (Chapter 5) discuss structuration at a more detailed level.

The articles that focus specifically on IT and organizational change form an important subcategory. Some of them explore the theories of and methods for analyzing and understanding the relationship between technology and organizational change. For example, Markus and Robey (1988) review the theories about the impact of technology on organizations with an explicit intent of improving the qualities of these theories. Orlikowski and Robey (1991), in turn, suggest a theoretical framework (based on Giddens' structuration theory) for guiding studies on two areas that are closely linked to IT-enabled and induced organizational change: systems development and the effects of IT use on organizations. Orlikowski (1993b) takes a closer look at the role of specific systems development tools (CASE tools) in organizational change processes. In another widely cited paper, she (Orlikowski, 1996) provides a detailed look at how small, frequently repeated actions by organizational actors while making sense of a new IT and applying it to their everyday work lead to organizational change.

In this volume, Markus discusses the complex relationship between IT and organizational structure (Chapter 67). Jarvenpaa's chapter on open innovation management illustrates the very significant impact IT has had on the development of innovation management, an essential organizational capability (Chapter 68).

Returning back to the highly cited articles, two papers in the IT and organizational change subcategory take quite a different approach from the ones discussed earlier. Both Henderson and Venkatraman (1993) and Venkatraman (1994) are more prescriptive in their approach, explicitly providing guidance regarding the ways IT can be used to enable certain types of organizational transformation processes. Henderson and Venkatraman (1993) also suggest an important mechanism for understanding and guiding the alignment between IT and business strategies (strategic integration) and the fit between business and IT infrastructures (functional or operational integration). Venkatraman (1994) presents a five-level model to describe different types of IT-enabled business transformation, starting from localized exploitation and moving through internal integration, business process redesign, and business network redesign to business scope redefinition.

In this volume, several chapters discuss related issues, including Luftman on alignment (Chapter 65), and both Rossi & Lindman and Devadoss the relationship between organizational transformation processes and the design, implementation, and use of large-scale enterprise systems (Chapters 23 and 24, respectively).

1.2.1.3 Methodology

The next most widely cited main category of IS research articles consists of those providing guidance on *methodology* to the community. The most influential articles have covered case research (Benbasat et al., 1987; Walsham, 1995), interpretive field studies (Myers, 1997; Klein and Myers, 1999), structural equation modeling (Chin, 1998; Gefen et al., 2000; Chin et al., 2003), and instrument validation (Straub, 1989).

1.2.1.4 Media and Organizational Communication

The list of frequently cited article categories continues with *media and organizational communication*. In this category, by far the most influential paper is Daft and Lengel's (1986) study on the reasons why organizations process information and the relationship between structure, media characteristics (specifically, media richness), and organizational information processing. A common general theme within this category is the impact of communication media on organizational communication effectiveness. The emergence of e-mail and its impact on managerial communication has been of particular interest (Sproull and Kiesler, 1986), together with media selection (Daft et al., 1987). Markus (1987) and Yates and Orlikowski (1992) have produced widely used general analyses of interactive media and genres of organizational communication, respectively.

1.2.1.5 Knowledge Management

Knowledge management is the next highly influential category. In this area, a review article by Alavi and Leidner (2001) is clearly the most widely cited contribution to the literature, demonstrating the usefulness of high-quality integrative summaries (see also Alavi and Leidner, 1999). Alavi and Leidner also articulate the key positions of the longstanding debate regarding the relationship between data, information, and knowledge, an essential question when the focus is on understanding how knowledge management differs from and depends on information and data management. Davenport et al. (1998) have contributed another broadly used (although significantly more practically focused) review article, which emphasizes the importance of knowledge management projects and explores their success factors. The factors that motivate individuals within an organization to share their knowledge and contribute it to the relevant organizational knowledge management systems have been a specific question of interest, explored by Wasko and Faraj (2005) and Bock et al. (2005).

In this volume, Newell and Marabelli (Chapter 17) provide a comprehensive overview of the current status of knowledge management research and practice.

1.2.1.6 IS Success

The meaning, role, and importance of *IS success* is the focus of four highly influential articles that have explored the question of what it means to achieve success with IS. The seminal article in this area is DeLone and McLean (1992), which was later updated by the same authors 10 years later (DeLone and McLean, 2003). Seddon's respecification and extension of the original model (Seddon, 1997) has also reached a strong position as a widely cited article. The original DeLone and McLean paper identified six interdependent IS success constructs: system quality, information quality, use, user satisfaction, individual impact, and organizational impact. These constructs are organized into a process model with organizational impact as the ultimate IS success construct affected either directly or indirectly by the others. The updated 2003 model added a third quality dimension, service quality, recognized the connection between intention to use and use, and replaced all impact measures with a single net benefit construct.

DeLone, McLean, and Sedera continue to build on their earlier work, evaluate its impact, and propose new directions for future work in Chapter 70 of this volume.

1.2.1.7 IT and Business Performance

Another significant area of interest in IS research is the relationship between *IT and business performance*, often explored using the methods and approaches of economics. Two highly influential papers in this category (Mata et al., 1995; Bharadwaj, 2000) have adopted the resource-based view of the firm as the underlying perspective for exploring the role of IT as a source of competitive advantage. Mata and colleagues concluded that of the five factors that they analyzed conceptually, only one (managerial IT skills) is likely to be a source of competitive advantage; the others (access to capital, proprietary technology, technical IT skills, and customer switching costs) do not satisfy the criteria that would grant them this role. Bharadwaj (2000) categorizes IT resources into three groups: IT infrastructure, human IT skills (managerial and technical), and IT-enabled intangibles (such as customer orientation, knowledge assets, and synergy). Based on an empirical analysis, she comes to a different conclusion compared to Mata et al., stating that "IT capability is rent generating resource that is not easily imitated or substituted" (p. 186), attributing the ability to sustain superior performance to time compression diseconomies, resource connectedness, and social complexity.

Two authors who have contributed very significantly to the discussion regarding the relationship between IT and business performance are Brynjolfsson and Hitt. In this analysis, four of their articles were included among the most highly cited ones: Brynjolfsson and Hitt (1996, 2000), Hitt and Brynjolfsson (1996), and Brynjolfsson (1993). Two of them (Brynjolfsson, 1993; Brynjolfsson and Hitt, 1996) analyzed the widely cited productivity paradox of computing (which refers to the invisibility of the effects of IT in aggregate output statistics). By analyzing firm-level data, the authors discovered that IT spending (both capital and labor) had a significant impact on firm output and concluded that at least in their sample the paradox had disappeared. Brynjolfsson and Hitt (2000, p. 45) emphasized the role of IT as a "general purpose technology" that enables complementary innovations, such as new business processes and new organizational and industry structures. The authors further emphasized the importance of performing the analysis at the firm level instead of evaluating only economies or industries at the aggregate level.

In this volume, Chapter 71 by Hoadley and Kohli and Chapter 72 by Mithas and Lucas explore the questions related to the business value of IS/IT investments, each with their own distinctive perspective.

1.2.1.8 Trust

IS research on *trust* has achieved a highly visible status, gaining it the eighth place in this analysis. This category has two clearly different elements: trust within virtual teams (Jarvenpaa and Leidner, 1998; Jarvenpaa et al., 1998) and trust in the context of consumer electronic commerce or electronic markets (Jarvenpaa et al., 1999; Gefen, 2000; McKnight et al., 2002; Jøsang et al., 2007). These two streams are connected through the trust construct itself, but the contexts and core issues are quite different. In the former, the key questions are related to the role of trust (or the lack thereof) in the context of globally distributed teams that are conducting work without the benefit of establishing personal relationships in a face-to-face environment. In the latter case, the focus is on the mechanisms through which virtual marketplaces and the organizations maintaining them can create trust that has the potential to alleviate the users' concerns regarding the risks associated with the marketplaces. In both cases, the scholarly work has improved our understanding of how trust is formed and maintained in environments where physical presence is missing.

This volume recognizes the importance of trust in the design and use of IS by including a dedicated chapter related to this topic area (Chapter 41 by Li).

1.2.1.9 Electronic Marketplaces

Moving forward in the list of categories of influential, highly cited articles, the next one focuses on *electronic marketplaces*. Malone et al. (1987) wrote a pre-electronic business (as we know it now) analysis of electronic markets and hierarchies, stating insightfully "We should not expect electronically interconnected world of tomorrow to be simply a faster and more efficient version of the world we know today. Instead, we should expect fundamental changes in how firms and markets organize the flow of goods and services in our economy" (p. 497). This has definitely been the case, and the transformation process is not over yet. A few years after Malone et al. (1987) and Bakos (1991) analyzed the potential strategic impact of electronic marketplaces, focusing on key dimensions of the impact of the digitalization of business: reduction of search costs, network externalities, technological uncertainty, switching costs, and economies of scale and scope (p. 308). Bakos (1997) takes a closer look at the reduction of buyer search costs and their impact on electronic marketplaces. Brynjolfsson and Smith (2000) compare conventional and Internet retailers to each other, with a specific focus on the impact on Internet retailing on pricing.

1.2.1.10 Adoption and Diffusion of IT Innovations

Adoption and diffusion of IT innovations has been an influential area of research, exemplified in our group of highly cited articles by Cooper and Zmud (1990), Moore and Benbasat (1991), and Karahanna et al. (1999). Cooper and Zmud studied the adoption (allocation of resources to acquire the innovation) and infusion (achieving high levels of effectiveness with the innovation) of an organizational innovation (material requirements planning systems, precursors of current ERPs), finding significant differences in factors that affect adoption compared to those affecting infusion. Moore and Benbasat focused on individual adoption of an IT, developing an instrument for measuring perceptions that might have an impact on the adoption process. Moore and Benbasat's work was strongly based on Rogers's theory of innovation diffusion (Rogers, 2003) and the five attributes of innovations identified by Rogers (relative advantage, compatibility, complexity, observability, and trialability). Both the instrument introduced in this article and the development process have been widely used. One of the adopters of the instrument was Karahanna et al. (1999), who examined at the individual-level pre-adoption and post-adoption beliefs and attitudes, demonstrating the fundamental importance of perceptions regarding usefulness and enhancing one's image post-adoption.

In the context of this volume, issues of technology adoption are most visible in Chapter 38 on individual-level technology adoption research by Venkatesh et al. and in Chapter 45 in which Grudin explores factors affecting the adoption of new communication and collaboration technologies in organizations. These chapters do not, however, incorporate the innovation diffusion perspective.

1.2.1.11 Group Support Systems

The next topic was prominently featured in the IS literature in late 1980s and early 1990s: *group (decision) support systems (GSS) and other collaborative systems*. The most highly cited paper in this category, DeSanctis and Poole (1994), could also have been categorized based on the way it introduces the adaptive structuration theory as an approach to study the role of technology in the context of organizational change. The empirical context for the paper is, however, GSS (or GDSS, as the authors call it). The article presents seven propositions of the adaptive structuration theory and explores the applicability of the theory to the study of GSS use in a small group context. The key issue it highlights (in addition to the theoretical contribution) is that the introduction of the same technology may lead to widely different organizational outcomes and that the effectiveness of technology use is dependent on how it is appropriated at a specific organizational context. An earlier article by DeSanctis (DeSanctis and Gallupe, 1987) presents a framework for the study of GSS, demonstrating the importance of group size, the proximity of the group members, and the nature of the task as important contingency variables.

DeSanctis represents a GSS research tradition that emerged from the University of Minnesota; another very significant center for this research was the University of Arizona, where the work by Nunamaker and his research team led to a long stream of research contributions and software products in this area. The most highly cited of these articles is Nunamaker et al. (1991), which presents a summary of research conducted at the Arizona research program on electronic meeting systems (EMS). In addition to the framework for the study of these systems, the article gives a good overview of the Arizona approach that combines the design and implementation of technology artifacts with research on them both in the laboratory and in the field.

The current state of GSS research and the broader area of research on enhancing team-level problem solving and decision making is reviewed by de Vreede et al. in Chapter 44 of this volume. In addition, Poltrock provides a comprehensive overview of a closely related area, computer-supported cooperative work, in Chapter 43.

1.2.1.12 User Satisfaction

Before the emergence of the TAM research and focus on perceptions of perceived ease of use and usefulness, several papers were published on *user satisfaction*. Bailey and Pearson (1983) described the process of developing an instrument for measuring "computer user satisfaction." Despite the name, the construct itself has little to do with computers per se. Instead, the instrument appears to capture at a very broad level users' satisfaction with the services that the IT (EDP) department provides (the instrument includes questions related to, among others, top management involvement, charge-back method for payment for services, and relationship with the EDP staff with 36 other categories). This is understandable given the time when these data were collected (around the time when first PCs were introduced).

Ives et al. (1983) fine-tuned the Bailey and Pearson instrument and developed a short form of it. The scope of the issues that the two instruments measure is equally broad; interestingly, the authors suggest that the instrument can be used to "evaluate an IS or general systems development effort in an organization." Doll and Torkzadeh (1988) reported on the development of a different type of a satisfaction instrument, which was intended to focus on the users' satisfaction with the outcomes and ease of use of a specific application (instead of, say, the general satisfaction with the IS department). The resulting instrument was designed to capture data on five factors: content, accuracy, format, ease of use, and timeliness. Even though the seeds of the ideas underlying TAM are clearly included in this article, it is interesting to note that the work was parallel with that of Davis (whose dissertation was published in 1985).

1.2.1.13 Design Science

There are only two articles in the following category, *design science*, but particularly the more highly cited of them, Hevner et al. (2004), has already had a very impressive impact on research within the field despite its recent publication date. The popularity and influence of this article can be at least partially explained by its focus on an issue that is fundamental from the perspective of the identity of the IS field. It gave revived legitimacy for the design science paradigm in IS research at a time when many thought that only the behavioral research paradigm would survive. Hevner et al. built on earlier work, such as Nunamaker et al. (1990) and March and Smith (1995), the latter of which is the other highly cited article in this category). Hevner et al. (2004) articulate a conceptual framework that positions IS research in the context of both the environment within which the accumulated body of knowledge is applied and the knowledge base that the research develops. They also specify a set of prescriptive guidelines for conducting design research.

Hevner provides an in-depth introduction to design science research and discusses its achievements and future in Chapter 22 of this volume.

1.2.1.14 IT Implementation

The final category with more than two articles is labeled *IT implementation*. The three articles in this category, Markus (1983), Kwon and Zmud (1987), and Orlikowski (1993b), all explore questions related to IS implementation from a broad perspective that acknowledges the richness and complexity of the factors which affect the successfulness of IT implementation in organizations. In one of the early classics of the IS literature, Markus (1983) compares three theories (people-determined, system-determined, and the political variant of interaction theory) regarding the causes of resistance against the implementation of a new IS and finds support only for the political variant of interaction theory. Kwon and Zmud (1987) present a stage model for IT implementation activities, and Orlikowski (1993) develops a theoretical framework that helps us understand a variety of "organizational issues" in the context of the adoption and use of CASE tools. The key lesson from all of these is that technical issues are seldom (if ever) the sole reason for difficulties or failures of IS implementation processes.

1.2.1.15 Other Categories

The remaining 11 categories of widely cited articles each have only one or two articles (in most cases one) and have a citation index value less than 1000. To keep the length of the material reasonable, these categories will not be discussed at a more detailed level.

1.2.2 Mapping of Categories of Interest to Prior Models

This section will briefly analyze the relationship between the categories identified in this chapter and discussed earlier (1–14) and those specified in two recent citation analysis–based articles that evaluate the status and development of the field of IS: Sidorova et al. (2008) and Taylor et al. (2010). Using different approaches, these articles provide a useful and interesting analysis of the identity of the field based on an analysis of published literature. Both analyses track the change of the field over time, demonstrating how the key areas of interest have changed.

As Table 1.2 shows, the mapping between Sidorova et al. (2008) and the categories identified in Section 1.2.1 is straightforward, particularly after two of our categories (*IS Success* and *Adoption and Diffusion of IT Innovations*) were divided into two new ones based on the level of analysis (individual vs. organizational). The most significant insight from this analysis is the dearth of articles on IS development among the most highly cited IS studies—the only category that fits at least partially with IS development is design science (in which both articles advocate for more research on design science and provide guidelines for it, instead of presenting empirical evidence). Based on our analysis, most highly cited articles in the field of IS do not cover topics that are related to analysis, design, or coding of system solutions. Overall, IT and organizations and IT and individuals dominate; IT and markets and IT and groups have two and one subcategories associated with them, respectively.

Mapping to Taylor et al. (2010) was significantly more difficult than completing the same process in the context of Sidorova et al. (2008), at least partially because of the broad and somewhat amorphous boundaries of Taylor's IS Thematic Miscellany and Qualitative Methods Thematic Miscellany categories. The results of the matching process are included in Table 1.3. Not surprisingly, IS Thematic Miscellany covers the largest number of subcategories, while the rest are distributed equally over the other categories.

TABLE 1.2   Mapping of Categories Identified in Our Citation Analysis with Sidorova et al. (2008)

Sidorova et al. (2008)	Current Chapter
IT and markets	• IT and business performance
	• Electronic marketplaces
IT and organizations	• Impact of IT on organizations
	• Knowledge management
	• IT implementation
	• IS success—Organizational
	• Adoption and diffusion of IT innovations—Organizational
IT and groups	• Group support systems
IT and individuals	• User acceptance and use of IT
	• User satisfaction
	• IS success—Individual
	• Adoption and diffusion of IT innovations—Individual
IS development	• Design science
[Across or outside Sidorova categories]	• Methodology
	• Trust
	• Media and organizational communication

TABLE 1.3   Mapping of Current Chapter Categorization with Taylor et al. (2010)

Taylor et al. (2010)	Current Chapter
Inter-business systems	• Electronic marketplaces
IS strategy	• IS success
	• IT and business performance
IS thematic miscellany	• IT implementation
	• Design science
	• User acceptance and use of IT
	• User satisfaction
	• Adoption and diffusion of IT innovations
Qualitative methods thematic miscellany	• Impact of IT on organizations
Group work and decision support	• Group support systems
	• Knowledge management
[Across Taylor categories]	• Methodology
	• Trust

1.2.3 Shift of Focus on Research Categories over Time

One interesting question worth exploring is the shifts in the topics on which the field has produced highly cited articles. Table 1.4 shows for each of the 5 year periods starting from 1980 and ending in 2009, eight (or fewer, if there total number of categories represented was smaller than eight) most highly cited categories and the total number of articles and the sum of the citation metric counts from that time period.

There are a few categories that have stayed highly influential throughout the timeframe that is being evaluated. The consistency with which *User acceptance and use of IT* has maintained its top position is remarkable: it held the top or the second position throughout the 1985–2004 time period, dropping to the second place only once. *Impact of IT on organizations* was also included in the top eight categories for each of the time periods in 1985–2004, but with much modest rankings (except 1990–1994 when it held the top position). Another topic area with a consistent high-visibility presence throughout the observation period is *IS success*, although its positions at the top have been more periodic, punctuated

TABLE 1.4   Shifts in Research Categories over Time

Time Period	Most Significant Categories during the Time Period	Total of Citation Index Values	Total Number of Articles
1980–1984	User satisfaction	3,548	4
	IT implementation		
	IS success		
1985–1989	User acceptance and use of IT	23,735	17
	Media		
	Methodology		
	Markets		
	GSS		
	User satisfaction		
	Impact of IT on organizations		
	IT implementation		
1990–1994	Impact of IT on organizations	21,607	24
	User acceptance and use of IT		
	Adoption		
	GSS		
	IS success		
	Coordination		
	IT and business performance		
	Media		
1995–1999	User acceptance and use of IT	23,518	30
	Methodology		
	Trust		
	Self-efficacy		
	IT and business performance		
	Knowledge management		
	Impact of IT on organizations		
	Enterprise systems		
2000–2004	User acceptance and use of IT	27,089	26
	Knowledge management		
	IT and business performance		
	Trust		
	Methodology		
	Design science		
	IS success		
	Impact of IT on organizations		
2005–2009	Artificial intelligence	3,451	4
	Knowledge management		
	Trust		

by highly influential articles in 1980–1984, 1990–1994, and 2000–2004. There has also been a consistent stream of influential articles on various aspects of *Methodology*.

Other categories have gained a high-impact status later and then maintained it. *IT and business performance* emerged in 1990–1994 and has stayed as a top category ever since. Both *Knowledge management* and *Trust* have stayed as highly influential areas of research since they entered the top topic rankings in 1995–2009.

Research on *User satisfaction* and *IT implementation* peaked in 1980s. *Media and organizational communication* and *Group support systems* were in the center of the discipline between 1985 and 1995;

research in these areas has not regained the same type of prominence later, although it will be interesting to see what impact a recent article by Dennis et al. (2008) will have, given that it builds heavily on the earlier media research tradition. In addition, several topics earned a visible position once based on one or two highly influential articles. These include *Electronic marketplaces* in 1985–1989 (Malone et al., 1987), *Adoption* in 1990–1994 (Cooper and Zmud, 1990; Moore and Benbasat, 1991), *Coordination* in 1990–1994 (Malone and Crowston, 1994), *Self-efficacy* in 1995–1999 (Compeau and Higgins, 1995; Compeau et al., 1999), *Design science* in 2000–2004 (Hevner et al., 2004), and *Artificial intelligence* in 2005–2009 (Adomavicius and Tuzhilin, 2005). These are obviously all important research areas, but still much less consistently prominent than those that have maintained top positions throughout the time period.

1.2.4 Summary

The analysis of the most highly cited articles in the field of IS demonstrates unequivocally the dominance of the research on *User acceptance and use of IT*. About 30% of the total citation index values to these top articles referred to this category, within which research based on the TAM and its derivatives (such as Unified Theory of Acceptance and Use of Technology or UTAUT; Venkatesh et al., 2003) dominates. As Venkatesh et al. describe both in this volume (Chapter 38) and in Venkatesh et al. (2007), TAM has been replicated in a variety of contexts to show its generalizability, its predictive validity has been demonstrated through a number of studies, it has been shown to perform better than competing models, and it has been expanded to study the antecedents of its key constructs. Despite (or maybe because of) the dominant position of TAM, the question about its value and future is controversial in the field of IS, as demonstrated, for example, by the *Journal of Association of Information Systems* special issue in 2007 entitled "Quo Vadis TAM—Issues and Reflections on Technology Acceptance Research" (Hirschheim, 2007).

While technology acceptance research focuses on individuals and their decisions regarding technology, the second highly influential research category, *The Impact of IT on Organizations*, specifically emphasizes issues at the organizational level. In this research, the key questions are related to the impact of IT on how organizations can and do change in the context of technology introductions and interventions. Particularly interesting are the unintended consequences that emerge when users find ways to appropriate the technology in ways that aligns with their goals, sometimes in a way that is very different from the designer's or management's intent.

Overall, the analysis in this chapter illustrates both the broad diversity of the research in the field and the dominance of some of the approaches. These results suggest the IS discipline strives to understand questions as diverse as the following (and naturally many more):

- Why do users choose to accept some technologies readily and others not at all?
- What are the mechanisms through which IT affects organizations and can be used to transform them?
- How do people make decisions regarding communication options using computer-based technologies?
- What structures, incentives, and technologies should organizations have in place to ensure success in knowledge management?
- How do we define and measure the success of IS in organizations?
- Do IT-based solutions really have a measurable impact on business performance?
- What impact do various technologies have on the formation and maintenance of trust in virtual environments?
- What are the special characteristics of electronic marketplaces and how do they operate?
- What are the factors that determine whether or not an individual or an organization adopts an IT solution and how an IT innovation spreads among individuals and organizations?

It is interesting to note that our analysis of the most influential papers in IS research included very few that are related to areas that many faculty members in the field consider to be its core and that form the key elements of the IS curriculum recommendations (see, e.g., Topi et al., 2010): data, database, and information management; systems analysis and design (the analysis, design, and implementation of systems that utilize computing technology); or IT infrastructure (design and implementation of computing artifact that provide either processing or communication capabilities).

1.3 Key Conversations within the Field

During its history, the IS discipline has engaged in a number of highly intensive conversations reflecting on its own identity and position in academia in which many of the leading authors of the field have participated. All these discussions consist of a number of connected journal articles and/or book chapters. There is no formal way of identifying these conversations in an exhaustive way and thus, while this chapter covers a few well-known representative examples of these dialogues, it makes no claims regarding completeness. The purpose is to use this as another mechanism to illustrate topics that have had in a central role within the IS community.

1.3.1 Rigor and Relevance

One of the most persistent topics discussed by the IS community is the question about the relevance of IS research and the relationship between relevance and scientific rigor. This conversation has been ongoing throughout the history of the field, punctuated by a number of highly intensive periods. One such period was the publication of a number of Issues and Opinions articles on this topic in the March 1999 issue of *MIS Quarterly*. Benbasat and Zmud (1999) made, without hesitation, a claim that most IS research "lacks relevance to practice," identified this as a significant weakness of the field, and suggested approaches for changing the situation. In an interesting response to Benbasat and Zmud (1999), Davenport and Markus (1999) agree that the problem exists but they believe that an even more profound change to the field is needed—instead of moving further in our desire to emulate other management disciplines, IS should use medicine and law as our role models. In another response, Lyytinen (1999) calls for large and heterogeneous research teams that are able to attack large-scale problems from multiple perspectives. He also suggests that this debate looks quite different depending on the geographic context (e.g., the United States vs. Europe).

The next major event in this ongoing discussion was a large number of opinion pieces published in CAIS in 2001 as a "special volume on relevance," following a particularly vigorous discussion on IS world (an electronic distribution list for the IS community). Given that the total number of articles in this particular conversation episode is close to 30, it is not possible to address all the issues covered at a detailed level or give justice to individual contributions. In his introduction to the special volume, Gray (2001) summarizes the outcomes of the conversation by identifying as common themes that IS research should be relevant and that practitioners do not read academic literature. The authors do, however, provide a wide variety of explanations regarding why this might be the case and what the field should do about the situation.

Over the past several years, this topic has been frequently discussed in articles published in top IS journals. For example, Rosemann and Vessey (2008) lamented the lack of progress in achieving a higher level of relevance despite the active conversations and provided specific guidance regarding the ways in which the IS community could improve the relevance of its research. They suggest a multilevel approach, which would address the problem at the institutional, project governance, and research process levels. They advocate for the use of the "applicability check" method as an integral part of the research process as a way to ensure that a research project achieves its relevance objectives. In another *MIS Quarterly* article, Klein and Rowe (2008) proposed modifications to the doctoral programs that would make them more attractive to professionally qualified students and potentially help them publish in forms and outlets that are interesting to practitioners. Also in *MISQ*, Gill and Bhattacherjee (2009) found that the ability of the IS discipline to inform its constituents has declined, not improved, despite the high level of attention given to the question. They present five very specific proposals regarding actions the field should take in order to improve its relevance, emphasizing strongly the importance of these actions to ensure the ability of the field to survive.

Every episode of the rigor vs. relevance discussion tends to end with a prediction that the conversation will again lift its head in a few years. This is, indeed, likely to continue to be the case: despite the widely recognized ongoing challenge with relevance and the prescriptive guidance given, little seems to have changed. At the same time, demonstrating the relevance of the field for various stakeholder groups, such as practitioners, students, and public funding sources, has become increasingly important. This is clearly an area where much more work and specific action are needed, particularly if the field desires to be recognized among the practitioners as a source of valuable new practices, fresh ideas, and competent graduates.

1.3.2 Legitimacy and Core of the Discipline

Another very intensive debate within the field of IS has been related to the feeling of anxiety about the legitimacy of the discipline and the associated search for the disciplinary core. Lyytinen and King (2004) provide an excellent concise description of the history of what they call "anxiety discourse," starting from Dearden's (1972) *MIS is a Mirage* and including strongly worded concerns by acknowledged leaders such as Culnan and Huff (1986), Benbasat and Weber (1996), and Straub (1999). Since its beginning, the field has been unsure about itself, to the extent that—paraphrasing Weber—it cannot even agree whether or not it suffers from an identity crisis (Weber, 2003). Lyytinen and King (2004) identify the lack of a strong theoretical core as one possible reason underlying the anxiety (although they do not believe that to be the case). Not surprisingly, there have been many related debates regarding whether or not the field should have a one strong focused core and if it should, what it should be.

The key articles of this conversation and authors' commentaries on their own work a few years later are captured in an important book edited by King and Lyytinen (2006). A comprehensive review of the individual articles or the issues covered in this dialogue is not possible within the scope of this chapter; instead, we will provide a short summary of the key questions and strongly encourage the interested reader to read the King and Lyytinen (2006) book.

One of the key questions debated in this context is whether or not the IS field should have a strong shared identity in the first place. Some very senior scholars within the field have over the years expressed a strong need to identify and articulate the identity of the field so that it can be more unified and defend its legitimacy more easily (e.g., Benbasat and Weber, 1996; Benbasat and Zmud, 2003; Weber, 2006). For example, Benbasat and Weber (1996) very clearly state that "[w]e run the risk, therefore, that diversity will be the miasma that spells the demise of the discipline" (p. 397). Particularly Benbasat and Zmud (2003) led to a number of contributions from other leading academics in the field to defend an opposite perspective that highly values diversity without articulated boundaries (DeSanctis, 2003; Galliers, 2003; Robey, 2003; Ives et al., 2004; see also Robey, 1996; Lyytinen and King, 2006). No true compromise or integrated position has emerged, although seeds of that can be seen already in Benbasat and Weber (1996), who recognize the need for both focusing on the core and achieving diversity around it. According to Taylor et al. (2010), the field has done this by maintaining a "polycentric core" while allowing for a diversity of methods, topics, and research contexts. Whinston and Geng (2004) suggest that allowing "strategic ambiguity" is important to allow continuous innovation within IS research, while still maintaining the centrality of the IT artifact.

Another frequently revisited topic addressed by the most senior members of the IS community is whether or not the field is in crisis. Markus (1999) made an important contribution to this discussion in a book chapter with a provocative title "Thinking the Unthinkable—What Happens if the IS Field as we

Know it Goes Away." Her insightful views regarding the changes needed to the field are still important. Strong concerns regarding enrollments throughout the early 2000s and funding challenges caused by the great recession of 2007–2009 have strengthened the calls for quick action. For example, Hirschheim and Klein (2003) use the question related to the existence of the crisis as a framework to present their own suggestions regarding the state of the field and the direction it should take (interestingly, they end up not taking a stand regarding the existence of the crisis). For Benbasat and Zmud (2003), the crisis is specifically an identity crisis, with potentially very serious effects on the field. Galliers (2003) adopts a very different position, explicitly "deny[ing] that we are at a crossroads in the field" (p. 338). No consensus has emerged regarding this issue, either.

Yet another widely debated topic is the role of the "IT artifact" within the IS discipline. Orlikowski and Iacono (2001) call the IS discipline to engage more deeply in the process of understanding the IT artifact, which they defined as "those bundles of material and cultural properties packaged in some socially recognizable form such as hardware and/or software" (p. 121). After articulating five different views of technology (tool, proxy, ensemble, computational, and nominal), they demonstrate how the great majority of IS articles published in *Information Systems Research* in 1990s are either ignoring the IT artifact entirely or view it from perspectives that hide the significant interdependencies between system characteristics and the social and organizational context in which the systems are used. Orlikowski and Iacono (2001, p. 133) present a beautiful articulation of the special contribution the IS discipline can make at the intersection of technology and the contexts in which it is used, if it takes seriously its role in understanding the IT artifact in a deeper way:

> However, none of these groups attempts to understand the complex and fragmented emergence of IT artifacts, how their computational capabilities and cultural meanings become woven in dense and fragile ways via a variety of different and dynamic practices, how they are shaped by (and shape) social relations, political interests, and local and global contexts, and how ongoing developments in, uses of, and improvisations with them generate significant material, symbolic, institutional, and historical consequences. Yet, this is precisely where the IS field—drawing as it does on multiple disciplines and different types of analyses—is uniquely qualified to offer essential insights and perspectives.

Building on Orlikowski and Iacono (2001), Weber (2003) and Benbasat and Zmud (2003) make a strong call for the centrality of the IT artifact and "its immediate nomological net" (Benbasat and Zmud, 2003, p. 186) within the IS discipline. For Benbasat and Zmud, the IT artifact is "the application of IT to enable or support some task(s) embedded within a structure(s) that itself is embedded within a context(s)" (p. 186) Others, such as Galliers (2003) and DeSanctis (2003), strongly advocate for the need to let the field move without constraints created by the need to link the research to the IT artifact, however it is defined. Recent reviews of the discipline's areas of focus (Sidorova et al., 2008; Taylor et al., 2010) both suggest that in practice, the field continues to embraces a broad perspective without constraints and at best a polycentric core with extensive surrounding diversity.

1.3.3 Reference Disciplines

The final broad discussion topic that the field of IS engages in at regular intervals is that of the relationship between IS and other disciplines, particularly from the perspective of identifying which discipline is serving as a reference discipline to the other. From the early days of the IS discipline, the field has been characterized as one that has its roots in a number of reference disciplines, such as computer science, economics, and psychology (Keen, 1980), management and organization science in addition to computer science (Culnan and Swanson, 1986) or, increasingly, marketing in addition to economics and less computer science (Grover et al., 2006). In the early days of the field, many faculty members had backgrounds in these fields, which (and many others, such as sociology, social psychology, philosophy, and organizational behavior) also were the source of key theories, conceptual frameworks, and method-ological innovations for IS. This continues to be the case (Grover et al., 2006).

The key question more recently has been whether or not IS has reached a status of being a reference discipline itself, contributing to other fields. This is another area where an intense debate appears to continue without a conclusion: some authors (Baskerville and Myers, 2002; Grover et al., 2006) firmly believe that there is strong evidence that IS has emerged as a real reference discipline for related fields whereas others disagree or at least skeptical (e.g., Wade et al., 2006; Gill and Bhattacherjee, 2009). In a recent editorial, Straub (2012) points out that at least one requirement of IS serving as a reference disci-pline is fulfilled: over its relatively brief history, IS has been able to develop native theories, such as TAM and UTAUT (Section 1.2.1.1 of this chapter), task-technology fit theory (Section 1.2.1.1), adaptive struc-turation theory (Section 1.2.1.11), and the information success model (Section 1.2.1.6). Regardless of the extent to which IS serves as a reference discipline to other areas of study, the discipline has advanced very significantly from the times when it was solely depending on other fields in terms of its theories and methodologies.

1.4 Summary and Conclusions

This chapter has taken us through the most influential categories of IS research during the past 30 years and a number of ongoing or frequently reemerging scholarly conversations within the field. This last section of the chapter will discuss some of the implications of the findings described earlier.

In Section 1.2.4, the following questions were used to illustrate in practical terms the diverse nature of the most widely cited IS research:

- Why do users choose to accept some technologies readily and others not at all?
- What are the mechanisms through which IT affects organizations and can be used to transform them?
- How do people make decisions regarding communication options using computer-based technologies?
- What structures, incentives, and technologies should organizations have in place to ensure suc-cess in knowledge management?
- How do we define and measure the success of IS in organizations?
- Do IT-based solutions really have a measurable impact on business performance?
- What impact do various technologies have on the formation and maintenance of trust in virtual environments?
- What are the special characteristics of electronic marketplaces and how do they operate?
- What are the factors that determine whether or not an individual or an organization adopts an IT solution?

Level of analysis varies from individual to societal, and research methods are equally varied. Philosophically IS researchers view these questions through multiple lenses. The field does not have a shared view regarding whether or not this diversity is good (as indicated in Section 1.3.2). Even if the field came to the conclusion that this diversity should somehow be harnessed, finding practical mecha-nisms for doing so is difficult (except through very specific editorial policies in our top journals). Still, it is important to keep emphasizing the importance of building cumulative results (Keen, 1980) that will continue to build toward the discipline's own Body of Knowledge.

As discussed earlier, the discipline continues to suffer from an uncertainty regarding its own aca-demic legitimacy and whether or not the field is in crisis. Self-reflective evaluations are important for the development of the field; at the same time, it is essential that an unnecessarily large share of the discipline's top resources is not used on introspective processes. To thrive, the field needs to develop a shared understanding and, indeed, an identity that gives its members a sense of belonging and a basis on which to build the inter- and transdisciplinary collaboration that will be increasingly important.

One surprising result that emerged from the analysis presented in this chapter is the very insignificant role that articles focusing on core IS education topics have among the most influential IS research articles. One can legitimately wonder why research on systems analysis and design, data management, IT infrastructure, enterprise architecture or even IT strategy, management, governance, and acquisition has no visibility among the most highly cited works (together with IS project management, these form the core topics in the latest IS model curriculum, IS 2010; Topi et al., 2010). Maybe even more surprisingly, where are information-related topics? Given the recent emergence of iSchools (sometimes under the informatics title, sometimes information science, or simply information) as serious research enterprises, is the strongest expertise related to information and related topics any more within the field of IS—or has it ever been there? How long can the field make strong claims regarding its true contributions to the practice of IT if its most influential research is very far from these topics?

At the same time, this volume of the Computing Handbook includes a large number of excellent examples of outstanding scholarly work in areas that have been traditionally considered the core in IS (at least in IS education), demonstrating that this work exists and is thriving. It has just not been cited (i.e., used by others) at the same level of intensity as the research that emerged on the top in this analysis. Citations per se are not of importance, but they have an increasingly significant role as a core metric within the research enterprise. In addition, they tell at least a partial story about which research gets noticed and which does not. Also, one has to wonder what these results suggest about the possibilities for integration of research and teaching in the IS field, given that the most highly visible research topics are not aligned with what we teach.

If our focus in education will increasingly move toward general business topics such as, for example, creativity, business process design, service design, or organizational innovation, what differentiates us from other fields within business that have these same interests (such as operations management or marketing)? It appears that Orlikowski and Iacono's (2001) call to move toward a deeper understanding of and stronger theorizing about the IT artifact is as relevant as it was more than a decade ago. Shouldn't IS find its legitimacy and focus in the areas where a true understanding of information and IT is integrated with a true understanding of a domain that wants to benefit from effective and efficient use of information and IT resources? This essential space is there for us to keep and nurture—this is the area that we are uniquely qualified to occupy.

1.A Appendix

	Article	Total Citation Index	Category
1.	Davis (1989) (MISQ)	3237	User acceptance and use
2.	Davis et al. (1989) (ManSci)	2004	User acceptance and use
3.	Daft and Lengel (1986) (ManSci)	1510	Media
4.	Venkatesh et al. (2003) (MISQ)	1404	User acceptance and use
5.	DeLone and McLean (1992) (ISR)	1286	IS success
6.	Venkatesh and Davis (2000) (ManSci)	1183	User acceptance and use
7.	Alavi and Leidner (2001) (MISQ)	1067	Knowledge management
8.	Moore and Benbasat (1991) (ISR)	941	Adoption
9.	Taylor and Todd (1995) (ISR)	905	User acceptance and use
10.	Malone et al. (1987) (CACM)	774	Markets
11.	Hevner et al. (2004) (MISQ)	729	Design science
12.	Delone and McLean (2003) (JMIS)	706	IS success
13.	Orlikowski (1992) (OrgSci)	701	Impact of IT on organizations
14.	Adomavicius and Tuzhilin (2005) (KDE)	701	AI

(continued)

(continued)

	Article	Total Citation Index	Category
15.	Malone and Crowston (1994) (CSUR)	647	Coordination
16.	Compeau and Higgins (1995) (MISQ)	636	Self-efficacy
17.	DeSanctis and Poole (1994) (OrgSci)	602	GSS
18.	Benbasat et al. (1987) (MISQ)	600	Methods
19.	Gefen et al. (2003) (MISQ)	599	User acceptance and use
20.	Davenport (1998) (HBR)	595	Enterprise systems
21.	Mathieson (1991) (ISR)	593	User acceptance and use
22.	Klein and Myers (1999) (MISQ)	588	Methods
23.	Sproull and Kiesler (1986) (ManSci)	547	Media
24.	DeSanctis and Gallupe (1987) (ManSci)	541	GSS
25.	Davenport et al. (1998) (SMR)	531	Knowledge management
26.	Orlikowski (2000) (OrgSci)	530	Impact of IT on organizations
27.	Adams et al. (1992) (MISQ)	515	User acceptance and use
28.	Jarvenpaa and Leidner (1998) (JCMC)	512	Trust
29.	Bharadwaj (2000) (MISQ)	508	IT and business performance
30.	Orlikowski and Baroudi (1991) (ISR)	506	Impact of IT on organizations
31.	Venkatesh (2000) (ISR)	505	User acceptance and use
32.	Bailey and Pearson (1983) (ManSci)	504	User satisfaction
33.	Goodhue and Thompson (1995) (MISQ)	494	User acceptance and use
34.	Markus (1983) (CACM)	475	IT implementation
35.	Cooper and Zmud (1990) (ManSci)	474	Adoption
36.	Brynjolfsson and Smith (2000) (ManSci)	470	Markets
37.	Davis et al. (1992) (JASP)	467	User acceptance and use
38.	Karahanna et al. (1999) (MISQ)	466	Adoption
39.	Rockart (1979) (HBR)	458	Managerial IT use
40.	Ives et al. (1983) (CACM)	445	User satisfaction
41.	Gefen et al. (2000) (CAIS)	443	Methods
42.	Bakos (1997) (ManSci)	432	Markets
43.	Venkatesh and Morris (2000) (MISQ)	430	User acceptance and use
44.	Henderson and Venkatraman (1993) (IBMSJ)	430	Impact of IT on organizations
45.	Doll and Torkzadeh (1988) (MISQ)	426	User satisfaction
46.	Nunamaker et al. (1991) (CACM)	425	GSS
47.	Chin et al. (2003) (ISR)	418	Methods
48.	Brynjolfsson and Hitt (2000) (JEP)	412	IT and business performance
49.	McKnight et al. (2002) (ISR)	411	Trust
50.	Davis (1993) (IJMMS)	410	User acceptance and use
51.	Brynjolfsson and Hitt (1996) (ManSci)	407	IT and business performance
52.	Orlikowski (2002) (OrgSci)	406	Knowledge management
53.	Daft et al. (1987) (MISQ)	404	Media
54.	Venkatesh and Davis (1996) (Decis. Sci.)	403	User acceptance and use
55.	Markus and Robey (1988) (ManSci)	400	Impact of IT on organizations
56.	Davis (1985) (Diss)	399	User acceptance and use
57.	Walsham (1995) (EJIS)	397	Methods
58.	Agarwal and Karahanna (2000) (MISQ)	392	User acceptance and use
59.	Mata et al. (1995) (MISQ)	387	IT and business performance
60.	Iacovou et al. (1995) (MISQ)	383	Impact of IT on organizations

(continued)

	Article	Total Citation Index	Category
61.	Straub (1989) (MISQ)	375	Methods
62.	Bhattacherjee (2001) (MISQ)	365	IS continuance
63.	Wasko and Faraj (2005) (MISQ)	364	Knowledge management
64.	Chin (1998) (MISQ)	360	Methods
65.	Orlikowski and Iacono (2001) (ISR)	356	Discipline identity
66.	Josang et al. (2007) (DSS)	352	Trust
67.	Gold et al. (2001) (JMIS)	350	Knowledge management
68.	Ives and Olson (1984) (ManSci)	350	IS success
69.	Brynjolfsson (1993) (CACM)	349	IT and business performance
70.	Jarvenpaa et al. (1999) (JCMC)	336	Trust
71.	Taylor and Todd (1995) (MISQ)	333	User acceptance and use
72.	Koufaris (2002) (ManSci)	329	User acceptance and use
73.	Gefen and Straub (1997) (MISQ)	325	User acceptance and use
74.	Kwon and Zmud (1987) (CIIS)	323	IT implementation
75.	Orlikowski and Robey (1991) (ISR)	321	Impact of IT on organizations
76.	Ba and Pavlou (2002) (MISQ)	316	Trust
77.	Wang and Strong (1996) (JMIS)	316	Data quality
78.	Gefen (2000) (Omega)	315	Trust
79.	Thompson et al. (1991) (MISQ)	313	User acceptance and use
80.	Compeau et al. (1999) (MISQ)	310	Self-efficacy
81.	Orlikowski (1996) (ManSci)	310	Impact of IT on organizations
82.	Jarvenpaa et al. (1998) (JMIS)	310	Trust
83.	Bock et al. (2005) (MISQ)	308	Knowledge management
84.	Dellarocas (2003) (ManSci)	304	Media
85.	Ackoff (1967) (ManSci)	303	Managerial IT use
86.	Melville et al. (2004) (MISQ)	301	IT and business performance
87.	March and Smith (1995) (DSS)	298	Design science
88.	Pavlou (2003) (IJEC)	296	User acceptance and use
89.	Orlikowski (1993b) (MISQ)	293	Impact of IT on organizations
90.	Agarwal and Prasad (1999) (Decis. Sci.)	292	User acceptance and use
91.	Yates and Orlikowski (1992) (AMR)	291	Media
92.	Szajna (1996) (ManSci)	289	User acceptance and use
93.	Seddon (1997) (ManSci)	286	IS success
94.	Bakos (1991) (MISQ)	286	Markets
95.	Grudin (1994) (CACM)	275	GSS
96.	Pitt et al. (1995) (MISQ)	256	IS effectiveness
97.	Hitt and Brynjolfsson (1996) (MISQ)	255	IT and business performance
98.	Orlikowski and Gash (1994) (TOIS)	253	Impact of IT on organizations
99.	Alavi and Leidner (1999) (CAIS)	239	Knowledge management
100.	Markus (1987) (CR)	236	Media
101.	Venkatraman (1994) (SMR)	233	Impact of IT on organizations
102.	Myers (1997) (MISQ)	226	Methods
103.	Lacity and Hirschheim (1993) (SMR)	224	Outsourcing
104.	Orlikowski (1993) (TIS)	221	IT implementation
105.	Jarvenpaa and Todd (1996) (IJEC)	173	Electronic commerce

Acknowledgments

I gratefully acknowledge the thoughtful feedback I received from Wendy Lucas, Allen Tucker, and Ryan Wright during the preparation of this chapter. The work on this chapter was made possible by a sabbatical leave and a summer research grant from Bentley University, both of which I also acknowledge with gratitude. Finally, I want to thank Aalto University School of Business for the opportunity to work on this chapter while I was at Aalto as Visiting Scholar and Radford University for a few important days of peace and quiet in the middle of an international move.

References

Ackoff, R.L. 1967. Management misinformation systems. *Management Science* 14(4) (December 1): B147–B156.

Adams, D.A., R.R. Nelson, and P.A. Todd. 1992. Perceived usefulness, ease of use, and usage of information technology: A Replication. *MIS Quarterly* 16(2) (June 1): 227–247.

Adomavicius, G. and A. Tuzhilin. 2005. Toward the next generation of recommender systems: A survey of the state-of-the-art and possible extensions. *IEEE Transactions on Knowledge and Data Engineering* 17(6) (June): 734–749.

Agarwal, R. and E. Karahanna. 2000. Time flies when you're having fun: Cognitive absorption and beliefs about information technology usage. *MIS Quarterly* 24(4) (December): 665–694.

Agarwal, R. and J. Prasad. 1999. Are individual differences germane to the acceptance of new information technologies? *Decision Sciences* 30(2): 361–391.

Alavi, M. and D. Leidner. 1999. Knowledge management systems: Issues, challenges, and benefits. *Communications of the Association for Information Systems* 1(1) (February 22). http://aisel.aisnet.org/cais/vol1/iss1/7

Alavi, M. and D.E. Leidner. 2001. Review: Knowledge management and knowledge management systems: Conceptual foundations and research issues. *MIS Quarterly* 25(1) (March): 107–136.

Ba, S. and P. Pavlou. 2002. Evidence of the effect of trust building technology in electronic markets: Price premiums and buyer behavior. *SSRN eLibrary*. http://papers.ssrn.com/sol3/papers.cfm?abstract_id = 951734

Bailey, J.E. and S.W. Pearson. 1983. Development of a tool for measuring and analyzing computer user satisfaction. *Management Science* 29(5): 530–545.

Bakos, J.Y. 1991. A strategic analysis of electronic marketplaces. *MIS Quarterly* 15(3): 295–310.

Bakos, J.Y. 1997. Reducing buyer search costs: Implications for electronic marketplaces. *Management Science* 43(12): 1676–1692.

Baskerville, R.L. and M.D. Myers. 2002. Information systems as a reference discipline. *MIS Quarterly* 26(1) (March 1): 1–14.

Benbasat, I. and H. Barki. 2007. Quo Vadis TAM? *Journal of the Association for Information Systems* 8(4) (April 1). http://aisel.aisnet.org/jais/vol8/iss4/16

Benbasat, I., D.K. Goldstein, and M. Mead. 1987. The case research strategy in studies of information systems. *MIS Quarterly* 11(3): 369–386.

Benbasat, I. and R. Weber. 1996. Research commentary: Rethinking 'Diversity' in information systems research. *Information Systems Research* 7(4) (December): 389.

Benbasat, I. and R.W. Zmud. 1999. Empirical research in information systems: The practice of relevance. *MIS Quarterly* 23(1) (March 1): 3–16.

Benbasat, I. and R.W. Zmud. 2003. The identity crisis within the IS discipline: Defining and communicating the discipline's core properties. *MIS Quarterly* 27(2) (June 1): 183–194.

Bharadwaj, A.S. 2000. A resource-based perspective on information technology capability and firm performance: An empirical investigation. *MIS Quarterly* 24(1): 169–196.

Bhattacherjee, A. 2001. Understanding information systems continuance: An expectation-confirmation model. *MIS Quarterly* 25(3) (September 1): 351–370.

Bock, G.-W., R.W. Zmud, Y.-G. Kim, and J.-N. Lee. 2005. Behavioral intention formation in knowledge sharing: Examining the roles of extrinsic motivators, social-psychological forces, and organizational climate. *MIS Quarterly* 29(1): 87–111.

Brynjolfsson, E. 1993. The productivity paradox of information technology. *Communications of the ACM* 36(12) (December): 66–77.

Brynjolfsson, E. and L. Hitt. 1996. Paradox lost? Firm-level evidence on the returns to information systems spending. *Management Science* 42(4): 541–558.

Brynjolfsson, E. and L.M. Hitt. 2000. Beyond computation: Information technology, organizational transformation and business performance. *The Journal of Economic Perspectives* 14(4): 23–48.

Brynjolfsson, E. and M.D. Smith. 2000. Frictionless commerce? A comparison of internet and conventional retailers. *Management Science* 46(4): 563–585.

Chin, W.W. 1998. Commentary: Issues and opinion on structural equation modeling. *MIS Quarterly* 22(1): vii–xvi.

Chin, W.W., B.L. Marcolin, and P.R. Newsted. 2003. A partial least squares latent variable modeling approach for measuring interaction effects: Results from a Monte Carlo simulation study and an electronic-mail emotion/adoption study. *Information Systems Research* 14(2) (June): 189–217.

Compeau, D.R. and C.A. Higgins. 1995. Computer self-efficacy: Development of a measure and initial test. *MIS Quarterly* 19(2): 189–211.

Compeau, D., C.A. Higgins, and S. Huff. 1999. Social cognitive theory and individual reactions to computing technology: A longitudinal study. *MIS Quarterly* 23(2): 145–158.

Cooper, R.B. and R.W. Zmud. 1990. Information technology implementation research: A technological diffusion approach. *Management Science* 36(2): 123–139.

Culnan, M.J. and S. Huff. 1996. Back to the future: Will there be an ICIS in 1996? In *Proceedings of the 7th International Conference on Information Systems*, p. 352. San Diego, CA.

Culnan, M.J. and E.B. Swanson. 1986. Research in management information systems, 1980–1984: Points of work and reference. *MIS Quarterly* 10(3) (September 1): 289–302.

Daft, R.L. and R.H. Lengel. 1986. Organizational information requirements, media richness and structural design. *Management Science* 32(5): 554–571.

Daft, R.L., R.H. Lengel, and L.K. Trevino. 1987. Message equivocality, media selection, and manager performance: Implications for information systems. *MIS Quarterly* 11(3): 355–366.

Davenport, T.H. 1998. Putting the enterprise into the enterprise system. *Harvard Business Review* 76(4) (August): 121–131.

Davenport, T.H., D.W. De Long, and M.C. Beers. 1998. Successful knowledge management projects. *Sloan Management Review* 39(2): 43–57.

Davenport, T.H. and M.L. Markus. 1999. Rigor vs. relevance revisited: Response to Benbasat and Zmud. *MIS Quarterly* 23(1) (March 1): 19–23. doi:10.2307/249405.

Davis, F.D. 1985. A technology acceptance model for empirically testing new end-user information systems: Theory and results. John C. Henderson, Sloan School of Management, Cambridge, MA, http://hdl.handle.net/1721.1/15192

Davis, F.D. 1989. Perceived usefulness, perceived ease of use, and user acceptance of information technology. *MIS Quarterly* 13(3): 319–340.

Davis, F.D. 1993. User acceptance of information technology: System characteristics, user perceptions and behavioral impacts. *International Journal of Man-Machine Studies* 38(3) (March): 475–487.

Davis, F.D., R.P. Bagozzi, and P.R. Warshaw. 1989. User acceptance of computer technology: A comparison of two theoretical models. *Management Science* 35(8): 982–1003.

Davis, F.D., R.P. Bagozzi, and P.R. Warshaw. 1992. Extrinsic and intrinsic motivation to use computers in the workplace. *Journal of Applied Social Psychology* 22(14): 1111–1132.

Dearden, J. 1972. MIS is a mirage. *Harvard Business Review* 50(1): 90–99.

Dellarocas, C. 2003. The digitization of word of mouth: Promise and challenges of online feedback mechanisms. *Management Science* 49(10) (October 1): 1407–1424.

DeLone, W.H. and E.R. McLean. 1992. Information systems success: The quest for the dependent variable. *Information Systems Research* 3(1) (March 1): 60–95.

DeLone, W.H. and E.R. McLean. 2003. The DeLone and McLean model of information systems success: A ten-year update. *Journal of Management Information System* 19(4) (April): 9–30.

Dennis, A., R. Fuller, and J.S. Valacich. 2008. Media, tasks, and communication processes: A theory of media synchronicity. *Management Information Systems Quarterly* 32(3) (October 27): 575–600.

DeSanctis, G. 2003. The social life of information systems research: A response to Benbasat and Zmud's call for returning to the IT artifact. *Journal of the Association for Information Systems* 4(1) (December 12). http://aisel.aisnet.org/jais/vol4/iss1/16

DeSanctis, G. and R.B. Gallupe. 1987. A foundation for the study of group decision support systems. *Management Science* 33(5) (May 1): 589–609.

DeSanctis, G. and M.S. Poole. 1994. Capturing the complexity in advanced technology use: Adaptive structuration theory. *Organization Science* 5(2) (May 1): 121–147.

Doll, W.J. and G. Torkzadeh. 1988. The measurement of end-user computing satisfaction. *MIS Quarterly* 12(2) (June 1): 259–274. doi:10.2307/248851.

Galliers, R.D. 2003. Change as crisis or growth? Toward a trans-disciplinary view of information systems as a field of study: A response to Benbasat and Zmud's call for returning to the IT artifact. *Journal of the Association for Information Systems* 4(1) (November 20). http://aisel.aisnet.org/jais/vol4/iss1/13

Gefen, D. 2000. E-commerce: The role of familiarity and trust. *Omega* 28(6): 725–737.

Gefen, D., E. Karahanna, and D.W. Straub. 2003. Trust and TAM in online shopping: An integrated model. *MIS Quarterly* 27(1) (March 1): 51–90.

Gefen, D. and D.W. Straub. 1997. Gender differences in the perception and use of e-mail: An extension to the technology acceptance model. *MIS Quarterly* 21(4) (December 1): 389–400. doi:10.2307/249720.

Gefen, D., D. Straub, and M.-C. Boudreau. 2000. Structural equation modeling and regression: Guidelines for research practice. *Communications of the Association for Information Systems* 4(1) (October 24). http://aisel.aisnet.org/cais/vol4/iss1/7

Gill, G. and A. Bhattacherjee. 2009. Whom are we informing? Issues and recommendations for MIS research from an informing science perspective. *Management Information Systems Quarterly* 33(2) (June 1): 217–235.

Gold, A.H., A. Malhotra, and A.H. Segars. 2001. Knowledge management: An organizational capabilities perspective. *Journal of Management Information Systems* 18(1): 185–214.

Goodhue, D.L. and R.L. Thompson. 1995. Task-technology fit and individual performance. *MIS Quarterly* 19(2) (June 1): 213–236.

Gray, P. 2001. Introduction to the special volume on relevance. *Communications of the Association for Information Systems* 6(1) (March 12). http://aisel.aisnet.org/cais/vol6/iss1/1

Grover, V., R. Gokhale, J. Lim, J. Coffey, and R. Ayyagari. 2006. A citation analysis of the evolution and state of information systems within a constellation of reference disciplines. *Journal of the Association for Information Systems* 7(5) (May 25). http://aisel.aisnet.org/jais/vol7/iss5/13

Grudin, J. 1994. Groupware and social dynamics: Eight challenges for developers. *Communication of the ACM* 37(1) (January): 92–105.

Harzing, A.W. 2007. Publish or Perish, available from http://www.harzing.com/pop.htm

Henderson, J.C. and N. Venkatraman. 1993. Strategic alignment: Leveraging information technology for transforming organizations. *IBM Systems Journal* 32(1) (January): 4–16.

Hevner, A.R., S.T. March, J. Park, and S. Ram. 2004. Design science in information systems research. *MIS Quarterly* 28(1) (March): 75–105.

Hirschheim, R. 2007. Introduction to the special issue on 'Quo Vadis TAM—Issues and reflections on technology acceptance research'. *Journal of the Association for Information Systems* 8(4) (April 1). http://aisel.aisnet.org/jais/vol8/iss4/18

Hirschheim, R.A. and H.K. Klein. 2003. Crisis in the IS field? A critical reflection on the state of the discipline. *Journal of the Association for Information Systems* 4(1) (October 1). http://aisel.aisnet.org/jais/vol4/iss1/10

Hirschheim, R. and H. Klein. 2012. A glorious and not-so-short history of the information systems field. *Journal of the Association for Information Systems* 13(4) (April 30). http://aisel.aisnet.org/jais/vol13/iss4/5

Hirschheim, R., C. Saunders, and D. Straub. 2012. Historical interpretations of the IS discipline: An introduction to the special issue. *Journal of the Association for Information Systems* 13(4) (April 30). http://aisel.aisnet.org/jais/vol13/iss4/4

Hitt, L.M. and E. Brynjolfsson. 1996. Productivity, business profitability, and consumer surplus: Three different measures of information technology value. *MIS Quarterly* 20(2): 121–142.

Hu, P.J.-H. and H. Chen. 2011. Analyzing information systems researchers' productivity and impacts: A perspective on the H index. *ACM Transactions on Management Information System* 2(2) (July): 7:1–7:8.

Iacovou, C.L., I. Benbasat, and A.S. Dexter. 1995. Electronic data interchange and small organizations: Adoption and impact of technology. *MIS Quarterly* 19(4) (December 1): 465–485.

Ives, B. and M.H. Olson. 1984. User involvement and MIS success: A review of research. *Management Science* 30(5): 586–603.

Ives, B., M.H. Olson, and J.J. Baroudi. 1983. The measurement of user information satisfaction. *Communications of the ACM* 26(10) (October): 785–793.

Ives, B., M.S. Parks, J. Porra, and L. Silva. 2004. Phylogeny and power in the IS domain: A response to Benbasat and Zmud's call for returning to the IT artifact. *Journal of the Association for Information Systems* 5(3) (March 1). http://aisel.aisnet.org/jais/vol5/iss3/4

Jarvenpaa, S.L., K. Knoll, and D.E. Leidner. 1998. Is anybody out there? Antecedents of trust in global virtual teams. *Journal of Management Information Systems* 14(4) (March): 29–64.

Jarvenpaa, S.L. and D.E. Leidner. 1998. Communication and trust in global virtual teams. *Journal of Computer-Mediated Communication* 3(4): 0–0.

Jarvenpaa, S.L. and P.A. Todd. 1996. Consumer reactions to electronic shopping on the world wide web. *International Journal of Electronic Commerce* 1(2): 59–88.

Jarvenpaa, S.L., N. Tractinsky, and L. Saarinen. 1999. Consumer trust in an internet store: A cross-cultural validation. *Journal of Computer-Mediated Communication* 5(2): 0–0.

Jøsang, A., R. Ismail, and C. Boyd. 2007. A survey of trust and reputation systems for online service provision. *Decision Support Systems* 43(2) (March): 618–644.

Karahanna, E., D.W. Straub, and N.L. Chervany. 1999. Information technology adoption across time: A cross-sectional comparison of pre-adoption and post-adoption beliefs. *MIS Quarterly* 23(2) (June 1): 183–213.

Keen, P.G.W. 1980. MIS research: Reference disciplines and a cumulative tradition. In *Proceedings of the First International Conference on Information Systems*, pp. 9–18. Philadelphia, PA.

King, J.L. and K. Lyytinen. 2006. *Information Systems: The State of the Field.* Chichester, England: Wiley.

Klein, H.K. and M.D. Myers. 1999. A set of principles for conducting and evaluating interpretive field studies in information systems. *MIS Quarterly* 23(1) (March 1): 67–93.

Klein, H. and F. Rowe. 2008. Marshaling the professional experience of doctoral students: A contribution to the practical relevance debate. *Management Information Systems Quarterly* 32(4) (November 5): 675–686.

Koufaris, M. 2002. Applying the technology acceptance model and flow theory to online consumer behavior. *Information Systems Research* 13(2) (June 1): 205–223.

Kwon, T.H. and R.W. Zmud. 1987. Unifying the fragmented models of information systems implementation. In *Critical Issues in Information Systems Research*, eds. R.J. Boland, Jr. and R.A. Hirschheim, pp. 227–251. New York: John Wiley & Sons, Inc. http://dl.acm.org/citation.cfm?id = 54905.54915

Lacity, M.C. and R. Hirschheim. 1993. The information systems outsourcing bandwagon. *Sloan Management Review* 35(1): 73–86.

Lyytinen, K. 1999. Empirical research in information systems: On the relevance of practice in thinking of IS research. *MIS Quarterly* 23(1) (March 1): 25–27.

Lyytinen, K. and J.L. King. 2004. Nothing at the center? Academic legitimacy in the information systems field. *Journal of the Association for Information Systems* 5(6) (June 21). http://aisel.aisnet.org/jais/vol5/iss6/8

Lyytinen, K. and J. King. 2006. The theoretical core and academic legitimacy: A response to professor Weber. *Journal of the Association for Information Systems* 7(10) (October 31). http://aisel.aisnet.org/jais/vol7/iss10/27

Malone, T.W. and K. Crowston. 1994. The interdisciplinary study of coordination. *ACM Computing Survey* 26(1) (March): 87–119.

Malone, T.W., J. Yates, and R.I. Benjamin. 1987. Electronic markets and electronic hierarchies. *Communications of the ACM* 30(6) (June): 484–497.

March, S.T. and G.F. Smith. 1995. Design and natural science research on information technology. *Decision Support Systems* 15(4) (December): 251–266.

Markus, M.L. 1983. Power, politics, and MIS implementation. *Communications of the ACM* 26(6) (June): 430–444.

Markus, M.L. 1987. Toward a 'Critical Mass' theory of interactive media universal access, interdependence and diffusion. *Communication Research* 14(5) (October 1): 491–511.

Markus, M.L. 1999. Thinking the unthinkable-What happens if the IS field as we know it goes away. In *Rethinking MIS,* eds. W.L. Currie and R. Galliers, Oxford, UK: Oxford University Press.

Markus, M.L. and D. Robey. 1988. Information technology and organizational change: Causal structure in theory and research. *Management Science* 34(5) (May 1): 583–598.

Mata, F.J., W.L. Fuerst, and J.B. Barney. 1995. Information technology and sustained competitive advantage: A resource-based analysis. *MIS Quarterly* 19(4) (December 1): 487–505.

Mathieson, K. 1991. Predicting user intentions: Comparing the technology acceptance model with the theory of planned behavior. *Information Systems Research* 2(3) (September 1): 173–191.

McKnight, D.H., V. Choudhury, and C. Kacmar. 2002. Developing and validating trust measures for e-commerce: An integrative typology. *Information Systems Research* 13(3) (September 1): 334–359.

Meho, L.I. and K. Yang. 2007. Impact of data sources on citation counts and rankings of LIS faculty: Web of science versus scopus and google scholar. *Journal of the American Society for Information Science and Technology* 58(13): 2105–2125.

Melville, N., K. Kraemer, and V. Gurbaxani. 2004. Review: Information technology and organizational performance: An integrative model of it business value. *MIS Quarterly* 28(2) (June): 283–322.

Moore, G.C. and I. Benbasat. 1991. Development of an instrument to measure the perceptions of adopting an information technology innovation. *Information Systems Research* 2(3): 192–222.

Myers, M.D. 1997. Qualitative research in information systems. *MIS Quarterly* 21(2) (June 1): 241–242.

Nunamaker, J.F. Jr., M. Chen, and T.D.M. Purdin. 1990. Systems development in information systems research. *Journal of Management Information System* 7(3) (October): 89–106.

Nunamaker, J.F., A.R. Dennis, J.S. Valacich, D. Vogel, and J.F. George. 1991. Electronic meeting systems. *Communications of the ACM* 34(7) (July): 40–61.

Orlikowski, W.J. 1992. The duality of technology: Rethinking the concept of technology in organizations. *Organization Science* 3(3) (August 1): 398–427.

Orlikowski, W.J. 1993a. Learning from notes: Organizational issues in groupware implementation. *The Information Society* 9(3): 237–250.

Orlikowski, W.J. 1993b. CASE tools as organizational change: Investigating incremental and radical changes in systems development. *MIS Quarterly* 17(3) (September): 309–340.

Orlikowski, W.J. 1996. Improvising organizational transformation over time: A situated change perspective. *Information Systems Research* 7(1) (March 1): 63–92.

Orlikowski, W.J. 2000. Using technology and constituting structures: A practice lens for studying technology in organizations. *Organization Science* 11(4) (July): 404.

Orlikowski, W.J. 2002. Knowing in practice: Enacting a collective capability in distributed organizing. *Organization Science* 13(3) (May 1): 249–273.

Orlikowski, W.J. and J.J. Baroudi. 1991. Studying information technology in organizations: Research approaches and assumptions. *Information Systems Research* 2(1) (March 1): 1–28.

Orlikowski, W.J. and D.C. Gash. 1994. Technological frames: Making sense of information technology in organizations. *ACM Transactions of Information System* 12(2) (April): 174–207.

Orlikowski, W.J. and C.S. Iacono. 2001. Research commentary: Desperately seeking the 'IT' in IT research—A call to theorizing the IT artifact. *Information Systems Research* 12(2) (June 1): 121–134.

Orlikowski, W.J. and D. Robey. 1991. Information technology and the structuring of organizations. *Information Systems Research* 2(2) (June 1): 143–169.

Pavlou, P.A. 2003. Consumer acceptance of electronic commerce: Integrating trust and risk with the technology acceptance model. *International Journal of Electronic Commerce* 7(3): 101–134.

Pitt, L.F., R.T. Watson, and C.B. Kavan. 1995. Service quality: A measure of information systems effectiveness. *MIS Quarterly* 19(2) (June 1): 173–187.

Robey, D. 1996. Research commentary: Diversity in information systems research: Threat, promise, and responsibility. *Information Systems Research* 7(4) (December): 400.

Robey, D. 2003. Identity, legitimacy and the dominant research paradigm: An alternative prescription for the IS discipline: A response to Benbasat and Zmud's call for returning to the IT artifact. *Journal of the Association for Information Systems* 4(1) (December 10). http://aisel.aisnet.org/jais/vol4/iss1/15

Rockart, J.F. 1979. Chief executives define their own data needs. *Harvard Business Review* 57(2) (April): 81–93.

Rogers, E.M. 2003. *Diffusion of Innovations, 5th Edition.* New York: Free Press.

Rosemann, M. and I. Vessey. 2008. Toward improving the relevance of information systems research to practice: The role of applicability checks. *Management Information Systems Quarterly* 32(1) (October 10): 1–22.

Seddon, P.B. 1997. A respecification and extension of the DeLone and McLean model of IS success. *Information Systems Research* 8(3) (September 1): 240–253.

Sidorova, A., N. Evangelopoulos, J.S. Valacich, and T. Ramakrishnan. 2008. Uncovering the intellectual core of the information systems discipline. *MIS Quarterly* 32(3): 467–482.

Sproull, L. and S. Kiesler. 1986. Reducing social context cues: Electronic mail in organizational communications. *Management Science* 32(11) (November 1): 1492–1512.

Straub, D.W. 1989. Validating instruments in MIS research. *MIS Quarterly* 13(2) (June 1): 147–169.

Straub, D. 2012. Editor's comments (Does MIS have native theories?). *MIS Quarterly* 36(2) (June 1): iii–xii.

Szajna, B. 1996. Empirical evaluation of the revised technology acceptance model. *Management Science* 42(1) (January 1): 85–92.

Taylor, H., S. Dillon, and M.V. Wingen. 2010. Focus and diversity in information systems research: Meeting the dual demands of a healthy applied discipline. *MIS Quarterly* 34(4): 647–667.

Taylor, S. and P. Todd. 1995a. Assessing IT usage: The role of prior experience. *MIS Quarterly* 19(4) (December 1): 561–570.

Taylor, S. and P.A. Todd. 1995b. Understanding information technology usage: A test of competing models. *Information Systems Research* 6(2) (June 1): 144–176.

Thompson, R.L., C.A. Higgins, and J.M. Howell. 1991. Personal computing: Toward a conceptual model of utilization. *MIS Quarterly* 15(1) (March 1): 125–143.

Topi, H., J.S. Valacich, R.T. Wright, K. Kaiser, J.F. Nunamaker, Jr, J.C. Sipior, and G.J. de Vreede. 2010. IS 2010: Curriculum guidelines for undergraduate degree programs in information systems. *Communications of the Association for Information Systems* 26(18). http://aisel.aisnet.org/cais/vol26/iss1/18

Triandis, H.C. 1980. Values, attitudes, and interpersonal behavior. In *Nebraska Symposium on Motivation. 1979: Beliefs, Attitudes and Values*, Vol. 27, pp. 195–259. Lincoln, NE: University of Nebraska Press. http://www.ncbi.nlm.nih.gov/pubmed/7242748

Venkatesh, V. 2000. Determinants of perceived ease of use: Integrating control, intrinsic motivation, and emotion into the technology acceptance model. *Information System Research* 11(4) (December): 342–365.

Venkatesh, V. and F.D. Davis. 1996. A model of the antecedents of perceived ease of use: Development and test. *Decision Sciences* 27(3): 451–481.

Venkatesh, V. and F.D. Davis. 2000. A theoretical extension of the technology acceptance model: Four longitudinal field studies. *Management Science* 46(2): 186–204.

Venkatesh, V., F.D. Davis, and M.G. Morris. 2007. Dead or alive? The development, trajectory and future of technology adoption research. *Journal of the Association for Information Systems* 8(4) (April): 268–286.

Venkatesh, V. and M.G. Morris. 2000. Why don't men ever stop to ask for directions? Gender, social influence, and their role in technology acceptance and usage behavior. *MIS Quarterly* 24(1) (March 1): 115–139.

Venkatesh, V., M.G. Morris, G.B. Davis, and F.D. Davis. 2003. User acceptance of information technology: Toward a unified view. *MIS Quarterly* 27(3): 425–478.

Venkatraman, N. 1994. IT-enabled business transformation: From automation to business scope redefinition. *Sloan Management Review* 35(2): 73–87.

Wade, M., M. Biehl, and H. Kim. 2006. Information systems is not a reference discipline (and what we can do about it). *Journal of the Association for Information Systems* 7(5) (May 25). http://aisel.aisnet.org/jais/vol7/iss5/14

Walsham, G. 1995. Interpretive case studies in IS research: Nature and method. *European Journal of Information Systems* 4(2) (May): 74–81.

Wang, R.W. and D.M. Strong. 1996. Beyond accuracy: What data quality means to data consumers. *Journal of Management Information Systems* 12(4): 5–33.

Wasko, M.M. and S. Faraj. 2005. Why should I share? Examining social capital and knowledge contribution in electronic networks of practice. *MIS Quarterly* 29(1) (March 1): 35–57.

Weber, R. 2003. Editor's comments (Still desperately searching for the IT artifact). *MIS Quarterly* 27(2) (June 1): iii–xi.

Weber, R. 2006. Reach and grasp in the debate over the IS core: An empty hand? *Journal of the Association for Information Systems* 7(10) (November 3). http://aisel.aisnet.org/jais/vol7/iss10/28

Whinston, A.B. and X. Geng. 2004. Operationalizing the essential role of the information technology artifact in information systems research: Gray area, pitfalls, and the importance of strategic ambiguity. *MIS Quarterly* 28(2) (June 1): 149–159.

Yates, J. and W.J. Orlikowski. 1992. Genres of organizational communication: A structurational approach to studying communication and media. *The Academy of Management Review* 17(2) (April 1): 299–326.

2

Discipline of Information Technology: History and Development

Barry M. Lunt
Brigham Young University

Han Reichgelt
Southern Polytechnic State University

2.1 Introduction

Since the introduction of computers several decades ago, there have always been those who spend some of their time solving problems with computers. This became particularly true after the advent of the personal computer, as in this new era the majority of computer users were not trained in computers, and, indeed, did not want to be. This was intensified as the World Wide Web became ubiquitous. These trends gave rise to the need for professionals with a unique skill set, which, as we will discuss later in this chapter, were not being provided by the traditional computing programs.

In response to these needs, new 4 year computing degree programs emerged at several institutions in the United States; the earliest was introduced in 1992. These programs, which were typically titled "information technology" (IT), grew out of the need to produce professionals with the skills necessary to support the use of the newly ubiquitous computing tools. In the past decade, the number of these programs has grown. In 2012, there are at least 50 such programs, of various flavors, in the United States.

The objectives of this chapter are to further chronicle the emergence of this new computing discipline of IT and its maturation, and to draw a contrast between it and the other computing disciplines.

2.1.1 Emergence of the IT Discipline

The underlying principles of IT have grown out of its history. Historically, those who have worked closely with computers were well educated in computing (starting with those whose academic background is in the disciplines of computer science (CS) and (IS)). This was also mostly true of those who used comput-ers. And if someone outside the "computing cognoscenti" wanted to use computers, it was necessary for them to immerse themselves in this knowledge domain.

This began to change with the introduction of the personal computer, and with further system soft-ware releases (particularly the graphical user interfaces of the Mac OS and Windows). Computers were soon in the hands of people who wanted to harness their capabilities, but did not want to have to learn all the details of how they worked. Thus, the need arose for a computing discipline whose responsibility was to solve computer application problems for this new computing generation. From this history grew the broad goals of IT programs[1]:

1. Explain and apply appropriate information technologies and employ appropriate methodologies to help an individual or organization achieve its goals and objectives
2. Function as a user advocate
3. Manage the IT resources of an individual or organization
4. Anticipate the changing direction of IT and evaluate and communicate the likely utility of new technologies to an individual or organization
5. Understand and, in some cases, contribute to the scientific, mathematical, and theoretical foun-dations on which information technologies are built
6. Live and work as a contributing, well-rounded member of society

In 2001, several of the institutions with 4 year programs in IT met to formally define IT as a separate academic discipline. The outgrowth of this meeting and the many meetings that followed included a formally accepted model curriculum,[2] accreditation guidelines and criteria,[3] and a special-interest group of the ACM: SIGITE—Special Interest Group for Information Technology Education.[4]

The IT model curriculum, the first draft of which was published in 2005 and which was formally accepted by the Education Council of the ACM in 2008, defines the academic discipline of IT as follows:

> "IT, as an academic discipline, is concerned with issues related to **advocating for users** and meeting their needs within an organizational and societal context through the **selection, cre-ation, application, integration and administration** of computing technologies."[5] (emphasis in original).

This same document further lists the five pillars of IT: programming, networking, human–computer interaction, databases, and web systems, with information assurance and security being a "pervasive theme," which, with other pervasive themes, are "woven-like threads throughout the tapestry of the IT curriculum."[6] The other pervasive themes are as follows:

- User centeredness and advocacy
- The ability to manage complexity through abstraction and modeling, best practices, patterns, standards, and the use of appropriate tools
- Extensive capabilities for problem solving across a range of information and communication
- Technologies and their associated tools
- Adaptability
- Professionalism (lifelong learning, professional development, ethics, responsibility)
- Interpersonal skills

The IT Model Curriculum also points to another important characteristic of the IT discipline, namely, its breadth. As mentioned in the IT model curriculum: "The depth of IT lies in its breadth…,"[7] meaning that an identifying characteristic of the IT discipline is that it is very broad—anything having to do with the applications of computers lies within the domain of IT.

2.1.2 Unique Nature of IT

Until fairly recently, the only computing programs that had had formally defined model curricula were CS, whose first model curriculum was promulgated in 1968, and IS, whose first model curriculum was accepted in 1973. For many years, CS and IS were the only widely recognized computing programs. However, in the early part of the 2000s, the relatively well-established computing disciplines of software engineering (SWE) and computer engineering (CE) started formulating new model curricula, as did the newer computing discipline of IT. These efforts took place primarily under the auspices of the ACM, the Association for Information Systems (AIS), and The Computer Society of the Institute for Electrical and Electronic Engineers (IEEE-CS). It also soon became clear that there was a need to describe the similarities and differences among the different computing disciplines.

In 2003, the ACM invited representatives of all five of these computing disciplines to write a document that would form a guide to undergraduate degree programs in computing. With oversight from the ACM, AIS, and IEEE-CS, two representatives from each of these disciplines met on several occasions. This effort resulted in 2005 in a report entitled "Computing Curricula 2005—The Overview Report." This document became very influential in helping to define the different computing disciplines with respect to each other. It provides, for each of these computing disciplines, insights into their history, descriptions, and graphical views, a table of the comparative weights of computing and non-computing topics (Table 2.1), and a table of the relative performance capabilities of computing graduates.

The Overview Report describes IT in the following way:

"In the previous section, we said that Information Systems focuses on the information aspects of information technology. Information Technology is the complement of that perspective: its emphasis is on the technology itself more than on the information it conveys. IT is a new and rapidly growing field that started as a grassroots response to the practical, everyday needs of business and other organizations. Today, organizations of every kind are dependent on information technology. They need to have appropriate systems in place. These systems must work properly, be secure, and upgraded, maintained, and replaced as appropriate. Employees throughout an organization require support from IT staff who understand computer systems and their software and are committed to solving whatever computer-related problems they might have. Graduates of information technology programs address these needs.

Degree programs in information technology arose because degree programs in the other computing disciplines were not producing an adequate supply of graduates capable of handling these very real needs. IT programs exist to produce graduates who possess the right combination of knowledge and practical, hands-on expertise to take care of both an organization's information technology infrastructure and the people who use it. IT specialists assume responsibility for selecting hardware and software products appropriate for an organization, integrating those products with organizational needs and infrastructure, and installing, customizing, and maintaining those applications for the organization's computer users. Examples of these responsibilities include the installation of networks; network administration and security; the design of web pages; the development of multimedia resources; the installation of communication components; the oversight of email systems; and the planning and management of the technology lifecycle by which an organization's technology is maintained, upgraded, and replaced."[8]

As indicated in the second paragraph of this quote, the needs of the rapidly growing computing community were not being fully met by the existing computing disciplines in academia. The desire to supply these needs was the major motivating factor that gave rise to the IT programs in academia.

On page 20 of The Overview Report is a depiction of the problem space of computing, specifically depicting the space occupied by IT, as shown in Figure 2.1.

This figure shows that IT is strongly focused on the application, deployment, and configuration of computing systems, and is concerned with four of the five domains identified on the left side of the figure.

TABLE 2.1   Computing and Non-Computing Topics and Performance Capabilities Which Identify IT

Computing Topics

Programming fundamentals
Integrative programming
Operating systems configuration and use
Net centric principles and design
Net centric use and configuration
Platform technologies
Human–computer interaction
Information management (DB) practice
Legal/professional/ethics/society
Analysis of technical requirements
Security: implementation and management
Systems administration
Systems integration
Digital media development
Technical support
Non-computing topics
Risk management (project, safety risk)
Mathematical foundations
Interpersonal communications
Performance capabilities
Provide training and support for users of productivity tools
Create a software user interface
Train users to use information systems
Maintain and modify information systems
Model and design a database
Select database products
Configure database products
Manage databases
Train and support database users
Develop computer resource plan
Schedule/budget resource upgrades
Install/upgrade computers
Install/upgrade computer software
Design network configuration
Select network components
Install computer network
Manage computer networks
Manage communication resources
Manage mobile computing resources
Manage an organization's web presence
Configure and integrate e-commerce software
Develop multimedia solutions
Configure and integrate e-learning systems
Evaluate new forms of search engine

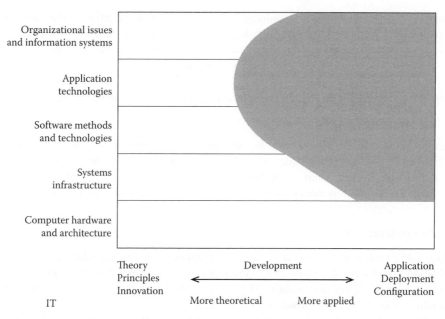

FIGURE 2.1 The problem space occupied by IT. (From p. 20 of The Overview Report, The Joint Task Force for Computing Curricula 2005, a cooperative project of ACM, AIS, IEEE-CS, http://www.acm.org/education/education/curric_vols/CC2005-March06Final.pdf)

It should be noted that since this figure was created and the field has matured, IT has developed a stronger focus in the area of systems infrastructure, which would be depicted in this figure by extending to the left the area of the gray shape in that fourth domain. For example, the IT model curriculum, which was still in draft form when the Overview Volume was being formulated and which was formally accepted in 2008, contains 10 IT-specific knowledge areas, of which four (platform technologies, networking, system integration and architecture, and system administration and maintenance) are most directly concerned with the systems infrastructure and one (information assurance and security) has a direct bearing on this layer.

Commenting on the information depicted in Figure 2.1, and similar figures for CE, CS, IS, and SWE, Agresti,[9] in his 2011 paper, points out that these figures give each of the disciplines a "disciplinary theme," and these themes are:

CE = Hardware
CS = Theory
IS = Organizations
IT = Deployment
SE = Development

In explaining his choice of "deployment" for the IT disciplinary theme, Agresti states:

"[The figures depicting computing space] portray IT as anchored in the world of deployment, structuring and configuring computing artifacts. As new technologies appear, they need to be integrated with existing systems to have any beneficial effect for users." (p. 260)

Thus, one of the main underlying principles of IT is that students in this discipline, compared to the other computing disciplines, should be the broadest in their knowledge of computing. They should know how to work with people of any discipline who know practically nothing about computers. They should know how to speak the language of those who do not care (or need) to know the details of how computers work or how to fix them. They should know how to get disparate computing systems to work together. They should be very knowledgeable about each of the IT pillars of databases, human–computer interaction, networking, programming and web systems, and the pervasive themes from the IT model curriculum. The depth of this discipline is not in a single or a few knowledge areas, but in the breadth required. This goes along well with Agresti's comment:

"Integration of technologies and interactions among them should be major focus areas of the IT discipline."[10]

Another insight provided in this Overview Report is the data summarized in Table 2.1, consisting of the computing topics, non-computing topics, and relative performance capabilities of graduates of IT programs. In this table, the only topics or capabilities included are those for which IT programs were rated at a 4 or a 5 (out of 5), meaning that IT programs considered them an essential part of the identity of IT.[11]

2.1.3 Accreditation

In 2003, SIGITE produced the criteria by which IT programs could be accredited by the Computing Accreditation Commission (CAC) of ABET. These criteria were accepted by CSAB Inc. and ABET in 2004, and in 2005 ABET began accrediting IT programs, along with programs in other major computing disciplines which ABET had already been accrediting.

At the time the IT criteria were created, the IT discipline was the only one of the three disciplines accredited by ABET CAC that had adopted an outcome-based approach to accreditation criteria.

TABLE 2.2   Program Criteria for Information Technology and Similarly Named Computing Programs

Student Outcomes

    a. An ability to apply knowledge of computing and mathematics appropriate to the discipline

    b. An ability to analyze a problem, and identify and define the computing requirements appropriate to its solution

    c. An ability to design, implement, and evaluate a computer-based system, process, component, or program to meet desired needs

    d. An ability to function effectively on teams to accomplish a common goal

    e. An understanding of professional, ethical, legal, security, and social issues and responsibilities

    f. An ability to communicate effectively with a range of audiences

    g. An ability to analyze the local and global impact of computing on individuals, organizations, and society

    h. Recognition of the need for and an ability to engage in continuing professional development

    i. An ability to use current techniques, skills, and tools necessary for computing practice

    j. *An ability to use and apply current technical concepts and practices in the core information technologies

    k. *An ability to identify and analyze user needs and take them into account in the selection, creation, evaluation and administration of computer-based systems

    l. *An ability to effectively integrate IT-based solutions into the user environment

    m. *An understanding of best practices and standards and their application

    n. *An ability to assist in the creation of an effective project plan

Curriculum

    1. Coverage of the fundamentals of

        a. The core information technologies of human–computer interaction, information management, programming, networking, web systems, and technologies

        b. Information assurance and security

        c. System administration and maintenance

        d. System integration and architecture

    2. Advanced course work that builds on the fundamental course work to provide depth

The outcomes preceded with an ∗ are those which are uniquely associated with IT programs.

TABLE 2.3 Universities with IT Programs Accredited by CAC of ABET, as of February 2012, Ordered Alphabetically

Academic Institution	Location
Brigham Young University	Provo, UT
Capella University	Minneapolis, MN
Drexel University	Philadelphia, PA
East Tennessee State University	Johnson City, TN
George Mason University	Fairfax, VA
Georgia Southern University	Statesboro, GA
Indiana University-Purdue University Indianapolis	Indianapolis, IN
Macon State College	Macon, GA
Purdue University at West Lafayette	West Lafayette, IN
Rochester Institute of Technology	Rochester, NY
Slippery Rock University	Slippery Rock, PA
Southern Polytechnic State University	Marietta, GA
United States Military Academy	West Point, NY
United States Naval Academy	Annapolis, MD
University of Cincinnati	Cincinnati, OH
University of Missouri-Kansas City	Kansas City, MO
University of North Florida	Jacksonville, FL
University of South Alabama	Mobile, AL

In other words, IT was the first of the three computing disciplines that formulated its criteria primarily in terms of the skills and knowledge that graduates from accredited IT programs could expect to have acquired, rather than in terms of, for example, the courses students needed to complete. This was probably a direct consequence of the fact that ABET itself had moved to an outcome-based approach shortly before the IT community formulated its accreditation criteria.

However, with ABET's greater emphasis on outcomes, CS and IS also reformulated their criteria in terms of outcomes. Moreover, since ABET CAC wanted to be able to accredit computing programs that were not necessarily named "CS," "IT," or "IS," the criteria were divided into general criteria, which any program in computing must meet to be accreditable, and program-specific criteria for specific computing disciplines. As part of this effort, a distinction was made between attributes that any graduate from a computing program should possess, and those attributes that are specific to graduates from IT programs. Table 2.2 lists the attributes for IT. Today there are 18 IT programs which have been accredited by the CAC, as shown in Table 2.3.

2.2 Distinguishing Characteristics of the IT Discipline

Because of the emergence of the computing discipline in response to a workforce development need, it can be argued that one of the distinguishing characteristics of IT is its highly practical nature. Computing has permeated a myriad of activities and disciplines, and graduates from IT programs are expected to be able to support computing tools and their users in a variety of settings. Because of this, most IT programs have an expectation that students develop some knowledge of a particular application domains, be it geology, health care, crime investigation, archeology, entomology, etc.

A good example of this is provided by the IT program at Georgia Southern University that originally required all IT students to include a so-called second discipline in their program of study, essentially a directed minor, with the explicit aim of making sure that students developed sufficient knowledge of an application domain to be able to develop and support IT applications for that domain.[1] As a result, the main impact that IT has on practice lies in the application domains where it is deployed, and upon which each domain has become increasingly dependent. A few examples will help make the point.

2.2.1 Applications in Health Care

All health-care providers have the need to have access to accurate and current patient records. These records should be secure, convenient, thorough, and readily available wherever they or their patients may be. This is an IT issue.

Drugs for health care must constantly be under inventory control, to keep supplies adequate for needs and to monitor for theft or abuse. The supply chain for these drugs must be documented to assure their authenticity and purity. Much of this is an IT issue.

Over the past few decades, billions of patient health-care records have been kept on computers. Such records often include data such as the demographics of the patient, the symptoms they experienced, the treatments provided, and the results of the treatments. This information, if fully searchable, constitutes a vast database which could yield dramatic insights into health-care treatments and their effectiveness. Making such a database readily available across all platforms (the Internet, desktops, laptops, tablets, and smart phones) is another research area with a large IT component.

2.2.2 Applications in Law Enforcement

Perpetrators usually have a characteristic method of operation (MO), which can often be used to recognize a group of crimes which may have all been committed by the same person. If there were a readily accessible and secure database which could be easily searched by law-enforcement workers, it would greatly enhance their ability to solve crimes. There has been a lot of progress in this area, and this progress is sufficient to convince practically anyone that further effort in this direction will yield great improvements. Such a database would be even more valuable if it were worldwide and available across all platforms. Since the Internet is nearly ubiquitous, it is probably only a matter of time and effort before this can be accomplished. Much of the wide deployment of this crime database is an IT issue.

Each person who has a criminal record has specific conditions of their position, whether incarcerated, on parole, in a half-way house, or free. Law-enforcement officers are greatly aided when they are quickly able to determine what these conditions are for each individual. Much of this is an IT problem.

Crime-scene investigators are greatly aided by rapid analysis of forensic evidence. And today, much of that evidence is digital, including cell-phone records and contents, laptop and pad-type computers, and GPS records. Accessing this evidence and amalgamating it with all the other crime-scene evidence is largely an IT issue.

2.2.3 Applications in Manufacturing

Manufacturing of a typical integrated circuit (IC) involves approximately 500 steps, each of which must be precisely controlled in the materials used, the temperature, the pressure, the time, and other parameters, which together are the "recipe" for each step and for each type of IC being produced. If any of these ingredients or conditions in any step of the recipe is not exactly controlled, or if any of the 500 steps is performed in the wrong order, hundreds or even thousands of ICs will be faulty and would have to be scrapped. Solutions in this domain always involve people from varied backgrounds, needing to access multiple data sources and integrate multiple applications. Thus, a very significant need is to have professionals able to work in nearly all areas of computing. These are IT professionals.

For most products today, each step in the manufacturing process is carefully monitored and data are gathered. These data, as with health-care and law-enforcement databases, become a rich store of data which can be mined to find ways to improve the manufacturing process. Much of this is an IT problem.

2.2.4 Applications in Education

Only a few decades ago, students registered for college classes with pen-and-paper forms which they took around to classes and had their professors sign. Problems and conflicts were many, especially in classes for which the demand was particularly high. Today, students at most institutions of higher education can register based on their class standing (with graduate students getting first pick, then seniors, juniors, etc.). They can see how many openings there are in every class they wish to register for. And they can do it all from their mobile platform of choice or a fixed computer. This dramatic improvement in the registration process has been primarily an IT solution because it has involved the integration of databases, hardware, networks, human–computer interfaces, web systems, security, storage, applications, and integrative programming.

College professors today have access to a very wide array of materials, including video clips, audio clips, photos, simulations, questionnaires, forms, equations, and applications. Many of these are available on the Internet. But there are concerns with copyrights, licenses, controlling access, and availability. At many institutions of higher education, such materials are made available through "courseware," which is basically software that pulls together solutions for appropriate delivery of all such materials. This has greatly enriched the educational experience for all teachers and students with such access. Applying such courseware across an institution and all its varied courses is primarily an IT issue for the same reason as mentioned in the previous paragraph—because it involves the integration of so many computing areas and is deeply technical.

2.2.5 Summary: Impact on Practice

The preceding examples are illustrative only; many more examples could be provided in dozens of application domains. Graduates from computing disciplines such as CS and SE have done much to produce software for these applications. Graduates from CE have done much to provide hardware specific for these domains. And graduates from IS have done much to define how the software and hardware must perform in their specific organization and how information is processed to benefit the organization. And when the software and hardware are both deployed, there are always some changes that need to be made to make the system more effective in specific applications—that is the part most often played by the IT professionals. They must be able to see the entire situation as a system, consisting of multiple entities, each of which must be understood, and their interactions also known and designed for. In each of the scenarios described previously, the unique contribution provided by IT professionals is their view of the system as a whole. Thus IT professionals must be broad, able to deal with problems in the infrastructure, the software, the hardware, the networking, the web, the database, the human–computer interaction, the security, or any other part of the computing system.

2.3 IT Literature

The more practice-oriented nature of IT is also reflected in its relationship to the so-called practitioner literature, including white papers and case studies (some published by consulting firms), and the articles that appear in more practice-oriented publications, such as *Educause Review* and *CIO Magazine*. This material often reflects valuable insights from practice, but lacks the scientific rigor that is typically demanded by academic researchers. At the same time, the practitioner literature tends to have a greater influence on practice than the academic literature and since IT programs strongly prepare their students for practice, the discipline has to pay greater attention to this literature than some of the other computing disciplines do.

A good example that illustrates this is the emergence of the IT Infrastructure Library (ITIL), a set of best practices for IT service delivery. While there is relatively little academic literature on ITIL, the practitioner literature is replete with examples of implementations of ITIL and claims about its ability to improve the quality of IT service delivery. IT service organizations are starting to widely adopt the ITIL framework. As a reaction to this trend, IT programs are therefore starting to include coverage of ITIL.[12]

As Reichgelt et al. (Chapter 4 in this volume) argue, the fact that the practitioner literature lacks the rigor demanded by academia opens up a viable stream of possible research topics for the academic IT community, namely, projects to find solid evidence for the claims made by practitioners using the traditional methods used in the various computing disciplines.

2.4 Research Issues

This volume contains a chapter on the topic of research in IT (see Chapter 4); the reader is advised to study that chapter for further depth in this area.

The academic discipline of IT is centered on the scholarship of application and the scholarship of integration, as defined in Boyer.[13] This means that research in IT is most likely to be concerned with applications of IT in solving problems in many fields of endeavor. A bottom-up study of IT research embodied in IT masters theses[14] shows that of the 70 theses studied, they could be classified into 5 categories:

1. Development projects (case studies)
2. Education and IT for the education domain
3. Information assurance, security, and forensics
4. Project management and IT for the project management domain
5. Technology evaluation and modeling

Subsequent research[15,16] has reinforced these categories. Taken together, these categories demonstrate the point that IT is focused on application and integration. It is important to recall that the third category embraces the most central of the IT pervasive themes, which further emphasizes the point that the domain of IT is best defined by application and integration.

2.5 Future of Information Technology

There are two trends that are likely to influence the further evolution of the discipline of IT.

The first is that we are likely to see further growth in IT programs around the world. For example, the shift toward cloud computing in the way information services are delivered will result in a growing number of data centers, a trend that is likely to be exacerbated by the continued growth in the use of computing to support organizations and individuals. There will therefore be a growing need for graduates with the type of skills instilled by IT programs.

The growth of IT is likely to manifest itself in three ways. First, it is likely that there will be more programs calling themselves "IT" emerging at more institutions. However, there is also a second, more subtle, way in which IT is likely to become more widespread, in that programs will emerge that may not call themselves "IT" but that cover a large part of the IT model curriculum. Examples might include programs in information security or network technologies. Third, there is likely to be a shift in existing computing programs toward a greater coverage of aspects of the IT model curriculum. For example, more and more CS programs are introducing courses on information security that place a much greater emphasis on the more practical way in which these topics are covered in the IT model curriculum than the more theoretical way in which related topics are covered in the CS model curriculum. For example, such courses place much less emphasis on encryption algorithms and more emphasis on the more practical aspects of hardening the IT infrastructure against attacks.

All of this raises the question of how an IT program is defined. Rowe et al. have developed a series of metrics based on the IT model curriculum to determine how close a program is to being an IT program, independently of how it decides to designate itself.[17] Using this metric, over 900 BS programs in computing in the United States were evaluated and over 300 of these programs were scored using these metrics. The results of this analysis produced a list of 4 year IT programs in the United States, as well as a list of names by which these programs call themselves.[18] The point of this is that there is a great deal of diversity in computing programs, and in programs that appear to be IT programs. It is very likely that this diversity will continue to increase, as it seems to have done over the past few decades.

A second likely development is the greater emphasis on research. There is a growing realization within the academic IT community that if IT is to thrive, it needs to gain more credibility as a serious academic discipline. This is especially important in light of the fact that high-quality IT programs typically rely on more expensive equipment than, say, CS, IS, or SWE. For example, given the growing importance of cloud computing, and given the need to allow students to develop knowledge and skills of the type of data centers that support cloud computing, IT programs may wish to build scaled-down versions of such data centers. However, even scaled-down data centers are considerably more expensive than the more traditional computing labs to support programs in the other computing disciplines, and, at a time where resources are scarce in academia, it is often to make the case for a discipline that lacks academic credibility.

As an example, SIGITE, the ACM Special Interest Group for IT Education and an organization heavily involved in the formulation of both the IT model curriculum and the ABET CAC accreditation criteria for IT, has recently expanded its annual conference to include a joint conference on Research in Information Technology (RIIT). We are also seeing a growing number of papers calling for the need for an IT research agenda (see, e.g., [13], or the chapter by Reichgelt et al. in this volume).

The greater emphasis on research is also likely to result in a greater number of graduate programs in IT. An interesting issue that is likely to arise in this context is whether the traditional PhD program, with its much greater emphasis on theory, is best for IT, or whether the more professionally focused doctoral programs, such as Ed.D (Doctorate of Education) or DBA (Doctorate of Business Administration) programs provide a more appropriate model.

2.6 Summary

IT is both a very broad term and a much narrower term. The broad meaning of IT is anything having to do with computers. The much narrower meaning is the academic discipline of IT, which first arose in about 1992, and which was formally defined starting in 2003 and ending in 2008. As an academic discipline in computing, it has now taken its place alongside the CS and IS veterans and the CE and SE relative newcomers.

IT is primarily focused on integration and system-type thinking in computing. Its purview includes anything in computing, but it is generally not deep in any single domain of computing. IT prepares computing generalists, whose job includes the selection, application, creation, integration, and administration of computing solutions for an organization.

IT has accreditation criteria which have been used by ABET to accredit 18 4 year IT programs in the United States. IT also has a model curriculum which is of great value to any institution looking to establish or modify an IT program. There are at least 50 4 year IT programs in the United States today, and that number appears to be growing.

Glossary

Forensics: For IT, this term refers to the efforts necessary to preserve evidence found in all types of IT artifacts, including phones, computers of all kinds, flash memory, optical disks, hard-disk drives, and magnetic tape. This also includes analyzing, recovering, identifying, and presenting summary information of the evidence found.

Information assurance and security: This field is known by at least three terms: information assurance; security; and cybersecurity. In this document, we have chosen to use information assurance and security for this field, as is done in the IT model curriculum.

Model curriculum: Each of the five major computing disciplines has defined a model curriculum (see http://www.acm.org/education/curricula-recommendations). This formally defined and approved document represents the best thinking of professionals in IT education and describes the core of what a 4-year program in IT should include.

Further Information

Reference #1, cited several times, is definitive and should be consulted for further information. For comparing the various computing programs with formally defined model curricula, the reader should consult reference #8, which is also definitive in that regard.

References

1. Lunt, B. M. et al., Information technology 2008: Curriculum guidelines for undergraduate degree programs in information technology, ACM, http://www.acm.org//education/curricula/IT2008%20 Curriculum.pdf (accessed on March 29, 2013), p. 10, 2008.
2. Lunt, B. M. et al., Information technology 2008: Curriculum guidelines for undergraduate degree programs in information technology, ACM, http://www.acm.org//education/curricula/IT2008%20 Curriculum.pdf (accessed on March 29, 2013), 2008.
3. Program criteria for information technology and similarly named computing programs, ABET, http://www.abet.org/computing-criteria-2012-2013/ (accessed on March 29, 2013).
4. Special Interest Group for Information Technology Education, SIGITE. http://www.sigite.org/ (accessed on March 29, 2013).
5. Lunt, B. M. et al., Information technology 2008: Curriculum guidelines for undergraduate degree programs in information technology, ACM, http://www.acm.org/education/curricula/IT2008 Curriculum.pdf (accessed on March 29, 2013), p. 9, 2008.
6. Ibid, p. 33.
7. Ibid, p. 18.
8. The Overview Report, The Joint Task Force for Computing Curricula 2005, a cooperative project of ACM, AIS, and IEEE-CS, http://www.acm.org/education/education/curric_vols/CC2005-March06Final.pdf (accessed on March 29, 2013), pp. 14 –15, 2005.
9. Agresti, W. W., Toward an IT Agenda, *Communications of the Association for Information Systems*: Vol 28, Article 17, 2011.
10. Ibid, p. 261.
11. The Overview Report, The Joint Task Force for Computing Curricula 2005, a cooperative project of ACM, AIS, IEEE-CS, http://www.acm.org/education/education/curric_vols/CC2005-March06Final.pdf (accessed on March 29, 2013), p. 24, 25, 28.
12. Van Bon, J., de Jong, A., Kolthof, A., Peiper, M., Tjassing, R., van der Veen, A., Verheijen, T. *Foundations of IT Service Management Based on ITIL V3*. Zaltbommel, NL: Van Haren Publishing, 2007.
13. Boyer, E. L., *Scholarship Reconsidered*, San Francisco, CA: Jossey-Bass, 1997.
14. Cole, C., Ekstrom, J. J., and Helps, R., Collecting IT Scholarship: the IT Thesis Project, *Proceedings of the 2009 ACM SIGITE Conference*, Fairfax, VA, Oct 22-24, 2009.
15. Ekstrom, J. J. et al., A research agenda for information technology: Does research literature already exist? *Proceedings of the 2006 ACM SIGITE Conference*, Minneapolis, MN, Oct 19-21, 2006.
16. Cole, C., and Ekstrom, J. J., The IT Thesis Project: A slow beginning, *Proceedings of the 2010 ACM SIGITE Conference*, Midland, MI, Oct 7-9, 2010.
17. Rowe, D., Lunt, B. M., and Helps, R. G., An assessment framework for identifying information technology bachelor programs, *Proceedings of the 2011 ACM-SIGITE Conference*, West Point, NY, Oct 19-22, 2011.
18. Lunt, B. M., Bikalpa, N., Andrew, H., and Richard, O., Four-Year IT programs in the USA, *Proceedings of the 2012 ACM SIGITE Conference*, Calgary, AL, Oct 11-13, 2012.

3

Information Systems as a Practical Discipline

Juhani Iivari
University of Oulu

3.1 Introduction

The purpose of this chapter is to discuss information systems (IS) as a practical discipline or science, "which is conceived in order to make possible, to improve, and to correct a definite kind of extra-scientific praxis" (Strasser 1985, p. 59). Referring to Hassan (2011) but contrary to Strasser (1985), this chapter prefers to speak about a discipline rather than a science in the context of IS.

The practical relevance of disciplines is often discussed as a dilemma between relevance and rigor (Keen 1991, Benbasat and Zmud 1999). A dilemma more specific to IS is how to ensure its practical relevance without condescending into commercial faddism. Much of the IS research agenda still continues to be determined by consultants and gurus, and various fads are part of our scientific language (Baskerville and Myers 2009). Following the commercial culture of the fashion setters, many IS researchers tend to introduce each hype as a radical or disruptive innovation, instead of seeking invariants and continuities between ideas and technologies.

Partly because of this commercial faddism, it is hard to see a greater cumulative tradition in IS research than 25 years ago when Banville and Landry (1989) characterized it as a "fragmented adhocracy." Due to the technological development, IS research problems and topics have continued to proliferate, theoretical frameworks applied have become more numerous and diverse, and research methods more pluralistic. Precisely because of these factors, IS researchers should, however, resist the temptation to artificially reinforce and amplify fashion waves.

Benbasat and Zmud (2003) attempted to bring some order into this fragmentation. According to them, the core of IS is determined by the information technology (IT) artifact and its immediate nomological net. They discuss their proposal in terms of errors of exclusion and errors of inclusion in IS research. Errors of exclusion refer to studies that do not include the IT artifact or elements of its immediate nomological net. Errors of inclusion, on the other hand, refer to studies that include non-IT constructs—that is, those not included in the nomological net—especially if they are causally distant from those of the nomological net.

Their proposal has received exceptional attention and also severe criticism (see Agarwal and Lucas 2005, King and Lyytinen 2006), much from the viewpoint of the academic legitimacy of IS. The academic legitimacy of IS is hardly a serious problem, at least in Scandinavia or elsewhere in Europe, and according to Mason (2006) not even in North America. Thus, this point of criticism is not very relevant, although Benbasat and Zmud (2003) themselves raise the legitimacy issue as a major motivation for their proposal.

On the contrary, the thesis of the present chapter is that the intellectual core of IS is crucial for the practical relevance. Because of errors of exclusion, IS research has suffered from a weak attention to the IT artifact (Orlikowski and Iacono 2001). Theories are mainly borrowed from reference disciplines. These theories are often totally void of any IT-specific substance. All this has likely diminished the practical relevance of IS research.

Furthermore, the present chapter is based on the conviction that seeking invariants and continuities between ideas and technologies is a more effective way of ensuring the practical relevance of IS than adhering to commercial faddism. There are a number of reasons for this. First, there is no point competing with consultants and gurus by just emulating their behavior. Second, helping practitioners to understand that a new fad is a mutated version of an older idea or just old wine in a new bottle helps them to make better sense of the fad in question. Third, a better understanding of predecessors makes it possible to transfer knowledge and experience from the past and possibly to avoid repeating mistakes. All these can be expected to increase the credibility of IS research at least in the longer run.

The structure of this chapter is as follows. The next section defines the territory of IS among the disciplines of computing, suggesting that the special focus of IS lies on IT applications rather than on IT artifacts in general. After it, the external praxes or practices to be supported by IS research are discussed, identifying three major areas—IT management, IT application development, and IT application usage. Focusing on IT application development, the following section suggests a three-tier model that includes design by vendor, design by client collective, and design by users. Continuing to focus on design, the question of how IS research can effectively support the identified external practices is discussed next. The section suggests that IS research should focus on actionable and, more specifically, on designable qualities of IT applications so that the research outcomes can inform design at different stages of design in the three-tier model. Finally, the implications of the article are discussed.

3.2 IT Artifacts and IT Applications

All disciplines of computing—computer engineering, computer science, software engineering, IT, and IS (Shackelford et al. 2005)—are interested in IT artifacts. IT artifacts form the *raison d'être* of their existence. This common interest in IT artifacts makes disciplines of computing—especially the last three of them—sister disciplines. There is a considerable overlap between them especially at the level of base technologies such as data communication, software, databases, and user interfaces, as well as in the case of system development methods, techniques, and tools.

Iivari (2007) suggests that the distinctive focus of IS among disciplines of computing lies in IT applications rather in IT artifacts in general. As a consequence, IS can be conceived as a discipline of computing that specifically focuses on IT applications, on their development, use, and impact at individual, group, organizational, community, society, and global levels.

IT applications are IT artifacts that have human beings as users and have functionality/capability to produce direct services to their users.* The concept of IT application is close to "application software," but not exactly the same. First, an IT application does not comprise only software but also the physical device(s) required for the operation and use of the system. Second, it may also include content that is not software by nature (e.g., information content). Third, an IT application may be an instantiation

---

* The word "direct" attempts to exclude so-called systems software from the category of IT applications. The services provided by systems software are indirect in the sense that they are required for providing the services of IT applications.

TABLE 3.1   Ideal Types of IT Applications

Intended Purpose	Examples
To automate	Many embedded systems
	Many transaction processing systems
To augment	Many personal productivity tools, computer-aided design tools
To mediate	Computer-mediated communication (CMC): Internet phone calls, e-mail, instant messaging, chat rooms, etc.
To informate	Information systems (proper)
To entertain	Computer games (Aarseth 2001, Davis et al. 2009)
To artisticize[a]	Computer art (Noll 1995, Edmonds et al. 2004, Oates 2006)
To accompany	Digital (virtual and robotic) pets (Kusahara 2001, Friedman et al. 2003)
To fantasize	Digital fantasy world applications (Davis et al. 2009)

[a] The word "artisticize" is introduced to refer to subjecting a person to an artistic experience as a consumer or possibly also as an interactive creator of a piece of (computer) art.

of application software (e.g., an enterprise resource planning (ERP)-based information system is an instantiation of ERP software, but ERP software is not an application as itself).

Iivari (2007) also introduces an open-ended typology of IT applications that, in its current version, includes eight types depicted in Table 3.1. The proposed typology is not based on any reference theory. Instead it is a Type I theory in the framework of Gregor (2006)—"theory for analysing" that is specific to IT applications, based on the analysis of their major purposes.

The first four archetypes are close to "technology as a labor substitution tool," "technology as a productivity tool," "technology as a social relations tool," and "technology as an information processing tool" in Orlikowski and Iacono (2001). So, essentially Table 3.1 extends Orlikowski and Iacono (2001) by introducing four additional types. Computer games illustrate the capability of IT applications to entertain. IT applications may also attempt to arouse artistic experience, and one can easily imagine a new sort of art that is essentially built on the dynamic and interactive character of computer technology. IT artifacts such as digital pets can accompany human users. Finally, computers allow users to build digital fantasy worlds and have fantasy-like experiences. Table 3.1 identifies some references to these non-traditional application types.

As ideal types, the IT applications of Table 3.1 may not occur in their pure forms in practice, but one can identify examples that are close to them. Yet, many real IT applications are combinations of the ideal types for several reasons. First, a single IT application may be deliberately designed to include several purposes in terms of Table 3.1. For example, an information system (proper) may be designed to include game features to motivate the use of the system and to support learning in the case of educational IS as exemplified by Age of Empires II: The Age of Kings (http://www.microsoft.com/games/age2/). Second, an IT application of one archetype may include auxiliary functions of another archetype to support of the use of the system. For example, e-mail with the original function of communicating messages includes mailboxes that allow one to build a directory to informate about previous communications. Thus, it can be developed into a fairly sophisticated information system about one's electronically mediated social network and one's electronic communication within that network. Third, Zuboff (1988) claimed that to automate also allows one to informate. This can obviously be extended to cover other uses of IT applications, so that computer games, for example, could at least in principle collect information about their use and about users' actions and reactions during playing. A final point is that applications of one archetype can be used to implement applications of another archetype. For example, spreadsheet software as an augmenting tool can be used to implement a specific information system.

Referring to the proliferation of IT applications, one should note that a sound typology of ideal types also enables to address this increased variety of IT applications. The eight ideal types alone can be combined in 255 ways. If one adds different application domains (e.g., business, government, health-care, education, library, and geography), one could have thousands of different application types just in terms of the eight ideal types. It is this variety of IT applications that makes the ideal types significant. Recognizing that many practical IT applications are combinations of ideal types, research can focus on much fewer ideal types, assuming that a deeper understanding of ideal types—their similarities and differences within and across application domains—also informs about the hybrid applications in different application domains.

Related to fashion waves, the first question to be asked in the case of each new (type of) IT application is whether or not it really is so new that it does not have any predecessors and cannot be viewed as a subset or an instance of a more general type of IT applications of Table 3.1. Because IT applications represent artificial reality, it may well be that fundamentally new IT applications will be invented that cannot be interpreted in terms of the eight archetypes of Table 3.1. That is why the typology of Table 3.1 is open-ended. Yet, situations like this are likely quite rare events.

If one accepts that IS is a discipline of computing that specifically focuses on IT applications in Table 3.1, it is clear that the territory of IS goes beyond the workplace use of utilitarian IT artifacts. Although non-utilitarian use is occasionally recognized in the IS literature (e.g., Davis et al. 2009, Lin and Bhattacherjee 2010), much of IS research still continues to assume utilitarian use in the workplace (e.g., Benbasat and Zmud 2003, Benbasat and Zmud 2006, Burton-Jones and Straub 2006, Barki et al. 2007).

3.3 External Practices to Be Supported by IS Research

The idea of IS as a practical discipline leads to the need to identify those practices that IS attempts to enable, to improve, and to correct. One can distinguish three interacting focus areas with associated actor groups depicted in Figure 3.1.*

Among three subfields of IT management—IT project management, management of IT function, and IT governance—the last one is introduced to emphasize three points.[†] First, it can be practiced at the national, regional, and global levels in addition to the organizational level as normally assumed. Second, IT governance is exercised by actors external to IT such as senior management in organizations, top government officials, and political leaders at the national, regional, and global levels. Third, IT governance practiced by these external actors points out that the core of IS should not be interpreted as exclusive. If IS researchers are able to inform these decision-makers in the issues of IT governance, the relevance of their work is likely very high, even though may include some errors of inclusion.

The three subfields of IT application usage are inspired by Barki et al. (2007), who distinguish three behaviors—technology interaction behaviors, task-technology adaptation behaviors, and individual adaptation behaviors—in the case of individual usage. The idea of Figure 3.1 is, however, that IT application usage can be viewed at different levels such as individuals, groups, organizations, communities, and societies. Therefore, the three behaviors suggested by Barki et al. (2007) are renamed aiming at more generality, pointing out that in addition to IT application use, the application may be redesigned in the adopting unit while used, and also the adopting unit may adapt or be adapted to reflect the IT application and its use.

Figure 3.1 identifies three subfields of IT application development: design, implementation, and evaluation. Design is here interpreted to cover also analysis activities needed in design. IS implementation refers to the process of adoption, appropriation, acceptance, instantiation, and/or institutionalization of the designed IT application in the adopting unit in question. Adapting Ammenwerth et al. (2004),

---

* The actor groups in Figure 3.1 may be interested in several areas, but for simplicity they are associated with one.
† IT Governance Institute (ITGI 2003, p. 10) defines IT governance as "the responsibility of the board of directors and executive management. It is an integral part of enterprise governance and consists of the leadership and organizational structures and processes which ensure that the organization's IT sustains and extends the organization's strategy and objectives."

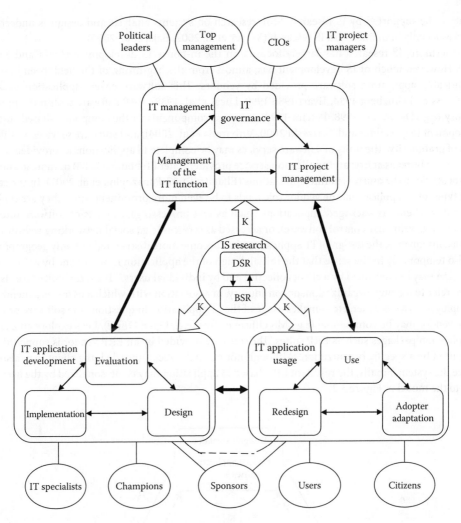

FIGURE 3.1 Extra-scientific practices supported by IS research.

evaluation is interpreted here as an act of measuring or exploring qualities (such as worth, value, benefits, cost, and impact) of an IT application and the related changes, the result of which informs a decision to be made concerning that application in a specific context. Evaluation studies in health informatics (Kaplan and Shaw 2004) illustrate the potential of evaluation studies to inform the practice. General IS research has also a similar potential, if IT constructs are incorporated more strongly in behavioral science research (BSR) models—as implied by the nomological net of Benbasat and Zmud (2003). In that case much of descriptive BSR would actually evaluate IT applications or types of IT application.

The dotted arrow in Figure 3.1 reminds that depending on the perspective redesign during usage may be viewed as a part of IT application usage and a part of design activity in IT application development. The following section elaborates this by distinguishing a three-tier model of IT application design.

3.4 A Three-Tier Model of IT Application Design

Benbasat and Zmud (2003, p. 191) claim that "our focus should be on how to best design IT artifacts and IS systems to increase their compatibility, usefulness, and ease of use or on how to best manage and support IT or IT-enabled business initiatives." If one accepts this view, design of IT applications is a key activity to be supported by IS research. Still, research on systems analysis and design is underrepresented especially in mainstream IS research (Vessey et al. 2002, Bajaj et al. 2005).

Furthermore, IS research has largely focused on the in-house IS development (Light and Sawyer 2007). However, much of IS development has, almost from the beginning of the field, been based on prefabricated application software provided by vendors. This software covers application packages (e.g., Gross and Ginzberg 1984, Iivari 1986/1990, Lucas et al. 1988), ERP software within the previous category (e.g., Davenport 1998, Pereira 1999), business components in the component-based software development (e.g., Fellner and Turowski 2000, Vitharana et al. 2004), and software services at different levels of granularity. The units of software services may be complete IS applications as provided by software service bureaus at first and application service providers later (Tebboune 2003) or more granulated units of service to be assembled into applications (Elfatatry 2007, Papazoglou et al. 2007). In the case of other types of IT application, the vendor role tends to be even more prominent, since they are typically acquired by clients as packaged applications (such as text processing), as parts of software-intensive products (e.g., vehicular control software), or accessed as services (e.g., social networking websites).

As a consequence, the design of IT applications has become more distributed not only geographically but also temporarily in the sense that the initial design of the application is often done by a vendor and the design may be continued by a client collective and by individual users.* The term "collective" is used here to refer to a group, organization, or community, in association with which a user—as a member, as an employee, as a customer, etc.—makes usage of the IT application in question. The software provided by the vendor may be commercial (e.g., MS Office) or free (e.g., Open Office). Let's explore an example of application packages such as ERP suites. The software provided by an ERP vendor is supposed to be configured by a specific organization. Although not normally recommended, the client may also customize the system. Finally, the resultant ERP-based IS application may be personalized by the user. This leads to the model of Figure 3.2.

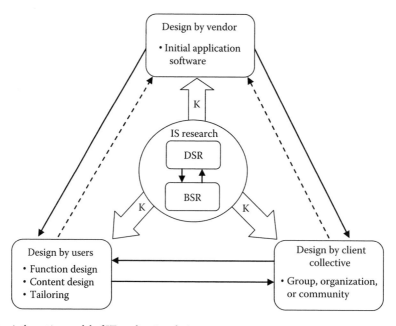

FIGURE 3.2 A three-tier model of IT application design.

---

* In each case there may be agents involved. The vendor may use contractors, the collective may have outsourced part of its
   software design, and the user may use technical experts to help. The present article does not delve into this complication.

Design by users in Figure 3.2 corresponds to "secondary design" in Germonprez et al. (2011) who discuss it at the functional layer and at the content layer. Figure 3.2 also includes tailoring by users as a third example, since "secondary design" does not necessarily concern the functionality of the system or its content, but may be limited to the redesign of the user interface, for example.

Figure 3.2 expands the distinction between primary design and secondary design in Germonprez et al. (2011) into a three-tier model. Design between users and design by client collective do not assume a predefined order, but can proceed either way or interactively. For example, a collective such as an organization may design a standard for web pages of its employees and after that employees are assumed to design their web pages so that they comply with the standard. On the other hand, a group or community may design its web pages bottom-up so that members design their web pages first and after that they are linked together to form the group's website.

From the viewpoint of the practical relevance of IS the three-tier model is significant in two respects. First, if IS research wishes to influence the quality of IT applications in practice, it is pivotal to pay attention to the initial design of those applications by vendors. It is obvious that the quality of this initial design significantly determines the quality of the final IT application, after possible configuration, customization, and user tailoring. Second, all this implies that we should pay much more attention to the vendor side in addition to the client side and user side as is traditionally done, although a better understanding of the latter ones may inform the vendor side, too.

3.5 How Can Information Systems Support Practice?

Figure 3.1 as well as Figure 3.2 assume that IS research—whether following the BSR orientation or the design science research (DSR) orientation—may enable, improve, or correct the related practices by providing relevant knowledge (K). This is the case also in the context of DSR, since the essence of IT artifacts as DSR contributions is knowledge (Iivari 2007).

Although DSR in IS has a potential to increase the practical relevance of IS research (Iivari 2007, Sein et al. 2011), the present chapter does not specifically focus on this research orientation, since it is widely discussed in the recent IS literature (e.g., Hevner et al. 2004).

Instead, this section focuses on how to make descriptive nomothetic research more design-oriented. Much of recent nomothetic IS research has been based on the TAM (Davis et al. 1989; see also Chapter 38 by Venkatesh et al. in this volume). Benbasat and Barki (2007) criticize it in that it has been weakly linked with design. Indeed, it has focused on contextual aspects such as computer anxiety, end user support, experience, gender, personal innovativeness, and self-efficacy (Lee et al. 2003) rather than on designable qualities of IT applications. Perceived ease of use, which—when operationalized as in Davis (1989)—closely corresponds to the core usability (Preece et al. 1994), is the only fairly designable quality of IT applications that has been systematically analyzed in TAM research. Other designable qualities such as system quality, information quality ($\approx$output quality) are only occasionally included (Venkatesh and Davis 2000, Wixom and Todd 2005).*

Benbasat and Zmud (2003) recommend that IS researchers should focus on IS aspects of the phenomena instead of treating IT artifacts as a "black box." In line with their recommendation, this section suggests a set of designable qualities of IT artifacts to be considered in nomological empirical research into IT applications.

Each IT application type in Table 3.1 has essential designable qualities (see Table 3.2). Since automating applications often aim at high autonomy of the system execution, system quality—such as reliability, security, and safety—is very important in their case (e.g., embedded vehicular software systems). Augmenting IT applications, on the other hand, typically are highly interactive, underlining usability in the sense of ease of use and ease of learning. Mediating IT applications provide a medium for social

---

* One should note that perceived usefulness in TAM is a more dependent variable. In terms of Table 3.2, it can be expected to be dependent on functional quality, information quality, usability ($\approx$perceived ease of use), and so on.

TABLE 3.2   Designable Qualities of IT Applications

Ideal Type	Designable Quality	Comments
Automating	System quality	System quality in the sense of DeLone and McLean (1992), excluding aspects of usability
Augmenting	Usability	Usability in the sense of ease of use and ease of learning (Preece et al. 1994)
Mediating	User-to-user interactivity	User-to-user interactivity supported by qualities such as reciprocity, multimodality, and responsiveness (McMillan 2002, Johnson et al. 2006)
	User identifiability	User identifiability covering user anonymity as a special case
Informating	Information quality	Information quality in the sense of DeLone and McLean (1992)
Entertaining	Funniness	Funniness of the characters, story, sound, graphics, interface, etc. (Davis and Carini 2005)
	User-to-system interactivity	User-to-system interactivity supported by qualities such as user control of the interaction, multimodality, and system responsiveness of the user–system the interaction (McMillan 2002, Johnson et al. 2006).
Artisticizing	Artistic quality	Artistic quality comprising technical skill and originality (Kozbelt 2004)
	Aesthetic quality	Aesthetic quality referring to the beauty of IT artifacts, covering classical aesthetics and expressive aesthetics (Lavie and Tactinsky 2004)
Accompanying	Emotional quality	The quality of emotions built in the IT artifact (Bates 1994, Fujita 2001)
Fantasizing	Identity constructability	Identity construction by choosing nicknames, wearing different embodiments (e.g., Avatars), and talking, discussing, and negotiating about various identities (Talamo and Ligorio 2001).
All	Functional quality	Does the system provide the expected or desired functionality to clients and its users or user groups?
		Does the system exclude undesired or not allowed functionality from clients, users, or user groups?

interaction between human beings. Therefore, user-to-user interactivity—such as reciprocity, multimodality, and responsiveness—is particularly significant in their case. Because information content is an essential part of IS proper, information quality is an essential characteristic of IS.

It seems obvious that computer games should be funny in some sense (Davis and Carini 2005) so that they are attractive and engaging. Just as in the case of any piece of art, "artistic quality"—that is, the technical skills and originality exhibited by the piece of art in question—is very essential in the context of computer art. In the case of digital pets, users are assumed to build a personal attachment with the pet. The pet also provides companionship (Friedman et al. 2003). Therefore, emotional quality as the quality of emotions built into a digital pet and exhibited by it is very essential in the case of this type of application.[*][†] Although one can claim that all communication includes identity construction

---

[*] Emotional quality as used in the present article is a topic of affective computing (Picard 1997), although affective computing is more interested in how computers can recognize, interpret, and process users' emotions than in producing artificial emotions in computers.

[†] Emotional quality differs from affective quality as defined by Zhang and Li (2005). Affective quality in Zhang and Li (2005) refers to the ability of an IT artifact to cause a change in the (core) affect of the user. Obviously, emotional quality of an IT artifact may influence its affective quality, but to my knowledge this is an unexplored territory. Sun and Zhang (2006) review the role of affect in IS research, but do not specifically address artificial emotions built in the IT artifacts.

(Hermans 2004), it is the most conspicuous in the case of virtual fantasy world applications where a user can typically choose and/or construct his/her identity on will or have a number of identities (Talamo and Ligorio 2001, Yee and Bailenson 2007). Identity constructability includes self-presentability as a special case, if the user wishes to present his/her real identity.

Table 3.2 summarizes the qualities presented earlier including a few additional ones. The final row identifies "functional quality" as a generic quality characteristic that is common to all software-based systems. ISO 9126 standard for software quality characteristics defines functionality by a question: "Are the required functions available in the software." Table 3.2 does not consider required functions or functional requirements as a reference of functional quality, but expectations and desires of clients, users, or user groups. Required functions or functional requirements as a baseline artifact in software development may more or less correspond to these expectations and desires. Functional quality also covers the question if the system excludes undesired or not allowed functionality from clients, users, or user groups. This aspect is close to compliance that is one subcharacteristic of functionality in the ISO 9126 standard (ISO 1991).

Table 3.2 identifies an application type where the relevance of the designable quality in question is most obvious and justifies its inclusion in that way. It does not, however, claim that the designable quality is relevant only in the case of that application type. As discussed earlier, IT applications in real life tend to be hybrids of several ideal types in Table 3.1. Therefore, when considering relevant designable qualities of a real application, it is justified to consider all the 12 qualities—whether or not they are relevant in the case of the IT application at hand.

Of course, one cannot claim that Table 3.2 is an exhaustive list of designable qualities of IT artifacts. For example, one could consider complementing Table 3.2 with some innovation characteristics such as complexity, compatibility, result demonstrability, and trialability (Rogers 1995). Although they can partly be influenced by design, generally they are less designable than the qualities in Table 3.2, and above all they are more general without any IT-specific substance.

The AIS website of theories used in IS research (http://istheory.byu.edu/wiki/Main_Page, accessed March 15, 2012) identifies 14 theories that have originated in IS. Of these, DeLone and McLean (1992, 2003) models and Media Synchronicity Theory (Dennis et al. 2008) are strongest in addressing designable qualities. Since the latter is very specific focusing CMC applications only, let us compare Table 3.2 with the DeLone and McLean models, which are more general. DeLone and McLean (1992) explicitly include only system quality and information quality and implicitly usability as an aspect of system quality as designable qualities. One should note, however, that the core usability differs quite a lot from more technical aspects of system quality such as reliability of the system, its maintainability (flexibility), and efficiency (resource utilization). Therefore, it is justifiable to separate it—in particular since users are the best experts to assess usability, while IT specialists are the best experts to assess the technical aspects of system quality. DeLone and McLean (2003) additionally include service quality, acknowledging that the latter may be more relevant in the case of IS functions than in the case of individual applications. Table 3.2 incorporates functional quality that resembles the "technical quality" aspect of service quality in Grönroos (1984).*

So, despite the fact that Table 3.2 is tentative, it provides a far more systematic and comprehensive list of designable qualities to be included compared with the DeLone and McLean models. As a consequence, one can imagine research inspired by Wixom and Todd (2005), for example, in which one has the qualities of Table 3.2 as independent variables and TAM/UTAUT constructs (Davis et al. 1989, Venkatesh et al. 2003) as intervening and/or dependent variables. Inclusion of all or most of the qualities of Table 3.2 as independent variables reduces the risk of specification errors. If they are

---

* Grönroos (1984) distinguishes two aspects in service quality: technical quality, which refers to what the customer is actually receiving from the service; and functional quality, which involves the manner in which the service is delivered. As a consequence, "technical quality" in the sense of Grönroos (1984) is close to functional quality in Table 3.2, whereas "functional quality" in the sense of Grönroos (1984) covers aspects of systems quality and usability.

systematically included, one can expect a more cumulative research tradition on designable qualities that significantly affect the success of IT applications.

3.6 Discussion and Conclusions

To summarize, the present chapter argues that IS is as a practical discipline of computing that specifically focuses on IT applications, on their development, use, and impact at individual, group, organizational, community, society, and global levels.

The major ideas of the chapter can be summarized in four points. First, the chapter encourages IS researchers to seek technological continuities and invariants in order to foster a cumulative research tradition in this world of technological change. The first question to be asked in the case of each new technology is whether or not it really is so new that it does not have any predecessors and cannot be viewed as an instance of a more general genre of technologies. If not, then one can ask how prior research into its possible predecessors and more general genre of technologies may inform the focal new technology, how the focal technology differs from existing comparable technologies, and what implications these differences may have.

To illustrate, although ERP-based IS are examples of application package-based IS, it is not a norm in the ERP literature to emphasize this continuity and to carefully review how earlier research on the implementation of application package-based IS could inform ERP implementation; "implementation" in this context covering design by the client organization (configuration and possible customization) and the final implementation. This has likely resulted in omission of relevant earlier research.*

Second, the chapter advices IS researchers to identify the extra-scientific practices (Strasser 1985) IS research aims at supporting and the associated actor groups to be informed. When suggesting practical or managerial implications of their research, IS researchers could be much more explicit in this regard. One reason is that what is actionable in practice depends on the actor groups. For example, top management support is actionable to managers but not to individual users, users' age and gender are not actionable to users and actionable to managers through recruiting at most.

Third, the chapter proposes that design of IT applications should be a central topic of IS research, because it is the key practice to be supported by IS research and because design of IT applications essentially affects the quality of IT applications and their success. Recognizing that software for most IT applications is acquired as packaged software or as prefabricated components and services, we should also pay more attention to the whole chain of design from the initial design by vendors and further design by client collectives and individual users.

Finally, the chapter suggests that IT applications can be conceptualized in terms of designable qualities. Focusing on designable qualities helps to build more IS-specific theories for explaining and predicting (Gregor 2006) than theories adopted from reference disciplines, which tend to be void of any IT-specific substance. Also many theories, which have originated in IS, tend to be fairly weak in addressing IT. For example, it is almost paradoxical that the major theoretical achievement of IS—Technology Acceptance Model (Davis et al. 1989)—is so general that it is totally empty of any IT-specific substance.† As a consequence, it can be applied in the case of almost any technology (such as farming technologies, Flett et al. 2004). One could expect that more IT-specific theories would be more informative to practitioners.

---

* As an example, Iivari (1986/1990) suggested a theoretical model that explains implementation challenges of application package-based information systems in terms of originality, complexity, radicalness, and divisibility of the systems. It seems obvious that these four dimensions are particularly relevant in the case of ERP-based information systems, which tend to be complex, may imply radical changes in business processes, and prefer borrowing (configuration) rather than adaptation (customization). Overall, all these three factors accentuate implementation challenges, only divisibility (i.e., modularity) of ERP systems mitigating them. Yet, the four factors and their interactions have not been systematically discussed in the ERP literature.

† The IT substance is just a matter operationalization of perceived ease of use and perceived usefulness.

Furthermore, modeling application types and specific applications in terms of variables such as designable qualities in Table 3.2 allows us to conduct more general and time-invariant research, since application types typically include several specific applications that vary in terms of designable qualities just as specific IT applications do during their evolution. For example, we have currently numerous social networking websites. When researching them, it would be advisable to include a number of sites in a study rather than a single one in order to increase "objective" variance of each designable quality. Assuming that there are standard measurement instruments for designable qualities, independent studies investigating different websites or a single website at different times are also more easily comparable.

But above all, designable qualities constitute parameters for design. Therefore, when conducting nomothetic research, this chapter suggests that the incorporation of designable qualities is primary and the inclusion of non-designable and non-actionable contextual factors only secondary. Although the latter may significantly increase the variance explained and therefore may be intellectually interesting to include, interpreting IS as a practical discipline underlines that it is more important to understand how designable qualities explain the success of IT applications and in this way to help practitioners to design better systems.

References

Aarseth, E., Computer game studies, Year One, *The International Journal of Computer Game Research* [online] 1(1), 2001, (http://www.gamestudies.org/0101/editorial.html)

Agarwal, R. and Lucas, H.C. Jr., The information systems identity crisis: Focusing on high-visibility and high-impact research, *MIS Quarterly*, 29(3), 2005, 381–398.

Ammenwerth, E., Brender, J., Nykänen, P., Prokosch, H.U., Rigby, M., and Talmon, J., Visions and strategies to improve evaluation of health information systems: Reflections and lessons based on the HIS-EVAL workshop in Innsbruck, *International Journal of Medical Informatics*, 73, 2004, 479–491.

Bajaj, A., Batra, D., Hevner, A., Parsons, J., and Siau, K., Systems analysis and design: Should we be researching what we teach? *Communications of the Association for Information Systems*, 15, 2005, 478–493.

Banville, C. and Landry, M., Can the field of MIS be disciplined? *Communications of the ACM*, 32(1), 1989, 48–60.

Barki, H., Titah, R., and Boffo, C., Information system use-related activity: An expanded behavioral conceptualization of individual-level information system use, *Information Systems Research*, 18(2), 2007, 173–192.

Baskerville, R.L. and Myers, M.D., Fashion waves in information systems research and practice, *MIS Quarterly*, 33(4), 2009, 647–662.

Bates, J., The role of emotion in believable agents, *Communications of the ACM*, 37(7), 1994, 122–125.

Benbasat, I. and Barki, H., Quo vadis, TAM? *Journal of the Association for Information Systems*, 8(4), 2007, 211–218.

Benbasat, I. and Zmud, R.W., Empirical research in information systems: The practice of relevance, *MIS Quarterly*, 23(1), 1999, 3–16.

Benbasat, I. and Zmud, R.W., The identity crisis within the discipline: Defining and communicating the discipline's core properties, *MIS Quarterly*, 27(2), 2003, 183–194.

Benbasat, I. and Zmud, R.W., Further reflections on the identity crisis, in King, J.L. and Lyytinen, K. (eds.), *Information Systems: The State of the Field*, John Wiley & Sons, Chichester, England, 2006, pp. 300–306.

Burton-Jones, A. and Straub, D.W., Conceptualizing system usage: An approach and empirical test, *Information Systems Research*, 17(3), 2006, 228–246.

Davenport, T.H., Putting the enterprise into the enterprise system, *Harvard Business Review*, 76(4), July–August 1998, 121–131.

Davis, F.D., Perceived usefulness, perceived ease of use, and user acceptance of information technology, *MIS Quarterly*, 13(3), 1989, 319–339.

Davis, F.D., Bagozzi, R.P., and Warshaw, P.R., User acceptance of computer technology: A comparison of two theoretical models, *Management Science*, 35(8), 1989, 982–1003.

Davis, S. and Carini, C., Constructing a player-centred definition of fun for video games design, *People and Computers XVIII—Design for Life*, Springer-Verlag, London, U.K., 2005, pp. 117–132.

Davis, A., Murphy, J., Owens, D., Khazanchi, D., and Zigurs, I., Avatars, people, and virtual worlds: Foundations for research in metaverses, *Journal of the Association for Information Systems*, 10(2), 2009, 90–117.

DeLone, W.H. and McLean, E.R., Information systems success: The quest for the dependent variable, *Information Systems Research*, 3(1), 1992, 60–95.

DeLone, W.H. and McLean, E.R., The DeLone and McLean model of information systems success: A ten-year update, *Journal of Management Information Systems*, 19(4), 2003, 9–30.

Dennis, A.R., Fuller, R.M., and Valacich, J.S., Media, tasks, and communication processes: A theory of media synchronicity, *MIS Quarterly*, 32(3), 2008, 575–600.

Edmonds, E., Turner, G., and Candy, L., Computer human interface: Approaches to interactive art systems, *Proceedings of the 2nd International Conference on Computer Graphics and Interactive Techniques in Australasia and South East Asia*, Singapore, 2004, pp. 113–117.

Elfatatry, A., Dealing with change: Components versus services, *Communications of the ACM*, 50(8), 2007, 35–39.

Fellner, K.J. and Turowski, K., Classification framework for business components, *HICSS'2000, Proceedings of the 33rd Hawaii International Conference on System Sciences*, Maui, HI, 2000, pp. 1–10.

Flett, R., Alpass, F., Humphries, S., Massey, C., Morris, S., and Long, N. The technology acceptance model and use of technology in New Zealand dairy farming, *Agricultural Systems*, 80(2), 2004, 199–211.

Friedman, B., Kahn, P.H. Jr., and Hagman, J., Hardware companions? What online AIBO discussion forums reveal about the human-robotic relationship, *Proceedings of the SIGCHI Conference on Human Factors in Computing Systems (CHI 2003)*, Ft. Lauderdale, FL, 2003, pp. 273–280.

Fujita, M., AIBO: Toward the era of digital creatures, *The International Journal of Robotics Research*, 20(10), 2001, 781–794.

Germonprez, M., Hovorka, K., and Gal, U., Secondary design: A case of behavioral design science research, *Journal of Association for Information Systems,* 12(10), 2011, 662–683.

Gregor, S., The nature of theory in information systems, *MIS Quarterly*, 30(3), 2006, 611–642.

Grönroos, C., A service quality model and its marketing implications, *European Journal of Marketing*, 18(4), 1984, 36–44.

Gross, P.H.B. and Ginzberg, M.J., Barriers to the adoption of application software packages, *Systems, Objectives, Solutions*, 4(4), 1984, 211–226.

Hassan, N.R., Is information systems a discipline? Foucauldian and Toulminian insights, *European Journal of Information Systems*, 20, 2011, 456–476.

Hermans, H.J.M., Introduction: The dialogical self in a global and digital age, *Identity: An International Journal of Theory and Research*, 4(4), 2004, 297–320.

Hevner, A.R., March, S.T., Park, J., and Ram, S., Design science in information systems research, *MIS Quarterly*, 28(1), 2004, 75–105.

Iivari, J., Implementability of in-house developed vs. application package based information systems, in *ICIS'1986, Proceedings of the Seventh International Conference on Information Systems*, San Diego, CA, 1986, pp. 67–80 (reprinted in *Data Base*, 21(1), 1990, pp. 1–10).

Iivari, J., A paradigmatic analysis of information systems as a design science, *Scandinavian Journal of Information Systems*, 19(2), 2007, 39–63.

ISO (International Organization for Standardization), ISO/IEC: 9126 Information technology-software product evaluation-quality characteristics and guidelines for their use, 1991. (www.cse.dcu.ie/essiscope/sm2/9126ref.html) (accessed February 17, 2012).

ITGI, Board briefing on IT governance, 2003. Available online: www.itgi.org (accessed January 22, 2012).

Johnson, G.J., Bruner, G.C., and Kumar, A., Interactivity and its facets revisited, theory and empirical test, *Journal of Advertising*, 35(4), 2006, 35–52.

Kaplan, B. and Shaw, N.T., Future directions in evaluation research: People, organizational, and social issues, *Methods of Information in Medicine*, 43(3), 2004, 215–231.

Keen, P., Relevance and rigor in information systems research, in Nissen, H.-E., Klein, H.K., and Hirschheim, R. (eds.), *Information Systems Research: Contemporary Approaches and Emergent Traditions*, Elsevier Publishers, Amsterdam, the Netherlands, 1991, pp. 27–49.

King, J.L. and Lyytinen, K. (eds.), *Information Systems, The State of the Field*, John Wiley & Sons, Chichester, U.K., 2006.

Kozbelt, A., Originality and technical skills as components of artistic quality, *Empirical Studies of the Arts*, 22(2), 2004, 157–170.

Kusahara, M., The art of creating subjective reality: An analysis of Japanese digital pets, *Leonardo*, 34(4), 2001, 299–302.

Lavie, T. and Tractinsky, N., Assessing dimensions of perceived visual aesthetics of web sites, *International Journal Human-Computer Studies*, 60, 2004, 269–298.

Lee, Y., Kozar, K.A., and Larsen, K.R.T., The technology acceptance model: Past, present, and future, *Communications of the Association for Information Systems*, 12(50), 2003, 752–780.

Light, B. and Sawyer, S., Locating packaged software in information systems research, *European Journal of Information Systems*, 16, 2007, 527–530.

Lin, C.-P. and Bhattacherjee, A., Extending technology usage models to interactive hedonic technologies: A theoretical model and empirical test, *Information Systems Journal*, 20(2), 2010, 163–181.

Lucas, H.C., Walton, E.J., and Ginzberg, M.J., Implementing packaged software, *MIS Quarterly*, 1988, 12(4), 537–549.

Mason, R., Comments on the Weber commentary and Lyytinen/King response, *Journal of the Association for Information Systems*, 7(11), 2006, 722–724.

McMillan, S.J., Exploring models of interactivity from multiple research traditions: Users, documents, and systems, in Liverow, L. and Livingstone, S. (eds.), *Handbook of New Media*, Sage, London, U.K., 2002, pp. 162–182.

Noll, A.M., The beginnings of computer art in the United States: A memoir, *Computers and Graphics*, 19(4), 1995, 495–503.

Oates, B.J., New frontiers for information systems research: Computer art as an information system, *European Journal of Information Systems*, 15, 2006, 617–626.

Orlikowski, W.J. and Iacono, C.S., Research commentary: Desperately seeking the "IT" in IT research— A call theorizing the IT artifact, *Information Systems Research*, 12(2), 2001, 121–134.

Papazoglou, M.P., Traverso, P., Dustdar, S., and Leymann, F., Service-oriented computing: State of the art and research challenges, *IEEE Computer*, 40(11), 2007, 38–45.

Pereira, R.E., Resource view theory analysis of SAP as a source of competitive advantage for firms, *The Data Base for Advances in Information Systems*, 30(1), 1999, 38–46.

Picard, R.W., *Affective Computing*, MIT Press, Cambridge, MA, 1997.

Preece, J., Rogers, Y., Sharp, H., Benyon, D., Holland, S., and Carey, T., *Human-Computer Interaction*, Addison-Wesley, Workingham, U.K., 1994.

Rogers, E.M., *Diffusion of Innovations*, 4th edn., The Free Press, New York, 1995.

Sein, M., Henfridsson, O., Purao, S., Rossi, M., and Lindgren, R., Action design research, *MIS Quarterly*, 35(1), 2011, 37–56.

Shackelford, R., Cross, J., Davies, G., Impagliazzo, J., Kamali, R., LeBlanc, R., Lunt, B., McGettrick, A., Sloan, R., and Topi, H., *Computing Curricula 2005—The Overview Report*, ACM/IEEE Computer Society, 2005, (http://www.acm.org/education/education/curric_vols/CC2005-March06Final.pdf) (accessed on March 29, 2013).

Strasser, S., *Understanding and Explanation Basic Ideas Concerning the Humanity of the Human Sciences*, Duquesne University Press, Pittsburg, PA, 1985.

Sun, H. and Zhang, P., The role of affect in information systems research: A critical survey and research model, in Zhang, P. and Galletta, D. (eds.), *Human-Computer Interaction and Management Information Systems: Foundations*, M.E. Sharpe, Armonk, NY, 2006, pp. 295–329.

Talamo, A. and Ligorio, B., Strategic identities in cyberspace, *Cyberpsychology and Behavior*, 4(1), 2001, 109–122.

Tebboune, D.E.S., Application service provision: Origins and development, *Business Process Management Journal*, 9(6), 2003, 722–734.

Venkatesh, V. and Davis, F.D., A theoretical extension of the technology acceptance model: Four longitudinal field studies, *Management Science*, 46(2), 2000, 186–204.

Venkatesh, V., Morris, M.G., Davis, G.B., and Davis, F.D., User acceptance of information technology: Toward a unified view, *MIS Quarterly*, 27(3), 2003, 425–478.

Vessey, I., Ramesh, V., and Glass, R.L., Research in information systems: An empirical study of diversity in the discipline and its journals, *Journal of Management Information Systems*, 19(2), 2002, 129–174.

Vitharana, P., Jain, H., and Zahedi, F.M., Strategy-based design of reusable business components, *IEEE Transactions on Systems, Man, and Cybernetics—Part C: Applications and Reviews*, 34(4), 2004, 460–474.

Wixom, B.H. and Todd, P.A., A theoretical integration of user satisfaction and technology acceptance, *Information Systems Research*, 16(1), 2005, 85–102.

Yee, N. and Bailenson, J., The proteus effect: The effect of transformed self-representation on behavior, *Human Communication Research*, 33, 2007, 271–290.

Zhang, P. and Li, N., The importance of affective quality, *Communications of the ACM*, 48(9), 2005, 105–108.

Zuboff, S., *In the Age of the Smart Machine, The Future of Work and Power*, Heineman, Oxford, U.K., 1988.

4

Information Technology: Principles, Methods, and Theory

Han Reichgelt
*Southern Polytechnic
State University*

Joseph J. Ekstrom
Brigham Young University

Art Gowan
Georgia Southern University

Barry M. Lunt
Brigham Young University

4.1 Introduction

Information technology (IT), in the narrow sense in which the term is used in particular in North America, is the newest computing discipline to emerge over the past 20 years or so. The Computing Accreditation Commission of ABET, the primary accreditation agency for programs in computing, engineering, and technology in the United States, and increasingly beyond, first promulgated separate accreditation criteria for programs in IT in 2002. The Association for Computing Machinery (ACM) and Computer Society of the Institute of Electrical and Electronic Engineers (IEEE-CS) published its first model curriculum for programs in IT in 2008 (Lunt et al., 2008) although IT was mentioned already in the 2005 overview volume (JTFCC, 2005) in which an ACM, IEEE-CS, and Association for Information Systems (AIS) joint task force compared the various computing disciplines (http://www.acm.org/education/education/curric_vols/CC2005-March06Final.pdf).

Before discussing research in IT, it is good to make a distinction between the broader use of the term "information technology" and the narrower sense in which the term will be used in this chapter. In the broader sense, the term "information technology," which was first coined by Leavitt and Whisler (1958), refers to the whole of computing and includes all the computing disciplines, including computer engineering, computer science, information systems (IS), software engineering, and the myriad of emerging computing disciplines. In this broader sense, "IT" is therefore more or less synonymous with the terms "computing" and "informatics."

In the narrower sense, the term "information technology" is defined as follows by Lunt et al. (2008, p. 9):

> IT, as an academic discipline, is concerned with issues related to advocating for users and meeting their needs within an organizational and societal context through the selection, creation, application, integration and administration of computing technologies.

With mathematics serving as the foundation, modern computer science was essentially born with the advent of the digital computer, proceeded in theory in 1931 by Kurt Gödel's incompleteness theorem and 1936 Alan Turing's and Alonzo Church's formalization of the concept of an algorithm. The first computerized business applications of the 1950s, often called electronic data processing, gave birth to the field of IS. Other scientific and engineering computing applications gave rise to software and computer engineering. By the 1990s, with the growing complexity of computing technologies, especially with the wide scale of integration provided by the Internet, IT emerged as a separate computing discipline primarily because of the workforce development needs (Lunt et al., 2008; Gowan and Reichgelt, 2010) to support the infrastructure. IT programs evolved because institutions heard from potential employers of graduates from computing programs that computer science programs often did not impart the more practice-oriented system and network administration skills in their graduates, focusing instead on the more theoretical aspects of computing. Programs in IS, on the other hand, often did not instill the required technical knowledge in their graduates, concentrating instead on issues concerning the management and use of IT in organizations.

4.2 Identity of the Field of IT

So, unlike many other new disciplines, the field of IT did not emerge out of a need to answer novel research questions, or out of a need to approach existing research questions through new methodological approaches. The discipline of IS arose out of the realization that questions regarding the use of computing technologies within organizations were not addressed sufficiently well in, for example, social and/or organizational psychology, sociology, or computer science. And one can argue that the discipline of cognitive science arose out of the realization that there was a greater need for theory building regarding and modeling of cognition than was possible in the main other discipline studying cognition, cognitive psychology. At the time of the emergence of cognitive science, psychology was still dominated by the behaviorist mind-set and attempted to explain every cognitive behavior simply in terms of the stimuli the individual received and the responses produced. Any speculation about the internal processes that mediated between stimuli and responses was frowned upon. Cognitive science was less beholden to this tradition and used the new methodology of computer modeling to approach the study of cognition.

The origin of the discipline of IT to a large extent explains the fact that it has not as yet coalesced around as clear a set of research questions as the other computing disciplines. Benbasat and Zmud (2003), refining work by Aldrich (1999) on organizational identity, argue that for a discipline to establish its own recognized identity within an existing environment, it must resolve two issues, namely, a legitimacy issue and a learning issue. The legitimacy issue itself consists of two subissues, namely, sociopolitical legitimacy and cognitive legitimacy. Sociopolitical legitimacy is conferred on a discipline when it is recognized as a separate discipline by key stakeholders, including the general public, governmental agencies, and, in the case of academic disciplines, academic administrators. IT fairly quickly obtained sociopolitical legitimacy. For example, shortly after its emergence, the Computing Accreditation Commission of ABET, the lead organization for accrediting programs in computing in the United States, and increasingly elsewhere, invited the IT community to formulate its own set of accreditation criteria. IT was also included in the efforts by the ACM, the AIS, and the IEEE-CS to formulate model curricula for the different computing disciplines.

However, the same cannot be said for cognitive legitimacy. Cognitive legitimacy is conferred on a discipline if it is recognized by others within the academic community as addressing a set of questions that are not addressed by any other discipline (such as IS did when it emerged) or when it is recognized as addressing questions that are addressed by other disciplines in novel ways (such as cognitive science). IT has not achieved cognitive legitimacy as yet. For example, there are very few PhD programs in North America in IT in the narrow sense in which the term is used here and very few research-oriented journals that focus on IT as a discipline.

However, an academic discipline is unlikely to achieve cognitive legitimacy until it can resolve the other issue highlighted by Benbasat and Zmud, namely, the learning issue. The learning issue essentially involves the community establishing a coherent set of issues to study and/or developing its own set of methodologies, theories, and concepts to approach these issues. The IT community appears to be aware of the importance of resolving the learning issue, and there are a number of papers that attempt to discuss what unique contribution the discipline of IT might make to research in computing and what an IT agenda might look like (e.g., Reichgelt, 2004; Ekstrom et al., 2006; Agresti, 2011).

Many of the papers attempting to resolve the learning issue for IT contrast the discipline with other computing disciplines, in particular the field of IS. This may reflect the fact that the field of IS, more so than the other computing disciplines, seems to periodically engage in the type of meta-level discussions that, to date, have characterized many of the research-oriented discussions in the IT community (see, e.g., Benbasat and Zmud, 2003; Myers et al., 2011). It may also reflect the fact that the field of IS is the discipline that is most closely related to IT. For example, the most recent model curriculum for IS states that IS programs sometimes go under the name "information technology" (Topi et al., 2010, p. 12).

Moreover, when trying to draw a contrast between IT and the other computing disciplines, authors tend to rely on information about undergraduate programs. For example, early in the emergence of IT as a new computing discipline, a number of studies compared undergraduate programs in computer science, IS, and IT (e.g., Hislop, 2003; Reichgelt et al., 2004; Lunt et al., 2005), both in an attempt to legitimize IT as a separate computing discipline and to show how it differed from the other computing disciplines. More recent attempts to draw a contrast between IT and the other computing disciplines (e.g., Agresti, 2011) have relied on the so-called overview volume. The overview volume was published by a combination of professional organizations in an effort to draw a contrast between the computing disciplines of computer engineering, computer science, IS, IT, and software engineering (JTFCC, 2005). One can, of course, argue as to whether the structure of undergraduate programs should be used in trying to contrast the research agendas of the discipline. In the absence of any other sources, this is as good a place to start as any.

Three themes emerge from these discussions, namely, the central role of what has been called the IT infrastructure, its deployment, and the use of integration in its creation.

4.3 Underlying Themes for the Discipline of IT

The first theme that is central to many of the discussions on how to resolve the learning issue for IT is that of the IT infrastructure. The term "IT infrastructure" is defined by van Bon et al. (2007, p. 338) as "all the hardware, software, networks, facilities etc., that are required to develop, test, deliver, monitor or support IT services." It then goes on to define an IT service as a service that is based on the use of the IT [defined as the use of technology for the storage, communication, or processing information] and supporting the customer's business processes. The IT infrastructure is a subset of the "IT artifact." The latter has figured prominently in discussions on the nature of the field of IS and is defined by Benbasat and Zmud (2003) as "the application of IT to enable and support task(s) within a structure(s) that is itself embedded within a context(s)." While there have been calls for improved conceptualizations of the IT artifact (e.g., Orlikowski and Iacono, 2001), the term, as defined by Benbasat and Zmud, encompasses not only the IT infrastructure but also the tasks, the task structure, and the task context in which the IT infrastructure is used. While there is considerable debate within the IS community as to the centrality of this concept to the discipline, with, for example, Benbasat and Zmud (2003) arguing that it had to have a central place in defining the discipline, with others disagreeing (e.g., Alter, 2003; DeSanctis, 2003; Ives et al., 2004), there is little debate within the IT community that the IT infrastructure has to play a central role in the discipline. For example, the IT model curriculum (Lunt et al., 2008) contains the following 13 knowledge areas:

1. Programming Fundamentals [PF]
2. Information Technology Fundamentals [ITF]
3. Information Management [IM] (covering essentially database management)
4. Human Computer Interaction [HCI]
5. Social and Professional Issues [SP]
6. Platform Technologies [PT] (covering essentially hardware)
7. Networking [N]
8. System Integration and Architecture [SIA]
9. Integrative Programming and Technologies [IPT]
10. Web Systems and Technologies [WEB]
11. Mathematics and Statistics for IT [MATH]
12. Information Assurance and Security [IAS]
13. System Administration and Maintenance [SAM]

If we ignore those knowledge areas that either cover basic material or focus on general education topics (PF, ITF, SP, MATH), five of the remaining nine (IM, PT, N, IPT, WEB) directly relate to the IT infrastructure.

Clearly, the field of IT is not the only computing discipline to cover portions of these knowledge areas. For example, computer engineering covers knowledge areas such as platform technologies and networking, while software engineering and computer science would both claim to cover information management, programming (PF and IPT in the IT model curriculum), web technologies, human computer interaction, and at least some aspects of networking. IM and HCI both have a central role in the IS discipline.

However, and here is where the second theme emerges, as Agresti (2011) points out, the way in which these topics are covered in IT differs significantly from the way in which they are covered in the other computing disciplines. Agresti argues that each of the major computing disciplines has a separate theme. Thus, computer science, in its purest form, focuses on the theory underlying computing and seeks to formulate the mathematical and logical foundations of computing. Computer engineering focuses on the hardware necessary to physically realize computation, while software engineering focuses on the development of particularly the applications to run on the hardware. IS's focus is on the use of IT (used in the broader sense) to create business value and to provide support to the organization. IT, finally, focuses on the deployment of the technologies and their integration with existing systems. Agresti summarizes his analysis as follows:

Discipline	Focus
Computer Science	Theory
Computer Engineering	Hardware
Software Engineering	Development
Information Systems	Organizations
Information Technology	Deployment

The model curriculum for IT programs includes knowledge areas that are directly relevant to deployment, in particular SIA and SAM. One could also argue that the knowledge areas of IAS and HCI are related to the issue of the deployment of the IT infrastructure.

A third, closely related theme, in the IT literature is the theme of integration. There are two different ways in which this theme emerges.

The first is most directly related to the deployment theme and concerns the integration of the IT infrastructure into the organization, including users of the IT infrastructure. Clearly, this is very similar to what the field of IS considers to be its primary domain of research, and there is a significant overlap between the two disciplines in this area. However, whereas IT tends to focus more on the individual users, as witnessed by the inclusion of HCI among the knowledge areas in its model curriculum, IS places a greater emphasis on broader organizational issues. This is not to say that IT completely ignores the organizational focus, or that IS ignores the individual user. Indeed, HCI is a thriving area of research within IS and, indeed, computer science. The difference is primarily one of emphasis. Going back to the Zmud and Benbasat definition of the IT artifact as "the application of IT to enable and support task(s) within a structure(s) that is itself embedded within a context(s)," one could argue that IT approaches the issue of HCI primarily by focusing on the application of IT to support tasks, while IS is also concerned about the organizational structures and the context in which these structures are embedded.

The second way in which the theme of integration emerges in the field of IT concerns the construction of the IT infrastructure. Whereas the fields of computer engineering and software engineering tend to focus on one element of the IT infrastructure, hardware and software, respectively, the field of IT is primarily concerned with how one can integrate existing components into a new IT infrastructure to the benefit of the organization. Ekstrom and Lunt (2003) argue that IT programs should not require students to develop a deep knowledge of how individual technology components are to be implemented. Rather, such programs should instill in their students a deep understanding of the interfaces between the component technologies. This sentiment is also reflected in the IT model curriculum, where the more advanced programming knowledge area is called "IPT."

4.4 Formulating Research Questions for IT

Having distinguished IT from the other computing disciplines, a number of questions arise when it comes to the issue of research, namely,

1. Are there a series of research questions that are unique to the field of IT and that are not covered by any other computing discipline?
2. Are there areas that are currently "claimed" by one of the other computing disciplines that could be "claimed" by IT?
3. For those research questions for which there is a significant overlap with other disciplines, is there a unique approach that IT brings and that is likely to uncover insights that are not provided by the other computing disciplines?

The questions in some way parallel the three strategies that Agresti (2011) suggests the IT community may follow in formulating a research agenda.

Agresti's first proposed strategy is to seize on "judicial vacancies," i.e., to lay claim on research questions that were never considered by other disciplines, or that are used to be but are no longer considered by another discipline. Clearly, this strategy parallels our first and third questions.

Agresti's second proposed strategy is to build a research agenda by looking at current practice and abstracting frameworks and theories from these. This particular strategy would align well with the research that is done by many IT practitioners often resulting in white papers and case studies. Whereas such material often provides valuable insights into practice, there are few attempts to extract insights from these that are widely applicable and even fewer attempts to empirically or otherwise validate the more general conclusions from these studies. Agresti's suggestion of building theoretical frameworks from practice parallels our third question. We return to this point in the next section.

Agresti's third strategy is to establish a stronger tie with a reference discipline. Agresti argues that one of the reasons for the cognitive legitimacy of some of the other computing disciplines is their strong tie to well-established reference disciplines. Thus, computer science has as its reference discipline mathematics and derives some of its cognitive legitimacy from this link. Agresti argues that IT would do well to strengthen its tie with a well-established reference discipline and suggests that it looks to systems science (Bailey, 2005) as its reference discipline.

It is of course debatable whether systems science is as appropriate a reference discipline for IT as, say, mathematics is for computer science, primarily because the discipline is not yet as widely established. However, the general notion that IT looks at insights coming out of systems science and systems engineering does seem appropriate.

4.5 Research Questions Emanating from the Underlying Themes for IT

Given the absence of a well-established research tradition in IT, many of the answers to the three questions posed on the previous section are likely not to be unanimously agreed to by the IT community. In many ways, this chapter is, therefore, more intended as a catalyst for further discussion on IT research both within the IT community and within the larger computing research community. However, given the themes of the central role of the IT infrastructure, its deployment, and the use of integration in its creation, we can start formulating some tentative answers.

The use of integration in the creation of the IT infrastructure in particular is likely to lead to a series of research questions that are unique to IT. The following examples, in particular, are relevant:

First, is it possible to develop a framework to predict some of the emergent properties of the IT infrastructure, based on the value of that property in the components that were integrated and the way in which they were integrated to create the IT infrastructure? For example, given some value of the level of security provided by each of the components integrated into a particular IT infrastructure and the methods used to integrate them, is it possible to make some predictions about the level of security provided by the resulting system? One can ask similar questions, of course, about availability, cost of ownership, and so on. Moreover, those questions can also be asked about more dynamic system performance-related issues, such as throughput and power consumption.

The framework question also allows one to draw a cleaner distinction between the questions one might pursue as an IT researcher and related questions one might ask as a researcher in a different computing discipline. For example, computer scientists and computer engineers, in particular network engineers, are likely to be able to provide measures of security, availability, etc., as far as individual components are concerned. Thus, a computer scientist or software engineer will be able to answer questions about how vulnerable a particular software application is to malicious attacks, while a computer engineer may be able to answer questions about how easy it is to hack into network traffic on a particular type of network. Similarly, an IS researcher is very likely to be able to answer questions about the integration of the IT infrastructure into the organization and what factors need to be taken into account to ensure that security risks are minimized. However, the emphasis on the IT infrastructure itself is unique to IT.

Clearly, once developed, the framework should also have the ability to predict how the integration of a new component into an existing IT infrastructure is likely to affect the properties of the overall infrastructure. How will security be affected by the integration of some cloud-based service into the IT infrastructure? What effect will it have on availability, access, reliability, etc. Again, what sets IT apart in this respect from, in particular, IS is the emphasis on the IT infrastructure. The field of IS has interesting insights to offer on how the introduction of a new aspect may affect organizational performance, but its focus is exclusively on the organization and on the individuals within it. It does not concern itself with the IT infrastructure as an entity in its own right.

A second related question concerns issues to do with trade-offs between different aspects of the IT infrastructure. For example, it is possible to provide top-quality user support by making available a personal assistant to every user of the IT infrastructure, but clearly this is not feasible from a financial point of view. Similarly, it is possible to fully secure the IT infrastructure by denying access to any user, but this also is impractical. The question that follows from this is whether it is possible to arrive at a high-level framework for determining the optimum point on various trade-offs. For example, what is the most appropriate point on the security–availability trade-off, what factors determine the appropriate point on this trade-off, and how are these factors weighted in a given organization?

Again, the question about trade-offs is not unique to IT. For example, the space–time trade-off is a well-known principle in computer science and computer engineering. Similarly, concerns about the cost/benefits trade-offs of IT applications underlie IT business value analysis, a subfield of IS. The main question of IT business value analysis is how IT impacts organizational performance of firms both at the intermediate process level and at the organizational level, and both in terms of (internal) efficiency and (external) in terms of competitiveness (Melville et al., 2004).

Similar concerns about trade-offs between costs and otherwise desirable properties of software applications, such as maintainability and loose coupling, also underlie Boehm and Sullivan's proposal that software engineering research pay greater attention to software economics (Boehm and Sullivan, 2000). Software economics proposes to bring economic reasoning about product, process, program, portfolio, and policy issues to software design and engineering. There is an obvious similarity here with IT in that one part of the IT research agenda is aimed at bringing reasoning about processor, process, program, portfolio and policy issues to the creation, implementation, and management of the IT infrastructure.

However, IT again brings a unique perspective on questions related to trade-offs and significantly broadens these concerns. The trade-offs considered in IT concern all aspects of the IT infrastructure and, as a result, concern a larger number of trade-offs than those considered in other disciplines. Thus, computer science and computer engineering restrict themselves to the space–time trade-off, whereas IT business value and software economics tend to concentrate on the trade-off between certain aspects of the IT infrastructure and value, where value is typically defined in financial terms. While the financial and organizational aspects are relevant to IT as well, IT also considers more technical trade-offs between different properties of the IT infrastructure, including cost and value, security, availability, usability, and scalability.

There are also a series of research questions that follow from the theme of deployment of the IT infrastructure. Whereas other computing disciplines have contributions to make to questions related to the creation of components of the IT infrastructure and/or to questions related to the deployment within an organizational framework, IT uniquely considers deployment from a technical point of view.

The theme of deployment is most closely aligned with Agresti's suggestion that IT attempts to build theoretical insights from practice. As we stated earlier, there is a lot of anecdotal evidence on the value of certain processes, procedures, and best practices in deploying (aspects) of an IT infrastructure, often in the form of case studies and white papers, and often published by IT practitioners. While this material is extremely valuable, the research reported in it often does not meet the standards of rigor that is expected in academic research. In addition, almost no attempt is made to generalize from these case studies. As a result, such studies are often rejected by academic researchers.

However, the fact that the research as it stands does not meet the standards set by the academic community does not mean that it should be dismissed out of hand. As an academic discipline, IT might try to take some of this anecdotal information and turn it into scientifically validated theories. For example, there are a large number of studies available from IT consulting firms demonstrating the usefulness of adopting best practices as documented in the IT Infrastructure Library (ITIL) to improve the quality of the IT service delivery. Such papers provide valuable insight into practice and a thorough analysis of such papers might lead to the type of insights regarding the IT infrastructure that are unlikely to be obtained by following the more traditional scientific methodologies adopted by the other computing disciplines, in particular IS, which involve formulating a hypothesis based on a more widely adopted theory and testing that hypothesis. One can, of course, create similar research projects also around other IT best practice frameworks, such as CoBIT.

A related deployment-related series of questions concerns the derivation of theoretical insights, including best practices, from anecdotal information about other deployment-related issues, including business continuity and virtualization and cloud computing. In all cases, the thrust would be from documented case studies to general insights.

The theme of deployment and integration of the IT infrastructure within a structure of tasks to be performed by an end user also provides another avenue for IT research. While the various issues identified earlier do not appear to have been researched in a more abstract way, there is a growing literature on the use of the IT infrastructure within a particular application domain. A good example is provided by the health care industry where there is a growing number of well-established and well-regarded scientific journals on the use of IT to support health care delivery, including, for example, the *BMC Medical Informatics and Decision Making, Journal of the American Medical Informatics Association*, and the *Journal of Biomedical Engineering and Computing*. Moreover, one often finds articles related to the deployment of the IT infrastructure in journals that appear to be more geared to health care providers. For example, there are a number of articles on the inappropriateness of the interfaces to many electronic health records in such journals as *Critical Care Medicine* (Ahmed et al., 2011), the *Journal of Oncology Practice* (Corrao et al., 2010), and *Current Oncology Reports* (Markman, 2011). Health care delivery is but one example. One can find similar examples for many other domains, including education, criminal justice, and the military.

However, just as there typically is no attempt to generalize from the case studies published by IT practitioners, there typically is no attempt to generalize the lessons learned about the IT infrastructure, its deployment, and the use of integration from one domain to another domain. For example, are the issues of information assurance and security and the consequences that it has for the design of the IT infrastructure that arise in the health sector significantly different from the way these issues arise in the field of education? If so, why; what specific features of these application domains lead to these differences? Again, the thrust would be from domain-specific insights to more general insights and theories.

4.6 Conclusion

The discipline of IT, uniquely among the computing disciplines, primarily emerged in reaction to workforce development needs, rather than as reaction to a need for finding answers to new research questions, or to a need to employ new research methodologies to existing questions. As a result, it has not yet developed the strong research tradition that the other computing disciplines have developed. Thus, there are very few PhD programs in IT, and there are no recognized avenues for publishing high-quality IT research.

However, the IT community is acutely aware of this need to address this issue. It knows that in order to obtain legitimacy in the academic community, it must solve the learning issue for the discipline and arrive at a set of the research questions and/or research methodologies that it can legitimately claim as its own. One can expect IT research to emerge organically, especially as the consensus within the community about what sets IT apart from the other computing disciplines. However, such organic growth will take time and will, in any case, be encouraged through a meta-level research discussion about what an IT research agenda might look like.

This chapter has argued that the unique nature of IT as focusing on the IT infrastructure, the use of integration in its creation, and its deployment raises a series of questions to which IT can bring a unique focus. It has also argued that a practice-driven formulation of theories provides a unique methodological approach. The hope is that the points made in this chapter will help fertilize and expedite the organic growth of IT research that the IT community is so keenly looking forward to.

References

Agresti, W. (2011) Towards an IT agenda. *Comm AIS*, 28, Article 17.

Ahmed, A., Chandra, S., Herasevich, V., Gajic, O., Pickering, B. (2011) The effect of two different electronic health record user interfaces on intensive care provider task load, errors of cognition, and performance. *Critical Care Medicine*, 39(7), 1626–1634.

Aldrich, H. (1999) *Organizations Evolving*. Sage Publications, Thousand Oaks, CA.

Alter, S. (2003) Sorting out issues about the core, scope and identity of the IS field. *Comm AIS*, 12, 607–628.

Bailey, K. (2005) Fifty years of systems science: Further reflections. *Systems Research and Behavioral Science*, 22, 355–361.

Benbasat, I. and Zmud, R. (2003) The identity crisis within the IS discipline: Defining and communicating the discipline's core properties. *MIS Quarterly*, 27, 183–194.

Boehm, B. and Sullivan, K. (2000) Software economics: A roadmap. *22nd ICSE: Future of Software Engineering*, Limerick, Ireland, pp. 319–343.

Corrao, N., Robinson, A., Swiernik, M., and Naeim, A. (2010) Importance of testing for usability when selecting and implementing an electronic health or medical record system. *Journal of Oncology Practice*, 6(13), 121–124.

DeSanctis, G. (2003) The social life of information systems research: A response to Benbasat and Zmud's call for returning to the IT artifact. *Journal of the AIS*, 4, 360–376.

Ekstrom, J. J., Dark, M. J., Lunt, B. M., and Reichgelt, H. (2006). A research agenda for information technology: Does research literature already exist? *Proceedings of SIGITE 06*, Houston, TX, pp. 19–24.

Ekstrom, J. and Lunt, B. (2003) Education at the seams: Preparing students to stitch systems together: Curriculum and issues for 4-year IT programs. *Proceedings of SIGITE 03*, West Lafayette, IN, pp. 196–200.

Gowan, A. and Reichgelt, H. (2010) Emergence of the information technology discipline. *IEEE Computer*, 43(7), 79–81.

Hislop, G. (2003) Comparing undergraduate degrees in information technology and information systems. *Proceedings of SIGITE-03*, West Lafayette, IN, pp. 9–12.

Ives, B., Parks, S., Porra, J., and Silva, L. (2004) Phylogeny and power in the IS domain: A response to Benbasat and Zmud's call for returning to the IT artifact. *Journal of the AIS*, 5(3), 108–124.

JTFCC (Joint Task Force for Computing Curricula). (2005) Computing curricula 2005: The overview report covering undergraduate degree programs in Computer Engineering, Computer Science, Information Systems, Information Technology, Software Engineering, ACM, AIS and IEEE-CS, Available from http://www.acm.org/education/curricula-recommendations, retrieved March 5, 2012.

Leavitt, H. and Whisler, T. (1958) Management in the 1980's. *Harvard Business Review*, 11, 41–48.

Lunt, B., Ekstrom, J., Gorka, S., Hislop, G., Kamali, R., Lawson, E., LeBlanc, R., Miller, J., and Reichgelt, H. (2008) Information Technology 2008: Curriculum guidelines for undergraduate degree programs in Information Technology. ACM and IEEE Computer Society. Available from http://www.acm.org/education/curricula-recommendations, retrieved March 5, 2012.

Lunt, B., Ekstrom, J., Lawson, E., Kamali, R., Miller, J., Gorka, S., and Reichgelt, H. (2005) Defining the IT curriculum: The results of the past 3 years. *Journal of Issues in Informing Science and Information Technology*, 2, 259–270.

Lunt, B., Ekstrom, J., Reichgelt, H., Bailey, M., and LeBlanc, R. (2008) IT 2008: The history of a new computing discipline. *Communications of the ACM*, 53(12), 133–141.

Markman, M. (2011) Information overload in oncology practice and its potential negative impact on the delivery of optimal patient care. *Current Oncology Reports*, 13(4), 249–251.

Melville, N., Kraemer, K., Gurbanaxi, V. (2004) Review: Information technology and organizational performance: An integrative model of IT business value. *MIS Quarterly*, 28, 283–322.

Myers, M., Baskerville, R., Gill, G., and Ramiller, N. (2011) Setting our research agendas: Institutional ecology, informing sciences, or management fashion theory. *Comm AIS*, 28, Article 23.

Orlikowsky, W. and Iacono, C. (2001) Research commentary: Desperately seeking the "IT" in IT research—A call to theorizing the IT artifact. *Information Systems Research*, 12(2), 183–186.

Reichgelt, H. (2004) Towards a research agenda for information technology. *Proceedings of SIGITE 04*, Salt Lake City, UT, pp. 248–254.

Reichgelt, H., Lunt, B., Ashford, T., Phelps, A., Slazinski, E., and Willis, C. (2004) A comparison of baccalaureate programs in information technology with baccalaureate programs in computer science and information systems. *Journal of Information Technology Education*, 3, 19–34.

Topi, H., Valacich, J., Wright, R., Kaiser, K., Nunamaker, F., Jr., Sipior, J., and de Vreede, G. (2010) IS 2010: Curriculum guidelines for undergraduate degree programs in information systems. ACM and AIS. Available from http://www.acm.org/education/curricula-recommendations, retrieved March 5, 2012.

Van Bon, J., de Jong, A., Kolthof, A., Peiper, M., Tjassing, R., van der Veen, A., and Verheijen, T. (2007) *Foundations of IT Service Management Based on ITIL V3*. Zaltbommel, the Netherlands: Van Haren Publishing.

5

Sociotechnical Approaches to the Study of Information Systems

Steve Sawyer
Syracuse University

**Mohammad
Hossein Jarrahi**
Syracuse University

5.1 Sociotechnical Premise

Through this chapter, we introduce and explain the sociotechnical premise relative to the study of information systems (IS). The sociotechnical premise can be articulated as (1) the mutual constitution of people and technologies (and, specifically, digital technologies*), (2) the contextual embeddedness of this mutuality, and (3) the importance of collective action. Some readers will value this chapter for its breadth of coverage. Established sociotechnical scholars will likely thirst for more advanced discussions than what we provide here. Some readers will value the material in this chapter for identifying particular debates, current themes, or emerging approaches. We see this as a special opportunity and focus on these topics at the chapter's end.

---

* For this chapter, we use technology to mean digital technologies, information and communication technologies (ICT), information technologies, and computer technologies. So, technology here is more narrowly circumscribed (than all kinds of technologies) but inclusive (of digital technologies) conceptual shorthand.

We begin by noting the widespread ethos within the IS research community that it is sociotechnical by definition. As Allen Lee notes:

> Research in the information systems field examines more than just the technological system, or just the social system, or even the two side by side; in addition, it investigates the phenomena that emerge when the two interact. This embodies both a research perspective and a subject matter that differentiate the academic field of information systems from other disciplines. In this regard, our fields so called reference disciplines are actually poor models for our own field. They focus on the behavioral or the technological, but not on the emergent sociotechnical phenomena that set our field apart. For this reason, I no longer refer to them as reference disciplines, but as contributing disciplines at best (Lee 2001, p. iii).

Our goal in writing this chapter is to encourage scholars to move beyond this rhetorically pleasant articulation of sociotechnical thinking toward more deliberate conceptual development, increased empirical activity, and greater methodological capacity.

5.1.1 Mutual Constitution

Sociotechnical research is premised on the interdependent and inextricably linked relationships among the features of any technological object or system and the social norms, rules of use, and participation by a broad range of human stakeholders. This mutual constitution of social and technological is the basis of the term sociotechnical. Mutual constitution directs scholars to consider a phenomenon without making *a priori* judgments regarding the relative importance or significance of social or technological aspects (e.g., Bijker 1987, 1995; Latour 1999).

Mutual constitution differs from social determinism as social determinists see information and communication technologies (ICTs) as being caused by or created for the organizational needs and peoples' decisions about how to meet those needs (albeit, imperfectly at times). In particular, socially deterministic researchers contend there cannot be any specific "effects of technology" attributable to the material qualities of technology. That is, the "cause" and the nature of "particular effects" are always social interpretations (Grint and Woolgar 1997).

Mutual constitution stands apart from technological determinism, which regards the technology as the main cause of organizational change. Technological determinists characterize ICT as an independent (and often the primary) variable. The essence of this cause-and-effect relationship is conveyed by the word "impact." Technology is seen as an exogenous, independent, and material force that determines certain behavior of individual and organizations, producing predictable changes in organizational traits such as structure, size, decision-making, work routines, and performance (e.g., Leavitt and Whisler 1958; Pfeffer and Leblebici 1977).

Mutual constitution implies (1) both humans and technologies may have some sort of agency (some ability to act) in a given situation and (2) these actions are not deterministic (actions are not independent of surrounding events). The underlying premise of mutual constitution is coevolution among that which is technological and that which is social. The focus on interdependency among technology and human organization is done by attending to material triggers, actions of social groups, pressures from contextual influences, and the complex processes of development, adoption, adaptation, and use of new (digital) technologies in people's social worlds (Jones and Orlikowski 2007). Directionality is a property of the situation, not inherent to a relation.

Why is this important? Barley (1988) argues a singular focus on the material aspects of technology leads researchers into inappropriate materialism. On the other hand, a focus on technology as a solely social production has led to an overreliance on social orders as primary drivers, potentially leading to social determinism. In contrast, the sociotechnical perspective, and in particular the principle of mutual constitution, speaks directly to the complex and dynamic interactions among technological capacities, social histories, situated context, human choices, and actions rather than looking for simplified causal agency.

5.1.2 View of Context

The second element of the sociotechnical premise is that all technologies are socially situated. Any IS or ICT is embedded into a social context that both adapts to, and helps to reshape, social worlds through the course of their design, development, deployment, and uses (e.g., Kling 1980; Land and Hirschheim 1983; Orlikowski 1992; Walsham 1993; Avgerou 2001). This situated and mutually adaptive conceptualization stands in contrast to much of the current IS research that is "... de-problematized <of> time, space and the uses of ICT" (Lee and Sawyer 2009, p. 3), which is self-limiting in its failure to recognize the depth of social engagement. The desire for such an a-contextual perspective on IS (and ICT) continues to fascinate scholars (e.g., Brynjolfsson and Hitt 1996). However, this discourse, through its own framing, fails to account for social changes within which technological innovations unfold (e.g., Fleck 1994).

Adherents to de-contextualized models of IS pay too little attention to the environment of the organization and temporal dimension of technological innovation. For example, the literature focused on investigating the strategic implications of IS has sought to provide guidance for how to best harness strategic potential of new ICT/IS as a source of competitive advantage (e.g, Earl 1989; Morton 1991; Hammer and Champy 1993; Porter and Millar 1985). However, the depiction of an organization's environment does not typically delve into the complicated social processes that embed technological innovation within organizational or social contexts. In addition, these appraoches tend to ignore the temporal dimension of ICT. More than 20 years ago, Orlikowski and Baroudi (1991) found nearly 90% of IS research represents a single-snapshot data collection method, which does not include observations and data collection over time. This continues today (e.g., Avital 2000; Pollock and Williams 2009).

In contrast to a-contextualized and detemporalized approaches, the sociotechnical perspective is premised on the embedding of the ICT/IS into the more complex world of situated action: a world that is tightly tied to the characteristics of where the actions occur. In their analysis, sociotechnical scholars focus on situating work and seek to examine all contextual factors. This type of inquiry leads to a holistic view of context: one that does not diminish or remove contextual elements, even those with limited influence. In this situated view, context is not taken as fixed or delineable but is defined dynamically. Sociotechnical approaches focus on building situational and temporal conditions directly into their theories, relating these to conceptualizations of ICT/IS (e.g., Dourish 2004).

5.1.3 Collective Action

The third element of the sociotechnical premise is collective action: the pursuit of goal(s) by two or more interested parties (Merton 1968). Collective action undergirds the concept of organization (and all sociality) without implying positive or negative outcomes: there are shared pursuits. So (and typically), multiple parties pursue different goals, creating conflict. The underlying premise is the joint pursuit of one or more shared goals by two or more parties, focusing on the design, development, deployment, and uses of a particular ICT or IS, is both shaped by, and shapes, the nature of collective action.

Collective action means that joint interests and multiple goals are intertwined with both the context and the technological elements. Contemporary sociotechnical research focuses less on individual workers and simplified social situations—such as worker/manager differences—to attend more complex problematizations of social settings and tensions of multiple parties such as distributed online organizing (Sawyer and Eschenfelder 2002; Kling 2007).

5.1.4 Sociotechnical Premise in Contemporary IS Research

Sociotechnical approaches differ from what Kling and Lamb (2000) call "the standard model" of an IS. Standard models of IS focus on the technological aspects and pay far less attention to social roles and structures.* Second, standard models of IS decontextualize the work in search of generalities and best practices. Third, standard models of IS emphasize the cognitive and behavioral aspects of people's involvement with technologies.

The sociotechnical approach eschews simplifying rationales that seek a single or dominant cause of change. Instead, sociotechnical approaches foreground both the complexity and the uncertainty involved in the process of technologically involved change. In contrast to socially or technologically deterministic views, sociotechnical approaches require a detailed understanding of dynamic organizational processes and the occurrence of events over time in addition to knowledge about the intention of actors (situated rationality) and the features of technologies. As a result, studies in this stream of work construct neither independent nor dependent variables but instead adopt process logic to investigate the reciprocity and coevolution of the contextual interactions and outcomes (Barrett et al. 2006).

Sociotechnical researchers focus attention to the heterogeneous networks of institutions, people, and technological artifacts that together play roles in the design, development, deployment, take-up, and uses of any particular IS (Kling et al. 2003). So, as Michel Callon and John Law put it, any distinction between ICT (or IS) and society as context is an oversimplification that obscures the complex processes where human and technologies jointly construct sociotechnical entities (Callon and Law 1989).

5.2 Historical Roots and Conceptual Developments

In this section, we outline the historical roots of sociotechnical research.

5.2.1 Tavistock Tradition

The term "sociotechnical" was coined by researchers at the Tavistock Institute of Human Relations in England (Mumford 2000). The Tavistock Institute was founded in London in 1946 under the auspices of the Rockefeller Foundation. Its initial mission was to weave together social and psychological sciences in order to benefit a society damaged by the effects of the Second World War. Tavistock researchers, including therapists and a wide variety of consultants, strove to formulate techniques that could help to rehabilitate war-damaged soldiers (Jones 2004). Some Tavistock scholars came to believe that the same techniques could be employed in support of a more humane (and human-centered) organization of work in industry. These scholars suspected the Tavistock techniques would be applicable to the work of lower-rank employees who spent most of their time on routine and simple tasks without any clear prospect for job satisfaction or personal development (Mumford 2000).

This extension of the Tavistock agenda to the workplace showcased two attributes of sociotechnical scholarship that unfolded in quite different ways over time. The first, and most prominent, attribute is the close association between the technological and the social (sub-) systems of organizations. The technological elements, perceived as machines and associated work practices by Tavistock's researchers, were not meant to be the sole controlling factor when a new technological system was implemented. Tavistock scholars advocated equal attention should be paid to providing a satisfactory work environment for employees. In this regard, the main innovation of the Tavistock research was the design of technology-supported work arrangements that could enrich work practices using multiskilled jobs with workers organized into teams. The sociotechnical approach was, in this way,

---

* There exists an insightful body of IS literature focusing on social analyses of IS and ICT (Avgerou et al. 2004). These social analyses of IS and ICT emphasize the social and behavioral activities which frame and underpin the development, deployment, and uses of any ICT or IS. Social analysts take a strong social constructivist view on the sociotechnical relationships. Many social analyses draw on sociotechnical theories. But they take a "weak constructionist" view of the ICT. That is, the technological elements are seen primarily as an outcome of social practices and social forces, with the technology having little or no agency or impact. Our articulation of the sociotechnical premise differs primarily in the elevation of the importance of the ICT's and IS's roles—we assign some agency or the potential for direct effects to be of the ICT's doing. In all other ways the sociotechnical perspective is very similar to the social analytic perspective.

a rebellion against the evolution of work-design practices of the time (that took an instrumental view of work and the workforce).

The second attribute of the Tavistock approach was the importance of worker involvement. At its core, Tavistock's sociotechnical approach was interventionist and activist. As we note later, this orientation to worker's interests and activism underlies the action-research orientation of Enid Mumford's Effective Technical and Human Implementation of Computer-based Systems (ETHICS) and Peter Checkland's Soft Systems Method (SSM) (Mumford and Weir 1979; Checkland 1995), the participatory design principles that characterize the Scandinavian and Nordic scholarship, and perhaps some of the more contemporary design-centric approaches to IS.

The Tavistock researcher's fundamental goal was to humanize jobs through redesigning work practices and workplace technologies while propagating democracy at work. This led to the formulation of theories that entailed concepts like "quality of working life" (Kling and Lamb 2000). These theories postulated that employees who were involved in the work system should be given a voice in the design process to determine how the new system could improve the quality of their work. In addition, the practical side of the sociotechnical approach sought to give equal weight to both technical and social aspects in the design process.

5.2.1.1 Early Developments of Sociotechnical Approaches in IS

The Tavistock approach to sociotechnical design inspired a number of researchers in the nascent field of IS. Enid Mumford, greatly influenced by her association with the Tavistock Institute, is considered to be the most influential researcher to initiate sociotechnical research within IS (Davenport 2008). She, along with her peers such as Frank Land and students (Land 2000), voiced concerns that the bulk of IS research and professional know-how were limited to engineering approaches that centered on the effective construction of reliable technical artifacts.

Mumford's work placed the social context and human activities/needs at the center of IS design. The essence of the early sociotechnical discourse in IS is found in the proceedings of the "Human Choice and Computers" conference (Sackman 1975). The overall tone was critical of the perceived computer's impact on social institutions like the social order of in the workplace.

Findings from projects across the 1960s and 1970s were consolidated by Mumford and her colleagues and students and gave rise to an information systems development methodology called ETHICS (Mumford and Weir 1979). This methodology drew upon sociotechnical principles and involved a double-design effort: the design of IT-based systems/IS and the design of work processes. Initially, the two design efforts were conducted separately. The design of IS followed the technical system analysis method, whereas the design of work processes was aimed at the elicitation of "job satisfaction" requirements of workers. The latter analysis involved the application of work quality principles such as multi-skilled jobs. The two streams of design were brought together to achieve a "sociotechnical optimization" (Mumford 2006).

In ETHICS, the starting point was work design, rather than system design, and the methodology placed emphasis on the interaction of technologies and people.* The main objective of the method was to develop IS that are both technically viable and entail social qualities that would lead to high worker satisfactions (Mumford and Weir 1979). To this end, an IS designed to solely meet technical requirements is "likely to have unpredictable human consequences" (Mumford and Weir 1979, p. 13). Therefore, ETHICS uncompromisingly elevated socially-centered approaches to system design and called attention to scoring the quality of working life in terms of fit between personal achievement and organizational goals.†

---

* More recently, Alter's work systems (Alter 2006) approach has rekindled interest in this approach, though his conceptual frame is not rooted in Tavistock or sociotechnical thinking.

† At about the same time, Peter Checkland advanced SSM as an IS design approach. Similar to ETHICS in its attention to human and technological subsystems, the conceptual framing builds from systems theory (see Checkland 1998, 1995).

5.2.1.2 Industrial Engineering/Human Factors

Scholars from other intellectual communities also drew from the Tavistock work, translating the sociotechnical principles into new intellectual communities and, in turn, impacting practice. One of the most successful of these translators was Albert Cherns (1976, 1987). Cherns was an associate of the Tavistock Institute and published his summation of sociotechnical design principles in *Human Relations* (see Table 5.1). The design principles Cherns laid out quickly gained prominence in the intellectual communities concerned with ergonomics, human factors, and large-scale systems design and usability.

For IS, Cherns' sociotechnical principles were translated and introduced through two articles by Bostrom and Heinen (1977a,b).* This translation emphasized the computer-based IS as the focal technological subsystem and centered attention on its design rather than on the Tavistock principles of quality of working life. Bostrom and Heinen further problematized, and constrained, the social subsystem in two ways. First, they characterized the social aspects of an IS as emphasizing the tension between worker's interests and manager's interests. Second, Bostrom and Heinen defined workers (but not managers!) as users.

TABLE 5.1 Chern's Principles of Sociotechnical Systems

Compatibility	The design process should be compatible with its objectives. If the design is intended to foster democracy in work situations, the design processes themselves must be democratic.
Minimal critical specification	No more should be specified than is absolutely essential. However, the designers should ascertain what is essential.
The sociotechnical criterion	Variances, as deviations from expected standards, must be kept as close to their point of origin as possible. In other words, solution to problems should be devised by the groups that directly experience them, not by supervisory groups.
Multifunctionality principle	In order for groups to respond to the changing work environment, they need a variety of skills. These include skills that go beyond what day-to-day production activities require.
Boundary location	Boundaries exist where work activities pass from one group to another and where a new set of skills is required. However, boundaries should facilitate knowledge sharing. All groups should be able to learn from one another despite the existence of the boundaries.
Information	Information must reside where it is principally needed for action. A sociotechnical design gives the control authority to the groups whose efficiency is being monitored.
Support congruence	A social support system must be in place to enjoin the desired social behaviors.
Design and human values	High-quality work involves 1. Jobs to be reasonably demanding 2. Opportunity to learn 3. An area of decision-making 4. Social support 5. The opportunity to relate work to social life 6. A job that leads to a desirable future
Incompletion	Practitioners must recognize the fact that the design is an iterative process. It never stops. The new changes in environment require continual revisions of objectives and structures.

---

* See also the more recent work of Clegg (2000).

Bostrom and Heinen's problematization of the social system as workers versus managers, and their definition of workers as users, likely reflects contemporary management thinking of the 1970s in the United States. Contemporary scholars, such as Avgerou et al. (2004), who have built from and extended the Tavistock approaches to sociotechnical scholarship, are consistently critical of the conceptual limitations of this managerialist reframing and the diminished emphasis on humane design. And, while the Bostrom and Heinen papers are cited, their work has had a lesser impact on methods and concepts in IS than might be expected given the rhetorical enthusiasm for sociotechnical claims of IS.

5.2.1.3 Participatory Design and Worker Involvement

The importance of worker participation, fundamental to the Tavistock approach, was widely adopted in practice (Land et al. 1979; Land and Hirschheim 1983). In its original sense, user participation encouraged all intended users to be involved in all IS development tasks and stages. However, across the latter part of the twentieth century, user involvement was implemented in a much more limited way: Users were typically consulted primarily to learn about the tasks and the technical systems that support them, then not involved in the rest of the design process (Land et al. 1979).

However, the sociotechnical ideals of the Tavistock Institute found fertile ground in Scandinavian countries. In the late 1960s, "the Norwegian Industrial Democracy Projects" introduced the principle that technology innovation should improve work practices along with productivity measures (Thorsrud 1970). This was meant to empower employees to organize their own jobs. In the 1970s, figures like Kristen Nygaard—and more recently Bo Dahlbom, Pelle Ehn, Erik Stolterman, and their students—pioneered the Scandinavian approaches to the social analyses of computing. This approach reflects a strong orientation toward worker involvement in designing IS and attends to both the quality of working life and the potentially humanizing power of ICTs.

Perhaps the Tavistock approach garnered attention in Scandinavia because it resonated with the sociopolitical context grounded in deep appreciation for workers' right (Sawyer and Tapia 2007). This tendency was also reflected in the principles of participatory design (PD) and more recently activity theory-centric approaches (Kuutti 1996).* In addition, the information system development methodology suggested by Mumford in the United Kingdom has been tried in Denmark and Sweden (Bjorn-Andersen and Eason 1980; Hedberg 1980). To this day, PD and social-theoretically inspired analyses of computing underpin much of information system research in the Scandinavian and Nordic countries (Iivari and Lyytinen 1998).

Despite its promising principles, sociotechnical design in IS failed to proliferate outside of the United Kingdom, Scandinavia, and the Nordic countries. During the 1990s, economic, business, and technological arenas witnessed dramatic changes; the consequences turned out to be frustrating for advocates of sociotechnical design. In the harsh competitive environment, corporations were forced to embark on methods like lean production and business process reengineering that took little consideration of employees' needs or their quality of working life (Kling and Lamb 2000). As Carr (2008) notes, pointedly, the focus in corporate IT over the past 15 years has been to routinize, automate, and outsource. And, despite its progress in places like Scandinavia and the Nordic countries, few contemporary organizations were interested in adopting PD and sociotechnical design approaches.

Workplace trends across the 1990s had companies creating flatter hierarchies and encouraged people to see innovative companies as requiring highly skilled groups to work together as members of high-performance teams. While these trends expanded these organizational member's responsibility and autonomy, this also created a more divided workplace as only high-status groups gained these benefits.

---

* An entire intellectual community, centered on participatory design, has grown up and meets every other year (see http://pdc2012.org/about.html) to advance sociotechnical design principles. Many of the leading scholars are known to the IS community but are more likely to publish in computer–human interaction (CHI) or computer-supported cooperative work (CSCW) venues.

Lower-status positions (e.g., service, support, and administrative) were often outsourced or transferred to temporary employment and given far less voice or status in the contemporary firm. This segmentation of the workforce further undermined sociotechnical approach, which targeted broad groups of organizational members (Mumford 2006).

5.2.2 Sociological Perspective on Sociotechnical Systems

Independent (at first) of the Tavistock efforts, in the late 1960s, some scholars in sociology began a stream of theorizing on the roles of ICT (e.g., Joerges 1990). These scholars typically focused on ICT relative to social structures such as groups, communities, organizations, social stratifications/societies, and changes in power and equity (e.g., Latham and Sassen 2005). The focal interest of sociologists, however, has led them to foregrounding macro-sociological interests relative to the roles of ICT in and of society. So, computing continues to be a tertiary topic for contemporary sociologists.*

The sociological perspective on sociotechnical systems unfolds in discourses that Kling (1980) sought to capture and articulate as belonging to the rhetoric of either "systems rationalists" or "segmented institutionalists". Systems rationalists focused on the positive, intended, and often technologically inspired opportunities from computing and automation of the workplace. Segmented institutionalists focused on the importance of social power, conflict, and pursuit of noncommensurate goals as the undergirding of any computerization activity.

And, as we turn to later, Kling and his colleagues thundered against the naïve simplifications of systems rationalists. In doing so, Kling and colleagues advanced—through empirical, methodological, and analytical efforts—a range of segmented institutionalist approaches to the sociological analysis of computing (Wellman and Hiltz 2004). This approach was not pursued by many computer scientists or IS scholars and even fewer sociologists (as Kling [2003] himself rued in a posthumously published reflection on his work). Many sociologists continue to grapple with the roles of ICT in social life, albeit with little attention to the work of Tavistock scholars, the IS community in Scandinavia, the Nordic countries, the United Kingdom, or the Kling-inspired colleagues in North America.

In the 1990s, new research interests in the social dimensions of IS emerged. These were directed to the relationships among IS development, uses, and resultant social and organizational changes. The new stream of research has offered fresh insights into the emerging role of ICT within differing organizational contexts (e.g., Kling and Scacchi 1982; Orlikowski 1992; Walsham 1993; Hirschheim et al. 1996). Drawing directly on sociological theories of institutions, this wave of sociotechnical research has informed, if not directly shaped, IS scholarship (Avgerou 2002). The sociological theories leveraged by the new stream of sociotechnical research had flourished in parallel with the early sociotechnical approaches, while they similarly dispensed with technological determinism. Even though not explicitly using the term "sociotechnical," these sociological theories offered a solid basis upon which emerging sociotechnical research built. In the following section, we discuss institutional theories, structuration theory, social network perspective, and social philosophy that proved to be instrumental in the creation of this scholarly space.

5.2.2.1 Institutional Theories

Institutional theories in organization studies emerged as a counterpoint to organizational theories that saw people as rational actors. In contrast, institutional theories focused on cultural and normative explanations for organizational phenomena (DiMaggio and Powell 1991). Barley and Tolbert (1997, p. 93) note: "… organizations, and the individuals who populate them, are suspended in a web of values, norms, rules, beliefs, and taken-for-granted assumptions, that are at least partially of their own making."

---

* With this noted, there is a thriving community of sociologists interested in online communities and the nature of social networks (see Wellman et al. 2001 among others).

The institutional perspective is premised on the concept of "institution" that denotes the importance of authoritative, established, and rule-like procedures that provide order in society.

A key aspect of institutions is their interest in continuing to exist. As institutions become taken for granted, people are likely to believe that there is a functional rationale for their existence, and therefore, their validity is not questioned by social actors. By employing the concept of "institution," new intuitional theorists build a conceptual basis for delineating contextual influences and the broader influences of organizations' social environments.

From an institutional perspective, ICT is viewed not as a set of material features functioning according to the functional rules that are inscribed into their physical components but as a product of embeddedness into social institutions (Avgerou 2002). In this way, ICT can be considered as an institution in its own right, one that interacts with other institutions in modern society. As an institution, ICT can be characterized as emergent, embedded, evolving, fragmented, and connected to an ephemeral social presence that is shaped as much by other institutional and contextual forces as by technical and economic rationales (Orlikowski and Barley 2001). Examining the interactions of ICT with other intuitions, Swanson and Ramiller (1997) conclude that institutional forces are significant in shaping the perception about the organizational potentials of ICT. Avgerou (2000), building from the institutional perspective, captures the process of IT-enabled organizational change with both the ICT innovation and the organizational practices considered to be institutions, each having its own mechanisms and legitimating elements. In this way, the interaction between these two types of institutions is theorized as the dual processes of institutionalization of IT and deinstitutionalization of established organizational structures and practices.

While institutional theory provides a powerful framework for understanding the context of organizational change, its pillars rest upon the assumptions of stability, persistence, and the limited scope for change. By overstressing stability and the robustness of institutions, this often underestimates the role of human agency and ignores their potentials in transforming institutional patterns (Barley and Tolbert 1997; Dacin et al. 2002). By the same token, ICT artifacts also retreat to a secondary position in institutional analyses. In fact, the institutional perspective does not inquire into the nature of the technological artifact and does not address whether IT itself can represent any material influences independent of institutional forces.

5.2.2.2 Structuration

Many scholars have theorized about the mutual constitutions of individuals and society (e.g., Bourdieu 1977; Urry 1982; Bhaskar 1998); however, it was the work of Anthony Giddens that introduced the concept of structuration to a wide range of social sciences (Bryant et al. 2001, p. 43). Through the concept of structuration, Giddens (1986) rejects the traditional characterizations of social phenomena as determined by either objective social structures or totally autonomous human agents. In his view, social phenomena are neither structure nor agency but are continuously constituted in their duality. Giddens adopts an unconventional definition of structure as "rules and resources, organized as properties of social systems" that exists only as structural properties (Giddens 1989, p. 258). Human agents draw on social structures in their actions, and at the same time, these actions produce or reproduce social structures. In this sense, action and structures are in a recursive relationship, and this is the meaning imparted by the term "structuration."

Structuration's dynamic conception of structure, as being recurrently produced and reproduced through situated interactions of people, facilitates studying technological change (Orlikowski 2000). The fundamental duality of actions and structures as featured in the theory helped IS researchers to break away from both technological and social determinism (Markus and Robey 1988). Early champions of structuration theory in IS research employed the theory to link the context and process of change, examining the ways ICT contributes to the structuring of organizations (Orlikowski and Robey 1991; Walsham and Han 1991; Orlikowski 1992). DeSanctis and Poole (1994) developed an IS-specific version of the theory called Adaptive Structuration Theory (AST). AST focuses on (1) structures that are embedded in the technologies and (2) structures that emerge as human actors interact with those technologies. Scholars have noted the inconsistency of AST as: Giddens theorized that structure cannot be inscribed or embedded in technology since they do not exist separately from the practices of social actors (Jones 1999).

Rejecting the hypothesis of structure as embedded in technology, subsequent interpretations of structuration theory in IS gave more weight to agency, focusing attention on improvisation, enactment, and the emergent nature of ICT-enabled organizational practices (Orlikowski 1996; Orlikowski and Hoffman 1997; Weick 1998). The shift from technology as an embedded structure toward agency of humans underpins the practice lens (Orlikowski 2000; Schultze and Orlikowski 2004). She proposes the notion of technology-in-practice, which refers to the structure of technology use enacted by social actors while they interact recurrently with a particular technology artifact. Seen this way, technology-in-practice is emergent and enacted, not embodied or appropriated.

Despite its potential for explaining the social processes through which ICT and organizations interconnect, structuration theory has limitations for empirical IS research. For instance, Giddens makes almost no references to IS or ICT in his writings (Jones and Karsten 2008). In structurational analyses, technologies can have a material influence on human actions, but the effects are contingent upon how social actors engage with them through their practices. Thus, "… as they do things in relation to machines and so forth, these are the stuff out of which structural properties are constructed" (Giddens and Pierson 1998, p. 83). What this "relation to machines" might be, and how it affects social actors' practices, however, is not elaborated in structuration theory, and analysis of these properties remains largely underdeveloped (Jones and Orlikowski 2007). Moreover, structuration theory deals with social phenomena at a high level of abstraction. This leads to it being seen as a "metatheory": a way of thinking about the world, rather than an empirically testable explanation of organizational practices.

5.2.2.3 Social Network/Structure

The social network perspective first blossomed in the 1920s at Harvard University's Department of Sociology. Since then this approach has been taken up and advanced by scholars in sociology, anthropology, psychology, and organization studies, to name a few (Scott 2000). In organization studies, scholars have explored different elements of social networks relative to embeddedness of organizational routines, informal information sharing, holes and bridges in knowledge flow, and the transfer of social norms and tacit knowledge (e.g., Merton 1957; Granovetter 1973; Wellman and Berkowitz 1988; Burt 1992). In IS, social network concepts have been used to study the behaviors of teams, organizations, industries, and ICT-enabled communities (Nohria and Eccles 2000; Barabasi 2003; Monge and Contractor 2003; Christakis and Fowler 2009).

The social network perspective is useful for studying some of the emerging forms of social or organizational arrangements and the roles of ICT (and more recently, social media). A social network is a social structure made up of individuals called "nodes" who are "connected" by an interdependency like friendship, kinship, common interest, financial exchange, and knowledge or prestige. Networks of relations are enacted by sharing of information or resources among nodes (people) via ties between them (their interdependency). These simple concepts continue to offer scholars " …a surprisingly fruitful way to analyze how social formations organize, change, and grow" (Oinas-Kukkonen et al. 2010, p. 62).

Concepts of network analysis such as social capital (as resources derived from networks of relationships) provide a vehicle for examining digitally enabled social networks. For example, Robert et al. (2008) use social capital as the lens to study the differences between lean digital networks and face-to-face interactions. In addition, via emerging computational capabilities, researchers are also enabled to conduct complex network analysis such as ICTs (such as social media) that can enhance interactions across social networks (Agarwal et al. 2008). In this regard, building from the theory and methods of social network analysis, Kane and Alavi (2008) studied how multiple users interact with multiple ICTs within healthcare groups.

5.2.2.4 Social Philosophy

Scholars in IS have drawn on social philosophers to gain insight into contemporary and complicated dimensions of IS design, management, and theory. These dimensions include but are not limited to politics, power, cognition, and rationality of IS (Fitzgerald 1997). The philosophy-oriented conceptual thinking came about in part as a challenge to the conventions of systems rationalism that pervaded much of the early approaches to studying IS. This work also provided a more conceptually nuanced view of the world, shifting attentions from systems and technology to the interplay of people and their complex universe.

Most notably, IS researchers have levered philosophical approaches with a significant input from critical social theorists to address IS sociotechnical problems. Arguably, critical social theory in these studies is mostly rooted in the work of Jurgen Habermas and Michel Foucault. These critical social theorists posit that social structures are not independent of people: they are produced and reproduced by them. And people's ability to transform their social and economic circumstances is constrained by multiple cultural, social, and political forces. To this end, critical research seeks to outline the restrictive dimensions of the status quo and carry forward an emancipatory agenda by focusing on conflicts and contradictions in contemporary society (Myers 1997).

Other IS researchers have drawn on critical social theories to study freedom, social control, and power with regard to the development, use, and role of IT in organizations and society at large. Critical social theory has allowed these researchers to account for ethical and moral aspects of IS, by focusing on emancipatory approaches and the betterment of people's lives (Ngwenyama and Lee 1997). Critical approaches critique taken-for-granted assumptions, directing attention to the potentials of people in changing their social and material circumstances. Germane to this view is the thesis that people who interact with IS need not be constrained by their social circumstances (Orlikowski and Baroudi 1991). As an example, Hirschheim and Klein (1994) applied critical social theory to study IS development. Their critical approach discards oversimplistic notion of users and the dynamics of context by paying attention to how different user groups are affected by system development and may react to it. It also explicates how the system may fit into organizational context and how other organizational dynamics such as power and politics come into play.

While sharing some of the same philosophical roots, recent sociotechnical scholarship has shifted focus from railing against prescriptive IS development methodologies toward a broader view on the roles of ICT in contemporary organizations. In particular, the new sociotechnical approach that emerged in the 1990s is strongly inspired by the broader stock of social science theory (i.e., theories from science and technology studies [STS]) in order to address the multiple substantive issues that are associated with ICT (Dunlop and Kling 1991). These theoretical insights are critical in contrast to the classic sociotechnical approaches that had some distinctive "blind spots" in dealing with rapidly changing contingencies of workplaces and malleable technologies that lend themselves to user improvisation (Malone et al. 1995; Ciborra and Lanzarra 1999).

5.2.3 Science and Technology Studies

Scholars in the intellectual community known as STS* are concerned with the reciprocal relationships among social, political, and cultural structures, science/scientific research, and technological innovation. Scholars of STS bring a broad range of perspectives to focus on the relationships among technology and society. So, historians, communications, sociology, computer science, political science/policy studies, and information studies scholars are found in central roles in STS.

Given the variety of models, conceptual frameworks, and domains of study, there is no agreed upon definition of what constitutes or qualifies as studies of technology within STS (Van House 2003).

---

* For more about the STS community, see Hackett et al. (2008) and http://www.4sonline.org

This intellectual ambiguity serves STS well as it makes the community open to exploring new technologies. And, by the 1980s, technology studies began to proliferate in STS. Two seminal works on social shaping of technology (SST) (MacKenzie and Wajcman 1985) and the social construction of technological systems (Pinch and Bijker 1987) signaled what Steve Woolgar was to call the "turn to technology" (Woolgar 1991).

Despite its ontological and epistemological heterogeneity, STS scholars pursue common themes. Virtually, all STS researchers are united in their aim to dispense with the predominance of technological determinism. They believe that the simplicity offered by such a perspective fails to recognize the dynamic and complex process through which technologies interact with society (Bijker 1995; Latour 2005). STS researchers also unanimously argue that the "black box of technology" should be opened up for sociological analysis (Law and Bijker 2000). To do so, researchers must pay attention to the process and content of technology itself. Shifting away from "the impact" of technology, this body of research tends to highlight how technology is constructed during research, development, and innovation phases and how structural and political circumstances of its development are reflected in technology. Over time, STS scholars have embraced several theoretical approaches. Three of the most prominent are the social construction of technology (SCOT), focusing on constructs like interpretive flexibility and relevant actors; the SST drawing on concepts like configuration and trajectories; and actor–network theory (ANT) that introduces networks, enrollment, translation, and irreversibility.

5.2.3.1 Social Construction of Technology and Social Shaping of Technology

As the name implies, SCOT conceptualizes technology as not determining human action: humans socially construct technology (Bijker 1997). Social constructionists, the advocates of SCOT, downplay self-evident explanations of effects stemming from the material attributes of a technology: constraining and enabling effects of technologies are matters of interpretative practices of people in the social context. The process through which technology is constructed is described through a number of stages. The first stage involves the concept of interpretive flexibility to explain how a technology is socially constructed. The second stage explains how technology reaches stabilization, a state where the "relevant social groups" have their problems resolved and desires manifested in the artifact. In the final stage, the technological content of the artifact is linked to the social, through considering the meanings that are assigned to the artifact by the relevant social groups: it comes to "closure."

The conceptual pillars of SCOT are found in a number of IS studies. One example is Orlikowski and Gash's (1994) study of people's interpretation of Lotus Notes. Building on Bijker's (1987) conception of "technological frames," their study stressed the specific uses to which the technology is put in a given setting and how the context of use influences the users' interaction with the IS. The lack of uniformity among the technological frames of disparate social groups is also attributed to the fact that "technologies are social artifacts, their material form and function will embody their sponsors' and developers' objectives, values and interests, and knowledge of that technology" (Orlikowski and Gash 1994).

Some scholars see SCOT's notion of closure as privileging design over use as this diminishes the recurrent reinterpretations of technological artifacts by different groups of users (Wajcman 2000). In response, STS scholars advanced the SST perspective. The SST perspective shares many commonalties with SCOT but avoids the problems that lead SCOT theorists to implicitly assume that technological structures will become something external to human actions during use. This noted, early SST work was rebutted for giving undue attentions to technology developers and technology design (Russell 1986).

Through a number of revisions, SST researchers (notably Russell and Williams 2002) fine-tuned their theorization to open up the concept of design and development, portraying technology development as an open and indeterminate activity, through concepts like "innofusion" and "configuration." Innofusion is the process through which innovation extends to implementation, consumption, and use (Fleck 1988). The technological artifact emerges and remerges "through a complex process of action and interaction between heterogeneous players"—what Fleck calls "learning by trying" (Fleck 1994). Along the same line, "configuration" advances the notion of open design and implementation, by appealing to technology in cases where a constellation of heterogeneous components is locally incorporated into some kind of working orders. In this sense, technology developers only preconfigure their products, which later become subject to various reconfigurations in the local site of use (Williams 2000).

The "strong constructivism" that underpins both SCOT and SST however, can lead analysts to discount any trace of, or role for, technological agency. Hutchby (2001, p. 450) argues strong constructivist approaches showcase "… humans are capable of interpreting the capacities of technologies [such as a bridge or an airplane] in varying ways." But strong constructivists do not address the central question of "… does the aeroplane lend itself to the same set of possible interpretations as the bridge; and if not, why not?" (Hutchby 2001, p. 447).

5.2.3.2 Actor–Network Theory

In contesting technological determinism, both SCOT and SST have been accused of falling into a form of social determinism: overstressing social choice at the expense of technological considerations. By contrast, ANT seeks to avoid determinism. While ANT shares with social constructivists the central premise that social structures and practices cannot be realized solely through an account of the material properties of technologies, ANT distances itself from social determinism by making no analytical distinction between the social and technical (Latour 1987, 1999). Rooted in a "… ruthless application of semiotics" (Latour 1999, p. 3), ANT's first premise is that entities have no inherent qualities: they acquire their form and functionality only through their relations with other entities.*

The second premise of ANT is "symmetry" among humans and technological artifacts, arguing against any *a priori* distinction between what is technical and what is not (Bloomfield and Vurdubakis 1997). As Latour (1991, p. 129) maintains: "… rather than assuming that we are dealing with two separate, but related, ontological domains—technology and organizations—we propose to regard them as but phases of the same essential action." Building from the concept of symmetry, both technological and social entities are explained as "actants" (Akrich 1992).

An actor–network is constructed through the enrollment of actants (both human and nonhuman actants) into a network of relations by means of negotiations. This process is explicated by the "sociology of translation" that aims to describe, rather than explain, the transitions and negations that take place as the network is configured or "translated" (Callon 1986). The translation process is political in nature and begins with a certain "problematization" when one actor identifies a problem that is shared by others and starts to convince other actors that the problem is significant enough to dedicate resources for its solutions.

Actor–networks reach stability when they become irreversible. Irreversibility is when it would be either too costly to reverse the relationships or doing so becomes improbable. Reaching network stability requires (1) successfully negotiating the enrollment of participants, followed by the (2) translation of an (3) obligatory passage point (when the sets of relations and those enrolled become (4) irreversible) (Latour 1987). Mobilization of network members ensues as a result of irreversibility and stability where social investment in the network reaches a point at which withdrawal would be unthinkable. The durability of a network is a matter of the robustness of the translation. Networks collapse or undergo changes if the translation processes that brought the networks to their current state can revert or if the networks cannot resist alternative translations.

To date, ANT's conceptual vocabulary and methodological demands have been used eclectically in IS research (Walsham 1997; e.g., Pouloudi and Whitley 2000). Numerous case studies inspired by ANT serve as a means to improve the understanding of IS researchers of the design and use of ICT, which is of

---

* One implication of this relational premise is that concepts of context cannot be distinct from an actor–network—what one might characterize as "context" or "aspects of the situation" matter in ANT only if there is a relationship developed among entities of interest. That is, ANT is neither contextual nor a-contextual: the actor–network is both local and distant and the relationships between entities of interest, action, and location are the focus.

significance for IS (Hanseth et al. 2004). And there exists a prominent and meticulous adoption of ANT in the literature on information infrastructure. The size and complexity of infrastructural technologies such as groupware, and the characteristic that they generally build upon existing technologies, make researchers direct their focus from ICT or isolated technological artifacts to a more complex notion of IT infrastructure (Hanseth et al. 1996). Several researchers have drawn upon ANT to account for the sociotechnical nature of the information infrastructure, which includes not only artifact but also human habits, norms, and roles that may prove its most intractable elements (Jackson et al. 2007). For example, building on ANT's conceptual vocabulary, Hanseth et al. (1996) investigate how any given elements of information infrastructure constrain others and how these elements inscribe certain patterns of use. To do so, they identify explicit anticipations of use by various actors during use and the way these anticipations are translated and inscribed into standards.

5.3 Contemporary Sociotechnical Theories

There has been, over the past 15 years, substantial sociotechnical theorizing. Many of these contemporary approaches to sociotechnical theorizing draw together concepts from several of these sociotechnical sources and combine both empirical and theoretical works to advance sociotechnical thinking, as we outline later.

5.3.1 Genres

The concept of "genre of organizational communication" was introduced to examine organizational communication with a particular focus on the mediating role of communicative technologies. Rooted in ancient Greek word, *genos*, genre denotes races, kinds, and classes (Zimmerman 1994). The academic use perhaps dates back to the 1950s and these early formulations were mostly taxonomy oriented, seeking to classify the form and content of written and spoken communicative actions. Contemporary efforts have shifted toward the classification of communicative or rhetorical practices (Østerlund 2007).

Yates and Orlikowski (1992), building on Millers (1984), define a genre as "… a typified communicative action invoked in response to a recurrent situation" (p. 301). The recurrent situation encompasses the history and nature of established practices, social relations, and communication media within organizations. In this view, genres are meaningless without the daily practices into which they are situated. Central to this theorization of genres is a focus on community. That is, the communicative purpose of a genre has to be shared within a community (i.e., organization) by more than one person.

The structural features, communicative technologies, and symbol systems of a genre are all made sense of within the community (Yates and Orlikowski 1992). For example, an e-mail is not an instance of communicative genres, while a memo or a meeting agenda, which may be mediated by the use of e-mail, is (Päivärinta 2001). With this distinction, genre theory dispenses with information richness theory, which defines a medium rich or lean due to its intrinsic characteristics. Based on genre theory, the same communicative medium can convey messages using different genres, just as the same genre can ensue using different communicative media. For instance, e-mail can carry messages in the form of business letter or informal personal notes that are essentially considered disparate forms of genres (Lee 1994). Genres are dynamic: introducing and using a new medium may transform daily practices and changing communication practices. For example, Davidson (2000) examined how computerized clinical order system changed interactions among multiple groups of organizational actors (e.g., physicians, nurses, and pharmacists).

5.3.2 Boundary Objects

A boundary object is something that is simultaneously "rigid" enough to be commonly understood among different social worlds and "plastic" enough to be understood within each social world (Star 1989;

Star and Griesemer 1989). Some scholars have focused on how ICTs serve as boundary objects among different groups (Fleischmann 2006a,b; Gal et al. 2008). Others have focused on the negotiations that occur around boundary objects (Lee 2007; Lutters and Ackerman 2007; Rajao and Hayes 2012). Still others have focused on how ICTs, acting as boundary objects, facilitate the flow of information in collaborative work (Osterlund and Boland 2009; Tyworth et al. 2012). Still others have focused on how to design ICT to function as effective boundary objects (Akoumianakis et al. 2011).

5.3.3 Biographies of Artifacts

The biographies of artifacts (BoA) approach has emerged from the work of a small group of scholars at Edinburgh University (Pollock et al. 2003). Building on the series of deep studies of various broad-scale commercial technologies, the BoA conceptualization is that an ICT or an IS often exists at one or more places (installed in more than one location) and varies by both the version (temporal) and the location (spatial). To understand and study such diverse, distributed, and popular (or common) IS requires one to pursue a biographical approach to analysis, one that focuses on the evolution of the particular artifact over time, its changes and evolution based on uses and events at various sites, and the players in these various arenas (Pollock et al. 2003). Building, as it does, from the SST tradition and from cultural studies (where the BoA concept first emerged), BoA provides a framework to approach large-scale technological analysis. Its emphasis on trajectories and social shaping, and the demands of maintaining a dual focus—on both common events and unique events, as both can shape the direction of the artifact's evolution—make it a demanding approach at present (e.g., Pollock and Williams 2009).

5.3.4 Domestication

The domestication of ICTs is premised on the processes of consumption and focuses on the ways in which a new artifact, concept, or object of interest is brought into an existing social world (Berker et al. 2006). Consumption is more than use as it encompasses the mutual adaptations of social structures, processes, and meanings as people bring the object of interest into their world and learn to live with it (or not). Much of the original domestication scholarship focused on the take-up and uses of the media and digital technologies in the home (Silverstone and Haddon 1996). Domestication scholars broadened their focus of inquiry and the conceptual approach soon could be found in IS (e.g., Frissen 2000).

Domestication focuses attention to the ongoing and mutually adaptive processes of bringing ICTs into existing social worlds (often the home or people's lives). Domestication scholars see ICTs as arriving with preformed meanings constructed by advertising, design features, and both informal discourse (e.g., word-of-mouth and personal networks) and formal media messages. Once acquired individuals invest the ICT with their own significance, create and adapt strategies to manage, use, and value these, and deal with the changes their presence, uses, and meanings entail (e.g., Stewart 2003). Haddon (2006) makes clear domestication focus on consumption, rather than use, highlights processes of negotiation in bringing ICT into the social world. This process of making something (like a commodity technology or a particular artifact) personal is a "taming" or training that takes place over time. This may not lead to success (or a desired outcome) and domestication scholarship often highlights the churning or seemingly cyclical patterns of nonprogress toward full acceptance/belonging. In focusing on this process, domestication emphasizes the relationships among the object, individuals, and the larger social milieu in which this process unfolds (and often grants agency to the media messages and opinion leaders for framing the roles, uses, and expectations of ICT).

5.3.5 Computerization Movements

The computerization movements approach extends concepts of social movements in two ways. A social movement, broadly defined, is coordinated collective action that relies on informal social units to pursue institution building toward shared goals. These goals can either be aimed at a specific and narrow policy or be more broadly aimed at cultural change. Often this institution building creates conflict with existing institutions. Institution building requires leaders, visions, resources, followers, and events. Adapting this broad area of sociological inquiry, Kling and Iacono (1988) articulated a computerization movement as a specific form of social movement, one focused on institutionalizing a particular ICT or IS. Moreover, a computerization movement emphasizes the public discourses around a new ICT, the ways in which various adherents and opponents frame the computerization efforts, and the practices through which these frames and discourses unfold. Elliott and Kraemer (2008) compiled a volume of computerization studies as a tribute to the late Rob Kling. Hara and Rosenbaum (2008) have extended computerization movements, developing a typology based on an empirical analysis of published work in this area.

5.3.6 Sociomateriality

Derived in part from the sociology of science, sociomateriality posits social practices as intrinsically conjoined with material things—and for our interests, this tends to focus on the material aspects of ICT and IS (Orlikowski and Scott 2008). Sociomateriality does not assign agency to people or technology but views the social and technological to be ontologically inseparable (Suchman 2007). It hence provides a means for understanding how social meanings and technological actions are inextricably related and, together, shape social practices.

The sociomaterial perspective considers knowledge to be enacted—every day and over time—in and through people's practices. Practices are "… recurrent, materially bounded and situated action engaged in by members of a community" (Orlikowski 2002, p. 256). Sociomaterial scholars treat knowledge and practice as mutually constitutive: knowing is inseparable from knowledge practices and is constituted through those actions. Central to this view is the thesis that technological affordances are achieved in practice and can only be understood by focusing on their material performance that is always enacted by humans. In this way, the performativity of ICT and IS is not given *a priori*; they are temporally emergent and enacted (Orlikowski and Scott 2008).

5.3.7 Social Informatics, Social Actors, and Sociotechnical Interaction Networks

Inspired by constructivist approaches to studying technology found in STS and sociology, social informatics (SI) is particularly concerned with computerization movements: The transformation of human social arrangements and activities that follows from the implementation, use, and adoption of computers in different types of organizations and social systems (Horton et al. 2005). SI is a perspective, a way of framing, the particular dynamics of IS and ICT in social and organizational worlds. SI scholars seek to theorize ICT and IS as evolving crystallizations of interests, activities, structures, and artifacts that are constructed over time in response to local and institutional conditions (Davenport 2008).

SI scholarship arose in response to the research in computer science's failure to adequately explain or even examine the changes in social structures and organizational processes that arise as new ICT and IS are taken up and used. Kling (2007) notes that SI is "a body of research that examines the social aspects of computerization". A more formal definition is "the interdisciplinary study of the design(s), uses, and consequences of information technology that takes into account their interaction with institutional and cultural contexts."

Kling is recognized as the foremost proponent of SI during his lifetime (Day 2007) and hailed as the man who "brought computing and sociology together" (Wellman and Hiltz 2004). Kling, as a prominent member of a research group at the University of California, Irvine, began to explore IS in local government across the United States in the early 1970s. Soon, Kling and Scacchi (1982) developed an explanatory framework, "the web of computing," in which they asserted that organizational computing must be viewed as an ensemble of equipment, applications, and techniques with identifiable information processing capabilities. This ensemble or web stands as an alternative to "engineering models" that focused on the technological equipment and their information processing capabilities.

As part of the perspective, SI scholars have advanced a set of findings: insights drawn from many studies that, together, can be taken as common or expected outcomes from any computerization effort (see Kling 1996; Kling et al. 2005). First, the uses of any ICT lead to multiple and sometimes paradoxical effects. Second, the uses of ICT shape thought and action in ways that benefit some groups more than others. The design, development, and uses of ICT will reshape access in unequal and often ill-considered ways. Third, the differential effects of the design, implementation, and uses of ICT often have moral and ethical consequences. This finding is so often (re)discovered in studies across the entire spectrum of ICT, and across various levels of analysis, that ignorance of this point borders on professional naiveté. Fourth, the design, implementation, and uses of ICT have reciprocal relationships with the larger social context: it is not possible to isolate an ICT or its effects. Finally, effects of ICT will vary by the level of analysis, making it appear as if there are different phenomena at work across these levels.

Two characteristics of SI distinguish it from the broader sociotechnical research literature. First, SI provides an intellectual framework that can guide methodological practices of IS researchers, one focusing specifically on ICT (and not more generically on "technology"). In the "web of computing" paper (Kling and Scacchi 1982) and later work, Kling (1987) presented IS researchers with a methodological guide for studying institutional uses of ICTs with a focus on the institutional level. A second differentiating characteristic of SI is the greater level of agency provided to the ICT and IS. This stands in contrast to the more common strong constructivism found in SST and SCOT, where technical capacities are not fixed but essentially indeterminate and open to interpretive flexibility (for an overview of this debate, see Grint and Woolgar 1992; Kling 1992a,b).

The social aspects of the web of computing were advanced by Lamb and Kling (2003). They emphasized that the focus on workers and others as users diminished their larger and more complex roles as "social actors" and that only a portion (and often a remarkably small portion) of one's work is as a user. The social actor model further articulated several important social relations relative to the ways in which ICT and IS are taken up and brought into the social worlds of these actors (e.g., Rowlands 2009).

Several SI scholars have also extended the web of computing model, advancing sociotechnical interaction networks (STIN) as a more fully developed version (Kling et al. 2003). Inspired by the symmetrical treatment of humans and nonhumans in ANT, Kling defines STIN as a network that "includes people (also organizations), equipment, data, diverse resources (money, skill, and status), documents and messages, legal arrangements and enforcement mechanisms, and resource flows" (Kling et al. 2003, p. 48). STIN distinguishes itself from ANT's conceptual and abstract vocabulary by providing a methodological heuristic for identifying the network's members and boundaries (e.g., Meyer 2006).

5.4 Emerging Scholarly Spaces

Beyond the substantial intellectual activity surrounding the development and conceptual advances in contemporary sociotechnical theorizing as described in Section 5.3, we provide an overview of two emerging scholarly spaces—multidimensional networks and economic sociology—whose members are focusing on the roles of ICT and are reaching out to sociotechnical scholars for guidance. The first space reflects a subcommunity of social network scholars who are grappling with the empirical and conceptual issues of various ICTs being involved in or interacting with people's social networks. The second space is a distinct intellectual subcommunity that is generating attention in both economics and sociology.

5.4.1 Multidimensional Networks

Many scholars are drawing on concepts from social networks to study the roles and effects of ICTs in social and organizational life. Relative to this, two major views coexist. In the first view, ICT is considered an exogenous variable that can shape social networks. For example, the work of scholars like Barry Wellman and colleagues has investigated how ICTs can substitute for communication networks and how it can reconfigure these human-centric relationships (i.e., Wellman and Gulia 1999; Wellman 2001; Wellman and Haythornthwaite 2002).

Recently, a second view on the role of ICTs in social networks is emerging. This view rejects any separation between ICT and the network, treating ICT as an endogenous variable (e.g., Monge and Contractor 2003). These scholars advance the concept of a multidimensional network to help researchers capture the entangled and multifaceted relationships between individuals and technologies. By definition, multidimensional network is both multimodal and multiplex. Multimodal networks contain nodes of different types, and multiplex networks are comprised of multiple types of relationships among nodes (Contractor et al. 2011).* These simple concepts help sociotechnical researchers focus on the structure and dynamics of networks involving different types of players (people and technology) and different types of relations among multiple people and technologies.

5.4.2 Economic Sociology

Economic sociology has emerged in the past 20 years as an intellectual community that focuses attention to the socially embedded nature of economic activity (see Swedberg 1994). Social embeddedness is at the core of economics sociology and has three characteristics. First, an economic transaction is done by and for people. As such, this transaction reflects sociality in the ways it is conducted, in the sets of assumptions regarding the behaviors of the participants (and the behaviors of the nonparticipants), and the roles that these transactions play in the larger social world (their structuring potential). A second characteristic is that the motives of the participants are influenced by their social relations, norms, and structures: participants are social agents (e.g., Swedberg and Granovetter 2001). In this rejection of "homo-economicus," economic sociologists go beyond Simon's concepts of satisficing and reflect the fundamentals of bias, anchoring, and attention to social activity that Tversky and Kahneman (1974) theorized on in their Nobel-winning prospect theory. The third characteristic of economic sociology is the importance of institutions and their construction as sets of shared interests and social relations that are so important to the structuring of society that these interests and relations are often encoded as laws and regulating social norms (e.g., Rauch and Casella 2001).

The empirical and conceptual premise of economic sociology is being advanced by economists, sociologists, and political scientists. However, they have, collectively, done little to conceptualize the roles or effects of ICT. That is, the roles of ICT as both a participant in economic life and social life, and the potential effects of the presence and uses of ICT on social relations and social structure, remain anomalous. With this noted, there has been some recent work to advance the role of technology (more broadly) in economic life by building on the flexible conceptualizations of technological materiality (e.g., Pinch and Swedberg 2008).

5.5 Conceptualizing the ICT Artifact

We began this chapter by noting that IS scholarship focuses on the relationships among ICT and organizational actions, processes, structures, and changes (Leavitt and Whisler 1958; Markus and Robey 1988; Benbasat and Zmud 2003; Sidorova et al. 2008). As these scholars and others have noted, conceptualizing the ICT artifact, IS more broadly, and their effects remains both a core mission and a difficult task for IS

---

* This discourse has similarities with the sociomateriality discussion (Section 5.3.5). The most distinctive difference between the two is the explicit embedding of a technological artifact into a network of social relationships that multimodal networks demand. Concepts of networks and embedding may also hearken ANT. But the multimodal conceptualizations are not premised on the conceptual foundations of symmetry, enrolment, negotiation, translation, and irreversibility on which ANT relies.

scholars (Orlikowski and Iacono 2001; King 2011). As we have outlined here, the sociotechnical perspective and its underlying premises provide IS scholars with a range of conceptual tools to advance our empirical bases, theoretical understanding, and design interventions relative to IS in organizations and society.

At the core of the sociotechnical perspective are (1) the characterization of ICT and IS and (2) the conceptual focus on technological agency. Relative to the characterization of ICT and IS, sociotechnical approaches provide a range of conceptualizations: the strong social constructionism of SCOT, the more limited social constructionism of SST, the conceptual symmetry of ANT, and the weaker constructionism of STIN and sociomateriality. Relative to technological agency, sociotechnical approaches from this are mutually interdependent with social and organizational aspects of the situated phenomena. This provides a useful alternative to both the standard models of IS and other emerging approaches seen in IS.

Over the past decade, scholars have begun advancing design science and design theory as another approach to IS research (e.g., Gregor 2002; Pries-Heje and Baskerville 2008). Given the design focus and interventionist orientation of the Tavistock traditions of sociotechnical research, it may be that design research is one vehicle for sociotechnical scholars to pursue (e.g., Markus et al. 2002; Baskerville and Pries-Heje 2010). More generally, given the broad scope of IS scholar's research interests and the equally broad range of conceptual approaches taken, the sociotechnical perspective provides a set of conceptual tools and empirical insights to advance the state of IS research (King 2011; Sawyer and Winter 2011). That is, sociotechnical approaches to studying ICT and IS provide useful intellectual guidance to advance our theorizing on technological artifacts and how people's work practices and organizational arrangements are afforded by technological resources and inhibited by technological constraints.

Acknowledgments

Thanks to Brian Butler, Sean Goggins, and Heikki Topi for their comments on earlier versions of this chapter. Thanks to students of the Syracuse iSchool's "sociotech reading group," and particularly Gabe Mugar, Matt Willis, Andreas Kuehn, and Janet Marsden, for comments and questions on formative parts of the work presented here.

References

Agarwal, R., A.K. Gupta, and R. Kraut. 2008. Editorial overview—The interplay between digital and social networks. *Information Systems Research* 19 (3):243–252.

Akoumianakis, D., G. Vidakis, D. Vellis, G. Kotsalis, A. Milolidakis, A. Akrivos Plemenos, and D. Stefanakis. 2011. Transformable boundary artifacts for knowledge-based work in cross-organization virtual communities spaces. *Intelligent Decision Technologies* 5:65–82.

Akrich, M. 1992. The de-scription of technical objects. In *Shaping Technology/Building Society: Studies in Socioteclmical Change*, ed. J. Law. Cambridge, MA: MIT Press, pp. 205–224.

Alter, S. 2006. *The Work System Method: Connecting People, Processes, and IT for Business Results*. Larkspur, CA: Work System Press.

Avgerou, C. 2000. IT and organizational change: An institutionalist perspective. *Information Technology and People* 13 (4):234–262.

Avgerou, C. 2001. The significance of context in information systems and organizational change. *Information Systems Journal* 11 (1):43–63.

Avgerou, C. 2002. The socio-technical nature of information systems innovation. In *Information Systems and Global Diversity*, ed. C. Avgerou. New York: Oxford University Press.

Avgerou, C., C. Ciborra, and F. Land. 2004. *The Social Study of Information and Communication Technology*. Oxford, U.K.: Oxford University Press.

Avital, M. 2000. Dealing with time in social inquiry: A tension between method and lived experience. *Organization Science* 11:665–673.

Barabasi, A.L. 2003. *Linked: How Everything Is Connected to Everything Else and What It Means.* New York: Penguin Group.

Barley, S.R. 1988. Technology, power, and the social organization of work: Towards a pragmatic theory of skilling and deskilling. *Research in the Sociology of Organizations* 6:33–80.

Barley, S.R. and P.S. Tolbert. 1997. Institutionalization and structuration: Studying the links between action and institution. *Organization Studies* 18 (1):93.

Barrett, M., D. Grant, and N. Wailes. 2006. ICT and organizational change: Introduction to the special issue. *Journal of Applied Behavioral Science* 42 (1):6.

Baskerville, R. and J. Pries-Heje. 2010. Explanatory design theory. *Business and Information Systems Engineering* 2 (5):271–282.

Benbasat, I. and R.W. Zmud. 2003. The identity crisis within the IS discipline: Defining and communicating the discipline's core properties. *MIS Quarterly* 27 (2):183–194.

Berker, T., M. Hartmann, and Y. Punie. 2006. *Domestication of Media and Technology.* Berkshire, U.K.: Open University Press.

Bhaskar, R. 1998. *The Possibility of Naturalism.* Hoboken, NJ: Routledge.

Bijker, W.E. 1987. *The Social Construction of Bakelite: Towards a Theory of Invention. The Social Construction of Technological Systems*, eds. W.E. Bijker, T.P. Hughes, and T. Pinch. Cambridge, MA: The MIT Press, pp. 159–187.

Bijker, W.E. 1995. Sociohistorical technology studies. In *Handbook of science and technology studies*, eds. S. Jasanoff, G.E. Markle, J.C. Petersen, and T. Pinch. London, U.K.: Sage, pp. 229–256.

Bijker, W.E. 1997. *Of Bicycles, Bakelites, and Bulbs: Toward a Theory of Sociotechnical Change.* Cambridge, MA: The MIT Press.

Bjorn-Andersen, N. and K. Eason. 1980. Eason. KD, Myths and realities of information systems contributions to organizational rationality. In *Human Choice and Computers*, ed. A. Mowshowitz. Amsterdam, the Netherlands: North-Holland, pp. 97–109.

Bloomfield, B.P. and T. Vurdubakis. 1997. Visions of organization and organizations of vision: The representational practices of information systems development. *Accounting, Organizations and Society* 22 (7):639–668.

Bostrom, R.P. and J.S. Heinen. 1977a. MIS problems and failures: A socio-technical perspective. Part I: The causes. *MIS Quarterly* 1 (3):17–32.

Bostrom, R.P. and J.S. Heinen. 1977b. MIS Problems and failures: A socio-technical perspective. Part II: the application of socio-technical theory. *MIS Quarterly* 1 (4):11–28.

Bourdieu, P. 1977. *Outline of a Theory of Practice* (R. Nice, Trans.). Cambridge, U.K.: Cambridge University.

Bryant, C.G.A., D. Jary, and W. Hutton. 2001. *The Contemporary Giddens: Social Theory in a Globalizing Age.* New York: Palgrave.

Brynjolfsson, E. and L. Hitt. 1996. Paradox lost? Firm-level evidence on the returns to information systems spending. *Management Science* 42 (4):541–558.

Burt, R.S. 1992. *Structural Holes.* Chicago, IL: University of Chicago Press.

Callon, M. 1986. The Sociology of an Actor-Network: The Case of the Electric Vehicle. In *Mapping the Dynamics of Science and Technology. Sociology of Science in the Real World*, eds. M. Callon, H. Law, and A. Rip. London, U.K.: Macmillan, pp. 19–34.

Callon, M. and J. Law. 1989. On the construction of sociotechnical networks: Content and context revisited. *Knowledge and Society* 8:57–83.

Carr, N.G. 2008. *The Big Switch: Rewiring the World, from Edison to Google.* New York: WW Norton & Company.

Checkland, P. 1995. Model validation in soft systems practice. *Systems Research* 12 (1):47–54.

Checkland, P. 1998. *Systems Thinking, Systems Practice.* New York: John Wiley & Sons.

Chems, A. 1976. The principles of sociotechnical design. *Human Relations* 29 (8):783–792.

Cherns, A. 1987. Principles of sociotechnical design revisited. *Human Relations* 40 (3):153.

Christakis, N.A. and J.H. Fowler. 2009. *Connected: The Surprising Power of Our Social Networks and How They Shape Our Lives.* Boston, MA: Little, Brown.

Ciborra, C. and G.F. Lanzarra. 1999. A Theory of Information Systems Based on Improvisation. In *Rethinking Management Information Systems*, eds. W. Currie and R. Galliers. Oxford, U.K.: Oxford University Press, pp. 136–155.

Clegg, C.W. 2000. Sociotechnical principles for system design. *Applied Ergonomics* 31 (5):463–477.

Contractor, N., P.R. Monge, and P. Leonardi. 2011. Multidimensional networks and the dynamics of sociomateriality: Bringing technology inside the network. *International Journal of Communication* 5:682–720.

Dacin, M.T., J. Goodstein, and W.R. Scott. 2002. Institutional theory and institutional change: Introduction to the special research forum. *Academy of Management Journal* 45 (1):43–56.

Davenport, E. 2008. Social informatics and sociotechnical research—A view from the UK. *Journal of Information Science* 34 (4):519.

Davidson, E.J. 2000. Analyzing genre of organizational communication in clinical information systems. *Information Technology and People* 13 (3):196–209.

Day, R.E. 2007. Kling and the critical: Social informatics and critical informatics. *Journal of the American Society for Information Science* 58 (4):575–582.

DeSanctis, G. and M.S. Poole. 1994. Capturing the complexity in advanced technology use: Adaptive structuration theory. *Organization Science* 5 (2):121–147.

DiMaggio, P.J. and W.W. Powell. 1991. *The New Institutionalism in Organizational Analysis*. Chicago, IL: Chicago University Press.

Dourish, P. 2004. What we talk about when we talk about context. *Personal and Ubiquitous Computing* 8 (1):19–30.

Dunlop, C. and R. Kling. 1991. *Computerization and Controversy: Value Conflicts and Social Choices*. San Diego, CA: Academic Press Professional, Inc.

Earl, M.J. 1989. *Management Strategies for Information Technology*. Upper Saddle River, NJ: Prentice-Hall, Inc.

Elliott, M.S. and K.L. Kraemer. 2008. *Computerization Movements and Technology Diffusion: From Mainframes to Ubiquitous Computing*. Medford, NJ: Information Today Inc.

Fitzgerald, G. 1997. Foreword. In *Philosophical Aspects of Information Systems*, eds. R. Winder and I.A. Beeson. London, U.K.: Taylor & Francis Group, pp. 1–4.

Fleck, J. 1988. *Innofusion or Diffusation? The Nature of Technological Development in Robotics*. Edinburgh University, Department of Business Studies, Working Paper No. 7.

Fleck, J. 1994. Learning by trying: The implementation of configurational technology. *Research Policy* 23 (6):637–652.

Fleischmann, K.R. 2006a. Boundary objects with agency: A method for studying the design-use interface. *Information Society* 22 (2):77–87.

Fleischmann, K.R. 2006b. Do-it-yourself information technology: Role hybridization and the design-use interface. *Journal of the American Society for Information Science and Technology* 57 (1):87.

Frissen, V.A.J. 2000. ICTs in the rush hour of life. *The Information Society* 16 (1):65–75.

Gal, U., K.J. Lyytinen, and Y. Yoo. 2008. The dynamics of IT boundary objects, information infrastructures, and organisational identities: The introduction of 3D modelling technologies into the architecture, engineering, and construction industry. *European Journal of Information Systems* 17 (3):290–304.

Giddens, A. 1986. *The Constitution of Society: Outline of the Theory of Structuration*. Berkeley, CA: University of California Press.

Giddens, A. 1989. A reply to my critics. In *Social Theory of Modern Societies: Anthony Giddens and His Critics*, eds. D. Held and J.B. hompson. Cambridge, U.K.: Cambridge University Press, pp. 249–301.

Giddens, A. and C. Pierson. 1998. *Conversations with Anthony Giddens: Making Sense of Modernity*. Stanford, CA: Stanford University Press.

Granovetter, M.S. 1973. The strength of weak ties. *American Journal of Sociology* 78 (6):1360.

Gregor, S. 2002. Design theory in information systems. *Australasian Journal of Information Systems* (December), pp. 14–22.

Grint, K. and S. Woolgar. 1992. Computers, guns, and roses: What's social about being shot? *Science, Technology, and Human Values* 17 (3):366–380.

Grint, K. and S. Woolgar. 1997. *The Machine at Work: Technology, Work, and Organization*. Cambridge, U.K.: Polity Press.

Hackett, E.J., O. Amsterdamska, M. Lynch, and J. Wajcman. 2008. *The Handbook of Science and Technology Studies* (3rd edn.). Cambrdige, MA: The MIT Press.

Haddon, L. 2006. The contribution of domestication research to in-home computing and media consumption. *Information Society* 22 (4):195–203.

Hammer, M. and J. Champy. 1993. *Reengineering the Corporation*. New York: Harper Business.

Hanseth, O., M. Aanestad, and M. Berg. 2004. Guest editors' introduction: Actor-network theory and information systems. What's so special? *Information Technology and People* 17 (2):116–123.

Hanseth, O., E. Monteiro, and M. Hatling. 1996. Developing information infrastructure: The tension between standardization and flexibility. *Science, Technology and Human Values* 21 (4):407.

Hara, N. and H. Rosenbaum. 2008. Revising the conceptualization of computerization movements. *Information Society* 24 (4):229–245.

Hedberg, B. 1980. Using computerized information systems to design better organizations and jobs. In *The Human Side of Information Processing*, eds. N. BjornAndersen, Amsterdam: North-Holland, pp. 15–45.

Hirschheim, R. and H.K. Klein. 1994. Realizing emancipatory principles in information systems development: The case for ETHICS. *MIS Quarterly* 18 (1) :83–109.

Hirschheim, R., H.K. Klein, and K. Lyytinen. 1996. Exploring the intellectual structures of information systems development: A social action theoretic analysis. *Accounting, Management and Information Technologies* 6 (1–2):1–64.

Horton, K., E. Davenport, and T. Wood-Harper. 2005. Exploring sociotechnical interaction with Rob Kling: Five "big" ideas. *Information Technology and People* 18 (1):50–67.

Hutchby, I. 2001. Technologies, texts and affordances. *Sociology* 35 (02):441–456.

Iivari, J. and K. Lyytinen. 1998. Research on information systems development in Scandinavia—Unity in plurality. *Scandinavian Journal of Information Systems* 10 (1):135–185.

Jackson, S.J., P.N. Edwards, G.C. Bowker, and C.P. Knobel. 2007. Understanding infrastructure: History, heuristics, and cyberinfrastructure policy. *First Monday* 12 (6).

Joerges, B. 1990. Images of technology in sociology: Computer as butterfly and bat. *Technology and Culture* 31 (2):203–227.

Jones, M. 1999. Structuration theory. In *Rethinking Management Information Systems*, eds. W. Currie and R. Galliers. New York, NY: Oxford University Press, pp. 103–135.

Jones, E. 2004. War and the practice of psychotherapy: The UK experience 1939–1960. *Medical History* 48 (4):493.

Jones, M.R. and H. Karsten. 2008. Giddens's structuration theory and information systems research. *MIS Quarterly* 32 (1):127–157.

Jones, M. and W.J. Orlikowski. 2007. Information technology and the dynamics of organizational change. In *The Oxford Handbook of Information and Communication Technologies (Oxford Handbooks in Business and Management)*, eds. R. Mansell, C. Avgerou, D. Quah, and R. Silverstone. Oxford, U.K.: Oxford University Press.

Kane, G.C. and M. Alavi. 2008. Casting the net: A multimodal network perspective on user-system interactions. *Information Systems Research* 19 (3):253–272.

King, J.L. 2011. CIO: Concept is over. *Journal of Information Technology* 26 (2):129–138.

Kling, R. 1980. Social analyses of computing: Theoretical perspectives in recent empirical research. *ACM Computing Surveys (CSUR)* 12 (1):61–110.

Kling, R. 1987. Defining the boundaries of computing across complex organizations. In *Critical Issues in Information Systems Research*, eds. R. Boland and R. Hirschheim. London, U.K.: John Wiley, pp. 307–362.

Kling, R. 1992a. Audiences, narratives, and human values in social studies of technology. *Science, Technology, and Human Values* 17 (3):349–365.

Kling, R. 1992b. When gunfire shatters bone: Reducing sociotechnical systems to social relationships. *Science, Technology, and Human Values* 17 (3):381–385.

Kling, R. 1996. *Computerization and Controversy: Value Conflicts and Social Choices* (2nd edn.). San Diego, CA: Academic Press.

Kling, R. 2003. Critical professional education about information and communications technologies and social life. *Information Technology and People* 16 (4):394–418.

Kling, R. 2007. What is social informatics and why does it matter? *Information Society* 23 (4):205–220.

Kling, R. and S. Iacono. 1988. The mobilization of support for computerization: The role of computerization movements. *Social Problems* 35 (3):226–243.

Kling, R. and R. Lamb. 2000. IT and organizational change in digital economies: A socio-technical approach. Understanding the digital economy—Data, tools and research. Cambridge, MA: MIT Press.

Kling, R., G. McKim, and A. King. 2003. A bit more to it: Scholarly communication forums as socio-technical interaction networks. *Journal of the American Society for Information Science and Technology* 54 (1):47–67.

Kling, R., H. Rosenbaum, and S. Sawyer. 2005. *Understanding and Communicating Social Informatics: A Framework for Studying and Teaching the Human Contexts of Information and Communication Technologies.* Medford, NJ: Information Today Inc.

Kling, R. and W. Scacchi. 1982. The web of computing: Computer technology as social organization. *Advances in Computers* 21 (1):90.

Kuutti, K. 1996. Activity theory as a potential framework for human-computer interaction research. In *Context and consciousness: Activity theory and human-computer interaction*, ed. B. Nardi. Cambridge, MA: MIT Press, pp. 17–44.

Lamb, R. and R. Kling. 2003. Reconceptualizing users as social actors in information systems research. *MIS Quarterly* 27 (2):197–236.

Land, F. 2000. Evaluation in a socio-technical context. In *Organizational and Social Perspectives on Information Technology*, eds. R. Basskerville, J. Stage, and J. DeGross. Boston, MA: Kluwer Academic Publishers, pp. 115–126.

Land, F. and R. Hirschheim. 1983. Participative systems design: Rationale, tools and techniques. *Journal of Applied Systems Analysis* 10 (10):15–18.

Land, F., E. Mumford, and J. Hawgood. 1979. *Training the Systems Analyst for the 1980s: Four New Design Tools to Assist the Design Processs.* The Information Systems Environment, North Holland.

Latham, R. and S. Sassen. 2005. *Digital Formations: IT and New Architectures in the Global Realm.* Princeton, NJ: Princeton University Press.

Latour, B. 1987. *Science in Action: How to Follow Scientists and Engineers through Society.* Cambridge, MA: Harvard University Press.

Latour, B. 1991. Technology is society made durable. In *A Sociology of Monsters: Essays on Power, Technology and Domination*, ed. J. Law. London, U.K.: Routledge, pp. 103–131.

Latour, B. 1999. On recalling ANT. In *Actor Network Theory and After*, eds. J. Law and J. Hassard. Oxford, U.K.: Blackwell, pp. 15–25.

Latour, B. 2005. *Reassembling the Social: An Introduction to Actor-Network-Theory.* Oxford, U.K.: Oxford University Press.

Law, J. and W.E. Bijker. 2000. *Shaping Technology/Building Society: Studies in Sociotechnical Change.* Cambridge, MA: MIT Press.

Leavitt, H.J. and T.L. Whisler. 1958. Management in the 1980's. *Harvard Business Review* 36:165.

Lee, A.S. 1994. Electronic mail as a medium for rich communication: An empirical investigation using hermeneutic interpretation. *MIS Quarterly* 18 (2):143–157.

Lee, A. 2001. Editorial. *MIS Quarterly* 25 (1):iii–vii.

Lee, C.P. 2007. Boundary negotiating artifacts: Unbinding the routine of boundary objects and embracing chaos in collaborative work. *Computer Supported Cooperative Work* 16:307–339.

Lee, H. and S. Sawyer. 2009. Conceptualizing time, space and computing for work and organizing. *Time and Society* 19 (3):293–317.

Lutters, W.G. and M.S. Ackerman. 2007. Beyond boundary objects: Collaborative reuse in aircraft technical support. *Computer Supported Cooperative Work* 16 (3):341–372.

MacKenzie, D. and J. Wajcman. 1985. *The Social Shaping of Technology*. Buckingham, U.K.: Open University Press.

Malone, T.W., K.Y. Lai, and C. Fry. 1995. Experiments with oval: A radically tailorable tool for cooperative work. *ACM Transactions on Information Systems (TOIS)* 13 (2):177–205.

Markus, M.L., A. Majchrzak, and L. Gasser. 2002. A design theory for systems that support emergent knowledge processes. *MIS Quarterly* 26 (3):179–212.

Markus, M.L. and D. Robey. 1988. Information technology and organizational change: Causal structure in theory and research. *Management Science* 34 (5):583–598.

Merton, R.K. 1957. *Social Theory and Social Structure*. New York: The Free Press of Glencoe.

Merton, R.K. 1968. *Social Theory and Social Structure*. New York: The Free Press of Glencoe.

Meyer, E.T. 2006. Socio-technical interaction networks: A discussion of the strengths, weaknessesand future of Kling's STIN model. In *IFIP International Federation for Information Processing, Social Informatics: An Information Society for All? In Remembrance of Rob Kling*, eds. J. Berleur, M.I. Numinen, and J. Impagliazzo. Boston, MA: Springer, pp. 37–48.

Miller, C.R. 1984. Genre as social action. *Quarterly Journal of Speech* 70 (2):151–167.

Monge, P.R. and N.S. Contractor. 2003. *Theories of Communication Networks*. New York: Oxford University Press.

Morton, M.S.S. 1991. *The Corporation of the 1990s: Information Technology and Organizational Transformation*. New York: Oxford University Press.

Mumford, E. 2000. Socio-technical design: An unfulfilled promise or a future opportunity. In *Organizational and Social Perspectives on Information Technology*, eds. R. Baskerville, J. Stage, and J. DeGross. Boston, MA: Kluwer Academic Publishers, pp. 33–46.

Mumford, E. 2006. The story of socio-technical design: Reflections on its successes, failures and potential. *Information Systems Journal* 16 (4):317.

Mumford, E. and M. Weir. 1979. *Computer Systems in Work Design: The ETHICS Method*. New York: Wiley.

Myers, M.D. 1997. Qualitative research in information systems. *MIS Quarterly* 21 (2):241–242.

Ngwenyama, O.K. and A.S. Lee. 1997. Communication richness in electronic mail: Critical social theory and the contextuality of meaning. *MIS Quarterly* 21 (2):145–167.

Nohria, N. and R. Eccles. 2000. Face-to-face: Making network organizations work. In *Technology, Organizations and Innovation: Towards 'Real Virtuality'?* 1659 eds. I. McLoughlin, D. Prece, and P. Dawson. New York, NY: Taylor & Francis, pp. 1659–1682.

Oinas-Kukkonen, H., K. Lyytinen, and Y. Yoo. 2010. Social networks and information systems: Ongoing and future research streams. *Journal of the Association for Information Systems* 11 (2):3.

Orlikowski, W.J. 1992. The duality of technology: Rethinking the concept of technology in organizations. *Organization Science* 3 (3):398–427.

Orlikowski, W.J. 1996. Improvising organizational transformation over time: A situated change perspective. *Information Systems Research* 7 (1):63–92.

Orlikowski, W.J. 2000. Using technology and constituting structures: A practice lens for studying technology in organizations. *Organization Science* 11 (4):404–428.

Orlikowski, W.J. 2002. Knowing in practice: Enacting a collective capability in distributed organizing. *Organization Science* 13 (3):249–273.

Orlikowski, W.J. and S.R. Barley. 2001. Technology and institutions: What can research on information technology and research on organizations learn from each other? *MIS Quarterly* 25 (2):145–165.

Orlikowski, W.J. and J.J. Baroudi. 1991. Studying information technology in organizations: Research approaches and assumptions. *Information Systems Research* 2 (1):1–28.

Orlikowski, W.J. and D.C. Gash. 1994. Technological frames: Making sense of information technology in organizations. *ACM Transactions on Information Systems (TOIS)* 12 (2):174–207.

Orlikowski, W. and D. Hoffman. 1997. An imporvisational model for change managment: The case of groupware technologies. *Sloan Management Review* 38 (2):11–21.

Orlikowski, W. J. and C. S. Iacono. 2001. Research commentary: Desperately seeking the "IT" in IT research-A call to theorizing the IT artifact. *Information Systems Research* 12 (2):121–134.

Orlikowski, W.J. and D. Robey. 1991. Information technology and the structuring of organizations. *Information Systems Research* 2 (2):143–169.

Orlikowski, W.J. and S.V. Scott. 2008. Sociomateriality: Challenging the separation of technology, work and organization. *Academy of Management Annals* 2 (1):433–474.

Østerlund, C. 2007. Genre combinations: A window into dynamic communication practices. *Journal of Management Information Systems* 23 (4):81–108.

Osterlund, C. and R.J. Boland. 2009. Document cycles: Knowledge flows in heterogeneous healthcare information system environments. Paper Read at *System Sciences, 2009. HICSS '09. 42nd Hawaii International Conference*. January 5–8, 2009, Big Island, HI, pp. 1–11.

Päivärinta, T. 2001. The concept of genre within the critical approach to information systems development. *Information and Organization* 11 (3):207–234.

Pfeffer, J. and H. Leblebici. 1977. Information technology and organizational structure. *The Pacific Sociological Review* 20 (2):241–261.

Pinch, T.J. and W.E. Bijker. 1987. The social construction of facts and artifacts. Cambridge, MA: The MIT Press, pp. 17–50.

Pinch, T. and R. Swedberg. 2008. *Living in a Material World*. Cambridge, MA: MIT Press.

Pollock, N. and R. Williams. 2009. *Software and Organizations: The Biography of the Enterprise-Wide System or How SAP Conquered the World*. Vol. 5. London, U.K.: Routledge.

Pollock, N., R. Williams, and R. Procter. 2003. Fitting standard software packages to non-standard organizations: The 'biography' of an enterprise-wide system. *Technology Analysis and Strategic Management* 15 (3):317–332.

Porter, M.E. and V.E. Millar. 1985. How information gives you competitive advantage. *Harvard Business Review*, 63 (4):149–160.

Pouloudi, A. and E.A. Whitley. 2000. Representing human and non-human stakeholders: On speaking with authority. In *Organizational and Social Perspectives on Information Technology*, eds. R. Basskerville, J. Stage, and J. DeGross. Boston, MA: Kluwer Academic Publishers, pp. 339–354.

Pries-Heje, J. and R. Baskerville. 2008. The design theory nexus. *MIS Quarterly* 32 (4):731.

Rajao, R. and N. Hayes. 2012. Boundary objects and blinding: The contradictory role of GIS in the protection of the amazon rainforest. In *European Conference on Information Systems (ECIS)*, Paper 76, Barcelona, Spain.

Rauch, J.E. and A. Casella. 2001. *Networks and Markets*. New York: Russell Sage Foundation.

Robert, L.P., A.R. Dennis, and M.K. Ahuja. 2008. Social capital and knowledge integration in digitally enabled teams. *Information Systems Research* 19 (3):314–334.

Rowlands, B. 2009. A social actor understanding of the institutional structures at play in information systems development. *Information Technology and People* 22 (1):51–62.

Russell, S. 1986. The social construction of artefacts: A response to Pinch and Bijker. *Social Studies of Science* 16 (2):331–346.

Russell, S. and R. Williams. 2002. Social shaping of technology: Frameworks, findings and implications for policy with glossary of social shaping concepts. In *Shaping Technology, Guiding Policy: Concepts, Spaces and Tools*, eds. K.H. Sorensen, and R. Williams. Cheltenham, U.K.: Edward Elgar, pp. 37–131.

Sackman, H. 1975. *Human Choice and Computers*, eds. E. Mumford and H. Sackman. London, U.K.: North-Holland.

Sawyer, S. and K.R. Eschenfelder. 2002. Social informatics: Perspectives, examples, and trends. *Annual Review of Information Science and Technology (ARIST)* 36:427–465.

Sawyer, S. and A. Tapia. 2007. From findings to theories: Institutionalizing social informatics. *Information Society* 23 (4):263–275.

Sawyer, S. and S. Winter. 2011. Special issue on the future of information systems: Prometheus unbound. *Journal of Information Technology* 26 (4):4–9.

Schultze, U. and W.J. Orlikowski. 2004. A practice perspective on technology-mediated network relations: The use of internet-based self-serve technologies. *Information Systems Research* 15 (1):87.

Scott, J. 2000. *Social Network Analysis: A Handbook* (2nd edn.). London,U.K.: Sage Publications.

Sidorova, A., N. Evangelopoulos, J.S. Valacich, and T. Ramakrishnan. 2008. Uncovering the intellectual core of the information systems discipline. *MIS Quarterly* 32 (3):467–482.

Silverstone, R. and L. Haddon. 1996. Design and the domestication of information and communication technologies: Technical change and everyday life. In *Communication by design: The politics of information and communication technologies,* eds. R. Mansell and R. Silverstone, Oxford, U.K.: Oxford University Press, pp. 44–74.

Star, S.L. 1989. The structure of Ill-structured solutions: Boundary objects and heterogeneous distributed problem solving. In *Readings in Distributed Artificial Intelligence*, eds. M. Kuhn and L. Gasser. Menlo Park, CA: Morgan Kaufman, pp. 37–54.

Star, S.L. and J.R. Griesemer. 1989. Institutional ecology, 'Translations' and boundary objects: Amateurs and professionals in Berkeley's museum of vertebrate zoology, 1907–1939. *Social Studies of Science* 19 (3):387–420.

Stewart, J. 2003. The social consumption of information and communication technologies (ICTs): Insights from research on the appropriation and consumption of new ICTs in the domestic environment. *Cognition, Technology and Work* 5 (1):4–14.

Suchman, L.A. 2007. *Human-Machine Reconfigurations: Plans and Situated Actions*. Cambridge, U.K.: Cambridge University Press.

Swanson, E.B. and N.C. Ramiller. 1997. The organizing vision in information systems innovation. *Organization Science* 8 (5):458–474.

Swedberg, R. 1994. Markets as social structures. In *Handbook of Economic Sociology*, eds. R. Smelser and R. Swedberg. Princeton, NJ: Russell Sage Foundation.

Swedberg, R. and M.S. Granovetter. 2001. *The Sociology of Economic Life*. Boulder, CO: Westview Press.

Thorsrud, E. 1970. A strategy for research and social change in industry: A report on the industrial democracy project in Norway. *Social Science Information* 9 (5):64–90.

Tversky, A. and D. Kahneman. 1974. Judgment under uncertainty: Heuristics and biases. *Science* 185 (4157):1124–1131.

Tyworth, M., N. Giacobe, V. Mancuso, C. Dancy, and E. McMillan. 2012. The distributed nature of cyber situation awareness. In *SPIE Conference on Defense, Security and Sensing 2012*, Baltimore, MD, pp. 174–178.

Urry, J. 1982. Duality of Structure: Some critical questions. *Theory, Culture and Society* 1:121–154.

Van House, N.A. 2003. Science and technoloav studies and information studies. *Annual Review of Information Science and Technology* 38, pp. 3–86.

Wajcman, J. 2000. Reflections on gender and technology studies: in what state is the art? *Social Studies of Science* 30 (3):447–464.

Walsham, G. 1993. *Interpreting Information Systems in Organizations*. New York: John Wiley & Sons, Inc.

Walsham, G. 1997. Actor-Network Theory and IS Research: Current status and future prospects. In *Information Systems and Qualitative Research*, eds. A. Lee, J. Liebenau, and J.I. DeGross. London, U.K.: Chapman & Hall, pp. 466–480.

Walsham, G. and C.K. Han. 1991. Structuration theory and information systems research. *Journal of Applied Systems Analysis* 18:77–85.

Weick, K.E. 1998. Introductory essay: Improvisation as a mindset for organizational analysis. *Organization Science* 9 (5):543–555.

Wellman, B. 2001. Computer networks as social networks. *Science* 293 (5537):2031.

Wellman, B. and S.D. Berkowitz. 1988. *Social Structures: A Network Approach.* Cambridge, U.K.: Cambridge University Press.

Wellman, B. and M. Gulia. 1999. Virtual communities as communities. In *Communities in Cyberspace,* eds. M.A. Smith, and P. Kollock. New York, NY: Routledge, pp. 167–194.

Wellman, B., A.Q. Haase, J. Witte, and K. Hampton. 2001. Does the internet increase, decrease, or supplement social capital? *American Behavioral Scientist* 45 (3):436–455.

Wellman, B. and C.A. Haythornthwaite. 2002. *Internet in Everyday Life.* Malden, MA: Wiley-Blackwell.

Wellman, B. and S.R. Hiltz. 2004. Sociological Rob: How Rob Kling brought computing and sociology together. *Information Society* 20 (2):91–95.

Williams, R. 2000. Public choices and social learning: The new multimedia technologies in Europe. *Information Society* 16 (4):251–262.

Woolgar, S. 1991. The turn to technology in social studies of science. *Science, Technology and Human Values* 16 (1):20.

Yates, J.A. and W.J. Orlikowski. 1992. Genres of organizational communication: A structurational approach to studying communication and media. *Academy of Management Review* 17 (2):299–326.

Zimmerman, E. 1994. On definition and rhetorical genre. In *Genre and the New Rhetoric,* eds. A. Freedman and P. Medway. London, U.K.: Taylor & Francis Group, pp. 125–132.

6

IT and Global Development

Erkki Sutinen
*University of
Eastern Finland*

6.1 Introduction

The increasing role of information technology (IT) in global development has a fascinating history, especially from the viewpoint of information systems (IS). In the following, we will briefly follow two parallel threads. The first one has transformed the interpretation of the concept of global development from a set of universal indicators to the one emphasizing local potential and strengths, conceptualized as the term of bottom of the pyramid. The other thread portrays the changing self-image of IS from that of analyzing the efficiency of a given information system from the perspective of the surrounding business or governance structure toward one that makes a positive change in the surrounding societal, even global, context.

Interestingly, the two threads have recently encountered each other at the crossroads of IT that—as an artifact—has remained the commonly accepted (although also widely debated, see Chapter 1) research focus for the discipline of IS.

The definition of global development depends on the factors that are used to measure it. In addition to the economically oriented gross national product, indicators such as human development index (HDI) that measures the level of well-being, based on factors such as education, life expectancy, and standard of living (Human Development Reports, 2012), are also used. In addition to objective approaches based on government statistics, interview-based barometers address development from the viewpoint of individual, subjective experiences, such as the World Happiness Report (Word Happiness Report, 2012) by researchers from Columbia University.

HDI and WHR are examples of indices that measure the development from a human perspective. Sustainable development, however, takes also into account the environmental aspect. Encyclopedia of Earth (2012) as a member-administered website shows the manifold of ecologically oriented indicators for sustainable development.

Indicators and definitions are keys for analyzing the pace of global development. However, the crucial challenge is to make development take place. This requires setting goals of which the UN-defined Millennium Development Goals (2012) are the best known at the moment. The recent Rio + 20 Conference promoted the process toward sustainable development goals.

Complementary to the needs-focused MDGs, the bottom-of-the-pyramid approach (Prahalad, 2004) emphasizes the potential of the poor communities as a consuming market as well as a producer. This is a particularly interesting viewpoint from the IS perspective, since one of the main products for poor communities are mobile services—which also can market the communities' own production to the global markets.

The other thread, parallel to the emerging theme and variations of global development, is the redefining scope of IS as a discipline. As coarse, objective and global development indicators are fabricated into more focused and detailed ones for limited contexts, to achieve better accuracy, IS research is getting more focused on improving the life of an individual user or a user community. For example, the generalizability of research outcomes seems to give way for their transferability, from one user or context to another one.

The traditional agenda of IS research has been to improve business and governance processes by technology (see, e.g., Nilsson et al., 1999). Over the years, the agenda reduced to analyzing the usability of off-the-shelf solutions in given user scenarios. The research has followed the positivist approach for explaining and predicting the usefulness of technology for the users' performance and applied quantitative methods.

As neat and controlled the laboratory-oriented approach of IS might be, applying the natural science paradigm to the observed human users of technology, the agenda easily leaves IS in a disciplinary isolation, trapped by its own methodological dogmatism. Recently, this trend has been criticized by, for example, Walsham (2012) and Niederman and March (2012). Whereas Walsham calls for an interdisciplinary openness that allows IS to see whether ICTs make the world a better place, Niederman and March argue for a clear role for design science and action research that allows IS research to make a difference for the user communities.

The rebirth of IS as a world changing discipline—that would also inspire young talent and attract research funding—is built on the dual roots of the discipline. American pragmatism by Dewey and others emphasized the complementarity of thought and actions: what works is true. The continental critical theory, dating back to Marx, Nietzsche, and Freud, differentiates what is wrong from what is right. This academic heritage is the background of a discipline for a better, developing world, to be combined with the recent trends in IS research for design science and action research.

What many IS researchers (but not all; see the debates reported in Chapter 1) seem to agree upon is the role of IT artifacts as a focus of IS research. An IT artifact refers not only to a technical implementation, like product, service, application, or system, but can also be a construct, a model, or a method (Niederman and March, 2012). The consensus paves way to the very encounter of global development and IS.

For pragmatically oriented IS research, IT does not remain an observable artifact, but becomes an artifact for innovation. This is where IS meets with the challenges of global development: an innovation changes the existing practices, routines, and processes (Denning and Dunham, 2010). The agenda is also aligned with the expectations of and fears from IT in the context of developing countries. Critical approach is required to distinguish between the threats and opportunities where the affordances and limitations of IT take their users and the surrounding communities, even nations. For instance, the role of SMSs for improved communication is different before and after a revolution. For the oppressed and the ones in power, a simple SMS can carry a completely different meaning—even if the person in question is the same.

The global stage for the encounter of IS and universal development challenges is scaffolded by the recently emerged penetration of mobile technology, especially in the world's poor communities. The poorest one billion people at the bottom of the pyramid are not only using innovations imported from outside, but actively bringing about novel applications for changing their everyday. The potential of the whole new user group is also a crowd for sourcing innovations (Howe, 2006). The innovations from the grassroots might not only be incremental but also radically change the conditions of life. From the organizational learning point of view, we talk about double-loop learning instead of single-loop learning (Argyris and Schön, 1978).

From the viewpoint of an IT user, development always takes place within an immediate, physically or virtually local context. This observation sets the background for the current article. Therefore, rather than portraying the global image of IT users (Dutta and Mia, 2011), we focus on how IS research can develop an ordinary life.

Following the newly emphasized interdisciplinary orientation of IS, we will deliberately move from the IS-centric discussion to the dialogue between IT and global development. This is the *de facto* agenda of the renewing, interdisciplinary IS: to analyze the encounter of people with IT and have technology make a positive change to the lives of these people.

In many ways, the concepts of global development and IT seem slightly outdated from the mainstream of contemporary discussion. Development was on the global agenda at the turn of the millennium when the United Nations accepted the list of Global Development Goals (MDGs), to be reached by 2015. At about the same time, the expectations from IT were at their peak years. Within the cabinets of political decision makers, the challenges of global development and the opportunities of IT are still at the status as high as that of their politically correct promoters, but early adopters and change makers have already chosen alternative paths of uncertainty and new sources of inspiration, as the Arab spring has shown.

The intent of the chapter is to show that when juxtaposed with or cross-fertilizing one another, global development and IT can refresh, if not transform each other. In fact, global development, among other vast challenges of the humankind, offers a particularly interesting call for information technologists. But to comprehend this opportunity, one must go to the roots of both concepts, like for any renaissance.

What does global development mean from the viewpoint of IT? Who is the client of IT solutions that can launch the first step toward positive change and progress that the term development refers to? Is he/she a street kid sleeping in a wasteland next to a five star hotel? Or is he/she a president of a developing country? Or is it an anonymous group of bottom-of-the-pyramid customers that buy airtime for their text messages? From the IT perspective, there is nothing truly global in development, as there are very few, if any, universal solutions to all seven billion people. Development, although global as an abstract phenomenon and challenge, is local in contents and concrete customer requirements. Relatively, a poor community at the outskirts of a rich city may benefit from IT far more than a well-off landowner of a developing countryside. Therefore, rather than global development or a developing country, we will use terms such as *local development* or *developing context*, even a *developing individual*. From the IT perspective, a developing context is a challenge that technology can meet with information.

What, then, is IT from the aspect of a developing context? Customers in a developing context want to improve their lives by relevant, just-in-time, and functional information that requires technology. Far too often, these expectations are latent, unknown to those in the need of information, but likewise to potential solution designers. But when a street kid connects to an online supporter and can get regular food coupons as a text message, he can experience that the information works for him. Or when the president can promote equal access to learning basic literacy and numeracy skills, his/her nation can break the schooling barrier. And when a group of handicapped children in a remote Tanzanian city of Iringa get smart blocks for their rehabilitation (Lund, 2012), the information hits their bodies, not just minds. From the point of view of a developing context, IT is an opportunity beyond expectation.

The whole spectrum of information is relevant for a developing context: the information needs to exceed that of administrative databases or network protocols. Sensory data can be used to analyze the fertility of a soil or the muscular movements of a child with cerebral palsy; climate information can be refined to expand the knowledge of a small farmer toward that of an expert. In most developing contexts, the technologies should cope with and even make use of uncertain information. The concept of awareness with respect to information can be understood technically as in context-aware applications for unconsciously used data: a mobile device is aware of its user's position and the user does not need to pay attention to that awareness. But awareness can also be used along the tradition of Gestalt therapy, as in tools that promote HIV and AIDS explicit awareness for the user—far from the unconsciousness-related connotation of the term. Digital storytelling, a natural platform for presenting information especially among illiterate users, would even require technologies that convey emotions. The diversity and originality of the demands for IT in developing contexts are indeed apparent already within the concept of information: its types, representation, visualization, and reliability. This starting point provides challenges for IT researchers at the fundamental level of information itself.

Table 6.1 exemplifies state-of-the-art research approaches in the intersection of IT and global development, categorized by the type of the IT artifact studied and whether the motivation is on global development as a challenge or IT as an opportunity. For implementations, or instantiations of constructs, models, and methods, the division between challenge and opportunity loses meaning, because the application-oriented instantiations are based on an explicit challenge, but portray either a design or an analysis (or both) of an IT opportunity that meets with the challenge. The table also illustrates the diversity of the research area. This is why this chapter sketches a scaffold for the research area, based on a conceptualization of computing as a context-oriented discipline.

Classically, computer science has been positioned in the crossroads of mathematics, natural science, and engineering (Wegner, 1976). Methodologically, it has inherited one main principle from each of the scientific parents, namely, abstraction, modeling, and design, respectively. When the area of computing has expanded from strict computer science, computer engineering, and software engineering, the human and social aspects have increasingly gained relevance also in the methods of computing. This has had consequences for identifying user scenarios, developing user experiences, understanding usability issues from a cultural viewpoint, and developing ethical principles for IT (for cultural usability, see, e.g., Kamppuri, 2011). The interplay of the diverse scientific traditions of IT becomes very challenging in developing contexts, which differ significantly from those that computing grew within. For an illiterate farmer, climate information on a mobile platform might be represented as a story. A customer will transform the traditionally Western interpretations of abstraction, modeling, design, and user experience.

TABLE 6.1 Examples of Research Approaches on IT and Global Development by the Type of IT Artifact

Type of IT Artifact	Focus on Global Development as a Challenge	Focus on IT as an Opportunity
Construct	Ethnocomputing (Tedre et al., 2006)	Emergence of ICT4D as a research area (Unwin, 2009)
Model	A posteriori analysis	Future-oriented agenda of IS (Walsham, 2012)
		ICT4D business models (World e-Business Initiative http://www.worldebusiness.org)
Method	Design-reality gap (Heeks, 2002)	Improving mobile interfaces (Medhi et al., 2011)
Implementation	Interventions in application areas such as health, governance, education, agriculture, banking, tourism, business	

With its cross-cutting nature, IT is raising obvious expectations in developing countries. This far, the encounter between global development and IT has taken place in very large-scale projects. The approach goes back to the heroic, hall-of-fame type of understanding of global development and IT. Identify a major problem and attack it with well-trained troops carrying massive weaponry. Typical examples are extensive networking projects, teaching laboratories with hundreds of computers, or national programs for establishing community multimedia centers. A techno hero, trained outside the battlefield, is expected to plan and establish an extensive technical infrastructure to combat ignorance. But ignorance prevails as long as activities or services on the top of the infrastructure are not available. Plain walls or cables do not advance anyone else's progress than that of their constructor's. The projects are mostly funded by massive development aid programs whose benefits are getting more and more controversial and debatable (Moyo, 2009).

Recent concretizations of IT-flavored megaprojects are various programs with a focus on information or knowledge society and innovation; that is, services on the top of the technical infrastructure. The almost miraculous success of the global north, in particular natural resource poor Nordic countries, has been understood to be based on innovative designs and uses of IT. The Nordic road is an attractive progress story that most developing countries want to learn from and adapt to. In essence, national innovation system programs in developing countries aim at a level more advanced than building a plain technical infrastructure: they are building a national system of innovation or a whole knowledge society. While this goal setting looks at the first sight a bit more elaborated than just an extensive setup of material infrastructure, typical components of the innovation system might not take an intended beneficiary, like an individual farmer of a poor community, very far in their concrete development expectations. In order to succeed, incubation or funding services need to address the contents and links to the citizens' everyday life.

Like the devil is told to dwell in details, the challenge of twisting—or, indeed, distorting, see later—IT into a vehicle for progress is in individual encounters, at the very local level, at the bottom or base of the society's pyramid; although, interestingly, a truly networked society has a topology with neither top nor base. The contents and relevance in innovative IT designs for development can only be born at the grassroots.

Interestingly, the metaphors of the current IT streams reflect the ideological change from comprehensive, one-size-fits-all mega solutions toward personalized ones. At the time of ecosystems, cloud services, or grid computing, we are most excited of the individual components vitalizing the systems. Supercomputers do not raise the excitement of ubiquitous technology. In terms of the solutions, the perspective is changing from delivery-oriented electronic malls to piazzas where everyone can set up their small digital apartments, stages, and garages. With facebooks, twitters, and microcredits, individuals can make their expressions known, their chirps heard, and their talents realized. This is exactly what developing contexts need.

Developing communities can make use of IT-based innovations. However, it is essential to distinguish between various genres of innovation. The distinction between aiming at incremental or radical innovations (Henderson and Clark, 1990) has consequences for how IT takes up its role as an agent of change. The traditional mega(lomaniac) approach usually ends at incremental innovation in IT uses. It orients toward conservative changes: users live as earlier but with improvements. Text messages help life, but do not necessarily give value for money in terms of new challenges or alternative jobs. A radical innovation changes the rules of the game, transforms living conditions. A mobile endoscopic surgery unit with specialists consulting from overseas saves lives. A radical innovation, very often, is born in the periphery, behind the infrastructural services and out of the mainstream. We can talk about *pockets of development* by *pocket technologies*, in local and limited contexts.

A radical innovation at the grassroots—"radical" comes from the Latin word radix, for root—makes a creative use of IT as a source of innovation. But individual radical innovations need a way to the mainstream. The recent discussion on innovation emphasizes this dynamics between grassroots and periphery on one side and the mainstream on the other. Open innovation (Hagel and Brown,

2008) refers to facilitating a movement where finding solutions to individual and open challenges can be outsourced to crowds: we talk about crowdsourcing (Howe, 2006). IT offers extensive opportunities for crowdsourcing by giving a face and voice to individuals, independently of their backgrounds. The potential of the young people's energy and fresh ideas can become the radical solution to many of the globe's problems.

One of the main lessons that every foreign expert in a developing context understands sooner or later is that a solution is viable only if it brings benefit to all those engaged in the project or initiative. The principle easily brings about jealousy, misunderstanding, and mistrust, especially in well-funded mega projects. Accusations of corruption or waste of resources start to emerge. However, the whole idea of technology is the benefit for people, and the participatory approach should treat funders, decision makers, designers, and users at an equal basis. At the same time, the nature of the expected benefits might vary significantly between diverse partners and stakeholders. However, demanding benefit for all is a useful starting point for IT development. An individual community in Chiure, northern Mozambique, built their multimedia center with their own money. In order to increase their benefit, they need connectivity, that is, infrastructure. The government will get goodwill for responding to this need. The benefit for all can only happen when all join in the effort, but everyone also gains.

Developing contexts offer several examples, such as mobile banking and e-health, of bending available technology into novel tools that have brought about prosperity and wealth by information and its communication. Whether already success stories or yet weak signals, they challenge IT professionals to familiarize themselves in the dynamics of development, so that they can proactively recognize latent demands for information within developing contexts. Innovative solutions, with their origin in developing contexts, will reshape the field of computing that has far too long become so conventional that IT studies face recruitment problems among the globe's brightest young minds.

6.2 Underlying Principles

This chapter portrays the common ground for a developing context and IT that is designed and used in the context. IT is a cross-cutting technology that has potential to boost diverse areas of a developing society. The contextual challenges may *divert* the face of IT from the image that it is known of in developed countries. We will analyze the diverted image of IT from the four viewpoints of modeling, design, abstraction, and human aspects that we will call *encounter*. Although these perspectives were originally identified to position the discipline of computer science onto the map of sciences, they work well in setting up the stage for IT in a developing context.

In a developing context, the principles of construction and ethnocomputing (Tedre et al., 2006) will shape the four perspectives on computing and, hence, IT artifacts (Figure 6.1). Unlike the mainstream of the ICT4D (ICT for development) (Unwin, 2009) movement that is mostly interested in analyzing the impacts of the transfer of available IT into developing contexts, we take seriously the constructive agenda of computer science (Sutinen and Tedre, 2010). This means that the theoretical emphasis shifts from a conservative and controllable evaluation toward radical and risk taking art of creative design, development, construction, and implementation. A developing context will divert its challenges to artifacts, including products and services that are not yet even dreamed of among the IT users or developers of the global north.

In order to construct, a developer needs to find a comfort zone in his/her workshop. In developing contexts, not only the information technologies but also their related concepts are of alien origin. Similar problems have aroused in other fields, like learning mathematics. A set of ethnosciences or ethnoarts have answered to the question: ethnomathematics, ethnomusic, ethnobotany, and recently ethnocomputing. Whereas ethnomathematics (Gerdes, 2007) is interested in different representations of mathematical concepts, ethnocomputing emphasizes that everyone should be able to enter the world of IT from his/her own background. For example, programming courses

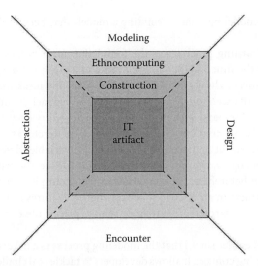

FIGURE 6.1 The four perspectives of computing focus on constructing IT artifacts by ethnocomputing. The ethnocomputing approach ensures that the context tunes the construction process of a relevant IT artifact.

might start from contextually important problems rather than from the conceptology of variables and statements, based on the Western science.

6.2.1 Constructive Modeling in Context

The modeling perspective to IT refers to its roots in natural sciences. The key idea of a model is a set of variables and a function that maps the variables to predict the phenomenon that the model is assumed to portray. In IT, the predicted phenomena cover a wide spectrum of automated processes from the behavior of algorithms up to the user experience of applications. Estimating the performance of an algorithm is an example of a quantitative model, explaining the experience of using, say, a social media platform might require a qualitative approach. For IT artifacts in the developing contexts, the most interesting phenomena to be explained, predicted, or interpreted are the *functionality* and *relevance* of the systems and services in physically, culturally, economically, or politically demanding conditions. Functionality denotes the extent to which the evaluated technology works according to its expectations that can vary throughout the modeling process. Relevance measures issues like the benefit from, added value, cost efficiency, or robustness of, and improved quality of life promoted by the technology in a given context.

Along the lines of modern science education, the empirical paradigm calls for modeling that is a learning process based on observations in the perceived surrounding reality, rather than previously derived, existing theories. The contemporary pedagogical approach is contradictory to the earlier one where a lesson started from the teacher's solo introduction to a theory that the students needed to learn by heart; the students' own activities with applications and examples were secondary to the teacher's presentation. The old approach believes in localizing the commonly approved standards in students' minds, the modern one in students building their own insights from their own observations. When measuring the learning outcomes, the first approach emphasizes global standards, the latter the contextual relevance of what has been learnt.

Often, the transfer of information technologies to a developing context is analogous to the outdated pedagogical model. Technologists seem to believe largely in massifying, delivering, or generalizing the so-called best practices: technologies that have shown to work in a similar context are assumed to be ready for others to adapt to, or learn from. This understanding materializes in countless seminars, workshops, conferences, and courses that call policy makers to learn by heart IT systems, services, or solutions that have shown to be functional or relevant by the models developed elsewhere. One of the key ideas of computing—modeling, that is, building a model—has been replaced by imitating, that is, following the given model.

In fact, an empirical learning process sets the foundation for a process of building a contextual model for understanding the functionality and relevance of a given IT in a given context. In essence, an empirical modeling process always involves trial and error: the measures to understand the functionality and relevance of the technology are iterated by the feedback from the context. The iterative modeling process goes hand in hand with its parallel of *participatory design*. Whereas the design process results in artifacts, also in iteration, the modeling process iterates the criteria against which the functionality and relevance of the artifacts are measured. The variables determining the functionality and relevance of the technology vary from context to context. Hence, the criteria for functionality and relevance cannot be given beforehand, or standardized. For example, a mobile climate service for a user group of illiterate farmers in an electricity-secure context requires criteria different from those of a similar service to industrial farmers in extremely remote places: the criteria for both functionality and relevance are different.

The analysis mentioned earlier shows that the modeling process is a natural part of the construction of an IT in a given, developing context. It allows developers to tackle real challenges by making use of the opportunities of IT in a bottom-up, empirical, organic way. The modeling process turns feedback from observations into criteria for improving the functionality and relevance of the systems and services to be designed.

Somewhat surprisingly, robotics gives an opportunity to derive a potential user's own understanding of a computational process in a concrete way in a developing context. For instance, Lego robots have been used in IT education to promote self-confidence with technology. Programming robots opens the black box of IT in a way that sparks the creativity of local designers (Sutinen and Vesisenaho, 2006).

Quality assurance is a standard part of the modern design and construction process of an IT system or service. However, in many cases a certified production process is only assessed by superficial quantitative factors or, even worse, by the outcomes that are called performance indicators. For example, in an educational process a certified primary schooling process might count the percentage of children that have taken IT lessons at school, but pay no attention to the competences they have at the time of graduation, or the way they are using computers 1 year after finishing the school. The quality assurance factors and indicators are normally set before the process, thus resembling the outdated theory-driven science education approach.

The definition of quality as the relevance of the contents (Machili, 2012) shifts the focus from the process to the contents at its each step. While the empirically derived, emerging model of an IT system or service focuses on functionality and relevance issues, each modeling iteration improves the quality of the system or service as understood as the relevance of the contents of the system or service.

Modeling answers to the question "Why does a system or service function in a relevant way?" "Why" is also the main question of any natural science inquiry to the observed phenomena in the surrounding physical world.

6.2.2 Constructive Design in Context

The design approach in computing stems from the engineering legacy of the discipline. In developing contexts, engineers have typically been invited as constructors and builders implementing designs that have their origins outside the context. Hence, from the design point of view, engineering has been neither constructive nor contextual. This applies also to most IT systems and services, with exceptions like mobile banking where Kenya is among the global pioneers of novel designs.

Participatory design is a movement within design that calls for ordinary users to join a design process. Culturally, the approach is of Nordic origin: it is a design incarnation of the political egalitarianism of the Nordic countries. In this approach, participation does not separate designers from the rest of the society into an own class of ivory tower experts, but invites them to learn from and struggle with laymen for innovative designs. Participatory design can also be understood as a consumer movement: consumers are seen the best experts of not only the everyday realities but also the criteria for the functionality and relevance of the solutions to be designed. As a bottom-up or base-of-the pyramid-up movement, we can call participatory design *contextualized* or *contextualizing design*.

The inherent dilemma of participatory design—obvious particularly in developing contexts—is that the users are not aware of the opportunities of technology and for this or other reasons do not have imagination for novel, out-of-everyday designs. This is also reflected in Steve Jobs's comment that radical designs can never be achieved by listening to users' needs. This is certainly true if the users' imagination is not empowered. But designers can also be intellectually lazy or tied to their own ethnocentrism. A latent obstacle making designers reserved for a contextualized approach is the common attitude that developing contexts should use affordable and tested second-hand technologies. This attitude is based on the understanding that every technology has its predestined sequence of steps that every culture should go through. Hence, typewriters should be used before text processing, landline telephones before mobile phones, dull phones before smart phones, and smart phones ahead of tablets. As we know of several examples of developing contexts leapfrogging the predestined steps, the idea of predetermined sequences of technical developments does not hold.

Because of the challenges described previously, participatory design requires close dialogue between professional and laymen designers, as to discover the latent demands for relevant designs and to match them with the opportunities that the advancing, not-yet-available or not-yet-designed technology provides. A design dialogue is a reciprocal learning process between the professionals and the laymen. The professionals need to learn the resources and requirements of the future users of the to-be-designed technologies. The laymen have to learn the principles of technology and get hands-on experience with it to the extent that they can use their imagination to ideate what is possible. A fundamental prerequisite for a mutual learning process—that might take several years—is trust among the professional and laymen designers, because the users might be shy to explicate their still very vague ideas. A fast and politically correct exchange of system specification forms is but a remote image of a committed dialogue for sustainable designs.

One of the ancestors of the participatory or contextual design approach is the Bauhaus movement from the Germany of the 1920s. The Bauhaus movement was a counter reaction to the success story of industrial design of the contemporary England. As a movement, Bauhaus emphasized interdisciplinary and explorative design communities for functional solutions. A concurrent parallel of the Bauhaus communities are living laboratories that accommodate technology developers to implement and analyze a design process *in situ*, within a real-life context. As their best, living laboratories are pockets of development that radiate their successes and enthusiasm to the surrounding region that—particularly now at the time of social media—might easily extend its physical limitations. A living laboratory might be a street studio that gathers street vendors to share the ideas of their business strategies with designers that are creating context-aware applications to city dwellers. A creative design milieu is an environment that features the characteristics for a novel and fresh design.

Unlike the bottom-up approach to constructive design in context, many international development aid funders are stuck in the outdated top-down ideology, leading to strictly hierarchical planning. The planning scheme with its follow-up monitoring mechanisms is called the *logical framework* (AusGuideline, 2005), logframe for short, and it is, in theory, based on a temporal logic model for understanding the dependencies between various components of a project. While the logframe can be used for projects that follow a linear sequence of predefined steps, like those of building a house, it is unsuitable for bottom-up technology design projects with—hopefully—several simultaneous and unexpected inventions, stages, and iterations. The many simultaneous threads of a contextual design project make it impossible to represent the project as a hierarchy. Structured as a hierarchy, the logframe does not allow nor support links between the leaves of the tree-typed structure. However, for an innovative design project of an IT artifact, the most exciting opportunities are based on combining components distant from each other.

Despite all its weaknesses, logframe can be used as a structure for planning a technology design project. The reason is that it is closely aligned with needs-based design. An expatriate designer, whether an individual or a group of experts, is tempted to understand a user community from the viewpoint of their assumed hierarchy of needs: the needs hierarchies can be easily identified outside the user community as a desk research exercise or plainly copied from sources like the MDGs. By the logframe approach, the community can easily be imposed by a structure completely unfamiliar to them. The serious shortcoming of the logframe approach is that it does not fit into a reality with a diverted structure, different from but more valid than its Western, hierarchical representation. Indeed, the diversion is in the viewers' eyes: the hierarchical representation might be equally diverted from a viewer representing a developing context as a developing context can seem chaotic to a Western professional designer or planner. In any case, the logframe is unable to accommodate unexpected, emerging threads that a contextual design process is known of. We conclude that as a design tool, the logframe is based on the outdated waterfall model. The modern agile development should replace it especially in the developing contexts.

In essence, participatory design is aligned with the strengths-based rather than needs-based orientation for design expectations. Unlike the needs-based approach based on the undisputable lists of given targets, the strength-based approach starts from the given resources. While the needs-based approach leads to a simplistic and linear design process consisting of a sequence of predetermined steps, the strengths-based approach requires explorative and creative screening of resources that are not included in any given wish list of ideal contents: the resources might be hidden and unexpected. Whereas the needs-based approach is based on the negatively flavored list of what a community does not have, the strength-based approach is positive and based on what is available.

As a constructive design approach in context, the participatory design process makes IT an agent of change in the design community and beyond it. The design process makes its milieu a cradle for fresh designs. Because of its designed products, the cradle starts to attract people from outside. What was a laid back community from the viewpoint of technology has a potential to transform to an origin of truly new technologies.

A properly organized participatory design process is an answer to the How question of constructing functional and relevant IT.

6.2.3 Constructive Abstraction in Context

The role of abstraction as a key approach in computing goes back to the mathematical roots of the discipline. Mathematics forms the foundation of any automation and is seen in the formal representation and language of computational processes, and the interrelations and functional dependencies of the formal entities. While many of the formal structures of computing, such as database theory, live their own, independent mathematical life, they originate in the real world problems and their formal representation, like that of occupying flights or arranging a census. As a reminder of their real-life background, formal structures have physically descriptive names such as trees for certain types of data structures or divide and conquer for a class of algorithms.

While modeling and design as threads of computing were almost straightforward to interpret as underlying principles of constructing IT artifacts in a developing context, abstraction requires more elaboration. The most natural approach is to analyze abstraction from the viewpoints of representation and language.

Ethnostudies and ethnoarts, in this case particularly ethnocomputing, refer to the ways that culturally central principles, ideas, or forms of expression have been used as entry points to understanding or practicing a certain field of science or art. Ethnocomputing suggests that the language of computing can be learned from one's own culture, not necessarily following the standard steps of a universal curriculum. In practice, an ethnocomputing approach to learning programming in a developing context can take place by constructing programs for gadgets made of physical blocks. Ron Eglash's monograph

African Fractals (1999) gives an example on how fractal, self-iterative, or recursive expressions repeat in West African architecture and how they can be used in (learning) computing.

Contrary to the common belief, mathematics is not *per se* a discipline of a theoretical or deductive character, at least when created. According to Polya (1957), mathematicians work as inductive experimenters, not deductive theorists, and this should be more transparent in how mathematics is taught. From the viewpoint of abstraction in IT, Polya's message translates to IT explorations, which are inspired also by cultural contexts, along the lines of mathematical adventures by de Guzman (1986).

The idea of abstraction calls for unprejudiced theory and concept formation in the developing contexts earlier unexposed to IT, based on identifying the computational interpretations and formulations of their real-life problems. Coping with tolerance, uncertainty, and diversity, whether in natural languages or cultural representations, can form starting points for indigenous abstraction. In many contexts, storytelling might be an appropriate representation for even the core of computing, like programming.

Indigenous abstraction can serve as a counterattack to the conceptual counterpart of technology transfer, namely, concept transfer. Almost any IT conference in developing countries seems to repeat the global slogans, such as the quest for community multimedia centers or integrating IT in school curricula. While innocent and sometimes even beneficial initiatives in themselves, transferred concepts seldom have carried their original meaning to the audiences of their importers.

Abstraction names and positions the inner mechanics of IT. Thus it generates the navigation or mapping system for IT. Each cultural context might need to (and, indeed, needs to) identify IT using its own representations. Abstraction answers to the What, Where, and When questions related to a technology that needs to be in the comfort zone of its users.

6.2.4 Constructive Encounter in Context

We call the human perspective to computing an *encounter*, to emphasize that the development and use of IT takes place in an encounter that always involves people, whether as intellectuals searching for inventions, nerds enjoying the fascination of computational devices, or ordinary users eager to improve their everyday living.

Much of the research and practice of the users' encounter with IT in developing contexts sees people as objects using the technology. This is called an evaluative approach of ICT4D. The objects need to be understood, their needs analyzed, and IT solutions either modestly bent accordingly or, what happens also in developed countries, the encounter simplified by training the objects to use technologies that are far from their everyday needs and unnecessarily hard to learn and master. Rather than empowering people in their everyday life, the technology makes them its slaves. We call the evaluative view of IT encounter an *objectifying* view of users.

A constructive encounter between IT and its users in their context sees people as subjects who want to influence on their situation and make a difference. It is thus aligned with the participatory design approach. However, instead of engineering solutions, the encounter approach focuses on the *human rights perspective*. It promotes aspects like digital rights, ownership, access, and collaboration within diverse user groups. We call the constructive view of IT encounter a *subjectifying* view of users.

Whereas the objectifying view evaluates the extent to which users can *access* information by technological systems and services, the subjectifying view is interested in the opportunities that people can use technology for creating, delivering, and promoting the full spectrum of information: data, knowledge, and awareness. Access of information is usually identified with digital rights. The latter view emphasizes the *ownership* and sharing of information and tools that allow people to process, generate, elaborate, and modify information. Let us consider an example from the area of e-government. The objectifying view evaluates the e-government services from the perspective of government as a service provider and taxpayers as service users. The subjectifying view is interested in the construction of digital platforms that facilitate the citizens to make their voice heard.

From the human rights perspective, the objectifying view is interested in how human rights set standards to information services and access thereof. Hence, the services are analyzed against the given human rights. According to the subjectifying view, the focus is on how people can use IT to promote human rights, even those yet unheard of. Awareness is of a particular interest to human rights, especially from the viewpoint of a developing context. Awareness has two connotations. Awareness can be unconscious and latent or conscious and explicated. As latent, awareness can be understood as a smooth and transparent access to information, and it is based on human rights in the objectifying sense. Explicated awareness is close to emancipation by information, and thus, it promotes human rights in the subjectifying sense. For instance, an emancipatory HIV and AIDS awareness campaign allows its participants to share their stories of coping with the disease. Interestingly, as awareness, also context has both the unconscious or transparent and explicit or manifest connotations. In order for IT to contribute to the development of a context, the members of the context need to be fully aware of that they belong to it.

As an indication that MDGs are aligned with the objectifying view of IT users, the MDGs do not refer to ICTs promoting human rights while they cover, for example, environmental sustainability and global partnership that can set standards for evaluating IT access from the human rights perspective. The UN declaration of human rights (2012) is equally vague in how IT can promote human rights, see relevant articles 17 (right for one's property), 19 (information), and 21 (public service). In general, in developing contexts IT has a significant role in promoting dialogue between conflictive human rights, for instance, those of environmental awareness and the rights of the owner of a mine (http://en.wikipedia.org/wiki/Human_rights#Human_rights_and_the_environment).

The Arab Spring in 2011 showed that IT is in particular a human rights tool for the young population. The challenge is in how IT can help them to join forces to tackle development challenges in their contexts.

The encounter aspect of IT answers to the Who questions—who, by whom, and whose—of IT.

6.2.5 Toward Life Computing

The underlying principles of IT and global development answer the fundamental challenges of making IT a tool for development: Why does it work (modeling)? How does it work (design)? What works, where and when (abstraction)? Who is the subject of IT (encounter)? The answers set up a scene where a diversity of experts and users make use of observation and measuring instruments for modeling; shared workbench for design, various languages, representations, and navigation tools for multilayered understanding; and refresh their human rights for active encounters. As understood in this way, IT is in the center of human activity for a better world, and it can be called life computing.

6.3 Impact on Practice

Even if the landscape of IT within developing contexts is complex and sporadic, it is easy to observe several instances where it has dramatically changed not only the way the discipline of computing is understood by professionals but also the conditions of its users. An obvious example is the mobile phone revolution among the poor people: people at the grassroots are using technology for everything from getting a job up to making a revolution.

Due to the diversity of IT uses in developing contexts, this section can but portray pockets of the impacts that IT has had on people's lives. The perspective is that of individual technologies, their designers, and users.

6.3.1 Impact on the Discipline

We will go through the foreseen impact of IT and global development on the discipline of computing by the underlying principles of modeling, design, abstraction, and encounter.

Modeling emphasizes the empirical orientation in the context. The interdisciplinary orientation that is required for IT to make a difference in a given context strengthens the role of humanities in the discipline.

Participatory *design* generates unconventional crosses between disciplines. Environmental challenges in developing contexts and the demand for renewable local energy connect IT to forestry. Opportunities of modern endoscopic surgery allow specialized pediatric surgeons to use highly advanced technology even in remote areas: this sets demands for using IT for learning modern surgery in real contexts. IT can support a farming community throughout the lifetime of a crop from purchasing the seeds until selling the harvest: successful mobile service design requires collaboration between meteorology, agriculture, anthropology, sociology, and IT.

In the area of *abstraction*, digital storytelling generates novel representations for IT. The contextual orientation transforms usability toward cultural usability, and replaces the idea of localizing universal services to contextualization that starts from the requirements of a given context.

Encounter emphasizes ethics and an anthropological orientation throughout the discipline. For instance, the anthropological concept of awareness transforms standard user expectations to those promoting the metacognition of the user's behavior. The subjectifying orientation of the actor calls for open platforms, for multiple contributions, on an equal basis. This is foreseen in the open source movement that paved the way to the culture of tweets and micromovies.

The general trend common to all the impacts above leads toward a more contextualized discipline that is capable to absorb inspiration from the challenges. The trend poses requirements for an adaptive and flexible curriculum that interconnects to other disciplines.

6.3.2 Impact on the Profession

We will go through the foreseen impact of IT and global development on the profession of computing by the underlying principles of modeling, design, abstraction, and encounter.

Empirical *modeling* as an integrated thread of a software design and implementation process requires that the criteria for functionality and relevance are iterated throughout the process. This demands that IT professionals are able to replace given criteria lists by those created together with the users.

The participatory *design* projects in developing contexts change the landscape of IT consulting. Emerging areas cover areas like mobile banking, mobile learning, IT for health, and IT education that take place in the context. An IT professional changes from a person supporting available practices by conservative solutions toward a change maker. For example, mobile learning focuses on informal learning settings rather than conventional schooling.

Abstraction calls IT professionals to listen to the local voices at the bottom of the pyramid, instead of following the traditional top-down approach of various needs-lists. IT professionals turn into language experts and need to interpret the hidden demands with novel IT representations.

For *encounter*, IT professionals need to design technologies that promote their users' human rights. These cover applications that support interfaith and intercultural dialogue, conflict resolution, and crisis management. Natural language processing and text tools form key techniques for these applications.

In general, the impact from IT in the context of global development has transformed the IT profession toward diverse job profiles. The competences required from IT professionals reflect increased sensitivity and ethical orientation.

6.4 Research Issues

The impact of IT for global development is still at its infancy. At the same time, IT can significantly gain from novel challenges and solutions within the developing contexts. Therefore, it is crucial that IT is developed within developing contexts where an interdisciplinary and multi-stakeholder collaboration renews and rejuvenates both technology and the development that it shall serve. This means that mere evaluation of existing technology in various developing contexts is not enough; proactive and engaging design and implementation are required.

In a way, the relation between technology and its users in developing contexts reminds that of digital natives and digital immigrants in other contexts. There is a difference, though: instead of differentiating between *users* from the viewpoint of their technology use history, we can measure the extent to which *technology* itself has been exposed to or rooted in a particular user context. Technology can be either *context native*, that is, born—designed and implemented—in a given context, or *context immigrant*, that is, transferred from its origin to an alien place to serve. Using this terminology, most of the technologies used in developing contexts are context immigrants. The key research challenge is how to design context native IT.

Like the impact on IT discipline and profession, the research challenges can be categorized in modeling, design, abstraction, and encounter. Contrary to the issues of a practical nature, research looks, though, for radical rather than incremental changes. In the area of language learning, an incremental innovation allows students to learn Portuguese by their mobile phones during their holidays in Lisbon. A radical innovation lets Mozambican secondary school students to earn airtime by chatting with students that are becoming teachers in the Portuguese language. The role of radical innovation and creativity is crucial for the research-based creation of context native technology, which can also be called indigenous technology.

Empirical modeling requires that the measures for the functionality and relevance of context native technology be rethought. This results in out-of-box, fresh *models* as criteria for successful technologies. To measure the success of context immigrant technologies, it is normally enough to adjust the weights of the measured variables, like in the case of the mobile Portuguese learning environment those related to the user behavior. A set of completely new variables needs to be introduced for evaluating the functionality and relevance of the user scenarios of children teaching adults from the global south to the global north. To measure the impact of context native technology requires qualitative indicators, and thus, methodologies based on the formative approach, like artifact contests and focus user group barometers.

Strength-based or resource-based *design* of context native technologies can be easily juxtaposed with the needs-based design of context immigrant technologies. On one hand, the first is based on how to use the existing resources in a creative way for matching the demands and expectations of the context; both the resources and the demands need to be explored without prejudices or available lists. Needs-based design, on the other hand, makes use of existing, straightforward lists of what is lacking. Research on creative design milieu is an example of a task that results in context native technologies. This requires designing a relevant IT curriculum for the particular developing context.

From the perspective of *abstraction*, research in non-linear information representations and non-structured media is of a particular importance for context native technologies. In a developing context, digital storytelling requires workbenches and studios for diverted stories. An interesting trade-off takes place between abstraction and concretization: a multilayered representation supports both of them.

As of the research for enriching the *encounters* between users and their context native technologies, a completely new research area in the intersection of human rights and information technologies is envisaged. The key question is how IT can promote human rights by strengthening the ownership in information in all its forms, including awareness.

An example of a possible research initiative is microstudies on context native IT. Microstudies is an equivalent of microcredits in the domain of education. Microstudies enable a bottom-up study process where each study—performed using a mobile device—gives a narrow competence in a user-requested area. The acquired competences are stored as a digital portfolio. When a student has reached a critical set of studies, she can apply for a certificate.

The rationale of microstudies is in the massive and fast-changing educational landscape of a developing context. Educational demands are hard to forecast, mobile technology is available, and people need fast track for IT competences required at jobs. At the same time, the culture recognizes traditional, formal aspects of education, like certificates.

The exemplified research project will be carried out following the strength-based design in a given context. Representation of the learning process is a digital, interactive, multimedia-based learning story, shared by social mobile media. A focus group representing potential users identifies indicators for the functionality and relevance of microstudies.

We can summarize the challenge of context native technology with the alternative concepts of garage computing, bottom-of-the-pyramid computing or life computing. The research toward context native technology takes place in developing technology pockets, that is, a periphery far way from centers of conventional technology.

6.5 Summary

The era of fast global development sets an attractive stage for IT to return to its passionate and radical roots. Developing context native technologies will require young talents that the developing contexts are full of. The emphasis is on designing fascinating solutions that only later might find their satisfied customers. The carrying force of the challenge is the belief that fresh ideas grow in peripheries.

Acknowledgments

In addition to the review of the key academic and industry literature, this chapter reflects the lessons learned by the author during his 2 years' work as the Chief Technical Advisor in STIFIMO, a program of cooperation in Science, Technology, and Innovation between Finland and Mozambique in 2010–2014. The job offered a unique opportunity to discuss issues related to IT and global development with diverse stakeholders and also with people who did not have a clue of how science, technology, or innovation could benefit individual Mozambicans—mostly, the latter were expatriates. The opinions do not necessarily reflect those of the Ministry of Science and Technology of Mozambique or of the Finnish Ministry for Foreign Affairs. For the research part of the work, the author is indebted to the partial support from the Academy of Finland.

Further Information

http://ictlogy.net
http://www.ict4d.org.uk, for links http://www.gg.rhul.ac.uk/ict4d/links.html
e-Government for Development: http://www.egov4dev.org/

References

Argyris C. and Schön D. (1978). *Organizational Learning: A Theory of Action Perspective*. Reading, MA: Addison Wesley.

AusGuideline (2005). 3.3 The logical framework approach. Available at http://www.ausaid.gov.au/ausguide/Documents/ausguideline3.3.pdf (accessed on March 29, 2013).

Denning P.J. and Dunham R. (2010). *The Innovator's Way: Essential Practices for Successful Innovation*. Cambridge, MA: The MIT Press.

Dutta S. and Mia I. (eds.) (2011). The global information technology report 2010-11. Transformations 2.0. World Economic Forum. Available at http://www3.weforum.org/docs/WEF_GITR_Report_2011.pdf (accessed on March 29, 2013).

Eglash R. (1999). *African Fractals: Modern Computing and Indigenous Design*. New Brunswick, NJ: Rutgers University Press.

Encyclopedia of Earth (2012). Expert-reviewed information about the Earth, ed. Cleveland, C.J. (Washington, D.C. Environmental Information Coalition, National Council for Science and the Environment). Available at http://www.eoearth.org (accessed on March 29, 2013).

Gerdes P. (2007). *Etnomatemática. Reflexões sobre Matemática e Diversidade Cultural.* Ribeirão, Portugal: Edições Humus.

de Guzman M. (1986). *Aventuras Matemáticas (Mathematical Adventures).* Barcelona, Spain: Labor.

Hagel J. III and Brown J.S. (2008). Creation nets: Harnessing the potential of open innovation. *Journal of Service Science* 1(2), 27–40.

Heeks R. (2002). e-Government in Africa: Promise and practice. *Information Polity* 7(2), 97–114.

Henderson R.M. and Clark K.B. (1990). Architectural innovation: The reconfiguration of existing product technologies and the failure of established firms. *Administrative Science Quarterly* 35, 9–30.

Howe J. (2006). The Rise of crowdsourcing. *Wired Magazine* 14(6), 1–4. Available at http://www.wired.com/wired/archive/14.06/crowds.html (accessed on March 29, 2013).

Human Development Reports (2012). Available at http://hdr.undp.org (accessed on March 29, 2013).

Kamppuri M. (2011). Theoretical and methodological challenges of cross-cultural interaction design. Publications of the University of Eastern Finland Dissertations in Forestry and Natural Sciences No 29. Available at http://epublications.uef.fi/pub/urn_isbn_978-952-61-0407-2/urn_isbn_978-952-61-0407-2.pdf

Lund H.H. (2012). A concept for a flexible rehabilitation tool for sub-Saharan Africa. In: *IST-Africa 2012 Conference Proceedings,* Dar es Salaam, Tanzania, Cunningham P. and Cunningham M. (eds.), IIMC International Information Management Corporation, Dublin, Ireland.

Machili C. (2012). Personal communication.

Medhi I., Patnaik S., Brunskill E., Gautama S.N.N., Thies W., and Toyama K. (2011). Designing mobile interfaces for novice and low-literacy users. *ACM Transactions on Computer-Human Interaction* 18(1), 1–28.

Millennium Development Goals (2012). A Gateway to the UN System's Work on the MDGs. Available at http://www.un.org/millenniumgoals/ (accessed on March 29, 2013).

Moyo D. (2009). *Dead Aid: Why Aid Is Not Working and How There Is Another Way for Africa.* New York: Farrar, Straus and Giroux.

Niederman F. and March S.T. (2012). Design science and the accumulation of knowledge in the information systems discipline. *ACM Transactions on Management Information Systems (TMIS)* 3(1), Article No. 1, 1–15.

Nilsson A.G., Tolis C., and Nellborn C. (eds.) (1999). *Perspectives on Business Modelling: Understanding and Changing Organisations.* Berlin, Germany: Springer-Verlag.

Polya G. (1957). *How to Solve It,* 2nd edn. London, U.K.: Penguin Books Ltd.

Prahalad C.K. (2004). *The Fortune at the Bottom of the Pyramid.* Upper Saddle River, NJ: Wharton School Publishing.

Sutinen E. and Tedre M. (2010). ICT4D: A computer science perspective. In: Elomaa, T., Mannila, H., and Orponen, P. (eds.), *Algorithms and Applications, Lecture Notes in Computer Science (LNCS) 6060.* Berlin, Germany: Springer, pp. 221–231.

Sutinen E. and Vesisenaho M. (2006). Ethnocomputing in Tanzania: Design and analysis of a contextualized ICT course. *Research and Practice in Technology Enhanced Learning* 1(3), 239–267.

Tedre M., Sutinen E., Kähkönen E., and Kommers P. (2006). Ethnocomputing: ICT in cultural and social context. *Communications of the ACM* 49(1), 126–130.

The UN Declaration of Human Rights (2012). Available at http://www.un.org/en/documents/udhr/ (accessed on March 29, 2013).

Unwin T. (ed.) (2009). *ICT4D, Information and Communication Technology for Development.* Cambridge, U.K.: Cambridge University Press.

Walsham W. (2012). Are we making a better world with ICTs? Reflection on a future agenda for the IS field. *Journal of Information Technology* 27, 87–93.

Wegner P. (1976). Research paradigms in computer science. In: *Proceedings of the 2nd International Conference on Software Engineering,* San Francisco, CA, October 13–15, pp. 322–330.

World Happiness Report (2012). Available at http://www.earth.columbia.edu/sitefiles/file/Sachs%20Writing/2012/World%20Happiness%20Report.pdf (accessed on March 29, 2013).

Initiatives

Bridges.org: http://www.bridges.org
Computing Community Consortium on Global Development: http://www.cra.org/ccc/globaldev.php
Global e-Schools and Communities Initiative: http://www.gesci.org
Mobile Technology for Social Impact: http://mobileactive.org
One Laptop per Child: http://one.laptop.org
World e-Business Initiative: http://www.worldebusiness.org

Journals

Information Technologies & International Development: http://itidjournal.org/itid
The *Electronic Journal of Information Systems in Developing Countries*: http://www.ejisdc.org/ojs2/ index.php/ejisdc

Conferences

IFIP WG 9.4 Social Implications of Computers in Developing Countries: http://www.ifipwg94.org/ conference-tracks
International Conference on Information and Communication Technologies and Development: http:// ictd2012.org
ACM Annual Symposium on Computing for Development: http://dev2012.org
International Development Informatics Association: http://www.developmentinformatics.org
InfoDev: http://www.infodev.org
World Summit on the Information Society (WSIS): http://www.itu.int/wsis/index.html

7

Using ICT for Development, Societal Transformation, and Beyond: Closing the Digital Divide in Developing Countries: Case of Egypt*

Sherif Kamel
The American
University in Cairo

7.1 Overview

Developing countries around the world face various challenges across different sectors. These challenges vary depending on infrastructure availability, level of education, and economic conditions among other reasons. During the past few decades, new challenges have been imposed on the developing world. These challenges were introduced following the massive introduction and diffusion of information and communications technology (ICT) with different societies experiencing digital inclusion in some sectors but also digital exclusion and gaps in many more areas. The digital divide

* This is an updated and an amended version of "Information and Communication Technology for Development: Building the Knowledge Society in Egypt," published in *Access to Knowledge in Egypt New Research on Intellectual Property, Innovation and Development*, edited by Nagla Rizk and Lea Shaver, London, U.K.: Bloomsbury Academic, 2010, ISBN 978-1-84966-008-2, pp. 174–204.

has been described and defined in a variety of ways; it is fundamentally the gap between those who have access to and control of technology and those who do not (Tinio, 2003). In other words, the term describes the difference between the haves and the have-nots when it comes to information access and knowledge dissemination (Loch et al., 2003).

One of the important factors in bridging the digital divide is the availability of ICT and Internet skills and capacities coupled with the spread of the ICT infrastructure investment (Opesade, 2011). Unless that is realized, ICT could be the platform through which gaps widen between societies not only inter (between) but also intra (within) communities and countries alienating remote and underprivileged communities and leading to the widening of the socioeconomic gap resulting from minimal information access and dissemination as well as knowledge acquisition opportunities.

Different countries experience different reasons for which readiness of infrastructure for digital inclusion is realized. For example, in the case of South Africa, technology infrastructure and human skills exist; however, there is a need for more determination and political will from the government (Parker, 2011). Information-poor and information-rich societies are still present in different countries around the world reflecting the case of intrasociety digital divides. In the case of Egypt, closing the divide is handled by a public–private partnership where the government enabled the regulatory environment, the private sector made the bulk of the investment and the civil society had an invaluable role to play in education and learning. The diffusion of ICT is not following the conventional wisdom in most developing nations where technology is only diffused in the capital, but it is also in the remote areas, villages, and towns as well as in different cities around Egypt. According to Opesade (2011), "investment in ICT alone cannot bridge the digital divide, but a value-based investment targeted at specific developmental issues will go a long way in achieving this." It needs a holistic view for comprehensive societal development.

While the World Bank classifies nations based on gross national income (GNI) per capita, in the ICT age, introduction, diffusion and access to information and knowledge, technology platforms, and the percentage of digital literacy are important factors for societal development and growth. Moreover, these elements are used to rank nations and represent important factors attracting companies and foreign direct investment (FDI). Penetration rates of ICT, Internet, and mobility are vital indicators that help minimizing the digital gap (Kamel and Hussein, 2002). Basic needs, food, shelter, and housing, are no more the same; both information and technology have become of primary importance. According to the World Bank (2002), "the knowledge gap in many developing countries is a contributory factor to poverty, and there is no better way to bridge this divide than through the use of ICTs." For emerging economies and developing nations, one of the primary important issues that need to be addressed is a clearly defined information technology policy that is critical to the development of a comprehensive information technology infrastructure (Aqili and Moghaddam, 2008). These ideas were raised throughout the past decade and highlighted during the World Summit of the Information Society (WSIS) both in Geneva in 2003 and Tunis in 2005 (World Summit on the Information Society, 2008). One of the challenges that were discussed at both summits was to avail the ecosystem required to provide "a society where everyone can access and share information and knowledge that can enable individuals and communities to achieve their full potential in promoting their development and improving their quality of life, and by that the digital divide can only be closed through a collective multidimensional and multisectoral approach" (Rao, 2005). The ICT policies should be framed in conjunction with a more holistic ecosystem that serves ICT in the wider-national scale. This includes investment in education, ICT infrastructure, creating an environment conducive to ICT development and growth, and fostering innovation and entrepreneurial start-ups (Quibria et al., 2003).

The digital divide comes in different forms within countries as well as between countries. Wealthy people have better access than poor people, younger people usually have better access than older people and men usually have more access than women and usually access is better in cities rather than in rural areas (Busch, 2010). There is also a relation to the issue of classes across the community that leads to a social hierarchy and having unequal access to resources (Lindquist, 2002). According to Illich (1973), "a convivial society would be the result of social arrangements that guarantee for each member the most ample and free access to the ICT tools of the community and limit this freedom only in favor of another member's equal freedom." Information, in itself, is a resource of immense social and economic value, and it is important as an invaluable tool for a productive, effective, and efficient economy since information impact different sectors and help rationalize the decision-making process when they are timely and accurate (Cawkell, 2001).

7.2 Introduction

Since ancient history, Egypt has witnessed massive information flows through different means. This included inscription on Rosetta stones and papyrus papers and the establishment of the library of Alexandrina, the world's first and most famous library and the gateway for knowledge creation and accessibility (Kamel, 1998a). During the middle ages, Arabic manuscripts became one of the most common means for information and knowledge dissemination. In the modern age, paper printing and publishing started in Egypt during the nineteenth century, witnessing the publishing of the first journal in Egypt in 1826. A few years later in 1830, Egypt witnessed the establishment of the first national archive system (Kamel, 1998b).

However, in the twentieth century prior to 1985, a number of characteristics identified the status of information in Egypt. The country was rich in data but poor in information, known for accumulated bureaucracy through red tape, computers were viewed as ends and not means, and there were islands of innovation with no bridges (Kamel, 1998b, 1999). From a government perspective, the focus was more on technical issues and not decision outcomes; there was poor multisector coordination and no synergy between information and socioeconomic development strategies.

Given this reality, even as ICT was increasingly becoming a necessity for socioeconomic development (Press, 1999), the government of Egypt recognized the need to take proactive measures and build the required information infrastructure. The strategy deployed followed a two-tier approach, inviting society with its different stakeholders to contribute in shaping the infostructure, which, in turn, would effectively contribute to the socioeconomic development (World Bank, 2006). Between 1985 and 2007, the government announced nine major policy initiatives to promote the development of Egypt's information society (Kamel, 2007). The policy initiatives were amongst the early attempts to close the nationwide digital divide in Egypt that was not just caused by the lack of ICT penetration rates but also by the level of illiteracy rates among the population. Such efforts were followed by a tenth policy initiative introduced in 2011 post Egypt's uprising, where ICT and universal Internet access across the nation had an invaluably effective impact in its successful realization to meet its initial targeted objective of the population as shown in Table 7.1. The tenth initiative was mainly targeting availing better and universal broadband across Egypt's different provinces. Since the uprising the Internet access grew to reach 32.4% of the population, mobility reached 105%, and Facebook 13.2% (Egypt ICT Indicators, 2012).

Since the early 1990s, Egypt has been undergoing a liberalization program of its public sector. The government has announced that it would invest in its human capital, encourage foreign direct investment (FDI), and emphasize innovative ICT as a platform for business and socioeconomic development (Kamel, 2005a). The government in collaboration with the private sector through a variety of public–private sector partnerships has announced the restructuring of sectors such as education and health as well as working on closing the digital divides and promoting social inclusion using ICT tools and applications. Egypt's population is growing at 1.9% annually with over 58% under the age of 25 years. Since 2011, over 20 million of the population have successfully capitalized on the effective use and spread of ICT through social networking and managed to voice their opinion and introduce initial changes to the political scene in Egypt. The digital inclusion of many had effective and concrete outcomes on

TABLE 7.1   ICT-Related Policy Initiatives and Programs

Year	Policy Initiatives/Programs
1985	Economic Reform Program
1985	Information Project Cabinet of Ministers (IPCOM)
1986	Information and Decision Support Program (IDSC)
1989	National Information and Administrative Reform Initiative
1994	Egypt Information Highway
1999	Ministry of Communications and Information Technology (MCIT)
2000	National Information and Communications Technology Master Plan
2003	Egypt Information Society Initiative (EISI)
2007	Egypt ICT Strategy 2007–2010
2011	Broadband Initiative

engaging large segments in the community in the socioeconomic and political debates taking place in Egypt (Ghonim, 2012). Further examples include the huge increase of bloggers and online debates post January 2011, leading to being one of the top countries in the world in terms of active bloggers per capita. With the growing penetration rates in ICT-related tools and mobility, the digital divide is gradually closing across the nation, and many are gradually being digitally and socially included in the cyberspace. Many called it the emergence of Egypt 2.0.

This chapter examines the introduction and diffusion of ICT on Egyptian society and the role it played in providing awareness and access to knowledge to different clusters in the community. In January 2011, ICT applications—mainly social networks—proved to have had a direct impact on society at large being a platform enabler for Egypt's uprising that was spearheaded by the nation's youth and that successfully mobilized a nation and led to major transformation in the political landscape in Egypt. The chapter addresses four main research questions. These include: whether and how did ICT become a vehicle for development and a platform to access knowledge and minimize the digital divide? How did ICT for development policy and strategy formulation and infrastructure deployment evolve? What were the challenges and the lessons learnt from efforts aimed at using ICT for socioeconomic development? What are the implications of diffusing ICT on access to knowledge in the society in Egypt?

7.3 Building Egypt's Information Society

The idea of ICT4Development (ICT4D) came into vogue in the early 1980s when Egypt was faced with the chronic challenges of developing nations, such as debt, economic reform, public sector reform, a balance of payment deficit, a high illiteracy rate, poor technological and telecommunications infrastructure, constrained financial resources, unemployment, environmental protection, and cultural heritage preservation. During that period, Egypt was striving to implement a nationwide strategy toward its socioeconomic development objectives and ICT was identified as a catalyst for that process. Therefore, the government of Egypt adopted a set of information-based projects leading to the establishment in 1985 of the Information Project Cabinet of Egypt (IPCOM), a project that was formulated to help introduce ICT into government and public administration in Egypt. With the growing interest in what IPCOM can offer to the government and decision makers, it was transformed into the Information and Decision Support Center (IDSC), a think tank affiliated with the cabinet of Egypt. The IDSC's objective was to develop and implement, using a supply-push strategy, large informatics projects to achieve socioeconomic development using state-of-the-art ICT (El-Sherif and El-Sawy, 1988).

During the 1990s, technological innovation and economic and social organization became more tightly linked than ever. This was spreading across the government institutions, the public and private sectors as well as the civil society. Continuous innovation in ICTs geared industry and society toward information acquisition and knowledge dissemination (Branscomb, 1994). Consequently, innovation has transformed the activities and relationships of individuals and organizations into a new information society, or *knowledge society*, in which ICT services pose both challenges and opportunities (Shapiro and Varian, 1999). The knowledge society refers to a second-generation information society. Whereas an information society aims to make information available and invent the technology necessary for this, a knowledge society aims to generate knowledge, create a culture of sharing and develop applications that operate via emerging ICTs like the Internet (ESCWA, 2005).

The knowledge society is now a force for fundamental global change (Garito, 1996). Knowledge and ICT innovation are becoming important values for business, socioeconomic development, and wealth creation with implications at the macro and micro levels (ESCWA, 2005). The changeover is complex, requiring new forms of partnership and cooperation between public and private sector organizations (Kamel and Wahba, 2002). Such partnerships were intended to engage different constituencies in the society to help leverage the role and impact of key sectors in the economy including education. This is best achieved through collaborative strategies that diffuse best practices and develop ICT applications, with the primary objectives of promoting growth and strengthening competitiveness. For many emerging economies such as Egypt, formulating strategies and policy frameworks to support the growing knowledge society could significantly accelerate development. In the context of Egypt, the goal of joining the knowledge society is to fill societal needs, create wealth, and sustainably enhance the quality of life of the community (Kamel, 2009). With a young and fresh population, the potential of a knowledgeable society will set the pace for years to come. For Egypt, intellectual power and contribution will be a crucial determining factor in the development process of the nation.

For millennia, the basic needs of humankind have been food, clothing, and shelter. Now it is time to add information to this list. Information and knowledge are nowadays the drivers in the global society, much more than land, capital, or labor. The capacity to manage knowledge-based intellect is the critical skill of this era; a firm and/or a society with a strong base of knowledge can leverage that base to create further knowledge, increasing its advantage over its competitors (Kamel and Wahba, 2002). ICT innovations are making a growing impact on business and socioeconomic development by introducing and diffusing the concepts of knowledge sharing, community development, and equality. These impacts are felt at the individual, organizational, and societal levels. ICT is not an end in itself, however, but a means of reaching broader policy objectives. The main objective of ICTs should be to improve the everyday lives of community members, fight poverty, and advance the Millennium Development Goals (MDGs). This is to a great extent related to addressing the different aspects that can help closing the digital divide. In this respect, ICT is delivering the key productivity gains that enable lives of material comfort for many around the world that would have been unthinkable only two centuries ago. Expanding access to knowledge through these new channels and tools creates emerging opportunities for learning and employment with strong implications for social and economic development. Moreover, in a country like Egypt, its young and growing human resources represents for the country "the oil of the 21st century," an invaluable resource that could transform the nation with a strong economy and an educated society that includes over 16 million children in schools, 1.2 million students, and around 300,000 university graduates every year (Kamel, 2010).

The technology innovations could have remarkably positive implications for developing nations, if they are properly introduced and managed. However, if they are not well supported, or if ICT is marginalized in the development process, inequities may increase between the developed and developing world, leading to widening the digital divide. It is a commonplace in development literature that the developing world lacks access to ICT, leading to such a divide. Nevertheless, it is important to note that such a divide exists *within* nations, both developed and developing as well. This internal digital divide, also referred to as a gap between "haves" and "have-nots," relates the possession of ICT resources by individuals, schools, and libraries to variables such as income level, age, ethnicity, education, gender, and rural–urban residence (Kamel, 2005b). It also addresses the illiteracy in computing as opposed to the conventional definition of illiteracy that addresses reading and writing.

The usual causes of this divide include, but are not limited to, expensive personal computers that are unaffordable for most developing country citizens, poor or limited telecommunications infrastructure especially in remote locations, and high illiteracy rates and poor educational systems (Kamel and Tooma, 2005). However, the major obstacle is the ICT ecosystem, including the complexity of the necessary operational details that need cultural adaptation and localization; this included the use of the Arabic language in what is widely known as *Arabization*. For societies to develop, grow, and benefit from the ICT revolution, nationwide introduction, adoption, diffusion, and adaptation of ICT should occur. ICT utilization is a living proof of the generation gap between different users and the case of Egypt's uprising is one of the most recent examples. However, this is rarely seen in developing nations where most of the ICT implementations and infrastructure are focused in capitals and major cities and are mainly related to conventional utilization of ICT. Universal access is becoming widely advocated where ICT should reach all the remote and underprivileged communities. This could lead to leverage educational, cultural, and societal capacities. The power and outreach of ICT could help emerging economies like Egypt leapfrog in their quest for development and growth.

The knowledge society promises to capitalize on emerging ICTs to create economic and social benefits. It encompasses ways in which various high-technology businesses, including ICT, universities, and research institutions can contribute to the economy of a nation while enabling economic sectors to operate more efficiently and effectively. In this context, Egypt has made efforts toward adapting to the changing global and technological conditions while catering to local markets. The ongoing restructuring of the ICT sector to serve development is liberalizing the telecommunications sector and opening the market to new competition. This restructuring has involved designing laws and regulations related to telecommunications, electronic commerce, intellectual copyrights and industry development; investing in human resources and promoting innovation and research and development; and promoting entrepreneurial initiatives in the marketspace in a world increasingly becoming positively affected from the removal of time and distance barriers.

It is important to note that, ICT is not the only enabler of the knowledge society. The European Commission (2003) defined the knowledge society as characterized by a number of interrelated trends, including major advances in diffusing and using ICTs, increased emphasis on innovation in the corporate and national context, the development of knowledge-intensive business service economies and knowledge management, in addition to trends toward globalization and economic restructuring. The most highly valued and profitable assets in a knowledge society are intellectual: knowledge and expertise acquired by workers. Culture is one of the most important factors in formulating the knowledge society: universities, education and training institutions from both the public and the private sectors will also need to cooperate to realize the knowledge society paradigm. This should be coupled with massive efforts to introduce and diffuse the notion of ICT and lifelong learning as a medium for leapfrogging and realizing multiplier effects on the community.

As the impact of the availability of information on socioeconomic development became apparent, governments around the globe started to invest in national information infrastructure (Petrazzini and Harindranath, 1996). Egypt, too, has heavily invested in its technology and infostructure to become the platform for the economy's development and growth (Kamel, 2005a). During the period 1985–1995, a government–private sector partnership had a remarkable impact on the build-up of Egypt's infostructure (Kamel, 1995, 1997). Hundreds of informatics projects and centers were established in different government, public, and private sector organizations targeting socioeconomic development (Kamel, 1998b). These projects included human, technological, and financial infrastructure development. They also included projects promoting entrepreneurship and transforming youth-led ideas into start-ups. Such elements represented the major building blocks necessary to establish a full-fledged infostructure capable of keeping pace with the developments taking place globally (American Chamber of Commerce in Egypt, 2001).

In 1999, ICT was identified as a priority at the highest policy level and a new cabinet office was established, namely, the Ministry of Communications and Information Technology (MCIT) (Kamel, 2005b).

MCIT was charged with the task of creating an information society, which started with the preparation of the national ICT plan. MCIT has articulated a strong vision and strategy on development and infrastructure deployment since its first national plan in 1999. The emphasis was on supporting the private sector so that it can take a leading role in transforming an ICT sector where the government could play the role of the enabler of a platform that could have various positive implications on different sectors and industries. Due to changes in global and local markets, both vision and strategy were amended in 2000 and 2004 (Kamel, 2009). MCIT took concrete steps like establishing the National Telecommunications Authority (NTRA) in 2003 and the IT Industry Development Agency (ITIDA) in 2004, and radically modernizing Egypt National Postal Organization (ENPO) in 2002. The partnership between these institutions and the ICT private sector accelerated ICT growth during the period 2006–2008, which reached 20% in 2006 and surpassed 25% in 2007 (American Chamber of Commerce in Egypt, 2007). The ICT sector is driving the gross domestic product (GDP) growth in many nations, and Egypt is no exception according to Tarek Kamel, Egypt's Minister of ICT in 2011, "we are positively contributing to the treasury through a steady growth in our services to the local community and through outsourcing." This is expected to grow given a more extensive liberalization of the sector and increasing competition with an emerging roster of local and private ICT companies in Egypt. Moreover, in 1993, the government established the Information Technology Institute (ITI) focused on providing high-end IT training for government and public and private sector employees. Almost, two decades later, ITI graduates represent the backbone and the invaluable workforce of the IT workforce in Egypt with hundreds of young promising human resource capacities trained every year. This has been the plan since the inception of ITI where the objective was to close the gap between the needs of the market in Egypt and the qualifications of Egyptian university graduates. Another divide is created by the quality of education when compared to the skills and capacities expected by different employers. "The mission of ITI is to create, shape, nurture and empower the Egyptian IT community by developing and disseminating state-of-the-art training processes," according to the minister of ICT, Mohamed Salem (Press, 2010).

The government of Egypt has formulated various initiatives to promote ICTs and pave the way for an electronically ready community that can benefit from public, universal access to knowledge. Other factors, however, will also be critical to closing the digital divide and promoting social inclusion within the digital economy: the legal and regulatory environment, awareness and capacity development, and mechanisms needed for collaboration between different sectors in the economy. The experiences vary based on the access and sophistication of the use of computers and information technology in general (Selwyn, 2006). This underscores the importance of developing national ICT strategies that recognize the role of ICT in enabling access to knowledge. In the final analysis, the challenge is to leverage ICT as a platform for knowledge dissemination in the community, and the focus should always be on outcome assessment for this strategy. The use of computers as a gateway to knowledge accessibility plays an invaluable role in such setting.

7.4 ICT4Development: How to Close the Digital Divide: Cases from Egypt

The evolution in the knowledge society heralds a new socioeconomic order. This era is witnessing the emergence of knowledge-based economies, with traditional economic, industrial, and business activities moving toward more knowledge-driven processes and the progressive transformation of advanced economies into knowledge-based, technology-driven, services-dominated economies. These shifts are increasingly laying emphasis on economic activities with intellectual content and knowledge, enabled by the development and exploitation of new information and communications technologies within all spheres of human endeavor. In the marketspace, ICT represents a massive opportunity for small- and medium-sized enterprises (SMEs), where the differential between large corporations and small companies is relatively small. ICT can provide various opportunities for small entrepreneurs with ideas that can have impacts on employment opportunities but also develop solutions that could benefit the community.

Against that background, Egypt's government has announced efforts aiming at developing its information and knowledge base through investments in ICT and human capacity development, improving and broadening universal access to higher and quality education, and training with an emphasis on lifelong learning and creating digital content accessible to the society. Efforts in Egypt for ICT development are government-led in collaboration with the private sector and civil society. In that respect, Egypt has developed a number of policies and strategies to facilitate socioeconomic development and accelerate the transformation of the nation's economy and society to become information-rich and knowledge-based.

In May 2007, MCIT released its 2007–2010 national ICT strategy (MCIT, 2007b). The plan paved the way for the Egyptian Information Society Initiative (EISI), which represented the vision of the ICT strategy translated into specific initiatives and programs to diffuse ICT connectivity (MCIT, 2005a,b). EISI is structured around seven major tracks, each designed to help bridge the digital divide and progress Egypt's evolution into an information society (MCIT, 2005b). The theme of the strategy is "closing the digital divide and promoting social inclusion" (Table 7.2).

TABLE 7.2   Seven Tracks of the Egyptian Information Society Strategy

eReadiness *"Equal Access for All"*	**eLearning** *"Nurturing Human Capital"*
• Enabling all citizens with easy and affordable access to new technologies • Developing a crucial robust communication infrastructure	• Promoting the use of ICT in education • Shaping a new generation of citizens who understand ICT and are comfortable with its use in their daily lives
eGovernment *"Government Now Delivers"*	**eBusiness** *"A New Way of Doing Business"*
• Delivering high-quality government services to the public in the format that suits them • Reaching a higher level of convenience in government services • Offering citizens the opportunity to share in the decision-making process	• Creating new technology-based firms • Improving workforce skills • Using electronic documents • Developing ePayment infrastructure • Using ICT as a catalyst to increase employment, create new jobs, and improve competitiveness
eHealth[a] *"Increasing Health Services Availability"*	**eCulture** *"Promoting Egyptian Culture"*
• Improving citizens' quality of life and healthcare workers' work environment • Using ICT through to reach remote populations • Providing continuous training for doctors • Developing the tools for building a national health network	• Documenting Egyptian cultural identity by using ICT tools to preserve manuscripts and archives and index materials • Offering worldwide access to cultural and historical materials • Generating and promoting interest in Egyptian cultural life and heritage
ICT export initiative *"Industry Development"*	
• Fostering the creation of an export-oriented ICT industry • Developing an ICT industry that will be a powerful engine for export growth and job creation	

[a] eHealth through its growing digital infrastructure attempts to provide an information/knowledge repository that caters to patients scattered throughout Egypt's 27 provinces and who are geographically dispersed. The initiative's primary objective is to decrease the health gaps for Egyptians in remote locations and villages when compared to others who live in the capital and major cities. Empowering marginalized citizens with information access on different health issues would have long-term positive effects on the economy at large and also on closing the digital divide enabling a healthier population and a society more equipped to develop and grow (Gilbert et al., 2007).

Following is an evaluation of some of the initiatives and programs implemented by the government to capitalize on emerging ICT and disseminate knowledge to the society with all its segments, sectors, and communities. Rather than attempt to cover all seven elements of the EISI, the discussion will emphasize efforts in the eReadiness, eLearning, and eCulture areas considered of greatest relevance to the access to knowledge theme.

7.4.1 Electronic Readiness: ICT for All

Electronic readiness is a complex process that harnesses social, political, economic, infrastructure, and policy issues to bridge the digital divide. Most countries adopt the same path where an assessment of eReadiness is followed by a strategy and an action plan. Therefore, digital inclusion and equality are becoming integral factors in the electronic readiness of different societies (Kamel, 2007). Therefore, the government of Egypt announced that it was launching efforts aiming at universal, easy, affordable, and fast access to ICT for all citizens while raising awareness of the potential in ICT tools and techniques. The MCIT has implemented different programs promoting computer literacy and encouraging the use of ICT across the nation. One of these programs is "ICT for all," also known as the electronic readiness (eReadiness) building block of EISI. Recognizing universal access to ICT as key to socioeconomic development, the program is devised with two main objectives. First, it aims to assist the government policy to integrate ICT in government and public services by (a) increasing ICT penetration; (b) fostering inclusion in the knowledge society and better public services and quality of life; and (c) expanding the use of post offices to provide public services. Second and more obviously, this strategy aims to facilitate ICT access for all citizens by (a) increasing PC penetration, (b) expanding the reach of Internet connectivity and broadband to all communities, (c) raising youth employability through ICT training and (d) encouraging government employees to attain international accreditation in ICT skills. The process of ICT for all is obviously perceived to have multiple positive implications on the community including job creation and community engagement. Following is a description of three selected projects aimed at promoting electronic readiness: the Free Internet Initiative, Egypt PC 2010, and IT clubs. All three employ an implementation model of public–private partnerships in which the government's role is to articulate policy and regulatory frameworks for the private sector and civil society to implement. At the close of this section, the impact of these projects will be examined through the lens of representative indicators and secondary analyses, to assess the nation's progress in achieving its eReadiness goals.

7.4.1.1 Free Internet Initiative (Internet and Broadband Connectivity)

The Internet was first introduced to Egypt in 1993 by the Egyptian Universities Network of the Supreme Council of Egyptian Universities, originally serving 2000 users (Kamel and Hussein, 2002). In 1994, in an effort to diffuse Internet usage among the broader society, the Cabinet of Egypt's Information and Decision Support Center (IDSC) in collaboration with the Regional Information Technology and Software Engineering Center (RITSEC) began providing free Internet access on a trial basis to public and private organizations. The vertical and horizontal penetration of Internet was intending to minimize the digital gap from the outset. This was done with financial support from the government, in an attempt to boost global exposure of the local market and pave the way for commercialization of Internet services.

The free access formula was credited with accelerating the growth of Internet users, particularly within small and medium-sized enterprises and industry professionals (Kamel, 1998b). In 1996, the government replaced its free access policy with an open access policy: commercial Internet services were privatized, and a dozen Internet service providers (ISPs) began operation (Mintz, 1998). By December 2001, more than 600,000 Egyptians were online, but only 77,000 were paid subscribers, served by 51 private ISPs. Such limited growth was perceived as hindering the development of the knowledge society and creating a divide between the haves and the have-nots. Therefore, in January 2002, MCIT launched

TABLE 7.3   Electronic Readiness in Egypt

Indicators	October 1999	December 2002	December 2004	December 2006	December 2008	December 2011
Internet subscribers	300,000	1.2 million	3.6 million	6 Million	11.4 million	29.8 million
ADSL subscribers	N/A	N/A	N/A	206,150	593,042	1.65 million
Internet penetration per 100 inhabitants	0.38%	2.53%	5.57%	8.25%	15.59%	34.83%
Mobile phones	654,000	4.5 million	7.6 million	18 Million	38.06 million	78.99 million
Mobile phones penetration per 100 inhabitants	0.83%	5.76%	9.74%	23.07%	50.7%	97.93%
Fixed lines	4.9 million	7.7 million	9.5 million	10.8 million	11.4 million	8.96 million
Fixed lines penetration per 100 inhabitants	6.2%	9.8%	12.1%	13.8%	15.2%	11.98%
Public pay phones	13,300	48,000	52,700	56,449	58,002	23,664
IT clubs	30	427	1,055	1,442	1,751	2,163
ICT companies	870	1,533	1,870	2,211	2,621	4,250
IT companies	266	815	1,374	1,970	2,012	3,599
Communications companies	59	75	152	244	265	295
Services companies	88	121	148	211	242	356
IT economic indicators						
Number of employees in the ICT sector[a]	48,090	85,983	115,956	147,822	174,478	212,260

[a]   There are also 26,000 indirect workers in both IT clubs and Internet cafés.

a new initiative providing free nationwide Internet access to all citizens (Kamel and Ghaffar, 2003). This has contributed to the fast-growing use, with the percentage of the population online rising from 5.5% (3.9 million users) in 2004 to 15.6% (11.4 million users) in 2008, still rising at a rate of 16.7% annually (MCIT, 2008, p. 1). Table 7.3 demonstrates the current demographics of the ICT sector in Egypt. It is important to note that huge increase in all indicators was realized post Egypt's uprising in January 2011. This was caused by an overwhelming engagement of Egypt's youth in learning, communicating, and exchanging ideas and starting-up companies as part of what has been widely known since the uprising as Egypt 2.0 and the creation of an environment that promotes freedom of expression, the private sector, and more engagement of different constituencies in the society.

Egypt's free Internet initiative has made connectivity affordable to most citizens by enabling access on all fixed phone lines without additional monthly fees. The cost of dial-up access is the same as a local telephone call, less than 5 U.S. dollars per month (MCIT, 2008, p. 4). Dial-up modems, however, are generally capable of a maximum speed of only 56 kbps and occupy the telephone line. To enhance the Internet experience, broadband (ADSL) connectivity, which supplies at least 256 kbps and does not disrupt telephone use, has been offered since 2004. The continued expansion of broadband service may be expected to positively affect access to knowledge on the web due to its reliability, stability, and capacity compared to the dial-up option. In 2011, as a reaction to Egypt's uprising, a new initiative was introduced focused on diffusing broadband across the nation's 1 million km². This would help reach out to the remote locations spread across the country and could be effective in one of the priority sectors in Egypt post January 2011: education.

Today, however, the higher cost of broadband connectivity still remains a challenge to its acquisition by more households. The broadband tariff initiated in 2004 has been revisited twice and in March 2009 was reduced again to 17 US dollars per month for a 256 kb speed. This cost may be further reduced by sharing a connection across multiple households. More than 1.6 million Egyptians now subscribe to the web through broadband service, yet this is small compared with more than 29 million total Internet subscribers; this encouraged the formulation of the broadband initiative in 2011. Both figures likely reflect the use of superior connections in Internet cafés, as well as sharing of a single broadband subscription across households (MCIT, 2008, p. 4). According to a survey by Arab Advisors Group released in April 2008, 63.4% of households in Egypt with ADSL subscription reported sharing the ADSL with neighbors, and 81.9% of those share it with more than three neighboring households (Arab Advisors Group, 2008).

7.4.1.2 Computer in Every Home: Egypt PC 2010

Egypt PC 2010 is an initiative to bring Egypt as a nation online. It is an amended version of the 2002 PC for every home initiative launched in collaboration with Telecom Egypt (TE) (MCIT, 2007a). The first initiative offered locally assembled PCs with bank credit for up to 3 years, using ownership of a landline telephone as loan collateral. PCs could be bought on hire–purchase terms by anyone with a TE telephone line, with the periodic loan repayments included in the phone bill. In the original initiative, a limited variety of PC models and specifications were offered, which rendered the product unaffordable to many. Additionally, only TE customers could participate, which limited the project's scope to urban communities. These two issues hindered the success of the program among the community. Five years after the launch of the 2002 initiative, the Internet penetration rate had increased to only 7% (MCIT, 2007a). Since, as noted earlier, dial-up service is available to any home with a telephone landline at no additional charge, the slow Internet penetration rate must be attributed to the continuing difficulties to acquire home computers.

The PC 2010 initiative implemented several lessons learned. The new program offers local and international brand PCs, from simple models for beginners to high-end desktops and laptops. Participants no longer need to be TE customers to be eligible for the extended payment terms because financing banks offer the required loans through facilitated retail banking procedures. Under the new scheme, the PC can be purchased on installments for as little as 8.50 U.S. dollars per month, which comes to just over 100 U.S. dollars per year. Compared with Egypt's average GDP per capita of approximately 4337 U.S. dollars per year (UNDP, 2007, p. 231), this rate is quite affordable. Moreover, the new initiative focuses on improving PC distribution in all provinces, with an emphasis on serving underprivileged communities through partnerships with civil society organizations. This is likely to change with the broadband initiative that is currently being prepared with plans to cover Egypt across its different provinces. The importance of PC 2010 could be seen in the context that access to computers reflects the extent to which technology use enables individuals to participate and be part of society versus being social excluded through the barrier of not having access to a growing network of connected communities of interest (Berman and Philips, 2001; Haddon, 2000).

7.4.1.3 Integrated Computing and Training through IT Clubs

The digital divide is often created by the remote and demographic location of some communities that prohibits the access and use of the Internet in the first place (Katz and Aspden, 1997; Katz and Rice, 2002). These are the underprivileged societies located in rural settings especially in developing nations. This has been the case in Egypt which led to the emergence of a network of IT clubs across Egypt's 27 provinces that provide citizens with access to information technology tools and applications for minimal cost, many services are for free. Their primary objective is to open the global eSociety to Egyptian youth and rural and underprivileged communities by offering an affordable site for Internet access and training. The initial vision for the program was to open 300 such IT clubs (MCIT, 2001, p. 2). The model proved highly replicable; today there are over 2163 IT clubs across the nation, a figure still growing by approximately 13.5% annually (http://www.egyptictindicators.com). This number is likely to increase with a growing number of business plan competitions and enabling platforms for young entrepreneurs from different provinces who are eager to start their own businesses. The American University in Cairo (AUC) School of Business Entrepreneurship and Innovation Program (EIP), the American Chamber of Commerce in Egypt, Google Egypt, Sawari Venture's Flat6 initiative, and others are among a number of organizations that have been growing regularly since January 2011 with a targeted objective to support young entrepreneurs with innovative ICT ideas that could have various impacts on the community. Most of the start-up ideas are IT services and solutions. There is a massive need for training and lifelong learning for those young and potential entrepreneurs who want to make a difference for their society and for the future of Egypt. These ICT clubs are perfectly positioned to play such much needed role in disseminating knowledge in ICT-related issues and help prepare those future private sector leaders creating jobs at the local level. An example of AUC School of Business EIP efforts is helping over 3500 seniors and fresh graduates from 11 different provinces train on taking their ICT ideas to the next level and starting their own companies. The key is to create jobs throughout the nation; this could be the most effective way to close the holistic divide that includes among its implications the digital divide.

At this time, more IT clubs are connected and many of these are remote in the sense that new IT clubs that are touring the nation in the form of buses were introduced so that those living in remote and underprivileged communities also benefit from such effort. The remote and poor villages benefited the most from this project, especially those who distantly located from major cities and towns. The case of the IT clubs diffusion in remote locations reflects the notion that it is difficult to adequately account for the effective uses of ICTs without accounting for the availability of appropriate and current ICT infrastructure (Grabill, 2003).

7.4.2 Assessing the Impact of Egypt's eReadiness Initiatives

Through the efforts exerted as described previously, significant progress has been made in achieving Egypt's eReadiness goals. As a reflection of this point, Egypt ranked 76th out of 134 economies surveyed for the 2009 Networked Readiness Index (WEF, 2009, p. xvii).* Egypt also earned recognition as "an emerging outsourcing gateway in the Middle East" (WEF, 2009, p. xiii), in part due to its competitive Internet usage charges (WEF, 2009, p. 116). An alternative eReadiness index puts Egypt even higher, at 57th out of 70 nations, noting an upward momentum due to improvements in connectivity (Economist Intelligence Unit, 2008a, p. 3).†

However, it is obvious that additional work is still needed. The slow increase in fixed line density rates, in particular, has been identified as an element slowing down Egypt's eReadiness. An external source reports that "Despite the moderate growth which Egypt's fixed-line market has continued to experience, the market has failed to keep pace with the country's expanding population" (Business Monitor International, 2008). The report predicted a shrink in the sector in 2009, with more customers relying on mobile phones in place of fixed lines. A chance to move forward again was predicted,

---

* The World Economic Forum (WEF) and INSEAD develop the NRI. The index is based on hard data produced by organizations such as the World Bank, the International Telecommunication Union and the United Nations, and survey data generated from the Executive Opinion Survey that is annually conducted by WEF. The three components of the index include the ICT environment created by the government, the readiness of the community's key stakeholders including government, businesses and individuals and the usage of ICT amongst those stakeholders. The index ranks Egypt 60th in terms of market environment and 70th in terms of IT infrastructure; while Egypt ranks 51st in government readiness, it is still lingering at 97th rank in individual readiness.

† The Economist Intelligence Unit's eReadiness ranking is based on a set of quantitative and qualitative criteria that include connectivy and technology infrastructure, the business environment, the social and cultural environment, the legal and policy environment, the government's policy line and business adoption. The data is sourced from such institutions as the World Bank, the World Intellectual Property Organization, and the Economist's network of national experts and economists. In 2008, Egypt held an eReadiness score of 4.81 out of 10 as opposed to 4.26 in 2007 (The Economist Intelliedence Unit, 2008b).

however, only if competition enters the market. "The arrival of a new fixed-line operator sometime in the next 2 years could result in a new round of growth for the sector, particularly if the new entrant started providing fixed wireless services," according to the former minister of ICT. Given the recent developments in Egypt, this plan has been put on hold. However, it is perceived that in the coming few years, a new fixed line operator will be introduced into the market in Egypt, which will be in favor of market competition.

Taking the long view, significant progress on eReadiness has unquestionably been made. The ICT infrastructure witnessed in the past decade massive developments in international links for telephony and the Internet backbone in addition to disseminating Internet across Egypt's 27 provinces. The ultimate goal, however, is not to achieve increased ICT adoption for its own sake. As the name "eReadiness" implies, these efforts merely lay the foundation to take advantage of ICT tools for development ends. With this in mind, the following section examines Egypt's efforts to date in leveraging ICT development for education.

7.4.3 Electronic Learning: ICT for Education

Education and lifelong learning are central drivers of socioeconomic development and growth, and have particularly relevant implications for access to knowledge. Boosting performance on these measures, however, has historically been a challenge for Egypt. The country's adult literacy rate stands at only 71.4% (UNDP, 2007, p. 231), indicating that a substantial proportion of the country must overcome barriers beyond the merely technological in order to take advantage of the Internet. Egypt's education system has been fully subsidized by the government for decades, yet challenges with regard to infrastructure and quality persist. In the public school system, class sizes of 70–80 pupils are common, teachers are poorly qualified, and the emphasis is on rote memorization rather than problem-solving (Kozma, 2004, pp. 14–15). Since 2003, the country's former ruling National Democratic Party has declared education reform to be a key priority (Essam El-Din, 2003); nothing has materialized in a sector that has been suffering since the coup d'état of 1952 that always managed the education sector based on a fire fighting approach rather than developing a holistic approach that could transform a sector highly needed for socioeconomic development. This has been one of the key elements for the creation of the knowledge and digital gaps in the community. Education has been and will always be the determining platform of whether these gaps diminish or grow.

One major component of such reform is embedding ICT in education to promote information acquisition and knowledge dissemination. The objectives of deploying ICT for education include optimizing ICT investments to avail the required infrastructure that promotes education and lifelong learning, satisfying the ICT industry training requirements, creating an open learning environment by connecting the education community through broadband, and increasing the efficiency and effectiveness of education institutions and embedding ICT in the curriculum. Based on the events of January 2011, the growing access to ICT tools and applications and the exposure of Egypt's youth were benefiting from enabling the unconventional use of the technological platforms in changing the political landscape of the nation.

Toward those ends, MCIT is supposed to work closely and strategically with the Ministry of Education (MOE) and Ministry of Higher Education and Scientific Research (MOHE); where in the context of developing nations, ICT deployment is invaluable for educational improvement and spread across the society. Accordingly, a number of projects were devised; most notably the Smart Schools Network, the Egyptian Education Initiative (EEI), and ICT for Illiteracy Eradication. These programs share the common target of increasing ICT awareness and promoting education and lifelong learning. Their strategy is meant to capitalize on the potential of ICT to provide universal access to knowledge and education to all constituencies in Egypt, irrespective of socioeconomic group, gender, age, or background. Egypt offers a model for how rising economies can develop a skilled IT workforce and a healthy IT sector that are closely aligned with each other.

7.4.4 Assessing the Impact of Egypt's eLearning Initiatives

Although Egypt has made significant progress in achieving its eReadiness objectives, its eLearning programs are generally still at the pilot stage. According to a regional report, Egypt's ICT for education implementation was ranked at a maturity level of two out of four, indicating a number of sporadic projects and initiatives that had concrete impacts, but lacking the consistency and long-term vision for successful implementation and sustainability (ESCWA, 2007). It is important to note that of the three-eLearning programs, the Egyptian Education Initiative has had the greatest impact, training over 100,000 teachers in using ICT. The Smart Schools Network, although still quite small, holds the potential to build on this success to use ICT in a way that truly transforms the educational experience. Finally, the ICT for Illiteracy Eradication project demonstrates that eLearning can be leveraged to achieve results in lifelong learning, even for those at the greatest educational disadvantage. To achieve a significant impact upon access to knowledge in Egypt, however, these programs must be greatly scaled up. This will require broader advancements in access to computers and the Internet (eReadiness) than has already been achieved. Once this infrastructure is in place, however, the eLearning programs may benefit from the economics of easily reproduced open source software to scale up with an efficiency of resources.

7.5 Social Networks Impact: 2011 and Beyond

In recent years, many case studies indicated that ICT tools and applications including mobile phones, short messaging systems (SMS), and the Internet have had a variety of impacts on democratic freedoms (Shirazi et al., 2010). The provision of such tools and applications are central to information access and participation in social and political life (Becker, 2001; Bennett and Fielding, 1997; Drezner and Farrell, 2008; Harwit and Clark, 2001; Snellen, 2001). These mechanisms help facilitate the diffusion of information and in many ways help in closing the digital divide and in improving social inclusion (Norris, 1999). Moreover, some people think that ICT diffusion help create a new type of technology adopter called the "digital citizen" (Katz and Aspden, 1997). According to Balkin (2004), "the digital revolution brings features of freedom of expression to the forefront of our concern and makes possible for widespread cultural participation and interaction."

ICTs provided the platform to enable citizens the participation in the transformation and democratization process as well as mobilization of resources, debating issues, influencing decisions, and leveraging civic engagement, all contributing to minimizing the digital divide (Bennett and Fielding, 1997; Dertouzos, 1997 and Sussman, 1997; Gilbreth and Otero, 2001; Norris, 2001; Suarez, 2006; Weber et al., 2003).

The impact of ICT with an emphasis on social networking was optimized in Egypt in the past few years. This was spread across the community. It also enabled an effective platform to oppose the convention wisdom that in the Middle East women face unequal access to the Internet. Social networking phenomenal growth, from 6 to 20 million in 18 months, allowed social inclusion through the digital world for different groups including women, youth, and underprivileged communities in remote areas in Egypt. Since 2008, Egypt's youth (58% of the population totaling 90 million in March 2012) realized the power of sharing news, knowledge, and momentum using emerging ICT tools and applications, social networking. Applications such as Facebook were growing in popularity in the Middle East region for a few years now recording, as a region, one of the highest growth rates in the world. The tools were perceived as a platform to share ideas, empower them, as well as a mechanism that can help transform the society. This is made clear in Ghonim's statement, as one of the young Egyptians who contributed in Egypt's uprising "the power of the people is greater that the people in power" (Ghonim, 2012). The experience witnessed in Egypt in early 2011 for the duration of 18 days demonstrated the power ICT can provide and how access to information and knowledge in a nation that is one million square kilometers can be effective with concrete impacts. As an attempt to stop the uprising, one of the strategies used by the government of Egypt was cutting-off cellular phones and Internet connectivity; a clear statement of how effective communication devices were influential, even in a country that rates high in illiteracy. In the case of Egypt, the Internet has been instrumental in setting the stage for a revival of Egypt's political scene for the first time in 60 years. There is no doubt that ICT and social networks made public exposure inevitable (Ghonim, 2012). In the build-up to January 2011 and to date, the impact on the growth of Internet users and mobile holders was phenomenal as indicated earlier. Mobile phones have increasingly and effectively helped in closing the digital divide among the different constituencies of the community irrespective their social class, income, educational background, skills, and capacities (James, 2011). In many ways, the likes of Facebook and Twitter in Egypt became the primary news channel of communication for many Egyptians both in Egypt and with the over 6 million Egyptians who are living around the world but very much connected to their homeland. Over a year after Egypt's uprising, the transformation process continues in different models using various ICT tools and techniques supported by several ICT-based business plan competitions and initiatives that target several start-ups aiming at addressing socioeconomic and political issues and developing solutions and suggestions to approach them by different constituencies.

7.6 Conclusion

Over the past decade, Egypt has made significant progress toward realizing the vision of the knowledge society through ICT. The developments of the initial phase addressed legal, technical, and business fundamentals, enabling the ICT industry to develop significantly (IDSC, 2005). These have been reflected positively in the overall growth of the sector, which exceeded 20% during the period 2006–2008 and contributed to overall GDP growth by more than 7% (Fayed, 2009). In the words of the former ICT minister, the ICT sector has transformed itself "from a sector looking for support and subsidies to a sector contributing tangibly and intangibly to the economy with a total of 5.2 billion US dollars received by the treasury since early 2006" (Kamel, 2008). The ICT sector has also served as a role model for other sectors of reform and liberalization, capitalizing on a free market economy and catering to different social groups and interests.

The National ICT Action Plan 1999–2009 was realized in many ways over the past decade, although not in its full capacity as envisioned in 1999. This action plan, set shortly after the establishment of MCIT, aimed to build a knowledge-based society that can boost socioeconomic development and entice economic growth. As originally conceived, the plan identified eight goals: *completing the ICT infrastructure build-up* to avail universal interconnectivity among all 27 provinces including 520 local administrations and over 8000 cities and villages; *realizing infostructure interconnectivity* among value-added information networks in government, private sector, and civil society organizations; *linking Egypt locally and globally* within the growing global digital marketspace; *investing in human capital* through lifelong learning programs and serving different segments of the community; *building an electronically ready community* capable of engaging in the global information society; *updating Egypt's information infrastructure* as a step in building the nation's information highway; *encouraging an ICT export industry* by promoting and supporting innovation, creativity, and research and development in ICT-related areas; and *collaborating through public–private partnerships* engaging different stakeholders in high-tech projects with business and socioeconomic implications.

Although none of these objectives have been fully attained, the achievements to date of the plan lie in gradually helping Egypt to bridge the nation's digital divide and in sharpening its competitive edge on the global ICT scene. Penetration rates are gradually increasing for infrastructure like Internet access, PCs, and mobile and fixed phone lines. The liberalization of the telecom sector created competitive forces that are working for the best interest of the consumer; more is expected in the years to come. The action plan also helped avail an ecosystem that is empowered by deregulation policies, which laid the foundations of the ICT sector's continuing development. The MCIT has gained valuable experience through a number of public–private partnership initiatives that can be expanded and improved upon to achieve fully the vision of the knowledge society over the coming decade.

The development of the knowledge society cannot be left to market forces; it deserves and needs the attention of the highest political decision-makers with a vision to expanding access and contribution to knowledge. Nations like Egypt should prioritize information needs for business and socioeconomic development, just as they do already for sectors such as industry, agriculture, education, and health. Governments are responsible for taking a strategic approach to the demands of an information-intensive global environment. This approach should include creating a shared vision of the knowledge society, intensifying the process of information acculturation, generating the necessary human capacities, accelerating the development and deployment of ICT infrastructure, and building an electronically ready community.

A critical issue in the information age will be developing a win–win partnership between the government and the private sector. According to the former MCIT Minister Tarek Kamel, "institutional build up is important for the optimization of effective impacts of ICT on the society" Transparency is important. Adherence to law is important. But I believe that the most crucial issue is genuine public–private partnership including all the various stakeholders in the dialogue and in the development process (Atallah, 2008). The private sector is now seen as a major stakeholder in the progress toward the knowledge society. Use of public–private partnerships will continue to be instrumental for the government's strategy. The nature of this partnership will be determined by the answer to this question: how will governance be exercised in the information-based world? While the framework is not yet defined, the private sector will probably provide information-based services while governments construct a supporting regulatory framework based on the greater public participation and consensus essential for a knowledge society.

The knowledge society requires not just an intricate web of legal measures but also a strong, comprehensive infrastructure, a human resource investment plan, good education, and concrete incentives for local and foreign investments. Moreover, it requires full transparency in the transfer and use of data within an environment that encourages creativity and innovation. Information is power, and it is a factor in the manipulation of discourse about socioeconomic reform (Stiglitz, 2002). Historically in Egypt, the government has dominated the supply of information. The process of information sharing and dissemination was orchestrated by a number of public and private sector organizations led by the Central Agency for Public Mobilization and Statistics (CAPMAS), established in 1964 and considered the official source of data collection in the nation (El-Mikawy and Ghoneim, 2005). This strategy has been gradually changing since the mid-1980s, when the government opted for a relatively more transparent strategy by collaborating with the private sector and by allowing research entities to conduct market studies, sharing findings and outcomes, and generally contributing in the build-up of the knowledge society. This promising trend has opened venues for information sharing to the public, empowered the society, and disclosed opportunities for business and socioeconomic development.

Access to knowledge will not reach all segments of the society across all provinces, however, until further efforts are expended. Despite significantly increasing ICT penetration rates, too many Egyptians are still excluded from the opportunity to participate in the knowledge society. Not only technological expansion is needed, but also educational opportunities and the economic resources for Egypt's people to avail themselves of the opportunities the new technologies provide. Moreover, a political revisiting and reform for media freedom will be conducive to an improved access to information, which is currently still perturbed by a series of legal and extra-legal restrictions. A critical leveraging of the potential of open source software and open-licensed content is also in order to expand access to eLearning and eCulture. This is becoming on-the-top priorities of the agenda especially with Egypt's uprising and the growing role of ICT, the Internet, and social networking in information sharing, acquisition, and dissemination.

In this respect, access to knowledge emerges as an invaluable platform for development and growth in the global marketplace of the twenty-first century. With the increasing competition taking place around the world, investing in human capacities and disseminating knowledge through multiple channels is integral to business and socioeconomic development. Expanding access to ICT plays a pivotal role in this effort. Over the next decade, Egypt should further develop its ICT policies and programs within an overall ecosystem that entices knowledge sharing and collaborative work, and which is guided by the notion that access to knowledge is the path to societal development and growth.

The developing world must understand the importance of ICT in order to optimize the associated benefits. ICT investments and infrastructure should be considered as a platform, a tool for development; only then ICT can help solve societal problems and realize its targeted objectives. ICT, through connectivity, timeliness, currency, and accessibility, is an enabling environment for digital inclusion and can have positive and concrete impacts on different sectors such as education, health, and more. Bridging the digital divide goes beyond the influence of ICT penetration. Closing the gap caused by the digital divide is a precondition to many more influences on the community including reducing poverty, improving economic conditions, supporting education diffusion, and achieving sustainable markets. While the digital divide reflects the gap in information and knowledge access, it also relates to gaps that relate to different elements including economic, social, political, and cultural aspects. In addition, it is important to note the role of income and education as significant contributors to access to computers and the Internet (Grabill, 2003). Access to information and the closing of the digital divide within the community is becoming as important as the creation of wealth, jobs, and prosperity for citizens.

References

American Chamber of Commerce in Egypt. 2001. *Information Technology in Egypt.* Cairo, Egypt: Business Studies and Analysis Center.

American Chamber of Commerce in Egypt. 2007. *Information Technology in Egypt.* Cairo, Egypt: Business Studies and Analysis Center.

Aqili, S. V. and Moghaddam, A. I. 2008. Bridging the digital divide: The role of librarians and information professionals in the third millennium. *The Electronic Library* 26(1), 226–237.

Arab Advisors Group. 2008. ICT Statistics Newslog—63.4% of Egyptian households with ADSL, share the ADSL connection with neighbors. News related to ITU Telecommunication/ICT Statistics. http://www.itu.int/ITUD/ict/newslog/634+Of+Egyptian+Households+With+ADSL+Share+The+ADSL+Connection+With+Neighbors.aspx (accessed: April 14, 2009)

Attalah, L. 2008. Interview with Tarek Kamel. *Talk* 7, Fall 2008.

Balkin, M. J. 2004. Digital speech and democratic culture: A theory of freedom of expression for the information society. *New Work University Law Review* 79, 1.

Becker, T. 2001. Rating the impact of new technologies on democracy. *Communications of the ACM* 44(1), 39–43.

Bennett, D. and Fielding, P. 1997. *The Net Effect: How Cyber Advocacy is Changing the Political Landscape.* Merrifield, VA: e-Advocate Press.

Berman, Y. and Phillips, D. 2001. Information and social quality. *Aslib Proceedings* 53(5), 179–188.

Branscomb, A. W. 1994. *Who Owns Information.* New York: Basic Books.

Busch, T. 2010. Capabilities in, capabilities out: Overcoming digital divides by promoting corporate citizenship and fair ICT. *Ethics and Information Technology* 13, 339–353.

Business Monitor International. 2008. Egypt telecommunications report Q1 2009. Document ID: BMI2066607. December 31, 2008. http://www.marketresearch.com/product/print/default.asp?g = 1&productid = 2066607

Cawkell, T. 2001. Sociotechnology: The digital divides. *Journal of Information Science* 27(1), 55–60.

Dertouzos, M. 1997. *What Will Be: How the Information Marketplace Will Change Our Lives.* San Francisco, CA: Harper.

Drezner, D. W. and Farrell, H. 2008. The power and politics of blogs. *Public Choice* (134), pp. 15–30.

Economist Intelligence Unit (EIU). 2008a. *E-readiness Rankings 2008, Maintaining Momentum: A White Paper from the Economist Intelligence Unit.* http://a330.g.akamai.net/7/330/25828/20080331202303/graphics.eiu.com/upload/ibm_ereadiness_2008.pdf

Economist Intelligence Unit (EIU). 2008b. http://graphics.eiu.com/upload/ibm_ereadiness_2008.pdf

Economic and Social Commission for Western Asia-ESCWA. 2005. Measuring the information society, Geneva, Switzerland, ESCWA, February 7–9.

Economic and Social Commission for Western Asia-ESCWA. 2007. The information society from declaration to implementation, Geneva, Switzerland, ESCWA, May 21.

Egypt ICT Indicators. 2012. www.egyptictindicators.com (accessed: May 15, 2012)

El-Mikawy, N. and Ghoneim, A. 2005. *The Information Base, Knowledge Creation and Knowledge Dissemination in Egypt.* Center for Development Research, University of Bonn, Bonn, Germany.

El-Din, E. G. "Education in Flux," *Al-Ahram Weekly Online,* 649, July 31–August 6, 2003. http://weekly.ahram.org.eg/2003/649/eg2.htm

El-Sherif, H. and El-Sawy, O. 1988. Issue-based decision support systems for the cabinet of Egypt. *MIS Quarterly* 12(4), 551–569.

European Commission. 2003. Commission of the European Community. *Building the Knowledge Society: Social and Human Capital Interactions.* May 28, 2003.

Fayed, S. 2009. The net exporter. *Business Monthly* 26(1), 54–58.

Garito, M. A. 1996. The creation of the euro-mediterranean information society. *Proceedings of the European Union Meeting on the Creation of the Information Society,* Rome, Italy.

Ghonim, W. 2012. *Revolution 2.0.* London, U.K.: HarpersCollins Publishers.

Gilbert, M., Masucci, M., Homko, C., and Bove, A. A. 2007. Theorizing the digital divide: Information and communication technology use frameworks among poor women using a telemedicine system. *Geoforum* 39, 912–925.

Gilbreth, C. and Otero, G. 2001. Democratization in Mexico: The Zapatista uprising and civil society. *Latim American Perspectives* 28(4), 7–29.

Grabill, J. T. 2003. Community computing and citizens' productivity. *Computers and Composition* 20, 131–150.

Haddon, L. 2000. Social exclusion and information and communication technologies: Lessons from studies of single parents and the young elderly. *New Media and Society* 2(4), 387–408.

Harwitt, E. and Clark, D. 2001. Shaping the internet in China: Evolution of political control over network infrastructure and content. *Asian Survey* 41(3), 377–408.

Illich, I. 1973. *Tools for Conviviality,* New York. http://preservenet.com/theory/illich/illich/tools.html

Information and Decision Support Center (IDSC). 2005. Annual report on Egypt. The Cabinet of Egypt Information and Decision Support Center, Egypt.

James, J. 2011. Sharing mobile phones in developing countries: Implications for the digital divide. *Technological Forecasting and Social Change* 78, 729–735.

Kamel, S. 1995. Information superhighways, a potential for socio-economic and cultural development. Managing information and communications in a changing global environment. *Proceedings of the 6th International IRMA Conference on Managing Information and Communications in a Changing Global Environment,* Khosrowpour, M. (ed.). Atlanta, GA, May 21–24, pp. 115–124.

Kamel, S. 1997. DSS for strategic decision-making. In: *Information Technology Management in Modern Organizations,* Khosrowpour, M. and Liebowitz, J. (eds.). Hershey, PA: Idea Group Publishing, pp. 168–182.

Kamel, S. 1998a. Building the African information infrastructure. In: *Business Information Technology Management: Closing the International Divide,* Banerjee, P., Hackney, R., Dhillon, G., and Jain, R. (eds.). New Delhi, India: Har-Anand Publications, pp. 118–144.

Kamel, S. 1998b. Building an information highway. *Proceedings of the 31st Hawaii International Conference on System Sciences,* Maui, HI, January 6–9, pp. 31–41.

Kamel, S. 1999. Information technology transfer to Egypt. *Proceedings of the Portland International Conference on Management of Engineering and Technology (PICMET). Technology and Innovation Management: Setting the Pace for the Third Millennium,* Kocaoglu, D.F. and Anderson, T. (eds.), Portland, Oregon, July 25–29, pp. 567–571.

Kamel, S. 2005a. Assessing the impacts of establishing an internet cafe in the context of a developing nation. *Proceedings of the 16th International IRMA Conference on Managing Modern Organizations with Information Technology*, San Diego, CA, May 15–16, pp. 176–181.

Kamel, S. 2005b. The evolution of information and communication technology infrastructure in Egypt. In: *Information Systems in an e-World*, Hunter, G. and Wenn, A. (eds.). Washington, DC: The Information Institute, pp. 117–135.

Kamel, S. 2007. The evolution of the ICT industry in Egypt. In: *Science, Technology and Sustainability in the Middle East and North Africa*, Ahmed, A. (ed.). Brighton, U.K.: Inderscience Enterprises Limited, pp. 65–79.

Kamel, S. 2008. The use of ICT for social development in underprivileged communities in Egypt. *Proceedings of the International Conference on Information Resources Management (Conf-IRM) on Information Resources Management in the Digital Economy*, Niagara Falls, ON, Canada, May 18–20.

Kamel, T. 2009. *Together towards a Better Tomorrow*. Presentation to the American Chamber of Commerce in Egypt, Washington, DC, June 23.

Kamel, S. 2010. Egypt: A growing ICT hub in the Middle East–North Africa (MENA) region. *The Journal of Information Technology Management, Cutter IT Journal*, 23(7), 223–236. Special Issue on IT's Promise for Emerging Markets.

Kamel, S. and Ghaffar, H. A. 2003. Free internet in the lands of the pharaohs—A study of a developing nation on a mission to narrow its digital divide. *Proceedings of the 14th Information Resource Management Association International Conference (IRMA) on Information Technology and Organizations: Trends, Issues, Challenges and Solutions*, Philadelphia, PA, May 19–21, pp. 228–229.

Kamel, S. and Hussein, M. 2002. The emergence of eCommerce in a developing nation—The case of Egypt. *Benchmarking—An International Journal*, 2, 146–153. Special Issue on Electronic Commerce: A Best Practice Perspective.

Kamel, S. and Tooma, E. 2005. Exchanging debt for Development: Lessons from the Egyptian debt-for-development swap experience, Economic Research Forum and Ministry of Communications and Information Technology, September.

Kamel, S. and Wahba, K. 2002. From an information Island to a knowledge society—The case of Egypt. In: *New Perspective on Information Systems Development—Theory, Methods and Practice*, Harindranath, G., Wojtkowski, W.G., Zupancic, J., Rosenberg, D., Wojtkowski, W., Wrycza, S., and Sillince, J.A. (eds.). New York: Kluwer Academic/Plenum Publishers, pp. 71–82.

Katz, J. E. and Aspden, P. 1997. Motives, hurdles and dropouts. *Communications of the ACM* 40, 7–102.

Katz, J. E. and Rice, R. E. 2002. *Social Consequences of Internet Use: Access, Involvement and Interaction*. Cambridge, MA: MIT.

Kozma, R. B. 2004. Technology, economic development and education reform: Global challenges and Egyptian response. Partners for a Competitive Egypt. http://robertkozma.com/images/kozma_egyptian_report.pdf

Lindquist, J. 2002. *A Place to Stand: Politics and Persuasion in a Working Class Bar*. Oxford, U.K.: Oxford University Press.

Loch, K., Straub, D., and Kamel, S. 2003. Use of the internet: A study of individuals and organizations in the Arab world. *Proceedings of the 1st Global Information Technology Management Association Conference (GITMA)*, Memphis, TN, June 11–13, 2003, p. 191.

Ministry of Communications and Information Technology (MCIT). 2001. *Navigating Youth Toward an E-Future: The Information Technology Club (ITC) Model*. http://www.mcit.gov.eg/Publications.aspx?realName = IT_Clubs_Navigating_Youth_Toward_E-Future20077915323.pdf

Ministry of Communications and Information Technology (MCIT). 2005a. *Egypt Information Society Initiative*, 4th edn.

Ministry of Communications and Information Technology (MCIT). 2005b. *Building Digital Bridges: Egypt's Vision of the Information Society*. http://www.wsis-egypt.gov.eg/wsis2003/assets/OffDocuments/Building%20Digital%20Bridges.doc

Ministry of Communications and Information Technology (MCIT). 2007a. *Egypt's ICT Annual Book.*

Ministry of Communications and Information Technology (MCIT). 2007b. *Egypt's ICT Strategy 2007–2010.* May 2007. http://www.mcit.gov.eg/Publications.aspx?realName = Egypt-ICT-Strategy.pdf.

Ministry of Communications and Information Technology (MCIT). 2008. *Information and Communications Technology Indicators Bulletin: December 2008 Quarterly Issue.* http://www.mcit.gov.eg/ Publications.aspx?realName = ICT_Indicators_Bulletin_Q3_08_Eng2009223113215.pdf.

Mintz, S. 1998. The Internet as a tool for Egypt's economic growth, An International Development Professionals Inc. Report, October.

Norris, P. 2001. *Digital Divide: Civic Engagement, Information Poverty and the Internet in Democratic Societies.* Cambridge, U.K.: Cambridge University Press.

Opesade, A. 2011. Strategic, value-based ICT investment as a key factor in bridging the digital divide. *Information Development.* Thousands Oaks, CA: Sage Publications, pp. 100–108.

Parker, S. 2011. The digital divide is still with us. *Information Development.* Thousands Oaks, CA: Sage Publications, pp. 83–84.

Petrazzini, B. and Harindranath, G. 1996. Information infrastructure initiatives in emerging economies: The case of India. *The National Information Infrastructure Initiatives,* Kahin, B. and Wilson, E. (eds.). Cambridge, MA: Massachusetts Institute of Technology Press, pp. 217–260.

Press, L. 1999. Connecting Villages: The Grameen Bank Success Story. *OnTheInternet.* 5(4), July/August, 32–37.

Press, G. 2010. Accelerating IT development in Egypt, *On Life in Information,* 2.

Quibria, M. G., Ahmed, S. N., Tschang, T., and Reyes-Mascasaquit, M. 2003. Digital divide: Determinants and policies with special reference to Asia. *Journal of Asian Economics* 13, 811–825.

Rao, S. S. 2005. Bridging digital divide: Efforts in India. *Telematics and Informatics* 22, 361–375.

Selwyn, N. 2006. Digital division or digital inclusion? A study of non-users and low-users of computers. *Poetics* 34, 273–292.

Shapiro, C. and Varian, H. 1999. *Information Rules.* Boston, MA: Harvard Business School Press.

Shirazi, F., Ngwenyama, O., and Morawczynski, O. 2010. ICT expansion and the digital divide in democratic freedoms: An analysis of the impact of ICT expansion, education and ICT filtering on democracy. *Telemetics and Informatics* 27, 21–31.

Snellen, I. 2001. ICTs, bureaucracies and the future of democracy. *Communications of the ACM.* 44(1), 45–48. doi:10.1145/357489.357504.

Suarez, L. S. 2006. Mobile democracy: Text messages, voters turnout, and the 2004 Spanish general election. *Representation* 42(2), 117–128.

Sussman, G. 1997. *Communication, Technology and Politics in the Information Age.* Thousands Oaks, CA: Sage.

Stiglitz, J. 2002. Transparency in Government. In: *The Right to Tell: The Role of Mass Media in Economic Development,* Islam, R. (ed.), Washington, DC: World Bank Publications, pp. 27–44.

Tinio, V. L. 2003. ICT in Education. Available at www.apdip.net/publications/iespprimers/eprimer-edu.pdf.

United Nations Development Program (UNDP). 2007. *Human Development Report 2007/2008: Fighting Climate Change: Human Solidarity in a Divided World.* http://hdr.undp.org/en/media/ HDR_20072008_EN_Complete.pdf.

Weber, M. L., Loumakis, A., and Bergman, J. 2003. Who participated and why? *Social Science Computer Review* 21(1), 26–42.

World Bank. 2002. Telecommunications and information services to the poor. Towards a strategy for universal access. Available at: http://rru.worldbank.org/douments/paperslinks/1210.pdf

World Bank. 2006. *Information and Communication Technology for Development—Global Trends and Policies.* Washington, DC: World Bank Publications.

World Economic Forum (WEF). 2009. Global Information technology report. http://www3.weforum.org/ docs/WEF_GITR_Report_2010.pdf

World Summit on the Information Society (WSIS). 2008. http://www.wsis-online.net (accessed: March 2).

II

Technical Foundations of Data and Database Management

8

Data Models

Avi Silberschatz
Yale University

Henry F. Korth
Lehigh University

S. Sudarshan
*Indian Institute of
Technology Bombay*

Underlying the structure of a database is the concept of a *data model*. Modeling is done at three levels of abstraction: physical, logical, and view. The physical level describes how data are stored at a low level and captures implementation details that are not covered in this chapter. The logical level describes the real-world entities to be modeled in the database and the relationships among these entities. The view level describes only a part of the entire database and serves the needs of users who do not require access to the entire database. Typically in a database, the same data model is used at both the logical and the view levels. Data models differ in the primitives available for describing data and in the amount of semantic detail that can be expressed.

Database system interfaces used by application programs are based on the logical data model; databases hide the underlying implementation details from applications. This conceptual separation of physical organization from logical structure allows the implementation of database applications to be independent of the underlying implementation of the data structures representing the data. The low-level implementations affect system performance and there has been considerable effort to design physical organizations that take advantage of how certain applications access data and how the underlying computer architectures store and manipulate data.

In this chapter, we focus on logical data models. In Section 8.1 we cover the relational data model. In Section 8.2 we cover object-based data models including the E-R model (in Section 8.2.1), the object-oriented model (in Section 8.2.2), and the object-relational model (in Section 8.2.3). The E-R model is a representative of a class of models called semantic data models, which provide richer data structuring and modeling capabilities than the record-based format of the relational model.

In Section 8.3, we describe two models for representing nested structures, namely, XML (in Section 8.3.1) and JSON (in Section 8.3.2). These models are widely used for data exchange between applications. In Section 8.5 we outline models for streaming data.

Our coverage of data models is intended as an introduction to the area. In Section 8.6 we provide references for further reading on the topics covered in this chapter, as well as on related topics such as data modeling issues related to knowledge representation.

8.1 Relational Model

The *relational model* is today the primary data model for commercial data-processing applications. It has attained its primary position because of its simplicity, which eases the job of the programmer, as compared to earlier data models.

A relational database consists of a collection of *tables*, each of which is assigned a unique name. An *instance* of a table storing customer information is shown in Table 8.1 The table has several rows, one for each customer, and several columns, each storing some information about the customer. The values in the *customer-id* column of the *customer* table serve to uniquely identify customers, while other columns store information such as the name, street address, and city of the customer.

The information stored in a database is broken up into multiple tables, each storing a particular kind of information. For example, information about accounts and loans at a bank would be stored in separate tables. Table 8.2 shows an instance of the *loan* table, which stores information about loans taken from the bank.

In addition to information about "entities" such as customers or loans, there is also a need to store information about "relationships" between such entities. For example, the bank needs to track the relationship between customers and loans. Table 8.3 show the *borrower* table, which stores information indicating which customers have taken which loans. If several people have jointly taken a loan, the same loan number would appear several times in the table with different customer-ids (e.g., loan number L-17). Similarly, if a particular customer has taken multiple loans, there would be several rows in the table with the customer-id of that customer (e.g., 019-28-3746), with different loan numbers.

TABLE 8.1 The *Customer* Table

Customer-Id	Customer-Name	Customer-Street	Customer-City
019-28-3746	Smith	North	Rye
182-73-6091	Turner	Putnam	Stamford
192-83-7465	Johnson	Alma	Palo Alto
244-66-8800	Curry	North	Rye
321-12-3123	Jones	Main	Harrison
335-57-7991	Adams	Spring	Pittsfield
336-66-9999	Lindsay	Park	Pittsfield
677-89-9011	Hayes	Main	Harrison
963-96-3963	Williams	Nassau	Princeton

TABLE 8.2 The *Loan* Table

Loan-Number	Amount
L-11	900
L-14	1500
L-15	1500
L-16	1300
L-17	1000
L-23	2000
L-93	500

TABLE 8.3 The *Borrower* Table

Customer-Id	Loan-Number
019-28-3746	L-11
019-28-3746	L-23
244-66-8800	L-93
321-12-3123	L-17
335-57-7991	L-16
555-55-5555	L-14
677-89-9011	L-15
963-96-3963	L-17

8.1.1 Formal Basis

The power of the relational data model lies in its rigorous mathematical foundations and a simple user-level paradigm for representing data. Mathematically speaking, a *relation* is a subset of the Cartesian product of an ordered list of domains. For example, let E be the set of all employee identification numbers, D the set of all department names, and S the set of all salaries. An employment relation is a set of 3-tuples (e,d,s) where $e \in E$, $d \in D$, and $s \in S$. A tuple (e,d,s) represents the fact that employee e works in department d and earns salary s.

At the user level, a relation is represented as a table. The table has one column for each domain and one row for each tuple. Each column has a name, which serves as a column header, and is called an *attribute* of the relation. The list of attributes for a relation is called the *relation schema*. The terms table and relation are used synonymously, as are row and tuple, as also column and attribute.

Data models also permit the definition of *constraints* on the data stored in the database. For instance, *key constraints* are defined as follows. If a set of attributes L is specified to be a *super-key* for relation r, in any consistent ("legal") database, the set of attributes L would uniquely identify a tuple in r; that is, no two tuples in r can have the same values for all attributes in L. For instance, *customer-id* would form a super-key for relation *customer*. A relation can have more than one super-key, and usually one of the super-keys is chosen as a *primary key*; this key must be a minimal set, that is, dropping any attribute from the set would make it cease to be a super-key.

Another form of constraint is the *foreign key* constraint, which specifies that for each tuple in one relation, there must exist a matching tuple in another relation. For example, a foreign key constraint *from borrower referencing customer* specifies that for each tuple in *borrower*, there must be a tuple in *customer* with a matching *customer-id* value.

Users of a database system may query the data, insert new data, delete old data, or update the data in the database. Of these tasks, the task of querying the data is usually the most complicated. In the case of the relational data model, since data are stored as tables, a user may query these tables, insert new tuples, delete tuples, and update (modify) tuples. There are several languages for expressing these operations.

The tuple relational calculus and the domain relational calculus are nonprocedural languages that represent the basic power required in a relational query language. Both of these languages are based on statements written in mathematical logic. We omit details of these languages.

The relational algebra is a procedural query language that defines several operations, each of which takes one or more relations as input, and returns a relation as output. For example:

- The *selection* operation is used to get a subset of tuples from a relation, by specifying a predicate. The selection operation $\sigma_p(r)$ returns the set of tuples of r that satisfy the predicate P.
- The *projection* operation, $\Pi_L(r)$, is used to return a relation containing a specified set of attributes L of a relation r, removing the other attributes of r.
- The *union*, operation $r \cup s$ returns the union of the tuples in r and s. The *intersection* and *difference* operations are similarly defined.

- The *natural join* operation ⋈ is used to combine information from two relations. For example, the natural join of the relations *loan* and *borrower*, denoted *loan* ⋈ *borrower*, would be the relation defined as follows. First, match each tuple in *loan* with each tuple in *borrower* that has the same values for the shared attribute *loan-number*; for each pair of matching tuples, the join operation creates a tuple containing all attributes from both tuples; the join result relation is the set of all such tuples.

   For instance, the natural join of the *loan* and *borrower* tables in Tables 8.2 and 8.3 contains tuples (L-17, 1000, 321-12-3123) and (L-17, 1000, 963-96-3963), since the tuple with loan number L-17 in the *loan* table matches two different tuples with loan number L-17 in the *borrower* table.

The relational algebra has other operations as well, for instance, operations that can aggregate values from multiple tuples, for example, by summing them up, or finding their average.

Since the result of a relational algebra operation is itself a relation, it can be used in further operations. As a result, complex expressions with multiple operations can be defined in the relational algebra.

Among the reasons for the success of the relational model are its basic simplicity, representing all data using just a single notion of tables, as well as its formal foundations in mathematical logic and algebra.

The relational algebra and the relational calculi are terse, formal languages that are inappropriate for casual users of a database system. Commercial database systems have, therefore, used languages with more "syntactic sugar." Queries in these languages can be translated into queries in relational algebra.

8.1.2 SQL

The SQL language has clearly established itself as *the* standard relational database language. The SQL language has a data definition component for specifying schemas, and a data manipulation component for querying data as well as for inserting, deleting, and updating data.

We illustrate some examples of queries and updates in SQL. The following query finds the name of the customer whose customer-id is 192-83-7465:

```
select customer.customer-name
from customer
where customer.customer-id = '192-83-7465'
```

Queries may involve information from more than one table. For instance, the following query finds the amount of all loans owned by the customer with customer-id 019-28-3746:

```
select loan.loan-number, loan.amount
from borrower, loan
where borrower.customer-id = '019-28-3746' and
borrower.loan-number = loan.loan-number
```

If the previous query were run on the tables shown earlier, the system would find that the loans L-11 and L-23 are owned by customer 019-28-3746, and would print out the amounts of the two loans, namely, 900 and 2000.

The following SQL statement adds an interest of 5% to the loan amount of all loans with amount greater than 1000.

```
update loan
set amount = amount * 1.05
where amount > 10000
```

Over the years, there have been several revisions of the SQL standard. The most recent is SQL:2011.

QBE and Quel are two other significant query languages. Of these, Quel is no longer in wide-spread use. Although QBE itself is no longer in use, it greatly influenced the QBE-style language in Microsoft Access.

8.1.3 Relational Database Design

The process of designing a conceptual-level schema for a relational database involves the selection of a set of relation schemas. There are several approaches to relational database design. One approach, which we describe later, in Section 8.2.1, is to create a model of the enterprise using a higher-level data model, such as the entity-relationship model, and then translate the higher-level model into a relational database design.

Another approach is to directly create a design, consisting of a set of tables and a set of attributes for each table. There are often many possible choices that the database designer might make. A proper balance must be struck among three criteria for a good design:

1. Minimization of redundant data
2. Ability to represent all relevant relationships among data items
3. Ability to test efficiently data dependencies that require certain attributes to be unique identifiers

To illustrate these criteria for a good design, consider a database of employees, departments, and managers. Let us assume that a department has only one manager, but a manager may manage one or more departments. If we use a single relation *emp-info1(employee, department, manager)*, then we must repeat the manager of a department once for each employee. Thus we have *redundant* data.

We can avoid redundancy by *decomposing* (breaking up) the above relation into two relations *emp-mgr(employee, manager)* and *emp-dept(manager, department)*. However, consider a manager, Martin, who manages both the sales and the service departments. If Clark works for Martin, we cannot represent the fact that Clark works in the service department but not the sales department. Thus we cannot represent all relevant relationships among data items using the decomposed relations, such a decomposition is called a *lossy-join decomposition*. If instead, we chose the two relations *emp-dept(employee, department)* and *dept-mgr(department, manager)*, we would avoid this difficulty, and at the same time avoid redundancy. With this decomposition, joining the information in the two relations would give back the information in *emp-info1*, such a decomposition is called a *lossless-join decomposition*.

There are several types of data dependencies. The most important of these are *functional dependencies*. A functional dependency is a constraint that the value of a tuple on one attribute or set of attributes determines its value on another. For example, the constraint that a department has only one manager could be stated as "department functionally determines manager." Because functional dependencies represent facts about the enterprise being modeled, it is important that the system check newly inserted data to ensure no functional dependency is violated (as in the case of a second manager being inserted for some department). Such checks ensure that the update does not make the information in the database inconsistent. The cost of this check depends on the design of the database.

There is a formal theory of relational database design that allows us to construct designs that have minimal redundancy, consistent with meeting the requirements of representing all relevant relationships, and allowing efficient testing of functional dependencies. This theory specifies certain properties that a schema must satisfy, based on functional dependencies. For example, a database design is said to be in a *Boyce–Codd normal form* if it satisfies a certain specified set of properties; there are alternative specifications, for instance, the *third normal form*. The process of ensuring that a schema design is in a desired normal form is called *normalization*.

More details can be found in standard textbooks on databases; Abiteboul et al. [AHV95] provide a detailed coverage of database design theory.

8.1.4 History

The relational model was developed in the late 1960s and early 1970s by E. F. Codd. For this work, Codd won the most prestigious award in computing, the ACM A. M. Turing Award. The 1970s saw the development of several experimental database systems based on the relational model and the emergence of a formal theory to support the design of relational databases. The commercial application of relational databases began in the late 1970s, but was limited by the poor performance of early relational systems. During the 1980s, numerous commercial relational systems with good performance became available. The relational model has since established itself as the primary data model for commercial data processing applications.

Earlier generation database systems were based on the *network data model* or the *hierarchical data model*. Those two older models are tied closely to the data structures underlying the implementation of the database. We omit details of these models since they are now of historical interest only.

8.2 Object-Based Models

The relational model is the most widely used data model at the implementation level; most databases in use around the world are relational databases. However, the relational view of data is often too detailed for conceptual modeling. Data modelers need to work at a higher level of abstraction.

Object-based logical models are used in describing data at the conceptual level. The object-based models use the concepts of *entities* or *objects* and relationships among them rather than the implementation-based concepts of the record-based models. They provide flexible structuring capabilities and allow data constraints to be specified explicitly. Several object-based models are in use; some of the more widely known ones are

- The entity-relationship model
- The object-oriented model
- The object-relational model

The entity-relationship model has gained acceptance in database design and is widely used in practice. The object-oriented model includes many of the concepts of the entity-relationship model, but represents executable code as well as data. The object-relational data model combines features of the object-oriented data model with the relational data model.

8.2.1 Entity-Relationship Model

The E-R data model represents the information in an enterprise as a set of basic objects called *entities*, and *relationships* among these objects. It facilitates database design by allowing the specification of an *enterprise schema*, which represents the overall logical structure of a database. The E-R data model is one of several *semantic data models*; that is, it attempts to represent the meaning of the data. In addition to providing an easily understood conceptual view of a database design, an E-R design can be converted algorithmically to a relational design that often requires little further modification.

8.2.1.1 Basics

There are three basic notions that the E-R data model employs: entity sets, relationship sets, and attributes. An *entity* is a "thing" or "object" in the real world that is distinguishable from all other objects. For example, each person in the universe is an entity.

Each entity is described by a collection of features, called *attributes*. For example, the attributes *account-number* and *balance* may describe one particular account in a bank, and they form attributes of the *account* entity set. Similarly, attributes *customer-name*, *customer-street* address, and *customer-city* may describe a *customer* entity.

The values for some attributes may uniquely identify an entity. For example, the attribute *customer-id* may be used to uniquely identify customers (since it may be possible to have two customers with the same name, street address, and city). A unique customer identifier must be assigned to each customer.

An entity may be concrete, such as a person or a book, or it may be abstract, such as a bank account, or a holiday, or a concept.

An *entity set* is a set of entities of the same type that share the same properties (attributes). The set of all persons working at a bank, for example, can be defined as the entity set *employee*, and the entity John Smith may be a member of the *employee* entity set. Similarly, the entity set *account* might represent the set of all accounts in a particular bank. An E-R design thus includes a collection of entity sets, each of which contains any number of entities of the same type.

Attributes are descriptive properties possessed by all members of an entity set. The designation of attributes expresses that the database stores similar information concerning each entity in an entity set; however, each entity has its own value for each attribute. Possible attributes of the *employee* entity set are *employee-name*, *employee-id*, and *employee-address*. Possible attributes of the *account* entity set are *account-number* and *account-balance*. For each attribute, there is a set of permitted values called the *domain* (or *value set*) of that attribute. The domain of the attribute *employee-name* might be the set of all text strings of a certain length. Similarly, the domain of attribute *account-number* might be the set of all positive integers.

Entities in an entity set are distinguished based on their attribute values. A set of attributes that suffices to distinguish all entities in an entity set is chosen, and called a *primary key* of the entity set. For the *employee* entity set, *employee-id* could serve as a primary key; the enterprise must ensure that no two people in the enterprise can have the same employee identifier.

A *relationship* is an association among several entities. Thus, an *employee* entity might be related by an *emp-dept* relationship to a *department* entity where that employee entity works. For example, there would be an *emp-dept* relationship between John Smith and the bank's credit department if John Smith worked in that department. Just as all *employee* entities are grouped into an *employee* entity set, all *emp-dept* relationship instances are grouped into an *emp-dept relationship set*. A relationship set may also have descriptive attributes. For example, consider a relationship set *depositor* between the *customer* and the *account* entity sets. We could associate an attribute *last-access* to specify the date of the most recent access to the account. The relationship sets *emp-dept* and *depositor* are examples of a binary relationship set, that is, one that involves two entity sets. Most of the relationship sets in a database system are binary.

The overall logical structure of a database can be expressed graphically by an *E-R diagram*. There are multiple different notations used for E-R diagrams; we describe in the following one of the widely used notations, based on the one proposed by Chen [Che76]. Such a diagram consists of the following major components:

- *Rectangles*, which represent entity sets
- *Ellipses*, which represent attributes
- *Diamonds*, which represent relationship sets
- *Lines*, which link entity sets to relationship sets, and link attributes to both entity sets and relationship sets

An entity-relationship diagram for a portion of our simple banking example is shown in Figure 8.1. The primary key attributes (if any) of an entity set are shown underlined.

Composite attributes are attributes that can be divided into subparts (that is, other attributes). For example, an attribute *name* could be structured as a composite attribute consisting of *first-name*, *middle-initial*, and *last-name*. Using composite attributes in a design schema is a good choice if a user wishes to refer to an entire attribute on some occasions, and to only a component of the attribute on other occasions.

The attributes in our examples so far all have a single value for a particular entity. For instance, the *loan-number* attribute for a specific loan entity refers to only one loan number. Such attributes are said

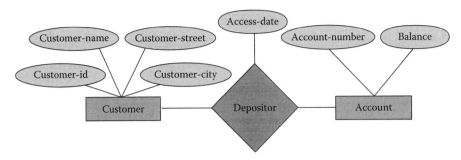

FIGURE 8.1   E-R diagram.

to be *single valued*. There may be instances where an attribute has a set of values for a specific entity. Consider an *employee* entity set with the attribute *phone-number*. An employee may have zero, one, or several phone numbers, and hence this type of attribute is said to be *multivalued*.

Suppose that the *customer* entity set has an attribute *age* that indicates the customer's age. If the *customer* entity set also has an attribute *date-of-birth*, we can calculate *age* from *date-of-birth* and the current date. Thus, *age* is a *derived attribute*. The value of a derived attribute is not stored, but is computed when required.

Figure 8.2 shows how composite, multivalued, and derived attributes can be represented in the E-R notation. Ellipses are used to represent composite attributes as well as their subparts, with lines connecting the ellipse representing the attribute to the ellipse representing its subparts. Multivalued attributes are represented using a double ellipse, while derived attributes are represented using a dashed ellipse.

Figure 8.3 shows an alternative E-R notation, based on the widely used UML class diagram notation, which allows entity sets and attributes to be represented in a more concise manner. The figure shows the *customer* entity set with the same attributes as described earlier, the *account* entity set with attributes *account-number* and *balance*, and the *depositor* relationship set with attribute *access-date*. For more details on this alternative E-R notation, see [SKS11].

Most of the relationship sets in a database system are *binary*, that is, they involve only two entity sets. Occasionally, however, relationship sets involve more than two entity sets. As an example, consider the entity sets *employee*, *branch*, and *job*. Examples of *job* entities could include manager, teller, auditor, and so on. Job entities may have the attributes *title* and *level*. The relationship set *works-on* among

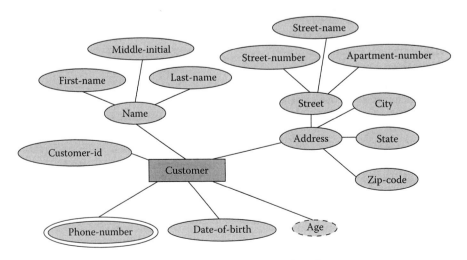

FIGURE 8.2   E-R diagram with composite, multivalued, and derived attributes.

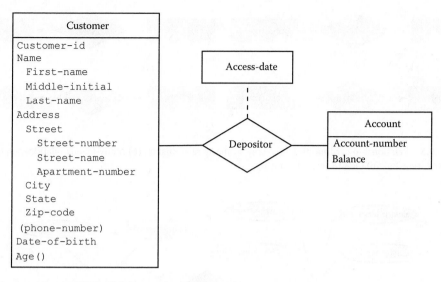

FIGURE 8.3 Alternative E-R diagram notation based on UML.

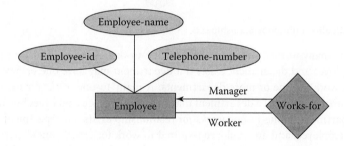

FIGURE 8.4 E-R diagram with role indicators.

employee, *branch*, and *job* is an example of a *ternary relationship*. A ternary relationship among Jones, Perryridge, and manager indicates that Jones acts as a manager at the Perryridge branch. Jones could also act as auditor at the Downtown branch, which would be represented by another relationship. Yet another relationship could be among Smith, Downtown, and teller, indicating Smith acts as a teller at the Downtown branch.

Consider, for example, a relationship set *works-for* relating the entity set *employee* with itself. Each employee entity is related to the entity representing the manager of the employee. One employee takes on the role of *worker*, whereas the second takes the role of *manager*. Roles can be depicted in E-R diagrams as shown in Figure 8.4.

Although the basic E-R concepts can model most database features, some aspects of a database may be more aptly expressed by certain extensions to the basic E-R model. Commonly used extended E-R features include specialization, generalization, higher- and lower-level entity sets, attribute inheritance, and aggregation. The notion of specialization and generalization are covered in the context of object-oriented data models in Section 8.2.2. A full explanation of the other features is beyond the scope of this chapter; we refer readers to [SKS11] for additional information.

8.2.1.2 Representing Data Constraints

In addition to entities and relationships, the E-R model represents certain constraints to which the contents of a database must conform. One important constraint is *mapping cardinalities*, which express the number of entities to which another entity can be associated via a relationship set. Therefore, relationships

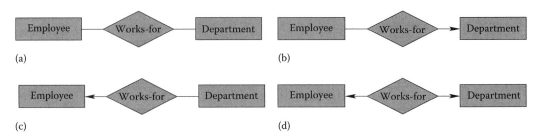

(a)   (b)

(c)   (d)

FIGURE 8.5   Relationship cardinalities. (a) Many-to-many relationship. (b) Many-to-one relationship. (c) One-to-many relationship. (d) One-to-one relationship.

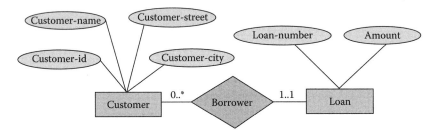

FIGURE 8.6   Cardinality limits on relationship sets.

can be classified as many-to-many, many-to-one, or one-to-one. A many-to-many *works-for* relationship between employee and department exists if a department may have one or more employees and an employee may work for one or more departments. A many-to-one *works-for* relationship between *employee* and *department* exists if a department may have one or more employees but an employee must work for only department. A one-to-one *works-for* relationship exists if a department were required to have exactly one employee, and an employee required to work for exactly one department. In an E-R diagram, an arrow is used to indicate the type of relationship as shown in Figure 8.5. The arrow indicates "exactly one."

E-R diagrams also provide a way to indicate more complex constraints on the number of times each entity participates in relationships in a relationship set. An edge between an entity set and a binary relationship set can have an associated minimum and maximum cardinality, shown in the form $l..h$, where l is the minimum and h the maximum cardinality. A maximum value of 1 indicates that the entity participates in at most one relationship, while a maximum value * indicates no limit.

For example, consider Figure 8.6. The edge between *loan* and *borrower* has a cardinality constraint of 1..1, meaning the minimum and the maximum cardinality are both 1. That is, each loan must have exactly one associated customer. The limit 0..* on the edge from *customer* to *borrower* indicates that a customer can have zero or more loans. Thus, the relationship *borrower* is one to many from *customer* to *loan*.

It is easy to misinterpret the 0..* on the edge between *customer* and *borrower*, and think that the relationship *borrower* is many to one from *customer* to *loan*—this is exactly the reverse of the correct interpretation. In this case, the earlier arrow notation is easier to understand.

If both edges from a binary relationship have a maximum value of 1, the relationship is one-to-one. If we had specified a cardinality limit of 1..* on the edge between *customer* and *borrower*, we would be saying that each customer must have at least one loan.

8.2.1.3 Use of E-R Model in Database Design

A high-level data model, such as the E-R model, serves the database designer by providing a conceptual framework in which to specify, in a systematic fashion, the data requirements of the database users and how the database will be structured to fulfill these requirements. The initial phase of database design, then, is to characterize fully the data needs of prospective database users. The outcome of this phase will be a *specification of user requirements*. The initial specification of user requirements may be based on interviews with the database users and the designer's own analysis of the enterprise. The description that arises from this design phase serves as the basis for specifying the logical structure of the database.

By applying the concepts of the E-R model, the user requirements are translated into a conceptual schema of the database. The schema developed at this *conceptual design* phase provides a detailed overview of the enterprise. Stated in terms of the E-R model, the conceptual schema specifies all entity sets, relationship sets, attributes, and mapping constraints. The schema can be reviewed to confirm that all data requirements are indeed satisfied and are not in conflict with each other. The design can also be examined to remove any redundant features. The focus at this point is on describing the data and their relationships rather than on physical storage details.

A fully developed conceptual schema also indicates the functional requirements of the enterprise. In a *specification of functional requirements*, users describe the kinds of operations (or transactions) that will be performed on the data. Example operations include modifying or updating data, searching for and retrieving specific data, and deleting data. A review of the schema for meeting functional requirements can be made at the conceptual design stage.

The process of moving from a conceptual schema to the actual implementation of the database involves two final design phases. In the *logical design* phase, the high-level conceptual schema is mapped onto the implementation data model of the database management system (DBMS). In the next section, we discuss how a relational schema can be derived from an E-R design. The resulting DBMS-specific database schema is then used in the subsequent *physical design* phase, in which the physical features of the database are specified. These features include the form of file organization and the internal storage structures.

Because the E-R model is extremely useful in mapping the meanings and interactions of real-world enterprises onto a conceptual schema, a number of database design tools draw on E-R concepts. Further, the relative simplicity and pictorial clarity of the E-R diagramming technique may well account in large part for the widespread use of the E-R model.

8.2.1.4 Deriving a Relational Database Design from the E-R Model

A database that conforms to an E-R diagram can be represented by a collection of relational tables. For each entity set and each relationship set in the database, there is a unique table that is assigned the name of the corresponding entity set or relationship set. Each table has a number of columns that, again, have unique names. The conversion of database representation from an E-R diagram to a table format is the basis for deriving a relational database design.

The column headers of a table representing an entity set correspond to the attributes of the entity, and the primary key of the entity becomes the primary key of the relation. The column headers of a table representing a relationship set correspond to the primary key attributes of the participating entity sets, and the attributes of the relationship set. Rows in the table can be uniquely identified by the combined primary keys of the participating entity sets. For such a table, the primary keys of the participating entity sets are *foreign keys* of the table. The rows of the tables correspond to individual members of the entity or relationship set.

Tables 8.1 through 8.3 show instances of tables that correspond, respectively, to the *customer* and *loan* entity sets, the *borrower* relationship set, of Figure 8.6.

The full set of steps to convert an E-R design to a well-designed set of tables is presented in [SKS11].

8.2.2 Object-Oriented Model

The object-oriented data model is an adaptation of the object-oriented programming paradigm to database systems. The object-oriented approach to programming was first introduced by the language Simula 67, which was designed for programming simulations and advanced further by Smalltalk.

Today, C++ and Java have become the most widely known object-oriented programming languages, and most application development is done in these or other object-oriented languages. The model is based on the concept of encapsulating data, and code that operates on those data, in an object.

In this section, we present an overview of the concepts of object-oriented application development and how they are used in database systems.

8.2.2.1 Basics

Like the E-R model, the object-oriented model is based on a collection of objects. Entities, in the sense of the E-R model, are represented as *objects* with attribute values represented by *instance variables* within the object. The value stored in an instance variable may itself be an object. Objects can contain objects to an arbitrarily deep level of nesting. At the bottom of this hierarchy are objects such as integers, character strings, and other data types that are built into the object-oriented language and serve as the foundation of the object-oriented model. The set of built-in object types varies among languages.

Usually, the data within an object are private to that object and manipulated only by procedures (usually called methods) within that object. Therefore, the internal representation of an object's data need not influence the implementation of any other object; all access to data within an object is via publicly accessible methods for that object. This encapsulation of code and data has proven useful in developing higher modular systems.

Unless the developer decides to make some parts of the object structure public, the only way in which one object can access the data of another object is by invoking a method of that other object. Thus, the call interface of the methods of an object defines its externally visible part. The internal parts of the object—the instance variables and method code—are not visible externally. The result is two levels of data abstraction.

To illustrate the concept, consider an object representing a bank account. Such an object contains instance variables *account-number* and *account-balance*, representing the account number and account balance. It contains a method *pay-interest*, which adds interest to the balance. Assume that the bank had been paying 4% interest on all accounts but now is changing its policy to pay 3% if the balance is less than $1000 or 4% if the balance is $1000 or greater. Under most data models, this would involve changing code in one or more application programs. Under the object-oriented model, the only change is made within the *pay- interest* method. The external interface to the object remains unchanged.

8.2.2.2 Classes

Objects that contain the same types of values and the same methods are grouped together into *classes*. A class may be viewed as a type definition for objects. This combination of data and code into a type definition is similar to the programming language concept of abstract data types. Thus, all *employee* objects may be grouped into an *employee* class. Classes themselves can be grouped into a hierarchy of classes; for example, the *employee* class and the *customer* classes may be grouped into a *person* class. The class *person* is a *superclass* of the *employee* and *customer* classes, since all objects of the *employee* and *customer* classes also belong to the *person* class. Superclasses are also called *generalizations*. Correspondingly, the *employee* and *customer* classes are *subclasses* of *person*; subclasses are also called *specializations*.

The hierarchy of classes allows sharing of common methods. It also allows several distinct views of objects: an employee, for an example, may be viewed either in the role of person or employee, whichever is more appropriate.

8.2.2.3 Object-Oriented Database Programming Languages

There are two approaches to creating an object-oriented database language: the concepts of object orientation can be added to existing database languages, or existing object-oriented languages can be extended to deal with databases by adding concepts such as persistence and collections. Object-relational database systems take the former approach. Persistent programming languages follow the latter approach.

Any persistent programming language must have a mechanism for making objects persist; those not designated persistent are *transient* and vanish when the program terminates. Among the approaches to persistence are

- *Persistence by class.* A class may be declared persistent, making all objects of such classes persistent. Objects of all other classes are transient. This approach is simple but lacks flexibility.
- *Persistence by creation.* At the time of object creation, the program may choose to declare the object persistent; otherwise it is transient.
- *Persistence by marking.* All objects are transient at the time they are created. However, at any point prior to program termination, an object can be marked as persistent. This approach improves on persistence by creation in that the decision on persistence can be deferred beyond creation time. It has the disadvantage of making it less clear to the developer which objects are in fact persistent.
- *Persistence by reachability.* Objects may be declared to be persistent. Not only are those objects persistent but also those objects reachable from those objects by a sequence of object references. This makes it easy to create a persistent data structure simply by making its root persistent, but it makes it more expensive for the system to determine which objects are persistent.

The Object Data Management Group (ODMG) developed standards for integrating persistence support into several programming languages such as C++ and Java. The Java Database Objects (JDO) standard provided a similar functionality, with tighter integration with Java. However, both these standards are no longer extensively used.

Instead, a new approach called *Object-Relational Mapping* (ORM) has found widespread acceptance. In this approach, a schema designer provides an object-oriented data model, along with a specification of how the data are to be mapped to a relational model. Library functions provided by ORM systems convert data from the relational representation used for storage to an object-oriented representation stored in memory. Updates to the in-memory object-oriented representation are translated back into updates on the underlying relational model. ORM systems provide an API to fetch objects based on a primary key; in addition, when the program traverses a reference from a previous fetched object, the referenced object is transparently fetched from the database in case it had not been fetched earlier. These features are ideally suited for online transaction processing systems. ORM systems also typically provide a query language for bulk data processing, with queries expressed in terms of the object-oriented data mode. The Hibernate ORM system is currently the most widely used ORM implementation.

The alternative to persistent programming languages is to leave persistence to the database system and provide ways for standard programming language applications to access the database. The most popular of these are ODBC for C and C++ programs and JDBC for Java.

8.2.3 Object-Relational Data Models

Object-relational data models extend the relational data model by providing a richer type system that allows representation of instance variables and methods. In many implementations, stored procedures serve in place of methods.

The rigid structure of relations arose historically from the use of fixed length records, first on punched cards and later in early file systems. A more general model of relations would allow set-valued and relation-valued attributes. Such concepts were proposed in the 1980s [RK87:Design], but have only more recently appeared in practice. SQL has implemented this concept through *structured types*. Constructs are added to relational query languages such as SQL to deal with the added data types. The extended type systems allow attributes of tuples to have complex types including non-atomic values such as nested relations. Such extensions attempt to preserve the relational foundations, in particular the declarative access to data, while extending the modeling power.

As an example, consider a *person* entity whose attributes include name, which consists of first- and last-name) and address, which consists of street, city, and postal code. This can be expressed as

```
create type Name as
   (firstname varchar(20),
   lastname varchar(20))
   final;
create type Address as
   (street varchar(20),
   city varchar(20),
   zipcode varchar(9))
   not final;
create table Person (
   name Name,
   address Address,
   dateOfBirth date);
```

Observe that the *Person* table has two attributes, *Name* and *Address* that themselves have substructures. Such data types were specifically disallowed in early relation systems and, indeed, *first normal form* was defined to specify that relations have only atomic, indivisible values. Clearly, then, relations such as *Person* are non-first-normal-form relations.

In addition to allowing tuples within tuples as in our example, SQL allows attributes to hold arrays and multisets (sets in which values may occur multiple times). These features now give SQL the full modeling capability of earlier nested relational model proposals.

Data types, like *Address* in our example, that have the *not final* specification can have *subtypes* defined under them, allowing the definition of subclasses. Finally, data types may have methods defined on them that, such as methods in an object-oriented language, operate on an instance (object) of these types. In addition to object-oriented data-modeling features, SQL supports an imperative extension of the basic SQL query language, providing features such as for and while loops, if-then-else statements, procedures, and functions. SQL also supports references to objects; references must be to objects of a particular type, which are stored as tuples of a particular relation. Objects, however, do not have an independent existence, they correspond to tuples in a relation.

Relations are allowed to form an inheritance hierarchy; each tuple in a lower level relation must correspond to a unique tuple in a higher-level relation that represents information about the same object. Inheritance of relations provides a convenient way of modeling roles, where an object can acquire and relinquish roles over a period of time.

SQL support this via table inheritance; if *r* is a subtable of *s*, the type of tuples of *r* must be a subtype of the type of tuples of *s*. Every tuple present in *r* is implicitly (automatically) present in *s* as well. A query on *s* would find all tuples inserted directly to *s* as well as tuples inserted into *r*; however, only the attributes of table *s* would be accessible, even for the *r* tuples. Thus subtables can be used to represent specialization/generalization hierarchies. However, while subtables in SQL can be used to represent disjoint specializations, where an object cannot belong to two different subclasses of a particular class, they cannot be used to represent the general case of overlapping specialization.

The combination of these features gives SQL a rich set of object-oriented features. In practice, however, the exact syntax of these constructs varies from the published standard. The reason for this is that, prior to object-relational features being added to the SQL standard, many commercial database systems incorporated their own proprietary object-relational extensions into their SQL implementations. While these all offer similar computational and modeling capabilities, the syntactic distinctions create problems when porting database designs across database systems. As a result, object-relational extensions of SQL are not widely used in practice.

Object-relational database systems (that is, database systems based on the object-relation model) provide a convenient migration path for users of relational databases who wish to use object-oriented features. Complex types such as nested relations are useful to model complex data in many applications. Object-relational systems combine complex data based on an extended relational model with object-oriented concepts such as object identity and inheritance.

8.3 Modeling Nested Structures

We have already seen how SQL has been extended over the years to allow representation of nested relational structures. There are other models in wide use for representing nested structures, including XML and JSON. These models are particularly useful when several applications need to share data, but do not all use the same database. In the section, we show how these models can be used in conjunction with a database system, but we do not go into the full feature set of these models.

8.3.1 XML

Extensible Markup Language (XML) was not originally conceived as a database technology. In fact, like the *Hyper-Text Markup Language (HTML)* on which the world wide web is based, XML has its roots in document management. However, unlike HTML, XML can represent database data, as well as many other kinds of structured data used in business applications.

The term *markup* in the context of documents refers to anything in a document that is not intended to be part of the printed output. For the family of markup languages that includes HTML and XML, the markup takes the form of *tags* enclosed in angle-brackets, <>. Tags are used in pairs, with <tag> and </tag> delimiting the beginning and the end of the portion of the document to which the tag refers.

Unlike HTML, XML does not prescribe the set of tags allowed, and the set may be specialized as needed. This feature is the key to XML's major role in data representation and exchange, whereas HTML is used primarily for document formatting.

For example, in our running banking application, account and customer information can be represented as part of an XML document as in Figure 8.7. Observe the use of tags such as `account` and `account-number`. These tags provide context for each value and allow the semantics of the value to be identified. The contents between a start tag and its corresponding end tag is called an *element*.

Compared to storage of data in a database, the XML representation is inefficient, since tag names are repeated throughout the document. However, in spite of this disadvantage, an XML representation has significant advantages when it is used to exchange data:

- First, the presence of the tags makes the message *self-documenting*; that is, a schema need not be consulted to understand the meaning of the text.
- Second, the format of the document is not rigid. For example, if some sender adds additional information, such as a tag `last-accessed` noting the last date on which an account was accessed, the recipient of the XML data may simply ignore the tag. The ability to recognize and ignore unexpected tags allows the format of the data to evolve over time, without invalidating existing applications.
- Third, elements can be nested inside other elements, to any level of nesting. Figure 8.8 shows a representation of the bank information from Figure 8.7, but with `account` elements nested within `customer` elements.
- Finally, since the XML format is widely accepted, a wide variety of tools are available to assist in its processing, including browser software and database tools.

Just as SQL is the dominant *language* for querying relational data, XML is becoming the dominant *format* for data exchange.

```
<bank>
  <account>
    <account-number> A-101 </account-number>
    <branch-name> Downtown </branch-name>
    <balance> 500 </balance>
  </account>
  <account>
    <account-number> A-102 </account-number>
    <branch-name> Perryridge </branch-name>
    <balance> 400 </balance>
  </account>
  <account>
    <account-number> A-201 </account-number>
    <branch-name> Brighton </branch-name>
    <balance> 900 </balance>
  </account>
  <customer>
    <customer-name> Johnson </customer-name>
    <customer-street> Alma </customer-street>
    <customer-city> Palo Alto </customer-city>
  </customer>
  <customer>
    <customer-name> Hayes </customer-name>
    <customer-street> Main </customer-street>
    <customer-city> Harrison </customer-city>
  </customer>
  <depositor>
    <account-number> A-101 </account-number>
    <customer-name> Johnson </customer-name>
  </depositor>
  <depositor>
    <account-number> A-201 </account-number>
    <customer-name> Johnson </customer-name>
  </depositor>
  <depositor>
    <account-number> A-102 </account-number>
    <customer-name> Hayes </customer-name>
  </depositor>
</bank>
```

FIGURE 8.7   XML representation of bank information.

In addition to elements, XML specifies the notion of an *attribute*. For example, the type of an account is represented in the following as an attribute named acct-type.

```
    <account acct-type = "checking">
        <account-number> A-102 </account-number>
        <branch-name> Perryridge </branch-name>
        <balance> 400 </balance>
    </account>
    ...
```

The attributes of an element appear as *name = value* pairs before the closing ">" of a tag. Attributes are strings, and do not contain markup. Furthermore, an attribute name can appear only once in a given tag, unlike subelements, which may be repeated.

Note that in a document construction context, the distinction between subelement and attribute is important—an attribute is implicitly text that does not appear in the printed or displayed document.

```
<bank-1>
  <customer>
    <customer-name> Johnson </customer-name>
    <customer-street> Alma </customer-street>
    <customer-city> Palo Alto </customer-city>
    <account>
      <account-number> A-101 </account-number>
      <branch-name> Downtown </branch-name>
      <balance> 500 </balance>
    </account>
    <account>
      <account-number> A-201 </account-number>
      <branch-name> Brighton </branch-name>
      <balance> 900 </balance>
    </account>
  </customer>
  <customer>
    <customer-name> Hayes </customer-name>
    <customer-street> Main </customer-street>
    <customer-city> Harrison </customer-city>
    <account>
      <account-number> A-102 </account-number>
      <branch-name> Perryridge </branch-name>
      <balance> 400 </balance>
    </account>
  </customer>
</bank-1>
```

FIGURE 8.8   Nested XML representation of bank information.

However, in database and data exchange applications of XML, this distinction is less relevant, and the choice of representing data as an attribute or a subelement is often arbitrary.

The *document type definition* (DTD) is an optional part of an XML document. The main purpose of a DTD is much like that of a schema: to constrain and type the information present in the document. However, the DTD does not in fact constrain types in the sense of basic types like integer or string. Instead, it only constrains the appearance of subelements and attributes within an element. The DTD is primarily a list of rules for what pattern of subelements appears within an element.

The *XMLSchema* language plays the same role as DTDs, but is more powerful in terms of the types and constraints it can specify. The *XPath* and *XQuery* languages are used to query XML data. The XQuery language can be thought of as an extension of SQL to handle data with nested structure, although its syntax is different from that of SQL.

Many database systems store XML data by mapping them to relations. Unlike in the case of E-R diagram to relation mappings, the XML to relation mappings are more complex and done transparently. Users can write queries directly in terms of the XML structure, using XML query languages.

8.3.2 JSON

JavaScript Object Notation (JSON), such as XML, is a text-based format for data exchange. Whereas XML is based on a pairing of tags and values, JSON is based on a nested collection of pairings of *names* and values. A simple name–value pair takes the form

$$n: v$$

where
   n is a name
   v is a value

```
"bank-1" : {
   "customer" : {
      "customer-name" : Johnson,
      "customer-street" : Alma,
      "customer-city" : Palo Alto,
      "account" : {
         "account-number" : A-101,
         "branch-name" : Downtown,
         "balance" : 500 },
      "account" : {
         "account-number" : A-201,
         "branch-name" : Brighton,
         "balance" : 900 },
   },
   "customer" : {
      "customer-name" : Hayes,
      "customer-street" : Main,
      "customer-city" : Harrison,
      "account" : {
         "account-number" : A-102,
         "branch-name" : Perryridge,
         "balance" : 400 }
   }
}
```

FIGURE 8.9   JSON representation of bank information.

The name *n* is a quoted string (in double quotes), and the value *v* is one of the JSON basic types (number, string, Boolean), or a composite type (array elements listed in square brackets separated by commas), or object (set of name–value pairs, listed within set brackets and separated by commas). The bank information we listed in XML in Figure 8.7 appears in JSON format in Figure 8.9. It is easy to note the similarity between the JSON and the XML representations of our sample bank data. JSON's representation is somewhat more compact.

Because JSON syntax is compatible with the popular scripting language JavaScript, JSON data are easily manipulated in JavaScript. JSON Schema permits specification of a schema for JSON data. Tools are available to test whether input JSON data are compatible with a given schema.

8.4 Temporal Data

Suppose we retain data in our bank showing not only the address of each customer but also all former addresses of which the bank is aware. We may then ask queries such as "Find all customers who lived in Princeton in 1981." In this case, we may have multiple addresses for customers. Each address has an associated start and end date, indicating when the customer was resident at that address. A special value for the end date, e.g., null, or a value well into the future such as 9999-12-31, can be used to indicate that the customer is still resident at that address.

In general, *temporal data* are data that have an associated time interval during which they are *valid*. (There are other models of temporal data that distinguish between *valid time* and *transaction time*, the latter recording when a fact was recorded in the database. We ignore such details for simplicity.) We use the term *snapshot* of data to mean the value of the data at a particular point in time. Thus a snapshot of customer data gives the values of all attributes, such as address, of customers at a particular point in time.

Modeling temporal data is a challenging problem for several reasons. For example, suppose we have a *customer* entity with which we wish to associate a time-varying address. To add temporal information to an address, we would then have to create a multivalued attribute, each of whose values is a composite value containing an address and a time interval. In addition to time-varying attribute values, entities may themselves have an associated valid time. For example, an account entity may have a valid time from the date it is opened to the date it is closed. Relationships too may have associated valid times. For example, the *depositor* relationship between a customer and an account may record when the customer became an owner of the account. We would thus have to add valid time intervals to attribute values, entities, and relationships. Adding such detail to an E-R diagram makes it very difficult to create and to comprehend. There have been several proposals to extend the E-R notation to specify in a simple manner that an attribute or relationship is time-varying, but there are no accepted standards.

Consider the constraint that a department has only one manager, which could be stated as "department functionally determines manager." Such a constraint is expected to hold at a point in time, but surely cannot be expected to hold across time, since a department may have different managers at different points in time. Functional dependencies that hold at a particular point in time are called temporal functional dependencies; data in every snapshot of the database must satisfy the functional dependency.

8.5 Stream Data

The discussion of data models to this point is based on the assumption that there is a collection of data to model and store. Another view of data is that of a *stream*, in which data arrive continuously and in a sufficiently large volume that all of the data cannot be stored in the system. When new data arrive in the stream, the system uses that input to update a set of pre-defined computations. These could be simple aggregates (such as sum) or more complicated queries expressed in a SQL-like language designed for streams. These are sometimes referred to as *continuous queries*. A continuous query has a set of associated tuples at any point in time; and whenever the set of tuples changes, the system outputs either a tuple insertion or tuple deletion event. Thus, the output of a continuous query on a data stream is itself a data stream.

Because all prior data entering from the stream cannot be stored in the system, any computation done by the system is usually restricted to a limited amount of stream data. That restricted amount of data is called a *window* and stream query languages permit users to specify the maximum size of a window and how window boundaries are computed.

Stream data are becoming increasing common. Examples include telecommunication billing systems in which each billable event (call, text message, etc.) requires a price calculation and bill update, sensor networks in which a large volume of deployed sensors transmit data to a collection site, and ground stations for satellite data that need to collect and process data in real time as they arrive from the satellite.

The data in a stream could then be modeled, for example, using the relational data model, and one could view handling of data streams as a lower-level physical design issue. However, designers of streaming data systems have found it beneficial to model streaming data using a stream data model, which is an extension of the relational model, and to provide operators that convert data between the stream data model and the relational model.

The stream data model extends the relational model by associating a sequence number, typically a timestamp, with each tuple in the data stream. A window operation operates on a data stream, and creates multiple windows, with data in each window being modeled as a relation. There are also operators that take the result of a relational query using a window relation, and create an output stream whose timestamp/sequence number is defined by a timestamp/sequence number associated with the window. More recently, there have been proposals to model streaming data as having a beginning and an end timestamp; the timestamps of a particular tuple corresponding to the period in which the tuple is considered valid.

Further Readings

1. *The Relational model*

   The relational model was proposed by E. F. Codd of the IBM San Jose Research Laboratory in the late 1960s [Cod70]. Following Codd's original paper, several research projects were formed with the goal of constructing practical relational database systems, including System R at the IBM San Jose Research Laboratory (Chamberlin et al. [CAB⁺81]), Ingres at the University of California at Berkeley (Stonebraker [Sto86b]), and Query-by-Example at the IBM T. J. Watson Research Center (Zloof [Zlo77]).

   General discussion of the relational data model appears in most database texts, including Date [Dat00], Ullman [Ull88], ElMasri and Navathe [EN11], Ramakrishnan and Gehrke [RG02], and Silberschatz et al. [SKS11]. Textbook descriptions of the SQL-92 language include Date and Darwen [DD97] and Melton and Simon [MS93].

   Textbook descriptions of the network and hierarchical models, which predated the relational model, can be found on the website http://www.db-book.com (this is the website of the text by Silberschatz et al. [SKS11]).

2. *The object-based models*

   a. The entity-relationship model: The entity-relationship data model was introduced by Chen [Che76]. Basic textbook discussions are offered by ElMasri and Navathe [EN11], Ramakrishnan and Gehrke [RG02], and Silberschatz et al. [SKS11]. Various data manipulation languages for the E-R model have been proposed, though none is in widespread commercial use. The concepts of generalization, specialization, and aggregation were introduced by Smith and Smith [SS77].

   b. *Object-oriented models*: Numerous object-oriented database systems were implemented as either products or research prototypes. Some of the commercial products include ObjectStore, Ontos, Orion, and Versant. More information on these may be found in overviews of object-oriented database research, such as Kim and Lochovsky [KL89], Zdonik and Maier [ZM90], and Dogac et al. [DOBS94]. The ODMG standard is described by Cattell [Cat00]. Information about the Hibernate object-relational mapping system can be found at http://www.hibernate.org.

   c. *Object-relational models*: The nested relational model was introduced in [Mak77] and [JS82]. Design and normalization issues are discussed in [OY87], [RK87], and [MNE96]. POSTGRES ([SR86] and [Sto86a]) was an early implementation of an object-relational system. Commercial databases such as IBM DB2, Informix, and Oracle support various object-relational features of SQL:1999. Refer to the user manuals of these systems for more details.

   Melton et al. [MSG01] and Melton [Mel02] provide descriptions of SQL:1999; [Mel02] emphasizes advanced features, such as the object-relational features, of SQL:1999. Date and Darwen [DD00] describes future directions for data models and database systems.

3. *Nested structures*

   The World Wide Web Consortium (W3C) acts as the standards body for web-related standards, including basic XML and all the XML-related languages such as XPath, XSLT, and XQuery. A large number of technical reports defining the XML-related standards are available at http://www.w3c.org.

   A large number of books on XML are available in the market. These include [CSK01], [CRZ03], and [E⁺00].

   The JSON data interchange format is defined in IETF RFC 4627, available online at http://www.ietf.org/rfc/rfc4627.

4. *Stream data*

   Several systems have been built to handle data streams, including Aurora, StreamBase, Stanford STREAMS, and Oracle CEP. The Aurora data stream management system is described in [ACc⁺03].

[MWA+03] provides an overview of many issues in managing data streams, in the context of the Stanford STREAMS system. The CQL query language, described in [ABW06], was developed as part of the Stanford STREAMS project and is used in the Oracle Complex Event Processing (CEP) system. The StreamBase system (http://www.streambase.com) implements an SQL-based query language, which has some similarities to the CQL language, but differs in several respects. [JMS+08] is a step toward a standard for SQL on data streams.

5. *Knowledge representation models*

The area of knowledge representation, which has been widely studied in the AI community, aims at modeling not only data but also knowledge in the form of rules. A variety of knowledge representation models and languages have been proposed; Brachman and Levesque [BL04] provide detailed coverage of the area of knowledge representation and reasoning. We note that one of the knowledge representation models, called Resource Description Framework, or RDF for short, has been widely used in recent years for modeling semistructured data. RDF is an example of a graph-based data model, where nodes represent concepts, entities, and values, while labeled edges specify relationships between the concepts/entities/values. See http://www.w3.org/RDF for more information on RDF.

Glossary

Attribute: (1) A descriptive feature of an entity or relationship in the entity-relationship model; (2) the name of a column header in a table, or, in relational-model terminology, the name of a domain used to define a relation.

Class: A set of objects in the object-oriented model that contain the same types of values and the same methods; also, a type definition for objects.

Data model: A data model is a collection of conceptual tools for describing the real-world entities to be modeled in the database and the relationships among these entities.

Element: The contents between a start tag and its corresponding end tag in an XML document.

Entity: A distinguishable item in the real-world enterprise being modeled by a database schema.

Foreign key: A set of attributes in a relation schema whose value identifies a unique tuple in another relational schema.

Functional dependency: A rule stating that given values for some set of attributes, the value for some other set of attributes is uniquely determined. X functionally determines Y if whenever two tuples in a relation have the same value on X, they must also have the same value on Y.

Generalization: A super-class; an entity set that contains all the members one or more specialized entity sets.

Instance variable: Attribute values within objects.

Key: (1) A set of attributes in the entity relationship model that serves as a unique identifier for entities. Also known as *super-key*. (2) a set of attributes in a relation schema that functionally determines the entire schema. (3) *candidate key*—a minimal key; that is, a super-key for which no proper subset is a super-key. (4) *primary key*—a candidate key chosen as the primary means of identifying/accessing an entity set, relationship set, or relation.

Message: The means by which an object invokes a method in another object.

Method: Procedures within an object that operate on the instance variables of the object and/or send messages to other objects.

Normal form: A set of desirable properties of a schema. Examples include the Boyce–Codd normal form and the third normal form.

Object: Data and behavior (methods) representing an entity.

Persistence: The ability of information to survive (persist) despite failures of all kinds, including crashes of programs, operating systems, networks, and hardware.

Relation: (1) A subset of a Cartesian product of domains. (2) informally, a table.

Relation schema: A type definition for relations consisting of attribute names and a specification of the corresponding domains.

Relational algebra: An algebra on relations; consists of a set of operations, each of which takes as input one or more relations and returns a relation, and a set of rules for combining operations to create expressions.

Relationship: An association among several entities.

Subclass: A class that lies below some other class (a superclass) in a class inheritance hierarchy; a class that contains a subset of the objects in a superclass.

Subtable: A table such that (a) its tuples are of a type that is a subtype of the type of tuples of another table (the supertable), and (b) each tuple in the subtable has a corresponding tuple in the supertable.

Specialization: A subclass; an entity set that contains a subset of entities of another entity set.

References

[ABW06] A. Arasu, S. Babu, and J. Widom. The CQL continuous query language: Semantic foundations and query execution. *VLDB Journal*, 15(2):121–142, 2006.

[ACc⁺03] D. J. Abadi, D. Carney, U. Çetintemel, M. Cherniack, C. Convey, S. Lee, M. Stonebraker, N. Tatbul, and S. B. Zdonik. Aurora: A new model and architecture for data stream management. *VLDB Journal*, 12(2):120–139, 2003.

[AHV95] S. Abiteboul, R. Hull, and V. Vianu. *Foundations of Databases*. Addison Wesley, Reading, MA, 1995.

[BL04] R. Brachman and H. Levesque. *Knowledge Representation and Reasoning*. Morgan Kaufmann Series in Artificial Intelligence, Boston, MA, 2004.

[CAB⁺81] D. D. Chamberlin, M. M. Astrahan, M. W. Blasgen, J. N. Gray, W. F. King, B. G. Lindsay, R. A. Lorie et al. A history and evaluation of system R. *Communications of the ACM*, 24(10):632–646, October 1981.

[Cat00] R. Cattell, (ed). *The Object Database Standard: ODMG 3.0*. Morgan Kaufmann, Boston, MA, 2000.

[Che76] P. P. Chen. The entity-relationship model: Toward a unified view of data. *ACM Transactions on Database Systems*, 1(1):9–36, January 1976.

[Cod70] E. F. Codd. A relational model for large shared data banks. *Communications of the ACM*, 13(6):377–387, June 1970.

[CRZ03] A. B. Chaudhri, A. Rashid, and R. Zicari. *XML Data Management: Native XML and XML-Enabled Database Systems*. Addison Wesley, Boston, MA, 2003.

[CSK01] B. Chang, M. Scardina, and S. Kiritzov. *Oracle9i XML Handbook*. McGraw Hill, New York, 2001.

[Dat00] C. J. Date. *An Introduction to Database Systems*, 7th edn. Addison Wesley, Reading, MA, 2000.

[DD97] C. J. Date and G. Darwen. *A Guide to the SQL Standard*, 4th edn. Addison Wesley, Boston, MA, 1997.

[DD00] C. J. Date and H. Darwen. *Foundation for Future Database Systems: The Third Manifesto*, 2nd edn. Addison Wesley, Reading, MA, 2000.

[DOBS94] A. Dogac, M. T. Ozsu, A. Biliris, and T. Selis. *Advances in Object-Oriented Database Systems*, Vol. 130. Springer Verlag, New York, 1994. Computer and Systems Sciences, NATO ASI Series F.

[E⁺00] K. Williams (ed.) et al. *Professional XML Databases*. Wrox Press, Birmingham, U.K., 2000.

[EN11] R. Elmasri and S. B. Navathe. *Fundamentals of Database Systems*, 6th edn. Addison Wesley, Boston, MA, 2011.

[JMS⁺08] N. Jain, S. Mishra, A. Srinivasan, J. Gehrke, J. Widom, H. Balakrishnan, U. Çetintemel, M. Cherniack, R. Tibbetts, and S. B. Zdonik. Towards a streaming sql standard. *PVLDB*, 1(2):1379–1390, 2008.

[JS82] G. Jaeschke and H. J. Schek. Remarks on the algebra of non first normal form relations. In *Proceedings of the ACM Symposium on Principles of Database Systems*, Los Angeles, CA, pp. 124–138, 1982.

[KL89] W. Kim and F. Lochovsky (ed.). *Object-Oriented Concepts, Databases, and Applications*. Addison Wesley, Reading, MA, 1989.

[Mak77] A. Makinouchi. A consideration of normal form on not-necessarily normalized relations in the relational data model. In *Proceedings of the International Conference on Very Large Databases*, Tokyo, Japan, pp. 447–453, 1977.

[Mel02] J. Melton. *Advanced SQL: 1999—Understanding Object-Relational and Other Advanced Features*. Morgan Kaufmann, Boston, MA, 2002.

[MNE96] W. Y. Mok, Y.-K. Ng, and D. W. Embley. A normal form for precisely characterizing redundancy in nested relations. *ACM Transactions on Database Systems*, 21(1):77–106, March 1996.

[MS93] J. Melton and A. R. Simon. *Understanding The New SQL: A Complete Guide*. Morgan Kaufmann, Boston, MA, 1993.

[MSG01] J. Melton, A. R. Simon, and J. Gray. *SQL: 1999—Understanding Relational Language Components*. Morgan Kaufmann, Boston, MA, 2001.

[MWA+03] R. Motwani, J. Widom, A. Arasu, B. Babcock, S. Babu, M. Datar, G. S. Manku, C. Olston, J. Rosenstein, and R. Varma. Query processing, approximation, and resource management in a data stream management system. In *CIDR*, Pacific Grove, CA, 2003.

[OY87] G. Ozsoyoglu and L. Yuan. Reduced MVDs and minimal covers. *ACM Transactions on Database Systems*, 12(3):377–394, September 1987.

[RG02] R. Ramakrishnan and J. Gehrke. *Database Management Systems*, 3rd edition, McGraw Hill, New York, 2002.

[RK87] M. A. Roth and H. F. Korth. The design of 1nf relational databases into nested normal form. In *Proceedings of the ACM SIGMOD Conference on Management of Data*, San Francisco, CA, pp. 143–159, 1987.

[SKS11] A. Silberschatz, H. Korth, and S. Sudarshan. *Database System Concepts*, 6th edn. McGraw Hill, New York, 2011.

[SR86] M. Stonebraker and L. Rowe. The design of POSTGRES. In *Proceedings of the ACM SIGMOD Conference on Management of Data*, Washington, DC, 1986.

[SS77] J. M. Smith and D. C. P. Smith. Database abstractions: Aggregation and generalization. *ACM Transactions on Database Systems*, 2(2):105–133, March 1977.

[Sto86a] M. Stonebraker. Inclusion of new types in relational database systems. In *Proceedings of the International Conference on Data Engineering*, Los Angeles, CA, pp. 262–269, 1986.

[Sto86b] M. Stonebraker (ed). *The Ingres Papers*. Addison Wesley, Reading, MA, 1986.

[Ull88] J. D. Ullman. *Principles of Database and Knowledge-Base Systems*, Vol. 1. Computer Science Press, Rockville, ML, 1988.

[Zlo77] M. M. Zloof. Query-by-example: A data base language. *IBM Systems Journal*, 16(4):324–343, 1977.

[ZM90] S. Zdonik and D. Maier. *Readings in Object-Oriented Database Systems*. Morgan Kaufmann, San Mateo, CA, 1990.

9

Tuning Database Design for High Performance

Philippe Bonnet
IT University of Copenhagen

Dennis Shasha
New York University

9.1 Introduction

In fields ranging from arbitrage to sensor processing, speed of access to data can determine success or failure. Database tuning is the activity of making a database system run faster. Like optimization activities in other areas of computer science and engineering, database tuning must work within certain constraints. Just as compiler optimizers, for example, cannot directly change the underlying hardware but can change register allocations, database tuners cannot change the underlying database management system software, but can make use of the options it offers.

The database administrator in charge of tuning can, for example, modify the design of tables, select new indexes, rearrange transactions, tamper with the operating system, or buy hardware. The goals are to increase throughput and reduce response time. In our personal experience, tuning efforts can have dramatic effects, e.g., reduce a query time from nine hours to 15 seconds.

Further, interactions between database components and the nature of the bottlenecks change with technology. For example, hard drives have been used as secondary storage since the advent of relational database systems; nowadays, only solid state drives (SSD) offer the high-performance throughput needed to match the speed of processing units in a balanced system. In SSDs, sequential IOs are not so much faster than random IOs (which is the case for disks).

Tuning, then, is for well-informed generalists. This chapter introduces a principled foundation for tuning, focusing on principles that have been robust for years and promise to still hold true for years to come. We include experiments from specific commercial and free database systems, which will become obsolete much faster.

9.2 Underlying Principles

To understand the principles of tuning, you must understand the two main kinds of database applications and what affects performance.

9.2.1 What Databases Do

At a high level of abstraction, databases are used for two purposes: on-line transaction processing and decision support. **On-line transaction processing** typically involves access to a small number of records, generally to modify them. A typical such transaction records a sale or updates a bank account. These transactions use indexes to access their few records without scanning through an entire table. **E-commerce** applications are modern examples of OLTP applications. It seems that potential e-customers will abandon a site if they have to wait more than for a google search.

Decision support queries, by contrast, read many records often from a **data warehouse**, compute an aggregate result, and sometimes apply that aggregate back to an individual level. Typical decision support queries are "find the total sales of widgets in the last quarter in the northeast" or "calculate the available inventory per unit item." Sometimes the results are actionable as in "find frequent flyer passengers who have encountered substantial delays in their last few flights and send them free tickets and an apology."

Data mining is, in practice, best done outside the database management system, though it may draw samples from the database. In so doing, it issues decision support queries.

9.2.2 Performance Spoilers

Having divided the database applications into two broad areas, we can now discuss what slows them down.

1. *Random vs. sequential disk accesses*: On hard disk drives, sequential disk bandwidth is between one and two orders of magnitude (10–100 times) greater than random-access disk bandwidth. On SSDs, there is no intrinsic difference between the throughput of sequential and random reads because read performance depends only on the degree of parallelism achieved within the SSD. Depending on the SSD model, the relative performance of sequential and random reads will vary, but not much. Thus on hard drives, many sequential IOs might be much faster than few random IOs; while on SSDs fewer IOs is always better. Concretely, index accesses tend to be random whereas scans are sequential. Thus, on hard disks, removing an index may sometimes improve performance, because the index used performs random reads and behaves poorly. While on SSDs, indexed accesses are better than scans provided they result in fewer IOs.

2. *Lack of parallelism*: Modern computer systems support parallelism in processing (e.g., multicore and/or graphics processors), storage (controllers, disks, and SSDs). Poor layout and operating system choices may inhibit parallelism.

3. *Imprecise data searches*: These occur typically when a selection retrieves a small number of rows from a large table, yet must search the entire table to find those data. Establishing an index may help in this case, though other actions, including reorganizing the table, may also have an effect.

4. *Many short data interactions, either over a network or to the database*: This may occur, for example, if an object-oriented application views records as objects and assembles a collection of objects by accessing a database repeatedly from within a "for" loop rather than as a bulk retrieval.

5. Delays due to lock conflicts.

These occur either when update transactions execute too long or when several transactions want to access the same datum, but are delayed because of locks. A typical example might be a single variable that must be updated whenever a record is inserted. In the following example, the COUNTER table contains the next value which is used as a key when inserting values in the ACCOUNT table.

```
begin transaction
    NextKey: = select nextkey from COUNTER;
    insert into ACCOUNT values (nextkey, 100, 200);
    update COUNTER set nextkey = NextKey + 1;
end transaction
```

When the number of such transactions issued concurrently increases, COUNTER becomes a bottleneck because all transactions read and write the value of next key. This problem is only amplified by the need to achieve a high degree of parallelism.

As mentioned in the introduction, avoiding performance problems requires changes at all levels of a database system. We will discuss tactics used at several of these levels and their interactions—hardware, concurrency control subsystem, indexes, and conceptual level. There are other levels such as recovery and query rewriting that we mostly defer to Ref. [4].

9.3 Best Practices

Understanding how to tune each level of a database system requires understanding the factors leading to good performance at that level. Each of the following subsections discusses these factors before discussing tuning tactics.

9.3.1 Tuning Hardware

Each processing unit consists of one or more processors, one or more disks, and some memory. Assuming a high-end 180-GIPS (billion instructions per second) processor, the CPU will become the bottleneck for on-line transaction-processing applications when it is attached to 1 high-end SSDs, 10 mid-range SSDs, or 7,200 hard disks (counting 500,000 instructions per random IO issued by the database system and 400,000 random IO per seconds on a high-end SSD, 40,000 random IO per second on a mid range SSD, and 50 random IO per second on a high-end hard disk).

Decision-support queries, by contrast, often entail massive scans of a table. In theory, a 180-GIPS processor is saturated when connected to 1 high-end SSD, 5 mid-range SSDs, or 36 hard disks (counting 500,000 instruction per sequential IO and 1 million sequential IO per second for high-end SSD, 100,000 sequential IO per second for mid-range SSDs, and 10,000 IO per second for hard disks). A few years ago, the system bus was the bottleneck when reading sequentially from several disks. Nowadays, a commonplace PCIe system bus of 5 GT/s (Giga Transfers per second) is not saturated by the aggregated bandwidth of 10 SSDs. The problem is to find enough connectors on the system bus.

Note that a 180 GIPS processor relies on tens of hardware threads to deliver such throughput. Each core delivers typically 20–30 GIPS. Also the figures mentioned earlier assume that the IO submission rate is high enough to leverage the internal parallelism of the SSDs. Achieving a high IO submission rate is the key objective when tuning a database system on SSD.

Summing up, decision-support sites may need fewer disks per processor than transaction-processing sites for the purposes of matching aggregate disk bandwidth to processor speed.* Note that storing 1 TB of data on hard disk is an order of magnitude cheaper than storing 1 TB of data on a SSD.

---

* This point requires a bit more explanation. There are two reasons you might need more disks: (1) for disk bandwidth (the number of bytes coming from the disk per second); or (2) for space. Disk bandwidth is usually the issue in on-line transaction processing. Decision support applications tend to run into the space issue more frequently, because scanning allows disks to deliver their optimal bandwidth.

All applications will benefit from a mix of solid state drives for high-performance access and hard disks for inexpensive storage.

Random access memory (RAM) obviates the need to go to disk. Database systems reserve a portion of RAM as a *buffer*. In all applications, the buffer usually holds frequently accessed pages (*hot* pages, in database parlance) including the first few levels of indexes. Increasing the amount of RAM buffer tends to be particularly helpful in on-line transaction applications where disks are the bottleneck, particularly for smaller tables.

The read **hit ratio** in a database is the portion of database reads that are satisfied by the buffer. Hit ratios of 90% or higher are common in on-line transaction applications, but less common in decision support applications. Even in transaction processing applications, hit ratios tend to level off as you increase the buffer size if there is one or more tables that are accessed unpredictably and are much larger than available RAM (e.g., sales records for a large department store).

The decreasing cost of RAM makes it economical to attach 256 GB of RAM to a CPU. As a consequence, mid-sized databases can be manipulated entirely in RAM centrally on a single server, or in a distributed manner across a cluster. From a tuning point of view, the main goal is then to minimize the number of buffer reads on each server and the amount of communication across servers.

9.3.2 Tuning the Virtual Machine

It is today commonplace to rely on virtual machines to manage the hardware resources in a company. As a result, database systems often run within a virtual machine. Tuning the virtual machine to match the needs of the database system has become an important task. While configuring the number of CPUs, the size of the RAM or the number of disks associated to a virtual machine is rather straightforward. The configuration of the disk controllers and of the disk characteristics is a bit more subtle:

- *No host caching*: The database system expects that the IOs it submits are transferred to disk as directly as possible. It is thus critical to disable file system caching on the host for the IOs submitted by the virtual machine.
- *Hardware supported IO virtualization*: A range of modern CPUs incorporate components that speed up access to IO devices through direct memory access and interrupt remapping (i.e., Intel's VT-d and AMD's AMD-vi). Such a feature is important to allow databases executing within a virtual machine to leverage the performance characteristics of high-end SSDs.
- *Statically allocated disk space*: When creating a virtual disk, it is possible to reserve the space physically on disk (static allocation). Alternatively, the virtual disk grows dynamically as required. Static allocation is preferable for database systems. First, the overhead of space reservation is paid only once at disk creation time, not while executing transactions. Second, static allocation guarantees that a disk is mapped on contiguous physical space. This is especially important when using hard disks, because contiguous allocation guarantees that the sequential IOs submitted by a database system on a virtual disk are actually mapped onto sequential IOs on disk.

9.3.3 Tuning the Operating System

The operating system provides the storage and processing abstractions to the database system. More specifically, file system and thread management impact database performance.

9.3.3.1 Thread Management

Processor affinity refers to binding of a given software thread to a specific hardware thread. It allows efficient cache reuse, and avoids context switches. For transactional workloads, the log write can be assigned to a specific hardware thread to guarantee that it can continuously write to the log.

Modern database systems allow a database administrator to dynamically change the priority of the different threads running in the system. Also, the database system implements mechanisms that avoid the priority inversion problem, where a high-level transaction waits for a low-level transaction to release a resource it is waiting for.

9.3.4 Tuning Concurrency Control

As the chapter on Concurrency Control and Recovery in this handbook explains, database systems attempt to give users the illusion that each transaction executes in isolation from all others. The ANSI SQL standard, for example, makes this explicit with its concept of degrees of isolation [3,5]. Full isolation or **serializability** is the guarantee that each transaction that completes will appear to execute one at a time *except that its performance may be affected by other transactions*. This ensures, for example, that in an accounting database in which every update (sale, purchase, etc.) is recorded as a double-entry transaction, any transaction that sums assets, liabilities, and owners' equity will find that assets equal the sum of the other two. There are less stringent notions of isolation that are appropriate when users do not require such a high degree of consistency.

The concurrency-control algorithm in predominant use is two-phase locking, sometimes with optimizations for data structures. Two-phase locking has **read** (or *shared*) and **write** (or *exclusive*) locks. Two transactions may both hold a shared lock on a datum. If one transaction holds an exclusive lock on a datum, however, then no other transaction may hold any lock on that datum; in this case, the two transactions are said to **conflict**. The notion of datum (the basic unit of locking) is deliberately left unspecified in the field of concurrency control, because the same algorithmic principles apply regardless of the size of the datum, whether a page, a record, or a table. The performance may differ, however. For example, record-level locking works much better than table-level locking for on-line transaction processing applications.

Snapshot isolation avoids read locks as it gives each transaction the illusion that it accesses, throughout its execution, the state of the database that was valid when it started. With snapshot isolation read transactions do not conflict with write transactions [9]. However, snapshot isolation does not ensure serializability. For example, consider a transaction T1 that reads value x and writes that value into y. Suppose that T2 reads y and writes that value into x. If x is initially 3 and y is initially 17, then any serial execution will guarantee that x and y are the same at the end. Under snapshot isolation T1 may read x at the same time that T2 reads y, in which case x will have the value 17 at the end and y will have the value 3.

9.3.4.1 Rearranging Transactions

Tuning concurrency control entails trying to reduce the number and duration of conflicts. This often entails understanding application semantics. Consider, for example, the following code for a purchase application of item i for price p for a company in bankruptcy (for which the cash cannot go below 0):

PURCHASE TRANSACTION (p,i)

```
1  BEGIN TRANSACTION
2  if cash < p then roll back transaction
3  inventory(i): = inventory(i) + p
4  cash: = cash − p
5  END TRANSACTION
```

From a concurrency-control-theoretical point of view, this code does the right thing. For example, if the cash remaining is 100, and purchase P1 is for item i with price 50, and purchase P2 is for item j with price 75, then one of these will roll back.

From the point of view of performance, however, this transaction design is very poor, because every transaction must acquire an exclusive lock on cash from the beginning to avoid deadlock.

(Otherwise, many transactions will obtain shared locks on cash and none will be able to obtain an exclusive lock on cash.) That will make cash a bottleneck and have the effect of serializing the purchases. Since inventory is apt to be large, accessing inventory(i) will take at least one disk access, taking about 5 ms. Since the transactions will serialize on cash, only one transaction will access inventory at a time. This will limit the number of purchase transactions to about 50 per second. Even a company in bankruptcy may find this rate to be unacceptable.

A surprisingly simple rearrangement helps matters greatly:

REDESIGNED PURCHASE TRANSACTION (p,i)

1 BEGIN TRANSACTION
2 inventory(i): = inventory(i) + p
3 if cash < p then roll back transaction
4 else cash: = cash − p
5 END TRANSACTION

Cash is still a hot spot, but now each transaction will avoid holding cash while accessing inventory. Since cash is so hot, it will be in the RAM buffer. The lock on cash can be released as soon as the commit occurs.

Advanced techniques are available that "chop" transactions into independent pieces to shorten lock times further. We refer interested readers to Ref. [4].

9.3.4.2 Living Dangerously

Many applications live with less than full isolation due to the high cost of holding locks during user interactions. Consider the following full-isolation transaction from an airline reservation application:

AIRLINE RESERVATION TRANSACTION (p,i)

1 BEGIN TRANSACTION
2 Retrieve list of seats available.
3 Reservation agent talks with customer regarding availability.
4 Secure seat.
5 END TRANSACTION

The performance of a system built from such transactions would be intolerably slow, because each customer would hold a lock on all available seats for a flight while chatting with the reservations agent. This solution does, however, guarantee two conditions: (1) no two customers will be given the same seat, and (2) any seat that the reservation agent identifies as available in view of the retrieval of seats will still be available when the customer asks to secure it.

Because of the poor performance, however, the following is done instead:

LOOSELY CONSISTENT AIRLINE RESERVATION TRANSACTION (p,i)

1 Retrieve list of seats available.
2 Reservation agent talks with customer regarding availability.
3 BEGIN TRANSACTION
4 Secure seat.
5 END TRANSACTION

This design relegates lock conflicts to the secure step, thus guaranteeing that no two customers will be given the same seat. It does allow the possibility, however, that a customer will be told that a seat is available, will ask to secure it, and will then find out that it is gone. This has actually happened to a particularly garrulous colleague of ours.

9.3.5 Indexes

Access methods, also known as **indexes**, are discussed in another chapter. Here we review the basics, then discuss tuning considerations. An **index** is a data structure plus a method of arranging the data tuples in the table (or other kind of collection object) being indexed. Let us discuss the data structure first.

9.3.5.1 Data Structures

Two data structures are most often used in practice: B-trees and Hash structures. Of these, B-trees are used the most often (one vendor's tuning book puts it this way: "When in doubt, use a B-tree"). Here, we review those concepts about B-trees most relevant to tuning.

A **B-tree** (strictly speaking a B + tree) is a balanced tree whose nodes contain a sequence of key–pointer pairs [2]. The keys are sorted by value. The pointers at the leaves point to the tuples in the indexed table).

B-trees are self-reorganizing through operations known as splits and merges (though occasional reorganizations for the purpose of reducing the number of seeks do take place). Further, they support many different query types well: equality queries (find the employee record of the person having a specific social security number), min–max queries (find the highest-paid employee in the company), and range queries (find all salaries between $70,000 and $80,000).

Because an access to disk secondary memory costs about 5 ms if it requires a seek (as index accesses will), the performance of a B-tree depends critically on the number of nodes in the average path from root to leaf. (The root will tend to be in RAM, but the other levels may or not be, and the farther down the tree the search goes, the less likely they are to be in RAM.) The number of nodes in the path is known as the number of levels. One technique that database management systems use to minimize the number of levels is to make each interior node have as many children as possible (1000 or more for many B-tree implementations). The maximum number of children a node can have is called its *fanout*. Because a B-tree node consists of key–pointer pairs, the bigger the key is, the lower the fanout.

For example, a B-tree with a million records and a fanout of 1000 requires three levels (including the level where the records are kept). A B-tree with a million records and a fanout of 10 requires 7 levels. If we increase the number of records to a billion, the numbers of levels increase to 4 and 10, respectively. This is why accessing data through indexes on large keys is slower than accessing data through small keys on most systems.

Hash structures, by contrast, are a method of storing key–value pairs based on a pseudorandomizing function called a *hash function*. The hash function can be thought of as the root of the structure. Given a key, the hash function returns a location that contains either a page address (usually on disk) or a directory location that holds a set of page addresses. That page either contains the key and associated record or is the first page of a linked list of pages, known as an *overflow chain* leading to the record(s) containing the key. (You can keep overflow chaining to a minimum by using only half the available space in a hash setting.)

In the absence of overflow chains, hash structures can answer equality queries (e.g., find the employee with Social Security number 156-87-9864) in one disk access, making them the best data structures for that purpose. The hash function will return arbitrarily different locations on key values that are close but unequal, e.g., Smith and Smythe. As a result, records containing such close keys will likely be on different pages. This explains why hash structures are completely unhelpful for range and min–max queries.

9.3.5.2 Clustering and Sparse Indexes

The data structure portion of an index has pointers at its leaves to either data pages or data records.

- If there is at most one pointer from the data structure to each data page, then the index is said to be **sparse**.
- If there is one pointer to each record in the table, then the index is said to be **dense**.

If records are small compared to pages, then there will be many records per data page and the data structure supporting a sparse index will usually have one less level than the data structure supporting a dense index. This means one less disk access if the table is large. By contrast, if records are almost as large as pages, then a sparse index will rarely have better disk access properties than a dense index.

The main virtue of dense indexes is that they can support certain read queries within the data structure itself in which case they are said to **cover** the query. For example, if there is a dense index on the keywords of a document retrieval system, a query can count the records containing some term, e.g., "derivatives scandals," without accessing the records themselves. (Count information is useful for that application, because queriers frequently reformulate a query when they discover that it would retrieve too many documents.) A secondary virtue is that a query that makes use of several dense indexes can identify all relevant tuples before accessing the data records; instead, one can just form intersections and unions of pointers to data records.

A **clustering index** on an attribute (or set of attributes) X is an index that puts records close to one another if their X-values are *near* one another. What "near" means depends on the data structure. On B-trees, two X-values are near if they are close in their sort order. For example, 50 and 51 are near, as are Smith and Sneed. In hash structures, two X-values are near only if they are identical.

Some systems such as Oracle and InnoDB use an implicit form of clustering called an **index organized table**. This is a table that is clustered on its primary key (or a system-generated row id, if there is no primary key).

Sparse indexes must be clustering, but clustering indexes need not be sparse. In fact, clustering indexes are sparse in some systems (e.g., SQL Server, ORACLE hash structures) and dense in others (e.g., ORACLE B-trees, DB2). Because a clustering index implies a certain table organization and the table can be organized in only one way at a time, there can be at most one clustering index per table.

A *nonclustering index* (sometimes called a *secondary* index) is an index on an attribute (or set of attributes) Y that puts no constraint on the table organization. The table can be clustered according to some other attribute X or can be organized as a heap, as we discuss later. A nonclustering index must be dense, so there is one leaf pointer per record. There can be many nonclustering indexes per table.

A **heap** is the simplest table organization of all. Records are ordered according to their time of entry. That is, new insertions are added to the last page of the data structure. For this reason, inserting a record requires a single page access. Reading a record requires a scan.

A table and the indexes associated with it might be partitioned. Each partition is associated to a file and each file is associated to a disk. This way a table might be accessed in parallel. Indexes are defined on each partition. The dispatching of tuples into partitions is either based on round-robin, on ranges defined on a partitioning attribute, or on a hash function applied on a partitioning attribute.

Nonclustering indexes are useful if each query retrieves significantly fewer records than there are pages in the file. We use the word "significant" for the following reason: a table scan can often save time by reading many pages at a time, provided the table is stored on contiguous tracks. Therefore, on hard disk, if the scan and the index both read all the pages of the table, the scan may complete 100 times faster than if it read one page at a time. On SSD, this ratio is much smaller.

In summary, nonclustering indexes work best if they cover the query. Otherwise, they work well, specially on SSDs, if the average query using the index will access fewer records than there are data pages. Large records and high selectivity both contribute to the usefulness of nonclustering indexes. On hard disks, scans are hard to beat.

9.3.5.3 Data Structures for Decision Support

Decision support applications often entail querying on several, perhaps individually unselective, attributes. For example, "Find people in a certain income range who are female, live in California, buy climbing equipment, and work in the computer industry." Each of these constraints is unselective in itself, but together form a small result. The best all-around data structure for such a situation is the bitmap.

A bitmap is a collection of vectors of bits. The length of each such "bit vector" equals the length of the table being indexed and has a 1 in position i if the ith record of the table has some property. For example a bitmap on state would consist of 50 bit vectors, one for each state. The vector for California would have a 1 for record i if record i pertains to a person from California. In our experiments, bitmaps outperform multidimensional indexes by a substantial margin.

Some decision support queries compute an aggregate, but never apply the result of the aggregate back to individuals. For example, you might want to find the approximate number of Californian women having the properties mentioned earlier. In that case, you can use approximate summary tables as a kind of indexing technique. The Aqua system [1], for example, proposes an approximation based on constructing a database from a random sample of the most detailed table T (sometimes known as the fact table in data warehouse parlance) and then joining that result with the reference tables R1, R2, ..., Rn based on foreign key joins.

9.3.5.4 Final Remarks Concerning Indexes

The main point to remember is that the use of indexes is a two-edged sword: we have seen an index reduce the time to execute a query from hours to a few seconds in one application, yet increase batch load time by a factor of 80 in another application. Add them with care.

9.3.6 Tuning Table Design

Table design is the activity of deciding which attributes should appear in which tables in a relational system. The Conceptual Database Design chapter discusses this issue, emphasizing the desirability of arriving at a **normalized** schema. Performance considerations sometimes suggest choosing a nonnormalized schema, however. More commonly, performance considerations may suggest choosing one normalized schema over another or they may even suggest the use of redundant tables.

9.3.6.1 To Normalize or Not to Normalize

Consider the normalized schema consisting of two tables: *Sale*(sale_id, customer_id, product, quantity) and *Customer*(customer_id, customer_location).

If we frequently want sales per customer location or sales per product per customer location, then this table design requires a join on customer_id for each of these queries. A denormalized alternative is to add customer_location to *Sale*, yielding *Sale*(sale_id, customer_id, product, quantity, customer_location) and *Customer*(customer_id, customer_location). In this alternative, we still would need the *Customer* table to avoid anomalies such as the inability to store the location of a customer who has not yet bought anything.

Comparing these two schemas, we see that the denormalized schema requires more space and more work on insertion of a sale. (Typically, the data-entry operator would type in the customer_id, product, and quantity; the system would generate a sale_id and do a join on customer_id to get customer_location.) On the other hand, the denormalized schema is much better for finding the products sold at a particular customer location.

The tradeoff of space plus insertion cost vs. improved speeds for certain queries is the characteristic one in deciding when to use a denormalized schema. Good practice suggests starting with a normalized schema and then denormalizing sparingly.

9.3.6.2 Redundant Tables

The previous example illustrates a special situation that we can sometimes exploit by implementing wholly redundant tables. Such tables store the aggregates we want. For example:

Sale(sale_id, customer_id, product, quantity) *Customer*(customer_id, customer_location) *Customer_Agg* (customer_id, totalquantity) *Loc_Agg* (customer_location, totalquantity). This reduces the query time, but imposes an update time as well as a small space overhead. The tradeoff is worthwhile in situations where many aggregate queries are issued (perhaps in a data warehouse situation) and an exact answer is required.

9.3.6.3 Tuning Normalized Schemas

Even restricting our attention to normalized schemas without redundant tables, we find tuning opportunities because many normalized schemas are possible. Consider a bank whose *Account* relation has the normalized schema (account_id is the key):

- *Account*(account_id, balance, name, street, postal_code)

Consider the possibility of replacing this by the following pair of normalized tables:

- *AccountBal*(account_id, balance)
- *AccountLoc*(account_id, name, street, postal_code)

The second schema results from **vertical partitioning** of the first (all nonkey attributes are partitioned). The second schema has the following benefits for simple account update transactions that access only the id and the balance:

- A sparse clustering index on account_id of *AccountBal* may be a level shorter than it would be for the *Account* relation, because the name, street, and postal_code fields are long relative to account_id and balance. The reason is that the leaves of the data structure in a sparse index point to data pages. If *AccountBal* has far fewer pages than the original table, then there will be far fewer leaves in the data structure.
- More account_id–balance pairs will fit in memory, thus increasing the hit ratio. Again, the gain is large if *AccountBal* tuples are much smaller than *Account* tuples.

On the other hand, consider the further decomposition:

- *AccountBal*(account_id, balance)
- *AccountStreet*(account_id, name, street)
- *AccountPost*(account_id, postal_code)

Though still normalized, this schema probably would not work well for this application, since queries (e.g., monthly statements, account update) require both street and postal_code or neither. Vertical partitioning, then, is a technique to be used for users who have intimate knowledge of the application.

In recent years, database systems have been designed as **column-stores**, where vertical partitioning is the rule and efficient mechanisms are provided for improving hit ratio both in the database cache and in CPU cache lines, and for minimizing IOs based on efficient column compression. Column stores are well suited for decision support applications in which tables may have hundreds of columns, but each query uses only a few. The benefit then is that only the few relevant columns need to be processed, saving both input/output time and processing time. Further, storing columns separately gives greater opportunities to achieve high compression.

9.4 Tuning the Application Interface

A central tuning principle asserts *start-up costs are high; running costs are low*. When applied to the application interface, this suggests that you want to transfer as much necessary data as possible between an application language and the database per connection. Here are a few illustrations of this point.

9.4.1 Assemble Object Collections in Bulk

Object-oriented encapsulation allows the implementation of one class to be modified without affecting the rest of the application, thus contributing greatly to code maintenance. Encapsulation sometimes is interpreted as "the specification is all that counts." Unfortunately, that interpretation can lead to horrible performance.

The problem begins with the fact that the most natural object-oriented design on top of a relational database is to make records (or sometimes fields) into objects. Fetching one of these objects then translates to a fetch of a record or a field. So far, so good.

But then the temptation is to build bulk fetches from fetches on little objects (the "encapsulation imperative"). The net result is to execute many small queries, each of which goes across the programming language to database system boundary, instead of one large query.

Consider, for example, a system that delivers and stores documents. Each document type (e.g., a report on a customer account) is produced according to a certain schedule that may differ from one document type to another. Authorization information relates document types to users. This gives a pair of tables of the form:

    authorized(user, documenttype)
    documentinstance(id, documenttype, documentdate)

When a user logs in, the system should say which document instances he or she can see. This can easily be done with the join:

```
select documentinstance.id, documentinstance.documentdate
from documentinstance, authorized
where documentinstance.documenttype = authorized.documenttype
and authorized.user = <input user name>
```

But if each document type is an object and each document instance is another object, then one may be tempted to write the following code:

```
Authorized authdocs = new Authorized();
authdocs.init(<input user name>);
for (Enumeration e = authdocs.elements(); e.hasMoreElements();)
{
    DocInstance doc = new DocInstance();
    doc.init(e.nextElement());
    doc.print();
}
```

This application program will first issue one query to find all the document types for the user (within the init method of Authorized class):

```
select documentinstance.documenttype
from authorized
where authorized.user = <input user name>
```

and then for each such type t to issue the query (within the init method of DocInstance class):

```
select documentinstance.id, documentinstance.documentdate
from documentinstance
where documentinstance.documenttype = t
```

This is much slower than the previous SQL formulation. The join is performed in the application and not in the database server.

The point is not that object-orientation is bad. Encapsulation contributes to maintainability. The point is that programmers should keep their minds open to the possibility that accessing a bulk object (e.g., a collection of documents) should be done directly rather than by forming the member objects individually and then grouping them into a bulk object on the application side.

9.4.2 Art of Insertion

We have spoken so far about retrieving data. Inserting data rapidly requires understanding the sources of overhead of putting a record into the database:

1. As in the retrieval case, the first source of overhead is an excessive number of round trips across the database interface. This occurs if the batch size of your inserts is too small. In fact up to 100,000 rows, increases in the batch size improve performance on most systems.
2. The second reason has to do with the ancillary overhead that an insert causes: updating all the indexes on the table. Even a single index can hurt performance.
3. Finally, the layers of software within a database system can get in the way. Database systems provide bulk loading tools that achieve high performance by bypassing some of the database layers (mostly having to do with transactional recovery) that would be traversed if single row INSERT statements were used. For instance, *SQL* Loader* is a tool that bulk loads data into Oracle databases. It can be configured to bypass the query engine of the database server (using the direct path option).

The SQL Server BULK INSERT command and SQL*Loader allow the user to define the number of rows per batch or the number of kilobytes per batch. The minimum of the two is used to determine how many rows are loaded in each batch. There is a tradeoff between the performance gained by minimizing the transaction overhead in the omitted layers and the work that has to be redone in case a failure occurs.

9.5 Monitoring Tools

When your system is slow, you must figure out where the problem lies. Is it a single query? Is some specific resource misconfigured? Is there insufficient hardware? Most systems offer the following basic monitoring tools [6,7,8]:

1. Time-spent monitors capture for a transaction, or a session the time spent in the different database components. This information allows to identify for a transaction, which component can be tuned to significantly improve performance.
2. Event monitors (sometimes known as Trace Data Viewer or Server Profiler) capture usage measurements (processor usage ratio, disk usage, locks obtained, etc.) throughout the system at the end of each query. It is sometimes hard to differentiate the contribution of the different transactions to the indicators. Also, it is hard to evaluate how a given indicator impacts the performance of a specific transaction.
3. If you have found an expensive query, you might look to see how it is being executed by looking at the query plan. These Plan Explainer tools tell you which indexes are used, when sorts are done and which join ordering is chosen.
4. If you suspect that some specific resource is overloaded, you can check the consumption of these resources directly using operating system commands. This includes the time evolution of processor usage, disk queuing, and memory consumption. Dynamic tracing frameworks are allowed to collect detailed information even in the context of production systems.

9.6 Tuning Rules of Thumb

Often, tuning consists in applying the techniques cited earlier, such as the selection and placement of indexes or the splitting up of transactions to reduce locking conflicts. At other times, tuning consists in recognizing fundamental inefficiencies and attacking them.

1. Simple problems are often the worst. We have seen a situation where the database was very slow because the computer supporting the database was also the mail router. Offloading nondatabase applications is often necessary to speed up database applications.
2. Another simple problem having a simple solution concerns locating and rethinking specific queries. The authors have had the experience of reducing query times by a factor of 10 by the judicious use of outer joins to avoid superlinear query performance.
3. The use of triggers can often result in surprisingly poor performance. Since procedural languages for triggers resemble standard programming languages, bad habits sometimes emerge. Consider, for example, a trigger that loops over all records inserted by an update statement. If the loop has an expensive multitable join operation, it is important to pull that join out of the loop if possible. We have seen another 10-fold speedup for a critical update operation following such a change.
4. There are many ways to partition load to avoid performance bottlenecks in a large enterprise. One approach is to distribute the data across sites connected by wide-area networks. This can result, however, in performance and administrative overheads unless networks are extremely reliable. Another approach is to distribute queries over time. For example, banks typically send out 1/20 of their monthly statements every working day rather than send out all of them at the end of the month.

9.7 Summary and Research Results

Database tuning is based on a few principles and a body of knowledge. Some of that knowledge depends on the specifics of systems (e.g., which index types each system offers), but most of it is independent of version number, vendor, and even data model (e.g., hierarchical, relational, or object-oriented). This chapter has attempted to provide a taste of the principles that govern effective database tuning.

Various research and commercial efforts have attempted to automate the database tuning process [10]. Among the most successful is the tuning wizard offered by Microsoft's SQL server. Given information about table sizes and access patterns, the tuning wizard can give advice about index selection among other features. Tuners would do well to exploit such tools as much as possible. Human expertise then comes into play when deep application knowledge is necessary (e.g., in rewriting queries and in overall hardware design) or when these tools do not work as advertised (the problems are all NP-hard).

Diagnosing performance problems and finding solutions may not require a good bedside manner, but good tuning can transform a slow and sick database into one full of pep.

9.8 Information

Whereas the remarks of this chapter apply to most database systems, each vendor will give you valuable specific information in the form of tuning guides or administrator's manuals. The guides vary in quality, but they are particularly useful for telling you how to monitor such aspects of your system as the relationship between buffer space and hit ratio, the number of deadlocks, the input/output load, and so on.

Our book *Database Tuning: Principles, Experiments, and Troubleshooting Techniques*, published by Morgan Kaufmann goes into greater depth regarding all the topics in this chapter. Our website http://www.databasetuning.org contains a repository of experiments and results obtained on current database systems.

Glossary

B-tree: The most used data structure in database systems. A B-tree is a balanced tree structure that permits fast access for a wide variety of queries. In virtually all database systems, the actual structure is a B+ tree in which all key–pointer pairs are at the leaves.

Clustering index: A data structure plus an implied table organization. For example, if there is a clustering index based on a B-tree on last name, then all records with the last names that are alphabetically close will be packed onto as few pages as possible.

Column-oriented store: In such a storage layout, tables are laid out column-wise, so all the employee ids are contiguous, all the salaries are contiguous, etc.

Conflict (between locks): An incompatibility relationship between two lock types. Read locks are compatible (nonconflicting) with read locks, meaning different transactions may have read locks on the same data item s. A write lock, however, conflicts with all kinds of locks.

Covering index: An index whose fields are sufficient to answer a query.

Data mining: The activity of finding actionable patterns in data.

Decision support: Queries that help planners decide what to do next, e.g., which products to push, which factories require overtime, and so on.

Denormalization: The activity of changing a schema to make certain relations denormalized for the purpose of improving performance (usually by reducing the number of joins). Should not be used for relations that change often or in cases where disk space is scarce.

Dense index: An index in which the underlying data structure has a pointer to each record among the data pages. Clustering indexes can be dense in some systems (e.g., ORACLE). Nonclustering indexes are always dense.

E-commerce applications: Applications entailing access to a website and a back end database system.

Hash structure: A tree structure whose root is a function, called the hash function. Given a key, the hash function returns a page that contains pointers to records holding that key or is the root of an overflow chain. Should be used when selective equality queries and updates are the dominant access patterns.

Heap: In the absence of a clustering index, the tuples of a table will be laid out in their order of insertion. Such a layout is called a heap. (Some systems, such as RDB, reuse the space in the interior of heaps, but most do not.)

Hit ratio: The number of logical accesses satisfied by the database buffer divided by the total number of logical accesses.

Index: A data organization to speed the execution of queries on tables or object-oriented collections. It consists of a data structure, e.g., a B-tree or hash structure, and a table organization.

Index organized table: A table clustered based on its primary key or on a row id if no primary key is defined.

Locking: The activity of obtaining and releasing read locks and write locks (see corresponding entries) for the purposes of concurrent synchronization (concurrency control) among transactions.

Nonclustering index: A dense index that puts no constraints on the table organization, also known as a secondary index. For contrast, see **clustering index**.

Normalized: A relation R is normalized if every functional dependency "X functionally determines A," where A and the attributes in X are contained in R (but A does not belong to X), has the property that X is the key or a superset of the key of R. X functionally determines A if any two tuples with the same X values have the same A value. X is a key if no two records have the same values on all attributes of X.

On-line transaction processing: The class of applications where the transactions are short, typically 10 disk I/Os or fewer per transaction, the queries are simple, typically point and multipoint queries, and the frequency of updates is high.

Read lock: If a transaction T holds a read lock on a data item x, then no other transaction can obtain a write lock on x.

Seek: Moving the read/write head of a disk to the proper track.

Serializability: The assurance that each transaction in a database system will appear to execute in isolation of all others. Equivalently, the assurance that a concurrent execution of committed transactions will appear to execute in serial order as far as their input/output behaviors are concerned.

Solid State Drive (SSD): A storage device that is composed of tens of flash chips connected to a micro-controller. The current generation of SSD exposes a block device interface similar to the magnetic hard drive interface. The micro-controller runs a firmware called flash translation layer (or FTL) that maps operations on logical block addresses onto operations on physical blocks on flash. SSDs can be directly installed on a server's memory bus (PCI bus), or they can be accessed as IO devices (via a SATA interface).

Sparse index: An index in which the underlying data structure contains exactly one pointer to each data page. Only clustering indexes can be sparse.

Track: A narrow ring on a single platter of a disk. If the disk head over a platter does not move, then a track will pass under that head in one rotation. The implication is that reading or writing a track does not take much more time than reading or writing a portion of a track.

Transaction: A program fragment delimited by Commit statements having database accesses that are supposed to appear as if they execute alone on the database. A typical transaction may process a purchase by increasing inventory and decreasing cash.

Two-phase locking: An algorithm for concurrency control whereby a transaction acquires a write lock on x before writing x and holds that lock until after its last write of x; acquires a read or write lock on x before reading x and holds that lock until after its last read of x; and never releases a lock on any item x before obtaining a lock on any (perhaps different) item y. Two-phase locking can encounter deadlock. The database system resolves this by rolling back one of the transactions involved in the deadlock.

Vertical partitioning: A method of dividing each record (or object) of a table (or collection of objects) so that some attributes, including a key, of the record (or object) are in one location and others are in another location, possibly another disk. For example, the account id and the current balance may be in one location and the account id and the address information of each tuple may be in another location.

Write lock: If a transaction T holds a write lock on a datum x, then no other transaction can obtain any lock on x.

References

1. Acharya S., Gibbons, P.B., Poosala, V., and Ramaswamy, S. 1999. The aqua approximate query answering system. *Proceedings of the SIGMOD Conference 1999*, Philadelphia, PA.
2. Comer, D. 1979. The ubiquitous B-tree. *ACM Comput. Surveys* 11(2):121–137.
3. Gray, J. and Reuter, A. 1993. *Transaction Processing: Concepts and Techniques*. Morgan Kaufmann, San Mateo, CA.
4. Shasha, D. and Bonnet, P. 2002. *Database Tuning: Principles, Experiments, and Troubleshooting Techniques*. Morgan Kaufmann, San Mateo CA. Experiments may be found in the accompanying web site: http://www.distlab.dk/dbtune/ (accessed on March 29, 2013).
5. Weikum, G. and Vossen, G. 2001. *Transactional Information Systems: Theory, Algorithms, and Practice of Concurrency Control and Recovery*. Morgan Kaufmann, San Francisco, CA.
6. Oracle. http://otn.oracle.com/ (accessed on March 29, 2013).
7. DB2. http://www.ibm.com/software/data/db2/ (accessed on March 29, 2013).
8. SQL Server. http://www.microsoft.com/sql/ (accessed on March 29, 2013).
9. Thomasian, A. and Ryu, K. 1991. Performance analysis of two-phase locking. *IEEE Trans. Software Eng.* 17(5):68–76.
10. Weikum, G., Hasse, C., Moenkeberg, A., and Zabback, P. 1994. The comfort automatic tuning project. *Inf. Systems* 19(5):381–432.

10

Access Methods

Apostolos N.
Papadopoulos
Aristotle University
of Thessaloniki

Kostas Tsichlas
Aristotle University
of Thessaloniki

Anastasios
Gounaris
Aristotle University
of Thessaloniki

Yannis
Manolopoulos
Aristotle University
of Thessaloniki

10.1 Introduction

We are witnessing a tremendous growth in the size of the data gathered, stored, and processed by various kinds of information systems. Therefore, no matter how big memories become, there is always the need to store data in secondary or even tertiary storage to facilitate access. Even if the data set can fit in main memory, there is still a need to organize data to enable efficient processing. In this chapter, we discuss the most important issues related to the design of efficient access methods (i.e., indexing schemes), which are the fundamental tools in database systems for efficient query processing. For the rest of the discussion, we are going to use the terms access method and index interchangeably.

Take, for example, a large data set containing information about millions of astronomical objects (e.g., stars, planets, comets). An astronomer may require some information out of this data set. Therefore, the most natural way to proceed is to store the data set in a database management system (DBMS) in order to enjoy SQL-like query formulation. For example, a possible query in natural language is: "show me all stars which are at most 1000 light-years away from the sun." To answer such a query efficiently, one should avoid the exhaustive examination of the whole data set. Otherwise, the execution of each query will occupy the system for a long period of time, which is not practical and leads to performance degradation.

For the rest of the discussion, we are mainly interested in **disk-based access methods**, where the data set as well as the auxiliary data structures to facilitate access reside on magnetic disks. The challenge in this case is to perform as few **disk accesses** as possible, because each random access to the disk (i.e., reading or writing a block) costs about 5–8 ms, which is significantly slower than processing in main memory. Moreover, we assume that our data are represented by records of the form $<a_1; a_2,..., a_m>$, where each a_i denotes an attribute value. Attribute values may be simple, an integer for example, or may correspond to more complex objects such as points in 3D space or other geometric shapes. When needed, we are going to make clear the kind of data supported by each access method. For example, some access methods are good in organizing unidimensional objects (e.g., price, salary, population), whereas others have been specifically designed to handle points or rectangles in 2D or 3D space, text, DNA sequences, time-series, to name a few.

10.2 Underlying Principles

In contrast to memory-resident data structures, handling large data collections requires the corresponding access method (or at least a large part of it) to reside on secondary storage. Although flash memories are currently widely used, the magnetic disk continues to be the predominant secondary storage medium, used extensively by large information systems. The fundamental disk limitation is that accessing data on a disk is hundreds of times slower than accessing it in main memory. In fact, this limitation was the driving force underlying the development of efficient access methods trying to reduce the impact of this limitation as much as possible. In this section, we discuss briefly some key issues in access methods.

10.2.1 Blocks and Records

The fundamental characteristic of an access method is that the data are accessed in chunks called **pages** or **blocks**. In particular, whenever a data item x is requested, instead of fetching only x, the system reads a whole set of data items that are located near item x. In our context, near means within the same block. Each block can accommodate a number of data items. Usually, all blocks are of the same size B. Typical block sizes are 4 Kb, 8 Kb, 16 Kb, or larger. Obviously, the larger the block size, the more data items can fit in every block. Moreover, the number of items that fit in each block depends also on the size of each data item. One of the primary concerns in the design of efficient access methods is storing data items that are likely to be requested together in the same block (or nearby blocks). Thus, the target is to fetch into memory more useful data by issuing only one block access. Since this is not feasible in all cases, a more practical goal is to reduce the number of accesses (reads or writes) as much as possible.

There are two fundamental types of block accesses that are usually supported by access methods: a **random access** involves fetching a randomly selected block from the disk, whereas a **sequential access** just fetches the next block. Usually, a random access is more costly because of the way magnetic disks operate. To facilitate a random access, the **disk heads** must be positioned right on top of the track that contains the requested block, thus requiring a significant amount of time, called **seek time**. In fact, seek time is the predominant cost of an I/O disk operation. On the other hand, a sequential access just reads blocks one by one in a get-next fashion, thus requiring less seek time. However, to facilitate sequential access blocks must be located in nearby positions on the disk to minimize the required seek operations. In general, sequential access is more restrictive and easier to obtain than random access. In addition, random access is more useful because of the flexibility offered to access any block any time.

Next, we describe briefly how records are organized inside each block. We limit our discussion for the case where records are of **fixed size** denoted as R. Therefore, the maximum number of records that can fit in a block of size B is simply $\lfloor B/R \rfloor$. There are two basic alternatives we may follow to organize these records inside the block. The first approach is to force that all free space will be placed at the end of the block. This means that whenever we delete a record, its place will be taken by the last record in the block. The second alternative is to use a small index in the block header recording information about which record slot is occupied and which one is free. The second alternative avoids moving records inside the block, but reduces the capacity of the block because of the index used. In case **variable size** records are allowed, we expect that less storage will be required but processing time may increase due to some extra bookkeeping required to locate each record.

10.2.2 Fundamental Operations

Although access methods have different capabilities depending on their design and the problem they call to support, they all are primarily built to support a set of fundamental operations. The most significant one is **searching**. In the simplest case, a search operation takes as input a value and returns some information back to the caller. For example: "What is the perimeter of the Earth?," "Display the ids of the customers located in Greece," "What is the salary of John Smith?" or "Is Jack Sparrow one of our customers?" All these **queries** can be formulated as simple search operations. However, to support these queries as efficiently as possible, the corresponding access method should provide access to the appropriate record attributes. For example, to find the perimeter of the Earth, the access method must be searchable by the name of an astronomical object. Otherwise, the only way to spot the answer is to resort to sequential scanning of the whole data set. Similarly, to display the customers residing in Greece, our access method must be able to search by the name of the country.

There are other search-oriented queries that are clearly much more complex. For example, "find the names of the cities with a population at least 1 million and at most 5 millions." Clearly, this query involves searching in an interval of populations rather than focusing on a single population value. To support such a query efficiently, the access method must be equipped with the necessary tools. Note also, that depending on the application, searching may take other forms as well. For example, if the access method organizes points in the 2D space, then we may search by a region asking for all points falling in the region of interest. In any case, to facilitate efficient search, the access method must be organized in such a way that queries can be easily handled, avoiding scanning the whole database.

Two operations that change the contents of an access method are the insertion of new objects and the deletion of existing ones. If the access method does not support these operations, it is characterized as **static**; if both are supported, it is called **dynamic**; whereas if only insertions are supported, then it is called **semi-dynamic**. In the static case, the access method will be built once and there is no need to support insertions/deletions. The dynamic case is the most interesting and challenging one, since most of the real-life applications operate over data sets that change continuously, and potentially quite rapidly. Thus, insertions and deletions must be executed as fast as possible to allow for efficient maintenance of the access method.

In some cases, there is a need to build an access method when the corresponding data set is known in advance. The simplest solution is to just perform many invocations of the insertion operation. However, we can do much better because the data set is known, and therefore with appropriate preprocessing the index may be built much faster than by using the conservative one-by-one insertion approach. The operation of building the index taking into consideration the whole data set is called **bulk loading**.

10.3 Best Practices

In this section, we study some important indexing schemes that are widely used both in academia and industry. First, we discuss about the **B-tree** and **hashing** which are the predominant access methods for 1D indexing. Then, we center our focus to spatial access methods and discuss the **R-tree** and briefly some of its variations.

10.3.1 Fundamental Access Methods

The two dominant categories of fundamental external memory indexing methods are tree-based methods and hash-based methods. For tree-based methods, the dominant example is the *B*-tree [5], while for hash-based methods linear [38] and extendible [18] hashing are the most common ones. In the following, we briefly present both methods and their variants/extensions.

10.3.1.1 *B*-Tree

The *B*-tree [5] is a ubiquitous data structure when it comes to external memory indexing, and it is a generalization of balanced binary search trees. The intuition behind this generalization is that reading a block should provide the maximum information to guide the search. The use of binary trees may result in all nodes on a search path to reside in distinct blocks, which incurs an $O(\log_2 n)$ overhead while our goal is to impose that the number of blocks to read in order to find an element is $O(\log_B n)$ (i.e., logarithmic with respect to the data set cardinality). The *B*-tree with parameters k (corresponding to the internal node degree) and c (corresponding to leaf capacity) is defined as follows:

1. All internal nodes v have degree $d(v)$ such that $\ell \cdot k \leq d(v) \leq u \cdot k$, where $u > \ell > 0$. The only exception is the root, whose degree is lower-bounded by 2.
2. All leaves lie at the same level; that is, the depth of all leaves is equal.
3. All leaves l have size $|l|$ such that $\ell' \cdot c \leq |l| \leq c$, where $0 < \ell' < 1$.

The maximum height of the tree is $\lfloor \log_{\ell \cdot k}(n/\ell'c) + 1 \rfloor$ while its minimum height is $\lceil \log_{u \cdot k}(n/c) \rceil$. Depending on the satellite information of each element, the size of the pointers, as well as any other needed information within each block, we may set accordingly constants k, c, ℓ, ℓ', and u. Additionally, these parameters are also affected by the desired properties of the tree. Henceforth, for simplicity and without loss of generality, we assume that $k = c = B$, $u = 1$, and $\ell = \ell' = 1/2$.

In the classic *B*-tree, internal nodes may store elements (records) apart from pointers to children. This results in the decrease of parameter k and thus the height of the tree is increased. In practice, and to avoid this drawback, the B^+-trees [15] are extensively used. These trees store elements only at leaves while internal nodes store routing information related to the navigation during search within the tree. In this way, the parameter k is increased considerably and the height of the tree is reduced.

To perform a search for an element x, the search starts from the root and moves through children pointers to other internal nodes toward the leaves of the tree. When a leaf l is found, it is brought into main memory and a sequential or binary search is performed to find and return the element x or to report failure. One can also return the predecessor or the successor of the element x (which is x if it exists in the tree), but one more I/O may be needed. To support range search queries, the B^+-tree is usually changed so that all leaves constitute a linked list. As a result, for the range query $[x_1, x_2]$, first the leaf is located that contains the successor of x_1, and then a linear scan of all leaves whose value is within the range $[x_1, x_2]$ is performed with the help of the linked list of leaves. Before moving to update operations, we first briefly discuss the rebalancing operations. The B^+-tree (in fact all such trees) can be restored after an update operation by means of splits, fusions, and shares.

A node v is split when there is no available space within the node. In this case, half the information contained in v (pointers and routing information in internal nodes or elements at leaves) is transferred to a new node v' and thus both nodes have enough free space for future insertions. However, the father of v has its children increased by one, which means that there may be a cascading split possibly reaching even the root. A node v requires fusion with a sibling node v' when the used space within v is less than $\frac{1}{2} B$ due to a deletion. In this case, if the combined size of v and v' is $>B$, then some information is carried over from v' to v. In this case we have a share operation which is a terminal rebalancing operation in the sense that the number of children of the father of v and v' remains the same. Otherwise, all information of v is transferred to v' and v is deleted. The father of v' has its children reduced by one and so there may be cascading fusions toward the root.

An insertion operation invokes a search for the proper leaf l in which the new element must be inserted to respect the sorted order of elements in the B^+-trees. If l has available space, then the new element is inserted and the insertion terminates. If it does not have available space, then the leaf is split into two leaves l and l' and the new element is inserted to either l or l', depending on its value. Then, based on whether the father of l and l' needs split, the process continuous until an internal node with free space is reached or until the root is split. A deletion operation is similar with the exception that it invokes a fusion operation for rebalancing. In Figure 10.1 an example of a B^+-tree is depicted.

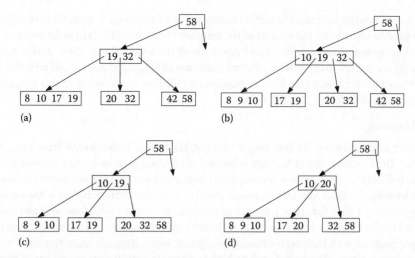

FIGURE 10.1 An example of update operations for the B^+-tree is depicted. Internal nodes contain only routing information. (a) An example of a B^+-tree ($B = 4$). (b) Adding 9 and performing a split. (c) Removing 42 and performing a fusion. (d) Removing 19 and performing a share.

A variant of the B^+-tree is the B^*-tree [15], which balances adjacent internal nodes to keep them more densely packed. This variant requires that $l = l' = \frac{2}{3}$ instead of $\frac{1}{2}$. To maintain this, instead of immediately splitting up a node when it gets full, its elements are shared with an adjacent node. As soon as the adjacent node gets full, then the two nodes are split into three nodes.

In total, search and update operations can be carried out in $O(\log_B n)$ I/Os while a range search query can be carried out in $O(\log_B n + (t/B))$ I/Os, where t is the number of reported elements.

10.3.1.2 *B*-Tree Variations and Extensions

There are numerous variants and extensions of B-trees, and we only report some of them. Most of these, if not all, are quite complicated to implement and can be used in practice only in particular scenarios. The Lazy B-tree [33] support updates with $O(1)$ worst-case rebalancings (not counting the search cost) by carefully scheduling these operations over the tree. The ISB-tree [33] uses interpolation search in order to achieve a $O(\log_B \log n)$ expected I/Os for searching and updating, provided that the distribution of the elements belongs to a large family of distributions with particular properties. It is simple to extend basic B^+-trees to maintain a pointer to the parent of each node as well as to maintain that all nodes at each level are connected in a doubly linked list. Applying these changes, the B^+-tree can support efficiently finger searches [11] such that the number of I/Os for searching becomes $O(\log_B d)$, where d is the number of leaves between a leaf l designated as the finger and the leaf l' we are searching for. When searching for nearby leaves, this is a significant improvement over searching for l' from the root. One can also combine B-trees with hashing [43] to speed-up operations.

There are numerous extensions of B-trees that provide additional functionalities and properties. The weight-balanced B-tree [4] has the weight property that normal B-trees lack. For an internal node v let $w(v)$ be its weight, which is the number of elements stored in the subtree of v. The weight property states that an internal node v is rebalanced only after $\Theta(w(v))$ updates have been performed at its subtree since the last update. This property is very important to reduce complexities when the B-tree has secondary structures attached to internal nodes [4]. A partial persistent B-tree [7] is a B-tree that maintains its history attaining the same complexities with normal B-trees. Efficient fully persistent B-trees have very recently been designed [12] and allow updates in the past instances of the B-tree giving rise to different history paths. String B-trees [19] have been designed to support efficiently search and update operations on a set of strings (and its suffixes) in external memory. Cache-oblivious B-trees [9] have been designed that do not need to know basic parameters of the memory hierarchy (like B) in order to attain the same complexities as normal B-trees, which make heavy use of the knowledge of these parameters. Finally, buffer trees [3] are B-trees that allow for efficient execution of batch updates (or queries). This is accomplished by lazily flushing toward the leaves all updates (or queries) in the batch. In this way, each update can be supported in $O((1/B)\log_{M/B}(n/B))$ I/Os.

10.3.1.3 Hashing

Another much used technique for indexing is hashing. Hashing is faster than B-trees by sacrificing in functionality. The functionality of hashing is limited when compared to B-trees because elements are stored unsorted and thus there is no way to support range queries or find the successor/predecessor of an element. In hashing, the basic idea is to map each object to a number corresponding to the location inside an array by means of a hash function. It is not our intention to describe hash functions and thus we refer the interested reader to [16] and the references therein for more information on practical hash functions.

The largest problem with hashing is collision resolution, which happens when two different elements hash to the same location. The hash function hashes elements within buckets, which in this case is a block. When the bucket becomes full, then either a new overflow bucket is introduced for the same hashed value or a rehashing is performed. All buckets without the overflowing buckets constitute the primary area. Two well-known techniques are **linear hashing** [38] and **extendible hashing** [18]. The first method extends the hash-table by one when the fill factor of the hash-table (the number of elements divided by the size of the primary area) goes over a critical value. The second one extends the primary area as soon as an overflowing bucket is about to be constructed.

In linear hashing, data are placed in a bucket according to the last k bits or the last $k + 1$ bits of the hash function value of the element. A pointer *split* keeps track of this boundary. The insertion of a new element may cause the fill factor to go over the critical value, in which case the bucket on the boundary corresponding to k bits is split into two buckets corresponding to $k + 1$ bits. The number of buckets with k bits is decreased by one. When all buckets correspond to $k + 1$ bits, another expansion is initiated constructing new buckets with $k + 2$ bits. Note that there is no relationship between the bucket in which the insertion is performed and the bucket that is being split. In addition, linear hashing uses overflow buckets although it is expected that these buckets will be just a few. The time complexity for a search and update is $O(1)$ expected. In Figure 10.2 an example of an insertion is depicted.

On the other hand, extendible hashing does not make use of overflow buckets. It uses a directory that may index up to 2^d buckets, where d, the number of bits, is chosen so that at most B elements exist in each bucket. A bucket may be pointed by many such pointers from the directory since they may be indexed

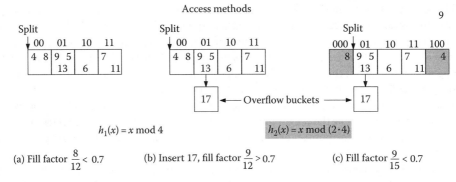

FIGURE 10.2 Assuming that the critical fill factor is 0.7, then when 17 is inserted in (a), an overflow bucket is introduced (b) since the bucket corresponding to bits 01 has no space. After the insertion, the fill factor is >0.7 and as such we construct a new hash function $h_2(x)$ and introduce a new bucket (c). $h_2(x)$ is applied to gray buckets (3 bits) while $h_1(x)$ is applied to white buckets (2 bits) only.

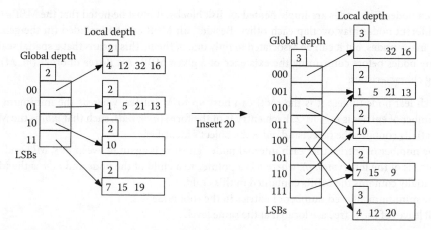

FIGURE 10.3 An example of re-organization for extendible hashing. After the insertion of 20 the first bucket overflows and we construct a new table of global depth 3; that is, with 2^3 entries. Only the overflowing bucket is split and has local depth 3, thus needing 3 bits to navigate. The buckets that did not overflow need only 2 bits and thus 2 pointers point to them.

by using less than d bits. An insertion either causes the directory to double or some of entries of the directory are changed. For example, a bucket pointed by eight entries will be split and both new buckets are pointed by four entries. In case a bucket is pointed only by one entry and needs to be split, then the directory needs to be doubled. The advantage of extendible hashing is that it never has more than two disk accesses for any record since there are no overflow buckets to traverse. The main problems with this variation on hashing are total space utilization and the need for massive reorganization (of the table). In Figure 10.3, an example of such a reorganization for an insertion is depicted.

There are many other hashing schemes with their own advantages and disadvantages. Cuckoo hashing [45] is one such scheme which is very promising because of its simplicity. The basic idea is to use two hash functions instead of only one and thus provide two possible locations in the hash-table for each element. When a new element is inserted, it is stored in one of the two possible locations provided by the hash functions. If both of these locations are not empty, then one of them is kicked out. This new displaced element is put into its alternative position and the process continues until a vacant position has been found or until many such repetitions have been performed. In the last case, the table is rebuilt with new hash functions. Searching for an element requires inspection of just two locations in the hash-table, which takes constant time in the worst case.

10.3.2 Spatial Access Methods

The indexing schemes discussed previously support only one dimension. However, in applications such as Geographic Information Systems (GIS), objects are associated with spatial information (e.g., latitude/longitude coordinates). In such a case, it is important to organize the data taking into consideration the spatial information. Although there are numerous proposals to handle spatial objects, in this chapter we will focus on the *R*-tree index [24], which is one of the most successful and influential spatial access methods, invented to organize large collections of rectangles for VLSI design.

10.3.2.1 *R*-Tree

R-trees are hierarchical access methods based on B^+-trees. They are used for the dynamic organization of a set of d-dimensional geometric objects representing them by the minimum bounding d-dimensional rectangles (for simplicity, MBRs in the sequel). Each node of the *R*-tree corresponds to the MBR that bounds its children. The leaves of the tree contain pointers to the database objects instead of pointers to children nodes. The nodes are implemented as disk blocks. It must be noted that the MBRs that surround different nodes may overlap each other. Besides, an MBR can be included (in the geometrical sense) in many nodes, but it can be associated to only one of them. This means that a spatial search may visit many nodes before confirming the existence of a given MBR. An R-tree of order (m, M) has the following characteristics:

- Each leaf node (unless it is the root) can host up to M entries, whereas the minimum allowed number of entries is $m \leq M/2$. Each entry is of the form (mbr, oid), such that mbr is the MBR that spatially contains the object and oid is the object's identifier.
- The number of entries that each internal node can store is again between $m \leq M/2$ and M. Each entry is of the form (mbr, p), where p is a pointer to a child of the node and mbr is the MBR that spatially contains the MBRs contained in this child.
- The minimum allowed number of entries in the root node is 2.
- All leaves of the R-tree are located at the same level.

An R-tree example is shown in Figure 10.4 for the set of objects shown on the left. It is evident that MBRs R_1 and R_2 are disjoint, whereas R_3 has an overlap with both R_1 and R_2. In this example we have assumed that each node can accommodate at most three entries. In a real implementation, the capacity of each node is determined by the block size and the size of each entry which is directly related to the number of dimensions.

The R-tree has been designed for dynamic data sets and, therefore, it supports insertions and deletions. To insert a new object's MBR, the tree is traversed top-down, and at each node a decision is made to select a branch to follow next. This is repeated until we reach the leaf level. The decision we make at each node is based on the criterion of area enlargement. This means that the new MBR is assigned to the entry which requires the least area enlargement to accommodate it. Other variations of the R-tree, such as the R^*-tree [8], use different criteria to select the most convenient path from the root to the leaf level. Upon reaching a leaf L, the new MBR is inserted, if L can accommodate it (there is at least one available slot). Otherwise, there is a node overflow and a **node split** occurs, meaning that a new node L' is reserved. Then, the old entries of L (including the new entry) are distributed to two nodes L and L'. Note that a split at a leaf may cause consecutive splits in the upper levels. If there is a split at the root, then the height of the tree increases by one.

The split operation must be executed carefully to maintain the good properties of the tree. The primary concern while splitting is to keep the overlap between the two nodes as low as possible. This is because the higher the overlap, the larger the number of nodes that will be accessed during a search operation. In the original R-tree proposal, three split policies have been studied, namely, exponential, quadratic, and linear.

Exponential split: All possible groupings are exhaustively tested and the best one, with respect to the minimization of the MBR enlargement, is chosen.

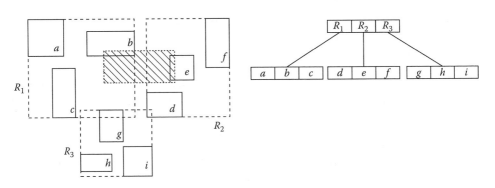

FIGURE 10.4　*R*-tree example.

Algorithm RANGESEARCH (**Node** *N*, **Region** *Q*)
Input: root mode *N*, query region *Q*
Output: answer set *A*

---

1.   **if** (*N* is not a leaf node)
2.   examine each entry *E* of *N* to find those *E*.mbr that intersect *Q*
3.   foreach such entry *E* call RANGESEARCH(*E.ptr,Q*)
4.   **else** // *N* is a leaf node
5.   examine all entries *E* and find those for which *E.mbr* intersects *Q*
6.   add these entries to the answer set *A*
7.   **Endif**

FIGURE 10.5   The *R*-tree range search algorithm.

Quadratic split: Choose two objects as seeds for the two nodes, where these objects if put together create as much empty space as possible. Then, until there are no remaining objects, insert the object for which the difference of dead space if assigned to each of the two nodes is maximized in the node that requires less enlargement of its respective MBR.

Linear split: Choose two objects as seeds for the two nodes, where these objects are as far apart as possible. Then, consider each remaining object in a random order and assign it to the node requiring the smallest enlargement of its respective MBR.

The quadratic split policy is the best choice that balances efficiency and effectiveness and, therefore, it is widely used in *R*-tree implementations.

In a deletion, an entry is removed from the leaf level of the tree, which may cause a node underflow. In this case, **re-insertion** of entries is applied to reduce the space requirements of the tree. In general, a deletion has the opposite effect than that of an insertion. In both cases, the path from the root to the leaf level that was affected by the operation must be adjusted properly to reflect the changes performed in the tree. Thus, the deletion of an entry may cause a series of deletions that propagate up to the root of the *R*-tree. If the root has only one child, then it is removed and the height of the tree decreases by one.

In the sequel, we examine briefly how search is performed in an *R*-tree. We will center our focus to range queries, where the user defines a query region *Q* and the answer to the query contains the spatial objects that intersect *Q*. For example, in Figure 10.4, the query region is shown filled, whereas the answer to the query is composed of the objects *b* and *e*. The outline of the algorithm that processes range queries in an *R*-tree is given in Figure 10.5. For a node entry *E*, *E.mbr* denotes the corresponding MBR and *E.p* the corresponding pointer to the next level. If the node is a leaf, then *E.p* denotes the corresponding object identifier (*oid*). We note that the rectangles that are found by range searching constitute the candidates of the filtering step. The actual geometric objects intersected by the query rectangle have to be found in a refinement step by retrieving the objects of the candidate rectangles and testing their intersection.

10.3.2.2 R-Tree Bulk Loading

Recall that bulk loading is the process of building an index by taking into consideration the data set which is known in advance. Thus, usually this operation is applied for static data sets or when insertions and deletions are rare. In most of the cases, when bulk loading is applied, the leaf level of the tree is created first. Then, the upper tree levels can be built one by one until we reach the root.

The first bulk-loading algorithm for *R*-trees proposed in [47] first sorts the data objects by using the *x* coordinate of their center. If objects are points rather than rectangles, then the *x* coordinate of the point is used. By using the sorted order, the leaves may be formed by placing the first *M* entries in the first leaf, until no more data are available. This way, all leaves will be 100% full, except maybe of the last leaf which may contain less.

Another contribution to this problem is reported in [30]. The algorithm is similar to that of [47] in that again a sorting is performed in order to build the leaf level of the tree. Sorting is performed by using the Hilbert value of the data objects' centroids. According to the performance evaluation given in [30], this approach shows the best overall performance with respect to the cost of performing queries.

STR (Sort-Tile-Recursive) is a bulk-loading algorithm for R-trees proposed by Leutenegger et al. [32]. Let n be a number of rectangles in 2D space. The basic idea of the method is to tile the address space by using V vertical slices, so that each slice contains enough rectangles to create approximately $\sqrt{n/M}$ nodes, where M is the R-tree node capacity. Initially, the number of leaf nodes is determined, which is $L = \lceil n/M \rceil$. Let $V = \sqrt{L}$. The rectangles are sorted with respect to the x-coordinate of the centroids, and V slices are created. Each slice contains $V.M$ rectangles, which are consecutive in the sorted list. In each slice, the objects are sorted by the y-coordinate of the centroids and are packed into nodes (placing M objects in a node). Experimental evaluation performed in [32] has demonstrated that the STR method is generally better than previously proposed bulk-loading methods. However, in some cases the Hilbert packing approach performs marginally better.

10.3.2.3 R-Tree Variations and Extensions

Several R-tree variations have been proposed in the literature to improve the performance of queries. Here we discuss briefly three successful variations that show better performance than the original proposal by Guttman. These variations differ in several aspects like the way insertions and deletions are performed, the optimization criteria being used, the split policy applied, and the storage utilization.

The R^+-tree: R^+-trees were proposed as an alternative that avoids visiting multiple paths during point location queries, aiming at the improvement of query performance [52]. Moreover, MBR overlapping of internal modes is avoided. This is achieved by using the clipping technique. In other words, R^+-trees do not allow overlapping of MBRs at the same tree level. In turn, to achieve this, inserted objects have to be divided in two or more MBRs, which means that a specific object's entries may be duplicated and redundantly stored in several nodes. Therefore, a potential limitation of R^+-trees is the increased space requirements due to redundancy.

The R^*-tree: R^*-trees [8] were proposed in 1990 but are still very well received and widely accepted in the literature as a prevailing performance-wise structure that is often used as a basis for performance comparisons. As already discussed, the R-tree is based solely on the area minimization of each MBR. On the other hand, the R^*-tree goes beyond this criterion and examines the following: (1) minimization of the area covered by each MBR, (2) minimization of the overlap between MBRs, (3) minimization of MBR margins (perimeters), and (4) maximization of storage utilization. The R^*-tree follows an engineering approach to and the best possible combinations of the aforementioned criteria. This approach is necessary, because the criteria can become contradictory. For instance, to keep both the area and the overlap low, the lower allowed number of entries within a node can be reduced. Therefore, storage utilization may be impacted. Also, by minimizing the margins so as to have more quadratic shapes, the node overlapping may be increased.

The Hilbert R-tree: The Hilbert R-tree [31] is a hybrid structure based on the R-tree and the B^+-tree. Actually, it is a B^+-tree with geometrical objects being characterized by the Hilbert value of their centroid. The structure is based on the Hilbert space-filling curve. It has been shown in [42] that the Hilbert space-filling curve preserves well the proximity of spatial objects. Entries of internal tree nodes are augmented by the largest Hilbert value of their descendants. Therefore, an entry e of an internal node is a triple of the form $< mbr, H, p >$ where mbr is the MBR that encloses all the objects in the corresponding subtree, H is the maximum Hilbert value of the subtree, and p is the pointer to the next level. Entries in leaf nodes are exactly the same as in R-trees, R^+-trees, and R^*-trees and are of the form $< mbr, oid >$, where mbr is the MBR of the object and oid the corresponding object identifier.

10.3.3 Managing Time-Evolving Data

Time information plays a significant role in many applications. There are cases where in addition to the data items per se, the access method must maintain information regarding the time instance that a particular event occurred. For example, if one would like to extract statistical information regarding the sales of a particular product during the past 5 years, the database must maintain historical information. As another example, consider an application that tracks the motion patterns of a specific species. To facilitate this, each location must be associated with a timestamp. Thus, by inspecting the location in consecutive timestamps, one may reveal the motion pattern of the species. For the rest of the discussion, we assume that time is discrete and each timestamp corresponds to a different time instance.

One of the first access methods that was extended to support time information is the *B*-tree and its variations. The most important extensions are the following:

The time-split B-tree [39]: This structure is based on the write-once *B*-tree access method proposed in [17], and it is used for storing **multi-version data** on both optical and magnetic disks. The only operations allowed are insertions and searches, whereas deletions are not supported. The lack of deletions in conjunction with the use of two types of node splits (normal and time-based) is the main reason for the structure's space and time efficiency. However, only exact-match queries are supported efficiently.

The fully persistent B+-tree [36]: In contrast to the time-split *B*-tree, the fully persistent *B+*-tree supports deletions in addition to insertions and searches. Each record is augmented by two fields t_{start} and t_{end}, where t_{start} is the timestamp of the insertion and t_{end} is the timestamp when the record has been deleted, updated, or copied to another node. This way, the whole history can be recorded and queries may involve the past or the present status of the structure.

The multi-version B-tree [6]: This access method is asymptotically optimal and allows insertions and deletions only at the last (current) timestamp, whereas exact-match and range queries may be issued for the past as well. The methodology proposed in [6] may be used for other access methods, when there is a need to transform a simple access method to a multi-version one.

There are also significant research contributions in providing time-aware spatial access methods. An index that supported space and time is known as **spatiotemporal** access method. Spatiotemporal data are characterized by changes in location or shape with respect to time. Supporting time increases the number of query types that can be posed by users. A user may focus on a specific time instance or may be interested in a time interval. Spatiotemporal queries that focus on a single time instance are termed **time-slice queries**, whereas if they focus on a time interval, they are termed **time interval queries**. If we combine these choices with spatial predicates and the ability to query the past, the present, or the future, spatiotemporal queries can be very complex, and significant effort is required to process them.

A large number of the proposed spatiotemporal access methods are based on the well-known *R*-tree structure. In the following, we discuss briefly some of them.

The 3D R-tree [54]: In this index, time is considered as just another dimension. Therefore, a rectangle in 2D becomes a box in 3D. The 3D *R*-tree approach assumes that both ends of the interval $[t_{start}, t_{end})$ of each rectangle are known and fixed. If the end time t_{end} is not known, this approach does not work well, due to the use of large MBRs leading to performance degradation with respect to queries. In addition, conceptually, time has special characteristic, i.e., it increases monotonically. This suggests the use of more specialized access methods.

The partially persistent R-tree [34]: This index is based on the concept of partial persistency [35]. It is assumed that in spatiotemporal applications, updates arrive in time order. Moreover, updates can be performed only on the last recorded instance of the database, in contrast to general bitemporal data. The partially persistent *R*-tree is a directed acyclic graph of nodes with a number of root nodes, where each root is responsible for recording a subsequent part of the ephemeral *R*-tree evolution. Object records are stored in the leaf nodes of the PPR-tree and maintain the evolution of the ephemeral *R*-tree data objects. As in the case of the fully persistent *B*+-tree, each data record is augmented to include the two lifetime fields, t_{start} and t_{end}. The same applies to internal tree nodes, which maintain the evolution of the corresponding directory entries.

The multi-version 3D R-tree [53]: This structure has been designed to overcome the shortcomings of previously proposed techniques. It consists of two parts: a multi-version *R*-tree and an auxiliary 3D *R*-tree built on the leaves of the former. The multi-version *R*-tree is an extension of the multi-version *B*-tree proposed by Becker et al. [6]. The intuition behind the proposed access method is that time-slice queries can be directed to the multi-version *R*-tree, whereas time-interval queries can be handled by the 3D *R*-tree.

10.4 Advanced Topics

In this section, we discuss some advanced topics related to indexing. In particular, we focus on three important and challenging issues that attract research interest mainly due to their direct impact on performance and because they radically change the way that typical indexing schemes work. Namely, these topics are (1) cache-oblivious indexing, (2) on-line indexing, and (3) adaptive indexing.

10.4.1 Cache-Oblivious Access Methods and Algorithms

In 1999, Frigo et al. [20] introduced a new model for designing and analyzing algorithms and data structures taking into account memory hierarchies. This is the **cache-oblivious model** which, as its name implies, is oblivious to the parameters as well to some architectural characteristics of the memory hierarchy. Data structures designed in this model do not know anything about the memory hierarchy, but have performance which is comparable to data structures that have been designed with knowledge of the particular memory hierarchy. In a nutshell, this is accomplished by laying out the data structure over an array which is cache-oblivious by default. The main problem is how to design this layout and how to maintain it when it is subjected to update operations. The main advantage of these indexing schemes (an example is the cache oblivious *B*-tree [9]) is not only their simplicity and portability among different platforms but also their innate ability to work optimally in all levels of the memory hierarchy.

10.4.2 On-Line and Adaptive Indexing

Traditional physical indexing building follows an off-line approach to analyzing workload and creating the data structures to enable efficient query processing. More specifically, typically, a sample workload is analyzed with a view to selecting the indices to be built with the help of auto-tuning tools (e.g., [2,14,56]). This is an expensive process, especially for very large databases. On-line and adaptive indexing have introduced a paradigm shift, where indices are created on the y during query processing. In this way, database systems avoid a complex and time-consuming process, during which there is no index support, and can adapt to dynamic workloads. We draw a distinction between on-line and adaptive indexing following the spirit of [28]. According to that distinction, on-line indexing monitors the workload and creates (or drops) indices on-line during query execution (e.g., [13,40]). Adaptive indexing takes one step further: in adaptive indexing the process of index building is blended with query execution through extensions to the logic of the operators in the query execution plan (e.g., [23,25]). In the sequel, we discuss briefly these two approaches which differ significantly from the typical bulk-loading process for index building.

On-line Indexing: The paradigm of on-line indexing solutions departs from traditional schemes in that the index tuning mechanism is always running. A notable example of such category is described in [13], where the query engine is extended with capabilities to capture and analyze evidence information about the potential usefulness of indices that have not been created thus far. Another interesting approach is soft-indices [40], which build on top of an index-tuning mechanism that continuously collects statistical information and periodically solves the NP-hard problem of index selection. Similarly to adaptive indexing, the new decisions are enforced during query processing but without affecting the operator implementation.

Adaptive Indexing: The most representative form of adaptive indexing is **database cracking** [25], which is mostly tailored to (in-memory) column-stores [1]. Database cracking revolves around the concept of continuous physical re-organization taking place at runtime. Such physical re-organization is automated, in the sense that does not involve human involvement, does not contain any off-line preparatory phase, and may not build full indices. The latter means that the technique is not only adaptive but also relies on partial indexing that is refined during workload execution in an incremental manner. To give an example, suppose that a user submits a range query on the attribute *R.A* for the first time and the range predicate is *value*1 < *R.A* < *value*2. According to database cracking, during execution, a copy of *R.A* will be created, called the *cracker column* of *R.A*. The data in that cracker column is physically re-organized and is split in three parts: (1)*value*1 ≤ *R.A*; (2) *value*1 < *R.A* < *value*2; and (3) *R.A* ≤ value2. Moreover, an AVL tree, called *cracker index*, is created to maintain the partitioning information. This process of physical re-organization continues with every query. Overall, each query may benefit from the cracking imposed by previous queries and the new splits introduced may be of benefit for subsequent queries. In other words, the access methods are created on the y and are query-driven, so that they eventually match the workload. Database cracking is characterized by low initialization overhead, is continuously refined, and may well outperform techniques that build full indices, because the latter need a very large number of queries in order to amortize the cost of full index building to be outweighed. Cracking can be extended to support updates [26] and complex queries [27].

Nevertheless, database cracking may converge slowly and is sensitive to query pattern. Such limitations are mitigated by combining traditional indexing techniques (*B*-trees) with database cracking, resulting in the so-called adaptive merging [23]. The key distinction between the two approaches is that adaptive merging relies on merging rather than on partitioning, and that adaptive merging is applicable to disk-based data as well.

10.5 Research Issues

The field of access methods continues to be one of the most important topics in database management systems, mainly because of its direct impact on query performance. Although the field has been active for several decades, modern applications pose novel challenges that must be addressed carefully toward efficient processing. Here, we discuss briefly some of these challenges.

High-dimensional data: One of the most important challenges emerges when the number of attributes (dimensions) increases significantly. *R*-trees and related indexing schemes perform quite well for dimensionalities up to 15 or 20. Above this level, the dimensionality curse renders indexing difficult mainly due to concentration of measure effect. Unfortunately, the cases where high dimensionalities appear are not few. For example, in multimedia data management, images are often represented by feature vectors, each containing hundreds of attributes. There are several proposals in the literature to improve performance in these cases, trying to reduce the effects of dimensionality curse. For example, the X-tree access method proposed in [10] introduces the concept of **supernode** and avoids node splits if these cannot lead to a good partitioning. As another example, Vector Approximation File [55] is an approximation method that introduces error in query processing. For static data sets, dimensionality reduction may also be applied to first decrease the number of dimensions in order to be easier for access methods to organize the data.

Modern hardware: Hardware is evolving, and although the magnetic disk is still the prevailing secondary storage medium, new types of media have been introduced, such as **solid-state drives**.

Also, processors enjoy a dramatic change by the ability to include multiple cores in the same chip. These changes in hardware bring the necessity to change the way we access and organize the data. For example, solid-state drives do not suffer from the seek time problem and, therefore, the impact of I/O time on the performance becomes less significant. This means that operations that are I/O-bounded with magnetic disks may become CPU-bounded with modern hardware, due to the reduction in I/O time. The indexing problem becomes even more challenging as new processing and storage components like GPUs and FPGAs are being used more often. The new access methods must take into account the specific features of new technology.

Parallelism and distribution: The simultaneous use of multiple instances of resources such as memory, CPU, and disks brings a certain level of flexibility to query processing, but it also creates significant challenges in terms of overall efficiency. By enabling concurrent execution of tasks, we are facing the problem of coordinating these tasks and also the problem of determining where data reside. Therefore, parallel and distributed access methods are needed that are able to scale well with the number of processors. In addition to scalability, these techniques must avoid bottlenecks. For example, if we simply take an access method and place each block to a different processor, the processor that hosts the root will become a hot spot. Although there are many significant contributions for parallel/distributed indexing (e.g., distributed hash-tables), modern programming environments such as MapReduce on clusters or multi-core CPUs and GPUs call for a reconsideration of some concepts to enable the maximum possible performance gains and scalability.

On-line and adaptive indexing: On-line and adaptive indexing are technology in evolution; thus multiple aspects require further investigation. Among the most important ones, we highlight the need of investigation of concurrency control in database cracking techniques, the most fruitful combination of off-line, on-line, and adaptive techniques, and the application of database cracking to row stores. Finally, as new indexing methodologies are being proposed, we need a comprehensive benchmark in order to assess the benefits and weaknesses of each approach; a promising first step toward this direction is the work in [29].

10.6 Summary

Access methods are necessary toward efficient query processing. The success of an access method is characterized by its ability to organize data in such a way that locality of references is enhanced. This means that data that are located in the same block are likely to be requested together. In this chapter, we discussed some fundamental access methods that enjoy a wide-spread use by the community due to their simplicity and their excellent performance. Initially, we discussed some important features of access methods and then we described *B*-trees and hashing which are the prevailing indexing schemes for 1D data and *R*-trees, which is the most successful family for indexing spatial and other types of multidimensional data. For *R*-trees we also described briefly bulk-loading techniques, which is an important operation for index creation when the data set is available. Finally, we touched the issues of on-line and adaptive indexing, which enjoy a growing interest due to the ability to adapt dynamically based on query workloads. Access methods will continue to be a fundamental research topic in data management. In the era of big data, there is a consistent need for fast data management and retrieval, and thus indexing schemes are the most important tools in this direction.

Glossary

B-tree: A tree-based index that allows queries, insertions, and deletions in logarithmic time.
Bulk loading: The insertion of a known before-hand sequence of objects into a set of objects.
Cache-oblivious model: A model that supports the design of algorithms and data structures for memory hierarchies without knowing its defining parameters.

Cuckoo hashing: Hashing that uses two hash functions to guarantee worst-case $O(1)$ accesses per search.

Database cracking: A form of continuous physical re-organization taking place at runtime.

Disk access: Reading (writing) a block from (to) the disk.

Disk-based access method: An access method that is designed specifically to support access for disk-resident data.

Disk head: The electronic component of the disk responsible for reading and writing data from and to the magnetic surface.

Dynamic access method: An access method that allows insertions, deletions, and updates in the stored data set.

Extendible hashing: A hashing method that doubles the index size to avoid overflowing.

Fixed-size record: A record with a predetermined length which never changes.

Hashing: An index method that uses address arithmetic to locate elements.

Linear hashing: A hashing method that extends the main index by one each time the fill factor is surpassed.

Multi-version data: Data augmented by information about the time of their insertion, deletion, or update.

Node split: The operation that takes place when a node overflows, i.e., there is no room to accommodate a new entry.

On-line indexing: An indexing scheme where the index tuning mechanism is always running.

R-tree: A hierarchical access method that manages multidimensional objects.

Random access: The operation of fetching into main memory a randomly selected block from disk or other media.

Re-insertion: The operation of re-inserting some elements in a block. Elements are first removed from the block and then inserted again in the usual manner.

Semi-dynamic access method: An access method that supports queries and insertions of new items; it does not support deletions.

Seek time: Time required for the disk heads to move to the appropriate track.

Solid-state drive: A data storage device that uses integrated circuits solely to store data persistently.

Spatiotemporal data: Data containing both spatial and temporal information.

Static access method: An access method that does not support insertions, deletions, and updates.

Supernode: A tree node with a capacity which is a multiple of the capacity of a regular node.

Time-slice query: A query posed on a specific timestamp.

Time-interval query: A query that refers to a time interval defined by two timestamps.

Variable-size record: A record whose length may change due to to changes in the size of individual attributes.

Further Information

There is a plethora of resources that the interested reader may consult to grasp a better understanding of the topic. Since access methods are related to physical database design, the book of Lightstone et al. [37] may be of interest to the reader. For a thorough discussion of *B*-trees and related issues, we recommend the article of Graefe in [22]. Also, almost every database-oriented textbook contains one or more chapters devoted to indexing. For example, Chapters 8–11 of [46] cover indexing issues in detail, in particular *B*-trees and hashing, in a very nice way.

For a detailed discussion of spatial access methods, the reader is referred to two books from Samet, [50] and [51]. In [50] the reader will find an in-depth examination of indexing schemes that are based on space portioning (e.g., linear quadtrees). Also, in [51], in addition to the very detailed study of spatial access methods, the author presents metric access methods, where objects reside in a metric space rather than in a vector space. For a thorough discussion of the *R*-tree family, the reader is referred to the book of Manolopoulos et al. [41]. We also mention a useful survey paper of Gaede and Günther [21], which covers indexing schemes that organize multidimensional data.

In many applications, time information is considered very important and thus it should be recorded. For example, in a fleet management application, we must know the location of each vehicle and the associated timestamp. There are many approaches to incorporate time information into an index. Some of them were briefly discussed in Section 10.3.3. The reader who wants to cover this issue more thoroughly may consult other more specialized resources such as the survey of Salzberg and Tsotras [49] as well as the article of Nguyen-Dinh et al. [44]. In addition to the indexing schemes designed to support historical queries, some access methods support future queries. Obviously, to answer a query involving the near future, some information about the velocities of moving objects is required. The paper by Šaltenis et al. [48] describes such an index.

The previous resources cover more or less established techniques. For modern research issues in the area, the best sources are the proceedings of leading database conferences and journals. In particular, the proceedings of the major database conferences *ACM SIGMOD/PODS*, *VLDB*, *IEEE ICDE*, *EDBT/ ICDT*, *ACM CIKM*, and *SSDBM* contain many papers related to access methods and indexing. Also, we recommend the journals *ACM Transactions on Database Systems*, *IEEE Transactions on Knowledge and Data Engineering*, *The VLDB Journal*, and *Information Systems* where, in most cases, the published articles are enhanced versions of the corresponding conference papers.

References

1. DJ. Abadi, PA. Boncz, and S. Harizopoulos. Column oriented database systems. *PVLDB*, 2(2):1664–1665, 2009.
2. S. Agrawal, S. Chaudhuri, L. Kollár, AP. Marathe, VR. Narasayya, and M. Syamala. Database tuning advisor for Microsoft SQL server 2005. In *Proceedings of VLDB*, Toronto, ON, Canada, pp. 1110–1121, 2004.
3. L. Arge. The buffer tree: A technique for designing batched external data structures. *Algorithmica*, 37(1):1–24, 2003.
4. L. Arge and JS. Vitter. Optimal external memory interval management. *SIAM Journal on Computing*, 32(6):1488–1508, 2003.
5. R. Bayer and EM. McCreight. Organization and maintenance of large ordered indices. *Acta Informatica*, 1:173–189, 1972.
6. B. Becker, S. Gschwind, T. Ohler, B. Seeger, and P. Widmayer. On optimal multiversion access structures. In *Proceedings of the 3rd International Symposium on Advances in Spatial Databases*, Singapore, pp. 123–141, 1993.
7. B. Becker, S. Gschwind, T. Ohler, B. Seeger, and P. Widmayer. An asymptotically optimal multiversion b-tree. *The VLDB Journal*, 5(4):264–275, 1996.
8. N. Beckmann, H-P. Kriegel, R. Schneider, and B. Seeger. The r*-tree: An efficient and robust access method for points and rectangles. In *Proceedings of ACM SIGMOD*, Atlantic City, NJ, pp. 322–331, 1990.
9. MA. Bender, ED. Demaine, and M. Farach-Colton. Cache-oblivious b-trees. *SIAM Journal on Computing*, 35(2):341–358, 2005.
10. S. Berchtold, DA. Keim, and H-P. Kriegel. The x-tree: An index structure for high-dimensional data. In *Proceedings of VLDB*, Bombay, India, pp. 28–39, 1996.
11. GS. Brodal, G. Lagogiannis, C. Makris, A. Tsakalidis, and K. Tsichlas. Optimal finger search trees in the pointer machine. *Journal of Computer and System Sciences*, 67(2):381–418, 2003.
12. GS. Brodal, K. Tsakalidis, S. Sioutas, and K. Tsichlas. Fully persistent b-trees. In *SODA*, pp. 602–614, 2012. Available at http://siam.omnibooksonline.com/2012SODA/data/papers/498.pdf (accessed on April 15, 2013).
13. N. Bruno and S. Chaudhuri. An online approach to physical design tuning. In *ICDE*, Orlando, FL, pp. 826–835, 2007.

14. S. Chaudhuri and VR. Narasayya. Self-tuning database systems: A decade of progress. In *Proceedings of VLDB*, Vienna, Austria, pp. 3–14, 2007.

15. D. Comer. Ubiquitous b-tree. *ACM Computing Surveys*, 11(2):121–137, 1979.

16. M. Dietzfelbinger. Universal hashing and k-wise independent random variables via integer arithmetic without primes. In *Proceedings of the 13th Annual Symposium on Theoretical Aspects of Computer Science, STACS'96*, Grenoble, France, pp. 569–580, 1996.

17. MC. Easton. Key-sequence data sets on indelible storage. *IBM Journal of Research and Development*, 30(3):230–241, 1986.

18. R. Fagin, J. Nievergelt, N. Pippenger, and HR. Strong. Extendible hashing: A fast access method for dynamic files. *ACM Transactions on Database System*, 4(3):315–344, 1979.

19. P. Ferragina and R. Grossi. The string b-tree: A new data structure for string search in external memory and its applications. *Journal of the ACM*, 46(2):236–280, 1999.

20. M. Frigo, CE. Leiserson, H. Prokop, and S. Ramachandran. Cache-oblivious algorithms. In *IEEE FOCS*, New York, pp. 285–298, 1999.

21. V. Gaede and O. Günther. Multidimensional access methods. *ACM Computing Surveys*, 30(2):170–231, 1998.

22. G. Graefe. Modern b-tree techniques. *Foundations and Trends in Databases*, 3(4):203–402, 2011.

23. G. Graefe and HA. Kuno. Self-selecting, self-tuning, incrementally optimized indexes. In *EDBT*, Lausanne, Switzerland, pp. 371–381, 2010.

24. A. Guttman. R-trees: A dynamic index structure for spatial searching. In *Proceedings of ACM SIGMOD*, Boston, MA, pp. 47–57, 1984.

25. S. Idreos, ML. Kersten, and S. Manegold. Database cracking. In *CIDR*, Asilomar, CA, pp. 68–78, 2007.

26. S. Idreos, ML. Kersten, and S. Manegold. Updating a cracked database. In *Proceedings of ACM SIGMOD*, Beijing, China, pp. 413–424, 2007.

27. S. Idreos, ML. Kersten, and S. Manegold. Self-organizing tuple reconstruction in column-stores. In *Proceedings of ACM SIGMOD*, Providence, RI, pp. 297–308, 2009.

28. S. Idreos, S. Manegold, and G. Graefe. Adaptive indexing in modern database kernels. In *EDBT*, Berlin, Germany, pp. 566–569, 2012.

29. I. Jimenez, J. LeFevre, H. Polyzotis, H. Sanchez, and K. Schnaitter. Benchmarking online index-tuning algorithms. *IEEE Data Engineering Bulletin*, 34(4):28–35, 2011.

30. I. Kamel and C. Faloutsos. On packing r-trees. In *CIKM*, Washington, DC, pp. 490–499, 1993.

31. I. Kamel and C. Faloutsos. Hilbert r-tree: An improved r-tree using fractals. In *Proceedings of VLDB*, Santiago, Chile, pp. 500–509, 1994.

32. I. Kamel and C. Faloutsos. Str: A simple and efficient algorithm for r-tree packing. In *ICDE*, Sydney, New South Wales, Australia, pp. 497–506, 1999.

33. A. Kaporis, C. Makris, G. Mavritsakis, S. Sioutas, A. Tsakalidis, K. Tsichlas, and C. Zaroliagis. Isb-tree: A new indexing scheme with efficient expected behaviour. *Journal of Discrete Algorithms*, 8(4):373–387, 2010.

34. G. Kollios, VJ. Tsotras, D. Gunopoulos, A. Delis, and M. Hadjieleftheriou. Indexing animated objects using spatiotemporal access methods. *IEEE Transactions on Knowledge and Data Engineering*, 13(5):441–448, 2001.

35. A. Kumar, VJ. Tsotras, and C. Faloutsos. Designing access methods for bitemporal databases. *IEEE Transactions on Knowledge and Data Engineering*, 10(1):1–20, 1998.

36. S. Lanka and E. Mays. Fully persistent b+-trees. In *Proceedings of ACM SIGMOD*, Denver, CO, pp. 426–435, 1991.

37. SS. Lightstone, TJ. Teorey, and T. Nadeau. *Physical Database Design: The Database Professional's Guide to Exploiting Indexes, Views, Storage, and More*. Morgan Kaufmann, San Fransisco, CA, 2007.

38. W. Litwin. Linear hashing: A new tool for file and table addressing. In *Proceedings of VLDB, VLDB'80*, Montreal, Quebec, Canada, pp. 212–223, 1980.

39. D. Lomet and B. Salzberg. Access methods for multiversion data. In *Proceedings of ACM SIGMOD*, Portland, OR, pp. 315–324, 1989.

40. M. Lühring, K-U. Sattler, K. Schmidt, and E. Schallehn. Autonomous management of soft indexes. In *ICDE Workshops*, Istanbul, Turkey, pp. 450–458, 2007.

41. Y. Manolopoulos, A. Nanopoulos, AN. Papadopoulos, and Y. Theodoridis. *R-Trees: Theory and Applications*. Springer-Verlag, New York, 2005.

42. B. Moon, HV. Jagadish, C. Faloutsos, and JH. Saltz. Analysis of the clustering properties of the hilbert space-filling curve. *IEEE Transactions on Knowledge and Data Engineering*, 13(1):124–141, 2001.

43. MK. Nguyen, C. Basca, and A. Bernstein. B+hash tree: Optimizing query execution times for on-disk semantic web data structures. In *Proceedings of the 6th International Workshop on Scalable Semantic Web Knowledge Base Systems*, SSWS'10, Shanghai, China, pp. 96–111, 2010.

44. LV. Nguyen-Dinh, WG. Aref, and MF. Mokbel. Spatiotemporal access methods: Part2 (2003–2010). *IEEE Data Engineering Bulletin*, 33(2):46–55, 2010.

45. R. Pagh and FF. Rodler. Cuckoo hashing. In *Proceedings of the 9th Annual European Symposium on Algorithms*, ESA'01, Berlin, Germany, pp. 121–133, 2001.

46. R. Ramakrishnan and J. Gehrke. *Database Management Systems*. McGraw-Hill, Boston, MA, 2002.

47. N. Roussopoulos and D. Leifker. Direct spatial search on pictorial databases using packed r-trees. In *Proceedings of ACM SIGMOD*, Austin, TX, pp. 17–31, 1985.

48. S. Šaltenis, CS. Jensen, ST. Leutenegger, and MA. Lopez. Indexing the positions of continuously moving objects. In *Proceedings of ACM SIGMOD*, Dallas, TX, pp. 331–342, 2000.

49. B. Salzberg and VJ. Tsotras. Comparison of access methods for time-evolving data. *ACM Computing Surveys*, 31(2):158–221, 1999.

50. H. Samet. *The Design and Analysis of Spatial Data Structures*. Addison-Wesley, Reading, MA, 1990.

51. H. Samet. *Foundations of Multidimensional and Metric Data Structures*. Morgan Kaufmann, San Francisco, CA, 2005.

52. T. Sellis, N. Roussopoulos, and C. Faloutsos. The r+-tree: A dynamic index for multi-dimensional objects. In *Proceedings of VLDB*, Brighton, U.K., pp. 507–518, 1987.

53. Y. Tao and D. Papadias. Mv3r-tree: A spatio-temporal access method for timestamp and interval queries. In *Proceedings of VLDB*, Rome, Italy, pp. 431–440, 2001.

54. Y. Theodoridis, M. Vazirgiannis, and T. Sellis. Spatiotemporal indexing for large multimedia applications. In *Proceedings of 3rd IEEE International Conference on Multimedia Computing and Systems*, Hiroshima, Japan, pp. 441–448, 1996.

55. R. Weber, H-J. Schek, and S. Blott. A quantitative analysis and performance study for similarity-search methods in high-dimensional spaces. In *Proceedings of VLDB*, New York, pp. 194–205, 1998.

56. DC. Zilio, J. Rao, S. Lightstone, GM. Lohman, AJ. Storm, C. Garcia-Aranello, and S. Fadden. Db2 design advisor: Integrated automatic physical database design. In *Proceedings of VLDB*, Toronto, Ontario, Canada, pp. 1087–1097, 2004.

11

Query Optimization

Nicolas Bruno
Microsoft Corporation

11.1 Introduction

Database management systems, or DBMSs for short, let users specify queries using high-level declarative languages such as SQL. In such languages, we write *what* we want but not *how* to obtain the desired results. The latter task is delegated to the *query optimizer*, a component in the DBMS responsible for finding an efficient execution plan to evaluate the input query. The query optimizer searches a large space of alternatives and chooses the one that is expected to be least expensive to evaluate. Query optimization is absolutely necessary in a DBMS because the difference in runtime among alternatives and thus the overhead of a bad choice can be arbitrarily large. In fact, for queries of medium complexity, execution plans producing the same result can run from milliseconds to several hours. To answer a SQL query, a typical DBMS goes through a series of steps, illustrated in Figure 11.1*:

1. The input query is *parsed* and transformed into an algebraic tree. This step performs both syntactic and semantic checks over the input query, rejecting all invalid requests.
2. The algebraic tree is *optimized* and turned into an execution plan. A query execution plan indicates not only the operations required to evaluate the input query but also the order in which they are performed and the specific algorithms used at each step.
3. The execution plan is either compiled into machine code or interpreted by the execution engine, and results are sent back to the user.

---

* Note that DBMSs can cache execution plans for repeating input queries and skip the first two steps until the cached plan becomes invalid due to changes in the database design (e.g., an index deletion) or in the data distribution itself.

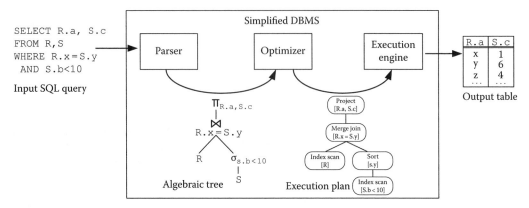

FIGURE 11.1　Executing SQL queries in a relational database system.

Although implementation details vary among specific DBMSs, virtually all optimizers share the same high-level conceptual structure and conduct the search of execution plans in a *cost-based* manner. More specifically, optimizers assign each candidate plan its *estimated* cost and choose the plan with the least expected cost for execution. Alternative formulations attempt to minimize other related metrics, such as the time to obtain the first tuple. Query optimization can therefore be viewed as a complex search problem defined by three interacting components, illustrated in Figure 11.2:

1. Search space: The search space characterizes the set of execution plans that can be considered by the optimizer. For scalability, some optimizers restrict the set of alternatives that can be used to evaluate queries.
2. Cost model: The cost model assigns a cost value to every plan in the search space by estimating the resources consumed by the plan. The quality of the resulting plan is only as good as the underlying cost model.
3. Enumeration strategy: The search space can be characterized without specifying how to generate all candidate plans. The enumeration strategy serves this purpose and dictates how to traverse all interesting plans in the search space efficiently.

The task of an optimizer is nontrivial given the large number of execution plans for an input query, the large variance in response time of the plans in the search space, and the difficulty of accurately estimating costs. A desirable optimizer is one for which (i) the search space includes low-cost plans, (ii) the cost model is accurate, and (iii) the enumeration strategy is efficient.

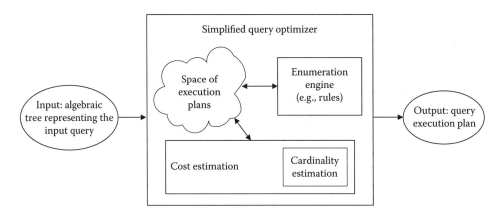

FIGURE 11.2　Architecture of a query optimizer in a DBMS.

The area of query optimization is very rich within the database field. It has been studied in a great variety of contexts and from many different angles, giving rise to diverse solutions in each case. The purpose of this chapter is to primarily discuss the core architecture of modern query optimizers and only touch upon the wealth of results that exist beyond that. Likewise, we make no attempt to provide a complete survey of the literature, in most cases providing only a few example references. More extensive surveys can be found elsewhere [1,2]. The rest of the chapter is organized as follows. Section 11.2 discusses some choices that exist in the space of execution plans and the restrictions usually imposed by current optimizers to make the whole process more manageable. Section 11.3 explains the problem of estimating the cost of execution plans and describes cardinality estimation in more detail, since it is a crucial component in any optimizer cost model. Section 11.4 focuses on enumeration strategies that have been proposed in the context of commercial query optimizers. Section 11.5 touches upon several advanced types of query optimization that have been proposed to address hard problems in the area and raises some open questions in the field. Finally, Section 11.6 concludes the chapter.

11.2 Search Space

The search space of an optimizer depends on the set of physical operators supported by the execution engine and the set of algebraic equivalences among execution subplans. The main source of diversity within an optimizer arises from the available equivalence-preserving transformations.

11.2.1 Operator Reordering

An important class of transformations exploits the commutativity and associativity properties among operators. We next describe three examples of such transformations.

11.2.1.1 Join Reordering

Given a set of relations to be combined in a query, the set of all alternative join trees is determined by two algebraic properties of join: commutativity ($R1 \bowtie R2 \equiv R2 \bowtie R1$), which determines which relations will be inner and outer in the join execution, and associativity [$(R1 \bowtie R2) \bowtie R3 \equiv R1 \bowtie (R2 \bowtie R3)$], which determines the order in which joins will be executed. The alternative join trees that are generated by commutativity and associativity are very large, up to $O(N!)$ for N relations.

Thus, optimizers generally restrict the allowed sequences of join operations to limit the search space and therefore improve optimization time. For example, some optimizers only allow linear sequences of join operations.* Of course, the sequence of joins in an operator tree does not need to be linear. For example, a query joining relations R_1, R_2, R_3, and R_4 can be algebraically represented and evaluated as $(R_1 \bowtie R_2) \bowtie (R_3 \bowtie R_4)$. Such query trees are called bushy and typically require materializing intermediate relations. While bushy trees may result in a cheaper query plan, they considerably increase the cost of enumerating the search space. In general, most optimizers focus on linear join sequences and at most some restricted subsets of bushy join trees.

Another common restriction is to defer Cartesian products until after all joins are processed. This restriction speeds up optimization but may also miss opportunities in some decision-support queries where a single big table joins with multiple small tables. Performing some Cartesian products among these small tables before joining them with the big table can result in a significant reduction in cost.

11.2.1.2 Group-by and Join Clauses

In a conceptual evaluation of a SQL query block, the processing of the join precedes that of the group-by. Similarly, a group-by clause in a sub-query in the FROM clause is conceptually evaluated before the joins in the outer query block. Some algebraic transformations enable commuting group-by and join

---

* In a linear sequence of joins, at least one operand in each join is a base relation.

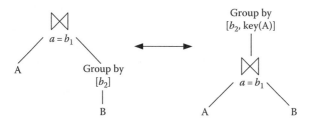

FIGURE 11.3　Reordering joins and group-by clauses.

operators [3,4]. The evaluation of a group-by operator can potentially result in a significant reduction in the number of tuples, since only one tuple is generated for every partition of the relation induced by the group-by. At the same time, joins might eliminate large portions of an input relation, resulting in fewer tuples to subsequently group. Therefore, we can significantly reduce the cost of a query by reordering join and group-by clauses.

As an example of such transformations, a group-by clause can be pulled up above a join, as long as (i) we add a key of the other join relational input to the set of grouping columns and (ii) the join predicate is not defined over an aggregated column, or else the resulting join is not well-formed. Figure 11.3 illustrates this transformation, assuming that b_2 includes b_1 so that the original join is well-formed. Conversely, a group-by clause can be pushed below a join whenever certain conditions are satisfied.

11.2.1.3 Outer Joins

One-sided outer joins are asymmetric operators in SQL that preserve all of the tuples of one relation. For instance, a left outer join between tables R and S, denoted $R \rightarrow S$, is similar to the join $R \bowtie S$, but additionally returns all rows in R that do not match any row in S filling with NULL values all columns in S. Symmetric outer joins preserve both the input relations. Unlike regular joins, a sequence of outer joins and joins do not freely commute. However, the following algebraic equivalence allows to pull outer joins above a block of joins:

$$R \bowtie (S \rightarrow T) \equiv (R \bowtie S) \rightarrow T$$

After outer joins have been delayed in this manner, joins may be freely reordered. There is no guarantee as to which alternative is better in all cases, and thus the use of this transformation needs to be cost-based [5].

11.2.2　Access Path Selection

Earlier query optimizers allowed at most one index for accessing tuples of each table in a given query. For queries with complex single-table predicates, more advanced strategies can manipulate multiple indexes simultaneously to efficiently obtain the qualifying tuples. Consider, as an example, a SQL query that returns information on engineers that earn less than $50,000

```
SELECT *
FROM Emp
WHERE Title = 'Engineer' AND Salary < 50000
```

and suppose that we have single-column indexes $E1 = Emp(Title)$ and $E2 = Emp(Salary)$. Traditionally, the alternative plans that the optimizer considers for the query are either (i) a clustered index scan filtering each tuple by both predicates or (ii) an index seek using either $E1$ or $E2$, followed by look-ups to the clustered index to obtain the remaining columns, ended by a filter on the predicate that was not evaluated by the index. By using both indexes simultaneously, we can sometimes obtain better execution plans. In this case, we could perform what is called an *index intersection* by (i) using index *E*1 to obtain all record identifiers (RIDs) of tuples satisfying `Title = 'Engineer,'` (ii) using index *E*2 to obtain all RIDs of tuples that satisfy `Salary < 50000`, (iii) intersecting both lists of RIDs, and (iv) fetching from the clustered index the resulting records, which satisfy both predicates.

This alternative can greatly reduce execution times because the individual indexes *E*1 and *E*2 are in general smaller than the clustered index, and the intersected list of RIDs can be much smaller than either of the original lists, resulting in much fewer RID lookups.

In general, we can intersect several indexes to answer complex single-table predicates, although at some point the benefit from additional index intersections is outweighed by its cost [6]. Additionally, there are analogous strategies to handle disjunctive predicates using RID unions and negation predicates using RID differences or two seek operations (e.g., employees that are not engineers). Considering these alternatives certainly increases the search space considerably, so some optimizers use heuristics similar to those used for join reordering to limit the number of such index-based alternative plans.

11.2.3 Other Algebraic Transformations

There are several additional transformations that exploit special cases or common patterns, some of which we briefly discuss in the following text:

Query decorrelation: The traditional strategy to process nested queries evaluates the inner sub-query *for each tuple* of the outer sub-query, in what sometimes is called *tuple iteration semantics*. Much work in the literature has identified techniques to decorrelate or *unnest* such queries by "flattening" them into a single block that can be executed more efficiently [7,8]. The complexity of the unnesting problem depends on the structure of the query. The simplest scenarios can be directly unnested using semijoins. The problem is more complex when aggregates are present in the nested sub-query, since unnesting requires pulling up the aggregation without violating the semantics of the nested query, which can be especially tricky in presence of duplicates and NULL values.

Partial preaggregation: In some scenarios, rather than fully pushing a group-by below a join, we can unfold a group-by clause into a local, more fine-grained group-by clause, which is pushed below the join, and a global group-by clause, which performs the final aggregation [9]. Such staged computation may still be useful in reducing the join costs because of the data reduction effect of the local aggregate, but requires the aggregate function to satisfy some additional algebraic properties.

Materialized views: Materialized views are results of queries that are stored and used by the optimizer transparently to answer other queries. Given a set of materialized views and a query, the problem is to optimize the query while leveraging available materialized views as much as possible. This problem introduces the challenge of identifying potential reformulations of the query so that it uses one or more of the materialized views [10,11].

Common subexpressions: When intermediate results are defined multiple times in a query, there are opportunities for avoiding multiple identical computations. Several specialized techniques leverage this fact to produce efficient plans that spool and reuse common subexpressions [12].

11.2.4 Implementation Choices

In addition to all the algebraic equivalences discussed so far, the search space also contains alternative implementations for each operator. For instance, joins can be implemented by different algorithms (e.g., hash-based, merge-based, and nested-loop-based), aggregates can be hash- or stream-based, supporting data structures can be built on the fly, and other implementation characteristics of this sort. The optimizer needs to consider different implementations for the same operator because depending on data and environment characteristics, any such alternative can be the most efficient one.

11.3 Cost Model

As we discussed so far, there are many logically equivalent algebraic expressions associated to a given query and multiple ways to implement each of these expressions. Even if we ignore the computational complexity of enumerating the space of possibilities, there remains the question of deciding which of the operator trees consumes the least resources. Resources may be CPU time, I/O cost, memory, or a combination of these. Therefore, given a partial or full operator tree, being able to accurately and efficiently evaluate its cost is of fundamental importance. Cost estimation must be *accurate*, since optimization is only as good as its cost estimates, and *efficient*, since it is in the inner loop of query optimization. The basic framework for estimating costs is based on the following recursive approach:

1. Collect statistical summaries of stored data.
2. Given an operator in the execution plan and statistical summaries for each of its subplans, determine (i) statistical summaries of the output and (ii) the estimated cost of executing the operator, which involves formulas that account the corresponding actions in the operator.

The procedure can be applied to an arbitrary tree to derive the costs of each operator. The estimated cost of a plan is then obtained by combining the costs of each of its operators. The resources needed to execute a query plan, and therefore its cost, is a function of the sizes of the intermediate query results. For that reason, cost estimation heavily depends on cardinality estimates of subplans generated during optimization. The following example illustrates how sizes of intermediate results may significantly change the chosen plan. Consider the following query template, where @C is a numeric parameter:

```
SELECT * FROM R,S
WHERE R.x = S.y and R.a < @C
```

Figure 11.4 shows optimal execution plans when @C is 20, 200, and 2000. Although the three query instances are syntactically very similar, the resulting plans are rather different. In Figure 11.4a, the number of tuples in *R* satisfying R.a < 20 is very small. Consequently, using a non-clustered index over R.a, the resulting plan first retrieves the RIDs of all tuples in *R* that satisfy R.a < 20. Then, using lookups against table *R*, it lookups the actual tuples that correspond to those RIDs. Finally, it performs a nested-loop join between the subset of tuples of *R* calculated before, and table *S*, which is sequentially scanned. For the case @C = 2000 in Figure 11.4c, the number of tuples of *R* satisfying R.a < 2000 is rather large. The resulting plan therefore scans both tables sequentially, discards on the fly tuples from *R* that do not satisfy the condition R.a < 2000, and then performs a hash join to obtain the result. (Note that in this scenario, the lookups of the previous plan would have been too numerous and therefore too expensive.) Finally, Figure 11.4b shows the execution plan when the number of tuples of *R* satisfying the predicate is neither too small nor too large. In this case, table *S* is scanned in increasing order of S.y using a

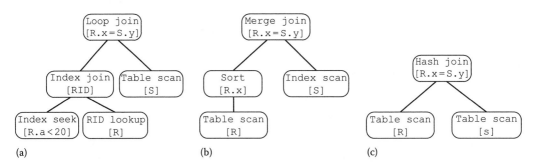

FIGURE 11.4   Query plans for various instances of a template query (a) R.a < 20, (b) R.a < 200, (c) R.a < 2000.

clustered index, and table R is first scanned sequentially discarding invalid tuples as before and then sorted by $R.x$. Finally, a merge join is performed on the two intermediate results.

11.3.1 Statistical Summaries of Data

For every table, statistical information usually includes the number of tuples, the average size of each tuple, the number of physical pages used by the table, and the number of distinct tuples in the table, among others. Statistical information on table columns, if available, is helpful for estimating the cardinality of range or join predicates. A large body of work studies the representation of statistics on a given column or combination of columns. Such alternatives include histograms [13], wavelets [14], sketches [15], sampling [16,17], curve-fitting/parametric information [18], and many others [13]. In most commercial systems, information on the data distribution on a column is provided by histograms, since they can be built and maintained efficiently.

Histograms divide the values of a column or set of columns into a number of buckets, and associate with each bucket some aggregated information. The number of buckets not only influences the accuracy of the histogram, but also affects memory usage, since relevant histograms are loaded into memory during optimization. Both the particular procedure used to select bucket boundaries and the aggregated information associated with each bucket lead to different families of histograms [13]. A popular example is equi-depth histograms, which divide the domain of a column in a table into ranges so that each range has approximately the same number of rows. Other histogram variants, illustrated in Figure 11.5, are equi-width (where the data domain is divided into ranges of equal size), max-diff (where bucket boundaries are placed between pairs of consecutive values that have the largest *frequency gap*), and end-biased (where separate counts are maintained for a small number of very frequent values, and the remaining values are modeled using an equi-depth histogram). Although single-dimensional histograms are widely used, multidimensional histograms are relatively rare in commercial DBMSs.

11.3.2 Cardinality Estimation

Histograms and other statistical structures provide a compressed representation of the underlying column distribution. When manipulating such compact data structures, we generally make a number of simplifying assumptions. For instance, we assume that the tuples in a histogram bucket are distributed uniformly in the bucket domain. In absence of multidimensional statistics, we also assume that predicates in different columns are independent. Using these assumptions, there are simple procedures

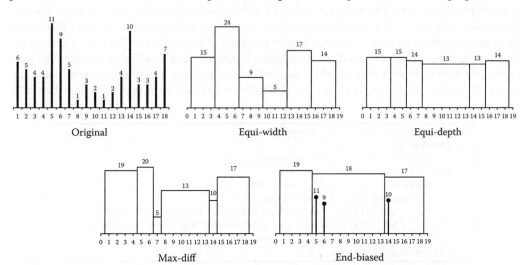

FIGURE 11.5 Histograms approximate value distributions of table columns.

to estimate the number of output tuples for each operator. For instance, to estimate the cardinality of single-table range predicates we use interpolation over the histogram buckets. Join estimation requires that we combine information of two histograms, and involves aligning buckets under certain simplifying assumptions on data distributions. Group-by queries leverage the number of distinct values, stored either globally at the column level, or inside each histogram bucket. Of course, these assumptions only approximate the true cardinality values, and the aforementioned techniques are prone to very large errors in the worst case [19]. To mitigate these problems, some approaches directly store statistics on intermediate expressions and are able to estimate cardinality values without simplifying assumptions whenever the input subexpressions match such intermediate statistics [20].

11.4 Enumeration Strategy

An enumeration algorithm must pick an efficient execution plan for an input query by effectively exploring the search space. An important consideration while designing an enumeration algorithm is to allow it to gracefully adapt to changes in the search space or cost model. Optimization architectures built with this paradigm are called *extensible optimizers*. Building an extensible optimizer does not just involve designing a better enumeration algorithm but also providing an infrastructure for evolution of the optimizer design. In this section, we focus on three representative examples of such extensible optimizers. Rather than providing a detailed description of all the features in these enumeration architectures, we give instead a high-level overview of the frameworks and focus on the aspects that are relevant to our discussion.

11.4.1 Dynamic Programming

System-R was a seminal project that influenced much of the subsequent work in query optimization [21]. The join enumeration algorithm in System-R uses dynamic programming, and it assumes that the cost model satisfies the *principle of optimality*. Specifically, we assume that in order to obtain an optimal plan for a query consisting of k joins, it suffices to consider only the optimal plans for query subexpressions that consist of fewer joins and extend those plans with additional joins. In other words, the optimal plan is obtained by combining optimal subplans.

The dynamic programming algorithm, illustrated in Figure 11.6, views the input as a set of relations $\{R_1, ..., R_n\}$ to be joined and works in a bottom-up manner. We assume there is an associative array bestPlan which, or each subset of tables $\mathcal{R}$, returns the best calculated plan so far for $\mathcal{R}$. First, we generate plans for every table in the query (lines 1 and 2). Such access path selection is encapsulated in function bestAccessPath and consists of scans, seeks, or optionally, more complex plans using index intersections and unions. Then, the dynamic programming algorithm performs $n - 1$ iterations in lines 3–8. At the end of the *i-th* iteration, we produce the optimal plans for all subexpressions that join i tables. Line 4 considers each subset $\mathcal{R}$ of i tables, and line 5 attempts to further partition each such subset into $\mathcal{R}'$ and $\mathcal{R} - \mathcal{R}'$. For each pair of table subsets, line 6 obtains the best plan that joins tables in $\mathcal{R}'$

```
JoinEnumeration ({R₁, …, Rₙ}:input tables)
1    foreach Rᵢ ∈ {R₁, …, Rₙ}
2        bestPlan({Rᵢ}) = bestAccessPath(Rᵢ)
3    for i = 2 to n
4        foreach R ⊂ {R₁, …, Rₙ} such that |R|=i
5            foreach R' ⊂ R such that R' ≠ ∅
6                candPlan = bestJoin(bestPlan(R'), bestPlan(R-R'))
7                if (cost(candPlan) < cost(bestPlan(R)) // cost(null)=∞
8                    bestPlan(R) = candPlan
9    return bestPlan({R₁, …, Rₙ})
```

FIGURE 11.6   Dynamic programming algorithm to enumerate joins.

and tables in $\mathcal{R} - \mathcal{R}'$. For that purpose, the procedure bestJoin tries different join implementations (e.g., hash-, merge-, and index-based joins). Note that the input to bestJoin is bestPlan($\mathcal{R}'$) and bestPlan($\mathcal{R} - \mathcal{R}'$), therefore leveraging the principle of optimality. Lines 7 and 8 update bestPlan for the current subset $\mathcal{R}$ based on the cost model. Suboptimal plans are discarded and never considered again. Finally, line 9 returns the best plan for the whole set of tables, which corresponds to the optimal join reordering for the input query. We now supplement the description of the dynamic programming algorithm for join reordering with a couple of important refinements.

First, consider a query that represents the join among $\{R_1, R_2,$ and $R_3\}$ with the predicates $R_1.a = R_2.a = R_3.a$. Let us also assume that the cost of a hash-based join plan for $R_1 \bowtie R_2$ is smaller than that of a merge-based join alternative. In such a case, bestPlan($\{R_1, R_2\}$) would only keep the hash-based join alternative discarding the suboptimal merge-based join. However, note that if merge-based join alternative is used in $R_1 \bowtie R_2$, the result of the join is sorted on columns $R_1.a$ and $R_2.a$. The sorted order may significantly reduce the cost of the subsequent join with R_3. Thus, pruning the merge join alternative for $R_1 \bowtie R_2$ can result in suboptimality of the global plan. The problem arises because the result of the merge join between R_1 and R_2 has an ordering of tuples in the output stream that is useful in a subsequent join. However, the hash-based join alternative, while cheaper locally, does not have such ordering. To address this problem, the dynamic programming algorithm is extended to keep in bestPlan an optimal plan for every choice of an *interesting* sort order. Two plans are then compared if they represent the same expression and have the same interesting order.

Second, note that the algorithm in Figure 11.6 enumerates all bushy plans, including those that perform early Cartesian products. The algorithm can easily be adapted to enumerate through different search spaces by suitably modifying line 5. To consider only (left) linear join trees, line 5 is changed to:

5 **foreach** $\mathcal{R}' \subset \mathcal{R}$ such that $|\mathcal{R}'| = 1$

The strategy discussed in this section was extended several times, culminating in a technique in [22] that efficiently enumerates all bushy trees with no cross products and additionally considers outerjoins.

11.4.2 Rule-Based Systems

In the context of extensible DBMSs, several rule-based strategies have been proposed. Rules are defined on how plans can be constructed or modified, and the enumeration strategy follows the rules to explore the search space. The most representative efforts are those of Starburst [23] and Volcano/Cascades [24,25]. In this section, we briefly describe the Cascades framework, used as the foundation for both academic and industrial query optimizers. This framework works by manipulating *operators*, which are the building blocks of *operator trees* and are used to describe both the input declarative queries and the output execution plans. Consider the simple SQL query in the following text:

```
SELECT *
FROM R, S, T
WHERE R.x = S.x AND S.y = T.y
```

Figure 11.7a shows a tree of logical operators that specify, in an almost one-to-one correspondence, the relational algebra representation of the query in the preceding text. In turn, Figure 11.7c shows a tree of physical operators that corresponds to an execution plan for the aforementioned query. The goal of the optimization process is to transform the original logical operator tree into an efficient physical operator tree. For that purpose, Cascades-based optimizers rely on two components: the MEMO data structure, which keeps track of the explored search space, and *optimization tasks*, which guide the enumeration strategy.

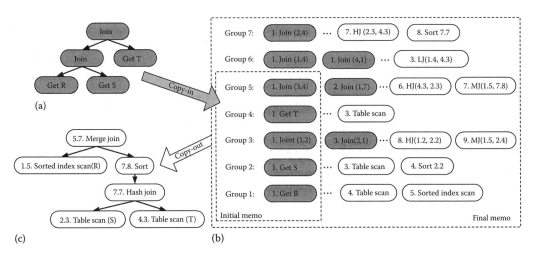

FIGURE 11.7   The MEMO structure in a Cascades optimizer (a) Logical operator tree, (b) MEMO, and (c) physical operator tree.

The MEMO data structure provides a compact representation of the search space of plans by relying on memoization, which is a variant of dynamic programming. A MEMO consists of two key data structures, called *groups* and *groupExpressions*. A *group* represents all equivalent operator trees producing the same output. To reduce memory requirements, a *group* does not explicitly enumerate all its operator trees. Instead, it represents all the operator trees by using *groupExpressions*. A *groupExpression* is an operator having other *groups* as children (rather than other operators). As an example, consider Figure 11.7b, which shows a MEMO for the simple query in the previous section, where logical operators are shaded and physical operators use white background. In the figure, *group* 1 represents all equivalent expressions that return the contents of table *R*. Some operators in *group* 1 are logical (e.g., *Get R*) and some are physical (e.g., *Table Scan* and *Sorted Index Scan*). Likewise, *group* 3 contains all equivalent expressions for *R* ⋈ *S*. Note that *groupExpression* 3.1, (i.e., *Join(1,2)*), represents all operator trees whose root is *Join*, first child is in *group* 1, and second child is in *group* 2. In this way, a MEMO compactly represents a very large number of operator trees.

The enumeration algorithm in Cascades is divided into several *tasks*. Initially, the optimizer schedules the optimization of the *group* corresponding to the root of the original query tree (*group* 5 in the figure). This task in turn triggers the optimization of smaller and smaller operator subtrees and eventually returns the most efficient execution plan for the input query. At the heart of the optimization tasks there are transformation rules. A transformation rule is a pair of (i) an antecedent expression to match in the MEMO and (ii) a consequent expression to generate and introduce back in the MEMO. On one side, *exploration* rules transform logical operator trees into equivalent logical operator trees, and can range from simple rules like join reordering to more complex ones like pushing aggregates below joins. On the other side, *implementation* rules transform logical operator trees into hybrid logical/physical trees by introducing physical operators into the MEMO. Implementation rules range from simple ones like transforming a logical join into a physical hash join to very complex ones that generate index strategies.

Applying a rule can be broken down into four steps. First, all bindings for the rule antecedent are identified and iterated over (note there can be different ways of matching the antecedent of the rule with operator trees in the current *group*). Second, the rule is applied to each binding generating one or more new expressions. Third, the resulting expressions are integrated back into the MEMO, possibly creating new *groups* (e.g., applying join associativity to expression 5.1 in the figure results in *groupExpression* 5.2, which points to a newly created *group* 7). Finally, each new *groupExpression* triggers new tasks that continue exploring logical equivalences or implement physical plans.

11.4.3 Heuristic Enumeration Strategies

The enumeration strategies discussed so far exhaustively explore the plan search space. Their memory requirements and running time thus grow exponentially with query size in the worst case. For that reason, many modern systems have a limit on the size of queries that can be exhaustively optimized (usually around 15 joins). Beyond these limits, query optimizers apply heuristics to enumerate a fraction of the search space or use faster specialized strategies for special cases. As an example of the latter approach, it is known that for chain queries, the classical dynamic programming algorithm works in polynomial time [26]. Also, if the join graph is acyclic and the cost model satisfies the adjacent sequence interchange (ASI) property [27], the IKKBZ technique in [28] returns the optimal plan in $\mathcal{O}(N^2)$ where N is the number of tables.

To effectively address more complex scenarios, heuristic enumeration strategies are proposed, which explore only a fraction of the search space. One line of work adapts randomized techniques and combinatorial heuristics to address this problem, and considers the space of plans as points in a high-dimensional space, that can be *traversed* via transformations (e.g., join commutativity and associativity). Such strategies include iterative improvement, simulated annealing, and genetic algorithms [29] and can be seen as heuristic variations of transformation-based exhaustive enumeration algorithms. Another line of work implements heuristic variations of dynamic programming. Some notable approaches include iteratively performing dynamic programming on a subset of tables and replacing intermediate results with a new *virtual* table [30], simplifying the initial join graph by disallowing non-promising join edges and exhaustively searching the resulting simpler problem [31], and greedily building execution trees one operator at a time [32,33].

11.5 Advanced Optimization Topics

In this section, we discuss some advanced types of optimization that researchers have proposed over the past few years, and for which fundamental questions remain open. The descriptions are based on examples only; further details may be found in the references provided. Furthermore, there are several research challenges that we do not discuss at all due to space limitations, although much interesting work has been and remains to be done on them, such as query optimizer generators; query optimizer validation; and domain-specific query optimization, such as object-oriented, heterogeneous, recursive, XML-based, stream-based, and probabilistic-based, among others.

11.5.1 Parallel/Distributed Optimization

A parallel DBMS is typically an individual system controlling multiple processors that are in the same location. When multiple processors are available, each operator in the query tree can be parallelized by partitioning the input among available processors and later combining results using intraoperator parallelism. Additionally, there are multiple choices to place operators into groups that should be executed simultaneously by the available processors, called interoperator parallelism and further divided into pipelining and independent parallelism. Parallelism can reduce the overall time to process a query, but is associated with much larger search spaces over traditional sequential environments. For that reason, most systems and research prototypes adopt heuristics to avoid dealing with a very large search space. One popular approach first identifies the optimal sequential plan using conventional techniques and then identifies the optimal parallelization/scheduling of that plan [34]. Other approaches only consider schedules that process memory-resident right-deep segments of possibly bushy query plans one at a time with no independent interoperator parallelism [35,36]. More recent strategies integrate parallel plan optimization mode deeply into the query optimizer, instead of being done during postoptimization [37]. With recent growing interest in large-scale computation, parallel optimization remains an active research area.

Distributed Databases are composed of semiautonomous processing sites that are connected via a network that could be spread over a large geographic area. Many prototypes of distributed DBMSs have been implemented [38,39], and several commercial systems offer distributed versions of their products as well. The main differences between centralized and distributed query optimization is in the search space, which additionally contains various processing strategies and opportunities for transmitting data among sites. The most well-known opportunity is to preprocess joins use semijoins in order to only transmit tuples that would certainly contribute to join results [38]. An extension of that idea is using Bloom filters, which are bit vectors that approximate join columns and are transferred across sites to determine which tuples might participate in a join so that only these may be transmitted [39].

11.5.2 Parametric Query Optimization

Many commercial applications rely on precompiled parameterized queries to interact with a database. The DBMS optimizes and caches a parameterized query the first time it is submitted and then reuses the cached execution plan when the query is resubmitted with different parameter values. This approach is especially useful for relatively simple queries, as each optimization call may be a significant fraction of the total execution time. Unfortunately, executing a cached query with a set of parameters different from those used at compilation may be suboptimal. Parametric query optimization addresses this problem by exhaustively determining the optimal plans in each point of the parameter space at compile time. A parametric optimizer determines a set of plans such that, for each point in the parameter space, there is at least one plan in the set that it is optimal. At runtime, these techniques use the actual parameter values and simply pick the plan that was found optimal for them with little or no overhead. Parametric optimization proposals often assume that the cost formulas of physical plans are linear or piecewise linear with respect to the cost parameters or that the regions of optimality are connected and convex. Various approaches to parametric query optimization include dynamic plans [40], randomized algorithms [41], and non-intrusive procedures on top of traditional query optimizers [42]. Recent work provides interesting and actionable diagrams of the plan choices of the optimizer over a space of input parameters [43]. Although much work has been done in the area of parametric optimization, there is still significant room for improvement, especially in the context of industry-strength database systems.

11.5.3 Adaptive Query Processing

Traditional approaches to query processing can be crisply divided into query optimization (which produces an execution plan) and execution proper (which evaluates the plan and obtains results). The quality of the resulting execution plan depends, to a large extent, on (i) the cost model and underlying cardinality estimations being accurate and (ii) the absence of variability in the environment variables used during optimization (e.g., available memory). In real scenarios, however, both these assumptions are not valid, and thus execution plans can be highly suboptimal. This problem motivates *intra-query adaptivity*, which essentially attempts to somehow interleave query optimization and execution, allowing the system to correct bad choices as soon as possible [44]. A simple approach uses execution feedback to progressively improve the statistical information used during optimization [18,45]. More dynamic approaches to address this problem insert checkpoints in the execution plan that monitor actual cardinality values during execution. If the actual intermediate results are very different from the estimated ones, the system suspends the execution of the query, reoptimizes it using the new information, and chooses whether to continue with the original plan or switch to a new one [46,47]. An extreme approach continuously reorders operators in a query plan as it runs, conceptually considering different plans for each routed tuple [48]. Adaptive query processing is a topic that received much attention lately, but has not yet fully found its way into commercial systems.

11.5.4 Multiquery Optimization

So far, this chapter focused on optimizing individual queries. Quite often, however, multiple queries become available for optimization at roughly the same time. Instead of optimizing each query separately, we may obtain a *global* plan that, although suboptimal for each individual query, is optimal for the execution of all of them as a group [49]. Consider the following queries:

```
Q1 =    SELECT Emp.Name          Q2 =    SELECT Dept.floor, Emp.Name
        FROM Emp, Dept                   FROM Emp, Dept
        WHERE ON Emp.id = Dept.id        WHERE ON Emp.id = Dept.id
        AND Dept.Budget > 1M             AND Emp.Title = 'Programmer'
```

Depending on the sizes of the Emp and Dept relations and the selectivities of the predicates, it may well be that computing the entire join once and then applying the two predicates are more efficient than doing the join twice, each time taking into account the corresponding predicate.

11.5.5 Semantic Optimization

Semantic query optimization is a form of optimization that uses integrity constraints defined in the database to rewrite a given query into semantically equivalent ones [50,51]. These can then be optimized as regular queries, and the most efficient plan among all can be used to answer the original query. Examples of semantic optimization include introduction of inferred predicates, removal of joins under foreign key constraints when only columns from the primary table are used, and inference about distinct values for grouping queries. As a simple example, consider an integrity constraint on table Emp of the form Title = "Programmer" $\Rightarrow$ salary > 100K and the following query:

```
SELECT name, floor
from Emp JOIN Dept ON Emp.id = Dept.id
WHERE Title = "Programmer"
```

Using this constraint, the query can be rewritten into an equivalent one that includes an extra predicate salary > 100 K, which results in an efficient plan if the only available index is on Emp.salary.

```
SELECT name, floor
from Emp JOIN Dept ON Emp.id = Dept.id
WHERE Title = "Programmer" AND salary > 100K
```

Semantic optimization is a relatively old topic that periodically finds renewed interest in the research community.

11.5.6 Physical Database Design

DBMSs provide physical data independence, so that queries return the same result independently of the set of available indexes. Of course, the *performance* of evaluating queries will significantly vary depending on index availability. Together with the capabilities of the execution engine and query optimizer, the physical design of a database determines how efficiently a query is ultimately executed. For this reason, there has been a great amount of interest in automatically deciding, for a given query workload, what would be a reasonably good index configuration to materialize and maximize query processing performance [52]. Physical database design is a complex search problem over the space of available indexes, and current solutions iterate over different index configurations evaluating the benefit of such index subsets. These solutions require evaluating the expected cost of a query under a given candidate index configuration in the search space. To do so in a scalable manner, optimizers are typically extended with a new interface that *simulates* a hypothetical index configuration in the DBMS and optimize queries without actually materializing new indexes, a process called *what-if optimization*. What-if optimization is a cornerstone of automated physical database design, which enabled virtually all commercial tuners. It is also one of the first scenarios in which the query optimizer itself is made into the inner loop of an even larger application. Relying on a query optimizer to recommend physical design changes is a novel research direction that found significant interest in the industry, and is currently being actively refined and generalized in the research community.

11.6 Conclusion

To a large extent, the success of a DBMS lies in the quality, functionality, and sophistication of its query optimizer. In this chapter, we gave a bird's eye view of query optimization, presented an abstraction of the architecture of a query optimizer, and focused on the techniques currently used by most commercial systems. In addition, we provided a glimpse of advanced issues in query optimization, whose solutions are still partial and have not completely found their way into commercial systems. Although query optimization has existed as a field for more than 30 years, it is very surprising how fresh it remains in terms of being a source of research problems, as illustrated in the previous section. At the same time, in virtually every component of the traditional optimization architecture of Figure 11.2, there are crucial questions for which we do not have complete answers. When is it worthwhile to consider bushy trees instead of just linear trees? How can buffering and other runtime variables be effectively modeled in the cost formulas? Which search strategy can be used for complex queries with reliability and confidence? We believe that the future in query optimization will be as active as the past and will bring many advances to optimization technology, changing many of the approaches currently used in practice. Despite its age, query optimization remains an exciting field.

Acknowledgment

I thank Ravi Ramamurthy for many useful comments that improved the presentation and content of this chapter. Portions of this chapter also appeared in earlier papers and book chapters [52,53].

References

1. S. Chaudhuri. An overview of query optimization in relational systems. In: *Proceedings of the ACM Symposium on Principles of Database Systems (PODS)*, Seattle, WA, 1998.
2. M. Jarke and J. Koch. Query optimization in database systems. *ACM Computing Surveys*, 16(2), 1984.
3. S. Chaudhuri and K. Shim. An overview of cost-based optimization of queries with aggregates. *IEEE Data Engineering Bulleting*, 18(3), 3–9, 1995.
4. W.P. Yan and P.-Å. Larson. Performing group-by before join. In: *Proceedings of the International Conference on Data Engineering (ICDE)*, Houston, TX, 1994.
5. C.A. Galindo-Legaria and A. Rosenthal. Outerjoin simplification and reordering for query optimization. *ACM Transactions on Database Systems*, 22(1), 43–74, 1997.
6. C. Mohan, D.J. Haderle, Y. Wang, and J.M. Cheng. Single table access using multiple indexes: Optimization, execution, and concurrency control techniques. In: *Proceedings of the International Conference on Extending Database Technology (EDBT)*, Venice, Italy, 1990.
7. M. Elhemali, C. Galindo-Legaria, T. Grabs, and M. Joshi. Execution strategies for sql subqueries. In: *Proceedings of the ACM International Conference on Management of Data (SIGMOD)*, Beijing, China, 2007.

8. M. Muralikrishna. Improved unnesting algorithms for join aggregate SQL queries. In: *Proceedings of the International Conference on Very Large Databases (VLDB)*, Vancouver, British Columbia, Canada, 1992.

9. P.-Å. Larson. Data reduction by partial preaggregation. In: *Proceedings of the International Conference on Data Engineering (ICDE)*, San Jose, CA, 2002.

10. S. Chaudhuri, R. Krishnamurthy, S. Potamianos, and K. Shim. Optimizing queries with materialized views. In: *Proceedings of the International Conference on Data Engineering (ICDE)*, Taipei, Taiwan, 1995.

11. J. Goldstein and P.-Å. Larson. Optimizing queries using materialized views: A practical, scalable solution. In: *Proceedings of the ACM International Conference on Management of Data (SIGMOD)*, Santa Barbara, CA, 2001.

12. J. Zhou, P.-Å. Larson, J.-C. Freytag, and W. Lehner. Efficient exploitation of similar subexpressions for query processing. In: *Proceedings of the ACM International Conference on Management of Data (SIGMOD)*, Beijing, China, 2007.

13. Y.E. Ioannidis. The history of histograms (abridged). In: *Proceedings of the International Conference on Very Large Databases (VLDB)*, Berlin, Germany, 2003.

14. Y. Matias, J.S. Vitter, and M. Wang. Wavelet-based histograms for selectivity estimation. In: *Proceedings of the ACM International Conference on Management of Data (SIGMOD)*, Seattle, WA, 1998.

15. F. Rusu and A. Dobra. Sketches for size of join estimation. *ACM Transactions on Database Systems*, 33, 1–46, 2008.

16. F. Olken and D. Rotem. Random sampling from database files: A survey. In: *Statistical and Scientific Database Management, 5th International Conference (SSDBM)*, Charlotte, NC, 1990.

17. R.J. Lipton, J.F. Naughton, and D.A. Schneider. Practical selectivity estimation through adaptive sampling. In: *Proceedings of the ACM International Conference on Management of Data (SIGMOD)*, Atlantic City, NJ, 1990.

18. C.-M. Chen and N. Roussopoulos. Adaptive selectivity estimation using query feedback. In: *Proceedings of the ACM International Conference on Management of Data (SIGMOD)*, Minneapolis, MN, 1994.

19. Y. Ioannidis and S. Christodoulakis. On the propagation of errors in the size of join results. In: *Proceedings of the ACM International Conference on Management of Data (SIGMOD)*, Denver, Co, 1991.

20. N. Bruno and S. Chaudhuri. Conditional selectivity for statistics on query expressions. In: *Proceedings of the ACM International Conference on Management of Data (SIGMOD)*, Paris, France, 2004.

21. P.G. Selinger et al. Access path selection in a relational database management system. In: *Proceedings of the ACM International Conference on Management of Data (SIGMOD)*, Boston, MA, 1979.

22. G. Moerkotte and T. Neumann. Dynamic programming strikes back. In: *Proceedings of the ACM International Conference on Management of Data (SIGMOD)*, Vancouver, British Columbia, Canada, 2008.

23. L.M. Haas et al. Extensible query processing in Starburst. In: *Proceedings of the ACM International Conference on Management of Data (SIGMOD)*, Portland, OR, 1989.

24. G. Graefe. The Cascades framework for query optimization. *Data Engineering Bulletin*, 18(3), 19–29, 1995.

25. G. Graefe and W. McKenna. The Volcano optimizer generator: Extensibility and efficient search. In: *Proceedings of the International Conference on Data Engineering (ICDE)*, Vienna, Austria, 1993.

26. K. Ono and G.M. Lohman. Measuring the complexity of join enumeration in query optimization. In: *Proceedings of the International Conference on Very Large Databases (VLDB)*, Brisbane, Queensland, Australia, 1990.

27. T. Ibaraki and T. Kameda. On the optimal nesting order for computing n-relational joins. *ACM Transactions on Database System*, 9(3), 483–502, 1984.

28. R. Krishnamurthy, H. Boral, and C. Zaniolo. Optimization of nonrecursive queries. In: *Proceedings of the International Conference on Very Large Databases (VLDB)*, Kyoto, Japan, 1986.

29. M. Steinbrunn, G. Moerkotte, and A. Kemper. Heuristic and randomized optimization for the join ordering problem. *The VLDB Journal*, 6(3), 1997.

30. D. Kossmann and K. Stocker. Iterative dynamic programming: A new class of query optimization algorithms. *ACM TODS*, 25(1), 43–82, 2000.

31. T. Neumann. Query simplification: Graceful degradation for join-order optimization. In: *Proceedings of the ACM International Conference on Management of Data (SIGMOD)*, Providence, RI, 2009.

32. N. Bruno, C. Galindo-Legaria, and M. Joshi. Polynomial heuristics for query optimization. In: *Proceedings of the International Conference on Data Engineering (ICDE)*, Long Beach, CA, 2010.

33. A. Swami. Optimization of large join queries: Combining heuristics and combinatorial techniques. *SIGMOD Record*, 18(2), 367–376, 1989.

34. W. Hong and M. Stonebraker. Optimization of parallel query execution plans in XPRS. In: *Proceedings of the First International Conference on Parallel and Distributed Information Systems (PDIS)*, Miami Beach, FL, 1991.

35. E. Shekita, H. Young, and K.-L. Tan. Multi-join optimization for symmetric multiprocessors. In: *Proceedings of the International Conference on Very Large Databases (VLDB)*, Dublin, Ireland, 1993.

36. D. Schneider and D. DeWitt. Tradeoffs in processing complex join queries via hashing in multiprocessor database machines. In: *Proceedings of the International Conference on Very Large Databases (VLDB)*, Brisbane, Queensland, Australia, 1990.

37. J. Zhou, P.-Å. Larson, and R. Chaiken. Incorporating partitioning and parallel plans into the SCOPE optimizer. In: *Proceedings of the International Conference on Data Engineering (ICDE)*, Long Beach, CA, 2010.

38. P.A. Bernstein, N. Goodman, E. Wong, C.L. Reeve, and J.B. Rothnie Jr. Query processing in a system for distributed databases (SDD-1). *ACM Transactions on Database Systems*, 6(4), 602–625, 1981.

39. L. Mackert and G. Lohman. Validation and performance evaluation for distributed queries. In: *Proceedings of the International Conference on Very Large Databases (VLDB)*, Kyoto, Japan, 1986.

40. R. Cole and G. Graefe. Optimization of dynamic query evaluation plans. In: *Proceedings of the ACM International Conference on Management of Data (SIGMOD)*, Minneapolis, MN, 1994.

41. Y. Ioannidis, R. Ng, K. Shim, and T. Sellis. Parametric query optimization. In: *Proceedings of the International Conference on Very Large Databases (VLDB)*, Vancouver, British Columbia, Canada, 1992.

42. A. Hulgeri and S. Sudarshan. Anipqo: Almost non-intrusive parametric query optimization for nonlinear cost functions. In: *Proceedings of the International Conference on Very Large Databases (VLDB)*, Berlin, Germany, 2003.

43. N. Reddy and J.R. Haritsa. Analyzing plan diagrams of database query optimizers. In: *Proceedings of the International Conference on Very Large Databases (VLDB)*, Trondheim, Norway, 2005.

44. A. Deshpande, Z.G. Ives, and V. Raman. Adaptive query processing. *Foundations and Trends in Databases*, 1(1), 1–140, 2007.

45. N. Bruno, S. Chaudhuri, and L. Gravano. STHoles: A multidimensional workload-aware histogram. In: *Proceedings of the ACM International Conference on Management of Data (SIGMOD)*, Santa Barbara, CA, 2001.

46. N. Kabra and D.J. DeWitt. Efficient mid-query re-optimization of sub-optimal query execution plans. In: *Proceedings of the ACM International Conference on Management of Data (SIGMOD)*, Seattle, WA, 1998.

47. V. Markl, V. Raman, D.E. Simmen, G.M. Lohman, and H. Pirahesh. Robust query processing through progressive optimization. In: *Proceedings of the ACM International Conference on Management of Data (SIGMOD)*, Paris, France, 2004.

48. R. Avnur and J. Hellerstein. Eddies: Continuously adaptive query processing. In: *Proceedings of the ACM International Conference on Management of Data (SIGMOD)*, Dallas, TX, 2000.

49. P. Roy, S. Seshadri, S. Sudarshan, and S. Bhobe. Efficient and extensible algorithms for multi query optimization. In: *Proceedings of the ACM International Conference on Management of Data (SIGMOD)*, Dallas, TX, 2000.

50. Q. Cheng et al. Implementation of two semantic query optimization techniques in DB2 Universal Database. In: *Proceedings of the International Conference on Very Large Databases (VLDB)*, Edinburgh, U.K., 1999.

51. J. King. Quist: A system for semantic query optimization in relational databases. In: *Proceedings of the International Conference on Very Large Databases (VLDB)*, Cannes, France, 1981.

52. N. Bruno. *Automated Physical Database Design and Tuning.* CRC Press, Boca Raton, FL, 2011.

53. A.B. Tucker, (ed.). *The Computer Science and Engineering Handbook.* CRC Press, Boca Raton, FL, 1997.

12

Concurrency Control and Recovery

Michael J. Franklin
University of California at Berkeley

12.1 Introduction

Many service-oriented businesses and organizations, such as banks, airlines, catalog retailers, and hospitals, have grown to depend on fast, reliable, and correct access to their "mission-critical" data on a constant basis. In many cases, particularly for global enterprises, 7×24 access is required, that is, the data must be available 7 days a week, 24 h a day. *Database management systems* (DBMSs) are often employed to meet these stringent performance, availability, and reliability demands. As a result, two of the core functions of a DBMS are (1) to protect the data stored in the database and (2) to provide correct and highly available access to those data in the presence of concurrent access by large and diverse user populations, despite various software and hardware failures. The responsibility for these functions resides in the *concurrency control* and *recovery* components of the DBMS software. *Concurrency control* ensures that individual users see consistent states of the database even though operations on behalf of many users may be interleaved by the database system. *Recovery* ensures that the database is fault-tolerant, that is, that the database state is not corrupted as the result of a software, system, or media failure. The existence of this functionality in the DBMS allows applications to be written without explicit concern for concurrency and fault tolerance. This freedom provides a tremendous increase in programmer productivity and allows new applications to be added more easily and safely to an existing system.

For database systems, correctness in the presence of concurrent access and/or failures is tied to the notion of a *transaction*. A transaction is a unit of work, possibly consisting of multiple data accesses and updates that must *commit* or *abort* as a single atomic unit. When a transaction *commits*, all updates that it performed on the database are made permanent and visible to other transactions. In contrast, when a transaction *aborts*, all of its updates are removed from the database, and the database is restored (if necessary) to the state it would have been in if the aborting transaction had never been executed.

Informally, transaction executions are said to respect the *atomicity, consistency, isolation, and durability (ACID) properties* (Gray and Reuter 1993):

Atomicity: This is the "all-or-nothing" aspect of transactions discussed earlier—either all operations of a transaction complete successfully or none of them do. Therefore, after a transaction has completed (i.e., committed or aborted), the database will not reflect a partial result of that transaction.

Consistency: Transactions preserve the consistency of the data—a transaction performed on a database that is internally consistent will leave the database in an internally consistent state. Consistency is typically expressed as a set of declarative *integrity constraints*. For example, a constraint may be that the salary of an employee cannot be higher than that of his or her manager.

Isolation: A transaction's behavior is not impacted by the presence of other transactions that may be accessing the same database concurrently. That is, a transaction sees only a state of the database that could occur if that transaction were the only one running against the database and produces only results that it could produce if it was running alone.

Durability: The effects of *committed* transactions survive failures. Once a transaction commits, its updates are guaranteed to be reflected in the database even if the contents of volatile (e.g., main memory) or nonvolatile (e.g., disk) storage are lost or corrupted.

Of these four transaction properties, the concurrency control and recovery components of a DBMS are primarily concerned with preserving *atomicity, isolation*, and *durability*. The preservation of the *consistency* property typically requires additional mechanisms such as compile-time analysis or run-time triggers in order to check adherence to integrity constraints.* For this reason, this chapter focuses primarily on the A, I, and D of the ACID transaction properties.

Transactions are used to structure complex processing tasks, which consist of multiple data accesses and updates. A traditional example of a transaction is a money transfer from one bank account (say account A) to another (say B). This transaction consists of a withdrawal from A and a deposit into B and requires four accesses to account information stored in the database: a read and write of A and a read and write of B. The data accesses of this transaction are as follows:

```
TRANSFER ()
  01 A_bal: = Read(A)
  02 A_bal: = A_bal - $50
  03 Write(A, A_bal)
  04 B_bal: = Read(B)
  05 B_bal: = B_bal + $50
  06 Write(B, B_bal)
```

The value of A in the database is read and decremented by $50, and then the value of B in the database is read and incremented by $50. Thus, TRANSFER preserves the invariant that the sum of the balances of A and B prior to its execution must equal the sum of the balances after its execution, regardless of whether the transaction commits or aborts.

Consider the importance of the atomicity property. At several points during the TRANSFER transaction, the database is in a temporarily inconsistent state. For example, between the time that account A is updated (statement 3) and the time that account B is updated (statement 6), the database reflects the decrement of A but not the increment of B, so it appears as if $50 has disappeared from the database. If the transaction reaches such a point and then is unable to complete (e.g., due to a failure or an unresolvable conflict), then the system must ensure that the effects of the partial results of the transaction (i.e., the update to A) are removed from the database—otherwise the database state will be

---

* In the case of triggers, the recovery mechanism is typically invoked to abort an offending transaction.

incorrect. The durability property, in contrast, only comes into play in the event that the transaction successfully commits. Once the user is notified that the transfer has taken place, he or she will assume that account B contains the transferred funds and may attempt to use those funds from that point on. Therefore, the DBMS must ensure that the results of the transaction (i.e., the transfer of the $50) remain reflected in the database state even if the system crashes.

Atomicity, consistency, and durability address correctness for *serial execution* of transactions where only a single transaction at a time is allowed to be in progress. In practice, however, DBMSs typically support *concurrent execution*, in which the operations of multiple transactions can be executed in an interleaved fashion. The motivation for concurrent execution in a DBMS is similar to that for multi-programming in operating systems, namely, to improve the utilization of system hardware resources and to provide multiple users a degree of fairness in access to those resources. The *isolation* property of transactions comes into play when concurrent execution is allowed.

Consider a second transaction that computes the sum of the balances of accounts A and B:

```
REPORTSUM()
  01 A_bal: = Read(A)
  02 B_bal: = Read(B)
  03 Print(A_bal + B_bal)
```

Assume that initially, the balance of account A is $300 and the balance of account B is $200. If a REPORTSUM transaction is executed on this state of the database, it will print a result of $500. In a database system restricted to *serial execution* of transactions, REPORTSUM will also produce the same result if it is executed after a TRANSFER transaction. The atomicity property of transactions ensures that if the TRANSFER aborts, all of its effects are removed from the database (so REPORTSUM would see A = $300 and B = $200), and the durability property ensures that if it commits, then all of its effects remain in the database state (so REPORTSUM would see A = $250 and B = $250).

Under concurrent execution, however, a problem could arise if the isolation property is not enforced. As shown in Table 12.1, if REPORTSUM were to execute after TRANSFER has updated account A but before it has updated account B, then REPORTSUM could see an inconsistent state of the database. In this case, the execution of REPORTSUM sees a state of the database in which $50 has been withdrawn from account A but has not yet been deposited in account B, resulting in a total of $450—it seems that $50 has disappeared from the database. This result is not one that could be obtained in any serial execution of TRANSFER and REPORTSUM transactions. It occurs because in this example, REPORTSUM accessed the database when it was in a temporarily inconsistent state. This problem is sometimes referred to as the *inconsistent retrieval* problem. To preserve the isolation property of transactions, the DBMS must prevent the occurrence of this and other potential anomalies that could arise due to concurrent execution. The formal notion of correctness for concurrent execution in database systems is known as *serializability* and is described in Section 12.2.

Although the transaction processing literature often traces the history of transactions back to antiquity (such as Sumerian tax records) or to early contract law (Gray 1981, Gray and Reuter 1993,

TABLE 12.1 An Incorrect Interleaving of TRANSFER and REPORTSUM

TRANSFER	REPORTSUM	
01 A_bal: = Read (A)		
02 A_bal: = A_bal - $50		
03 Write(A, A_bal)		
	01 A_bal: = Read (A)	/* value is $250 */
	02 B_bal: = Read (B)	/* value is $200 */
	03 Print (A_bal + B_bal)	/* value is $450 */
04 B_bal: = Read (B)		
05 B_bal: = B_bal + $50		
06 Write (B, B_bal)		

Korth 1995), the roots of the transaction concept in information systems are typically traced back to the early 1970s and the work of Bjork (1973) and Davies (1973). Early systems such as IBM's IMS addressed related issues, and a systematic treatment and understanding of ACID transactions was developed several years later by members of the IBM System R group (Gray et al. 1975, Eswaran et al. 1976) and others (e.g., Lomet 1977, Rosenkrantz et al. 1977). Since that time, many techniques for implementing ACID transactions have been proposed, and a fairly well-accepted set of techniques has emerged. The remainder of this chapter contains an overview of the basic theory that has been developed as well as a survey of the more widely known implementation techniques for concurrency control and recovery. A brief discussion of work on extending the simple transaction model is presented at the end of the chapter.

It should be noted that issues related to those addressed by concurrency control and recovery in database systems arise in other areas of computing systems as well, such as file systems and memory systems. There are, however, two salient aspects of the ACID model that distinguish transactions from other approaches. First is the incorporation of both isolation (concurrency control) and fault tolerance (recovery) issues. Second is the concern with treating arbitrary groups of *write* and/or *read* operations on multiple data items as atomic, isolated units of work. While these aspects of the ACID model provide powerful guarantees for the protection of data, they also can induce significant systems implementation complexity and performance overhead. For this reason, the notion of ACID transactions and their associated implementation techniques have remained largely within the DBMS domain, where the provision of highly available and reliable access to "mission-critical" data is a primary concern.

12.2 Underlying Principles

12.2.1 Concurrency Control

12.2.1.1 Serializability

As stated in the previous section, the responsibility for maintaining the isolation property of ACID transactions resides in the concurrency control portion of the DBMS software. The most widely accepted notion of correctness for concurrent execution of transactions is *serializability*. Serializability is the property that an (possibly interleaved) execution of a group of transactions has the same effect on the database, and produces the same output, as some serial (i.e., noninterleaved) execution of those transactions. It is important to note that serializability does not specify any *particular* serial order, but rather, only that the execution is equivalent to *some* serial order. This distinction makes serializability a slightly less intuitive notion of correctness than transaction initiation time or commit order, but it provides the DBMS with significant additional flexibility in the scheduling of operations. This flexibility can translate into increased responsiveness for end users.

A rich theory of database concurrency control has been developed over the years (see Papadimitriou 1986, Bernstein et al. 1987, Gray and Reuter 1993), and serializability lies at the heart of much of this theory. In this chapter, we focus on the simplest models of concurrency control, where the operations that can be performed by transactions are restricted to *read*(x), *write*(x), *commit*, and *abort*. The operation *read*(x) retrieves the value of a data item from the database, *write*(x) modifies the value of a data item in the database, and *commit* and *abort* indicate successful or unsuccessful transaction completion, respectively (with the concomitant guarantees provided by the ACID properties). We also focus on a specific variant of serializability called *conflict serializability*. Conflict serializability is the most widely accepted notion of correctness for concurrent transactions because there are efficient, easily implementable techniques for detecting and/or enforcing it. Another well-known variant is called *view serializability*. View serializability is less restrictive (i.e., it allows more legal schedules) than conflict serializability, but it and other variants are primarily of theoretical interest because they are impractical to implement. The reader is referred to Papadimitriou (1986) for a detailed treatment of alternative serializability models.

12.2.1.2 Transaction Schedules

Conflict serializability is based on the notion of a *schedule* of transaction operations. A schedule for a set of transaction executions is a partial ordering of the operations performed by those transactions, which shows how the operations are interleaved. The ordering defined by a schedule can be partial in the sense that it is only required to specify two types of dependencies:

1. All operations of a given transaction for which an order is specified by that transaction must appear in that order in the schedule. For example, the definition of REPORTSUM in the preceding text specifies that account A is read before account B.
2. The ordering of all *conflicting operations* from different transactions must be specified. Two operations are said to conflict if they both operate on the same data item and at least one of them is a *write()*.

The concept of a schedule provides a mechanism to express and reason about the (possibly) concurrent execution of transactions. A *serial* schedule is one in which all the operations of each transaction appear consecutively. For example, the serial execution of TRANSFER followed by REPORTSUM is represented by the following schedule:

$$r_0[A] \to w_0[A] \to r_0[B] \to w_0[B] \to c_0 \to r_1[A] \to r_1[B] \to c_1 \tag{12.1}$$

In this notation, each operation is represented by its initial letter, the subscript of the operation indicates the *transaction number* (tn) of the transaction on whose behalf the operation was performed, and a capital letter in brackets indicates a specific data item from the database (for read and write operations). A tn is a unique identifier that is assigned by the DBMS to an execution of a transaction. In the example earlier, the execution of TRANSFER was assigned tn 0 and the execution of REPORTSUM was assigned tn 1. A right arrow ($\to$) between two operations indicates that the left-hand operation is ordered before the right-hand one. The ordering relationship is transitive; the orderings implied by transitivity are not explicitly drawn.

For example, the interleaved execution of TRANSFER and REPORTSUM shown in Table 12.1 would produce the following schedule:

$$r_0[A] \to w_0[A] \to r_1[A] \to r_1[B] \to c_1 \to r_0[B] \to w_0[B] \to c_0 \tag{12.2}$$

The formal definition of serializability is based on the concept of equivalent schedules. Two schedules are said to be *equivalent* ($\equiv$) if

1. They contain the same transactions and operations.
2. They order all conflicting operations of non-aborting transactions in the same way.

Given this notion of equivalent schedules, *a schedule is said to be serializable if and only if it is equivalent to some serial schedule*. For example, the following concurrent schedule is serializable because it is equivalent to schedule 12.1:

$$r_0[A] \to w_0[A] \to r_1[A] \to r_0[B] \to w_0[B] \to c_0 \to r_1[B] \to c_1 \tag{12.3}$$

In contrast, the interleaved execution of schedule 12.2 is *not* serializable. To see why, notice that in any serial execution of TRANSFER and REPORTSUM either *both* writes of TRANSFER will precede *both* reads of REPORTSUM or vice versa. However, in schedule 12.2 $w_0[A] \to r_1[A]$ but $r_1[B] \to w_0[b]$. schedule 12.2, therefore, is not equivalent to any possible serial schedule of the two transactions so it is not serializable. This result agrees with our intuitive notion of correctness, because recall that schedule 12.2 resulted in the apparent loss of $50.

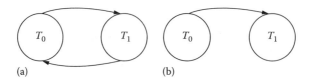

FIGURE 12.1   Precedence graphs for (a) nonserializable and (b) serializable schedules.

12.2.1.3 Testing for Serializability

A schedule can easily be tested for serializability through the use of a *precedence graph*. A precedence graph is a directed graph that contains a vertex for each *committed* transaction execution in a schedule (noncommitted executions can be ignored). The graph contains an edge from transaction execution T_i to transaction execution T_j ($i \neq j$) if there is an operation in T_i that is constrained to precede an operation of T_j in the schedule. A schedule is serializable if and only if its precedence graph is *acyclic*. Figure 12.1a shows the precedence graph for schedule 12.2. That graph has an edge $T_0 \rightarrow T_1$ because the schedule contains $w_0[A] \rightarrow r_1[A]$ and an edge $T_1 \rightarrow T_0$ because the schedule contains $r_1[B] \rightarrow w_0[b]$. The cycle in the graph shows that the schedule is nonserializable. In contrast, Figure 12.1b shows the precedence graph for schedule 12.1. In this case, all ordering constraints are from T_0 to T_1, so the precedence graph is acyclic, indicating that the schedule is serializable.

There are a number of practical ways to implement conflict serializability. These and other implementation issues are addressed in Section 12.3. Before discussing implementation issues, however, we first survey the basic principles underlying database recovery.

12.2.2 Recovery

12.2.2.1 Coping with Failures

Recall that the responsibility for the atomicity and durability properties of ACID transactions lies in the recovery component of the DBMS. For recovery purposes, it is necessary to distinguish between two types of storage: (1) *volatile storage*, such as main memory, whose state is lost in the event of a system crash or power outage and (2) *nonvolatile storage*, such as magnetic disks or tapes, whose contents persist across such events. The recovery subsystem is relied upon to ensure correct operation in the presence of three different types of failures (listed in order of likelihood):

1. *Transaction failure*: When a transaction that is in progress reaches a state from which it cannot successfully commit, all updates that it made must be removed from the database in order to preserve the atomicity property. This is known as *transaction rollback*.
2. *System failure*: If the system fails in a way that causes the loss of volatile memory contents, recovery must ensure that (1) the updates of all transactions that had committed prior to the crash are reflected in the database and (2) all updates of other transactions (aborted or in progress at the time of the crash) are removed from the database.
3. *Media failure*: In the event that data are lost or corrupted on the nonvolatile storage (e.g., due to a disk-head crash), then the online version of the data is lost. In this case, the database must be restored from an archival version of the database and brought up to date using operation logs.

In this chapter, we focus on the issues of rollback and crash recovery, the most frequent uses of the DBMS recovery subsystem. Recovery from media crashes requires substantial additional mechanisms and complexity beyond what is covered here. Media recovery is addressed in the recovery-related references listed at the end of this chapter.

12.2.2.2 Buffer Management Issues

The process of removing the effects of an incomplete or aborted transaction for preserving atomicity is known as *UNDO*. The process of reinstating the effects of a committed transaction for durability is known as *REDO*. The amount of work that a recovery subsystem must perform for either of these functions depends on how the DBMS buffer manager handles data that are updated by in-progress and/or committing transactions (Haerder and Reuter 1983, Bernstein et al. 1987). Recall that the buffer manager is the DBMS component that is responsible for coordinating the transfer of data between main memory (i.e., volatile storage) and disk (i.e., nonvolatile storage). The unit of storage that can be written atomically to nonvolatile storage is called a *page*. Updates are made to copies of pages in the (volatile) buffer pool, and those copies are written out to nonvolatile storage at a later time. If the buffer manager allows an update made by an *uncommitted* transaction to overwrite the most recent committed value of a data item on nonvolatile storage, it is said to support a *STEAL* policy (the opposite is called *NO-STEAL*). If the buffer manager ensures that all updates made by a transaction are reflected on nonvolatile storage before the transaction is allowed to commit, then it is said to support a *FORCE* policy (the opposite is *NO-FORCE*).

Support for the STEAL policy implies that in the event that a transaction needs to be rolled back (due to transaction failure or system crash), UNDOing the transaction will involve restoring the values of any nonvolatile copies of data that were overwritten by that transaction back to their previous committed state. In contrast, a NO-STEAL policy guarantees that the data values on nonvolatile storage are valid, so they do not need to be restored. A NO-FORCE policy raises the possibility that some committed data values may be lost during a system crash because there is no guarantee that they have been placed on nonvolatile storage. This means that substantial REDO work may be required to preserve the durability of committed updates. In contrast, a FORCE policy ensures that the committed updates *are* placed on nonvolatile storage, so that in the event of a system crash, the updates will still be reflected in the copy of the database on nonvolatile storage.

From the earlier discussion, it should be apparent that a buffer manager that supports the combination of NO-STEAL and FORCE would place the fewest demands on UNDO and REDO recovery. However, these policies may negatively impact the performance of the DBMS during normal operation (i.e., when there are no crashes or rollbacks) because they restrict the flexibility of the buffer manager. NO-STEAL obligates the buffer manager to retain updated data in memory until a transaction commits or to write those data to a temporary location on nonvolatile storage (e.g., a swap area). The problem with a FORCE policy is that it can impose significant disk write overhead during the critical path of a committing transaction. For these reasons, many buffer managers support the STEAL and NO-FORCE (*STEAL/NO-FORCE*) policies.

12.2.2.3 Logging

In order to deal with the UNDO and REDO requirements imposed by the STEAL and NO-FORCE policies, respectively, database systems typically rely on the use of a *log*. A log is a sequential file that stores information about transactions and the state of the system at certain instances. Each entry in the log is called a *log record*. One or more log records are written for each update performed by a transaction. When a log record is created, it is assigned a *log sequence number* (LSN), which serves to uniquely identify that record in the log. LSNs are typically assigned in a monotonically increasing fashion so that they provide an indication of relative position in the log. When an update is made to a data item in the buffer, a log record is created for that update. Many systems write the LSN of this new log record into the page containing the updated data item. Recording LSNs in this fashion allows the recovery system to relate the state of a data page to logged updates in order to tell if a given log record is reflected in a given state of a page.

Log records are also written for transaction management activities such as the commit or abort of a transaction. In addition, log records are sometimes written to describe the state of the system at certain periods of time. For example, such log records are written as part of the *checkpointing* process. Checkpoints are taken periodically during normal operation to help bound the amount of recovery work that would be required in the event of a crash. Part of the checkpointing process involves the writing of one or more *checkpoint records*. These records can include information about the contents of the buffer pool and the transactions that are currently active. The particular contents of these records depend on the method of checkpointing that is used. Many different checkpointing methods have been developed, some of which involve quiescing the system to a consistent state, while others are less intrusive. A particularly nonintrusive type of checkpointing is used by the ARIES recovery method (Mohan et al. 1992) that is described in Section 12.3.

For transaction update operations, there are two basic types of logging: *physical* and *logical* (Gray and Reuter 1993). Physical log records typically indicate the location (e.g., position on a particular page) of modified data in the database. If support for UNDO is provided (i.e., a STEAL policy is used), then the value of the item prior to the update is recorded in the log record. This is known as the *before image* of the item. Similarly the *after image* (i.e., the new value of the item after the update) is logged if REDO support is provided. Thus, physical log records in a DBMS with STEAL/NO-FORCE buffer management contain both the old and new data values of items. Recovery using physical log records has the property that recovery actions (i.e., UNDOs or REDOs) are *idempotent*, meaning that they have the same effect no matter how many times they are applied. This property is important if recovery is invoked multiple times, as will occur if a system fails repeatedly (e.g., due to a power problem or a faulty device).

Logical logging (sometimes referred to as *operational logging*) records only high-level information about operations that are performed, rather than recording the actual changes to items (or storage locations) in the database. For example, the insertion of a new tuple into a relation might require many physical changes to the database such as space allocation, index updates, and reorganization. Physical logging would require log records to be written for all of these changes. In contrast, logical logging would simply log the fact that the insertion had taken place, along with the value of the inserted tuple. The REDO process for a logical logging system must determine the set of actions that are required to fully reinstate the insert. Likewise, the UNDO logic must determine the set of actions that make up the inverse of the logged operation.

Logical logging has the advantage that it minimizes the amount of data that must be written to the log. Furthermore, it is inherently appealing because it allows many of the implementation details of complex operations to be hidden in the UNDO/REDO logic. In practice, however, recovery based on logical logging is difficult to implement because the actions that make up the logged operation are not performed atomically. That is, when a system is restarted after a crash, the database may not be in an *action consistent* state with respect to a complex operation—it is possible that only a subset of the updates made by the action had been placed on nonvolatile storage prior to the crash. As a result, it is difficult for the recovery system to determine which portions of a logical update are reflected in the database state upon recovery from a system crash. In contrast, physical logging does not suffer from this problem, but it can require substantially higher logging activity.

In practice, systems often implement a compromise between physical and logical approaches that has been referred to as *physiological logging* (Gray and Reuter 1993). In this approach, log records are constrained to refer to a single page, but may reflect logical operations on that page. For example, a physiological log record for an insert on a page would specify the value of the new tuple that is added to the page, but would not specify any free-space manipulation or reorganization of data on the page resulting from the insertion; the REDO and UNDO logic for insertion would be required to infer the necessary operations. If a tuple insert required updates to multiple pages (e.g., data pages plus multiple index pages), then a separate physiological log record would be written for each page updated. Physiological logging avoids the action consistency problem of logical logging, while reducing, to some extent, the amount of logging that would be incurred by physical logging. The ARIES recovery method is one example of a recovery method that uses physiological logging.

12.2.2.4 Write-Ahead Logging

A final recovery principle to be addressed in this section is the *write-ahead logging* (WAL) protocol. Recall that the contents of volatile storage are lost in the event of a system crash. As a result, any log records that are not reflected on nonvolatile storage will also be lost during a crash. WAL is a protocol that ensures that in the event of a system crash, the recovery log contains sufficient information to perform the necessary UNDO and REDO work when a STEAL/NO-FORCE buffer management policy is used. The WAL protocol ensures that

1. All log records pertaining to an updated page are written to nonvolatile storage before the page itself is allowed to be overwritten in nonvolatile storage
2. A transaction is not considered to be committed until all of its log records (including its commit record) have been written to stable storage

The first point ensures that UNDO information required due to the STEAL policy will be present in the log in the event of a crash. Similarly, the second point ensures that any REDO information required due to the NO-FORCE policy will be present in the nonvolatile log. The WAL protocol is typically enforced with special support provided by the DBMS buffer manager.

12.3 Best Practices

12.3.1 Concurrency Control

12.3.1.1 Two-Phase Locking

The most prevalent implementation technique for concurrency control is locking. Typically, two types of locks are supported, *shared* (S) locks and *exclusive* (X) locks. The compatibility of these locks is defined by the *compatibility matrix* shown in Table 12.2. The compatibility matrix shows that two different transactions are allowed to hold S locks simultaneously on the same data item, but that X locks cannot be held on an item simultaneously with any other locks (by other transactions) on that item. S locks are used for protecting *read* access to data (i.e., multiple concurrent readers are allowed), and X locks are used for protecting *write* access to data. As long as a transaction is holding a lock, no other transaction is allowed to obtain a conflicting lock. If a transaction requests a lock that cannot be granted (due to a lock conflict), that transaction is *blocked* (i.e., prohibited from proceeding) until all the conflicting locks held by other transactions are released.

S and X locks as defined in Table 12.2 directly model the semantics of conflicts used in the definition of conflict serializability. Therefore, locking can be used to enforce serializability. Rather than testing for serializability after a schedule has been produced (as was done in the previous section), the blocking of transactions due to lock conflicts can be used to prevent nonserializable schedules from *ever* being produced.

A transaction is said to be *well formed* with respect to *reads* if it always holds an S or an X lock on an item while reading it, and well formed with respect to *writes* if it always holds an X lock on an item while writing it. Unfortunately, restricting all transactions to be well formed is not sufficient to guarantee serializability. For example, a nonserializable execution such as that of schedule 12.2 is still possible using well-formed transactions. Serializability can be enforced, however, through the use of *two-phase locking (2PL)*. Two-phase locking requires that all transactions be well formed and that they respect the following rule: *Once a transaction has released a lock, it is not allowed to obtain any additional locks.* This rule results in transactions that have two phases:

1. A *growing phase* in which the transaction is acquiring locks
2. A *shrinking phase* in which locks are released

TABLE 12.2 Compatibility Matrix for S and X Locks

	S	X
S	*y*	*n*
X	*n*	*n*

The two-phase rule dictates that the transaction shifts from the growing phase to the shrinking phase at the instant it first releases a lock.

To see how 2PL enforces serializability, consider again schedule 12.2. Recall that the problem arises in this schedule because $w_0[A] \to r_1[A]$ but $r_1[B] \to w_0[B]$. This schedule could not be produced under 2PL, because transaction 1 (REPORTSUM) would be blocked when it attempted to read the value of A because transaction 0 would be holding an X lock on it. Transaction 0 would not be allowed to release this X lock before obtaining its X lock on B, and thus it would either abort or perform its update of B before transaction 1 is allowed to progress. In contrast, note that schedule 12.1 (the serial schedule) would be allowed in 2PL. 2PL would also allow the following (serializable) interleaved schedule:

$$r_1[A] \to r_0[A] \to r_1[B] \to c_1 \to w_0[A] \to r_0[B] \to w_0[B] \to c_0 \qquad (12.4)$$

It is important to note, however, that 2PL is sufficient but not necessary for implementing serializability. In other words, there are schedules that are serializable but would not be allowed by 2PL. Schedule 12.3 is an example of such a schedule.

In order to implement 2PL, the DBMS contains a component called a *lock manager*. The lock manager is responsible for granting or blocking lock requests, for managing queues of blocked transactions, and for unblocking transactions when locks are released. In addition, the lock manager is also responsible for dealing with *deadlock* situations. A deadlock arises when a set of transactions is blocked, each waiting for another member of the set to release a lock. In a deadlock situation, none of the transactions involved can make progress. Database systems deal with deadlocks using one of two general techniques: avoidance or detection. Deadlock avoidance can be achieved by imposing an order in which locks can be obtained on data, by requiring transactions to predeclare their locking needs, or by aborting transactions rather than blocking them in certain situations.

Deadlock detection, on the other hand, can be implemented using *timeouts* or explicit checking. Timeouts are the simplest technique; if a transaction is blocked beyond a certain amount of time, it is assumed that a deadlock has occurred. The choice of a timeout interval can be problematic, however. If it is too short, then the system may infer the presence of a deadlock that does not truly exist. If it is too long, then deadlocks may go undetected for too long a time. Alternatively the system can explicitly check for deadlocks using a structure called a *waits-for graph*. A waits-for graph is a directed graph with a vertex for each active transaction. The lock manager constructs the graph by placing an edge from a transaction T_i to a transaction T_j ($i \neq j$) if T_i is blocked waiting for a lock held by T_j. If the waits-for graph contains a cycle, all of the transactions involved in the cycle are waiting for each other, and thus they are deadlocked. When a deadlock is detected, one or more of the transactions involved is rolled back. When a transaction is rolled back, its locks are automatically released, so the deadlock will be broken.

12.3.1.2 Isolation Levels

As should be apparent from the previous discussion, transaction isolation comes at a cost in potential concurrency. Transaction blocking can add significantly to transaction response time.[*] As stated previously, serializability is typically implemented using 2PL, which requires locks to be held at least until all necessary locks have been obtained. Prolonging the holding time of locks increases the likelihood of blocking due to data contention.

In some applications, however, serializability is not strictly necessary. For example, a data analysis program that computes aggregates over large numbers of tuples may be able to tolerate some inconsistent access to the database in exchange for improved performance. The concept of *degrees of isolation* or *isolation levels* has been developed to allow transactions to trade concurrency for consistency in a controlled manner (Gray et al. 1975, Gray and Reuter 1993, Berenson et al. 1995). In their 1975 paper,

---

[*] Note that other, non-blocking approaches discussed later in this section also suffer from similar problems.

Gray et al. defined four degrees of consistency using characterizations based on locking, dependencies, and anomalies (i.e., results that could not arise in a serial schedule). The degrees were named degrees 0–3, with degree 0 being the least consistent and degree 3 intended to be equivalent to serializable execution.

The original presentation has served as the basis for understanding relaxed consistency in many current systems, but it has become apparent over time that the different characterizations in that paper were not specified to an equal degree of detail. In this section, we focus on the locking-based definitions of the isolation levels, as they are generally acknowledged to have "stood the test of time" (Berenson et al. 1995). However, the definition of the degrees of consistency requires an extension to the previous description of locking in order to address the *phantom problem*.

An example of the phantom problem is the following: assume a transaction T_i reads a set of tuples that satisfy a query predicate. A second transaction T_j inserts a new tuple that satisfies the predicate. If T_i then executes the query again, it will see the new item, so that its second answer differs from the first. This behavior could never occur in a serial schedule, as a "phantom" tuple appears in the midst of a transaction; thus, this execution is anomalous. The phantom problem is an artifact of the transaction model, consisting of reads and writes to *individual* data that we have used so far. In practice, transactions include *queries* that dynamically define sets based on predicates. When a query is executed, all of the tuples that satisfy the predicate at that time can be locked as they are accessed. Such individual locks, however, do not protect against the later addition of further tuples that satisfy the predicate.

One obvious solution to the phantom problem is to lock predicates instead of (or in addition to) individual items (Eswaran et al. 1976). This solution is impractical to implement, however, due to the complexity of detecting the overlap of a set of arbitrary predicates. Predicate locking can be approximated using techniques based on locking clusters of data or ranges of index values. Such techniques, however, are beyond the scope of this chapter. In this discussion, we will assume that predicates can be locked without specifying the technical details of how this can be accomplished (see Gray and Reuter (1993) and Mohan et al. (1992) for detailed treatments of this topic).

The locking-oriented definitions of the isolation levels are based on whether or not read and/or write operations are well formed (i.e., protected by the appropriate lock), and if so, whether those locks are *long duration* or *short duration*. Long-duration locks are held until the end of a transaction (i.e., when it commits or aborts); short-duration locks can be released earlier. Long-duration write locks on data items have important benefits for recovery, namely, they allow recovery to be performed using *before images*. If long-duration write locks are not used, then the following scenario could arise:

$$w_0[A] \rightarrow w_1[A] \rightarrow a_0 \tag{12.5}$$

In this case, restoring A with T_0's before image of it will be incorrect because it would overwrite T_1's update. Simply ignoring the abort of T_0 is also incorrect. In that case, if T_1 were to subsequently abort, installing it before image would reinstate the value written by T_0. For this reason and for simplicity, locking systems typically hold long-duration locks on data items. This is sometimes referred to as *strict locking* (Bernstein et al. 1987).

Given these notions of locks, the degrees of isolation presented in the SQL-92 standard can be obtained using different lock protocols. In the following, all levels are assumed to be well formed with respect to writes and to hold long-duration *write* (i.e., exclusive) locks on updated data items. Four levels are defined (from weakest to strongest:)*

---

* It should be noted that two-phase locks can be substituted for the long-duration locks in these definitions without impacting the consistency provided. Long-duration locks are typically used, however, to avoid the recovery-related problems described previously.

READ UNCOMMITTED: This level, which provides the weakest consistency guarantees, allows transactions to read data that have been written by other transactions that have not committed. In a locking implementation, this level is achieved by being ill formed with respect to reads (i.e., not obtaining read locks). The risks of operating at this level include (in addition to the risks incurred at the more restrictive levels) the possibility of seeing updates that will eventually be rolled back and the possibility of seeing some of the updates made by another transaction but missing others made by that transaction.

READ COMMITTED: This level ensures that transactions only see updates that have been made by transactions that have committed. This level is achieved by being well formed with respect to reads on individual data items but holding the read locks only as short-duration locks. Transactions operating at this level run the risk of seeing *nonrepeatable* reads (in addition to the risks of the more restrictive levels). That is, a transaction T_0 could read a data item twice and see two different values. This anomaly could occur if a second transaction were to update the item and commit in between the two reads by T_0.

REPEATABLE READ: This level ensures that reads to individual data items are repeatable, but does not protect against the phantom problem described previously. This level is achieved by being well formed with respect to reads on individual data items, and holding those locks for long duration.

SERIALIZABLE: This level protects against all of the problems of the less restrictive levels, including the phantom problem. It is achieved by being well formed with respect to reads on *predicates* as well as on individual data items and holding all locks for long duration.

A key aspect of this definition of degrees of isolation is that as long as all transactions execute at the READ UNCOMMITTED level or higher, they are able to obtain at least the degree of isolation they desire without interference from any transactions running at lower degrees. Thus, these degrees of isolation provide a powerful tool that allows application writers or users to trade off consistency for improved concurrency. As stated earlier, the definition of these isolation levels for concurrency control methods that are not based on locking has been problematic. This issue is addressed in depth in Berenson et al. (1995).

It should be noted that the discussion of locking so far has ignored an important class of data that is typically present in databases, namely, *indexes*. Because indexes are auxiliary information, they can be accessed in a non-two-phase manner without sacrificing serializability. Furthermore, the hierarchical structure of many indexes (e.g., B-trees) makes them potential concurrency bottlenecks due to high contention at the upper levels of the structure. For this reason, significant effort has gone into developing methods for providing highly concurrent access to indexes. Pointers to some of this work can be found in the Further Information section at the end of this chapter.

12.3.1.3 Hierarchical Locking

The examples in the preceeding discussions of concurrency control primarily dealt with operations on a single granularity of data items (e.g., tuples). In practice, however, the notions of conflicts and locks can be applied at many different granularities. For example, it is possible to perform locking at the granularity of a page, a relation, or even an entire database. In choosing the proper granularity at which to perform locking, there is a fundamental tradeoff between potential concurrency and locking overhead. Locking at a fine granularity, such as an individual tuple, allows for maximum concurrency, as only transactions that are truly accessing the same tuple have the potential to conflict. The downside of such fine-grained locking, however, is that a transaction that accesses a large number of tuples will have to acquire a large number of locks. Each lock request requires a call to the lock manager. This overhead can be reduced by locking at a coarser granularity, but coarse granularity raises the potential for *false conflicts*. For example, two transactions that update different tuples residing on the same page would conflict under page-level locking but not under tuple-level locking.

The notion of hierarchical or multigranular locking was introduced to allow concurrent transactions to obtain locks at different granularities in order to optimize the above tradeoff (Gray et al. 1975).

TABLE 12.3 Compatibility Matrix for Regular and Intention Locks

	IS	IX	S	SIX	X
IS	y	y	y	y	n
IX	y	y	n	n	n
S	y	n	y	n	n
SIX	y	n	n	n	n
X	n	n	n	n	n

TABLE 12.4 Hierarchical Locking Rules

To Get	Must Have on All Ancestors
IS or S	IS or IX
IX, SIX, or X	IX or SIX

In hierarchical locking, a lock on a granule at a particular level of the granularity hierarchy implicitly locks all items included in that granule. For example, an S lock on a relation implicitly locks all pages and tuples in that relation. Thus, a transaction with such a lock can read any tuple in the relation without requesting additional locks. Hierarchical locking introduces additional lock modes beyond S and X. These additional modes allow transactions to declare their *intention* to perform an operation on objects at lower levels of the granularity hierarchy. The new modes are IS, IX, and SIX for *intention shared*, *intention exclusive*, and *shared with intention exclusive*, respectively. An IS (or IX) lock on a granule provides no privileges on that granule, but indicates that the holder intends to obtain S (or X) lock on one or more finer granules. An SIX lock combines an S lock on the entire granule with an IX lock. SIX locks support the common access pattern of scanning the items in a granule (e.g., tuples in a relation) and choosing to update a fraction of them based on their values.

Similarly to S and X locks, these lock modes can be described using a compatibility matrix. The compatibility matrix for these modes is shown in Table 12.3. In order for transactions locking at different granularities to coexist, all transactions must follow the same hierarchical locking protocol starting from the root of the granularity hierarchy. This protocol is shown in Table 12.4. For example, to read a single record, a transaction would obtain IS locks on the database, relation, and page, followed by an S lock on the specific tuple. If a transaction wanted to read all or most tuples on a page, then it could obtain IS locks on the database and relation, followed by an S lock on the entire page. By following this uniform protocol, potential conflicts between transactions that ultimately obtain S and/or X locks at different granularities can be detected.

A useful extension to hierarchical locking is known as *lock escalation*. Lock escalation allows the DBMS to automatically adjust the granularity at which transactions obtain locks, based on their behavior. If the system detects that a transaction is obtaining locks on a large percentage of the granules that make up a larger granule, it can attempt to grant the transaction a lock on the larger granule so that no additional locks will be required for subsequent accesses to other objects in that granule. Automatic escalation is useful because the access pattern that a transaction will produce is often not known until run time.

12.3.1.4 Other Concurrency Control Methods

As stated previously, 2PL is the most generally accepted technique for ensuring serializability. Locking is considered to be a *pessimistic* technique because it is based on the assumption that transactions are likely to interfere with each other and takes measures (e.g., blocking) to ensure that such interference does not occur. An important alternative to locking is *optimistic concurrency control*. Optimistic methods (e.g., Kung and Robinson 1981) allow transactions to perform their operations without obtaining any locks. To ensure that concurrent executions do not violate serializability, transactions must perform a *validation phase* before they are allowed to commit. Many optimistic protocols have been proposed. In the algorithm of Kung and Robinson (1981), the validation process ensures that the reads and writes performed by a validating transaction did not conflict with any other transactions with which it ran concurrently. If during validation it is determined a conflict had occurred, the validating transaction is aborted and restarted.

Unlike locking, which depends on *blocking* transactions to ensure isolation, optimistic policies depend on transaction *restart*. As a result, although they do not perform any blocking, the performance of optimistic policies can be hurt by data contention (as are pessimistic schemes)—a high degree of data contention will result in a large number of unsuccessful transaction executions. The performance tradeoffs between optimistic and pessimistic have been addressed in numerous studies (see Agrawal et al. 1987). In general, locking is likely to be superior in resource-limited environments because blocking does not consume CPU or disk resources. In contrast, optimistic techniques may have performance advantages in situations where resources are abundant, because they allow more executions to proceed concurrently. If resources are abundant, then the resource consumption of restarted transactions will not significantly hurt performance. In practice, however, resources are typically limited, and thus concurrency control in most commercial database systems is based on locking.

Another class of concurrency control techniques is known as *multiversion concurrency control* (e.g., Reed 1983). As updating transactions modify data items, these techniques retain the previous versions of the items online. Read-only transactions (i.e., transactions that perform no updates) can then be provided with access to these older versions, allowing them to see a consistent (although possibly somewhat out-of-date) snapshot of the database. Optimistic, multiversion, and other concurrency control techniques (e.g., timestamping) are addressed in further detail in Bernstein et al. (1987).

12.3.2 Recovery

The recovery subsystem is generally considered to be one of the more difficult parts of a DBMS to design for two reasons: First, recovery is required to function in failure situations and must correctly cope with a huge number of possible system and database states. Second, the recovery system depends on the behavior of many other components of the DBMS, such as concurrency control, buffer management, disk management, and query processing. As a result, few recovery methods have been described in the literature in detail. One exception is the ARIES recovery system developed at IBM (Mohan et al. 1992). Many details about the ARIES method have been published, and the method has been included in a number of DBMSs. Furthermore, the ARIES method involves only a small number of basic concepts. For these reasons, we focus on the ARIES method in the remainder of this section. The ARIES method is related to many other recovery methods such as those described in Bernstein et al. (1987) and Gray and Reuter (1993). A comparison with other techniques appears in Mohan et al. (1992).

12.3.2.1 Overview of ARIES

ARIES is a refinement of the WAL protocol. Recall that the WAL protocol enables the use of a STEAL/ NO-FORCE buffer management policy, which means that pages on stable storage can be overwritten at any time and that data pages do not need to be forced to disk in order to commit a transaction. As with other WAL implementations, each page in the database contains an LSN, which uniquely identifies the log record for the latest update that was applied to the page. This LSN (referred to as the *pageLSN*) is used during recovery to determine whether or not an update for a page must be redone. LSN information is also used to determine the point in the log from which the REDO pass must commence during restart from a system crash. LSNs are often implemented using the physical address of the log record in the log to enable the efficient location of a log record given its LSN.

Much of the power and relative simplicity of the ARIES algorithm is due to its REDO paradigm of *repeating history*, in which it redoes updates for *all* transactions—including those that will eventually be undone. Repeating history enables ARIES to employ a variant of the *physiological logging* technique described earlier: it uses *page-oriented REDO* and a form of *logical UNDO*. Page-oriented REDO means that REDO operations involve only a single page and that the affected page is specified in the log record. This is part of physiological logging. In the context of ARIES, logical UNDO means that the operations performed to undo an update do not need to be the exact inverses of the operations of the original update.

In ARIES, logical UNDO is used to support fine-grained (i.e., tuple-level) locking and high-concurrency index management. For an example of the latter issue, consider a case in which a transaction T1 updates an index entry on a given page P1. Before T1 completes, a second transaction T2 could split P1, causing the index entry to be moved to a new page (P2). If T1 must be undone, a physical, page-oriented approach would fail because it would erroneously attempt to perform the UNDO operation on P1. Logical UNDO solves this problem by using the index structure to find the index entry, and then applying the UNDO operation to it in its new location. In contrast to UNDO, page-oriented REDO can be used because the repeating history paradigm ensures that REDO operations will always find the index entry on the page referenced in the log record—any operations that had affected the location of the index operation at the time the log record was created will be replayed before that log record is redone.

ARIES uses a three-pass algorithm for restart recovery. The first pass is the *analysis* pass, which processes the log forward from the most recent checkpoint. This pass determines information about dirty pages and active transactions that is used in the subsequent passes. The second pass is the *REDO* pass, in which history is repeated by processing the log forward from the earliest log record that could require REDO, thus ensuring that all logged operations have been applied. The third pass is the *UNDO* pass. This pass proceeds backward from the end of the log, removing from the database the effects of all transactions that had not committed at the time of the crash. These passes are shown in Figure 12.2. (Note that the relative ordering of the starting point for the REDO pass, the endpoint for the UNDO pass, and the checkpoint can be different than that shown in the figure.) The three passes are described in more detail in the subsequent sections.

ARIES maintains two important data structures during normal operation. The first is the *transaction table*, which contains status information for each transaction that is currently running. This information includes a field called the *lastLSN*, which is the LSN of the most recent log record written by the transaction. The second data structure, called the *dirty-page table*, contains an entry for each "dirty" page. A page is considered to be dirty if it contains updates that are not reflected on stable storage. Each entry in the dirty-page table includes a field called the *recoveryLSN*, which is the LSN of the log record that caused the associated page to become dirty. Therefore, the *recoveryLSN* is the LSN of the earliest log record that might need to be redone for the page during restart. Log records belonging to the same transaction are linked backward in time using a field in each log record called the *prevLSN* field. When a new log record is written for a transaction, the value of the *lastLSN* field in the transaction-table entry is placed in the *prevLSN* field of the new record and the new record's LSN is entered as the *lastLSN* in the transaction-table entry.

During normal operation, checkpoints are taken periodically. ARIES uses a form of fuzzy checkpoints which are extremely inexpensive. When a checkpoint is taken, a checkpoint record is constructed which includes the contents of the transaction table and the dirty-page table. Checkpoints are efficient, since no operations need be quiesced and no database pages are flushed to perform a checkpoint. However, the effectiveness of checkpoints in reducing the amount of the log that must be maintained is limited in

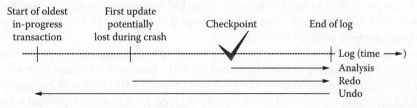

FIGURE 12.2 The three passes of ARIES restart.

part by the earliest *recoveryLSN* of the dirty pages at checkpoint time. Therefore, it is helpful to have a background process that periodically writes dirty pages to nonvolatile storage.

12.3.2.2 Analysis

The job of the analysis pass of restart recovery is threefold: (1) it determines the point in the log at which to start the REDO pass, (2) it determines which pages could have been dirty at the time of the crash in order to avoid unnecessary I/O during the REDO pass, and (3) it determines which transactions had not committed at the time of the crash and will therefore need to be undone.

The analysis pass begins at the most recent checkpoint and scans forward to the end of the log. It reconstructs the transaction table and dirty-page table to determine the state of the system as of the time of the crash. It begins with the copies of those structures that were logged in the checkpoint record. Then, the contents of the tables are modified according to the log records that are encountered during the forward scan. When a log record for a transaction that does not appear in the transaction table is encountered, that transaction is added to the table. When a log record for the commit or the abort of a transaction is encountered, the corresponding transaction is removed from the transaction table. When a log record for an update to a page that is not in the dirty-page table is encountered, that page is added to the dirty-page table, and the LSN of the record which caused the page to be entered into the table is recorded as the *recoveryLSN* for that page. At the end of the analysis pass, the dirty-page table is a conservative (since some pages may have been flushed to nonvolatile storage) list of all database pages that could have been dirty at the time of the crash, and the transaction table contains entries for those transactions that will actually require undo processing during the UNDO phase. The earliest *recoveryLSN* of all the entries in the dirty-page table, called the *firstLSN*, is used as the spot in the log from which to begin the REDO phase.

12.3.2.2.1 REDO

As stated earlier, ARIES employs a redo paradigm called *repeating history*. That is, it redoes updates for *all* transactions, committed or otherwise. The effect of repeating history is that at the end of the REDO pass, the database is in the same state with respect to the logged updates that it was in at the time that the crash occurred. The REDO pass begins at the log record whose LSN is the *firstLSN* determined by analysis and scans forward from there. To redo an update, the logged action is reapplied and the *pageLSN* on the page is set to the LSN of the redone log record. No logging is performed as the result of a redo. For each log record, the following algorithm is used to determine if the logged update must be redone:

- If the affected page is not in the dirty-page table, then the update does *not* require redo.
- If the affected page is in the dirty-page table, but the *recoveryLSN* in the page's table entry is *greater than* the LSN of the record being checked, then the update does *not* require redo.
- Otherwise, the LSN stored on the page (the *pageLSN)* must be checked. This may require that the page be read in from disk. If the *pageLSN* is *greater than or equal to* the LSN of the record being checked, then the update does *not* require redo. Otherwise, the update *must* be redone.

12.3.2.2.2 UNDO

The UNDO pass scans backwards from the end of the log. During the UNDO pass, all transactions that had not committed by the time of the crash must be undone. In ARIES, undo is an *unconditional* operation. That is, the *pageLSN* of an affected page is not checked, because it is always the case that the undo must be performed. This is due to the fact that the *repeating of history* in the REDO pass ensures that all logged updates have been applied to the page.

When an update is undone, the undo operation is applied to the page and is logged using a special type of log record called a *compensation log record* (CLR). In addition to the undo information, a CLR contains a field called the *UndoNxtLSN*. The *UndoNxtLSN* is the LSN of the next log record that must be undone for the transaction. It is set to the value of the *prevLSN* field of the log record being undone. The logging of CLRs in this fashion enables ARIES to avoid ever having to undo the effects of an undo

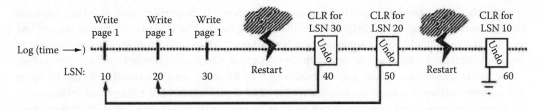

FIGURE 12.3 The use of CLRs for UNDO.

(e.g., as the result of a system crash during an abort), thereby limiting the amount of work that must be undone and bounding the amount of logging done in the event of multiple crashes. When a CLR is encountered during the backward scan, no operation is performed on the page, and the backward scan continues at the log record referenced by the *UndoNxtLSN* field of the CLR, thereby jumping over the undone update and all other updates for the transaction that have already been undone (the case of multiple transactions will be discussed shortly). An example execution is shown in Figure 12.3.

In Figure 12.3, a transaction logged three updates (LSNs 10, 20, and 30) before the system crashed for the first time. During REDO, the database was brought up to date with respect to the log (i.e., 10, 20, and/or 30 were redone if they were not on nonvolatile storage), but since the transaction was in progress at the time of the crash, they must be undone. During the UNDO pass, update 30 was undone, resulting in the writing of a CLR with LSN 40, which contains an *UndoNxtLSN* value that points to 20. Then, 20 was undone, resulting in the writing of a CLR (LSN 50) with an *UndoNxtLSN* value that points to 10. However, the system then crashed for a second time before 10 was undone. Once again, history is repeated during REDO, which brings the database back to the state it was in after the application of LSN 50 (the CLR for 20). When UNDO begins during this second restart, it will first examine the log record 50. Since the record is a CLR, no modification will be performed on the page, and UNDO will skip to the record whose LSN is stored in the *UndoNxtLSN* field of the CLR (i.e., LSN 10). Therefore, it will continue by undoing the update whose log record has LSN 10. This is where the UNDO pass was interrupted at the time of the second crash. Note that no extra logging was performed as a result of the second crash.

In order to undo multiple transactions, restart UNDO keeps a list containing the next LSN to be undone for each transaction being undone. When a log record is processed during UNDO, the *prevLSN* (or *UndoNxtLSN*, in the case of a CLR) is entered as the next LSN to be undone for that transaction. Then the UNDO pass moves on to the log record whose LSN is the most recent of the next LSNs to be redone. UNDO continues backward in the log until all of the transactions in the list have been undone up to and including their first log record. UNDO for *transaction rollback* works similarly to the UNDO pass of the restart algorithm as described above. The only difference is that during transaction rollback, only a single transaction (or part of a transaction) must be undone. Therefore, rather than keeping a list of LSNs to be undone for multiple transactions, rollback can simply follow the backward chain of log records for the transaction to be rolled back.

12.4 Research Issues and Summary

The model of ACID transactions that has been described in this chapter has proven to be quite durable in its own right, and serves as the underpinning for the current generation of database and transaction processing systems. This chapter has focused on the issues of concurrency control and recovery in a centralized environment. It is important to note, however, that the basic model is used in many types of distributed and parallel DBMS environments and the mechanisms described here have been successfully adapted for use in these more complex systems. Additional techniques, however, are needed in such environments. One important technique is *two-phase commit*, which is a protocol for ensuring that all participants in a distributed transaction agree on the decision to commit or abort that transaction.

While the basic transaction model has been a clear success, its limitations have also been apparent for quite some time (e.g., Gray 1981). Much of the ongoing research related to concurrency control and recovery is aimed at addressing some of these limitations. This research includes the development of new implementation techniques, as well as the investigation of new and extended transaction models.

The ACID transaction model suffers from a lack of flexibility and the inability to model many types of interactions that arise in complex systems and organizations. For example, in collaborative work environments, strict isolation is not possible or even desirable (Korth 1995). Workflow management systems are another example where the ACID model, which works best for relatively simple and short transactions, is not directly appropriate. For these types of applications, a richer, multilevel notion of transactions is required.

In addition to the problems raised by complex application environments, there are also many computing environments for which the ACID model is not fully appropriate. These include environments such as mobile wireless networks, where large periods of disconnection are possible, and loosely coupled wide-area networks (the Internet as an extreme example) in which the availability of systems could be low. The techniques that have been developed for supporting ACID transactions must be adjusted to cope with such highly variable situations. New techniques must also be developed to provide concurrency control and recovery in nontraditional environments such as heterogeneous systems and dissemination-oriented environments.

A final limitation of ACID transactions in their simplest form is that they are a general mechanism, and hence, do not exploit the semantics of data and/or applications. Such knowledge could be used to significantly improve system performance. Therefore, the development of concurrency control and recovery techniques that can exploit application-specific properties is another area of active research.

As should be obvious from the preceding discussion, there is still a significant amount of work that remains to be done in the areas of concurrency control and recovery for database systems. The basic concepts, however, such as serializability theory, 2PL, and WAL, will continue to be a fundamental technology, both in their own right and as building blocks for the development of more sophisticated and flexible information systems.

Glossary

Abort: The process of rolling back an uncommitted transaction. All changes to the database state made by that transaction are removed.

ACID properties: The transaction properties of atomicity, consistency, isolation, and durability that are upheld by the DBMS.

Checkpointing: An action taken during normal system operation that can help limit the amount of recovery work required in the event of a system crash.

Commit: The process of successfully completing a transaction. Upon commit, all changes to the database state made by a transaction are made permanent and visible to other transactions.

Concurrency control: The mechanism that ensures that individual users see consistent states of the database even though operations on behalf of many users may be interleaved by the database system.

Concurrent execution: The (possibly) interleaved execution of multiple transactions simultaneously.

Conflicting operations: Two operations are said to conflict if they both operate on the same data item and at least one of them is a *write*().

Deadlock: A situation in which a set of transactions is blocked, each waiting for another member of the set to release a lock. In such a case, none of the transactions involved can make progress.

Log: A sequential file that stores information about transactions and the state of the system at certain instances.

Log record: An entry in the log. One or more log records are written for each update performed by a transaction.

LSN: A number assigned to a log record, which serves to uniquely identify that record in the log. LSNs are typically assigned in a monotonically increasing fashion so that they provide an indication of relative position.

Multiversion concurrency control: A concurrency control technique that provides read-only transactions with conflict-free access to previous versions of data items.

Nonvolatile storage: Storage, such as magnetic disks or tapes, whose contents persist across power failures and system crashes.

Optimistic concurrency control: A concurrency control technique that allows transactions to proceed without obtaining locks and ensures correctness by validating transactions upon their completion.

Recovery: The mechanism that ensures that the database is fault-tolerant, that is, the database state is not corrupted as the result of a software, system, or media failure.

Schedule: A schedule for a set of transaction executions is a partial ordering of the operations performed by those transactions, which shows how the operations are interleaved.

Serial execution: The execution of a single transaction at a time.

Serializability: The property that an (possibly interleaved) execution of a group transactions has the same effect on the database, and produces the same output, as some serial (i.e., noninterleaved) execution of those transactions.

STEAL/NO-FORCE: A buffer management policy that allows committed data values to be overwritten on nonvolatile storage and does not require committed values to be written to nonvolatile storage. This policy provides flexibility for the buffer manager at the cost of increased demands on the recovery subsystem.

Transaction: A unit of work, possibly consisting of multiple data accesses and updates, that must commit or abort as a single atomic unit. Transactions have the ACID properties of *atomicity, consistency, isolation,* and *durability.*

Two-phase locking (2PL): A locking protocol that is a sufficient but not a necessary condition for serializability. Two-phase locking requires that all transactions be well formed and that once a transaction has released a lock, it is not allowed to obtain any additional locks.

Volatile storage: Storage, such as main memory, whose state is lost in the event of a system crash or power outage.

Well formed: A transaction is said to be well formed with respect to reads if it always holds a shared or an exclusive lock on an item while reading it, and well formed with respect to writes if it always holds an exclusive lock on an item while writing it.

Write-ahead logging: A protocol that ensures all log records required to correctly perform recovery in the event of a crash are placed on nonvolatile storage.

Acknowledgment

Portions of this chapter are reprinted with permission from Franklin, M., Zwilling, M., Tan, C., Carey, M., and DeWitt, D., Crash recovery in client-server EXODUS. In: *Proc. ACM Int. Conf. on Management of Data (SIGMOD'92)*, San Diego, CA, June 1992. © 1992 by the Association for Computing Machinery, Inc. (ACM).

Further Information

For many years, what knowledge that existed in the public domain about concurrency control and recovery was passed on primarily though the use of multiple-generation copies of a set of lecture notes written by Jim Gray in the late seventies ("Notes on Database Operating Systems" in *Operating Systems: An Advanced Course* published by Springer–Verlag, Berlin, Germany, 1978). Fortunately, this state of affairs has been supplanted by the publication of *Transaction Processing: Concepts and Techniques* by

Jim Gray and Andreas Reuter (Morgan Kaufmann, San Mateo, CA, 1993). This latter book contains a detailed treatment of all of the topics covered in this chapter, plus many others that are crucial for implementing transaction processing systems.

An excellent treatment of concurrency control and recovery theory and algorithms can be found in *Concurrency Control and Recovery in Database Systems* by Phil Bernstein, Vassos Hadzilacos, and Nathan Goodman (Addison–Wesley, Reading, MA, 1987). Another source of valuable information on concurrency control and recovery implementation is the series of papers on the ARIES method by C. Mohan and others at IBM, some of which are referenced in this chapter. The book *The Theory of Database Concurrency Control* by Christos Papadimitriou (Computer Science Press, Rockville, MD, 1986) covers a number of serializability models.

The performance aspects of concurrency control and recovery techniques have been only briefly addressed in this chapter. More information can be found in the recent books *Performance of Concurrency Control Mechanisms in Centralized Database Systems* edited by Vijay Kumar (Prentice Hall, Englewood Cliffs, NJ, 1996) and *Recovery in Database Management Systems*, edited by Vijay Kumar and Meichun Hsu (Prentice Hall, Englewood Cliffs, NJ, in press). Also, the performance aspects of transactions are addressed in *The Benchmark Handbook: For Database and Transaction Processing Systems* (2nd ed.), edited by Jim Gray (Morgan Kaufmann, San Mateo, CA, 1993).

Finally, extensions to the ACID transaction model are discussed in *Database Transaction Models*, edited by Ahmed Elmagarmid (Morgan Kaufmann, San Mateo, CA, 1993). Papers containing the most recent work on related topics appear regularly in the ACM SIGMOD Conference and the International Conference on Very Large Databases (VLDB), among others.

References

Agrawal, R., Carey, M., and Livny, M. 1987. Concurrency control performance modeling: Alternatives and implications. *ACM Trans. Database Systems* 12(4), 609–654, December.

Berenson, H., Bernstein, P., Gray, J., Melton, J., Oneil, B., and Oneil, P. 1995. A critique of ANSI SQL isolation levels. In: *Proc. ACM SIGMOD Int. Conf. on the Management of Data*, San Jose, CA, June.

Bernstein, P., Hadzilacos, V., and Goodman, N. 1987. *Concurrency Control and Recovery in Database Systems*. Addison–Wesley, Reading, MA.

Bjork, L. 1973. Recovery scenario for a DB/DC system. In: *Proc. ACM Annual Conf.*, Atlanta, GA.

Davies, C. 1973. Recovery semantics for a DB/DC system. In: *Proc. ACM Annual Conf.*, Atlanta, GA.

Eswaran, L., Gray, J., Lorie, R., and Traiger, I. 1976. The notion of consistency and predicate locks in a database system. *Commun. ACM* 19(11), 624–633, November.

Gray, J. 1981. The transaction concept: Virtues and limitations. In: *Proc. Seventh Int. Conf. on Very Large Databases*, Cannes, France.

Gray, J., Lorie, R., Putzolu, G., and Traiger, I. 1975. Granularity of locks and degrees of consistency in a shared database. In: *IFIP Working Conf. on Modelling of Database Management Systems, Freudenstadt, Germany*.

Gray, J. and Reuter, A. 1993. *Transaction Processing: Concepts and Techniques*. Morgan Kaufmann, San Mateo, CA.

Haerder, T. and Reuter, A. 1983. Principles of transaction-oriented database recovery. *ACM Comput. Surveys* 15(4), 287–317.

Korth, H. 1995. The double life of the transaction abstraction: Fundamental principle and evolving system concept. In: *Proc. Twenty-First Int. Conf. on Very Large Databases*, Zurich, Switzerland.

Kung, H. and Robinson, J. 1981. On optimistic methods for concurrency control. *ACM Trans. Database Systems* 6(2), 213–226.

Lomet, D. 1977. Process structuring, synchronization and recovery using atomic actions. *SIGPLAN Notices* 12(3), 128–137, March.

Mohan, C., Haderle, D., Lindsay, B., Pirahesh, H., and Schwarz, P. 1992. ARIES: A transaction method supporting fine-granularity locking and partial rollbacks using write-ahead logging. *ACM Trans. Database Systems* 17(1), 94–162, March.

Papadimitriou, C. 1986. *The Theory of Database Concurrency Control.* Computer Science Press, Rockville, MD.

Reed, D. 1983. Implementing atomic actions on decentralized data. *ACM Trans. Comput. Systems* 1(1), 3–23, February.

Rosenkrantz, D., Sterns, R., and Lewis, P. 1977. System level concurrency control for distributed database systems. *ACM Trans. Database Systems* 3(2), 178–198.

13

Distributed and Parallel Database Systems

M. Tamer Özsu
University of Waterloo

Patrick Valduriez
INRIA and LIRMM

13.1 Introduction

The maturation of database management system (DBMS) technology [1] has coincided with significant developments in distributed computing and parallel processing technologies. The end result is the development of *distributed DBMSs* and *parallel DBMSs* that are now the dominant data management tools for highly data-intensive applications. With the emergence of cloud computing, distributed and parallel database systems have started to converge.

A parallel computer, or multiprocessor, is itself a distributed system made of a number of nodes (with processors, memories, and disks) connected by a high-speed network within a cabinet. Distributed database technology can be naturally extended to implement *parallel database systems*, that is, database systems on parallel computers [2,3]. Parallel database systems exploit the parallelism in data management in order to deliver high-performance and high-availability database servers.

In this chapter, we present an overview of the distributed DBMS and parallel DBMS technologies, highlight the unique characteristics of each, and indicate the similarities between them. We also discuss the new challenges and emerging solutions.

13.2 Underlying Principles

The fundamental principle behind data management is data independence, which enables applications and users to share data at a high conceptual level while ignoring implementation details. This principle has been achieved by database systems that provide advanced capabilities such as schema management, high-level query languages, access control, automatic query processing and optimization, transactions, and data structures for supporting complex objects.

A distributed database [1] is a collection of multiple, logically interrelated databases distributed over a computer network. A distributed database system is defined as the software system that permits the management of the distributed database and makes the distribution transparent to the users. Distribution transparency extends the principle of data independence so that distribution is not visible to users.

These definitions assume that each site logically consists of a single, independent computer. Therefore, each site has the capability to execute applications on its own. The sites are interconnected by a computer network with loose connection between sites that operate independently. Applications can then issue queries and transactions to the distributed database system that transforms them into local queries and local transactions and integrates the results.

The database is physically distributed across the data sites by fragmenting and replicating the data. Given a relational database schema, for instance, fragmentation subdivides each relation into partitions based on some function applied to some tuples' attributes. Based on the user access patterns, each of the fragments may also be replicated to improve locality of reference (and thus performance) and availability. The use of a set-oriented data model (like relational) has been crucial to define fragmentation, based on data subsets.

The functions provided by a distributed database system could be those of a database system (schema management, access control, query processing, transaction support, etc.). But since they must deal with distribution, they are more complex to implement. Therefore, many systems support only a subset of these functions. When the data and the databases already exist, one is faced with the problem of providing integrated access to heterogeneous data. This process is known as *data integration*, which consists of defining a global schema over the existing data and mappings between the global schema and the local database schemas. Data integration systems have received, over time, several other names such as federated database systems, multidatabase systems (MDBMS), and mediator systems. Standard protocols such as Open Database Connectivity (ODBC) and Java Database Connectivity (JDBC) ease data integration using standard query language (SQL). In the context of the web, mediator systems allow general access to autonomous data sources (such as files, databases, and documents) in read-only mode. Thus, they typically do not support all database functions such as transactions and replication.

When the architectural assumption of each site being a (logically) single, independent computer is relaxed, one gets a parallel database system, that is, a database system implemented on a tightly coupled multiprocessor or a cluster. The main difference with a distributed database system is that there is a single operating system which facilitates implementation, and the network is typically faster and more reliable. The objective of parallel database systems is high performance and high availability. High performance (i.e., improving transaction throughput or query response time) is obtained by exploiting data partitioning and query parallelism, while high availability is obtained by exploiting replication. Again, this has been made possible by the use of a set-oriented data model, which eases parallelism—in particular, independent parallelism between data subsets.

There are three forms of parallelism inherent in data-intensive application workloads. *Inter-query parallelism* enables the parallel execution of multiple queries generated by concurrent transactions. *Intra-query parallelism* makes the parallel execution of multiple, independent operations (e.g., select operations) possible within the same query. Both inter-query and intra-query parallelism can be obtained by using *data partitioning*, which is similar to horizontal fragmentation. Finally, with *intra-operation parallelism*, the same operation can be executed as many sub-operations using *function partitioning* in addition to data partitioning. The set-oriented mode of database languages (e.g., SQL) provides many opportunities for intra-operation parallelism.

In a distributed or parallel environment, it should be easier to accommodate increasing database sizes or increasing performance demands. Major system overhauls are seldom necessary; expansion can usually be handled by adding more processing and storage power to the system.

Ideally, a parallel DBMS (and to a lesser degree a distributed DBMS) should demonstrate two advantages: *linear scaleup* and *linear speedup*. Linear scaleup refers to a sustained performance for a linear increase in both database size and processing and storage power. Linear speedup refers to a linear increase

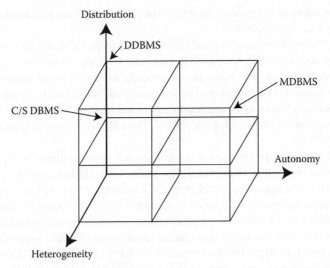

FIGURE 13.1 Distributed DBMS architectures. (Adapted from Özsu, T.M. and Valduriez, P., *Principles of Distributed Database Systems*, 3rd edn., Springer, 2011.)

in performance for a constant database size and a linear increase in processing and storage power. Furthermore, extending the system should require minimal reorganization of the existing database.

There are a number of possible ways in which a distributed DBMS may be architected. We use a classification (Figure 13.1) that organizes the systems as characterized with respect to three dimensions: (1) the autonomy of local systems, (2) their distribution, and (3) their heterogeneity. Autonomy, in this context, refers to the distribution of control, not of data. It indicates the degree to which individual DBMSs can operate independently. Whereas autonomy refers to the distribution (or decentralization) of control, the distribution dimension of the taxonomy deals with the physical distribution of data over multiple sites (or nodes in a parallel system). There are a number of ways DBMSs have been distributed. We distinguish between client/server (C/S) distribution and peer-to-peer (P2P) distribution (or full distribution). With C/S DBMS, sites may be clients or servers, thus with different functionality, whereas with homogeneous P2P DBMS (DDBMS in Figure 13.1), all sites provide the same functionality. Note that DDBMS came before C/S DBMS in the late 1970s. P2P data management struck back in the 2000 with modern variations to deal with very-large-scale autonomy and decentralized control. Heterogeneity refers to data models, query languages, and transaction management protocols. MDBMS deal with heterogeneity, in addition to autonomy and distribution.

13.3 Distributed and Parallel Database Technology

Distributed and parallel DBMSs provide the same functionality as centralized DBMSs except in an environment where data are distributed across the sites on a computer network or across the nodes of a parallel system. As discussed earlier, the users are unaware of data distribution. Thus, these systems provide the users with a *logically integrated* view of the *physically distributed* database. Maintaining this view places significant challenges on system functions. We provide an overview of these new challenges in this section. We assume familiarity with basic database management techniques.

13.3.1 Query Processing and Optimization

Query processing is the process by which a declarative query is translated into low-level data manipulation operations. SQL is the standard query language that is supported in current DBMSs.

Query optimization refers to the process by which the "best" execution strategy for a given query is found from among a set of alternatives.

In centralized DBMSs, the process typically involves two steps [1]: *query decomposition* and *query optimization*. Query decomposition takes an SQL query and translates it into one expressed in relational algebra. In the process, the query is analyzed semantically so that incorrect queries are detected and rejected as easily as possible, and correct queries are simplified. Simplification involves the elimination of redundant predicates, which may be introduced as a result of query modification to deal with views, security enforcement, and semantic integrity control. The simplified query is then restructured as an algebraic query.

For a given SQL query, there are usually more than one possible algebraic equivalents. Some of these algebraic queries are "better" than others. The quality of an algebraic query is defined in terms of expected performance. The traditional procedure is to obtain an initial algebraic query by translating the predicates and the target statement into relational operations as they appear in the query. This initial algebraic query is then transformed, using algebraic transformation rules, into other algebraic queries until the "best" one is found.* The query optimizer is usually seen as three components: a search space, a cost model, and a search strategy. The *search space* is the set of alternative execution plans to represent the input query. These plans are equivalent, in the sense that they yield the same result but they differ on the execution order of operations and the way these operations are implemented. The *cost model* predicts the cost of a given execution plan. To be accurate, the cost model must have accurate knowledge about the parallel execution environment. The *search strategy* explores the search space and selects the best plan. It defines which plans are examined and in which order.

In distributed DBMSs, two additional steps are involved between query decomposition and query optimization: *data localization* and *global query optimization* [1].

The input to data localization is the initial algebraic query generated by the query decomposition step. The initial algebraic query is specified on global relations irrespective of their fragmentation or distribution. The main role of data localization is to localize the query's data using data distribution information. In this step, the fragments that are involved in the query are determined, and the query is transformed into one that operates on fragments rather than on global relations. As indicated earlier, fragmentation is defined through fragmentation rules that can be expressed as relational operations (horizontal fragmentation by selection, vertical fragmentation by projection). A distributed relation can be reconstructed by applying the inverse of the fragmentation rules. This is called a *localization program*. The localization program for a horizontally (vertically) fragmented query is the union (join) of the fragments. Thus, during the data localization step, each global relation is first replaced by its localization program, and then the resulting fragment query is simplified and restructured to produce another "good" query. Simplification and restructuring may be done according to the same rules used in the decomposition step. As in the decomposition step, the final fragment query is generally far from optimal; the process has only eliminated "bad" algebraic queries.

The input to global query optimization step is a fragment query, that is, an algebraic query on fragments. The goal of global query optimization is to find an execution strategy for the query which is close to optimal. Remember that finding the optimal solution is computationally intractable. An execution strategy for a distributed query can be described with *relational algebra operations* and *communication primitives* (send/receive operations) for transferring data between sites. The previous layers have already performed some optimizations—for example, by eliminating redundant expressions. However, this optimization is independent of fragment characteristics such as cardinalities. In addition, communication operations are not yet specified. By permuting the ordering of operations within one fragment query, many equivalent query execution plans may be found.

---

* The difference between an optimal plan and the best plan is that the optimizer does not, because of computational intractability, examine all of the possible plans.

In a distributed environment, the cost function, often defined in terms of time units, refers to computing resources such as disk space, disk I/Os, buffer space, central processing unit (CPU) cost, and communication cost. Generally, it is a weighted combination of I/O, CPU, and communication costs. To select the ordering of operations, it is necessary to predict execution costs of alternative candidate orderings. Determining execution costs before query execution (i.e., static optimization) is based on fragment statistics and the formulas for estimating the cardinalities of results of relational operations. Thus the optimization decisions depend on the available statistics on fragments. An important aspect of query optimization is *join ordering*, since permutations of the joins within the query may lead to improvements of several orders of magnitude. One basic technique for optimizing a sequence of distributed join operations is through the use of the semijoin operator. The main value of the semijoin in a distributed system is to reduce the size of the join operands and thus the communication cost. However, more recent techniques, which consider local processing costs as well as communication costs, do not use semijoins because they might increase local processing costs. The output of the query optimization layer is an optimized algebraic query with communication operations included on fragments.

Parallel query optimization exhibits similarities with distributed query processing [4,5]. It takes advantage of both intra-operation parallelism, which was discussed earlier, and inter-operation parallelism.

Intra-operation parallelism is achieved by executing an operation on several nodes of a multiprocessor machine. This requires that the operands have been previously partitioned, that is, horizontally fragmented, across the nodes [6]. The way in which a base relation is partitioned is a matter of physical design. Typically, partitioning is performed by applying a hash function on an attribute of the relation, which will often be the join attribute. The set of nodes where a relation is stored is called its *home*. The *home of an operation* is the set of nodes where it is executed. The home of an operation must also be the home of its operand relations in order for the operation to access its operands. For binary operations such as join, this might imply repartitioning one of the operands [7]. The optimizer might even sometimes find that repartitioning both operands is useful. Parallel optimization to exploit intra-operation parallelism can make use of some of the techniques devised for distributed databases.

Inter-operation parallelism occurs when two or more operations are executed in parallel, either as a dataflow or independently. We designate as *dataflow* the form of parallelism induced by *pipelining*. *Independent* parallelism occurs when operations are executed at the same time or in an arbitrary order. Independent parallelism is possible only when the operations do not involve the same data.

13.3.2 Concurrency Control

Whenever multiple users access (read and write) a shared database, these accesses need to be synchronized to ensure database consistency. The synchronization is achieved by means of *concurrency control algorithms*, which enforce a correctness criterion such as *serializability*. User accesses are encapsulated as *transactions* [8], whose operations at the lowest level are a set of read and write operations to the database. Transactions typically have four properties: *atomicity, consistency, isolation,* and *durability,* which are collectively known as *ACID properties*. Concurrency control algorithms enforce the *isolation* property of transaction execution, which states that the effects of one transaction on the database are isolated from other transactions until the first completes its execution.

The most popular concurrency control algorithms are *locking*-based. In such schemes, a lock, in either shared or exclusive mode, is placed on some unit of storage (usually a page) whenever a transaction attempts to access it. These locks are placed according to lock compatibility rules such that *read–write, write–read,* and *write–write* conflicts are avoided. It is a well-known theorem that if lock actions on behalf of concurrent transactions obey the following simple rule, then it is possible to ensure the serializability of these transactions: "No lock on behalf of a transaction should be set once a lock previously held by the transaction is released." This is known as *two-phase locking* [9], since transactions go through a growing phase when they obtain locks and a shrinking phase when they release locks.

In general, releasing of locks prior to the end of a transaction is problematic. Thus, most of the locking-based concurrency control algorithms are *strict* in that they hold on to their locks until the end of the transaction.

In distributed DBMSs, the challenge is to extend both the serializability argument and the concurrency control algorithms to the distributed execution environment. In these systems, the operations of a given transaction may execute at multiple sites where they access data. In such a case, the serializability argument is more difficult to specify and enforce. The complication is due to the fact that the serialization order of the same set of transactions may be different at different sites. Therefore, the execution of a set of distributed transactions is serializable if and only if

1. The execution of the set of transactions at each site is serializable.
2. The serialization orders of these transactions at all these sites are identical.

Distributed concurrency control algorithms enforce this notion of *global serializability*. In locking-based algorithms, there are three alternative ways of enforcing global serializability: centralized locking, primary copy locking, and distributed locking.

In *centralized locking*, there is a single lock table for the entire distributed database. This lock table is placed under the control of a single lock manager at one of the sites. The lock manager is responsible for setting and releasing locks on behalf of transactions. Since all locks are managed at one site, this is similar to centralized concurrency control, and it is straightforward to enforce the global serializability rule. These algorithms are simple to implement, but suffer from two problems. The central site may become a bottleneck, both because of the amount of work it is expected to perform and because of the traffic that is generated around it; and the system may be less reliable since the failure or inaccessibility of the central site would cause system unavailability.

Primary copy locking is a concurrency control algorithm that is useful in replicated databases where there may be multiple copies of a data item stored at different sites. One of the copies is designated as a primary copy, and it is this copy that has to be locked in order to access that item. The set of primary copies for each data item is known to all the sites in the distributed system, and the lock requests on behalf of transactions are directed to the appropriate primary copy. If the distributed database is not replicated, primary copy locking degenerates into a distributed locking algorithm.

In *distributed* (or *decentralized*) *locking*, the lock management duty is shared by all the sites in the system. The execution of a transaction involves the participation and coordination of lock managers at more than one site. Locks are obtained at each site where the transaction accesses a data item. Distributed locking algorithms do not have the overhead of centralized locking ones. However, both the communication overhead to obtain all the locks and the complexity of the algorithm are greater.

One side effect of all locking-based concurrency control algorithms is that they cause *deadlocks*. The detection and management of deadlocks in a distributed system is difficult. Nevertheless, the relative simplicity and better performance of locking algorithms make them more popular than alternatives such as *timestamp-based algorithms* or *optimistic concurrency control*. Timestamp-based algorithms execute the conflicting operations of transactions according to their timestamps, which are assigned when the transactions are accepted. Optimistic concurrency control algorithms work from the premise that conflicts among transactions are rare and proceed with executing the transactions up to their termination at which point a validation is performed. If the validation indicates that serializability would be compromised by the successful completion of that particular transaction, then it is aborted and restarted.

13.3.3 Reliability Protocols

We indicated earlier that distributed DBMSs are potentially more reliable because there are multiples of each system component, which eliminates single point of failure. This requires careful system design and the implementation of a number of protocols to deal with system failures.

In a distributed DBMS, four types of failures are possible: *transaction failures*, *site (system) failures*, *media (disk) failures*, and *communication line failures*. Transactions can fail for a number of reasons. Failure can be due to an error in the transaction caused by input data, as well as the detection of a present or potential deadlock. The usual approach that is followed in cases of transaction failure is to abort the transaction, resetting the database to its state prior to the start of the database.

Site (or system) failures are due to a hardware failure (e.g., processor, main memory, and power supply) or a software failure (bugs in system or application code). The effect of system failures is the loss of main memory contents. Therefore, any update to the parts of the database that are in the main memory buffers (also called *volatile database*) is lost as a result of system failures. However, the database that is stored in secondary storage (also called *stable database*) is safe and correct. To achieve this, DBMSs typically employ *logging protocols*, such as *Write-Ahead Logging* [9], which record changes to the database in system logs and move these log records and the volatile database pages to stable storage at appropriate times. From the perspective of distributed transaction execution, site failures are important since the failed sites cannot participate in the execution of any transaction.

Media failures refer to the failure of secondary storage devices that store the stable database. Typically, these failures are addressed by duplexing storage devices and maintaining archival copies of the database. Media failures are frequently treated as problems local to one site and therefore are not specifically addressed in the reliability mechanisms of distributed DBMSs.

The three types of failures described earlier are common to both centralized and distributed DBMSs. Communication failures, on the other hand, are unique to distributed systems. There are a number of types of communication failures. The most common ones are errors in the messages, improperly ordered messages, lost (or undelivered) messages, and line failures. Generally, the first two of these are considered to be the responsibility of the computer network protocols and are not addressed by the distributed DBMS. The last two, on the other hand, have an impact on the distributed DBMS protocols and, therefore, need to be considered in the design of these protocols. If one site is expecting a message from another site and this message never arrives, this may be because (a) the message is lost, (b) the line(s) connecting the two sites may be broken, or (c) the site which is supposed to send the message may have failed. Thus, it is not always possible to distinguish between site failures and communication failures. The waiting site simply timeouts and has to assume that the other site is unable to communicate. Distributed DBMS protocols have to deal with this uncertainty. One drastic result of line failures may be *network partitioning* in which the sites form groups where communication within each group is possible but communication across groups is not. This is difficult to deal with in the sense that it may not be possible to make the database available for access while at the same time guaranteeing its consistency.

Two properties of transactions are maintained by reliability protocols: *atomicity* and *durability*. Atomicity requires that either all the operations of a transaction are executed or none of them are (all-or-nothing). Thus, the set of operations contained in a transaction is treated as one atomic unit. Atomicity is maintained even in the face of failures. Durability requires that the effects of successfully completed (i.e., committed) transactions endure subsequent failures.

The enforcement of atomicity and durability requires the implementation of *atomic commitment protocols* and *distributed recovery protocols*. The most popular atomic commitment protocol is *two-phase commit* (2PC). The recoverability protocols are built on top of the local recovery protocols, which are dependent upon the supported mode of interaction (of the DBMS) with the operating system [10,11].

2PC is a very simple and elegant protocol that ensures the atomic commitment of distributed transactions. It extends the effects of local atomic commit actions to distributed transactions by insisting that all sites involved in the execution of a distributed transaction agree to commit the transaction before its effects are made permanent (i.e., all sites terminate the transaction in the same manner). If all the sites agree to commit a transaction, then all the actions of the distributed transaction take effect; if one of the sites declines to commit the operations at that site, then all of the other sites are required to abort the transaction. Thus, the fundamental 2PC rule states

1. If even one site rejects to commit (which means it votes to abort) the transaction, the distributed transaction has to be aborted at each site where it executes.
2. If all the sites vote to commit the transaction, the distributed transaction is committed at each site where it executes.

The simple execution of the 2PC protocol is as follows. There is a *coordinator* process at the site where the distributed transaction originates and *participant* processes at all the other sites where the transaction executes. There are two rounds of message exchanges between the coordinator and the participants (hence the name 2PC protocol): in the first phase, the coordinator asks for votes regarding the transaction from the participants and determines the fate of the transaction according to the 2PC rule, and, in the second phase, it informs the participants about the decision. In the end, all of the participants and the coordinator reach the same decision. Two important variants of 2PC are the *presumed abort 2PC* and *presumed commit 2PC* [12]. These are important because they reduce the message and I/O overhead of the protocols. Presumed abort protocol is included in the X/Open XA standard and has been adopted as part of the International Standards Organization (ISO) standard for Open Distributed Processing.

One important characteristic of 2PC protocol is its *blocking* nature. Failures can occur during the commit process. As discussed earlier, the only way to detect these failures is by means of a time-out of the process waiting for a message. When this happens, the process (coordinator or participant) that timeouts follows a *termination protocol* to determine what to do with the transaction that was in the middle of the commit process. A non-blocking commit protocol is one whose termination protocol can determine what to do with a transaction in case of failures under any circumstance. In the case of 2PC, if a site failure occurs at the coordinator site and one participant site while the coordinator is collecting votes from the participants, the remaining participants cannot determine the fate of the transaction among themselves, and they have to remain blocked until the coordinator or the failed participant recovers. During this period, the locks that are held by the transaction cannot be released, which reduces the availability of the database.

13.3.4 Replication Protocols

In replicated distributed databases,* each logical data item has a number of physical instances. For example, the salary of an employee (*logical data item*) may be stored at three sites (*physical copies*). The issue in this type of a database system is to maintain some notion of consistency among the copies. The most discussed consistency criterion is *one copy equivalence*, which asserts that the values of all copies of a logical data item should be identical when the transaction that updates it terminates.

If replication transparency is maintained, transactions will issue read and write operations on a logical data item x. The replica control protocol is responsible for mapping operations on x to operations on physical copies of x ($x_1, \ldots, x_n$). A typical replica control protocol that enforces one copy serializability is known as *Read-Once/Write-All* (ROWA) protocol. ROWA maps each read on x [Read(x)] to a read on one of the physical copies x_i [Read(x_i)]. The copy that is read is insignificant from the perspective of the replica control protocol and may be determined by performance considerations. On the other hand, each write on logical data item x is mapped to a set of writes on *all* copies of x.

ROWA protocol is simple and straightforward, but it requires that all copies of all logical data items that are updated by a transaction be accessible for the transaction to terminate. Failure of one site may block a transaction, reducing database availability.

---

* Replication is not a significant concern in parallel DBMSs because the data are normally not replicated across multiple processors. Replication may occur as a result of data shipping during query optimization, but this is not managed by the replica control protocols.

A number of alternative algorithms have been proposed, which reduce the requirement that all copies of a logical data item be updated before the transaction can terminate. They relax ROWA by mapping each write to only a subset of the physical copies.

This idea of possibly updating only a subset of the copies, but nevertheless successfully terminating the transaction, has formed the basis of quorum-based voting for replica control protocols. The majority consensus algorithm can be viewed from a slightly different perspective: It assigns equal votes to each copy and a transaction that updates that logical data item can successfully complete as long as it has a majority of the votes. Based on this idea, an early *quorum-based voting algorithm* [13] assigns a (possibly unequal) vote to each copy of a replicated data item. Each operation then has to obtain a *read quorum* (V_r) or a *write quorum* (V_w) to read or write a data item, respectively. If a given data item has a total of V votes, the quorums have to obey the following rules:

1. $V_r + V_w > V$ (a data item is not read and written by two transactions concurrently, avoiding the read–write conflict)
2. $V_w > V/2$ (two write operations from two transactions cannot occur concurrently on the same data item, avoiding write–write conflict)

The difficulty with this approach is that transactions are required to obtain a quorum even to read data. This significantly and unnecessarily slows down read access to the database. An alternative quorum-based voting protocol that overcomes this serious performance drawback [14] has also been proposed. However, this protocol makes unrealistic assumptions about the underlying communication system. It requires that failures that change the network's topology are detected by all sites instantaneously, and that each site has a view of the network consisting of all the sites with which it can communicate. In general, communication networks cannot guarantee to meet these requirements. The single copy equivalence replica control protocols are generally considered to be restrictive in terms of the availability they provide. Voting-based protocols, on the other hand, are considered too complicated with high overheads. Therefore, these techniques are not used in current distributed DBMS products. More flexible replication schemes have been investigated where the type of consistency between copies is under user control. A number of *replication servers* have been developed or are being developed with this principle.

13.4 New Challenges and Emerging Solutions

The pervasiveness of the web has spurred all kinds of data-intensive applications and introduced significant challenges for distributed data management [15]. New data-intensive applications such as social networks, web data analytics, and scientific applications have requirements that are not met by the traditional distributed database systems in Figure 13.1. What has changed and made the problems much harder is the scale of the dimensions: very-large-scale distribution, very high heterogeneity, and high autonomy. In this section, we discuss some of these challenges and the emerging solutions to these challenges.

13.4.1 Cloud Data Management

Cloud computing is the latest trend in distributed computing and has been the subject of much debate. The vision encompasses on demand, reliable services provided over the Internet (typically represented as a cloud) with easy access to virtually infinite computing, storage, and networking resources. Through very simple web interfaces and at small incremental cost, users can outsource complex tasks, such as data storage, system administration, or application deployment, to very large data centers operated by cloud providers. Thus, the complexity of managing the software/hardware infrastructure gets shifted from the users' organization to the cloud provider. From a technical point of view, the grand challenge is to support in a cost-effective way the very large scale of the infrastructure, which has to manage lots of users and resources with high quality of service.

However, not all data-intensive applications are good candidates for being supported in a cloud [16]. To simplify, we can classify between the two main classes of data-intensive applications: Online Transaction Processing (OLTP) and Online Analytical Processing (OLAP). OLTP deals with operational databases of average sizes (up to a few terabytes), is write-intensive, and requires complete ACID transactional properties, strong data protection, and response time guarantees. OLAP, on the other hand, deals with historical databases of very large sizes (up to petabytes), is read-intensive, and thus can accept relaxed ACID properties. Furthermore, since OLAP data are typically extracted from operational OLTP databases, sensitive data can be simply hidden for analysis (e.g., using anonymization) so that data protection is not as crucial as in OLTP.

OLAP is more suitable than OLTP for cloud primarily because of two cloud characteristics [16]: elasticity and security. To support elasticity in a cost-effective way, the best solution that most cloud providers adopt is a shared-nothing cluster, which has a fully distributed architecture, where each node is made of processor, main memory, and disk and communicates with other nodes through message passing. Shared-nothing provides high scalability but requires careful data partitioning. Since OLAP databases are very large and mostly read-only, data partitioning and parallel query processing are effective [4]. However, it is much harder to support OLTP on shared-nothing systems because of ACID guarantees that require complex concurrency control. For these reasons and because OLTP databases are not very large, shared disk, where any processor has direct access to any disk unit through the interconnection network, is the preferred architecture for OLTP.

The second reason that OLTP is not so suitable for cloud is that highly sensitive data get stored at an untrusted host (the provider site). Storing corporate data at an untrusted third-party, even with a carefully negotiated Service Level Agreement with a reliable provider, creates resistance from some customers because of security issues. However, this resistance is much reduced for historical data, with anonymized sensitive data.

There is much more variety in cloud data than in scientific data since there are many different kinds of customers (individuals, small- and medium-size enterprises, large corporations, etc.). However, we can identify common features. Cloud data can be very large, unstructured (e.g., text-based) or semistructured, and typically append-only (with rare updates). Cloud users and application developers may be in high numbers, but not DBMS experts.

Generic data management solutions (e.g., relational DBMS) that have proven effective in many application domains (e.g., business transactions) are not efficient at dealing with emerging cloud applications, thereby forcing developers to build ad hoc solutions that are labor-intensive and that cannot scale. In particular, relational DBMSs have been lately criticized for their "one-size-fits-all" approach. Although they have been able to integrate support for all kinds of data (e.g., multimedia objects, XML documents, and new functions), this has resulted in a loss of performance and flexibility for applications with specific requirements because they provide both too much and too little. Therefore, it has been argued that more specialized DBMS engines are needed. For instance, column-oriented DBMSs, which store column data together rather than rows as in traditional row-oriented relational DBMSs, have been shown to perform more than an order of magnitude better on OLAP workloads. The "one-size-does-not-fit-all" counterargument generally applies to cloud data management as well.

Therefore, current data management solutions for the cloud have traded consistency for scalability, simplicity, and flexibility. As alternative to relational DBMS (which use the standard SQL language), these solutions have been recently quoted as "Not Only SQL" (NoSQL) by the database research community. Many of these solutions exploit large-scale parallelism, typically with shared-nothing clusters. Distributed data management for cloud applications emphasizes scalability, fault-tolerance, and availability, sometimes at the expense of consistency or ease of development. We illustrate this approach with three popular solutions: Bigtable, PNUTS, and MapReduce.

Bigtable is a database storage system initially proposed by Google for a shared-nothing cluster [17]. It uses the Google File System (GFS) [18] for storing structured data in distributed files, with fault-tolerance and availability. It also uses a form of dynamic data partitioning for scalability. There are also open source implementations of Bigtable, such as Hadoop Hbase, as part of the Hadoop project of the Apache foundation (http://hadoop.apache.org), which runs on top of Hadoop Distributed File System, an open source implementation of GFS. Bigtable supports a simple data model that resembles the relational model, with multi-valued, timestamped attributes. It provides a basic application programming interface (API) for defining and manipulating tables, within a programming language such as C++, and various operators to write and update values and to iterate over subsets of data, produced by a scan operator. There are various ways to restrict the rows, columns, and timestamps produced by a scan, as in a relational select operator. However, there are no complex operators such as join or union, which need to be programmed using the scan operator. Transactional atomicity is supported for single row updates only. To store a table in GFS, Bigtable uses range partitioning on the row key. Each table is divided into partitions, called *tablets*, each corresponding to a row range.

PNUTS is a parallel and distributed database system for cloud applications at Yahoo! [19]. It is designed for serving web applications, which typically do not need complex queries, but require good response time, scalability, and high availability and can tolerate relaxed consistency guarantees for replicated data. PNUTS supports the relational data model, with arbitrary structures allowed within attributes of Blob type. Schemas are flexible as new attributes can be added at any time even though the table is being queried or updated, and records need not have values for all attributes. PNUTS provides a simple query language with selection and projection on a single relation. Updates and deletes must specify the primary key. PNUTS provides a replica consistency model that is between strong consistency and eventual consistency, with several API operations with different guarantees. Database tables are horizontally partitioned into tablets, through either range partitioning or hashing, which are distributed across many servers in a cluster (at a site).

Both Bigtable and PNUTS provide some variation of the relational model, a simple API or language for manipulating data, and relaxed consistency guarantees. They also rely on fragmentation (partitioning) and replication for fault-tolerance. Thus, they capitalize on the well-known principles of distributed data management.

MapReduce [20] is a good example of generic parallel data processing framework, on top of a distributed file system (GFS). It supports a simple data model (sets of (key, value) pairs), which allows user-defined functions (map and reduce). MapReduce was initially developed by Google as a proprietary product to process large amounts of unstructured or semistructured data, such as web documents and logs of web page requests, on large shared-nothing clusters of commodity nodes and produce various kinds of data such as inverted indices or uniform resource locators (URLs) access frequencies.

MapReduce enables programmers to express in a simple, functional style their computations on large data sets and hides the details of parallel data processing, load balancing, and fault-tolerance. The programming model includes only two operations, `map` and `reduce`, which we can find in many functional programming languages such as Lisp and ML. The Map operation is applied to each record in the input data set to compute one or more intermediate (key, value) pairs. The Reduce operation is applied to all the values that share the same unique key in order to compute a combined result. Since they work on independent inputs, map and reduce can be automatically processed in parallel, on different data partitions using many cluster nodes.

Different implementations of MapReduce are now available such as Amazon MapReduce (as a cloud service) or Hadoop MapReduce (as open source software). There is also much research going on improving the performance of the MapReduce framework, which performs full scans of data sets. For instance, Hadoop++ [21] introduces noninvasive, DBMS-independent indexing and join techniques to boost the performance of Hadoop MapReduce.

13.4.2 Scientific Data Management

Scientific data management has become a major challenge for the database and data management research community [22]. Modern science such as agronomy, bio-informatics, physics, and environmental science must deal with overwhelming amounts of experimental data produced through empirical observation and simulation. Such data must be processed (cleaned, transformed, and analyzed) in all kinds of ways in order to draw new conclusions, prove scientific theories, and produce knowledge. However, constant progress in scientific observational instruments (e.g., satellites, sensors, and Large Hadron Collider) and simulation tools (that foster in silico experimentation, as opposed to traditional in situ or in vivo experimentation) creates a huge data overload. For example, climate modeling data are growing so fast that they will lead to collections of hundreds of exabytes expected by 2020.

Scientific data are also very complex, in particular, because of heterogeneous methods used for producing data, the uncertainty of captured data, the inherently multi-scale nature (spatial scale and temporal scale) of many sciences, and the growing use of imaging (e.g., satellite images), resulting in data with hundreds of attributes, dimensions, or descriptors. Processing and analyzing such massive sets of complex scientific data are, therefore, a major challenge since solutions must combine new data management techniques with large-scale parallelism in cluster, grid, or cloud environments [23].

Furthermore, modern science research is a highly collaborative process, involving scientists from different disciplines (e.g., biologists, soil scientists, and geologists working on an environmental project), in some cases from different organizations distributed in different countries. Since each discipline or organization tends to produce and manage its own data, in specific formats, with its own processes, integrating distributed data and processes gets difficult as the amounts of heterogeneous data grow.

Another major difficulty that is inherent with science is that knowledge and understanding may keep evolving, making it sometimes very hard to model data. Thus, we need to better understand the fundamental aspects of the scientific data management problem, in relationship with the main users, that is, scientists.

Despite their variety, we can identify common features of scientific data [24]: massive scale; manipulated through complex, distributed workflows; typically complex, for example, multidimensional or graph-based; with uncertainty in the data values, for example, to reflect data capture or observation; important metadata about experiments and their provenance; heavy floating-point computation; and mostly append-only (with rare updates). For reasons similar to those discussed for cloud data, relational DBMS are not efficient for dealing with most scientific data and NoSQL can be useful for scientific data.

The goal of scientific data management is to make scientific data easier to access, reproduce, and share by scientists of different disciplines and institutions. In recent international interdisciplinary workshops and conferences (e.g., http://www-conf.slac.stanford.edu/xldb), the following key requirements for scientific data management (that cannot be supported by current technology) have been identified [25]:

- Built-in support for managing and processing uncertain data (e.g., inaccurate data generated by faulty sensors or by imprecise observations) in distributed environments
- Rich representation of scientific data with an extensible data model that features multidimensional arrays, graph structures, sequences, etc.
- Distributed and parallel workflow execution involving large numbers of distributed processes and large amounts of heterogeneous data, with support of data provenance (lineage) to understand result data
- Scalability to hundreds of petabytes and thousands of nodes in high-performance computing environments (e.g., very large clusters), with high degrees of tolerance to failures
- Efficient data and metadata management, in particular, with semantics (ontologies), in order to help integrating data coming from different sources, with different formats and semantics
- Open source software in order to foster a community of contributors (from many different laboratories and institutes) and to insure data independence from proprietary systems

Addressing the aforementioned challenges and requirements is now on the agenda of a very active research community composed of scientists from different disciplines and data management researchers. For instance, the SciDB organization (http://www.scidb.org) is building an open source database system for scientific data analytics. SciDB supports an array data model, which generalizes the relational model, with array operators. User-defined functions also allow for more specific data processing. SciDB will be certainly effective for similar applications for which the data is well understood (with well-defined models). However, to avoid that the one-size-fits-all argument applies to SciDB as well, the key question is: How generic should scientific data management be, without hampering application-specific optimizations? For instance, to perform scientific data analysis efficiently, scientists typically resort to dedicated indexes, compression techniques, and specific algorithms. Thus, generic techniques, inspired from the database (DB) research community should be able to cope with these specific techniques.

Genericity in data management encompasses two dimensions: data model, which provides data structures (captured by the data model), and data processing (inferred by the query language). Relational DBMS have initially provided genericity through the relational data model (that subsumes earlier data models) and a high-level query language (SQL). However, successive object extensions to include new data structures such as lists and arrays and support user-defined functions in a programming language have resulted in a yet generic, but more complex data model and language for the developers. Therefore, emerging NoSQL solutions tend to rely on a more specific data model (e.g., to deal with graphs, arrays, or sequences) with a simple set of operators easy to use from a programming language with a simple API.

In emerging solutions, it is interesting to witness the development of algebras to raise the level of abstraction for the programmer and provide automatic optimization. For example, Pig Latin [26] is an alternative data management solution to MapReduce with an algebraic query language. Another example is the approach proposed in [27] to deal with the efficient processing of scientific workflows that are computational and data-intensive, that is, manipulating huge amounts of data through specific programs and files and requiring execution in large-scale parallel computers. The authors propose an algebraic approach (inspired by relational algebra) and a parallel execution model that enable automatic optimization of scientific workflows. With the algebra, data are uniformly represented by relations and workflow activities are mapped to operators that have data-aware semantics. The execution model is based on the concept of activity activation, inspired from data activations proposed for parallel DBMS [28], which enables transparent distribution and parallelization of activities.

13.4.3 Stream Applications and Stream Data Management

The database systems that we have discussed until now consist of a set of unordered objects that are relatively static, with insertions, updates, and deletions occurring less frequently than queries. They are sometimes called *snapshot databases* since they show a snapshot of the values of data objects at a given point in time. Queries over these systems are executed when posed, and the answer reflects the current state of the database. In these systems, typically, the data are persistent and queries are transient.

However, the past few years have witnessed an emergence of applications that do not fit this data model and querying paradigm. These applications include network traffic analysis, financial tickers, online auctions, and applications that analyze transaction logs (such as web usage logs and telephone call records). In these applications, data are generated in real time, taking the form of an unbounded sequence (stream) of values. These are referred to as the *data stream* applications, and the systems that handle these types of data are known as *data stream management systems* (DSMS) (or stream data management systems) [29,30].

A data stream is commonly modeled as an append-only sequence of timestamped items that arrive in some order. Stream items may contain explicit source-assigned timestamps or implicit timestamps assigned by the DSMS upon arrival. In either case, the timestamp attribute may or may not be part of the stream schema, and therefore may or may not be visible to users. Stream items may arrive out of order (if explicit timestamps are used) and/or in preprocessed form. For instance, rather than propagating the header of each IP packet, one value (or several partially pre-aggregated values) may be produced to summarize the length of a connection between two IP addresses and the number of bytes transmitted.

A fundamental assumption of the data stream model is that new data are generated continually and in fixed order, although the arrival rates may vary across applications from millions of items per second (e.g., Internet traffic monitoring) down to several items per hour (e.g., temperature and humidity readings from a weather monitoring station). The ordering of streaming data may be implicit (by arrival time at the processing site) or explicit (by generation time, as indicated by a *timestamp* appended to each data item by the source). As a result of these assumptions, DSMSs face the following novel requirements:

1. Much of the computation performed by a DSMS is push-based or data-driven. Newly arrived stream items are continually (or periodically) pushed into the system for processing. On the other hand, a DBMS employs a mostly pull-based or query-driven computation model, where processing is initiated when a query is posed.

2. As a consequence of the preceding text, DSMS queries are *persistent* (also referred to as *continuous*, *long-running*, or *standing queries*) in that they are issued once, but remain active in the system for a possibly long period of time. This means that a stream of updated results must be produced as time goes on. In contrast, a DBMS deals with one-time queries (issued once and then "forgotten"), whose results are computed over the current state of the database.

3. The system conditions may not be stable during the *lifetime* of a persistent query. For example, the stream arrival rates may fluctuate and the query workload may change.

4. A data stream is assumed to have unbounded, or at least unknown, length. From the system's point of view, it is infeasible to store an entire stream in a DSMS. From the user's point of view, recently arrived data are likely to be more accurate or useful.

5. New data models, query semantics, and query languages are needed for DSMSs in order to reflect the facts that streams are ordered and queries are persistent.

The applications that generate streams of data also have similarities in the type of operations that they perform. We list a set of fundamental continuous query operations over streaming data as follows:

- *Selection*: All streaming applications require support for complex filtering.
- *Nested aggregation*: Complex aggregates, including nested aggregates (e.g., comparing a minimum with a running average) are needed to compute trends in the data.
- *Multiplexing and demultiplexing*: Physical streams may need to be decomposed into a series of logical streams, and conversely, logical streams may need to be fused into one physical stream (similar to group-by and union, respectively).
- *Frequent item queries*: These are also known as *top-k* or *threshold* queries, depending on the cut-off condition.
- *Stream mining*: Operations such as pattern matching, similarity searching, and forecasting are needed for on-line mining of streaming data.
- *Joins*: Support should be included for multi-stream joins and joins of streams with static metadata.
- *Windowed queries*: All of the aforementioned query types may be constrained to return results inside a window (e.g., the last 24 h or the last one hundred packets).

Proposed data stream systems resemble the abstract architecture shown in Figure 13.2. An input monitor regulates the input rates, perhaps by dropping items if the system is unable to keep up. Data are typically stored in three partitions: temporary working storage (e.g., for window queries that will be discussed shortly), summary storage for stream synopses, and static storage for metadata (e.g., physical location of each source). Long-running queries are registered in the query repository and placed into groups for shared processing, though one-time queries over the current state of the stream may also be posed. The query processor communicates with the input monitor and may re-optimize the query plans in response to changing input rates. Results are streamed to the users or temporarily buffered. Users may then refine their queries based on the latest results.

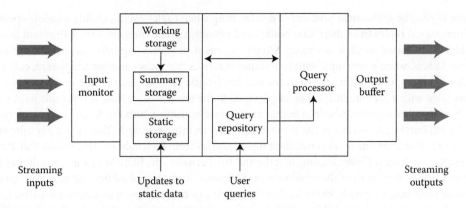

FIGURE 13.2 Abstract reference architecture for a DSMS. (Adapted from Özsu, T.M. and Valduriez, P., *Principles of Distributed Database Systems*, 3rd edn., Springer, 2011.)

Unbounded streams cannot be stored locally in a DSMS, and only a recent excerpt of a stream is usually of interest at any given time. In general, this may be accomplished using a *time-decay model*, and the most common way to achieve this is by means of *window models* where items within the window are given full consideration and items outside the window are ignored. There are a number of ways to classify windows, but the most common are the *logical*, or *time-based* windows, which are defined in terms of a time interval, and *physical*, or *count-based* windows, which are defined in terms of the number of tuples in the window.

Query languages developed for DSMSs fall into three classes: declarative, object-based, and procedural. Declarative languages have SQL-like syntax, but stream-specific semantics, as described earlier. Similarly, object-based languages resemble SQL in syntax, but employ DSMS-specific constructs and semantics, and may include support for streaming abstract data types and associated methods. Finally, procedural languages construct queries by defining data flow through various operators.

While the streaming languages discussed earlier may resemble standard SQL, their implementation, processing, and optimization present novel challenges. Some relational operators are blocking. For instance, prior to returning the next tuple, a nested loop join algorithm may potentially scan the entire inner relation and compare each tuple therein with the current outer tuple. Since data streams are unbounded, it is not possible to scan them, requiring nonblocking implementations. Some operators have non-blocking counterparts; in general, unblocking a query operator may be accomplished by re-implementing it in an incremental form, restricting it to operate over a window (more on this shortly), and exploiting stream constraints. However, there may be cases where an incremental version of an operator does not exist or is inefficient to evaluate, where even a sliding window is too large to fit in main memory, or where no suitable stream constraints are present. In these cases, compact stream summaries may be stored, and approximate queries may be posed over the summaries. This implies a trade-off between accuracy and the amount of memory used to store the summaries. An additional restriction is that the processing time per item should be kept small, especially if the inputs arrive at a fast rate.

Since windowed processing is crucial in DSMSs, we briefly discuss window operators. Sliding window operators process two types of events: arrivals of new tuples and expirations of old tuples. The actions taken upon arrival and expiration vary across operators. A new tuple may generate new results (e.g., join) or remove previously generated results (e.g., negation). Furthermore, an expired tuple may cause a removal of one or more tuples from the result (e.g., aggregation) or an addition of new tuples to the result (e.g., duplicate elimination and negation). Moreover, operators that must explicitly react to expired tuples (by producing new results or invalidating existing results) perform state purging eagerly (e.g., duplicate elimination, aggregation, and negation), whereas others may do so eagerly or lazily (e.g., join).

There is also the orthogonal problem of determining when tuples expire; sliding window operators must remove old tuples from their state buffers and possibly update their answers. Expiration from an individual time-based window is simple: A tuple expires if its timestamp falls out of the range of the window. That is, when a new tuple with timestamp ts arrives, it receives another timestamp, call it exp, that denotes its expiration time as ts plus the window length. In effect, every tuple in the window may be associated with a lifetime interval of length equal to the window size. Now, if this tuple joins with a tuple from another window, whose insertion and expiration timestamps are ts' and exp', respectively, then the expiration timestamp of the result tuple is set to min (exp,exp'). That is, a composite result tuple expires if at least one of its constituent tuples expires from its windows. This means that various join results may have different lifetime lengths and furthermore, the lifetime of a join result may have a lifetime that is shorter than the window size. Moreover, as discussed earlier, the negation operator may force some result tuples to expire earlier than their exp timestamps by generating negative tuples. Finally, if a stream is not bounded by a sliding window, then the expiration time of each tuple is infinity.

In a count-based window, the number of tuples remains constant over time. Therefore, expiration can be implemented by overwriting the oldest tuple with a newly arrived tuple. However, if an operator stores state corresponding to the output of a count-based window join, then the number of tuples in the state may change, depending upon the join attribute values of new tuples. In this case, expirations must be signaled explicitly using negative tuples.

Window sizes can be set system-wide or by each application (or a combination of both). It is possible to use P2P techniques to dynamically increase the size of the sliding window [31].

In addition to the implementation of operators, there is the larger question of optimizing and processing queries, specifically continuous queries. A DSMS query optimizer performs tasks that are similar to a DBMS, but it must use an appropriate cost model and rewrite rules. Additionally, DSMS query optimization involves adaptivity, load shedding, and resource sharing among similar queries running in parallel, as summarized in the subsequent sections.

Traditional DBMSs use selectivity information and available indices to choose efficient query plans (e.g., those which require the fewest disk accesses). However, this cost metric does not apply to (possibly approximate) persistent queries, where processing cost per unit time is more appropriate. Alternatively, if the stream arrival rates and output rates of query operators are known, it may be possible to optimize for the highest output rate or to find a plan that takes the least time to output a given number of tuples. Finally, quality-of-service metrics such as response time may also be used in DSMS query optimization.

A DSMS query optimizer rewrites queries as usual, but with particular attention to the commutativity rules that arise in data stream systems. For example, selections and time-based sliding windows commute, but not selections and count-based windows. Furthermore, the notion of adaptivity is important in query rewriting; operators may need to be re-ordered on-the-fly in response to changes in system conditions. In particular, the cost of a query plan may change for three reasons: change in the processing time of an operator, change in the selectivity of a predicate, and change in the arrival rate of a stream. Initial efforts on adaptive query plans include mid-query re-optimization and query scrambling, where the objective is to preempt any operators that become blocked and schedule other operators instead. To further increase adaptivity, instead of maintaining a rigid tree-structured query plan, the Eddy approach [32] performs scheduling of each tuple separately by routing it through the operators that make up the query plan. In effect, the query plan is dynamically re-ordered to match current system conditions. This is accomplished by tuple routing policies that attempt to discover which operators are fast and selective, and those operators are scheduled first. A recent extension adds queue length as the third factor for tuple routing strategies in the presence of multiple distributed Eddies [33]. There is, however, an important trade-off between the resulting adaptivity and the overhead required to route each tuple separately. Adaptivity involves online reordering of a query plan and may therefore require that the internal state stored by some operators be migrated over to the new query plan consisting of a different arrangement of operators.

The stream arrival rates may be so high that not all tuples can be processed, regardless of the (static or run-time) optimization techniques used. In this case, two types of load shedding may be applied—random or semantic—with the latter making use of stream properties or quality-of-service parameters to drop tuples believed to be less significant than others. For an example of semantic load shedding, consider performing an approximate sliding window join with the objective of attaining the maximum result size. The idea is that tuples that are about to expire or tuples that are not expected to produce many join results should be dropped (in case of memory limitations), or inserted into the join state but ignored during the probing step (in case of CPU limitations). Note that other objectives are possible, such as obtaining a random sample of the join result.

In general, it is desirable to shed load in such a way as to minimize the drop in accuracy. This problem becomes more difficult when multiple queries with many operators are involved, as it must be decided where in the query plan the tuples should be dropped. Clearly, dropping tuples early in the plan is effective because all of the subsequent operators enjoy reduced load. However, this strategy may adversely affect the accuracy of many queries if parts of the plan are shared. On the other hand, load shedding later in the plan, after the shared sub-plans have been evaluated and the only remaining operators are specific to individual queries, may have little or no effect in reducing the overall system load.

One issue that arises in the context of load shedding and query plan generation is whether an optimal plan chosen without load shedding is still optimal if load shedding is used. It has been shown that this is indeed the case for sliding window aggregates, but not for queries involving sliding window joins.

Note that instead of dropping tuples during periods of high load, it is also possible to put them aside (e.g., spill to disk) and process them when the load has subsided. Finally, note that in the case of periodic re-execution of persistent queries, increasing the re-execution interval may be thought of as a form of load shedding.

13.4.4 Web Data Management

The World Wide Web ("WWW" or "web" for short) has become a major repository of data and documents. For all practical purposes, the web represents a very large, dynamic, and distributed data store, and there are the obvious distributed data management issues in accessing web data [34]. The web consists of "pages" that are connected by hyperlinks, and this structure can be modeled as a directed graph that reflects the hyperlink structure. In this graph, commonly referred to as the *web graph*, static HTML web pages are the nodes, and the links between them are represented as directed edges. Studying the web graph is obviously of interest to theoretical computer scientists, because it exhibits a number of interesting characteristics, but it is also important for studying data management issues since the graph structure is exploited in web search [35–37], categorization and classification of web content, and other web-related tasks. The important characteristics of the web graph are the following [38]:

- It is quite volatile. We already discussed the speed with which the graph is growing. In addition, a significant proportion of the web pages experience frequent updates.
- It is sparse. A graph is considered sparse if its average degree is less than the number of vertices. This means that the each node of the graph has a limited number of neighbors, even if the nodes are in general connected. The sparseness of the web graph implies an interesting graph structure that we discuss shortly.
- It is "self-organizing." The web contains a number of communities, each of which consists of a set of pages that focus on a particular topic. These communities get organized on their own without any "centralized control," and give rise to the particular subgraphs in the web graph.
- It is a "small-world network." This property is related to sparseness—each node in the graph may not have many neighbors (i.e., its degree may be small), but many nodes are connected through intermediaries. Small-world networks were first identified in social sciences where it was noted

that many people who are strangers to each other are connected by intermediaries. This holds true in web graphs as well in terms of the connectedness of the graph.

- It is a power law network. The in- and out-degree distributions of the web graph follow power law distributions. This means that the probability that a node has in- (out-) degree i is proportional to $1/i^\alpha$ for some $\alpha > 1$. The value of α is about 2.1 for in-degree and about 7.2 for out-degree.

The management of the very large, dynamic, and volatile web graph is an important issue. Two alternatives have been proposed for this purpose. The first is compressing the web graph, and the second is to develop special storage structures that allow efficient storage and querying (e.g., S-nodes [39]).

Web search is arguably the most common method for accessing the web. It involves finding "all" the web pages that are relevant (i.e., have content related) to keyword(s) that a user specifies. Naturally, it is not possible to find all the pages, or even to know if one has retrieved all the pages; thus the search is performed on a database of web pages that have been collected and indexed. Since there are usually multiple pages that are relevant to a query, these pages are presented to the user in ranked order of relevance as determined by the search engine.

A search engine consists of a number of components. One component is the *crawler*, which is a program used by a search engine to scan the web on its behalf and collect data about web pages. A crawler is given a starting set of pages—more accurately, it is given a set of URLs that identify these pages. The crawler retrieves and parses the page corresponding to that URL, extracts any URLs in it, and adds these URLs to a queue. In the next cycle, the crawler extracts a URL from the queue (based on some order) and retrieves the corresponding page. This process is repeated until the crawler stops. A control module is responsible for deciding which URLs should be visited next. The retrieved pages are stored in a page repository.

The *indexer module* is responsible for constructing indexes on the pages that have been downloaded by the crawler. While many different indexes can be built, the two most common ones are *text indexes* and *link indexes*. In order to construct a text index, the indexer module constructs a large "lookup table" that can provide all the URLs that point to the pages where a given word occurs. A link index describes the link structure of the web and provides information on the in-link and out-link state of pages.

The *ranking module* is responsible for sorting the large number of results so that those that are considered to be most relevant to the user's search are presented first. The problem of ranking has drawn increased interest in order to go beyond traditional information retrieval (IR) techniques to address the special characteristics of the web—web queries are usually small and they are executed over a vast amount of data.

There have also been attempts to perform declarative querying of web data and the efficient execution of these queries. Although these technologies are well developed for relational DBMSs, there are difficulties in carrying over traditional database querying concepts to web data. Perhaps the most important difficulty is that database querying assumes the existence of a strict schema. As noted earlier, it is hard to argue that there is a schema for web data similar to databases.* At best, the web data are *semistructured*—data may have some structure, but this may not be as rigid, regular, or complete as that of databases, so that different instances of the data may be similar but not identical (there may be missing or additional attributes or differences in structure). There are, obviously, inherent difficulties in querying schema-less data.

A second issue is that the web is more than the semistructured data (and documents). The links that exist between web data entities (e.g., pages) are important and need to be considered. Similar to search that we discussed in the previous section, links may need to be followed and exploited in executing web queries. This requires links to be treated as first-class objects.

A third major difficulty is that there is no commonly accepted language, similar to SQL, for querying web data. As we noted in the previous section, keyword search has a very simple language, but this is not

---

* We are focusing on the "open" web here; deep web data may have a schema, but it is usually not accessible to users.

sufficient for richer querying of web data. Some consensus on the basic constructs of such a language has emerged (e.g., path expressions), but there is no standard language. However, a standardized language for XML has emerged (XQuery), and as XML becomes more prevalent on the web, this language is likely to become dominant and more widely used.

One way to approach querying the web data is to treat it as a collection of semistructured data. Then, models and languages that have been developed for this purpose can be used to query the data. Semistructured data models and languages were not originally developed to deal with web data; rather they addressed the requirements of growing data collections that did not have as strict a schema as their relational counterparts. However, since these characteristics are also common to web data, later studies explored their applicability in this domain.

There have also been special purpose web query language developments. These directly address the characteristics of web data, particularly focusing on handling *links* properly. Their starting point is to overcome the shortcomings of keyword search by providing proper abstractions for capturing the content structure of documents (as in semistructured data approaches) as well as the external links. They combine the content-based queries (e.g., keyword expressions) and structure-based queries (e.g., path expressions).

13.4.5 P2P Data Management

In contrast to C/S DBMS, P2P systems [40] adopt a completely decentralized approach to distributed data sharing. By distributing data storage and processing across autonomous peers in the network, they can scale without the need for powerful servers. Popular examples of P2P systems such as BitTorrent and eMule have millions of users sharing petabytes of data over the Internet. Although very useful, these systems are quite simple (e.g., file sharing), support-limited functions (e.g., keyword search) and use simple techniques (e.g., resource location by flooding), which have performance problems. To deal with the dynamic behavior of peers that can join and leave the system at any time, they rely on the fact that popular data get massively duplicated. Furthermore, they are single-application systems and focus on performing one task, and it is not straightforward to extend them for other applications/functions [41].

To provide proper database functionality (schema, queries, replication, consistency, availability, etc.) over P2P infrastructures, the following data management issues must be addressed:

- Data location: Peers must be able to refer to and locate data stored in other peers.
- Data integration: When shared data sources in the system follow different schemas or representations, peers should still be able to access that data, ideally using the data representation used to model their own data.
- Query processing: Given a query, the system must be able to discover the peers that contribute relevant data and efficiently execute the query.
- Data consistency: If data are replicated or cached in the system, a key issue is to maintain the consistency between these duplicates.

Data location depends on the underlying infrastructure. P2P networks can be of two general types: pure and hybrid. *Pure P2P networks* are those where there is no differentiation between any of the network nodes—they are all equal. In *hybrid P2P networks*, on the other hand, some nodes are given special tasks to perform. Hybrid networks are commonly known as *super-peer systems*, since some of the peers are responsible for "controlling" a set of other peers in their domain. The pure P2P networks can be further divided into structured and unstructured networks. *Structured networks* tightly control the topology and message routing, for instance using a distributed hash table (DHT) or a tree structure, whereas in *unstructured networks*, each node can directly communicate with its neighbors and can join the network by attaching themselves to any node.

Structured P2P networks have emerged to address the scalability issues faced by unstructured P2P networks, which rely on flooding to locate data. They achieve this goal by tightly controlling the network topology and the placement of resources. Thus, they achieve higher scalability at the expense of lower autonomy as each peer that joins the network allows its resources to be placed on the network based on the particular control method that is used.

Data integration raises issues that are more difficult than for designing database integration systems. Due to specific characteristics of P2P systems, for example, the dynamic and autonomous nature of peers, the approaches that rely on centralized global schemas no longer apply. The main problem is to support decentralized schema mapping so that a query on one peer's schema can be reformulated in a query on another peer's schema. The approaches, which are used by P2P data management systems for defining and creating the mappings between peers' schemas, can be classified as follows: pairwise schema mapping, common agreement mapping, and schema mapping using IR techniques. With pairwise schema mapping [42], each user defines the mapping between the local schema and the schema of any other peer that contains data that is of interest. Relying on the transitivity of the defined mappings, the system tries to extract mappings between schemas, which have no defined mapping. In common agreement mapping [43], the peers that have a common interest agree on a common schema description (CSD) for data sharing. Given a CSD, a peer schema can be specified using views. This is similar to the local-as-view approach in data integration systems, except that queries at a peer are expressed in terms of the local views, not the CSD. Another difference is that the CSD is not a global schema, that is, it is common to a limited set of peers with common interest. The last approach extracts the schema mappings at query execution time using IR techniques by exploring the schema descriptions provided by users. PeerDB [41] follows this approach for query processing in unstructured P2P networks.

P2P networks provide basic techniques for routing queries to relevant peers and this is sufficient for supporting simple, exact-match queries. For instance, a DHT provides a basic mechanism for efficient data lookup based on a key value. However, supporting more complex queries in P2P systems, particularly in DHTs, is difficult and has been the subject of much recent research. The main types of complex queries, which are useful in P2P systems, are top-k queries, join queries, and range queries.

With a top-k query, the user can specify a number k of the most relevant answers to be returned by the system. The degree of relevance (score) of the answers to the query is determined by a scoring function. An efficient algorithm for top-k query processing in centralized and distributed systems is the threshold algorithm (TA) [44]. TA assumes a general model based on lists of data items sorted by their local scores and is applicable for queries where the scoring function is monotonic, that is, any increase in the value of the input does not decrease the value of the output. The stopping mechanism of TA uses a threshold which is computed using the last local scores seen under sorted access in the lists. Many of the popular aggregation functions such as min, max, and average are monotonic. TA has been the basis for several TA-style algorithms in P2P systems. There are many database instances over which TA keeps scanning the lists although it has seen all top-k answers. Thus, it is possible to stop much sooner. Based on this observation, best position algorithms that execute top-k queries much more efficiently than TA have been proposed [45].

Structured P2P systems, in particular, DHTs are very efficient at supporting exact-match queries (i.e., queries of the form "$A = value$") but have difficulties with range queries. The main reason is that hashing tends to destroy the ordering of data which is useful to find ranges quickly. There are two main approaches for supporting range queries in structured P2P systems: extend a DHT with proximity or order-preserving properties [46], or maintain the key ordering with a tree-based structure [47].

The most efficient join algorithms in distributed and parallel databases are hash-based. Thus, the fact that a DHT relies on hashing to store and locate data can be naturally exploited to support join queries efficiently. A basic solution [48] is a variation of the traditional parallel hash join algorithm. The joined relations and the result relations have a namespace that indicates the nodes, which store horizontal fragments of the relation. Then it makes use of the DHT put method for distributing tuples onto a set of peers based on their join attribute so that tuples with same join attribute values are stored at the same peers. Then, joins can be performed locally at each peer, for example, using a hash join algorithm.

To increase data availability and access performance, P2P systems replicate data. However, different P2P systems provide very different levels of replica consistency. The earlier, simple P2P systems such as Gnutella and Kazaa deal only with static data (e.g., music files), and replication is "passive" as it occurs naturally as peers request and copy files from one another (basically, caching data). In more advanced P2P systems where replicas can be updated, there is a need for proper replica management techniques. Most of the work on replica consistency has been done only in the context of DHTs.

To improve data availability, most DHTs rely on data replication by storing data items at several peers by, for example, using several hash functions. If one peer is unavailable, its data can still be retrieved from the other peers that hold a replica. Some DHTs provide basic support for the application to deal with replica consistency, for example, Tapestry [49].

Although DHTs provide basic support for replication, the mutual consistency of the replicas after updates can be compromised as a result of peers leaving the network or concurrent updates. For some applications (e.g., agenda management), the ability to get the current data is very important. Supporting data currency in replicated DHTs requires the ability to return a current replica despite peers leaving the network or concurrent updates. The problem can be partially addressed by using data versioning, where each replica has a version number that is increased after each update. To return a current replica, all replicas need to be retrieved in order to select the latest version. However, because of concurrent updates, it may happen that two different replicas have the same version number, thus making it impossible to decide which one is the current replica. A more complete solution considers both data availability and data currency in DHTs, using distributed timestamping [50].

Replica reconciliation goes one step further than data currency by enforcing mutual consistency of replicas. Since a P2P network is typically very dynamic, with peers joining or leaving the network at will, eager replication solutions are not appropriate. The major solution is lazy distributed replication as used in OceanStore [51] and P-Grid [52]. APPA provides a general lazy distributed replication solution, which assures eventual consistency of replicas [53].

13.5 Summary

The impact of distributed and parallel data management technology on practice has been very high as all commercial DBMSs today have distributed and parallel versions. Parallel database systems are used to support very large databases (e.g., hundreds of terabytes or petabytes). Examples of applications that deal with very large databases are e-commerce, data warehousing, and data mining. Very large databases are typically accessed through high numbers of concurrent transactions (e.g., performing online orders on an electronic store) or complex queries (e.g., decision-support queries). The first kind of access is representative of OLTP applications while the second is representative of OLAP applications. Although both OLTP and OLAP can be supported by the same parallel database system (on the same multiprocessor), they are typically separated and supported by different systems to avoid any interference and ease database operation.

To address the requirements of new data-intensive applications (e.g., scientific applications or cloud), emerging solutions tend to rely on a more specific data model (e.g., Bigtable, which is a variant of the nested relational model) with a simple set of operators easy to use from a programming language. For instance, to address the requirements of social network applications, new solutions rely on a graph data model and graph-based operators. To address the requirements of scientific applications, SciDB [54] supports an array data model, which generalizes the relational model, with array operators. User-defined functions also allow for more specific data processing. In emerging solutions, it is interesting to witness the development of algebras, with specific operators, to raise the level of abstraction in a way that enables optimization, while making it easy to manipulate data from a programming language with a simple API.

Distributed and parallel database technology can now be used in two complementary large-scale distributed contexts: P2P and cloud. A P2P solution is appropriate to support the collaborative nature of many applications, for example, scientific applications, as it provides scalability, dynamicity, autonomy, and decentralized control. Peers can be the participants or organizations involved in collaboration and may share data and applications while keeping full control over their (local) data sources. But for very-large-scale data analysis or to execute very large workflow activities, cloud computing is the right approach as it can provide virtually infinite computing, storage, and networking resources.

References

1. T. M. Özsu and P. Valduriez. *Principles of Distributed Database Systems*, 3rd edn. Springer, New York, 2011.
2. D. J. DeWitt and J. Gray. Parallel database systems: The future of high-performance database systems. *Communications of the ACM*, 35(6):85–98, 1992.
3. P. Valduriez. Parallel database management. In: L. Liu and M. T. Özsu, eds., *Encyclopedia of Database Systems*, pp. 2026–2029. Springer, New York, 2009.
4. E. Pacitti. Parallel query processing. In: L. Liu and M. T. Özsu, eds., *Encyclopedia of Database Systems*, pp. 2038–2040. Springer, New York, 2009.
5. H. Zeller and G. Graefe. Parallel query optimization. In: L. Liu and M. T. Özsu, eds., *Encyclopedia of Database Systems*, pp. 2035–2038. Springer, New York, 2009.
6. P. Valduriez. Parallel data placement. In: L. Liu and M. T. Özsu, eds., *Encyclopedia of Database Systems*, pp. 2024–2026. Springer, New York, 2009.
7. G. Graefe. Parallel hash join, parallel merge join, parallel nested loops join. In: L. Liu and M. T. Özsu, eds., *Encyclopedia of Database Systems*, pp. 2029–2030. Springer, New York, 2009.
8. J. Gray. The transaction concept: Virtues and limitations. In: *Proceedings of the 7th International Conference on Very Large Data Bases*, Cannes, France, pp. 144–154, 1981.
9. J. N. Gray. Notes on data base operating systems. In: R. Bayer, R. M. Graham, and G. Seegmüller, eds., *Operating Systems: An Advanced Course*, pp. 393–481. Springer-Verlag, New York, 1979.
10. P. A. Bernstein, V. Hadzilacos, and N. Goodman. *Concurrency Control and Recovery in Database Systems*. Addison-Wesley, Reading, MA, 1987.
11. G. Weikum and G. Vossen. *Transactional Information Systems*. Morgan-Kaufmann, San Francisco, CA, 2001.
12. C. Mohan and B. Lindsay. Efficient commit protocols for the tree of processes model of distributed transactions. In: *Proceedings of ACM SIGACT-SIGOPS 2nd Symposium on the Principles of Distributed Computing*, Montreal, Quebec, Canada, pp. 76–88, 1983.
13. D. K. Gifford. Weighted voting for replicated data. In: *Proceedings of the 7th ACM Symposium on Operating System Principles*, Pacific Grove, CA, pp. 50–159, 1979.
14. A. E. Abbadi, D. Skeen, and F. Cristian. An efficient, fault-tolerant protocol for replicated data management. In: *Proceedings of ACM SIGACT-SIGMOD Symposium on Principles of Database Systems*, Portland, OR, pp. 215–229, 1985.
15. P. Valduriez. Principles of distributed data management in 2020? In: *Proceedings of the International Conference on Database and Expert Systems Applications*, Toulouse, France, pp. 1–11, 2011.
16. D. J. Abadi. Data management in the cloud: Limitations and opportunities. *IEEE Data Engineering Bulletin*, 32(1):3–12, 2009.
17. F. Chang, J. Dean, S. Ghemawat, W. C. Hsieh, D. A. Wallach, M. Burrows, T. Chandra, A. Fikes, and R. E. Gruber. Bigtable: A distributed storage system for structured data. *ACM Transactions on Computer Systems*, 26(2), 2008.
18. S. Ghemawat, H. Gobioff, and S.-T. Leung. The google file system. In: *Proceedings of the 19th ACM Symposium on Operating System Principles*, Bolton Landing, NY, pp. 29–43, 2003.
19. B. F. Cooper, R. Ramakrishnan, U. Srivastava, A. Silberstein, P. Bohannon, H.-A. Jacobsen, N. Puz, D. Weaver, and R. Yerneni. Pnuts: Yahoo!'s hosted data serving platform. *Proceedings of VLDB*, 1(2):1277–1288, 2008.

20. J. Dean and S. Ghemawat. MapReduce: Simplified data processing on large clusters. In: *Proceedings of Symposium on Operating System Design and Implementation*, San Francisco, CA, pp. 137–150, 2004.

21. J. Dittrich, J.-A. Quiané-Ruiz, A. Jindal, Y. Kargin, V. Setty, and J. Schad. Hadoop++: Making a yellow elephant run like a cheetah (without it even noticing). *Proceedings of VLDB*, 3(1):518–529, 2010.

22. A. Ailamaki, V. Kantere, and D. Dash. Managing scientific data. *Communications of the ACM*, 53(6):68–78, 2010.

23. E. Pacitti, P. Valduriez, and M. Mattoso. Grid data management: Open problems and new issues. *Journal of Grid Computing*, 5(3):273–281, 2007.

24. J. Gray, D. T. Liu, M. A. Nieto-Santisteban, A. S. Szalay, D. J. DeWitt, and G. Heber. Scientific data management in the coming decade. *CoRR*, abs/cs/0502008, 2005.

25. M. Stonebraker, J. Becla, D. J. DeWitt, K.-T. Lim, D. Maier, O. Ratzesberger, and S. B. Zdonik. Requirements for science data bases and scidb. In: *Proceedings of the 4th Biennial Conference on Innovative Data Systems Research*, Asilomar, CA, 2009.

26. C. Olston, B. Reed, U. Srivastava, R. Kumar, and A. Tomkins. Pig Latin: A not-so-foreign language for data processing. In: *Proceedings of ACM SIGMOD International Conference on Management of Data*, Vancouver, British Columbia, Canada, pp. 1099–1110, 2008.

27. E. Ogasawara, D. De Oliveira, P. Valduriez, D. Dias, F. Porto, and M. Mattoso. An algebraic approach for data-centric scientific workflows. *Proceedings of VLDB*, 4(11):1328–1339, 2011.

28. L. Bouganim, D. Florescu, and P. Valduriez. Dynamic load balancing in hierarchical parallel database systems. In: *Proceedings of the 22nd International Conference on Very Large Data Bases*, Mumbai, India, pp. 436–447, 1996.

29. L. Golab and M. T. Özsu. Issues in data stream management. *ACM SIGMOD Record*, 32(2):5–14, 2003.

30. L. Golab and M. T. Özsu. *Data Stream Systems*. Morgan & Claypool Publishers, San Rafael, CA, 2010.

31. W. Palma, R. Akbarinia, E. Pacitti, and P. Valduriez. DHTJoin: Processing continuous join queries using DHT networks. *Distributed and Parallel Databases*, 26(2–3):291–317, 2009.

32. L. Adamic and B. Huberman. The nature of markets in the world wide web. *Quarterly Journal of Electronics Communication*, 1:5–12, 2000.

33. F. Tian and D. DeWitt. Tuple routing strategies for distributed Eddies. In: *Proceedings of the 29th International Conference on Very Large Data Bases*, Berlin, Germany, pp. 333–344, 2003.

34. S. Abiteboul, I. Manolescu, P. Rigaux, M.-C. Rousset, and P. Senellart. *Web Data Management*. Cambridge University Press, Cambridge, U.K., 2011.

35. J. M. Kleinberg, R. Kumar, P. Raghavan, S. Rajagopalan, and A. Tomkins. The Web as a graph: Measurements, models, and methods. In: *Proceedings of the 5th Annual International Conference on Computing and Combinatorics*, Tokyo, Japan, pp. 1–17, 1999.

36. S. Brin and L. Page. The anatomy of a large-scale hypertextual web search engine. *Computer Networks*, 30(1–7):107–117, 1998.

37. J. M. Kleinberg. Authoritative sources in a hyperlinked environment. *Journal of the ACM*, 46(5):604–632, 1999.

38. A. Bonato. *A Course on the Web Graph*. American Mathematical Society, Providence, RI, 2008.

39. S. Raghavan and H. Garcia-Molina. Representing web graphs. In: *Proceedings of the 19th Conference on Data Engineering*, Bangalore, India, pp. 405–416, 2003.

40. P. Cudré-Mauroux. Peer data management system. In: L. Liu and M. T. Özsu, eds., *Encyclopedia of Database Systems*, pp. 2055–2056. Springer, 2009.

41. B. Ooi, Y. Shu, and K.-L. Tan. Relational data sharing in peer-based data management systems. *ACM SIGMOD Record*, 32(3):59–64, 2003.

42. I. Tatarinov, Z. G. Ives, J. Madhavan, A. Y. Halevy, D. Suciu, N. N. Dalvi, X. Dong, Y. Kadiyska, G. Miklau, and P. Mork. The piazza peer data management project. *ACM SIGMOD Record*, 32(3):47–52, 2003.

43. R. Akbarinia, V. Martins, E. Pacitti, and P. Valduriez. Design and implementation of atlas p2p architecture. In: R. Baldoni, G. Cortese, and F. Davide, eds., *Global Data Management*, pp. 98–123. IOS Press, Amsterdam, Netherlands, 2006.

44. R. Fagin, J. Lotem, and M. Naor. Optimal aggregation algorithms for middleware. *Journal of Computer and System Sciences*, 66(4):614–656, 2003.

45. R. Akbarinia, E. Pacitti, and P. Valduriez. Best position algorithms for top-k queries. In: *Proceedings of the 33rd International Conference on Very Large Data Bases*, Vienna, Austria, pp. 495–506, 2007.

46. A. Gupta, D. Agrawal, and A. El Abbadi. Approximate range selection queries in peer-to-peer systems. In: *International Conference on Innovative Data Systems Research (CIDR)*, Asilomar, CA, pp. 141–151, 2003.

47. H. V. Jagadish, B. C. Ooi, and Q. H. Vu. Baton: A balanced tree structure for peer-to-peer networks. In: *VLDB*, Trondheim, Norway, pp. 661–672, 2005.

48. R. Huebsch, J. Hellerstein, N. Lanham, B. Thau Loo, S. Shenker, and I. Stoica. Querying the internet with pier. In: *Proceedings of the 29th International Conference on Very Large Data Bases*, Berlin, Germany, pp. 321–332, 2003.

49. B. Y. Zhao, L. Huang, J. Stribling, S. C. Rhea, A. D. Joseph, and J. D. Kubiatowicz. Tapestry: A resilient global-scale overlay for service deployment. *IEEE Journal on Selected Areas in Communications*, 22(1):41–53, January 2004.

50. R. Akbarinia, E. Pacitti, and P. Valduriez. Data currency in replicated dhts. In: *Proceedings of ACM SIGMOD International Conference on Management of Data*, Beijing, China, pp. 211–222, 2007.

51. J. Kubiatowicz, D. Bindel, Y. Chen, S. Czerwinski, P. Eaton, D. Geels, R. Gummadi, S. Rhea, H. Weatherspoon, W. Weimer, C. Wells, and B. Zhao. Oceanstore: An architecture for global-scale persistent storage. In: *ACM International Conference on Architectural Support for Programming Languages and Operating Systems*, Cambridge, MA, pp. 190–201, 2000.

52. K. Aberer, P. Cudré-Mauroux, A. Datta, Z. Despotovic, M. Hauswirth, M. Punceva, and R. Schmidt. P-grid: A self-organizing structured p2p system. *ACM SIGMOD Record*, 32(3):29–33, 2003.

53. V. Martins, E. Pacitti, M. El Dick, and R. Jimenez-Peris. Scalable and topology-aware reconciliation on p2p networks. *Distributed and Parallel Databases*, 24(1–3):1–43, 2008.

54. M. Stonebraker, J. Becla, D. J. DeWitt, K.-T. Lim, D. Maier, O. Ratzesberger, and S. B. Zdonik. Requirements for science data bases and SciDB. In: *Proceedings of the 4th Biennial Conference on Innovative Data Systems Research*, Asilomar, CA, 2009.

14

Multimedia Databases: Description, Analysis, Modeling, and Retrieval

Vincent Oria
New Jersey Institute of Technology

Ying Li
IBM T.J. Watson Research Center

Chitra Dorai
IBM T.J. Watson Research Center

Michael E. Houle
National Institute of Informatics

14.1 Introduction

With rapidly growing collections of images, news programs, music videos, movies, digital television programs, and training and education videos on the Internet and corporate intranets, new tools are needed to harness digital media for different applications, ranging from image and video cataloging, to media archival and search, multimedia authoring and synthesis, and smart browsing. In recent years, we have witnessed the growing momentum in building systems that can query and search video collections efficiently and accurately for desired video segments, just in the manner text search engines on the web have enabled easy retrieval of documents containing a required piece of text located on a server anywhere in the world. Digital video archival and management systems are also important to postproduction houses, broadcast studios, stock footage houses, and advertising agencies working with large videotape and multimedia collections, to enable integration of content in their end-to-end business processes. Further, since the digital form of videos enables rapid content editing, manipulation, and synthesis, there is burgeoning interest in building cheap, personal desktop video production tools.

An image and video content management system must allow archival, processing, editing, manipulation, browsing, and search and retrieval of image and video data for content repurposing, new program production, and other multimedia interactive services. Annotating or describing images and videos manually through a preview of the material is extremely time consuming, expensive, and unscalable. A content management system, for example, in a digital television studio serves different sets of users, including the program producer who needs to locate material from the studio archive, the writer who needs to write a story about the airing segment, the editor who needs to edit in the desired clip, the librarian who adds and manages new material in the archive, and the logger who annotates the material in terms of its metadata such as medium ID, production details, and other pertinent information about the content that allows it to be located easily. Therefore, a content management tool must be scalable and highly available, ensure integrity of content, and enable easy and quick retrieval of archived material for content reuse and distribution. Automatic extraction of image and video content descriptions is highly desirable to ease the pain of manual annotation and to provide a consistent language of content description when annotating large video collections.

In order to answer user queries during media search, it is crucial to define a suitable representation for the media, their metadata, and the operations to be applied to them. The aim of a data model is to introduce an abstraction between the physical level (data files and indices) and the conceptual representation, together with some operations to manipulate the data. The conceptual representation corresponds to the conceptual level of the American National Standards Institute (ANSI) relational database architecture [106], where algebraic optimization and algorithm selections are performed. Optimization at the physical level (data files and indices) consists of defining indices and selecting the right access methods to be used in query processing.

This chapter surveys techniques used to extract descriptions of multimedia data (mainly image and video) through automated analysis and current database solutions in managing, indexing, and querying of multimedia data. The chapter is divided into two parts: multimedia data analysis and database techniques for multimedia. The multimedia data analysis part comprises Section 14.2, which presents common features used in image databases and the techniques to extract them automatically, and Section 14.3, which discusses audio and video content analysis, extraction of audiovisual descriptors, as well as the fusion of multiple media modalities. The second part comprises Section 14.4, which presents an example of database models defined for multimedia data, and Section 14.5, which discusses similarity search in multimedia. Finally, Section 14.6 concludes the chapter.

14.2 Image Content Analysis

Existing work in image content analysis could be coarsely categorized into two groups based on the features employed. The first group indexes an image based on low-level features such as color, texture, and shape, while the second group attempts to understand the semantic content of images by using mid- to high-level features, as well as by applying more complex analysis models. This section surveys representative work in both groups.

14.2.1 Low-Level Image Content Analysis

Research is this area proposes to index images based on low-level features, which are easy to extract and fast to implement. Some well-known content-based image retrieval (CBIR) systems such as query by image content (QBIC) [39], multimedia analysis and retrieval system (MARS) [54], WebSEEK [119], and Photobook [102] have employed these features for image indexing, browsing, and retrieval, with reasonable performance achieved. However, due to the low-level nature of these features, there still exists a gap between the information revealed by these features and the real image semantics. Obviously, more high-level features are needed to truly understand the image content. In the following, we review some commonly used image features, including color, texture, and shape.

14.2.1.1 Color

Color is one of the most recognizable elements of image content. It is widely used as a feature for image retrieval due to its invariance to image scaling, translation, and rotation. Key issues in color feature extraction include the selection of color space and the choice of color quantization scheme.

A color space is a multidimensional space in which different dimensions represent different color components. Theoretically, any color can be represented by a linear combination of the three primary colors, red (R), green (G), and blue (B). However, the RGB color space is not perceptually uniform, and equal distances in different areas and along different dimensions of this space do not correspond to equal perception of color dissimilarity. Therefore, other color spaces, such as CIELAB and CIEL*u*v*, have been proposed. Other widely used color spaces include YCbCr, YIQ, YUV, HSV, and Munsell spaces. Readers are referred to [20] for more detailed descriptions on color spaces. The MPEG-7 standard [85], formally known as the "multimedia content description interface," has adopted the RGB, YCbCr, and HSV color spaces, as well as linear transformation matrices with reference to RGB.

Quantization is used to reduce the color resolution of an image. Since a 24 bit color space contains 2^{24} distinct colors, using a quantized map can considerably decrease the computational complexity of color feature extraction. The most commonly used color quantization schemes are uniform quantization, vector quantization, tree-structured vector quantization, and product quantization. MPEG-7 supports linear and nonlinear quantization, as well as quantization via lookup tables.

Three widely used color features for images (also called color descriptors in MPEG-7) are global color histogram, local color histogram, and dominant color. The global color histogram captures the color content of the entire image while ignoring information on spatial layout. Specifically, a global color histogram represents an image I by an N-dimensional vector $H(I) = [H(I, j), j = 1, 2, ..., N]$, where N is the total number of quantized colors and $H(I, j)$ is the number of pixels having color j.

In contrast, the local color histogram representation considers the position and size of each individual image region so as to describe the spatial structure of the image color. For instance, Stricker and Dimai [123] segmented each image into five nonoverlapping spatial regions, from which color features were extracted and subsequently used for image matching. In [96], a scalable blob histogram was proposed, where the term blob denotes a group of pixels with a homogeneous color. This descriptor is able to distinguish images, containing objects of different sizes and shapes, without performing color segmentation.

The dominant color representation is considered to be one of the major color descriptors due to its simplicity and its association with human perception. Various algorithms have been proposed to extract this feature. For example, Ohm and Makai took the means of color clusters (the cluster centroids) as dominant colors of images [97]. Considering that human eyes are more sensitive to changes in smooth regions than changes in detailed ones, Deng et al. proposed the extraction of dominant colors by hierarchically merging similar and unimportant color clusters while leaving distinct and more important clusters untouched [34].

Other commonly used color feature representations include color correlogram [52] and color moments [123]. Specifically, the color correlogram captures the local spatial correlation of pairs of colors and is a second-order statistic on the color distribution. It is rotation, scale, and (to some extent) viewpoint invariant. Moreover, it is designed to tolerate moderate changes in appearance and shape due to viewpoint changes, camera zoom, noise, and compression. Therefore, color correlogram representations have been widely used in video copy detection [92] and video concept modeling and detection [140]. On the other hand, color moments are mainly adopted to overcome undesirable color quantization effects and are defined as the moments of the distribution formed by RGB triplets in a given image. Note that by using a proper combination of moments it is possible to normalize against photometric changes. Such combinations are called color moment invariants. Color moments and color moment invariants have been applied to applications such as CBIR.

14.2.1.2 Texture

Texture refers to visual patterns with properties of homogeneity that do not result from the presence of only a single color or intensity. Tree bark, clouds, water, bricks, and fabrics are some examples of textures. Typical texture features include contrast, uniformity, coarseness, roughness, frequency, density, and directionality, which contain important information about the structural arrangement of surfaces as well as their relationship to the surrounding environment. So far, much research effort has been devoted to texture analysis due to its usefulness and effectiveness in applications such as pattern recognition, computer vision, and image retrieval.

There are two basic types of texture descriptors: statistical and transform based. The statistical approach explores the gray-level spatial dependence of textures and extracts meaningful statistics as the texture representation. For example, Haralick et al. [47] proposed the representation of textures using co-occurrence matrices that capture gray-level spatial dependence. They also performed a statistical analysis of the spatial relationships of lines, as well as the properties of their surroundings. Interestingly, Tamura et al. addressed this topic from a totally different viewpoint [125]: based on psychological measurements, they claimed that the six basic textural features should be coarseness, contrast, directionality, line-likeness, regularity, and roughness. Two well-known CBIR systems, QBIC and MARS, have adopted this representation. Other texture representations may involve a subset of the aforementioned six features, such as contrast, coarseness, and directionality.

Two other commonly used statistical texture features are edge histograms [85] and histograms of oriented gradient (HOG) [29]. Edge histograms describe the distribution of edges in terms of both frequency and directionality of the brightness gradients within an image. HOG extends the edge orientation histogram or histogram of gradient directions by means of computation over a dense grid of uniformly spaced cells. Furthermore, improved performance is often achieved within overlapping local contrast normalization. Edge histograms and HOG features have been widely applied in image and video object detection and recognition [45].

In recent years, local binary pattern (LBP) feature has gradually gained in popularity, especially in computer vision applications. Specifically, the LBP operator is defined as a grayscale invariant texture measure, derived from a general definition of texture in a local neighborhood. Its properties of invariance under monotonic gray-level changes, together with its computational simplicity, have made it a very good feature for applications such as facial expression recognition [151].

Commonly used transforms for texture extraction include the discrete cosine transform (DCT), the Fourier–Mellin transform, the polar Fourier transform, the Gabor transform, and the wavelet transform. Alata et al. [5] proposed the classification of rotated and scaled textures using a combination of the Fourier–Mellin transform and a parametric 2D spectrum estimation method (Harmonic Mean Horizontal Vertical). In [130], Wan and Kuo reported their work on texture feature extraction for Joint Photographic Experts Group (JPEG) images based on the analysis of DCT–AC coefficients. Chang and Kuo [23] presented a tree-structured wavelet transform, which provided a natural and effective way to describe textures that have dominant middle- to high-frequency subbands. Readers are referred to [20] for a detailed description on texture feature extraction.

Building upon Gabor features, Oliva and Torralba proposed the GIST feature for the characterization of scenes [98]. GIST features concatenate spatially pooled Gabor filter responses in different scales and orientations in different spatial blocks of the image. They have been successfully applied to directly represent, understand, and recognize image scenes, bypassing the segmentation and processing of individual objects or regions. Natseve et al. have also successfully applied GIST (along with other features) for video copy detection and video event detection in their text retrieval conference—video retrieval evaluation (TRECVID) 2010 effort [92].

14.2.1.3 Shape

Compared to color and texture, the representation of shape is inherently much more complex. Two major steps are required to extract a shape feature: object segmentation and shape representation.

Although object segmentation has been studied for decades, it remains a very difficult research topic in computer vision. Existing image segmentation techniques include the global threshold-based approach, the region growing approach, the split and merge approach, the edge-detection-based approach, the color- and texture-based approach, and the model-based approach. Generally speaking, it is difficult to achieve perfect segmentation due to the complexity of individual object shapes, as well as the existence of shadows and noise.

Existing shape representation approaches can be categorized into the following three classes: boundary-based representation, region-based representation, and their combination. Boundary-based representations emphasize the closed curve that surrounds the shape. Numerous models have been proposed to describe this curve, including chain codes [42], polygons [28], circular arcs [38], splines [33], explicit and implicit polynomials [61], boundary Fourier descriptors, and Universidade Nova de Lisboa (UNL) descriptors [16]. Region-based representations, on the other hand, emphasize the area within the closed boundary. Various descriptors have been proposed to model interior regions, such as moment invariants [77], Zernike moments [62], morphological descriptor [81], and pseudo-Zernike moments. Generally speaking, region-based moments are invariant to image's affine transformations. Readers are referred to [147] for more details.

Recent work in shape representation includes the finite element method (FEM), the turning function, and the wavelet descriptor. Besides the earlier work in 2D shape representation, research effort has been devoted to 3D shape representation. Readers are referred to [20] for more detailed discussions on shape features.

14.2.2 Key Points and Bag of Visual Words

14.2.2.1 Key Point Extraction

Key points (or points of interest) are used to describe local features of images. The most popular technique is the scale invariant feature transform (SIFT) proposed by Lowe in 2004 [82]. SIFT is designed to be scale, rotation, and shift invariant so that it can be reliably applied to object recognition and image matching. The four basic steps in extracting SIFT features include scale-space extrema detection, key point localization, orientation assignment, and key point descriptor formation.

Based upon SIFT, several other variants have been subsequently proposed, including the speeded up robust features (SURFs) [10], the principal component analysis-based representation of SIFT features (PCA-SIFT) [60], and the rotation-invariant feature transform (RIFT) [66]. SURF is a high-performance scale- and rotation-invariant interest point detector/descriptor, which has been claimed to approximate or even outperform SIFT with respect to repeatability, distinctiveness, and robustness. PCA-SIFT differs from SIFT in terms of calculation of the image gradients and the definition of gradient regions. It also applies PCA to further reduce the feature dimension. Finally, RIFT is a rotation-invariant generalization of SIFT, constructed using circular normalized patches, which are divided into concentric rings of equal width. A gradient orientation histogram is then computed within each ring.

Key point extraction methods such SIFT represent an image as a large collection of feature vectors, each of which describes a specific key point in the image. Visual words are obtained by grouping similar key points into clusters. Hence, a visual word represents a specific visual pattern and the set of visual words represent a visual word vocabulary. An image can then be represented as a collection (or "bag") of visual words. The entire process from the key point extraction to the final visual word vectors is illustrated by Figure 14.1.

14.2.3 Mid- to High-Level Image Content Analysis

Research in this area attempts to index images based on semantics content in the form of salient image objects. To achieve this goal, various mid- to high-level image features, as well as more complex analysis models, have been proposed. An example reported in [34] uses a low-dimensional color indexing scheme based on homogeneous image regions. After first applying a color segmentation method (JSEG) to obtain homogeneous regions, the colors within each region are quantized and grouped into a small number of clusters. Finally, color centroids together with their relative weights were used as feature descriptors.

Some approaches attempt to understand image content by learning its semantic concepts. For example, Minka and Picard [88] developed a system that first generated segmentations or groups of image

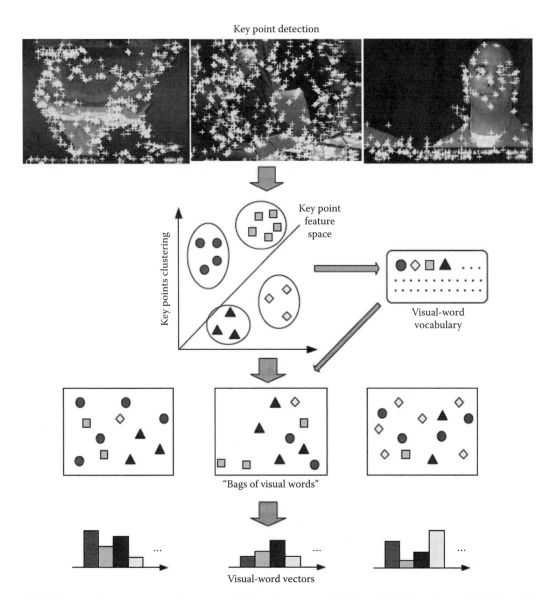

FIGURE 14.1   From key points to bag of words. (From Yang, L. et al., *Multimedia Inform. Retrieval (MIR)*, 1, 265, September 2007.)

regions using various feature combinations. The system was later improved by learning from the user input to decide which combinations best represent predetermined semantic categories. The system, however, requires supervised training over various parts of the image. In contrast, Li et al. proposed the detection of detect salient image regions based on segmented color and orientation maps without any human intervention [74].

Targeting the construction of a moderately large lexicon of semantic concepts, Naphade et al. proposed a support vector machine (SVM) learning system for detecting 34 visual concepts, including 15 scene concepts (e.g., outdoors, indoors, landscape, cityscape, sky, beach, mountain, and land) and 19 object concepts (e.g., face, people, road, building, tree, animal, text overlay, and train) [91]. Using the TREC 2002 benchmark corpus for training and validation, this system has achieved reasonable performance with moderately large training samples.

With the proliferation of photo sharing services and general purpose sites with photo sharing capabilities, such as Flickr and Facebook, automated image annotation and tagging have become a problem of great interest [35,107,115,129]. In fact, image annotation has been extensively studied by several research communities, including the image retrieval, computer vision community, multimedia and web, and human–computer interface communities. Several benchmark evaluation campaigns such as Pattern Analysis, Statistical Modelling and Computational Learning (PASCAL) and TRECVID have been launched to encourage and evaluate algorithms and systems developed along this line [56,117,128]. In [50], Houle et al. proposed a knowledge propagation algorithm as a neighborhood-based approach to effectively propagate knowledge associated to a few objects of interest to an entire image database. In [114,121], an ontology is utilized for image annotation. The keywords are organized in a hierarchical structure to help remove semantic ambiguities. Photo sharing services such as Facebook and Flickr contain community-contributed media collections, and applying social knowledge can help rerank, filter, or expand image tags. Both [43] and [115] were Flickr image tag recommendation systems developed based on tag concurrence and intertag aggregation. Liu et al. further incorporated image similarity into tag co-occurrence and evaluated the system with subjective labeling of tag usefulness [79]. In [135], an end-to-end image tagging system was presented along with a detailed user evaluation of the accuracy and value of the machine-generated image tags. Several important issues in building an image tagging system were addressed, including the design of tagging vocabulary, taxonomy-based tag refinement, classifier score calibration for effective tag ranking, and selection of valuable tags as opposed to accurate ones.

Finally, there is a recent trend to adopt a new computing paradigm, known as data-intensive scalable computing (DISC), to keep up with the increasingly high-volume multimedia modeling and analysis. DISC proposes the use of a data parallel approach to processing Big Data, which are large volumes of data in the order of terabytes or petabytes [19]. DISC has defined a new research direction, and with the emergence of Big Data, research effort has been devoted to large-scale data processing. For example, in [140], Yan et al. proposed an algorithm called robust subspace bagging (RB-SBag) for large-scale semantic concept modeling, along with its MapReduce implementation (MapReduce [30–32] was proposed by Google as a leading example of DISC). RB-SBag combines both random subspace bagging and forward model selection into a unified approach. Evaluations conducted on more than 250K images and several standard TRECVID benchmark data sets have showed that the RB-SBag approach can achieve a more than ten-fold speedup over baseline SVM-based approaches.

14.3 Video Content Analysis

Video content analysis, which consists of visual content analysis, audio content analysis, as well as multimodality analysis, has attracted an enormous interest in both academic and corporate research communities. This research appeal, in turn, encourages areas that are primarily built upon content analysis modules such as video abstraction, video browsing, and video retrieval to be actively developed. In this section, a comprehensive survey of these research topics will be presented.

14.3.1 Visual Content Analysis

The first step in video content analysis is to extract its content structure, which could be represented by a hierarchical tree exemplified in Figure 14.2 [72]. As shown, given a continuous video bitstream, we first segment it into a series of cascaded video shots, where a shot contains a set of contiguously recorded image frames. Because the content within a shot is always continuous, in most cases, one or more frames (known as key frames) can be extracted to represent its underlying content. However, while the shot forms the building block of a video sequence, this low-level structure does not directly correspond to the video semantics. Moreover, this processing often leads to a far too fine segmentation of the video data. That risks the obscuration of its semantic meaning.

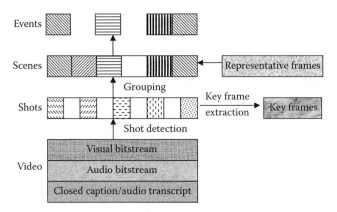

FIGURE 14.2 A hierarchical representation of video content.

Most recent work attempts to understand the video semantics through the extraction of the underlying video scenes, where a scene is defined as a collection of semantically related and temporally adjacent shots that depicts and conveys a high-level concept or story. A common approach to video scene extraction is to group semantically related shots into a scene.

Nevertheless, not every scene can be associated with a meaningful theme. For example, in feature films, there are certain scenes that are used only to establish story environment without involving any thematic topics. Therefore, it is necessary to find important scenes that contain specific thematic topics, such as dialogue or sports highlight. In the area of surveillance video analysis, users are likely to be interested only in video segments that contain specific or abnormal patterns or behaviors. In this chapter, such video units or segments are referred to as events.

Finally, it is also important to automatically annotate videos or detect specific video concepts, so as to better understand video content and facilitate video browsing, search, and navigation at a semantic level. This section reviews previous work on the detection of video shots, scenes, and events, as well as video annotation.

14.3.1.1 Video Shot Detection

A shot can be detected by identifying the camera transitions, which may be either abrupt or gradual. Abrupt transition, also called a camera break or cut, occurs when the content changes significantly between two consecutive frames. In contrast, a gradual transition is usually associated with special effects such as a dissolve, wipe, fade-in, and fade-out, where a smooth change in content is observed over a consecutive sequence of frames.

Existing work in shot detection can be generally categorized into the following five classes: pixel based, histogram based, feature based, statistics based, and transform based. The pixel-based approach detects the shot change by counting the number of pixels, which have changed from one frame to the next. However, while this approach may be the simplest way to detect the content change between two frames, a pixel-based method may be too sensitive to object and camera motion. Consequently, the histogram-based approach, which detects content change by comparing the histograms of neighboring frames, has gained more popularity as histograms are invariant to image rotation, scaling, and transition. In fact, it has been reported that histogram methods can achieve good trade-offs between the accuracy and speed [78].

The feature-based approach applies features such as intensity edge [146] and visual rhythm [26] to identify the shot boundary. Other technologies such as image segmentation and object tracking have also been employed. On the other hand, information such as mean and standard deviation of pixel intensities are exploited by statistics-based approach. Finally, transform-based methods use hidden Markov model (HMM) or Markov random field (MRF) to model shot transitions.

The topic of shot detection has matured in recent years, as the current research literature contains approaches that can perform reasonably well on real-world applications [145].

14.3.1.2 Video Scene and Event Detection

Existing scene detection approaches can be classified into two: model based and model-free. With the former category, specific structure models are constructed for up specific video applications by exploiting scene characteristics, discernible logos, or marks. For example, in [148], temporal and spatial structures were defined for the parsing of TV news, where the temporal structure was modeled by a series of shots, including anchorperson shots, news shots, commercial break shots, and weather forecast shots. Meanwhile, the spatial structure was modeled by four frame templates, with each containing either two anchorpersons, one anchorperson, one anchorperson with an upper-right news icon, or one anchorperson with an upper-left news icon. Some other work in this direction has tried to integrate multiple media cues such as visual, audio, and text (closed captions or audio transcripts) to extract scenes from real TV programs.

Compared to the model-based approach, which has very limited applications, the model-free approach can be applied to very generic situations. Work in this area can be categorized into three classes according to the use of visual, audio, or audiovisual cues. In visual-based approaches, color or motion information is utilized to locate the scene boundary. For example, Yeung et al. proposed to detect scenes by grouping visually similar and temporally close shots [142]. Moreover, they also constructed a transition graph to represent the detected scene structure. Compressed video sequences were used in their experiments. Other work in this area has involved the cophenetic dissimilarity criterion or a set of heuristic rules to determine the scene boundary.

Pure audio-based work was reported in [150], where the original video was segmented into a sequence of audio scenes such as speech, silence, music, speech with music, song, and environmental sound based on low-level audio features. In [89], sound tracks in films and their indexical semiotic usage were studied based on an audio classification system that could detect complex sound scenes as well as the constituent sound events in cinema. Specifically, it has studied the car chase and the violence scenes for action movies based on the detection of their characteristic sound events such as horns, sirens, car crashes, tires skidding, glass breaking, explosions, and gunshots.

However, due to the difficulty of precisely locating the scene boundaries based on audio cues alone, more recent research has concentrated on the integration of multiple media modalities for more robust results [37,53,120]. For example, three types of media cues including audio, visual, and motion were employed by [53] to extract semantic video scenes from broadcast news. They considered two types of scenes, N-type (normal scenes) characterized by a long-term consistency of chromatic composition, lighting conditions, and sound and M-type (montage scenes) characterized by visual features for such attributes as lighting conditions, location, and time of creation. N-type scenes were further classified into pure dialogue scenes (with a simple repetitive visual structure among the shots), progressive scenes (linear progression of visuals without repetitive structure), and hybrid scenes (dialogue embedded in a progressive scene) [124]. An integration of audio and visual cues was reported in [73], where audio cues including ambient noise, background music, and speech were cooperatively evaluated with visual feature extraction in order to precisely locate the scene boundary. Special movie editing patterns were also considered in this work.

Compared to that of a scene, the concept of an event is more subjectively defined as it can take on different meanings for different applications many assign. For example, an event could be the highlight of a sports video or an interesting topic in a video document. In [84], a query-driven approach was presented for the detection of topics of discussion by using image and text contents of query foil presentation. While multiple media sources were integrated in their framework, identification results were mainly evaluated in the domain of classroom presentations due to the special features adopted. In contrast, work on sports highlight extraction mainly focuses on detecting the announcer's speech, the audience ambient speech noise, the game-specific sounds (e.g., the sounds of batting in baseball), and various other background noises (e.g., the cheering of audiences). Targeting the analysis of movie content, Li et al. proposed the detection of three types of events, namely, 2-speaker dialogues, multispeaker dialogues, and hybrid events, by exploiting multiple media cues and special movie production rules [75,95].

In the computer vision community, event detection usually refers to abnormal event detection or abnormal object activity detection/recognition in surveillance videos. For example, in [8], Basharat et al. first built the trajectories of objects by tracking them using an appearance model and then, based on the measurement of the abnormality of each trajectory, detecting anomalous behaviors such as unusual path taken by pedestrians or moving object with abnormal speed and size. Similar ideas were also proposed in [104], where a nonparametric object trajectory representation was employed to detect abnormal events. In [63], a space–time MRF model was proposed to detect abnormal activities in videos. The nodes in the MRF graph corresponded to a grid of local regions in the video frames, and neighboring nodes in both space and time were associated with links. Experimental results on several surveillance videos showed that the MRF model could not only localize atomic abnormal activities in a busy video scene but also capture global-level abnormalities caused by irregular interactions between local activities. Also aiming at modeling both spatial and temporal activities, Xiang and Gong developed an HMM to incrementally and adaptively detect abnormal object behaviors [134].

In contrast with the detection of moving objects, Tian et al. proposed a framework for the robust efficient detection of abandoned and removed objects in complex surveillance videos based on background subtraction and foreground analysis complemented by tracking [127]. Three Gaussian mixture models (GMMs) were applied to model the background, as well as to detect static foreground regions. Context information of the foreground masks was also exploited to help determine the type of static regions. Finally, a person-detection process was integrated to distinguish static objects from stationary people. The proposed method has been tested on a large set of real-world surveillance videos and has demonstrated its robustness to quick changes in lighting and occlusions in complex environments.

A very comprehensive and scalable object and event detection, indexing, and retrieval system, the IBM Smart Surveillance Solution, was presented in [41]. The system allows user to define criteria for alerts with reference to a specific camera view. The criteria can be parked car detection, trip wire, etc. Moreover, it automatically generates descriptions for the events that occur in the scene and stores them in an indexed database to allow users to perform rapid search. Analytics supported by this system include moving object (such as people and vehicle) detection and tracking, object classification with calibration, object behavior analysis, color classification, and license plate recognition. The system has been successfully deployed in several large cities such as Chicago and New York City for public safety monitoring in urban environments.

In recent years, with the proliferation of videos on social media sites such as YouTube, detecting real-world events from user-uploaded videos and monitoring and tracking their usage and life cycle have attracted increasing interests from researchers in both multimedia analysis and social media mining communities. Cha et al. conducted a large-scale YouTube measurement study to characterize content category distributions, as well as to track exact duplicates of popular videos [22]. In [136], Xie et al. developed a large-scale event monitoring system for YouTube videos based on proposed visual memes. A meme is a short video segment frequently remixed and reposted by different authors. Once memes were extracted from tens of thousands of YouTube videos, the authors built a graph model to model the interactions among them, with people and content as nodes and meme postings as links. Then using this model they were able to derive graph metrics that capture content influence and user roles. The authors observed that over half of the event videos on YouTube contained remixed content, that about 70% of video creators participated in video remixing, and that over 50% of memes were discovered and reposted within 3 h after their first appearance.

14.3.1.3 Video Annotation

The automatic annotation of videos at the semantic concept level has recently emerged as a very active topic within the multimedia research community. Video annotation entails a multilabeling process whereby a video clip is associated with one or more labels. For example, a video clip can be tagged as "garden," "plant," "flower," and "lake" simultaneously. The concepts of interest in this case usually include a wide range of categories describing scenes, objects, events, and certain named entities.

Current research on video annotation follows three general paradigms: individual-concept detection, context-based concept fusion (CBCF), and integrated multilabeling [105]. Research involving the first paradigm generally uses binary classification to detect each individual concept, thus treating a collection of concepts as isolated points. A typical approach in this case is to build a binary detector for each concept and create either a presence or absence label based on video content modeling and analysis. An alternative is to stack a set of binary detectors into a single discriminative classifier [113]. The left branch in Figure 14.3 illustrates a typical process flow.

Solutions proposed according to the individual-concept detection paradigm have only achieved limited success, as they do not model the inherent correlations between real-world concepts. For example, the presence of "sky" often occurs together with the presence of "cloud," whereas "ocean" normally does not co-occur with "truck." Furthermore, while simple concepts can be directly modeled from low-level features, it could be rather difficult to learn individual models for such complex concepts as "political protest." Instead, complex concepts could be inferred from other more primitive concepts based on correlation. For example, if the concepts of "people," "walking," and "banners" are detected, then they likely indicate the presence of the concept "political protest."

In contrast, studies of context-based concept fusion add a second step after individual-concept detection by fusing multiple concepts together. A CBCF-based active learning approach was proposed in [123] where user annotations were solicited for a small number of concepts from which the remainder of the concept could be inferred. In [132], Wu et al. proposed an ontology-based multiclassification learning method or video concept detection. Each concept was first independently modeled by a classifier, and then a predefined ontology hierarchy was applied to improve the detection accuracy of individual classifiers. An alternative fusion strategy was proposed in [49], where logistic regression (LR) was applied to fuse individual detections. In [118], Smith et al. described a two-step discriminative model fusion (DMF) approach to derive unknown or indirect relationship among concepts by constructing model vectors based on detection scores of individual classifiers. An SVM was then trained to refine the detection results of these classifiers. The typical flow of this process is illustrated by the center branch in Figure 14.3.

While it is intuitive to exploit the contextual relationship among concepts to improve their detection accuracy, CBCF approaches do not always provide stable performance improvement. The reasons are twofold. First, detection errors in individual classifiers can be propagated to the fusion step, thereby degrading the overall performance. Second, the training data for the conceptual fusion may be insufficient, leading to overfitting. Models learned in this way usually lack generalization capability.

Recently, Qi et al. proposed a third paradigm that simultaneously classifies concepts and models their correlations using a so-called correlative multilabel (CML) framework [105]. This approach, illustrated by the right-most branch (Figure 14.3), follows the principle of least commitment [86] in the sense that both learning and optimization are performed in a single step, thus avoiding the error propagation problem of CBCF. Moreover, since the entire training data are simultaneously used for modeling both individual concepts and their correlations, the risk of overfitting is significantly reduced. Experiments conducted on the benchmark TRECVID 2005 data sets have demonstrated its superior performance over other state-of-the-art approaches.

Xie et al. proposed probabilistic visual concept trees for modeling large visual semantic taxonomies and demonstrated their effective use in visual concept detection in a limited setup [137]. They considered three salient attributes for a real-world visual taxonomy: concept semantics, image appearance, and data statistics. They proposed a multifaceted concept forest structure within which these attributes are expressed in the form of parent–child relationships, mutually exclusive relationships, and multiple aspect labelings. They then used the proposed visual concept tree model to encode the multifaceted concept forest under uncertainty. The probabilistic visual concept trees were designed to extend the functionalities of several predecessors, such as the Bayes Net and tree-structured taxonomy [138].

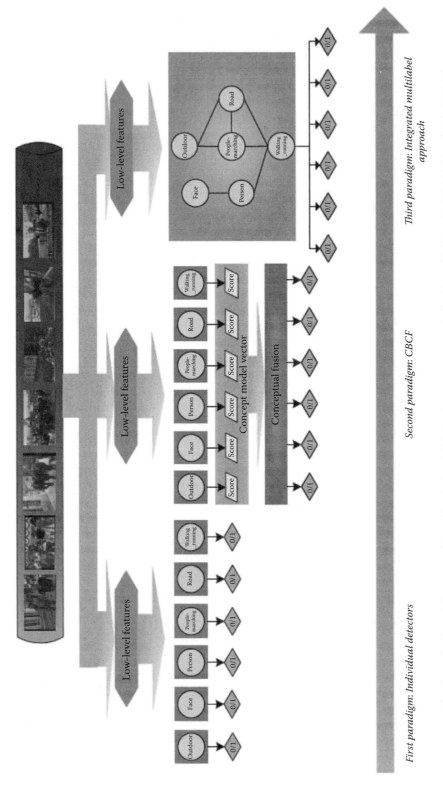

FIGURE 14.3   Flowcharts for three different paradigms of typical multilabel video annotation methods. (From Qi, G. et al., *ACM Multimedia*, 1, 17, 2007.)

They are also capable of supporting inference with the junction tree algorithm. Evaluations conducted over 60K web images with a taxonomy of 222 concepts have shown that probabilistic visual concept trees are capable of effective representation of visual semantic taxonomies, including the encoding of hierarchy, mutually exclusion, and multifaceted concept relationships under uncertainty.

14.3.2 Audio Content Analysis

Existing research on content-based audio data analysis is still quite limited and can be categorized into the following two classes.

14.3.2.1 Audio Segmentation and Classification

One basic problem in audio segmentation is the discrimination between speech and music, which are the two most important audio types. In general, a relatively high accuracy can be achieved when distinguishing speech from music, since their respective signals are significantly different in their spectral distributions as well as their temporal change patterns. A general solution is to first extract various audio features such as the average zero-crossing rate (ZCR) and the short-time energy from the signals, then distinguish the two sound types based on the feature values. For example, 13 audio features calculated in time, frequency, and cepstrum domains were employed in [112] for the purpose of classification. The authors also examined and compared several popular classification schemes, including the multidimensional Gaussian maximum a posteriori estimator, the GMM, a spatial partitioning scheme based on k-d-trees, and a nearest neighbor classifier.

State-of-the-art classification algorithms are usually based more on sound. For example, Wyse and Smoliar classified audio signals into three types, including "music," "speech," and "others" [133]. Music was first detected based on the average length of the interval during which peaks were within a narrow frequency band; speech was then separated out by tracking the pitches. Research in [103] was devoted to analyze the signal's amplitude, frequency, and pitch. Simulations were also conducted on human audio perception, the results of which were utilized to segment the audio data and to recognize the music component. More recently, Zhang and Kuo presented an extensive feature extraction and classification system for audio content segmentation and classification purposes [149]. Five audio features including energy, average ZCR, fundamental frequency, and spectral peak tracks were extracted to fulfill the task. A two-step audio classification scheme was proposed in [83], where in the first step, speech and nonspeech were discriminated using k-nearest neighbor (K-NN) and line spectral pair vector quantization (LSPVQ) classification schemes. In the second step, the nonspeech signals were further classified into music, environment sounds, and silence using a feature thresholding scheme. In [70], an SVM-based audio classification mechanism was proposed to assist the content analysis of instructional videos such as presentations and lectures.

In [143], Yu et al. proposed the combination of local summarization and multilevel locality-sensitive hashing (LSH) for multivariant audio retrieval. Based on spectral similarity, they divide each audio track into multiple continuously correlated periods of variable length. This removes a significant part of the redundant information but provides support for more compact but yet accurate descriptions. They then compute weighted mean chroma features for each of these periods. Then, by exploiting the characteristics of the content description, they adapted a two-level LSH scheme for efficiently delineating a narrow relevant search region. In [144], Yu et al. proposed MultiProbe histograms as global summaries of audio feature sequences that retain local temporal acoustic correlations by concatenating major bins of adjacent chroma features. They then exploited the order statistics of the MultiProbe histograms to more efficiently organize and probe LSH tables. The resulting approximate retrieval method is faster than exact K-NN retrieval while still achieving useful accurate rates.

14.3.2.2 Audio Analysis for Video Indexing

Five different video classes, including news report, weather report, basketball, football, and advertisement, were distinguished in [80] using both multilayer neural networks (MNNs) and HMMs.

Features such as the silence ratio, the speech ratio, and the subband energy ratio were extracted to fulfill this task. It was shown that while MNN worked well in distinguishing among reports, games, and advertisements, it had difficulty in classifying different types of reports or games. On the contrary, the use of HMM increased the overall accuracy, but it could not achieve good classification of the five video types. In [101], features such as the pitch, the short-time average energy, the band energy ratio, and the pause rate were first extracted from the coded subband of an MPEG audio clip; these features were then integrated to characterize the clip as either silence, music, or dialogue. Another approach to video indexing based on music and speech detection was proposed in [87], where image processing techniques were applied to the spectrogram of the audio signals. The spectral peaks of music were recognized by applying an edge-detection operator and the speech harmonics were detected with a comb filter.

Li et al. [69] proposed to the detection of discussion scenes in instructional videos by identifying scenes that contain narrations and dialogues using on an audio classification scheme.

14.3.3 Fusion of Multiple Modalities

Recently, researchers have begun to focus their attention on the analysis of video content by integrating multiple modalities including audio, visual, and textual cues. Multiple media sources usually complement each other to convey the complete semantics of the content. Consequently, by taking all sources into account, we can achieve better understanding of the content.

A general solution for the fusion of multiple modalities is to first perform individual visual, audio, or textual content analysis to obtain separate sets of analysis results and then to combine them based on fusion rules, perhaps under a probabilistic framework. Another popular way of media integration is to employ different information sources at different processing stages. For example, we can first employ visual cues to generate coarse-level results, then introduce audio cues to refine the results. However, when and how to efficiently and effectively integrate multiple media sources still remain an open issue in need of further study.

In [36,71], a system called metadata automated generation for instructional content (MAGIC) was presented, which aimed to assist content authors and course developers in generating metadata for learning materials and information assets, so as to enable wider reuse of these objects across departments and organizations. The MAGIC metadata generation environment consisted of a set of text and video processing tools to recover the underlying narrative structures of learning materials, including text analysis components for extracting titles and keywords, a taxonomy and classification system for automatically assigning documents to specific taxonomy, and audio and video segmentation and analysis components for segmenting learning videos. Various video analysis algorithms and machine-learning techniques have been applied to both visual features extracted from the video track and aural features computed from the audio track. The segments were further annotated with text features extracted from time-stamped closed caption text based on advanced text analysis. Experiments conducted on a midsized set of videos from various Department of Homeland Security agencies have demonstrated the advantage of integrating multiple modalities in video content analysis to assist more intuitive and effective video browsing, search, and authoring.

An atomic topical segment detection framework was proposed in [76] for instructional videos to facilitate topic-based video navigation and offer efficient video authoring. A comprehensive text analysis component was first developed to extract informative text cues such as keyword synonym sets and sentence boundary information from video transcripts. These text cues were then integrated with various audiovisual cues, such as silence/music breaks and speech similarity, to identify topical segments. Encouraging user feedback has been gathered regarding the usefulness of such system.

Aiming at the detection of affective content in movies, Xu et al. exploited cues from multiple modalities to represent emotions and emotional atmosphere [139]. They first generated midlevel representations from low-level features by integrating audio sounds, dialogue, and subtitles with the visual analysis and then applied them to infer the affective content. Experiments carried out on different movie genres have validated the effectiveness of their approach.

A comprehensive study and large-scale tests of web video categorization methods were reported in [141]. Considering that web videos are characterized by a much higher diversity of quality, subject, style, and genres compared with traditional video programs, the authors concluded that multiple modalities should be employed. In addition to the application of low-level visual and audio features, they proposed two new modalities: a semantic modality and a surrounding text modality. The semantic modality was represented by three types of features: concept histogram, visual word vector model, and visual word latent semantic analysis (LSA). Meanwhile, the text modality was represented by titles, descriptions, and tags of web videos. A comprehensive experiment on evaluating the effectiveness of proposed feature representations with three different classifiers, including SVM, GMM, and manifold ranking, was conducted over 11K web videos, and a mean average precision of 0.56 has been achieved.

An earlier work on classifying news videos using multimodal information was reported in [24]. The authors first applied Fisher's linear discriminant (FLD) to select a smaller number of discriminative features from a large set of features extracted from image, audio, transcript, and speech; they then concatenated the projections into a single synthesized feature vector to represent the scene content. Using SVM as the classifier, they have achieved good performance on anchor and commercial detection in the TREC 2003 video track benchmark. A later development of this work was described in [48], wherein a hybrid approach to improving semantics extraction from news video was proposed. Through extensive experiments, the authors demonstrated the value of careful parameter tuning, exploitation of multiple feature sets and multilingual linguistic resources, applying text retrieval approaches for image features, as well as establishing synergies between multiple concepts through undirected graphical models.

A novel visual feature-based machine-learning framework called IBM multimedia analysis and retrieval system (IMARS) was presented in [6,93], which aimed to make digital photos and videos searchable through large-scale semantic modeling and classification. IMARS consists of two important components. First, the multimedia analysis engine applies machine-learning techniques to model semantic concepts in images and videos from automatically extracted visual descriptors. It then automatically assigns labels to unseen content to improve its searching, filtering, and categorization capabilities. Second, the multimedia search engine combines content-based, model-based, semantic concept-based, and speech-based text retrieval for more effective image and video searching.

14.3.4 Video Abstraction

Video abstraction, as the name implies, generates a short summary of a long video. A video abstract is a sequence of still or moving images, which represent the video essence in a very concise way. Video abstraction is primarily used for video browsing and is an indispensable part of any video indexing and retrieval system.

There are two fundamentally different kinds of video abstracts: still- and moving-image abstracts. The still-image abstract, also known as a static storyboard or video summary, is a small collection of key frames extracted or generated from the underlying video source. The moving-image abstract, also known as a moving storyboard or video skim, consists of a collection of image sequences, as well as the corresponding audio abstract extracted from the original sequence. It is itself a video clip but of considerably shorter length than the original.

There are basically two types of video skim: the summary sequence and the highlight. A summary sequence is used to provide users an impression about the entire video content, while a highlight generally only contains a relatively small number of interesting video segments. A good example of a video highlight is the movie trailer, which only shows a few very attractive scenes without revealing the story's ending.

Compared to video skimming, video summarization has attracted much more research interest. Based on the way the key frames are extracted, existing work in this area can be categorized into the following three classes: sampling based, shot based, and segment based.

FIGURE 14.4   A mosaic image generated from a panning sequence.

Most of the earlier work on summarization work was sampling based, where key frames were either randomly chosen or uniformly sampled from the original video. Sampling is the simplest way to extract key frames, yet it often fails to adequately represent the video content. More sophisticated methods thus tend to extract key frames by adapting to the dynamic video content using shots. Because a shot is taken within a continuous capture period, a natural and straightforward approach is to extract one or more key frames from each shot. Features such as color and motion have been applied to find the optimal key frames in this case. Mosaics have also been proposed for the representation of shots. One such example is shown in Figure 14.4 [85].

A major drawback of using one or more key frames for each shot is that it does not scale well for long video sequences. Therefore, researchers have proposed to work with an even higher-level video units or segments. A video segment could be a scene, an event, or even the entire sequence.

14.4 Modeling and Querying Multimedia Data

Most research prototypes and applications in multimedia directly organize and search for the vectors obtained from content analysis (see Section 14.5). However, a few database models for multimedia have been proposed in the 1990s and early 2000s. A database model defines the theoretical foundation by which the data are stored and retrieved. The models were proposed as extensions of existing database models (relational, object-oriented, and semistructured models).

14.4.1 Multimedia Database Model as an Extension of Object-Relational Models

Defining a multimedia data model as an extension of the relational or object-relational model is appealing, as the database world is dominated by the relational database model. In [110], an image is stored in a table T (h: *Integer*, x_1: X_1, ..., x_n: X_n), where h is the image identifier and x_i is an image feature attribute of domain (or type) X_i (note that classical attributes can be added to this minimal schema). The tuple corresponding the image k is indicated by $T[k]$. Each tuple is assigned a real-valued score (ζ) such that $T[k]$. ζ is a distance between the image k and the current query image. The value of ζ is assigned by a scoring operator $\Sigma_T(s)$ given a scoring function s: $\Sigma_T(s)[k] \cdot \zeta = s(T[k] \cdot x_1, ..., T[k] \cdot x_n)$.

Since many image queries are based on distance measures, a set of distance functions (d: $X \times X \rightarrow [0,1]$) is defined for each feature type X. Given an element x: X and a distance function d defined on X, the scoring function s assigns $d(x)$, a distance from x to every element of X. In addition, a set of score combination operators $\Diamond$: $[0,1] \times [0,1] \rightarrow [0,1]$ is defined. New selection and join operators defined on the image table augmented with the scores allow the selection of the n images with lowest scores, the images whose scores are less than a given score value ρ and the images from a table $T = T(h$: *Integer*, x_1:X_1, ..., x_n: X_n) that match images from a table $Q = Q(h$: *Integer*, y_1: Y_1, ..., y_n: Y_n), based on score combination functions as follows:

- K-NNs: $\sigma_k^\#(\Sigma_T(s))$ returns the k rows of the table T with the lowest distance from the query.
- Range query operator: $\sigma_\rho < (\Sigma_T(s))$ returns all the rows of the table T with a distance less than ρ.
- $\Diamond$ Join: $T \bowtie Q$ joins the tables T and Q on their identifiers h and returns the table $W = W(h$: *Integer*, x_1: X_1,...,x_n: X_n,y_1: Y_1,...,y_n: Y_n). The distance in the table W is defined as $W \cdot d = T \cdot d \Diamond Q \cdot d$.

In [111], the same authors proposed a design model with four kinds of feature dependencies that can be exploited for the design of efficient search algorithms. For multimedia data models defined as the object-relational model, an extension of SQL called SQL/MM was proposed as a standard query language [122].

14.4.2 Multimedia Database Model as Extension of Object-Oriented Models

Multimedia content description is application dependent. The vectors and related similarity functions are defined for specific applications. Because they allow the user to provide methods for specific classes, object-oriented models have been extended to support multimedia data [68,99]. The distributed image database management system (DISIMA) model [99] is an example of a multimedia data model defined as an extension of an object-oriented data model. In the DISIMA model, an image is composed of physical salient objects (regions of the image) whose semantics are given by logical salient objects representing real-world objects. Both images and physical salient objects can have visual properties. The DISIMA model uses an object-oriented concept and introduces three new types: image, physical salient objects, and logical salient objects as well as operators to manipulate them.

Images and related data are manipulated through predicates and operators defined on images; physical and logical salient objects are used to query the images. They can be directly used in calculus-based queries to define formulas or in the definition of algebraic operators. Since the classical predicates $\{=, <, \leq, >, \geq\}$ are not sufficient for images, a new set of predicates were defined for use on images and salient objects:

- Contain predicate: Let i be an image and o be an object with a behavior *pso* that returns the set of physical salient objects associated with o: $contains(i,o) \Leftrightarrow \exists p \in o \cdot pso \wedge p \in i \cdot pso$.
- Shape similarity predicates: Given a shape similarity metric d_{shape} and a similarity threshold ε_{shape}, two shapes s and t are similar with respect to d_{shape} if $d_{shape}(s,t) \leq \varepsilon_{shape}$. In other words, *shape_similar* $(s,t,\varepsilon_{shape}) \Leftrightarrow d_{shape}(s,t) \leq \varepsilon_{shape}$.
- Color similarity predicates: Given two color representations (c_1,c_2) and a color distance metric d_{color}, the color representations c_1 and c_2 are similar with respect to d_{color} if $d_{color}(c_1,c_2) \leq \varepsilon_{color}$.

Based on the previously defined predicates, several operators are defined: *contains* or *semantic join* (to check whether a salient object is found in an image), the *similarity join* that is used to match two images or two salient objects with respect to a predefined similarity metric on some low-level features (color, texture, shape, etc.), and *spatial join* on physical salient objects:

- Semantic join: Let S be a set of semantic objects of the same type with a behavior *pso* that returns, for a semantic object, the physical salient objects it describes. The *semantic join* between an image class extent I and the semantic object class extent S, denoted by $I \bowtie_{contains} S$, defines the elements of $I \times S$ where for $i \in I$ and $s \in S$, $contains(i,s)$ is true.
- Similarity join: Given a similarity predicate *similar* and a threshold $\in$, the *similarity join* between two sets R and S of images or physical salient objects, denoted by $R \bowtie_{similar(r.i,s.j,\in)} S$ for $r \in R$ and $s \in S$, is the set of elements from $R \times S$ where the behaviors i defined on the elements of R and j on the elements of S return some compatible metric data type T and $similar(r \cdot i, s \cdot j)$ (holds true the behaviors i and j) can be the behaviors that return color, texture, or shape.
- Spatial join: The spatial join of the extent of two sets R and S, denoted by $R \bowtie_{r.i\theta s.j} S$, is the set of elements from $R \times S$ where the behaviors i defined on the elements of R, and j on the elements of S, return some spatial data type, θ is a binary spatial predicate, and $R \cdot i$ stands in relation θ to $S \cdot j$ (θ is a spatial operator like north, west northeast, intersect).

The previously defined predicates and the operators are the basis of the declarative query languages MOQL [67] and Visual MOQL [100]. The DISIMA model was later extended to support video data [25]. MOQL is an extension of the Object Query Language (OQL) [21]. Most extensions introduced to OQL by MOQL involve the where clause, in the form of four new predicate expressions:

spatial expression, temporal expression, contains predicate, and similarity expression. The spatial expression is a spatial extension that includes spatial objects, spatial functions, and spatial predicates. The temporal expression deals with temporal objects, functions, and predicates for videos. The *"contains"* predicate is used to specify the salient objects contained in a given media object. The similarity predicate checks if two media objects are similar with respect to a given metric. Visual MOQL uses the DISIMA model to implement the image facilities of MOQL.

14.4.3 Example of Semistructured Model

MPEG-7, the multimedia content description interface, is an ISO metadata standard defined for the description of multimedia data [44]. The MPEG-7 standard aims at helping with searching, filtering, processing, and customizing multimedia data through specifying its features in a "universal" format. MPEG-7 does not specify any applications, but rather the representation format for the information contained within the multimedia data is represented, thereby supporting descriptions of multimedia made using many different formats. The objectives of MPEG-7 include creation of methods to describe multimedia content, manage data flexibly, and globalize data resources.

MPEG-7 essentially provides two tools: the description definition language (MPEG-7 DDL) [1] for the definition of media schemes and an exhaustive set of media description schemes (DSs) mainly for media low-level features. The predefined media DSs are composed of visual feature descriptor schemes [2], audio feature descriptor schemes [3], and general multimedia DSs [4]. Media description through MPEG-7 is achieved through three main elements: descriptors (Ds), DSs, and a DDL. The descriptors essentially describe a feature or distinctive aspect of the multimedia data. An example of a descriptor would be the camera angle used in a video document. The DS organizes the descriptions, specifying the relationship between descriptors. DSs, for example, would represent how a picture or a movie would be logically ordered. The DDL is used to specify the schemes, as well as to allow modifications and extension to the schemes.

The MPEG-7 DDL is a superset of XML Schema [126], the W3C schema definition language for XML documents. The extensions to XML Schema comprise support for array and matrix data types as well as additional temporal data types. Because MPEG-7 media descriptions are XML documents that conform to the XML Schema definition, it is natural to suggest XML database solutions for the management of MPEG-7 documents, as proposed in [64]. Current XML database solutions are text oriented, and MPEG-7 encodes nontextual data. Directly applying current XML database solutions to MPEG-7 would lower its expressive power, as only textual queries will be allowed.

The models presented in this section are representative of the solutions proposed for multimedia data; however, all have significant limitations. Object or object-relational models assume that multimedia data are well structured, with their content clearly defined. This is not always the case. For example, the content description of an image as it relates to salient objects relies on segmentation and object recognition, both of which are still active research topics. Multimedia content is often represented as a collection of vectors with specific similarity functions, which cannot be represented using the semistructured data models that are commonly applied to text data. Multimedia data are complex data whose representations can involve structured data (multimedia metadata), text (multimedia semantics), as well as vectors (multimedia content). A simple extension of existing data models is not sufficient, especially as the retrieval of multimedia data involves browsing and search as well as classical text-based queries.

14.5 Similarity Search in Multimedia Databases

The aim of a similarity query is to select and return a subset of the database with the greatest resemblance to the query object, all according to a measure of object-to-object similarity or dissimilarity. Data dissimilarity is very commonly measured in terms of distance metrics such as the Euclidean distance, other Lp metrics, the Jaccard distance, or the vector angle distance (equivalent to cosine similarity);

many other measures exist, both metric and nonmetric. The choice of measure generally depends on the types of descriptors used in modeling the data.

Similarity queries are of two main types: range queries and K-NN queries. Given a query object $q \in \mathcal{U}$, a data set $S \subseteq \mathcal{U}$, and a distance function $d: \mathcal{U} \times \mathcal{U} \rightarrow \mathbb{R}_0^+$, where $\mathcal{U}$ is a data domain, range queries and K-NN queries can be characterized as follows:

- Range queries report the set $\{v \in S \mid d(q,v) \leq \varepsilon\}$ for some real value $\varepsilon \geq 0$.
- K-NN queries report a set of k objects $W \subseteq S$ satisfying $d(q,u) \leq d(q,v)$ for all $u \in W$ and $v \in S \setminus W$.

Note that a K-NN query result is not necessarily unique, whereas a range query result is. Of the two types of similarity queries, K-NN queries are arguably more popular, perhaps due to the difficulty faced by the user in deciding a range limit when the similarity values are unknown or when they lack an interpretation meaningful to the user.

Most similarity search strategies seek to reduce the search space directly by means of an index structure. Spatial indices are among the first such structures; a great many examples exist, including k-d-trees [12], B-trees [11], R-trees [46], X-trees [13], SR-trees [59], and A-trees [108]. Such structures explicitly rely on knowledge of the data representation for their organization, which makes them impractical when the representational dimension is very high. Moreover, they usually can handle only queries with respect to Euclidean or other Lp distance measures. Later, indices were developed for use with general metric data, such as multi-vantage-point trees [17], geometric near-neighbor access trees (GNAT) [18], and M-trees [27]. Unlike spatial indices, metric indices rely solely on pairwise distance values to make their decisions. Many more examples of spatial and metric data structures can be found in [109].

Both spatial and metric search structures do not in general scale well due to an effect known as the curse of dimensionality, so-called due to the tendency of the query performance of indexing structures to degrade as the data dimensionality increases. With respect to a given query item, as the data dimensionality increases, the distribution of distances from a given query object to the data set objects changes: these distances tend to concentrate more tightly around their mean value, leading to a reduction in the ability of the similarity measure to discriminate between relevant and nonrelevant data objects [14]. Evidence shows that as the data dimension approaches 20, the performance of such search structures as the R^*-tree, SS-tree, and SR-tree typically degrades to that of a sequential scan of the entire data set [14]. However, the dimensionality of multimedia feature vectors is usually much higher than that which can ordinarily be dealt with using spatial or metric indices. For example, color histograms typically span at least 64 dimensions and sometimes many more. Applications using bag-of-words modeling use vectors of many thousands of dimensions to represent the data [141].

More recently, many other approaches have been developed for both exact and approximate K-NN search, in an attempt to partially circumvent the curse of dimensionality. For exact search, Koudas et al. [65] proposed a method that used a bitmap data representation to facilitate the search process. Jagadish et al. [57] developed iDistance, in which high-dimensional objects are transformed into 1D values, which are then indexed for K-NN search using a B+-tree. Beygelzimer et al. [15] introduced the cover tree index, in which the search is performed by identifying a set of nodes whose descendants are guaranteed to include all nearest neighbors.

For approximate search, Andoni et al. [7] developed the MedScore technique, in which the original high-dimensional feature space is transformed into a low-dimensional feature space by means of random projections. For each dimension in the new feature space, a sorted list is created. The search is done by aggregating these sorted lists to find the objects with the minimum median scores. The well-known threshold algorithm [40] is used here for the aggregation. Indyk and Motwani [55] proposed a popular hashing-based search framework called LSH, in which pairs of similar objects have higher probabilities of mapping to the same hash location than pairs of dissimilar objects. Many variants and improvements upon LSH have subsequently been proposed, such as spectral hashing [131], where the hash bits are determined by thresholding a subset of eigenvectors of the Laplacian of the similarity graph on the data objects.

All of these recently proposed methods attempt to improve upon the performance of their predecessors on high-dimensional data, through the use of domain reduction techniques: via the direct reduction of either the data set (sampling) or attribute set (projection) or via hashing or other data domain transformations. Generally speaking, these methods can cope with data of up to several hundred dimensions before succumbing to the effects of the curse of dimensionality. Further improvements in performance can then be sought through the use of parallelism, as is possible using MapReduce [30], or other forms of distributed processing [9]. Ensembles of similarity indices have also been considered. The fast library for approximate nearest neighbors (FLANN) uses an ensemble of randomized *k-d*-tree indices [116] and hierarchical *K*-means trees [94], along with automatic parameter tuning of these indices for optimized performance [90]. It has been shown to achieve very good practical performance for many data sets with up to approximately 200 dimensions.

Recent advances have been made as to the indexability of multimedia data sets of very large scale, in both data set size and data set dimensionality. For data sets of extreme size, a product quantization-based approach for approximate nearest neighbor search has recently been developed that has been successfully applied to the indexing of two billion SIFT vectors of dimension 128, taken from a set of one million images [58]. After decomposing the space into a Cartesian product of low-dimensional subspaces, each subspace is quantized separately. The Euclidean distance between two vectors can be efficiently estimated from short codes composed of the vectors' quantization indices. Inverted list techniques are then used to assemble a final query result.

For data sets of extreme dimensionality, Houle and Sakuma [51] proposed an approximate search index called a spatial approximation sample hierarchy (SASH), a multilevel structure recursively constructed by building a SASH on a large random data sample and then connecting the remaining objects to several of their approximate nearest neighbors from within the samples. For lower-dimensional data (on the order of 200 dimensions or less), the overheads associated with their sampling strategy generally render the SASH index less competitive than ensemble methods such as FLANN [90] or the heavily optimized product quantization method of [58]. However, the use of near neighbors for distance discrimination can often overcome the effects of the curse of dimensionality for sparse data sets of extremely high dimension (with both the number of vectors and the dimensionality in the millions), as sometimes arise in applications where the data are modeled using the bag-of-words representation.

14.6 Conclusion

In this chapter, we have surveyed multimedia databases in terms of data analysis, data querying, and indexing. In multimedia databases, raw multimedia data are unfortunately of limited use due to its large size and lack of interpretability. Consequently, they are usually coupled with descriptive data (including both low-level features and mid- to high-level semantic information) obtained by analyzing the raw media data. Various types of features and concepts related to image, video, and audio data are thus required along with some popular techniques for extracting them. Some latest work on integrating multiple media modalities to better capture media semantics has also been reviewed.

Retrieving a specific image, video, or song that a user has in mind remains a challenging task. Currently, the search relies more on metadata (structured data or textual description) than on the media content. The general problem with images and videos is that their digital representations do not convey any meaningful information about their content. Often to allow semantic search, an interpretation must be added to the raw data, either manually or automatically. Manual annotation of images and video is tedious, and automatic inference of the semantics of images and videos is not always accurate. Multimedia data annotation is still an active research topic, and existing approaches motivate users either to annotate images by simplifying the procedure [35,107,129] or to annotate images in an automatic or semiautomatic way [50,123,132]. In [115], Flickr generates candidate tag lists derived from user-defined tags of each image by computing tag co-occurrence and then recommending new tags by aggregating and ranking the candidates. The ideal solution in which the user does not intervene or manually annotates objects of interest is yet to be proposed.

A related and pressing issue is the lack of a complete data model for multimedia data. Although the idea of extending existing multimedia data models is an appealing one, models defined for classical data cannot effectively handle multimedia data. Metadata such as names, dates, and locations can indeed be stored in a relational database; however, for other types, a relational database does not suffice. For example, the content of an image can also be described using text, and high-dimensional vectors can be described using SIFT descriptors. Across differing real-life applications, different media types are used to describe the same information. This is the case for news applications, where an individual event can be covered by TV channels using video, by radio stations using audio, and by newspapers using text and images. Any successful multimedia data model must be sufficiently general so as to be able to address each of these issues.

References

1. ISO/IEC JTC 1/SC 29/WG 11. Information technology multimedia content description interface part 2: Description definition language. International Organization for Standardization/International Electrotechnical Commission (ISO/IEC)ISO/IEC Final Draft International Standard 15938-2:2001, International Organization for Standardization/International Electrotechnical Commission, September 2001.
2. ISO/IEC JTC 1/SC 29/WG 11. Information technology multimedia content description interface part 3: Visual. International Organization for Standardization/International Electrotechnical Commission (ISO/IEC)ISO/IEC Final Draft International Standard 15938-2:2001, International Organization for Standardization/International Electrotechnical Commission, September 2001.
3. ISO/IEC JTC 1/SC 29/WG 11. Information technology multimedia content description interface part 4: Audio. International Organization for Standardization/International Electrotechnical Commission (ISO/IEC)ISO/IEC Final Draft International Standard 15938-4:2001, International Organization for Standardization/International Electrotechnical Commission, June 2001.
4. ISO/IEC JTC 1/SC 29/WG 11. Information technology multimedia content description interface part 5: Multimedia description schemes. International Organization for Standardization/International Electrotechnical Commission (ISO/IEC)ISO/IEC Final Draft International Standard 15938-5:2001, International Organization for Standardization/International Electrotechnical Commission, October 2001.
5. O. Alata, C. Cariou, C. Ramannanjarasoa, and M. Najim. Classification of rotated and scaled textures using HMHV spectrum estimation and the Fourier-Mellin Transform. In *ICIP '98*, Vol. 1, Chicago, IL, pp. 53–56, 1998.
6. IBM AlphaWorks. http://www.alphaworks.ibm.com/. IBM multimedia analysis and retrieval system (IMARS), accessed March 27, 2013.
7. A. Andoni, R. Fagin, R. Kumar, M. Patrascu, and D. Sivakumar. Corrigendum to Efficient similarity search and classification via rank aggregation by R. Fagin, R. Kumar, and D. Sivakumar (Proc. SIGMOD '03). In *SIGMOD Conference*, Vancouver, Canada, pp. 1375–1376, 2008.
8. A. Basharat, A. Gritai, and M. Shah. Learning object motion patterns for anomaly detection and improved object detection. In *IEEE Conference on Computer Vision and Pattern Recognition (CVPR)*, Anchorage, AK, Vol. 1, pp. 1–8, 2008.
9. M. Batko, F. Falchi, C. Lucchese, D. Novak, R. Perego, F. Rabitti, J. Sedmidubsky, and P. Zezula. Building a web-scale image similarity search system. *Multimedia Tools and Applications*, 47:599–629, 2010. DOI: 10.1007/s11042-009-0339-z.
10. H. Bay, T. Tuytelaars, and L. Gool. SURF: Speeded up robust features. *European Conference on Computer Vision (ECCV)*, Graz, Austria, May 2006.
11. R. Bayer and E. M. McCreight. Organization and maintenance of large ordered indices. *Acta Informatica*, 1:173–189, 1972.
12. J. L. Bentley. Multidimensional binary search trees used for associative searching. *Communications of the ACM*, 18(9):509–517, September 1975.

13. S. Berchtold, D. A. Keim, and H.-P. Kriegel. The x-tree : An index structure for high-dimensional data. In T. M. Vijayaraman, A. P. Buchmann, C. Mohan, and N. L. Sarda, eds., VLDB '96, *Proceedings of 22nd International Conference on Very Large Data Bases*, Bombay, India, pp. 28–39, September 3–6, 1996, Burlington, MA, 1996, Morgan Kaufmann.

14. K. S. Beyer, J. Goldstein, R. Ramakrishnan, and U. Shaft. When is "nearest neighbor" meaningful? In *ICDT'* 1999, Jerusalem, Israel, pp. 217–235, 1999.

15. A. Beygelzimer, S. Kakade, and J. Langford. Cover trees for nearest neighbor. In *International Conference on Machine Learning (ICML)*, pp. 97–104, 2006.

16. M. Bober. MPEG-7 visual shape descriptors. *IEEE Transactions on Circuits and Systems for Video Technology*, 11(6):716–719, 2001.

17. T. Bozkaya and Z. M. Özsoyoglu. Indexing large metric spaces for similarity search queries. *ACM Transactions on Database System*, 24(3):361–404, 1999.

18. S. Brin. Near neighbor search in large metric spaces. In *Proceedings of the 21st International Conference on Very Large Data Bases*, Zurich, Switzerland, pp. 574–584, September 1995.

19. R. Bryant. Data-intensive supercomputing: The case for DISC. Technical report, School of Computer Science, Carnegie Mellon University, Pittsburgh, PA, 2007.

20. V. Castelli and L. D. Bergman. *Image Databases-Search and Retrieval of Digital Imagery*. John Wiley & Sons, Inc., New York, 2002.

21. R. G. G. Cattell, D. Barry, D. Bartels, M. Berler, J. Eastman, S. Gamerman, D. Jordan, A. Springer, H. Strickland, and D. Wade, eds. *The Object Database Standard: ODMG 2.0*. Morgan Kaufmann, Burlington, MA, 1997.

22. M. Cha, H. Kwak, P. Rodriguez, Y. Ahn, and S. Moon. I tube, you tube, everybody tubes: Analyzing the world's largest user generated content video system. *ACM SIGCOMM Conference on Internet Measurement*, 1:1–14, 2007.

23. T. Chang and C.-C. Kuo. Texture analysis and classification with tree-structured wavelet transform. *IEEE Transactions on Image Processing*, 2(4):429–441, 1993.

24. M. Chen and A. Hauptmann. Multi-modal classification in digital news libraries. In *International Conference on Acoustics, Speech, and Signal Processing (ICASSP)*, Montreal, Quebec, Canada, Vol. 1, pp. 212–213, 2004.

25. S. Chen, M. Shyu, C. Zhang, and R. Kashyap. Video scene change detection method using unsupervised segmentation and object tracking. In *ICME '01*, Tokyo, Japan, Vol. 1, pp. 56–59, 2001.

26. M. Chung, H. Kim, and S. Song. A scene boundary detection method. In *ICIP '00*, Vol. 1, Vancouver, British Columbia, Canada, pp. 933–936, 2000.

27. P. Ciaccia, M. Patella, and P. Zezula. M-tree: An efficient access method for similarity search in metric spaces. In *Proceedings of the 23rd International Conference on Very Large Data Bases*, Athens, Greece, pp. 426–435, 1997.

28. S. D. Cohen and L. J. Guibas. Partial matching of planar polylines under similarity transformation. In *Proceedings of the 8th Annual ACM-SIAM Symposium on Discrete Algorithms*, New Orleans, LA, pp. 777–786, 1997.

29. N. Dalal and B. Triggs. Histograms of oriented gradients for human detection. In *IEEE Conference on Computer Vision and Pattern Recognition (CVPR)*, San Diego, CA, Vol. 1, pp. 886–893, 2005.

30. J. Dean and S. Ghemawat. Mapreduce: Simplified data processing on large clusters. In *OSDI*, San Francisco, CA, pp. 137–150, 2004.

31. J. Dean and S. Ghemawat. Mapreduce: simplified data processing on large clusters. *Communications of the ACM*, 51(1):107–113, 2008.

32. J. Dean and S. Ghemawat. Mapreduce: A flexible data processing tool. *Communications of the ACM*, 53(1):72–77, 2010.

33. A. Delbimbo and P. Pala. Visual image retrieval by elastic matching of user sketches. *IEEE Transactions on Pattern Analysis and Machine Intelligence*, 19(2):121–132, 1997.

34. Y. Deng, C. Kenney, M. Moore, and B. Manjunath. Peer group filtering and perceptual color quantization. *IEEE International Symposium on Circuits and Systems*, 4:21–24, 1999.

35. C. Desai, D. V. Kalashnikov, S. Mehrotra, and N. Venkatasubramanian. Using semantics for speech annotation of images. In *International Conference on Data Engineering (ICDE)*, pp. 1227–1230, 2009.

36. C. Dorai, R. Farrell, A. Katriel, G. Kofman, Y. Li, and Y. Park. MAGICAL demonstration: System for metadata automated generation for instructional content. In *ACM Multimedia*, Vol. 1, pp. 491–492, 2006.

37. L.-Y. Duan, M. Xu, Q. Tian, C.-S. Xu, and J. S. Jin. A unified framework for semantic shot classification in sports video. *IEEE Transactions on Multimedia*, 7(6):1066–1083, 2005.

38. J. P. Eakins, K. Shields, and J. Boardman. ARTISAN-a shape retrieval system based on boundary family indexing. *Proceedings of SPIE*, 2670:17–28, 1996.

39. J. J. Ashley et al. Automatic and semiautomatic methods for image annotation and retrieval in query by image content (QBIC). *Proceedings of SPIE*, 2420:24–35, March 1995.

40. R. Fagin, A. Lotem, and M. Naor. Optimal aggregation algorithms for middleware. In *PODS*, Santa Barbara, CA, 2001.

41. R. Feris, A. Hampapur, Y. Zhai, R. Bobbitt, L. Brown, D. Vaquero, Y.-L. Tian, H. Liu, and M.-T. Sun. *Case Study: IBM Smart Surveillance System*. CRC Press, Intelligent Video Surveillance: Systems and Technology, Boca Raton, FL, 2009.

42. H. Freeman. Computer processing of line drawing images. *Computer Surveys*, 6(1):57–98, 1974.

43. N. Garg and I. Weber. Personalized tag suggestion for Flickr. In *International World Wide Web Conference (WWW)*, Beijing, China, Vol. 1, pp. 1063–1064, 2008.

44. MPEG Requirements Group. MPEG-7 context, objectives and technical roadmap. Doc. ISO/MPEG N2861, *MPEG Vancouver Meeting*, Vancouver, British Columbia, Canada, July 1999.

45. W. Guan, N. Haas, Y. Li, and S. Pankanti. A comparison of FFS+LAC with AdaBoost for training a vehicle localizer. *First Asian Conference on Pattern Recognition*, pp. 42–46, Beijing, China, November 2011.

46. A. Guttman. R-trees: A dynamic index structure for spatial searching. In *Proceedings of the ACM SIGMOD 1984 Annual Meeting*, Boston, MA, pp. 47–57, June 1984.

47. R. M. Haralick, K. Shanmugam, and I. Dinstein. Texture features for image classification. IEEE Transactions on Systems, Man, and Cybernetics, SMC-3(6):1345–1350, 1973.

48. A. Hauptmann, M. Chen, M. Christel, W. Lin, and J. Yang. A hybrid approach to improving semantic extraction of news video. In *IEEE International Conference on Semantic Computing (ICSC)*, Irvine, CA, Vol. 1, pp. 79–86, 2007.

49. A. G. Hauptmann, M. Chen, and M. Christel. Confounded expectations: Informedia at TRECVID 2004. *TREC Video Retrieval Evaluation (TRECVID)*, Gaithersburg, MA, 2004.

50. M. E. Houle, V. Oria, S. Satoh, and J. Sun. Knowledge propagation in large image databases using neighborhood information. In *ACM Multimedia*, pp. 1033–1036, 2011.

51. M. E. Houle and J. Sakuma. Fast approximate similarity search in extremely high-dimensional data sets. In *International Conference on Data Engineering (ICDE)*, Tokyo, Japan, pp. 619–630, 2005.

52. J. Huang, S. Kumar, M. Mitra, W. Zhu, and R. Zabih. Spatial color indexing and applications. *International Journal of Computer Vision*, 35(3):602–607, December 1999.

53. Q. Huang, Z. Liu, and A. Rosenberg. Automated semantic structure reconstruction and representation generation for broadcast news. *Proceedings of SPIE*, 3656:50–62, January 1999.

54. T. S. Huang, S. Mehrotra, and K. Ramachandran. Multimedia analysis and retrieval system (MARS) project. In *Proceedings of 33rd Annual Clinic on Library Application of Data Processing-Digital Image Access and Retrieval*, Urbana-Champaign, IL, Vol. 1, pp. 1–15, March 1996.

55. P. Indyk and R. Motwani. Approximate nearest neighbors: Towards removing the curse of dimensionality. In *ACM Symposium on the Theory of Computing (STOC)*, Dallas, TX, pp. 604–613, 1998.

56. National Institute of Standards and Technology (NIST). TREC video retrieval evaluation. http://www-nlpir.nist.gov/projects/trecvid, accessed March 27, 2013.

57. H. V. Jagadish, B. C. Ooi, K.-L. Tan, C. Yu, and R. Zhang. Idistance: An adaptive b+-tree based indexing method for nearest neighbor search. *ACM Transactions on Database Systems*, 30(2):364–397, 2005.

58. H. Jegou, M. Douze, and C. Schmid. Product quantization for nearest neighbor search. *IEEE Transactions on Pattern Analysis and Machine Intelligence*, 33:117–128, 2011.

59. N. Katayama and S. Satoh. The SR-tree: An index structure for high-dimensional nearest neighbor queries. In *Proceedings of the ACM SIGMOD International Conference on Management of Data*, Tucson, AZ, pp. 369–380, May 1997.

60. Y. Ke and R. Sukthankar. PCA-SIFT: A more distinctive representation for local image descriptors. *IEEE Conference on Computer Vision and Pattern Recognition (CVPR)*, 2:506–513, 2004.

61. D. Keren, D. B. Cooper, and J. Subrahmonia. Describing complicated objects by implicit polynomials. *IEEE Transactions on Pattern Analysis and Machine Intelligence*, 16:38–54, 1994.

62. H. Kim, J. Kim, D. Sim, and D. Oh. A modified Zernike moment shape descriptor invariant to translation, rotation and scale for similarity-based image retrieval. In *International Conference on Multimedia and Expo*, Vol. 1, New York, pp. 307–310, 2000.

63. J. Kim and K. Grauman. Observe locally, infer globally: A space-time MRF for detecting abnormal activities with incremental updates. In *IEEE Conference on Computer Vision and Pattern Recognition (CVPR)*, Miami, FL, Vol. 1, pp. 2921–2928, 2009.

64. H. Kosch. MPEG-7 and multimedia database systems. *ACM SIGMOD Record*, 31(2):34–39, 2002.

65. N. Koudas, B. C. Ooi, H. T. Shen, and A. K. H. Tung. LDC: Enabling search by partial distance in a hyper-dimensional space. In *International Conference on Data Engineering* (ICDE), Boston, MA, pp. 6–17, 2004.

66. S. Lazebnik, C. Schmid, and J. Ponce. Semi-local affine parts for object recognition. *Proceedings of the British Machine Vision Conference*, 1:959–968, 2004.

67. J. Z. Li, M. T. Özsu, D. Szafron, and V. Oria. MOQL: A multimedia object query language. In *Proceedings of the 3rd International Workshop on Multimedia Information Systems*, Como, Italy, pp. 19–28, September 1997.

68. W.-S. Li, K. S. Candan, K. Hirata, and Y. Hara. SEMCOG: An object-based image retrieval system and its visual query language. In *Proceedings of ACM SIGMOD International Conference on Management of Data*, Tucson, AZ, pp. 521–524, May 1997.

69. Y. Li and C. Dorai. Detecting discussion scenes in instructional videos. In *IEEE Conference on Multimedia and Expo (ICME)*, Vol. 2, Taipei, Taiwan, pp. 1311–1314, 2004.

70. Y. Li and C. Dorai. SVM-based audio classification for instructional video analysis. *IEEE Conference on Acoustics, Speech, and Signal Processing (ICASSP)*, 5:897–900, 2004.

71. Y. Li, C. Dorai, and R. Farrell. Creating MAGIC: System for generating learning object metadata for instructional content. *ACM Multimedia*, 1:367–370, 2005.

72. Y. Li and C.-C. Kuo. *Video Content Analysis Using Multimodal Information: For Movie Content Extraction, Indexing and Representation*. Kluwer Academic Publishers, Norwell, MA, 2003.

73. Y. Li and C.-C. Kuo. A robust video scene extraction approach to movie content abstraction. *International Journal of Imaging Systems and Technology*, 13(5):236–244, 2004, Special Issue on Multimedia Content Description and Video Compression.

74. Y. Li, Y. Ma, and H. Zhang. Salient region detection and tracking in video. In *ICME '03*, Vol. 1, Baltimore, MD, pp. 269–272, 2003.

75. Y. Li, S. Narayanan, and C.-C. Kuo. Content-based movie analysis and indexing based on audiovisual cues. *IEEE Transactions on Circuits and Systems for Video Technology*, 14(8):1073–1085, 2004.

76. Y. Li, Y. Park, and C. Dorai. Atomic topical segments detection for instructional videos. *ACM Multimedia*, 1:53–56, 2006.

77. Z. Li, Q. Zhang, Q. Hou, L. Diao, and H. Li. The shape analysis based on curve-structure moment invariants. *International Conference on Neural Networks and Brain*, 2:1139–1143, 2005.

78. R. W. Lienhart. Comparison of automatic shot boundary detection algorithms. In *Electronic Imaging '99*, San Jose, CA, pp. 290–301, International Society for Optics and Photonics, 1998.

79. D. Liu, X. Hua, L. Yang, M. Wang, and H. Zhang. Tag ranking. In *International World Wide Web Conference (WWW)*, Madrid, Spain, Vol. 1, pp. 351–360, 2009.

80. Z. Liu, J. Huang, and Y. Wang. Classification of TV programs based on audio information using hidden Markov model. *Proceedings of IEEE 2nd Workshop on Multimedia Signal Processing*, St. Thomas, Virgin Islands, pp. 27–32, December 1998.

81. A. Loui, A. Venetsanopoulos, and K. Smith. Two-dimensional shape representation using morphological correlation functions. *International Conference on Acoustics, Speech, and Signal Processing*, 4:2165–2168, 1990.

82. D. Lowe. Distinctive image features from scale-invariant keypoints. *International Journal of Computer Vision*, 60(2):91–110, 2004.

83. L. Lu, H. Jiang, and H. Zhang. A robust audio classification and segmentation method. In *ACM Multimedia*, Vol. 1, Ottawa, Ontario, Canada, pp. 203–211, 2001.

84. T. S. Mahmood and S. Srinivasan. Detecting topical events in digital video. *ACM Multimedia '00*, Marina del Rey, Los Angeles, CA, pp. 85–94, November 2000.

85. B.S. Manjunath, P. Salembier, T. Sikora, and P. Salembier. *Introduction to MPEG 7: Multimedia Content Description Language*. John Wiley & Sons, Inc., New York, June 2002.

86. D. Marr. *Vision: A Computational Investigation into the Human Representation and Processing of Visual Information*. W. H. Freeman, New York, 1982.

87. K. Minami, A. Akutsu, H. Hamada, and Y. Tonomura. Video handling with music and speech detection. *IEEE Multimedia*, 5(3):17–25, 1998.

88. T. P. Minka and R. W. Picard. Interactive learning using a society of models. *Pattern Recognition*, 30(4):565–581, 1997.

89. S. Moncrieff, C. Dorai, and S. Venkatesh. Detecting indexical signs in film audio for scene interpretation. In *ICME '01*, Vol. 1, Tokyo, Japan, 2001.

90. M. Muja and D. G. Lowe. Fast approximate nearest neighbors with automatic algorithm configuration. In *VISAPP International Conference on Computer Vision Theory and Applications*, pp. 331–340, 2009.

91. M. R. Naphade, C. Lin, A. Natsev, B. Tseng, and J. Smith. A framework for moderate vocabulary semantic visual concept detection. In *ICME '03*, Vol. 1, Baltimore, MD, pp. 437–440, 2003.

92. A. Natsev, J. Smith, M. Hill, G. Hua, B. Huang, M. Merler, L. Xie, H. Ouyang, and M. Zhou. IBM Research TRECVID-2010 video copy detection and multimedia event detection system. In *TRECVID 2010 Workshop*, Gaithersburg, MD, 2010.

93. A. Natsev, J. Smith, J. Tesic, L. Xie, and R. Yan. IBM multimedia analysis and retrieval system. In *ACM International Conference on Image and Video Retrieval (CIVR)*, Niagara Falls, Canada, Vol. 1, pp. 553–554, 2008.

94. D. Nistér and H. Stewénius. Scalable recognition with a vocabulary tree. In *IEEE Conference on Computer Vision and Pattern Recognition (CVPR)*, New York, Vol. 2, pp. 2161–2168, June 2006.

95. N. Nitta, N. Babaguchi, and T. Kitahashi. Generating semantic descriptions of broadcasted sports videos based on structures of sports games and TV programs. *Multimedia Tools and Applications*, 25(1):59–83, 2005.

96. Sharp Laboratories of America. Scalable blob histogram descriptor. *MPEG-7 Proposal 430, MPEG-7 Seoul Meeting*, Seoul, Korea, March 1999.

97. J. R. Ohm, F. Bunjamin, W. Liebsch, B. Makai, and K. Mller. A set of descriptors for visual features suitable for MPEG-7 applications. *Signal Processing: Image Communication*, 16(1–2):157–180, 2000.

98. A. Oliva and A. Torralba. Modeling the shape of the scene: A holistic representation of the spatial envelope. *International Journal of Computer Vision*, 42(3):145–175, 2001.

99. V. Oria, M. T. Özsu, X. Li, L. Liu, J. Li, Y. Niu, and P. J. Iglinski. Modeling images for content-based queries: The DISIMA approach. In *Proceedings of 2nd International Conference of Visual Information Systems*, San Diego, CA, pp. 339–346, December 1997.

100. V. Oria, M. T. Özsu, B. Xu, L. I. Cheng, and P.J. Iglinski. VisualMOQL: The DISIMA visual query language. In *Proceedings of the 6th IEEE International Conference on Multimedia Computing and Systems*, Vol. 1, Florence, Italy, pp. 536–542, June 1999.

101. N. V. Patel and I. K. Sethi. Audio characterization for video indexing. In *Proceedings of SPIE: Storage and Retrieval for Image and Video Databases IV*, San Jose, CA, February 1996.

102. A. Pentland, R. W. Picard, W. Rosalind, and S. Sclaro. Photobook: Tools for content-based manipulation of image databases. *Proceedings of SPIE*, 2368:37–50, 1995.

103. S. Pfeiffer, S. Fischer, and W. Effelsberg. Automatic audio content analysis. *ACM Multimedia '96*, Boston, MA, pp. 21–30, November 1996.

104. C. Piciarelli, C. Micheloni, and G. Foresti. Trajectory-based anomalous event detection. *IEEE Transactions on Circuits and Systems for Video Technology*, 18(11):1544–1554, 2008.

105. G. Qi, X. Hua, Y. Rui, J. Tang, T. Mei, and H. Zhang. Correlative multi-label video annotation. *ACM Multimedia*, 1:17–26, 2007.

106. R. Ramakrishnan and J. Gehrke. *Database Management Systems*, 2nd edn. McGraw Hill, New York, 2000.

107. B. C. Russell, A. Torralba, K. P. Murphy, and W. T. Freeman. Labelme: A database and web-based tool for image annotation. *International Journal of Computer Vision*, 77(1–3):157–173, 2008.

108. Y. Sakurai, M. Yoshikawa, S. Uemura, and H. Kojima. The A-tree: An Index Structure for High-Dimensional Spaces Using Relative Approximation. *In Proceedings of the 26th International Conference on very Large Data Bases*, San Francisco, CA, pp. 516–526, 2000.

109. H. Samet. *Foundations of Multidimensional and Metric Data Structures*, The Morgan Kaufmann Series in Computer Graphics and Geometric Modeling. Morgan Kaufmann Publishers Inc., San Francisco, CA, 2005.

110. S. Santini and A. Gupta. An extensible feature management engine for image retrieval. In *Proceedings of SPIE Storage and Retrieval for Media Databases*, San Jose, CA, Vol. 4676, pp. 86–97, 2002.

111. S. Santini and A. Gupta. Principles of schema design in multimedia databases. *IEEE Transactions on Multimedia Systems*, 4(2):248–259, 2002.

112. E. Scheirer and M. Slaney. Construction and evaluation of a robust multifeature speech/music discrimination. In *ICASSP '97*, Munich, Germany, p. 4, 1997.

113. X. Shen, M. Boutell, J. Luo, and C. Brown. Multi-label machine learning and its application to semantic scene classification. In *International Symposium on Electronic Imaging*, San Jose, CA, 2004.

114. R. Shi, C.-H. Lee, and T.-S. Chua. Enhancing image annotation by integrating concept ontology and text-based Bayesian learning model. In *ACM Multimedia*, Augsburg, Germany, pp. 341–344, 2007.

115. B. Sigurbjörnsson and R. Zwol. Flickr tag recommendation based on collective knowledge. In *International World Wide Web Conference (WWW)*, Beijing, China, pp. 327–336, 2008.

116. C. S.-Anan and R. Hartley. Optimised kd-trees for fast image descriptor matching. In *Computer Vision and Pattern Recognition (CVPR)*, Vol. 1, Anchorage, AL, pp. 1–8, 2008.

117. A. Smeaton, P. Over, and W. Kraaij. Evaluation campaigns and TRECVid. In *ACM International Workshop on Multimedia Information Retrieval*, Vol. 1, Santa Barbara, CA, pp. 321–330, 2006.

118. J. Smith and M. Naphade. Multimedia semantic indexing using model vectors. In *IEEE International Conference on Multimedia and Expo*, Baltimore, MD, Vol. 2, pp. 445–448, 2003.

119. J. R. Smith and S. F. Chang. Image and video search engine for the World Wide Web. *Proceedings of SPIE*, 3022:84–95, 1997.

120. C. G. M. Snoek and M. Worring. Multimodal video indexing: A review of the state-of-the-art. *Multimedia Tools and Applications*, 25(1):5–35, 2005.

121. M. Srikanth, J. Varner, M. Bowden, and D. I. Moldovan. Exploiting ontologies for automatic image annotation. In *ACM Special Interest Group on Information Retrieval (SIGIR)*, Salvador, Bahia, Brazil, pp. 552–558, 2005.

122. K. Stolze. SQL/MM spatial the standard to manage spatial data in a relational database system. In *GI Symposium Database Systems for Business, Technology and Web* (*BTW*), Leipzig, Germany, pp. 247–264, 2003.

123. M. Stricker and A. Dimai. Color indexing with weak spatial constraints. *Proceedings of SPIE*, 2670:29–40, 1996.

124. H. Sundaram and S. F. Chang. Determining computable scenes in films and their structures using audio-visual memory models. *ACM Multimedia '00*, Marina Del Rey, Los Angeles, CA, November 2000.

125. H. Tamura, S. Mori, and T. Yamawaki. Texture features corresponding to visual perception. *IEEE Transactions on Systems, Man, and Cybernetics*, SMC-8(6):780–786, 1978.

126. H. Thompson, D. Beech, M. Maloney, and N. Mendelsohn. XML schema part 1: Structures. *W3C Recommendation, World Wide Web Consortium* (*W3C*), May 2001. http://www.w3.org/TR/2001/REC-xmlschema-1-200105021 (accessed on May 16, 2013).

127. Y. Tian, R. Feris, H. Liu, A. Hampapur, and M. Sun. Robust detection of abandoned and removed objects in complex surveillance videos. *IEEE Transactions on Systems, Man, and Cybernetics—Part C: Applications and Reviews*, 41(5):565–576, 2010.

128. The PASCAL visual object classes homepage. http://pascallin.ecs.soton.ac.uk/challenges/voc/. March 27, 2013.

129. L. von Ahn and L. Dabbish. Labeling images with a computer game. In *Proceedings of Conference on Human Factors in Computing Systems* (*CHI*), Vienna, Austria, pp. 319–326, 2004.

130. X. Wan and C. C. Kuo. Image retrieval based on JPEG compressed data. *Proceedings of SPIE*, 2916:104–115, 1996.

131. Y. Weiss, A. Torralba, and R. Fergus. Spectral hashing. In *Neural Information Processing Systems* (*NIPS*), British Columbia, Canada, pp. 1753–1760, 2008.

132. Y. Wu, B. Tseng, and J. Smith. Ontology-based multi-classification learning for video concept detection. In *IEEE International Conference on Multimedia and Expo*, Vol. 2, Taipei, Taiwan, pp. 1003–1006, 2004.

133. L. Wyse and S. Smoliar. Towards content-based audio indexing and retrieval and a new speaker discrimination technique. Downloaded from http://www.iss.nus.sg/People/lwyse/lwyse.html, Institute of Systems Science, National University of Singapore, Singapore, March 27, 2013.

134. T. Xiang and S. Gong. Incremental and adaptive abnormal behaviour detection. *Journal of Computer Vision and Image Understanding*, 111(1):59–73, 2008.

135. L. Xie, A. Natsev, M. Hill, J. Smith, and A. Philips. The accuracy and value of machine-generated image tags—Design and user evaluation of an end-to-end image tagging system. In *ACM International Conference on Image and Video Retrieval* (*CIVR*), Vol. 1, Xi'an, China, pp. 58–65, 2010.

136. L. Xie, A. Natsev, J. Kender, M. Hill, and J. Smith. Visual memes in social media: Tracking real-world news in YouTube videos. *ACM Multimedia*, 1:53–62, 2011.

137. L. Xie, R. Yan, J. Tesic, A. Natsev, and J. Smith. Probabilistic visual concept trees. *ACM Multimedia*, 1:867–870, 2010.

138. L. Xie, R. Yan, and J. Yang. Multi-concept learning with large-scale multimedia lexicons. In *IEEE International Conference on Image Processing* (*ICIP*), San Diego, CA, Vol. 1, pp. 2148–2151, 2008.

139. M. Xu, S. Luo, J. Jin, and M. Park. Affective content analysis by midlevel representation in multiple modalities. In *First International Conference on Internet Multimedia Computing and Service*, Vol. 1, Kunming, China, pp. 201–207, 2009.

140. R. Yan, M. Fleury, M. Merler, A. Natsev, and J. Smith. Large-scale multimedia semantic concept modeling using robust subspace bagging and MapReduce. In *First ACM Workshop on Large-Scale Multimedia Retrieval and Mining*, Beijing, China, pp. 35–42, 2009.

141. L. Yang, J. Liu, X. Yang, and X. Hua. Multi-modality web video categorization. *Multimedia Information Retrieval* (*MIR*), 1:265–274, September 2007.

142. M. Yeung, B. Yeo, and B. Liu. Extracting story units from long programs for video browsing and navigation. In *IEEE Proceedings of Multimedia Computing and Systems*, Hiroshima, Japan, pp. 296–305, 1996.

143. Y. Yu, M. Crucianu, V. Oria, and L. Chen. Local summarization and multi-level LSH for retrieving multi-variant audio tracks. In *ACM Multimedia*, Beijing, China, pp. 341–350, 2009.

144. Y. Yu, M. Crucianu, V. Oria, and E. Damiani. Combining multi-probe histogram and order-statistics based LSH for scalable audio content retrieval. In *ACM Multimedia*, Firenze, Italy, pp. 381–390, 2010.

145. J. Yuan, H. Wang, L. Xiao, W. Zheng, J. Li, F. Lin, and B. Zhang. A formal study of shot boundary detection. *IEEE Transactions on Circuits and Systems for Video Technology*, 17(2):168–186, 2007.

146. R. Zabih, J. Miller, and K. Mai. A feature-based algorithm for detecting and classifying scene breaks. In *ACM Multimedia '95*, Vol. 1, San Francisco, CA, pp. 189–200, 1995.

147. D. Zhang and G. Lu. Review of shape representation and description techniques. *Pattern Recognition*, 37:1–19, 2004.

148. H. J. Zhang, S. Y. Tan, S. W. Smoliar, and G. Y. Hong. Automatic parsing and indexing of news video. *Multimedia Systems*, 2(6):256–266, 1995.

149. T. Zhang and C.-C. Kuo. Audio content analysis for on-line audiovisual data segmentation. *IEEE Transactions on Speech and Audio Processing*, 9(4):441–457, 2001.

150. T. Zhang and C.-C. J. Kuo. Audio-guided audiovisual data segmentation, indexing and retrieval. *Proceedings of SPIE*, 3656:316–327, 1999.

151. G. Zhao and M. Pietikainen. Boosted multi-resolution spatiotemporal descriptors for facial expression recognition. *Pattern Recognition Letters*, 30(12):1117–1127, 2009.

III

Data, Information, and Knowledge Management

15

Building Conceptual Modeling on the Foundation of Ontology

Andrew
Burton-Jones
*The University of
Queensland*

Ron Weber
Monash University

15.1 Introduction

Conceptual models provide a representation, often graphical, of some features of a real-world domain. Frequently, information system professionals (e.g., system analysts) prepare these models to achieve two outcomes (e.g., Fettke, 2009). First, during the system development process, conceptual models facilitate communications among stakeholders. For instance, they may be used by analysts to check their understanding of the domain to be supported by the system being developed with likely users of the system. Second, after a system has been implemented, conceptual models assist users of the system or information systems professionals who must maintain and operate the system to undertake their work. For instance, they may be employed by managers to help formulate queries to interrogate the system's database or programmers to understand the system so they can undertake maintenance work.

Conceptual models are generated using some kind of conceptual modeling technique. For instance, an early technique was entity–relationship modeling (Chen, 1976), which provided its users with three constructs—entities, relationships, and attributes—to represent real-world phenomena that were of interest to them. When entity–relationship modeling is used, any real-world phenomenon of interest has to be mapped to one of these three constructs. Figure 15.1 shows an entity–relationship model of a simple domain.

FIGURE 15.1   Simple entity–relationship model.

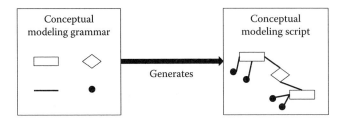

FIGURE 15.2   Generating conceptual modeling scripts via a conceptual modeling grammar.

More formally, a conceptual modeling technique can be conceived as a type of *grammar* that is used to generate *scripts* (Figure 15.2) (Wand and Weber, 1993). The conceptual modeling grammar provides various constructs to model different types of phenomena in the real world. The scripts generated via the grammar contain strings of *instances* of constructs in the grammar. These instances provide representations of specific real-world phenomena that are of interest to stakeholders in the system.

Historically, many different conceptual modeling techniques have been proposed. Indeed, at one stage of their development, the term "YAMA" was coined (Oei et al., 1992, p. 2), which stood for "yet another modeling approach." This term manifested the frustration many researchers felt at the time with the proliferation of existing modeling techniques, the ongoing emergence of still more techniques, and the paucity of rigorous research that evaluated their relative strengths and weaknesses. Moreover, the tendency of researchers and practitioners to propose new techniques *without* strong theory or empirical evaluation remains:

> "Even though hundreds of modelling methods are in existence today, practitioners and researchers are zealously 'producing' new modelling methods. ... Many of these methods lack theoretical foundations and empirical evidence to demonstrate their worthiness" (Siau and Rossi, 2011, p. 249).

The conceptual modeling techniques that have been proposed over the years tend to fall into one of two categories. The first category primarily provides constructs to model *substance and form* in the real world—specifically, things and their properties (including how things are associated with other things and how properties of things are related). These techniques are often called "data modeling techniques" or "semantic modeling techniques." Contrary to their name, however, often they are *not* used to model "data" in the real world. Instead, they are used to model certain real-world phenomena directly rather than the data that in turn might be chosen to represent the phenomena. Moreover, while the techniques can be used to model certain real-world semantics, they are capable of modeling only a *subset* of the real-world semantics that might interest stakeholders.

The second category of conceptual modeling techniques that have been developed primarily provides constructs to model *possibility and change* in the real world—specifically, the state spaces and event spaces that things in the real world might traverse and the transformations these things might undergo. These techniques are often called "process modeling techniques." This name is more appropriate than the names chosen for the first category of techniques. Process modeling techniques indeed provide constructs to model how changes occur in a domain.

Interestingly, historically few conceptual modeling techniques have been proposed that have a comprehensive focus on *both* substance and form as well as possibility and change. The reasons are unclear.

The so-called data or semantic modeling techniques developed first, perhaps because early designers of conceptual modeling techniques found the real-world semantics of substance and form easier to understand and thus simpler to model. Process modeling techniques developed somewhat later, perhaps because designers of conceptual modeling techniques found the real-world semantics of possibility and change harder to understand and thus more difficult to model. Moreover, historically data and semantic modeling techniques have been linked with database development, whereas process modeling techniques have been linked with application development. A recent trend is to advocate using *collections* of techniques to overcome the limitations of individual techniques. For instance, the unified modeling language (UML) includes some techniques more focused on substance and form (such as class diagrams) and other techniques more focused on possibilities and change (such as activity diagrams and statecharts) (Rumbaugh et al., 1999).

Motivated by the importance of conceptual modeling and the lack of theory and evidence to inform conceptual modeling practice, in the late 1980s, Wand and Weber began to develop an approach to evaluating conceptual modeling techniques that relied on the philosophical field of ontology* (e.g., Wand and Weber, 1993, 1995). This approach was an innovation at that time because previously conceptual modeling researchers and practitioners had not relied on theories of ontology. Wand and Weber noted that ontological researchers are concerned with building theories about the nature of existence—that is, theories about the nature of the fundamental types of phenomena that occur in the real world. Furthermore, they noted that, in essence, conceptual modelers are also concerned with building models of some subset of phenomena in the real world. Specifically, their work can be conceived as developing *specialized ontologies*—models of a particular domain in the real world.

Given that philosophical researchers had been working on theories of ontology for many hundreds of years, Wand and Weber believed these theories could be used to inform evaluations of the quality of both conceptual modeling grammars and conceptual modeling scripts. Specifically, they argued that ontological theories could be used to make predictions about the strengths and weaknesses of conceptual modeling grammars and scripts. These predictions could then be tested empirically.

Wand and Weber's proposed approach differed substantially from previous approaches used to evaluate conceptual modeling grammars and scripts. These prior approaches had relied primarily on case studies in which different conceptual modeling techniques were used to model some domain (e.g., Floyd, 1986). The strengths and weaknesses of the resulting models were then compared. Unfortunately, the case study approach often proved unsatisfactory as a way of resolving disputes. For instance, one group of protagonists inevitably made allegations that the case study chosen had been selected deliberately to downplay the strengths and highlight the weaknesses of the particular technique they favored. In the absence of theory, it was difficult to see how disputes might be resolved.

Building conceptual modeling on the foundation of ontology is now an accepted approach in the conceptual modeling field (Siau and Rossi, 2011). The purpose of this chapter, therefore, is to provide an overview of this approach, describe two theories used by researchers that are based on ontology, and illustrate the kinds of the results that have emerged from empirical investigations of these theories. We first discuss the underlying principles that have motivated theoretical analyses of and empirical research on conceptual modeling practice. Next, we describe some examples of the research done (based on these principles) and how the findings have implications for conceptual modeling practice. Subsequently, we consider some significant research opportunities that exist in the application of ontological principles to conceptual modeling practice. Finally, we present a brief summary of our analyses and some conclusions.

---

* Unfortunately, the term "ontology" now has multiple meanings in the information technology literature. It is sometimes used as the generic term that refers to the branch of philosophy that deals with the nature of existence in the real world. It is also used to refer to a particular scholar's theory about the nature of existence in the real world (e.g., Bunge's, 1977, 1979 ontology). More recently, it is sometimes used to refer to a description of a particular domain.

15.2 Underlying Principles

The idea that conceptual modelers in essence are building specialized ontologies motivates the conclusion that the modeling grammars they use must therefore be capable of representing the major types of phenomena that exist in the real world. This conclusion has underpinned the development of two theories about conceptual modeling grammars—one to account for the strengths and weaknesses of individual grammars and the other to account for how multiple grammars might be selected for conceptual modeling purposes. The following two subsections explain the nature and uses of both theories.

15.2.1 Theory of Ontological Expressiveness

The foundation for much ontologically based work on conceptual modeling is Wand and Weber's (1993) *theory of ontological expressiveness* (TOE). The TOE was formulated to address the question of how well conceptual modeling grammars are able to model real-world phenomena.

TOE is based upon a mapping between two sets of constructs: (1) a set of ontological constructs developed to describe various phenomena in the real world and (2) a set of grammatical constructs developed to generate "scripts" that describe various phenomena in the real world. The ontological constructs should be derived from a coherent theory about the nature and form of phenomena in the real world. The grammatical constructs should be derived from a careful specification of the grammar's capabilities.

Wand and Weber propose a conceptual modeling grammar that is *ontologically expressive* when the mapping between ontological constructs and grammatical constructs is one-to-one and onto (isomorphic)—that is, each and every ontological construct maps to one and only one grammatical construct, and each and every grammatical construct is the "image" (or representation) of one and only one ontological construct. Effective visual reasoning with software engineering notations also seems to depend on having an isomorphic mapping between semantic constructs and visual symbols (syntax) (Moody, 2009, pp. 762–763).

In TOE, an isomorphic pattern exists when four conditions hold (Figure 15.3):

1. Each ontological construct maps to one and only one grammatical construct.
2. Each grammatical construct maps to one and only one ontological construct.

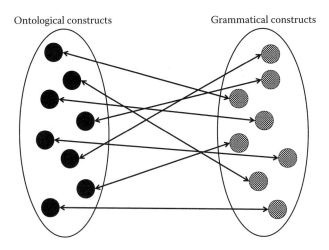

FIGURE 15.3   An ontologically expressive conceptual modeling grammar.

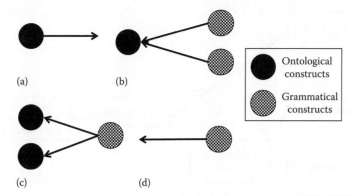

FIGURE 15.4 Factors that undermine the ontological expressiveness of a conceptual modeling grammar. (a) Construct deficit, (b) construct redundancy, (c) construct overload, and (d) construct excess.

3. All ontological constructs are "covered" by a grammatical construct—in other words, every ontological construct maps to a grammatical construct.
4. All grammatical constructs are "covered" by an ontological construct—in other words, every grammatical construct maps to an ontological construct.

When a mapping is undertaken between the two sets of constructs, four situations can arise that Wand and Weber argue undermine the ontological expressiveness of a conceptual modeling grammar (Figure 15.4). The first situation—construct deficit—undermines the *ontological completeness* of a modeling grammar. The next three situations—construct redundancy, construct overload, and construct excess—undermine the *ontological clarity* of a grammar. The four situations are defined as follows:

1. *Construct deficit*: Construct deficit arises when no grammatical construct exists that maps to a particular ontological construct. The consequence is that some aspect of the real world cannot be represented in scripts generated using the grammar.
2. *Construct redundancy*: Construct redundancy arises when two or more grammatical constructs map to a single ontological construct (m:1 mapping). The consequence is that different grammatical constructs can be used to represent the same real-world phenomenon. As a result, users of a script may become confused about what instances of the grammatical constructs represent when they try to interpret a script. They may try to interpret a script using the assumption that instances of different grammatical constructs are meant to represent different types of real-world phenomena.
3. *Construct overload*: Construct overload arises when a grammatical construct maps to two or more ontological constructs (1:m mapping). The consequence is that users of a script may be unable to understand the meaning of a real-world phenomenon represented by an instance of the construct.
4. *Construct excess*: Construct excess arises when a grammatical construct exists that does not correspond to any ontological construct. The consequence is that users of a script may be unable to understand what an instance of the construct is intended to represent.

Figure 15.5 provides an overview of TOE. It shows that user task effectiveness and efficiency will be higher when a modeling grammar has higher levels of ontological expressiveness. The level of a grammar's ontological expressiveness will be higher when it has lower levels of ontological deficit and higher levels of ontological clarity. Its ontological clarity will be undermined when it has higher levels of construct redundancy, overload, and excess.

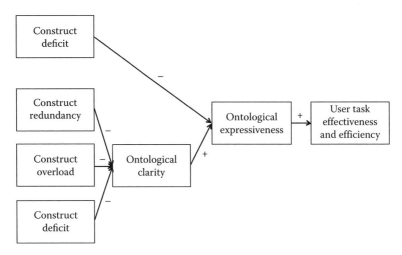

FIGURE 15.5 Theory of ontological expressiveness (TOE).

Ontological completeness and clarity are relative concepts. In other words, whether a conceptual modeling grammar is deemed ontologically complete and/or clear depends on the particular ontology used to undertake the evaluation (the "reference" ontology). For instance, some ontologies make a distinction between "things" and "properties of things" (e.g., Bunge, 1977). Others make no such distinction and have only the concept of an "entity" or "thing" (e.g., Cocchiarella, 1972). A conceptual modeling grammar that provides only a single construct to model both "things" and "properties of things" will be *ontologically unclear* under the former ontology. (It will possess ontological overload because a single grammatical construct maps to two ontological constructs.) It will be ontologically clear, however, under the latter ontology.

Similarly, an ontology may have a single construct to represent the real-world phenomenon of an event. A conceptual modeling grammar may have several constructs, however, to represent different *types* of events (e.g., initial events, intermediate events, and final events). Providing the real-world meanings of the different event constructs in the modeling grammar do not overlap, the grammar does *not* have construct redundancy. Rather, it is based on a more specialized ontology than the ontology used to undertake the evaluation for completeness and clarity.

Whether ontological completeness and clarity (or lack thereof) have any practical impact is an empirical issue (Gemino and Wand, 2005, p. 307). Predictions about the effects of an ontologically complete and clear grammar or an ontologically incomplete or unclear grammar need to be tested in the context of users' performance with scripts generated via the grammar. For instance, a conceptual modeling grammar might be deemed ontologically unclear because it provides only a single construct to represent "things" and "properties of things." Nonetheless, lack of ontological clarity has no import unless the performance of stakeholders who engage with the grammar or scripts generated via the grammar is undermined in some way. Similarly, the performance of stakeholders who engage with an ontologically clear conceptual modeling grammar might be poor because the ontology used to evaluate ontological clarity is deficient.

15.2.2 Theory of Multiple Grammar Selection

Because stakeholders often engage with conceptual modeling grammars that have construct deficit, they may conclude they must use multiple grammars to meet their modeling needs. Based on TOE, Green (1996) proposed a theory to account for how stakeholders will select the grammars to use from a set of

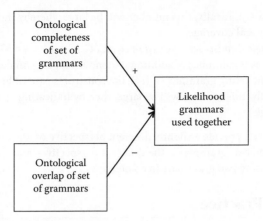

FIGURE 15.6 Theory of multiple grammar selection (TMGS).

alternatives—the theory of multiple grammar selection (TMGS) (Figure 15.6). He proposed that stakeholders will seek to satisfy two objectives:

1. *Maximum ontological coverage (MOC)*: Stakeholders will choose a combination of grammars to afford them maximum coverage of the phenomena they wish to model. In other words, they will seek to minimize problems caused by construct deficit.
2. *Minimum ontological overlap (MOO)*: Stakeholders will choose a combination of grammars to afford them minimum ontological redundancy. In other words, they will seek to mitigate problems arising from having multiple grammatical constructs covering the same ontological construct.

Green argued the primary objective in selecting a combination of modeling grammars is to achieve MOC. This objective has to be moderated, however, by problems that might arise as the level of ontological overlap among the grammars increases. Moreover, he argued stakeholders will seek to achieve parsimony in their selection of grammars. They will select the smallest number of grammars that allow them to achieve their MOC and MOO objectives satisfactorily.

Mindful of the MOC and MOO objectives, Green et al. (2007) and zur Muehlen and Indulska (2010) have proposed five metrics to evaluate when selecting a combination of modeling grammars to use. In the context of two grammars, G_1 and G_2, these metrics are based on the following sets:

- *Union of ontological constructs in G_1 and G_2*: $G_1 \cup G_2 = x \in G_1 \vee x \in G_2$: The number of elements in this set is the number of ontological constructs that can be represented by using *both* modeling grammars. Ideally, this number will equal the number of constructs in the reference ontology.
- *Symmetric difference of ontological constructs covered in G_1 and G_2*: $G_1 \triangle G_2 = (G_1 - G_2) \cup (G_2 - G_1)$: The number of elements in this set is the number of constructs in the reference ontology that G_1 and G_2 represent *distinctly* (i.e., free from overlap). Relative to the number of constructs in the reference ontology, ideally the number in the symmetric difference will be *large* (thereby indicating little overlap exists in the ontological constructs covered by the two grammars).
- *Intersection of ontological constructs covered in G_1 and G_2*: $G_1 \cap G_2 = x \in G_1 \wedge x \in G_2$: The number of elements in this set is the number of ontological constructs that can be represented in *both* modeling grammars. Relative to the number of constructs in the reference ontology, ideally the number in the intersection will be *small* (thereby indicating little overlap exists in the ontological constructs covered by the two grammars).
- *Difference of ontological constructs covered in G_1 and G_2*: $G_1 - G_2 = x \in G_1 \wedge x \notin G_2$: The number of elements in this set is the number of additional (nonoverlapping) ontological constructs that can be represented in modeling grammar G_1 (i.e., the constructs cannot be represented in

modeling grammar G_2). Ideally, this number will be large, thereby indicating G_1 adds substantially to G_2's ontological coverage.

* *Difference in ontological constructs covered in G_2 and G_1*: $G_2 - G_1 = x \in G_2 \wedge x \notin G_1$: The number of elements in this set is the number of additional (nonoverlapping) ontological constructs that can be represented in modeling grammar G_2 (i.e., the constructs cannot be represented in modeling grammar G_1). Ideally, this number will be large, thereby indicating G_2 adds substantially to G_1's ontological coverage.

Note the last four metrics all provide a slightly different perspective on the extent to which ontological overlap exists between the two grammars. The different perspectives may prove helpful, however, in appraising the merits of using two grammars in combination.

15.3 Impact on Practice

The two theories we have outlined (TOE and TMGS) essentially stem from the same general argument, which is that the ontological "quality" of a grammar or grammars affects the use of, or consequences of using, the grammar or grammars. While the outcome construct in TOE concerns the consequences of using grammars (see Figure 15.5), second-order consequences might arise (e.g., poor system design). Likewise, while the outcome construct in TMGS concerns the selection of grammars (see Figure 15.6), the guiding logic is that people will choose to use grammars that yield beneficial consequences. Some of the literature we review in this section makes precisely such predictions.

According to Moody (2005), most conceptual modeling techniques have had little impact on practice. As a result, the purpose of developing theories such as TOE and TMGS is to help stakeholders in the conceptual modeling community explain and predict the use and consequences of using conceptual modeling techniques in practice. In this vein, we subscribe to a view, often attributed to Lewin (1945), that "nothing is quite so practical as a good theory."

> Lewin's [view]…is as important today as it was in Lewin's time. Good theory is practical precisely because it advances knowledge in a scientific discipline, guides research toward crucial questions, and enlightens the profession… (Van de Ven, 1989, p. 486).

As noted earlier, TOE and TMGS are ontology-agnostic. Nevertheless, in this section, we examine how both theories have been used with a *particular* theory of ontology—namely, Bunge's (1977) theory of ontology—to derive predictions about and recommendations for conceptual modeling practice. We have chosen Bunge's theory of ontology because for two reasons it has informed much ontologically based conceptual modeling research*: (a) it has been articulated carefully and formally; and (b) it uses constructs employed widely within the information technology field.

To show how Bunge's theory of ontology can be used in conjunction with TOE and TMGS to inform conceptual modeling practice, we describe examples of research studies that have first derived analytical results and then evaluated these results empirically. We have chosen these studies because, like Moody (2005) and Siau and Rossi (2011), we believe empirical results add weight to analytical results. Nonetheless, the results obtained from the combined analytical and empirical studies mirror those obtained from the purely analytical studies—for instance, analytical studies of the ontological expressiveness of the entity–relationship grammar (Wand and Weber, 1993), UML grammar (Evermann and Wand, 2005a,b), NIAM grammar (Weber and Zhang, 1996), object-modeling grammar (Opdahl and Henderson-Sellers, 2004), and ARIS grammar (Fettke and Loos, 2007), and the ways in which various process modeling and rule modeling grammars should be used in conjunction with one another (zur Muehlen and Indulska, 2010).

---

* We reference some of this research in the sections and subsections that follow.

In the following two subsections, note we have not sought to provide a comprehensive review of prior studies. Rather, our goal is to illustrate how Bunge's theory of ontology, TOE, and TMGS have been employed in prior research to derive outcomes that can be used to inform practice.

15.3.1 Conceptual Modeling Practice and the TOE

Several research studies have now been conducted that use TOE and Bunge's theory of ontology as the reference ontology to make predictions about the strengths and weaknesses of different conceptual modeling grammars and to test these predictions empirically. Based on the results obtained, some conclusions can be drawn for conceptual modeling practice.

15.3.1.1 Construct Deficit

Recall, construct deficit arises in a conceptual modeling grammar when a construct exists in the reference ontology for which no corresponding construct exists in the grammar. In the presence of construct deficit, TOE motivates a prediction that instances of real-world phenomena represented by instances of the ontological construct cannot be represented in scripts generated via the grammar. The outcome is that the scripts provide an incomplete description of the real-world domain of interest.

Recker et al. (2009) studied 12 process modeling grammars (ANSI flowcharts, data flow diagrams, IDEF method 3 process description capture method, ISO/TC97, Merise, EPC, BPML, WSCI, ebXML, WS-BPEL, Petri nets, and BPMN) from the perspective of ontological completeness and clarity. As their reference ontology, they used the so-called Bunge–Wand–Weber (BWW) ontology, which is an extension and adaptation of Bunge's theory of ontology by Wand and Weber (1990).*

In terms of ontological completeness, they found all the grammars had construct deficit according to the BWW ontology. As a measure of construct deficit, for each grammar, they counted the number of constructs in the BWW ontology that were not covered by a grammatical construct and divided this number by the total number of constructs defined in the BWW ontology. The average measure of construct deficit across the 12 grammars was 59.21% (the range was 9.10%–81.80%). In other words, on average, approximately 59% of the constructs in the BWW ontology could not be represented via constructs in the grammars.

In light of their analysis, Recker et al. (2009) predicted users of the grammars they studied would encounter a number of difficulties. For instance, they concluded

- Good process decompositions could not always be achieved because some of the grammars had no constructs to model phenomena such as things, classes of things, and properties of things, which are needed to design good decompositions (Burton-Jones and Meso, 2006).
- Some of the grammars had no constructs to model lawful and conceivable state spaces. As a result, they provided limited support for the identification of unlawful states and the design of exception handling processes.
- The ability to define and refine business rules was limited because some of the grammars had no constructs to model the history of things (and thus the sequence of states things traversed).
- Most of the grammars had no constructs to model the environment of a system. As a result, the process models they could represent were decoupled from their environment. The design of context-aware process models became difficult, if not impossible.

Recker et al. (2011) then undertook a web-based survey of users' experiences with one of the grammars they analyzed—namely, BPMN. They chose BPMN because it is used extensively to undertake business process modeling. In their survey instrument, they asked BPMN users about the extent to which they perceived BPMN has certain kinds of construct deficit (e.g., "BPMN does not provide sufficient symbols to represent business rules in process models").

---

* Wand and Weber (1990) sought to show how Bunge's (1977) theory of ontology could be applied to the information technology domain, to extend Bunge's theory to include additional constructs that were useful in the information technology domain, and to correct certain errors that they believed existed in Bunge's theory.

Recker et al. (2011) obtained 528 usable responses from BPMN users around the world. They found a negative association between the extent to which these users perceived BPMN had certain kinds of construct deficit (based on their experience with BPMN) and the extent to which they perceived BPMN was *useful* in achieving their modeling objectives. Moreover, Recker et al. (2011) found these users' perceptions of BPMN's usefulness declined as a function of the number of different types of construct deficit they had encountered.

Recker et al. (2010) also interviewed 19 business process modeling practitioners in six Australian organizations about their experiences with BPMN. Again, they tested the extent to which construct deficit in BPMN undermined these practitioners' work. As with their survey results, for the most part they found the forms of construct deficit they investigated caused difficulties for the practitioners in the work they undertook.

When considering the acquisition of a grammar for conceptual modeling purposes, practitioners can use Bunge's theory of ontology and TOE to analyze the grammar for construct deficit. To the extent egregious instances of construct deficit exist in the candidate grammar, other grammars should be considered as alternatives. In some cases, practitioners may be unable to find a single grammar that covers the ontological constructs important to their modeling needs. As a result, multiple grammars might be chosen that provide *complementary* capabilities. In other words, the grammars are selected so that constructs that are missing in one grammar will be present (covered by) in another grammar. The recognition that multiple modeling grammars are needed to address construct deficit problems that practitioners confront motivated the articulation of the TMGS (which we discuss further in the subsequent sections).

15.3.1.2 Construct Redundancy

Recall, construct redundancy arises in a conceptual modeling grammar when two or more constructs in the grammar map to a single construct in the reference ontology. In the presence of construct redundancy, TOE motivates a prediction that users of the grammar will encounter difficulties when they have to make decisions about which of the redundant constructs to choose in modeling an instance of the ontological phenomenon in the real world. Moreover, stakeholders who have to interpret scripts that contain instances of the redundant constructs will encounter difficulties interpreting the meaning of instances of the constructs.

In their study of 12 process modeling grammars (see preceding text), Recker et al. (2009) were able to assess the level of construct redundancy in *seven* of the grammars they examined—namely, EPC, BPML, WSCI, ebXML, WS-BPEL, Petri nets, and BPMN. They measured construct redundancy by counting the number of redundant grammatical constructs (based on the BWW ontology) and dividing this number by the total number of constructs in the grammar. The average measure of construct redundancy across the seven grammars was 26.93% (the range was 0%–51.30%). In other words, on average, approximately 27% of the constructs in the grammars were redundant.

In light of their analysis, Recker et al. (2009) predicted users of the grammars they studied would encounter a number of difficulties. For instance, they concluded

- For some of the grammars, the high level of construct redundancy that existed in relation to several ontological constructs was likely to cause confusion among users of the grammars.
- Construct redundancy in the grammars sometimes seemed to have occurred because of specialization (subtyping) of a BWW ontological construct. The motivation for the specialization appeared to have been particular objectives that designers of the grammars were seeking to achieve (e.g., assisting users to undertake validation of scripts generated using the grammar). Whether such specializations were likely to be successful was unclear.
- Some types of construct redundancy appeared to represent variations on the same ontological construct with the goal of better meeting the needs of different user communities who have different process modeling purposes (e.g., compliance management versus enterprise systems configuration).

In data obtained from their web-based survey of global BPMN users (see preceding text), Recker et al. (2011) found a negative association between the extent to which these users perceived BPMN had certain kinds of construct redundancy (e.g., "I often have to choose between a number of BPMN symbols to represent one kind of real-world object in a process model") and the extent to which they perceived BPMN was *easy to use*. They also found these users' perceptions of BPMN's ease of use declined as a function of the number of different types of construct redundancy they had encountered.

Based on interview responses obtained from 19 business process modeling practitioners in six Australian organizations (see preceding text), however, Recker et al. (2010) found the existence of construct redundancy in BPMN for the most part did *not* undermine the practitioners' work. While the practitioners recognized construct redundancy existed in BPMN, they found ways to mitigate its effects. For instance, one way in which users of process modeling grammars cope with construct redundancy is to confine their use of the grammar to certain constructs and not others (zur Muehlen and Recker, 2008).

In short, based on Recker et al.'s (2009) study, the evidence about the impact of construct redundancy on the users of conceptual modeling grammars is mixed. Nonetheless, there is a straightforward way in which stakeholders might seek to cope with problems that might arise from construct redundancy in a conceptual modeling grammar. Specifically, they can choose to use only one of the redundant constructs associated with a particular ontological construct and proscribe use of the other redundant construct(s).

15.3.1.3 Construct Overload

Recall, construct overload arises in a conceptual modeling grammar when a grammatical construct maps to two or more constructs in the reference ontology. In the presence of construct overload, TOE motivates a prediction that users of the grammar will encounter difficulties when they have to make decisions about how to model instances of the different ontological constructs. Moreover, stakeholders who have to interpret scripts that contain instances of the overloaded construct may fail to notice important differences in the meaning of different instances of the overloaded construct. They may also encounter difficulties interpreting the meaning of different instances of the overloaded construct.

Not all conceptual modeling grammars provide separate constructs to model classes of things (entities) and properties (attributes) in general of these classes of things. For instance, the object role modeling (ORM) grammar uses only a single construct (an entity, shown graphically via a named ellipse or rounded rectangle) to represent both things and properties (Halpin, 2008). Interestingly, users of ORM are told to focus on "facts," which are stated in terms of *objects* having *properties* or multiple objects participating in a *relationship*. In the scripts generated via ORM, however, objects and properties are both represented using the same grammatical construct. For instance, Figure 15.7 shows an ORM script in which two entities (warehouse and inventory) and two attributes of these entities (inventory item name and retail price of inventory item) are all represented by named ellipses.

Bunge's theory of ontology sustains a clear distinction between "things" and "properties of things." In the context of TOE, therefore, ORM has construct overload because a single grammatical construct (an ellipse) is used to represent two ontological constructs—namely, classes of things and properties in general of classes of things. As a result, the theory motivates a prediction that users of ORM scripts will become confused when they have to interpret instances of ellipse constructs in the scripts.

Weber (1996) provides evidence to suggest the organization of human memory relies on a distinction being sustained between "things" and "properties of things." He undertook a multi-trial free-recall experiment in which participants where first shown NIAM diagrams (NIAM is an earlier version of ORM). Participants were then asked to recall the diagrams once the diagrams were removed. Across multiple trials of the experiment, participants' recall protocols indicated they had a propensity to recall ellipses that designated "things" first followed by ellipses that designated "properties of things." Moreover, having recalled a "thing" ellipse, they then recalled ellipses that designated "properties of things" in clusters where the clusters related to the thing they had first recalled. In light of his experimental results,

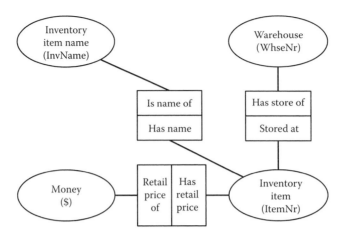

FIGURE 15.7   Object role model of a simple domain.

Weber (1996) concluded users of NIAM diagrams imposed a structure of "things" and "properties of things" on the diagrams even though this structure was not represented explicitly in the diagrams. In short, his results supported the proposition that the object construct in ORM was overloaded because it represented both a class of things and a property in general of a class of things.

In subsequent research, Shanks et al. (2010) conducted an experiment in which the participants were 80 individuals who had work experience but little or no experience with conceptual models. The participants were presented with four different versions of an entity–relationship model of a sales order domain. Based on Bunge's (1977) ontology, the first version used three different grammatical constructs to represent classes of things, mutual properties in general, and intrinsic properties in general. The second version represented classes of things and mutual properties in general via the same grammatical construct (an entity), but it still represented intrinsic properties in general via a different grammatical construct (an attribute). The third version represented classes of things, mutual properties in general, and some intrinsic properties in general (but not all) via the same grammatical construct (an entity). The fourth version represented all classes of things, mutual properties in general, and intrinsic properties in general via a single grammatical construct (an entity). Thus, the four versions of the conceptual model contained increasing levels of construct overload.

Shanks et al. (2010) had four different groups of participants perform three tasks: answer comprehension questions about the domain; solve problems relating to the domain; and check the correspondence of the conceptual model to a textual description of the domain. They found performance differences for the comprehension task but not the problem-solving and discrepancy-testing tasks. In particular, participants who received the conceptual model that showed a clear distinction between classes of things, mutual properties in general, and intrinsic properties in general outperformed participants who received the conceptual model that did not maintain a distinction among these three constructs. Similar results were then obtained in a protocol study that Shanks et al. (2010) undertook with 12 participants, all of whom had at least 3 years' industry experience. Again, the entity construct caused problems for participants when it was overloaded.

In their study of 12 process modeling grammars (see preceding text), Recker et al. (2009) were able to assess the level of construct overload in *seven* of the grammars they examined—namely, EPC, BPML, WSCI, ebXML, WS-BPEL, Petri nets, and BPMN. They measured construct overload by counting the number of grammatical constructs that mapped to more than one BWW construct and dividing this number by the total number of constructs in the grammar. The average measure of construct overload across the seven grammars was 15.04% (the range was 0%–42.90%). In other words, on average, approximately 15% of the constructs in the grammars mapped to more than one ontological construct.

In light of their analysis of construct overload, Recker et al. (2009) predicted users of the grammars they studied would encounter a number of difficulties. For instance, they concluded

- Some grammars achieve higher levels of ontological completeness only by allowing their constructs to be overloaded. Prior research on process models suggests, however, that the benefits of ontological completeness are quickly undermined when the meaning of instances of overloaded constructs is difficult to interpret.
- Use of an overloaded construct often requires process modelers to provide additional explanations to assist readers of process models to interpret the meaning of instances of the construct. As a result, any benefits from allowing more modeling flexibility via overloaded constructs dissipate quickly.

In data obtained from their web-based survey of global BPMN users (see preceding text), Recker et al. (2011) found a negative association between the extent to which these users perceived BPMN had certain kinds of construct overload (e.g., "I often have to provide additional information to clarify the context in which I want to use the Pool symbol in a process model") and the extent to which they perceived BPMN was *easy to use*. They also found these users' perceptions of BPMN's ease of use declined as a function of the number of different types of construct overload they encountered.

In their interviews with 19 business process modeling practitioners in six Australian organizations (see preceding text), Recker et al. (2010) also found the existence of construct overload in BPMN undermined the work on these practitioners. The practitioners confirmed the real-world meaning of the overloaded constructs was unclear. As a result, the overloaded constructs were often used inconsistently in the business process models they encountered.

One way in which stakeholders might seek to cope with problems that arise from construct overload in a conceptual modeling grammar is to create variations of the overloaded construct so that a one-to-one mapping exists between each variation and one of the reference ontological constructs. For instance, instances of the grammatical construct might be annotated, shaded, or colored in some way to indicate the particular ontological construct they represent.

15.3.1.4 Construct Excess

Recall, construct excess arises in a conceptual modeling grammar when a grammatical construct does not map to any construct in the reference ontology. In the presence of construct excess, TOE motivates a prediction that users of the grammar may become confused about the nature and purposes of the excess construct (at least from a conceptual modeling perspective). Moreover, stakeholders who have to interpret scripts containing instances of the excess construct may be confused about the real-world meaning to be ascribed to instances of the construct.

At first glance, the notion of construct excess might seem curious. After all, surely each construct in a conceptual modeling grammar must refer to *something*. As with construct deficit, overload, and redundancy, however, construct excess is determined relative to a particular ontological benchmark. In this light, a grammar may have construct excess according to one ontological benchmark but not another one. For any given benchmark, instances of excess may occur because the ontology is silent on its capacity to represent a particular type of phenomenon or because it expressly states a particular type of phenomenon does not exist in the real world.

Consider the case of optional attributes. Many conceptual modeling grammars have a construct that can be used to show whether an attribute of an entity is mandatory or optional. For instance, the closed (shaded) circle on the arc between the entity "PhD student" and the attribute "advisor's name" in Figure 15.8 indicates all PhD students must have an advisor. The open (unshaded) circle on the arc between the entity PhD student and the attribute "thesis title," however, indicates not all PhD students have finalized a thesis title.

When optional attributes are used in a conceptual model, the meaning intended is that only some instances of things (entities or objects) in a type or class of things possess the attribute. Conversely, some

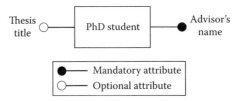

FIGURE 15.8   Entity with mandatory and optional attributes.

instances of things in the type or class possess a "negative" or "null" attribute—an attribute they do *not* have. For instance, in Figure 15.8, some instances of PhD student possess an attribute "does not have a thesis title."

In Bunge's theory of ontology, all properties of things must be mandatory—he brooks no notion of an optional or "negative" property (a property a thing does not possess). Specifically, he argues: "We certainly need negation to understand reality and argue about it or anything else, but external reality wears only positive traits" (Bunge, 1977, p. 60). We might conclude *logically* that something does not possess a property, but this conclusion does not mean the thing possesses the property of not possessing the property. The "negation" falls within the domain of logic and not the domain of ontology.

In the context of TOE, the optional attribute construct constitutes an example of construct excess. It is a grammatical construct that maps to no ontological construct. As a result, based on Bunge's theory of ontology, TOE motivates a prediction that users of scripts containing optional attributes will become confused when they have to interpret the scripts.

At first glance, it may not be clear why this outcome will occur. Problems become apparent, however, when having to reason about a thing that has two or more optional attributes. For instance, in Figure 15.9, two optional attributes of PhD students are "has a confirmed thesis title" and "has passed comprehensive examinations." Consider, now, the following question: Must all PhD students who have a confirmed thesis title also have passed their comprehensive examinations? The information contained in the conceptual model in Figure 15.9 does not allow us to answer this question. Figure 15.9's unclear semantics can be resolved, however, by using a subclass that has mandatory properties only. In this regard, Figure 15.10 shows PhD students who have a confirmed thesis title must also have passed their comprehensive examinations.

In three experiments, Bodart et al. (2001) investigated the impact of using optional attributes on users' understanding of conceptual models. Based on two theories of human memory—semantic network theory (Collins and Quillian, 1969) and spreading activation theory (Anderson and

FIGURE 15.9   Conceptual model with unclear semantics.

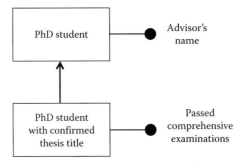

FIGURE 15.10   Conceptual model with clear semantics.

Pirolli, 1984)—Bodart et al. predicted optional attributes would assist users to undertake tasks where only a surface-level understanding of a conceptual model was needed. For instance, conceptual models with optional attributes might be satisfactory for senior management who are interested in obtaining only a broad understanding of a domain. Subclasses that have mandatory attributes only might clarify the semantics of a conceptual model, but this outcome is attained often at the cost of the model being more complex. Where users of conceptual models have to undertake tasks that require a deep-level understanding of a domain, however, Bodart et al. (2001) predicted use of subclasses with mandatory attributes only would be beneficial. For instance, database designers need to have a precise understanding of a domain's semantics when they design the normal-form relations to be used in a database. The experimental results obtained by Bodart et al. (2001) for the most part supported their predictions.

Gemino and Wand (2005) also investigated the impact of using optional versus mandatory attributes in conceptual models on users' understanding of a domain. Based on Mayer's (2001) cognitive theory of multimedia learning, they predicted users who were given a conceptual model that had subclasses with mandatory attributes only would have greater comprehension and understanding of the domain represented by the conceptual model than users who were given a conceptual model that had optional attributes. They found support for their predictions in the results obtained from a laboratory experiment they conducted.

Bowen et al. (2006) studied the impact of using optional versus mandatory attributes in conceptual models on users' performance in formulating SQL queries. On the one hand, they hypothesized use of mandatory attributes only in a conceptual model would assist users to formulate more accurate and more complete SQL queries because the domain's semantics were clearer. On the other hand, they hypothesized use of mandatory attributes only in a conceptual model would undermine users' query performance because they had to deal with a greater number of entities in the conceptual model (because more subclasses were present). In an experimental test of their hypotheses, overall they found users who received conceptual models with mandatory attributes only showed the highest levels of query performance. Nonetheless, detailed analyses of errors that their participants made revealed those who received conceptual models with mandatory attributes only made a greater number of certain types of errors—specifically, those where they had to use JOIN operations to assemble data from different relations (an outcome of the need for more subclasses in conceptual models with mandatory attributes only).

In their study of 12 process modeling grammars (see preceding text), Recker et al. (2009) were able to assess the level of construct excess in *seven* of the grammars they examined—namely, EPC, BPML, WSCI, ebXML, WS-BPEL, Petri nets, and BPMN. They measured construct excess by counting the number of grammatical constructs that did not map to any BWW construct and dividing this number by the total number of constructs in the grammar. The average measure of construct excess across the seven grammars was 22.09% (the range was 0%–42.90%). In other words, on average, approximately 22% of the constructs in the grammars did not represent an ontological construct.

In light of their analysis of construct excess, Recker et al. (2009) predicted users of the grammars they studied would encounter a number of difficulties. For instance, they concluded

- Some instances of construct excess arose because the grammars contained a mixture of conceptual, logical, and physical process modeling constructs. For instance, constructs such as "parameters" and "activity instance state" supported process implementation and execution rather than conceptual modeling of processes in a domain. As a result, the grammars were unnecessarily complex. Moreover, they confused the functions of analysis, design, implementation, and execution in process modeling.
- Some instances of construct excess arose because the grammars contained constructs to support the mechanics or "act" of modeling. For instance, constructs such as "off-page connector" or "text annotation" had the purpose of linking different parts of a model or providing supplementary descriptions for different components of the model. These constructs made the grammars more complex and would be better provided by tools developed to support use of the grammars.

In data obtained from their web-based survey of global BPMN users (see preceding text), Recker et al. (2011) found a negative association between the extent to which these users perceived BPMN had certain kinds of construct excess (e.g., "[t]he Basis Event symbol does not have any real-world meaning in a process model") and the extent to which they perceived BPMN was *easy to use*. They also found these users' perceptions of BPMN's ease of use declined as a function of the number of different types of construct excess the users had encountered.

Based on interview responses obtained from 19 business process modeling practitioners in six Australian organizations (see preceding text), however, Recker et al. (2010) found the impact of construct excess in BPMN on the practitioners' work was somewhat mixed. Certain excess constructs in BPMN were simply ignored by the practitioners, while others were deemed important to the "act of modeling" even though they have no real-world correspondence.

One way in which stakeholders might seek to cope with problems arising from construct excess in a grammar is to proscribe use of the excess constructs for conceptual modeling purposes. In some cases (e.g., optional attributes), the merits of the excess construct for any kind of purpose are debatable. In other cases (e.g., activity instance state), the excess construct might be useful for logical (design) and/or implementation modeling purposes. In these latter cases, proscribing their use during conceptual modeling will force modelers to clearly differentiate conceptual modeling activities from logical (design) and implementation modeling activities. The outcome should be that higher-quality work is done for each of the three kinds of modeling activities.

15.3.2 Conceptual Modeling Practice and the TMGS

Relative to TOE, only a small number of research studies have so far been conducted using TMGS. Moreover, most have used TMGS for analytical purposes (e.g., Green et al., 2007). Few studies have tested their analytical results empirically.

In an extensive study of stakeholders' use of a computer-aided software engineering tool, Green (1996) found the stakeholders acted strategically in the ways they chose from the set of modeling grammars provided in the tool to address problems caused by construct deficit in individual modeling grammars. First, they chose grammars to achieve MOC. Green found that, either consciously or subconsciously the stakeholders selected a set of grammars to try to ensure all constructs in the reference ontology he used (the BWW ontology) were covered by at least one of the grammars. Second, the stakeholders chose grammars to achieve MOO. Either consciously or unconsciously they selected a set of grammars to try to minimize construct redundancy. They seemed to recognize that having redundant grammatical constructs would cause them difficulties.

Green's (1996) research shows the MOC and MOO strategies are a good way for practitioners to deal with problems caused by construct deficit. Practitioners first need to select a reference ontology—one they feel is congruent with the ways they would like to model domains of interest to them. They then need to select a set of conceptual modeling grammars that appear capable of modeling these domains. Next, they need to analyze the ontological expressiveness of these grammars and examine alternative combinations of them to see how well they meet the MOC and MOO objectives. To achieve parsimony, they should choose the combination that includes the smallest number of grammars that achieves the MOC and MOO objectives.

15.4 Research Issues

The use of ontology as a theoretical underpinning for conceptual modeling phenomena is still a new field of research. Much work remains to be done before the merits of this approach as a means of improving our knowledge of conceptual modeling practice can be better understood. In this section, therefore, we briefly discuss some research opportunities that might be pursued.

15.4.1 Refining the TOE

The TOE (Wand and Weber, 1993) needs to be refined if its explanatory and predictive powers are to be improved. Several types of refinement might be undertaken:

- TOE motivates a prediction that construct deficit, redundancy, overload, and excess undermine stakeholders' ability to effectively and efficiently use conceptual modeling grammars and the scripts generated via these grammars. The ways in which stakeholders' performance will be undermined, however, are unclear. A taxonomy of stakeholders and the tasks they perform through engaging with conceptual modeling grammars and scripts needs to be articulated. The impact of grammar and scripts that lack ontological expressiveness on the different types of stakeholders and the different tasks they perform then needs to be enunciated. As noted earlier, for example, Bodart et al. (2001) predicted construct excess would be more problematical when readers had to engage in tasks that required them to have a deep rather than a shallow understanding of scripts. Depth of understanding required is just one way to differentiate tasks, however, and a more detailed taxonomy would assist future research.
- The nature of the associations between the four types of factors that undermine ontological expressiveness (construct deficit, redundancy, overload, and excess) and different outcome factors needs to be articulated more precisely. Currently, TOE simply indicates a negative effect on stakeholder performance will occur. A refined form of TOE would indicate the functional form of this effect. For instance, does the impact of increasing levels of construct overload have a negative exponential effect on stakeholder performance (i.e., increasing sharply and then leveling off)? Or, does the impact of increasing levels of construct overload follow a sigmoid shape (i.e., initially a slow increase, then a sharp increase, then a leveling off)? Stronger tests of theories are possible when the functional form of associations between their constructs are specified more precisely (Edwards and Berry, 2010; Weber, 2012).
- The relative strength of the effects of construct deficit, redundancy, overload, and excess on different types of stakeholders and different types of tasks has not been articulated. Moreover, the ways in which these factors interact is unknown. For instance, do higher levels of construct excess exacerbate the effects of construct overload because instances of excess constructs that seemingly have no real-world domain meaning compound the ambiguity surrounding instances of overloaded constructs in a script?

To some extent, refinements of TOE depend upon obtaining more empirical evidence from tests of TOE. For instance, evidence from such tests might indicate the relative weightings that should be assigned to different factors that undermine ontological expressiveness in terms of their impact on users of modeling grammars and the tasks in which they engage.

15.4.2 Refining the TMGS

Several types of refinement might be undertaken to the TMGS (Green, 1996):

- TMGS motivates a prediction that a higher level of ontological completeness among a set of conceptual modeling grammars is associated with a higher likelihood that the grammars will be used together for modeling purposes. The level of ontological completeness is measured in terms of the number of constructs in the reference ontology covered by constructs in the modeling grammars. For particular stakeholders and particular tasks, however, certain ontological constructs and thus certain grammatical constructs may be more important than others. A more refined measure of ontological completeness would therefore involve more than a simple count of the number of ontological constructs covered by the modeling grammars. Instead, it would incorporate a weighting of individual ontological constructs to reflect their relative importance for different stakeholders and different tasks.

- TMGS motivates a prediction that a higher level of ontological overlap among a set of conceptual modeling grammars will be associated with a lower likelihood that the grammars will be used together. The level of ontological overlap is measured in terms of the number of constructs in the reference ontology covered by more than one construct in the modeling grammars. As with the level of ontological completeness among a set of grammars, however, certain kinds of construct overlap may undermine stakeholder performance in different ways (perhaps as a function of the type of stakeholder and the type of task). A more refined measure of ontological overlap would take into account that certain kinds of overlap are more problematic than others.
- As TMGS currently stands, the likelihood of certain grammars being selected together for modeling purposes is not impacted by an interaction effect between the grammars' level of ontological completeness and level of ontological overlap. Such an interaction effect might be important, however, in the selection of grammars to use together. For instance, if ontological completeness can be achieved only with more ontological overlap, the benefits of ontological completeness may be diminished.

As with TOE, to some extent, refinements to TMGS depend on obtaining further empirical evidence from tests of TMGS. For instance, the results of such tests might provide insights into the relative weightings that should be assigned to different types of ontological completeness, and overlap in terms of their impact on the likelihood grammars will be used together for modeling purposes.

15.4.3 Testing Alternative Ontologies

Recall, TOE and TMGS are not tied to a particular theory of ontology. Nonetheless, as we have discussed earlier, many scholars have selected the BWW ontology as the reference ontology because it has been articulated rigorously and it contains constructs that seem especially relevant to the information technology field. Moreover, the BWW ontology has provided the basis for a number of predictions about conceptual modeling practice that have been supported empirically.

Nonetheless, some scholars believe that the BWW ontology provides a flawed foundation for evaluating conceptual modeling grammars and scripts (e.g., Guizzardi et al., 2006; Wyssusek, 2006). Unfortunately, while criticisms have been made of the BWW ontology, little work has been done to articulate alternative reference ontologies, adapt these ontologies to the information technology domain, evaluate the ontological expressiveness of conceptual modeling grammars using these ontologies, and empirically test the conclusions generated on the basis of such evaluations. Until such work is undertaken, the merits of alternative reference ontologies remain uncertain.

Comparative *analytic* evaluations of alternative ontologies (e.g., Chisholm, 1996) that could be used as a reference ontology for TOE or TMGS are likely to have merits. We believe, however, that better research outcomes will occur if predictions based upon the alternatives chosen are made and tested *empirically*. In particular, the most interesting predictions are likely to be those that *contradict* each other. By testing competing predictions empirically, the relative merits of alternative reference ontologies will become more apparent.

15.4.4 Improving the Design and Execution of Empirical Tests

Historically, for two reasons, undertaking valid and reliable empirical tests of conceptual modeling grammars and scripts has proved challenging. First, it is often difficult to mitigate threats to construct validity, internal validity, and external validity (Burton-Jones et al., 2009). Second, as discussed earlier, the theory that underpins the tests is imprecise and thus the ways the tests should be designed and the constructs measured are unclear.

To illustrate some of the difficulties encountered, consider Figures 15.5 and 15.6, which show two outcome constructs: user task effectiveness and efficiency and the likelihood of grammars being used together. It is unlikely that these two constructs cover the full set of outcomes that might arise from having greater ontological expressiveness in a single grammar or combination of grammars. For instance, in a group setting, one outcome of interest might be the development of a shared understanding among users of a grammar (or set of grammars). Intuitively, clear conceptual models should help users reach a shared understanding. To the best of our knowledge, however, this outcome construct—shared understanding—has never been examined.

Even if we focus only on the outcome constructs shown in Figures 15.5 and 15.6, the best way to measure them is unclear. For instance, as a measure of task effectiveness and efficiency, researchers have used recall time, recall accuracy, comprehension scores, problem-solving scores, scores on fill-in-the-blank tests, cognitive breakdowns, and perceptual variables indicating users' ease of understanding conceptual models. The results have varied considerably depending on the particular measure chosen (Bodart et al., 2001; Burton-Jones and Meso, 2008; Gemino, 1998). More research is needed, therefore, to discern the best measures to use.

15.4.5 Action Research to Assist the Designers of Conceptual Modeling Grammars

There is little evidence to suggest the design of conceptual modeling grammars is informed by high-quality theory. Rather, the design of many current grammars seems to have been based on earlier grammars that were developed, perceptions about stakeholders' experiences with these grammars, the beliefs of influential individuals about appropriate features for a grammar, and the deliberations of and compromises reached by committees (Siau and Rossi, 2011).

Through action research (e.g., Davison et al., 2004), TOE could be used as the basis for designing new conceptual modeling grammars or refining and enhancing existing grammars. For new grammars, a set of constructs should be chosen so a one-to-one mapping exists between each construct and each construct in the reference ontology. For existing grammars, their constructs should be modified and/or enhanced to achieve a one-to-one mapping with ontological constructs. Moreover, for existing grammars, through the identification of construct excess (grammatical constructs for which no corresponding ontological construct exists), TOE may help clarify which, if any, grammatical constructs are focused on logical (design) modeling or implementation (physical) modeling. The grammatical constructs can then be partitioned and advice provided to users of the grammar about which constructs to employ for conceptual modeling and which constructs to employ for logical and implementation modeling. Such work hopefully would lead to better conceptual modeling practice as well as refinement and enhancement of TOE.

15.4.6 Action Research to Assist the Users of Conceptual Modeling Grammars

Through action research, TOE and TMGS could be used as the basis for assisting stakeholders to select conceptual modeling grammars best suited to their needs. Stakeholders first must choose a reference ontology. They then must identify which ontological constructs are important for their modeling needs (possibly all will be deemed necessary). Using TOE and the ontological constructs they have chosen, they can then evaluate the relative levels of ontological expressiveness possessed by the grammars they are considering. If the goal is to select only a single grammar for use, stakeholders will have to compare the costs of construct deficit, redundancy, overload, and excess among the grammars they are evaluating. If stakeholders are willing to engage with multiple grammars, a choice might be made by seeking to achieve the goals of MOC and MOO.

TOE might also be used as a basis for deciding how a conceptual modeling grammar (or grammars) should be used. For instance, in communicating the semantics of a domain to senior managers, some ontological constructs might not be deemed important. Senior managers might be interested in the major things (objects) in a domain and how they are related, but they might not be interested in the intrinsic properties things possess or the nature of the state spaces or event spaces associated with things. An organization might put in place a set of standards and protocols that dictate what grammatical constructs will be used to generate scripts for particular stakeholders and particular tasks.

Where the conceptual modeling grammar used has construct redundancy, overload, or excess, an organization might also put in place a set of standards and protocols to mitigate any detrimental effects that might arise. For instance,

- In the case of construct redundancy, a standard might require that only one of the redundant grammatical constructs should ever be used to represent the reference ontological construct.
- In the case of construct overload, several versions of the grammatical construct might be created and somehow differentiated from each other so that a one–one mapping is achieved between the different versions of the grammatical construct and the reference ontological constructs.
- In the case of construct excess, a standard might require the excess constructs never be used to generate conceptual modeling scripts.

Via action research, stakeholder's experiences in using both TOE and TMGS can be documented and evaluated. Hopefully, stakeholder's performance in using conceptual modeling grammars and scripts can be improved and TOE and TMGS can be refined and enhanced.

One promising basis for future action research relates to the development of modeling rules or guidelines. The basic idea is to encapsulate the implications of theories such as TOE and TMGS in concrete rules or guidelines, whether for modeling domains in general (e.g., Evermann and Wand, 2005a) or modeling domains to support specific tasks (e.g., Bera et al., 2011). These rules or guidelines might then be incorporated in software to help analysts follow them more easily and accurately (Wand et al., 2008). Bera et al. (2011, p. 904) explain the logic underlying this stream of work:

> …the domain [should be] presented in a way that fits the task. This implies that aspects of the domain that are most relevant to the task…should be made salient in the representation. … This is where principles from philosophical ontology can provide guidance as to how the formal modeling grammar can be used to make the relevant domain aspects more salient in the script. We refer to this guidance…as "representational rules." In short, the approach we have followed is a method for creating representations that have ontologically informed fit for a given task.

Action researchers could test the benefit of these modeling rules and guidelines by first training people to use them or training them to use the software that implements them and then observing whether they improve modeling outcomes.

15.5 Summary

In this chapter, we have examined some ways in which the philosophical field of ontology has informed practice and research on conceptual modeling. Specifically, we have discussed how two theories that rely on ontological foundations have been used to predict, explain, and understand conceptual modeling phenomena. The first is the TOE, which allows the ontological completeness and clarity of conceptual modeling grammars to be evaluated. The second is the TMGS, which assists stakeholders to decide which combinations of conceptual modeling grammars they should use to undertake conceptual modeling work. We have examined the nature of these theories, some ways in which they have been used to generate recommendations for conceptual modeling practice, and some examples of research that has been undertaken to test predictions about conceptual modeling phenomena they have motivated.

Studying the ways in which ontological theories can be used to inform conceptual modeling practice and research is a relatively new field of endeavor. Much theoretical and empirical work remains to be done. For instance, existing theories that rely on ontological theories, such as TOE and TMGS, need refinement and enhancement. Moreover, additional empirical research needs to be undertaken to evaluate these theories. Based on work done already, it is also clear that improved measures and procedures are needed to increase the validity and reliability of empirical work. Opportunities also exist to develop new, innovative ways in which ontological theories can be used to study conceptual modeling phenomena.

Glossary

Conceptual model: A representation, often graphical, of some features of a real-world domain; assists with the design, implementation, maintenance, and use of an information system.

Conceptual modeling grammar: Comprises a set of constructs to represent specific types of phenomena in a domain and a set of rules for combining the constructs to show how phenomena in the domain are related.

Conceptual modeling script: A string of instances of constructs in a conceptual modeling grammar that provides a representation of a specific domain.

Construct deficit: An attribute of a conceptual modeling grammar that arises when a construct in a theory of ontology cannot be represented by a construct in the conceptual modeling grammar.

Construct excess: An attribute of a conceptual modeling grammar that arises when a construct in the conceptual modeling grammar has no counterpart in a theory of ontology.

Construct overload: An attribute of a conceptual modeling grammar that arises when a construct in the conceptual modeling grammar is used to represent two or more constructs in a theory of ontology.

Construct redundancy: An attribute of a conceptual modeling grammar that arises when a construct in a theory of ontology can be represented by two or more constructs in the conceptual modeling grammar.

Grammatical construct: A component in a conceptual modeling grammar that is used to represent a specific type of phenomenon in the real world.

Ontology: A theory about the nature of and types of phenomena that exist in the real world.

Ontologically clear: An attribute of a conceptual modeling grammar that arises when the grammar has no instances of construct redundancy, overload, and excess.

Ontologically complete: An attribute of a conceptual modeling grammar that arises when the grammar has no instances of construct deficit.

Ontological construct: A component in an ontological theory that is used to represent a specific type of phenomenon in the real world.

Ontological coverage: The extent to which two or more conceptual modeling grammars have constructs that cover all constructs in a theory of ontology; analogous to ontological completeness for a single conceptual modeling grammar.

Ontologically expressive: An attribute of a conceptual modeling grammar that describes the extent to which the grammar represents the constructs in a theory of ontology in a complete and clear way.

Ontological overlap: The extent to which two or more conceptual modeling grammars have constructs that represent the same construct in a theory of ontology; analogous to construct redundancy for a single conceptual modeling grammar.

Acknowledgment

The preparation of this paper was supported, in part, by Australian Research Council Discovery Grant DP110104386 to both authors.

References

Anderson, J.R. and Pirolli, P.L. (1984). Spread of activation. *Journal of Experimental Psychology: Learning, Memory, and Cognition*, 10, 791–798.

Bera, P., Burton-Jones, A., and Wand, Y. (2011). Guidelines for designing visual ontologies to support knowledge identification. *MIS Quarterly*, 25(4), 883–908.

Bodart, F., Patel, A., Sim, M., and Weber, R. (2001). Should optional properties be used in conceptual modeling? A theory and three empirical tests. *Information Systems Research*, 12(4), 384–405.

Bowen, P. L., O'Farrell, R. A., and Rhode, F. L. (2006). Analysis of competing data structures: Does ontological clarity produce better end user query performance. *Journal of the Association for Information Systems*, 7(8), 514–544.

Bunge, M. (1977). *Treatise on Basic Philosophy: Volume 3: Ontology I: The Furniture of the World*. Dordrecht, the Netherlands: D. Reidel Publishing Company.

Bunge, M. (1979). *Treatise on Basic Philosophy: Volume 4: Ontology II: A World of Systems*. Dordrecht, the Netherlands: D. Reidel Publishing Company.

Burton-Jones, A. and Meso, P. (2006). Conceptualizing systems for understanding: An empirical test of decomposition principles in object-oriented analysis. *Information Systems Research*, 17(1), 38–60.

Burton-Jones, A. and Meso, P. (2008). The effects of decomposition quality and multiple forms of information on novices' understanding of a domain from a conceptual model. *Journal of the Association for Information Systems*, 9(12), 748–802.

Burton-Jones, A., Wand, Y., and Weber, R. (2009). Guidelines for empirical evaluations of conceptual modeling grammars. *Journal of the Association for Information Systems*, 10(6), 495–532.

Chen, P. P.-S. (1976). The entity-relationship model—Toward a unified view of data. *ACM Transactions on Database Systems*, 1(1), 9–36.

Chisholm, R. M. (1996). *A Realistic Theory of Categories: An Essay on Ontology*. Cambridge, U.K.: Cambridge University Press.

Cocchiarella, N. B. (1972). Properties as individuals in formal ontology. *Noûs*, 6(3), 165–187.

Collins, A. M. and Quillian, M. R. (1969). Retrieval times from semantic memory. *Journal of Verbal Learning and Verbal Behavior*, 8, 240–247.

Davison, R. M., Martinsons, M. G., and Kock, N. (2004). Principles of canonical action research. *Information Systems Journal*, 14(1), 65–86.

Edwards, J. R. and Berry, J. W. (2010). The presence of something or the absence of nothing: Increasing theoretical precision in management research. *Organizational Research Methods*, 13(4), 668–689.

Evermann, J. and Wand, Y. (2005a). Ontology based object-oriented domain modelling: Fundamental concepts. *Requirements Engineering*, 10(2), 146–150.

Evermann, J. and Wand, Y. (2005b). Toward formalizing domain modeling semantics in language syntax. *IEEE Transactions on Software Engineering*, 31(1), 21–37.

Fettke, P. (2009). How conceptual modeling is used. *Communications of the Association for Information Systems*, 25, 571–592.

Fettke, P. and Loos, P. (2007). Ontological evaluation of Scheer's reference model for production planning and control systems. *International Journal of Interoperability in Business Information Systems*, 2(1), 9–28.

Floyd, C. (1986). A comparative analysis of system development methods. In: Olle, T. W., Sol, H. G., and Verrijn-Stuart, A. A. (Eds.), *Information System Design Methodologies: Improving the Practice* (pp. 19–54). Amsterdam, the Netherlands: Elseveier Science Publishers, North-Holland.

Gemino, A. (1998). To be or may to be: An empirical comparison of mandatory and optional properties in conceptual modeling. Paper presented at the *Annual Conference of the Administrative Sciences Associaton of Canada*, Information Systems Division, Saskatoon, Saskatchewan, Canada.

Gemino, A. and Wand, Y. (2005). Complexity and clarity in conceptual modeling: Comparison of mandatory and optional properties. *Data & Knowledge Engineering, 55*(3), 301–326.

Green, P. (1996). An ontological analysis of ISAD grammars in upper CASE tools. PhD, The University of Queensland, Brisbane, Queensland, Australia.

Green, P., Rosemann, M., Indulska, M., and Manning, C. (2007). Candidate interoperability standards: An ontological overlap analysis. *Data & Knowledge Engineering, 62*(2), 274–291.

Guizzardi, G., Masolo, C., and Borgo, S. (2006). In defense of a trope-based ontology for conceptual modeling: An example with the foundations of attributes, weak entities and datatypes. Paper presented at the *Conceptual Modeling—ER 2006: 25th International Conference on Conceptual Modeling*, Tuscon, AZ. 10.1007/11901181.

Halpin, T. A. (2008). *Information Modeling and Relational Databases*, 2nd edn. San Francisco, CA: Morgan Kaufman.

Lewin, K. (1945). The research center for group dynamics at Massachusetts Institute of Technology. *Sociometry, 8*, 126–135.

Mayer, R. E. (2001). *Multimedia Learning*. Cambridge, MA: Cambridge University Press.

Moody, D. L. (2005). Theoretical and practical issues in evaluating the quality of conceptual models: Current state and future directions. *Data & Knowledge Engineering, 55*, 243–276.

Moody, D. L. (2009). The "physics" of notations: Toward a scientific basis for constructing visual notations in software engineering. *IEEE Transactions on Software Engineering, 35*(6), 756–779.

zur Muehlen, M. and Indulska, M. (2010). Modeling languages for business processes and business rules: A representational analysis. *Information Systems, 35*(4), 379–390.

zur Muehlen, M. and Recker, J. (2008). How much language is enough? Theoretical and practical use of the Business Process Modeling Notation. Paper presented at the *Advanced Information Systems Engineering—CAiSE 2008*, Montpellier, France.

Oei, J. L. H., van Hemmen, L. J. G. T., Falkenberg, E. H., and Brinkkemper, S. (1992). The meta model hierarchy: A framework for information systems concepts and techniques. Technical report 92–17. Department of Information Systems, University of Nijmegen, Nijmegen, the Netherlands.

Opdahl, A. L. and Henderson-Sellers, B. (2004). A template for defining enterprise modelling constructs. *Journal of Database Management, 15*(2), 39–73.

Recker, J., Indulska, M., Rosemann, M., and Green, P. (2010). The ontological deficiencies of process modeling in practice. *European Journal of Information Systems, 19*(5), 501–525.

Recker, J., Rosemann, M., Green, P., and Indulska, M. (2011). Do ontological deficiencies in modeling grammars matter? *MIS Quarterly, 35*(1), 57–79.

Recker, J., Rosemann, M., Indulska, M., and Green, P. (2009). Business process modeling—A comparative analysis. *Journal of the Association for Information Systems, 10*(4), 333–363.

Rumbaugh, J., Jacobson, I., and Booch, G. (1999). *The Unified Modeling Language Reference Manual*. Reading, MA: Addison-Wesley.

Shanks, G., Moody, D. L., Nuridini, J., Tobin, D., and Weber, R. (2010). Representing classes of things and properties in general in conceptual modelling: An empirical evaluation. *Journal of Database Management, 21*(2), 1–25.

Siau, K. and Rossi, M. (2011). Evaluation techniques for systems analysis and design modelling methods—A review and comparative analysis. *Information Systems Journal, 21*, 249–268.

Van de Ven, A. H. (1989). Nothing is quite so practical as a good theory. *Academy of Management Review, 14*(4), 486–489.

Wand, Y. and Weber, R. (1990). An ontological model of an information system. *IEEE Transactions on Software Engineering, 16*(11), 1282–1292.

Wand, Y. and Weber, R. (1993). On the ontological expressiveness of information systems analysis and design grammars. *Journal of Information Systems, 3*(4), 217–237.

Wand, Y. and Weber, R. (1995). On the deep structure of information systems. *Information Systems Journal, 5*, 203–223.

Wand, Y., Woo, C., and Wand, O. (2008). Role and request based conceptual modeling—A methodology and a CASE tool. Paper presented at the *Proceedings of the CAiSE'08 Forum*, Montpellier, France.

Weber, R. (1996). Are attributes entities? A study of database designers' memory structures. *Information Systems Research*, *7*(2), 137–162.

Weber, R. (2012). Evaluating and developing theories in the information systems discipline. *Journal of the Association for Information Systems*, *13*(1), 323–346.

Weber, R. and Zhang, Y. (1996). An analytical evaluation of NIAM's grammar for conceptual schema diagrams. *Information Systems Journal*, *6*(2), 147–170.

Wyssusek, B. (2006). On ontological foundations of conceptual modelling. *Scandinavian Journal of Information Systems*, *18*(1), 63–80.

16

Data and Information Quality Research: Its Evolution and Future

Hongwei Zhu
University of Massachusetts

Stuart E. Madnick
Massachusetts Institute of Technology

Yang W. Lee
Northeastern University

Richard Y. Wang
Massachusetts Institute of Technology

16.1 Introduction

Organizations have increasingly invested in technology and human resources to collect, store, and process vast quantities of data. Even so, they often find themselves stymied in their efforts to translate this data into meaningful insights that they can use to improve business processes, make smart decisions, and create strategic advantages. Issues surrounding the quality of data and information that cause these difficulties range in nature from the technical (e.g., integration of data from disparate sources) to the nontechnical (e.g., lack of a cohesive strategy across an organization ensuring the right stakeholders have the right information in the right format at the right place and time).

Although there has been no consensus about the distinction between data quality and information quality, there is a tendency to use *data quality* (DQ) to refer to technical issues and *information quality* (IQ) to refer to nontechnical issues. In this chapter, we do not make such distinction and use the term *data quality* to refer to the full range of issues. More importantly, we advocate interdisciplinary approaches to conducting research in this area. This interdisciplinary nature of research demands that we integrate and introduce research results, regardless of technical and nontechnical in terms of the nature of the inquiry and focus of research. In fact, much of the research introduced here has a mixed focus, just as problems of data and information quality reveal themselves.

As a focused and established area of research, data and information quality began to attract the research community's attention in the late 1980s. To address data quality concerns, researchers at Massachusetts Institute of Technology (MIT) started investigating issues such as inter-database instance identification (Wang and Madnick 1989) and data source tagging (Wang and Madnick 1990). In 1992, the MIT's Total Data Quality Management (TDQM) program was formally launched to underscore data quality as a research area (Madnick and Wang 1992). The pioneering work at the MIT TDQM program and later MIT Information Quality (MITIQ) program laid a foundation for data quality research and attracted a growing number of researchers to conduct cutting-edge research in this emerging field. The substantial output of this research community has been a primary driver for the creation of related conferences, workshops, and a journal dedicated to the data and information quality, the *ACM Journal of Data and Information Quality (JDIQ)* (Madnick and Lee 2009a,b, 2010a,b).

This chapter provides an overview of the current landscape of data quality research and discusses key challenges in the field today. We do not attempt to provide a comprehensive review of all—or even most—prior work in the field. Instead, for each topic and method, we introduce representative works to illustrate the range of issues addressed and methods used in data quality research. The cited works also serve as pointers for interested researchers to other relevant sources in the literature.

The rest of the chapter is organized as follows. In Section 16.2, we briefly review some of the pioneering work in data quality. In Section 16.3, we present a framework for characterizing data quality research. In Section 16.4, we describe various research topics in data quality research and cite a sample of works to exemplify issues addressed. In Section 16.5, we review data quality research methods and show how they have been used to address a variety of data quality issues. Finally, in Section 16.6, we conclude the chapter with a brief discussion on the challenges that lie ahead in data quality research.

16.2 Evolution of Data Quality Research

Early data quality research focused on developing techniques for querying multiple data sources and building large data warehouses. The work of Wang and Madnick (1989) used a systematic approach to study related data quality concerns. Their research identified and addressed entity resolution issues that arose when integrating information from multiple sources with overlapping records. These researchers explored ways to determine whether separate records actually corresponded to the same entity. This issue has become known by terms such as *record linkage*, *record matching*, and more broadly, *data integration* and *information integration*.

Later, Wang and Madnick (1990) developed a *polygen* (*poly* for multiple and *gen* for source) model to consider the processing of data source tags in the query processor so it could answer data quality-related questions such as "Where is this data from?" and "Which intermediary data sources were used to arrive at this data?" Follow-up research included the development of a modeling method (known as the *quality entity–relationship* (ER) model) to systematically capture comprehensive data quality criteria as metadata at the conceptual database design phase (Wang et al. 1993; Storey and Wang 1998) and used an extended relational algebra to allow the query processor to process hierarchical data quality metadata (Wang et al. 1995a). This stream of research has led to impacts on modern database research and design such as data provenance and data lineage (Buneman et al. 2001) and other extensions to relational algebra for data security and data privacy management. More importantly, these early research efforts motivated researchers to embark on the systematic inquiry of the whole spectrum of data quality issues, which in turn led to the inauguration of the MIT TDQM program in the early 1990s and later the creation of the MITIQ.

16.2.1 TDQM Framework as the Foundation

Early research at the TDQM program developed the TDQM framework, which advocates continuous data quality improvement by following the cycles of *define, measure, analyze,* and *improve* (Madnick and Wang 1992). The framework extends and adapts the total quality management framework for quality improvement in the manufacturing domain (Deming 1982; Juran and Godfrey 1999) to the domain of data. A key insight was that although data is, in fact, a product (or by-product) *manufactured* by most organizations, it was not treated nor studied as such. Subsequent research developed theories, methods, and techniques for the four components of the TDQM framework, which we briefly describe next.

Define. A major breakthrough was to define data quality from the consumer's point of view in terms of *fitness for use* and to identify dimensions of data quality according to that definition via a systematic multistage survey study (Wang and Strong 1996). Prior to this research, data quality had been characterized by attributes identified via intuition and selected unsystematically by individual researchers. Key data quality dimensions were uncovered using a factor analysis on more than 100 data quality attributes identified systematically by the empirical survey study. These dimensions have been organized into four data quality categories: intrinsic, contextual, representational, and accessibility. Intrinsic DQ (e.g., accuracy and objectivity) denotes that data have quality in their own right. Contextual DQ (e.g., timeliness and completeness) emphasizes that data quality must be considered within the context of the task at hand. Both representational DQ (e.g., interpretability) and accessibility (e.g., access security) concern the role of systems and tools that enable and facilitate the interactions between users (including user applications) and data (Wang and Strong 1996).

Measure. A comprehensive data quality assessment instrument was developed for use in research as well as in practice to measure data quality in organizations (Lee et al. 2002). The instrument operationalizes each dimension into four to five measurable items, and appropriate functional forms are applied to these items to score each dimension (Pipino et al. 2002). The instrument can be adapted to specific organizational needs.

Analyze. This step interprets measurement results. Gap analysis techniques (Lee et al. 2002) reveal perceptual and experiential differences between data dimensions and data roles about the quality of data (Strong et al. 1997). The three major roles in and across most organizations are: data collectors, data custodians, and data consumers (Lee and Strong 2004). The knowledge held by different data roles reveals the different aspects and levels of knowledge held by different groups in and across organizations roles (Lee and Strong 2004). Analysis also identifies the dimensions that most need improvement and root causes of data quality problems.

Improve. In this step, actions are taken either to change data values directly or, often more suitably, to change processes that produce the data. The latter approach is more effective as discussed in Ballou et al. (1998) and Wang et al. (1998) where steps toward managing information as a product are provided. In addition, technologies mentioned earlier such as polygen and quality ER model (Storey and Wang 1998) can be applied as part of the continuous improvement process. When data quality software tools are embedded in the business processes, the strength and the limitation of the tools and the business processes need to be clarified (Lee et al. 2002). Experienced practitioners in organizations solve data quality problems by reflecting on and explicating knowledge about contexts embedded in, or missing from, data. These practitioners break old rules and revise actionable dominant logic embedded in work routines as a strategy for crafting rules in data quality problem solving (Lee 2004).

16.2.2 Establishment and Growth of the Field and Profession

In addition to developing and enriching the TDQM framework, the TDQM program and MITIQ program made significant efforts toward solidifying the field of data quality, broadening the impact of research, and promoting university–industry–government collaborations via publications, seminars, training courses, and the annual International Conference on Information Quality (ICIQ) started in 1996. With the help of the TDQM and the MITIQ programs, the University of Arkansas at Little Rock has established the first-of-its-kind Master's and PhD data quality degree programs in the United States to meet the increasing demand for well-trained data quality professionals and to prepare students for advanced data quality research (Lee et al. 2007b).

Today, data quality research is pursued by an ever-widening community of researchers across the globe. In addition to ICIQ, other professional organizations have organized focused workshops on various areas within the field of data quality (e.g., SIGMOD Workshop on Information Quality in Information Systems, CAiSE Workshop on Information Quality, and SIGIQ Workshop on Information Quality). On the industry side of the data quality field, major software vendors have begun to implement data quality technologies in their product and service offerings. In government, data quality has become an important component in many e-government and enterprise architecture (EA) initiatives (OMB 2007). In the private sector, organizations have adopted variations on the TDQM methodology. An increasing number of companies have appointed a chief data officer (CDO) or senior executives with responsibilities similar to the CDO to oversee data production processes and manage data improvement initiatives (Lee et al. 2012). Some groups have started to use the title information strategists to signify that data quality has critical and compelling applicability for an organization's strategies.

In the meantime, data quality research faces new challenges that arise from ever-changing business environments, regulatory requirements, increasing varieties of data forms/media, and Internet technologies that fundamentally impact how information is generated, stored, manipulated, and consumed. Data quality research that started two decades ago has entered a new era where a growing number of researchers actively enhance the understanding of data quality problems and develop solutions to emerging data quality issues.

16.3 Framework for Characterizing Data Quality Research

An early framework for characterizing data quality research was presented in Wang et al. (1995b). It was adapted from ISO 9000 based on an analogy between physical products and data products. The framework consisted of seven elements that impact data quality: (1) management responsibilities, (2) operation and assurance costs, (3) research and development, (4) production, (5) distribution, (6) personnel management, and (7) legal function. Data quality research in 123 publications up to 1994 was analyzed using this framework. Although the framework was comprehensive, it lacked a set of intuitive terms for characterizing data quality research and thus was not easy to use. Furthermore, the seven elements do not provide sufficient granularity for characterization purposes.

To help structure our overview of the landscape of data quality research, we have developed a framework that is easier to use. We took a pragmatic approach to develop this framework based on two principles. First, the types of research topics continue to evolve. Instead of developing distinct or orthogonal categories, we selected and combined commonly known categories from various research communities to encourage multidisciplinary research methods. Second, research methods known and used by researchers have evolved over time and continue to be used in different disciplinary areas. Some methods overlap with others, but the methodological nomenclature offers a cue for researchers in corresponding research areas. Thus the framework has two dimensions, *topics* and *methods*, and is derived from a simple idea: Any data quality research project addresses certain issues (i.e., topics) using certain research methods. For each dimension, we have chosen a small set of terms (i.e., keywords) that have intuitive meanings and should encompass all possible characteristics along the dimension. These keywords are listed in Table 16.1, and their detailed explanations are provided in the next two sections. These topic and method keywords are also used to categorize papers submitted for publication in the *ACM JDIQ*.

For ease of use, we have chosen intuitive and commonly used keywords, such as *organizational change* and *data integration* for the topics dimension and *case study* and *econometrics* for the methods dimension. The methods and the topics are not necessarily orthogonal. We have grouped the topics into

TABLE 16.1 Topics and Methods of Data Quality Research

Topics		Methods
1. Data quality impact	1.	Action research
1.1 Application area (e.g., CRM, KM, SCM, and ERP)	2.	Artificial intelligence
1.2 Performance, cost/benefit, and operations	3.	Case study
1.3 IT management	4.	Data mining
1.4 Organization change and processes	5.	Design science
1.5 Strategy and policy	6.	Econometrics
2. Database-related technical solutions for data quality	7.	Empirical
2.1 Data integration and data warehouse	8.	Experimental
2.2 Enterprise architecture and conceptual modeling	9.	Mathematical modeling
2.3 Entity resolution, record linkage, and corporate householding	10.	Qualitative
2.4 Monitoring and cleansing	11.	Quantitative
2.5 Lineage, provenance, and source tagging	12.	Statistical analysis
2.6 Uncertainty (e.g., imprecise and fuzzy data)	13.	System design and implementation
3. Data quality in the context of computer science and IT	14.	Survey
3.1 Measurement and assessment	15.	Theory and formal proofs
3.2 Information systems		
3.3 Networks		
3.4 Privacy		
3.5 Protocols and standards		
3.6 Security		
4. Data quality in curation		

four major categories. For the research methods, which are listed in alphabetical order, we have included terms with varying levels of specificity. For example, *econometrics* is more specific than *quantitative* method. This framework gives users the flexibility to choose a preferred level of specificity in characterization based on the tradition used in one's disciplinary background. When using the framework to characterize a particular piece of research, the researcher would choose one or more keywords from each dimension. For example, the paper "AIMQ: a methodology for information quality assessment" (Lee et al. 2002) addresses the *measurement* and *assessment* topic and uses a particular *qualitative* method (i.e., field study interviews and survey) along with a *quantitative* method (i.e., statistical analysis of data and analysis instrument).

We can also view the framework as a 2D matrix where each cell represents a topic–method combination. We can place a research paper in a particular cell according to the topic addressed and the method used. It is possible to place one paper in multiple cells if the paper addresses more than one issue and/or uses more than one method. A paper that uses a more specific method can also be placed in the cell that corresponds to a more general method. Obviously, some cells may be empty or sparsely populated, such as cells corresponding to certain combinations of technical topics (e.g., data integration) and social science methods (e.g., action research). Researchers are encouraged to consider employing more than one research method, including one or more quantitative methods with one or more qualitative methods.

In the next two sections, we use the framework to describe the landscape of data quality research. We also provide descriptions of keywords and illustrate their uses by citing relevant literature.

16.4 Research Topics

Data quality is an interdisciplinary field. Existing research results show that researchers are primarily operating in two major disciplines: management information systems and computer science (CS). We encourage researchers in other areas also to engage in data quality research, and we encourage researchers in one field to borrow theoretical and methodological traditions from other disciplines as well. As a result of its interdisciplinary nature, data quality research covers a wide range of topics. In the subsequent sections, we provide a categorization scheme of data quality research topics. The scheme is broad enough to encompass topics addressed in existing research and those to be explored in future research. The scheme includes four major categories, each having a number of subcategories. A particular research activity can be categorized into multiple categories if it addresses multiple issues or multiple aspects of a single issue.

16.4.1 Data Quality Impact

Research in this area investigates impacts of data quality in organizations, develops methods to evaluate those impacts, and designs and tests mechanisms that maximize positive impacts and mitigate negative ones. There are five subcategories.

16.4.1.1 Application Area

Research in this category investigates data quality issues related to specific application areas of information systems such as customer relationship management (CRM), knowledge management, supply chain management, and enterprise resource management (ERP). For example, Mikkelsen and Aasly (2005) reported that patient records often contain inaccurate attribute values. These inaccuracies make it difficult to find specific patient records. In another study, Xu et al. (2002) developed a framework for identifying data quality issues in implementing ERP systems. Nyaga et al. (2011) explored drivers for information quality in contemporary inter-organizational supply chain relations. Heinrich et al. (2009) provided a procedure for developing metrics for quantifying the currency of data in the context of CRM.

16.4.1.2 Performance, Cost/Benefit, and Operations

Research in this area investigates the impact of data quality on the performance of organizational units (including individuals), evaluates the costs and benefits of data quality initiatives, and assesses the impact of data quality on operations and decision making. As suggested by Redman (1998), poor data quality can jeopardize the effectiveness of an organization's tactics and strategies. Poor data quality can be a factor leading to serious problems (Fisher and Kingma 2001). The impact of data quality and information about data quality on decision making has been investigated in several studies (Chengular-Smith et al. 1999; Raghunathan 1999; Fisher et al. 2003; Jung et al. 2005). Preliminary research has assessed the impact of data quality on firm performance (Sheng and Mykytyn 2002). Another study (Lee and Strong 2004) investigated whether a certain mode of knowledge, or *knowing-why*, affects work performance and whether knowledge held by different work roles matters for work performance. A recent study has shown evidence that the relationship between information quality and organizational outcomes is systematically measurable and the measurements of information quality can be used to predict organizational outcomes (Slone 2006). Still more research is needed to assess the impact of data quality on entities as diverse as individual firms and the national economy.

16.4.1.3 Information Technology Management

Research in this area investigates interactions between data quality and information technology (IT) management (e.g., IT investment, chief information officer (CIO) stewardship, and IT governance). The "fitness-for-use" view of data quality positions data quality initiatives as critical to an organization's use of IT in support of its operations and competitiveness. Organizations have begun to move from reactive to proactive ways of managing the quality of their data (Otto 2011). We expect to see more empirical studies that gauge their effectiveness and uncover other effects of proactive data quality management.

16.4.1.4 Organization Change and Processes

Ideally, data should be treated as a product, which is produced through a data manufacturing process. As suggested in prior research, data quality improvement often requires changes in processes and organizational behaviors. Research in this area investigates interactions between data quality and organizational processes and changes. For example, Lee et al. (2004) investigated data quality improvement initiatives at a large manufacturing firm, which iteratively adapted technical data integrity rules in response to changing business processes and requirements. A longitudinal study builds a model of data quality problem solving (Lee 2004). The study analyzes data quality activities portrayed by practitioners' reflection-in-action at five organizations via a 5 year action research study. The study finds that experienced practitioners solve data quality problems by reflecting on and explicating knowledge about contexts embedded in, or missing from, data. The study also specifies five critical data quality contexts: role, paradigm, time, goal, and place. Cao and Zhu (2013) investigated inevitable data quality problems resulting from the tight coupling effects and the complexity of ERP-enabled manufacturing systems in their case study.

16.4.1.5 Strategy and Policy

Research in this area investigates strategies and policies for managing and improving data quality at various organizational and institutional levels. For example, Kerr (2006) studied strategies and policies adopted by the healthcare sector in New Zealand. The study shows that the adoption of a data quality evaluation framework and a national data quality improvement strategy provides clear direction for a holistic way of viewing data quality across the sector and within organizations as they develop innovations through locally devised strategies and data quality improvement programs. Data quality strategies and policies at a firm level are laid out in Lee et al. (2006). Weber et al. (2009) studied a data governance model, including data quality roles, decision areas, and responsibilities.

16.4.2 Database-Related Technical Solutions for Data Quality

Research in this area develops database technologies for assessing, improving, and managing data quality. It also develops techniques for reasoning with data quality and for designing systems that can produce data of high quality. There are six subcategories.

16.4.2.1 Data Integration and Data Warehouse

Information systems within and between organizations are often highly distributed and heterogeneous. For analytical and decision-making purposes, there is a need to gather and integrate data from both internal and external sources (e.g., trading partners, data suppliers, and the Internet). Integration can be enabled via a flexible query answering system that accesses multiple sources on demand or via a data warehouse that pre-assembles data for known or anticipated uses. For example, Fan et al. (2001) provided ways to integrate numerical data. Data integration improves the usability of data by improving consistency, completeness, accessibility, and other dimensions of data quality. It is still an active research area after more than two decades of extensive study.

Goh et al. (1999) and Madnick and Zhu (2006) present a flexible query answering system, named COIN for COntext INterchange, which employs knowledge representation, abductive reasoning coupled with constraint solving, and query optimization techniques. The system allows users to query data in multiple sources without worrying about most syntactic or semantic differences in those sources. In practice, many alleged "data quality" problems actually have been "data misinterpretation" problems. By understanding the contexts of both data sources and data consumers, *COIN* attempts to overcome data misinterpretation problems. It converts data, when necessary, to forms users prefer and know how to interpret.

Two other issues addressed by data integration research are entity resolution (sometimes known as record linkage or record deduplication, which is discussed later) and schema matching. Schema matching research (Rahm and Bernstein 2001; Doan and Halevy 2005) develops techniques to automatically or semiautomatically match data schemas. The results can be used for a query answering system to rewrite queries using one schema to query against other matched schemas. The results can also be used to construct a global schema (Batini et al. 1986) for a data warehouse.

A data warehouse is often built via *extract, transform, load* (ETL) processes and provides tools to quickly interrogate data and obtain multidimensional views (e.g., sales by quarter, by product line, and by region). A framework for enhancing data quality in data warehouses is presented in Ballou and Tayi (1999). The data warehouse quality project (Jarkes et al. 1999) has produced a set of modeling tools to describe and manage ETL processes to improve data quality (Vassiliadis et al. 2001). In addition to various data cleansing tools designed specifically for ETL, a flexible query answering system such as COIN can be used as a transformation engine in ETL processes.

16.4.2.2 Enterprise Architecture and Conceptual Modeling

EA (Zachman 1987; Schekkerman 2004; OMB 2007) is a framework for understanding the structure of IT elements and how IT is related to business and management processes. EA allows an organization to align its information systems with its business objectives. This alignment is often accomplished by documenting, visualizing, and analyzing relationships between systems and organizational needs. EA methods have been widely used. For example, federal agencies in the United States are required to adopt a set of federal enterprise architecture methods in IT operations, planning, and budgeting (OMB 2007). Research in this area develops technologies to inventory, visualize, analyze, and optimize information systems and link their functionality to business needs.

Conceptual modeling is primarily used for database and system design. It is also useful for modeling EA. The ER model (Chen 1976) and its extensions are the most prevalent data modeling techniques. One important extension is to add data quality characteristics to an ER model. As illustrated in (Wang et al. 1993; Storey and Wang 1998), this extension captures data quality requirements as metadata at the cell level. Furthermore, the querying system can be extended to allow for efficient processing of data quality metadata (Wang et al. 1995a). Further research in this area aims to develop modeling extensions and query answering mechanisms to accommodate the need to manage data quality-related metadata such as quality metrics, privacy, security, and data lineage (Naumann 2002; Karvounarakis et al. 2010).

16.4.2.3 Entity Resolution, Record Linkage, and Corporate Householding

An entity, such as a person or an organization, often has different representations in different systems, or even in a single system. Entity resolution (Wang and Madnick 1989; Talburt et al. 2005), also known as record linkage (Winkler 2006) and object identification (Tejada et al. 2001), provides techniques for identifying data records pertaining to the same entity. These techniques are often used to improve completeness, resolve inconsistencies, and eliminate redundancies during data integration processes.

A corporate entity is often composed of multiple sub-entities that have complex structures and intricate relationships. There are often differing views about the structures and relationships of the sub-entities of a corporate entity. For example, the answer to "What was the total revenue of IBM in 2008?" depends on the purpose of the question (e.g., credit risk assessment or regulatory filing). The purpose would determine if revenues from subsidiaries, divisions, and joint ventures should be included or excluded. This phenomenon sometimes is known as the corporate household problem. In certain cases, it can be modeled as an aggregation heterogeneity problem (Madnick and Zhu 2006). More corporate household examples can be found in (Madnick et al. 2005). Actionable knowledge about organizations and their internal and external relationships is known as corporate household knowledge (Madnick et al. 2001). Corporate householding research develops techniques for capturing, analyzing, understanding, defining, managing, and effectively using corporate household knowledge. Preliminary results of using context mediation for corporate householding management can be found in (Madnick et al. 2004).

16.4.2.4 Monitoring and Cleansing

Certain data quality problems can be detected and corrected either online as data comes in or in batch processes performed periodically. Research in this area develops techniques for automating these tasks. For example, a technique for detecting duplicate records in large datasets is reported in (Hernandez and Stolfo 1998). The AJAX data cleansing framework has a declarative language for specifying data cleansing operations (Galahards et al. 2001). This declarative approach allows for separation of logical expression and physical implementation of data transformation needed for data cleansing tasks. The framework has been adapted for data cleansing needs in biological databases (Herbert et al. 2004).

16.4.2.5 Lineage, Provenance, and Source Tagging

Data lineage and data provenance information, such as knowledge about sources and processes used to derive data, is important when data consumers need to assess the quality of the data and make appropriate use of that data. Early research in this area (Wang and Madnick 1990) developed a data model that tags each data element with its source and provides a relational algebra for processing data source tags. A more general model was developed later (Buneman et al. 2001); it can be applied to relational databases as well as to hierarchical data such as XML. While much prior work focused on developing theories, an effort at Stanford University (Widom 2005) has developed a database management system to process data lineage information as well as uncertainties in data. A method of evaluating data believability using data provenance is developed in (Pratt and Madnick 2008).

16.4.2.6 Uncertainty

From a probabilistic viewpoint, there is a certain degree of uncertainty in each data element, or conversely, an attribute can probabilistically have multiple values. Numeric values also have a precision. Research in this area develops techniques for storing, processing, and reasoning with such data (Dalvi and Suciu 2007). For example, Benjelloun et al. (2006) present a novel extension to the relational model for joint processing of uncertainty and lineage information. While certain tasks require data with high precision and low uncertainty, other tasks can be performed with data that are less precise and more uncertain. Thus there is a need to effectively use data of differing levels of precision and uncertainty to meet a variety of application needs. Kaomea and Page (1997) present a system that dynamically selects different imagery data sources to produce information products tailored to different user constraints and preferences. In other cases, trade-offs need to be made between certainty or precision and other metrics of data quality. A mechanism of optimizing the accuracy–timeliness trade-off in information systems design is given in (Ballou and Pazer 1995).

16.4.3 Data Quality in the Context of Computer Science and Information Technology

Research in this area develops technologies and methods to manage, ensure, and enhance data quality. There are six subcategories.

16.4.3.1 Measurement and Assessment

To manage data quality, an organization first needs to evaluate the quality of data in existing systems and processes. Given the complexity of information systems and information product manufacturing processes, there are many challenges to accurate and cost-effective assessments of data quality. Research in this area develops techniques for systematic measurement of data quality within an organization or in a particular application context. The measurement can be done periodically or continuously. Lee et al. (2002) present a data quality assessment and improvement methodology that consists of a questionnaire to measure data quality and gap analysis techniques to interpret the data quality measures. Useful functional forms used for processing the questionnaire results are discussed in Pipino et al. (2002).

Data quality can also be assessed using other methods. For example, Pierce (2004) suggests the use of control matrices for data quality assessment. Data quality problems are listed in the columns of the matrix, quality checks and corrective processes form the rows, and each cell is used to document the effectiveness of the quality check in reducing the corresponding data quality problem. To improve the computation efficiency of data quality assessments in a relational database, Ballou et al. (2006) developed a sampling technique and a method of estimating the quality of query results based on a sample of the database. As more information is being generated on the web and through user contribution, numerous methods for measuring the quality of semi-structured and unstructured information have been developed (Gertz et al. 2004; Agichtein et al. 2008; Caro et al. 2008).

16.4.3.2 Information Systems

In the broad field of information systems, data quality research identifies data quality issues in organizations, investigates practices that enhance or deteriorate data quality, and develops techniques and solutions for data quality management in an organizational setting. For example, taking a *product* view of information (Wang et al. 1998), Wang and his colleagues (Wang 1998; Shankaranarayan et al. 2003) developed a modeling technique, called IPMap, to represent the manufacturing process of an information product (IP). Using a similar modeling technique, Ballou et al. (1998) illustrated how to model an information product manufacturing system and presented a method for determining quality attributes of information within the system. Lee et al. (2007a) developed a context-embedded IPMap to explicitly represent various contexts of information collection, storage, and use. In a 5 year longitudinal study of data quality activities in five organizations, Lee (2004) investigated how practitioners solved data quality problems by reflecting on and explicating knowledge about contexts embedded in, or missing from, data, and the contexts of data connected with otherwise separately managed data processes (i.e., collection, storage, and use).

16.4.3.3 Networks

There are a multitude of networks that connect various parts of a system and multiple systems. Networks can consist of physical communications networks, logic and semantic linkages between different systems, connections between systems and users, or even connections among users, such as social networks. Research into such networks can provide insights into how data are used and how the quality of data changes as data travel from node to node. These insights can be used to optimize network topology and develop tools for analyzing and managing networks. For example, O'Callaghan et al. (2002) proposed a single-pass algorithm for high-quality clustering of streaming data and provided the corresponding empirical evidence. Marco et al. (2003) investigated the transport capacity of a dense wireless sensor network and the compressibility of data.

16.4.3.4 Protocols and Standards

Data quality can be affected by protocols and standards. Research in this area develops protocols and standards to improve the quality of data exchanged among multiple organizations or within a single organization. Data standards improve data quality in dimensions such as consistency, interpretability, accuracy, etc. However, when data standards are too cumbersome, users may circumvent the standards and introduce data that deviate from those standards (Zhu and Wu 2011a). Thus research in this area also needs to study how protocols and standards impact data quality and how organizations can promote user compliance. In addition, the quality of the protocols or standards is also subject to quality evaluation. For example, Bovee et al. (2002) evaluated the quality of the eXtensible Business Reporting Language standard to see if its vocabulary is comprehensive enough to support the needs of financial reporting. Methods have also been developed to measure the quality of data standards (Zhu and Fu 2009; Zhu and Wu 2011b).

16.4.3.5 Privacy

Certain systems contain private information about individuals (e.g., customers, employees, and patients). Access to such information needs to be managed to ensure only authorized users view such data and only for authorized purposes. Privacy regulations in different jurisdictions impose different requirements about how private data should be handled. Violating the intended privacy of data would represent a failure of data quality. Although there have been commercial tools for creating privacy rules and performing online auditing to comply with regulations, there are still many challenges in developing expressive rules and efficient rule enforcement mechanisms. Recent research also addresses privacy preservation issues that arise when certain data must be disclosed without other private information being inferred from the disclosed data. Such research has focused on developing algorithms to manipulate the data to prevent downstream users from inferring information that is supposed to be private (Li and Sarkar 2006; Xiao and Tao 2006).

Privacy concerns engender multiple requirements. For example, one aspect of privacy, called *autonomy* is the right to be left alone. The "do not call" list in the United States is an example of legal protection for an autonomy requirement of privacy. As modes of communication with customers evolve, future research needs to develop effective solutions for describing the various privacy requirements and designing systems to meet these requirements. Further complicating privacy issues, some requirements such as those for data provenance can simultaneously increase quality while compromising privacy.

16.4.3.6 Security

Data security has received increasing attention. Research in this area develops solutions for secure information access, investigates factors that affect security, and develops metrics for assessing overall information security across and between organizations. A recent study (Ang et al. 2006) extends the definition of information security in three avenues: (1) *locale* (beyond the boundary of an enterprise to include partner organizations), (2) *role* (beyond the information custodians' view to include information consumers' and managers' views), and (3) *resource* (beyond technical dimensions to include managerial dimensions). This research attempts to develop an instrument for assessing information security based on this extended definition.

16.4.4 Data Quality in Curation

Digital curation is an emerging area of study originated in the fields of library and information science. It involves selecting, preserving, and managing digital information in ways that promote easy discovery and retrieval for both current and future uses of that information. Thus digital curation needs to consider current as well as future data quality issues. Consider the accessibility dimension of data quality: data preserved on 8 and 5 in. floppy disks have become nearly inaccessible because it is difficult to find a computer that is equipped with a compatible floppy drive. There are other technical and nontechnical issues that need to be considered. For example, implicit contextual information that is known today (and often taken for granted) and necessary to interpret data may become unknown to future generations requiring explicit capture now to ensure future interpretability of curated data.

Standards and policies can improve data curation processes and strategies. A collection of curation related standards can be found at http://www.dcc.ac.uk/diffuse/, a site maintained by the Digital Curation Centre in the United Kingdom.

In addition to database-related concerns, there are issues inherent in curation processes such as manually added annotations. Such manual practices provide challenges for data provenance requirements. Buneman et al. (2006) developed a technique to track provenance information as the user manually copies data from various sources into the curated database. The captured provenance information can be queried to trace the origins and processes involved in arriving at the curated data.

16.5 Research Methods

Just as there is a plethora of research topics, there is a wide range of research methods suitable for data quality research. We identify 15 high-level categories of research methods.

16.5.1 Action Research

Action research is an empirical and interpretive method used by researchers and practitioners who collaboratively improve the practices of an organization and advance the theory of a certain discipline. It differs from consultancy in its aim to contribute to theory as well as practice. It also differs from the case study method in its objective to intervene, not simply to observe (Baskerville and Wood-Harper 1996). An example of this research method can be found in Lee et al. (2004), which studied how a global manufacturing company improved data quality as it built a global data warehouse.

16.5.2 Artificial Intelligence

The field of artificial intelligence was established more than 50 years ago and has developed a set of methods that are useful for data quality research. For example, knowledge representation and automatic reasoning techniques can be used to enable semantic interoperability of heterogeneous systems. As demonstrated in Madnick and Zhu (2006), the use of such techniques can improve the interpretability and consistency dimensions of data quality. Agent technologies can be used to automate many tasks such as source selection, data conversion, predictive searches, and inputs that enhance system performance and user experience.

16.5.3 Case Study

The case study is an empirical method that uses a mix of quantitative and qualitative evidence to examine a phenomenon in its real-life context (Yin 2002). The in-depth inquiry of a single instance or event can lead to a deeper understanding of *why* and *how* that event happened. Useful hypotheses can be generated and tested using case studies (Flyvbjerg 2006). This method is widely used in data quality research. For example, Davidson et al. (2004) reported a longitudinal case study in a major hospital on how information product maps were developed and used to improve data quality.

16.5.4 Data Mining

Evolving out of machine learning of artificial intelligence and statistical learning of statistics, data mining is the science of extracting implicit, previously unknown, and potentially useful information from large datasets (Frawley et al. 1992). The data mining approach can be used to address several data quality issues. For example, data anomaly (e.g., outlier) detection algorithms can be used for data quality monitoring, data cleansing, and intrusion detection (Dasu and Johnson 2003; Petrovskiy 2003; Batini and Scannapieco 2006). Data mining has also been used in schema matching to find one-to-one matches (Doan et al. 2001) as well as complex matching relationships (He et al. 2004). While many data mining algorithms are robust, special treatment is sometimes necessary when mining data with certain known data quality issues (Zhu et al. 2007).

16.5.5 Design Science

There is an increasing need for better design of information systems as many organizations have experienced failed IT projects and the adverse effects of bad data. A systematic study of design science has been called for in the information systems community. With an artifact-centric view of design science,

Hevner et al. (2004) developed a framework and a set of guidelines for understanding, executing, and evaluating research in this emerging domain. As more artifacts such as quality ER (Wang et al. 1993; Storey and Wang 1998) and IPMap (Shankaranarayan et al. 2003) are created to address specific issues in data quality management, it is important that they are evaluated using appropriate frameworks, such as the one suggested in Hevner et al. (2004).

16.5.6 Econometrics

A field in economics, econometrics develops and uses statistical methods to study and elucidate economic principles. A comprehensive economic theory for data quality has not been developed, but there is growing awareness of the cost of poor quality data (Øvretveit 2000) and a large body of relevant literature in the economics of R&D (Dasgupta and Stiglitz 1980) and quality (De Vany and Saving 1983; Thatcher and Pingry 2004). As we continue to accumulate empirical data, there will be econometric studies to advance economic theory and our overall understanding of data quality practices in organizations.

16.5.7 Empirical

The empirical method is a general term for any research method that draws conclusions from observable evidence. Examples include the survey method (discussed later) and methods discussed earlier such as action research, case study, statistical analysis, and econometrics.

16.5.8 Experimental

Experiments can be performed to study the behavior of natural systems (e.g., physics), humans and organizations (e.g., experimental psychology), or artifacts (e.g., performance evaluation of different algorithms). For example, Jung et al. (2005) used human subject experiments to examine the effects of contextual data quality and task complexity on decision performance. Klein and Rossin (1999) studied the effect of error rate and magnitude of error on predictive accuracy. Li (2009) proposed a new approach for estimating and replacing missing categorical data. Applying the Bayesian method, the posterior probabilities of a missing attribute value belonging to a certain category are estimated. The results of this experimental study demonstrate the effectiveness of the proposed approach.

16.5.9 Mathematical Modeling

Mathematical models are often used to describe the behavior of systems. An example of this research method can be found in Ballou et al. (1998) where a mathematical model is used to describe how data quality dimensions such as timeliness and accuracy change within an information manufacturing system. System dynamics, a modeling technique originated from systems and control theory, has been used to model a variety of complex systems and processes such as software quality assurance and development (Abdel-Hamid 1988; Abdel-Hamid and Madnick 1990), which are closely related to data quality.

16.5.10 Qualitative

Qualitative research is a general term for a set of exploratory research methods used for understanding human behavior. Qualitative research methods suitable for data quality research include action research, case study, and ethnography (Myers 1997). Examples of data quality research that used action research and case study have been discussed earlier. Ethnography is a research method where the researcher is immersed in the environment of the subjects being studied to collect data via direct observations and interviews. The method was used in Kerr (2006), which studied data quality practices in the health sector in New Zealand.

16.5.11 Quantitative

Quantitative research is a general term for a set of methods used for analyzing quantifiable properties and their relationships for certain phenomena. Econometrics and mathematical modeling are examples of quantitative methods suitable for data quality research. See the discussions earlier for comments and examples of these method types.

16.5.12 Statistical Analysis

Statistical analysis of data is widely used in data quality research. For example, factor analysis was used in Wang and Strong (1996) to identify data quality dimensions from survey data. Furthermore, statistics is the mathematical foundation of other quantitative methods such as data mining and econometrics.

16.5.13 System Design and Implementation

This research method draws upon design methodology in software engineering, database design, data modeling, and system architecture to design systems that realize particular data quality solutions. Using this method, trade-offs in the feature space can be evaluated systematically to optimize selected objectives. Researchers often use this method to design and implement proof-of-concept systems. The COIN system (Goh et al. 1999) was developed using this research method.

16.5.14 Survey

Survey studies often use questionnaires as instrument to collect data from individuals or organizations to discover relationships and evaluate theories. For example, surveys were used in Wang and Strong (1996) to identify data quality dimensions and the groupings of those dimensions. A survey with subsequent statistical analysis was also used in Lee and Strong (2004) to understand the relationship between modes of knowledge held by different information roles and data quality performance and in Slone (2006) to uncover the relationship between data quality and organizational outcomes.

16.5.15 Theory and Formal Proofs

This method is widely used in theoretical CS research such as developing new logic formalism and proving properties of computational complexity. The method is useful in theoretical data quality research. For example, Shankaranarayan et al. (2003) applied graph theory to prove certain properties of IPMap. Fagin et al. (2005) formalized the data exchange problem and developed the computational complexity theory for query answering in data exchange contexts.

16.6 Challenges and Conclusion

Data quality research has made significant progress in the past two decades. Since the initial work performed at the TDQM program (see web.mit.edu/tdqm) and later the MITIQ program (see mitiq.mit.edu) at MIT, a growing number of researchers from CS, information systems, and other disciplines have formed a community that actively conducts data quality research. In this chapter, we introduced a framework for characterizing data quality research along the dimensions of topic and method. Using this framework, we provided an overview of the current landscape and literature of data quality research.

Looking ahead, we anticipate that data quality research will continue to grow and evolve. In addition to solving existing problems, the community will face new challenges arising from ever-changing technical and organizational environments. For example, most of the prior research has focused on the quality of structured data. In recent years, we have seen a growing amount of semi-structured and unstructured data as well as the expansion of datasets to include image and voice. Research is needed to develop techniques for managing and improving the quality of data in these new forms. New ways of delivering information have also emerged. In addition to the traditional client–server architecture, a service-oriented architecture has been widely adopted as more information is now delivered over the Internet to traditional terminals as well as to mobile devices. As we evolve into a pervasive computing environment, user expectations and perceptions of data quality will also change. We feel that the "fitness for use" view of data quality has made some of the early findings extensible to certain issues in the new computing environment. Other issues are waiting to be addressed by future data quality research.

Much current research focuses on individuals and organizations. A broader perspective at a societal or group level can also be pursued. New research can also address issues that face inter-industry information sharing in this "big data" era, ushered in by increasingly diverse data consumers around the world in a social networking and networked environment. Researchers can collaborate across continents to uncover new insights into how data quality shapes global business performance and collaborative scientific endeavors, looking inward and outward. As seen in other areas, we envision an evolving set of topics and methods to address new sets of research questions by new generations of researchers and practitioners.

References

Abdel-Hamid, T.K. 1988. The economics of software quality assurance: A simulation-based case study. *MIS Quarterly 12*, 3, 395–411.

Abdel-Hamid, T.K. and Madnick, S.E. 1990. *Dynamics of Software Project Management*. Prentice-Hall, Englewood Cliffs, NJ.

Agichtein, E., Castillo, C., Donato, D., Gionis, A., and Mishne, G. 2008. Finding high-quality content in social media. In: *Proceedings of the International Conference on Web Search and Web Data Mining (WSDM '08)*. ACM, New York, pp. 183–194.

Ang, W.H., Lee, Y.W., Madnick, S.E., Mistress, D., Siegel, M., Strong, D.M., Wang, R.Y., and Yao, C. 2006. House of security: Locale, roles, resources for ensuring information security. In: *Proceedings of the 12th Americas Conference on Information Systems*. Acapulco, Mexico, August 4–6, 2006.

Ballou, D.P., Chengalur-Smith, I.N., Wang, R.Y. 2006. Sample-based quality estimation of query results in relational database environments. *IEEE Transactions on Knowledge and Data Engineering 18*, 5, 639–650.

Ballou, D. and Pazer, H. 1995. Designing information systems to optimize accuracy-timeliness trade-off. *Information Systems Research 6*, 1, 51–72.

Ballou, D. and Tayi, G.K. 1999. Enhancing data quality in data warehouse environments. *Communications of ACM 41*, 1, 73–78.

Ballou, D., Wang, R.Y., Pazer, H., and Tayi, G.K. 1998. Modeling information manufacturing systems to determine information product quality. *Management Science 44*, 4, 462–484.

Baskerville, R. and Wood-Harper, A.T. 1996. A critical perspective on action research as a method for information systems research. *Journal of Information Technology 11*, 235–246.

Batini, C., Lenzerini, M., and Navathe, S. 1986. A comparative analysis of methodologies for database schema integration. *ACM Computing Survey 18*, 4, 323–364.

Batini, C. and Scannapieco, M. 2006. *Data Quality: Concepts, Methodologies, and Techniques*. Springer Verlag, New York.

Benjelloun, O., Das Sarma, A., Halevy, A., and Widom, J. 2006. ULDBs: Databases with uncertainty and lineage. In: *Proceedings of the 32nd VLDB Conference*, Seoul, Korea, September 2006, pp. 935–964.

Bovee, M., Ettredge, M.L., Srivastava, R.P., and Vasarhelyi, M.A. 2002. Does the year 2000 XBRL taxonomy accommodate current business financial-reporting practice? *Journal of Information Systems 16*, 2, 165–182.

Buneman, P., Chapman, A., and Cheney, J. 2006. Provenance management in curated databases. In: *Proceedings of ACM SIGMOD International Conference on Management of Data,* Chicago, IL, pp. 539–550.

Buneman, P., Khanna, S., and Tan, W.C. 2001. Why and where: A characterization of data provenance. In: Jan Van den B. and Victor V., Eds., *International Conference on Database Theory*. Springer, New York, LNCS 1973, pp. 316–330.

Cao, L., and Zhu, H. 2013. Normal accident: Data quality problems in ERP-enabled manufacturing. *ACM Journal of Data and Information Quality,* 4, 3, Article 12.

Caro, A., Calero, C., Caballero, I., and Piattini, M. 2008. A proposal for a set of attributes relevant for Web portal data quality. *Software Quality Journal 6,* 4, 513–542.

Chen, P.P. 1976. The entity–relationship model: Toward a unified view of data. *ACM Transactions on Database Systems 1,* 1, 1–36.

Chengular-Smith, I., Ballou, D.P., and Pazer, H.L. 1999. The impact of data quality information on decision making: An exploratory analysis. *IEEE Transactions on Knowledge and Data Engineering 11,* 6, 853–865.

Dalvi, N. and Suciu, D. 2007. Management of probabilistic data: Foundations and challenges. *ACM Symposium on Principles of Database Systems (PODS),* Beijing, China, pp. 1–12.

Dasgupta, P. and Stiglitz, J. 1980. Uncertainty, industrial structure, and the speed of R&D. *The Bell Journal of Economics 11,* 1, 1–28.

Dasu, T. and Johnson, T. 2003. *Exploratory Data Minding and Data Cleaning*. John Wiley & Sons, Hoboken, NJ.

Davidson, B., Lee, Y.W., and Wang, R. 2004. Developing data production maps: Meeting patient discharge data submission requirements. *International Journal of Healthcare Technology and Management 6,* 2, 223–240.

De Vany, S. and Saving, T. 1983. The economics of quality. *The Journal of Political Economy 91,* 6, 979–1000.

Deming, W.E. 1982. *Out of the Crisis*. MIT Press, Cambridge, MA.

Doan, A., Domingos, P., and Halevy, A. 2001. Reconciling schemas of disparate data sources: A machine learning approach. *ACM SIGMOD,* Santa Barbara, CA, pp. 509–520.

Doan, A. and Halevy, A.Y. 2005. Semantic-integration research in the database community: A brief survey. *AI Magazine 26,* 1, 83–94.

Fagin, R., Kolaitis, P.G., Miller, R., and Popa, L. 2005. Data exchange: Semantics and query answering. *Theoretical Computer Science 336,* 1, 89–124.

Fan, W., Lu, H., Madnick, S.E., and Cheung, D.W. 2001. Discovering and reconciling data value conflicts for numerical data integration. *Information Systems 26,* 8, 635–656.

Fisher, C., Chengular-Smith, I., and Ballou, D. 2003. The impact of experience and time on the use of data quality information in decision making. *Information Systems Research 14,* 2, 170–188.

Fisher, C. and Kingma, B. 2001. Criticality of data quality as exemplified in two disasters. *Information and Management 39,* 109–116.

Flyvbjerg, B. 2006. Five misunderstandings about case study research. *Qualitative Inquiry 12,* 2, 219–245.

Frawley, W.J., Piateksky-Shapiro, G., and Matheu, S.C.J. 1992. Knowledge discovery in databases: An overview. *AI Magazine 13,* 3, 57–70.

Galahards, H., Florescu, D., Shasha, D., Simon, E., and Saita, C.A. 2001. Declarative data cleaning: Language, model and algorithms. In: *Proceedings of the 27th VLDB Conference*. Rome, Italy, pp. 371–380.

Gertz, M., Ozsu, T., Saake, G., and Sattler, K.-U. 2004. Report on the Dagstuhl seminar "Data Quality on the Web." *SIGMOD Record 33,* 1, 127–132.

Goh, C.H., Bressan, S., Madnick, S.E., and Siegel, M.D. 1999. Context interchange: New features and formalisms for the intelligent integration of information. *ACM Transactions on Information Systems 17,* 3, 270–293.

He, B., Chang, K.C.C., and Han, J. 2004. Mining complex matchings across Web query interfaces. *Proceedings of the 9th ACM SIGMOD Workshop on Research Issues in Data Mining and Knowledge Discovery,* Paris, France, pp. 3–10.

Heinrich, B. Klier, M., and Kaiser, M. 2009. A procedure to develop metrics for currency and its application in CRM, 2009. *ACM Journal of Data and Information Quality 1*, 1, 5/1–5/28.

Herbert, K.G., Gehani, N.H., Piel, W.H., Wang, J.T.L., Wu, C.H. 2004. BIO-AJAX: An extensible framework for biological data cleaning. *SIGMOD Record 33*, 2, 51–57.

Hernandez, M. and Stolfo. 1998. Real-world data is dirty: Data cleansing and the merge/purge problem. *Journal of Data Mining and Knowledge Discovery 2*, 1, 9–37.

Hevner, A.T., March, S.T., Park, J., and Ram, S. 2004. Design science in information systems research. *MIS Quarterly 28*, 1, 75–105.

Jarke, M., Jeusfeld, M.A., Quix, C., and Vassiliadis, P. 1999. Architecture and quality in data warehouse: An extended repository approach. *Information Systems 24*, 3, 229–253.

Jung, W., Olfman, L., Ryan, T., and Park, Y. 2005. An experimental study of the effects of contextual data quality and task complexity on decision performance. In: *Proceedings of IEEE International Conference on Information Reuse and Integration*, August 15–17, Las Vegas, NY, pp. 149–154.

Juran, J. and Goferey, A.B. 1999. *Juran's Quality Handbook*. 5th edn. McGraw-Hill, New York.

Kaomea, P. and Page, W. 1997. A flexible information manufacturing system for the generation of tailored information products. *Decision Support Systems 20*, 4, 345–355.

Karvounarakis, G., Ivies, Z.G., and Tannen, V. 2010. Querying data provenance. In: *Proceedings of the 2010 International Conference on Management of Data (SIGMOD '10)*. ACM, New York, pp. 951–962.

Kerr, K. 2006. The institutionalisation of data quality in the New Zealand health sector. PhD dissertation, The University of Auckland, New Zealand.

Klein, B.D. and Rossin, D.F. 1999. Data quality in neural network models: Effect of error rate and magnitude of error on predictive accuracy. *Omega 27*, 5, 569–582.

Lee, Y.W. 2004. Crafting rules: Context-reflective data quality problem solving. *Journal of Management Information Systems 20*, 3, 93–119.

Lee, Y.W., Chase, S., Fisher, J., Leinung, A., McDowell, D., Paradiso, M., Simons, J., Yarawich, C. 2007a. CEIP maps: Context-embedded information product maps. In: *Proceedings of Americas' Conference on Information Systems*, August 15–18, Denver, CO, Paper 315.

Lee, Y.W., Chung, W.Y., Madnick, S., Wang, R., Zhang, H. 2012. *On the rise of the chief data officers in a world of big data*, SIM/MISQE Workshop, Orlando, FL, Society for Information Management.

Lee, Y.W., Pierce, E., Talburt, J., Wang, R.Y., and Zhu, H. 2007b. A curriculum for a master of science in information quality. *Journal of Information Systems Education 18*, 2, 233–242.

Lee, Y.W., Pipino, L.L., Funk, J.F., and Wang, R.Y. 2006. *Journey to Data Quality*. The MIT Press, Cambridge, MA.

Lee, Y.W., Pipino, L., Strong, D., and Wang, R. 2004. Process embedded data integrity. *Journal of Database Management 15*, 1, 87–103.

Lee, Y. and Strong, D. 2004. Knowing-why about data processes and data quality. *Journal of Management Information Systems 20*, 3, 13–39.

Lee, Y., Strong, D., Kahn, B., and Wang, R. 2002. AIMQ: A methodology for information quality assessment. *Information & Management 40*, 133–146.

Li, X.B. 2009. A Bayesian approach for estimating and replacing missing categorical data. *ACM Journal of Data and Information Quality 1*, 1, 3/1–3/11.

Li, X.B. and Sarkar, S. 2006. Privacy protection in data mining: A perturbation approach for categorical data. *Information Systems Research 17*, 3, 254–270.

Madnick, S. and Lee, Y. 2009a. Editorial for the inaugural issue of the *ACM Journal of Data and Information Quality*, *ACM Journal of Data and Information Quality 1*, 1, 1/1–1/6.

Madnick, S. and Lee, Y. 2009b. Where the JDIQ articles come from: Incubating research in an emerging field. *ACM Journal of Data and Information Quality 1*, 3, 13/1–13/5.

Madnick, S. and Lee, Y. 2010a. Editorial: In search of novel ideas and solutions with a broader context of data quality in mind. *ACM Journal of Data and Information Quality 2*, 2, 7/1–7/3.

Madnick, S. and Lee, Y. 2010b. Editorial notes: Classification and assessment of large amounts of data: Examples in the healthcare industry and collaborative digital libraries. *ACM Journal of Data and Information Quality 2*, 3, 12/1–12/2.

Madnick, S. and Wang, R.Y. 1992. Introduction to total data quality management (TDQM) research program. TDQM-92-01, Total Data Quality Management Program, MIT Sloan School of Management.

Madnick, S.E., Wang, R.Y., Dravis, F., and Chen, X. 2001. Improving the quality of corporate household data: Current practices and research directions. In: *Proceedings of the Sixth International Conference on Information Quality*, Cambridge, MA, November 2001, pp. 92–104.

Madnick, S.E., Wang, R.Y., Krishna, C., Dravis, F., Funk, J., Katz-Hass, R., Lee, C., Lee, Y., Xiam, X., and Bhansali, S. 2005. Exemplifying business opportunities for improving data quality from corporate household research. In: Wang, R.Y., Pierce, E.M., Madnick, S.E., and Fisher, C.W., Eds. *Information Quality*, M.E. Sharpe, Armonk, NY, pp. 181–196.

Madnick, S.E., Wang, R.Y., and Xian, X. 2004. The design and implementation of a corporate household-ing knowledge processor to improve data quality. *Journal of Management Information Systems 20*, 3, 41–69.

Madnick, S.E. and Zhu, H. 2006. Improving data quality with effective use of data semantics. *Data and Knowledge Engineering 59*, 2, 460–475.

Marco, D., Duate-Melo, E., Liu, M., and Neuhoffand, D. 2003. On the many-to-one transport capacity of a dense wireless sensor network and the compressibility of its data. In: Goos, G., Hartmanis, J., and van Leeuwen, J. Eds., *Information Processing in Sensor Networks, Lecture Notes in Computer Science,* 2634, Springer, Berlin, Germany, p. 556.

Mikkelsen, G. and Aasly, J. 2005. Consequences of impaired data quality on information retrieval in electronic patient records. *International Journal of Medical Informatics 74*, 5, 387–394.

Myers, M.D. 1997. Qualitative research in information systems. MISQ Discovery, June 1997. http://www.misq.org/discovery/MISQD_isworld/index.html, retrieved on October 5, 2007.

Naumann, F. 2002. *Quality-Driven Query Answering for Integrated Information Systems*. Springer, Berlin, Germany.

Nyaga, G., Lee, Y., and Solomon, M. 2011. Drivers of information quality in supply chains, SIGIQ Workshop, *Quality Information, Organizations, and Society*. Shanghai, China, December 3, 2011.

O'Callaghan, L., Mishira, N., Meyerson, A., Guha, S., and Motwaniha, R. 2002. *Proceedings of the 18th International Conference on Data and Engineering*, San Jose, CA, pp. 685–694.

OMB (Office of Management & Budget). 2007. FEA reference models. http://www.whitehouse.gov/omb/egov/a-2-EAModelsNEW2.html, retrieved on October 5, 2007.

Otto, B. 2011. Data governance. *Business & Information Systems Engineering 3*, 4, 241–244.

Øvretveit, J. 2000. The economics of quality—A practical approach. *International Journal of Health Care Quality Assurance 13*, 5, 200–207.

Petrovskiy, M. I. 2003. Outlier detection algorithms in data mining systems. *Programming and Computing Software 29*, 4, 228–237.

Pierce, E.M. 2004. Assessing data quality with control matrices. *Communications of the ACM 47*, 2, 82–86.

Pipino, L., Lee, Y., and Wang, R. 2002. Data quality assessment. *Communications of the ACM 45*, 4, 211–218.

Pratt, N. and Madnick, S. 2008. Measuring data believability: A provenance approach. In: *Proceedings of 41st Annual Hawaii International Conference on System Sciences,* January 7–10, Big Island, HI.

Raghunathan, S. 1999. Impact of information quality and decision-making quality on decision quality: A theoretical model. *Decision Support Systems 25*, 4, 275–287.

Rahm, E. and Bernstein, P. 2001. On matching schemas automatically. *VLDB Journal 10*, 4, 334–350.

Redman, T.C. 1998. The impact of poor data quality on the typical enterprise. *Communications of the ACM 41*, 2, 79–82.

Schekkerman, J. 2004. *How to Survive in the Jungle of Enterprise Architecture Frameworks: Creating or Choosing an Enterprise Architecture Framework*. Trafford Publishing, Victoria, Canada.

Shankaranarayan, G., Ziad, M., and Wang, R.Y. 2003. Managing data quality in dynamic decision environment: An information product approach. *Journal of Database Management 14*, 4, 14–32.

Sheng, Y. and Mykytyn, P. 2002. Information technology investment and firm performance: A perspective of data quality. In: *Proceedings of the 7th International Conference on Information Quality*, Cambridge, MA, pp. 132–141.

Slone, J.P. 2006. Information quality strategy: An empirical investigation of the relationship between information quality improvements and organizational outcomes. PhD dissertation, Capella University, Minneapolis, MN.

Storey, V. and Wang, R.Y. 1998. Modeling quality requirements in conceptual database design. *Proceedings of the International Conference on Information Quality*, Cambridge, MA, November, 1998, pp. 64–87.

Strong, D., Lee, Y.W., and Wang, R.Y. 1997. Data quality in context. *Communications of the ACM 40*, 5, 103–110.

Talburt, J., Morgan, C., Talley, T., and Archer, K. 2005. Using commercial data integration technologies to improve the quality of anonymous entity resolution in the public sector. *Proceedings of the 10th International Conference on Information Quality (ICIQ-2005)*, MIT, Cambridge, MA, November 4–6, 2005, pp. 133–142.

Tejada, S., Knoblock, C., and Minton, S. 2001. Learning object identification rules from information extraction. *Information Systems 26*, 8, 607–633.

Thatcher, M.E. and Pingry, D.E. 2004. An economic model of product quality and IT value. *Information Systems Research 15*, 3, 268–286.

Vassiliadis, P., Vagena, Z., Skiadopoulos, S., Karayannidis, N., and Sellis, T. 2001. ARKTOS: Towards the modeling, design, control and execution of ETL processes. *Information Systems 26*, 537–561.

Wang, R.Y. 1998. A product perspective on Total Data Quality Management. *Communications of the ACM 41, 2*, 58–65.

Wang, R.Y., Kon, H.B., and Madnick, S.E. 1993. Data quality requirements analysis and modeling. In: *Proceedings of the 9th International Conference of Data Engineering*, Vienna, Austria, pp. 670–677.

Wang, R.Y., Lee, Y., Pipino, L., and Strong, D. 1998. Managing your information as a product. *Sloan Management Review, Summer 1998*, 95–106.

Wang, R.Y. and Madnick, S.E. 1989. The inter-database instance identification problem in integrating autonomous systems. In: *Proceedings of the 5th International Conference on Data Engineering*, Los Angeles, CA, pp. 46–55.

Wang, R.Y. and Madnick, S.E. 1990. A polygen model for heterogeneous database systems: The source tagging perspective. In: *Proceedings of the 16th VLDB Conference*, Brisbane, Australia, pp. 519–538.

Wang, R.Y., Reddy, M., and Kon, H. 1995a. Toward quality data: An attribute-based approach. *Decision Support Systems 13*, 3–4, 349–372.

Wang, R.Y., Storey, V.C., and Firth, C.P. 1995b. A framework for analysis of data quality research. *IEEE Transactions on Knowledge and Data Engineering 7*, 4, 623–640.

Wang, R.Y. and Strong, D.M. 1996. Beyond accuracy: What data quality means to data consumers. *Journal of Management Information Systems 12*, 4, 5–34.

Weber, K., Otto, B., and Osterle, H. 2009. One size does not fit all-a contingency approach to data governance. *ACM Journal of Data and Information Quality 1*, 1, 5/1–5/28.

Widom, J. 2005. Trio: A system for integrated management of data, accuracy, and lineage. In: *Proceedings of the Second Biennial Conference on Innovative Data Systems Research (CIDR '05)*, Pacific Grove, CA, January 2005.

Winkler, W.E. 2006. Overview of record linkage and current research directions. Technique report, U.S. Census Bureau, Statistics #2006-2.

Xiao, X. and Tao, Y. 2006. Anatomy: Simple and effective privacy preservation. In: *Proceedings of the 32nd VLDB Conference*, Seoul, Korea.

Xu, H., Nord, J.H., Brown, N., and Nord, G.G. 2002. Data quality issues in implementing an ERP. *Industrial Management & Data Systems 102*, 1, 47–58.

Yin, R. 2002. *Case Study Research: Design and Methods*, 3rd edn. Sage Publications, Thousand Oaks, CA.

Zachman, J.A. 1987. A framework for information systems architecture. *IBM Systems Journal 26*, 3, 276–292.

Zhu, X., Khoshgoftaar, T., Davidson, I., and Zhang, S. 2007. Editorial: Special issue on mining low-quality data. *Knowledge and Information Systems 11*, 2, 131–136.

Zhu, H. and Fu, L. 2009. Towards quality of data standards: Empirical findings from XBRL. *30th International Conference on Information System (ICIS'09)*, December 15–18, Phoenix, AZ.

Zhu, H. and Wu, H. 2011a. Interoperability of XBRL financial statements in the U.S. *International Journal of E-Business Research 7*, 2, 18–33.

Zhu, H. and Wu, H. 2011b. Quality of data standards: Framework and illustration using XBRL taxonomy and instances. *Electronic Markets 21*, 2, 129–139.

17

Knowledge Management

Susan Newell
Bentley University
University of Warwick

Marco Marabelli
Bentley University

Knowledge management (KM) refers to the multiple processes that support the creation and use of knowledge in an organization, or as March (1991) describes it, the exploration and exploitation of knowledge. KM is relevant to the information systems (IS) discipline, because information and communication technologies (ICT) are important tools involved in managing knowledge, especially given the increasingly distributed nature of organizational activity. However, despite the marketing of various ICT packages as "KM tools" by vendors, the IS discipline has recognized that ICT tools alone (i.e., the ICT artifact) are not sufficient. The combination of people and technology, in a particular context will influence how knowledge is produced and reproduced. This (re)production can be either purposive (e.g., when knowledge is integrated in order to develop a new method for dealing with customer enquiries or when knowledge is shared so that a new process in one geographical location can be reproduced in another location) or an emergent and unintended outcome of practice (e.g., when people start to use their e-mail to save important files for themselves rather than to communicate with other people). How the ICT, people, and context interrelationships are theorized will vary, of course, depending on the ontological and epistemological assumptions of the author (e.g., we will contrast an epistemology of possession and an epistemology of practice, Cook and Brown (1999), later in this chapter) but simplistic notions of ICT as the causal or determining factor in managing knowledge have long since been exploded, at least in the IS discipline. In other words, we now recognize that even though an organization might adopt ICT for some purpose (e.g., to improve knowledge sharing), simply introducing technology will not, on its own, lead to the desired change. Indeed, the introduction of ICT can sometimes have the opposite outcomes to those anticipated, for example, when the introduction of an intranet reduces knowledge sharing across organizational units rather than improving knowledge sharing as intended (Newell et al., 2001). Thus, we do not agree with Spender (2006, p. 128) when he writes that: "IT or management information theorists and economists normally treat knowledge as separable from the people who generate and use it in their decision-making."

It is perhaps unfortunate that there is still a focus, even if only in practice and not our theorizing, on the ICT tools in KM. When things go wrong, the tools and the people who create the tools are then blamed, rather than recognizing the people–technology–context interactions that are actually at the root of any unsuccessful (as well as successful) initiative that is aimed to improve knowledge production and reproduction in an organizational context. This fallacy of blaming "the ICT" and information technology (IT) specialists for any KM initiative problem is captured nicely in the following joke that has done its rounds on the Internet:

A man piloting a hot air balloon discovers he has wandered off course and is hopelessly lost. He descends to a lower altitude and locates a man down on the ground. He lowers the balloon further and shouts "Excuse me, can you tell me where I am?"

The man below says: "Yes, you're in a hot air balloon, about 30 feet above this field."

"You must work in Information Technology," says the balloonist.

"Yes I do," replies the man. "And how did you know that?"

"Well," says the balloonist, "what you told me is technically correct, but of no use to anyone."

The man below says, "You must work in management."

"I do," replies the balloonist, "how did you know?"

"Well," says the man, "you don't know where you are, or where you're going, but you expect my immediate help. You're in the same position you were before we met, but now it's my fault!"

The skit nicely sums up how ICT (and those who work in the IT function) can become the scapegoat for KM initiatives. In this chapter, we will examine how a more sophisticated conceptualization of the interrelationships between people, technology, and context can help us to better understand how to create a working KM system for the production and reproduction of knowledge. We begin, though, with a discussion about "what is knowledge" because this is essential for our understanding of attempts to manage knowledge in an organization. We use production and reproduction of knowledge (rather than creation and use), because, as we will argue, knowledge is not transferred between individuals or collectives "ready-to-wear" or use directly. It is rather reproduced (or translated) in each new context of application, indeed, from a practice perspective, in each new moment of time.

Following the discussion on the nature of knowledge, the rest of the chapter is as follows: The next section provides an overview of knowledge (and KM) with respect to the epistemology of possession and the epistemology of practice; the following section (17.3) highlights how technology supports organizational memory (repository view), which is in line with the epistemology of possession. In contrast, Section 17.4 illustrates the "network view," which is rooted in the epistemology of practice. Section 17.5 provides an overview of KM systems (KMS), while Section 17.6 adopts the lens of absorptive capacity to highlight two contrasting aspects of knowledge: exploration and exploitation. The last section (17.7) draws some conclusions.

17.1 What Is Knowledge?

We watch detectives on TV piece together evidence, using a variety of tools, talking with colleagues, following up leads, interviewing suspects, and finally working out "who did it." The novice detective marvels at the demonstrated ability, having been unable himself/herself to make sense of all the various cues that point to the guilty suspect. We also watch doctors, ask others to do particular kinds of tests, conduct their own examination of a patient using various pieces of medical equipment, discuss the case with other specialists, and finally determine "what's wrong" and "what's the cure." Again, the trainee doctor can only stand and watch, unable to work out how all the different symptoms add up to a particular diagnosis. We recognize these people as having (or not having) knowledge and acting (or not acting) knowledgeably. The main feature that distinguishes between the expert and the novice in these situations is their ability to differentiate within and between the patterns of complex information and data that they can draw upon in the particular context in which they are operating. As Tsoukas and Vladimirou (2001) put it, knowledge is "the individual ability to draw distinctions within a collective domain of action, based on appreciation of context or theory or both" (p. 979). The ability to discriminate and so draw meaning in a particular context, then, is the essence of knowledge. The next question is how do we gain knowledge?

Philosophers have long discussed and argued about this question. It has carried over into the social sciences generally today and into the area of KM very specifically. While there are multiple views, a helpful distinction in terms of the nature of knowledge and how knowledge is acquired is provided by Cook and Brown (1999). These authors distinguish between an epistemology of possession (that treats knowledge as something individuals and groups have or own based on prior experience but separable from that experience) and an epistemology of practice (that treats knowing as something people do that is context-dependent and socially situated). Essentially, this distinction differentiates between the idea that people "have knowledge" and that people "act knowledgeably."

The possession perspective focuses on cognitions (mental processes) since it is the human mind that is viewed as the carrier or engine of knowledge. Knowledge is, thus, a personal property of an individual knower (or potentially also a collective, Spender (1996)), with mental processes being the key to conferring meaning from data and information, which exist "out there" (i.e., outside the head) in the world. These mental processes are the product of past experiences, perceptions, and understandings, which create a "frame of reference" that allows an individual to infer particular things (i.e., to make discriminations) from some data or information. So, an individual with prior training in physics can infer meaning from the equation—$e = mc^2$—that someone without this possessed knowledge will be unable to. This possessed knowledge includes both tacit and explicit knowledge (Polanyi, 1967). Explicit knowledge is that which can be written down or articulated in language or some other symbolic form. Tacit knowledge is knowledge that is impossible or certainly hard to write down and, even if written down, does not express the knowledge adequately. The example that is often used is knowledge to ride a bike. This includes both explicit knowledge (you must sit on the saddle and hold on to the handle bars, which might seem obvious to you but if you had never seen a bike before you would be unlikely to know that it was something you could sit on and peddle) and tacit knowledge (you must balance to stay upright). The knowledge of balance will always remain tacit—you can be told to shift weight if you are falling to one side, but what this "shifting weight" actually means cannot be explicitly shared, although once it is mastered, the individual will possess it as tacit knowledge. It is important to remember in Cook and Brown's classification that both explicit and tacit knowledge are possessed knowledge (tacit knowledge is not the practice element), albeit the epistemology of practice is needed, in their view, to generate this possessed knowledge.

The epistemology of practice, then, sees knowledge, or better knowing, as intrinsic to localized social situations and practices where people perform or enact activities with a variety of others (both human and nonhuman) such that acting knowledgeably emerges from this practice and cannot be separated from this practice. Knowledge and practice are thus, immanent; knowledge is not something that stands outside of practice but is rather constantly (re)produced as people and their tools work together with certain consequences (which may be more or less intentional or purposive and which may demonstrate more or less "acting knowledgeably"). The collective, or community, within which practice is undertaken is characterized by a particular set of stories, norms, representations, tools, and symbols, which together produce the knowledge-related outcomes. These outcomes include the development of shared identities as well as shared beliefs, which underpin being a knowledgeable actor in any particular setting. Moreover, these outcomes are always emergent—never completely predictable.

Cook and Brown (1999) see these two epistemologies not as alternative or opposing views, but rather as complementary, with knowledge (possessed) being a tool of knowing. Possessed and practiced knowledge, thus, work together in what they describe as a "generative dance." Gherardi (2006) refers to this as the mutual constitution view of knowledge and she contrasts this with the containment and the radical practice views—all practice perspectives but differing slightly in their orientation.

The containment view focuses on how practice is developed within a specific community—of nurses, flute makers, photocopy technicians, tailors, radiologists, academics, etc. All the literature on communities of practice (Lave and Wenger, 1991) fits within this containment perspective and is focused on how newcomers learn "to be" nurses, flute makers, etc., through gradual participation in the situation of doing. A good example of such a containment perspective is demonstrated in the work of Julian Orr (1996) who examined the work of photocopy technicians and showed how learning to act knowledgeably occurred socially, often emerging through storytelling as well as through physical interactions with machines-in-context. In terms of storytelling, Orr observed that the technicians would meet up for coffee between site visits where they were repairing or maintaining copy machines. At these informal (and unofficial) meetings, they would tell "war stories" of machines they were or had been working on. These stories were crucial vehicles for the sharing of know-how that contributed to their skillful practice. Unfortunately, management saw the informal meetings rather differently—as unnecessary detours from their next scheduled visit and sought to outlaw the meetings. Ironically, as Orr relates, in doing so they reduced the ability of the technicians to learn their skills. In terms of interactions with machines-in-context, Orr points out that technicians needed to observe the copy machine in-context because its situated use would provide vital cues as to what might be the problem. Such cues could never be part of the training manual because there were so many aspects of the context that might be relevant. Learning to act knowledgeably, then, is socially situated within a particular community of practice.

The radical view, on the other hand, broadens this perspective, focusing on the interconnectedness of practices within any social setting or as Nicolini (2011) refers to it, in a "site of knowing." From this perspective, it is not individual cognitions that are important but the discursive practices, artifacts, and spaces that form a nexus for knowledgeable action in a particular context. For example, in a hospital, there may be a variety of communities of practice (nurses, oncologists, radiologists, and even patients with a particular disease), and to understand practice in this setting, it is important to look at the inter-relationships between these communities as well as the relationships between people and objects and spaces. The knowing of the oncologist is intimately tied up with the tools that she uses and the relationships she has with a variety of other healthcare professionals (and the patient) in this setting. Without these relationships, she could not act knowledgeably. Within this context, or site of knowing, people and objects do not "carry knowledge" but instead are mediators that are actively translated (Latour, 2005) either "by contact" or "at distance" (Nicolini, 2011). Given this process of translation, practice (and so knowing) is always emergent—"pursing the same thing, necessarily implies doing something different" (Nicolini, 2007, p. 893–894). Translation captures the idea that knowledge is (re)produced rather than transferred, so that certain elements will be foregrounded or take on different meaning depending on the context and the mix of actors (human and nonhuman) that are present. Knowing is in these relationships, rather than in the heads of the individuals involved.

While the details of the differences between these practice views can be important, the key point that we wish to make in this chapter is that it is crucial to see knowledge as something that does not simply reside in individuals' minds, but rather recognize how knowing is a practical accomplishment. A chef cannot become a master chef simply by reading about cooking or even by practicing in isolation; rather she becomes a master through working in a particular kitchen where certain norms and values, together with particular culinary tools and ingredients, produce and re-produce her ability to act knowledgeably, as an emergent accomplishment of the social setting.

Moreover, the distinctions between knowledge as possession and practice and the different variants of the practice perspective are important because they influence how one views the interrelationships between people, technology, and context. We will next discuss two different images of the role of technology and the associated views of people and contexts in relation to KM: the repository view (that we link to the possession view of knowledge and the socio-technical school in IS) and the network view (that we link to the practice perspective of knowledge and the sociomateriality school in IS). The repository versus network views have been previously discussed (Alavi and Tiwana, 2002). More recently, McAfee (2006) has distinguished between the platform and channel views of KM, which is very similar. These authors, however, make these distinctions based on developments in IT (from Web 1.0 to Web 2.0 and beyond), while we think it is important to see these different views of KM not as a product simply of technological advancement, but in terms of the epistemological perspective on knowledge and the ontological assumptions about the people–technology relationships that underpins their use.

17.2 Repository View

The repository view is probably the best established view of KM and is linked to the knowledge as possession perspective discussed earlier. From this perspective, individuals own knowledge, but they can also transfer knowledge to others, just as if I own a book I can give it to someone else, if I so choose. The idea behind the repository then is that individuals are encouraged to make explicit their knowledge (typically through a written document but this could also be through other media, like a film). This knowledge is then stored in a repository of some kind (typically a database using a document management system of some kind that can be accessed through an intranet that includes a search engine). Others can then search the repository when they need certain knowledge that they lack, and once they find what they are looking for, they can then apply this knowledge in their own context. It is a supply and demand model—those who have knowledge provide it for those who need it.

The repository view does not assume that putting in place a searchable database will alone be sufficient to improve knowledge sharing in an organization. In other words, it is recognized that the repositories on their own will not improve KM across an organization, in the absence of attention to the social aspects of organizing (Connelly et al., 2012). We can describe this as a socio-technical perspective (Kling and Courtright, 2006) on KM (Davenport and Prusak, 1998). A socio-technical perspective recognizes how the social and the technical influence each other so that what may seem like a perfect technical solution may not operate as such, because it undermines some aspect of the social, like motivation. The technical and the social are seen as having separate ontological status but are recognized to influence each other. It is, therefore, important from a socio-technical perspective to ensure that the social and the technical are aligned in relation to the goals that one is trying to pursue. For example, implementing a lessons learned database for project managers to store their experiences from which others can learn may not be effective if project managers are very time-pressured and/or not recognized for adding to the lessons learned database. It is also not very helpful if project managers are encouraged to share their knowledge if, at the same time, they are not also encouraged to look for and reuse knowledge from others (the demand side must be addressed as well as the supply side, an issue which has often been ignored in KM initiatives; Newell and Edelman (2008)). A number of factors, therefore, are recognized as being crucial to the success of such an initiative. We discuss some of those issues next.

17.2.1 Sticky Knowledge

While the repository view assumes that knowledge can be transferred from place to place, it also recognizes that some knowledge, especially tacit knowledge, can be more difficult to transfer because as Szulanski (1996) notes, knowledge can be "sticky." Nonaka (1994) suggests that this problem can be overcome by converting tacit knowledge to explicit knowledge. However, as already hinted at, this misunderstands the essential nature of tacit knowledge, which, as Polanyi (1961) indicates, will always remain tacit. Tsoukas (1996) summarizes this nicely when he refers to tacit and explicit knowledge as "mutually constituted," meaning that in any knowledgeable act, there will always be both tacit and explicit knowledge at work.

Sticky knowledge that is tacit may, then, simply not be transferable through a repository, but this still leaves a lot, from the possession perspective that can be shared. Stickiness, however, resides not just in the tacitness of knowledge but also in the fact that people have to be persuaded to both share their knowledge at the supply end and then use knowledge from elsewhere at the demand end. This is where the "people aspect" of KM is often discussed. Incentives may be necessary to encourage people to share their knowledge, because knowledge is often a source of power in an organization (Bartol and Srivastava, 2002). Knowledge hoarding can, thus, be a major barrier to the success of a KM repository initiative (Lam and Chua, 2005). This means that incentives may be needed either to discourage hoarding or because it is time-consuming to "make one's knowledge available to others" (Alavi et al., 2006). Incentives may be financial but they can also be symbolic, as when knowledge workers share knowledge to increase their professional reputation (McLure-Wasko and Faraj, 2005).

Moreover, the possession perspective would also recognize that an individual's mind is not open to receive any kind of new knowledge—there is a path dependency to knowledge accumulation (see section on absorptive capacity). So there will be barriers to knowledge sharing, for example, across departments where different syntax is used, or where meanings are different (Carlile, 2004) (see following section on knowledge boundaries).

17.2.2 Leaky Knowledge

While from the repository view the stickiness of knowledge has to be overcome, it is also the case that because knowledge can be so readily transferred, it also has to be protected since it is the main source of sustainable advantage (Trkman and Desouza, 2012), as per the resource-based (Barney, 1991) or knowledge-based view of the firm (Grant, 1996). Technical solutions to this problem are addressed to securing databases and other ICT from unwanted predators. In fact, knowledge leaks associated with security breaches are not infrequent (Desouza and Vanapalli, 2005); for instance, Fitzgerald (2003) documents a number of cases of loss of intellectual property and sensitive data due to weak security measures. However, it is also recognized that a focus on the social is important. This can be in the form of contractual obligations, for example, with stipulating that an individual leaving an organization is not allowed to join a competitor within a certain amount of time, or it can take the form of training to encourage employees to have secure passwords (not sticky-taped to their computer) or to know what information is propriety and so should not be shared. Leaky knowledge is, in effect, the opposite of absorptive capacity (Cohen and Levinthal, 1990; Zahra and George, 2002) discussed below. Absorptive capacity is about building the capacity to recognize, process, and use external knowledge, while in this section, we have been discussing the importance of protecting organizational knowledge from external parties.

17.2.3 Organizational Culture

Given these issues of both sticky and leaky knowledge, scholars have identified organizational culture as key to effective KM practice. People need to be encouraged to share their knowledge and reuse others' knowledge through developing a culture or climate that encourages such activity (Alavi et al., 2006), or the culture needs to prevent such sharing where this is important. Culture refers to the basic assumptions and values that exist within an organization (Schein, 1992). There can be multiple cultures across an organization (Meyerson and Martin, 1987), but the key is that culture has been shown to be either an enabler (Alavi et al., 2006) or barrier (Connelly et al., 2012) to KM success. The cultivation of a culture that supports knowledge sharing has thus been described as an essential facilitator of the ability to exploit a KM repository (Donate and Guadamillas, 2010). Hackett (2000) describes this as a knowledge-sharing culture, which should include a shared vision, value-based leadership at all levels, openness and continuous communication, and rewards and recognition.

17.3 Network View

The network view has been linked to the emergence of social networking sites that encourage social exchanges in a much more dynamic and personalized way than repositories can. Social media are not simply platforms that individuals can communicate from (to one or many) but instead provide tools for ongoing dynamic exchanges using multiple channels, including video, text, chat, and discussion forums. The network view is linked to the communities of practice perspective on knowledge sharing (Wenger et al., 2002) and to the practice perspective of knowledge more generally (Gherardi, 2006). Since knowing is an emergent outcome that is negotiated through social and material interactions, repositories that contain knowledge that is divorced from context are seen to be less helpful. However, ICT that can promote dialogue and shared experiences (albeit virtual) can play a role in the social learning process that underpins becoming a knowledgeable actor. Moreover, ICT is not just a pipe through which knowledge flows, rather ICT is seen to be part of the very practice itself, scaffolding that practice along with a host of other actors (material and human). We can describe this as a sociomaterial perspective (Feldman and Orlikowski, 2011).

From a sociomaterial perspective, it is not simply that the social and the technical are inter related, rather they are seen as "constitutively entangled in everyday life" (Orlikowski, 2007, p. 1437). As Pickering (1993) suggests, subjects and objects are mangled in the flow of practice, which can be seen as a series of resistances and subsequent accommodations in an ongoing flow of activity that is directed toward some goal (in the case he describes, the creation of a bubble chamber by a physicist). In the ongoing flow of practice, it is not possible or useful to distinguish between the social and the technical/material because they are entangled together in a form of "absorbed coping" (Dreyfus, 1995). Only when something breaks down, are we able to separate (artificially) the social and the material elements of the practice (Sandberg and Tsoukas, 2011).

From the practice perspective, ICT is a potential mediator, which will provide access to information that will be translated-in-context. For example, Nicolini (2011), examining telemonitoring in a hospital, describes how the ICT, infrastructures, programs, and interfaces that allow the doctors to collect information from a patient's home are all mediators of knowledge that is translated in different settings. For instance, Nicolini (2011) quotes a dialogue (phone call) between a patient (at home) and a nurse (in the hospital). The nurse first gathers a lot of information from the patient (i.e., the patient was sick the previous night, his daily pressure reading is fine, and he made a decision to stop taking a drug). Then the nurse needs to reach the doctor (in the example, a cardiologist), refer everything to him/her, and make sure that the doctor listens carefully to the whole conversation and makes the right decisions that will be then be reported to the patient (as Nicolini highlights, from the nurses perspective, "often doctors listen to you only briefly and support your decisions"—p. 7). In this example, the nurse acts as a mediator "at distance" that *translates* knowledge across contexts.

Given that, from this perspective, the social and the technical are seen as inseparable, it would not be sensible to think about ICT independent of its use in practice. Nevertheless, there are a number of important points to recognize from this perspective, which we discuss next.

17.3.1 Emergent (Unintended) Use

The material (including ICT) may scaffold the social (Orlikowski, 2007), but the scaffold is not fixed in design but rather unfolds through practice. So, the video conferencing facility that was designed to support virtual teamwork may be used by those involved in the project to watch movies together, rather than to hold joint team working sessions. In other words, from the practice perspective, there is always some emergence that may not be predicted but which managers need to be aware of if they hope to promote knowledgeable work. In saying this, we are not suggesting that managers should be trying to constrain this emergence—a distributed team using their video connection to watch films together (rather than, or at least as well as, to conference call) may actually promote strong team bonds that can facilitate project work (as with Orr's photocopy technicians taking time out to meet and chat between customer visits). Managers rather should be aware of why emergence happens and work with the users to create technical scaffolds that can support the work that they are doing, in ways that suit the particular context, recognizing that this may differ across time and space.

17.3.2 Role of Boundaries

The role of boundaries in managing knowledge is crucial for both the possession and the practice perspectives; boundaries exist between groups of any kind that are needing to share knowledge. Carlile (2004) distinguishes between syntactic, semantic, and pragmatic boundaries that increase in importance as the novelty of the collaboration increases, making knowledge sharing increasingly difficult:

Syntactic boundaries: When novelty is low and there are few differences between those sharing knowledge because they have worked on similar problems before, the main boundary is a syntactic one. Developing and maintaining a common lexicon can resolve this boundary problem. Once a common lexicon is in place, knowledge can be *transferred* relatively unproblematically (Carlile, 2004, p. 558).

Semantic boundaries: As novelty increases, differences between the collaborating groups become greater since they are working on new problems so that there are likely to be more ambiguities because there will be different meaning for the same word or convention, for example, about what to call something. Different workers might use conventions that are not clear to those outside a "circle," leading to different interpretations. In these situations, ambiguous terms and expressions (semantics) need to be *translated* to ensure consistent use of words and conventions.

Pragmatic boundaries: In situations where novelty is high, differences between collaborating groups are likely to cause the most problems for knowledge sharing. In these situations, it is not simply that different groups have different meaning systems, but that there are political issues associated with whose knowledge is to be most highly valued. Thus, we are "invested" in our knowledge, and if a change threatens our knowledge, we will resist it. For example, in healthcare, a new medical breakthrough may potentially move the treatment of a particular disease like cancer from surgeons (who cut out the cancerous cells) to radiologists (who kill the cells using radioactive treatments), and in this situation, we might expect the surgeons to resist the change (Swan and Scarbrough, 2005). In these situations, knowledge needs to be *transformed* through negotiations that may eventually change the frame of reference, dominant values, and the practices of the participating parties.

Linking this work on boundaries to the different perspectives on knowledge, from the possession perspective, the major challenges related to sticky knowledge are addressed by overcoming syntactic and semantic boundaries. However, from the practice perspective, a more formidable challenge is to overcome pragmatic boundaries that arise because we are invested in our practices (Carlile, 2004). Thus, while Nonaka (1994) defines knowledge as "justified true belief" what is justified depends on ones' practice history. In this sense, it is the boundaries between communities that create problems for knowledge sharing with community members "blinded" by their particular worldview and taken-for-granted sociomaterial practices. Since most action in organizations occurs in the context of multiple communities "tied together by interconnected practices" (Gherardi and Nicolini, 2002), the socially situated and contextually based nature of knowing poses as many problems as opportunities. Recognizing pragmatic boundaries between diverse epistemic communities (Knorr-Cetina, 1999) indicates the importance of considering the role of boundary objects (Star, 1989) and mediators (Latour, 2005) for helping the translation (or transformation in Carlile's terms) of knowledge across communities and contexts.

17.4 Technology for Support of KM Processes

While earlier we have differentiated between the repository and network views of KM and suggested that these are the product of different epistemological and ontological assumptions, in reality, most organizations combine the repository and network perspectives in developing their KMS. This is consistent with our mutual constitution view that recognizes both possessed knowledge and practice-based knowing. A variety of technical systems are thus introduced, together with social systems that can support the use of these technical systems. Quaddus and Xu (2005), for example, identified a number KMS, which are reported in Table 17.1 (the % in the table indicates the popularity of the different technologies).

TABLE 17.1   Technology-Based Knowledge Management Systems

E-mail (92%)	Video conferencing (43%)	Electronic bulletin boards (29%)	Best practice database (22%)	Extranet (17%)
Internet (90%)	Online discussion systems (40%)	Electronic meeting systems (26%)	Corporate yellow pages (22%)	Issue management systems (16%)
Databases (86%)	Workflow systems (39%)	Learning tools (25%)	Online analytical processing systems (21%)	Knowledge directories (15%)
Intranet (80%)	Data warehousing/ mining (37%)	People information archive (23%)	Knowledge repositories (21%)	Expert systems (8%)
Document management systems (60%)	Search and retrieval tools (36%)	Decision support systems (23%)	Knowledge portals (19%)	Artificial intelligence (5%)
Customer management systems (48%)	Executive information systems (34%)	Groupware (22%)	Lessons-learnt databases (18%)	

TABLE 17.2   Knowledge Management System and Knowledge Processes

Knowledge Process	Knowledge Creation	Knowledge Storage	Knowledge Transfer	Knowledge Application
KMS	–E-learning systems	–Knowledge repositories (data warehousing and data mining)	–Communication support systems (e-mail)	–Expert systems
	–Collaboration support systems		–Enterprise information portals (intranets/ internets)	–Decision support systems

Alavi and Tiwana (2002) categorize these different KMS in terms of the knowledge processes that they aim to enhance (knowledge creation, storage, transfer, and application) as depicted in Table 17.2.

This is helpful in the sense that it recognizes that different types of ICT will be more or less useful for different knowledge processes. At the same time, the Quaddus and Xu (2005) study demonstrates that it is the storage and transfer technologies (based on the repository view) that have been most popular in terms of the types of KMS that are used in practice. However, unfortunately this was a 2005 published paper and may not reflect the current state of play.

17.5 Knowledge Exploration and Exploitation

Most of the discussion so far has considered knowledge at the individual or local level—in particular, sites of knowing. However, we are also interested in looking at KM from an organizational perspective. On the one hand, organizational knowledge can be described as "more than the sum of the parts" (Simon, 1962); on the other hand, "organizations know more than they can tell" (Polanyi, 1967). To some extent, these two sayings reflect the possession and the practice perspectives, respectively, and so from a mutual constitution perspective, both are, paradoxically, insightful even though they seem to be opposite ideas. Moreover, from the perspective of this section of the chapter, the issue of organizations knowing more than they can tell relates to the problems surrounding exploitation, while the issue of organizational knowledge being more than the sum of the individual parts relates to the opportunities for exploration.

It is widely argued that organizational knowledge is a key asset for any organization and can lead to success (Grant, 1996). It is also suggested that a balance between knowledge exploration and knowledge exploitation is positively associated with performance (Cohen and Levinthal, 1990; Levinthal and March, 1993). Among others, March (1991) indicates that achieving a balance between exploitation and exploration will affect an organization's survival and prosperity. In this section, we aim to deepen the issues associated with organizational knowledge, knowledge exploration, and knowledge exploitation. In so doing, we will apply the lens of absorptive capacity (Cohen and Levinthal, 1990), that is, the capability to capture external knowledge and transform it into organizational assets for innovation. In particular, we will focus on Zahra and George's (2002) reconceptualization of the construct. In their paper, they highlight that knowledge should be "explored" (potential absorptive capacity) and then "exploited" (realized absorptive capacity), and they also provide a process view of knowledge absorption that fits with the view of managing knowledge taken in this chapter.

17.5.1 Absorptive Capacity (a la Zahra and George)

Absorptive capacity was originally defined by Cohen and Levinthal (1990) as "the ability of a firm to recognize the value of new, external information, assimilate it, and apply it to commercial ends" (p. 128). Cohen and Levinthal (1990) also suggest that organizational absorptive capacity (1) builds on prior investments in building absorptive capacity in individuals; (2) develops cumulatively and tends to be path-dependent (i.e., what has been learnt in the past sets limits on what can be learnt in the future); and (3) depends on the organization's ability to share knowledge internally. Thus, since individual learning is cumulative, learning performance is greater when the object of learning (the new knowledge to be identified, assimilated, and applied) is related to what individuals already know. In sum, absorptive capacity is a mix of capabilities that allow individuals and organizations to capture external knowledge, process it, and apply it to innovation and therefore to organizational performance.

Zahra and George (2002) revised the absorptive capacity construct and highlighted that it has two main states, potential and realized: "potential capacity comprises knowledge acquisition and assimilation capabilities, and realized capacity centers on knowledge transformation and exploitation" (p. 185). This suggests that potential absorptive capacity captures processes of (external) knowledge exploration; that is, organizational actors scan the external environment with the aim to identify knowledge that, potentially, can be processed and used within the organization. However, potential absorptive capacity "does not guarantee the exploitation of this knowledge" (Zahra and George, 2002, p. 190). Realized absorptive capacity reflects the firm's capability to elaborate and exploit the knowledge that has been identified. Therefore, knowledge exploration and knowledge exploitation are complementary (March, 1991). Moreover (as discussed in the next sections), processes and mechanisms that support potential absorptive capacity incorporate characteristics of the epistemology of possession while processes and mechanisms that support realized absorptive capacity incorporate characteristics of the epistemology of practice. This twofold lens of analysis of absorptive capacity (possession vs. practice perspective) and the assumption that potential and realized absorptive capacity are complementary follow the mutual constitution view of knowledge that was introduced in the first part of this chapter. The next paragraphs discuss this in more detail.

17.5.2 Potential Absorptive Capacity and the Epistemology of Possession

Potential absorptive capacity includes two main capabilities: knowledge acquisition and knowledge assimilation (Zahra and George, 2002).

Knowledge acquisition is the firm's capability to identify and acquire only the external knowledge that is relevant for its business; that is, not all external knowledge should be taken into consideration, instead, a selection process should take place. Relevant knowledge is the knowledge that might contribute to innovation and, therefore, organizational performance (Cohen and Levinthal, 1990; Jansen et al., 2005). Moreover, in order for a firm to gain and maintain a competitive advantage, it is crucial that external knowledge is acquired quickly—or, at least, quicker than other competitors.

The ICT industry is very rich with "cases" of exploration and exploitation of technology-related knowledge (and associated issues). In this chapter, we would like to use the example of Facebook's recent acquisition of Instagram and the processes that supported the implementation of Instagram's functionalities into the Facebook portal and led to Facebook's users exploiting the new software (i.e., the software from Instagram).

Instagram, before being acquired by Facebook, was a start-up company that provided a set of tools to automatically upload and share photos on personal websites. Some of Instagram's features help users to graphically modify photos so that even those users who are not very familiar with photo-editing software (such as Photoshop®) are able to create amazing graphical effects.

In April 2012, Facebook acquired Instagram. From an absorptive capacity perspective, we can say that Facebook boundary spanners identified the start-up company whose business was tightly related to one of the main functionalities of Facebook—uploading personal photos to one's profile. After the identification of Instagram as a potential acquisition, communication between the board and the Facebook CEO (Mark Zuckerberg) took place. Shortly after, the decision to acquire Instagram was made and took less than a week to complete. Of course, at Facebook, both the boundary spanners and the board (in particular, Mr. Zuckerberg, who has considerable decision power in Facebook) had a background knowledge that allowed them to evaluate the external knowledge. That is, knowledge acquisition happened through boundary spanners, and decisions were made by powerful people in the organization (CEO and the board). How successful this acquisition will be in the long term is difficult to judge, but the case does illustrate a very swift process of (knowledge) acquisition, which is associated by some with competitive advantage (Argote and Ingram, 2000).

Knowledge assimilation includes processes of analysis and interpretation of external knowledge (Sun and Anderson, 2010). Knowledge assimilation can occur informally (through social networks) or formally (with the use of coordinators) (Zahra and George, 2002). Drawing again from the Facebook example, after the acquisition Instagram's technology was incorporated into Facebook's portal, moreover, Facebook users were "trained" on how to use the new service and connect it to their Facebook webpage. At Facebook, knowledge assimilation occurred once the technicians were able to make the new software available to the users; moreover, the marketing department identified effective ways to "sell" Instagram to Facebook users. Thus, while knowledge identification involved just senior management, knowledge assimilation involved technical and nontechnical processes and business units to make the new knowledge available for the users (with Facebook "customers" (i.e., the users) seen as part of the same—at large—community where absorptive capacity assimilation takes place).

In sum, both knowledge acquisition and assimilation are driven by prior knowledge; consequently, potential absorptive capacity is path-dependent (Cohen and Levinthal, 1990; Lane et al., 2006). This suggests that potential absorptive capacity is a combination of exploratory processes that can be effectively analyzed using the epistemology of possession. In fact, it is emphasized (in line with Lane and Lubatkin, 1998) that once the external knowledge is sufficiently similar to the knowledge of the potential recipient, identification and assimilation occur relatively easily. Moreover, it is quite common that the recognition and assimilation of knowledge are undertaken by individual actions rather than collective efforts, as per the Facebook example (i.e., boundary spanners and upper management identified new external knowledge and some technicians made the knowledge available for users).

Finally, the boundaries between Facebook and the environment (in the case of knowledge identification) and between Facebook and the users (in the case of knowledge assimilation) were syntactic and semantic. Syntactic boundaries were (easily) overcome in the case of knowledge identification: The knowledge related to Instagram (and its functionalities) was straightforwardly transferred to the Facebook technicians. On the other hand, syntactic boundaries were overcome when the functionalities of Instagram were made available to the worldwide Facebook users (i.e., with basic communication to the users and the translation in multiple languages of essential to-do lists on how to use the new photo-editing software and on how to manage and automatically upload the photos on one's Facebook's profile). Moreover, semantic boundaries existed (and were overcome) when the technicians were able to incorporate the characteristics of Instagram into the Facebook portal to make the new service easily accessible by the users. As previously highlighted, once a relevant amount of prior knowledge overlaps the external knowledge, syntactic and semantic boundaries can be overcome without major challenges. Lane and Lubatkin's (1998) study on "relative absorptive capacity"—that is, knowledge transfer between two firms, one "learner" organization (the donator of knowledge) and one "student" organization (the recipient)—suggests that the more the overlapping knowledge between two firms, the more the process of knowledge transfer is straightforward.

17.5.3 Issues Associated with Knowledge Acquisition and Assimilation (Potential Absorptive Capacity)

From the preceding text, it is clear that knowledge acquisition and assimilation (potential absorptive capacity) are affected by path dependency. That is, an organization is likely to acquire knowledge that is similar to the knowledge it already knows. Once the acquisition of some particular knowledge starts, the organization is likely to keep trying to acquire similar knowledge in a particular area. In the following text, we discuss some of the most common issues that can occur within the process of knowledge acquisition and assimilation (potential absorptive capacity).

Issue 1: Context specificity of knowledge. Knowledge identification and assimilation can be difficult when external knowledge incorporates context-specific characteristics (Carlile, 2004) that often prevent outsiders from understanding and replicating such knowledge; for instance, as noted earlier in this chapter, Szulanski (1996) calls it "stickiness."

Issue 2: Complementary assets. Identifying and assimilating external knowledge can be difficult when the value of knowledge is associated with the existence of complementary assets that might not be available in the recipient firm (Teece, 1981; Zahra and George, 2002).

Issue 3: Prior knowledge and incremental innovation. One could argue that the very close relationship between prior knowledge and external knowledge might represent a hindrance to the implementation of disruptive innovation (Christensen, 1997). Namely, along with the idea that organizations are good when pursuing incremental innovation (Facebook improved an existing functionality of the platform—in fact, it was already possible to upload photos, even if with a more complex process), managers are less likely to identify knowledge that creates breakthrough innovation.

Issue 4: Power. Another issue that can be associated with potential absorptive capacity and, therefore, with the identification and assimilation capabilities is power. The idea that power affects managing knowledge is not new (Argote et al., 2003; Marshall and Brady, 2001; Pfeffer, 1981). While, in the previous sections, we highlighted that knowledge can lead to power (Bartol and Srivastava, 2002), here we also show how power can affect knowledge absorption (Easterby-Smith et al., 2008). In the literature on absorptive capacity, for instance, Todorova and Durisin (2007) focus on why some firms develop more absorptive capacity than others and suggest that power affects the extent to which organizations are able to explore external knowledge. This is interesting because it suggests that (few) people exercise power in organizations regarding decision-making processes that involve the introduction of technological innovation based on external knowledge. For instance, Marabelli and Newell (2009) document the case of an enterprise resource planning (ERP) implementation that initially was led by some managers from the adopting firm's sales department: they decided which system to implement, when to start the implementation, and with which characteristics (a lot of customization, in their case). One negative consequence of the exercise of power by just a few organizational actors was that while the recognition process was conducted very quickly, the assimilation phase—actual introduction of the system in the organization—encountered a number of technical and nontechnical problems. Differently from Facebook and Instagram, in this example, (Marabelli and Newell, 2009) power acted as a barrier (rather than a source of competitive advantage) to knowledge identification and assimilation.

In sum, we can conclude that, although recognizing and assimilating external knowledge can be seen as straightforward, a number of obstacles can occur: External knowledge can be associated with other invisible assets that are needed to fully understand new knowledge; since external knowledge is often very similar to the knowledge that an organization already has, radical innovation is limited; and power relationships can also affect potential absorptive capacity. In terms of power, the effects can be either positive or negative and this probably depends on the core competences of the leaders who exercise power: In the case of Facebook, Mr. Zuckerberg is a visionary innovator; in the case of Alpha Co. (the example in Marabelli and Newell, 2009), the management was probably not so skilled to make decisions regarding how to introduce a new ERP system.

17.5.4 Realized Absorptive Capacity

Within realized absorptive capacity, two main capabilities are identified: knowledge transformation and knowledge exploitation (Zahra and George, 2002).

Knowledge transformation is the ability to combine existing knowledge and the newly acquired and assimilated knowledge. This can occur by adding or deleting knowledge or by interpreting existing in-firm knowledge in light of external inputs. Zahra and George, drawing from Koestler (1966), depict knowledge transformation as a process where organizations recognize two apparently incongruous sets of information (external knowledge and in-firm knowledge) and are able to combine them to arrive at a new schema. Knowledge transformation capabilities are needed for strategically managing organizational change. During the transformation phase of absorptive capacity, knowledge has to be shared internally since business units' coordination, cross-division communication, and integration processes facilitate knowledge flows (Van de Bosch et al., 1999). Zahra and George's conceptualization of knowledge transformation (at least the way it is interpreted in this chapter) is very similar to Carlile's discussion on pragmatic boundaries (he also refers to knowledge transformation) and to Latour's idea of knowledge translation—that is, knowledge needs to be, to some extent, recreated or reproduced by those who are practicing it.

Bringing up again the case of Facebook (and Instagram), we can identify processes that can be linked to knowledge transformation. For instance, once the main technical problems were solved (the incorporation of Instagram's functionalities into the Facebook portal), and once a marketing campaign was conducted to inform the Facebook users about the availability of a new photo-editing/uploading tool, a more challenging issue emerged: The external knowledge (e.g., Instagram, and how photo digital editing can work for "dummies") had to be acknowledged by the users as interesting, appealing, and helpful. The literature on the introduction of new technology highlights that if the users are aware of its benefits they are also more likely to adopt it (Ferneley and Sobreperez, 2006; Hwang and Thron, 1999). And this is what happened at Facebook: In the phase of knowledge assimilation, Facebook put a lot of effort into informing the users of the benefits of Instagram for uploading photos on their personal pages. However, having the users adopt the technology is not an automatic process (Easterby-Smith, 2008). In fact, while tasks such as the Marketing department promoting innovation can be relatively straightforward, the actual adoption of technology requires more complex elaboration of external knowledge (Elbanna, 2006) that, as we argue in this chapter, requires in-context practice. This means that it is not just that the users need to be aware that some new functionality is available; they must be able to incorporate the novelty into their current routines. So, basic processes such as taking a picture with the IPhone and uploading it using the "photos" functionality available in the IPhone–Facebook "app" is replaced with taking the picture and using the "Instagram" functionality (still available as an iPhone app). Initially, moving from a familiar photo-editing and upload system to a more innovative system will be perceived as a more complex way to have photos on one's personal page; users need to become familiar with a different way to process pictures and this can take time and needs practice. The driver that makes the change happen is the user's belief that a process, which seems relatively complex, will be more "productive" in the long term (i.e., it is more powerful and adds new functionalities such as for-dummies image editing features).

Moreover, the new knowledge (again, Instagram) is shared among groups within social networks, based on virtual word-of-mouth (Gruhl et al., 2004). The way the Facebook portal is designed allows others to see which application was used to create a photo that was shared, stimulating people's curiosity and creating the willingness to try a new, potentially interesting service. Users are then able to evaluate the new "tool" through practicing it. For users, moving to Instagram does not imply that they have to use it every time they want to upload a photo. Therefore, users may perceive that it is worth trying the tool. Online discussions can be seen as virtual communities of practice (Dube' et al., 2005) where new external knowledge is collectively combined in an online forum of practice.

Knowledge exploitation is the capability that allows organizations to actually apply the previously transformed knowledge for commercial purposes (Cohen and Levinthal, 1990); successful knowledge exploitation includes refinement, extension, and leverage of existing competences and creation of new ones by incorporating external knowledge into existing organizational processes and procedures (Tiemessen et al., 1997).

The successful knowledge exploitation at Facebook occurred when new users started to use Instagram. Knowledge exploitation can take time; in fact, as previously highlighted, Facebook users needed to practice the new tool for a while before deciding that it was better than the old system to upload photos. Moreover, at the time of writing, knowledge exploitation processes are still ongoing at Facebook.

Nevertheless, some conclusions can be drawn: First, effective knowledge exploration can lead to successful knowledge exploitation. Even if we do not argue that knowledge acquisition, assimilation, transformation, and exploitation can be represented as a linear series of processes, it is evident that only the acquisition of relevant knowledge can lead to its assimilation (because the firm has the skills to implement it into existing routines and processes; Zahra and George, 2002). Moreover, only effective assimilation (i.e., knowledge integration) leads people to (try to) practice new external knowledge with the aim to translate it and, consequently, to exploit it for commercial use (Cohen and Levinthal, 1990).

The preceding text suggests that during transformation and exploitation processes, knowledge should be collectively "practiced." Consequently, an epistemology of practice should be used to investigate realized absorptive capacity. Moreover, in the case of knowledge transformation, knowledge-sharing boundaries are pragmatic: As indicated in Carlile (2004), shared artifacts and methods play an important role in achieving the capacity to negotiate interests and transform knowledge. This is particularly evident in the Facebook example where the users have the possibility to see others' activities (i.e., somebody who was an early adopter of Instagram) and send positive feedback ("I like it" or "I don't like it") through the online portal (Facebook).

17.5.5 Issues Associated with Knowledge Transformation and Exploitation (Realized Absorptive Capacity)

Although effective knowledge identification and assimilation can lead to successful knowledge transformation and exploitation, realized absorptive capacity can encounter problems. In the following text, we list common issues related to the effective transformation and exploitation of new external knowledge.

Issue 1: Barriers to knowledge sharing. Knowledge transformation and exploitation are negatively affected by (potential) barriers to knowledge sharing. Since knowledge transformation happens through practice, people can benefit from the possibility of interacting with and discussing the new knowledge. It is important that users' experiences are shared among other users so that social processes of knowledge reproduction take place. Moreover, knowledge exploitation is the collective effort of implementing novelty into organizational routines so that the whole organization benefits from external knowledge (Zahra and George, 2002). Therefore, cross-unit barriers need to be overcome and clear common objectives identified. Pragmatic boundaries can make this difficult.

Issue 2: Long-term processes. Knowledge transformation and exploitation often require long-term practice (Lane et al., 2001) for the external knowledge to be recreated in the form of working knowledge. It is arguable that transformation capabilities are achieved in particular organizational settings such as communities of practice where frequent member interactions (as happens between Facebook members, Ellison et al., 2007; Wilson et al., 2009) allow both the appropriation of the new knowledge and the creation of new knowledge using ideas from external inputs. However, short-term objectives and the rush to transform inputs (i.e., external knowledge) into tangible outputs (i.e., innovation, therefore performance) do not always promote knowledge transformation processes.

Issue 3: Power. Power affects knowledge transformation and exploitation. Given that both capabilities should be developed through practice, Nicolini (2011) points to the importance of addressing power issues. Moreover, Newell et al. (2009) noted that, although ideally organizational actors will channel their efforts to successfully share (and eventually transform) knowledge, this is quite a simplistic view of organizations. Assuming that common goals are always clear to employees is a very functionalist view of organizations (Burrell and Morgan, 1979). On the contrary, empirical results (Perez-Bustamante, 1992; Wagner et al., 2002) indicate that very often encouraging employees to surrender their knowledge for the benefit of the organization has the result of disempowering them. This is one reason why individuals might choose to "hoard" rather than share knowledge. Knowledge exploration too can be (often negatively) affected by reflecting interests of powerful groups. For instance, in large-scale projects of ERP implementation, it is not infrequent that decisions of powerful coalitions on how to exploit external knowledge result in a failure of the overall implementation of the external knowledge (the ERP system) (Kholeif et al., 2007; Yusuf et al., 2004).

As with potential absorptive capacity, then, realized absorptive capacity also presents issues because although effective knowledge exploration *can* lead to successful knowledge exploitation, this is not necessarily a straightforward process. Realized absorptive capacity requires practice for external knowledge to be transformed, and this practice is likely to throw up various problems because it can conflict with previous ways of doing things. This indicates that power issues may emerge during this period of transformation. Power relationships, thus, affect both potential and realized absorptive capacity. Our examples indicate that while power and successful potential absorptive capacity are mediated by the skills and experience of powerful decision makers (i.e., Mr. Zuckerberg), for realized absorptive capacity, the transformation and application phases are less likely to be fully controlled by dominant coalitions. Instead, emergent interactions among users (in this case, Facebook users) take place, which influence whether and how the knowledge is transformed and exploited. This is also consistent with the unpredictable nature of knowledge, spelled out by scholars who theorize from the practice perspective (Chia, 2003; Feldman and Orlikowski, 2011; Sandberg and Tsoukas, 2011).

In conclusion, we believe that the absorptive capacity construct is appropriate to provide a theoretical lens to clearly identify processes of knowledge exploration and exploitation. We acknowledge that absorptive capacity depicts these processes as being very linear: knowledge exploration, therefore, knowledge exploitation. This might not be what actually happens in organizations when they try to deal with processes of transformation of external knowledge into innovation. For example, the literature on innovation shows that a number of iterations, failures, and lessons learned might be needed (Dougherty and Dunne, 2011). However, we argue that the absorptive capacity construct is useful to graphically represent some knowledge processes that might occur when organizations strive to pursue innovation. Drawing from the Facebook example, we propose a synthesis of these processes in light of absorptive capacity, which are depicted in Figure 17.1.

Figure 17.1 highlights that there are different types of external knowledge and that the knowledge that is identified is likely to be similar to the knowledge that organizations already know (recognition capability); once some specific knowledge is identified as "potential," the organization needs to (quickly) implement it into existing routines. Firstly, technical implementation occurs (assimilation), and secondly, organizational actors (the "users" of the new knowledge) need to actually adopt the innovation and, we argue, this happens through practice (transformation); practicing the new knowledge might result in the creation of new "working" knowledge which is, eventually, fully exploited (i.e., the new knowledge becomes part of daily routines).

The absorptive capacity construct and, in particular, the tension between potential and realized absorptive capacity is also linked to other knowledge "elements" and metaphors that were discussed previously in this chapter (i.e., the repository view vs. the network view). Therefore, the aim of the next

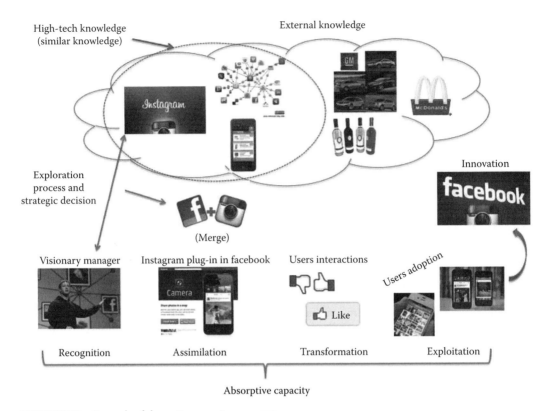

FIGURE 17.1 Example of absorptive capacity processes.

section is to synthesize the aforementioned and to suggest a comprehensive framework of knowledge that highlights the tension between exploration and exploitation under the umbrella of the epistemology of possession and the epistemology of practice.

17.5.6 Combining Potential and Realized Absorptive Capacity: The Mutual Constitution View

It is now clear that both exploration and exploitation processes coexist within a firm. Interpreting knowledge exploration and exploitation with the lens of absorptive capacity also suggests that while potential absorptive capacity can be interpreted with the lens of the epistemology of possession (emphasis on individual decision makers who own knowledge and emphasis on knowledge that is transferrable), realized absorptive capacity is better interpreted with the lens of the epistemology of practice. On the one hand, the epistemology of possession suggests that knowledge is owned by individuals and can be easily transferrable using ICT artifacts such as centralized databases and electronic archives (repository view); this is particularly true for potential absorptive capacity, as per our examples where the processes of knowledge *recognition* and *assimilation* were straightforward because syntactic and semantic boundaries can be easily overcome. On the other hand, the epistemology of practice indicates that some knowledge is also embedded in everyday practices; the main consequence is that it might not be sufficient to "store" knowledge in some repository to have it transferred to others. Instead, in this second case (practice perspective), knowledge is equated to practice (Nicolini, 2011) and, therefore, knowledge is *transformed* through negotiations (pragmatic boundaries) in order for the knowledge to be fully *exploited*. The network view and the realized absorptive capacity are, therefore, more appropriate concepts to represent these processes.

TABLE 17.3   Managing Knowledge: A Comprehensive Framework

Epistemology	Views	Ways of Sharing	Boundaries	AC View	Articulations
Possession	Repository	Transfer	Syntactic	Exploration	Acquisition
			Semantic		Assimilation
Practice	Network	Transformation ((re)production)	Pragmatic	Exploitation	Transformation
					Exploitation

Table 17.3 synthesizes the earlier theorizing and examples that are collapsed into a comprehensive framework of knowledge that considers epistemologies, views, ways of sharing it, and absorptive capacity capabilities.

From Table 17.3, it emerges that knowledge can be seen both as a possession (knowledge can be "moved about" within and between organizations) and as a practice. This is in line with the mutual constitution view that was illustrated in the first part of this chapter.

Interestingly, the mutual constitution view requires multilevel analyses. The possession view discusses both individual and organizational knowledge. For instance, Nonaka's (1994) SECI (Socialization, Externalization, Combination, and Internalization) process of knowledge conversion (from tacit to explicit) discusses the individual as well as the organizational level. However, the underlying assumption is that knowledge is "exchanged" among organizational actors (using socialization processes) and, therefore, the very nature of knowledge lies in individuals. The practice view discusses both individual and organizational knowledge. In fact, even if the emphasis is on collective knowledge that originates with practice, individuals can affect knowledge processes (being part of a community of practice, if we adopt the containment view, or being "actors that act in practice" (D'Adderio, 2006) if we adopt the radical view). This is highlighted by concepts such as consequentiality (Feldman and Orlikowski, 2011): An action is considered to be a practice when it is consequential for the development of the activity. Therefore, it is implicitly acknowledged that individuals can "actively" affect practices. Another concept from the practice view that highlights the role of individuals is the "nexus of practices," which indicates that a link between (individual) actions exists and that some actions might "matter more" than other actions (i.e., power relationships play a role in action). However, although these two examples highlight that the practice epistemology of knowledge considers individuals, the main focus of this epistemology is on collective processes. Thus, even while individuals matter from the practice epistemology, knowledge does not lie in an individual's mind but rather in their practices, which are always undertaken in conjunction with other actors (whether human and/or material actors). In conclusion, combining the epistemology of possession and the epistemology of practice is helpful for exploring knowledge processes from a multilevel perspective which, in turn, is helpful for fully exploring knowledge management dynamics (Hannah and Lester, 2009).

17.6 Future of Knowledge Management

Some have argued that KM is a fad, just like there have been many other fads related to ICT adoption in organizations (Swan et al., 1999). We would argue that while the term KM may not survive, the focus on improving ways to explore and exploit knowledge in organizations will not diminish. Indeed, while traditionally organizations were either designed to maximize efficiency or flexibility (Thompson, 1967), today organizations are exhorted to be "ambidextrous" (Adler et al., 1999; Newell et al., 2003; O'Reilly and Tushman, 2007; Tushman and O'Reilly, 1996). Being ambidextrous requires a focus on both exploitation (to constantly increase efficiency of what the organization is currently doing) and exploration (to increase flexibility through innovating in products, services, or organizational processes). In the past, as we have seen, much KM effort has been on the exploitation aspect, with organizations introducing document management systems, searchable databases, and intranets so that people can more effectively share and so exploit the knowledge that already exists within an organization. In the future, we anticipate that a similar level of effort will be placed on exploration, using social networking and other multimedia technologies that can help to connect people. Such social network-based online platforms (Facebook-like technologies for use within an organization) that focus on collaboration and cross-fertilization among different business units can help organizations develop informal ways to share information and knowledge to promote innovation. These types of KMS can help to stimulate the development of (virtual) communities of practice that bring together people from different locations, and so take advantage, potentially, of talent that is globally distributed (Dube' et al., 2006).

It has been demonstrated that innovation leads to long-term competitive advantage and that interactions (i.e., knowledge exchanges) among people with different backgrounds and skills can help to promote innovative and disruptive ideas (Desrochers, 2001; Hanna and Walsh, 2002). Therefore, in the past, a lot of attention was given to organizational design, structuring processes so that departments would communicate at least for specific initiatives, for example, product innovation involving marketing, sales, and R&D, and cross-unit process innovation involving different business units (Hansen and Birkinshaw, 2007; Martin and Eisenhardt, 2010). Moreover, strategic links between organizational actors such as project managers, process managers, product managers, and so on were put in place. Nowadays, organizations are less bureaucratic and more network-based (Borgatti and Foster, 2003; Capaldo, 2007), that is, they do not try and undertake all activities but rather work with other organizations who may be better placed to provide particular types of service, as with outsourced IT service provision. This network will typically be globally distributed and rely heavily on ICT for information and knowledge sharing. Along with dealing with syntactic and semantic boundaries (i.e., different languages and therefore difficulties in communicating to exchange basic information; time change differences; cultural and background differences), overcoming pragmatic boundaries are likely to be the most challenging issue KM (and KMS) face. That is, worldwide dispersed units of a networked organization need to be able to collaborate online and create synergies that are typical of localized communities of practice. While, from a technical point of view, effective online platforms that allow video meetings/conference calls already exist, people are now learning how to use ICT as a substitute for face-to-face relationships. While such virtual relationships can be very seductive (as evidenced by the amount of time people spend on sites such as Facebook and Twitter, as well as in online gaming sites), questions remain about differences between online and face-to-face relationships and their respective impact on our social world.

17.7 Conclusions and Implications

In this chapter, we aimed to provide an overview of knowledge and its exploratory and exploitative nature (*What is knowledge?*) with respect to two different epistemologies that relate to different approaches to managing knowledge: the epistemology of possession (*repository view*) and the epistemology of practice (*network view*). We then highlighted that, today, technology plays a role in supporting knowledge processes (*technology for support of KM processes*), and we suggested that the absorptive capacity construct could be adopted to highlight knowledge exploration and exploitation processes (*knowledge exploration and exploitation*).

We concluded arguing that the mutual constitution view (a "generative dance," Cook and Brown, 1999) is an effective representation of KM processes and can be linked to potential and realized absorptive capacity. Moreover, we noted that adopting a mutual constitution view allows multilevel analysis: Individuals possess knowledge generated by engaging and investing in (collective) practices. Finally, adopting the mutual constitution view of knowledge has implications for technology: For instance, while the possession perspective indicates that data sharing and centralization (e.g., using ERP architectures) has several advantages for creating an organizational memory (Vandaie, 2008), the practice perspective highlights that technology can be fully exploited only if the "users" actually adopt and use IT artifacts in their practices (Feldman and Orlikowski, 2011).

References

Adler P. S., Goldoftas B., and Levine D. I., 1999. Flexibility versus efficiency? *A Case Study of Model Changeovers in the Toyota Production System*, Vol. 10(1), pp. 43–68.

Alavi M., Kayworth T. R., and Leidner D. E., 2006. An empirical examination of the influence of organizational culture on knowledge management practices, *Journal of Management Information Systems*, 22(3), 191–224.

Alavi M. and Tiwana A., 2002. Knowledge integration in virtual teams, *Journal of the American Society for Information Science and Technology*, 53(12), 1029–1037.

Argote L. and Ingram P., 2000. Knowledge transfer: A basis for competitive advantage, *Organizational Behavior and Human Decision Processes*, 82(1), 150–169.

Argote L., McEvily B., and Reagans R., 2003. Managing knowledge in organizations: An integrative framework and review of emerging themes, *Management Science*, 49(4), 571–582.

Barney J., 1991. Firm resources and sustained competitive advantage, *Journal of Management*, 17, 99–120.

Bartol K. M. and Srivastava A., 2002. Encouraging knowledge sharing: The role of organizational renewal systems, *Journal of Leadership and Organizational Studies*, 9(1), 64–76.

Borgatti S. P. and Foster P. C., 2003. The network paradigm in organizational research: A review and typology, *Journal of Management*, 29(6), 991–1013.

Burrell, G. and Morgan G., 1979. *Sociological paradigms and organizational analysis*. London: Heinemann Educational Books.

Capaldo A., 2007. Network structure and innovation: The leveraging of a dual network as a distinctive relational capability, *Strategic Management Journal*, 28, 585–608.

Carlile P. R., 2004. Transferring, translating, and transforming: An integrative framework for managing knowledge across boundaries, *Organization Science*, 15(5), 555–568.

Chia R., 2003. From knowledge-creation to the perfecting of action: Tao, Basho and pure experience as the ultimate ground of knowing, *Human Relations* 56(8), 953–981.

Christensen C. M., 1997. *The Innovator's Dilemma: When New Technologies Cause Great Firms to Fail*. Boston, MA: Harvard Business School Press.

Cohen W. M. C. and Levinthal D. A., 1990. Absorptive capacity: A new perspective on learning and innovation, *Administrative Science Quarterly*, 35, pp. 128–152.

Connelly C. E., Zweig D., Webster J., and Trougakos J. P., 2012. Knowledge hiding in organizations, *Journal of Organizational Behavior*, 33(1), pp. 64–88.

Cook S. D. N. and Brown, J. S., 1999. Bridging epistemologies: The generative dance between organizational knowledge and organizational knowing, *Organization Science* 10(4), 381–400.

D'Adderio L., 2006. The performativity of routines: Theorizing the influence of artifacts and distributed agencies on routines dynamics, *Research Policy* 37(5), 769–789.

Davenport T. H. and Prusak L., 1998. *Working knowledge: How organizations manage what they know*. Boston, MA: Harvard Business School Press.

Desouza K. C. and Vanapalli G. K., 2005. Securing knowledge assets and processes: Lessons from the defense and intelligence sector, *Proceedings of the 38th Hawaii International Conference on System Sciences (HICSS)*, Island of Hawaii, HI.

Desrochers P., 2001. Local diversity, human creativity, and technology innovation, *Growth and Change*, 32, 369–394.

Donate M. J. and Guadamillas F., 2010. The effect of organizational culture on knowledge management practices and innovation, *The Journal of Corporate Transformation*, 17(2), 82–94.

Dougherty D. and Dunne D. D., 2011. Organizing ecologies of complex innovation. *Organization Science*, 22(5), 1214–1223.

Dreyfus H. L., 1995. *Being-in-the-World: A Commentary on Heidegger's Being and Time, Division I*. Cambridge, MA: MIT Press.

Dube' L., Bourhis A., and Jacob R., 2005. The impact of structural characteristics on the launching of intentionally formed virtual communities of practice. *Journal of Organizational Change Management*, 18(2), 145–166.

Dube' L., Bourhis A., and Jacob R., 2006. Towards a typology of virtual communities of practice, *Interdisciplinary Journal of Information, Knowledge, and Management*, 1, 69–92.

Easterby-Smith M., Lyles M. A., and Tsang E. W. T., 2008. Inter-organizational knowledge transfer: Current themes and future prospects, *Journal of Management Studies*, 45(4), 677–690.

Elbanna A. R., 2006. The validity of the improvisation argument in the implementation of rigid technology: The case of ES Systems, *Journal of Information Technology*, 21, 165–175.

Ellison N. B., Steinfield C., and Lampe C., 2007. The benefits of Facebook "Friends": Social capital and college students' use of online social network sites, *Journal of Computer-Mediated Communication*, 12(4), 1143–1168.

Feldman M. S. and Orlikowski W., 2011. Theorizing practice and practicing theory, *Organization Science*, 22, 1240–1253.

Ferneley E. H. and Sobreperez P., 2006. Resist, comply or workaround? An examination of different facets of user engagement with information systems, *European Journal of Information Systems*, 15, 345–356.

Fitzgerald M., 2003. At Risks Offshore, *CIO Magazine*, November.

Gherardi S., 2006. *Organizational Knowledge: The Texture of Workplace Learning*. Oxford, U.K.: Blackwell.

Gherardi S. and Nicolini D., 2002. Learning in a constellation of inter-connected practices: Canon or dissonance? *Journal of Management Studies*, 39(4), 419–436.

Grant R. M., 1996. Towards a knowledge-based theory of the firm, *Strategic Management Journal*, 17(Winter Special Issue), 109–122.

Gruhl D., Guha R., Liben-Nowell D., and Tomkins A., 2004. Information diffusion through blogspace, *Proceedings of the 13th International Conference on World Wide Web*, New York, pp. 491–501.

Hackett B., 2000. *Beyond Knowledge Management: New Ways to Work and Learn*, New York: Conference Board Report.

Hanna V. and Walsh K., 2002. Small firm networks: A successful approach to innovation, *R&D Management*, 32(3), 201–207.

Hannah S. T. and Lester P. B., 2009. A multilevel approach to building and leading learning organizations, *The Leadership Quarterly*, 20(1), 34–48.

Hansen M. T. and Birkinshaw J., 2007. The innovative value chain, *Harvard Business Review*, 85(6), 339–360.

Hwang M. I. and Thron R. G., 1999. The effect of user engagement on system success: A meta-analytical integration of research findings, *Information and Management*, 35(4), 229–236.

Jansen J. J. P., Van de Bosch F. A. J., and Volberda H. W., 2005. Managing potential and realized absorptive capacity: How do organizational antecedents matter? *The Academy of Management Journal*, 48(6), 999–1015.

Kholeif A. O. R., Abdel-Kadel M., and Sheren M., 2007. ERP customization failure: Institutionalized accounting practices, power relations and market forces, *Journal of Accounting & Organizational Change*, 3(3), 250–269.

Kling R. and Courtright C., 2006. Group behavior and learning in electronic forums: A sociotechnical approach, *The Information Society: An International Journal*, 19(3), 221–235.

Knorr-Cetina K., 1999. *Epistemic Cultures: How the Sciences Make Knowledge*, Cambridge, MA: Harvard University Press.

Koestler A., 1966. *The Act of Creation*. London, U.K.: Hutchinson.

Lam W. and Chua A., 2005. Knowledge management projects abandonment: An exploratory examination of root causes, *Communication of the ACM*, 16, 723–743.

Lane P. J., Koka B. R., and Pathak S., 2006. The reification of absorptive capacity: A critical review and rejuvenation of the construct. *Academy of Management Review*, 31(4), 833–863.

Lane P. J. and Lubatkin M., 1998. Relative absorptive capacity and interorganizational learning, *Strategic Management Journal*, 19, 461–477.

Lane P. J., Salk J. E., and Lyles M. A., 2001. Absorptive capacity, learning, and performance in international joint ventures, *Strategic Management Journal*, 22, 1139–1161.

Latour B., 2005. *Reassembling the Social: An Introduction to Action-Network Theory.* Oxford, NY: Oxford University Press.

Lave J. and Wenger E., 1991. *Situated Learning. Legitimate Peripheral Participation.* Cambridge, U.K.: Cambridge University Press.

Levinthal D. A. and March J. G., 1993. The myopia of learning, *Strategic Management Journal,* 14(S2), 95–112.

Marabelli M. and Newell S., 2009. Organizational learning and absorptive capacity in managing ERP implementation processes, *ICIS 2009 Proceedings,* paper 136, Phoenix, AZ.

March J. G., 1991. Exploration and exploitation in organizational learning, *Organization Science,* 2(1), 71–87.

Markus M. L., Tanis C., and Fenema P. C., 2000. Multisite ERP implementation, *Communications of the ACM,* 43, 42–46.

Marshall N. and Brady T., 2001. Knowledge management and the politics of knowledge: Illustrations from complex products and systems, *European Journal of Information Systems,* 10(2), 99–112.

Martin J. A. and Eisenhardt K. M., 2010. Rewiring: Cross-business-unit collaborations in multibusiness organizations, *The Academy of Management Journal,* 53(2), 265–301.

McAfee A. P., 2006. Enterprise 2.0: The dawn of emergent collaboration, *MIT Sloan Management Review,* 47(3), 21–28.

McLure-Wasko M. and Faraj S., 2005. Why should I share? Examining social capital and knowledge contribution in electronic networks of practice, *MIS Quarterly,* 29(1), 35–57.

Meyerson D. and Martin J., 1987. Cultural change: An integration of three different views, *Journal of Management Studies,* 24(6), 623–647.

Newell S., Huang J. C., Galliers R. D., and Pan S. L., 2003. Implementing enterprise resource planning and knowledge management systems in tandem: Fostering efficiency and innovation complementarity. *Information and Organization,* 13(1), 25–52.

Newell, S., Robertson, M., Scarbrough, H., and Swan, J., 2009. Introducing knowledge work: processes, purposes, and context. In: *Managing Knowledge, Work, and Innovation,* 2nd ed. Basingstoke, U.K.: Palgrave MacMillan.

Newell S., Swan J., and Scarbrough H., 2001. From global knowledge management to internal electronic fences: Contradictory outcomes of intranet development, *British Journal of Management,* 12(2), 97–112.

Newell S. M. and Edelman L. F., 2008. Developing a dynamic project learning and cross-project learning capability: Synthesizing two perspectives, *Information Systems Journal,* 18(6), 567–591.

Nicolini D., 2007. Stretching out and expanding work practices in time and space: The case of telemedicine, *Human Relations,* 60(6), 889–920.

Nicolini D., 2011. Practice as site of knowing: Insights from the field of telemedicine, *Organization Science,* 22(3), 602–620.

Nonaka I., 1994. A dynamic theory of organizational knowledge creation, *Organization Science,* 5(1), 14–37.

O'Reilly C. A. and Tushman M. L., 2007. Ambidexterity as a dynamic capability: Resolving the innovator's dilemma, *Research in Organizational Behavior,* 28, 1–60.

Orlikowski W. J., 2007. Sociomaterial practices: Exploring technology at work, *Organization Studies,* 28(9), 1435–1448.

Orr J., 1996. *Talking About Machines: An Ethnography of a Modern Job.* Ithaca, NY: Cornell University Press.

Pérez-Bustamante G., 1992. Knowledge management in agile innovative organizations, *Journal of Knowledge Management,* 3(1), 6–17.

Pfeffer J., 1981. *Power in Organizations.* Cambridge, MA: Ballinger.

Pickering A., 1993. The mangle of practice: Agency and emergence in the sociology of science, *The American Journal of Sociology,* 99(3), 559–589.

Polanyi M., 1961. Knowing and Being. *Mind,* 70(280), 458–470.

Polanyi M., 1967. *The Tacit Dimension.* London, U.K.: Routledge and Kegan Paul.

Quaddus J. and Xu M., 2005. Adoption and diffusion of knowledge management systems: Field studies of factors and variables, *Knowledge-Based Systems,* 18(2–3), 107–115.

Sandberg J. and Tsoukas H., 2011. Grasping the logic of practice: Theorizing through practical rationality, *Academy of Management Review*, 36(2), 338–360.

Schein E., 1992. *Organizational Culture and Leadership*. Hoboken, NJ: John Wiley & Sons, Inc.

Simon H. A., 1962. The architecture of complexity, *Proceeding of the American Philosophical Society*, 106(6), 467–482.

Spender J. C., 1996. Organizational knowledge, learning, and memory: three concepts in search of a theory, *Journal of Organizational Change and Management*, 9(1), 63–78.

Star S. L., 1989. The structure of ill structured solutions: Boundary objects and heterogeneous distributed problem solving. In: Huhns M. and Gasser L., (Eds.) *Readings in Distributed Artificial Intelligence*. Menlo Park, CA: Morgan Kaufman, 37–54.

Sun P. Y. T. and Anderson M. H., 2010. An examination of the relationship between absorptive capacity and organizational learning, and a proposed integration, *International Journal of Management Review*, 12(2), 130–150.

Swan J. and Scarbrough H., 2005. The politics of networked innovation, *Human Relations*, 58(7), 913–943.

Swan J., Scarbrough H., and Preston J., 1999. Knowledge management—The next fad to forget people? *Proceedings of ECIS 1999*, Copenhagen, Denmark.

Szulanski G., 1996. Exploring internal stickiness: Impediments to the transfer of best practice within the firm, *Strategic Management Journal*, 17, 27–43.

Teece D. J., 1981. The multinational enterprise: Market failure and market power considerations. *Sloan Management Review*, 22(3), 3–17.

Thompson J. D., 1967. *Organizations in action: Social science bases of administrative theory*. New York: McGraw-Hill.

Tiemessen I., Lane H. W., Crossan M., and Inkpen A. C., 1997. Knowledge management in international joint ventures. In: Beamish P. W. and Killing J. P. (Eds.) *Cooperative Strategies: North American Perspective*. San Francisco, CA: New Lexington Press, 370–399.

Todorova G. and Durisin B., 2007. Absorptive capacity: Valuing a reconceptualization, *Academy of Management Review*, 32(3), 774–786.

Trkman P. and Desouza K. C., 2012. Knowledge risks in organizational networks: An exploratory framework, *Journal of Strategic Information Systems*, 21(1), 1–17.

Tsoukas H., 1996. The firm as a distributed knowledge system: A social constructionist approach. *Strategic Management Journal*, 17(Winter special issue), 11–25.

Tsoukas H. and Vladimirou E., 2001. What is organizational knowledge? *Journal of Management Studies*, 38(7), 973–999.

Tushman M. L. and O'Reilly C. A., 1996. Ambidextrous organizations: Managing evolutionary and revolutionary change, *California Management Review*, 38(4), 8–30.

Van de Bosch F. A., Volberda H. W., and De Boer M., 1999. Coevolution of firm absorptive capacity and knowledge environment: Organizational forms and combinative capabilities, *Organization Science*, 10(5), 551–568.

Vandaie R., 2008. The role of organizational knowledge management in successful ERP implementation projects, *Knowledge-Based Systems*, 21, 920–926.

Wagner C., Cassimjee N., and Nel H., 2002. Some key principles in implementing knowledge management: The views of employees in a small software company, *SA Journal of Industrial Psychology*, 28(2), 49–54.

Wenger E., McDermott R., and Snyder W. M., 2002. *Cultivating Communities of Practice*. Boston, MA: Harvard Business School Press.

Wilson C., Boe B., Sala A., Puttaswamy K. P. N., and Zhao B. Y., 2009. User interactions in social networks and their implications, *Proceedings of the 4th ACM European Conference of Computer Systems*, Nuremberg, Germany, 205–218.

Yusuf Y., Gunasekaran A., and Abthorpe M. S., 2004. Enterprise information systems project implementation: A case study of ERP in Rolls-Royce, *International Journal of Production Economics*, 87(3), 251–266.

Zahra S. A. and George G., 2002. Absorptive capacity: A review. Reconceptualization, and extension, *Academy of Management Review*, 27(2), 185–203.

18

Digital Libraries

Edward A. Fox
Virginia Tech

Monika Akbar
Virginia Tech

Sherif Hanie El
Meligy Abdelhamid
Virginia Tech

Noha Ibrahim
Elsherbiny
Virginia Tech

Mohamed Magdy
Gharib Farag
Virginia Tech

Fang Jin
Virginia Tech

Jonathan P. Leidig
*Grand Valley State
University*

Sai Tulasi Neppali
Virginia Tech

18.1 Theory

In his book "Libraries of the Future," J.C.R. Licklider predicted that a world network of automated libraries would together support global information requirements, supported by a hodgepodge of different technologies, since there was no integrated theory (Licklider 1965). After more than 40 years, it has become clear that advances in understanding and technology have allowed a theory-based approach to this field that should aid those interested in computer science, information systems, and/or information technology to better grasp the key concepts regarding digital libraries (DLs).

18.1.1 DL Definitions

Though storing and retrieving information with the help of computers dates back at least to the 1950s (Taube 1955), those focused on what has emerged as the field of DLs did not select that phrase until about 1991, giving it preference over terms such as electronic library or virtual library.

Accordingly, some have been interested in defining what constitutes a DL as a step in clarifying its particular place in the digital world. Three illustrative examples of such definitions follow:

1. "A digital library is an assemblage of digital computing, storage and communications machinery together with the content and software needed to reproduce, emulate and extend the services provided by conventional libraries, based on paper and other material means of collecting, cataloging, finding and disseminating information" (Gladney et al. 1994).

2. "A set of electronic resources and associated technical capabilities for creating, searching and using information. In this sense they are an extension and enhancement of information storage and retrieval systems that manipulate digital data in any medium (text, images, sounds; static or dynamic images) and exist in distributed networks. The content of digital libraries includes data, metadata that describe various aspects of the data (e.g., representation, creator, owner, reproduction rights) and metadata that consist of links or relationships to other data or metadata, whether internal or external to the digital library" (Borgman 1996).

3. "Digital libraries are a focused collection of digital objects, including text, video and audio along with methods for access and retrieval and for selection, organization and maintenance of the collection" (Witten and Bainbridge 2002).

18.1.2 Digital Library Curriculum

By 2005, many DLs were in widespread use, providing increasingly popular means of collecting and finding information. It became clear that educational and training support was needed for the emerging community of digital librarians. Toward this goal, a team at Virginia Tech and the University of North Carolina at Chapel Hill collaborated to develop a curriculum, founded in theory and practice, that could be used to teach about DLs in library schools, information science departments, or computer science programs (Pomerantz et al. 2006). Figure 18.1 gives one perspective on how some of the modules that have been developed could be used in a two course sequence; it also demonstrates many of the key topics associated with DLs. An increasing number (already more than 40) of learning modules, each supporting either self-study, tailored training, or roughly a week of coverage in a course, can be accessed both through Wikiversity and our website (Fox et al. 2012b).

18.1.3 Formal Approaches

In computing, rapid advances often follow the development of a theory that enables a formal approach to practical problems, for example, the use of the relational model for databases (Codd 1970). In the DL field, early efforts ignored the use of formalisms, so the availability and features associated with services were haphazard. Clearly, the community of DL producers can benefit from formal definitions of existing and proposed DLs, along with the services they provide. Formalisms within the field of DLs provide a theoretical foundation. This can save work for designers and implementers, for example, by enabling the reuse of formally and programmatically defined services, and the deployment of existing service modules in the generation of proposed DLs for a specific context. The two major efforts toward a theoretical framework include the 5S formal framework (Goncalves et al. 2004; Fox et al. 2012a) and the DELOS reference model (Candela et al. 2011).

18.1.3.1 5S

The 5S (streams, structures, spaces, scenarios, and societies) framework builds on five precisely defined yet intuitively understandable basic constructs, allowing a high level foundation to be used to describe all key DL concepts (Fox et al. 2012a). Streams encompass sequences of information that represent content. Streams include audio, video, bit streams on disk or transmitted over a network, and character strings that constitute text. Structures represent any type of organization (e.g., a database, taxonomy, thesaurus,

Core topics

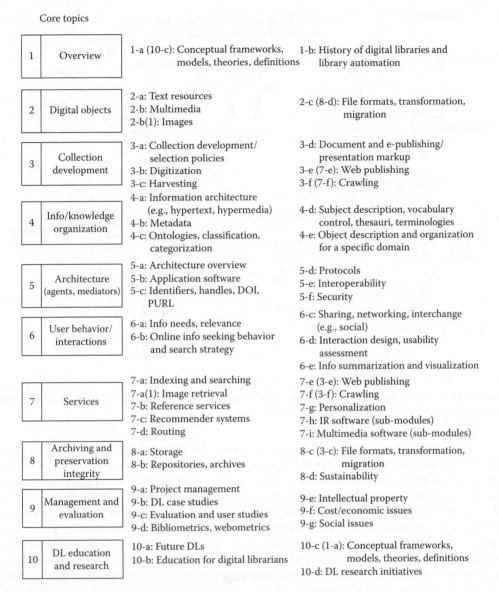

1	Overview	1-a (10-c): Conceptual frameworks, models, theories, definitions 1-b: History of digital libraries and library automation
2	Digital objects	2-a: Text resources 2-b: Multimedia 2-b(1): Images 2-c (8-d): File formats, transformation, migration
3	Collection development	3-a: Collection development/ selection policies 3-b: Digitization 3-c: Harvesting 3-d: Document and e-publishing/ presentation markup 3-e (7-e): Web publishing 3-f (7-f): Crawling
4	Info/knowledge organization	4-a: Information architecture (e.g., hypertext, hypermedia) 4-b: Metadata 4-c: Ontologies, classification, categorization 4-d: Subject description, vocabulary control, thesauri, terminologies 4-e: Object description and organization for a specific domain
5	Architecture (agents, mediators)	5-a: Architecture overview 5-b: Application software 5-c: Identifiers, handles, DOI, PURL 5-d: Protocols 5-e: Interoperability 5-f: Security
6	User behavior/ interactions	6-a: Info needs, relevance 6-b: Online info seeking behavior and search strategy 6-c: Sharing, networking, interchange (e.g., social) 6-d: Interaction design, usability assessment 6-e: Info summarization and visualization
7	Services	7-a: Indexing and searching 7-a(1): Image retrieval 7-b: Reference services 7-c: Recommender systems 7-d: Routing 7-e (3-e): Web publishing 7-f (3-f): Crawling 7-g: Personalization 7-h: IR software (sub-modules) 7-i: Multimedia software (sub-modules)
8	Archiving and preservation integrity	8-a: Storage 8-b: Repositories, archives 8-c (3-c): File formats, transformation, migration 8-d: Sustainability
9	Management and evaluation	9-a: Project management 9-b: DL case studies 9-c: Evaluation and user studies 9-d: Bibliometrics, webometrics 9-e: Intellectual property 9-f: Cost/economic issues 9-g: Social issues
10	DL education and research	10-a: Future DLs 10-b: Education for digital librarians 10-c (1-a): Conceptual frameworks, models, theories, definitions 10-d: DL research initiatives

FIGURE 18.1 Curriculum modules by semester.

dictionary, ontology, data structure, record, file in a particular format, or any digital object following some type of schema) and can be described using labeled graphs. Spaces are sets of items along with rules and operations valid for each set of items. In addition to the usual 1D, 2D, and 3D spaces, there are vector, probabilistic, and topological spaces. Typical user interfaces are implemented on a 2D space, while many search engines are based on a vector space model. Scenarios are sequences of events that lead to an outcome. All services follow scenarios, as do workflows, procedures, and algorithms. Societies are made up of entities, either human or computing-based, with relationships among them. Thus, they encompass the set of end users, groups engaged in collaboration, social networks, and other aggregates studied in psychology, the social sciences, or fields like law or economics. As can be seen in Figure 18.2, each of the five constructs can be used alone or in combination with the others to define every one of the key constructs we consider to be required parts of even the most minimal DLs.

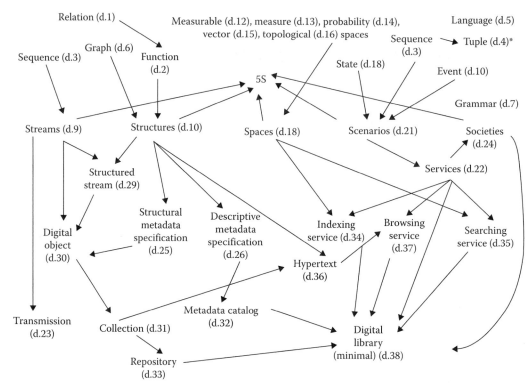

FIGURE 18.2 Formal definition dependencies for the constructs of a minimal digital library. (Reprinted from Goncalves et al., *ACM Trans. Inform. Syst.*, 22(2), 270, 2004.)

The 5S formal framework also may be used to encode descriptions of DLs and related concepts. Common services required by a range of DLs were initially defined, for example, indexing, searching, browsing, and annotation (Goncalves et al. 2008). The framework has since been extended to describe concepts found in a number of different domain-specific DLs (Murthy et al. 2007). Such extensions allow for emerging classes of DLs to be formalized, for example, for content-based image retrieval (Murthy et al. 2011) and scientific workflow systems (Leidig et al. 2010). Definitions of new services or aspects to DLs may build on top of existing definitions, and producing formal descriptions of DLs is facilitated through reuse of the existing set of definitions. Additional research connected with 5S has produced tools to aid in developing DLs, involving quality evaluations (5SQual) (Moreira et al. 2007), a description language (5SL) (Goncalves and Fox 2002), and graphical representations (5SGraph) (Zhu 2002). The 5S framework has been used in multiple DLs including Networked Digital Library of Theses and Dissertations (NDLTD), electronic tools and ancient near east archives (ETANA), Ensemble, and scientific DLs (Fox 1999; Ravindranathan et al. 2004; Fox et al. 2006, 2010; Leidig et al. 2011).

18.1.3.2 DELOS

The DELOS digital library reference model aims to provide a foundation for describing, understanding, and evolving the field of DLs. The model describes the DL universe in terms of digital library management systems (DLMS), digital library systems (DLS), and DLs. A DL instance is a practically running system that is perceived and used by end users. A DLS is the software implementation of the required functionality of a class of DLs. A DLMS addresses a conceptual problem, provides a software solution, and implements a DL approach as given in formal definitions of architecture, content, functionality, policy, quality, and users. The reference model has been repeatedly updated to provide concept maps, definitions, and conformance evaluation checklists (Candela et al. 2011).

We believe that mappings could be defined between concepts in the reference model and the 5S framework, so either approach might be employed, depending on the situation at hand. However, we prefer (1) a bottom-up approach to theory development, with everything precisely defined; (2) starting from a minimalist perspective; and (3) growing the framework incrementally as needed. Hence, we focus on 5S and use it in the rest of this chapter.

18.2 Data, Information, and Knowledge

Content is an essential part of a DL. The computing-related fields generally view content from varied perspectives based on decomposition into three levels of abstraction or complexity: data, information, and knowledge. This is particularly useful since it encourages use of tailored services inside a DL, such as database management systems, search engines (sometimes based on information retrieval toolkits), and knowledge management systems (e.g., for managing ontologies). Yet, DLs also have simplified work with content by shifting focus to a different set of abstractions, discussed in the subsequent sections.

18.2.1 Digital Objects

In a DL, the simplest abstraction for an element of content is to consider it as a digital object. Clearly, with the help of computers, much content is created in a digital form, like this chapter. Nevertheless, there are myriad objects in the real world that are not in digital form, such as physical books, paintings, statues, buildings, chemical or biological samples, and mathematical constructs. Fortunately, building on practices in linguistics and information science, we can discuss such objects if they are assigned a name, identifier, descriptor, and/or surrogate (e.g., a photograph or image prepared by a scanner). Thus, we can simplify the discussion of the content of a DL if we assume that anything of interest has some associated digital representation, allowing us to focus solely on such digital objects.

18.2.1.1 Atomic/Complex

There are many types of digital objects that can be stored in a DL. Some basic types include text, image, audio, and video. Each of these types has different formats and structures, for example, text formats include web pages, PDF, XML, or any file that contains a stream of characters. Digital objects also can be categorized based on the complexity of the structure of the object as either atomic or complex. Atomic objects are the basic unit of storage in a DL. All the basic types can be considered atomic objects. Basic types cannot be decomposed into smaller objects. Complex objects are objects that are composed of an aggregation of atomic objects into one unit with a certain structure that records the relationships between the different components (Kozievitch et al. 2011). Complex objects also can be constructed using parts of several other objects. Extending this notion further, we refer to the concept of extracting a part of an object (e.g., a sub-document) and treating it as a whole unit by itself, together with connecting together a set of such units, as superimposed information (Murthy et al. 2004). This approach provides a foundation for any type of annotation system and leads to a class of DLs that is particularly useful for a wide variety of scholarly activities (Murthy 2011).

18.2.1.2 Streams, Structures, and Spaces

These three constructs were briefly introduced earlier in Section 18.1.3.1. Their utility in describing the content in DLs needs further elaboration, however.

Streams are sequences of objects, for example, bits, bytes, characters, numbers, images, audio samples, and clicks. Thus, streams can be constructed from objects with different granularity levels. For example, a text is based on characters, while a book is based on pages. Streams may be static, as when a sequence of characters makes up a name or an abstract. Or, they may be dynamic, as in the stream of data that comes from a sensor or surveillance camera, or that is recorded into a log. In part to support all such dynamic streams, data stream management systems (Arasu et al. 2004) have been developed.

Structures also are essential in DLs. Understanding data structures, related algorithms, and their use (Shaffer 1997) is an essential basis for working with structures in DLs. This also aids the use of special languages used to describe structures, such as schema for relational databases, XML (Sperberg-McQueen and Thompson 2000) files/databases, or resource description framework (RDF) (Brickley and Guha 2004) (triple) stores. Of particular importance are record structures, commonly used for metadata (see Section 18.2.5). DLs support other important types of structures, such as classification systems, taxonomies, dictionaries, ontologies, databases, and hyperbases. Furthermore, when combined with streams, we have structured streams, which lead to constructs like documents, protocols (e.g., the organization of the streams connecting clients and servers), and files (that generally have some format or type, which specifies how the bytes are organized to carry meaning).

Spaces support representations that can help in organizing, retrieving, and presenting the content in a DL. One broad class relates to user interfaces that often adopt some spatial metaphor and connect some abstract spatial view with a concrete spatial instantiation (e.g., a window system). Another broad class relates to geographic information systems, often with time added as another dimension, supporting maps, timelines, global positioning system (GPS) data, mobile applications, studies of earth/atmosphere/astronomy, and combinations like mashups. A third broad class relates to multidimensional representations using various features that leads to particular implementations using vector or probability spaces, suitable for indexing, searching (especially approximate, nearest neighbor, cluster based, or range), retrieval, and ranking. An interesting example is an index that often is essential to ensure efficient operations of DLs; such an index generally combines streams of characters (for entries in the index), structures (to allow rapid response for key operations), and spaces (as in the vector space model).

18.2.1.3 Media

It is common for DLs to handle text documents with embedded images. But to address the broad needs of users around the world, DLs must support all types of content, including multimedia and hypermedia (multimedia with links), with sufficient quality of service, through speedy connections.

Multimedia has become an effective information sharing resource. DLs should provide direct support, including content-based searching, summarization, annotation, and specialized linking, as well as adequate storage (amounting to petabytes), for video, speech, audio, images, graphics, and animations. This is important, since web users expect services like those found with Flickr, Pandora, Shutterstock, YouTube, and Webmusic.IN. Sources of multimedia include libraries, museums, archives, universities, and government agencies, as well as individuals comfortable with uploading from cameras, smartphones, and camcorders.

Specialized DLs also support oral histories, recordings of dying languages (and translations), reuse of video segments, searching for similar images/videos, geo-coding, analysis of health-related images and data, support for bioinformatics, management of chemical information, access to tables and figures in documents, and preservation of virtual environments. Since formats keep changing in specialty areas, long-term preservation of such content is a particular challenge.

18.2.2 Collections

In a DL, a collection is generally an assembled, cataloged, and easily accessed digital information aggregation, whose resource data are usually provided through underlying database and storage structures (Unsworth 2000). Within a collection, storage preserves many kinds of digital information (such as documents and media files), which have been cataloged (see Section 18.2.5) and indexed for users' easy access. Thus, a collection is a set of digital objects, which may be atomic or complex, which may represent a variety of media and content types, and which should be accessible, such as by browsing (if in the form of a hyperbase) and searching (through an index and catalog). Within a collection, the format and class of digital objects generally is consistent in structure and representation. As an example, a book collection generally has associated descriptive information for each item in the collection, such as "title," "publication information," "material type," "call no," "location," "status," and "ISBN."

A DL may be made up of one or several collections. Minimal support for DLs that manage multiple collections involves supporting each one independently. Integration challenges must be addressed if a unified view is to be afforded across the collections. This can be further complicated when the DL provides portal services while storing key information for digital objects that are distributed (Fox et al. 2010).

18.2.3 Linking and Linked Data

As can be seen in Figure 18.2, a DL includes a hypertext collection, or, more generally, if there is multimedia content, a hyperbase. This links objects in the collection together and facilitates browsing. Additional capabilities may be provided, depending on the nature of the hyperbase and on related services; we can understand this more clearly in the context of the World Wide Web (WWW).

The WWW is a huge source of data that is linked into what is called a web graph. The nodes in the graph are web pages, while the edges (arcs) are the hyperlinks between those pages. The WWW is considered a dynamically linked graph, where the graph is rapidly increasing in size and the links are continuously being created and changed. Hyperlinks connect web pages for purposes of navigation, connectivity, and reuse. The relationships between linked web pages are unspecified, and their nature can only be guessed by studying the anchoring texts.

Hyperbases can be much richer and supported by more powerful services. For example, links can be bidirectional, and stored in a database rather than in objects themselves. Furthermore, links can connect either sub-documents or documents, so the nature of each of the origin and destination can be anywhere in the range from highly specific to very general. When the hyperbase includes content fitting with the RDF (W3C 2004b), a node in the hyperbase can be as small as a string that names something, while when multimedia is included, a node can refer to a large video. Since the WWW uses general addresses, a node can even be a website or database.

With linked data, the granularity often focuses on the small side. In PlanetMath (Krowne 2005), there is automatic linking from many of the words in each encyclopedic entry, so the number of links is very high, and connectivity is strong (Krowne 2003). Linking web pages based on the semantic concepts and relations between them is considered in the Semantic Web (Berners-Lee et al. 2001) approach. Examples of works with extensive linking include Wikipedia, DBpedia, and ontologies.

18.2.4 Semantic Digital Library

Incorporating semantics (machine-oriented representations of meaning) in DLs can be done at different levels and using different formats. Semantics can help enhance the organization of data, aiding interoperability, facilitating integration of different collections, enhancing and enriching the services provided, and providing a domain model for each collection managed by the DL. In keeping with the recommendations of the World Wide Web Consortium (W3C) and fitting with plans for the Semantic Web, at least two types of representations should be supported. The first is RDF (W3C 2004a,b), typically managed with some type of triplestore, as is used in the Cyberinfrastructure for Network Science (CINET) project (see Section 18.4.1). The second involves ontologies, as represented by the web ontology language (OWL) (W3C 2004a) or other schemes. One of the goals of the Ensemble system (see Section 18.5.3) (Ensemble 2009) is to use ontologies to aid access by stakeholders. Thus, instructors in business, teaching about databases, can make use of connections in the ontology to identify database-oriented educational resources in an Ensemble collection developed by computer scientists.

18.2.5 Metadata and Catalogs

Metadata provides information about one or more aspects of the digital objects stored in the DL, such as: how the data are created, the purpose of the data, time and date of creation, creator or author of data, location on a computer network where the data resides, and standards enforced on the data. Many current DLs at least include content for each of the 15 specified metadata elements that constitute the Dublin Core metadata standard, which was developed to enable easy description of web content (Dublin-Core-Community 2002).

Metadata are valuable for information retrieval. If digital objects themselves cannot be accessed and indexed, the metadata may be the only information usable to enable exploration. Even when full text is available, often the subject description and other elements in the metadata description can be used to enhance retrieval. Accordingly, the choice of metadata scheme can affect how well the system can satisfy the user's information needs and greatly affects its long-term ability to maintain its digital objects.

Metadata records are stored in a catalog, so that every digital object has at least one metadata record in the catalog (or more if several metadata formats are supported). In some cases, the catalog is called a union catalog, when digital objects come from a variety of sources. A representative example is the union catalog of the NDLTD (2012).

18.2.6 Repositories

Repositories are a leading way of collecting, organizing, archiving, preserving, and providing access to intellectual properties. One important role is to have collections of metadata.

18.2.6.1 Institutional Repository

Institutions producing scholarly work often collect, store, and provide access to intellectual properties generated within the institution. These repositories often allow universal access to resources including theses and dissertations, courseware, and technical reports. Services of these repositories include cataloging, organizing, indexing, browsing, and searching. One of the main goals of these repositories is to archive and preserve the contents while providing long-term access to the material.

A number of software packages are available to help institutions set up and maintain digital repositories. DSpace, introduced by the DSpace Foundation (DSpace 2008), is an open-source software package used by a number of institutional repositories, in part, since it is easy to install and deploy. Other systems include Fedora (Staples et al. 2003), ePrints (EPrints.org 2002), and various open-source content management systems (CMSs).

18.2.6.2 Open Archives Initiative

Interoperability between different repositories is important for linking and cataloging resources across these repositories. Many DLs participate in the Open Archives Initiative (OAI) in order to ensure interoperability (Van de Sompel and Lagoze 2000). The OAI Protocol for Metadata Harvesting (PMH) is a standard for gathering metadata across DLs. Under OAI-PMH, actors fall into one of two roles, the data provider and the service provider. A data provider has a collection of metadata that it exposes through the protocol. A service provider then can provide any of a number of services (e.g., searching and browsing) from its local collection that is built through a series of requests for metadata (using PMH), sent to data providers. A number of open-source software packages such as jOAI (Weatherley 2012) and the collection workflow integration system (CWIS) (Internet Scout Project 2012) are available to allow repositories to expose their metadata following PMH.

The OAI also has aided interoperability through another initiative, OAI-ORE (Lagoze and Van de Sompel 2007), which focuses on object reuse and exchange. This operates in the context of the Semantic Web to enable interoperability involving groupings of information items (see Section 18.2.4). It addresses compound objects, such as those described using RDF, so those aggregations can be transferred between repositories, and used in each location.

18.2.7 Life Cycle

Content management involves more than the storage and retrieval of content. Policies and procedures are required to support content throughout its life cycle. A traditional life cycle describes stages like generation, management, use, change, and deletion (see Figure 18.3). Long-term aspects of content management include curation, archiving, and preservation. Curation involves deciding what content to maintain, under what conditions, and for how long. The following subsections discuss further the subsequent stages of archiving and presentation.

18.2.7.1 Archives

Archiving is the process of preparing, representing, and storing content to be accessed later. Key issues to consider include storage media, format representations, and access tools. A large number of physical and digital archives exist in a broad range of domains. Many countries have national archives, and there are special graduate programs for archivists.

18.2.7.2 Preservation

Preservation is the process of ensuring that the format, meaning, and accessibility of content are not lost over time. Various preservation operations may be performed on collections, for example, reformatting, copying, converting, migrating, and transforming. Physical media and content representations are changed periodically to keep up with the progression of technology. If the meaning and integrity

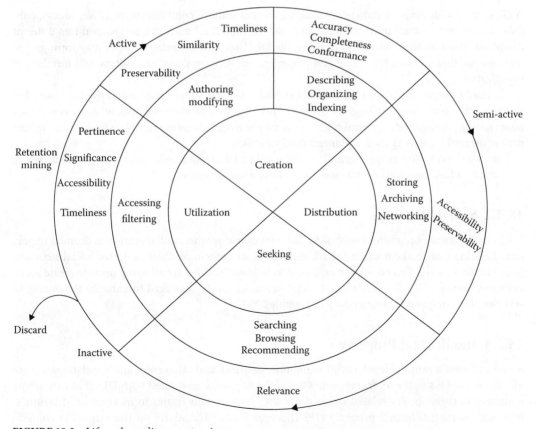

FIGURE 18.3 Life cycle quality on outer ring.

of content are not preserved, then content may be lost or may become inaccessible. There are a host of external threats to digital collections (Rosenthal et al. 2005). These include media failure, hardware failure, software failure, network problems, media and hardware obsolescence, software obsolescence, human error, natural disaster, attack by hackers, lack of funding to continue maintenance, and the demise of the hosting institution. Technology obsolescence complicates preservation services as computers and software are updated frequently and lead to the inability to access data. Technology fragility leads to inaccessible data when byte streams are changed or corrupted. Content should be examined periodically to mitigate some of these factors and ensure that bits have not changed, for example, using MD5 checksums. Duplication and replication techniques also are used to ensure an acceptable copy is available if a given copy is corrupted or lost, for example, using lots of copies keep stuff safe (LOCKSS) (Reich and Rosenthal 2001). Accessibility is provided through emulation, simulation, and translation from older software and operating systems.

The full information life cycle shown in Figure 18.3 includes the phases discussed earlier, along with others. For each phase, as can be seen in the outer ring, it is important to ensure that quality is maintained, so users interested in information at every stage have suitable services that address their information needs. This leads to a broader discussion of users, or, in 5S terms, societies, in the next section.

18.3 Societies

18.3.1 Users, Roles, Collaborators, and Communities

A DL serves a wide range of audiences including content authors, contributors, reviewer, editors, publishers, managers, content users, viewers, and readers. Individual users can be grouped into different categories based on their interest and use of the DL. Users with similar interests may form groups and communities within a DL. These groups can include reviewers, authors, editors, and members of organizations.

The need for access control on DL content can lead to systems supporting various roles for users. For example, while some users can be given the "Edit" privilege (i.e., an advanced role), other users may only have viewing privilege (i.e., a general role). A user may be assigned one or multiple roles based on factors such as ownership, activity level, and membership status.

User interactions such as tagging and ratings within a DL create a collaborative environment where community-added information enhances the value of existing content.

18.3.2 Social Networks

Social networks are becoming essential in different domains where collaboration is deemed important. Existing tools make it easier for DL users to share content on their preferred social networks (e.g., Twitter, G+, and Facebook). Systems used to manage DLs also are allowing users to create social networks within a DL. The existence of social networks can be leveraged to enhance the quality of services like rating and recommending (Ensemble 2009).

18.3.3 Intellectual Property

Societies follow a range of legal, social, economic, political, and other rules and regulations—these all connect to DLs in the 5S framework. One of the key areas associated with DLs that cuts across a number of these society-related issues deals with how content relates to its creators, disseminators, and users. Intellectual property (IP) concerns focus particularly on the origins of content, the ownership of economic interests, who has access (and under what conditions), and how all this is managed.

18.3.3.1 Open Access

Some DLs provide access to their content online, free of charge, and free of most copyright and licensing restrictions. Such libraries may be called open-access DLs. These are common in the case of institutional repositories and of other sites that are part of the OAI. However, sometimes access must be controlled, as is explained in the following section.

18.3.3.2 Digital Right Management, Security, and Privacy

Security is important to consider when designing certain types of DLs. Security weaknesses in DLs can be exploited to access confidential information, result in loss of integrity of the data stored, or cause the whole DL system to become unavailable for use. This can have a damaging effect on the trust afforded to publishers or other content providers, can cause embarrassment or even economic loss to DL owners, and can lead to pain and suffering or other serious problems if urgently needed information is unavailable (Fox and ElSherbiny 2011).

There are many security requirements to consider because of the variety of different actors working with a DL. Each of these actors has different security needs, which are sometimes in conflict; this can make the security architecture of a DL more complex (Chowdhury and Chowdhury 2003). For example, a DL content provider might be concerned with protecting IP rights and the terms of use of content, while a DL user might be concerned with reliable access to content stored in the DL.

Some content stored in a DL may not be provided for free. Many content providers are concerned with protecting their IP rights. They may employ digital rights management (DRM), which refers to the protection of content from logical security attacks and the addressing of issues relating to IP rights and authenticity. DRM provides protection by encrypting content and associating it with a digital license (Tyrväinen 2005). The license identifies the users allowed to view the content, lists the content of the product, and states the rights the users have to the resource, in a computer readable format, using a digital rights expression language or extensible rights markup language that also describes constraints and conditions.

18.4 Systems

Building upon the discussion of formal approaches in Section 18.1.3, we can explain how systems fit with DLs. According to the DELOS reference model (Candela et al. 2011), this can be considered in levels (see Section 18.1.3.2). Further details are given in the following subsections, beginning with a broader discussion to provide context.

18.4.1 Cyberinfrastructure

Cyberinfrastructure (CI) refers to a broad area of software and systems deployment to aid societies with computing solutions including automation and provision of desired services (Atkins et al. 2003; NSF 2007). DLs generally fit into this either as standalone or embedded systems. A standalone DL can be utilized to support information satisfaction in a web portal, provide an underlying storage system, perform logical control and organizational functions, and conduct domain-specific tasks. The AlgoViz case study discussed in Section 18.5.2 explains such a standalone DL with its own portal and management system. On the other hand, a DL can operate at the middleware layer, embedded in a larger environment, for example, to provide a variety of data management and complex services to support scientific tasks. This makes sense, since eScience systems are complex and contain a variety of computing components. In general, CI often involves user interfaces, communication protocols, distributed high-performance computing resources, instrumentation, software applications, repositories, content collections, and multiple societies and user types. The CINET case study in Section 18.5.6 explains about a DL that is fully integrated within such a larger CI.

18.4.2 Architecture

In the software world, the structural aspects of a system are referred to as architecture, which in the case of DLs can be quite varied (Fox et al. 2005), including whether the system is a monolithic entity or an interconnection of components or whether it is centralized or distributed. Varying approaches can be employed, such as using agents (Birmingham 1995) or having a service-oriented architecture (Petinot et al. 2004).

DLs from multiple institutions and domains can be combined into a single federated system (Leazer et al. 2000), even when constituent DLs span multiple organizations with related but often conflicting purposes, goals, content, formats, collections, users, and requirements. As such, integration and interoperability is needed between federated DLs. These actually are easier if harvesting replaces federated search (Suleman and Fox 2003). This use of harvesting was one of the initial contributions of the OAI, discussed earlier in Section 18.2.6.2. Thus, harvesting is used in the National Science Digital Library (NSDL, see Section 18.5.4), which connects multiple DLs serving a variety of science-related domains.

18.4.3 Building DLs

In general, the different phases that go into building a DL are (Dawson 2004)

Planning
Requirements gathering and analysis
Standards and policy making
Collection development
Architecture design
User interfaces
Metadata collection and enhancement
Classification and indexing
Implementation of services
Access management
Security/authenticity
Preservation
User studies
Evaluation

Many DLs have been built from scratch, or from general components, as we did with the Computing and Information Technology Interactive Digital Educational Library (CITIDEL 2001), Multiple Access Retrieval of Information with ANnotations (MARIAN) (France et al. 2002), and Ensemble (2009). Such custom efforts are important in large-scale DLs catering to a very specific audience. More common, however, especially in recent years, is to make use of readily available software products, toolkits, and generic systems like Greenstone (Witten et al. 2000) for developing DLs. Many such solutions, often open-source, now exist for building DLs. The most widely used ones are presented in the following text.

The Greenstone DL software was born out of the New Zealand Digital Library Project at the University of Waikato and is one of the most extensively used open-source multilingual software packages for building DLs (Witten et al. 2001, 2009). It is a comprehensive system for constructing and presenting collections of millions of documents including text, images, audio, and video. The runtime system of the software has two parts: the receptionist, which takes requests from clients, and the collection server, which is the communication channel between the receptionist and the database. Collections can be built by specifying the Uniform Resource Locators (URLs) to be crawled. The user interface can be customized by the user.

Other widely used software for DL construction include Fedora Commons and DSpace, both of which are managed by a nonprofit organization called DuraSpace. DSpace is a "turnkey institutional repository" that is used for creating open-access collections of all types of media. It is used most commonly in academic institutions. Fedora Commons is a software platform for managing, storing, and accessing digital content. Due to its modularity, it can easily be integrated with other database systems. DSpace, being an out-of-the-box solution, is easy to use in small repositories, while Fedora Commons is a more robust suite of services for an advanced, completely customizable repository (Bagdanov et al. 2009).

18.4.4 Services

DLs carry out many scenarios, and, thus, provide varied services to societies. We discuss this broadly in the 5S context in (Fox et al. 2012a), including providing a comprehensive taxonomy of services and describing which services we believe should be built into any system that is called a DL. The key minimal services are discussed later in Sections 18.4.4.1, 18.4.4.2.1, and 18.4.4.2.2; the following discussion also sets the context for understanding additional DL services. Some services, like indexing (see Section 18.4.4.1), relate to having a suitable DL infrastructure, while others address societal needs (see Section 18.4.4.2).

18.4.4.1 Indexing

DLs contain a vast amount of data and information. Indexing mechanisms allow users to search the information space efficiently. Most DLs offer search services based on text indexing, which parses the sentences to build an index of words. Search engine technologies such as Lucene are widely used to repeatedly index and search through text files. Apache Solr, an open-source search platform, uses Lucene as its search engine and adds additional features including hit highlighting, faceted search, and dynamic clustering. Indexing of multimedia content such as images and video also is incorporated in some DLs (Wei 2002; Orio 2008) to allow content-based information retrieval.

18.4.4.2 User Needs

Societies, in general, and users, in particular, have need for information. Accordingly, DLs provide a range of services to help users satisfy their needs; key among those are the ones discussed in the following subsections.

18.4.4.2.1 Search

Searching is one of the most common activities of DL users. DLs offer a capability to search content at varied levels of object granularity. Generally, two types of searching can be supported: full-text search and metadata search. The searching process consists of two steps, sometimes integrated: retrieving and ranking. Supported by the indexing service (see Section 18.4.4.1), the retrieving step accesses the indexing data structure using the query terms entered by the user. Documents that (approximately) match the query terms are retrieved and given as output to the ranking process. The purpose of the ranking process is to find the best order of the documents for the user.

Faceted search is a powerful feature that provides a combination of searching and browsing (see subsequent sections), where the searching process is constrained using one or more of the attributes of the objects.

18.4.4.2.2 Browse

A browsing service provides another means for users to explore information in DLs. Browsing is based on the logical or physical organization of objects in the DL. Often, a rich organization like a hyperbase (resulting from linking texts or multimedia, as in hypertext and hypermedia) supports browsing, but simpler structures, like lists of dates or author names, also are widely used.

Browsing and searching can be considered as dual operations (Fox et al. 2012a). One can have the same results of searching using browsing and also the same results of browsing using searching. In the first case, query terms are used to determine the nodes in the navigation tree (graph) that match them. The navigation path can be built by traversing the navigation tree backward from the matched nodes to the root of the tree. In the second case, one issues multiple queries that produce the same results as those from following the navigational path during browsing. Each query corresponds to one step in the navigational path that restricts more results than those of the previous query.

18.4.4.2.3 Visualize

Visualizing provides the DL users with the results of their requests, like searching and analyzing, through an interactive representation. The representation of the results depends on the kind of the operations producing the results, the context where the results are presented, and the type of data. Search results can be represented as a list or as a grid of items (Fox et al. 1993). Analysis results can be represented by graphs, charts, timelines, or maps.

18.4.4.2.4 Personalize

Personalization of information and services within a DL can be achieved in many different forms. Services can be personalized to meet the specific needs of a group or community of users. Notification, subscription, and tagging are some examples of personalized services for individual users. Along with services, information can be personalized based on user profile, interest, and activity within the DL (Neuhold et al. 2003), for example, in content recommender systems.

18.4.5 Integration

Within DLs, integration is required to bring people together and get coordinating processes to work. Content, storage structures, file formats, metadata schemas, policies, and software are often diverse within collaborations and federated systems. The integration of databases, content collections, and metadata structures remains a difficult task. One approach, used in DLs, involves accessing diverse underlying collections through Application Programming Interfaces (APIs) to conduct search over each system. DLs also provide a means of implementing a standard set of services and user interfaces that allow for integrated discovery and management in non-standardized collections. The goal of integration is to make the differences between systems, content, and services to appear seamless to end users, despite differences and conflicts in underlying components.

18.4.6 Interoperability

DLs can be considered to be information management systems that deal with different kinds of digital objects and provide different kinds of services that facilitate the manipulation of these objects and the extraction of information from them. Thus, there are many collections of objects in either digital form or real life that can be accessed from a DL through one interface. Accordingly, DLs should support interoperability, to overcome the differences and difficulties of interfacing such sources of information.

Collections of differing objects vary regarding organization (structure) and metadata. These syntactic-level differences sometimes aid when dealing with even more complicated semantic differences, which users expect will be addressed in a comprehensive interoperability solution. Two such types of interoperability that should be supported in DLs include data interoperability and services interoperability. Data interoperability focuses on providing access to different collections with different structures. One of the approaches developed for supporting data interoperability is through OAI (see Section 18.2.6.2).

18.4.7 Quality

DLs manage different types of objects and support services for different types of users. Assessing, evaluating, and improving the performance of a DL often is based on identifying the different concepts included in a DL and the dimensions of quality that are associated with these concepts. Table 18.1 shows

TABLE 18.1 Digital Library Concepts and Quality Dimensions

Digital Library Concept	Dimensions of Quality
Digital object	Accessibility
	Pertinence
	Preservability
	Relevance
	Similarity
	Significance
	Timeliness
Metadata specification	Accuracy
	Completeness
	Conformance
Collection	Completeness
	Impact factor
Catalog	Completeness
	Consistency
Repository	Completeness
	Consistency
Services	Composability
	Efficiency
	Effectiveness
	Extensibility
	Reusability
	Reliability

the main concepts that are needed to build a DL management system and what quality aspects we measure with respect to each concept. For example, for each digital object in the DL, we need to consider the relevance of the object to the user needs, whether the object can be preserved, how similar the object is to other objects, etc.

Information in DLs goes through several steps, starting from adding a raw digital object and ending by presenting the information to satisfy a user need, according to the "Information Life Cycle" (Borgman 1996). We can evaluate a DL by measuring the quality of information in each step of the information life cycle. Figure 18.3 shows the different steps included in the information life cycle in a DL, the services provided in each step, and the quality criteria that are used in evaluating those services.

Efficiency, effectiveness, completeness, relevance, compliance, timeliness, and similarity are examples of such evaluation criteria. For each criterion, one has to specify some measures (quality indicators). For example, efficiency of interactive services in a DL can be evaluated by measuring response time. Likewise, the relevance of the objects in the DL can be evaluated by measuring precision and recall.

18.5 Case Studies

There have been thousands of DLs used over the past decades, so understanding the DL field is aided by considering how the concepts explained earlier relate to a representative set of case studies.

18.5.1 ETANA

The ETANA-DL was built in support of Near Eastern archaeology (Shen et al. 2008). It involved collection of information from a number of archaeological sites, development of schema for each, mapping the schema into a global schema, harvesting metadata from sites using OAI-PMH, transforming that metadata according to the schema mapping to be stored in a union metadata catalog, constructing an ontology for the domain with the aid of experts, and providing a broad range of services for archaeologists and others with interest. The browsing, searching, and visualizing were integrated, as was the whole system, based on the 5S framework, into a union DL.

18.5.2 AlgoViz

The AlgoViz project (AlgoViz 2011) is one of the primary content contributors to Ensemble. AlgoViz contains a catalog of visualizations that can be used to demonstrate various aspects of different algorithms. AlgoViz also has a bibliography collection related to algorithm visualizations (AVs) and hosts the field report collection—a forum like place where educators share their experience on using any particular AV. AlgoViz implements an OAI-PMH data provider service that maps catalog entry content to tags in the Dublin Core format. The metadata are thereby made available to Ensemble and other potential OAI service providers.

18.5.3 Ensemble

Ensemble (Ensemble 2009; Fox et al. 2010), the computing pathway project in the NSDL (see subsequent sections), is a distributed portal that collects and provides access to a wide range of computing education materials while preserving the original collections. It also hosts a community space to support groups of people with different interests related to computing education. Furthermore, there is a "Technologies" section that lists educational tools. Ensemble also shares key metadata with NSDL.

18.5.4 NSDL

The NSDL supports learning in Science, Technology, Engineering, and Mathematics (STEM) by providing access to a vast collection of related educational resources (NSDL 2001). Since 2000, the U.S. National Science Foundation has funded many groups in a variety of projects (e.g., Ensemble) to collect and manage quality metadata covering all key STEM areas (NSF 2012). NSDL uses harvesting and federated search, social networks, and a broad range of services, some centralized and some distributed, in one of the largest DL efforts yet undertaken.

18.5.5 CTRnet

The Crisis, Tragedy, and Recovery (CTR) network (CTRnet 2011) is a project aimed at building a human and digital network for collecting all information related to CTR events. A DL has been developed for managing the collected data and for providing services for communities that would help them in recovering from these events and for researchers to analyze data and extract useful information.

The CTRnet DL collects all types of data like reports, webpages, images, and videos. In collaboration with the Internet Archive (Internet_Archive 2000), many collections of websites have been prepared for further analysis and display to users. Integration with social media is also supported by collecting tweets related to CTR events. Tweet collections are analyzed for topic identification and opinion leader detection. The URLs in tweets are extracted, expanded, and stored for further use as seeds for archive building using local focused crawlers (around Heritrix) or the Internet Archive crawling service. The CTRnet DL provides communities and stakeholders with services like searching, browsing, visualizing, sharing, and analyzing data related to CTR events. Figure 18.4 shows key related data flows.

To aid in organizing the collected information, an ontology has been developed for the domain that includes all types of disasters and CTR events. The CTR ontology began as a merger of three disaster databases and is iteratively enhanced through the concepts extracted from collections archived and tweets collected using the yourTwapperKeeper (TwapperKeeper 2011) toolkit.

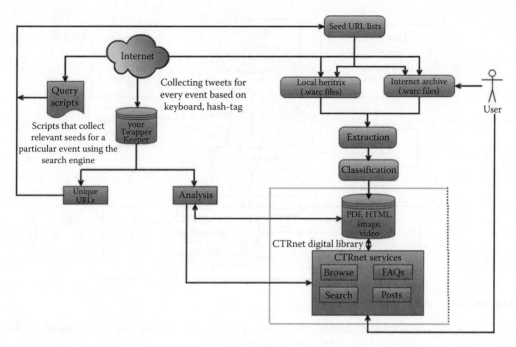

FIGURE 18.4 CTRnet data flow.

18.5.6 CINET

Educators teach about graphs and networks in multiple academic domains. In computer science, these topics are taught in courses on discrete structures. Graduate network science courses are based on this type of content, supporting education and research. These courses require real-world large graphs, scalable algorithms, and high-performance computing resources. CI including a portal has been developed to provide management for network science content and algorithms, including teaching advanced courses.

The CINET project provides educators, students, and researchers with access to a cloud computing cyber environment. This resource may be utilized to submit, retrieve, simulate, and visualize graph and network resources. It builds on our simulation-supporting digital library (SimDL) (Leidig 2012), which manages graphs, simulation datasets, large-scale analysis algorithms, results, and logging information.

Educators can use the DL as a repository for class projects, lecture materials, and research involving preexisting content and software. High-value educational content has been added to the repository; the DL includes the key datasets referred to in textbooks within this domain. Figure 18.5 details the system components in the CINET infrastructure. The DL provides core services to store and retrieve network graphs and algorithms. It also provides higher level services related to user interfaces, simulation applications, and computing resources.

This class of DLs provides services not commonly found in full-text collections. Thus, the DL automatically processes computer-generated content, extracts metadata from files, builds indexes of distributed scientific content, and provides search over highly structured numeric files. Infrastructure independent communication brokers are used to send and receive data in the DL and invoke DL services through automated workflow processes. This architectural design allows for generated scientific information to be captured, indexed, managed, curated, and accessed without human intervention. Management of scientific content in this manner provides a standardized management system for later efforts to support interoperability within a domain. Additional simulation-specific services aim to support advanced scientific workflow functions. These services include incentivizing researchers to contribute content, memorizing (i.e., caching) and reusing existing results, scientific workflow provenance

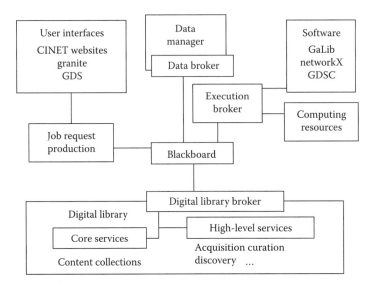

FIGURE 18.5   CINET system components.

tracking, curating of large-scale content, and scientific document searching. SimDL and its related services represent an emerging class of DLs applied in the sciences.

18.5.7 Publisher DLs

By the late 1980s, it was becoming clear that academic, professional, and commercial publishers were in the midst of an electronic revolution that would directly affect their activities (Fox 1988). This has led to the adoption of DL technologies to cover much of the publishing world. Automated systems are commonly used to receive submissions, handle reviewing, manage editorial work, collect final versions of documents, archive publications, and provide access (often through subscription or per-item payment schemes) (ACM 2000; IEEE-CS 2004).

18.5.8 WDL

The World Digital Library (WDL 2012) aims to promote international understanding and cross-cultural awareness by providing free multilingual access to online metadata and images of cultural riches of United Nations for Education, Science and Culture Organization (UNESCO) member states and others that have become partners. The types of items include books, journals, manuscripts, maps, motion pictures, newspapers, prints/photographs, and sound recordings.

In June 2005, Dr. James H. Billington, U.S. Librarian of Congress, proposed the establishment of the WDL to UNESCO. In December 2006, key stakeholders from around the world attended an Experts Meeting sponsored by UNESCO and the Library of Congress, deciding to establish working groups focused on standards and content selection. By October 2007, a prototype of WDL was presented at the UNESCO General Conference. In April 2009, WDL was officially launched and opened to the public. WDL has an emphasis on quality, with quantity also desired but not at the expense of quality. Regarding the development of WDL, some of the key features and aspects are as follows:

1. Metadata: Each item in WDL is described, indicating its significance as well as what it is. Curators and experts provide context to stimulate the curiosity of students and public interest in cultural heritage. There is consistent coverage of topical (based on the Dewey Decimal System), geographical, and temporal information.

2. Languages: Access to metadata, navigational aids, and supporting content like curator videos is enabled for seven languages: Arabic, Chinese, English, French, Portuguese, Russian, and Spanish.
3. Collaborative networks: Technical and programmatic networks connect partners, stakeholders, and users—all in a spirit of openness. There are contributions, including from libraries and archives, of: content collections, technology, finances, and involvement in working groups.
4. Digitization: The Library of Congress and partners in Brazil, Egypt, Iraq, and Russia have digital conversion centers to produce high-quality digital images, as can be found in the WDL. WDL intends to work with UNESCO to encourage the establishment of additional digital conversion centers around the world.

18.6 Future

The DL field—and its "children" like repositories, CMSs, and web portals—will continue to have broad impact as people around the globe utilize specialized systems to work with particular types of content, providing tailored assistance that goes beyond the general services of search engines. Supporting research and development efforts will continue, sometimes referred to using "DLs," but also carried out by those whose first loyalty is database processing, information retrieval, knowledge management, human–computer interaction, hypertext, electronic publishing, or other related domains. DLs are the result of synergies informed by all these fields, so conceptual and practical integration and interoperability is perhaps the most important type of DL research required. Establishing a firm formal foundation, supporting transparency across data/information/knowledge collections, developing consensus regarding quality measures and indicators, and shifting toward modular Semantic Web–oriented implementations atop a service-oriented architecture, are among the key challenges now faced. Funding and collaboration are required to address these and other problems explained earlier, so users working in a particular domain can benefit from improved services (e.g., support for collection security, integrated text and data analysis workflows, and enhanced information visualization). The discussion earlier should help you understand this important area, so you can more effectively use DLs and/or can assist in their further unfolding.

Acknowledgment

Some of this material is based upon work supported by the National Science Foundation (NSF) under Grant Nos. CCF-0722259, DUE-9752190, DUE-9752408, DUE-0121679, DUE-0121741, DUE-0136690, DUE-0333531, DUE-0333601, DUE-0435059, DUE-0532825, DUE-0840719, IIS-9905026, IIS-9986089, IIS-0002935, IIS-0080748, IIS-0086227, IIS-0090153, IIS-0122201, IIS-0307867, IIS-0325579, IIS-0535057, IIS-0736055, IIS-0910183, IIS-0916733, ITR-0325579, OCI-0904844, OCI-1032677, and SES-0729441. Any opinions, findings, and conclusions or recommendations expressed in this material are those of the authors and do not necessarily reflect the views of the National Science Foundation.

This work also has been partially supported by NIH MIDAS project 2U01GM070694-7, DTRA CNIMS Grant HDTRA1-07-C-0113, and R&D Grant HDTRA1-0901-0017.

References

ACM. 2000. *ACM Digital Library* [cited March 24, 2013]. Available from http://www.acm.org/dl/

AlgoViz. 2011. *AlgoViz.org, The Algorithm Visualization Portal* [cited March 24, 2013]. Virginia Tech Available from http://algoviz.org

Arasu, A., Babcock, B., Babu, S., Cieslewicz, J., Datar, M., Ito, K., Motwani, R., Srivastava, U., and Widom, J. 2004. *STREAM: The Stanford Data Stream Management System*. Technical Report. Stanford Infolab, Stanford University, Palo Alto, CA.

Atkins, D. E., Droegemeier, K. K., Feldman, S. I., Garcia-Molina, H., Klein, M. L., Messerschmitt, D. G., Messina, P., Ostriker, J. P., and Wright, M. H. 2003. *Revolutionizing Science and Engineering Through Cyberinfrastructure: Report of the National Science Foundation Blue-Ribbon Advisory Panel on Cyberinfrastructure.* Arlington, VA: NSF.

Bagdanov, A., Katz, S., Nicolai, C., and Subirats, I. 2009. Fedora Commons 3.0 Versus DSpace 1.5: Selecting an enterprise-grade repository system for FAO of the United Nations. In: *4th International Conference on Open Repositories.* Atlanta, GA: Georgia Institute of Technology.

Berners-Lee, T., Hendler, J., and Lassila, O. 2001. The Semantic Web: A new form of web content that is meaningful to computers will unleash a revolution of new possibilities. *Scientific American* 284 (5):34–43.

Birmingham, W. P. 1995. An agent-based architecture for digital libraries. *D-Lib* 1 (7), http://dx.doi.org/10.1045/july 95-birmingham.

Borgman, C. L. 1996. Social aspects of digital libraries. In: *DL'96: Proceedings of the 1st ACM International Conference on Digital Libraries*–Bethesda, MD. ACM. pp. 170–171.

Brickley, D. and Guha, R. V. 2004. *RDF Vocabulary Description Language 1.0: RDF Schema: W3C Recommendation 10 February 2004.* ed. B. McBride. Cambridge, MA: W3C.

Candela, L., Athanasopoulos, G., Castelli, D., El Raheb, K., Innocenti, P., Ioannidis, Y., Katifori, A., Nika, A., Vullo, G., and Ross, S. 2011. *Digital Library Reference Model* Version 1.0. Deliverable D3.2b. Pisa, Italy: DL.org, ISTI-CNR, pp. 1–273.

Chowdhury, G. and Chowdhury, S. 2003. *Introduction to Digital Libraries.* London, U.K.: Facet Publishing.

CITIDEL. 2001. Computing and information technology interactive digital educational library, CITIDEL Homepage [Web site]. Virginia Tech [cited March 24, 2013]. Available from http://www.citidel.org

Codd, E. F. 1970. A relational model for large shared data banks. *Communications of the ACM* 13 (6):377–387.

CTRnet. 2011. Crisis, tragedy, and recovery network website [cited March 24, 2013]. Available from http://www.ctrnet.net/

Dawson, A. 2004. Building a digital library in 80 days: The Glasgow experience. In: *Digital Libraries: Policy, Planning, and Practice*, eds. Law, D. G. and Andrews, J. Aldershot. Hants, England: Ashgate Pub. Ltd.

DSpace. 2008. *Manakin.* Cambridge, MA: DSpace Federation.

Dublin-Core-Community. 2002. *Dublin Core Metadata Element Set.* Dublin, OH: OCLC.

Ensemble. 2009. Ensemble distributed digital library for computing education homepage [cited March 24, 2013]. Available from http://www.computingportal.org

EPrints.org. 2002. E-Prints. University of Southampton [cited March 24, 2013]. Available from http://www.eprints.org/

Fox, E. A. 1988. ACM press database and electronic products: New services for the information age. *Communications of the ACM* 31 (8):948–951.

Fox, E. A. 1999. The 5S framework for digital libraries and two case studies: NDLTD and CSTC. In: *Proceedings NIT'99 International Conference on New Information Technology.* Taipei, Taiwan: NIT, pp. 115–125.

Fox, E. A., Chen, Y., Akbar, M., Shaffer, C. A., Edwards, S. H., Brusilovsky, P., Garcia, D. D. et al. 2010. Ensemble PDP-8: Eight Principles for Distributed Portals. In: *Proceeding of the 10th Annual Joint Conference on Digital libraries* (JCDL). Gold Coast, Australia: ACM, pp. 341–344.

Fox, E. A. and ElSherbiny, N. 2011. Security and digital libraries. In: *Digital Libraries—Methods and Applications*, ed. Huang, K. H. Rijeka, Croatia: InTech.

Fox, E. A., Goncalves, M. A., and Shen, R. 2012a. *Theoretical Foundations for Digital Libraries: The 5S (Societies, Scenarios, Spaces, Structures, and Streams) Approach.* San Francisco, CA: Morgan & Claypool Publishers.

Fox, E. A., Hix, D., Nowell, L., Brueni, D., Wake, W., Heath, L., and Rao, D. 1993. Users, user interfaces, and objects: Envision, a digital library. *Journal of American Society Information Science* 44 (8):480–491.

Fox, E. A., Suleman, H., Gaur, R. C., and Madalli, D. P. 2005. Design architecture: An introduction and overview. In: *Design and Usability of Digital Libraries: Case Studies in the Asia Pacific*, eds. Theng, Y.-L. and Hershey, S. F. Hershey, PA: Idea Group Publishing, pp. 298–312.

Fox, E. A., Yang, S., and Ewers, J. 2012b. Digital libraries curriculum development. Digital Library Research Laboratory, Virginia Tech, 2012 [cited March 24, 2013]. Available from http://curric.dlib.vt.edu/

Fox, E. A., Yang, S., and Kim, S. 2006. ETDs, NDLTD, and Open Access: A 5S perspective. *Ciencia da Informacao* 35 (2): 75–90.

France, R. K., Goncalves, M. A., and Fox, E. A. 2002. MARIAN digital library information system (home page). Virginia Tech [cited March 24, 2013]. Available from http://www.dlib.vt.edu/products/marian.html

Gladney, H., Fox, E. A., Ahmed, Z., Ashany, R., Belkin, N. J., and Zemankova, M.. 1994. Digital library: Gross structure and requirements: Report from a March 1994 workshop. In: *Digital Libraries '94, Proc. 1st Annual Conf. on the Theory and Practice of Digital Libraries*, eds. Schnase, J., Leggett, J., Furuta, R., and Metcalfe, T. College Station, TX.

Goncalves, M. A. and Fox, E. A. 2002. 5SL—A language for declarative specification and generation of digital libraries. In: *Proc. JCDL'2002, Second ACM/IEEE-CS Joint Conference on Digital Libraries, July 14–18*, ed. Marchionini, G. Portland, OR: ACM.

Goncalves, M. A., Fox, E. A., and Watson, L. T. 2008. Towards a digital library theory: A formal digital library ontology. *International Journal of Digital Libraries* 8 (2):91–114.

Goncalves, M. A., Fox, E. A., Watson, L. T., and Kipp, N. A. 2004. Streams, structures, spaces, scenarios, societies (5S): A formal model for digital libraries. *ACM Transactions on Information Systems* 22 (2):270–312.

IEEE-CS. 2004. IEEE Computer Society Digital Library. IEEE-CS [cited March 24, 2013]. Available from http://www.computer.org/publications/dlib/

Internet_Archive. 2000. Internet archive [cited March 24, 2013]. Available from http://www.archive.org

Internet Scout Project. 2012. CWIS: Collection workflow integration system [cited March 24, 2013]. Available from https://scout.wisc.edu/Projects/CWIS/

Kozievitch, N. P., Almeida, J., da Silva Torres, R., Leite, N. A., Goncalves, M. A., Murthy, U., and Fox, E. A. 2011. Towards a formal theory for complex objects and content-based image retrieval. *Journal of Information and Data Management* 2 (3):321–336.

Krowne, A. P. 2003. An architecture for collaborative math and science digital libraries. Masters thesis, Computer Science, Virginia Tech, Blacksburg, VA.

Krowne, A. 2005. PlanetMath. Virginia Tech [cited March 24, 2013]. Available from http://planetmath.org/

Lagoze, C. and Van de Sompel, H. 2007. Compound information objects: The OAI-ORE perspective. Open archives initiative [cited March 24, 2013]. Available from http://www.openarchives.org/ore/documents/CompoundObjects-200705.html

Leazer, G. H., Gilliland-Swetland, A. J., and Borgman, C. L. 2000. Evaluating the use of a geographic digital library in undergraduate classrooms: ADEPT. In: *Proceedings of the Fifth ACM Conference on Digital Libraries: DL 2000*. San Antonio, TX: ACM Press.

Leidig, J. 2012. *Epidemiology Experimentation and Simulation Management through Scientific Digital Libraries, Computer Science*. Blacksburg, VA: Virginia Tech.

Leidig, J., Fox, E. A., Hall, K., Marathe, M., and Mortveit, H. 2011. SimDL: A Model Ontology Driven Digital Library for Simulation Systems. In: *Proceedings of the 11th Annual International ACM/IEEE Joint Conference on Digital libraries (JCDL'11)*. Ottawa, Ontario, Canada: ACM.

Leidig, J., Fox, E., Marathe, M., and Mortveit, H. 2010. Epidemiology experiment and simulation management through schema-based digital libraries. In: *2nd DL.org Workshop at 14th European Conference on Digital Libraries (ECDL)*, Glasgow, Scotland, pp. 57–66.

Licklider, J. C. R. 1965. *Libraries of the Future*. Cambridge, MA: MIT Press.

Moreira, B. L., Goncalves, M. A., Laender, A. H. F., and Fox, E. A. 2007. Evaluating Digital Libraries with 5SQual. In *Research and Advanced Technology for Digital Libraries, ECDL 2007, LNCS 4675*. eds. L. Kovács, N. Fuhr, and C. Meghini. Berlin/Heidelberg, Germany: Springer, pp. 466–470.

Murthy, U. 2011. Digital libraries with superimposed information: Supporting scholarly tasks that involve fine grain information. Dissertation, Dept. of Computer Science, Virginia Tech, Blacksburg, VA.

Murthy, U., Gorton, D., Torres, R., Gonçalves, M., Fox, E., and Delcambre, L. 2007. Extending the 5S Digital Library (DL) Framework: From a Minimal DL towards a DL Reference Model. In *First Digital Library Foundations Workshop,* ACM/IEEE-CS Joint Conference on Digital Libraries (JDCL 2007). Vancouver, British Columbia, Canada: ACM.

Murthy, U., Li, L. T., Hallerman, E., Fox, E. A., Perez-Quinones, M. A., Delcambre, L. M., and da Silva Torres, R. 2011. Use of subimages in fish species identification: A qualitative study. In: *Proceedings of the 11th Annual International ACM/IEEE Joint Conference on Digital Libraries.* Ottawa, Ontario, Canada: ACM.

Murthy, S., Maier, D., Delcambre, L., and Bowers, S. 2004. Putting integrated information into context: Superimposing conceptual models with SPARCE. In: *Proceedings of the First Asia-Pacific Conference of Conceptual Modeling.* Denedin, New Zealand.

NDLTD. 2012. Find ETDs: NDLTD union catalog [cited March 24, 2013]. Available from http://www.ndltd.org/find

Neuhold, E., Niederée, C., and Stewart, A. 2003. Personalization in digital libraries—An extended view. In: *Digital Libraries: Technology and Management of Indigenous Knowledge for Global Access, Lecture Notes in Computer Science V. 2911,* eds. by Sembok, T., Zaman, H., Chen, H., Urs, S., and Myaeng, S.-H. Berlin/Heidelberg, Germany: Springer.

NSDL. 2001. NSDL (National Science Digital Library) homepage. NSF [cited March 24, 2013]. Available from http://www.nsdl.org

NSF. 2007. *Cyberinfrastructure Vision for 21st Century Discovery, NSF 07–28.* Arlington, VA: National Science Foundation Cyberinfrastructure Council.

NSF. 2012. *National STEM Education Distributed Learning (NSDL).* NSF. Available from http://www.nsf.gov/funding/pgm_summ.jsp?pims_id=5487

Orio, N. 2008. Music indexing and retrieval for multimedia digital libraries. In: *Information Access through Search Engines and Digital Libraries, Vol. 22 in The Information Retrieval Series,* ed. Agosti, M. Berlin/Heidelberg, Germany: Springer.

Petinot, Y., Giles, C. L., Bhatnagar, V., Teregowda, P. B., Han, H., and Councill, I. G. 2004. A service-oriented architecture for digital libraries. In: *Proceedings of the Second International Conference on Service Oriented Computing,* New York. ACM. pp. 263–268.

Pomerantz, J., Wildemuth, B. M., Yang, S., and Fox, E. A. 2006. Curriculum Development for Digital Libraries. In: *Proc. JCDL 2006.* Chapel Hill, NC.

Ravindranathan, U., Shen, R., Goncalves, M. A., Fan, W., Fox, E. A., and Flanagan, J. W. 2004. Prototyping digital libraries handling heterogeneous data sources—The ETANA-DL case study. In: *Research and Advanced Technology for Digital Libraries: Proc. 8th European Conference on Digital Libraries, ECDL2004,* eds. Heery, R. and Bath, L. L. Berlin, Germany: Springer-Verlag GmbH.

Reich, V. and Rosenthal, D. S. H. 2001. LOCKSS: A permanent web publishing and access system. *D-Lib Magazine* 7 (6), http://dx.doi.org/10.1045/june2001-reich.

Rosenthal, D., Robertson, T., Lipkis, T., Reich, V., and Morabito, S. 2005. Requirements for digital preservation systems: A bottom-up approach. *D-Lib Magazine* 11 (11), http://dx.doi.org/10.1045/november2005-rosenthal

Shaffer, C. A. 1997. *A Practical Introduction to Data Structures and Algorithm Analysis, C++ Version.* Englewood Cliffs, NJ: Prentice Hall.

Shen, R., Vemuri, N. S., Fan, W., and Fox, E. A. 2008. Integration of complex archaeology digital libraries: An ETANA-DL experience. *Information Systems* 33 (7–8):699–723.

Sperberg-McQueen, C. M., and Thompson, H. S. XML schema. W3C 2000. Available from http://www.w3.org/XML/Schema

Staples, T., Wayland, R., and Payette, S. 2003. The Fedora project—An Open-Source Digital Object Repository Management System. *D-Lib Magazine* 9 (4), http://dx.doi.org/10.1045/april2003-staples

Suleman, H. and Fox, E. 2003. Leveraging OAI harvesting to disseminate theses. *Library Hi Tech* 21 (2):219–227.

Taube, M. 1955. Storage and retrieval of information by means of the association of ideas. *American Documentation* 6 (1):1–18.

TwapperKeeper. 2011. Your twapperkeeper archive your own tweets [cited March 24, 2013]. Available from http://your.twapperkeeper.com/

Tyrväinen, P. 2005. Concepts and a design for fair use and privacy in DRM [cited March 24, 2013]. *D-Lib Magazine* (2), http://www.dlib.org/dlib/february05/tyrvainen/02tyrvainen.html.

Unsworth, J. 2000. The Scholar in the Digital Library, University of Virginia [cited March 24, 2013] April 6, 2000. Available at http://people.lis.illinois.edu/~unsworth/sdl.html

Van de Sompel, H. and Lagoze, C. 2000. Open archives initiative [WWW site]. Cornell University [cited March 24, 2013]. Available from http://www.openarchives.org

W3C. 2004a. OWL web ontology language overview. W3C, 10 February 2004a. Available from http://www.w3.org/TR/owl-features/

W3C. 2004b. Resource description framework (RDF), http://www.w3.org/RDF. W3C [cited March 24, 2013]. Available from http://www.w3.org/RDF/

WDL. 2012. World digital library [cited March 24, 2013]. Available from http://www.wdl.org/en/

Weatherley, J. 2012. jOAI: Open archives initiative protocol for metadata harvesting (OAI-PMH) data provider and harvester tool that runs in Tomcat. UCAR [cited March 24, 2013]. Available from https://wiki.ucar.edu/display/nsdldocs/jOAI

Wei, J. 2002. Color object indexing and retrieval in digital libraries. *IEEE Transactions on Image Processing* 11 (8):912–922.

Witten, I. H. and Bainbridge, D. 2002. *How to Build a Digital Library*. San Francisco, CA: Morgan Kaufmann.

Witten, I. H., Bainbridge, D., and Nichols, D. C. 2009. *How to Build a Digital Library*. 2nd edn. San Francisco, CA: Morgan Kaufmann.

Witten, I. H., Loots, M., Trujillo, M. F., and Bainbridge, D. 2001. The promise of DLs in developing countries. *CACM* 44 (5):82–85.

Witten, I. H., McNab, R. J., Boddie, S. J., and Bainbridge, D. 2000. Greenstone: A comprehensive Open-Source Digital Library Software System. In *Proceedings of the 5th ACM Conference on Digital Libraries*, San Antonio, TX. ACM. pp. 113–121.

Zhu, Q. 2002. 5SGraph: A modeling tool for digital libraries. Master's thesis, Virginia Tech—Department of Computer Science, Blacksburg, VA.

19

Knowledge Discovery and Data Mining

Jennifer Jie Xu
Bentley University

19.1 Introduction

In today's information age, the competitive advantage of an organization no longer depends on their information storage and processing capabilities (Carr 2003) but on their ability to analyze information and discover and manage valuable knowledge. Knowledge discovery and data mining, which is often referred to as *knowledge discovery from data* (KDD), is a multidisciplinary field bringing together several disciplines, including computer science, information systems, mathematics, and statistics (Han et al. 2011). The field emerged in response to the demand for knowledge management and decision support based on large volumes of data in business, medicine, sciences, engineering, and many other domains. Since the first KDD conference in 1989 (Fayyad et al. 1996b), KDD research has made significant progress and had a far-reaching impact on many aspects in our lives.

The primary goal of knowledge discovery and data mining is to identify "valid, novel, potentially useful, and ultimately understandable patterns in data" (Fayyad et al. 1996c, p. 30). Huge amounts of data are gathered daily and stored in large-scale databases and data warehouses: sales transactions, flight schedules, patient medical records, news articles, financial reports, blogs, and instant messages, among many others. Valuable knowledge can be extracted from these data to generate important intelligence for various purposes. For example, credit card companies can analyze their customers' credit card transaction records and profile each customer's purchasing patterns (e.g., the average spending amount, the average monthly balance, and the time and locations of regular purchases). Such knowledge can, for example, be used to detect credit card fraud when some unusual, suspicious transactions occur (e.g., unusually high spending amount) (Bhattacharyya et al. 2011).

Many KDD techniques have been developed and applied to a wide range of domains (e.g., business, finance, and telecommunication). However, KDD still faces several fundamental challenges, such as

- *Data volume*: Because of the rapid advances in information technology, the volumes of data that can be used for knowledge discovery purposes grow exponentially: companies collect transactional data in their daily business, research institutions and laboratories record and make available results from their scientific experiments, and healthcare providers and hospitals monitor patients' health conditions and changes. Therefore, the KDD techniques must be sufficiently efficient and scalable to handle large volumes of data in a timely manner. Especially for real-time applications such as computer network intrusion detection, the KDD technology must be quick enough to detect and respond to any intrusion to prevent it from causing devastating damages to computer systems and networks (Ryan et al. 1998).
- *Data types and formats*: Data can be of many different types and formats: numerical, text, multimedia, temporal, spatial, and object-oriented. Numerical data (e.g., product price, order total, blood pressure, and weather temperature) are also called structured data and can be organized easily using relational databases. Other types of data are often called unstructured data because they may contain text and other data types that do not fit in some predefined data models. Because of the variety and complexity in data types and formats, the KDD techniques must be able to handle the specific characteristics of a data type. For example, text mining techniques must be able to recognize terms and phrases in free-text documents, parse sentences into syntactical components, and extract meaningful syntactical and semantic patterns from texts (Feldman and Sanger 2006).
- *Data dynamics*: In many applications, the data are constantly changing. For example, in traffic monitoring and control systems, the data collected using video cameras and sensors installed along streets are in the form of continuous sequences of digitalized signals. This poses a significant challenge on the storage and processing power of the systems.
- *Data quality*: The quality of data can substantially affect the quality and value of the patterns extracted from the data. When the data contain many errors, noise, and missing values, the extracted patterns based on such data may be inaccurate, incorrect, misleading, or useless. Therefore, the KDD techniques must be robust enough to handle low-quality data to ensure the quality of the results.

The KDD community has kept developing innovative techniques and methods to address these challenges. This chapter will provide an overview of the principles, techniques, applications, and practical impacts of KDD research. The next section will introduce the KDD process and the underlying principles for major data mining techniques. We will then review the applications and impacts of KDD technologies. The research issues and trends of future KDD research will be discussed. The last section will summarize this chapter.

19.2 KDD Principles and Techniques

As introduced earlier, KDD is a multidisciplinary field, which is related to several different research areas and fields—statistics, machine learning, artificial intelligence, neural networks, databases, information retrieval, and linguistics—which all have contributed to the development and evolution of the KDD field. There have been several views regarding the fundamental principles of KDD: data reduction view, data compression view, probabilistic view, microeconomic view, and inductive database view (Mannila 2000).

The data reduction view treats KDD as a process to reduce the data representation. In the data compression view, KDD is a way to compress given data into certain patterns or rules. The probabilistic view regards KDD as a statistic problem and the extracted knowledge as hypotheses about the data (Mannila 2000). The microeconomic view considers KDD as a problem-solving process intended to find useful knowledge to support decision making (Kleinberg et al. 1998). The inductive database view assumes that the patterns are already embedded in the data and the task of KDD is to query the database and find the existing patterns (Imielinski and Mannila 1996).

These views and theories have together built the foundation of the KDD field. In reality, knowledge discovery is not a simple, one-step task but a complex process involving multiple stages and activities. With different contexts, goals, applications, and techniques, the underlying principles may not be the same.

19.2.1 KDD Process

Although many people use "knowledge discovery" and "data mining" interchangeably, the two terms actually refer to different things. More specifically, data mining is a core step in the multistage process of knowledge discovery (Han et al. 2011). Figure 19.1 presents an overview of a typical knowledge discovery process:

- *Cleaning and integration*: This preprocessing step includes such operations as removing noise, correcting errors, resolving inconsistencies, handling missing and unknown values, and consolidating data from multiple sources. For example, when mining sales data, abandoned or invalid transactions can be removed from the dataset. This step is critical to ensure that the preprocessed data meet the quality requirements for the later steps in the process.
- *Selection and transformation*: Sometimes the original dataset contains features (or fields and attributes in database terms) that are not relevant to a particular KDD application or task. By selecting only relevant features, the number of variables under investigation can be reduced and the efficiency of the KDD process can be improved. In addition, the original data may need to be transformed, summarized, or aggregated before they can be used for pattern extraction. For example, in social network mining applications, the data about social relationships between people may not be directly available but have to be inferred from other types of data such as group membership (Chau and Xu 2007) and paper coauthorship (Newman 2004a).
- *Data mining*: This is the most important step in the entire KDD process. In this step, effective and efficient algorithms, methods, and techniques are applied to the data to extract patterns, rules, or models. We will discuss this step in greater detail in the following sections.

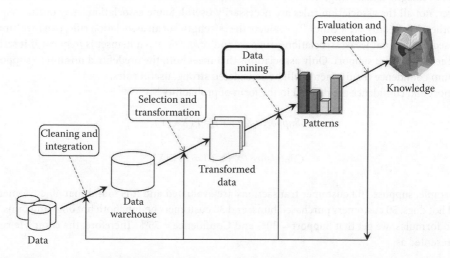

FIGURE 19.1 Overview of the KDD process. (Adapted from Han J. et al., *Data Mining: Concepts and Techniques*, 3 edn., Morgan Kaufmann, Amsterdam, the Netherlands, 2011.)

- *Evaluation and presentation*: The patterns extracted from the data must be subject to evaluation (e.g., by domain experts) to determine its quality, validity, interestingness, and value. Patterns that are trivial or not useful for the particular application will not be considered knowledge. Finally, the valid, novel patterns extracted from the data must be presented in a way that they can be interpreted and understood relatively easily by users or decision makers because the ultimate goal of KDD is to provide "actionable" knowledge that supports decision making (Han et al. 2011). Various presentation and visualization techniques may be used in this step to accomplish this goal.

Because data mining is the key step in the KDD process, we will focus in the following sections on data mining techniques, applications, and impacts. Depending on the goals of KDD tasks, traditional data mining techniques can be categorized into four types: *association mining, classification and prediction, clustering,* and *outlier analysis*. We will discuss the basic principles for major algorithms and methods of these types in this section. We will also briefly introduce a few relatively new mining types including text mining, web mining, and network mining.

19.2.2 Association Mining

Association mining identifies frequently occurring relationships among items in a dataset (Agrawal et al. 1993). A widely used example of association mining is the market basket analysis, which is intended to find items that customers often purchase together. For example, a market basket analysis on customer grocery shopping transactions may reveal that when customers buy hot dogs, very likely they will also buy buns to go with the hot dogs. Such findings may help stores better design the floor layout and plan the shelf space (e.g., placing buns close to hot dogs) so as to encourage sales of both items. Market basket analysis has also been used in online stores. The online retailer, Amazon.com, has long been using customers' "co-purchasing" information to promote sales of related items.

An association relationship is often represented as a rule, $X \Rightarrow Y$, where X and Y are two itemsets, to indicate that itemset X and itemset Y are associated. Moreover, the rule implies that the occurrence of X will lead to the occurrence of Y. Note that the reverse of this rule may not necessarily be true. In the aforementioned example about hot dogs and buns, the rule can be represented as {Hot dog} $\Rightarrow$ {Bun}, which means that customers who buy hot dogs will also buy buns. An itemset consisting of k items is called a k-itemset. In this example, both {Hot dog} and {Bun} are 1-itemsets.

It can be imagined that a huge number of association rules can be mined out of a transactional dataset. However, not all the association rules are necessarily useful. Some associations may occur frequently, while others occur only a few times. To evaluate the "strength" of an association rule, two measures have been used to filter out weak associations: *support* and *confidence*. An itemset is *frequent* if it satisfies a predefined minimum support. Only associations that meet both the predefined minimum support and minimum confidence requirements will be considered strong, useful rules.

Support and confidence are defined in the form of probability:

$$Support\ (X \Rightarrow Y) = P(X \cup Y),$$

$$Confidence\ (X \Rightarrow Y) = P(Y|X).$$

For example, suppose 100 customer transactions are evaluated and it is found that 40 customers purchased hot dogs, 50 customers purchased buns, and 30 customers bought both hot dogs and buns. Using the two formulas, we get that Support = 30% and Confidence = 75%. Therefore, the complete rule will be represented as:

$$\{\text{Hot dog}\} \Rightarrow \{\text{Bun}\}\ [Support = 30\%,\ Confidence = 75\%].$$

This means that this association rule is supported by 30% of the transactions under consideration (i.e., hot dogs and buns co-occur in 30% of the transactions), and 75% of the customers who purchased hot dogs also purchased buns.

Researchers have proposed many methods and algorithms for finding association rules in databases. Among these methods, the Apriori algorithm (Agrawal and Srikant 1994) is the most widely used. The fundamental principle on which this algorithm is based is that there exists the prior knowledge of the property of a frequent itemset: all nonempty subsets of a frequent itemset must also be frequent. This property is used to search from data for all frequent itemsets, which have at least the minimum support. The frequent itemsets are then further examined to find strong association rules that also have minimum confidence.

The search for frequent itemsets is a progressive, iterative process in which the frequent k-itemsets are found using $(k - 1)$-itemsets. The algorithm has two steps in each iteration: joining and pruning. The algorithm starts with the initial scan of the dataset and finds the set S_1, which contains all 1-itemsets that meet the minimum support requirement. S_1 will be joined with itself to find the candidate set, S_2, for frequent 2-itemsets. Based on the prior knowledge about the property of frequent itemsets, all nonfrequent subsets of S_2 (i.e., the subsets are not contained in S_1) will be removed from S_2 in the pruning step. These two steps are repeated for each k until no frequent itemset can be found for $k + 1$.

After all the frequent itemsets are identified, association rules can be generated by finding the subsets of each itemset and calculate their confidences. Rules that do not satisfy the minimum confidence will be discarded. The output is a number of strong association rules.

The Apriori algorithm is intuitive and easy to implement. However, as the data volume increases, it becomes quite slow because it requires scanning the dataset repeatedly. Many variations of the algorithm have been proposed to increase the efficiency and scalability to handle large volumes of data, including the hash-based methods (Park et al. 1995), transaction reduction, partitioning (Savasere et al. 1995), and sampling (Toivonen 1996). An approach called frequent pattern growth (FP-Growth) that compresses the original dataset to avoid time-consuming candidate itemset generation has also been proposed (Han et al. 2000). All these new techniques have improved the performance of association rule mining in large databases.

19.2.3 Classification and Prediction

Classification is a type of data mining technology used to map records into one of several predefined categories based on attribute values of the records. For example, attributes in a patient's medical records such as his/her blood sugar level, age, weight, and family medical history can be used to predict the patient's risk level of having diabetes in the future (low vs. high) (Prather et al. 1997).

Classification is closely related to machine learning, pattern recognition, and statistics (Han et al. 2011). It usually consists of a training step and a testing step. During the training step, a classification algorithm reads records with their known category labels and generates a classification model (i.e., the classifier) based on the training data. In the testing step, the known category labels of the testing records are removed and the algorithm will predict the labels based on the learned model. These classifier-assigned labels are then compared against the known labels of the testing records to determine the performance of the classification. If the performance exceeds a predefined threshold, the algorithm, together with the learned model, can then be used to classify new data whose category labels are unknown. Because classification involves the training step in which the category labels are given, classification is also called supervised learning. Examples of classification applications include fraud detection (Bhattacharyya et al. 2011), computer and network intrusion detection (Ryan et al. 1998), corporate failure prediction (Eksi 2011), and image categorization (Zaiane et al. 2001).

Classifiers are used to generate discrete-valued category labels (e.g., low risk vs. high risk, success vs. failure). For applications dealing with continuous variables (e.g., yearly revenue), a traditional statistical method, regression analysis, can be used. Because regression analysis has long been studied and used

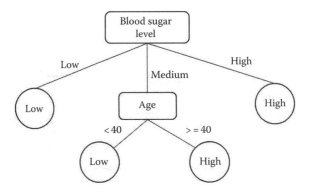

FIGURE 19.2   An example of a decision tree.

in statistics and many textbooks and literature that introduce and explain regression analysis can be found, we focus in this chapter on classifiers.

The model generated by a classifier can be represented as decision rules (if-then statements), decision trees, mathematical formulas, or neural networks. In a decision tree, for example, each internal node represents a test on the value of an attribute, each branch specifies an outcome from the test, and each leaf node corresponds to a category label. Figure 19.2 presents an example of a hypothetical decision tree for predicting the risk level (low vs. high) of diabetes for patients.

There have been many classification techniques including induction decision tree-based methods (Breiman et al. 1984; Quinlan 1986, 1993), naïve Bayesian classifiers (Weiss and Kulikowski 1991), Bayesian belief networks (Heckerman 1996), neural networks (Rumelhart and McClelland 1986), support vector machines (SVM) (Burges 1998; Vapnik 1998), and genetic algorithms (Goldberg 1989), among many others. Different classification methods are based on different principles. We briefly introduce a few methods as follows.

Decision tree induction methods such as ID3 (Quinlan 1986), C4.5 (Quinlan 1993), and classification and regression trees (CART) (Breiman et al. 1984) are based on the principle of *information entropy* in information theory (Shannon 1951). Information entropy generally measures the uncertainty and randomness in the "information content" of messages. The higher the uncertainty is, the higher the information entropy is. These classification methods take a divide-and-conquer strategy. The process begins with the entire set of attributes, and the attribute set is recursively partitioned into smaller subsets while the decision tree is built. The attribute that can reduce the information entropy the most will be selected as the splitting attribute for a partition. Each partition corresponds to an internal node of the decision tree.

The underlying principle of naïve Bayesian classification is Bayes' theorem of posterior probability, along with the assumption that the effects of different attributes on the categories are independent of each other (Weiss and Kulikowski 1991). Based on this theorem, the posterior probability, $P(C_i|R)$, for a record R belonging to a category C_i given the attribute values of R, can be calculated using the posterior probability $P(R|C_i)$ and prior probabilities $P(R)$ and $P(C_i)$:

$$P(C_i|R) = P(R|C_i)P(C_i)/P(R).$$

The record R is assigned to a specific category C_i if $P(C_i|R) > P(C_j|R)$, for all $j \neq i$.

Neural network classifiers learn a classification model by simulating the ways in which neurons process information (Rumelhart and McClelland 1986). A neural network usually consists of several layers of nodes (neurons). The nodes at the first layer (the input layer) correspond to the record attributes and the nodes at the last layer (the output layer) correspond to the categories with associated labels. There may be one or more layers between the first and the last layer. The nodes at a layer are connected with nodes at the adjacent layers and each connection has an associated weight. During the training step, the network receives the values of the attributes from the input layer. In order to generate the category labels that match the given labels at the output layer, the network dynamically adjusts the weights of the connections between layers of nodes. During the testing step, the learned weights are used to predict the labels of testing records. The biggest disadvantage of neural networks is that the learned classification model is hard to interpret because it is encoded in the weights of the connections.

Genetic algorithms incorporate the principles of natural evolution by imitating the genetic crossover and mutation operations during reproduction processes of living beings (Bäck 1996). The training stage involves multiple iterations. The initial set of decision rules, which are represented by strings of bits, are randomly generated. In each iteration, the rules are updated by swapping segments of the strings (crossover) and inverting randomly selected bits (mutation). The fitness of rules, which is usually measured by accuracy, is assessed in each iteration. The process stops when all rules' fitness scores exceed a predefined threshold.

The performance of a classification method can be measured by efficiency and effectiveness (accuracy, precision, and recall) in general. Studies have shown that decision tree induction methods are quite accurate but slow, limiting their applicability to large datasets. The naïve Bayesian classifiers are comparable to decision tree induction methods in effectiveness. Both neural networks and genetic algorithms can achieve high effectiveness in some domains but often require long training time (Han et al. 2011).

19.2.4 Clustering

Clustering is used to group similar data items into clusters without the prior knowledge of their category labels (Jain et al. 1999). In this sense, clustering is a type of exploratory, unsupervised learning. The basic principle of clustering is to maximize within-group similarity while minimizing between-group similarity (Jain et al. 1999). Figure 19.3a presents an example of clustering analysis. Except for the two data points p and q, all points fall into either one of the clusters represented by the dashed circles. Points within a circle are closer to each other than to points in the other circle. Clustering has been used in a variety of applications including image segmentation (Jain and Flynn 1996), gene clustering (Getz et al. 2000), and document categorization (Roussinov and Chen 1999).

Two types of clustering methods have been widely used: hierarchical and partitional. Hierarchical methods group data items into a series of nested clusters, forming a hierarchy of partitions. Partitional methods, in contrast, generate only one partition of the entire dataset.

Hierarchical methods are further categorized into agglomerative and divisive methods. Agglomerative methods such as variations of the single-link algorithm (Sibson 1973) and complete-link algorithm (Defays 1977) take a bottom-up approach and grow the clusters progressively. At the beginning, each data item is treated as a cluster. Using distance as the measure for the similarity (or dissimilarity) between data items, the two closest data items are merged into a new cluster and the distances between this new cluster and other clusters are updated. The algorithm then searches for another pair of clusters

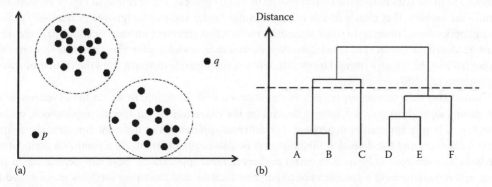

(a)   (b)

FIGURE 19.3 (a) An example of a clustering analysis and (b) an example of a dendrogram.

whose distance is the smallest among all between-cluster distances and merges them together. This process is repeated until all clusters are merged into a single cluster. Divisive methods reverse the clustering process of agglomerative methods by taking a top-down approach. The initial cluster includes all data items and the algorithms progressively split the clusters into smaller ones.

The pattern (i.e., nested partitions) generated by hierarchical methods is represented as a dendrogram in which each distance level corresponds to a merger of two clusters. The dendrogram can be cut at any level to generate a particular partition of the dataset. Figure 19.3b presents an example of a dendrogram.

Partitional clustering methods require the prior knowledge of the number of clusters, k. A partitional algorithm groups the data items into k clusters by maximizing an objective function. The most commonly known partitional method is the k-means algorithm (Lloyd 1957). At the beginning, the algorithm randomly selects k data items and treats them as the centers of the k clusters. The remaining items are then assigned to their closest clusters. The new center of each cluster is recalculated after receiving new items. All the data items are then reassigned to clusters based on the updated centers. The process terminates if the cluster centers do not change. Because k-means algorithms often terminate at the local optimum, they may need to be run several times to find the global optimum.

Both hierarchical and partitional methods have their advantages and disadvantages. The number of clusters does not need to be prespecified for hierarchical methods. However, it is often difficult to determine the level at which the dendrogram should be cut to yield a meaningful partition. Partitional methods require the prior knowledge regarding the number of clusters but can generate the desired partition of the data (Jain and Dubes 1988). In terms of efficiency and scalability, partitional methods are faster and more scalable for large datasets than hierarchical methods are.

Other types of clustering techniques have also been developed. For example, density-based methods can be used to cluster spatial data (Ester et al. 1996). Self-organizing map (SOM) (Kohonen 1995), which is a neural network-based clustering technique, can be used to directly distribute multidimensional data items into regions (clusters) in a two-dimensional space (Chen et al. 1996). In network data mining applications, various clustering methods have been proposed for detecting communities, which can be viewed as densely knit clusters of nodes in networks. For instance, Girvan and Newman (2002) proposed a divisive algorithm that progressively removes links to break a connected network into communities. However, the algorithm is rather slow. A few alternative methods, such as the modularity-based algorithm (Clauset et al. 2004), have been proposed to provide better efficiency.

19.2.5 Outlier Analysis

Most data mining tasks, such as association mining, classification, and clustering, are intended to search for commonality among data and to seek patterns that occur frequently, regularly, or repeatedly. Outliers, which deviate substantially from the rest of the data items, are often treated as noise and removed from the data during the cleaning stage in a KDD process. For example, in Figure 19.3a, points p and q are outliers. They clearly do not belong to either cluster and may be ignored or removed in clustering applications. However, in some situations, outliers that represent unusual or abnormal behaviors must be identified. Credit card fraud detection, for instance, is looking for abnormal transactions that do not fall into the range of normal transactions or are significantly different from the regular purchasing patterns of credit card owners.

Unlike other data mining types such as classification and clustering, in which the development of the mining algorithms may not have to depend on the characteristics of specific applications, outlier detection is largely application dependent. For different applications in different domains, the definitions for outliers and the detection methods may be drastically different. As a result, no dominating methods have emerged to be used to detect outliers in most applications. Basically, depending on the goals and requirements of a specific application, classification and clustering methods may be used to detect outliers (Han et al. 2011).

For example, if the data sample contains records that are labeled by domain experts as outliers, classification methods can be used to detect future outliers that are similar to the known ones. For data without category labels, unsupervised methods such as clustering algorithms can be used to identify data items that are exceptionally dissimilar with the rest of the data.

In addition, statistical approaches have also been proposed for outlier detection in situations where the "normal behavior" of the data is known (Abraham and Box 1979; Agarwal 2006). Such approach is under the assumption that the normally behaved data are all generated by the same stochastic mechanism and outliers are those generated by different mechanisms form the rest of the data.

19.2.6 Other Data Mining Types

In addition to the four types of traditional data mining, there have been several new types of data mining including text mining, web mining, network mining, spatiotemporal data mining, stream data mining, and visual data mining, among many others. We briefly introduce the first three types here:

1. *Text mining*: Text mining has been employed in a wide range of applications such as text summarization (Fattah and Ren 2009), text categorization (Xue and Zhou 2009), named entity extraction (Schone et al. 2011), and opinion and sentiment analysis (Pang and Lee 2007). Text mining is closely related to computational linguistics, natural language processing, information retrieval, machine learning, and statistics (Feldman and Sanger 2006). Text mining requires a great deal of preprocessing in which the text (e.g., news articles) must be decomposed (parsed) into smaller syntactical units (e.g., terms and phrases). Sometimes, the text data may also need to be transformed into other types. For example, in some text mining applications, terms extracted from the documents in the entire corpus are treated as features and documents are treated as records. Thus, each document can be represented as a Boolean vector in which a true (or false) value for a feature indicates the presence (or absence) of the corresponding term in the document. During the mining stage, depending on the requirements of the specific applications, various data mining methods such as association mining, classification, and clustering may be used to find patterns in the text. For example, classification and clustering are frequently used in text categorization applications (Feldman and Sanger 2006).
2. *Web mining*: The web has provided a vast amount of publicly accessible information that could be useful for knowledge discovery. Web mining techniques can be categorized into three types (Kosala and Blockeel 2000): content mining, structure mining, and usage mining. Web content mining extracts useful information from the text, images, audios, and videos contained in web pages (Zamir and Etzioni 1998; Pollach et al. 2006). Web structure mining examines hyperlink structures of the web. It usually involves the analysis of in-links and out-links of a web page and has been used for search engine result ranking and other web applications (Brin and Page 1998; Kleinberg 1999). Web usage mining analyzes search logs or other activity logs to find patterns of users' navigation behavior or to learn user profiles (Srivastava et al. 2000; Nasraoui et al. 2008; Malik and Rizvi 2011).
3. *Network mining*: Unlike traditional data mining that extracts patterns based on individual data items, network mining is used to mine patterns based on the relationships between data items. Network mining is grounded on three theoretical foundations: graph theory from mathematics and computer science (Bollobás 1998), social network analysis from sociology (Wasserman and Faust 1994), and topological analysis from statistical physics (Albert and Barabási 2002). Network mining is intended to find various structural patterns, such as structural positions that have special roles (e.g., leaders, gatekeepers) (Freeman 1979), communities (Clauset et al. 2004; Newman 2004b; Bagrow 2008), and future links (Liben-Nowell and Kleinberg 2007). The techniques include descriptive measures (e.g., centrality measures, network diameters), statistical approaches (e.g., degree distributions), and clustering analysis (e.g., modularity-based algorithms), among many others. Network mining is a young, fast-growing area and has great potential for various applications.

19.3 KDD Applications and Impacts

KDD technologies have made huge impact on every aspect of our lives whether we realize it or not. When we go to a retail store to shop, for example, we may be able to find related items easily on the shelves due to the store's effective use of the results from market basket analyses. When we search for information using a search engine (e.g., Google), by entering just a few query terms, we can find the information we need on the web pages returned by the search engine based on web content and structure mining results. Many credit card companies and banks constantly watch for suspicious, fraudulent transactions to protect us from identify theft crimes. These benefits and the associated convenience have been provided by advanced technologies developed in the KDD research.

The KDD research, to a large extent, is application-driven. Many new methods and techniques are developed in response to the demand for knowledge discovery applications in various domains. In this section, we review a few examples of the domains in which KDD technologies have made significant impacts.

19.3.1 Finance

KDD technologies have long been used in the financial industry to support decision making in various applications including fraud detection, stock market forecasting, corporate distress and bankruptcy prediction, portfolio management, and financial crime (e.g., money laundering) investigation (Zhang and Zhou 2004; Kovalerchuk and Vityaev 2010). In these applications, multiple types of data mining methods often are combined and integrated to achieve high performance, quality, and interpretability of the results:

- *Fraud detection*: Fraud detection is used to identify fraudulent transactions, behaviors, and activities that deviate from the regular patterns of behavior. Phua et al. (2010) categorized frauds into four types: internal frauds (e.g., fraudulent financial reporting) (Kirkos et al. 2007; Glancy and Yadav 2011; Ravisankar et al. 2011), insurance frauds (Bentley 2000; Viaene et al. 2004), credit transaction frauds, and telecommunication frauds (Alves et al. 2006). In the financial context, the most frequently studied type is credit card fraud detection. Because fraudulent transactions are usually outliers that occur infrequently and irregularly, classification-based outlier analysis is often employed in such applications. For example, Paasch (2007) used neural networks together with genetic algorithms to detect fraudulent credit card transactions in a real dataset that contained 13 month's worth of 50 million credit card transactions. Using the same dataset, Bhattacharyya et al. (2011) combined SVM and logistic regression models for fraud detection. Many other data mining methods have also been used in credit card fraud detection such as Bayesian classifiers (Panigrahi et al. 2009), hidden Markov models (Srivastava et al. 2008), and association rules (Sánchez et al. 2009).
- *Stock market forecasting*: Stock market forecasting attempts to predict the future prices or returns of stocks or other securities such as bonds. The input data usually are time-series data (e.g., stock prices) that are measured at successive time points at equal time intervals. There have been a large number of studies on stock market forecasting using neural network-based methods (Trippi and Desieno 1992; Grudnitski and Osburn 1993; Wood and Dasgupta 1996; Wong and Selvi 1998). Neural networks have been also integrated with other methods to generate better prediction accuracy. For example, Hadavandi et al. (2010) proposed to integrate genetic fuzzy systems and artificial neural networks to build a stock price forecasting expert system. This approach has been tested and used in the prediction of the stock prices of companies in the IT and airline sectors.
- *Corporate distress and bankruptcy prediction*: Firms and corporations may face difficult economic and financial conditions. Financial distress sometimes may lead to bankruptcy. Accurate and timely prediction of financial distress and bankruptcy can help the important stakeholders of

the firms take appropriate strategies to reduce or avoid possible financial losses. Chen and Du (2009) proposed to use neural networks to construct a distress model based on several financial ratios. Similarly, neural network-based techniques are used in other studies to predict financial distresses facing firms and corporations (Coats and Fant 1991–1992, 1993; Altman et al. 1994). The prediction of bankruptcy is also formulated as a classification problem using various features. In addition to the classic bankruptcy prediction model, the Z-score model, which is based on discriminant analysis* (Altman 1968), a variety of classification methods have been used including genetic algorithms (Shin 2002), neural networks (Fletcher and Goss 1993; Zhang et al. 1999), logistic regression and discriminant analysis (Back et al. 1996), and the hybrid approach combining several classifiers (Lee et al. 1996). Sung et al. (1999) distinguished crisis economic conditions from normal conditions under which a firm is facing the possibility of bankruptcy. The interpretive classification model they used identified different factors that should be used to predict bankruptcy under different conditions.

19.3.2 Business and Marketing

Business is a well-fit domain for knowledge discovery and data mining. Companies around the world are capturing large volumes of data about their customers, sales, transactions, goods transportation and delivery, and customer reviews of products and services. Using these data, KDD technologies have been playing a critical role in supporting various business functions such as customer relationship management (CRM), customer profiling, marketing, supply chain management, inventory control, demand forecasting, and product and service recommendation (Ghani and Soares 2006):

- *CRM and customer profiling*: CRM is aimed at helping businesses understand and profile the needs and preferences of individual customers and manage the interaction and relationships with their customers. As a business expands and the size of its customer base increases, it becomes increasingly difficult to learn customer profiles using manual approaches. Data mining technologies provide a great opportunity to serve the purposes of CRM in terms of customer identification, attraction, retention, and development (Ngai et al. 2009; Chopra et al. 2011). Customer profiles may include not only the demographics of the customers but also their purchasing history and patterns such as frequency and size of purchases and customer lifetime values (Shaw et al. 2001). Association mining-based market basket analysis, as reviewed in the previous section, can provide a collection of frequent itemsets and association rules that represent such patterns. In addition, neural networks, decision trees, and clustering are also widely used in CRM (Ngai et al. 2009). For example, neural networks and genetic algorithms were combined for customer targeting (Kim and Street 2004), a Bayesian network classifier was used to model the changes in the customer lifecycles (Baesens et al. 2004), multiple classifiers were integrated to predict customer purchasing patterns (Kim et al. 2003), and k-means algorithms were used to identify groups of customers motivated by the importance of stores in shopping centers (Dennis et al. 2001).
- *Direct marketing and viral marketing*: As today's markets become more competitive and fast changing, mass marketing, which relies on mass media such as newspaper, television, and radio to advertise products and services to the general public, becomes less effective. In contrast, direct marketing selects and targets individual customers that are predicted to be more likely to respond to promotions and marketing campaigns (Ling and Li 1998). Direct marketing is closely related to CRM as it is also based on the concept of customer differentiation and profiling. In addition, due to advanced web technologies, online social networks have become important channels for information disseminations and diffusion. As a result, network mining techniques have been used in

---

* Discriminant analysis is a type of statistical analysis used to express a categorical-dependent variable using a linear combination of a set of independent variables.

viral marketing programs (Domingos and Richardson 2001; Watts and Dodds 2007). By identifying and targeting the most influential customers in a company's online customer networks, the company may be able to spread the information about promotions of new products and services quickly through these customers, achieving high effectiveness with low costs. In addition, by analyzing the community structure in customer networks, companies may also obtain insights into customers' opinions and attitudes toward their products and services (Chau et al. 2009).

- *E-commerce*: KDD technologies have brought great opportunities for e-commerce. In addition to the patterns that can be mined using traditional data mining techniques such as association rules, classification, and clustering analysis, new patterns that are unique to the ways that customer use the websites can be extracted using web usage mining (Srivastava et al. 2000). For example, the paths along which customers navigate among web pages can help design personalized websites so as to provide better online shopping convenience and experience and to increase the chances of customer retention. In addition, web usage patterns can also help generate better recommendations of products using content-based approaches (i.e., recommending items similar to the items purchased in the past) or collaborative approaches (i.e., recommending items that are purchased by other customers).

19.3.3 Healthcare and Biomedicine

KDD has become increasingly popular in the domain of healthcare and biomedicine because of the availability of large-size databases and data warehouses for clinical records, gene sequences, and medical literature. However, the "information overload" problem also poses challenges to healthcare professionals (e.g., doctors, nurses, hospital officials, insurers, and pharmaceutical companies) and medical researchers. It is nearly impossible for anyone to process, analyze, and digest such large volumes of healthcare data to make correct, timely decision using manual approaches. Data mining technologies enable healthcare professionals and researchers to leverage healthcare data for the purposes of diagnosing diseases, developing effective treatments, drugs, nurse care plans, and novel hypotheses (Koh and Tan 2005), and detecting infectious disease outbreaks (Chapman et al. 2004; Zeng et al. 2005):

- *Disease diagnosis*: Data mining techniques, especially classification methods, have been used to help with the diagnosis and treatments of various diseases. For example, Breault et al. (2002) used the decision tree algorithm, CART, to classify over 30,000 records of diabetic patients in a large medical data warehouse and found that younger age was the most important variable associated with bad glycemic control. Similarly, Kaur and Wasan (2006) applied classification methods to the early diagnosis of type-I diabetes in children. Classification algorithms have also been used in the diagnosis of cancers and tumors (Ball et al. 2002; Delen et al. 2005), skin lesions (Dreiseitl et al. 2001), preterm birth (Prather et al. 1997), and neurological disorders (Xu et al. 2011).
- *Gene expression and microarray data analysis*: The advances in new biomedical technologies for genome sequencing and protein identification have brought tremendous opportunities for mining gene expression[*] and microarray data[†] (Getz et al. 2000; Sturn et al. 2002) to help develop personalized drugs and treatment for patients (Debouck and Goodfellow 1999), or find new biological solutions to medical problems. Clustering analysis is widely used on microarray data to identify tissue similarity or groups of genes sharing similar expression patterns (Matsumoto et al. 2003). Examples of these applications include using hierarchical clustering for identifying genome-wide expression patterns (Eisen et al. 1998), k-means for clustering complementary CDNA fingerprinting data (Herwig et al. 1999), hierarchical self-organizing methods for discovering cancer classes

---

[*] Gene expression is a process in which the genetic information encoded in a DNA stretch is used to synthesize a functional protein.

[†] A microarray is a two-dimensional array on a glass slide or silicon thin film that records the activities and interactions of a large amount of biological entities (e.g., genes, proteins, and antibodies) using high-throughput screening methods.

and identifying marker genes (Hsu et al. 2003), and neural network-based methods for finding gene groups that are consistent with those defined by functional categories and common regulatory motifs (Sawa and Ohno-Machado 2003).

- *Biomedical text analysis*: Text mining techniques have been used in recent years to analyze biomedical texts in the form of research articles, laboratory reports, clinical documents, and patient records to discover relationships between biological entities (e.g., genes, proteins, drugs, and diseases) that are potentially useful but previously unnoticed. For example, Kazama et al. (2002) used SVM to extract biological entities in the GENIA corpus. Tanabe and Wilbur (2002) used the part-of-speech tagging and a Bayesian model to extract genes and proteins in biomedical text. Biological relationships (e.g., gene regulations, metabolic pathways, and protein interactions) are identified using various text mining and data mining methods (Song and Chen 2009; Steele et al. 2009).

19.3.4 Security and Intelligence

In response to the tragic events of 9/11 and the following series of terrorist attacks around the world, there has been a pressing demand for advanced technologies for helping intelligence communities and law enforcement agencies combat terrorism and other crimes. A new interdisciplinary field called *Intelligence and Security Informatics (ISI)* (Chen et al. 2003) has emerged to leverage technologies and knowledge from different disciplines to assist crime investigation and help detect and prevent terrorist attacks. KDD is a core component in the ISI technology collection (Chen et al. 2003):

- *Crime investigation*: A number of data mining techniques have been employed in crime investigation applications. For example, classification and clustering techniques have been applied to detect various types of crimes including computer crimes. Adderley and Musgrove (2001) and Kangas et al. (2003) employed the SOM approach to cluster crime incidents based on a number of offender attributes (e.g., offender motives and racial preferences) to identify serial murders and sexual offenders. Ryan et al. (1998) developed a neural network-based intrusion detection method to identify unusual user activities based on the patterns of users' past system command usage. In addition, spatial pattern analysis and geographical profiling of crimes play important roles in solving crimes (Rossmo 1995). Brown (1998) proposed a k-means and nearest neighbor approach to cluster spatial data of crimes to find "hot spot" areas in a city (Murray and Estivill-Castro 1998). Koperski and Han (1995) used spatial association-rule mining to extract cause–effect relations among geographically referenced crime data to identify environmental factors that attract crimes (Estivill-Castro and Lee 2001).

- *Counterterrorism*: Terrorism and terrorist activities substantially threaten national security and have considerable negative impact on the society. However, because terrorist groups are covert organizations, the data about their members and activities are extremely difficult to gather. This has made knowledge discovery for counterterrorism a rather challenging task. Nonetheless, network mining techniques have been used to analyze and extract patterns from terrorist networks such as the Global Salafi Jihad networks (Sageman 2004) that were already destroyed by intelligence agencies. The structural characteristics and evolutionary dynamics have been discovered using techniques such as community detection and topological analysis (Krebs 2001; Xu and Chen 2008; Xu et al. 2009). Although these findings were based on historical data, they enhanced our understanding of the organization, behavior, and characteristics of terrorist groups, and such understanding will help develop better strategies to prevent and fight future terrorist attacks (Krebs 2001; van Meter 2001; Dombroski and Carley 2002). In addition, because the web has been widely used by terrorist groups as a communication medium (Burris et al. 2000; Gerstenfeld et al. 2003), web mining techniques have been employed to discover patterns in the contents and structures of terrorist groups' websites (Chen 2012).

In addition to the application domains we have reviewed, KDD technologies have also made great impacts in a number of other domains such as sciences (Fayyad et al. 1996a), engineering (Grossman et al. 2001), and manufacturing (Harding et al. 2006), among many others.

19.4 Research Issues

Research on knowledge discovery and data mining has made considerable progress in recent years (Han et al. 2011). However, many issues still remain challenging. In addition, with the rapid innovation and development of technologies in database management, data collection, computer memory, and computational power, new trends and directions keep emerging in the field of KDD:

- *Diversity of data types*: The diversity of data types will continue to be both a challenge and an opportunity for KDD research. On one hand, more new data types have become available for mining purposes including web data (e.g., hyperlinks, web pages, and web usage data), multimedia data (e.g., videos, autos, and images), sequence data (e.g., time series, symbolic sequences, DNA sequences, and protein sequences), network data (e.g., social networks, computer networks, and information networks), spatiotemporal data, and data streams (e.g., sensor data). These new data types may pose unique challenges on traditional data management and mining technologies. For example, sensor data usually arrive rapidly and continuously, and traditional database management systems are not designed for loading such continuous streams. In addition, many mining algorithms can only access data via a few linear scans and generally cannot perform repeated random access on data streams (Yates and Xu 2010). On the other hand, these new data types also expose the KDD research to new areas and application domains. For example, image mining and spatiotemporal data mining can be used in mobile phones, global positioning system (GPS) devices, weather services, and satellites (Han et al. 2011). Mining on DNA sequence and microarray data may provide new insights and directions to the development of new drugs and medicine (Debouck and Goodfellow 1999).
- *High-performance mining techniques*: As the number of KDD applications continues to grow, the demand for developing new mining techniques with high performance remains strong. In addition to the improvement of general-purpose techniques such as classification and clustering, the development of application-specific algorithms is becoming a trend. Many applications have their unique goals, constraints, and characteristics, which require that the mining techniques be tailored to address the unique problems in the specific applications. For example, in network mining applications, traditional clustering algorithms based on similarity or distance measures could not be used to uncover community structure in unweighted networks since all links are of the same length (distance). As a result, non-distance-based methods need to be developed to address this problem (Clauset et al. 2004). Moreover, as the volumes of data continue to grow, the mining algorithms and methods must be scalable enough to mine large-scale datasets effectively and efficiently.
- *User interaction and domain knowledge*: The process of KDD should be made interactive so that users and analysts can set parameters, select techniques, specify constraints and conditions, and incorporate their domain knowledge to guide the mining process (Fayyad et al. 1996c). However, most existing data mining technologies are not interactive. To many users, the mining process is a black box over which they do not have any control. Another issue is concerning the presentation of KDD results. That is, the patterns generated should be presented in a way that users and analysts, who may not understand the technology, could easily interpret and comprehend them. This requires the development of effective knowledge presentation and visualization techniques to help users and analysts understand the discovered patterns and make correct, timely decisions based on the knowledge.
- *Evaluation and validation of mining results*: Since knowledge discovery is a process to extract novel patterns that are previously unknown, many data mining technologies are exploratory in nature. Unlike statistical results, which can be evaluated using standard tests, results generated

by data mining algorithms may not have standard tests available for assessing their "significance." The patterns yielded by hierarchical clustering algorithms, for example, are dendrograms, which can be cut at any similarity (or distance) level to produce a partition of the data. Except for domain experts' subjective evaluation, there is no "gold" standard for evaluating the quality or utility of a particular partition (Jain et al. 1999). Therefore, while developing new techniques and methods is important, it is also critical to develop appropriate evaluation and validation methods, in addition to existing methods and metrics (e.g., multiple-fold cross validation, accuracy, precision and recall), to ensure the quality of the extracted patterns.

- *Privacy concerns*: A recent trend in KDD research is on the privacy-preserving data mining due to the increasing concerns with data privacy and security. Especially in domains such as healthcare, medicine, and security and intelligence, confidential data (e.g., patients' medical history and criminals' identities) must be handled with great caution. There are established rules, regulations, and laws to protect privacy and confidentiality of data. The 1996 Health Insurance Portability and Accountability Act (HIPAA), for example, set the rules for handling patient records in electronic forms. Violations of such regulations may lead to serious social, ethical, or even legal consequences. Privacy-preserving data mining, therefore, is designed to extract knowledge from data without threatening the privacy and confidentially of records by using various transformation approaches such as data de-identification (Cios and Moore 2002), perturbation (Li and Sarkar 2006), or reconstruction (Zhu et al. 2009).

19.5 Summary

KDD is a fast-growing, interdisciplinary field. More than two decades of research and practice have resulted in significant progress in this field. This chapter introduces the principles and major techniques of data mining techniques including association mining, classification, clustering, and outlier analysis. Relatively new data mining topics such as text mining, web mining, and network mining are also discussed. Examples of application domains in which KDD has made tremendous impact include finance, business, marketing, healthcare, biomedicine, security, and intelligence. Although various issues remain to be challenging, the KDD field will keep growing and thriving with emerging new trends and directions.

Further Information

1. The annual ACM special interest group on knowledge discovery and data mining conference (SIGKDD): http://www.kdd.org.
2. Han, J., Kamber, M., and Pei, J. 2011. *Data Mining: Concepts and Techniques*, 3rd edn. Morgan Kaufmann. ISBN: 978-0-12-381479-1.
3. Maimon, O. and Rokach, L. (eds.). 2010. *Data Mining and Knowledge Discovery Handbook*, 2nd edn. New York: Springer. ISBN: 978-0-38-709822-7.
4. Hand, D., Mannila, H., and Smyth, P. 2001. *Principles of Data Mining*. A Bradford Book. ISBN: 978-0-26-208290-7.

References

Abraham, B. and Box, G. E. P. 1979. Bayesian analysis of some outlier problems in time series. *Biometrika* 66:229–248.

Adderley, R. and Musgrove, P. B. 2001. Data mining case study: Modeling the behavior of offenders who commit serious sexual assaults. Paper read at the *7th ACM SIGKDD International Conference on Knowledge Discovery and Data Mining*, San Francisco, CA.

Agarwal, D. 2006. Detecting anomalies in cross-classified streams: A Bayesian approach. *Knowledge and Information Systems* 11(1):29–44.

Agrawal, R., Imielinski, T., and Swami, A. 1993. Mining association rules between sets of items in large databases. Paper read at *1993 ACM-SIGMOD International Conference on Management of Data (SIGMOD'93)*, May 1993, Washington, DC.

Agrawal, R. and Srikant, R. 1994. Fast algorithms for mining association rules. Paper read at *1994 International Conference on Very Large Data Bases (VLDB'94)*, September 1994, Santiago, Chile.

Albert, R. and Barabási, A.-L. 2002. Statistical mechanics of complex networks. *Reviews of Modern Physics* 74(1):47–97.

Altman, E. 1968. Financial roatios, discriminant analysis, and the prediction of corporate bankruptcy. *Journal of Finance* 23:589–609.

Altman, E. I., Marco, G., and Varetto, F. 1994. Corporate distress diagnosis: Comparisons using linear discriminant analysis and neural networks (the Italian experience). *Journal of Banking and Finance* 18:505–529.

Alves, R., Ferreira, P., Belo, O., Lopes, J., Ribeiro, J., Cortesao, L., and Martins, F. 2006. Discovering telecom fraud situations through mining anomalous behavior patterns. Paper read at *the ACM SIGKDD Workshop on Data Mining for Business Applications*, New York.

Bäck, T. 1996. *Evolutionary Algorithms in Theory and Practice: Evolution Strategies, Evolutionary Programming, Genetic Algorithms*. Oxford, U.K.: Oxford University Press.

Back, B., Laitinen, T., Sere, K., and Wezel, M. V. 1996. Choosing bankruptcy predictors using discriminant analysis, logit anaysis, and genetic algorithms. Paper read at the *First International Meeting of Artificial Intelligence in Accounting, Finance, and Tax*, Huelva, Spain.

Baesens, B., Verstraeten, G., Dirk, V. D. P., Michael, E. P., Kenhove, V. K., and Vanthienen, J. 2004. Bayesian network classifiers for identifying the slope of the customer-lifecycle of long-life customers. *European Journal of Operational Research* 156:508–523.

Bagrow, J. P. 2008. Evaluating local community methods in networks. *Journal of Statistical Mechanics: Theory and Experiment* May 2008 (5): P05001.

Ball, G., Mian, S., Holding, F., and Ro, A. 2002. An integrated approach utilizing artificial neural networks and SELDI mass spectrometry for the classification of human tumors and rapid identification of potential biomakers. *Bioinformatics* 18:395–404.

Bentley, P. 2000. "Evolutionary, my dear Watson" Investigating committee-based evolution of fuzzy rules for the detection of suspicious insurance claims. Paper read at *Genetic and Evolutionary Computation Conference (GECCO)*, Las Vegas, NV.

Bhattacharyya, S., Jha, S., Tharakunnel, K., and Westland, J. C. 2011. Data mining for credit card fraud: A comparative study. *Decision Support Systems* 50(3):602–613.

Bollobás, B. 1998. *Modern Graph Theory*. New York: Springer-Verlag.

Breault, J. L., Goodall, C. R., and Fos, P. J. 2002. Data mining a diabetic data warehouse. *Artificial Intelligence in Medicine* 26(1):37–54.

Breiman, L., Friedman, J., Olshen, R., and Stone, C. 1984. *Classification and Regression Trees*. Belmont, CA: Wadsworth International Group.

Brin, S. and Page, L. 1998. The anatomy of a large-scale hypertextual web search engine. Paper read at the *7th International Conference on World Wide Web*, April 1998, Brisbane, Queensland, Australia.

Brown, D. E. 1998. The regional crime analysis program (RECAP): A framework for mining data to catch criminals. Paper read at the *1998 International Conference on Systems, Man, and Cybernetics*, San Diego, CA.

Burges, C. J. C. 1998. A tutorial on support vector machines for pattern recognition. *Data Mining and Knowledge Discovery* 2:121–168.

Burris, V., Smith, E., and Strahm, A. 2000. White supremacist networks on the Internet. *Sociological Focus* 33(2):215–235.

Carr, N. G. 2003. IT doesn't matter. *Harvard Business Review* 81(5):41–49.

Chapman, W. W., Dowling, J. N., and Wagner, M. M. 2004. Fever detection from free-text clinical records for biosurveillance. *Journal of Biomedical Informatics* 37:120–127.

Chau, M. and Xu, J. 2007. Mining communities and their relationships in blogs: A study of hate groups. *International Journal of Human-Computer Studies* 65:57–70.

Chau, M., Xu, J., Cao, J., Lam, P., and Shiu, B. 2009. Blog mining: A framework and example applications. *IEEE IT Professional* 11(1):36–41.

Chen, H. 2012. Dark Web: Exploring and mining the dark side of the web. Paper read at the *International Conference of Formal Concept Analysis (ICFCA)*, Leuven, Belgium, May 7–10.

Chen, W.-S. and Du, Y.-K. 2009. Using neural networks and data mining techniques for the financial distress prediction model. *Expert Systems with Applications* 36(2):4075–4086.

Chen, H., Miranda, R., Zeng, D. D., Demchak, C., Schroeder, J., and Madhusudan, T., eds. 2003. *Intelligence and Security Informatics: Proceedings of the 1st NSF/NIJ Symposium on Intelligence and Security Informatics*. Tucson, AZ: Springer.

Chen, H., Schuffels, C., and Orwig, R. 1996. Internet categorization and search: A self-organizing approach. *Journal of Visual Communication and Image Representation* 7(1):88–102.

Chopra, B., Bhambri, V., and Krishan, B. 2011. Implementation of data mining techniques for strategic CRM issues. *International Journal of Computer Technology and Applications* 2(4):879–883.

Cios, K.J. and Moore, G. W. 2002. Uniqueness of medical data mining. *Artificial Intelligence in Medicine* 26(1–2):25–36.

Clauset, A., Newman, M. E. J., and Moore, C. 2004. Finding community structure in very large networks. *Physical Review E* 70:066111.

Coats, P. K. and Fant, L. F. 1991–1992. A neural network approach to forecasting financial distress. *Journal of Business Forecasting* 10(4):9–12.

Coats, P. K. and Fant, L. F. 1993. Recognizing financial distress patterns using a neural network tool. *Financial Management* 22(3):142–155.

Debouck, C. and Goodfellow, P. N. 1999. DNA microarrays in drug discovery and development. *Nature Genetics* 21(1 Suppl):48–50.

Defays, D. 1977. An efficient algorithm for a complete link method. *The Computer Journal* 20(4):364–366.

Delen, D., Walker, G., and Kadam, A. 2005. Predicting breast cancer survivability: A comparison of three data mining methods. *Artificial Intelligence in Medicine* 34(2):113–127.

Dennis, C., Marsland, D., and Cockett, T. 2001. Data mining for shopping centrescustomer knowledge management framework. *Journal of Knowledge Management* 5:368–374.

Dombroski, M. J. and Carley, K. M. 2002. NETEST: Estimating a terrorist network's structure. *Computational & Mathematical Organization Theory* 8:235–241.

Domingos, P. and Richardson, M. 2001. Mining the network value of customers. Paper read at the *7th ACM SIGKDD International Conference on Knowledge Discovery and Data Mining*, August 26–29, San Francisco, CA.

Dreiseitl, S., Ohno-Machado, L., Kittler, H., Vinterbo, S., Billhardt, H., and Binder, M. 2001. A comparison of machine learning methods for the diagnosis of pigmented skin lesions. *Journal of Biomedical Informatics* 34:28–36.

Eisen, M., Spellman, P., Brown, P., and Bostein, D. 1998. Cluster anlaysis and display of genome-wide expression patterns. *Proceedings of the National Academy of Sciences* 95:14863–14868.

Eksi, I. H. 2011. Classification of firm failure with classification and regression trees. *International Research Journal of Finance and Economics* 75:113–120.

Ester, M., Kriegel, H.-P., Sander, J., and Xu, X. 1996. A density-based algorithm for discovering clusters in large spatial databases. Paper read at *International Conference on Knowledge Discovery and Data Mining (KDD'96)*, August 1996, Portland, OR.

Estivill-Castro, V. and Lee, I. 2001. Data mining techniques for autonomous exploration of large volumes of geo-referenced crime data. Paper read at the *6th International Conference on GeoComputation*, September 24–26, Brisbane, Queensland, Australia.

Fattah, M. A. and Ren, F. 2009. GA, MR, FFNN, PNN and GMM based models for automatic text summarization. *Computer Speech & Language* 23(1):126–144.

Fayyad, U., Haussler, D., and Stolorz, P. 1996a. Mining scientific data. *Communications of the ACM* 39(11):51–57.

Fayyad, U., Piatetsky-Shapiro, G., and Smyth, P. 1996b. From data mining to knowledge discovery: An overview. In: *Advances in Knowledge Discovery and Data Mining*, eds. Fayyad, U. M., Piatetsk-Shapiro, G., Smyth, P., and Uthurusamy, R. Menlo Park, CA: AAAI Press/The MIT Press.

Fayyad, U., Piatetsky-Shapiro, G., and Smyth, P. 1996c. The KDD process for extracting useful knowledge from volumes of data. *Communications of the ACM* 39(11):27–34.

Feldman, R. and Sanger, J. 2006. *The Text Mining Handbook*. New York: Cambridge University Press.

Fletcher, D. and Goss, E. 1993. Forecasting with neural networks: An application using bankruptcy data. *Information & Management* 24(3):159–167.

Freeman, L. C. 1979. Centrality in social networks: Conceptual clarification. *Social Networks* 1:215–240.

Gerstenfeld, P. B., Grant, D. R., and Chiang, C.-P. 2003. Hate online: A content analysis of extremist internet sites. *Analyses of Social Issues and Public Policy* 3:29.

Getz, G., Levine, E., and Domany, E. 2000. Coupled two-way clustering analysis of gene microarray data. *Proceedings of the National Academy of Sciences* 97(22):12079–12084..

Ghani, R. and Soares, C. 2006. Data mining for business applications. *SIGKDD Explorations* 8(2):79–81.

Girvan, M. and Newman, M. E. J. 2002. Community structure in social and biological networks. *Proceedings of the National Academy of Sciences* 99:7821–7826.

Glancy, F. H. and Yadav, S. B. 2011. A computational model for financial reporting fraud detection. *Decision Support Systems* 50(3):595–601.

Goldberg, D. E. 1989. *Genetic Algorithms in Search Optimization and Machine Learning*. Reading, MA: Addison Wesley.

Grossman, R. L., Kamath, C., Kegelmeyer, P., Kumar, V., and Namburu, R., eds. 2001. *Data Mining for Scientific and Engineering Applications*. Boston, MA: Springer.

Grudnitski, G. and Osburn, L. 1993. Forecasting S and P and gold futures prices: An application of neural networks. *The Journal of Futures Markets* 13(16):631–643.

Hadavandi, E., Shavandi, H., and Ghanbari, A. 2010. Integration of genetic fuzzy systems and artificial neural networks for stock price forecasting. *Knowledge-Based Systems* 23(8):800–808.

Han, J., Kamber, M., and Pei, J. 2011. *Data Mining: Concepts and Techniques*, 3 edn. Amsterdam, the Netherlands: Morgan Kaufmann.

Han, J., Pei, J., and Yin, Y. 2000. Mining frequent patterns without candidate generation. Paper read at *2000 ACM-SIGMOD International Conference on Management of Data (SIGMOD'00)*, May 2000, Dallas, TX.

Harding, J. A., Shahbaz, M., Srinivas, S., and Kusiak, A. 2006. Data mining in manufacturing: A review. *Journal of Manufacturing Science and Engineering* 128:969–976.

Heckerman, D. 1996. Bayesian networks for knowledge discovery. In: *Advances in Knowledge Discovery and Data Mining*, eds. Fayyad, U. M., Piatetsky-Shapiro, G., Smyth, P., and Uthurusamy, R. Cambridge, MA: MIT Press.

Herwig, R., Poustka, A., Muller, C., Bull, C., Lehrach, H., and O'Brien, J. 1999. Large-scale clustering of cDNA fingerprinting data. *Genome Research* 9:1093–1105.

Hsu, A. L., Tang, S., and Halgamuge, S. K. 2003. An unsupervised hierarchical dynamic self-organizing approach to cancer class discovery and marker gene identification in microarray data. *Bioinformatics* 19(16):2131–2140.

Imielinski, T. and Mannila, H. 1996. A database perspective on knowledge discovery. *Communications of the ACM* 39(11):58–64.

Jain, A. K. and Dubes, R. C. 1988. *Algorithms for Clustering Data*. Upper Saddle River, NJ: Prentice Hall.

Jain, A. K. and Flynn, P. J. 1996. Image segmentation using clustering. In: *Advances in Image Understanding: A Festschrift for Azriel Rosenfeld*, eds. Ahuja, N. and Bowyer, K. Piscataway, NJ: IEEE Press.

Jain, A. K., Murty, M. N., and Flynn, P. J. 1999. Data clustering: A review. *ACM Computing Surveys* 31(3):264–323.

Kangas, L. J., Terrones, K. M., Keppel, R. D., and La Moria, R. D. 2003. Computer aided tracking and characterization of homicides and sexual assaults (CATCH). In: *Investigative Data Mining for Security and Criminal Detection*, ed. Mena, J. Amsterdam, the Netherlands: Butterworth Heinemann.

Kaur, H. and Wasan, S. K. 2006. Empirical study on applications of data mining techniques in healthcare. *Journal of Computer Science* 2(2):194–200.

Kazama, J., Maino, T., Ohta, Y., and Tsujii, J. 2002. Tuning support vector machines for biomecial named entity recognition. Paper read at the *Workshop on Natural Language Processing in the Biomedical Domain*, July 2002, Philadelphia, PA.

Kim, E., Kim, W., and Lee, Y. 2003. Combination of multiple classifiers for the customer's purchase behavior prediction. *Decision Support Systems* 34:167–175.

Kim, Y. S. and Street, W. N. 2004. An intelligent system for customer targeting: A data mining approach. *Decision Support Systems* 37:215–228.

Kirkos, E., Spathis, C., and Manolopoulos, Y. 2007. Data mining techniques for the detection of fraudulent financial statements. *Expert Systems with Applications* 32(4):995–1003.

Kleinberg, J. 1999. Authoritative sources in a hyperlinked environment. *Journal of the ACM* 46(5):604–632.

Kleinberg, J., Papadimitriou, C., and Raghavan, P. 1998. A microeconomic view of data mining. *Data Mining and Knowledge Discovery* 2(4):311–324.

Koh, H. C. and Tan, G. 2005. Data mining applications in healthcare. *Journal of Healthcare Management* 19(2):64–72.

Kohonen, T. 1995. *Self-Organizing Maps*. Berlin, Germany: Springer.

Koperski, K. and Han, J. 1995. Discovery of spatial association rules in geographic information databases. Paper read at the *4th International Symposium on Large Spatial Databases*, August 6–9, Portland, ME.

Kosala, R. and Blockeel, H. 2000. Web mining research: A survey. *SIGKDD Explorations* 2:1–15.

Kovalerchuk, B. and Vityaev, E. 2010. Data mining for financial applications. In: *Data Mining and Knowledge Discovery Handbook*, 2nd edn., eds. Maimon, O. and Rokach, L. New York: Springer.

Krebs, V. E. 2001. Mapping networks of terrorist cells. *Connections* 24(3):43–52.

Lee, K. C., Han, I., and Kwon, Y. 1996. Hybrid neural network models for bankruptcy predictions. *Decision Support Systems* 18:63–72.

Li, X.-B. and Sarkar, S. 2006. Tree-based data perturbation approach for privacy-preserving data mining. *IEEE Transactions on Knowledge and Data Engineering* 18(9):1278–1283.

Liben-Nowell, D. and Kleinberg, J. 2007. The link prediction problem for social networks. *Journal of the American Society for Information Science and Technology* 58(7):1019–1031.

Ling, C. X. and Li, C. 1998. Data mining for direct marketing: Problems and solutions. Paper read at the *Fourth International Conference on Knowledge Discovery and Data Mining*, August 27–31, 1998, New York.

Lloyd, S. P. 1957. Least squares quantization in PCM. *IEEE Transactions on Information Theory* 28:128–137.

Malik, S. K. and Rizvi, S. 2011. Information extraction using web usage mining, web scrapping and semantic annotation. Paper read at *2011 International Conference on Computational Intelligence and Communication Networks (CICN)*, October 2011, Gwalior, India.

Mannila, H. 2000. Theoretical frameworks for data mining. *SIGKDD Explorations* 1(2):30–32.

Matsumoto, I., Emori, Y., Nakamura, S., Shimizu, K., Arai, S., and Abe, K. 2003. DNA microarray cluster analysis reveals tissue similarity and potential neuron-specific genes expressed in cranial sensory ganglia. *Journal of Neuroscience Research* 74(6):818–828.

van Meter, K. M. 2001. Terrorists/liberators: Research and dealing with adversary social networks. *Connections* 24(3):66–78.

Murray, A. T. and Estivill-Castro, V. 1998. Cluster discovery techniques for exploratory spatial data analysis. *International Journal of Geographical Information Science* 12:431–443.

Nasraoui, O., Soliman, M., Saka, E., Badia, A., and Germain, R. 2008. A web usage mining framework for mining evolving user profiles in dynamic Web sites. *IEEE Transactions on Knowledge and Data Engineering* 20(2):202–215.

Newman, M. E. J. 2004a. Coauthorship networks and patterns of scientific collaboration. *Proceedings of the National Academy of Sciences* 101:5200–5205.

Newman, M. E. J. 2004b. Detecting community structure in networks. *European Physical Journal B* 38:321–330.

Ngai, E. W. T., Xiu, L., and Chau, D. C. K. 2009. Application of data mining techniques in customer relationship management: A literature review and classification. *Expert Systems with Applications* 36:2592–2602.

Paasch, C. 2007. *Credit Card Fraud Detection Using Artificial Neural Networks Tuned by Genetic Algorithms.* Hong Kong, China: Hong Kong University of Science and Technology.

Pang, B. and Lee, L. 2007. Opinion mining and sentiment analysis. *Foundations and Trends in Information Retrieval* 2(1):1–135.

Panigrahi, S., Kundu, A., Sural, S., and Majumdar, A. K. 2009. Credit card fraud detection: A fusion approach using Dempster–Shafer theory and Bayesian learning. *Information Fusion* 10(4):354–363.

Park, J. S., Chen, M. S., and Yu, P. S. 1995. An effective hash-based algorithm for mining association rules. Paper read at *1995 ACM-SIGMOD International Conference on Management of Data (SIGMOD'95)*, May 1995, San Jose, CA.

Phua, C., Lee, V., Smith, K., and Gayler, R. 2010. A comprehensive survey of data mining-based fraud detection research. Available at http://arxiv.org/pdf/1009.6119v1.pdf (accessed on March 29, 2013).

Pollach, I., Scharl, A., and Weichselbraun, A. 2006. Web content mining for comparing corporate and third-party online reporting: A case study on solid waste management. *Business Strategy and the Environment* 18(3):137–148.

Prather, J. C., Lobach, D. F., Goodwin, L. K., Hales, J. W., Hage, M. L., and Hammond, W. E. 1997. Medical data mining: Knowledge discovery in a clinical data warehouse. *Proceedings of the AMIA Annual Fall Symposyum* 101(5):101–105.

Quinlan, J. R. 1986. Induction of decision trees. *Machine Learning* 1:81–106.

Quinlan, J. R. 1993. *C4.5: Programs for Machine Learning.* San Francisco, CA: Morgan Kaufmann.

Ravisankar, P., Ravia, V., Raghava Rao, G., and Bose, I. 2011. Detection of financial statement fraud and feature selection using data mining techniques. *Decision Support Systems* 50(2):491–500.

Rossmo, D. K. 1995. Overview: Multivariate spatial profiles as a tool in crime investigation. *Crime Analysis Through Computer Mapping.* eds. C. R. Block, M. Dabdoub, and S. Fregly. Washington, DC: Police Executive Research Forum.

Roussinov, D. G. and Chen, H. 1999. Document clustering for electronic meetings: An experimental comparison of two techniques. *Decision Support Systems* 27:67–79.

Rumelhart, D. E. and McClelland, J. L. 1986. *Parallel Distributed Processing.* Cambridge, MA: MIT Press.

Ryan, J., Lin, M.-J., and Miikkulainen, R. 1998. Intrusion detection with neural networks. In: *Advances in Neural Information Processing Systems*, eds. M. I. Jordan, M. J. Kearns, and Solla S. A. Cambridge, MA: MIT Press.

Sageman, M. 2004. *Understanding Terror Networks.* Philadelphia, PA: University of Pennsylvania Press.

Sánchez, D., Vila, M. A., Cerda, L., and Serrano, J. M. 2009. Association rules applied to credit card fraud detection. *Expert Systems with Applications* 36(2):3630–3640.

Savasere, A., Omiecinski, E., and Navathe, S. 1995. An efficient algorithm for mining association rules in large databases. Paper read at *International Conference on Very Large Data Bases (VLDB'95)*, September 1995, Zurich, Switzerland.

Sawa, T. and Ohno-Machado, L. 2003. A neural network-based similarity index for clustering DNA microarray data. *Computers in Biology and Medicine* 33(1):1–15.

Schone, P., Allison, T., Giannella, C., and Pfeifer, C. 2011. Bootstrapping multilingual relation discovery using English Wikipedia and Wikimedia-induced entity extraction. Paper read at the *23rd IEEE International Conference on Tools with Artificial Intelligence*, November 2011, Boca Raton, FL.

Shannon, C. E. 1951. Prediction and entropy of printed English. *The Bell System Technical Journal* 30:50–64.

Shaw, M. J., Subramaniam, C., Tan, G. W., and Welge, M. E. 2001. Knowledge management and data mining for marketing. *Decision Support Systems* 31:127–137.

Shin, K. S. 2002. A genetic algorithm application in bankruptcy prediction modeling. *Expert Systems with Applications* 23(3):321–328.

Sibson, R. 1973. SLINK: An optimally efficient algorithm for the single-link cluster method. *The Computer Journal* 16(1):30–34.

Song, Y.-L. and Chen, S.-S. 2009. Text mining biomedical literature for constructing gene regulatory networks. *Interdisciplinary Sciences: Computational Life Science* 1(3):179–186.

Srivastava, J., Cookey, R., Deshpande, M., and Tan, P.-N. 2000. Web usage mining: Discovery and applications of usage patterns from web data. *SIGKDD Explorations* 1(2):12–23.

Srivastava, A., Kundu, A., Sural, S., and Majumdar, A. K. 2008. Credit card fraud detection using hidden Markov model. *IEEE Transactions on Dependable and Secure Computing* 5(1):37–48.

Steele, E., Tucker, A., 't Hoen, P. A. C., and Schuemie, M. J. 2009. Literature-based priors for gene regulatory networks. *Bioinformatics* 25(14):1768–1774.

Sturn, A., Quackenbush, J., and Trajanoski, Z. 2002. Genesis: Cluster analysis of microarray data. *Bioinformatics* 18(1):207–208.

Sung, T. K., Chang, N., and Lee, G. 1999. Dynamics of modeling in data mining: Interpretive approach to bankruptcy prediction. *Journal of Management Information Systems* 16(1):63–85.

Tanabe, L. and Wilbur, W. J. 2002. Tagging gene and protein names in biomedical text. *Bioinformatics* 18(8):1124–1132.

Toivonen, H. 1996. Sampling large databases for association rules. Paper read at *International Conference on Very Large Data Bases (VLDB'96)*, September 1996, Bombay, India.

Trippi, R. R. and Desieno, D. 1992. Trading equity index futures with a neural network. *The Journal of Portfolio Management* 19(1):27–33.

Vapnik, V. N. 1998. *Statistical Learning Theory*. New York: John Wiley & Sons.

Viaene, S., Derrig, R., and Dedene, G. 2004. A case study of applying boosting naive Bayes to claim fraud diagnosis. *IEEE Transactions on Knowledge and Data Engineering* 16(5):612–620.

Wasserman, S. and Faust, K. 1994. *Social Network Analysis: Methods and Applications*. Cambridge, U.K.: Cambridge University Press.

Watts, D. J. and Dodds, P. S. 2007. Influentials, networks, and public opinion formation. *Journal of Consumer Research* 34:441–458.

Weiss, S. M. and Kulikowski, C. A. 1991. *Computer Systems That Learn: Classification and Prediction Methods from Statistics, Neural Nets, Machine Learning, and Expert Systems*. San Francisco, CA: Morgan Kaufmann.

Wong, B. K. and Selvi, Y. 1998. Neural network applications in finance: A review and analysis of literature (1990–1996). *Information & Management* 34(3):129–139.

Wood, D. and Dasgupta, B. 1996. Classifying trend movements in the MSCI USA capital market index: A comparison of regression, ARIMA and neural network methods. *Computers and Operations Research* 23(6):611–622.

Xu, X., Batalin, M. A., Kaiser, W. J., and Dobkin, B. 2011. Robust hierarchical system for classification of complex human mobility characteristics in the presence of neurological disorders. Paper read at *2011 International Conference on Body Sensor Networks (BSN)*, May 2011, Dallas, TX.

Xu, J. and Chen, H. 2008. The topology of dark networks. *Communications of the ACM* 51(10):58–65.

Xu, J., Hu, D., and Chen, H. 2009. Dynamics of terrorist networks: Understanding the survival mechanisms of Global Salafi Jihad. *Journal of Homeland Security and Emergency Management* 6(1):Article 27.

Xue, X.-B. and Zhou, Z.-H. 2009. Distributional features for text categorization. *IEEE Transactions on Knowledge and Data Engineering* 21(3):428–442.

Yates, D. and Xu, J. 2010. Sensor field resource management for sensor network data mining. In: *Intelligent Techniques for Warehousing and Mining Sensor Network*, ed. A. Cuzzocrea. Hershey, PA: IGI Global.

Zaiane, O. R., Maria-Luiza, A., and Coman, A. 2001. Application of data mining techniques for medical image classification. Paper read at the *2nd International Workshop on Multimedia Data Mining*, August 26, 2001, San Francisco, CA.

Zamir, O. and Etzioni, O. 1998. Web document clustering: A feasibility demonstration. Paper read at *21st Annual International ACM SIGIR Conference on Research and Development in Information Retrieval*, August 1998, Melbourne, Victoria, Australia.

Zeng, D., Chen, H., Lynch, C., Eidson, M., and Gotham, I. 2005. Infectious disease informatics and outbreak detection. In: *Medical Informatics: Knowledge Management and Data Mining in Biomedicine*, eds. Chen, H., Fuller, S. S., Friedman, C., and Hersh, W. New York: Springer.

Zhang, G., Hu, M. Y., Patuwo, B. E., and Indro, D. C. 1999. Artificial neural networks in bankruptcy prediction: General framework and cross-validation analysis. *European Journal of Operational Research* 116:16–32.

Zhang, D. and Zhou, L. 2004. Discovering golden nuggets: Data mining in finance application. *IEEE Transactions on Systems, Man, and Cybernetic—Part C: Applications and Reviews* 34(4):513–522.

Zhu, D., Li, X.-B., and Wu, S. 2009. Identity disclosure protection: A data reconstruction approach for privacy-preserving data mining. *Decision Support Systems* 48(1):133–140.

20

Big Data*

Stephen Brobst
Teradata Corporation

Bill Franks
Teradata Corporation

20.1 Introduction to Big Data

Perhaps nothing will have as large an impact on advanced analytics in the coming years as the ongoing explosion of new and powerful data sources. When analyzing customers, for example, the days of relying exclusively on demographics and sales history are past. Virtually every industry has at least one completely new data source coming online soon, if it is not here already. Some of the data sources apply widely across industries; others are primarily relevant to a very small number of industries or niches. Many of these data sources fall under a new term that is receiving a lot of buzz: big data.

20.1.1 What Is Big Data?

There is not a consensus in the marketplace as to how to define big data, but there are a couple of consistent themes. Two sources have done a good job of capturing the essence of what most would agree big data is all about. The first definition is from Gartner's Merv Adrian in an early 2011 Teradata Magazine article. According to Adrian: "Big data exceeds the reach of commonly used hardware environments and software tools to capture, manage, and process it within a tolerable elapsed time for its user population" [1]. Another good definition is from a paper by the McKinsey Global Institute in May 2011: "Big data refers to data sets whose size is beyond the ability of typical database software tools to capture, store, manage and analyze" [2].

---

* Portions of this chapter have been adapted from: Taming the Big Data Tidal Wave: Finding the Opportunities in Huge Data Streams with Advanced Analytics, Bill Franks. Copyright ©2012 Bill Franks. Used with permission of John Wiley & Sons, Inc.

These definitions imply that what qualifies as big data will change over time as technology advances. What was big data historically or what is big data today will not be big data tomorrow. This aspect of the definition of big data is one that some people find unsettling. The preceding definitions also imply that what constitutes big data can vary by industry, or even organization, if the tools and technologies in place vary greatly in capability.

A couple of interesting facts reported in the McKinsey paper help bring into focus how much data is out there today:

1. $600 today can buy a disk drive that will store all of the world's recorded music.
2. There are 30 billion pieces of information shared on Facebook each month.
3. Fifteen of 17 industry sectors in the United States have more data per company on average than the U.S. Library of Congress [3].

Despite these impressive statistics, big data is not just about the size of the data in numbers of bytes. According to Gartner, the "big" in big data also refers to several other characteristics of a big data source [4]. These aspects include not just increased volume but increased velocity and increased variety. These factors, of course, lead to extra complexity as well. What this means is that you are not just getting a lot of data when you work with big data. It is also coming at you fast, it is coming at you in complex formats, and it is coming at you from a variety of sources.

The analytic techniques, processes, and systems within organizations will be strained up to, or even beyond, their limits. It will be necessary to develop additional analysis techniques and processes utilizing updated technologies and methods in order to analyze and act upon big data effectively. We will discuss these techniques later in the chapter.

20.1.2 How Is Big Data Different?

There are some important ways that big data is different from traditional data sources. Not every big data source will have every feature that follows, but most big data sources will have several of them.

First, big data is often automatically generated by a sensor or a machine. If you think about traditional data sources, there was always a person involved. Consider retail or bank transactions, telephone call detail records, product shipments, or invoice payments. All of those involve a person doing something in order for a data record to be generated. Somebody had to deposit money, or make a purchase, or make a phone call, or send a shipment, or make a payment. In each case, there is a person who is taking action as part of the process of new data being created. This is not so for big data in many cases. A lot of sources of big data is generated without any human interaction at all. A sensor embedded in an engine, for example, spits out data about its surroundings even when nobody touches it or asks it to.

Second, big data is typically an entirely new source of data. It is not simply an extended collection of existing data. For example, with the use of the Internet, customers can now execute a transaction with a bank or retailer online. But the transactions they execute are not fundamentally different transactions from what they would have done traditionally. They have simply executed the transactions through a different channel. An organization may capture web transactions, but they are really just more of the same old transactions that have been captured for years. However, actually capturing browsing behaviors as customers execute a transaction creates fundamentally new data, which represent a transition from analyzing only transactions to extending analysis to understand the interactions that lead up to (and follow) the transactions.

Sometimes, "more of the same" can be taken to such an extreme that the data becomes something new. For example, your power meter has probably been read manually each month for years. An argument can be made that automatic readings every 15 min by a smart meter is more of the same. It can also be argued that it is so much more of the same and that it enables such a different, more in-depth level of analytics that such data is really a new data source.

Third, many big data sources are not designed to be friendly. In fact, some of the sources are not designed at all! Take text streams from a social media site. There is no way to ask users to follow certain standards of grammar, or sentence ordering, or vocabulary. You are going to get what you get when people make a posting. It is almost always difficult to work with such data at best and very, very ugly at worst. Most traditional data sources were designed up-front to be friendly. Systems used to capture transactions, for example, provide data in a clean, preformatted template that makes the data easy to load and use. This was driven, in part, by the historical need to be highly efficient with space. There was no room for excess fluff. Traditional data sources were very tightly defined up-front. Every bit of data had a high level of value or it would not be included. With the cost of storage space becoming almost negligible, big data sources are not always tightly defined up-front and typically capture everything that may be of use. This can lead to having to wade through messy, junk-filled data when performing an analysis.

Finally, large swaths of big data streams may not have much value. In fact, much of the data may even be close to worthless. Within a web log, there is information that is very powerful. There is also a lot of information that does not have much value at all. It is necessary to weed through and pull out the valuable and relevant pieces. Traditional data sources were defined up-front to be 100% relevant. This is because of the scalability limitations that were present. It was far too expensive to have anything included in a data feed that was not critical. Not only were data records predefined, but every piece of data in them was of high value. Storage space is no longer a primary constraint. This has led to the default with big data being to capture everything possible and worry later about what matters. This ensures nothing will be missed but also can make the process of analyzing big data more painful.

Big data will change some of the tactics analytic professionals use as they do their work. New tools, methods, and technologies will be added alongside traditional analytic tools to help deal more effectively with the flood of big data. Complex filtering algorithms will be developed to siphon off the meaningful pieces from a raw stream of big data. Not only SQL (NoSQL) programming methods for incorporating parallelization of procedural capabilities alongside traditional analytic tools, late-binding techniques for deferring the imposition of structure on top of nontraditional data, and polymorphic file systems for handling multi-structured data are key technologies that have emerged to exploit big data. Modeling and forecasting processes are being updated to include big data inputs on top of currently existing inputs. These tactical changes do not fundamentally alter the goals or purpose of analysis, or the analysis process itself.

20.1.3 Structure of Big Data

As you research big data, you will come across a lot of discussion on the concept of data being structured, unstructured, semi-structured, or even multi-structured. Big data is often described as unstructured and traditional data as structured. The lines are not as clean as such labels suggest, however. Most traditional data sources are fully in the structured realm. This means traditional data sources come in a clear, predefined format that is specified in detail. There is no variation from the defined formats on a day-to-day or update-to-update basis. For a stock trade, the first field received might be a date in a MM/DD/YYYY format, next might be an account number in a 12-digit numeric format, next might be a stock symbol that is a three- to five-digit character field, and so on. Every piece of information included is known ahead of time, comes in a specified format, and occurs in a specified order. This makes it easy to work with.

Unstructured data sources are those that you have little or no control over. You are going to get what you get. Text data, video data, and audio data all fall into this classification. Although all of these data types have structure, they do not fit well into traditional file systems or relational database technologies. For example, a picture has a format of individual pixels set up in rows, but how those pixels fit together to create the picture seen by an observer is going to vary substantially in each case. Trying to store a picture into a traditional file system using anything other than a binary large object (BLOB) format is difficult.

There are sources of big data that are truly unstructured such as those described earlier. However, most data is at least semi-structured. Semi-structured data has a logical flow and format that can be understood, but the format is not user-friendly. Sometimes semi-structured data is referred to as

96.255.99.50 - - [01/Jun/2013:05:28:07 +0000] "GET /origin-log.enquisite.com/d.js?id=a1a3af-ly61645&referrer=http://www.google.com/search?hl=en&q=budget+planner&aq=5&aqi=g10&aql=&oq=budget+&gs_rfai=&location=https://money.strands.com/content/simple-and-free-monthly-budget-planner&ua=Mozilla/4.0 (compatible; MSIE 7.0; Windows NT 6.0; SLCC1; .NET CLR 2.0.50727; .NET CLR 3.0.30618; .NET CLR 3.5.30729; InfoPath.2)&pc=pgys63w0xgn102in8ms37wka8 quxe74e&sc=cr1kto0wmxqik1wlr9p9weh6yxy8q8sa&r=0.07550191624904945 HTTP/1.1" 200 380 "-" "Mozilla/4.0 (compatible; MSIE 7.0; Windows NT 6.0; SLCC1; .NET CLR 2.0.50727; .NET CLR 3.0.30618; .NET CLR 3.5.30729; InfoPath.2)" "ac=bd76aad174480000679a044cfda00e005b130000"

FIGURE 20.1   Example of raw web log data.

multi-structured data. There can be a lot of noise or unnecessary data intermixed with the nuggets of high value in such a feed. Reading semi-structured data to analyze them is not as simple as specifying a fixed file format. To read semi-structured data, it is necessary to employ complex rules that dynamically determine how to proceed after reading each piece of information.

Web logs are a good example of semi-structured data. Web logs are pretty ugly when you look at them; however, each piece of information does, in fact, serve a purpose of some sort. Whether any given piece of a web log serves your purposes is another question (see Figure 20.1 for an example of a raw web log).

There is logic to the information in the web log even if it is not entirely clear at first glance. There are fields, there are delimiters, and there are values just like in a structured source. However, they do not follow each other consistently or in a set way. The log text generated by a click on a website right now can be longer or shorter than the log text generated by a click from a different page 1 min from now. In the end, however, it is important to understand that semi-structured data has an underlying logic. It is possible to develop relationships between various pieces of it. It simply takes more effort than structured data.

20.1.4 Filtering Big Data Effectively

The biggest challenge with big data may not be the analytics you do with it, but the extract, transform, and load (ETL) processes you have to build to get it ready for analysis. ETL is the process of taking a raw feed of data, reading it, and producing a usable set of output. The data is extracted (E) from whatever source it is starting from. The data is next transformed (T) through various aggregations, functions, and combinations to get it into a usable state. Finally, the data is loaded (L) into whatever environment will be leveraged to analyze the data. That is the process of ETL.

When you are drinking water out of a hose, you do not really care which parts of the stream of water get in your mouth. With big data, you care very much about which parts of the data stream get captured. It will be necessary to explore and understand the entire data stream first. Only then can you filter down to the pieces that you need. This is why the up-front effort to tackle big data can take so long. The goal is to sip the right amount of data out of the data stream as it flows past, not to try and drink it all. By focusing on the important pieces of the data, it makes big data easier to handle and keeps efforts focused on what is important.

Analytic processes may require filters on the front end to remove portions of a big data set when it first arrives in order to manage the economics of huge data streams. There will be additional filters along the way as the data is processed. The complexity of the rules and the magnitude of the data being removed or kept at each stage will vary by data source and by business problem. The load processes and filters that are put on top of big data is absolutely critical. Without getting those correct, it will be very difficult to succeed. Traditional structured data does not require as much effort in these areas since it is specified, understood, and standardized in advance. With big data, it is necessary to specify, understand, and standardize it as part of the analysis process in many cases.

Deferring the imposition of specific structure on the big data until the time of analysis provides a lot of flexibility in organizing the data in accordance with the needs of each analytic undertaking. Moreover, avoiding the high cost of ETL as an up-front prerequisite to data processing can lead to much greater agility in analytics. This "NoETL" approach to data processing means that there is a higher burden placed onto the analyst to extract the structure required from data at execution time for a query. This late-binding technique for big data analytics is particularly useful when content of the data changes quickly and is not well aligned to relational modeling structures.

20.2 Big Data Use Cases

If Web data is not the original big data, it is probably the most widely used and recognized source of big data. But there are many other sources of big data as well, and they all have their own valuable uses. Some are fairly well known and some are relatively obscure. The following sections provide a representative cross section of big data sources that go beyond web log data. The purpose of these examples is to illustrate the breadth and types of big data available, as well as the breadth of analysis that the data enables.

The big data sources we will cover include

- Auto insurance: The value of telematics data
- Multiple industries: The value of text data
- Multiple industries: The value of time and location data
- Retail and manufacturing: The value of radio-frequency identification (RFID) data
- Utilities: The value of smart-grid data
- Gaming: The value of casino chip tracking data
- Industrial engines and equipment: The value of sensor data
- Telecommunications and other industries: The value of social network data

20.2.1 Auto Insurance: The Value of Telematics Data

Telematics has started to receive serious attention in the auto insurance industry. Telematics involves putting a sensor, or black box, into a car to capture information about what is happening with the car. This black box can measure any number of things depending on how it is configured. It can monitor speed, mileage driven, or if there is a pattern of heavy braking. Telematics data help insurance companies better understand customer risk levels and set insurance rates. With telematics, it is possible to keep track of everywhere a car was driven, when it was there, how fast it was going, and what features of the car were in use.

There are insurance companies pursuing telematics-based insurance in many countries across the globe—and the number is growing. Early programs focus on collecting minimal information from cars. What the early programs do track is how far a car is driven, what time of day it is driven, and if speeding occurs. This is fairly basic information with limited privacy concerns, which is intentional. Using this basic information allows discounts to be granted for safer driving behaviors—incenting good driving habits and retaining lowest risk driving insurance customers.

Telematics data is taking hold initially as a tool to help consumers and companies get better, more effective auto and fleet insurance. Over time, telematics devices may end up being present in a large number of vehicles, and uses for telematics data outside of insurance will emerge. There are already onboard computers managing systems within an automobile, but a telematics device can take it to an entirely new level. There are some very interesting uses for telematics data.

Imagine that a critical mass of millions or tens of millions of cars end up with telematics devices within your country. Let's also imagine that a third-party research firm arranges with consumers to collect very detailed telematics data from their cars in an anonymous fashion. As opposed to the limited data collected for insurance purposes, the data in this case has minute-by-minute or second-by-second updates on speed, location, direction, and other useful information.

This data feed will provide information on thousands of cars in any given traffic jam on any given day. Researchers will know how fast each car was moving along the way. They will know where traffic started, where it ended, and how long it lasted. This is an amazingly detailed view of the reality of traffic flow. Imagine the impact on the study of traffic jams and the planning of road systems! In fact, much of this data can be collected from location-enabled smart phones in geographies where these devices are highly penetrated.

The wealth of possibilities for telematics data is a terrific example of putting big data to use in a way that was not initially foreseen. Often, the most powerful uses for a given data source will be something entirely different from why it was created. Be sure to consider alternative uses for every big data stream encountered. Once researchers have access to thousands of cars in every rush hour, every day, in every city, they will have the ability to diagnose traffic causes and effects in immense detail. They will be able to pinpoint the answers to questions such as

- How does a tire in the road impact traffic?
- What happens if a left lane gets blocked?
- When a traffic light gets out of sync, what are the effects?
- Which traffic intersections are poorly timed, even if they are acting the way they were programmed to act?
- How fast does a backup in one lane spread to other lanes?

It has traditionally been very difficult to effectively study such questions today, outside of very focused and expensive testing. It is possible to physically send people out to monitor a given stretch of road and record information, to put down sensors to count cars that go by, or to install a video camera. But those options are very limited in practice due to costs.

It is a traffic engineer's dream to have access to the telematics information outlined here. If telematics devices do become common, any location populated enough to have traffic can be studied. The changes that are made to roads and traffic management systems, as well as the plans for how roads are built in the first place, will provide huge benefits to all of us. Telematics got its start as a mechanism to assist in insurance pricing. But it may well revolutionize how we manage our highway systems and improve our lives by reducing the stress and frustration we experience when sitting in traffic.

20.2.2 Multiple Industries: The Value of Text Data

Text is one of the highest volume and most common sources of big data. Just imagine how much text is out there. There are e-mails, text messages, tweets, social media postings, instant messages, real-time chats, and audio recordings that have been translated into text. Text data is one of the least structured and largest sources of big data in existence today.

Luckily, a lot of work has been done already to tame text data and utilize it to make better business decisions. Text analytics typically starts by parsing text and assigning meaning to the various words, phrases, and components that comprise it. This can be done by simple frequency counts or more sophisticated methods. The discipline of natural language processing (NLP) comes into play heavily in such analytics.

Text mining tools are available as part of major analytic tool suites, and there are also stand-alone text mining packages available. Some of these text analysis tools focus on a rules-based approach where analysts have to tune the software to identify the patterns that they are interested in. Others use machine learning and advanced algorithms that help to find patterns within the data automatically.

Once the parsing and classification phases for preparing text are completed, the results of those processes can be analyzed. The output of a text mining exercise is often an input to other analytic processes. For example, once the sentiment of a customer's e-mail is identified, it is possible to generate a variable that tags the customer's sentiment as negative or positive. That tag is now a piece of structured data that can be fed into an analytic process. Creating structured data out of unstructured text is often called information extraction.

For another example, assume that we have identified which specific products a customer commented about in his or her communications with our company. We can then generate a set of variables that identify the products discussed by the customer. Those variables are again metrics that are structured and can be used for analysis purposes. These examples illustrate how it is possible to capture pieces of unstructured data and create relevant and structured data from them.

Text analysis with NLP techniques is an excellent example of taking purely unstructured data, processing it, and creating structured data that can be used by traditional analytics and reporting processes. One major part of effectively exploiting big data is getting creative in the ways that unstructured and semi-structured data is made usable in this way.

Interpreting text data is actually quite difficult. The words we say change meaning based on which words we emphasize and also the context in which we state them. When looking at pure text, you will not know where the emphasis was placed, and you often will not know the full context. This means that an analyst must make some assumptions when deriving meaning from pure text data. In this way, text analysis is both an art and a science, and it will always involve a level of uncertainty. When performing text analysis, there will be issues with misclassification as well as issues with ambiguity. This is to be expected. If a pattern can be found within a set of text that enables a better decision to be made, then it should be used—even if the results are not perfect. The goal of text analysis is improvement of the decisions being made, not perfection. Text data can easily cross the bar of improving decisions and providing better information than was present without it. This is true even given the noise and ambiguity that text data contains.

A popular use of text analysis is "sentiment" analysis. Sentiment analysis looks at the general direction of opinion across a large number of people to provide information on what the market is saying, thinking, and feeling about a brand, product, or action associated with an organization. It often uses data from social media sites. Examples include the following:

- What is the "buzz" around a company or product?
- Which corporate initiatives are people talking about?
- Are people saying good or bad things about an organization and the products and services it offers?

One tough part of text analysis is that words can be good or bad depending on the context. It will be necessary to take that into account, but across a lot of individuals, the direction of sentiment should become clear. Getting a read on the trends of what people are saying across social media outlets or within customer service interactions can be immensely valuable in planning what to do next.

If an organization captures sentiment information at an individual customer level, it will provide a view into customer intent and attitudes. Similar to how it is possible to use web data to infer intent, knowing whether a customer's general sentiment about a product is positive or negative is valuable information. This is particularly true if the customer has not yet purchased that product. Sentiment analysis will provide information on how easy it is going to be to convince a specific customer to purchase a specific product.

Another use for text data is pattern recognition. By sorting through complaints, repair notes, and other comments made by customers, an organization will be more quickly able to identify and fix problems before they become bigger issues. As a product is first released and complaints start to come in, text analysis can identify the specific areas where customers are having problems. It may even be possible to identify a brewing issue in advance of a wave of customer service calls coming in. This will enable a much faster, more proactive reaction. The corporate response will be better both in terms of putting in place a fix to address the problem in future products and also in what can be done to reach out to customers and mitigate the issues that they are experiencing today.

Fraud detection is also a major application for text data. Within property and casualty or disability insurance claims, for example, it is possible to use text analysis to parse out the comments and justifications that have been submitted to identify patterns that are associated with fraud. Claims can be analyzed and scored based on the text content to flag high risk of fraud. Claims with higher risk patterns can be checked much more carefully. On the flip side, it is possible to enhance automation of claims adjudication. If there are patterns, terms, and phrases that are associated with clean, valid claims, those claims can be identified as low risk and can be expedited through the system, while claim examiners are more appropriately focused on the claims that have a higher risk.

The legal profession also benefits from text analysis. In a legal case, it is routine that e-mail or other communication histories are subpoenaed. The messages are then examined in bulk to identify statements that may contain information tied to the case at hand. For example, which e-mails have potential insider information in them? Which people made fraudulent statements as they interacted with others? What is the specific nature of threats that were made? Applying such analytics in a legal setting is often called eDiscovery. All of the preceding analytics can lead to successful prosecutions. Without text analysis, it would be almost impossible to manually scan all the documents required. Even if an effort was made to manually scan them, there would be a good chance of missing key information due to the monotonous nature of the task.

Text data has the potential to impact almost every industry. It is one of the most widely used forms of big data.

20.2.3 Multiple Industries: The Value of Time and Location Data

With the advent of global positioning systems (GPS), personal GPS devices, and cellular phones, time and location information is a growing source of data. A wide variety of services and applications including foursquare, Google Places, and Facebook Places are centered on registering where a person is located at a given point in time. Cell phone applications can record your location and movement on your behalf. Cell phones can even provide a fairly accurate location using cell tower signals if a phone is not formally GPS-enabled.

There are some very novel ways that consumer applications use this information, which leads to individuals allowing it to be captured. For example, there are applications that allow you to track the exact routes you travel when you exercise, how long the routes are, and how long it takes you to complete the routes. If you carry a cell phone, you can keep a record of everywhere you have been. You can also open up that data to others if you choose. As more individuals open up their time and location data more publicly, very interesting possibilities start to emerge.

Many organizations are starting to realize the power of knowing "when" their customers are "where" and are attempting to get permission to collect such information from their customers. Today, organizations are coming up with compelling value propositions to convince customers to release time and location information to them.

Time and location data is not just about consumers, however. The owner of a fleet of trucks is going to want to know where each is at any point in time. A pizza restaurant will want to know where each delivery person is at any given moment. Pet owners want to be able to locate pets if they get out of the house. A large banquet facility wants to know how efficiently servers are moving around and covering patrons in all areas of the facility.

As an organization collects time and location data on individual people and assets, it starts to get into the realm of big data quickly. This is especially true if frequent updates to that information are made. It is one thing to know where every truck ends up at the start and end of every day. It is another thing to know where every truck is every second of every day. Time and location data is going to continue to grow in adoption, application, and impact.

One application of these data that is only beginning to be leveraged is the development of time and location sensitive offers. It is no longer just about what offer to develop for a customer today or this week but rather what offer is best for that customer based on when they will be where. Today, this is typically achieved by having customers check in and report where they are so they can receive an offer. Eventually, organizations may track the whereabouts of customers continually and react as necessary (with permission of the customer, of course).

Time and location data not only helps understand consumers' historical patterns but also allows accurate prediction where consumers will be in the future. This is especially true for consumers who stick to a regular schedule. If you know where a given person is and where they are heading, you can predict where they might be in 10 min or an hour based on that information. By looking at where consumers were going historically when on the same route, you can make an even more educated prediction as to where they are going now. At minimum, you can greatly narrow down the list of possibilities. This enables better targeting.

There is likely to be an explosion in the use of time and location data in the coming years. Opt-in processes and incentives for consumers will begin to mature. This will enable messaging to become even more targeted and personal than it is today. The idea of getting offers that are not targeted to the here and now may well be considered old-fashioned in the not-too-distant future.

20.2.4 Retail and Manufacturing: The Value of Radio-Frequency Identification Data

An RFID tag is a small tag placed on objects like shipping pallets or product packages. It is important to note that an RFID tag contains a unique serial number, as opposed to a generic product identifier like a universal product code (UPC) code. In other words, it does not just identify that a pallet contains some Model 123 computers. It identifies the pallet as being a specific, unique set of Model 123 computers.

When an RFID reader sends out a signal, the RFID tag responds by sending information back. It is possible to have many tags respond to one query if they are all within range of the reader. This makes accounting for a lot of items easy. Even when items are stacked on top of one another or behind a wall, as long as the signals can penetrate, it will be possible to get a response. RFID tags remove the need to manually log or inventory each item and allow a census to be taken much more rapidly.

Most RFID tags used outside of very-high-value applications are known as passive. This means that the tags do not have an embedded battery. The radio waves from a reader create a magnetic field that is used to provide just enough power to allow a tag to send out the information embedded within it. While RFID technology has been around for a long time, costs were prohibitive for most applications. Today, a passive tag costs just a few cents and prices continue to drop. As prices continue to drop, the feasible uses will continue to expand. There are some technical issues with today's RFID technology. One example is that liquids can block signals. As time progresses, these issues should be solved with updates to the technologies used.

There are uses of RFID today that most people will have come in contact with. One use is the automated toll tags that allow drivers to pass through a toll booth on a highway without stopping. The way it works is that the card provided by the toll authority has an RFID tag in it. There are also readers placed on the road. As a car drives through, the tag will transmit back the car's data so that the fact that you went through the toll can be registered.

Another major use of RFID data is asset tracking. For example, an organization might tag every single PC, desk, or television that it owns. Such tags enable robust inventory tracking. They also enable alerts if items are moved outside of approved areas. For example, readers might be placed by exits. If a corporate asset moves through the door without having been granted prior approval, an alarm can be sounded and security can be alerted. This is similar to how the item tags at retail stores sound an alarm if they have not been deactivated.

One of the biggest uses for RFID today is item and pallet tracking in the manufacturing and retail spaces. Each pallet a manufacturer sends to a retailer, for example, may have a tag. It makes it easy to take stock of what is in a given distribution center or store. Eventually, it is possible that every individual product in a store that is above a trivial price point will end up being tagged with an RFID chip or an updated technology that serves the same purpose. Let's look at some examples of how RFID data can be used [5].

One application where RFID can add value is in identifying situations where an item has no units on the shelf in a retail environment. If a reader is constantly polling the shelves to identify how many of each item remains, it can provide an alert when restocking is needed. RFID enables much better tracking of shelf availability because there is a key difference between being out of stock and having shelf availability. It is entirely possible that the shelf in a store has no product on it, yet simultaneously there are five cases in the storage room in the back. In such a scenario, any traditional out-of-stock analysis is going to show that there is plenty of stock remaining and nothing to worry about. When sales start to drop, people are going to wonder why. If products have an RFID tag, it is possible to identify that there are five units in the back, yet no product on the shelf. As a result, the problem can be fixed by moving product from the back room to the shelf. There are some challenges in terms of cost and technology in this example today, but work is being done to overcome them.

RFID can also be a big help for tracking the impact of promotional displays. Often, during a promotion, product may be displayed in multiple locations throughout a store. From traditional point-of-sale data, all that will be known is that an item on promotion sold. It is not possible to know which display it came from. Through RFID tags, it is possible to identify which products were pulled from which displays. That makes it possible to assess how different locations in the store impact performance.

As RFID is combined with other data, it gets even more powerful. If a company has been collecting temperature data within a distribution center, product spoilage can be traced for items that were present during a specific power outage or other extreme event. Perhaps the temperature of a section of a warehouse got up to 90° for 90 min during a power outage. With RFID, it is possible to know exactly which pallets were in that part of distribution center at exactly that point in time and appropriate action can be taken. The warehouse data can then be matched to shipment data. A targeted recall can be issued if the products were likely to be damaged or retailers can be alerted to double-check their product as it arrives.

There are operational applications as well. Some distribution centers may tend to be too rough with merchandise and cause a high level of breakage. Perhaps this is true only for specific work teams or even specific workers. A human resources (HR) system will report who was working at any point in time. By combining that data with RFID data that show when product was moved, it is possible to identify employees who have an unusually high rate of breakage, shrinkage, and theft. The combination of data allows stronger, better quality action.

A very interesting future application of RFID is tracking store shopping in a similar fashion as web shopping. If RFID readers are placed in shopping carts, it is possible to know exactly what customers put into their carts and exactly what order that they added those items. Even if individual items are not tagged, the cart's path can still be identified. Many of the advantages of analyzing web data in an online environment can be re-created in a store environment through such a use of RFID.

RFID has the potential to have huge implications within the manufacturing and retail industries in the years to come. It has had a slower adoption rate than many expected. But as tag prices continue to drop and the quality of the tags and readers continues to improve, it will eventually make financial sense to pursue wider adoption.

20.2.5 Utilities: The Value of Smart-Grid Data

Smart grids are the next generation of electrical power infrastructure. A smart grid is much more advanced and robust than the traditional transmission lines all around us. A smart grid has highly sophisticated monitoring, communications, and generation systems that enable more consistent service and better recovery from outages or other problems. Various sensors and monitors keep tabs on many aspects of the power grid itself and the electricity flowing through it.

One aspect of a smart grid is what is known as a smart meter. A smart meter is an electric meter that replaces traditional meters. On the surface, a smart meter will not look much different than the meters we have always had. A smart meter is, however, much more functional than a traditional meter. Instead of a human meter reader having to physically visit a property and manually record consumption every few weeks or months, a smart meter automatically collects data on a regular basis, typically every 15 min to every hour. As a result, it is possible to have a much more robust view of power usage both for every household or business individually, as well as across a neighborhood or even the entire grid.

While we will focus on smart meters here, the sensors placed throughout a smart grid deserve a mention. The data that utilities capture from unseen sensors placed throughout the smart grid dwarf smart meter data in size. Synchrophasors that take 60 readings per second across the power system and home area networks recording each appliance cycling on or off are just two examples. The average person will not have any idea that most of these other sensors exist, but they will be crucial for the utilities. Such sensors will capture a full range of data on the flow of power and the state of equipment throughout the power grid. The data generated will be very, very big.

Smart-grid technology is already in place in some parts of Europe and the United States of America. Virtually, every power grid in the world will be replaced with smart-grid technology over time. The amount of data on electricity usage that utility companies will have available to them as a result of smart grids is going to grow exponentially.

From a power management perspective, the data from smart meters will help people to better understand demand levels from customers all the way up the chain. But the data can also benefit consumers. An individual homeowner, for example, will be able to explicitly test how much power various appliances use by simply turning them on while holding other things constant and then monitoring the detailed power usage statistics that flow from their smart meter.

Utilities around the world are already aggressively moving to pricing models that vary by time of day or demand, and the smart grid will only accelerate that. One of the primary goals of the utility firms is to utilize new pricing programs to influence customer behavior and reduce the demand during peak times. It is the peaks that require additional generation to be built, which drives significant costs and environmental impacts. If the cost of power can be flexibly applied by time of day and measured by the meter, customers can be incented to change their behavior. Lower peaks and more even demand equate to fewer new infrastructure requirements and lower costs.

The power company, of course, will be able to identify all sorts of additional trends through the data provided by smart meters. Which locations are drawing power on off-peak cycles? What customers have a similar daily or weekly cycle of power needs? A utility can segment customers based on usage patterns and develop products and programs that target specific segments. The data will also enable identification of specific locations that appear to have very unusual patterns that might point to problems needing correction.

In effect, power companies will have the ability to do all of the customer analytics other industries have been able to do for years. Imagine a phone company knowing your month-end total bill, but none of your calls. Consider a retailer knowing only your total sales, but none of your purchase details. Think about a financial institution knowing only your month-end balances, but none of the movements of money and funds throughout the month. In many ways, power companies have been dealing with data that are equally poor for understanding their customers. They had a simple month-end total usage, and even that month-end figure was often an estimated, not an actual, usage amount.

With smart meter data, a whole variety of new analytics will be enabled that will benefit all. Consumers will have customized rate plans based on their individual usage patterns, similar to how telematics enables individualized rates for auto insurance. A customer who is using power during peak periods is going to be charged more than nonpeak users. Such differentiated pricing will encourage consumers to shift usage patterns once they are able to see the incentives to do so. Perhaps consumers will run the dishwasher early in the evening instead of right after lunch, for example.

Utilities will have much better forecasts of demand as they are able to identify where demand is coming from in more detail. They will know what types of customers are demanding power at what points in time. The utility can look for ways to drive different behaviors to even out demand and lower the frequency of unusual spikes in demand. All of this will limit the need for expensive new-generation facilities.

Each household or business will gain the capability through smart meter data to better track and proactively manage its energy usage. This will not only save energy and make the world more green, but it is going to help everyone save money. After all, if you are able to identify where you are spending more than you intended, you will adjust as needed. With only a monthly bill, it is impossible to identify such opportunities. Smart meter data makes it simple.

In some cases, big data will literally transform an industry and allow it to take the use of analytics to a whole new level. Smart-grid data in the utility industry is an example. No longer limited by monthly meter readings, information on usage will be available at intervals measured in seconds or minutes. Add to that the sophisticated sensors throughout the grid, and it is a whole different world from a data perspective. The analysis of these data will lead to innovation in rate plans, power management, and more.

20.2.6 Gaming: The Value of Casino Chip Tracking Data

Earlier, we discussed RFID technology as it is applied to the retail and manufacturing industries. However, RFID technology has a wide range of uses, many of which lead to big data. An interesting use for RFID tags is to place them within the chips at a casino. Each chip, particularly high-value chips, can have its own embedded tag so that it can be uniquely identified via the tag's serial number.

Within a casino environment, slot machine play has been tracked for many years. Once you slide your frequent-player card in the machine, every time you pull the handle or push a button, it is tracked. Of course, so are the amounts of your wagers and any payouts that you receive. Robust analysis of slot patterns has been possible for years. However, casinos did not traditionally have the capacity to capture such details from table games. By embedding tags within gaming chips, they are now evolving to have it.

One obvious benefit of casino chip tags is the ability to precisely track each player's wagers. This ensures a player gets full credit in a frequent-player program, but no more or less. This benefits players and the casino. For the casino, resources will be allocated more precisely to the correct players. Over-rewarding the wrong players and under-rewarding the right players both lead to suboptimal allocation of limited marketing resources. Players, of course, always want their credits to be accurate.

Wager data collected across players will allow casinos to better segment players and understand betting patterns. Who typically bets $5 at a time, yet every once in a while jumps up to a $100 bet? Who bets $10 every time? Players can be segmented based on these patterns. The betting patterns can also point to those who are card counting in blackjack as there are certain patterns of wagers that will become clear when a player is using card counting techniques.

With chip tracking, it is also a lot harder to purposely defraud a casino or even for a dealer to make a mistake. Since bets and payouts can be tracked to the chip, it is easy to go back and compare video of the results of a blackjack hand and the payouts that were made. Even if arms or heads obscure what chips were put down or picked up, the RFID readings will provide the details. It will allow casinos to identify errors or fraud that took place. One example is when a player puts down extra chips after the fact when the dealer is looking the other way.

Analysis over time can identify dealers or players involved in an unusual number of mistakes. This will lead to either addressing fraudulent activity or providing additional training for a dealer who just happens to make a lot of innocent mistakes. Errors will also be lowered in the counting of chips in the casino cage. Counting large stacks of chips of different denomination is monotonous, and people can make mistakes. RFID allows faster, more accurate tallying.

Taking the preceding example further, the tracking of individual chips is a strong deterrent for thieves. If a stack of chips is stolen, the RFID identifiers for those chips can be flagged as stolen. When someone comes in to cash the chips, or even sits at a table with the chips, the system can realize it and alert security. If thieves remove or alter the chips so that they cannot be read, that will be a flag. The casino will know exactly what chip IDs exist and will expect each chip to report a valid ID. When a chip does not report an ID, or when the ID reported is not valid, they can take action.

As with any business, the more a casino can stop fraud and ensure that appropriate payouts are being made, the less risk it has. This will lead to the ability to provide both better service and better odds to players since there will be fewer expenses to cover. It can be a win for both casinos and their players.

Retailers and manufacturers leverage RFID technology. So do casinos. How they use RFID is totally different in many ways and has similarities in others. The interesting part is that a single technology can be used by different industries to create their own distinct sources of big data. These examples illustrate that some of the same underlying technologies will enable different big data streams that are similar in nature but completely different in scope and application. One fundamental technology can have different and completely distinct uses that generate multiple forms of big data in multiple industries.

20.2.7 Industrial Engines and Equipment: The Value of Sensor Data

There are a lot of complex machines and engines in the world. These include aircraft, trains, military vehicles, construction equipment, drilling equipment, and others. Keeping such equipment running smoothly is absolutely critical given how much it costs. In recent years, embedded sensors have begun to be utilized in everything from aircraft engines to tanks in order to monitor the second-by-second or millisecond-by-millisecond status of the equipment.

Monitoring may be done in immense detail, particularly during testing and development. For example, as a new engine is developed, it is worth capturing as much detailed data as possible to identify if it is working as expected. It is quite expensive to replace a flawed component once an engine is released, so it is necessary to analyze performance carefully up-front. Monitoring is also an ongoing endeavor. Perhaps not every detail is captured for every millisecond on an ongoing basis, but a large amount of detail is captured in order to assess equipment lifecycles and identify recurring issues.

Consider an engine. A sensor can capture everything from temperature, to revolutions per minute, to fuel intake rate, to oil pressure level. These data can be captured as frequently as desired. All of these data get massive quickly as the frequency of readings, number of metrics being read, and number of items being monitored in such a fashion increases.

Engines are very complex. They contain many moving parts, must operate at high temperatures, and experience a wide range of operating conditions. Because of their cost, they are expected to last many years. Stable and predictable performance is crucial, and lives often quite literally depend on it. For example, taking an aircraft out of service for maintenance will cost an airline or a country's air force a lot of money, but it must be done if a safety issue is identified. It is imperative to minimize the time that aircraft and aircraft engines, as well as other equipment, are out of service.

Strategies for minimizing down time include holding spare parts or engines that can be quickly swapped for the asset requiring maintenance, creating diagnostics to quickly identify the parts that must be replaced, and investing in more reliable versions of problem parts. All three of these strategies depend on data for effective implementation. Data is used to create diagnostic algorithms and as input to the algorithms to diagnose a specific problem. Engineering organizations can use sensor data to pinpoint the underlying causes of failure and design new safeguards for longer, more dependable operation. These considerations apply whether the engine is in an aircraft, watercraft, or ground-based equipment.

By capturing and analyzing detailed data on engine operations, it is possible to pinpoint specific patterns that lead to imminent failures. Patterns over time that lead to lower engine life and/or more frequent repair can also be identified. The number of permutations of the various readings, especially over time, makes analysis of these data a challenge. Not only does the process involve big data, but the analysis that must be developed is complex and difficult. Some examples of the types of questions that can be studied are as follows:

- Does a sudden drop in pressure indicate imminent failure with near certainty?
- Does a steady decrease in temperature over a period of a few hours point to other problems?
- What do unusual vibration levels imply?
- Does heavy engine revving upon start-up seriously degrade certain components and increase the frequency of maintenance required?
- Does a slightly low fuel pressure over a period of months lead to damage to some of the engine's components?

When major problems arise, it is extremely helpful to go back and examine what was taking place in the moments up until the problem manifested itself. In this case, sensors act similarly to how well-known airplane black boxes help diagnose the cause of an accident. The data from the sensors in an engine also provide data that can be leveraged for diagnostic and research purposes. The sensors being discussed here are conceptually a more sophisticated form of the telematics devices discussed in the auto insurance example. The use of data from sensors that are continually surveying their surroundings is a recurrent theme in the world of big data. While we have focused on engines here, there are countless other ways that sensors are also being used today, and the same principles discussed here apply to those uses.

If the sensor data capture process is repeated across a lot of engines and a long period of time, it leads to a wealth of data to analyze. Analyze it well, and it is possible to find glitches in equipment that can be proactively fixed. Weak points can also be identified. Then, procedures can be developed to mitigate the problems that result from those findings. The benefits are not just added safety but also lowered cost. As sensor data enables safer engines and equipment that remain in service a higher percentage of the time, it will enable both smoother operations and lower costs.

Sensor data offers a difficult challenge. While the data collected is structured and its individual data elements are well understood, the relationships and patterns between the elements over time may not be understood at all. Time delays and unmeasured external factors can add to the problem's difficulty. The process of identifying the long-term interactions of various readings is extremely difficult given all the information to consider. Having structured data does not guarantee a highly structured and standardized approach to analyzing it.

20.2.8 Telecommunications and Other Industries: The Value of Social Network Data

Social network data qualifies as a big data source even though in many ways it is more of an analysis methodology against traditional data. The reason is that the process of executing social network analysis requires taking data sets that are already large and using them in a way that effectively increases their size by orders of magnitude.

One could argue that the complete set of cell phone calls or text message records captured by a cellular carrier is big data in and of itself. Such data is routinely used for a variety of purposes. Social network analysis, however, is going to take it up a notch by looking into several degrees of association instead of just one. That is why social network analysis can turn traditional data sources into big data.

For a modern phone company, it is no longer sufficient to look at all callers and analyze them as individual entities. With social network analysis, it is necessary to look at whom the calls were between and then extend that view deeper. You need to know not only whom I called but also whom the people I called in turn called, and whom those people in turn called, and so on. To get a more complete picture of a social network, it is possible to go as many layers deep as analysis systems can handle. The need to navigate from customer to customer and call to call for several layers makes the volume of data become multiplied. It also increases the difficulty of analyzing it, particularly when it comes to traditional tools.

The same concepts apply to social networking sites. When analyzing any given member of a social network, it is not that hard to identify how many connections the member has, how often she posts messages, how often she visits the site, and other standard metrics. However, knowing how wide a network a given member has when including friends, friends of friends, and friends of friends of friends requires much more processing.

One thousand members or subscribers are not hard to keep track of. But they can have up to one million direct connections between them and up to one billion connections when "friends of friends" are considered. That is why social networking analysis is a big data problem. The analysis of such connections has a number of applications being pursued today. Moreover, social network analysis will typically use graph structures and sophisticated graph algorithms for calculation of social network metrics. While graphs can be represented by casting them into relational structures, a native representation of graphs with the ability to parallelize graph operators is very useful for this type of analysis.

Social network data, and the analysis of it, has some high-impact applications. One important application is changing the way organizations value customers. Instead of solely looking at a customer's individual value, it is now possible to explore the value of his or her overall network. The example we are about to discuss could apply to a wide range of other industries where relationships between people or groups are known, but we will focus on wireless phones since that is where the methods have been most widely adopted.

Assume a wireless carrier has a relatively low-value customer as a subscriber. The customer has only a basic call plan and does not generate any ancillary revenue. In fact, the customer is barely profitable, if the customer generates profit at all. A carrier would traditionally value this customer based on his or her individual account. Historically speaking, when such a customer calls to complain and threaten to leave, the company may have let the customer go. The customer just is not worth the cost of retention.

With social network analysis, it is possible to identify that the few people our customer does call on that inexpensive calling plan are very heavy users who have very wide networks of friends. In other words, the connections that customer has are very valuable to the organization. Studies have shown that once one member of a calling circle leaves, others are more likely to leave and follow the first. As more members of the circle leave, it can catch like a contagion and soon circle members are dropping fast. This is obviously a bad thing.

A very compelling benefit of social network data is the ability to identify the total revenue that a customer influences rather than just the direct revenue that he or she provides. This can lead to drastically different decisions about how to invest in that customer. A highly influential customer needs to be coddled well beyond what his or her direct value indicates if maximizing a network's total profitability takes priority over maximizing each account's individual profitability.

Using social network analysis, it is possible to understand the total value that the customer in our example influences for the organization rather than only the revenue directly generated. This allows a completely different decision regarding how to handle that customer. The wireless carrier may overinvest in the customer to make sure that they protect the network the customer is a part of. A business case can be made to provide incentives that dwarf the customer's individual value if doing so will protect the wider circle of customers that he or she is a part of.

This is an excellent example of how the analysis of big data can help provide new contexts in which decisions that would have never happened in the past now make perfect sense. Without big data, the customer would have been allowed to cancel and the wireless carrier would not have seen the avalanche of losses that soon came as the customer's friends followed. The goal shifts from maximizing individual account profitability to maximizing the profitability of the customer's network.

Identification of highly connected customers can also pinpoint where to focus efforts to influence brand image. Highly connected customers can be provided free trials and their feedback can be solicited. Attempts can be made to get them active on corporate social networking sites through incentives to provide commentary and opinion. Some organizations actively recruit influential customers and shower them with perks, advance trials, and other goodies. In return, those influential customers continue to wield their influence, and it should be even more positive in tone given the special treatment they are receiving.

Within social networking sites like LinkedIn or Facebook, social network analysis can yield insights into what advertisements might appeal to given users. This is done by considering not just the interests customers have personally stated. Equally important, it is based on knowing what their circle of friends or colleagues has an interest in. Members will never declare all of their interests on a social networking site, and every detail about them will never be known. However, if a good portion of a customer's friends have an interest in biking, for example, it can be inferred that the customer does as well, even though the customer might never have stated that directly.

Law enforcement and antiterrorism efforts can leverage social network analysis. Is it possible to identify people who are linked, even if indirectly, to known trouble groups or persons? Analysis of this type is often referred to as link analysis. It could be an individual, a group, or even a club or restaurant that is known to attract a bad element. If a person is found to be hanging out with a lot of these elements in a lot of these places, he or she might be targeted for a deeper look. While this is another analysis that raises privacy concerns, it is being used in real life situations today.

Online video gaming is another area where such analysis can be valuable. Who plays with whom? How does that pattern change across games? It is possible to identify a player's preferred partners on a game-by-game basis. Telemetry data makes it possible for players to be segmented based on individual playing style. Do players of similar styles team up when playing together? Or do players seek a mix of styles? Knowing such information can be valuable to know if a game producer wants to suggest groups for players to team up with (e.g., providing suggestions as to which group, out of the many options available in a list, a player might prefer to join in with when he or she logs in to play).

There have also been interesting studies on how organizations are connected. It starts by looking at the connections established through e-mail, phone calls, and text messages within an organization. Are departments interacting as expected? Are some employees going outside typical channels to make things happen? Who has a wide internal influence and would be a good person to take part in a study on how to better improve communications within the organization? This type of analysis can help organizations better understand how their people communicate.

Social network analysis will continue to grow in prevalence and impact. One of the interesting features it has is that it is always going to make a data source much bigger than it began due to the exponentially expanding nature of the analysis process. Perhaps the most appealing feature is how it can provide insights into a customer's total influence and value, which can completely change how that customer is viewed by an organization.

20.3 Big Data Technologies

It goes without saying that the world of big data requires new levels of scalability. As the amount of data organizations process continues to increase, the same old methods for handling data just will not work anymore. Organizations that do not update their technologies to provide a higher level of scalability will quite simply choke on big data. Luckily, there are multiple technologies available that address different aspects of the process of taming big data and making use of it in analytic processes. Three key technologies that are essential components of most big data deployment strategies are as follows: (1) massively parallel processing (MPP) systems, (2) cloud computing, and (3) MapReduce.

20.3.1 Massively Parallel Processing Systems

MPP systems have been around for decades. While the specifics of individual vendor implementations may vary, MPP is the most mature, proven, and widely deployed mechanism for storing and analyzing large amounts of data. So what is an MPP, and why is it special?

An MPP system allocates data into a collection of partitions and assigns them to independent servers using a shared-nothing organization of storage and central processing unit (CPU) resources. A shared-nothing organization of the data means that a given partition of data is "owned" by a single server and there can be no read or write contention for data blocks across multiple servers. This approach delivers superior scalability for large amounts of data because there is no need for locking or other data management coordination between servers that are executing software for reading and writing data into a file system or database. Because there are no coordination costs for database buffer cache coherency or logging across the multiple servers in an MPP environment, it is straightforward to scale this type of architecture to thousands—and even tens of thousands—of servers executing a single (or multiple) workload.

Let's look at an example in the big data world. The nonparallel execution of a query will scan a large table or file one block at a time. If there are a billion records in a database table or file, this could take a very long time. However, idealized parallel execution using a shared-nothing MPP approach with 1000 servers (often referred to as nodes) would reduce the time for execution of the query by a factor of (nearly) 1000. The billion records of data would be divided up among the 1000 servers in the MPP system (usually through use of a hashing algorithm against a DBA-assigned partitioning key for each file or table). Each server would then execute in parallel against the (roughly) one million rows assigned to it (Figure 20.2).

There are two critical success factors for obtaining linear scalability in a shared-nothing MPP system for data management:

1. Data must be divided up evenly across the many servers in the MPP because all servers must wait for the slowest one (or one with the most work) to finish.
2. Assembly of results from across the multiple servers must be done as much in parallel as possible.

Dividing data evenly across the MPP servers is critical because the time that it takes to execute a query will be dictated by the slowest server (or the one with the most work). All servers must finish their portion of the work in order for a query to complete. If any server is lagging behind because it has more data to process than the others, the overall query execution time can suffer dramatically. Applying a hashing algorithm to the partitioning key for each record in a file allows placement of data across servers in an MPP system in a relatively balanced way so as to ensure efficiency in parallel processing. Studies have shown that scale-out parallelism with many commodity servers working together will outperform the scale-up computer systems using traditional symmetric multiprocessing (SMP) architectures using a shared memory or nonuniform memory access (NUMA) implementation [6]. However, since there are

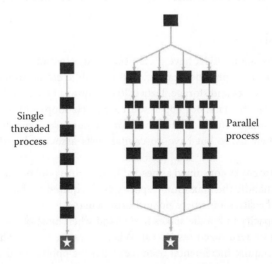

Single threaded process

Parallel process

FIGURE 20.2 Serial query execution versus MPP query execution.

many hardware and software components in a large MPP system, failures are inevitable. MPP systems are architected with built-in redundancy so that data is stored in more than one location to make recovery easy in cases where there is a hardware or software error associated with a node.

20.3.2 Cloud Computing

As with any new and emerging technology, there are conflicting definitions describing the concept of cloud computing. We will consider two that serve as a solid foundation for our discussion. The first comes from a McKinsey and Company paper published in 2009 [7]. The paper listed three criteria for a cloud environment:

1. Enterprises incur no infrastructure or capital costs, only operational costs. Those operational costs will be incurred on a pay-per-use basis with no contractual obligations.
2. Capacity can be scaled up or down dynamically and immediately. This differentiates clouds from traditional hosting service providers where there may have been limits placed on scaling.
3. The underlying hardware can be anywhere geographically. The architectural specifics are abstracted from the user. In addition, the hardware will run in multi-tenancy mode where multiple users from multiple organizations can be accessing the exact same infrastructure simultaneously.

A key concept behind this definition is that cloud infrastructure takes away concern for resource constraints. Users can get whatever they need whenever they need it. They will get charged, of course, but only for what they use. No more fighting with system administrators over resources.

Another definition comes from the National Institute of Standards and Technology (NIST), which is a division of the U.S. Department of Commerce. NIST lists five essential characteristics of a cloud environment [8]:

1. On-demand self-service
2. Broad network access
3. Resource pooling
4. Rapid elasticity
5. Measured service

Each of those criteria has to come into play. The similarities between the McKinsey definition and the NIST definition are easy to see. You can learn more about the work the NIST has done on the cloud on their website [9]. There are two primary types of cloud environments: (1) public clouds and (2) private clouds.

20.3.2.1 Public Clouds

Public clouds, such as Amazon's EC2, have gotten the most attention in the marketplace and align most closely to the definitions provided earlier. With a public cloud, users are basically "renting" access to a large farm of servers and storage. Data is loaded onto a host system and resources are allocated as required to process the data. Both open source (e.g., Hadoop) and commercial (e.g., Teradata) software packages are available in the public cloud for the purpose of big data processing. Users get charged according to their usage but have no capital costs associated with hardware or software acquisition.

Because processing capacity is consumed as needed from a huge pool of resources, it is not necessary to buy a system sized to handle the maximum capacity ever required for big data analytics. The public cloud approach avoids the situation where nonuniform demand for computation creates the need to buy large amounts of capacity to handle spikes in demand when most of the time much less capacity is required (resulting in underutilized resources). When there are bursts where lots of processing is needed, it is possible to acquire incremental resources from the public cloud without actually owning those resources. The user is simply billed for the extra resources on a pay-for-use basis.

A huge advantage of cloud computing for big data analytics is fast ramp up because end users are leveraging preexisting infrastructure with little to no setup time. Once granted access to the cloud environment, users load their data and start analyzing. It is easy to share data with others regardless of their geographic location since a public cloud is accessible over the Internet and is outside of corporate firewalls. Anyone can be given permission to log on to the environment created. Of course, the advantage of sharing needs to be weighed against the potential security exposures associated with public cloud infrastructure.

As a result of the scale of resources available in a large public cloud infrastructure, computing throughput is generally quite scalable and relatively high performance. But, like any shared infrastructure, there can be peak periods of demand when performance cannot be guaranteed. One solution to this issue is "reserve" resources in a public cloud through use of virtual private cloud allocation. By "leasing" an allocation of resources from the public cloud, a user can be protected from variations in demand for resources during the leasing period.

20.3.2.2 Private Clouds

A private cloud has many of the same features as a public cloud, but it is owned exclusively by one organization and typically housed behind a corporate firewall. An important implication of the private cloud approach is that the owning organization incurs the capital costs of acquiring the hardware and software for enabling the cloud. A private cloud serves the same function as a public cloud but just for the people or teams within a given organization (see Figure 20.3 for an illustration of a public versus a private cloud).

One huge advantage of a private cloud is that the owning organization will have complete control over the data and system security. Data is not being deployed outside the corporate firewall so there is no concern about violating policies (or government regulations) related to geographic placement of data or reliance on security implementations put in place by third parties. The data is at no more risk than it is on any other internal system.

A significant downside of a private cloud is that it is necessary to purchase and own the entire cloud infrastructure before allocating it out to users. This takes time and money. Capital outlay this year will increase this year's expense relative to a public cloud. In the years that follow, there will generally be a lower cost for allocating resources from the (already paid for) private cloud than if resources are acquired from a public cloud.

The overall total cost of ownership will depend on economies of scale and efficiency of an organization's data center operations. Private cloud computing rarely makes sense for a small organization

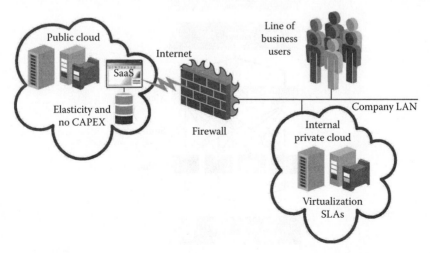

FIGURE 20.3 Public cloud versus private cloud.

because lack of buying power and lack of scale for creating world class data center capabilities means that public cloud computing will almost always be lower total cost of ownership.

However, for large organizations with world class data center operational capability, it is likely that private cloud computing will be more cost effective than public cloud computing. Organizations such as eBay, Facebook, Yahoo!, Google, and many others use private cloud computing for their big data processing. In the long run, if a lot of big data analysts will be using a cloud environment, it can be much cheaper to have a private cloud with good workload management than pay fees for accessing resources delivered via a public cloud. However, the initial capital outlay and setup will cause higher costs in the short term when implementing a private could. Over time, the balance will shift. This is especially true if large volumes of data would need to have been moved from enterprise systems into a public cloud (versus localized movement within the data centers of the enterprise).

20.3.3 MapReduce

MapReduce is a parallel programming framework popularized by Google to simplify data processing across massive data sets [10,11]. MapReduce consists of two primary processes that a programmer builds: the "map" steps for structuring and assigning data across the parallel streams of processing and the "reduce" steps for computation and aggregation. The MapReduce framework coordinates the execution of the map and reduce logic in parallel across a set of servers in an MPP environment. The logic for mapping and reducing can be written in traditional programming languages, such as Java, C++, Perl, and Python.

Assume that we have 20 TB of data and 20 MapReduce server nodes on which to run our analysis. The first step is to distribute 1 TB to each of the 20 nodes using a simple file copy process. The data should be distributed prior to the MapReduce process being started. Data is in a file of some format determined by the user. There is no standard format like in a relational database. Next, the programmer submits two programs to the scheduler. One is a map program; the other is the reduce program. In this two-step processing, the map program finds the data on disk and executes the mapping logic. This occurs independently on each of the 20 servers in our example. The results of the map step are then passed to the reduce process to summarize and aggregate the final answers (see Figure 20.4 for a visual description of the MapReduce process). MapReduce programming breaks an analytic job into many tasks and runs them in parallel.

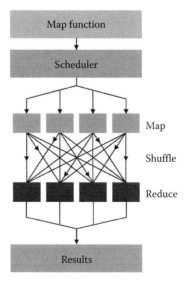

FIGURE 20.4   The MapReduce process.

20.3.3.1 Scheduler

Let's look at an example where an organization has a large volume of text flowing in from online customer service chats taking place on its website. This is the classic example to show how the MapReduce programming framework parallelizes large-scale data processing. In this example, text paragraphs are provided as input. A programmer creates a map step to parse out each word present in the chat text. In this example, the map function will simply find each word, parse it out of its paragraph, and associate a count of one with it. The end result of the map step is a set of value pairs such as "<my, 1>," "<product, 1>," "<broke, 1>." When each worker task is done executing the map step, it lets the scheduler know.

Once the map step is complete, the reduce step is started. At this point, the goal is to figure out how many times each word appeared. What happens next is called shuffling. During shuffling the answers from the map steps are distributed through hashing so that the same words end up on the same reduce task. For example, in a simple situation, there would be 26 reduce tasks so that all the words beginning with A go to one task, all the Bs go to another, all the Cs go to another, and so on.

The reduce step will simply get the count by word. Based on our example, the process will end up with "<my, 10>," "<product, 25>," "<broke, 20>," where the numbers represent how many times the word was found. There would be 26 files emitted (one for each reduce node) with sorted counts by word. Note that another process is needed to combine the 26 output files. Multiple MapReduce processes are typically required to get to a final answer set.

Once the word counts are computed, the results can be fed into an analysis. The frequency of certain product names can be identified. The frequency of words like "broken" or "angry" can be identified. The point is that a text stream that was "unstructured" has been structured in a fashion that allows it to be easily analyzed. The use of MapReduce is often a starting point in this way. The output of MapReduce can be an input to another analysis process (Figure 20.5).

There can be thousands of map and reduce tasks running across thousands of machines. When there are large volume streams of data and the processing that needs to be executed against the data can be partitioned into separate streams, the performance of the MapReduce programming framework will be very scalable. If one worker does not need to know what is happening with another worker to execute effectively, it is possible to achieve linear scalability in the parallel processing of large data sets. The essence of the MapReduce framework is to allow many servers to share the burden of working through a large amount of data. When the logic required can be run independently on different subsets of the data, the shared-nothing nature of the MapReduce framework makes processing very scalable.

The overall MapReduce word count process

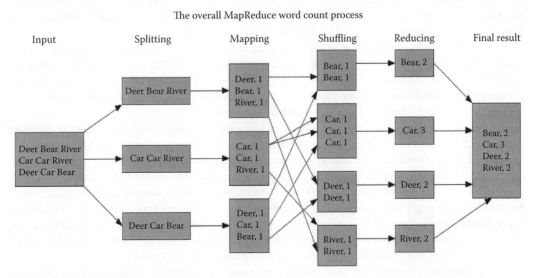

FIGURE 20.5 Simple word count with MapReduce.

In our example, each word can be parsed by itself, and the contents of the other words are irrelevant for a given map worker task.

Recall that well-designed MPP systems spread data out across nodes that can then be queried with efficient use of parallelism. The MapReduce framework makes use of a technique called "data parallelism" whereby each worker runs the same code against its portion of the data. However, the parallel workers are independent and do not interact or even have knowledge of each other. If there is a steady stream of web logs coming in, chunks of data would be allocated to the various workers so that they can process in parallel. A simple method would be a round robin procedure where chunks of web logs are passed to nodes sequentially over and over. Hashing algorithms are also a common method for distributing data across workers. In this case, records are passed to workers based on a mathematical formula so that records with the same partitioning key get sent to the same worker. For example, hashing on customer ID will send all records for a given customer to the same worker. The hashing method can dramatically increase efficiency if analysis by customer ID is planned.

MapReduce offers some significant advantages for big data analytics that go beyond what relational database management systems provide. First, for data that is nontraditional (e.g., not in record format), the lack of imposed data structure from the programming model means that there is a lot more flexibility in handling non-relational data such as web logs, text, graphs, and rich media. Structure can be overlaid on top of file system data at the time of analytics rather than at the time of loading. Thus, MapReduce provides a quite suitable programming framework to handle analytics with multi-structured data. A polymorphic file system is capable of efficiently storing data in many different structures and is a foundational component of a big data processing environment.

Second, MapReduce facilitates the use of programming models that extend beyond the traditional SQL processing capabilities available on an American National Standards Institute (ANSI) standard relational database management system. As a set processing language, there are certain capabilities that are not efficiently executed within SQL. For example, time series analysis is generally quite awkward to express in SQL because set processing is not well aligned to ordered analytics. Market basket analytics or execution of graph algorithms have been implemented on top of relational databases for many years, but MapReduce offers the opportunity to simplify and improve performance on these types of more sophisticated analytics.

Hadoop is the most widely used open source implementation of the MapReduce programming framework and is used widely in the dot-com industry. Hadoop is available at no cost via the Apache open source project. Hadoop was originally started as an open source search engine project by the name of "Nutch" in 2004 and was later spun off and adopted by Yahoo! in 2006. The principle design goals for Hadoop were centered around providing extreme scalability for batch processing of analytics on top large volumes of (unstructured) text data from web pages and their links.

The NoSQL movement within the big data analytics community will sometimes position traditional relational databases and the SQL programming language as obsolete. The idea is that MapReduce on top of the Hadoop Distributed File System (HDFS) can do everything that SQL on top of a relational database can do—and more—so why not replace all Relational Database Management System (RDBMS) implementations with Hadoop? The reality is not quite so simple.

There is quite a bit of controversy surrounding the relative strengths and weaknesses of Hadoop versus relational databases [12,13]. Relational databases are far more mature than the Hadoop environment in the areas of optimizer technology, schema abstraction, security, indexes, interactive query and reporting tools, performance management, and many other areas. The reality is that Hadoop is largely deployed in highly technical environments where computer scientists are directly involved in analysis of big data. In order to get wider adoption of big data analytics, enterprises have a strong desire to deploy technologies that can be used by data scientists who are not necessarily computer programmers. The open source community has a number of initiatives to close the usability gap that exists with Hadoop deployments through use of higher level data access languages such as Pig and Hive.

At the same time, commercial vendors have begun extending traditional relational database products with MapReduce capabilities. Major vendors in the analytics marketplace such as EMC, IBM, Oracle, and Teradata have either already deployed relational databases with MapReduce extensions or have initiatives along these lines well underway. We would propose that a more rational realization of the NoSQL movement would result in "Not Only" SQL deployments. Sometimes SQL is a very appropriate language for big data analytics and sometimes something more is required. MapReduce is an important extension to the analytic capabilities of traditional databases that allows much more effective big data processing than SQL on its own.

The open source implementation of Hadoop makes it a very attractive storage area for capturing large volumes of data. The use of commodity storage and Just a Bunch of Disk (JBOD) storage along with zero-cost software licensing creates a low total cost of ownership value proposition. Increasingly, organizations are using Hadoop to stage high-volume data for analytic purposes. Preprocessing and filtering big data using batch jobs in preparation for discovery analytics is performed on the Hadoop staging platform. The data prepared on Hadoop is then loaded into next-generation relational databases implemented with MapReduce extensions to the SQL programming language. The extended relational platforms are better suited for efficient execution of complex workloads and providing interactive analytics with big data, while the Hadoop platform provides excellent total cost of ownership characteristics for large-scale data storage and more simplistic scanning, filtering, and aggregation of data.

20.4 Conclusions

The tools and technologies associated with big data processing are still emerging in their capabilities and adoption. Using Geoffrey Moore's framework for new technology introduction into the marketplace [14], we can observe that big data analytics is well past the incubation and early adopters stages and is just at the cusp of being embraced by the early majority pragmatists within the Fortune 1000.

The early hype associated with big data technologies that implied that enterprise data warehouse infrastructures were in imminent danger of being replaced by the new-generation NoSQL tools for all analytics was clearly overstated. However, it is clear that there is high value-added analytic capability provided by these tools. In the early days, big data ecosystems primarily consisted of the MapReduce programming model implemented on top of the HDFS. The ecosystem has evolved significantly over the past several years to include better support for management and administration (ZooKeeper, Karmasphere, Cloudera), query scripting and programming language support (Hive, Pig, Jaql, Mahout), query tools (Pentaho, Datameer), data sourcing (chuckwa), workflow optimization and scheduling (Oozie), metadata management (Hortonworks), and so on.

Core investments, being made by both the open source community and commercial vendors, are targeted at making the big data analytics platforms more scalable, reliable, and performant. Moreover, integration with other data platforms (relational databases, data integration tools, data mining tools, etc.) is addressing mainstream market requirements for adoption of big data technologies. It is clear that big data technologies are here to stay and that the open source and commercial communities will continue to evolve capabilities in this area. Eventually, like object technologies in the late part of the previous millennium, big data technologies will become part of the standard capabilities expected of all participants in the area of enterprise data management.

References

1. M. Adrian. Big data, *Teradata Magazine*, First Quarter, 2011, http://www.teradatamagazine.com/v11n01/Features/Big-Data/ (accessed on March 29, 2013).
2. McKinsey Global Institute. *Big Data: The Next Frontier for Innovation, Competition, and Productivity*, May 2011. Available at http://www.mckinsey.com/insights/business_technology/big_data_the_next_frontier_for_innovation (accessed on April 22, 2013).
3. Ibid.

4. *CEO Advisory: "Big Data" Equals Big Opportunity*, Gartner, March 31, 2011. Available at http://www.gartner.com/id=1614215 (accessed on April 22, 2013).

5. S. Brobst and R. Beaver. The value of RFID is the data, *Teradata Magazine*, Third Quarter, 2006, 26–31.

6. M. Michael, J. E. Moreira, D. Shiloach, and R. W. Wisniewski. Scale-Up x Scale-Out: A Case Study Using Nutch/Lucene. In *Parallel and Distributed Processing Symposium, 2007. IPDPS 2007. IEEE International*, 1–8. doi:10.1109/IPDPS.2007.370631.

7. McKinsey and Company, Clearing the air on cloud computing, March, 2009.

8. National Institute of Standards and Technology, Draft. NIST working definition of cloud computing, 2009, version 15. Available at http://csrc.nist.gov/publications/drafts/800-146/Draft-NIST-SP800-146.pdf (accessed on August 8, 2009).

9. National Institute of Standards and Technology. http://www.nist.gov/itl/cloud/index.cfm (accessed on March 29, 2013).

10. J. Dean and S. Ghemawat. MapReduce: Simplified data processing on large clusters. *Communications of the* ACM, 51(1)(January):107–113, 2008.

11. J. Dean and S. Ghemawat. Mapreduce: A flexible data processing tool. *Communications of the ACM*, 53(1):72–77, 2010.

12. D. J. DeWitt and M. Stonebraker. Mapreduce: A major step backwards. *Blog*, 2008. Available at http://craig-henderson.blogspot.com/2009/11/dewitt-and-stonebrakers-mapreduce-major.html (accessed on March 29, 2013).

13. M. Stonebraker, D. Abadi, D. J. DeWitt, S. Madden, E. Paulson, A. Pavlo, and A. Rasin. Mapreduce and parallel dbmss: Friends or foes? *Communications of the ACM*, 53(1):64–71, 2010.

14. G. Moore. *Crossing the Chasm*. October, 2000, New York: HarperBusiness.

21

Governance of Organizational Data and Information

Walid el Abed
Global Data Excellence

Paul Brillant
Feuto Njonko
*University of
Franche-Comté*

21.1 Introduction to the Notion of Governance

Before diving into the specific case of data governance (DG), it is important to outline the notion of governance in the broader sense because DG is but one form of governance. A governance program is used to define the decision-making processes and authority around a specific domain. In other words, a governance program is about deciding how to decide in order to be able to handle complex situations or issues in the future in the most controlled and efficient way possible.*

Over the past decade, corporate governance has been defined as a set of relationships between a company's management, its board, its shareholders, and other stakeholders that provide a structure for determining organizational objectives and monitoring performance, thereby ensuring that corporate objectives are attained (Khatri and Brown, 2010). Under this broad umbrella that sets enterprise-wide policies and defined standards, it is necessary to implement different governance areas to achieve the goal of a good and responsible corporate governance structure. Weill and Ross propose a framework

---

* http://www.element61.be/e/resourc-detail.asp?ResourceId=6

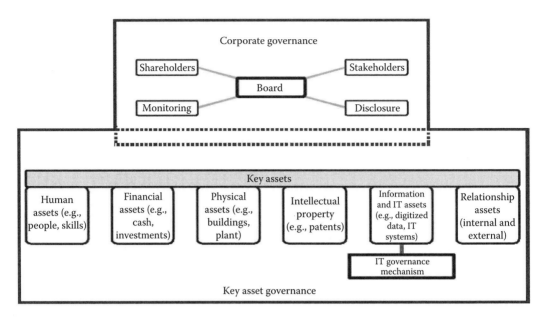

FIGURE 21.1   Corporate and key asset governance.

for linking corporate governance with six key assets of a company, which are the basis of generating business value and for realizing the company's strategies (Weill and Ross, 2004). Figure 21.1 is based on Weill and Ross' understanding of corporate and key asset governance and depicts the dependencies between these areas.

Since information technology (IT) is an essential part within a company's structure, quality management has to ensure corporate governance compliance of its IT systems. Thus, the discipline IT governance has been established and concentrates its activities on handling of technology and its usage within the company. For Weill and Ross, the IT governance has the goals in specification of decision rights and accountability frameworks to encourage a desirable behavior in the IT usage. In addressing of the following three questions, effective IT governance will be set up:

- What decisions regarding the use of IT must be made?
- Who should make these decisions?
- What are the characteristics of the decision and monitoring processes?

In this context, IT governance relates to the management of IT life cycles and IT portfolio management. This scope does not include the information itself, its value, and its life cycle management. The necessary examination of the relevance and the value chain of information should be part of an information and DG. Unlike Weill and Ross, many researchers emphasize that data and information governance are not a subset of IT governance. Accomplishing corporate data quality requires close collaboration among IT and business professionals who understand the data and its business purpose; in addition, both have to follow corporate governance principles. DG broadens the scope of IT governance by considering data quality aspects and processes especially defined to clarify the life cycle of data. According to Khatri and Brown, IT assets refers to technologies (computers, communication, and databases) that help support the automation of well-defined tasks, while information assets (or data) are defined as facts having value or potential value that are documented. Data Governance Institute (DGI)* makes the difference between IT and DG by analogy with the plumbing system: IT is

---

* http://www.datagovernance.com

like the pipes and pumps and storage tanks in a plumbing system. Data (information) are like the water flowing through those pipes.

21.1.1 Scope of Data (Information) Governance in the Overall Corporate Governance

DG is an essential part of an overall corporate governance strategy as an equal subdiscipline alongside IT governance. DG is a cross-organization and organization-wide initiative that requires that any barriers between IT and business are brought down and that they are replaced with well-defined roles and responsibilities for both the business areas and the technical areas of the organization. The *"who does what and when"* question is much more important than the *"where"* question. Its goal is to govern and manage data and information by organizing the proper usage of data and defining the responsibilities for data and respective data quality.

Furthermore, DG should be clearly distinguished from data management and data quality management (DQM) (Brown and Grant, 2005) because the most common definitional mistake companies make is to use DG synonymously with data management. DG is the decision rights and policy making for corporate data, while data management is the tactical execution of those policies. In other words, DG is not about what specific decisions are made. That is management. Rather, it is about systematically determining who makes each type of decision (a decision right), who has input to a decision (an input right), and how these people (or groups) are held accountable for their role. DG requires a profound cultural change demanding leadership, authority, control, and allocation of resources. DG is the responsibility of the board of directors and executive management and more focused on corporate environment and strategic directions, including other areas beyond DQM, such as Data Security and Privacy and Information Life-Cycle Management (Lucas, 2010). With DG, organizations are able to implement corporate-wide accountabilities for DQM, encompassing professionals from both business and IT units. Figure 21.2, extended from Brown and Grant (2005), illustrates the scope of DG in the corporate governance and the position between governance and management.

In this chapter, in line with most data or information governance publications, we have used synonymously the terms "data" and "information" although they may have different meanings in some contexts.

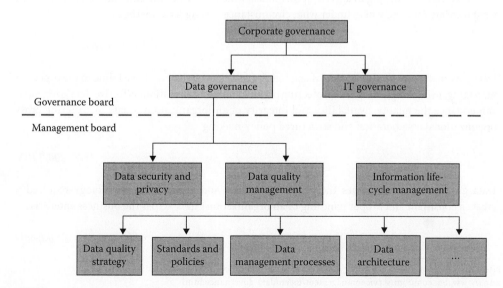

FIGURE 21.2 Position of DG in the corporate governance and difference between governance and management.

21.2 Data Governance

21.2.1 Introduction

Nowadays, volumes of data are steadily growing up, and organizations are seeking new ways to exploit their data as enterprise assets in order to fuel their growth. However, data of high quality are a prerequisite for fulfilling these objectives. Moreover, understanding how the data are being used throughout the business is vital for ensuring that legal, ethical, and policy-based requirements are met. Over the past few years, many organizations adopted DQM focusing on collecting, organizing, storing, processing, and presenting high-quality data. The organizational issues that must be addressed, such as defining accountabilities, managing expectation and compliance, avoiding scope creep, and handling political issues, push DQM into an area of conflict between business and IT. On the one hand, DQM has to provide many stakeholders (e.g., CEOs, sales, controlling, procurement, IT, business units, customers, public authorities) with high-quality corporate data. On the other hand, these stakeholders have different interests (e.g., local/regional/company-wide differences), which do not necessarily accumulate to the best achievable result for the company as a whole. Because of these particularities of DQM, large multi-business companies are likely to have difficulties with institutionalizing DQM, that is, defining accountabilities, assigning people accountable for DQM within the organizational structure, and enforcing DQM mandates throughout the company.

DG addresses these particular issues by implementing corporate-wide accountabilities for DQM that encompass professionals from both business and IT. It specifies the framework for decision rights and accountabilities to encourage desirable behavior in the use of data. It establishes organization-wide guidelines and standards for DQM and assures compliance with corporate strategy and laws governing data.

21.2.2 Definition

DG is an emerging discipline with an evolving definition. There are varying definitions of DG because each practitioner tailors DG to its needs and abilities. Here are some definitions that cover almost all the components and goals of DG:

> Data Governance is a system of decision rights and accountabilities for information-related processes, executed according to agreed-upon models which describe who can take what actions with what information, and when, under what circumstances, using what methods.

> *The Data Governance Institute**

> Data governance is a quality control discipline for adding new rigor and discipline to the process of managing, using, improving and protecting organizational information. Effective data governance can enhance the quality, availability and integrity of an [organization's] data by fostering cross-organizational collaboration and structured policy-making

> *[IBM, 2007] IBM†*

> Data governance encompasses the people, processes, and information technology required to create a consistent and proper handling of an organization's data across the business enterprise.

> *Wikipedia‡*

---

* http://www.datagovernance.com/adg_data_governance_definition.html (accessed on 31 May 2012)
† http://www.ibm.com/ibm/servicemanagement/us/en/data-governance.html
‡ http://en.wikipedia.org/wiki/Data_governance

The formal orchestration of people, processes, and technology to enable an organization to leverage data as an enterprise asset.

*Customer Data Integration Institute**

All these definitions help us understand the aim and principle of DG, which is to help maximize the value of an organization's data and information assets. In the absence of academic definitions of DG, the working definition used in this chapter was adapted from research on IT governance definition (Weill and Ross, 2004). Thus, DG specifies the framework for decision rights and accountabilities to encourage desirable behavior in the use of data; it encompasses people, policies, procedures, and technologies to enable an organization to leverage data as an enterprise asset.

21.3 Designing Effective Data Governance

The scope of a DG program can be very wide or specific due to the wide variety of backgrounds, motivations, and expectations of DG practitioners. However, a good DG framework ensures that organizational data are formally managed throughout the enterprise as an asset and provides efficient help to the organization to achieve one of the following business drivers:

- Increasing revenue and value
- Lowering costs
- Reducing risks

The following are few descriptions of goals, focus areas, and principles of a DG program. Due to the limited academic research, most of sources come from practitioners on DG, namely, reports by analysts, white papers by consulting companies, vendors, or software manufacturers such as IBM, GDE, Gartner, and Oracle.

21.3.1 Goals of Data Governance

DG is not meant to solve all business or IT problems in an organization. Its goal is to continually produce high-quality data while lowering cost and complexity and supporting risk management and regulatory compliance. The main goals and objectives of DG include the following points:

- Ensure data meets the business needs
- Protect, manage, and develop data as a valued enterprise asset
- Lower the costs of managing data and increase the revenue
- Enable decision making and reduce operational friction
- Train management and staff to adapt common approaches
- Build and enforce conformance to data policies, standards, architecture, and procedures
- Oversee the delivery of data management projects and services
- Manage and resolve data-related issues

21.3.2 Focus of Data Governance

DG frameworks can have a focus on the following areas:

- Policy, standards, and strategy
  The focus on data policies, data standards, and overall data strategies are usually the first step when an organization initiates a DG program. The DG program contributes to the definition of a data strategy and will ensure that consistent policies and standards are used within the

---

* http://www.tcdii.com

organization. It will steer the creation of new policies and standards and work on the alignment of the existing ones. It will finally monitor that those policies and standards are correctly used with the organization. The main activities of a DG program with a focus on policy, standards, and strategy include the following:

- Establish, review, approve, and monitor policy
- Collect, choose, approve, and monitor standards
- Establish enterprise data strategies
- Align sets of policies and standards
- Contribute to business rules
- Data quality

The desire to improve the quality and usability of the data is the main driver for this type of DG program. It will set the direction for the data quality program. It will help the organization to define and monitor the quality of its data and help to establish ownership, decision rights, and accountabilities to ensure the quality, integrity, and usability of the data.

- Privacy, compliance, and security

  Depending on the type of the organization, this area might be more or less critical. Companies within the financial services and health-care industries are main adopters of these programs for apparent reasons. In any case, the DG program will help to protect sensitive data and enforce the legal, contractual, or architectural compliance requirements. This will help the organization to control and limit the risks on its data.

- Architecture and integration

  In this domain, the DG program will mainly have a support role by bringing cross-functional attention to integration challenges. It will support initiatives like metadata or MDM programs and help in the definition of architectural policies and standards in order to improve systems integration.

- Data warehousing and business intelligence

  The DG program will help to enforce rules and standards in the various source systems of the organization, which will affect the data warehouses and business intelligence data by providing better quality data. It will also help to clarify and promote the value of the data-related projects in general.

- Management alignment

  The DG program will promote and measure the value of any data-related projects and initiatives toward management. It will help management to align the different initiatives and monitor for them the data-related projects.

Independently of the area of focus, the main activity should be to identify stakeholders, establish decision rights, and clarify accountabilities.

21.3.3 Data Governance Principles

Here are some DG guiding principles that help stakeholders come together to resolve the types of data-related conflicts that are inherent in companies:

- Integrity

  DG participants will practice integrity with their dealings with each other; they will be truthful and forthcoming when discussing drivers, constraints, options, and impacts for data-related decisions.

- Transparency

  DG and stewardship processes will exhibit transparency; it should be clear to all participants and auditors how and when data-related decisions and controls were introduced into the processes.

- Audit ability
  Data-related decisions, processes, and controls subject to DG will be auditable; they will be accompanied by documentation to support compliance-based and operational auditing requirements.
- Accountability
  DG will define accountabilities for cross-functional data-related decisions, processes, and controls.
- Stewardship

DG will define accountabilities for stewardship activities that are the responsibilities of individual contributors, as well as accountabilities for groups of data stewards.

- Checks-and-balances
  DG will define accountabilities in a manner that introduces checks-and-balances between business and technology teams as well as between those who create/collect information, those who manage it, those who use it, and those who introduce standards and compliance requirements.
- Standardization
  DG will introduce and support standardization of enterprise data.
- Change management
  DG will support proactive and reactive change management activities for reference data values and the structure/use of master data and metadata

21.3.4 Benefits of Data Governance

The scope of a DG program can be very wide and needs to be carefully defined in order to provide the benefits the organization is looking for. It also requires specific expertise to implement DG initiatives in a successful way in order to transform the data in the organization into a real asset. Effective DG does not come together all at once, but it is an ongoing process. Before adopting an approach, it is important to assess the current state maturity of the DG capability. Usually, DG program appears in an organization when it reaches a certain maturity level. In that case, it will be necessary to align the existing rules and processes of different groups by using the following iterative approach:

- *Creates and aligns rules*: the governance program will help in the definition of global rules by coordinating and balancing the need of all the stakeholders. It will define the decision process as well as the roles and responsibilities of all the stakeholders involved.
- *Enforces rules and resolves conflicts*: the governance program will then make sure that those rules and processes are used and applied correctly by all the stakeholders. If necessary, they will be enforced. The governance program will also help to prevent or resolve conflicts by providing ongoing services.
- *Provide ongoing services*: the governance program must support the stakeholders in the application of the processes and rules and by identifying any new opportunities to create new rules or to align or adapt existing ones.

Thus, DG program can provide the following benefits to the organization:

- *Reduce costs and complexity*: reduce duplicate data and management processes and reduce likelihood of errors due to the lack of understanding of data or poor-quality data
- *Support compliance*: quicker access to authoritative data in order to achieve compliance goals and avoid cost of noncompliance penalties
- *Support impact analysis*: increase ability to do useful impact analysis (by providing authoritative business rules, system of record information, and data lineage metadata) and provide a capability to assess cross-functional impacts of data-related decisions

- *Help align efforts*: assist business teams to articulate their data-related business rules and requirements to IT, architecture, and data management teams; consider requirements and controls in an integrated fashion; craft cross-functional accountabilities; and develop common data definitions to be used across various systems increasing integration and cross-flow of data
- *Improve data repositories*: provide accountability and support for improving the quality of data in the repository so it can become an authoritative source of information, reduce likelihood of architectural decisions that limit the organization's ability to analyze its information, and increase ability to find authoritative information quickly
- *Improve confidence in data*: increase confidence in data-related decisions and output reports, increase ability to make timely data-related decisions, and increase confidence in data strategy by providing a cross-functional team to weigh in on key decisions

21.3.5 Key Data Governance Components

DG is a cross-organization and organization-wide initiative where people work together to establish standards and enforce policies (or rules) over data processes. Figure 21.3 illustrates that an effective DG involves four key components:

Standards and policies: A key function of DG is to establish standards and enforce policies that guide DG program implementation. These include vision and goals of the organization regarding data standards, data management processes, decision-making jurisdiction, responsibilities, enforcement, and controls. The DG approach should be consistent with the organization's overall mission and stakeholders' expectations. Program goals should be clearly stated, and it should be made clear how these goals address data content needs, what outcomes will be considered a success, and how the progress will be measured. Furthermore, an organization should evaluate the resources required for the long-term sustainability of the program to ensure it can sustain necessary levels of data quality and security over the entire data lifecycle.

Processes: Establishing and enforcing processes around the creation, development, and management of data is the foundation of an effective DG practice. Companies need to define data and data-related business rules, control access to and delivery of data, establish ongoing monitoring and measurement mechanisms, and manage changes to data.

People: Arguably the most important issue that companies must address when launching a DG initiative is how to manage people in order to design the organizational structure. Companies need to define the

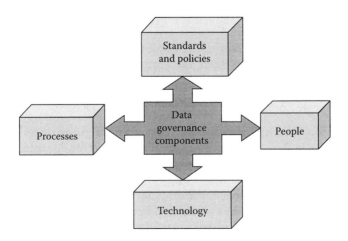

FIGURE 21.3   Four components of DG.

roles and responsibilities within the organization that are accountable for the data. The organization may include several different roles at different levels, involving both business and IT personnel from executive councils to day-to-day implementers, such as data stewards and data analysts. Addressing training and organizational change management issues is also critical if DG programs are to succeed.

Technology: Hypothetically, companies could embark on a DG initiative without an underlying technology infrastructure. Indeed, many organizations launch their initial DG programs using manual tools (spreadsheets, Word documents, etc.) to capture data definitions and document processes. However, most quickly realize that this kind of manual approach is severely limited. It is difficult to ensure high data quality and availability, security is at risk given the ad hoc nature of the approach, and maintaining detailed documentation is an almost insurmountable task. Indeed, it is nearly impossible to achieve the ultimate goals of DG using a manual approach. Thus, technologies can help automate and scale the development and enforcement of DG standards, policies, and processes.

21.4 State of the Art in Data Governance

Both researchers and practitioners consider DG as a promising approach for companies to improve and maintain the quality of corporate data, which is seen as critical for being able to meet strategic business requirements. It receives increasing interest and has recently been given prominence in many leading conferences, such as The Data Warehousing Institute (TDWI) World Conference, the Excellence Quality Information* Association, the Data Management Association (DAMA) International Symposium, the Data Governance Annual Conference, and the MDM Summit. Recently, Gartner has published the 2012 list of Cool Vendors in Information Governance and MDM (Bill et al. 2012). A recent survey conducted by TDWI highlights the increasing number of organizations investing in formal DG initiatives. Eight percent of the 750 responding organizations have deployed a DG initiative; 17% were in the design or implementation phase and 33% are considering it (Russom, 2006). The results also make clear that DG is an emerging practice area, most governance efforts are in the early phases, and many organizations are still working to build the business case for DG. However, only few scientific findings have been produced so far (Otto, 2010) and academic research on DG is still in its infancy. Research on IT governance is more advanced with the first publication released in 1997 (Wende, 2007) and constitutes the starting point for research in DG.

IT governance follows a more flexible approach for the assignment of accountabilities. Early research distinguished two IT governance models: in centralized models, corporate IT performs all IT functions, whereas in decentralized models, business units' IT performs these tasks. Subsequent research specified more precise IT governance models, acknowledging several IT functions and more than one organizational level. Finally, Weill and Ross (2004) proposed five IT functions, three organizational units, and a distinction between decision and input rights. The combination of these three dimensions resulted in six feasible IT governance models. In conclusion, IT governance research proposes three elements that compose an IT governance model: roles, major decisions areas, and assignment of accountabilities. This idea of IT governance models has been the starting point for research on DG models.

21.4.1 Academic Research on Data Governance

In the academic research area, Wende has proposed a model for DG similar to IT governance models to document the company-specific decision-making framework of DQM (Wende, 2007). His model, shown in Table 21.1, outlines the three components of such a framework, namely, DQM roles, decision areas, and responsibilities. For the components, he identifies typical data quality roles and decision areas and proposes a method to assign responsibilities. The approach respects the fact that each company needs a specific DG configuration.

---

* http://www.exqi.asso.fr/excellence-qualite-information/

TABLE 21.1   Draft of a DG Model

	Roles					
Decision Areas	Executive Sponsor	Data Governance Council	Chief Steward	Business Data Steward	Technical Data Steward	...
Plan data quality initiatives	A	R	C	I	I	
Establish a data quality review process	I	A	R	C	C	
Define data producing processes		A	R	C	C	
Define roles and responsibilities	A	R	C	I	I	
Establish policies, procedures and standards for data quality	A	R	R	C	C	
Create a business data dictionary		A	C	C	R	
Define information systems support		I	A	C	R	
...						

Source: Wende, K., A model for data governance—Organizing accountabilities for data quality management, In *18th Australasian Conference on Information Systems*, 2007.
    R: Responsible; A: Accountable; C: Consulted; I: Informed

The three components are arranged in a matrix, the columns of the matrix indicate the roles in DQM, the rows of the matrix identify the key decision areas and main activities, and the cells of the matrix are filled with the responsibilities, that is, they specify degrees of authority between roles and decision areas. A company outlines its individual DG configuration by defining data quality roles and decision areas and responsibilities and by subsequently arranging the components into the model. This configuration is unique for each company. However, this model only focused on the DQM accountability aspect of DG and did not address other areas of DG such as compliance and privacy. Similarly, Khatri and Brown (2010) proposed a framework, which inherits an existing IT governance framework with a set of five data decision domains. For each decision domain, they create a DG matrix, which can be used by practitioners to design their DG model. The framework places special interests in data principles and data life cycle in addition to data quality, data access, and metadata. Boris Otto has developed a morphology of DG organization for answering the question: What are the aspects that need to be considered in order to capture the entirety of DG organization? Based on deductive analysis, the morphology constitutes an analytic theory, that is, it identifies and structures the basic concepts of DG organization (Otto, 2010).

Figure 21.4 describes the morphology of DG organization developed in Otto's research and identifies 28 organizational dimensions.

The first organizational dimension (1) relates to an organization's goals divided into formal goals (1A) and functional goals (1B). Whereas the former measure an organization's performance and maintain or raise the value of a company's data assets, the latter refer to the tasks an organization has to fulfill represented by the decision rights defined. The first set of formal goals specified by the morphology comprises a number of business goals or focus areas of DG. The second set of formal goals specified by the morphology is a two-piece set of information system (IS)-/IT-related goals: "increase data quality" and "support IS/IT integration." Functional goals exclusively relate to the decision areas for which DG specifies certain rights and responsibilities.

The second organizational dimension (2) is the organizational structure. It comprises three aspects, namely, the positioning of decision-making power within the hierarchical structure of an organization

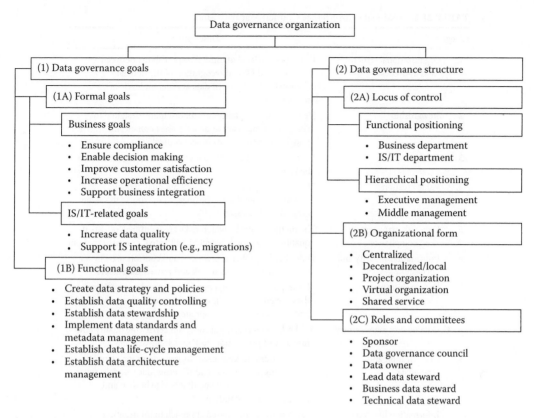

FIGURE 21.4 Morphology of DG organization.

(2A, "locus of control"), the division of labor and the organizational form (2B) resulting from this, and the allocation of tasks to roles and committees (2C).

21.4.2 Data Governance from Practitioners

Numerous DG frameworks and maturity models have been proposed by different organizations with an attempt to provide guidance in designing and developing effective DG approaches. The DG maturity model is a tool to assess your organization's current state of DG awareness and effectiveness. These instruments need to be used judiciously and with regard to particular organizational environments (Chartres, 2012). Each organization will set its own goals and methods based on individual business scenarios, plans, and needs. This leads different organizations to have their own focus on specific aspects of DG. Despite this, they still share some common decision domains, such as data quality at least. The following are a few examples (not comprehensive) of DG frameworks from DG organizations.

21.4.2.1 IBM Data Governance Council Maturity Model

The IBM Data Governance Council is an organization formed by IBM consisting of companies, institutions, and technology solution providers with the stated objective to build consistency and quality control in governance, which will help companies better protect critical data. This common forum, where practitioners explore challenges and solutions, has been instrumental in developing benchmarks, best practices, and guides to successful DG. This is known as the Data Governance Council Maturity Model, shown in Table 21.2. The model has 11 categories, where each category is both a

TABLE 21.2   IBM Data Governance Council Maturity Model

Category	Description
1. Organizational structures and awareness	Describes the level of mutual responsibility between business and IT, and recognition of the fiduciary responsibility to govern data at different levels of management
2. Stewardship	Stewardship is a quality control discipline designed to ensure custodial care of data for asset enhancement, risk mitigation, and organizational control
3. Policy	Policy is the written articulation of desired organizational behavior
4. Value creation	The process by which data assets are qualified and quantified to enable the business to maximize the value created by data assets
5. Data risk management and compliance	The methodology by which risks are identified, qualified, quantified, avoided, accepted, mitigated, or transferred out
6. Information security and privacy	Describes the policies, practices, and controls used by an organization to mitigate risk and protect data assets
7. Data architecture	The architectural design of structured and unstructured data systems and applications that enable data availability and distribution to appropriate users
8. Data quality management	Methods to measure, improve, and certify the quality and integrity of production, test, and archival data.
9. Classification and metadata	The methods and tools used to create common semantic definitions for business and IT terms, data models, types, and repositories. Metadata that bridge human and computer understanding
10. Information life cycle management	A systematic policy-based approach to information collection, use, retention, and deletion
11. Audit information, logging and reporting	The organizational processes for monitoring and measuring the data value, risks, and efficacy of governance

Source: IBM, The IBM data governance council maturity model: Building a roadmap for effective data governance, Technical report, IBM Software Group, 2007.

starting place for change and a component in a larger plan. The categories are grouped into four major DG domains (IBM, 2007):

1. *Outcomes:* data risk management and compliance; value creation
2. *Enablers:* organizational structures and awareness; policy; stewardship
3. *Core disciplines:* DQM; information life cycle management; information security and privacy
4. *Supporting disciplines:* data architecture; classification and metadata; audit information logging and reporting

21.4.2.2 Data Governance Institute Framework

The DGI framework* provided by the DGI is a logical structure for classifying, organizing, and communicating complex activities involved in making decisions about and taking action on enterprise data. Figure 21.5 describes the architecture of the framework with the 10 universal components:

1. Mission and vision
2. Goals, governance metrics and success measures, and funding strategies

---

* http://www.datagovernance.com/fw_the_DGI_data_governance_framework.html

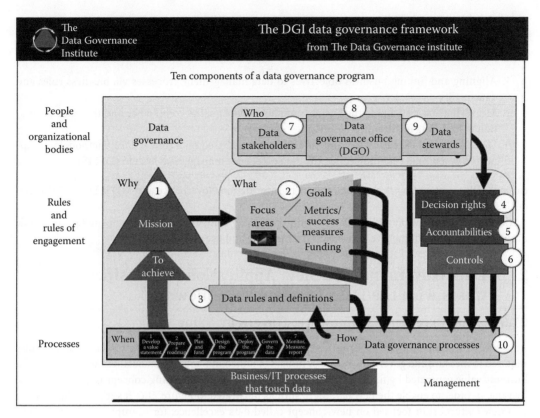

FIGURE 21.5 DGI framework.

3. Data rules and definitions
4. Decision rights
5. Accountabilities
6. Controls
7. Data stakeholders
8. A DG office
9. Data stewards
10. Proactive, reactive, and ongoing DG processes

21.4.2.3 Data Excellence Framework

The Data Excellence Framework (DEF) provided by Global Data Excellence (GDE)* describes the methodology, processes, and roles required to generate business value while improving business processes using data quality and business rules (El Abed, 2009b). The framework supports the creation of a new cultural shift focused on data excellence, motivating the broader team and supporting collaboration between the stakeholders. DEF is more than improving data quality and reducing costs; it is about taking a sustainable approach to an organization's data assets to drive value from the data. A key difference about the DEF is that it is focused on generating value in comparison to most initiatives relating to data, which are only focused on reducing costs. The key success of DEF is that it is grounded on both academic research (Feuto and El Abed, 2012) and practical experiences.

---

* http://www.globaldataexcellence.com

Thus, the data excellence maturity model helps organizations worldwide successfully move from chaos to data excellence stage.

Innovative capabilities in maximizing the value of enterprise data through more effective ways in

1. Aligning and linking business objectives to data management processes via business rules and data quality
2. Measuring and visualizing the business impact of data quality compliance and the value generation of high data quality
3. Organizing and executing sustainable DG processes based on a proven DEF methodology supported by a product solution called the data excellence management system (DEMS)

earned GDE a spot on Gartner's Cool Vendors in Information Governance and MDM for 2012 list (Bill et al. 2012).*

Through the Cool Vendors list, Gartner shines light on companies that use information to enable business outcomes and improve and enrich the meaning and use of master data.

21.5 Data Governance Journey: Application and Business Cases with the Data Excellence Framework

21.5.1 Data Excellence Framework: Govern by Value

As already mentioned in the previous section, DEF describes the methodology, processes, and roles required to measure and govern the business impact of noncompliant data on business activities and the business value generated by fixing the data (El Abed, 2009a,b). The leading concept is the governance by value in order to enable the new paradigm shift as illustrated in Figure 21.6. It supports the creation of a new cultural shift focused on new concept called data excellence, for example, elevating data to

FIGURE 21.6   Govern by value: the new paradigm shift.

---

* http://www.gartner.com/id=1984515

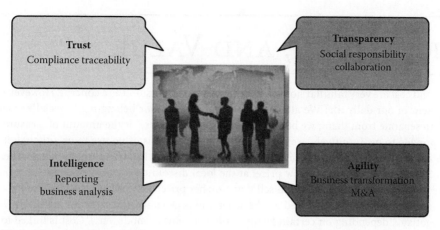

FIGURE 21.7 The value pillars of DEF.

business excellence, motivating the broader team, and supporting collaboration between the business stakeholders, the data management team, and IT. Data excellence arises from the natural evolution of organizations and society in the information age and represents a golden opportunity to enable growth, profit, and flawless execution.

DEF is based on four value pillars (shown in Figure 21.7) that are essential to the survival of any organization or enterprise in the information age: *agility, trust, intelligence*, and *transparency*.

These characteristics are fundamental value pillars to enable business sustainability and support economic growth:

- *Agility* is needed to react to external and internal changes and ensure prompt and successful integration that supports fast business: how fast can you change a process, introduce a new product, or merge a new business.
- *Trust* is associated with the integrity of the data, which induces confidence in enterprise image, brands, and products; empowers business partners, employees, and stakeholders; and ensures legal and compliance traceability.
- *Intelligence* at all levels of the enterprise leads to better and flawless execution, operational efficiency, and accurate financial consolidation based on just-in-time quality data from reporting systems and applications.
- *Transparency* is critical to the organization's performance as it increases visibility and collaboration across and outside the enterprise: not about showing everything but controlling what should be communicated, for example, social responsibility.

These fundamental value pillars result in new ways of working and will lead to maximizing the value and lowering the cost of using data.

21.5.2 Managing Data as a Company Asset

A key difference about the DEF is that it is focused on generating value in comparison to most initiatives related to data that are only focused on reducing costs. Data are an asset, and an asset has a value. The main difference between a tangible asset and the data asset is that the value of data is dynamic and will depend on the context of its usage. Out of the context of usage, the value of data is zero. Moreover, it has a cost, and when the value is not demonstrated, the organization tends to cut this cost.

PRICE, COST, AND VALUE

Price and cost are very familiar notions, but value is less understood as a concept. However, value is present in our daily life: We attach sentimental value to our belongings especially when we have to separate from them; we buy goods according to the use or the amount of pleasure they will provide to us. Intuitively, we don't attach value to what is of no use to us although it may have a price or a cost. If you have a fridge filling your needs, it is of no use to buy another one even if there is bargain (e.g., a low price) at the local discount store: therefore, the item has no value for you (except if you think to sell it at a higher price). The price is the monetary amount required to buy an item. It is fixed by the seller and is generally the same for all buyers. It can be negotiated depending on certain factors (volume, delivery mode, etc.). Cost is linked to cost price, which is the total amount of ownership required for the producing and the utilization of an item. Taking the fridge case, its price is the posted price to buy it and cost is the price plus the total expense to run it through its use: electricity, repair, maintenance, etc. Cost is also objective (same for all users).

However, value is subjective and contextual depending on the use we make of it. A bottle of water in the desert can be worth your life (i.e., a lot of money if your life depends on this water), but the same bottle has little value in your daily life as you have plenty of water on the tap. Hence, the value is contextual and instantiate itself at transaction time. Value is therefore subjective as it can vary for different individuals and vary in time for the same person.

Since Frederick Taylor invented his theory of scientific management to improve economic efficiency in the industrial age, enterprises have separated the value generation and the cost management. The former was given to the "business" and the latter to the support functions whose mission is to rationalize and optimize processes (the means). This approach has proved efficient and effective in increasing volumes and reducing costs in an industrial world. However, since we have entered the information age, this separation has disconnected the value creation from the means supporting it and has, in a way, "devaluated" the support functions like IT or data management.

21.5.2.1 What Are the Three Barriers to Manage Data as an Enterprise Asset?

The main barriers are as follows (Figure 21.8):

- *Insufficient alignment between business managers, data management, and IT*: data are noncompliant and are not fit for purpose, and as a result, transactions fail. Despite investing millions in infrastructure and architecture, organizations have not invested in the data. However, data are a key differentiator for organizations when all organizations have similar ERP systems implemented by the same consulting firms. The business, the data management team, and IT are not aligned, resulting in problematic decision outcomes.
- *Inability to demonstrate the business impact of bad data management*: nobody knows the cost of poor data quality and nobody is able to quantify the potential benefit of fixing the data. Worse, nobody knows which data are critical and what the impact of this bad data on the business is. Building a business case to invest in data quality is always difficult, and as a result, the problem still exists. People have lots of anecdotes but find it difficult to quantify their thoughts.

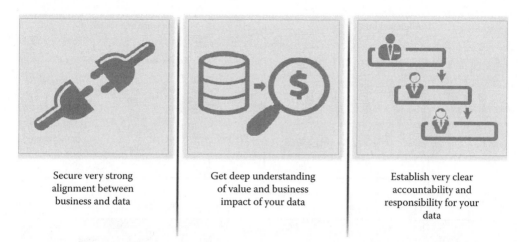

Secure very strong alignment between business and data

Get deep understanding of value and business impact of your data

Establish very clear accountability and responsibility for your data

FIGURE 21.8 Barriers to manage data as an asset.

- *Unclear roles and responsibilities to govern data as an asset*: most organizations are struggling to maintain momentum to achieve sustainable business process improvement and flawless execution through data quality improvement processes. This is because nobody feels accountable or responsible for data quality. There is no understanding of the roles or processes needed to sustain high-quality data to fulfill the goals of a DG process. There is no easy way to motivate or sustain the data management team or IT to support the business goals. The DEF provides a method to lower each of these three barriers.

21.5.3 Data Excellence Framework Methodology

The DEF's methodology emphasizes a value-driven approach for enabling business excellence through data excellence. Data excellence becomes an imperative to unlock the enterprise potential and enable sustainable value generation. Thus, data are made visible so that they can become resources for the enterprise to make informed decisions.

In order to lower the three barriers to manage data as a company asset and to maximize the business value of enterprise data, DEF proposes a three-step approach presented at Figure 21.9.

21.5.3.1 Align and Link: Business Excellence Requirements

The purpose here is to align and link business objectives to data management processes via business rules and data quality (see example in Figure 21.10). We introduce the concept of business excellence requirements (BERs): the BER is a prerequisite, business rule, standard, policy, or best practice that business processes, transactions, and data should comply with in order to have the business goals flawlessly executed to generate value. Defining BERs is defining compliance on data quality. The role of the business rules concept implementation enables to action the data quality key performance indicators (KPIs) at the record level and to address the root cause (El Abed, 2008a).

21.5.3.2 Measure and Visualize: DEI and KVI

We measure and visualize the business impact of data quality compliance and the value generation of high data quality. To do so, we use a data excellence index (DEI) measuring the ratio of compliant data and a key value indicator (KVI) measuring the value of compliant data and the value of noncompliant

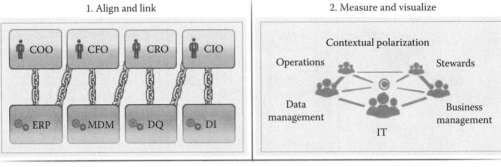

FIGURE 21.9   The three steps to maximize the business value of enterprise data.

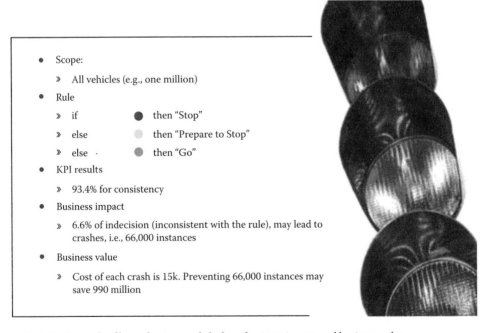

FIGURE 21.10   Example of how a business rule links to business impact and business value.

data (the impact). In compliant and noncompliant data DEIs and KVIs, value and impact can be visualized through different angles (geography, business unit, process, business rule, role, etc.) using the contextual polarization. This helps organizations in

- Providing global visibility of data quality to the whole organization at all levels.
- Publishing the data KPIs monthly on the operation site on the intranet at all levels.

- Presenting the data KPIs at the top management level and at the market management level each month (El Abed, 2008b).
- Measuring the business value through KVIs (aggregated sum and number for real business impact) enables the business users to visualize how DG can help them get there fast (Michellod and El Abed, 2010).

21.5.3.3 Organize and Execute: Accountability and Responsibility

In step 3, the business impact of poor-quality data must be reduced by fixing the wrong data and identifying and analyzing the root causes. In order to manage it efficiently and effectively, new DG must be established including the business, the data management, and IT. The business managers who are accountable and responsible of the value generated in the enterprise will take accountability for the BERs and the associated data, DEIs and KVIs. They will decide which data should be corrected to generate more value by reducing the impact. They will also define accountability and responsibility around the business rules KVIs with business stakeholders (Orazi and Gessler, 2010). Data managers, in charge of the data management of the BER-associated data, will take responsibility for these data. They will correct data and optimize processes upon business order. IT will provide and support the systems necessary to manage these data. The Global DEF is modifying the role of the MDM manager, giving the role higher visibility and strategic positioning (Delez, 2010). The higher the person is in the hierarchy, the faster he or she understands how powerful and useful the DEF to achieve their objectives, and people can no longer question the importance of data errors—with the KVIs, the business impact and value generation are now obvious (Michellod and El Abed, 2010).

In order to facilitate the implementation of this three-step approach, a process including eight questions is provided as shown in Figure 21.11. A simplified view of DEMS, which is a collaborative and evolutionary system to fully support the implementation of DEF capabilities, is presented in Figure 21.12.

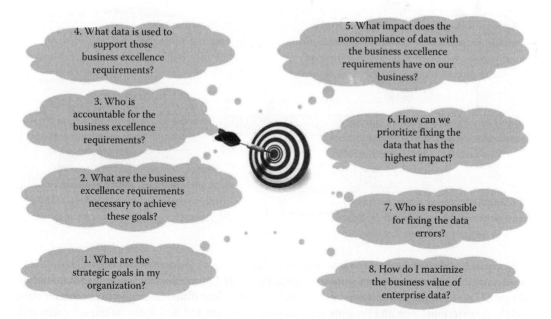

FIGURE 21.11 The eight critical questions linking your business and your data.

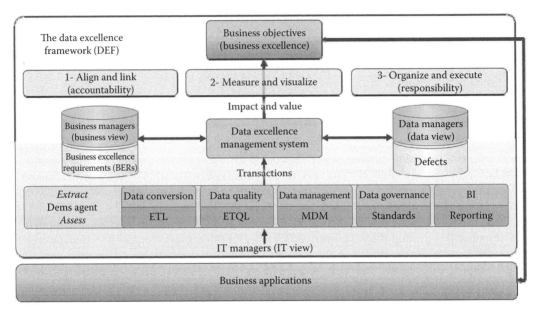

FIGURE 21.12   DEMS.

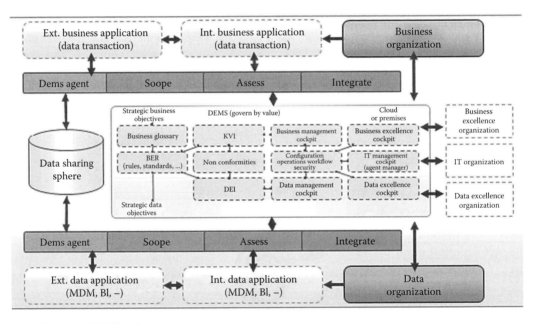

FIGURE 21.13   DEMS in the enterprise environment.

Figure 21.12 shows the general architecture of DEMS. However, it does not show how DEMS could be integrated in the enterprise environment. Thus, Figure 21.13 provides a detailed functional architecture of DEMS in the enterprise environment. It makes a clear difference between the master data and the transactional data and presents the data sharing sphere and how it can be implemented and governed.

21.5.4 Business Case

We are now going to illustrate the application of the DEF through a simple business case in financial institution chosen from many possibilities: the allocation of loans by a bank to its customers.

BUSINESS CASE: ALLOCATION OF LOANS BY A BANK TO ITS CUSTOMERS

SmartBank is a retail bank who has two brands: BankA and BankB both operating in two regions, West and East.

Strategic goal: to govern the risk linked to the distribution of loans and maximize their value

Scope: all loan applications

Data: customer data

21.5.4.1 Scope

The scope is the set of data objects and business processes that are defined before starting the exercise. This scope is dynamic and can be extended or reduced according to the business critical requirements.

In this example, the scope is the set of loan applications at BankA and BankB in the East and West regions. This scope could be reduced to BankA in the East region or extended to another brand of SmartBank (BankC) or to all regions.

The business process is the loan validation process and data objects are the loans (transactions) and customer data.

21.5.4.2 Business Excellence Requirements

The BERs, a unique value proposition representing the backbone of the DEF, are a prerequisite business rule, standard, policy, or best practice that business processes, transactions, and data should comply with in order to have the business goals flawlessly executed to generate value.

21.5.4.2.1 Example

In this example, the SmartBank loan business manager has defined three BERs for the loan allocation process:

BER 1: "a customer record of default must be provided for any loan application if the amount is over or equal to 10000€." This record states whether the customer had defaulted on a loan repayment in the past.

BER 2: "no loan is granted to minor customers (e.g., under 18)."

BER 3: "each application must be accompanied with a payslip."
It is the responsibility of each branch account manager to make sure that any loan application is compliant with these three BERs before they are examined by the SmartBank loan committee for approval.

21.5.4.3 Data Excellence Index

21.5.4.3.1 Definition and Role

The DEI is the instrument used to measure the compliance of data records with the BERs. It consists of the following:

- The percentage of data records compliant with a collection of BERs
- The list of compliant and noncompliant records

Each data record impacts the DEI (positively or negatively) only once regardless of the number of BER applied to this record.

The DEI results are used to evaluate the value and impact of data on business processes and transactions. The BER concept makes the DEI actionable at the record level allowing finest root cause analysis and surgical DG.

21.5.4.3.2 Example

In this example, the three BERs have to be applied to the two data objects: the loans and the clients included in the case scope (see preceding text). The extraction of data from BankA and BankB ISs yields the following data:

Records	Région	Bank	Account Manager	Customer Name	Customer First Name	Age	Income	Loan Amount	Pay slip Provided	Default Record Provided
Record 1	East	BankA	Claus French	Dupont	Jean	75	30,000	9,000	No	
Record 2	East	BankB	Claus French	Proust	Marcel	50	40,000	11,000	Yes	Yes
Record 3	East	BankA	Julian Wood	Flaubert	Gustave	35	25,000	11,000	Yes	Yes
Record 4	East	BankB	Gilbert Sullivan	Monet	Claude		22,000	5,000	Yes	
Record 5	East	BankA	Claus French	Courbet	Gustave	57	60,000	25,000	Yes	Yes
Record 6	East	BankA	Julian Wood	Dupont	Jean	75	31,000	5,000	Yes	Yes
Record 7	West	BankA	Julian Wood	Hugo	Victor	52	58,000	17,000	Yes	
Record 8	West	BankB	Gilbert Sullivan	Cezanne	Paul	40		1,500	Yes	Yes
Record 9	West	BankA	Claus French	Duchamp	Marcel	32	15,000	2,500	Yes	
Record 10	West	BankB	Gilbert Sullivan	Lupin	Arsene	XXX	XXXXX	XXXXX	XXXXXX	XXXXXX
Record 11	West	BankA	Julian Wood							

There are two records with specific features:

1. Record 10 could not be retrieved for security reasons (no access rights).
2. Record 11 was not found, although account manager, Julian Wood, reported it in the CRM.

A simple check of each record against each BER yields the following results:

Records	Customer Name	Customer First Name	Age	Income	Loan Amount	Pay Slip Provided	Default Record Provided	BER 1	BER 2	BER 3	All BERs
Record 1	Dupont	Jean	75	30,000	9,000	No		Y	Y	N	N
Record 2	Proust	Marcel	50	40,000	11,000	Yes	Yes	Y	Y	Y	Y
Record 3	Flaubert	Gustave	35	25,000	11,000	Yes	Yes	Y	Y	Y	Y
Record 4	Monet	Claude		22,000	5,000	Yes		Y	N	Y	N
Record 5	Courbet	Gustave	57	60,000	25,000	Yes	Yes	Y	Y	Y	Y
Record 6	Dupont	Jean	75	31,000	5,000	Yes	Yes	Y	Y	Y	Y
Record 7	Hugo	Victor	52	58,000	17,000	Yes		N	Y	Y	N
Record 8	Cezanne	Paul	40		1,500	Yes	Yes	Y	Y	Y	Y
Record 9	Duchamp	Marcel	32	15,000	2,500	Yes		Y	Y	Y	Y
Record 10	Lupin	Arsene	XXX	XXXXX	XXXXX	XXXXXX	XXXXXX	N	N	N	N
Record 11								N	N	N	N
						Compliant records		8	8	8	6
						# records		11	11	11	11
						DEI		72.73%	72.73%	72.73%	54.55%

The DEI is the ratio of compliant records against the total number of records. Noncompliant records for all three BERs are records #1, #4, #7, #10, and #11. Action in the further steps will be focused on these records (see subsequent sections).

It should be noted that

- Although each BER have a DEI of 72.7%, the global DEI is lower (54.5%), which shows that this model is different from the traditional aggregation models (OLAP).
- Each record counts only for one in the final calculation of DEI (records 10 and 11) although they can be noncompliant to several BERs.

At this stage, we have a level of compliance (DEI) similar to a KPI. However, we have no information regarding the impact in value of noncompliant transactions (records).

21.5.4.3.3 Views

Using the same method, other DEIs can be calculated according to different views: per region, per bank, per account manager, per customer, etc. It is easy to check that the DEI by region yields the following results:

Regions	BER 1	BER 2	BER 3	All
East	100.00%	83.33%	83.33%	66.67%
West	40.00%	60.00%	60.00%	40.00%
All regions	72.73%	72.73%	72.73%	54.55%

Although DEIs by regions (and for other views) are different, the global DEI stays unchanged as the whole set of records is identical and noncompliant records remain the same.

21.5.4.4 Data Excellence Dimensions

Another view called data quality view according to data quality dimensions (called data excellence dimensions (DEDs) in the framework) can be calculated. DEDs are as follows:

- *Uniqueness*: The uniqueness dimension is to allow the identification in a deterministic way of an entity, a relationship, or an event instance within a specific context to execute a specific business process (e.g., the business process must be able to identify the unique address for sending the invoice for a specific customer). In the example, Records #1 and #6 raise an issue regarding the uniqueness dimension.

- *Completeness*: The completeness dimension is to ensure that the data required for a successful execution of a process in a specific domain and context are present in the database (e.g., to pay your supplier, you need its bank account number). In the example, records #4 and #8 are not complete.
- *Accuracy*: The accuracy dimension is to ensure that the data reflect a real-world view within a context and a specific process. In the example, either record #1 or #6 may not be accurate.
- *Non-obsolescence*: The non-obsolescence dimension ensures that the data required to execute a specific process in a specific context are up to date. Either record #1 or #6 may be obsolete.
- *Consistency*: The consistency dimension is to ensure that the data values are delivered consistently across all the databases and systems for the execution of a specific business process in a specific context. In the example, records #1 and #6 may not be consistent.
- *Timeliness*: The timeliness dimension is to enable first-time right delivery of data required to enable flawless execution of business processes and fulfilling the service-level agreements. In the example, record #11 has not been delivered on time.
- *Accessibility*: The accessibility dimension is to ensure that people, systems, or processes have access to data according to their roles and responsibilities. In the example, record #10 is not accessible.

Each DED provides a DEI, and the DEI resulting from all DED is called the GDE index.

21.5.4.5 KVIs

21.5.4.5.1 Definition and Role

A KVI is a measurement of the value and impact of the DEI on the business operations. It consists of

- *Value*: related to the collection of data elements of the DEI that comply with the BERs
- *Impact*: related to the collection of defective data elements of the DEI that do not comply with the BERs

A fundamental KVI represents the key business value and impact affected by a specific BER. A BER could impact multiple KVIs and a specific KVI can be affected by multiple BERs.

The KVI is a fundamental deliverable of the DEF: Through the KVI, the DEF enables multifocal governance linking business management and data management. The business transaction is, therefore, considered as a key component that enables managing data as a company asset. The DEI becomes the pivot that links the BER, the KVI, and the data elements.

21.5.4.5.2 KVI's Guiding Principles

In order to be efficient and effective, the KVI must fill the following requirements:

- *Tangible value*: a KVI must have tangible value to be considered as an asset linked to the data through the DEI.
- *Stewardship*: each KVI must have a clear steward in the business function and at all organizational levels. The steward from the function must practice their accountability striving for the improvement of the said KVI.
- *BERs*: each KVI must be based on defined BERs.
- *Adequate*: the KVI should reflect real and important business factors to enable control or analysis/evaluation of business operations and to enable sharp decision support.
- *Simple and intuitive*: the KVIs should convey their meaning in a compelling way. Their impact on business should be clearly understandable using common sense.
- *Precise and unambiguous*: the measures should provide no room for misinterpretation or result negotiation.
- *Comprehensive*: they should describe all aspects that are relevant to achieving expected results.

- *Manageable:* calculation of indicators and their presentation to interested parties (reporting) must be straightforward and must not involve personnel interaction.
- *Limited in number:* the whole set of measures should be looked at when evaluating performance.

21.5.4.5.3 Example

Back to our loan application management example, the loan business manager of SmartBank can decide to define the KVI as the amount of the loan for each application. The rationale behind it is that the risk for the bank is the amount of money lent to its customers; it is also linked to the risk to lose a transaction if some data are missing for a loan application. More widely, a specific KVI linked to the loss of customer derived to a bad loan application management could be considered.

From the DEI calculation, we have derived that records #1, #4, #7, #10, and #11 are noncompliant: the following table shows the KVI and impact according to the KVI definition for each of the three BERs, all BERs.

The value and impact per BER are calculated by summing, respectively, the loan amount of all compliant records and noncompliant records. For example, for BER1 only, record #7 is noncompliant. The impact is the loan amount of record #7, that is, 17,000€. The KVI (value) is the sum of the loan amount of the compliant records, that is, 70,000€. The total value is 87,000€. If the ratio KVI/total value of 80.9% is slightly better that the DEI (72.7%), the impact can be zeroed by fixing record #7 by providing the information related to the customer default record. The value impact % column in the table is the value impact distribution across noncompliant records. It can be seen that fixing record #7 and record #1 (surgical approach to data quality) would cut the impact by 84%, which may be sufficient for the business.

Records	Customer Name	Customer First Name	Age	Income	Loan Amount	Pay Slip Provided	Default Record Provided	BER 1	BER 2	BER 3	All BERs	Value Impact %
Record 1	Dupont	Jean	75	30,000	9,000	No		Y	Y	N	N	29.03%
Record 2	Proust	Marcel	50	40,000	11,000	Yes	Yes	Y	Y	Y	Y	
Record 3	Flaubert	Gustave	35	25,000	11,000	Yes	Yes	Y	Y	Y	Y	
Record 4	Monet	Claude		22,000	5,000	Yes		Y	N	Y	N	16.13%
Record 5	Courbet	Gustave	57	60,000	25,000	Yes	Yes	Y	Y	Y	Y	
Record 6	Dupont	Jean	75	31,000	5,000	Yes	Yes	Y	Y	Y	Y	
Record 7	Hugo	Victor	52	58,000	17,000	Yes		N	Y	Y	N	54.84%
Record 8	Cezanne	Paul	40		1,500	Yes	Yes	Y	Y	Y	Y	
Record 9	Duchamp	Marcel	32	15,000	2,500	Yes		Y	Y	Y	Y	
Record 10	Lupin	Arsene	XXX	XXXXX	XXXXX	XXXXXX	XXXXXX	N	N	N	N	
Record 11								N	N	N	N	
						Compliant records		8	8	8	6	
						# records		11	11	11	11	
						DEI		72.73%	72.73%	72.73%	54.55%	
						Total value		87,000	87,000	87,000	87,000	
						Impact		17,000	5,000	9,000	31,000	
						KVI		70,000	82,000	78,000	56,000	
						Ratio KVI/Total value		80.46%	94.25%	89.66%	64.37%	

Note that if DEIs for the three different BERs are identical (72.7%), the KVIs are different because each record does not represent the same business value. This is the main reason why the management by KPIs (DEIs) of the industrial age is inadequate as a good KPI may hide a terrible KVI leading to a severe impact for the business. Alternatively, a bad KPI could generate expensive process improvements, which could prove useless with regards to high KVIs.

In the information age, enterprises must shift from a KPI-driven organization to a KVI-driven one. For emphasizing the point, Patrick Kabac has concluded in Givaudan MDM governance initiative that: "*Business understands facts, not theory*" (Kabac, 2010).

21.5.5 Data Excellence Process

> **Data and information governance mission**: *The establishment of accountability and responsibility at all levels to maximize the business value of enterprise data*
> **Accountability**
> *The willingness and commitment of business executives to be accountable for the definition and management of the BERs of their area and related KVI targets*
> **Responsibility**
> *The willingness and commitment of data managers to be responsible for individual data records assuring the compliance with BER*
> DG can be operationalized through collaborative governance networks where data stewards can play a pivotal role between business, data management, and IT.

21.5.5.1 Roles

We have defined the key roles at the center of the DEF:

- *Data accountables*: they have BER responsibility. They are part of line of business managers responsible for processes and applications. They define, approve BER and KVI, follow the results, and drive priorities of actions. There is a tree of data accountability called the governance tree. In our example, the loan account manager is data accountable for the loan application process of SmartBank. He or she has responsibility over the local account managers in BankA and BankB in all regions for the loan application process: this defines the governance tree.
- *Data responsibles*: they are the data managers. Part of the local data management, they are responsible for individual source data records for each BER and defect management.

In our example, we suppose that for each bank (BankA and BankB), there is a data responsible per region for customer and loan data. Each data responsible in each bank reports to the bank's global data responsible reporting to the SmartBank's data responsible: this defines the responsibility tree. Note that there could be different data responsibility for customer data and loan application data.

- *Data stewards:* link the business, data management, and IT. They coordinate the setup of the DG structure, DEI, KVIs, and BERs with data accountable. They collaborate with data responsibles and IT.
  In our example, the SmartBank loan account manager has named a data steward dedicated to the loan application management.
- *IT*: part of the data integration team, they support the implementation and development of DEIs and KVIs.

In our example, IT is a central team in the headquarters of SmartBank (Figure 21.14).

21.5.5.2 Description

This chart is a summary of the continuous data excellence process described just before. It demonstrates explicitly how in the DEF, data and information are governed and driven from the business requirements, by the business through data stewards into data management. The data accountable will define the context, BERs and KVIs. By setting target for the KVIs, he or she will give mandate to the data steward to meet it by fixing data issues that impact the value.

21.5.5.3 Example

In our example, the loan account manager has given instruction to the data steward to have a KVI ratio above 94% (i.e., 81,780€). The data steward notes that the current ratio is 64.37%: by fixing records #1 and #7,

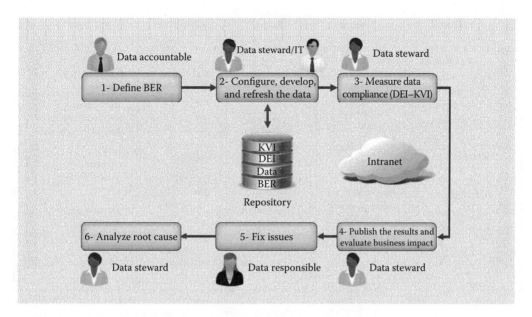

FIGURE 21.14 The continuous data excellence process: Role-based process flow.

he will add 26,000€ to the KVI (56,000€) and reach a 94.25% ratio. Record #1 is managed by the data responsible of BankA in the East, and record #7 is managed by the data responsible of BankA in the West. The data steward will send a requirement to both data managers of BankA to fix those records. Record #1 will be fixed by asking the account manager of BankA in the East region to request the payslip for Jean Dupont and update the loan application system. Record #7 will be fixed by requesting the default record of the customer.

If the root cause analysis shows that there is a defect in a process in BankA, some action might be decided to fix it in order to avoid further defects and value impact.

21.5.5.4 Conclusion

This example shows that the DEF is not a traditional scientific total data quality approach driven by KPIs, but a pragmatic contextual data quality approach driven by value (KVI).

21.5.6 Governance Model

This new approach to cultivate the data assets requires a totally different mind-set as shown in Figure 21.15.

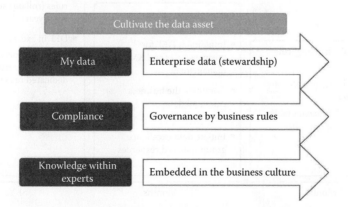

FIGURE 21.15 The governance of the data asset: Change management proposition.

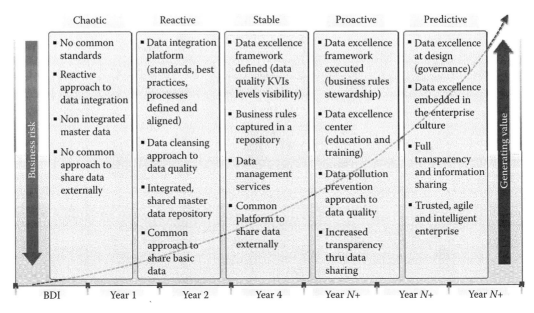

FIGURE 21.16   The data excellence maturity model: From chaos to data excellence.

Shifting from

- Data as a personal power asset to data as an enterprise asset: moving from traditional hierarchal organization to collaborative networks organization
- Compliance data management to a BER- and KVI-driven governance: moving from top-down standardization and MDM approach to "fit for purpose" value-driven DG
- Knowledge within experts to knowledge embedded in the business culture: moving the paradigm from "having the knowledge is the power" to "sharing the knowledge is the power"

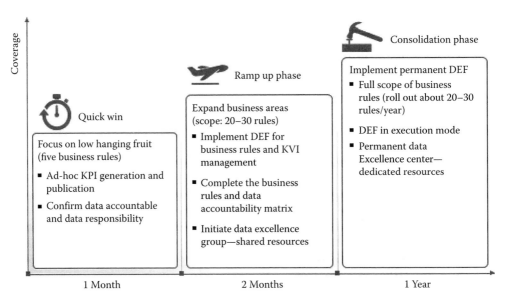

FIGURE 21.17   The DEF: Example of progressive approach.

21.5.7 Maturity Model

Obviously, a KPI-driven organization will not move overnight to the value-driven management. This is a progressive step-by-step evolution illustrated by the data excellence maturity model (Figure 21.16), which tracks organization's progress from the early stages of DG (which is described as "chaotic") to the most mature stages where data are utilized as a core enterprise asset (described as "predictive"; Figure 21.16).

This can be implemented smoothly with short-term returns and small investments for the first phases into a virtuous circle that will lead the enterprise into the data excellence culture (Figure 21.17).

You can achieve DG and high data quality without earthquake in the organization and without costly initiatives. Be progressively incremental with clear communicated business objectives and value of each business rule to be implemented (Figure 21.17). For instance, DG initiative at Nestlé was started with one (1) business rule (El Abed, 2007).

21.6 Conclusion and Future Research on Data Governance

The industrial age was focused on "how" to improve and optimize processes: this approach has proven successful for increasing volumes and lowering costs. However, in the information age characterized by a deluge of data and rules, processing data in a scientific way proves to be time consuming and very costly. In the twenty-first century, business is driven by data and better governance of data means better business. Thus, more organizations start to recognize that they need better data management processes capable of combining business-driven and technical perspectives in order to respond to strategic and operational challenges demanding high-quality corporate data. Both researchers and practitioners consider DG as a promising approach for companies to improve and maintain the quality of corporate data, which is seen as critical for being able to meet strategic business requirements, such as compliance or integrated customer management.

DG specifies the framework that encompasses professionals from both business and IT for decision rights and accountabilities to control and to protect the investments made in data in order to transform them in a real asset for the company. It helps an organization define itself and its strategy and then embed those definitions into core business data ensuring that all of its business processes, systems, projects, and strategies are aligned to create trust and enable consistent executive decision making. DG is to the organization what breathing is to the body. When it is working right, you hardly notice, but when it is not, everything quickly becomes more difficult and labored.

We have been discussing in this chapter the implications of DG for both academics and practitioners and presented the state of the art in the domain. Both sides have proposed a number of DG models and frameworks, which can assist organizations to take off with a DG program. Even though slight differences exist between these models, they all share the basic principle of what a DG program should do: "Defining how to deal with data-related issues."

However, a number of limitations need to be considered. Many models are based on organizational structures that should fit all companies alike. They have, thereby, neglected the fact that each company requires a specific DG configuration that fits a set of contingencies. Furthermore, knowledge has been transferred from IT governance research to DG, and the research on contingencies influencing IT governance models is used as starting point for the contingency research on DG. So far, the proposed contingencies and their impact lack validation in the context of DG. DG model might be characterized by additional parameters, such as a time dimension, which respects the fact that the configuration might evolve over time. In the literature, these issues are investigating and the future direction of DG model is to look toward a contingency approach for DG. Also, one key is to shift the approach from "data-driven" to "value-driven" mind-set meaning that the "business transaction" must be put at the center of DG rather than "data" as advocated by data management trends. This approach for defining and implementing the DG model would help companies to align with shareholders accountability, to assure a sustainable just-in-time data quality enabling the realization of the ultimate "business transaction" between the shareholders and the customers. The purpose of the "enterprise" is to realize the "business transaction" between the shareholders and the customers and exchange the expected value for each. Therefore, DG must be a mean to enable flawless execution of business strategy maximizing the enterprise potential and revealing the business value of the ultimate transaction.

References

Brown E. A. and Grant G. Framing the frameworks: A review of IT governance research, *Communication of the Association for Information Systems*, 15, 696–712, 2005.

Chartres N. *Data Governance, Thought Leadership Series*, Vol. 1. North Melbourne, VIC, Australia: Health Informatics Society of Australia Ltd., 2012.

Delez T. Data governance driven master data management, developing and optimizing the MDM function within business operations, *Firmenich Business Case for the Proceedings of the Master Data Management Conference*, Berlin, Germany, September 2010.

El Abed W. Data quality framework at Nestlé, *Proceedings for the Data Management and Information Quality Conference Europe*, London, U.K., October 2007.

El Abed W. Data governance at Nestlé, *Proceedings for the Data Governance Conference Europe*, London, U.K., February 2008a.

El Abed W. Data excellence framework from vision to execution and value generation in global environment, *Proceedings for the Information and Data Quality Conference*, San Antonio, TX, 2008b.

El Abed W. Mergers and acquisitions: The data dimension, a White Paper, 2009a.

El Abed W. Data governance: A business value-driven approach, a White Paper, November 2011, 2009b. Available at http://community.informatica.com/mpresources/docs/GDE_Data_Governance_March_Nov_2009.pdf (accessed on March 29, 2013).

Feuto P. B. and El Abed W. From natural language business requirements to executable models via SBVR, *Proceedings of the 2012 IEEE International Conference on Systems and Informatics (ICSAI 2012)*, Yantai, China, May 19–20, 2012.

International Business Machines (IBM). The IBM data governance council maturity model: Building a roadmap for effective data governance. Technical report. IBM Software Group, 2007.

Kabac P. MDM governance initiative at Givaudan. In *Givaudan Case Study Presentation*. Geneva, Switzerland, June 2010.

Khatri V. and Brown C. Designing data governance, *Communications of the ACM*, 53(1), 148–152, 2010.

Lucas A. Corporate data quality management in context, *Proceedings of the 15th International Conference on Information Quality*, Little Rock, AR, 2010.

Michellod E. and El Abed W. Data governance case study, introduction of the GDE data governance framework at GROUPE MUTUEL GMA SA Switzerland, *GMA Case Study for the Proceedings of the Data Governance Conference*, San Diego, CA, June 2010.

O'Kane B., Radcliffe, J., White, A., Friedman, T., and Casonato, R. Cool vendors in information governance and master data management, Gartner Research, ID Number: G00226098, April 2012.

Orazi L. and Gessler B. How to create and measure business value applying data governance practices? Supply chain efficiency improvement through product data quality, *Alcatel-Lucent Business Case for the Proceedings of the Master Data Management Conference*, Berlin, Germany, September 2010.

Otto B. A morphology of the organisation of data governance, *Proceedings of the 19th European Conference on Information Systems (ECIS 2011)*, Helsinki, Finland, 2010.

Russom P. Taking data quality to the enterprise through data governance, TDWI Report Series, March 2006.

Weill P. and Ross, J. W. *IT Governance: How Top Performers Manage IT Decision Rights for Superior Results*. Harvard Business School Press, Boston, MA, 2004, 269pp, ISBN 1-59139-253-5.

Wende K. A model for data governance—Organizing accountabilities for data quality management, *18th Australasian Conference on Information Systems*, Toowoomba, Queensland, 2007.

IV

Analysis, Design, and Development of Organizational Systems

22

Design Science Research

Alan R. Hevner
University of South Florida

The proper study of mankind is the science of design.

Herbert Simon

Engineering, medicine, business, architecture and painting are concerned not with the necessary but with the contingent—not with how things are but with how they might be—in short, with design.

Herbert Simon

22.1 Introduction

The design science research (DSR) paradigm has its roots in engineering and the sciences of the artificial (Simon 1996). It is fundamentally a problem-solving paradigm. DSR seeks to enhance human knowledge with the creation of innovative artifacts. These artifacts embody the ideas, practices, technical capabilities, and products through which information and computing technology and systems (ICTS) can be efficiently developed and effectively used. Artifacts are not exempt from natural laws or behavioral theories. To the contrary, their creation relies on existing laws and theories that are applied, tested, modified, and extended through the experience, creativity, intuition, and problem-solving capabilities of the researcher. Thus, the results of DSR include both the newly designed artifact and a fuller understanding of the theories of why the artifact is an improvement to the relevant application context.

Design activities are central to most applied disciplines. Research in design has a long history in many fields including architecture, engineering, education, psychology, and the fine arts (Cross 2001). The ICTS field since its advent in the late 1940s has appropriated many of the ideas, concepts, and methods of design science that have originated in these other disciplines. However, ICTS as composed of inherently mutable and adaptable hardware, software, and human interfaces provide many unique and challenging design problems that call for new and creative ideas.

The DSR paradigm is highly relevant to ICTS research because it directly addresses two of the key issues of the discipline: the central role of the ICTS artifact (Orlikowski and Iacono 2001) and the importance of professional relevance of ICTS research (Benbasat and Zmud 1999). Design science, as conceptualized by Simon (1996), follows a pragmatic research paradigm that calls for the creation of innovative artifacts to solve real-world problems and to address promising opportunities. DSR in ICTS fields involves the construction of a wide range of *socio-technical* artifacts such as decision support systems (DSSs), modeling tools, governance strategies, methods for software systems development and evaluation, and system change interventions. Thus, DSR in ICTS combines a focus on the ICTS artifact with a high priority on relevance in the application domain.

This chapter begins with a concise introduction to the basic concepts and principles of DSR. The key to appreciating the core ideas and goals of DSR is a clear understanding of how DSR relates to human knowledge. The appropriate and effective consumption and production of knowledge are related issues that researchers should consider throughout the research process—from initial problem selection to the use of sound research methods, to reflection, and to communication of research results in journal, magazine, and conference articles. The chapter concludes with a discussion of ongoing challenges in DSR.

22.2 Design Science Research Concepts and Principles

ICTS artifacts are implemented within an application context (e.g., a business organization) for the purpose of improving the effectiveness and efficiency of that context. The utility of the artifact and the characteristics of the application—its work systems, its people, and its development and implementation methodologies—together determine the extent to which that purpose is achieved. Researchers in the ICTS disciplines produce new ideas to improve the ability of human organizations to adapt and succeed in the presence of changing environments. Such new ideas are then communicated as knowledge to the relevant ICTS communities. In this section, the basic concepts and principles of DSR are surveyed with a focus on its application in ICTS contexts.

22.2.1 Cycle between Design Science Research and Behavioral Science Research

Acquiring ICTS knowledge involves two complementary but distinct paradigms, behavioral science and design science (March and Smith 1995). The behavioral science paradigm has its roots in natural science research methods. It seeks to develop and justify theories (i.e., principles and laws) that explain or predict organizational and human phenomena surrounding the analysis, design, implementation, and use of ICTS. Such theories ultimately inform researchers and practitioners of the interactions among people, technology, and organizations that must be managed if an information system (IS) is to achieve its stated purpose, namely, improving the effectiveness and efficiency of an organization. These theories impact and are impacted by design decisions made with respect to the system development methodology used and the functional capabilities, information contents, and human interfaces implemented within the IS.

Technological advances are the result of innovative, creative DSR processes. If not "capricious," they are at least "arbitrary" (Brooks 1987) with respect to business needs and existing knowledge. Innovations—such as database management systems, high-level languages, personal computers, software components, intelligent agents, object technology, the Internet, and the World Wide Web (WWW)—have had dramatic and at times unintended impacts on the way in which ICTS are conceived, designed, implemented, and managed.

A key insight here is that there is a complementary research cycle between design science and behavioral science to address fundamental problems faced in the productive application of ICTS

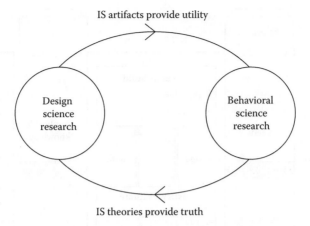

FIGURE 22.1 Complementary nature of design science and behavioral science research.

(see Figure 22.1). Technology and behavior are not dichotomous in an IS. They are inseparable. They are similarly inseparable in ICTS research. These arguments draw from a pragmatist philosophy that argues that truth (justified theory) and utility (artifacts that are effective) are two sides of the same coin and that scientific research should be evaluated in light of its practical implications. In other words, the practical relevance of the research result should be valued equally with the rigor of the research performed to achieve the result.

22.2.2 Design Science Research in ICTS

To achieve a true understanding of and appreciation for the DSR paradigm, an important dichotomy must be faced. Design is both a process (set of activities) and a product (artifact)—a verb and a noun (Walls et al. 1992). It describes the world as acted upon (*processes*) and the world as sensed (*artifacts*). This view of design supports a problem-solving paradigm that continuously shifts perspective between design processes and designed artifacts for the same complex problem. The design process is a sequence of expert activities that produces an innovative product (i.e., the design artifact). The evaluation of the artifact then provides feedback information and a better understanding of the problem in order to improve both the qualities of the product and the design process. This build-and-evaluate loop is typically iterated a number of times before the final design artifact is generated. During this creative process, the design science researcher must be cognizant of evolving both the design process and the design artifact as part of the research.

March and Smith (1995) identify two design processes and four design artifacts produced by DSR in ICTS. The two processes are *build* and *evaluate*. The artifacts are *constructs*, *models*, *methods*, and *instantiations*.

Purposeful artifacts are built to address relevant problems. They are evaluated with respect to the utility provided in solving those problems. Constructs provide the language in which problems and solutions are defined and communicated. Models use constructs to represent a real-world situation—the design problem and its solution space. Models aid problem and solution understanding and frequently represent the connection between problem and solution components enabling exploration of the effects of design decisions and changes in the real world. Methods define processes. They provide guidance on how to solve problems, that is, how to search the solution space. These can range from formal, mathematical algorithms that explicitly define the search process to informal, textual descriptions of "best practice" approaches, or some combination. Instantiations show that constructs, models, or methods can be implemented in a working system. They demonstrate feasibility, enabling concrete assessment

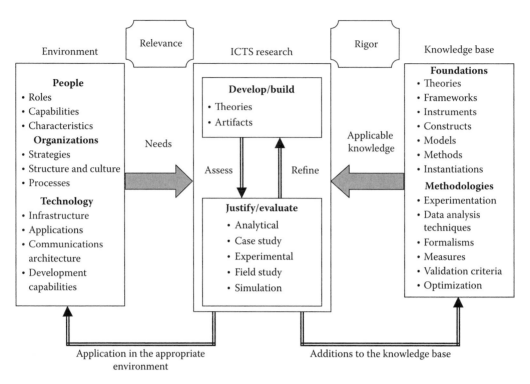

FIGURE 22.2   ICTS research framework. (Adapted from Hevner, A., et al., *MIS Quarterly* 28(1), 75, 2004.)

of an artifact's suitability to its intended purpose. They also enable researchers to learn about the real world, how the artifact affects it, and how users appropriate the design artifact in a real-world context.

Figure 22.2 presents a conceptual framework for understanding, executing, and evaluating research combining behavioral science and design science paradigms (Hevner et al. 2004). The environment defines the problem space in which the phenomena of interest reside. For ICTS research, it is composed of people, organizations, and existing or planned technologies. In it are the goals, tasks, problems, and opportunities that define needs as they are perceived by stakeholders within the organization. Needs are assessed and evaluated within the context of organizational strategies, structure, culture, and existing work processes. They are positioned relative to existing technology infrastructure, applications, communication architectures, and development capabilities. Together these define the "research problem" as perceived by the researcher. Framing research activities to address real stakeholder needs assures research relevance.

Given such articulated needs, IS research is conducted in two complementary phases. Behavioral science addresses research through the *development* and *justification* of theories that explain or predict phenomena related to the identified needs. Design science addresses research through the *building* and *evaluation* of artifacts designed to meet the identified needs. The knowledge base provides the raw materials from and through which ICTS research is accomplished. The knowledge base is composed of foundations and methodologies. Prior research and results from reference disciplines provide foundational theories, frameworks, instruments, constructs, models, methods, and instantiations used in the develop/build phase of a research study. Methodologies provide guidelines used in the justify/evaluate phase. Rigor is achieved by appropriately applying existing foundations and methodologies. In behavioral science, methodologies are typically rooted in data collection and empirical analysis techniques. In design science, computational and mathematical methods are primarily used to evaluate the quality and effectiveness of artifacts; however, empirical techniques may also be employed.

TABLE 22.1 DSR Guidelines

Guideline	Description
Guideline 1: Design as an artifact	Design science research must produce a viable artifact in the form of a construct, a model, a method, or an instantiation
Guideline 2: Problem relevance	The objective of design science research is to develop technology-based solutions to important and relevant business problems
Guideline 3: Design evaluation	The utility, quality, and efficacy of a design artifact must be rigorously demonstrated via well-executed evaluation methods
Guideline 4: Research contributions	Effective design science research must provide clear and verifiable contributions in the areas of the design artifact, design foundations, and/or design methodologies
Guideline 5: Research rigor	Design science research relies upon the application of rigorous methods in both the construction and evaluation of the design artifact
Guideline 6: Design as a search process	The search for an effective artifact requires utilizing available means to reach desired ends while satisfying laws in the problem environment
Guideline 7: Communication of research	Design science research must be presented effectively both to technology-oriented as well as management-oriented audiences

As discussed earlier, DSR is inherently a problem-solving process. The fundamental principle is that knowledge and understanding of a design problem and its solution are acquired in the building and application of an artifact. This principle leads to the statement of seven basic guidelines for the performance of DSR (Hevner et al. 2004). That is, DSR requires the creation of an innovative, purposeful artifact (Guideline 1) for a specified and important problem domain (Guideline 2). Because the artifact is "purposeful," it must yield utility for the specified problem. Hence, thorough evaluation of the artifact is crucial (Guideline 3). Novelty is similarly crucial since the artifact must be "innovative," solving a heretofore unsolved problem or solving a known problem in a more effective or efficient manner (Guideline 4). In this way, DSR is differentiated from the practice of design. The artifact itself must be rigorously defined, formally represented, coherent, and internally consistent (Guideline 5). The process by which it is created, and often the artifact itself, incorporates or enables a search process whereby a problem space is constructed and a mechanism posed or enacted to find an effective solution (Guideline 6). Finally, the results of the DSR project must be communicated effectively (Guideline 7) both to a technical audience (researchers who will extend them and practitioners who will implement them) and to a managerial audience (researchers who will study them in context and practitioners who will decide if they should be implemented within their organizations). Researchers, reviewers, and editors must use their creative skills and judgment to determine when, where, and how to apply each of the guidelines in a specific research project. However, each of these guidelines should be addressed in some manner for DSR to be complete. Table 22.1 summarizes the seven guidelines.

22.2.3 Three DSR Cycles

Key insights can be gained by identifying and understanding the existence of three DSR cycles in any design research project as shown in Figure 22.3 (Hevner 2007). This figure borrows the IS research framework found in (Hevner et al. 2004) and overlays a focus on three inherent research cycles. The *relevance cycle* bridges the contextual environment of the research project with the design science activities. The *rigor cycle* connects the design science activities with the knowledge base of scientific foundations, experience, and expertise that informs the research project. The central *design cycle* iterates between the

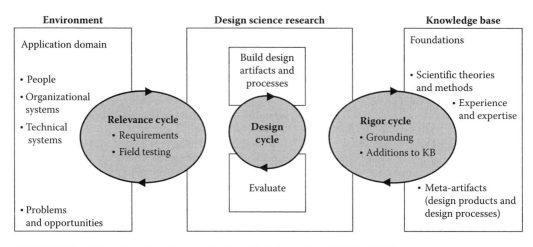

FIGURE 22.3 DSR cycles. (From Hevner, A., *Scand. J. Inform. Syst.*, 19(2), 87, 2007.)

core activities of building and evaluating the design artifacts and processes of the research. These three cycles must be present and clearly identifiable in a DSR project. The following sections briefly expand on the definitions and meanings of each cycle.

22.2.3.1 Relevance Cycle

The relevance cycle initiates DSR within an application context that not only provides the requirements for the research (e.g., the opportunity/problem to be addressed) as inputs but also defines acceptance criteria for the ultimate evaluation of the research results. Does the design artifact improve the environment? How can this improvement be measured? The output from DSR must be returned into the environment for study and evaluation in the application domain. The field study of the artifact can be executed by means of appropriate technology transfer methods such as action research (Sein et al. 2011).

The results of the field testing will determine whether additional iterations of the relevance cycle are needed in this DSR project. The new artifact may have deficiencies in functionality or in its inherent qualities (e.g., performance, usability) that may limit its utility in practice. Another result of field testing may be that the requirements input to the DSR project were incorrect or incomplete with the resulting artifact satisfying the requirements but still inadequate to the opportunity or problem presented. Another iteration of the relevance cycle will commence with feedback from the environment from field testing and a restatement of the research requirements as discovered from actual experience.

22.2.3.2 Rigor Cycle

DSR draws from a vast knowledge base of scientific theories and engineering methods that provides the foundations for rigorous research. Equally important, the knowledge base also contains two types of additional knowledge:

- The experiences and expertise that define the state of the art in the application domain of the research
- The existing artifacts and processes found in the application domain

The rigor cycle provides past knowledge to the research project to ensure its innovation. Consideration of rigor in DSR is based on the researcher's skilled selection and application of the appropriate theories and methods for constructing and evaluating the artifact. DSR is grounded on existing ideas drawn from the domain knowledge base. Inspiration for creative design activity can be drawn from many different sources to include rich opportunities/problems from the application environment, existing artifacts, analogies/metaphors, and theories (Iivari 2007). This list of design inspiration can be expanded to include additional sources of creative insights (Csikszentmihalyi 1996).

Additions to the knowledge base as results of DSR will include any additions or extensions to the original theories and methods made during the research, the new artifacts (design products and processes), and all experiences gained from performing the iterative design cycles and field testing the artifact in the application environment. It is imperative that a DSR project makes a compelling case for its rigorous foundations and contributions lest the research be dismissed as a case of routine design. Definitive research contributions to the knowledge base are essential to selling the research to an academic audience just as useful contributions to the environment are the key selling points to a practitioner audience.

22.2.3.3 Design Cycle

The internal design cycle is the heart of any DSR project. This cycle of research activities iterates rapidly between the construction of an artifact, its evaluation, and subsequent feedback to refine the design further. Simon (1996) describes the nature of this cycle as generating design alternatives and evaluating the alternatives against requirements until a satisfactory design is achieved. As discussed earlier, the requirements are input from the relevance cycle and the design and evaluation theories and methods are drawn from the rigor cycle. However, the design cycle is where the hard work of DSR is done. It is important to understand the dependencies of the design cycle on the other two cycles while appreciating its relative independence during the actual execution of the research.

During the performance of the design cycle, a balance must be maintained between the efforts spent in constructing and evaluating the evolving design artifact. Both activities must be convincingly based in relevance and rigor. Having a strong grounded argument for the construction of the artifact, as discussed earlier, is insufficient if the subsequent evaluation is weak. Artifacts must be rigorously and thoroughly tested in laboratory and experimental situations before releasing the artifact into field testing along the relevance cycle. This calls for multiple iterations of the design cycle in DSR before contributions are output into the relevance cycle and the rigor cycle.

22.2.4 DSR Process Models

The growing interest in performing DSR projects has led to several proposed process models for scheduling and coordinating design activities. Seminal thinking in this area was achieved by Nunamaker and his research group at the University of Arizona (Nunamaker et al. 1990–1991). They claim that the central nature of systems development leads to a multi-methodological approach to ICTS research that consists of four research strategies: theory building, experimentation, observation, and systems development. Systems development corresponds closely to DSR and consists of five stages: conceptual design, constructing the architecture of the system, analyzing the design, prototyping (may include product development), and evaluation. The framework is shown in Figure 22.4.

More recently, Vaishnavi and Kuechler (2008) have extended the general design cycle model of Takeda et al. (1990) to apply specifically to DSR as illustrated in Figure 22.5. In this model, all design begins with awareness of problem. Here, you not only identify the problem but also define it. The next stage is a preliminary suggestion for a problem solution that is drawn from the existing knowledge or theory base for the problem area. Once a tentative design is decided, the next stage is actual development. This is a creative stage where the design is further refined and an actual artifact is produced through many iterations. Once an implementation (or prototype) is ready, it is evaluated according to the functional specification implicit or explicit in the suggestion. Empirical methods are often used in evaluation. There are iterations and feedback involved in these stages cited as circumscription. Finally, a project is terminated and concluded.

Peffers et al. (2008) propose and develop a design science research methodology (DSRM) for the production and presentation of DSR as shown in Figure 22.6. This DSR process includes six steps (problem identification and motivation, definition of the objectives for a solution, design and development, demonstration, evaluation, and communication) and four possible entry points (problem-centered initiation, objective-centered solution, design and development–centered initiation, and client/context initiation). A brief description of each DSR activity follows.

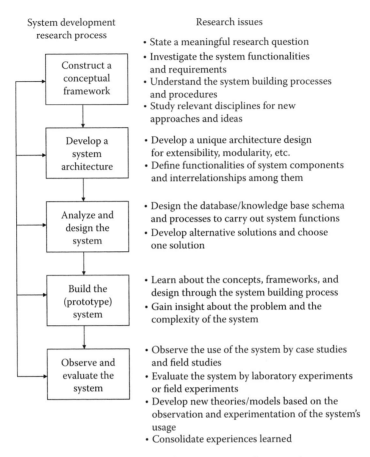

System development
research process

Research issues

Construct a
conceptual
framework

- State a meaningful research question
- Investigate the system functionalities
  and requirements
- Understand the system building processes
  and procedures
- Study relevant disciplines for new
  approaches and ideas

Develop a
system
architecture

- Develop a unique architecture design
  for extensibility, modularity, etc.
- Define functionalities of system components
  and interrelationships among them

Analyze and
design the
system

- Design the database/knowledge base schema
  and processes to carry out system functions
- Develop alternative solutions and choose
  one solution

Build the
(prototype)
system

- Learn about the concepts, frameworks, and
  design through the system building process
- Gain insight about the problem and the
  complexity of the system

Observe and
evaluate the
system

- Observe the use of the system by case studies
  and field studies
- Evaluate the system by laboratory experiments
  or field experiments
- Develop new theories/models based on the
  observation and experimentation of the system's
  usage
- Consolidate experiences learned

FIGURE 22.4　Systems development research model. (From Nunamaker, J. et al., *J. Manage. Inform. Syst.*, 7(3), 89, 1990–1991.)

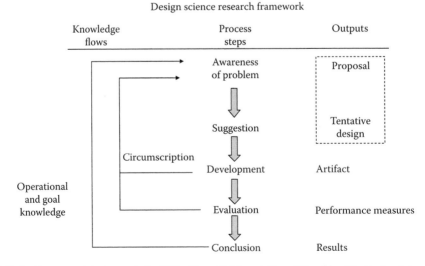

Design science research framework

Knowledge
flows

Process
steps

Outputs

Awareness
of problem

Proposal

Suggestion

Tentative
design

Circumscription

Development

Artifact

Operational
and goal
knowledge

Evaluation

Performance measures

Conclusion

Results

FIGURE 22.5　The general design cycle for DSR. (From Vaishnavi, V. and Kuechler, W., *Design Science Research Methods and Patterns: Innovating Information and Communication Technology*, Auerbach Publications, Boston, MA, 2008.)

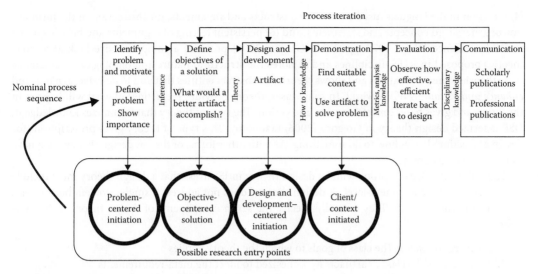

FIGURE 22.6 DSRM process model. (From Peffers, K. et al., *J. MIS*, 24(3), 45, 2008.)

Activity 1: Problem identification and motivation. This activity defines the specific research problem and justifies the value of a solution. Justifying the value of a solution accomplishes two things: It motivates the researcher and the audience of the research to pursue the solution and it helps the audience to appreciate the researcher's understanding of the problem. Resources required for this activity include knowledge of the state of the problem and the importance of its solution.

Activity 2: Define the objectives for a solution. The objectives of a solution can be inferred from the problem definition and knowledge of what is possible and feasible. The objectives can be quantitative, for example, terms in which a desirable solution would be better than current ones, or qualitative, for example, a description of how a new artifact is expected to support solutions to problems not hitherto addressed. The objectives should be inferred rationally from the problem specification.

Activity 3: Design and development. An artifact is created. Conceptually, a DSR artifact can be any designed object in which a research contribution is embedded in the design. This activity includes determining the artifact's desired functionality and its architecture and then creating the actual artifact.

Activity 4: Demonstration. This activity demonstrates the use of the artifact to solve one or more instances of the problem. This could involve its use in experimentation, simulation, case study, proof, or other appropriate activity.

Activity 5: Evaluation. The evaluation measures how well the artifact supports a solution to the problem. This activity involves comparing the objectives of a solution to actual observed results from use of the artifact in context. Depending on the nature of the problem venue and the artifact, evaluation could take many forms. At the end of this activity, the researchers can decide whether to iterate back to step three to try to improve the effectiveness of the artifact or to continue on to communication and leave further improvement to subsequent projects.

Activity 6: Communication. Here, all aspects of the problem and the designed artifact are communicated to the relevant stakeholders. Appropriate forms of communication are employed depending upon the research goals and the audience, such as practicing professionals.

22.2.5 Design Theory

In the natural and social sciences, well-developed or mature theory refers to a cohesive body of knowledge (BOK) that has certain distinguishing characteristics. "Scientific" knowledge, as opposed to common-sense knowledge, has (1) explanations of why statements and beliefs hold, (2) delimitation of the boundaries within which beliefs hold, (3) a logically consistent set of statements and beliefs,

(4) precision in the language used to specify constructs and statements, (5) abstraction in the formulation of generalized concepts and statements, and (6) persistent testing of arguments and beliefs against available evidence (Nagel 1979, pp. 2–14). Gregor (2006) provides a comprehensive look at various theories proposed in ICTS disciplines and explores the structural nature and ontological character of these theories. She summarizes and shows theories as abstract entities that aim to *describe, explain,* and *enhance understanding* of the world. In some cases, theory provides *predictions* of what will happen in the future and gives bases for intervention and action. The type of theory that formalizes knowledge in DSR is termed design theory in Gregor's (2006) taxonomy. This type of theory gives prescriptions for design and action: It says how to do something. As with other forms of theory, design theory must have some degree of generality.

Walls et al. (1992), early proponents of design theory in IS, stress that a design theory must include components that are at a meta-level. A design theory "does not address a single problem but rather a class of problems" (p. 42). They formally specify the seven components of design theory in ICTS as follows:

1. *Meta-requirements*: The class of goals to which the theory applies
2. *Meta-design*: The class of artifacts hypothesized to meet the meta-requirements
3. *Kernel* design product theories*: Theories from natural and social sciences that govern design requirements
4. *Testable design product hypotheses*: Statements required to test whether the meta-design satisfies meta-requirements
5. *Design method*: A description of the procedures for constructing the artifact
6. *Kernel design process theories*: Theories from natural or social sciences that inform the design process
7. *Testable design process hypotheses*: Statements required to test whether the design method leads to an artifact that is consistent with meta-design

Extending this stream of thought, Gregor and Jones (2007) propose that a DSR project should produce a design theory or provide steps toward a design theory. Their view, as expressed similarly in other branches of science, is that all research should result in a contribution to knowledge in the form of partial theory, incomplete theory, or even some particularly interesting and perhaps surprising empirical generalization in the form of a new design artifact (Merton 1968; Sutton and Staw 1995).

Gregor and Jones (2007) provide arguments to show that well-developed design knowledge can satisfy these criteria for scientific knowledge and also the criteria for theory that Dubin (1978) specifies. They argue that any DSR design theory should include as minimum criteria the following core components:

1. *Purpose and scope*: "What the system is for," the set of meta-requirements or goals that specifies the type of artifact to which the theory applies and in conjunction also defines the scope, or boundaries, of the theory
2. *Constructs*: Representations of the entities of interest in the theory
3. *Principles of form and function*: The abstract "blueprint" or architecture that describes an IS artifact, either product or method/intervention
4. *Artifact mutability*: The changes in state of the artifact anticipated in the theory, that is, what degree of artifact change is encompassed by theory
5. *Testable propositions*: Truth statements about the design theory
6. *Justificatory knowledge*: The underlying knowledge or theory from the natural or social or design sciences that gives a basis and explanation for the design (kernel theories)

---

* The term "kernel theories" is used by Walls et al. (1992) to denote descriptive theories from natural and social sciences that are used to support and explain design theories. In general, kernel theory refers to the descriptive theory that provides the underpinning justificatory knowledge that informs artifact construction and explains why artifacts work (Gregor and Jones 2007).

Additional design theory components may be

8. *Principles of implementation*: A description of processes for implementing the theory (either product or method) in specific contexts

9. *Expository instantiation*: A physical implementation of the artifact that can assist in representing the theory both as an expository device and for purposes of testing

The decision whether to present DSR results as design theory will depend on the levels of artifact abstraction (see Section 22.3.3) and desired generality in the research. An artifact that is presented with a higher degree of abstraction can be generalized to other situations and is more interesting than a simple descriptive case study of what happened in one situation. What distinguishes design theory, however, is that it includes kernel theory to explain why the artifact works and testable propositions. If research can be expressed in these terms, with more explanation, more precision, more abstraction, and more testing of beliefs facilitated, then there is a move toward a more mature and well-developed BOK (Nagel 1979). However, it is important for researchers to clearly understand what contributions can be claimed in a DSR project. Researchers should not force results into a design theory description if such a presentation is not appropriate or useful. Often, DSR results can be presented effectively as artifact representations and rigorous evaluations of the artifact in use. Only if expression of these results in a design theory provides a useful generalization for extending knowledge in the problem or solution domains should such a design theory be presented.

22.3 Knowledge Contributions of Design Science Research

Contributing to human knowledge is seen as the key criterion for the credibility and publication of DSR. The appropriate and effective consumption and production of knowledge are fundamental issues that researchers should consider throughout the research process—from initial problem selection to the use of sound research methods, to reflection, and to communication of research results in journal and conference articles. Experience shows that many authors, reviewers, and editors struggle to present DSR work well with a clear demonstration of how their research contributes to knowledge. The difficulties here likely arise from a combination of factors, which include the relative youth of the ICTS disciplines and the comparatively recent recognition of DSR as a distinct, yet legitimate, research paradigm. The potential impact of rigorous DSR is lost or marginalized when knowledge contributions are inadequately positioned and presented.

22.3.1 Types of Knowledge

Drawing from the economics of knowledge field, useful DSR knowledge can be divided into two distinct types (see Mokyr 2002). *Descriptive knowledge* (denoted Ω or omega) is the "what" knowledge about natural phenomena and the laws and regularities among phenomena. *Prescriptive knowledge* (denoted Λ or lambda) is the "how" knowledge of human-built artifacts. Figure 22.7 shows that both Ω knowledge and Λ knowledge comprise a knowledge base for a particular DSR application context.

22.3.1.1 Descriptive Knowledge

Descriptive knowledge has two primary forms. The descriptions of natural, artificial, and human-related phenomena are composed of observations, classifications, measurements, and the cataloging of these descriptions into accessible forms. Additionally, we discover knowledge of the sense-making relationships among the phenomena. This sense-making is represented by natural laws, principles, regularities, patterns, and theories. Together, phenomena and their sense-making relationships provide the natural, artificial, and human (social) science bases for the world in which we live. An addition of knowledge to Ω is a *discovery* of new facts or laws that have always existed but have not been understood and described until now. Over time, the Ω knowledge base accumulates the "BOK" surrounding a natural, artificial, social, or human phenomenon. Such knowledge resides in people's minds or in external storage devices (e.g., data repositories).

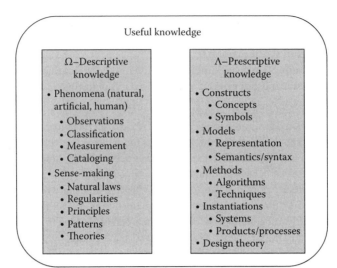

FIGURE 22.7 The DSR knowledge base. (From Gregor, S. and Hevner, A., *Management Information Systems Quarterly*, 2013.)

22.3.1.2 Prescriptive Knowledge

Prescriptive knowledge concerns artifacts designed by humans to improve the natural world. Simon (1996) labels such knowledge as belonging to the sciences of the artificial. March and Smith (1995) define four types of prescriptive knowledge: constructs, models, methods, and instantiations. Design theories are also forms of presenting prescriptive knowledge, so the Λ knowledge base includes the following:

- *Constructs*, which provide the vocabulary and symbols used to define and understand problems and solutions; for example, the constructs of "entities" and "relationships" in the field of information modeling. The correct constructs have a significant impact on the way in which tasks and problems are conceived, and they enable the construction of models for the problem and solution domains.
- *Models*, which are designed representations of the problem and possible solutions. For example, mathematical models, diagrammatical models, and logic models are widely used in the ICTS field. New and more useful models are continually being developed. Models correspond to "principles of form" in the Gregor and Jones (2007) taxonomy: the abstract blueprint of an artifact's architecture, which show an artifact's components and how they interact.
- *Methods*, which are algorithms, practices, and recipes for performing a task. Methods provide the instructions for performing goal-driven activities. They are also known as techniques (Mokyr 2002) and correspond to "principles of function" in the Gregor and Jones (2007) taxonomy and Bunge's (1998) technological rules.
- *Instantiations*, which are the physical systems that act on the natural world, such as an IS that stores, retrieves, and analyzes customer relationship data. Instantiations can embody design knowledge, possibly in the absence of more explicit description. The structural form and functions embodied in an artifact can be inferred to some degree by observing the artifact.
- *A design theory*, which is an abstract, coherent body of prescriptive knowledge that describes the principles of form and function, methods, and justificatory theory that are used to develop an artifact or accomplish some end (Gregor and Jones 2007).

Adding knowledge to Λ concerns *inventions*, that is, things that would not exist except for human creativity.

22.3.2 Roles of Knowledge in DSR

Figure 22.8 illustrates the various relationships and interactions of Ω knowledge and Λ knowledge in the performance of DSR. DSR begins with an important opportunity, challenging problem, or insightful vision/conjecture for something innovative in the application environment (Hevner et al. 2004; Hevner 2007; Iivari 2007). Research questions typically center on how to increase some measure of operational utility via new or improved design artifacts. To study the research questions, the first enquiry is: What do we know already? From what existing knowledge can we draw? Both Ω and Λ knowledge bases are investigated for their contributions to the grounding of the research project. Such investigations are contingent on researchers having ready and efficient access to both knowledge bases.

From the Ω base, the researcher appropriately draws relevant descriptive and propositional knowledge that informs the research questions. Relevant knowledge may be drawn from many different elements in Ω. This would include existing justificatory (i.e., kernel) theories that relate to the goals of the research. At the same time, from the Λ base, the researcher investigates known artifacts and design theories that have been used to solve the same or similar research problems in the past. The objective is to provide a baseline of knowledge on which to evaluate the novelty of new artifacts and knowledge resulting from the research. In many cases, the new design research contribution is an important extension of an existing artifact or the application of an existing artifact in a new application domain. The success of a DSR project is predicated on the research skills of the team in appropriately drawing knowledge from both Ω and Λ bases to ground and position the research; the teams' cognitive skills (e.g., creativity, reasoning) in designing innovative solutions; and the teams' social skills in bringing together all the individual members' collective and diverse intelligence via effective teamwork.

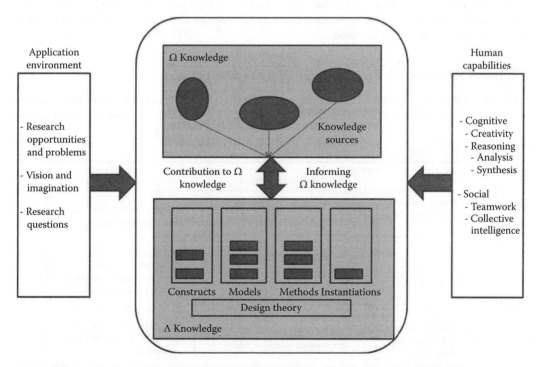

FIGURE 22.8 The roles of knowledge in DSR. (From Gregor, S. and Hevner, A., *Management Information Systems Quarterly*, 2013.)

DSR activities include the building of innovative artifacts, evaluating the utility of the artifacts, and reflecting on what has been learned in the research process. The researcher adds new knowledge to the Λ base with the design of new constructs, models, methods, and instantiations and with the development of new design theories or the more complete development of nascent design theories. The results of the evaluations then may also be a contribution to Ω knowledge in the form of propositions such as "System A outperforms System B in Environment C by Measurement D." Thus, we assert that rigorous DSR produces both Λ and Ω knowledge as research contributions.

22.3.3 ICTS Artifact

Information technology (IT) artifact plays a key role in knowledge contribution. In general, the term artifact refers to a thing that has, or can be transformed into, a material existence as an artificially made object (e.g., model, instantiation) or process (e.g., method, software). Many IT artifacts have some degree of abstraction but can be readily converted to a material existence; for example, an algorithm converted to operational software. In contrast, a theory is more abstract, has a nonmaterial existence, and contains knowledge additional to the description of a materially existing artifact. The construction of an artifact and its description in terms of DSR concepts and principles can be seen as steps in the process of developing more comprehensive bodies of knowledge or design theories.

Table 22.2 shows three maturity levels of DSR artifact types with examples of each level (Purao 2002; Gregor and Hevner 2013). A specific DSR research project can produce artifacts on one or more of these levels ranging from specific instantiations at Level 1 in the form of products and processes to more general (i.e., abstract) contributions at Level 2 in the form of nascent design theory (e.g., constructs, design principles, models, methods, technological rules), to well-developed design theories about the phenomena under study at Level 3. Note that "artifact or situated implementation" (Level 1) is considered a knowledge contribution, even in the absence of abstraction or theorizing about its design principles or architecture because the artifact can be a research contribution in and of itself. Demonstration of a novel artifact can be a research contribution that embodies design ideas and theories yet to be articulated, formalized, and fully understood.

22.3.4 A DSR Knowledge Contribution Framework

Gregor and Hevner (2013) posit a DSR knowledge contribution framework as an effective way to understand and position a DSR project's research contributions. Clearly identifying a knowledge contribution

TABLE 22.2 DSR Artifact Contribution Types

	Contribution Types	Example Artifacts
More abstract, complete, and mature knowledge	Level 3: Well-developed design theory about embedded phenomena	Design theories (mid-range and grand theories)
$\updownarrow\ \updownarrow\ \updownarrow\ \updownarrow$	Level 2: Nascent design theory—knowledge as operational principles/architecture	Constructs, methods, models, design principles, and technological rules.
More specific, limited, and less mature knowledge	Level 1: Situated implementation of artifact	Instantiations (software products or implemented processes)

Source: Gregor, S. and Hevner, A., *Management Information Systems Quarterly*, 2013.

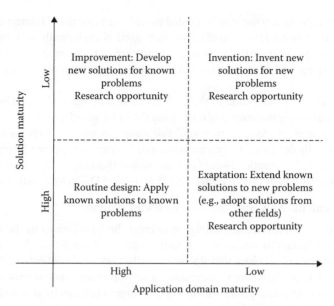

FIGURE 22.9 DSR knowledge contribution framework. (From Gregor, S. and Hevner, A., *Management Information Systems Quarterly*, 2013.)

is often difficult in DSR because it depends on the nature of the designed artifact (as seen in Table 22.2), the state of the field of knowledge, the audience to whom it is to be communicated, and the publication outlet. In addition, the degree of knowledge contribution can vary: there might be incremental artifact construction or only partial theory building, but this may still be a significant and publishable contribution. The size of the knowledge increase could be offset by the practical impact in a knowledge area.

A fundamental issue is that nothing is really "new." Everything is made out of something else or builds on some previous idea. When is something really novel or a significant advance on prior work? This problem is highly relevant when considering patents and intellectual property. A DSR project has the potential to make different types and levels of research contributions depending on its starting points in terms of *problem maturity* and *solution maturity*. Figure 22.9 presents a 2 × 2 matrix of research project contexts and potential DSR research contributions. The x-axis shows the maturity of the problem context from high to low. The y-axis represents the current maturity of artifacts that exist as potential starting points for solutions to the research question, also from high to low.

For each quadrant, research contributions are made in terms of different levels of artifact/theory in DSR and also in terms of contributions to and flows between Ω and Λ knowledge bases.

22.3.4.1 Invention: New Solutions for New Problems

True invention is a radical breakthrough—a clear departure from the accepted ways of thinking and doing. The invention process can be described as an exploratory search over a complex problem space that requires the cognitive skills of curiosity, imagination, creativity, insight, and knowledge of multiple realms of inquiry to find a feasible solution. While this process of invention is perhaps ill-defined, invention activities can still be considered DSR when the result is an artifact that can be applied and evaluated in a real-world context and when new knowledge is contributed to the Ω and/or Λ knowledge bases.

DSR projects in this quadrant will entail research in new and interesting applications where little current understanding of the problem context exists and where no effective artifacts are available as solutions. In fact, so little may be known about the problem that research questions may not even have been raised before. Research contributions in this quadrant result in recognizably novel artifacts or inventions. In this category, the idea of the artifact itself is new, for example, the first bicycle or the first DSS. Here, a recognized problem may not necessarily exist and the value of a solution may be unclear.

As Simon (1996) says, the researcher may be guided by nothing more than "interestingness." In part, a key contribution is the conceptualization of the problem itself. Knowledge flows in the invention quadrant are typically from prescriptive to descriptive. The new artifact is invented and then other researchers see it employed in use and begin to formulate descriptive knowledge about its use in context (in a different quadrant).

A well-known exemplar of invention in the IT field is found in Agrawal et al. (1993). This research developed the first full conceptualization of mining databases for association rules as well as an efficient method for discovering them. As an "invention," this paper has generated and influenced whole new fields of research broadly described as data analytics and business intelligence (Davenport and Harris 2007). Other examples of IT inventions would include the first thinking on DSSs by Scott-Morton (1967) and the subsequent evolution of DSS into executive IS and group DSSs (Nunamaker et al. 1991).

22.3.4.2 Improvement: New Solutions for Known Problems

The goal of DSR in the improvement quadrant is to create better solutions in the form of more efficient and effective products, processes, services, technologies, or ideas. Researchers must contend with a known application context for which useful solution artifacts either do not exist or are clearly suboptimal. Researchers will draw from a deep understanding of the problem environment to build innovative artifacts as solutions to important problems. The key challenge in this quadrant is to clearly demonstrate that the improved solution genuinely advances on previous knowledge. Several examples of improvement DSR projects are listed below.

- Better data mining algorithms for knowledge discovery (extending the initial ideas invented by Agrawal et al. (1993)), for example, (Fayyad et al. 1996; Zhang et al. 2004; Witten et al. 2011).
- Improved recommendation systems for use in e-commerce, for example, (Herlocker et al. 2004; Adomavicius and Tuzhilin 2005).
- Better technologies and use strategies for saving energy in IT applications, for example, (Donnellan et al. 2011; Watson and Boudreau 2011).
- Improved routing algorithms for business supply chains, for example, (van der Aalst and Hee 2004; Liu et al. 2005).

Improvement DSR is judged first on its ability to clearly represent and communicate the new artifact design. The presentation will show how and why the new solution differs from current solutions. The reasons for making the improvement should be formally grounded in kernel theories from the Ω knowledge base. Once the design improvement is described, then the artifact must be evaluated to provide convincing evidence of its improvement over current solutions. Improvement may be in the form of positive changes in efficiency, productivity, quality, competitiveness, market share, and other quality measures, depending on the goals of the research.

In the improvement quadrant, DSR projects make contributions to the Λ knowledge base in the form of artifacts at one or more levels as described in Table 22.2. Situated instantiations (Level 1) are often constructed to evaluate the level of improvements in comparison with instantiations of the existing solution artifacts. As appropriate, more general artifacts (Level 2) in the form of constructs, methods, models, and design principles are proposed as research improvements. In addition, new Λ knowledge may be formulated as midrange design theory (Level 3) as a result of improved understanding of the problem and solution spaces. Furthermore, the evaluations of the improved artifact may lead to knowledge contributions to the Ω knowledge base in the form of expanded understanding of the kernel theories or the development of new behavioral theories of the artifact in use.

22.3.4.3 Exaptation: Known Solutions Extended to New Problems

Original ideas often occur to individuals who have training in multiple disciplines of thought. Such training allows interconnections and insights among the fields to result in the expropriation of artifacts in one field to solve problems in another field. Thus, there may exist a research situation in which existing artifacts are available but are suboptimal. However, effective artifacts may exist in related problem areas that may be adapted or, more accurately, exapted* to the new problem context. In this quadrant are contributions where design knowledge that already exists in one field is extended or refined so that it is extendable to some new application area. In exaptation, one needs to justify that the extension of known design knowledge into a new field is nontrivial and interesting.

This type of research is common in ICTS fields, where new technology advances inspire new applications and a consequent need to test or refine prior ideas. Often, these new advances open opportunities for the exaptation of theories and artifacts to new fields. Several classic exemplars of exaptation in ICTS research are as follows:

- Codd's exaptation of relational mathematics to the problem of database systems design, which led to relational database concepts, models, methods, and instantiations (Codd 1970, 1982).
- Berners-Lee's (2000) original concept of the WWW was one of simply sharing research documents in a hypertext form among multiple computers. Soon, however, many individuals saw the potential of this rapidly expanding interconnection environment to exapt applications from old platforms to the WWW platforms. These new Internet applications were very different from previous versions and added many new artifacts to Λ knowledge.
- Berndt et al. (2003) research on the CATCH data warehouse for health-care information. Well-known methods of data warehouse development (e.g., Inmon 1992) were exapted to new and interesting areas of health-care systems and decision-making applications.

In the exaptation quadrant, similar to the improvement quadrant, DSR can make contributions to the Λ knowledge base in the form of artifacts at all three levels, as appropriate, to the research project goals. Ω-knowledge contributions may also be produced via a greater understanding of the new artifacts in use.

22.3.4.4 Routine Design: Known Solutions for Known Problems

Routine design occurs when existing knowledge for the problem area is well understood and when existing artifacts are used to address the opportunity or question. Research opportunities are less obvious, and these situations rarely require DSR to solve the given problem. In this quadrant is work that would not normally be thought of as contributing to research because existing knowledge is applied in familiar problem areas in a routine way. However, routine work may in some cases lead to surprises and discoveries, but, in such cases, these discoveries will likely involve moving the research to one of the other quadrants.

It is important that high-quality professional design or system building be clearly distinguished from DSR. Professional design is the application of existing knowledge to organizational problems, such as constructing a financial or marketing IS using "best practice" artifacts (constructs, models, methods, and instantiations) that exist in the knowledge base. The key differentiator between professional design and DSR is the clear identification of contributions to the Ω and Λ knowledge bases and the communication of the contribution to the stakeholder communities.

In the early stages of a discipline, or with significant changes in the environment, each new artifact created for that discipline or changed environment is "an experiment" that "poses a question to nature" (Newell and Simon 1976, p. 114). Existing knowledge is used where appropriate; however, often the requisite knowledge is nonexistent. In other words, the knowledge base is inadequate. Reliance on creativity and trial-and-error searches are characteristic of such research efforts. Because DSR results are codified in the knowledge base, they have the potential to become "best practices." Professional design and system building then become the routine application of the knowledge base to known problems.

---

* In biological evolution, *exaptation* is the adaptation of a trait for a different purpose than its original purpose. The classic example, featured in Gould and Vrba (1982), is the exaptation of bird feathers to the purposes of flight from the original purposes of temperature regulation.

22.4 Summary and Design Science Research Challenges

DSR is a widely practiced form of enquiry due to the natural desire of researchers to improve things. For some it is not enough to study and understand why nature is as it is, but they want to know how to improve the way it is. DSR attempts to focus creativity into the design and construction of artifacts that have proven utility in application contexts.

Design as a research paradigm focuses on the construction and evaluation of novel artifacts that enable the solution of important problems for which extant theory (descriptive knowledge) and current designs (prescriptive knowledge) are inadequate. It utilizes theory and design knowledge but is fundamentally a creative activity in which new knowledge is acquired through the building and use of novel problem-solving artifacts. That knowledge must be tested through the evaluation of the produced artifact. Rigorous testing results in a demonstration that the design can be utilized to solve real problems.

Designs have no special dispensation from the laws of nature. Hence, natural science research methods are utilized in a natural research cycle (Figure 22.1) to gain an understanding of why a design works and to specify contingencies upon it. The resultant theories provide principles that can then become part of the "best practice" and new research opportunities kick off a new cycle of DSR.

As the ICTS fields move forward to achieve a fuller appreciation and use of DSR, there exist a number of exciting challenges to be faced. To conclude this chapter, a few of these challenges are summarized.

22.4.1 DSR and Wicked Problems

DSR in ICTS addresses what are considered to be *wicked problems* (Rittel and Webber 1984; Brooks 1987), that is, those problems characterized by

- Unstable requirements and constraints based on ill-defined environmental contexts
- Complex interactions among subcomponents of the problem
- Inherent flexibility to change design processes as well as design artifacts (i.e., malleable processes and artifacts)
- A critical dependence upon human cognitive abilities (e.g., creativity) to produce effective solutions
- A critical dependence upon human social abilities (e.g., teamwork) to produce effective solutions

Thus, the execution of effective DSR calls upon refined cognitive and social skills in the research team. DSR requires environments that support and enhance such human skills.

22.4.2 Design as Research versus Researching Design

DSR has been interpreted as including two distinctly different classes of research—"design as research" and "researching design." While this chapter focuses on the former class of research, it is important to recognize the existence and importance of both types of research.

Design as research encompasses the idea that doing innovative design that results in clear contributions to the knowledge base constitutes research. Knowledge generated via design can take several forms including constructs, models, methods, instantiations, and design theory. DSR projects are often performed in a specific application context, and the resulting designs and research contributions may be clearly influenced by the opportunities and constraints of the given application domain. Additional research may be needed to generalize the research results to broader domains. Design as research, thus, provides an important strand of research that values research outcomes that focus on improvement of an artifact in a specific domain as the primary research concern and, then, seeks a broader, more general, understanding of theories and phenomena surrounding the artifact as an extended outcome.

Researching design shifts the focus to a study of designs, designers, and design processes. The community of researchers engaged in this mode of research was organized under the umbrella of the Design

Research Society starting as early as the mid-1960s. Because of their focus on methods of designing, they have been able to articulate and follow the goal of generating domain-independent understanding of design processes. Their investigations have been focused largely in the fields of architecture, engineering, and product design. Although it is difficult to provide unambiguous and universally accepted definitions of design processes, working definitions suggest designing is an iterative process of planning, generating alternatives, and selecting a satisfactory design. Examples of work from this stream, therefore, include use of representations and languages (Oxman 1997), use of cognitive schemas (Goldschmidt 1994), and theoretical explorations (Love 2002).

Although similarities are many, the two fields of design study have been different in their focus and trajectory. Of the differences, three are most visible. First, design as research emphasizes the domain in which the design activity will take place, placing a premium on innovativeness within a specific context. In contrast, researching design emphasizes increased understanding of design methods often independent of the domain. Second, the domains of study for the first subfield have typically been the ICTS disciplines as opposed to architecture and engineering for the second. Finally, the closest alliances from the design as research have been formed with disciplines such as computer science, software engineering, and organization science. Researching design is more closely allied with cognitive science and professional fields such as architecture and engineering.

22.4.3 Innovation and DSR

An emerging focus on design thinking has driven transformations in organizations to achieve competitive advantages in the marketplace via the introduction of new products and services (Brown 2009; Martin 2009; Verganti 2009). The need for continual innovation to support future growth and sustainability has led to organizations devoting significant attention and resources to innovation activities.

An intriguing question is how the research fields of DSR and innovation are related. Both disciplines continue to evolve rapidly, with research agendas driven by researchers with deep expertise in either of the two fields. However, little attention has been given to identifying how the concepts of DSR and innovation impinge on each other. A recent investigation based on a case study of the innovation process implemented in Chevron suggests that there are key insights that can be drawn from the DSR guidelines that can potentially impact and improve organizational innovation processes (Anderson et al. 2011). A future objective is to also explore the impacts going in both directions: How can successful innovation processes inform and improve DSR activities? How can successful DSR processes inform and improve innovation activities?

22.4.4 Presentation of DSR Contributions

The DSR community lags the behavioral science community in having useful templates for communicating knowledge contributions. Helpful advice for preparing behavioral research for publication is provided by numerous authors including Bem (2003) for psychology, Perry (1998) for marketing, and Neumann (2006) for the social sciences in general.

Some basic guidance on presenting DSR knowledge contributions is given in work including Hevner et al. (2004), Hevner and Chatterjee (2010), Peffers et al. (2008), Vaishnavi and Kuechler (2008), Zobel (2005), and Sein et al. (2011). This guidance is not, however, highly detailed. In addition, the treatment of publication schema for DSR is scant. In part, this problem arises because much of the work on DSR sees the development of the artifact itself as the whole point and, accordingly, there has been little emphasis on what it means to make a contribution to generalized knowledge. Gregor and Hevner (2013) address this gap in a recent paper on the positioning and presenting of DSR with a focus on knowledge contributions.

A particular issue is adequately describing a design artifact in a scientific journal. Describing the complexities of an artifact within the confines of a journal article is not easy. The level of design detail in a journal paper will vary based on the application domain, the designed artifact, and the audience to which the presentation is made. The increased use of on line appendices, data repositories, and executable systems to supplement DSR presentations in journals will become more pervasive for these presentation challenges.

22.4.5 DSR Is Perishable

Rapid advances in ICTS technology can invalidate DSR results before they are implemented effectively in the business environment or, just as importantly to managers, before adequate payback can be achieved by committing organizational resources to implementing those results. Just as important to researchers, ICTS design results can be overtaken by new technology advances before they even appear in the research literature due to the delays of academic publication.

Similarly, social and organizational environments also change rapidly. Thus, "design theories" are also perishable. That is, they are subject to change as the social reality changes. There may be, in fact, no immutable "laws of organizational design" to be discovered and codified. The significance of this observation for academic researchers is that they must constantly challenge the assumptions that have characterized research in the management disciplines. They must also recognize that creativity, change, innovation, and qualitative novelty are a significant part of the phenomena they study.

22.4.6 Artifact Utility Evaluation in DSR

Current thinking in DSR defines *utility* of an artifact as the primary research goal. In this context, the close relationship of utility to practical *usefulness* is emphasized. The choice of usefulness as the preeminent dependent variable for DSR is well established in ICTS literature (DeLone and McLean 1992, 2003). Given these strong connections to existing well-established research streams, does it even make sense to question if usefulness should *always* be our central criteria for evaluating design?

A recent research direction questions this accepted view. Gill and Hevner (2011) state that the search for the dependent variable in DSR requires some rethinking. They propose a pair of alternative dependent variables: *design fitness* and *design utility*. In the case of fitness, the focus is on its biological meaning—the ability of an entity to reproduce itself and evolve from generation to generation. In the case of utility, rather than viewing it as being roughly equivalent to usefulness, the focus is on its meaning in fields such as economics and decision sciences, where it serves as the basis for ranking decision alternatives. Naturally, usefulness plays an important role in determining both fitness and utility. Neither of these variables, however, is solely determined by usefulness. Thus, a new understanding of the relationship between the three variables via a new fitness-utility model would complement current thinking and provide important insights into the nature of DSR.

22.4.7 Invention and DSR

A genuine new invention is a difficult goal for DSR research projects and we can expect few contributions to fall in the invention quadrant. However, exploration for new ideas and artifacts should be encouraged regardless of the hurdles. Invention comes about when the existing knowledge base is insufficient for design purposes and designers must rely on intuition, experience, and trial-and-error methods. A constructed artifact embodies the designer's knowledge of the problem and solution. In new and emerging applications of technology, the artifact itself represents an experiment. In its execution, we learn about the nature of the problem, the environment, and the possible solutions—hence the importance of developing and implementing prototype artifacts (Newell and Simon 1976). Thus, editors and reviewers should be more open to claims of "newness" in regard to problem and solution domains. However, rigorous surveys of existing knowledge bases must be done before any claims of inventions in DSR are made.

References

Adomavicius, G. and Tuzhilin, A. 2005. Toward the next generation of recommender systems: A survey of the state-of-the-art and possible extensions, *IEEE Transactions of Knowledge and Data Engineering* (17:6), 734–749.

Agrawal, R., Imielinski, T., and Swami, A. 1993. Mining association rules between sets of items in large databases, *Proceedings of the 1993 ACM SIGMOD Conference*, Washington, DC, May.

Anderson, J., Donnellan, B., and Hevner, A. 2011. Exploring the relationship between design science research and innovation: A case study of innovation at Chevron, *Proceedings of the European Design Science Symposium (EDSS 2011)*, Leixlip, Ireland, October.

Bem, D. 2003. Writing the empirical journal article, in: *The Compleat Academic: A Practical Guide for the Beginning Social Scientist* (2nd edn.), Darley, J. M., Zanna, M. P., and Roediger III, H. L. (eds.), Washington, DC: American Psychological Association.

Benbasat, I. and Zmud, R. 1999. Empirical research in information systems: The practice of relevance, *MIS Quarterly* (23:1), 3–16.

Berndt, D., Hevner, A., and Studnicki, J. 2003. The CATCH data warehouse: Support for community health care decision making, *Decision Support Systems* (35), 367–384.

Berners-Lee, T. 2000. *Weaving the Web: The Original Design and Ultimate Destiny of the World Wide Web*, New York: Harper Business, Inc.

Brooks, F. P. Jr. 1987. No silver bullet: Essence and accidents of software engineering, *IEEE Computer* (20:4), April, 10–19.

Brown, T. 2009. *Change by Design*, New York: HarperCollins Publishers.

Bunge, M. 1998. *Philosophy of Science, Vol. 2: From Explanation to Justification*, New Brunswick, NJ: Transaction Publishers.

Codd, E. F. 1970. A relational model of data for large shared data banks, *Communications of the ACM* (13:6), 377–387.

Codd, E. F. 1982. Relational database: A practical foundation for productivity (The 1981 Turing Award Lecture), *Communications of the ACM* (2:25), 109–117.

Cross, N. 2001. Designerly ways of knowing: Design discipline vs. design science, *Design Issues* (17:3), 49–55.

Csikszentmihalyi, M. 1996. *Creativity: Flow and Psychology of Discovery and Invention*, New York: HarperCollins.

Davenport, T. and Harris, J. 2007. *Competing on Analytics: The New Science of Winning*, Cambridge, MA: Harvard Business School Publishing.

DeLone, W. and McLean. E. 1992. Information systems success: The quest for the dependent variable, *Information Systems Research* (3:1), 60–95.

DeLone, W. and McLean, E. 2003. The DeLone and McLean model of information systems success: A ten-year update, *Journal of Management Information Systems* (19:4), 9–30.

Donnellan, B., Sheridan, C., and Curry, E. 2011. A capability maturity framework for sustainable information and communication technology, *IEEE IT Professional* (13:1), 33–40.

Dubin, R. 1978. *Theory Building* (Rev. edn.), New York: The Free Press.

Fayyad, U., Piatetsky-Shapiro, G., and Smyth, P. 1996. From data mining to knowledge discovery in databases, *AI Magazine* (17:3), 37–54.

Gill, G. and Hevner, A. 2011. A fitness-utility model for design science research, *Proceedings of the Design Science Research in Information Systems and Technology (DESRIST 2011)*, Milwaukee, WI, May.

Goldschmidt, G. 1994. On visual thinking: The vis kids of architecture, *Design Studies* (15:2), 158–174.

Gould, S. and Vrba, E. 1982. Exaptation—A missing term in the science of form, *Paleobiology* (8:1), 4–15.

Gregor, S. 2006. The nature of theory in information systems, *MIS Quarterly* (30:3), 611–642.

Gregor, S. and Hevner, A. 2013. Positioning and presenting design science research for maximum impact, *Management Information Systems Quarterly*, (37:2), 337–355.

Gregor, S. and Jones, D. 2007. The anatomy of a design theory, *Journal of the Association of Information Systems* (8:5), 312–335.

Herlocker, J., Konstan, J., Terveen, L., and Riedl, J. 2004. Evaluating collaborative filtering recommender systems, *ACM Transactions on Information Systems* (22:1), 5–53.

Hevner, A. 2007. A three cycle view of design science research, *Scandinavian Journal of Information Systems* (19:2), 87–92.

Hevner, A. and Chatterjee, S. 2010. *Design Research in Information Systems*, New York: Springer Publishing.

Hevner, A., March, S., Park, J., and Ram, S. 2004. Design science in information systems research, *MIS Quarterly* (28:1), 75–105.

Iivari, J. 2007. A paradigmatic analysis of information systems as a design science, *Scandinavian Journal of Information Systems* (19:2), 39–64.

Inmon, W. 1992. *Building the Data Warehouse*, New York: Wiley and Sons Publishers.

Liu, J., Zhang, S., and Hu, J. 2005. A case study of an inter-enterprise workflow-supported supply chain management system, *Information and Management* (42:3), 441–454.

Love, T. 2002. Constructing a coherent cross-disciplinary body of theory about designing and designs: Some philosophical issues, *International Journal of Design Studies* (23:3), 345–361.

March, S. and Smith, G. 1995. Design and natural science research on information technology, *Decision Support Systems* (15), 251–266.

Martin, R. 2009. *The Design of Business: Why Design Thinking is the Next Competitive Advantage*, Cambridge, MA: Harvard Business Press.

Merton, R. 1968. *Social Theory and Social Structure* (Enlarged edn.), New York: The Free Press.

Mokyr, J. 2002. *The Gifts of Athena: Historical Origins of the Knowledge Economy*, Princeton, NJ: Princeton University Press.

Nagel, E. 1979. *The Structure of Science Problems in the Logic of Scientific Explanation*, Indianapolis, IN: Hackett Publishing Co.

Neumann, W. L. 2006. *Basics of Social Research: Qualitative and Quantitative Approaches* (2nd edn.), Columbus, OH: Pearson.

Newell, A. and Simon, H. 1976. Computer science as an empirical inquiry: Symbols and search, *Communications of the ACM* (19:3), 113–126.

Nunamaker, J., Chen, M., and Purdin, T. 1990–1991. Systems development in information systems research, *Journal of Management Information Systems* (7:3), 89–106.

Nunamaker, J., Dennis, A., Valacich, J., Vogel, D., and George, J. 1991. Electronic meetings to support group work, *Communications of the ACM* (34:7), 40–61.

Orlikowski, W. and Iacono, C. 2001. Research commentary: Desperately seeking the 'IT' in IT research— A call to theorizing the IT artifact, *Information Systems Research* (12:2), June, 121–134.

Oxman, R. 1997. Design by re-representation: A model of visual reasoning in design, *Design Studies* (18:4), 329–347.

Peffers, K., Tuunanen, T., Rothenberger, M., and Chatterjee, S. 2008. A design science research methodology for information systems research, *Journal of MIS* (24:3), 45–77.

Perry, C. 1998. A structured approach for presenting theses, *Australasian Journal of Marketing* (6:1), 63–85.

Purao, S. 2002. Design research in the technology of information systems: Truth or dare, *GSU Department of CIS Working Paper*, Atlanta, GA: Georgia State University.

Rittel, H. and Webber, M. 1984. Planning problems are wicked problems, in: *Developments in Design Methodology*, N. Cross (ed.), New York: John Wiley & Sons.

Scott-Morton, M. 1967. Computer-driven visual display devices—Their impact on the management decision-making process, Unpublished PhD thesis, Harvard Business School.

Sein, M., Henfredsson, O., Purao, S., Rossi, M., and Lindgren, R. 2011. Action design research, *Management Information Systems Quarterly* (35:1), 37–56.

Simon, H. 1996. *The Sciences of the Artificial* (3rd edn.), Cambridge, MA: MIT Press.

Sutton, R. I. and Staw, B. M. 1995. What theory is not, *Administrative Sciences Quarterly* (40:3), 371–384.

Takeda, H., Veerkamp, P., Tomiyama, T., and Yoshikawam, H. 1990. Modeling design processes, *AI Magazine*, (Winter), 37–48.

Vaishnavi, V. and Kuechler, W. 2008. *Design Science Research Methods and Patterns: Innovating Information and Communication Technology*, Boston, MA: Auerbach Publications.

Van der Aalst, W. and van Hee, K. 2004. *Workflow Management: Models, Methods, and Systems*, Cambridge, MA: MIT Press.

Verganti, R. 2009. *Design-Driven Innovation: Changing the Rules of Competition by Radically Innovating What Things Mean*, Cambridge, MA: Harvard Business Press.

Walls, J., Widemeyer, G., and El Sawy, O. 1992. Building an information system design theory for vigilant EIS, *Information Systems Research* (3:1), 36–59.

Watson, R. and Boudreau, M. 2011. *Energy Informatics*, Athens; GA: Green ePress.

Witten, I., Frank, E., and Hall, M. 2011. *Data Mining: Practical Machine Learning Tools and Techniques* (3rd edn.), Burlington, MA: Morgan Kaufmann.

Zhang, H., Padmanabhan, B., and Tuzhilin, A. 2004. On the discovery of significant statistical quantitative rules, in *Proceedings of ACM International Conference on Knowledge Discovery and Data Mining*, Seattle, WA, pp. 374–383.

Zobel, J. 2005. *Writing for Computer Science* (2nd edn.), London, U.K.: Springer-Verlag.

23

Identifying Opportunities for IT-Enabled Organizational Change

Matti Rossi
Aalto University
School of Business

Juho Lindman
Hanken School of
Economics

23.1 Introduction

Process modeling and designing transformative information technology (IT) were once seen as panacea for implementing changes in organizations when they modernize their operations (Williams, 1967). Manager's role was seen primarily as modeling of processes and creating interventions to systems that resulted in measurable impact (Hammer and Champy, 1993) or finding ways to increase the amount and transformative capability of IT in the organization (Venkatraman, 1994). The problems of failed and misused IT were identified as a strategic management issue, including issues such as poor alignment between business strategy, IT strategy, organizational infrastructure and processes, and IT infrastructure and processes (Henderson and Venkatraman, 1993). However, the causal link often implied in these theories that explains coexistence of IT implementations and organizational change is deeply problematic to the degree that the direction of the causal link and the identities of the actors are at times questioned (Markus and Robey, 1988). One way of addressing this issue is the situational practice lens, which studies the transformation of IT and organizational structure and practice together as a process that involves institutional environment and thus is complex and difficult to predict (Orlikowski, 1996).

When systems have grown larger, it has become clear that many integrated enterprise systems (ESs) act more like organizational cement (Davenport, 2000b). Systems that were implemented and maintained to allow for change have actually become harnesses that slow down the change. Complex integrated systems develop institutional inertia, a legacy, on their own (Hanseth and Lyytinen, 2010). This is natural; the systems have grown so large that they are the core infrastructure of modern multinational organizations. However, organizations are facing increasing pressures for change and, thus, it is not optimal that the main information systems are inhibitors rather than contributors to the change (Markus, 2004). This state of affairs is increasingly unsustainable, and there could be ways of changing it for the better.

Implementing change in organizations is notoriously difficult. After failed efforts for organizational change, it is not unusual to introduce a new information system as a way of introducing process changes. This is a way to externalize the organizational problems into a project, where the costs of the organizational change can be attributed to the management of the project. If this is the starting point for an IT project, the process can easily become an uphill battle. However, if the modernization of information systems is seen as and is planned to be an enabler of future change, the road ahead is simpler (Upton and Staats, 2008). Maintaining shared set of beliefs underlying the implementation project and its rationality while aiming at inclusiveness of the different interest groups in the organization is a critical issue in delivering the desired organizational change (Willis and Chiasson, 2007). Unlike in the past, most employees, customers, and business partners will be users of the corporate systems, and thus, their input on the requirements and to the organizational change is critical (Mathiassen, 2002). There are a few indicators suggesting that future users will be different in terms of skills and expectations. In general, a likely further increase in specialization and professionalization will be a reaction to the growing complexity organizations face from their internal and external stakeholders. More and more professionalization will include using specific software systems, such as human resource, customer relationship management, or supply chain management packages. Also, many future users are digital natives so that dealing with software will be natural for them (Vodanovich et al., 2010). Hence, they can be expected to not only accept but even demand for the extensive use of elaborate software systems in their work environment. At the same time, future users can be expected to be more demanding. This can become a challenge, if the systems fail to provide the promised functionality, as often seems to be the case in the early phases of any implementation process. Users may require smart—and fancy—user interfaces as consumer products feature them. Therefore, they will probably more profoundly resist against being bothered with information they do not need—or against performing steps that are redundant, meaningless, or a threat to their autonomy.

Alignment is a process where business requirements, organizational and cultural elements, and technological solutions are brought together. Conventional wisdom approached alignment and building the applications using the waterfall model: First (real world) business requirements are mapped out and then the technical solutions are designed and implemented to those problems. This might require several iterations. New approaches highlight the variety in the uses of these IT systems such as workarounds, conflicts in the requirements for the systems, and improvisation and creativity (Macredie and Sandom, 1999). Business architecture is a concept that ideally connects high-level strategy and business models to the digital execution platforms. Both practitioners (Ross et al., 2006; Upton and Staats, 2008) and researchers (Hanseth and Lyytinen, 2010) see a flexible architecture as a key enabler of change in future digital organizations.

In this chapter, we look at the ways to allow for what Markus (2004) calls technochange to happen more often. Technochange means that instead of limiting scope of change to process and technology modification, organizations should link the technical elements to the social, personal, and political aspects of the units and individuals who are undergoing the change. First, we outline ways of making the core ESs of an organization to work. Then we describe how the business architecture based on the ESs can be used to change work processes and customer facing processes through the use of mobile systems and other lighter interfaces.

23.2 Enterprise IT and Organizational Change

The relationship between IT and organizational change has been widely recognized (Boudreau and Robey, 1999). Changing roles and responsibilities or changed distribution of power, resulting from the new information system, may cause resistance to change in an organization (Markus, 1983).

Large integrated system implementations change the status quo of the target organization: Some departments and individuals gain more than others. Engaging and coordinating different individual and group interests require often subtle political maneuvering. The scoping of the project is a critical concern here: Large projects gather more inertia, and the costs of enrollment are much higher than in the smaller and more incremental projects.

ESs have been defined in many ways in prior research. We consider ESs to include the enterprise resource planning (ERP) system functions and all the other applications providing an integrated information system for most functions of a company (Davenport, 2000a). According to Davenport (Davenport, 1998), an ES seamlessly integrates all the information flowing through the company (e.g., financial, accounting, human resources, supply chain management, and customer information) into a centralized database. Thus, an ES provides a technology platform that enables a company to integrate and coordinate business processes and to share information across the company. ESs also allow allocation and coordination of resources across time zones and geographical locations while keeping the data available and centralized. In Zuboff's words, "activities, events, and objects are translated into and made visible by information" (Zuboff, 1988). Thus, an ES aims to organize information processing in an organization around standardized processes or best practices and offers uniform tools to access the data (Davenport, 1998). Best of breed solutions may also enforce standardized processes that come directly from the vendor company. Informational visibility of the objects also changes organizational information accessibility, if it is used, for example, in performance monitoring.

Although ESs are seen to increase organizational efficiency, this depends on the context and whether the associated changes in business processes can be achieved. Light et al. (2001) argue that the benefits may realize only after some advanced modules, such as customer resource management, are implemented. Organizational change either before or during the implementation is seen as beneficial for the implementation (Davenport et al., 2004). In this chapter, we view ES in use as a change agent and a way of enforcing the changed practices.

During a company's organizational transformation, ESs may adopt different roles. ES should become the digital infrastructure (or architecture) of the company (Tilson et al., 2010) and allow for changes to strategy and organizational structure. In the following section, we assume that ES is developed through an open architecture, which allows external systems to use the ES data and processes.

23.3 How to Seek for Opportunities?

Markus states that "a reason that managers use IT to drive organizational change is that starting major cross-functional changes without an IT focus does not work in many organizational cultures" (Markus, 2004). Tools for cultural change [traditionally called critical success factors (Al-Mashari et al., 2003)] include senior management support, user involvement in system design, breaking of departmental silos, and improve staff motivation and managerial decision making (Jackson and Philip, 2010). These kinds of tactics are by themselves not enough (ibid), but they should be both accompanied with technological change and ongoing attention to issues that typically emerge when cultural and technological changes occur.

In what follows, we will note some interesting general technological developments that we consider important. We complete the chapter with specific examples of the interplay of organizations and two technological changes taking place due to (1) increased mobility and (2) increased openness.

Radio frequency identification tags (Curtin et al., 2007; Weinstein, 2005) provide a new means for data retrieval and storage. Their main usage is related to surveillance and identification of items remotely from a short range. The price of sensors and tags has gone down, and their usage for different kinds of automated tracking scenarios has increased vastly. Examples include integrated tasks such as workplace monitoring (Kaupins and Minch, 2005), supply chain management (Chow et al., 2006), or parcel tracking. New electronic and mobile integrated payment systems offer more real-time transactions and monitoring of funds and investments (Mallat et al., 2004, 2009). In addition, new devices provide a means for secure physical identification, which then enables organizations and individual's ways to conduct legal transactions (e.g., payment or electronic signature), which are not constrained by the paper trail. Cloud-based access to organizations resources calls into question conventional truths about control and management of organizations databases as well as their content and different kinds of digital assets.

23.4 Possibilities Based on Specific New Technologies

In this section, we briefly outline the possibilities of two specific technologies to act as change agents. First is a process technology, mobile information systems, and second is what is called the steam engine of the twenty-first century, open data. Both have possibilities for radically altering the current ways of working in private and public enterprises and affecting the ways that people work and use services.

23.4.1 Mobile Systems as Change Agents

One clear possibility for searching new opportunities is through augmenting ES with mobile technology. Mobile technologies are expected to enhance business efficiency by distributing information to the workforce and by offering new communication channels with customers (Leung and Antypas, 2001). In earlier studies, mobile applications have been seen suitable for truck drivers, service teams, traveling salesmen, and so forth (Berger et al., 2002; Bruegge and Bennington, 1995). There are, however, limitations in each case, related to the technology-based change. These limitations are resources, timetable, the urgency of the need, current infrastructure, such as IT equipment and software, development of mobile technology, the pace of the change, organizational culture, and, in particular, old habits and resistance to change. Furthermore, mobile fieldwork makes it challenging for management to coordinate sales people, particularly in terms of information transfer (Watson-Manheim and Bélanger, 2002). Recent advances in mobile technology, such as LTE networks and handheld tablet computers utilizing these networks, are rapidly expanding the use of mobile devices in different types of enterprises. Examples of uses that have reached a critical mass are the aforementioned mobile payments, the use of tablets in cockpit (Murphy, 2011), and health care (Basole et al., 2012).

23.4.2 Open Data as a Change Agent

Increased availability of data about organizations and society holds untapped potential for change (Shadbolt et al., 2006). Open data refer to a specific subset of all data—data that are "openly" made available. Usually created by various levels of government, open data are based on the ideas that if more data are made available, better and more transparent decisions are made in organizations. Open data are defined as follows: "A piece of content or data is open if anyone is free to use, reuse, and redistribute it— subject only, at most, to the requirement to attribute and/or share-alike." (http://opendefinition.org/). The idea is that the organization, which releases the data, intends that the data are used as a basis for new applications and services. The ideology underlying open data is similar to that of open source (Lindman and Tammisto, 2011). Open source is about the applications in a society, open data are about data. These two phenomena also have strong links historically, culturally, and technically, although many open data activists do not endorse this comparison. Work of Kuk and Davies (2011) identifies several artifacts and stages in the service development, which builds on open data sets.

Now let's turn from viewing the organizations as a consumer of data and focus on organizations as a source of open data. *Private open data* means data collected and owned by a commercial actor that are made available for reuse over the Internet. What could these type of data be about? The objects described in the data sets are different but can be divided into (1) data about the organization in question and (2) data about the environment of the organization. The question becomes important in the discussion about what data sets are created and how. When we take the service design perspective, we need to know what the benefits offered by opening the data to its ultimate users are. This, in turn, requires theoretical understanding of the relation between the digital data artifacts and organizations. Especially of interest here are the organizational goals that lead to opening up the datasets.

Another concern is the tension between organizational requirements of data and the interests and claims to the authority of individual users. This means that the increased openness of data challenges the current ways of controlling information resources. The availability of more an open digital archive of (organizational) transactions makes certain issues more visible and hides others [for a thorough discussion, see Marton (2011)]. Radical redistribution of organizational knowledge might occur if the digital trail of the actors is available to more users in an organization. This redistribution has its winners and losers but could in correct circumstances result in more transparent decision making and thus better or at least more deliberative decisions. Unfortunately, it could also result in more technological surveillance (Fuchs, 2011; Zuboff, 1988). Surveillance theory ties in the organizational authority structures, new competence requirements, and the digital trail of the actors archived in the databases. One of the main sources of this information would be accounting systems, data warehouses, and integrated ERP solutions (Zuboff, 1988).

Another epistemological question is related to the malleability of the data. How do we know whether the data we are relying are trustworthy? One answer to this question is to revert to the old open source saying "more eyeballs make bugs shallow" and rely on the wisdom of the crowds. Additionally, if the processes that bring about the data are transparent, the likelihood of the data also being valid and trustworthy increases.

Data are also an interesting kind of digital object in its own right as it represents something, but often this representation is not trivial. Data do not necessarily include a description of a process by which they came about, but their provenance is still tied to the data gathering, storage system, retrieval classification, and organizational usage. One industry facing substantial changes in regard to data management is the newspaper media.

23.4.3 Concrete Case Example: Augmenting ERP with Mobile Technology

In an action research study (Rossi et al., 2007), we studied the re-engineering of field sales processes of Amer, a Finnish tobacco company operating in Estonia until the end of 2004, when the change of the owner of Amer's business unit in Estonia took place. The importance of mobilizing the workforce, adopting mobile technologies, and providing mobile information access integrated into the enterprise IT systems as well as applications for field force professionals is highlighted by Lyytinen and Yoo (2002), who outline the evolution from traditional enterprise IT toward ubiquitous IT. The case company employed mobile field sales personnel that worked in remote locations most of the time; some of them needed to come to the central office only once in a month.

To enhance customer support, the company decided to launch a project to solve the problems of work and document flows and to empower the field sales staff. It was decided early on that the work processes on the field sales have to be modeled so that the processes could be supported by advanced IT. The main objective was to create a solution to document production in the field. This would result in a faster document flow while allowing the dispersion of work in terms of time and place. The initial approach taken was based on process re-engineering utilizing mobile communications technologies. There were two main reasons for this: First, the mobility of the workers and the need to cut down the document cycles forced the company to seek solutions that can be used anywhere. Second, a real-time process was seen essential to overcome the process inefficiencies that interfered with client service.

The goal of this project was to streamline and shorten the ordering and billing cycles. Efficiency gains were sought through recording information at its source and delivering it immediately to the back-office systems of the company. The starting point was to be able to use mobile technologies by mobile employees in such a way that the data created in the field could be entered in real time to the ERP system of the company, thus replacing several error-prone manual steps with one data entry procedure.

23.4.3.1 The Case Company: Amer Tobacco

Amer Tobacco AS (hereinafter Amer) was a part of Amer Tobacco Ltd. that belonged to the Finnish concern Amer Group until the end of 2004. Amer had started its operations in Estonia in 1995, and it conducted its business through active field sales and deliveries of products to retailers and wholesalers.

Amer's clients in Estonia were small retail food businesses. Forty percent of sales were generated from wholesale and contract customers who were administered by key account managers. The rest (60%) of sales were generated from field sales. Amer's field sales personnel handled sales to stores, kiosks, and restaurants. The company served altogether around 1500 customers, two-thirds of which by the field sales representatives. For the most important store chains, deliveries were made from the central warehouse.

The operations of Amer were organized into business functions. Key business functions included sales, marketing, administration, and warehousing. Efforts had been made to simplify administration to the largest possible degree, and the same staff members often carried out several phases in different processes.

All Amer's operations were managed by an ERP system called Hansa. The Hansa modules in use were accounting, inventory, ledgers, purchase and sales orders, invoicing and cash register, and client management. Centralized data management in the Hansa system made working regimes more structured and reporting more standardized and thus more unambiguous. Extra reporting for the management purposes was prepared mainly with Microsoft's Office tools.

23.4.3.2 Old Field Sales Processes

All business processes in Amer evolved around the customer and customer service teams, to which the field sales representatives belonged. These field sales representatives operated in different parts of Estonia and were responsible for their clientele in the respective areas. They visited 15–25 customers a day, negotiated with customers with the objective of getting an immediate order, delivered the products to the customer, and invoiced the sale. The functions of the field sales representatives included keeping record of receivables from his or her customers. In case there had not been a payment, the objective of the visit was no longer sales, but collection of receivables.

Ordering, invoicing, and delivery from a van took place during the same customer contact. Invoicing was completed by writing an invoice by hand on a carbon copy paper pad. After closing the deals, the sales representatives sent invoices and other documents required for reporting purposes to the Amer office by mail once or twice a week. Invoicing data were handled and entered in the office to an ERP system by the office secretary.

23.4.3.3 Problems with the Old Processes

The main problem of the old order-and-delivery process was its long flow time. Firstly, because of the delay between the date of the invoice and the date it was entered to the ERP system, and because, for several clients, the payment term was short (typically 7 days), the company had often received payments before the invoice was even entered to the system. Collection of overdue receivables could start only after the keying in into the system was completed. A delay in the collection could further slow down the continuous business with the same customer, thus weakening the quality of customer service.

Secondly, entering sales information was time-consuming for both the field sales representatives as well as for the office staff. Furthermore, the same sales information was recorded twice: First, a sales representative issued a hand-written receipt to the customer and then, an office employee entered the information in the system. Entering the same data twice also created increased possibilities for errors. Delivering these hand-written receipts to the office created another factor weakening the flow. The fact that the receipts were delivered to the office only twice a week created a significant delay in recording information.

The process flow regularly contained operating irregularities that weakened the level of service quality. For instance, sales needed to be put on hold while taking inventory to ensure validity of the results, and reporting was even more delayed if the paper receipts were lost in transportation. It was not unusual for the check pads to be late from the printing house, and invoicing was interrupted. In the course of time, additional activities had been developed to correct these irregularities. An example of this was checking information on customer payments. Because of a delay in recording, sales representatives were forced to check payment information directly from the customer.

After receiving the invoice data from sales representatives, an office staff member entered it to the ERP system as a collection run. The same person also handled possible error messages before the confirmation run, after which the stock inventory information was updated in the EPR system.

23.4.3.4 Developing a Mobile Field Sales System

The solution needed middleware for efficient data transfer between the mobile terminals and the Amer Estonia office in Tallinn. Gateway software supplied by Klinkmann Ltd. and a GSM modem that connected the office PC into the GSM network and the terminal devices of sales representatives were installed. PC that handles data communication at the office was connected to the company's IT network and the ERP package.

The selected GSM control software was compatible with Microsoft Excel, which served as a bridge to the ERP system. The bridge allowed automatic securing of wireless communication messages (sent/received) and indexing of messages, which allowed handling of messages in case of operational malfunctions. If and when there were changes in the order–delivery process or its functions, configurations to the solution would have been relatively easy to complete.

23.4.3.5 The Mobile Sales Solution

The chosen mobile terminals, Nokia Communicators (NCs), include a standard user interface based on Excel. NCs were chosen over portable computers because of their smaller size and corresponding significantly better portability, as well as for their greater battery capacity.

For enquiries and orders, the NC had a set of Excel-based forms to be filled out by the sales representatives. After printing an invoice, the field sales representative sent a copy of it to the main office. Control gateway software transferred the Excel spreadsheet data as SMS messages over the GSM network. Only the data that had been changed was transferred, limiting the length and number of the SMS messages. To keep the message short, the SMS message included only coded information: client number, invoice number, price list code, date, payment time, product code and quantity, and the sender's ID code. This was important for keeping the costs down. The server software combined the individual messages and updated the predefined cells in the Excel spreadsheet with data received. Information from the Excel spreadsheets was then imported into Amer's ERP system.

In the new system, during a customer sales process, the order could be stored directly on the NC terminal of the field sales representative, eliminating the need to make notes separately on a paper. Changing invoicing details was also made simpler and quicker. Dispatch of invoicing information remained double: After the invoice was printed out, the invoice data was sent to the office in real time as an SMS message through the mobile GSM network. The printed invoice document signed by the customer, however, was still sent by mail to the office afterward, as a delivery confirmation.

23.4.3.6 Business Impacts of the Project

In terms of event flow, the new and old processes were very similar. As the modifications in the process flow were not radical, the change was well accepted and employed by the staff quickly. This made it possible to complete the pilot project in the planned schedule of 5 months. The implemented IT solution removed a lot of inefficient and overlapping functions.

The new solution significantly reduced document handling at the back-office (see Table 23.1). The only remaining manual work was archiving, since the invoice information was entered directly in real time by the mobile users and transferred to the back-end systems automatically. This significantly reduced typing errors as well as the amount of clerical work needed. Another enhancement was the single direction of the

TABLE 23.1   Measures of the Project Success

Measure	Before	After
Invoice verification	7–9 days	1 day
Billing cycle	Week	3 days

document flow. Even though invoice copies still needed to be sent to the office, the sales representatives now had more flexibility in choosing the most convenient time of a week for the delivery. The information about the orders and invoices was entered into the ERP system in real time. Thus, the sales and stock reports were updated practically in real time, and the representatives could review the regional stock levels anytime throughout the day. Furthermore, since the office assistants no longer needed to manually enter the data, they could concentrate their efforts in collection of receivables as well as other duties.

23.4.3.7 Evaluation and Learning from the Project

The order–delivery process was the most important and critical process for the case company. The reported process re-engineering project took into account process risks, and the old model was maintained in case the results of the re-engineering would turn out to be unsuccessful. The focus in the project was solely on increasing the efficiency of invoice documentation, but it also enhanced the process transparency and overall performance. This is largely in line with what (Schierholz et al., 2007) propose with their method for mobilization of CRM: focus on the analyses of business processes and the corporate benefits, which can be realized by their mobilization.

23.4.3.8 Lessons Learned Regarding Process Change

The new process model improved the level of customer service quite noticeably. Credit monitoring in real time reduced credit risk, because the field sales representatives now received payment information on real time, making sure that all customer visits focused on sales.

In addition, the new process and solution increased the efficiency of collecting receivables and handling temporary interruptions in deliveries. The increase in the quality of process output reduced the need of phone inquiries for information from the customers and field sales representatives on payment details. Stock management and all sales information in the ERP system were available in real time. Freeing working time from paperwork routines and clarification of errors was seen as a factor that motivated both the field personnel as well as the office staff.

Mobile technology together with process re-engineering brought noticeable efficiency gains to Amer, both in terms of significantly improved process flow time and reduced delay time. Mobility also enabled true independence from time and place: A field sales representative was now able to send the documentation immediately after completing sales deal, irrespective of his or her location and office hours. It is notable that the changes to the processes were designed to cause minimal overhead and learning curve to the users. We believe that this made the change resistance far easier to handle and minimized implementation risks. In this case, we can claim that the process can be made simpler and less error prone and at the same time customer service can be enhanced.

In several cases, companies have integrated wireless technology with their business processes in an attempt to enhance efficiency and improve service lead times. This can be argued to be a profound change into how companies use IT and change into more service oriented enterprises (Grönroos, 2008).

23.5 Conclusion and Implications

In this chapter, we identified the possibilities IT creates in organizational transformation processes. It was found that IT has two roles in the transformation: it helps to implement change into the core operations of the company through ESs and architecture, and it helps by allowing future change through improvising at the edges of the systems.

23.5.1 Theoretical Implications

ESs play a dual role of enabling the change and reinforcing the new organizational structures, and thus there is a reciprocal relationship between the IT and the organizational change (Mattila et al., 2011). While the IT enables and helps sustain the change, it is also getting shaped by the organizational change, and in some cases, it is an inhibitor of change (Alter, 1999).

The advent of mobile enterprise and open data challenge the dominance of monolithic ESs. This calls for research on how more agile enterprises can be supported by more agile enterprise architectures and how we change from process centric to connection and data centric enterprise architectures. Organizational theory has been struggling to better capture the role of the new availability of information and system artifacts in organizations concerning organizations strategies, boundaries, environment, etc.

23.5.2 Practical Implications

ESs should be used in such a way that they help the change process and create new business opportunities instead of hindering them. The management expectation that systems could act as change agents is theoretically problematic and practically much challenged. For the change to succeed, knowledge workers need to understand and be well motivated to use the system according to the new processes. If the system does not support the users in their daily routines, it is unlikely they will utilize it due to this task misfit (Markus, 2004), thus rendering the sought management coordination obsolete.

23.5.3 Future Directions

IT developments are often difficult to forecast. The social and institutional implications of the IT technologies are even more elusive. However, several continuing trends seem prevalent. Trends such as the increased computational capabilities, improvements of sensor technologies, and communications infrastructure open up organizations to further increasing virtualization of workplace, changing practices of knowledge work, rapid internalization of certain activities, and in general, acquisition of contribution outside the traditional institutional boundaries of organizations. These advances require new designs for systems and probably also more modular architectures of the new systems (Ross et al., 2006). We foresee that the problems caused by complexity of these systems and marketplace operations of the IT system vendors likely create risks of more nonfunctional, non-compliant, and non-interoperable systems. Larger patterns in society's infrastructures such as the emergence of social networking sites, for example, Facebook, challenge and offer new possibilities for these systems. The immediate impact of, for example, enterprise social networking to system design, support, and work practice seems difficult to foresee, but we argue that smaller systems working together is a more likely scenario than ever larger monolithic systems.

References

Al-Mashari, M., Al-Mudimigh, A., and Zairi, M. (2003). Enterprise resource planning: A taxonomy of critical factors. *European Journal of Operational Research, 146*(2), 352–364.

Alter, S. (1999). A general, yet useful theory of information systems. *Communications of the AIS, 1*(3es), 3.

Basole, R.C., Braunstein, M.L., and Rouse, W.B. (2012). Enterprise transformation through mobile ICT: A framework and case study in healthcare. *Journal of Enterprise Transformation, 2*(2), 130–156.

Berger, S., Lehmann, H., and Lehner, F. (8–9 July, 2002). Mobile B2B applications—A critical appraisal of their utility. *Paper presented at the M-Business 2002*, Athens, Greece.

Boudreau, M.-C. and Robey, D. (1999). Organizational transition to enterprise resource planning systems: Theoretical choices for process research. *Paper presented at the 20th International Conference on Information Systems (ICIS)*, Charlotte, NC.

Bruegge, B. and Bennington, B. (1995). Applications of mobile computing and communication. *IEEE Journal on Personal Communications, 3*(1), 64–71.

Chow, H.K.H., Choy, K.L., Lee, W.B., and Lau, K.C. (2006). Design of a RFID case-based resource management system for warehouse operations. *Expert Systems with Applications, 30*(4), 561–576.

Curtin, J., Kauffman, R.J., and Riggins, F.J. (2007). Making the 'MOST'out of RFID technology: A research agenda for the study of the adoption, usage and impact of RFID. *Information Technology and Management, 8*(2), 87–110.

Davenport, T. (1998). Putting the enterprise into the enterprise system. *Harvard Business Review*, July/August, 121–131.

Davenport, T. (2000a). The future of enterprise system-enabled organizations. *Information Systems Frontiers; 2*(2), 163–180, special issue of on the future of enterprise resource planning systems frontiers.

Davenport, T. (2000b). *Mission Critical: Realizing the Promise of Enterprise Systems.* Boston, MA: Harvard Business School Press.

Davenport, T., Harris, J.G., and Cantrell, S. (2004). Enterprise systems and ongoing process change. *Business Process Management Journal, 10*(1), 16–26.

Fuchs, C. (2011). Towards an alternative concept of privacy. *Communication and Ethics in Society, 9*(4), 220-237.

Grönroos, C. (2008). Service logic revisited: Who creates value? And who co-creates? *European Business Review, 20*(4), 298–314.

Hammer, M. and Champy, J. (1993). *Business Process Reengineering.* London, U.K.: Nicholas Brealey.

Hanseth, O. and Lyytinen, K. (2010). Design theory for dynamic complexity in information infrastructures: The case of building internet. *Journal of Information Technology, 25*(1), 1–19.

Henderson, J.C. and Venkatraman, N. (1993). Strategic alignment: Leveraging information technology for transforming organizations. *IBM Systems Journal, 32*(1), 4–16.

Jackson, S. and Philip, G. (2010). A techno-cultural emergence perspective on the management of technochange. *International Journal of Information Management, 30*(5), 445–456.

Kaupins, G. and Minch, R. (2005). Legal and ethical implications of employee location monitoring. *Proceedings of the 38th Annual Hawaii International Conference on System Sciences*, 3–6 January, 2005.

Kuk, G. and Davies, T. (2011). The Roles of Agency and Artifacts in Assembling Open Data Complementarities. In *Proceedings of 32nd International Conference on Information Systems*, Shanghai, China.

Leung, K. and Antypas, J. (2001). Improving returns on m-commerce investments. *Journal of Business Strategy, 22*(5), 12–13.

Light, B., Holland, C.P., and Wills, K. (2001). ERP and best of breed: A comparative analysis. *Business Process Management Journal, 7*(3), 216–224.

Lindman, J. and Tammisto, Y. (2011). Open Source and Open Data: Business perspectives from the frontline. Hissam, S.A., Russo, B., de Mendonca Neto, M.G., Kon, F. (eds.), In OSS 2011. IFIPAICT, vol. 365, 330–333. Heidelberg: Springer.

Lyytinen, K. and Yoo, Y. (2002). Ubiquitous computing. *Communications of the ACM, 45*(12), 63–65.

Macredie, R.D. and Sandom, C. (1999). IT-enabled change: Evaluating an improvisational perspective. *European Journal of Information Systems, 8*(4), 247–259.

Mallat, N., Rossi, M., and Tuunainen, V.K. (2004). Mobile electronic banking. *Communications of the ACM, 47*(5), 42–46.

Mallat, N., Rossi, M., Tuunainen, V.K., and Öörni, A. (2009). The impact of use context on mobile services acceptance: The case of mobile ticketing. *Information & Management, 46*(3), 190–195. doi: doi:10.1016/j.im.2008.11.008.

Markus, M.L. (1983). Power, politics, and MIS implementation. *Communications of the ACM, 26*(6), 430–444.

Markus, M.L. (2004). Technochange management: Using IT to drive organizational change. *Journal of Information Technology, 19*(1), 4–20.

Markus, M.L. and Robey, D. (1988). Information technology and organizational change: Causal structure in theory and research. *Management Science, 34*(5), 583–598.

Marton, A. (2011). *Forgotten as Data—Remembered through Information Social Memory Institutions in the Digital Age: The Case of the Europeana Initiative.* London, U.K.: London School of Economics.

Mathiassen, L. (2002). Collaborative practice research. *Information Technology and People, 15*(4), 321–345.

Mattila, M., Nandhakumar, J., Hallikainen, P., and Rossi, M. (2011). Role of enterprise system in organizational transformation. *Engineering Management Journal, 23*(3), 8–12.

Murphy, K. (2011). The paperless cockpit, *New York Times,* p. 1. Retrieved from http://www.nytimes.com/2011/07/05/business/05pilots.html?pagewanted=all (accessed on March 21, 2013).

Orlikowski, W. (1996). Improvising organizational transformation over time: A situated change perspective. *Information Systems Research, 7*(1), 63–92.

Ross, J.W., Weill, P., and Robertson, D. (2006). *Enterprise Architecture as Strategy: Creating a Foundation for Business Execution.* Boston, MA: Harvard Business Press.

Rossi, M., Tuunainen, V.K., and Pesonen, M. (2007). Mobile technology in field customer service—Big improvements with small changes. *Business Process Management Journal, 13*(6), 853–865.

Schierholz, R., Kolbe, L.M., and Brenner, W. (2007). Mobilizing customer relationship management: A journey from strategy to system design. *Business Process Management Journal, 13*(6), 830–852.

Shadbolt, N., Hall, W., and Berners-Lee, T. (2006). The Semantic Web Revisited, IEEE *Intelligent Systems, 21*(3), 96–101.

Tilson, D., Lyytinen, K., and Sørensen, C. (2010). Research commentary—Digital infrastructures: The missing IS research agenda. *Information Systems Research, 21*(4), 748–759.

Upton, D.M. and Staats, B.R. (2008). Radically simple IT. *Harvard Business Review, 86*(3), 118–124.

Venkatraman, N. (1994). IT-Enabled Business Transformation: From automation to business scope redefinition. *Sloan Management Review, 35*(2), 73–87.

Vodanovich, S., Sundaram, D., and Myers, M. (2010). Research commentary—Digital natives and ubiquitous information systems. *Information Systems Research, 21*(4), 711–723.

Watson-Manheim, M.B. and Bélanger, F. (2002). An in-depth investigation of communication mode choises in distributed teams. *Paper presented at the Twenty-Third International Conference on Information Systems,* Barcelona, Spain.

Weinstein, R. (2005). RFID: A technical overview and its application to the enterprise. *IT Professional, 7*(3), 27–33.

Williams, S. (1967). Business process modeling improves administrative control. *Automation,* 44–50.

Willis, R. and Chiasson, M. (2007). Do the ends justify the means?: A Gramscian critique of the processes of consent during an ERP implementation. *Information Technology & People, 20*(3), 212–234.

Zuboff, S. (1988). *In the Age of the Smart Machine: The Future of Work and Power.* New York: Basic Books.

24

Deconstructing Enterprise Systems: Emerging Trends in the Next Generation of ES Applications

Paul Devadoss
Lancaster University

24.1 Introduction

Enterprise systems (ESs) are packaged systems serving a wide range of functions in organizations. Such enterprise-wide systems are designed to achieve scalable intra-unit integration and rapid dissemination of information (Ross and Vitale, 2000). ES evolved from inventory control (IC) and materials requirement planning since the 1970s and 1980s to their current potentially all-encompassing nature in business information systems (IS). ESs now represent an entire category of IS that has gained momentum and widespread use since the 1990s.

ES applications are industry-specific, customizable software packages that integrate information and business processes in organizations (Markus and Tanis, 2000; Rosemann and Watson, 2002). ESs are encoded with "best practices" in organizational business processes. "Best practices" are identified through academic research on industry practices and shared understanding of acceptable efficient processes among practitioners. In any ES implementation, organizations may choose a single package or a combination of ES modules that suit their particular functional needs from several packages and implement them accordingly. They may also need to implement middleware (applications that bridge across software systems). ES packages are often customized, although extensive customization is inadvisable due to cost, time, and complexity (Markus and Tanis, 2000). In addition, since they are modeled after "best practices," modification and customization would seem unnecessary and detrimental. Companies often view such ES as a panacea for fragmented information, incompatible legacy systems, and outdated inefficient processes.

Organizations spent about USD $45.5 billion on ES in 2011 (Forrester Research, 2011). IT spend specifically in this segment has grown at about 11% over the past 6 years. However, despite extensive use in several industries, ES implementation is still regarded as a highly risky process. It is often reported that a large number of ES projects fall short of targets in terms of costs, time, or functionality (Esteves and Bohorquez, 2007). Despite the intuitively obvious benefits of ES to an organization, there are several aspects of implementation and use of ES that can and have gone wrong (e.g., Scott and Vessey, 2002). Several studies on ES implementations have identified project failures and difficulties in ES implementation (Esteves and Bohorquez, 2007). Such dramatic and, often public, failures have perpetuated a lingering message of difficulty with such systems. Financial analyses do, however, suggest that there are positive returns from ES-related investments if an organization is willing to work at it (e.g., Anderson et al., 2003). Due to the scope, complexity, and risks associated with ES, they are fundamentally different from other IS, requiring significant organizational investments in terms of technology, time, training, and human resources.

In this chapter, we look at the structure and implementation of ES in organizations to understand the nature of business processes and their reflection in ES. The chapter also looks at some modern developments in IT to suggest future directions for enterprise wide computing applications.

24.2 ES Components and Infrastructure

ESs have evolved from internal standard IC packages to include material requirements planning (MRP) and manufacturing resource planning in the late 1980s. They have since grown to include other enterprise processes, primarily classified as finance and control, materials management, sales and distribution, and human resource management. Several other fundamental organizational functions have since been added, including some add-on modules that have now become critical due to increased modern day usage, such as business intelligence. Early forms of these systems were often locally developed or assembled through a combination of developed and packaged systems. Currently, ESs are typically packaged, off-the-shelf software solutions that embed industry-wide best practices to integrate business functions and processes into a single, comprehensive framework (Markus and Tanis, 2000).

The size and complexity of ESs have meant that implementations of such systems are extremely complex, requiring a blend of technical and functional expertise (Gallagher et al., 2012). They are difficult to deploy requiring extensive resources and are complex to use (Devadoss and Pan, 2007). Implementation projects require extensive organizational participation and clearly defined goals with adequate resources to successfully complete a project. Consequently, ES literature has accumulated a large number of studies on implementation. Such studies have identified ES selection frameworks, project management techniques, success and failure factors, and lifecycle management frameworks, among several other aspects of ES implementation (Esteves and Bohorquez, 2007).

Organizations face several risks during ES implementation. Such projects often run behind schedule, fail to deliver on project goals, or over-run on costs. Organizations also potentially experience a period of low productivity during and immediately after ES implementations due to resources absorbed by implementation projects. Some ill-timed projects may even contribute to bankrupting an organization (Hitt et al., 2002; Scott and Vessey, 2002). Much of the underestimation of ES implementation complexities may perhaps be attributed to a failure in recognizing the significant differences between ES and traditional IS (Markus and Tanis, 2000).

ESs, by design, support process-oriented view of the enterprise and its business functions (Nah and Lau, 2001). They streamline the flow of information across traditional business functions by utilizing a common IT infrastructure (Newell et al., 2003). Traditional IS generally support functional units independently, potentially with disparate platforms, and, hence, are not integrated. If they are to be integrated, they may require another layer of software to interpret among various systems (known as middleware). In contrast, ESs offer tight interdependencies among a firm's functional units, bringing together all activities so that they operate as a whole with real-time data (Klaus et al., 2000; O'Leary, 2000; Robinson and Wilson, 2001; Ross and Vitale, 2000).

Due to the integrative nature of ES, implementing ES is a complex task with complications and risks that go beyond those associated with traditional IS (Bunker, 2000; Markus and Tanis, 2000; Sumner, 2000). For instance, integration means standardization of data across functional units. Furthermore, an ES imposes "its own logic on a company's strategy, organization and culture" (Davenport, 1998, p. 121). Hence, ES users must not only be trained to use the software but also educated on the overall logic of the system. Due to such complexity, ES implementations often end up late, over budget, or in failure (Davenport, 1998; Gibson et al., 1999; Scott and Vessey, 2000). Furthermore, as new business models evolve in modern globalized marketplaces, ESs evolve to include more functions and processes among their modules, making such systems far more complex.

In their approach, ESs have been closely linked with business process redesign, although there is some debate as to their sequence in implementation, thus affecting users with process and system changes (Bancroft et al., 1998; Esteves et al., 2002). Processes in ES, by virtue of the best practices embedded in them, are often potentially in conflict with, or, at best, different from an organization's existing practices (Krumbholz and Maiden, 2001). This conflict can be resolved either by tailoring the enterprise resource planning (ERP) software to suit the existing business processes or by reengineering the organization's processes to adopt best practices in the software, with minimal customization of the latter (Al-Mudimigh et al., 2001). Minimal customization is often considered beneficial to avoid high cost of customization and complexity (Bancroft et al., 1998; Holland et al., 1999). Recent technical developments such as componentization and enterprise frameworks are attempts to mask some of this complexity by allowing ES to be tailored across different areas of the business in a more effective manner (Fan et al., 2000). However, modern ESs have attempted to simplify implementation by further tightly integrating modules. Such efforts reduce implementation time, but hide greater complexity should an organization pursue customization. A significant proportion of ES projects also fail due to an underestimation of changes necessary in the organization (Appleton, 1997).

According to Davenport (2000), reengineering results in changes in an organization's strategy, structure, and culture, and the behavior of its workers. Davenport (1998) states that such changes tend to be paradoxical: ERP promotes more flexible and flatter organizations through increased information sharing, while centralization of control over such information and standardization of processes (according to the software) tend to promote command-and-control organizational structures that stifle innovation. ESs thus impact employee roles by rendering tasks more broad-based, necessitating a shift in the nature of skills and knowledge possessed by the individual (Davenport, 2000). Some studies illustrate how individual knowledge becomes more divergent due to ES as subject specialists need to know more about other business areas (Baskerville et al., 2000; Sia et al., 2002).

Despite implementation complexity, efficiency is a major motivation for ES use (Davenport, 1998; Nah and Lau, 2001; Newell et al., 2003). ESs are expected to reduce operating costs and improve productivity through greater information exchange and redesign of business processes. Another stream of literature reports that primary incentives for an organization should be strategic in nature, providing competitive advantage (e.g., Davenport, 2000). Other motivations include the organizational need for a common technology platform, improved data visibility, enhanced decision-making capabilities, and better customer responsiveness through unified view of the organization (Markus and Tanis, 2000; Ross and Vitale, 2000). Visibility of standardized real-time data across the entire organization is a significant advantage offered by ES.

A smaller proportion of ES studies have examined issues in the use of ES. Yet, it is during use that many ES problems are manifested, even if they might have originated from implementation (Markus and Tanis, 2000). During implementation, initial patterns of usage emerge. As early patterns of usage are known to congeal and become resistant to change (Tyre and Orliwkoski, 1994), they form an interesting subject for researchers seeking to understand ES use in general. Typical issues around ES use relate to lack of visibility to understand purpose of data collected in processes, complexity of systems, or lack of awareness or participation in ES implementation, consequently leading to user apathy (Devadoss and Pan, 2007).

In the context of earlier discussions in literature, this chapter shall deliberately avoid discussions that dominate ES literature, such as ES implementation projects and success/failure factors.

Instead, this chapter will discuss three key aspects of ES that are necessary to understand the nature of ES and their impact on organizations. First, we begin by understanding ES design based on business processes and consequent issues in implementation projects. The next section will follow up with a discussion on participants in an ES project and the nature of ES market ecosystem. This section further identifies some issues that impact the deployment of such complex IS in organizations. Finally, with our understanding of design problems in the existing generation of ES, we look at modern technologies that are likely to address issues identified in these discussions and influence the next generation of ES. We then discuss some implications for research and practice.

Several discussions on ES have often focused on outcomes observed during and post-implementation of such systems. However, it is necessary to deconstruct the underlying design of ES to understand the source of its complexity and the nature of issues that appear during implementation projects. Toward this purpose, in the next section, we begin by understanding how work is structured as a process and its consequent appearance in IS as discrete tasks. This section also lays the foundation for our understanding of business processes as incomplete representation of work in organizations.

24.3 Understanding Business Processes (or Deconstructing ES)

ESs are designed around what is recognized as "best practices" in the industry and academia to perform business tasks. It relies on users and their interaction with inscribed business processes within organizations. Consequently, we begin by understanding the nature of tasks in organizations. While tasks are represented through discrete, logical business procedures, the execution of a task entails much more than just a series of steps, within or outside a software system.

The guiding influences of Frederick Taylor's (1911) ideas are evident in the modern definition of a business process. Taylor's ideas suggested developing scientific methods for each stage of a work process and training workers in each of these stages to achieve a completed task with the requisite skill. Davenport and Short (1990, p. 12) define a business process as "the logical organization of people, materials, energy, equipment, and procedures into work activities designed to produce a specified end result (work product)." They further state that "A set of processes forms a business system – the way in which a business unit, or a collection of units, carries out its business." In this view, an organization is defined as a collection of externalized business processes, which are executed in order to reach predefined organizational goals.

Suchman (1983, p. 321) notably discusses the "general relationship of any normative rules to the actual occasions of their use." This view suggests that executing any business process requires some additional and possibly significant work, in order to produce a procedure that "works"—something that is seen as meaningful and skilled by both employees and their customers. This work (of situating normative rules in actual use) involves selecting appropriate procedures and making sense of them in the organizational context. "Practical action," as Suchman calls it, refers to the "work of finding the meaning of organizational plans in actual cases" (Suchman, 1983, p. 321). Her "practical action" provides an alternate conceptualization of a business process, with a number of important implications for ES implementation and business process design.

The modern definition of a business process often excludes the important aspect of business process construction and execution in an organization—"practical action." We consider this view of practical action a direct challenge to the current view of ES implementation and business process. At a minimum, Suchman's (1983) "practical action" expands our view to the numerous and potentially infinite variety of activities beyond the discrete steps in executing a task. The focus on ES implementation and the best practices represented ostensively in an ES distracts and detracts from this important organizational and social reality of situating a process in use.

Related to this challenge, other researchers have explored the importance and primacy of "practical action" in organizational life. One such perspective is "sensemaking," introduced by Weick (1995). These two ideas suggest that people make sense of various organizational inputs in an open system in order to create organizing order (Weick et al., 2005). These individuals then make sense of tasks, and resources available to them, and such sensemaking informs their action.

Discussing sensemaking in the context of pervasive electronic contexts in organizations, Weick (1985) suggests that electronic data force users to abandon social cues that preserve properties of events, since such social cues are not captured electronically. He argues that human operators benefit from their ability to step outside a situation to make sense of any set of circumstances. Human reliance on electronic terminal affects this ability to step outside a situation and constrains them by reducing their ability to comprehend the information in its entirety. There are several reasons that such information conveyed through electronic terminals is constrained (Weick, 1985). These factors constraining human cognition of information represented in electronic screens are also reflected in modern ES, centralizing control and reducing user access to the "bigger picture" for data relevant to their business processes (Devadoss and Pan, 2007). As a result, due to sheer inability to convey comprehensive information through electronic screens, sensemaking often suffers with consequent failures within organizations. Narrow focus on individual business processes without making sense of the large organizational issues can also lead to failure of the organization as a whole (Devadoss and Pan, 2007).

Central to sensemaking (in the context of using IS) is that people need to make sense of a situation beyond the limitations of the electronic representation of the data. The current influence of ES thinking about business processes, however, falls short in recognizing a need for sensemaking. A reformed view would argue that technologies might help to trigger sensemaking by users (Griffith, 1999). Such a view would require organizations to provide users with the resources to act on information in a meaningful way. This is in contrast to the current generation of ES that are often rigid systems that often confine users' role even within a particular business process.

24.3.1 Representing Business Processes in ES

ES can be distinguished from other information technologies, based on their highly integrated and packaged nature (Devadoss and Pan, 2007). ES imposes a particular informational logic on the organization (Davenport, 1998). Furthermore, such highly integrated technologies can impose restrictions on their interpretive flexibility during use as well (Devadoss and Pan, 2007). Complexity of ES can also serve to unquestioningly preserve institutional influences (Gosain, 2004). Consequently, ES implementation and use has often been a mindless instead of a "mindful" choice grounded in the organizational context (Swanson and Ramiller, 2004).

Despite extensive developments in ES since its MRP roots, ESs suffer the same limitations as any other IS: they can only represent an abstraction of the continuously changing needs of an organization (Hirschheim et al., 1995). Even as memory systems, ES can only capture a part of the knowledge of tasks embedded in the system (Stijn and Wensley, 2001). Changing the organization to suit the system or customizing the system to the organization may address such knowledge gaps (Stijn and Wensley, 2001).

Given the complexity of ES and their impact on constraining human action in an organization (e.g., Gosain, 2004), the task-technology fit of ES is an issue of importance (Gattiker and Goodhue, 2004). Bridging this gap too appears to be down to either of two options: change the system or change practice, and in some circumstances, both these options may be impossible (Gattiker and Goodhue, 2004).

The essence of the earlier discussion in terms of tasks embedded in ES brings us back to our discussions of "practical action" and the "performative" aspect of routines. Pervasive adoption of IS designed on Tayloristic influences on tasks as procedural processes has led to two important outcomes. First, as a procedural workflow, business processes are meant to capture aspects (or perhaps abstractions) of task performance for various reasons such as institutional or organizational requirements. Performing a task is more about achieving a certain output rather than the means of achieving that task. In other words, a task can be performed in more ways than one predesignated abstraction of work, and the accomplishment of that task is more important than ensuring repetitive completion of individual steps in a business process (as encoded in any IS). Feldman and Pentland (2003) cite the example of police routinely selecting suitable facts to fill in forms after arrests to comply with regulations.

A second important implication is individual control over data in performing a task. Individuals in an organization have lost control over how tasks are executed. In some sense, choices on how their input is processed have been predetermined. In the case of ES, such predetermination is most likely to have occurred outside the organizational context (Gosain, 2004). In the case of ESs that do closely suit some tasks, other tasks in the same system are at risk of not suiting an organization completely (or at all). Such predetermined design and implementation leads to lack of visibility and lack of control over data in organizations using ES, leaving users finding workarounds that may even exclude the ES (Robey et al., 2002). Such lack of control is accentuated in organizations that are functionally organized, leaving task performers few cues toward fully comprehending the data as suggested by Weick (1985). ES literature often reports on lack of visibility, and consequent disinterest among users in completing data that they do not directly rely on. Such users often have no knowledge of consequent impacts elsewhere in the interdependent systems (e.g., Eriksen et al., 1999).

In many work-related situations, with or without ES, tasks are complex and adaptable to numerous nuances in the context of use. Suchman (1983) discusses the example of an accounting department in an office, where processing invoices is not necessarily orderly in practice, though procedures cast work in orderly sequences. In another case involving ES, numerous pieces of material from a manufacturing plant were to be reused (Gattiker and Goodhue, 2004), but because each reusable piece needed a unique identification number per item in the ES, it was infeasible to use the implemented ES. Instead, employees used an Excel spreadsheet, and a clerk was added to the workforce. In this and many other situations, ESs failed to capture nuanced work practices and situations. The two examples illustrate different issues with the systems view of work that relies on sequentially accomplishing tasks and information technology support for such procedures. A Tayloristic and structured view of work common in ES implementations fails to capture the notion that "procedures is an outcome to which practitioners orient their work – it is not the work itself" (Suchman, 1983, p. 327).

Alternatively, Feldman and Pentland (2003) put forward a theory that identifies the need to discuss the knowledge of a task and the execution of the task based on routines. They define an organizational routine as "a repetitive, recognizable pattern of interdependent actions, involving multiple actors" (ibid, p. 96). Routines are often portrayed as consisting of fixed patterns. Organizational support for these patterns, for example, through ES, may result in people choosing some patterns more frequently over others. They identify two problems with routines: lack of agency and variability of routines. They proposed a theory of routine as a duality, which is defined by both the *ostensive* and *performative*. The ostensive aspect is essentially an abstraction of a routine, and is nonspecific. The performative aspect is the action taken by people performing a routine in an organization and it includes situation-specific interpretations of the rules in executing routines. Hence, the performative aspect includes the patterns of action used to accomplish a task. Consider the example of a travel itinerary to identify these two aspects of a routine. The ostensive aspect is the travel itinerary itself, listing travel "details" such as flight details and destination. The performative aspect includes the actual details of the trip, such as getting to the airport, completing related formalities, security checks, and waiting.

Like Feldman and Pentland (2003), we argue that the act of accomplishing a task in an organization involves more than following an ostensive procedure. If this were true, a business system could be reduced to a book of instructions, or perhaps a fully automated IS that would require no manual intervention. Instead, it is argued that we should look beyond these implicit and Tayloristic views of task performance as a series of steps and information in an ES and consider a business process as a performative and sensemaking act that is informed by the IS, but not restricted to it.

As a result, the representation of best practice in an ES omits the numerous forms of *practical action* required to successfully execute a specific business process. Such practical action involves all of the expert and nuanced work required before and during the use of an ES, which invokes ostensive routines in order to satisfy a specific business need. Following that, a focus on only the IS can only be performative where human expertise is no longer required, and a rigid rule can encapsulate all performative knowledge in a rigid and singular routine. In considering all best practices as embodied in an ES, this mechanistic thinking is carried to an extreme, where the ostensive functionality in the ES is considered best practice, and all human expertise of practical action is considered idiosyncratic and haphazard.

In summary, the aforementioned discussion traces problems with ES usage to ES design as a "business system." Since tasks in an organization can be seen to have two aspects, as in the theory of routines (Feldman and Pentland, 2003), or as including "practical action" (Suchman, 1983), ESs need to be simple enough to allow flexible use of the system to empower users and account for the need for internal flexibility following external change. However, despite whatever ability ES may have to empower, it will also be down to individual intentions in allowing such empowerment, or scuttling it with more control (e.g., Sia et al., 2002).

Three key issues have been discussed in this section. Primarily, business processes are incomplete representations of work in organizations. Consequently, relying on business processes as accurate abstract representations further distances the technologies that are inscribed with these processes from the reality of task performance in an organization. Finally, this design problem manifests itself in various forms during implementation and use of ES. With this understanding, we now look at participants in an ES implementation project, and the overall ES ecosystem with its services providers.

24.4 An ERP Project and EcoSystem

ES products are an aggregation of business processes encoded in IS, as discussed earlier. Since business practices vary due to arrangements within industries of a kind, these products are differentiated primarily by industry verticals. Within each vertical, they are inscribed with options to perform similar tasks in some typical and popular methods to perform business functions in an organization. They also include a variety of business functions required within an organization. Due to advancements in technologies and programming techniques, these products are also integrated tightly across various functional components, making them easier to deploy in shorter implementation time frames. This has consequently lead to more "standard" implementations and less customization. All these developments have facilitated the growth of ES around the world in large and medium sized organizations. Large organizations have gone through several cycles of implementation and upgrades since the early years of MRP. ES providers have now expanded their reach to SMEs, providing products and solutions that address key concerns in terms of pricing and speed of deployment. This, however, does not make the process less difficult, as it involves several players who need to collaborate to successfully complete the project. In this section, we illustrate a typical ES project and the key players in the industry that provide ES-related services.

There are several discussions on implementation models and its participants (e.g., Somers and Nelson, 2004). However, consider the typical case of an organization that decides to implement an ERP system. In practice, it may need to engage a consultant to study and identify its key organizational requirements, mindful of what is offered in ERP systems in the market, and transfer knowledge to the client (Ko et al., 2005). Such requirements will reflect in the Request for Proposal (RFP) issued to vendors. The proposal is then picked up by ES package vendors and passed on to potential implementation partners. Implementation partners are companies with the expertise and required manpower to deploy ES packages from vendors. The implementation partners work along with ES vendors and respond to an organization's RFP with a suggested solution.

The solution generally identifies an ES product, modules and any customizations that may be required to implement the solution. Such responses also include suggestions and workarounds to prevent cost escalations through customizations. In typical implementations, other third-party applications may be required to deliver some of the functionality required by the client organization (Markus

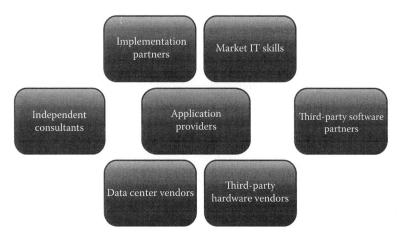

FIGURE 24.1   The enterprise systems ecosystem.

and Tanis, 2000). Implementation partners may then take on the responsibility of delivering third-party solutions as well and will manage their delivery projects. If projects are large, several implementation partners may need to be managed to deliver various segments of the project for manageability and audit.

During this phase, the client organization and potential vendors attempt to understand with some level of detail the requirements of the organization and the capabilities of the system. Considering the time and effort required to delve into details of multiple processes and identify quirks in individual systems and modules, this is often constrained to process level detail. Furthermore, such investigations are also often restricted to key processes in major functional areas (and/or functional areas of high impact as decided by the organization). In some cases, where the management considers key business processes their competitive advantage, some important business processes may be explored in minute detail along with business users. However, unless an impossible technical situation arises, implementers are often willing to find workarounds to deliver solutions to the client. The real cause for concern at this stage is the inability to pursue the suitability of any proposed solution in fine detail, and different styles of decision making (Bernroider and Koch, 2001).

The client organization then enters a period of negotiations with the vendors and implementation partners to finalize licensing costs, implementation costs, manpower requirements, project scope, time, and effort. Implementation schedules, penalties, and potential causes for concern are identified and a negotiated understanding emerges. A hardware partner is then selected to supply the hardware requirements for the chosen system. Hardware choices are made based on client organization's choice of platforms and ERP vendor/implementation partners' recommendations (Figure 24.1).

The ES product market is made of a few large vendors and several smaller vendors. Smaller product vendors specialize in specific functions in an organization. Larger providers have expanded their product offering by acquiring popular smaller systems to grow the range of functions integrated in their enterprise-wide systems. With each acquisition, larger systems have become more capable in delivering wider range of solutions to any client, while also increasing complexity in deploying solutions.

24.5 ES Implementation

A result of the high growth in the ES market is that the ecosystem for ES products is well developed with several products and services providers. Key participants in the ecosystem are vendors of ES, implementers, and consultants who assist the client organization. Since choosing and implementing a system requires appropriate level of detail, these participants are crucial to the success of any project. This is an often-neglected topic of research, though recent literature has investigated the complexity of the ecosystem and its value creation (Sarker et al., 2012).

Every organization should investigate the suitability of a product prior to implementation. However, considering the time and cost involved in investigating minute details, decisions are often made on the basis of top-level functions and business processes supported by such systems. In any typical implementation process, the organization undergoes a period where processes are identified, documented, and mapped. This process is known as blueprinting. It identifies the fundamental details such as fields, tables, and labels required to set up various processes that will be encoded for the final system.

While deploying these systems, several details are addressed by making appropriate changes to the system or the organizations business processes. Consequently, this process requires exceptional dedication and participation from users at all levels of an organization. Furthermore, each stage of implementation is interdependent on the others to effect a successful implementation (Velcu, 2010). Hence, these individual process level details are factors that affect project outcomes (Holland and Light, 1999). Users are needed to provide relevant domain expertise on processes that are being inscribed in the new system, and senior managers are required to assist in making decisions about any changes in business processes as inscribed in the new system. Incompatibility among processes in practice and processes inscribed in ERP systems can lead to delays in deployment. Often, such contradictions are left unresolved and can later lead to expensive delays in projects.

Furthermore, knowledge level of an implementation team varies according to the size and scope of the project (Ko et al., 2005). Implementation partners often vary the skill level of consultants in an implementation team to ensure balance of new and experienced consultants. The balance allows them opportunities to manage their costs and career development commitments to newer consultants in their organizations. Consequently, the composition of implementation teams and their management throughout the project can contribute to effectiveness of the implementation. Apart from these observations, implementation projects are intensive projects with several teams of diverse expertise, and hence such projects need to be managed closely to monitor and dynamically manage quality of deployment (Markus and Tanis, 2000). Inadequate experience in an implementation team can result in less optimal solutions being inscribed in the system. There are often several choices that implementers need to make in terms of choosing appropriate modules/processes/variables, etc. Experience with the system and appropriate support systems from the implementation partner are crucial to implementing optimal solutions for the client (Ko et al., 2005).

24.6 Evolving Technological Changes

ES technologies have evolved along with the progress of technologies over the past two decades. Large centralized systems delivered in cryptic terminal-based user interfaces have now migrated to reasonably informative web browser–based user interfaces. Backend technologies have become sophisticated enough to manage large volumes of data. They also include complex computing capabilities. However, the integrated nature of the system has inevitably increased the complexity in system architecture, leading to a setup that requires layers of database and applications over hardware management systems. In addition, since organizations now possess large volumes of data about their products and services, new data mining applications assist in identifying improvements and opportunities in existing business processes (Moller, 2006). Such applied usage, one of the fundamental reasons for the introduction of ES in organizations, has also benefitted from advances in technologies, such as cloud computing (Figure 24.2). Cloud computing has made it possible for organizations to deliver applications to its distributed workforce across supply chains and geographical distances (Marston et al., 2010).

Several new technological developments have an impact on modern day usage of computing within organizations, increase in use of modern computing devices has led the growth of the phenomenon known as "consumerisation," where user expectations and computing usage patterns are affected by the growth in consumer electronics (such as the use of smartphones and tablet devices influencing similar expectations in work settings). Consequently, users are now familiar with more advanced devices and applications with advanced user interfaces to accomplish work-related computing tasks, promoting

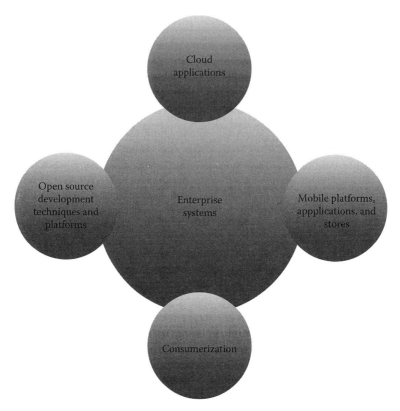

FIGURE 24.2 Erosion of traditional ES models.

enterprise mobility (Consoli, 2012). This is the opposite of what happened in the early years of computing in organizations, when computer self-efficacy was a significant factor in IT use (Agarwal and Sambamurthy, 2000). As a result, business applications now have to adapt to delivering applications over devices and platforms that users find familiar. Smart phones and tablet devices are commonplace in modern day organizations, as are "application" stores to deliver applications over these devices. Since these modern devices are smaller, with less intensive computing power, and operate on platforms that are stripped down versions of operating systems, cloud computing could become an important part of such IT infrastructure for organizations. There are now several mobile device platforms, applications, developers, and application stores (Giessmann et al., 2012).

As a consequence of the "consumerization" phenomenon, there is also a growing workforce of programmers who can deliver targeted applications that focus on specific business processes. Currently, such applications are made available in corporate private application stores for enterprise users. Such business applications on mobile platforms can interact with cloud architectures to store, retrieve, and perform complex processes outside the capabilities of such devices, thus delivering complex information processing support for organizational processes (Giessmann et al., 2012). As a result of such applications and platforms, two key issues are fast gaining traction in corporate use: focused business applications that deliver specific functionality, and abundance of low cost programming solutions for organizations.

The tremendous growth in such technical capabilities could lead to the availability of greater variety of applications that can together deliver important functions for any organization. This will result in the ability to design and develop third-party solutions (or even open source solutions) that can be deployed without the complexity of implementing expensive and time consuming and complex integrated ES. A "swarm" of such applications can replace our current understanding of ES.

Enterprise application stores for modern application platforms are now available in large numbers (Giessmann, 2012). However, further research is necessary to identify application requirements engineering techniques, and coordination mechanisms among these process focused applications to integrate and deliver functionality in organizational settings. Since large organizations often operate their own stores, they need to evolve mechanisms for coherent requirements engineering and development. Third-party stores for smaller organizations may need other protocols to coordinate development and delivery of applications. This exciting new development in ES requires significant research from IS researchers to identify organizational practices and mechanisms for absorbing these stores into organizational use.

24.7 Key Issues Affecting Current ES

The discussion in this chapter highlights some key issues around ES design and technological changes potentially affecting ES. We now look at a brief summary of three key issues from our discussion and suggest directions for further research and suggest what the next generation of ES may deliver in organizations given the developments in technologies discussed in this chapter.

The ES literature has been obsessed with implementation issues because, as discussed earlier in this chapter, implementing such complex and expensive systems does put organizations at severe risk. Hence, the attention in research has been justified. Implementation studies have contributed to our extensive knowledge of the complexity and difficult in the design and deployment of such systems. The fundamental issue stems from the nature of business processes in modern day organizations. IS and most business practices continue to draw on principles drawn from industrial practices in the manufacturing sector. However, nature of work in organizations has shifted from manufacturing-based practices to "services sector"—based practices. Most modern global economies have a large share of their gross domestic product contributed by the services sector. However, this shift in the nature of any economy is not adequately reflected in our systems designs and research activities. Along with the shift to services economy, developments in IT have changed the nature of knowledge creation and its consumption, necessitating a rethink in systems design (Chesbrough and Spohrer, 2006). This is far removed from the dominant manufacturing sector-based economies of the past, where several of our current organizational systems designs were modeled. Hence, ES studies need to reexamine the needs of modern enterprises to design and develop systems that reduce complexity, devolve skills, and decision making to knowledge workers, who can act as process owners.

Much research is needed to understand true knowledge intensive activities in organizational settings and the role information technologies can play in facilitating task performance. In knowledge intensive activities, knowledge creation and application can be devolved to task owners rather than centralizing tasks through complex enterprise applications. In other words, modern ESs have preferred to centralize control and limit access and visibility to its users (Devadoss and Pan, 2007). However, most modern work environments may require the opposite, with devolved control and greater visibility for users of their own tasks and others'. Such systems will then be able to provide the focus on knowledge workers and productivity, as necessary in a modern knowledge economy (Drucker, 1999).

Advances in information technologies should assist in creating simple and uncluttered business processes that can interact with other business processes. This should decentralize business process ownership to assist knowledge users in delivering higher productivity through effective IS, while providing fully functional ability to audit a knowledge workers performance. This implies oversight on tasks and performance for managerial purposes, but decentralization of control and knowledge application to individual users. Simplified user interfaces, consumer friendly devices and greater control over task performance are key issues that will contribute significantly to the growth of ES adoption and its use in organizations. For example, these issues in the context of knowledge management have shown great promise for organizations (Massa and Testa, 2009). In the context of modern and mobile technologies and growing numbers of enterprise application stores, more research is needed to identify the nature of task performance and the role of IT in assisting knowledge workers. Furthermore, given the growth in mobile applications stores for enterprises, further investigation is necessary to understand their design, development, delivery, and application in organizational use (Giessmann et al., 2012).

Computing ideas from the past have explored environments where users had greater control over design and use of objects in organizations, such as Lotus Notes. End-user computing may have a different context in organizations in modern days, with better-equipped workforce that has greater computer self-efficacy, and also has the need for greater control and flexibility over task performance (Fischer et al., 2004). Furthermore, open source software solutions and associated contributors were expected to revolutionize computing solutions. However, large projects with independent developers, each with varying independent visions for their products have been significant barriers in organizational adoption (Nagy et al., 2010). Though they have made vital contributions in some segments such as Internet applications, open source ES solutions have been difficult due to the design issues in business processes (as discussed in this chapter) and architectures used in such solutions. There is a market for open source ES, though advanced functionality does not come free. As discussed previously, modern day computing solutions are now capable of breaking down solution complexity and facilitating implementation of less complex solutions that focus on specific business process level details. The cost of developing and deploying such solutions are less in terms of time, cost, and risk. This is an emerging area of interest that needs more attention in future, and will have a significant impact on ES applications in organizations given the growing numbers of enterprise application stores (Giessmann et al., 2012). The emergence of mobile platforms and enterprise application stores should also encourage greater development from distributed programming resources and leave design and implementation to process experts instead of technology experts, as suggested by Fischer et al. (2004). The next generation of ES should cater to a new kind of knowledge worker who is equipped with appropriate ES to deploy their expertise to execute their tasks.

References

Agarwal, R. and Sambamurthy, V. (2000). The evolving relationship between general and specific computer self-efficacy—An empirical assessment. *Information Systems Research*, 11(4), 418–430.

Al-Mudimigh, M., Zairi, M., and Al-Mashari, M. (2001). ERP software implementation: An integrative framework. *European Journal of Information Systems*, 10(2), 216–226.

Anderson, M. C., Banker, R. D., and Ravindran, S. (2003). The new productivity paradox. *Communications of the ACM*, 46(3), 91–95.

Appleton, E. (1997). How to survive ERP. *Datamation*, 43(3), 50–53.

Bancroft, N. H., Seip, H., and Sprengel, A. (1998). *Implementing SAP R/3*, 2nd edn, Manning Publications Company, Greenwich, CT.

Baskerville, R., Pawlowski, S., and McLean, E. (2000). Enterprise resource planning and organizational knowledge: Patterns of convergence and divergence. In: *Proceedings of the 21st International Conference on Information Systems*, Brisbane, Queensland, Australia, December 10–13, 2000, pp. 396–406.

Bernroider, E. and Koch, S. (2001). ERP selection process in midsized and large organizations. *Business Process Management Journal*, 7(3), 251–257.

Bunker, D. (2000). Enterprise resource planning (ERP) system tools: The context of their creation and use within the technology transfer process. In: *Proceedings of the Americas Conference on Information Systems*, Long Beach, CA, pp. 1533–1536.

Chesbrough, H. and Spohrer, J. (2006). A research manifesto for services science. *Communications of the ACM*, 49(7), 35–40.

Consoli, D. (2012). An advanced platform for collaborative and mobile enterprise 2.0. *Journal of Mobile, Embedded and Distributed Systems*, IV(2), 121–133.

Davenport, H. (1998). Putting the enterprise into the enterprise system. *Harvard Business Review*, 76(4), 121–131.

Davenport, T. H. (2000). *Mission Critical: Realizing the Promise of Enterprise Systems*. Harvard Business School Press, Boston, MA, 2000.

Davenport, T. H. and Short, J. E. (1990). The new industrial engineering: Information technology and business process redesign. *Sloan Management Review*, 31(4), 11–27.

Devadoss, P. and Pan, S.-L. (2007). Enterprise systems use: Towards a structurational analysis of enterprise systems induced organizational transformation. *Communications of the Association for Information Systems*, 19, 352–385.

Drucker, P. (1999). Knowledge-worker productivity: The biggest challenge. *California Management Review*, 41(2), 79–94.

Eriksen, L. B., Axline, S., and Markus, M. L. (1999). What happens after 'going live' with ERP systems? Competence centers can support effective institutionalization. In: *Proceedings of the Fifth Americas Conference on Information Systems*, Milwaukee, WI, August 13–15, University of Wisconsin-Milwaukee, pp. 776–778.

Esteves, J. and Bohorquez, V. (2007). An updated ERP systems annotated bibliography: 2001–2005. *Communication of the Association for Information Systems*, 19(1), 386–446.

Esteves, J., Pastor, J., and Casanovas, J. (2002). Monitoring business process redesign in ERP implementation projects. In: *Proceedings of the Eighth Americas Conference on Information Systems*, Dallas, TX, August 9–11, 2002, pp. 865–873.

Fan, M., Stallaert, J., and Whinston, A. B. (2000). The adoption and design methodologies of component-based enterprise systems. *European Journal of Information Systems*, 9(1), 25–35.

Feldman, M. S. and Pentland, B. T. (2003). Reconceptualizing organizational routines as a source of flexibility and change. *Administrative Science Quarterly*, 48(1), 94–118.

Fischer, G., Giaccardi, E., Ye, Y., Sutcliffe, A. G., and Mehandjiev, N. (2004). Meta-design: A manifesto for end-user development. *Communications of the ACM*, 47(9), 33–37.

Forrester Research. (2011). The State of ERP in 2011: Customers have more options inspite of market consolidation. http://www.pioneerconsulting.co.uk/erp-market-to-grow-to-50.3bn-in-2015-forrester (accessed on August 30, 2012).

Gallagher, K. P., "Jamey" Worrell, J. L., and Mason, R. M. (2012). The negotiation and selection of horizontal mechanisms to support post-implementation ERP organizations. *Information Technology & People*, 25(1), 4–30.

Gattiker, T. F. and Goodhue, D. L. (2004). Understanding the local-level costs and benefits of ERP through organizational information processing theory. *Information & Management*, 41(4), 431–443.

Gibson, N., Holland, C. P., and Light, B. (1999). Enterprise resource planning: A business approach to systems development. In: *Proceedings of the 32nd Hawaii International Conference on System Sciences*, Maui, HI.

Giessmann, A., Stanoevska-Slabeva, K., and de Bastiaan, V. (2012). Mobile enterprise applications—Current state and future directions. In Proceedings of the *45th Hawaii International Conference on System Sciences*, Maui, HI, pp. 1363–1372.

Gosain, S. (2004). Enterprise information systems as objects and carriers of institutional forces: The new iron cage? *Journal of the Association for Information Systems*, 5(4), 151–182.

Griffith, T. L. (1999). Technology features as triggers for sensemaking. *The Academy of Management Review*, 24(3), 472–488.

Hirschheim, R., Klein, H. K., and Lyytinen, K. (1995). *Information Systems Development and Data Modeling: Conceptual and Philosophical Foundations*. Cambridge University Press, Cambridge, U.K.

Hitt, L., Wu, D. J., and Zhou, X. (2002). Investment in enterprise resource planning: Business impact and productivity measures. *Journal of Management Information Systems*, 19(1), 71–98.

Holland, C. R. and Light, B. (1999). A critical success factors model for ERP implementation. *IEEE Software*, 16(3), 30–36.

Holland, C. P., Light, B., and Gibson, N. (1999). A critical success factors model for enterprise resource planning systems. In: *Proceedings of the 7th European Conference on Information Systems*, Copenhagen, Denmark, June 23–25, 1999, pp. 273–297.

Klaus, H., Rosemann, M., and Gable, G. G. (2000). What is ERP? *Information Systems Frontiers*, 2(2), 141–162.

Ko, D. G. Kirsch, L. J., and King, W. R. (2005). Antecedents of knowledge transfer from consultants to clients in enterprise system implementations. *MIS Quarterly*, 29(1), 59–85.

Krumbholz, M. and Maiden, N. (2001). The implementation of enterprise resource planning packages in different organizational and national cultures. *Information Systems*, 26(3), 185–204.

Markus, M. L. and Tanis, C. (2000). The enterprise system experience—From adoption to success. In: Zmud, R. W. (Ed.), *Framing the Domains of IT Management: Projecting the Future through the Past*. PinnaFlex Educational Resources, Cincinnati, OH, pp. 173–207.

Marston, S., Li, Z., Bandhyopadhyay, S., Zhang, J., and Ghalsasi, A. (2010). Cloud computing—The business perspective. *Decision Support Systems*, 51, 176–189.

Massa, S. and Testa, S. (2009). A knowledge management approach to organizational competitive advantage: Evidence from the food sector. *European Journal of Information Systems*, 27(2), 129–141.

Moller, C. (2006). ERP II: A conceptual framework next-generation enterprise systems? *Journal of Enterprise Information Management*, 18(4), 483–497.

Nagy, D., Yassin, M. A., and Bhattacherjee, A. (2010). Organizational adoption of open source software: Barriers and remedies. *Communications of the ACM*, 53(3), 148–151.

Nah, F. F. and Lau, J. L. (2001). Critical success factors for successful implementation of enterprise systems. *Business Process Management Journal*, 7, 285–296.

Newell, S., Huang, J. C., Galliers, R. D., and Pan, S. L. (2003). Implementing enterprise resource planning and knowledge management systems in tandem: Fostering efficiency and innovation complementarity. *Information and Organization*, 13, 25–52.

O'Leary, D. E. (2000). *Enterprise Resource Planning Systems: Systems, Life Cycle, Electronic Commerce, and Risk*. Cambridge University Press, New York.

Robey, D., Ross, J. W., and Boudreau, M. (2002). Learning to implement enterprise systems: An exploratory study of the dialectics of change. *Journal of Management Information Systems*, 19(1), 17–46.

Robinson, B. and Wilson, F. (Fall 2001). Planning for the market? Enterprise resource planning systems and the contradictions of capital. *The Database for Advances in Information Systems*, 32(4), 21–33.

Rosemann, M. and Watson, E. E. (2002). Special issue on the AMCIS 2001 workshops: Integrating enterprise systems in the University Curriculum. *Communications of the Association for Information Systems*, 8, 200–218.

Ross, J. W. and Vitale, M. R. (2000). The ERP revolution: Surviving vs. thriving. *Information Systems Frontiers*, 2(2), 233–241.

Sarker, S., Sarker, S., Sahaym, A., and Bjorn-Andersen, N. (2012). Exploring value co-creation in relationships between an ERP vendor and its partners: A revelatory case study. *MIS Quarterly*, 36(1), 317–338.

Scott, J. E. and Vessey, I. (2000). Implementing enterprise resource planning systems: The role of learning from failure. *Information Systems Frontiers*, 2, 213–232.

Scott, J. and Vessey, I. (2002). Managing risks in enterprise systems implementation. *Communications of the ACM*, 45(4), 74–81.

Sia, S. K., Tang, M., Soh, C., and Boh, W. F. (2002). Enterprise resource planning (ERP) systems as a technology of power: Empowerment or panoptic control? *Database for Advances in Information Systems*, 33(1), 23–37.

Somers, M. T. and Nelson, G. K. (2004). A taxonomy of players and activities across the ERP project life cycle. *Information & Management*, 41(3), 257–278.

Stijn, E. V. and Wensley, A. (2001). Organizational memory and completeness of process modeling in ERP systems. *Business Process Management Journal*, 7(3), 181–194.

Suchman, L. (1983). Office procedures as practical action. *ACM Transactions on Office Information Systems*, 1, 320–328.

Sumner, M. (2000). Risk factors in enterprise—Wide/ERP projects. *Journal of Information Technology*, 15(4), 317–328.

Swanson, E. B. and Ramiller, N. C. (December 2004). Innovating mindfully with information technology. *MIS Quarterly*, 28(4), 553–583.

Taylor, F. W. (1911). *The Principles of Scientific Management*, Harper Bros, New York.

Tyre, M. J. and Orlikowski, W. J. (February 1994). Windows of opportunity: Temporal patterns of technological adaptation in organizations. *Organization Science*, 5(1), 98–118.

Velcu, O. (2010). Strategic alignment of ERP implementation stages: An empirical investigation. *Information & Management*, 27(3), 158–166.

Weick, K. E. (1985). Cosmos vs. chaos: Sense and nonsense in electronic contexts. *Organizational Dynamics*, 14(2), 51–64.

Weick, K. E. (1995). Sensemaking in organizations, Vol. 3. Sage Publications, Incorporated.

Weick, K. E., Sutcliffe, K. M., and Obstfeld, D. (2005). Organizing and the process of sensemaking. *Organization Science*, 16(4), 409–421.

25

Enterprise Architecture

Martin Meyer
Dublin City University

Markus Helfert
Dublin City University

25.1 Introduction

Since the inceptions of information technology (IT) and information systems (IS), companies have experienced a constant increase in size, scope, and complexity of enterprise IS. To manage and organize these systems, logical constructions and representations in the form of models were needed, and an architecture approach to IS was developed in response to these challenges. Architectures are used commonly in IT and IS domains to construct blueprints of an enterprise for organizing system components, interfaces, processes, and business capabilities, among others. Architectures are discussed extensively, describing various aspects of IS. Architectures in the wider context of IS are often used to model aspects of a system, especially computer, network, software, application, service-oriented, business, and project-development architectures, among others. Thus, the term architecture plays a central role in IS and IT and can be defined generally as *"the fundamental organization of a system embodied in its components, their relationships to each other, and to the environment, and the principles guiding its design and evolution"* (IEEE Computer Society, 2007).

Combined with increased importance of IS in the 1990s (Morton 1991), information system architectures—and later enterprise architectures (EA)—became important. Yet, this is still a broad and multifaceted domain with various viewpoints. Over the years, many EA frameworks, modeling concepts, and tools were proposed, including the Zachman framework (Zachman 1987), Department of Defense Architecture Framework (Chief Information Officer U.S. Department of Defense 2010), Federal Enterprise Architecture Framework (Office of Management and Budget 2012), The Open Group Architecture Framework (TOGAF) (The Open Group 2011), and Architecture of Integrated

Information Systems (ARIS) (Scheer 1992). The purpose of this chapter is to describe EA from a business and IS viewpoint as well as review prominent EA frameworks. A common concept found among all frameworks is the conceptualization of strategic business decisions in systems design choices, in which EA aims to support and enable transformations within an organization while aligning business and IT. To reduce complexity, many frameworks use views to describe elements of architectural content (e.g., strategy, information, process, service, and technology). Each view is meaningful to various stakeholder groups. The intent of EA is to determine—from various stakeholder viewpoints—how an organization can most effectively achieve current and future objectives in the form of transformations. Strategic options are often translated into systems design options, and EA helps manage this process carefully.

The terms enterprise architecture (EA) and enterprise architecture management (EAM) are often used interchangeably (Ahlemann et al. 2012b), though the scopes of the two are different. EA is concerned with creation in the form of design and development of architectural blueprints and artifacts; EAM is concerned with the sustained management of such artifacts using lifecycles. Consequently, EAM is a process to plan, control, govern, and evolve an enterprise's organizational structure, business processes/services, and technical infrastructure in accord with a vision and strategy. We view EAM as part of EA, and hence, refer only to the EA discipline as whole. EA is concerned with business process integration and standardization (Ross et al. 2006), and thus, business processes are at the core of EA. We view EA from a business process and modeling perspective, though EA often includes a large variety of IS-related viewpoints and approaches in practice.

25.2 Positioning Enterprise Architecture

EA evolved into its own research domain, and various ways of positioning and organizing EA emerged (Ahlemann et al. 2012b; Dern 2009; Hanschke 2010; Op 't Land et al. 2009; Ross et al. 2006; Scheer et al. 2000; Schmidt et al. 2011; The Open Group 2011). Early work on architectures model aspects of systems, often referred to as IS or system architectures. These architectures do not emphasize strategic directions or business aspects strongly (Kim et al. 1994). Given a need for IT and business alignment and coming from early system-oriented concepts, a widely adopted approach emerged that organizes EA along various organizational layers and emphasizes an IT business alignment paradigm. The purpose is to relate strategic aspects to application and technology. Layering helps researchers and managers understand and describe the scope and function of EA as they relate to boundary points from business strategy to technical infrastructure. Figure 25.1 illustrates both architectural layers and the contexts with which they interact. Usually, strategic planning serves as input for EA, especially when EA is viewed as a strategy-driven enterprise function. Core layers represent business architectures (BAs), application architectures (AAs) (including data/information architectures), and technology architecture, all used with EA processes and services to design, develop, govern, and manage them. EA produces some form of content or output such as artifacts (Winter et al. 2006) and principles (Greefhorst et al. 2011). In addition, EA uses frameworks and tools.

We view EA from a service perspective; each layer offers a service to the business, and the service-oriented paradigm occurs in each architectural layer. For example, the technology architecture offers infrastructure services in the form of hardware and networks. The application architecture provides services centered on software applications and data. The BA is concerned with business processes and services. The BA, through business processes, emphasizes the dynamic aspects of workflows and activities supported by application components and infrastructures. Examples in the literature present simple, three-layered frameworks with which to view EA (Hasselbring 2000; The Open Group 2012) to multilayered EA frameworks (Winter et al. 2006). Approaches that include enterprise strategy as a separate layer are also documented (Godinez et al. 2010). Whether or not strategy is a dedicated EA layer,

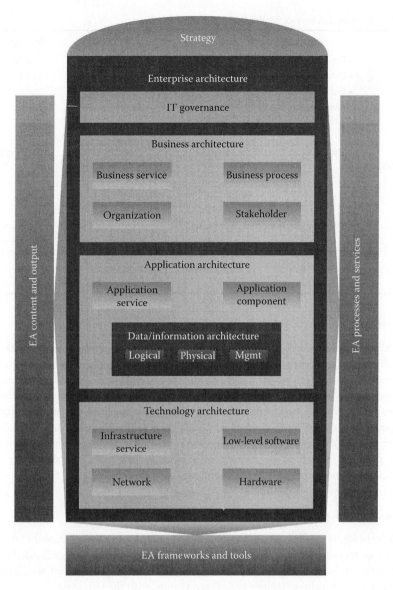

FIGURE 25.1 EA overview.

it initiates a discussion of whether EA is solely an IT or also a business function. Due to the strategic importance of business–IT alignment (BITA), we argue EA is a hybrid enterprise function (Ahlemann et al. 2012b). In the following sections, we describe EA processes and services, the role of IT governance, and core architectural layers.

25.2.1 EA Processes and Services

EA services provide benefits to stakeholders within an organization and may consist of several EA processes. EA service model (Robertson 2008) can be introduced to document the service delivery of EA. An example is described in Table 25.1. Notably, there are more processes in relation to these fundamental EA services, and EA services need to be adapted to specific enterprise contexts.

TABLE 25.1   Example EA Service Model

EA Services	Design	Development	Management	Training and Support
EA processes	• Define standards • Define guidelines • Define principles	• Build • Implement	• Governance • Planning	• Consulting • Training • Support
Roles	• Business architect	• Enterprise architect	• Business architect • Enterprise architect	• All
Outcomes (deliverables)	• Standards • Guidelines • Principles	• Strategic architecture • Solution architecture • Artifacts	• Reports	• Training material
Benefits	• Standardization	• Reduced complexity • Improved integration	• Improved decision support • Improved risk management • Reduced costs	• Common understanding • Improved skills • Improved knowledge

25.2.2 Role of IT Governance

In the context of EA, the role of IT governance is important since it provides a means for management to direct and govern IT toward desired outcomes (IT Governance Institute). IT governance is a tool that provides necessary means to achieve strategies and goals. EA is crucial for effective IT governance because it provides methods to measure success (Weill et al. 2004). However, a holistic model that describes IT governance in an enterprise context has not surfaced due to the complexities and diversity of perspectives inherent in the domain (Rüter et al. 2006). EA governance is a subdiscipline of IT governance that focuses on the principles, decision rights, rules, and methods that drive architecture development and alignment in an organization (Greefhorst et al. 2011). EA governance is concerned with standardization and compliance with current and future practices.

25.2.3 Business Architecture

BA, the top layer of EA, typically contains business processes/services, organizational structures (including roles and responsibilities), and value drivers, which are aligned to a strategy divided into goals and objectives (The Open Group 2011; Versteeg et al. 2006). BA marks the foundation of effective BITA, and sample scenarios in a BA context include (The Open Group 2010) the following:

- Mergers and acquisitions planning and developments
- Business unit consolidations
- New product and service rollouts
- Introduction of new lines of business
- Consolidating suppliers across supply chains
- Outsourcing business functions
- Divesting business lines
- Management changes
- Regulatory compliances
- Operation cost reductions
- Federated architecture alignments in government
- Business transformations
- Entering international markets

25.2.4 Application Architecture

AAs deal with combined logical capabilities, developing specifications for individual applications including interactions and relationships to business processes. Applications offer services to other business units. Within AA, service-oriented architectures (SOA) (Erl 2007; Krafzig et al. 2007) and cloud computing (Mahmood 2011) are discussed often. These emerged as recent EA paradigms, besides classical enterprise application integration challenges.

25.2.5 Data/Information Architecture

We argue data and information architectures (DA/IA) are situated as part of the AA layer, due to its importance and relation to applications. This does not mean this sub-layer is unimportant or can be seen in isolation; in fact, many view it as the core of an enterprise relating to multiple layers of EA, referred to as the data-centric view. Information consists of business objects such as orders and production plans, and includes raw information gathered throughout the company by means of performance measures (Neely 2004) and translating it into meaningful information as a key objective of EA (Godinez et al. 2010). In this way, DA/IA relates to business processes and translates information into a business context. Consequently, it is an important EA aspect because it has significant impacts on the business, and quality of data and information is particularly important (Borek et al. 2012). It comprises the logical, physical, and management views that provide sufficient information to form a basis for communication, analysis, and decision-making (Allen et al. 1991; Xie et al. 2011). For example, EA is concerned with data models, logical and physical models, data flows and transformations, metadata, and standards on how information should be stored and exchanged.

25.2.6 Technology Architecture

The technical layer of EA is concerned with an underlying infrastructure (i.e., hardware and system-level software). Hardware includes servers and network infrastructures, and all of their components. Applications are built using appropriate infrastructures, and hence, this layer provides required infrastructure services. An example service at this layer is the provisioning of a server. Technology architecture offers infrastructure services to applications; notably a database is a type of infrastructure.

25.3 Reasons for Enterprise Architecture

To provide a review of reasons to employ EA, we examine various EA drivers and key EA applications, and we describe benefits and common outputs of EA.

25.3.1 EA Drivers

EA drivers are the primary reasons for enterprises to employ EA, and one of the most important drivers is BITA (Schöenherr 2008). Critical aspects are described in detail when aligning business and IT (Henderson et al. 1993; Luftman 2003; Mahr 2010; Pereira and Sousa 2005). Cost reductions and managing complexities are other important EA drivers. For example, dismantling legacy systems and components reduces costs, and EA reduces complexity by providing a streamlined and manageable system landscape described through various models. In addition to these important and frequently mentioned drivers, some external drivers do not originate within the enterprise. An example is regulatory compliance requirements issued by legal organizations or governments to which the enterprise must adhere.

TABLE 25.2 Benefits of EA

Technology-related benefits	
IT costs	When reducing non-value-added variations of technologies, a company can reduce IT operations unit and application maintenance costs.
IT responsiveness	Through standardization, decision-makers spend less time choosing the right technologies or dealing with recurring errors; thus, development time is reduced, increasing overall IT responsiveness.
Risk management	IT infrastructure cleanup improves manageability, and thus, contributes to reduced business risk, improved regulatory compliance, increased disaster tolerance, and reduced security breaches.
Business-related benefits	
Shared business platforms	Through data and process standardization, greater data sharing and integrated process standards emerge.
Managerial satisfaction	Although subjective, satisfaction indicates the confidence of business executives in the ability of IT to deliver business value; an increase in senior management satisfaction and business unit IT leadership results from effective EA.
Strategic business impacts	EA enables: • Operational excellence • Customer intimacy • Product leadership • Strategic agility

25.3.2 EA Benefits and Value Contributions

Outcomes and benefits depend on context and the extent to which an enterprise uses EA. Usual EA output is described in the form of principles, models, and architecture views. Less tangible outputs in the form of common understanding and improved communications are benefits of EA. In the literature, a number of researchers discuss EA contributions in terms of direct benefits and business value (Ahlemann et al. 2012b; Meyer et al. 2012; Ross et al. 2006; van Steenbergen et al. 2011). We summarize some of the typical EA benefits in Table 25.2 (Ross et al. 2005).

25.4 EA Frameworks

A number of EA frameworks emerged over the past few decades (Schekkerman 2006), and several studies of evaluations and analyses of select EA frameworks exist, based usually on varying foci (Leist and Zellner 2006; Tang et al. 2004). To provide an overview of typical frameworks, we describe well-known and widely used ones in the following section. Common to these frameworks is reducing enterprise complexities by considering disparate viewpoints and organizing various aspects in ways that make an enterprise understandable. Another important purpose is support for an architecture development process. The EA process provides directions and guidelines to architects to design a model or even an entire architecture.

25.4.1 Zachman EA Framework

Seminal work on the Zachman framework (Zachman 1987) was a key contribution to EA. As IT became more complex, a need arose for a method to organize complex IS. For this reason, primitive interrogatives of *what*, *how*, *when*, *who*, *where*, and *why* identify aspects of an enterprise. The reification or transformation of an abstract idea into an instantiation is labeled *Identification*, *Definition*, *Representation*, *Specification*, *Configuration*, and *Instantiation*. Intersections among the interrogatives and transformations result in 6 by 5 matrix. Another important aspect of this framework is consideration of roles (i.e., stakeholders) related to the various perspectives.

The Zachman framework provides a schema for describing the structure of an enterprise; it represents ontology since it provides a structured set of an object's essential components. It is a meta-model, without implying how to establish architecture. In its current version 3.0, the Zachman framework is an 8 by 8 matrix, depicted in Figure 25.2. It is important to note that this framework was not conceived originally as a methodology since it did not provide a process to instantiate an object (i.e., implement the enterprise). The limited focus on the meta-model design may not warrant consistency among layers and views. Nevertheless, there exist some methodological approaches for applying this framework (e.g., Marques Pereira et al. 2004).

25.4.2 Generalized Enterprise Reference Architecture and Methodology

To provide a conceptual basis for IS architectures, the Generalized Enterprise Reference Architecture and Methodology (GERAM) framework was developed by the IFAC/IFIP (IFIP IFAC Task Force on Architectures for Enterprise Integration 2003), building on (Bernus et al. 1996, 1997). One advantage of using GERAM is that it is not a domain-specific architecture; it is a general framework of concepts for designing and maintaining any type of enterprise. GERAM is a framework in which all methods, models, and tools necessary to build and maintain an integrated enterprise are defined; however, it is limited in providing guidelines or reference models for EA in practice. The framework applies to all types of enterprises, including part of an organization, a single organization, or a network of organizations.

Illustrated in Figure 25.3, GERAM is based on the lifecycle concept and identifies three dimensions for defining the scope and content of enterprise modeling:

- *Lifecycle dimensions*, which provide for a controlled modeling process of enterprise entities according to lifecycle activities
- *Genericity dimensions*, which provide for a controlled instantiation process from generic and partial to particular
- *View dimensions*, which provide for a controlled visualization of views of the enterprise entity

To facilitate efficient modeling by enhancing the modeling process, GERAM uses one important concept; it provides various partial enterprise models with an underlying generic enterprise modeling concept. Instead of developing new models within the process, models and concepts are adopted and configured for each situation. Since these concepts and models play an important role in integrating heterogeneous systems, we focus on partial enterprise models (i.e., reference models), including their supported generic modeling concepts. Partial enterprise models are reusable models of human roles, processes, and technologies that capture characteristics common to many enterprises within and across one or more industries. They include various enterprise entities such as products, projects, and companies and may represent each from various viewpoints. To describe multiple levels of abstraction along lifecycle phases, partial models exist at various levels, including concept, requirements, and design levels. Generic enterprise modeling concepts define enterprise modeling constructs. These concepts may be defined using natural language (e.g., glossaries), meta-models (e.g., entity relationship meta-schema), or ontological theories. Bernus et al. (2010) took the GERAM meta-model to the next level by incorporating standards such as IEEE Computer Society (2007) and ISO/IEC (2000), though one of the older frameworks never found a place in practice because of its overwhelming complexity.

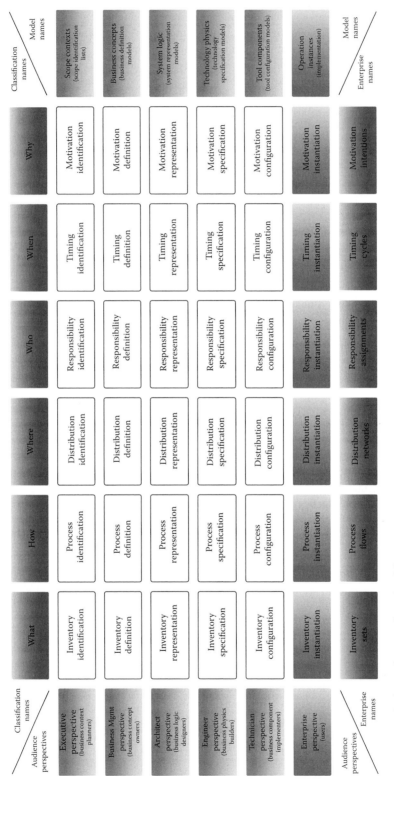

FIGURE 25.2   Zachman framework version 3.0.

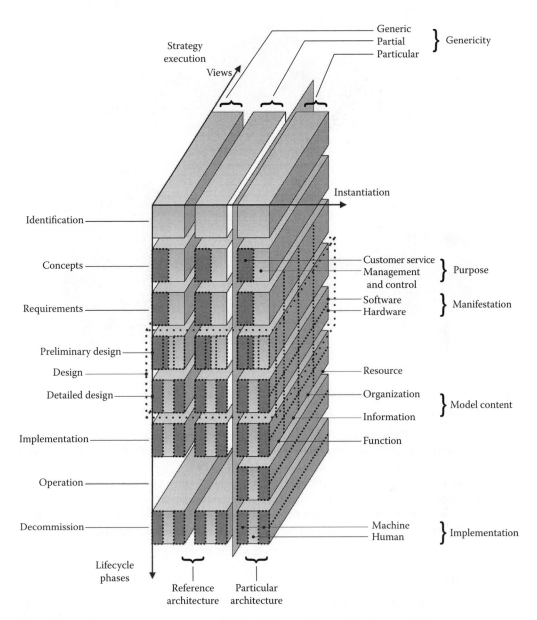

FIGURE 25.3 GERAM cube.

25.4.3 ARIS

The ARIS framework is a prominent architectural framework often used in practice that represents various aspects of information system design (Scheer 1992). The framework has its roots in the 1980s, with a trend toward corporate data models and material requirements planning. The ARIS house (Figure 25.4) is an architecture that describes business processes, providing modeling methods and meta-structures combined with suggestions for modeling techniques. As a modeling technique and methodology, ARIS is a model for comprehensive, computer-aided business process management, although only representing selected EA views and layers. The concept is also incorporated in a toolset

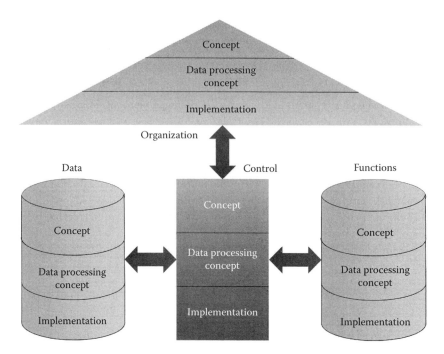

FIGURE 25.4   ARIS house.

to support modeling. To reduce complexity, the model is divided into individual views that represent discrete design aspects:

- *Data view*: Conditions and events form the data view. Events such as order, order reception, and production release are information objects represented by data. Reference field conditions such as customer status and article status are also represented by data
- *Function view*: Functions to be performed and relationships among them
- *Organization view*: Structure and relationships among staff members and organizational units
- *Control view*: An essential component that restores and retains relationships among the various views. By structuring a process, it incorporates a data, function, and organizational view

25.4.4 TOGAF

TOGAF (The Open Group 2011) is a comprehensive methodology, including a set of tools to develop an EA. It is based on an iterative process model supported by best practices and a reusable set of existing architecture assets. TOGAF offers clear definitions for various terms, including basics such as architecture(s), views, and stakeholders. TOGAF is organized into three primary sections: *Architecture Development Method* (ADM), *Enterprise Continuum*, and *Architecture Repository*. Additionally, TOGAF provides a comprehensive content meta-model that describes entities and their relationships to ensure common understanding of framework mechanics.

The ADM is a generic method used to realize an organization-specific EA in accord with business requirements. It is organized into several phases, each of which contains various steps to achieve the desired output at the current stage of development. Combined, all phases mark the complete development lifecycle. The *Enterprise Continuum* represents a view of the repository containing all architecture assets, including architecture descriptions, models, building blocks, patterns, and viewpoints, in addition to other architecture and solution artifacts. Architecture and solution artifacts can be internal or external. Internal artifacts are extant architecture assets such as models that are ready for reuse. External artifacts

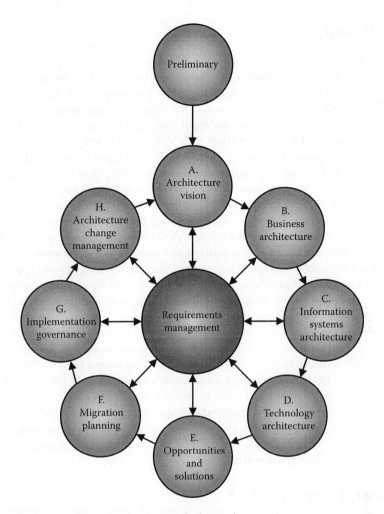

FIGURE 25.5 TOGAF architecture development method (ADM).

comprise industry reference models and architecture patterns that are applicable to aspects of IT and industry and are constantly emerging and evolving. The *Architecture Repository* supports the Enterprise Continuum by storing classes of architectural output at varying levels of abstraction. The ADM generates this output during several phases. The ADM and its phases are depicted in Figure 25.5.

25.4.5 Other Frameworks

While there exist a number of dedicated EA frameworks (Schekkerman 2006), there are also some related IT frameworks, occasionally working with them, such as IT Infrastructure Library (ITIL) (Office of Government Commerce 2007) and Control Objectives for Information and Related Technology (COBIT) (IT Governance Institute 2007). ITIL is the standard for IT service management. Control objectives for Information and Related Technology (COBIT) began as an IT governance framework, but it has evolved since to a comprehensive IT framework comprised of control objectives, reference processes, management guidelines, and maturity models. Both frameworks mention EA explicitly and demonstrate how they consider EA's position. COBIT considers EA an explicit IT function, concerned with IT processes with goals derived from business goals as input. ITIL follows a more traditional domain or layer approach, introducing domains such as service, environmental, management, and product architectures.

25.5 EA Modeling

EA modeling is an essential part of contemporary EA practice. Since the EA discipline is concerned with developing a blueprint for an enterprise, practitioners use a variety of models such as data, business interaction, and organizational models (Olivé 2007). An example of an EA modeling language is ArchiMate (The Open Group 2012), based loosely on Unified Modeling Language (UML). ArchiMate supports three layers for modeling—business, application, and technology—each of which contains various modeling elements. The language aligns fully with TOGAF, and each layer connects to ADM phases. Looking at the content meta-model of ArchiMate (The Open Group 2012), each layer offers services. The technology layer offers infrastructure services to the application layer, which in turn offers application services to the business layer. The latter provides the actual business service, consumed either within the enterprise or by external clients. We reflect the SOA paradigm in this language.

Chen et al. (2009) present a domain-modeling approach based on the Zachman framework. UML is one of the most important languages in enterprise modeling. With its great diversity of diagrams, it offers views and models for many situations both at a high level of abstraction when it comes to conceptualization and at a low level for comprehensive technical documentation of inner application aspects. Although the primary foci and applications of UML are software architectures, many enterprise architects use UML (or some derivative) to accomplish the modeling part of their work (i.e., to produce EA artifacts based on models) (Kruchten et al. 2006).

At the center of BAs are business processes, which capture dynamic aspects of the business element in an enterprise. Regarding modeling, there are many process and workflow modeling languages such as Business Process Modeling Notation (Business Process Management Initiative 2011), Event Driven Process Chains (Scheer 1999), Business Process Execution Language (OASIS 2007), and others (List and Korherr 2006). Business Process Management System is a generic software system driven by explicit process designs to enact and manage operational business processes (Weske 2007). It offers a platform that allows both modeling and execution, and is an important part of enterprise modeling and, hence, EA (Grunninger 2003; Scheer et al. 2000). Regardless of models or modeling languages, it is critical that stakeholders understand them (Kaisler et al. 2005).

25.6 Practical Aspects of EA

In larger enterprises, a team of enterprise architects usually conducts the EA function (Murer et al. 2011). In smaller companies, EA is often part of the IT department, with no separate EA unit. Most companies use some of the common EA frameworks even though they are, in many cases, adapted or customized. However, employing a framework alone does not provide an EA; architects are a necessary component. Depending on the company, there are a few architecture job descriptions with varying skill sets such as solution, data, and enterprise architects. We must be concise about who participates in the EA function and what their function is. The Open Group (2011) details a presentation of EA roles and associated skills, and Ahlemann et al. (2012a) outline another explanation of what types of architects are needed. Another practical aspect is EA maturity, indicating how much the function has evolved in terms of various aspects such as architecture planning, value, and stakeholder communication. All EA processes should be supported by a variety of EA tools, and we examine that support in terms of EA processes and services.

25.6.1 EA Maturity

Maturity is a measure that describes advancement and progress in business processes or EA. At each stage, maturity indicates whether requirements are met. Ross (2003) outlines a general description of EA maturity and its stages. The four stages are characterized by management practices to design and protect the architecture. In the first stage, we find business silos, and within them, business cases and project methodologies. In stage two, standardized technologies and structures are in place, including an IT steering committee, a formal compliance process, centralized handling of enterprise applications, and architects on a project team, among other practices. An optimized core marks the third stage where process owners, EA principles, business leadership of project teams, and IT program managers are in place. The last maturity stage is called business modularity, characterized by an EA graphic, post-implementation assessments, technology research, adoption processes, and a full-time EA team.

There exists a variety of EA maturity frameworks, and Meyer et al. (2011) present an overview and analysis. All of these frameworks possess classifications and characteristics of an EA and a method to assess it. The IT Capability Maturity Framework (IT-CMF) (Curley 2006, 2009) offers five maturity levels for nine capability building blocks. It comprises aspects of planning, practices in the form of governance, processes and value contributions, and organizational and communication aspects. IT-CMF offers detailed assessment for each capability building block.

25.6.2 EA Tools

Companies have broad choices when using EA tools, and these tools help enterprise architects and other stakeholders plan, model, develop, and monitor architectures. As with most business software, companies are not always dependent on vendor solutions, and may choose to develop their own. Most EA tools offer

- EA planning
- EA modeling
- EA simulation and improvement
- EA monitoring and reporting
- Shared artifact repository
- Reusability support

An overview of EA tools and a selection guide is available online from the Institute For Enterprise Architecture Developments (2012), and Wilson and Short (2011) provides a magic quadrant for EA tools.

25.7 Critical Aspects

EA should reduce structural and operational complexity within an organization. Conducted properly, EA delivers projected benefits as discussed earlier, but it can be an investment without any significant return. One challenge with EA is that it is not always obvious what enterprise architects do, how architects generate business value, and whether investments in EA are necessary (Meyer et al. 2012; Robertson 2008). A first step is to generate an EA service model. Mentioned earlier, EA has come a long way over the past few decades, but there is still no single definition or unifying framework. GERAM provides a useful meta-framework, however, with less practical applicability. The Zachman framework and ARIS offers great flexibility and is often used in practice, but is limited with its meta-model and methodology. TOGAF together with ArchiMate supports modeling in conjunction with an ADM but can be challenging to adapt in a practical setting. In practice, often a combination of several frameworks can be found. Also, several studies of evaluations and analyses of select EA frameworks exist (e.g., Leist and Zellner 2006; Tang et al. 2004). Furthermore, distinctions among processes such as EA development and EA management do not appear in the literature. Although Schöenherr (2008) points out that common terminology is important for the EA discipline, we are unable to offer concise descriptions of what EA is and does, and Kaisler et al. (2005) mention critical problems with EA that are barely solved after years of research.

25.8 Summary

Since its origins, described as IS Architecture in the 1980s and 1990s, EA evolved into an extensive domain with numerous approaches and frameworks. A core characteristic of EA is that it enables and supports constant transformations of an enterprise from a current to a target state. EA supports this transformation with tools and frameworks to manage complex enterprise systems. We outline prominent frameworks and modeling approaches and describe the importance of practical tools. EA provides directions and guidelines for high-level enterprise evolvement by encapsulating and relating business aspects, application portfolios, information and data, and technology. Related to strategic management, alignment of business and IT is one of the most important drivers of EA. An effective EA function yields a number of benefits such as reduced complexity, reduced risks, reduced costs, and improved business–IT alignment (Lankhorst 2009; Op 't Land et al. 2009; Ross et al. 2005, 2006; van Steenbergen et al. 2011). Enterprises evolve constantly; producing drivers for employing an EA function—with all its capabilities—make it an ever-emerging discipline. New developments and innovative modeling approaches, tools, and model-based EA frameworks can be expected, combined with practical methods to employ and value EA in enterprises. These are the foci of current research and development.

References

Ahlemann, F., Legner, C., and Schäfczuk, D. EAM governance and organization, in: *Strategic Enterprise Architecture Management—Challenges, Best Practices, and Future Developments*, Ahlemann, F., Stettinger, E., Messerschmidt, M., and Legner, C. (eds.), Berlin Heidelberg, Germany, Springer, 2012a, pp. 81–110.

Ahlemann, F., Legner, C., and Schäfczuk, D. Introduction, in: *Strategic Enterprise Architecture Management—Challenges, Best Practices, and Future Developments*, Ahlemann, F., Stettinger, E., Messerschmidt, M., and Legner, C. (eds.), Berlin Heidelberg, Germany, Springer, 2012b, pp. 1–34.

Allen, B. R. and Boynton, A. C. Information architecture: In search of efficient flexibility, *MIS Quarterly* (15:4) 1991, 435–445.

Bernus, P. and Nemes, L. A framework to define a generic enterprise reference architecture and methodology, *Computer Integrated Manufacturing Systems* (9:3) 1996, 179–191.

Bernus, P. and Nemes, L. Requirements of the generic enterprise reference architecture and methodology, *Annual Reviews in Control* (21) 1997, 125–136.

Bernus, P. and Noran, O. A *Metamodel for Enterprise Architecture, Enterprise Architecture, Integration and Interoperability*, Berlin Heidelberg, Germany, Springer, 2010, pp. 56–65.

Borek, A., Helfert, M., Ge, M., and Parlikad, A. K. IS/IT Resources and Business Value: Operationalization of an Information Oriented Framework, *EIS LNBIP 102*, Berlin Heidelberg, Springer, 2012, pp. 420–434.

Business Process Management Initiative. Business Process Modeling Notation (BPMN 2.0), 2011.

Chen, Z. and Pooley, R. Domain modeling for enterprise information systems—Formalizing and extending Zachman framework using BWW ontology, in: *WRI World Congress on Computer Science and Information Engineering 2009*. Los Alamitos, CA, IEEE Computer Society, 2009, pp. 634–643.

Chief Information Officer U.S. Department of Defense. DoD Architecture Framework 2.02, 2010.

Curley, M. An IT value based capability maturity framework, in: *CISR Research Briefing*. Cambridge, MIT Sloan CISR, 2006.

Curley, M. Introducing an IT capability maturity framework. Berlin Heidelberg, Springer, 2009, pp. 63–78.

Dern, G. *Management von IT-Architekturen*. Wiesbaden, Vieweg+Teubner, 2009.

Erl, T. *SOA: Principles of Service Design*. Upper Saddle River, Prentice Hall, 2007.

Godinez, M., Hechler, E., Koenig, K., Lockwood, S., Oberhofer, M., and Schroeck, M. *The Art of Enterprise Information Architecture*. Upper Saddle River, IBM Press, 2010.

Greefhorst, D. and Proper, E. *Architecture Principles*. Berlin Heidelberg, Germany, Springer, 2011.

Grunninger, M. *Enterprise Modelling*. Berlin Heidelberg, Germany, Springer, 2003.

Hanschke, I. *Strategisches Management der IT-Landschaft*, (2nd edn.). München, Hanser, 2010.

Hasselbring, W. Information systems integration, *Communications of the ACM* (43:6) 2000, 32–38.

Henderson, J. C. and Venkatraman, H. Strategic alignment: Leveraging information technology for transforming organizations, *IBM Systems Journal* (38), 1993, pp. 472–484.

IEEE Computer Society. Std 1471-2000: Systems and software engineering, in: *Recommended Practice for Architectural Description of Software-Intensive Systems*, New York, IEEE, 2007.

IFIP IFAC Task Force on Architectures for Enterprise Integration *GERAM. The Generalized Enterprise Reference Architecture and Methodology*. Berlin Heidelberg, Germany, Springer, 2003.

Institute for Enterprise Architecture Developments. *Enterprise Architecture Tools Overview*. Amersfoort, The Netherlands, IAFD, 2012.

ISO/IEC. ISO 15704:2000 Industrial automation systems—Requirements for enterprise-reference architectures and methodologies, 2000.

IT Governance Institute. Control objectives for information and related technology (COBIT) 4.1, 2007.

Kaisler, S. H., Armour, F., and Valivullah, M. Enterprise architecting: Critical problems, *Proceedings of the 38th International Conference on System Sciences*, Hawaii, IEEE, 2005.

Kim, Y.-G. and Everest, G. C. Building an IS architecture: Collective wisdom from the field, *Information & Management* (26:1) 1994, 1–11.

Krafzig, D., Banke, K., and Slama, D. *Enterprise SOA: Service-Oriented Architecture Best Practices*. Upper Saddle River, Prentice Hall PTR, 2007.

Kruchten, P., Obbink, H., and Stafford, J. The past, present, and future for software architectures, *IEEE Software* (23:2) 2006, 22–30.

Lankhorst, M. *Enterprise Architecture at Work*, (2nd edn.). Berlin Heidelberg, Germany, Springer, 2009.

Leist, S. and Zellner, G. Evaluation of current architecture frameworks. *Proceedings of the 2006 ACM symposium on Applied computing*, Dijon, France, ACM, 2006, pp.1546–1553. 2006.

List, B. and Korherr, B. An evaluation of conceptual business process modelling languages. SAC'06, ACM, 2006.

Luftman, J. N. *Competing in the Information Age: Align in the Sand*. New York, Oxford University Press, 2003.

Mahmood, Z. Cloud computing for enterprise architectures: Concepts, principles and approaches, in: *Cloud Computing for Enterprise Architectures*, Mahmood, Z. and Hill, R. (eds.), Berlin Heidelberg, Germany, Springer, 2011, pp. 3–19.

Mahr, F. *Aligning Information Technology, Organization, and Strategy—Effects on Firm Performance*. Wiesbaden, Gabler, 2010.

Marques Pereira, C. and Sousa, P. A method to define an enterprise architecture using the Zachman framework, ACM Symposium of Applied Computing, Nicosia, Cyprus, 2004.

Meyer, M., Helfert, M., Donnellan, B., and Kenneally, J. *Applying Design Science Research for Enterprise Architecture Business Value Assessments*, DESRIST 2012, Berlin Heidelberg, Germany, Springer, 2012, pp. 108–121.

Meyer, M., Helfert, M., and O'Brien, C. *An Analysis of Enterprise Architecture Maturity Frameworks*, BIR 2011 LNBIP 90, Berlin Heidelberg, Springer, 2011, pp. 167–177.

Morton, M. S. S. Introduction, in: *The Corporation of the 1990s—Information Technology and Organizational Transformation*, Morton, M.S.S. (ed.), New York, Oxford University Press, 1991, pp. 3–23.

Murer, S., Bonati, B., and Furrer, F. J. *Managed Evolution—A Strategy for Very Large Information Systems*. Berlin Heidelberg, Germany, Springer, 2011.

Neely, A. *Business Performance Measurement: Theory and Practice*. Cambridge, Cambridge University Press, 2004.

OASIS. Web Services Business Process Execution Language Version 2.0, 2007.

Office of Government Commerce. IT infrastructure library 3.0, The Stationary Office, 2007.

Office of Management and Budget. Federal enterprise architecture (FEA), 2012.

Olivé, A. *Conceptual Modeling of Information Systems*. Berlin Heidelberg, Germany, Springer, 2007.

Op 't Land, M., Proper, E., Waage, M., Cloo, J., and Steghuis, C. *Enterprise Architecture*. Berlin Heidelberg, Springer, 2009.

Pereira, C. M. and Sousa, P. Enterprise architecture: Business and IT alignment, in: *Proceedings of the 2005 ACM symposium on Applied computing*, New York, ACM, 2005, pp. 1344–1345.

Robertson, B. Organize your enterprise architecture effort: Services, *Gartner Research* G00160689, 2008, pp 1–7.

Ross, J. W. Creating a strategic IT architecture competency: Learning in stages. *MISQ Executive* (2:1) (March) 2003, 31–43.

Ross, J. W. and Weill, P. Understanding the Benefits of Enterprise Architecture, in: *CISR Research Briefing*. Cambridge, MIT Sloan CISR, 2005.

Ross, J. W., Weill, P., and Robertson, D. C. *Enterprise Architecture as Strategy*. Boston, Havard Business Press, 2006.

Rüter, A., Schröder, J., Göldner, A., and Niebuhr, J. *IT-Governance in der Praxis—Erfolgreiche Positionierung der IT im Unternehmen. Anleitung zur erfolgreichen Umsetzung regulatorischer und wettbewerbsbedingter Anforderungen*. Berlin Heidelberg, Germany, Springer, 2006.

Scheer, A.-W. *Architecture of Integrated Information Systems: Foundations of Enterprise Modelling*. Berlin Heidelberg, Springer, 1992.

Scheer, A.-W. *ARIS—Business Process Frameworks*. Berlin Heidelberg, Germany, Springer, 1999.

Scheer, A.-W. and Nüttgens, M. *ARIS Architecture and Reference Models for Business Process Management*, Business Process Management, Models, Techniques, and Empirical Studies, Berlin Heidelberg, Springer, 2000, pp. 301–304.

Schekkerman, J. *How to Survive in the Jungle of Enterprise Architecture Frameworks: Creating or Choosing an Enterprise Architecture Framework*. Victoria, Trafford, 2006.

Schmidt, C. and Buxmann, P. Outcomes and success factors of enterprise IT architecture management: Empirical insight from the international financial services industry. *European Journal of Information Systems* (20), 2011, pp. 168–185.

Schöenherr, M. Towards a common terminology in the discipline of enterprise architecture, in: *LNCS 5472*, Feuerlicht, G. and Lamersdorf, W. (eds.), Springer, Berlin Heidelberg, Germany, 2008, pp. 400–413.

van Steenbergen, M., Mushkudiani, N., Brinkkemper, S., Foorthuis, R., Bruls, W., and Bos, R. Achieving enterprise architecture benefits: What makes the difference?, *15th International Enterprise Distributed Object Computing Conference Workshops*, Helsinki, Finland, 2011, pp. 350–359.

Tang, A., Han, J., and Chen, P. A comparative analysis of architecture frameworks, *Proceedings of the 11th Asia-Pacific Software Engineering Conference* (APSEC'04), 2004.

The Open Group. Business architecture scenarios. 2010.

The Open Group. *The Open Group Architecture Framework (TOGAF) Version 9.1*, 2011.

The Open Group. ArchiMate 2.0 Specification, 2012.

Versteeg, G. and Bouwman, H. Business architecture: A new paradigm to relate business strategy to ICT. *Information Systems Frontiers* (8:2), 2006, 91–102. Berlin Heidelberg, Germany, Springer.

Weill, P. and Ross, J. *IT Governance*. Boston, Harvard Business School Press, 2004.

Weske, M. *Business Process Management: Concepts, Languages, Architectures*. Berlin Heidelberg, Germany, Springer, 2007.

Wilson, C., and Short, J. Gartner Magic Quadrant for Enterprise Architecture Tools. *Gartner Research* G00219250, 2011.

Winter, R. and Fischer, R. Essential layers, artifacts, and dependencies of enterprise architecture, in: *10th IEEE International Enterprise Distributed Object Computing Conference Workshops (EDOCW'06)*, Hong Kong, China, IEEE Computer Society, 2006.

Xie, S. and Helfert, M. An architectural approach to analyze information quality for inter-organizational service, *Proceedings of the 13th International Conference on Enterprise Information Systems (ICEIS 2011)*, Beijing, China, 2011, pp. 438–443.

Zachman, J. A. A framework for information systems architecture. *IBM Systems Journal* (26:3) 1987, 276–292.

26

Business Process Management and Business Process Analysis

Jan vom Brocke
University of Liechtenstein

Christian
Sonnenberg
University of Liechtenstein

26.1 Introduction

"Panta rhei"—"everything flows"—an aphorism leading back to the thoughts of Heraclitus provides a metaphor for the *dynamic nature of physical existence*. Humans have been conceiving sets of actions ever since to deal with and survive in a dynamic environment. Sets of consecutive actions that are directed to some end are commonly referred to as a *process*. In fact, human societies are organized around such generic processes as crafting tools, producing energy, producing food, producing goods, building structures, making astronomical observations, communicating messages, or celebrating rituals. Process can also be perceived on the level of an individual like preparing a meal, buying a car, or simply completing a to-do list. For processes that require the participation of multiple individuals and that are intended to be executed on a continuing basis are usually embedded within an *organization*.

Consequently, an organization can be best described in terms of *what it does and can do* (cf. Rescher, 1996) and processes—in a very simple sense—are *what companies do* (cf. vom Brocke and Rosemann, 2013). The capability and potential performance of an organization is thus built into its processes. Processes do happen, regardless whether we manage them or not. Managing processes, therefore, intends to make sure that the organization meets the requirements in operating processes that support the strategic corporate objectives. Over the past decades, business process management (BPM) emerged as an academic discipline that intersects with both information systems and management research. BPM represents an integrated, process-oriented management approach utilizing tools and methods for business process analysis and improvement (cf. Rosemann and vom Brocke, 2010).

The BPM discipline, as it is understood today, evolved around two main strands (Hammer, 2010):

- *Continuous change*: Earlier studies focused on continuous or incremental improvement of existing processes. Examples of this approach have been total quality management (TQM), lean management, or kaizen. Basic methodological principles have been contributed by Deming, for example, on statistical process control, dealing with a systematic analysis of processes by means of both quantitative and qualitative criteria (Deming, 1986).
- *Radical change*: By introducing the concept of business process reengineering, Hammer and Champy (1993) advocated an approach, which is based on the fundamental rational to challenge all existing processes in an organization. Here, the goal is to rethink and, if deemed appropriate, redesign business processes from scratch (so-called green field approach). Information and communication technology is considered to be the main facilitator and driver of such radical process redesigns (Davenport, 1993). The focus is less on fine grained analytical methods but more on discussing the design of "large-scale, truly end-to-end processes" (Hammer, 2010, p. 4).

Today, BPM aims to combine both approaches (Hammer, 2010), recommending that there should be an alternation of activities of continuous improvement and radical change (vom Brocke and Rosemann, 2013).

26.2 Core Elements of BPM

In order to enable the management of processes in an effective and efficient way, companies need to develop a wide range of capabilities. From research on maturity models (de Bruin et al., 2005), six central core elements for BPM were derived (Rosemann and vom Brocke, 2010). They are shown in Table 26.1 and are briefly explained in the following.

- *Strategy*: Process management needs to be guided by strategic organizational goals. The main question to answer: How can each process be assessed in order to be able to determine its specific value?
- *Governance*: Process management needs to be rooted in the organizational structure. The main question to answer: Who is supposed to be responsible for which process management task?
- *Methods*: Process management needs methods for process design supporting different phases of process development. The main question to answer: What methods are being used by the organization?
- *Technology*: Process management needs to be supported by technology, particularly information technology (IT), as the basis for process design. The main question to answer: What systems are available in the organization in order to be able to execute different process management tasks?
- *People*: Process management needs certain capabilities and skills on the part of an organization's staff. The main question to answer: What capabilities and skills are available and what must be done to develop missing capabilities?
- *Culture*: Process management needs a common value system in order to effect certain changes required. The main question to answer: Is the culture of an organization and the basic attitude of its people supportive of change and innovation?

Research on BPM has shown that no core element should be neglected when BPM is to be implemented successfully in an organization (e.g., Bandara et al., 2007; Kettinger and Grover, 1995; Trkman, 2010). The challenge is to develop each core element adequately while at the same time matching all core elements with each other in a reasonable way. The context an organization is embedded in plays a crucial role here. There is no "out-of-the BPM approach" that can be readily applied to any organization whatsoever.

TABLE 26.1 Core Elements of BPM

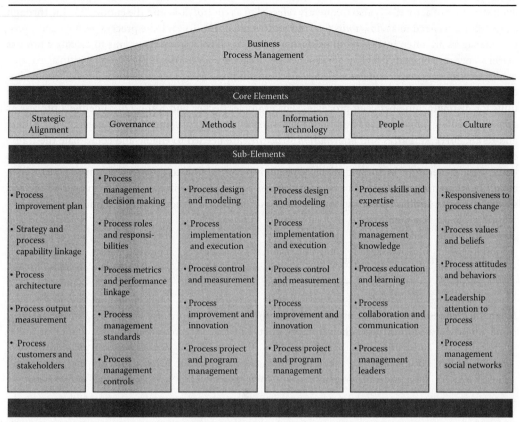

Business Process Management					
Core Elements					
Strategic Alignment	Governance	Methods	Information Technology	People	Culture
Sub-Elements					
• Process improvement plan • Strategy and process capability linkage • Process architecture • Process output measurement • Process customers and stakeholders	• Process management decision making • Process roles and responsibilities • Process metrics and performance linkage • Process management standards • Process management controls	• Process design and modeling • Process implementation and execution • Process control and measurement • Process improvement and innovation • Process project and program management	• Process design and modeling • Process implementation and execution • Process control and measurement • Process improvement and innovation • Process project and program management	• Process skills and expertise • Process management knowledge • Process education and learning • Process collaboration and communication • Process management leaders	• Responsiveness to process change • Process values and beliefs • Process attitudes and behaviors • Leadership attention to process • Process management social networks

Source: Rosemann, M. and vom Brocke, J., The six core elements of business process management, In: *Handbook on Business Process Management*, Vol. 1, ed. vom Brocke, J. and Rosemann, M., Springer, Berlin, Germany, pp. 107–122, 2010.

Every organization has to analyze and decide very precisely what their particular BPM approach with regard to each of the six core elements should look like.

The particular BPM approach taken is dependent not only on the actual maturity level attained within the six core elements but also on the organization-specific perception of the business process concept. There are several views about the nature of business processes that require BPM activities to be approached differently depending on the view adopted. The subsequent section briefly sketches four possible views on business processes and outlines how BPM activities—in particular process analysis tasks—can be organized within a BPM life cycle. It will also be discussed how different views both enable and limit specific approaches to business process analysis.

26.3 On the Nature of Business Processes and Process Analysis

26.3.1 Common Business Process Definitions

A typical definition of a business process as presented in a large number of BPM literature regards business processes as a *sequence of activities generating a specific value for a customer* (e.g., Davenport, 1993; Hammer and Champy, 1993). The activities contained in a business process are assumed to transform some *input* of suppliers into *outputs* for customers, whereby the transformation activities can be hierarchically decomposed into sub-processes (vom Brocke and Rosemann, 2013). *Activities* in a process are not always arranged in a strict sequence but they are rather linked though network structures.

These network structures determine the order in which entities may flow through a process. The network structure of activities is also frequently referred to as *control flow*, and the entities flowing through a process are referred to as *flow units* (cf. Anupindi et al., 2012). To evaluate process performance, process managers are often interested in analyzing the paths of individual flow units to calculate process performance measures such as *flow time*, *inventory* (total number of flow units present within a process), *quality*, *cost*, or *value-added* (determined in terms of added or changed flow unit attributes that customers consider important). In this regard, some process definitions account for the significance of flow units for distinguishing between processes. For example, Becker and Kahn (2011) define a process as a "timely and logical sequence of activities necessary to handle a business relevant object" (cf. Becker and Kahn, 2011). *Resources* and *information structures* are needed to enable and coordinate the transformation of inputs into outputs. *Resources* (like raw materials, manufacturing facilities, equipment, or employees) are required by individual activities within control flow to process flow units. An appropriate *information structure* supports managerial decision making within a process on an operational and strategic level. In summary, five aspects are central for managing processes (cf. Anupindi et al., 2012):

1. Inputs and outputs
2. Flow units
3. Activities and control flow structures
4. Resources
5. Information structure

Figure 26.1 illustrates how these aspects relate to the concept of a process. Figure 26.2 extends on this conceptualization by illustrating the central characteristics of a business process.

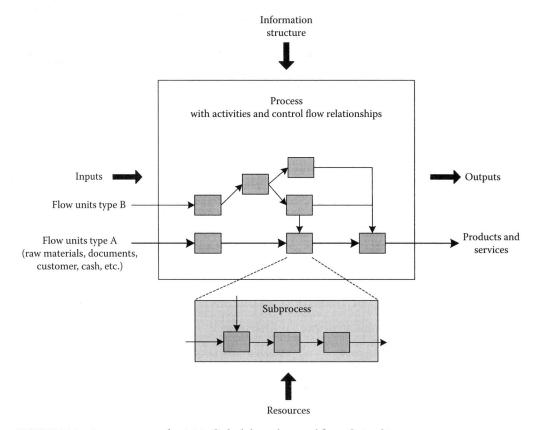

FIGURE 26.1   A process as set of activities linked through control flow relationships.

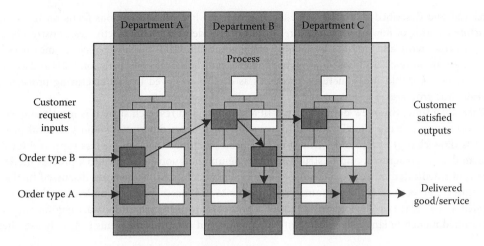

FIGURE 26.2 Business process as cross-functional process.

A *business process* is commonly regarded as a particular process type which is characterized as having *interfaces to a customer or a supplier* and which contains the activities that represent the *core business* of an organization (cf. Becker and Kahn, 2011). Business processes are thus designed and perceived as *end-to-end processes* (see Figure 26.2). An example of a typical end-to-end business processes is an "Order-to-Cash" process, which comprises all activities for generating and fulfilling a customer order and to collect the payments for fulfilling an order. These activities are usually divided into sub-business processes like "Sales," "Purchasing," "Production," "Shipment," and "Delivery."

Ultimately, the application of BPM in organizations requires a "thinking in processes." Process thinking emphasizes the significance of inter-departmental, cross-functional coordination of business activities with the aim to better serve customer needs (cf. Hammer, 2010). Traditional "functional thinking" on the contrary would rather emphasize the "local optimization" of activities within a particular function by exploiting resource efficiency gains through specialization. However, this local optimization does not necessarily yield an overall optimal coordination of all the cross-functional activities involved in a business process. "Process thinking," as mandated by BPM, stimulates optimization across an organization along defined process structures and to effectively and efficiently coordinate activities in the interest of customer needs.

26.3.2 Views on Business Processes

While the general understanding of the business process concept as outlined earlier is a useful common denominator for BPM, real-world processes are multifaceted and show great diversity according to their specific nature. BPM thus has to account for this diversity by allowing business processes to be approached from different perspectives. Melão and Pidd (2000) present four different views on business processes. According to them, processes can be viewed as (1) a *deterministic machine*, (2) a *complex dynamic system*, (3) an *interacting feedback loop*, or (4) a *social construct*. These views emphasize different important features of a business process and also resemble different assumptions about organizational life (cf. Walsham, 1991). In the following paragraphs, these views are briefly outlined along with their implications on BPM practice and research.

26.3.2.1 Business Process as a Deterministic Machine

According to this view, a process is regarded as a "fixed sequence of well-defined activities or tasks performed by 'human machines' that convert inputs into outputs in order to accomplish clear objectives" (see discussion earlier) (Melão and Pidd, 2000). This view holds that a process can be objectively perceived and described in rational and technical terms. Process descriptions focus on the *process structure* (routing of flow units), *procedures* (logical dependencies between activities, constraints, and required resources), and *goals*. Goodness of a process is evaluated in terms of *efficiency* measures like time, costs, or productivity. This view has its roots in scientific management (Taylor, 1911) and assumes the presence of stable and structured processes as can be observed in manufacturing processes or bureaucratic processes.

For a long time, this view dominated the BPM practice and research since it is mostly concerned with capturing process knowledge within semi-formal process descriptions and graphical process models (flow charts) of all kinds. This "machine view" assumes that any process type, technical or human driven, structured, and unstructured, can be unambiguously mapped to a process model by means of a dedicated modeling formalism. Exemplary modeling formalisms are discussed further in the subsequent sections and are summarized in Table 26.2. Process modeling under a machine view is applied in a wide variety of domains for various purposes, for example, in software engineering (e.g., the unified modeling language (UML) with its activity and sequence diagrams is heavily used here),

TABLE 26.2 Summary of Different Perspectives on Business Processes

	Deterministic Machine	Complex Dynamic System	Interacting Feedback Loop	Social Construction
Main assumptions	Processes can be described and designed in a rational way	Processes can be described and designed in a rational way	Magnitude of impact of policies on process performance is easily quantifiable through equations	Process change or process design can be negotiated in presence of conflicting interests
Consideration of dynamics	Static snapshot (of as-is or to-be situation)	Interaction between process components	Interaction between process structure and policies	Dynamics are considered implicitly as part of the subjective constructions of a business process
A process design is good if…	…the parts work *efficient*	…the process as a whole is *effective*	…the process is *effective* given a set of *policies* to be obeyed	…the perceived process exhibits "*cultural feasibility*"
BPM techniques	Flow charting and workflow modeling	Process simulation, quality management (statistical process control), business activity monitoring, and process mining	Process simulation and system dynamics modeling	Soft systems methodology
Related BPM activities	Process documentation, application development, certification, and benchmarking	"What if" analysis, discovery of previously unknown process behavior, and process control	"What if" analysis and gaining qualitative insight about process behavior	Learning about processes, engagement in unstructured process design problems, and management of "creative" business processes.
Main limitations	Lack of a time dimension Neglect of sociopolitical issues	Neglect of interaction between process structures and policies. Neglect of sociopolitical issues	Difficult to apply since the magnitude of impact of policies is hard to quantify Neglect of sociopolitical issues	Focus on "cultural feasibility" may impede radical process designs Not accessible to objective and quantitative assessment of process performance

the modeling of automated processes in the domain of workflow management, or for certification (e.g., quality management according to the ISO 9000 requires a documentation of business processes). Moreover, process models could be used to customize enterprise systems according to individual needs of an organization.

Process modeling is the main focus in the machine view, and research in this area increasingly focuses on the understandability and usability of particular modeling notations (see, e.g., zur Muehlen and Recker, 2008). Other research activities additionally study capabilities of modeling notations to map existing business realities onto a process model (e.g., see work on modeling grammars, as in Recker et al., 2011).

The main limitations of this deterministic machine view on a business processes are its assumption that business processes can be objectively perceived and modeled in a rational way (like engineering machines). It therefore neglects the sociopolitical dimension of BPM initiatives. Moreover, business processes as seen from a machine-view represent static snapshots of an as-is or a to-be situation and thus neglect the dynamic nature of processes, which are likely to change over time. However, as long as the main goal is to understand and clarify structural features of a process, then this view can still be useful.

26.3.2.2 Business Process as a Complex Dynamic System

In this view, a process can be understood as a system that dynamically adapts to a changing environment. From a systems perspective, a process has inputs, transformational capabilities, outputs, and a boundary to its environment. A process can also be hierarchically decomposed into subsystems (people, tasks, organizational structure, and technology) and elements (e.g., individual entities of each subsystem). Subsystems and elements interact with each other (internal relationships) and with the system's environment (external relationships) to achieve some superordinate objective (i.e., the system's objective). As opposed to the mechanistic perspective, this view emphasizes dynamic behavior of processes and interactions with an external environment (cf. Melão and Pidd, 2000). In particular, instead of analyzing the efficiency of its parts, this view encourages process managers to analyze process behavior holistically. Consequently, goodness of a process is more likely measured in terms of *effectiveness* (e.g., by employing quality and service level measures).

Besides the mechanistic view, this view on business processes as dynamic complex system is the predominant view today in both BPM research and practice. Many techniques and tools have been developed within the past decades to analyze and control complex process behavior. For example, TQM and techniques of *statistical process control* like *Six Sigma* have been developed and are applied under this particular view. Moreover, complex dynamic interactions within a process can be modeled and analyzed by means of *discrete-event simulation*. Many information systems (so-called process-aware information systems, cf. Dumas et al., 2005) keep track of user activities in so-called *event logs* enabling the analysis of process executions (PEs) through *business activity monitoring* (zur Muehlen and Shapiro, 2010) and also the discovery of "hidden" or formerly unknown process structures through *process mining* (van der Aalst, 2011). Furthermore, *execution semantics* of process modeling languages are defined in this view (see detailed discussion in this chapter on the business process modeling notation (BPMN) language in the subsequent sections). By means of such semantics, it can be analyzed if a given process model is "sound," that is, if it is able to be executed and if there is a danger of deadlocks in a process. Formalisms like Petri nets allow for mathematically proving process soundness. Many tools that support process modeling have a built-in functionality to animate the flow of flow units through a process and to automatically check process soundness.

In the past years, BPM research has widened its focus from solely analyzing process structures toward understanding interrelationships between process structures and other subsystems. For example, a research stream on value-oriented process modeling (see discussion in the subsequent sections and vom Brocke et al., 2010) explores the relations of process structures and the *value system* of an organization. Within this stream, important research contributions have been made with regard to goal-oriented process modeling linking the goal-systems of an organization with process designs (e.g., Neiger et al., 2008). Goal-oriented process modeling can be applied to facilitate a strategic alignment of business processes with organizational goals or to support risk management activities in BPM. Another contribution in the value-oriented research stream has been made with regard to linking process modeling with the financial system of an organization (vom Brocke and Grob, 2011; vom Brocke et al., 2011) in order to analyze the impact of process design decisions on economic performance measures like the total cost of ownership, the return on investment (ROI), or the net present value (NPV).

While this complex dynamic systems view facilitates analyzing and understanding dynamic process behavior, it still shares critical assumptions with the mechanistic view. It still bears the risk to neglect the sociopolitical dimensions of a business process and it still implicitly assumes that process design and analysis can be approached in logical and rational terms (cf. Melão and Pidd, 2000).

Moreover, business processes as viewed from a complex dynamic systems perspective are assumed to have *no intrinsic control*, that is, the behavior of the system's elements is controlled by exogenous input variables and not by internal system states. In particular, policies that might "regulate" a system's behavior are not considered. Such systems are called *open loop systems*. Figure 26.3 illustrates an example of a complex dynamic system. The system under study is an airport, which might consist of subsystems like groups of passengers, IT resources, employees, and of course, a process system that determines the behavior of the "airport system." In particular, Figure 26.3 considers the process of "boarding a plane" from the perspective of the airport management. Airport management is interested in analyzing process performance in terms of passenger throughput, passenger queue length, or passenger waiting times (i.e., the relevant flow unit here are passengers). Since policies for conducting a passenger boarding are not considered, the process behavior at time t is determined by the arrival rates of passengers at time t (exogenous variable), the passenger queue length at time t, and the process structure (which in this case might be a sequence of a "check boarding pass" and a "complete boarding" activity). In the extreme event that all passengers arrive at the same time in front of the gate and are allowed to board the plane in the order they have queued (implicit first-in-first-out policy) process performance is likely to decline as time evolves since seat numbers among passengers are arbitrarily distributed. This might cause congestions within the airplane since passengers who have seats in the front of the plane temporarily block passengers seated in the back of the plane when trying to occupy a seat (anyone who has ever boarded a plane might have experienced such a situation). Figure 26.3 depicts the situation at three points in time. Initially, 150 passengers are waiting in front of the gate. After 3 min, 10 passengers have already found a seat while 20 passengers are waiting in front of an airplane. Additionally, 10 passengers arrive at the gate. After 10 min, already 50 passengers have found their seats. As is indicated in Figure 26.3, the number of passengers having found their seats in a particular row is randomly distributed and follows no particular pattern.

However, the existence of a policy, that is, the ability of the process to exert internal control over the arrival rate of passengers could lead to a better process performance since seats can be occupied according to particular patterns. An exemplary policy could be that "if the number of passengers in front of a gate G exceeds a threshold of x then only passengers in seat rows $[Y..Z]$ of a plane are allowed to board first." Such a policy would increase the probability that passengers find their seats more conveniently and thus congestion (see waiting queue in front of airplane after 10 min) could be avoided. This policy would interact with the first activity (e.g., "check boarding pass") and requires to model (information) feedback loops from "inside the airplane" activities to the boarding activity (so that the boarding activity is updated if boarding is complete for particular rows). The inability of an open loop systems view to account for the additional dynamics arising from internal system control (e.g., policies and information feedback loops) is the main limitation of this particular view. In fact, policies and feedback loops determine the behavior of many real-world processes (Melão and Pidd, 2000) and are thus worth to be considered properly. Remember that information structures have also been considered as an important aspect of the process concept in the preceding text (see Figure 26.1).

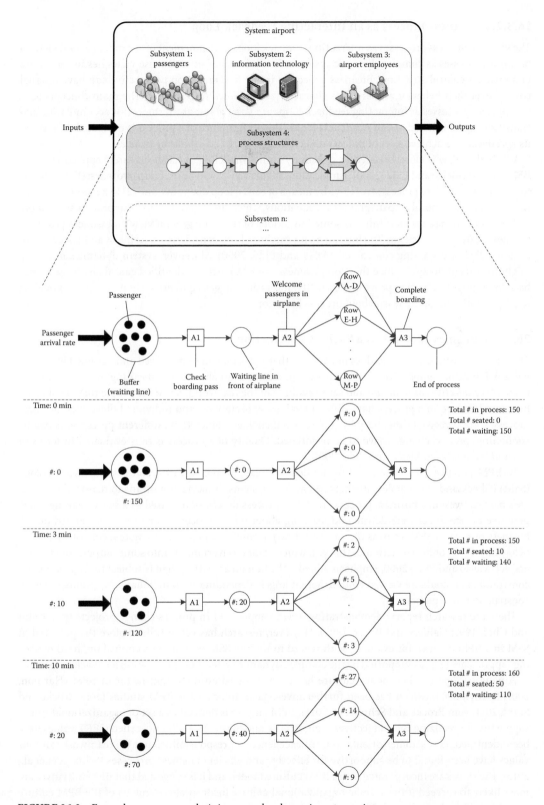

FIGURE 26.3 Exemplary process analysis in a complex dynamic systems view.

26.3.2.3 Business Process as an Interacting Feedback Loop

The view on business processes as interacting feedback loops is a refinement of the more general view on business processes as complex dynamic systems. The difference between these views lies in the nature of a system's control structure. Business processes in the feedback loop perspective can have internal control over their behavior and are thus characterized as *closed loop systems*. The main concepts used to represent processes as interacting feedback loops are *flows of resources*, *stocks* (for accumulating and transforming items), and *causal (feedback) loops*. The performance of a process is evaluated in terms of its *effectiveness* to achieve a set of process objectives given a set of *regulating policies*.

While this view is closely related to system dynamics modeling, it has been rarely applied within a BPM context (Melão and Pidd, 2000). Applications of this view could be useful in two ways: (1) as a means to *qualitatively* explore process structures by means of causal loop diagrams, and (2) in a *quantitative* way that translates the causal loop diagrams into equations, which then serve as a basis for process simulation.

The system dynamics view inherits some limitations of the more general view on business processes as complex dynamic systems. The human factor might only be considered in terms of an instrument to be controlled or exercising control (cf. Melão and Pidd, 2000). Moreover, system dynamics modeling might be difficult to apply since the "completeness" and "impact" of identified causal chains and feedback loops might be hard to prove and quantify. It is rather suggested to use system dynamics modeling in a "qualitative mode" to learn about process behavior.

26.3.2.4 Business Process as a Social Construction

This view on business processes does not assume that a process can be perceived and designed in a rational way but rather emphasizes that a business processes is an abstraction, meaning, or judgment as a result of a *subjective construction* and sense-making of the real world. Since the people that are affected by or are involved in a process have different values, expectations, and individual objectives, business processes are perceived differently by different stakeholders. Since these different perceptions can be conflicting, process changes have to be negotiated. Quality of a process is thus evaluated in terms of *cultural feasibility* (cf. Melão and Pidd, 2000).

In BPM practice, this view could be applied by means of the so-called soft systems methodology (SSM) (Checkland and Scholes, 1990). SSM employs systems thinking as a means to reason about people's perspectives on a business process. That is, a process description is used as a sense-making interpretative device to generate debate and learning about how a process is or should be carried out. A business process in SSM terms is thus a "would-be *purposeful human activity system* consisting of a set of logically interconnected activities through which actors convert inputs into some outputs for customers" (cf. Melão and Pidd, 2000, emphasis added). The human activity system is subject to environmental constraints and could be viewed from different angels depending on individual perceptions or social constructions of a process.

There are research reports demonstrating how to apply SSM in process change projects (e.g., Cahn and Choi, 1997; Galliers and Baker, 1995). However, research has yet to fully explore the potential of SSM in a BPM context, for example, with regard to linking SSM with more technical or "hard modeling" approaches (e.g., as supported by the complex dynamic systems or information feedback loop view).

Among sociopolitical elements, culture has been reported an important factor in BPM (Harmon, 2007; Spanyi, 2003), which has been further investigated in academic BPM studies (vom Brocke and Sinnl, 2011; vom Brocke and Schmiedel, 2011). BPM culture is defined as a set of organizational values supportive for realizing BPM objectives (vom Brocke and Sinnl, 2011). The so-called CERT values have been identified: (a) customer orientation, (b) excellence, (c) responsibility, and (d) teamwork; all four values have been found to be supportive for effective and efficient business processes (Schmiedel et al., 2013). The values are incorporated into a BPM culture model, and it is suggested that BPM initiatives are more likely to succeed if the specific organizational culture incorporates elements of the BPM culture model (vom Brocke and Sinnl, 2011).

A main limitation of this view on business processes is its sole reliance on *cultural feasibility*, which may impede more radical process designs and changes (cf. Melão and Pidd, 2000). Moreover, while this view is capable of capturing sociopolitical issues, it does not disclose how such issues could be dealt with. More recent research in BPM, however, does seem to overcome this limitation by designing tools for measuring, for example, the cultural fitness of an organization for BPM to very early identify barriers for process change and take preventive actions (like skill development and culture development) accordingly (Schmiedel et al., 2012).

26.3.3 Integrating the Multiple Perspectives on Business Processes

As can be seen from the aforementioned discussion, business processes can be viewed from several perspectives, which provide an account for the multifaceted nature of reality (cf. Melão and Pidd, 2000). While each of the discussed views has its limitations, they might well serve as a starting point for pluralistic approaches to BPM. For example, if a process under study is not well understood and bears potentially problematic implications for a group of people, then soft approaches can be used initially to gain consensus about what the process under study process looks like. A static approach as suggested by a mechanistic view on business processes can then be employed to document and communicate a process structure. In addition to such temporal combinations of the views, different process context can also call for specific process views to be predominantly applied. The mechanistic view on processes, for instance, is useful for analyzing and improving technical, well-defined processes, while the interacting feedback loop view or the social construction view might be more appropriate to study less structured business areas. Davenport (2005), for instance, discusses how the management of knowledge work can benefit from a process-oriented view. Also, the primary objective of process management can have an influence on the right view: If the analysis and improvement of process behavior is of interest, then the process analysis can be complemented by an approach that employs systems thinking. In particular, if (unpredictable) complex interactions within a process are not well understood, then approaches to disclose the process dynamics might prove to be useful, like process simulation, process mining, or business activity monitoring. Finally, if a business process is assumed to have feedback loops or some intrinsic control structures, then the systems view on processes should be extended to also consider the dynamics of policies.

Except for the social construction view on business processes, all views considerably assume that processes can be objectively perceived and process performance is rather independent of the people who participate within a process. Taken to an extreme, these views would suggest that processes can be optimized (even automatically) based on some quantifiable attributes. For complex and less technical processes, this assumption, however, has proven problematic already during the reengineering wave. Many re-engineering projects failed due to an overemphasis on designing technically sound processes at the cost of fostering social, cultural, and political resistance to change (Willcocks and Smith, 1995). Thus, a technically feasible and sophisticated process is a necessary but by no means a sufficient condition for the success of process change projects. Contemporary BPM therefore needs to follow a more holistic picture particularly integrating both technical and social elements of organizations as socio-technical systems (Bostrom and Heinen, 1977). BPM is, therefore, considered an integrated management discipline (Rosemann and vom Brocke, 2010) that seeks to build organizational capabilities to continuously improve and innovate business processes according to their very specific nature and given objectives and context factors.

For successful BPM initiatives, the problem, therefore, is not to favor a particular view on business processes. Instead, the contrary is the case. In order to capture the full richness and complexity of a situation, business process managers benefit from approaching the analysis and improvement of business processes in a multifaceted way (cf. Davenport and Perez-Guardo, 1999). Each view has its merits and limitations with regard to the analysis and improvement objectives to be achieved in a particular situation. Table 26.2 summarizes the different views on business processes as discussed earlier with its assumptions, limitations, and possible areas of application.

26.3.4 Process Analysis in the Context of a BPM Life Cycle

Having clarified the multifaceted nature of business processes, we now briefly discuss how BPM activities are affected by particular views and why process analysis is a significant task in an overall BPM initiative. Figure 26.4 depicts a BPM life cycle, which sketches the relationships between the activities to be conducted while managing business processes.

The BPM cycle starts with designing, documenting, and implementing a process. As opposed to the assumptions of the reengineering movement, process implementations do not always require the use of IT. For example, implementations of human-driven processes would require some simple procedures or policies to be obeyed by employees or other process participants. Once a process is up and running, a process manager might be interested in how the process performs with regard to some performance measures and with regard to the some competitors. Once a process performance gap is identified, corrective actions have to be planned in order to change the current process and transform it into a process with a more desirable performance. Once the process has been redesigned, it is again subject to performance measurement, and the cycle starts over again.

What can be seen from the description and from Figure 26.4 is that process analysis is a prerequisite to almost all BPM activities. Designing and documenting a process necessitates the initial analysis of existing process structures. Identifying performance gaps necessitates an analysis of the process behavior. Overcoming a performance gap and redesigning a process necessitates an understanding of the current process behavior and also an understanding of the to-be process as well as of the potentials to be unlocked by a new process design (cf. vom Brocke et al., 2009). Process analysis is thus central for gaining transparency about processes (in terms of structure and behavior).

In this regard, process modeling languages constitute an important methodological basis for BPM. Examples of such modeling languages are the event-driven process chain (EPC) or the UML activity diagrams. Selecting the "right" modeling language and the "right" view on a process in a given situation is becoming more and more difficult as BPM initiatives become more complex. This chapter provides a brief overview of some more widely used modeling languages and then introduces one specific modeling language that meanwhile has become an international industry standard: the BPMN. We employ a

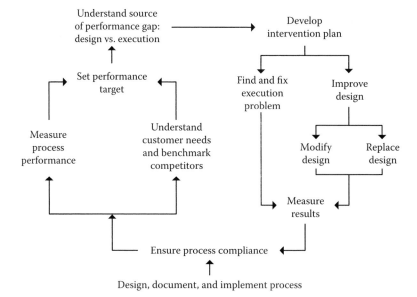

FIGURE 26.4 BPM life cycle. (From Hammer, M., What is business process management, In: *Handbook on Business Process Management*, Vol. 1, ed. vom Brocke, J. and Rosemann, M., p. 5, Springer, Berlin, Germany, 2010.)

complex dynamic systems view on business processes. We focus on the capturing of process structures under this view and the analysis of a particular kind of process behavior: *economic value creation*.

This particular type of behavior addresses an important question in BPM, which has so far received little attention. Process managers frequently have to choose a process (design) alternative (out of the large number of alternatives available in total) by judging which alternative appears to be the most beneficial one. Thus, this question is critical to the successful management of business processes. This chapter, therefore, particularly takes a look at this important dimension and presents methods that allow their users to compare and determine the economic value of different process reorganization alternatives. The methods for creating structural process transparency and process value transparency as presented in this chapter build upon each other in order to achieve maximum synergies between the modeling and the economic assessment of processes. Furthermore, the chapter demonstrates how useful financial key indicators, such as total cost of (process) ownership or ROI (of process redesign), can be calculated. Finally, the chapter presents a real-world example and briefly discusses its results.

26.4 Creating Transparency about Process Structures

26.4.1 Process Modeling Languages: A Brief Overview

In order to be able to create transparency about process structures, process modeling languages are used, which allow to describe the sequence of activities (in terms of time and logical order) within a process. As far as the development and use of such process modeling languages is concerned, a trade-off has to be made regarding the formalism to be employed. On one extreme, processes can be described in natural language (text-based modeling). On the other extreme, formal modeling languages can be used that restrict the use of the "modeling vocabulary" (modeling based on modeling grammars, cf. Recker et al., 2011). Natural language-based modeling languages are quite easy to understand. However, they are characterized by a large degree of freedom regarding the representation of process structures, which may be a reason not to use them for certain analyses as they might be too inexact. Formal modeling languages, on the other hand, do support such analyses, yet they are understood by relatively few experts only. The use of methods for process management therefore is closely connected with methods for semiformal specification of business processes.

For example, the EPC (Keller et al., 1992) is a frequently applied process modeling language and is used for customizing the SAP® enterprise resource planning system. Nowadays, due to the considerable number of available modeling formalisms, process modeling languages are subject to standardization initiatives, for example, the UML (OMG-UML, 2011) or the BPMN (OMG-BPMN, 2011, as of Version 2.0 called business process modeling and notation). Especially the development of BPMN has been influenced a lot by some related and already widely used modeling languages (cf. OMG-BPMN, 2011, p. 1). Besides the EPC and UML activity diagrams, Integrated DEFinition for Process Description Capture Method (IDEF3; cf. Mayer et al., 1995) and ebXML Business Process Specification Schema (ebXML BPSS, cf. OASIS, 2006) have been considered throughout development of the BPMN. In addition, the BPMN execution semantics explicitly builds upon the formally defined semantics of Petri nets (Petri, 1962), which in turn is based on the concepts of "tokens" (see explanations further in the subsequent sections). Apart from BPMN, also EPC and UML activity diagrams each incorporate the Petri nets semantics.

The process modeling languages described in Table 26.3 address different *modeling purposes*. EPC and IDEF3 are primarily used for specifying the functional requirements of business processes, mainly for designing *organizational structures* (in order to be able to meet requirements regarding documentation, certification, benchmarking, or process-oriented reorganization) (cf. Becker et al., 2011). UML activity diagrams and Petri nets, on the other hand, are predominantly used for designing process-oriented *application systems*. Some Petri net variants are used for modeling *workflows* (i.e., processes of

TABLE 26.3 Overview of Process Modeling Languages

Modeling Language	Short Description
EPC (Keller et al., 1992)	EPC is a semi-formal modeling language for describing processes. EPC represents processes as *sequences of events and functions*. The EPC does not distinguish between start events, intermediate events, and end events. The architecture of integrated information systems (ARIS) (Scheer, 2000) uses EPCs for describing the *process/control view*, which integrates of the four other ARIS views (*organization, function, data*, and *output*). The EPC is particularly suited for specifying the functional requirements of business processes (i.e., the modeling of business processes from a business perspective).
IDEF3 (Mayer et al., 1995)	Similar to BPMN and EPC, IDEF3 was developed with the intention to be easily learned and used by practitioners. IDEF3 is domain neutral and primarily supports the documentation of *process knowledge* (Mayer et al., 1995). IDEF3 process models are predominantly used for *process analysis* aiming at process re-organization. IDEF3 supports two strategies of documentation (*process centered* and *object centered*), for which two *views* (*schematics*) are provided (*process schematics* and *object schematics*). An IDEF3 process description may comprise one or more process and object schematics.
UML activity diagrams (OMG-UML, 2011)	UML activity diagrams represent one a particular UML diagram type. They are used for describing the behavior of *application systems*. Activity diagrams illustrate the dynamic aspects of single use cases, showing start and end events but not intermediate events. Although activity diagrams just like BPMN 2.0 are based on *Petri net semantics*, the execution semantics of the two modeling languages differ widely with regard to certain aspects. For example, a process described by an activity diagram ends when at least one final node has been marked with a token (see the discussion on the token concept in the subsequent text). In contrast, a BPMN process remains active as long as there are any tokens in a process.
	When specifying the functional requirements of business processes, practitioners often combine activity diagrams with other UML diagram types in order to be able to separately illustrate functional requirements and technical requirements in a *use case diagram* or to describe relations between business relevant process object types in *class diagrams*.
Petri nets (Petri, 1962)	Petri nets come with a simple graphical process modeling language that can be used both for describing the behavior of application systems and for modeling business processes. Petri nets are based on a concept of *tokens, places* (i.e., *events*), and *transitions* (i.e., *functions/activities*). Petri nets do not allow the expression of different views of a business process. By means of the token concept, Petri nets provide an *instrument for abstraction* that allows to describe and *formally analyze the semantics of control flows*. A lot of the more recent modeling languages use Petri nets semantics, which allows analyzing processes with regard to cycles, deadlocks, or inaccessible paths.
BPEL (OASIS, 2007)	The OASIS is standard "business process execution language for web services (BPEL4WS or BPEL)" is an *XML-based, machine-readable* process description language. A BPEL process can be understood as a *web service* that interacts with other web services. BPEL can be used to describe both *inner-organizational and cross-organizational processes*. The BPMN defines *rules* for transforming graphical BPMN process models into BPEL models.
ebXML BPSS (OASIS, 2006)	The ebXML business process specification schema (BPSS) is an *XML standard* maintained by OASIS that allows describing processes in a textual, *machine-readable* way. ebXML-based process descriptions specify which business partners, roles, collaborations, choreographies, and documents are involved in the execution of a process. The focus is on describing *cross-organizational exchange of business documents*, i.e., the modeling of cross-organizational processes.

fine granularity) where transitions of activities are controlled by application systems (a *process engine*). Process descriptions used as input for such application systems are thus required to be machine readable. Such descriptions can be provided by non-graphical, XML-based process modeling languages (such as BPEL or ebXML BPSS), which allow to generate technical, machine-readable process descriptions that can be interpreted and executed by a process engine. The graphical representation of process models is not a primary aspect of these process modeling languages. However, there are a few languages that allow to seamlessly map graphical process descriptions onto technical, machine-readable process descriptions in order to close the gap between process modeling from a business perspective (functional requirements of a business process) and the actual PE. Sample notations that aim at linking a business perspective with a technical perspective on business processes are the BPMN (OMG-BPMN, 2011), the activity and sequence diagrams of the UML (OMG-UML, 2011), or yet another workflow language (ter Hofstede et al., 2010).

In recent years, BPMN has developed into an industry standard for process modeling. BPMN aims to provide a notation for graphical description of business processes that allows to consistently connect both *process design* and *process implementation* (cf. OMG-BPMN, 2011, p. 1). BPMN process descriptions are supposed to be understood by both technical developers and business analysts. Furthermore, the BPMN aims at adequately visualizing technical, XML-based process descriptions in order to be interpretable by human actors. In general, BPMN is supposed to facilitate the end-to-end modeling of processes, starting with process descriptions from a functional perspective, which are then more and more refined until a detailed BPMN model is developed from which a technical process description (e.g., in BPEL) can be generated.

Version 1.0 of BPMN was launched in 2004. It was developed by the Business Process Management Initiative Notation Working Group, which then was merged with the Object Management Group (OMG) in 2005. Since then the OMG has been responsible for maintaining the BPMN. In 2011, the OMG published Version 2.0 of the BPMN, following Version 1.2. Among the new features are two additional diagram types: conversation diagrams and choreography diagrams. As of Version 2.0, it is also possible to extend BPMN with own notation elements.

Taking into account certain element types that had already been part of earlier versions of BPMN, Version 2.0 was extended primarily with elements that focus very much on the modeling of executable process models. Thus, BPMN is focusing more on the *design of workflows*. While there are a large number of BPMN notational elements available, in practice only a relatively small number of elements are actually required and used for the modeling and design of business processes (the so-called BPMN core, cf. zur Muehlen and Recker, 2008). The following section presents a selection of these core elements and concepts of BPMN.

26.4.2 Business Process Modeling and Notation

The presentation of central concepts and elements of BPMN refers to literature covering the fundamentals of BPMN (Havey, 2005; OMG-BPMN, 2011; Weske, 2007). The concepts and elements to be presented in the following sections have been selected according to their significance for getting a basic understanding of BPMN models. After providing a short overview of the BPMN diagram types, one specific diagram type, the *Business Process Diagram* (BPD), together with its most important notational elements is presented in detail. Subsequently, one section is dedicated to the analysis of the behavior of BPMN process models by discussing the BPMN execution semantics.

26.4.2.1 BPMN Diagram Types

As has already been mentioned, the BPMN supports the modeling of sequences of activities (in terms of time and logical order). Prior to Version 2.0, the BPMN considered one diagram type only: the *BPD*. Unlike other process modeling languages (such as EPC or IDEF3), the BPMN did not provide separate views on process models (e.g., value chain view, organizational structures, or data view).

As of Version 2.0, however, the BPMN offers the possibility to model a "global view" of the sequence of interactions taking place between several collaboration partners (i.e., of inter-organizational processes). Therefore, the BPMN has been extended with two additional diagram types: *conversation diagrams*, which specify conversations taking place in an interaction scenario (who is involved, what messages are exchanged), and *choreography diagrams*, which specify the course of a conversation. Both diagram types establish an external, cross-organizational view on the interplay between collaboration partners. The details of affected internal processes of conversation participants are then specified in BPDs.

For multi-perspective process modeling, the BPD has to be extended with additional diagram types, which are not part of BPMN (such as UML class diagrams for specifying complex organizational and data structures). Because of the central role of the BPD, its most important elements are presented in the following section.

26.4.2.2 BPD: Core Elements

The notational elements associated with the BPD represent so-called object types used to describe business processes (cf. OMG-BPMN, 2011, p. 27 ff.). The BPMN makes a distinction between *flow objects*, *connecting objects*, *swimlanes*, *artifacts*, and *data objects* (see Figure 26.5). The BPMN comprises much more elements than shown in Figure 26.5. The elements presented here are considered to be core elements of BPMN and the ones that are most frequently used in BPMN-based process modeling projects (cf. zur Muehlen and Recker, 2008).

Flow objects are used to describe the sequence of activities (*tasks*). Basically, the BPMN distinguishes atomic activities and activities that can be refined into sub-processes. That way it is possible to describe hierarchical process structures. Besides the tasks, *events* and *gateways* represent central notational elements of the BPMN. Events can initiate a process (start events), occur as external events during PE, be generated internally as a result of the execution of a task (intermediate events), or terminate a process instance (end events). When events occur, a process engine (i.e., an application system that coordinates and monitors the execution of the process) saves process-related data. The more a process description focuses on the technical implementation of a process, the more important it is to actually model events. Process modeling from a functional perspective often focuses on start events and end events only, including events related to state changes of relevant process objects (e.g., sending or requesting messages, creating or modifying documents). Besides simple, un-typed events, the BPMN provides multiple event types, of which *message events* are frequently used in process modeling. *Gateways*, finally, represent a third core notational element. They are used to illustrate decision points within a process.

Flow objects can be related to each other by means of *connecting objects*. There are three types of connecting objects provided by the BPMN. *Sequence flows* connect tasks with each other and specify the order in which they are supposed to be executed. *Message flows* connect tasks contained in different pools (see Figure 26.6) in order to be able to model a communication channel for exchanging message objects. *Associations* are used to relate flow objects to artifacts in order to be able to integrate additional information into the process description (e.g., to add information about input and output conditions for single tasks).

Swimlanes enable the specification of responsibilities regarding the execution of tasks. In particular, swimlanes allow the description of basic organizational structures. The BPMN divides swimlanes into *pools* and *lanes*. Self-contained processes are modeled within pools. Lanes are used to group tasks according to responsibilities for their execution.

Artifacts are used to integrate context specific information into a process model. The most frequently used artifact type is *annotation*. As of BPMN Version 2.0, it is possible to integrate additional custom notational elements (i.e., which are not defined by the BPMN standard) as artifacts in order to be able to adapt the BPMN to individual process modeling requirements (cf. OMG-BPMN, 2011, p. 8). Artifacts can be connected with flow objects and sequence flows.

Data objects represent all information and documents used or produced within a process. Unlike the EPC, for example, the BPMN does not provide the possibility to represent descriptions of

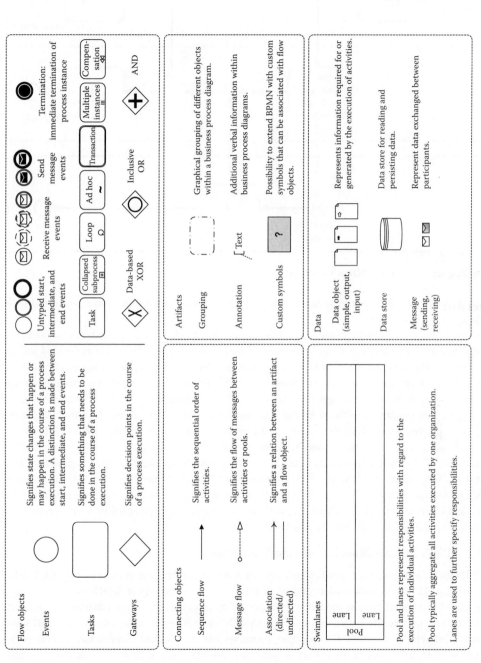

FIGURE 26.5 BPMN core elements. (From OMG-BPMN, Business process modeling notation (BPMN) information. http://www.omg.org/spec/BPMN/2.0/PDF (accessed March 7, 2011), p. 27 ff., 2011).

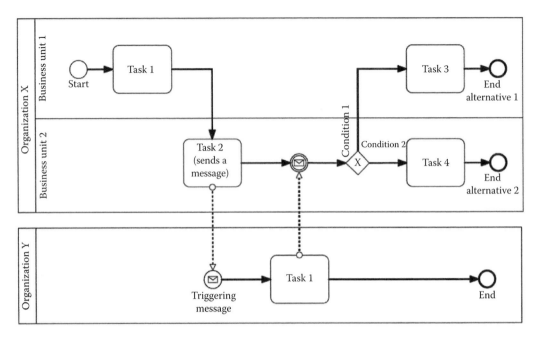

FIGURE 26.6   Example of a BPMN process model.

complex data structures within a process model or a separate view. Data objects are connected with flow objects via associations.

Figure 26.6 shows an example of an abstract BPMN process model. The graphic shows two internal processes (one for "Organization X" and one for "Organization Y"), each of which is embedded within a separate pool. The depicted situation typically applies in inter-organizational settings (such as, when a credit card check service is used prior to booking a flight). Processes contained in separate pools are connected with each other by message flows only (so-called choreography). In the example in Figure 26.6, such a choreography is initiated by the internal process of "Organization X." After "Business Unit 1" has executed the first activity (Task 1), the process continues in "Business Unit 2" (Task 2), which sends a message to "Organization Y." Upon receipt of this message, a process is triggered in "Organization Y." Following an activity (Task 1) in "Organization Y," a new message is generated and sent back to "Organization X." Meanwhile, the process of "Organization X" waits until the message from "Organization Y" has been received. When the message has been received by "Organization X," a gateway decides which one of two alternative activities is triggered and executed in "Organization X," depending on predefined conditions (i.e., business rules).

26.4.2.3 BPMN Process Model Execution Semantics

The BPMN execution semantics specifies how a process behaves at the time of its execution (cf. OMG-BPMN, 2011, p. 425 ff.). A basic understanding of the execution semantics of BPMN process descriptions is useful to prevent process modelers from designing processes that do not behave as intended. Without considering execution semantics process models that are syntactically correct may still show unexpected behavior at runtime. For example, processes may be blocked (deadlocks), process paths could not be executed (inaccessible paths), or process areas repeatedly executed accidentally (cycles). Such malfunction of processes should be taken into account already during the phase of process modeling. For this reason, the BPMN 2.0 standard explicitly defines execution semantics of BPMN processes. These semantics have been derived from the Petri nets semantics and is based on the concept of *tokens*.

The token concept allows validating process structures in both the modeling phase as well as during PEs at runtime. Tokens are assumed to flow through a process instance along the control

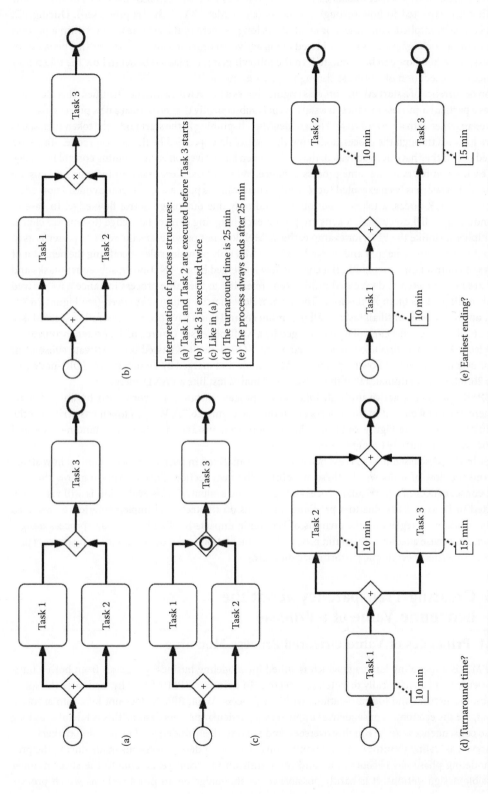

FIGURE 26.7 Examples of BPMN execution semantics. (a) Synchronization; (b) No Synchronization; (c) Synchronization (despite OR join); (d) Turnaround time?; (e) Earliest ending?

flow structures. In this sense, tokens can be interpreted as an abstraction of instances of single flow units that are expected to flow through a process (e.g., order "XYZ" that is processed). During PE, tokens can be multiplied, consumed, or created. As long as there is at least one token within a process instance, the associated process is considered being active. In order to validate the runtime behavior of a process, the token flow can be simulated. In the following, it is explained in detail how the token flow influences the execution of a process through a process engine.

Upon occurrence of a start event, a process engine creates a token with a unique identifier. This identifier signifies a particular process instance. If a start event is subsequently triggered, then a new process instance is created together with a new token and token identifier. Beginning at the start node, the token then starts to move forward through the process following the control flow specified by the process model. Tasks are executed only if they have been activated, that is, if a token has arrived on each incoming control flow edge and if each token refers to the same process instance. When a task is executed, all incoming tokens are consumed. If a task has been executed successfully, a new token is put on each outgoing control flow edge.

In case of XOR-nodes, a token is put on one and only one outgoing control flow edge. In case of AND-nodes (parallel execution), tokens are put on each outgoing control flow edge (tokens are cloned or multiplied). Unlike the EPC, for example, the BPMN does not provide specific rules or patterns as to how control flows are to be split and merged again. BPMN only provides rules specifying the behavior of mergers of control flows during PE. If a control flow is merged by an AND-node, subsequent tasks and events are not triggered and executed until tokens referring to the same process instance have arrived at all incoming edges of an AND-node. The control flow thereby gets *synchronized* (see Figure 26.7a), that is, one has to wait until all tasks of all incoming control flows have been successfully executed (see Figure 26.7d). If a parallel control flow is merged by a XOR-node (see Figure 26.7b), no synchronization takes place. As soon as one token has arrived at the XOR-node, it is passed on to activate subsequent tasks or events. If a control flow is split by an AND-node and merged by an outage risk (OR)-node (see Figure 26.7c), BPMN stipulates that the OR-merge behaves just like an AND-merge.

A BPMN process is active until all tokens of a process instance are consumed by an end node (i.e., there are no tokens left in the process any more, see Figure 26.7e). When a token has arrived at the "termination" node (see Figure 26.5), all tokens of a process instance are deleted instantaneously, and the process is immediately brought to an end.

After having discussed how process behavior can be analyzed in terms of process control flow structures and the flow of units within these structures, the next section aims at exploring how the economic consequences of a particular process design can be evaluated. This evaluation is still very often neglected in BPM practice due to a predominant focus on the technical implementation of specified process structures. Neglecting the analysis of economic impacts implied by a certain process design, however, may cause a technically sound process implementation to not deliver a satisfying process performance in terms of economic performance measures.

26.5 Creating Transparency about the Economic Value of a Process

26.5.1 Principles of Value-Oriented Process Modeling

The BPMN is a modeling language which is suited for modeling business processes from both a business-oriented and a technically oriented perspective. In doing so, BPMN basically facilitates the specification of structural and logical relations within a process. What BPMN does not take into account, however, are the economic consequences implied by a particular process design. This is tolerable as long as these consequences can be roughly overseen and are intuitively understood by decision makers.

However, assessing intuitively the economic consequences of design decisions made during the process modeling phase has become more and more difficult in recent years. Due to the sheer number of possible design options, it is hardly possible to reliably judge on an intuitive basis which process alternative is superior to another and under what conditions this is the case. Because of that, it is important to explicitly analyze the economic consequences (i.e., economic value contribution) of design alternatives already at the time of process modeling (vom Brocke, 2007). In particular, decision makers are often interested in the financial performance resulting from the execution of business processes. When taking this aspect into consideration, it becomes obvious that transparency about process structures alone is not sufficient. What is needed in addition is transparency about the economic value created or consumed by a process. Thus, the process of business process modeling needs to be complemented by an accompanying process of calculating the value contribution of a process (vom Brocke, 2007) enabling a continuous and integrated assessment of the economic consequences pertinent to different business process alternatives (vom Brocke et al., 2009).

The analysis of economic consequences can be approached from different perspectives (cf. vom Brocke et al., 2010). A reasonable approach seems to consider economic value in terms of the degree of fulfillment of specific economic goals. The question to be answered is what economic goals are considered to be relevant in each individual case and how different goal dimensions need to be weighed. Stakeholder theory (as originally proposed by Freeman, 1984) provides a conceptual framework here to balance interests and goals of different groups that have a vested interest in a process or an organization. The problem of balancing different interests and assessing different goal dimensions properly has been reflected in various controlling (e.g., the balanced scorecard; cf. Kaplan and Norton, 1992). Process management frequently refers to different goal dimensions, such as cost, quality, and time (the magic triangle). More recent studies have extended this view beyond economic goals, integrating other goal dimensions in order to be able to assess also the ecological and social effects of the execution or reorganization of processes (Hailemariam and vom Brocke, 2010).

The present chapter aims to illustrate how economic consequences of process designs can be assessed by means of methods for capital budgeting. We chose to refer to this specific perspective since it is considered to be of particular importance in a capitalist economy. This perspective is also important since benefits of BPM initiatives are ultimately judged by means financial performance measures (cf. vom Brocke et al., 2010). Moreover, the approach to be presented may serve as a pattern for developing other assessment perspectives. A crucial prerequisite for the presented approach is the existence of process models. The overall goal of the approach is to overcome the methodological divide—that still exists both in theory and in practice—between process modeling and process controlling and evaluation in order to be able to improve both the efficiency and the quality of decisions taken in BPM and process controlling.

26.5.2 Constructs for Value-Oriented Process Modeling

In many practical cases, the question arises whether reorganizing a certain process pays off. Two aspects are decisive here: the benefits to be expected from a process reorganization and its price. Both aspects should always be assessed as early as possible in order to be able to ponder different design alternatives and make decisions that are as rational as possible. A lot of studies on process controlling (also often referred to as process monitoring or business activity monitoring) have proposed to use key performance indicators for controlling a process' value contribution at process run-time (for an overview on this see zur Muehlen and Shapiro, 2010). This proposal, however, is of limited use for assessing planned processes as opposed to evaluate running or finished processes. Studies have shown that the potential profitability of processes is significantly determined already during the phase of process modeling and less during the run-time phase of a process (cf. vom Brocke et al., 2010). What is remarkable though is that hardly any methods for such process evaluation during the phase of process design have been developed so far. Figure 26.8 shows a calculation scheme that can be applied to assess the economic value implied by a particular process design. This calculation scheme assumes that economic value is analyzed in monetary terms and follows a capital budgeting approach. The approach is explained in detail in the subsequent text.

The *benefit* of a process reorganization or a process (re-)design can be ascertained by comparing the *total payments of process ownership (TPPO)* for the case with reorganization (process p) and the case

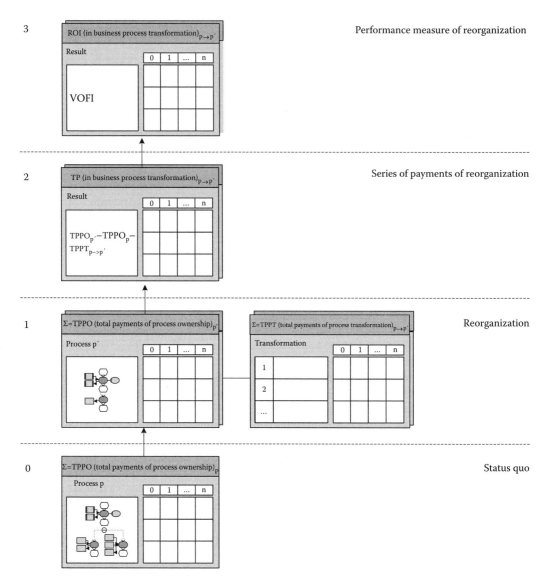

FIGURE 26.8 Calculation scheme (return on process transformation).

without reorganization (process p' = status quo) (level 1 and level 0 in Figure 26.8). The benefit resulting from a process reorganization is expressed by a positive difference between the $TPPO_{p'}$ of the new process p' and the $TPPO_p$ of the same process at the status quo level (process p). The *price* of a process reorganization is the sum of the payments that need to be made for the transformation of the process (level 1). These payments are referred to as *total payments of process transformation (TPPT)* in the following. TPPT typically comprise payments for procurement of new IT, for the development of process knowledge, or for training employees affected by a (new) process design.

In order to be able to measure the TPPO and the TPPT, long-term economic consequences of the process reorganization need to be taken into account, which is why the planning horizon for the payments needs to span a substantial period of time (e.g., 5 years). By netting the series of payments ($TPPO_{p'}$ – $TPPO_p$) and the investment in the process transformation (TPPT), the total expected payments resulting from the process reorganization can be calculated (level 2). This sequence of direct payments provides the basis for taking into accounting further financial consequences, including indirect (derived) payments (such as interest and tax payments). To calculate the derived payments, various standard methods for investment controlling can be applied. Apart from applying classical methods for capital budgeting (such as the NPV or the internal rate of return), the method of visualization of financial implications (Grob, 1993) provides a suitable means to make the financial consequences of a particular investment transparent. It allows to make transparent how direct and indirect payments that are related by different financial conditions lead to a particular economic performance measure. This transparence has already proved to be very beneficial for process evaluation purposes (vom Brocke and Grob, 2011) as process design decisions could be directly related to expected payments and specific conditions of funding.

Assessing the TPPO on level 0 and on level 1 can be made at different degrees of detail. The financial planning approach allows both to specify in advance net payments to be expected for each planning period and to further differentiate the calculation of net payments without explicitly linking the payments to a particular process structure. Another interesting approach uses BPDs (see Section 26.3) as a basis for the assessment. To do so, BPDs must be complemented with information about economic consequences, as shown in Figure 26.9 with the BPMN.

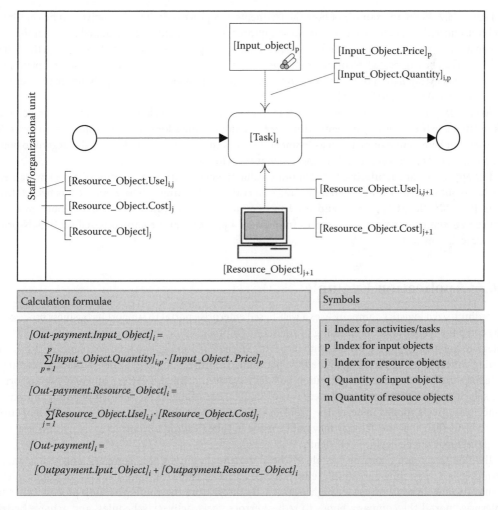

FIGURE 26.9 Detailed scheme for calculating out-payments on an activity level. (From vom Brocke, J. and Grob, H. L., Profitability of business processes, In: *Process Management: A Guide for the Design of Business Processes*, 2nd edn., eds. Becker, J., Kugeler, M., and Rosemann, M., pp. 421–446, Springer, Berlin, Germany, 2011.)

	0	1	2	...	5
− Integration infrastructure					
− Development of a wrapper service	−20.000 €	−700 €	−700 €		−700 €
− Development phase	−20.000 €	0 €	0 €		0 €
- Requirements analysis	−8.000 €	0 €	0 €		0 €
- Implementation	−7.500 €	0 €	0 €		0 €
- Testing	−5.500 €	0 €	0 €		0 €
+ Operating phase	0 €	200 €	200 €		200 €
+ Adaptation phase	0 €	500 €	500 €		500 €
+ Human resource development	−1.500 €	−1.200 €	−600 €		−200 €
Payments (total)	−21.500 €	−1.900 €	−1.300 €		−900 €

FIGURE 26.10　Example template for determining the TPPT.

Figure 26.9 shows an example of how the payments of a process design alternative can be captured and consolidated through an *activity-oriented* approach. Out-payments are calculated based on the use or consumption of input objects and resources objects. All payments calculated for each activity (or task) can then be aggregated according to the process structure and extrapolated over several planning periods. This extrapolation of payments can be done either individually or by applying trend rates (cf. vom Brocke and Grob, 2011; vom Brocke et al., 2010).

To calculate the price (TPPT) of a process reorganization, certain calculation templates can be used for putting together typical payments with representational values for certain transformation tasks. It is important here as well to make an assessment that is as complete as possible, with due regard to both time and logical order. An example is given in Figure 26.10.

The approach for calculating the economic value in synch with process modeling, as presented earlier, is subsequently illustrated by means of a real-world example. The example has been taken from (vom Brocke et al., 2009) and used to demonstrate how transparency about both process structured and economic consequences implied by a process can be gained based on BPMN process descriptions.

26.6 Application Example

A medium-sized logistics company uses a web-based enterprise portal to support its business processes. The company management is thinking of conducting a process reorganization project with the aim to integrate the route planning process into its portal. In route planning, two kinds of planning types are distinguished: detailed planning and adhoc planning. In order to be able to determine the planning type for each delivery, order prioritization policies have been defined. Delivery orders of high priority are subject to detailed planning. If there is not enough time available for detailed planning, however, adhoc planning is applied instead. The problem with adhoc planning is that resulting routes may turn out to be inefficient as the delivery may not be made in time (leading to contractual penalties), and the truck fleet may not be deployed efficiently.

The fact that route planning has always been done manually by this company and that manual planning is a very time-consuming effort, has led to a drastic overrepresentation of adhoc route plans, also for delivery orders of high priority. By integrating the route planning process into the enterprise portal the company hopes to reduce errors, meet delivery schedules, and achieve better deployment of resources. The technical implementation of the solution is to be done on the basis of a service-oriented architecture. Prior to the implementation, different design alternatives need to be assessed that particularly impact the effectiveness and efficiency of the planning process:

1. *GlobalRoutePlanning*: An IT solution by means of which route plans can be created over an online interface and saved to the company's database. Using the service requires specific information, such as delivery orders, truck fleet capacity, order prioritization, delivery addresses, and delivery dates. Using this solution, the process of route planning is fully "out-tasked."

2. *GeoDataForLogistics*: An in-house solution by means of which internal routing rules and customer data get enriched by external route information provided by a special geographic map service particularly suited to the needs of logistics companies. While this service is able to substantially reduce the planning effort, it also requires the development of a number of data services (wrappers) in-house.

3. *IntelligentRouting*: A web service by means of which fully fledged route plans can be created (similar to *GlobalRoutePlanning*, but only for a particular geographic region). As the geographical data of this service is very much up to date (providing information on construction sites or blocked roads, for example), the planning quality is very likely to be significantly improved by this solution.

The internal route planning process is represented by means of a BPMN process model (see Figure 26.11). In order to be able to depict process alternatives and specify the quantity structure, company-specific notational elements (BPMN artifacts) have been integrated (see explanation of symbols in Figure 26.11). Calculation of payments is basically done according to the scheme illustrated in Figure 26.9. Two resource object types are relevant here: organizational unit (here: dispatching/scheduling) and the services to be integrated. The quantity structure relevant to the calculation of the use of resources is specified within the BPD by means of tables (proprietary artifact).

The process design alternatives are considered implicitly in Figure 26.11. In order to be able to integrate the services into the process, various infrastructure requirements are needed (e.g., purchase and maintenance of an enterprise service bus (ESB), implementation of interfaces, and in-house developments). Deciding in favor or against a certain service in this specific example is expected to have a local impact only, as there are no structural or institutional interdependencies with other process elements.

The BPD illustrates the design alternatives as they were given in the concrete case. Selection of a *to-be* model here is made on the basis of a comparison of alternatives taking into account financial key indicators. This calculation is shown in Figure 26.12 and illustrated in more detail for the case of the *IntelligentRouting* web service.

As the impact of the design decisions is only local, the calculation of the differences between alternatives can be made by means of a partial analysis for determining the direct payments. When calculating the difference, it has to be determined what the expected *additional payments*—compared to the status quo level (*as-is*)—of each design alternative are (*to-be*).

Using the service for the route planning process promises both a higher number of detailed plans potentially possible and a better quality of realized detailed plans (*rDP*). It is expected that the processing time of the activity "Create detailed plan" is reduced from 10 to 5 min. It is also expected that the error ratio is reduced (from 20% to 3% in case of the *IntelligentRouting* web service). Taking into account an available capacity of 1,825 working hours per period, a number of *PEs* of 18,250, and an OR of the web service of 3%, direct cost savings of 22,356 € compared to the status quo level can be expected in period 1 (calculation is based on the assumption of an average advantage of 2.50 € for creating a detailed plan compared to creating an ad-hoc plan).

For the *IntelligentRouting* web service, a transaction-based pricing model is assumed, with an average calculation rate of 0.25 € per transaction. The payments per period to be calculated on a service level are to be calculated on the basis of the execution frequency expected for the task

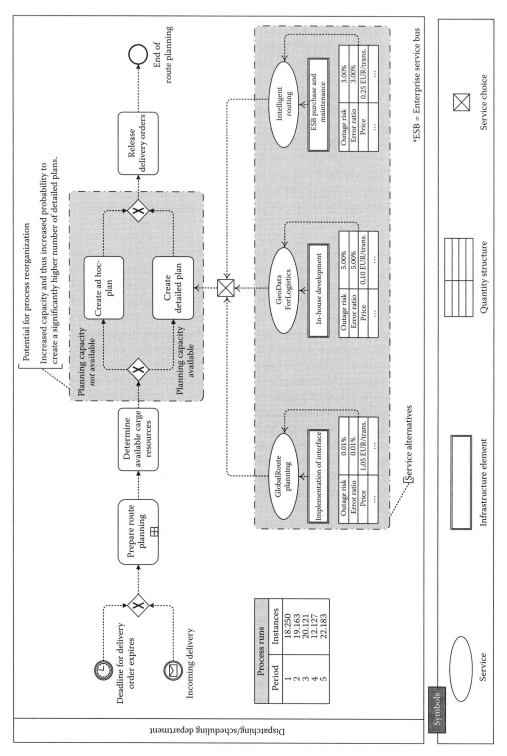

FIGURE 26.11   Business process diagram for process (route planning).

Direct payments					
Period	0	1	2	...	5
− *IntelligentRouting*	−25.000 €	11.431 €	12.422 €	...	16.463 €
− Activity level		22.356 €	24.569 €	...	29.614 €
− Savings		22.356 €	24.569 €	...	29.614 €
(1) CA		1.825	1.825		1.825
(2) PE		18.250	19.163		22.183
(3) $PT_{as\text{-}is}$		10	10		10
(4) $PT_{to\text{-}be}$		5	5		5
(5) $ER_{as\text{-}is}$		0,2	0,2		0,2
(6) $ER_{to\text{-}be}$		0,03	0,03		0,03
(7) $OR_{to\text{-}be}$		0,03	0,03		0,03
(8) $pDP_{as\text{-}is} = (1)/(3)$		10.950	10.950	...	10.950
(9) $pDP_{to\text{-}be} = [(1)/(4)]\cdot[1-(7)]$		21.243	21.243		21.243
(10) $rDP_{as\text{-}is} = MIN\,[(2)\,;\,(8)\cdot\,[1-(5)]$		8.670	8.670		8.670
(11) $rDP_{to\text{-}be} = MIN\,[(2)\,;\,(9)\cdot\,[1-(6)]$		17.703	18.588		20.606
(12) $\Delta rDP = (10)-(11)$		8.943	9.828		11.846
· Savings per detailed plan (€)		2,50	2,50		2,50
= Total savings (€)		22.356	24.569		29.614
− Infrastructure level	−25.000 €	−6.500 €	−7.500 €	...	−8.000 €
− Template [payments ESB]	−25.000 €	−6.500 €	−7.500 €	...	−8.000 €
Initial expenditure	−25.000 €				
−Maintenance		4.000 €	4.000 €		4.000 €
−Adaptation		2.500 €	3.500 €	...	4.000 €
Depreciations		5.000 €	5.000 €		5.000 €
− Service level	0 €	−4.426 €	−4.647 €	...	−5.151 €
− Service payments	0 €	−4.426 €	−4.647 €	...	−5.151 €
Service charge rate (price per transaction)		0,25 €	0,25 €		0,25 €
· $rDP_{to\text{-}be}$		17.702	18.588	...	20.606
= Service payments		−4.426 €	−4.647 €		−5.151 €
− *GlobalRoutePlanning*	−1.000 €	3.060 €	5.183 €	...	9.349 €
+ Activity level		23.720 €	26.001 €	...	32.839 €
+ Infrastructure level	−1.000 €	−1.500 €	−700 €	...	−500 €
+ Service level		−19.161 €	−20.119 €	...	−22.990 €
+ *GeoDataForLogistics*	−32.500 €	11.770 €	13.229 €	...	16.814 €

Formulae for evaluating the economic potentials

$$pDP_{as\text{-}is} = \frac{CA}{PT_{as\text{-}is}}$$

$$mDP_{to\text{-}be} = \frac{CA}{PT_{to\text{-}be}} \cdot (1-OR)$$

$$rDP_{as\text{-}is} = MIN\,(PE\,;\,pDP_{as\text{-}is}) \cdot (1-ER_{as\text{-}is})$$

$$rDP_{to\text{-}be} = MIN\,(PE\,;\,pDP_{tp\text{-}be}) \cdot (1-ER_{to\text{-}be})$$

$$\Delta rDP = rDP_{to\text{-}be} - rDP_{as\text{-}is}$$

Symbols

- CA Capacity (hours/period)
- PE Process executions (#)
- PT Processing time (min)
- ER Error ratio (%)
- OR Outage risk (%)
- pDP Possible detail plans (#)
- rDP Realized detail plans (#)

FIGURE 26.12 Calculation of direct payments of the process reorganization.

"Create detailed plan" ($\approx rDP_{to\text{-}be}$). In period 1, for example, for 17,702 detailed plans created, the service payments amount to –4,426 €. If the route planning process is out-tasked, the calculation rate—at a lower OR—is 1.05 € per transaction, so that the expected payments rise accordingly (–19,161 €).

Apart from the activity-based payments ($TPPO_p$ and $TPPO_{p'}$), the payments for the process transformation (TPPT) also need to be taken into consideration. In this case, most of these payments primarily result from investments in the technical infrastructure. Using the *IntelligentRouting* web service requires the implementation of an ESB solution. In the case example, it is assumed that the company pursues an incremental implementation strategy. Therefore, a decision in favor of the *IntelligentRouting* web service brings about all payments for purchase of technical infrastructure (25,000 €) as well as all follow-up payments for maintenance and adaptation that occur periodically. In case *GlobalRoutePlanning* is used, with activities for detailed planning being out-tasked, payments for the technical infrastructure are substantially lower. If *GeoDataForLogistics* is used, more payments for the technical infrastructure are expected due to the relatively high implementation and development effort required.

With direct payments being consolidated by means of methods for capital budgeting, the future values of investing into the reorganization of the process are 30,379 € (*IntelligentRouting*), 25,424 € (*GlobalRoutePlanning*), and 26,235 € (*GeoDataForLogistics*). Compared to the future value of opportunity of 11,425 € (own equity compounded with an interest rate of 6 per cent), implementing any service can be considered beneficial to the company. The *IntelligentRouting* web service is the design alternative which generates the highest additional future value.

26.7 Conclusion and Outlook

This chapter aimed at providing an introduction into the basic concepts of BPM. It focused on phenomena pertinent to the task of business process analysis, as this task very central for BPM. While incrementally developing a multifaceted perspective on business processes and the task of business process analysis, this chapter particularly calls for integrating process modeling and process evaluation. We think the integration of process modeling with an economic analysis represents a meaningful contribution to the field of BPM. Current approaches for process modeling do not allow for assessing the economic value of process reorganization projects. Approaches for process evaluation or controlling (such as activity-based costing), on the other hand, have no differentiated understanding of process structures and take no advantage of process descriptions already available in practice.

As a consequence, we call for an integrated analysis of the structure of a process and its economic value contribution using an integrated set of methods and instruments. As both, established methods for process modeling (e.g., BPMN) and for capital budgeting are to be applied, the main challenge is to create interfaces between these two domains. How such interfaces can be realized on the basis of BPMN process descriptions has been one of the foci of this chapter. In particular, the chapter presented a procedure model to calculate the *return on process transformation*, which integrates both a process structure perspective as well as a process value perspective.

Applying these methods in practice has shown that it is not always necessary to calculate at a high degree of detail. Using the set of methods presented in the article allows to simplify the evaluation procedures by starting to capture process structures in their entirety on a higher level of abstraction in order to be able to conduct a kind of "pre-assessment" first. Evaluations on a higher level may take into account only a few approximate values at the beginning, which can later be refined successively according to what is deemed necessary. Following this principle, one-day or two-day workshops often are a very good instrument for developing a basic understanding of which direction to go when thinking of reorganizing a certain business process.

More generally, we see the BPM discipline further establishing into a comprehensive management discipline. While contributions over the past decades were mainly focused on modeling, designing, and implementing selected processes of interest, BPM in the future will focus much more on developing capabilities inside organizations to continuously improve and innovate their business. With this, the view of processes as social constructions in particular will play an increasingly important role. While the relevance of the social factor in BPM is often pointed at, little contributions are yet available to facilitate the management of processes as social constructions. Evaluating the financial performance of alternative process design is one approach to facilitate communicating process innovation and process transformation to different stakeholder groups, such as shareholders, in particular. Such approaches can help to strengthen capability areas such as strategic alignment and governance. Further areas or future research include people and culture as two distinct capability areas dedicated to the human factor in BPM.

References

van der Aalst, W. M. P. 2011. *Process Mining: Discovery, Conformance and Enhancement of Business Processes*. Berlin, Germany: Springer.

Anupindi, R., Chopra, S., Deshmukh, S. D., van Mieghem, J. A., and Zemel, E. 2012. *Managing Business Process Flows. Principles of Operations Management*, 3rd edn. Boston, MA: Prentice Hall.

Bandara, W., Indulska, M., Chong, S., and Sadiq, S. 2007. Major issues in business process management: An expert perspective. In: *Proceedings of the 15th European Conference on Information Systems (ECIS)*, pp. 1240–1251. St. Gallen, Switzerland.

Becker, J. and Kahn, D. 2011. The process in focus. In: *Process Management: A Guide for the Design of Business Processes*, 2nd edn., eds. Becker, J., Kugeler, M., and Rosemann, M., pp. 3–14. Berlin, Germany: Springer.

Becker, J., Kugeler, M., and Rosemann, M. 2011. *Process Management: A Guide for the Design of Business Processes*. Berlin, Germany: Springer.

Bostrom, R. P., and Heinen, J. S. 1977. MIS problems and failures: A sociotechnical perspective, Part I: the Cause. *MIS Quarterly*, 1(3):17–32.

vom Brocke, J. 2007. Service portfolio measurement. Evaluating financial performance of service-oriented business processes. *International Journal of Web Services Research (IJWSR)* 4:1–32.

vom Brocke, J. and Grob, H. L. 2011. Profitability of business processes. In: *Process Management: A Guide for the Design of Business Processes*, 2nd edn., eds. Becker, J., Kugeler, M., and Rosemann, M., pp. 421–446. Berlin, Germany: Springer.

vom Brocke, J., Recker, J., and Mendling, J. 2010. Value-oriented process modeling: Integrating financial perspectives into business process re-design. *Business Process Management Journal* 16:333–356.

vom Brocke, J. and Rosemann, M. 2013. Business process management. In: *Encyclopedia of Management Information Systems* (Vol. 7), eds. Straub, D. and Welke, R., in the *Wiley Encyclopedia of Management*, 3rd edn. Chichester, U.K.: Wiley.

vom Brocke, J. and Schmiedel, T. 2011. Towards a conceptualisation of BPM culture: Results from a literature review. In: *Proceedings of the 15th Pacific Asia Conference on Information Systems (PACIS)*, Brisbane, Queensland, Australia.

vom Brocke, J. and Sinnl, T. 2011. Culture in business process management: A literature review. *Business Process Management Journal* 17: 357–377.

vom Brocke, J., Sonnenberg, C., and Baumoel, U. 2011. Linking accounting and process-aware information systems—Towards a generalized information model for process-oriented accounting. In: *Proceedings of the 19th European Conference on Information Systems (ECIS)*, Helsinki, Finland.

vom Brocke, J., Sonnenberg, C., and Simons, A. 2009. Value-oriented information systems design: The concept of potentials modeling and its application to service-oriented architectures. *Business & Information Systems Engineering*, 1: 223–233.

de Bruin, T., Rosemann, M., Freeze, R., and Kulkarni, U. 2005. Understanding the main phases of developing a maturity assessment model. In: *Proceedings of the 16th Australasian Conference on Information Systems (ACIS)*, Sydney, New South Wales, Australia.

Cahn, S. and Choi, C. 1997. A conceptual and analytical framework for business process re-engineering. *International Journal of Production Economics*, 50: 211–223.

Checkland, P. and Scholes, J. 1990. *Soft Systems Methodology in Action*. Chichester, U.K.: John Wiley.

Davenport, T. H. 1993. *Process Innovation: Reengineering Work through Information Technology*. Boston, MA: Harvard Business School Press.

Davenport, T. H. 2005. *Thinking for a Living: How to get Better Performances and Results from Knowledge Workers*. Boston, MA: Harvard Business Press.

Davenport, T. and Perez-Guardo, M. 1999. Process ecology: A new metaphor for reengineering-oriented change. In: *Business Process Engineering: Advancing the State of the Art*, eds. Elzinga, J., Gulledge, T., and Chung-Yee, L., pp. 25–41. Norwell, MA: Kluwer Academic Publishers.

Deming, W. E. 1986. *Out of the Crisis*. Cambridge, MA: MIT Press.

Dumas, M., van der Aalst, W. M. P., and ter Hofstede, A. H. 2005. *Process Aware Information Systems: Bridging People and Software through Process Technology*. Hoboken, NJ: Wiley.

Freeman, R. E. 1984. *Strategic Management: A Stakeholder Approach*. Boston, MA: Pitman.

Galliers, R. and Baker, B. 1995. An approach to business process re-engineering: The contribution of socio-technical and soft OR concepts. *Information* 33:263–277.

Grob, H. L. 1993. *Capital Budgeting with Financial Plans, an Introduction*. Wiesbaden, Germany: Gabler.

Hailemariam, G. and vom Brocke, J. 2010. What is sustainability in business process management? A theoretical framework and its application in the public sector of ethiopia. In: *Proceedings of the 8th Business Process Management Conference*. Hoboken, NJ.

Hammer, M. 2010. What is business process management. In: *Handbook on Business Process Management*, (Vol. 1), eds. vom Brocke, J. and Rosemann, M., pp. 3–16. Berlin, Germany: Springer.

Hammer, M. and Champy, J. 1993. *Reengineering the Corporation: A Manifesto for Business Revolution*. London, U.K.: Nicholas Brealey Publishing.

Harmon, P. 2007. *Business Process Change. A Guide for Business Managers and BPM and Six Sigma Professionals*, 2nd revised edition. San Francisco, CA: Morgan Kaufmann.

Havey, M. 2005. *Essential Business Process Modeling*. Sebastopol, CA: O'Reilly Media.

ter Hofstede, A. H. M., van der Aalst, W. M. P., Adams, M., and Russell, N. 2010. *Modern Business Process Automation—YAWL and Its Support Environment*. Berlin, Germany: Springer.

Kaplan, R. S. and Norton, D. P. 1992. The balanced scorecard—Measures that drive performance. *Harvard Business Review* 70(1):71–79.

Keller, G., Nüttgens, M., and Scheer, A.-W. 1992. Semantische Prozeßmodellierung auf der Grundlage, Ereignisgesteuerter Prozeßketten (EPK). In: *Veröffentlichungen des Instituts für Wirtschaftsinformatik Scheer*, ed. Scheer, A.-W. Saarland University, Saarbrücken, Germany.

Kettinger, W. J. and Grover, V. 1995. Toward a theory of business process change management. *Journal of Management Information Systems* 12:9–30.

Mayer, R. J., Menzel, C. P., Painter, M. K., de Witte, P. S., Blinn, T., and Perakath, B. 1995. Information integration for concurrent engineering (IICE) IDEF3 process description capture method report. http://www.idef.com/pdf/Idef3_fn.pdf (accessed March 7, 2011).

Melão, N. and Pidd, M. 2000. A conceptual framework for understanding business processes and business process modeling. *Information Systems Journal* 10:105–129.

zur Muehlen, M. and Recker, J. 2008. How much language is enough? Theoretical and practical use of the business process modeling notation. In: *Proceedings of the 20th International Conference on Advanced Information Systems Engineering (CAiSE)*, eds. Bellahsène, Z. and Léonard, M., pp. 465–479. Berlin, Heidelberg: Springer.

zur Muehlen, M. and Shapiro, R. 2010. Business process analytics. In: *Handbook on Business Process Management* (Vol. 2), eds. vom Brocke, J. and Rosemann, M., pp. 137–158. Berlin, Germany: Springer.

Neiger, D., Churilov, L., and Flitman, A. 2008. *Value-Focused Business Process Engineering: A Systems Approach: with Applications to Human Resource Management*. New York: Springer.

OASIS 2006. ebXML Business process specification schema technical specification v2.0.4. http://docs. oasis-open.org/ebxml-bp/2.0.4/OS/spec/ebxmlbp-v2.0.4-Spec-os-en.pdf (accessed March 7, 2011).

OASIS 2007. Web services business process execution language. Version 2.0 OASIS standard. http://docs. oasis-open.org/wsbpel/2.0/OS/wsbpel-v2.0-OS.pdf (accessed March 7, 2011).

OMG-BPMN. 2011. Business process modeling notation (BPMN) information. http://www.omg.org/ spec/BPMN/2.0/PDF (accessed March 7, 2011).

OMG-UML. 2011. Object management group—UML. http://www.uml.org/ (accessed March 7, 2011).

Petri, C. A. 1962. Kommunikation mit Automaten. In: *Schriften des Instituts für Instrumentelle Mathematik der Universität Bonn*. Bonn, Germany.

Recker, J., Rosemann, M., Green, P. F., and Indulska, M. 2011. Do ontological deficiencies in modeling grammars matter? *MIS Quarterly* 35:57–79.

Rescher, N. 1996. *Process Metaphysics: An Introduction to Process Philosophy*. Albany, NY: State University of New York Press.

Rosemann, M. and vom Brocke, J. 2010. The six core elements of business process management. In: *Handbook on Business Process Management* (Vol. 1), eds. vom Brocke, J. and Rosemann, M., pp. 107–122. Berlin, Germany: Springer.

Scheer, A.-W. 2000. *ARIS—Business Process Modeling*. Berlin, Germany: Springer.

Schmiedel, T., vom Brocke, J., and Recker, J. 2012. Is your organizational culture fit for business process management? *BPTrends* 9 (May):1–5.

Schmiedel, T., vom Brocke, J., and Recker, J. 2013. Which cultural values matter to business process management? Results from a global Delphi study. *Business Process Management Journal* 19(2).

Spanyi, A. 2003. *Business Process Management Is a Team Sport: Play It to Win*. Tampa, FL: Anclote Press.

Taylor, F. W. 1911. *The Principles of Scientific Management*. New York: Harper & Brothers.

Trkman, P. 2010. The critical success factors of business process management. *International Journal of Information Management* 30:125–134.

Walsham, G. 1991. Organizational metaphors and information systems research. *European Journal of Information Systems* 1:83–94.

Weske, M. 2007. *Business Process Management. Concepts, Languages, Architectures*. Berlin, Germany: Springer.

Willcocks, L. and Smith, G. 1995. IT-enabled business process re-engineering: Organizational and human resource dimensions. *Journal of Strategic Information Systems* 4:279–301.

27

Information Requirements Determination

Glenn J. Browne
Texas Tech University

27.1 Introduction

Information requirements determination (IRD) is the process of eliciting, representing, and verifying the functional and non functional needs for an information system (Browne and Ramesh 2002; Davis 1982; Hickey and Davis 2004; Larsen and Naumann 1992; Vitalari 1992). More broadly, IRD is a form of needs analysis, an activity required in any designed artifact, ranging from consumer products to software to industrial processes (Simon 1996).

The primary IRD process occurs early in the development of an information system, generally following the project planning process. IRD has traditionally been conceptualized within the traditional systems development life cycle (SDLC) (Kendall and Kendall 2010) as an early stage in the process. However, IRD is also utilized in its basic form in new approaches to systems development, including most "agile" development methods (Baskerville and Pries-Heje 2004; Lee and Xia 2010).

The importance of IRD to systems development is difficult to overstate. Because the IRD process occurs early in development and determines the needs for the system, all remaining activities in development, from modeling to design to coding to implementation, depend on specifying requirements that are as accurate and complete as possible. Elegantly designed systems that do not meet user requirements will not be used. Queries to databases that do not contain information users need will not be made. Thus, IRD is arguably the central activity in systems development.

Computer information systems serve a wide range of purposes in society. Regardless of whether the system is fully automated, such as a control system (e.g., a system operating temperature controls for a building) or a system that acts as a replacement for human decision makers (such as a loan application system), or is a system designed for interactive decision support (e.g., web-based shopping assistants), the requirements for the system must be determined by the systems analyst and developer. For this reason, requirements determination is also key to the computing profession generally.

27.2 Underlying Principles

The full information determination requirements process is often conceptualized as the development of a strategy for the elicitation of the requirements, their representation, and their verification (Larsen and Naumann 1992; Vitalari 1992). Development of a strategy was described well by Davis (1982), but little recent research has addressed the subject. Elicitation, or gathering or capturing of requirements, can be accomplished in at least four ways: asking, deriving from an existing system, synthesis from characteristics of the utilizing system, and discovering from experimentation from an evolving system (prototyping) (Davis 1982). Although all these sources are used at various times, this chapter will focus on the asking strategy, as it is the primary strategy utilized.

The asking strategy for elicitation is performed primarily through discussions with people who use or will use the system. Two other methods also important in elicitation include (1) examination of (a) internal documents such as organizational forms, reports, procedures manuals, policy documents, strategic plans, training materials, documents concerning existing systems (whether computerized or manual), and other company records and (b) external documents such as books, scholarly articles, white papers, and web-based materials and (2) observations of people performing tasks relevant to the system being developed. The focus of this chapter will be primarily on elicitation of requirements from users, but examining documents and observation will also be discussed briefly.

Representation of requirements is accomplished by analysts using informal models (such as flow charts, conceptual maps, knowledge maps, and cognitive maps) and semi-formal models (such as entity-relationship diagrams, data-flow diagrams, and Unified Modeling Language diagrams) (Browne and Ramesh 2002; Montazemi and Conrath 1986; Robertson and Robertson 2006). Verification is conducted by discussing the analyst's understanding of the requirements with the user, utilizing models, diagrams, written notes, and/or any other representational or verbal means available. Although the full process is critical to the systems development process, as noted, this chapter will focus on the elicitation of requirements from users.

Information system development can proceed in several different ways. The traditional SDLC model (usually termed the "waterfall" model or the "planned" model) is still used in a large majority of systems development efforts (Light 2009; Norton 2008) as it brings numerous significant advantages to the development process. However, the lack of quick user feedback on design efforts in the waterfall model has led to the creation of several alternative methods for systems development over the past 15 years or so. These methods are usually grouped together under the term "agile." Agile methods rely on short cycle times, small work modules, and rapid user feedback to improve system development outcomes (Baskerville and Pries-Heje 2004; Lee and Xia 2010). A set of basic requirements is needed even in agile methodologies, and so the discussion of IRD in this chapter is relevant to all methodologies. Nonetheless, there are differences in IRD between the two general approaches, and these differences are discussed in the *Impact on Practice* section.

27.2.1 Individual Level Principles

Asking users to specify information requirements relies on several fundamental cognitive and behavioral principles. Knowledge elicitation utilizes an interactive process in which the analyst stimulates the user to describe his needs in terms of inputs, processes, and outputs in the business process. Users are often encouraged not just to state "facts" but also to tell stories that reveal process activities that may otherwise remain unsaid (e.g., Alvarez 2002; Davidson 2002; Urquhart 2001). Stimulating users to evoke their knowledge requires that the analyst ask the right questions (Browne and Rogich 2001; Marakas and Elam 1998; Moody et al. 1998; Wetherbe 1991) to cause the user to access the correct "path" in memory. The more substantively knowledgeable (in the application domain) and procedurally knowledgeable (in systems development) the analyst, the more likely he will be to stimulate the user in appropriate ways. Failure to stimulate the correct path for the user can result in serious (although unintentional) "misinformation" or lack of information elicited from the user and inaccurate requirements. Thus, the analyst's role in eliciting knowledge from users is usually both critical and challenging.

Analysts (and users) are limited in their information processing capabilities (Reisberg 2009; Simon 1996), and this constrains their ability to gather accurate and reasonably complete requirements (note that it can be persuasively argued that requirements are never "complete"). Perception is vastly limited; humans perceive a tiny fraction of information available in the environment at any point in time. Similarly, attention is also significantly limited. Most of what is perceived is not attended to and therefore does not enter working memory. Working memory is the mechanism through which everything we refer to as "thinking" takes place (e.g., reasoning, judgment, decision making, reflection, and introspection). Information in working memory that is elaborated upon and that has sufficient time to be stored ("consolidation") enters long-term memory. Working memory is severely constrained both by speed of processing (generally speaking, we are serial processors and can think about only one thing at a time) and capacity of processing (we are generally limited to four to nine pieces of information in working memory at one time, although the more complex the piece of information, the lower the number of pieces that can be considered). Long-term memory is also constrained. Because working memory processes so little information so slowly, most information to which we attend either does not enter long-term memory or enters only as fragments. We remember the "gist," or essence, of events but not the details (Brainerd and Reyna 1990). When asked to recall an event, a person must fill in the details of the event that were not stored. So, for example, when a user is asked by an analyst to recall an event that occurred, he will have to fill in the details (i.e., invent them) because he has no memory of those details. This can, of course, mislead the analyst. The other major problem with long-term memory is that people forget information. Nonetheless, it is generally believed that nothing is ever truly forgotten, and thus the goal for analysts is finding the right stimulus material to help the user locate the correct "path" to the memory (Reisberg 2009). This effort can be more or less successful.

Researchers have investigated numerous issues related to user and analyst cognition in IRD. As one example, recent research in requirements determination has investigated specific issues in long-term memory that affect the accuracy of requirements. Appan and Browne (2010) investigated a phenomenon referred to as "retrieval-induced forgetting." This psychological phenomenon shows that the act of recalling certain information (e.g., in response to a question from an analyst) causes related but currently unrecalled information to be suppressed in memory and therefore to be unavailable for subsequent recall from the same or similar stimulus. Thus, the first relevant information an analyst receives from a user concerning a particular issue is also likely to be the last; if the analyst returns to the issue in subsequent IRD sessions, he is likely to receive the same information from the user (Appan and Browne 2010; see also MacLeod 2002). This problem reduces the amount of information an analyst will learn about user needs and suggests the need to use multiple users at multiple points in time for IRD.

A second unrelated study concerning long-term memory, also by Appan and Browne (2012), investigated the impact of the misinformation effect on IRD. The "misinformation effect" occurs when false or misleading information is suggested to someone (e.g., a user) following the occurrence of an event and the person recalls the misinformation rather than the information he originally observed during the event. Appan and Browne (2012) demonstrated empirically the existence of the misinformation effect during IRD and extended previous findings to show that the effect exists even with factual information learned over a long period of time. The latter finding is particularly troubling for IRD, since the results show that if an analyst suggests some misinformation contrary to users' long-held beliefs or factual knowledge that they have developed through their own experiences, those users are likely to report in response to subsequent questions the information provided by the analyst (which may be arbitrary or be due to ulterior motives) rather than their own knowledge or beliefs. The potential impact of the misinformation effect on eliciting accurate user requirements and on ultimate system success is therefore of significant concern.

Beyond specific problems with long-term memory, the overall limitations on information processing mean that systems analysts, and humans in general, are not cognitively equipped to gather all possible information in any but the simplest of problems. Requirements determination is not a simple problem. Therefore, analysts employ heuristics to help them narrow the problem space. A *heuristic* is a short-cut, or rule-of-thumb, for achieving satisfactory (and perhaps better) results in solving a problem (Simon 1996). People evolve heuristics for solving problems based on experience; the more experience a person has with a type of problem, the better his heuristics typically are. Heuristics allow humans to circumvent their cognitive limitations, particularly those associated with working memory, but they come at a cost. Because they are not optimization techniques (people do not gather all available information and do not integrate it optimally), heuristics result in errors (Tversky and Kahneman 1974). Some of the errors created are random and some are systematic. The systematic errors are usually referred to as cognitive biases. A *bias* is a systematic deviation from an accurate answer; accuracy in this case can mean normatively accurate information or a target standard resulting from consensus.

In the elicitation of requirements from users, numerous biases have been theorized and several have been investigated empirically. Basic issues in cognitive biases in requirements determination have been discussed by Browne and Ramesh (2002), Ralph (2011), and Stacy and MacMillan (1996). Prompting strategies for overcoming cognitive biases in requirements elicitation have been suggested by Browne and Rogich (2001), and a list of cognitive biases and strategies for overcoming them is provided in Pitts and Browne (2007).

The influence of cognitive biases on human behavior, including requirements elicitation, is extensive, and a full discussion is beyond the scope of this chapter. It is known that people are affected by the ease with which information can be recalled, by insensitivity to base rates (i.e., the actual rate of occurrence of events in populations), by vividness of stimuli, by giving inappropriate weight to small samples, by misconceptions of chance events (e.g., believing that certain patterns in data cannot occur without a causal explanation), by seeking primarily confirmatory evidence for conclusions, by anchoring on initial values (or existing procedures), and by overwhelming overconfidence, among many other biases (see, e.g., Bazerman and Moore 2013; Kahneman 2011). These biases all impact the information elicited in requirements determination as they affect both the users and the analysts. As one example, important research concerning the anchoring bias in requirements modeling has been performed by Parsons and Saunders (2004) and Allen and Parsons (2010) and in requirements elicitation by Jayanth et al. (2011). The interested reader is referred generally to Bazerman and Moore (2013) and Kahneman (2011) and, for examples specific to IRD and systems development, to Browne and Ramesh (2002) and Stacy and MacMillan (1996). In general, the study of cognitive biases is an under-researched area in the information systems field.

A second type of bias that affects IRD is motivational biases. Motivational biases result from perceived incentives in the decision-making environment or from internal desires or preferences (see Bazerman and Moore 2013 for a general discussion). Examples of motivational biases include self-serving attributions (i.e., taking credit for successes and blaming failures on external sources beyond the person's control), reporting false facts or beliefs to curry the favor of influential others (e.g., bosses), the illusion of control (i.e., the belief that one can control random events), real and anticipated regret, and deliberately escalating commitment to a failing project. A large body of literature in the information systems discipline concerning escalation of commitment has been developed by Keil (1995) and Mähring and Keil (2008). Most other motivational biases have not received attention, however. Motivational biases generally (although not universally) arise in IRD because requirements determination involves numerous parties in a social process, and both analysts and users are motivated to portray themselves in a positive light, to hide errors, and to avoid blame.

27.2.2 Group Level Principles

When analysts elicit requirements from users, the parties are engaged in a social process. In addition to the motivational biases just discussed, social influence concerns such as conformity, persuasion, and trying to please the analyst (the "demand effect" or "getting along effect") all are potential problems for the analyst gathering requirements (see Cialdini and Trost 1998). The process of IRD should not involve the analyst imposing his will on the user or attempting to persuade him to agree to certain requirements (Alvarez 2002). The process should be one of open inquiry, in which the analyst attempts to gain an accurate understanding of the user's needs. Thus, the analyst should approach the process with an open mind, ignoring pre-conceived notions about what the user may want or need and without thinking ahead to implementation details. The analyst should be aware that his role typically gives him a type of perceived legitimate power over the user (Alvarez 2002; Tan 1994), which may lead the user to attempt to subtly assess what the analyst wants and then report or agree to those "requirements" to conform and/ or to try to please the analyst. In such cases, the analyst may be pleased, but the requirements provided are likely to be less accurate than desired.

When IRD occurs in group settings, in information systems development, the interactions are typically organized as joint application development, or JAD, sessions (Liou and Chen 1994). JAD sessions are requirements determination sessions involving one or more analysts, several or many users, and often additional stakeholders in the system (Dennis et al. 1999). In such cases, in addition to the individual issues involved, analysts must be aware of group dynamics. Issues such as company politics, power and hierarchy (e.g., Milne and Maiden 2012), impression management, attempts at ingratiation, coalitions, group polarization, illusions of agreement (e.g., groupthink), conformity, roles, and norms in the organization all must be taken into account (see, e.g., Guinan et al. 1998). Analysts are often poorly prepared to manage group dynamics, and skilled meeting facilitators are often employed to ensure that the purpose of the session, that is, requirements for the system, is adequately met.

A major difficulty that arises because the asking strategy in IRD is a social process is communication issues between analysts and users (Davern et al. 2012). Much research has investigated communication issues (e.g., Bostrom 1989; Davis et al. 2006; Valusek and Fryback 1985). Attempts to overcome or mitigate user-analyst communication issues have been made, particularly by increasing user participation in the systems development process. However, the impact of user participation has been mixed in empirical studies (e.g., Gallivan and Keil 2003; Hunton and Beller 1997; McKeen et al. 1994), and further research is necessary to determine the amount and type of user participation that can result in the best requirements determination results.

27.2.3 Examination of Internal and External Documentation

As noted, a second method usually used to accompany the asking strategy is for the analyst to examine internal and external documents (e.g., Davis 1982; Kotonya and Summerville 1998; Maiden and Rugg 1996). Analysts must decide which documents contain valuable and relevant information for the firm and for the users and should be incorporated into the proposed information system and which documents are redundant, not used, or not valuable and should be omitted (Robertson and Robertson 2006; Watson and Frolick 1993). For example, Watson and Frolick (1993) discussed requirements determination for executive information systems and noted the importance of examining memos, e-mails, and various internal company documents as well as external documents such as articles, books, marketing research information, and government and industry publications. Company forms and reports often determine the desired inputs and outputs, respectively, of systems and are therefore a natural starting point for analysts. However, with all existing documents, analysts must be concerned about improper anchoring on current data and processes (Davis 1982). Simply because internal documentation exists or external data seem to support requirements does not mean they should be included in the system (Robertson and Robertson 2006). Because people prefer concrete examples to abstract concepts, there is a strong temptation to utilize an existing document, design, interface, etc. as a starting point. This initial anchoring point, however, typically precludes other classes of potential alternatives, including more innovative ideas. Oftentimes, business process reengineering using outside experts (with no pre-conceived notions or company "baggage") may be required to be certain that existing documentation reflects requirements that are useful and appropriate for the system currently under design.

27.2.4 Observation of Users and Organizational Task Performance

Observation of users as they perform organizational tasks is also an important method for eliciting requirements (Alvarez 2002; Byrd et al. 1992; Kendall and Kendall 2010; Leifer et al. 1994; Robertson and Robertson 2006). Users often have difficulty verbalizing how they perform their everyday tasks. As experience increases, people typically lose conscious access to the steps they take in performing a task (Leifer et al. 1994; Simon 1979). Observation can help analysts either as a first step in understanding work processes or in filling in gaps in user descriptions of how they perform their jobs (Kendall and Kendall 2010). Observation can be performed either silently or interactively. Silent observation requires the analyst to observe the user performing his task without interruption. The advantage of this method is that the user's typical behaviors are not (ideally) altered because of interruptions. The main disadvantage of this method is that the analyst can observe only behavior, not cognition, and therefore may not understand how or why an action is being performed. Interactive observation permits the analyst to interrupt the user's task performance when an action or the reasons for it are not clear (Robertson and Robertson 2006). This allows the analyst to understand how and why the action is being performed, assuming the user can articulate the reasons. The main disadvantage is that the user's behavior and thought processes are interrupted, which may lead to task performance that is not typical. This could cause the analyst to model the requirements inappropriately. Based on the strengths and weaknesses of the two techniques, using silent observation first and following that technique with interactive observation may be the best approach.

A disadvantage of observation generally is the Hawthorne effect, in which users act differently because they know they are being observed (Parsons 1992). There is no perfect cure for the Hawthorne effect, but analysts should assure users that they are not being judged and that the results of their individual task performance will not be reported (only aggregated results), if true. If the organization is unwilling to promise anonymity for individual performance, it is recommended that the analyst utilize a different technique.

Another method for observing task performance is through protocol analysis (Byrd et al. 1992; Ericsson and Simon 1993). Protocol analysis requires users to think aloud while they perform a task, and their thoughts (sometimes accompanied by movements) are recorded on tape, video, or digital media. The resulting vocalized thoughts can be analyzed to trace task performance and to capture certain aspects of behavior and cognition that both silent and interactive observation may miss. When the correct procedures are utilized, results from protocol analysis can be valid and highly valuable sources of information (see Ericsson and Simon 1993).

27.3 Impact on Practice

Requirements determination is arguably the principal job of systems analysts. Regardless of whether the proposed system is intended to automate or provide decision support for an existing manual system, upgrade an existing computerized information system, or tailor a purchased ("off-the-shelf") software program for a company's needs, user and organizational requirements must be determined. Because information systems are now the lifeblood of organizations (and, increasingly, individuals), requirements determination is a critical and ubiquitous activity in companies today. Thus, all of the issues discussed in the previous section are important and relevant for IRD practice.

Faulty requirements are generally thought to be the most common cause of information system failure in organizations (Bostrom 1989; Byrd et al. 1992; Hofmann and Lehner 2001; Wetherbe 1991). Underutilization, in which systems are used but not to their full functionality, is also a problem that results from inaccurate or incomplete requirements. The cost of information systems failures to businesses is estimated in the hundreds of billions or even trillions of dollars worldwide each year.* Thus, the impact of improving information systems on organizational performance is difficult to overestimate.

Requirements determination research over the years has directly influenced systems analysis practice. For instance, analysts are now trained in multiple elicitation techniques and are often warned about the influence of various problems in elicitation (Kendall and Kendall 2010). For example, analysts ideally should provide little input during early stages of the interview process, stimulating the user to talk and probing using only neutral questions. This allows the user to express his views without any priming or undue influence from the analyst. At later stages of the process, the analyst can probe to elicit specific knowledge not yet revealed (Appan and Browne 2012).

Research demonstrating the advantages of involving many stakeholders and multiple points of view has led to the use of JAD sessions, noted earlier. JAD sessions allow for informed discussion of various potential needs both between users and analysts and between users themselves. The assignment of devil's advocates in JAD sessions to improve and polish ideas is also a result of research on group decision making (see Browne and Ramesh 2002). Although JAD sessions have weaknesses, such as wasted time during group brainstorming (Brown and Paulus 2002; see also Sandberg 2006), they bring significant strengths to the IRD process, including user involvement and increased "buy-in" by stakeholders (Davidson 1999).

One practical implication of research in IRD that has yet to appear on a large-scale basis is training analysts to recognize the heuristics they use and the biases and other cognitive challenges to which they are subject. Articles in the practitioner literature have appeared (e.g., Stacy and MacMillan 1996; West 2008), but they have not been widespread and there is no evidence of systematic training. Such training has the potential to make a profound contribution to the effectiveness and efficiency of requirements elicitation, perhaps more than any other set of findings from the research literature. Motivational issues are also critical, but they are generally more easily recognized and remedied. Cognitive biases, recall issues, attentional deficits, and other cognitive problems operate primarily subconsciously and are largely hidden from view. This makes their effects particularly pernicious and difficult to discern.

As noted in the *Underlying Principles* section mentioned earlier, there are differences in requirements determination in the SDLC approach and agile approaches to systems development. In the SDLC, requirements are gathered primarily in one of the earliest stages of the process. After they are modeled and verified, users often "sign off" on the requirements, and system designers then proceed with development of the system. There are feedback loops in the SDLC, and analysts and designers often must return to users for additional requirements and/or clarifications. However, the initial stage subsumes the majority of the requirements determination effort. Research has demonstrated that one of the weaknesses of this approach is that users often do not know what they need during the early stages of the development of a system (especially a new system) and that their "sign off" is of dubious validity (Wetherbe 1991). Additionally, analysts may not understand the proposed system well enough to ask appropriate questions at this point, resulting in an inadequate determination of requirements (Davis 1982). A second weakness is that requirements often change during development (e.g., Maruping et al. 2009), and the "frozen" requirements document with which designers work when using the SDLC can therefore make the resulting system inappropriate or out of date by the time it is completed. These findings led to the agile methods discussed earlier, including such methods as scrum and extreme programming, which rely on short bursts of IRD followed by mock-ups or prototypes to which users can quickly react. It is likely that such agile methods will continue to be used and perhaps increase in popularity as the techniques are improved and the advantages, particularly quick user feedback, are fully utilized. It is unlikely, however, that an initial requirements gathering will disappear from agile methods that survive. Absent an initial substantial investment in requirements elicitation, the first prototypes will reflect what

---

* For one estimate in the trillions of dollars, and a critique of that number, see http://www.zdnet.com/blog/projectfailures/worldwide-cost-of-it-failure-62-trillion/7627?tag=content;siu-container and http://brucefwebster.com/2009/12/28/the-sessions-paper-an-analytical-critique/.

the analyst thinks he knows about the system or what he thinks should be included. There is no more fertile ground for the types of cognitive and motivational biases and other problems discussed in this chapter than this type of approach to systems development.

27.4 Research Issues

Despite the large body of research that has developed concerning requirements determination over the past 30 years, there is no shortage of work remaining. Because the activity involves individual and group behavior, and all the cognitive and motivational issues associated with such behavior, IRD progresses and improves largely as our understanding of human behavior progresses in the social sciences. This section describes a limited subset of research issues that are currently of interest in the field or will be important in the near future.

One important set of issues is the enduring topic of cognition in IRD. As we gain a better understanding of judgment and reasoning, and how decisions are made, we can improve the efficiency of IRD and elicit and model requirements more accurately. This process includes a better understanding of how analysts allocate their attention, how they reason and make judgments and choices, and what information they commit to long-term memory. Additionally, the recall of information from long-term memory, and the problems that have been shown to occur, remain issues that deserve further research. As noted in the *Underlying Principles* section in the preceding text, recent research in the information systems field has addressed some of these issues. In addition, neuroscience techniques such as functional magnetic resonance imaging and electroencephalography are allowing information systems researchers to develop an understanding of the brain at the neural level (e.g., Dimoka et al. 2011) that will aid immeasurably in our understanding of cognition and behavior in IS development.

A related concern is the issue of eliciting requirements that help analysts directly with decision making. Decisions about requirements require reasoning, judgment, and choices by systems analysts, and the direct elicitation of user preferences about requirements can aid immeasurably in these decisions. Some research in IRD has addressed direct elicitation of user preferences (e.g., Liaskos et al. 2011) and decision making more generally (e.g., Alenljung and Persson 2008). However, much more research from decision theory could be incorporated to improve these outcomes. Preference elicitation, probability elicitation, and many other decision analytic techniques can be incorporated in IRD to provide a stronger foundation for the choice and prioritization of requirements by analysts (see, e.g., Goodwin and Wright 2004; Jain et al. 1991; Keeney and Raiffa 1993; von Winterfeldt and Edwards 1986).

Another important issue for systems analysts in IRD is when to stop eliciting and modeling requirements. In unstructured problems with nearly limitless information, such as IRD, analysts can conceivably elicit information and model requirements indefinitely. However, time, cost, cognitive, and motivational considerations prevent them from doing so. Research has revealed that analysts (unsurprisingly) do not gather all available information in requirements elicitation settings and has demonstrated cognitive stopping rules that analysts employ to cease requirements determination efforts (Browne and Pitts 2004; Pitts and Browne 2004). Additional research has investigated stopping rules in information search behavior more generally (Browne et al. 2007). Stopping rules are particularly important in IRD because of the costs of underacquisition and overacquisition of information. Overacquisition results in wasted time and money and may lead to unnecessary features in software. Underacquisition results in underspecified requirements, leading to systems that likely do not meet user needs. Much additional work is necessary to understand both cognitive and motivational reasons that analysts stop during all activities in IRD, including elicitation, representation, and verification of requirements.

In addition to basic cognitive issues, an improved understanding of analysts' use of heuristics and the biases that result can have a profound impact on IRD. As noted earlier, human decision making is often biased in predictable ways. Both analysts and users are biased in their provision and interpretation of information and in the decisions they make based on that information. Therefore, an improved understanding of the conscious and subconscious heuristics that analysts use to elicit, model, and verify information, and of the biases that are present in both analysts and users, has the potential to improve IRD and the remainder of the systems development process significantly.

Another important issue is that software development is increasingly performed by people who are not colocated geographically. This means that requirements determination must be performed by people in different places, perhaps of different cultures, who likely have never met. The telecommunications challenges of such interactions have largely been solved, but the lack of face-to-face communication may have significant impacts on the process and outcomes of IRD. Research in communication theory and organizational behavior has demonstrated the differences between face-to-face communication, computer-mediated communication, and other forms of communication. Although some excellent research concerning IRD has been performed both in this area and the related area of off-shoring (e.g., Aranda et al. 2010; Dibbern et al. 2008; Evaristo et al. 2005), much more work is necessary to understand the full implications of distributed requirements determination.

Another important future research issue concerns requirements determination for purchased software. Much software is now purchased rather than produced "in-house," and requirements determination can be a complicated process with such software. Ideally, requirements are determined before software is purchased to ensure that the software meets user needs. However, software can be purchased for a variety of reasons that are unrelated to requirements, including political reasons, purchasing from a vendor with which the company already has a relationship, better pricing, etc. If requirements are determined beforehand, systems analysts and designers will still need to integrate the system into the company's environment. If requirements are not determined a priori, or if only a cursory attempt is made, the challenges upon the arrival of the software can be daunting. Analysts will need to determine user requirements and then assess how (and if) the software can be tailored to those requirements. Some research has addressed these issues (e.g., Parthasarathy and Ramachandran 2008). However, as software is increasingly purchased off-the-shelf or in large-scale company-wide packages (e.g., enterprise resource planning systems), analysts' challenges in determining requirements escalate dramatically. In such cases, arguments have been made that it is easier to change user processes rather than the software (e.g., Davenport 1998; but for a contrary view, see Hong and Kim 2002). In many cases, dramatic IT implementation failures have resulted from systems that could not be properly adapted to company requirements after purchase (see, e.g., Chen et al. 2009). More research is needed to help analysts manage the challenges associated with IRD in these contexts.

Another important research question concerns how users can be trained and motivated to provide better and more complete requirements. Users are the domain content experts in most systems development efforts, and systems analysts (and the organization) rely on them to provide requirements that will lead to successful systems. However, users suffer from the same biases as analysts. There is also often inertia among users who do not want to change how they perform work, which is nearly inevitable with the introduction of a new information system (Gallivan and Keil 2003; Markus and Benjamin 1997). Thus, users may not provide necessary information or may deliberately introduce misinformation (Appan and Browne 2012). Additional research investigating requirements elicitation issues from the user side would be highly valuable.

Specific content areas are also important. Although most research has focused on methods and tools for IRD that can be generalized across content domains, some tailoring is often necessary depending on the domain. For example, Schneider et al. (2012) note that many systems are now security-related, and requirements determination expertise for such systems is often lacking. They state that "Identifying security-relevant requirements is labor-intensive and error-prone" (p. 35) and suggest organizational learning tools to aid analysts in determining requirements for security-related software. Similarly, Vivas et al. (2011) point out the difficulties in designing security-related systems that ensure privacy, trust, and identity management. Fabian et al. (2010) provide a conceptual framework for understanding requirements elicitation for security-related applications. Considering the continuing threat to system security, further research into this content area, with methods and tools for appropriately determining system requirements, is crucial. Other content areas may similarly benefit from IRD research directed specifically toward them.

Convincing practitioners to use the methods and tools suggested by research is another topic worthy of further investigation. Sikora et al. (2012) point out that many methods generated by academics and supported by empirical research do not make their way into common usage among systems analysts. The authors attribute this adoption problem in part to a lack of understanding among academics concerning the tools practitioners need (ironically, perhaps, a requirements determination problem), but a lack of reading of the research literature and general suspicion of academics by practitioners also likely play a role. Better dissemination of research results and better connections with practicing systems analysts by academics have the potential to improve this situation, but more research into the full nature and causes of the problem would be useful.

Requirements determination is a critical and enduring topic in systems development. Many other issues are important for research, and the discussion in this section should be considered a starting point for topics worthy of investigation.

27.5 Summary

IRD is central to systems development because it occurs early in the process and directly influences all activities that follow it. Much research has been performed in the area, but because it is critical to all systems development it is an enduring research topic in computer information systems. Both successes and difficulties in eliciting requirements have been discussed in this chapter, as well as methods for mitigating the problems that occur. IS practice is directly impacted by research in the area because systems analysts are always gathering requirements and are under increasing competitive pressure to produce systems quickly with appropriate, and often innovative, features. Excellent requirements determination is like creative problem solving, with innovations flowing directly from requirements users may not have even imagined before talking with a systems analyst. Needless to say, technology is ubiquitous throughout world societies today. Technology directly improves productivity and standards of living. Advancements and innovations in technology can only be achieved through an improved understanding of users' needs and desires and through the proper management of these requirements during the entire systems development process.

Further Informations

Kendall, K.E. and Kendall, J.E. *Systems Analysis and Design*. Upper Saddle River, NJ: Prentice Hall, 2010.
Robertson, S. and Robertson, J. *Mastering the Requirements Process*, 2nd edn. Upper Saddle River, NJ: Addison-Wesley, 2006.
Valacich, J.S., George, J., and Hoffer, J.A. *Essentials of Systems Analysis and Design*, 5th edn. Upper Saddle River, NJ: Prentice Hall, 2011.

References

Alenljung, B. and Persson, A. Portraying the practice of decision-making in requirements engineering: A case of large scale bespoke development. *Requirements Engineering*, 13(4), 2008, 257–279.
Allen, G. and Parsons, J. Is query reuse potentially harmful? Anchoring and adjustment in adapting existing database queries. *Information Systems Research*, 21, 2010, 56–77.
Alvarez, R. Confessions of an information worker: A critical analysis of information requirements discourse. *Information and Organization*, 12(2), 2002, 85–107.
Appan, R. and Browne, G.J. Investigating retrieval-induced forgetting during information requirements determination. *Journal of the Association for Information Systems*, 11(5), 2010, 250–275.
Appan, R. and Browne, G.J. The impact of analyst-induced 'Misinformation' on the requirements elicitation process. *MIS Quarterly*, 36, 2012, 85–106.

Aranda, G.N., Vizcaíno, A., and Piattini, M. A framework to improve communication during the require-ments elicitation process in GSD projects. *Requirements Engineering*, 15(4), 2010, 397–417.

Baskerville, R. and Pries-Heje, J. Short cycle time systems development. *Information Systems Journal*, 14(3), 2004, 237–264.

Bazerman, M.H. and Moore, D. *Judgment in Managerial Decision Making*, 8th edn. New York: Wiley & Sons, 2013.

Bostrom, R.P. Successful application of communication techniques to improve the systems development process. *Information & Management*, 16(5), 1989, 279–295.

Brainerd, C.J. and Reyna, V.F. Gist is the grist: Fuzzy-trace theory and the new intuitionism. *Developmental Review*, 10(1), 1990, 3–47.

Brown, V.R. and Paulus, P.B. Making group brainstorming more effective: Recommendations from an associative memory perspective. *Current Directions in Psychological Science*, 11, 2002, 208–212.

Browne, G.J. and Pitts, M.G. Stopping rule use during information search in design problems. *Organizational Behavior and Human Decision Processes*, 95, 2004, 208–224.

Browne, G.J., Pitts, M.G., and Wetherbe, J.C. Cognitive stopping rules for terminating information search in online tasks. *MIS Quarterly*, 31, 2007, 89–104.

Browne, G.J. and Ramesh, V. Improving information requirements determination: A cognitive perspec-tive. *Information & Management*, 39, 2002, 625–645.

Browne, G.J. and Rogich, M.B. An empirical investigation of user requirements elicitation: Comparing the effectiveness of prompting techniques. *Journal of Management Information Systems*, 17, 2001, 223–249.

Byrd, T.A., Cossick, K.L., and Zmud, R.W. A synthesis of research on requirements analysis and knowl-edge acquisition techniques. *MIS Quarterly*, 16(1), 1992, 117–138.

Chen, C.C., Law, C.C.H., and Yang, S.C. Managing ERP implementation failure: A project management perspective. *IEEE Transactions on Engineering Management*, 56(1), 2009, 157–170.

Cialdini, R.B. and Trost, M.R. Social influence: Social norms, conformity, and compliance. In: Gilbert, D.T., Fiske, S.T., and Lindzey, G. (eds.), *The Handbook of Social Psychology*. New York: McGraw-Hill, 1998, pp. 151–192.

Davern, M., Shaft, T., and Te'eni, D. Cognition matters: Enduring questions in cognitive IS research. *Journal of the Association for Information Systems*, 13, 2012, 273–314.

Davenport, T.H. Putting the enterprise into the enterprise system. *Harvard Business Review*, July–August, 1998, 121–131.

Davidson, E.J. Joint application design (JAD) in practice. *Journal of Systems and Software*, 45, 1999, 215–223.

Davidson, E.J. Technological frames and framing: A socio-cognitive investigation of requirements deter-mination. *MIS Quarterly*, 26(4), 2002, 329–358.

Davis, G.B. Strategies for information requirements determination, *IBM Systems Journal*, 21(1), 1982, 4–30.

Davis, C.J., Fuller, R.M., Tremblay, M.C., and Berndt, D.J. Communication challenges in requirements elicitation and the use of the repertory grid technique. *Journal of Computer Information Systems*, 46(5), 2006, 78–86.

Dennis, A.R., Hayes, G.S., and Daniels, R.M. Business process modeling with group support systems. *Journal of Management Information Systems*, 15(4), 1999, 115–142.

Dibbern, J., Winkler, J., and Heinzl, A. Explaining variations in client extra costs between software proj-ects offshored to India. *MIS Quarterly*, 32(2), 2008, 333–366.

Dimoka, A., Pavlou, P.A., and Davis, F. NeuroIS: The potential of cognitive neuroscience for information systems research. *Information Systems Research*, 22(4), 2011, 687–702.

Ericsson, K.A. and Simon, H.A. *Protocol Analysis: Verbal Reports as Data*. Cambridge, MA: MIT Press, 1993.

Evaristo, R., Watson-Manheim, M.B., and Audy, J. e-Collaboration in distributed requirements determination. *International Journal of e-Collaboration*, 1(2), 2005, 40–56.

Fabian, B., Gürses, S., Heisel, M., Santan, T., and Schmidt, H. A comparison of security requirements engineering methods. *Requirements Engineering*, 15(1), 2010, 7–40.

Gallivan, M.J. and Keil, M. The user-developer communication process: A critical case study. *Information Systems Journal*, 13(1), 2003, 37–68.

Goodwin, P. and Wright, G. *Decision Analysis for Management Judgment*, 3rd edn. Chichester, U.K.: Wiley & Sons, 2004.

Grünbacher, P., Halling, M., Biffl, S., Kitapci, H., and Boehm, B.W. Integrating collaborative processes and quality assurance techniques: Experiences from requirements negotiation. *Journal of Management Information Systems*, 20(4), 2004, 9–29.

Guinan, P.J., Cooprider, J.G., and Faraj, S. Enabling software development team performance during requirements definition: A behavioral versus technical approach. *Information Systems Research*, 9(2), 1998, 101–125.

Hickey, A. and Davis, A. A unified model of requirements elicitation. *Journal of Management Information Systems*, 20(4), 2004, 65–85.

Hofmann, H.F. and Lehner, F. Requirements engineering as a success factor in software projects. *IEEE Software*, 18(4), 2001, 58–66.

Hong, K.-K. and Kim, Y.-G. The critical success factors for ERP implementation: An organizational fit perspective. *Information & Management*, 40, 2002, 25–40.

Hunton, J.E. and Beeler, J.D. Effects of user participation in systems development: A longitudinal field experiment. *MIS Quarterly*, 21(4), 1997, 359–388.

Jain, H., Tanniru, M., and Fazlollahi, B. MCDM approach for generating and evaluating alternatives in requirement analysis. *Information Systems Research*, 2(3), 1991, 223–239.

Jayanth, R., Jacob, V.S., and Radhakrishnan, S. Vendor and client interaction for requirements assessment in software development: Implications for feedback process. *Information Systems Research*, 22(2), 2011, 289–305.

Kahneman, D. *Thinking, Fast and Slow*. New York: Farrar, Straus, and Giroux, 2011.

Keeney, R.L. and Raiffa, H. *Decisions with Multiple Objectives: Preferences and Value Tradeoffs*. Cambridge, U.K.: Cambridge University Press, 1993.

Keil, M. Pulling the plug: Software project management and the problem of project escalation. *MIS Quarterly*, 19(4), 1995, 421–447.

Kendall, K.E. and Kendall, J.E. *Systems Analysis and Design*. Upper Saddle River, NJ: Prentice Hall, 2010.

Kotonya, G. and Sommerville, I. *Requirements Engineering*. New York: Wiley & Sons, 1998.

Larsen, T.J. and Naumann, J.D. An experimental comparison of abstract and concrete representations in systems analysis. *Information & Management*, 22(1), 1992, 29–40.

Lee, G. and Xia, W. Towards agile: An integrated analysis of quantitative and qualitative field data on software development agility. *MIS Quarterly*, 34(1), 2010, 87–114.

Leifer, R., Lee, S., and Durgee, J. Deep structures: Real information requirements determination. *Information & Management*, 27(5), 1994, 275–285.

Liaskos, S., McIlraith, S.A., Sohrabi, S., and Mylopoulos, J. Representing and reasoning about preferences in requirements engineering. *Requirements Engineering*, 16(3), 2011, 227–249.

Light, M. How the waterfall methodology adapted and whistled past the graveyard. Gartner Research Report #G00173423, December 18, 2009.

Liou, Y.I. and Chen, M. Using group support systems and joint application development for requirements specification. *Journal of Management Information Systems*, 10(3), 1994, 25–41.

MacLeod, M.D. Retrieval-induced forgetting in eyewitness memory: Forgetting as a consequence of remembering. *Applied Cognitive Psychology*, 16, 2002, 135–149.

Mähring, M. and Keil, M. Information technology project escalation: A process model. *Decision Sciences*, 39(2), 2008, 239–272.

Maiden, N. and Rugg, G. ACRE: Selecting methods for requirements acquisition, *Software Engineering Journal*, 11(5), 1996, 183–192.

Marakas, G.M. and Elam, J.J. Semantic structuring in analyst acquisition and representation of facts in requirements analysis. *Information Systems Research*, 9(1), 1998, 37–63.

Markus, M.L. and Benjamin, R.I. The magic bullet theory of IT-enabled transformation. *Sloan Management Review*, 38(2), 1997, 55–68.

Maruping, L.M., Venkatesh, V., and Agarwal, R. A control theory perspective on agile methodology use and changing user requirements. *Information Systems Research*, 20(3), 2009, 377–399.

McKeen, J.D., Guimaraes, T., and Wetherbe, J.C. The relationship between user participation and user satisfaction: Investigation of four contingency factors. *MIS Quarterly*, 18(4), 1994, 27–45.

Milne, A. and Maiden, N. Power and politics in requirements engineering: Embracing the dark side? *Requirements Engineering*, 17(2), 2012, 83–98.

Montazemi, A.R. and Conrath, D.W. The use of cognitive mapping for information requirements analysis. *MIS Quarterly*, 10(1), 1986, 45–56.

Moody, J.W., Blanton, J.E., and Cheney, P.W. A theoretically grounded approach to assist memory recall during information requirements determination. *Journal of Management Information Systems*, 15(1), 1998, 79–98.

Norton, D. The current state of agile method adoption. Gartner Research Report #G00163591, December 12, 2008.

Parsons, H.M. Hawthorne: An early OBM experiment. *Journal of Organizational Behavior Management*, 12, 1992, 27–43.

Parsons, J. and Saunders, C. Cognitive heuristics in software engineering: Applying and extending anchoring and adjustment to artifact reuse. *IEEE Transactions on Software Engineering*, 30(12), 2004, 873–888.

Parthasarathy, S. and Ramachandran, M. Requirements engineering method and maturity model for ERP projects. *International Journal of Enterprise Information Systems*, 4(4), 2008, 1–14.

Pitts, M.G. and Browne, G.J. Stopping behavior of systems analysts during information requirements elicitation. *Journal of Management Information Systems*, 21, 2004, 213–236.

Pitts, M.G. and Browne, G.J. Improving requirements elicitation: An empirical investigation of procedural prompts. *Information Systems Journal*, 17, 2007, 89–110.

Ralph, P. Toward a theory of debiasing software development. In: Wrycza, S. (ed.), *Research in Systems Analysis and Design: Models and Methods. Proceedings of the 4th SIGSAND/PLAIS Eurosymposium 2011*, Gdansk, Poland: Springer, 2011, pp. 92–105.

Reisberg, D. *Cognition: Exploring the Science of the Mind*, 4th edn. New York: W.W. Norton & Co., 2009.

Robertson, S. and Robertson, J. *Mastering the Requirements Process*, 2nd edn. Upper Saddle River, NJ: Addison-Wesley, 2006.

Sandberg, J. Brainstorming works best if people scramble for ideas on their own. *The Wall Street Journal*, June 13, 2006.

Schneider, K., Knauss, E., Houmb, S., Islam, S., and Jürjens, J. Enhancing security requirements engineering by organizational learning. *Requirements Engineering*, 17(1), 2012, 35–56.

Sikora, E., Tenbergen, B., and Pohl, K. Industry needs and research directions in requirements engineering for embedded systems. *Requirements Engineering*, 17(1), 2012, 57–78.

Simon, H.A. Information processing models of cognition. *Annual Review of Psychology*, 30, 1979, 363–396.

Simon, H.A. *The Sciences of the Artificial*, 3rd edn. Cambridge, MA: MIT Press, 1996.

Stacy, W. and MacMillan, J. Cognitive bias in software engineering. *Communications of the ACM*, 38, 1995, 57–63.

Tan, M. Establishing mutual understanding in systems design: An empirical study. *Journal of Management Information Systems*, 10(4), 1994, 159–182.

Tversky, A. and Kahneman, D. Judgment under uncertainty: Heuristics and biases. *Science*, 185(4157), 1974, 1124–1131.

Urquhart, C. Analysts and clients in organisational contexts: A conversational perspective, *Journal of Strategic Information Systems*, 10(3), 2001, 243–262.

Valusek, J.R. and Fryback, D.G. Information requirements determination: Obstacles within, among, and between participants. In: *Proceedings of the Twenty-First Annual Conference on Computer Personnel Research*. New York: ACM Press, 1985, pp. 103–111.

Vitalari, N.P. Structuring the requirements analysis process for information systems: A propositional viewpoint. In: Cotterman, W.W. and Senn, J.A. (eds), *Challenges and Strategies for Research in Systems Development*. New York: Wiley & Sons, 1992, pp. 163–179.

Vivas, J.L., Agudo, I., and López, J. A methodology for security assurance-driven system development. *Requirements Engineering*, 16(1), 2011, pp. 55–73.

Watson, H.J. and Frolick, M.N. Determining information requirements for an EIS. *MIS Quarterly*, 17(3), 1993, 255–269.

West, R. The psychology of security. *Communications of the ACM*, 51(4), 2008, 34–40.

Wetherbe, J.C. Executive information requirements: Getting it right. *MIS Quarterly*, 15(1), 1991, 51–65.

von Winterfeldt, D. and Edwards, W. *Decision Analysis and Behavioral Research*. Cambridge, U.K.: Cambridge University Press, 1986.

28

From Waterfall to Agile: A Review of Approaches to Systems Analysis and Design

Joey F. George
Iowa State University

28.1 Introduction

While there will always be some debate about when and where the first information system for business was developed, it is generally agreed that the first such system in the United Kingdom was developed at J. Lyons & Sons. In the United States, the first administrative information system was General Electric's payroll system, developed in 1954. Now, 60 years later, the people who designed and built those systems would scarcely recognize the approaches and tools currently used to develop information systems. Systems development, which will be used in this chapter synonymously with systems analysis and design, has changed quite a bit in the past six decades.

It is the premise of this chapter that systems development has evolved through three distinct phases (Figure 28.1). Initially, development was a craft, where each system was unique and not replicable. Changes in technology and in attitude, and concerns from organizational management, led to the second phase, where development was seen as a feat of engineering. Planning, structure, and documentation governed the process. But, according to critics, the approach grew too big, too rigid, and too slow. New approaches began to emerge, including the rational unified process (RUP) and participatory design. Eventually, a new phase dawned, and the focus shifted from plan-based to small, fast, and agile. Today, the plan-based and agile approaches coexist side-by-side. This coexistence is not tension free, however, and this uneasy state of affairs will no doubt be followed by another phase, although what

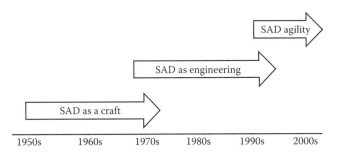

FIGURE 28.1   Three phases of systems development approaches.

that will be is not so clear. In this chapter, we will present an overview of the three phases of systems development approaches, and we will close with some conjecture on what the next step might be. Each phase will be presented with reference to specific methodologies and tools that emerged during that phase, but in each case, the list is only representative and not comprehensive. A complete list of all of the development methodologies that have been created and used in the past 60 years is far beyond the scope of this chapter.

28.2 Systems Development as Craft

In the early years, especially in the time before early third-generation languages such as COBOL, software development more closely resembled an art form than a production business process. Early systems development often occurred in a chaotic and haphazard manner, depending on the skills and experience of the people performing the work (Green and DiCaterino, 1988). In this era, developers were technically trained to work on particular hardware systems, which were incompatible with systems manufactured by other vendors. Many capitalized on peculiarities of a particular machine, using their proprietary machine and assembly languages, for writing and optimizing code. Developers had little understanding of the business, and user needs were poorly understood, if considered at all. There were no development methodologies. "The approach programmers took to development was typically individualistic, often resulting in poor control and management of projects" (Avison and Fitzgerald, 2003, p. 79). Schedules, budgets, and functionality were all unpredictable, and maintenance was a nightmare, as documentation was largely non-existent. When the system designer left the firm, few of those remaining had clear ideas about how to maintain the systems left behind. As technology improved, with the development of procedural programming languages, database management systems, and as management grew tired of the chaos and waste, this era started to come to an end. Calls for structured programming (Dijkstra, 1968) were accompanied by calls for a more structured and predictable approach to systems development.

28.3 Systems Development as Engineering

The structured approach emerged in the 1970s and continued to dominate systems development well into the 1990s. At the heart of the structured approach was a focus on designing and building systems the same way other artifacts, such as bridges and buildings, are designed and built. Like these artifacts, systems would start as concepts, which would then be modeled using several different tools, and these logical models would be converted to physical plans that would guide the construction of the system. The overall process would become an engineering process, and systems developers would become software engineers. Development would become systematic, predictable, repetitive and repeatable, measurable, heavily documented, and reliable. As a result, the productivity of programmers and developers would increase, as would the quality of the resulting systems.

Although many have been associated with it, maybe the best known advocate for the structured approach was Edward Yourdon. Yourdon (1989, p. 3) listed the major components of the structured approach as:

Structured analysis: A collection of graphical modeling tools that allows the systems analyst to replace the classical functional specification (text based) with a model that the users can actually understand….

Top-down design and implementation: The strategy of designing a system by breaking it into major functions, and then breaking those into smaller subfunctions, and so on, until the eventual implementation can be expressed in terms of program statements.

Structured design: A set of guidelines and techniques to help the designer distinguish between "good" and "bad" designs. Proper use of structured design will lead to a system composed of small, highly independent modules, each of which is responsible for carrying out one small single-purpose function.

Structured programming: An approach to programming that constructs all program logic from combinations of three basic forms (sequence, loops, and conditional statements).

The graphical models that structured analysis relied on included data flow diagrams and entity relationship diagrams. Structured design relied on tools such as structure charts and on concepts such as modularization, coupling, and cohesion. Structured programming made use of flow charts and decision tables. Despite the intuitive appeal of bringing order to chaos that structured analysis and design promised, adoption by organizational software development groups was generally slow (Fichman and Kemerer, 1993). For many, it was difficult to accept discipline where it had been missing before. The learning curve for mastering the many diagrams and other tools that supported structured analysis and design was substantial.

28.3.1 Systems Development Life Cycle

At the core of structured analysis and design was the systems development life cycle (SDLC). The SDLC embodies the key ideas of any product life cycle. At some point, a need is recognized; the need is investigated and a solution is designed, built, and released. Eventually, the solution reaches its maximum usefulness and begins to decline. Eventually, it must be replaced and a new solution must be created. One cycle has ended and a new one begins.

Although no two people in the IS field, either academics or practitioners, seem to agree about the exact form and contents of the SDLC, it does constitute a useful model of the systems development process. Figure 28.2 shows a generic SDLC model, typically referred to as a waterfall model.

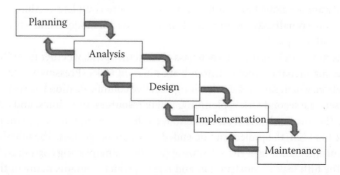

FIGURE 28.2 A waterfall systems development life cycle.

Although different versions of the waterfall SDLC have different numbers of phases, the generic model here has five phases that capture the basic structure of the systems development process. The first phase, planning, represents all of the activities that result in the identification of a need for a system. Once the need is recognized, it is important to gain a clear understanding of the problem or opportunity that sparked the need. The existing system should be analyzed, and any additional requirements should be determined in this phase, analysis. Useful tools for analysis include joint application design (JAD) and prototyping (explained in more detail later in this chapter). Once the existing situation is understood, it can be compared to where things should or ought to be. At that point, a solution can be designed, and this is what occurs in the design phase. Design involves both high-level logical design and physical design tied to a particular computing platform and operating environment. Once designed, the solution can be constructed, tested, and installed or released, the primary activities of implementation. Once the solution has been installed in an organization, the maintenance phase begins. In maintenance, the work centers on keeping the solution viable, including making any changes required due to the changing business environment and changing regulations and legal conditions. In maintenance, programmers also work to correct errors in the system and to optimize system performance. At some point, the solution is no longer viable and a new solution must be sought, ending one cycle and beginning another.

The end of each phase is a major development project milestone. As a phase ends, a decision must be made to stop or to continue. If the decision is made to continue, the next phase can begin. Notice that there are two sets of arrows connecting the phases in Figure 28.2. One set goes downhill, connecting the phases in the order in which they were described previously. Yet, the systems development process is rarely this cleanly structured. Early waterfall models contained only the downhill set of arrows, and this was the source of much early criticism of the model. In practice, it is much more typical for developers to repeat a phase, or to return to an earlier phase, as system requirements change and as work completed earlier needs to be reexamined or even redone. Iterations within the life cycle are very common. The second set of arrows, the ones going backward and uphill, has been included in Figure 28.2 to represent the possibility and likelihood of iterations within the life cycle. The flow of work is not linear, as would be the case in a true waterfall, but instead it is cyclical. The uphill arrows allow the work flow to move from one phase to an earlier phase, and in conjunction with the downhill arrows, allow for numerous iterations between phases.

The name waterfall refers to the shape of the model, in that the work flow cascades from one phase down to the next until the end is reached. As indicated previously, there are many variations of the standard waterfall SDLC. Techniques such as rapid application development (RAD) (Martin, 1991) used modified versions of the waterfall SDLC. RAD will be discussed in a later section.

Another way to visualize the SDLC is the spiral model (see Boehm, 1988), shown in Figure 28.3. The spiral model is also referred to as an evolutionary model. The spiral model combines the iterative nature of development with the systematic, stepwise approach of the waterfall model. Software is seen as being developed in a series of incremental releases. Early releases might be prototypes, but increasingly, the versions of the software developed are more complete. As more complete versions of the software are released, what was conceptualized as maintenance in the waterfall model is conceptualized here as subsequent passes around the spiral.

According to the model in Figure 28.3, each pass around the spiral, whether for initial iterative development or for later maintenance, moves through six different tasks (Pressman, 1997). The first is customer communication, which involves establishing effective communication between the developer and the customer or user. The second task is planning, where resources, time-lines, and other project information is defined. The third task is risk assessment. Here, technical and management risks are assessed, and if the risks are too high, the project can be ended at the go-no-go axis. The fourth task is engineering, which represents the building of the software application. Engineering corresponds to design in the waterfall model. The fifth task is construction and release, which contains many of the same activities as were included in the implementation phase earlier: construction, testing, installation, and providing

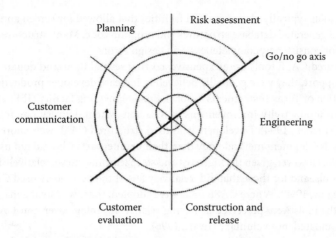

FIGURE 28.3 A spiral systems development life cycle.

user support. The sixth task is customer evaluation. The evaluation process elicits feedback on the engineering and construction and release tasks.

While the waterfall model is a simple and easy to understand representation of the systems development process, the spiral model presents a more complex and hence more realistic view. The spiral model adds iteration, risk assessment, and a more sophisticated view of maintenance to the basic waterfall model. Many of the basic tasks central to system development are the same in both models, however, as would be expected. The spiral model has also had great success in practice. Boehm (1988) cites adherence to the model as the cause of productivity improvements as high as 50% in development projects at TRW (an aerospace and automotive company, purchased by Northrop Grumman in 2002), where Boehm's particular version of the model was developed and used. The spiral model is not suited to all software development efforts, however. It best fits large-scale development projects. Another limitation in practice is its demands for considerable risk assessment expertise in order to succeed, an issue of particular concern to both Boehm and Pressman.

28.3.2 Structured Tools and Techniques

As the structured approach to systems development became more widely adopted, various tools and techniques were developed to enrich the approach and to make it easier to follow. People began to automate the graphical tools that supported the approach, and eventually the automated tools were combined with a rigorous methodology in computer-assisted software engineering, or CASE, tools. Others borrowed techniques from engineering, such as prototyping, where a smaller working model of an artifact was built as a way to help system users articulate the system's requirements. Other techniques created to assist the requirements determination process included JAD, a method that brought together the key stakeholders (or their representatives) in one place at one time, to help clarify what was desired in the new or revamped system. Each of these tools and techniques, all of which strongly supported the structured approach to systems development, will be discussed in more detail in the paragraphs that follow.

CASE tools were developed as a way to use the power of information technology to support the software development process. CASE tools are bundles of software tools that make analysis, design, documentation, and coding consistent, complete, and easier to accomplish. CASE packages built for the structured approach included tools for creating and maintaining data flow diagrams, entity-relationship diagrams, and structure charts, among others. The diagramming tools were linked in that each object created in one of the tools was entered into the central repository around which the CASE product was built. The repository stored data about all of the different elements that make up the system analysis and design. CASE tools typically also contained facilities that allowed for screen and report design, as well as facilities that generated database structures and program code. Many structured CASE tools also allowed checking for consistency, completeness, and design errors.

Although structured CASE tools were expensive, costing several thousand dollars per copy, including training and support, they were generally believed to increase developer productivity. A case study of the implementation of Texas Instruments' information engineering facility (IEF) at a British firm in 1989–1990 reported productivity improvements of 85% and system delivery rate increases of around 200% (Finlay and Mitchell, 1994). Developers who used structured CASE were more likely to follow a structured systems development methodology than their counterparts who did not use CASE (Lending and Chervany, 1998). However, given the high cost and steep learning curve, relatively few firms adopted structured CASE tools, and for those that did, relatively few of the developers used CASE (Iivari, 1996; Lending and Chervany, 1998). Where CASE tools were adopted, there was strong management support for the approach, the tools were perceived as having relative advantage over non-CASE methods, and use tended to be mandated, not voluntary (Iivari, 1996).

Prototyping was a process borrowed from engineering, in which a scale model of an object or system was developed, both as a test-of-concept and as a device to facilitate communication between designer and client (Gavurin, 1991; Naumann and Jenkins, 1982). The software tools used for prototyping could be as simple as paint programs or as complex as CASE tools that allowed users to enter data and navigate between screens. The main idea behind prototyping as a technique was the provision of a means by which developers could transform user requirements into objects that users could see, touch, and use. Users could then provide feedback to developers about the prototype, about what worked and what did not, and about what else they would have liked to see. The developer would then take that feedback and use it to modify the prototype, beginning another round of iteration between user and developer. At some point, it was decided to either keep the prototype as the basis for the production system (evolutionary prototyping) or to throw it away (sometimes called mock-up design prototyping), while using the knowledge gained in the prototyping process to construct the production system. Prototyping became very popular in system development efforts, with over 70% of firms surveyed reporting the adoption of prototyping by 1995 (Hardgrave, 1995). Studies of prototyping revealed that, compared to an SDLC approach that did not use it, prototyping was associated with higher levels of both user and developer satisfaction, higher levels of user commitment, improved communication with users, earlier detection of errors, better code, and improvements in the development process in terms of shorter development time and less effort for developers (Beyton-Davies et al., 1999; Cerpa and Verner, 1996; Mahmood, 1987). By its very nature, prototyping encouraged meaningful and frequent interaction between user and developer. Prototyping was not without problems, however. Prototyping was associated with difficulties in managing user expectations, enticing users to commit to fully participate in the prototyping effort, and with project management generally (Beyton-Davies et al., 1999).

JAD was a technique developed primarily to address the issues involved in determining user requirements for a system (Wood and Silver, 1995). Traditionally, a developer determined requirements by interviewing users, studying existing systems, perusing input form, reports, and other documents, and observing users while working. User interviews were an important source of information for what the system being developed should do and look like. Interviews, however, were difficult to schedule, as users had their own regular work to complete while assisting the developer. Contradictions between users also had to be reconciled, which required follow-up interviews. The process could be very time consuming and frustrating.

JAD was developed to deal with the difficulties in scheduling interviews and reconciling the information collected from them. In JAD, key users met with management representatives and systems developers, sometimes for a day, sometimes for an entire week, to determine requirements for a system. The process was very structured, typically led by a trained JAD facilitator whose primary role it was to keep the process moving. JAD essentially allowed for many interviews to be conducted at once and for contradictions to be reconciled on the spot (although not always possible). A JAD session was typically held off-site to minimize the distractions participants would experience if working at the office. There has been little empirical research into the effectiveness of JAD. One study reported that JAD was superior to traditional methods, with respect to the quality of user–designer interactions, effectiveness of consensus management, and user acceptance of design specifications (Purvis and Sambamurthy, 1997). Other studies combined JAD with other approaches for managing group work, from electronic group support to the nominal group technique. JADs run with electronic group support had more equal participation among group members but lacked the session discipline and conflict resolution common to traditional JAD (Carmel et al., 1995). JAD combined with the nominal group technique was found to be more effective than traditional JAD, although just as efficient, with higher levels of user satisfaction and better system requirements determination (Duggan and Thachenkary, 2004).

28.3.3 Other Structured Approaches

Information engineering (IE) is a structured approach to systems analysis and design originally developed by Clive Finkelstein (http://www.ies.aust.com/cbfindex.htm). He later worked with James Martin, who developed his own version of the methodology in the mid-1980s (Martin, 1986). As with many other structured approaches to systems analysis and design, IE depended on sets of related and interconnected diagramming models. These models included entity–relationship diagrams, data flow diagrams, decision trees (called action diagrams), and report and screen layout tools. IE also stressed the importance to the development process of automated tools, which allowed high levels of user participation, as well as high levels of horizontal integration (between systems) and of vertical integration (linking systems to top management goals and corporate strategy). IE had four stages: (1) information strategy planning; (2) business area analysis; (3) system design; and (4) construction. The planning phase focused on the entire enterprise, while a different business area analysis would be done for each different business area. These two processes together took from 9 to 18 months to complete. The latter stage focused on identifying the processes and data needed for each business area. Design was conducted via automated development tools, and construction was aided by code generators and other tools. At all stages of the IE process, knowledge was stored in an encyclopedia, or central repository.

Martin later worked with Texas Instruments on instantiating the methodology into a CASE tool called IEF, for information engineering facility. IEF was released in 1987. In 1997, Texas Instruments sold IEF to Sterling Software, which was later acquired by Computer Associates.

RAD was a development methodology also credited to James Martin (1991). The idea behind RAD was captured in its name: dramatically shorten the time necessary for systems to be developed. Complex business systems typically took years to develop using structured analysis and design processes, but with the dynamic pace of business in an expanding global economy, firms could not wait for systems that, once completed, might no longer be adequate models of their business processes they modeled. According to Martin, a system that would take 2 years to complete following the traditional approach might instead be completed in 6 months following the RAD approach.

RAD was purported to save time and other related resources, like money, because of how it was designed to work. First RAD made heavy use of JAD and of CASE tools to support prototyping. The first JAD meetings in a RAD effort might involve prototyping very early in the process, before a more traditional JAD would. The prototype developed also tended to become the basis for the production system, rather than being thrown away, as discussed previously. Second, the prototyping process in a RAD required more intensive participation from users in a true partnership with developers. Users might also have become involved in the design process itself instead of ending their participation after requirements determination was complete, which was more typically the case using traditional approaches. Martin (1991) also argued that developers had to be trained in the right skills and methodologies, and management had to lend its complete support, for RAD to be successful.

As good as it might sound, RAD was not without its problems. Sometimes, due to the speed with which systems were developed, some of the basics of traditional systems development were overlooked. These included interface consistency across the system, programming standards such as documentation and data naming standards, upward scalability to serve larger numbers of more diverse users, and planning for system administration chores such as database maintenance and organization, backup and recovery, and so on (Bourne, 1994).

28.3.4 Alternative Development Approaches and Practices

Even though it was not universally adopted, the structured approach to systems development dominated development work for many years. Before the transition to agile methods that occurred in the late 1990s and the early part of the twenty-first century, there existed several alternatives to the structured approach. One, the RUP, was created to guide object-oriented systems development efforts. Another, participatory design, focused more on people and creative hands-on techniques for development rather than focusing on the technology and using it to support development. A third approach, reuse, is really not so much a separate methodology as it is a practice created to make any development process more efficient through reusing existing software code and components. Each of these approaches and practices are discussed in more detail in the sections that follow.

RUP was a development methodology that emerged from the object-oriented perspective on analysis and design. It had its origins in the work of Jacobson et al. (1999), although the process now is owned by IBM (http://www-01.ibm.com/software/awdtools/rup/). The approach was built around iteration, with a focus on being risk-driven (Aked, 2003; Ambler, 2005). The risk-driven approach refers to the practice of tackling head on concerns that could adversely affect the development project, rather than dealing with easy issues first and dealing with risk later (Aked, 2003). The methodology crossed four phases (inception, elaboration, construction, and transition) with nine disciplines, which described the different types of work to be done. These nine disciplines are: (1) business modeling; (2) requirements; (3) analysis and design; (4) implementation; (5) test; (6) deployment; (7) configuration and change management; (8) project management; and (9) environment. The phases and disciplines are often shown in what is called a "hump diagram," where the relative effort devoted to each discipline within each phase is shown (Figure 28.4 shows only the first six disciplines). Each phase was also divided into a number of iterations. At the end of each phase, there was a go/no go decision, much as was the case in an SDLC from the structured approach.

The inception phase included defining scope, determining the feasibility of the project, understanding user requirements, and preparing a software development plan. In the elaboration phase, detailed

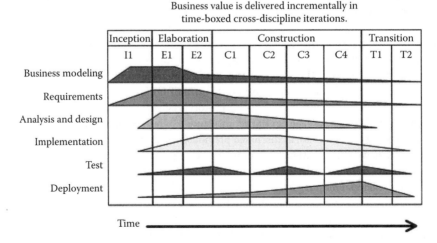

FIGURE 28.4 The rational unified process depicted in a hump diagram. (This figure is in the public domain and is used by permission of its author Dutchguilder. http://en.wikipedia.org/wiki/File:Development-iterative.gif)

user requirements and a baseline systems architecture were developed. Analysis and design activities constituted the bulk of the elaboration phase. In the construction phase, the software was coded, tested, and documented. In the transition phase, the system was deployed and the users were trained and supported. The construction phase was generally the longest and the most resource intensive. The elaboration phase was also long but less resource intensive, relatively speaking. The transition phase was resource intensive but short. The inception phase was short and the least resource intensive.

The nine disciplines were one way to package the skills and processes needed for successful systems development. The first six, from business modeling to deployment, corresponded to a typical development process, from planning through analysis, design, testing, and implementation. Configuration and change management corresponded to the ongoing maintenance process, which follows implementation. Both project management and environment ran more or less throughout the development process. While the former is self-explanatory, the latter refers to making sure the appropriate guidance and tools are available to the development (and maintenance) team throughout the process (Ambler, 2005). As RUP came out of the object-oriented perspective, for implementation it relied heavily on the unified modeling language (UML) (Booch et al., 2005). RUP and UML were captured in the Rational Rose CASE tool, which is now sold by IBM as the Rational series of tools. Stories of RUP use in organizations and the resulting successes and problems were typically told in case study form (see Hanssen et al., 2005, and Lee, 2006, for examples).

Participatory design was a very different approach to systems development than anything presented so far. Developed in Northern Europe, participatory design focused primarily on the system's users and how the system would affect their work lives. In some cases, the entire user community might play an active role in the systems development process, while in others, users might select representatives to represent them. Typically, system developers worked for the users. Management and outside consultants provided advice but did not control the process. Participatory design efforts included a repertoire of flexible practices and general guidelines. Running through the repertoire were the twin themes of mutual reciprocal learning, in which users and developers teach each other about work and technology, respectively, and design by doing, which resembles prototyping but does not necessarily require sophisticated computing to implement (Carmel et al., 1993). While well known in Northern Europe, especially in the Nordic countries, participatory design was rarely followed elsewhere.

Reuse was the use of previously written software resources in new applications. As so many bits and pieces of applications are relatively generic across applications, it seems intuitive that great savings could be achieved in many areas if those generic bits and pieces did not have to be written anew each time they were needed. Reuse should increase programmer productivity, as being able to use existing software for some functions meant they could perform more work in the same amount of time. Reuse should also decrease development time, minimizing schedule overruns. Because existing pieces of software have already been tested, reusing them should also result in higher quality software with lower defect rates, which is easier to maintain.

Although reuse could conceivably apply to many different aspects of systems, typically it was most commonly applied to object-oriented code and components. In object-oriented code, both data and function are combined in one item. For example, an employee object would contain both the data about employees and the instructions necessary for calculating payroll for a variety of job types. The object could be used in any application that dealt with employees, but if changes had to be made in calculating payroll for different type employees, the changes would only have to be made to the object and not to the various applications that used it. By definition, using the employee object in more than one application constitutes reuse. As with CASE tools and structured development, developers were slow to adopt object-orientation (Fichman and Kemerer, 1993), although that has changed dramatically in the past few years. Today, four of the five most used programming languages are object-oriented (Java, C#, C++, and Objective-C, with the other being C) (TIOBE, 2012). Components are general-purpose pieces of software that can be used interchangeably in many different programs. Components can be as small as objects or as large as pieces of software that handle single business functions, such as currency conversion. The idea behind component-based development is the assembly of an application from many different components at many different levels of complexity and size. Reuse of objects and components has proven to be effective, resulting in increased productivity, reduced defect density, and reduced rework (Basili et al., 1996). Reuse has it problems, however. In a study of 24 European companies that undertook reuse projects, one-third of the projects failed (Maurizio et al., 2002). The reasons for failure were not technical—they were managerial. How changes to work processes that govern reuse are implemented and managed are key to its success (Sherif et al., 2006).

28.4 Agile Development

Although the engineering approach remains in use today, to many critics, it has become bloated and slow, no longer as useful in a global economy that runs on Internet time. The convergence of the object-oriented approach and the Internet economy set the stage for the current major phase in systems development. Relying on object-orientation and the need for speed, the new approach sacrifices the milestones and multiple phases of the engineering approach in favor of close cooperation between developers and clients, combining many life cycle phases into few phases, and multiple rapid releases of software. Although there are many individual methods that reflect the new approach, collectively they are called the agile methodologies.

Many different individual methodologies come under the umbrella of agile methodologies. In February 2001, many of the proponents of these alternative approaches to systems analysis and design met in Utah, USA, to reach a consensus on many of the underlying principles their various approaches contained. This meeting let to a document they called "The Agile Manifesto" (Beck et al., 2001). The agile methodologies share 12 principles (http://www.agilemanifesto.org/principles.html), which can be summarized in three key points: (1) a focus on adaptive rather than predictive methodologies, (2) a focus on people rather than roles, and (3) a focus on self-adaptive processes (Valacich et al., 2012). Agile adherents argue that software development methodologies adapted from engineering generally are not a good fit for the reality of developing software. In the engineering disciplines, such as civil engineering, requirements tend to be well understood. Once the creative and difficult work of design is completed, construction becomes very predictable (Fowler, 2005). In addition, construction may account for as much as 90% of the total project effort. For software, on the other hand, requirements are rarely understood well, and they change continually during the lifetime of the project. Construction may account for as little as 15% of the total project effort, leaving design to constitute as much as 50% of the project effort (McConnell, 1996), depending on the project type. Applying techniques that work well for predictable, stable projects, like building a bridge, tend not to work well for fluid, design-heavy projects like writing software, say the agile methodology proponents. One mechanism for dealing with a lack of predictability, which all agile methodologies share, is iterative development. Iterative development focuses on the frequent production of working versions of a system that have a subset of the total number of required features. Iterative development provides feedback to customers and developers alike.

Second, the focus on people in agile methodologies is a focus on individuals rather than on the roles that people perform. The roles that people fill, of systems analyst or tester or manager, are not as important as the individuals who fill those roles. Fowler (2005) argues that the focus on engineering principles applied to systems development has resulted in a view of people as interchangeable units instead of a view of people as talented individuals, each of which has something unique to bring to the development team. Third, the agile methodologies also promote a self-adaptive software development process. As software is developed, the process used to develop it should be refined and improved. Development teams can do this through a review process, often associated with the completion of iterations.

28.4.1 eXtreme Programming

One of the best known and most written about agile methodologies is called eXtreme programming. Developed by Kent Beck in the late 1990s (Beck, 2000), eXtreme programming illustrates many of the central philosophies of this new approach to systems development. eXtreme programming is distinguished by its short development cycles, its incremental planning approach, its focus on automated tests written by programmers and customers to monitor the process of development, and its reliance on an evolutionary approach to development that lasts throughout the lifetime of the system. One of the key emphases of eXtreme programming is having a customer on-site during the development process. Another is its use of two-person programming teams.

Under this approach, users generate stories that serve as the basis for system functions and features. These user requirements are not consolidated into a single, detailed and complete design specification, as would be expected in an engineering-based development effort. Instead, requirements serve as the basis for a design, which is captured in code and then tested. Coding and testing are intimately related parts of the same process. The programmers who write the code also write the tests. The emphasis is on testing those things that can break or go wrong, not on testing everything. The overall philosophy behind eXtreme programming is that code will be integrated into the system it is being developed for and tested within a few hours after it has been written. If all the tests run successfully, then development proceeds. If not, the code is reworked until the tests are successful. Pair programming makes the code-and-test process work smoothly. All coding and testing is done by two people working together, writing code and writing tests. The programmers work together on the problem they are trying to solve, exchanging information and insight and sharing skills.

To continually improve the quality of the design, eXtreme programming relies on a practice called simple design. Simple design is what it sounds like: keeping the design simple. According to Beck (2000), it is common for developers and programmers to look ahead and to try to anticipate changes in how the system will work and to design accordingly. Many times, those anticipated future conditions never materialize, and the time and effort that went into designing for the uncertain future is wasted. Under the eXtreme programming approach, design should focus on solving the immediate problem, not on solving future problems that may or may not occur.

28.4.2 Scrum

Scrum originated in 1995 and was developed by Jeff Sutherland and Ken Schwaber (Schwaber and Sutherland, 2011). Scrum represents a framework that includes Scrum teams and their associated roles, events, artifacts, and rules. Each team consists of three roles: the product owner, the development team, and the Scrum master. The owner is essentially accountable for the product and the work that produces it. The development team is small, within the preferred range of 3–9. The Scrum master is there to teach and enforce the rules (which vary by team and by product). Scrum is designed for speed and for multiple functional product releases. The primary unit is the sprint, a 1 month (or less) unit of effort. Each sprint is a complete project in and of itself. It starts with an 8 hour sprint planning meeting, which focuses on two questions: What will need to be delivered by the end of the sprint and how will the work needed to do that be accomplished? The sprint goal provides guidance for the team for the duration of the sprint. During the sprint, there is a daily Scrum, a 15 min meeting held to essentially evaluate what progress has been made within the past 24 hours and what still needs to be done. At the end of the sprint, there are two other meetings, the sprint review (4 hours) and the sprint retrospective (3 hours). While the review focuses on the product, what has been accomplished, and what needs to be done in the next sprint, the retrospective is more broad. It focuses as well on the performance of the team and how it can improve in the next sprint. There are three primary artifacts in the Scrum process. The first is the product backlog. This is an ordered list of everything that might be included in the product, that is, a list of potential requirements. The list includes "all features, functions, requirements, enhancements and fixes (Schwaber and Sutherland, 2011, p. 12)" that make up all the changes to be made to the product. The sprint backlog is a subset of the product backlog, consisting of only those items to be addressed in a particular sprint. Finally, the increment is the sum of all the product backlog items completed during a sprint. Each increment must be in complete enough form to be useable, whether or not the product owner decides to release it. It is called an increment because it represents an increment of total functionality for the product. Each increment is thoroughly tested, not only as a standalone but also in conjunction with all prior increments.

28.4.3 Agile Performance

Several studies have investigated agile methods in practice. One study, a survey of over 100 agile projects, found three primary critical success factors for agile development (Chow and Cao, 2008). The first is delivery strategy, which refers to the continuous delivery of working software in short time scales. The second is following agile software engineering practices. That means managers and programmers must continually focus on technical excellence and simple design. The third critical success factor is team capability, which refers to the agile principle of building projects around motivated individuals.

Another study found that, once implemented, agile methods can lead to improved job satisfaction and productivity on the part of programmers (Dyba and Dingsoyr, 2008). They can also lead to increased customer satisfaction, even though the role of on-site customer representative can be tiring and so not sustainable for very long. Agile methods tend to work best in small projects. In some instances, it may make sense to combine them with traditional development methods.

A detailed study of one agile development effort showed that some of the key principles of agile development had to be modified to help ensure success (Fruhling and DeVreede, 2006). For example, in the agile project studied, pair programming was not always used, especially when resources were needed elsewhere. Second, the process of writing the test case first and then the code was followed until the system became too complex. Third, the customer was not located in the same place as the programmers. Instead, the customer stayed in contact through regular meetings and continual e-mail and phone communication. Even with these modifications, the resulting system was considered a success—fewer than 10 updates were issued due to errors and none were issued due to implementing the wrong functionalities. Working together, the users and developers were able to clarify system requirements and create a user interface that was easy to learn and use.

28.5 Reconciliation and the Next Phase

Although the agile approach to systems development has many adherents who believe it is the only viable approach, others have found that agile works best for certain types of development projects. Traditional, or plan-based, approaches work best for a different set of development projects. For example, Boehm and Turner (2004) argue that agile methods are better for projects that involve small teams and small products; that agile methods are untested on safety-critical products; that the approach works best in a dynamic as opposed to stable environment; that agile projects require a critical mass of scarce agility experts; and that the approach works best where the participants do well in a work environment that is openly defined and even chaotic. Traditional approaches, then, work best for large, complex, and critical products where the product environment is stable and relatively few experts are needed during the course of development and where people are most comfortable in a work environment where their roles are well-defined and practices and procedures are clear. Additionally, agile methods are thought to work best where participants are collocated, while plan-based methods can function just as well where participants are physically distributed. Meanwhile, there is some evidence that agile methods may be used successfully in environments and for products that do not fit the constraints outlined earlier (Abrahamsson et al., 2009). Similarly, some of the practices thought central to agility, such as having an on-site customer and pair programming, may not contribute to agility as much as previously thought (Conboy, 2009).

In fact, there seems to be a movement toward using mixed approaches for systems development, where some aspects of plan-based and agile approaches are used together. Port and Bui (2009) showed through simulation that hybrid approaches yielded better results than either traditional or agile approaches alone. Results were measured by the value and the cost of the requirements implemented during the simulation from both pure and hybrid approaches. In 2008, Baskerville et al. studied three software development companies where parts of the development process were supported through Scrum, while other parts were carried out using traditional plan-based methods, including a heavy emphasis on project management techniques (Baskerville et al., 2009). The people they interviewed stressed the need for alignment between the two approaches, where alignment was "described as necessary to ensure that the work carried out by the Scrum teams [wa]s in line with the overall scope document, budget, and project plan (Baskerville et al., 2009, p. 551)." The Scrum process and teams existed within a larger wrapper consisting of the traditional project management structure. Baskerville et al. (2009) speculated that the next stage in the overall march of systems development processes, which they called "post-agility," will blend the goals of agility and the plan-based approach through new ways of organizing, managing and communicating, and through new methods and software tools. Agility would no longer be a separate approach but would be deeply incorporated into all systems development methodologies. As they pointed out, the post-agility era would also be the post-plan-driven era.

28.6 Future Research Opportunities

The study of information systems development—of the methodologies, methods, and tools that have been created over the years to facilitate and support the process—was once the primary topic of research in the information systems field (Sidorova et al., 2008). Today, the topic is less studied, but it remains an important part of the field. There are still several interesting questions related to systems development that need to be investigated. This chapter suggests at least three such areas: (1) comparisons of different approaches to systems development, (2) investigations into the practices and processes currently used in industry, and (3) speculative studies of what will constitute the next stage in the evolution of system development approaches.

In the Management Information Systems (MIS) field, there has been a long tradition of studies that compare two or more approaches to systems development (see, for example, Howard et al., 1999). These studies have typically been experimental in nature, where different groups or individuals have used competing methods or tools to address a specific systems development problem. While many of these studies have compared different aspects of the structured plan-driven approach, relatively few have compared the structured approach to the object-oriented approach. Fewer still have compared plan-driven and agile approaches. This is the case even though agile methods have existed for over a decade, as of this writing. There is a clear need for systematic experimental studies that compare different approaches to systems development to each other. For both research and practical reasons, we should strive to understand the relative strengths and weaknesses of the various approaches.

A second area of research that has great potential is the study of how organizations actually approach systems development. While agile methods have received much attention in the trade press, to what extent have organizations actually adopted them? We know from past research that industry is slow to adopt new approaches to development (Fichman and Kemerer, 1993). Is that also the case for agile methods today? To what extent do organizations adopt agile methods for some projects but plan-driven approaches for others? And on what basis do they choose which approach to use? With the widespread practice of organizations outsourcing many aspects of their systems development, how do they reconcile the need for planning apparent in outsourcing with the need for agility? Also, to what extent do companies use RUP and/or follow reuse policies? These are just some of the many questions that can be answered only through field work in organizations that develop systems, either for sale or for their own use. Industry-based research will help identify best practices that can benefit many different organizations as they seek to improve their systems development efforts.

Finally, as Baskerville et al. (2009) suggested, we are on the verge of a new, fourth approach to systems development. What will this post-agility, post-plan-driven era look like? What should it look like? How can its advent be hastened? We are in a unique position vis-a-vis those who studied the transitions between the previous approaches to systems development—few of them anticipated the previous transitions. Study of the transitions, and the new approaches they produced, came only after they had already occurred. Meanwhile, we know the next transition is coming, and we have some idea of what the next approach will look like. Not only can we anticipate the change and study it as it occurs, but we can also influence the form it eventually takes and its underlying philosophy.

28.7 Conclusions

Administrative information systems have been a part of organizational life for almost 60 years. During that time, information systems technology, organizations, the business environment, and the global economy have all changed dramatically. Information systems development methods have evolved to better deal with and take advantage of changes in technology as well as in the various external environments that affect it. Initially, when the technology was relatively primitive, systems development followed a craft approach. This was suitable for an era without standards, where computers varied considerably by vendor and by model, and where there were no standard programming languages, much less standard development approaches. Although the systems developed during this era worked, and some were truly innovative and worked amazingly well, the lack of predictability and documentation and the costs involved became too much to bear. Toward the end of this phase, standardized development methodologies were developed, just as standardized procedural programming languages had been developed. With standardization came control: over budgets, over schedules, over functionality, and over information systems personnel. Planning became paramount, and artists became engineers. While there was some resistance to this transformation, the plan-based approach soon became the norm. But in a dynamic human enterprise such as systems development, things did not stay the same for long. The plan-based approach had its own problems. To its critics, it was too slow to keep up with a dynamic business environment, it was too bloated, and it was too big. Such critiques were becoming common even before the Internet and the concept of "Internet time" exploded on the scene. Eventually, a new approach, agility, emerged. It was slow to spread, but eventually agility became just as popular as the plan-based approach. Today, these two approaches coexist, and there is some evidence to suggest that aspects of the two perspectives are slowly merging into a new approach. We can call this the "post-agility" phase, although a new and more descriptive name will be invented as the approach itself becomes better articulated. We know from the history of systems development approaches that the next phase will take some time yet to fully emerge. It may be 5 years from now, or it may be 10. If there is a revision of this chapter 10 years from now, the new phase will almost certainly be included. While there is still much uncertainty about the next phase, what it will encompass, and what it will be called, what is certain is that there will eventually be a fourth phase and that it will not be the last.

References

Abrahamsson, P., Conboy, K., and Wang, X. (2009). Lots done, more to do: The current state of agile systems development research. *European Journal of Information Systems*, 18, 281–284.

Aked, M. (2003). Risk reduction with the RUP phase plan. Available at http://www.ibm.com/developer-works/rational/library/1826.html#N100E4, Accessed January 13, 2012.

Ambler, S.W. (2005). A manager's introduction to the rational unified process (RUP). Available at http://www.ambysoft.com/unifiedprocess/rupIntroduction.html, Accessed January 13, 2012.

Avison, D.E. and Fitzgerald, G. (2003). Where now for development methodologies? *Communications of the ACM*, 46(1), 79–82.

Basili, V.R., Briand, L.C., and Melo, W.L. (1996). How reuse influences productivity in object-oriented systems. *Communications of the ACM* 39(10), 104–116.

Baskerville, R., Pries-Heje, J., and Madsen, S. (2009). Post-agility: What follows a decade of agility? *Information and Software Technology*, 53, 543–555.

Beck, K. (2000). *eXtreme Programming eXplained*. Upper Saddle River, NJ: Addison-Wesley.

Beck, K. et al. (2001). Manifesto for Agile Software Development. Available at http://www.agilemanifesto. org/ (accessed on March 25, 2013).

Beyton-Davies, P., Tudhope, D., and Mackay, H. (1999). Information systems prototyping in practice. *Journal of Information Technology*, 14, 107–120.

Boehm, B.W. (1988). A spiral model of software development and enhancement. *Computer* 21(5), 61–72.

Boehm, B. and Turner, R. (2004). *Balancing Agility and Discipline*. Boston, MA: Addison-Wesley.

Booch, G., Rumbaugh, J., and Jacobson, I. (2005). *Unified Modeling Language User Guide*, 2nd edn. Reading, MA: Addison-Wesley.

Bourne, K.C. (1994). Putting rigor back in RAD. *Database Programming and Design* 7(8), 25–30.

Carmel, E., George, J.F., and Nunamaker, J.F., Jr. (1995). Examining the process of electronic-JAD. *Journal of End User Computing*, 7(1), 13–22.

Carmel, E., Whitaker, R., and George, J.F. (1993). PD and joint application design: A transatlantic comparison. *Communications of the ACM*, 36(6), 40–48.

Cerpa, N. and Verner, J. (1996). Prototyping: Some new results. *Information and Software Technology*, 38, 743–755.

Chow, T. and Cao, D.-B. (2008). A survey study of critical success factors in agile software projects. *The Journal of Systems and Software*, 81, 961–971.

Conboy, K. (2009). Agility from first principles: Reconstructing the concept of agility in information systems development. *Information Systems Research*, 20(3), 329–354.

Dijkstra, E.W. (1968). Letter to the editor: Goto statement considered harmful. *Communications of the ACM*, 11(3), 147–148.

Duggan, E.W. and Thachenkary, C.S. (2004). Integrating nominal group technique and joint application development for improved systems requirements determination. *Information & Management*, 41, 399–411.

Dyba, T. and Dingsoyr, T. (2008). Empirical studies of agile software development: A systematic review. *Information and Software Technology*, 50, 833–859.

Fichman, R.G. and Kemerer, C.F. (1993). Adoption of software engineering process innovations: The case of object orientation. *Sloan Management Review*, winter, 7–22.

Finlay, P.N. and Mitchell, A.C. (1994). Perceptions of the benefits from the introduction of CASE: An empirical study. *MIS Quarterly*, 18(4), 353–370.

Fowler, M. (2005). The new methodology. April. Available at martinfowler.com/articles/newMethodology. html, Accessed January 12, 2012.

Fruhling, A. and De Vreede, G.J. (2006). Field experiences with eXtreme programming: Developing an emergency response system. *Journal of MIS*, 22(4), 39–68.

Gavurin, S.L. (1991). Where does prototyping fit in IS development? *Journal of Systems Management*, 42(2), 13–17.

Green, D. and DiCaterino, A. (1988). A survey of system development process models. SUNY Albany. Available at http://www.ctg.albany.edu/publications/reports/survey_of_sysdev, Accessed January 12, 2012.

Hanssen, G.K., Westerheim, H., and Bjørnson, F.O. (2005). Using rational unified process in an SME—A case study. In: Richardson, I. et al. (Eds.). *EuroSPI 2005, LNCS 3792*, Springer-Verlag, Berlin, Germany, pp. 142–150.

Hardgrave, B.C. (1995). When to prototype: Decision variables used in industry. *Information and Software Technology*, 37(2), 113–118.

Howard, G.S., Bodnovich, T., Janicki, T., Liegle, J., Klein, S., Albert, P., and Cannon, D. (1999). The efficacy of matching information systems development methodologies with application characteristics—An empirical study. *Journal of Systems and Software*, 45, 177–195.

Iivari, J. (1996). Why are CASE tools not used? *Communications of the ACM*, 39(10), 94–103.

Jacobson, I., Booch, G., and Rumbaugh, J. (1999). *The United Development Software Process*. Reading, MA: Addison-Wesley.

Lee, K. (2006). An empirical development case of a software-intensive system based on the rational unified process. In: Gavrilova, M. et al. (Eds.). *ICCSA 2006, LNCS 3984*, Springer-Verlag, Berlin, Germany, pp. 877–886.

Lending, D. and Chervany, N. (1998). *The Use of CASE Tools*. Computer Personnel Research 98, Boston, MA, pp. 49–58.

Mahmood, M.A. (1987). System development methods: A comparative investigation. *MIS Quarterly*, 11(3), 293–311.

Martin, J. (1986). *Information Engineering: The Key to Success in MIS*. Carnforth, U.K.: Savant Research Studies.

Martin, J. (1991). *Rapid Application Development*. New York: Macmillan Publishing Company.

Maurizio, M., Ezran, M., and Tully, C. (2002). Success and failure factors in software reuse. *IEEE Transactions on Software Engineering*, 28(4), 340–357.

McConnell, S. (1996). *Rapid Development*. Redmond, WA: Microsoft Press.

Naumann, J.D. and Jenkins, A.M. (1982). Prototyping: The new paradigm for systems development. *MIS Quarterly*, 6(3), 29–44.

Port, D. and Bui, T. (2009). Simulating mixed agile and plan-based requirements prioritization strategies: Proof-of-concept and practical implications. *European Journal of Information Systems*, 18, 317–331.

Pressman, R.S. (1997). Software engineering. In: Dorfman, M. and Thayer, R.H. (Eds.). *Software Engineering*. Los Alamitos, CA: IEEE Computer Society Press, 57–74.

Purvis, R. and Sambamurthy, V. (1997). An examination of designer and user perceptions of JAD and the traditional IS design methodology. *Information & Management*, 32(3), 123–135.

Schwaber, K. and Sutherland, J. (2011). The Scrum guide. Available at http://www.scrum.org/scrumguides, Accessed January 12, 2012.

Sherif, K., Zmud, R.W., and Browne, G.J. (2006). Managing peer-to-peer conflicts in disruptive information technology innovations: The case of software reuse. *MIS Quarterly*, 30(2), 339–356.

Sidorova, A., Evangelopoulos, N., Valacich, J.S., and Ramakrishnan, T. (2008). Uncovering the intellectual core of the information systems discipline. *MIS Quarterly*, 32(3), 467–482.

TIOBE Software. (2012). TIOBE Programming Community Index for March 2012. Available at http://www.tiobe.com/index.php/content/paperinfo/tpci/index.html (accessed on March 23, 2012).

Valacich, J.S., George, J.F., and Hoffer, J.A. (2012). *Essentials of Systems Analysis & Design*, 5th edn. Upper Saddle River, NJ: Prentice Hall.

Wood, J. and Silver, D. (1995). *Joint Application Design*, 2nd edn. New York: John Wiley & Sons.

Yourdon, E. (1989). *Managing the Structured Techniques*, 4th edn. Englewood Cliffs, NJ: Prentice Hall.

Further Reading

Agility. Available at http://www.agilealliance.org.

Bryce, T. (2006). A short history of systems development, Available at http://it.toolbox.com/blogs/irm-blog/a-short-history-of-systems-development-8066 (accessed on January 12, 2012).

James, M. Available at http://www.jamesmartin.com.

UML. Available at http://www.uml.org.

Yourdon, E. Available at http://yourdon.com/about/.

29

Human-Centered System Development

Jennifer Tucker
U.S.Department
of Agriculture

29.1 Introduction

In the previous edition of this handbook, this chapter opened with the declaration, "The last decade has seen a significant change in the information technology landscape, resulting in a fundamental shift in the activities and skills required of technology professionals and teams." Those words were typed at about the same time that Facebook was being coded in a Harvard dorm room, 3 years before the first iPhone app launched, and 4 years before the Android source code was released. In only a few years, the landscape of information technology (IT) and systems development has again fundamentally changed.

With the evolution of computer languages, target platforms, and development methods, the skills demanded of IT professionals also evolve. In addition to maintaining and expanding technical skills, most IT professionals must communicate with an increasing technologically literate user base, navigate faster development cycles, and work in complex interdisciplinary teams. There is, of course, still the mystical archetype of the lone programmer huddled in the basement coding through the night with pizza boxes piled high. Today's zeitgeist, however, also recognizes that this same programmer might become the next tech company's Chief Executive Officer; and the project that begins as a loner's experiment might become a venture capitalist's target.

The success stories of today's social web, where every programmer is a potential millionaire, paint a positive view of the state of systems and software development. Statistics across the industry, however, raise more concern. While some question the study methods and criteria, the often-quoted Standish CHAOS reports signal ongoing industry dysfunction. The CHAOS 2000 report found that 23% of the projects studied essentially failed (Standish Group, 2001). The CHAOS Summary 2009 (Standish Group, 2010) concluded that just 32% of all projects were delivered on time, on budget, with required features and functions; that 44% of projects were "challenged" (e.g., they were considered late, over budget, and/or with less than the required features and functions); and that 24% failed (e.g., were canceled or just never used) (Standish Group, 2009).

These and other studies highlight different failure rates for different project types, but importantly also highlight the common factors contributing to these failures. These reasons often include lack of user involvement, poor project planning and execution, lack of executive support, stakeholder politics, an overemphasis on process activities, and poor communication or lack of shared understanding between business and technology groups (Standish Group, 2001, 2010; Schneider, 2002; Charette, 2005; Dyba, 2005; Kelly, 2007; Krigsman, 2009; McCafferty, 2010). Even with our best development methodologies and technological sophistication, the human dimension of systems and software development remains the key element to success.

At the most basic level, the output from any development project emerges from the conversations and collaboration among individuals working together over time. As such, success can be influenced by the ability to manage the following types of human dynamics:

- *Individual development*: Today's technology professional must be a lifelong learner of different application development frameworks, platforms, programming and markup languages, and development tools. Individuals, therefore, must have self-awareness about personal learning style and make strategic choices about whether and where to specialize and generalize. Like all professionals, technologists should also have a sense of their strengths, weaknesses, and preferences in how they work with others. Do they prefer working in a structured, predictable environment, or do they like flexibility and change? Do they prefer to work alone or with others? Too often, technology professionals are drawn to the technology aspects of the work, without fully considering the people-oriented dimension of the career choices being made. Furthermore, technical mentors may not always help their technical protégées navigate these more personal questions of self-development.
- *Team collaboration*: The range of talents required of an IT development team today is both broad and deep. Today's development teams include functional experts, business analysts, architects, programmers, content strategists and writers, data modelers and scientists, network and systems administrators, and project managers. In one study of IT-focused organizations, only 12% of IT professionals surveyed reported their role as being programmer or developer (Tucker et al., 2004). The result is that IT teams, which once shared a common technical base for building relationships, are coming together from separate specialties and backgrounds within their own fields. This diversity creates profound opportunities for collaboration if team members are able to transfer knowledge internally and communicate effectively, but it can also lead to miscommunication and misfires if not managed well.
- *User communication*: Technical individuals and teams rarely work in a vacuum. Even in teams where a project manager or business analyst serves as a primary point person for customers, developers need to interpret requirements and understand how their work will meet the user's or customer's needs. Self-perception, self-expression, interpersonal skills, decision making, and stress management are all factors of emotional intelligence (EQ) that shape the ability to work effectively with customers (Stein and Book, 2006), and motivational states and cognitive preferences can dramatically shape understanding and interpretation. Deciding how users are to be treated and engaged in a systems development process, and then following through even when conflict is encountered, is a critical factor in the design of a technology program.

These realities and dimensions of the systems development experience call us to again reexamine IT and systems development work from a human dynamics perspective. This chapter considers the underlying principles informing human-centered systems development and considers applications and directions for future research.

29.2 Principles and Applications

Understanding human-centered systems development begins with defining the term itself. For the purpose of this chapter, human-centered systems development considers the people side of systems development: the personality dynamics of IT professionals, the people-related dynamics of IT teams, and the interpersonal elements of the systems development life cycle.

29.2.1 Personality Dynamics of IT Professionals

Many research studies have used psychometric tools to describe the personality dynamics of IT professionals. While these studies have ranged in design, they collectively reveal that certain personality characteristics tend to be more and less prevalent among IT specialists than the general population (Mastor and Ismael, 2004; Tucker et al., 2004; Williamson et al., 2005; Lounsbury et al., 2007). These distinctive characteristics include

- A higher preference for logical, impersonal, and objective analysis, sometimes captured as tough-mindedness
- Higher levels of intrinsic motivation, personal control, and emotional resilience
- Higher levels of independence
- High desire for supervisor support, but without the desire for increased levels of control by those same supervisors (consistent with independence)
- Lower levels of rules orientation and conformity, which in some psychological tests are characterized as conscientiousness
- Lower levels of political savvy and interest, sometimes known as image management
- Lower needs for interpersonal connection and inclusion in other people's social activities

The practical impact of this research is that understanding these personality characteristics both reveals the common strengths that information technologists might bring to their work and points to possible systemic weaknesses or blind spots that may occur when strengths are overdone. The strengths of logic, objectivity, independence, and control can yield powerful technology solutions, but they also make it understandable why the mythos of the lone, independent, even awkward coder is so persistent and so often encountered in real life.

When overdone, the strengths most commonly associated with IT professionals can be perceived as a lack of empathy for the user, a lack of concern for relationship management aspects of the development process, or a bias toward working on interesting individual tasks rather than attending to more routine team needs.

Independence and personal control over work are often repeated themes when referencing technology professional preferences. Developing high-quality software requires attention to detail and clear controls. Controlling quality, however, is different from controlling the people developing the code, particularly when those people value high levels of independence and personal autonomy. Structured project management addresses software quality because it introduces controls and testing check points that prevent low quality code from being committed and released. Too much management, however, can stifle innovation and creativity among the people creating that code. An awareness of these human dynamics is vital for leaders seeking ways to control the code without overly controlling the coders.

Interpersonal and relationship management skills are vital in any systems development effort, but while increasingly referenced in curriculum models, are not central in the formal education of most technology professionals. Once in the workplace, many technology professionals tend to avoid training that focuses on the development of interpersonal and communication skills, believing that such investment of time and resources is not as beneficial as learning a new technical skill.

At the same time, personal experience suggests that many technical specialists have at least occasionally experienced conflict with team members, misunderstandings with the boss, or confusion with end users or clients. Many of these interpersonal issues can be traced to fundamental style or communication differences rather than a technical deficiency. Technical professionals who recognize these root causes are better prepared to manage problems effectively, leading to better technical work. Understanding what one brings to the team from a human dynamics standpoint, and how that might mesh or clash with coworkers, is a key starting point.

29.2.2 IT Team Dynamics

The increasing specialization of roles in systems development brings together teams of engineers, business analysts, architects, developers, content strategists, testers, and project managers to guide technology projects to fruition. Larger team sizes lead to more complexity, and while many acknowledge that project success is often linked to the "people side" of the equation, there are no agreed upon algorithms for project success. Project success in the real world, in fact, comes from carefully balancing a variety of factors. Just as overdone personality strengths become weaknesses, an overemphasis on any particular project or team factor becomes a liability as well.

Human-centered systems development requires understanding factors and models from a variety of fields, including psychology, process improvement, project management, and organization development. This section reviews a number of factors important to the success of technology teams (Ahern et al., 2001; Rutledge and Tucker, 2006; Basri and O'Connor, 2011).

29.2.2.1 Mission and Goal Clarity

A shared understanding of the goal of any particular project keeps both the project team and user base focused on the end state and provides a stable benchmark when tough trade-offs and decisions must be made. When different segments of a project team hold different understandings of the mission, disconnects and conflicts may occur without people actually realizing their cause.

For example, drawing from a true case study, picture a large-scale software radio development program with teams distributed across a number of sites. While the program had an overarching and rather lengthy mission statement, people generally reported either not knowing what it was, or, not really understanding what it meant. To explore this, two teams in the program were asked to develop a statement summarizing the overarching mission of the project. One team responded that the goal was to create a low-cost piece of integrated software and hardware that would require little maintenance once fielded. The second team responded that the goal was to develop cutting-edge technology that would advance the software radio field and spark new technology investments. There had been a variety of conflicts between these two teams, and integration between systems had consistently encountered difficulties and delays. Unfortunately, the teams had not ever discovered their fundamental disconnect, leading to conflicts that were ultimately based on foundationally different ideas of the project mission.

Any factor that is overemphasized can, of course, lead to blind spots. A shared understanding of a mission provides teams with a target against which to mutually align and a common goal around which to rally. When there is too much emphasis on mission achievement, however, the team can miss shifts in the environment that led to the initial need, leading to possible obsolescence in the mission itself. Additionally, an overemphasis on goal orientation can lead to high levels of team stress if too much pressure is imposed.

29.2.2.2 Structure and Leadership

Structure and leadership includes such formal elements as role clarity, reporting relationships, and organizational structure, and informal elements such as power dynamics, personal accountability, and empowerment. Structure provides information about hierarchies, governance, and team organization. Leadership can be either formal as managers give direction, make decisions, and allocate resources; or informal and emergent as different team members lead each other as the project unfolds (Howell et al., 1986; Carter, 2004; Rutledge and Tucker, 2006; Senge, 2006).

Structure and leadership often evolve as organizations and teams grow. The team that begins as two programmers grows into a small team and beyond. Technical questions become supplemented by organizational questions of centralization versus decentralization, and self-directed teams versus matrix relationships. These are important questions about how the team functions and also reveal dynamics about how a technology team engages with others outside the group.

For example, in formal organizations, examining an organization chart (if there is one), can reveal the team's relationship with its users. Who is the person or group responsible for ensuring that the users' voices are heard? Where does this role reside on the chart, and how much power is inherent in the position?

The same teams described earlier that struggled with mission clarity also struggled with user relationships and senior stakeholder buy-in. The team responsible for customer relationships was organizationally located in an entirely different team than the engineering and testing teams, and there were few formal or even informal connections between them. As such, the team responsible for relationship management was listening to the user base, but the structure of the large program did not facilitate information flow from one part of the program to another. One key consequence of this was that the customer felt heard, but when elements of the product were delivered, there were significant usability gaps between what the customer had asked for and what was delivered.

Over time, informal connections between project team members built informal pathways for questions and feedback, but leaders did not make easy transmission of customer feedback a priority, causing inefficiency in communication and a great deal of invisible labor behind the scenes. The informal leadership of team members ultimately evolved into an unwritten "shadow org chart" for information sharing. The degree of match between formal and informal power structures on a team reveals a great deal about the efficiency and effectiveness of team operations, and the degree to which formal leadership is respected and followed.

A very different case study can be seen in the structure and leadership of the open source community that develops the operating system Linux. This is a model where structure and leadership are highly interdependent. The Linux organizational structure is centered on a person rather than a role; Linus Torvalds, the "benevolent dictator of Planet Linux," (Rivlin, 2003) and a small set of key deputies hold decision-making authority for all elements of a Linux release. Outside this structure, there are loose networks of developers and contributors but no formal organization charts, reporting relationships, or hierarchies.

In this model, a small core of formal leadership is at the center, making the institutional decisions that define the product. In the absence of formal institutional structures, technical and people leadership from Torvalds is cited as a defining factor in the success of the overall enterprise (Rivlin, 2003; Vaughan-Nichols, 2010). Complementing this, informal and emergent leadership is exercised as individual developers choose what features to work on and who to follow; it is this grassroots leadership that ultimately provides the assets from which the formal leadership core can select. While Torvalds may be the central figure, the relationships across this community form an informal network that provides intellectual capital and becomes its own type of organizational structure.

29.2.2.3 Process and Communication

How much team processes are known, agreed on, communicated, and acted upon greatly impacts team effectiveness and morale. While structure defines roles and relationships, process defines how work flows between these roles to lead to decisions and deliverables. Software development models, business processes, workflow, and decision matrices all fall under the category of process. While process and communication are often treated separately in process management, they are fundamentally linked; communication between team members, whether technically or interpersonally mediated, is generally the mechanism by which process is expressed.

In the past, the technology industry attacked the dynamic problems of system development with an increased focus on process, creating a plethora of standards, protocols, and maturity models. As noted previously, however, any strength maximized can become a liability; and process management taken too far can result in an overemphasis on project reporting and process documentation, compromising functional delivery.

This is an area where the pendulum may be swinging, or at least diversifying. In the previous edition of this handbook, this chapter declared, "Every complex endeavor needs an organized plan to keep it on track." Since that edition, however, the success of emergent start-ups and open source projects has called this project management assumption into question. For example, Facebook proudly proclaims its preference for "The Hacker Way" through the mantra: "Done is better than perfect." Facebook has embraced a development process centered upon small iterative releases, with constant testing both to correct errors and to identify the next steps (Zuckerberg, 2012). Free and open source software development also usually does not follow formal process models or plan-driven approaches because it is generally based on volunteer collaboration; yet, these projects can generate high-quality products quite quickly (Magdalenoa et al., 2012).

These development processes are very different from the structured waterfall methods of the past and require different communication approaches and styles. The absence of plan-driven process does not, of course, mean the absence of process. The question is: How do the formal and informal processes and communication pathways within a team support or detract from the team's success, however that is measured, and how do the processes encourage or discourage the participation of the technical professionals most qualified to do the work? An emphasis on structured, plan-driven process and communication appeals to many professionals. It is predictable, clear, and when managed effectively, efficient. Other professionals are more attracted to the "Hacker Way," where the product emerges from the unpredictability of trying out new ideas. It is innovative, flexible, and when managed effectively, efficient.

Different process approaches have different impacts that can be both positive and negative. These impacts may also be perceived differently by people with different personalities. What is a positive process practice or communication approach for one person may be problematic for another. Just as engineers must understand the impact of different architecture and design choices, those working in a human-centered systems development context must anticipate how different process and communication approaches are likely to impact different people and plan accordingly. This may mean recruiting people that either enhance or counter-balance different team characteristics, customizing communication for different people based on individual values and styles, and identifying and monitoring the risks that may emerge when specific approaches are too far over-emphasized.

29.2.2.4 Interpersonal Skills

The unequaled mourning over the death of Steve Jobs sent tremors of paradox through the leadership development community. After years of advocating the importance of interpersonal skills in leading organizations and teams, many struggled with the sense of cognitive dissonance that a man so feared would also be so celebrated for his ability to inspire other people. How could this leader, known for his angry outbursts and harsh feedback, be so revered?

A review of leadership articles and blogs published shortly after Jobs' death reveals a common theme: Jobs' leadership style worked because of his unique blend of technical and design intelligence, because of his insistence on excellence from everyone around him, and because he was inspirational in how he communicated his vision to others (Isaacson, 2012). Many articles, however, also carried a warning echoed by leadership coaches far and wide: "Don't try this at home" (Allen, 2011; Nocera, 2011; Riggio, 2012). Jobs' ability to lead despite his difficult interpersonal style was cited as many as an anomaly; something "mere mortals" should not attempt.

From a practical viewpoint, it is more useful, perhaps, to reframe the situation as a matter of fit. Jobs had a specific style that attracted certain people to him, but there are likely many that self-selected away from him as a result of that same style. Investing in developing interpersonal skills provides the flexibility needed to work well with a range of other people, increasing the opportunities for both leadership and followership.

Interpersonal skills include the ability to listen and demonstrate understanding of others' views, display trust, build relationships, and support others. Closely linked with communication and leadership, interpersonal skills allow people to manage conflict effectively, negotiate, advocate their views compellingly, and use power to achieve desired ends. Interpersonal skills are not dependent on a particular set of personality preferences and styles. Read biographies of both Linus Torvald and Steve Jobs, and you will read portraits of very different men. Interpersonal skills result from self-awareness, the ability to read and empathize with others, and the ability to work with others toward a common goal. While "code may win arguments" in hacker culture, it is the combination of both technical and interpersonal skills that brings people together for the argument in the first place.

Psychological type, popularized through the Myers-Briggs Type Indicator (MBTI), is a popular model in understanding a set of different dimensions of personality that can contribute to interpersonal skills (Myers and Myers, 1980, 1995; Kroeger et al., 2002). Like technical skills, interpersonal skills are teachable, and the psychological type model has helped many professionals understand how their personalities impact behavior and their relationships with others (Peslak, 2006).

The MBTI assesses four dimensions of personality: your source of energy (internal or external); your preference for data gathering (detailed and specific, or abstract and global); your preference for decision making (logical and objective, or subjective and values driven); and your orientation to the world (whether you prefer to show others your decisions or your perceptions) (Kroeger et al., 2002).

Preferences often manifest themselves in behaviors, which can either support or irritate relationships with others. As an example, imagine a developer who is energized by the internal world (rather than engaging with others), and when in the external world, prefers a sense of order, structure, and closure (rather than openness and flexibility). Because this developer prefers to communicate final thoughts in a decision-based format, she expects that others generally operate this way as well. Our developer is in a prototype review session with a user who is energized by the external world, and who prefers to bounce ideas around to explore and learn in real time. This user may view interaction periods as open brainstorming sessions, designed to explore new ideas, cover the possibilities, and think out loud.

What might happen as an outcome of this prototype review session? From the developer's perspective, the meeting resulted in a series of decisions, and a long list of new requirements. From the user's perspective, the discussion was simply the sharing of ideas and options, and alternatives for consideration, no strings attached. This scenario, which is neither exaggerated nor uncommon, is an example of how personality type can impact relationships. It also reframes the enduring problem of requirement creep, a key issue facing the IT community. Requirements creep is usually described as a problem at the system or project level; however, every need ever articulated for a system came from a human user expressing an individual thought. How that thought is heard and managed is ultimately what leads to the project-level problem of requirement creep. This is only one example. Consider the full range of preference combinations between developers and users, and the complexity and potential impact of human dynamics in the development process becomes even more dramatic.

The MBTI is, of course, only one psychological assessment used to study personality and team style and preference, and it is by no means universally accepted. Some, for example, reject the idea of personality testing altogether. Others reject the MBTI based upon reviews of its psychometric characteristics, such as construct validity and applicability (Pittenger, 1993; Capraro and Capraro, 2002).

Other models with associated assessments that provide personality-related insights across different scales include the "Big Five," a generic term for assessments that study the five personality characteristics of extraversion, openness, agreeableness, conscientiousness, and emotional stability (McCrae and John, 1992; Grucza and Goldberg, 2007); EQ indicators, which assess a variety of factors related to self-awareness, self-management, social awareness, and relationship management (Petrides and Furnham, 2000; Stein and Book, 2006); and reversal theory, which assesses motivational states, changeability, and emotion (Apter, 2007; Tucker and Rutledge, 2007).

What is the best practice that can be gained from using psychometric tools? Investing the time and effort to learn about personality preferences, and learning how to recognize them in others yields better relationship management skills and less frustration, because having a language with which to talk about differences makes them easier to resolve. This conclusion is also supported by evidence from a study of 40 software development teams, which concluded "how people work together is a stronger predictor of performance than the individual skills and abilities of team members" (Sawyer, 2001).

29.2.3 Systems Development Life Cycle Models

The previous section touched upon process as a critical element of team dynamics. Process includes systems and software development models. Given their importance, this section considers different systems development life cycle models or methodologies, through the lens of the human-centered demands that they place on a development professional. While there are many overlaps between methods, the presentation here is based upon the type of people-driven factors that tend to be called upon the most for that method type.

29.2.3.1 Formal Structured Methods

Formal structured methods, such as the classic waterfall methodology and the risk-based spiral model, are methods that are generally governed by a plan-driven predictable and repeatable process (McConnell, 1996; Boehm, 1998; Ahren, 2001). In the formal structured methods, development occurs through a series of sequential phases, with a major milestone review marking the end of each phase or iteration. While phases differ across projects and different variations of the development model, they generally cover the following types of activities: concept development or needs assessment, requirements analysis, high-level or general design, detailed design, coding/unit test/debugging, system testing, and implementation. Spiral models add an iterative element to the structure, allowing the team to learn between iterations and the user to see some outcomes, while still maintaining the discipline of structured methods.

Formal structured methods work well for low-speed and low-change projects, in which there is a very clear and stable definition of the requirements, the product, and the supporting technology. It is also used for large projects where high quality, low error outputs must be combined with other engineered systems, or for long-term expensive projects where budget and resource planning is vital to keep the project funded and staffed.

Ultimately, the strengths of any method also hold the source of its weaknesses. The discipline of formal methods can lead to a project that is inflexible to changing needs, and an overemphasis on process documentation can take time away from actual project development. Structured development models are likely to appeal to developers and users who have lower tolerance for ambiguity, enjoy the stability of long-term projects, do not mind formal process and documentation, prefer concrete milestones and metrics, and enjoy the process of creating high-quality code. It can be frustrating for users and developers who want to see tangible outcomes more quickly.

29.2.3.2 Agile and Other Adaptive Methods

Agile, adaptive methods are defined here primarily by their orientation toward user involvement and the degree of change and flexibility encouraged in the development process. Agile methods emphasize interpersonal interactions, communication, customer collaboration, rapid production of code, and frequent feedback (Agile Alliance, 2001; Vinekar et al., 2006; Moe et al., 2010; Stettina and Heijstek, 2011; Thibodeau, 2011). In contrast to formal structured methods, agile development methods have been praised for bringing together developers and users and for allowing fluidity in the development process as participants co-learn and evolve their understanding of the true business needs and requirements.

While agile methods have become widely adopted over the past few years, there are critiques as well (Boehm and Turner, 2004; Dyba and Dingsøyr, 2008; Barlow et al., 2011; Gualtieri, 2011). The bias toward rapid development of code can lead to software modules that solve an immediate problem but that are not built with an eye to the customer organization's business strategies and need for larger-scale integration points. Furthermore, the method is heavily dependent on the presence of knowledgeable users to iterate business requirements in real time. This can lead to solutions that represent the local needs of users that participated in the process, but that may not sufficiently represent the needs of a diverse user base across a larger enterprise.

Agile, adaptive methods require significant interpersonal skills for those engaging frequently with users and comfort with ambiguity, change, and real-time problem solving. It can be frustrating for those wanting more structure and a broader scope or planning horizon, or for those that are uncomfortable working with users or receiving uncensored feedback in real-time application development environments.

29.2.3.3 Hackathon Method

Some may bristle at the classification of the hackathon as a method, seeing it as an extension of the iterative methods discussed earlier; however, these problem-focused coding approach and events have made a sufficient mark on the programming landscape and culture to be considered separately here. Described as "a contest to pitch, program, and present," (Leckart, 2012) hackathons bring together programmers and systems engineers to work on specific software problems or to compete for funding for further development. A time-bound variation on the long-standing "code and fix" process, hackathons are characterized by their emphasis on real-time, intensive, outcome-driven coding processes, with developers often using preexisting software resources and toolkits to generate a usable product (such as a new app, or new application programming interface) in a time-boxed period.

Hackathons have become a normal activity for many software firms and programs and have even become a practice in government agencies. In the initial public offering documentation for Facebook, Mark Zuckerberg (2012) described the nature of hacking: "Hacking is also an inherently hands-on and active discipline. Instead of debating for days whether a new idea is possible or what the best way to build something is, hackers would rather just prototype something and see what works." At hackathon events, a problem or goal is presented, and a time span defined in which to work. At the end, the best solution is generally selected with some kind of acknowledgement or award.

Hackathons require personality factors such as comfort and skill with working informally with new people in a small team, the ability to quickly extend and extrapolate preexisting code and tools to solve particular problems, the ability to work intensely under time pressure, and the ability to present one's work to others in a compelling way in a short time. Not everyone thrives in these conditions, but these events have proven to be a valuable mechanism by which new software components can be created quickly.

29.3 Research Opportunities and Conclusions

No technical team sets out to develop a system that does not meet user needs. Moreover, no user purposefully sets out to confuse or muddy the requirements that he or she is trying to articulate. In today's environment of mobile apps and social software, human-centered software is user-driven software, and it is becoming common to refer to interactions with software as "experiences." Given this, "experience creators" must be increasingly trained to be aligned with both the tangible and emotional needs of an increasingly technologically savvy user base. Research and education are two vital paths to maximize this alignment.

29.3.1 Research Landscape and Opportunities

To better assess the research opportunities in human-centered systems development, it is useful to review the type of research generally conducted across selected different technologically oriented disciplines interested in this area. Primary fields of interest here include information systems, systems engineering, and human–computer interaction.

Information systems is a broad and interdisciplinary academic field, covering a range of topics related to the processing and management of information: technologically, organizationally, and cognitively. Information systems covers such domains as data and information management, enterprise architecture, project management, infrastructure, systems analysis and design, information strategy, cognitive science, and informatics (Topi et al., 2010; Buckland, 2011). The field of information systems provides a broad social and systems level perspective of the intersection between people, technology, and information.

Systems engineering research is also a vital resource for those interested in human-centered systems development. It focuses on the theory, practice, and methods associated with the development of often large socio-technical systems. In the realm of systems engineering, case-based studies concerning personality and team performance is the research that most closely relates to the content of this chapter.

Human–computer interaction research also addresses topics at the intersection of people and technology, but generally looks at this intersection from the users' perspective, not the developers (Helander, 1997). For example, a search on "human computer interaction emotion" yields multiple articles about the simulation of emotion by computers, and methods for invoking emotions among users, but includes only isolated references to the emotions of information technologists themselves.

Outside these technology-oriented fields, researchers interested in human-centered systems development might look to the fields of organization development, management, and organization and industrial psychology for both theoretical and applied research. Subfields of leadership development, technology innovation, and team dynamics also can yield valuable insights that can be applied to human-centered systems development.

Unfortunately, in many real-world environments, the work of organizational development professionals and IT professionals remains quite distinct, even though they share the same foundational goals of supporting organization change. Organizational development professionals tend to approach problems from the people perspective; IT professionals tend to approach problems from the technical or information flow perspective. The different language and tools used by each type of professional makes cross-fertilization less common in professional settings than one might hope.

Examples of research that would further support the principles and applications presented in this chapter are as follows:

- *Empirically based case studies*: More empirical research using validated psychological assessments is required to solidify the connection between effective human dynamics, communication, and project success. Too often, empirical studies rely not on published and controlled validated assessments (such as the MBTI), but rather, on publically available "spin-offs," which while grounded in the same theories, lack the validity to support confidence in study results.

- *Cross-team comparison studies*: Unfortunately, the lack of standardization between existing studies of personality and team project success makes their results difficult to generalize or to usefully compare findings across projects with different characteristics. More studies using similar methods across well-defined but distinct environments are needed to better understand the intersection of personality preferences, development models, and product outcomes and successes over time.

- *Ethnographic research*: Observational studies of development teams at work, including their interaction with each other and with users, would provide additional ground-level evidence that could inform future practice. For example, comparing the actual content of joint applications development sessions or agile-based user-developer discussions against the final specifications of a completed product could yield useful insights about the types of elicitation techniques or conversational arcs that are most effective in yielding useful requirements and efficient user feedback.

Research helps to support the work of practitioners by providing insight into human-centered systems development and its processes, and cross-disciplinary research helps bridge the existing gaps between computer science and the social sciences. The key, however, is to continue to work to bridge the boundaries between academic communities and industry and government so that the research generated in the academic world is communicated in a way that is practical and useful for nonacademic professionals fighting the fires of delayed and at-risk projects in real time. Collaborative research efforts with industry can also to help unite today's curriculum and students with the industry challenges and opportunities of tomorrow.

29.3.2 Teaching Methods

A focus on human-centered systems development points to several themes for educators to consider in developing curricula for technical professionals. Key competencies to develop in these efforts include

- Self-awareness and learning
- User interaction and communication
- Team dynamics
- Innovation
- Working with ambiguity and uncertainty
- Process management and quality improvement

Encourage a broad curriculum: Educators are in a unique position to help technology professionals experience a broad curriculum that includes interpersonal skills development and learning projects that focus on team-based work (Crawley et al., 2007). Teaching skills such as critical thinking and analysis, writing, strategies to learn new development tools and languages as they emerge, and helping students develop the habits of self-awareness and introspection will serve them well in any professional setting.

Balance theory and practice: Personality type theorists argue that how people perceive or gather information is the most important factor in determining how people learn. Some people learn best when given a theoretical framework within which to place new information. Others learn best when presented with practical applications or when they are able to interact with tools that allow practice and interaction with the new knowledge. Balancing theory and practice develops concrete applications knowledge, while also providing a broad theoretical foundation from which future applications can be derived. Practice also reminds hands-on students why they self-selected into a field in the first place, particularly important for students who selected into the field because of its practical applications and potential for impact.

Incorporate human uncertainty into assignments: Many educators use simulated exercises or case studies to allow programmers to practice programming skills. These methods are even more useful when they integrate real users and user needs into the equation. Projects being completed under the humanitarian free and open source software program are a good example of instances where educators are providing students with real life software development opportunities. These opportunities are not just isolated exercises, but are full life cycle experiences completed over the course of a semester, giving students exposure to a wide range of activities and interactions that they will encounter in the future as professionals (Morelli et al., 2010).

These projects benefit both students and nonprofit organizations; the students gain invaluable exposure and experience, and the users gain a piece of software that supports a humanitarian-focused nonprofit mission. Research related to the long-term impact that these experiences have on students as they enter their careers would be a useful step to validate the usefulness of these experiences over the long term.

29.3.3 Conclusions

Recognizing the human-centered elements of systems development, including personality preferences, team dynamics, and development models, may help technical professionals select projects or roles that support personal values and areas of personal strength. Some people simply would rather work alone to generate a product that later can be joined with the efforts of others. Other people enjoy the shared nature of actively collaborative efforts. Some developers enjoy the iterative nature of evolutionary prototyping; others find the constant change and uncertainty more frustrating than freeing.

Technology professionals should take the time and be given the experiences that allow them to recognize their personal strengths, preferences, and potential blind spots when selecting jobs or projects. If a developer, for example, chafes against the requirements for structure and documentation that come with working on a project governed by structured methods and capability maturity models, it is better to know this in advance. Research provides useful context for asking new questions, but in the end, self-knowledge is the foundational key to making more effective decisions: decisions that will impact the individual, the development team, and ultimately the end user.

References

Agile Alliance. 2001. The agile manifesto. http://www.agilealliance.org/the-alliance/the-agile-manifesto/ (accessed on May 29, 2012).

Ahern, D.M., Turner, R., and Clouse, A. 2001. *CMMI(SM) Distilled: A Practical Introduction to Integrated Process Improvement*. Reading, MA: Addison-Wesley.

Allen, F.E. 2011. Steve Jobs broke every leadership rule. Don't try it yourself. *Forbes*. August 27, 2011. http://www.forbes.com/sites/frederickallen/2011/08/27/steve-jobs-broke-every-leadership-rule-dont-try-that-yourself/

Apter, M.J. 2007. *Reversal Theory: The Dynamics of Motivation, Emotion and Personality*, 2nd edn. Oxford, U.K.: Oneworld Publications.

Barlow, J.B., Giboney, J.S., Keith, M.J., Wilson, D.W., Schuetzler, R.M., Lowry, P., and Vance, A. 2011. Overview and guidance on agile development in large organizations. *Communications of the Association for Information Systems*. 29:2, 25–44.

Basri, S. and O'Connor, R.V. 2011. A study of software development team dynamics in Software Process Improvement. In: *18th European Software Process Improvement Conference*. June 27–29, 2011. Roskilde, Denmark.

Boehm, B.W. 1998. A spiral model of software development and enhancement. *IEEE Computer*. 21:61–72.

Boehm, B.W. and Turner, R. 2004. *Balancing Agility and Discipline: A Guide for the Perplexed*. Boston, MA: Addison-Wesley. Appendix A, pp. 165–194.

Buckland, M. 2011. What kind of science can information science be? *Journal of Information Science and Technology*. 63(1), 1–7. Pre-published version. http://people.ischool.berkeley.edu/~buckland/whatsci.pdf

Capraro, R.M. and Capraro, M.M. 2002. Myers-Briggs Type Indicator score reliability across: Studies a meta-analytic reliability generalization study. *Educational and Psychological Measurement (SAGE Publications)*. 62:4, 590–602.

Carter, L.L. 2004. *Best Practices in Leadership Development and Organization Change*. San Francisco, CA: Jossey Bass.

Charette, R.N. 2005. Why software fails. *IEEE Spectrum*, 42:9, 42–49.

Crawley, E., Malmquist, J., Ostlund, S., and Brodeur, D. 2007. *Rethinking Engineering Education: The Conceiving, Designing, Implementing, and Operating (CDIO) Approach*. New York: Springer.

Dyba, T. 2005. An empirical investigation of the key factors for success in software process improvement. *Software Engineering, IEEE Transactions*. 31:5, 410–424.

Dyba, T. and Dingsøyr, T. 2008. Empirical studies of agile software development: A systematic review. *Information and Software Technology*. 50:9–10, 833–859.

Grucza, R.A. and Goldberg, L.R. 2007. The comparative validity of 11 modern personality inventories: Predictions of behavioral acts, informant reports, and clinical indicators. *Journal of Personality Assessment*. 89:2, 167–187.

Gualtieri, M. 2011. Agile software is a cop-out; Here's what's next. *Forrester Blogs*. October 12, 2011. http://blogs.forrester.com/mike_gualtieri/11-10-12-agile_software_is_a_cop_out_heres_whats_next

Helander, H. 1997. *Handbook of Human-Computer Interaction*, 2nd edn. Amsterdam, the Netherlands: Elsevier.

Howell, J.P., Dorfman, P.W., and Kerr, S. 1986. Moderator variables in leadership research. *The Academy of Management Review*. 11:1, 88–102.

Isaacson, W. 2012. The real leadership lessons of Steve Jobs. *Harvard Business Review*. 90:4, 92–100.

Kelly, N. 2007. High failure rate hits IT projects. *Computing*. 20 August 2007. http://www.computing.co.uk/ctg/news/1829160/high-failure-rate-hits-it-projects (accessed on March 25, 2013).

Krigsman, M. 2009. CRM failure rates: 2001–2009. *Technology News, Analysis, Comments and Product Reviews for IT Professionals*. ZDNet, 3 August 2009. http://www.zdnet.com/blog/projectfailures/crm-failure-rates-2001-2009/4967 (accessed on March 25, 2013).

Kroeger, O., Thuesen, J., and Rutledge, H. 2002. *Type Talk at Work: How the 16 Personality Types Determine Your Success on the Job*. New York: Dell Publishing.

Leckart, S. 2012. The hackathon is on: Pitching and programming the next killer App. *Wired*. March 2012.

Lounsbury, J.W., Moffitt, L., Gibson, L.W., Drost, A.W., and Stevenson, M.W. 2007. An investigation of personality traits in relation to the job and career satisfaction of information technology professionals. *Journal of Information Technology*. 22, 174–183.

Magdalenoa, A.M., Wernera, C.M.L., and de Araujob, R.M. 2012. Reconciling software development models: A quasi-systematic review. *Journal of Systems and Software*. 85:2, 351–369.

Mastor, K.A. and Ismael, A.H. 2004. Personality and cognitive style differences among matriculation engineering and information technology students. *World Transactions on Engineering and Technology Education*. 3:1, 101–105.

McCafferty, D. 2010. Why IT projects fail. *CIO Insight*. June 10, 2010.

McConnell, S. 1996. *Rapid Development*. Redmond, WA: Microsoft Press.

McCrae, R.R. and John, O.P. 1992. An introduction to the five-factor model and its applications. *Journal of Personality*. 60:2, 175–215.

Moe, N.B., Dingsøyr, T., and Dyba, T. 2010. A teamwork model for understanding an agile team: A case study of a Scrum project. *Information and Software Technology*. 52:5, 480–491.

Morelli, R., Tucker, A., and de Lanerolle, T. 2010. The humanitarian FOSS project. *Open Source Business Resource*. Talent First Network, Ottawa, Ontario, Canada, December 2010.

Myers, I.B. and Myers, P.B. 1980, 1995. *Gifts Differing: Understanding Personality Type*. Mountain View, CA: Davies-Black Publishing.

Nocera, J. 2011. What makes Steve Jobs great. *New York Times: Opinion Pages*. August 26, 2011. http://www.nytimes.com/2011/08/27/opinion/nocera-what-makes-steve-jobs-great.html?_r=1

Peslak, A.R. 2006. The impact of personality on information technology team projects. *SIGMIS-CPR 2006*, 13–15 April 2006, Claremont, CA.

Petrides, K.V. and Furnham, A. 2000. On the dimensional structure of emotional intelligence. *Personality and Individual Differences*. 29, 313–320.

Pittenger, D.J. 1993. Measuring the MBTI…and coming up short. *Journal of Career Planning and Employment*. 54:1, 48–52.

Riggio, R.E. 2012. Why Steve Jobs is a leadership nightmare. *Psychology Today*. February 7, 2012. http://www.psychologytoday.com/blog/cutting-edge-leadership/201202/why-steve-jobs-is-leadership-nightmare

Rivlin, G. 2003. Leader of the free world. *Wired Magazine*. Issue 11.11, November 2003.

Rutledge, D.H. and Tucker, J. 2006. Transforming cultures: A new approach to assessing and improving technical programs. *CrossTalk—Department of Defense Journal of Software Engineering*. January 2006. http://www.crosstalkonline.org/storage/issue-archives/2006/200601/200601-Rutledge.pdf (accessed on March 25, 2013).

Sawyer, S. 2001. Effects of intra-group conflict on packaged software development team performance. *Information Systems Journal*. 11:155–178.

Schneider, K. 2002. Non-technical factors are key to ensuring project success. *Computer Weekly*. February 28, 2002, 40.

The Standish Group, 2001. *Extreme Chaos*. Boston, MA: The Standish Group International, Inc. 1–12.

The Standish Group, 2009. *CHAOS Summary 2009*. Boston, MA: The Standish Group International, Inc. 1–4.

The Standish Group, 2010. *CHAOS Summary for 2010*. Boston, MA: The Standish Group International, Inc. 1–26.

Senge, P.M. 2006. *The Fifth Discipline: The Art and Practice of a Learning Organization*. New York: Doubleday/Currency.

Stein, S.J. and Book, H. 2006. *The EQ Edge: Emotional Intelligence and Your Success*. Jossey-Bass Leadership Series. Ontario, Canada: John Wiley and Sons.

Stettina, C.J. and Heijstek, W. 2011. Five agile factors: Helping self-management to self-reflect. *Proceedings of the 18th European System Software Process Improvement and Innovation Conference EUROSPI 2011*. Roskilde, Denmark, pp. 84–96.

Thibodeau, P. 2011. It's not the coding that's hard, it's the people. *Computerworld*. October 6, 2011.

Topi, H., Valacich, J.S., Wright, R.T., Kaiser, K., Nunamaker, J.F., Jr., Sipior, J.C., and de Vreede, G.J. 2010. IS 2010: Curriculum guidelines for undergraduate degree programs in information systems. *Communications of the Association for Information Systems*. 26:18.

Tucker, J., Mackness, A., and Rutledge, D.H. 2004. The human dynamics of IT teams. *CrossTalk—Department of Defense Journal of Software Engineering*. February 2004. Hill AFB, Utah: U.S. Department of Defense Software Technology Support Center, pp. 15–19. http://www.crosstalkonline.org/storage/issue-archives/2004/200402/200402-Tucker.pdf

Tucker, J. and Rutledge, D.H. 2007. Motivation and emotion in technology teams. *CrossTalk—Department of Defense Journal of Software Engineering*. November 2007. Hill AFB, Utah: U.S. Department of Defense Software Technology Support Center, pp. 10–13. http://www.crosstalkonline.org/storage/issue-archives/2007/200711/200711-0-issue.pdf

Vaughan-Nichols, S.J. 2010. Do open-source projects need strong leaders? *Computerworld*. November 2, 2010.

Vinekar, V., Slinkman, C.W., and Nerur, S. 2006. Can agile and traditional systems development approaches coexist? An ambidextrous view. *Information Systems Management*. 23:3, 31–42.

Williamson, J.W., Pemberton, A.E., and Lounsbury, J.W. 2005. An investigation of career and job satisfaction in relation to personality traits of information professionals. *Library Quarterly*. 75:2, 122–141.

Zuckerberg, M. 2012. Letter from Mark Zuckerberg. U.S. Securities and Exchange Commission: Form S-1: Registration Statement: Facebook, Inc. p. 67. February 1, 2012.

30

Developing and Managing Complex, Evolving Information Infrastructures

Ole Hanseth
University of Oslo

30.1 Introduction

30.1.1 The Growth of Complexity in ICT Solutions, the Emergence of Information Infrastructures, and the Need for a New Paradigm

Increased processing power and higher transmission and storage capacity have made it possible to build increasingly integrated and versatile information technology (IT) solutions with dramatically increased complexity (BCS/RAE 2004, Hanseth and Ciborra 2007, Kallinikos 2007). The complexity of IT solutions has also been continuously growing as existing systems, new and old, have been increasingly integrated with each other. Complexity can be defined here as the dramatic increase in the number and heterogeneity of included components, relations, and their dynamic and unexpected interactions in IT solutions (c.f. Hanseth and Lyytinen 2010).

The Software Engineering Institute (SEI) at Carnegie-Mellon University describes this trend as the emergence of ultra-large-scale (ULS) systems (Northrop 2006). The report argues that these ULS systems will push far beyond the size of today's systems and systems of systems by every measure:

- Number of technological components of various kinds
- Number of people and organizations employing the system for different purposes

- Number of people and organizations involved in the development, maintenance, and operations of the systems
- Amount of data stored, accessed, manipulated, and refined
- Number of connections and interdependencies among the elements involved

The report argues further that the sheer scale of ULS systems will change everything: ULS systems will necessarily be decentralized in a variety of ways, developed and used by a wide variety of stakeholders with conflicting needs that it will be evolving continuously, and constructed from heterogeneous parts. Further, people will not just be users of a ULS system; they will be elements of the system. The acquisition of a ULS system will be simultaneous with its operation and will require new methods for control. These characteristics are, according to the report, emerging in today's systems of systems; in the near future, they will dominate.

The SEI report states that the scale of ULS systems presents challenges that are unlikely to be addressed adequately by incremental research within the established paradigm. Rather, *they require a broad new conception of both the nature of such systems and new ideas for how to develop them.* We will need to look at them differently, not just as systems or systems of systems, but as socio-technical ecosystems.

The growth in complexity has brought to researchers' attention novel mechanisms to cope with complexity, such as architectures and modularity and standards (Parnas 1972, Schmidt and Werle 1998, Baldwin and Clark 2000). Another, more recent, line of research has adopted a more holistic, socio-technical and evolutionary approach, putting the growth in the combined social and technical complexity at the centre of an empirical scrutiny (see, e.g., Edwards et al. 2007). These scholars view these complex systems as new types of IT artifacts and denote them with a generic label of information infra-structures (IIs) (e.g., Star and Ruhleder 1996, Hanseth and Lyytinen 2010, Tilson et al. 2010).

Hanseth and Lyytinen (2010) define an II, consistent with the aforementioned characterization of ULS, as a shared, open (and unbounded), heterogeneous, and evolving socio-technical system (called installed base) consisting of a set of IT capabilities and their users, operations, and design communities. Typical examples of IIs are the Internet, solutions supporting the interaction among manufacturers along a supply chain, and portfolios of integrated applications in organizations.

30.1.2 Key Characteristics of IIs

The most distinctive feature of IIs and ULSs is their overall complexity. Managing IIs, then, is about understanding and managing complexity. More specifically: IIs, like all complex systems, are evolving and not designed from scratch. So managing IIs means managing their evolution.

IIs are radically different from how information and software systems are presented in the literature. Infrastructures have no life cycle—they are "always already present." This is strictly true of some infrastructures, such as our road infrastructure that has evolved through modifications and extensions ever since animals created the first paths. IT infrastructures certainly have a much shorter history, but an II like the Internet has now been around and constantly evolving for roughly 40 years. The same is the case for IIs like portfolios of integrated information system (IS) (often numbering in the thousands) in larger organizations. Developing IIs, then, requires approaches that are different from the traditional "design from scratch" ones (Edwards et al. 2007, Hanseth and Lyytinen 2010, Tilson et al. 2010). It requires somewhat opposite approaches—it is about modifying (changing and extending) the installed base, existing IIs, so that the installed base evolves as far as possible toward what is desired (user requirements).

The evolution of IIs involves regularly a large number of actors. All these actors cannot be strictly controlled from one single point (like, for instance, a manager at the top of a hierarchically structured project organization). They will often act independently. In the case of the Internet, there are thousands, if not millions, of actors developing new services of the top of the Internet or adding new features to lower level services, like, for example, quality of service mechanisms. This means that even though there are many institutions (like ICANN, IETF, etc.) that are involved in the governance of the Internet, the Internet is not evolving in strictly planned or controlled way. Its evolution is mostly the aggregated result of the various autonomous actors' actions. Institutions having responsibility for the governance of the Internet can have impact and shape its evolution in a way similar to how we may influence the growth of an organism or a piece of land. The same is the case for large application portfolios. The individual applications and their relations (degree of integration) will change continuously, and no single actor will have a total overview of these changes. Accordingly, we call our approach to how we can shape the evolution of an II "*installed base cultivation*" (Hanseth and Lyytinen 2010).

If infrastructures are "always already present," are new infrastructures never emerging? Was not the Internet a new infrastructure at some point in time? Yes, of course. New IIs are indeed emerging. It happens primarily in two ways. One, a system may be growing in terms of number of users and along that path gradually changing from being a system (or application) of limited reach and range into a large scale II. E-mail, for instance, was introduced into the Internet (or rather the Arpanet) at a time when the net consisted of only four computers. Those computers (and the e-mail service) did not constitute an II, but a distributed system of limited complexity. But as the net was growing in terms of computers connected, services, developers, and users, it was taking on more and more the character of an infrastructure. The other way an II may emerge has been seen in most large organizations. Over time, the number of applications has continuously been growing at the same time as old and new applications have been increasingly more integrated. During this process, the application portfolio is growing from the first stand-alone one to a few loosely integrated ones and toward an increasingly more complex II.

The next section will present theories of complexity that help us understand the dynamics of IIs followed by a section giving a brief outline of what I see as the key challenges related to the management of the evolution of IIs and what kind of "tools" that are available for this task. Finally, I will present three examples of II evolution and how the choices of strategy and management tools have shaped the evolution of these IIs.

30.2 Socio-Technical Complexity and Theories of Complexity

In this section, I will present three theories of complexity that help us understand the core issues related to IIs.

30.2.1 Actor Network Theory

Actor network theory (ANT) was originally developed as a "toolbox" for studying processes from individual experiments to a situation where scientific facts or theories are universally accepted as such. This "toolbox" consists of a set of concepts for describing how relations between elements of various kinds are established in a process where heterogeneous networks are constructed. Central concepts in early ANT research are closure, enrolment, and alignment. Specifically, closure indicates a state where consensus emerges around a particular technology. Closure stabilizes the technology by accumulating resistance against change. It is achieved through a negotiation process and by enrolling actors/elements of various kinds into a network and translating (re-interpreting or changing) them so that the elements are aligned, that is, fit together, in a way that supports the designers' intentions.

The early ANT studies can be said to have focused on complexity in the sense that they spelled out the rich and complex relations between the scientific and the technological on the one hand, and the social on the other, related to the making of scientific theories and technological solutions. ANT has been used, for instance, in research on the negotiation of IS standards and the embedding of their local context of development and use (Star and Ruhleder 1996, Hanseth and Monteiro 1997, Timmermans and Berg 1997, Bowker and Star 1999, Fomin et al. 2003).

Since their emergence in the early 1980s, ANT and ANT research have evolved beyond their (so-called) "managerial" approach, which focuses on how a single actor-network is aligned by a dominating central actor (Law 2003b). Complexity has been addressed more explicitly as the focus has turned to the dynamics unfolding when independent actors try to align different but intersected actor-networks (Latour 1988, Star and Griesemer 1989, Law and Mol 2004, Law and Urry 2004, Law 2003a). This has happened as attention has moved toward more complex cases where order and universality cannot be achieved in the classical way.* These cases are described as "worlds," which are too complex to be closed and ordered according to one single mode or logic. There will only be partial orders, which are interacting in different ways, or interconnected and overlapping subworlds, which are ordered according to different logics. The interconnectedness of the subworlds means that when one is trying to make order in one subworld by imposing a specific logic, the same logic is making disorder in another—an order also has its disorder (Berg and Timmermans 2000, Law 2003b). Rather than alignment, stabilization, and closure, the keywords are now *multiplicities, inconsistencies, ambivalence, ambiguities, and fluids* (Law and Mol 2002, Law 2003a). Mastering this new world is not about achieving stabilization and closure, but rather about more ad hoc practices—"ontological choreography" of an ontological patchwork (Cussins 1998). This approach has been applied to studies of cases such as train accidents (Law 2003a), a broad range of high-tech medical practices (Mol and Berg 1998), and interdisciplinary research (Star and Griesemer 1989). This approach to complexity has also been applied to analyzing the challenges, not to say impossibility, of achieving closure and stabilization in relation to IIs and ICT standards (Aanestad and Hanseth 2000, Alphonse and Hanseth 2010).

30.2.2 Complexity Science

Just like the ANT community, the emergence of order has been in focus within the field of complexity science. However, while the first has focused on order making where humans are actively involved, the latter has had a focus on how order emerges without any "intelligent designer" involved like, for instance, the order found in complex organisms and "animal societies" like bee hives and ant heaps. But important contributions to complexity science have been made from studies of more social phenomena, in particular within the economy, like financial markets and also, an issue central to this book, standardization (David 1986, Arthur 1994). Complexity science is made up of a broad range of disciplines such as chaos theory and complex adaptive systems (CAS). CAS is concerned with the dynamic with which complex systems evolve through adaptation and is increasingly used in organizational studies, for example, in health care. CAS is made up of semiautonomous agents with the inherent ability to change and adapt to other agents and to the environment (Holland 1995). Agents can be grouped, or aggregated into meta-agents, and these can be part of a hierarchical arrangement of levels of agents. Agents can respond to stimuli—they behave according to a set of rules (schema).

Adaptation is the process whereby the agent fits into the environment and the agent as well as the CAS undergoes change. Adaptation—and creativity and innovation—is seen as being optimal at "the edge of chaos" (Stacey 1996), or more generally, adaptation occurs within the zone of complexity, which is located between the zone of stasis and the zone of chaos (Eoyang 1996). Dooley (1996) suggests that CAS behave according to three principles: order is emergent, the system's history is irreversible, and the system's future is unpredictable.

Overall, complexity science investigates systems that *adapt* and *evolve* as they *self-organize* through *time* (Urry 2003). Central to the emergence of orders are *attractors*, that is, a limited range of possible states within which the system stabilizes. The simplest attractor is a single point. There are also attractors with specific shapes, which are called "strange attractors," that is, unstable spaces to which the trajectory of dynamical systems are attracted through millions of iterations" (Capra 1996).

---

* John Law and Annamarie Mol (2002, p. 1) define complexity as follows: "There is complexity if things relate but don't add up, if events occur but not within the process of linear time, and if phenomena share a space but cannot be mapped in terms of a single set of three-dimensional coordinates." This definition is very brief and rather abstract, but is in perfect harmony with Cillier's definition presented earlier.

Orders emerge around attractors through various *feedback* mechanisms, including *increasing returns*, and through *path-dependent* processes of many small steps that may end in *lock-in* situations (David 1986). Some steps may be crucial in the sense that they may force the process in radically different (unexpected) directions. Such points are called tipping or bifurcation points (Akerlof 1970). The existence of such points makes the evolution of complex systems *nonlinear* in the sense that small changes in a system at one point in time may lead to hugely different consequences at a later point in time. Many of these concepts have been presented to the IS community by Shapiro and Varian (1999) under the label information economy. They will be explained in the subsequent sections.

30.2.2.1 Increasing Returns and Positive Feedback

Increasing returns mean that the more a particular product is produced, sold, or used, the more valuable or profitable it becomes. Infrastructure standards are paradigm examples of products having this characteristic. The development and diffusion of infrastructural technologies are determined by "the overriding importance of standards and the installed base compared to conventional strategies concentrating on programme quality and other promotional efforts" (Grindley 1995: 7).

A communication standard's value is to a large extent determined by the number of users using it— that is, the number of users you can communicate with if you adopt the standard. This phenomenon is illustrated by well-known examples such as Microsoft Windows and the rapid diffusion of the Internet. Earlier examples are the sustainability of FORTRAN and COBOL far beyond the time when they had become technologically outdated.

The basic mechanism is that the large installed base attracts complementary products and makes the standard cumulatively more attractive. A larger base with more complementary products also increases the credibility of the standard. Together these make a standard more attractive to new users. This brings in more adoptions, which further increases the size of the installed base, and so on (Grindley 1995: 27).

Further, there is a strong connection between increasing-returns mechanisms and *learning* processes. Increased production brings additional benefits: producing more units means gaining more experience in the manufacturing process, achieving greater understanding of how to produce additional products even more cheaply. Moreover, *experience* gained with one product or technology can make it easier to produce new products incorporating similar or related technologies. Accordingly, Shapiro and Varian (1999) see positive feedback as the central element in the information economy, defining information as anything that may be digitized. An information good involves high fixed costs but low marginal costs. The cost of producing the first copy of an information good may be substantial, but the cost of producing (or reproducing) additional copies is negligible. They argue that the key concept in the network economy is positive feedback.

30.2.2.2 Network Externalities

Whether real or virtual, networks have a fundamental economic characteristic: The value of connecting to a network depends on the number of other people already connected to it. This fundamental value proposition goes under many names: Network effects, network externalities, and demand-side economies of scale. They all refer to essentially the same point: Other things being equal, it is better to be connected to a bigger network than a smaller one (Shapiro and Varian 1999.)

Externalities arise when one market participant affects others without compensation being paid. In general, network externalities may cause negative as well as positive effects. The classic example of negative externalities is pollution: My sewage ruins your swimming or drinking water. Positive externalities give rise to positive feedback (Shapiro and Varian 1999).

30.2.2.3 Path Dependency

Network externalities and positive feedback give rise to a number of more specific effects. One such is *path dependence*. Path dependence means that past events will have large impacts on future development, and events that appear to be irrelevant may turn out to have tremendous effects (David 1986).

For instance, a standard that builds up an installed base ahead of its competitors becomes cumulatively more attractive, making the choice of standards "path dependent" and highly influenced by a small advantage gained in the early stages (Grindley 1995: 2). The classical and widely known example illustrating this phenomenon is the development and evolution of keyboard layouts, leading to the development and *de facto* standardization of QWERTY (David 1986).

We can distinguish between two forms of path dependence:

1. Early advantage in terms of numbers of users leads to victory
2. Early decisions concerning the design of the technology will influence future design decisions

The first one has already been mentioned earlier. When two technologies of a kind where standards are important—such as communication protocols or operating systems—are competing, the one getting an early lead in terms of number of users becomes more valuable for the users. This may attract more users to this technology, and it may win the competition and become a *de facto* standard. The establishment of Microsoft Windows as the standard operating system for PCs followed this pattern. The same pattern was also followed by the Internet protocols during the period they were competing with OSI protocols.

The second form of path dependence concerns the technical design of a technology. When, for instance, a technology is established as a standard, new versions of the technology must be designed in a way that is compatible (in one way or another) with the existing installed base. This implies that design decisions made early in the history of a technology will often live with the technology as long as it exists. Typical examples of this are various technologies struggling with the backward compatibility problem. Well-known examples in this respect are the different generations of Intel's microprocessors, where all later versions are compatible with the 8086 processor, which was introduced into the market around 1982.

Early decisions about the design of the Internet technology, for instance, have had a considerable impact on the design of new solutions both to improve existing services and to add new ones to the Internet. For example, the design of the TCP/IP protocol constrains how improved solutions concerning real-time multimedia transfer can be designed and how security and accounting services can be added to the current Internet.

30.2.2.4 Lock-In: Switching Costs and Coordination Problems

Increasing return may lead to yet another effect: *lock-in*. Lock-in means that after a technology has been adopted, it will be very hard or impossible to develop competing technologies. "Once random economic events select a particular path, the choice becomes locked-in regardless of the advantages of alternatives" (Arthur 1994). In general, lock-in arises whenever users invest in multiple complementary and durable assets specific to a particular technology. We can identify different types of lock-in: contractual commitments, durable purchases, brand-specific training, information and databases, specialized suppliers, search costs, and loyalty programmes (Shapiro and Varian 1999). We can also say that lock-ins are caused by the huge switching costs or by coordination problems (or a combination of these) that would be incurred when switching from one standardized technology to another.

Switching costs and lock-ins are ubiquitous in ISs, and managing these costs is very tricky both for sellers and buyers. For most of the history of computers, customers have been in a position where they could not avoid buying (more or less) all their equipment and software from the same vendor. The switching costs of changing computer systems could have been astronomical—and certainly so high that no organization did. To change from one manufacturer (standard) to another would imply changing all equipment and applications at the same time. This would be very expensive—far beyond what anybody could afford. But it would also be an enormous waste of resources, because the investments made have differing economic lifetimes, so there is no easy time to start using a new, incompatible system. As a result, others face switching costs, which effectively lock them into their current system or brand (Shapiro and Varian 1999).

Switching costs also go beyond the amount of money an organization has to pay to acquire a new technology and install it. Since many software systems are mission critical, the risks in using a new vendor, especially an unproven one, are substantial. Switching costs for customers include the risk of a severe disruption in operations.

Lock-in is not only created by hardware and software. Information itself—its structures in databases as well as the semantics of the individual data elements—is linked together into huge and complex networks that create lock-ins. One of the distinct features of information-based lock-in is that it proves to be so durable: equipment wears out, reducing switching costs, but specialized databases live on and grow, increasing lock-in over time (Shapiro and Varian 1999).

The examples of lock-ins and switching costs mentioned so far are all related to infrastructures that are seen as local to one organization. As infrastructures and standards are shared across organizations, lock-in problems become even more challenging.

Network externalities make it virtually impossible for a small network to thrive. But every network has to start from scratch. The challenge to companies introducing new but incompatible technology into the market is to build a network size that overcomes the collective switching costs—that is, the combined switching costs of all users. In many information industries, collective switching costs are the biggest single force working in favor of incumbents. Worse yet for would-be entrants and innovators, switching costs work in a nonlinear way: Convincing 10 people connected in a network to switch to your technology is more than 10 times as hard as getting one customer to switch. But you need all 10, or most of them: No one will want to be the first to give up the network externalities and risk being stranded. Precisely because various users find it so difficult to coordinate a switch to an incompatible technology, control over a large installed base of users can be the greatest asset you can have.

But lock-in is more than cost. As the community using the same technology or standard grows, switching to a new technology or standard becomes an increasingly larger *coordination* challenge. The lock-in represented by QWERTY, for instance, is most of all a coordination issue. It is shown that the individual costs of switching are marginal (David 1986), but, as long as we expect others to stick to the standard, it is best that we do so ourselves as well. There are too many users (everybody using a typewriter or PC/computer). It is impossible to bring them together so that they could agree on a new standard and commit themselves to switch.

30.2.2.4.1 Inefficiency

The last consequence of positive feedback we mention is what is called *possible inefficiency*. This means that the best solution will not necessarily win. An illustrative and well-known example of this phenomenon is the competition between the Microsoft Windows operating system and Macintosh. Macintosh was widely held to be the best technology—in particular from a user point of view—but Windows won because it had early succeeded in building a large installed base.

30.2.3 World Risk Society and Reflexive Modernization

Ulrich Beck provides an extensive analysis of the essence and impact of globalization processes. IIs, such as the Internet, are key elements in globalization processes at the same time as globalization generates increased demands for more globally distributed and integrated IIs. The key element in Beck's argument is that globalization radically changes our possibilities for controlling processes in nature and society, and then also the conditions and possibilities for managing IIs. The reason for this is, in short, the growth in complexity produced by globalization.

Beck builds his theory of (World) Risk Society on a distinction between what he calls first and second modernity. First modernity is based on nation-state societies, where social relations, networks, and communities are understood in a territorial sense. He argues further that the collective patterns of life, progress and controllability, full employment, and exploitation of nature that were typical in this first modernity have now been undermined by five interlinked processes: globalization, individualization, gender revolution, underemployment, and global risks. The kind of risks he primarily focuses on is those related to the ecological crises (climate change) and global financial markets.

At the same time, as the risks are becoming global, their origin is also globalization (Beck 1986, 1999, Giddens 1990, Beck et al. 1994). And, further, both on-going globalization processes as well as the outcome—that is, increased risks—are fundamentally related to modernity. Globalization is the form modernization has taken today, and risks are increasing because of the changing nature of modernity. The very idea of controllability, certainty, and security—which was so fundamental in the first modernity—collapses in the transfer to second modernity (Beck 1999, p. 2). Modernity's aim has been to increase our ability to control processes in nature and society in a better way through increased knowledge and improved technology. The concept of "world risk society" draws our attention to the limited controllability of the dangers we have created for ourselves. The reflexive modernization argument (at least in Beck's version) says that while it has been the case that increased modernization implied increased control, in second modernity, modernization, that is, enhanced technologies and an increase in bodies of knowledge, may lead to less control, that is, higher risks.

This is what lies at the heart of the "reflexivity" argument. In particular, the theory of reflexive modernization contradicts the instrumental optimism regarding the predetermined controllability of uncontrollable things: "the thesis that more knowledge about social life... equals greater control over our fate is false" (Giddens 1990: 43), and "the expansion and heightening of the intention of control ultimately ends up producing the opposite" (Beck et al. 1994: 9).

This shift, which may appear contradictory, can be explained by *the ubiquitous role of side effects*. Modernization means integration. At the same time, all changes and actions—new technologies introduced, organizational structures and work procedures implemented, and so on—have unintended side effects. The more integrated the world becomes, the longer and faster side effects travel, and the heavier their consequences. Globalization, then, means globalization of side effects. In Beck et al.'s (1994: 175, 181) own words: "It is not knowledge but rather non-knowledge that is the medium of reflexive modernization... we are living in the age of side effects... The side effect, not instrumental rationality, is becoming the motor of social change."

The theories of Risk Society and Reflexive Modernization can be seen as complexity theories in the sense that complexity is at their core. Seeing globalization as an integration process, for instance, can very well lead to it also being seen as a process of making the world more complex exactly through this integration process. The role attributed to side effects also links naturally to complexity. The role of side effects is increasing exactly because the "system" is becoming so complex. Because of its complexity, we cannot know how all components are interacting; accordingly, the outcomes of those interactions will be more unpredictable.

30.3 Design Challenges and Management "Tools"

30.3.1 Design Dilemmas

30.3.1.1 Tension between Standards and Flexibility

Infrastructures are made up of a huge number of components. Accordingly, standards defining the interfaces between components are essential features of IIs, and the specification and implementation of standards are important activities in the establishment of IIs. Standards are closely associated with stability. This is the case partly because keeping a standard stable is required for various reasons (e.g., data standards that make it possible to store data today and read them at a later time). But standards are also stable because widely diffused standards are so hard to change when we have to (because of the lock-in problem). However, standards need the change over time. Just like ISs, IIs need to change and adapt to changing user requirements. And some such changes mean that implemented standards need to change, too. At the same time, the scaling and growth of an II can generate need for changes in standards even though user requirements stay unchanged (Hanseth et al. 1996, Tilson et al. 2010).

Accordingly, IIs will evolve as a dynamic driven by a tension between standards (stability and uniformity) and flexibility (change and heterogeneity). Managing this tension is a key to the management of IIs (Hanseth et al. 1996).

30.3.1.2 Bootstrapping—Adaptability

The dynamic complexity of IIs poses a chicken–egg problem for the II designer that has been largely ignored in the traditional approaches. On one hand, IT capabilities embedded in IIs gain their value, as explained earlier, by being used by a large number of users demanding rapid growth in the user. Therefore, II designers have to come up early on with solutions that persuade users to adopt while the user community is non-existent or small. This requires II designers to address head on the needs of the very first users before addressing completeness of their design or scalability. This can be difficult, however, because II designers must also anticipate the completeness of their designs. This defines the *bootstrap problem of II design*. On the other hand, when the II starts expanding by benefitting from the network effects, it will switch to a period of rapid growth. During this growth, designers need to heed for unforeseen and diverse demands and produce designs that cope technically and socially with these increasingly varying needs. This demands infrastructural flexibility in that the II adapts technically and socially. This defines the *adaptability problem of II design* (Edwards et al. 2007, Hanseth and Lyytinen 2010). Clearly, these two demands contradict and generate tensions at any point of time in II design (Edwards et al. 2007).

30.3.2 Management "Tools"

So what kinds of "tools" are available for managing the evolution of IIs—or for installed base cultivation? The answer given here is: *process strategy, architecture,* and *governance regime*. The rationale behind the focus on these three aspects is, first of all, the fact that these three "tools" are what development efforts are all about: the steps to be taken to develop some new technology (bottom-up or top-down, incremental/ iterative or "big-bang," evolutionary and learning driven or specification driven, etc.); the architecture and overall design of the technology (the modularization of a system determines how and how easy it may be maintained and modified); and how to govern, manage, and organize the effort. In addition, these "tools" have also been at the centre of extensive discussions and research on the evolution of the Internet. These three factors have been three aspects as key factors behind its success: an experimental bottom-up development strategy (which includes a strategy and rules for bottom-up development and settlement of standards), the end-to-end architecture, and distributed control and governance structures combined with open source software licenses (Hanseth et al. 1996, Benkler 2006, Zittrain 2006).

Traditionally, the management of ICT has focused on the management of projects developing ICT solutions. Such projects are typically organized as a hierarchy of subprojects each with a subproject manager. Each manager has the rights to make decisions within the domain of the subprojects and give instructions to the managers at the level below. The management of such a project organization is normally supported by various management tools such as the detailed plans and establishment of milestones. And further, the production of detailed plans and the monitoring of the progress made in the project—if it is progressing according to the plan or not—are supported by various computer based project management tools. Together this package of project organization, decision rights, and management tools is an example of what I call governance regime. Governing the complexity of IIs requires new and different governance regimes. In the case of the Internet, for example, its successful evolution has been shaped by a governance regime consisting of a few central institutions like IETF and ICANN. Another important part of this regime has been the fact that most of its technology has been distributed based on open source license (like the GNU Public License). But may be the most important elements of the governance regime has been the organizing of the development activities as a loosely connected network of individuals that coordinate their work through extensive use of e-mail and by making all software and relevant information publicly available on FTP and Web servers.

Software and IS development often takes place by following specific methodologies. The central element of such methodologies is a specification of which steps to be taken, and in which sequence, to develop a specific IS solution. This is what I call a process strategy. The complexity of IIs requires process strategies different of those prescribed for traditional IS development efforts. In particular, we need process strategies that address the role of network externalities and path dependence; that is, we need strategies that address the bootstrap and adaptability problems.

The architecture of an IS is traditionally considered important for its maintenance. In general, modularization is an important strategy for coping with complexity. And in the case of IIs, the architecture plays a crucial role. This is illustrated with the role attributed to the Internet's architecture in explaining its successful evolution. The Internet's so-called end-to-end architecture (the functionality is located in the ends of the network, i.e., in the computers connected to the Internet, and not in the network itself, which has been the case within traditional telecommunication networks) has made the Internet extremely flexible in the sense that anybody having a computer connected to the Internet could develop and provide new services.

The management of IIs, then, requires process strategies, architectures, and governance regimes that in combination make an II evolve along the desired path. Exactly which combinations that is appropriate for specific IIs is still a major research issue.

30.4 Three Examples of Evolving IIs

I will here present three examples of IIs. They are all from the health care sector. They are presented as individual and separate IIs, but in reality, they will be linked together and with others and, accordingly, be parts of larger scale IIs in the health care sector. These examples illustrate the variety among IIs and, most importantly, how specific process strategies, architectures, and governance regimes are interacting and shaping the evolution of IIs.

30.4.1 Electronic Patient Records

30.4.1.1 Case Presentation

The introduction of electronic patient record (EPR) systems in hospitals is intended to improve the quality of patient care by replacing the existing fragmented and often unavailable paper-based patient records by an electronic one which would make any information instantly available to anybody, anywhere, and anytime. In 1996, the Medakis project was established with the aim to develop and implement an EPR system in all the five largest hospitals in Norway. Siemens was involved to take care of the software development. The system was given the name DocuLive. We focus here on the implementation at "Rikshospitalet" (see Hanseth et al. 2006), where DocuLive was intended to serve several ambitions: It should include all clinical patient information, covering the needs of all users; it should be built as one single integrated system; it should enable better collaboration and coordination of patient treatment and care through electronic information sharing and exchange. Finally, a more general and important aim, in the contexts of the arguments developed in this paper, was that the system should be a standard EPR solution for all Norwegian hospitals. This was an important goal for two reasons: enable information exchange and making it possible for health care personnel to use the EPR system without training when they are moving from one hospital to another. The deadline set for the delivery of the final system was the end of 1999. The project started with the best intentions of involving users, acknowledging current work practices, and favoring a bottom-up development strategy. However, as we illustrate in the subsequent sections, the difficulties associated with the standardization of both technology and work practices were dramatically underestimated.

Shortly after it began, project members became aware that within Siemens EPR projects were also underway in Sweden, the United Kingdom, Germany, and India. Moreover, they realized that the Norwegian project was not at the top of Siemens' priorities since Norway represented the smallest market.

This meant there was a high risk of overrun as Siemens prioritized more profitable markets. As a consequence, the project members decided to make moves toward "internationalizing" the project, first to the Scandinavian level and later to the European one. However, this decision weakened the consortium's position with respect to Siemens, since now the requirements from all projects from across countries would need to be merged and a new architecture designed.

In 1999, Siemens acquired a large U.S. software company developing software solutions for health care. As a consequence, Siemens' medical division's headquarters was moved from Europe to the United States, and the project's scope became global. As the number of involved users grew, large-scale participatory development became unmanageable. After a few years, only a small number of user representatives from each hospital continued to actively participate in the development. Moreover, the need to continuously find common agreements between the hospitals turned the intended bottom-up approach into a top-down one.

Also, the efforts aimed at solving the fragmentation problem with a complete and smoothly integrated EPR system turned out to be more challenging than foreseen. Paradoxically, the volume of paper records increased, and the patient record became even more fragmented for a variety of reasons. First, this was because new laws on medical documentation required detailed records from professional groups not previously obliged to maintain a record (such as nurses, physiotherapists, and social workers). Second, for both practical and legal reasons, the hospital had to keep updated versions of the complete record. As long as lots of information only existed on paper, the complete record had to be paper-based. Thus, each time a clinical note was written in the EPR, a paper copy was also printed and added to the paper record. Printout efficiency was not a design principle for the current EPR, causing non-adjustable print layouts that could result in two printed pages for one electronic page form. Third, multiple printouts of preliminary documents (e.g., laboratory test results) were often stored in addition to final versions. The result was that the volume of paper documents increased. This growth created a crisis at the paper record archival department. The hospital had moved into new facilities designed with a reduced space for the archive as it was supposed to handle electronic records only. In 2003, the archive was full, and more than 300 shelf meters of records were lying on the floors. This situation also affected the time needed to find records, and often requests failed to be satisfied.

When the implementation of DocuLive started, five local systems containing clinical patient information existed. The plan was to replace these with DocuLive so as to have the EPR as one single integrated IS. In spite of this, the number of local systems was growing at an accelerating speed, based on well-justified needs of the different medical specialties and departments. For example, the in vitro fertilization clinic needed a system that allowed them to consider a couple as a unit, as well as allow tracking of information from both semen and egg quality tests through all procedures involved, up to the birth of the child. The intensive care unit acquired a system that allowed them to harvest digital data from a vast array of medical equipment and thus eliminate the specialized paper forms previously used to document events and actions. Moreover, new digital instruments in use in many different departments include software components with medical record functionality. The number of such specialized systems had grown from 5 in 1996 to 135 in 2003.

The original plan said that the final system should be delivered in 1999. Four years later, toward the end of 2003, the version of DocuLive in use included information types covering between 30% and 40% of a patient record. This meant that the paper records were still very important, but unfortunately more fragmented and inaccessible than ever. The increased fragmentation was partly due to not only the large volume of paper caused by DocuLive printouts but also the high number of specialized systems containing clinical patient information.

At Rikshospitalet they decided to change their strategy and approach complexity in quite different way in 2004. They realized that the idea of one complete EPR system had failed. Instead, they decided to "loosely couple" the various systems containing clinical patient information underneath a "clinical portal" giving each user group access to the relevant information in a coherent way.

30.4.1.2 Another EPR Case

I will here briefly present another EPR case as a contrast to the aforementioned one. Around the same time as the Medakis project started, a small and simple application supporting a few work tasks for a couple of medical doctors was developed by a software developer at a smaller hospital in Northern Norway. The users were very happy with the system. After a period of use, they asked for some added functionality, which was implemented. At the same time, other users became aware of the system and wanted to try using it as a tool supporting some of their work tasks. This was the first iteration of a long evolutionary process. The functionality, number of users, and developers involved started to grow. After a few years, the software and the developers were taken over by a company set up for this task. The company and the software product was given the name DIPS. It has continuously been growing and is today standard EPR system in Norway (Ellingsen and Monteiro 2008). In September 2012, the Oslo University Hospital (which Rikshospitalet is a part of) decided to replace DocuLive with DIPS. This means that four of the five hospitals involved in the Medakis project will have abandoned DocuLive in favor of DIPS.

30.4.1.3 Case Analysis

How can the project trajectory of DocuLive at Rikshospitalet be analyzed? Key issues are the complexities involved and the handling of these complexities. The complexity of the Medakis project and the DocuLive system was made up of a mix of technical, organizational, and medical issues—that is, a very complex actor-network. Further, this case demonstrates the limitations of traditional strategies for managing software development projects and how these strategies lead to actions triggering lots of side effects. These side effects are propagating throughout the actor-network, creating domino- and boomerang effects, which are returning to the origin of the initial actions so that the final results became opposite to what one aimed for. In this way, the case demonstrates the relevance of the theory of reflexive modernization in the sense that strategies aiming at control of the development process leads to less control.

The primary complexity involved is that of the work practices related to patient treatment and care. By trying to make one integrated system that should cover the needs of all hospitals in Norway, one brings together the total complexity of these practices. The hospitals realized that dealing with this was demanding. Accordingly, they concluded that they had to involve a supplier that was strong both financially and in terms of ICT competence. So Siemens was chosen. But then, the complexity of Siemens was merged with that of the hospitals. The medical division within Siemens is large, with a traditional base within medical imaging technologies. The imagining instruments had become digital, and supplementary software systems had been built. As the EPR development activities were increasing within Siemens, it became more and more important to align and integrate the EPR strategy and product(s) with other Siemens products and strategies.

Within this world, Norway become marginal, as the appetite for larger markets escalated in a self-feeding process. A side effect of the expansion of ambitions and scope was the increased complexity: the larger the market Siemens was aiming at, the more diverse the user requirements, and accordingly, the more complex the system had to be in order to satisfy them. This implied that the development costs were growing, which again implied that a larger market was required to make the whole project profitable. The project went through this spiral of self-reinforcing and escalating complexity several times until it collapsed.*

The failure of DocuLive can be explained in more theoretical terms as a failure in attempting to control complexity. Arguably, the main mistake was to follow a "traditional" IS development approach—typical

---

* At Rokshospitalet, they decided in 2003 to change their strategy quite dramatically. In this new strategy, the central element was loose coupling of the various systems through a portal, which was giving various user groups coherent interfaces to the systems they needed to access. This strategy has been much more successful—but not without its own challenges.

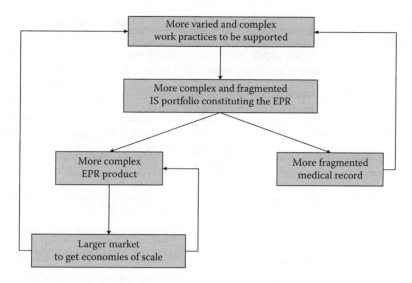

FIGURE 30.1   The reflexive standardization process.

for (first) modernity, that is, overemphasizing criteria of universality, uniformity, and centralization of control to achieve alignment, stabilization, and closure. In line with the more recent developments within actor network theory, our case data suggest that the complexity defines II development as the emergence of multiplicities, inconsistencies, ambivalence, and ambiguities (Law and Mol 2002, Law 2003b). Ironically, what happened became the opposite of the initial aims. When actors tried to stabilize the requirement and design specification by enrolling more actors, this made it less stable. Attempts to improve fragmented records by means of one integrated EPR made the records more fragmented. The complexity of DocuLive turned out to be one where the ordering efforts created disorders. The side effects triggered new ones, which again were reflected back on the origin—the standardization process turned out to be reflexive and self-destructive. The dynamics of reflexive processes at work are summarized in Figure 30.1.

The concept of reflexivity offers thus an interpretation of the dynamics of the case. The theory of "high" modernity helps to observe how the logics of the "first" industrial modernity find their limits (Beck et al. 1994). The intensified interconnectedness of social practices with technical artifacts on the one hand, and the need to align geographically dispersed actors on the other hand, effectively undermines the reductionist approach to control complexity. The weaknesses of such approach become visible when the control itself reflexively reproduced the complexity—thus creating the immanent paradox of modernity. This fact highlights the need to develop alternative standardization approaches that better overcome the paradoxes to deal with complexity.

The case is also an example of self-reinforcing processes driving the dynamics of complex systems. In this case, what the most salient self-reinforcing process is that of the escalation of the complexity.

I will now zoom in a bit more on the specific "management tools" chosen and their impact on the evolution of the DocuLive solution and the Medakis project. Most important is the principal architectural decision made: The solution should be a tightly integrated (or rather monolithic) one. On this basis, it was decided to organize the development activities as a classical hierarchical organization controlled from the top by a project manager and to develop the solution according to a top-down, specification-driven process. The choice of process strategy and governance regime was largely given by the choice of architecture. The important role played by the architecture is illustrated by Rikshospitalet's decision to switch to the almost exact opposite strategy in 2004, that is, an architecture allowing the implementation and use of a large number of applications loosely coupled through a portal. This strategy, then, implied a more evolutionary approach (i.e., process strategy) to the development of individual applications as well as the overall II, and this evolutionary process was managed by means of a network oriented governance structure in terms of a number of loosely coupled projects. This strategy turned out to be much more successful, and the paper-based patient record was replaced within a couple of years.

The history of DIPS is almost the exact opposite of the Medakis project and DocuLive. It started out as a very small and simple solution design to support a very narrow range of works tasks and very few users. The very first step taken was successful. Learning from and capitalizing on this, the solution has been growing into a national Norwegian EPR II. This process strategy illustrates very well how IIs may bootstrap: as the installed base of an II grows, the use value of the II increases which, in turn, attracts more users in a self-reinforcing spiral. The architecture of DIPS, however, shares its basic features with DocuLive: it is a tightly coupled system. But the bottom-up evolutionary process strategy chosen made it possible to make the solution over time grow into one satisfying the required range of user requirements of a national II. The governance regimes have some common features (proprietary software delivered by a commercial company), but are also different in important areas. The Medakis project was set up as a national and hierarchical organization including representatives from each of the five hospitals. The DIPS governance regime had a more entrepreneurial flavor. The strategy for how a DIPS-based II should grow toward a national one has been determined by the company. User organizations have primarily been involved as customers and as individual hospitals (later regional companies) related to individual implementation projects at each hospital.

30.4.2 Information Exchange between Health Care Institutions

30.4.2.1 Case Presentation

The second case presented is an example of an II supporting information exchange between health care institutions involved in diagnosing and treatment of patients. We will look at one such project, called the Elin project, which aimed at establishing electronic co-operation between the general practitioners (GPs) and other actors in the healthcare sectors, such as hospitals, pharmacies, and welfare authorities.

The aim of the project was first to develop comprehensive requirements specifications as a basis for user friendly, standardized solutions for electronic healthcare-related communications for GPs. The main aim was "better communication and collaboration, and not just development of technical solutions for message exchange." This included the development of solutions for exchange of admission and discharge letters, laboratory orders and reports, illness and doctor's declarations, prescriptions, and communication with patients. The project had a strong focus on standards representing the main information objects, like laboratory orders and reports, admission and discharge letters, and prescriptions.

The project was split into three phases. In the first, the focus was on exchange of discharge letters between GPs and hospital departments and outpatient clinics. In the second phase, the focus was on exchange of discharge letters between medical specialists' offices and GPs, exchange of orders and reports between radiology laboratories to GPs, and information exchange with patients. The third phase focused on improving and piloting the technical solutions. The II's architecture was derived from what we can call the "EDI paradigm." That means that the overall II would be built by adding functionality to the applications the users were using, like the GPs' patient record systems for sending and receiving the specified messages. This means that all vendors of applications within the use area need to be involved and agree on the specifications and implementing them.

A number of challenges surfaced. For example, the existing EDIFACT standards did not fit well with the defined requirements, and for some messages, standards were not yet available. Another challenge was that the exchange of the messages took too long time for various (and often mysterious) reasons. Accordingly, new standards and messages had to be specified based on a web services model but within the framework of the EDI paradigm and its architecture.

The project has played a major role in the development of user requirement for ICT solutions supporting communication between GPs and other healthcare institutions, which are well aligned with user needs and requirements from healthcare authorities. The project also specified a standardized architecture for solutions for information exchange between GPs and hospitals (Aksnes 2006). It was quite successful in establishing strong, enthusiastic collaborative networks of users and suppliers. However, the implementation and diffusion of solutions have definitely been very, very slow. In spite of this, the approach taken by the Elin project has been considered a great leap forward compared to previous efforts within the same domain because it had a stronger focus on user requirements and less on specification of technical standards only. For this reason, it triggered a series of follow-up projects in related areas. But the success of these projects in terms of developing solutions that are adopted by the intended users is modest.

30.4.2.2 Case Analysis

The overall approach taken by the Elin projects is similar to Medakis, and so are the outcomes. But there are also differences. The overall complexity of the work practices one aimed at supporting is significantly lower. But the most important difference is the complexity of the organization developing the Elin IIs, that is, the overall complexity of the actor-network created by the chosen architecture. The organization of the development work was a direct side effect of the choice of architecture.

Even though the number of people involved was rather small, the project involved many different and totally independent vendor organizations. The way the development work was organized was a direct implication of the architecture chosen, that is, the EDI model where the information exchange is implemented as add-ons to and tightly integrated with the applications in use. This work can hardly be done by others that the applications vendors. So all of them needed to be involved, they had to agree upon quite complicated technical specifications, and each of them needed to implement the specifications "correctly." The experience is that this requires tight coordination between a large number of independent actors—in fact, tighter coordination that what is possible.

30.4.3 Telemedicine in Ambulances

30.4.3.1 Case Presentation

The last case is the implementation of a telemedicine solution in ambulances. This is a successful case—largely due to the bootstrapping strategy followed (Hanseth and Aanestad 2003).

Østfold is the south-easternmost county in Norway, with a population of around 250,000 on an area of 3600 km². Until January 1, 1998, there were five independent public hospitals in the county, each with their own fully equipped emergency care unit (ECU). In a process of rationalization and centralization, the hospitals were merged into one organization. The number of ECUs was reduced to two, which created public concern over increased transport time and possible loss of lives in emergency situations. Also time spent within the hospital before treatment is started ("door-to-needle-time") is generally longer in large than in small hospitals. For example, for myocardial infarction, treatment should preferably be given within an hour ("the golden hour"). Usually myocardial infarction is treated with thrombolytic agents that dissolve the blood clots; however, this should not be given in certain cases as it may cause massive bleedings. The procedure was thus to wait with thrombolytic agents until the patient was brought form the ECUs to the heart intensive care unit and a doctor had verified the diagnosis of myocardial infarction using electrocardiography (ECG) equipment. This means that in many cases the patients do not reach doctors that can give this important treatment within an hour—unfortunately with serious consequences.

In 1996, prior to the merger, one doctor had heard of MobiMed, a telemedicine systems that facilitated transmission of text and ECGs from ambulances to a receiver (e.g., in a hospital). He and some ambulance drivers went to look at the system in use in Sweden. When approaching the county's health administration, the doctor did not manage to convince them to support this financially, but he got a permission to try out the system. The vendor lent the equipment, and in February 1998, two senders were installed in two ambulances, and a receiver was installed in the cardiology ward in the hospital where these ambulances would bring their patients. The first aim was to transmit ECG to the cardiology ward for interpretation by a doctor. A verified diagnosis of myocardial infarction would allow the ambulance personnel to bypass the ECU and bring the patient directly to the heart ICU.

The ambulances in this town were selected because there was here a rather high likelihood that there would be an anesthesia nurse available for emergency trips. Anesthesia nurses were trained in using the equipment, and the practical testing started. During 1998, it was used on 166 patients; of these, 16 had an infarction. In these cases, the ECU was bypassed and the "door-to-needle time" was reduced between 25 and 30 min. The ambulance personnel were also taught how to use an ECG recorder, and they practiced on each other during quiet time periods. After some time, they were also allowed to use the equipment when there were no nurses in the ambulances.

In January 1999, the ambulance nurses were allowed to administer the thrombolytic medication (i.e., giving the patient the shot) after the diagnosis was verified by the doctor on duty at the hospital, and the total time from the patient became ill to the medication was given ("call-to-needle-time") was further reduced by 25–30 min. Sometime later the ambulance personnel were also trained and allowed to administer the medication.

Based on the success with the first two ambulances and the benefits and savings that they could demonstrate, the county's health authorities decided to support the purchase of the equipment also for ambulances from the other towns that did not have an ECU. Here, the equipment was installed in April 2000, and the transition from just bypassing the ECU to also give medication in the car came about in just 6 months (October 2000), as compared to about 1 year in the first case.

During 1999, the Norwegian health authorities were approving the practice where nurses were administering the medication after a verified diagnosis. Up to that time, they could do it as a task delegated by the doctor setting the diagnosis. Such an approval from the authorities was necessary to scale up the activities. Later on, a similar authorization was given to the ambulance personnel and how to do it was included into their regular education.

By December 2001, it was decided to purchase senders also for the rest of the ambulances, including the towns *with* an ECU. The aim was then not to administer medication in the ambulance but to bypass the ECU when appropriate. Similar projects were started in several other counties and the extension of the system for also supporting the diagnosis of brain stroke was under discussion.

30.4.3.2 Case Analysis

This case demonstrates, first of all, that substantial benefits may be obtained with limited resources when an appropriate strategy is chosen. In the two previous cases, the chosen strategy generated an unmanageable complexity. Like the DIPS case, this one followed a strategy keeping the complexity at lower level. But also this case was complex in the sense that it included a rather complex mix (actor-network) of technological, medical, and legal issues. First of all, this case demonstrates the importance of an evolutionary and learning centered approach, which balances short-term gains and long-term strategic aims. This evolutionary strategy was crucial in order to prove that it was possible for first nurses and later on other ambulance personnel to give the shots to patients in a safe way. This was achieved within a framework of collaborative arrangements between personnel in the ambulances and cardiologists in the hospital that were developed over time in combination with the overall II that was established. These collaborative arrangements were developed in tandem with new rules and regulations that made these arrangements legal at the same time as they both (regulations and collaborative arrangements) were based on solid practical experience proving them safe.

The evolutionary process strategy chosen was made possible by key features of the technological architecture: a very simple technological solution, which was loosely coupled to the larger infrastructure in terms of the existing equipment in the ambulances and the EPR solution and overall IT infrastructure in the hospitals. Lastly, the strategy was also supported by the entrepreneurial governance regime:

The II evolution was primarily driven by a few enthusiastic doctors. This lean governance regime also contributed to making the smooth evolution of the II possible.

30.5 Conclusion

Overall the complexity of ICT solutions, individually and through integration with others, has reached a level of complexity that demands radically new approaches to the way we develop and manage them. Such new approaches also need to be built on new concepts. In this chapter, the concept of IIs is used in order to capture the complexities ICT developers and managers are confronted with, the key characteristics of these complexities, and the challenges they raise regarding their development and management. The chapter also points to some ways of coping with these challenges.

The key characteristic of IIs is, of course, their complexity. Complex systems are evolving in a way largely driven by side effects. So managing IIs needs to focus on these side effects—how to avoid creating them but at the same time draw upon them. In addition to their complexities, the key characteristic of IIs is the fact that they are not designed from scratch and that they do not have a life cycle—they just evolve. So approaches to coping with their complexities need to focus on how to shape their evolution—*cultivating the installed base*. In this chapter, three "tools" are presented, which may help us in managing complex IIs: process strategies, architectures, and governance regimes. Managing IIs, then, requires the choice of an appropriate combination of these tools. We have illustrated how some such combinations shape the evolution of three IIs within the healthcare sector. But we have still a long way to go before we have the "toolbox" required for proper management of IIs. Unfortunately, we might never reach that stage because as we are developing better "tools," the complexity—and the challenges for management—is growing. Substantial research into this critical issue is desperately needed. In my view, what is most in demand is research on how specific combinations of the management "tools" mentioned here in interaction shape the evolution of IIs. Based on such research, better advice can be given on how specific "tools" should be selected for specific IIs.

References

Aanestad, M. and Hanseth, O. (2000). Implementing open network technologies in complex work practices. A case from telemedicine. In: *Proceedings from IFIP WG 8.2 International Conference. The Social and Organizational Perspective on Research and Practice in Information Technology*, June 10–12, 2000, Aalborg, Denmark.

Akerlof, G. A. (1970). The Market for "Lemons": Quality Uncertainty and the Market Mechanism. *The Quarterly Journal of Economics*, 84(3), 488–500.

Aksnes, B. (2006). Samhandlingsarkitektur for helsesektoren. City: KITH: Trondheim, Norway. http://www.kith.no/upload/2949/Samhandlingsarkitektur.pdf (accessed on March 26, 2013).

Alphonse, J. and Hanseth, O. (2010). Knowledge, technology and fluids-developing a model to view practices of knowledge and technology in the telecom industry. In: *Proceedings of the Ninth International Conference on Mobile Business and 2010 Ninth Global Mobility Roundtable (ICMB-GMR), Athens, Greece, 2010*. June 13–15, 2010.

Arthur, W. B. (1994). Increasing returns and path dependence in the economy. Ann Arbor, Michigan: University of Michigan Press.

Baldwin, C. and Clark, K. (2000). *Design Rules*, MIT Press, Cambridge, MA.

BCS/RAE (2004). The challenges of complex IT projects, British Computer Society and Royal Academy Engineering Project [www document], http://www.bcs.org/upload/pdf/complexity.pdf (accessed August 2009).

Beck, U. (1986). *Risk Society: Towards Another Modernity*, Routledge, London, U.K.

Beck, U. (1999). *World Risk Society*, Polity Press, Cambridge, U.K.

Beck, U., Giddens, A., and Lash, S. (1994). *Reflexive Modernization*, Polity Press, Cambridge, U.K.

Benkler, Y. (2006). *The Wealth of Networks. How Social Production Transforms Markets and Freedom*, Yale University Press, New Haven, CT.

Berg, M. and Timmermans, S. (2000). Orders and their others: On the construction of universalities in medical work. *Configurations* 8, 31–61.

Bowker, G. and Star, S. L. (1999). *Sorting Things Out. Classification and Its Consequences*, MIT Press, Cambridge, MA.

Capra, F. (1996). *The Web of Life: A New Scientific Understanding of Living Systems*, Harper Collins, London, U.K.

Cussins, C. (1998). Ontological Choreography: Agency for women patients in an infertility clinic. In *Differences in Medicine: Unravelling Practices, Techniques and Bodies*, M. Berg and A. Mol (eds.), Duke University Press, Durham, NC, 166–201.

David, P. A. (1986). Understanding the economics of QWERTY. In: Parker, W. N. (ed.) *Economic History and the Modern Economist*, Basil Blackwell, London, U.K.

Dooley, K. (1996). Complex adaptive systems: A nominal definition, http://www.public.asu.edu/~kdooley/papers/casdef.PDF (accessed August 13, 2009).

Edwards, P., Jackson, S., Bowker, G., and Knobel, C. (2007). Report of a workshop on history and theory of infrastructures: Lessons for new scientific infrastructures, School of Information, University of Michigan, Ann Arbor, MI, http://www.si.umich.edu/InfrastructureWorkshop/documents/UnderstandingInfrastructure2007.pdf (accessed March 15, 2007).

Ellingsen, G. and Monteiro, E. (2008). The organizing vision of integrated health information systems. *Health Informatics Journal* 14(3), 223–236.

Eoyang, G. (1996). "Complex? Yes! Adaptive? Well, maybe..." *Interactions* 3(1), 31–37.

Fomin, V., Keil, T., and Lyytinen, K. (2003). Theorising about standardization: Integrating fragments of process theory in light of telecommunication standardization wars. *Sprouts: Working Papers on Information Environments, Systems and Organizations*.

Giddens, A. (1990). *Consequences of Modernity*, Polity Press, Cambridge, MA.

Grindley, P. (1995). *Standards, Strategy, and Politics. Cases and Stories*, Oxford University Press, New York.

Hanseth, O. and Aanestad, M. (2003). Bootstrapping networks, infrastructures and communities. *Methods of Information in Medicine* 42, 384–391.

Hanseth, O. and Ciborra, C. (2007). *Risk, Complexity and ICT*, Edward Elgar Publishing, London, U.K.

Hanseth, O., Jacucci, E., Grisot, M., and Aanestad, M. (2006). Reflexive standardization: Side effects and complexity in standard making. *MIS Quarterly* 30(2), 563–581.

Hanseth, O., and Lyytinen, K. (2010). Design theory for dynamic complexity in information infrastructures: the case of building internet. *Journal of Information Technology*, 25(1), 1–19.

Hanseth, O. and Monteiro, E. (1997). Inscribing behaviour in information infrastructure standards. *Accounting Management & Information Technology* 7(4), 183–211.

Hanseth, O., Monteiro, E., and Hatling, M. (1996). Developing information infrastructure: The tension between standardization and flexibility. *Science, Technology and Human Values* 21(4), 407–426.

Holland, J. (1995). *Hidden Order*, Addison-Wesley, Reading, MA.

Kallinikos, J. (2007). Technology, contingency and risk: The vagaries of large-scale information systems. In: Hanseth, O. and Ciborra, C. (eds.) *Risk, Complexity and ICT*, Edward Elgar Publishing, London, U.K., 46–74.

Latour, B. (1988). *Science in Action*, Harvard University Press, Cambridge, MA.

Law, J. (December, 2003a). Landbroke grove, or how to think about failing systems, http://www.lancs.ac.uk/sociology/papers/law-ladbroke-grove-failing-systems.pdf, (accessed on March 26, 2013).

Law, J. (November, 2003b). Traduction/trahison: Notes on ANT, http://cseweb.ucsd.edu/~goguen/courses/175/stslaw.html (accessed on March 26, 2013).

Law, J. and Mol, A. (2002). *Complexities: Social Studies of Knowledge Practices*, Duke University Press, Durham, NC.

Law, J. and Urry, J. (2004) Enacting the social. *Economy and Society*, 33(3), 390–410.

Mol, A. and Berg, M. (1998). Introduction, In: *Differences in Medicine: Unravelling Practices, Techniques and Bodies*, Berg, M. and Mol, A. (eds.), Duke University Press, Durham, NC, pp. 1–12.

Northrop, L., Feiler, P., Gabriel, R.P., Goodenough, J., Linger, R., Longstaff, T., Kazman, R., Klein, M., Schmidt, D., Sullivan, K., and Wallnau, K. (2006). *Ultra-Large-Scale Systems: The Software Challenge of the Future*, Carnegie Mellon Software Engineering Institute. http://www.sei.cmu.edu/library/assets/ ULS_Book20062.pdf (accessed on March 29, 2013).

Parnas, D. L. (1972). A technique for software module specification with examples. *Communications of the ACM* 15(5), 330–336.

Schmidt, S. K. and Werle, R. (1998). *Coordinating Technology. Studies in the International Standardization of Telecommunications*, MIT Press, Cambridge, MA.

Shapiro, C. and Varian, H. R. (1999). *Information Rules: A Strategic Guide to the Network Economy*, Harvard Business School Press, Boston, MA.

Stacey, R. D. (1996). *Complexity and Creativity in Organisations*, Berrett-Koehler, San Francisco, CA.

Star, S. L. and Griesemer, J. R. (1989). Institutional ecology, "Translations" and boundary objects: Amateurs and professionals in Berkeley's Museum of Vertebrate Zoology, 1907–39. *Social studies of Science* 19, 387–420.

Star, S. L. and Ruhleder, K. (1996). Steps toward an ecology of infrastructure: Design and access for large information spaces. *Information Systems Research* 7(1), 111–134.

Tilson, D. and Lyytinen, K., and Sørensen, C. (2010). Research commentary-digital infrastructures: The missing IS research agenda. *Information Systems Research* 21(4): 748–759. ISSN 1047-7047.

Timmermans, S. and Berg, M. (1997). Standardization in action: Achieving local universality through medical protocols. *Social Studies of Science* 27, 273–305.

Urry, J. (2003). *Global Complexities*. Polity Press, Cambridge, U.K.

Zittrain, J. (2006). The generative Internet. *Harward Law Review* 119, 1974–2040.

31

Impact of Culture on Information Systems Design and Use: A Focus on E-Business

Dianne Cyr
Simon Fraser University

31.1 Introduction

For many years, the impact of culture on information systems (IS) design and use has been recognized. Based on cross-cultural research, Ein-Dor et al. (1993) assess that IS are influenced by socio-psychological, demographic, and economic factors. The various factors are integrated into a research framework in which the most important finding is that culture has the potential to impact technical and procedural elements of IS greatly. Accordingly, there are numerous authors who suggest that a viable agenda for research links prior literature on international business and IS (Deans and Ricks, 1993; Niederman et al., 2012; Sagi et al., 2004).

   A recent article by Niederman et al. (2012) is useful for assessing key topics in global information management. Based on a decade of research published in the *Journal of Global Information Management*, the authors designate 11 categories or topics that evolve from their investigation. These topics include: (1) adoption and diffusion of technology; (2) cultural variations; (3) e-commerce; (4) e-government; (5) information technology (IT) in developing countries; (6) IT in multinational firms to facilitate global operations and decision-making; (7) knowledge management; (8) managerial and worker actions, beliefs, and values; (9) national infrastructure; (10) offshoring of IS tasks; and (11) IS support of virtual teams involving multiple cultures.* While these topics are not exhaustive, they do represent a snapshot of the field as evidenced by research published in a journal exclusively focused on global information management. Niederman et al. further note that there are overlapping concerns of international business and IS and that the relationship between international business thinking and global information management is complex. Each of these 11 topics is expansive and inclusion of them all within the scope of this chapter is not possible. However, extracting from these themes, the following pages are focused on areas designated by Niederman et al. (2012) as most significant and where there is greatest representation of research. These areas include e-commerce (or e-business as the terminology used in this chapter) and culture.

Studies of e-business consider characteristics of consumer use of IS functions and website offerings, user differences across cultures, as well as the impact of website design on e-business use. More specifically, this chapter will emphasize website design as the medium of communication in e-business. For several years, researchers have examined the effectiveness of website design related to culture, and the subsequent impact on user trust and e-loyalty in e-business (Belanche et al., 2012; Cyr, 2008a; Flávian et al., 2006; Sia et al., 2009; Smith et al., 2011). These investigations are a reasonable proxy for the impact of culture on IS design and use more generally. Therefore, this chapter aims to contribute to understanding website design elements, which facilitate e-business success in diverse cultural settings. Since a primary goal of e-business vendors is to solicit trust of online users, this topic is likewise discussed.

The chapter begins with a brief overview of e-business and national culture. This is followed by a discussion and research findings about culture and website design. More specifically, we focus on how elements of information design, navigation design, and visual design are influenced by culture. Since trust is central to e-business, this topic also receives attention in a cross-cultural context. In addition to the focus on research, the chapter covers the impact of culture and website design on practice. The chapter concludes with suggestions for future research challenges and new directions for research.

31.2 The Global E-Business Phenomenon

The rapid increase of Internet users globally is staggering. Equally impressive is the surge in the number of people shopping or searching for product information online. As of December 31, 2011, Internet users are over 2.2 billion strong, which represents 528.1% growth since 2000 (Internet World Stats, 2011). Of these Internet users, the majority reside in Asia (44.8%), followed by Europe (22.1%), North America (12.0%), Latin America/Caribbean (10.4%), Africa (6.2%), the Middle East (3.4%), and Oceania/Australia (1.1%). Remarkable statistics are the increase in penetration and growth in certain regions. For instance, in Africa the growth of Internet users is 2988% and in the Middle East growth is 2244% in the period from 2000 to 2011. The relative number of Internet users in North America is declining as the number of users in other geographical locations grows.

In 2008, over 875 million consumers had shopped online, up 40% from 2006. Among users with Internet access utilized for shopping, the highest percentage of online shoppers is found in South Korea (99%), United Kingdom (97%), Germany (97%), Japan (97%), and the United States (94%). The most popular and purchased items are: books (41%), clothing/accessories/shoes (36%), videos/DVDs/games (24%), airline tickets (24%), and electronic equipment (23%) (Multilingual Search, 2008). Based on *Internet Retailer*'s 2011 Top 500 Guide, one trend is evident—web-only merchants took business away from the rest of the market in 2010. Combined revenue for 87 merchants indicated that annual web sales increased 32.9% to $46.53 billion in 2010 from $35 billion in 2009. In comparison, according to the National Retail Federation, total retail sales grew year over year about 3% to $2.4 trillion in 2010 from $2.33 trillion in 2009 (Internet Retailer, 2011).

Finally, it was reported earlier that the largest segment of Internet users reside in Asia. With respect to global e-business, China is by far the world's largest online market. A September 2011 study of online

---

* For an elaboration of each of these topics, as well as observed relationships between them, please see Niederman et al. (2012).

buyers conducted by PricewaterhouseCoopers determined that 86% of China's nearly 200 million online shoppers consider themselves as experts in online shopping. This is compared to 72% of online shoppers in the United States and 70% in the United Kingdom. Chinese shoppers make 8.4 purchases online per month, compared to 5.2 purchases in the United States, or 2.4 in France and the Netherlands (Multilingual Search, 2012). There is no question that e-business is a growing global phenomenon, representing unique cultural requirements.

31.3 What Is Culture?

Culture has implications for Internet usage and affects e-commerce trust (Gefen and Heart, 2006; Jarvenpaa et al., 1999), B2C attitudes (Tan et al., 2007), website development (Junglas and Watson, 2004; Sun, 2001), information and communication technology adoption (Erumban and Jong, 2006), and Internet marketing (Tian and Emery, 2002). There are differences in online communication strategies for target markets between Japan, Spain, and the United States (Okayazaki and Rivas, 2002). Further, differences exist between cultures concerning web interface acceptance and preferences for design features (Evers and Day, 1997; Vyncke and Brengman, 2010). Specific to the IS field, there have been calls for research that integrates IS and national culture (Ford et al., 2003; Gefen and Heart, 2006). In the past several years, this call has been heeded by numerous researchers who have examined culture related to website characteristics. For a comprehensive listing of studies in which culture and e-commerce research are investigated, please see Vyncke and Brengman (2010).

Numerous definitions of culture exist, as do multiple levels of culture (Karahanna et al., 2005; McCoy et al., 2005). Within a given national culture, there are individual, team, or organizational differences—but it is expected there will be a dominant set of shared cultural values, attributes, beliefs, and behaviors (Erumban and Jong, 2006; Matsumoto, 1994). Culture is denoted as "a system of values and norms that are shared among a group of people and that when taken together constitute a design for living" (Doney et al., 1998, p. 67). While a single definition of culture is impossible, most frequently researchers have used nation state as a loose categorization for national culture (Doney et al., 1998).

For more than 30 years, researchers have relied on Hofstede's (1980) work to make comparisons based on country affiliation. Hofstede's dimensions of individualism–collectivism, power distance, uncertainty avoidance, or masculinity–femininity have been frequently used in previous research in IS, and more specifically in e-commerce (Vyncke and Brengman, 2010).* Although there have been questions regarding the validity of using Hofstede's findings, results of his work are supported quantitatively and qualitatively by numerous studies in various disciplines (Sondergaard, 1990; Straub, 1994). Researchers often use Hofstede's classifications to study social psychological phenomena (Dawar et al., 1996; Gefen and Heart, 2006; Yamagishi and Yamagishi, 1994), computer self-efficacy (Srite et al., 2008), and specifically cross-cultural differences in an e-business context (Cyr et al., 2009a; Jarvenpaa et al., 1999; Simon, 2001).

In addition to Hofstede, other cultural work related to website design relies on Hall's (1976, 1990) two dimensions of culture for context and time (Smith et al., 2013 Vyncke and Brengman, 2010). Regarding context, Hall contends that meaning is formed based on how information is perceived, and that the form and function of information varies between cultures. Certain high context cultures like Japan rely on implicit information embedded in a context of social cues such as body language, eye movement, silence, or other nuances. Alternately, in low context cultures such as Canada, there is reliance on explicit forms of communication as found in text or speech. In addition, Hall categorizes cultures based on how members of that culture manage and perceive time. In monochronic cultures like Canada, time is seen as linear and in fixed segments that can be quantified and measured. Members of monochronic

---

* It expected most readers are familiar with Hofstede's categorizations so they will not be elaborated here. However, for a thorough review and comprehensive discussion of cultural theories (including Hofstede's work), as well as culture related to IT, the reader is referred to Corbitt et al. (2004).

societies tend to focus on a single task within a specified period of time and work on it until completion. In polychronic societies, time is viewed as continuous and unstructured, and the completion of the task is more important then the schedule of achieving it. Latin countries are examples of polychronic cultures when time is flexible.

In addition to an examination of users who differ by country, a useful but less utilized approach is to investigate differences and similarities between countries based on Clustering Theory (Hartigan, 1975). According to Clustering Theory, prediction occurs in two ways. First, if a group (country) is classified into a given cluster (country cluster), then information about behaviors or values of other members of that cluster would serve as predictors of expected behavior of the classified group. Second, measures (such as website design elements) that demonstrate a cultural affinity would be predictive of other members of another given group as long as they belong to the same cluster. For example, user perceptions of website design in Canada would be aligned to the United States since users are within the same cultural cluster. This approach is consistent with cultural researchers (i.e., Ronen and Shenkar, 1985) who adhere to a "culture cluster" model.

An alternate perspective concerning IT use does not focus on the impact of culture, but rather that IT and knowledge management systems may have a suppression effect (Mason, 2003). In this case, technology has a homogenizing effect on national preferences, including IT design and use. However, based on the majority of work related to website design in e-business, culture does matter (Cyr and Trevor-Smith, 2004; Smith et al., 2011). As an additional caveat, while research has mostly focused on culture and IT, an emerging and interesting avenue of consideration is how *socio-cultural* perspectives have a role to play in how users think about and utilize technology. More specifically, Cyr et al. (2009b) examined how socio-cultural values of masculinity and femininity were expressed not only across different cultural groups but also between men and women. The findings of this study will be elaborated in a later section of this chapter.

In sum, culture is pervasive and influences how users perceive ITs. Culture has an impact on the design and use of technology, and in this chapter, the spotlight is on website design. In the following sections, key elements of website design in an e-business context are elaborated, together with the impact of culture on the design–use relationship.

31.4 Culture and Website Design

Online consumers are attracted and engaged by effective website design (Agarwal and Venkatesh, 2002; Cyr, 2008a; Fogg and Tseng, 1999; Fogg et al., 2002; Hoffman et al., 1999; Nielsen, 2001). According to Gommans et al. (2001, p. 51), "A website has to be designed for a targeted customer segment … Local adaptation should be based on a complete understanding of a customer group's culture." The merging of culture and usability is known as "culturability"—when cultural elements in website design are expected to affect the way a user interacts with the site directly (Barber and Badre, 2001). When websites are culturally appropriate, or "localized," then users are more likely to visit them, spend time on the site, and to revisit them in the future (Barber and Badre, 2001; Evers and Day, 1997). Localization is the process of adapting a product or service to a particular language, culture, and desired local "look and feel." In localizing a product, in addition to language translation, details such as currency, color sensitivities, product or service names, images, gender roles, and geographic examples are considered.

Many corporations with international operations operate multiple country websites, and many of these websites are designed to exhibit localized characteristics. For example, in April 2012, Coca-Cola operated 157 websites for different countries. This is an increase of 32 websites from June 2009 (as reported in Vyncke and Brengman, 2010). Related to the earlier discussion of Internet usage and e-business, Coca-Cola segments its markets into six geographical regions with multiple websites in each: North America (4), Latin America (22), Europe (35), Eurasia (32), Africa (52), and Asia Pacific (12). Recall that some of the highest growth of Internet use was in Africa, with 2988% increase from 2000 to 2011.

It is interesting to note that Coca-Cola has 52 websites for Africa. In the North American market, two websites exist for the United States (English and Spanish) and two more for Canada (French and English). For the American sites, besides language, there are multiple other differences including use of color, design features such as animation, navigation, and type of information provided. These design differences, therefore, localize the website and make the site more appealing to local users.

Further, based on research investigations, different user preferences are found when design characteristics are considered across cultures (Cyr et al., 2009a; del Galdo and Nielsen, 1996; Marcus and Gould, 2000). In a study that systematically compared domestic and Chinese websites for 40 American-based companies, significant differences in cultural characteristics were found for all major categories tested (Singh et al., 2003). Cyr and Trevor-Smith (2004) examined design elements using 30 municipal websites in each of Germany, Japan, and the United States. Use of symbols and graphics, color preferences, site features (links, maps, search functions, and page layout), language, and content were examined, and significant differences were uncovered in each design category. In other research in which color (Cyr et al., 2010) or human images (Cyr et al., 2009a) were specifically investigated, cultural differences were likewise noted across culturally diverse groups.

Vyncke and Brengman (2010) aimed to determine if "culturally congruent" websites are more effective than websites when culture is not taken into account. A culturally congruent website is one that is localized (as defined earlier) and therefore matches the norms, values, and expectations of the user. Cultural congruency is measured on a variety of dimensions as identified by Singh and his colleagues (Singh and Baack, 2004; Singh and Boughton, 2005) based on the degree to which a website was localized. Websites range from standardized, when there is no customization across country locations, to highly localized. In the Vyncke and Brengman study, website effectiveness was determined based on whether or not users perceive a website to be useful, easy to use, entertaining, and facilitating positive attitudes and behaviors that result in users wanting to return to the website in the future. In total, 27 research studies published in 16 different journals are evaluated for cultural congruency related to effectiveness. This investigation provides strong empirical support for the positive impact of cultural congruency on performance measures.

31.5 A Suggested Taxonomy of Design Characteristics

Overall, design features have been broad and diffusely defined. However, in the realm of user experience design, Garrett (2003) suggests a number of viable design categorizations including information design, visual design, and navigation design. These categories encompass much of the research conducted in the website design area. They are used and experimentally validated by Cyr and her colleagues over several years from 2004 to 2012. It is expected that information design, visual design, and navigation design provide a useful framework for considering websites generally as well as culturally embedded websites in e-business. These categories are elaborated in the subsequent sections.

Information design refers to website elements that convey accurate or inaccurate information about products or services to a user. "[C]ustomers dissatisfied with web site information contents will leave the site without making a purchase" (McKinney et al., 2002, p. 308). Information is considered an important prerequisite to trust (Flavián et al., 2006; Yoon, 2002). Furthermore, there are differences in the type and amount of information that is considered appropriate across cultures. In North America, substantial amounts of product information are considered desirable, while in other cultures, the same level of information would be considered inappropriate as outlined in the subsequent sections (Cyr, 2002, p. 172).

> [O]n the [customer] support side, there's a lot of pride in some European countries. In France they have a long history of what they're doing, and status comes from the knowledge and expertise acquired. So it's very important to only tell customers information about products which they assume it's reasonable not to have … otherwise, it's like trying to tell them how to do their job.

Research comparing user preferences for perceived access and presentation of product information in Canada, the United States, Germany, and Japan uncovered few significant differences between the United States, Canada, and Germany but significant differences ($p < 0.01$) between these countries and highly collectivist Japan (Cyr et al., 2005). Based on qualitative comments from the study, there appears a desire on the part of Canadians, Americans, and Germans for utility—at least as far as obtaining site information is concerned. As one Canadian user elaborates:

> For a first glance I like the first ten bullet points, the ten most important things. But if I'm looking for detail information I want it to be there. For example, the sizes and dimensions or something like that (Ibid, p. 41).

Elements of *Visual design* deal with balance, emotional appeal, aesthetics, and uniformity of the website overall graphical look. This includes colors, photographs, shapes, or font type (Garrett, 2003). Users from collectivist cultures such as China have a strong preference for visuals, whereas users from more individualistic cultures like Germany prefer a logical and structured page layout (Szymanski and Hise, 2000). As an example, in a study by Cyr et al. (2005), which includes Japanese participants and their perception of various website elements, these users noted they preferred pictures, bright colors, and animation. This sentiment is captured by a Japanese user who notes: "I say … use more pictures, more drawings to appeal to Japanese people … Japanese people like the emotional approach" (Ibid, p. XX). Alternately, an American respondent indicates: "… Banners drive me crazy, they are very distracting actually, when I got deeper into the site, there was a flashy thing over here, it is very distracting" (Ibid, p. 41). Color is a common differentiator by culture and connotes different meaning (Barber and Badre, 2001; Singh et al., 2003). Red means happiness in China but danger in the United States.

Navigation design refers to the navigational scheme used to help or hinder users as they access different sections of a website (DeWulf et al., 2006; Garrett, 2003). "No matter how thorough the information content of a site is, a customer who has difficulty in searching and getting the needed information is likely to leave the site" (McKinney et al., 2002, p. 308). User preferences for the form of navigational scheme are expected to vary by culture (Marcus and Gould, 2000). Germans who are moderately high on uncertainty avoidance "feel anxiety about uncertain or unknown matters" (Marcus and Gould, 2000, p. 39), and therefore prefer "navigation schemes intended to prevent users from becoming lost" (Ibid, p. 41). Simon (2001) found that individualistic Europeans and North Americans prefer navigation that enhances movement and makes the site simpler to use. Alternately, Asian/Latin and South Americans (generally collectivists) desire navigation aids to change the appearance of the site without particular concern for movement.

It is known that a website design effective in information design, visual design, and navigation design can support website trust (Cyr, 2008a,b). Further, the comparative degree to which these elements are important to users and engender trust varies across cultures, as further outlined in the subsequent sections.

31.6 Website Trust and Culture

Whether or not a website is trustworthy is an important consideration for potential online buyers. Unlike vendor–shopper relationships, as established in traditional retail settings, the primary communication interface between the user and the vendor is an IT artifact—the website. As such, trust is generally more difficult to establish. A lack of trust is one of the most frequently cited reasons for consumers not purchasing from Internet vendors (Grabner-Krauter and Kaluscha, 2003).

Corritore et al. (2001) provide a definition of online trust that includes cognitive and emotional elements. Hence, trust encompasses "an attitude of confident expectation in an online situation or risk that one's vulnerabilities will not be exploited" (Ibid, p. 740). Website trust implies consumer confidence in a website and "willingness to rely on the seller and take actions in circumstances where such action makes the consumer vulnerable to the seller" (Jarvenpaa et al., 1999, p. 4). Establishing consumer trust in the website is fundamental to online loyalty, which includes user intentions to revisit an online vendor or to buy from it in the future (Cyr, 2008b; Flavián et al., 2006; Gefen, 2000; Yoon, 2002).*

Trust is a highly researched topic in e-business. Antecedents to website trust vary and include website design characteristics (Cyr, 2008b; Cyr et al., 2009a; Ou and Sia, 2010). Other factors that engender trust are social presence (Cyr et al., 2007; Gefen et al., 2003; Hassanein et al., 2009); perceived vendor reputation (Jarvenpaa et al., 1999; Koufaris, 2002); clear and trustworthy privacy policies (Reichheld and Schefter, 2000); online transaction security (Palmer et al., 2000); and information privacy (Hoffman and Novak, 1996), among others.

Cultural differences prevail related to user propensity to trust. For instance, individualists are more optimistic than collectivists concerning benevolence from strangers (Inglehart et al., 1998; Yamagishi and Yamagishi, 1994). Further, Kim and Son (1998) measured levels of distrust between highly individualist Americans and highly collectivist Koreans and find that 59% of Americans trust members of a different ethnic group in their society, and 57% trust people from a different country. For Koreans, the average responses are 23% and 18%, respectively. According to Yamagishi and Yamagishi (1994), exchange relationships outside a cultural group only occur when there are substantial institutional safeguards such as strong cultural norms or legal sanctions.

Prior research on website trust and culture compared Internet trust in collectivist and individualist cultures (Jarvenpaa et al., 1999). The researchers expect that consumers from individualist cultures would exhibit higher trust levels in an Internet store than consumers from collectivist cultures. Contrary to this hypothesis, no strong cultural effects are found for trust. Alternately, Simon (2001) finds that for information provided on American and European websites, Asians are most trusting (83% positive), while Europeans (46% positive) and North Americans (42% positive) exhibit substantially lower levels of trust toward the websites.

In related research, Cyr et al. (2005) conducts a study to investigate whether or not local websites engender higher levels of trust for web users than a foreign website of the same vendor (Samsung in this case). With reference to earlier work by Yamagishi and Yamagishi (1994) and others, it is expected that users from individualistic cultures such as Canada or the United States would be least likely to trust the local website, and most likely to trust the foreign website than moderately individualistic German users and collectivist Japanese users. When comparing the level of trust between countries for the *local* website, almost no differences are reported between the Canadians, Americans, and Germans. However, there are large differences between the Japanese with Americans, Canadians, or Germans. Contrary to expectations, Japanese respondents trust their local website least, while Germans trust their local site most. Similar results are found for users viewing the *foreign* version of the website.

Closely related to trust and information design is how legitimacy of the vendor, products, or services is conveyed to website visitors. According to Chen and Dhillon (2003), "Since transactions [on the Internet] occur without personal contact, consumers are generally concerned with legitimacy of the vendor and authenticity of products or services" (p. 1). In a study of website users in the United States, Canada, Germany, and Japan, all country groups score highly regarding the need to trust an Internet store with a well-known reputation, and regarding concern for the legitimacy of the online sales contact. As expected, Japanese are most concerned with online risk and security when buying on the Internet (Cyr, 2011). Canadians note they are aware of security problems when using the Internet, but feel the benefits outweigh the risks. As one Canadian describes,

> You realize that some of the concerns the market has, or some of the perception that people have with security are unfounded ... The likelihood that someone is going to intercept the transmission between your computer and a website, and decipher it, is very low. (Ibid)

---

* A thorough review of trust in offline and online settings is not feasible within the scope of this chapter. However, the reader may wish to refer to Rousseau et al. (1998) for a critique of offline trust and Gefen et al. (2003) for a summary of online trust. Also of interest may be how trust differs from distrust in website design (Ou and Sia, 2010).

In research by Cyr (2008a,b), information design, visual design, and navigation design are modeled to determine if a statistical relationship exists between these design elements and trust. A total of 571 participants located in Canada, Germany, or China completed an experimental task and online survey (N = 230 in Canada; 118 in Germany; and 223 in China). Results of the investigation indicate that navigation design results in trust for Canada and China, visual design results in trust for China only, and information design results in trust for Canada only. It is clear there are distinct design proclivities between the countries in this study. Also of interest is that the three design characteristics explain trust better in Canada and China than in Germany. In the case of Germany, it appears that other characteristics not captured in this study also contribute to online trust. This may include the company name and reputation, or perceived security of information.

Hence, while it was already established that affinity to trust varies between cultures, studies that focus on website design characteristics indicate that design elements have the capability to elicit user trust—and that differing design elements are more important in certain cultures than in others.

31.7 Emotion, Website Design, and Culture

Online shopping is recognized to encompass both utilitarian and hedonic components (Childers et al., 2001; Kim et al., 2007; Lim and Cyr, 2009, 2010; Sun and Zhang, 2006; Tractinsky, 2004). Utilitarian-focused shopping is aimed to achieve predetermined and cognitively oriented goals. In terms of website design, effective information design or navigation design assists the online shopper to achieve such goals. On both a theoretical and a practical level, utilitarian outcomes of online shopping are expected to result in perceived usefulness or effectiveness as extensively investigated in IS using the technology acceptance model (TAM) (Davis, 1989).

Alternately, much less researched are hedonic website elements used to create an experience for the online consumer which is positive, enjoyable, and that provides emotive and sensory pleasure (Bruner and Kumar, 2003; Cyr et al., 2006; Kim et al., 2007; van der Heijden, 2003). Recently, researchers examine how vividness (i.e., the richness and in media and information presented to users) influences user emotional responses in e-commerce (Sheng and Joginapelly, 2012). In other work, website "socialness" leads to enjoyment and influenced user intentions to use online products (Wakefield et al., 2011). Finally, the effective use of images or colors on a website can contribute to emotional appeal for users (Garrett, 2003; Rosen and Purinton, 2004).

Further, in the context of hedonic consumer interactions, researchers investigate perceived social presence as an important antecedent to online consumer enjoyment (Cyr et al., 2007; Hassanein and Head, 2007). Social presence implies a psychological connection with the user who perceives the website as "warm," personal, sociable, thus creating a feeling of human contact (Yoo and Alavi, 2001). Hassanein and Head (2007) demonstrate that emotive text and pictures of humans result in higher levels of perceived social presence for websites. As such, social presence straddles the areas of information design and visual design. In addition to social presence resulting in online enjoyment, it has implications for website trust (Cyr et al., 2007; Gefen and Straub, 2003; Hassanein and Head, 2007), website involvement (Kumar and Benbasat, 2002; Witmer et al., 2005), and utilitarian outcomes such as perceived usefulness or effectiveness (Hassanein and Head, 2004, 2006, 2007).

While the foray to investigate hedonic elements of website design is relatively less studied than utilitarian design elements, the amount of research that examines these topics from a cultural perspective is more scarce. Only a few studies have been conducted in which hedonic website design features are systematically modeled across diverse cultures. More recently, research in this area is beginning to emerge. For instance, Tsai et al. (2008) examine emotion and visual information uncertainty for websites using a Taiwanese sample with differing levels of uncertainty avoidance (one of Hofstede's cultural dimensions). Using a three country sample (Canada, Germany, and China), Cyr (2008) discovered that visual design resulted in satisfaction for users in all three countries and trust in China only.

Hassanein et al. (2009) aimed to determine if perceived social presence is culture specific or universal in online shopping settings. These researchers find that for Chinese and Canadian users' social presence led to similar levels of usefulness and enjoyment, but trust only for Canadians. Cyr and her colleagues conducted two separate research investigations with Canadian, German, and Japanese users regarding their reaction to visual design website elements. In one study, survey data indicate that human images universally result in image appeal and perceived social presence; while interviews and eye-tracking data suggested participants from different cultures experience the design images differently (Cyr et al., 2009a). In the second study, website color appeal is found to be a significant determinant of website trust and satisfaction, with differences across cultures (Cyr et al., 2010). Collectively, these research findings highlight unique perceptions and outcomes of users in different countries based on different cultural orientations.

The widely used TAM, which represents utilitarian outcomes for usefulness, ease of use, and behavioral intention (to use the technology) has also received attention in a cultural context (i.e., McCoy et al., 2005; Rose and Straub, 1998; Straub et al., 1997). Although not specific to design elements in an e-business setting, a landmark study by McCoy et al. (2005) examines the viability of TAM with almost 4000 students representing 25 countries. Results of the investigation reveal that TAM is not universal, and in particular does not apply for individuals who are low on uncertainty avoidance, high on power distance, high on masculinity, and high on collectivism.

Further, in a cross-cultural examination of online shopping behavior in Norway, Germany, and the United States (Smith et al., 2013), patterns for cognitive (i.e., utilitarian) and affective (i.e., hedonic) involvement with a website are examined using a TAM framework across the cultures. The key finding is that the TAM model does not apply for Norway and Germany, although it does for the United States where it was developed. Additional findings reveal that as expected, cognitive involvement is an antecedent to perceived usefulness and perceived ease of use in all three countries. However, for affective involvement, there is a relationship to behavioral intention for the U.S. and Norwegian samples, but not for Germans.

In one additional study, Cyr et al. (2009b) uses TAM to examine gender and socio-cultural differences in samples located in Canada and the Netherlands. Canada scores highly on Hofstede's masculinity dimension (i.e., a focus on material success and achievement), while the Netherlands scores very high on the femininity dimension (i.e., a focus on relationships and cooperation). In the study, website design elements are used to create online social presence, which was then modeled to perceived usefulness and behavioral intention (operationalized as e-loyalty). Of particular interest is whether masculinity and femininity impact social presence, perceived ease of use, and trust. Masculinity was further hypothesized to influence perceived usefulness. The results show that while biological sex affects only perceived ease of use, the effects of femininity extend also to trust, and those of masculinity also to trust and social presence. Therefore, masculinity and femininity are relevant predictors in the research model and suggest a need to reexamine the relationship between IS constructs, on the one hand, and biological sex and masculinity or femininity, on the other. Of interest, nationality has a direct effect on masculinity, while femininity is influenced by an interaction of nationality and biological sex. This study further emphasizes the subtle relationships that exist concerning IS design and use across cultures.

31.8 Additional Applications for IS Design and Use

This chapter has been focused on the interplay of culture and IS with a focus on e-business. However, as indicated in the Introduction, there are numerous other related areas in which the lessons learned regarding e-business are also applicable. More specifically, e-business is a technology supported by design characteristics in the realm of information design, visual design, and navigation design. These design elements are likewise relevant for e-government, virtual teams, online corporate training programs, management of IS, or e-health, among other areas.

As one example, for e-government, the same technologies that assist to build trust and loyalty among users are useful as in e-business. Dynamic websites are required that are culturally appropriate, and which engage the user. The main difference for e-government websites is the expectations of the users who view information in the public domain, and who have differing requirements from e-business users. In one study in which trust is evaluated in the e-government tax system of Singapore (Lim et al., 2012), one respondent indicates:

> There is excellent service at the e-file helpline. Also, there are user-friendly fields with help buttons. The entire set up in e-filing is meant to be easy for taxpayers. The system has user-friendly navigation, simple interface, easy for anyone to get tax filing done without adding more grief to the already undesirable task. (Taxpayer)

When design elements meet user expectations and facilitate the completion of online tasks, then trust results. Although the study by Lim et al. does not examine cultural elements, the appropriate use of symbols, colors, etc. for the e-government website in the Singaporean context will contribute to user satisfaction and loyalty. The same can be said for the other topics mentioned above. Further elaboration beyond e-business of design use and applications appears in the sections on practice in the subsequent text.

31.9 Impact on Practice

Website localization carries a substantial cost to companies, particularly those companies which operate in multiple locations (e.g., in 2012 Coca-Cola operated in 157 countries). However, based on much of the research presented in this chapter, effective localization of web content and design features to specific country users is necessary and has the potential to further engender user trust. Despite this finding, many international companies fail to localize their websites.

While in an ideal world, companies will localize their websites for each country or cultural group, it is known that certain countries cluster together regarding user perceptions of website design. For example, users within North American or Latin American clusters are very similar (Cyr, 2013). Adaptation of website design in these clusters may only require minimal modifications. Further, in Canada while language localization (French and English) is essential, other forms of website adaptation are less necessary. It is noteworthy that "cluster localization" serves the purpose of providing some degree of cultural adaptation to user groups, but with a substantial cost savings to online vendors over deep localization procedures.

The importance of creating websites in the local language of the user is obvious, and perceived usability is increased when a website is originally conceived in the native language of the users. Even though translation may be of an excellent quality, there is still a level of cultural distance, which impacts a user's evaluation of a website (Nantel and Glaser, 2008). Hence quality of the language is important, along with culturally embedded metaphors of the target group. Subsequently, managers for e-business sites will ideally test the usability of their sites for various international locations with respect to language and redesign them as necessary.

In terms of information design, information quality should be high, with online product information complete, detailed, and trustworthy. Information design is especially important to some users (e.g., Canadians), and it is statistically related to trust. This is a beacon to online vendors to pay special attention to effective presentation of information on the website. Type and amount of information varies by country and should also be tailored to particular users. For example, the majority of users note they prefer few product details when first entering a website, and these details should be easily accessible. Detailed information can be embedded at the next level of the website (Cyr et al., 2005). Generally speaking, users in lower context cultural locations especially desire clarity of information and larger amounts of it.

Visual design preferences likewise vary by culture. For instance, Cyr (2008) find visual design to be especially important to Chinese users. Among this group, high-quality visual design, in turn, resulted in website trust. This was not the case for Canadians and Germans who were included in the same investigation. As such, website designers should pay particular attention to the colors, images, shapes, and overall graphical look of websites for high context, collectivist users who prefer a more "emotional approach." More specifically, brighter colors such as red are preferred, along with animation.

Still in the realm of visual design, based on survey data, users universally prefer websites with images including facial features over human images without facial features, or no human images. Human images result in image appeal and perceived social presence, eventually resulting in website trust (Cyr et al., 2009a). In this same study, interview data reveals different categorizations as to how images are perceived. This includes (1) aesthetics—pleasantness of attractiveness of the website; (2) symbolism—the implied meaning of design elements (i.e., that an image of a man and a little girl are a representation of father and daughter); (3) affective property—emotion eliciting qualities; and (4) functional property—website structure as organized or distracting. Of interest, perceptions of images vary by culture between Canada, Germany, and Japan. For example, in the condition when images include facial features, for Canada these are perceived as aesthetic and affective; for Germany as function, affective, and symbolic; and for Japan as affective and symbolic. While images are expected to be perceived differently across cultures, this work suggests a fine-grained analysis of differences and that images must be carefully chosen to be localized to user preferences. An additional finding from the work by Cyr and her colleagues is that partial human images (i.e., a torso only) generally appear unnatural and unexpected to users and therefore should be avoided.

Color appeal has the potential for emotional impact leading to website trust and satisfaction, and subsequently to online loyalty (Cyr et al., 2010). In this investigation, Canadians have a stronger preference for a grey color scheme for an electronics e-business website compared to Japanese and Germans. Germans prefer a blue color scheme for the same website. Similar to the subtleties that occur when using variations of images for website design, when examining colors across cultures user preferences emerged from interview data. For example, the color blue is universally one of the most popular and trusted colors. The cultural differences are still quite significant: for Canadians, this color elicits a wide range of attributes including being perceived as aesthetic, affective, functional, harmonious, and appropriate. For Germans, blue is perceived as functional, affective, and appropriate. For Japanese, it is affective and functional. In sum, from a practical perspective, website designers need to pay attention to color for specific cultural groups—which in turn results in loyalty wherein users revisit the website in the future.

The third design element that is discussed in this chapter is navigation design. While elements of navigation vary, consistency of page layout and quick access to navigational features are universally desirable. Navigation design is highly related to trust in some countries (e.g., for Canadians and Chinese), which suggests that users from these countries expect websites that are clear and transparent (Cyr, 2008a). Navigation itself has cultural nuances. For instance, Canadians expect utilitarian websites that enhance movement and are easy to use. Generally speaking, cultures high in uncertainty avoidance will prefer clear and detailed platforms for how to navigate a website.

Expanding beyond e-business, it would make sense that information design, visual design, and navigation design be tailored to other areas of website design in various cultural contexts. This may include websites used by virtual teams, e-learning-training sites as used for constituents from diverse cultures, or for cross subsidiary IT management committees. For instance, trust is important for virtual team performance (Edwards and Sridhar, 2005) and for distributed teams (Dinev at al., 2006). Based on the material in this chapter, it is proposed that such trust can be developed through effectively designed and localized websites.

As website design becomes more sophisticated, and with emphasis on hedonic elements and how to develop warm and sociable websites, design features for enhancement of social presence in website design become increasingly of interest. As already noted, social presence relies primarily on human images and emotive text—and as such includes elements of both visual design and information design.

Since social presence typically results in enjoyment, trust (Cyr et al., 2007; Hassanein and Head, 2007), involvement (Kumar and Benbasat, 2002; Witmer et al., 2005), perceived usefulness, and effectiveness (Hassanein and Head, 2004, 2006, 2007), then website designers will do well to incorporate images and emotive text as an important element of design—both within and across cultures. However, usability testing will need to be conducted to determine exactly the best method for how this will occur.

The addition of social presence elements is receiving attention in e-health and shows promise to ensure that users remain on health websites or revisit them in the future. Crutzen et al. (2012) experimentally test the addition of social presence elements on a hepatitis website. While social presence elements are of interest to users, there is no significant difference between website viewing when social presence elements are present versus absent. It appears that for e-health websites, information design is paramount, and that users first require trustworthy information; visual design elements are secondary. However, as the authors point out, personalization is an important strategy for e-health websites. As such, communication is individualized for a user, with the expectation that this will result in a more positive experience (Hawkins et al., 2008). In a similar way, social presence elements on e-health websites may require tailoring or personalization—a technique which is commonly used and found to be effective for the dissemination of information on intervention websites (Krebs et al., 2010). Across cultures, differences are likewise found related to masculine and feminine value orientations in website design, and these differences deserve attention. For instance, in higher masculinity cultures, interface design elements are focused on traditional gender/family/age distinctions; work tasks, roles, and mastery; navigation oriented to exploration and control; attention gained through games and competitions; and graphics, sound, and animation used for utilitarian purposes. Feminine cultures emphasize blurring of gender roles; mutual cooperation, exchange, and support rather than mastery and winning; and attention gained through poetry, visual aesthetics, and appeals to unifying values (Zahedi et al., 2006).

Applying the masculinity–femininity concept to international users, multinational companies that market to a variety of countries might do well to align the values of the designers to the users for whom they are designing. That is, rather than have a team of North American web designers localize websites for all country constituents, designers with values of either masculinity or femininity may be matched to the overarching cultural values present within a country. For North America, more masculine-oriented designers would be appropriate, while in feminine cultures such as the Netherlands, designers with feminine-oriented value systems would be employed.

In sum, there are numerous ways in which websites can be adapted to accommodate the preferences of users in different cultures or country locations. Based on the preceding, in addition to more obvious modifications such as language translation, there are numerous subtleties that exist related to information design, visual design, and navigation design, among other design features. While there is a cost attached for companies who localize websites, I propose that this is a necessary component of doing international online business if online shoppers are to trust the website and be loyal to the vendor in the future. In addition, the same lessons for website design in e-business find application in other areas of IS use and design such as e-government, e-health, virtual teams, or other settings.

31.10 Research Issues

While there have been several rigorous studies in which the impact of website design as an IT medium has been examined in e-business across cultures, there is considerable scope for future research in alignment with practical considerations as outlined in the previous section.

One such research area is to move beyond cultural groups as the defining characteristic for the study of website localization. Although there is merit in using Hofstede (1980) or other cultural theorists to determine broad cultural categorizations, an alternative point of view is to utilize a wider variety of country characteristics as they impact user perceived value of websites for e-business. For instance, in a huge study involving 8886 consumers from 23 countries and 30 websites, Steenkamp and Geyskens (2006) considered country characteristics such as "rule of law" and the regulatory system, national identity, and individualism as influencing users' perceived value in website design. They found that the effects of privacy/security protection (which would be related to trust) on perceived value is stronger for users in countries with weak regulatory systems. Further, users in countries high on national identity find it of particular importance that there is cultural congruity between themselves and the website—and therefore that website localization occurs. It would be of interest to investigate this further as it applies in e-government. Finally, users from more individualistic countries find pleasure or hedonic outcomes to be essential, along with privacy/security protection, and customization of website design compared to users in collectivist countries.

In this vein, in addition to Hofstede's dimension of uncertainty avoidance, Cyr (2013) uses clustering theory (i.e., wherein countries with similar values and beliefs are clustered together) to determine preferences for website design across cultures.* As already noted, this resulted in some remarkable similarities between certain country groupings such as for North America or Latin America. In this same investigation, country indicators such as Internet connectivity and infrastructure, the digital and legal environment, and consumer and business electronic adoption were used to discuss user reactions to an e-business website. In future research, it is suggested that such additional parameters be used more frequently related to website design and use. Such clustering might be used for the adoption and diffusion of various IS technologies such as adoption of large-scale applications (e.g., ERP systems and Intranet facilities) or in more specific adaptations of training procedures or software development for certain cultural groups. In each case, cultural localization can occur, but for similar cultural clusters rather than on a country specific basis.

The collection of individual-level data is an additional alternative to the use of established cultural categorizations for the study of IT use, and website design more particularly. Collection of individually based or "espoused" user data has been advocated by scholars (i.e., McCoy et al., 2005; Srite and Karahanna, 2006), and has significant advantages to determine culturally specific user values. For example, in Canada there are not only significant value differences for French- or English-speaking users but also for numerous other cultural groups. To elaborate, in Vancouver on the West coast of Canada, there is a predominant Chinese community—and even within this "cultural" group, it is expected there are widely varying cultural values related to whether a user is Chinese born in China or indigenous Chinese born in Canada. Collecting data in this way leads researchers into the area of learning more about socio-cultural value systems, in which more specific values are accessed compared to those represented in established global cultural groupings. As research into cultural values and IT use matures, such fine grained comparisons will be useful—if not imperative. The results of such investigations will be useful for better understanding IS employee values and attitudes, and how IS managers can best facilitate successful implementations and strategies based on cultural proclivities. As one example, it is known that in different countries, MIS project managers perceive projects and risks differently (Choi and Choi, 2003; Peterson and Kim, 2003). Additional research on risk based on socio-cultural values will potentially provide more focused and relevant information than relying on country-based cultural groupings.

Research on the topic of IS design and use, including website design in e-business has typically relied on surveys with single or multiple item scales. However, as measurement techniques expand, there is opportunity to delve into new methods that more deeply and comprehensively attend to what users are experiencing. More specifically, neurophysiological techniques such as functional magnetic resonance imaging (fMRI) have been used for testing reactions to product packaging (Reimann et al., 2010), and there is merit to pursue these alternative methodologies as they inform website design communities. To this end, human images in website design (Cyr et al., 2009a) as well as user reactions to different colors on websites (Cyr et al., 2010) have been measured using eye-tracking equipment to determine exactly where and for how long users look at elements of design. Coupled with interviews to determine why users look where they do, these methods offer a systematic analysis of website elements.

---

* The results of the GLOBE project appear in a series of papers in the *Journal of World Business* by House et al. (2002).

Most recently, a paper published in the top IS journal, *MIS* Quarterly, by Dimoka et al. (2012) charts a research agenda for the use of neurophysiological tools in IS research. The use of methodologies such as eye-tracking and fMRI are part of an evolving research agenda and are well applicable to study IS design and the cognitive and affective outcomes for users related to their reactions to website design principles. In this regard, Djamasbi (2011) examine online viewing and aesthetic preferences using an eye-tracking device, while Sheng and Joginapelly (2012) explore website atmospherics. These methodologies offer precise insights into why users respond as they do—which in turn serves to develop or elaborate design theory.

With respect to the topics as outlined in this chapter, the realm of emotion and hedonic responses by users to various design features is generally under-researched and thus presents opportunities for additional study. As previously indicated, one particular area of future research concerns the application of personalization through social presence elements in e-health. Further study could also be devoted to social presence in website design across cultures, where to date these appears to be only a very few investigations on this topic (i.e., Cyr et al., 2009b; Hassanein et al., 2009). Finally, the impact of design aesthetics in mobile commerce is a contemporary topic worthy of future investigations. Visual design aesthetics significantly impact perceived usefulness, ease of use, and enjoyment of mobile services (Cyr et al., 2006), although how mobile designs are perceived by diverse cultural users is mostly unknown. With the high usage of mobile devices, it would be interesting to know how cultural users perceive various design characteristics—on both utilitarian and hedonic dimensions.

31.11 Summary

The impact of culture on IS design and use is pervasive and very broad in scope. In this chapter, emphasis has been on website design features in e-business, and in particular information design, visual design, and navigation design. These all have potential to influence user trust toward a website, and all have a varied impact related to culture of the user. From a practical perspective, the impact of website localization (and the attainment of cultural congruity) has the potential to turn web browsers into online purchasers, and the implications for vendor profits are substantial as a result. According to Reichheld and Schefter (2000), an increase in customer retention rates by only 5% can increase profits by 25%–95%. Therefore, the development of loyal customer behavior is a valued goal for managers, marketers, and strategists. Further, from a research perspective, there are numerous topics that merit investigation, and there are novel and interesting methodologies that enable investigators to gain new and deeper insights into how cultural variables impact IS design and use.

Keywords

National culture, information systems design, e-business, website design, localization, information design, visual design, navigation design, and website trust.

References

Agarwal, R. and Venkatesh, V. (2002). Assessing a firm's web presence: A heuristic evaluation procedure for measurement of usability. *Information Management Research*, 13(2), 168–121.

Barber, W. and Badre, A.N. (2001). Culturability: The merging of culture and usability. *Proceedings of the 4th Conference on Human Factors and the Web*. Basking Ridge, NJ.

Belanche, D., Casalo, L., and Guinaliu, M. (2012). Website usability, consumer satisfaction and the intension to use a website: The moderating effect of perceived risk. *Journal of Retailing and Consumer Services*, 19, 124–132.

Bruner, G. and Kumar, A. (2003). Explaining consumer acceptance of handheld internet devices. *Journal of Business Research*, 58, 115–120.

Chen, S.C., and Dhillon, G.S. (2003). Interpreting dimensions of consumer trust in e-commerce. *Information Management and Technology*, 4, 2–3 (2003), 303–318.

Childers, T., Carr, C., Peck J., and Carson, S. (2001). Hedonic and utilitarian motivations for online retail shopping. *Journal of Retailing*, 77(4), 511–535.

Choi, H.-Y. and Choi, H. (2003). An exploratory study and design of cross-cultural impact of information systems managers' performance, Job satisfaction and managerial value. *Journal of Global Information Management*, 12(1), 60–67.

Corbitt, B.J., Peszynski, K.J., Inthanond, S., Hill, B., and Thanasankit, T. (2004). Culture differences, information and codes. *Journal of Global Information Management*, 12(3), 65–85.

Corritore, C.L., Kracher, B., and Wiedenbeck, S. (2001). Trust in online environments. In: Smith, M.J., Salvenjy, G., Harris, D., Koubek, R.J. (Eds.), *Usability Evaluation and Interface Design: Cognitive Engineering, Intelligent Agents and Virtual Reality*. Erlbaum, Mahway, NJ, pp. 1548–1552.

Crutzen, R., Cyr, D., Larios, H., Ruiter, R., and de Vries, K. N. (2012). Social presence and use of internet-delivered interventions: A multi-method approach. *Working paper Department of Health Promotion*, Maastricht University/CAPHRI, the Netherlands.

Cyr, D. (2002). CreoScitex: The next step. In: Cyr, D., Dhaliwal, J., and Persaud, A. (Eds.), *E-Business Innovation: Cases and Online Readings*, Prentice-Hall, Toronto, Canada, pp. 164–173.

Cyr, D. (2008). Modeling website design across cultures: Relationships to trust, satisfaction and e-loyalty. *Journal of Management Information Systems*, 24(4), 47–72.

Cyr, D. (2011). Website design and trust across cultures. In: Douglas, I. and Liu, Z. (Eds.), *Global Usability, Human–Computer Interaction Series*, Springer, New York, pp. 39–55.

Cyr, D. (2013). Website Design and Trust: An Eight Country Investigation. *Electronic Commerce Research and Applications*.

Cyr, D., Bonanni, C., Bowes, J., and Ilsever, J. (2005). Beyond trust: Website design preferences across cultures. *Journal of Global Information Management*, 13(4), 24–52.

Cyr, D., Gefen, D., and Walczuch, R. (2009b). Masculinity-femininity values and biological sex: New perspectives related to TAM, online trust and social presence. *Proceedings for the Pre-ICIS Cross-cultural Workshop*, Phoenix, AZ.

Cyr, D., Hassanein, K., Head, M., and Ivanov, A. (2007). The role of social presence in establishing loyalty in e-service environments. *Interacting with Computers*. Special Issue on "Moving Face-to-Face Communication to Web-based Communication", 19(1), 43–56.

Cyr, D., Head, M., and Ivanov, A. (2006). Design aesthetics leading to M-loyalty in mobile commerce. *Information and Management*, 43(8), 950–963.

Cyr, D., Head, M., and Larios, H. (2010). Colour appeal in website design within and across cultures: A multi-method evaluation. *International Journal of Human Computer Studies*, 68(1–2), 1–21.

Cyr, D., Head, M., Larios, H., and Pan, B. (2009a). Exploring human images in website design: A multi-method approach. *MIS Quarterly*, 33(3), 539–566.

Cyr, D. and Trevor-Smith, H. (2004). Localization of web design: An empirical comparison of German, Japanese, and U.S. Website characteristics. *Journal of the American Society for Information Science and Technology*, 55(13), 1–10.

Davis, F. D. (1989). Perceived usefulness, perceived ease of use, and user acceptance of information technology. *MIS Quarterly*, 13(3), 319–333.

Dawar, N., Parker, P., and Price, L. (1996). A cross-cultural study of interpersonal information exchange. *Journal of International Business Studies*, 27(3), 97–516.

Deans, P. and Ricks, D. (1993). An agenda for research linking information systems and international business: Theory methodology and application. *Journal of Global Information Management*, 1(1), 6–19.

Del Galdo, E. and Neilson, J. (1996). *International User Interfaces*. John Wiley & Sons, New York.

DeWulf, K., Schillewaert, N., Muylle, S., and Rangarajan, D. (2006). The role of pleasure in web site success. *Information and Management*, 43, 434–446.

Dimoka, A., Banker, R.D., Benbasat, I., Davis, F.D., Dennis, A.R., Gefen, D., Gupta, A. et al. (2012). On the use of neurophysiological tools in IS research: Developing a research agenda for NeuroIS. *MIS Quarterly*, 36(3), 679–702.

Dinev, T., Bellotto, M., Hart, P.L., Russo, V., Serra, I., and Colautti. C. (2006). Internet users' privacy concerns and beliefs about government surveillance: An exploratory study of differences between Italy and the United Sates. *Journal of Global Information Management*, 14(4), 57–93.

Djamasbi, S., Siegel, M., Skorinko, J., and Tullis, T. (2011). Online viewing and aesthetic preferences of generation Y and baby boomers: Testing user website experience through eye tracking. *International Journal of Electronic Commerce*, 15(4), 121–158.

Doney, P.M., Cannon, J.P., and Mullen, M.R. (1998). Understanding the influence of national culture on the development of trust. *Academy of Management Review*, 23(3), 601–620.

Edwards, H.K. and Sridhar, V. (2005). Analysis of software requirements engineering exercises in a global virtual team setup. *Journal of Global Information Management*, 13(2), 21–41.

Ein-Dor, P., Segev, E., and Orgad, M. (1993). The Effect of national culture on IS: Implications for international information systems. *Journal of Global Information Management*, 1(1), 33–45.

Erumban, A. and de Jong, S. (2006). Cross-country differences in ICT adoption: A consequence of culture? *Journal of World Business*, 41(4), 303–314.

Evers, V. and Day, D. (1997). The role of culture in interface acceptance. In: Howard, S., Hammond, J., and Lindegaard, G. (Eds.), *Proceedings of Human Computer Interaction, INTERACT '97*, Chapman and Hall, London, U.K, pp. 260–267.

Flávian, C., Guinalíu, M., and Gurrea, R. (2006). The role played by perceived usability, satisfaction and consumer trust on website loyalty. *Information & Management*, 43(1), 1–14.

Fogg, B.J., Soohoo, C., and Danielson, D. (2002). *How People Evaluate a Web Site's Credibility? Results from a Larger Study*. Persuasive Technology Lab, Stanford University, Stanford, CA.

Fogg, B.J. and Tseng, S. (1999). Credibility and computing technology. *Communications of the ACM*, 14(5), 39–87.

Ford, D.P., Connelly, C.E., and Meister, D.B. (2003). Information systems research and Hofstede's culture's consequences: An uneasy and incomplete partnership. *IEEE Transactions on Engineering Management*, 50(1), 8–25.

Garrett, J.J. (2003). *The Elements of User Experience: User-Centered Design for the Web*. New Riders, Indianapolis, IN.

Gefen, D. (2000). E-commerce: The role of familiarity and trust. *The International Journal of Management Science*, 28, 725–737.

Gefen, D. and Heart, T. (2006). On the need to include national culture as a central issue in e-commerce trust beliefs. *Journal of Global Information Management*, 14(4), 1–30.

Gefen, D., Karahanna, E., and Straub, D.W. (2003). Trust and TAM in online shopping: An integrated model. *MIS Quarterly*, 27(1), 51–90.

Gefen, D. and Straub, D. (2003). Managing user trust in B2C e-services. *e-Service Journal*, 2, 2(2003), 7–24.

Gommans, M., Krishan, K.S., and Scheddold, K.B. (2001). From brand loyalty to e-loyalty: A conceptual framework. *Journal of Economic and Social Research*, 3(1), 51.

Grabner-Krauter, S. and Kaluscha, E. (2003). Empirical research in online trust: A review and critical assessment. *International Journal of Human-Computer Studies*, 58, 783–812.

Hall, R.M. (1976). *Beyond Culture*. Doubleday, New York.

Hall, R.M. (1990). *Understanding Cultural Difference*. Intercultural Press Inc., Boston, MA.

Hartigan, J. (1975). *Clustering Algorithms*, Wiley, New York.

Hassanein, K. and Head, M. (2004). Building online trust through socially rich web interfaces. *Proceedings of the 2nd Annual Conference on Privacy, Security and Trust*, Fredericton, New Brunswick, Canada, pp. 15–22.

Hassanein, K. and Head, M. (2006). The impact of infusing social presence in the web interface: An investigation across different products. *International Journal of Electronic Commerce*, 10(2), 31–55.

Hassanein, K. and Head, M. (2007). Manipulating social presence through the web interface and its impact on attitude towards online shopping. *International Journal of Human-Computer Studies*, 65(8), 689–708.

Hassanein, K., Head, M., and Ju, C. (2009). A cross-cultural comparison of the impact of social presence on website trust, usefulness and enjoyment. *International Journal of Electronic Business*, 7(6), 625–641.

Hawkins, R.P., Kreuter, M.W., Resnicow, K., Fishbein, M., and Dijkstra, A. (2008). Understanding tailoring in communicating about health. *Health Education Research*, 23, 454–466.

van der Heijden, H. (2003). Factors influencing the usage of websites: The case of a generic portal in the Netherlands. *Information & Management*, 40(6), 541–549.

Hoffman, D.L. and Novak, T.P. (1996). Marketing in hypermedia computer-mediated environments: Conceptual foundations. *Journal of Marketing*, 60, 50–68.

Hoffman, D.L., Novak, T.P., and Peralta, M.A. (1999). Building consumer trust online. *Communications of the ACM*, 42(4), 80–85.

Hofstede, G.H. (1980). *Culture's Consequences, International Differences in Work-Related Values*. Sage, Beverly Hills, CA.

House, R., Javidan, M., Hanges, P., and Dorfman, P. (2002). Understanding cultures and implicit leadership theories across the globe: An introduction to project GLOBE. *Journal of World Business*, 37, 3–10.

Inglehart, R., Basanez, M., and Moreno, A. (1998). *Human Values and Beliefs: A Cross-cultural Sourcebook*. University of Michigan Press, Ann Arbor, MI.

Internet Retailer. (2011). Accessed April 23, 2012 at: http://www.internetretailer.com/2011/02/01/top-500-analysis-it-was-web-only-merchants-and-then-everyo.

Internet World Stats. (2011). Accessed April 23, 2012 at: http://www.internetworldstats.com/stats.htm

Jarvenpaa, S., Tractinsky, N., Saarinen, L., and Vitale, M. (1999). Consumer trust in an internet store: A cross-cultural validation. *Journal of Computer Mediated Communication*, 5. WWW document. http://jcmc.indiana.edu/vol5/issue2/jarvenpaa.html (accessed on April 8, 2013).

Junglas, I. and Watson, R. (2004). National culture and electronic commerce: A comparative study of U.S. and German web sites, *E-Service Journal*, 3(34), 3–34.

Karahanna, E., Evaristo, J.R., and Srite, M. (2005). Levels of culture and individual behavior: An integrative perspective. *Journal of Global Information Management* 13(2), 51–20.

Kim, H., Chan, H.C., and Chan, Y.P. (2007). A balanced thinking-feelings model of information systems continuance. *International Journal of Human-Computer Studies*, 65, 511–525.

Kim, Y.H., and Son, J. (1998). Trust, co-operation and social risk: A cross-cultural comparison. *Korean Journal*, 38, (spring), 131–153.

Koufaris, M. (2002). Applying the technology acceptance model and flow theory to online consumer behavior. *Information Systems Research*, 13(2), 205–223.

Krebs, P., Prochaska, J.O., and Rossi, J.S. (2010). A meta-analysis of computer-tailored interventions for health behavior change. *Preventive Medicine*, 51, 214–221.

Kumar, N. and Benbasat, I. (2002). Para-social presence and communication capabilities of a website. *E-Service Journal*, 1(3), 5–24.

Lim, E.T.K. and Cyr, D. (2010). Modeling utilitarian consumption behaviors in online shopping. An expectation disconfirmation perspective. *Proceedings 16th Americas Conference on Information Systems (AMCIS)*, Lima, Peru.

Lim, E.T. K. and Cyr, D. (2009). Modeling hedonic consumption behaviors in online shopping. *Proceedings for the Eighth Pre-ICIS HCI Research in MIS Workshop (HCI/MIS'08)*, Phoenix, AZ.

Lim, E., Tan, C.W., Cyr, D., Pan, S., and Xiao, B. (2012). Advancing public trust relationships in electronic government: The Singapore e-filing journey. *Information Systems Research*, 23(4), 1110–1130.

Marcus, A. and Gould, E.W. (2000). Cultural dimensions and global web user interface design. *Interactions*, July/August, 7, 33–46.

Mason, R.M. (2003). Culture-free or culture-bound? A boundary spanning perspective on learning in knowledge management systems. *Journal of Global Information Management*, 11(4), 20–37.

Matsumoto, D. (1994). *Psychology from a Cultural Perspective*, Brookes/Cole, Pacific Grove, CA.

McCoy, S., Galletta, D.F., and King, W.R. (2005). Integrating national culture into IS research: The need for current individual-level measures. *Communications of the Association for Information Systems*, 15, 211–224.

McKinney, V., Yoon, K., and Zahedi, F.M. (2002). The measurement of web-customer satisfaction: An expectation and disconfirmation approach. *Information Systems Research*, 13(3), 308.

Multilingual Search. (2008). World statistics on the number of Internet shoppers. Available at http://www.multilingual-search.com/world-statistics-on-the-number-of-internet-shoppers/ (accessed on April 23, 2012).

Multilingual Search. (2012). Global e-Commerce statistics: How avid are chinese online shoppers? Available at http://www.multilingual-search.com/global-ecommerce-statistics-how-avid-are-chinese-online-shoppers/17/04/2012/ (accessed on April 23, 2012).

Nantel, J. and Glaser, E. (March–June 2008). The impact of language and culture on perceived website usability. *Journal of Engineering and Technology Management*, 25(1–2), 112–122.

Niederman, F., Alhorr, H., Park, Y., and Tolmie, C.R. (2012). Global information management research: What have we learned in the past decade? *Journal of Global Information Management*, 20(1), 18–56.

Nielsen, J. (2001). *Designing for Web Usability*. New Riders Publications, Indianapolis, IN.

Okayazaki, S. and Rivas, J.A. (2002). A content analysis of multinationals' web communication strategies: Cross-cultural research framework and pre-testing. *Internet Research: Electronic Networking Applications and Policy*, 12(5), 380–390.

Ou, C.X. and Sia, C.L. (December, 2010). Consumer trust and distrust: An issue of website design. *International Journal of Human-Computer Studies*, 68(12), 913–934.

Palmer, J.W., Bailey, J.P., Faraj, S., and Smith, R.H. (2000). The role of intermediaries in the development of trust on the WWW: The use and prominence of trusted third parties and privacy statements. *Journal of Computer Mediated Communication,* 5(3). On-line journal. Available at http://jcmc.indiana.edu/vol5/issue3/palmer.html (accessed on April 8, 2013).

Peterson, K.K. and Kim, C. (2003). Perceptions on IS risks and failure types: A comparison of designers from the United States, Japan and Korea. *Journal of Global Information Management*, 11(3), 19–38.

Reichheld, F.F. and Schefter, P. (2000). E-loyalty: Your secret weapon on the web. *Harvard Business Review*, 78(4), 105–114.

Reimann, M., Zaichkowsky, J., Neuhaus, C., Bender, T., and Weber, B. (2010). Aesthetic package design: A behavioral, neural, and psychological investigation. *Journal of Consumer Psychology*, 20(4), 431–441.

Ronen, S. and Shenkar, O. (1985). Clustering countries on attitudinal dimensions: A review and synthesis. *Academy of Management Review*, 10, 435–454.

Rose, G. and Straub, D. (1998). Predicting general IT use: Applying TAM to the Arabic world. *Journal of Global Information Management*, 6(3), 39–46.

Rosen, D. E. and Purinton, E. (2004). Website design: Viewing the web as a cognitive landscape. *Journal of Business Research*, 57(7), 787–794.

Rousseau, D.M., Sitkin, S.M., Burt, R.S., and Camerer, C. (1998). Not so different after all: A cross-discipline view of trust. *Academy of Management Review*, 23(3), 393–404.

Sagi, J., Carayannis, E., Dasgupta, S., and Thomas, G. (Jul-Sept, 2004). ICT and business in the new economy: Globalization and attitudes towards eCommerce. *Journal of Global Information Management*, 12(3), 44–64.

Sheng, H. and Joginapelly, T. (2012). Effects of web atmospheric cues on users' emotional responses in e-commerce. *Transactions on Human-Computer Interaction*, 4(1), 1–24.

Sia, C., Lim, K.H., Leung, K., Lee, M.O., Huang, W., and Benbasat, I. (2009). Web strategies to promote internet shopping: Is cultural-customization needed? *MIS Quarterly*, 33(3), 491–512.

Simon, S.J. (2001). The impact of culture and gender on web sites: An empirical study. *The Data Base for Advances in Information Systems*, 32(1), 18–37.

Singh, N. and Baack, D.W. (2004). Web site adaptation: A cross-cultural comparison of U.S. and Mexican web sites. *Journal of Computer-Mediated Communication*, 9(4). Available at http://www.ascusc.org/jcmc/vol9/issue4/singh_baack.html (accessed on March, 2012).

Singh, N. and Boughton, P.D. (2005). Measuring website globalization: A cross-sectional country and industry level analysis. *Journal of Website Promotion*, 1(3), 3–20.

Singh, N., Xhao, H., and Hu, X. (2003). Cultural adaptation on the web: A study of American companies' domestic and Chinese websites. *Journal of Global Information Management*, 11(3), 63–80.

Smith, R., Deitz, G., Royne, M., Hansen, J., Grünhagen, M., and Witte, C. (2013). Cross-cultural examination of online shopping behavior: A comparison of Norway, Germany, and the United States. *Journal of Business Research*, 66(3), 328–335.

Sondergaard, M. (1990). Hofstede's consequences: A study of reviews, citations and replications. *Organization Studies*, 15(3), 447–456.

Srite, M. and Karahanna, E. (2006). The role of espoused national cultural values in technology acceptance. *MIS Quarterly*, 30(3), 679–704.

Srite, M., Thatcher, J.B., and Galy, E. (2008). Does within-culture variation matter? An empirical study of computer usage. *Journal of Global Information Management*, 16(1), 1–25.

Steenkamp, J. and Geyskens, I. (2006). How country characteristics affect the perceived value of web sites. *Journal of Marketing*, 70, 136–150.

Straub, D.W. (1994). The effect of culture on IT diffusion: E-mail and fax in Japan and the US. *Information Systems Research*, 5(1), 23–47.

Straub, D., Keil, M., and Brenner, W. (1997). Testing the technology acceptance model across cultures: A three country study. *Information and Management*, 33, 1–11.

Sun, H. (2001). Building a culturally-competent corporate web site: An explanatory study of cultural markers in multilingual web design. *SIGDOC '01*, Los Alamitos, CA, October 21–24, pp. 95–102.

Sun, H. and Zhang, P. (2006). The role of affect in IS research: A critical survey and a research model. In: Zhang, P. and Galletta, D. (Eds.), *Human-Computer Interaction and Management Information Systems: Foundations*. M E Sharpe Inc Armonk, New York, Chapter 14, 295–329.

Szymanski, D.A. and Hise, R.T. (2000). E-satisfaction: An initial examination. *Journal of Retailing*, 76(3), 309–322.

Tan, F., Yan, L., and Urquhart, C. (2007). The effect of cultural differences on attitude, peer influence, external influence, and self-efficacy in actual online shopping behavior. *Journal of Information Science and Technology*, 4(1), 3–26.

Tian, R. and Emery, C. (2002). Cross-cultural issues in internet marketing. *The Journal of American Academy of Business*, 1(2), 217–224.

Tractinsky, N. (2004). Toward the study of aesthetics in information technology. In: *Proceedings for the Twenty-Fifth International Conference on Information Systems*, Washington, DC, pp. 771–780.

Tsai, U., Chang, T., Chuang, M., and Wang, D. (2008). Exploration in emotion and visual information uncertainty of websites in culture relations. *International Journal of Design*, 2(2), 55–66, ISSN 1991–3761.

Vyncke, F. and Brengman, M. (2010). Are culturally congruent websites more effective? An overview of a decade of empirical evidence. *Journal of Electronic Commerce Research*, 11, 14–29.

Wakefield, R.L., Wakefield, K.L., Baker, J., and Wang, L.C. (2011). How website socialness leads to website use. *European Journal of Information Systems*, 20, 118–132. Published online September 21, 2010.

Witmer, B.G., Jerome, C.J., and Singer, M.J. (2005). The factor structure of the presence questionnaire. *Presence*, 14(3), 298–312.

Yamagishi, T. and Yamagishi, J. (1994). Trust and commitment in the United States and Japan. *Motivation and Emotion*, 18(2), 129–165.

Yoo, Y. and Alavi, M. (2001). Media and group cohesion: Relative influences on social presence, task participation, and group consensus. *MIS Quarterly*, 25(3), 371–390.

Yoon, S.-J. (2002). The antecedents and consequences of trust in online-purchase decisions. *Journal of Interactive Marketing*, 16(2), 47–63.

Zahedi, F.M., Van Pelt, W., and Srite, M. (2006). Web documents' cultural masculinity and femininity. *Journal of Management Information Systems*, 23(1), 87–128.

V

Human–Computer Interaction and User Experience

32

Usability Engineering

John M. Carroll
*The Pennsylvania
State University*

Mary Beth Rosson
*The Pennsylvania
State University*

32.1 Introduction

The essence of usability as a software development goal is that tools and systems can be designed to facilitate and enhance desirable experiences for their users—ease, transparency, engagement, fun, usefulness, self-efficacy, satisfaction—as well as to mitigate undesirable experiences, such as confusion, frustration, anger, disappointment, lack of confidence, and so forth. Usability is a *nonfunctional* requirement in that it describes the criteria for assessing the quality of a system beyond merely specifying the functions that the system performs, its *functional* requirements.

As a supporting framework for achieving usable systems, the insights, guiding principles, and methods of usability engineering are deceptively simple and straightforward. The first is that ensuring the usability of interactive systems is necessarily a *process*: usability cannot be reduced to fixed laws, scripts, or guidelines. The usability process must be initiated at the beginning of the system development process, and continue throughout and beyond the traditional scope of system development. This is a painfully important insight, because it is far more difficult and costly to manage a process than to follow a rule.

The second principle is that usability can indeed be *engineered*: to a considerable extent, usability can be systematically managed through replicable and teachable methods to meet explicit and measurable objectives (often called usability requirements). Usability can be empirically tracked and iteratively improved throughout the system development process to ensure better outcomes. In other words, usability is not a matter of trial and error in development, followed by empirical checking at the end to see if things worked out.

The third principle is that usability is an *open concept*: conceptions of usability have evolved throughout the past 40 years, and continue to evolve. One obvious reason for this is that the nature of interactive systems has evolved very rapidly. The concepts appropriate to understand and manage the usability of a programming environment in 1977, a PC database in 1987, a website in 1997, or a mobile personal system in 2007 are all quite distinct. But, another reason for the continuing evolution of usability is that our understanding of usability and our arsenal of usability engineering tools and methods have improved continually. The trajectory of usability through the past four decades is that it has become far more articulated, far more nuanced, and far more comprehensive.

These three principles are obviously interdependent and contingent. Thus, the scope and rate of change in interactive systems is a key factor in the dynamic nature of usability itself, which, in turn, is part of why usability must be managed as a process. It is difficult to see any narrowing or slowing in the evolution of interactive technologies and applications. Change is clearly accelerating.

32.2 Emergence of Usability Engineering

32.2.1 Software Psychology

The concept of usability and the practices of usability engineering began to take shape in the context of 1970s' mainframe/terminals system configurations with line-oriented displays, batch processing for many interactions, and purely textual command-driven interfaces (sometimes with cosmetic menus). Early usability research struggled against two persistent fallacies of system development. The first is the tendency of programmers and designers to assume that their own experiences are valid indicators of what other people will experience in using their software. This seems naively optimistic when stated in the abstract, but it is surprisingly seductive. One still sees it regularly among students in introductory software design classes, but in the 1970s, it was a pervasive attitude among software professionals. The second fallacy is that usability flaws can usually be addressed through adding help and training support to the overall system. In fact, serious usability issues cannot be addressed by help and training. Indeed, designers must always recognize that adding help and training is ipso facto adding complexity to the overall system design.

Throughout the 1970s, usability was investigated as part of a project called "software psychology" (Shneiderman, 1980). This effort tried to confront the fallacies of usability with scientific studies of usability phenomena. Of course, in the 1970s, ordinary people did not use computers, and there were no "users" in the sense we understand that term now. Rather, there were a variety of data processing professionals who designed, implemented, maintained, and used mainframe systems and applications, for the most part, as their full-time jobs. The vision of software psychology was to apply concepts and methods of experimental psychology to better understand the problems of professional programmers, including the then-emerging ranks of application programmers. Three examples of early software psychology contributions are with respect to commands and command languages, design problem solving, and collaborative work; these will be discussed in the following paragraphs.

There are many fairly direct mappings of phenomena from experimental psychology into programming. Thus, especially in the 1970s, programs were long lists of expressions, and command interfaces were long lists of command-function pairs. Psychologists had demonstrated severe limitations of human working memory, including distinctive patterns such as primacy and recency, the tendency in memorizing lists to recall items early and late in the list better than items in the middle of the list. It was also known that people spontaneously impose sophisticated structuring on verbal material. Thus, if a command "down" moved the cursor ahead one record, people automatically assume there will be a command "up" and that it will move the pointer back one record. From this, it could be expected that memory failures, including particular patterns such as primacy and recency, and confusions created by badly designed command languages would be pervasive and debilitating in a text-based computing paradigm. And, indeed they were (e.g., Card et al., 1983; Carroll, 1985; Grudin and Barnard, 1984; Norman, 1988).

The psychology of problem solving had demonstrated that humans often decompose complex problems into subproblems and lose track of the dependencies across a subproblem, that they rely on metaphors and analogies in simplifying complex problems, that they frequently underanalyze problems, that they mistake merely familiar problem elements as being more important elements, and that they rarely optimize solutions (Carroll, 1987; Kahneman and Tversky, 1979; Reitman, 1965; Simon, 1972).

Of course, people are also still the best problem-solving engines we know of, and regularly generate deeply creative insights into complex problems, often "solving" problems that were too complex to even be stated precisely. Technologies can be crafted to address the inherent flaws of human problem solving, for example, representing problem analyses and decompositions in ways that help to expose missing parts (Norman, 1988). Technologies can also strengthen the strengths of human problem solving, for example, suggesting criteria for evaluating metaphors and analogies (Carroll and Kellogg, 1989). The technologies of the 1970s were poorly matched to the characteristics of human problem solvers, and this became more obvious as the decade progressed.

Finally, the psychology of programming identified issues and challenges in the coordination of work in teams. Among the most ambitious software projects of the 1960s was the development of IBM's Operating System 360 (Brooks, 1975). The size and schedule of this project required a large team, and a standard conception of the time was the "man-month," in this case, the work a single programmer could contribute in a month. However, reflections on the project revealed that adding team members did not increase team productivity linearly. Indeed, there was a severe drop in productivity owing to the need for additional team members to spend much more time coordinating their individual efforts. In this case, the analysis of software development experience slightly led the basic psychology of collective effort that is its foundation. But since the 1960s, social psychologists have extensively investigated the issues in optimizing collective efforts (West, 2012).

32.2.2 Early Human–Computer Interaction

Software psychology primarily focused on characterizing usability issues with respect to a scientific foundation of psychological concepts and methods. It was largely descriptive and reflective. In the early 1980s, computing was broadly transformed by the emergence of the personal computer. In the course of only a few years, the problems of professional programmers receded as the focus of research on usability. Software psychology was absorbed into a larger area called Human–Computer Interaction (HCI). HCI expanded the software psychology agenda to be more proactive with respect to new concepts, tools, and methods, including a fundamental rethinking of the software development process. This was good timing; the "waterfall" conceptions of software development had failed even for 1970's conceptions of users and applications. In the 1980s, developers were eagerly seeking new paradigms that could help them design for the growing PC world, where the focus was the user interface and the user. HCI took on this challenge, and through the next decade transformed computer science, establishing what is now taken for granted: software that is not highly usable is just a failed software.

Early HCI embraced three distinctive goals. All three contrast with the goals of software psychology, and all three positioned HCI as a focal area of challenge and change in computing. First, HCI was conceived of as an area in which new science and theory would emerge and develop, not merely as an application area for existing basic knowledge. Initially, HCI was an area of cognitive science, and later it also sought to make fundamental contributions to sociology and anthropology. Second, HCI was conceived of as an area in which new models and techniques for software design and development would emerge and develop, not merely as a project to enrich or improve existing software development models. Third, HCI was conceived of as a technology area in which new user interface software and new applications, new software architectures and tools, and even entirely new types of software would be developed and investigated. Remarkably, all of these goals were achieved in the next couple of decades.

In the area of science and theory, HCI adapted descriptive models from many realms of cognitive and social science, including visual and auditory perception, models of routine cognitive skill, mental models, common ground, distributed cognition, activity theory, social psychology, and sociological studies of work and of community (Carroll, 2003). HCI also was a primary incubator for new theoretical perspectives and technical achievements that have influenced basic sciences, including integrated information processing models (Card et al., 1983), analyses of situated actions (Suchman, 1987), and conceptions of distributed cognition (Hollan et al., 2000), activity (Kaptelinin and Nardi, 2012), and experience

(McCarthy and Wright, 2004). This incredibly varied science base is used in many ways in usability engineering, including design techniques (such as metaphorical design), design and evaluation tools (Bellamy et al., 2011), theoretically grounded design rationales (Carroll and Rosson, 2003), and conceptual frameworks for critiquing designs (McCarthy and Wright, 2004).

In the area of models and techniques for software development, HCI framed the principle of iterative development, the claim that systems can never be successfully specified without extensive prototyping and refinement (Brooks, 1975; Carroll and Rosson, 1985; Gould and Lewis, 1985). The principle of iterative development entrained several further paradigmatic ideas that have shaped the development process for interactive systems. The paradigm of user-centered design requires that system design goals be aligned with the goals of the eventual users of the system. This is sometimes achieved through empirical requirements development, empirical studies of user activities and preferences that are then used to guide envisionment of future systems. Sometimes, it is achieved through direct user participation in the design team, called participatory design.

Iterative, user centered, and participatory design alter the system development process in fundamental ways. These paradigmatic commitments entailed a wide variety of specific usability engineering concepts and techniques to gather data from or about users and to represent and interpret that data. For example, creating scenario descriptions of user activity—essentially stories—is now a core technique in interactive system design, often called scenario-based design (Carroll, 2000; Rosson and Carroll, 2002). As late as in the 1980s, this was seen as a radical approach to software development; today, not doing it would be radical.

In the area of software technology, HCI research produced software architectures and tools that modularized graphical user interfaces and their many successors among personal devices, embedded systems, immersive systems, etc. The initial design strategy was to isolate user interface functionality in a User Interface Management System (UIMS), to make it easier to adjust characteristics of the user interface through the course of design iterations, without entailing disruption to underlying system functionality. Although such a clean separation proved to be intractable, UIMS work helped to establish a software requirement of malleability to enable iterative design.

32.2.3 Managing Experience Design

HCI is inescapably a design discipline, but the term *design* has a variety of different meanings. HCI is a design in the sense that creating interactive systems is always an ill-structured problem; a problem that can only be stated precisely after it is solved (Reitman, 1965). But, interactive systems and applications have expanded into every arena of human activity, and as the richness of user experiences afforded and evoked by interactive systems has expanded, HCI has also become design in the sense that it shapes our interaction with the social and material world, such as consumer product designs, mass media designs, and so forth. Experience design, as it is sometimes called, is now an important focus in HCI and usability engineering (Diller et al., 2005).

This entails a further expansion of what usability and usability engineering are about. The user's experience is no longer just an issue of understanding what is going on well enough to achieve a specific task goal. That conception of usability is still valid, still necessary, but it is no longer sufficient for understanding the user experience. It can be argued that the user experience was always much more complicated than merely easy-to-learn/easy-to-use. As early as the mid-1980s, it was argued that usable systems should be fun to interact with (Carroll and Thomas, 1988). One way to see this is that until the field had developed far enough to effectively and broadly address the simplest conceptions of usability, such as ease of learning and use, the much richer issues that are currently in focus could not get to center stage.

Throughout the past four decades, the concepts and practices of usability engineering have expanded to manage increasingly richer and more nuanced conceptions of usability. The field has been shaped by external factors, a technology context of new devices and networking infrastructures, new systems, services, and applications, new user interface designs, and new types of users with an ever-widening range of interests and motivations, domain knowledge and knowledge of computing, etc. But it has also been shaped by an astounding diversity of skills and sensitivities among its researchers and practitioners. There is now a solid core of usability engineering tools and practices, as well as an active and rapidly advancing frontier of new concepts and techniques growing year by year.

32.3 Core Methods of Usability Engineering

The tools and practices of usability engineering are often applied within a life cycle framework, for instance, scenario-based usability engineering (Rosson and Carroll, 2002), contextual development (Beyer and Holtzblatt, 1997), or the wheel model (Hartson and Pyla, 2012). The goal of such frameworks is to recommend and demonstrate the use of a variety of techniques at different points, often in a repeated and iterative fashion, through the development of a system.

A high-level summary of how core usability engineering activities might be integrated within a scenario-based development (SBD) life cycle appears in Table 32.1 (the hypothetical example concerns a new e-commerce website). In the SBD framework, scenarios and associated design rationale serve as a central representation that assures a pervasive focus on the task goals and experiences a design team envisions for users. The scenarios also help team members with differing expertise (e.g., social vs. computational science) to establish and maintain common ground as they discuss usability engineering trade-offs. At different points through the cycle, the usability engineering team may use a range of methods, and in some cases the same basic method may be used in support of different usability development goals.

32.3.1 Direct Observation

The most straightforward way to learn about and respond to human needs and preferences is to observe them as they carry out activities that rely on the system or design concept under development. The types of methods and analyses selected for these activities depend on many factors, including the goals of the observational studies, the resources available to the evaluation team, and the nature of the interactive system being developed. Depending on where a team is in the product development life cycle,

TABLE 32.1 Usability Engineering Methods for Phases of Scenario-Based Design

Life Cycle Phase	Examples of Usability Engineering Methods
Problem analysis: Scenarios synthesizing current practices with associated trade-offs	• Contextual inquiry of administrative employees using BuyOnlineNow.com to order office supplies • Convenience survey of current users of amazon.com • Focus groups discussing personas and scenarios synthesized from contextual inquiry and survey
Design: Increasingly detailed scenarios of new activities with associated rationale	• Claims analysis and reasoning about trade-offs in problem scenarios to guide redesign • Participatory design sessions with administrative employees mocking up rough design scenarios for online purchasing tasks
Evaluation: Formative evaluation that guides redesign, complemented by summative evaluation to measure success	• Think-aloud study of administrative employees working with a low-fidelity prototype • Performance study measuring time and errors for several benchmark tasks (e.g., locating item, completing purchase) • Convenience survey of users of a new system once fielded to gather subjective reactions, loyalty, etc.

there may be different evaluation goals in focus (Scriven, 1967). For example, early in the life cycle, *formative* methods are an important mechanism for producing rich and often qualitative findings that can guide iterative design activities. Toward the end of the life cycle, or at critical milestones, the team may instead use *summative* methods to characterize whether and how well usability objectives have been met. In this sense, some of the very early observational studies may include analysis of precursor systems that the target users already employ for similar tasks.

A usability evaluation team may have the luxury of a relatively open-ended mission, spanning many months and having access to a diverse team of usability experts and target users, but more commonly their activities are constrained by time and other resources available. Because of such constraints, the team may need to address their studies to a small set of features or usability questions that imply the use of specific observational methods. For instance, answering a question about two different dialog designs for an e-commerce system implies a performance-oriented observation, perhaps complemented by user satisfaction. In contrast, a more exploratory study of a novel information structure for the same e-commerce site might be best served by a study gathering users' verbal protocols. In the following, we summarize several common observational methods found in usability engineering, ordered roughly by the goals and resources available to the team.

32.3.1.1 Workplace Ethnography

A workplace ethnography would typically be carried out either in the early phases of a project, or perhaps after a project has been completed and fielded. Its goals are to study the established use of some system(s) of interest, without any intervention or structure provided by the observers. In fact, the goal of an ethnographer is to "disappear" into the background of a workplace setting. The primary source of data is the analyst's field notes about what he or she is observing; evaluators may introduce more comprehensive records like audio- or video-recordings, but generally would do so only if these recordings can be made in an unobtrusive fashion. In a pure ethnography, the evaluator would attempt to record everything that is happening, perhaps with a special focus on technology usage episode, but to do this in as unbiased as fashion as possible. Because such evaluations rely entirely on natural work settings and activity rhythms to reveal user tasks or reactions "of interest," they generally require long time periods—weeks or even months—of regular observation. A classic example of workplace ethnography that was used to fuel design thinking can be found in the study of Hughes et al. (1992) of an air traffic control room.

The ethnographic evaluation that is more commonly found in usability engineering contexts is more brief and necessarily more focused in its observation, sometimes referred to as "quick and dirty" or rapid ethnography (Hughes et al., 1995). The general concepts of rich observation and analyst-centered interpretation are still in place, but the abbreviated length and reduced scope assume that an expert observer can enter a situation where design-relevant behavior is underway, and in a relatively short time extract key practices, preferences, and obstacles from workers' behaviors. In the e-commerce example, a quick and dirty ethnographic study may involve observations of shoppers in a store as well as interviews with people who have used existing e-commerce systems; it might also involve the collection and analysis of web interaction logs and context-specific artifacts such as a "filled" online shopping cart, a shopper profile, recommendations made by the system, and so on.

32.3.1.2 Contextual Inquiry

Contextual inquiry provides a more focused approach to field observation; these methods were first described and implemented by Hugh Beyers and Karen Holtzblatt (Beyer and Holtzblatt, 1995) as a centerpiece of their contextual design framework. Like ethnographic methods, this observational technique requires evaluators to collect data from the field as workers engage in their everyday activities. It is particularly useful during the requirements analysis conducted in the early phases of interactive product design, because it produces a number of analysis documents that encode important artifacts currently in use, including technology, physical, social, or organizational factors that impinge on current practices, and so on. The goals of this technique are similar to rapid ethnography, but the data collection includes more intervention from the observer.

A typical setting for contextual inquiry might include an observer who sits near the workstation or desk of a worker engaged in everyday tasks. The observer watches and takes notes (perhaps also digital recordings), as would an ethnographer. However, the observer's presence is more salient to the worker (e.g., because of proximity) and she is expected to ask questions at times when something of "interest" takes place. This requires the observer to begin the session with a particular focus, perhaps even a set of task-related questions, and to inject these concerns when as relevant. Early on in a system design process, the data collection may be guided by very generic prompts, for example, "Why did you do that?" or "What just happened?" but at times it may be appropriate to make these more specific, for instance, "How did you know it was time to press the Submit button just now?" In our e-commerce example, a contextual inquiry could be directed at either (or both) shoppers in a store setting (e.g., shadowing and observing but also asking shoppers about their search and decision making) or shoppers who interact with an existing e-commerce site.

32.3.1.3 Think Aloud Observations

When usability evaluators want to probe the goals, expectations, and reactions of a user to a particular set of user interface features or controls, they may conduct a session in which they ask users to "think out loud" about their experiences while attempting a task. Such sessions could be done in a field setting (e.g., in conjunction with contextual inquiry visits), but more commonly take place in a laboratory setting where participants are recruited to work with a system while their mental experiences are vocalized and recorded. The resulting auditory file is typically synchronized with a log of their system interactions, and is often termed a verbal protocol (e.g., as in the methods pioneered by cognitive psychologists studying problem solving [Ericsson and Simon, 1980]).

In a typical think aloud session, a representative user will be asked to attempt a specified task (e.g., locate and order *The Girl with the Dragon Tattoo* using Amazon.com), while also saying out loud her moment-to-moment interpretations of what she sees, her plans for what to do next, and her reactions to the system's responses at each step. The evaluator's goal in this is to infer important aspects of the user's *mental model* (Carroll et al., 1987; Payne, 2003), that is, her current knowledge of what an interactive system can do for her and how to instruct it to meet these goals. These results are particularly useful if the software being designed relies on novel or unusual information abstractions (e.g., new vocabulary or conceptual relationships), or if it relies on user interface controls that have not been seen or used in similar systems. These studies are most useful when the primary objective for the system is that it requires little, if any, learning for new users. Because a think-aloud requirement creates a secondary—and often distracting—task with respect to the primary usage task, these methods would normally be used only to serve formative evaluation goals.

32.3.1.4 Human Performance Studies

The most common form of observational study—and what most usability practitioners mean when they talk about doing a user test—is a controlled study of representative users' performance with an interactive system. The method is straightforward: a set of typical tasks is designed to exercise the aspects of a system needing feedback, and users with characteristics similar to those of the project's target users are recruited to carry out these tasks (Card et al., 1983).

For instance, continuing with the e-commerce example, a team might design four to five product search or purchasing tasks that involve different subdialogs within a high-fidelity prototype; perhaps some rely on recommendations or reviews, whereas others are more specific examples of item location or purchase decisions. The data collected in these cases would be objective in form and might include: (1) the number of steps required to complete the task; (2) the actual path followed; (3) the number of missteps taken; (4) the time required to complete the task, and so on. These data would be summarized for comparison to preestablished task performance expectations, perhaps based on competitive analysis, on pilot studies with an earlier version of this system, or mathematical human performance models. In a formative evaluation setting, such data would likely be complemented by other methods such as those described earlier, but for summative evaluation, they might be used on their own to determine relative success or failure of the system with respect to user performance.

32.3.2 User Expectations and Preferences

Beyond direct observation of behavior, usability engineers normally seek to gain insight into users' expectations, beliefs, or experiences with respect to the system or design concept. One challenge in this process is assuring that test users report their actual feelings as accurately as they are able. There is a general tendency to exhibit socially desirable behavior, and in a system test, this may bias participants to be more positive than their experiences warrant. Another challenge is to evoke reactions that are nuanced and specific, rather than "Sure, I like this!" or "There is no way I would ever use this system!" Finally, usability engineers must be careful to frame users' reports of their expectations and preferences as subjective data that can be affected by a wide variety of individual differences and contextual factors.

32.3.2.1 Focus Groups

Usability engineers may conduct focus group sessions in the early phases of a software development project; group discussions are particularly useful for exploratory data collection. A group of representative users (or perhaps other stakeholders, for instance, decision makers) is gathered together and guided through a discussion about issues relevant to the current phase of system development. Depending on the level of design and development, the discussion may be anchored by demonstration of a design storyboard or even an early prototype. In SBD, scenarios are often communicated informally as textual narratives that revolve around personas (Cooper, 1997), but with a bit more effort, they can be elaborated using sketches or screen shots that capture key aspects of the imagined interaction sequence.

Early on in design, before system functionality or user interactions have been proposed, a focus group may be shown other existing technologies that are related to the design ideas being considered; in our e-commerce example, a session may start out with a few videotapes of people using amazon.com, so as to ground the discussion on particular aspects of current approaches. Depending on the time and resources available, the facilitator may rely instead on a semistructured script of questions or topics. Focus group discussions would normally be recorded for later review; a video of the participants interacting with each other can also be helpful in later analysis of when and how different ideas were generated. A typical result from a focus group would be a summary of ideas raised, perhaps organized by the question or topic that evoked the ideas. If the group has discussed specific design features, there may be a more detailed analysis of the positive and negative comments that were made, or the suggestions for changes.

32.3.2.2 Interviews

An interview is similar to a focus group, but the questions are directed to an individual. As with focus groups, the interview may be more or less structured and may include demonstrations. Typically, the interview is semistructured such that it includes a set of high-level questions but the evaluator is left free to pursue interesting details that come up, or even to engage in an entirely new topic that seems to be a rich source of information from a given participant.

In an interview, a person answers questions individually, so it is important to establish rapport with her quickly so that she feels comfortable enough to expand on her personal experiences and attitudes. One technique for establishing rapport is to begin the questioning with some background discussion about the participant's current work (or leisure, education, etc.) setting. General background information is simple and easy to provide, and has the added advantage of providing participant-specific context that the interviewer can later inject into questions or follow-up probes. For example, in the e-commerce setting, the usability engineers might ask questions about how often and for what sorts of purposes shopping trips are made, as well as what the interviewee most enjoys or finds irritating about shopping as an activity. Another useful technique when interviewing is to prompt for examples or illustrations of points being made; by asking the participant to remember and describe specific cases, the interviewer can both clarify what a participant is trying to convey and, in general, evoke a richer experiential report.

32.3.2.3 Questionnaires and Surveys

Usability engineers often develop questionnaires or surveys to gather users' expectations or reactions as a complement to open-ended methods. The analysis of qualitative data gathered during direct observation (e.g., in a think aloud study) can be tedious and time-consuming. In contrast, a well-designed survey can gather probe a number of specific user experiences using rating scales or other item types that are simple for users to complete and straightforward to summarize. In fact, usability professionals have adapted the psychometric methods of sociology and psychology to generate and disseminate several validated scales for usability assessment. For example, the QUIS instrument (Questionnaire for User Interface Satisfaction) was developed in the late 1980s as a research project in the HCI lab at the University of Maryland (Chin et al., 1988); QUIS has been validated and refined over the years and is available for licensing to practicing usability engineers (http://lap.umd.edu/quis).

Existing and validated scales are useful in positioning users' general reactions or preferences. However, by their nature, such instruments are conceptualized at a relatively general level. QUIS includes subscales for general usability objectives such as screen design, vocabulary use, and support for learning. However, usability engineers who need to evaluate novel design features (e.g., a small touch screen or natural language input) will need to extend such instruments with items specifically oriented to assess these features. These additional items will be particularly critical in the formative phases of system development, when the design team is deciding which new ideas to pursue and how to refine them.

When used in tandem with direct observation studies, usability engineers typically prepare a brief background survey that is administered before the tasks are attempted; this survey can be used to gather participants' expectations or other preexisting knowledge that may (1) influence how they approach the task; or may (2) change as a result of working with the new system. A posttask survey may repeat some of these items, enabling an analysis of *changes* in users' expectations or preferences, contrasting their responses before and after their usage experience. For example, in the hypothetical e-commerce project that includes a novel interface for finding and reading product reviews, the background survey might ask users about the perceived cost of finding or absorbing product reviews, and compare ratings on this question to the same item presented after the users try out the new approach.

In addition to the development or adaptation of usability-related rating scales, usability questionnaires and surveys often include open-ended questions (e.g., describe the three things you liked the most about the system you just used). As for other qualitative methods, these items are particularly useful early on in the development process when the goal is to develop as rich an understanding as possible of users' experiences. The open-ended items may also be used to probe a participant's baseline comprehension of specific concepts or procedures that are incorporated into the system design. Responses to open-ended questions may be summarized or coded in a fashion similar to transcripts of interviews or focus groups.

Survey methods are also used for requirements gathering in advance of designing a system, or for field-testing of new software after it has been launched. In these cases, considerable effort should be spent in developing and refining a set of questions that addresses the trade-off between information gathering and survey length, simply because in general participants may not be willing to complete a survey that requires more than 10–15 min of their time.

32.3.2.4 Participatory Design

Focus groups, interviews, and surveys can be used to gather preferences or reactions to an existing or proposed interactive system. In these cases, the design team has the responsibility to make sense of the information gathered from users, determining whether and how best to refine the team's current concepts or prototype system. A more direct way of injecting users' preferences or concerns into a design project is to engage them as participants on the design team. Participatory design activities can take place at specific points in the life cycle (Muller, 2007) or as a long-term evolutionary process (Carroll et al., 2000; Greenbaum and Kyng, 1991). The general concept is similar to a focus group; however, the goal is not to explore concerns and ideas but rather to generate new design concepts, often represented physically using mock-ups of varying levels of fidelity.

In projects that incorporate participatory design, a single set of representative users often meets repeatedly with the design team and collaborates directly on extending or refining the existing analysis or design representations. For example, early on the process they may help to analyze data collected from field studies, such as snippets from videos collected from earlier fieldwork (Chin et al., 1997). Once a problem situation has been analyzed, the group of prospective users may work with the designers to develop or refine user descriptions or task scenarios that envision new or revised activities (Carroll et al., 2000). As the design ideas become more concrete, the users may also help to sketch out user interface displays or interaction sequences (Muller, 1991). Finally, if the software is being constructed in an open and iterative fashion, and if appropriate end user development tools are provided, the user representatives may even operate directly on the software under construction to make changes to a prototype (Costabile et al., 2004).

32.3.3 Analytic Methods

Software projects do not always have the time or resources to recruit and involve representative users or their surrogates in the development process. Alternatively, they may be working within a task domain where detailed performance studies and optimizations are crucial to the success of the project (e.g., safety- or time-critical problem domains). In these situations, a usability engineer may choose to apply analytic methods—techniques that rely on the usability expertise of professionals to identify and manage usability concerns. Like empirical methods, analytic approaches vary in their precision and the corresponding cost of application.

32.3.3.1 Heuristic Evaluation

One class of analytic methods draws from heuristics or high-level guidelines to structure an analyst's review of a design. The interactive software concepts under review might be preliminary (e.g., initial ideas that have been documented through scenarios or storyboards), or it may be a functioning prototype. The heuristics may be a list of best practices that have been summarized across a number of different projects—perhaps within a "family" of interactive applications such as the user interface guidelines used by Apple or Microsoft. They might also be drawn from expert usability practitioners (e.g., http://www.useit.com/papers/heuristic/heuristic_list.html) or HCI researchers (e.g., the guidelines found in Shneiderman et al., 2009). For instance, one common heuristic is to "Speak the user's language," implying a careful focus on the terms and phrasing used by the system's user interface controls and other explanatory text. An analytic evaluation that used this guideline as a heuristic would result in a detailed critique of vocabulary-based issues, perhaps broken down into different segments of the user population. Of course, most such evaluations would be addressing multiple guidelines or heuristics at once, leading to an extensive report at the end of the analysis. Large companies often maintain internal guidelines that address usability issues; depending on the market for the product, one or more national or international standards may also be consulted (see e.g., http://zing.ncsl.nist.gov/iusr/documents/CISU-R-IR7432.pdf). Thus, a professional usability engineer is expected to gain expertise with many forms of standards and guidelines.

The process of applying usability heuristics and guidelines may range from an ongoing background activity (e.g., an e-commerce usability engineer would always be considering issues of web page accessibility by vision-impaired users) to concrete usability inspection episodes. For instance, an expert might conduct a usability inspection of the e-commerce application in our example by receiving a specified set of tasks for product browsing, searching, purchasing, and so on from the design team, and then enacting a careful step-by-step inspection of these tasks in light of the usability heuristics or guidelines provided by the organization. The inspection might be conducted by an individual expert, by a pair of experts, or by multiple experts conducting inspections independently. The general goal is to provide a report of usability problems that are evoked by the heuristics; the problems are often further classified by severity, the price of addressing the problem, or other company priorities.

32.3.3.2 Design Rationale

The analysis of a designed artifact's design rationale draws from the precepts of ecological science originally pioneered by Gibson (1966, 1979) and Brunswick (1956). Brunswick showed how some of the central perceptual phenomena identified and investigated with line drawings, such as visual illusions, could be explained in terms of the ways that objects align and occlude in the physical world. Gibson showed that perception in movement permits direct recognition of higher-order optical properties such as the expansion pattern one sees in approaching a textured scene or surface. Gibson concluded that real-world perception typically consists of direct recognition of complex properties, which he called "affordances." This contradicted the most fundamental assumption of perceptual psychology, namely, that higher-order properties, such as being an object or being something that can be poured, are inferred on the basis of piecemeal recognition of local features such as contours, edges, and angles. In HCI, the use of design rationale leverages affordances, arguing that a useful analysis for usability is to consider in detail the affordances—or consequences—of an artifact's features for its use in the world. Typically, such analyses produce trade-off descriptions, because almost every system feature, one can imagine, implies a mix of usability pros and cons for different users in different settings.

One approach to developing and maintaining usability design rationale is *claims analysis*, a method used to hypothesize features of an artifact in use (e.g., as illustrated by a usage scenario) that have causal relations with a set of positive and negative consequences for end users (Carroll, 2000; Rosson and Carroll, 2002). Claims analysis was developed to articulate the causal relations inhering in artifacts-as-theory (Carroll and Kellogg, 1989). It has been incorporated as a central reasoning and documentation method in scenario-based design: throughout the evolution of the design scenarios, the design team reasons about the claims, attempting to envision new features that can address the negative consequences hypothesized, while also maintaining or enhancing the positives (Rosson and Carroll, 2002).

Claims development is a form of analytic evaluation. At the same time, the process can set up a structure for claims-based empirical evaluations—this becomes a form of *mediated* usability evaluation that synthesizes analytic and empirical evaluation methods (Carroll and Rosson, 1995; Scriven, 1967). The claims that are created through an analytic process embody hypotheses about what users will expect, attempt, or experience when engaged in a particular activity context (scenario). Such hypotheses can be tested specifically (e.g., in a controlled study that engages users in the relevant activity) or indirectly (e.g., through a generalization process that leads to broader hypotheses that are evaluated through a series of tests).

In the context of claims analysis, design artifacts under construction (assuming some context of use) always have one or more design features in focus; the analysis helps a design team to reflect on and learn from their design discussions and experimentation. Examples of claims analysis as a usability engineering method can be found in case studies of collaborative learning software (Chin, 2004), community networking software and activities (Carroll and Rosson, 2003; 2008), and developmental online communities (Rosson and Carroll, 2013).

32.3.3.3 Cognitive and Computational Models

Another class of analytic methods has its foundations in mathematical models of human information processing (Card et al., 1983). Rather than inspecting a piece of software to identify problematic features, the cognitive modeling approach requires one to build a formal model of the knowledge and procedures needed to complete a task. For example, in the GOMS method, these models require the decomposition of a user interaction task into goals, operators, methods, and selection rules. The cognitive model produced in this fashion can subsequently be translated into a computational model;

performance parameters can be estimated for the model's primitive operations (e.g., pressing a key, moving a mouse, recognizing a word), and the resulting program can be executed to produce performance estimates (Hornoff and Halverson, 2003). Once a model has been created, it can be revised and re-executed to analyze whether specific changes in task design may lead to performance gains. The use of analytic methods at this level of detail is time-consuming, but may be cost-effective when the tasks being optimized are routine behaviors that will be executed hundreds of times per day by thousands of users (Gray et al., 1993; John and Kieras, 1996).

32.3.4 Rapid Prototyping and Iteration

The most direct way to learn about and respond to human needs and preferences is to observe them as they carry out activities related to the system or design concept under development; in other words, to use one or more of the empirical methods described in Sections 32.3.1 and 32.3.2. A key enabler of this approach is to build prototypes for testing, generally with the assumption that these prototypes will be built rapidly and in a way that enables iteration in response to feedback from users or other stakeholders.

As with other usability engineering methods, rapid prototyping can be conducted at differing levels of detail and veridicality with respect to the design vision. For instance, early in a project, one or more prototypes may be created using everyday office supplies such as cardboard, paper, and colored markers. This phase of usability engineering is often referred to as "paper prototyping." Practitioners have documented a range of useful techniques for constructing these low-fidelity prototypes (Snyder, 2003; http://www.paperprototyping.com).

A low-fidelity prototype might be created to represent key display screens or dialog steps, and users can be recruited to respond and "act upon" the design concepts; the problems and successes they have would be noted and used to guide refinement of the underlying concepts and interaction techniques. In a focus group setting, the team might arrive with storyboards or multimedia animations of current design scenarios, presenting these to the group of users for reactions and suggestions. As the design concepts become more stable, the team's software developers will begin to build software prototypes; this may be done in phases using different authoring tools with increasing fidelity, or in an evolutionary fashion where a single system is gradually expanded and refined until it is fully operational.

A number of usability engineering practitioners have debated and studied the impact of prototype fidelity in evaluating users' performance and subjective reactions to a system under development (e.g., Nielsen, 1990; Sauer et al., 2008; Virzi et al., 1996). Many of these studies have shown that low-fidelity prototypes can evoke results from user tests very similar to those carried out with more veridical systems, particularly with respect to detecting and classifying the most severe problems that a team would almost certainly want to address (Sauer and Sonderegger, 2009). Interestingly, users seem to adjust their aesthetic evaluations to some extent when presented with "rough" prototypes, perhaps recognizing that these surrogates should not be judged according to the same criteria (Sauer and Sonderegger, 2009). Practitioners might be expected to prefer lower fidelity methods, when possible, simply because of the reduced cost of creating them and the simplicity of iteration (e.g., redrawing a screen). However, it is clear that when realistic and comprehensive measures of user experience and performance are required, a fully functioning prototype must be provided. This is unlikely to be available until relatively late in the design and development life cycle. Rudd et al. (1996) discuss the practical trade-offs between low versus high fidelity.

32.3.5 Cost-Justifying Usability in Real-World Settings

As was implied in the discussion of analytic methods and the trade-offs in low- versus high-fidelity prototyping, the costs and benefits of doing usability engineering is an important issue in the real-world settings of companies doing software development. At the most extreme, if the target user population will accept and adopt a piece of software that has horrible usability characteristics (i.e., because it fulfills an essential function not available in any other way), why would a company spend time and money to analyze and support user interaction requirements? Even when it is clear that a product's usability will influence people's decision to buy and use it, how can a company justify an empirical study of its usability when the team's usability inspection can yield a rich number of issues to consider and address? Why build a high-fidelity prototype at all?

Because of such concerns, researchers and practitioners over the years have attempted to weigh the costs and benefits of conducting usability evaluations. Jakob Nielsen pioneered this thread of work with his proposals for "discount usability engineering" and his development of the heuristic evaluation methods (Nielsen, 1994; Nielsen and Molich, 1990). He has provided a number of analyses that contrast different usability engineering methods, estimates for how many test users are needed to discover critical problems, the role of analyst expertise, and so on. Several detailed discussions can be found on the website he maintains as part of his usability consulting business (useit.com). For example, he summarizes the observed usability improvement (as indicated by task time and errors) across the iterative design of four different case studies, estimating that each new version led to an improvement of about 40% in usability (useit.com/iterative_design). However, Nielsen does not provide the corresponding costs of iterating the designs, nor does he consider the variability in improvement one would expect across different design projects and teams.

One of the first practitioners to describe the use of cost–benefit analysis for usability engineering methods was Claire-Marie Karat (1990). Such a calculation depends on first estimating the benefits, for example, the seconds saved by future users who carry out tasks with a piece of software that has been improved through user testing (an average savings of 4 s per e-commerce transaction for a task carried out one time per week by 100 K users can quickly add up to many dollars of saved customer time). These benefits are then compared to the costs of conducting the iterative design: the costs of one or more usability engineers to analyze tasks, develop test materials, recruit users and conduct test sessions, and so on; the cost of the test users' time and associated testing equipment; the cost of responding to the user feedback, and so on. Of course, any such estimation process depends greatly on the assumptions made concerning the usability engineering methods to be used and their effectiveness in guiding redesign. A number of usability cost–benefit analyses developed by Karat and others can be found in the books edited by Bias and Mayhew (1994, 2005).

32.4 Current Challenges for Usability Engineering

Most of the methods for usability engineering emerged during the 1980s and 1990s, as computers became increasingly pervasive and important as personal productivity tools. During that era, the conventional view of a "user test" was a single user sitting at a single computer with single screen, working on a single-threaded task. Such a setting was quite amenable to engineering methods, because it could be structured, observed, and analyzed with considerable precision. However, the variety in interactive software and user interaction settings has increased enormously in the past two decades, creating tremendous challenges for usability engineers. We turn now to a discussion of trends in both technology and usage settings, along with corresponding implications for usability engineers working in these new arenas.

32.4.1 New Technology Contexts for Usability

Users no longer work alone. Increasingly the presumption is that people are connected to others through their technology and that often their day-to-day activities depend on those connections. As computing has become more ubiquitous in our everyday lives, the devices and associated input and output channels that people use for technology interactions have expanded. Finally, user interaction dialogs have become more intelligent, largely as a result of expanding access to large datasets and a concomitant growth in computational power.

32.4.1.1 The Web and Network-Based Systems

The World Wide Web has become a pervasive element of many technology-mediated activities. For some tasks, the web provides critical support for communication with other people or institutions; for others, it is the broad access to information that is most important. Even software that is designed to operate without a network connection will often rely on the web for delivery of regular updates or other maintenance services. As a result, the networked information and services available through the web must be recognized and analyzed when usability engineers are studying whether and how users will interact with a given software system.

A particularly common way for application designers to leverage the web is in user help and information services (Lee and Lee, 2007). While it is possible to package a core set of information with the software itself, a much larger and more dynamic body of information can be provided through the web. Web-based reference information and other documentation can be maintained and expanded in a centralized and convenient fashion as the application is used and evolves, making the help available more extensive. An application's users may also contribute content, perhaps by asking questions that become an index into the help provided, or by answering questions posed by other less-experienced users (Ackerman and Malone, 1990). More broadly, users regularly turn to the web at large for tips or guidance on how to solve a problem, even when the application offers an extensive help system of its own. Surveys of users' preferences for learning about novel technologies often report strong ratings for either asking a colleague or finding an example solution (Rosson et al., 2005; Zang and Rosson, 2009).

One challenge faced by usability engineers is that the web and its services are extremely dynamic and may result in rather different experiences for different users; the dynamic nature of the web also gives designers significantly less control over the user experience. Although the designers of a web application may carefully architect its content, look, and feel, the boundary between the designed application and the rest of the web may still be quite blurred in users' experiences. Considering our e-commerce example, it would be typical for the service to invite comments or reviews from customers, often with little oversight in the details of what and how comments are posted. It is also quite common for Internet applications to include links to other network content or applications, and it is not clear whether and how users integrate across applications when they follow such links.

32.4.1.2 Ubiquitous Computing

The mechanisms through which users interact with software have also expanded in the past two decades. While the most common context for use continues to be interaction with a personal computer using a mouse and keyboard, other options are increasingly common. Computing devices have decreased in size, and wireless connectivity to the Internet is available throughout the world through telephone or Wi-Fi networks.

As the variety in location and activity contexts of use have broadened, so have the options available for interacting with devices. Users who are moving around in the world may find it difficult to look at a small screen or to interact with mouse and keyboard. As a result, technologies that enable voice or touch input are becoming more common (Ishi, 2008; Vertanen and MacKay, 2010). Sensors built into the devices (e.g., a Global Positioning System for tracking location) can also provide useful information to an application that the individual no longer needs to input. However, adding an input channel that is beyond the user's control can raise a new set of usability issues, for example, problems with maintaining user engagement and appropriate awareness of information collected over the external channel (Brown and Laurier, 2012; Leshed et al., 2008). Because ubiquitous use implies interactions with different devices or modalities depending on the context, system support for synchronizing personal data and application preferences across these different contexts has also become critical.

Ubiquitous computing has usability consequences that are similar to that of pervasive network access; that is, it becomes very difficult to predict exactly where, when, and how an application will be activated to carry out a task. Using a mobile application while driving has very different usability requirements and consequences than using the same functionality in one's home or office (Williamson et al., 2011). Reading from a display that is outside in natural sunlight is a very different perceptual process than reading the same display under normal office lighting. These real-world, constantly changing usage contexts create considerable challenges for lab-based testing; so usability engineers must rely increasingly on fieldwork to evaluate the effectiveness of their designs.

32.4.1.3 Intelligent User Interfaces

A third general trend for interactive computing is the level of intelligence that software contains. With pervasive access to the Internet, an application can access and process a huge volume of data to "assist" a user with a task. This has become commonplace in e-commerce activities, where customer reviews and ratings are omnipresent, and where the bargains, prices, and even the presentation of products may be finely tuned to the person's prior shopping history. Many online retailers provide social recommendations—guesses as to which items a user might find of interest based on the behavior of other users judged to be similar—to guide users' shopping decisions. In this context, usability experts have worried about the types and amounts of information that is being tracked for individual users (Karat and Karat, 2010), even though most users count on these sorts of guidance.

Designers of learning support systems have long been interested in intelligent user interfaces. These include tutors that build and evolve a model of the learner's current knowledge state so that the learning activities and feedback can be carefully matched to the person's needs (London, 1992). But, like online recommendations, systems that adapt to a given user have become pervasive—even an application as commonplace as a web browser will now adapt to users' location and language; personal mail clients "learn" about what is and is not junk mail; and cell phone battery apps adjust to the owner's typical usage rhythms for energy-consuming features over a 24-hour period. Most of these intelligent adaptations are designed to improve usability and are usually easy to turn on and off. It is, however, still an open question whether an end user would be interested and willing to tune the adaptive model to make it even more effective (Kulesza et al., 2012).

Perhaps, the most salient example of intelligent user interfaces is the wide variety of robots that increasingly are being developed for special purpose activities such as cleaning (Marrone et al., 2002), information provision (Lee et al., 2010), or dangerous work environments like war zones or firefighting (Motard et al., 2010). Often, robots are designed as autonomous systems, but enough of them involve human interaction that an entirely new subarea within HCI has developed: human–robot interaction (HRI; see e.g., the Conference on Human–Robot Interaction series at http://humanrobotinteraction.org/category/conference/). Many of the usability issues in HRI are similar to those studied for years in HCI, for example, the impact and relative desirability of anthropomorphism, or the trade-offs in building affective features into a robot whose job is to answer questions. Thus, although robots may seem qualitatively different from an "interactive application," in principle, HRI should follow the same usability life cycle used by other software projects—fieldwork that assesses how a robot might fit into or enhance current practices, prototyping and formative evaluation to envision and build the system, and summative evaluation to assess whether and how it meets stated requirements (Lee et al., 2009).

32.4.2 New Usage Contexts for Usability

In parallel with changing technologies for user interface development, the activity contexts in which computing takes place have changed significantly in the past two decades. Two aspects that are very relevant for usability engineering are the degree of collaboration and multitasking present in many computing activities.

32.4.2.1 Collaborative Computing

Because users are often connected to a network while they engage in computing, there is a collaborative background to their work or play, even when the application does not explicitly support or encourage interaction with other people—for instance, simply sending a document or image as an attachment through email represents a loosely coupled form of collaborative computing. When examining current practices as part of a design process, the consideration of an activity's social or cultural context has become a prime concern along with the more traditional human factors stemming from users' physical, perceptual, or cognitive capacities (Beyer and Holtzblatt, 1997; Rosson and Carroll, 2002).

When explicitly designing for collaborative use, usability engineers must address a broad range of factors that interact with the nature and types of interaction that will take place among collaborators. For example, an asynchronous collaborative activity that transpires over different locations and times has rather different needs than a synchronous interaction that connects remote collaborators in real time (Neale et al., 2004). The different options for computer-mediated communication bring different affordances for the style and quality of the exchange (Clark and Brennan, 1991), with text-based communication being adequate for simple information sharing but richer audiovisual channels important for collaborations that rely on evaluation and negotiation of options for decision-making (Monk, 2008). To further complicate the issue, users' communication may change over time, as collaborators construct mental representations of one another that can guide and simplify their interactions (Convertino et al., 2011). Beyond the general concern for matching communication channels to collaborators' needs is the more general requirement for helping users to stay aware of what collaborators have been doing, whether in real time or not (Carroll et al., 2006). For software intended to support global collaboration contexts, a special usability challenge is the conflicts that may arise when global teams use a mix of face-to-face (i.e., within a locale) and remote (i.e., across locales) interactions (Ocker et al., 2009).

32.4.2.2 Multitasking

Modern operating systems are designed to support multithreaded activities, allowing users to launch and interact with multiple applications at the same time. In a work setting, this might mean transitioning in and out of a mail client, a document editor, an analysis package, and so on. Sometimes, the decision to transition is initiated by the user, but at other times, a user is "called to attention" by a notification of an application's status change. In general, the presence of multitasking does not raise new requirements for user interface design and usability engineering. Multitasking increases the importance of features that have long been in focus—for example, providing feedback during and after a task or subtask has been completed, and providing status information that conveys whether and how an interrupted task can be resumed (Norman, 1988). When designers support multitasking directly with complementary input and output channels, the situation becomes even more complex, because users must not only be able to track multiple goal–action sequences in parallel, but may also be required to orient to different configurations of displays or input devices for these different task contexts.

There are many usability engineering challenges that ensue from the complications of web-based platforms, pervasive connectivity and collaborative tasks, and the general trend toward greater and greater multitasking and heterogeneity in devices. To some extent, the challenges are simply those of greater complexity. For example, a meaningful reference task for evaluation may involve multiple parallel tasks, a group of users, or tasks that take place over extended periods of time and multiple locations (Neale et al., 2004). However, the new contexts also foreground new usability issues, for example, issues of *privacy* related to the persistent collection, analysis and repurposing of users' personal data or online behavior, and issues of perceptual and motor "transfer" among rather different input and output devices used for the same tasks.

32.4.3 New Concepts and Methods for Usability

Early HCI focused on software systems to support work activity, typically the activity of single users in the workplace. Today, HCI addresses learning and education, leisure, and entertainment; it addresses people individually, in groups, and as mass society. HCI is now about coping with everyday life, citizenship and civic participation, and very much about satisfaction as a consumer. This diversification has evoked new problems in usability engineering, required new concepts and approaches, and has drawn new people and skills into the field. The fundamental goals however remain the same: to design and refine software systems to meet human needs. This diversification is a natural consequence of growth and success, though it has created microcosms of specialized usability engineering work that are not always well coordinated with one another or with a broader conception. Nevertheless, these initiatives are driving usability engineering forward.

32.4.3.1 Universal Usability

Many of the most important innovations in display technologies or input devices have emerged from user interface design research directed at meeting the needs of people with disabilities. While the interest in accommodating people with special needs—whether with respect to visual, motor behavior, cognitive, or other capacities—has a long history, the resources and scientific interest aimed at these questions increased significantly in the 1990s, at least partly due to significant pieces of national legislation and policies (e.g., the Americans with Disabilities Act of 1990, the Telecommunications Act of 1996) that mandated accommodations in some work settings. The rapidly expanding population of elderly users in the United States has also helped to fuel research interest in these topics, with annual conferences such as *ACM ASSETS* and *ACM Transactions on Accessible Computing* created to enhance visibility of emerging findings and technology innovations.

A focus on universal usability—sometimes also called universal design—often produces findings or solutions that yield better designs for everyone. For example, the continued progress on voice commands helps not only the motor-impaired but also users who are in a hands-busy environment such as driving or working with mechanical equipment (Schütz et al., 2007). Alternate input devices such as joysticks or foot pedals are often integrated into custom HCI settings like games or training simulations (Zon and Roerdink, 2007). The many years of research on systems for text-to-speech translation not only support low-sight individuals (e.g., using screen readers like JAWS) but also is now omnipresent in voice-based telephone dialogs.

32.4.3.2 Interaction Design

As computer technologies have diversified, approaches to usability have broadened to consider the role and impacts of technology in consumer product design, for example, including furniture (Streitz et al., 1998), kitchen and other household tools (Chi et al., 2007), toys (Tomitsch et al., 2006), electronic books and games (Druin et al., 2009), and even clothing and accessories (Cho et al., 2009). Many such efforts fall under the umbrella of interaction design—often abbreviated as IxD—that differs from usability engineering in its pronounced emphasis on creative design thinking. That is, whereas in usability engineering the designers typically expect to begin with a detailed analysis of users' current practices and associated requirements, IxD designers specialize in out-of-the-box imagination and exploration of how things might be without the constraints of the current world (Löwgren, 2008).

In the IxD vision, the aesthetic characteristics of a design concept are of equal importance to its technical or functional elements, and these characteristics are often tied to physical form of the product. For instance, Apple has developed a consistent and successful design aesthetic through its i-devices; people choose to purchase and use these products not just because they satisfy their task needs but because they like the way the objects look and feel. Because of the importance of the overall aesthetic, sketches and other tangible artifacts are significant means for communicating design ideas. Learning to create and elaborate these early design prototypes is an essential skill for IxD professionals (Buxton, 2007).

32.4.3.3 Experience Design

Similar to IxD, the emerging discipline of user experience design (UX) places great value on the subjective experience of people who will be interacting with technology. UX recognizes that this subjective experience is embedded in a complex network of related concerns, for example, the "family" of products associated with a particular offering, how a product's users feel about the company that produces it, how often and by what means the product will be upgraded, its price relative to competitors, and so on (Diller et al., 2005).

The UX design process centers on the mechanisms for engagement. Design representations such as personas (Cooper, 1999; Grudin and Pruitt, 2002) are particularly popular within this design community, because a well-crafted persona offers a means of vividly evoking subjective experiences. Knowing that 32-year-old Jane Pelham loves to watch professional sports on TV, is an avid cook, and is carefully watching her weight may seem incidental to discussions about a new feature for an office product, but this relatively vivid depiction can help the product designers imagine how Jane would react to a design decision involving support for a touch-screen interface (e.g., how might she use it in the living room vs. in the kitchen). In this sense, UX can be seen as a more holistic version of the SBD process overviewed earlier—it relies centrally on representing and reasoning about individual user experiences rather than generic population statistics, but unlike SBD, the process does not entrain any systematic analysis and representation of design trade-offs and rationale.

32.5 Summary and Conclusions

To appropriate Robert Browning, usability engineering is an area of technical endeavor where our reach must necessarily exceed our grasp. New technology infrastructures, new applications, and new human aspirations drive usability engineering practices toward an expanding frontier of new concepts and techniques. Like many topics in computer science and engineering, we are not likely to ever close the book on usability engineering practices.

Why would we want to? Science and technology are most exciting at the frontier. The primary constant in usability engineering through four decades has been a developmental trajectory. Usability engineering has constantly become richer, more broadly applicable, more diverse conceptually and methodologically, and much more effective. The need for this trajectory to continue is unabated, and the prospects for further innovation and growth, driven by both external technological factors and human imagination, remain strong.

References

Ackerman, M.S. and Malone, T.W. 1990. Answer Garden: A tool for growing organizational memory. In *Proceedings of Conference on Organizational Computing: COCS 1990* (pp. 31–39). New York: ACM.

Bellamy, R., John, B.E., and Kogan, S. 2011. Deploying Cogtool: Integrating quantitative usability assessment into real-world software development. In *Proceedings of the 33rd International Conference on Software Engineering: ICSE '11* (pp. 691–700). New York: ACM.

Beyer, H. and Holtzblatt, K. 1995. Apprenticing with the customer. *Communications of the ACM*, 38(5), 45–52.

Beyer, H. and Holtzblatt, K. 1997. *Contextual Design: Defining Customer-centered Systems*. San Francisco, CA: Morgan Kaufmann.

Bias, R.G. and Mayhew, D.J. 1994. *Cost-Justifying Usability*. San Francisco, CA: Morgan Kaufmann.

Bias, R.G. and Mayhew, D.J. 2005. *Cost-justifying Usability: An Update for the Internet Age*. San Francisco, CA: Morgan Kaufmann.

Blythe, M., Overbeeke, K., Monk, A., and Wright, P. (Eds.) 2005. *Funology: From Usability to Enjoyment*. New York: Springer.

Brooks, F.P. 1975. *The Mythical Man-Month*. Reading, MA: Addison-Wesley.

Brown, B. and Laurier, E. 2012. The normal natural troubles of driving with GPS. In *Proceedings of Human Factors in Computing Systems: CHI 2012* (pp. 1621–1630). New York: ACM.

Brunswick, E. 1956. *Perception and the Representative Design of Psychological Experiments*. Berkeley, CA: University of California Press.

Buxton, B. 2007. *Sketching the User Interface: Getting the Design Right and Getting the Right Design*. San Francisco, CA: Morgan Kaufmann.

Card, S.K., Moran, T.P., and Newell, A. 1983. *The Psychology of Human-Computer Interaction*. Hillsdale, NJ: Lawrence Erlbaum Associates.

Carroll, J.M. 1985. *What's in a Name? An Essay in the Psychology of Reference*. New York: W.H. Freeman.

Carroll, J.M. 2000. *Making Use: Scenario-based Design of Human-Computer Interactions*. Cambridge, MA: MIT Press.

Carroll, J.M. (Ed.) 2003. *HCI Models, Theories and Frameworks: Toward a Multidisciplinary Science*. San Francisco, CA: Morgan Kaufman.

Carroll, J.M., Chin, G., Rosson, M.B., and Neale, D.C. 2000. The development of cooperation: Five years of participatory design in the virtual school. In *Proceedings of Designing Interactive Systems: DIS 2000* (pp. 239–252). New York: ACM.

Carroll, J.M. and Kellogg, W.A. 1989. Artifact as theory-nexus: Hermeneutics meets theory-based design. In *Proceedings of Human Factors in Computing Systems: CHI 1989* (pp. 7–14). New York: ACM.

Carroll, J.M., Olson, J.R., and Anderson, N.S. 1987. *Mental Models in Human-Computer Interaction: Research Issues about What the User of Software Knows*. Washington, DC: National Research Council, Committee on Human Factors.

Carroll, J.M. and Rosson, M.B. 1985. Usability specifications as a tool in iterative development. In Hartson, H.R. (Ed.), *Advances in Human-Computer Interaction*. Norwood, NJ: Ablex, pp. 1–28.

Carroll, J.M. and Rosson, M.B. 1995. Managing evaluation goals for training. *Datamation*, 30, 125–136.

Carroll, J.M. and Rosson, M.B. 2003. Design rationale as theory. In Carroll, J.M. (Ed.), *HCI Models, Theories and Frameworks: Toward a Multidisciplinary Science*. San Francisco, CA: Morgan Kaufman, pp. 431–461.

Carroll, J.M. and Rosson, M.B. 2008. Theorizing mobility in community networks. *International Journal of Human-Computer Studies*, 66, 944–962.

Carroll, J.M., Rosson, M.B., Convertino, G., and Ganoe, C. 2006. Awareness and teamwork in computer-supported collaborations. *Interacting with Computers*, 18(1), 21–46.

Carroll, J.M. and Thomas, J.C. 1988. Fun. *ACM SIGCHI Bulletin*, 19(3), 21–24.

Chi, P., Chao, J., Chu, H., and Chen, B. 2007. Enabling nutrition-aware cooking in a smart kitchen. *Extended Abstracts of CHI 2007* (pp. 2333–2338). New York: ACM.

Chin, J.P., Diehl, V.A., and Norman, K.L. 1988. Development of an instrument measuring user satisfaction of the human-computer interface. In *Proceedings of SIGCHI '88* (pp. 213–218). New York: ACM.

Chin, G., Rosson, M.B., and Carroll, J.M. 1997. Participatory analysis: Shared development of requirements from scenarios. In *Proceedings of Human Factors in Computing Systems: CHI'97* (pp. 162–169). New York: ACM.

Cho, G., Lee, S., and Cho, J. 2009. Review and reappraisal of smart clothing. *International Journal of Human-Computer Interaction*, 25(6), 582–617.

Clark, H.H. and Brennan, S.E. 1991. Grounding in communication. In Resnick, L.B., Levine, J.M., and Teasley, J.S. (Eds.), *Perspectives on Socially Shared Cognition*. Washington, DC: American Psychological Association, pp. 127–149.

Convertino, G., Mentis, H.M., Slakovic, A., Rosson, M.B., and Carroll, J.M. 2011. Supporting common ground and awareness in emergency management planning: A design research project. *ACM Transactions on Computer-Human Interaction*, 18(4), Article 22, 34.

Cooper, A. 1999. *The Inmates Are Running the Asylum*. Indianapolis, IN: SAMS.

Costabile, M.F., Fogli, D., Mussio, P., and Piccinno, A. 2004. End-user development: The software shaping workshop approach. In Lieberman, H., Paterno, F. and Wulf, V. (Eds.), *End User Development—Empowering People to Flexibly Employ Advanced Information and Communication Technology*. Dordrecht, the Netherlands: Kluwer Academic Publishers, pp. 183–205.

Diller, S., Shedroff, N., and Rhea, D. 2005. *Making Meaning: How Successful Businesses Deliver Meaningful Customer Experiences*. Upper Saddle River, NJ: New Riders Press.

Druin, A., Bederson, B.B., Rose, A., and Weeks, A. 2009. From New Zealand to Mongolia: Co-designing and deploying a digital library for the world's children. *Children, Youth, and Environment: Special Issue on Children in Technological Environments*, 19(1), 34–57.

Ericsson, K.A. and Simon, H.A. 1980. Verbal reports as data. *Psychological Review*, 87(3), 215–251.

Forlizzi, J. and Battarbee, K. 2004. Understanding experience in interactive systems. In *Proceedings of the 5th Conference on Designing Interactive Systems: Processes, Practices, Methods, and Techniques: DIS '04* (pp. 261–268). New York: ACM.

Gibson, J.J. 1966. *The Senses Considered as Perceptual Systems*. Boston, MA: Houghton Mifflin.

Gibson, J.J. 1979. *The Ecological Approach to Visual Perception*. Boston, MA: Houghton Mifflin.

Gould, J.D. and Lewis, C. 1985. Designing for usability: Key principles and what designers think. *Communications of the ACM*, 28(3), 300–311.

Gray, W.D., John, B.E., and Atwood, M.E. 1993. Project Ernestine: Validating a GOMS analysis for predicting and explaining real-world task performance. *Human-Computer Interaction*, 8, 237–309.

Greenbaum, J. and Kyng, M. (Eds.) 1991. *Design at Work: Cooperative Design of Computer Systems*. Hillsdale, NJ: Lawrence Erlbaum Associates.

Grudin, J. and Barnard, P. 1984. The cognitive demands of learning and representing command names for text editing. *Human Factors*, 26(4), 407–422.

Grudin, J. and Pruitt, J. 2002. Personas, participatory design and product development: An infrastructure for engagement. In *Proceedings of Participatory Design Conference: PDC 2002* (pp. 144–161). Malmo, Sweden, June 2002.

Hartson, R. and Pyla, P.S. 2012. *The UX Book: Process and Guidelines for Ensuring a Quality User Experience*. San Francisco, CA: Morgan Kaufmann.

Hollan, J., Hutchins, E., and Kirsh, D. 2000. Distributed cognition: Toward a new foundation for human-computer interaction research. *ACM Transactions on Computer-Human Interaction*, 7(2), 174–196.

Hornoff, A.J. and Halverson, T. 2003. Cognitive strategies and eye movements for searching hierarchical computer displays. In *Proceedings of Human Factors in Computing Systems: CHI 2003* (pp. 249–256). New York: ACM.

Hughes, J., King, V., Rodden, T., and Andersen, H. 1995. The role of ethnography in interactive systems design. *Interactions*, April 2(2), 56–65.

Hughes, J.A., Randall, D., and Shapiro, D. 1992. Faltering from ethnography to design. In *Proceedings of CSCW 1992* (pp. 115–122). New York: ACM.

Ishi, H. 2008. The tangible user interface and its evolution. *Communications of the ACM*, 51(6), 32–37.

John, B.E. and Kieras, D.E. 1996. The GOMS family of user interface analysis techniques: Comparison and contrast. *ACM Transactions on Computer-Human Interaction*, 3(4), 320–351.

Kahneman, D. and Tversky, A. 1979. Prospect theory: An analysis of decision under risk. *Econometrica*, 47(2), 263–292.

Kaptelinin, V. and Nardi, B. 2012. Activity theory in HCI: Fundamentals and reflections. *Synthesis Lectures on Human-Centered Informatics*, 5(1), 1–105.

Karat, C.-M. 1990. Cost-benefit analysis of usability testing. In *Proceedings of INTERACT '90: Third International Conference on Human-Computer Interaction* (pp. 351–356). Cambridge, U.K.

Karat, C.-M. and Karat, J. 2010. *Designing and Evaluating Usability Technology in Industrial Research: Three Case Studies*. San Francisco, CA: Morgan Claypool Synthesis Series.

Kulesca, T., Stumpt, S., Burnett, M., and Kwan, I. 2012. Tell me more? The effects of mental model soundness on personalizing an intelligent agent. In *Proceedings of Human Factors in Computing Systems: CHI 2012* (pp. 1–10). New York: ACM.

Lee, M.K., Forlizzi, J., Rybski, P.E., Crabbe, F., Chung, W., Finkle, J., Glaser, E., and Kiesler, S. 2009. The Snackbot: Documenting the design of a robot for long-term human-robot interaction. In *Proceedings of Human-Robot Interaction: HRI 2009* (pp. 7–14). New York: ACM.

Lee, M.K., Kiesler, S., and Forlizzi, J. 2010. Receptionist or information kiosk: How do people talk with a robot? In *Proceedings of Computer-Supported Cooperative Work: CSCW 2010* (pp. 31–40). New York: ACM.

Lee, K.C. and Lee, D.H. 2007. An online help framework for web applications. In *Proceedings of ACM Design of Communication: SIGDOC 2007* (pp. 176–180). New York: ACM.

London, R.B. 1992. Student modeling to support multiple instructional approaches. *User Modeling and User-Adapted Interaction*, 2(1–2), 117–154.

Löwgren, J. 2008. Interaction design. In Soegaard, M. and Dam, R.F. (Eds.), *Encyclopedia of Human-Computer Interaction*. Available at http://www/interaction-design.org/books/hci.html

Leshed, G., Veldon, T., Rieger, O., Kot, B., and Sengers, P. 2012. In-car GPS navigation: Engagement with and disengagement from the environment. In *Proceedings of Human Factors in Computing Systems: CHI 2012* (pp. 1621–1630). New York: ACM.

Marrone, F., Raimondi, F.M., and Strobel, M. 2002. Compliant interaction of a domestic service robot with a human and the environment. In *Proceedings of the 33rd International Symposium on Robotics: ISR 2002*, Stockholm, Sweden.

McCarthy, J. and Wright, P. 2004. *Technology as Experience*. Cambridge, MA: MIT Press.

Monk, A. 2008. *Common Ground in Electronically-Mediated Conversation*. Synthesis Lectures on Human-Computer Interaction. San Rafael, CA: Morgan & Claypool Publishers.

Motard, E., Naghsh, A., Roast, C., Arancon, M.M. and Marques, L. 2010. User interfaces for human robot interactions with a swarm of robots in support to firefighters. In *Proceedings of International Conference on Robotics and Automation: ICRA 2010* (pp. 2846–2851), May 3–7, 2010, Anchorage, AK.

Muller, M.J. 1991. PICTIVE: An exploration in participatory design. In *Proceedings of Human Factors in Computing Systems: CHI 1991* (pp. 225–231). New York: ACM.

Muller, M.J. 2007. Participatory design: The third space. In Jacko, J. and Sears, A. (Eds.), *The Human-Computer Interaction Handbook* (pp. 1051–1068). Hillsdale, NJ: Lawrence Erlbaum Associates.

Neale, D.C., Carroll, J.M., and Rosson, M.B. 2004. Evaluating computer-supported cooperative work: Models and frameworks. In *Proceedings of Computer-supported Cooperative Work: CSCW 2004* (pp. 112–121). New York: ACM Press.

Nielsen, J. 1990. Paper versus computer implementations as mockup scenarios for heuristic evaluation. In *Proceedings of the IFIP TC13 Third International Conference on Human–Computer Interaction: INTERACT '90* (pp. 315–320). Cambridge, U.K.: North-Holland.

Nielsen, J. 1994. Heuristic evaluation. In Nielsen, J. and Mack, R.L. (Eds.), *Usability Inspection Methods* (pp. 25–62). New York: John Wiley & Sons.

Nielsen, J. and Molich, R. 1990. Heuristic evaluation of user interfaces. In *Proceedings of Human Factors in Computing Systems: CHI '90* (pp. 249–256). New York: ACM.

Norman, D.A. 1988. *The Psychology of Everyday Things*. New York: Basic Books.

Ocker, R., Zhang, J., Hiltz, S.R., and Rosson, M.B. 2009. Determinants of partially distributed team performance: A path analysis of socio-emotional and behavioral factors. In *Proceedings of AMCIS 2009*, paper 707. San Francisco, CA, August 2009.

Payne, S.J. 2003. Users' mental models: The very ideas. In Carroll, J.M. (Ed.), *HCI Models, Theories and Frameworks: Toward a Multidisciplinary Science* (pp. 135–156). San Francisco, CA: Morgan Kaufmann.

Reitman, W.R. 1965. *Cognition and Thought: An Information Processing Approach*. New York: John Wiley & Sons.

Rosson, M.B., Ballin, J., and Rode, J. 2005. Who, what and why? A survey of informal and professional web developers. In *Proceedings of Visual Languages and Human-Centric Computing 2005* (pp. 199–206). New York: IEEE.

Rosson, M.B. and Carroll, J.M. 2013. Developing an online community for women in computer and information science: A design rationale analysis. *AIS Transactions on Human-Computer Interaction* 5(1), 6–27.

Rudd, J., Stern, K., and Isensee, S. 1996. Low vs. high-fidelity prototyping debate. *Interactions*, 3(1), 76–85.

Sauer, J., Franke, H., and Ruttinger, B. 2008. Designing interactive consumer products: Utility of low-fidelity prototypes and effectiveness of enhanced control labeling. *Applied Ergonomics,* 39, 71–85.

Sauer, J. and Sonderegger, A. 2009. The influence of prototype fidelity and aesthetics of design in usability tests: Effects on user behavior, subjective evaluation and emotion. *Applied Ergonomics*, 40, 670–677.

Schütz, R., Glanzer, G., Merkel, A.P, Wießflecker, T., and Walder, U. 2007. A speech-controlled user interface for a CAFM-based disaster management system. *Cooperative Design, Visualization, and Engineering* (pp. 80–87). Lecture Notes in Computer Science, 4674, New York: Springer.

Scriven, M. 1967. The methodology of evaluation. In Tyler, R.W., Gagne, R.M., and Scriven, M. (Eds.), *Perspectives of Curriculum Evaluation* (pp. 39–83). Chicago, IL: Rand McNally.

Shneiderman, B.A. 1980. *Software Psychology: Human Factors in Computer and Information Systems*. Cambridge, MA: Winthrop.

Shneiderman, B.A., Plaisant, C., Cohen, M.S., and Jacobs, S.M. 2009. *Designing the User Interface: Strategies for Effective Human–Computer Interaction*, 5th edn. New York, NY: Addison Wesley.

Simon, H.A. 1972. Theories of bounded rationality. In McGuire, C.B. and Radner, R. (Eds.), *Decision and Organization* (pp. 161–176). Amsterdam, the Netherlands: North Holland.

Snyder, C. 2003. *Paper Prototyping: The Fast and Easy Way to Design and Refine User Interfaces*. San Francisco, CA: Morgan Kaufmann.

Streitz, N.A., Geißler, J., and Holmer, T. 1998. Roomware for cooperative buildings: Integrated design of architectural spaces and information spaces. In Streitz, N.A., Konomi, S., and Burkhardt, H.-J. (Eds.), *Cooperative Buildings—Integrating Information, Organization and Architecture* (pp. 4–21). Lecture Notes in Computer Science 1370, Heidelberg, Germany.

Suchman, L.A. 1987. *Plans and Situated Action: The Problem of Human-Machine Communication*. New York: Cambridge University Press.

Tomitsch, M., Grechenig, T., Kappel, K., and Koltringer, T. 2006. Experiences from designing a tangible musical toy for children. In *Proceedings of the Conference on Interaction Design for Children: IDC 2006* (pp. 169–170). New York: ACM.

Vertanen, K. and MacKay, D.J.C. 2010. Speech dasher: Fast writing using speech and gaze. In *Proceedings of Human Factors in Computing Systems: CHI 2010* (pp. 1–4). New York: ACM.

Virzi, R.A., Sokolov, J.L., and Karis, D. 1996. Usability problem identification using both low- and high-fidelity prototypes. In *Proceedings of Human Factors in Computing Systems: CHI '96* (pp. 236–243). New York: ACM.

West, M.A. 2012. *Effective Teamwork: Practical Lessons from Organizational Research*. Chichester, U.K.: Wiley.

Williamson, J.R., Crossan, A., and Brewster, S. 2011. Multimodal mobile interactions: Usability studies in real world settings. In *Proceedings of International Conference on Multimodal Interaction: ICMI 2011* (pp. 361–368). New York: ACM.

Zang, N. and Rosson, M.B. 2009. Playing with data: How end users think about and integrate dynamic data. In *Proceedings of Visual Languages and Human-Centric Computing: VL/HCC 2009* (pp. 85–92). Piscataway, NJ: IEEE.

Zon, R. and Roerdink, M. 2007. HCI testing in flight simulator: Set up and crew briefing procedures. In *Proceedings of the 7th International Conference on Engineering Psychology and Cognitive Ergonomics: EPCE 2007* (pp. 867–876). Berlin, Germany: Springer-Verlag.

33

Task Analysis and the Design of Functionality

David Kieras
University of Michigan

Keith A. Butler
University of Washington

33.1 Introduction

In *human–computer interaction* (*HCI*), *task analysis* is a family of methods for understanding the user's task thoroughly enough to help design a computer system that will effectively support users in doing the task. By *task* is meant the user's job or goal-driven work activities, what the user is attempting to accomplish; an individual user's task is not just to interact with the computer, but to get a job done. Thus, understanding the user's task involves understanding the user's *task domain* and the user's larger job goals.

By *analysis* is meant a relatively systematic approach to understanding the user's task that goes beyond unaided intuitions or speculations and attempts to document and describe exactly what the task involves. At the level of large military systems and business processes, the analysis can span multiple users, machines, and organizations; this can still be termed task analysis, but other labels, such as business process analysis, tend to be used. A large-scale analysis is necessary to determine the context in which individual users or systems do their work, and a finer-grain analysis is required to determine how the individual users do their work within that larger context.

The design of *functionality* is a stage of the design of computer systems in which the user-accessible functions of the computer system are chosen and specified. The basic thesis of this chapter is that the successful design of functionality requires a task analysis early enough in the system design to enable the developers to create a system that effectively supports the user's task. Thus, the proper goal of the design of functionality is to choose functions that are both *useful* in the user's task and which together with a good *user interface* result in a system that is *usable*, being easy to learn and easy to use. If the scale is expanded to a larger work context, effective design of functionality becomes the design of work processes, deciding what parts of the work are going to be done by humans and what by machines.

The purpose of this chapter is to provide some historical and theoretical background, and beginning "how to" information on how to conduct a task analysis and how to approach the design of functionality.

Although they are closely intertwined, for purposes of exposition, task analysis and the design of functionality are presented in two separate major sections in that order. Each section will start with background principles and then present some specific methods and approaches.

33.2 Task Analysis

33.2.1 Principles

A task analysis presupposes that there already exists some method or approach for carrying out the work, involving some mixture of human activity and possible machine activity. This method may have existed for a long time, or might be a new, even hypothetical, method based on new technology. The goal of the initial task analysis is to describe how the work is currently being done in order to understand how computing may improve it.

33.2.1.1 Role of Task Analysis in Development

33.2.1.1.1 Development of Requirements

This stage of development is emphasized in this chapter. A task analysis should be conducted before developing the system requirements to guide the choice and design of the system functionality; the ultimate, upper boundary of usability of the system is actually determined at this stage. The goal of task analysis at this point is to find out what the user needs to be able to accomplish, and how they will do it, so that the functionality of the system can be designed so that the tasks can be accomplished more easily, with better quality, etc. While some later revision is likely to be required, these critical choices can be made before the system implementation or user interface is designed.

Designers may believe they understand the user's task adequately well without any task analysis. This belief is often not correct; even ordinary tasks are often complex and poorly understood by developers. However, many economically significant systems are intended for expert users, and understanding their tasks is absolutely critical and well beyond normal software developer expertise. For example, a system to assist a petroleum geologist must be based on an understanding of the knowledge and goals of the petroleum geologist. To be useful, such a system will require functions that produce information useful for the geologist; to be usable, the system will have to provide these functions in a way that the frequent and most important activities of the geologist are well supported.

33.2.1.1.2 User Interface Design and Evaluation

Task analysis is also needed during interface design to effectively design and evaluate the user interface. Most applications of task analysis in software projects are focused on user interface design (Diaper and Stanton 2004a). During interface design, valuable usage scenarios can be chosen that are properly representative of user activities. Task analysis results can be used to choose benchmark tasks for user testing that will represent important uses of the system. A task analysis will help identify the portions of the interface that are most important for the user's tasks. Once an interface is designed and is undergoing evaluation, the original task analysis can be revised to specify or document how the task would be done with the proposed interface. This can suggest usability improvements from either modifying the interface or improving the fit of the functionality to the more specific form of the user's task entailed by a proposed interface.

33.2.1.1.3 Follow-Up after Installation

Task analysis can be conducted on fielded or in-place systems to compare systems or identify potential problems or improvements. These results could be used to compare the demands of different systems, identify problems that should be corrected in a new system, or to determine properties of the task that should be preserved in a new system.

33.2.1.2 Contribution of Human Factors

Task analysis was developed in the discipline of *Human Factors*, which has a long history of concern with the design process and how human-centered issues should be incorporated into system design. Task analysis methods have been applied to a broad range of systems, covering handheld radios, radar consoles, whole aircraft, chemical process plant control rooms, and very complex multiperson systems such as a warship combat information center. The breadth of this experience is no accident. Historically, Human Factors is the leading discipline with an extended record of involvement and concern with human-system design, long predating the newer field of HCI. The resulting task-analysis methods have been presented in comprehensive collections of task analysis techniques (Beevis et al. 1992; Diaper and Stanton 2004a; Kirwan and Ainsworth 1992), and it has been an active area of research within Human Factors and allied fields (Annett and Stanton 2000a; Schraagen et al. 2000). One purpose of this chapter is to summarize some of these techniques and concepts.

Precomputer technologies usually had interfaces consisting of fixed physical displays and controls in a spatial arrangement. So, the original forms of task analysis focused on how the operator's task was related to the fixed physical "knobs and dials" interface. However, the user interfaces provided by computer-based systems are potentially much more flexible, resulting in a need for task analysis to identify what display and control functions should be present to support the operator. In other words, the critical step in computer-based system design is the *choice of functionality*. Once the functions are chosen, the constraints of computer interfaces mean that the procedural requirements of the interface are especially prominent; if the functions are well chosen, the procedures that the operator must follow will be simple and consistent. The importance of this combination of task analysis, choice of functionality, and the predominance of the procedural aspects of the interface is the basis for the recommendations provided later in this chapter.

33.2.1.3 Contributions of Human–Computer Interaction

The field of HCI is a relatively new and highly interdisciplinary field and still lacks a consensus on scientific, practical, and philosophical foundations. Consequently, a variety of ideas have been discussed concerning how developers should approach the problem of understanding what a new computer system needs to do for its users. While many original researchers and practitioners in HCI had their roots in Human Factors, several other disciplines have had a strong influence on HCI theory and practice.

There are roughly two groups of these disciplines. The first is ***cognitive psychology***, which is a branch of scientific psychology that is concerned with human cognitive abilities, such as comprehension, problem-solving, and learning. The second is a mixture of ideas from the social sciences, such as social-organizational psychology, ethnography, and anthropology. While the contribution of these disciplines has been important in developing the scientific basis of HCI, they either have little experience with humans in a work context, as is the case with cognitive psychology, or no experience with practical system design problems, as with the social sciences. On the other hand, Human Factors is almost completely oriented toward solving practical design problems in an ad hoc manner, and is generally atheoretic in content. Thus, the disciplines with a broad and theoretical science base lack experience in solving design problems, and the discipline with this practical knowledge lacks a comprehensive scientific foundation.

The current state of task analysis in HCI is thus rather confused (Diaper and Stanton 2004b); there has been a tendency to reinvent task analysis under a variety of guises, as each theoretical approach and practice community presents its own insights about how to understand a work situation and design a system to support it. Apparently, most task analysis methods are tapping into the same underlying ontology and differ mainly in what aspects are represented or emphasized, suggesting that they could be merged them into a single analysis (van Welie et al. 1998), but in the meantime, the most successful ad hoc approaches address a much wider set of requirements than the user interface. Contextual Design (Beyer and Holtzblatt 1998) is a collection of task-analytic techniques and activities, which will lead the developer through the stages of understanding, what users do, what information they need, and what system functions and user interfaces will help them do it better.

33.2.1.3.1 Contributions from Cognitive Psychology

The contribution of cognitive psychology to HCI is both more limited and more successful within its limited scope. Cognitive psychology treats an individual human being as an information processor who is considered to acquire information from the environment, transform it, store it, retrieve it, and act on it. This information-processing approach, also called the computer metaphor, has an obvious application to how humans would interact with computer systems. In a cognitive approach to HCI, the interaction between human and computer is viewed as two interacting information-processing systems with different capabilities, and in which one, the human, has goals to accomplish, and the other, the computer, is an artificial system which should be designed to facilitate the human's efforts. The relevance of cognitive psychology research is that it directly addresses two important aspects of usability: how difficult it is for the human to learn how to interact successfully with the computer, and how long it takes the human to conduct the interaction. The underlying topics of human learning, problem-solving, and skilled behavior have been intensively researched for decades in cognitive psychology.

The systematic application of cognitive psychology research results to HCI was first proposed by Card et al. (1983). Two levels of analysis were presented: the lower-level analysis is the Model Human Processor, a summary of about a century's worth of research on basic human perceptual, cognitive, and motor abilities in the form of an engineering model that could be applied to produce quantitative analysis and prediction of task execution times; the higher-level analysis was the *GOMS model*, which is a description of the *procedural knowledge* involved in doing a task. The acronym GOMS can be explained as follows: The user has Goals that can be accomplished with the system. Operators are the basic actions such as keystrokes performed in the task. Methods are the procedures, consisting of sequences of operators that will accomplish the goals. Selection rules determine which method is appropriate for accomplishing a goal in a specific situation. In the Card et al. formulation, the new user of a computer system will use various problem-solving and learning strategies to figure out how to accomplish tasks using the computer system, and then with additional practice, these results of problem-solving will become GOMS procedures that the user can routinely invoke to accomplish tasks in a smooth, skilled manner. The properties of the procedures will thus govern both the ease of learning and ease of use of the computer system. In the research program stemming from the original proposal, approaches to representing GOMS models based on cognitive psychology theory have been developed and validated empirically, along with the corresponding techniques and computer-based tools for representing, analyzing, and predicting human performance in HCI situations (see John and Kieras 1996a,b for reviews).

The significance of the GOMS model for task analysis is that it provides a way to describe the task procedures in a way that has a theoretically rigorous and empirically validated scientific relationship to human cognition and performance. Space limitations preclude any further presentation of how GOMS can be used to express and evaluate a detailed interface design (see John and Kieras 1996a,b; Kieras 1997, 2004). Later in this chapter, a technique based on GOMS will be used to couple task analysis with the design of functionality.

33.2.2 Best Practices: How to Do a Task Analysis

The basic idea of conducting a task analysis is to understand the user's activity in the context of the whole system, either an existing or a future system. While understanding human activity is the subject of scientific study in psychology and the social sciences, the conditions under which systems must be designed usually preclude the kind of extended and intensive research necessary to document and account for human behavior in a scientific mode. Thus, a task analysis for system design must be rather more informal, and primarily heuristic in flavor compared to scientific research. The task analyst must do his or her best to understand the user's task situation well enough to influence the system design given the limited time and resources available. This does not mean that a task analysis is an easy job; large amounts of detailed information must be collected and interpreted, and experience in task analysis is valuable even in the most structured methodologies (see Annett 2004).

33.2.2.1 Role of Formalized Methods for Task Analysis

Despite the fundamentally informal character of task analysis, many formal and quasiformal systems for task analysis have been proposed and have been widely recommended. Several will be summarized later. It is critical to understand that these systems do not in themselves analyze the task or produce an understanding of the task. Rather, they are ways to structure the task analysis process and notations for representing the results of task analysis. They have the important benefit of helping the analyst observe and think carefully about the user's actual task activity, both specifying what kinds of task information are likely to be useful to analyze, and providing a heuristic test for whether the task has actually been understood. That is, a good test for understanding something is whether one can represent it or document it, and constructing such a representation can be a good approach to trying to understand it. Relatively formal representations of a task further this goal because well-defined representations can be more easily inspected, criticized, and developed with computer-based tools. Furthermore, they can support computer simulations or mathematical analyses to obtain quantitative predictions of task performance. But, it must be understood that such results are no more correct than the original, and informally obtained, task analysis underlying the representation.

33.2.2.2 An Informal Task Analysis Is Better Than None

Most of the task analysis methods to be surveyed require significant time and effort; spending these resources would usually be justified, given the near-certain failure of a system that fails to meet the actual needs of users. However, the current reality of software development is that developers often will not have adequate time and support to conduct a full-fledged task analysis. Under these conditions, what can be recommended? As pointed out by sources such as Gould (1988) and Grudin (1991), perhaps the most serious problem to remedy is that the developers often have no contact with actual users. Thus, if nothing more systematic is possible, the developers should spend some time in informal observation of real users actually doing real work. The developers should observe unobtrusively, but ask for explanation or clarification as needed, perhaps trying to learn the job themselves. They should not, however, make any recommendations, and do not discuss the system design. The goal of this activity is simply to try to gain some experience-based intuitions about the nature of the user's job, and what real users do and why (see Gould 1988 for additional discussion). Such informal, intuition-building contact with users will provide tremendous benefits at relatively little cost. Approaches such as Contextual Design (Beyer and Holzblatt 1998) and the more elaborate methods presented here provide more detail and more systematic documentation, and will permit more careful and exact design and evaluation than casual observation, but some informal observation of users is infinitely better than no attempt at task analysis at all.

33.2.2.3 Collecting Task Data

The data collection methods summarized here are those that have been found to produce useful information about tasks (see Gould 1988; Kirwan and Ainsworth 1992); the task-analytic methods summarized in the next section are approaches that help analysts perform the synthesis and interpretation.

33.2.2.3.1 Observation of User Behavior

In this fundamental family of methods, the analyst observes actual user behavior, usually with minimal intrusion or interference, and describes what has been observed in a thorough, systematic, and documented way.

The setting for the user's activity can be the actual situation (e.g., in the field) or a laboratory simulation of the actual situation. Either all of the user's behavior can be recorded, or it could be sampled periodically to cover more time while reducing the data collection effort. The user's activities can be categorized, counted, and analyzed in various ways. For example, the frequency of different activities can be tabulated, or the total time spent in different activities could be determined. Both such measures contribute valuable information on which task activities are most frequent or time-consuming, and thus important to address in the system design. Finer grain recording and analysis can provide information on the exact timing and sequence of task activities, which can be important in the detailed design of the interface. Video recording of users supports both very general and very detailed analysis at low cost; consumer-grade equipment is often adequate.

A more intrusive method of observation is to have users "think aloud" about a task while performing it, or to have two users discuss and explain to each other how to do the task while performing it. The verbalization can disrupt normal task performance, but such *verbal protocols* are believed to be a rich source of information about the user's mental processes such as inferences and decision-making. The pitfall for the inexperienced is that the protocols can be extremely labor intensive to analyze, especially if the goal is to reconstruct the user's cognitive processes. The most fruitful path is to transcribe the protocols, isolate segments of content, and attempt to classify them into an informative set of categories.

A final technique in this class is *walkthroughs* and *talkthroughs*, in which the users or designers carry out a task and describe it as they do so. The results are similar to a think-aloud protocol, but with more emphasis on the procedural steps involved. An important feature is that the interface or system need not exist; the users or designers can describe how the task would or should be carried out.

33.2.2.3.2 Critical Incidents and Major Episodes

Instead of attempting to observe or understand the full variety of activity in the task, the analyst chooses incidents or episodes that are especially informative about the task and the system, and attempts to understand what happens in these; this is basically a case-study approach. Often, the critical incidents are accidents, failures, or errors, and the analysis is based on retrospective reports from the people involved and any records produced during the incident. An important extension on this approach is the *critical decision method* (Wong 2004), which focuses on understanding the knowledge involved in making expert-level decisions in difficult situations. However, the critical "incident" might be a major episode of otherwise routine activity that serves especially well to reveal the problems in a system. For example, observation of a highly skilled operator performing a very specialized task revealed that most of the time was spent doing ordinary file maintenance; understanding why led to major improvements in the system (Brooks, personal communication).

33.2.2.3.3 Questionnaires

Questionnaires are a fixed set of questions that can be used to collect some types of user and task information on a large scale quite economically. The main problem is that the accuracy of the data is unknown compared to observation, and can be susceptible to memory errors and social influences. Despite the apparent simplicity of a questionnaire, designing and implementing a successful one is not easy, and can require an effort comparable to interviews or workplace observation.

33.2.2.3.4 Structured Interviews

Interviews involve talking to users or domain experts about the task. Typically, some unstructured interviews might be done first, in which the analyst simply seeks any and all kinds of comments about the task. Structured interviews can then be planned; a series of predetermined questions for the interview is prepared to ensure a more systematic, complete, and consistent collection of information.

33.2.2.4 General Issues in Representing Tasks

Once the task data is collected, the problem for the analyst is to determine how to represent the task data, which requires a decision about what aspects of the task are important and how much detail to represent. The key function of a representation is to make the task structure visible or apparent in some way that supports the analyst's understanding of the task. By examining a task representation, an analyst hopes to identify problems in the task flow, such as critical bottlenecks, inconsistencies in procedures, excessive workloads, and activities that could be better supported by the system. Traditionally, a graphical representation, such as a flowchart or diagram, has been preferred, but as the complexity of the system and the operating procedures increase, diagrammatic representations lose their advantage.

33.2.2.4.1 Task Decomposition

One general form of task analysis is often termed task decomposition (Kirwan and Ainsworth 1992). This is not a well-defined method at all, but merely reflects a philosophy that tasks usually have a complex structure, and a major problem for the analyst will be to decompose the whole task situation into subparts for further analysis, some of which will be critical to the system design, and others possibly less important. Possible decompositions include controls and their arrangement or display coding. But an especially important approach is to decompose a task into a hierarchy of subtasks and the procedures for executing them, leading to a popular form of analysis called (somewhat too broadly) *Hierarchical Task Analysis* (HTA). However, another approach would be to decompose the task situation into considerations of how the controls are labeled, how they are arranged, and how the displays are coded. This is also task decomposition, and might also have a hierarchical structure, but the emphasis is on describing aspects of the displays in the task situation. Obviously, depending on the specific system and its interface, some aspects of the user's task situation may be far more important to analyze than others. Developing an initial task decomposition can help identify what is involved overall in the user's task, and thus allows the analyst to choose what aspects of the task merit intensive analysis.

33.2.2.4.2 Level of Detail

The question of how much detail to represent in a task analysis is difficult to answer. At the level of whole tasks, Kirwan and Ainsworth (1992) suggest a *probability times cost* rule: if the probability of inadequate performance multiplied by the cost of inadequate performance is low, then the task is probably not worthwhile to analyze. But, even if a task has been chosen as important, the level of detail at which to describe the particular task still must be chosen. Some terminology must be clarified at this point: task decompositions can be viewed as a standard inverted tree structure, with a single item, the overall task, at the top, and the individual actions (such as keystrokes or manipulating valves) or interface objects (switches, gauges) at the bottom. A high-level analysis deals only with the low-detail top parts of the tree, while a low-level analysis includes all of the tree from the top to the high-detail bottom.

The cost of task analysis rises quickly as more detail is represented and examined, but on the other hand, many critical design issues appear only at a detailed level. For example, at a high enough level of abstraction, the Unix operating system interface is essentially just like the Macintosh operating system interface; both interfaces provide the functionality for invoking application programs and copying, moving, and deleting files and directories. The notorious usability problems of the Unix command line relative to the Mac OS X GUI only appear at a level of detail that the cryptic, inconsistent, and clumsy command structure and generally poor feedback come to the surface. "The devil is in the details." Thus, a task analysis capable of identifying usability issues in an interface design typically involves working at a low, fully detailed level that involves individual commands and mouse selections. The opposite consideration holds for the design of functionality, as will be discussed later. When choosing functionality, the question is how the user will carry out tasks using a set of system functions, and it is important to avoid being distracted by the details of the interface.

33.2.2.5 Representing What the User Needs to Know

Another form of task analysis is to represent what knowledge the human needs to have in order to effectively operate the system. Clearly, the human needs to know how to operate the equipment; such procedural knowledge is treated under its own heading in the next section. But, the operator might need additional procedural knowledge that is not directly related to the equipment, and additional nonprocedural conceptual background knowledge. For example, a successful fighter aircraft pilot must know more than just the procedures for operating the aircraft and the onboard equipment; he or she must have additional procedural skills such as combat tactics, navigation, and communication protocols, and an understanding of the aircraft mechanisms and overall military situation is valuable in dealing with unanticipated and novel situations.

Information on what the user needs to know is clearly useful for specifying the content of operator training and operator qualifications. It can also be useful in choosing system functionality in that large benefits can be obtained by implementing system functions that make it unnecessary for users to know concepts or skills that are difficult to learn; such simplifications typically are accompanied by simplifications in the operating procedures as well. Aircraft computer systems that automate navigation and fuel conservation tasks are an obvious example.

In some cases where the user knowledge is mostly procedural in content, it can be represented in a straightforward way, such as decision–action tables that describe what interpretation should be made of a specific situation, as in an equipment troubleshooting guide. However, the required knowledge can be extremely hard to identify if it does not have a direct and overt relationship to "what to do" operating procedures. An example is the petroleum geologist, who after staring at a display of complex data for some time, comes up with a decision about where to drill, and probably cannot provide a rigorous explanation for how the decision was made. Understanding how and why users make such decisions is difficult, because there is very little or no observable behavior prior to producing the result; it is all "in the head," a *purely cognitive task.*

Analyzing a purely cognitive task in complete detail is essentially a cognitive psychology research project, and so is not usually practical in a system design context. Furthermore, to the extent that a cognitive task can be completely characterized, it becomes a candidate for automation, making the design of the user interface moot. Expert systems technology, when successfully applied using "knowledge acquisition" methods (see Boose 1992), is an example of task analyses carried out in enough detail that the need for a human performer of the task is eliminated.

However, most cognitive tasks seem neither possible nor practical to analyze this thoroughly, but there is still a need to support the human performer with system functions and interfaces and training programs and materials that improve performance in the task even if it is not completely understood. For example, Gott (1988) surveys cases in which an intensive effort to identify the knowledge required for tasks can produce large improvements in training programs for highly demanding cognitive tasks such as electronics troubleshooting.

During the 1990s, there were many efforts to develop methods for cognitive task analysis (CTA). Cognitive task analysis emphasizes the knowledge required for a task, and how it is used in decision-making, situation-recognition, or problem-solving rather than just the procedures involved (see Chipman et al. 2000 for an overview, and Dubois and Shalin 2000 and Seamster et al. 2000 for useful summaries of some important methods). Most CTA methods involve some form of interview techniques for eliciting a subset of the most critical knowledge involved in a task, often focusing on critical incidents (Crandall et al. 2006; Militello and Hutton 2000; Wong 2004).

33.2.2.6 Representing What the User Has to Do

A major form of task analysis is describing the actions or activities carried out by the human operator while tasks are being executed. Such analyses have many uses; the description of how a task is currently conducted, or would be conducted with a proposed design, can be used for prescribing training, assisting in the identification of design problems in the interface, or as a basis for quantitative or simulation modeling to obtain predictions of system performance. Depending on the level of detail chosen for the analysis, the description might be at very high level, or might be fully detailed, describing the individual valve operations or keystrokes needed to carry out a task. The following are the major methods for representing procedures.

33.2.2.6.1 *Operational Sequence Diagrams*

Operational sequence diagrams and related techniques show the sequence of the operations (actions) carried out by the user (or the machine) to perform a task, represented graphically as a flowchart using standardized symbols for the types of operations. Such diagrams are often partitioned, showing the user's actions on one side, and machine's on the other, to show the pattern of operation between the user and the machine.

33.2.2.6.2 Hierarchical Task Analysis

Hierarchical task analysis involves describing a task as a hierarchy of tasks and subtasks, emphasizing the procedures that operators will carry out, using several specific forms of description. The term "hierarchical" is somewhat misleading, since many forms of task analysis produce hierarchical descriptions; a better term might be "procedure hierarchy task analysis." The results of an HTA are typically represented either graphically, as a sort of annotated tree diagram of the task structure similar to the conventional diagram of a function-call hierarchy in programming, or in a more compact tabular form. This is the original form of systematic task analysis (Annett et al. 1971) and still the single most heavily used form of task analysis (see Annett 2004 for additional background and procedural guide).

HTA descriptions involve *goals, tasks, operations,* and *plans.* A goal is a desired state of affairs (e.g., a chemical process proceeding at a certain rate). A task is a combination of a goal and a context (e.g., get a chemical process going at a certain rate given the initial conditions in the reactor). Operations are activities for attaining a goal (e.g., procedures for introducing reagents into the reactor, increasing the temperature, and so forth). Plans specify which operations should be applied under what conditions (e.g., which procedure to follow if the reactor is already hot). Plans usually appear as annotations to the tree-structure diagram that explain which portions of the tree will be executed under what conditions. Each operation in turn might be decomposed into subtasks, leading to a hierarchical structure. The analysis can be carried out to any desired level of detail, depending on the requirements of the analysis.

33.2.2.6.3 GOMS Models

GOMS models (John and Kieras 1996a,b), introduced earlier, are closely related to HTA, and in fact Kirwan and Ainsworth (1992) include GOMS as a form of HTA. GOMS models describe a task in terms of a hierarchy of goals and subgoals, methods which are sequence of operators (actions) that when executed will accomplish the goals, and selection rules that choose which method should be applied to accomplish a particular goal in a specific situation. However, both in theory and in practice, GOMS models are different from HTA. The concept of GOMS models grew out of research on human problem-solving and cognitive skills, whereas HTA appears to have originated out of the pragmatic commonsense observation that tasks often involve subtasks, and eventually involve carrying out sequences of actions. Because of its more principled origins, GOMS models are more disciplined than HTA descriptions. The contrast is perhaps most clear in the difficulty HTA descriptions have in expressing the flow of control: the procedural structure of goals and subgoals must be deduced from the plans, which appear only as annotations to the sequence of operations. In contrast, GOMS models represent plans and operations in a uniform format using only methods and selection rules. An HTA plan would be represented as simply a higher-order method that carries out lower-level methods or actions in the appropriate sequence, along with a selection rule for when the higher-order method should be applied.

33.2.2.7 Representing What the User Sees and Interacts With

The set of information objects and controls that the user interacts with during task execution constrain the procedures that the user must follow (Zhang and Norman 1994, 1997). Consequently, a full procedural description of the user's task must refer to all objects in the task situation that the user must observe or manipulate. However, it can be useful to attempt to identify and describe the relevant objects and event independently of the procedures in which they are used. Such a task analysis can identify some potential serious problems or design issues quite rapidly. For example, studies of nuclear power plant control rooms (Woods et al. 1987) found that important displays were located in positions such that they could not be read by the operator. A task decomposition can be applied to break the overall task situation down into smaller portions, and the interface survey technique mentioned earlier can then determine the various objects in the task situation. Collecting additional information, for example, from interviews or walkthroughs, can then lead to an assessment of whether and under what conditions the individual controls or displays are required for task execution. There are then a variety of guidelines in human factors for determining whether the controls and displays are adequately accessible.

33.2.2.8 Representing What the User Might Do Wrong

Human factors practitioners and researchers have developed a variety of techniques for analyzing situations in which errors have happened, or might happen. The goal is to determine whether human errors will have serious consequences, and to try to identify where they might occur and how likely they are to occur. The design of the system or the interface can then be modified to try to reduce the likelihood of human errors, or mitigate the consequences of them. Some key techniques can be summarized.

33.2.2.8.1 Event Trees

In an event tree, the possible paths, or sequences of behaviors, through the task are shown as a tree diagram; each behavior outcome is represented either as success/failure, or a multiway branch, for example, for the type of diagnosis made by an operation in response to a system alarm display. An event tree can be used to determine the consequences of human errors, such as misunderstanding an alarm. Each path can be given a predicted probability of occurrence based on estimates of the reliability of human operators at performing each step in the sequence these estimates are controversial; see Reason1990 for discussion).

33.2.2.8.2 Failure Modes and Effects Analysis

The analysis of human failure modes and their effects is modeled after a common hardware reliability assessment process. The analyst considers each step in a procedure, and attempts to list all the possible failures an operator might commit, such as to omit the action, perform it too early, too late, too forcefully, and so forth. The consequences of each such failure "mode" can then be worked out, and again a probability of failure could be predicted.

33.2.2.8.3 Fault Trees

In a fault tree analysis, the analyst starts with a possible system failure, and then documents the logical combination of human and machine failures that could lead to it. The probability of the fault occurring can then be estimated, and possible ways to reduce the probability can be determined.

33.2.2.8.4 Application to Computer User Interfaces

Until recently, these techniques had not been applied in computer user interface design to any visible extent. At most, user interface design guides contained a few general suggestions for how interfaces could be designed to reduce the chances of human error. Promising work such as Stanton and Baber (2005) and Wood (1999), use task analysis as a basis for systematically examining how errors might be made and how they can be detected and recovered from. Future work along these lines will be an extraordinarily important contribution to system and interface design.

33.3 Design of Functionality

33.3.1 Principles

Given a task analysis that describes how the work is done currently, can a new system be designed that will enable the work to be done more quickly, efficiently, reliably, or accurately? Such improvement usually involves changes in *function allocation*—assigning functions to the human and the computer—in ways different from how it is currently being done. For example, if an existing process for selling and shipping merchandise involves only paper documents, and a computer-supported process is desired, then it is necessary to choose a role for the computer in terms of its functions. For example, these could merely allow the same forms to be filled out on the computer screen and then printed out and handled manually, or the completed forms could be handled electronically, or the work process redesigned so that no forms need to be filled out at all. Each allocation of machine functions implies a corresponding and complementary allocation of human user functions.

The design of functionality is closely related to task analysis. One must start with an analysis of the existing tasks, considered as the whole work process, and then propose new machine functionality that will improve the existing task or work process; this new version of the task must then be analyzed to determine if there has been a net benefit, and whether additional benefits could be obtained by further redesign of the process.

33.3.1.1 Problems with the Current Situation

In many software development organizations, some groups, such as a marketing department or government procurement agents, prepare a list of requirements for the system to be developed. Such requirements specify the system functions at least in part, typically in the form of lists of desired features. The designers and developers then further specify the functions, possibly adding or deleting functions from the list, and then begin to design an implementation for them. Problems with misdefined functionality are common, because the typical processes for preparing requirement specifications often fail to include a substantial task analysis.

Typically, only after this point is the user interface design begun. The design of the user interface may use appropriate techniques to arrive at a usable design, but these techniques normally are based on whatever conception of the user and the user's tasks have already been determined, and the interface is designed in terms of the features or functions that have already been specified. Thus, even when a usability process is followed, it might well arrive at only a local optimum defined by an inadequate characterization of the user's needs and the corresponding functions (cf. Buxton 2007).

That is, the focus of usability methods tends to be on relatively low-level questions (such as menu structure) and on how to conduct usability tests to identify usability problems; the problem is posed as developing a usable interface to functions that have already been chosen. Typically, the system requirements have been prepared by a group which then "throws it over the wall" to a development group who arrives at the overall design and functional implementation, and then throws it over the wall to a usability group to "put a good interface on it."

The problem is that if the initial requirements and system functions are poorly chosen, the rest of the development will probably fail to produce a usable product. It is a truism in HCI that if customers need the functionality, they will buy and use even a clumsy, hard-to-use product. If the functionality is poorly chosen, no amount of effort spent on user interface design will result in a usable system, and it might not even be useful at all. This is by no means a rare occurrence; it is easy to find cases of poorly chosen functionality that undermine whatever usability properties the system otherwise possesses. Some classic examples are as follows.

33.3.1.1.1 Interface Is Often Not the Problem

An important article with this title by Goransson et al. (1987) presents several brief case studies that involve failures of functionality design masquerading as usability problems. The most painful is a database system in a business organization that was considered to be too difficult to use. The interface was improved to make the system reasonably easy to use, but then it became clear that nobody in the organization needed the data provided by the system! Apparently, the original system development did not include an analysis of the needs of the organization or the system users. The best way to improve the usability of the system would have been to simply remove it.

33.3.1.1.2 Half a Loaf Is Worse Than None

The second major version of an otherwise easy-to-use basic word processing application included a multiple-column feature; however, it was not possible to mix the number of columns on a page. Note that documents having a uniform layout of two or three columns throughout are rare in the real world; rather, real multicolumn documents almost always mix the number of columns on at least one page. For example, a common pattern is a title page with a single column for the title that spans the page, followed by the body of the document in two-column format. The application could produce such a document only if two separate documents were prepared, printed, and then physically cut-and-pasted together! In other words, the multiple-column feature of this second version was essentially useless for preparing real documents. A proper task analysis would have determined the kinds and structures of multiple-column documents that users would be likely to prepare. Using this information during the product development would have led to either more useful functionality (like the basic page-layout features in the third major release of the product), or a decision not to waste resources on a premature implementation of incomplete functionality.

33.3.1.1.3 Why Doesn't It Do That?

A first-generation handheld "digital diary" device provided calendar and date book functions equivalent to paper calendar books, but included no clock, no alarms, and no awareness of the current date, although such functions would have been minor additions to the hardware. In addition, there was no facility for scheduling repeating meetings, making such scheduling remarkably tedious. The only shortcut was to use a rather clumsy copy–paste function, but it did not work for the meeting time field in the meeting information. A task analysis of typical user's needs would have identified all of these as highly desirable functions. Including them would have made the first generation of these devices much more viable.

33.3.1.1.4 Progress Is Not Necessarily Monotonic

The second version of an important Personal Digital Assistant also had a problem with recurring meetings. In the first version, a single interface dialog was used for specifying recurring meetings, and it was possible to select multiple days per week for the repeating meeting. Thus, the user could easily specify a weekly repeating meeting schedule of the sort common in academics, for example, scheduling a class that meets at the same time every Monday, Wednesday, and Friday for a semester. However, in the second version, which attempted many interface improvements, this facility moved down a couple of menu levels and became both invisible in the interface (unless a certain option was selected) and undocumented in the user manual. If any task analysis was done in connection with either the original or second interface design, it did not take into account the type of repeating meeting patterns needed in academic settings, a major segment of the user population that includes many valuable "early adopter" customers.

33.3.1.1.5 Missing Functions Can Be Hazardous to Your Health

The aforementioned are examples of missteps made long ago. However, poorly designed functionality is still common, even in the currently active and important field of health information systems. For example, at a major eye surgery clinic, a key part of the patient record is sketches on a paper form that are hand-drawn by the retina specialists during slit-lamp ophthalmoscopy examinations. These sketches show internal features of the patient's eyes that even modern imaging techniques cannot adequately convey. A new computer-based record system being installed has no provision for input or storage of such sketches. How will this be handled without degrading patient care, requiring additional personnel, or disrupting the workflow of these medical experts?

33.3.2 Best Practices: How to Design Functionality

33.3.2.1 Task Analysis at the Whole-System Level

When large systems are being designed, an important component of task analysis is to consider how the system, consisting of all the machines, and all the humans, is supposed to work as a whole in order to accomplish the overall system goal.

In the military systems, often analyzed heavily in Human Factors, the overall system goal is normally rather large-scale and well above the level of concerns of the individual humans operating the machines. For example, in designing a new naval fighter aircraft, the system goals might be stated in terms such as "enable air superiority in naval operations under any conditions of weather and all possible combat theaters through the year 2020." At this level, the "users" of the system as a whole are military strategist and commanders, not the pilot, and the system as a whole will involve not just the pilot, but other people in the cockpit (such as a radar intercept operator), and also maintenance and ground operations personnel. Thus, the humans involved in the system will have a variety of goals and tasks, depending on their role in the whole system.

At first glance, this level of analysis would appear to have little to do with computer systems; we often think of computer users as isolated individuals carrying out their tasks by interacting with their individual computers. However, when considering the needs of an organization, the mission level of analysis is clearly important; the system is supposed to accomplish something as a whole, and the individual humans all play roles defined by their relationship with each other and with the machines in the system. HCI has begun to consider higher levels of analysis, as in the field of computer-supported collaborative work, but perhaps the main reason why the mission level of analysis is not common parlance in HCI is that HCI has a cultural bias that organizations revolve around humans, with the computers playing only a supporting role. Such a bias would explain the movement mentioned earlier toward incorporating more social-science methodology into system design. In contrast, in military systems, the human operators are often viewed as "parts" in the overall system, whose ultimate "user" is the commanding officer, leading to a concern with how the humans and machines fit together. Only after due regard for this whole-system perspective, which provides the context for the activities of individual humans, is it then possible to properly consider what goals the individual operators will need to accomplish and how their machines will support them.

This kind of very high-level analysis can be done even with very large systems, such as military systems involving multiple machines and humans. The purpose of the analysis is to determine what role in the whole system the individual human operators will play. Various methods for whole-system analysis have been in routine use for some time. Briefly, these are as follows (see Beevis et al. 1992; Kirwan and Ainsworth 1992).

33.3.2.1.1 Mission and Scenario Analysis

Mission and scenario analysis is an approach to starting the system design from a description of what the system has to do (the mission), especially using specific concrete examples, or scenarios (see Beevis et al. 1992 for related discussion).

33.3.2.1.2 Function-Flow Diagrams

Function-flow diagrams are constructed to show sequential or information-flow relationships of the functions performed in the system. Beevis et al. (1992) provide a set of large-scale examples, such as naval vessels.

33.3.2.1.3 Function Allocation

Function allocation is a set of fairly informal techniques for deciding which system functions should be performed by machines, and which by people. Usually mentioned in this context is the *Fitts list* that describes what kinds of activities can be best performed by humans versus machines (Beevis et al. 1992). Two examples for humans are the ability to perceive patterns and make generalizations about them, and to react to unexpected low-probability events. Two examples for machines are performing repetitive and routine tasks, and doing many different things at one time. However, according to surveys described in Beevis et al. (1992), this classic technique is rarely used in real design problems since it is simply not specific enough to drive design decisions. Rather, functions are typically allocated in an ad hoc manner, often simply maintaining whatever allocation was used in the predecessor system, or following the rule that whatever can be automated should be, even though it is known that automation often produces safety or vigilance problems for human operators (Kantowitz and Sorkin 1987). There have been some attempts to make function allocation more systematic (see Sheridan, Woods, Pew, and Hancock 1998 for an overview), but at least when it comes to computer system design, methods both new and still under development must be relied upon.

As argued earlier, the key feature of computer-based tasks is the dominance of procedural aspects of the task; this means that function allocation approaches that are based on the task procedures are mostly likely to be valuable in the design of functionality. Thus, function allocation decisions are best driven by consideration of the task procedures; a good function allocation will be characterized by fast, simple, and efficient procedures for the user to follow with a minimum of overhead imposed by the use of the machine. The next sections describe some developing approaches to function allocation and functionality design.

33.3.2.2 Using GOMS Task Analysis in Functionality Design

The task analysis techniques described previously work well enough to have been developed and applied in actual system design contexts. However, they have mainly developed in the analysis of traditional interface technologies rather than computer user interfaces. While as discussed earlier, the general concepts of task analysis hold for computer interfaces, there are some key differences and a clear need to address computer user interfaces more directly. In summary, the problems with traditional task representations are as follows:

1. Computer interface procedures tend to be complicated, repetitious, and hierarchical. The primarily graphical and tabular representations traditionally used for procedures become unwieldy when the amount of detail is large.
2. They use different representation approaches for procedures that differ only in the level of analysis. For example, in HTA, a plan is represented differently from a procedure, even though a plan is simply a kind of higher-order procedure.
3. To a great extent, Human Factors practice uses different representations for different stages of the design process (see Beevis et al. 1992; Kirwan and Ainsworth 1992). It would be desirable to have a single representation that spans these stages even if it only covers part of the task analysis and design issues.

This section describes how GOMS models could be used to represent a high-level task analysis that can be used to help choose the desirable functionality for a system. Because GOMS models have a programming-language-like form, they can represent large quantities of procedural detail in a uniform notation that works from a very high level down to the lowest level of the interface design.

33.3.2.3 High-Level GOMS Analysis

Using high-level GOMS models is an alternative to the conventional requirements development and interface design process discussed in the introduction to this chapter. The approach is to drive the choice of functionality from the high-level procedures for doing the tasks, choosing functions that will produce simple procedures for the user. By considering the task at a high level, these decisions can be made independently of, and prior to, the interface design, thereby improving the chances that the chosen functionality will enable a highly useful and usable product once a good interface is developed. The analyst can then begin to elaborate the design by making some interface design decisions and writing the corresponding lower-level methods. If the design of the functionality is sound, it should be possible to expand the high-level model into a more detailed GOMS model that also has simple and efficient methods. If desired, the GOMS model can be fully elaborated down to the keystroke level of detail that can produce usability predictions (see John and Kieras 1996a,b; Kieras 1997).

GOMS models involve goals and operators at all levels of analysis, with the lowest level being the so-called keystroke level, of individual keystrokes or mouse movements. The lowest level goals will have methods consisting of keystroke-level operators, and might be basic procedures such as moving an object on the screen, or selecting a piece of text. However, in a high-level GOMS model, the goals may refer only to parts of the user's task that are independent of the specific interface, and may not specify operations in the interface. For example, a possible high-level goal in creating a document would be *Add a footnote*, but not *Select insert footnote from edit menu*. Likewise, the operators must be well above the keystroke level of detail, and not be specific interface actions. The lowest level of detail an operator may have is to invoke a system function, or perform a mental decision or action such as choosing which files to delete or thinking of a file name. For example, an allowable operator would be *Invoke the database update function*, but not *Click on the update button*.

The methods in a high-level GOMS model describe the order in which mental actions or decisions, sub-methods, and invocations of system functions are executed. The methods should document what information the user needs to acquire in order to make any required decisions and invoke the system functions, and also should represent where the user might detect errors and how they might be corrected with additional system functions. All too often, support for error detection and correction by the user is either missing, or is a clumsy add-on to a system design; by including it in the high-level model for the task, the designer may be able to identify ways in which errors can be prevented, detected, and corrected, early and easily.

33.3.2.3.1 An Example of High-Level GOMS Analysis

33.3.2.3.1.1 Task Domain The domain for this example is electronic circuit design and computer-aided design systems for electronic design (ECAD). A task analysis of this domain would reveal many very complex activities on the part of electronic circuit designers. Of special interest are several tasks for which computer support is feasible: After a circuit is designed, its correct functioning must be verified, and then its manufacturing cost estimated, power and cooling requirements determined, the layout of the printed circuit boards designed, automated assembly operations specified, and so forth.

This example involves computer support for the task of verifying the circuit design by using computer simulation to replace the traditional slow and costly "breadboard" prototyping. For many years, computer-based tools for this process have been available and undergoing development, based on the techniques for simulating the behavior of the circuit using an abstract mathematical representation of the circuit. For purposes of this example, attention will be limited to the somewhat simpler domain of digital circuit design. Here, the components are black-box modules (i.e. integrated circuits, "chips") with known behaviors, and the circuit consists simply of these components with their terminals interconnected with wires. Thus, the basic task goal for this example is to arrive at a correctly functioning circuit—a configuration of interconnected chips that performs the required function.

However, the user can make errors in performing the task; in this domain, the user errors can be divided into *semantic* errors, in which the specified circuit does not do what it is supposed to do, and *syntactic* errors, in which the specified circuit is invalid, regardless of the function. The semantic errors can only be detected by the user evaluating the behavior of the circuit, and discovering that the idea for the circuit was incorrect or incorrectly specified. In fact, the basic concept of the ECAD simulator is to support finding and correcting the semantic errors, so these are already being accounted for. The syntactic errors in the digital circuit domain are certain connection patterns that are incorrect just in terms of what is permissible and meaningful with digital logic circuits. Two common cases are disallowed connections (e.g., shorting an output terminal to ground) or missing connections (e.g., leaving an input terminal unconnected). It is important to correct syntactic errors prior to running the simulator, because their presence will cause the simulator to fail or produce invalid results.

The high-level functionality design of an actual system will be illustrated; then to show how the analysis can lead to design improvements, a better design will be illustrated.

33.3.2.3.1.2 High-Level GOMS Model The basic functionality concept is using a simulation of a circuit to verify the circuit design. This concept is described in the *top-level method*, which is the first method shown in Figure 33.1, which accomplishes the goal *Verify circuit with ECAD system*. This first method needs some explanation of the notation, which is the "Natural" GOMS Language (NGOMSL) described by Kieras (1997) for representing GOMS models in a relatively nontechnical readable format. The first line introduces a method for accomplishing the top-level user goal. It will be executed whenever the goal is asserted, and terminates when the return with goal accomplished operator in Step 4 is executed. Step 1 represents the user's "black-box" mental activity of thinking up the original idea for the circuit design; this think-of operator is just a placeholder; no attempt is made to represent the extraordinarily complex

```
Method for goal: Verify circuit with ECAD system
     Step 1. Think-of circuit idea.
     Step 2. Accomplish Goal: Enter circuit into ECAD system.
     Step 3. Accomplish Goal: Proofread drawing.
     Step 4. Invoke circuit simulation function and get results.
     Step 5. Decide: If circuit performs correct function,
                   then return with goal accomplished.
     Step 6. Think-of modification to circuit.
     Step 7. Make modification with ECAD system.
     Step 8. Go to 3.

Method for goal: Enter circuit into ECAD system
     Step 1. Invoke drawing tool.
     Step 2. Think-of object to draw next.
     Step 3. If no more objects, then Return with goal accomplished.
     Step 4. Accomplish Goal: draw the next object.
     Step 5. Go to 2.

Selection rule set for goal: Drawing an object
     If object is a component, then accomplish Goal: draw a component.
     If object is a wire, then accomplish Goal: draw a wire.
     ...
     Return with goal accomplished

Method for goal: Draw a component
     Step 1. Think-of component type.
     Step 2. Think-of component placement.
     Step 3. Invoke component-drawing function with type and placement.
     Step 4. Return with goal accomplished.

Method for goal: Draw a wire
     Step 1. Think-of starting and ending points for wire.
     Step 2. Think-of route for wire.
     Step 3. Invoke wire-drawing function with starting point,
                   ending point, and route.
     Step 4. Return with goal accomplished
```

FIGURE 33.1   High-level methods for ECAD system.

cognitive processes involved. Step 2 asserts a subgoal to specify the circuit for the ECAD system by entering the circuit into the system; the method for this subgoal appears next in Figure 33.1. Step 3 of the top-level method asserts a subgoal to proofread the entered circuit to find and correct the syntactic errors. Step 4 is a high-level operator for invoking the functionality of the circuit simulator and getting the results; no commitment is made at this point to how this will be done; the method merely asserts that there is a function to do this. Step 5 documents that at this point the user will decide whether the job is complete or not based on the output of the simulator. If not, Step 6 is another placeholder for a complex cognitive process of deciding what modification to make to the circuit. Step 7 invokes the functionality to modify the circuit, which would probably be much like that involved in Step 2, but which will not be further elaborated in this example. Finally, the loop at Step 8 shows that the top-level task is iterative.

At this point, the top-level method shows the basic structure of how the ECAD system will work: the user inputs a circuit specification, checks the specification for entry errors, runs the simulation, interprets the results, and modifies the circuit and repeats if necessary. Notice that the iteration in the top-level method provides for detecting and correcting semantic errors in the circuit concept. This basic design is motivated by a function allocation: only the human user knows how the circuit must behave, and what candidate circuit might accomplish that behavior, so these parts of the process are performed by the human. However, the computer can simulate the behavior of the circuit much faster and more reliably than a human could, so it performs this function. This basic function allocation insight is the motivation for developing the system.

The methods must be expanded to show how the top-level steps will get done. First, consider the method for entering a circuit into the ECAD system. In this domain, schematic circuit drawings are the conventional representations for a circuit, so a good choice of functionality for entering the candidate circuit information is a tool for drawing circuit diagrams that also produces the data representation needed for the simulation. This is reflected in the method for the goal *Enter circuit into ECAD system.*

This method starts with invoking the drawing tool, and then has a simple iteration consisting of thinking of something to draw, accomplishing the goal of drawing it, and repeating until done. The methods get more interesting when the goal of drawing an object is considered, because in this domain there are some fundamentally different kinds of objects, and the information requirements for drawing them are different. Only two kinds of objects will be considered here.

A selection rule is needed in a GOMS model to choose what method to apply depending on the kind of object, and then a separate method is needed for each kind of object; the selection rule set in Figure 33.1 thus accomplishes a general goal by asserting a more specific goal, which then triggers the corresponding method. The method for drawing a component requires a decision about the type of component (e.g., what specific multiplexer chip should be used) and where in the diagram the component should be placed to produce a visually clear diagram; the drawing function for a component creates the appropriate graphical representation of the component. Drawing a connecting wire requires deciding which two points in the circuit the wire should connect, and then choosing a route for the wire to produce a clear appearance in the diagram.

At this point, the analysis has revealed some possibly difficult and time-consuming user activities; candidates for additional system functions to simplify these activities could be considered. For example, the step of thinking of an appropriate component might be quite difficult, due to the huge number of integrated circuits available, and likewise thinking of the starting and ending points for the wire involves knowing which input or output function goes with each of the many pins on the chips. Some kind of online documentation or database to provide this information in a form that meshes well with the task might be valuable. Likewise, a welcome function might be some automation to choose a good routing for wires.

Next to be considered is how the goal of proofreading the circuit would be accomplished. In the actual design presented here, the functionality decisions were apparently motivated by simple after-the-fact "add-ons." As shown in Figure 33.2, the method for proofreading the drawing first invoked a checking

---

```
Method for goal: Proofread drawing
    Step 1. Invoke checking function and get list of error messages.
    Step 2. Look at next error message.
    Step 3. If no more error messages, Return with goal accomplished.
    Step 4. Accomplish Goal: Process error message.
    Step 5. Go to 2.

Method for goal: Process error message
    Step 1. Accomplish Goal: Locate erroneous point in circuit.
    Step 2. Think-of modification to erroneous point.
    Step 3. Make modification to circuit.
    Step 4. Return with goal accomplished.

Method for goal: Locate erroneous point in circuit
    Step 1. Read type of error, netlist node name from error message.
    Step 2. Invoke identification function with netlist node name.
    Step 4. Locate highlighted portion of circuit.
    Step 5. Return with goal accomplished.
```

FIGURE 33.2 Diagram of proofreading methods for an actual ECAD system.

function, which was designed to produce a series of error messages that the user would process one at a time. For ease in implementation, the checking function did not work in terms of the drawing, but in terms of the abstract circuit representation, the "netlist," and so reported the site of the syntactically illegal circuit feature in terms of the name of the "node" in the netlist. However, the only way the user had to examine and modify the circuit was in terms of the schematic diagram. So, the method for processing each error message requires first locating the corresponding point in the circuit diagram, and then making a modification to the diagram. To locate the site of the problem on the circuit diagram, the user invoked an identification function and provided the netlist node name; the function then highlights the corresponding part of the circuit diagram, which the user can then locate on the screen. Only then can the user think of what the problem is and how to correct it. In summary, this set of functionality design decisions means that in order to check the diagram for errors, the user must first complete the entire diagram, and then invoke a function, which produces output for each error that must be manually transferred into another function to identify the location of the error on the diagram.

Although no specific details of the interface have yet been specified, the high-level methods clearly point to a usability problem in the making: a clumsy iterative process must be performed just to make syntactic errors visible in a natural way. This design could be dismissed as a silly example, except for the fact that at least one major vendor of ECAD software used exactly this design.

Are there alternative functionality designs that would simplify the proofreading process? Certainly, combining the checking and identification functions would remove the need to manually transfer the netlist node name between the two device functions; the user could view the error message and the highlighted part of the circuit diagram simultaneously. However, a bit more analysis suggests that a different function allocation might be much better. In particular, rather than allow the user to complete a diagram that contains errors which are then detected and corrected, it would be desirable to allow errors to be detected and corrected during the drawing; the proofreading functionality would be moved from the top-level method into the circuit-entry method and its submethods. In this domain, disallowed connections appear at the level of individual wires, while missing connections show up at the level of the entire diagram. Thus, the functions to assist in detecting and correcting disallowed connections should be operating while the user is drawing wires, and those for missing connections while the user is working on the whole drawing. Figure 33.3 shows the corresponding revised methods. In this version, the function for drawing a wire checks for disallowed connections and thus enables their correction immediately; proofreading the entire diagram has been moved from the top-level method to the method for entering the circuit, and involves detecting missing connections and correcting them by drawing a wire.

At this point in the design of functionality, the analysis has documented what information the user needs to get about syntactic errors in the circuit diagram, and where this information could be used to detect and correct the error immediately. Some thought can now be given to what interface-level functionality might be useful to support the user in finding syntactically missing and disallowed connections. One obvious candidate is to use color-coding; for example, perhaps unconnected input terminals could start out as red in color on the display, and then turn green when validly connected. Likewise, perhaps as soon as a wire is connected at both ends, it could turn red if it is a disallowed connection, and green if it is legal. This use of color should work very well, since properly designed color-coding is known to be an extremely effective way to aid visual search. In addition, the color-coding calls the user's attention to the problems but without forcibly interrupting the user. This design rationale for using color is an interesting contrast to the actual ECAD system display, which generally made profligate use of color in ways that lacked any obvious value in performing the task. Figure 33.4 presents a revision of the methods to incorporate this preliminary interface design decision.

33.3.2.3.1.3 Implications The previous example of how the design of functionality can be aided by working out a high-level GOMS model of the task seems straightforward and unremarkable. A good design is usually intuitively "right," and once presented, seems obvious. However, at least the first few generations of ECAD tools did not implement such an "intuitively obvious" design at all, probably because nothing

```
Method for goal: Verify circuit with ECAD system
    Step 1. Think-of circuit idea.
    Step 2. Accomplish Goal: Enter circuit into ECAD system.
    Step 3. Invoke circuit simulation function and get results.
    Step 4. Decide: If circuit performs correct function, then
               return with goal accomplished.
    Step 5. Think-of modification to circuit.
    Step 6. Make modification with ECAD system.
    Step 7. Go to 3.

Method for goal: Enter circuit into ECAD system
    Step 1. Invoke drawing tool.
    Step 2. Think-of object to draw next.
    Step 3. Decide: If no more objects, then go to 6.
    Step 4. Accomplish Goal: draw the next object.
    Step 5. Go to 2.
    Step 6. Accomplish Goal: Proofread drawing.
    Step 7. Return with goal accomplished.

Method for goal: Proofread drawing
    Step 1. Find missing connection in drawing.
    Step 2. Decide: If no missing connection, return with goal
            accomplished.
    Step 3. Accomplish Goal: Draw wire for connection.
    Step 4. Go to 1.

Method for goal: Draw a wire
    Step 1. Think-of starting and ending points for wire.
    Step 2. Think-of route for wire.
    Step 3. Invoke wire drawing function with starting point,
               ending point, and route.
    Step 4. Decide: If wire is not disallowed, return with goal
            accomplished.
    Step 5. Correct the wire.
    Step 6. Return with goal accomplished.
```

FIGURE 33.3 Revised methods for error detection and correction steps during diagram entry.

```
Method for goal: Proofread drawing
    Step 1. Find a red terminal in drawing.
    Step 2. Decide: If no red terminals, return with goal accomplished.
    Step 3. Accomplish Goal: Draw wire at red terminal.
    Step 4. Go to 1.

Method for goal: Draw a wire
    Step 1. Think-of starting and ending points for wire.
    Step 2. Think-of route for wire.
    Step 3. Invoke wire drawing function with starting point,
               ending point, and route.
    Step 4. Decide: If wire is now green, return with goal accomplished.
    Step 5. Decide: If wire is red, think-of problem with wire.
    Step 6. Go to 1.
```

FIGURE 33.4 Methods incorporating color-coding for syntactic drawing errors.

like a task and functionality analysis was performed. Rather, a first version of the system was probably designed and implemented in which the user draws a schematic diagram in the obvious way, and then runs the simulator on it. However, once the system was in use, it became obvious that errors could be made in the schematic diagram that would cause the simulation to fail or produce misleading results. The solution was minimal add-ons of functionality, leaving the original flawed design of functionality intact.

In summary, the high-level GOMS model clarifies the difference between the two designs by showing the overall structure of the interaction. Even at a very high level of abstraction, a poor design of functionality can result in task methods that are obviously inefficient and clumsy. Thus, high-level GOMS models can capture critical insights from a task analysis to help guide the initial design of a system and its functionality. As the design is fleshed out, the high-level GOMS models can be elaborated into methods for a specific proposed interface, so that a good design of both functionality and interface can be developed prior to usability testing, providing a seamless pathway from initial design concept to final detailed design.

33.3.2.4 Work-Centered Design

High-level GOMS models show how a specific set of proposed functions would be used by a user to accomplish tasks. Intuitively, a good choice of functionality would be revealed by the high-level GOMS methods being simple and apparently easy for users to learn and execute. Rather than propose some functionality and then see how well it plays out in detail for the user, a better approach would be to design both the functionality and the user methods concurrently; that is, the design of the top-level interaction of the user and system should be the explicit goal of the system design. This approach was first presented by Butler et al. (1999), followed by a specific application reported by Butler et al. (2007). The approach is called work-centered design (WCD), because the focus has moved upward and outward from the user to the entire work process.

WCD is an approach for systematically designing the functionality and user processes as two complementary parts of a complex system; the name reflects a change in orientation—rather than *user-centered design* which focuses on the users and their needs, it focuses instead on the nature of the work to be done, and seeks to determine the combination of system and user activity that will enable the work to get done most effectively. It thus directly addresses the problem that most system development is *feature-based*; as mentioned earlier, the requirements for a system typically consist of a list of features, whose relation to the work to be done is rarely made explicit, and is often only haphazardly related to the user's tasks. In fact, users normally must *infer* how to use the system features to get the work done; they are often left in the position of being responsible for the "leftovers"—whatever bits of the process did not happen to be allocated to the computer systems. Typically, the resulting overly complex systems do not support the work very well. WCD proposes that instead, the system development process should deliberately focus on the design of the *Top-Level Algorithm* (TLA), the basic pattern of work division between the system and the user.

WCD combines ideas from business process modeling, industrial engineering, software engineering, mission-level task analysis, and the function allocation concept. WCD treats the user(s) and machine(s) of HCI as different types of agent resources, which perform their respective procedures. Taken together, their procedures must constitute a TLA that is capable of transforming the entity (product) of work to its goal state. The key idea is to perform a deliberate design of the top-level interaction between the user and the system in order to allocate functions to the user and system in a way that will optimize the overall workflow, while minimizing overhead tasks for users. The basic steps in WCD process are as follows:

1. *Characterize the work domain*: Similar to the high-level GOMS analysis, WCD takes a whole-system view. It begins by asking the purpose of the system—what has to get done? More specifically, what is the entity of work that the system must produce, independently of how the system will do it? WCD introduces conceptual knowledge as a complement to the procedural knowledge

in task analysis. Adapting a concept from industrial engineering, WCD explicitly specifies the entity of work that the system must produce. Similar to Contextual Research, WCD acknowledges the way the context in which work is performed constrains the procedures of humans and machines. For example, factors such as laws or organizational rules, or spatial separation of agents, must be taken into account if the procedures are to succeed.

2. WCD then asks what operations must be performed to produce that final result? In WCD, transforming the work entity to its goal state is a fundamental requirement for the TLA. For example, consider the problem of mission and maintenance scheduling for an Air Force Transport Squadron presented by Butler et al. (2007). The basic objects in the work entity are aircraft that must be maintained, and missions that need to be performed. The final product of the work is a combined schedule for flight operations and maintenance actions that as much as possible enables all the missions to be flown and all the maintenance to be performed in a timely way. The basic operation in the work domain is assigning time slots for planes when they can be assigned to missions or they will be assigned to maintenance. The assignments are constrained to satisfy all mission and maintenance activities within their needed time windows. The basic constraints are the mission time requirements, the maintenance schedule time requirements, and the maintenance resources available. If these constraints conflict, only designated humans are authorized to make exceptions to the mission or schedule requirements.

   The basic methodology for collecting this information is the task-analysis methods described earlier. But since the goal of the design effort is to produce a good human–machine function allocation and interaction, the emphasis is on characterizing the procedural aspects of the task, such as describing the workflow with a process-modeling tool such as BPMN (White and Miers 2008), a flowchart-like notation that shows the activities and interactions of each actor or system involved. This is similar to the form of operational sequence diagrams in which each actor/system has a "swim lane" and the interaction between these actors/systems is made explicit.

   In addition, it is essential to characterize the entity or entities that are being modified by the process. In the squadron-scheduling domain, a form of ontology model was useful to characterize the kinds of entities, how they could be modified, and the properties of a solution to the scheduling problem. Ideally, this initial characterization of the work to be done can avoid commitments to what part is done by humans versus machines even in the case of an existing system.

3. *Allocate the functions and design the TLA*: Given the operations that must be performed to transform the entity to its goal state, determine which parts of them should be done by humans, and which by machines, and how the work process will sequence human and machine functions to accomplish the work. What is feasible for the machine to do? What must be reserved for the human? If the process is already in place with allocated functions, should it be changed? Are there nonessential, overhead tasks that can be eliminated or reduced?

   Continuing the squadron-scheduling example, the first major step is that the computer system will automatically assign aircraft to missions and maintenance operations as long as both mission requirements and maintenance constraints can be met. However, if they cannot, then the second major step is that humans must decide in a socially determined chain of command whether maintenance actions can be postponed in order to meet the mission requirements, or whether the mission requirements can be modified to meet the maintenance requirements. The third major step is to input these decisions into the machine, which will then finalize the schedule. The sequence, or workflow, in which the machine and human functions interact is described by the TLA.

4. *Assess the quality of the TLA*: How effective is the proposed combination of process and function allocation? A *correct* TLA ensures that the work goals are accomplished—the correct work products are produced. If not, the process and functions must be corrected. An *efficient* TLA has an optimum human–machine function allocation that minimizes the overhead (unnecessary work) and that humans are not performing work that could be done by the machine. This goes beyond simple record-keeping and automation of complex functions such as optimization. For example,

well-designed displays will require less cognitive effort to interpret, and thus eliminate unnecessary work; but the functionality behind computing the required content must be designed. Other overhead issues show up in procedures that are complex, difficult to learn, or impose memory load. If the low-level details of the user interface are well-designed, an efficient TLA will be apparent in usability testing.

But, designing a user interface for a poor TLA is wasted effort; so, some means of evaluating the quality of the TLA prior to interface development is valuable. Task network simulations or high-level GOMS models can provide rough predictions of task performance times or user workloads. If the TLA has problems, then return to Step 2 and attempt to devise a different process or function allocation.

For example, at one point in the development of the squadron scheduling system, a high-level GOMS analysis showed that there was a high overhead user function involving multiple inspections of the individual aircraft information to determine the final schedule. Once seen, it was easy to assign this particular function to the optimization software, resulting in the disappearance of a large amount of user overhead effort. A less formal example was the realization that the human authorized to make changes to the maintenance or mission requirements was not a direct user of the system; so the TLA had to be modified to include using a report-generating function to produce a concise description of the schedule conflict to be given to the decision-maker.

WCD changes the focus from user interface design to the design of effective and efficient work procedures. Once a good TLA has been designed, techniques and guidelines for good user interface design can be applied to complete the design of the user's side of the system. In the case of the squadron scheduling system, the resulting design in its first user testing iteration enabled a single user to successfully complete in less than a half-hour the work that previously required 3 days of work by three experienced users (Butler et al. 2007). Thus, WCD techniques for task analysis and design of functionality can lead to system designs whose quality is high on the first testable version, a great advantage in the development of complex systems. Work is underway to further develop the approach and supporting tools for designing health information systems (Butler et al. 2011).

33.4 Research Issues and Concluding Summary

The major research problem in task analysis is attempting to bring some coherence and theoretical structure to the field. While psychology as a whole, rather seriously, lacks a single theoretical structure, the subfields most relevant to human–system interaction are potentially unified by work in cognitive psychology on cognitive architectures, which are computational modeling systems that attempt to provide a framework for explaining human cognition and performance (see Byrne 2003 for an overview). These architectures are directly useful in system design in two ways: First, because the architecture must be "programmed" to perform the task with task-specific knowledge, the resulting model contains the content of a full-fledged task analysis, both the procedural and cognitive components. Thus, constructing a cognitive-architectural model is a way to represent the results of a task analysis and verify its completeness and accuracy. Second, because the architecture represents the constants and constraints on human activity (such as the speed of mouse pointing movements and short-term memory capacity), the model for a task is able to predict performance on the task, and so can be used to evaluate a design very early in the development process (see Kieras 2012 for an overview).

The promise is that these comprehensive cognitive architectures will encourage the development of coherent theory in the science base for HCI and also provide a high-fidelity way to represent how humans would perform a task. While some would consider such predictive models as the ultimate form of task analysis (Annett and Stanton 2000b), there is currently a gap between the level of detail required to construct such models and the information available from a task analysis, both in principle and in practice (Kieras and Meyer 2000); there is no clear pathway for moving from one of the well-established task analysis methods to a fully detailed cognitive-architectural model. It should be possible to bridge this gap, because GOMS models, which can be viewed as a highly simplified form of cognitive-architectural model (John and Kieras 1996a,b), are similar enough to HTA that it is easy to move from this most popular task analysis to a GOMS model. Future work in this area should result in better methods for developing high-fidelity predictive models in the context of more sophisticated task analysis methods.

Another area of research concerns the analysis of team activities and team tasks. A serious failing of conventional psychology and social sciences in general is a gap between the theory of humans as individual intellects and actors and the theory of humans as members of a social group or organization. This leaves HCI as an applied science without an articulated scientific basis for moving between designing a system that works well for an individual user and designing a system that meets the needs of a group. Despite this theoretical weakness, task analysis can be done for whole teams with some success, as shown by Essens et al. (2000), Klein (2000), and Zachary et al. (2000). What is less clear at this point is how such analyses can be used to identify an optimum team structure or interfaces to optimally support a team performance. One approach will be to use computational modeling approaches that take individual human cognition and performance into account as the fundamental determiner on the performance of a team (Kieras and Santoro 2004).

While there are serious practical problems in performing task analysis, the experience of Human Factors shows that these problems can be overcome, even for rather large and complex systems. The numerous methods developed by Human Factors for collecting and representing task data are ready to be used and adapted to the problems of computer interface design. The additional contributions of cognitive psychology have resulted in procedural task analyses that can help evaluate designs rapidly and efficiently. System developers thus have a powerful set of concepts and tools already available, and can anticipate even more comprehensive task analysis methods in the future.

Already developing is one set of ideas about the design of functionality, in which rather than the unsystematic choice of features which the user somehow infers how to apply, the top-level pattern of user–machine function interaction is the target of basic system design. This approach links conventional concepts of software development more directly to task analysis to enable systems that are both effective and highly usable.

Glossary

Cognitive psychology: A branch of psychology that is concerned with rigorous empirical and theoretical study of human cognition, the intellectual processes having to do with knowledge acquisition, representation, and application.

Cognitive task analysis: A task analysis that emphasizes the knowledge required for a task and its application, such as decision-making and its background knowledge.

Functionality: The set of user-accessible functions performed by a computer system; the kinds of services or computations performed that the user can invoke, control, or observe the results of.

GOMS model: A theoretical description of human procedural knowledge in terms of a set of Goals, Operators (basic actions), Methods, which are sequences of operators that accomplish goals, and Selection rules that select methods appropriate for goals. The goals and methods typically have a hierarchical structure. GOMS models can be thought of as programs that the user learns and then executes in the course of accomplishing task goals.

Human factors: Originated when psychologists were asked to tackle serious equipment design problems during World War II, this discipline is concerned with designing systems and devices so that they can be easily and effectively used by humans. Much of Human Factors is concerned with psychological factors, but important other areas are biomechanics, anthropometrics, work physiology, and safety.

Task: This term is not very well defined, and is used differently in different contexts, even within Human Factors and HCI. Here, it refers to purposeful activities performed by users, either a general class of activities, or a specific case or type of activity.

Task domain: The set of knowledge, skills, and goals possessed by users that is specific to a kind of job or task.

Usability: The extent to which a system can be used effectively to accomplish tasks; a multidimensional attribute of a system, covering ease of learning, speed of use, resistance to user errors, intelligibility of displays, and so forth.

User interface: The portion of a computer system that the user directly interacts with, consisting not just of physical input and output devices, but also the contents of the displays, the observable behavior of the system, and the rules and procedures for controlling the system.

Further Information

The reference list contains useful sources for following up this chapter. Landauer (1995) provides excellent economic arguments on how many systems fail to be useful and usable. The most useful sources on task analysis are Kirwan and Ainsworth (1992), Diaper and Stanton (2004a), and Beevis et al. (1992). John and Kieras (1996a,b) and Kieras (1997, 2004) provide detailed overviews and methods for GOMS analysis.

Acknowledgment

The concept of high-level GOMS analysis was developed in conjunction with Ruven Brooks, of Rockwell Automation, who also provided helpful comments on the first version of this chapter.

References

Annett, J. 2004. Hierarchical task analysis. In *The Handbook of Task Analysis for Human-Computer Interaction*, eds., D. Diaper and N. A. Stanton, pp. 67–82. Mahwah, NJ: Lawrence Erlbaum Associates.

Annett, J., Duncan, K.D., Stammers, R.B., and Gray, M.J. 1971. *Task Analysis*. London, U.K.: Her Majesty's Stationery Office.

Annett, J. and Stanton, N.A. (eds.) 2000a. *Task Analysis*. London, U.K.: Taylor & Francis Group.

Annett, J. and Stanton, N.A. 2000b. Research and development in task analysis. In *Task Analysis*, eds., J. Annett and N.A. Stanton, pp. 3–8. London, U.K.: Taylor & Francis Group.

Beevis, D., Bost, R., Doering, B., Nordo, E., Oberman, F., Papin, J-P. I., Schuffel, H., and Streets, D. 1992. Analysis techniques for man-machine system design. (Report AC/243(P8)TR/7). Brussels, Belgium: Defense Research Group, NATO HQ.

Beyer, H. and Holtzblatt, K. 1998. *Contextual Design: Defining Customer-centered Systems*. San Francisco, CA: Morgan Kaufmann.

Boose, J. H. 1992. Knowledge acquisition. In *Encyclopedia of Artificial Intelligence* (2nd edn.), pp. 719–742. New York: Wiley.

Butler, K.A., Esposito, C., and Hebron, R. 1999. Connecting the design of software to the design of work. *Communications of the ACM*, 42(1), 38–46.

Butler, K.A., Haselkorn, M., Bahrami, A., and Schroeder, K. 2011. Introducing the MATH method and toolsuite for evidence-based HIT. *AMA-IEEE 2nd Annual Medical Technology Conference*, Boston, MA, October 16–18, 2011, http://ama-ieee.embs.org/wp-content/uploads/2011/10/Butler-AMA-IEEE-final.2.pdf

Butler, K.A., Zhang, J., Esposito, C., Bahrami, A., Hebron, R., and Kieras, D. 2007. Work-centered design: A case study of a mixed-initiative scheduler. In *Proceedings of CHI 2007*, San Jose, CA, April 28–May 3, 2007. New York: ACM.

Byrne, M.D. 2003. Cognitive architecture. In *The Human-Computer Interaction Handbook,* eds., J.A. Jacko and A. Sears, pp. 97–117. Mahwah, NJ: Lawrence Erlbaum Associates.

Chipman, S.F., Schraagen, J.M., and Shalin, V.L. 2000. Introduction to cognitive task analysis. In *Cognitive Task Analysis*, eds., J.M. Schraagen, S.F. Chipman, and V.L. Shalin, pp. 3–24. Mahwah, NJ: Lawrence Erlbaum Associates.

Crandall, B., Klein, G., Hoffman, R.R. 2006. *Working Minds: A Practitioner's Guide to Cognitive Task Analysis*. New York: Bradford.

Diaper, D. and Stanton, N.A. (eds.) 2004a. *The Handbook of Task Analysis for Human-Computer Interaction*. Mahwah, NJ: Lawrence Erlbaum Associates.

Diaper, D. and Stanton, N.A. 2004b. Wishing on a sTAr: The future of task analysis. In *The Handbook of Task Analysis for Human-Computer Interaction*, eds., D. Diaper and N.A. Stanton, pp. 603–619. Mahwah, NJ: Lawrence Erlbaum Associates.

Dubois, D. and Shalin, V.L. 2000. Describing job expertise using cognitively oriented task analysis. In *Cognitive Task Analysis*, eds., J.M. Schraagen, S.F. Chipman, and V.L. Shalin, pp. 41–56. Mahwah, NJ: Lawrence Erlbaum Associates.

Essens, P.J.M.D., Post, W.M., and Rasker, P.C. 2000. Modeling a command center. In *Cognitive Task Analysis*, eds., J.M. Schraagen, S.F. Chipman, and V.L. Shalin, pp. 385–400. Mahwah, NJ: Lawrence Erlbaum Associates.

Goransson, B., Lind, M., Pettersson, E., Sandblad, B., and Schwalbe, P. 1987. The interface is often not the problem. In *Proceedings of CHI+GI 1987*. New York: ACM.

Gott, S.P. 1988. Apprenticeship instruction for real-world tasks: The coordination of procedures, mental models, and strategies. In *Review of Research in Education*, ed., E.Z. Rothkopf. Washington, DC: AERA.

Gould, J.D. 1988. How to design usable systems. In *Handbook of Human-Computer Interaction*, ed. M. Helander, pp. 757–789. Amsterdam, the Netherlands: North-Holland.

Grudin, J. 1991. Systematic sources of suboptimal interface design in large product development organizations. *Human-Computer Interaction*, 6, 147–196.

John, B.E. and Kieras, D.E. 1996a. Using GOMS for user interface design and evaluation: Which technique? *ACM Transactions on Computer-Human Interaction*, 3, 287–319.

John, B.E. and Kieras, D.E. 1996b. The GOMS family of user interface analysis techniques: Comparison and contrast. *ACM Transactions on Computer-Human Interaction*, 3, 320–351.

Kantowitz, B.H. and Sorkin, R.D. 1987. Allocation of functions. In *Handbook of Human Factors*, ed. G. Salvendy, pp. 355–369. New York: Wiley.

Kieras, D.E. 1997. A guide to GOMS model usability evaluation using NGOMSL. In *Handbook of Human-Computer Interaction* (2nd edn.), eds., M. Helander, T. Landauer, and P. Prabhu, pp. 733–766. Amsterdam, the Netherlands: North-Holland.

Kieras, D.E. 2004. GOMS models for task analysis. In *The Handbook of Task Analysis for Human-Computer Interaction*, eds., D. Diaper and N.A. Stanton, pp. 83–116. Mahwah, NJ: Lawrence Erlbaum Associates.

Kieras, D.E. 2012. Model-based evaluation. In *Human-Computer Interaction Handbook* (3rd edn.), ed. J.A. Jacko, pp. 1295–1341. New York: CRC Press.

Kieras, D.E. and Meyer, D.E. 2000. The role of cognitive task analysis in the application of predictive models of human performance. In *Cognitive Task Analysis*, eds., J.M. Schraagen, S.F. Chipman, and V.L. Shalin, pp. 237–260. Mahwah, NJ: Lawrence Erlbaum.

Kieras, D.E. and Santoro, T.P. 2004. Computational GOMS modeling of a complex team task: Lessons learned. In *Proceedings of CHI 2004: Human Factors in Computing Systems*, Vienna, Austria, pp. 97–104.

Kirwan, B. and Ainsworth, L.K. 1992. *A Guide to Task Analysis*. London, U.K.: Taylor & Francis Group.

Klein, G. 2000. Cognitive task analysis of teams. In *Cognitive Task Analysis*, eds., J.M. Schraagen, S.F. Chipman, and V.L. Shalin, pp. 417–430. Mahwah, NJ: Lawrence Erlbaum Associates.

Landauer, T. 1995. *The Trouble with Computers: Usefulness, Usability, and Productivity*. Cambridge, MA: MIT Press.

Militello, L.G. and Hutton, R.J.B. 2000. Applied cognitive task analysis (ACTA): A practitioner's toolkit for understanding cognitive task demands. In *Task Analysis*, eds., J. Annett and N.A. Stanton, pp. 90–113. London, U.K.: Taylor & Francis Group.

Reason, J. 1990. *Human Error*. Cambridge, MA: Cambridge University Press.

Schaafstal, A. and Schraagen, J.M. Training of troubleshooting: A structured task analytical approach. In *Cognitive Task Analysis*, eds., J.M. Schraagen, S.F. Chipman, and V.L. Shalin, pp. 57–70. Mahwah, NJ: Lawrence Erlbaum Associates.

Schraagen, J.M., Chipman, S.F., and Shalin, V.L. (eds.) 2000. *Cognitive Task Analysis*. Mahwah, NJ: Lawrence Erlbaum Associates.

Seamster, T.L., Redding, R.E., and Kaempf, G.L. 2000. A skill-based cognitive task analysis framework. In *Cognitive Task Analysis*, eds., J.M. Schraagen, S.F. Chipman, and V.L. Shalin, pp. 135–146. Mahwah, NJ: Lawrence Erlbaum Associates.

Sheridan, T.B., Van Cott, H.P., Woods, D.D., Pew, R.W, and Hancock, P.A. 1998. Allocating functions rationally between humans and machines. *Ergonomics in Design*, 6, 20–25.

Stanton, N.A. and Baber, C. 2005. Validating task analysis for error identification: reliability and validity of a human error prediction technique. *Ergonomics*, 48(9), 1097–1113.

Van Welie, M., Van der Veer, G.C., and Eliëns, A. 1998. An ontology for task world models. In *Design, Specification and Verification of Interactive Systems, DSV-IS 98*, eds., P. Markopoulos and P. Johnson, pp. 57–70. Springer Lecture Notes on Computer Science, Vienna.

White, S. and Miers, D. 2008. *BPMN Modeling and Reference Guide*. Lighthouse Point, FL: Future Strategies Inc.

Wong, W. 2004. Critical decision method data analysis. In *The Handbook of Task Analysis for Human-Computer Interaction*, eds., D. Diaper and N.A. Stanton, pp. 327–346. Mahwah, NJ: Lawrence Erlbaum Associates.

Wood, S.D. 1999. The application of GOMS to error-tolerant design. *In Proceeding of 17th International System Safety Conference*, Orlando, FL. Unionville, VA: System Safety Society Publications.

Woods, D.D., O'Brien, J.F., and Hanes, L.F. 1987. Human factors challenges in process control: The case of nuclear power plants. In *Handbook of Human Factors*, ed., G. Salvendy. New York: Wiley.

Zachary, W.W., Ryder, J.M., and Hicinbotham, J.H. 2000. Building cognitive task analyses and models of a decision-making team in a complex real-time environment. In *Cognitive Task Analysis*, eds., J.M. Schraagen, S.F. Chipman, and V.L. Shalin, pp. 365–384. Mahwah, NJ: Lawrence Erlbaum Associates.

Zhang, J. 1997. The nature of external representations in problem solving. *Cognitive Science*, 21(2), 179–217.

Zhang, J. and Norman, D.A. 1994. Representations in distributed cognitive tasks. *Cognitive Science*, 18, 87–122.

34

Designing Multimedia Applications for User Experience

Alistair Sutcliffe
University of Manchester

34.1 Introduction

Multimedia is rapidly becoming the default in most applications apart from transaction processing systems where numbers and text dominate, although multimedia has also been used extensively in task-based applications for process control and safety-critical systems (Hollan et al., 1984; Alty, 1991). With the advent of the Web 2.0 and beyond, interactive multimedia is a continuing design challenge. Note the "interactive" in multimedia; this is important since the objective is not only to deliver information through a variety of media (text, speech, image, video, music, etc.), but also to engage the user in simulating and exciting interaction. The view that the user interface (UI) design goes beyond functional and usability into user experience (UX) is well established in the Human–Computer Interaction (HCI) community (Hassenzahl and Tractinsky, 2006). In this chapter, the author takes a UX-oriented view of multimedia, so interaction will be treated as one medium in the multimedia design space.

This chapter has four aims:

1. Describe the properties of media resources and how they are used in design.
2. Propose design guidelines for conveying information in multimedia.
3. Explain the concept of UX and its relationship to interactive multimedia.
4. Propose multimedia design guidelines for UX.

The first part of the chapter will cover the design process for multimedia applications; starting with an information analysis, it then progresses to deal with issues of media selection and integration. A more detailed description of the process and information modeling techniques is given in Sutcliffe (2003, 2012). Multimedia design involves several special areas that are technical subjects in their own right. For instance, design of text is the science (or art) of calligraphy that has developed new fonts over many years; visualization design encompasses the creation of images, either drawn or captured as photographs. Design of moving images, cartoons, video, and film are further specializations, as are musical composition and design of sound effects. Multimedia design lies on an interesting cultural boundary between the creative artistic community and science-based engineering. One implication of this cultural collision is that space precludes "within-media" design being dealt with in depth in this chapter, i.e., guidelines for design of one particular medium.

The second part of the chapter broadens multimedia design to cover UX and design of applications where the goal is to persuade, entertain, or educate the user. Design in this section considers interaction design, aesthetics, and emotional effects of multimedia. UX design is a complex subject, so the background to this topic will only be reviewed briefly. For a more in-depth treatment of UX design, the reader is referred to Sutcliffe (2009).

34.2 Media Properties and Design Principles

We perceive multimedia using either our visual sense (text, photographs, drawings, animations), audio (speech, music, natural and artificial sounds), or both (film with a sound track). Haptic (touch), olfactory (smell), and gustatory (taste) senses may be more important in the near future; however, for the present, visual and audio senses dominate. How we perceive and understand multimedia is limited by cognitive constraints inherent in the way our brain processes information. These are summarized as follows; more in-depth explanation of these issues is given in Sutcliffe (2012):

- Selective attention: We can only attend to one input on one sensory channel at a time. Visually, this is a consequence of the way our eyes focus on images, on only one small area at a time. For audio, we can only pick out one message from the background noise; two or more separate sounds interfere, and we have to selectively attend to one or the other. We can overcome this limitation by time-slicing our attention between different images or sounds, but at the penalty of forgetting.
- Working memory: Information from our senses is perceived rapidly, but the understanding phase encounters a very limited bottleneck, working memory. Working memory is a very limited scratch pad or cache memory containing about five facts/ideas or concepts (chunks in the psychological jargon) at any one time. As information is received, it overwrites the current content of working memory, so we can only actively process a small quantity of information at once.
- Knowledge integration: We make sense of the world by seeing or hearing patterns, and those patterns are easier to understand when we can associate them with things we already know, our memory. This allows us to overcome some of the selective attention and working memory limitations. For example, when moving images and an audio sound track are integrated in a film, we can process information on two channels without difficulty.
- Arousal and affect: Any change in images or audio automatically demands our attention (see selective attention); however, change also alerts us, and this increases our arousal. Change in multimedia is therefore more interesting and possibly exciting. Content of multimedia can connect with our emotions as we react instinctively to scary images or loud noises.

The links between different types of media and these psychological implications are summarized in Table 34.1.

The psychological implications also motivate multimedia design principles (ISO, 1997, 1998). The principles are high-level concepts, which are useful for general guidance, but they have to be interpreted

TABLE 34.1 Psychological Constraints and Implications for Multimedia Design

Medium	Selective Attention	Working Memory	Arousal and Affect
Image: Photograph	Eye scanning depends on salient features on images	No detail of image persists; only a few facts can be extracted at one time	Highlighting components increase arousal; other effects depend on content
Text	Eye scanning is set by text format and reading order	Only a few (four to five) facts can be extracted at once	Depends on content
Moving image: Film, cartoons	Attention is automatic; it is difficult to ignore dynamic media	Rapid overwriting; only high-level "gist" facts can be extracted	Dynamic media are more arousing than static media; content can be very effective in emotional reactions
Audio: Speech	Attention is automatic. Speech dominates over other sound	Rapid overwriting; only high-level "gist" facts can be extracted	Dynamic media are more arousing; content and delivery (voice tone) can be used for emotional effects
Audio: Other sounds	Attention is automatic	Facts depend on recognition, i.e., matching sound to memory	Audio has a range of arousal and emotional effects

in a context to give more specific advice. The following design principles are cross-referenced to the corresponding psychological constraints:

- *Thematic congruence.* Messages presented in different media should be linked together to form a coherent whole. This helps comprehension as the different parts of the message make sense by fitting together. Congruence is partly a matter of designing the content so that it follows a logical theme (e.g., the script or story line makes sense and does not assume too much about the user's domain knowledge), and partly a matter of attentional design to help the user follow the message thread across different media (knowledge integration, selective attention).

- *Manageable information loading.* Messages presented in multimedia should be delivered at a pace that is either under the user's control or at a rate that allows for effective assimilation of information without causing fatigue. The rate of information delivery depends on the quantity and complexity of information, the effectiveness of the design in helping the user extract the message from the media, and the user's domain knowledge and motivation. Some ways of reducing information overload are to avoid excessive use of concurrent dynamic media and give the user time to assimilate complex messages. This guards against overloading user's working memory. Since our working memory has a small capacity and gets continually overwritten by new input, controlling the pace and quantity of input can reduce overloading.

- *Compatibility with the user's understanding.* Media should be selected that convey the content in a manner compatible with the user's existing knowledge, for example, the radiation symbol and road sign icons are used to convey hazards and dangers to users who have the appropriate knowledge and cultural background. The user's ability to understand the message is important for designed image media (diagrams, graphs) when interpretation is dependent on the user's knowledge and background (knowledge integration).

- *Complementary viewpoints.* Similar aspects of the same subject matter should be presented on different media to create an integrated whole. Showing different aspects of the same object, for example, a picture and a design diagram of a ship, can help memorization by developing richer schema and better memory cues (knowledge integration).

- *Consistency.* It helps users learn an interface by making the controls, command names, and layout follow a familiar pattern. People recognize patterns automatically, so operating the interface becomes an automatic skill. Consistent use of media to deliver messages of a specific type can help by cueing users with what to expect (selective attention, knowledge integration).
- *Interaction and engagement.* These help understanding and learning by encouraging the user to solve problems. Memory is an active process. Interaction increases arousal, and this makes the user's experience more vivid, exciting, and memorable (arousal and affect).
- *Reinforce messages.* Redundant communication of the same message on different media can help learning. Presentation of the same or similar aspects of a message helps memorization by the frequency effect. Exposing users to the same thing in a different modality also promotes rich memory cues (knowledge integration; see also complementary viewpoints).

Many of the principles are motivated by knowledge integration, which is one of the reasons for the "multi" in multimedia.

The following section provides some guidelines to address these issues.

34.3 Media Selection and Combination

Designers have to solve two problems:

1. Which media resources to select (or create) to convey information?
2. How to integrate media resources to convey information more effectively and improve UX?

Media resources may be classified as either static (do not change, e.g., text, images) or dynamic in the sense that they change during presentation (e.g., speech, all audio media, moving image media such as films, cartoons). Another distinction is also useful: realistic/not realistic, reflecting the designer's involvement in creating the medium. For instance, a photograph or video may capture an image of the real world; alternatively, the hand of the designer may be more overt in drawings, sketches, and diagrams. Note that it is the content that determines the media resource rather than the presentation medium; for example, a diagram is still a diagram whether it has been photographed or drawn on paper and scanned. The design of media resources is important, because abstract concepts can only be conveyed by language (text/speech), designed audio (music), or by designed images (diagrams, cartoons). More fine-grained taxonomic distinctions can be made, for instance, between different signs and symbolic languages (see Bernsen, 1994), but richer taxonomies increase specification effort. Media resources can be classified using the three criteria in the decision tree illustrated in Figure 34.1.

The approach to classifying media uses a decision tree "walkthrough" with the following questions that reflect the facets of the classification:

- Is the medium perceived to be realistic or not? Media resources captured directly from the real world will usually be realistic, for example, photographs of landscapes, sound recordings of bird song, etc. In contrast, nonrealistic media are created by human action. However, the boundary case category that illustrates the dimension is a realistic painting of a landscape.
- Does the medium change over time or not? The boundary case here is the rate of change, particularly in animations where some people might judge 10 frames/s to be a video but 5 slides in a minute shown by a PowerPoint presentation to be a sequence of static images.
- Which modality does the resource belong to? Most media are visual (image/text) or audio (music/speech), although one resource may exhibit two modalities. For example, a film with a sound track communicates in both visual and audio modalities.

The walkthrough process informs selecting appropriate media resources to convey information content. Information content may be apparent from the resources provided, such as diagrams, text, etc; however, in some cases, only high-level user requirements may be given, such as "present

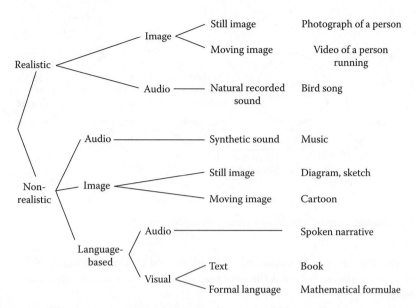

FIGURE 34.1 Decision tree for classifying media resources.

information to influence people to adopt a healthy diet, give up smoking, or purchase a product from an e-commerce website." The designers will have to source the content if it is not provided, and then further media resources may have to be designed to amplify the message. For example, education materials to explain the working of a car engine may be simple diagrams and text. To really convey how an engine works requires an interactive animation so the student can see the sequence of movements and relate these to a causal model of the internal combustion engine. Multimedia design, therefore, often involves transforming and augmenting information content for effective communication.

Recommendations for selecting media have to be interpreted according to the users' task and design goals. If information provision is the main design goal, for example, in a tourist kiosk information system, then persistence of information and drawing attention to specific items is not necessarily as critical as in tutorial applications. Task and user characteristics influence media choice; for instance, verbal media are more appropriate to language-based and logical reasoning tasks; visual media are suitable for spatial tasks involving moving, positioning, and orienting objects. Some users may prefer visual media, while image is of little use for blind users.

34.3.1 Information Architecture

This activity consists of several activities, which will differ according to the type of application. Some applications might have a strong task model; for instance, a multimedia process control application where the tasks are monitoring a chemical plant, diagnosing problems, and supporting the operator in controlling plant operation. In goal-driven applications, information requirements are derived from the task model. In information provision applications such as websites with an informative role, information analysis involves categorization, and the architecture generally follows a hierarchical model. In the third class of explanatory or thematic applications, analysis is concerned with the story or argument, that is, how the information should be explained or delivered. Educational multimedia and websites with persuasive missions fall into the last category.

In information-provision applications, classification of the content according to one or more user views defines the information architecture; for example, most university departments have an information structure with upper-level categories for research, undergraduate courses, postgraduate courses, staff interests, departmental organization, mission, and objectives. For explanatory applications,

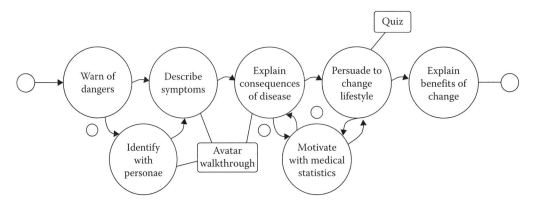

FIGURE 34.2   Thematic map for a healthcare promotion application.

a theme or story line needs to be developed. This will depend on the application's objectives and the message the owner wishes to deliver. A thematic map from a health awareness application is illustrated in Figure 34.2.

In the case described in Figure 34.2, the requirement is to convince people of the dangers of heart disease. The theme is a persuasive argument that first tries to convince people of the dangers from smoking, poor diet, stressful lifestyles, etc. It then explains how to improve their lifestyle to prevent heart disease, followed by reinforcing the message with the benefits of a healthy lifestyle such as lower health insurance, saving money, longer life, etc. Subthemes are embedded at different points, so users can explore the facts behind heart disease, the statistics and their exposure, how to get help, etc. Information is then gathered for each node in the thematic map. How this architecture will be delivered depends on interaction design decisions: it could, for example, become an interactive story to explore different lifestyle choices, combined with a quiz. The outcome of information architecture analysis will be an information-enhanced task model, a thematic map, or a hierarchy/network to show the structure and relationships of information categories. The next step is to analyze the information content by classifying it by types.

Information types are amodal, conceptual descriptions of information components that elaborate the content definition. Information components are classified into one or more of the following:

- Physical items relating to tangible observable aspects of the world
- Spatial items relating to geography and location in the world
- Conceptual abstract information, facts, and concepts related to language
- Static information that does not change: objects, entities, relationships, states, and attributes
- Dynamic, or time-varying information: events, actions, activities, procedures, and movements
- Descriptive information, attributes of objects, and entities
- Values and numbers
- Causal explanations

Information is often complex or composite, so one component may be classified with more than one type; for instance, instructions on how to get to the railway station may contain procedural information (the instructions "turn left, straight ahead," etc.), and spatial or descriptive information (the station is in the corner of the square, painted blue). The information types are "tools for thought," which can be used either to classify specifications of content or to consider what content may be necessary. To illustrate, for the task "navigate to the railway station," the content may be minimally specified as "instructions how to get there," in which case the information types prompt questions in the form "what sort of information does the user need to fulfill the task/user goal?" Alternatively, the content may be specified as a scenario narrative of directions, waymarks to recognize, and description of the target. In this case, the types classify components in the narrative to elucidate the deeper structure of the content. The granularity of components is a matter of the designer's choice and will depend on the level of detail demanded by the application. To illustrate the analysis:

Communication goal: Explain how to assemble a bookshelf from ready-made parts.

Information component 1:

  Parts of the bookshelf, sides, back, shelves, connecting screws

  Mapping to information types:
  Physical-Static-Descriptive: parts of the bookshelf are tangible, do not change, and need to be described
  Physical-Static-Spatial: dimensions of the parts, the way they are organized
  Physical-Static-Relationship: this type could also be added to describe which parts fit together

Information component 2:

  How to assemble parts instructions?

  Mapping to information types:
  Physical-Dynamic-Discrete action (each step in the assembly)
  Physical-Dynamic-Procedure (all the steps so the overall sequence is clear)
  Physical-Static-State (to show final assembled bookshelf)

The mapping physical information at the action and then the procedure level is to improve integration of information and hence understanding, following the "reinforce messages" principle. First, each step is shown, then the steps in a complete sequence (the procedure), and finally the end state of the completed assembly.

34.3.2 Matching Information to Media

The following heuristics are supplemented by more detailed examples in Sutcliffe (2012) and ISO (1998).

- To convey detail, use static media, for example, text for language-based content, diagrams for models, or still image for physical detail of objects (Booher, 1975; Faraday and Sutcliffe, 1998).
- To engage the user and draw attention, use dynamic media, for example, video for physical information, animation, or speech.
- For spatial information, use diagrams, maps, with photographic images to illustrate detail and animations to indicate pathways (Bieger and Glock, 1984; May and Barnard, 1995).
- For values and quantitative information, use charts and graphs for overviews and trends, supplemented by tables for detail (Bertin, 1983; Tufte, 1997).
- Abstract concepts, relationships, and models should be illustrated with diagrams explained by text captions and speech to give supplementary information.
- Complex actions and procedures should be illustrated as a slideshow of images for each step, followed by a video of the whole sequence to integrate the steps. Text captions on the still images and speech commentary provide supplementary information (Hegarty and Just, 1993). Text and bullet points summarize the steps at the end, so choice trade-offs may be constrained by cost and quality considerations.
- To explain causality, still and moving image media need to be combined with text (Narayanan and Hegarty, 1998), for example, the cause of a flood is explained by text describing excessive rainfall with an animation of the river level rising and overflowing its banks. Causal explanations of physical phenomena may be given by introducing the topic using linguistic media, showing cause and effect by a combination of still image and text with speech captions for commentary; integrate the message by moving image with voice commentary and provide a bullet point text summary.

34.4 UX and Multimedia

The term User EXperience (UX) grew out of concerns that traditional concepts of usability (ISO, 1997; Shneiderman and Plaisant, 2004) did not cover the more aesthetic aspects of design. Traditionally, usability has emphasized ease of use, ease of learning, and effective operation, in other words, the "drivability of an interface," and how well it fits the user's tasks and goals. Norman (2004) questioned the traditional view of usability in his book on emotion in design and pointed to the importance of aesthetic aspects in UIs and users' emotional responses to well-designed products. Researchers in HCI began to question how aesthetic design might be related to usability, led by the pioneering studies of Tractinsky (1997) and Lavie and Tractinsky (2004) who experimentally manipulated usability and aesthetic qualities of a design to coin the now well-known aphorism "what is beautiful is usable."

UX also evolved from marketing concepts such as "consumer experience" (Thomas and Macredie, 2002), which refers to the totality of product experience including sales, setup, use, support during use, and maintenance. Other influences came from research into enjoyment and fun (Blythe et al., 2004) and application areas such as games and entertainment where traditional usability appeared to be less appropriate. When excitement and amusement are the major design goals, interaction, metaphor, and aesthetics become important concerns. Hence, UX generally refers to a wider concept of design beyond functional products, which encompasses interaction, flow (Csikszentmihalyi, 2002), and aesthetic design. The content of media as well as the design of interaction in multimedia systems contribute to UX.

The criteria which may influence UX, and hence user satisfaction and use of IT products, are summarized in Figure 34.3.

Content and functionality are related to users' goals or requirements, so this aspect is dealt with in processes for requirement analysis (Sutcliffe et al., 2006) and user-centered design. Multimedia represents content to improve UX compared to monomedia (e.g., text or image alone) and is used to implement interaction using graphical metaphors, characters, speech, and audio. Content and services are closely related to task goals, so these criteria will be more important in work-related applications. In contrast, engagement and aesthetics may gain importance in entertainment and leisure applications.

Customization interacts with content and services since users are rarely a homogenous group, and different subgroups often require subsets of the overall content and functions. Customization may be at the subgroup or individual level (personalization), to increase the fit between the application and the user's needs, which, in turn, increases the user's commitment to a design. UX is enhanced not only through more stimulating interaction and better fit to the user's task and abilities (Sutcliffe et al., 2005), but also through the sense of ownership. Personalizing the choice of media and content within images, videos, etc. are examples of customization.

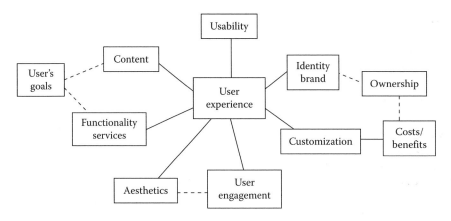

FIGURE 34.3 Components of UX.

Brand and identity interact with content and the users' background. If users are familiar with product brands, they are naturally going to be well disposed toward an application. Similarly, familiar content can be motivating as it is easier to assimilate. However, the users' reaction to content is complex; in some domains, for example, education and entertainment, challenging and unfamiliar content may be at a premium for stimulating learning and excitement, while in traditional information processing, familiar patterns of content make task performance more effective. At a high level, content and the tone of its expression in language may be familiar or not in a cultural sense, for example, the expression of material in English, American English, or the user's native language in international UIs.

Usability is always important, but the degree of ease of use may depend on the domain, varying between high-quality usability when content and functionality are priorities, to sufficient usability when user engagement and fun are the main criteria. Aesthetics or the "look and feel" of a multimedia application may be important in domains where design quality and brand image are at a premium (especially on websites). The debate about the relative influence of aesthetics is complex, where aesthetics is often restricted to the perception or "look" of a UI, separate from interaction or the "feel" of users' interfaces, which focuses on making interaction exciting and motivating for the user while also delivering effective content, and functionality that meets the users' goals.

34.5 Components of User Engagement

User engagement synthesizes several influences to promote a sense of flow and fluid interaction leading to satisfying arousal and pleasurable emotions of curiosity, surprise, and joy. The contributing influences are summarized in Figure 34.4.

The three main components of user engagement are *interaction*, *multimedia*, and *presence*. Interaction describes how the user controls the computer. The default is standard WIMP (windows, icons, mouse, pointer) interaction on graphical UIs. Multimedia not only deals with the representation of content, but also the interactive environment and how the user and the means of interaction are represented, ranging from simple cursors to icons and interactive avatars with fly-through interactive paradigms. Presence is determined by the representation of the user, which ranges from a cursor in a 2D interactive surface to an avatar in a more elaborate 3D interactive world. Presence is also related to, and augmented by, interaction. Acting in a 3D world and interaction with objects all increase the sense of "being there" and arousal. Flow is the key concept for understanding interaction in terms of the pace of action, complexity of actions, and the rate of change.

Flow, presence, and immersion are concepts which influence the choice of media for interaction; so in this case it is UI rather than content which is the subject of multimedia design. Text dominates

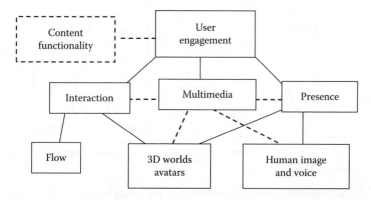

FIGURE 34.4   Issues related to user engagement and multimedia.

traditional form-filling and menu interfaces, although graphics in the form of icons and structures (windows, sliders, menu bars) are now standard UI components. In the emerging and future UIs, new media will appear, as the user interacts not just via a cursor but also through an avatar, situated in a graphical virtual world, occupied by animated objects, with speech and audio. Multimedia interaction improves UX through flow, presence, and immersion.

34.5.1 Interaction and Flow

Flow is a finely tuned balancing act between the user's abilities and skills, and the challenges provided by learning new interactions, and then responding to events. It is the sense of engagement and being absorbed in an interactive experience. The concept involves optimal arousal produced by a "sweet spot" trade-off between challenge and difficulty on one hand and ease of operation and achievement on the other. If operating a UI is too difficult, users will get frustrated and discouraged and may give up, leaving them with negative emotions and adverse memory of the experience. In contrast, if operating the UI is too easy, they become bored, excitement (or arousal) decreases, and they turn their attention to more interesting things. The trick is to keep interaction in the flow zone (see Figure 34.5), an intuition appreciated by game designers. Games need to maintain the pace of change with unpredictable events, while not overwhelming the user with too much change that exceeds their capabilities. They rapidly become used to patterns of events leading to decreased arousal as the unfamiliar becomes familiar.

34.5.2 Presence

The origins of presence are in virtual reality (VR), in which the user is represented by an avatar or virtual character. The avatar places the user inside a 3D graphic world, so interaction becomes very different from operating an interface through a cursor. Presence is the sense of "being there" inside a virtual world as a representation of oneself (embodied or immersed interaction). Embodied interaction is more complex than standard 2D interfaces, since the user can control movement directly, manipulate objects in an almost natural manner, and even feel objects if haptic feedback is provided. The sense of immersion and control is enhanced via natural movement. Virtual worlds become engaging because they invoke curiosity and arousal. Interaction becomes transparent, i.e., the user is not aware of the computer, instead feeling immersed and becoming absorbed in the virtual graphical world (see Figure 34.6).

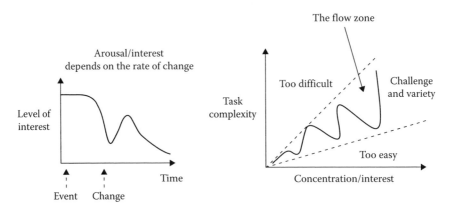

FIGURE 34.5 Change in arousal as events become familiar. The concept of flow.

FIGURE 34.6 Virtual world illustrating the user's presence represented as two virtual hands.

34.5.3 Engagement and Social Presence

Social presence theory (Short et al., 1976) argues that different communication channels and representations promote more or less sense of social presence, i.e., awareness of the identity, location, and personalities of other people. The theory does not give a formal classification or model of social presence; although it can be reinterpreted in terms of a model of communication which describes degradations from the ideal of face-to-face, colocated communication. E-mail is the least engaging since the time gap between turns destroys the sense of a dialog, whereas instant messaging (IM) approaches synchronous exchange and becomes more engaging. In IM, the pace of exchange approaches a conversation, and the sense of presence increases. Adding video or even still images improves presence by providing more information about the other person (see Figure 34.7). Contextual information gives us the background to an ongoing conversation. The context may be provided by a user profile, a room or whatever has been constructed in the virtual world for interpreting the conversation.

FIGURE 34.7 *SecondLife*, illustrating avatars that represent the characters people adopt and the location with others, both of which promote social presence.

Strange as it may seem, it takes very little to create an illusion of presence. Our powerful imaginations just need a few hints (priming or framing effects in psychology) to conjure up a perceived reality, as Pinter* and other proponents of minimalist theater have demonstrated. In a series of experiments, Reeves and Nass (1996) showed how we treat computer-based media as if they corresponded to real people. Their computer as social actor (CSA) paradigm explains how we treat computers as virtual people even when we are presented with limited cues, such as a photograph of a person or human voice. Indeed, the image can be artificial, cartoon-like, with little correspondence to reality; the same applies to the voice. Chatterbots, avatars on the web equipped with simple semi-intelligent scripts for responding to human conversations, are treated like real characters, and some people actually form relationships with these virtual characters (De Angeli and Brahnam, 2008).

34.6 Media Design for User Engagement

Multimedia design is frequently motivated by the need to attract users' attention and to make the UX interesting and engaging. These considerations may contradict some of the earlier guidelines, because the design objective is to please users and capture their attention rather than deliver information effectively. First, a health warning should be noted: the old saying "beauty is in the eye of the beholder" has a good foundation. Judgments of aesthetic quality suffer from considerable individual differences. A person's reaction to a design is a function of their motivation (Brave and Nass, 2008), individual preferences, knowledge of the domain, and exposure to similar examples, to say nothing of peer opinion and fashion.

In human–human conversation, we modify our reactions according to our knowledge (or assumptions) about the other person's role, group identification, culture, and intention (Clark, 1996). For example, reactions to a military mannequin will be very different from those to the representation of a parson. Male voices tend to be treated as more authoritative than female voices. Simple photographs or more complex interactive animations (talking heads or full body mannequins) have an attractive effect; however, the effectiveness of media representing people depends on the characters' appearance and voice (see Figure 34.8).

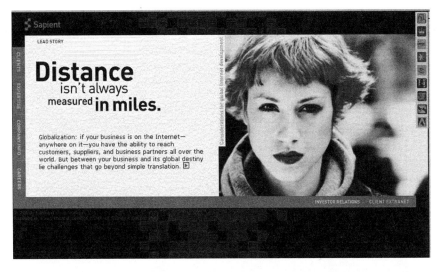

FIGURE 34.8 Effective use of human image for attraction. The picture attracts by the direction of gaze to the user as well as by the appearance of the individual.

---

* http://en.wikipedia.org/wiki/Harold_Pinter

Use of human-like forms is feasible with prerecorded videos and photographs; however, the need depends on the application. Video representation of the lecturer can augment presentations, and video communication helps interactive dialog. A good speaker holds our attention by a variety of tricks, such as maintaining eye contact, varying the voice tone, using simple and concise language, as well as delivering an interesting message. These general effects can be reinforced by projected personality. Friendly people are preferred over colder, more hostile individuals. TV announcers who tend to be middle-aged, confident, but avuncular characters have the attention-drawing power of a dominant yet friendly personality. Both sexes pay attention to extrovert, young personalities, while the male preference for beautiful young women is a particularly strong effect. These traits have long been exploited by advertisers. There are lessons here for multimedia designers as the web and interactive TV converge, and when we want media to convey a persuasive message (Reeves and Nass, 1996; Fogg et al., 2008). Media selection guidelines for motivation and persuasion, adapted from Reeves and Nass (1996), can be summarized as follows:

- Human image and speech invokes the CSA effect to facilitate motivation and persuasion.
- Photographs of people attract attention especially when the person is looking directly at the user.
- Faces that represent the norm in a population (Mr./Ms. Average) and young children are more attractive. We are very susceptible to the large-eyes effect in young animals, as exploited by Disney cartoons.
- Polite praise: Use of please, thank you, and simple compliments such as "that was an excellent choice" increase people's tendency to judge the computer as pleasant and enjoyable.
- Short compelling argument such as the well-known British World War I recruiting poster featuring General Kitchener gazing directly at the viewer with the caption "your country needs you."

34.6.1 Media for Emotional Effects

Media design for affect (emotional response and arousal) involves both choice of content and interaction. Arousal is increased by interactive applications, surprising events during interaction, use of dynamic media, and challenging images. In contrast, if the objective is to calm the users, arousal can be decreased by choice of natural images and sounds, and soothing music. The most common emotional responses that designers may want to invoke are pleasure, anxiety and fear, and surprise. Pleasure, anxiety, and fear usually depend on our memory of agents, objects, and events (Ortony et al., 1988), so content selection is the important determinant. Anxiety can be evoked by uncertainty in interaction and cues to hidden effects, while emotional response of fear or pleasure will depend on matching content to the user's previous experience. Some guidelines to consider are as follows:

- *Dynamic media*, especially video, have an arousing effect and attract attention; hence, video and animation are useful in improving the attractiveness of presentations. However, animation must be used with care as gratuitous video that cannot be turned off quickly offends (Spool et al., 1999).
- *Speech* engages attention because we naturally listen to conversation. Choice of voice depends on the application: female voices for more restful and information effects, male voices to suggest authority and respect (Reeves and Nass, 1996).
- *Images* may be selected for mood setting, for example, to provide a restful setting for more important foreground information (Mullet and Sano, 1995). Backgrounds in half shades and low saturation color provide more depth and interest in an image.
- *Music* has an important emotive appeal, but it needs to be used with care. Classical music may be counterproductive for a younger audience, while older listeners will not find heavy metal pop attractive. Music can set the appropriate mood, for example, loud strident pieces arouse and excite, whereas romantic music calms and invokes pleasure, etc.
- *Natural sounds* such as running water, wind in trees, bird song, and waves on a seashore have restful properties and hence decrease arousal.

- *Dangerous and threatening episodes*, for example, being chased by a tiger, gory images (mutilated body), and erotic content all increase arousal and invoke emotions ranging from fear to anger, whereas pleasant images (e.g., flowers, sunset) tend to decrease it, having calming effects and producing pleasurable emotional responses.
- *Characters* can appear threatening or benevolent depending on their appearance or dress. For example, disfigured people appear threatening and evoke emotions ranging from fear to disgust. Characters familiar from popular culture can be used for emotional effect.
- *Spoken dialog* is probably the most powerful tool for creating emotional responses, from threats to empathy. Emotional effects are additive; so choice of character with a threatening appearance, complemented by a menacing voice tone and an aggressive dialog, all reinforce the emotions of anxiety and fear.

Media integration rules may be broken for emotive effects. For example, use of two concurrent video streams might be arousing for a younger audience, as Music TV (MTV) and pop videos indicate. Multiple audio and speech tracks can give the impression of complex, busy, and interesting environments.

34.6.2 Multimedia and Aesthetic Design

Judging when aesthetics may be important will be set by the design goals of the owner of the application. For example, in e-commerce applications with high-value, designer-label products, aesthetic presentation is advisable; similarly, when selling to a design-oriented audience. Aesthetic design primarily concerns graphics and visual media. Evaluation questionnaires can assess aesthetics and more creative aspects of visual design (Lavie and Tractinsky, 2004); however, these measure user reaction to general qualities, such as "original," "fascinating," "clear," and "pleasant." The following heuristics provide more design-directed guidance, but they may also be employed for evaluation (Sutcliffe, 2002; Sutcliffe and De Angeli, 2005).

- *Judicious use of color*: Color use should be balanced, and low saturation pastel colors should be used for backgrounds. Designs should not use more than two to three fully saturated intense colors. Yellow is salient for alerting, red/green have danger/safety positive/negative associations, and blue is more effective for background. Low-saturated colors (pale shades with white) have a calming effect and are also useful for backgrounds. Color is a complex subject in its own right; for more guidance, refer to Travis (1991).
- *Depth of field*: Use of layers in an image stimulates interest and can attract by promoting curiosity. Use of background image with low-saturated colors provides depth for foreground components. Use of layers in an image and washed-out background images stimulate curiosity and can be attractive by promoting a peaceful effect.
- *Use of shape*: Curved shapes convey an attractive visual style, in contrast to blocks and rectangles that portray structure, categories, and order in a layout.
- *Symmetry*: Symmetrical layouts, for example, bilateral, radial organization, can be folded over to show the symmetrical match.
- *Simplicity and space*: Uncluttered, simple layouts that use space to separate and emphasize key components.
- *Design of unusual or challenging images* that stimulate the users' imagination and increase attraction; unusual images often disobey normal laws of form and perspective.
- *Visual structure and organization*: Dividing an image into thirds (Right, Center, Left; or Top, Middle, Bottom) provides an attractive visual organization, while rectangular shapes following the golden ratio (height/width = 1.618) are aesthetically pleasing. Use of grids to structure image components promotes consistency between pages.

Although guidelines provide ideas that can improve aesthetic design and attractiveness of interfaces, there is no guarantee that these effects will be achieved. Design is often a trade-off between ease of use and aesthetic design; for instance, use of progressive disclosure to promote flow may well be perceived by others as being difficult to learn. Visual effects often show considerable individual differences and learning effects; so, a well-intentioned design might not be successful. The advice, as with most design, is to test ideas and preliminary designs with users to check interpretations, critique ideas, and evaluate their acceptability. There are several sources of more detailed advice on aesthetics and visual design (Mullet and Sano, 1993; Kristof and Satran, 1995; Lidwell et al., 2003).

34.6.3 Metaphors and Interaction Design

While task and domain analysis can provide ideas for interaction design, this is also a creative process. Interaction design is essentially a set of choices along a dimension from simple controls, such as menus and buttons where the user is aware of the interface, to embodiment in which the user becomes involved as part of the action by controlling an avatar or other representation of their presence. At this end of the dimension, multimedia interaction converges with virtual reality. Interactive metaphors occupy the middle ground.

Some interactive metaphors are generally applicable, such as timelines to move through historical information, the use of a compass to control the direction of movement in an interactive space, controls based on automobiles (steering wheels) or ships (rudders). Others will be more specific, for example, selecting and interacting with different characters (young, old, male, female, overweight, fit, etc.) in a health-promotion application. Design of interaction also involves creating the microworld within which the user moves and interactive objects that can be selected and manipulated.

Interaction via characters and avatars can increase the user's sense of engagement first by selecting or even constructing the character. In character-based interaction, the user can either see the world from an egocentric viewpoint, i.e., from their character's position, or exocentric when they see their character in the graphical world. Engagement* is also promoted by surprise and unexpected effects; so, as the user moves into a particular area, a new subworld opens up, or system-controlled avatars appear. These techniques are well known to games programmers; however, they are also applicable to other genres of multimedia applications. The design concepts for engagement can be summarized as follows (see Sutcliffe, 2009, for more detail):

- *Character-driven interaction*: It provides the user with a choice of avatars or personae they can adopt as representations of themselves within the interactive virtual world (see Figure 34.6). Avatar development tools enable virtual characters to be designed and scripted with actions and simple speech dialogs. Most sophisticated, semi-intelligent "chatterbots" (e.g., Alice http://alice.pandorabots.com/& Jabberwacky http://www.jabberwacky.com/) use response-planning rules to analyze user input and generate naturally sounding output; however, it is easy to fool these systems with complex natural language input.
- *Tool-based interaction*: This places the tools in the world which users can pick up; the tool becomes the interface, for example, a virtual mirror magnifies, a virtual helicopter flies (Tan et al., 2001).
- *Collaborative characters*: In computer-mediated communication, these characters may represent other users; in other applications, system-controlled avatars appear to explain, guide, or warn the user.
- *Surprise effects*: Although conventional HCI guidelines should encourage making the affordances and presence of interactive objects explicit, when designing for engagement, hiding and surprise are important.

Interaction design for an explanatory/tutorial application is illustrated in Figure 34.9. This is an interactive microworld in which the user plays the role of a dinosaur character, illustrating the use of the engagement concepts. A compass navigation metaphor allows the user to act as a dinosaur moving around the landscape, as illustrated in the figure. The user is given feedback on the characteristics of other predators and prey in the vicinity, and has to decide whether to attack or avoid them.

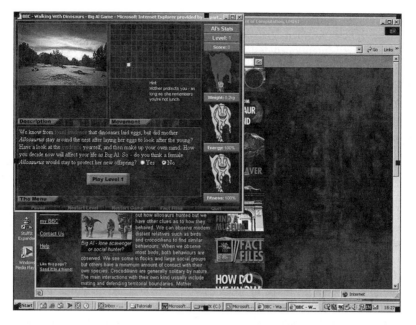

FIGURE 34.9 Interactive microworld: Big Al game (http://www.bbc.co.uk/sn/). The user plays the dinosaur role by navigating with the compass metaphor. The photograph updates with each move and the user is given a choice of attacking or avoiding other dinosaurs in the virtual world.

Other controls that might be added to such interactive microworlds could be the settings to change the environment, for example, add more predators, change the weather, etc.

34.7 A Brief Example

A healthcare example illustrates the use of some of the UX-multimedia guidelines. The objective is to persuade people through a web application to adopt healthy diets to tackle the obesity problems. Current designs (e.g., http://www.nhs.uk/livewell/; http://www.nhs.uk/Change4Life) use image and text with limited video to provide advice on the types of food to eat, what to avoid, portion size, etc. However, interactive functions are limited to simple quizzes on diet choices, which are analyzed to show the calories consumed. More adventurous designs might employ an interactive "day in your life" metaphor, where the user controls an avatar while progressing through a typical day making lifestyle choices about exercise (e.g., walk to work or drive), what to eat for each meal, choice of portion sizes, etc. The interactive game-style application provides feedback on choices, approving sensible ones or disapproving poor ones. The CSA guidelines would suggest feedback with praise for good choices given by a sympathetic character. Long-term tracking of diet and weight loss or gain could employ emotional effects to encourage (empathy, praise, empowerment), and to admonish poor progress with warnings about the dangers of obesity. The potential for multimedia design can be seen in the increasing number of mobile applications which use interactive games–style interaction with monitoring to encourage fitness.

34.8 Conclusions

Multimedia still poses many issues for further research. The design method described in this chapter coupled with user-centered design can improve quality; however, there is still a need for experts to create specific media resources, for example, film/video, audio. Furthermore, considerable research is still necessary before we fully understand the psychology of multimedia interaction. Design for UX in multimedia is still poorly understood, for example, some people prefer avatars and rich graphical metaphors, while others find graphical interfaces annoying and prefer simpler designs. Although guidelines can improve design with advice that is based on psychological knowledge, the reaction to designs often depends on the individual; so, one-size-fits-all solutions will not work for everyone. As with most design, testing with users is always advisable, and where possible personalizing and customizing designs for the user or giving them the choice can improve the UX.

Current research frontiers are unpacking the role of interactivity in UX, as multimedia converges with VR and 3D interactive worlds, such as Google street view technology, become commonplace. Immersion and presence increase user engagement with games (Jennet et al., 2009); however, application of virtual worlds to other domains, such as education, does not always improve UX (Sutcliffe and Alrayes, 2012). Although interactive features do appear to have a positive appeal in many websites (Cry et al., 2009), the debate continues about the relative importance of interactive and aesthetic design on overall UX (Bargas-Avila and Hornbæk, 2011; Lindgaard et al., 2011). On balance, it appears that "pragmatic" design considerations such as functionality and usability outweigh "hedonic" factors such as aesthetics (Diefenbach and Hassenhalz, 2009), but the contribution of interaction design remains unclear.

In the future, language and multimodal communication will change our conception of multimedia from its current CD-ROM or web-based form into interfaces that are conversational and multisensory. Multimedia will become part of wearable and ubiquitous UIs where the media is part of our everyday environment. Design for multisensory communication will treat media and artifacts (e.g., our desks, clothes, walls in our homes) as a continuum, while managing the diverse inputs to multimedia from creative design, technology, and usability engineering will be one of the many interesting future challenges.

References

Alty, J. L. (1991). Multimedia: What is it and how do we exploit it? In D. Diaper and N. V. Hammond (Eds), *Proceedings of HCI'91: People and Computers VI* (pp. 31–41). Cambridge, U.K.: Cambridge University Press.

Bargas-Avila, J.A. and Hornbæk, K. (2011). Old wine in new bottles or novel challenges? A critical analysis of empirical studies of user experience. In *Proceedings CHI-11* (pp. 2689–2698). New York: ACM Press.

Bernsen, N. O. (1994). Foundations of multimodal representations: A taxonomy of representational modalities. *Interacting with Computers,* 6(4), 347–371.

Bertin, J. (1983). *Semiology of Graphics.* Madison, WI: University of Wisconsin Press.

Bieger, G. R. and Glock, M. D. (1984). The information content of picture-text instructions. *Journal of Experimental Education,* 53, 68–76.

Blythe, M. A., Overbeeke, K., Monk, A. F., and Wright, P. C. (2004). *Funology: From Usability to Enjoyment.* Boston, MA: Kluwer.

Booher, H. R. (1975). Relative comprehensibility of pictorial information and printed word in proceduralized instructions. *Human Factors,* 17(3), 266–277.

Brave, S. and Nass, C. (2008). Emotion in human computer interaction. In A. Sears and J. A. Jacko (Eds), *The Human-Computer Interaction Handbook: Fundamentals, Evolving Technologies and Emerging Applications,* 2nd edn. (Chapter 4, pp. 77–92). New York: CRC Press/Taylor & Francis.

Clark, H. H. (1996). *Using Language.* Cambridge, U.K.: Cambridge University Press.

Csikszentmihalyi, M. (2002). *Flow: The Classic Work on How to Achieve Happiness.* (Revised edn.). London, U.K.: Rider.

Cyr, D., Head, M., and Ivanov, A. (2009). Perceived interactivity leading to e-loyalty: Development of a model for cognitive-affective user responses. *International Journal of Human-Computer Studies,* 67, 850–869.

De Angeli, A. and Brahnam, S. (2008). I hate you! Disinhibition with virtual partners. *Interacting with Computers, 20,* 302–310.

Diefenback, S. and Hassenzahl, M. (2009). The "beauty dilemma": Beauty is valued but discounted in product choice. In *Proceedings CHI 2009* (pp. 1419–1426). New York: ACM Press.

Faraday, P. and Sutcliffe, A. G. (1998). Making contact points between text and images. In *Proceedings ACM Multimedia 98 of 6th ACM International Multimedia Conference.* New York: ACM Press.

Fogg, B. J., Cueller, G., and Danielson, D. (2008). Motivating, influencing and persuading users: An introduction to Captology. In A. Sears and J. A. Jacko (Eds), *The Human-Computer Interaction Handbook: Fundamentals, Evolving Technologies and Emerging Applications,* 2nd edn. (pp. 133–146). New York: CRC Press/Taylor & Francis.

Heller, R. S. and Martin, C. (1995). A media taxonomy. *IEEE Multimedia, Winter,* 36–45.

Hochberg, J. (1986). Presentation of motion and space in video and cinematic displays. In K. R. Boff, L. Kaufman, and J. P. Thomas (Eds), *Handbook of Perception and Human Performance, 1: Sensory Processes and Perception.* New York: Wiley.

Hollan, J. D., Hutchins, E. L., and Weitzman, L. (1984). Steamer: An interactive inspectable simulation-based training system. *AI Magazine,* 5(2), 15–27.

ISO. (1997). *ISO 9241: Ergonomic Requirements for Office Systems with Visual Display Terminals (VDTs).* International Standards Organisation. Geneva, Switzerland.

ISO. (1998). *ISO 14915 Multimedia UI Design Software Ergonomic Requirements, Part 1: Introduction and Framework; Part 3: Media Combination and Selection.* International Standards Organisation, Geneva, Switzerland.

Jennett, C., Cox, A. L., Cairns, P., Dhoparee, S., Epps, A., Tijs, T., and Walton, A. (2008). Measuring and defining the experience of immersion in games. *International Journal of Human-Computer Studies,* 66(9), 641–661. doi:10.1016/j.ijhcs.2008.04.004.

Kristof, R. and Satran, A. (1995). *Interactivity by Design: Creating and Communicating with New Media.* Mountain View, CA: Adobe Press.

Lavie, T. and Tractinsky, N. (2004). Assessing dimensions of perceived visual aesthetics of web sites. *International Journal of Human-Computer Studies,* 60(3), 269–298.

Lidwell, W., Holden, K., and Butler, J. (2003). *Universal Principles of Design.* Gloucester, MA: Rockport.

Lindgaard, G., Dudek, C., Sen, D., Sumegi, L., and Noonan, P. (2011). An exploration of relations between visual appeal, trustworthiness and perceived usability. *ACM Transactions on Computer-Human Interaction,* 18(1), 1–30.

May, J. and Barnard, P. (1995). Cinematography and interface design. In K. Nordbyn, P. H. Helmersen, D. J. Gilmore, and S. A. Arnesen (Eds), *Proceedings: Fifth IFIP TC 13 International Conference on Human-Computer Interaction,* Lillehammer, Norway, June 27–29, 1995 (pp. 26–31). London, U.K.: Chapman & Hall.

Mullet, K. and Sano, D. (1995). *Designing Visual Interfaces: Communication Oriented Techniques.* Englewood Cliffs, NJ: SunSoft Press.

Narayanan, N. H. and Hegarty, M. (1998). On designing comprehensible interactive hypermedia manuals. *International Journal of Human-Computer Studies,* 48, 267–301.

Norman, D. A. (2004). *Emotional Design: Why We Love (or Hate) Everyday Things.* New York: Basic Books.

Ortony, A., Clore, G. L., and Collins, A. (1988). *The Cognitive Structure of Emotions.* Cambridge, U.K.: Cambridge University Press.

Reeves, B. and Nass, C. (1996). *The Media Equation: How People Treat Computers, Television and New Media Like Real People and Places.* Stanford, CA/Cambridge, U.K.: CLSI/Cambridge University Press.

Shneiderman, B. and Plaisant, C. (2004). *Designing the UI: Strategies for Effective Interaction,* 4th edn. Reading, MA: Addison-Wesley.

Short, J., Williams, E., and Christie, B. (1976). *The Social Psychology of Telecommunications.* Chichester, PA: Wiley.

Spool, J. M., Scanlon, T., Snyder, C., Schroeder, W., and De Angelo, T. (1999). *Web Site Usability: A Designer's Guide*. San Francisco, CA: Morgan Kaufmann.

Sutcliffe, A. G. (2002). Assessing the reliability of heuristic evaluation for website attractiveness and usability. In *Proceedings HICSS-35: Hawaii International Conference on System Sciences*, Big Island, HI, January 7–10, 2002 (pp. 1838–1847). Los Alamitos, CA: IEEE Computer Society Press.

Sutcliffe, A. G. (2003). *Multimedia and Virtual Reality: Designing Multisensory User Interfaces*. Mahwah, NJ: Lawrence Erlbaum Associates.

Sutcliffe, A. G. (2009). Designing for user engagement: Aesthetic and attractive user interfaces. In J. M. Carroll (Ed), *Synthesis Lectures on Human Centered Informatics*. San Rafael, CA: Morgan Claypool.

Sutcliffe, A. G. (2012). Multimedia user interface design. In J. A. Jacko (Ed), *The Human-Computer Interaction Handbook: Fundamentals, Evolving Technologies and Emerging Applications*, 2nd edn. (pp. 387–404). New York: CRC Press/Taylor & Francis.

Sutcliffe, A. G. and Alrayes, A. (2012). Investigating user experience in Second Life for collaborative learning. *International Journal of Human Computer Studies*, 70(7), 508–525.

Sutcliffe, A. G. and De Angeli, A. (2005). Assessing interaction styles in web user interfaces. In *Proceedings of Human Computer Interaction—Interact 2005*, Rome, Italy. Berlin, Germany: Springer.

Sutcliffe, A. G., Fickas, S., and Sohlberg, M. (2005). Personal and contextual requirements engineering. In *Proceedings: 13th IEEE International Conference on Requirements Engineering*, Paris, France, August 29–September 2, 2005 (pp. 19–28). Los Alamitos, CA: IEEE Computer Society Press.

Sutcliffe, A. G., Fickas, S., and Sohlberg, M. M. (2006). PC-RE: A method for personal and contextual requirements engineering with some experience. *Requirements Engineering*, 11, 157–163.

Tan, D. S., Robertson, G. R., and Czerwinski, M. (2001). Exploring 3D navigation: Combining speed coupled flying with orbiting. In J. A. Jacko, A. Sears, M. Beaudouin-Lafon, and R. J. K. Jacob (Eds), *CHI 2001 Conference Proceedings: Conference on Human Factors in Computing Systems*, Seattle, WA, March 31–April 5, 2001 (pp. 418–425). New York: ACM Press.

Thomas, P. and Macredie, R. D. (2002). Introduction to the new usability. *ACM Transactions on Computer-Human Interaction*, 9, 69–73.

Tractinsky, N. (1997). Aesthetics and apparent usability: Empirically assessing cultural and methodological issues. In S. Pemberton (Ed.), *Human Factors in Computing Systems: CHI 97 Conference Proceedings*, Atlanta, GA, May 22–27, 1997 (pp. 115–122). New York: ACM Press.

Travis, D. (1991). *Effective Colour Displays: Theory and Practice*. Boston, MA: Academic Press.

Tufte, E. R. (1997). *Visual Explanations: Images and Quantities, Evidence and Narrative*. Cheshire, CN: Graphics Press.

35

Applying International Usability Standards

Tom Stewart
System Concepts

35.1 Introduction

Usability has become recognized as a major factor in determining the acceptance of information systems. Poor usability has been identified as a significant cause of system failure—from large-scale government systems to small bespoke developments. The expanding use of information technology in all types of consumer products from music players and mobile phones to websites and washing machines has brought usability into even sharper focus.

In the past, poor usability was either ignored or blamed on lack of user training. The video cassette recorder (VCR) was one of the most recognizable unusable consumer products employing information technology (albeit relative simple timer technology). The number of such products with clocks "flashing 1200" after a power failure was testament to the difficulty of making a simple change to the clock setting. This in itself did not, however, appear to reduce VCR sales or adoption significantly, and customers only discovered the problems when they had already bought the machines. So, few manufacturers bothered to invest in usability.

The growth of the World Wide Web changed the information technology scene dramatically. Generally, users are offered so much choice (too much to understand properly, some might argue) that they no longer tolerate poor usability. For online retailers, their competitors are "simply a click away," and few would now launch a website without some user testing. Similarly, the mobile phone industry has recognized that the wide availability of competing products means that good usability design pays dividends. Many with a technical background find Apple products frustrating because they build a layer between the user and technology. The limitations are not technical but commercial. This so-called walled garden approach means that Apple can control almost all aspects of the user interface and the experience of the user. For many users, this seamless (albeit restricted) interaction not only highly satisfies but also apparently justifies a significant premium on price.

Apple has long made strong claims about the "intuitiveness" of its user interfaces. Watching young children quickly use pinch and swipe on an iPad, it is hard to argue. However, the computer operating system (OS) is a different matter. Some of the ideas of point and click and direct manipulation are indeed relatively natural to many people, but such concepts as dragging a floppy disk icon to the "trash can" to eject it are far from natural. Where Apple has been successful with its limited and close system is policing a consistent look and feel on its interfaces (even as these have developed).

As an organization, they have been able to retain the control over their part of the industry which no one else has achieved. However, they are not alone in trying to ensure that users can actually use their systems and can achieve results effectively, efficiently, and even find the process satisfying. There are international usability standards that any developer or user organization can use to make their systems usable. By usable, we mean effective, efficient, and satisfying (the definition of usability in one of the international standards ISO 9241-11: 1998 Guidance on usability). ISO/TC159/SC4 is the subcommittee (SC) that deals with human–system interaction.

These standards have been developed over nearly 30 years with new and updated standards being published continuously. The purpose of this chapter is to describe how to apply international usability standards to improve the design of user interfaces and the experience of the users of information systems and technology.

Before describing the standards in detail and their impact on practice, it is important to understand the underlying principles of standardization, especially international standardization.

35.2 Underlying Principles

International standards (ISs) are powerful tools for improving the quality, effectiveness, and safety of goods and services; they are also the facilitators of world trade. International usability standards play an important role in information systems and information technology as they encompass widely accepted best practices. However, one of the most common problems people experience when trying to use these standards (and usability standards in particular) is that they do not understand the limitations and constraints that apply to the standards development (Smith 1986; Potonak 1988; Stewart 2000). This section describes the principles underlying ISs in general and international usability standards in particular.

35.2.1 Purpose of Standards

There are two main purposes for standards. First, an important purpose in some areas is to ensure what is called interoperability. For example, there are standards which define screw threads, the prime purpose of which is to ensure that nuts fit bolts. In the information and communications technology arena, there are many standards concerned with interoperability, such as standards for encoding images JPEG, which allow pictures to be shared easily.

Second, there are standards that enshrine best practice and aim to ensure that products, systems, or services meeting these standards perform at some specified level. For example, there are standards for life jackets that aim to ensure that they provide an appropriate level of buoyancy that ensures that people who have to abandon vessels can be kept in a position that allows them to breathe (i.e., face up out of the water) even if they were unconscious. In the usability standards area, there are hardware standards which aim to ensure that computer displays meeting them provide an appropriate level of visibility that ensures that people who use them will be able to read them in typical office environments. There are software usability standards which aim to ensure that dialog screens are designed to match the users and the tasks they are performing and take account of relevant user psychology and cognitive processes.

35.2.1.1 Best Practice

International usability standards are primarily concerned with providing definitive statements of good practice. The growing field of usability and more recently user experience (UX) is filled with experts offering their own, often conflicting, views of good practice. There are those who argue that user interface design should be regarded as a science and that systematic user research and usability testing techniques are essential in order to design a usable interface. There are others, the most notable example being the late Steve Jobs who famously said in a magazine interview that "You can't just ask the customers what they want and then try to give that to them. By the time you get it built, they'll want something new."

The truth probably lies somewhere between these two extremes. Simply relying on user research will not result in the kind of highly innovative interfaces for which Apple has been famous. However, simply relying on creatively guessing user needs is not enough, and even Apple has conducted extensive testing of its ideas and prototypes with many not making it to market.

In this volatile arena, international usability standards can provide independent and authoritative guidance. International standards are developed slowly, by consensus, using extensive consultation and development processes. This has its disadvantages in such a fast-moving field as user interface design, and some have criticized any attempts at standardization as premature. However, there are areas where a great deal is known about good design practices which can be made accessible to designers through appropriate standards (Dzida et al. 1978; Shneiderman 1987; Sperandio 1987; Newman and Lamming 1995). There are approaches to user interface standardization, based on human characteristics, which are relatively independent of specific design solutions or technologies (e.g., display technology).

The practical discipline of having to achieve international consensus helps moderate some of the wilder claims of user interface enthusiasts and helps ensure that the resulting standards do indeed represent good practice. Although the slow development process means that usability standards can seldom represent the leading edge of design, nonetheless, when properly written, they need not inhibit creativity. Indeed, as discussed later, the development of process standards avoids the trap of limiting standards to current technology. Another approach—which not only does not inhibit creativity but actually enhances it—involves the development of user-based test methods (as exemplified in ISO 9241-304:2008). In this standard, the check for conformance involves comparing the performance of a user executing a standard task on a reference display (which meets the standard) and the new display. If the performance on the new display is as good as or better than the reference display, the new display passes. Since the task involves reading and recognizing characters on the display, it does not matter what technology is used. The test simply checks the performance of the user viewing the display. Similar tests have also been agreed for keyboards and other input devices. Far from limiting creativity, these test standards encourage designers to develop even more innovative technical solutions that work well for their users.

35.2.1.2 Common Understanding

Another problem in the field of usability and user experience is that there are many views not only about the potential interface solutions but also about the terminology involved at all stages. Even what the main players involved in designing systems for people should be called can become a point of argument—are they UX designers? Information designers or architects? Interaction or interface designers? What happened to system designers, developers, software designers, and so on. Apart from potentially overlapping or conflicting roles, good usability design involves a number of different perspectives and viewpoints. One of the benefits of international usability standards is to provide a framework for different parties to share a common understanding when specifying interface quality in design, procurement, and use. International usability standards allow:

- Users to set appropriate procurement requirements and to evaluate competing suppliers' offerings
- Suppliers to check their products during design and manufacturing and provide a basis for making claims about the quality of their products
- Regulators to assess quality and provide a basis for testing products against national and international requirements

35.2.1.3 Consistency

Anyone who uses computers knows only too well the problems of inconsistency between applications and often even within the same application. Inconsistency, even at the simplest level, can cause problems. On the World Wide Web, inconsistency is very common. Even something as straightforward as a hypertext link may be denoted by underlining on one site, by a mouseover on a second site, and by nothing at all on a third site.

Usability standards can help organizations develop their own internal user interface standards and style guides to address this problem. Such documents can provide a consistent reference across design teams or across time to help avoid time-consuming and error-inducing inconsistency in the user interface.

35.2.1.4 Raising the Profile of User Interface Issues

One of the most significant benefits of international usability standardization is that it places user interface issues squarely on the agenda. Standards are serious business, and whereas many organizations may choose to pay little regard to research findings, few organizations can afford to ignore standards. Indeed in Europe, and increasingly in other parts of the world, compliance with relevant standards is a mandatory requirement in major contracts (Earthy et al. 2001).

35.2.2 Use of Language

It is important to understand that although ISs are written in English, French, or Russian (the three official languages of ISO, the International Organization for Standardization), there are extensive drafting rules, which mean that certain terms have quite specific meanings. For example, in the English versions of the standards, the use of the word "shall" means that something is necessary in order to meet the requirements in the standard. Meeting all the "shalls" in the standard is necessary in order to be able to claim that something "conforms" to the standard.

In the field of usability, there are few circumstances where there is only one solution and therefore where a "shall" is required. One of the frustrations that information systems designers experience is that when they ask a usability expert for advice, the answer usually start with "it depends … ." The reason for this is not that usability people share with politicians an unwillingness to give straight answers but rather a recognition that the "correct answer" depends on the context. So, for example, an apparently simple question such as "what is the maximum number of items that can be presented as a menu?" does not have a simple answer. If the menu is a simple drop-down list which the user will scroll down using a mouse or arrow key, then according to ISO 9241-14:2000 Menu Dialogues, that number is generally 7 ± 2 corresponding to short-term memory. However, if the menu is an alphabetic listing of every song on an iPod which will be accessed by the user touching a list which allows scrolling to be sped up or slowed down, then the answer is as many songs as the user possesses.

In practice, therefore, usability standards generally provide "guidance" and use the word "should" (in English versions) to reflect that this is a "recommendation" rather than a "requirement."

35.2.3 Conformance

International standards are generally voluntary, that is, it is up to the user of the standard to decide whether to follow a standard or not. In practice, however, especially in procurement conformance with standards may be required in order to satisfy the procuring organization. Conformance means complying with the mandatory requirements set out in the specific standard, usually indicated by the word "shall" in the text and a description on how to assess compliance in an annex (a test method).

As discussed earlier, many international usability standards do not contain many requirements and instead provide guidance, that is, "shoulds," which depend on the context. However, it is still possible to claim conformance to such standards. One way is simply to declare that "all relevant recommendations in standard x have been followed." It is then up to the party to whom such a claim is made to assess whatever evidence is provided and determine whether they agree that the recommendations have been followed. Many user experience consultancies offer such evaluation services to their clients—both to suppliers wishing to make such claims about their products and to user organizations wishing to have suppliers' offerings assessed.

In some cases, the usability standards contain what is known as conditional compliance clauses. These standards do contain a "shall," that is, a mandatory requirement. That requirement typically stipulates that users of the standard shall determine which clauses and recommendations apply to their product, system, or service, and then detail how they determined that they met that requirement. For example, clause 4.3 in ISO 9241-14:1997 Menu Dialogues states, "If a product is claimed to have met the applicable recommendations in this part of ISO 9241, the procedure used in establishing requirements for, developing, and/or evaluating the menus shall be specified. The level of specification of the procedure is a matter of negotiation between the involved parties. Users of this part of ISO 9241 can either utilize the procedures provided in Annex A, or develop another procedure tailored to their particular development and/or evaluation environment."

35.2.4 How ISO Develops International Standards

The ISO is the world's largest developer and publisher of ISs. Its members are the National Standards Bodies (NSBs) of 162 countries. It is supported by a Central Secretariat based in Geneva, Switzerland. The principal deliverable of ISO is the IS.

Standards development work is conducted by technical committees (TCs) and subcommittees (SCs) which meet every year or so and are attended by formal delegations from participating members of that committee. In practice, the technical work takes place in working groups (WGs) of experts, nominated by national standards committees but expected to act as independent experts. Most NSBs set up "mirror committees" to coordinate their input to the international work.

- The work starts when a new work item (NWI) is proposed identifying a suitable topic for standardization. This proposal document is circulated for vote to the appropriate TC or SC.
- If approved, it is passed to a WG to be developed into a working draft (WD), which is a partial or complete first draft of the text of the proposed standard. The work is usually done by a project editor with members of the WG.
- The first formal stage is to produce and circulate a committee draft (CD) and circulate it for comment and approval within the committee and the national mirror committees. It is not unusual for a second CD to be required if there is significant disagreement over the first CD.
- The first public stage is a draft international standard (DIS), which is circulated widely for public comment via the NSBs.
- Once this is approved and any relevant comments addressed, it is issued as final DIS. At this stage, no further substantive comments are permitted, except a final go/no go vote.
- Eventually (under ISO rules, each stage takes a minimum of 3–6 months to allow for full review in each country), a final IS is published.

35.3 Impact on Practice

ISO usability standards have become increasingly important in the design of information systems and technology. They are specified in many procurement contracts, and most major software developers rely on them to demonstrate the quality of their products. Their history stretches back over more than 30 years.

35.3.1 Early History of ISO Usability Standards

Usability standards have been under development by ISO since 1983 when a NWI with the title "Visual Information Processing" was approved. This work item became the responsibility of ISO TC159 SC4 Ergonomics of human–system interaction, and after several years work resulted in a 17-part standard ISO 9241 entitled "Ergonomic requirements for office work with visual display terminals." The focus on office work reflected concerns at the time about the impact of VDU use on people's health, but many of the parts of ISO 9241 dealt with broader usability and software interface issues. It has been an influential standard across the world and is referenced in many countries' regulations, including the United Kingdom's Health and Safety (display screen equipment) Regulations 1992 (Stewart 1992). Thirty five countries take part in the standards work and many go on to adopt the ISs as national standards.

When it came to revising this standard, the committee wanted both to retain the ISO 9241 "brand" which had become a recognized benchmark and also to broaden its scope. A new title "The Ergonomics of Human–System Interaction" reflected this ambition, and the multipart structure allowed the committee to integrate other usability standards.

The structure reflected the numbering of the original ISO 9241 standard; for example, displays were originally Part 3, and became the 300 series. In each section, the "hundred" is an introduction to the section, for example, Part 100 gives an introduction to the software ergonomics parts. The number of parts that will be developed in each section varies depending on the complexity of the area and the need for specific standards.

Only three part numbers have been retained (following revision) from the original ISO 9241 structure, as these address issues that apply across all other parts of ISO 9241:

Part 1 Introduction
Part 2 Job design
Part 11 Hardware and software usability

An additional part that also applies across all of ISO 9241 is

Part 20 Accessibility and human–system interaction

The remaining parts are structured in "hundreds" as follows:

100 series Software ergonomics
200 series Human–system interaction processes
300 series Displays and display-related hardware
400 series Physical input devices—ergonomics principles
500 series Workplace ergonomics
600 series Environment ergonomics
700 series Application domains—control rooms
900 series Tactile and haptic interactions

35.3.2 How Do You Apply Usability Standards in Practice?

During the development of ISO 9241, it became clear that user interface technology was developing so quickly that the traditional approach to standards could not keep up in all areas. Two solutions were adopted for this problem.

First, standards concerned with hardware including displays and input devices contained what were known as "user performance test methods" as well as conventional hardware test methods. These user performance test methods allowed a new device to be used under controlled conditions by a defined group of participants performing specific tasks. The performance of the participants was then compared with their performance on a test device, which met the standard. If the participants' performance was as good or better on the test device, then this device was deemed to pass. This approach means that any technology can be tested if it is designed to allow users to perform suitable tasks, for example, read text on a screen, or enter characters into a system.

Second, it was realized that in many cases the best way to ensure that a design was usable was to follow a human-centered design process. Human-centered design is an approach to developing and acquiring software that improves system effectiveness and the efficiency and satisfaction of users (usability). Following this systematic framework helps reduce the risk that new software is unusable or fails to work properly for its intended users. Across all industries and governments, major software development projects are notorious for being delivered late and overbudget. Worse still, they often fail to deliver the service promised (Clegg et al. 1997). In hindsight, the problems are often due to a failure to engage fully with all stakeholders, especially those intended to operate and use the systems. Furthermore, excessive ambition which overlooks the reality of existing working practices, and design and development methods which are rigid, inflexible, and unable to adapt to emerging business needs and changes in the environment are also reasons for failure. In 1999, ISO TC159 SC4 published a new standard ISO 13407 on Human-Centered Design for Interactive Systems. The standard provides a systematic framework, which directly addresses these and other problems in delivering usable software, products, and services. It is aimed at Project Managers and describes how they can apply human-centered design processes to procure or develop technical software that will deliver value to the users and the business. This standard presents four principles of human-centered design, including the iteration of design solutions and the active involvement of users. It also describes key activities including the need to understand and specify the context of use, produce designs and prototypes of potential solutions, and evaluate these solutions against the human-centered criteria. The term human-centered design was used in preference to user-centered design, because the process includes all stakeholders, not just those usually considered as users.

Both approaches have been adopted widely, and the new parts of ISO 9241 that deal with displays and input devices both contain user performance test methods. Human-centered design has been recognized as the best way to develop usable systems, and the standard has been endorsed by such bodies as the Usability Professionals Association and the International Ergonomics Association.

The human-centered design approach is complementary to existing design methodologies—it does not describe all the processes that are necessary to design and develop a system. It is characterized by the following:

- Systems are designed to take account of all the people who will use them and other stakeholder groups, including those who might be affected (directly or indirectly) by their use. Constructing systems based on an inappropriate or incomplete understanding of user needs is one of the major sources of systems failure.
- Users are involved throughout design and development. This provides knowledge about the context of use, the tasks, and how users are likely to work with the future system. User involvement should be active, whether by participating in design, acting as a source of relevant data, or by evaluating solutions.
- The design is driven and refined by user-centered evaluation. Early and continuous feedback from users minimizes the risk that the system does not meet user and organizational needs (including those requirements that are hidden or difficult to specify explicitly). Such evaluation ensures preliminary design solutions are tested against "real world" scenarios, with the results being fed back into progressively refined solutions. User-centered evaluation should also take place as part of final acceptance of the product to confirm that requirements have been met. Feedback from users during operational use identifies long-term issues and provides input to future design.
- The process is iterative. For complex systems, it is impossible to specify completely and accurately every aspect of the interaction at the beginning of development. Initial design solutions rarely satisfy all the user needs. Many users and stakeholders need only emerge during development, as the designers refine their understanding of the users and their tasks, and as users provide feedback on potential solutions.

- The design addresses the whole user experience. This involves considering, where appropriate, organizational impacts, user documentation, on-line help, support and maintenance (including help desks and customer contact points), training, and long-term use.
- The design team reflects a range of skills and perspectives. The team does not have to be large, but should be sufficiently diverse to understand the constraints and realities of the various disciplines involved. This can have other benefits too. For example, technical experts can become more sensitized to user issues, and the users themselves can become more aware of technical constraints.

35.3.3 Using Standards to Support Human-Centered Design

Although, as explained earlier, the human-centered design standard ISO 13407 was developed in parallel with ISO 9241, many parts of ISO 9241 were developed with an implicit human-centered design process in mind. These parts can also be used to support specific design activities whether or not a full human-centered design process is being followed. Therefore, the next section describes the human-centered design process, originally specified in ISO 13407, and explains how the various parts of ISO 9241 can be used to support this process.

As part of the overall restructuring of ISO 9241 mentioned earlier, the opportunity was taken when ISO 13407 was undergoing its systematic review (a key part of the ISO process to ensure continued relevance) to incorporate it within the new ISO 9241 structure as ISO 9241-210:2010 Human-centered design for interactive systems. The revised standard remained largely unchanged, with one major exception. The four key human-centered design activities were no longer just recommendations; they became "requirements." This meant that people could now claim conformance, that is, that they followed and adhered to the ISO 9241-210 human-centered design process. To do so, they have to ensure that they

- Understand and specify the context of use (including users, tasks, environments)
- Specify the user requirements in sufficient detail to drive the design
- Produce design solutions which meet these requirements
- Conduct user-centered evaluations of these design solutions and modify the design taking account of the results

Note that the requirement is to provide evidence to show that they have done these successfully. There is no requirement to produce lots of documents to conform to the standard. The onus is on the organization claiming conformance to provide whatever level of detailed evidence the recipient requires.

In addition to these four activities, the standard provides guidance and recommendations on how to plan the human-centered design process and how to follow-up to ensure that the system is implemented properly. These six main steps are described in the following sections, and the other parts of ISO 9241, which support these steps, are explained and described briefly.

Standards development is an ongoing process, and standards can change radically during the various stages. The tables therefore only reference ISO standards that have been published up to February 2012.

35.3.3.1 Planning When and How the Human-Centered Design Activities Fit in the Program

The Project Manager should prepare the plan, which shows how the human-centered activities integrate into the overall project plan. Where the software is being developed in-house within the organization, the Project Manager creates the plan directly ensuring due time and resources are scheduled to allow the human-centered activities to be undertaken properly and any results fed back into the overall development process.

Where software is being purchased from an external supplier, there are two options:

1. The software is already available "off the shelf."
2. Significant customizing and development are required.

In option 1, the Project Manager only has to plan any organization-specific activities involved in selecting, evaluating, or implementing the software.

In option 2, the Project Manager requires the supplier to provide the plan, taking full account of the time and resources, which may be required from the organization, to participate in user research, requirements gathering, user testing, and any other user-based activities.

Underestimating the extent of user interaction required is a common feature of projects that do not plan human-centered design appropriately. For example, a system may be initially intended to be fully automated but ends up requiring significant user interaction, which then has to be designed in a rush without a sound understanding of the context of use.

The level of detail in the plan will vary depending on the scale of the project, the degree of user interaction expected, and the risks associated with poor usability or user rejection of the system.

The plan

- Identifies appropriate methods and resources for the human-centered activities
- Defines how these and their outputs fit with other system development activities
- Identifies the individuals and the organization(s) responsible for the human-centered design activities and the range of skills and viewpoints they provide
- Describes how user feedback and communication are collected and communicated
- Sets appropriate milestones and timescales for human-centered activities within the overall plan (allowing time for user feedback, and possible design changes, to be incorporated into the project schedule)

The plan for human-centered design is subject to the same project disciplines (e.g., responsibilities, change in control) as other key activities.

Time should also be allocated within the plan for communication among design team participants. Time spent resolving usability issues early in the project will deliver significant savings at later stages (when changes are, inevitably, more costly).

Human-centered design activities should start early in the project (e.g., at the initial concept stage) and continue throughout the life of the project (Table 35.1).

35.3.3.2 Understand and Specify the Context of Use (Including Users, Tasks, Environments)

The characteristics of the users, their tasks, and the organizational, technical and physical environment define the context in which the system is used (the context of use).

TABLE 35.1 Parts of ISO 9241 and Associated Standards Relevant to Planning

Relevant parts of ISO 9241	Brief Description
ISO 9241-210:2010 Human-centered design for interactive systems (supersedes ISO 13407 1999)	This part provides guidance on and requirements for the human-centered design process and is aimed at those responsible for managing the design of interactive systems.
ISO/TR 18529:2000 Human-centered life cycle process descriptions	Uses the ISO standard format for process models to describe the processes necessary for ensuring human-centered design content in systems strategy.

TABLE 35.2   Parts of ISO 9241 Relevant to Understanding and Specifying the Context of Use

Relevant Part of ISO 9241	Brief Description
ISO 9241-11:1998 Guidance on usability	Defines usability as "Extent to which a product can be used by specified users to achieve specified goals with effectiveness, efficiency, and satisfaction in a specified context of use," and provides guidance on how to address usability in design projects.

The context of use description includes details of the following:

- The users and other stakeholder groups.
- The user profile in terms of relevant knowledge, skill, experience, training, physical attributes (e.g., any disabilities), preferences, and capabilities of the users.
- The environment profile in which the system will operate the technical environment, including the hardware, software, and materials; the relevant characteristics of the physical environment, including thermal conditions, lighting, spatial layout, and furniture.
- The user task profile, including the frequency and duration of tasks and interdependencies (Table 35.2).

The user, environment, and task profiles should be described in sufficient detail to support the requirements, design, and evaluation activities. They are working documents that start as outlines, and are reviewed, maintained, extended, and updated during the design and development process.

The next step is to turn the profiles into concrete, detailed usage "scenarios" and "personnas" (Dzida 1998). Scenarios are realistic stories that describe the use of the software from the user's perspective and are combinations of individual tasks in the right sequence to achieve an overall goal. Personnas are rich descriptions of target users.

The scenarios and personnas provide a high-level summary of the key goals, which the software is intended to support, in user-centered language.

Where the software is being custom-developed (whether in-house or by an external supplier), the scenarios and personnas help communicate how the users will expect to use the software and what they will want to achieve from it. Designers find these kinds of descriptions useful in maintaining a focus on the user when designing specific functionality.

Where software is being purchased "off the shelf" from an external supplier, the scenarios help the software selectors understand the user's context and communicate this to potential suppliers.

In both situations, the scenarios provide a consistent framework for the subsequent evaluation of emerging or proposed designs.

35.3.3.3 Specify the User Requirements in Sufficient Detail to Drive the Design

Identifying user needs and specifying the functional and other requirements for the system are major software development activities. When following the human-centered design approach, these activities are extended to create an explicit specification of user requirements in relation to the context of use and the objectives of the system (Table 35.3).

The user requirements specification should be

- Stated in terms that permit subsequent testing
- Verified by the relevant stakeholders
- Internally consistent
- Updated as necessary, during the life of the project

This specification is then used to drive the design and provide agreed quality measures used to evaluate the software.

TABLE 35.3 Parts of ISO 9241 Relevant to Specifying the User Requirements

Relevant Parts of ISO 9241	Brief Description
ISO 9241-11:1998 Guidance on usability	Defines usability as "Extent to which a product can be used by specified users to achieve specified goals with effectiveness, efficiency, and satisfaction in a specified context of use," and provides guidance on how to address usability in design projects.
ISO 9241-20:2009 Accessibility guidelines for information/ communication technology equipment and services	A high-level overview standard covering both hardware and software. It covers the design and selection of equipment and services for people with a wide range of sensory, physical, and cognitive abilities, including those who are temporarily disabled, and the elderly.

These quality measures should

- Capture the key success factors for the product and what value it will deliver for users.
- Summarize who will be doing what and under what circumstances.
- Identify the type of measures to be used (e.g., performance data, user preference data, or conformance to standards).
- Specify the level that the software must achieve to be released to users. Scores below that level mean that the products must be improved. Scores above that level may mean that resources have been wasted.

35.3.3.4 Produce Design Solutions That Meet These Requirements

There are several chapters in this handbook that provide guidance for the design of usable software. Within the ISO 9241 series of standards, there are several standards that provide detailed guidance, but the overall process involves six main steps in designing usable software.

1. Structure solutions around key tasks and workflows based on the context of use and user requirements specification. A key issue concerns establishing an appropriate "allocation of function"—deciding which system tasks are to be automated and which should be under user control. Table 35.4 lists standards relevant to designing jobs and tasks.
2. Design the interaction, interface, and navigation from the user's perspective and keep them consistent. Be aware of other software that the user is likely to be using, and be cautious about introducing different styles of interface. Table 35.5 lists standards relevant to designing dialogs and navigation.
3. Define the navigation and keep it consistent. If different teams are developing separate parts of the software, maintain consistency by establishing an agreed style guide at the beginning of the development.
4. Follow interface design best practice. Various parts of ISO 9241 offer guidance and are based on internationally agreed best practice for achieving usable hardware and software (see Tables 35.6 through 35.8). The HCI Bibliography website lists more than 53 journals, many peer-reviewed, relevant to usability and human–computer interaction (HCI Bibliography 2012).

TABLE 35.4 Parts of ISO 9241 Relevant to Designing Jobs and Tasks

Relevant Part of ISO 9241	Brief Description
ISO 9241-2:1992 Guidance on task requirements deals with the design of tasks and jobs involving work with visual display terminals	Provides guidance on how task requirements may be identified and specified within individual organizations and how task requirements can be incorporated into the system design and implementation process.

TABLE 35.5   Parts of ISO 9241 Relevant to Designing User–System Dialogs and Interface Navigation

Relevant Parts of ISO 9241	Brief Description
ISO/TR 9241-100:2010 Introduction to standards related to software ergonomics	Provides an introduction to the "100 series" of ISO 9241 software parts. Published as a TR, which has a shorter development time than a full IS. This will allow it to be updated regularly to reflect the current content and structure of ISO 9241.
ISO 9241-110:2006 Dialog principles (supersedes ISO 9241 10:1996)	Sets out seven dialog principles and gives examples. The dialog should be suitable for the task (including the user's task and skill level); self-descriptive (it should be obvious what to do next); controllable (especially in pace and sequence); conform to user expectations (i.e., consistent); error-tolerant and forgiving; suitable for individualization and customizable; and should support learning.
ISO 9241-129:2010 Guidance on software individualization	This part provides ergonomics guidance on individualization within interactive systems, including recommendations on where individualization might be appropriate or inappropriate and how to apply individualization. The standard provides general guidance on individualization rather than specific implementations of individualization mechanisms. It is intended to be used along with other parts of ISO 9241—not in isolation.
ISO 9241-14:2000 Menu dialogues	Recommends best practice for designing menus (pop-up, pull-down, and text-based menus). Topics include menu structure, navigation, option selection, and menu presentation (including placement and use of icons). One of the annexes contains a 10-page checklist for determining compliance with the standard.
ISO 9241-15:1998 Command dialogues	This part provides recommendations for the ergonomic design of command languages used in user–computer dialogs. The recommendations cover command language structure and syntax, command representations, input and output considerations, and feedback and help. Part 15 is intended to be used by both designers and evaluators of command dialogs, but the focus is primarily toward the designer.
ISO 9241-16:1999 Direct manipulation dialogues	This part provides recommendations for the ergonomic design of direct manipulation dialogs, and includes the manipulation of objects, and the design of metaphors, objects, and attributes. It covers those aspects of "Graphical User Interfaces," which are directly manipulated, and not covered by other parts of ISO 9241. Part 16 is intended to be used by both designers and evaluators of command dialogs, but the focus is primarily toward the designer.
ISO 9241-143:2012 Forms	ISO 9241-143:2012 provides requirements and recommendations for the design and evaluation of forms, in which the user fills in, selects entries for, or modifies labeled fields on a "form" or dialog box presented by the system. It contains guidance on the selection and design of interface elements relevant to forms. The requirements and recommendations can be used during design, as a basis for heuristic evaluation, as guidance for usability testing, and in the procurement process. It replaces ISO 9241-17:1998 Form-filling dialogs
ISO 9241-151:2008 Guidance on World Wide Web user interfaces	Sets out detailed design principles for designing usable websites—these cover high-level design decisions and design strategy; content design; navigation; and content presentation.
ISO9241-171:2008 Guidance on software accessibility	Aimed at software designers and provides guidance on the design of software to achieve as high a level of accessibility as possible. Replaces the earlier Technical Specification ISO TS 16071:2003 and follows the same definition of accessibility—"usability of a product, service, environment, or facility by people with the widest range of capabilities." Applies to all software, not just web interfaces.

TABLE 35.6 Parts of ISO 9241 Relevant to Designing or Selecting Displays

Relevant Parts of ISO 9241	Brief Description
ISO 9241-300:2008 Introduction to electronic visual display requirements (The ISO 9241-300 series supersedes ISO 9241, parts 3, 7, and 8)	A very short (four pages) introduction to the ISO 9241-300 series, which explains what the other parts contain.
ISO 9241-302:2008 Terminology for electronic visual displays	Definitions, terms, and equations that are used throughout ISO 9241:300 series.
ISO 9241-303:2011 Requirements for electronic visual displays	Sets general image quality requirements for electronic visual displays. The requirements are intended to apply to any kind of display technology.
ISO 9241-304:2008 User performance test methods for electronic visual displays	Unlike the other parts in the subseries which focus on optical and electronic measurements, this part sets out methods which involve testing how people perform when using the display. The method can be used with any display technology.
ISO 9241-305:2008 Optical laboratory test methods for electronic visual displays	Defines optical test methods and expert observation techniques for evaluating a visual display against the requirements in ISO 9241-303. Very detailed instructions on taking display measurements.
ISO 9241-306:2008 Field assessment methods for electronic visual displays	Provides guidance on how to evaluate visual displays in real-life workplaces.
ISO 9241-307:2008 Analysis and compliance test methods for electronic visual displays	Supports ISO 9241-305 with very detailed instructions on assessing whether a display meets the ergonomics requirements set out in part 303.
ISO/TR 9241-308:2008 Surface-conduction electron-emitter displays	Technical report on a new "ecofriendly" display technology called "Surface-Conduction Electron-Emitter Displays."
ISO/TR 9241-309:2008 Organic light-emitting diode displays	Technical report on another new display technology called "organic light emitting diode displays," which are better for fast-moving images than LCDs.
ISO/TR 9241-310:2010 Visibility, aesthetics, and ergonomics of pixel defects	Pixel defects are controversial. Users expect none, but would not be willing to pay the extremely high price that would be charged to achieve this. This TR explains the current situation and gives guidance on the specification of pixel defects, visibility thresholds, and aesthetic requirements for pixel defects.
ISO 9241-12:1999 Presentation of information	This part contains specific recommendations for presenting and representing information on visual displays. It includes guidance on ways of representing complex information using alphanumeric and graphical/symbolic codes, screen layout, and design as well as the use of windows.

5. Produce sketches and mock-ups early to test assumptions. Using scenarios and simple sketches or mock-ups enables the designers to communicate the proposed design to users and other stakeholders to obtain feedback before design decisions are finalized.
6. Keep testing emerging solutions with users until the quality criteria are met. Human-centered evaluation can take place throughout the design process from initial concepts to final (signed-off) designs.

One example of the kind of high-level guidelines in ISO 9241 is reproduced in simplified form in Table 35.9, which is based on ISO 9241-110:2006 Dialogue Principles.

See Table 35.10 for standards relevant to selecting usability methods for design and evaluation.

TABLE 35.7   Parts of ISO 9241 Relevant to Designing or Selecting Keyboards and Other Input Devices

Relevant Parts of ISO 9241	Brief Description
ISO 9241-4:1998 Keyboard requirements (some clauses in this standard have been superseded by ISO 9241-400 and ISO 9241-410	This part specifies the ergonomics design characteristics of an alphanumeric keyboard, which may be used comfortably, safely, and efficiently to perform office tasks.
ISO 9241-400:2007 Principles and requirements for physical input devices	Sets out the general ergonomics principles and requirements, which should be taken into account when designing or selecting physical input devices.
ISO 9241-410:2008 Design criteria for physical input devices (supersedes ISO 9241-9: 1998)	Describes ergonomics characteristics for input devices, including keyboards, mice, pucks, joysticks, trackballs, touchpads, tablets, styli, and touch-sensitive screens. The standard is aimed at those who design such devices and is very detailed.
ISO 9241-420:2011 Selection of physical input device	Provides ergonomics guidance for selecting input devices for interactive systems. It describes methods for evaluating a wide range of devices from keyboards and mice to pucks, joysticks, trackballs, trackpads, tablets and overlays, touch-sensitive screens, styli, and light pens. It encourages user organizations and systems integrators to consider the limitations and capabilities of users and the specific tasks and context of use when selecting input devices.
ISO 9241-910:2011 Framework for tactile and haptic interaction	Provides a framework for understanding and communicating various aspects of tactile/haptic interaction. It defines terms, describes structures and models, and also serves as an introduction to the other parts of the ISO 9241 "900" subseries. It provides guidance on how tactile/haptic interaction can be applied to a variety of user tasks. It does not specifically cover gesture-based interfaces, although it does offer some relevant guidance for understanding such interactions.
ISO 9241-920:2009 Guidance on tactile and haptic interactions	Gives recommendations for tactile and haptic hardware and software interactions. It provides guidance on the design and evaluation of hardware, software, and combinations of hardware and software interactions. It does not provide recommendations specific to Braille, but can apply to interactions that make use of Braille.

TABLE 35.8   Parts of ISO 9241 Relevant to Designing or Workplaces for Display Screen Users

Relevant Parts of ISO 9241	Brief Description
ISO 9241-5:1999 Workstation layout and postural requirements	This part specifies the ergonomics requirements for a visual display terminal workplace, which will allow the user to adopt a comfortable and efficient posture.
ISO 9241-6:2000 Guidance on the work environment	This part specifies the ergonomics requirements for the visual display terminal working environment, which will provide the user with comfortable, safe, and productive working conditions.

TABLE 35.9 Usability Principles (Based on ISO 9241-110)

1. Fit with user's task

The interface should match the way users perform their tasks

1.1 Are screens designed to provide users with the relevant information they need to complete tasks?

1.2 Is only necessary information displayed?

1.3 Does the system accept input and produce output in useful and appropriate formats?

1.4 Do input fields contain useful default values?

1.5 Is the sequence of actions required optimum? (i.e., all that is needed, but no more)

1.6 If paper documents are needed for the task, do the screens match their format?

2. Sign posting

The users should be able to tell from the screens, where they are in the task and what they can do next

2.1 Do the screens guide the user what to do next?

2.2 Can most users work without needing to refer to manuals?

2.3 Is it obvious when input is required or about to be required?

2.4 Is it clear what input is required?

2.5 Is it clear how to interact with the different system components?

2.6 Do input fields indicate what format and units are required (for example, dd/mm/yyyy)?

3. Intuitiveness

The system should respond to the task context and follow accepted conventions for the target users

3.1 Is the terminology appropriate to the target users?

3.2 Is there timely feedback on users' actions?

3.3 Are users given useful feedback on response times?

3.4 Do the data structures correspond to users' expectations?

3.5 Is there sufficient and useful feedback to users?

3.6 Are the behavior and appearance of the system consistent throughout?

3.7 Does the input focus default appropriately?

3.8 Are system messages useful and constructive?

3.9 Is the interface suitable for users who do not have English as their first language?

4. Learnability

The system should support learning

4.1 Does the system provide explanations which help users learn?

4.2 Are there prompts and reminders to help users learn?

4.3 Is sufficient support available to help infrequent users?

4.4 Does the system help users build a mental model of what is going on?

4.5 Can users see the steps needed to complete a task?

4.6 Does the system minimize the amount of input required (e.g., using autocomplete)?

4.7 Does the system encourage users to explore (if appropriate) without negative consequences?

5. User control

The user should be able to control the interaction

5.1 Can the user control the pace of the interaction?

5.2 Can the user control the sequence of steps?

5.3 Does the system allow the user to restart their task if necessary?

5.4 Does the system offer an "undo" where possible?

5.5 Can the user control the amount of data displayed at a time?

5.6 Does the system support alternative input/output devices?

5.7 Can users set appropriate default values?

5.8 If appropriate, can users "track changes" and view original data after a modification?

(continued)

TABLE 35.9 (continued)   Usability Principles (Based on ISO 9241-110)

6. Error tolerance

The system should minimize the likelihood and severity of errors

6.1 Does the system help minimize input errors?

6.2 Is the system robust?

6.3 Are error messages helpful for correcting the error?

6.4 Does the system actively support error recovery?

6.5 If automatic error correction is provided, can users override this functionality?

6.6 Can users continue (where appropriate) and defer correcting an error until a more suitable time?

6.7 Can users get additional help for errors?

6.8 Does the system validate input data before processing?

6.9 Has the number of steps for error correction been minimized?

6.10 Does the system require confirmation for potentially critical actions?

7. Customizing

Users should be able to customize the system to suit their skills and preferences

7.1 Can the system be tailored to suit different user profiles?

7.2 Can the user customize how data is displayed?

7.3 Can the help information be tailored for users with different skills?

7.4 Can users customize system speed and sensitivity?

7.5 Can users choose different interaction styles, for example, command line versus GUI?

7.6 Can users choose from a variety of interaction methods (e.g., keyboard vs. mouse)?

7.7 Can users customize how input/output data is displayed, to a degree that meets their needs?

7.8 Can users customize the interface or functionality to suit their task requirements and preferences?

7.9 Can users modify names of objects and actions to suit their requirements?

7.10 Can default settings be restored?

TABLE 35.10   Standards Relevant to Selecting Usability Methods for Design and Evaluation

Relevant Standard	Brief Description
ISO/TR 16982:2002 Usability methods supporting human-centered design	Provides information on human-centered usability methods, which can be used for design and evaluation. It details the advantages, disadvantages, and other factors relevant to using each usability method.

35.3.3.5 Conduct User-Centered Evaluations of These Design Solutions and Modify the Design, Taking Account of the Results

User-centered evaluation (evaluation based on the users' perspective) is at the heart of human-centered design. It occurs throughout the design life cycle from initial testing of concepts to final testing to confirm that the software meets the quality criteria.

There are two main approaches:

- Inspection-based evaluation
- User-based testing

35.3.3.5.1 Inspection-Based Evaluation

Inspection-based evaluation helps eliminate major issues before user testing (and hence makes the user testing more cost-effective). It is ideally performed by usability experts, with broad experience of problems encountered by users, working with domain experts who understand the software and the problem domain. Inspection can be based on task scenarios or on guidelines and standards, for example, the Web Content Accessibility Guidelines (WCAG).

During the development process, or where there are no users already familiar with the software, inspection-using scenarios can be used. This involves the assessor putting themselves in the position of the user working through the scenarios agreed in the requirements stage. This can be done with working prototypes or with partial prototypes or even nonfunctioning mock-ups of the software. At the very early stages in design, these mock-ups can be simple sketches on paper, or at a slightly later stage, PowerPoint screenshots with hotspots to simulate functionality. The assessor works through the tasks in the scenario and notes any issues, for example, points in the sequence where it is not clear what to do next or where the result is unexpected. The objective is to identify breaks in flow and incomplete or unclear screens and dialogs at an early enough stage for them to be changed quickly and cheaply. Delaying assessing early designs is not only wasteful, but also tends to inhibit assessors who may be reluctant to recommend changes, which they think may be time-consuming or expensive. Starting assessment early also provides a useful benchmark, which shows how the design has improved, as it is modified, based on user-centered feedback.

Inspection-based evaluation can also be conducted using guidelines and standards. Software accessibility, the degree to which software can be used by people with disabilities, is one area where it can be necessary to check compliance with published external standards. In particular, the WCAG, published by the World Wide Web Consortium (W3C), are the widely accepted accessibility standards for web-based applications. More information on available standards to support inspection is shown in Tables 35.5 through 35.10.

35.3.3.5.2 User Testing

User testing can be undertaken at any stage of the design and development cycle. At an early stage, users can be presented with sketches or mock-ups of design concepts and asked to evaluate them by working through how they would carry out key tasks. This provides much more information than simply showing or demonstrating them to users. For instance, scenarios could be used to provide a useful framework for structuring this walkthrough of the software. Later in development, user testing is used to confirm that the software meets the agreed quality criteria established earlier.

User testing is best conducted in a usability laboratory, which allows user behavior including detailed mouse and keyboard use to be observed and recorded for subsequent analysis. This can be supplemented with eye-tracking technology, which shows exactly where on the screen the users have been looking during the tasks. It can also be conducted more simply at the user's normal workstation by observing and video recording their behavior as they work through the tasks.

In addition to direct observations or measurements of user behavior (e.g., task completion times), user testing can involve users vocalizing their thought process while they perform the tasks. Further feedback can be obtained through questionnaires or interviews and checklists (before, during, or after the tasks).

The procedure for conducting user-based testing involves the following:

- Writing a test plan, which uses the scenarios to test the software against the release criteria, collects the relevant user behavior information, and identifies appropriate test participant profiles, for example, experienced user of competing product
- Ensuring that all stakeholders review and agree the plan in advance
- Recruiting the participants to the agreed profile
- Preparing the software, computer hardware, usability lab or other equipment, and questionnaires and checklists
- Conducting a pilot study to ensure that the testing is likely to produce the appropriate data for establishing whether the release criteria 3 have been met

- Conducting the formal test with the requisite number of participants (typically a minimum of three to four per unique user group) ensuring that the test conditions are randomized or balanced to avoid confounding learning effects or bias
- Analyzing and reporting the results in an agreed format with usability issues prioritized in terms of severity (see following paragraphs)
- Working with the design team to review the usability defects and decide on the basis of their severity, number of users affected, cost to fix, and how and when to fix them

The severity of usability defect is often classified as follows:

- Critical—the user is unable to use the software
- High—the user is severely restricted in using the software
- Medium—the user can circumvent the problem and use the software with only moderate inconvenience
- Low—the user can use the software with limitations, but these do not restrict its use

The Project Manager should ensure that the test plan is designed to allow the agreed quality criteria to be tested. Testers should be competent in conducting usability tests and follow the Usability Professionals Association or equivalent code of conduct. Where the user testing is being conducted by a third party, for example, independent usability specialists or usability teams within the software supplier, then the Project Manager may require the test reporting to follow ISO/IEC 25062:2006 software engineering—software products quality and requirements evaluation (SQuaRE)—common industry format (CIF) for usability tests.

When the testing is conducted internally, the worksheet presented in Table 35.11 (which is based on the full standard) may be used for reporting usability test results.

35.3.3.6 Implement the System with Appropriate Training and Support

The final stage in deploying a new software, whether it is developed in-house or purchased "off the shelf," is implementation.

Implementation involves a large number of technical processes to ensure that the new software (and/or hardware) operates properly in its environment. This includes tasks such as installation, configuration, testing, and making necessary changes before the software is finally deployed.

Human-centered design is complementary to and does not replace existing standards or processes for implementing the software.

Following a human-centered design process means that users are already prepared for the new software through their earlier involvement in requirements gathering and user testing. The scenarios identified earlier provide a useful basis for structuring any training required and also for other training and support materials, such as on-screen help and user manuals. Any usability issues that could not be fixed on time can be addressed through work-arounds and specific "frequently asked questions" (FAQs). While users find it convenient to ask more knowledgeable colleagues or dedicated support staff for help, one of the benefits of following a human-centered design process is that users require less support, even at implementation. This reduces the burden on colleagues and the dependence on costly on-site support.

Once the software is fully deployed, the final human-centered design activity is to track real world usage, not only to ensure that the original objectives were achieved but also to provide input to further development. Help desk data, log files, defect reports, user satisfaction surveys, and requests for changes, all provide valuable data. Formal follow-up evaluation can be carried out within a specific period, for example, 6 months to 1 year after system installation to test system performance and collect data on whether the user requirements were met.

Table 35.12 identifies standards relevant to reporting usability test results.

TABLE 35.11   Worksheet for Reporting Usability Test Results

Report title page

Name of the product and version that was tested

When the test was conducted?

Date the report was prepared

Contact name(s) for questions and/or clarifications

Background

Product name, purpose, and version

What parts of the product were evaluated?

The user population for which the product is intended

Test objectives

Description of context of use (note any difference between test context and real context)

Methods

Experimental design

Participant characteristics

Tasks and scenarios used

Test facilities

Computer configuration, including model, OS version, required libraries, or settings

Display devices, including screen size, resolution

Input devices, including make and model

Measures

Checklists, scoring sheets or questionnaires used—attach copies in the Appendix

Procedure including time limits on tasks and whether participants "thought aloud"

Participant instructions including task instructions—attach in Appendix

Metrics for effectiveness, completion rates, errors, and assists

Metrics for efficiency, completions rates/time on task

Metrics for satisfaction

Data analysis including any statistical analysis

Performance results

Satisfaction results

Appendices

Checklists, scoring sheets, and questionnaires used

Participant task instructions

Based on ISO/IEC 25062:2006 Software engineering: SQuaRE—Common Industry Format (CIF) for usability tests.

Check that the user test report contains at least the listed information to a sufficient level of detail to allow the validity and reliability of the findings to be assessed (✓ to confirm).

TABLE 35.12   Standards Relevant to Reporting Usability Test Results

Relevant Standard	Brief Description
ISO/IEC 25062:2006 Software product Quality Requirements and Evaluation (SQuaRE)— Common Industry Format (CIF) for usability test reports	Provides a standard method for reporting usability test findings. The format is designed for reporting results of formal usability tests in which quantitative measurements were collected, and is particularly appropriate for summative/ comparative testing.

35.4 Research Issues

The relationship between international usability standards and research confuses some people. There are two major issues. First, although standards should be based on research evidence, research results themselves are not usually sufficient to determine what the standard should recommend or require. So, for example, in the standards concerned with identifying a suitable level of office illumination for work with visual display screens, research can show how individual sensitivity and acuity are affected by overall illumination level, image contrast, and so on. What such data show is that people vary in their sensitivity, and while some may be able to tolerate high ambient illumination, others will complain that their visual performance is significantly impaired. What the data itself cannot determine is where to set limits in the standard. The decision on where to set such limits involves a judgment on the part of the standards maker. Perhaps, this is more obvious in the area of workplace design. Human dimensions vary enormously, and so it is not realistic to design a single item of equipment to take account of the full range of human variability. In practice, many standards aim to address the middle 90% of the population and accept that the lower 5th percentile and the higher 95th percentile may not be accommodated. Such practicalities may be difficult for purist researchers to accept.

Second, much usability practice is closer to a craft than to a science. Usability practitioners regularly test systems with very small samples of participants, often as few as five. Anyone skilled in psychophysics knows that such small numbers make it almost impossible to draw reliable conclusions about the user population. However, even such small-scale testing can reveal useful and valuable data about some of the problems that real users may experience with the final product. And, there is a substantial body of practical data that can be used to guide user interface designers to improve the usability of their systems. But, the uncertainties associated with this body of practical experience mean that usability standards, especially international usability standards, can seldom be as prescriptive as traditional industry standards. Subtle differences in context can completely change the user experience, and emerging technology develops faster than the standards.

Both of these issues pose major problems for researchers wishing to support international usability standards. However, there is a need both for sound human-centered design and evaluation methods and for empirical data on new and emerging technologies. Therefore, data from well-constructed research studies will always be welcomed by the relevant standards committees.

In terms of future developments in usability standards, the path is not entirely clear. Some of the standards discussed here are based on human characteristics, and these are unlikely to change much. Although ISO standards go through a formal review process every 5 years, there seems little point in making major updates to standards which deal with human basics, such as visual perception. However, even if the principles are the same, there is evidence that readers and users of such standards are unlikely to trust them if the examples are out of date. There is, therefore, a need to continue to "refresh" such standards to ensure that they not only remain relevant but also appear to be most up-to-date.

In terms of future standardization efforts, new technological developments are likely to continue to generate new requirements. As discussed earlier, some usability standards are better written as process standards rather than product standards. However, experience shows that designers much prefer specific product standards, provided that they remain relevant. So, for example, work is currently underway in ISO TC159 SC4 WG2 to develop a technical report (TR) dealing with autostereoscopic displays, which can appear three-dimensional without the need for the viewer to wear glasses. Specific technical guidance in this chapter will help display designers avoid the many pitfalls in this emerging technology.

Ever more innovative interfaces are likely to emerge, requiring standard makers to remain agile and innovative in their efforts to ensure that all users can benefit from such developments in technology.

35.5 Summary

This chapter introduced the reader to international usability standards from the Human–System Interaction Committee of the ISO. It explained how standards are developed and the implications for how they can be used. It then went on to describe the human-centered design process as contained in ISO 9241, part 210 and explained how different standards can be used to support these human-centered design activities.

Further Information

Any readers who wish to get involved in the international usability standardization processes described in this chapter should contact their national standards body, which belongs to ISO and ask if it participates in ISO TC159 SC4. If it does, it will usually be able to put the reader in touch with a National Mirror Committee. If it does not, then it may be willing to become involved if there is sufficient support in that country.

References

International Standards

All parts of ISO 9241 have a series title "Ergonomics of Human System Interaction" as well as a part title as shown below.
ISO 9241-1:1997/Amd 1:2001 General Introduction (supersedes ISO 9241-1:1993)
ISO 9241-2:1992 Guidance on task requirements
ISO 9241-4:1998 Keyboard requirements (some clauses in this standard have been superseded by ISO 9241-400 and ISO 9241-410)
ISO 9241-5:1999 Workstation layout and postural requirements
ISO 9241-6:2000 Guidance on the work environment
ISO 9241-11:1998 Guidance on usability
ISO 9241-12:1999 Presentation of information
ISO 9241-13:1999 User guidance
ISO 9241-14:2000 Menu dialogues
ISO 9241-15:1998 Command dialogues
ISO 9241-16:1999 Direct manipulation dialogues
ISO 9241-20:2009 Accessibility guidelines for information/communication technology (ICT) equipment and services
ISO/TR 9241-100:2010 Introduction to standards related to software ergonomics
ISO 9241-110:2006 Dialogue principles (supersedes ISO 9241-10:1996)
ISO 9241-129:2010 Guidance on software individualization
ISO 9241-143:2012 Forms
ISO 9241-151:2008 Guidance on World Wide Web user interfaces
ISO 9241-171:2008 Guidance on software accessibility
ISO 9241-210:2010 Human-centered design for interactive systems (supersedes ISO 13407 1999)
ISO 9241-300:2008 Introduction to electronic visual display requirements
(The ISO 9241-300 series supersedes ISO 9241, parts 3, 7, and 8)
ISO 9241-302:2008 Terminology for electronic visual displays
ISO 9241-303:2011 Requirements for electronic visual displays
ISO 9241-304:2008 User performance test methods for electronic visual displays
ISO 9241-305:2008 Optical laboratory test methods for electronic visual displays
ISO 9241-306:2008 Field assessment methods for electronic visual displays

ISO 9241-307:2008 Analysis and compliance test methods for electronic visual displays

ISO/TR 9241-308:2008 Surface-conduction electron-emitter displays (SED)

ISO/TR 9241-309:2008 Organic light-emitting diode (OLED) displays

ISO/TR 9241-310:2010 Visibility, aesthetics, and ergonomics of pixel defects

ISO 9241-400:2007 Principles and requirements for physical input devices

ISO 9241-410:2008 Design criteria for physical input devices (supersedes ISO 9241-9:1998)

ISO 9241-420:2011 Selection of physical input device

ISO 9241-910:2011 Framework for tactile and haptic interactions

ISO 9241-920:2009 Guidance on tactile and haptic interactions

ISO/TR 16982:2002 Usability methods supporting human-centered design

ISO/IEC 25062:2006 Software product quality requirements and evaluation (SQuaRE)—common industry format (CIF) for usability test reports provides a standard method for reporting usability test findings

ISO/TR 18529:2000 Human-centered life cycle process descriptions

Other References

Clegg, C., Axtell, C., Damodaran, L., Farbey, B., Hull, R., Lloyd-Jones, R., Nicholls, J., Sell, R. and Tomlinson, C. 1997. Information technology: A study of performance and the role of human and organizational factors. *Ergonomics*, 40(9), 851–871.

Dzida, W. and Freitage, R. 1998. Making use of scenarios for validating analysis and design. *IEEE Transactions on Software-Engineering*, 24(12), 1182–1196.

Dzida, W., Herda, S. and Itzfield, W.D. 1978. User perceived quality of interactive systems. *IEEE Transactions on Software-Engineering*, SE(4), 270–276.

Earthy, J., Sherwood Jones, B. and Bevan, N. 2001. The improvement of human-centered processes—Facing the challenge and reaping the benefit of ISO 13407. *International Journal of Human Computer Studies*, 55(4), 553–585.

Newman, W.M. and Lamming, M.G. 1995. *Interactive System Design*. Addison-Wesley, Boston, MA.

Potonak, K. 1988. What's wrong with standard user interfaces? *IEEE Software*, 5(5), 91–92.

Shneiderman, B. 1987. *Designing the User Interface: Strategies for Effective Human-Computer Interaction*. Addison Wesley, Boston, MA.

Smith, S.L. 1986. Standards versus guidelines for designing users interface software. *Behaviour and Information Technology*, 5(1), 47–61.

Sperandio, L.J. 1987. Software ergonomics of interface design. *Behaviour and Information Technology*, 6, 271–278.

Stewart, T.F.M. 1992. The role of HCI standards in relation to the directive. *Displays*, 13, 125–133.

Stewart, T. 2000. Ergonomics user interface standards: are they more trouble than they are worth? *Ergonomics*, 43(7), 1030–1044.

Web Link

HCI Bibiography (2012) http://hcibib.org/hci-sites/JOURNALS.html (accessed on April 8, 2013).

36

Designing Highly Usable Web Applications

Silvia Abrahão
Universitat Politècnica de València

Emilio Insfran
Universitat Politècnica de València

Adrian Fernandez
Universitat Politècnica de València

36.1 Introduction

Usability is considered to be one of the most important quality factors for Web applications, along with others such as reliability and security (Offutt 2002). It is not sufficient to satisfy the functional requirements of a Web application in order to ensure its success. The ease or difficulty experienced by users of these applications is largely responsible for determining their success or failure. Usability evaluations and technologies that support the usability design process have therefore become critical in ensuring the success of Web applications.

Today's Web applications deliver a complex amount of functionality to a large number of diverse groups of users. The nature of Web development forces designers to focus primarily on time-to-market issues in order to achieve the required short cycle times. In this context, model-driven engineering (MDE) approaches seem to be very promising, since Web development can be viewed as a process of transforming a model into another model until it can be executed in a development environment. MDE has been successfully used to develop Web-based interactive applications. Existing approaches usually comprise high-level models (e.g., a presentation model, task model) that represent interactive tasks that are performed by the users of the Web application in a way that is independent from platforms and interaction modalities. From these models, the final Web application can be generated by applying model transformations.

Model-driven engineering approaches can play a key role in helping Web designers to produce highly usable Web applications. The intermediate models that comprise the Web application can be inspected and corrected early in the development process, ensuring that a product with the required level of usability is generated. Although a number of model-driven Web development processes have emerged, there is a need for usability evaluation methods that can be properly integrated into these types of processes (Fernández et al. 2011). This chapter addresses this issue by showing how usability evaluations are integrated into a specific model-driven development method to ensure the usability of the Web applications developed.

36.2 Underlying Principles

Traditional Web development approaches do not take full advantage of the usability evaluation of interactive systems based on the design artifacts that are produced during the early stages of the development. These intermediate software artifacts (e.g., models and sketches) are usually used to guide Web developers in the design but not to perform usability evaluations. In addition, since the traceability between these artifacts and the final Web application is not well-defined, performing usability evaluations by considering these artifacts as inputs may not ensure the usability of the final Web application.

This problem may be alleviated by using a MDE approach due to its intrinsic traceability mechanisms that are established by the transformation processes. The most well-known approach to MDE is the model-driven architecture (MDA) defined by the object management group—OMG (2003). In MDA, platform-independent models (PIMs) such as task models or navigation models may be transformed into platform-specific models (PSMs) that contain specific implementation details from the underlying technology platform. These PSMs may then be used to generate the source code for the Web application (Code Model—CM), thus preserving the traceability among PIMs, PSMs, and source code.

A model-driven development approach therefore provides a suitable context for performing early usability evaluations. Platform-independent (or platform-specific) models can be inspected during the early stages of Web development in order to identify and correct some of the usability problems prior to the generation of the source code. We are aware that not all usability problems can be detected based on the evaluation of models since they are limited by their own expressiveness and, most importantly, they may not predict the user behavior or preferences. However, studies such as that of Hwang and Salvendy (2010) claim that usability inspections, applying well-known usability principles on software artifacts, may be capable of finding around 80% of usability problems. In addition, as suggested by previous studies (Andre et al. 2003), the use of inspection methods for detecting usability problems in product design (PIMs or PSMs in this context) can be complemented with other evaluation methods performed with end users before releasing a Web application to the public.

36.3 State of the Art in Usability Evaluation for Web Applications

Usability evaluation methods can be mainly classified into two groups: empirical methods and inspection methods. Empirical methods are based on observing, capturing, and analyzing usage data from real end users, while inspection methods are performed by expert evaluators or designers, and are based on reviewing the usability aspects of Web artifacts (e.g., mock-ups, conceptual models, user interfaces [UIs]), which are commonly UIs, with regard to their conformance with a set of guidelines.

The employment of usability evaluation methods to evaluate Web artifacts was investigated through a systematic mapping study in a previous work (Fernandez et al. 2011). This study revealed various findings:

1. There is lack of usability evaluation methods that can be properly integrated into the early stages of Web development processes.
2. There is a shortage of usability evaluation methods that has been empirically validated.

Usability inspection methods have emerged as an alternative to empirical methods as a means to identify usability problems since they do not require end user participation and they can be employed during the early stages of the Web development process (Cockton et al. 2003). There are several proposals based on inspection methods to deal with Web usability issues, such as the cognitive walkthrough for the Web—CWW (Blackmon 2002) and the web design perspectives—WDP (Conte et al. 2007). Cognitive walkthrough for the Web assesses the ease with which a user can explore a website by using semantic algorithms. This method extends and adapts the heuristics proposed by Nielsen and Molich (1990) with the aim of drawing closer to the dimensions that characterize a Web application: content, structure, navigation, and presentation. However, these kinds of methods tend to present a considerable degree of subjectivity in usability evaluations. In addition, this method only supports ease of navigation.

Other works present Web usability inspection methods that are based on applying metrics in order to minimize the subjectivity of the evaluation, such as the WebTango methodology (Ivory and Hearst 2002) and the Web quality evaluation method—WebQEM (Olsina and Rossi 2002). The WebTango methodology allows us to obtain quantitative measures, which are based on empirically validated metrics for UIs, to build predictive models in order to evaluate other UIs. Web quality evaluation method performs a quantitative evaluation of the usability characteristics proposed in the ISO 9126-1 (2001) standard, and these quantitative measures are aggregated in order to provide usability indicators.

The aforementioned inspection methods are oriented toward application in the traditional Web development context; they are therefore principally employed in the later stages of Web development processes. As mentioned earlier, model-driven Web development offers a suitable context for early usability evaluations since it allows models, which are applied in all the stages, to be evaluated. This research line has emerged recently, and only a few works address Web usability issues, such as those of Atterer and Schmidt (2005), Abrahão and Insfran (2006), and Molina and Toval (2009).

Atterer and Schmidt (2005) proposed a model-based usability validator prototype with which to analyze models that represent enriched Web UIs. This approach takes advantage of models that represent the navigation (how the website is traversed) and the UI of a Web application (abstract properties of the page layout).

Abrahão and Insfran (2006) proposed a usability model to evaluate software products that are obtained as a result of model-driven development processes. Although this model is based on the usability subcharacteristics proposed in the ISO 9126 standard, it is not specific to Web applications and does not provide specific metrics. The same model was used by Panach et al. (2008) with the aim of providing metrics for a set of attributes that would be applicable to the conceptual models that are obtained as a result of a specific model-driven Web development process. Molina and Toval (2009) presented an approach to extend the expressivity of models that represent the navigation of Web applications in order to incorporate usability requirements. It improves the application of metrics and indicators to these models.

Nevertheless, to the best of our knowledge, there is no generic process for integrating usability evaluations into model-driven Web development processes. In this chapter, we address this issue by showing how usability evaluations can be integrated into a specific model-driven development method.

36.4 Early Usability Evaluation in Model-Driven Web Development

Currently, there are several Web development methods (e.g., WebML [Ceri et al. 2000], OO-H [Gomez et al. 2011]), environments (e.g., Teresa [Mori et al. 2004]), and languages (e.g., UsiXML [Limbourg et al. 2004]) that exploit MDE techniques to develop Web applications.

These approaches usually comprise high-level models that represent tasks that are performed by the Web application users in a way that is independent from the technology used.

The usability of a software product (e.g., a Web application) obtained as a result of a transformation process can be assessed at several stages of a model-driven Web development process. In this chapter, we suggest the use of a **Web Usability Model** that contains a set of usability attributes and metrics that can be applied by the Web designer in the following artifacts obtained in a MDA-based Web development process:

- PIM, to inspect the different models that specify the Web application independently of platform details (e.g., screen flow diagrams, screen mock-ups, screen navigation diagrams)
- PSM, to inspect the concrete design models related to a specific platform, and
- CM, to inspect the UI generated or the Web application source code (see Figure 36.1)

In order to perform usability evaluations, the Web designer should select the set of relevant usability attributes and metrics from the Web Usability Model according to the purpose of the evaluation, the development phase, and the type of artifacts (models) to be inspected. There are some usability attributes that can be evaluated independently of the platform (at the PIM level), other attributes that can only be evaluated on a specific platform and taking into account the specific components of the Web application interface (at the PSM level), and finally some attributes that can be evaluated only in the final Web application (CM).

The usability evaluations performed at the PIM level produce a *platform-independent usability report* that provides a list of usability problems with recommendations to improve PIM (Figure 36.1—1A). Changes in PIM are reflected in CM by means of model transformations and explicit traceability between models (PIM to PSM and PSM to CM). This prevents usability problems from appearing in the Web application generated.

Evaluations performed at PSM produce a *platform-specific usability report*. If PSM does not allow a Web application with the required level of usability to be obtained, this report will suggest changes to correct the following: the PIM (Figure 36.1—2A), the transformation rules that transform PIM into PSM (Figure 36.1—2B), and/or the PSM itself (Figure 36.1—3B). Nevertheless, the evaluations at the PIM or PSM level should be done in an iterative way until these models allow the generation of a Web application with the required level of usability. This allows usability evaluations to be performed at early stages of a Web development process.

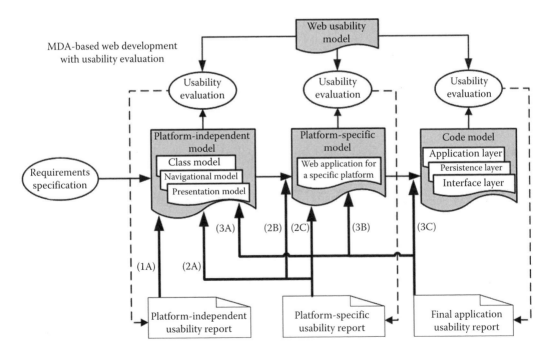

FIGURE 36.1   Integrating usability evaluations in Model-driven Web development.

Finally, evaluations performed at the CM level produce a *final application usability report*. Rather than suggesting changes to improve the final Web application, as is usual in other approaches, this usability report suggests changes to correct the PIM (Figure 36.1—3A), the PSM (Figure 36.1—3B), and/or the transformation rules that transforms the PSM into the final Web application (Figure 36.1—3C).

Our evaluation strategy is aimed at providing support to the intrinsic usability of the Web application generated by following a model-driven development process, and to the notion of usability proven by construction (Abrahão et al. 2007). Usability by construction is analogous to the concept of correctness by construction (Hall and Chapman 2002) introduced to guarantee the quality of a safety-critical system. In this development method, the authors argue that to obtain software with almost no defects (0.04% per thousand lines of code; KLOC), each step in the development method should be assessed with respect to correctness. If we can maintain proof of the correctness of a software application from its inception until its delivery, it would mean that we can prove that it is correct by construction. Similarly, if we can maintain proof of the usability of a Web application from its model specification until the source code generation, it would mean that we can prove it is usable by construction. Of course, we can only hypothesize that each model may allow a certain level of usability in the generated Web application to be reached. Therefore, we may predict the global usability of an entire Web application by estimating the relative usability levels that the models and transformations involved in a specific model-driven Web development method allow us to accomplish. We cannot prove that a Web application is totally usable (i.e., it satisfies all the usability attributes for all the users), but we can prove that it is usable at a certain level.

In the following section, we provide a brief overview of our Web usability model and show how it can be used to evaluate the usability of a Web application, early in the Web development process.

36.5 Web Usability Model

The objective of the Web usability model is to help Web designers and developers to achieve the required Web application usability level through the definition of usability characteristics and attributes, measurement of usability attributes, and evaluation of the resulting usability.

The Web usability model (Fernandez et al. 2009) is an adaptation and extension of the usability model for model-driven development processes, proposed by Abrahão and Insfran (2006). The model was adapted to be compliant with the ISO/IEC 25010 (2011) standard, also known as SQuaRE (software product quality requirements and evaluation). This standard was created for the purpose of providing a logically organized, enriched, and unified series of standards covering two main processes: software quality requirements specification and software quality evaluation. Both of these processes are supported by a software quality measurement process. SQuaRE replaces the previous ISO/IEC 9126 (2001) and ISO/IEC 14598 (1999) standards.

In order to define the Web usability model, we have paid special attention to the SQuaRE quality model division (ISO/IEC 2501n), where three different quality models are proposed: the software product quality model, the system quality in use model, and the data quality model. Together, these models provide a comprehensive set of quality characteristics relevant to a wide range of stakeholders (e.g., software developers, system integrators, contractors, and end users). In particular, the software quality model defines a set of characteristics for specifying or evaluating software product quality; the data quality model defines characteristics for specifying or evaluating the quality of data managed in software products; and the quality in use model defines characteristics for specifying or evaluating the quality of software products in a particular context of use.

The goal of the Web usability model is to extend the software quality model proposed in SQuaRE, specifically the usability characteristic, for specifying, measuring, and evaluating the usability of Web applications that are produced throughout a model-driven development process from the end user's perspective.

In the following, we introduce a brief description of the main subcharacteristics, usability attributes, and metrics of our Web usability model. The entire model, including all the subcharacteristics, usability attributes, and their associated metrics is available at http://www.dsic.upv.es/~afernandez/WebUsabilityModel.

36.5.1 Usability Attributes

SQuaRE decomposes usability into seven high-level subcharacteristics: appropriateness recognizability, learnability, operability, user error protection, accessibility, UI aesthetics, and compliance. However, these subcharacteristics are generic and need to be further broken down into measurable usability attributes. For this reason, our proposed Web usability model breaks down these subcharacteristics into other subcharacteristics and usability attributes in order to cover a set of Web usability aspects as broadly as possible. This breakdown has been done by considering the ergonomic criteria proposed by Bastien and Scapin (1993) and other usability guidelines for Web development (Lynch and Horton 2002; Leavit and Shneiderman 2006).

The first five subcharacteristics are related to user performance and can be quantified using objective metrics.

Appropriateness recognizability refers to the degree to which users can recognize whether a Web application is appropriate for their needs. In our Web usability model, this subcharacteristic was broken down by differentiating between those attributes that enable the optical *legibility* of texts and images (e.g., font size, text contrast, position of the text), and those attributes that allow information *readability*, which involves aspects of information grouping cohesiveness, information density, and pagination support. In addition, it also includes other subcharacteristics such as *familiarity*, the ease with which a user recognizes the UI components and views their interaction as natural; *workload reduction*, which is related to the reduction of user cognitive effort; *user guidance*, which is related to message availability and informative feedback in response to user actions; and *navigability*, which is related to how easily the content is accessed by the user.

Learnability refers to the degree to which a Web application facilitates learning about its employment. In our model, this subcharacteristic was broken down into other subcharacteristics such as *predictability*, which refers to the ease with which a user can determine the result of his/her future actions; *affordance*, which refers to how users can discover which actions can be performed in the next interaction steps; and *helpfulness*, which refers to the degree to which the Web application provides help when users need assistance.

Several of the aforementioned concepts were adapted from the affordance term that has been employed in the Human–Computer Interaction field in order to determine how intuitive the interaction is (Norman 1988). These subcharacteristics are of particular interest in Web applications. Users should not spend too much time learning about the Web application employment. If they feel frustrated when performing their tasks, it is likely that they may start finding other alternatives.

Operability refers to the degree to which a Web application has attributes that make it easy to operate and control. In our model, this subcharacteristic was broken down into other subcharacteristics related to the technical aspects of Web applications such as *compatibility* with other software products or external agents that may influence the proper operation of the Web application; *data management* according to the *validity of input data* and its *privacy*; *controllability* of the action execution such as cancel and undo support; *capability of adaptation* by distinguishing between *adaptability*, which is the Web application's capacity to be adapted by the user; and *adaptivity*, which is the Web application's capacity to adapt to the users' needs (i.e., the difference is in the agent of the adaptation); and *consistency* in the behavior of links and controls.

User error protection refers to the degree to which a Web application protects users against making errors. In the ISO/IEC 9126-1 (2001) standard, this subcharacteristic was implicit in the operability term. However, the ISO/IEC 25010 (SQuaRE) standard made it explicit since it is particularly important to achieve freedom from risk. In our model, this subcharacteristic was broken down into other subcharacteristics related to *error prevention* and *error recovery*.

Accessibility refers to the degree to which a Web application can be used by users with the widest range of characteristics and capabilities. Although the concept of accessibility is so broad that it may require another specific model, the SQuaRE standard added this new subcharacteristic as an attempt to integrate usability and accessibility issues. In our model, this subcharacteristic was broken down into usability attributes by considering not only a range of human disabilities (e.g., blindness, deafness), but also temporary technical disabilities (e.g., element unavailability, device dependency). The usability attributes include *magnifier support*, which indicates that the text of a Web page must be resized regardless of the options offered by the browser for this action; *device independency*, which indicates that the content should be accessible regardless of the type of input device employed (mouse, keyboard, voice input); and *alternative text support*, which indicates that the multimedia content (images, sounds, animations) must have an alternative description to support screen readers and temporary unavailability of these elements.

The last two usability subcharacteristics are related to the perception of the end user (UI aesthetics) or evaluator (compliance) using the Web application. This perception is mainly measured using subjective metrics.

User interface aesthetics refers to the degree to which UI enables pleasing and satisfying interaction for the user. This definition evolved from the attractiveness characteristic proposed in the ISO/IEC 9126 (2001) standard. Although this subcharacteristic is clearly subjective and can be influenced by many factors in a specific context of use, it is possible to define attributes that might have a high impact on how users perceive the Web application.

In our model, this subcharacteristic was broken down into other subcharacteristics related to the *uniformity* of the elements presented in UI (e.g., font, color, position); *interface appearance customizability*, which should not be confused with the subcharacteristic *capability of adaption*, since it is not related to user needs, but to aesthetic preferences; and *degree of interactivity*, whose definition was proposed by Steuer (1992) as "the extent to which users can participate in modifying the form and content of a media environment in real time." This is a concept that has recently become increasingly important owing to collaborative environments and social networks through the Web.

Compliance refers to how consistent the Web application is with regard to rules, standards, conventions, and design guidelines employed in the Web domain. In our model, this subcharacteristic was broken down in other subcharacteristics such as the degree of fulfillment to the ISO/IEC 25010 (2011) standard and other relevant usability and Web design guidelines.

36.5.2 Usability Metrics

Once the subcharacteristics and usability attributes have been identified, metrics are associated to the measurable attributes in order to quantify them. The values obtained from the metrics (i.e., measures), and the establishment of thresholds for these values, will allow us to determine the degree to which these attributes help to achieve a usable Web application. A measure is therefore a variable to which a value is assigned as the result of measurement (ISO/IEC 9126 2001).

The metrics included in this usability model were extracted and adapted from several sources: a survey presented by Calero et al. (2005), the ISO/IEC 25010 standard (2011), and the Web design and accessibility guidelines proposed by the W3C Consortium (2008). The metrics selected for our model were mainly the ones that were theoretically and/or empirically validated. Each metric was analyzed taking into account the criteria proposed in SQuaRE (e.g., its purpose, its interpretation, its measurement method, the measured artifact, the validity evidence). If the Web metric was applied to the final UI (CM), we also analyzed the possibility of adapting it to the PIM and/or PSM levels. Due to space constraints, we have only illustrated some examples of how we defined and associated metrics to the attributes of the Web usability model. Table 36.1 shows some examples of metrics for some selected usability attributes. It also shows at what level of abstraction the metric can be applied.

TABLE 36.1   Examples of Usability Metrics

Subcharacteristic/ Usability Attribute	Metric	Definition	Model
Learnability/ Predictability/Icon/ Link Title Significance	Proportion of titles chosen suitably for each icon/title	Ratio between the total number of titles that are appropriate to the link icon they are associated and the total number of titles associated with existing links or icons (scale: $0 < X \leq 1$)	PIM, PSM, or CM
Appropriateness recognizability/User guidance/Quality of messages	Proportion of meaningful messages (error, advise, and warning messages)	Ratio of error messages explaining an error in a proper way and the total contexts where this error may occur (scale: $0 < X \leq 1$)	PIM, PSM, or CM
Appropriateness recognizability/ Navigability/ Reachability	Breadth of the internavigation (BiN)	Level of breadth in the user navigation. It represents the different paths that can be selected by the user in a certain context of the user navigation (i.e., homepage, internal sections) (scale: Integer > 0)	PIM
	Depth of the navigation (DN)	Level of depth in the user navigation. It indicates the longest navigation path (without loops), which is needed to reach any content or feature (scale: Integer > 0)	PIM

36.6 Operationalizing the Web Usability Model

The operationalization of the Web usability model consists of the instantiation of the model to be used in a specific model-driven Web development method. This is done by defining specific measurement functions (calculation formulas) for the generic metrics of the usability model, taking into account the modeling primitives of the PIMs and PSMs of the Web development method.

36.6.1 Selecting the Web Development Process

In this chapter, we use the Object-Oriented Hypermedia method—OO-H (Gómez et al. 2001) as an example of operationalization of the Web usability model. OO-H provides the semantics and notation for developing Web applications. The PIMs that represent the different concerns of a Web application are the class model, the navigational model, and the presentation model. The *class model* is UML-based and specifies the content requirements; the *navigational model* is composed of a set of navigational access diagrams (NADs) that specify the functional requirements in terms of navigational needs and users' actions; and the *presentation model* is composed of a set of abstract presentation diagrams (APDs), whose initial version is obtained by merging the former models, which are then refined in order to represent the visual properties of the final UI. The PSMs are embedded into a model compiler, which automatically obtains the source code (CM) from the Web application by taking all the previous PIMs as input.

36.6.2 Applying the Web Usability Model in Practice

We applied the Web usability model (operationalized for the OO-H method) to a Task Manager Web application designed for controlling and monitoring the development of software projects in a software company.

A Web development project involves a series of *tasks* that are assigned to a *user*, who is a programmer in the company. For each task, the start date, the estimated end date, priority, etc. are recorded. The project manager organizes the tasks in folders according to certain criteria: pending tasks, critical tasks, and so on. Additionally, it is possible to attach external *files* to a task (e.g., requirements documents, models, code). Programmers can also add *comments* to the tasks and send *messages* to other programmers. Every day, programmers can generate a report (*Daily Report*), including information related to the tasks they are working on. Finally, the customers (i.e., stakeholders) of the project are recorded as *Contacts* in the Web application.

Due to space constraints, we focus only on an excerpt of the Task Manager Web application: the user access to the application, the task management, and the registration of customer's (contact's) functionalities. Figure 36.2a shows a fragment of the NAD that represents the user access to the application, whereas Figure 36.2b shows the abstract presentation model generated from the class model and the navigational model.

The NAD has a unique *entry point* that indicates the starting point of the navigation process. This diagram is composed of a set of navigational elements that can be specialized as navigational nodes or navigational links. A *navigational node* represents a view in the UML class diagram. A navigational node can be a navigational target, a navigational class, or a collection. A *navigational target* groups elements of the model (i.e., navigational classes, navigational links, and collections) that collaborate in the coverage of a user navigational requirement. A *navigational class* represents a view over a set of attributes (navigational attributes) and operations (navigational operations) of a class from the UML class diagram. A *collection* is a hierarchical structure that groups a set of navigational links. Navigational links define the navigation paths that the user can follow through UI. There are two types of links: *source links* when new information is shown in the same view (depicted as an empty arrow); and *target links* when new information is shown in another view (depicted as a filled arrow).

In Figure 36.2a, NAD shows the programmer (*User*) accessing the Web application, starting at the *User Entry Point*. The next node in the diagram is the *User* navigational class, which corresponds to a data entry form through which the programmer can log into the application. This process requires *profile* information, a *username*, and a *password*. If the user exists, which is represented by the navigational link *LI4*, the system displays a menu (*restricted home*) with links to each of the three possible navigational destinations (*Tasks, Contacts,* and *Reports*). If the user does not exist, the system shows an error message (*LI2*) returning to the initial data entry form (LI6). The *restricted home* menu is represented by a collection node, which includes the navigational links LI63, LI19, LI75, and LI28. In addition, the label *connected as* shows the log in information (LI92).

In Figure 36.2b, the abstract presentation model shows the three abstract pages generated from the previous NAD. The *User* navigational class produces a Web page showing the abstract access to the application (see Figure 36.2—b1). The navigational link *LI2* generates the error message represented by the *error* collection in a different abstract Web page (see Figure 36.2—b2). Finally, the *restricted home* collection, which is accessed through the navigational link *LI4*, generates the user's home page in a different abstract Web page (see Figure 36.2—b3), representing the possible navigational targets (*Tasks, Reports,* and *Contacts*). However, as the label *connected as,* represented by the navigational link *LI96*, is a *source link* (empty arrow), the log in information is shown in the same user's generated home page (see Figure 36.2—b3).

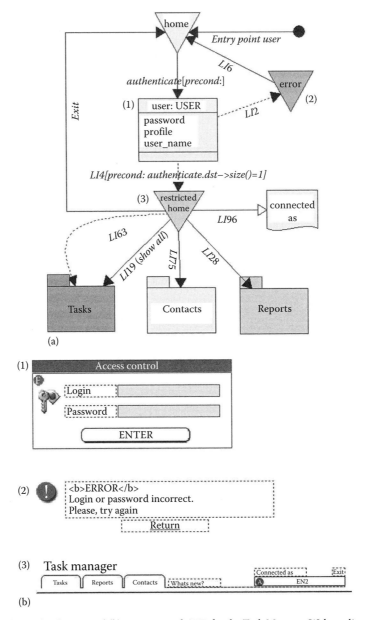

FIGURE 36.2 (a) First level NAD and (b) its associated APD for the Task Manager Web application.

The following usability metrics can be applied to the Navigational Model in Figure 36.2a:

- *Breadth of the internavigation (see Table 36.1)*: This metric can only be operationalized in those NADs that represent the first navigation level (i.e., internavigation). The calculation formula to be applied is *BiN (NAD) = Number of output target links from collections connected to navigational targets*. The thresholds* defined for the metric are [BiN = 0]: critical usability problem; [1 ≤ BiN ≤ 9]: no usability problem; [10 ≤ BiN ≤ 14]: low usability problem; and [15 ≤ BiN ≤ 19]: medium

---

* These thresholds were established considering hypertext research works such as those of Botafogo et al. (1992), and usability guidelines such as those of Leavit and Shneiderman (2006) and Lynch and Horton (2002).

usability problem. Therefore, the breadth of the internavigation for NAD shown in Figure 36.2a is 4 since there are four first-level navigational targets (i.e., Home, Tasks, Contacts, and Reports). This signifies that no usability problem was detected.

- *Depth of the navigation (see Table 36.1)*: This metric considers the number of navigation steps from the longest navigation path. Therefore, the formula to be applied is *DN (NAD) = Number of navigation steps from the longest navigation path,* where "navigation step" is when a target link exists between two nodes and "longest navigation path" is the path with the greatest number of navigation steps, which begins in the first navigational class or collection where navigation starts, and which ends in the last navigational class or service link, from which it is not possible to reach another previously visited modeling primitive. The thresholds* defined for this metric are $[1 \leq DN \leq 4]$: no usability problem; $[5 \leq DN \leq 7]$: low usability problem; $[8 \leq DN \leq 10]$: medium usability problem; and $[DN \geq 10]$: critical usability problem. Therefore, the depth of the navigation for the NAD shown in Figure 36.2a is 2, because the length of the longest navigational path between the root of the navigational map and its leaves (Tasks, Contacts, and Reports—navigational targets) passes through two navigational links. This means that no usability problem was detected since obtained values are within the threshold $[1 \leq BaN \leq 9]$.

The values obtained through these metrics are of great relevance to the navigability of a Web application. If the navigational map of the Web application is too narrow and deep, users may have to click too many times and navigate through several levels to find what they are looking for. However, if the navigational map is too wide and shallow, users may be overwhelmed due to the excessive amount of information they can access. In this context, for large Web applications that have a hierarchical structure, the recommended value is usually a *depth* of less than 5 levels and an *amplitude* of less than 9 levels. Since the navigational map of our Task Manager Web application obtains values below these thresholds, this Web application has a good navigability.

Regarding the abstract presentation model in Figure 36.2b, the following metrics, which are closer to the final UI, may be applied:

- *Proportion of meaningful messages (see Table 36.1)*: There is only one error message "Log in or password incorrect. Please try again." As the message properly explains what is wrong, the value of the metric is $1/1 = 1$. This result is very positive because values closer to 1 indicate that the error messages are very significant, thus guiding the user toward the correct use of the Web application.
- *Proportion of titles chosen suitably for each icon/title (see Table 36.1)*: As an example, we focus on the titles of the links of the APD shown in Figure 36.2b: *Enter, Return, What's new,* and *Exit.* Since all the titles make us intuitively predict the target of these links, the value for this metric is $4/4 = 1$. This result is very positive because values closer to 1 indicate that the selected titles are appropriate to the icons or links, allowing the user to predict what actions will take place.

Obviously, these metrics can be applied to each one of the abstract Web pages of an application to determine to which extent they provide proper labels and meaningful error messages. In this running example, we have used only a fragment of the Task Manager Web application. However, the Web usability model is fully applicable to a complete Web application that has been developed following a MDE approach. Although the usability model has been operationalized in OO-H, it can also be used with other model-driven Web development methods, for example, WebML (Ceri et al. 2000) or UWE (Kraus et al. 2006). The key is to relate the attributes of the Web usability model with the modeling primitives of the PIMs, PSMs, or the final code of the Web application. To establish these relationships, it is essential to understand the purpose of the generic usability metrics (to understand what concepts they are intended to measure) and then to identify which elements of the models provide the semantics needed to support them.

---

* These thresholds were established considering hypertext research works such as those of Botafogo et al. (1992), and usability guidelines such as those of Leavit and Shneiderman (2006) and Lynch and Horton (2002).

36.7 Summary

This chapter introduced a usability inspection strategy that can be integrated into specific model-driven Web development methods to produce highly usable Web applications. This strategy relies on a usability model that has been developed specifically for the Web domain and which is aligned with the SQuaRE standard to allow the iterative evaluation and improvement of the usability of Web-based applications at the design level. Therefore, this strategy does not only allow us to perform usability evaluations when the Web application is completed, but also during the early stages of its development in order to provide feedback to the analysis and design stages.

The inherent features of model-driven development (e.g., well-defined models and the traceability between models and source code established by model transformations) provide a suitable context in which to perform usability evaluations. This is due to the fact that the usability problems that may appear in the final Web application can be detected early and corrected at the model level. The model-driven development also allows the automation of common usability evaluation tasks that have traditionally been performed by hand (e.g., generating usability reports that include recommendations for improvement).

From a practical point of view, our usability inspection strategy enables the development of more usable Web-based applications by construction, meaning that each step of the Web development method (PIM, PSM, Code) satisfies a certain level of usability, thereby reducing the effort of fixing usability problems during the maintenance phase. This is in line with research efforts in the computer science community toward raising the level of abstraction not only of the artifacts used in the development process but also in the development process itself.

In this context, there are several open research challenges, mainly due to the highly evolving nature of the Internet and Web applications, and also because of the concept of usability. Although there are some underlying usability concepts that may remain immutable (e.g., the concept of navigation), there are new devices, platforms, services, domains, technologies, and user expectations that must be taken into consideration when performing usability inspections. This may lead us to determine the families of Web applications and to identify the most relevant usability attributes for these families. In addition, there is a need for the study of aggregation mechanisms to merge values from metrics in order to provide scores for usability attributes that will allow different Web applications from the same family to be compared.

36.8 Future Research Opportunities

The challenge of developing more usable Web applications has promoted the emergence of a large number of usability evaluation methods. However, most of these methods only consider usability evaluations during the last stages of the Web development process. Works such as those of Juristo et al. (2007) claim that usability evaluations should also be performed during the early stages of the Web development process in order to improve user experience and decrease maintenance costs.

This is in line with the results of a systematic review that we performed to investigate which usability evaluation methods have been used to evaluate Web artifacts and how they were employed (Fernández et al. 2011). The study suggests several areas for further research, such as the need for usability evaluation methods that can be applied during the early stages of the Web development process, methods that evaluate different usability aspects depending on the underlying definition of the usability concept, the need for evaluation methods that provide explicit feedback or suggestions to improve Web artifacts created during the process, and guidance for Web developers on how the usability evaluation methods can be properly integrated at relevant points of a Web development process.

Glossary

Model-driven engineering (MDE): A software development approach that focuses on creating and exploiting domain models (abstract representations of the knowledge and activities that govern a particular application domain), rather than on the computing (or algorithmic) concepts.

Platform-independent model (PIM): A view of a system that focuses on the operation of a system while hiding the details necessary for a particular platform.

Platform-specific model (PSM): A view of a system from the platform-specific point. A PSM combines the specifications in the PIM with the details that specify how that system uses a particular type of platform.

Usability by construction: Proof of the usability of a software application from its model specification to its generated final code.

Usability problem: The problem that users will experience in their own environment, which affects their progress toward goals and their satisfaction.

Acknowledgments

This research work is funded by the MULTIPLE project (MICINN TIN2009-13838) and the FPU program (AP2007-03731) from the Spanish Ministry of Science and Innovation.

References

Abrahão, S. and Insfran, E. 2006. Early usability evaluation in model-driven architecture environments. In *6th IEEE International Conference on Quality Software (QSIC'06)*, Beijing, China, pp. 287–294.

Abrahão, S., Iborra, E., and Vanderdonckt, J. 2007. Usability evaluation of user interfaces generated with a model-driven architecture tool. In Law, E. L-C., Hvannberg, E., and Cockton, G. (Eds.), *Maturing Usability: Quality in Software, Interaction and Value*, Springer, New York, pp. 3–32.

Andre, T.S., Hartson, H.R., and Williges, R.C. 2003. Determining the effectiveness of the usability problem inspector: A theory-based model and tool for finding usability problems. *Human Factors* 45(3): 455–482.

Atterer, R. and Schmidt, A. 2005. Adding usability to Web engineering models and tools. In *5th International Conference on Web Engineering (ICWE 2005)*, Sydney, Australia, Springer, New York, pp. 36–41.

Bastien, J.M. and Scapin, D.L. 1993. Ergonomic criteria for the evaluation of human-computer interfaces. Technical Report n.156. INRIA, Rocquencourt, France.

Blackmon, M.H., Polson, P.G., Kitajima, M., and Lewis, C. 2002. Cognitive walkthrough for the Web. In *Proceedings of the ACM CHI'02*, Minneapolis, MN, pp. 463–470.

Botafogo, R.A., Rivlin, E., and Shneiderman, B. 1992. Structural analysis of hypertexts: Identifying hierarchies and useful metrics. *ACM Transactions of Information Systems* 10(2): 142–180.

Calero, C., Ruiz, J., and Piattini, M. 2005. Classifying Web metrics using the Web quality model. *Online Information Review* 29(3): 227–248.

Ceri, S., Fraternali, P., and Bongio, A. 2000. Web modeling language (WebML): A Modeling language for designing Web sites. In *9th WWW Conference (2000)*, Amsterdam, the Netherlands, pp. 137–157.

Cockton, G., Lavery, D., and Woolrychn, A. 2003. Inspection-based evaluations. In Jacko, J.A. and Sears, A. (Eds.), *The Human-Computer Interaction Handbook*, 2nd edn., L. Erlbaum Associates, Hillsdale, NJ, pp. 1171–1190.

Conte, T., Massollar, J., Mendes, E., and Travassos, G.H. 2007. Usability evaluation based on Web design perspectives. In *1st International Symposium on Empirical Software Engineering and Measurement (ESEM 2007)*, pp. 146–155, Spain.

Fernandez, A., Insfran, E., and Abrahão, S. 2009. Integrating a usability model into a model-driven web development process. In *10th International Conference on Web Information Systems Engineering (WISE 2009)*, Poznan, Poland, pp. 497–510, Springer-Verlag, New York.

Fernandez, A., Insfran, E., and Abrahão, S. 2011. Usability evaluation methods for the web: A systematic mapping study. *Information & Software Technology* 53(8): 789–817.

Gómez, J., Cachero, C., and Pastor, O. 2001. Conceptual modeling of device-independent Web applications. *IEEE Multimedia* 8(2): 26–39.

Hall, A. and Chapman, R. 2002. Correctness by construction: Developing a commercial secure system. *IEEE Software* 19(1): 18–25.

Hwang, W. and Salvendy, G. 2010. Number of people required for usability evaluation: The 10 ± 2 rule. *Communications of the ACM* 53(5): 130–133.

ISO/IEC 14598. 1999. Information technology, Software product evaluation.

ISO/IEC 9126-1. 2001. Software engineering, Product quality—Part 1: Quality model.

ISO/IEC 25010. 2011. Systems and software engineering—Systems and software quality requirements and evaluation (SQuaRE)—System and software quality models.

Ivory, M. Y. and Hearst, M. A. 2002. Improving Web site design. *IEEE Internet Computing* 6(2): 56–63.

Juristo, N., Moreno, A., and Sanchez-Segura, M.I. 2007. Guidelines for eliciting usability functionalities. *IEEE Transactions on Software Engineering* 33(11): 744–758.

Leavit, M. and Shneiderman, B. 2006. Research-based Web design & usability guidelines. U.S. Government Printing Office, Washington, DC. http://usability.gov/guidelines/index.html

Limbourg, Q., Vanderdonckt, J., Michotte, B., Bouillon, L., and López-Vaquero, V. 2004. USIXML: A language supporting multi-path development of user interfaces. In Bastide, R., Palanque, P. A., and Roth, J. (Eds.), *EHCI/DS-VIS*, Vol. 3425 of LNCS, Springer, New York, pp. 200–220.

Lynch, P. J. and Horton, S. 2002. *Web Style Guide: Basic Design Principles for Creating Web Sites*, 2nd edn., Yale University Press, London, U.K.

Molina, F. and Toval, J.A. 2009. Integrating usability requirements that can be evaluated in design time into model driven engineering of Web information systems. *Advances in Engineering Software* 40(12): 1306–1317.

Mori, G., Paterno, F., and Santoro, C. 2004. Design and development of multidevice user interfaces through multiple logical descriptions. *IEEE TSE* 30(8): 507–520.

Nielsen, J. and Molich, R. 1990. Heuristic evaluation of user interfaces. In *Proceedings of ACM CHI'90 Conference*, Seattle, WA, pp. 249–256.

Norman, D.A. 1988. *The Psychology of Everyday Things*, Basic Books, New York.

Object Management Group (OMG). The MDA guide version 1.0.1. omg/2003–06–01. http://www.omg.org/cgi-bin/doc?omg/03–06–01 (accessed June 18, 2012).

Offutt, J. 2002. Quality attributes of Web software applications. *IEEE Software* 19(2): 25–32.

Olsina, L. and Rossi, G. 2002. Measuring Web application quality with WebQEM. *IEEE Multimedia* 9(4): 20–29.

Panach, I., Condori, N., Valverde, F., Aquino, N., and Pastor, O. 2008. Understandability measurement in an early usability evaluation for model-driven development: An empirical study. In *International Empirical Software Engineering and Measurement (ESEM 2008)*, Kaiserslautern, Germany, pp. 354–356.

Steuer, J. 1992. Defining virtual reality: Dimensions determining telepresence. *Journal of Communication* 42(4): 73–93.

Web Content Accessibility Guidelines 2.0 (WCAG 2.0). 2008. In Caldwell, B., Cooper, M., Guarino Reid, L., Vanderheiden, G. (Eds.), www.w3.org/TR/WCAG20 (accessed June 18, 2012).

37

Transforming HCI: The Art, Science, and Business of User Experience Design

William M.
Gribbons
Bentley University

Roland Hübscher
Bentley University

Over the past 30 years, our approach to designing technology has evolved in response to a dramatically changing marketplace. Many of these changes were a reaction to an ever-increasing shift in people's expectations for the use of technology. The result is a sure-to-continue cycle where technological innovation and better design elevate user expectations, further driving future design and innovation. In response to these changes, the human computer interaction (HCI) discipline must reexamine past research and professional practices and make appropriate accommodations to better serve new technologies, conditions, and market pressures. The need for this chapter should not be viewed as a failure of past efforts. Instead, this is simply a continuation of the discipline's ongoing evolution from industrial psychology to human factors and then to HCI.

This chapter examines changes in the marketplace for technology products over the past 30 years and proposes transforming our traditional HCI framework based on an expanded perspective offered by user experience (UX) research and design. This chapter will begin by examining the conditions that drove this change and then offer a more detailed review of the human elements relevant to emerging technologies as well as research necessary to meet the more sophisticated demands of the evolving technology and marketplace. The chapter concludes by proposing a process that supports each of these elements, thereby increasing the probability of a successful product design.

37.1 Changing Market Conditions

If the HCI discipline has learned anything over the last three decades, it is that we can rarely force change without the need first existing in the user community. Of course, there will always be an exception for extraordinary technological breakthroughs, but even in these cases you can typically trace the success of a given innovation to the fulfillment of an unmet human need or the delivery of unique value.

Factors influencing the success of new technology offerings have evolved dramatically over the past three decades. During the 1980s and 1990s, success was determined almost exclusively by the power, stability, and functionality of new technologies. This was the correct strategy at the time, given technology delivered value by lessening the drudgery of a previously manual activity or through previously unobtainable levels of productivity, quality, or safety. However, the competitive advantage offered by this focus alone diminished over time. In response to pressures from the marketplace, the technology community continued to produce innovative technologies with an ever-increasing focus on the goals, values, and abilities of the end user (Raskin 2000). Along this journey, the HCI discipline shifted from a role of enhancing design late in development, to influencing the early design, and finally to a strategic role of defining the design space. In an issue of *Interactions* dedicated to reimagining HCI, Panu Korhonen (2011) of Nokia captured this transformation of the profession nicely:

> In the early days, HCI people were told "Please evaluate our interface and make it easy to use." That gave way to "Please help us design this user interface so that it is easy to use." That in turn, led to a request "Please help us find what the users really need so that we know how to design this user interface." And now, the engineers are pleading with us "to look at this area of life and find us something interesting."

Increasingly, a deep and systematic focus on the user is emerging as a key competitive differentiator in the most progressive product development groups and for leading business service providers, particularly for those operating in hypercompetitive markets.

Naturally, a close consideration of the user in product design does not minimize the importance of the technological components of product development; it simply suggests that those qualities alone are rarely sufficient to ensure success in a world where powerful technology has become the norm. Figure 37.1 depicts the evolution of product design over the past three decades from the early days of a pure engineering focus to the one focused on usability and now the UX. In this model, each stage is a prerequisite for the next. Accompanying that shift is a progressively deeper integration of the UX strategy in the development and business culture. In the 1980s, the success of a product was determined almost exclusively by the power, performance, functionality, and its ability to enable work efficiently. Once these critically important goals were met and became commonplace, users came to expect this as a standard and then demanded that a given technology become more useful and usable (Veryzer and Borja de Mozota 2005). Usability was soon defined by ease-of-learning, transfer-of-learning, and a close mapping to users' abilities and usefulness focused on users' goals and what they valued (Dumas and Redish 1999). We then saw the pattern repeat again. Once usability and usefulness became best practices in the marketplace, users then demanded a

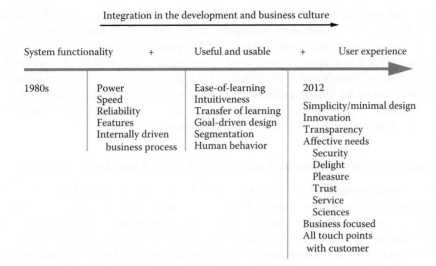

FIGURE 37.1   Evolution of product design over the past three decades.

deeper, far-reaching, and carefully orchestrated UX. A component of that stage was the move toward minimal design (simplicity) as seen in the lean start-up movement (Maeda 2006; Ries 2011), reducing the load imposed by the system, and by focusing on those aspects of the product that users value most (Dixon et al. 2010). This approach to product design stands in stark contrast to the bloated, feature-rich development strategy of the past where we looked to incremental feature enhancements to sustain a product through the market life cycle. That was a time characterized by an incomplete understanding of the requirements of the user community leaving us with no alternative but "just-in-case-design." Finally, UX considers the affective component of the experience as well as how the experience can be extended beyond the technology itself to every touchpoint with the customer (Khalid and Helander 2006; Norman 2003; Walters 2011).

Three dominant market forces contributed significantly to this evolution:

1. The evolving expectations of users
2. The expanding diversity of the user community
3. The growing application of technology in all aspects of our lives

37.1.1 Evolving Expectations of Users

A key characteristic of human behavior is that our expectations evolve based on a continually changing base of experiences. Each new experience that delivers greater value, pleasure, performance, or efficiency creates a new and elevated expectation. This behavior governs all aspects of our human experience (Eysenck 2012). In the case of technology, this change is partially a product of constantly improving technology. In a changing marketplace, user expectations evolve based on the enabling qualities of each new technology, interaction design potential, and with each increase in product usability. In short, what was once a successful "standard" of excellence in the 1990s is no longer acceptable to today's more demanding user, a trajectory that will continue in perpetuity. The lesson learned in the marketplace is clear: technology producers who recognize and respond in a timely fashion to this new product life cycle will thrive, while those who do not will fail (Christenson 1997; Martin 2009).

37.1.2 Expanding Diversity of the User Community

Further contributing to the changing marketplace is the significantly expanding diversity of users (Langdon et al. 2007). Gone are the days when we designed for a homogeneous community of well-educated, highly motivated, technologically savvy, and adaptable users. In the past, this community was required to "bend" to the demands of powerful systems that were often misaligned with their needs and abilities. Fortunately, this population was willing and able to make the required accommodations based on highly developed abilities and motivated by the value derived from the system. Today, the system must increasingly bend, to varying degrees based on context, to the user while still serving the larger purpose of the system. Today's user might be less technologically savvy, illiterate, unmotivated, cognitively or physically disabled, elderly, or living in a developing country. The topic of universal accessibility is addressed later in this chapter. User experience for "all" has raised the art and science of system and interaction design to previously unknown levels of sophistication.

37.1.3 Growing Integration of Technology in All Aspects of Our Lives

Finally, users were more capable and willing to "bend" when a limited number of technologies required this accommodation. This too has changed. In many areas of the global marketplace, technology has "penetrated" each and every aspect of our lives: our homes, communication, social communities, transportation systems, work, education, entertainment, commerce, government, and healthcare.

Our understanding of the human information processing system has long confirmed that we possess finite resources for interacting with our environment (Wickens and Hollands 1999).

As ubiquitous technology places ever-increasing demands on people, their capacity to manage this workload is challenged. If technology is to continue enriching and improving our lives, new technology must impose a lighter load by more closely aligning with the values and abilities of the user. This clearly requires designers to develop an increasingly deeper understanding of user's interactions with technology and the alignment of that interaction with the strengths and limitations imposed by the human information processing system.

37.2 Common Framework

No technology sector is insulated from these market forces. Separately, the pressures of this change have been felt in business information technology, commercial software/hardware, consumer electronics, government systems, games, social media, healthcare, entertainment, and transportation. Rather than approaching each of these sectors as a *unique* challenge requiring a *unique* framework for understanding users, a common approach can be identified and appropriate support found in the merged UX and HCI disciplines. A fractured strategy, focused on individual sectors, will ultimately weaken the larger cause and fail to fully leverage a shared research agenda.

Moving forward, the merged HCI–UX discipline can take great comfort in knowing that the traditional behavioral and perceptual foundations of the HCI discipline continue to serve us well. However, the range of behaviors and perceptual inputs will expand, and the complexity of the interaction across subvariables will increase. This is a challenge the discipline has met in response to past innovation. Expansion of the research agenda is particularly necessary in the social, emotional, aesthetic, and cultural elements of the UX. On the interaction design front, we will continue our progress to full integration of multimodal interaction—audible, haptic, and auditory (Heim 2007). Here, we must examine how a particular mode best supports an interaction requirement and manage the complex and difficult-to-predict interactions across these interaction modes.

While we work to identify a common research and practice framework to accommodate diverse industries, it remains useful to recognize that unique challenges do exist within industry sectors. For example, making the case for focusing on the customer is relatively clear in the commercial marketplace for hardware, software, and consumer electronics. More challenging is the case for business IT and government systems—these are difficult, but not impossible. In the case of IT, our past focus on business rules, work processes, productivity, and the enabling qualities of the technology often occurred at the expense of the users responsible for implementing, supporting, and using the system (Babaian et al. 2010). This often resulted in high failure rates, partial implementations, slow and incomplete adoption of the system, and unexpectedly high total life cycle costs. In the case of government systems, we have witnessed a slow and steady march to making government and healthcare available to the public through information and communication technologies (Becker 2005; McClelland 2012; Sainfort et al. 2012). Here too, acceptance and use of these systems is critically dependent on the ease of use and quality of the UX. E-government and e-health possess tremendous potential for improving people's lives; however, we will only realize this potential when the supporting systems are accessible to the larger population (Borycki and Kushniruk 2005; Croll and Croll 2007; Gribbons 2012). Any solution focused narrowly on the enabling ICT without a clear consideration of usability and the UX is likely to fail.

37.3 From HCI to UX

Building on a traditional HCI foundation, the UX promotes the careful alignment of human behaviors, needs, and abilities with the core value delivered through a product or service. Depending on context, this experience may have psychological, social, cultural, physiological, and emotional components—most likely, a combination of all five. In contrast to traditional HCI, UX broadens the scope of the human experience considered in interaction design including the emotional, social, internationalization, and accessibility. Similar to traditional HCI, the UX discipline defines the optimum experience through the detailed study and assessment of the user in the appropriate use environment. Even here we see a need to move from less formal modes of inquiry to more rigorous methods such as ethnographic inquiry. Ultimately, UX design applies these theoretical principles and research practices with a goal of exercising greater control over users' response to a design. In other words, they strive to produce predictable outcomes for the system consistent with a set of desirable, defined, and measurable goals.

37.4 UX and a Holistic View of Product Design

UX also embraces a more holistic view of product design. In the past, user requirements were often defined in a vacuum, most often assuming ideal conditions. Learning from the shortcomings of past practice, UX identifies, manages, and accommodates the various business and development forces in play that might necessarily or needlessly influence the finished product design and eventually compromise the UX.

As depicted in Figure 37.2, user goals are compared to business goals, and where the conflict between these two cannot be resolved, tensions are documented and their effects minimized in the design.

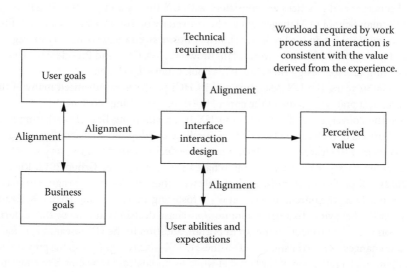

FIGURE 37.2 UX and product design model.

As goals are moved to implementation, the UX practitioner then assesses the impact of technical and regulatory requirements on the UX. Given that the regulatory requirements must be supported, tensions between those requirements and the abilities and expectations of the user must be noted and accommodated. Under ideal circumstances, technical requirements would specify leading-edge technologies that align with and support the user's expectations and abilities. In many cases—sometimes driven by business goals—older technologies are employed, existing code is repurposed, and requirements for compatibility with other systems drive technical requirements. Tensions then arise with users' expectations and abilities. In the final analysis, UX design recognizes that unavoidable tensions can exist within a system; yet, it can still produce acceptable results if these tensions are properly managed and when the perceived value of the system exceeds the workload required by the suboptimum interaction experience. The converse is true for a misaligned system that delivers marginal value at high interaction costs.

37.5 UX Meets Business

In contrast to traditional views of HCI, UX balances business and user needs given its growing contribution to competitive differentiation. As noted earlier, the "differentiation value" of the UX has grown as many technologies and business services become commoditized in hypercompetitive markets (Pine and Gilmore 1999). In this new role, we must deeply integrate UX theories and practices in the development process as well as in the strategic framing, pricing, and competitive positioning of products and services. Finally, we must recognize the contributions of UX in nontechnology-based areas such as process design, customer support, or invoicing—further expanding the influence of the discipline.

In the past, one would rarely hear discussions of HCI's contribution within the ranks of senior executives or see HCI play a prominent role in the producing organization's marketing or product positioning. This is not the case with UX where we now find the discipline prominently displayed in competitive positioning and advertising—in the most extreme cases, the experience becomes the brand as seen in companies such as Apple, Facebook, and Amazon. Recognizing that strategic value, many development organizations have subsequently promoted UX professionals to senior executive ranks.

User experience's journey up the value chain in the business organization could not have come at a more opportune time. Paralleling the advancement of the UX profession is a growing need in business to continuously innovate products, services, and processes as well as innovation's critical role in sustaining long-term competitive advantage. While the business community often frames this discussion in the context of "design thinking," a careful examination of that "movement" reveals that many of its foundational concepts and practices are consistent with UX theory and practice. Roger Martin (2009), Dean of the Rotman School of Management at the University of Toronto, builds a compelling case for the value of design thinking in his book *The Design of Business*, characterizing this strategy as the next competitive advantage in the business world. Tim Brown (2009), CEO and President of IDEO, makes an equally compelling case for design thinking in his book *Change by Design*.

For more than 30 years, the UX discipline and its HCI predecessor advanced many of the foundational principles and practices found at the core of the design thinking movement.

We also see the convergence of business and UX in an emerging line of marketing research that examines the features of technology that drive a purchase or adoption decision. Emerging research in this area suggests that the features that drive a purchase decision are often the very same features that compromise the user's experience with the purchased product (George Group 2012; Rust et al. 2006; Zhao et al. 2012). UX professionals must work closely with their marketing colleagues to strike an artful balance between driving the purchase decision while fostering a positive long-term UX. Examining the tension between the behaviors that drive adoption/purchase decisions and those that determine long-term satisfaction with the product's use is an important addition to the UX research agenda.

Each of these changes—UX to business and business to UX—is a recognition of the previously discussed market conditions and a reflection of the fact that while technological innovation alone may be adequate for the initial success of a product it is rarely adequate to sustain long-term success in the marketplace.

37.6 Ethics and Unintended Consequences

As the UX discipline extends its reach beyond the early challenges of enabling work, improving usability, extending productivity, and expanding interaction design capabilities, the discipline is increasingly likely to encounter far more complex and unexpected issues resulting from people's use of technology and a deeper understanding of the depth and breadth of the UX. Clearly, new technologies have always produced unintended secondary consequences as noted by historians (Kranzberg and Pursell 1967). What has changed is the pervasiveness of these consequences and their severity. As a consequence, the UX discipline must consider and debate its contribution to and responsibility for the unattended consequences of new technologies, some of them with a clear ethical dimension. Issues examined within this section include

- Accommodating and influencing user's behavior
- Human cost of automation and devaluing work
- Deskilling
- Erosion of privacy
- Dangers of distraction

While fueling an ongoing discussion within the profession, these issues will also contribute to the discipline's research agenda moving forward.

37.6.1 Accommodating and Influencing User's Behavior

As we deepen our understanding of human behavior, we have increased our ability to deliberately influence and control users' behavior through system design. While design has always had this effect to some degree, the implications moving forward are potentially more far-reaching and significant based on the potential of emerging technologies. A case in point is the growing desire to subtly alter the irrational behavior regularly exhibited by people.

Ariely (2008) demonstrates convincingly the endless number of occasions where people make poor choices in critical aspects of their lives, thus challenging the traditional view that people exhibit logical, rational behavior in important situations. Ariely suggests that this irrational behavior is so predictable that we need to question our traditional view of what constitutes rational behavior.

Contributing to this behavior are various emotions and biases that direct our actions in irrational ways. In response to the predictability of this suboptimum behavior, some suggest that we should direct this behavior to a more positive and beneficial outcome. Sunstein and Thaler (2008) correctly propose, based on a deep understanding of human behavior, that we can design "choice environments" that make it "easier" for people to choose what is best for them, their families, and society, thereby minimizing the possibility of irrational responses, biases, and emotions. The authors are careful to frame this proposal in a positive fashion with careful safeguards to protect one's freedom of choice while producing more positive outcomes in terms of health, financial security, and personal happiness.

Such thinking has interesting implications for UX design since it would be the designer who frames the experience—both the interaction and the information design—to produce an outcome some authority deems "in the best interest of the user." In an ideal world, this all seems well and good if there were some objective measures of which outcome was in the best interest of the user. Despite the designer's best intentions, they must be cognizant that the selections they make in creating the optimum "choice environment" are subject to the very same biases they are trying to minimize and that these biases operate outside of their conscious control. In the worst case, UX designers must also consider that for every product with the "best intention" there will be another that deliberately "nudges" the user to ends not in the user's best interest, but to those of a different stakeholder.

This creates a vexing ethical conundrum for UX designers. On the one hand, they recognize that human behavior often results in suboptimum choices and actions. On the other hand, they recognize that they have the potential, through design, to affect that behavior in other ways—positive and negative. At the very least, this begs the question: "how do UX professionals define their ethical responsibilities as they subconsciously influence users' decisions or actions?" The case of producing negative outcomes is clear, but less clear is who determines what is "positive" and the recognition that the line between the two is often not well defined.

As the UX profession gains a deeper understanding of human behavior and its implication for interaction design, they have begun to question past practices. For example, system designers typically create products assuming a rational user. Based on the goal of supporting a logical, rational response, they conduct research to determine appropriate information support, an optimum sequence of actions, optimum feedback, and ideal outcomes. This research is likely to validate the initial assumptions because the research activity engaged the conscious, "rational side" of the user's mind. This information is then captured as user requirements and scenarios that are implemented and supported through the system design. The designer is then surprised when the user—now guided by natural, emotional behaviors outside their conscious control—interacts with the system in an unexpected, suboptimum fashion (Kahneman 2011). Traditionally, interaction designers would characterize this behavior as "carelessness," "laziness," or "human error." This raises yet another challenging question: can the UX designer predict and accommodate this behavior without falling into the manipulation trap described previously?

37.6.2 Human Cost of Automation and Devaluing Work

So much of the HCI discipline's early effort was driven by the desire to improve human performance and productivity while reducing errors. Few questioned the value of these gains—gains achieved by optimizing the system design, augmenting human ability, and automating various components of the supported activity. At the same time, some in the discipline have examined the human cost of automation (Parasuraman and Mouloua 1996), a discussion not limited to the loss of jobs. While most applaud automation when it eliminates dangerous, repetitive, or tedious work, other forms of automation often come at the cost of diminishing the intellectual and emotional value of that work. In some cases, such as the level of automation found in fast food restaurants or warehouse fulfillment centers (McClelland 2012), it has reached a state where work is nearly dehumanized, opportunities for worker growth diminished, and the value of rewarding work stripped away. Given the UX professional's role in designing that experience, what responsibility does the profession have for these undesirable outcomes?

37.6.3 Deskilling

Over the past two decades, there have been tremendous advances in the development of powerful intelligent support systems that augment human intelligence in the most demanding environments. For example, modern aircraft have grown so complex that pilots can no longer operate the aircraft without an "intelligent" assistant, particularly in situations of catastrophic failure. The positive benefits of this technology enable the use of sophisticated technologies and improve safety (Charette 2010). At the same time, the UX research agenda must examine the possible "deskilling" effects of automation on the operator and evaluate the ability and need of the operator to respond in the absence of intelligent support (Zuboff 1982). Similar discussions are also present in the medical community where medical professionals, including doctors, are increasingly employing powerful diagnostic systems with the same benefits and concerns raised in aviation (Berner 2008). Moving forward, UX professionals must go beyond the positive, intended outcomes of the technologies they design and consider the unattended consequences of new technologies for the user. In some cases, the gains will clearly outweigh the negative consequences and the loss accepted. In others, the level of support and automation might warrant reconsideration. Whatever be the outcome, it is critical that UX designers initiate this conversation.

37.6.4 Erosion of Privacy

The case of remote monitoring technologies for the elderly and children provides another interesting case of unintended consequences. As life expectancies increase in many societies, so too has the attention directed to improving the quality of life for this population. With the best of intentions, technologies have been developed and employed to remotely monitor the daily activities of independently living elders—their every movement, what and how much they eat, and when they take their prescribed medications. Such advances have enabled a high-value goal of elders to live independently in a place of their choosing. While the benefits are clear, an unintended consequence is the loss of privacy and dignity—also highly valued by the population—as a consequence of monitoring beyond their control. Similar technologies are also employed by parents to monitor their children's every movement, how fast they are driving, and their location at any moment in time. Product designers create each of these technologies focused on the positive intended consequence; however, each technology is accompanied by an unintended outcome of the loss of privacy, dignity, and perhaps independence.

37.6.5 Dangers of Distraction

Finally, the convergence of technologies in many use environments produces an attention load that threatens the very limits of human capabilities. A case in point is the ever-increasing integration of communication, navigation, and entertainment technologies in automotive design. While these technologies deliver unquestionable value and pleasure to the driver and passenger, they divide the operator's attention in ways that distract from the primary driving task—a situation creating life-threatening situations for all who travel our roads and highways (Beede and Kass 2006). What responsibility do UX professionals have in situations such as these where we might focus on only the positive outcome of a technology and either intentionally or inadvertently fail to recognize possible negative consequences? The ever-increasing likelihood of distraction and its consequences should become an area of intense focus in the discipline's research agenda.

At the end of the day, UX professionals must increasingly consider where their responsibilities lie, with the product organization that reaps financial gains from the technology they sell or with the user who will suffer possible negative or potentially life-threatening consequences of these products. Because the occurrence of these events has become more frequent and the consequences more severe, the profession can no longer avoid this question.

37.7 UX for All Users

Moving forward, the UX discipline must embrace all users, not simply those of a certain ability, economic class, or region of the world. The discipline must avoid the past narrow focus on capable, economically desirable populations and instead address the needs of an increasingly diverse user community, for example, those of a poor illiterate farmer in a developing nation using ICTs to better manage the planting, harvesting, and marketing of his crops. A serious mistake was made in the past where the discipline focused on an artificial designation of the "typical user" and "typical behavior" and relegated accessibility and internationalization to the periphery—factors the designer addresses once the core produced is created. This approach communicated that each of these activities was optional and virtually eliminated them from common practice. Instead, we must seamlessly integrate universal design (Shneiderman 2000) and internationalization (Marcus and Gould 2012) into the core design in all but the most regionalized product directed to homogenous populations.

It is easy to understand why the previous approach was adopted. First, the foundational psychological research focused primarily on western subjects. Second, the cognitive/perceptual research most often assumed an optimally functioning individual. Third, the early marketplace for technology and the HCI profession's connection to this marketplace initiated and perhaps perpetuated this bias. Because technology now touches users of unimaginable diversity in every corner of the world, the discipline's past approach is no longer acceptable either in theory or practice.

37.7.1 Global

The UX movement must become a movement for all users, not simply an experience based on the values, beliefs, and behaviors of one geographic region or the other. The discipline has long understood that the sensory component of the human experience is the same in almost every practical way. In other words, the register of colors, sounds, and touch sensations is essentially universal (Ware 2000). In contrast, how these signals are encoded and acted upon during higher levels of cognitive processing will vary across cultures. In some cases, there will be significant differences (Matsumoto and Juang 2008). Over the past several decades, research in the field of intercultural psychology has offered invaluable insight into cultural difference in areas such as communication, creative thinking (Paletz and Peng 2008), collaboration, learning (Li 2003), evaluation, decision making (Briley et al. 2000), and the like (Berry et al. 2011; Kamppuri et al. 2006). This research has implications not only for the internationalization of the interaction experience but also for the research methods we use to study users and how we interpret those findings.

37.7.2 Universal Design

As discussed earlier, a distinctive characteristic of today's technology marketplace is the incredible diversity of the user community, with an ever-increasing likelihood that the user might be illiterate, cognitively disabled (Gribbons 2012), physically handicapped (Jacko et al. 2012), or elderly (Czaja and Lee 2008). The UX academy and profession must be at the vanguard of creating an inclusive environment where all users fully realize the remarkable benefits of technology. This is especially important given technology is interwoven into the very fabric of our lives at home, work, civic engagement, wellness, and play. For years, the HCI profession led the crusade for addressing the needs of special populations, first those with physical disabilities, then the unique needs of the elderly, and finally those with cognitive disabilities. This crusade eventually coalesced under the universal design moniker. The UX movement must advance this cause by more deeply and seamlessly integrating universal design principles into the core tenets of the discipline.

The first principle of universal design dictates "provide the same means of use for all users: identical if possible—equivalent when not" (Story 1998). When executed properly, the required accommodations are transparently embedded in the system, interface, and interaction design and are unlikely to affect development costs or create obstacles for the nondisabled. It is widely accepted in the accessibility community that the required accommodations typically improve the usability of the product for every user (Dickinson et al. 2003; Gribbons 2012). If all people are to have equal opportunities to become contributing members of society and fully participate in civic activities, the UX discipline must ensure that technology is accessible to all. This goal is especially critical in the areas of e-health, e-government, and the economic benefits afforded by technology in the workplace.

37.8 User Experience Design

As noted earlier in this chapter, HCI has evolved from a rather narrow testing-focused human factors (HF) discipline to the much broader design approach of UX. New technologies, novel user interfaces, and the expanding number of relevant contexts require an increasing set of design solutions and research that informs them. Information technology developed many innovations, with some emerging as "game changers" for the field. This section will examine some of these innovations and consider how they contributed to the emerging UX discipline. Innovations reviewed include

- Commercialization of the web
- Social networks
- Mobile computing
- Multimodal input
- Applications to apps
- Games

In the mid-1980s, computers were stand-alone machines used mainly for office-related tasks like text processing, creating simple graphics, and computing with spreadsheets (Gentner and Nielsen 1996). They were accompanied by a set of manuals often just as large and heavy as the computers themselves. Relatively few games were available and, since neither the necessary advanced network infrastructure nor the World Wide Web (WWW) was yet developed, almost nobody was connected to the Internet.

In such a context, where the focus was on individual users performing a task efficiently with minimal errors, HF methods were appropriate. These methods could be used to measure whether the necessary functionality was present and accessible to the user, and whether the tasks could be executed efficiently. Naturally, the focus was on the psychological and physiological characteristics of the user and the functionality and efficiency of the computer interface.

Interestingly, as simple as these interfaces may appear in the light of today's advanced ubiquitous computing devices, we cannot claim that the design of these so-called simple interfaces was ever mastered. A brief look at the interfaces in our daily surroundings will illustrate very quickly that the field still struggles to get even the simple interfaces right.

37.9 World Wide Web

The WWW, itself enabled by the existence of the Internet, was more of an enabler than a game changer. Websites started to sprout here and there, but the mostly academic users did not do much more than share documents. The interfaces with their static pages were rather a step backward, and this minimal interaction did not require HCI to do anything special (Myers 1998). This was all changed by the commercialization of the WWW and social networks.

37.9.1 Commercialization of the Web

As early as 1994, one "could even order pizza" on the web as Press (1994) somewhat incredulously stated. Online businesses slowly realized that websites had to appeal to customers in ways similar to brick-and-mortar stores. The shopping experience had to become appealing, or at least not upsetting. This included making sure that transactions were not only safe using secure web protocols, but that the customers trusted the businesses enough to hand over their credit card numbers. Furthermore, many of these online businesses had to deal with highly diverse customers that became increasingly global over time. Many sadly learned the bitter lesson that an attractive pitch to one group of customers may be offensive to another.

When is an interface like a website trustworthy (Hampton-Sosa and Koufaris 2005; Kim and Benbasat 2010)? This was an uncharted territory for HCI given its previous narrow focus on human–machine interactions. Furthermore, too often, trust and risk are predominantly addressed from a technology perspective (Camp 2000) even though that approach alone is problematic. Users need to decide whether they trust the people behind the technology. Websites had to reduce the appearance of risk, build long-term relationships, and build the reputation of the e-commerce business (Lanford and Hübscher 2004; Lynch 2010). These are issues not easily approached with traditional HF methods alone. Social interactions between individuals and businesses became very important.

The trustworthiness of a site cannot be easily tested in a lab setting because, for instance, people will spend their own money very differently than the experimenter's fake money.

37.9.2 Social Networks

The WWW also enabled online communities (Erickson 2011), typically focused on specific issues or goals, and social networks, which are much more focused on individuals (Howard 2010). Commercial sites tend to automate as much as possible and use as few as possible people on the business side of the interaction. Thus, whereas the interaction between people is often at the center of social networks, between a business and its customers, the interaction is a necessary but only secondary design focus. As a result, person-to-person interaction on business sites is often less than optimally designed, sometimes to such a degree that it discourages communication altogether.

In social networks, UX no longer focuses on individual users alone, but must now be designed for social interaction between individuals. Some individuals know each other, some do not. Privacy issues become important as it may not be clear to all users what information is accessible to whom. Real-world analogies are often severely flawed and may result in incorrect expectations, complicating these issues (Gentner and Nielsen 1996; Neale and Carroll 1997). For example, it is very easy to be linked—maybe indirectly—to people one is not aware of. These indirect connections often have no consequences in the real world, but this may not be the case in the virtual space. Do users realize this, given that the conceptual model of these linkages is often less than clear? Interestingly, some interfaces may even take advantage of these confusions. However, this could be considered unethical and not consistent with the goals of UX design.

With the incredible growth of online communities and networks into millions of users, the diversity of the users has also increased (Wade 2011). Differences across these users are no longer limited to interests and languages but also include different alphabets, social norms, and educational and literacy levels, to name just a few. As discussed earlier, accommodating such multidimensional diversity is a huge problem, considering that designing for a narrowly and well-defined audience tends to be difficult enough.

Because of such web-enabled interactions, UX needs to deal more with interactions between people including conceptual models of complex social interactions as they are defined by the various social networks. Privacy issues have also become increasingly more important given their legal implications.

37.10 Mobile Computing

Mobile computing was enabled during the mid-1990s by the availability of wireless networks and was initially mostly restricted to simple cell phones (Pierre 2001). Nevertheless, cell phones were soon commonplace all over the world including third-world countries. With the arrival of smart phones, their access to all the content on the WWW and many other services on the Internet through apps, the cell phone morphed into a general computing device enabling true ubiquitous computing. For many users, making phone calls has become a secondary function of these devices.

37.10.1 Integration in Daily Life

Over 87% of the world population had a mobile subscription for a cell phone in 2011 (Ekholm and Fabre 2011). The availability of smart phones is much lower at the moment at 17% of the population but growing quickly, and already two thirds of Americans between the ages of 24 and 35 own a smart phone (Nielsen 2012). Besides smart phones, many other interfaces use mobile computer technology and are neither considered computers including car dashboards, TVs, thermostats, and watches. Computational interfaces have become part of many people's daily life so transparently that it sometimes takes a power outage to realize they are now an inseparable component of our lives.

37.11 Multimodal Input

Frequently, new technologies like the WWW or mobile computing are the driving or enabling forces of new interfaces resulting in a challenge within the UX community to find appropriate ways to deal with them. Such technologies consist of hardware like ambient displays, augmented reality, haptic interfaces, and affective computing, as well as software like adaptive interfaces and recommender systems. Early implementations of new interfaces based on technological progress alone are typically rather awkward to use.

It is important that UX does not simply react to new technologies, hardware or software. From the very beginning of the innovation and design process, UX must be part of shaping how to use them and how to design optimum UXs enabled by new technologies (Sproll et al. 2010). User experience design, as the name implies, must be integrated early in the design process and not limited to testing before release. Thus, the UX community must be part of the development of new interfaces. The members of this community must help discover and adopt new interaction methods and technologies to improve the existing UX for a more diverse group of users who are demanding innovative interfaces, not just incremental improvements.

Computer input in the past was limited to keyboards and mice. However, this too is changing rapidly because they are too large and cumbersome to support ubiquitous computing. Many tablet computers and smart phones avoid the use of physical keyboards and rely heavily on gestures, voice input, and virtual keyboards. Most of these alternatives to physical keyboards and mice, like voice and gestures input, are often ambiguous and imprecise and therefore, cannot be taken literally. They need to be interpreted often based on the user's input history, his/her preferences, the physical location, time of day, and so on.

Users have also adapted to the new, somewhat limited interfaces. For instance, they have developed new languages because of message-length restrictions in SMS messaging on cell phones, microblogging, and online chat applications (Honeycutt and Herring 2009).

The old interaction paradigms need to be completely revised by shifting the tedious operations to the computer. Intelligent support can provide the necessary methods for multimodal user interfaces. Direct manipulation requires the user to implement the tasks using mouse and keyboard, but this is no longer necessary. For instance, take the remote control. We used to have to "program" them so that we could watch the show later, then this got reworded to "recording" a show, and finally, the universal remote managed to figure out for us which devices to turn on if we want to listen to a CD or watch a TV show. A goal-oriented approach lets us state our goal, and the system will take care of the rest (Faaborg and Lieberman 2006). For instance, telling the system that we want to watch the news will result in different choices, for instance, depending on the person asking for it and the time of day.

Intelligent approaches are necessary to support emerging interfaces. Intelligent agents interpret imprecise and ambiguous inputs; they reason about what actions need to be executed proactively; they execute certain tasks autonomously; and they can create individualized experiences by adapting the interface to the user. This approach is in contrast to the typical direct-manipulation approach, where the user has to tediously implement an action one simple step after another. Adaptive interfaces are already individualizing experiences in educational applications including museum exhibition guides and in many recommender systems used in e-commerce sites (Aroyo 2011; Brusilovsky 2012; Brusilovsky et al. 2010).

However, many of these intelligent user interfaces require a user model consisting not only of information explicitly provided by the user but also behavior data collected by the interface. Related to the earlier discussion of unintended consequences, this raises serious privacy issues that need to be taken into consideration during the design process; they cannot be treated simply as an afterthought.

New technologies open up completely new ways to interact with computational devices as well as with the real world via computational devices. Considering how tricky it can be to develop a "normal" user interface, it is not surprising that applications of new interface technologies are often initially suboptimal. This should not keep us from using them. However, it is critical that the UX community actively participates in helping to shape the use of new technologies in ways consistent with the abilities and goals of people.

The UX with many of these new technologies can be aptly described as reality-based interaction (Jacob et al. 2008). The goal is to go beyond the stale WIMP (window, icon, menu, and pointing device) interfaces. Ubiquitous computing implies that we can access the Internet from anywhere at any time, which we often do with laptops or smart phones. However, any computational device can take advantage of this anywhere/anytime access of information. With tangible interaction (Hornecker 2009; Shaer and Hornecker 2010), information is not just displayed on a screen but can be integrated into the real-world interaction. Leveraging the infinite variety found in nature for displaying and interacting with information is important, that is, the designers need to consider existing natural and artificial objects instead of adding yet another rectangular screen. Why is a classroom wall just a wall to hold paint and separate groups of learners? Why is everything in stores, museums, and office buildings labeled with static paper labels? To find innovative answers to such questions, UX designers need to consider the whole body of a person and also physical spaces, thus moving into the realm of architecture (Alexander et al. 1977). The arts also need to be included when designing large public displays and, especially, ambient displays. Ambient displays show information in a nonintrusive way, allowing the user to perceive it preattentively with little cognitive load (Mankoff et al. 2003). This is quite different from the usual design elements in more traditional user interfaces, and therefore, we cannot simply apply the same design and evaluation heuristics (Mankoff et al. 2003; Pousman and Stasko 2006). Ambient displays are sometimes used to change the behavior of the user (Jafarinaimi et al. 2005), something that is often explicitly avoided in more traditional systems where the business process is to be supported, not disrupted.

For a long time, user interfaces have been rather restrictive: it is easy for the device to interpret a key press or a mouse click; however, it is not trivial for a user to translate some complex goal into a lengthy sequence of digital actions. As the computer becomes more and more invisible (Norman 1998), multi-modal interaction will become more important, enabling design to become really user-centered for all users, no matter what their specific needs are, and in all social and physical contexts.

37.12 From Applications to Apps

With mobile computing, the traditional focus on large software applications with ever-expanding functionality has shifted to much smaller, highly focused apps. This move has a major impact on the design, development, and marketing of software.

Apps typically focus on a narrow, well-defined set of functions normally targeting an audience with similar goals. As a consequence, the interface can target the audience better, and the conceptual model of an app can be more coherent. The functionality needs to be restricted and highly focused on the tasks targeted by the specific app (Wroblewski 2011). As a result, the relative simplicity of apps, as compared to the highly complex applications used on desktop and laptop computers, allows small teams, even individuals, to design, develop, and market a software product, potentially to a worldwide market.

Due to the narrow focus of apps, users need only buy the functionality they are interested in, which is then reflected in a lower and more appropriate pricing model than when paying for lots of functionality that is of no interest to the user (Muh et al. 2011). For instance, some text processors on the desktop contain every functionality imaginable, plus the kitchen sink. On a tablet or smart phone, this functionality is separated out into specific apps providing a dictionary, an outliner, or a bibliography database, to name a few. The new pricing model, where the user pays proportionately to use, is terribly disruptive to traditional design and pricing models. However, to sell an app, the interface needs to market itself within minutes or even seconds to the potential buyers, as they frequently make a purchase decision not based on a long evaluation but based on word-of-mouth and a quick-and-dirty evaluation at purchase time. Thus, it is important for apps to be easy-to-use with a very flat learning curve supported transparently within the application. Learning is seamlessly embedded, similar to games, and the interaction is using gestures more or less based on physical interactions in the real world. However, such natural user interfaces are not necessarily the panacea for all mobile interactions, at least not in its current state (Norman and Nielsen 2010). Most gestures have to be learned and differ from culture to culture.

37.13 Games

Although computer games are as old as computers themselves, they have interesting characteristics not found, unfortunately, in many other computational user interfaces. Computer games used to be text-based but soon evolved into complex, sometimes massively multiuser, games with highly interactive, graphical user interfaces. Today, games are everywhere, part of social networks, mobile platforms, but also in the physical world where the boundaries between physical and computation world start to blur.

A wide range of games are played on smart phones by people all over the world with many different sociocultural backgrounds. Yet, most of these users do not seem to have serious problems learning to play these games, even though the interfaces are frequently rather complex. Although the learning curves may be just as steep as in other complex software, there seems to be a much better support of the user to overcome this obstacle. The immersive character of games often results in users trying harder at mastering the software application than when having to solve an unauthentic, imposed problem (Rooney 2007). Furthermore, in games, users are typically scaffolded (Puntambekar and Hübscher 2005) with context-sensitive support built-in and not separate from using the software to accomplish a task (Wroblewski 2011).

When analyzing tasks supported by an interface, it is important to also understand the users' goals and why they want to complete that task. While the goals for playing games are often intrinsic, the ones for doing many work-related tasks are not. They are imposed from the outside, and as a result, it may not be surprising that gamers may be more willing to spend some effort to learn a nontrivial user interface. Furthermore, the social interaction between the users is often characteristic in helping newcomers to get more proficient at playing. These observations should not be used as an excuse for missing support for novice users of a product, but as a suggestion of where to look for improved ways of helping the user to master an interface.

Although most UX designers do not design games, we should attempt to appropriately apply lessons learned within the game-design community to a broader class of interfaces. The mechanics-of-games guidelines (Sicart 2008) that drive an enjoyable game-playing experience can be adapted to the design of nongame user interaction (Zichermann and Cunningham 2011). Unlike most software that requires training or frequent trips to online documentation, games rarely require users to go through a tutorial. Support is seamlessly integrated as part of the game-playing activities. Such scaffolding needs to be added to all nontrivial interfaces, not just educational ones. Furthermore, many educational applications are not very interactive, and students rarely become as immersed in them as in games.

37.14 UX Theory and Practice: Looking Forward

The pace of new technological developments and their integration in the lives of an increasingly diverse user community will continue to accelerate. User experience research and design must continue to evolve accordingly. Again, the need to change should not be viewed as a failure of past practice, but rather seen favorably as a discipline's response to the ever-shifting demands of new technologies, enhanced interaction possibilities, and expanded user communities. This future produces challenges for the discipline and profession on a number of different levels as considered in this chapter:

- Address the needs of new technologies, user populations, and an expanded array of human behaviors in academic programs in HCI, human factors, and UX design.
- Expand the research agenda to address issues surrounding a more diverse user population and newly encountered behaviors. Most especially, this expansion should include the seamless integration of a more global research base and the fundamental tenets of universal design.
- Appreciate the intersection of UX with the larger business enterprise including, but not limited to marketing and competitive advantage in increasingly competitive markets.

- Understand more deeply how this strategic role shapes the research agenda moving forward and informs the practice.
- Expand the guiding research for new modes of interaction and their possible effects, both positive and negative, in the UX. From a purely interaction design perspective, the interaction experience has finally realized its full multimodal potential involving virtually all of the human senses. In practice, the UX profession must develop a systematic approach to optimizing the multimodal interaction experience, delivering highest possible value with minimal negative consequences.
- Consider the profession's ethical role and responsibilities related to the consequences of the technologies they design. There is no easy or clear answer here; the important thing is that the profession continues to consider these issues and not blindly focus on the intended outcomes.

This agenda should not be viewed as negative or threatening to the past focus of the HCI discipline. Instead, this expanded view has provided exciting new research opportunities and increased recognition of the value the discipline offers the user as well as the product design organization.

In practice, a successful product or service must encompass each of the elements discussed in this chapter. User experience professionals must understand the relevance of market conditions, appreciate how those conditions shift over time, and appropriately coordinate a response. They must recognize and leverage the implications of game-changing technologies, appreciate the expanding array of cognitive and perceptual factors associated with those technologies, weigh these elements appropriately based on an actual user population, use context and value realized. And finally, the UX professional must balance the human element against business, regulatory, marketing, and technical requirements. Figure 37.3 frames a process that considers each of these elements.

Beginning with a deep knowledge of human behavior and ability traditionally captured in HCI, the UX professional complements this knowledge with research gathered through user research identifying appropriate contextual factors. From this effort, they glean insights that lead to design innovation and implementation. Finally, they anticipate and accommodate the changing demands of the marketplace through a deliberate and systematic focus on sustaining innovation.

In this model, UX is seen as an art, science, and business. Through this model, the discipline will define their research agenda and professional practice moving forward. By deeply integrating and managing this process, they can avoid the narrow focus of the past. Instead, the discipline will greatly increase the probability of a carefully orchestrated UX, ultimately delivering value to both user and business.

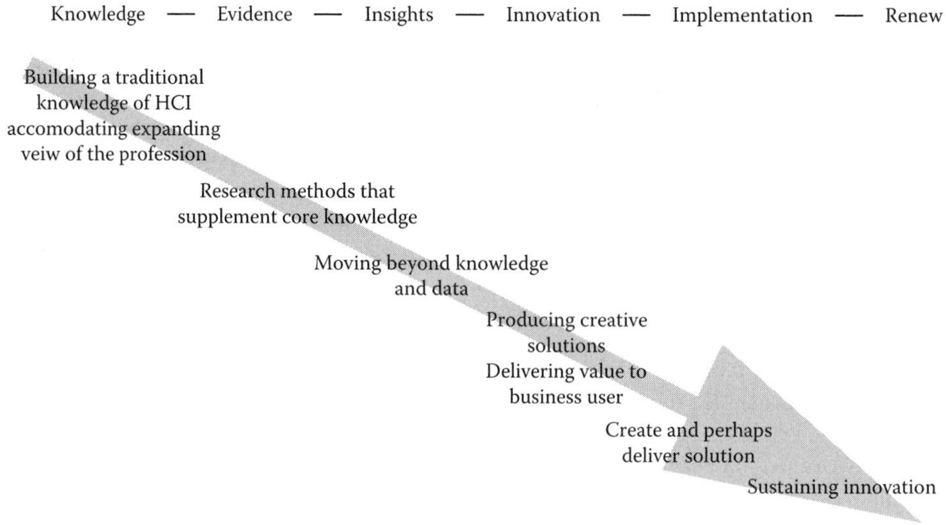

Knowledge — Evidence — Insights — Innovation — Implementation — Renew

Building a traditional
knowledge of HCI
accomodating expanding
veiw of the profession

Research methods that
supplement core knowledge

Moving beyond knowledge
and data

Producing creative
solutions
Delivering value to
business user

Create and perhaps
deliver solution

Sustaining innovation

FIGURE 37.3   UX research and design process.

References

Alexander, C., S. Ishikawa, M. Silverstein 1977. *A Pattern Language: Towns, Buildings, Construction*. London, U.K.: Oxford University Press.

Ariely, D. 2008. *Predictably Irrational*. New York: Harper Collins.

Aroyo, L., F. Bohnert, T. Kuflik, J. Oomen 2011. Personalized access to cultural heritage (PATCH 2011). Paper presented at the meeting of the *Proceedings of the 16th International Conference on Intelligent User Interfaces*, New York.

Babaian, T., W. Lucas, J. Xu, H. Topi 2010. Usability through system-user collaboration: Design principles for greater ERP usability. Paper presented at the meeting of the *Proceedings of the 5th International Conference on Global Perspectives on Design Science Research*, St. Gallen, Switzerland.

Becker, S. A. 2005. E-government usability for older adults. *Communications of ACM* 48(2): 102–104.

Beede, K. E., S. J. Kass 2006. Engrossed in conversation: The impact of cell phone on simulated driving performance. *Accident Analysis and Prevention* 38(2): 415–421.

Berner, E. 2008. Ethical and legal issues in the use of health information technology to improve patient safety. *HEC Forum* 20(3): 243–258.

Berry, J., Y. Poortinga, S. Breugelmans, A. Chasiotis 2011. *Cross Cultural Psychology: Research and Applications*. Cambridge, MA: Cambridge University Press.

Borycki, E., A. Kushniruk 2005. Identifying and preventing technology induced error using simulations: Application of usability engineering techniques. *HealthCare Quarterly* 8: 99–105.

Briley, D., M. Morris, I. Simonson 2000. Reasons as carriers of culture: Dynamic versus dispositional models of cultural influence on decision making. *Journal of Consumer Research* 27: 157–178.

Brown, T. 2009. *Change by Design: How Designing Thinking Transforms Organizations and Inspires Innovation*. New York: Harper Business.

Brusilovsky, P. 2012. Adaptive hypermedia for education and training. In *Adaptive Technologies for Training and Education*, eds. P. J. Durlach and A. M. Lesgold, pp. 46–68. Cambridge, MA: Cambridge University Press.

Brusilovsky, P., J.-W. Ahn, E. Rasmussen 2010. Teaching information retrieval with Web-based interactive visualization. *Journal of Education for Library and Information Science* 51(3): 187–200.

Camp, L. J. 2000. *Trust and Risk in Internet Commerce*. Cambridge, MA: MIT Press.

Charette, R. N. 2010. Automated to death. *IEEE Spectrum*. http://aerosrv.cls.calpoly.edu/dbiezad/ Aero_Courses/Aero_420/2012%20ASSIGNMENTS%20420/Automated_to_Death_Spectrum.pdf (accessed on May 21, 2012).

Christensen, C. 1997. *The Innovator's Dilemma*. Cambridge, MA: Harvard Business School Press.

Croll, P., J. Croll 2007. Investigating risk exposure in e-Health systems. *International Journal of Medical Informatics* 76(5): 460–465.

Czaja, S., C. Lee 2008. Older adults and information technology: Opportunities and challenges. In *The Human Computer Interaction Handbook*, eds. J. Jacko and A. Sears. Mahwah, NJ: Lawrence Erlbaum.

Dickinson, A., R. Eisma, P. Gregor 2003. Challenging interfaces/redesigning users. Paper presented at the meeting of the *Proceedings of the 2003 Conference on Universal Usability*, New York.

Dixon, M., K. Freeman, N. Toman 2010. Stop trying to delight your customers. *Harvard Business Review* 88: 116–122.

Dumas, J., J. Redish 1999. *A Practical Guide to Usability Testing*. Portland, OR: Intellect.

Ekholm, J., S. Fabre 2011. Gartner says worldwide mobile connections will reach 5.6 billion in 2011 as mobile data services revenue totals $314.7 billion. http://www.gartner.com/it/page.jsp?id = 1759714 (accessed on March 22, 2012).

Erickson, T. 2011. Social computing. In *Encyclopedia of Human-Computer Interaction*, eds. M. Soegaard and R. F. Dam. Aarhus, Denmark: The Interaction-Design.org Foundation.

Eyseneck, M. 2012. *Fundamentals of Cognition*. Hove, U.K.: Psychology Press.

Faaborg, A., H. Lieberman 2006. A goal-oriented web browser. Paper presented at the meeting of the *Proceedings of the SIGCHI Conference on Human Factors in Computing Systems*, New York.

Gentner, D., J. Nielsen 1996. The anti-Mac interface. *Communications of ACM* 39(8): 70–82.

George Group 2012. Unraveling complexity in products and services. http://knowledge.wharton.upenn.edu/index.cfm?fa = SpecialSection&specialId = 45 (accessed on March 21, 2012).

Gribbons, W. M. 2012. Universal accessibility and low literacy populations: Implications for HCI, design, and research methods. In *The Human Computer Interaction Handbook*, eds. J. Jacko and A. Sears, pp. 913–932. Mahwah, NJ: Lawrence Erlbaum.

Hampton-Sosa, W., M. Koufaris 2005. The effect of Web site perceptions on initial trust in the owner company. *International Journal of Electronic Commerce* 10(1): 55–81.

Heim, S. 2007. *The Resonant Interface*. Reading, MA: Addison Wesley.

Honeycutt, C., S. C. Herring 2009. Beyond microblogging: Conversation and collaboration via Twitter. Paper presented at the meeting of the *Proceedings of the 42nd Hawai'i International Conference on System Sciences (HICSS-42)*, Big Island, HI.

Hornecker, E. 2009. Tangible interaction. http://www.interaction-design.org/encyclopedia/tangible_interaction.html (accessed on March 22, 2012).

Howard, T. 2010. *Design to Thrive: Creating Social Networks and Online Communities That Last*. San Francisco, CA: Morgan Kaufmann Publishers Inc.

Jacko, J. A., V. K. Leonard, M. A. McClellan, I. U. Scott 2012. Perceptual impairments. In *The Human Computer Interaction Handbook*, eds. J. Jacko and A. Sears, pp. 893–912. Mahwah, NJ: Lawrence Erlbaum.

Jacob, R. J., A. Girouard, L. M. Hirshfield, M. S. Horn, O. Shaer, E. T. Solovey, J. Zigelbaum 2008. Reality-based interaction: A framework for post-WIMP interfaces. Paper presented at the meeting of the *Proceedings of the 26th Annual SIGCHI Conference on Human Factors in Computing Systems*, New York.

Jafarinaimi, N., J. Forlizzi, A. Hurst, J. Zimmerman 2005. Breakaway: An ambient display designed to change human behavior. Paper presented at the meeting of the *CHI'05 Extended Abstracts on Human Factors in Computing Systems*, New York.

Kahneman, D. 2011. *Thinking Fast and Slow*. New York: Farrar, Straus and Giroux.

Kamppuri, M., R. Bednarik, M. Tukiainen 2006. The expanding focus of HCI: Case culture. Paper presented at the meeting of the *Proceedings of NordiCHI 2006: Changing Roles*, Oslo, Norway.

Khalid, H., M. Helander 2006. Customer emotional needs in product design. *Concurrent Engineering* 14(3): 197–206.

Kim, D., I. Benbasat 2010. Designs for effective implementation of trust assurances in Internet stores. *Communications of ACM* 53(2): 121–126.

Korhonen, P., L. Bannon 2011. Reimagining HCI: Toward a more human-centered perspective. *Interactions* 18(4): 50–57.

Kranzberg, M., C. Pursell 1967. *Technology in Western Civilization*. London, U.K.: Oxford University Press.

Lanford, P., R. Hübscher 2004. Trustworthiness in E-commerce. Paper presented at the meeting of the *Proceedings of the 42nd Annual ACM Southeast Conference*, Huntsville, AL, pp. 315–319.

Langdon, P., T. Lewis, J. Clarkson 2007. The effects of prior experience on the use of consumer products. *Universal Access in the Information Society* 6(2): 179–191.

Li, J. 2003. U.S. and Chinese cultural beliefs about learning. *Journal of Educational Psychology* 95(2): 258–267.

Lynch, P. J. 2010. Aesthetics and trust: Visual decisions about web pages. Paper presented at the meeting of the *Proceedings of the International Conference on Advanced Visual Interfaces*, New York, pp. 11–15.

Maeda, J. 2006. *The Laws of Simplicity*. Cambridge, MA: MIT Press.

Mankoff, J., A. K. Dey, G. Hsieh, J. Kientz, S. Lederer, M. Ames 2003. Heuristic evaluation of ambient displays. Paper presented at the meeting of the *Proceedings of the SIGCHI Conference on Human Factors in Computing Systems*, New York, pp. 169–176.

Marcus, A., E. W. Gould 2012. Globalization, localization, and cross-cultural user interface design. In *The Human Computer Interaction Handbook*, eds. J. Jacko and A. Sears, pp. 341–366. Mahwah, NJ: Lawrence Erlbaum.

Martin, R. 2009. *The Design of Business: Why Design Thinking is the Next Competitive Advantage*. Cambridge, MA: Harvard Business School Press.

Matsumoto, D., L. Juang 2008. *Culture and Psychology*. Belmont, CA: Thompson.

McClelland, M. 2012. I was a warehouse wage slave. http://www.motherjones.com/politics/2012/02/mac-mcclelland-free-online-shipping-warehouses-labor (accessed on March 18, 2012).

Muh, F., A. Schmietendorf, R. Neumann 2011. Pricing mechanisms for cloud services: Status quo and future models. Paper presented at the meeting of the *Proceedings of the 2011 Fourth IEEE International Conference on Utility and Cloud Computing*, Washington, DC, pp. 202–209.

Myers, B. A. 1998. A brief history of human-computer interaction technology. *Interactions* 5(2): 44–54.

Neale, D. C., J. M. Carroll 1997. The role of metaphors in user interface design. In *Handbook of Human-Computer Interaction*, eds. M. Helander, T. K. Landauer and P. Prabhu, pp. 441–458. Amsterdam, the Netherlands: Elsevier Science.

Nielsen, J. 2011. Nielsen survey: New U.S. smartphone growth by age and income (electronic document), February 20, 2012. (Date retrieved: March 22, 2012).

Norman, D. A. 1998. *The Invisible Computer: Why Good Products Can Fail, the Personal Computer Is So Complex, and Information Appliances Are the Solution*. Cambridge, MA: MIT Press.

Norman, D. A., 2003. *Emotional Design: Why We Love (or Hate) Everyday Things*. New York: Basic Books.

Norman, D. A., J. Nielsen 2010. Gestural interfaces: A step backward in usability. *Interactions* 17(5): 46–49.

Paletz, S., K. Peng 2008. Implicit theories of creativity across cultures. *Journal of Cross-Cultural Psychology* 39: 286–302.

Parasuraman, R., M. Mouloua 1996. *Automation and Human Performance*. New York: Psychology Press.

Pierre, S. 2001. Mobile computing and ubiquitous networking: Concepts, technologies and challenges. *Telematics and Informatics* 18(2–3): 109–131.

Pine, J., J. Gilmore 1999. *The Experience Economy*. Cambridge, MA: Harvard Business School Press.

Pousman, Z., J. Stasko 2006. A taxonomy of ambient information systems: Four patterns of design. Paper presented at the meeting of the *Proceedings of the Working Conference on Advanced Visual Interfaces*, New York, pp.67–74.

Press, L. 1994. Commercialization of the Internet. *Communications of the ACM* 37(11): 17–21.

Puntambekar, S., R. Hübscher 2005. Tools for scaffolding students in a complex learning environment: What have we gained and what have we missed? *Educational Psychologist* 40(1): 1–12.

Raskin, J. 2000. *The Humane Interface: New Directions for Designing Interactive Systems*. Boston, MA: Addison Wesley.

Ries, E. 2011. *The Lean Startup*: How today's entrepreneurs use continuous innovation to create radically successful business. New York: Crown Business.

Rooney, P. 2007. Students @ play: Serious games for learning in higher education. Paper presented at the meeting of the *Proceedings of the International Technology, Education and Development*, Valencia, Spain.

Rust, R., D. Thompson, R. Hamilton 2006. Why do companies keep adding features to already complex products? *Harvard Business Review* 98–107.

Sainfort, F., J. A. Jacko, M. A. McClellan, P. J. Edwards 2012. Human computer interaction in healthcare. In *The Human Computer Interaction Handbook*, eds. J. Jacko and A. Sears, pp. 701–724. Mahwah, NJ: Lawrence Erlbaum.

Shaer, O., E. Hornecker 2010. *Tangible User Interfaces*. Delft, the Netherlands: Now Publishers Inc.

Shneiderman, B. 2000. Universal usability. *Communications of the ACM* 43(5): 84–91.

Sicart, M. 2008. Defining game mechanics. *The International Journal of Computer Game Research* 8(2). http://gamestudies.org/0802/articles/sicart (accessed on May 27, 2013).

Sproll, S., M. Peissner, C. Sturm 2010. From product concept to user experience: Exploring UX potentials at early product stages. Paper presented at the meeting of the *Proceedings of the 6th Nordic Conference on Human-Computer Interaction: Extending Boundaries*, New York.

Story, M. 1998. Maximizing usability: The principles of universal design. *Assistive Technology* 10(1): 4–12.

Sunstein, C., C. Thaler 2008. *Nudge: Improving Decisions about Health, Wealth, and Happiness.* New Haven, CT: Yale University Press.

Veryzer, R., B. Borja de Mozota 2005. The impact of user-oriented design on new product development: An examination of fundamental relationships. *Journal of Product Innovation Management* 22(2): 128–143.

Wade, V. 2011. Challenges for the multi-lingual, multi-dimensional personalised web. Paper presented at the meeting of the *Proceedings of the First Workshop on Personalised Multilingual Hypertext Retrieval*, New York.

Walters, A. 2011. *Designing for Emotion.* New York: Basic Books.

Ware, C. 2000. *Information Visualization.* San Francisco, CA: Morgan Kaufmann Publishers Inc.

Wickens, C., J. Hollands 1999. *Engineering Psychology and Human Performance.* New York: Prentice Hall.

Wroblewski, L. 2011. *Mobile First.* New York: A Book Apart.

Zhao, S., R. Meyer, J. Han 2012. A tale of two judgments: Biases in prior valuation and subsequent utilization of novel technological product attributes. http://knowledge.wharton.upenn.edu/papers/download/Prior_Valuation_Bias.pdf (accessed on February 17, 2012).

Zichermann, G., C. Cunningham 2011. *Gamification by Design: Implementing Game Mechanics in Web and Mobile Apps.* ed. M. Treseler and O'Reilley Sebastopol, CA: O'Reilly Media.

Zuboff, S. 1982. New worlds of computer-mediated work. *Harvard Business Review* 60: 142–152.

VI

Using Information Systems and Technology to Support Individual and Group Tasks

38

Individual-Level Technology Adoption Research: An Assessment of the Strengths, Weaknesses, Threats, and Opportunities for Further Research Contributions

Viswanath
Venkatesh
University of Arkansas

Michael G. Morris
University of Virginia

Fred D. Davis
University of Arkansas

38.1 Introduction

Imagine the following conversation between two researchers at a research university located anywhere in the world:

Doctoral student/new researcher: "I've been reading about how and why people choose to use new technologies and am very interested in building on some of the recent technology adoption models."

Doctoral advisor/senior researcher: "I don't recommend it. There has been so much work done in that area that it is virtually impossible to demonstrate a contribution and get a paper on technology acceptance accepted at any top journal anymore. You would be much better off spending your time working on… ."

Although the aforementioned dialog is imagined, the issue at the core of the conversation is not. How does a researcher interested in the broad phenomenon of individual-level technology adoption make a significant contribution? There can be little doubt that technology adoption has been one of the most dominant, if not *the* most dominant, streams of research in information systems research

(Benbasat and Barki 2007; Venkatesh et al. 2007), and papers from this stream of research are among the most cited in all of business and economics (ScienceWatch.com 2009). A quick examination of the titles and author-supplied keywords indicates that at least 7 of the 10 most highly cited papers in *MIS Quarterly* (*MISQ*) and 11 of the top 20 are related to individual "user acceptance" of information technologies (http://www.misq.org/skin/frontend/default/misq/pdf/MISQStats/MostCitedArticles.pdf). Moreover, the technology acceptance model (TAM), with Venkatesh et al. (2003) as the most influential paper in this stream, has been identified as one of the four fastest growing research fronts in business and economics (Sciencewatch.com 2009). Reflecting on our work in this stream, some of the most-cited papers in various journals are related to technology adoption: the two papers that originally intro-duced TAM—i.e., Davis (1989) and Davis et al. (1989)—are the most-cited papers in *MIS Quarterly* and *Management Science*, respectively; the paper that introduced unified theory of acceptance and use of technology (UTAUT)—i.e., Venkatesh et al. (2003)—is the second most-cited paper in *MIS Quarterly;* the paper that introduced TAM 2—i.e., Venkatesh and Davis (2000)—is the fifth most-cited paper in *Management Science;* papers that studied the antecedents of TAM constructs—i.e., Venkatesh and Davis (1996) and Venkatesh (2000)—are among the most-cited papers published in *Decision Sciences* (most cited) and *Information Systems Research* (second most cited), respectively. Thus, it seems clear that tech-nology adoption is and is likely to remain an area of importance and interest to business researchers. If so, identifying answers to the question of how to identify and make a significant theoretical contribu-tion to this very rich research stream is of vital importance to the academic community.

The maturity of individual-level technology adoption research has fueled much debate about whether this area of research is so mature that there are no more interesting and unanswered ques-tions related to technology adoption left to be answered—this, in fact, is the central issue tackled in this chapter. This debate has revolved primarily around TAM (Davis et al. 1989). A special issue of the *Journal of AIS* in 2007 was dedicated to the discussion about TAM and related models (hereafter referred to as *JAIS-TAM*). The articles in the *JAIS-TAM* largely reflected the *subjective* reactions of sev-eral senior scholars to questions ranging from whether there is a future in TAM research to whether the model has even been good for the field at all. Criticisms of TAM range from its simplicity, to research simply replicating TAM in new contexts or making minor refinements to the model, to TAM hinder-ing the development of more complex theories and detracting from the study of more relevant prob-lems (see Bagozzi 2007; Benbasat and Barki 2007; see also Lee et al. 2003). In light of prior research explaining up to 70% of the variance in intention to use a system (Venkatesh et al. 2003), articles in the *JAIS-TAM* ranged from strongly hinting to explicitly stating that individual-level technology adoption research as we currently know it has almost no future.

Nonetheless, from a practical perspective, there is little doubt that technology adoption continues to be a significant business challenge as information system projects continue to fail at an alarmingly high rate (see Davis and Venkatesh 2004; Devadoss and Pan 2007; Morris and Venkatesh 2010). Despite apparent progress on the scientific front, if project success rates in practice have not improved, there are two possible explanations: (1) practice is disconnected from science—i.e., the message has not reached the practitioners or they have not applied the "lessons learned" from scientific research and/or (2) sci-ence is disconnected from practice—i.e., researchers have failed to provide meaningful actionable guid-ance (see Benbasat and Zmud 1999, 2003; Sambamurthy 2005; Straub and Ang 2008; Weber 2003). Our belief is that technology adoption research remains vital and is one important vehicle for addressing both of these broad gaps.

One of the sentiments conveyed in the *JAIS-TAM* is that although TAM and the associated body of research have been useful and served to integrate disparate bodies of research in the 1980s, it may have well outlived its useful life (Benbasat and Barki 2007). Benbasat and Barki (2007) further note that TAM is, at best, a middle range theory and, to advance scientific inquiry, it is important to move beyond TAM. Most of the other articles in the *JAIS-TAM* concurred with this basic idea and suggested that now, about 20 years after the publication of the original TAM papers, was an appropriate time to further our thinking about technology adoption research. We concur that significant progress has been made (see Venkatesh et al. 2007 for a review) and that more replications and minor extensions are not likely to be helpful in moving the boundaries of the domain forward. By definition, such approaches will only advance our knowledge in a very limited way. Thus, we agree with many scholars who have suggested (in the *JAIS-TAM* and elsewhere) that technology adoption research is stagnant in many ways, consistent with the normal science phase of a stream of research (Kuhn 1996; see also Silva 2007). In suggesting potential future research directions, the collection of papers in the *JAIS-TAM* presented some broad ideas about how one might make a theoretical contribution in this stream. Of course, each of the papers brought its own ideas (and biases) to the debate, and there was no attempt to group or coordinate the suggestions. The potential directions can be grouped into the following general categories:

1. *Determinants*: Benbasat and Barki (2007) point to the need to focus on determinants specifically tied to the IT artifact (i.e., system design characteristics). Similarly, Venkatesh et al. (2007) issued a call for additional determinants based on the fact that work in the reference disciplines has examined a broader set of determinants. Most of the work in this stream to date has primarily studied only technology characteristics and technology-related individual characteristics as determinants.

2. *Consequences/outcomes*: The various papers in the *JAIS-TAM* similarly called for richer theorizing around the outcomes and consequences of technology adoption and use (see Lucas et al. 2007; Venkatesh et al. 2007). Researchers have begun to probe this area more deeply (Barki et al. 2007; Burton-Jones and Straub 2006; Venkatesh et al. 2008); however, work in this area remains in its infancy.

3. *Alternative theoretical mechanisms*: Many have called for richer theorizing by examining alternative mechanisms that can better explain technology use. Bagozzi (2007) proposes a different theoretical perspective on technology adoption, grounded in much of his previous work. Citing Kuhn's thoughts on scientific revolution, Benbasat and Barki (2007) have called for pushing the boundaries of our thinking with regard to theoretical explanations. They cite Wixom and Todd (2005) as an exemplar of such work. Lucas et al. (2007) call for research on multilevel models that can lead to a richer understanding of the phenomenon, and Schwarz and Chin (2007) call for a rethinking of the very meaning of acceptance—habit is used as an example of an alternative theoretical explanation (see Kim 2009; Kim et al. 2005; Polites and Karahanna 2012; Venkatesh et al. 2012). Venkatesh et al. (2007) note the scientific progress made in related reference discipline in examining alternative theoretical mechanisms and draw parallels to the work on technology adoption.

4. *Analytical issues*: Straub and Burton-Jones (2007) are most forceful in raising concerns related to common method bias, and others have recently echoed similar concerns about the research in this area (Sharma et al. 2009). Another example of a potential advance by leveraging more recent analytical techniques is presented by Lucas et al. (2007), who suggested the use of hierarchical linear modeling as a mechanism for providing new theorizing and analysis across levels that have largely been ignored in the work to date. Similarly, Kang et al. (2012) suggested multilevel analyses for the adoption of collaborative technologies as a fruitful step for additional work.

38.2 Strengths, Weaknesses, and Threats of Current Technology Adoption Research

As noted, although many authors have offered important and valid critiques of technology adoption research, it is first worth noting that the prior research has a number of important strengths. First and foremost, prior work has provided a mechanism for integrating disparate bodies of research. With its roots in social psychology, particularly the theory of reasoned action (TRA; Fishbein and Ajzen 1975) and the theory of planned behavior (TPB; Ajzen 1991), models of technology adoption tested in workplace settings have been used to explore individual use of technologies, ranging from personal productivity tools used in organizations (Davis et al. 1989) to electronic commerce applications (Duan et al. 2009), and everything in between. Clearly, the generality of the various technology adoption models has helped create a strong, cumulative tradition, with a rich set of empirical results. For example, a meta-analysis examining the role of voluntariness included a sample of 71 studies on technology acceptance (Wu and Lederer 2009). As indicated earlier, studies on technology adoption are among the most highly cited in all of business and economics and, thus, continue to exert a strong influence on research in various fields within business and economics.

Unfortunately, many of these strengths, when turned around, contribute to the stream's most visible weaknesses. For example, although the extant literature provides a number of influential models of adoption capable of helping scholars understand a vast array of technologies, their generality precludes including any direct design guidance (Venkatesh et al. 2003). This criticism has been recurrent within this stream of research itself for quite some time (Benbasat and Barki 2007; Venkatesh and Davis 1996). Further, although the models are easy to apply, they are heavily reliant on perceptions as predictors of use rather than performance itself (Nielsen and Levy 1994; Venkatesh et al. 2003). Likewise, most technology adoption research starts and ends with the question on what drives technology use itself. This is an important and appropriate starting point for understanding the implementation of information systems in organizations. However, use (in and of itself) offers few advantages to organizations or managers; rather, the consequences of use are factors that are most relevant to researchers and managers interested in improving productivity and performance (see Burton-Jones and Straub 2006; Morris and Venkatesh 2010).

The series of identifiable strengths and weaknesses gives rise to a series of potential threats or concerns with the current state of technology adoption research. Kuhn (1996) argues that scientific thought in any given area proceeds from "prescience" (that lacks a central unifying paradigm), followed by "normal science" that is typified by a program of research on a phenomenon that is guided by foundational work acknowledged by the community as the basis for further investigation and practice. Kuhn (1996) charts the progress of dozens of scientific domains that follow a similar pattern of evolution. One of the main threats associated with this period of "normal science" is that it crowds out or suppresses inquiry that might lead to more revolutionary developments or a "paradigm shift." Given the volume and impact of TAM and related follow-on models (e.g., TAM2, UTAUT, TAM3, UTAUT2), one can argue that by Kuhn's criteria, the existing research on technology adoption provides an almost textbook example of a dominant theoretical paradigm that is in a period of "normal science" (see Silva 2007; Straub 2009a).

As opposed to a series of "one off" recommendations, one of our goals in providing a more organized, comprehensive framework in this chapter is to help structure future research in order to break out of the dominant theoretical paradigm and open new avenues of thought. In so doing, many researchers interested in this area of study might help create one or more revolutionary theoretical perspectives that lead to advances that might otherwise have been missed or ignored. A similar line of reasoning holds for the primary methodological techniques used in much prior work in this domain. Just as one can argue that there is a dominant theoretical paradigm in this stream, based on the methods, measures, and analytical techniques used in the vast majority of the influential papers in the stream, one can argue that there is a dominant methodological paradigm as well that results in a biased set of conclusions (Sharma et al. 2009; Straub and Burton-Jones 2007; Straub et al. 1995). Like a dominant theoretical paradigm, a dominant methodological paradigm can also have a stifling effect on creativity and scientific innovation.

We thus acknowledge and agree with many of our colleagues that demonstrating a meaningful contribution within the technology adoption and research stream can be a daunting task. The central question of this chapter is to determine what to do about it. On the one hand, the community can adopt a skeptical stance (as the fictional senior researcher in the opening paragraph) and back away from the challenge, essentially saying that the field has "been there, done that," and it is time to move on to other important areas of research. Or, on the other hand, the community can look for ways to break free from the shackles of the existing dominant theoretical paradigm by supporting creative research that builds upon the legacy of work in what still appears to be an ongoing phenomenon

TABLE 38.1 Strengths, Weaknesses, Opportunities, and Threats for Current Technology Adoption Research

Strengths	• Remains a highly influential body of work within the IS discipline
	• Investigates a problem important to researchers and managers
	• Integrates disparate bodies of research
	• Applies to many different technology contexts
	• Provides a well-understood model
	• Explains a high percentage of variance in intentions and use
	• Is easily applied and modified
Weaknesses	• Does not offer direct design guidance for improving system acceptability
	• Emphasizes the importance of perceptions over actual behavior
	• Focuses on use as the primary dependent variable of interest
Threats	• Creating and reinforcing a dominant theoretical paradigm
	• Creating and reinforcing a dominant methodological paradigm
	• Propagating the perception that there is nothing new to be learned
	• Stifling scientific creativity and innovation
	• Finding ways to demonstrate a significant contribution
	• Becoming stuck in the normal science phase
Opportunities	• See framework (Figure 38.1)

of interest. Table 38.1 briefly summarizes the strengths, weaknesses, opportunities, and threats that we have outlined in the preceding analysis. In the sections that follow, we outline what we (and others) believe constitutes a theoretical contribution and offer our ideas for how to craft work that can make a strong contribution to this research stream.

38.3 Making a Contribution: Promising Future Research Directions

Before presenting a framework outlining future research directions, we first highlight some of the critical elements deemed essential for making a theoretical contribution drawn from the comments of Benbasat and Zmud (1999), Sutton and Staw (1995), Weber (2003), and Whetten (1989), among others. Many of the points made by these scholars have been reinforced in the editorial statements of the editors at *MIS Quarterly* (Saunders 2005; Straub 2009a; Weber 2003) and *Information Systems Research* (Sambamurthy 2005).

Weber (2003), in his editorial comments, stated that theory "is an *account* that is intended to explain or predict *phenomena* that we perceive in the world" (p. iv, emphasis in original). His articulation of the various ways in which theoretical contributions can be made is consistent with those described in Whetten's (1989) influential remarks. Whetten (1989) noted and discussed four building blocks of theory development: *What* (constructs), *How* (relationships), *Why* (justification), and *Who, Where, When* (conditions). *Why* is likely the "…most fruitful, but also most difficult avenue for theory development" (Whetten 1989, p. 493). Weber's (2003) commentary echoes this point by suggesting that theory should be able to account for phenomena. Also, he notes that the constructs and relationships serve as bases for building richer explanations for phenomena of interest. *Who, Where,* and *When* also possess potential if the conditions being examined are qualitatively different from prior research and/or if the nature of the relationships and outcomes are substantially different from prior research that helps understand the limits and boundaries of existing knowledge, thus furthering our understanding of the phenomenon at hand. As theorists have often noted, good theory is usually laced with a set of convincing and logically interconnected arguments. It can have implications that we might not be able to see with our naked (or theoretically unassisted) eye. It may have implications that run counter to our commonsense. As Weick (1995) put it succinctly, "a good theory explains, predicts, and delights" (p. 378).

In Straub (2009a), the four required elements for a paper to be accepted at *MIS Quarterly* or other premier journals were identified to be (1) blue ocean ideas,* (2) research questions that are nontrivial, (3) popular themes, and (4) the use and development of theory. Both the notion of blue ocean and the importance of research questions ultimately boil down to contribution (Straub 2009a). As he goes on to note, "I believe that it is the unbeatable combination of a reasonably well applied theory answering interesting, novel questions in a well-known thematic stream of work that leads to blue oceans" (p. vii). In his example of nontrivial research questions, he notes that one of his papers (Gefen and Straub 1997) had a "simple enough" research question about how gender affected the causal relationships in TAM but given that the research question had not been studied before, the review team felt that the question was nontrivial. We leverage Straub's (2009b) articulation of "blue ocean ideas" to propose our framework. Researchers can use the framework as a starting point to break out of the normal science phase that seems to currently typify technology adoption research (Kuhn 1996; Silva 2007; Straub 2009a).

The previously mentioned suggestions and guidelines for a theoretical contribution suggest four elements that are important motivators for future technology adoption research. After making a few observations about the relevance of each element for future technology adoption research, we then present a more comprehensive framework that can be used to motivate research that we believe has the potential to make a substantive contribution to the stream.

1. *Answer unanswered research questions related to important and interesting phenomena*: The pursuit of unanswered research questions should be considered by keeping in mind the need to focus on important and relevant business management phenomena (see Benbasat and Zmud 1999). Triviality of phenomena or findings is a fatal flaw for which papers should be rejected (see Saunders 2005; Zmud 1996). In other words, simply because a question has not been previously studied does not make it worthy of study. For example, even though it may not have been studied yet, research applying existing adoption models to the adoption of the next Windows operating system does not make it important or interesting. To this end, we concur with some of the ideas expressed by Benbasat and Barki (2007) by suggesting that one useful area within the technology adoption stream that has the potential to specifically address unanswered research questions—even within the context of existing models—is the exploration of determinants related to the IT artifact and theories that directly tap into properties of specific classes of technologies.

2. *Question/challenge prior theoretical perspectives and consider new theoretical perspectives*: Technology adoption research is dominated by research from the same paradigm of research (i.e., social psychology) from within which TAM was originally developed. We believe that questioning and challenging this dominant theoretical paradigm is essential to developing new theory and going beyond the boundaries of what is firmly established (see Sambamurthy 2005; Saunders 2005; Weber 2003; Zmud 1996). Thus, technology adoption research must leverage alternative theoretical perspectives to enrich our understanding, make progress, and provide innovative solutions to technology implementation and diffusion challenges. This can be achieved either by employing entirely new theories or by integrating previously unused theories in this context. Within technology adoption, examples of employing truly *new theories* include those offered by Kim and Kankanhalli (2009) using a *status quo* perspective and Malhotra et al. (2008) using organismic integration theory; unfortunately, although such examples exist, they are all too rare. Perhaps not surprisingly, there are a few more examples of integrating *previously unused theories* into the technology adoption context, including those by Sykes et al. (2009) for social networking, Devaraj et al. (2008) for personality, and Wixom and Todd (2005) for technology adoption and satisfaction.

---

* Drawing from marketing, Straub (2009b) describes blue ocean ideas to be those that are extremely innovative ideas that make competition irrelevant.

3. *Relative to prior research, develop richer and more comprehensive explanations*: A natural by-product of questioning and challenging prior theory is that the process of thinking through such new theoretical perspectives and executing research that builds on new theory will almost certainly lead to richer and more comprehensive explanations of underlying phenomena (see Sambamurthy 2005). Although most new theoretical perspectives, as discussed earlier, can provide some insights into technology adoption processes, we believe there will be particular benefit and richness from spanning across levels of analysis. We must emphasize that we are *not* suggesting that technology adoption research should simply apply theories from other disciplines; rather, it should use/develop theories in ways that can help explain the unique elements of technology diffusion that will likely call for new constructs, relationships, explanations, and/or boundary conditions (see Herold et al. 2007; Johns 2006). Alvesson and Karreman (2007) note that examining general theories in specific contexts is essential to identify a theory's boundary conditions in order to create the opportunity to extend existing theories. Identifying relevant boundary conditions and contingencies has been far too underrated in technology adoption research (see Weber 2003). We believe that a compelling demonstration of how relationships may change under new conditions—for example, gender as a moderator of key relationships (Venkatesh and Morris 2000; Venkatesh et al. 2000) or the implicit moderation of relationships when different types of systems were investigated (Van der Heidjen 2004)—can represent an important contribution if they are anchored strongly in theory.

4. *Expand the nomological network of relationships*: In most prior research on technology adoption, the nomological network of relationships starts with technology perceptions (on the predictor side), and with few exceptions extends only as far as technology use (on the outcomes side). Building on the ideas of Sutton and Staw (1995), we suggest the need to study broader organizational and/or social phenomena where technology plays a role. We call for research to fully explicate and understand the determinants of technology-related constructs and behaviors and appreciate the impact of technology on outcomes that go beyond use as the primary dependent variable of interest. For example, it is surprising that little or no research on job satisfaction in organizational behavior or psychology has considered technology in any significant way despite the fact that technology has, without question, had a pervasive role in organizational life for well over two decades. A similar problem exists when examining research on technology acceptance and use at the individual level—researchers have simply not ventured far beyond technocentric constructs. Recent examples that guide us in this direction include Morris and Venkatesh (2010) who study job satisfaction and Raghunathan et al. (2008) who study stress. These exemplars illustrate why it behooves us to extend the reach and scope of our nomological networks (see Sambamurthy 2005) by bringing fresh new perspectives to the individual technology adoption phenomenon.

Against the backdrop of these four suggestions, we present a framework that we believe can help guide researchers interested in technology adoption topics (broadly construed) make a substantive contribution to the stream. For example, starting with a generic model of acceptance as the core layer in our framework, we have presented ideas for how researchers might begin to address unanswered questions and probe boundary conditions (points 1 and 4). From this starting point, we have constructed what we call "layers" that provide some specific suggestions and—more importantly—offer researchers a structure for finding ways to contribute meaningfully to the technology adoption literature. Working out from the core layer, our framework includes technology, individual, organizational, societal, and methodological/analytical layers. We believe that, as researchers explore ideas within and across the layers in the framework, it may help illuminate ideas that can be used to exploit potential new theoretical perspectives (point 2) and expand the nomological network of relationships (point 3) to new areas that have not been previously addressed or that are just starting to emerge in the literature.

Within the framework, we offer a number of specific suggestions that we believe have a high potential for making a contribution. This list is not exhaustive, but instead, it is meant to be illustrative of how the framework might be applied. Implicit in the framework and building on the suggestions of other researchers is the notion that research agendas must be theory-driven. Simply because a construct or an idea appears in the framework does not, in and of itself, mean to imply that we believe all research on that focal construct or idea constitutes a contribution by definition. Rather, it means that we believe that research that focuses on that particular element or dimension of the framework has the *potential* to offer a contribution. In other words, it is incumbent upon researchers interested in studying the phenomenon to motivate their work based on the theory within the relevant dimension of the framework, not simply use the framework as the sole motivator for their work.

The first of the elements of the framework include technology adoption determinants (e.g., technology characteristics, individual characteristics, and job characteristics); technology adoption outcomes (e.g., job satisfaction, job performance); intra-individual alternative theoretical mechanisms (e.g., non-productive use or continuance); and extra-individual alternative theoretical mechanisms (e.g., teams, social networks, culture). In addition to theoretically driven research directions, the framework presents research directions that involve novel analytical techniques, such as polynomial modeling and agent-based modeling (ABM). Research that leverages new analytical techniques can span the theoretical ideas presented earlier and may, in fact, reveal new constructs or phenomena that were previously unaccounted. We hope that this framework can help scholars pursue important questions and, perhaps equally important, help reviewers and editors make fewer "type II" errors (see Straub 2008) when evaluating work in this stream.

Figure 38.1 and Table 38.2 present our framework and summarize our future research directions. We elaborate on this next.

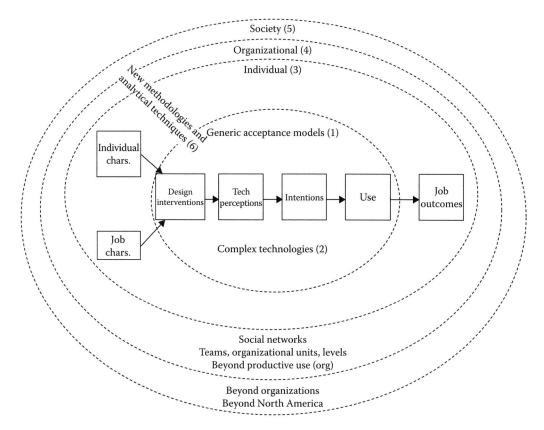

FIGURE 38.1   Framework for technology adoption research contributions.

TABLE 38.2 Potential New Research Directions

Framework Layer	Concept	Illustrative Focal Directions
1. Existing models	Interventions	• Actionable design characteristics
		• Training mechanisms
	Reconceptualizing constructs	• Nonproductive use
		• Deep structure use
		• Various technology adoption constructs
2. Technological	New or complex technologies	• Enterprise systems
		• Workflow management
		• Decision aids/agents
3. Individual	Individual characteristics	• Personality
		• Other theoretically driven demographic or psychographic constructs
	Job characteristics	• Task significance, skill variety, task identity, autonomy, feedback
		• Job demand, job control
	Job outcomes	• Job satisfaction
		• Organizational commitment
		• Turnover intention
		• Job performance
4. Organizational	Social networks and teams	• Size, shape, centrality, density
		• Team variables (e.g., cohesion)
	Organizational units or levels	• Multilevel analysis of technology use
		• Nonproductive use (team or org-level)
5. Societal	Beyond organizations	• Environment
		• Public policy
		• Culture
	Beyond North America	• Culture
		• Societal norms
		• Questions of generalizability
		• Questions of omitted variables
6. Methodological and analytical	New methodologies and analytical techniques	• Curvilinear effects
		• Polynomial modeling
		• Common method bias
		• Multilevel analysis (see organizational layer)
		• Simulation
		• Network analysis (see organizational layer)

Note: These are illustrative areas that we see as important, but, of course, do not represent an exhaustive list of possible directions that could constitute a contribution.

38.3.1 Existing Models' Layers

38.3.1.1 Interventions

In order for research on technology adoption to have meaningful and maximal implications for system designers and managers, it is essential to design and test interventions that can be used in practice. Thus, looking at specific technology characteristics is a particularly promising avenue for future work. The interventions that can be studied range from specific design characteristics (Cyr et al. 2009; Lee et al. 2012; Lohse and Spiller 1999; Sun 2012) to training (Venkatesh 1999) to other types of managerial interventions (see Leonard-Barton and Deschamps 1988). The pervasive interest in the technology artifact (see Benbasat and Weber 1996; Benbasat and Zmud 2003; King and Lyytinen 2004; Orlikowski and Iacono 2001) certainly suggests that the design, development, theorizing, and evaluation of interventions, particularly tied to specific system design characteristics, are important. With respect to technology adoption in particular, Benbasat and Barki (2007) state that "…theorizing the effect of system characteristics on TAM variables is definitely harder" but go on to advocate this type of work as a valuable contribution to the discipline. In addition, the integration of technology adoption research with human–computer interaction research on usability is a noteworthy step in the direction of tying design interventions to known determinants of adoption (Agarwal and Venkatesh 2002; Palmer 2002; Zhang et al. 2002). It is important to note that we are not advocating a cycle of never-ending elaboration, incremental testing, and documentation of a range of outcomes associated with the use of various design characteristics, such as those specified by "design gurus" or best practices. Rather, we are calling for theory-driven examinations of design options or other interventions, grounded in research in software engineering, human–computer interaction, information visualization, and IS (among others). Such an exploration of key design interventions offers the potential for answering interesting questions, extending the nomological net of relationships, and providing a richer explanation for the adoption of new systems in ways that are practically relevant and keep the core properties of technologies as the appropriate focal point of study (Benbasat and Zmud 1999, 2003).

38.3.1.2 Reconceptualizing Constructs

For obvious reasons, technology adoption research has had a near-solitary focus on what could be termed "productive use." Inherent in this work is one critical assumption: more use is better. This assumption implies that benefits derived from technologies will be maximized when the use of technologies is also maximized (Agarwal 2000; Devaraj and Kohli 2003; Venkatesh and Davis 2000). The broader implication of this assumption is that maximizing use always leads to more positive outcomes (in a linear fashion) and that there are no significant negative consequences associated with such use. In contrast, the trade press in the past decade or so has been rich with stories of several negative consequences of technology use, such as excessive Internet use, often leading to dysfunctional outcomes including decreased social involvement and increased depression (Kraut et al. 1998a; Young 2004). Although technology adoption research has long advocated use as a positive outcome that drives other positive outcomes, such as increased productivity at work and improved home life, psychologists have begun to focus on the potential negative consequences of excessive technology use. In addition to negative impact that technology can have on users in homes, technology use can result in loss of productivity due to employees using technologies to engage in nonwork-related tasks—e.g., chatting, browsing, and playing games—without being noticed. Interesting streams of research from psychology, sociology, and organizational behavior on addictions and withdrawal behaviors can be leveraged to help further our understanding of wasteful technology use, its determinants, its consequences, potential contingencies, and helpful interventions (see Tarafdar et al. 2010; Turel et al. 2011). However, this first calls for a fundamental rethinking of the use construct in order to broaden the scope of technology adoption research, paralleling the way the reach of technologies themselves has broadened.[*]

Burton-Jones and Straub (2006) challenged the traditional conceptualization of system use in favor of richer conceptualizations. Specifically, they proposed cognitive absorption and deep structure use as viable alternative dependent variables. They provided preliminary evidence that these conceptualizations of use were more strongly related to task performance. Burton-Jones and Gallivan (2007) and Barki et al. (2007) have also made suggestions about how to move forward with better conceptualizations of this key construct. Yet, much work remains to be done. What are the predictors of these conceptualizations of use? What are its consequences? How well do existing models of technology adoption

---

[*] Although beyond the immediate focus of this discussion, other areas that could be of interest along these lines include rethinking the conceptual foundations and role in the nomological network for various other constructs, such as service quality (see Wixom and Todd 2005).

predict these conceptualizations? What are the temporal dynamics and other contingencies involved in predicting these conceptualizations of use? These are but a few questions, that if pursued, would result in the emergence of the blue ocean ideas to which Straub (2009b) refers. We note that some recent work has started to examine different types of use and even its consequences (see Sykes and Venkatesh in press; Venkatesh et al. 2008), which we view as a positive step in helping open up new productive areas of research in the broad theoretical domain.

In addition to use, others have already made contributions to the stream by reconceptualizing other adoption-related constructs. Examples of this work include Schwarz and Chin (2007) who called for researchers to rethink the very idea of acceptance and Karahanna et al. (2006) who called for a reconceptualization of the compatibility construct. Likewise, Compeau et al. (2007) refined the conceptualization of perceived characteristics of innovations from Moore and Benbasat (1991). Taken together, this is an important and fruitful direction as it helps sharpen the definitions and measurements of constructs known to be important within the stream of research. Not only will this in turn result in better prediction of key outcomes, but it also offers the potential to provide greater richness and clarity in our explanations.

38.3.2 Technological Layer

Pushing beyond the inner layer of existing models in our framework, we believe that, at the technological layer, theorizing about newer, more complex types of systems is another fruitful avenue for new research. As an example, one type of system that researchers have successfully studied to date with a strong theoretical grounding is recommendation agents (see review in Xiao and Benbasat 2007). The collection of works examines various design characteristics that are unique to the context of recommendation agents and their impact on key evaluation criteria (Hevner et al. 2004; Komiak and Benbasat 2006; Wang and Benbasat 2005). Other such investigations have included collaborative systems (Brown et al. 2010), mobile data services (Hong and Tam 2006; Thong et al. 2011; Venkatesh et al. 2012), and protective ITs (Dinev and Hu 2007; Johnston and Warkentin 2010; Liang and Xu 2009).

Further expanding the idea of the need to look at specific technology characteristics, it is certainly true that one of the most significant changes in the last several years with respect to technology itself lies in the sheer complexity of modern technological systems. Interestingly, many of the technological systems studied in early technology adoption research were fairly simple, individual-level productivity tools. More recently, the implications for the use of more complex enterprise-wide systems that can span business units within the organization and even across organizational boundaries have been studied (Grewal et al. 2007; Karimi et al. 2009). Although researchers have begun to study the impacts of such systems, they are still typically conceptualized as a "black box" (Beaudry and Pinnsoneault 2005; Morris and Venkatesh 2010), suggesting the need for more detailed theorizing around the nature of new technologies themselves. Clearly, these new systems have the potential to push the limits of our existing technology adoption theories. Although some elements of prior research findings may indeed hold in the context of complex systems—e.g., perceived usefulness will likely continue to be a driver of intention to use in these systems—it is also likely that complex systems will provide an opportunity for deeper theorizing about technologies, their implementation in organizations, and challenges associated with deploying such complex systems (see Boudreau and Robey 2005). As complex technologies will increase the interdependence among users both for information *in* the system and also for help *using* the system, they may increase the range of relevant independent and dependent variables of interest. This, in turn, will give rise to the need to use richer theoretical perspectives, such as social networks and multiple levels of analysis, which are discussed later in this chapter. Finally, complex technologies that integrate various business processes are in keeping with current phenomena and will extend the boundaries of existing models and may call for more sophisticated models (see Sambamurthy 2005; Weber 2003).

38.3.3 Individual Layer

38.3.3.1 Individual and Job Characteristics

Working outward in the framework to move beyond the existing models and technological layers, in organizational settings, it is clear that understanding the influence of both individual character-istics and job characteristics as technology adoption determinants also holds great promise (Morris and Venkatesh 2010). Specifically, moving beyond "technocentric" outcomes, such as use, expands the nomological network of relationships and creates a broader understanding of the implications of tech-nology adoption. Such work is responsive to the suggestions of Delone and McLean (1992, 2003) who have specifically cited the importance of considering other metrics of system success, such as individual and organizational impacts. We extend their idea by suggesting that, within our framework, a similar case could be made on the predictor side as well.

Although research in psychology predicts many beliefs and behaviors using personality (Funder 2001), only limited technology adoption research has studied such constructs in much depth beyond technology-related personality traits, such as computer playfulness, computer self-efficacy, computer anxiety, and personal innovativeness with information technology (see Agarwal and Prasad 1998; Compeau and Higgins 1995a,b; Venkatesh 2000; Webster and Martocchio 1992). One notable and important exception is Devaraj et al. (2008). We believe that further work that studies the role of personality both as an antecedent and as a moderator could greatly enhance the predictive validity of existing models by helping organizations identify individuals with specific personality profiles that can help promote the use, success, and diffusion of technologies. Such work might include, but is not limited to, the big five (see Costa and McCrae 1992), locus of control (see Rotter 1954), and goal orientation (see Dweck 1986).

Just as personality would provide a better representation of the person, consideration of job characteristics can provide a greater understanding of the situational circumstances of employees. Surprisingly, few (if any) technology adoption studies have considered job characteristics or other job-related factors in predicting technology adoption decisions (see Venkatesh et al. 2003) as a means to advance our understanding. The job characteristics model (JCM; Hackman and Oldham 1976) and the demand–control model (Karasek 1979) offer useful ways to consider new theoretical perspec-tives that can be leveraged to study the impacts on technology adoption decisions (see Morris and Venkatesh 2010).

38.3.3.2 Job Outcomes

On the "downstream" side of the framework, outcomes, such as job satisfaction, organizational commitment, job performance—particularly objectively rated performance that avoids the common method variance concerns articulated in Straub and Burton-Jones (2007)—and other job outcomes that are of interest to employees and organizations should be integrated into technology adoption models to extend our understanding of the influence of use within the organization. Such research also has the potential to influence organizational behavior research by refining models of constructs like job satisfaction to encompass technology-related constructs as predictors that are especially important given the central role that technologies play in most jobs today (Herold et al. 2007). This would be particularly pertinent to organizational change interventions where both new business processes and new technologies are introduced concurrently as is often the case with technolo-gies, such as enterprise resource planning systems (see discussion under the technological layer of the framework). In sum, we call for studies to help build a comprehensive nomological network of relationships that incorporate a richer, broader integration of upstream and downstream constructs that are related—yet distinct—from the traditional determinants and outcomes of technology adop-tion decisions.

38.3.4 Organizational Layer

38.3.4.1 Social Networks

Moving outward in the framework beyond internal, individualized theoretical layers to "extra-individual factors" is a potential avenue for further exploration. Given that the space of theoretical perspectives is vast, we provide one illustration of a promising avenue (among many possible avenues)—namely, social networks.* Traditional models of adoption and use have often (but not always) included a construct meant to capture the impact of social factors. Sometimes called subjective norm or social influence, this construct captures the degree to which referent others influence an individual's intention to use (or not) the system in question (Taylor and Todd 1995; Venkatesh and Davis 2000; Venkatesh et al. 2003, 2012). However, despite its conceptual importance, the role of others in influencing individuals to adopt new technologies has been inconsistent. Researchers have identified a series of moderating influences, such as gender, age, voluntariness, and experience (Venkatesh et al. 2003) that might clarify the role that social influence plays. The next steps in this area should focus on examining the problem using an extra-individual theoretical lens, such as that available via social networks theory.

Social networks have been used to explain a wide array of behaviors in organizations including early promotion, career mobility, adaptability to change, and supervisor ratings of performance (Burt 1992; Carroll and Teo 1996; Gargiulo and Benassi 2000; Podolny and Baron 1997). Further, within a more technology-centric context, research on social networks has found that centrality in a workflow network is positively associated with supervisor ratings in high-technology companies (Mehra et al. 2001). It also has discovered that individuals are typically more reliant on individual relationships for finding new information (e.g., about technologies) rather than organizational databases or other media including Intranets or the Internet itself (see also Borgatti and Cross 2003; Davis et al. 2009).

Despite the positive individual (and by extension, organizational) performance benefits associated with the appropriate leveraging of social networks, however, the role of social networks, including their size, structure, and nature of the relational ties has not been fully leveraged within the technology adoption research stream. One recent paper that offers a promising look at ways to incorporate social network theory, method, and analysis is Sykes et al. (2009), where network constructs were shown to explain significant additional variance in system use beyond the typical predictors studied in technology adoption research. Other forthcoming research leveraging social networks to better understand the impacts of an implementation include Sykes et al. (in press) and Sykes and Venkatesh (in press).

We contend that social networks represent one theoretical paradigm that may be useful to understanding the role of others in influencing use in a general way as well as gaining a deeper understanding of *which* others inside or outside the organization are the most important, and perhaps even *why* they are important. Once again, we must note that we are not advocating a straight application of social network theory and findings to technology adoption, but rather, we hope researchers will draw on it as a new lens for developing a deeper understanding of behavior and as a mechanism for considering more deeply the *what, how, why, who, where,* or *when* questions that remain relevant for technology adoption and use in today's organizations.

38.3.4.2 Organizational Units†

There are other potential avenues for productive technology adoption research that mandate theorizing beyond the individual level of analysis. For example, Lucas et al. (2007) call for multilevel approaches

---

* Here, we use the term to refer to the scientific domain around social network theory, method, and analysis techniques—this should not be confused with the emergent popular applications, such as Facebook. Applications such as Facebook can and are used to develop and maintain social networks.

† The authors thank Dr. Likoebe Maruping for his advice on the integration of technology acceptance research with research on teams.

that account for the influence of factors across different levels of analysis. Although there has been research at different levels, they have seldom been integrated. Individuals' adoption and use of technology are often embedded in a social context and, thus, are open to influence by factors that are not currently accounted by an individual level of analysis (see Sarker and Valacich 2010; Sarker et al. 2005). As organizational researchers have noted, microphenomena in organizations and other social systems are embedded within macrocontexts (Kozlowski and Klein 2000). Hence, *meso* theories, which cross multiple levels of analysis, are necessary in order to advance our understanding of organizational phenomena (House et al. 1995). Unit-level factors (e.g., organizational, subdivision, functional, group) could and should be incorporated into models of technology adoption. For instance, within organizations, job types, skills, and the centrality or necessity of technology to the work differ across functional units. Multilevel models can capture the nature of these differences and how they affect individual adoption of organization-wide systems, whereas more sophisticated statistical techniques (which will be explicated in more detail in the methodological layer discussion) can be employed to help analyze such multilevel data (Bryk and Raudenbush 1992; Hofmann 1997).

Once again, to provide an illustrative example of how our framework might be applied in this context, one area that could benefit from a multilevel perspective on technology adoption is research on virtual teams. Much of the research on virtual teams has focused on how information and communication technologies (ICTs) support specific team processes (Dennis and Kinney 1998; Dennis et al. 2008; Nicholson et al. 2006; Powell et al. 2004; Sarker et al. 2005). Virtual teams need to be deliberate in their selection of technology to facilitate task completion (Maruping and Agarwal 2004); however, there is still a need to understand the processes through which the adoption of specific ICTs occurs. Maruping and Agarwal (2004) point to a need for future research to examine how individual differences (e.g., computer anxiety, remote work efficacy) affect virtual teams' choice of ICTs. Reciprocal cross-level models, such as those suggested by Griffin (1997), could be used to model bottom-up processes fostered by individual differences as well to examine how the resulting team interactions influence individual intention to adopt a specific ICT. In sum, the development and testing of such models would provide a richer understanding of the dynamics affecting the adoption of technology in situations that extend beyond the scope of the individual.

38.3.5 Societal Layer

38.3.5.1 Beyond Organizations

There has been a good bit of recent research on nonorganizational (e.g., household) use of technologies (Brown and Venkatesh 2005; Venkatesh and Brown 2001). Other focal areas of study for technologies typically used outside of the organizational context in recent years have been e-commerce (Pavlou 2003; Pavlou and Fygenson 2006; Pavlou et al. 2007) and m-commerce (Mathew et al. 2004; Sarker and Wells 2003; Venkatesh and Ramesh 2006) and other societal-level effects (Hsieh et al. 2008, 2010). The importance of extra-organizational use is only likely to increase in the coming decades. The emergence of social media (e.g., Facebook) makes our need to reexamine the application of existing theory and to consider new theoretical perspectives quite important (Gallaugher and Ransbotham 2010; Kane 2011; Kane et al. 2009).

38.3.5.2 Beyond North America

Clearly, most contemporary technology adoption models have been developed, tested, and subsequently applied in North America. Despite new technologies reaching developing and even underdeveloped countries, our understanding of their adoption, use, and impact in these countries is remarkably limited. Although a few studies in IS have examined cultural differences, several practical and theoretical questions regarding the influence of culture on technology adoption and use remain unanswered (Karahanna et al. 2000, 2005; Straub et al. 2002). Culture, environment, and public policy are among the factors that will almost surely play a significant role in adoption; yet, our understanding of their influence is limited. Importantly, there have been a few studies that have laid the groundwork for future investigations of culture and adoption (Bajwa et al. 2008; Cyr 2008; Gefen and Straub 1997; Sia et al. 2009; Srite and Karahanna 2006). In light of the spread and infusion of technology to new contexts, studies that further explore the boundary conditions around current models of adoption will be particularly useful. For example, some of the questions that are worthy of additional study may include: Is there cross-cultural generalizability for the models developed primarily in North America? Are there any omitted variables as we move the models to non-North American settings?

Complicating matters further, there are several specific methodological idiosyncrasies in cross-cultural research that are largely different from intracultural research that could potentially change the nature of results obtained in prior studies. For example, there is a high level of construct, method, and item nonequivalence across different cultures that raise questions about the generalizability of prior findings to new settings (see Johns 2006 for a detailed discussion about how different contexts, such as culture, can cause various changes to our theory and empirical findings). There are also several regional, ethnic, religious, and linguistic differences within each culture suggesting that examining culture as a phenomenon in its own right is required. As before, we would like to note that simply examining any question or model in a new cultural setting does not fit the bill of a substantive contribution. Rather, such investigations must start begin from a strong theoretical foundation, grounded in the unique and important psychological, environmental, or other social mechanisms relevant to the cultural context of study (see Venkatesh et al. 2010).

38.3.6 Methodological and Analytical Layer

38.3.6.1 New Method: Agent-Based Modeling

Despite the progress in technology adoption research, there are several puzzles and inconsistencies that remain that may be an artifact of the methodological approaches used in contemporary adoption research. Why do some technologies that would appear to an external observer to be useful and easy to use so often fail to garner the acceptance of target users (Markus and Keil 1994)? Why do some technologies lacking true usefulness so often become widely accepted (Davis and Kottemann 1995)? Why do some promising technologies gain initial adoption, only to fall by the wayside over time (Shapiro and Varian 1999)? More generally, why do technologies sometimes exhibit oscillating patterns of adoption (Maienhofer and Finholt 2002)? Technology adoption research to date has been unable to solve these puzzles.

To overcome these limitations, we propose the use of ABM methodologies that offer key complementary strengths (see Railsback and Grimm 2011) to the narrative, mathematical theories, and methods traditionally employed in technology adoption. Research on organizational adoption of innovations, such as total quality management (Repenning 2002), quality circles (Strang and Macy 2001), downsizing (Rosenkopf and Abrahamson 1999), and production technologies (Bullnheimer et al. 1998), have all found ABM valuable in providing insights into the dynamics of adoption beyond those afforded by more traditional methods (see also Latane 1996). A major reason for the insights is that ABM was used to incorporate both rational and social forces within a common modeling paradigm. Rational and social explanations have often been regarded as competing accounts for technology adoption (Kraut et al. 1998b; Schmitz and Fulk 1991). Even though neither is able to single-handedly explain technology discontinuance decisions, when combined, they can account for fad-like waves of adoption and rejection (Strang and Macy 2001). These initial applications of ABM to the study of organizational adoption illustrate how the complex interplay of various social influences give rise to *unintuitive* emergent global behavior across a population of firms. Similar advantages are expected to accrue through the use of ABM to specifically study individual-level adoption of technology. In a similar fashion, the dynamic complexity revealed by these organizational adoption studies portend that ABM can be used to solve similar challenges in studying technology adoption at the individual and group layers of the framework.[*] ABM allows for simulations that have varying time granularity as well as detailed models of influence, thus allowing for a rich understanding of emergent phenomena that extends well beyond what exists in current models.

38.3.6.2 New Analytical Technique: Polynomial Modeling and Response Surface Analysis

In addition to new theoretical approaches and new methods, new analytical techniques might be utilized in order to bring new insights or perspectives to technology adoption and use. One approach—common in the management literature—that has not been employed to its full potential in technology adoption research is polynomial modeling and associated response surface modeling. Rather than being limited to the use of "best fit" linear models, the use of polynomial models allows researchers to capture more complex and dynamic behavior that, in the past, has been difficult to explore analytically. Although an in-depth discussion of polynomial modeling is beyond the scope of this chapter, the use of polynomial models and the related use of response surfaces can be used to plot two independent variables in relation to a dependent variable on a three-dimensional graph (see Edwards 1995; Edwards and Harrison 1993; Edwards and Parry 1993 for an extended discussion on polynomial models and response surface analysis). Such a technique is better able to capture the complex and nonlinear effects of change (e.g., the shift between pre-implementation and post-implementation attitudes) and avoids many of the methodological problems associated with difference scores that are often used such as weak reliability, conceptual ambiguity, and encapsulation of confounding effects of the independent variables on the components of the difference (Edwards 1995).

With respect to technology adoption and use, polynomial response surface modeling might be applied in a number of ways. First, although some have recognized the potentially important nature of changes in perception or use over time, most existing models have either attempted to control for prior perceptions (e.g., perceptions of usefulness at time 1 are used as a control on those at time 2), or have conceptualized the change as a single (new) construct meant to capture the dynamics of change. Either technique, although understandable given prior analytical constraints, reduces potentially nonlinear, multivariate relationships into a linear, univariate one, thereby suppressing a number of potentially interesting elements that may contribute to changing patterns of adoption and use over time.

Two examples of how polynomial response surface modeling might be applied to technology adoption and use may be helpful to illustrate its benefits. First, in studying some of the downstream consequences of use, one might examine how both pre-adoption and post-adoption perceptions combine in novel ways to influence user attitudes and/or performance. Such an approach would plot pre- and post-adoption measures on the X and Y axes, respectively, with the dependent variable of interest (say, performance) on the Z (vertical) axis. Another approach, consistent with the theoretical underpinnings of expectation–confirmation models, would be to measure pre-implementation expectations and post-implementation beliefs together (X and Y axes) to predict downstream perceptions or use (Z axis), rather than relying on a single construct, such as confirmation or disconfirmation, that typically had to be used in prior research to capture the differences (Bhattacherjee 2001; Bhattacherjee and Premkumar 2004). Venkatesh and Goyal (2010) and Titah and Barki (2009) are exemplars that use this approach to model nonlinear effects as a means to advance our understanding of technology adoption (see also Brown et al. 2012; Brown et al. in press). New work could build on these recent examples in order to identify opportunities for deeper theorizing about how shifts in constructs, rather than absolute levels themselves, might influence important outcomes such as performance or satisfaction, while simultaneously challenging the assumptions of linearity that underlie most current theorizing.

---

[*] Note that the methodological and analytical layer spans the other layers of the framework in Figure 38.1.

38.3.6.3 Qualitative Research

We believe that qualitative research methods and associated approaches to interpret data hold particular promise for breaking the shackles of normal science in which technology adoption is currently situated. Qualitative research blends induction with search for novel insights that hypothetico-deductive approaches that have dominated this stream lack (see Locke 2007). Qualitative research has a rich tradition in business management research, with many well-known papers that present guidelines on how to do qualitative research (Benbasat et al. 1987; Lee 1989). Within the technology adoption stream, there have been some innovative papers that have brought to the fore new ideas, concepts, mechanisms, and even problems that were previously undiscovered (Sarker and Lee 2002, 2003, 2006; Sutanto et al. 2008; Wagner and Newell 2007). Taking the cue from these exemplars, we believe that the qualitative approach is underrepresented in technology adoption research and that researchers in the stream can further leverage these methodologies across each of the layers of our framework (Figure 38.1) to help identify new constructs, develop new theory, and create blue ocean ideas, consistent with the ideas of Straub (2009b). Further, the employment of multiple methods can lead to rich insights that any single method cannot provide (see Venkatesh et al. 2013).

38.4 Conclusions

In this chapter, we have made a case for the continued relevance and vitality of the technology adoption research stream. In so doing, we acknowledge many of the criticisms that have recently been levied at research in the area and agree with the broad conclusion that this important stream of research runs the danger of becoming stagnated unless we, as a community of researchers, can harness our creative talents to reinvigorate research in the domain. Although challenging, the problem remains critically important to research and practice, and we have a responsibility to conduct value-added studies that bring new scientific knowledge into the fold. Within the context of our critique and assessment of research on technology adoption, we have offered a new framework that we hope serves as a useful starting point for motivating additional work in the area that pushes all of us interested in the area out of our "comfort zone," beyond "normal science," and toward a period of research that pushes the boundaries of our existing knowledge through perhaps a scientific revolution that comes from breaking away from existing paradigms. In doing so, rather than declaring technology adoption research "dead" and having this chapter serve as its epitaph, we hope the thoughts and commentary here instead bring to mind the famous adages from such disparate personalities as Mark Twain and Steve Jobs indicating that accounts of one's death—in this case, the technology adoption research stream—are greatly exaggerated. Indeed, we hope that our ideas here might serve a useful role in revitalizing one of the most mature streams in business management research. In doing so, we hope to bring an end to the debate on the merits of TAM and its derivatives per se and, instead, motivate proponents and critics of the current state of the research stream (and we count ourselves as both) to extend technology adoption theory to new frontiers of knowledge.

References

Agarwal, R. 2000. Individual adoption of new information technology. In *Framing the Domains of IT Management: Glimpsing the Future through the Past*, ed. R.W. Zmud, pp. 85–104. Cincinnati, OH: Pinnaflex.

Agarwal, R. and Prasad, J. 1998. A conceptual and operational definition of personal innovativeness in the domain of information technology. *Information Systems Research* 9(2):204–215.

Agarwal, R. and Venkatesh, V. 2002. Assessing a firm's web presence: A heuristic evaluation procedure for the measurement of usability. *Information Systems Research* 13(2):168–186.

Ajzen, I. 1991. The theory of planned behavior. *Organizational Behavior and Human Decision Processes* 50(2):179–211.

Alvesson, M. and Karreman, D. 2007. Constructing mystery: Empirical matters in theory development. *Academy of Management Review* 32(4):1265–1281.

Bagozzi, R. P. 2007. The legacy of the technology acceptance model and a proposal for a paradigm shift. *Journal of the Association for Information Systems* 8(4):244–254.

Bajwa, D. S., Lewis, L. F., Pervan, G., Lai, V. S., Munkvold, B. E., and Schwabe, G. 2008. Factors in the global assimilation of collaborative information technologies: An exploratory investigation in five regions. *Journal of Management Information Systems* 25(1):131–165.

Barki, H., Titah, R., and Boffo, C. 2007. Information system use-related activity: An expanded behavioral conceptualization of individual-level information system use. *Information Systems Research* 18(2):173–192.

Beaudry, A. and Pinnsoneault, A. 2005. Understanding user responses to information technology: A coping model of user adaptation. *MIS Quarterly* 29(3):493–524.

Benbasat, I. and Barki, H. 2007. Quo vadis, TAM? *Journal of the Association for Information Systems* 8(4):212–218.

Benbasat, I., Goldstein, D. K., and Mead, M. 1987. The case research strategy in studies of information systems. *MIS Quarterly* 11(3):369–386.

Benbasat, I. and Weber, R. 1996. Research commentary: Rethinking "diversity" in information systems research. *Information Systems Research* 7(4):389–399.

Benbasat, I. and Zmud, R. W. 1999. Empirical research in information systems: The practice of relevance. *MIS Quarterly* 23(1):3–16.

Benbasat, I. and Zmud, R. W. 2003. The identity crisis within the IS discipline: Defining and communicating the discipline's core properties. *MIS Quarterly* 27(2):183–194.

Bhattacherjee, A. 2001. Understanding information systems continuance: An expectation-confirmation model. *MIS Quarterly* 25(3):351–370.

Bhattacherjee, A. and Premkumar, G. 2004. Understanding changes in belief and attitude toward information technology usage: A theoretical model and longitudinal test. *MIS Quarterly* 28(2):229–254.

Borgatti, S. and Cross, R. 2003. A social network view of organizational learning: Relational and structural dimensions of "know who." *Management Science* 49(4):432–445.

Boudreau, M. C. and Robey, D. 2005. Enacting integrated information technology: A human agency perspective. *Organization Science* 16(1):3–18.

Brown, S. A., Dennis, A., and Venkatesh, V. 2010. Predicting collaboration technology use: Integrating technology adoption and collaboration research. *Journal of Management Information Systems* 27(2):9–53.

Brown, S. A. and Venkatesh, V. 2005. Model of adoption of technology in households: A baseline model test and extension incorporating household life cycle. *MIS Quarterly* 29(3):399–426.

Brown, S. A., Venkatesh, V., and Goyal, S. Expectation confirmation in IS research: A test of six competing models. *MIS Quarterly*, forthcoming.

Brown, S. A., Venkatesh, V., and Goyal, S. 2012. Expectation confirmation in technology use. *Information Systems Research* 23(2):474–487.

Bryk, A. S. and Raudenbush, S. W. 1992. *Hierarchical Linear Models: Applications and Data Analysis Methods*. Thousand Oaks, CA: Sage.

Bullnheimer, B., Dawid, H., and Zeller, R. 1998. Learning from own and foreign experience: Technological adaptation by imitating firms. *Computational and Mathematical Organization Theory* 4(3):267–282.

Burt, R. S. 1992. *Structural Holes*. Cambridge, MA: Harvard University Press.

Burton-Jones, A. and Gallivan, M. J. 2007. Toward a deeper understanding of system usage in organizations: A multi-level perspective. *MIS Quarterly* 31(4):657–679.

Burton-Jones, A. and Straub, D. W. 2006. Reconceptualizing system usage: An approach and empirical test. *Information Systems Research* 17(3):228–246.

Carroll, G. and Teo, A. 1996. On the social networks of managers. *Academy of Management Journal* 39(2):421–440.

Compeau, D. R. and Higgins, C. A. 1995a. Application of social cognitive theory to training for computer skills. *Information Systems Research* 6(2):118–143.

Compeau, D. R. and Higgins, C. A. 1995b. Computer self-efficacy: Development of a measure and initial test. *MIS Quarterly* 19(2):189–211.

Compeau, D. R., Meister, D. B., and Higgins, C. A. 2007. From prediction to explanation: Reconceptualizing and extending the perceived characteristics of innovating. *Journal of the Association for Information Systems* 8(8):409–439.

Cyr, D. 2008. Modeling web site design across cultures: Relationships to trust, satisfaction, and e-loyalty. *Journal of Management Information Systems* 24(4):47–72.

Cyr, D., Head, M., Larios, H., and Pan, B. 2009. Exploring human images in website design: A multi-method approach. *MIS Quarterly* 33(3):539–566.

Davis, F. D. 1989. Perceived usefulness, ease of use, and user acceptance of information technology. *MIS Quarterly* 13(3):319–340.

Davis, F. D., Bagozzi, R. P., and Warshaw, P. R. 1989. User acceptance of computer technology: A comparison of two theoretical models. *Management Science* 35(8):982–1003.

Davis, F. D. and Kottemann, J. E. 1995. Determinants of decision rule use in a production planning task. *Organizational Behavior and Human Decision Processes* 63(2):145–157.

Davis, F. D. and Venkatesh, V. 2004. Toward preprototype user acceptance testing of new information systems: Implications for software project management. *IEEE Transactions on Engineering Management* 51(1):31–46.

Davis, J. M., Kettinger, W. J., and Kunev, D. G. 2009. When users are IT experts too: The effects of joint IT competence and partnership on satisfaction with enterprise-level systems implementation. *European Journal of Information Systems* 18(1):26–37.

DeLone, W. H. and McLean, E. R. 1992. Information systems success: The quest for the dependent variable. *Information Systems Research* 3(1):60–95.

DeLone, W. H. and McLean, E. R. 2003. The DeLone and McLean model of information systems success: A ten-year update. *Journal of Management Information Systems* 19(4):9–30.

Dennis, A. R., Fuller, R. M., and Valacich, J. S. 2008. Media, tasks, and communication processes: A theory of media synchronicity. *MIS Quarterly* 32(3):575–600.

Dennis, A. R. and Kinney, S. T. 1998. Testing media richness theory in the new media: The effects of cues, feedback, and task equivocality. *Information Systems Research* 9(3):256–274.

Devadoss, P. and Pan, S. 2007. Enterprise systems use: Towards a structurational analysis of enterprise systems induced organizational transformation. *Communications of the Association for Information Systems* 19:352–385.

Devaraj, S., Easley, R. F., and Crant, J. M. 2008. How does personality matter? Relating the five-factor model to technology acceptance and use. *Information Systems Research* 19(1):93–105.

Devaraj, S. and Kohli, R. 2003. Performance impacts of information technology: Is actual usage the missing link? *Management Science* 49(3):273–289.

Dinev, T. and Hu, Q. 2007. The centrality of awareness in the formation of user behavioral intention toward protective information technologies. *Journal of the Association for Information Systems* 8(7):386–408.

Duan, W., Gu, B., and Whinston, A. B. 2009. Informational cascades and software adoption on the internet: An empirical investigation. *MIS Quarterly* 33(1):23–48.

Dweck, C. S. 1986. Motivational processes affecting learning. *American Psychologist* 41:1040–1048.

Edwards, J. R. 1995. Alternatives to difference scores as dependent variables in the study of congruence in organizational research. *Organizational Behavior and Human Decision Processes* 64(3):307–324.

Edwards, J. R. and Harrison, R. V. 1993. Job demands and worker health: Three-dimensional reexamination of the relationship between person-environment fit and strain. *Journal of Applied Psychology* 78(4):628–648.

Edwards, J. R. and Parry, M. E. 1993. On the use of polynomial regression equations as an alternative to difference scores in organizational research. *Academy of Management Journal* 36(6):1577–1598.

Fishbein, M. and Ajzen, I. 1975. *Belief, Attitude, Intention and Behavior: An Introduction to Theory and Research*. Reading, MA: Addison-Wesley.

Funder, D. C. 2001. Personality. *Annual Review of Psychology* 52:197–221.

Gallaugher, J. and Ransbotham, S. 2010. Social media and customer dialog at Starbucks. *MIS Quarterly Executive* 9(4):197–212.

Gargiulo, M. and Benassi, M. 2000. Trapped in your own net? Network cohesion, structural holes, and the adaptation of social capital. *Organization Science* 11(2):183–196.

Gefen, D. and Straub, D. W. 1997. Gender differences in the perception and use of email: An extension to the technology acceptance model. *MIS Quarterly* 21(4):389–400.

Grewal, R., Johnson, J., and Sarker, S. 2007. Crises in business markets: Implications for interfirm linkages. *Journal of the Academy of Marketing Science* 35(3):398–416.

Griffin, M. A. 1997. Interaction between individuals and situations: Using HLM procedures to estimate reciprocal relationships. *Journal of Management* 23(6):759–773.

Hackman, J. R. and Oldham, G. R. 1976. Motivation through design of work: Test of a theory. *Organizational Behavior and Human Performance* 16(2):250–279.

Herold, D. M., Fedor, D. B., and Caldwell, S. D. 2007. Beyond change management: A multilevel investigation of contextual and personal influences on employees' commitment to change. *Journal of Applied Psychology* 92(4):942–951.

Hevner, A. R., March, S. T., Park, J., and Ram, S. 2004. Design science in information systems research. *MIS Quarterly* 28(1):75–105.

Hofmann, D. A. 1997. An overview of the logic and rationale of hierarchical linear models. *Journal of Management* 23(6):723–744.

Hong, S. J. and Tam, K. Y. 2006. Understanding the adoption of multipurpose information appliances: The case of mobile data services. *Information Systems Research* 17(2):162–179.

House, R., Rousseau, D. M., and Thomas-Hunt, M. 1995. The meso paradigm: A framework for integration of micro and macro organizational research. In *Research in Organizational Behavior*, eds. L. L. Cummings and B. Staw, pp. 71–114. Greenwich, CT: JAI Press.

Hsieh, J. J. P.-A., Rai, A., and Keil, M. 2008. Understanding digital inequality: Comparing continued use behavioral models of the socio-economically advantaged and disadvantaged. *MIS Quarterly* 32(1):97–126.

Hsieh, J. J. P.-A., Rai, A., and Keil, M. 2010. Addressing digital inequality for the socioeconomically disadvantaged through government initiatives: Forms of capital that affect ICT utilization. *Information Systems Research* 22(2):233–253.

Johns, G. 2006. The essential impact of context on organizational behavior. *Academy of Management Review* 31(2):386–408.

Johnston, A. C. and Warkentin, M. 2010. Fear appeals and information security behaviors: An empirical study. *MIS Quarterly* 34(3):549–566.

Kane, G. 2011. A multimethod study of Wiki-based collaboration. *ACM Transactions on Management Information Systems* 2(1):1–16.

Kane, G., Fichman, R. G., Gallaugher, J., and Glaser, J. 2009. Community relations 2.0: With the rise of real-time social media, the rules about community outreach have changed. *Harvard Business Review* 87(11):45–50.

Kang, S., Lim, K. H., Kim, M. S., and Yang, H.-D. 2012. Research note: A multilevel analysis of the effect of group appropriation on collaborative technologies use and performance. *Information Systems Research* 23(1):214–230.

Karahanna, E., Agarwal, R., and Angst, C. M. 2006. Reconceptualizing compatibility beliefs in technology acceptance research. *MIS Quarterly* 30(4):781–804.

Karahanna, E., Evaristo, R., and Srite, M. 2000. Methodological issues in MIS cross-cultural research. *Journal of Global Information Management* 10(1):48–55.

Karahanna, E., Evaristo, R., and Srite, M. 2005. Levels of culture and individual behavior: An investigative perspective. *Journal of Global Information Management* 13(2):1–20.

Karasek, R. A. 1979. Job demands, job decision latitude, and mental strain: Implications for job redesign. *Administrative Science Quarterly* 24(2):285–311.

Karimi, J., Somers, T. M., and Bhattacherjee, A. 2009. The role of ERP implementation in enabling digital options: A theoretical and empirical analysis. *International Journal of Electronic Commerce* 13(3):7–42.

Kim, S. S. 2009. The integrative framework of technology use: An extension and test. *MIS Quarterly* 33(3):513–537.

Kim, H. W. and Kankanhalli, A. 2009. Investigating user resistance to information systems implementation: A status quo bias perspective. *MIS Quarterly* 33(3):567–582.

Kim, S. S., Malhotra, N. K., and Narasimhan, S. 2005. Two competing perspectives on automatic use: A theoretical and empirical comparison. *Information Systems Research* 16(4):418–432.

King, J. L. and Lyytinen, K. 2004. Reach and grasp. *MIS Quarterly* 28(4):539–551.

Komiak, S. Y. X. and Benbasat, I. 2006. The effects of personalization and familiarity on trust and adoption of recommendation agents. *MIS Quarterly* 30(4):941–960.

Kozlowski, S. W. J. and Klein, K. J. 2000. A multilevel approach to theory and research in organizations: Contextual, temporal, and emergent processes. In *Multilevel Theory, Research, and Methods in Organizations: Foundations, Extensions, and New Directions*, eds. K. J. Klein and S. W. J. Kozlowski, pp. 3–90. San Francisco, CA: Josey-Bass.

Kraut, R. E., Patterson, M., Lundmark, V., Kiesler, S., Mukopadhyay, T., and Scherlis, W. 1998a. Internet paradox: A social technology that reduces social involvement and psychological well-being? *American Psychologist* 53(9):1017–1031.

Kraut, R. E., Rice, R. E., Cool, C., and Fish, R. S. 1998b. Varieties of social influence: The role of utility and norms in the success of a new communication medium. *Organization Science* 9(4):437–453.

Kuhn, T. S. 1996. *The Structure of Scientific Revolutions*. Chicago, IL: University of Chicago Press.

Latane, B. 1996. Dynamic social impact: The creation of culture by communication. *Journal of Communication* 46(4):13–25.

Lee, A. S. 1989. A scientific methodology for MIS case studies. *MIS Quarterly* 13(1):33–50.

Lee, Y., Chen, A. N. K., and Ilie, V. 2012. Can online wait be managed? The effects of filler interfaces and presentation modes on perceived waiting time online. *MIS Quarterly* 36(2):365–394.

Lee, Y., Kozar, K. A., and Larsen, K. 2003. The technology acceptance model: Past, present, and future. *Communications of the Association for Information Systems* 12(50):752–780.

Leonard-Barton, D. and Deschamps, I. 1988. Managerial influence in the implementation of new technology. *Management Science* 34(10):1252–1265.

Liang, H. and Xue, Y. 2009. Avoidance of information technology threats: A theoretical perspective. *MIS Quarterly* 33(1):71–90.

Locke, E. A. 2007. The case for inductive theory building. *Journal of Management* 33(6):867–890.

Lohse, G. L. and Spiller, P. 1999. Internet retail store design: How the user interface influences traffic and sales. *Journal of Computer-Mediated Communication* 5(2), available at: http://onlinelibrary.wiley.com/doi/10.1111/j.1083-6101.1999.tb00339.x/full (accessed on April 8, 2013).

Lucas, H. C., Swanson, E. B., and Zmud, R. W. 2007. Implementation, innovation, and related themes over the years in information systems research. *Journal of Association for Information Systems* 8(4):206–210.

Maienhofer, D. and Finholt, T. 2002. Finding optimal targets for change agents: A computer simulation of innovation diffusion. *Computational and Mathematical Organization Theory* 8(4):259–280.

Malhotra, Y., Galletta, D. F., and Kirsch, L. J. 2008. How endogenous motivations influence user intentions: Beyond the dichotomy of extrinsic and intrinsic user motivations. *Journal of Management Information Systems* 25(1):267–300.

Markus, M. L. and Keil, M. 1994. If we build it, they will come: Designing information systems that people want to use. *Sloan Management Review* 35(4):11–25.

Maruping, L. M. and Agarwal, R. 2004. Managing team interpersonal processes through technology: A task-technology fit perspective. *Journal of Applied Psychology* 89(6):975–990.

Mathew, J., Sarker, S., and Varshney, U. 2004. M-commerce services: Promises and challenges. *Communications of the Association for Information Systems* 14(1):1–16. http://aisel.aisnet.org/cais/vol14/iss1/26/

McCrae, R. R., Costa, P. T., and Martin, T. A. 2005. The NEO-PI-3: A more readable revised NEO personality inventory. *Journal of Personality Assessment* 84(3):261–270.

Mehra, A., Kilduff, M., and Brass, D. J. 2001. The social networks of high and low self-monitors: Implications for workplace performance. *Administrative Science Quarterly* 46(1):121–146.

MIS Quarterly. 2010. MIS Quarterly's most cited articles. December 2010, available at: http://misq.org/skin/frontend/default/misq/pdf/MISQStats/MostCitedArticles.pdf (accessed on February 20, 2011).

Moore, G. C. and Benbasat, I. 1991. Development of an instrument to measure the perceptions of adopting an information technology innovation. *Information Systems Research* 2(3):192–222.

Morris, M. G. and Venkatesh, V. 2010. Job characteristics and job satisfaction: Understanding the role of enterprise resource planning system implementation. *MIS Quarterly* 34(1):143–161.

Neilsen, J. and Levy, J. 1994. Measuring usability: Preference vs. performance. *Communications of the ACM* 37(4):66–75.

Nicholson, D., Sarker, S., Sarker, S., and Valacich, J. S. 2006. Determinants of effective leadership in information systems development teams: An exploratory study of cross-cultural and face-to-face contexts. *Journal of Information Technology Theory & Applications* 8(3):39–56.

Orlikowski, W. J. and Iacono, C. S. 2001. Desperately seeking the "IT" in IT research: A call to theorizing the IT artifact. *Information Systems Research* 12(2):121–134.

Palmer, J. W. 2002. Web site usability, design, and performance metrics. *Information Systems Research* 13(2):151–167.

Pavlou, P. A. 2003. Consumer acceptance of electronic commerce: Integrating trust and risk with the technology acceptance model. *International Journal of Electronic Commerce* 7(3):101–134.

Pavlou, P. A. and Fygenson, M. 2006. Understanding and predicting electronic commerce adoption: An extension of the theory of planned behavior. *MIS Quarterly* 30(1):115–143.

Pavlou, P. A., Liang, H. G., and Xue, Y. J. 2007. Understanding and mitigating uncertainty in online exchange relationships: A principal-agent perspective. *MIS Quarterly* 31(1):105–136.

Podolny, J. and Baron, J. 1997. Resources and relationships: Social networks and mobility in the workplace. *American Sociological Review* 62:673–693.

Polites, G. L. and Karahanna, E. 2012. Shackled to the status quo: The inhibiting effects of incumbent system habit, switching costs, and inertia on new system acceptance. *MIS Quarterly* 36(1):21–42.

Powell, A., Piccoli, G., and Ives, B. 2004. Virtual teams: A review of current literature and directions for future research. *Database for Advances in Information Systems* 35(1):6–36.

Raghu-Nathan, T. S., Tarafdar, M., and Raghu-Nathan, B. S. 2008. The consequences of technostress for end users in organizations: Conceptual development and empirical validation. *Information Systems Research* 19(4):417–433.

Railsback, S. F. and Grimm, V. 2011. *Agent-Based and Individual-Based Modeling: A Practical Introduction.* Princeton, NJ: Princeton University Press.

Repenning, N. P. 2002. A simulation-based approach to understanding the dynamics of innovation implementation. *Organization Science* 13(2):109–127.

Rosenkopf, L. and Abrahamson, E. 1999. Modeling reputational and informational influences in threshold models of bandwagon innovation diffusion. *Computational and Mathematical Organization Theory* 5(4):361–384.

Rotter, J. B. 1954. *Social Learning and Clinical Psychology.* New York: Prentice Hall.

Sambamurthy, V. 2005. Editorial notes. *Information Systems Research* 16(1):1–5.

Sarker, S. and Lee, A. S. 2002. Using a positivist case research methodology to test three competing practitioner theories-in-use of business process redesign. *Journal of the Association for Information Systems* 2(7):1–72.

Sarker, S. and Lee, A. S. 2003. Using a case study to test the role of three key social enablers in ERP implementation. *Information & Management* 40(8):813–829.

Sarker, S. and Lee, A. S. 2006. Does the use of computer-based BPC tools contribute to redesign effectiveness? Insights from a hermeneutic study. *IEEE Transactions on Engineering Management* 53(1):130–145.

Sarker, S., Sarker, S., Joshi, K. D., and Nicholson, D. 2005. Knowledge transfer in virtual systems development teams: An empirical examination of key enablers. *IEEE Transactions on Professional Communications* 48(2):201–218.

Sarker, S. and Valacich, J. S. 2010. An alternative to methodological individualism: A non-reductionist approach to studying technology adoption by groups. *MIS Quarterly* 34(4):779–808.

Sarker, S., Valacich, J. S., and Sarker, S. 2005. Technology adoption by groups: A valence perspective. *Journal of the Association for Information Systems* 6(2):37–72.

Sarker, S. and Wells, J. D. 2003. Understanding mobile wireless device use and adoption. *Communications of the ACM* 46(12):35–40.

Saunders, C. 2005. Looking for diamond cutters. *MIS Quarterly* 29(1):iii–iv.

Schmitz, J. and Fulk, J. 1991. Organizational colleagues, media richness, and electronic mail: A test of the social influence model of technology use. *Communication Research* 18(4):487–523.

Schwarz, A. and Chin, W. 2007. Looking forward: Toward an understanding of the nature and definition of IT acceptance. *Journal of the Association for Information Systems* 8(4):232–243.

Science Watch. 2009. Tracking trends and performance in basic research. ScienceWatch.com, August 2009, available at: http://sciencewatch.com/dr/tt/2009/09-augtt-ECO/ (accessed on February 20, 2011).

Shapiro, C. and Varian, H. R. 1999. *Information Rules: A Strategic Guide to the Network Economy*. Boston, MA: Harvard Business School Press.

Sharma, R., Yetton, P., and Crawford, J. 2009. Estimating the effect of common method variance: The method-method pair technique with an illustration from TAM research. *MIS Quarterly* 33(3):473–490.

Sia, C. L., Lim, K. H., Leung, K., Lee, M. K. O., Huang, W. W., and Benbasat, I. 2009. Web strategies to promote internet shopping: Is cultural-customization needed? *MIS Quarterly* 33(3):491–512.

Silva, L. 2007. Post-positivist review of technology acceptance model. *Journal of the Association for Information Systems* 8(4):255–266.

Srite, M. and Karahanna, E. 2006. The role of espoused national cultural values in technology acceptance. *MIS Quarterly* 30(3):679–704.

Strang, D. and Macy, M. W. 2001. In search of excellence: Fads, success stories, and adaptive emulation. *American Journal of Sociology* 107(1):147–182.

Straub, D. W. 2008. Type II reviewing errors and the search for exciting papers. *MIS Quarterly* 32(2):v–x.

Straub, D. W. 2009a. Why top journals accept your paper. *MIS Quarterly* 33(3):iii–x.

Straub, D. W. 2009b. Creating blue oceans of thought via highly citable articles. *MIS Quarterly* 33(4):iii–vii.

Straub, D. W. and Ang, S. 2008. Reliability and the relevance versus rigor debate. *MIS Quarterly* 32(4):iii–xiii.

Straub, D. W. and Burton-Jones, A. 2007. Veni, vidi, vici: Breaking the TAM logjam. *Journal of the Association for Information Systems* 8(4):224–229.

Straub, D. W., Limayem, M., and Karahanna, E. 1995. Measuring system usage: Implications for IS theory testing. *Management Science* 41(8):1328–1342.

Straub, D. W., Loch, K., Evaristo, R., Karahanna, E., and Srite, M. 2002. Toward a theory-based measurement of culture. *Journal of Global Information Management* 10(1):13–23.

Sun, H. 2012. Understanding user revisions when using information system features: Adaptive system use and triggers. *MIS Quarterly* 36(2):453–478.

Sutanto, J., Kankanhalli, A., and Tay, J. 2008. Change management in interorganizational systems for the public. *Journal of Management Information Systems* 25(3):133–175.

Sutton, R. I. and Staw, B. M. 1995. What theory is not. *Administrative Science Quarterly* 40(3):371–384.

Sykes, T. A. and Venkatesh, V. Explaining post-implementation employee system use and friendship, advice and impeding social ties. *MIS Quarterly* in press.

Sykes, T. A., Venkatesh, V., and Gosain, S. 2009. Model of acceptance with peer support: A social network perspective to understand employees' system use. *MIS Quarterly* 33(2):371–393.

Sykes, T. A., Venkatesh, V., and Johnson, J. L. Enterprise system implementation and employee job performance: Understanding the role of advice networks. *MIS Quarterly*, forthcoming.

Tarafdar, M., Tu, Q., and Ragu-Nathan, T. S. 2010. Impact of technostress on end-user satisfaction and performance. *Journal of Management Information Systems* 27(3):303–334.

Taylor, S. and Todd, P. A. 1995. Understanding information technology usage: A test of competing models. *Information Systems Research* 6(2):144–176.

Thong, J. Y. L., Venkatesh, V., Xu, X., Hong, S. J., and Tam, K. Y. 2011. Consumer acceptance of personal information and communication technology services. *IEEE Transactions on Engineering Management* 58(4):613–625.

Titah, R. and Barki, H. 2009. Nonlinearities between attitude and subjective norms in information technology acceptance: A negative synergy? *MIS Quarterly* 33(4):827–844.

Turel, O., Serenko, A., and Giles, P. 2011. Integrating technology addiction and use: An empirical investigation of online auction users. *MIS Quarterly* 35(4):1043–1061.

Van der Heijden, H. 2004. User acceptance of hedonic information systems. *MIS Quarterly* 28(4):695–704.

Venkatesh, V. 1999. Creation of favorable user perceptions: Exploring the role of intrinsic motivation. *MIS Quarterly* 23(2):239–260.

Venkatesh, V. 2000. Determinants of perceived ease of use: Integrating control, intrinsic motivation, and emotion into the technology acceptance model. *Information Systems Research* 11(4):342–365.

Venkatesh, V., Bala, H., and Sykes, T. A. 2010. Impacts of information and communication technology implementations on employees' jobs in service organizations in India: A multi-method longitudinal field study. *Production and Operations Management* 19(5):591–613.

Venkatesh, V. and Brown, S. A. 2001. A longitudinal investigation of personal computers in homes: Adoption determinants and emerging challenges. *MIS Quarterly* 25(1):71–102.

Venkatesh, V., Brown, S. A., and Bala, H. 2013. Bridging the qualitative-quantitative divide: Guidelines for conducting mixed methods research in information systems. *MIS Quarterly*, 37(1), 21–54.

Venkatesh, V., Brown, S. A., Maruping, L. M., and Bala, H. 2008. Predicting different conceptualizations of system use: The competing roles of behavioral intention, facilitating conditions, and behavioral expectation. *MIS Quarterly* 32(3):483–502.

Venkatesh, V. and Davis, F. D. 1996. A model of the antecedents of perceived ease of use: Development and test. *Decision Sciences* 27(3):451–481.

Venkatesh, V. and Davis, F. D. 2000. A theoretical extension of the technology acceptance model: Four longitudinal field studies. *Management Science* 46(2):186–204.

Venkatesh, V., Davis, F. D., and Morris, M. G. 2007. Deadoralive? The development, trajectory and future of technology adoption research. *Journal of the Association for Information Systems* 8(4):268–286.

Venkatesh, V. and Goyal, S. 2010. Expectation disconfirmation and technology adoption: Polynomial modeling and response surface analysis. *MIS Quarterly* 34(2):281–303.

Venkatesh, V. and Morris, M. G. 2000. Why don't men ever stop to ask for directions? Gender, social influence, and their role in technology acceptance and usage behavior. *MIS Quarterly* 24(1):115–139.

Venkatesh, V., Morris, M. G., and Ackerman, P. L. 2000. A longitudinal field investigation of gender differences in individual technology adoption decision-making processes. *Organizational Behavior and Human Decision Processes* 83(1):33–60.

Venkatesh, V., Morris, M. G., Davis, G. B., and Davis, F. D. 2003. User acceptance of information technology: Toward a unified view. *MIS Quarterly* 27(3):425–478.

Venkatesh, V. and Ramesh, V. 2006. Web and wireless site usability: Understanding differences and modeling use. *MIS Quarterly* 30(1):181–206.

Venkatesh, V., Thong, J. Y. L., and Xu, X. 2012. Consumer acceptance and use of information technology: Extending the unified theory of acceptance and use of technology. *MIS Quarterly* 36(1):157–178.

Wagner, E. L. and Newell, S. 2007. Exploring the importance of participation in the post-implementation period of an ES project: A neglected area. *Journal of the Association for Information Systems* 8(10):508–524.

Wang, W. and Benbasat, I. 2005. Trust in and adoption of online recommendation agents. *Journal of the Association for Information Systems* 6(3):72–101.

Weber, R. 2003. Still desperately seeking the IT artifact. *MIS Quarterly* 27(2):iii–xi.

Webster, J. and Martocchio, J. J. 1992. Microcomputer playfulness: Development of a measure with workplace implications. *MIS Quarterly* 16(2):201–226.

Weick, K. E. 1995. What theory is not, theorizing is. *Administrative Science Quarterly* 40(3):385–390.

Whetten, D. A. 1989. What constitutes a theoretical contribution. *Academy of Management Review* 14(4):490–495.

Wixom, B. H. and Todd, P. A. 2005. A theoretical integration of user satisfaction and technology acceptance. *Information Systems Research* 16(1):85–102.

Wu, J. M. and Lederer, A. 2009. A meta-analysis of the role of environment-based voluntariness in information technology acceptance. *MIS Quarterly* 33(2):419–432.

Xiao, B. and Benbasat, I. 2007. E-commerce product recommendation agents: Use, characteristics, and impact. *MIS Quarterly* 31(1):137–209.

Young, K. S. 2004. Internet addiction: A new clinical phenomenon and its consequences. *American Behavioral Scientist* 48(4):402–415.

Zhang, P., Benbasat, I., Carey, J., Davis, F. D., Galletta, D., and Strong, D. 2002. AMCIS 2002 panels and workshops 1: Human-computer interaction research in the MIS discipline. *Communications of the Association of the Information Systems* 9(20):334–355.

Zmud, R. W. 1996. On rigor and relevancy. *MIS Quarterly* 20(3):xxvii–xxxix.

39

Computer Self-Efficacy

George M. Marakas
Florida International University

Miguel I.
Aguirre-Urreta
DePaul University

Kiljae Lee
University of Kansas

39.1 Introduction

Every year, both organizations and their employees invest a significant amount of resources in training and development programs, in the hope that these will have an important impact on employee growth and ultimately on organizational performance. Among the many skill sets, computer skills are the most frequent type of training provided by organizations (Yi and Davis, 2003). Rooted in Social Cognitive Theory (SCT) (Bandura, 1997), computer self-efficacy (CSE), generally defined as a "judgment of one's capability to use a computer" (Compeau and Higgins, 1995b), has been repeatedly identified as a key outcome of training, mediating the effects of a number of influences on performance, such as training treatments, past experience and demographic variables (Marakas et al., 1998), or personality characteristics and other individual differences (Johnson, 2005), and affecting performance both directly and through its impacts on different motivational and affective mechanisms. Individuals displaying high levels of CSE are expected to be more focused and persistent, put more effort into their endeavors and be more committed to achieving their goals, be more able to cope with negative feedback, and be generally less anxious about completing the task (Marakas et al., 1998).

In addition, there is an important literature base, both in information systems and in other reference and related disciplines, bearing on the issue of self-efficacy modification and development. Among those examined, programs based on behavior modeling, i.e., observing someone perform a target behavior and then attempting to reenact it, have been shown to be particularly effective (Johnson and Marakas, 2000). Further, such research has become the focus of more detailed examinations into the processes by which learning is affected (Yi and Davis, 2003). Research has also focused on collaborative learning, either by itself or in combination with other approaches, such as behavior modeling (Davis and Yi, 2004; Keeler and Anson, 1995). Overall, this rich and growing stream of research has made CSE an attractive target for the implementation of treatments and interventions.

In general, it is then possible to build a theoretical chain of reasoning connecting computer training interventions with ultimate task performance. Indeed, researchers from both the academic and applied communities often use this presumed state of affairs as justification for the importance of furthering our understanding in this realm. For example, Compeau and Higgins (1995a) argued that user training was a widely recognized factor contributing to productive use of information systems in organizations. Yi and Im (2004) highlighted computer task performance as a major contributor to end-user productivity, while Yi and Davis (2003) remarked that effective computer training is a major contributor to organizational performance.

Given the ubiquitous nature of computers in the workplace, the home, and our daily lives, it may be tempting to assume that because we are immersed in using computers and their associated technologies, we all possess the appropriate level of CSE—that is to say, "we all have it." As such, this temptation results in a declaration that while CSE was important 15–20 years ago, it is far less so today. It is our belief that as computers, and IT in general, become more and more pervasive, CSE stands to become more, and not less, important in the future. Both the academic and applied communities are far from being done with their focus on CSE. Both stand at a threshold where academic research can now begin to bring useful understanding to the goal of achieving a high computer literacy and skill set in the workplace and our society.

39.2 Underlying Principles

The primary foundation for the CSE construct is rooted in SCT (Bandura, 1997) and its major theoretical concepts and relationships. Given the very rich and complex nature of this framework, it is unlikely that this brief summary will do it justice. Thus, the reader is referred to the work of Bandura (1986a, 1997) himself for a full account. SCT and its central variable, *self-efficacy*, have proven to be an extremely popular basis for research in a number of distinct areas, such as academic and health behaviors, management and organizational psychology, leadership, training, negotiation, and career development, to name just a few. By one account, more than 10,000 investigations involving this theory have been conducted since it was first formulated more than 25 years ago (most notably, in 2004 alone, an average of 1.67 articles per day were published on self-efficacy) (Judge et al., 2007).

In contrast to unidirectional models of causality that attempt to explain human behavior as being controlled either by environmental influences or by internal dispositions, SCT explains functioning in terms of a *triadic reciprocal causation* between behavior, cognitive and other personal factors, and the external environment. This model of reciprocal determinism is depicted in Figure 39.1.

The model is argued to be deterministic in the sense of the production of effects by events, rather than suggesting that actions are completely determined by a prior sequence of causes independent of the individual and his or her actions. Given that behavior is jointly determined by a number of factors operating interactively, effects produced by particular events are considered to be probabilistic, rather than inevitable, in this conceptual system.

Another important aspect of Figure 39.1 is the postulated reciprocality between the *environment*, *cognitive and personal factors*, and *behavior*. While the underlying logic states that these three factors influence each other in a bidirectional relationship, it neither does assign equal strength to the different sources, nor does state that influences occur simultaneously. Rather, *time* is involved in the activation of causal factors, their influence on others, and the emergence of reciprocal influences. Given this bidirectionality, individuals are considered to be both the products and the producers of their environment (Wood and Bandura, 1989b).

Self-efficacy beliefs are thus concerned with an individual perception of capability to mobilize the motivational and cognitive resources and courses of action needed to exert control over events of personal relevance. This complements true skills and capabilities, in that it stimulates their use under difficult and stressing circumstances. Individuals with the same level of skill may then perform poorly, adequately, or extraordinarily depending on the effects their personal efficacy beliefs have on their motivation and problem-solving efforts (Wood and Bandura, 1989b). Thus defined, perceived self-efficacy is

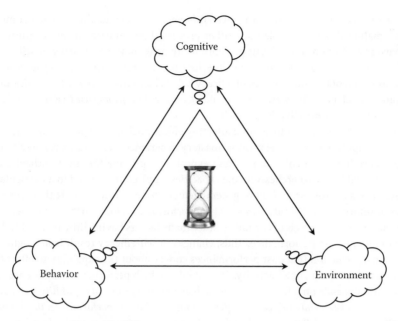

FIGURE 39.1   Triadic reciprocal determinism.

a generative capability in which a number of subskills involving cognitive, social, emotional, and behavioral aspects of functioning must be organized and orchestrated effectively to serve certain purposes.

39.2.1 Dimensions of Self-Efficacy Beliefs

All self-efficacy beliefs vary along three important dimensions (Bandura, 1997). First, they differ in *level*; efficacy beliefs of individuals may be limited to simple task demands, extend to moderately complex ones, or encompass the most taxing performance demands within a particular domain of functioning. Thus, efficacy beliefs are not decontextualized traits upon which situational demands operate. Rather, they represent the actual performance requirements for the context under consideration against which efficacy is judged.

Second, efficacy beliefs vary in *generality*. Individuals may judge themselves to be efficacious across a wide range of activities and domains or only within certain specific activities and domains of human functioning. This assessment of generality itself can vary along different dimensions, such as the degree of perceived similarity present in activities, the mode in which capabilities are expressed (e.g., cognitive, affective, behavioral), qualitative features of different situations, and also characteristics of individuals toward whom behaviors are directed.

Finally, self-efficacy beliefs also vary in *strength*; whereas weak beliefs are easily negated by discomforting experiences, individuals possessing strong convictions will persevere in the face of obstacles and difficulties, and will not be easily overwhelmed by adversity. Strength of self-efficacy, however, is not linearly related to choice behavior, since it is postulated that a certain threshold level must be crossed in order to even attempt a certain course of action, and higher levels of self-efficacy would also result in the same behavior being attempted. Strength of efficacy beliefs manifests itself in the greater perseverance, and thus likelihood of success, in the chosen activity.

39.2.2 Primary Sources of Efficacy Beliefs

Following SCT, beliefs of personal efficacy are constructed from four main sources of information relevant for the formation of a judgment about capability for performing a task. Moreover, these sources are not exclusive of each other, and thus any given influence may operate through one or more of them. However, information that is available but neither processed nor integrated into cognitive thought is unlikely to have any effect on self-referent perceptions. In general, the processing of self-efficacy information depends on two separable functions. The first one relates to the *types of information people attend to and use*; each of the four sources of efficacy beliefs has a distinctive set of indicators and cues. The second function refers to *the rules, processes and heuristics that people use to weight and integrate the different sources of information* (Bandura, 1997).

The most influential source of information affecting efficacy beliefs is *enactive mastery*, e.g., repeated performance accomplishments. While positive mastery experiences serve to increase self-efficacy, negative ones are more likely to debilitate it. These experiences provide the most authentic evidence of whether an individual is able to orchestrate the capacities required to succeed in a particular endeavor. Enactive mastery has a stronger and more generalized effect on efficacy beliefs than any of the other three sources of information. Despite this strength, performance alone does not provide oneself with enough information to judge levels of capability, since many factors having little to do with the latter can have an important effect on performance. Thus, changes in self-efficacy result from cognitive processing of diagnostic information that past performances convey about the capability to effect them rather than the mere act of carrying out an activity. The extent to which performance experiences alter beliefs of self-efficacy also depends on a host of factors such as *preconceptions of capability, perceptions of task difficulty, the amount of external aid received, the circumstances of the performance, the temporal pattern of failures and successes,* and *the process by which these experiences are organized and reconstructed in memory* (Bandura, 1997).

Second, people not only do rely on past experience as the only source of information about their performance capabilities, but also are influenced by *vicarious experiences* obtained from various modeled attainments. Models build beliefs of efficacy by conveying and exemplifying effective strategies to manage a variety of situations. Social comparison processes, such as watching a similar person perform the same or similar task under consideration, also play a role in the effects modeling obtains on self-efficacy, given that human beings partially judge their capabilities in comparison with others. Given that, for most activities, there are no objective measures of adequacy, referential comparisons become very salient. In some cases, performance or social standards are available such that individuals may judge where they stand in comparison with some expected outcome. In others, people seek to compare themselves with similar others, such as coworkers, classmates, competitors, etc., to inform their efficacy beliefs.

Wood and Bandura (1989b) commented extensively on different mechanisms governing modeling processes and on their primary importance for the development of competencies through targeted training interventions. Indeed, a significant portion of research involving self-efficacy, and CSE in particular, as reviewed later, has been conducted in the context of training programs. In reference specifically to vicarious experience, four different processes govern this source of efficacy belief formation.

First, *attentional processes* play an important role in determining what people selectively observe in the large variety of modeling influences available to them and what information is extracted from those that are observed. The logic is that people cannot be influenced by observed performances if those cannot be remembered. Next, *representational processes* involve the transformation and restructuring of remembered information about events into the form of rules and conceptions. Retention of this information is greatly improved when individuals symbolically transform the modeled information into mental codes and then rehearse the coded information. During the third component of behavior modeling, *behavioral production processes*, those symbolic conceptions are now translated into appropriate courses of action, through a matching process in which guided patterns of behavior are enacted and the adequacy of these actions are contrasted against the conceptual model, leading to an iterative process by which individuals adjust their behavior in order to achieve correspondence between actions and conceptions. Finally, *motivational processes* involve three major types of incentive motivators. Individuals are more likely to enact modeled strategies that produce valued outcomes and set aside those with unrewarding or punishing effects. People are also motivated by the success of similar others, but are discouraged from pursuing behaviors that have been known to result in negative consequences. Personal, self-produced standards provide yet another source of performance motivation.

Verbal persuasion represents the third important source of information involved in the development of efficacy belief, but lags behind enactive mastery and vicarious experiences with regard to the strength by which it is able to do so. While verbal persuasion alone may have a limited effect in creating enduring efficacy beliefs, faith in one's capabilities as expressed by significant others, particularly during times of doubt or despair, can have a substantial effect with respect to the mobilization and maintenance of effort. Persuasion has its greater impact on those individuals that, for whatever reason, already believe they can produce certain effects through their actions. The raising of efficacy beliefs to unrealistic levels, however, is likely to result in a failure that would not only undermine future conceptions of self-efficacy, but also discredit the source of information. The degree of appraisal disparity, i.e., how different one's own beliefs are from what people are told, is an important contingency on the effects of social persuasion. While social appraisals that differ markedly from judgments of current capability may be considered believable in the long run, they are unlikely to be acted upon on the short term. Persuasive appraisals are more likely to be believed when they are only moderately beyond what individuals can do at the time.

Lastly, people rely on the information provided by physiological and emotional states in assessing their capabilities. They read their *emotional arousal* and tension as signs of vulnerability to poor performance. Even before CSE had become the subject of focused research, computer anxiety (Heinssen et al., 1987) had already attracted the interest of the information systems community. Because high arousal can debilitate performance, people are more inclined to expect success when they are not undermined by aversive arousal. Treatments that reduce emotional reactions to subjective threats through mastery experiences can help increase beliefs in coping efficacy and correspondingly improvements in performance. Thus, the fourth major way of altering efficacy beliefs is to enhance physical status, reduce stress levels and negative emotional influences, or alter dysfunctional interpretations of bodily states.

39.2.3 Cognitive Processes

Efficacy beliefs affect thought patterns that can either enhance or undermine performance, and these influences take various forms. People with a high sense of efficacy tend to take a future time perspective in structuring their lives, and since most courses of action take initial shape in thought, these cognitive constructions then serve as guides for future actions. Strong efficacy beliefs also affect perceptions of situations as presenting realizable opportunities.

Those who judge themselves inefficacious, on the other hand, construe uncertain situations as risky and visualize scenarios involving failure. Perceived self-efficacy, and this process of cognitive simulation, also affects each other bidirectionally, positive cognitive constructions in turn strengthening efficacy beliefs. A major function of thought is to enable people to both predict the likely outcomes of different courses of action and create the means to exert some degree of control over those (Bandura, 1997).

People draw on their existing knowledge in order to construct and weigh options, integrate predictive factors into rules, test and revise judgments against the results of actions, and also remember which factors were tested and how well those performed. A strong sense of efficacy exerts a powerful influence on self-regulatory cognitive processes and supports the ability to remain task oriented in the face of causal ambiguity, personal and situational demands, and judgment failures that can potentially have important repercussions for the individual. Results from a series of experiments provide strong support for the diverse set of influences that alter efficacy beliefs, and those in turn influence performance attainments through their effects on goals and efficiency of analytic thinking.

39.2.4 Motivational Processes

Bandura (1997) argued that most human motivation is cognitively generated, e.g., people motivate themselves and guide their actions through exercise of anticipation and forethought. Human beings form beliefs about what they can and cannot do, anticipate the likely positive or negative consequences of different courses of action, and then set goals and make plans to realize valued outcomes and avoid the unpleasant ones. It is possible to distinguish three different bodies of theory on motivational processes, built around *causal attributions, outcome expectancies,* and *cognized goals.* Efficacy beliefs play a central role in all of them.

Retrospective judgments of the causes of past performances are postulated to have a motivational effect according to *attribution theory.* Individuals who attribute past successes to personal capability and failure to insufficient effort will undertake more difficult tasks and persist longer in the face of negative feedback, since they perceive outcomes as being a function of how much effort is spent in pursuing them. In contrast, those who ascribe failures to situational determinants will display reduced interest and abandon courses of action in the face of difficulties. Results from extensive research indicate that causal attributions can indeed influence future performance, but that the effect is fully mediated by their influences on beliefs of self-efficacy. For instance, attributions of past success to one's ability heighten beliefs of personal efficacy, which in turn influence future attainments.

SCT, however, presents a more comprehensive picture of the mechanisms by which past performance influences self-efficacy than does attribution theory, which is generally limited to considerations of effort, ability, task difficulty, and chance. People also consider the situation under which they performed, the amount of assistance received, and the rate and pattern of past successes as valuable information that is integrated into judgments of capability. In addition, efficacy beliefs can also bias causal attributions. Thus, efficacious individuals are more likely to attribute failures to insufficient effort or environmental impediments, whereas those with a low sense of efficacy attribute them as arising from a lack of ability. Performance feedback that is inconsistent with perceived self-efficacy is dismissed as less accurate and more likely to be attributed to extraneous factors.

Individuals also motivate themselves by considering the *outcomes* they expect to accrue from following a given course of behavior. Expectancy-value theory essentially predicts that motivation to perform an activity is the result of expecting that doing so would secure specific outcomes and that those outcomes are highly valued by the person considering the performance. Bandura (1997) notes, however, that people are less systematic in their consideration of potential courses of action and in their appraisal of likely outcomes than expectancy-value models would suggest and argues that individuals act on their beliefs about what they can do as well as on their beliefs about the likely effects of their actions. A diminished sense of efficacy can then eliminate the potential allure of certain outcomes if individuals believe that they cannot successfully perform the actions that would lead to them. In activities where outcomes depend on the quality of the performance, efficacy beliefs thus determine which outcomes will be foreseen, and expected outcomes contribute little to future performance when efficacy beliefs are statistically controlled for in research models.

Lastly, behavior is also motivated and directed by *goals* that result from forethought and self-regulatory mechanisms. Cognitive motivation, based on the pursuit of goals or standards, is further mediated by different types of self-influences—affective reactions to performance. Such self-influences include anticipated satisfaction from fulfilling valued standards or self-dissatisfaction with poor performance, perceived self-efficacy for goal attainment (which influences which challenges to undertake), how much effort to spend in the endeavor and how long to persevere in the face of difficulties, and adjustment of standards in light of past attainments (which depend on the construal of the pattern and level of progress being made).

39.2.5 Affective Processes

Efficacy beliefs also play a pivotal role in the regulation of affective states, impacting the nature and intensity of emotional experiences through the exercise of personal control on thought, action, and affect. In the first case, efficacy beliefs create attention biases and influence whether and how events are construed, represented, and retrieved in ways that are emotionally benign or perturbing. Also perceived cognitive abilities influence the control of negative trains of thought that intrude in the flow of consciousness and distract attention from the task and situation at hand. In the action-oriented mode of influence, efficacy beliefs support courses of action that transform the environment in ways that improve its emotional potential. The affective mode influences self-efficacy to negatively affect aversive emotional states once they have been aroused.

39.3 General versus Specific Computer Self-Efficacy

Consistent with the root theory, SCT, upon which the CSE construct is based, we recognize that CSE is a multilevel construct. CSE can be, and has been, operationalized at both the general computing behavior level and at the specific computer application level. Thus, GCSE refers to *an individual's judgment of computing capability across multiple computer application domains* whereas CSE refers to *an individual's perception of efficacy in performing specific computer-related tasks within the domain of general computing.* Research has shown that these two constructs are correlated but distinct (Johnson and Marakas, 2000).

Figure 39.2 presents a theoretical model describing the relationship between GCSE and CSE as well as how the two constructs uniquely affect both task-specific and overall computing performance. Each CSE perception is the major determinant of its associated task performance, and GCSE is the major predictor of general computer performance. Each computer user can have a different set of CSE perceptions related to multiple domain-related experiences that affect the formation of GCSE and can apply different weights to each of them in forming their GCSE perception.

The level of CSE formed at one time further influences the level of subsequent CSE perceptions. Thus, the model explains why two individuals can have the same level of GCSE when their CSEs are different in an application-specific computer domain. It also explains why two individuals can have different levels of GCSE while having similar levels of CSE at the task level. From this, we expect an increase in GCSE as an individual gains experience with different computer applications and computer-related tasks and experiences. We also expect an increasing initial level of CSE for future application-specific tasks over time. Furthermore, the model suggests that particularized measures of CSE will surpass general measures of GCSE in explanatory and predictive power for specific performance as suggested

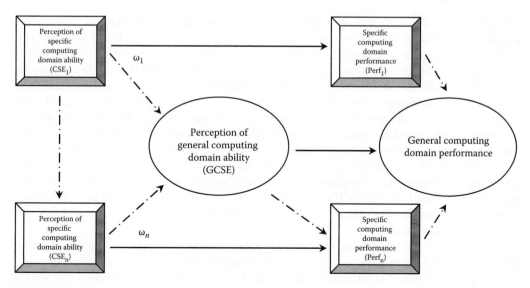

FIGURE 39.2 General computer self-efficacy.

by Bandura (1986a) and Gist (1987) but not for general computing performance. This implies that CSE is a more cost-effective and powerful mechanism to make an impact on user's specific computer performance whereas GCSE is a less malleable but more enduring mechanism over time. This has significant implications for organizational training programs that use CSE formations to assist in performance enhancement within the computing domain.

From an academic perspective, it is important to note that out of the 59 IS papers that used self-efficacy published in the top 4 IS journals from 1985 to 2011, 40 articles chose to employ the GCSE construct while only 19 articles used the CSE construct. This suggests the original premise advanced by Marakas et al. (1998) regarding the need for specific task-related measures of CSE versus simply measuring GCSE has yet to be adopted by the IS research community at large. Additional research is needed to establish the true boundaries of GCSE perception formation such that a generalizable, and reliable, measure of the GCSE construct can be developed.

39.3.1 CSE and Task-Specific Self-Efficacy

Although Bandura (1997) convincingly argued that multiple types of self-efficacy are relevant to performance in any given domain, few studies have examined the joint effects of more than one type on performance (Tierney and Farmer, 2002). Complex organizational tasks generally involve the use of particular means in order to achieve the desired level of performance. However, a mismatch between self-efficacy and performance occurs when only judgments of capability for performing the prescribed means are involved as predictors of performance, since the latter depends also on the particular influence of the selected means and how they relate to task accomplishment (Stajkovic and Luthans, 1998). In particular, Bandura (1997) stated:

> When personal efficacy to perform the prescribed means is used as a predictor in the hypothesized causal model, one is testing not only the predictive power of efficacy beliefs but also the validity of the posited influence of the prescribed means on attainments in the causal model... When the means that produce certain behavioral attainments are only partially understood or have not been adequately verified, efficacy beliefs should be measured at two levels of specificity: efficacy to execute the prescribed means successfully and efficacy to achieve different levels of performance attainments by whatever means people choose to do so (p. 63).

It seems reasonable, given the earlier text, to consider the appropriate positioning of CSE, which is argued to represent the *prescribed means of attainment* (e.g., performing a certain task with the use of computer technologies), in relationship to task self-efficacy (TSE), which represents the more *general level of specificity*. In line with these arguments, Marakas et al. (2007) noted the need to develop complex models of performance that incorporate efficacy estimations of both domain and technical skills, involving the use of multiple instruments. However, there is essentially little existing research (with the singular exception of the work of Looney et al. 2006) that sheds light on the form of the relationship between domain and computing skills and its effects on task performance. In other words, when considering CSE as it relates to a computer-related task, we must not only consider a person's perception of their ability to use a computer to perform the task, but we also simultaneously consider the person's perception of their ability to perform the task. People do not simply use a computer—rather, they use a computer to do something specific. If a person is faced with using an Excel spreadsheet to construct a budget, we must consider their perceived ability to use Excel (CSE) simultaneous with their perceived ability to construct a budget (TSE).

Given this, it is argued that the more specific perception of self-efficacy, one that captures the use of technologies to perform the task, will be most closely related to performance. The next step, then, is to theorize how CSE is related to this intervening construct. Aguirre-Urreta (2008) positions CSE as moderating the effects of TSE on computer-specific task self-efficacy (CSTSE), "an individual's perception of efficacy in performing a particular task with the support of computer technology" (pp. 84–85). While judgments of capability in both realms are jointly needed to influence CSTSE, TSE is positioned as the main driver of the relationship. Paralleling creativity self-efficacy research, it is argued here that, even in the presence of strong CSE, it would be difficult for an individual to feel efficacious in performing a task when devoid of perceived functional capacity (Tierney and Farmer, 2002). On the other hand, weak computer skills are likely to erode the positive effects that TSE brings to performance.

It is expected, then, that feelings of efficacy to perform a task with the support of technology would be stronger when feelings of adequacy for both functional and technology skills are present. Conversely, when both TSE and CSE are low, it is expected that CSTSE would be low too. As theorized by Aguirre-Urreta (2008), the depiction of TSE as the main driver and CSE as the moderating influence would lead one to expect that in situations where TSE is high and CSE is low, CSTSE would be higher than that in those occurrences when the opposite occurs. Finally, when multiple self-efficacies come into play in predicting performance, it is expected that the more specific and proximal type would account for the direct effects on the dependent variable (Bandura, 1997). This positioning of the CSE construct as a moderating variable is a significant departure from previous research, which commonly positioned the CSE construct as an independent or mediating variable.

39.4 CSE and Performance

Of the myriad of antecedent and consequent relationships that have been empirically employed using the CSE construct, there is extensive support for the relationship between CSE and computer performance, in different contexts and for a variety of software applications (Compeau and Higgins, 1995a; Johnson, 2005; Johnson and Marakas, 2000; Marakas et al., 2007; Yi and Davis, 2003; Yi and Im, 2004). Given the objective of these efforts on isolating the performance effects of this central construct, dependent measures have focused exclusively on syntactic- and feature-based uses of the different software packages involved, most notably productivity suites. As such, these outcomes represent knowledge situated in the first three levels of the hierarchy developed by Sein et al. (1998): syntax (actual language through which a user interacts with the application), semantic (meaning of those commands), and functional (e.g., grouping commands into a task such as creating a document).

On the broader level, decades of research involving the self-efficacy construct and work-related behaviors have generated a rich and extensive literature that would be impossible to review in any comprehensive manner but that is, nonetheless, applicable to the more narrowly focused CSE construct.[*] Empirical findings have demonstrated the relationship with work performance in a wide variety of settings, such as *idea generation* (Gist, 1989), *newcomer adjustment* (Saks, 1995), *coping with career events* (Stumpf et al., 1987), *creativity in the workplace* (Tierney and Farmer, 2002), *skill acquisition* (Mitchell et al., 1994), and even *seasickness* (Eden and Zuk, 1995), along with the ever-present research regarding *computer-related behaviors* (Compeau and Higgins, 1995a; Gist et al., 1989; Johnson, 2005; Johnson and Marakas, 2000; Marakas et al., 2007; Martocchio, 1994; Martocchio and Judge, 1997; Yi and Davis, 2003; Yi and Im, 2004).

Continuing with the focal area of interest, some of the early work on CSE was conducted by Hill et al. (1987) and by Gist et al. (1989). In the first case, the authors postulated that efficacy beliefs with respect to computers would be important determinants of the decision to use computers in the future and thus a possible avenue through which adoption could be fostered. In particular, they investigated

---

[*] The reader is referred to the large-scale meta-analysis conducted by Stajkovic and Luthans (1998). Reviewing in excess of 100 different studies, the authors examined both the general performance effects of self-efficacy and the significant within-group heterogeneity of individual correlations. Overall results from the meta-analysis, exclusive of moderator considerations, indicated a positive and significant correlation of self-efficacy and performance of 0.38 ($p < 0.01$). Further analyses indicated the presence of an interaction between moderators, such that average correlations between self-efficacy and performance were strongest for low-complexity tasks performed in laboratory settings (0.50, $p < 0.01$), and weakest, but always significant, for highly complex tasks assessed in field settings (0.20, $p < 0.01$).

the importance of efficacy beliefs over and above beliefs about the instrumental value of using computers. The underlying rationale for their Study 1 was that perceptions of controllability would have an important impact on whether individuals decided to use computers, separately from whether they believed their use implied any particular positive or negative consequences. It has been noted that, in this regard, self-efficacy beliefs may not be altogether different from perceived behavioral control, as incorporated into Ajzen's Theory of Planned Behavior (see Ajzen, 2002; Terry and O'Leary, 1995). Their results showed that, for both men and women, behavioral intentions to use computers were significantly predicted by both instrumentality and efficacy beliefs, and that behavior, measured as enrollment in classes requiring the use of computers, was significantly predicted by behavioral intentions.

In a second study, Hill et al. (1987) predicted that the same relationships would hold, but after adding previous experience with computers as a covariate with both efficacy and instrumentality beliefs. In addition, the behavior of interest was expanded to incorporate the decision to use technological innovations in general. Using a different subject sample, their results confirmed those discussed earlier, even when controlling for previous experience, measured in terms of the number of times a computer had been used in the past, whether the participants had written a computer program, or used a packaged computer system.

Gist et al. (1989) studied the relative effectiveness of alternative training methods designed to foster self-efficacy and the relationship of these perceptions to performance on an objective measure of computer software mastery. The two chosen training approaches were modeling, where participants watched a video of a model illustrating specific steps needed to perform a task, and a tutorial setting, where participants in the study employed an individual training program that presented concepts similar to those shown in the modeling condition. The authors assessed CSE prior to training as a general variable that attempted to capture the confidence subjects brought to the training sessions, and software self-efficacy, post-treatment, as the focal variable of interest.

Results showed support for the hypotheses that trainees in the modeling condition would develop higher software self-efficacy than those in the tutorial condition and that participants initially low in CSE would report higher software self-efficacy in the modeling compared to tutorial training condition. While the relationship between software self-efficacy and performance was not examined, pretest computer efficacy was significantly related to training performance, thus providing additional support for the efficacy–performance relationship in a computer-related context. In addition, this study represents an early example of the separation between more specific and more general types of computer-efficacy perceptions.

Marakas et al. (1998) proposed a model of the relationship between CSE and performance, discussed previously, which included direct and indirect effects of the former on the latter, in some cases also involving moderated relationships.

Yi and Davis (2003) examined the relationship between self-efficacy and performance as part of their efforts to develop an improved training approach based on behavior modeling and observational learning processes such as attention, retention, production, and motivation. This new model was proposed in order to provide a more detailed account of the mechanisms by which modeling-based interventions enhanced training and thus provide a base for both the evaluation and the improvement of training techniques in the future. Software self-efficacy was measured prior to training, as part of the effort to control for pre-training individual differences, and after the training intervention as one of the main outcomes of the proposed approach, declarative knowledge being the other. Task performance was measured immediately after training and a second time with a 10 day delay in between. Results from PLS analysis show that pre-training self-efficacy is significantly related to the post-training construct. The latter had significant effects on both immediate and delayed task performance, even when declarative knowledge was also included in the analysis. This confirms the role of CSE as both one of the primary determinants of performance and one of the primary outcomes of training interventions.

Yi and Im (2004) studied the relationship between CSE, personal goals, and performance, using data from a software training program. Considering beliefs of efficacy and personal goals is important because they provide two different, even if complementary, answers to the question of why some individuals perform better on tasks when they are similar in ability and knowledge to others that do not perform as well on the same tasks. Individuals that display higher levels of self-efficacy believe that they can do more with their capacities than others that possess less confidence in their capability to perform a task. On the other hand, participants that set more challenging and meaningful goals for themselves are more likely to exert additional effort in order to achieve increased performance than those that set less challenging objectives for themselves. In order to account for other pre-training differences, the authors also included prior experience and age as determinants of computer task performance. Results from their structural model showed no significant direct effect of CSE on performance, but did show a significant, if rather small, indirect effect through personal goals. Prior experience had a strong positive effect and age a negative one on performance. Overall, the model proposed by Yi and Im (2004) accounted for almost 40% of the variance in the dependent variable.

Also within a training context, Johnson (2005) proposed two mediators, in addition to a direct effect, of the relationship between application-specific self-efficacy and performance: goal level and goal commitment. These two mechanisms have been shown to be primary determinants of the motivation with which individuals approach tasks and improve performance. Individuals who set challenging goals for themselves exert greater effort and maintain it for longer, and perform better than those who set lower goals. Commitment to these goals, however, is required for them to have any motivational impact on performance; personally, meaningless objectives are unlikely to motivate greater effort expenditure.

Indeed, these motivational processes have been shown to be two of the primary channels by which cognitive appraisals of capability (e.g., efficacy beliefs) operate on behavior and ultimately performance in the series of studies by Bandura and colleagues (Bandura and Jourden, 1991; Bandura and Wood, 1989; Cervone et al., 1991; Wood and Bandura, 1989a). Results from Johnson (2005) confirm these previous findings. Application-specific CSE has significant impacts on performance both directly and through its positive effects on self-set goals and goal commitment, for a combined standardized path coefficient of 0.2933. In addition, it explains 13% and 14% of the variance, respectively, in the two mediating constructs.

Marakas et al. (2007), although specifying CSE as a formative composite, also examined its relationship to performance in the context of analyzing the differential predictive validity of measures developed at different points in the evolution of the computing domain. In order to accomplish this, the authors compared the spreadsheet measure used by Gist et al. (1989) and the one developed by Johnson and Marakas (2000). While both measures significantly predicted spreadsheet performance, the more recent essentially tripled the proportion of variance explained in the dependent variable (it is not clear whether the scale developed by Gist et al., 1989, was also specified as a formative composite).

39.5 CSE and Technology Adoption

In addition to performance, researchers also investigated the role of CSE as a distal predictor of technology adoption behavior and as an antecedent to perceived ease of use (EOU). Although Davis (1989) originally considered self-efficacy as part of the root formulation of perceived EOU, later research positioned CSE in its current role. Venkatesh and Davis (1996) conducted a three-experiment study to test, among other relationships, the hypotheses that GCSE would be a strong predictor of perceived EOU, both before and after hands-on experience with the focal systems. Results from this research revealed significant and large effects of CSE on perceived EOU, both for two different systems and before and after hands-on use. The authors theorized, and empirically supported, that objective usability of the application would only be a factor after participants in their studies had the opportunity to experience the systems by themselves, but that in either case, subjects would still anchor their general perception of EOU on their individual level of self-efficacy.

Building on these results, Venkatesh (2000) used an anchoring and adjustment framework to propose that in forming system-specific EOU, individuals anchor on key individual and situational variables that relate to control, intrinsic motivation, and emotion. With increasing experience, individuals adjust their system-specific perceived EOU to reflect their interaction with the system. Understanding determinants of EOU, such as CSE and computer anxiety, become important from two standpoints: (1) the construct has a direct effect on intention to adopt and indirectly through perceived usefulness and (2) perceived EOU is argued to be an initial hurdle that users have to overcome for acceptance, adoption, and usage of a system. In the absence of direct hands-on experience with new systems, perceived EOU of systems is not distinct across different new systems, suggesting the existence of "common" set of determinants. In this light, CSE and computer anxiety are conceptualized as anchors of this perception. Through a multisite longitudinal (three measurement points) study, the author showed the consistent and strong effects of both constructs on perceived EOU, which did not significantly change as additional experience with the application was gained.

Agarwal et al. (2000) also examined the relationship between CSE and perceived EOU, but distinguished between general and specific levels of the former. In this research, both perceptions of efficacy and EOU were conceptualized at the level of particular applications: the general Windows environment and a spreadsheet package. In addition, a general measure of CSE, adapted from Compeau and Higgins (1995b), was used. The authors predicted that the specific self-efficacy measure would be a more proximal predictor of perceptions of EOU, given that it was a particularized judgment, as opposed to the more global feeling of confidence represented by the more general measure of computer efficacy. Their results show the specific efficacy measures having significant effects on their respective EOU judgments, and the more general measure having indirect and partially mediated direct effects on perceived EOU for Windows (through the specific Windows self-efficacy measure).

CSE was also included as part of the nomological network of the construct of cognitive absorption, developed and validated by Agarwal and Karahanna (2000). Cognitive absorption, defined by the authors as a "state of deep involvement with the software" (p. 673) was tested in relation to the two main technology acceptance beliefs, perceived usefulness and perceived EOU (Davis, 1989). The inclusion of CSE as an antecedent to these beliefs served to help establish the value of the new construct over and above already known influences. In addition to this purpose, beliefs of efficacy were found to have a significant and positive effect on perceived EOU, once more validating the important role CSE has on forming these perceptions and, by extension, in the overall process of technology adoption.

39.6 Impact on Practice

The study of both CSE and GCSE has had a significant impact on both the applied and academic communities. Given its known relationship to performance, the applied community has a powerful tool to effect improvements to computer-related task performance throughout its workforce.

The distinction between internal and external sources of efficacy was first formulated by Eden (2001) based on earlier work on Pygmalion-style leadership (Eden, 1988, 1990, 1992), conceptual distinctions between different sources of efficacy beliefs by Gist and Mitchell (1992), and earlier work on the subjective assessment of the adequacy of tools for job performance (Eden and Aviram, 1993). Though of relatively recent appearance in the management literature, there is a small but growing collection of empirical studies that provides support for the validity of its main propositions across a number of different contexts, such as *psychology* (Stirin et al., 2011), leadership (Walumbwa et al., 2008), *training* (Eden and Sulimani, 2002), and, most notably, *the introduction of new information technologies to the workplace* (Aguirre-Urreta, 2008; Eden et al., 2010).

The vast majority of research related to self-efficacy, however, has been conducted following the seminal work of Bandura (1986b, 1997) with its clear focus on self-efficacy as a subjective judgment of competence for performing specific actions or achieving specific goals. Indeed, Bandura (1997, p. 21) defines self-efficacy as "… a judgment of one's ability to organize and execute given types of performances." After more than 30 years of research in this area, there is overwhelming empirical evidence of the effects of self-efficacy on performance, in both the psychology and management literatures, which includes experimental evidence in both laboratory studies and field experiments (Bandura and Jourden, 1991; Bandura and Wood, 1989), that support the causality of those relationships. There is also ample literature on the determinants of self-efficacy (Bandura, 1997; Gist, 1989; Gist and Mitchell, 1992).

39.7 Implications for Future Research into CSE and GCSE

As discussed in the opening comments of this chapter, it may be easy to conclude that the pervasive nature of computing technologies and their associated skills has rendered further investigation in CSE a moot issue. We wish to be clear that we believe nothing could be further from the truth. As computing technologies continue to become a part of our daily lives, any research that advances our understanding with regard to how we mere mortals interact and perceive such technologies must be regarded as relevant. To that end, continued research focusing on the CSE construct at all levels of analysis and all generalizable contexts should be encouraged and welcomed by both the academic and applied communities alike.

Despite the rich empirical literature, across a wide variety of disciplines, into the CSE construct, there exists a significant portion of our body of knowledge in this area yet to be determined and validated. Extant research has largely ignored the temporal development of self-efficacy beliefs and the possibility of different individual patterns underlying their growth. While this is a rather common limitation of this literature, which has largely focused on cross-sectional and between-individual studies, rather than considering intra-individual variation, a wide variety of specific areas of focus and specific research questions remain to be addressed and answered. Future research should study whether the development of the self-efficacy construct occurs at different rates for different individuals. Further, is there a ceiling effect with regard to either the development of CSE perceptions or their effects on performance-related constructs? Such research becomes important as we begin to place greater emphasis on the development of computer literacy at the elementary education levels. Is there an optimal starting point for such skill development as it relates to raising levels of CSE or GCSE? Can we manipulate CSE at early stages of chronological and cognitive development to mitigate the known gender bias observed in much of the CSE research? Can we use the temporal development of CSE in individuals to predict future performance in computer-related domains? The answers to such research questions would serve to inform for the applied and academic communities and would become instrumental in the design and development of more rigorous and complex training programs and mechanisms.

Along these same lines, research should focus on whether these individual temporal patterns of growth and development of CSE beliefs can be reduced to a limited number of archetypes and can be, therefore, identified based on observable features such that targeted development interventions can be derived. This line of research would be of value not only to the information systems domain, but also to the general literature on efficacy beliefs and human development, and falls in line with recent calls to reconsider the importance of individual units of analysis in psychology (Molenaar, 2004).

Another fruitful area of research that has remained largely unexplored is the determination of exactly what the relationship is between the task-specific CSE and the GCSE level. Marakas et al. (1998) theorized a relationship between the two levels similar to that depicted previously in Figure 39.2. Despite this conceptualization rapidly approaching two decades old (along with the popular application of the GCSE construct in numerous research efforts), no known empirical work has been conducted to validate this conceptualization—or any other, for that matter. It seems reasonable to conclude that the level of interest with regard to the use of the two levels of the construct can be advanced only by an empirical investigation into their true relationship to each other. Is GCSE simply a weighted average of the sum of CSE formations as suggested by Marakas et al. (1998) or is there a different mechanism in place? Do all CSE formations contribute to a formation of perception regarding GCSE or is there a point where no further significant change in GCSE levels is observed despite continued exposure to new opportunities to form CSE perceptions at the task-specific level?

More recently, work by Marakas et al. (2007) specified CSE using a formative, as opposed to reflective, model. Up to this point, researchers using SCT, and even within the domain of computing, had always used the latter form to posit the relationships between latent and manifest variables. The authors argued for this new conceptualization of CSE by noting that indicators currently used to measure the construct did not necessarily follow any particular pattern of correlations, as required by the reflective mode of measurement. In particular, they proposed that a person answering positively to one item might not necessarily answer in a similar manner to a second one, e.g., "It is possible, even likely in many cases, a person responding to the instrument might be capable of installing software, but not capable of accurately describing how a computer works" (p. 21).

The road chosen by Marakas et al. (2007) follows a literature that has so far been mainly concerned with the issue of direction of causality between manifest and latent variables (Bollen and Lennox, 1991; Edwards and Bagozzi, 2000; Jarvis et al., 2003; Mackenzie et al., 2005). As argued by Diamantopoulos and Siguaw (2006), it is not clear why researchers seem to argue that, following a process of measurement construction, the same items should have been selected to comprise either formative or reflective specifications. Pushing the issue a little further, it is also not clear why Diamantopoulos and Siguaw (2006) believe that the issue boils down to selecting different items from the same pool, as opposed to creating separate pools of items altogether. In all this literature, however, there is a more fundamental question that has so far been left unanswered, and from which all other considerations follow directly, namely, why would a researcher choose a reflective or formative specification in the first place? Borsboom et al. (2003) propose that the answer depends on the ontological status of the latent variable being invoked in the research model. In addition, a number of other important issues arise from the choice of specification that extends beyond the mere direction of arrows in the graphical depiction of a causal model, such as the error-free nature of the manifest indicators. Recent research has also questioned the quantitative accuracy of studies implying that misspecification of formative specifications as reflective result in large estimation biases (Aguirre-Urreta and Marakas, 2012), and even others disagree with the choice of model specification by Marakas et al. (2007), as well (Hardin et al., 2008a,b). While the issue of model specification has implications that extend beyond the CSE construct, it has happened that this construct has become the subject matter on which many of these issues are discussed.

This area of focus, along with those previously identified, provides a rich and fertile opportunity for CSE research. Beyond the work conducted by Marakas et al. (2007), no specific focus on advancing our understanding of this debate has been identified. Future research should begin an organized effort to better understand the implications of specifying the construct as formative or reflective. Is it possible that each perspective provides a window into a better understanding of the construct or its direct implications to the applied community that the other does not? Can the two perspectives be reconciled such that the decision as to how to specify the construct lies with the convenience of the researcher? Does the level of analysis at which an investigation is conducted receive any benefit from one perspective over the other? The benefits derived from this area of future research will most assuredly extend beyond the CSE research community and, as such, potentially provide a much larger return, as measured by the breadth and depth of potential beneficiaries, than those discussed earlier. Regardless of the chosen area of focus, however, the need for continued and varietal research into the CSE construct appears to be both clear and unequivocal. Any thought of the sun setting on this stream of research is still far off in the distance, to be sure.

39.8 Summary

The CSE construct, and its related general level, has enjoyed, and continues to enjoy, a rich nomological net developed from a wide variety of disciplines and empirical studies. Further, it stands as one of the few academic foci that can be easily extended into the applied realm to effect material changes and value with regard to the continued development of computer-related skill sets.

Given its strong theoretical foundation, CSE represents a product developed within the academic realm, through the combined efforts of a wide variety of researchers, that has sustainable value in furthering our understanding of human behavior in a society where computing technologies are, and will continue to remain, pervasive.

References

Agarwal, R. and Karahanna, E. (2000). Time flies when you're having fun: Cognitive absorption and beliefs about information technology usage. *MIS Quarterly*, 24(4), 665–694.

Agarwal, R., Sambamurthy, V., and Stair, R. (2000). Research report: The evolving relationship between general and specific computer self-efficacy—An empirical assessment. *Information Systems Research*, 11(4), 418–430.

Aguirre-Urreta, M. (2008). *An Empirical Investigation into the Moderating Relationship of Computer Self-Efficacy on Performance in a Computer-Supported Task*. Lawrence, KS: University of Kansas.

Aguirre-Urreta, M. and Marakas, G. (2012). Revisiting bias due to construct misspecification: Different results from considering coefficients in standardized form. *MIS Quarterly*, 36(1), 123–138.

Ajzen, I. (2002). Perceived behavioral control, self-efficacy, locus of control, and the theory of planned behavior. *Journal of Applied Social Psychology*, 32, 665–683.

Bandura, A. (1986a). *Social foundations of thought and action: A social-cognitive theory*. Englewood Cliffs, NJ: Prentice-Hall.

Bandura, A. (1986b). The explanatory and predictive scope of self-efficacy theory. *Journal of Social and Clinical Psychology*, 4(3), 359–373.

Bandura, A. (1997). *Self-Efficacy—The Exercise of Control*. New York: H. W. Freeman and Company.

Bandura, A. and Jourden, F. (1991). Self-regulatory mechanisms governing the impact of social comparison on complex decision making. *Journal of Personality and Social Psychology*, 60(6), 941–951.

Bandura, A. and Wood, R. (1989). Effect of perceived controllability and performance standards on self-regulation of complex decision making. *Journal of Personality and Social Psychology*, 56(5), 805–814.

Bollen, K. and Lennox, R. (1991). Conventional wisdom on measurement: A structural equation perspective. *Psychological Bulletin*, 110(2), 305–314.

Borsboom, D., Mellenbergh, G., and van Heerden, J. (2003). The theoretical status of latent variables. *Psychological Review*, 110(2), 203–219.

Cervone, D., Jiwani, N., and Wood, R. (1991). Goal setting and the differential influence of regulatory processes on complex decision-making performance. *Journal of Personality and Social Psychology*, 61(2), 257–266.

Compeau, D. and Higgins, C. (1995a). Application of social cognitive theory to training for computer skills. *Information Systems Research*, 6(2), 118–143.

Compeau, D. and Higgins, C. (1995b). Computer self-efficacy: Development of a measure and initial test. *MIS Quarterly*, 19(2), 189–211.

Davis, F. (1989). Perceived usefulness, perceived ease of use, and user acceptance of information technology. *MIS Quarterly*, 13(3), 319–340.

Davis, F. and Yi, M. (2004). Improving computer skill training: Behavior modeling, symbolic mental rehearsal, and the role of knowledge structures. *Journal of Applied Psychology*, 89(3), 509–523.

Diamantopoulos, A. and Siguaw, J. (2006). Formative versus reflective indicators in organizational measure development: A comparison and empirical illustration. *British Journal of Management*, 17, 263–282.

Eden, D. (1988). Pygmalion, goal setting, and expectancy: Compatible ways to boost productivity. *Academy of Management Review*, 13(4), 639–652.

Eden, D. (1990). *Pygmalion in Management: Productivity as a Self-Fulfilling Prophecy*. Lexington, MA: Lexington Books.

Eden, D. (1992). Leadership and expectations: Pygmalion effects and other self-fulfilling prophecies in organizations. *Leadership Quarterly*, 3, 271–305.

Eden, D. (2001). Means efficacy: External sources of general and specific subjective efficacy. In M. Erez, U. Kleinbeck, and H. Thierry (Eds.), *Work Motivation in the Context of a Globalizing Economy*. Hillsdale, NJ: Lawrence Erlbaum.

Eden, D. and Aviram, A. (1993). Self-efficacy training to speed redeployment: Helping people to help themselves. *Journal of Applied Psychology*, 78, 352–360.

Eden, D., Ganzach, Y., Granat-Flomin, R., and Zigman, T. (2010). Augmenting means efficacy to improve performance: Two field experiments. *Journal of Management*, 36(3), 687–713.

Eden, D. and Sulimani, R. (2002). Pygmalion training made effective: Greater mastery through augmentation of self-efficacy and means efficacy. In B. Avolio and F. Yammarino (Eds.), *Transformational and Charismatic Leadership: The Road Ahead*, Vol. 2. Kidlington, Oxford, U.K.: JAI Press.

Eden, D. and Zuk, Y. (1995). Seasickness as a self-fulfilling prophecy: Raising self-efficacy to boost performance at sea. *Journal of Applied Psychology*, 80(5), 628–635.

Edwards, J. and Bagozzi, R. (2000). On the nature and direction of relationships between constructs and measures. *Psychological Methods*, 5(2), 155–174.

Gist, M.E. (1987). Self-Efficacy: Implications for organizational behavior and human resource management. *Academy of Management Review*, 12(3), 472–485.

Gist, M. (1989). The influence of training method on self-efficacy and idea generating among managers. *Personnel Psychology*, 42, 787–805.

Gist, M. and Mitchell, T. (1992). Self-Efficacy: A Theoretical analysis of its determinants and malleability. *Academy of Management Review*, 17(2), 183–211.

Gist, M., Schwoerer, C., and Rosen, B. (1989). Effects of alternative training methods on self efficacy and performance in computer software training. *Journal of Applied Psychology*, 74(6), 884–891.

Hardin, A., Chang, J., and Fuller, M. (2008a). Clarifying the use of formative measurement in the IS discipline: The case of computer self-efficacy. *Journal of the Association for Information Systems*, 9(9), 544–546.

Hardin, A., Chang, J., and Fuller, M. (2008b). Formative vs. reflective measurement: Comment on Marakas, Johnson, and Clay (2007). *Journal of the Association for Information Systems*, 9(9), 519–534.

Heinssen, R., Glass, C., and Knight, L. (1987). Assessing computer anxiety: Development and validation of the computer anxiety rating scale. *Computers in Human Behavior*, 3(1), 49–59.

Hill, T., Smith, N., and Mann, M. (1987). Role of efficacy expectations in predicting the decision to use advanced technologies: The case of computers. *Journal of Applied Psychology*, 72(2), 307–313.

Jarvis, C., Mackenzie, S., and Podsakoff, P. (2003). A critical review of construct indicators and measurement model misspecification in marketing and consumer research. *Journal of Consumer Research*, 30, 199–218.

Johnson, R. (2005). An empirical investigation of sources of application-specific computer self efficacy and mediators of the efficacy-performance relationship. *International Journal of Human-Computer Studies*, 62, 737–758.

Johnson, R. and Marakas, G. (2000). Research report: The role of behavioral modeling in computer skills acquisition—Toward refinement of the model. *Information Systems Research*, 11(4), 402–417.

Judge, T., Jackson, C., Shaw, J., Scott, B., and Rich, B. (2007). Self-efficacy and work-related performance: The integral role of individual differences. *Journal of Applied Psychology*, 92(1), 107–127.

Keeler, C. and Anson, R. (1995). An assessment of cooperative learning used for basic computer skills instruction in the college classroom. *Journal of Educational Computing Research*, 12(4), 379–393.

Looney, C., Valacich, J., Todd, P., and Morris, M. (2006). Paradoxes of online investing: Testing the influence of technology on user expectancies. *Decision Sciences*, 37(2), 205–246.

Mackenzie, S., Podsakoff, P., and Jarvis, C. (2005). The problem of measurement model misspecification in behavioral and organizational research and some recommended solutions. *Journal of Applied Psychology*, 90(4), 710–730.

Marakas, G., Johnson, R., and Clay, P. (2007). The evolving nature of the computer self efficacy construct: An empirical investigation of measurement construction, validity, reliability and stability over time. *Journal of the Association for Information Systems*, 8(1), 16–46.

Marakas, G., Yi, M., and Johnson, R. (1998). The multilevel and multifaceted character of computer self-efficacy: Toward clarification of the construct and an integrative framework for research. *Information Systems Research*, 9(2), 126–163.

Martocchio, J. (1994). Effects of conceptions of ability on anxiety, self-efficacy, and learning in training. *Journal of Applied Psychology*, 79(6), 819–825.

Martocchio, J. and Judge, T. (1997). Relationship between conscientiousness and learning in eTraining: Mediating influences of self-deception and self-efficacy. *Journal of Applied Psychology*, 82(5), 764–773.

Mitchell, T., Hopper, H., Daniels, D., George-Falvy, J., and James, L. (1994). Predicting self-efficacy and performance during skill acquisition. *Journal of Applied Psychology*, 79, 506–517.

Molenaar, P. (2004). A manifesto on psychology as idiographic science: Bringing the person back into scientific psychology, this time forever. *Measurement*, 2(4), 201–218.

Saks, A. (1995). Longitudinal field investigation of the moderating and mediating effects of self-efficacy on the relationship between training and newcomer adjustment. *Journal of Applied Psychology*, 80, 211–225.

Sein, M., Bostrom, R., and Olfman, L. (1998). Rethinking end-user training strategy: Applying a hierarchical knowledge-level model. *Journal of End User Computing*, 9, 32–39.

Stajkovic, A. and Luthans, F. (1998). Self-efficacy and work-related performance: A meta analysis. *Psychological Bulletin*, 124(2), 240–261.

Stirin, K., Ganzach, Y., Pazy, A., and Eden, D. (2011). The effect of perceived advantage and disadvantage on performance: The Role of external efficacy. *Applied Psychology*, 61(1), 81–96.

Stumpf, S., Brief, A., and Hartmann, K. (1987). Self-efficacy expectations and coping with career events. *Journal of Vocational Behavior*, 31, 91–108.

Terry, D. and O'Leary, J. (1995). The theory of planned behavior: The effects of perceived behavioural control and self-efficacy. *British Journal of Social Psychology*, 34, 199–220.

Tierney, P. and Farmer, S. (2002). Creative self-efficacy: Its potential antecedents and relationship to creative performance. *Academy of Management Journal*, 45(6), 1137–1148.

Venkatesh, V. (2000). Determinants of perceived ease of use: Integrating control, intrinsic motivation, and emotion into the technology acceptance model. *Information Systems Research*, 11(4), 342–365.

Venkatesh, V. and Davis, F. (1996). A model of the antecedents of perceived ease of use: Development and test. *Decision Sciences*, 27(3), 451–481.

Walumbwa, F., Avolio, B., and Zhu, W. (2008). How transformational leadership weaves its influence on individual job performance: The role of identification and efficacy beliefs. *Personnel Psychology*, 61, 793–825.

Wood, R. and Bandura, A. (1989a). Impact of conceptions of ability on self-regulatory mechanisms and complex decision making. *Journal of Personality and Social Psychology*, 56(3), 407–415.

Wood, R. and Bandura, A. (1989b). Social cognitive theory of organizational management. *Academy of Management Review*, 14(3), 361–384.

Yi, M. and Davis, F. (2003). Developing and validating an observational learning model of computer software training and skill acquisition. *Information Systems Research*, 14(2), 146–169.

Yi, M. and Im, K. (2004). Predicting computer task performance: Personal goal and self efficacy. *Journal of Organizational and End User Computing*, 16(2), 20–37.

40

Developing Individual Computing Capabilities

Saurabh Gupta
University of North Florida

40.1 Introduction

Individual computing capabilities development continues to be a rich area of investment for enhancing the productivity of individuals, with over 38% of all training directed toward IT training (ASTD 2008). With an increasing percentage of both large and small businesses using computer applications in their daily work, this trend is likely to continue. Recent surveys have found that computer literacy requirements have skyrocketed in almost every job category. For example, over 70% of the companies included in the previous referenced survey now require computer competency in their middle and senior management positions. In addition, research has also shown that development of computing capabilities has a significant effect on new technology adoption (Igbaria et al. 1995).

Considerable amount of research has been done investigating methods to develop individual computing capabilities over the years. Early research in this area focused on traditional training methods in a structured classroom environment. However, an increase in the requirements for continuous learning and growth in the geographic dispersion of trainees is creating a "demand-pull" for going beyond the classroom-based training methods. This, combined with the declining cost of hardware and increasing ubiquity of communication networks, is creating an incentive for many organizations to move toward technology-mediated training or e-learning.

Over the years, three critical frameworks have been proposed to explain the development of individual computing capabilities (Alavi and Leidner 2001; Gupta and Bostrom 2009; Olfman and Pitsatron 2000). These frameworks, however, do not provide a consistent picture of the computer training phenomenon. While two of these focus on the psychological process of learning (Alavi and Leidner 2001; Olfman and Pitsatron 2000), they ignore the role of technology-mediated learning as well as the role of individual interaction with the training method. On the other hand, the most recent framework ignores the psychological learning process, focusing instead on the role of learning systems and their interaction with the learners (Gupta and Bostrom 2009).

The goal of this chapter is to define the nature of individual computing capabilities, outline the key underlying principles based on an integrative framework, along with their practical implications, and identify key areas of future research.

40.2 Defining Individual Computing Capabilities

Individual computing capabilities represent an understanding of the principles by which a system can be applied to a business task (Garud 1997). Drawing from research in educational psychology and IS, individual computing capabilities can be classified into four categories: skill-based capabilities, cognitive capabilities, affective capabilities, and metacognitive capabilities (Gupta et al. 2010).

Skill-based capabilities focus on the ability to use the target system. It is based on two specific types of knowledge base: command-based and tool procedural knowledge. Command-based knowledge is the knowledge of the syntax (set of commands and the command structure) and semantics (meaning of the commands) of the target system. Tool procedural refers to grouping these individual commands to perform a function or task. Without this level of knowledge, users are unable to use the system or recover from errors.

Cognitive capabilities focus on the mental awareness and judgment of the user. Business procedural, tool conceptual, and business conceptual knowledge bases focus on cognitive capabilities. Business procedural or task-based knowledge is about applying tool procedures to business processes. Tool conceptual knowledge focuses on the big picture, that is, the overall purpose and structure of the target system. Acting as an advance organizer, this knowledge provides a basis for the ability to transfer learning to new situations. Business conceptual is the knowledge of the specific business processes supported or enabled by the target system. This knowledge is required to understand the interdependencies of actions in complex systems.

Affective capabilities focus on the emotional aspects of the user's behavior. A review of the literature shows that three affective capabilities have been analyzed in IT training area (Gupta et al. 2010). The first one, motivational knowledge, is the knowledge about what the target system can do for the user's job, the organization, etc., that is, the usefulness of the software to the organization. The second important affective outcome is satisfaction with the training process. Given the continuous need for training, training programs are designed to not only impart knowledge, but also provide a high level of satisfaction. The last affective learning outcome, perceived anxiety, has received much less attention in IS research. Perceived anxiety deals with a feeling of apprehension, tension, or uneasiness in the capabilities of using the target system.

Metacognitive or self-regulated learning knowledge refers to an individual's knowledge regarding their own learning and information-processing processes. Among the most commonly investigated metacognitive variables is self-efficacy or users' belief about their ability to perform a specific behavior (Bandura 1986). Self-efficacy has been shown to affect other training capabilities and to be a strong antecedent to post-training intention to use (Compeau et al. 2005).

40.3 Underlying Principles for Individual Computing Capability Development

As mentioned earlier, two distinct research perspectives have been proposed to explain the development of individual computing capabilities, especially for the contemporary environment. The first one, characterized by Alavi and Leidner (2001), models the impact of training method on training outcomes as mediated by the individual's psychological processes. It argues for a shift in our attention to understanding the relevant instructional, psychological, and environmental factors that enhance learning. The other model, by Gupta and Bostrom (2009), uses adaptive structuration theory (AST) as a basis. AST argues that the influence of these advanced information technologies is moderated by the actions of the actors (learners in this case) (Barley 1986; DeSanctis and Jackson 1994; DeSanctis

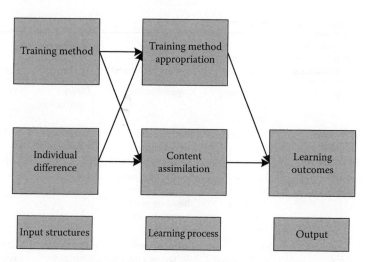

FIGURE 40.1   Individual computing capabilities development model.

and Poole 1994; Orlikowski 2000). AST makes both technology and human agents part of the system, accounting for the interplay between people and technology, as well as the full predictability of IS use in individuals, groups, and organizations (Fulk 1993). This allows AST to preserve the predictive potential of a deterministic perspective, while accounting for interpretive flexibility of the process perspective (Gouran 1989). Researchers have also argued that AST presents a good metatheory for IS research (Bostrom et al. 2009).

Using this as a basis, the model outlined by Gupta and Bostrom (2009) outlines the impact of training methods on training outcomes as moderated by the faithfulness of appropriation of training methods. These seemingly inconsistent models have one thing in common. Both stress the importance of learning process as well as the importance of individual differences. However, these models focus on different aspects of the learning process.

The model outlined in this chapter (see Figure 40.1) integrates the two models to provide a comprehensive overview of the development of individual computing capabilities through training. The combined model highlights five distinct constructs: training methods and individual differences as inputs into a training situation, appropriation of the training method and the psychological process as a part of the learning process, and the training outcomes as the final construct. The input structures and learning process are discussed in the following sections. We have already discussed individual computing capabilities earlier.

40.3.1 Input Structures

Training situations involve two critical inputs. The first one is the training method, as designed by the trainer, and the second is the individual trainees themselves. From an AST perspective, these are considered as input structures into a process.

40.3.1.1 Training Methods

A training method is defined as a combination of structures that guides individuals to achieve the learning outcomes. DeSanctis and Poole (1994) state that input structures can be described in terms of their features and spirit.

The spirit of a training method can be described in terms of the epistemological perspective the training method follows. Epistemology describes overarching beliefs about the nature of knowledge and about what it means to know something (Hannafin et al. 2004). It provides a design template for creating process and content structures embedded in the learning method. We utilize the most

TABLE 40.1 Epistemological Perspectives

Epistemological Perspectives	Assumption about the State of Knowledge	Learning Outcomes Focus On	Learning Happens By	Traditional Learning Technique	Technology-Based Learning Technique
Behaviorism	Objective	Conditioning learner behavior	Response to stimulus	Direct instruction	Drill
Cognitivism	Objective within context	The learner's thought process	Acquiring and reorganizing the cognitive structures	Behavioral modeling	Intelligent tutor
Constructivism	Constructed reality	The learner's ability to construct reality	Constructing new ideas or concepts based on prior knowledge and/or experience	Unguided case study	Self-regulated
Situationalism	Negotiated reality	Social construction by the learner	Participate and behave as a member of the community	Discussing, writing	Computer supported collaborative learning (CSCL)

common classifications in both IS and education literatures: behaviorism, cognitivism, constructivism, and situationalism (Leidner and Jarvenpaa 1995; Vries 2003). Table 40.1 describes these perspectives, their underlying assumptions, traditional learning techniques, and technology-based technique used to operationalize these perspectives.

The behaviorist view is based on the assumption that human behavior is predictable. Under this theory, learning takes place when new behaviors or changes in behaviors occur as the result of an individual's response to stimuli. Thus, the end goal is defined upfront, and each step necessary to achieve the goal is given to the learners (Burton et al. 2001). For example, direct instruction, which is the most popular form of learning method, is based on this perspective.

The cognitivist perspective holds that learning is a process that is dictated by the participant's cognitive structure and the presentation of the information to the participant. Under this perspective, learning is the change in the mental model as a result of the training. Thus, training methods under this theory have a predefined goal along with information necessary to reach the goal, but the process of cognition of the information is left to the learners (Winn and Snyder 2001). Several researchers have also compared training methods based on behaviorist and cognitive perspectives (Carroll et al. 1997; Olfman and Mandviwalla 1994; Santhanam and Sein 1994), as well as conceptual model or procedural model (Mayer 1981). Sein et al. (1989) found that the effectiveness of the conceptual model depends on the individual user characteristics interacting with the conceptual model and not on the type of model alone. More recently, Coulson et al. (2003) investigated the effectiveness of conceptual models for complex systems. When controlling for initial knowledge, the research found a significant positive effect for using a conceptual model.

The most prevalent theory used to understand participant learning in education as well as in IS is Social Cognitive Theory (Bandura 1986). This theory states that it is not just the exposure to a behavior, but learner action in exploring, manipulating, and influencing the environment that counts. Two kinds of observational learning methods have been differentiated in theory: (1) observation of others' actions, referred to as vicarious learning/modeling or behavior modeling and (2) observation of self-actions or enactive learning (Schunk 2004). Much of the end user training (EUT) literature has focused on vicarious modeling (VM) as a method of learning. Table 40.2 summarizes the literature in this area. Vicarious modeling treatment in previous research has been done by using an external actor to demonstrate actions (usually packaged in a video). Instructor-based treatment, on the other hand, uses the same content, but without demonstrations of the content being taught. A consistent finding is that VM yields better training outcomes than other methods such as instructor-based instruction or studying

TABLE 40.2 Vicarious Learning Literature in IT Training

Study/Target System	Training Intervention	Learning Outcomes	Findings
Gist (1988): Spreadsheet	VM vs. instruction-based training	Skill: Task performance	VM yielded higher task performance scores for both younger and older trainees
Compeau et al. (1995): Lotus 1-2-3 & WordPerfect	VM vs. instruction-based training	Skill: Task performance Metacognitive: Computer self-efficacy (CSE)	Subjects in the VM condition developed higher CSE and performed better than those in the instruction-based condition for spreadsheet program, but not for a word-processing program
Simon et al. (1996): MicroSnap II	Instruction exploration and VM	Skill: Task performance Cognitive: Comprehension Affective: End user satisfaction	VM outperformed the other two methods on all learning outcome measures
Johnson et al. (2000): Excel	Modeling vs. instruction-based training	Skill: Task performance Affective: Computer anxiety Metacognitive: CSE	Subjects in modeling treatment developed higher CSE and performed better than those in nonmodeling treatment. Computer anxiety was significantly related to CSE and task performance.
Bolt et al. (2001): Word and Excel	VM vs. instruction-based training when controlling for complexity	Skill: Task performance Metacognitive: CSE	VM outperformed nonmodeling when complexity was high
Yi et al. (2001): Excel	VM with practice vs. VM with retention enhancement vs. VM with retention enhancement and practice	Skill: Task performance Affective: Attitude	Subjects in the VM with retention enhancement and practice showed higher levels of learning outcomes when compared to the other groups
Yi et al. (2003): Excel	VM vs. VM with retention enhancement	Skill: Task performance Cognitive: Declarative knowledge Metacognitive: Self-efficacy	Subjects in the VM with retention enhancement showed higher levels of learning outcomes
Davis et al. (2004): Excel	VM vs. VM with symbolic mental rehearsal (SMR)	Skill: Task performance Cognitive: Declarative knowledge	VM with SMR was better than VM alone. Learning outcomes were mediated by the trainees' knowledge structures
Gupta (2006): Excel	VM with practice vs. VM with enactive learning (e-learning) vs. Collaborative learning in both treatments	Skill: Procedural knowledge Cognitive: Declarative knowledge Metacognitive: Self-efficacy, specific self-efficacy, satisfaction	Overall, VM with enactive learning performed better. The learning process was mediated by the attitude toward e-learning technology

from a manual. Current research has also tested four enhancements to VM: practice (Yi and Davis 2001), retention enhancement (Yi and Davis 2003), symbolic mental rehearsal (Davis and Yi 2004), and enactive learning (Gupta 2006). No significant impact of practice was found, but the later three enhancements have had a significant impact on learning outcomes.

The third perspective is the constructivist perspective, which states that individuals construct knowledge by working to solve realistic problems. Under this perspective, learning is the process whereby individuals construct new ideas or concepts based on prior knowledge and/or experience. A constructivist designer usually provides all the information necessary for learning, but allows the learner to absorb the materials and information in a way that is most comfortable and to arrive at their own conclusions (Duffy et al. 2001). Self-regulated learning to develop individual capabilities is an example of such a perspective; in recent studies, it has shown promise (Gravill and Compeau 2003; Kadlec 2008).

The situationist perspective on learning highlights the fact that meaning is not bound to the individual. Instead, meaning is socially constructed through negotiations among past and present members of the community involved in the domain. According to the proponents of this perspective, authentic activities, that is, the ordinary practices of the domain culture rather than traditional classroom activities, are needed for knowledge to be constructed. Computer-supported collaborative learning is a prime example of learning using this perspective. Drawing from the theory of personal relevance, Ross (1983) argued that motivation is enhanced in settings where personally relevant information can be processed and there exists a direct involvement in the message being evaluated. Olfman et al. (1991) compared training methods organized as application-based (personally relevant) or construct-based and found no significant difference in training outcomes, though there was some evidence that application-based training is best for novice users. Kang et al. (2003), investigating a similar phenomenon, suggested that training should include a broader context knowledge to be successful.

In summary, the most examined training methods in IT training are based on social cognitive theory. These methods, in general, have proven to be significantly better than traditional lectures. However, specific benefits have also been shown using training methods based on other epistemological perspectives. Thus, an important research theme would be to investigate what combination of epistemological perspectives, for example, combining demonstration within a specific situation, should be adopted in a given context: target system, individual differences, etc. In addition, how contemporary instructional technology can be incorporated in training continues to be a challenge. With the increase in the number of structures available for IT training like collaboration, learning objects, simulations, etc., there is also need for a good contingency theory in this area. Practitioners are struggling with how to create blended training methods. Systems design research (Schonberger 1980; Zhu 2002) and summary of various conceptualizations of fit (Venkatraman 1989) present a good foundation for a research stream in this area. Alternatively, design theories have argued for a theory-based design of training programs (Hevner et al. 2004; Reeves et al. 2004). Both of these areas present a very promising stream of research that will help explain inconsistent previous results, filling an important gap in the current body of knowledge.

40.3.2 Individual Structures

Individual differences have played an important role in IS (Sun and Zang 2005) as well as in education research (Lehtinen et al. 2001). In education, individual difference variables define the cognitive aspects of human activities that are often referred to as "learning ability." These variables influence training outcomes directly, by forming mental models, or indirectly, through interactions with training methods (Olfman and Pitsatron 2000). Table 40.3 summarizes the previous research with respect to these variables.

Generally, individual difference constructs are classified as traits or states (Bostrom et al. 1990). A trait is a distinguishing feature of an individual's nature and is relatively enduring, while psychological states are general emotion-based characteristics relating to the current context. States are dynamic; they change over time and from one situation to another.

The most examined individual differences are self-efficacy, motivation, and learning styles (see Table 40.3). More generally, in the case of both states and traits, IT training researchers have found a strong impact of individual differences on capability development. Theoretically, most IT training has focused on the direct impact of individual differences on individual capability development. This is consistent with the lower half of the model presented in Figure 40.1. The model states that individual differences affect the psychological process of content assimilation directly, which, in turn, affects learning.

TABLE 40.3 Individual Differences Research in IT Training

Study/Target System	Individual Difference	Learning Outcomes	Findings
Traits			
Sein et al. (1989); Bostrom et al. (1990): E-mail	Kolb learning style	Skill: Near transfer tests, far-transfer tests, efficiency	Abstract modelers performed better than concrete modeling subjects, especially in far-transfer tasks.
Sein et al. (1991): Email & Lotus 123		Affective: Satisfaction	Assimilators did not show a significant difference in learning outcomes between computer-based training (CBT) and lecture-based training.
Bohlen et al. (1997): Word-processing			
Gist (1988): Spreadsheet Webster et al. (1993): WordPerfect	Age	Skill: Test performance Affective: Post-training motivation to learn	Younger trainees performed better. Task labeling as play is better than task labeling as work for younger participants, but this was not true for older participants.
States			
Compeau et al. (1995): Lotus 123, WordPerfect	Self-efficacy	Skill: Task performance Metacognitive: Computer self-efficacy	Pretraining self-efficacy significantly influenced post-training learning outcomes.
Johnson et al. (2000); Marakas et al. (1998): Excel			
Martocchio et al. (1997): Windows 3.1			
Szajna et al. (1995): Basic IT skills	Computing attitude and achievement	Skill: Learning performance	Computing attitude and achievement are related to performance.
Szajna et al. (1995); Keller et al. (1995): Basic IT skills	Anxiety	Skill: Learning performance	Anxiety was not related to learning performance. Anxiety has significant interaction with training method.
Yi et al. (2003): Excel	Intrinsic motivation	Skill: Task performance Cognitive: Declarative knowledge Metacognitive: Self-efficacy	Positive correlation between intrinsic motivation and learning outcome.
Gravill (2004): Oracle 11i Navigator	Metacognitive + Self-regulated learning strategies + Goal orientation + Self-awareness + Task attention	Skill: Procedural knowledge Cognitive: Declarative knowledge Metacognitive: Self-efficacy	Learners who devoted effort toward the use of self-regulated learning strategies benefited in terms of higher learning outcomes.

Since, in most cases learning outcomes are used as a proxy for measuring content assimilation, it explains the direct effect that most of these studies found.

On the other hand, research in educational psychology (classified as aptitude treatment interaction) suggests that an individual's aptitude has an interaction effect with learning methods on learning outcomes (Ackerman et al. 1999). However, education researchers also noted that much of the learning process research has focused on post hoc analysis of results or on opinions of the researcher rather than theory-driven empirical analysis (Rohrbeck et al. 2003). In IT training literature, Szajna et al. (1995) found that computing aptitude and achievement are related to learning performance, whereas anxiety and preexisting experience are not. However, Khalifa et al. (1995) found a significant interaction effect for anxiety; students with higher initial anxiety did significantly better in a more supportive environment. Theoretically, this can be explained by the fact that individual differences have a direct impact on how a person appropriates the learning method, which in turn has an effect on individual capability development. However, in spite of the aforementioned theoretical arguments, the impact of individual differences on appropriation of training method has not been investigated in the IT training literature.

Overall, more work is needed in this area to clarify the role of individual differences in developing individual computing capabilities. In this regard, more IT-specific traits and states need to be examined further. In a learner-centric era, where the learner is responsible for learning, the impact of these IT-specific traits and states is likely to grow and create significant challenges needing further investigation (Salas and Cannon-Bowers 2001). For example, one salient trait found to apply in other IS contexts, yet not examined in EUT research, is personal innovativeness in IT (Agarwal and Karahanna 2000). In addition, researchers focusing on individual differences should examine both the direct effects on learning outcomes and the effect of individual differences on learning process. The model and arguments presented earlier provide a good framework for researchers to examine the effect of individual differences and assertions stated previously.

40.3.3 Process Mechanisms

The learning process is a series of actions or cognitive development that leads to a change in an individual's capability. Two critical activities happen during this process. First is the interaction of the individual with the elements of the training environment, and second is the development of mental models. Both are discussed later.

40.3.3.1 Structural Appropriation

The model outlined in Figure 40.1 provides AST as a basis for measuring the learning process. The learning process is viewed as an appropriation or structuration process where participants learn and adapt the learning method structures based on their interpretation of the designer's intent. Like all perceptions, this interpretation varies among learners. Although educational researchers do not have a concept similar to appropriation, they have focused a lot on scaffolding because of its importance (Chang et al. 2009; Ge and Land 2003). Scaffolding presupposes appropriation of structures/learning methods but focuses on how to guide or facilitate appropriation instead of structural impacts. The concepts of scaffolding and appropriation complement each other and would be useful for investigating use of learning method structures.

Assuming that the learning method reflects the values and assumptions of the epistemological perspective and the learning capabilities (i.e., for well-designed structures), a faithful appropriation occurs when participants' interaction is consistent with the spirit (Poole and DeSanctis 1992). Faithfulness is not necessarily concerned with the precise duplication of the procedures provided; rather, it is concerned with whether the structures are used in a manner consistent with the overarching intention in which the designer intended the system to be used. A participant's unique or innovative use of the structures may well be a faithful appropriation as long as their use is consistent with the spirit that the learning method intended to promote (Chin et al. 1997). Ironic appropriation occurs when the participants' interactions violate the spirit of the structure with or without abandoning the underlying learning method (Poole and DeSanctis 1990). In the case of well-designed learning methods, ironic appropriation could introduce internal contradictions within the structures governing interaction. Over time, these contradictions will cause tensions in interactions, which might lead to lower effectiveness of the structures. These contradictions must be addressed, detracting the participant(s) from the learning focus, leading to lower learning outcomes.

When analyzing technology appropriation, Poole and DeSanctis (1990) suggest three dimensions that indicate appropriation: faithfulness, attitudes, and level of consensus. That is, structures will only have their intended effect if the design principles are kept intact (faithfulness), if members do not react negatively to it (attitudes), and if members agree substantially over how structures are used (consensus). Considerable support for the proposition exists in AST literature, particularly the group support systems literature. Indirect support exists in the education literature, especially in the scaffolding literature. Most recently, in a follow-up to their conceptual model, Gupta and his colleagues studied appropriation as an influence on learning outcomes and found strong support (Gupta 2008).

40.3.3.2 Psychological Processes

The Alavi and Leidner (2001) model argues for a similar physiological or cognitive activity done by an individual as a part of the learning process. Their conceptualization of the learning process deals with how information is absorbed and stored in our minds, that is, content assimilation. Such a process deals with how a given content is captured to create new or modify existing mental models by the trainee. Analyzing the process, Davis and Yi (2004) found significant support for the importance of mental models on learning outcomes. Research in this area is often based on more general theories of content assimilation, such as Piaget's theory of cognitive development, Ausubel's theory of subsumption, or Mayer's theory of assimilation (Schunk 2004).

More specifically, educational psychologists have generally theorized three phases of learning: preparation for learning, acquisition and performance, and transfer of learning (Gagne 1985). Preparation for learning deals with the learners focusing on stimuli to the material and orienting themselves toward the goals. The main phase of learning is the acquisition and performance phase. This deals with recognizing the relevant stimulus features and transferring them to the working memory for processing. These are subsequently encoded, and new knowledge is transferred to create new/update existing mental models. Transfer of learning deals with developing models for new situations based on what has been learned. Only one study has examined this process in an IT training context (Yi and Davis 2003). This study, done in the context of social cognitive theory, found significant positive effect of the three aforementioned learning process phases on outcomes.

In addition to this, an individual can build a mental model of the system in three different ways: mapping via usage, mapping via analogy, and mapping via training. Only one research study on directly focusing on metal model building in an IT training context (focusing narrowly on mapping via training) has been conducted (Bostrom et al. 1990). While more studies are needed in understanding mapping via training, the other two methods also represent important areas in understanding how individuals build their computing capabilities.

In summary, the model in Figure 40.1 identifies four critical elements that influence the development of individual computing capability. The training methods, based on different epistemologies, and the trainees, with their individual differences, present the starting point. Individuals learn as they go through the process of content assimilation and training method appropriation. Key practical implications and future research areas are discussed next.

40.4 Practical Implications

The review of the literature presented in the previous paragraphs has many different implications in practice. First, individual capabilities need to be more broadly defined and assessed when compared to what has traditionally been done in practice. Most assessment methods focus only on skill-based capabilities, leaving out the other three. While skill-based capabilities are the foundation, the real exploitation of an information system is dependent on the levels of the other three capabilities.

Second, trainers need to shift their attention toward incorporating learning theories into their training methods. In addition, the literature clearly states that VM-based methods have the greatest impact on developing individual computing capabilities. Whether implemented in traditional environment or e-learning environment, trainers need to focus on how the components/features of the training method enhance critical dimensions of VM. Research also shows that certain enhancements, like the ability to take notes, work in simulations, work in teams, etc., can increase the effectiveness of VM.

The third implication deals with accommodating individual differences to personalized training. Personalization deals with the delivery of content that specifically meets the learner's needs and characteristics. Traditionally, the focus has been on the content of the training, and organizations have used job roles to define the content. However, with advances in technology, the focus is shifting to include the ability to personalize training methods using learning management systems (LMS). To personalize the learning method, a learner profile that captures individual differences should be mapped to different learning methods. Next, LMS can be used to link these methods, preferences, and competencies as needed (Mayer et al. 2004). As such, profiling and job-role competencies categorizations need to be developed and investigated, especially before their use in LMS.

However, this result of good training method and personalization is contingent on the faithful use of e-learning technology by the users. Higher levels of appropriation, especially a positive attitude, toward new training methods, such as e-learning, have a substantial impact on the extent of learning. A practical implementation of this is to encourage participants to continue with e-learning solutions for a longer period, as experience helps in enhancing attitudes toward e-learning. More importantly, it shows the importance of developing a positive attitude upfront. Additionally, trainers should use scaffolds to enhance overall faithfulness of the learning method appropriation.

Overall, this chapter provides a comprehensive view of designing contemporary training programs, in classrooms, as well as virtually. These outcomes can be used not only by corporate businesses/consultants/trainers but also by instructors in K-12 and universities.

40.5 Key Research Issues

The model outlined in Figure 40.1 points to a broad gamut of potential research questions. Most of the research issues deal with the input structures and the learning process. Some of these issues, such as personalization, combination of training methods, changes because of emerging technology, etc., were mentioned earlier in the chapter. In this section, we focus on four important research streams that have not yet been mentioned.

Although not specifically highlighted in research models, the creation of input structures requires a lot of pretraining activities (Olfman and Pitsatron 2000). These activities not only influence the development and design of the training method, but the rest of the training process as well. The study of how these components influence the design and development of the training method thus represents an important research topic for the future. This research will also provide an important feed into the "personalization" aspect of training—a critical input area for the training industry.

From a learning process perspective, there are two distinct ways of investigating the appropriation of training methods. First is the structural focus, concentrating on how well the training method structures are appropriated by the learner. Gupta (2006) found support for the effect of faithfulness of appropriation on learning outcomes. An important implication of this is that future researchers need to account for the level of appropriation in the interpretation of their studies. The second way of understanding the learning process, and a more enduring research theme, is the process focus involving the microlevel analysis of the learning process and the reciprocal causation phenomenon. Very limited research has been done in understanding the "moves" trainees make as a part of the learning process. "Move" studies are usually done at a microlevel analysis of the appropriation process. Poole et al. (Poole and DeSanctis 1992) have identified nine categories of such moves and provide a good starting point for the analysis. Such a research stream would provide a significant contribution to understanding the adaptive structuration process from the trainee's perspective. In all of the previous cases, researchers have suggested that the influence of structures changes over time and thus argue for longitudinal studies (Davis and Yi 2004; Gupta 2006). However, very limited knowledge currently exists on how appropriation changes over time and the causes for the changes.

Additionally, the learning process can also be manipulated through appropriation support or scaffolding for the training. There are various forms of process appropriation (procedural, metacognition, and strategic) support that need to be further studied. Procedural scaffolding helps learners make navigation decisions, such as how to utilize available resources and tools. Metacognitive scaffolds support individual reflection on learning, such as soliciting estimates of current understanding or cuing participants to identify prior related experiences they can reference. Strategic scaffolds support learners in anticipating their interactions with the learning method, such as analyzing, planning, and making tactical decisions. Different support forms are likely to have different influences on the learning process and would represent an important contribution to this stream of literature.

Finally, researchers also need to deal with the continuous and real-time nature of learning. Companies and vendors are incorporating learning technology into their applications and products. Given the nature of current technology evolution, IT training needs assessment will soon not be treated as a distinct event needing follow-up; rather, needs assessment using technology will be continuous, job-driven, and done on demand (Gruene 2005). In either case (whether needs assessment is done in real-time or not), an important research question is: How should organizations best assess the end user skill gap, especially using technology? In addition, as highlighted in the chapter, the capabilities of an EUT program are expanding. Different job roles using a variety of information systems need different kinds of knowledge (DeSousa 2004). Such a spread of information systems, coupled with the breadth of job roles associated with software, provides a compelling need for future study.

40.6 Concluding Remarks

Individual computing capabilities are very important productivity tools in today's digital economy. A particularly large number of job postings highlight this need. In this chapter, we have synthesized existing knowledge on the factors that could influence the development of individual computing capabilities, presented a nomological framework for organizing extant research, and provided areas of future research on ways to improve training effectiveness.

Glossary

Appropriation of learning method: Participants learn and adapt the learning method structures based on their interpretation of the designer's intent.

Collaborative learning: Groups of students work together in searching for understanding, meaning, solutions, or in creating a product.

Content assimilation: A process that deals with how a given content is captured to create new or modify existing mental models by the trainee.

E-collaboration technology: Technologies that offer a rich, shared, virtual workspace in which instructors and students can interact one-to-one, one-to-many, and many-to-many in order to learn together anytime and at any place.

Epistemological perspective or spirit: General intent with regard to values and goals underlying the choice of structure.

Individual computing capabilities: Individual computing capabilities represent an understanding of the principles by which a system can be applied to a business task.

IT training: IT training deals with the teaching of skills to effectively use computer applications to end users.

Learning method structures: Learning method structures are formal and informal procedures, techniques, skills, rules, and technologies embedded in a learning method, which organize and direct individual or group behavior.

Learning outcomes: Learning outcomes are the result of the learning process. These can be broadly classified into four dimensions, namely, skill, cognitive, affective, and metacognitive outcomes.

Learning process: A series of actions or cognitive development that leads to a change in an individual's capability.

Technology-mediated learning or e-learning: An environment in which the learner's interactions with learning materials, peers, and/or instructors are mediated through advanced information technology.

Training method: A training method is defined as a combination of structures that guides individuals to achieve the learning outcomes.

References

Ackerman, P. L., P. C. Kyllonen, and R. D. Roberts. 1999. *Learning and Individual Differences: Process, Trait, and Content Determinants.* Washington, DC: American Psychological Association.

Agarwal, R. and E. Karahanna. 2000. Time flies when you're having fun: Cognitive absorption and beliefs about information technology usage. *MIS Quarterly* 24(4):665–694.

Alavi, M. and D. E. Leidner. 2001. Research commentary: Technology-mediated learning—a call for greater depth and breadth of research. *Information Systems Research* 12(1):1–10.

ASTD. 2008. ASTD State of the Industry Report. Alexandria, VA: American Society for Training & Development.

Bandura, A. 1986. *Social Foundations of Thought and Action: A Social Cognitive Theory*, Prentice Hall Series in Social Learning Theory. Englewood Cliffs, NJ: Prentice Hall.

Barley, S. R. 1986. Technology as an occasion for structuring: Evidence from observations of CT scanners and the social order of radiology departments. *Administrative Science Quarterly* 31:78–108.

Bohlen, G. A. and T. W. Ferratt. 1997. End user training: An experimental comparison of lecture versus computer-based training. *Journal of End User Computing* 9(3):14–27.

Bolt, M. A., L. N. Killough, and H. C. Koh. 2001. Testing the interaction effects of task complexity in computer training using the social cognitive model. *Decision Sciences* 32(1):1–20.

Bostrom, R., S. Gupta, and D. Thomas. 2009. A meta-theory for understanding systems within sociotechnical systems. *Journal of Management Information Systems* 26(1):17–47.

Bostrom, R. P., L. Olfman, and M. K. Sein. 1990. The importance of learning style in end-user training. *MIS Quarterly* 14(1):101–119.

Burton, J. K., D. M. (Mike) Moore, and S. G. Magliaro. 2001. Behaviorism and instructional technology. In *Handbook of Research for Educational Communications and Technology*, ed. D. H. Jonassen. Mahwah, NJ: L. Erlbaum Associates.

Carroll, J. M. 1997. Toward minimalist training: Supporting the sense-making activities of computer users. In *Training for a Rapidly Changing Workplace: Applications of Psychological Research*, eds. M. A. Quiänones, and A. Ehrenstein. Washington, DC: American Psychological Association.

Chang, M.-Y., W. Tarng, and F.-Y. Shin. 2009. The effectiveness of scaffolding in a web-based, adaptive learning system. *International Journal of Web-Based Learning and Teaching Technologies* 4(1):1–15.

Chin, W., A. Gopal, and W. D. Salisbury. 1997. Advancing the theory of adaptive structuration: The development of a scale to measure faithfulness of appropriation. *Information Systems Research* 8(4):342–397.

Compeau, D., J. Gravill, N. Haggerty, and H. Kelley. 2005. Computer self-efficacy: A review. In *Human-Computer Interaction in Management Information Systems*, ed. D. Galletta and P. Zhang. Armonk, NY: M. E. Sharpe, Inc.

Compeau, D. R. and C. A. Higgins. 1995. Application of social cognitive theory to training for computer skills. *Information Systems Research* 6(2):118–143.

Coulson, A., C. Shayo, L. Olfman, and C. E. T. Rohm. 2003. ERP training strategies: Conceptual training and the formation of accurate mental models. Paper presented at *ACM SIGMIS Conference on Computer Personnel Research*, Philadelphia, PA.

Davis, F. D. and M. Y. Yi. 2004. Improving computer skill training: Behavior modeling, symbolic mental rehearsal, and the role of knowledge structures. *Journal of Applied Psychology* 89(3):509–523.

DeSanctis, G. and B. M. Jackson. 1994. Coordination of information technology management: Team-based structures and computer-based communication systems. *Journal of Management Information Systems* 10(4):85–110.

DeSanctis, G. and M. S. Poole. 1994. Capturing the complexity in advanced technology use: Adaptive structuration theory. *Organization Science* 5(2):121–147.

DeSousa, R. M. D. 2004. Complex information technology usage: Toward higher levels through exploratory use—The ERP systems case, *Management Information Systems*, University of Georgia, Athens, GA.

Duffy, T., D. J. Cunningham, and D. H. Jonassen. 2001. Constructivism: Implications for design and delivery of instruction. In *Handbook of Research for Educational Communications and Technology*. Mahwah, NJ: L. Erlbaum Associates.

Fulk, J. 1993. Social construction of communication technology. *Academy of Management Journal* 36(5):921–950.

Gagne, R. M. 1985. *The Conditions of Learning and Theory of Instruction*. New York: CBS College Publishing.

Garud, R. 1997. On the distinction between know-how, know-why and know-what in technological systems. In *Advances in Strategic Management*, ed. J. P. Walsh and A. S. Huff. Greenwich, CT: JAI Press.

Ge, X. and S. M. Land. 2003. Scaffolding students' problem-solving processes in an ill-structured task using question prompts and peer interactions. *Educational Technology, Research and Development* 51(1):21–38.

Gist, M. E. 1988. The influence of training method and trainee age on acquisition of computer skills. *Personnel Psychology* 41(2):255–265.

Gouran, D. S. 1989. Exploiting the predictive potential of structuration theory. *Communication Yearbook* 13:313–322.

Gravill, J. I. 2004. Self-regulated learning strategies and computer software training. PhD thesis, School of Business Administration, University of Western Ontario, London, Ontario, Canada.

Gravill, J. and D. R. Compeau. 2003. Self-regulated learning strategies and computer software training. Paper presented at *International Conference on Information Systems*, Seattle, WA.

Gruene, M., K. Lenz, and A. Oberweis. 2005. Pricing of learning objects in a workflow-based e-learning scenario. Paper presented at *Hawaii International Conference on System Sciences*, Big Island, HI.

Gupta, S. 2006. Longitudinal investigation of collaborative e-learning in an end user training context. PhD, MIS Department, University of Georgia, Athens, GA.

Gupta, S. 2008. *New Approaches to End-User Training*. Saarbrücken, Germany: VDM Verlag Dr. Mueller e.K.

Gupta, S. and R. P. Bostrom. 2009. Technology-mediated learning: A comprehensive theoretical model. *Journal of Association of Information Systems* 10(9):637–660.

Gupta, S., R. P. Bostrom, and M. Huber. 2010. End-user training methods: What we know, need to know. *SIGMIS Database for Advances in Information Systems* 41(4):9–39.

Hannafin, M. J., M. C. Kim, and H. Kim. 2004. Reconciling research, theory, and practice in web-based teaching and learning: The case for grounded design. *Journal of Computing in Higher Education* 15(2):3–20.

Hevner, A. R., S. T. March, J. Park, and S. Ram. 2004. Design science in information system. *MIS Quarterly* 28(1):75–105.

Igbaria, M., T. Guimaraes, and G. B. Davis. 1995. Testing the determinants of microcomputer usage via a structural equation model. *Journal of Management Information Systems* 11(4):87.

Johnson, R. D. and G. M. Marakas. 2000. Research report: The role of behavioral modeling in computer skills acquisition—toward refinement of the model. *Information Systems Research* 11(4):402–417.

Kadlec, C. A. 2008. Self-regulated learning strategies for the power user of technology, Management Information Systems, University of Georgia, Athens, GA.

Kang, D. and R. Santhanam. 2003. A longitudinal field study of training practices in a collaborative application environment. *Journal of Management Information Systems* 20(3):257–281.

Keeler, C. and R. Anson. 1995. An assessment of cooperative learning used for basic computer skills instruction in the college classroom. *Journal of Educational Computing Research* 12(4):379–393.

Lehtinen, E., K. Hakkarainen, L. Lipponen, M. Rahikainen, and H. Muukkonen 2001. Computer supported collaborative learning: A review. http://www.comlab.hut.fi/opetus/205/etatehtava1.pdf (accessed on May 26, 2011).

Leidner, D. E. and S. L. Jarvenpaa. 1995. The use of information technology to enhance management school education: A theoretical view. *MIS Quarterly* 19(3):265–291.

Marakas, G. M., M. Y. Yi, and R. D. Johnson. 1998. The multilevel and multifaceted character of computer self-efficacy: Toward clarification of the construct and an integrative framework for research. *Information Systems Research* 9(2):126–163.

Martocchio, J. J. and T. A. Judge. 1997. Relationship between conscientiousness and learning in employee training: Mediating influences of self-deception and self-efficacy. *Journal of Applied Psychology* 82(5):764–773.

Mayer, R. E. 1981. The psychology of how novices learn computer programming. *ACM Computing Surveys* 13(1):121–141.

Mayer, R. E., S. Fennell, L. Farmer, and J. Campbell. 2004. A personalization effect in multimedia learning: Students learn better when words are in conversational style rather than formal style. *Journal of Educational Psychology* 96(2):389–395.

Olfman, L. and R. P. Bostrom. 1991. End-user software training: An experimental comparison of methods to enhance motivation. *Journal of Information Systems* 1:249–266.

Olfman, L. and M. Mandviwalla. 1994. Conceptual versus procedural software training for graphical user interfaces: A longitudinal field experiment. *MIS Quarterly* 18(4):405–426.

Olfman, L. and P. Pitsatron. 2000. End-user training research: Status and models for the future. In *Framing the Domains of IT Management: Projecting the Future—through the Past*, ed. R. W. Zmud. Cincinnati, OH: Pinnaflex Education Resources Inc.

Orlikowski, W. J. 2000. Using technology and constituting structures: A practice lens for studying technology in organizations. *Organization Science* 11(4):404–428.

Poole, M. S. and G. DeSanctis. 1990. Understanding the use of group decision support systems: The theory of adaptive structuration. In *Organizations and Communication Technology*, ed. J. Fulk and C. W. Steinfield. Newbury Park, CA: Sage Publications.

Poole, M. S. and G. DeSanctis. 1992. Microlevel structuration in computer-supported group decision making. *Human Communication Research* 19(1):5–49.

Reeves, T. C., J. Herrington, and R. Oliver. 2004. A development research agenda for online collaborative learning. *Educational Technology Research and Development* 52(4):53–65.

Rohrbeck, C. A., M. D. Ginsburg-Block, J. W. Fantuzzo, and T. R. Miller. 2003. Peer-assisted learning interventions with elementary school students: A meta-analytic review. *Journal of Educational Psychology* 95(2):240–257.

Ross, S. M. 1983. Increasing the meaningfulness of quantitative material by adapting context to student background. *Journal of Educational Psychology* 75(4):519–529.

Salas, E. and J. A. Cannon-Bowers. 2001. The science of training: A decade of progress. *Annual Review of Psychology* 52(1):471–499.

Santhanam, R. and M. K. Sein. 1994. Improving end-user proficiency: Effects of conceptual training and nature of interaction. *Information Systems Research* 5(4):378–399.

Schonberger, R. J. 1980. MIS design: A contingency approach. *MIS Quarterly* 4(1):13–20.

Schunk, D. H. 2004. *Learning Theories: An Educational Perspective*. Upper Saddle River, NJ: Pearson/Merrill/Prentice Hall.

Sein, M. K. and R. P. Bostrom. 1989. Individual differences and conceptual models in training novice users. *Human Computer Interaction* 4(3):197–229.

Sein, M. K. and D. Robey. 1991. Learning style and the efficacy of computer-training methods. *Perceptual and Motor Skills* 72(1):243–248.

Simon, S. J. and J. M. Werner. 1996. Computer training through behavior modeling, self-paced, and instructional approaches: A field experiment. *Journal of Applied Psychology* 81(6):648–659.

Sun, H. and P. Zang. 2005. The role of affect in information systems research: A critical survey and a research model. In *Human-Computer Interaction in Management Information Systems*, ed. D. F. Galletta and P. Zhang. Armonk, NY: M. E. Sharpe, Inc.

Szajna, B. and J. M. Mackay. 1995. Predictors of learning performance in a computer-user training environment: A path-analytic study. *International Journal of Human-Computer Interaction* 7(2):167–185.

Venkatraman, N. 1989. The concept of fit in strategy research—toward verbal and statistical correspondence. *Academy of Management Review* 14(3):423–444.

Vries, E. 2003. Educational technology and multimedia from a cognitive perspective: Knowledge from inside the computer, onto the screen, and into our heads? In *Cognition in a Digital World*, ed. H. V. Oostendrop. Mahwah, NJ: Lawrence Erlbaum Associates.

Webster, J. and J. J. Martocchio. 1993. Turning work into play: Implications for microcomputer software training. *Journal of Management* 19(1):127–146.

Winn, W. and D. Snyder. 2001. Cognitive perspectives in psychology. In *Handbook of Research for Educational Communications and Technology*, ed. D. H. Jonassen. Mahwah, NJ: L. Erlbaum Associates.

Yi, M. Y. and F. D. Davis. 2001. Improving computer training effectiveness for decision technologies: Behavior modeling and retention enhancement. *Decision Sciences* 32(3):521–544.

Yi, M. Y. and F. D. Davis. 2003. Developing and validating an observational learning model of computer software training and skill acquisition. *Information Systems Research* 14(2):146–170.

Zhu, Z. 2002. Evaluating contingency approaches to information systems design. *International Journal of Information Management* 22(5):343–356.

41

Role of Trust in the Design and Use of Information Technology and Information Systems

Xin Li
University of South Florida

41.1 Introduction

Trust research has its long history in sociology, psychology, and business fields (Barber, 1983; Lewis and Weigert, 1985; Doney and Cannon, 1997). Mayer et al. (1995) defined trust as *"the willingness of a party to be vulnerable to the actions of another party based on the expectation that the other will perform a particular action important to the trustor, irrespective of the ability to monitor or control that other party"* (p. 712). The two parties involved in the trust relationship could be individuals or groups, such as organizations. Numerical research have suggested that trust is a key to understanding social relationships, as well as people's attitude and behaviors in the circumstances that uncertainty and/or risks are perceived (Mayer et al., 1995; Doney and Cannon, 1997; McKnight et al., 1998).

In the recent decades, the concept of trust has been introduced to information systems (IS) research. Specific research topics span from interpersonal trust in technology-enabled or technology-mediated environments (e.g., trust relationships in a virtual team or between online customers and web vendors) (Jarvenpaa et al., 2000; Piccoli and Ives, 2003) to trust in technological artifacts (e.g., trust in an online recommendation agent [RA] or trust in an inter-organizational system) (Ratnasingam, 2005; Wang and Benbasat, 2005). Trust has been widely found to play a crucial role in shaping information technology (IT)-related attitude and intended behaviors (Gefen et al., 2003; McKnight et al., 2011), especially behaviors of adopting a new, innovative, and emerging IT artifact (Luo et al., 2010) and of continuously using an IT artifact (Gefen et al., 2003). Trust research in IS has its roots in the traditional trust theories. Due to the new contexts, it also has its unique characteristics that traditional trust research may not address. We will discuss next the challenges IS researchers in this field have experienced in developing their own theories and literatures.

One challenge was the debate on the viability of the notion of trust in technology—whether trust, which usually requires moral capability and volitional control, can be extended to technological artifacts? It is not natural to label a technology as "honest" or "caring of others." But IS trust researchers stand on the shoulders of theories of social responses toward computing (Reeves and Nass, 1996; Nass and Moon, 2000), which posit that people treat computers and computer-based technologies as social actors and apply social rules to them. Recent IS research (Li et al., 2012) suggested that there could be multiple trusting objects involved in an IT/IS context. The technological artifact usually provides the functionality and reliability to complete a task, while the human actors (e.g., sponsors, designers, or other users) behind the technology contribute the volitional factors to the task performance and outcomes. Therefore, it is viable for people to trust technological artifacts in the sense that it is a dimension of the overall trust in the IT/IS context. Trust in technology has distinct psychological underpinning from traditional trust dimensions (e.g., trust in person), and they are complementary to one another to form trust outcomes. Researchers should not isolate trust in technology in their studies. Instead, they need to bridge it with other trust dimensions in the given context for a complete trust investigation.

Then how should a trust investigation be conducted in IT/IS contexts? Specifically, what trusting objects should be considered? How do people form trust toward different objects? How do the trust dimensions interact? What are their relative impacts on the trust outcomes? Numerous researchers have attempted to address these questions. However, their studies have adopted distinct focuses and lenses and used different terminologies. Also, trust is longitudinal in nature, and it evolves gradually along with the development of the relationships. All these reasons make it difficult to compare and contrast the research findings. There is no platform to establish an overview of the existing trust research in IS, to cumulate the research findings and build up the theories, or to provide systematic implications to the practice.

This chapter tries to close this gap by providing an overarching framework of trust in IT/IS contexts. The framework integrates various trust constructs and concepts studied in previous research, including interpersonal trust, technology trust, trust levels, trust developing stages, trust antecedents, and trust outcomes. It provides a comprehensive model for IS researchers to develop trust theories and a systematic guide to examine the role of trust in the system-design and system-use practices.

41.2 Underlying Principles

41.2.1 Trust in IT/IS

Beyond examining traditional, interpersonal trust in technology-mediated relationships, increasing research attention has been focused on technology trust. Technology trust, or trust in technology, refers to *the extent to which one is willing to depend on a technology because of its desirable attributes in performing certain tasks* (Pavlou and Ratnasingam, 2003; McKnight, 2005). When computer-based technologies have been widely used to complete individual or organizational tasks that were previously performed by other people, trust in these technologies becomes a key to understanding users' attitude and behavior in the contexts.

The notion of technology trust was not well accepted at first, as the traditional concept of trust requires volitional control for both the trustor (the party who grants trust) and the trustee (the party who receives trust). Based on theories of social responses toward computing (Reeves and Nass, 1996; Nass and Moon, 2000), IS trust researchers argued that computer-based technologies employed to substitute human roles and labor in completing certain tasks are usually seen and treated as social actors (Wang and Benbasat, 2005; Li et al., 2008). In human–technology interaction, all essential trust elements defined in traditional research (Mayer et al., 1995; Wang and Emurian, 2005) exist, and they work in the same way they do in interpersonal interactions. When using the technology to carry out a task, users develop the *expectations* on the technology-use process and the outcomes. They perceive *risks* in the situation too (e.g., what if the technology does not perform its responsibilities or reach the users' expectations). With all these considered, users make a trust decision—how much they are *willing to be vulnerable* to the technology. For example, to use an e-commerce system on a website for online shopping, customers would expect smooth, secure, and successful transactions, no less than what they would get at a brick-and-mortar storefront. Specifically, they would expect the e-commerce system to present rich and accurate product information, to process order placement with a few easy steps, and to provide customer services like gift wrapping, shipping choices, easy exchange and return, order tracking, etc. Customers are aware of the possibilities of delayed delivery, unsatisfactory products or privacy breaches. Based on their beliefs and expectations, they are, however, willing to give it a try. With all these elements working in a human–technology relationship, trust in technology is as viable as trust in a person.

As trust research in IS has its roots in the traditional trust literatures, some research has extended the traditional trust theories and models to operationalize technology trust. Specifically, trust in certain IT artifacts was measured using human-like attributes, including competence, benevolence, and integrity (Wang and Benbasat, 2005; Vance et al., 2008). This approach works fine when the technology of interest is more "human-like" (Tripp et al., 2011). For example, an online RA is usually seen as a "virtual advisor" that gives shopping recommendations to online customers, as a store assistant or a human expert would do. Thus, customers treat RA similarly to a human trustee (Wang and Benbasat, 2005). However, when it comes to a less human-like technology, or one that is not easy to personify (e.g., Microsoft Access [Tripp et al., 2011]), this operationalization method may not work well.

An information technology is usually a preprogrammed, intelligent tool that presents its unique characteristics in task performance. These technical characteristics may be more germane to make up the trustor's expectations of the IT artifacts in terms of task performance and outcomes. Prior studies have included various technical attributes in their trust models, such as site/system quality (McKnight et al., 2002; Vance et al., 2008), Technology Acceptance Model (TAM) attributes (i.e., perceived easy to use and perceived usefulness) (Gefen et al., 2003), and technical competence, reliability, and media understanding (Lee and Turban, 2001). In a recent study, McKnight et al. (2011) investigated technical trusting attributes in a more systematic way. They compared one's trusting beliefs in an IT artifact and those in a person. Similar to human trustees that are expected to do things in a competent way, technological trustees are expected to demonstrate needed functionality to complete a task. While IT artifacts cannot be benevolent and caring of the trustors, they are expected to provide effective help. Instead of having human integrity, a technical artifact can demonstrate its reliability and operate consistently. This study proposed and validated the approach of operationalizing trust in technology using three technical characteristics—functionality, helpfulness, and reliability. This approach does not include any volitional factors. But, without the conceptual difficulties of anthropomorphism, this approach may be easier to operate in most technology contexts.

Other studies have compared and contrasted the two operationalization approaches of technology trust. Tripp et al. (2011) measured technology trust using both human-like attributes and system-like (technical) attributes across three different technology contexts. They suggested that human-like trusting attributes (i.e., competence, benevolence, and integrity) had a strong influence on people's use of technologies with higher humanness (e.g., Facebook), while system-like trusting attributes (i.e., functionality, reliability, and helpfulness) were more strongly related to people's use of technologies with low humanness (e.g., Microsoft Access). However, there is not a clear definition of humanness or a standard criterion to categorize the humanness of a technology, which makes this suggestion less practical.

Given all this, what is the best approach to measure technology trust? In an IT/IS context, a specific IT artifact plays a critical role in completing certain tasks, but it never works alone. The task performance and outcomes are also affected by other factors. For example, the performance of an enterprise system is determined by both the technical attributes of the system and nontechnical factors, such as collaborative management, organizational support, system transition process, and user training and support. The social, environmental, and human factors should also be considered to form user trust toward technology use. Therefore, it should not be the question of whether volitional factors are required in technology trust, or whether we can ascribe volitional attributes to an IT artifact, but rather what role technology trust plays in building the associated trust outcomes in the context. In an IT/IS context, multiple trusting objects may exist and play distinct roles in completing certain tasks. The technological objects usually provide their functionality and reliability to ensure the completion of the tasks, while the organizational or human objects may contribute the volitional factors that affect the task performance and the outcomes. All these trusting objects should be considered to constitute the overall trust needed in forming the outcome attitude and behaviors.

41.2.2 Technology Trust and Interpersonal Trust

Although technologies have been used in various situations to replace human labor or to mediate human relationships, they do not totally replace the human role in the relationships. Specific technology is designed and employed by human actors for certain purposes, and the performance and outcomes of the technology may vary depending on the users and the procedures to use it. In order to use a technology, users need to trust both the technology and the human actors who exert impacts on the technology use. In contrary to technology trust, we call trust in the human actors as interpersonal trust. Li et al.'s (2012) study found that even in a pure e-commerce business where there is no physical storefront and where customers have no direct interaction with the e-vendor, trust in website and trust in e-vendor coexist. With different psychological underpinnings, technology trust and interpersonal trust complement each other and jointly affect user behaviors. To study trust in IT/IS contexts, one should not isolate technology trust from interpersonal trust.

Of course, researchers may pay different attention to interpersonal trust and technology trust, depending on their research interests and trust outcomes that they are concerned. If the researchers are interested in relationship building (e.g., initiating a new relationship or managing an existing relationship with the other human party), interpersonal trust may have a stronger and more meaningful impact on the model. On the other hand, if the technology-related behaviors (e.g., new technology acceptance or continued use) are concerned, researchers may find that technology trust has more significant impact on the research model. However, as discussed earlier, both technological artifacts and human actors play roles in building the overall trust in an IT/IS context, and thus IS trust researchers need to keep both types of trust in mind when predicting trust outcomes. For example, in this chapter, to explore the role of trust in design and use of IT/IS, technology trust may play a dominant role in the model, while the influence of interpersonal trust cannot be ignored.

We have discussed technology trust in the previous section. Interpersonal trust has been widely studied in the traditional trust research. In the IT/IS design and use processes, the human actors could be individual persons, like specific e-sellers in C2C e-commerce or certain IT specialists. In other cases, they are groups or organizations that render significant influences on the technology performance, such as IT department, software provider companies, senior management, etc. Trust established toward these individual or group human actors is based on their competence in performing their own responsibilities, and mostly their volitional attributes including benevolence and integrity.

The performance and outcomes of a technological artifact can be affected by different human actors. In organizations, for example, IS are used to support communication and collaboration between remotely distributed parties, to complete certain tasks that usually require human knowledge and labor, and to support integrated operations or processing. The human trustees in these contexts could be the collaborating partners, system designers and sponsors, and other users who have inputs to the systems. There could be a single human trustee considered in a context. For example, to use a personal service tool like a tax e-filling program, in addition to evaluating the trusting attributes of the program itself, a user may also research about the program providers' (e.g., TurboTax and TaxCut) reputation. In some other contexts, users may need to consider more than one human trustee. In the example of accepting and using an enterprise system, beyond the system itself, users have to trust the enterprise resource planning (ERP) vendor to provide a quality system and related services. They have to trust the in-house IT staff for technical support, and they have to trust all other users to provide the system with quality input. If any of these forms of trust is lacking, it may cause user resistance to the system.

To study trust in IT/IS contexts, researchers need to specify, case by case, the human trustees who affect the performance and outcomes of the technology.

In an IT/IS context, technology trust and interpersonal trust usually interact with each other. There are mutual relationships between the two dimensions of trust—one reinforces the other. For instance, users' trust in Microsoft, based on its reputation and market share, ensures their trust in the Office products; on the other hand, if a user loses trust in a software product because of a bad experience, it will impair his or her trust in the software provider too. The strength of the mutual impacts between technology trust and interpersonal trust may vary across contexts. Depending on the contexts and the research interests, researchers may be able to specify the primary and secondary influences between the mutual relationships.

41.2.3 Levels of Trust

Trust in IT/IS contexts is multidimensional as different technological, human, organizational, and environmental factors and attributes contribute to building the overall trust. The trust dimensions exert different impacts on the technology use. It is important to distinguish the levels of the trust dimensions in order to understand their relative influences and build a comprehensive understanding of trust in IT/IS contexts.

Barber (1983) articulated two levels of trust in social systems—general trust and specific trust. In a social relationship, general trust deals with the expectations of "persistence and fulfillment of the natural and moral social orders (p. 9)," which reduce complexity in life and allow effective and moral human actions to continue. The objects of general trust are usually foundational and environmental elements. In an IT/IS context, general trust include users' expectations on the technology infrastructure and those on the institutional structure and mechanisms. Established general trust upholds a supportive and dependable environment for all technology and system uses.

Specific trust was defined as expectations of "technically competent performance" and "fiduciary obligations and responsibilities" of a specific party with which a trustor interacts (Barber, 1983, p. 14). These expectations determine the trustor's attitudes and actions toward a specific party. In the IT/IS context, specific trust deals with users' expectations on the IT artifacts and human actors that they interact with and that influence the task performance and outcomes. It directly shapes users' attitudes and intended behaviors.

General and specific levels of trust are distinct in nature and serve different purposes. General trust reflects beliefs about the normality and stability of a situation, while specific trust is extended toward an actual partner in a social relationship (Barber, 1983). General trust is the precondition and foundation to establish special trust. Users who do not trust the technological and institutional environments would not trust any technology or system used in this environment. On the other hand, specific trust mediates the influences of general trust on the technology-related behaviors. In a trusted organizational and technological environment, users are more likely to trust the particular technology or systems, as well as the human actors involved in the technology use, which in turn, form the IT-related behaviors.

Specific trust dimensions in IT/IS contexts have been discussed in the previous section, including technology trust and interpersonal trust. As for general trust, two environmental trusting objects should be considered. They are IT infrastructure and institutional structure. *IT infrastructure* (consisting of networked computer hardware and software) is the foundation of all kinds of systems use. Users expect to see stable and secure infrastructure in place and proper protocols and up-to-date technical mechanisms and safeguards being used. When the users believe that the IT infrastructure can support fast and accurate computing, easy and secure data storage, and reliable transmission, it becomes simple for them to form specific trust in this technological environment.

Institutional structure refers to the organizational infrastructure as well as social rules, regulations, guidelines, and mechanisms that are followed in an organization to ensure the proper human procedures in the technology use. The effective use of these mechanisms and safeguards promotes users' trust in such organizational environment. When the users believe that there are effective ways to guide and regulate human actors' role, rights, and responsibilities in the technology use, they will trust the environment, and then are more likely to trust specific system used in the environment.

Trust in these two environmental objects can be called IT infrastructure trust and institutional trust. In the previous research, which mostly focuses on the specific trust, IT infrastructure trust and institutional trust were, if not overlooked, usually viewed and studied as antecedents of trust. Now with a broader view of trust in IT/IS contexts, they are users' trusting assessments on the bigger environment, and these trusting assessments contribute to the overall trust that users establish in the contexts. In addition, some empirical studies have operationalized and measured IT infrastructure trust and institutional trust with scales similar to those of specific trust (Pavlou and Gefen, 2004; Li et al., 2012). Thus, it is reasonable to see them as trust dimensions—the dimensions of general trust.

General and specific trust dimensions form differently. General trust is usually developed based on prior or related experience. In an organizational context, for example, if users have knowledge and experience with the technical and institutional mechanisms from using other systems or in other organizations, general trust can be easily and quickly extended to the current context. On the other hand, specific trust mostly relies on situational factors and requires current and more relevant information processing germane to the specific trusting objects. It is usually less extendable, which explains why most existent research interests are focused on studying specific trust. However, to investigate trust in IT/IS contexts, researchers should not overlook the impacts of general trust dimensions.

41.2.4 Trust Framework in IT/IS Contexts

To summarize the previously discussed trust concepts and constructs, Figure 41.1 presents an overarching framework of trust in IT/IS contexts. Trust in such contexts is multidimensional. Related trust dimensions can be classified in two levels. The general level of trust, including IT infrastructure trust and institutional trust, was built toward the overall environment of technology use. A stable, supportive and well-regulated environment promotes the specific level of trust—technology trust and interpersonal trust. These two specific trust dimensions interact and reinforce each other, and they jointly determine the trust outcomes of interest, such as relationship-building behaviors or IT-related behaviors. In this chapter, to explore the role of trust in the design and use of technologies and systems, the following discussion will focus on IT-related behaviors as the trust outcome, such as new technology acceptance or continued technology use.

This framework develops a comprehensive view of trust in IT/IS contexts. It integrates major trust concepts and constructs studied in previous research such as trusting beliefs and trusting behavior, trust antecedent and trust outcome, trust dimensions and trust levels, etc. It helps cumulate the existing research findings and guides the future trust investigations in IT/IS contexts.

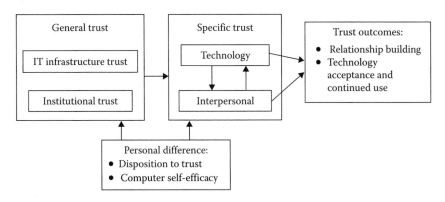

FIGURE 41.1   Framework of trust in IT/IS contexts.

The other constructs included in this framework are disposition to trust and computer self-efficacy. *Disposition to trust* refers to one's propensity to trust others (McKnight et al., 2002). *Computer self-efficacy* was defined as the self-judgment of one's capability to use computer technology (Compeau and Higgins, 1995). They have been included as antecedents to trust in many research studies, especially in the initial trust situations. Examining trust caused by personality traits and individual differences is interesting, but these constructs may provide limited implications to both the trust research and the practice that attempts to cultivate or manipulate trust in the design and use of IT and IS. In future trust research, they could be included in research models as control variables.

41.2.5 Longitudinal Nature of Trust

Trust also has its longitudinal nature. It evolves gradually along with the development of the relationship. Trust built in the early stage of a relationship, without much history and direct knowledge involved, is termed *initial trust* (McKnight et al., 1998). Over time, along with the relationship development, the trust evolves based on firsthand interactions and increasing knowledge between the two parties. It is called *developed trust*. This nature applies to trust in IT/IS contexts too. Initial trust is usually needed to justify new technology acceptance, while developed trust shaped the attitude and behavior of continued technology use.

To study IT/IS trust in initial and developed trust situations, researchers can expect the framework of trust as shown in Figure 41.1 to work differently. As mentioned earlier, in an initial situation, users may rely more on the peripheral trust dimensions to make their technology acceptance decision. Without direct knowledge and experience with the IT artifact, users may temporarily assume the specific trust based on their trusting perceptions of the institutional and technological environments. In addition, trust in related human actors may have a strong impact in promoting technology trust and dominantly determine users' technology acceptance behavior. However, once a technology has been adopted, along with the continuous interaction and increasing knowledge of the technology, many relationships in this framework may change. In a developed trust situation, users rely more and more on their trust judgment in the technology itself. The specific trust may depend less on the general trust, and the technology trust may depend less on the interpersonal trust. Users' real-time trust evaluation of the technology, which has been confirmed or denied in each and every use of the technology, becomes the key determinant in users' continued technology use (Bhattacherjee, 2001; Thatcher and McKnight, 2010; Li et al., 2012).

41.2.6 Uses of the Trust Framework

The framework of trust contributes to the IS trust research in different ways. It helps organize the existing literatures and provide a comprehensive guide for the future trust investigation in the IT/IS contexts.

First, the framework of trust provides a structure to map out the previous studies with distinct foci and lenses and to develop the streams of the IS trust research. For instance, studies that investigated virtual team trust (Jarvenpaa et al., 1998; Sarker et al., 2003; Robert et al., 2009) and customers' trust in online product or service providers (Jarvenpaa et al., 2000; McKnight et al., 2002; Gefen et al., 2003) mostly focused on interpersonal trust in the IT/IS contexts, while studies that examined trust in organizational IS (Li et al., 2009), inter-organizational systems (Ratnasingam, 2005), online RAs (Wang and Benbasat, 2008), and personal productivity software (McKnight et al., 2011) explored technology trust. Both streams have their foci and contributions to the field, but both of them only see one side of the overall trust in IT/IS contexts. A new stream of studies is being developed that models both interpersonal trust and technology trust in its trust investigation (Thatcher and McKnight, 2010; Li et al., 2012). They examine the interaction and the relative impacts of the two trust dimensions to the trust outcomes.

Another stream of the IS trust research is to explore the formation of the specific trust (i.e., interpersonal trust and technology trust), among which many studies examined IT infrastructure trust (Lee and Turban, 2001; Suh and Han, 2003) and institutional trust (Pennington et al., 2003; Pavlou and Gefen, 2004) and their influences in forming the overall trust in certain IT/IS contexts. Other streams are to explore the trust-inducing features from the interface design perspective (Wang and Emurian, 2005; Li et al., 2009). Features that hint the effectiveness of the infrastructure and institutional controls can strengthen the link between general trust and specific trust and induce trust for new technology adoption.

Second, this framework of trust also provides guidelines to the future IS trust research. To investigate trust in IT/IS contexts, researchers need to know the multidimensional nature of trust and adopt a broader view in examining the contexts. Their studies can focus on one or a few trust elements depending on their research purposes and interests. But, when designing the studies and interpreting the research findings, researchers need to keep a bigger picture in mind and control the impacts from the other trust elements. On the other hand, while most of prior research had scrutinized individual trust elements, the framework of trust suggests that additional research attention is needed to address the links among the trust dimensions. Their interactions are keys to improving our understanding of trust and manipulating trust to achieve desired outcomes in certain IT/IS contexts. Additional and detailed suggestions to the future research are discussed in Section 41.4.

41.3 Implications to the Practice

The trust framework presented in Figure 41.1 provides not only the base for trust scholars to cumulate the literatures and develop trust theories in IT/IS contexts, but it is also a platform to inspect the impact of IS trust research on the practice. Existent trust studies have provided valuable implications and useful directions to guide the design and use of IT and IS in an organizational environment.

41.3.1 Cultivating Trust in SDLC

Trust can be cultivated in the design process. To obtain a system, whether it is an in-house development, a purchased system, or an outsourced project, organizations need to understand and accommodate the trust-inducing elements in the systems development process. Specifically, additional attention on user involvement, help feature development, and documentation process in the system development life cycle (SDLC) can help cultivate user trust in the future system. Many of the trust-building techniques are also applicable to modern SDLC models, such as agile software development, rapid prototyping, etc.

User involvement in SDLC has been well known as an effective way to improve system quality. Users are invited to help with business requirements identification, to contribute opinions on desirable functions and features, and to participate in system testing. It narrows or closes the possible gap between the system capability and the organization's requirements. User involvement in SDLC also helps cultivate user trust in the final system.

First of all, early involvement of users in the system design and development process facilitates a positive and healthy user relationship with IT staff and other stakeholders of the system, and thus promotes interpersonal trust needed in future system use. The in-house IT staff, external IT consulting and support group, and other stakeholders, like system sponsors, senior management, and other users, could be the major human actors that affect users' perceptions of the future system use and performance. In many organizations, users may not have much coworking history or more friendly contexts to know IT staff other than in emerging troubleshooting situations. They may not even have a direct interaction or knowledge of other stakeholders in the system use either. A solid base for users to build trust in these human actors is missing. Early user involvement in SDLC provides opportunities for the users to meet and know the internal and external IT staff and system stakeholders in person before the actual system use. With a prolonged collaboration and personal interaction, it is more likely for them to establish a healthy and trusting relationship with these parties toward the future system use.

Second, user involvement in SDLC helps implant the desired trusting attributes into the system and thus promote users' technology trust. With a diverse knowledge background, system users may have different perceptions on the useful and important system functions and features than IT professionals.

They may also have their own judgment on system functionality, reliability, and helpfulness. It is important to facilitate early communication between IT professionals and system users and allow them to exchange their points of view. Specifically, users' opinions should be considered in the system defining and business requirement analysis stages to include user-desired trusting attributes of the system. In the meantime, IT professionals can help users shape their expectations and develop a realistic view of the system.

User involvement should also be encouraged in the system testing phase. A successful example is Microsoft, which expanded their testing of Office 2010 by preleasing a beta version to an expert user group and an experienced user group. They used the users' feedback to learn about potential difficulties and support needed in using the product. They also used the users' comments on the likes of the product in their market research and crafted a successful launch of the actual product the year after (Kroenke, 2011). User involvement in the system testing helps to engage external users and enhance user trust in the future systems.

In addition, user involvement in SDLC helps develop a feeling of ownership and a reasonable expectation of the final system. Previous research (Lewis and Weigert, 1985) has posited that trust has its affective base, which is complementary to the cognitive base. It is more likely for the users to grant trust to a system to which they have contributions and commitment in the development, than the one that they are simply given and asked to use. Also, user involvement in SDLC reduces the possibility of over- or underexpectations from third-party advertising and promotion, and helps users develop reasonable and accurate beliefs to the system in order to form trust.

To an outsourced project or a purchased system, it is beneficial to involve users in the vendor or contractor selection process. When users have a chance to participate in or just observe this process, they will have a better source to assess the quality of the system. On the other hand, it will increase users' affective trust in the future system, because it is developed and supported by the vendor or contractor they self-picked.

System designers and developers should understand the keys to users' technology trust and specifically address each of the trusting beliefs in the system. The importance of functionality and reliability of a system has been well understood and addressed in the system design and development process. Additional attention should be placed on the helpfulness features. Beyond the external help, like technical training and support, modern users with faster working pace and increasing computer self-efficacy have growing interest in using self-help features provided in the system or other self-serviced resources such as online user forums. System developers could enhance the self-help features and support the development of the peer-assisting services with the system. In systems interface design, help buttons should be placed at noticeable and convenient locations. Multiple ways to reach the help topics of interest should be provided (e.g., indexing contents, keyword search, and fuzzy search), and help should be provided in a more intuitive way with choices of elucidation methods (example analysis, video tutorials, etc.). User-generated content (UGC) is another way to improve helpfulness of a system. Online user forums that allow experience sharing and peer user assistance have become a popular and effective way for users to find help. System developers and sponsors could encourage user forums, and actively participate to monitor and guide the forum development.

In addition, system documentation is important in cultivating user trust. Users' expectations to the system are usually formed based on the predefined responsibilities of the system. Before and during the actual use, system responsibilities should be clearly stated and well communicated to all users in order to establish reasonable and realistic expectations. The system documentation, such as a handbook, training materials, etc. has the most authority in defining system responsibilities. Careful and thorough documentation and adequate communication will facilitate and guide a comprehensive and healthy user expectation development to form trust.

41.3.2 Promoting Trust During System Use

Once a system has been obtained and implemented, trust could be fostered to reduce user resistance and promote initial system acceptance and continued use. The current IT/IS trust literatures provide implications to the practice of fostering trust in system use too. Depending on the system use history and the trust development stages, the literatures suggest different focuses to promote user trust.

To promote a new system, organizations could place the attention on developing and advertising the general trust, as users usually lack the direct knowledge base to develop specific trust in such circumstances. Supportive and tech-savvy organizational culture, stable and well-regulated environment, and infrastructure are essential elements for a successful system acceptance. For example, to promote general trust in an organizational IS context, the organization needs to have well-defined technology-use and data policies, which should be thoroughly communicated and strictly enforced. Users should be well educated about organizational mission and goals and its technology strategies. These elements help construct a trustworthy environment for general technology uses.

Establishing such environment could be a long-term process that requires continuous inputs. In the meantime, particular technical safeguards and institutional mechanisms can be used as an immediate booster of users' trust in the environment. For example, recent research (Li et al., 2012) has suggested that third-party authorization and certification programs such as TRUSTe, BBB Online, and VeriSign could provide instantaneous assurance to customers' trust in the overall e-commerce society. New or small e-vendors could rely on these programs to promote customers' specific trust and encourage the first purchase. In different technical contexts, system designers and sponsors need to carefully identify effective technical and institutional mechanisms and use them to ensure initial trust and promote new technology acceptance.

After the initial system adoption, organizations should turn their attention to the specific trust dimensions in order to escort the continued system use. When users have started to use the system, they will turn to each and every interaction with the system and the outcomes of each system use to form trust. If their initial trusting perceptions get confirmed and the task outcomes are satisfactory in each of the system use, they will form developed trust, which suggests continued use of the system (Bhattacherjee, 2001). Of course, if their actual experiences suggest different or negative perceptions of the system, or the task outcomes are not as they expected, their developed trust will be low and will cause resistance to continue using the system. In the developed trust stage, users usually depend less on the environmental factors. It does not mean that the organizational and technological environments, as well as the technical and institutional safeguards, are not important anymore. They are still the precondition of the user trust, but they would have limited effects to increase the specific trust or promote continued system use. To foster or reinforce users' developed trust needed in such situations, organizations should focus their efforts on guiding and supporting each of user–technology interactions.

Specifically, IT professionals should work their best to bring productive and pleasant system experiences to each user. Thorough user training and real-time technical support are critical to ensure smooth user experiences and help users develop positive beliefs in the quality and usability of the system. Proper system maintenance and backup should be administered regularly to reduce the system downtime or recover quickly from system failures.

In addition, individual users may only use a small portion of the system in their jobs. Without a strategic and overall view of the system, they may form trusting beliefs based on fractional perceptions. In this case, system sponsors are responsible to educate users and help them see the complete and strategic value of the system to the entire organization. It may help individual users overcome their learning curve to continuously use the system.

41.4 Research Issues and Summary

This chapter reviewed the underlying principles of trust research in IT/IS contexts and presented an overarching framework of trust. The framework integrated major concepts and constructs in the existent trust literature. It is intended to help develop a comprehensive understanding of trust that facilitates trusting behaviors, especially technology acceptance or continued technology use in organizations. It provides a platform to cumulate research findings and develop theories in this field. This framework also helps identify several research issues and suggests future directions to the IS trust researchers.

First, the trust framework suggests a broad and comprehensive view to investigate trust in IT/IS contexts. With a broad view, researchers may be able to adopt different perspectives to answer a research question. Specifically, prior research has chosen the primary trust dimensions to study trust outcomes of interest. It will be interesting to explore the impacts from the other trust dimensions to the same trust outcomes. For example, prior studies mostly focused on technology trust when they investigated IT-related behaviors as the trust outcome. Now, based on the interaction between the two specific trust dimensions posited in the trust framework, it will be interesting to see how to manipulate interpersonal trust or how to facilitate the trust transference from trusted human actors in order to promote technology trust. What individual or organizational characteristics could induce users' trust toward a certain technology artifact? What features and information can be incorporated in the system design and use process to effectively transfer trust from a well-known and reputable software vendor or contractor to a new system? These sample research questions can bring new insights to the technology trust research.

The trust framework classified trust dimensions using general trust and specific trust, and suggested the interactions between these two trust levels. As discussed earlier, general trust provides the precondition and the foundation to build specific trust, especially in new technology acceptance situations. Prior research does not pay enough attention to general trust because it usually has less direct impacts on the trust outcomes. However, once a trustworthy environment is established, it will have long-term benefits to all technology use in the organization. Thus, future research could investigate general trust, including how to cultivate trustworthy organizational and social environments, and how to manipulate general trust to promote specific trust and to achieve the desired trust outcomes. A more applied research direction is to suggest effective procedures and practices and technical and institutional mechanisms to construct a trustworthy environment for system acceptance. Interface design research is also needed on how to cue the users about the established and trustworthy environment.

Recent trust research has introduced novel methodologies to the IS field. Dimoka (2010) and Riedl et al. (2010) used functional neuroimaging (fMRI) tools to explain how trust was formed biologically with various brain functions. This method provides a unique and novel perspective to study trust and explain trust formation. It can be used along with the trust framework to investigate trust in IT/IS contexts. For example, trust developed toward different objects in such contexts may differ in terms of the activation of the brain functions. It will be interesting to see the neural differences in forming the distinct trust dimensions, which may provide fresh implications in designing and using IT/IS in practice.

A thorough understanding of trust in IT/IS contexts provides a solid base in research endeavors that applies the concept of trust in larger nomological networks or in the different fields. Previous research has combined the trust model with the other technology theories, such as TAM (Gefen et al., 2003, Benamati et al., 2010). Other research studied trust with related concepts such as risk and expectation in a technology context (Luo et al., 2010). In many cases, trust is one psychological factor that plays certain role in a more complex situation. Thus, trust can be included in other theories and models to help improve the understanding in different domains. In these cases, it may not be appropriate to include a complete trust framework like what this chapter presents. But a thorough understanding of trust based on this framework can suggest the best parsimonious way to operationalize and measure trust in different research domains.

References

Barber, B. 1983. *The Logic and Limits of Trust*. Rutgers University Press, New Brunswick, NJ.

Benamati, J., Fuller, M. A., Serva, M. A. and Baroudi, J. 2010. Clarifying the integration of trust and TAM in e-commerce environments: Implications for systems design and management. *Engineering Management, IEEE Transactions* 57(3): 380–393.

Bhattacherjee, A. 2001. Understanding information systems continuance: An expectation-confirmation model. *MIS Quarterly* 25(3): 351–370.

Compeau, D. R. and Higgins, C. A. 1995. Computer self-efficacy: Development of a measure and initial test. *MIS Quarterly* 19(2): 189–211.

Dimoka, A. 2010. What does the brain tell us about trust and distrust? Evidence from a functional neuro-imaging study. *MIS Quarterly* 34(2): 373–396.

Doney, P. M. and Cannon, J. P. 1997. An examination of the nature of trust in buyer-seller relationships. *The Journal of Marketing* 61(2): 35–51.

Gefen, D., Karahanna, E. and Straub, D. W. 2003. Trust and TAM in online shopping: An integrated model. *MIS Quarterly* 27(1) (March): 51–90.

Jarvenpaa, S., Knoll, K. and Leidner, D. 1998. Is anybody out there? Antecedents of trust in global virtual teams. *Journal of Management Information Systems* 14(4): 29–64.

Jarvenpaa, S., Tractinsky, N. and Vitale, M. 2000. Consumer trust in an internet store. *Information Technology and Management* 1(1–2): 45–71.

Kroenke, D.M. 2011. *Using MIS*, 4th edition. Prentice Hall, Upper Saddle River, NJ.

Lee, M. K. O. and Turban, E. 2001. A trust model for consumer internet shopping. *International Journal of Electronic Commerce* 6(1): 75–91.

Lewis, J. D. and Weigert, A. 1985. Trust as a social reality. *Social Forces* 63(4): 967–985.

Li, X., Hess, T. J. and Valacich, J. S. 2008. Why do we trust new technology? A study of initial trust formation with organizational information systems. *The Journal of Strategic Information Systems* 17(1): 39–71.

Li, X., Rong, G. and Thatcher, J. B. 2012. Does technology trust substitute interpersonal trust? Examining technology trust's influence on individual decision-making. *Journal of Organizational and End User Computing* 24(2): 18–38.

Luo, X., Li, H., Zhang, J. and Shim, J. P. 2010. Examining multi-dimensional trust and multi-faceted risk in initial acceptance of emerging technologies: An empirical study of mobile banking services. *Decision Support Systems* 49(2): 222–234.

Mayer, R. C., Davis, J. H. and Schoorman, F. D. 1995. An integrative model of organizational trust. *The Academy of Management Review* 20(3): 709–734.

McKnight, D. H. 2005. Trust in information technology. In *The Blackwell Encyclopedia of Management, Vol. 7: Management Information Systems*, ed. G. B. Davis, pp. 329–331. Blackwell, Malden, MA.

McKnight, D. H., Carter, M., Thatcher, J. B. and Clay, F. P. 2011. Trust in a specific technology: An investigation of its components and measures. *ACM Transactions on Management Information Systems* 2(2): 12–32.

McKnight, D. H., Choudhury, V. and Kacmar, C. 2002. Developing and validating trust measures for e-commerce: An integrative typology. *Information Systems Research* 13(3): 334–359.

McKnight, D. H., Cummings, L. L. and Chervany, N. L. 1998. Initial trust formation in new organizational relationships. *The Academy of Management Review* 23(3): 473–490.

Nass, C. and Moon, Y. 2000. Machines and mindlessness: Social responses to computers. *Journal of Social Issues* 56(1): 81–103.

Pavlou, P. A. and Gefen, D. 2004. Building effective online marketplaces with institution-based trust. *Information Systems Research* 15(1): 37–59.

Pavlou, P. A. and Ratnasingam, P. 2003. Technology trust in B2B electronic commerce: Conceptual foundations. In *Business Strategies for Information Technology Management*, ed. K. Kangas, pp. 200–215. Idea Group Inc, Hershey, PA.

Pennington, R., Wilcox, H. D. and Grover, V. 2003. The role of system trust in business-to-consumer transactions. *Journal of Management Information systems* 20(3): 197–226.

Piccoli, G. and Ives, B. 2003. Trust and the unintended effects of behavior control in virtual teams. *MIS Quarterly* 27(3): 365–395.

Ratnasingam, P. 2005. Trust in inter-organizational exchanges: A case study in business to business electronic commerce. *Decision Support Systems* 39(3): 525–544.

Reeves, B. and Nass, C. 1996. *The Media Equation: How People Treat Computers, Television, and the New Media like Real People and Places*. Cambridge University Press, Stanford, CA.

Riedl, R., Hubert, M. and Kenning, P. 2010. Are there neural gender differences in online trust? An fMRI study on the perceived trustworthiness of eBay offers. *MIS Quarterly* 34(2): 397–428.

Robert, L. P., Dennis, A. R. and Hung, Y. T. C. 2009. Individual swift trust and knowledge-based trust in face-to-face and virtual team members. *Journal of Management Information Systems* 26(2): 241–279.

Sarker, S., Valacich, J. S. and Sarker, S. 2003. Virtual team trust: Instrument development and validation in an educational environment. *Information Resources Management Journal* 16(2): 35–56.

Suh, B. and Han, I. 2003. The impact of customer trust and perception of security control on the acceptance of electronic commerce. *International Journal of Electronic Commerce* 7(3): 135–161.

Thatcher, J. B., McKnight, D. H., Baker, E. W., Arsal, R. E. and Roberts, N. H. 2010. The role of trust in post-adoption IT exploration: An empirical examination of knowledge management systems." *IEEE Transactions on Engineering Management* 58(1): 56–71.

Tripp, J., McKnight, D. H. and Lankton, N. 2011. Degrees of humanness in technology: What type of trust matters? In the *Proceedings of the 17th America's Conference on Information Systems*, Detroit, MI.

Vance, A., Elie-Dit-Cosaque, C. and Straub, D. W. 2008. Examining trust in information technology artifacts: The effects of system quality and culture. *Journal of Management Information Systems* 24(4): 73–100.

Wang, T. D. and Emurian, H. H. 2005. An overview of online trust: Concepts, elements, and implications. *Computers in Human Behavior* 21(1): 105–125.

Wang, W. and Benbasat, I. 2005. Trust in and adoption of online recommendation agents. *Journal of the Association for Information Systems* 6(3): 72–101.

Wang, W. and Benbasat, I. 2008. Attributions of trust in decision support technologies: A study of recommendation agents for e-commerce. *Journal of Management Information systems* 24(4): 249–273.

42

Impacts of Information Systems on Decision-Making

Emre Yetgin
University of Oklahoma

Matthew L. Jensen
University of Oklahoma

Teresa Shaft
University of Oklahoma

42.1 Introduction

The promise of computer aids has captured the imagination of human inventors through numerous decades. Decision aids have the potential to free humans from mundane and tedious tasks, increase human productivity, and minimize human error. Decision aids appear in innumerable applications as varied as automobile manufacturing, airplane piloting, car navigation, corporate payroll management, and nuclear reactor management (Benbasat et al., 1993; Hayes-Roth and Jacobstein, 1994; Skitka et al., 2000). In fact, some researchers suggest that intelligent decision aids may be employed to assist human users in almost any environment imaginable (Russell and Norvig, 2003).

Since the early work on identifying the strengths and weaknesses of decision aids and their human operators (Fitts, 1951), there has been substantial literature regarding how people use information technology (IT) to improve their decisions. This literature spans decades and disciplines (e.g., psychology, computer science, management information systems [MIS]), and we cannot hope to offer a comprehensive review of all literature in this area in a single chapter. However, we will attempt to review key aspects of computer-aided decision-making with an emphasis on the impacts of decision aids on the decision-maker and focus especially on research published in MIS journals.

To examine current knowledge related to computer-aided decision-making, we first review key concepts in unassisted decision-making. Our approach is strongly anchored in the cognitive area of decision-making research. We acknowledge important work concerning decisions based on emotion and affective computing, but this work is outside the scope of this chapter. With this direction, we first describe a four-phase model of decision-making. We then discuss how decision aids affect decision-makers in each phase. Relevant findings are reviewed in terms of potential benefits and pitfalls for decision-makers.

42.2 Phases of Decision-Making and Technology Support

To understand how information systems impact human decision-making, we decompose decision-making into four simple steps: *information acquisition*, *information analysis*, *decision selection*, and *action implementation* (Parasuraman et al., 2000). These four steps are mostly analogous to those (i.e., intelligence, design, choice, and implementation) in the four-step decision-making model of Simon (1977). Although some cognitive psychologists may consider this four-phase model (Figure 42.1) a gross oversimplification, it is consistent with basic decision theory (Broadbent, 1958) and accurate enough to be useful for studying the role and design of IT support in human decision-making (Parasuraman, et al., 2000).

During each of the four phases of decision-making, IT support may play a significant role, a minor role, or no role at all. Parasuraman et al. (2000) have characterized the level of support IT can provide during each decision phase, as shown in Figure 42.2. Thus, a decision aid can be characterized based on the level of support it provides during each decision phase.

IT can support human information processing in all of these phases (Parasuraman, et al., 2000). We refer to such support as "decision support" in general, regardless of which phase it affects. Decision support can utilize IT to a varying extent such that the decision aid can decide everything without involving the user (i.e., full automation), or the user can make all decisions and take action with little assistance from the system. Parasuraman et al. (2000) treat these two extremes as the ends of a 10-level "automation" continuum. For the sake of simplicity, we assume three levels of automation: low, moderate, and high.

Information acquisition
Information analysis
Decision selection
Action implementation

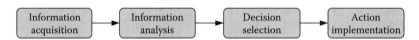

FIGURE 42.1 Four-phase model of decision-making.

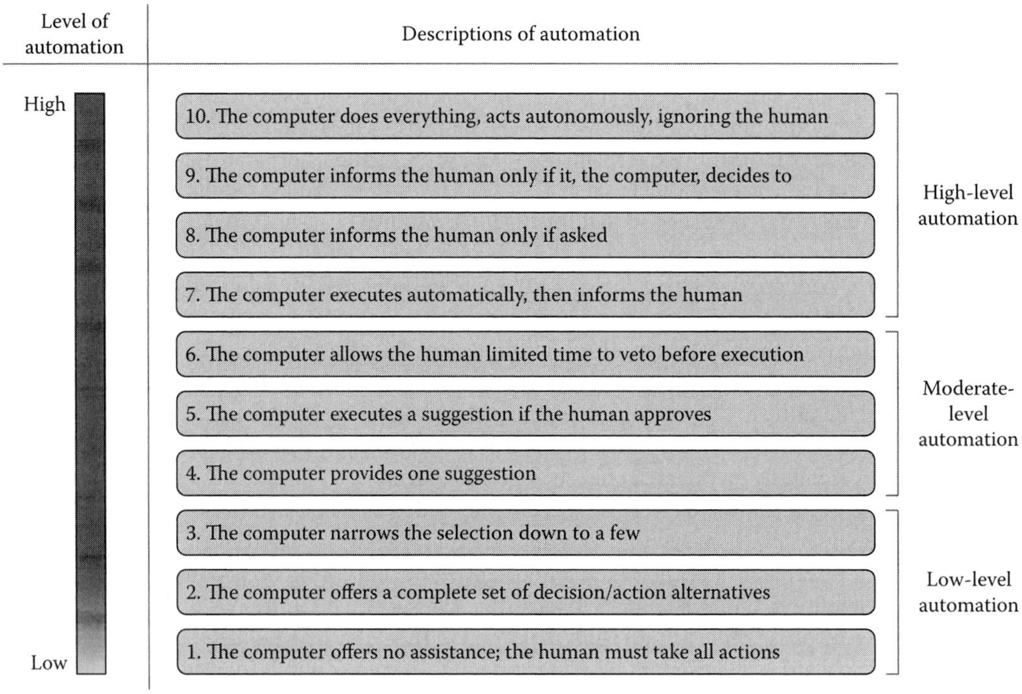

Level of automation
Descriptions of automation
High
Low
10. The computer does everything, acts autonomously, ignoring the human
9. The computer informs the human only if it, the computer, decides to
8. The computer informs the human only if asked
7. The computer executes automatically, then informs the human
6. The computer allows the human limited time to veto before execution
5. The computer executes a suggestion if the human approves
4. The computer provides one suggestion
3. The computer narrows the selection down to a few
2. The computer offers a complete set of decision/action alternatives
1. The computer offers no assistance; the human must take all actions
High-level automation
Moderate-level automation
Low-level automation

FIGURE 42.2 Descriptions of automation.

As one transitions from low to high levels of decision support through automation, the role of IT changes from assisting the user to advising or replacing him or her. Low decision support refers to situations in which the user performs the bulk or all of the decision-making. At this level, IT can provide support by acquiring, aggregating, and/or summarizing the information required for the decision or by providing several courses of action to choose from. Moderate levels of decision support imply more involvement of IT, such as suggesting decision alternatives. The system can provide the logic behind these alternatives and report their likely outcomes, assisting the user in making a better-informed decision. High levels of decision support require little or no input from the users. At this level, the decision aid can fully process the information and make a decision on its own. Users' involvement can be limited to selecting a decision logic or monitoring the decision aid's actions. In the case of complete automation, user involvement may not even be required before the system executes its decision. Examples of decision aids discussed in MIS research are summarized in Table 42.1. These examples span the four decision phases and the levels of automation.

TABLE 42.1 Examples of Decision Aids

Decision Phase	Level of Support	Reference	Task	System Description
Information acquisition	Low	Chervany and Dickson (1974)	Production, inventory, and workforce planning	The system provided statistical summaries of production costs, production amounts, inventory, and labor utilization
	Moderate	Poston and Speier (2005)	Designing a work plan for data modeling	The system enabled searching among existing work plans and provided detailed information and content ratings
	High	Price and Shanks (2011)	Rental property selection	The system enabled searching by attributes among properties, provided data quality tags, and sorted the results
Information analysis	Low	Hoch and Schkade (1996)	Evaluating loan applicants	The system approximated financial performance indicators (e.g., debt ratio and cash flow)
	Moderate	Goodwin (2005)	Adjusting production	The system provided statistical forecasts of future demands
	High	Nah and Benbasat (2004)	Financial analysis	The system (FINALYZER) performed several subanalyses in terms of liquidity, capital structure, and profitability and provided detailed explanations
Decision selection	Low	Wang and Benbasat (2009)	Choosing a digital camera	The system ranked cameras by attribute levels, depending on users' preferences
	Moderate	Eining and Dorr (1991)	Payroll evaluation	The system suggested a numerical evaluation of the internal control over payroll and displayed the rules for its decision
	High	Nissen and Sengupta (2006)	Supply chain management/procurement	The system (Intelligent Mall) analyzed quotations and recommended products and sources using analytical methods and heuristic rules
Action implementation	Low	Lerch and Harter (2001)	Mail sorting/planning	The system enabled users to assign mail to the sorting machines
	Moderate	Cummings and Mitchell (2008)	Unmanned air vehicle control	The system notified the user that it was going to take an action and gave him/her a limited time to veto the decision
	High	Nissen and Sengupta (2006)	Supply chain management/ordering	The system (Intelligent Mall) placed a purchase order and executed the final transaction

42.2.1 Information Acquisition

The first stage of the IT supported decision model, *information acquisition*, deals primarily with sensory processing as decision-makers gather information about their surrounding environment through sensory receptors. This information then serves as the input for the decision-making process.

It has long been known that computer-based decision aids can assist in the acquisition of information. In early considerations of the proper division of responsibilities between humans and machines, machines were thought to excel at data collection (Fitts, 1951). As early research on air traffic control evolved, the value of decision aids in the information acquisition was confirmed, and now decision aids play a crucial role in this labor-intensive task (Wickens et al., 1998). Over time, the ability to aggregate relevant information on which to base decisions has increased with some single decisions (e.g., hurricane forecasts) supported by terabytes of information (NCDC, 2010). A low-level decision aid can merely summarize the information required to make a decision (Chervany and Dickson, 1974), whereas a high-level aid can rank and sort this information, and enable parametric searches (Price and Shanks, 2011).

42.2.1.1 User Benefits

42.2.1.1.1 Better Sensing

MIS research has clearly indicated significant benefit from decision support during information acquisition. Much of this benefit has been achieved by extending the capabilities of human operators. For example, decision aids have clearly demonstrated value in gathering information through additional sensors or searching techniques that are unavailable to human operators. Examples include decision aids that use radars, infrared sensors, and robots with artificial visual and haptic sensors. Decision aids then supply this information to human decision-makers (Parasuraman et al., 2000).

42.2.1.1.2 Easier Information Gathering

Another way that decision aids facilitate information acquisition is through mimicking the actions of a human operator, but more quickly than a human can perform, or on a larger scale. In a recent example concerning computer-aided credibility assessment during a face-to-face interview, a decision aid successfully captured features of one person's behavior during the interview (Jensen et al., 2010). After the interaction was over, the decision aid reported relevant features (e.g., parts of speech, linguistic complexity, gestures, and head movement) it had gathered from the interaction to a third-party observer who had to evaluate the credibility of the interviewee. With sufficient time, the observer could count parts of speech, calculate linguistic complexity, and measure body movement, but this would have been very time consuming for the observer and would delay evaluation of the interview. Other examples can be seen in the procurement context (Nissen and Sengupta, 2006), where users would have to gather, analyze, and compare vendor information manually if not for an automated supply chain management system.

42.2.1.2 Pitfalls

42.2.1.2.1 Identifying Diagnostic Information

In order to be of use to a human operator, a decision aid must gather information that is relevant to the decision. The value of information is often described by how diagnostic the information is, or in other words, how clearly the information indicates an option during decision-making. If information gathered by the decision aid is highly diagnostic, then the decision aid will operate effectively, and the human user will likely benefit from the assistance. If the acquired information is not diagnostic, both the performance of the decision aid and the human user will suffer (Mudambi and Schuff, 2010).

The system's designers (usually in collaboration with the subject matter experts) bear the primary responsibility for determining what information is diagnostic and therefore should be gathered by the decision aid. With some decisions (e.g., a medical inventory alert system), identifying diagnostic information is straightforward, and system designers can successfully achieve a high degree of reliable automation (Chen et al., 2011). However, with other especially complex, naturalistic decisions (Orasanu and

Connolly, 1993) or decisions involving a great deal of ambiguity (Nissen and Sengupta, 2006), acquisition of diagnostic information plays a critical role, but is rarely clear cut. For example, in the case of computer-aided credibility assessment, cues evaluated by the decision aid are probabilistically linked to deception (DePaulo et al., 2003). In other words, the bits of information the decision aid gathers are not diagnostic individually. Only when the information is compiled and combined can they become diagnostic.

The determination of information diagnosticity is possible. When not clearly identified by subject matter experts, diagnostic information can also be uncovered via systematic examination of potentially relevant information. The process for doing so is beyond the scope of this chapter, but we direct interested readers to the work on signal detection (Green and Swets, 1966; Swets, 2000) and artificial intelligence (Russell and Norvig, 2003).

In cases where information diagnosticity is unclear or where resources are not available to determine it, one option that is often adopted is to gather as much information as possible and leave it to the user to determine what is useful and what should be ignored. This practice leads us to our next potential pitfall in information acquisition: information overload.

42.2.1.2.2 Information Overload

One need only type in some common search terms into a popular search engine to experience the sensation of information overload. A simple web search can yield millions of potentially relevant websites, few of which are consulted as part of the search. Information overload was identified in early MIS research as a potential pitfall. In his seminal work on potential pitfalls of decision aids, Ackoff (1967) suggested that information overload was a critical difficulty system users faced several decades ago. This issue is even more of a challenge today, as Paul and Nazareth (2010) demonstrate within the context of group support systems. Managers have no time to pore through all potentially relevant information when making their decisions. Similarly, consumers are not going to read all relevant product reviews before purchasing a product. Ackoff suggested that in countering the overabundance of information, the two most important features of an information system are filtration and condensation. However, filtering and condensing will function well only if they are built on a firm understanding of information diagnosticity. Otherwise, overfiltration and overcondensation of information may occur. In situations where all information could be important and relevant, such as financial fraud detection (Ngai et al., 2011), overfiltration and overcondensation of data could lead users to rely on inadequate information, and thus make incorrect decisions.

42.2.1.2.3 Codifying Errors

Incorrect information stored and codified during acquisition, either as a result of mistakes (Klein et al., 1997) or intentional deception (Biros et al., 2002), is exceptionally difficult for users to detect. These findings echo a well-known cliché in decision aid design: Garbage in, garbage out. However, it appears that once information is codified or stored in a decision aid, decision-makers rarely validate or verify the information. In other words, "garbage" potentially coming from a decision aid will rarely be identified.

To mitigate this pitfall, researchers have investigated the possibility of altering the analysis of information (Jiang et al., 2005) or providing an information quality indicator to inform users of potentially problematic information (Fisher et al., 2003).

42.2.2 Information Analysis

In the second stage, information analysis, information is selectively and consciously perceived, and then processed. Information analysis "involves cognitive functions such as working memory and inferential processes" (Parasuraman, et al., 2000) and refers to the processing of input data in order to make a decision. As with information gathering, there is a spectrum of automated support that an information system can offer a decision-maker. On the low end is simple presentation of relevant information that was gathered (and potentially filtered) as part of the information gathering stage. An example of a system that provides a low level of analysis capabilities is the one that is used for evaluating loan applicants (Hoch and Schkade, 1996), which calculates and presents several financial indicators. On the higher end of the analysis spectrum is another financial system, FINALYZER (Nah and Benbasat, 2004), which performs several complex subanalyses in terms of liquidity, capital structure, and profitability and provides its users with detailed explanations on these analyses.

The processing capabilities offered by decision support systems offer many obvious benefits. These benefits are widely acknowledged and are relied upon by countless individuals in their daily work and life. However, the processing capabilities of systems also carry some pitfalls, which if left unconsidered, may severely undermine effective computer-aided decision-making. Both benefits and potential pitfalls are addressed in the following sections.

42.2.2.1 Benefits

42.2.2.1.1 Analysis of Relevant Information

Perhaps the most obvious benefit from decision support is the capability to rapidly consider more information than is possible manually, and quickly apply analysis techniques that will yield a reasonable solution. This benefit of decision aids has long been supposed (Fitts, 1951) and is still being supported by current research. For example, a decision aid was able to determine optimal ordering qualities for small manufacturing companies (Choi et al., 2004). Thus, decision aids have the capacity to improve the accuracy of decisions through systematic analysis of relevant information.

Coupled with the potential of increased analysis capabilities, decision aids also offer the promise of decreasing mental effort. Effort reduction is perhaps as important as any gains in decision accuracy. Payne and colleagues (1993) have demonstrated that decision-makers often sacrifice increases in accuracy for a reduction in effort. Early decision theorists termed this tendency *satisficing* (Simon, 1956), where decision-makers adopt an acceptable (although not optimal) choice when they are weighing their options (Newell and Simon, 1972). MIS researchers have noted that computer-based decision aids simplify decisions and reduce the level of effort required during decision-making (Todd and Benbasat, 1991, 1992). Decision aids appear to provide the greatest benefit when users will not or cannot pay sufficient attention to the decision task to make a good choice (Reichenbach et al., 2011). Simply put, users of decision aids are likely to be able to analyze problems in more depth by using the expanded processing capabilities provided by such systems (Hoch and Schkade, 1996).

42.2.2.1.2 Presentation

One of the unique characteristics that systems offer decision-makers is the ability to visualize information in novel or unique ways. When task complexity is low, decision-makers most likely do not require advanced visual elements (Speier and Morris, 2003). However, when task complexity is higher, visualization techniques may assist the decision-maker in identifying links among the information to more easily come to a decision (Speier and Morris, 2003). For example, for complex decisions, a three-dimensional representation of information significantly improved decision performance (as measured by the time spent on analysis) (Kumar and Benbasat, 2004). This is consistent with previous results suggesting that adding visual elements to a decision aid's interface can assist decision-makers when task complexity is high (Speier and Morris, 2003). Other examples include geographic decision support systems (Mennecke et al., 2000), marketing information analysis (Lurie and Mason, 2007), and medical information presentation (Juraskova et al., 2008).

42.2.2.1.3 Feedback

During the information analysis phase, decision aids are in a unique position to offer the user feedback as they consider and weigh relevant information. This feedback may come in one of two forms: a response presented to the user after information is considered (feedback response) and a response to the user about actions that will need to take place (feedforward response) (Arnold et al., 2006; Dhaliwal and

Benbasat, 1996). Both forms of feedback are intended to inform the user about the decision model that the decision aid employs and engender the user's confidence in the decision aid (Dhaliwal and Benbasat, 1996; Gregor and Benbasat, 1999; Kayande et al., 2009). The decision aid's feedback may also instruct the user about analysis techniques that are transferable to other tasks (Arnold et al., 2006). However, in order for decision-makers to improve their performance, an estimate on upward potential (in addition to decision feedback) may be necessary (Kayande et al., 2009).

42.2.2.2 Pitfalls

42.2.2.2.1 Assumptions in Design

In the development of a decision aid, designers may make assumptions about the decision and the information to be analyzed to apply analysis techniques. For example, a designer may specify how to weigh a single piece of information when considering several criteria. Such an assumption may simplify the development of a decision aid and may even be applicable in the majority of cases. However, when assumptions are embedded into the design of a decision aid, problems in the decision aid's operation may surface when the assumptions are violated.

Although this pitfall may seem obvious, it becomes more insidious when one realizes that designers often do not recognize when they are inserting their own biases into the design (Korhonen et al., 2008). To make matters worse, decision-makers who deal with complex decisions are at times unaware of *their own* assumptions (Ackoff, 1967). This renders the task of extracting knowledge from subject matter experts, on whom designers rely to define decision parameters, especially difficult. The end result of unacknowledged assumptions in the design of a decision aid is improper recommendations where the decision aid may actually mislead the user, instead of assisting him or her. The impact of this pitfall will depend on the extent of assumptions and the consequences of the decision.

42.2.2.2.2 "Broken Leg" Phenomena

A special case of making assumptions in design merits singling out. This assumption is based on designers' beliefs that they have considered all necessary information to arrive at an optimal recommendation for the user. This is a common simplifying assumption that significantly reduces the design complexity of a decision aid. But, this assumption leaves a significant vulnerability against which the user must remain vigilant. As famously illustrated by Meehl (1954), assume that a person nearly always (say, 90% of the time) goes to the movies every Tuesday night. To build a decision aid predicting the person's appearance at the movie theater every Tuesday would be straightforward. But assume that the person just broke her leg. A broken leg would likely be outside the consideration of the decision aid but extremely relevant for the outcome. Therefore, the utility of the decision aid would substantially diminish in this case.

42.2.2.2.3 User and Task Characteristics

A great deal of research has explored how user characteristics impact how the user interacts with the decision aid. Expertise was among the first user characteristics considered by MIS researchers, and it was discovered that novices tended to request more information from the decision aid (Benbasat and Schroeder, 1977). In later works, other researchers have illustrated the importance of fit between the cognitive representations of users and capabilities of the decision aid (Goodhue and Thompson, 1995; Vessey and Galletta, 1991). In a review of relevant literature, Zmud (1979) suggested that cognitive behavior of future users should determine the system design. These findings suggest that if cognitive characteristics of users are not taken into account during the design, the decision aid could fall short of its potential to assist users. However, some researchers have suggested that cognitive characteristics of users are of minor importance (Huber, 1983), and other researches have shown no difference in decision improvement among those with high and low cognitive abilities (as measured by working memory) (Lerch and Harter, 2001). However, at this time, it appears that task characteristics are more predicative than characteristics of individual users (Davern et al., 2012).

42.2.3 Decision Selection

During the third stage, decision selection, a single option is selected from among the decision alternatives. The role of the decision aid in this stage of the decision process involves "varying levels of augmentation or replacement of human selection of decision options with machine decision making" (Parasuraman et al., 2000). A decision aid providing multiple decision alternatives is an example for low-level support. In this case, the user would have to pick a solution among the alternatives. For example, the recommendation agent in Wang and Benbasat (2009) provides the users with a list of products that it sorted according to users' preferences, and the users select one. A decision aid offering a moderate level of automated support can provide a recommendation and explain the logic behind it, for example, the expert system that suggests an evaluation of the internal control of a factory payroll (Eining and Dorr, 1991). The users can, in this case, view the explanation for the rules that the expert system used to determine this recommendation, and they can choose to agree with the recommended evaluation or create their own solution. A decision aid can also provide high-level support by following a series of rules and making a decision on its own. For instance, the fully automated supply chain management system, Intelligent Mall (Nissen and Sengupta, 2006), can analyze supplier quotes and recommend products and sources, without any requirement of human involvement.

When the information analysis stage of the decision produces an unambiguously superior option, decision selection may be a simple matter that can be rapidly carried out without much supervision. However, when information analysis does not produce a superior option or the information analysis and gathering is somehow deficient, decision selection becomes much less trivial. High levels of automation during decision selection carry several benefits and drawbacks that are briefly reviewed next.

42.2.3.1 Benefits

42.2.3.1.1 Rapid and Systematic Decision-Making

When the decision parameters are well defined (i.e., when the information gathered is diagnostic and sufficient for the decision), a decision aid featuring high decision selection automation has the capability to rapidly make correct decisions on a scale that a human decision cannot match. For example, many complex tools are available to investors that constantly poll and immediately react to price changes in securities. In a similar example, airline pilots rely heavily on rapid control changes from modern autopilot systems. In these examples, decisions are made by systematically considering a significant amount of information and rapidly coming to a conclusion. For well-defined decisions, a decision aid can substantially increase speed and accuracy of these decisions (Parikh et al., 2001).

42.2.3.1.2 Adaptable Decision Model

When a decision is not well defined, the decision model of the decision aid (i.e., how the aid decides on an option) is flexible and not hampered by bias or previous practice. For example, in early decision support research, the designers of MYCIN showed that their decision aid outperformed experienced doctors on a diagnosis task (Shortliffe, 1976). In its diagnoses, MYCIN considered known base rates that were adaptable to whatever the local and current prevalence of disease was. Findings in computer-aided credibility assessment (Jensen et al., 2010) and agent-based negotiation (Nissen and Sengupta, 2006) research suggested a similar conclusion. If a decision model in a decision aid needs to be altered, it is a simple matter. If a decision model in a human decision-maker needs to be altered, it may be more difficult (especially for very experienced decision-makers). Another benefit is that a decision aid may support several different decision models such as normative decision strategies (in place of elimination by aspects) (Wang and Benbasat, 2009) or recognition-primed decision-making (Hoch and Schkade, 1996).

42.2.3.2 Pitfalls

42.2.3.2.1 Complacency

According to Parasuraman et al. (2000), complacency occurs among decision-makers when a decision is highly, but not perfectly, reliable in decision selection. In a review on how decision-makers rely on decision aids, complacency is described as "over-trust" in the decision aid (Parasuraman and Riley, 1997). The decision-maker essentially abdicates decision-making responsibility to the decision aid. What results is a decision-maker who does not notice when a decision aid fails to perform as expected.

The tendency toward complacency is most evident in decision aids with high levels of automation where the user plays more a more supervisory than active role. Some research has suggested that the task of supervision substantially differs from decision-making, in terms of being vulnerable to the negative effects of complacency (Crocoll and Coury, 1990). On the one hand, the tendency toward complacency appears to be rational. After all, why would a user expend cognitive effort checking a decision aid that he or she expects to be right? Complacency becomes a problem however, when the costs of a bad decision are high. In considering an appropriate level of automation to build into a decision aid, Parasuraman et al. (2000) argue that the amount of expected complacency should be one of the primary considerations.

42.2.3.2.2 Reliance on Heuristics

A closely related issue to complacency is reliance on decision heuristics during decision-making. In making very complex decisions, individuals often are either not able to or not sufficiently motivated to come to a decision based on evidence and reasoning (Eagly and Chaiken, 1993). Instead, they rely on mental shortcuts such as "the decision aid recommends I should take this action, therefore I will do it." There is no serious consideration given to the reasons behind the aid's recommendation, but it is blindly adopted because investigating the "why" would require too many cognitive resources. Past research has shown that decision-makers depend on cues provided in support of heuristics (e.g., credibility or quality indicators) to a great degree (Fisher et al., 2003; Poston and Speier, 2005). If designers need to use such cues to assist decision-makers, they should take care to ensure the cues are as accurate as possible. Decision-makers will seek to reduce the amount of effort they expend during decision-making (Payne et al., 1993) and will use the cues even if they are far from perfect. Designers should also realize that quick decisions made based on heuristics may come with drawbacks. According to psychology research, such decisions are quickly made but can be transient and unstable (Petty and Cacioppo, 1986a,b). There appears to be a Goldilocks effect for the sustaining decision-makers' systematic processing when they are using a decision aid: not too much support, but not too little either (Tan et al., 2010).

42.2.3.2.3 Recommendation Rejection

After all the effort that goes into designing, developing, and verifying the function of decision aids, decision-makers may still not use the decision aid. The reluctance to use the decision aid may be especially salient among more experienced decision-makers (Jensen et al., 2010; Yang et al., 2012). It is clear that decision aids must create the impression in decision-makers that they are useful and pleasant to use (Kamis et al., 2008). Another way to entice decision-makers to engage more with the decision aid is to solicit their input and incorporate their input in the decision aid's recommendations (Arnold et al., 2006). However, the effects of providing this functionality are still unclear, and for some complex problems may be detrimental to decision performance (Jensen et al., 2011; Lee and Benbasat, 2011).

42.2.4 Action Implementation

Action implementation is the execution of the selected decision. Action implementation has not received as much attention in MIS research as the previous three stages of computer-aided decision-making. However, it deserves consideration because the action is the final outcome of decision-making.

In contrast to MIS research, other research disciplines (e.g., economics) focus solely on the actions. Among the little MIS research that does concern action implementation, there is a spectrum of support from low, for example, a system that enables users to assign mail to sorting machines (Lerch and Harter, 2001), to high, for example, a system that places a purchase order on its own and executes the transaction (Nissen and Sengupta, 2006).

42.2.4.1 Benefits

The primary benefit of high levels of automated support is the removal of the human in carrying out the action. In highly defined tasks such as monitoring (Aron et al., 2011), where the weaknesses of human operators (e.g., vigilance, speed) are the primary threat to successful action implementation, high levels of automation may be desirable and can significantly improve decision-making performance.

42.2.4.2 Pitfalls

The primary pitfall during action implementation is that actions might be carried out in a manner that is inconsistent with the wishes of the decision-maker. Such a situation would be most likely to arise where automation is on the higher end of the spectrum. From the perspective of the decision-maker, unexpected actions would likely be deemed as overreaching (at best) or just plain wrong (at worst). Such actions would also likely decrease users' trust in the decision aid and their intention to use it again in the future (Komiak and Benbasat, 2006). Other considerations for designers planning to implement a high level of automation include the costs and the likelihood of actions being implemented incorrectly. As automated decision aids are being developed for more and more complex problems (e.g., automated equity trading), the action implementation stage of a decision should attract more attention from MIS researchers.

42.3 Future Directions in Computer-Aided Decision-Making

To consider future directions for research in computer-aided decision-making is a little like gazing into a crystal ball. The field is ripe with creativity and innovation, and we are constantly amazed by new approaches that individuals and organizations develop for appropriately supporting decision processes. Nevertheless, several new trends have captured our attention, and we briefly discuss them next.

42.3.1 Decision Support Using Social Media

With the explosion of available information on the Internet and as users everywhere become individual content providers (through contributions to social media, community forums, blogs and microblogs, video sharing sites, etc.), designers of systems are devoting increasing attention to user-generated sources of information and developing tools to summarize and coalesce this information into useful knowledge (Hu and Liu, 2004). Examples include customer feedback and rating systems that online retailers develop (e.g., online product reviews), software performing sentiment analysis and data-driven market research, and location-aware computing. Individuals willingly contribute information about their own experiences and perceptions. This represents a significant opportunity for designers of decision aids as one of the most potent cues that individuals attend to in making decisions is the experience of others who have made the same or similar decisions (Mudambi and Schuff, 2010). For example, over 80% of holiday shoppers turn to online product reviews to assist with purchase decisions (Nielsen, 2008). Another example can be found in selecting a dining option. Combining location-aware computing with online reviews of dining options can effectively guide a decision-maker to a suitable dining option. The existing glut of user-contributed information will only continue to grow, and this presents a significant opportunity for designers of computer-based decision aids. If they can harness the experience of crowds, they will provide a powerful tool in support of decision-making.

42.3.2 Balancing Heuristics and Systematic Decision-Making

With the rising amount of information comes the responsibility of effectively managing it. Given the sheer magnitude of available information at decision-makers' fingertips, it is unreasonable to expect decision-makers to carefully and systematically review all relevant sources when making decisions. It is much more reasonable to expect decision-makers to rely on heuristics (e.g., helpfulness scores or star ratings) or recommendations (e.g., prescriptive advice) provided by a decision aid, because decision-makers likely lack the ability or motivation to examine all relevant sources (Cao et al., 2011). This tendency prompts a significant question that researchers must carefully deal with: Given a context and a decision aid, why do decision-makers reach the decision they do? In other words, what is the basis of a decision-maker's choice? This question is not new (Ackoff, 1967); however, it is growing in importance because the amount of information available to decision-makers makes the effort associated with even the simplest decisions substantial (e.g., Who offers the best Italian food in the city?), and decision-makers increasingly *are not able* to examine all relevant information and *must* rely on a decision aid.

Using heuristics to deal with complex information is actually a natural phenomenon. Even without a decision aid, people tend to rely on several cognitive heuristics, such as representativeness, availability of instances or scenarios, and adjustment from an anchor, in order to simplify complex decision-making tasks (Tversky and Kahneman, 1974). However, even though these heuristics reduce cognitive load and are usually effective, they can (and usually do) result in systematic errors. For example, people typically reach different estimates when rapidly multiplying ascending versus descending sequences of the same numbers. In addition to helping people interpret large amounts of data, decision aids can also help them avoid such cognitive biases and mitigate the resultant systematic errors (Bhandari et al., 2008).

The tendency (and even necessity) of decision-makers to rely on decision aids for assistance with decisions involving massive amounts of data places a great deal of responsibility on the designers of such systems and those who gather and contribute information that is later categorized by the decision aid. Individual decision-makers are unlikely to audit inputs to a system providing a recommendation, but they will likely be able to tell when they have relied upon an incorrect recommendation from a decision aid. Being wrong could carry with it substantial consequences for trust, use continuance and willingness to adopt future recommendations (Komiak and Benbasat, 2006).

42.3.3 Persuasive Technology

With increasingly sophisticated decision aids and the inability or unwillingness of decision-makers to consult relevant information during their decisions, there exists the potential for decision aids to unduly influence our opinions and attitudes (e.g., recommendation agents for commercial products, opinion, and social media aggregators). This possibility has been noted before (Fogg, 2002). But, as decision-makers increasingly become separated from information they would use to make decisions, the potential for manipulation increases.

42.3.4 "Big Data" Analytics

Recent research shows that top-performing organizations use data analytics more and are better at deriving value from analytics (LaValle et al., 2011). Given the increasing abundance of information, understanding and interpreting it becomes much more difficult yet critical, in order to gain a competitive advantage. Finding ways to correctly interpret and extract meaning from "big data" is a key challenge faced by scientists in many fields, and conventional tools may not be adequate to convey understanding of phenomena captured by "big data" (Frankel and Reid, 2008). The role of decision aids in discovering trends, patterns, and relationships in such data and transforming them into business value remains to be further scrutinized and improved, which should yield solid practical implications.

42.3.5 Action Implementation in Natural Environments

Finally, as noted earlier, there is little research on action implementation in technology-supported decision-making. In the MIS discipline, there has been a great deal of research investigating what individuals say they would do (i.e., their intentions). However, past MIS research has also shown that in terms of system use what individuals say they do and what they actually do are sometimes very different (Straub et al., 1995). This facet of supported decision-making deserves additional attention.

42.4 Conclusions

This chapter offers a brief glimpse into the potential benefits and pitfalls of using IT to facilitate decision-making. In countless cases, IT has improved and quickened decision-making. However, designers and decision-makers who rely on IT must also be aware of how decision aids can undermine the decision-maker. Our treatment of computer-aided decision-making has been largely limited to MIS research. However, we hope our treatment effectively illustrates the complexity of computer-aided decision-making. By being mindful of the trade-offs between benefits and potential pitfalls of using decision aids, our hope is that individuals will be able to appropriately rely on IT during decision-making and not misuse or abuse available decision aids.

References

Ackoff, R. L. (1967). Management misinformation systems. *Management Science, 14*(4), 147–156.

Arnold, V., Clark, N., Collier, P. A., Leech, S. A., and Sutton, S. G. (2006). The differential use and effect of knowledge-based system explanations in novice and expert judgment decisions. *MIS Quarterly, 30*(1), 79–97.

Aron, R., Dutta, S., Janakiraman, R., and Pathak, P. A. (2011). The impact of automation of systems on medical errors: Evidence from field research. *Information Systems Research, 22*(3), 429–446.

Benbasat, I., DeSanctis, G., and Nault, B. R. (1993). Empirical research in managerial support systems: A review and assessment. In C. W. Holsapple and A. B. Whinston (Eds.), *Recent Developments in Decision Support Systems*. New York: Springer-Verlag.

Benbasat, I. and Schroeder, R. G. (1977). An experimental investigation of some MIS design variables. *MIS Quarterly, 1*(1), 37–49.

Bhandari, G., Hassanein, K., and Deaves, R. (2008). Debiasing investors with decision support systems: An experimental investigation. *Decision Support Systems, 46*(1), 399–410.

Biros, D., George, J., and Zmud, R. (2002). Inducing sensitivity to deception in order to improve deception detection and task accuracy. *MIS Quarterly, 26*(2), 119–144.

Broadbent, D. E. (1958). *Perception and Communication*. London, U.K.: Pergamon.

Cao, Q., Duan, W., and Gan, Q. (2011). Exploring determinants of voting for the "helpfulness" of online user reviews: A text mining approach. *Decision Support Systems, 50*(2), 511–521.

Chen, Y.-D., Brown, S. A., Hu, P. J.-H., King, C.-C., and Chen, H. (2011). Managing emerging infectious diseases with information systems: Reconceptualizing outbreak management through the lens of loose coupling. *Information Systems Research, 22*(3), 447–468.

Chervany, N. L. and Dickson, G. W. (1974). An experimental evaluation of information overload in a production environment. *Management Science, 20*(10), 1335–1344.

Choi, H. R., Kim, H. S., Park, B. J., Park, Y. J., and Whinston, A. B. (2004). An agent for selecting optimal order set in EC marketplace. *Decision Support Systems, 36*(4), 371–383.

Crocoll, W. M. and Coury, B. G. (1990). Status or recommendation: Selecting the type of information for decision aiding. Paper presented at the *Proceedings of the Human Factors and Ergonomics Society Annual Meeting*, 34(19), 1524–1528. doi: 10.1177/154193129003401922.

Davern, M., Shaft, T., and Te'eni, D. (2012). Cognition matters: Enduring questions in cognitive IS research. *Journal of the Association for Information Systems, 13*(4), 273–314.

DePaulo, B., Lindsay, J., Malone, B., Muhlenbruck, L., Charlton, K., and Cooper, H. (2003). Cues to deception. *Psychological Bulletin, 129*(1), 74–118.

Dhaliwal, J. S. and Benbasat, I. (1996). The use and effects of knowledge-based system explanations: Theoretical foundations and a framework for empirical evaluation. *Information Systems Research, 7*(3), 342–362.

Eagly, A. H. and Chaiken, S. (1993). *The Psychology of Attitudes.* Fort Worth, TX: Harcourt Brace Jovanovich College Publishers.

Eining, M. M. and Dorr, P. B. (1991). The impact of expert system usage on experiential learning in an auditing setting. *Journal of Information Systems, 5*(1), 1–16.

Fisher, C. W., Chengalur-Smith, I. S., and Ballou, D. P. (2003). The impact of experience and time on the use of data quality information in decision making. *Information Systems Research, 14*(2), 170.

Fitts, P. M. (1951). *Human Engineering for an Effective Air Navigation and Traffic Control System.* Columbus, OH: Ohio State University Research Foundation Report.

Fogg, B. J. (2002). *Persuasive Technology: Using Computers to Change What We Think and Do.* Boston, MA: Morgan Kaufmann.

Frankel, F. and Reid, R. (2008). Big data: Distilling meaning from data. *Nature, 455*(7209), 30.

Goodhue, D. L. and Thompson, R. L. (1995). Task-technology fit and individual performance. *MIS Quarterly, 19*(2), 213–236.

Green, D. M. and Swets, J. A. (1966). *Signal Detection Theory and Psychophysics.* New York: Wiley.

Gregor, S. and Benbasat, I. (1999). Explanations from intelligent systems: Theoretical foundations and implications for practice. *MIS Quarterly, 23*(4), 497–530.

Hayes-Roth, F. and Jacobstein, N. (1994). The state of knowledge-based systems. *Communications of the ACM, 37*(3), 26–39.

Hoch, S. J. and Schkade, D. A. (1996). A psychological approach to decision support systems. *Management Science, 42*(1), 51–64.

Hu, M. and Liu, B. (2004). Mining and summarizing customer reviews. Paper presented at the *Proceedings of the Tenth International Conference on Knowledge Discovery and Data Mining (KDD '04),* New York.

Huber, G. P. (1983). Cognitive style as a basis for MIS and DSS designs: Much ado about nothing? *Management Science, 29*(5), 567–579.

Jensen, M. L., Burgoon, J. K., and Nunamaker, J. F. (2010). Judging the credibility of information gathered from face-to-face interactions. *ACM Journal of Data and Information Quality, 2*(1), 301–320. doi: 10.1145/1805286.1805289

Jensen, M. L., Lowry, P. B., Burgoon, J. K., and Nunamaker, J. F. (2010). Technology dominance in complex decision making: The case of aided credibility assessment. *Journal of Management Information Systems, 27*(1), 175–202.

Jensen, M. L., Lowry, P. B., and Jenkins, J. L. (2011). Effects of automated and participative decision support in computer-aided credibility assessment. *Journal of Management Information Systems, 28*(1), 203–236.

Jiang, Z., Mookerjee, V. S., and Sarkar, S. (2005). Lying on the web: Implications for expert systems redesign. *Information Systems Research, 16*(2), 131–148.

Juraskova, I., Butow, P., Lopez, A., Seccombe, M., Coates, A., Boyle, F., and Forbes, J. (2008). Improving informed consent: Pilot of a decision aid for women invited to participate in a breast cancer prevention trial (IBIS-II DCIS). *Health Expectations, 11*(3), 252–262.

Kamis, A., Koufaris, M., and Stern, T. (2008). Using an attribute-based decision support system for user-customized products online: An experimental investigation. *MIS Quarterly, 32*(1), 159–177.

Kayande, U., De Bruyn, A., Lilien, G. L., Rangaswamy, A., and van Bruggen, G. H. (2009). How incorporating feedback mechanisms in a DSS affects DSS evaluations. *Information Systems Research, 20*(4), 527–546.

Klein, B., Goodhue, D., and Davis, G. (1997). Can humans detect errors in data? Impact of base rates, incentives, and goals. *MIS Quarterly*, 21(2), 169–194.

Komiak, S. Y. X. and Benbasat, I. (2006). The effects of personalization and familiarity on trust and adoption of recommendation agents. *MIS Quarterly*, 30(4), 941–960.

Korhonen, P., Mano, H., Stenfors, S., and Wallenius, J. (2008). Inherent biases in decision support systems: The influence of optimistic and pessimistic DSS on choice, affect, and attitudes. *Journal of Behavioral Decision Making*, 21(1), 45–58.

Kumar, N. and Benbasat, I. (2004). The effect of relationship encoding, task type, and complexity on information representation: An empirical evaluation of 2D and 3D line graphs. *MIS Quarterly*, 28(2), 255–281.

LaValle, S., Lesser, E., Shockley, R., Hopkins, M. S., and Kruschwitz, N. (2011). Big data, analytics and the path from insights to value. *MIT Sloan Management Review*, 52(2), 21–32.

Lee, Y. E. and Benbasat, I. (2011). The influence of trade-off difficulty caused by preference elicitation methods on user acceptance of recommendation agents across loss and gain conditions. *Information Systems Research*, 22(4), 867–884.

Lerch, F. J. and Harter, D. E. (2001). Cognitive support for real-time dynamic decision making. *Information Systems Research*, 12(1), 63–82.

Lurie, N. H. and Mason, C. H. (2007). Visual representation: Implications for decision making. *Journal of Marketing*, 71(1), 160.

Meehl, P. E. (1954). *Clinical Versus Statistical Prediction: A Theoretical Analysis and a Review of the Evidence.* Minneapolis, MN: University of Minnesota Press.

Mennecke, B. E., Crossland, M. D., and Killingsworth, B. L. (2000). Is a map more than a picture? The role of SDSS technology, subject characteristics, and problem complexity on map reading and problem solving. *MIS Quarterly*, 24(4), 601–629.

Mudambi, S. M. and Schuff, D. (2010). What makes a helpful online review? A study of customer reviews on amazon.com. *MIS Quarterly*, 34(1), 185–200.

Nah, F. F. H. and Benbasat, I. (2004). Knowledge-based support in a group decision making context: An expert-novice comparison. *Journal of the Association for Information Systems*, 5(3), 125–150.

NCDC. (2010). Accomplishments Report of the National Oceanic and Atmospheric Administration's National Climatic Data Center. Retrieved from http://www.ncdc.noaa.gov/oa/about/2010-annual.pdf (accessed on March 12, 2012).

Newell, A. and Simon, H. A. (1972). *Human Problem Solving.* Englewood Cliffs, NJ: Prentice-Hall, Inc.

Ngai, E. W. T., Hu, Y., Wong, Y. H., Chen, Y., and Sun, X. (2011). The application of data mining techniques in financial fraud detection: A classification framework and an academic review of literature. *Decision Support Systems*, 50(11), 559–569.

Nielsen. (2008). 81 percent of online holiday shoppers read online customer reviews. Retrieved from http://www.nielsen-online.com/pr/pr_081218.pdf (accessed on April 10, 2010).

Nissen, M. E. and Sengupta, K. (2006). Incorporating software agents into supply chains: Experimental investigation with a procurement task. *MIS Quarterly*, 30(1), 145–166.

Orasanu, J. and Connolly, T. (1993). The reinvention of decision making. In G. A. Klein, J. Orasanu, R. Calderwood and C. E. Zsambok (Eds.), *Decision Making in Action: Models and Methods.* Norwood, NJ: Ablex Publishing, Corp.

Parasuraman, R. and Riley, V. (1997). Humans and automation: Use, misuse, disuse, abuse. *Human Factors*, 39(2), 230–254.

Parasuraman, R., Sheridan, T. B., and Wickens, C. D. (2000). A model for types and levels of human interaction with automation. *IEEE Transactions on Systems, Man, and Cybernetics*, 30(3), 286–297.

Parikh, M., Fazlollahi, B., and Verma, S. (2001). The effectiveness of decisional guidance: An empirical evaluation. *Decision Sciences*, 32(2), 303–332.

Paul, S. and Nazareth, D. L. (2010). Input information complexity, perceived time pressure, and information processing in GSS-based work groups: An experimental investigation using a decision schema to alleviate information overload conditions. *Decision Support Systems*, 49(1), 31–40.

Payne, J. W., Bettman, J. R., and Johnson, E. J. (1993). *The Adaptive Decision Maker*. Cambridge, MA: Cambridge University Press.

Petty, R. E. and Cacioppo, J. T. (1986a). *Communication and Persuasion: Central and Peripheral Routes to Attitude Change*. New York: Springer-Verlag.

Petty, R. E. and Cacioppo, J. T. (1986b). The elaboration likelihood model of persuasion. In L. Berkowitz (Ed.), *Advances in Experimental Social Psychology* (Vol. 19, pp. 123–205). New York: Academic Press.

Poston, R. S. and Speier, C. (2005). Effective use of knowledge management systems: A process model of content and credibility indicators. *MIS Quarterly*, 29(2), 221–244.

Price, R. and Shanks, G. (2011). The impact of data quality tags on decision-making outcomes and process. *Journal of the Association for Information Systems*, 12(4), 323–346.

Reichenbach, J., Onnasch, L., and Manzey, D. (2011). Human performance consequences of automated decision aids in states of sleep loss. *Human Factors: The Journal of the Human Factors and Ergonomics Society*, 53(6), 717–728.

Russell, S. and Norvig, P. (2003). *Artificial Intelligence: A Modern Approach*. Upper Saddle River, NJ: Prentice Hall.

Shortliffe, E. H. (1976). *Computer-Based Medical Consultations: MYCIN* (Vol. 388). New York: Elsevier.

Simon, H. A. (1977). *The New Science of Management Decision*. Englewood Cliffs, NJ: Prentice-Hall.

Skitka, L. J., Mosier, K., and Burdick, M. D. (2000). Accountability and automation bias. *International Journal of Human-Computer Studies*, 52(4), 701–717.

Speier, C. and Morris, M. G. (2003). The influence of query interface design on decision-making performance. *MIS Quarterly*, 27(3), 397–423.

Straub, D., Limayem, M., and Karahanna-Evaristo, E. (1995). Measuring system usage: Implications for IS theory testing. *Management Science*, 1328–1342.

Swets, J. A. (2000). Enhancing diagnostic decisions. In T. Connolly, H. R. Arkes and K. R. Hammond (Eds.), *Judgment and Decision Making: An Interdisciplinary Reader*. Cambridge, U.K.: Cambridge University Press.

Tan, C.-H., Teo, H.-H., and Benbasat, I. (2010). Assessing screening and evaluation decision support systems: A resource-matching approach. *Information Systems Research*, 21(2), 305–326.

Todd, P. and Benbasat, I. (1991). An experimental investigation of the impact of computer based decision aids on decision making strategies. *Information Systems Research*, 2(2), 87–115.

Todd, P. and Benbasat, I. (1992). An experimental investigation of the computer based DSS on effort. *MIS Quarterly*, 16(3), 373–393.

Tversky, A. and Kahneman, D. (1974). Judgment under uncertainty: Heuristics and biases. *Science*, 185, 1124–1131.

Vessey, I. and Galletta, D. (1991). Cognitive fit: An empirical study of information acquisition. *Information Systems Research*, 2(1), 63–84.

Wang, W. and Benbasat, I. (2009). Interactive decision aids for consumer decision making in ecommerce: The influence of perceived strategy restrictiveness. *MIS Quarterly*, 33(2), 293–320.

Wickens, C. D., Mavor, A., Parasuraman, R., and McGee, J. (1998). *The Future of Air Traffic Control: Human Operators and Automation*. Washington, DC: National Academy Press.

Yang, L., Su, G., and Yuan, H. (2012). Design principles of integrated information platform for emergency responses: The case of 2008 Beijing olympic games. *Information Systems Research*, 23(3), 761–786.

Zmud, R. W. (1979). Individual differences and MIS success: A review of the empirical literature. *Management Science*, 25(10), 966–979.

43

Computer-Supported Cooperative Work

Steven E. Poltrock
Bellevue, Washington

43.1 Introduction

Computer-supported cooperative work (CSCW) is a field of research addressing the intersection of collaborative behaviors and technology. CSCW includes collaboration among a few individuals, within teams, within and between organizations, and within online communities that may span the globe. CSCW addresses how technologies facilitate, impair, or simply change collaborative activities. It is a broader field than the four words comprising its name suggest. Although its primary focus is on computer-supported activities, it encompasses collaborative activities that have not yet been but could be computer supported. Its focus is on collaboration, but not necessarily achieved through intentional cooperation. Collaborators may compete with one another or may not even be aware of each other. The field's initial focus was on work-related activities, and this remains a topic of interest, but the scope includes collaboration in any area of human endeavor including games and social media.

43.2 Underlying Principles

Collaboration is the central issue in all CSCW research. The key challenge for both researchers and developers is to understand collaboration, how it is accomplished, and its impediments. Sometimes, these impediments include aspects of technologies intended to support it, potentially providing requirements for enhancements or suggesting new technologies.

43.2.1 Classes of Collaboration Context

How people collaborate depends on many aspects of their context. Classification models of collaboration context capture some of the features of this relationship. For example, people collaborate in real time and asynchronously, in the same location or different locations, and in small groups, organizations, or communities, as shown in Figure 43.1. Consider the collaborative activities likely to occur in each cell. A team or *small group* (typically less than seven people) may work together in a team room, collaborating in the *same location* at the *same time* by speaking to one another and by creating and sharing artifacts retained in the room. They are supported by technologies that facilitate small group, face-to-face interactions, such as digital projectors and meeting facilitation tools. Small groups may collaborate from *different locations* using collaboration technologies to communicate and share information with one another. A small group may work together at different times in a team room (*same location, different times*) by accessing or updating information posted in the room. A medical team providing patient care around the clock offers another example of collaboration in the *same location* at *different times*. A small group working together on a multiauthored paper produced by passing revisions back and forth is a familiar example of collaborating at different times and locations.

Larger groups form *organizations* to facilitate coordinated action across all members. Organizations typically contain many small groups that may collaborate in the ways described previously, but we should also consider how these groups collaborate with one another to achieve their common objectives. The collaborative activities of large organizations fall largely into two cells of the model in Figure 43.1. Organizational activities are rarely performed in a single location, because their size and missions demand geographic dispersion. Organizations are likely to engage in complex activities that require that groups work together in *different locations* at the *same time* and at *different times*, and coordinating these activities is the organizational challenge and their reason for adopting collaboration technologies. In a manufacturing organization, for example, collaboration focuses on the product, and group contributions to the product are coordinated by its development or construction plan and supporting technologies.

Communities are collections of people with common interests. Neighborhood communities are defined by their common location, but most communities are geographically distributed. Communities have no organizational structures, although they are often served by organizations that view the community as their marketplace. Collaboration within a community focuses on their common interest and requires little or no coordination. Consider, for example, the community of people interested in photography. Members of this community may seek or share photographs or information about photographic techniques at the *same time* (via presentations) or at *different times*. Organizations develop technologies that support sharing photographs while offering products and services to the photography community.

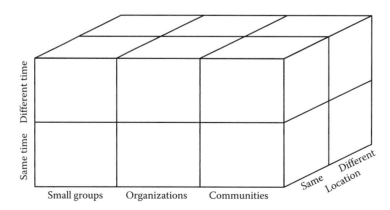

FIGURE 43.1   Classification model of three dimensions (time, place, and social unit) of collaboration context.

Each cell of Figure 43.1 constitutes a substantially different collaboration context, involving different collaborative activities and demanding different supporting technologies. This classification model describes three broad attributes of context (who, when, and where) that are relatively easily categorized. Other attributes of context such as the objectives of the collaboration are more difficult to categorize but no less important.

43.2.2 Categories of Collaboration Activities

The human activities involved in collaboration can be roughly described as communicating (e.g., speaking, writing, reading, listening), sharing information with one another (e.g., displaying a presentation, a map, a document, or other data), and coordinating (e.g., taking turns, sharing a resource, or performing different subtasks). Of course, these categories overlap and are tightly integrated; people may coordinate through speech and share information through both speech and written artifacts. People collaborating engage in all three categories of activity, shifting with little effort from one to another or combining them. Nonetheless, these categories are useful when thinking about technologies that support collaboration. Table 43.1 lists some well-known technologies or technology features that support each category of activities in both same-time and different-time contexts.

The temporal dimension shown in Table 43.1 is, of course, not the only aspect of context that influences use of these technologies. Communication is critically important within small groups, and most of the communication technologies listed in Table 43.1 are principally used by small groups, often just two people. Blogs and social networking sites are exceptions and are frequently used to communicate to a larger audience, perhaps a community of people with similar interests.

When small groups meet face-to-face, they may share information using meeting facilitation tools, a simple whiteboard, or a digital projector. When geographically distributed, small groups may share information using application-sharing technologies that allow everyone in the group to see and interact with the same information. Between meetings, they may store and access information managed by a document management system, shared folders, or a web-based team or project repository. Small groups may also use wikis to share information, but wikis are more widely known for their support of communities based on shared interests. Wikipedia is a prominent example of technological support for communities, not for small groups or organizations.

Small groups generally have little need for technologies that support coordination; they coordinate their activities by communicating and sharing information with one another. When geographically distributed, however, small groups may benefit from features of conferencing technologies that manage participation in a virtual meeting (session management) and limit participants' interactions

TABLE 43.1 Example Technologies Used in Collaborative Activity Categories

	Same Time	Different Time
Communication	• Telephone	• E-mail
	• Videoconferencing	• Voice mail
	• Instant messaging	• Blogs
	• Texting	• Social networking sites
Information sharing	• Whiteboard	• Document repositories
	• Application sharing	• Websites
	• Digital projector	• Team workspaces
	• Meeting facilitation	• Wikis
Coordination	• Floor control	• Workflow management
	• Session management	• Computer-aided software engineering tools
		• Project management
		• Calendar scheduling

(floor control). Social protocols accomplish these functions in face-to-face meetings, but in online conferences, these features prevent unwanted guests and facilitate turn taking. Small groups and communities have little need for technologies that help coordinate their different-time collaborations, but coordination is the principal challenge facing organizations. They invest in technologies that coordinate the collaborative work involved in designing, producing, and delivering their products and services such as computer-aided software engineering (CASE) tools, workflow management systems, and project management tools.

People collaborate for many purposes in different contexts and employing diverse collaborative activities. CSCW encompasses all these purposes, and collaboration technologies may be employed in any online collaborative activities. Small groups may use collaboration technologies to stay connected with distant team members, present ideas to a remote audience, generate ideas and make decisions in a group meeting, or play an online game. Organizations may use collaboration technologies to coordinate all aspects of their business, including production, delivery, and maintenance of their products. Communities may use technologies to discover others with shared interest, collect information about their interest, and provide community services.

43.2.3 Research Methods

The challenge in CSCW is to understand collaboration and how technologies can support it. Gaining this understanding is accomplished through research methods acquired from other disciplines that study human behavior such as psychology, sociology, and anthropology. Experiments in laboratories are of greatest value when designed to shed light on enduring properties of human behavior. For example, the experimental method has provided insights into the influence of visual display properties on trust and empathy during videoconferences (Nguyen and Canny, 2007, 2009) and the costs and benefits of communication between two collaborating people (Dabbish and Kraut, 2003, 2004).

The experimental method is less fruitful when studying complex collaborative activities such as software development, medical operations, or emergency responses. A researcher cannot randomly assign people to teams or treatments, and these contexts cannot easily be recreated in a laboratory setting. Instead, observational methods are employed, borrowing heavily from ethnography. Observers study and record team behavior, learning what team members do and how and why they do it in specific contexts. For example, Jackson et al. (2012) investigated maintenance and repair of the information and communication infrastructure in Nambia; Benford et al. (2012) investigated traditional Irish music making in pubs; and Pine (2012) investigated the coordination of care in the labor and delivery unit of a hospital. Observational research may identify opportunities for technological interventions, capture changes in behaviors during technology adoption, and reveal the effects (both positive and negative) of an implemented technology.

Both experimental and observational methods are inadequate for studies of large-scale online collaboration such as the construction and maintenance of Wikipedia. Participants in these collaborations are often anonymous, or at least unknown to the researcher, posing an obstacle to direct observation or controlled experimentation. These collaborations are, however, often prolific, and the products and traces of their production are accessible online. Methods for mining, analyzing, and interpreting the results of these collaborations are providing interesting insights. For example, Keegan et al. (2012) employed exponential random graph models (see Robins et al., 2007) to analyze and compare the coauthorship networks in Wikipedia of articles about breaking news and articles about contemporary and historical events. They found that the emergent organization of authors and editors of Wikipedia articles is influenced by attributes of both the editors and the article content, and editor experience is among the strongest influences. In addition to analyzing collaboration networks, mathematical methods may be employed to analyze content. For example, Al-Ani et al. (2012) employed topic modeling analysis (see Blei et al., 2003) to investigate the evolution of the content of blogs written in Egypt during the Egyptian revolution. By using mathematical models such as these, researchers can identify relationships in large-scale collaborations similar to those that ethnographers would find in smaller scale collaborations.

43.3 Impact on Practice

Collaborating with the support of computers is widespread today in small groups, organizations, and communities, and the practice of CSCW continues to accelerate. Despite obvious progress over the past 25 years, CSCW research offers some lessons, some warnings, and some advice for collaborators and for developers of collaboration technologies.

43.3.1 Grudin's Eight Challenges

In one of the most frequently cited articles in CSCW, Grudin (1994) listed eight challenges that continue to plague developers and users of collaboration technologies. These challenges are not technical problems; they arise from human psychology in social and organizational settings and may be avoided through thoughtful analysis of the collaborative context and careful design.

The first challenge is that users are reluctant to adopt a collaborative technology if the effort required outweighs the benefits they receive as individuals, regardless of the potential collective benefit. For example, employees have repeatedly resisted adopting time-tracking technologies despite their managers' belief that the information will yield greater overall work efficiency. Technologies that require more effort from a person should return a benefit to that person.

The second challenge is to achieve a critical mass of users. If some people involved in a collaboration decline to adopt a technology, the advantages of the technology for all other participants may be undermined. For example, a distributed group I observed using videoconferencing to support long-distance meetings abandoned the technology because one person was uncomfortable facing a camera.

The third challenge is to avoid interfering with complex and subtle social dynamics of a group. Our actions are guided by social conventions and awareness of people and priorities, awareness that computers generally do not share. Knowledge of the usage context may be required to design an application or to guide its adoption in a way that does not disrupt existing communication channels, create uncertainty about where to find information, or challenge existing authority structures.

The fourth challenge is to allow exceptions to standardized work practices. Technologies intended to coordinate work activities often incorporate and enforce models of the work processes and cannot easily accommodate deviations from these models. An important contribution of CSCW ethnographic studies (Suchman, 1983; Sachs, 1995) is the observation that handling exceptions and unanticipated events is normal behavior in the workplace, and strict adherence to standard processes is a rare exception.

The fifth challenge for technology designers is that the collaboration features of a technology must be readily accessible although they may be infrequently used. The challenge for the users of the technology is to understand how to use these potentially unfamiliar features while collaborating with other people. There is a risk that people may feel uncomfortable revealing to their friends or colleagues their lack of expertise with the technology.

The sixth challenge, as developer or user, is to evaluate the technology reliably. The methods developed for evaluating a single-user technology, such as walkthrough analysis or laboratory studies, are of limited value when assessing how well a team, organization, or community is supported by a technology. Users face the same challenge when choosing between technology alternatives. Evaluations of collaboration technologies generally take much longer to achieve a comparable level of reliability.

The seventh challenge is a breakdown in the intuitive decision-making process that normally guides product development and acquisition. These decisions are generally made by development or IT managers based on their informed intuitions regarding the strengths and weaknesses of alternatives.

Managers' expertise in an application domain may be a solid foundation for intuitive decision-making, but they are unlikely to have expertise in all the aspects of group dynamics that influence the success of a collaboration technology.

The eighth and final challenge is to manage the adoption of collaboration technologies. There is substantial risk of failure when a new collaboration technology is developed or acquired because of the pitfalls posed by the preceding seven challenges. Success requires that all the participants in a collaborative activity adopt the technology together, and this is unlikely to occur without careful planning.

43.3.2 Adoption of Collaboration Technologies

Grudin (1994) noted the challenge of managing adoption of collaboration technologies. The scope of this challenge and methods for addressing it depend on attributes of the collaboration context. For example, technologies intended to support small group, organizational, and community collaboration face very different challenges.

43.3.2.1 Adoption by Small Groups

Mark and Poltrock (2003, 2004) studied the adoption of desktop conferencing (sometimes called data conferencing) by geographically distributed teams in a large corporate setting. In this case, the technology was successfully adopted without a top–down management plan, but the diffusion process was slow and painful. Adoption was driven by users who recognized the benefits of the technology, demonstrated it to their coworkers, and helped each other learn to use it. Employees were often in one or more geographically distributed teams, and each such team constituted a social world. Information about the technology was communicated across social worlds, and people wanting to participate in a social world found that they needed the technology to be a successful contributor. The technology diffused rapidly across some parts of the company, but the managers of other parts considered this new technology to be a risk to network reliability and actively opposed its adoption. To function effectively, team members must uniformly adopt the technology, but people are members of multiple social worlds, and the active resistance of management in some social worlds created conflict and tensions. Eventually, the benefits of the technology were widely recognized, it was adopted by executive management, and resistance faded away, but the adoption process was long and painful for some individuals.

43.3.2.2 Adoption by Organizations

The primary challenge of organizational collaboration is the coordination of work activities. Workflow management systems, or product-focused tools with workflow management features, are often acquired for the purpose of coordinating the contributions of many distributed groups within an organization and sometimes across organizational boundaries. Acquisition and deployment of these systems are generally a substantial investment requiring top–down planning and integration with other tools and infrastructure. Unfortunately, these implementations are often unsuccessful (Bowers et al., 1995; Kingsbury, 1997) due to social and organizational factors that lead to adoption challenges.

Workflow management systems contain explicit representations of work processes and are intended to facilitate these processes by managing the transfer of work products from one person or group to another as tasks are completed. A promised benefit of these systems is that the defined work processes must be consistently followed, but the difficulty of defining stable work processes is a fundamental problem. Suchman (1983) asked members of an accounts payable organization to describe their processes, and she received a textbook explanation. When she observed their work, however, no single instance of their work followed the processes they had described. She distinguished between their plan (their documented process) and their procedures situated in the context of a particular process instance (Suchman, 1987). People engage in considerable problem-solving to ensure that the work complies with the expected results of the overall plan, but they accommodate the circumstances that are unique to each work instance.

Ishii and Ohkubo (1990) developed a workflow modeling and automation technology for office work and encountered this same problem. They successfully built a technology capable of managing the flow of work, but found that people did not consistently perform their work in accordance with the process they had described.

I participated in the design, deployment, and evaluation of several workflow management applications intended to support an engineering design organization. The organization employed a small staff of coordinators who acted as human workflow management engines; they collected the results of each task, delivered it to the next person or group, reminded people of incomplete tasks and deadlines, and kept track of each work instance. The workflow management applications were intended to perform much of this work, reducing the number of human coordinators required to manage the work. These coordinators were experts in the work processes, but nonetheless were unable to describe them accurately and completely.

A purported advantage of workflow management systems is the ease with which process models can be revised, achieving some degree of flexibility. Instead, the workflow applications that I helped deploy were inherently and unavoidably less flexible than the human-powered systems they replaced. Suppose the organization decided to revise a work process. A process facilitated by a human coordinator using e-mail and office documents can be changed by simply updating the process documentation and informing everyone about the revisions. The human coordinator would monitor the outcome, resolve any short-term problems, and propose additional revisions if any lasting problems are identified. But a workflow application is much more than a simple description of the process flow. Developing a workflow application requires modeling the data that will be managed in the workflow, defining the process rules, and developing the user interface required to view and edit the process-related data. It requires the same software development rigor as any other application development, and in this organization, all changes to software required thorough testing and release on a schedule coordinated with all other applications. If the revised processes create unexpected problems, there are no simple ways to make short-term adjustments, and further process revisions require another cycle of development and testing.

Even when processes remained stable, workflow management systems could not provide the flexibility offered by human coordinators. The workflow system sent e-mail messages to alert or remind people about tasks, and it provided lists of incomplete tasks. In contrast, coordinators alerted and reminded people in many ways, exploiting their knowledge of the individuals responsible for the tasks. An e-mail message was sufficient for some people, but coordinators knew who had an overflowing inbox and would be more receptive to a phone call or a personal visit. They knew when someone was on vacation and who was authorized as a substitute. To continue providing these services after the workflow management system was implemented, the coordinators continued to track all processes in spreadsheets, duplicating the information in the workflow system (Handel and Poltrock, 2011).

Achieving flexibility in workflow management applications remains an active area of research (Pesic and van der Aalst, 2006). Human coordinators could provide flexibility in many ways that exceeded the capacities of a workflow management system. Of course, human coordinators are expensive, and organizations may choose to adopt a workflow management system to avoid that cost, but these savings must be weighed against the costs of developing and maintaining a workflow management system.

43.3.2.3 Adoption by Communities

In striking contrast to the coordination demands of organizations, communities have little need for coordination technologies and primarily need tools for sharing information. An abundance of online social media services are aimed at addressing the needs of various communities. These include photograph (or other image) sharing sites, wikis devoted to topics of interest to specific communities, marketplaces where items can be bought and sold, and websites devoted to specific crafts, hobbies, or intellectual interests.

For the developer of such a technology, the first adoption challenge is to inform members of the target community about the opportunity offered by the technology. Early adopters may be found by visiting places where community members may gather in the real world or through existing organizations that support the community. The second challenge is to persuade potential adopters that the technology offers a value greater than its cost. Many community services address the cost side of this challenge by offering their services for free. The third challenge is to attract additional members of the community and engage them in active use of the technology such that it becomes indispensable. Social media technologies often address this challenge by encouraging users to invite their friends, often using their e-mail address books or other lists of contacts. Of course, such a technology service must have some source of income, and it may solicit charitable giving from the community or advertise products of interest to that specific community. Consider, for example, the community of postage stamp collectors. This community has constructed an excellent summary of stamp collecting in Wikipedia, including links to other online sources. One of these sources (learnaboutstamps.com) is sponsored by the U.S. Postal Service and the American Philatelic Society, among other organizations. This society's own website invites charitable contributions to support the community and includes an online marketplace where its members can buy and sell stamps.

These examples illustrate the challenges that both developers and users may encounter with technologies intended to support collaboration within small groups, organizations, or communities. Note that all the challenges have their roots in human behavior, not in the limitations of technologies. Collaboration is a social activity, and social practices can be leveraged to achieve extraordinary success or become obstacles that impede technology adoption.

43.4 Research Issues

CSCW research can be roughly divided into two large categories, one focused on understanding collaboration and how technology influences it, and the other focused on the computer science challenges involved in supporting collaboration. Jacovi et al. (2006) presented empirical evidence for this categorization. They analyzed the citation graph of all 465 papers presented at CSCW conferences between 1986 (the first conference) and 2004. They identified eight clusters, of which the two largest correspond roughly to social science and computer science. The social science cluster included papers about theories and models, ethnography, collocated teams, organizational memory, awareness, and user studies such as studies of adoption. The third largest cluster included meeting/decision support, shared media spaces, and conferencing. A fourth cluster included papers on instant messaging, presence, and social spaces. Smaller clusters included papers on the use of computer tools in the workplace, groupware design and workspace awareness, management of computing and information systems, and video-mediated communication and shared visual spaces.

Convertino et al. (2006) analyzed the same 465 research papers, categorizing each paper by type of institutional affiliation, author's geographical location, its level of analysis (individual, group, or organization), type of contribution (theory, design, or evaluation), and type of collaboration function investigated (communication, coordination, or cooperation). About 80% of the papers were about small group collaboration and nearly all the rest had an organizational focus. Research on community collaboration emerged as an important theme after the years from 1986 to 2004 covered by these analyses. Most papers contributed to either design (corresponding roughly to the computer science cluster of Jacovi et al., 2006) or evaluation (corresponding to the social science cluster). Contributions to theory were frequent at the first three conferences (about 30% of papers) but then declined to fewer than 10%. Grudin and Poltrock (2012) attributed this decline to the departure of the Management Information System (MIS) community from the conference, a departure driven by decisions to archive ACM conference proceedings and establish high-quality standards and high rejection rates for conferences. In many research communities, including MIS, researchers may present their work in progress at conferences and reserve their highest quality work for publication in journals. Indeed, the MIS community continued its active research in group support systems (e.g., research in collaboration engineering such as Kolfschoten et al. 2006) and organizational collaboration in business settings (e.g., business process management research). This research can be found in the *Proceedings of the Hawaii International Conference on Systems Science* and published in various Information Systems Journals.

This chapter does not address the challenging computer science issues in CSCW. It focuses, instead, on research seeking to understand collaboration and how technology influences it when collaboration takes place in small groups, organizations, and communities.

43.4.1 Small Group Collaboration—Videoconferencing

About 80% of the research presented at CSCW conferences between 1986 and 2004 explored aspects of small group collaboration (Convertino et al., 2006). Researchers have investigated ways to support small groups in face-to-face meetings and when they are geographically distributed. They have explored meeting or decision-support technologies, communication technologies, and ways to share information. This chapter focuses on videoconferencing, a form of small group communication that has been intensively studied in many ways over many decades.

Videoconferencing supports integrated and synchronized visual and audio communication. It is an old technology that has become dramatically more usable and affordable in recent years. In 1878, du Maurier created a cartoon of a Telephonoscope that would transmit both sound and images bidirectionally. In 1927, when Herbert Hoover was US Secretary of Commerce, he used a videoconference system in Washington to communicate with AT&T President Walter Gifford in New York. This, of course, was a prototype system, but AT&T continued pursuing the idea, developed the Picturephone in the 1950s, and launched the commercial Picturephone product in 1970. The technology was a commercial failure, but telephone companies continued to research and develop videoconferencing technology, and some of their research is included in the CSCW literature. Nonetheless, adoption of videoconferencing was surprisingly slow. Throughout the twentieth century, it was little used outside research labs and in some government and industry settings that required long-distance collaboration.

Improvements in videoconferencing and network technology have dramatically enhanced its quality and reduced its cost. This long history of improvements, including integration of videoconferencing with computers and their networks, has generated repeated resurgences of interest in videoconference systems. Engelbart and English (1968) demonstrated videoconferencing, including overlays of video and data, in their groundbreaking research exploring how computers could augment human intellect. Despite this steady stream of improvements, adoption of videoconferencing remained slow until recently. Today, many executives have personal videoconference systems on their desks, consumers use Skype's features to communicate with friends and family, and cameras are built into computers and mobile phones. The ability to communicate both visually and auditorially seems intuitively appealing, and its slow adoption despite improved performance and declining cost posed a puzzle that attracted many researchers.

At the second CSCW conference, Egido (1988) reviewed reasons for the failures of videoconferencing. She noted that forecasts in the early 1970s predicted that 85% of all business meetings would be electronically mediated by the end of that decade. But, when her paper was written in 1987, only about 100 videoconferencing facilities existed worldwide, and many of those were located in the offices of teleconferencing vendors and telecommunication companies. Egido argued that two factors were responsible for the low rate of videoconferencing adoption. The first factor was the inadequacy of needs-assessment methodologies, which were largely based on surveys at that time, with the consequence that technologies were not designed appropriately for the contexts in which they would be used. The second and closely related factor was the portrayal of videoconferencing as a direct replacement for face-to-face meetings. It does not provide the same opportunities for interpersonal interactions that occur in business meetings, and many years of research have found no decrease in face-to-face meetings when videoconferencing is available. Egido concluded that videoconferencing may be an attractive (if expensive) communication option in many situations, and research should focus on understanding communication tasks in the office environment and how such a technology could best support it.

At the same conference, Robert Root (1988) followed Egido's advice and proposed a system called CRUISER based on a thoughtful analysis of office communication patterns. Root observed the importance of frequent, low-cost face-to-face interactions in which both technical and personal information can be exchanged quickly and easily. He designed CRUISER to support social interaction in office environments that are geographically distributed, providing easy, convenient access to other people regardless of their physical location. Its users were able to stroll (metaphorically) down a virtual office hall, peek into virtual offices via videoconferencing, and stop to chat for a while if the office occupant was available. Note that this research involves an analysis of a social context, identification of the technology requirements to support collaboration in this context, and a proposed system intended to address these requirements. There is, of course, a big gap between a technology proposal such as this one and the development of a successful system that meets all the requirements and integrates seamlessly into the social environment.

Root and his colleagues implemented and evaluated CRUISER in a research and development laboratory (Fish et al., 1992). CRUISER was essentially a videoconferencing system with full-duplex audio, but it included three novel ways to initiate conversations: (1) *cruises* opened an immediate videoconference connection to a called party that timed out after 3 s unless the recipient explicitly accepted it, (2) *autocruises* initiated 3 s calls between selected users at random times that were extended only if both parties accepted it, and (3) *glances* opened a one-way video-only connection to the recipient's office that lasted only 1 s. Observations of how people used this system were enlightening about the challenges facing videoconferencing developers. CRUISER metrics indicated that the cruise feature was used much more than a glance, and autocruises were rarely used. Videoconferences initiated by the cruise feature were short, rarely lasting more than a few minutes, and were primarily used to schedule a face-to-face meeting or for status reporting. People frequently used it to see whether someone was present and available, and then arranged to visit them. But the sudden onset of the cruise or glance connection was perceived as intrusive, disruptive, and an invasion of privacy. Its developers had hoped to recreate the opportunities for informal social interaction that occur when walking between offices, but CRUISER's interaction methods were very different from the methods that people use when meeting one another face-to-face.

CRUISER was among the first of many systems developed to explore novel ways to establish videoconferences quickly and easily while respecting privacy (Cool et al., 1992; Tang et al., 1994). Researchers reasoned that persistent dissatisfaction with videoconferencing could, in part, be due to the clumsy process of starting and ending contact, a process that Tang (2007) termed *contact negotiation*. Members of a collocated team can easily see when other team members are busy or available, and they can easily negotiate contact with one another. Dourish and Bly (1992) developed Portholes to provide similar awareness of availability to distributed teams. Portholes displayed thumbnail images of team members from cameras in their offices, offering a quick and current view of who was present and available. Clicking on one of the thumbnail images in later versions of Portholes brought up a menu of communication options including videoconferencing (Lee et al., 1997). Systems such as Portholes facilitate awareness and contact negotiation but again raise concerns about privacy. Many people feel uncomfortable with the idea of being continuously visible via a camera image, even though they may be continuously visible to others in an open office environment. Tang (2007) reviewed research investigating alternative approaches to negotiating contact, concluding that the essential cognitive and social cues could be conveyed via text and icons instead of attempting to recreate sensory cues through video. His research demonstrates an approach that may successfully avoid the experience of an invasion of privacy.

While some researchers focused on simplifying contact negotiation, others questioned the benefits provided by video. The book *Video-Mediated Communication* (Finn et al., 1997) included 25 papers by noted researchers in the field exploring these benefits. To a great extent, these papers confirmed the findings of earlier studies that the addition of video to audio communication does not improve collaborative problem-solving. Furthermore, many people have stated informally that they do not want to see or be seen by the person with whom they are speaking while discussing business. This has been a disappointment to the telecommunication and network companies that hoped widespread adoption of videoconferencing by US industry would drive an increase in demand for bandwidth.

More recent research suggests that there are benefits of adding video to an audio communication channel, but the benefit is not better collaborative problem-solving. Williams (1997) found that video led to improved conflict resolution in stressful communication conditions, such as when two people with different native languages must negotiate in English. In three-way videoconferences (between three different locations), video showing both remote participants simultaneously (spatial video) helped people keep track of their three-way conversations and yielded higher quality conversations in the judgment of the participants (Inkpen et al., 2010). Video also provides a way that people can collaborate on a physical task, but in these situations the cameras are focused on the task and not the people (Kraut et al., 1996). Indeed, in some situations video can be a distraction and impair performance. Pan et al. (2009) presented English clips to Chinese students and tested their comprehension. The students achieved better comprehension with audio alone than with audio plus video.

Although the benefits of adding video to an audio conference are difficult to quantify, many people clearly enjoy the experience more. Even the studies that found no benefits (Finn et al., 1997) reported that participants liked seeing one another and believed that they had achieved a better solution, despite the objective evidence that they had not. This observation has led some researchers to explore emotional or affective consequences of videoconferencing.

One reason that people meet face to face is to establish a relationship and level of trust that will sustain their collaboration when geographically distributed. If videoconferencing is intended to replace face-to-face meetings, then its support for developing trust is important. Bos et al. (2002) studied the effects of four computer-mediated communication channels on trust development: face-to-face, audio, audio plus video, and text chat. They measured trust between three people using a game that poses a social dilemma. To cooperate, they must trust one another, and consequently their success in the game can be considered a measure of the trust within the group. Groups in the face-to-face condition began cooperating quickly; groups in the text chat condition never learned to cooperate; and both the audio groups and the audio plus video groups eventually learned to cooperate, but continued to experience defections throughout the game. These results suggest that communication mediated by technology does not support an awareness of the presence of others as effectively as face-to-face meetings.

The failure of videoconferencing to support development of trust posed a challenge to researchers in this field. Nguyen and Canny (2005, 2007) took up this challenge and investigated whether it was a social consequence of the breakdown of eye gaze awareness, which is disrupted in videoconferencing because camera lenses are not located where the eyes are displayed. They developed a group videoconferencing system that provides spatially faithful eye gaze awareness and studied the development of trust in three conditions: face-to-face meetings, their spatially faithful videoconference system, and a similar system with conventional nonfaithful displays. Like Bos et al. (2002), they indirectly measured trust via performance in a game that poses a social dilemma. They found that game performance and thus trust were essentially equal when people met face-to-face or when using the spatially faithful videoconference system, but was lower and more fragile when using the conventional system. These results suggest that eye gaze awareness and mutual gaze may contribute to the development of trust.

Many studies have found that face-to-face encounters are superior to videoconferences in establishing trust or feeling that a conversation is natural. Reconsidering the implications of their own research, Nguyen and Canny (2009) noted that even studies with mutual gaze awareness had found substantial and consistent differences between face-to-face interactions and videoconferencing. They hypothesized that eye gaze awareness was not the sole reason that they had found equal development of trust in face-to-face and video conditions. Instead, they proposed that video framing had played an important but unnoticed role. Videoconferencing equipment is usually configured to show only the heads or the heads and shoulders of participants, and most commercial videoconferencing systems (including laptop computers) are similarly configured. Nguyen and Canny's system, however, was configured to display the entire upper body at life size. In typical videoconferences, the main nonverbal channels of communication are facial expression and gaze, but their system had added three additional nonverbal communication channels: posture, gesture, and proxemics.

Nguyen and Canny (2009) conducted an experiment in which two participants talked for 20 min in one of three conditions: (1) face-to-face seated across a table from one another, (2) in a videoconference with a life size view of the other participant's upper body, or (3) in a videoconference with the same equipment but masked so that they could see only the face of the other participant. Near the end of the experiment, one of the two participants was asked to assist the experimenter by entering the room with the other participant and pretend to accidentally drop some pens. Nguyen and Canny reasoned that the other participant would be more likely to help pick up the pens if the two participants had indeed established empathy during their 20 min conversation. The experimenter measured whether the other participant helped pick up the pens and the time until the participant initiated that behavior. There were no significant differences in measures of empathy or the pen-drop experiment between the face-to-face meeting and the videoconference showing the whole upper body, but the two videoconferencing conditions were strikingly different. When participants viewed only the face, their ratings of empathy were lower; they were less likely to help pick up pens, and they were much slower to pick up pens. These results are strong evidence that a videoconference showing the whole upper body at life size will yield better nonverbal communication and more effectively support development of empathy than a videoconference showing only the face.

Today, video cameras are integrated in laptop computers and mobile telephones, and videoconferencing services that use these cameras are freely available. Although videoconferencing is freely available, its adoption remains much lower than other prominent communication technologies such as telephones, e-mail, and text messages. Researchers have identified key barriers to adoption and pointed the way to solutions. Nguyen and Canny's results tell us that videoconferences will be competitive with face-to-face meetings if they display the whole upper body and achieve good eye gaze awareness, and high-end or immersive commercial systems have adopted this framing recommendation. Contact negotiation in most commercial systems is modeled on a telephone calling interface, but researchers have evaluated other approaches that offer promising improvements in the videoconferencing experience.

43.4.2 Organizational Collaboration

Organizational collaboration occurs when two or more groups work together, requiring coordination of their mutual efforts. As described earlier, workflow management technologies provide coordination mechanisms intended to support organizational collaboration. Workflow management systems typically contain an explicit, editable model of work processes, and they pass work artifacts and assignments from person to person (or group to group) in accordance with this model.

Although workflow management is widely used in some industries, its value has been controversial in the CSCW research community. During the early years of CSCW, some researchers were optimistic about the potential for aiding collaboration through development and implementation of formal models that both described and guided the work. Winograd and Flores' (1986) explorations of the conversation-for-action analysis method inspired novel approaches for modeling work processes and using these models to coordinate work (Medina-Mora et al., 1992; De Michelis and Grasso, 1994). Suchman (1994) criticized these technologies on two grounds. First, the technologies embody a categorization scheme and impose this scheme on its user community, displacing their own preexisting categories for describing their work. Second, a workflow system that mediates all work activities amounts to an oppressive disciplinary regime and is politically repugnant. Winograd (1994) responded to these criticisms, explaining that the categorization schemes used in workflow management are not intended to serve as models of language but as structures that support unambiguous communication across organizational boundaries, and that workflow systems need not be oppressive. This debate essentially ended contributions to workflow management research within the CSCW community, with the exception of approaches intended to increase flexibility and give greater control of the work process to the user community (e.g., Dourish et al., 1996).

Research in organizational collaboration is principally found in management journals and conference proceedings, but some threads continue to find an audience within the CSCW community. Researchers have explored the flow of information within organizations, investigating properties of organizational memory (Ackerman and Halverson, 1998) and the role of social network systems in organizational acculturation, which is essential to successful organizational collaboration (Thom-Santelli et al., 2011). Balka and Wagner (2006) studied how technology is appropriated within organizations, focusing on the configurations required to make new technologies work successfully. One generally thinks about configuring the technology itself, but they noted that new technologies may require configuring organizational relations, space and technology relations, and the connectivity of people, places, and materials. Similarly, Lee et al. (2006) proposed expanding the concept of cyberinfrastructure to include the associated human infrastructure (people, organizations, and networks) as a way to bring into focus some of the factors that are critical to a successful cyberinfrastructure program.

Recent work in organizational collaboration has focused on domains or settings of special interest such as insurance, hospitals, and nonprofit enterprises. Painter (2002) studied the adoption of a new electronic claim file system in a health insurance program and its effects on collaboration. Abraham and Reddy (2008) studied coordination between clinical and nonclinical staff in the patient transfer process of a hospital, focusing on failures of information sharing and the reasons for these failures. Le Dantec and Edwards (2008) compared two nonprofit homeless outreach centers and found that differences in their organizational structures and composition were tightly coupled to differences in their use of information and communication technologies. Many workers in these settings are volunteers who do not receive much training in organizational processes or its technologies, offering both an opportunity and a challenge for coordinating technologies such as workflow management. Stoll et al. (2010) examined coordination across multiple nonprofit organizations seeking to help the victims of human trafficking. Their work highlights structural, technological, and organizational factors that challenge nonprofit organizations seeking to employ technologies that are widely used in the for-profit world.

43.4.3 Community Collaboration

In recent years, the CSCW research community has grown increasingly interested in collaborations that occur within communities, especially online communities. The participants in these collaborations are not members of the same teams or organizations. Often, they do not know one another. Nonetheless, online community collaborations have yielded impressive results, as exemplified by Wikipedia. Researchers seek to understand who participates in these communities, why they do it, how they become informed members of the communities, how their collaborations are structured or organized, and how their participation evolves over time.

Not surprisingly, the most frequently studied online community collaboration is the production and maintenance of the Wikipedia online encyclopedia. This research is important because the encyclopedia is massive and widely used. The Wikipedia system indirectly supports this research by saving all contributions by authors and editors online where researchers can readily access it.

Bryant et al. (2005) were among the first to study contributions to Wikipedia, and the first of several to focus on how people became active editors of Wikipedia. They interviewed nine people who frequently edit Wikipedia pages. They described their results using activity theory in which the building blocks are objects, subjects, community, division of labor, tools, and rules. They described the evolution from a novice editor to an expert Wikipedian as a transformation in each of these six areas. For example, editors develop a feeling of membership in a community while learning the community's rules and tools.

Of course, most users of Wikipedia only read articles and never contribute to an article. Antin and Cheshire (2010) ask why people contribute to Wikipedia when they could gain its benefits by just reading it without the work of writing and editing it, and then suggest that readers would also like to contribute but do not because they do not know enough about the system to feel ready to do it. They argue that reading is a form of legitimate peripheral participation, a category of behavior that provides a pathway to learn about the Wikipedia community and its environment. This is not, however, the only reason that people choose not to contribute. A survey of more than 40,000 users of Wikipedia explored reasons for not contributing, focusing on why so few women (less than 15%) are contributors (Collier and Bear, 2012). They found that women were less likely than men to feel confident in their expertise and the value of their contribution, they were more likely to feel uncomfortable with the high level of conflict involved in the editing process, and they were less likely to feel comfortable editing another person's work.

The distribution of contributions to Wikipedia follows a power law, with most people making very few edits. Antin et al. (2012) studied the factors that distinguish between people who make many edits and those who make few edits. They found that early diversification into multiple kinds of editing activities is strongly predictive of more editing behaviors later, people continue to engage in the same kinds of activity patterns over time, and early diverse editing behaviors are associated with later organizational and administrative behaviors.

Contributors to Wikipedia interact not only with the content but also with one another. Choi et al. (2010) studied socialization of new members in WikiProjects, which are topic-centric subgroups of Wikipedia contributors that improve coverage in a particular domain. They studied the messages written to newcomers by existing project members and identified types of socialization tactics. Using correlational analyses, they found that more socialization messages were associated with more editing behavior by the newcomers. They performed trace ethnography, a method that generates a rich account of interactions by combining a fine-grained analysis of automatically recorded editing traces with an ethnographically derived understanding of the sources of these traces.

Millions of people rely on Wikipedia as a source of information, and the quality of its content has been widely discussed. Researchers have explored how this high quality is achieved and how attempts to undermine its content, such as vandalism, are thwarted. Arazy and Nov (2010) found that communication and coordination between contributors increased quality and that editors who contribute to many topics in Wikipedia yield higher quality. Geiger and Ribes (2010) described the work involved in countering acts of vandalism in Wikipedia as an instance of distributed cognition in which the task is distributed across people and software tools. Keegan and Gergle (2010) studied the emergence of a new information-control role they call one-sided gatekeeping. They found that elite editors of the *In The News* section of Wikipedia were able to block inappropriate news items effectively, acting as gatekeepers, but their gatekeeping was one-sided because they were not more effective at promoting a proposed news item.

This research illustrates the questions and approaches asked about participants in online communities. Studies of Wikipedia have focused on contributors to its contents more than readers of the content, viewing the contributors as the most active members of the communities. The primary questions involve who is contributing, why some contribute more than others, and how content quality is achieved. Research methods include interviews and surveys, but new methods have emerged that integrate these traditional methods with large-scale statistical analyses of online data. The understanding that emerges from this research may suggest tools and methods for improved support to large-scale collaborations.

43.5 Summary

This chapter has selectively reviewed concepts and principles in the area of CSCW, focusing on research that contributes to our understanding of how people collaborate and the effects of technology on their collaboration. Context strongly influences how people collaborate, and categorization models describe some of these influences by categorizing who collaborates (groups, organizations, and communities), where they collaborate (same place or different places), and when they collaborate (same time or different times).

Our understanding of collaboration has implications for both developers and users of collaboration technologies. Developers must avoid well-established challenges that often befall collaboration technologies, and they should adopt potentially unfamiliar methods for establishing requirements and evaluating collaboration technologies. Technology adoption is especially challenging, because it often requires that all collaborators use the same technology, but this same factor may cause adoption to be extraordinarily rapid once critical mass is achieved.

Our evolving understanding of collaboration is illustrated with reviews of research on three prominent topics pertaining to small group, organizational, and community collaboration. Videoconferencing is of special interest because of its long history, wide appeal, yet remarkably slow adoption rate. Although the cost has plummeted and quality has risen, videoconferencing has not achieved its predicted popularity. Research suggests that the user interface is the principle obstacle to adoption. Contact negotiation remains awkward, and most videoconference devices fail to display important communicative behaviors. New technologies designed to benefit from these lessons may find greater success in the marketplace.

Workflow management is a prominent example of support for organizational collaboration, but the CSCW research community has been highly critical of both the practical value and social impact of this technology. More recent research has focused on the organizational collaboration challenges faced in domains of special interest such as health care and charities.

Interest in community collaboration, especially online communities, is growing. Online communities offer an attractive laboratory for studying collaboration because interaction data are saved automatically and vast quantities of data are available. Recent studies of Wikipedia reveal who participates, why, and how in the construction and maintenance of this online encyclopedia.

The CSCW field continues to evolve in response to changes in technologies, changes in ways people collaborate using technologies, and changes in the population of users. In its infancy, CSCW was constrained to studies of the workplace because few people had access to computers outside the workplace and networks were primitive. Today, mobile phones have greater computing power than those early personal computers, computing technologies are inexpensive, networks are ubiquitous, and information services are readily available. Other technology trends such as adoption of touch screens, tablet computers, and ubiquitous computing in the home, in the workplace, and in vehicles are all likely to influence how, where, and when people collaborate. Ongoing changes in technologies, usage contexts, and usage populations will pose new research questions and opportunities. For example, will the lessons learned about collaboration in the workplace prove applicable when children are collaborating in elementary schools? How will the ability to communicate and share information affect the lives of people living in poor rural areas? Researchers have been exploring collaboration among friends and within families, and this area of research will grow as people of all ages and incomes discover that collaboration technologies are both affordable and easy to use. Continuous communication availability may strengthen bonds within extended families, but it may also stress existing social norms. Access to shared information may empower people who have few economic resources, but the impacts on existing power and social structures may be unsettling.

Glossary

Awareness: Perception of relevant aspects of a communication context such as another person's availability for interaction or where another person is looking (eye gaze awareness).

Contact negotiation: The process of starting and ending contact in technology-mediated communication such as telephones and videoconferences.

Desktop conferencing: A category of technology that supports synchronous collaboration. All participants in a conferencing session see the same display with real-time updates, and some versions allow two or more people to interact with a display simultaneously.

Ethnography: A qualitative research approach for investigating what people do and how and why they do it. The data may include participant observation, field notes, interviews, surveys, and the work products of the people who are studied.

Legitimate peripheral participation: A theory describing how new members of a community are integrated into the community. Newcomers engage in simple, low-risk tasks that contribute to community goals.

Social world: A social world is a collection of people who have shared commitments to certain activities, share resources to achieve their goals, and build shared views by appropriate methods. Any type of collective unit can be a social world, including groups, organizations, and communities.

Workflow management: Coordination of sequences of work activities, assigning tasks to individuals, monitoring task progress, and, when a task is completed, forwarding relevant results to the person responsible for the next task. A workflow management system contains explicit models of the task sequences that comprise a work process and automatically manages the flow of work.

Further Information

CSCW research is presented and published in the proceedings of the annual *ACM Conference on Computer Supported Cooperative Work* (*CSCW*), the biannual *ACM International Conference on Supporting Group Work* (*GROUP*), the biannual *European Conference on Computer Supported Cooperative Work* (*ECSCW*), the *International Conference on Collaboration Technologies* (*CollabTech*), the *International Conference on Collaborative Computing* (*CollaborateCom*), and *Collaborative Technologies and Systems* (*CTS*). The principal journal in this field is *Computer Supported Cooperative Work: The Journal of Collaborative Computing*, published by Springer.

Influential early papers in the CSCW field can be found in Greif (1988) and Baecker (1993). Recent reviews include Olson and Olson (2011) and Grudin and Poltrock (2012).

References

Abraham, J. and Reddy, M. C. 2008. Moving patients around: A field study of coordination between clinical and non-clinical staff in hospitals. *Proceedings of the ACM Conference on Computer Supported Cooperative Work (CSCW 2008)*, San Diego, CA, pp. 225–228.

Ackerman, M. S. and Halverson, C. 1998. Considering an organization's memory. *Proceedings of the ACM Conference on Computer Supported Cooperative Work (CSCW'98)*, Seattle, WA, pp. 39–48.

Al-Ani, B., Mark, G., Chung, J. and Jones, J. 2012. The Egyptian blogosphere: A counter-narrative of the revolution. *Proceedings of the ACM Conference on Computer Supported Cooperative Work (CSCW 2012)*, Seattle, WA, pp. 17–26.

Antin, J. and Cheshire, C. 2010. Readers are not free-riders: Reading as a form of participation on Wikipedia. *Proceedings of the ACM Conference on Computer Supported Cooperative Work (CSCW 2010)*, Savannah, GA, pp. 127–130.

Antin, J., Cheshire, C. and Nov, O. 2012. Technology-mediated contributions: Editing behaviors among new wikipedians. *Proceedings of the ACM Conference on Computer Supported Cooperative Work (CSCW 2012)*, Seattle, WA, pp. 373–382.

Arazy, O. and Nov, O. 2010. Determinants of Wikipedia quality: The roles of global and local contribution inequality. *Proceedings of the ACM Conference on Computer Supported Cooperative Work (CSCW 2010)*, Savannah, GA, pp. 233–236.

Baecker, R. M. 1993. *Readings in Groupware and Computer-Supported Cooperative Work*. San Mateo, CA: Morgan Kaufmann.

Balka, E. and Wagner, I. 2006. Making things work: Dimensions of configurability as appropriation work. *Proceedings of the ACM Conference on Computer Supported Cooperative Work (CSCW 2006)*, Banff, Canada, pp. 229–238.

Benford, S., Tolmie, P., Ahmed, A. Y., Crabtree, A. and Rodden, T. 2012. Supporting traditional music-making: designing for situated discretion. *Proceedings of the ACM Conference on Computer Supported Cooperative Work (CSCW 2012)*, Seattle, WA, pp. 127–136.

Blei, D., Ng, A. and Jordan, M. 2003. Latent dirichlet allocation. *Journal of Machine Learning Research, 3,* 993–1022.

Bos, N., Olson, J., Gergle, D., Olson, G. and Wright, Z. 2002. Effects of four computer-mediated communications channels on trust development. *Proceedings of the ACM Conference on Computer Supported Cooperative Work (CSCW 2002)*, New Orleans, LA, pp. 135–140.

Bowers, J., Button, G. and Sharrock, W. 1995. Workflow from within and without: Technology and cooperative work on the print industry shopfloor. In H. Marmolin, Y. Sunblad, and K. Schmidt (Eds.), *Proceedings of European Conference on Computer-Supported Cooperative Work* Stockholm, Sweden, September 10–14, 1995. Dordrecht, the Netherlands: Kluwer, pp. 51–66.

Bryant, S. L., Forte, A. and Bruckman, A. 2005. Becoming Wikipedian: Transformation of participation in a collaborative online encyclopedia. *Proceedings of the ACM Conference on Supporting Group Work (GROUP'05)*, Sanibel Island, FL, pp. 1–10.

Choi, B., Alexander, K., Kraut, R. E. and Levine, J. M. 2010. Socialization tactics in Wikipedia and their effects. *Proceedings of the ACM Conference on Computer Supported Cooperative Work (CSCW 2010)*, Savannah, GA, pp. 107–116.

Collier, B. and Bear, J. 2012. Conflict, criticism, or confidence: An empirical examination of the gender gap in Wikipedia contributions. *Proceedings of the ACM Conference on Computer Supported Cooperative Work (CSCW 2012)*, Seattle, WA, pp. 383–392.

Convertino, G., Kannampallil, T. G. and Councill, I. 2006. Mapping the intellectual landscape of CSCW research. Poster presented at the *2012 ACM Conference on Computer Supported Cooperative Work (CSCW 2006)* Banff, Canada.

Cool, C., Fish, R. S., Kraut, R. E. and Lowery, C. M. 1992. Iterative design of video communication systems. *Proceedings of the ACM Conference on Computer Supported Cooperative Work (CSCW'92)*, Toronto, Canada, pp. 25–32.

Dabbish, L. A. and Kraut, R. E. 2003. Coordination communication: Awareness displays and interruption. *Extended Abstracts of the ACM Conference on Human Factors in Computing Systems (CHI 2003)*, Fort Lauderdale, FL, pp. 768–787.

Dabbish, L. A. and Kraut, R. E. 2004. Controlling interruptions: Awareness displays and social motivation for coordination. *Proceedings of the ACM Conference on Computer Supported Cooperative Work (CSCW 2004)*, Chicago, IL, pp. 182–191.

De Michelis, G. and Grasso, M. A. 1994. Situating conversations with the language/action perspective: The Milan conversation model. *Proceedings of the ACM Conference on Computer Supported Cooperative Work (CSCW'94)*, Chapel Hill, NC, pp. 89–100.

Dourish, P. and Bly, S. 1992. Portholes: Supporting awareness in a distributed work group. *Proceedings of the ACM Conference on Human Factors in Computing Systems (CHI'92)*, Monterey, CA, pp. 541–547.

Dourish, P., Holmes, J., MacLean, A., Marqvardsen, P. and Zbyslaw, A. 1996. Freeflow: Mediating between representation and action in workflow systems. *Proceedings of the ACM Conference on Computer Supported Cooperative Work (CSCW'96)*, Boston, MA, pp. 190–198.

Egido, C. 1988. Videoconferencing as a technology to support group work: A review of its failures. *Proceedings of the ACM Conference on Computer Supported Cooperative Work (CSCW'88)*, Portland, OR, pp. 13–24.

Engelbart, D. C. and English, W. K. 1968. A research center for augmenting human intellect. *Proceedings of the December 9–11, 1968 Fall Joint Computer Conference (AFIPS '68), Part I*, San Francisco, CA, pp. 395–410.

Finn, K. E., Sellen, A. J. and Wilbur, S. B. 1997. *Video-Mediated Communication.* Hillsdale, NJ: Erlbaum.

Fish, R. S., Kraut, R. E. and Root, R. W. 1992. Evaluating video as a technology for informal communication. *Proceedings of the SIGCHI Conference on Human Factors in Computing Systems (CHI'92)*, Monterey, CA, pp. 37–48.

Geiger, R. S. and Ribes, D. 2010. The work of sustaining order in Wikipedia: The banning of a vandal. *Proceedings of the ACM Conference on Computer Supported Cooperative Work (CSCW 2010)*, Savannah, GA, pp. 117–126.

Greif, I. 1988. *Computer-Supported Cooperative Work: A Book of Readings*. San Mateo, CA: Morgan Kaufmann.

Grudin, J. 1994. Groupware and social dynamics: Eight challenges for developers. *Communications of the ACM, 37*, 1, 92–105.

Grudin, J. and Poltrock, S. E. 2012. Taxonomy and theory in computer-supported cooperative work. In S. Kozlowski (Ed.), *Handbook of Industrial and Organizational Psychology*, Vol. 2, Oxford, U.K.: Oxford University Press, 1323–1348.

Handel, M. and Poltrock, S. 2011. Working around official applications: Experiences from a large engineering project. *Proceedings of the ACM Conference on Computer Supported Cooperative Work (CSCW 2011)*, Hangzhou, China, pp. 309–312.

Inkpen, K., Hegde, R., Czerwinski, M. and Zhang, Z. 2010. Exploring spatialized audio & video for distributed conversations. *Proceedings of the ACM Conference on Computer Supported Cooperative Work (CSCW 2010)*, Savannah, GA, pp. 95–98.

Ishii, H. and Ohkubo, M. 1990. Message-driven groupware design based on an office procedure model, OM-1. *Journal of Information Processing, 14*, 184–191.

Jackson, S. J., Pompe, A. and Krieschok, G. 2012. Repair Worlds: maintenance, repair, and ICT for development in rural Namibia. *Proceedings of the ACM Conference on Computer Supported Cooperative Work (CSCW 2012)*, Seattle, WA, pp. 107–116.

Jacovi, M., Soroka, V., Gilboa-Freedman, G., Ur, S., Shahar, E. and Marmasse, N. 2006. The chasms of CSCW: A citation graph analysis of the CSCW conference. *Proceedings of the ACM Conference on Computer Supported Cooperative Work (CSCW 2006)*, Banff, Canada, pp. 289–298.

Keegan, B. and Gergle, D. 2010. Egalitarians at the gate: One-sided gatekeeping practices in social media. *Proceedings of the ACM Conference on Computer Supported Cooperative Work (CSCW 2010)*, Savannah, GA, pp. 131–134.

Keegan, B., Gergle, D. and Contractor, N. 2012. Do editors or articles drive collaboration? Multilevel statistical network analysis of Wikipedia coauthorship. *Proceedings of the ACM Conference on Computer Supported Cooperative Work (CSCW 2012)*, Seattle, WA, pp. 427–436.

Kingsbury, N. 1997. Workflow in insurance. In P. Lawrence (Ed.), *Workflow Handbook*. New York: Wiley.

Kolfschoten, G. L., Briggs, R. O., de Vreede, J., Jaclobs, P. H. M. and Appelman, J. H. 2006. A conceptual foundation of the thinkLet concept for collaboration engineering. *International Journal of Human-Computer Studies, 64*, 611–621.

Kraut, R., Miller, M. and Siegal, J. 1996. Collaboration in performance of physical tasks: effects on outcomes and communication. *Proceedings of the ACM Conference on Computer Supported Cooperative Work (CSCW '96)*, Boston, MA, pp. 57–66.

Le Dantec, C. A. and Edwards, W. K. 2008. The view from the trenches: Organization, power, and technology at two nonprofit homeless outreach centers. *Proceedings of the ACM Conference on Computer Supported Cooperative Work (CSCW 2008)*, San Diego, CA, pp. 589–598.

Lee, C. P., Dourish, P. and Mark, G. 2006. The human infrastructure of cyberinfrastructure. *Proceedings of the ACM Conference on Computer Supported Cooperative Work (CSCW 2006)*, Banff, Canada, pp. 483–492.

Lee, A., Girgensohn, A. and Schlueter, K. 1997. NYNEX portholes: Initial user reactions and redesign implications. *Proceedings of the International ACM SIGGROUP Conference on Supporting Group Work (GROUP'97)*, Phoenix, AZ, pp. 385–394.

Mark, G. and Poltrock, S. 2003. Shaping technology across social worlds: Groupware adoption in a distributed organization. *Proceedings of the International ACM Conference on Supporting Group Work (GROUP '03)*, Sanibel, Island, FL, 284–293.

Mark G. and Poltrock, S. 2004. Groupware adoption in a distributed organization: Transporting and transforming technology through social worlds. *Information and Organization, 14*, 297–327.

du Maurier, G. 1878. Edison's telephonoscope (transmits light as well as sound). *Punch Magazine*, December 9.

Medina-Mora, R., Winograd, T., Flores, R. and Flores, F. 1992. The action workflow approach to workflow management technology. *Proceedings of the ACM Conference on Computer Supported Cooperative Work (CSCW'92)*, Toronto, Canada, pp. 281–288.

Nguyen, D. and Canny, J. 2005. MultiView: Spatially faithful group video conferencing. *Proceedings of the SIGCHI Conference on Human Factors in Computing Systems (CHI'05)*, Portland, OR, pp. 799–808.

Nguyen, D. and Canny, J. 2007. MultiView: Improving trust in group video conferencing through spatial faithfulness. *Proceedings of the SIGCHI Conference on Human Factors in Computing Systems (CHI'07)*, San Jose, CA, pp. 1465–1474.

Nguyen, D. T. and Canny, J. 2009. More than face-to-face: Empathy effects of video framing. *Proceedings of the SIGCHI Conference on Human Factors in Computing Systems (CHI'09)*, Boston, MA, pp. 423–432.

Olson, G. M. and Olson, J. S. 2011. Collaboration technologies. In J. Jacko (Ed.), *Handbook of Human-Computer Interaction*. Boca Raton, FL: CRC Press.

Painter, B. 2002. The electronic claim file: A case study of impacts of information technology in knowledge work. *Proceedings of the ACM Conference on Computer Supported Cooperative Work (CSCW 2002)*, New Orleans, LA, 276–285.

Pan, Y-X., Jiang, D-N., Picheny, M. and Qin, Y. 2009. Effects of real-time transcription on non-native speaker's comprehension in computer-mediated communications. *Proceedings of the SIGCHI Conference on Human Factors in Computing Systems (CHI'09)*, Boston, MA, pp. 2353–2356.

Pesic, M. and van der Aalst, W. M. P. 2006. A declarative approach for flexible business processes management. *Lecture Notes in Computer Science, 4103*, 169–180.

Pine, K. H. 2012. Fragmentation and choreography: Caring for a patient and a chart during childbirth. *Proceedings of the ACM Conference on Computer Supported Cooperative Work (CSCW 2012)*, Seattle, WA, pp. 887–896.

Robins, G., Pattison, P., Kalish, Y. and Lusher, D. 2007. An introduction to exponential random graph (p*) models for social networks. *Social Networks, 29*, 2, 173–191.

Root, R. W. 1988. Design of a multi-media vehicle for social browsing. *Proceedings of the ACM Conference on Computer Supported Cooperative Work (CSCW'88)*, Portland, OR, pp. 25–38.

Sachs, P. 1995. Transforming work: Collaboration, learning, and design. *Communications of the ACM, 38*, 9, 36–44.

Stoll, J., Edwards, W. K. and Mynatt, E. D. 2010. Interorganizational coordination and awareness in a nonprofit ecosystem. *Proceedings of the ACM Conference on Computer Supported Cooperative Work (CSCW 2010)*, Savannah, GA, pp. 51–60.

Suchman, L. 1983. Office procedures as practical action: Models of work and system design. *ACM Transactions of Office Information Systems, 1*, 4, 320–238.

Suchman, L. 1987. *Plans and Situated Actions: The Problem of Human-Machine Communication*. Cambridge, UK: Cambridge University Press.

Suchman, L. 1994. Do categories have politics? *Computer Supported Cooperative Work (CSCW), 2*, 177–190.

Tang, J. C. 2007. Approaching and leave-taking: negotiating contact in computer mediated communication. *Transactions on Computer-Human Interaction (TOCHI), 14*, 1–26.

Tang, J. C., Isaacs, E. A. and Rua, M. 1994. Supporting distributed groups with a montage of lightweight interactions. *Proceedings of the ACM Conference on Computer Supported Cooperative Work (CSCW'94)*, Chapel Hill, NC, pp. 23–34.

Thom-Santelli, J., Millen, D. R. and Gergle, D. 2011. Organizational acculturation and social networking. *Proceedings of the ACM Conference on Computer Supported Cooperative Work (CSCW 2011)*, Hangzhou, China, pp. 313–316.

Williams, G. 1997. Task conflict and language differences: Opportunities for videoconferencing? *Proceedings of the Fifth European Conference on Computer Supported Cooperative Work (ECSCW '97)*, Lancaster, U.K., pp. 97–108.

Winograd, T. 1994. Categories, disciplines, and social coordination. *Computer Supported Cooperative Work (CSCW), 2*, 191–197.

Winograd, T. and Flores, F. 1986. *Understanding Computers and Cognition: A New Foundation for Design.* Norwood, NJ: Ablex.

44

Information Technology for Enhancing Team Problem Solving and Decision Making

Gert-Jan de Vreede
University of Nebraska at Omaha

Benjamin Wigert
University of Nebraska at Omaha

Triparna de Vreede
University of Nebraska at Omaha

Onook Oh
University of Nebraska at Omaha

Roni Reiter-Palmon
University of Nebraska at Omaha

Robert Briggs
San Diego State University

44.1 Introduction

Fueled by a combination of decreasing costs and increasing capabilities, information technology (IT) has penetrated almost every successful workplace. Organizations have long focused on supporting the individual needs of their workers by offering applications such as word processors, spreadsheets, and database systems. However, with the advent of cost-effective communication technologies and ubiquitous Internet access, workplaces have become connected. As a result, the use of IT has shifted over the past 20 years from supporting pure computation activities toward coordination and collaboration activities within and between organizations. IT has become the essential vehicle for mediating interpersonal and group communication. As such, the nature and appearance of IT have changed accordingly. Information systems have taken the shape of socio-technical networks that are designed specifically to make information workers more effective at their jobs by means of making problem-solving and decision-making tasks more collaborative.

Collaboration technologies have enabled people to work together by sending, receiving, storing, retrieving, and co-creating information. This type of IT overcomes traditional collaboration constraints such as time, space, and hierarchy (Coleman and Levine, 2008; Lewin and Stephens, 1993). Although IT is often seen as an enabling technology, it can also be considered a *requirement* for effective organizational coordination and collaboration. Organizational work is increasingly becoming a group effort that requires flexibility and adaptiveness (Bunker and Alban, 2006; Ellinger, 2000; Hardy et al., 2005; Malone and Crowston, 1990; Tjosvold, 1988). Fundamentally, collaboration and coordination are the essential ingredients of organizational life: organizations form because people have to work together to create value that they cannot produce by themselves (Mintzberg, 1983). As coordination and collaboration require extensive information exchange, well-designed and usable systems are necessary to support the interdependencies between the actors in teams and organizations. Therefore, organizations require collaboration technologies that are capable of disseminating information fast, economically, reliably, and securely (Corbitt et al., 2000; Nunamaker et al., 2001).

This chapter addresses how organizations can enable teams to be more productive with collaboration technology. We first present a classification of the different types of collaboration technologies that organizations can use. Then, we introduce a model of team processes that describes the fundamental patterns of collaboration that teams can engage in, regardless of whether they use collaborative software applications or not. These patterns include *Generate, Converge, Organize, Evaluate,* and *Build consensus.* We describe key findings from past research, especially in the area of group support systems (GSSs), to illustrate how collaboration technology supports different types of team work in each of the five patterns. Next, we move from the team level to the organizational level where we discuss and illustrate the impact and added value of collaboration technology in an organization as a whole. We conclude this chapter by outlining three key future directions for collaboration technology research and practice.

44.2 Classification of Collaboration Technologies

This section examines the different types of collaboration technologies that are currently available. What began as an area with a few dozen professional applications in the early 1980s has grown to an industry for thousands of different systems, ranging from simple group texting apps to integrated work environments where teams communicate, deliberate, co-create, and store electronic artifacts (e.g., documents, models, and audio files). Taking a closer look at the typical capabilities that various collaboration technologies afford, Mittleman and colleagues classified them under four main categories according to their most fundamental capabilities: (1) jointly authored pages, (2) streaming tools, (3) information access tools, and (4) aggregated systems (Mittleman et al., 2009). The fourth category is for technologies that must integrate a mix of tools from the first three categories and optimize them to support work practices that cannot be achieved with a single technology. Each of these main categories can be subdivided into a number of subcategories based on the functions they are optimized to support. The complete classification scheme is presented in Table 44.1. Next, we discuss each of these categories in more detail.

44.2.1 Jointly Authored Pages

The most fundamental capability for all technologies in the jointly authored pages category is a digital page, defined as a single window to which multiple participants can contribute, often simultaneously. The pages might be able to represent text, graphics, numbers, or other digital objects. However, regardless of content, any contribution made by a participant will generally appear on the screens of the other participants who view the same page. A given technology based on jointly authored pages may provide a single page or multiple pages. Sometimes these tools allow for the creation of hierarchies or networks of pages. Jointly authored pages are the basis for several subcategories of collaboration technology including conversation tools, shared editors, group dynamics tools, and polling tools.

TABLE 44.1　Classification Scheme of Collaboration Technologies

Categories	Description
Jointly authored pages	*Technologies that provide one or more windows that multiple users may view, and to which multiple users may contribute, usually simultaneously*
Conversation tools	Optimized to support dialog among group members
Shared editors	Optimized for the joint production of deliverables like documents, spreadsheets, or graphics
Group dynamics tools	Optimized for creating, sustaining, or changing patterns of collaboration among people, making joint effort toward a goal (e.g., idea generation, idea clarification, idea evaluation, idea organization, and consensus building)
Polling tools	Optimized for gathering, aggregating, and understanding judgments, opinions, and information from multiple people
Streaming technologies	*Technologies that provide a continuous feed of changing data*
Desktop/application sharing	Optimized for remote viewing and/or control of the computers of other group members
Audio conferencing	Optimized for transmission and receipt of sounds
Video conferencing	Optimized for transmission and receipt of dynamic images
Information access tools	*Technologies that provide group members with ways to store, share, find, and classify data objects*
Shared file repositories	Provide group members with ways to store and share digital files
Social tagging systems	Provide means to affix keyword tags to digital objects so that users can find objects of interest and find others with similar interests
Search engines	Provide means to retrieve relevant digital objects from among vast stores of objects based on search criteria
Syndication tools	Provide notification of when new contributions of interest have been added to pages or repositories
Aggregated systems	*Technologies that combine other technologies and tailor them to support a specific kind of task*

Source: Mittleman, D.M. et al., Towards a taxonomy of groupware technologies. Groupware: Design, implementation, and use, *Proceedings of the Fourteenth 14th Collaboration Researchers International Workshop on Groupware*, pp. 305–317, Springer, Berlin, Germany, 2009.

Conversation tools are those primarily optimized to support dialog among group members. E-mail is a widely used conversation tool as well as short message service (SMS) (i.e., cell phone text messaging) that has become a ubiquitous application. According to the International Telecommunication Union, 6.1 trillion text messages were sent in 2010 (ITU, 2010). Other conversation tools include instant messaging, chat rooms, and blogs or threaded discussions. Instant messaging and chat rooms provide users with a single shared page to which they can make contributions to a chronologically ordered list. Participants may not move, edit, or delete their contributions. Instant messaging and chat rooms differ from one another only in their access and alert mechanisms. With instant messaging, an individual receives a pop-up invitation that another individual wishes to hold a conversation, while with chat rooms an individual browses to a website to find and join a conversation. Blogs (otherwise known as web logs) and threaded discussion tools are optimized for less-synchronous conversations. Users make a contribution, then come back later to see how others may have responded. Blogs and threaded discussions are typically persistent (i.e., their content remains even when users are not contributing), whereas chat rooms and instant messaging are usually ephemeral (i.e., when the last person exits a session, the session content disappears).

Shared editor tools are typically a jointly authored page optimized for the creation of a certain kind of deliverable by multiple authors. The content and affordances of these tools often match those of single-user office suite tools (e.g., word processing or spreadsheets). However, they are enhanced to accept contributions and editing by multiple simultaneous users. A wiki (the Hawaiian word for "fast") is another example of joint document authoring. Wikis are simple web pages that can be created directly through a web browser by any authorized user without the use of offline web development tools.

Group dynamics tools are optimized for creating, sustaining, or changing patterns of collaboration among individuals making a joint effort toward a goal. The patterns these tools support include generating ideas, establishing shared understanding of them, converging on those worth more attention, organizing and evaluating ideas, and building consensus (Briggs et al., 2006). These tools are often implemented as multiple layers of jointly authored pages such that each contribution on a given page may serve as a hyperlink to a subpage. The affordances of such tools are typically easily configurable, so at any given moment a group leader can provide team members with the features they need (e.g., view, add, and move) while blocking features they should not be using (e.g., edit and delete).

Polling tools are a special class of jointly authored pages, optimized for gathering, aggregating, and understanding judgments or opinions from multiple people. At a minimum, the shared pages of a polling tool must offer a structure of one or more ballot items, a way for users to record votes, and a way to display results. Polling tools may offer rating, ranking, allocating, or categorizing evaluation methods and may also support the gathering of text-based responses to ballot items.

44.2.2 Streaming Technologies

The core capability of all tools in the streaming technologies category is a continuous feed of dynamic data. Desktop sharing, application sharing, and audio/video conferencing are common examples of streaming technologies.

Desktop and application sharing tools allow the events displayed on one computer to be seen on the screens of other computers. With some application sharing tools, members may use their own mouse and keyboard to control the remotely viewed computer. This type of collaboration technology is often used in distributed training or support settings.

Audio conferencing tools provide a continuous channel for multiple users to send and receive sound while *video conferencing tools* allow users to send and receive sound and moving images. Typically all users may receive contributions in both types of tools. Systems may, however, vary in the mechanisms they provide for alerts and access control as well as by the degree to which affordances can be configured and controlled by a leader.

44.2.3 Information Access Technologies

Information access technologies provide ways to store, share, classify, and find data and information objects. Key examples in this category include shared file repositories, social tagging, search engines, and syndication tools.

Shared file repositories provide mechanisms for group members to store digital files where others in the group can access them. Some systems also provide version control mechanisms such as check-out/ check-in capabilities and version backups.

Social tagging allows users to affix keyword tags to digital objects in a shared repository. For example, the website del.icio.us allows users to store and tag their favorite web links (i.e., bookmarks) online so that they can access them from any computer. Users are not only able to access their own bookmarks by keyword, but bookmarks posted and tagged by others as well. More significantly, users can find other users who share an interest in the same content. Social tagging systems allow for the rapid formation of communities of interest and communities of practice around the content of the data repository. The data in a social tagging repository are said to be organized in a folksonomy, an organization scheme that emerges organically from the many ways that users think of and tag contributions, rather than in an expert-defined taxonomy.

Search engines use search criteria provided by team members to retrieve digital objects from among vast stores of such objects (e.g., the World Wide Web, the blogosphere, and digital libraries). Search criteria may include content, tags, and other attributes of the objects in the search space. Some search engines interpret the semantic content of the search request to find related content that is not an exact match for the search criteria.

Syndication tools allow a user to receive a notification when new contributions are made to pages or repositories they find interesting (e.g., blogs, wikis, and social networks). Users subscribe to receive update alerts from a feed on a syndicated site. Every time the site changes, the feed broadcasts an alert message to all its subscribers. Users view alerts using software called an aggregator. Any time a user opens their aggregator, they see which of their subscription sites has new contributions. Therefore, users do not need to scan all contents to discover new contributions.

44.2.4 Aggregated Technologies

Aggregated technologies integrate several technologies from the other three categories and optimize them to support tasks and processes that cannot be executed using a single technology. Aggregated technologies deliver value that could also be achieved with a collection of stand-alone tools. There are many examples of aggregated technologies, among them virtual workspaces, web conferencing systems, social networking systems, and GSS. Virtual workspaces often combine document repositories, team calendars, conversation tools, and other technologies that make it easier for team members to execute coordinated efforts (e.g., SharePoint). Remote presentation or web conferencing systems often combine application sharing and audio streams with document repositories and polling tools optimized to support one-to-many broadcast of presentations, with some ability for the audience to provide feedback to the presenter (e.g., WebEx or SameTime). Social networking systems (e.g., MySpace, Facebook, or Flickr) combine social tagging with elements of wikis, blogs, other shared page tools, and a search engine so that users can find and communicate with their acquaintances as well as establish new relationships based on mutual friends or mutual interests. Thus, aggregated technologies may combine any mix of shared-page, streaming, and information access technologies to support a particular purpose.

GSSs integrate collections of group dynamics tools to move groups seamlessly through a series of activities toward a goal, for example, by generating ideas in one tool, organizing them in another, and evaluating them in yet another (e.g., ThinkTank or MeetingWorks). Specifically, a GSS is a collection of collaborative meeting tools designed to improve creative co-creation, problem solving, and decision making by teams (de Vreede et al., 2003). A GSS is typically implemented on a collection of personal computers that are connected through a (wireless) LAN or through the Internet. Sometimes, participants work in meeting rooms especially designed for electronically supported meetings. Other times, the participants simply move laptop computers into a standard meeting room and begin their work. Still other times, participants used lean GSS applications asynchronously, especially since the emergence of web-based GSS applications in the mid-2000s.

A GSS permits all participants to contribute at once by typing their ideas into the system. The system immediately makes all contributions available to the other participants on their screens. Nobody has to wait for a turn to speak, so people do not forget what they want to say while waiting for the floor. People do not forget what has been said, because there is an electronic record of the conversation on their screens. GSSs also allow a team to enter ideas anonymously. During the generation phase of creativity, it is often useful to get every idea on the table, but people often hold back unconventional or unpopular ideas for fear of disapproval from peers or superiors. Anonymity allows ideas to surface without fear of repercussion.

In the remainder of this chapter, we will focus predominantly on GSSs as they represent the type of collaboration technology that probably has received most attention from collaboration researchers during the past three decennia. Especially from the late 1980s until the late 1990s, an extensive body of GSS research was published (for overviews, see Dennis et al., 2001; Fjermestad and Hiltz, 1998, 2000;

Nunamaker et al., 1997). Since the mid-2000s, GSS researchers have shifted from a predominant experimental focus to a focus on engineering and design science studies (see, e.g., Kolfschoten and de Vreede, 2009; de Vreede et al., 2009). The extensive scholarly attention for GSS is not surprising as this technology represents one of the few software applications that were originally developed by researchers in universities and subsequently became commercially available. Currently, there are a few dozen commercial GSS applications available in the market. Even though organizations have reported significant benefits from using GSS, sustained use of these applications has been challenging as will be discussed in a later section.

44.3 Types of Team Work: Patterns of Collaboration

Notwithstanding the plethora of collaboration technologies that are available for teams to facilitate their work, fundamentally teams need more than just technology to achieve productive collaboration efforts. Productive collaboration originates from the purposeful alignment of people, processes, information, and leadership with the technology the team uses (de Vreede et al., 2009). A critical component in this socio-technical setting concerns how teams move through their collaborative processes. Researchers have identified distinctly different ways in which a team's problem-solving and decision-making processes proceed, called *patterns of collaboration*. Patterns of collaboration are observable regularities of behavior and outcome that emerge over time when a team moves toward its goals (de Vreede et al., 2006). There are six different patterns of collaboration (Kolfschoten et al., in press; de Vreede et al., 2009):

- *Generate*: To move from having fewer concepts to having more concepts in the set of ideas shared by the group.
- *Reduce*: To move from having many concepts to a focus on fewer ideas deemed worthy of further attention.
- *Clarify*: To move from less to more shared understanding of the concepts in the set of ideas shared by the group.
- *Organize*: To move from less to more understanding of the relationships among concepts in the set of ideas shared by the group.
- *Evaluate*: To move from less to more understanding of the instrumentality of the concepts in the idea set shared by the group toward attaining group and private goals.
- *Build commitment*: To move from fewer to more group members who are willing to commit to a proposal for moving the group toward attaining its goal(s).

Most of the behaviors in which a team engages as it moves through a collaborative process can be characterized by these six patterns. For example, when a team moves through a product development activity, they may *generate* innovation ideas, *clarify* these ideas by re-stating them so that all members understand their meaning, *evaluate* the expected profitability and ease of implementation of each innovation idea, *reduce* the list to the innovations that are most promising, and *build commitment* on the actions to take to develop a complete business case for each innovation.

Researchers have studied phenomena relating to each of the six patterns of collaboration. With respect to the *Generate* pattern, for example, studies report the number of ideas a group produces (Connolly et al., 1990), or their originality, relevance, quality, effectiveness, feasibility, and thoroughness (Dean et al., 2006). People generate by creating new ideas (Reiter-Palmon et al., 1997), by gathering previously unshared ideas (Bock et al., 2005), or by elaborating on existing ideas with additional details (de Vreede et al., 2010). For the *Reduce* pattern, researchers address, for example, the number of ideas in the shared set, the degree to which a reduced idea set includes high-quality ideas and excludes low-quality ideas (Barzilay et al., 1999), and the degree to which reduction of idea sets yields reductions of actual and perceived cognitive load (Simpson and Prusak, 1995). Groups reduce idea sets through idea filtering (Chambless et al., 2005), generalizing ideas (Yeung et al., 1999), or selection (Rietzschel et al., 2006).

Researchers of the *Clarify* pattern focus on, among other things, reductions in ambiguity, reductions in the number of words required to convey meaning, and establishing mutual assumptions (Mulder et al., 2002). Among the phenomena of interest for research on the *Organize* pattern of collaboration are shared understandings of the relationships among concepts (Cannon-Bowers et al., 1993), cognitive load (Grisé and Gallupe, 2000), and the simplicity or complexity of the relationships among concepts (e.g., complex structures may signify sequence, hierarchy, and networks of relationships, which in turn may model, e.g., semantics of chronology, composition, heredity, or causation [Dean et al., 2000]). Research on the *Evaluate* pattern addresses projections of possible consequences of choices and the degree to which those consequences would promote or inhibit goal attainment (Westaby, 2002). Rating, ranking, and inclusion/exclusion are common means of evaluation (Gavish and Gerdes, 1997). Research on such techniques focuses, for example, on the degree to which participants can accurately project the likely outcomes of the proposals they consider (Laukkanen et al., 2002). Finally, phenomena of interest for the *Build Commitment* pattern pertain to the degree to which people are willing to contribute to the group's efforts (Montoya-Weiss et al., 2001). Issues of commitment arise in many phases of group work, starting with the formation of the group (Datta, 2007), and continuing through every proposed course of action and every choice group members make as they move through their activities (Saaty and Shang, 2007). More detailed findings of information systems research in each of these areas are described in the next section.

44.4 Supporting the Patterns of Collaboration

This section examines past research on technology support for each of the patterns of collaboration. For this purpose, the *reduce* and *clarify* patterns will be combined into a pattern called *converge* as research in this area often addresses both patterns simultaneously.

44.4.1 Generate

Idea generation (i.e., brainstorming) is the practice of gathering in a group and generating as many ideas for identifying or solving a problem as possible. Early brainstorming research conducted by Taylor et al. (1958) and later by Diehl and Stroebe (1987) indicated that nominal groups—groups in which ideas are pooled after individuals independently generate them—tended to generate more ideas than groups with team members who collectively interacted to generate ideas. This productivity loss in traditional lab teams occurred due to members being overwhelmed with information (i.e., cognitive load), social loafing, groupthink, and evaluation apprehension (Diehl and Stroebe, 1987). However, these experiments did not focus on teams that utilize strategic collaboration methods and technology. As such, later research showed that, under certain conditions, groups can improve key outcomes using collaboration strategies and technologies (for reviews, see Briggs et al., 1997; Dennis and Valacich, 1999; Fjermestad and Hiltz, 1998).

Nonetheless, effectively communicating pertinent instructions, monitoring participant contributions, dealing with conflict, and facilitating the flow of information/conversation can cause production losses even when a facilitator is charged with managing these issues. Fortunately, when combined appropriately, synergies between collaboration processes and technology can simplify and magnify the power of facilitation techniques. Collaboration technology, such as GSS, helps a facilitator simultaneously display tasks and instructions that group members need to understand in order to minimize the aforementioned challenges of collaboration and maximize the benefits of teamwork. In other words, this technology allows the facilitator to control what decision-making tasks group members see, how they can contribute to the group, and when they can contribute. Thus, it is easier for facilitators to breakdown the collaboration process into simpler, independent activities.

During the idea generation process, technology can also be used to make team members focus on generating ideas rather than evaluating them. When team members assess the merit of ideas or try to come to consensus while generating ideas, the quality, quantity, and originality of ideas tends to suffer (Basadur and Hausdorf, 1996). Further, technology allows team members to simultaneously contribute a massive number of ideas and transfer those ideas to subsequent collaboration activities, such as moving ideas into categories (i.e., idea organization) or voting on them (i.e., idea evaluation). Technology also can be used to greatly reduce interpersonal conflict that occurs during idea generation. For instance, GSS software can be configured so that participants contribute their ideas anonymously. This anonymity tends to improve collaborative problem-solving outcomes by reducing participants' social fears, such as being judged, retaliated against, or condemned for disagreeing with an authority figure (e.g., supervisor, boss, professor, and client) (McLeod, 2011).

Moreover, being the foundation of most collaborative problem-solving and decision-making efforts, idea generation is typically credited with initiating group creativity and innovation. Creativity is defined as a novel, useful, and socially valued product, idea, or service (Amabile, 1982). Creativity occurs when an ill-defined problem is assessed from multiple perspectives (Mumford and Gustafson, 1988). As such, idea generation is at the core of team creativity because this collaboration pattern requires participants to engage in divergent thinking. Frequently, participants see ideas generated by others that inspire them to make connections and recommendations that they would not have thought of by themselves. Here, ideas from teammates serve as new idea "seeds" that can inspire creativity. To test this assumption, de Vreede et al. (2010) divided large groups into subgroups that were forced to generate ideas in a serial manner by building on the ideas of previously engaged subgroups. This technique proved to be more effective at facilitating creativity than parallel group member idea generation. Other studies have provided further support for the notion that certain types of instructions or methods for generating ideas can be used to maximize group creativity (Sosik et al., 1998; Zurlo and Riva, 2009). These studies highlight the importance of effectively pairing collaboration techniques and technology.

While the aforementioned studies were conducted by means of synchronous collaboration, more and more collaboration is currently being conducted asynchronously online. Asynchronous collaboration overcomes temporal and spatial challenges faced by traditional synchronous collaboration. Although little published research addresses best practices for online asynchronous collaboration, a study by Michinov and Primois (2005) provides great insights. The researchers found that initiating social comparison between participants can improve the productivity and creativity of distributed teams. Specifically, Michinove and Primois learned that motivation to perform can be increased by simply labeling who contributes ideas to group brainstorming sessions. Based on these findings, team leaders and facilitators should consider the trade-offs of idea anonymity when collaboration is asynchronous. It may be the case that the performance-enhancing effects of social comparison may be more valuable to alleviating the "cold" nature of asynchronous collaboration than protecting participants from evaluation apprehension.

Despite the importance of the idea generation stage of collaboration, the best brainstorming interventions provide little utility to a group if generated ideas can cannot be understood, connected, and built upon. Therefore, the following sections describe how subsequent patterns of collaboration improve group problem solving and decision making.

44.4.2 Converge

After generating ideas for identifying or solving a problem, collaborators must *converge* on the shared knowledge necessary to address the problem. Convergence is a critical activity in group work that lays the foundation for shared understanding and the overall advancement of a group's task. To converge has been defined as "to move from having many concepts to a focus on and understanding of a few deemed worthy of further attention" (de Vreede and Briggs, 2005). Group work concerns convergence when a group deliberates on and reduces the amount of information they have to work with (de Vreede and Briggs, 2005).

Essentially, convergence lays the foundation for shared understanding and the overall advancement of the group's task (Davis et al., 2007). In doing so, a primary goal of convergence is to reduce a group's cognitive load in order to address all concepts, conserve resources, have less to think about, and achieve shared meaning of the concepts. However, convergence is time consuming and has been shown to be a painful process for groups (Chen et al., 1994; Easton et al., 1990). de Vreede and Briggs (2005) identify four aspects or subprocesses that can be combined in order to create useful variations in convergence activities. These include (1) judging or identifying which of the existing concepts merits further attention; (2) filtering or selecting a subset from a pool of concepts that will receive further attention; (3) generalizing or reducing the number of concepts under consideration through generalization, abstraction, or synthesis and then eliminating the lower-level concepts in favor of the more general concept; and (4) creating shared meaning or agreeing on connotation and establishing a shared meaning or understanding about the labels used to communicate various concepts.

To date, convergence has received limited scholarly attention. Researchers have argued that convergence activities in group work are complicated due to a variety of reasons, including information overload at the start of a convergent task, the cognitive effort that is required for convergent tasks, and the need for a higher granularity of meeting ideas to be stored (i.e., meeting memory) for future decision making and analysis (Chen et al., 1994). To address these issues, GSS researchers have argued about the most effective mode and means of communication for convergence issues. For example, Dennis and Valacich (1999) propose that verbal communication is most appropriate when a team needs to converge and establish shared meaning. Verbal communication is mostly recommended for efficiency, because it provides the fastest feedback. Other researchers argue for combining electronic and verbal communication modes during convergence: participants can benefit electronic tools during convergence due to the fact that key concepts can be identified and represented with a minimum of cognitive load on the group members (Briggs and de Vreede, 1997). These insights have been consolidated in media synchronicity theory (MST) that suggests that individuals would prefer highly synchronized communication media (i.e., face-to-face) for convergence communication, which relate to shared understanding, and less synchronized media (e.g., e-mail) for conveyance communication, which relates to facts or alternatives (DeLuca and Valacich, 2006; Dennis et al., 2008). Furthermore, MST argues that decision-making tasks require convergence before proceeding to the next task as well as the conveyance of specific information (e.g., facts or alternatives) (Dennis et al., 2001).

In contrast to the research that addresses convergence issues from the group members' perspective, recent research from a facilitator's perspective has shown that facilitators find convergence to be one of the least demanding patterns of collaboration, behind divergence (Hengst and Adkins, 2007). Yet, the same study found building consensus to be the most demanding pattern of collaboration, and the authors suggest that building consensus is most often done with convergence and organization subprocesses. Other researchers have proposed a set of performance criteria to provide a framework to compare and contrast different convergence techniques (Davis et al., 2007). These criteria include both result-oriented (e.g., level of shared understanding and level of reduction) and process- or experience-oriented criteria (e.g., ease of use for the team members and satisfaction with technique by the facilitator). Initial studies have applied some of these criteria (see, e.g., Badura et al., 2010), but clearly more research is required in this area.

In conclusion, a review of previous research suggests there is little structural attention on the topic of convergence. Moreover, there is very limited, detailed guidance available how to best structure convergence activities in groups (Briggs et al., 1997; Hengst and Adkins, 2007).

44.4.3 Organize

Groups cannot solve problems and innovate by means of simply compiling ideas and developing a shared understanding of them. As a result of idea generation and convergence, a large amount and complexity of the information shared between collaborators puts a burden on their time and attention.

In order to reduce demands on limited human resources and to overcome limitations on understanding, groups need to *organize* the information they generate. This section examines what is known about the organization stage of the collaborative problem-solving process and how this phenomenon has been conceptualized.

At the core of the organization stage of the decision-making process is the idea that collaborative organization activities are intended to facilitate greater understanding of the relationships among concepts. Little research has taken place in the general area of organizing information as team work. However, a special form of the organize pattern, *group modeling*, has received some focused attention (Renger et al., 2008). Broadly speaking, through collaborative modeling participants are able to create a shared understanding about a collection of concepts and their interrelationships. Often group model activities take place as part of an effort to represent a system design.

Early studies on IT-supported versus "pencil and paper" design modeling established that facilitation practices and IT can be leveraged to greatly improve collaborative design modeling processes and outcomes (Dean et al., 1995). For instance, the use of IDEF0 modeling was initially limited to collaboration in small groups because a group would first use flip charts, whiteboards, or transparencies to delineate information they were attempting to organize, then a facilitator would have to transcribe that information into a single-user tool that created graphical models. This process frequently led to misinterpretations of data and/or group members feeling that they had less control over the models than desired (Renger et al., 2008). Later group modeling systems integrated GSS and IDEF0-specific functionalities and allowed for active participation such that several participants would have direct control over how concepts are graphically, syntactically, and semantically modeled (Dean et al., 2000; Dennis et al., 1999). Other group modeling approaches that have been successfully employed in field settings include jointly understanding, reflecting, and negotiating strategy (JOURNEY Making) (Ackermann and Eden, 2005), which uses cognitive mapping techniques to support strategic planning and policy making, and group model building (Andersen et al., 2007), which refers to "a bundle of techniques used to construct system dynamics models working directly with client groups on key strategic decisions" (Andersen et al., 2007, p. 1).

The study of collaborative modeling has evolved with its use in practice, yet quantitative studies on specific comparative strengths and weaknesses of various technology-supported modeling methods are scant. A meta-analysis conducted by Renger et al. (2008) examined various outcomes in the collaborative modeling literature. It was found that across 22 group modeling studies, six studies yielded positive effectiveness results, three studies found high levels of participant satisfaction, and five studies discovered evidence of shared understanding resulting from certain collaborative design approaches. Thus, collaborative modeling has yielded effective outcomes when used to facilitate the organization stage of collaboration. Additional research is needed to identify context effects on collaborative modeling, as well as head-to-head comparisons of different modeling methods to determine which approaches are most effective under specific circumstances.

44.4.4 Evaluate

Another critical part of team work occurs when groups *evaluate* the merit of ideas. Admittedly, evaluation tends to happen at each stage of a collaboration process, but the most effective team efforts tend to prevent premature idea evaluation. That is, the generation, expansion, and linking of ideas tends to be disrupted when ideas are evaluated during or before convergence or organizing activities. However, once a group has converged on a set of ideas, this new found shared understanding must be leveraged to identify optimal options.

Two primary methods are used to evaluate the merit of ideas. First, ideas can be compared head-to-head using a rank-order or single criterion voting system. Second, a multi-criteria decision-making system can be utilized to develop a better understanding of the strengths and weaknesses of each idea. Dean et al. (2006) provide a thorough review of various criteria used to evaluate problem solutions in research studies conducted from 1990 to 2005. This study supports findings from MacCrimmon and Wagner (1994) that suggest that most decision-making criteria used in previous research tend to fall under one of four categories: novelty, workability, relevance, and specificity. Further, Dean et al. emphasized that when assessing the creativity of ideas, the constructs *originality* and *quality* succinctly embody the essential characteristics of creativity, as defined by Amabile (1996). Interestingly, Dean et al. conveyed that more accurate assessments of ideas occur when multi-criteria evaluations are kept separate, rather than combined into a summated or multiplicative unidimensional rating.

In all, ideas are relatively easy to compare head-to-head, but multi-criteria evaluations provide richer data. Further, many criteria have been utilized in previous research to effectively assess collaboratively generated and distilled ideas, but criteria need not be limited to these recommendations. Depending on the purpose or goal of collaborative decision making, criteria should be adjusted to assess customized considerations such as ease of implementation, cost, and anticipated impact.

44.4.5 Build Commitment

The fifth and final pattern of collaboration is building commitment. In the research literature, the terms building commitment and building consensus are often used interchangeably. Building commitment or consensus typically wraps up group decision-making efforts. For collaborative efforts to succeed, groups must find solutions that balance the differing requirements or needs of stakeholders, while acknowledging that it may not be possible to satisfy all needs of all stakeholders (Simon, 1996). Otherwise, individual stakeholders whose interests are thwarted by group processes may themselves be in a position to thwart the group's success. Thus, building consensus is a critical aspect of collaborative efforts.

To support decision making by teams and groups in consensus building, both rational (see, e.g., Yu and Lai, 2011) and social preferences (see e.g., Dong et al., 2010) should be considered. While some GSS have been developed to calculate optimal solutions with inclusion of both rational and social perspectives, these calculations are often complex and therefore less transparent. Thus, there is a possibility that the users may not trust them to represent team consensus. Even more importantly, GSSs aid in decision making, but to build consensus, groups also need to debate varying perspectives and overcome differences to arrive at a mutually agreeable decision (Er, 1988). Accordingly, one of the foci of GSS research has been to create the optimal conditions to achieve consensus. For example, Chen et al. (2012) suggested an adaptive consensus support model for group decision-making systems. Using this model, GSS can evaluate the comments of the contributors and provide suggestions for modification in order to enable progression toward consensus. Moreover, the system is also interactive so that the contributors can further modify these suggestions allowing the GSS to create an increasingly accurate model of consensus.

In addition to improving consensus building models, GSS research has also been extended to accommodate distributed decision making through the use of analytical algorithms to facilitate consensus. For example, Tavana and Kennedy (2006) used a strategic assessment model of consensus to enable distributed teams identify strengths, weaknesses, opportunities, and threats of the Cuban missile crisis situation. This model enabled consensus using various intuitive processes including subjective probabilities, an analytical hierarchy process, and environmental scanning to achieve consensus through focused interaction and negotiations among an international group of participants. However, it has been argued that these and other decision conferencing systems need to consider the cultural differences among international participants and adjust for it in order to achieve greater success (Quaddus and Tung, 2002).

Groups that use GSS tend to outperform their face-to-face counterparts on the aspects of information exchange and quality. In their study, Lam and Schaubroeck (2000) compared groups that were using GSS to groups that were involved in face-to-face decision making. They found that, in the absence of a pre-discussion consensus, the GSS groups were significantly more efficient in reaching a consensus about the superior hidden profile candidate than the face-to-face groups. In addition, being in a GSS facilitated group tended to shift the individual preferences toward group consensus more strongly.

The use of GSS may not only make the process of decision making more efficient, but it also offers an opportunity for researchers to evaluate the data in order to analyze patterns and recommend improvements to the process on a post hoc basis. On a simple level, initial levels of consensus can be statistically evaluated following a voting process by which participants use Likert-type scales (e.g., 1 = *strongly disagree* to 5 = *strongly agree*) to assess the merit of ideas (e.g., quality, novelty, and ease of implementation). Specifically, the average rating and standard deviation of each item are calculated. Then, the team typically agrees on ideas with the highest ratings and lowest standard deviations as their preferences. Moreover, a large standard deviation suggests participants do not agree on the merit of an idea. Such ideas should be further discussed, and then voted on again, to ensure the lack of consensus is due to a true disagreement, rather than a misunderstanding of the idea.

Other, more complex, methods and algorithms can also be used to assess group consensus. For example, a study by Ngwenyama et al. (1996) showed that the data that are collected by the means of GSS allow facilitators to identify the issues that have been most contentious as well as the individual and collective positions on all the issues raised. In addition, these data can also be useful to facilitators to analyze the group preferences and the available alternatives that support consensus formation. Similarly, in another study, a GSS-based consensus building approach was utilized to analyze a group's creative problem-solving and decision-making processes (Kato and Kunifuji, 1997). Using the GSS enabled the researchers to combine creative thinking methods with a consensus making process to solve complicated problems. The participants who used the combination of the two methods could construct the appropriate evaluation structure interactively and choose the optimal alternative in a rational manner.

Finally, the application of GSS has also moved beyond the laboratory and corporate decision making into the area of political disputes among conflicting parties. One such example can be found in the conflict resolution and consensus building effort that was successfully conducted by the water authorities in northwest Iran (Zarghami et al., 2008). In this context, a GSS was used to identify the criteria and rank the water resource projects in the area. This system guided the supervisor who was in charge of the group decision-making problem and allowed him to negotiate with the stakeholders effectively in order to resolve conflict and achieve consensus.

These studies are but a few that illustrate the significance of GSS in consensus building. The contribution of GSS to consensus building is far from saturated. With the advent of new technologies and the changing nature of groups and teams, GSSs have an opportunity to evolve and assimilate new ways of processing information and reaching consensus. For example, keeping up with the changing nature of technology, GSSs are evolving to a new level and may incorporate virtual reality and virtual interactive modeling (VIM) to assist with consensual decision making in groups (Jam et al., 2006). In their study, Jam et al. (2006) conducted a controlled experiment to evaluate the effectiveness of VIM in distributed groups. The results showed that the use of VIM leads to greater efficiency of decision making, greater satisfaction with the process, and improved group member attitude. Such visualization technologies can potentially help distributed groups integrate information easier and improve their communication.

44.5 Organizational Effects of Technology-Supported Collaboration

As the previous section shows, GSSs have been utilized in different ways and have yielded various outcomes in organizations. Early studies in university environments (e.g., Gallupe et al., 1988) were followed by studies at organizational sites (e.g., Grohowski et al., 1990). GSSs have since been commercialized and are present in an increasing number of domestic and international contexts (Briggs et al., 2003).

Interestingly, collaboration technologies such as GSS do not always render the effects that were antici-pated. As suggested by adaptive structuration theorists, organizations may utilize technologies in ways that were originally not intended (DeSanctis and Poole, 1994).

Yet, field research on GSS predominantly paints a positive picture (for an overview, see Fjermestad and Hiltz, 2000). Among field studies with GSS there are comparatively few that focus on organizational groups that used GSS in their own environment, that is, the organization of which they are part. Most studies on real groups report on visits that the group made to facilities outside their organization, most often on the premise of the researchers involved. It is important (and convenient) to do such studies in university contexts where variables can be more systematically explored and sufficient sample sizes be developed under more controlled circumstances. Exceptions to this practice include the use of software aided meeting management (SAMM) by the internal revenue service (IRS) in New York City (DeSanctis et al., 1992) and GroupSystems at the U.S. Navy ThirdFleet (Briggs et al., 1998).

Another limitation of GSS field studies is that most have been executed in North American settings, making it difficult to draw generalizations. Exceptions are reported in overviews in Nunamaker et al. (1997) and de Vreede et al. (1998). An interesting cross-cultural comparison of in situ GSS experiences was reported by de Vreede et al. (2003). Their work compared the experiences from four case studies, two in the United States and two in the Netherlands. Each case study used a GSS called GroupSystems:

1. *International Business Machines*: GroupSystems was introduced at IBM in 1987. A series of stud-ies at this site followed that demonstrated that GSS technology could be effectively introduced in organizational environments (Grohowski et al., 1990; Vogel et al., 1990). Based on success at the first facility, IBM installed the technology at six more sites over the following year and simi-larly expanded their internal facilitation support capabilities (Grohowski et al., 1990; Martz et al., 1992; Vogel et al., 1990). IBM continued expanding internally to 24 sites and beyond with the same format of use, that is, preplanned session agendas with facilitation support throughout the meeting process. The facilitation role was institutionalized with several generations of facilitators emerging from a wide variety of backgrounds and levels of experience with group and organiza-tional dynamics.

2. *Boeing Aircraft Corporation*: Encouraged by reports of IBM's success, Boeing Aircraft Corporation decided in 1990 to conduct a carefully controlled pilot test of GroupSystems in their organization. Boeing collected data so that a business case could be developed either in favor of or against the widespread use of GSS to support their projects. After 64 sessions, costs were evaluated. The project time, or number of calendar days required to produce the deliverables, was reduced by an average of 91%. The man-hour cost savings averaged 71%, or an average of $7,242 per session, for a total savings of $463,488 over the 64 sessions (Post, 1993). This was despite the fact that expense figures included the initial start-up of installing the meeting room technology, training facilita-tors, and collecting the measurement data.

3. *Nationale-Nederlanden (NN)*: Part of the ING Group, NN is the largest insurance firm in the Netherlands and one of the market leaders in Europe. NN was introduced to GroupSystems at Delft University of Technology in 1995. Based on early success, NN continued to use GSS and develop its own internal facilitation capabilities (de Vreede, 2001). Following the successful use at NN, other parts of the ING Group also adopted to use the technology.

4. *European Aeronautic Defense and Space Company, Military division (EADS-M)*: EADS-M is a cooperation of four European companies in producing the Eurofighter and other military air-crafts. EADS-M was first introduced to GSS by Delft University of Technology in 2001. Based on a successful pilot study on GSS, the company acquired a GroupSystems license and had a number of internal facilitators trained.

The cross-case findings from these four organizations are presented in Table 44.2 (all aspects, except savings, on a 5-point Likert scale, 5 being most positive). Overall, the results show that the application of GSS in each of the organizations can be considered successful, both in terms of the actual results that were

TABLE 44.2   Cross-Case Findings on Organizational Use of GSS

Aspect	IBM	Boeing	NN	EADS-M
Efficiency				
Session is efficient	3.9	4.0	4.0	4.3
Person-hours savings	55.5%	71.0%	53.0%	49.7%
Calendar time savings	92%/89%[a]	91.0%	57.7%	33.3%
Effectiveness				
GSS more effective than manual	4.2	—	4.0	4.1
GSS helps to achieve goals	4.1	4.0	3.9	4.0
Initiator's evaluation of outcome quality	4.4	4.1	3.9	4.3
User satisfaction				
Satisfaction with GSS process	4.1	—	4.1	4.3
Willingness to use GSS in similar projects/activities	4.2[b]	4.4	4.2	4.5

[a] Based on 11 and 59 sessions, respectively.
[b] For brainstorming activities.

achieved (e.g., time savings) and in terms of the participants' perceptions (e.g., meeting satisfaction). In each of the organizations, the expected savings would significantly outweigh the expenses for running the technology and hiring (technology) support staff (e.g., facilitators). These findings were particularly interesting given that the individual cases were conducted over a time span of 16 years in four very different organizations in terms of business focus and national/corporate culture.

What is even more interesting than the positive findings in the individual case studies is that although the results in each situation led to management decisions to implement the technology in the organization, the use of the GSS facilities slowed down over time and in three of the four cases the facilities were disbanded. This illustrates that the technology by itself may appear to represent a positive business proposition, but its actual sustained use over time is influenced by other factors. A stream of research in the area of Collaboration Engineering identified that the root cause behind the abandonment of seemingly successful GSS facilities concerned the need to use professional facilitators to let teams get value out of the technology. That is, the technology itself is too complex to configure and align with the team process for groups to do by themselves. Facilitators can be a costly option for an organization. Thus, many groups that could benefit from their services do not have access to them. Further, it turned out that it can be challenging for an organization to retain its facilitators because they are often either promoted to new positions or they leave the organization to establish consulting practices (see, e.g., Agres et al., 2005). Since then, researchers have pursued ways to reduce the reliance on professional facilitators to benefit from GSS-enabled collaboration. One particularly promising avenue has been to codify facilitator expertise in a collaboration design pattern language called ThinkLets (de Vreede et al., 2006, 2009). ThinkLets are like a Lego™ toolkit that allows collaboration engineers click together processes that can be easily transferred to domain experts in organizations to execute by themselves. A number of successful thinkLet-based interventions are described in de Vreede et al. (2009).

44.6 Future Directions for Research on Technology-Supported Collaboration

This final section presents three important directions for future research on technology-supported collaboration. The overview that has been presented in this chapter shows that, under the right circumstances, teams can dramatically benefit from the use of collaboration technology to support their problem-solving and decision-making activities. To unlock this potential, the key challenge for teams is twofold: First, they must purposefully align the people, process, information, and leadership specifics of a collaborative effort with the capabilities that the technology provides. Second, to keep reaping the benefits from the technology, they must find ways to create self-sustaining routines that they can use repeatedly to execute a technology-supported collaborative work practice.

The first of the future directions described addresses this challenge. The other two future directions illustrate that this challenge is even more important to address as team collaboration is moving into new technological environments that bring about a renewed focus on designing and sustaining productive technology-supported team efforts.

44.6.1 Task-Specific Collaboration Applications with Build-In Facilitation Guidance

To reduce the need to rely on professional facilitators to guide team efforts, researchers have begun to explore and prototype ways to build facilitation expertise into collaboration technology. The guiding research question is, *How can collaboration expertise be packaged with collaboration technology in a form that nonexperts can reuse with no training on either the collaboration tools or techniques?* To answer this question, researchers are developing a new class of software called Computer Assisted Collaboration Engineering system (CACE). CACE supports the design, development, and deployment of technology-supported collaborative work practices. An example of a CACE is the ActionCenters system (Briggs et al., 2010; Buttler et al., 2011; Mametjanov et al., 2011). ActionCenters supports the rapid design, configuration, and deployment of software applications custom tailored to support specific collaborative work practices. Each application presents the users with a series of activities for achieving their task. Each activity provides the users with just the tools, data, communication channels, and guidance they require for that activity, and nothing else. The applications can therefore appear very simple and obvious to the users, while providing powerful capabilities. The core of the ActionCenters environment is a collection of loosely coupled, highly configurable collaborative components like shared lists and outlines, shared graphics, polling tools, audio and video channels, and shared document repositories. In the ActionCenters CACE environment, the collaboration engineer snaps these elementary components together and configures them to behave as if they were a tightly integrated, task-specific system without having to write a new code. A CACE is fundamentally different from a GSS in that the GSS provides a generic, configurable toolkit for a plethora of collaboration processes whereas the system built in a CACE is a task-specific application that has all required configurations predefined and guides the users through each step in their process.

A CACE system can be seen as a design studio for creating technology support for repeatable group processes without writing any new code (see Figure 44.1). With a CACE system, collaboration engineers can embed different types of collaboration expertise into the collaborative tools in a format that enables practitioners to execute the technology-supported work practice with no training on either the tools or the collaboration techniques. The resulting collaborative application provides end users with a sequence of screens that represent the activities that together make up the team task. Each screen provides users only with the functionality that they require for that activity in the sequence. For example, Figure 44.2 depicts a screen at runtime in a requirements negotiation process. The banner heading at the top is implemented with an HTML control. The numbered list in the middle with its input panel is a configured outliner component. The caption-balloon icons to the right of each outline heading are links to a subordinate outliner component. When a caption balloon is clicked, the subordinate outline component pops up configured to appear and behave as a comment window, so users can comment on any heading in the main outline.

Pilot studies with ActionCenters for five technology-supported collaborative work practices have demonstrated that collaboration engineers could create complete, task-specific collaboration systems in hours or days that would have required several person-years of effort using conventional approaches. In a pilot study of a domain-specific creative problem-solving application, eight groups

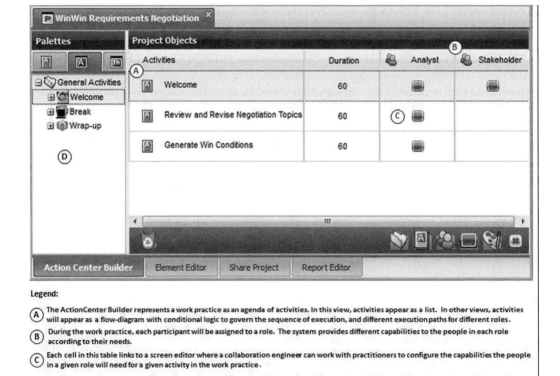

Legend:

(A) The ActionCenter Builder represents a work practice as an agenda of activities. In this view, activities appear as a list. In other views, activities will appear as a flow-diagram with conditional logic to govern the sequence of execution, and different execution paths for different roles.

(B) During the work practice, each participant will be assigned to a role. The system provides different capabilities to the people in each role according to their needs.

(C) Each cell in this table links to a screen editor where a collaboration engineer can work with practitioners to configure the capabilities the people in a given role will need for a given activity in the work practice.

(D) As a short-cut, the system provides palettes of pre-configured activities and techniques that can be dragged from the palette and dropped onto the agenda. The dropped objects instantiate on the agenda as one or more generic activities configured in advance with the capabilities to support a particular procedure. The collaboration engineer tailors the activities to suit the specifics of the task at hand.

FIGURE 44.1   Screenshot of the ActionCenters CACE studio.

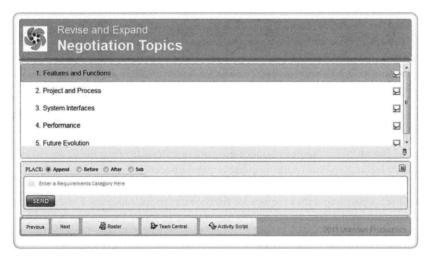

FIGURE 44.2   Run-time appearance of a screen in a requirements negotiation process step.

of practitioners with no prior experience on collaboration technology and no knowledge of the collaboration techniques embedded in the work practice nonetheless executed successfully a sequence of five techniques under the guidance of the ActionCenters application. Each group achieved its goals. Current efforts are under way to test the ActionCenters more rigorously in a program of laboratory experiments, which will be followed by a series of field trials.

44.6.2 Virtual Worlds

In the recent years, much excitement has built up around a new concept of work spaces called meta-verses or virtual three-dimensional (3-D) spaces. Metaverses are usually used for sharing information, creating personal virtual images (avatars), and manipulating virtual objects. They are immersive virtual worlds (VWs) where avatars interact with each other and with software agents, using the metaphor of the real world but without its physical limitations (Davis et al., 2009). VWs allow teams to move beyond simple electronic conversation and document distribution by providing an environment where people can also share a communal space with others.

Extrapolating from Media Richness Theory (Daft and Lengel, 1986) and Media Synchronicity Theory (Dennis et al., 2008), researchers suggest that users of VWs may be more engaged during VW collaboration than low-tech distributed collaboration because VW users experience co-presence—a feeling of shared social and physical space that adds an element of interpersonal realism to collaborative interactions. Further, the accessibility and multiple means of communication inherent to VWs makes it a powerful platform for supporting collaboration. Specifically, in VWs, users have many communication tools to their disposal, such as text-based messaging, video conferencing, object creation/sharing, audio, and even GSS. Research to date has shown that collaboration in VWs can be a motivating, engaging, productive, and satisfying experience for team members. Yet, teams face various challenges at the same time, including but not limited to the need to master the skills to smoothly operate their avatars, the risk of social loafing from team members, and the difficulty for team leaders to monitor participants' emotional states and establish rapport. For an overview of VW collaboration research to date, see Boughzala et al. (2012).

VWs are currently facing some major challenges to become a broadly accepted environment for team collaboration. In fact, VWs appear to be losing some popularity, especially as society is moving toward mobile technology that does not yet suitably supports VWs, such as cell phones and tablets. Also, researchers found that despite richer communication tools, the absence of face-to-face interaction makes convergence and consensus processes difficult as the body language, perceptions, and feelings of participants are difficult to assess in VWs (Wigert et al., 2012). Notably, much of the challenge of using VWs can be attributed to the difficulty faced when trying to effectively manipulate avatars to display appropriate body language. As such, future research is needed to determine how VW collaboration can be further enriched through advances that focus both on the technology itself and the processes that teams execute.

44.6.3 Crowdsourcing

The third and final direction for future research relates to collaboration through social media. Social media have expanded the capability of collaborative problem solving into social levels including online crowds of employees, professionals, amateurs, producers, and consumers. This has given rise to a new phenomenon, which is commonly known as *crowdsourcing* (Howe, 2008) that can be defined as a collaboration model enabled by social web technologies to solve organizational problems in partnership with online communities (Oh et al., 2012). Other popular terms include "mass collaboration" (Tapscott and Williams, 2006), "open collaboration," "collective intelligence" (Surowiecki, 2005), and "co-creation" (Kazman and Chen, 2009). All these terms put online users in the center of collaboration processes, while they have not been considered as important actors in traditional collaboration models.

In crowdsourcing, there are two modes of collaboration (closed collaboration versus open collaboration), which have distinctively different collaboration processes. Table 44.1 illustrates that "closed collaboration" focuses on *finding* the best solution out of many submitted ones, whereas "open collaboration" focuses on *facilitating collaboration* among undefined large number of online community members to solve problems (Table 44.3).

TABLE 44.3   Different Modes of Collaboration in Crowdsourcing

	Closed Collaboration	Open Collaboration
Types of problem	• Structured problem • *Organizations define a problem* and broadcast the defined problem to an online community so that community members can submit solutions to the organization Exemplary Problem: "A one-part adhesive is required that is activated at room temperature. The adhesive should have a minimum set-strength upon activation for gluing a fixed substrate on metals and synthetics (polymers) which can then be fully cured by other methods … Theoretical proposals (no verified method data) will be considered for a lesser reward." (*taken from InnoCentive.com*)	• Unstructured problem • Organizations broadcast an unstructured question to an online community so that the community can (1) identify and define specific problems in parallel ("brainstorming") and (2) clarify and evaluate each problem by commenting or voting to reduce the large number of suggested problems into a selection of the best few ideas worthy of more focused attention ("convergence") Exemplary Problem: "We've received many suggestions about closing select streets temporarily to create more opportunities for bicycling, walking, and events such as farmers markets and art walks. How often should we close streets for these types of activities?" (*taken from ideas. LA2B.org*)
Way of finding solution	The organization *selects* the best solution from a large number of solutions submitted by the online community. Collaboration between online community members is not allowed. Therefore, collaboration features such as voting, commenting, rating, and sharing in the web interface design are not important	The organization *facilitates collaboration* between online community members. The organization determines the best solution considering the results of both brainstorming and convergence activities. Therefore, collaboration features such as voting, commenting, rating, and sharing in the web interface design are important
Examples	*mTurk.com.com, InnoCentive.com, YourEncore.com, oDesk.com, freeLancer.com*	*Threadless.com, CambrianHouse.com MindMixer.com, CrowdCast.com, BrainReactions.net*

Crowdsourcing is based on two underlying assumptions: (1) a larger number of people can solve difficult or complex problems better than a small number of people and (2) a team or group that has high collective intelligence is more likely to excel in complex problem solving. Compared to traditional settings such as GSS, a distinctive feature of social media–driven collaboration is that it highlights the potential that a large online *amateur crowd* can be smarter than a handful of *organizational professional experts* in problem solving. To this end, Surowieck (2005, p.10) suggests four conditions that need to be met for online crowds to make wise decisions to solve problems collectively: (1) creating sufficient diversity of opinion, by which each individual involved has some private information, even if it is just an eccentric opinion; (2) maintaining independence, wherein each person's opinion or decision is not influenced by those around them; (3) enabling decentralization, through which individuals can specialize and tap into local sources of knowledge; and (4) facilitating aggregation, which stresses the importance of mechanisms for translating many private opinions or decisions into a collective decision. These notions are not entirely new. Apart from the fourth condition ("aggregation"), the first three conditions have been repeated research questions in collaboration science research in general and GSS research in particular. Although it is still an under-researched area, recent studies have begun to pursue the notion of "convergence," which is comparable to Surowiecki's notion of "aggregation." Finding ways to effectively create and sustain the conditions that Surowieck outlines provides an exciting opportunity for collaboration technology researchers in general and GSS researchers in particular.

44.7 Summary

The advent of ubiquitous computing power and access to the Internet has enabled team collaboration across boundaries of time, place, institutions, and culture. Modern collaboration processes have become increasingly reliant on advanced collaboration technologies and systems. Notwithstanding the clear value that these technologies bring, their use in organizations and teams is not without challenges. Collaboration researchers and practitioners have developed and leveraged new collaboration techniques and applications to address many challenges inherent to IT-supported collaboration. The aim of this chapter was to provide an overview of how IT in general and collaboration technologies in particular support team problem solving and decision making to improve their performance.

First, a classification scheme was presented to organize the different types of collaboration technologies teams can use in various situations. Next, some underlying principles of IT-supported collaboration were introduced in the form of six patterns of collaboration: *idea generation, reduction, clarification, organization, evaluation,* and *building commitment.* An overview was provided to show how past research has offered insights on supporting groups in each of these patterns through collaboration technology, with a special focus on GSS. Next, an account was given on the organizational value that organizations can derive in terms of cost savings and productivity gains through the use of GSS-supported team processes. Finally, three important directions for future research and practice were introduced: (1) the development of a new generation of collaboration technology in the form of CACE environments that enable collaboration engineers to snap together task-specific collaborative applications with embedded facilitation guidance without having to write software code, (2) the use of virtual worlds as an environment to provide a rich online collaboration experience through the use of avatar representations of team members, and (3) the emergence of crowdsourcing as a force for superior collaborative problem solving through the active involvement of large numbers of participants. Together these directions represent both an opportunity and challenge for researchers to transfer and expand insights from past research on technology-supported group work to future research efforts.

Glossary

Audio conferencing tools provide a continuous channel for multiple users to send and receive sound while video conferencing tools allow users to send and receive sound or moving images.

Convergence is the action of moving from having many concepts to a focus on and understanding of a few concepts deemed worthy of further attention.

Conversation tools are primarily optimized to support dialog among group members. E-mail is a widely used conversation tool as well as short message service (SMS) (i.e., cell phone text messaging), which is becoming increasingly common.

Crowdsourcing is used in order for the job of a specific agent or team to be outsourced to a larger, undefined group in an open call for solutions.

Desktop sharing tools allow the events displayed on one computer to be seen on the screens of other computers.

Enterprise analysis is a collaborative design model building approach that concentrates on collaboratively built models as a goal in itself rather than as a transitional object.

Group dynamics tools are optimized for creating, sustaining, or changing patterns of collaboration among individuals making a joint effort toward a goal.

Group model building refers to a bundle of techniques used to construct system dynamics models working directly with client groups on key strategic decisions.

Group support systems integrate collections of group dynamics tools to move groups seamlessly through a series of activities toward a goal, for example, by generating ideas in one tool, organize them in another, and evaluating them in yet another (e.g., GroupSystems or MeetingWorks).

Idea generation (i.e., brainstorming) is the practice of gathering in a group and generating as many ideas for identifying or solving a problem as possible.

Polling tools are a special class of jointly authored pages, optimized for gathering, aggregating, and understanding judgments, or opinions from multiple people.

Problem structuring methods (PSM) refer to a broad variety of methods and tools that have been developed mainly in the United Kingdom to cope with complexity, uncertainty, and conflict.

Search engines use search criteria provided by users to retrieve digital objects from among vast stores of such objects (e.g., the World Wide Web, the blogosphere, and digital libraries).

Shared editor tools are typically a jointly authored page optimized for the creation of a certain kind of deliverable by multiple authors.

Shared file repositories provide mechanisms for group members to store digital files where others in the group can access them.

Social tagging allows users to affix keyword tags to digital objects in a shared repository.

Syndication tools allow a user to receive a notification when new contributions to pages or repositories they deem to be of interest (e.g., blogs, wikis, and social networks).

Virtual worlds (VWs) are online virtual environments in which users interact using 3-D avatars in a shared space.

References

Ackermann, F. and Eden, C. (2005). Using causal mapping with group support systems to elicit an understanding of failure in complex projects: Some implications for organizational research. *Group Decision and Negotiation, 14*, 355–376.

Agres, A., de Vreede, G. J., and Briggs, R. O. (2005). A tale of two cities—Case studies on GSS transition. *Group Decision and Negotiation, 14*(4), 267–284.

Amabile, T. M. (1982). Social psychology of creativity: A consensual assessment technique. *Journal of Personality and Social Psychology, 43*, 997–1013. doi:10.1037/0022-3514.43.5.997.

Amabile, T. M. (1996). Attributions of creativity: What are the consequences? *Creativity Research Journal, 8*, 423–426. doi:10.1207/s15326934crj0804_10.

Andersen, D., Vennix, J., Richardson, G., and Rouwette, E. (2007). Group model building: Problem structuring, policy simulation and decision support. *Journal of the Operational Research Society, 58*(5), 691–694.

Badura, V., Read, A., Briggs, R. O., and de Vreede, G. J. (2010). Coding for unique ideas and ambiguity: The effects of convergence intervention on the artifact of an ideation activity. In *Proceedings of the 43rd Hawaiian International Conference on System Science*, Kauai, HI, IEEE Computer Society Press, Los Alamitos, CA.

Barzilay, R., McKeown, K. R., and Elhadad, M. (1999). Information fusion in the context of multi-document summarization. In Paper presented at the *Annual Meeting of the Association for Computational Linguistics on Computational Linguistics*, College Park, MD, 550–557.

Basadur, M. and Hausdorf, P. A. (1996). Measuring divergent thinking attitudes related to creative problem solving and innovation management. *Creativity Research Journal, 9*, 21–32. doi:10.1207/s15326934crj0901_3.

Bock, G. W., Zamud, W., Kim, Y., and Lee, J. (2005). Behavioral intention formation in knowledge sharing: Examining the roles of extrinsic motivators, social-psychological forces, and organizational climate. *Management Information Systems Quarterly, 29*, 87–111.

Boughzala, I., de Vreede, G. J., and Limayem, M. (2012). Team collaboration in virtual worlds. *Journal of the Association for Information Systems, 13*, 714–734.

Briggs, R. O., Adkins, M., Mittleman, D., Kruse, J., Miller, S., and Nunamaker, J. F. (1998). A technology transition model derived from field investigation of GSS use aboard the U.S.S. CORONADO. *Journal of Management Information Systems, 15*, 151–195.

Briggs, R., Kolfschoten, G., de Vreede, G. J., Albrecht, C., and Lukosch, S. (2010). Facilitator in a box: Computer assisted collaboration engineering and process support systems for rapid development of collaborative applications for high-value tasks. In *Proceedings of the 43rd Hawaii International Conference on System Sciences*, Kona, HI, pp. 1–10.

Briggs, R. O., Kolfschoten, G. L., de Vreede, G. J., and Dean, D. L. (2006). Defining key concepts for collaboration engineering. Irma Garcia, Raul Trejo (eds.). *Proceedings of the 12th Americas Conference on Information Systems*, AIS, Acapulco, Mexico. 121–128.

Briggs, R. O., Nunamaker, J. F., Jr., and Sprague, R. H. Jr., (1997). 1001 unanswered research questions in GSS *Journal of Management Information Systems, 14*(3), 3–21.

Briggs, R. O. and de Vreede, G. J. (1997). Meetings of the future: Enhancing group collaboration with group support systems. *Journal of Creativity and Innovation Management, 6*(2), 106–116.

Briggs, R. O., de Vreede, G. J., and Nunamaker, J. F. Jr. (2003). Collaboration engineering with ThinkLets to pursue sustained success with group support systems. *Journal of Management Information Systems, 19*, 31–63.

Bunker, B. B. and Alban, B. T. (2006). *The Handbook of Large Group Methods: Creating Systemic Change in Organizations and Communities.* Jossey-Bass, San Francisco, CA.

Buttler, T., Janeiro, J., Lukosch, S., and Briggs, R. O. (2011). In: Adriana S. Vivacqua, Carl Gutwin, Marcos R. S. Borges (eds.). Collaboration and Technology, *17th International Workshop, CRIWG 2011.* LNCS 6969/2011: 126–141. Springer.

Cannon-Bowers, J. A., Salas, E., and Converse, S. (1993). Shared mental models in expert team decision making. In *Individual and Group Decision Making*, Castellan NJ (ed.). Lawrence Erlbaum, Associates, Hillsdale, NJ, 221–246.

Chambless, P., Hasselbauer, S., Loeb, S., Luhrs, D., Newbery, T., and Scherer, W. (2005). Design recommendation of a collaborative group decision support system for the aerospace corporation. In *Systems and Information Engineering Design Symposium*, IEEE, pp. 183–191.

Chen, H., Hsu, P., Orwig, R., Hoopes, L., and Nunamaker, J. F. Jr., (1994). Automatic concept classification of text from electronic meetings, *Communications of the ACM, 37*(10), 56–72.

Chen, S.-M., Lee, L.-W., Yong, S.-W., and Sheu, T.-W. (2012). Adaptive consensus support model for group decision making systems. *Expert Systems with Applications: An International Jounral, 39*(16), 12580–12588.

Coleman, D. and Levine, S. (2008). *Collaboration 2.0: Technology and Best Practices for Successful Collaboration in a Web 2.0 World*, Happy About, Cupertino, CA.

Connolly, T., Jessup, L. M., and Valacich, J. S. (1990). Effects of anonymity and evaluative tone on idea generation in computer-mediated groups. *Management Science, 36*, 689–703.

Corbitt, G., Wright, L., and Christopolous, M. (2000). New approaches to business process redesign: A case study of collaborative group technology and service mapping. *Group Decision and Negotiation, 9*, 97–107.

Daft, R. L. and Lengel, R. H. (1986). Organizational information requirements, media richness, and structural design. *Management Science, 32*, 554–571.

Datta, D. (2007). Sustainability of community-based organizations of the rural poor: Learning from Concern's rural development projects, Bangladesh. *Community Development Journal, 42*, 1–47.

Davis, A., Murphy, J., Owens, D., Khazanchi, D., and Zigurs, I. (2009). Avatars, people, and virtual worlds: Foundations for research in metaverses. *Journal of the Association for Information Systems, 10*, 90–117.

Davis, A., de Vreede, G. J., and Briggs, R. O. (2007). Designing ThinkLets for convergence. In *Proceedings of the 13th Annual Americas Conference on Information Systems (AMCIS-13)*, Keystone, Colorado. http://aisel.aisnet.org/amcis2007/358 (accessed on April 10, 2013).

Dean, D. L., Hender, J. M., Rodgers, T. L., and Santanen, E. (2006). Identifying quality, novel, and creative ideas: Constructs and scales for idea evaluation. *Journal of Association for Information Systems, 7*, 649–699.

Dean, D., Lee, J., Orwig, R., and Vogel, D. (1995). Technological support for group process modeling. *Journal of Management Information Systems, 11,* 43–63.

Dean, D. L., Orwig, R. E., and Vogel, D. R. (2000). Facilitation methods for collaborative modeling tools. *Group Decision and Negotiation, 9,* 109–127.

DeLuca, D. and Valacich, J. S. (2006). Virtual teams in and out of synchronicity. *Information Technology and People, 19,* 323–344.

Dennis, A. R., Fuller, R. M., and Valacich, J. S. (2008). Media, tasks, and communication processes: A theory of media synchronicity. *MIS Quarterly, 32,* 575–600.

Dennis, A., Hayes, G., and Daniels, R. (1999). Business process modeling with group support systems. *Journal of Management Information Systems, 15,* 115–142.

Dennis, A. R. and Valacich, J. S. (1999). Research note. Electronic brainstorming: Illusions and patterns of productivity. *Information Systems Research, 10,* 375–377.

Dennis, A. R., Wixom, B. H., and Vandenberg, R. J. (2001). Understanding fit and appropriation effects in group support systems via meta-analysis. *MIS Quarterly, 25,* 2–16.

DeSanctis, G. and Poole, M. S. (1994). Capturing the complexity in advanced technology use: adaptive structuration theory. *Organization Science, 5*(2), 121–147.

DeSanctis, G., Poole, M. S., Lewis, H., and Desharnais, G. (1992). Using computing in quality team meetings: Some initial observations from the IRS-Minnesota project. *Journal of Management Information Systems, 8,* 7–26.

Diehl, M. and Stroebe, W. (1987). Productivity loss in idea-generating groups: Tracking down the blocking effect. *Journal of Personality and Social Psychology, 61,* 392–403.

Dong, Y., Zhanga, G., Hong, W. C., and Xua, Y. (2010). Consensus models for ahp group decision making under row geometric mean prioritization method. *Decision Support Systems, 49,* 281–289.

Easton, G. K., George, J. F., Nunamaker, J. F., Jr., and Pendergast, M. O. (1990). Using two different electronic meeting system tools for the same task: An experimental comparison. *Journal of Management Information Systems, 7,* 85–101.

Ellinger, A. E. (2000). Improving marketing/logistics cross-functional collaboration in the supply chain. *Industrial Marketing Management, 29,* 85–96.

Er, M. (1988). Decision support systems: A summary, problems, and future trends. *Decision Support Systems, 4,* 355–363.

Fjermestad, J. and Hiltz, S. R. (1998). An assessment of Group Support Systems experimental research: Methodology and results. *Journal of Management Information Systems, 15,* 7–149.

Fjermestad, J. and Hiltz, S. R. (2000). A descriptive evaluation of Group Support Systems case and field studies. *Journal of Management Information Systems, 17,* 35–86.

Gallupe, R., DeSanctis, G., and Dickson, G. W. (1988). Computer-based support for group problem-finding: An experimental investigation. *MIS Quarterly, 12,* 277–296.

Gavish, B. and Gerdes, J. H. (1997). Voting mechanisms and their implications in a GDSS environment. *Annals of Operation Research, 71,* 41–74.

Grisé, M. and Gallupe, R. B. (2000). Information overload: Addressing the productivity paradox in face-to-face electronic meetings. *Journal of Management Information Systems, 16,* 157–185.

Grohowski, R., McGoff, C., Vogel, D., Martz, B., and Nunamaker, J. (1990). Implementing electronic meeting systems at IBM: Lessons learned and success factors. *Management Information Systems Quarterly, 14,* 368–383.

Hardy, C., Lawrence, T. B., and Grant, D. (2005). Discourse and collaboration: The role of conversations and collective identity. *Academy of Management Review, 30,* 58–77.

Hengst, M. D. and Adkins, M. (2007). Which collaboration patterns are most challenging: A global survey of facilitators. In *Proceedings of the 40th Annual Hawaii International Conference on System Sciences*, Waikoloa, Big Island, HI, p. 17.

Howe, J. (2008). *Crowdsourcing: Why the Power of Crowd Is Driving the Future of Business.* Crown Business, New York.

ITU, (2010). *The world in 2010.* International Telecommunication Union, Geneva, Switzerland. http://www.itu.int/ITU-D/ict/material/FactsFigures2010.pdf (accessed: September 2012).

Jam, H. K., Ramamurthy, K. K., and Sundaram, S. (2006). Effectiveness of visual interactive modeling in the context of multiple-criteria group decisions. *IEEE Transactions On Systems, Man and Cybernetics: Part A, 36,* 298–318. doi:10.1109ITSMCA.2005.851296.

Kato, N. and Kunifuji, S. (1997). Consensus-making support system for creative problem solving. *Knowledge-Based Systems, 10,* 59–66. doi:10.1016/S0950–7051(97)00014–2.

Kazman, R. and Chen, H. M. (2009). The metropolis model: A new logic for development of crowd-sourced systems. *Communications of the ACM, 52*(7), 78–84.

Kolfschoten, G. L., Lowry, P. B., Dean, D. L., de Vreede, G. J., and Briggs, R. O. (in press). Patterns in collaboration. In Nunamaker, J. F. Jr., Romano, N. C. Jr., and Briggs, R. O. (Eds.) *AMIS Volume on Collaboration Science.*

Kolfschoten, G. L. and de Vreede, G. J. (2009). A Design approach for collaboration processes: A multi-method design science study in collaboration engineering. *Journal of Management Information Systems, 26*(1), 225–256.

Lam, S. K. and Schaubroeck, J. (2000). Improving group decisions by better pooling information: A comparative advantage of Group Decision Support Systems. *Journal of Applied Psychology, 85,* 565–573. doi:10.1037/0021–9010.85.4.565.

Laukkanen, S., Kangas, A., and Kangas, J. (2002). Applying voting theory in natural resource management: A case of multiple-criteria group decision support. *Journal of Environmental Management, 64,* 127–137. doi:10.1006/jema.2001.0511.

Lewin, A. and Stephens, C. (1993). Designing post-industrial organizations: Theory and practice. In Huber, G. and Glick, W. H. (Eds.) *Organization Change and Redesign: Combining Ideas and Improving Managerial Performance,* Oxford University Press, New York, pp. 393–409.

MacCrimmon, K. R. and Wagner, C. (1994). Stimulating ideas through creative software. *Management Science, 40,* 1514–1532.

Malone, T. and Crowston, K. (1990). What is coordination theory and how can it help design cooperative work systems? In *Proceedings of the Conference on Computer-Supported Cooperative Work,* Association for Computing Machinery, New York, pp. 357–370.

Mametjanov, A., Kjeldgaard, D., Pettepier, T., Albrecht, C., Lukosch, S., and Briggs, R. O. (2011). ARCADE: Action-centered rapid collaborative application development and execution. In Paper presented at the *44th Hawaii International Conference on System Sciences,* Poipu, HI, pp. 1–10.

Martz, B., Vogel, D., and Nunamaker, J. (1992). Electronic meeting systems: Results from the field. *Decision Support Systems, 8,* 155–192.

McLeod, P. (2011). Effects of anonymity and social comparison of rewards on computer-mediated group brainstorming. *Small Group Research, 42,* 475–503. doi:10.1177/1046496410397381.

Michinov, N. and Primois, C. (2005). Improving productivity and creativity in online groups through social comparison process: New evidence for asynchronous electronic brainstorming. *Computers in Human Behavior, 21,* 11–28. doi:10.1016/j.chb.2004.02.004.

Mintzberg, H. (1983). *Structures in Fives.* Prentice-Hall, Englewood cliffs, NJ.

Mittleman, D. M., Briggs, R. O., Murphy, J. D., and Davis, A. J. (2009). Towards a taxonomy of groupware technologies. Groupware: Design, implementation, and use. In *Proceedings of the Fourteenth 14th Collaboration Researchers International Workshop on Groupware,* Springer, Berlin, Germany, pp. 305–317.

Montoya-Weiss, M., Massey, P., and Song, M. (2001). Getting it together: Temporal coordination and conflict management in global virtual teams. *Academy of Management Journal, 44*(6), 1251–1262.

Mulder, I., Swaak, J., and Kessels, J. (2002). Assessing learning and shared understanding in technology-mediated interaction. *Educational Technology and Society, 5,* 35–47.

Mumford, M. D. and Gustafson, S. B. (1988). Creativity syndrome: Integration, application, and innovation. *Psychological Bulletin, 103,* 27–43. doi:10.1037/0033–2909.103.1.27.

Ngwenyama, O. K., Bryson, N., and Mobolurin, A. (1996). Supporting facilitation in group support systems: Techniques for analyzing consensus relevant data. *Decision Support Systems*, *16*, 155–168. doi:10.1016/0167-9236(95)00004-6.

Nunamaker, J. F. Jr., Briggs, R. O., Mittleman, D. D., Vogel, D. and Balthazard, P. A. (1997). Lessons from a dozen years of group support systems research: A discussion of lab and field findings. *Journal of Management Information Systems*, *13*, 163–207.

Nunamaker, J. F. Jr., Briggs, R. O., and de Vreede, G. J. (2001). From information technology to value creation technology. In Dickson, G. W. and DeSanctis, G. (Eds.) *Information Technology and the Future Enterprise: New Models for Managers*, Prentice-Hall, New York.

Oh, O., Nguyen, C., de Vreede, G. J., and Derrick, D. C. (2012, May). Collaboration science in the age of social media: A crowdsourcing view. In *Proceedings of Group Decision and Negotiation 2012*, Recife, Brazil.

Post, B. Q. (1993). A business case framework for group support technology. *Journal of Management Information Systems*, *9*, 7–26.

Quaddus, M. A. and Tung, L. L. (2002). Cultural differences explaining the differences in results in GSS: Implications for the next decade. *Decision Support Systems*, *33*, 177.

Reiter-Palmon, R., Mumford, M., Boes, J., and Runco, M. (1997). Problem construction and creativity: The role of ability, cue consistency, and active processing. *Creativity Research Journal*, *10*, 9–23.

Renger, M., Kolfschoten, G., and de Vreede, G. J. (2008). Challenges in collaborative modeling: A literature review and research agenda. *International Journal of Simulation and Process Modeling*, *4*, 248–263.

Rietzschel, E. F., Nijstad, T., and Stroebe, W. (2006). Productivity is not enough: A comparison of interactive and nominal brainstorming groups on idea generation and selection. *Journal of Experimental Social Psychology*, *42*, 244–251.

Saaty, T. L. and Shang, J. (2007). Group decision-making: Head-count versus intensity of preference. *Socio-Economic Planning Sciences*, *41*, 22–37.

Simon, H. A. (1996). *The Sciences of the Artificial*. MIT Press, Cambridge, MA.

Simpson, C. and Prusak, L. (1995). Troubles with information overload: Moving from quantity to quality in information provision. *International Journal of Information Management*, *15*(6), 413–425.

Sosik, J. J., Avolio, B. J., and Kahai, S. S. (1998). Inspiring group creativity: Comparing anonymous and identified electronic brainstorming. *Small Group Research*, *29*, 3–31. doi:10.1177/1046496498291001.

Surowiecki, J. (2005). *The Wisdom of Crowds*. Doubleday, New York.

Tapscott, D. and Williams, A. (2006). *Wikinomics: How Mass Collaboration Changes Everything*. Penguin Group, New York.

Tavana, M. and Kennedy, D. T. (2006). N-SITE: A distributed consensus building and negotiation support system. *International Journal of Information Technology and Decision Making*, *5*, 123–154.

Taylor, D. W., Berry, P. C., and Block, C. H. (1958). Does group participation when using brainstorming facilitate or inhibit creative thinking? *Administrative Science Quarterly*, *3*, 23–47.

Tjosvold, D. (1988). Cooperative and competitive interdependence: Collaboration between departments to serve customers. *Group and Organization Management*, *13*, 274–289.

Vogel, D., Nunamaker, J., Martz, B., Grohowski, R., and McGoff, C. (1990). Electronic meeting system experience at IBM. *Journal of Management Information Systems*, *6*, 25–43.

de Vreede, G. J. (2001). A field study into the organizational application of GSS. *Journal of Information Technology Cases and Applications*, *2*, 18–34.

de Vreede, G. J. and Briggs, R. O. (2005). Collaboration engineering: Designing repeatable processes for high-value collaborative tasks. In *The Proceedings of the 38th Hawaii International Conference on System Science*, Big Island, HI, IEEE Computer Society Press, Los Alamitos, CA.

de Vreede, G. J., Briggs, R. O., and Massey, A. (2009). Collaboration engineering: Foundations and opportunities. *Journal of the Association of Information Systems*, *10*, 121–137.

de Vreede, G. J., Briggs, R. O., and Reiter-Palmon, R. (2010). Exploring asynchronous brainstorming in large groups: A field comparison of serial and parallel subgroups. *Human Factors: The Journal of Human Factors and Ergonomics Society*, *52*(2), 189–202.

de Vreede, G. J., Davison, R., and Briggs, R. O. (2003). How a silver bullet may lose its shine—Learning from failures with group support systems. *Communications of the ACM*, Big Island, HI, *46*(8), 96–101.

de Vreede, G. J., Jones, N., and Mgaya, R. (1998). Exploring the application and acceptance of Group Support Systems in Africa. *Journal of Management Information Systems*, *15*, 197–234.

de Vreede, G. J., Kolfschoten, G. L., and Briggs, R. O. (2006). ThinkLets: A collaboration engineering pattern language. *International Journal of Computerized Applications in Technology*, *25*, 140–154.

de Vreede, G. J., Vogel, D., Kolfschoten, G., and Wien, J. (2003). Fifteen years of GSS in the field: A comparison across time and national boundaries. In *Proceedings of the Hawaiian Conference on System Sciences*, IEEE Computer Science Press, Big Island, HI.

Westaby, J. D. (2002). Identifying specific factors underlying attitudes toward change: Using multiple methods to compare expectancy-value theory to reasons theory. *Journal of Applied Social Psychology*, *32*, 1083–1104.

Wigert, B., de Vreede, G. J., Boughzala, I., and Bououd, I. (2012). Collaboration in virtual worlds: The role of the facilitator. *Journal of Virtual Worlds Research*, 5, 2, 1–18.

Yeung, A., Ulrich, O., Nason, S., and Von Glinow, M. (1999). *Organizational Learning Capability*. Oxford University Press, New York.

Yu, L. and Lai, K. K. (2011). A distance-based group decision-making methodology for multi-person multi-criteria emergency decision support. *Decision Support Systems*, *51*, 307–315.

Zarghami, M. M., Ardakanian, R. R., Memariani, A. A., and Szidarovszky, F. F. (2008). Extended OWA operator for group decision making on water resources projects. *Journal of Water Resources Planning and Management*, *134*, 266–275. doi:10.1061/(ASCE)0733-9496(2008)134:3(266).

Zurlo, R. and Riva, G. (2009). Online group creativity: The link between the active production of ideas and personality traits. *Journal of Cybertherapy and Rehabilitation*, *2*, 67–76.

45

Organizational Adoption of New Communication Technologies

Jonathan Grudin
Microsoft Research

45.1 Introduction

45.1.1 Why Today's Digital Technologies Are Different

Telegraph, telephone, photocopying, fax, courier mail, digital PBXs—organizations have steadily appropriated new communication technologies. In the digital era, e-mail came first. It had far fewer features than today, it was expensive, and adoption came slowly. Organizations had time to gauge the likely costs and benefits and to prepare.

This has changed. A wave of popular new communication technologies followed e-mail. Instant and text messaging, wikis, blogs and microblogs, social networking sites, video—many with free versions that employees can download, access via browsers, or bring in on smart phones. Many are used first for personal tasks. Work–life boundaries are blurring. The changed dynamic has benefits for an organization—acquisition and training costs are lower than they once were. It also creates challenges: Employees' preferences and behaviors can conflict with a planned deployment or produce a de facto deployment that takes management by surprise.

Historically, new technologies often find industrial uses prior to widespread domestic use—clocks, sewing machines, and telephones were initially too expensive for most consumers. Early computers were expensive, but because governments placed them in universities starting in the 1950s, many students entered industry experienced with e-mail, text editing, and other computer use. Abetted by

Moore's law and rapidly declining prices, students and other consumers continue to embrace many digital technologies before enterprises give them careful consideration.

This chapter focuses not on architecture and only a little on design; my concern is organizational use of messaging, wikis, weblogs, and social networking sites—communication technologies that support information sharing and coordination as well.

After a discussion of relevant terminology, I will briefly describe still-relevant high-level lessons from an earlier era. These inform two frameworks that I find invaluable in understanding seemingly contradictory outcomes that accompany widely used communication and collaboration systems. To motivate these frameworks, the first widely embraced digital communication technology, e-mail, is considered. The focus of this chapter then moves to studies of enterprise experiences with these new technologies that some employees first use spontaneously. The discussion and conclusion cover key points and forward-looking considerations based on the observations and analyses.

45.1.2 Terminology for Organizational Process

Information systems (IS) and computer science (CS) use the same terms to mean different things. IS generally has a longer-term perspective, from the first glimmer that something might be built or acquired to the system's retirement years or decades later. CS, even within software engineering, has a more limited and more technical focus.

The terminology of IS reflects the perspective of a manager or IT professional. Adapting the framework of Munkvold (2002), *initiation* is the identification of needs and possible solutions, *adoption* is the organizational sign-off on a plan, *adaptation* is the development or acquisition and installation of the technology, training plans, and polices governing use, and *acceptance* signals actual use by employees, which is followed by *operation and maintenance*. The entire process is called the organizational *implementation*. A CS perspective is that of a developer or user, with the sequence *requirements analysis*, *development* (also called *implementation*) or *acquisition*, and *adoption*, the latter representing actual use.

Why is a dry discussion of terms of interest? Partly because misunderstanding terms impede communication and exploration of the literatures. But a deeper difference is seen in the contrast between *acceptance* and *adoption* to identify system use. Historically, users of organizational systems had no discretion: Their task was to *accept* a system. CS focuses on professionals and consumers who have discretion—they can *adopt* a product or ignore it.

As suggested by this emphasis on the distinction between passive acceptance and active adoption, this chapter focuses on users' behaviors more than technical characteristics of technology. In organizations today, unlike the past, employee use of some technologies is initially discretionary. This aligns with the CS perspective, adoption, and can represent a change and challenge to IS managers and IT professionals.

Recapping, in the traditional systems perspective of IT professionals, once a technology is rolled out and found to work technically, a key process is complete, whether or not the technology is used at all. From a user perspective, the crucial adoption process *begins* at this point and focuses on whether the targeted individuals or groups *use* the technology. If included, a training perspective is in between: Developers of training focus on initial use and best practices.

45.2 Organizational Experiences in the Late Twentieth Century

For an existing organization, a new communication or collaboration system inevitably disrupts established communication channels, information repositories, and formally or informally designated authorities and responsibilities. The assumption that at some future point a more efficient and effective alignment will take shape is an assumption and is often not shared by everyone in the organization.

Reactions to the uncertainty of success take different forms. The role of evangelism is often stressed—encouraging people to suspend doubts and make a sincere effort to bring the organization to the promised land. At the other extreme is resistance, either complete opposition or an effort to limit or slow the deployment. Resistance can surface before or after a decision to adopt. In some circumstances, a full embrace or strong avoidance might be rewarded, but another option is to be as informed as possible to seek convincing evidence that there will be a productivity benefit—a measurable return on investment (but see the following caveats)—and identify best practices. When possible, it makes sense to mirror the practices of similar organizations that had a good outcome.

Two complications surface regularly in studies of organizational adoption. Diverging views of organizational stakeholders can be critical for a communication or collaboration technology intended for broad use. Opinions vary based on differences in personality, preferences, or past experiences, but there is a more systematic source of differences, based on the nature of different roles people occupy. The next section introduces Henry Mintzberg's framework of organizational parts, a wonderful aid in understanding studies of technology adoption. Mintzberg argues that almost all large or midsize organizations have several specific parts. Each part has a different structure. A new tool or process is viewed differently depending on where one sits. Failure to appreciate these differences and their consequences creates confusion and impedes adoption.

The second complication arises from the desire to minimize risk by focusing, often obsessively, on metrics and claims for a "return on investment." Proving a productivity benefit is a seductive goal for marketers and acquisition managers, but benefits are often impossible to prove or disprove, especially for communication and collaboration tools that are secondary to organizational production. A desperate quest for "return on investment" often leads to accepting hyperbolic claims and laboratory study outcomes that do not generalize to real-world situations. Joseph McGrath, a leader in research into team and group behavior, produced a framework for thinking about group activities that can widen our lens to bring into view a range of issues that are missed when we focus narrowly on performance.

45.2.1 E-Mail

E-mail is not a new digital communication technology, but a review of its reception by organizations helps motivate our consideration of the Mintzberg and McGrath frameworks.

Multitasking operating systems led to e-mail and real-time chat among users of the same machine. The top-down ARPANET and the grassroots unix-to-unix copy network (UUCPNET) appeared in the 1970s. In the 1980s, commercial e-mail arrived: IBM PROFS and various minicomputer office systems, followed by client–server networks of PCs and workstations.

Acquiring and managing an e-mail system was expensive, but even when in place, organizational acceptance was slow. In late 1983, my manager at a large minicomputer company advised me to ignore the e-mail system and write a formal paper memo, saying "e-mail is a way students waste time." Productivity benefits from e-mail use could never be proven. In the 1990s, leading analysts still argued that organizations would abandon e-mail once they understood the negative effects it had on productivity (Pickering and King, 1992).

Wasted time and the loss of confidential information were concerns. Some companies with internal e-mail systems blocked Internet access and restricted communication to work-related exchanges. Internet access was restricted until December 1995 at Microsoft. Even later, IBM considered it a bold step: "In 1997, IBM recommended that its employees get out onto the Net—at a time when many companies were seeking to restrict their employees' Internet access" (Snell, 2005).

Why was management skeptical? The anthropologist Constance Perin (1991) noted that e-mail was useful for individual contributors (ICs) who rely on rapid information exchange, but problematic for managers who must deal with security issues, rapid rumor propagation, and circumvention of hierarchy. Few managers then had keyboard skills, and interrupt-driven e-mail was not as useful for managers whose time was tightly structured and who dealt with formal documents more than informal discussion. Managerial resistance faced when the cost came down, familiarity based on home use rose, but most crucially, in the 1990s, it became possible to routinely send attachments—the documents, spreadsheets, and slide decks that are the lifeblood of managers. In some industries, management may yet rely on paper, but for much of the world, e-mail is mission-critical. Concerns about e-mail overload have led some enterprises to experiment with avoiding it, but its value is generally unquestioned. Employees use it to work more efficiently and effectively more than to waste time.

45.2.1.1 Assessing the E-Mail Experience

Most of the technologies we will examine reflect this e-mail progression:

1. Initial use by students and consumers, especially in North America
2. Resistance by organizations skeptical of productivity benefits
3. Young employees familiar with the technology see positive organizational uses
4. Organizational acceptance and adoption grows

Organizational resistance is not always irrational. There is a cost, a learning curve, and a disruption of the existing balance of communication channels, information repositories, and responsibilities. Utility varies across the organization, and benefits that are realized are not always those initially envisioned.

Passage of time has increased the speed with which a new technology can spread. Tools today cost less to introduce. Technophobia is gone. Executives and managers are hands-on users of technology at home and at work.

E-mail needed a quarter century to be widely embraced by organizations. Differential utility to people in different roles and the difficulty of proving an effect on productivity impeded adoption. Now we will consider this experience from the perspectives of Henry Mintzberg and Joseph McGrath.

45.2.2 Mintzberg's Typology of Parts

Henry Mintzberg's (1984) partitioning of an organization is shown in Figure 45.1. Executives (strategic apex), managers (middle line), ICs (operating core), the people formulating work processes (technostructure), and the support staff (everyone else) typically have different approaches, constraints, opportunities for action, and priorities.

Mintzberg observes that organizations can be grouped into five organizational forms, each marked by strong influence by a different part of the organization. In a start-up, the executives control directly; in a divisionalized company, middle management is empowered; in a professional bureaucracy such as a university or surgical ward of a hospital, the operating core (faculty, surgeons) has considerable

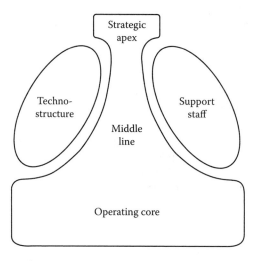

FIGURE 45.1 Five organizational parts.

authority; and so on. In addition, each part of an organization favors a distinct approach to collaboration and coordination: direct supervision (strategic apex), standardization of outputs (middle managers would like to be given a goal and left free to decide how to reach it), standardization of skills (faculty or doctors are certified and then assume they need not be strictly supervised), standardization of work processes (the focus of the technostructure), and mutual adjustment, in which all employees work out their contributions, the approach favored by people in support roles in organizations such as film companies that rely on them. Mintzberg's elegant framework illuminates the different perspectives that come naturally to people in different roles and different parts of an organization. It is no surprise that different features of a technology will appeal to or repel different stakeholders; indeed, this is found in many studies, often overlooked by the researchers, as we will see.

Mintzberg wrote before computers were used for communication and collaboration support in organizations. He did not focus on technology use, but we can extend his model. Consider the three central parts and corresponding roles: ICs, managers, and executives. They differ in the nature of their function and primary activities, their ability to delegate, and the sensitivity of their activities.

ICs who make up the operating core typically engage heavily in communication and are collocated to facilitate this. Managers focus more on sharing structured information—usually documents, which today include spreadsheets and slide decks. Executives of course also communicate and share information, but their primary job is to coordinate the activity of different groups. Communication, information sharing, and coordination—distinct emphases that affect their visions of what a new general-purpose technology will deliver.

Executives' calendars are heavily scheduled far into the future. Managers also have a lot of meetings. Most individual contributors are shielded from meetings so they can produce something—products, services, or documents for use by management.

The ability to delegate work correlates with one's level in the organization. Executives are more likely today to be hands-on users of a technology, but they may not use it long before assigning the task to a subordinate. Managers can delegate some work. The sensitivity to public disclosure of one's work activity also varies. What an IC does is often accessible to others in the organization, whereas an executive's meetings can be highly sensitive. Managers are in between—sensitivities trade off with the benefits in efficiency that can result when their peers know their availability and focus.

In sum, members of these three groups have different structures to their workdays, which impacts tool use. Differences affecting the technostructure and support staff have not been explored as thoroughly. Engaged in defining and improving work processes, technostructure members may advocate for the use of formal workflow systems that promise metrics. The difficulty of representing work activities in software with sufficient flexibility to accommodate exceptions that arise can create tension. IT professionals, members of the support staff, manifest yet another pattern of tool use and often focus on security and reliability, sometimes to the point of obstructing other employees' work.

In retrospect, many efforts to introduce technology are impeded or blocked by not anticipating or recognizing these differences. In Orlikowski's (1992) study of the deployment of the Lotus Notes shared workspace platform in a major consulting company, the strategic apex (senior partners) saw the potential benefits. The operating core (consultants) saw only drawbacks and did not use the system. The IT personnel deploying the system (support staff) had a different incentive structure and used the system very effectively.

Similarly, studies of shared calendar adoption showed sharp disagreements over features among executives, managers, and ICs (Palen, 1998; Palen and Grudin, 2002). When early desktop videoconferencing systems were installed in executive offices, the "direct-dial" calling that they required conflicted with admin-assisted executive meeting initiation (Poltrock and Grudin, 1995). In a study of web use recounted by Jones et al. (2001), the top managers unlike others delegated all save cursory web searching. Understanding this could inspire the design of features to support executive use. These and other examples are covered in more detail in Grudin (2004a,b).

TABLE 45.1   Functions and Modes of Team Activity

	Production	Group Well-Being	Member Support
Inception	Production demand and opportunity	Interaction demand and opportunity	Inclusion demand opportunity
Problem solving	Technical problem solving	Role network definition	Position and status achievements
Conflict resolution	Policy resolution	Power and payoff distribution	Contribution payoff distribution
Execution	Performance	Interaction	Participation

45.2.3 McGrath's Typology of Group Activities

McGrath (1991) described team behavior in terms of three functions and four modes (Table 45.1). Of particular interest are the functions, the columns. Production is the core driver of activity, the reason the group came together. But McGrath placed activities to promote group health and to ensure members get what they need at the same level, arguing they are necessary and in fact continually engaged in. They may go unnoticed—as social animals, we unconsciously attend to them, although not always adequately. The rows in Table 45.1 represent the modes in which a group may find itself at different times—taking on a new task, executing, and occasionally focusing on solving problems or resolving conflicts.

This typology may seem evident, yet it can be a revelation—because our attention is focused almost exclusively on the lower left cell, performance, the production function in execution mode. Management focuses on this cell. Researchers study it in the lab. The quest to show a *return on investment* is perpetual and can lead to narrow, short-term assessments of productivity.

Mysteries are solved by recognizing this myopic focus on performance. Why do some systems do well in lab studies but bomb in organizational settings? Why are others that show no effect in the lab embraced in practice? Dennis and Reinicke (2004) found that insufficient support for group and member well-being explains the lack of commercial success of "group support systems" despite their proven performance benefits in controlled studies. Anonymous brainstorming may work well in the lab, but in an enterprise setting, the identity of a speaker may be crucial, and credit for contributions may motivate participants. Another example is video. It showed no performance advantage over audio in decades of studies, but when other cells of McGrath's framework were examined, video was found to have significant effects in problem-solving and conflict-resolution tasks (Williams, 1997; Veinott et al., 2001). In addition, participants liked video. Greater positive affect could over time contribute to performance in subtle ways.

45.2.4 Reframing the E-Mail Experience

In an organization, e-mail is rarely a prime focus of work. It is used intermittently to support the central tasks. How can one measure the overall impact of getting an occasional quick answer to a question? Does a higher frequency of interaction improve a group's health? How much? How do we measure the effect on performance? Yet unless we believe that eventually a performance improvement will be realized, how do we justify the disruption and unknown outcome of adopting a new communication tool? Those resisting change can make a case.

The two frameworks provide a different perspective on the remark that "e-mail is a way students waste time." That perspective had support from the beginning. The ARPANET was planned as a file-sharing system, a natural vision for managers. E-mail became the tail that wagged the dog, a natural outcome for the ICs in different locations who had to keep the system running and were often its principal users.

A student's principal "production task" may be to learn, but a close second can be building and maintaining social networks that aid in studies and in reaching other career and life goals. E-mail supports activities that serve each of McGrath's functions. Bringing in Mintzberg, informal communication serves students—and ICs in organizations. Managers, not unlike instructors, rely more on formal communication and structured documents. What they perceive as wasted time might not be.

I observed a computer game on a public machine become popular in a competitive group comprising software engineers and a few often-marginalized employees in support roles—usability engineer and technical writer. The top 10 scores were displayed. The usability engineer began playing, culminating in a 16 h stretch during which he achieved by far the highest score. A waste of time? The respect he earned from the software engineers increased his influence. Playing the game may have yielded the highest return of anything he undertook that year. Employees can waste time, of course, but assessing what is ultimately productive is inevitably a judgment call.

45.2.5 Typologies as Guidance

Typologies are not theories, they are frameworks. They emphasize certain features. The test of a typology is whether or not it is useful. These typologies can be used as lenses for understanding empirical phenomena. Examples include those noted earlier and discussed later. Another use is as a checklist for a design or acquisition team, which can consider a system's impact or lack thereof on groups and activities represented by different cells in the typologies. Weiseth et al. (2006) constructed a comprehensive organizational process and technology framework drawing on these typologies. Their framework is centered on models of content, content lifecycle, and process integration. They identify 13 distinct activities in support of coordination, decision-making, or production, each a potential focus of technology support. They identify physical workspace, digital devices, and portals as technological components that bear consideration.

I emphasize that a challenge facing the use of these approaches is the difficulty of proving effects on productivity. When marketing a system, or when considering acquisition and deployment and the cost in terms of adjustment, there is an irresistible impulse to ask for evidence of productivity benefit. Those who claim that they can measure benefits are very popular, but the claims are often not well-founded. Effects, positive or negative, are often subtle.

45.3 A Decade of New Communication Tools in the Enterprise

For decades, as computer use spread, e-mail and bulletin boards or newsgroups were the primary digital communication tools. There were also some chat programs, and information sharing was supported by file-sharing systems built on the evolving file transfer protocol (FTP).

Prior to 2000, most groups—professional colleagues, a set of customers, hobbyists, classroom parents, a housing association, and so forth—could not assume all members were accessible via the Internet. As this changed and more groups become reliably connected, prospects improved for tools to support communication and information sharing. (I do not know of systematic studies, but a critical mass hit my community in the late 2000s—schools and other groups suddenly assumed parents and others had e-mail access, with telephone backup for emergencies.)

The tools that arrived included instant messaging (IM), text messaging, wikis, blogs, microblogs, photosharing sites, geolocation, and a range of social networking sites. Digital voice was built on the conventions of analog telephony; digital video reached and then surpassed familiar capabilities of analog video; the others for the most part broke new ground.

Widely adopted by students and consumers, most of these technologies generally crept into organizations under the radar of management and IT staff. IM and text messaging prospered among consumers in the late 1990s. After 2001, like a phoenix rising from the ashes of the Internet bubble, diverse new communication and information-sharing technologies were embraced. People had acquired technology and sought things to do with it. Wiki and weblog concepts dated back to the late 1990s, but significant levels of use appeared later.

45.3.1 Instant Messaging and Text Messaging

Basic messaging functionality dated to the early days of multitasking, but widespread use on the Internet arrived with clients such as ICQ (in 1996), AOL Instant Messenger (1997), Yahoo! Pager (1998), and Microsoft's MSN Messenger (1999). Text messaging had a similar history with a notable twist. Experiments in the early 1990s followed the advent of 2G cellular technology, but use mushroomed only after telecoms had worked out billing systems in the late. The twist was geographic: The other technologies discussed in this chapter reached prominence first in the United States, with most software and platforms developed there. Text messaging ran on mobile phones produced mainly in Europe and Asia; low messaging costs were set there and use followed.

By 2001, half of all U.S. Internet users, 60 million, had adopted IM. Many of them were part of the workforce, but IM (and text messaging) was overwhelmingly used by students and consumers. Contact lists comprised family and friends.

That said, IM clients were easily downloaded onto computers in most workplaces. IM slowly came into focus in organizational IT departments. In late 2001, an IM client intended for organizational use—no winks or emoticons—was included in Windows XP. Consulting companies began advising organizations on IM use. The advice might be summarized, "IM is a way students waste time." The expressions, "productivity drain," "communication quagmire," "worse than e-mail overload," "no data security," and "no enterprise management" appeared in Gartner reports over the next few years (Lovejoy and Grudin, 2003). Professional IM client uptake was not strong. Many employees were using IM, but did not want a second client, and those not using one for personal communication did not flock to adopt one for workplace use.

Table 45.2 lists the similarities between IM in the mid-2000s and e-mail 20 years earlier. In the 1980s, no directory of e-mail addresses existed, and addresses were highly unpredictable; people could contact acquaintances. Memory was too expensive to save e-mail messages, so it was an ephemeral, informal channel. All of that changed. Young people interviewed 20 years later praised IM and text messaging for being "informal, unlike e-mail." E-mail was by then routinely saved and used in court cases

TABLE 45.2   Parallels between E-mail and Instant Messaging

Email in 1983 *and 2003*	IM in 2003 *Was Evolving*
Used mostly by students	Used mostly by students
Used by everyone	*Use spreading rapidly*
Access limited to friends	Access limited to friends
Accessible to everyone	*Pressure to remove limits*
Clients not interoperable	Clients not interoperable
Complete interoperability	*Pressure for interoperability*
Conversations ephemeral	Conversations ephemeral
Conversations saved	*Recording is more common*
Chosen for informality	Chosen for informality
Became the formal option	*Becoming more formal*
Organizational distrust:	Organizational distrust:
Chit-chat? ROI?	Chit-chat? ROI?
Mission-critical technology	*Becoming more routine*

and government inquiries. A decade later, text messages are saved and appear in court cases; not quite formal, but moving down the same path.

Organizational use of IM built up under the radar. In 2003, a company marketing tools described an effective approach: Ask IT managers how much their employees used IM at work. The reply was often "none." They then installed measurement tools that revealed significant use of downloaded IM clients. They claimed that IT managers often then became interested in tools to monitor and control IM use (personal communication).

Workplace uses of IM were a major focus in studies of communication tools conducted in 2004 (Grudin et al., 2005). Use for quick questions and answers was reported by over 90% of IM users; for socializing by 43%. Many noted that IM was less expensive than the phone when on the road. Others used it during phone calls, teleconferences, and meetings—to get relevant information from people not present, to hold private conversations with others in the meeting, or to forward information to everyone present without interrupting the conversation. Some managers were as heavy IM users as ICs; for example, using it to do other work when in a large meeting that was focused on issues not relevant to them.

Other workplace uses addressed work–life balance issues. E-mail is less intrusive and therefore relied on to organize a school band concert or field trip, but when the event is imminent and quick responses are needed, work interruptions are tolerated and organizers sometimes shift temporarily to IM. In another example, a woman kept an IM window open on her computer at work so she could see from a status change that her son was safely home from school, since his first activity was to get on IM to interact with friends. Prior to using IM, she had interrupted work to telephone to confirm that he made it home.

Could the cumulative impact on productivity of these discrete, relatively low-frequency activities be measured? Laboratory studies do not reliably generalize to the low-frequency use typically found in workplaces. Yes, a quick answer to a question could be significant, but would not other channels have served equally well?

The IM client feature used most after messaging was file sharing. It took decades to fully integrate attachments into e-mail; its early presence in IM facilitated adoption in work settings.

Nevertheless, by the time IM was widely used in organizations, students and consumers were moving to web-based software.

45.3.1.1 Critical Mass Requirements for New Social Media

Web-based social media differ in terms of the scale of participation. A blog requires one author and an audience of variable size. A wiki requires group participation, but the group can be small. A social networking site such as MySpace, Facebook, LinkedIn, or Twitter requires larger numbers of participants to be effective. More specialized sites such as Foursquare and Pinterest require large scale with a narrower focus. These came into prominence in roughly that sequence.

45.3.2 Corporate or Employee Blogs

The term weblog was coined in 1997. The LiveJournal and Pyra Labs Blogger platforms appeared in 1999. By 2002, how-to-blog books were in bookstores. Visibility picked up in 2003 when Google bought Blogger, Wordpress was released, and LiveJournal hosted its millionth blog.

In late 2004, *Communications of the ACM* published a special issue on "the blogosphere." Bloggers were mostly teenagers or in their early 20s, maintaining online diaries on LiveJournal and elsewhere to be read by friends or family. A smaller but significant set of blogs sought larger readerships: essays or pronouncements on technology and other subjects by journalists, pundits, political candidates, and entrepreneurs. Some of the latter hesitated to call the former "bloggers," but some personal blogs were indexed by search engines and evolved to wider readership, whereas some pundits failed to find many readers. Early research examined both student blogs (e.g., Nardi et al., 2004) and the elite or "A-list" blogs (Bar-Ilan, 2004).

Technorati has published "state of the blogosphere" reports since 2004 (http://technorati.com/state-of-the-blogosphere/). "Corporate blog" initially referred to any whose author was clearly linked to an organization. Many were employees blogging about their work and lives. Today, a "corporate blog" is one authored by one or more people whose principal job responsibility is to blog on behalf of an organization. In this chapter, "employee blog" is used as a general term.

In late 2004, Technorati was tracking four million blogs, which produced four posts per second around the clock. Five thousand, one in 800, were employee blogs. Measurable numbers were from employees of companies such as Oracle, SAP, and Sun. Others were from companies marketing blogging tools and from mainstream media sites. These were more likely to include people paid to blog. The organization with the most employee bloggers identified by Technorati was Microsoft, which had close to 1000 or one-fifth of the world total.

In 2005, IBM formally encouraged its much larger employee population to blog publicly (Snell, 2005) and the picture changed. Gartner forecast that corporate blogging was losing its "hype" status and would be mainstream within 2 years.

A study conducted in the summer of 2005 identified complex reactions to the introduction of employee blogging at Microsoft (Efimova and Grudin, 2008) and is the basis for the case study that follows. Microsoft and IBM are not typical organizations, of course; many discourage most or all employees from blogging externally. However, many of the observations resonate with other accounts.

45.3.2.1 Case Study: Microsoft Adopts External Blogging (2003–2005)

Students with externally hosted weblogs, hired as interns or employees in 2000–2001, were the first Microsoft bloggers. They attracted negligible attention. By mid-2002, some employees manually hosted personal weblogs on their work machines.

There are three possibilities for employee blogging, two for externally facing blogs and one for internal blogs. By using a web-based blogging platform, any employee could manage an externally facing blog from his or her work computer. If a company is willing to dedicate a server to host external weblogs, it can facilitate, promote, and monitor employee blogging. It also makes it much easier for external readers to find a relevant employee blog—"one-stop shopping," at least for blogs on the server. Finally, a blog visible only inside the firewall would require an internal blog server.

Grassroots efforts by passionate employees to establish servers eventually brought blogs to the attention of low-level managers. An internal weblog server, maintained on an unused machine through volunteer efforts, hosted a few dozen weblogs by the end of 2002. Late in 2002, a list of externally visible employee weblogs was published by someone outside the organization. This helped create a sense of a community engaged in externally visible blogging. It also brought blogging to the attention of people in the company's legal and public relations groups for the first time. The subsequent meetings and reflection on practices led to some tensions but no actions.

By mid-2003, a server hosting externally visible weblogs was operating with budget support from a manager who perceived a benefit in using weblogs to communicate with customers. However, the company is decentralized enough that the wisdom of letting employees' blog was still being actively debated.

Blogging was still very much below most radar screens. In April 2003, one author came across a disgruntled employee's blog written for friends in which he described actions that would be grounds for termination. Later that year, a contractor was dismissed for a relatively minor disclosure in a blog. Many in the weblog community had made similar disclosures. This fed a sense that legal and public relations representatives who in meetings said "consult us when in doubt" wanted to shut blogging down. Mainstream media reported a Google employee fired for blogging about everyday life at work, and blog-related dismissals at Delta Airlines, ESPN, and elsewhere (e.g., Cone, 2005). IBM's public embrace of external employee blogging was still 18 months in the future.

Discussions among bloggers, human resources, legal, and public relations that followed the dismissal were regarded as mutually educational. Blogging picked up, leading to the 2004 Technorati report, but debate continued. A senior vice president had begun blogging in May 2003, and Bill Gates and Steve Ballmer spoke of blogging approvingly (Dudley, 2004). But another senior vice president in 2005 castigated bloggers as "self-appointed spokespeople" for projects, products, and the company.

By the summer of 2005, unofficial lists of employee bloggers for Amazon, Google, IBM, Sun, and other companies were on the web. Microsoft IT administrators responsible for two externally visible servers that carried 2,000 blogs estimated that 3,000 of the 60,000 employees were active, but a reliable employee survey found over 7,000 blogs, many hosted on web platforms. Some felt that external hosting promoted a perception of independence and gave them ownership of their writing. The internal server was still largely unmonitored and volunteer-run, to the disappointment of the 800 internal bloggers.

In interviews, senior attorneys and publication relations managers described a shift in attitude. Previously skeptical, they now saw value in employees connecting with developers, users, and others outside the company. Bloggers provided useful support and put a human face on the company. Two attorneys said that the 2003 contractor offense would no longer result in a punitive reaction. Avoiding disclosing of sensitive information was not a new concern; a blog might have a faster or broader impact than e-mail or a newsgroup post, but the rules were the same. Similarly, a senior public relations manager described his group's highly sophisticated approach to modeling how blogs, press releases, and the mainstream media interacted. Their goal had become to get in front of and work with employee blogging, not to suppress it.

Bloggers could be a source of joy or stress for middle managers. Many employees cleared their intention to blog with their direct manager, who knew and trusted them, but worried that higher-level managers would not approve. Tensions could arise when an employee's blog about a product or team effort developed a large external following. An employee might seek more time for blogging, irritate less visible but equally productive team members, lobby for changes by suggesting them in a visible forum, and create issues by taking a new position inside or outside the company, or otherwise discontinuing their blog.

Managers also had to assess whether time was being used productively. Bloggers who set out to post only about professional issues discovered that an occasional personal item led to a positive reader response. Authors believed it made them more effective; managers had to justify non-work-related posting in a company that had some skeptics in upper management.

An intense quest for metrics to assess the value of employee blogging did not get far. Employees who would never write a formal article for the company website blogged to provide information, share tips, and engage in discussions with customers or partners. But how is this value measured? Blogs humanized the organization, an intangible asset. Blogs were used to document and organize work. Employees reported following external blogs and discovering colleagues and ideas that they had not encountered by internal communication—again, how does one attach a numerical value to this?

In mid-2005, a few people were assigned to blog. Groups were experimenting with approaches to willfully creating a team or product blog, often contributed to and sometimes written by multiple people, with or without bylines. All involved editing, in some cases casual, in others more formal. Everyone recognized that immediacy and informality were keys to effective use of this medium.

It was at this time that IBM publicly encouraged its employees to blog, which helped validate the medium at Microsoft as well. Some full-time bloggers appeared, focused on image, marketing, and public relations.

In subsequent years, blogging has become more professional, regularly featured in newspapers. Multiauthored blogs are more prevalent. In some respects, blogs have been absorbed into the broader authored digital media, a little more spontaneous and unedited. Bloggers surveyed by Technorati aged. LiveJournal faded away. Although millions of blogs are active, focused on hobbies and other activities, the sense of immediacy passed to social network sites and microblogging (Twitter), discussed later.

45.3.3 Wiki Use in Organizations

Wikis are more constructional than conversational. The content of IM and text messages, blogs, and social network posts generally recedes untouched into the past; most wiki content is there to be edited. That said, wikis often contain conversational sections, social network sites blend persistent and conversational information, and generally there is a move toward hybrid applications.

A major organizational use of wikis is internal, as a dynamic information repository for project management or to communicate across groups. Wikis are also designed for interacting with partners, customers, or vendors (e.g., Wagner and Majchrzak, 2007). Finally, employees make use of Wikipedia and other public wikis. Published research into this is limited, but given that attorneys cite Wikipedia in legal articles and judicial opinions (People, 2009) and physicians consult it (Hughes, 2009), we can assume use is highly widespread.

Most published wiki research examines Wikipedia, followed by public wikis. Neither is the focus here, nor will findings generalize. Wikipedia is a special case, and any wiki on the web has about one million times as many potential viewers as a large organizational wiki. Vandalism is a major concern for public wikis but not inside a firewall. Information accuracy is a major issue for public wikis; in organizations, it plays out differently, focused mainly on staleness. Organizational wiki users typically also communicate via other channels (e-mail, face to face, etc.), whereas most Wikipedia discussion occurs through the tool itself. The "Academic Studies of Wikipedia" page in Wikipedia is a good starting point for those interested in that topic.

Another major focus of wiki research is on classroom use. Classes are about finding, organizing, and sharing knowledge. Students are often open to new tools. Nevertheless, results are mixed; educational institutions have routines that can create conflicts, notably around assessment (Forte, 2009). Virtually all classroom wikis are short duration, lasting one term. Classrooms bear some resemblance to start-up companies, but on the whole, they are not models for enterprise use.

Poole and Grudin (2010) classified organizational wikis. Our principal focus is on one category: topical or project-specific wikis contributed by multiple employees. A second category comprises "pedias" that span an entire organization, modeled on Wikipedia. They first appeared around 2005 when Wikipedia's success became evident. Most prominent in the media was Intellipedia, piloted in 2005 by the U.S. intelligence community and launched in 2006. MITREpedia, Pfizerpedia, and others followed. Careful studies have not been published, and recent years have seen little mass media attention to pedias (other than Wikipedia). The third wiki category comprises what are effectively personal intranet web pages. These are often started by an individual hoping to attract other contributors; when they do not materialize, either the wiki dies or the founder continues it as a solo endeavor, posting information deemed useful.

Published wiki research favors surveys of wiki users and papers written by wiki developers or evangelists in early stages of engagement. Majchrzak et al. (2006) received 165 survey responses from wiki-oriented listserv participants who felt wikis are sustainable and work better when directed at novel solutions or managed by credible sources. Interviews based on small numbers of participants and single sites identified problems familiar from earlier knowledge management system implementations: lack of management support, data that are obsolete or difficult to find, and usability problems (Ding et al., 2007; White and Lutters, 2007).

Egli and Sommerlad (2009) is an upbeat but objective report of a law firm's implementation of a wiki to support collaboration, information display, and personal use. Challenges they identified are covered later.

Phuwanartnurak (2009) describes wiki use in two short-term university IT department projects involving young, mostly wiki-savvy developers. The tool was used, though some disliked the slow updating and preferred IM for communication. This resembles the use of wikis in start-ups, also discussed later.

In a mixed survey and interview study covering thousands of wikis in several organizations, Grudin and Poole (2010) found that the vast majority of wikis were abandoned, often early. Many others never attracted a second contributor and became personal web pages on the intranet. Factors leading to abandonment and conditions conducive to success are discussed next.

With public wikis including Wikipedia, the bottom-up zeitgeist has meant that administrative structures, quality control practices, and reputation and incentive systems have emerged and evolved over time in response to the environment. The organization and reorganization of information also occur over time. Discussions happen through the wiki. This organic, self-contained process has slim chances of surviving the force fields of a dynamic organization.

Most organizations already have administrative structures, people responsible for certain topics, quality control processes, incentive and reputation systems, information repositories, and established communication channels. The existing structures cannot be grafted onto the wiki. If a known authority does not participate, others may be uncertain how to proceed. An employee may feel uncertain about editing the contribution of a senior person, whether or not that person encourages it. Should material in repositories elsewhere be duplicated in the wiki? Should communication happen in the wiki, in e-mail or distribution lists, or by convening meetings? Decisions on such matters do not bubble up and often remain unresolved.

Small start-ups are a sharp contrast. Lines of authority are ill-defined; everyone pitches in when they can. Employees are often younger, tech-savvy, and ready to try anything. There are few firmly established repositories, authorities, or communication channels. Literature cited earlier supports start-up ventures as promising for wiki use. Grudin and Poole (2010) reported successful wiki use in three start-ups—and also in two start-up-like efforts within a major corporation.

Information organization in a wiki is flexible when it is created, but often difficult to adjust later when information needs become better understood or change. Also, the formatting of content is often inflexible. These are sources of wiki abandonment. Changing a wiki in response to a group division or merger can be very difficult, as can the extraction of information for other uses. Wikis that use a global namespace hinder extension. Broken links and information that is obsolete or difficult to find are chronic problems. As with other file systems or document repositories, structure is critical, and the requirements are often not knowable or carefully worked out when a wiki is created.

Another major issue derives from Mintzberg: Even when they all agree the wiki will be a wonderful critter, executives, managers, and ICs have different views of exactly what it will be. Executives may envision the end of the elusive knowledge management quest to capture the knowledge or retiring employees for reuse by others. (This will not happen.) Managers may anticipate a "project dashboard" maintained by their team. (This is unlikely as well.) ICs may approach it as a resource for ad hoc problem solving, posting answers to frequently asked questions and other information useful to people like themselves. (This is the most likely success, if disillusioned executives and managers do not pull the plug.) Grudin and Poole (2010) found that enthusiasm could be strong at the top and bottom of organizations, with managers caught in the middle, trying to deal with the inevitable disruptions accompanying a new system.

Danis and Singer (2008) describe a 2-year-old enterprise wiki that had an exceptionally strong design team. It also benefited from an executive who was a strong proponent and no alternative to adoption: It replaced a system that was removed. The wiki was designed and deployed by an IBM research group for use in an annual project proposal review. The executive envisioned it being used year-round, but this did not materialize, and when he later retired, the wiki was retired as well (personal communication). Early stages in the migration of a document repository to a wiki are also described in Alquier et al. (2009).

45.3.4 Social Networking at Work

A decade after they began to attract notice, web-based social networking sites are being used by over 15% of the world's population. The prominent initial public offerings (IPOs) of LinkedIn and Facebook signal that social networking is considered reasonably mature. Their use by politicians, entertainers, and athletes helps legitimate them to organizations, which in any case have more difficulty keeping their use out than bringing it in.

Early terminal-based computers and the early Internet supported social activity. From 1995 to 1997, ICQ and AOL Instant Messenger were released and phone-based text messaging ramped up, acquainting more people with buddy lists and real-time or quasi-real-time digital communication. Classmates. com and Six Degrees were early efforts to use the web to provide a persistent device-independent online home for social activity.

Activity increased substantially 5 years later; perhaps the Internet bubble had to grow and burst, leaving many more people online and looking for something to do there. From 2002 to 2004, Cyworld, Friendster, Plaxo, Reunion.com, Hi5, LinkedIn, MySpace, Orkut, Facebook, and Microsoft Live Spaces were released and actively promoted. Different sites became prominent in different regions.

By 2013, the field has shaken out. Facebook, LinkedIn, and Twitter (which appeared in 2006) dominate outside China. Specialized new sites include Foursquare, Pinterest, Instagram, and Google+. Yammer is a prominent vendor of social networking that keeps exchanges within a corporate firewall. It was acquired by Microsoft in 2012.

Early use by students: Students were enthusiastic early adopters. Academic achievement and socializing are key student activities and social networking sites help with at least one of these. Students tend to start with less complex social networks, and early clients were not complex.

Colorful and chaotic MySpace was once the most active site. Studies of high school and college students (Hargitai, 2007; Boyd, 2008) found that students aiming at professional occupations tended to consider MySpace gaudy and not serious. Use declined in favor of Facebook.

Facebook famously began on college campuses. Student Facebook networks initially resembled IM or SMS buddy lists (Lampe et al., 2006, 2008). Many students felt faculty should be excluded, with women particularly concerned about privacy (Hewitt and Forte, 2006; Fogel and Nemad, 2009). Over time, fewer students allowed open access to their Facebook profiles, with women reporting more use of access control and long-time users less concerned (Tufekci, 2012). Attitudes to privacy depended on whether students networked primarily to communicate with existing friends or to find people with common interests.

Facebook was made available in anticipation-generating waves to elite universities, other universities, and high schools, and in 2006, employees of a few enterprises followed by general availability. When LinkedIn was launched in 2003, it asked for a person's occupation, employer, and past university, which discouraged student use. Its growth was deliberate but steady. It served job seekers, recruiters, and also professionals interested in a free, professional-looking web page and self-updating address book.

Use by the general population: The age, ethnicity, and socioeconomic profiles of the general population of social networking site users differ from most enterprise employee profiles. In addition, most studies rely on surveys of self-selected individuals. Enterprise use contexts differ in other ways; notably, anonymity is common in general web use but not on corporate intranets.

However, some studies of general use foreshadow results found in enterprise settings and discussed later. Naaman et al. (2010) found clusters of Twitter users: Some focused on posting about themselves and their activities, others primarily relaying information. Regrets expressed following Facebook posts remind us that technologies mature through trial and error (Wang et al., 2011). Similarly, flaming was a major topic in early e-mail use.

Media attention tends to focus on potential problems: students posting material they will regret, people fired for postings deemed inappropriate, and possible misuse of personal information by site owners. The growth in popularity of the medium indicates that this has not markedly affected uptake, but it could shape behavior in subtle ways.

Workplace studies: Many studies of enterprise use of social networking address internal systems confined to a company's employees. Most of these are prototype systems built and deployed at IBM, such as Beehive (later called SocialBlue), Blue Twit, Cattail, Dogear, and Timely.

A second set of studies examines employee use of sites such as Facebook, LinkedIn, and Twitter for work (and other) purposes. Some uses are directed outward—recruiting or public relations—and others inward—to build social capital, find answers to nonconfidential questions, and so forth.

Internal prototypes have advantages: With restricted membership, employees can discuss company-confidential information. Researchers can log and analyze use and conduct follow-up interviews. Drawbacks include limited participation—typically a few percent of enthusiastic employees who are willing to invest time in a system likely to be discontinued. This limits the generality of results, even within the enterprise deploying it.

45.3.4.1 Case Study: IBM Beehive (2007–2011)

In 2007, IBM launched a Facebook-like social networking system called Beehive. In 2008 and 2009, 15 papers on different aspects were published, with more to follow. At the time of the first published studies, fewer than 0.1% of IBM employees were using Beehive (DiMicco et al., 2009); toward the end 17% had profiles (Thom-Santelli et al., 2011), although the final study indicated that fewer than 1% of employees posted once in a 4-week period (Thom et al., 2012). The system was taken offline in early 2012, with some features incorporated into IBM Connections.

Nine months after deployment, employees used Beehive to share personal information, to promote themselves by describing skills and accomplishments, and to campaign for projects (DiMicco et al., 2008). It was not used to find information or get quick answers to questions. In a finding frequently reported, many people initially used it with close colleagues but later found it more useful for establishing weaker ties with other colleagues. They found new contacts through "friends-of-friends" and did not experience privacy concerns. Other popular uses included sharing structured lists (Geyer et al., 2008a).

Several efforts aimed to increase participation. A ratings-based incentive system had a short-term effect (Farzan et al., 2008, 2009); removing it had a strong negative effect (Thom et al., 2012). Recommender systems were explored in efforts to (1) increase participation by recommending expansion of profile information (Geyer et al., 2008b), (2) identify and suggest new contacts (Chen et al., 2009; Guy et al., 2009; Daly et al., 2010); (3) increase engagement (Freyne et al., 2009), and (4) direct people to items of interest (Freyne et al., 2010). More profile information correlated with more contacts (Dugan et al., 2008). An interesting finding was that personal network structures could be inferred more effectively from outward-facing social network information than from relatively private e-mail thread data (Guy et al., 2008).

A major theme was the use of Beehive to build social capital (Steinfield et al., 2009; Wu et al., 2010). There was evidence of utility for organizational acculturation (Thom-Santelli et al., 2011), though communication was largely within geographical region (Thom-Santelli et al., 2010). Employees expressed the wish that Beehive was integrated with other social networking sites (e.g., Facebook).

In early 2009, Zhang et al. (2010) studied one organization's use of Yammer, the Twitter-like tool for internal corporate use (with less constrained post length) launched in 2008. Yammer was used by 1.5% of employees to broadcast group or business unit status, ask and answer questions, send directed messages for real-time interaction, relay items of interest, and follow individual posters. Hashtag use to identify topic was rare. Almost all users were over 30 years old; about a quarter were also active Twitter users. With such limited adoption, even enthusiastic users had difficulty finding value.

Yammer reports use by employees of 200,000 companies, including over 400 of the Fortune 500, with around 4 million users or about 20 employees per company on average (http://www.yammer.com, 10/5/2012). It remains unclear how social networking inside a firewall will fare.

Higher uptake of internally developed social networking systems such as D Street at Deloitte is reported in the press. A complex social business platform at MITRE was the focus of a well-executed study employing logs, surveys, and interviews after 2 years of use (Holtzblatt et al., 2012). The Handshake system includes blog, wiki, file sharing, and social network features. Of the 60% of employees who tried it, benefits were largely reaped by the 18% who used it actively. Benefits were attributed to standard social networking features, the integration of capabilities that include heavier-weight collaboration around documents, and the ability to create groups that span the firewall to include customers. Focusing more on organizational and technical challenges, Rooksby and Sommerville (2012) describe internal and public site use in a large government agency. The appropriateness and risks of accessing public sites was a key issue.

Corporate uses of Facebook, Twitter, and LinkedIn: A quarter to a third of companies reportedly ban employee access to external social networking sites from the workplace (e.g., Purdy, 2012). However, employees can circumvent bans using smart phones. As with Internet, e-mail, and web use, prohibition is unlikely to persist. Past experience indicates that employers who can trust employees not to misuse new technologies benefit overall from employee use. Still, employee uncertainty about appropriateness affects behavior, as reported in Rooksby et al. (2012). Lampe et al. (2012), unexpectedly not finding Facebook used by university staff for work-related questions and answers, attributed it to such concerns.

Apart from the Microsoft case study provided later, research into enterprise uses of public social networking sites has focused on early adopters and heavy users. This provides a view of how sites *can* be used and some issues that are encountered but limits the generality of the findings, as does the rapid uptake of these technologies: Every year the landscape changes, with more users, new features and options, and higher levels of experience. Compare studies that were conducted in different years with caution. That said, some trajectories are evident.

In an early study of Facebook use in an enterprise, DiMicco and Millen (2007) identified three categories of Facebook profiles of young professionals moving from college to the workplace: "reliving the college days," comprising personal information, informal status messages, use of the Wall, and nonprofessional images; "dressed to impress," primarily job-related information with some personal information and formal images; and "living in the business world," consisting of limited profiles from new users as opposed to those who first adopted it as students.

Zhao and Rosson (2009) recruited 11 heavy users of Twitter at a large IT company in late 2008, using personal contacts and "snowball" referrals. Anticipating the Yammer findings described earlier, Twitter was used by these enthusiasts for "life updates" or personal status, timely sharing of information with friends or colleagues, and as "personal RSS feeds" to monitor trusted external sources of news.

In early 2009, Ehrlich and Shami (2010) compared Twitter and BlueTwit, a short-lived internal IBM tool that allowed posts of 250 characters. BlueTwit had been adopted by one-third of 1% of all employees. Thirty-four active users of both tools were identified. Fifty-seven percent of their posts were from five users who averaged 18 per day. On BlueTwit, they did less status posting and directed more information and comments to specific individuals than on Twitter. How to generalize from this sample of the 400,000 employees is unclear. A widely publicized Twitter-based event lasting a few days attracted stories conveying the company's culture (Thom and Millen, 2012). Although only about one-fifth of 1% of employees posted stories, it was considered a success.

Surveys of communication at a small company in mid-2008 and mid-2009 found Facebook, LinkedIn, and Twitter use increasing, with weekly use of Facebook and Twitter the norm (Turner et al., 2010). The authors anticipated that use of Twitter would thrive. Voida et al. (2012) is a study of the use of sites to communicate externally; they examined volunteer organizations and found them using sites to raise funds, publicize events, and recruit volunteers, but sites were not useful for volunteer coordinators, whose need was to focus on individuals and build strong ties.

45.3.4.2 Case Study: Cross-Sectional Trend Study of Microsoft (2008–2012)

Each April starting in 2008, a different random sample of 1,000 Microsoft employees (out of about 90,000) was invited to take a survey on communication practices. The survey explored their attitudes toward and behaviors around public social networking sites. Over 40% responded each year; their demographics matched those of the company as a whole. Employees varying in age, gender, role, level, geographic location, attitude toward social networking, and behavior were subsequently interviewed three of the years. Skeels and Grudin (2009) describes the first year, Archambault and Grudin (2012) the first four, and Zhang et al. (under review) the fifth year.

Microsoft is atypical but varied: 45% of the employees are in business groups, 45% are in sales and marketing, and 10% support operations. Almost half are in the Seattle area and the rest are distributed around the world. In 2008, 49% had Facebook profiles and 52% had LinkedIn profiles, the latter used less frequently. By 2012, over 80% of all age groups had Facebook profiles. Over 50% reported being daily users of Facebook, and 20% *daily* users of LinkedIn, primarily in sales and marketing divisions.

In 2008, a major issue was the tensions arising from having contacts from different facets of life or who spanned other boundaries—friends versus coworkers, family members, what do you do when your manager, vice president, or customer asks to be your friend? People tended to use just one site. Access control mechanisms were just appearing. By 2012, use was more sophisticated—people had set up different Facebook groups and used LinkedIn groups, used Facebook pages, and used Facebook, Twitter, and/or LinkedIn differentially. People wanted to be able to scan all their sites in one place; for example, on their phone as they waited in a queue. Concerns about privacy diminished as people felt they were establishing more control over their networks. One finding was that female employees made more use of access control features and expressed fewer concerns about the use of sites than male employees did.

The segmenting of online social worlds is discussed by many researchers, most incisively by Stutzman and Hartzog (2012) who studied a special population: adults with multiple identities on a single site. Although a highly atypical group, why and how their informants differentiated audiences and used their identities is consistent with the observations of Microsoft employees.

Are social networking sites useful for external professional networking? Many employees were unsure in 2008 but by 2012 agreed. For networking within the company, however, even in 2012, 20% disagreed and almost 30% were neutral. In 2008, managers were less likely to see internal utility; by 2011, managers were more likely than ICs to see the potential for internal use. For those seeing social networking as useful, strengthening ties with coworkers was from the start a strong rationale.

By 2011–2012, Facebook use had plateaued at 60% daily use and around 80% occasional use. Twitter use plateaued in 2010 with close to 10% of employees using it daily. Occasional Twitter use continued to grow, reaching 30% so far; people used it to track work-related and personal events and feeds and to publicize work-related developments. In conclusion, social networking was integrated into the personal and work lives of many, who showed no indication that they might abandon it. Instead, they took countermeasures when developments made them uneasy. Despite mass media stories of overload and burnout, social networking appears to be here to stay.

45.4 Discussion

When the Internet bubble burst in 2000 and 2001, it left behind a strong digital infrastructure. Businesses and homes had acquired computers and Internet access. Moore's law was not revoked, so capability grew and prices fell. The communication media discussed in this chapter capitalized on the infrastructure and critical mass of users remaining when loftier ambitions were swept away. This is one interpretation of the timing of the embrace of these innovations, along with Wikipedia, Delicious, Flickr, YouTube, and others.

The pace of experimentation and change is unlikely to be maintained. A vacuum was filled. The technologies are maturing. The high valuations of LinkedIn and Facebook indicate an expectation that they will prosper. Hundreds of millions of users are accumulating years of experience, finding benefits and comfort levels. Most new entries focus on niches—Instagram for pictures, Pinterest for web collages, and Foursquare for location check-in—rather than challenging Facebook in the way it challenged Friendster, Orkut, Bebo, MySpace, and others. Moore's law still has not been revoked. Smaller form factors are more important, but they too extend rather than replace existing practices.

45.4.1 Contagion

An innovation requires a critical mass willing to see through the initial disruption as effective practices are worked out and promulgated. Communication technologies have one advantage: Best practices can be shared through the technology itself.

A student cohort is ideal for a communication technology—a small group that communicates intensely, is young and willing to experiment, and has relatively few external ties and habits. One of the cohorts is excited about the possibilities of a new technology, and friends will give it a try. Start-ups share these qualities—an intense need to communicate, inherently experimental and youthful composition, a focus that blocks out other ties, few habits built up around other channels and repositories, and relatively little hierarchy.

For IM, weblogs, and social networking sites, achieving use was not the challenge. Consumer clients or platforms were quickly adopted. One challenge for enterprises is determining whether to embrace the technologies. If they opt to do so, they must either determine how to make use of the clients and platforms employees are already using, convince employees to shift to or also adopt corporate versions, or find a way to integrate the two.

Wikis are an exception. Most employees, even young ones, are not yet active wiki contributors, so it is a more traditional adoption context. Start-ups are promising candidates for the reasons listed. Adoption in mature organizations requires other approaches to achieving critical mass. A project that resembles a start-up—a new team or new assignments, no existing repositories, and preferred communication channels—is a good candidate.

On a national scale, the breakup of the Soviet bloc unleashed an entrepreneurial ferment among young professionals: ICQ IM and LiveJournal blogs, used overwhelmingly by students in the United States, established a strong presence in eastern European enterprises.

The importance of evangelism and executive backing in spurring adoption is often stressed for large enterprises in relatively individualistic or command-and-control cultures. Use may spread more contagiously in consensus-oriented cultures. This may be why some blogging and mobile technology adoption in Asia transcends age and occupation to a greater extent than in Europe and North America.

These emerging technologies benefited from two external developments. One is the uncertainty that surrounded the rise of the web. An anxious search for business models that would be viable in a digitally connected world opened doors for experimentation. Another is the successful use of the technologies by people in entertainment, government, and crisis situations. Examples are the use of Twitter by Oprah Winfrey, star athletes, and popular entertainers, and the complaints of the Obama team that they could not use familiar communication tools when entering the White House in January 2009—which revealed that they truly relied on these channels. Studies of use of blogging and social networking sites in Middle East crises (e.g., Al-Ani et al., 2012; Starbird and Palen, 2012) are evidence that these tools can be powerful in certain contexts.

45.4.2 Evolution

Over time, three aspects of the information in these channels changed: its formality, its progression from text to multimedia, and the sizes of intended audiences. Each presents a trade-off. Quick, informal information exchange can be useful; in other contexts, organized, formal documents can be essential. Text is economical and often sufficient; photos and other multimedia are engaging. A small targeted group was convenient and may be desirable, but some messages are intended for broad audiences, and we do not always know when and for whom one will be useful.

Information exchanged informally sometimes proves to be useful later, but when communication is routinely saved and could be forwarded, a medium becomes more formal. Spell checkers are added. It is subpoenaed in court cases. E-mail started informally and became formal. IM and text messaging moved into the informal communication niche; now they too are saved and subpoenaed.

For decades, most digital communication was text, but with websites came multimedia. At work as well as home, people love photos. Cell phone buddy lists and IM contacts were initially limited in number, but websites for maintaining lists of contacts and followers changed this.

These factors drove the progression from messaging to social network sites for some purposes: from limited social networks of friends to segmented audiences that can be useful for work. Many early bloggers addressed friends and family only or broadcasted to the public. Enterprises slowly engaged with blogging, internally or externally, seeking large but targeted audiences. Twitter feeds may be impacting heavy-weight corporate blogs, which require more time to construct and read. For most purposes, 140-character announcements of products or events are easier for everyone involved.

45.4.3 Approaches to Organizational Deployment

The single most important consideration when considering a new communication technology intended for wide use is the differing perspectives of ICs and managers. It is not that they are at loggerheads—their desired outcomes are usually shared. They differ in how they work, what will help them, and how they perceive each other working. They have different views of what a tool will do. Understand and address this during design and deployment and the odds of smooth adjustment should improve substantially.

Wikis are a clear example. On the surface, wikis appeal to both groups: Information can be structured and anyone can contribute. In practice, the structures a manager envisions may not align with the content individuals find useful to contribute. Manager and IC interests converge in posting answers to questions frequently asked by new employees streaming into a start-up. In other situations, lack of clear credit and accountability interfere with open editing. If a Wikipedia page has erroneous information, no one's job is on the line, but in an enterprise it can be different.

Executives may believe they know how managers should work and managers may believe they understand ICs, but understanding how a new technology will fit into one's own workflow is hard enough. Imagining it for someone in a different role is beyond reach. If employees can be trusted or lapses detected, letting people find the tools useful to them may be advisable, with high-level modeling, encouragement, and evangelism to get it bootstrapped.

The corollary is that an environment with diverse, interoperable tools available in one place is promising. Companies are building and experimenting with integrated platforms such as MITRE's Handshake and Deloitte's D Street. Experiences with an integrated system at Hewlett Packard are described in Brzozowski (2009) and Brzozowski et al. (2009). Commercial offerings are still taking shape, such as IBM Connections, a social networking capability that added blog, wiki, microblogging, forums, and communities over time, absorbing some of the functionality in research prototypes described earlier, and Sharepoint, Microsoft's document repository system that has also added blog, wiki, messaging, communities, and team sites.

Integrated internal systems have not yet fully proven themselves. The functionality is there, and organizations are trying them. The Holtzblatt et al. (2012) case study may be illustrative. After 2 years, 18% of employees reported benefits. This is significant, but short of the ultimate goal. Whether use will end up as widespread as e-mail is an empirical question.

45.4.4 Prospects for Enterprise Communication Technologies

In the consumer sphere, a decade of experimentation, exploration, and rapid change seems to have given way to a period of integration and consolidation. Organizations still face significant uncertainties, however. Should messaging remain a general platform or be embedded in different applications? Will an internal pedia or internal social networking platform be used, or should the focus be on making use of external tools? Can wiki and blog features address the chronic challenge of organizing information so that it can be retrieved with appropriate context? Other major challenges include addressing tensions and opportunities around work–life balance for individuals and cross-enterprise communication with suppliers, customers, and partners.

Fears that accompanied the arrival of individual communication tools are dissipating. Trepidation is reduced as workers become more technology-savvy. They may as likely to see unmet potential as they are to imagine the worst.

People are using more tools and consequently value the capability to select among information threads rapidly and efficiently. The solution is not to merge all communication into a single channel; it is to provide a single place in which to quickly skim over or select among feeds, drawing on different channels for different purposes. Prospects for a new service improve if it appears alongside familiar services. The "place" can be a sequence of phone app buttons, a desktop window that aggregates multiple sources, or a large persistent display sitting to the side of one's primary work displays. At the same time, highly relevant information should also appear in context.

People report finding useful information and contacts within their organization through reading public blogs or Twitter streams issued by their organization. Is this evidence of the inadequacy of internal tools, a strength of external tools that should be exploited, or both? By analogy, if a vendor finds that people encountering a problem with their product use a search engine to find solutions on the web and do not go to the vendor's site for help, should the vendor double down on their help site, focus on improving the prospects for search engine users, or let the crowd handle it?

An organization could focus on improving internal tools or assist employees using external sites. Microsoft took the latter approach with externally facing blog servers and their new Social Connector, which enables access to Facebook, Twitter, and LinkedIn from within Office. This could impede a shift to internal tools—but for what organizations should or will such a shift take place?

Information threads—streams or rivers might be more apt metaphors—include not only communication with other people, but also meta-data and compilations of traffic from people, organizations, and objects, including video, sensor data, and other sources. No one has time to track everything in their work and personal lives that is potentially of interest. As the proportion represented digitally grows and the costs of displays, bandwidth, and computation drop, we can monitor more of it. Tools will enable faster access and better display, and new interaction models will enable us to scan information quickly and expand it as desired.

In this time of uncertainty and opportunity, enterprises that find solutions that fit their needs could do very well. The choices may be difficult. For example, tools for monitoring employee activity are improving rapidly. They could help but can create dilemmas for managers or IT staff who use them. Casual banter in a workplace probably increases social capital and strengthens ties in ways that enable employees to work more effectively; it probably builds skills and expands contacts that can later be drawn on for work purposes, but it also distracts people from their principal tasks. How much should managers support? Benefits are difficult to measure. Managers rarely have detailed enough knowledge of their subordinates' jobs to confidently draw lines, being too strict could be counterproductive if it drives employees to use personal devices to communicate during work hours.

Organizations that for legal or competitive reasons must restrict communication among employees or with the outside world face greater challenges as technology continues to dissolve boundaries. Any website open to posting creates a simple bridge between people who might otherwise have had great difficulty finding one another.

Another thorny issue is information preservation. It is cheap to record and archive just about everything. The cost of archiving video communication may currently be noticeable, but the trend is toward effectively free storage. Archiving has benefits—it is not always possible to know what will later be of use. Information can be mined to find ways to help people work more effectively. It also has risks—not only clutter and difficulty finding what *is* useful, but also in potential personal or organizational liability for what is said or done, especially when retrieved out of context. For example, excerpts from years of recorded conversations and meetings could be used to build a case for poor performance or for a hostile work environment.

Historically organizations have preserved essential information in structured documents vetted by managers. Most communication was ephemeral. Today much more is saved, and it could all be saved. Tools are available that automatically delete information on a set date; we can anticipate struggles over where to draw boundaries among individual workers, mangers, legal staff, and IT staff. The decisions may differ according to the nature of the activities involved—the legal restrictions, potential for liability, and evolving case law.

Despite these complexities, on balance, new communication and collaboration technologies represent astonishing tools for living and working more productively and pleasantly. It is tempting to imagine that it will be easy to get there. It will not be. The record of the past decade is that we are moving forward; new tools are being picked up, tried, and improved. Many are proving useful in workplaces. A learning curve is behind us; the benefits, along with new challenges, lie ahead.

References

Al-Ani, B., Mark, G., Chung, J., and Jones, J. 2012. The Egyptian blogosphere: A counter-narrative of the revolution. *Proceedings of Computer Supported Cooperative Work 2012*, pp. 17–26. ACM Press, New York.

Alquier, L., McCormick, K., and Jaeger, E. 2009. knowIT, a semantic informatics knowledge management system. *Proceedings of the International Symposium on Wikis and Open Collaboration, WikiSym'09*, Article 20. ACM Press, New York.

Archambault, A. and Grudin, J. 2001. A longitudinal study of Facebook, LinkedIn, & Twitter use. *Proceedings of CHI 2012*, pp. 2741–2750. ACM Press, New York.

Bar-Ilan, J. 2004. An outsider's view on "topic-oriented blogging." *Proceedings of the WWW 2004*, pp. 28–34. ACM Press, New York.

Boyd, D. 2008. Taken out of context: American teen sociality in networked publics. PhD dissertation, University of California-Berkeley, Berkeley, CA.

Brzozowski, M.J. 2009. WaterCooler: Exploring an organization through enterprise social media. *Proceedings of GROUP 2009*, pp. 219–228. ACM Press, New York.

Brzozowski, M.J., Sandholm, T., and Hogg, T. 2009. Effects of feedback and peer pressure on contributions to enterprise social media. *Proceedings of GROUP 2009*, pp. 61–70. ACM Press, New York.

Chen, J., Geyer, W., Dugan, C., Muller, M., and Guy, I. 2009. Make new friends, but keep the old: Recommending people on social networking sites. *Proceedings of CHI 2009*, pp. 201–210. ACM Press, New York.

Cone, E. 2005. Rise of the Blog. *CIO Insight*, April 5, 2005. http://www.cioinsight.com/c/a/Past-News/Rise-of-the-Blog/ (accessed on March 26, 2013).

Daly, E.M., Geyer, W., and Millen, D.R. 2010. The network effects of recommending social connections. *Proceedings of the RecSys 2010*, pp. 301–304. ACM Press, New York.

Danis, C. and Singer, D. 2008. A wiki instance in the enterprise: Opportunities, concerns, and reality. *Proceedings of Computer Supported Cooperative Work 2008*, pp. 495–504. ACM Press, New York.

Dennis, A.R. and Reinicke, B.A. 2004. Beta versus VHS and the acceptance of electronic brainstorming technology. *MIS Quarterly*, 28(1), 1–20.

DiMicco, J., Geyer, W., Millen, D.R., Dugan, C., and Brownholtz, B. 2009. People sensemaking and relationship building on an enterprise social networking site. *Proceedings of HICSS 2009*. IEEE Computer Society, Washington, DC.

DiMicco, J.M. and Millen, D.R. 2007. Identity management: Multiple presentations of self in Facebook. *Proceedings of Group 2007*. ACM Press, New York.

DiMicco, J., Millen, D.R., Geyer, W., Dugan, C., Brownholtz, B., and Muller, M. 2008. Motivations for social networking at work. *Proceedings of Computer Supported Cooperative Work 2008*. ACM Press, New York.

Ding, X., Danis, C., Erickson, T., and Kellogg, W.A. 2007. Visualizing an enterprise wiki. *Proceedings of CHI'07*, pp. 2189–2194. ACM Press, New York.

Dudley, B. 2004. Bill Gates could join ranks of bloggers. *Seattle Times*, June 25, 2004. http://seattletimes.com/html/businesstechnology/2001964841_gatesblog25.html (accessed on March 26, 2013).

Dugan, C., Geyer, W., Muller, M., DiMicco, J., Brownholtz, B., and Millen, D.R. 2008. It's all 'about you': Diversity in online profiles. *Proceedings of Computer Supported Cooperative Work 2008*, pp. 703–706. ACM Press, New York.

Efimova, L. and Grudin, J. 2008. Crossing boundaries: Digital literacy in enterprises. In C. Lankshear and M. Knobel (Eds.), *Digital Literacies*, pp. 203–226. Peter Lang, New York.

Egli, U. and Sommerlad, P. 2009. Experience report—Wiki for law firms. *Proceedings of International Symposium on Wikis and Open Collaboration, WikiSym'09*, Article 19. ACM Press, New York.

Ehrlich, K. and Shami, N.S. 2010. Microblogging inside and outside the workplace. *Proceedings of the International AAAI Conference on Weblogs and Social Media 2010*. AAAI Press, Palo Alto, CA.

Farzan, R., DiMicco, J.M., and Brownholtz, B. 2009. Spreading the honey: A system for maintaining an online community. *Proceedings of GROUP 2009*, pp. 31–40. ACM Press, New York.

Farzan, R., DiMicco, J.M., Millen, D.R., Brownholtz, B., Geyer, W. and Dugan, C. 2008. Results from deploying a participation incentive mechanism within the enterprise. *Proceedings of CHI 2008*, pp. 563–572. ACM Press, New York.

Fogel, J. and Nemad, E. 2009. Internet social network communities: Risk taking, trust, and privacy concerns. *Computers in Human Behavior*, 25(1), 153–160.

Forte, A. 2009. Learning in public: Information literacy and participatory media. PhD dissertation, School of Interactive Computing, Georgia Institute of Technology, Atlanta, GA.

Freyne, J., Berkovski, S., Daly, E.M., and Geyer, W. 2010. Social networking feeds: Recommending items of interest. *Proceedings of RecSys 2010*, pp. 277–280. ACM Press, New York.

Freyne, J., Jacovi, M., Guy, I., and Geyer, W. 2009. Increasing engagement through early recommender intervention. *Proceedings of RecSys 2009*, pp. 85–92. ACM Press, New York.

Geyer, W., Dugan, C., DiMicco, J., Millen, D.R., Brownholtz, B., and Muller, M. 2008a. Use and reuse of shared lists as a social content type. *Proceedings of CHI 2008*, pp. 1545–1554. ACM Press, New York.

Geyer, W., Dugan, C., Millen, D.R., Muller, M., and Freyne, J. 2008b. Recommending topics for self-descriptions in online user profiles. *Proceedings of RecSys 2008*, pp. 59–66. ACM Press, New York.

Grudin, J. 2004a. Managerial use and emerging norms: Effects of activity patterns on software design and deployment. *Proceedings of HICSS-36*, 10pp. IEEE, New York. http://dl.acm.org/citation.cfm?id=962749.962778 (accessed on April 13, 2013).

Grudin, J. 2004b. Return on investment and organizational adoption. *Proceedings of Computer Supported Cooperative Work 2004*, pp. 274–277. ACM Press, New York.

Grudin, J. and Poole. E. 2010. Wikis at work: Success factors and challenges for sustainability of enterprise wikis. *Proceedings of International Symposium on Wikis and Open Collaboration, WikiSym 2010*, Article 5. ACM Press, New York.

Grudin, J., Tallarico, S., and Counts, S. 2005. As technophobia disappears: Implications for design. *Proceedings of GROUP 2005*, pp. 256–259. ACM Press, New York.

Guy, I., Jacovi, M., Meshulam, N., Ronen, I., and Shahar, E. 2008. Public vs. Private—Comparing public social network information with email. *Proceedings of Computer Supported Cooperative Work 2008*, pp. 393–402. ACM Press, New York.

Guy, I., Ronen, I., and Wilcox, E. 2009. Do you know?: Recommending people to invite into your social network. *Proceedings of IUI 2009*, pp. 77–86. ACM Press, New York.

Hargittai, E. 2007. Whose space? differences among users and non-users of social network sites. *Journal of Computer-Mediated Communication*, 13(1), Article 14. Wiley-Blackwell, Hoboken, NJ.

Hewitt, A. and Forte, A. 2006. Crossing boundaries: Identity management and student/faculty relationships on the Facebook. *Proceedings of Computer Supported Cooperative Work 2006 Poster*, Atlanta, GA.

Holtzblatt, L., Damianos, L.E., Drury, J.L., Cuomo, D.L., and Weiss, D. 2012. Evaluation of the uses and benefits of a social business platform. *LIFE-FORCE 2012 Extended Abstracts*, pp. 721–736. ACM Press, New York.

Hughes, B. 2009. Junior physician's use of Web 2.0 for information seeking and medical education: A qualitative study. *International Journal of Medical Informatics*, 78(10), 645–655.

Jones, W.P., Bruce, H., and Dumais, S.T. 2001. Keeping found things found on the web. *Proceedings of CIKM 2001*. ACM Press, New York, pp. 119–126.

Lampe, C., Ellison, N., and Steinfield, C. 2006. A Face(book) in the crowd: Social searching vs. social browsing. *Proceedings of Computer Supported Cooperative Work 2006*, pp. 167–170. ACM Press, New York.

Lampe, C., Ellison, N., and Steinfield, C. 2008. Changes in use and perception of Facebook. *Proceedings of Computer Supported Cooperative Work 2008*, pp. 721–730. ACM Press, New York.

Lampe, C., Vitak, J., Gray, R., and Ellison, N. 2012. Perceptions of Facebook's value as an information source. *Proceedings of CHI 2012*, pp. 3195–3204. ACM Press, New York.

Lovejoy, T. and Grudin, J. 2003. Messaging and formality: Will IM follow in the footsteps of email? *Proceedings of INTERACT 2003*, pp. 817–820. IOS Press, Amsterdam, the Netherlands.

Majchrzak, A., Wagner, C., and Yates, D. 2006. Corporate wiki users: Results of a survey. *Proceedings of International Symposium on Wikis and Open Collaboration*, *WikiSym'06*, pp. 99–104. ACM Press, New York.

McGrath, J.E. 1991. Time, interaction, and performance (TIP): A theory of groups. *Small Group Research*, 22(2), 147–174.

Mintzberg, H. 1984. A typology of organizational structure. In D. Miller and P.H. Friesen (Eds.), *Organizations: A Quantum View*, pp. 68–86. Prentice-Hall, Englewood Cliffs, NJ. Reprinted in R. Baecker (Ed.), *Readings in Computer Supported Cooperative Work and Groupware*. Morgan Kaufmann, San Mateo, CA, 1995.

Munkvold, B.E. 2002. *Implementing Collaboration Technologies in Industry: Case Examples and Lessons Learned*. Springer, New York.

Naaman, M., Boase, J., and Lai, C.H. 2010. Is it really all about me? Message content in social awareness streams. *Proceedings of Computer Supported Cooperative Work 2010*, pp. 189–192. ACM Press, New York.

Nardi, B.A., Schiano, D.J., and Gumbrecht, M. 2004. Blogging as social activity, or, would you let 900 million people read your diary? *Proceedings of Computer Supported Cooperative Work 2004*, pp. 222–231. ACM Press, New York.

Orlikowski, W. 1992. Learning from notes: Organizational issues in groupware implementation. *Proceedings of Computer Supported Cooperative Work 1992*, pp. 362–369. ACM Press, New York.

Palen, L. 1998. Calendars on the new frontier: Challenges of Groupware Technology. Dissertation, University of California, Irvine, CA.

Palen, L. and Grudin, J. 2002. Discretionary adoption of group support software: Lessons from calendar applications. In B.E. Munkvold (Ed.), *Organizational Implementation of Collaboration Technology*, pp. 68–86. Springer, New York.

Peoples, L.F. 2009–2010. The citation of Wikipedia in judicial opinions. *Yale Journal of Law and Technology*, 12(1), 38–39.

Perin, C. 1991. Electronic social fields in bureaucracies. *Communications of the ACM*, 34(12), 75–82.

Phuwanartnurak, A.J. 2009. Did you put it on the wiki? Information sharing through wikis in interdisciplinary design collaboration. *Proceedings of SIGDOC'09*, pp. 273–280. ACM Press, New York.

Pickering, J.M. and King, J.L. 1992. Hardwiring weak ties: Individual and institutional issues in computer mediated communication. *Proceedings of Computer Supported Cooperative Work 1992*, pp. 356–361. ACM Press, New York.

Poltrock, S. and Grudin, J. 2005. Videoconferencing: Recent experiments and reassessment. *Proceedings of HICSS'05*, 10pp. IEEE Computer Society, Washington, DC.

Poole, E. and Grudin, J. 2010. A taxonomy of wiki genres in enterprise settings. *Proceedings of International Symposium on Wikis and Open Collaboration, WikiSym 2010*, Article 14. ACM Press, New York.

Purdy, C. 2012. More than half of workers say their companies allow access to social networking, shopping, and entertainment sites. Monster. http://www.monsterworking.com/2012/09/21/more-than-half-of-workers-say-their-companies-allow-access-to-social-networking-shopping-and-entertainment-sites (accessed on March 26, 2013).

Rooksby, J. and Sommerville, I. 2012. The management and use of social network sites in a government department. *Computer Supported Cooperative Work*, 21(4–5), 397–415.

Skeels, M. and Grudin, J. 2009. When social networks cross boundaries: A case study of workplace use of Facebook and LinkedIn. *Proceedings of GROUP 2009*, pp. 95–104. ACM Press, New York.

Snell, J.A. 2005. blogging@ibm. Downloaded May 26, 2012, from https://www.ibm.com/developerworks/mydeveloperworks/blogs/jasnell/entry/blogging_ibm (accessed on March 26, 2013).

Starbird, K. and Palen, L. 2012. (How) will the revolution be retweeted?: Information diffusion and the 2011 Egyptian uprising. *Proceedings of Computer Supported Cooperative Work 2012*, pp. 7–16. ACM Press, New York.

Steinfield, C., DiMicco, J.M., Ellison, N.B., and Lampe, C. 2009. Bowling online: Social networking and social capital within the organization. *Proceedings of Communities and Technology 2009*. ACM Press, New York.

Stutzman, F. and Hartzog, W. 2012. Boundary regulation in social media. *Proceedings of Computer Supported Cooperative Work 2012*, pp. 769–778. ACM Press, New York.

Thom, J. and Millen. D.R. 2012. Stuff IBMers say: Microblogs as an expression of organizational culture. *Proceedings of the International AAAI Conference on Weblogs and Social Media 2012*, AAAI Press, Palo Alto, CA.

Thom, J., Millen, D.R., and DiMicco, J.M. 2012. Removing gamification from an enterprise SNS. *Proceedings of Computer Supported Cooperative Work 2012*, pp. 1067–1070. ACM Press, New York.

Thom-Santelli, J., Millen, D.R., and DiMicco, J.M. 2010. Characterizing global participation in and enterprise SNS. *Proceedings of ICIC 2010*, pp. 251–254. ACM Press, New York.

Thom-Santelli, J., Millen, D.R., and Gergle, D. 2011. Organizational acculturation and social networking. *Proceedings of Computer Supported Cooperative Work 2011*, pp. 313–316. ACM Press, New York.

Tufekci, Z. 2012. Youth and privacy in public networks. *Proceedings of the International AAAI Conference on Weblogs and Social Media 2012*, pp. 338–345. AAAI Press, Palo Alto, CA.

Turner, T., Qvarfordt, P., Biehl, J.Y., Golovchinsky, G., and Back, M. 2010. Exploring the workplace communication ecology. *Proceedings of CHI 2010*, pp. 841–850.

Veinott, E.S., Olson, J., Olson, G.M., and Fu, X. 2001. Video helps remote work: Speakers who need to negotiate common ground benefit from seeing each other. *Proceedings of CHI'01*, pp. 302–309. ACM Press, New York.

Voida, A., Harmon, E., and Al-Ani, B. 2012. Bridging between organizations and the public: Volunteer coordinators' uneasy relationship with social computing. *Proceedings of CHI 2012*, pp. 1967–1976. ACM Press, New York.

Wagner, C. and Majchrzak, A. 2007. Enabling customer-centricity using wikis and the wiki way. *Journal of Management Information Systems*, 23(3), 17–43.

Wang, Y., Norcie, G., Komandura, S., Acquisti, A. Leon, P.G., and Cranor, L. 2011. "I regretted the minute I pressed share": A qualitative study of regrets on Facebook. *Proceedings of SOUPS 2011*, Pittsburgh, PA. ACM Press, New York.

Weiseth, P.E., Munkvold, B.E., Tvedte, B., and Larsen, S. 2006. The wheel of collaboration tools: A typology for analysis within a holistic framework. *Proceedings of Computer Supported Cooperative Work 2006*, pp. 239–248. ACM Press, New York.

White, K.F. and Lutters, W.G. 2007. Midweight collaborative remembering: Wikis in the workplace. *Proceedings of CHIMIT'07*, Article 5. ACM Press, New York.

Williams, G. 1997. Task conflict and language differences: Opportunities for videoconferencing? *Proceedings of ECSCW'97*, pp. 97–108. Norwell, MA.

Wu, A., DiMicco, J., and Millen, D.R. 2010. Detecting professional versus personal closeness using an enterprise social network site. *Proceedings of CHI 2010*, pp. 1955–1964. ACM Press, New York.

Zhang, H., De Choudhury, M., and Grudin, J. Under review. Creepy but inevitable? Social networking approaches maturity.

Zhang, J., Qu, Y., Cody, J., and Wu, Y. 2010. A case study of micro-blogging in the enterprise: Use, value, and related issues. *Proceedings of CHI 2010*, pp. 123–132. ACM Press, New York.

Zhao, D. and Rosson, M.B. 2009. How and why people Twitter: The role that micro-blogging plays in informal communication at work. *Proceedings of GROUP 2009*, pp. 243–252. ACM Press, New York.

46

Social Media Use within the Workplace

Ester S. Gonzalez
California State University at Fullerton

Hope Koch
Baylor University

46.1 Introduction

Social media (hereinafter SM) use within the workplace has increased 50% since 2008. Forrester Research predicts that the sales of software to run corporate social networks will grow 61% a year, becoming a $6.4 billion dollar business by 2016 (Mullaney 2012). This growth is fueled by practitioners who feel they "have to do something with social media" (Pettit 2010), but are unsure about SM's capabilities (Deans 2011). Scholarship in this area can provide practitioners guidance on how to deploy SM strategically.

The purpose of this chapter is to provide a starting point for scholarship on SM within the workplace by discussing the current state of research in this area. Workplace SM research deals with how organizations implement SM to manage their internal operations such as employee relationships, communication, knowledge management, and innovation. This differs from external SM implementations, which deal with organizations using SM to manage customers, suppliers, and business partners. Both are a part of Enterprise 2.0, a term coined by McAfee in 2006, to describe organizational implementations of Web 2.0 technologies (McAfee 2006).

The term, Web 2.0, first appeared in 2004. It refers to technological capabilities that enable users to easily create and exchange web-based content (Kaplan and Haenlein 2010). High-speed Internet, cloud computing, clients, servers, application programming interfaces, open source software development, and increasingly mobile applications provide the infrastructure supporting SM (Falls 2010). Web 2.0 is the platform that spurred SM's popularization. College students began using MySpace and Facebook in 2004, and as of March 2012, more than 900 million people actively use Facebook (Facebook 2012). In addition to social networking sites like Facebook and LinkedIn, SM applications include collaborative projects (e.g., Wikipedia), blogs (e.g., Twitter), content communities (e.g., YouTube), virtual game worlds (e.g., *World of Warcraft*), and virtual social worlds (e.g., *Second Life*) (Kaplan and Haenlein 2010).

Given that college students initially adopted SM for recreational purposes, the initial SM research (Boyd and Ellison 2007; Ellison et al. 2007; Valenzuela et al. 2009) investigated the characteristics of students most likely to use SM, how students use SM, and the benefits they garner from its use. Given SM's proliferation in people's personal lives and desires to attract millennial generation new hires who expect to use these technologies in their working environment (Koch et al. 2012a; Majchrzak et al. 2009; Smith 2011), businesses began experimenting with SM in 2006.

The majority of research focuses on *external* SM use, describing how organizations like Wal-Mart, Dell, Starbucks, Ford, Burger King, and Zappos use online forums like Twitter, Facebook, blogs, and company-hosted forums (Culnan et al. 2010; Di Gangi et al. 2010). Primary SM uses include marketing, branding, customer service, and recruiting (Culnan et al. 2010; Gallaugher and Ransbotham 2010).

External SM research identified two key issues facing organizations: managing customer dialog and managing online communities. SM changes how organizations advertise their products from the historical megaphone strategy where companies broadcast what they want customers to know to strategies that incorporate customer-to-company and customer-to-customer communication (Gallaugher and Ransbotham 2010). Online community research identified four issues facing businesses: understanding the ideas posted in online communities, choosing the best ideas, balancing openness with competitiveness, and sustaining the community (Di Gangi et al. 2010).

Research on SM use *within the workplace* is more sparse but needed (Huysman 2011). Whereas workplace SM implementations were mostly experimental until 2011 (Rozwell 2011), in 2012, organizations began embracing SM applications to try and connect employees and increase their satisfaction and commitment despite geographic dispersion (Mullaney 2012). Organizations like Best Buy, IBM, and Kaiser Permanente are replacing their intranets with workplace SM sites (Bennett 2009). The most common software used to build these sites is Microsoft SharePoint. In fact, a survey of 1400 small, medium, and large organizations worldwide indicated that 70% of medium and large organizations build their workplace SM sites with Microsoft SharePoint (Ward 2012), which they customize to meet organizational needs (Koplowitz et al. 2011).

46.2 Underlying Principles

SM use within the workplace is based on three underlying principles: contribution, communication and collaboration, and social–work life integration. Table 46.1 highlights the theories that have guided research in this area.

TABLE 46.1   Theories Guiding Research on Social Media Use within the Workplace

Theory	Use in SM Workplace Research
Boundary theory (Ashforth et al. 2000; Sundaramurthy and Kreiner 2008)	Used to investigate SM's role integrating social life into the workplace (Koch et al. 2012a)
Social capital theory (Coleman 1988)	Used to investigate how workplace SM can help employees build capital to improve their working relationships (DiMicco et al. 2008)
Structural holes theory (Burt 1992; Granovetter 1973)	Used to investigate how SM promotes innovation and relationships by bringing together weak ties (Gray et al. 2011)
Theory of IT-culture conflict (Leidner and Kayworth 2006)	Used to investigate how organizations can resolve the conflict associated with implementing SM in the workplace (Koch et al. 2013)
Theory of positive emotions (Fredrickson 2000; Fredrickson and Branigan 2005)	Used to investigate why new hires used workplace SM and the benefits the SM initiative provided the new hires (Koch et al. 2012a; Leidner et al. 2010)

46.2.1 Contribution

Workplace SM value depends on employees frequently contributing small pieces of valuable knowledge that are easy to acquire, share, and use (Yates and Paquette 2011). A common misconception is that the crowd-sourcing capabilities of workplace SM will draw employees to adopt SM tools (Bradley and McDonald 2011). Unfortunately, organizations often struggle with employee adoption (Jackson et al. 2007; Koch et al. 2013). These struggles spawn from an interrelationship between the technology and the employees' workload. Some employees feel SM is a waste of time or that they are too busy to contribute, while others may not fully understand how their SM use may benefit the organization (Kiron 2012a; Koch et al. 2013). Many employees believe that if their coworkers see them contribute, they will think that they do not have enough work to do—a negative stigma (Koch et al. 2011).

To encourage employee participation in workplace SM, organizations have provided effort-investment and financial and reputational incentives. Effort-investment incentives allow employees to contribute to the SM with no extra effort. For example, corporate bookmarking applications allow employees to share the resources they consider valuable with coworkers simply by bookmarking resources in their own browser. Financial incentives include funding to develop popular ideas submitted over the SM, giving monetary awards to people and teams with the best ideas as well as incorporating workplace SM participation into performance reviews. For instance, Threadless t-shirt design system pays royalties to employees whose ideas are put into production (Malone et al. 2010). Reputational incentives include peer recognition, management recognition, and leadership opportunities. Tata consultants' leader board tracks the most prolific and highest-rated discussion board contributors (Kiron 2012c). Peer recognition and being on top of the leader board have led employees to respond to questions outside of their work group and helped Tata identify skills they did not know these employees possessed.

Even though organizations have adopted strategies to encourage employee contributions, the SM initiative's purpose and the organization's culture affect the success of workplace SM initiatives. Clearly, showing the purpose of the workplace SM by linking it to solving employee problems (Bradley and McDonald 2011) and monitoring outcomes rather than behaviors facilitate workplace SM success (Majchrzak et al. 2009). In addition, cultures characterized by youth, technical prowess, and enlightened leadership tend to successfully utilize workplace SM (Kiron 2012c). Employees want to understand why they should use the SM and feel confident that top management supports their SM use.

46.2.2 Communication and Collaboration

Both employees and managers adopt SM because SM encourages open communication and collaboration by providing forums where all employees can reach across time, space, interest, function, and hierarchy to seek and contribute information throughout the organization (Majchrzak et al. 2009). Using SM tools (e.g., wikis, blogs, and discussion boards), employees can see, use, reuse, augment, validate, critique, and rate one another's ideas (Bradley and McDonald 2011). This type of information sharing encourages transparency, informality, and democracy.

46.2.2.1 Employee Empowerment

SM's communication and collaboration principles have freed and empowered employees. Instead of responding to the same question repeatedly using private channels such as e-mail or phone calls, SM allows knowledge workers to publicly respond to questions (McAfee 2009). For example, a Kaiser Permanente team leader capitalized on SM's collaborative capabilities by posting employees' public relations plans on its SM site. This enabled all team members to comment on the proposals, without losing information in e-mails or offline conversations and documents (Kiron 2012b).

Empowered employees use SM to promote projects, advance their careers, and offer insight. Employees promote projects by driving traffic to project web pages and collecting comments that both shape the project and document support (DiMicco et al. 2008a,b; Majchrzak et al. 2009). SM supports career advancement by allowing employees to highlight their skills and interests to management and coworkers (DiMicco et al. 2008; Koch et al. 2012a). The openness embedded in SM applications allows employees to offer insights to problems outside their job responsibilities. USAA's new hires use the SM site to volunteer for projects that demonstrate their web development, graphic development, and project management skills to management (Leidner et al. 2010).

Management generally implements workplace SM to help employees address organizational problems including knowledge preservation, solution generation, and strategic insight. For example, IBM's Innovation Jam brings employees, clients, and partners together to discuss new business opportunities, whereas Scotland's Royal Bank uses a virtual world to gather employees' feedback on new banking environments (Birkinshaw et al. 2011).

46.2.2.2 Loss of Managerial Control

Given that SM promotes open communication, at times, employees may use SM to address problems that management would prefer employees ignore. In a notable example (da Cunha and Orlikowski 2008), petroleum company employees used an online discussion board to cope with organizational changes that were threatening their identity and livelihood. In SM postings, employees vented their anger and increased their solidarity by sharing painful and humiliating experiences. Their postings portrayed management as incompetent, insensitive, and unskilled while employees were portrayed as committed, competent, and willing to change. In another example, Telco's employees "hid behind the technology" to anonymously air grievances that they would not discuss when people knew their identity (Denyer et al. 2011).

As such, it is no wonder that management is still grappling with SM's empowerment capabilities. Organizational crowd-sourcing innovations often threaten innovation managers who feel a loss of control and protest about "having to spend their time sorting through the junk employees submit" (Kiron 2012c, p. 3). Middle managers have the most difficulty accepting workplace SM, possibly because SM threatens their role conveying information up and down the hierarchy (Kiron 2012c; Koch et al. 2012a; Leidner et al. 2010).

Many organizations have implemented workplace SM guidelines to address organizational fears about SM and help employees understand organizational-appropriate SM use (Bradley and McDonald 2011). These guidelines indicate what users can and cannot do with SM (Kaganer and Vaast 2010). However, most of the guidelines reflect a lack of understanding of SM's potential and instead see it as a risk or a way to convey management's message. The organizational policies address the risks by telling employees to consult authority if they are uncertain and that improper SM use could result in disciplinary actions. In more than one case (da Cunha and Orlikowski 2008; Denyer et al. 2011; Kiron 2012c), management has controlled information flow by disabling employees' ability to comment and instead appropriating SM as a megaphone to inform employees about policies and procedures.

46.2.3 Social and Work Life Integration

In addition to loss of managerial control, some organizations are concerned about the social aspect of workplace SM. In fact, organizations that implement workplace SM to encourage employees to make and maintain friends at work simultaneously struggle with whether the benefits of socializing outweigh potential productivity losses (Koch et al. 2012a).

Researchers have used boundary theory and the theory of positive emotions to investigate the SM principle of social and work life integration. Boundary theory (Ashforth et al. 2000) explains how people assume various roles (e.g., friend, coworker, supervisor, and family) that are determined by their task and social system (Katz and Kahn 1978; Perrone et al. 2003). When competing roles (e.g., supervisor and friend) interrupt each other, conflict occurs.

Contrary to boundary theory, an investigation of workplace SM use found that blurred work-social boundaries can create positive emotions (e.g., joy, contentment, and happiness) (Koch et al. 2012a) for SM users. This research combined boundary theory with the broaden and build theory of positive emotions. The broaden and build theory of positive emotions posits that positive emotional experiences can broaden an individual's thought–action repertoire, which can then lead to personal resource development. Personal resources include physical, social, intellectual, and psychological resources that individuals can draw upon when faced with negative situations (Fredrickson 2004).

Workplace SM use can increase employees' attachment to the workplace and improve working relationships by providing a forum for employees to make and maintain friends in the workplace (DiMicco et al. 2008; Koch et al. 2012a). Several SM applications support social–work life integration including employee profiles, activity feeds, and event planning (Koch et al. 2012a). Employee profiles encourage employees to share aspects of their personal lives including pictures and hobbies. Employees frequently scan coworker profiles to find a common ground with people they are seeking expertise from. Activity feeds allow employees to stay in touch with other employees whom they have met. Event planning features facilitate in-person social events where employees can meet one another and build relationships. Employees use these applications primarily to build stronger relationships with their acquaintances (i.e., weak ties) rather than reach out to people that they know well (DiMicco et al. 2009).

This relationship building helps build *social capital* (Coleman 1988) through trust, reciprocity, identification, and shared language (Tan et al. 2009). Social capital is difficult to see and may involve establishing purposeful relationships to generate benefits and offering short- and long-term kindness with the assumption that people will return this kindness (Coleman 1988; Nahapiet and Ghoshal 1998).

Combined SM applications engender bridging, bonding, and maintained social capital (Ellison et al. 2007). Bridging social capital involves weak ties between individuals who may provide useful information and new perspectives (Granovetter 1973). Bonding social capital is found in tightly knit emotionally close relationships such as family and close friends. It gives employees people to turn to in times of need. Employees experience maintained social capital when they can rely on their coworkers to do small favors (Ellison et al. 2007).

46.3 Impact on Practice

Most managers who champion workplace SM believe that better communication characterized by personal relationships, openness, and collaboration leads to positive outcomes for employees that transfer to the organization. Employees who use workplace SM may experience intrinsic motivations like job satisfaction, freedom, and fun (Tan et al. 2009) and engage in corporate citizenship behaviors (DiMicco et al. 2008, 2009; Skeels and Grudin 2009). This section highlights common SM applications and workplace SM initiatives that have proven effective.

46.3.1 Applications

Most workplace SM implementations encompass a common set of applications, which organizations integrate into customized portals. Table 46.2 describes these applications that have six characteristics (McAfee 2009): search, links, authoring, tags, extensions, and signals. Workplace SM applications may make searching and finding information more successful by allowing employees to share and post information in ways that they consider most beneficial (Kiron 2012c). Links may break down existing hierarchical data structures that have historically defined who can discuss certain topics as well as generate, share, and access information (Kiron 2012c). Links enable employees to define how the organization will organize its documents by promoting a self-imposed structure, which emerges out of everyday work practices. Authoring shifts corporate intranets from a top–down information flow reflecting management's voice to a tool that empowers employees to update information and provide commentary. Tags allow employees to keep up with useful web resources. Employees can assign descriptive reminders to the content they have accessed and browse their coworkers' content. Signals notify employees when new and relevant information is added such as when coworkers update documents of interest.

TABLE 46.2   Social Media Applications

Feature	Description
Blogs	Allows employees to author individually by posting thoughts and ideas in reverse, chronological order
Business intelligence	Analyzes e-mails, instant messaging, and virtual space usage to identify social networks and value-added networks
Collaborative projects	Enables employees to simultaneously create content
Content communities	Enables employees to share text, videos, photos, and PowerPoint presentations
Discussion boards	Forum for asking questions and leaving responses
Social bookmarks and tagging	Allows employees to bookmark and categorize frequently accessed documents and sites to share with colleagues
Social networking sites	Enables employees to construct a digital identity by developing a profile, articulating a list of users with whom they share a connection, viewing their list of connections, and viewing connections made by others within the system (Boyd and Ellison 2007). This is similar to public sites like Facebook and LinkedIn
Virtual worlds	Allows employees to replicate a three-dimensional environment that simulates real life
Wikis	Allows groups of employees to create and edit websites that are shared throughout the organization

Source: Kaplan, A.M. and Haenlein, M., *Business Horizons*, 53(1), 59, 2010; Majchrzak, A. et al., *MIS Quart. Exec.*, 8(2), 103, 2009.

46.3.2 Outcomes

While the business press is filled with stories of how workplace SM benefits organizations (Bradley and McDonald 2011; Li 2010; McAfee 2009), academic research highlights four areas where organizations have effectively deployed workplace SM. These include acclimating new employees, changing organizational culture, managing knowledge, and promoting innovation.

46.3.2.1 Acclimating New Hires

Organizations have leveraged workplace SM to acclimate new hires into the workplace (Koch et al. 2012a; Leidner et al. 2010). USAA implemented its workplace SM in 2009 in efforts to attract and retain college hires in the IT area. USAA's new hires use workplace SM throughout their new hire program. In the early stages of their program, these tools help new hires become familiar with their area, learn where to go for help, find people like them, make friends, and learn about social events. In the later stages of their program, the workplace SM allows more senior new hires to develop leadership skills by mentoring first-year new hires, developing content, and leading events.

While managers recognize several benefits from integrating SM into new hire acclimation programs including better retention, recruiting, morale, engagement, and productivity, concerns are still prevalent (Koch et al. 2012a). Since workplace SM shifts some of the burden of mentoring new hires away from middle managers while simultaneously exposing the new hires to upper management, middle managers can experience isolation and inequity. These emotions often manifest themselves in middle managers protesting that the SM encourages the new hires to socialize during the workday.

46.3.2.2 Changing Organizational Culture

In addition to making the workplace more social, organizations have implemented SM to also bring about information sharing cultural changes (Koch et al. 2012a; Majchrzak et al. 2009). SM promotes information sharing cultures in several ways. User profiles let employees know their coworkers' current projects and expertise. Document repositories provide central locations where employees can access documents such as contracts. Wikis and blogs allow employees to share meeting notes and frequently asked questions. Activity feeds keep project team members up to date on one another's project status. Organizations have leveraged these applications to promote employee health and safety and share best practices (Bradley and McDonald 2011).

SM can make the workplace more social by promoting social events where employees can physically get together and providing online forums for employees to learn about one another and nurture their personal relationships. These activities encourage employees to share their personal interests and hobbies. As part of its cultural change effort, a security company hosted a luncheon where employees brought and discussed pictures of themselves having fun (Koch et al. 2013). Subsequently, management encouraged the employees to post their pictures along with their hobbies on the workplace SM site.

Despite SM's potential to promote cultural changes, employees and management often challenge these efforts (Denyer et al. 2011; Koch et al. 2013). Both groups are concerned about socializing at work (Skeels and Grudin 2009), unnecessary distractions (O'Driscoll and Cummings 2009), and protecting confidential information from unauthorized employee access. While employees are sometimes reluctant to communicate openly in fear of management ramifications (Hewitt and Forte 2006; Skeels and Grudin 2009), management wants to control communication. SM can facilitate cultural change, when management highlights the need for organizational change and implements leadership-based, policy-based, and socialization-based integration mechanisms to align the SM with the organization's culture (Koch et al. 2013).

While SM-induced organizational change research is emerging, the theory of IT-culture conflict has proven useful in this area (Koch et al. 2013). This theory posits that IT implementations may encounter system conflict when the values embedded in the IT conflict with group member values (Leidner and Kayworth 2006). The theory suggests that organizations can resolve IT-culture system conflict by implementing integration mechanisms that bring group member values and the IT values closer together.

46.3.2.3 Managing Knowledge

Organizations can use SM to manage knowledge by facilitating knowledge sharing, knowledge transfer, and knowledge preservation (Bradley and McDonald 2011; McAfee 2006, 2009). In contrast to the knowledge management systems of the 1990s that required employees to preserve their knowledge by entering best practices and experiences in knowledge repositories, workplace SM can capture employee's knowledge as part of their routine work practices (McAfee 2006).

Workplace SM facilitates knowledge sharing by allowing employees to pose and answer questions. Employees who respond to the forums are often people with interest in an area rather than people assigned to an area, thus allowing organizations to tap into hidden skill resources. SM enables employees to find knowledge resources within the organization. For example, employees located at different worldwide divisions can locate their counterparts at other divisions. SM's social aspect (i.e., the profiles and the activity feeds) helps promote relationships that may ensure that the counterpart responds to the request for help.

SM facilitates knowledge transfer by allowing coworkers to stay abreast of one another's activities and facilitating succession. As an example of staying abreast of coworker activities, employees in the defense industry, who had historically e-mailed their managers weekly activity reports, began using SM to post their activity reports (Koch et al. 2013). This allowed their team members to monitor their project progress (Koch et al. 2013). By giving new team members access to their predecessor's portal, SM enables new employees to learn who their predecessor communicated with and the electronic resources they used.

Employees can use SM's knowledge transfer capabilities to gain broad-based or pulse knowledge (Jackson et al. 2007). Broad-based knowledge may include strategies, initiatives, policies, or events. For example, reading coworkers' blogs can help employees understand how changes to the companies' pension program affect them (Jackson et al. 2007). Pulse knowledge gives employees an insight to their coworkers' thoughts. Pulse knowledge can help executives assess organizational climate, assist human resources professionals to create a positive working environment, and allow newcomers to fit in.

46.3.2.4 Promoting Innovation

Research on workplace SM-enabled innovation includes business success stories, explanations of how organizations have used SM to promote innovation, and the relationship between SM tools and innovation. Business press authors like Charlene Li (2010), Andrew McAfee (Kiron 2012c; McAfee 2009), and Gartner Group executives (Bradley and McDonald 2011) share many case studies of organizations that have leveraged workplace SM to promote innovation. These case studies suggest that SM-enabled innovations have positively impacted efficiency, quality, culture, savings, and revenue. Examples range from a semiconductor design company that increased its engineering productivity by 25% to a healthcare company that created a collaborative community to provide better care for the elderly (Bradley and McDonald 2011).

With innovation becoming the entire organization's responsibility (Birkinshaw et al. 2011), workplace SM, like event planning systems, internal rating systems, and innovation forums, has the potential to involve all employees in the innovation process. Event planning systems promote innovation contests. USAA held a contest that encouraged its employees to develop a banking application for the Android mobile phone operating system before Android became mainstream. Innovation forums and internal rating systems allow employees to post, develop, and vote on innovative ideas. Threadless (Malone et al. 2010), IBM, and Scotland's Royal Bank (Birkinshaw et al. 2011) have leveraged ideas generated from workplace SM by creating connectors between idea generators and those that can make the idea come to life (Whelan et al. 2011).

Both blogging and social bookmarking can foster innovation by bringing together weak ties. Explicated in structural holes theory (Burt 1992; Granovetter 1973), weak ties describe people who are acquaintances and as a result of operating in different networks possess different information. Structural holes theory predicts that bridging gaps between unconnected individuals can expand one's sphere of influence, build knowledge resources, and foster innovation.

As an informal mechanism for linking weak ties, blogging has the potential to link disparate parts of the organization in an initial step toward promoting innovation through cross-functional collaboration (Jackson et al. 2007). As an example, a cement building and materials company credited its SM-enabled cross-functional collaboration with facilitating corporate process changes that helped decrease its emissions by raising its alternative fuel use (Bradley and McDonald 2011).

At the individual level, organizational social bookmarking applications can increase employee innovativeness when the process of reading coworkers' bookmarks provides employees novel information (Gray et al. 2011). For example, one employee attributed regularly reading his coworkers' bookmarks to staying abreast of developments in the visualization area and developing creative ideas for clients (Gray et al. 2011).

46.4 Future Research

Given the nascent stage of SM use within the workplace, additional research is necessary to better understand the phenomenon at both organizational and individual levels.

46.4.1 Organizational Level

Fruitful organizational level research topics include SM's impact, governance, and sustenance. Since SM often comes in a suite with a variety of applications including blogs and wikis, we need to understand the workplace impact of both particular SM applications (e.g., social networking and wikis) and SM initiatives as a whole.

Currently, practitioners have little guidance on what to do with SM or which SM applications may help solve workplace problems (Pettit 2010). To help practitioners, researchers should continue to investigate SM applications in the workplace, as part of a larger effort to build a workplace SM application taxonomy. Such a taxonomy would help practitioners understand which SM applications are best suited to address workplace needs like creating knowledge, fostering commitment, and promoting change. Further taxonomy development might investigate employee types that are most likely to benefit from particular SM applications. This research could investigate SM in different functional areas (e.g., accounting, marketing, information technology), levels (e.g., executive, middle management, entry level), and maybe even industries (e.g., retail, energy, distribution).

In addition to understanding how SM applications like wikis and social networking affect various processes in the organization, we need to understand how SM as a whole impacts the organization. This stream of research might bolster the case studies linking workplace SM use to positive organizational outcomes with quantitative evidence (Gray et al. 2011; Kiron 2012a). Research might measure how workplace SM initiatives affect collaboration, learning, information sharing, knowledge management and communication, and the resulting impact this has on key performance indicators such as turnover, innovation, and return on investment.

We also need to understand the dark side of SM in the workplace. While both the trade press (Cain et al. 2011) and academic articles (Denyer et al. 2011; Koch et al. 2013) document management's and employees' concerns about SM reducing productivity, we need quantifiable evidence about SM's negative effects in the workplace. Research in this area might investigate lost productivity, workplace disruption, and information overload.

These potential negative ramifications of workplace SM use might be addressed with SM governance. Governance refers to the management, policies, guidance, processes, information rights, and decision rights associated with an organization's SM initiatives (Weill and Ross 2004). Since many organizations implemented SM in an effort to experiment with this new medium and its open, collaborative, nonhierarchical values (Koch et al. 2013), we need to know more about effective SM governance. Research in this area might describe both effective and ineffective SM governance practices that address content management, information credibility, and employee rights. How do organizations balance the need for confidentiality vs. the need to share information, workplace surveillance vs. employee privacy, and encouraging collaboration vs. stifling gossip?

In addition, we know little about sustaining SM initiatives. Future research needs to address this along with how to integrate SM into existing work practices. As SM technologies continue to advance and incorporate Web 3.0, big data, predictive analytics, and mobile computing (Andriole 2010; King 2012), research might investigate mindful SM adoption (Swanson and Ramiller 2004).

46.4.2 Individual Level

Individual-level research will help us understand how workplace SM initiatives affect employees. How does SM impact communication processes like attention, persuasion, learning, and leadership? How does SM change coworker relationships? Some research areas include the effect of online relationships on offline relationships, social isolation, and weisure time. Weisure time refers to a time where employees cannot distinguish between where work ends and their social life begins (Luttenegger 2010). While we know something about how weisure time impacts employees from an organizational perspective (Koch et al. 2012a), we know less about the impact on the employees' personal life.

46.5 Conclusion

This chapter has provided a starting point for scholars interested in understanding SM use in the workplace. Contrary to SM use in the public domain (e.g., Facebook and LinkedIn) and organizational SM use for external relationship (e.g., customers and business partners), workplace SM use deals with employees. While these initiatives can empower employees, they can also result in loss of managerial control and social–work life integration. Case studies describing workplace SM uses such as acclimating new hires, changing organizational culture, managing knowledge, and promoting innovation characterize research in this area. Future research needs to investigate the quantifiable impacts of SM as well as its governance and sustenance. Some theories that may be useful to guide these inquiries are available on the IS theories page at http://istheory.byu.edu/wiki/Main_Page.

References

Andriole, S.J. 2010. Business impact of web 2.0 technologies. *Communications of the ACM* 53(12):67–79.

Ashforth, B.E., G.E. Kreiner, and M. Fugate. 2000. All in a day's work: Boundaries and micro role transitions. *Academy of Management Review* 25(3):472–491.

Bennett, J. 2009. Will social media kill off the intranet in years to come? http://www.internalcommshub.com/open/channels/whatsworking/intranetend.shtml (accessed April 28, 2012)

Birkinshaw, J., C. Bouquet, and J.L. Barsoux. 2011. The 5 myths of innovation. *MIT Sloan Management Review* 52(2): 42–50.

Boyd, D.M. and N.B. Ellison. 2007. Social network sites: Definition, history, and scholarship. *Journal of Computer Mediated Communication* 13(1):210–230.

Bradley, A. and M. McDonald. 2011. *The Social Organization: How to Use Social Media to Tap the Collective Genius of Your Customers and Employees*. Boston, MA: Harvard Business School Press.

Burt, R.S. 1992. *Structural Holes: The Social Structure of Competition*. Cambridge, MA: Harvard University Press.

Cain, M.W., J. Mann, M.A. Silver, M. Basso, A. Walls, and C. Rozwell. 2011. Predicts 2012: The rising force of social networking and collaboration services Gartner. Available from http://www.gartner.com/id = 1859216 (accessed February 6, 2012).

Coleman, J.S. 1988. Social capital in the creation of human capital. *Journal of Sociology* 94:S95–S120.

Culnan, M.J., P. McHugh, and J.I. Zubillaga. 2010. How large U.S. companies can use Twitter and other social media to gain business value. *MIS Quarterly Executive* 9(4):243–259.

da Cunha, J. and W.J. Orlikowski. 2008. Performing catharsis: The use of online discussion forums in organizational change. *Information and Organization* 18(2):132–156.

Deans, P.C. 2011. The impact of social media on C-level roles. *MIS Quarterly Executive* 10(4):187–200.

Denyer, D., E. Parry, and P. Flowers. 2011. "Social," "Open" and "Participative"? Exploring personal experiences and organizational effects of enterprise 2.0 use. *Long Range Planning* 44(2011):375–396.

Di Gangi, P.M., M. Wasko, and R. Hooker. 2010. Getting customers' ideas to work for you: Learning from dell how to succeed with online user innovation communities. *MIS Quarterly Executive* 9(4):213–228.

DiMicco, J.M., W. Geyer, D.R. Millen, C. Dugan, and B. Brownholtz. 2009. People sensemaking and relationship building on an enterprise social network site. Paper read at *42nd Annual Hawaii International Conference on System Sciences*, Waikoloa, HI.

DiMicco, J.M., D.R. Millen, W. Geyer, and C. Dugan. 2008b. Research on the use of social software in the workplace. Paper read at *CCSW' 08 Workshop, Social Networking in Organizations*, November 8–12, San Diego, CA.

DiMicco, J., D.R. Millen, W. Geyer, C. Dugan, B. Brownholtz, and M. Muller. 2008a. Motivations for social networking at work. Paper read at *CSCW'08*, November 8–12, San Diego, CA.

Ellison, N.B., C. Steinfield, and C. Lempe. 2007. The benefits of facebook "Friends": Social capital and college students' use of online social network sites. *Journal of Computer-Mediated Communication* 12(4):1143–1168.

FaceBook. 2012. Facebook statistics. http://newsroom.fb.com/content/default.aspx?NewsAreaId = 22 (accessed June 11, 2012).

Falls, J. 2010. What's next in social media ms now and mobile. Social media explorer. http://www. socialmediaexplorer.com/social-media-marketing/whats-next-in-social-media-is-now-and-mobile/ (accessed April 17, 2012).

Fredrickson, B.L. 2000. Cultivating positive emotions to optimize health and well-being. *Prevention and Treatment* 3(001a):1–25.

Fredrickson, B.L. 2004. The broaden-and-build theory of positive emotions. *Philosophical Transactions: Biological Sciences (The Royal Society of London)* 359(1449):1367–1377.

Fredrickson, B.L. and C. Branigan. 2005. Positive emotions broaden the scope of attention and thought-action repertoires. *Cognition and Emotion* 19(3):313–332.

Gallaugher, J. and S. Ransbotham. 2010. Social media and customer dialog management at Starbucks. *MIS Quarterly Executive* 9(4):197–212.

Granovetter, M.S. 1973. The strength of weak ties. *American Journal of Sociology* 78(6):1360–1380.

Gray, P., S. Parise, and B. Iyer. 2011. Innovation impacts of using social bookmarking systems. *MIS Quarterly* 35(3):629–649.

Hewitt, A. and A. Forte. 2006. Crossing boundaries: Identity management and student/faculty relationships on the facebook. Paper read at *Computer Supported Collaborative Work Poster*, November 4–8, Banff, Alberta, Canada.

Huysman, M. 2011. *Journal of Computer Mediated Communication Special Issue Call on Social Media and Communication in the Workplace*. Amsterdam, the Netherlands. Available from http://www.kinresearch.nl/research/call-for-papers/ (accessed April 17, 2012).

Jackson, A., A.Y. Jo, and W. Orlikowski 2007. Corporate blogging: Building community through persistent digital talk. In *40th Annual Hawaii International Conference on Systems Science*, Big Island, HI.

Kaganer, E. and E. Vaast. 2010. Responding to the (almost) unknown: Social representation and corporate policies on social media. In *31st International Conference on Information Systems*, St. Louis, MO.

Kaplan, A.M. and M. Haenlein. 2010. Users of the world, unite! The challenges and opportunities of social media. *Business Horizons* 53(1):59–68.

Katz, D. and R.L. Kahn. 1978. *The Social Psychology of Organizations*, 2nd edn. New York: John Wiley & Sons.

King, C. 2012. 30 Social media predictions for 2012 from the Pros. Available from http://www.socialmediaexaminer.com/30-social-media-predictions-for-2012-from-the-pros/ (accessed June 7, 2012).

Kiron, D. 2012a. The amplified enterprise: Using social media to expand organizational capabilities. *MIT Sloan Management Review* 53(2):1–6.

Kiron, D. 2012b. Social business at Kaiser Permanente: Using social tools to improve customer service, research and internal collaboration. *MIT Sloan Management Review* 53(3):1–3.

Kiron, D. 2012c. What sells CEOs on social networking. *MIT Sloan Management Review* 53(3):1–4.

Koch, H., E. Gonzalez, and D. Leidner. 2012a. Bridging the work/social divide: The emotional response to organizational social networking sites. *European Journal of Information Systems* advance online publication doi:10.1057/ejis.2012. 18:1–19.

Koch, H., D. Leidner, and E. Gonzalez. 2011. Resolving IT-culture conflict in enterprise 2.0 implementations. Paper read at *Americas Conference on Information Systems*, August 5–7, 2011, Detroit, MI.

Koch, H., D. Leidner, and E. Gonzalez. 2013. Digitally Enabling Social Networks: Resolving IT-Culture Conflict. *Information Systems Journal* 23(5).

Koplowitz, R., J.R. Rymer, M. Brown, J. Dang, and A. Anderson. 2011. SharePoint adoption: Content and collaboration is just the start. Available from http://www.forrester.com/SharePoint+Adoption+Content+And+Collaboration+Is+Just+The+Start/fulltext/-/E-RES60704 (accessed April 22, 2012).

Leidner, D., K. Hope, and E. Gonzalez. 2010. Assimilating generation Y IT new hires into USAA's workforce: The role of an enterprise 2.0 system. *MIS Quarterly Executive* 9(4):229–242.

Leidner, D.E. and T. Kayworth. 2006. Review: A review of culture in information systems research: Toward a theory of information technology culture conflict. *MIS Quarterly* 30(2):357–399.

Li, C. 2010. *Open Leadership: How Social Media Can Change the Way You Lead.* San Francisco, CA: Jossey-Bass.

Luttenegger, J. 2010. Smartphones: Increasing productivity, creating overtime liability. *The Journal of Corporate Law* 36(1):260–280.

Majchrzak, A., L. Cherbakov, and B. Ives. 2009. Harnessing the power of the crowds with corporate social networking tools: How IBM does IT. *MIS Quarterly Executive* 8(2):103–108.

Malone, T.W., R. Laubacher, and C. Dellarocas. 2010. The collective intelligence genome. *MIT Sloan Management Review* 51(3):21–31.

McAfee, A.P. 2006. Enterprise 2.0: The dawn of emergent collaboration. *MIT Sloan Management Review* Spring:21–28.

McAfee, A.P. 2009. *Enterprise 2.0.* Boston, MA: Harvard Business Press.

Mullaney, T. 2012. "Social Business" launched this burger. *USA Today*, May 17, 2012, 1A-1B.

Nahapiet, J. and S. Ghoshal. 1998. Social capital, intellectual capital, and the organizational advantage. *Academy of Management Review* 23(2):242–266.

O'Driscoll, T. and J. Cummings. 2009. Moving from attention to detail to detail to attention: Web 2.0 management challenges in red hat and the fedora project. Paper read at *SIM Enterprise and Industry Application for Web 2.0 Workshop*, Phoenix, AZ.

Perrone, V., A. Zaheer, and B. McEvily. 2003. Free to be trusted? Organizational constraints on trust in boundary spanners. *Organization Science* 14(4):422–439.

Pettit, A. 2010. The future of social media research. Available from http://www.scribemedia.org/2010/07/07/future-of-social-media-research/ (accessed May 27, 2012).

Rozwell, C. 2011. Roundup of social media research 2H10 to 1H11 gartner. Available from http://www.gartner.com/technology/core/home.jsp (accessed February 6, 2012).

Skeels, M.M. and J. Grudin. 2009. When social networks cross boundaries: A case study of workplace use of facebook and linkedin. Paper read at *GROUP'09*, May 10–13, 2009, Sanibel Island, FL.

Smith, A. 2011. Why Americans use social media. Available from http://www.pewinternet.org/Reports/2011/Why-Americans-Use-Social-Media/Main-report.aspx (accessed May 7, 2012).

Sundaramurthy, C. and G.E. Kreiner. 2008. Governing by managing identity boundaries: The case of family businesses. *Entrepreneurship Theory and Practice* 32(3):415–436.

Swanson, E.B. and N.C. Ramiller. 2004. Innovating mindfully with information technology. *MIS Quarterly* 28(4):553–583.

Tan, W.-K., K.K.O. Tha, T.T.D. Nguyen, and X. Yo. 2009. Designing groupware that fosters capital creation: Can facebook support global virtual teams? Paper read at *Americas Conference on Information Systems*, San Francisco, CA.

Valenzuela, S., N. Park, and K.F. Kee. 2009. Is there social capital in a social network site? Facebook use and college students' life satisfaction, trust, and participation. *Journal of Computer-Mediated Communication* 14(4):875–901.

Ward, T. 2012. The social intranet. Prescient Digital Media. Available from http://www.prescientdigital.com/articles/download-social-intranet-success-matrix (accessed April 22, 2012).

Weill, P. and J. Ross. 2004. *IT Governance: How Top Performers Manage IT Decision Rights for Superior Results.* Boston, MA: Harvard Business School Publishing.

Whelan, E., S. Parise, J. de Valk, and R. Aalbers. 2011. Creating employee networks that deliver open innovation. *MIT Sloan Management Review* 53(1):37–44.

Yates, D. and S. Paquette. 2011. Emergency knowledge management and social media technologies: A case study of the 2010 Haitian earthquake. *International Journal of Information Management* 31:6–13.

VII

Managing and Securing the IT Infrastructure and Systems

47

Virtualization of Storage and Systems

Yang Song
IBM Research—Almaden

Gabriel Alatorre
IBM Research—Almaden

Aameek Singh
IBM Research—Almaden

Jim Olson
IBM Global Technology Services

Ann Corrao
IBM Global Technology Services

All problems in computer science can be solved by another level of indirection.

David Wheeler

47.1 Overview of Virtualization

47.1.1 What Is Virtualization?

Virtualization is the abstraction of physical components, such as CPU, memory, and storage, into logical management entities. In other words, virtualization provides an additional layer of *indirection* that separates the actual resource pool from the management function, in order to achieve better system utilization with improved manageability (Figure 47.1). As an example, modern computer architecture requires a memory component to facilitate fast data access between CPU and storage devices. While outperforming hard disk drives (HDDs) in regard to the latency of data read/write, memory is usually much smaller in capacity than HDD and cannot satisfy the memory allocation requests from applications all the time. With the aid of virtualization, the host operating system (OS) can create a *virtual memory*, which essentially resides in the HDD, and allocate the virtual memory space for application memory requests if the physical memory is saturated. From the application perspective, the virtual memory appears the same as the physical memory, and each Input/Output (I/O) request will trigger a mapping, performed by the OS, from the virtual memory space to the actual location on the HDD. The complexity of indirection is transparent to the application and hence improved management flexibility and efficient system utilization can be achieved in a nonintrusive fashion. The concept of "additional layer of indirection" characterizes the

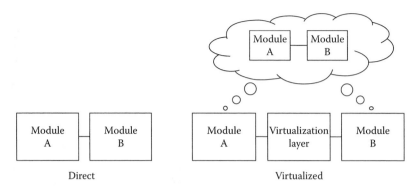

FIGURE 47.1   Virtualization as an additional layer of indirection.

essence of virtualization techniques that have been widely applied in many information technology (IT) applications, ranging from the inception of virtualization on IBM mainframes in the 1960s to the recent revitalization in emerging cloud computing paradigm shift.

47.1.2 Why Virtualization?

The key advantage of virtualization lies in the improved resource utilization and enhanced management flexibility. The virtualization layer presents a consistent interface to the system administrator while hiding the details of the underlying physical resources. The system administrator will have a logical view of the resource pool, and favorable features such as efficient resource allocation and dynamic control can be achieved. For example, a *virtual machine* (VM) can be created, which is a virtualized server with an isolated and complete OS (possibly different than the OS of the physical host server) and behaves exactly like a physical machine. Furthermore, multiple VMs can be consolidated into a single physical host machine to simplify the management overhead and reduce the operation cost such as electricity bills for cooling. By virtualizing all physical resources into a unified logical resource view, the overall IT infrastructure and system resources can be managed and shared more dynamically. For example, the disk space of one physical host can be shifted to another host that is running out of storage capacity, without interrupting the ongoing applications. Additional hardware such as HDD can be easily added to the system while keeping the logical resource view intact. Therefore, the overall IT infrastructure is more responsive and agile to the real-time environment changes, which facilitates better and faster business decisions. Also virtualization naturally supports multi-tenant IT infrastructure by virtualizing a separate environment for each tenant, and the overall system resources can be shared in an integrated fashion, where new value-added services can be deployed and delivered with less cost and complexity.

47.1.3 Types of Virtualization

In essence, virtualization provides a logical abstraction of the physical resources in a variety of forms such as host machine hardware, networking resources, and storage capacity, among many others. Next, we will introduce three common types of virtualization, i.e., platform virtualization, storage virtualization, and network virtualization, which are widely deployed in many IT scenarios such as cloud computing infrastructures.

47.1.3.1 Platform Virtualization

Traditional computers use specific hardware architectures with a host OS that is compatible with the underlying hardware platform. Therefore, the OS and the host hardware platform need to be developed side by side in order to interact properly. With platform virtualization, or sometimes referred to as hardware virtualization, multiple guest OSs (independent of the host OS) can run on the same physical host machine simultaneously and share the hardware in a coordinated fashion. The key enabling technique is a software, namely, *hypervisor*, which runs on the physical host and provides an integrated interface between the hardware platform and the guest OSs. Popular hypervisors include KVM [11], Xen [20], VMware ESX [18], Microsoft Hyper-V [12], and IBM PowerVM [7], among many others. By utilizing hypervisor and platform virtualization, multiple VMs are created on a physical server with their own set of virtual hardware. The interactions of VMs and the underlying host machine are intercepted and translated by the hypervisor. For example, a hypervisor can intercept the instructions of the guest OS of one VM and translate them to the native instruction set of the host machine. This enjoys great practical merit for enterprise IT administrators that need to execute legacy machine code on newly developed hardware platforms. In addition, the hypervisor can monitor the status of each VM, such as I/O rate and memory usage, and allocate the hardware resources on the host platform appropriately, in order to prioritize workloads and/or to achieve better system utilization. Therefore, hypervisors are also referred to as *virtual machine managers* (VMM). Due to its popularity, hypervisor architecture design and optimization have been studied extensively in the literature. For example, Barham et al. [23] and Menon et al. [32] discuss the overall architecture design of Xen and several performance enhancements to its I/O channel, respectively. In [24], a novel architecture of virtualization is designed and analyzed that enables a nested virtualization environment, i.e., a hypervisor can host other hypervisors with their own VMs, on existing x86-based systems. For more discussions on x86 virtualization, refer to [21].

According to the methods of interaction with host hardware platform, hypervisor can be categorized into two major classes, i.e., *bare metal* hypervisors and *hosted* hypervisors, as shown in Figure 47.2. Bare metal hypervisors interact with the physical hardware directly, e.g., KVM and ESX, while hosted hypervisors reside in a host OS, e.g., Oracle VM VirtualBox [15] installed on a Windows machine. Hypervisors usually can support virtualization in one of two modes, i.e., *full virtualization* and *paravirtualization*. In full virtualization mode, hypervisors provide a complete simulation of the underlying hardware to unmodified guest OSs. While in paravirtualization, hypervisors run modified guest OSs that leverage an Application Programming Interface (API) similar to that provided by the underlying hardware. Therefore, hypervisors with paravirtualization capabilities are less complex and provide faster interactions with the VMs due to reduced hardware and firmware emulation as well as less layers of indirection, as illustrated in Figure 47.3.

Platform virtualization provides significantly improved management flexibility compared to traditional IT architectures. For example, existing virtualization management tools, such as VMware's vMotion [19], are able to nondisruptively migrate a VM across multiple physical servers. VM migration facilitates improved server load-balancing, higher availability (e.g., moving VMs away from a failing or overloaded server), and easier management (e.g., moving VMs away from an old server scheduled for replacement). At the start of a migration, a copy of the VM, which resides in shared storage, is placed on the target server. Next, the active memory and execution state from the source server is transferred to the target until the physical location of the VM is updated.

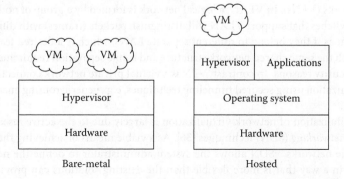

FIGURE 47.2 Bare metal hypervisors and hosted hypervisors.

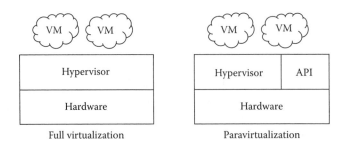

FIGURE 47.3   Full virtualization and paravirtualization.

47.1.3.2 Storage Virtualization

Storage virtualization is the abstraction and aggregation of physical storage capacity from multiple devices for efficient allocation and management in a coordinated way. By using storage virtualization, the complexity of physical storage devices is hidden from the management viewpoint, which is crucial for reducing the difficulty of management and improving performance. In the next section, we will elaborate on various techniques used in existing storage virtualization systems and the benefits of storage virtualization.

47.1.3.3 Network Virtualization

Network virtualization is the abstraction of network resources, such as routers and switches, as well as the physical network link bandwidth, into a logical *slice* of resources. With network virtualization, a single physical network appears to be multiple heterogeneous network architectures, and the overall network is shared among them. Similar to platform virtualization, network virtualization provides improved flexibility and enhanced network manageability. Each virtual network consists of a collection of virtual nodes and virtual links, and two virtual nodes can be connected via a virtual link regardless of their physical locations. In other words, network virtualization provides a shared yet isolated network environment that allows the coexistence of multiple overlay networks. Therefore, it is of particular interest to networking researchers since this enables deployment of heterogeneous network architectures and protocols on a shared physical infrastructure, where new networking techniques can be tested and evaluated. In [28], a systematic virtual network design that separates the network infrastructure provider and the network service providers is proposed. Such an architecture allows the traditional Internet Service Provider (ISP) to provide value-added services with customizable network protocol stacks. Bhatia et al. [25] propose a platform for virtualized networks based on commodity devices. Another example is the U.S. National Science Foundation's (NSF) support of the Global Environment for Network Innovations (GENI) project [17], which utilizes network virtualization techniques to provide a large-scale shared virtual test bed for networking research communities.

Traditional network virtualization techniques include virtual local area networks (VLANs) and virtual private networks (VPNs). In VLAN, a logical network is formed by a group of nodes with the same VLAN ID. The switches that support VLAN will distinguish packets (frames) with different VLAN IDs and forward them as if they belong to different separate LANs, which is a desired feature for multiple organizations with the same shared network structure and traffic separation requirements, possibly due to privacy and security reasons. In contrast, VPN is a virtual private network connecting multiple sites of the same organization using secured tunneling techniques, e.g., generic routing encapsulation (GRE) [9] and IPsec [10].

The recent revitalization of network virtualization is largely due to the active research activities on *software defined networking* (SDN) techniques [36]. As a viable means of achieving the general concept of "programmable networks," SDN allows the system administrator to define the network rules and shape the traffic in a way that is more flexible than the existing solutions can provide. For example, Open vSwitch [13] is an open source software stack that can virtualize a host as a programmable switch and enforce customized forwarding policies according to each packet's characteristics, e.g., source and destination address. Moreover, Open vSwitch supports OpenFlow [14], which is a communication protocol to allow switches to obtain control information, i.e., forwarding rules, from a central repository, in order to separate the data plane and control plane for efficient and programmable network utilization. As an example, in the emerging techniques of converging LAN/SAN (storage area networks), Ethernet is used to carry both regular data traffic and storage traffic, which was supported by Fibre Channel in traditional SANs, via either Fibre Channel over Ethernet (FCoE) or Internet Small Computer System Interface (iSCSI) protocols. In order to support lossless transmission for storage traffic, the system administrator can assign higher priorities for FCoE and iSCSI packets at the Ethernet switch, compared to regular data packets, to avoid packet losses due to network congestions. Such network-wide prioritizing policies can be configured conveniently via the OpenFlow protocol, with the aid of the centralized repository, without manual configurations on each router.

47.2 Storage Virtualization Explained

The Storage Networking Industry Association (SNIA) defines storage virtualization as follows [16]:

1. The act of abstracting, hiding, or isolating the internal functions of a storage (sub)system or service from applications, host computers, or general network resources, for the purpose of enabling application and network-independent management of storage or data.
2. The application of virtualization to storage services or devices for the purpose of aggregating functions or devices, hiding complexity, or adding new capabilities to lower-level storage resources.

The general structure of a storage system consists of host servers (usually the initiators of data read/write operations), disk subsystems (physical location of stored data), and the network connecting them, for example, SAN fabric based on Fibre Channel, or Ethernet IP networks. Storage virtualization can be implemented at many places along the I/O path, as illustrated in Figure 47.4, which should be determined according to the IT requirements and business needs. Next, we will introduce three common places where storage virtualization is implemented, i.e., server, disk subsystem, and SAN storage virtualization. It is worth noting that in practice, a combination of storage virtualization at different levels can be utilized according to specific IT requirements such as fault tolerance, scalability, ease of management, etc.

47.2.1 Server Storage Virtualization

As one of the earliest forms of storage virtualization, the server's OS utilizes a logical abstraction of storage capacity, in tandem with the file system, to manage the mapping between data and actual location on the disk. For example, a physical hard disk can be divided into multiple *partitions* to isolate the disk space into multiple parts. The system administrator may create a partition for Windows OS and another partition for Linux to prevent overlapping of different file systems and conflicts. A partition can consist of multiple disk *blocks*, where each block is the smallest operation unit between the OS and the

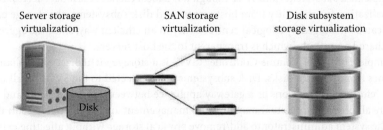

FIGURE 47.4 Storage virtualization on the I/O path.

hard disk. By virtualization, the system administrator can create a *logical* space on the disk, namely, a *volume*. Instead of physical partitions, the OS uses logical volumes to allocate storage spaces. In many modern OSs, such as Linux, there is a separate module, referred as the logical volume manager (LVM), to manage the mapping between physical disk blocks and logical blocks that can further aggregate to logical volumes. The LVM can resize each logical volume on the fly according to current system utilization and application demands. It is also common that multiple physical blocks are mapped to the same logical block for reliability concerns, e.g., mirroring.

47.2.2 Disk Subsystem Storage Virtualization

A disk subsystem is a specialized storage device optimized for fast I/O access, improved storage capacity utilization, and high availability, e.g., IBM System Storage DS8000 Series products [8] and EMC VMAX Family [5]. Physical disks are grouped together and presented as a logical disk using a storage virtualization technology known as redundant array of independent disks (RAIDs). Depending on its level, RAID provides striping for performance, mirroring for reliability, hot spares for resiliency, and error protection through the use of parity. These logical disks are in turn grouped together into a pool on which virtual disks are striped across. Greater parallelization (resulting in higher performance) and resiliency are achieved by striping a logical volume (i.e., virtual disk) across a group of RAIDs, which in turn stripe the data across multiple physical disks to achieve resiliency. This resource pooling abstraction also simplifies storage allocation by hiding the underlying physical storage device details from system administrators and avoids poor capacity utilization by allocating volumes efficiently and dynamically, with the appropriate amount of disk space (rather than entire disks) to host servers. In addition, modern disk subsystems leverage a virtualization technique named *thin provisioning* to further increase storage utilization by reducing the amount of allocated yet unused disk space within volumes. Thin provisioned volumes, also known as thin volumes, are volumes that present a virtual capacity to a host that is actually larger than its actual capacity. Thin volumes can have auto expand and auto shrink features to grow or reduce appropriately over time, in tune with the host server's application demands.

Another key advantage of virtualized storage disk subsystem is that due to the decoupling from physical location, data can be migrated from one volume pool to another. This aids in load balancing (e.g., alleviating overutilized devices) and attains better information lifecycle management (ILM) (e.g., moving frequently accessed data to a pool with faster devices such as solid-state disks). The subsystem executes data migration by making a copy of the data block in the target location and then updating its virtual to logical address mapping to redirect incoming I/O requests to the new location. Since the virtual address remains the same and the volume remains accessible the entire time, the data migration can be accomplished in a nondisruptive fashion.

47.2.3 SAN Storage Virtualization

SAN virtualization is able to reduce the cost and complexity of the overall storage network and improve performance and reliability. For example, a shared solid-state drive (SSD) can be placed in the SAN and used by multiple disk subsystems jointly, as though it is a local cache for fast data access. By shifting the storage virtualization functionality from host servers and disk subsystems, the storage capacity of the overall SAN can be managed by a logical central entity in an efficient way, since the aggregate resource pool can be shared seamlessly, which is transparent to the host servers.

As an example, IBM SAN Volume Controller (SVC) is a storage virtualization appliance that forms logical volumes from physical disks. Disk subsystems are connected to the SVC and all I/O access traverses it. Therefore, SVC functions as a gateway appliance between the host servers and the disk subsystems where all physical-to-virtual mapping and management are executed in a centralized fashion. This allows the system administrator to add/remove physical storage without affecting existing volumes since the virtual address remains unchanged.

SAN virtualization can be implemented in two ways, i.e., *in-band* and *out-of-band*. For in-band SAN virtualization, management, control, and data all share the same I/O path. In other words, all I/O data access traverses the management device, or SAN virtualization controller, such as IBM SVC appliance. In contrast, out-of-band solutions separate the SAN virtualization management from the data path, e.g., EMC Invista, which runs on SAN switches [4]. An additional software agent usually needs to be installed, and whenever an I/O request is initiated, the agent establishes a connection to the out-of-band SAN controller for data location lookups, and the actual data flow may go through a different path.

In-band and out-of-band SAN virtualization schemes are shown in Figures 47.5 and 47.6, respectively. In the in-band scenario, the centralized SAN controller will translate the logical location A to the physical location 1 and obtain the data on the server's behalf, while in the out-of-band case, the host agent will first inquire the logical–physical mapping from the out-of-band SAN controller and fetch the data next. Both SAN virtualization schemes are widely deployed in enterprise IT infrastructures. In-band solutions are usually easier to implement since no additional modifications are needed

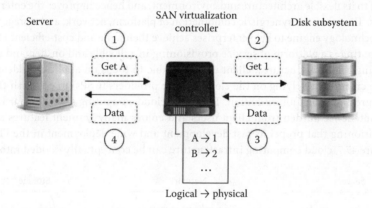

FIGURE 47.5 In-band SAN virtualization.

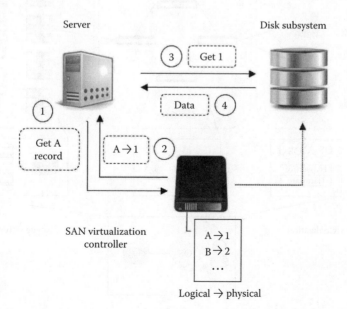

FIGURE 47.6 Out-of-band SAN virtualization.

on the host server devices. On the contrary, out-band solutions need a software agent installed on the host server whereas the SAN bandwidth is utilized more efficiently due to the separation of data and control flows.

47.3 Virtualization in Enterprise and Cloud Computing IT Systems

Enterprises utilize virtualization techniques extensively to build their scalable and reliable data centers. In traditional data centers, a single application usually resides in a designated server and the average server utilization is very low. Significant resource waste is due to the inflexibility of adjustment after the resources are allocated. In contrast, for modern data centers with virtualization, system efficiency can be remarkably enhanced by creating multiple VMs and consolidating them into a shared physical server to reduce space and maintenance cost (e.g., cooling). Meanwhile, improved performance can be achieved through the ease of resource sharing and portability across multiple VMs. Virtualization also enables faster IT solution deployment, e.g., hardware addition and replacement, and adaptive data migration, due to its flexible architecture and environment, and hence improves the enterprise business responsiveness. Therefore, the synergistic combination of platform, network, and storage virtualization serves as the technology engine to help enterprises achieve their agile and cost-efficient IT strategies.

Cloud computing is a platform of service provisioning in a flexible and on-demand manner. With cloud computing IT infrastructures, client users are able to focus on application development and business value creation by relying on the cloud service providers (CSPs) to establish the underlying IT infrastructure and actual job execution. Therefore, cloud computing shifts the hardware management and maintenance burden from client users and enjoys many eminent features such as elastic resource provisioning that propel its fast development and wide deployment in the IT industry. As shown in Figure 47.7, cloud computing infrastructure can be conceptually divided into compute and

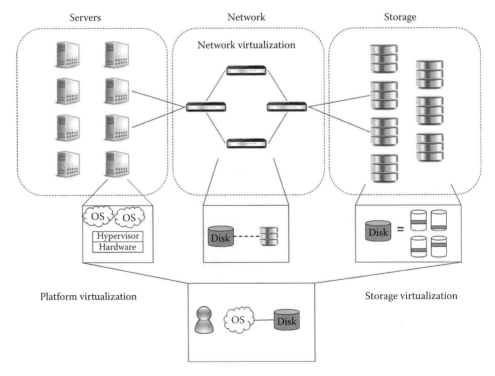

FIGURE 47.7   Virtualization in the cloud.

storage, which are interconnected via either Fibre Channel, IP networks, or gateway devices such as SVCs introduced in Section 47.2.

Cloud systems usually provide three types of on-demand services, i.e., Infrastructure-as-a-Service (IaaS), Platform-as-a-Service (PaaS), and Software-as-a-Service (SaaS). IaaS provides a pay-as-you-go model to utilize the cloud hardware resources, which allows users to specify the amount of CPU, memory, storage, and types of OSs, according to their application and business needs. For example, Amazon EC2 [2] provides a multi-tenant resource-sharing platform in the form of *VM instances*, and each instance is charged according to its CPU, memory, storage size, and the time duration it is leased. PaaS is an integration of both hardware and software infrastructures, e.g., Google App Engine [6] and Amazon Elastic MapReduce [3], which provide a convenient way for client users to develop and deploy their own business applications. SaaS usually provides an on-demand software service via a web browser, e.g., Google Gmail and Docs. From the cloud users' perspective, software development and upfront hardware cost can be saved and thus the CAPEX remains at a low level, compared to the conventional IT solutions.

Virtualization is *the* key enabling technique behind cloud computing systems to facilitate a flexible and scalable multi-tenant IT infrastructure. According to each user's application requirements, the cloud system can create multiple VMs, virtual disks, and connect them via a virtualized private network as desired. The resources allocated to multiple users are isolated while the scalability of each virtualized private environment is preserved. Cloud computing systems also leverage virtualization techniques for data migration and dynamic service provisioning to achieve both system resiliency and resource utilization efficiency.

47.4 Challenges and Research Issues

Despite the well-understood benefits and features, there are several remaining issues that need to be addressed to advance the virtualization techniques as well as the IT solutions based on them. We next discuss a few aspects of existing virtualization solutions as future research directions.

- *Virtual resource placement:*
  Due to the contention in cloud resource sharing, the overall system performance hinges largely on how virtual resources, such as VMs and virtual disks (VDisk), are placed on the physical servers. Obtaining the optimum resource allocation solution is usually computationally intensive and challenging due to its inherent combinatorial nature. In [31] and [34], the inter-VM traffic is considered and VMs with extensive mutual communication are allocated to nearby physical machines to alleviate network congestion. A joint VM and VDisk placement scheme is proposed in [35]. The authors of [27] consider the energy expenditure in VM placement where the resource allocation solution is adjusted based on power costs as well as application utilities. The chain effect of VM migrations is studied in [37], where a dynamic migration strategy is proposed for overcommitted cloud infrastructures. Specifically, the authors consider the statistical traffic patterns of VMs for multiplexing to avoid future VM migrations induced by the ongoing VM migration, and thus the number of migrations is reduced.

- *Multitenancy:*
  Most virtualization techniques require a centralized entity that provides the mapping between logical resources and physical resources. This inevitably introduces a single point of failure that triggers concurrent service disruptions for many clients in case of a failure. For example, during a reported outage of Amazon EC2 cloud [1], many customers such as Netflix, Instagram, Pinterest, and Heroku were affected simultaneously. In addition, while the system utilization can be remarkably improved, sharing application data of many different clients on the same infrastructure requires stringent security and privacy protection mechanisms. Therefore, how to share the physical resources in an efficient and trustworthy manner for multi-tenant virtualized systems remains an open question and deserves further investigation. For example, in [33], the

authors identify emerging security and privacy threats in a virtualized cloud environment due to the physical resource sharing among multiple VMs, and several mitigating solutions are proposed. A general framework based on an operational transformation model is designed in [29], which allows customers to collaborate securely on untrusted cloud servers.

- *Skyrocketing I/O demands:*
  Many virtualization solutions such as platform virtualization allow each physical system to host multiple VMs that share the same underlying hardware. However, such consolidation yields more intensive I/O demands that impose a remarkable burden on the I/O devices as well as the network and SAN switches along the I/O path. For example, if many VMs on the same server host database applications, voluminous I/O requests will be generated concurrently on the I/O path. Therefore, designing novel I/O techniques and architectures to accommodate large amount of (possibly bursty) I/O requests is imperative.

- *Software licensing:*
  Virtualization introduces new challenges from the software licensing perspective. With virtualized machines and platforms, the auditing and management of software license becomes more complex and difficult, especially due to the fact that VMs can be created and deleted at any time.

- *Root cause analysis:*
  As a cost of flexibility, virtualization adds a layer of indirection, and it is not uncommon that multiple layers of virtualization are implemented at different places within the overall IT system, which makes root cause analysis and network diagnosis more challenging. Information provenance and forensics in a virtualized environment are active research topics in the community, e.g., [26,30].

- *Chargeback policy:*
  In traditional IT infrastructures, the charge levied to each client can be easily computed by measuring the amount of allocated resources, e.g., the number of servers and the capacity of storage. However, in a shared virtualized environment, it is difficult to attribute the hardware expenditure using conventional methods due to many layers of abstraction and indirection. The actual hardware expense as well as the management cost should be charged to client users appropriately, which are only willing to pay for the amount of virtual resources that they requested. In [22], an integrated toolkit is designed to support fine-tuned cost models and charge policies of enterprise with shared IT infrastructure and heterogeneous users. Designing a flexible and profitable chargeback policy remains one of the key challenges for many CSPs with virtualized multi-tenant architectures.

47.5 Conclusions

Virtualization provides an additional layer of indirection by abstracting the physical resources into a logical resource pool, in order to attain efficient system utilization and flexible manageability. In this chapter, we provided an overview of various types of virtualization techniques utilized in current IT systems. Specifically, we introduced platform virtualization, network virtualization, storage virtualization, and their applications in cloud computing IT infrastructures. While virtualization has been widely deployed in many emerging IT systems, challenges such as multi-tenancy and software licensing issues remain as future research topics and need further investigation.

References

1. Sean Ludwig, "Amazon cloud outage takes down Netflix, Instagram, Pinterest, & more," June 29, 2012, http://venturebeat.com/2012/06/29/amazon-outage-netflix-instagram-pinterest/ (accessed on April 11, 2013).
2. Amazon Web Services, "Amazon Elastic Compute Cloud," 2013, http://aws.amazon.com/ec2/ (accessed on April 11, 2013).

3. Amazon Web Services, "Amazon Elastic MapReduce (Amazon EMR)," 2013, http://aws.amazon.com/elasticmapreduce/ (accessed on April 11, 2013).
4. Wikipedia, "EMC Invista," 2013, http://en.wikipedia.org/wiki/EMC_Invista (accessed on April 11, 2013).
5. EMC, "EMC VMAX Storage Family," 2013, http://www.emc.com/storage/symmetrix-vmax/symmetrix-vmax.htm (accessed on April 11, 2013).
6. Google, "Google App Engine," 2013, https://developers.google.com/appengine/ (accessed on April 11, 2013).
7. IBM, "IBM Power Systems," 2013, http://www-03.ibm.com/systems/power/ (accessed on April 11, 2013).
8. IBM, "IBM System Storage DS8000 Series," 2013, http://www-03.ibm.com/systems/storage/disk/ds8000/ (accessed on April 11, 2013).
9. S. Hanks et al., Generic Routing Encapsulation (GRE), IETF RFC 1701, October 1994, http://tools.ietf.org/html/rfc1701 (accessed on April 11, 2013).
10. S. Kent et al., Security Architecture for the Internet Protocol, IETF RFC 4301, http://tools.ietf.org/html/rfc4301 (accessed on April 11, 2013).
11. KVM, "Kernel Based Virtual Machine," 2013, http://www.linux-kvm.org/ (accessed on April 11, 2013).
12. Microsoft, "Microsoft Hyper-V Server," 2013, http://www.microsoft.com/en-us/server-cloud/hyper-v-server/ (accessed on April 11, 2013).
13. Open vSwitch, "Open vSwitch," 2013, http://openvswitch.org/ (accessed on April 11, 2013).
14. OpenFlow, "OpenFlow," 2013, http://www.openflow.org/ (accessed on April 11, 2013).
15. ORACLE, "Oracle VM VirtualBox," 2013, https://www.virtualbox.org/ (accessed on April 11, 2013).
16. SNIA, "The Storage Networking Industry Association," 2013, http://www.snia.org/ (accessed on April 11, 2013).
17. GENI, "The Global Environment for Network Innovations (GENI)," 2013, http://www.geni.net/ (accessed on April 11, 2013).
18. VMware, "VMware vSphere ESX and ESXi Info Center," 2013, http://www.vmware.com/products/vsphere/esxi-and-esx (accessed on April 11, 2013).
19. VMware, "VMware vMotion," 2013, http://www.vmware.com/products/datacenter-virtualization/vsphere/vmotion.html (accessed on April 11, 2013).
20. Xen, "Xen," 2013, http://www.xen.org/ (accessed on April 11, 2013).
21. K. Adams and O. Agesen. A comparison of software and hardware techniques for x86 virtualization. *International Conference on Architectural Support for Programming Languages and Operating Systems*, San Jose, CA, 2006.
22. S. Agarwala, R. Routray, and S. Uttamchandani. Chargeview: An integrated tool for implementing chargeback in it systems. *IEEE/IFIP Network Operations and Management Symposium (NOMS)*, Salvador, Brazil, 2008.
23. P. Barham, B. Dragovic, K. Fraser, S. Hand, T. Harris, A. Ho, R. Neugebauer, I. Pratt, and A. Warfield. Xen and the art of virtualization. *ACM Symposium on Operating Systems Principles (SOSP)*, Bolton Landing, NY, 2003.
24. M. Ben-Yehuda, M. D. Day, Z. Dubitzky, M. Factor, N. HarEl, A. Gordon, A. Liguori, O. Wasserman, and B.-A. Yassour. The turtles project: Design and implementation of nested virtualization. *USENIX Symposium on Operating Systems Design and Implementation (OSDI)*, Vancouver, British Columbia, Canada, 2010.
25. S. Bhatia, M. Motiwala, W. Muhlbauer, Y. Mundada, V. Valancius, A. Bavier, N. Feamster, L. Peterson, and J. Rexford. Trellis: A platform for building flexible, fast virtual networks on commodity hardware. *Workshop on Real Overlays and Distributed Systems (ROADS)*, Madrid, Spain, 2008.
26. D. Birk and C. Wegener. Technical issues of forensic investigations in cloud computing environments. *International Workshop on Systematic Approaches to Digital Forensic Engineering*, Oakland, CA, 2011.

27. M. Cardosa, M. Korupolu, and A. Singh. Shares and utilities based power consolidation in virtualized server environments. *IFIP/IEEE Integrated Network Management (IM)*, New York, 2009.

28. N. Feamster, L. Gao, and J. Rexford. How to lease the internet in your spare time. *ACM SIGCOMM Computer Communications Review*, 37(1), 61–64, 2007.

29. A. J. Feldman, W. P. Zeller, M. J. Freedman, and E. W. Felten. Sporc: Group collaboration using untrusted cloud resources. *USENIX Symposium on Operating Systems Design and Implementation (OSDI)*, Vancouver, British Columbia, Canada, 2010.

30. R. Lu, X. Lin, X. Liang, and X. S. Shen. Secure provenance: The essential of bread and butter of data forensics in cloud computing. *ACM Symposium on Information, Computer and Communications Security (ASIACCS)*, Beijing, China, 2010.

31. X. Meng, V. Pappas, and L. Zhang. Improving the scalability of data center networks with traffic-aware virtual machine placement. *IEEE INFOCOM*, San Diego, CA, 2010.

32. A. Menon, A. L. Cox, and W. Zwaenepoel. Optimizing network virtualization in xen. *USENIX Annual Technical Conference*, Boston, MA, 2006.

33. T. Ristenpart, E. Tromer, H. Shacham, and S. Savage. Hey, you, get off of my cloud: Exploring information leakage in third-party compute clouds. *ACM Conference on Computer and Communications Security*, Chicago, IL, 2009.

34. V. Shrivastava, P. Zerfos, K. won Lee, H. Jamjoom, Y.-H. Liu, and S. Banerjee. Application-aware virtual machine migration in data centers. *IEEE INFOCOM*, Shanghai, China, 2011.

35. A. Singh, M. Korupolu, and D. Mohapatra. Server-storage virtualization: Integration and load balancing in data centers. *IEEE/ACM Supercomputing (SC)*, Austin, TX, 2008.

36. K.-K. Yap, T.-Y. Huang, B. Dodson, M. S. Lam, and N. McKeown. Towards software-friendly networks. *ACM Asia-Pacific Workshop Systems*, New Delhi, India, 2010.

37. X. Zhang, Z.-Y. Shae, S. Zheng, and H. Jamjoom. Virtual machine migration in an over-committed cloud. *IEEE/IFIP Network Operations and Management Symposium (NOMS)*, Maui, HI, 2012.

48

Cloud Computing

Sharon E. Hunt
Pepperdine University

John G. Mooney
Pepperdine University

Michael L. Williams
Pepperdine University

48.1 Introduction

The concept of cloud computing has received widespread attention since its emergence in about 2007 and has become an important option for the provisioning of IS/IT infrastructure, platform, and applications. In 2012, the U.S. National Institute for Standards and Technology defined cloud computing as

> a model for enabling ubiquitous, convenient, on-demand network access to a shared pool of configurable computing resources (e.g., networks, servers, storage, applications, and services) that can be rapidly provisioned and released with minimal management effort or service provider interaction. (Mell and Grance 2011, p. 2).

Moreover, the NIST definition states that cloud computing, in its purest form, should exhibit the following "essential characteristics":

- *On-demand self-service*: users can access cloud computing capabilities as needed, without requiring human interaction.
- *Broad network access*: Cloud computing services should be available over the Internet and accessed through standard mechanisms, such as a web browser, which allows use by heterogeneous client devices including smartphones, tablets, laptops, and workstations.
- *Resource pooling*: Cloud computing capabilities are pooled to serve multiple users (a multi-tenancy model) and are assigned dynamically to users according to demand. As a result, users have no control or knowledge of the location of the underlying resources.

- *Rapid elasticity*: Computing capabilities can be instantly provisioned and released, typically automatically, to provide rapid and unlimited upward and downward scaling in response to computational needs.
- *Measured service*: Cloud computing services are metered according to service-appropriate metrics (e.g., storage, processing, bandwidth, number of active user accounts), with automatic monitoring of usage levels, and priced according to use.

While the combination of these characteristics is new, many of the underlying concepts of cloud computing go back to the origins of business computing in the 1950s. Unfortunately, a combination of semantic ambiguities and vendor opportunism has resulted in the term "cloud computing" being used very broadly to describe many variations of computing, storage, and network services across various "off-premise" deployment methods. From the business and user perspectives, cloud computing is an approach in which hardware, software, network, storage, and server resources can be accessed "as a service" from a third party, thereby eliminating the large up-front expense of purchasing the necessary components to build the required technical environment in-house. Additionally, cloud computing can refer to the virtualization of in-house resources, creating an internal infrastructure that maximizes the efficiency of proprietary hardware.

From a technical perspective, cloud computing is a style of computing that employs virtualization methods to create dynamically scalable IT resources. These can be offered to users as a service at a comparably affordable price as a consequence of the economies of scale that cloud service providers can achieve. Taken together, these business and technological perspectives have contributed to the rapid emergence of cloud providers over the past 5 years and the rapid adoption of cloud services. Cloud computing is becoming a computing style of choice by users, a dominant mode of computing within small- and medium-sized businesses, and a key component of the computing portfolio of large enterprises. Based on their extensive survey of cloud adoption practices at over 1000 firms, Willcocks et al. (2011) found that nearly two-third of executives viewed cloud computing as an enabling business service and IT delivery model that was driving innovation in organizations. In short, significant adoption of cloud computing has become inevitable (Seely Brown 2012) and is already well underway.

48.2 Underlying Dimensions

The breadth of resources and the variety of deployment models offered by cloud computing vendors contribute to its complexity. The following section lists the major categories of cloud computing commonly offered today with brief descriptions.

48.2.1 Domains of Cloud Computing

Infrastructure-as-a-Service (IaaS): "Infrastructure" typically refers to the tangible hardware components that are required in order to create a computing environment. In order to create computing and storage platforms, companies have traditionally built their own data centers that include the necessary servers, data storage units, power supplies, environmental controls, and other hardware components that provide the foundation for all their computing capabilities. In large companies, these are often large, dedicated centers that require full-time operations personnel and dedicated physical space. In smaller companies, this may simply be a "computer closet." Either way, infrastructure requires capital investment that must be kept operational and up-to-date in order to support the computing needs of the firm.

IaaS refers to the acquisition of these infrastructure capabilities as a "pay-for-use-service" on a contractual basis from a third party, in accordance with the essential characteristics as defined earlier. This eliminates the need for costly physical assets and operational expenses, as well as the need to hire experts to configure, operate, and maintain them. With IaaS, a company can host its enterprise applications on virtualized hardware that they never have to see, update, or troubleshoot.

Platform-as-a-Service (PaaS): In order to understand what is meant by PaaS cloud computing, it is first necessary to understand the concept of a computing platform. In software terminology, a platform is the underlying middle-layer code upon which applications and their interfaces are built. For example, the operating system on your laptop (such as Microsoft Windows or Mac OS) is an example of a computing platform. To simplify, we can think of a computing platform as simply an environment upon which software applications are launched and run.

When translated to cloud computing terminology, "platform" has been used to describe several different types of computing capabilities and uses, and this definition continues to evolve. PaaS refers to "renting" an operating platform, such as Force.com or Microsoft Azure, on which a firm can build customized software applications to meet its specific needs. Procuring a cloud-based platform also implies that the cloud vendor will be providing the infrastructure necessary to support the platform, so you can think of a cloud platform as the delivery of IaaS with an additional layer of functionality upon which a company can develop or purchase compatible software solutions. Cloud-based platforms are designed with software developers in mind and so provide the tools and support that developers need in order to create agile, scalable applications that typically are also mobile device and social media enabled. PaaS cloud computing remains one of the less developed areas of cloud computing. Over the coming years, its definition will likely evolve and become inclusive of more computing functionalities such as database management, business activity monitoring, and corporate governance (Mitchell Smith et al. 2012).

Software-as-a-Service: Cloud-based software-as-a-service offerings have been around for over a decade and have become the most commonly consumed cloud service genre across consumer and enterprise users. Essentially, SaaS delivers software functionality as a service, rather than as a packaged product that is licensed for use. The functionality is "bundled" with the underlying infrastructure needed to host and run an application and the operating platform on which the application is built. Examples of SaaS solutions in the end-user and consumer markets include web-hosted email providers (such as Gmail or Yahoo! Mail), search engines, Quicken, Pandora, and Netflix. Common enterprise SaaS offerings include Salesforce.com, Workday, and Radian6, an analytics tool used to gauge conversation and sentiment trends via social web conversation analysis. With SaaS, functionality is delivered remotely to registered users over the Internet via a browser. The service is paid for according to the terms of a contract (generally per user per month). Instead of consuming hard drive space and processing power on a native device such as desktop computer or laptop, SaaS solutions are provisioned remotely by a vendor's virtual infrastructure. Additionally, SaaS solutions are constantly being updated as its developers are fixing bugs and building/releasing new features and functionalities. These improvements to the application are automated for the end user and do not require a new version purchase or lengthy update downloads.

Business Process-as-a-Service: BPaaS remains the least mature, but potentially most business beneficial area of cloud computing. An extension of business process outsourcing (BPO) that incorporates cloud computing principles, BPaaS involves the provisioning of highly standardized end-to-end business processes that are offered through on-demand, pay-per-use, and self-service provisioning models. Currently, the most common process capabilities being sourced via this model include employee productivity, employee source to pay, payroll, customer contact to incident resolution, hire to retire, and order to cash (Karamouzis 2012). The underlying resources and capabilities used to provide the business processes are shared by multiple customers, with possible additional benefits that can emerge from pooling data across these multiple customers. Ried and Kisker (2011) suggest that "BPaaS in many ways represents the culmination of cloud based innovation" (p. 4). It is possible that future innovations in BPaaS services will evolve toward the concept of "Complex Adaptive Business Systems" as articulated by Tanriverdi et al. (2010).

48.2.2 Deployment Methods

Now equipped with a working knowledge of what cloud computing and the general categories of cloud services, the next step in fully understanding cloud technology is to compare and contrast the ways in which these solutions are delivered.

Public Cloud Platforms: The public cloud generally refers to a cloud service that is delivered over the public Internet and is shared by multiple users or "multitenant." Examples include Amazon Web Services, Google Docs, Facebook, and Salesforce.com. Generally, users subscribe to a public cloud service (such as an application or storage space) and then, instead of being supported by an internal virtualization network, users can access that account anywhere from which there is a public Internet connection.

Private Cloud Platforms: The private cloud differs from the public cloud much the same way that an intranet differs from the Internet. Essentially, a private cloud is created with a company's proprietary resources. However, it uses cloud computing techniques such as virtualization and data center automation in order to achieve the elasticity benefits that are one of the major advantages of public clouds but also the efficiency and productivity benefits associated with cloud-based provisioning. Generally, private clouds provide their owners with greater control over their capabilities and are often cited as providing a higher level of security (Accenture 2010). Companies can either invest in and create this cloud independent of any third-party vendor or utilize third-party services and the "pay-as-you-go" model with private cloud vendors. Examples of such a vendor are EMC's VMWare, a company that offers enterprise software virtualization products, and Pogoplug that offers products that allow a personal computer to become a web-based file sharing server, pooling storage capacity across all computers within a single company and creating a central location for file access.

The Chief Information Officer (CIO) of a global construction company that was an early adopter of cloud computing shared the view that the biggest difference between a public cloud and a private cloud is whose balance sheet includes the infrastructure asset. The idea behind this statement is that if the technology and capabilities of the cloud are the same or similar internally or externally, then there is no functional difference in regard to business processes (HBR Analytic Services 2011). Although this may be true for a company that does not have the need to create flexibility in its IT operations beyond what can be provided by a cloud vendor, for a company that competes in part through its ability to provide cutting-edge customer service, tracking, supply-chain management, or other functions that rely on increasingly improving IT capabilities, this may be less likely. Private clouds, although generally more expensive to build and maintain than public cloud services, can provide a company with the ability to create their own customized solutions above and beyond what is currently available in the vendor market.

Community or Virtual Private Cloud Platforms: A community cloud refers to a cloud that is hosted on the public Internet but is not fully multitenant. Rather, it is shared between a restricted community of firms or users, thus addressing the concerns of security and/or verifiability of the geographic location of the data. One major downfall of using a private cloud is the lack of economies of scale that arise from the efficiencies inherent in providing services to the large number of users that tap into public cloud solutions. In industries where similar businesses use private cloud computing capabilities, it is often advantageous to pool these resources in order to achieve some of the economies of scale that are lost in single-firm private clouds. Generally, these community clouds integrate numerous private clouds, using the unallocated resources from user machines to create a virtual data center.

Hybrid Cloud Platforms: Hybrid clouds refer to a type of cloud computing deployment that uses a combination of more than one delivery method, such as private and public cloud services. Typically, firms that are highly concerned with security issues and/or have legacy applications that would be too costly to move to public cloud computing services will maintain a portion of their computing capabilities within a private ecosystem and utilize public software and platform clouds to enable other business processes. Another reason hybrid clouds are popular is that they provide a means from which enterprises can remove data quickly from a public cloud, in response to either a pending security threat or complications with a vendor.

48.2.3 Location Types

There are three different ways that clouds can be located, each with its own set of capabilities and purposes:

On-site: An on-site cloud is a cloud that is located entirely on the premises of the user. The cloud can either be owned by the company itself or it could be the property of a third-party vendor that installs, maintains, and services the hardware. The latter option allows companies to have the benefits of private cloud computing without the hefty capital expenditure of purchasing the equipment. Hitachi Data Systems is a vendor that originally offered cloud computing services for telecommunication companies in the public cloud. However, having identified and understood the hesitation of executives in the telecommunication industry to move private consumer information into a public cloud, they began offering on-site, private clouds that they locate within the client company's own firewall.

Dedicated: Like an on-site cloud, a dedicated cloud serves only one client—however, the cloud is located in the provider's facilities. Companies can obtain dedicated infrastructure for their own use, instead of sharing with other "tenants." This allows a company to have greater control over the infrastructure—for example, if a company is utilizing a dedicated server to host its portfolio of corporate websites, then it would be able to select what operating system the website will use and what sort of security protection (such as encryption capabilities and firewalls) the server will provide. In general, a dedicated cloud gives the client more ability to customize the machine in accordance with its own needs (Loebbecke et al. 2012).

Shared: Shared clouds are the most commonly used type of cloud. These are located off-site (generally in the vendor's facilities) and allow multiple clients to utilize the same resources (Loebbecke et al. 2012). There is limited customization with this location method, but it is generally a less expensive option than on-site or dedicated clouds.

48.3 Impact on Business Practice: Applications of Cloud Computing and Their Outcomes

As more companies adopt cloud computing successfully to minimize cost, increase agility and standardization, and bring products to market faster, other firms across the spectrum of size, industry, and globalization are taking interest (Abokhodair et al. 2012). Although the potential benefits of cloud computing are compelling, managers must be aware of what objectives they are trying to accomplish with cloud technologies and develop effective management practices for eliciting and maintaining the intended benefits.

48.3.1 Benefits of Cloud Computing

The most commonly cited benefits offered by cloud computing are significantly reduced IT fixed costs and IT capital assets, increased speed of deployment of IT services, and greater flexibility or "elasticity" of IT services provisioning (Accenture 2012; Harris and Alter 2010; Iyer and Henderson 2012; McAfee 2011). Short-term variable cost reductions may also emerge from reductions in data center operational costs, IT operations, maintenance and support staff, electricity usage, and software licensing fees. Large, established companies can also take advantage of these savings by substituting internally developed or acquired assets with infrastructure, platform, and application services that are hosted in the cloud. Likewise, companies that are just entering new markets can experience decreased barriers to entry due to the fact that they have instant access to "world class" computing capabilities at affordable prices. Beyond cost, however, cloud computing offers other important areas of benefit for business organizations. Drawing from their research on early adopter companies, Iyer and Henderson (2012) identified six "benefit patterns" that companies had achieved from leveraging cloud computing capabilities: increased business focus, reusable infrastructure, collective problem solving, business model experimentation, orchestrating business partner dependencies and the "Facebook effect" of leveraging mobile devices, social networking, real-time data, and user-driven content generation. The aggregation of perceived business benefits across a wider range of adopters reveals the following primary areas of impact:

Flexibility: Cloud computing can offer flexible resources to companies that experience seasonal or sporadic increases in computing capacity needs. By offering these resources in such a way that is immediately elastic, companies need not invest in extra infrastructure just to support periodic bursts or suffer from overloading due to insufficient resources. These two factors not only add to the potential cost savings of a firm but also help to negate the risk of unsatisfactory service due to a lack of capacity. Due to the nature of cloud computing, new capabilities can be quickly implemented and deployed, increasing the speed in which a firm has access to technologies that assist in creating and delivering products, thereby increasing the speed in which these products get to market.

Standardization: Another compelling benefit of cloud computing is the ability for a large corporation to be able to standardize applications across previously segmented and siloed parts of the firm. Oftentimes, the apparent needs of different departments/regional locations within the same company give way to employing multiple legacy applications that are not capable of sharing information with one another, creating silos, communication difficulties, and compromising business intelligence. The task of extracting the data in these applications and moving them over to a new integrated system that attempts to solve these issues can be daunting and expensive. Cloud computing offers the possibility of deploying such a shared platform with greater speed, less expense, and less user resistance than creating an internal system.

Business intelligence and big data: Business insights and analytics tools powered by public and community clouds are becoming an increasingly popular means to access and analyze the vast datasets now available to business organizations. For example, some cloud analytics vendors create detailed geographic and demographic profiles by analyzing and interpreting the click behavior and interactions of millions of online consumers. By tapping into the resources generated by public clouds, firms are able to attain access to more granular (and therefore accurate) information than they would have otherwise. Some vendors, such as APTARE, leverage a community cloud in order to provide its members with analytical data regarding how well they are optimizing their data storage resources. The company, while keeping each individual community member's information private, is able to study and compare how well a firm is utilizing its resources compared to other members of the community with an infrastructure of similar size and scope.

Companies that develop their own internal clouds can use these capabilities to offer innovative products and services in addition to their core offerings, gaining a competitive advantage in the marketplace.

Enabling focus on core business competencies: In an ideal world, all IT systems would effectively act as enablers of core business functions, creating a more efficient and effective workforce. Too often, however, technology begins to create operational boundaries, limiting the very employees that it was implemented to help. Salespeople in the field are not able to provide customers with immediate information without access to their computers; IT professionals are distracted from their core duties when expanding technological capabilities to a new branch location; customer service agents cannot deliver speedy service when they are required to search through multiple, nonintegrated databases to gather basic information.

Cloud computing can offer solutions to such problems through multi-device access, anywhere anytime access, and standardization across all users. Salespeople can answer customer questions in real time via mobile cloud software services, branch openings are simplified by utilizing cloud computing vendor capabilities for immediate expansion, and SaaS offerings may provide integrated databases and immediate knowledge-sharing capabilities. In such scenarios, employees are empowered to focus on what matters most to the core business, whether that be customer service, knowledge expertise, or something else (Iyer and Henderson 2010).

48.3.2 Management Practices for Achieving Benefits from Cloud Computing

Although the possible business benefits of cloud computing are significant, managers of firms that intend to deploy cloud computing for business benefit must understand that cloud deployment is much like any other IT-enabled business process in that it must be carefully analyzed, planned, deployed, and monitored after implementation. Because cloud computing is an emerging technology with low barriers to adoption, it may be easy to fall into the trap of experimentation and speculation. Before moving to a cloud solution, managers must perform a comprehensive analysis of benefits, costs and risks, vendor due diligence, and develop compelling business cases built on appropriate, verifiable metrics (Herbert and Erickson 2011).

Although there are clear steps that should be taken before any business leader embarks on a journey to the cloud, research shows that managers who have had more than 5 years of experience with cloud computing are convinced that the benefits are greater than the risk (HBR Analytic Services 2011). Experience with cloud computing cultivates effective management capabilities, which in turn engenders confidence. Recent research showed that only 4% of senior managers in companies with more than five years of experience perceive that the risks of cloud computing outweigh the benefits (HBR Analytic Services 2011).

48.3.2.1 Understand the Market and Your Current IT Environment

Before undergoing a switch from traditional to cloud computing solutions, managers must understand thoroughly the rationale for this decision. In what areas does the firm need improvement? Is the market faced with an influx of new entrants that are more agile and more rapidly bringing products to market? Does the company need to standardize its processes in order to increase productivity and minimize the learning curve for cross-functional teams? Are legacy applications outdated and difficult to modernize?

A thorough understanding of the ways in which cloud computing is currently being used in your industry is necessary before coming to conclusions regarding how best to incorporate it into your company's current operations. This process is further aided by mapping out the company's IT infrastructure, platform, and application ecosystem. What business processes are supported by legacy applications? What is involved with moving these processes to cloud-based solutions? Is the cost of the transition sufficiently outweighed by the expected benefits gained from increased efficiency and productivity? How strong are the data used to support the decision?

Investigating industry best practices with the cloud and how conducive the firm's current technology management capabilities are to moving to a cloud environment will bring clarity to the adoption decision. Although the possible benefits of cloud computing are numerous, companies will need to have a clear vision for why they are moving to cloud computing and what challenges and/or synergies it will encounter with its current ecosystem.

48.3.2.2 Define and Manage the Intended Outcomes

Because of the extent of hype surrounding cloud computing, it may be easy to assume that moving a company over to cloud solutions will reduce costs and increase agility, but in order to get the most out of cloud-based solutions, managers must be specific regarding the intended processes they intend to improve, cost and revenue metrics, and other business outcomes. Managers cannot simply make the switch to cloud computing because it is trending in the industry or because they expect certain outcomes that are loosely defined. This goes beyond identifying areas of opportunity to creating a road map that describes hard objectives with useful measurements, the stakeholders involved, how the change will affect them, and the identification of who in the company will be responsible for ensuring that the objectives come to fruition (Peppard et al. 2007).

Involve stakeholders and stakeholder representatives in the formation of this plan in order to identify and address concerns that may not be apparent. Including those impacted by a change to a cloud computing solution in the decision itself will not only give managers greater clarity regarding the likely consequences of the change, but will also help to create buy-in from different user groups.

48.3.2.3 Define and Manage the Necessary Changes

If the current IT ecosystem is thoroughly understood and if that information aids in the selection of a cloud solution, deployment model, and vendor, then much of the legwork for defining what actual changes need to happen for a successful transition is already done. Examples of such changes could include redefining certain professional roles, training personnel on a new software solution, or redesigning a particular business process to be aligned with standardization goals. Whatever the change, managers must identify it and assign the management of it to a particular owner. These changes should be linked to their intended business outcome and should follow a plan that includes a timeline, milestones, and risk mitigation considerations.

In order to successfully create a realistic timeline and properly put into place mechanisms of risk mitigation, stakeholders must once again be consulted and considered. Who will be affected by this change? What pending projects are reliant on the legacy application that is being eradicated? What technological capabilities will be lost if all data centers are consolidated and an IaaS solution is deployed? What contingency plans are in place to help business processes continue in the event there is a hitch? What long-term cultural and organizational changes will result in this new business model?

48.3.3 Cloud Computing Risks

Although the possible benefits from switching to a business model that incorporates cloud computing has the potential to be dramatic, there are also many risks and challenges involved in the decision, which conscientious managers must address.

48.3.3.1 Confidentiality and Privacy of Data

What managers across the globe are concerned with is the threat of data privacy and confidentiality breaches (Accenture 2010). In addition, global companies may need to contend with differences in privacy laws depending on where they conduct business and prepare for subpoenas, searches, and seizures as permitted by local law. As we progress into the twenty-first century, the laws and regulations surrounding the use of cloud computing services will need to evolve with these developments. Firms will need to ensure that they remain under compliance and are protected. Before entering into new global territories, firms needed to closely scrutinize the liabilities that they may be subject to as users of cloud computing. They will also need to regularly audit their vendors to ensure that they are regularly iterating the development of their security measures to keep up with constantly evolving threats.

48.3.3.2 Market and Vendor Maturity

At this point in time, the cloud computing market is relatively immature, so there are constantly emerging vendors that are vying for market share. Although newer entrants may appear to offer better deals or other benefits, there is inherent value in a vendor that has proven its reliability as a trustworthy business partner. If a company's day-to-day operations rely on applications, stored data, or other capabilities that are powered by the cloud (as they likely will), then it could be potentially disastrous for that cloud to incur some sort of technical failure.

Relatively young firms and new entrants dominate the market space of cloud computing providers. This is particularly the case with the most innovative cloud service firms, many of which are new ventures that are dependent upon venture capital for financial stability, whose offerings are relatively new to market and are relatively unproven, and whose customer base is small when compared with the long-established vendors of comprehensive enterprise systems. The overall immaturity of this sector creates greater risks and concerns for adopting firms and will require robust due diligence in vendor evaluation and contracting (Phifer and Heiser 2012).

48.3.3.3 Integration with Existing On-Premise Systems

A significant challenge for many firms relates to the necessity to integrate data between traditional on-premise systems and cloud-based services in order to provide for data standardization and process coordination. One of the key benefits of tightly integrated enterprise resource planning (ERP) systems was their embedded support for data integration and master data management. The potential shift from, or augmentation of, these integrated ERP systems with specialized SaaS services creates potential risks of data silos and data inconsistencies (Radcliff 2012).

48.3.3.4 Low Maturity and Consistency of Local and Global Regulations

Another potential challenge for the adoption of cloud computing is the immaturity of regulations that govern various critical aspects of cloud computing services including vendor responsibilities, service-loss accountabilities, data loss prevention, and privacy protection. This sphere is further complicated by differences and inconsistencies in the prevailing policies and regulations across major regions of the world.

48.3.3.5 Reduced Control over IT Demand

Another potential risk associated with the availability of apparently more affordable, on-demand, direct access by users to unlimited computing resources is that demand and consumption of these services can grow exponentially, resulting in significant increases in overall computing costs. Contrast the likely behaviors when users must operate within a fixed storage allocation (e.g., mailbox, file share, or data archive quota) versus behaviors when offered apparently unlimited storage as a service. Combine this with the ability of business unit executives to independently contract for cloud services directly with cloud providers, and the potential for rapid escalation of demand and consumption of IT services emerges. For better or worse, IT units and traditional IT management approaches played a "demand management" role that served to control rising demand and consumption of IT services.

48.3.3.6 Vendor Reliability and Long-Term Viability

Any company that is considering moving critical computing capabilities to the cloud must thoroughly consider the risks of vendor failure. The disaster recovery and business contingency measures that a company should take to protect itself against possible vendor outages are not unlike those it would take to protect against failure of "on-premise" computing infrastructure. Data centers built internally are not immune to risk and must include contingency plans in the event of an unexpected interruption, whether it be from a natural disaster, faulty equipment, malevolent interruptions, or some other occurrence.

However, because of the media attention that cloud computing outages have received in the last couple of years, the actual risks associated with cloud computing reliability have been somewhat exaggerated. A cloud vendor's reputation and long-term viability are highly dependent on its reliability, and so cloud vendors have a string motivation to ensure that their services are backed by robust technologies, processes, and management practices. Although there are certainly examples of large failures such as Amazon E2C's 3 day outage in 2011, there is also evidence of solutions that have thus far proven to be more reliable than their in-house counterparts. For example, Google's Gmail-enabled email services achieved 99.984% uptime in 2010—making it approximately 32 times more reliable than the average enterprise email system (McAfee 2011).

This is not to say that a loss of service is not a serious threat for companies that rely on cloud computing capabilities. However, proper steps can and should be taken to mitigate the potential consequences of a system failure. Netflix is an example of a company that understands this principle and has truly integrated it into the DNA of its operations. When Amazon's cloud services famously went down in the spring of 2011, Netflix was able to continue to serve its customers uninterrupted because it had tested its systems for this contingency using an internal interference system it had built dubbed "Chaos Monkey." The system was designed to randomly attack subsets of the company's technology environment. These built-in attacks required all of Netflix's operative processes to develop adequate and immediate contingency plans, ensuring that the company was thoroughly prepared for an outside threat (McAfee 2011).

Beyond temporary services outages, a greater concern for cloud adopters is the long-term financial security and viability of highly innovative cloud service providers, many of which remain at the "early stage venture" of the business lifecycle, dependent upon investment funds for operating capital, and desperately striving to achieve the level of market success to ensure their survival and prosperity. Cloud adopters will need to conduct diligent analysis of the financial security and long-term viability of these highly innovative but immature cloud service providers.

48.3.3.7 Security

Skeptics of cloud computing often refer to the risks associated with off-premise cloud computing vendors maintaining the levels of security necessary to protect critical capabilities and proprietary data. These critics maintain that on-site data centers are superior in this regard due to the fact that the company itself controls them. This logic, however, is questionable at best. Cloud service providers are aware that their long-term viability is dependent on mitigating security breaches and consequently will typically dedicate far more resources and expertise to establishing and maintaining a highly secure environment than a typical client company would independently. That said, firms that have adopted cloud computing must manage the business risks associated with that decision and develop business continuity plans (backup plans) in the event the cloud service is inaccessible, or data are compromised, and develop appropriate data recovery processes. These plans should be reexamined periodically to ensure that the considerations remain relevant and complete.

In the absence of effective guarantees about service and compensatory damages from suppliers, and the absence of legal precedent in these areas, some firms have addressed the potential loss, risk and liability issues by taking out cyber-liability insurance (HBR Analytic Services 2011). Users can also manage this risk by dealing with well-established vendors, avoiding new vendors that have not yet proven reputations in the market. A clearly defined exit path should be developed to prepare for an event in which the vendor suffers some unexpected loss, i.e., bankruptcy or a natural disaster (Stroh 2009).

48.3.3.8 Vendor Lock-In

Firms need to pay special attention to the construction of their clouds, not just in the interest of building the most relevant and cost-effective solution, but also to mitigate the threat of too much vendor control. Vendor lock-in can come in a variety of forms: lengthy and binding service contracts, applications built with proprietary Application Programming Interfaces (APIs), voluminous data storage costs, and hefty switching costs resulting from an ecosystem dominated by one vendor. However, as the cloud computing market grows, firms have an ever-increasing pool of vendors to choose from and should leverage their buyer power to create flexible solutions to minimize these threats.

The first step for any firm that is considering deploying a cloud solution is to build a blueprint of the desired ecosystem. This blueprint should consider what services are necessary, required storage capacity, operating requirements, where user training would be needed, and what elements of design are required for enabling seamless portability of public or community cloud data to the private cloud (and vice versa). Once a firm has a holistic picture of its service needs, it should consider how best to transition from existing assets and resources to the proposed cloud platform.

After taking a comprehensive examination of how the firm intends to use the cloud to satisfy its immediate and longer-term needs, it will be prepared with the right set of questions to present to vendors. For example, if the firm intends to build custom applications, it should consider whether or not its PaaS vendor offers APIs that allow for app transfer to a new platform. In 2011, Google's App Engine enacted a sudden price increase, but because the applications built using this platform could not run anywhere else, firms had to weigh the cost of redevelopment against the cost of continuing to stay with App Engine. Users of PaaS offerings should also consider the type of code used to build applications, since some code is less expensive to alter than others (Gigaom 2011).

Considerable time should also be given to the negotiation of contracts, service agreements, and termination clauses. Of particular concern are the agreements around data portability. If possible, firms should diversify their cloud environment enough to limit reliability on any one vendor. Cloud vendors should be required to be transparent with their own ecosystem to ensure that customers are not unknowingly using third-party vendors that have not been appropriately vetted. As the market goes through a familiar pattern of mergers and acquisitions in the coming years, firms will need to dig deeper to uncover potentially problematic interdependencies in their proposed cloud ecosystems; otherwise, the diversification of vendors will not have the intended effect (White et al. 2010). By performing due diligence in this regard and leveraging buyer power to obtain preferential contracts, firms can hedge against multiple varieties of vendor lock-in.

48.3.3.9 Uncertainty around Return on Investment

It might be easy to get swept away in the hype around cloud computing. But like any major business decision, detailed analysis must precede any decision-making. First and foremost, there must be clearly defined benefits that are the intended outcome of the cloud migration. Without understanding *exactly why* a firm is considering a move to the cloud, a proper analysis (including opportunity costs and alternative options) cannot be performed. Once these intended benefits are articulated and included in the business case, however, firms should still rely as much as possible on hard metrics instead of speculative hunches. Tangible costs of the current environment should be compared with a holistic assessment of the costs of transition to cloud-based alternatives, including training, losses from temporary lapses in efficiencies, potential client attrition during the early stages, the added expense of hiring personnel to maintain vendor relationships, and the cost of the vendors themselves.

Depending on how a firm intends to use the cloud and the business benefits it hopes to gain, it will most likely need to develop hypotheses for the financial impact of these benefits. Where hard data do not exist, the hypotheses should be tested in microexperiments in order to avoid making larger, faulty assumptions. For example, if an insurance company believes that moving its customer data to the cloud would improve accessibility for its adjusters and therefore boost the speed of claim resolution and overall customer satisfaction, it could conduct a geographic experiment wherein a small subset of its covered locations are shifted to the cloud. This hypothesis could then be tested and proved (or disproved) before making a significant, company-wide transition (Harris and Alter 2010).

As hinted at in the previous example, return-on-investment, although a popular metric, may not be the best lens through which to view a cloud transition. For example, if one of the intended benefits is to improve employee collaboration or to become more agile in an increasingly dynamic marketplace, attaching a realistic Return on Investment (ROI) number may prove futile. However, the business impact of these improvements may be significant and should not be overlooked when making a case for cloud initiatives. Once again, experimentation and hypothesis testing can be very valuable for evaluating these assumptions and avoiding misguided or lofty predictions.

Finally, firms should ensure to carefully consider the various pricing models of the proposed vendors and evaluate the implications of these in the event of anticipated business growth and broader strategic goals. A thorough examination of available vendors should consider not only capabilities and lock-in potential, but whether or not their pricing models are aligned with the company's strategy (Stroh 2009).

48.3.3.10 Resistance from Executives, IT Managers, and Users

A thorough business case analysis should incorporate securing executive team support for moving critical capabilities to the cloud. However, studies show that IT professionals and business professionals tend to view the risks and benefits associated with moving to the cloud quite differently, and so the same pitch may not be effective across the board (Willcocks et al. 2011). Namely, IT professionals tend to see the threat of lock-in as much more looming than their business counterparts, who in turn are more skeptical of the potentially negative impacts of *not* moving to the cloud. In addition, IT managers tend to be much more conservative regarding the expected benefits of the cloud and the time horizon necessary for attaining these benefits. The gaps in risk perception and expectations between business and IT staff members can potentially cause tension during the shift to the cloud, causing undue strain on the transition (Willcocks et al. 2011).

An effective way to manage this likely divergence of views is to build a cloud strategy team that is cross-functional, including members from the executive team and multi-department managers, including business and IT representatives. Additional valued members of this team would include law experts, innovation specialists, and managers adept at instituting new processes. Such cross-functional teams can do much in terms of bridging the gaps between the business and technology sides of the firm, creating a holistic and realistic cloud strategy, and earning buy-in from key company players (Harris and Alter 2010).

48.4 Future Research Issues and Challenges

Given its recent emergence and relative maturity as an important alternative approach for the provisioning and consumption of computing resources, there is relatively little published academic IS research on cloud computing to date. However, a review of the research papers included within the leading IS academic conferences in 2012 (e.g., the 2012 European Conference on IS, 2012 Pacific-Asia Conference on IS and the 2012 Americas Conference on IS) clearly shows that much research on cloud computing is in process. That said, it is also clear that many opportunities for comprehensive academic and applied research studies on the technological, business, and managerial opportunities and challenges associated with cloud computing remain. The levels of adoption of cloud computing motivate a need to reevaluate much of the prevailing wisdom about computing across almost every chapter in this volume. Critical areas for which new insights will be needed include application services development and deployment methods for cloud computing environments; models and approaches for effective data management within distributed cloud-based systems; global regulations for data protection and data discovery, and management practices for adhering to these; enterprise architecture in the context of cloud-based services; services versioning and pricing; collaboration and collective intelligence in massively online cloud computing contexts; security and data privacy management; effective governance practices for cloud computing; and management practices for realizing value from cloud computing. Other areas of great importance are the implications of cloud computing for internal IT units and their requisite skills and capabilities for a model of business computing in which the dominant traditional competencies of design, development, and operation will be overshadowed by competencies in services architecture, brokering, and integration and in which users can assume much greater independence and control over their computing needs.

48.5 Summary

Cloud computing has emerged to offer important capabilities for corporate computing. For small- and medium-sized enterprises, the affordability, scalability, and simplicity of deployment of cloud computing are driving rapid adoption. Within these small- and medium-sized businesses, cloud computing is becoming the dominant approach for providing infrastructure services, application functionality, and business process capabilities. Cloud computing is diminishing many of the barriers to "enterprise-class computing resources" that have been previously available only to large firms with extensive financial resources and comprehensive internal IT management capabilities, and thus leveling the field for all players in the digital economy.

For large enterprises, cloud computing is providing compelling features and capabilities that augment and extend the accumulated portfolio of IT infrastructure and applications. Cloud-based approaches have been especially valuable for supporting the incorporation of social media interfaces, social network applications, tablet and smart phone devices, computing-intensive "deep analytics," and storage-intensive "big data" opportunities. For users, cloud computing has offered the opportunity to "untether" from desktops and embrace mobile computing devices with access to powerful on-demand applications and interactive information services.

Of great significance for the future of computing are the realities that almost all venture capital for software start-ups is being directed toward ventures that are embracing a cloud-based distribution model and that most established enterprise software companies are actively developing and executing a cloud strategy for their products. While some features of cloud computing are new, enabled by recent technology innovations, it also has to be realized that the emergence of cloud computing is entirely consistent with the evolving pattern of business computing and information systems provisioning that has been developing for the past 70 years. Fortunately, much of the knowledge about effective approaches for realizing value from computing that has been accumulated over this period is immediately applicable to cloud computing. However, consistent application of this knowledge to guide management practices remains a perennial challenge.

Further Information

Anthes, G. 2010. Security in the cloud. *Communications of the ACM* 53(11):16–18.

Armbrust, M., R. Griffith, A.D. Joseph et al. 2010. A view of cloud computing. *Communications of the ACM* 53(4): 50–58.

Babcock, C. 2010. *Management Strategies for the Cloud Revolution: How Cloud Computing Is Transforming Business and Why You Can't Afford to Be Left Behind.* New York: McGraw-Hill.

Backupify. 2012. The 2012 data liberation awards. http://blog.backupify.com/2012/07/02/infographic-first-annual-data-liberation-awards-the-rankings-are-in/ (accessed on November 2, 2012).

Benlian, A. and T. Hess. 2011. Opportunities and risks of software-as-a-service: Findings from a survey of IT executives. *Decision Support Systems* 52(1): 232–246.

Berman, S. et al. 2012. The power of cloud: Driving business model innovation. *IBM Institute for Business Value.* http://www.ibm.com/cloud-computing/us/en/assets/power-of-cloud-for-bus-model-innovation.pdf (accessed on October 3, 2012).

Brynjolfsson, E., P. Hofmann, and J. Jordan. 2010. Economic and business dimensions of cloud computing and electricity: Beyond the utility model. *Communications of the ACM* 53(5): 32–34.

Buyya, R., J. Broberg, and A. Goscinski. 2011. *Cloud Computing: Principles and Paradigms.* Hoboken, NJ: Wiley.

Cantara, M. 2012. BPM is critical for the adoption of applications and business processes in the cloud, Gartner Research Report.

Carr, N. 2005. The end of corporate computing. *MIT Sloan Management Review* 46(3): 67–73.

Carr, N. 2008. *The Big Switch: Rewiring the World, from Edison to Google.* New York, W. W. Norton & Co.

Chandrasekaran, N. 2012. The no-longer-hypothetical case for jumping on the cloud. *Tata Consultancy Services.* http://sites.tcs.com/cloudstudy/tag/computing#.UU9QG0JIKVs (accessed on September 22, 2012).

Computerworld. 2012. Pogoplug service turns your computers into private cloud. http://www.computerworld.com/s/article/9227018/Pogoplug_service_turns_your_computers_into_private_cloud (Accessed on August 2, 2012).

Creeger, M. 2009. CTO roundtable: Cloud computing. *Communications of the ACM* 52(8): 50–56.

Cusumano, M. 2010. Cloud computing and SaaS as new computing platforms. *Communications of the ACM* 53(4): 27–29.

Da Rold, C. and G. Tramacere. 2012. Chemical leader M&G achieves a fully-cloud-based, lean and low-cost IT. Gartner Research Note.

Etro, F. 2009. The economic impact of cloud computing on business creation, employment and output in Europe. *Review of Business and Economics* 54(2): 179–208.

Halpert, B. 2011. *Auditing Cloud Computing: A Security and Privacy Guide*. Hoboken, NJ: Wiley.

Henning, B. and H. Kemper. 2011. Ubiquitous computing—An application domain for business intelligence in the cloud? *Proceedings of the 2011 Americas Conference on IS*, Detroit, MI. http://aisel.aisnet.org/amcis2011_submissions/93 (Accessed on September 1, 2012).

Hirschheim, R., R. Welke, and A. Schwarz. 2010. Service-oriented architecture: Myths, realities, and a maturity model. *MIS Quarterly Executive* 9(1): 37–48.

Hoberg, P., J. Wollersheim, and H. Krcmar. 2012. The business perspective on cloud computing—A literature review of research on cloud computing. *Proceedings of the 2012 Americas Conference on ISs*, West Lafayette, IN, http://aisel.aisnet.org/amcis2012/proceedings/EnterpriseSystems/5 (Accessed on September 1, 2012).

Janssen, M. and A. Joha. 2011. Challenges for adopting cloud-based software as a service (SaaS) in the public sector. In *Proceedings of the 19th European Conference on Information Systems*. eds. V. Tuunainen, J. Nandhakumar, M. Rossi and W. Soliman. Helsinki, Finland.

Krutz, R. L. and R. D. Vines. 2010. *Cloud Security: A Comprehensive Guide to Secure Cloud Computing*, Indianapolis, IN: Wiley.

Lampe, U., O. Wenge, A. Müller, and R. Schaarschmidt. 2012. Cloud computing in the financial industry—A road paved with security pitfalls? *Proceedings of the 2012 Americas Conference on IS*, West Lafayette, IN, http://aisel.aisnet.org/amcis2012/proceedings/ISSecurity/4 (Accessed on September 1, 2012).

McKinsey Quarterly Global Survey Report. 2012. Minding your digital business.

Miller, C.C. 2011. Amazon cloud failure takes down web sites, *New York Times*, 21 April 2011.

Rai, A. et al. 2010. Transitioning to a modular enterprise architecture: Drivers, constraints, and actions. *MIS Quarterly Executive* 9(2): 83–94.

Reese, G. 2011. The AWS outage: The cloud's shining moment. O'Reilly Community.

Ryan, M. 2010. Viewpoint: Cloud computing privacy concerns on our doorstep. *Communications of the ACM* 54(1): 36–38.

Vaezi, R. 2012. Cloud computing: A qualitative study and conceptual model. *Proceedings of the 2012 Americas Conference on IS*, London, U.K., http://aisel.aisnet.org/amcis2012/proceedings/EndUserIS/3 (Accessed on September 1, 2012).

Ward, J. et al. 2008. Building better business cases for IT investments. *MIS Quarterly Executive* 7(1): 1–15.

West, D. 2010. *Saving Money through Cloud Computing*, Washington, DC: The Brookings Institution.

Wind, S., J. Repschlaeger, and R. Zarnekow. 2012. Towards a cloud computing selection and evaluation environment for very large business applications. *Proceedings of the 2012 Americas Conference on IS*, West Lafayette, IN, http://aisel.aisnet.org/amcis2012/proceedings/EnterpriseSystems/6 (Accessed on September 1, 2012).

References

Abokhodair, N., H. Taylor, J. Hasegawa, and S. J. Mowery. 2012. Heading for the clouds? Implications for cloud computing adopters. *Proceedings of the 2012 Americas Conference on IS*, Seattle, WA. http://aisel.aisnet.org/amcis2012/proceedings/ISSecurity/6 (Accessed on September 1, 2012).

Accenture. 2010. China's Pragmatic Path to Cloud Computing. http://www.accenture.com/us-en/Pages/insight-china-path-cloud-computing-summary.aspx (accessed on March 17, 2012).

Accenture. 2012. High performance IT insights: Five ways the cloud will change the way you run IT. http://www.accenture.com/us-en/Pages/insight-five-ways-cloud-change-run-it.aspx (accessed on July 16, 2012).

Gigaom. 2011. 5 Ways to protect against vendor lock-in in the cloud. http://gigaom.com/2011/09/24/5-ways-to-protect-against-vendor-lock-in-in-the-cloud/ (Accessed on September 17, 2012).

Harris, J. G. and A. E. Alter. 2010. Cloudrise: Rewards and risks at the dawn of cloud computing. Accenture Institute for Higher Performance Research Report.

Harvard Business Review Analytic Services. 2011. How the cloud looks from the top: Achieving competitive advantage in the age of cloud computing. http://download.microsoft.com/download/1/4/4/1442E796-00D2-4740-AC2D-782D47EA3808/16700%20HBR%20Microsoft%20Report%20LONG%20webview.pdf (accessed on January 5, 2012).

Herbert, L. and J. Erickson. 2011. The ROI of cloud apps. Forrester Research Report.

Iyer, B. and J. C. Henderson. 2010. Preparing for the future: Understanding the seven capabilities of cloud computing. *MIS Quarterly Executive* 11(1): 117–131.

Iyer, B. and J. C. Henderson. 2012. Business value from clouds: Learning from users. *MIS Quarterly Executive* 9(2): 51–60.

Karamouzis, F. 2012. Cloud business services: A closer look at BPaaS. Gartner Research Report G00229304.

Loebbecke, C., B. Thomas, and T. Ullrich. 2012. Assessing cloud readiness at continental AG. *MIS Quarterly Executive* 11(1): 11–23.

McAfee, A. 2011. What every CEO needs to know about the cloud. *Harvard Business Review* 89(11): 124–132.

Mell, P. and T. Grance. 2011. *The NIST Definition of Cloud Computing*, Gaithersburg, MD: National Institute of Standards and Technology, http://csrc.nist.gov/publications/nistpubs/800-145/SP800-145.pdf

Mitchell Smith, D. et al. 2011. Predicts 2012: Cloud computing is becoming a reality. Gartner Research Report G00226103.

Peppard, J. et al. 2007. Managing the realization of business benefits from IT investments. *MIS Quarterly Executive* 6(1): 1–11.

Phifer, G. and J. Heiser. 2012. Look before you leap into cloud computing. Gartner Research Report G00235058.

Radcliffe, J. 2012. The advent of master data management solutions in the cloud brings opportunities and challenges. Gartner Research Report.

Ried, S. and Kisker, H. 2011. Sizing the cloud, understanding and quantifying the future of cloud computing. Forrester Research Report.

Seely Brown, J. 2012. My take. *Tech Trends 2012: Elevate IT for Digital Business*. ed. Seely Brown, J, Chapters 5 and 33, Deloitte Consulting LLP, New York.

Stroh, S. et al. 2009. The cloud is ready for you are you ready for the cloud? Booz & Company Research Report.

Tanriverdi, H., A. Rai, and N. Venkatraman. 2010. Reframing the dominant quests of information systems strategy research for complex adaptive business systems. *Information Systems Research* 21(4): 822–834.

White, M. et al. 2010. There is no cloud. Deloitte Consulting, LLP Research Report.

Willcocks, L., W. Venters, and E. Whitley. 2011. Cloud and the future of business: From costs to innovation. *Accenture and the Outsourcing Unit at LSE*, London, U.K., http://outsourcingunit.org/publications/cloudPromise.pdf (Accessed on July 28, 2012).

World Economic Forum. 2010. Exploring the future of cloud computing: Riding the next wave of technology-driven transformation. http://www.weforum.org/reports/exploring-future-cloud-computing-riding-next-wave-technology-driven-transformation (Accessed on April 20, 2011).

49

Enterprise Mobility

Carsten Sørensen
*The London School
of Economics
and Political Science*

49.1 Mobilizing Interaction

49.1.1 From Telegraph to Mobile

The past 150 years signify unprecedented development in the ways organizations organize distributed activities. The invention of scientific management at the end of the nineteenth century along with technological innovations such as the telegraph, filing cabinets, and printers suddenly made it possible to manage large-scale organizational activities distributed across continents much more efficiently and effectively (Yates, 1989). A variety of physical information management technologies played an increasingly critical role, but the birth of the business computer in post-war 1950s marked the beginning of an era of rapid acceleration in the importance of complex digital information management for business across all sectors (Caminer et al., 1998). During the next 30 years, mainframe computers formed the backbone of organizational information management. The 1980s gave birth to the networked personal computer allowing further distribution of computing power (Hinds and Kiesler, 2002). Fast forward to the present and we witness the pulverization of computing access with myriads of mobile phones, digital recording devices, contact-less sensors, etc. (Yoo, 2010). Portable, handheld, and embedded client technologies are interconnected through personal, local, and global digital networks. The combination of powerful client technologies connected through high-speed networks to a variety of cloud services provides a powerful digital infrastructure, the consequences of which we have yet to fully grasp (Tilson et al., 2010).

49.1.2 Predictions

We tend to be rather poor at understanding the technological development when it is right underneath our noses. Several prominent predictions of the impact of information technology turned out to be off the mark (Sørensen, 2011, p. 20ff). IBM's Chairman Thomas Watson famously remarked in 1943, "I think there is a world market for maybe five computers." Digital Equipment Corporation cofounder Ken Olsen stated in 1977, "there is no reason for any individual to have a computer in his home." Even Bill Gates's proposition in the late 1980s that "Microsoft was founded with a vision of a computer on every desk, and in every home" now seems to greatly undershoot a reality littered with computation. Xerox Lab's Mark Weiser came close when he in 1991 stated that "The most profound technologies are those that disappear. They weave themselves into the fabric of everyday life until they are indistinguishable from it" and thus defined the foundation for ubiquitous information technology. Harold S. Osborne already in 1954 famously predicted a device similar to a mobile phone with a unique ID given to each person at birth and following him or her until death.

49.1.3 Researching Mobile Information Technology Practices

A variety of research has been conducted into mobile information technology practices, and this section briefly classifies the research into the four categories of studies: (1) the mobile phone applied in a general social context, (2) broad ranges of mobile and ubiquitous information technologies used in a social context, (3) the mobile phone used to support work activities, and (4) broad ranges of mobile and ubiquitous information technologies used for the purpose of work. The last category is labeled "enterprise mobility" and is the focus of the remainder of the chapter. Before leaping onto this important subject, the following will briefly discuss all four categories of mobile information technology research.

49.1.3.1 Mobile Phone in Society

It is not strange that a significant research effort has been put into understanding the role of one single technology—the mobile phone—in society. There are currently billions of mobile phones in use across the globe, and the use of mobile phones crosses geographical, social, religious, and income barriers. This has led to a number of books and articles within the field engaged in the social study of mobile communications. While significant and important, it is essential to note that this research field is mainly interested in studying social phenomena and seems less interested in understanding technological diversity. Good examples of this line of research are Agar's (2003) account of the history of the mobile phone; Baron's (2008) work on the coevolution of technological practices and language; and a large body of work exploring the new communicative practices and changes on social rituals emerging from mobile phone use by researchers such as Ling (2004, 2008), Harper et al. (2005), Licoppe and Heurtin (2001), Licoppe and Smoreda (2006), Licoppe (2004), Fortunati (2002, 2005), and Katz and Aakhus (2002), to name a few. The general and widespread global diffusion of mobile phone communication has been explored by Castells et al. (2007) and Funk (2004). Some research has also explored regional characteristics, for example, in Japan (Ito et al., 2005), in the Asia-Pacific region (Rao and Mendoza, 2005; Pertierra, 2007; Hjorth, 2009), and in developing countries in general (Horst and Miller, 2006; Donner, 2008).

49.1.3.2 Mobile and Ubiquitous Information Technologies in Society

Despite the immense global success of the mobile phone, there are good reasons for studying technological properties in use more broadly beyond this one incarnation. Mobile and ubiquitous technologies are used in conjunction with the mobile phone, and the mobile phone is integrating a number of previously distinct technologies. This shapes everyday life by what Yoo (2010) characterizes as "experiential computing," which is the everyday experience with a range of technologies embedding computational abilities, such as mobile phones, iPods, digital cameras, GPS navigators, digital photo frames, smart toasters, intelligent ovens, in-car entertainment systems, etc. Relatively little research beyond specific studies into the human–computer interaction aspects has been conducted within this area, which can be characterized as the social study of ubiquitous computing. Good example of such research is the work on design, intimacy, and environment as digital technologies disappear into everyday life by authors, such as Weiser (1991), Norman (1988, 1993, 1999), Dourish (2001), Mitchell (1995, 2003), and Mccullough (2004). Researchers have also studied changes in influence and power distribution as a result of mobile and ubiquitous technologies, for example, the consumer impact of radio frequency identification (RFID) technology (Albrecht and McIntyre, 2006) and wearable computing for citizen activism (Mann and Niedzviecki, 2002).

49.1.3.3 Mobile Phone at Work

There is a relatively small body of research specifically focused on the organizational use of the mobile phone at work, which is striking when considering the organizational proliferation of this technology. It is, however, often difficult to specifically isolate the mobile phone at work since it will often be an element in a larger portfolio of mobile and ubiquitous technologies now when the mobile phone is less associated with the isolated connectivity of voice calls and SMS messaging. The social study of the mobile phone at work has researched a diversity of issues, such as the role of the mobile phone in maintaining ongoing social contact while at work (Wajcman et al., 2008); the intensification of work (Bittman et al., 2009); work–life boundary maintenance (Golden and Geisler, 2007; Wajcman et al., 2008); and the organizational use of mobile e-mail (Mazmanian et al., 2006; Straus et al., 2010). Although not research based, Kellaway's (2005) amusing fictional account of mobile e-mail forms a highly informative image into contemporary organizational life with mobile e-mail.

49.1.3.4 Mobile and Ubiquitous Information Technologies at Work: Enterprise Mobility

The emergent organizational use of a range of diverse mobile and ubiquitous technologies led Lyytinen and Yoo (2002b) call for more research into this subject. This call has largely been ignored across the information systems field the passing decade (Sørensen, 2011). While a significant amount of research has explored how individual users in a social context use the mobile phone, there is little research exploring the organizational use of more complex portfolios of mobile and ubiquitous technologies (Sørensen, 2011). Some research, however, does explore the issue of enterprise mobility. Several edited collections broadly explore mobile and ubiquitous technologies at work and mobile working (Andriessen and Vartiainen, 2005; Sørensen et al., 2005; Basole, 2008; Hislop, 2008; Kourouthanassis and Giaglis, 2008). A range of books are aimed at practitioners engaged in implementing enterprise mobility within their organization, for example, Lattanzi et al. (2006), McGuire (2007), Darden (2009), Reid (2010), and Sathyan et al. (2012).

This chapter reports on such research into enterprise mobility. It focuses on the mobilization of interaction at work through the application of mobile and ubiquitous information technologies. The aim is to provide a conceptual map of the core challenges the mobilization of interaction pose to individuals, teams, and organizations. The chapter argues that the mobilization of interaction at work challenges existing arrangements and that the complex human–technology relationship must be more closely considered in order to understand how to manage unintended consequences.

49.1.4 Mobile Enterprise

Organizations have rapidly adopted a range of mobile and ubiquitous information technologies to support instant connectivity, to access corporate infrastructures, and for a range of other purposes. The first bulky mobile phones from the 1980s have been replaced with smartphones and tablets offering fast Internet access. Tiny microchips and associated sensor technologies can be used to track objects, vehicles, and people, using, for example, RFID, near-field communication (NFC), and the global positioning system (GPS). A great variety of portable and embedded devices beyond the ordinary mobile phone serve a range of purposes. These arrangements move far beyond establishing connectivity between people or granting access to information services. The technology also serves a variety of purposes implemented through machine-to-machine (M2M) connections.

Organizations, of course, constantly consider how emerging information technology can offer new possibilities, but equally, some technologies become important because they represent solutions to existing problems. This two-way bind of problems and solutions interacting through a mesh of technological and organizational arrangements is as old as information technology itself (Yates, 1989). Since the birth of the mainframe, organizations have engaged in an increasingly complex arrangement of locally and globally distributed activities (Urry, 2003). This has made it possible to geographically shift activities depending on changing needs, available skills, and cost levels, for example, extensively separating the design and production of goods and services. Apple's iPhone is designed in California and manufactured in China with components gathered from further distant countries. Its applications are downloaded from globally distributed server farms and are globally sourced from hundreds of thousands of small and large programming outfits. After-sales support is located wherever is most appropriate for the time of the day or the nature of the customer inquiry.

49.2 Capabilities and Performances

In order to understand this development in general and the organizational consequences of mobile and ubiquitous information technologies in particular, it can be helpful simply to consider the organizational application of mobile and ubiquitous information technologies as the technology performances carried out in the meeting of technological capabilities (see Figure 49.1) and the requirements of organizational activities. The diversity of capabilities is illustrated in Figure 49.1, and Figure 49.2 subsequently illustrates the relationships between capabilities, work requirements, and technology performances.

49.2.1 Mobile Capabilities

Let us first consider six categories of capabilities, or affordances, specifically important for the understanding of the unique characteristics of mobile and ubiquitous technologies at work.

49.2.1.1 Connectivity

The technological development has since the era of the telegraph signaled a process of radically increased connectivity, from the first transatlantic telegraph cable in 1866, over Bell's telephone patent in 1878 leading to ubiquity of telephones and subsequently fax machines, to Martin Cooper's first call on a

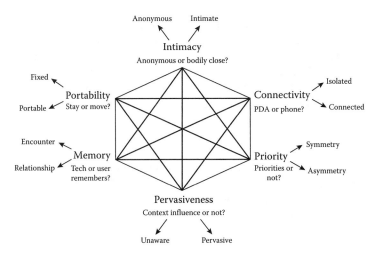

FIGURE 49.1 Overview of the six enterprise mobility capabilities.

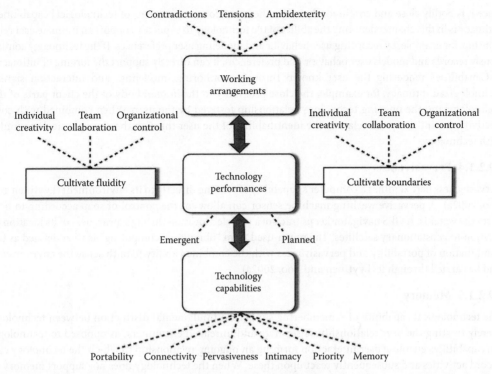

FIGURE 49.2 The cultivation of interactional fluidity and boundaries through emergent and planned technology performances when the contingencies of work meet technology capabilities.

handheld mobile phone in 1973 heralding the age of the mobile phone. Rapid development of cellular telecommunications networks followed from the first fully automated (1G) analog mobile network in Japan in 1979, over 2G networks from 1991, high data capacity (3G) networks from 2001, and to the emerging 4G networks. All of this has led to cheap global connectivity where in 2003 mobile phone subscriptions first exceeded fixed-line subscriptions. The year 2012 saw more mobile-connected devices than people on Earth, and it is estimated that 16 billion mobile devices will be interconnected in 2016 (Cisco, 2012).

49.2.1.2 Portability

The miniaturization of information technology has, during the past decades, produced increasingly portable, yet powerful, computational devices. This portability of the computing device forges closer relationships between the technology and the geographical movement of users, objects, and vehicles. Computation is not restricted to specific locations, but can follow organizational processes and members. The term "mobile technology" most frequently refers to the combination of a portable client connected to a network, for example, the mobile phone used for calls and SMS messages. However, in the context of this chapter, mobile technology in a much broader sense refers to technologies that can combine any of the six types of capabilities listed in Figure 49.1: portability, connectivity, intimacy, pervasiveness, memory, and priority.

49.2.1.3 Intimacy

One of the key assumptions is that the closer a technology is situated to the human body, the more critical the user–technology relationship becomes. The mainframe may be important, but the personal computer is highly personal, yet not as critical as the mobile phone, which takes up a position on par with keys and credit cards (Chipchase et al., 2004). The user may consider mobile and ubiquitous technologies as intimate since it is bodily close and continuously follow the user. However, in terms of technological capabilities, intimacy is in this chapter denoting the ability of the technology to explicitly support an intimate user relationship, for example, by recording user behavior and modeling user preferences. If the technology continuously records and models user behavior and preferences, it can directly support the forging of intimacy.

Capabilities rendering the user known through monitoring, modeling, and interaction signify technologized intimacy, for example, the close proximity to the human body of the client parts of the technology and the ongoing technology relationship fostered by the user. When combined with connectivity, the intimacy can relate to the identifiability of the user through the close physical proximity with technology.

49.2.1.4 Pervasiveness

Pervasiveness here signifies a computer's capability of relating directly to its environment (Lyytinen and Yoo, 2002a). A pervasive washing machine sensor can allow for the control of soap according to how dirty the water is. A GPS navigator keeps track of a vehicle's location through awareness of its location in relation to geostationary satellites. The much-used term "ubiquitous computing" is then defined as the combination of portability and pervasiveness with the combined ability to both sense the environment and be carried through it (Lyytinen and Yoo, 2002a).

49.2.1.5 Memory

The technological capability of remembering supports a fundamental distinction between technology merely treating the user relationship as separate and unrelated encounters, as opposed to technological capabilities enabling user interaction forming an ongoing relationship in which the technology can record activities and subsequently react upon these. When the technology does not support memory of an ongoing process, the user is solely responsible for constructing and maintaining memory of the process. When the technology explicitly maintains memory of aspects of the process, the technology can support by managing aspects of the process complexity (Carstensen and Sørensen, 1996). Technology-based memory is generally a prerequisite for comprehensive support of complex decision processes (Mathiassen and Sørensen, 2008).

49.2.1.6 Priority

In a similar manner to information technology supporting the remembering aspects of the interaction, it can also support the prioritization of interaction. While the user of course continually engages in assessments and prioritization when, for example, looking at a phone when called, some of the prioritization can also be supported explicitly by the technology. This can be implemented in the form of a variety of notification and filtering capabilities supporting rules stipulating how the interaction should unfold or directly supporting human choices through providing additional information about the interaction. Such prioritization explicitly treats the interaction as asymmetrical, whereas an absence of priorities implies the assumption of interaction symmetry. Mobile and ubiquitous technologies can directly support the prioritization of interaction through various filtering and awareness mechanisms, for example, by directly excluding or including specific interaction depending on information about content, caller identity, etc. (Oard, 1997; Ljungberg, 1999; Sørensen, 2010). The technological capability of supporting interaction prioritization relates directly to research into user awareness (Heath and Luff, 2000), interruptions (McFarlane, 1998; Rennecker and Godwin, 2005; Wiberg and Whittaker, 2005), and information overload (Hiltz and Turoff, 1985; Schultze and Vandenbosch, 1998; Eppler and Mengis, 2004; Iastrebova, 2006; Harper, 2010).

49.2.2 Complexity of Technology Performances

Enterprise mobility is, in short, the organizational application of complex combinations of the six capabilities discussed earlier embedded in mobile and ubiquitous information technologies. This application is governed by the emerging characteristics of the work situations the users find themselves in (see Figure 49.2). These situations are often conflicting and contradictory, and therefore calling for a combination of planned and emerging usage of the technology (Sørensen, 2011). Some activities are characterized by a very high degree of individual discretion regarding what needs to be done and how precisely this must be accomplished. Other activities are governed by next to no rules and procedures and thereby requiring extensive discretion on the part of the individual. A majority of activities share characteristics from both and will require some discretion and therefore also require both planned and emergent technology performances.

The complexity emerging when specific combinations of mobile information technology capabilities within a specific portfolio are evoked cannot easily be characterized in terms of straightforward linear causality between the aim of the activity and an associated performance. Rather, the negotiations of contradicting requirements emerging from the situation can lead to the user engaging different actions with the technology depending on the specific situation, the mood of the user, and a range of other aspects. As an example, the decision by an individual of whether or not to answer a call on their mobile phone is not necessarily straightforward and will depend on a range of factors, for example, who the caller is (if this can be ascertained a priori), what the recipient is currently doing, or what he or she guesses the caller wishes to discuss. A straightforward mapping can only be made between a set of technology performances by the user against a discrete set of situations if any possible individual discretion has been removed from the work conducted. Emerging aspects of work will in most cases make it necessary for the user to engage in emerging technology performances taking the specific situation into consideration (Sørensen, 2011).

49.3 Cultivating Fluidity and Boundaries at Work

The organizational impact of mobile and ubiquitous information technologies is difficult to document precisely given that the technology is applied in a great variety of roles, spanning work with very little individual discretion to work relying on extensive individual discretion. Within field force automation, it may be complicated, yet doable, to assess the benefits of increased use of mobile and ubiquitous information technologies as the work process can be codified and measured against a computer-calculated ideal process. The technology performances will in this case be treated similarly to the steps in an automated manufacturing process where the components are expected to behave in a certain manner at a certain point in time.

At the other end of the discretion spectrum, top executives or highly independent professionals are expected to largely govern their own activities and make choices they deem appropriate according to the specific situation. For this type of work, it can be almost impossible to devise simple measures for process quality, and simply measuring the outcome does relatively little to factor out the importance of particular technology performances. However, most work will be characterized by being governed by some organizational rules, while retaining some degree of discretion. Furthermore, it is often far from trivial from a distance to ascertain to what extent work can be conducted without discretionary decisions as unanticipated exceptions can call for emerging technology performances.

49.3.1 Anytime/Anywhere or Sometime/Somewhere?

The ability to engage in mobile technology performances anytime, anywhere, and with anyone is an often-cited practical implication of the application of mobile and ubiquitous information technologies. While this can technically be correct, such focus emphasizes the technological opportunities in Figure 49.2 over the organizational opportunities and constraints. While a mobile phone allows for interaction with virtually anyone on Earth, this does not reflect any organizational reality. Work is always situated in some context where certain issues, people, contexts, and time slots have more importance over others. Possessing a colleague's mobile phone number does not necessarily mean that the person can be called at any time. When combining the technological opportunities with the organizational realities, the technology performances are conducted as complex arrangements where interaction occurs sometimes, somewhere, and with someone (Wiberg and Ljungberg, 2001). The everyday mobile technology performances are ones hovering between being anytime and the sometime, and for the participants, this is in itself often a subject of negotiation.

49.3.2 Distance and Movement Matter

Olson and Olson (2000) argue that despite serious attempts to completely alleviate the importance of distance in collaboration, distance still matters. Collaboration across distances imposes a separate set of issues that must be overcome, even if the technology at the same time does facilitate highly distributed work (Olson and Olson, 2000; Dubé and Robey, 2009). In the use of networked personal computers, distance both matters as a barrier for interpersonal collaboration and, at the same time, matters as it also directly facilitates remote collaboration.

It can similarly be argued that the use of both mobile and ubiquitous technologies suddenly makes the physical location of the user more important than before—movement matters when mobile and ubiquitous technologies are deployed in an organization (Sørensen et al., 2012). The technology performances will frequently relate to human movement and therefore create a whole new set of issues related directly to the negotiation of interaction in movement. We can hardly assume that this does not create new challenges for the participants—especially if the movement cannot be made the subject of collective inquiry. The opening questions of "Where are you?" and "Did I disturb?" directly relate to matters of distance and movement.

However, for the organization, movement also matters when mobile technology helps loosen constraints for its members' locations and movements. When organizational members by circumstances are forced to work at their workstation, their movement is mostly a matter of ensuring sufficient number of working hours and to arrange around this constraint. When they are able to work from a range of locations while moving, their location and movement are no longer merely issues of personal preference. They become important operational, tactical, and even strategic concerns. Sales organizations with traveling salespeople have of course known this for a long time. However, mobile technology performances make it possible to engage in more fine-grained decisions regarding the distribution of work across time and location.

In the traditional workplace, roles could relatively easily be identifying those that required other arrangements that stationary working with no movement and being colocated with everyone else in a fixed office or by a machine (see Figure 49.3). In the mobile enterprise, roles are open for more flexible assignments. Some work may at times need to be organized as locally mobile within a specific location. At other times, work may require as arranged as remotely distributed from other colleagues, yet fixed at a desk. Yet at other times, work may need to be organized as mobile working, i.e., both remotely located and mobile.

49.3.3 Balancing Fluidity and Boundaries

As discussed earlier, mobile technology performances represent choices regarding the unfolding of activities and interactions across time and locations. With the networked personal computer, distance took on new meanings, and with mobile information technology, so does movement. The performances are localized individual or collective judgments regarding the combination of technological possibilities when faced with a possibly contradicting organizational reality. The performances can in these situations result in the removing of existing boundaries for interaction—establishing more fluid interaction. The use of mobile e-mail is a good example to the extent that members of the organization choose to use e-mail as a horizontal means of communication transcending previous lines of reporting and responsibility. However, mobile technology performances can also establish new or reinforce

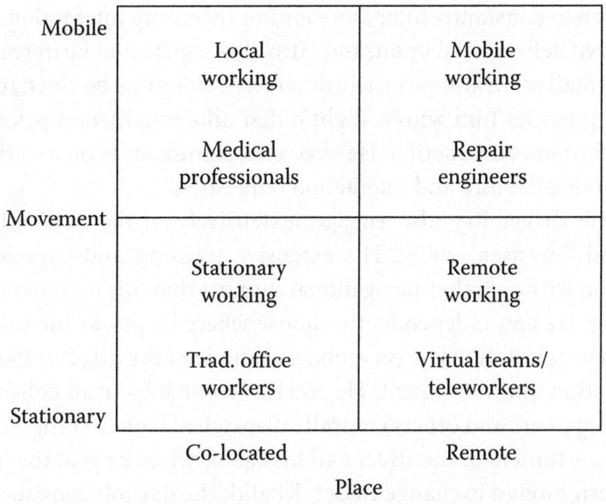

```
          Mobile ┌──────────────┬──────────────┐
                 │   Local      │   Mobile     │
                 │   working    │   working    │
                 │              │              │
                 │   Medical    │   Repair     │
                 │ professionals│  engineers   │
       Movement  ├──────────────┼──────────────┤
                 │  Stationary  │   Remote     │
                 │   working    │   working    │
                 │              │              │
                 │  Trad. office│ Virtual teams/│
                 │   workers    │  teleworkers │
       Stationary└──────────────┴──────────────┘
                   Co-located        Remote
                            Place
```

FIGURE 49.3 Separating remote distribution and movement, four analytical categories of working emerge: stationary, remote, local, and mobile working. (Adapted from Sørensen, C. *Enterprise Mobility: Tiny technology with global impact on work*. Technology, Work and Globalization Series, 2011.)

existing boundaries for interaction (see Figure 49.2). Not only can mobile performances set people free from their desks so they can move where it makes most sense to them, to the teams they work in, or to the organization as a whole. The performances can also bring organizational routines, procedures, and infrastructure along and through intimacy and pervasiveness support that specific procedures are enforced or that certain lines of command obeyed. It is, in other words, not a simple matter to ascertain what outcomes mobile performances can entail as mobile technology performances contain the inherent duality of both contributing to increased fluidity and boundaries in decision processes.

The following discusses the cultivation of fluidity and boundaries through mobile performances at three different levels of analysis: (1) individuals creatively seeking fluid performances by balancing availability and engaging boundaries for interaction, such as turning off the phone; (2) teams seeking to engage in transparent collaboration by balancing fluid internal team interaction with boundaries for how they interact with others outside the team; and (3) organizations seeking to cultivate interactional boundaries within the organization and at its boundaries.

49.3.3.1 Individual Creativity

The performance can support individuals in engaging in fluid working practices with few boundaries hindering interaction if this is deemed suitable and with the individual cultivation of barriers for interaction when needed, for example, through filters blocking out requests. The extensive use of mobile information technology over years renders the individual a highly experienced professional in engaging in such cultivation, within the technological constraints. For example, usage throughout a 40 h working week will within 5 years have accumulated the 10,000 h of practice cited as necessary for true expertise within a domain (Simon and Chase, 1973; Ericsson et al., 1993; Gladwell, 2008; Sørensen, 2011). We have found a number of very good examples of such expertise in creative cultivation of mobile performances in our studies (Sørensen, 2011, pp. 77–100).

The Japanese CEO, Hiro, for example, engages in highly complex performances when moving freely around Tokyo to get inspiration for new mobile services his small company possibly could build and launch (Kakihara and Sørensen, 2004). Hiro conducts "innovation by walking around" and extensively relies on advanced filtering features implemented on his mobile phone. This results in a number of profiles emitting different noises depending on the caller or the mobile e-mail sender. He prefers most interaction to be channeled to mobile e-mail so he can quickly scan the requests and prioritize which ones he acts upon when. This transforms synchronous and asynchronous interaction into the asynchronous e-mail, which allows Hiro to constantly filter and monitor incoming interaction. In this way, immediate requests can be prioritized before acted upon, and Hiro also argues that he in reality in this way uses the asynchronous mobile e-mail as an almost synchronous technology as he constantly checks e-mails. The arrangement, however, provides him with a slight buffer allowing instant prioritization—a task more difficult with truly synchronous interaction. He also relies significantly on a stationary administrator as a means of managing his availability and interaction requests.

The London black cab driver Ray also engages extensively in the cultivation of fluid interaction (Elaluf-Calderwood and Sørensen, 2008). His extensive training and experience in navigating the streets of Central London without other navigational support than his memory allows him to define his own rhythm of working. He can independently choose where he places the cab in order to maximize his income, and as such he is a stationary, yet mobile, example of the direct importance of discretionary choices concerning location and movement. He receives some jobs from colleagues via mobile phone when a sudden demand appears and others centrally dispatched from a computer cab system. However, most jobs appear when customers in the street hail his cab or when he is at the front of a taxi rank.

For the Middle Eastern foreign exchange trader, Khalid, the day job consists of office-based trading, but outside normal working hours, he engages in mobile trading using a mobile phone and a specialized mobile device providing market information (Sørensen and Al-Taitoon, 2008). Here, the balancing of home and working life is a major aspect of the balancing of fluidity and boundaries.

Common for these three examples is the extensive reliance on movement as a means of doing the work. For the CEO and the cab driver, the movement is in effect an integral part of the work, whereas for the trader, movement is an essential aspect of striking some balance between work and family requirements. When engaged in mobile working, all three are largely either independent or largely able to shape their own work. They can over time cultivate a sense of rhythms of working and of interaction supporting the cultivation of fluid working and the management of interruptions (Sørensen and Pica, 2005; Sørensen, 2011, pp. 94–100).

49.3.3.2 Team Collaboration

For teams engaged in collaboration and in the coordination of mutual interdependencies, the ability to obtain fluid and transparent negotiation of these interdependencies is a key concern. It is obviously both pregnant with opportunities and fraught with problems to obtain such fluidity when team members are both scattered and moving rather than when everyone is situated within the same office space.

For the police officers John and Mary, their interaction is largely conducted within their patrol vehicle, where they have continuous contact with the control room through an in-car terminal and radio system (Sørensen and Pica, 2005). When out of the vehicle and engaged in an incident, they rely heavily on the flexible means of a shoulder-mounted radio as a constant connection to the control room and to each other. The in-car system represents complex interaction boundaries between teams as it streamlines the coordination of officers accepting to attend incidents. The team of two officers constantly cultivates the rhythms of working both establishing fluid activities and cultivating interactional boundaries. In doing so, they rely extensively on a wide range of the technological capabilities discussed previously.

For Simon, a security guard doing rounds on industrial estates in Manchester, the coordination of work with the control room consists of little individual discretion (Kietzmann, 2008). His activities are almost automatically directed by a system whereby he swipes a RFID-reader mobile phone over an embedded RFID chip and the system informs him via SMS where to go next. Collaboration is here entirely codified through a specific set of stipulated performances.

One of the important aspects of managing the negotiation of mutual interdependencies is formed by decisions regarding interaction priority. If all interaction is prioritized equally, then the associated interactional symmetry will rely on the individual member to manually prioritize interaction. In both the cases mentioned earlier, systems embed capabilities supporting prioritization of interaction to facilitate more effective collaboration. Mobile technology performances also challenge traditional notions of individual versus collective working as the extensive use of mobile interaction can support the individualization of highly interdependent activities (Barley and Kunda, 2004; Felstead et al., 2005; Sørensen, 2011).

49.3.3.3 Organizational Control

From the perspective chosen here, an organization can be viewed as a specific arrangement of boundaries for interaction and facilitators for fluid interaction. Organizations will, from this perspective, spend considerable effort cultivating internal boundaries so they facilitate desirable fluid interaction and limit undesirable interaction. Similarly, much effort will be invested in cultivating the organizational boundary enabling some and hindering other forms of interaction.

For the restaurant supply delivery driver Jason, work is regularly challenging organizational arrangements (Boateng, 2010). When delivering goods to a restaurant he has, against procedure, provided the owners with his private mobile phone number instead of forcing customers to contact the call center who thereafter would contact Jason. He will also, for practical purposes, at times neglect procedures of not waiting in case a customer is not present at the restaurant. Such arrangements are obviously both efficient and desirable from Jason's perspective, yet ineffective from an organizational perspective. If and when Jason leaves his job, customers will be frustrated as they no longer can use the established fluid arrangement of just calling Jason's mobile. Also circumventing the call center results in order discrepancies not always being properly logged within the system. However, in the tension between centrally fixed boundaries and Jason's circumvention of these boundaries, Jason manages to carve out a little extra discretion for himself.

The hospital nurse Yin, who is further educating herself to a new role along with 15 other colleagues within the British Health Service, is caught up in a clash of arrangements of organizational fluidity and boundaries between the local hospital where she spends most of her time and the central London-based organization responsible for her further education, which she attends 1 week in 6 (Wiredu and Sørensen, 2006). Conflicting local and remote requirements cannot be reconsolidated through the use of mobile technology performances. The remote organization desires intense documentation of the remote work-integrated learning process. The local hospital does not see the importance of this documentation and establishes strict boundaries for interaction, for example, by not allowing her to use mobile technology in normal working hours. The end result is that the technology performances are abandoned for documentation purposes, and the London-based learning team instead relies on Yin's movement to resolve the issue. Every 6 weeks for a year, she attends 1 week sessions in London, and these sessions are deemed sufficient as means of documenting her learning. In this example, the organizational constraints and boundaries lead to human movement directly replacing mobile technology performances.

Table 49.1 summarizes the enterprise mobility capabilities discussed earlier and illustrates the diversity of capabilities across cases.

49.3.4 Enterprise Mobility and Organizational Complexity

Mobile technology performances can serve organizational purposes by providing both direct remote observation and indirect remote control. Such attempts can at the same time be subjected to a variety of counteractivities to serve individual or team purposes. For the organization, mobile technology performances represent both new opportunities for reengineering of traditional boundaries in order to obtain more effective decision flows and challenges as ineffective boundaries can lead to suboptimal interactional boundaries and much misdirected or wasted effort. For the Middle Eastern bank engaging in mobile foreign exchange trading after normal working hours, mobile technology performances represent the only viable solution to the problem that the bank wished to engage in 24 h trading without maintaining a trading floor around the clock. The solution of shifting trading across the globe following the sun was not feasible. Neither was home-based trading fixing traders by their personal computer all evening and night. However, although mobile trading clearly resulted in stress on the traders home life/ work life balance, it at least represented a solution acceptable to both the bank and the traders.

TABLE 49.1   Illustration of the Diversity of Capabilities across the Example Cases Discussed in Section 49.3.3.3

Name	Job	Technology	Enterprise Mobility Capability					
			Portability	Connectivity	Intimacy	Pervasiveness	Memory	Priority
Hiro	CEO	Mobile phone	Yes	Yes	Yes	No	Yes	Yes
Ray	Cabbie	Computer cab system	No	Yes	Yes	Some	No	Some
Ray	Cabbie	Mobile phone	Yes	Yes	No	No	No	No
Khalid	Trader	Terminal	Yes	Yes	Yes	No	Yes	Yes
John and Mary	Police	Mobile data terminal	No	Yes	Yes	No	Yes	Yes
John and Mary	Police	Personal radio	Yes	Yes	No	No	No	No
Simon	Security	RFID phone	Yes	Yes	Yes	Yes	Yes	Yes
Jason	Driver	Mobile phone	Yes	Yes	No	No	No	No
Yin	Nurse	Handheld assistant	Yes	No	No	No	No	No

Establishing much tighter and more interactive connections between individual workers and portable technology connected to communication networks allows for the contradicting purposes of both (1) tightening centralized control and codification of the work process and (2) providing increased decentralization and individual discretion through extensive access to the necessary corporate information. This development resembles the distinction between the use of information technology for either codifying or informating work discussed in the 1970s and 1980s when information technology was introduced in a range of sectors (Zuboff, 1988). However, this time it is personal as the technology is not an organizational anonymous entity but linked directly to individuals. The role of enterprise mobility and mobile technology performances can be viewed as one means of supporting organizational paradoxes.

Organizations experience the difficulties of balancing a variety of concerns, for example, balancing long-term exploration of opportunities with short-term exploitation of current abilities, being flexible and agile while securing stable structures for decision making, or balancing concerns for internal and external focus (Quinn and Rohrbaugh, 1983; Birkinshaw and Gibson, 2004; O'Reilly III and Tushman, 2004). The ability of an organization to be ambidextrous has been forwarded as a solution to managing organizational paradoxes where the organization needs to follow two seemingly contradicting logics. However, it has been argued that altering structural arrangements is not a long-term solution to this problem. Instead, the balancing of contradicting concerns must be delegated to individuals in the organization who then can make discretionary decisions depending on the actual situation and thereby enable contextual ambidexterity (Birkinshaw and Gibson, 2004). Enterprise mobility can serve as one of the means supporting organizations in achieving such contextual ambidexterity since it can ensure a much more consistent and comprehensive individual access to both the necessary information, access to the appropriate colleagues, while simultaneously support the detailed implementation of organizational procedures and control arrangements (Sørensen, 2011, pp. 49–52, pp. 158–167).

49.4 Enterprise Mobility Research Challenges

The research field studying enterprise mobility is still in its relative infancy despite the practical importance of the technology in many organizations. As an illustrative example, despite a prominent call for action to researchers in 2002 for the information systems field to investigate the organizational use of mobile and ubiquitous information technologies (Lyytinen and Yoo, 2002b), a decade later, only less than 4% of all papers in the top-eight journals even remotely explored the subject (Sørensen, 2011, p. 9).

49.4.1 Individual Intimacy and Organizational Trust

One of the primary challenges of understanding both opportunities and challenges in the organizational use of mobile and ubiquitous information technologies relates directly to the increased intimacy and pervasiveness offered by the technological capabilities. The technology affords intense investigation into individual micropatterns of behavior, thus making highly intimate aspects of everyday working life susceptible to intense scrutiny. It also affords the close and intimate connectivity between the work environment and technological properties, enabling the codification, mapping, monitoring, and calculation at an immense scale. Google's effort to map street views led to significant debate over privacy when the passing Google car has caught people on the images, but the combination of intimacy and pervasiveness will require a significant shift in our understanding of the complex relationships between technological opportunities and prevailing opinions of individual privacy. The success of advanced enterprise mobility capabilities will rely even more critically on interpersonal and organizational trust than other technological innovations in the enterprise, precisely because enterprise mobility connects technology and person even tighter than previous information technologies. The workers could quite literally end up having everywhere to go and nowhere to hide.

49.4.2 Optimizing Knowledge Supply Chains

From a broader organizational perspective of understanding the bigger picture, one of the significant challenges is to understand how to more flexibly manage the opportunities of fluidity and boundaries. Considering the advantages gained by computerizing the manufacturing supply chain, enterprise mobility represents the challenge of understanding how computerization of the knowledge or service supply chain through mobile information technology can lead to advances. However, while the nuts and bolts of highly distributed manufacturing processes do not care much about what is written about them in various databases, humans very much do. One of the challenges in this is to understand the balancing of individual privacy concerns with the organizational advantages of large-scale real-time big-data analysis. As technology and its application advance, increasing amounts of highly detailed data are collected concerning organizational and interorganizational processes and their outcomes (Benkler, 2006; Kallinikos, 2006; Ayres, 2007; The Economist, 2010). Such data represent the opportunity for the organization to flexibly define its internal and external behavior depending on the emerging contingencies and thereby achieve contextual ambidexterity. For individuals, teams, and organizations, the challenge is to understand how to flexibly configure portfolios of technological capabilities to support the balancing of interactional fluidity and boundaries as this increasingly is up for negotiation and not predetermined by the physical arrangement of workstations, offices, and buildings.

49.4.3 Rapid Technological Developments

It is a key concern to explain how enterprise mobility is not only similar to but also differs from other organizational technologies. It is clearly feasible to look at traditional implementation and work effectiveness issues with this particular technological intervention as with other technologies. However, the key challenge is to identify issues of particular importance for mobile and ubiquitous information technologies. As an example, the closer relationship and instant feedback loops between the user and the technology can perhaps create new types of dependencies (Mazmanian et al., 2005). Unpacking how the constantly changing technological opportunities and ongoing changes in working practices interrelate is a core concern and must relate directly to a theoretical understanding of essential changes in technological opportunities. As an example, the recent development of mobile smartphone and tablet application stores has significantly shifted the opportunities for organizations to more easily deploy complex mobile technology using worker self-service and relying on consumerized solutions. The unpacking of how rapidly changing technological circumstances and changing organizational practices interrelate is a complex endeavor but necessary for the further understanding of enterprise mobility. The aforementioned cases illustrate complex use of a variety of the technological capabilities presented in Figure 49.1: portability, connectivity, intimacy, pervasiveness, memory, and priority. The mere fact that the technology reaches widely from the user's hand over the built environment to global infrastructures makes the dynamics even further complex.

49.4.4 Paradoxical Technology Relationships

Much research of socio-technical phenomena tends to adopt the assumption that the user–technology relationships are straightforward, stable, and predictable. However, this is a questionable assumption, especially when the technology in question assumes a highly intimate relationship with the user (Arnold, 2003). For enterprise mobility, the user–technology relationship is highly interactive and relies on close physical proximity and constant connectivity, and the performance relating a user's goal and a specific technical capability cannot be described as linear and simple. Assuming that the user is bestowed some degree of individual discretion, the technology will often allow the same operation to be conducted applying different technology performances by selecting from a portfolio of opportunities. Furthermore, situational richness will likely imply that the user is not able to act as simply mapping discrete goals and technological opportunities. Rather, a range of concerns may concurrently compete as triggers of specific intent. When a person receives a mobile e-mail, the decisions regarding what to do will be informed by a range of situational factors. Also the outcome itself of the technology performances can influence future situations. The constant checking of mobile e-mail and instant replies also creates the need to further check and manage the environmental expectations of continuing to constantly reply to e-mails. This signals paradoxical technology relationships where the technology concurrently both fulfills and creates needs; simultaneously is public and private; supports mobility, yet enables a constantly fixed point of contact; and bestows the user with competence, yet also renders the user incompetent (Mick and Fournier, 1998; Jarvenpaa and Lang, 2005). While other technological innovations may also display such characteristics, mobile and ubiquitous information technologies will further challenge the notion of simple linear relationships between situations, technological capabilities, and technology performances. Research into mobile and ubiquitous technologies must seek to identify and explain the specific socio-technical complexities emerging when the technology is adapted and shaped by individuals, teams, and organizations. This chapter has sought to identify some of the unique aspects, for example, in terms of the importance of human movement. The chapter has also explored the organizational use of mobile and ubiquitous technologies beyond a simple logic of straightforward relationships between what the technology can offer and how it is used. This relates directly to the emerging notion of complex relationships between user behavior and available portfolios of organizational information services (Mathiassen and Sørensen, 2008). A promising research approach can be the detailed mapping of micro-coordination practices (Ling, 2004; Licoppe, 2010) using novel techniques seeking to uncover hidden patterns in complex nonlinear phenomena (Gaskin et al., 2010, www.orgdna.net).

49.5　Summary

Enterprise mobility—the organizational application of mobile and ubiquitous information technologies is a significant, yet underresearched, area of computing. This is despite a significant practical relevance of the subject. The challenges for practitioners and academics alike are to understand the specific opportunities and challenges imposed by this technological development and the associated organizational practices. This chapter has argued that the core challenges relate directly to the ability of the technology to intimately relate to and interact with the user, to manage, prioritize, and remember ongoing interaction processes with the immediate environment, while connecting through personal, local, and global communication networks. It is argued that these capabilities in nonlinear ways relate to user goals through planned and emergent technology performances. Such novel technology performances, it is argued, can intentionally or unintentionally challenge existing interactional boundaries at the individual, team, and organizational level. For the individual, mobile technology performances can both facilitate new inter-actional practices and require the creative management of these opportunities to establish fluid work-ing. For the team, the effective and transparent negotiation of highly distributed interdependencies also requires careful balancing of instant fluid access with boundaries for interaction. At the organizational level, fluid decision processes must be balanced with the cultivation of internal and external interactional boundaries. The introduction of mobile and ubiquitous technologies can mark a way of rendering (1) the individual more available while in charge of his or her own work; (2) teams more effective; and (3) organi-zations more flexible, yet better controlled. Such arrangements can be viewed as incremental changes to existing arrangements. However, the significant potential of enterprise mobility is of course to allow for new business models and disruptive organizational arrangements, for example, challenging assumptions about organizational membership and individual flexibility at work (Barley and Kunda, 2004).

Typically, an unintentional effect is increased interactional fluidity where individuals suddenly are given opportunities for more flexible and fluid access to information resources or colleagues. This can in turn result in both increased effectiveness if it allows for more meaningful connections, and also, as the examples demonstrate, tensions between individual and organizational concerns. It is, therefore, essential that individuals, teams, and organizations constantly cultivate the balance between technol-ogy performances increasing fluid organizational arrangements with those establishing interactional boundaries. Such efforts require significant improvements in our understanding of the complex and paradoxical relationships between the constantly developing diversity of technological capabilities and changing organizational practices.

Glossary

1G: First-generation analog cellular telecommunications networks
2G: Second-generation cellular telecommunications networks, which were digital
3G: Third-generation cellular telecommunications networks offering increased data transmission rates
4G: Fourth-generation cellular telecommunications networks with even faster data transmission rates
Cellular telecommunications network: Radio network that uses the geographical partitioning into cells as a means of reusing frequency bands
GPS: Global positioning system
M2M: Machine-to-machine
NFC: Near-field communication
RFID: Radio-frequency identification

Further Information

Mobility@lse: http://mobility.lse.ac.uk
Enterprise mobility forum: http://theemf.org
Mobile Enterprise Magazine: http://www.mobileenterprisemag.com
Enterprise mobility exchange network: http://www.enterprisemobilityexchange.com

References

Agar, J. (2003) *Constant Touch: A Global History of the Mobile Phone.* Cambridge, U.K.: Icon Books.
Albrecht, K. and McIntyre, L. (2006) *Spychips: How Major Corporations and Government Plan to Track Your Every Move with RFID.* Nashville, TN: Nelson Current.
Andriessen, J. H. E. and Vartiainen, M., eds. (2005) *Mobile Virtual Work: A New Paradigm?* Berlin, Germany: Springer.

Arnold, M. (2003) On the phenomenology of technology: The "Janus-Faces" of mobile phones. *Information and Organization*, 13:231–256.

Ayres, I. (2007) *Super Crunchers: Why Thinking-By-Numbers is the New Way to Be Smart*. New York: Bantam Books.

Barley, S. R. and Kunda, G. (2004) *Gurus, Hired Guns, and Warm Bodies: Itinerant Experts in a Knowledge Economy*. Princeton, NJ: Princeton University Press.

Baron, N. S. (2008) *Always on: Language in an Online and Mobile World*. New York: Oxford University Press.

Basole, R. C., ed. (2008) *Enterprise Mobility: Applications, Technologies and Strategies*. Tennenbaum Institute on Enterprise Systems, Vol. 2. Amsterdam: IOS Press.

Benkler, Y. (2006) *The Wealth of Networks*. New Heaven, CT: Yale University Press.

Birkinshaw, J. and Gibson, C. (2004) Building ambidexterity into an organization. *Sloan Management Review*, 45(4):47–55.

Bittman, M., Brown, J. E., and Wajcman, J. (2009) The mobile phone, perpetual contact and time pressure. *Work, Employment and Society*, 23(4):673–691.

Boateng, K. (2010) ICT-driven interactions: On the dynamics of mediated control. PhD Dissertation, London, U.K.: London School of Economics.

Caminer, D., Aris, J., Hermon, P., and Land, F. (1998) *L.E.O.—The Incredible Story of the World's First Business Computer*. London, U.K.: McGraw-Hill Education.

Carstensen, P. and Sørensen, C. (1996) From the social to the systematic: Mechanisms supporting coordination in design. *Journal of Computer Supported Cooperative Work*, 5(4, December):387–413.

Castells, M., Qiu, J. L., Fernandez-Ardevol, M., and Sey, A. (2007) *Mobile Communication and Society: A Global Perspective*. Cambridge, MA: The MIT Press.

Chipchase, J., Aarras, M., Persson, P., Yamamoto, T., and Piippo, P. (2004) Mobile essentials: Field study and concepting. In *CHI 2004*, Vienna, Austria: ACM,

Cisco (2012) Cisco visual networking index: Global mobile data traffic forecast update, 2011–2016. http://www.cisco.com/en/US/solutions/collateral/ns341/ns525/ns537/ns705/ns827/white_paper_c11-520862.html (accessed on March 21, 2013).

Darden, J. A., ed. (2009) *Establishing Enterprise Mobility at Your Company*. New York: Thomson Reuters/Aspatore.

Donner, J. (2008) Research approaches to mobile use in the developing world: A review of the literature. *The Information Society*, 24(3):140–159.

Dourish, P. (2001) *Where the Action Is: The Foundations of Embodied Interaction*. Cambridge, MA: MIT Press.

Dubé, L. and Robey, D. (2009) Surviving the paradoxes of virtual teamwork. *Information Systems Journal*, 19(1):3–30.

Elaluf-Calderwood, S. and Sørensen, C. (2008) 420 Years of mobility: ICT enabled mobile interdependencies in London Hackney cab work. In *Mobility and Technology in the Workplace*, ed. Hislop, D. London, U.K.: Routledge, pp. 135–150.

Eppler, M. J. and Mengis, J. (2004) The concept of information overload: A review of literature from organization science, accounting, marketing, MIS, and related disciplines. *The Information Society*, 20:325–344.

Ericsson, K. A., Krampe, R. T., and Tesch-Römer, C. (1993) The role of deliberate practice in the acquisition of expert performance. *Psychological Reviews*, 100:363–406.

Felstead, A., Jewson, N., and Walters, S. (2005) *Changing Places of Work*. London, U.K.: Palgrave Macmillan.

Fortunati, L. (2002) The mobile phone: Towards new categories and social relations. *Information, Communication and Society*, 5(4):513–528.

Fortunati, L. (2005) Mobile telephone and the presentation of self. In *Mobile Communications: Re-Negotiation of the Social Sphere*, eds. Ling, R. and Pedersen, P. E., London, U.K.: Springer.

Funk, J. L. (2004) *Mobile Disruption: The Technologies and Applications Driving the Mobile Internet*. Hoboken, NJ: Wiley-Interscience.

Gaskin, J., Schutz, D., Berente, N., Lyytinen, K., and Yoo, Y. (2010) The DNA of design work: Physical and digital Materiality in project-based design organizations. In Academy of Management Best Paper Proceedings. http://www.orgdna.net/wp-content/uploads/2011/10/voss-AOM1.pdf (accessed on March 21, 2013).

Gladwell, M. (2008) *Outliers: The Story of Success*. London, U.K.: Allen Lane.

Golden, A. G. and Geisler, C. (2007) Work–life boundary management and the personal digital assistant. *Human Relations*, 60:519–551.

Harper, R. (2010) *Texture: Human Expression in the Age of Communications Overload*. Cambridge, MA: The MIT Press.

Harper, R., Palen, L., and Taylor, A., eds. (2005) *The Inside Text: Social, Cultural and Design Perspectives on SMS*. Dordrecht, the Netherlands: Springer.

Heath, C. and Luff, P. (2000) *Technology in Action*. Cambridge, U.K.: Cambridge University Press.

Hiltz, S. R. and Turoff, M. (1985) Structuring computer-mediated communication systems to avoid information overload. *Communications of the ACM*, 28(7):680–689.

Hinds, P. J. and Kiesler, S., ed. (2002) *Distributed Work*. Cambridge, MA: MIT Press.

Hislop, D., ed. (2008) *Mobility and Technology in the Workplace*. London, U.K.: Routledge.

Hjorth, L. (2009) *Mobile Media in the Asia-Pacific: The Art of Being Mobile*. Oxford, U.K.: Routledge.

Horst, H. and Miller, D. (2006) *The Cell Phone: An Anthropology of Communication*. New York: Berg Publishers Ltd.

Iastrebova, K. (2006) Managers' information overload: The impact of coping strategies on decision-making performance. PhD Thesis, Rotterdam, the Netherlands: Erasmus University. http://repub.eur.nl/res/pub/7329/EPS2006077LIS_9058921115_IASTREBOVA.pdf

Ito, M., Okabe, D., and Matsuda, M., eds. (2005) *Persona, Portable, Pedestrian: Mobile Phones in Japanese Life*. Cambridge, MA: The MIT Press.

Jarvenpaa, S. L. and Lang, K. R. (2005) Managing the paradoxes of mobile technology. *Information Systems Management*, 22(4):7–23.

Kakihara, M. and Sørensen, C. (2004) Practicing mobile professional work: Tales of locational, operational, and interactional mobility. *INFO: The Journal of Policy, Regulation and Strategy for Telecommunication, Information and Media*, 6(3):180–187.

Kallinikos, J. (2006) *The Consequences of Information: Institutional Implications of Technological Change*. Cheltenham, U.K.: Edward Elgar.

Katz, J. E. and Aakhus, M., eds. (2002) *Perpetual Contact*. Cambridge, U.K.: Cambridge University Press.

Kellaway, L. (2005) *Martin Lukes: Who Moved My Blackberry?* London, U.K.: Penguin Books.

Kietzmann, J. (2008) Internative innovation of technology for mobile work. *European Journal of Information Systems*, 17(3):305–320.

Kourouthanassis, P. E. and Giaglis, G. M., eds. (2008) *Pervasive Information Systems*. New York: M.E. Sharpe.

Lattanzi, M., Korhonen, A., and Gopalakrishnan, V. (2006) *Work Goes Mobile: Nokia's Lessons from the Leading Edge*. Chichester, England: John Wiley & Sons.

Licoppe, C. (2004) 'Connected' presence: The emergence of a new repertoire for managing social relationships in a changing communication technoscape. *Environment and Planning D: Society and Space*, 22:135–156.

Licoppe, C. (2010) The "Crisis of the Summons" a transformation in the pragmatics of "Notifications," from phone rings to instant messaging. *The Information Society*, 26(4):288–302.

Licoppe, C. and Heurtin, J. P. (2001) Managing one's availability to telephone communication through mobile phones: A french case study of the development dynamics of mobile phone use. *Personal and Ubiquitous Computing*, 5:99–108.

Licoppe, C. and Smoreda, Z. (2006) Rhythms and ties: Towards a pragmatics of technologically-mediated sociability. In *Computers, Phones, and the Internet: Domesticating Information Technology*, eds. Kraut, R. E., Brynin, M., and Kiesler, S., New York: Oxford University Press.

Ling, R. (2004) *The Mobile Connection: The Cell Phone's Impact on Society*. Amsterdam, the Netherlands: Morgan Kaufmann.

Ling, R. (2008) *New Tech, New Ties: How Mobile Communication is Reshaping Social Cohesion*. Cambridge, MA: The MIT Press.

Ljungberg, F. (1999) Exploring CSCW mechanisms to realize constant accessibility without inappropriate interaction. *Scandinavian Journal of Information Systems*, 11(1):25–50.

Lyytinen, K. and Yoo, Y. (2002a) Issues and challenges in ubiquitous computing. *Communications of the ACM*, 45(12):62–65.

Lyytinen, K. and Yoo, Y. (2002b) The next wave of nomadic computing: A research Agenda for information systems research. *Information Systems Research*, 13(4):377–388.

Mann, S. and Niedzviecki, H. (2002) *Cyborg: Digital Destiny and Human Possibility in the Age of the Wearable Computer*. Toronto, Ontario, Canada: Doubleday.

Mathiassen, L. and Sørensen, C. (2008) Towards a theory of organizational information services. *Journal of Information Technology*, 23(4):313–329.

Mazmanian, M. A., Orlikowski, W. J., and Yates, J. (2005) Crackberries: The social implications of ubiquitous wireless e-mail devices. In *Designing Ubiquitous Information Environments: Socio-technical Issues and Challenges*, eds. Sørensen, C., Yoo, Y., Lyytinen, K., and DeGross, J. I., New York: Springer, pp. 337–343.

Mazmanian, M., Yates, J., and Orlikowski, W. (2006) Ubiquitous email: Individual experiences and organizational consequences of blackberry use. In *Proceedings of the 65th Annual Meeting of the Academy of Management*, Atlanta, GA.

Mccullough, M. (2004) *Digital Ground: Architecture, Pervasive Computing, and Environmental Knowing*. Cambridge, MA: The MIT Press.

McFarlane, D. C. (1998) Interruption of people in human-computer interaction. PhD thesis, The George Washington University. http://interruptions.net/literature/McFarlane-Dissertation-98.pdf (accessed on March 21, 2013).

McGuire, R. (2007) *The Power of Mobility: How Your Business Can Compete and Win in the Next Revolution*. Hoboken, NJ: John Wiley & Sons.

Mick, D. G. and Fournier, S. (1998) Paradoxes of technology: Consumer cognizance, emotions, and coping strategies. *Journal of Consumer Research*, 25:123–143.

Mitchell, W. J. (1995) *City of Bits: Space, Place and the Infobahn*. Cambridge, MA: The MIT University Press.

Mitchell, W. J. (2003) *Me++: The Cyborg Self and the Networked City*. Cambridge, MA: The MIT Press.

Norman, D. (1988) *The Psychology of Everyday Things*. New York: Basic Books.

Norman, D. A. (1993) *Things That Make Us Smart. Defending Human Attributes in the Age of the Machine*. Reading, MA: Addison-Wesley.

Norman, D. (1999) *The Invisible Computer: Why Good Products Can Fail, the Personal Computer Is So Complex, and Information Appliances are the Solution*. Cambridge, MA: The MIT Press.

O'Reilly III, C. A. and Tushman, M. L. (2004) The ambidextrous organization. *Harvard Business Review*, 4:74–81.

Oard, D. W. (1997) The state of the art in text filtering. *User Modeling and User-Adapted Interaction: An International Journal*, 7(3):141–178.

Olson, G. M. and Olson, J. S. (2000) Distance matters. *Human-Computer Interaction*, 15:139–178.

Pertierra, R., ed. (2007) *The Social Construction and Usage of Communication Technologies: Asian and European Experiences*. Quezon City, Philippines: The University of the Philippines Press.

Quinn, R. and Rohrbaugh, J. (1983) A spatial model of effectiveness criteria: Towards a competing values approach to organizational analysis. *Management Science*, 29(3):363–377.

Rao, M. and Mendoza, L., eds. (2005) *Asia Unplugged: The Wireless and Mobile Media Boom in Asia-Pacific*. New Delhi: Response Books. 0-7619-3272-0.

Reid, N. (2010) *Wireless Mobility*. New York: McGraw-Hill.

Rennecker, J. and Godwin, L. (2005) Delays and interruptions: A self-perpetuating paradox of communication technology use. *Information and Organization*, 15(3):247–266.

Sathyan, J., Anoop, N., Narayan, N., and Vallath, S. K. (2012) *A Comprehensive Guide to Enterprise Mobility*. San Francisco, CA: CRC Press.

Schultze, U. and Vandenbosch, B. (1998) Information overload in a groupware environment: Now you see it, now you don't. *Journal of Organizational Computing and Electronic Commerce*, 8(2):127–148.

Simon, H. A. and Chase, W. G. (1973) Skill in chess: Experiments with chess-playing tasks and computer simulation of skilled performance throw light on some human perceptual and memory processes. *American Scientist*, 61(4):394–403.

Sørensen, C. (2010) Cultivating interaction ubiquity at work. *The Information Society*, 26(4):276–287.

Sørensen, C. (2011) *Enterprise Mobility: Tiny Technology with Global Impact on Work*. Technology, Work and Globalization Series, New York: Palgrave.

Sørensen, C. and Al-Taitoon, A. (2008) Organisational usability of mobile computing: Volatility and control in mobile foreign exchange trading. *International Journal of Human-Computer Studies*, 66(12):916–929.

Sørensen, C. and Pica, D. (2005) Tales from the police: Rhythms of interaction with mobile technologies. *Information and Organization*, 15(3):125–149.

Sørensen, C., Yoo, Y., Lyytinen, K., and DeGross, J. I., eds. (2005) *Designing Ubiquitous Information Environments: Socio-Technical Issues and Challenges*. New York: Springer.

Straus, S. G., Bikson, T. K., Balkovich, E., and Pane, J. F. (2010) Mobile technology and action teams: Assessing blackBerry use in law enforcement units. *Computer Supported Cooperative Work (CSCW)*, 19(1):45–71.

The Economist (2010) Data, data everywhere: A special report on managing information. *The Economist*, February 27. http://www.economist.com/node/15557443

Tilson, D., Lyytinen, K., and Sørensen, C. (2010) Digital infrastructures: The missing IS research Agenda. *Information Systems Research*, 21(4):748–759.

Urry, J. (2003) *Global Complexity*. Cambridge, U.K.: Polity.

Wajcman, J., Bittman, M., and Brown, J. E. (2008) Families without borders: Mobile phones, connectedness and work-home divisions. *Sociology*, 42(4):635–652.

Weiser, M. (1991) The computer for the twenty-first century. *Scientific American*, 265(3):94–104.

Wiberg, M. and Ljungberg, F. (2001) Exploring the vision of "Anytime, Anywhere" in the context of mobile work. In *Knowledge Management and Virtual Organizations*, ed. Malhotra, Y., Hershey, PA: Idea Group Publishing, pp. 157–169.

Wiberg, M. and Whittaker, S. (2005) Managing availability: Supporting lightweight negotiations to handle interruptions. *ACM Transactions of Computer-Human Interaction*, 12(4):1–32.

Wiredu, G. and Sørensen, C. (2006) The dynamics of control and use of mobile technology in distributed activities. *European Journal of Information Systems*, 15(3):307–319.

Yates, J. (1989) *Control through Communication: The Rise of System in American Management*. Baltimore, MD: The Johns Hopkins University Press.

Yoo, Y. (2010) Computing in everyday life: A call for research on experiential computing. *MIS Quarterly*, 34(2):213–231.

Zuboff, S. (1988) *In the Age of the Smart Machine*. New York: Basic Books.

50

Sustainable IT

Edward Curry
*National University
of Ireland Galway*

Brian Donnellan
*National University of
Ireland Maynooth*

50.1 Introduction

Sustainable IT is the design, production, operation, and disposal of IT and IT-enabled products and services in a manner that is not harmful and may be positively beneficial to the environment during the course of its whole-of-life [1]. Sustainable IT requires the responsible management of resources (both IT and non-IT) encompassing environmental, economic, and social dimensions. The first wave of sustainable IT, *Greening of IT,* aims to reduce the 2% of global greenhouse gas (GHG) emissions for which information technology (IT) is responsible [2], by reducing the footprint of IT thought actions such as improving the energy efficiency of hardware (processors and disk drives) and reducing waste from obsolete hardware. The second wave of sustainable IT, *Greening by IT,* also called Green IT 2.0 [3], is shifting the focus toward reducing the remaining 98%, as illustrated in Figure 50.1, by focusing on the innovative use of IT in business processes to deliver positive sustainability benefits beyond the direct footprint of IT, such as monitoring a firm's emissions and waste to manage them more efficiently. The potential of *Greening by IT* to reduce GHG emissions has been estimated at approximately 7.8 Gt CO_2 of savings in 2020, representing a 15% emission cut in 2020 and 600 billion ($946.5 billion) of cost savings [2]. It is estimated that the use of IT for greening will play a key role in the delivery of benefits that can alleviate at least five times the GHG footprint of IT itself [4].

This chapter provides an overview of sustainable IT best practices followed by discussion of the challenges faced by IT organizations in delivering sustainable IT. Green for IT practices covered include how software should be designed to minimize its resource usage, how the energy efficiency of data centers (DCs) can be improved through the use of virtualization, air management, and cooling, and the life cycle assessment (LCA) process that is used to determine the environmental impact of a

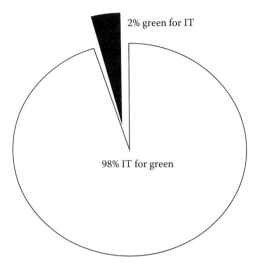

FIGURE 50.1   Target potential CO_2 reductions for sustainable IT.

product or service. Practices for IT for Green include how energy can be effectively managed with Green Information Systems (IS). Next, the chapter examines the challenges facing IT organizations as they put sustainable IT best practices into action, including how to benchmark the maturity of the sustainable IT. The chapter finishes with a brief look at future sustainable IT research directions and with a summary.

50.2 Two Faces of Sustainable IT

Sustainable IT is a broad topic that can have many motivations including:

- Self-interest (image, competitive advantage, innovation)
- Social, cultural, and political influence
- Regulatory and compliance requirements
- Environmental concerns
- Economic benefit

There are substantial inefficiencies in IT and its usage behavior that can be readily addressed. As illustrated in Figure 50.2, there are two faces to sustainable IT: Green for IT and IT for Green.

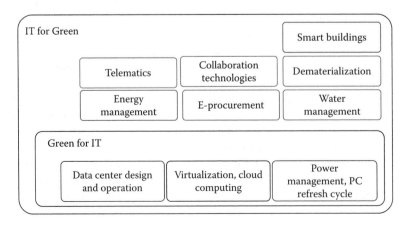

FIGURE 50.2   Two faces of sustainable IT.

50.2.1 Green for IT

Murugesan [5] details a holistic approach to *Green for IT* that comprehensively and effectively addresses the environmental impacts of IT using four key practices:

- *Green use of IT systems*: Reduce the energy consumption of computers and information systems and use them in an environmentally sound manner.
- *Green disposal of IT systems*: Refurbish and reuse old computers and properly recycle unwanted computers and other electronic equipment.
- *Green design of IT systems*: Design energy-efficient and environmentally sound components, computers, servers, and cooling equipment.
- *Green manufacturing of IT systems*: Manufacture electronic components, computers, and other associated subsystems with minimal or no impact on the environment.

By following these four practices, the environmental impacts of IT can be reduced to improve its sustainability throughout its entire life cycle. Examples of Green for IT include:

- Energy-efficient DC design and operation
- Virtualization and cloud computing
- Power management
- PC-refresh cycle
- Energy-efficient computing
- Responsible disposal and recycling of IT equipment
- Eco-labeling of IT products

50.2.2 IT for Green

There is substantial potential for sustainable IT to bring together business processes, resource planning, direct and indirect activities, and extended supply chains to effect positive changes across the entire activities of governments, organizations, and individuals. *IT for Green* is concerned with analyzing, designing, and implementing systems to deliver positive sustainability benefits beyond the direct footprint of IT, such as monitoring a firm's emissions and waste to manage them more efficiently. IT for Green provides an opportunity for the innovative use of IT in business processes to improve resource management. Examples of IT for Green are:

- Smart buildings
- Energy management
- E-procurement
- E-waste
- Telematics
- Water management
- Collaboration technologies
- Dematerialization

Understanding how to utilize sustainable IT requires understanding how IT can be used to improve the sustainability of an activity (both IT and non-IT). The following two sections discuss the application of IT to improve the sustainability of IT and then the application to a non-IT business activity.

50.3 Green for IT Practices

This section discusses Green for IT practices including Green Software, DC energy efficiency, and the LCA of products and services.

50.3.1 Green Software

Power-management features are becoming prevalent within hardware resulting in significantly improved hardware energy efficiency, with particularly high gains in mobile devices as a response to maximizing battery life. While hardware has been constantly improved to be energy efficient, software has not recorded a comparable track. The availability of increasingly efficient and cheaper hardware components has led designers up to now to neglect the energy efficiency of end-user software, which remains largely unexplored. While software does not consume power directly, software plays a critical role in the efficiency of the overall system as it interacts with power-consuming resources. Software causes the computations performed by the processor and is the root cause of all the consumption of the infrastructural layers within both a server and the DC (e.g., cooling, Uninterruptible Power Supply (UPS)).

The software development life cycle and related process management methodologies rarely consider energy efficiency as a design requirement. Software energy efficiency should be included in the initial design of software to optimize the efficiency of the software utilizing the hardware to satisfy a given set of functional requirements and workloads. Well-designed, efficient software should have good proportionality between the amount of useful work done and the amount of resources consumed [6]. In contrast, inefficient software demonstrates poor resource utilization to the amount of work done.

Effective resource management within software can contribute to the overall efficiency of the system. Where resources are over-allocated, or not efficiently used, the software can be responsible for lowering the efficiency of the entire system. The design of complex distributed software can affect the software energy efficiency. Well-known design issues that lower energy efficiency include sloppy programming techniques, excessive layering (deep inheritance trees leading to higher method invocation costs), code fragmentation (excessively small classes or small code objects inhibiting aggressive optimization), and overdesign (such as using databases to hold static configuration data) [7]. Software compilers cannot easily compensate for these design issues, resulting in lower energy efficiency [7]. An analysis [7] of the energy efficiency of software design has shown that:

- CPU is the component that absorbs most of the power (dependent on usage)
- Memory energy requirements are independent of use due to constant refreshing
- Hard disks consume most of the energy for their continuous spinning

The study concludes that an intelligent use of memory can shift the computational burden from the CPU to storage, possibly reducing the total energy consumption [7].

A concrete example of how software can be energy efficient is given by Steigerwald and Agrawal [8] for the efficiency of DVD playback software. When used on a notebook, is it better for the playback application to set up a large buffer and do less frequent optical drive reads or to keep the drive spinning? From an implementation perspective, there is not much difference between the two designs; however, the buffering strategy can add up to 20 min battery life. Within the analysis of the design of DVD applications, Steigerwald et al. [8] offer three key design considerations for energy-efficient software: (1) computational efficiency, (2) data efficiency, and (3) context awareness.

Computational efficiency: Reduce energy costs by improving application performance with respect to CPU utilization. The objective is to maximize the utilization of the CPU. The faster the workload can be completed, the sooner the CPU can return to idle and more energy can be saved. To achieve computational efficiency, use software techniques that achieve better performance such as efficient algorithms, multithreading, and vectorization.

Computational efficiency also extends to how the software interacts with the CPU. For example, a common design practice for threads waiting for a condition is to use a timer to periodically wake up to check if the condition has been satisfied. No useful work is being done as the thread waits, but each time it wakes up to check the condition, the CPU is forced to leave an idle power-managed state.

Ideally, applications should leverage an event-triggered mechanism to eliminate the need for time-based polling or, if possible, to move all periodic/polling activity to be batch processed.

Data efficiency: Reduce energy costs by minimizing data movement and using the memory hierarchy effectively. Data efficiency requires thinking about how an application reads and writes data (particularly I/O operations, such as read requests for a drive) and how it moves data around during execution. The DVD playback application is an example of data efficiency using I/O buffering to maximize I/O utilization by prefetching and caching. Other data efficiency techniques include:

- Software algorithms that minimize data movement
- Memory hierarchies that keep data close to processing elements
- Application software that efficiently uses cache memories

Data efficiency also covers the design of memory management. Data-efficient design will release memory that is no longer needed and watch for memory leaks. For example, a service that slowly leaks memory will over time allow its memory heap to grow to consume much of the system's physical memory. Even though the software does not actually need all this memory, it can restrict the opportunity to power manage memory since most of it has been allocated.

Context awareness: Reduce energy consumption by enabling applications to make intelligent decisions at runtime based on the current state of the system. Context-aware software knows the power state of the system and the current power policy, behaves appropriately, and responds to changes dynamically. These decisions are typically encapsulated in passive or active power policies. Passive power policies respond to a change in context by asking the user what action to take ("Switch to power-save mode?") or to acknowledge that the state change has occurred ("You have 10% battery left. OK?"). An active power policy would automatically take corrective actions, such as changing behavior to minimize energy consumption when a laptop is running on battery (i.e., dimming a screen when moving from AC to DC power sources).

Overall, the design of software can have a significant impact on its energy efficiency. If the software is less layered and more algorithmically efficient, it will consume less energy [7].

50.3.2 Data Center Energy Efficiency

With power densities of more than 100 times that of a typical office building, energy consumption is a central issue for DCs. Massive growth in the volumes of computing equipment, and the associated growth in areas such as cooling equipment, has led to increased energy usage and power densities within DCs. Trends toward cloud computing have the potential to further increase the demand for DC-based computing services [9].

The U.S. EPA estimates that servers and DCs are responsible for up to 1.5% of the total U.S. electricity consumption [10] or roughly 0.5% of U.S. GHG emissions for 2007. Power usage within a DC goes beyond the direct power needs of servers to include networking, cooling, lighting, and facilities management with power drawn for DCs ranging from a few kWs for a rack of servers to several tens of MWs for large facilities. While the exact breakdown of power usage will vary between individual DCs, Figure 50.3 illustrates an analysis of one DC where up to 88.8% of the power consumed was not used for computation; for every 100 W supplied to the DC, only 11.2 W was used for computation [10].

Air conditioners, power converters, and power transmission can use almost half of the electricity in the DC, and the IDC estimates that DC energy costs will be higher than equipment costs by 2015 [11]. The cost of operating a DC goes beyond just the economic bottom line; there is also an environmental cost. By 2020, the net footprint for DCs is predicted to be 259 $MtCO_2e$ [2]. There is significant potential to improve both the economic and environmental bottom line of DCs by improving their energy efficiency; however, a number of challenges exist.

The efficiency of a DC can be improved by using new energy-efficient equipment, improving airflow management to reduce cooling requirements, investing in energy management software, charging back

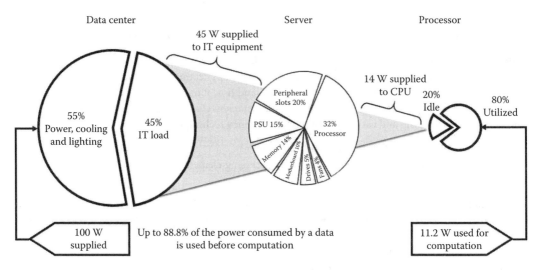

FIGURE 50.3   Example breakdown of power usage within a data center. (From Report to Congress on Server and Data Center Energy Efficiency Public Law 109–431, U.S. EPA, 2007.)

environmental impacts to consumers [12], and adopting environmentally friendly designs for DCs. In the remainder of this section, we will examine the most significant techniques for improving the energy efficiency within the DC, including virtualization, internal air management and cooling, and relevant metrics to understand the energy efficiency of a DC.

50.3.2.1 Virtualization

Virtualization is the creation of a virtual version (rather than actual physical version) of a computing resource such as a hardware platform, operating system, a storage device, or network resources. With virtualization, one physical server hosts multiple virtual servers as illustrated in Figure 50.4. Virtualization is a key strategy to reduce DC power consumption by consolidating physical server infrastructure to host multiple virtual servers on a smaller number of more powerful servers.

Virtualized DCs introduce additional degrees of freedom that are not available under traditional operating models where there is a hard binding between servers and the applications on which they run. Applications hosted in virtualized DCs run on virtualized operating systems. That is, the operating system does not run directly on the hardware, but it is mediated through a virtualization hypervisor. The hypervisor frees applications from a single physical host, allowing them to be moved around within a pool of servers to optimize the overall power and thermal performance of the pool. The loose binding between

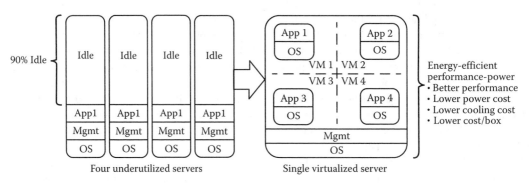

FIGURE 50.4   Saving achieved from virtualization of server resources.

applications and hosts allows treating a group of hosts as a pooled resources, allowing optimizations as a group that were not possible with individual machines, such as powering down servers during low demand.

50.3.2.2 Internal Air Management and Cooling

Air is the most common medium used to cool IT equipment. Moving air consumes energy; it is imperative that action is taken to ensure effective air management. How the air is supplied, how the exhaust heat is removed, what type of refrigeration cycle is utilized, and how the air circulates within the DC determine the energy efficiency of cooling [13,14].

Air segregation: Air mixing is an important issue with best practice to fully segregate hot and cold air. In the hot aisle, cold aisle configuration fans draw cool air from the front of the server, removing heat from the internal components and blowing the now heated air out through the back of the server chassis.

Air distribution strategies: Three common air cooling/distribution infrastructures are employed within the DC. (1) In *room-based (perimeter/peripheral) air-cooling*, cool air is delivered to the front of the server rack (cold aisle), typically through floor or overhead. The cool air is drawn through the servers and exhausted into the hot aisle. The hot air then propagates toward computer room air conditioners (CRACs) or computer room air handlers (CRAHs) at the peripheral of the room. (2) *Row-based* cooling is similar to perimeter cooling, but the individual CRACs/CRAHs are dedicated to a particular row of IT equipment as opposed to the room as a whole. (3) *Rack-based* cooling uses smaller CRAC/CRAH units associated with individual racks. In energy efficiency terms, row-based air distribution is typically an improvement on room-based cooling due to shortened air paths. However, rack-based cooling can be the most efficient given the reduced fan power required to move air within the confines of the rack itself.

Cooling: The main systems used to cool the DC include air-cooled direct expansion (DX) systems, glycol cooled systems, and water-cooled systems. Each utilizes a CRAC with an energy-hungry refrigeration cycle; a new practice is to use direct free cooling. Direct free cooling or airside economization is fast becoming an energy efficiency best practice. In climates that are suitable, air is supplied directly to the front of the IT equipment while the warmer exhaust air is ducted directly to the outside atmosphere. This essentially eliminates the refrigeration cycle providing what is typically called "free-cooling." In terms of redundancy and the percentage of the year when conditions are unfavorable, refrigeration-based backup is utilized.

50.3.2.3 Data Center Metrics

This section discusses a number of key metrics defined by the Green Grid* (a nonprofit group of IT enterprises formed in 2007) to understand the sustainability of a DC.

Power usage effectiveness/data center infrastructure efficiency: Power usage effectiveness (PUE) [15] is a measure of how efficiently a DC uses its power. PUE measures how much power the computing equipment consumes in contrast to cooling and other overhead uses. PUE is defined as follows:

$$PUE = \frac{Total\ Facility\ Power}{IT\ Equipment\ Power}$$

The reciprocal of PUE is data center infrastructure efficiency (DCiE) and is defined as follows:

$$DCiE = \frac{1}{PUE} = \frac{IT\ Equipment\ Power}{Total\ Facility\ Power} \times 100\%$$

---

* http://www.thegreengrid.org

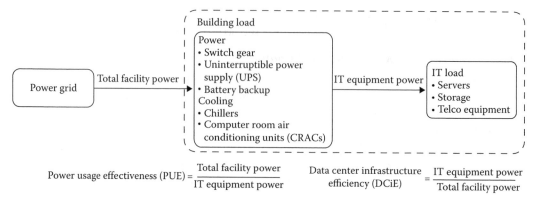

$$\text{Power usage effectiveness (PUE)} = \frac{\text{Total facility power}}{\text{IT equipment power}} \qquad \substack{\text{Data center infrastructure} \\ \text{efficiency (DCiE)}} = \frac{\text{IT equipment power}}{\text{Total facility power}}$$

FIGURE 50.5 PUE/DCiE calculation overview.

IT equipment power includes the load associated with all of the IT equipment, such as computing, storage, and network equipment, along with supplemental equipment used to monitor or otherwise control the DC including KVM switches, monitors, workstations, and laptops. *Total facility power* includes everything that supports the IT equipment load such as power delivery (UPS, generators, batteries, etc.), cooling system (chillier, computer room air conditioning units [CRACs], direct expansion air handler [DX] units, pumps, and cooling towers), compute, network, storage nodes, and other loads such as DC lighting.

Both PUE and DCiE, illustrated in Figure 50.5, are metrics that give an indication as to the use of power by supporting infrastructure of the DC. An ideal PUE would be 1.0; this means that no additional energy is consumed by the infrastructure that is supporting the IT load. The lower the PUE value the better, and conversely, the higher the DCiE value the better. If total power was 150 kW and IT power was 100 kW, then PUE would be 150/100 = 1.5; for DCiE, it would be 100/150 × 100 = 66%. This means that for every watt supplied to the IT equipment, half a watt is required to support it. A 2011 study by Koomey [16] estimates typical PUE between 1.93 and 1.83, with large DC operators such as Google, Yahoo, Facebook, and Microsoft reporting industry-leading PUE ratings in the 1.07–1.2 range.

While PUE/DCiE have proven to be an effective industry tool for measuring infrastructure energy efficiency, there is a need to measure the operational effectiveness of the DC. To this end, the Green Grid has defined a number of metrics [17] to measure dimensions including resource utilization and environmental impact. Water usage effectiveness (WUE) measures water usage to provide an assessment of the water used on-site for operation of the DC, while carbon usage effectiveness (CUE) measures DC-specific carbon emissions.

The xUE (PUE, WUE, CUE, etc.) family of consumption metrics does not tell the full story of the impacts of DCs on the environment. In order to understand the full environmental burden of a DC, a full life cycle analysis of the DC facilities and IT equipment is needed. These additional costs should not be underestimated. Take, for example, Microsoft's DC in Quincy, Washington, that consumes 48 MW (enough power for 40,000 homes) of power. In addition to the concrete and steel used in the construction of the building, the DC uses 4.8 km of chillers piping, 965 km of electrical wire, 92,900 m² of drywall, and 1.5 metric tons of batteries for backup power. Each of these components has its own impact that must be analyzed in detail; the means by which this is done is called an life cycle analysis (LCA).

50.3.3 Life Cycle Assessment

Understanding the impacts of an IT product or IT service requires an analysis of all potential impacts during its entire life cycle. An LCA, also known as life cycle analysis, is a technique to systematically identify resource flows and environmental impacts associated with all the stages of product/service provision [18]. The LCA provides a quantitative cradle-to-grave analysis of the product/services global environmental costs (i.e., from raw materials through materials processing, manufacture, distribution, use, repair and maintenance, and disposal or recycling). The demand for LCA data and tools has accelerated with the growing global demand to assess and reduce GHG emissions from different manufacturing and service sectors [18].

LCA can be used as a tool to study the impacts of a single product to determine the stages of its life cycle with most impact. LCA is often used as a decision support when determining the environmental impact of two comparable products or services.

50.3.3.1 Four Stages of LCA

The LCA standards, ISO 14040 and 14044 [19], follow a process with four distinct phases:

1. *Goal and scope definition*: It is important to ask the right question to ensure whether the LCA is successful. The first step in this process is the framing of the key questions for the assessment. Typical steps are to define the goal(s) of the project, determine what type of information is needed to inform decision makers, define functional units (environmental impact, energy efficiency, life span, cost per use, etc.), define the system boundaries, study perspective, allocation principles, environmental impact assessment categories, and level of detail.
2. *Inventory analysis*: The second phase involves data collection and modeling of the product/service system with process flow models and inventories of resource use and process emissions. The data must be related to the functional unit defined in the goal and scope definition and include all data related to environmental (e.g., CO_2) and technical (e.g., intermediate chemicals) quantities for all relevant unit processes within the study boundaries that compose the product system. Examples of inputs and outputs include materials, energy, chemicals, air emissions, water emissions, solid waste, radiation, or land use. This results in a life cycle inventory that provides verified information about all inputs and outputs in the form of elementary flows to and from the environment from all the unit processes involved in the study.
3. *Impact assessment*: The third phase evaluates the contribution to selected impact assessment categories, such as "Climate Change," "Energy Usage," and "Resource Depletion." Impact potential of the inventory is calculated and characterized according to the categories. Results can then be normalized across categories (same unit) and weighted according to the relative importance of the category.
4. *Interpretation*: The final phase involves interpretation of the results to determine the level of confidence and to communicate them in a fair, complete, and accurate manner. This is accomplished by identifying the data elements that contribute significantly to each impact category, evaluating the sensitivity of these significant data elements, assessing the completeness and consistency of the study, and drawing conclusions and recommendations based on a clear understanding of how the LCA was conducted and how the results were developed.

50.3.3.2 CRT Monitor vs LCD Monitor: Life Cycle Assessment

The U.S. Environmental Protection Agency's Design for the Environment program conducted a comprehensive environmental LCA of a traditional cathode ray tube (CRT) and a newer LCD monitor. The objective of the study [20] was to evaluate the environmental and human health life cycle impacts of functionally equivalent 17-in. CRT and 15-in. LCD monitors. The study assessed the energy consumption, resources input, and pollution produced over the lifetime of the equipment. The cradle-to-grave analysis was divided into three stages: (1) cradle-to-gate (manufacturing), (2) use, and (3) end-of-life (disposing or reusing). Each stage was assessed for the energy consumed, materials used in manufacturing along with associated waste. Components manufactured in different locations, where energy sources can differ due to the way local energy is produced, such as coal vs. nuclear, were taken into account. A sample of the results from a life cycle environmental assessment is presented in Table 50.1.

In summary, the LCA concluded that LCD monitors are about 10 times better for resource usage and energy use and 5 times better for landfill use. However, LCDs are only 15% better for global warming due to the fact that the LCD manufacturing process uses sulfur hexafluoride, a significant GHG.

TABLE 50.1 Life Cycle Analyses of CRT and LCD Monitors

	17″ CRT	15″ LCD
Total input material	21.6 kg	5.73 kg
Steel	5.16 kg	2.53 kg
Plastics	3 lb	1.78 kg
Glass	0.0 kg	0.59 kg
Lead-oxide glass	9.76 kg (0.45 lb of lead)	0.0 kg
Printed circuit boards (PCBs)	0.85 kg	0.37 kg
Wires	0.45 kg	0.23 kg
Aluminum	0.27 kg	0.13 kg
Energy (in manufacturing)	20.8 GJ	2.84 GJ
Power drawn	126 W	17 W
Energy (use—5 years' full power)	2.2 GJ	850 MJ

50.4 IT for Green Practices

50.4.1 Green Information Systems

As sustainable information is needed at both the macro and micro levels, it will require a multilevel approach that provides information and metrics that can drive high-level strategic corporate/regional sustainability plans, as well as low-level actions like improving the energy efficiency of an office worker. Relevant and accurate data, information, metrics, and Green Performance Indicators (GPIs) are important key to support sustainable practices, and the development of information systems that support this information need has led to the emergence of Green IS as a field in itself. There is substantial potential for Green IS to bring together business processes, resource planning, direct and indirect activities, and extended supply chains to effect positive changes across the entire activities of governments, organizations, and individuals. Green IS has been applied to a number of problems, from optimization of logistical networks [21], to buildings, DCs [12], and even cities. Within organizations, Green IS is the engine driving both the strategic and operational management of sustainability issues. Organizations pursuing a sustainability agenda will need to consider their Green IS to be a critical part of their operations [22]. This presents many challenges for Green IS research, and Melville [23] has outlined these under six themes:

1. *Context*: How do the distinctive characteristics of the environmental sustainability context, such as values and altruism, affect intention to use and usage of information systems for environmental sustainability?
2. *Design*: What design approaches are effective for developing information systems that influence human actions about the natural environment?
3. *Causality*: What is the association between information systems and organizational sustainability performance?
4. *New business models*: What is the association between IS and business models from an efficiency and environmental perspective?
5. *Systems approaches*: How can system approaches shed light on organizational and environmental outcomes that result from the use of IS for environmental sustainability?
6. *Models/metrics*: What are the multilevel models and metrics that encompass enterprise-wide sustainability initiatives?

50.4.1.1 Energy Informatics

The research field of energy informatics [22], a subfield of Green IS, recognizes the role that information systems can play in reducing energy consumption and thus CO_2 emissions. Energy informatics is concerned with analyzing, designing, and implementing systems to increase the efficiency of energy demand and supply systems. The core idea requires the collection and analysis of energy data to support optimization of energy. Watson [22] expresses this as:

$$\text{Energy} + \text{Information} < \text{Energy}$$

The Energy Informatics Framework [22], as illustrated in Figure 50.6, addresses the role of the information systems in the management and optimization of energy. The key components of the framework are as follows:

Supply and demand: There are two parties to any energy consumption transaction: a supplier and a consumer. Both sides have a common need for information to manage the flow of the resources they deliver and consume.

Energy system technologies: Three types of technology are present in an intelligent energy system:

1. *Flow network*: a set of connected transport components that supports the movement of continuous matter (e.g., electricity, oil, air, and water) or discrete objects (e.g., cars, packages, containers, and people).
2. *Sensor network*: a set of spatially distributed devices that reports the status of a physical item or environmental condition (e.g., air temperature, location of a mobile object) providing data that can be analyzed to determine the optimum use of a flow network.
3. *Sensitized object*: a physical good that a consumer owns or manages and has the capability to sense and report data about its use (i.e., GPS in a car). They provide information about the use of an object so that a consumer is better informed about the impact of the object on their finances and the environment. In addition, there needs to be remote control of the state of some sensitized objects so that suppliers and consumers can manage demand (i.e., smart appliances).

Information system: An information system ties together the various elements to provide a complete solution. It has several important functions from collecting data from the sensor network and feeding them into flow optimization algorithms, to transmitting data to automated controllers in the flow network to dynamically change a network based on the output of the optimization algorithms. The information system is also responsible for information provision to flow network managers, consumers, and governments about the consumption of resources. They can also enable consumers to automate or control object usage to reduce energy consumption.

Key stakeholders: The three most critical stakeholders in typical energy supply/demand systems are *suppliers* (provide energy/services, manage flow networks), *governments* (regulation), and *consumers* (user of the resource).

Eco-goals: The sustainability literature has identified three broad sustainability goals: *eco-efficiency* (ecological friendly competitively priced goods and services), *eco-equity* (social responsibility), and *eco-effectiveness* (working on the right products and services and systems).

The energy informatics framework provides a solid basis for managing different resource types with a Green IS (i.e., logistics [21]). Having discussed a number of sustainable IT practice, attention will now be turned to the challenges faced by IT organizations as they put sustainable IT best practice into action.

50.4.2 Sustainable IT in the Enterprise

To deliver on the promise of sustainable IT, IT organizations need to develop a sustainable IT capability to deliver benefits both internally and across the enterprise. However, due to the new and evolving nature of the field, few guidelines and guidance on best practices are available. Organizations face many challenges in developing and driving their overall sustainable IT strategies and programs as follows:

- The complexity of the subject and its rapid evolution
- The lack of agreed-upon and consistent standards

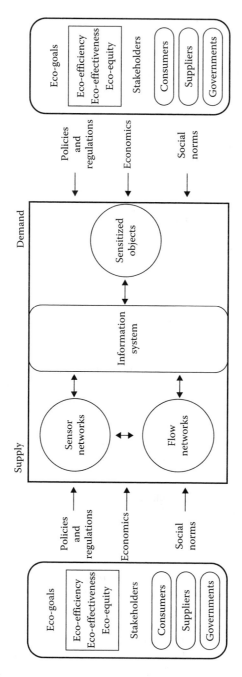

FIGURE 50.6　Energy informatics framework. (From Watson, R.T. et al., *MIS Quart.*, 34(1), 23, 2010.)

- Changing stakeholder expectations
- The lack of subject-matter expertise
- The need for new metrics and measures
- Evolving and increasing regulations and legislation around the world

Unfortunately, organizations often do not exploit IT's full potential in their efforts to achieve sustainability. Business and IT leaders frequently cannot find satisfactory answers to questions such as the following:

- Does the organization recognize IT as a significant contributor to its overall sustainability strategy?
- How is IT contributing to the organization's sustainability goals?
- What more could IT do to contribute to those goals?
- Are there clear measurable goals and objectives for sustainable IT?

IT departments face additional challenges specific to new IT methods and tools, industry metrics, and standards bodies as well as a general lack of relevant information such as power consumption quantifications. The challenge for IT departments is further complicated by the fact that sustainability is an enterprise-wide issue that spans the full value chain. The business is facing its own challenges in developing clear strategies and priorities to address a burning problem in such a dynamic and uncertain environment. It might lack the maturity to fully include sustainable IT in its efforts. This puts the onus on the IT organization to deliver sustainable IT benefits across the organization. In the remainder of this section, we examine two tools that can help an IT organization to develop a sustainable IT program: (1) a set of principles for Sustainable Enterprise IT and (2) the Sustainable Information Communication Technology-Capability Maturity Framework (SICT-CMF) from the Innovation Value Institute (IVI).

50.4.3 Sustainable IT Principles

In order to drive the adoption of sustainable IT, it is important for an IT organization to develop a clear set of IT sustainability principles for their activities. Sustainable IT principles are a set of aspirational statements to provide guidance on sustainable IT practices. An example of principles from the IT organization of Intel is described in Table 50.2 [24]. Within Intel IT, the principles play an important role in decision-making and are included in measurement models, standards, and processes. The criteria set out by the principles may also influence programs toward suppliers with sustainable business practices.

TABLE 50.2 Examples of Sustainable IT Principles from Intel IT

Principle	Example Actions
Consciously manage our capabilities	Include sustainability value and impacts in our proposals, measurements, and decision-making
Select sustainable suppliers	Work with our supplier managers to ensure our purchases represent and support sustainable business practices
Enable the organization to meet global sustainability compliance	Proactively monitor global regulations and requirements to ensure IT's and the organization's compliance
Measure, monitor, and optimize consumption	Reduce consumption and actively manage resources using sustainability metrics
Enable sustainable facilities	Work with Technology and Manufacturing Group (TMG) and Corporate Services (CS) to use IT capability to reduce resource consumption within TMG and CS IT capabilities to reduce resource consumption
Enable IT sustainability behavior	Create global awareness of IT sustainability that encourages IT employees to be corporate role models
Enable travel avoidance	Showcase all IT collaboration technologies across Intel
Promote IT's sustainability innovations across Intel and externally	Support sustainability innovations in our IT solutions and share those successes across Intel and externally

50.4.4 Capability Maturity Framework for Sustainable Information Communication Technology

The SICT-CMF gives organizations a vital tool to manage their sustainability capability [25,26]. The framework provides a comprehensive value-based model for organizing, evaluating, planning, and managing sustainable information communication technology (ICT) capabilities. The framework targets the broader scope of ICT to include communication technologies such as online collaboration tools, video conferencing, and telepresence, which can have positive impacts on sustainability (i.e., reduce business travel). Using the framework, organizations can assess the maturity of their SICT capability and systematically improve capabilities in a measurable way to meet the sustainability objectives. The SICT-CMF offers a comprehensive value-based model for organizing, evaluating, planning, and managing SICT capabilities, and it fits within the IVI's IT-CMF [27]. The SICT-CMF complements existing approaches for measuring SICT maturity, such as the G-readiness framework (which provides a benchmark score against SICT best practices [28,29]) or the Gartner Green IT Score Card (which measures corporate social responsibility compliance).

The SICT-CMF assessment methodology determines how SICT capabilities are contributing to the business organization's overall sustainability goals and objectives. This gap analysis between what the business wants and what SICT is actually achieving positions the SICT-CMF as a management tool for aligning SICT capabilities with business sustainability objectives.

The framework focuses on the execution of four key actions for increasing SICT's business value:

- Define the scope and goal of SICT
- Understand the current SICT capability maturity level
- Systematically develop and manage the SICT capability building blocks
- Assess and manage SICT progress over time

50.4.4.1 Defining the Scope and Goal

First, the organization must define the scope of its SICT effort. As a prerequisite, the organization should identify how it views sustainability and its own aspirations. Typically, organizational goals involve one or more of the following:

- Develop significant capabilities and a reputation for environmental leadership.
- Keep pace with industry or stakeholder expectations.
- Meet minimum compliance requirements and reap readily available benefits.

Second, the organization must define the goals of its SICT effort. It is important to be clear on the organization's business objectives and the role of SICT in enabling those objectives. Having a transparent agreement between business and IT stakeholders can tangibly help achieve those objectives. Significant benefits can be gained even by simply understanding the relationship between business and SICT goals.

50.4.4.2 Capability Maturity Levels

The framework defines a five-level maturity curve [26] for identifying and developing SICT capabilities (Figure 50.7):

1. *Initial*: SICT is ad hoc; there is little understanding of the subject and few or no related policies. Accountabilities for SICT are not defined, and SICT is not considered in the systems life cycle.
2. *Basic*: There is a limited SICT strategy with associated execution plans. It is largely reactive and lacks consistency. There is an increasing awareness of the subject, but accountability is not clearly established. Some policies might exist but are adopted inconsistently.

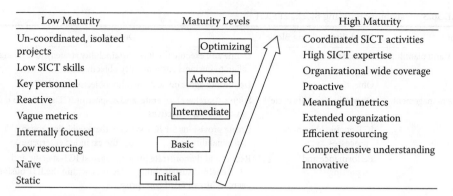

Low Maturity	Maturity Levels	High Maturity
Un-coordinated, isolated projects	Optimizing	Coordinated SICT activities
		High SICT expertise
Low SICT skills		Organizational wide coverage
Key personnel	Advanced	Proactive
Reactive		Meaningful metrics
Vague metrics	Intermediate	Extended organization
Internally focused		Efficient resourcing
Low resourcing	Basic	Comprehensive understanding
Naïve		Innovative
Static	Initial	

FIGURE 50.7 Comparison of low and high maturity of SICT.

3. *Intermediate*: An SICT strategy exists with associated plans and priorities. The organization has developed capabilities and skills and encourages individuals to contribute to sustainability programs. The organization includes SICT across the full systems life cycle, and it tracks targets and metrics on an individual project basis.

4. *Advanced*: Sustainability is a core component of the IT and business planning life cycles. IT and business jointly drive programs and progress. The organization recognizes SICT as a significant contributor to its sustainability strategy. It aligns business and SICT metrics to achieve success across the enterprise. It also designs policies to enable the achievement of best practices.

5. *Optimizing*: The organization employs SICT practices across the extended enterprise to include customers, suppliers, and partners. The industry recognizes the organization as a sustainability leader and uses its SICT practices to drive industry standards. The organization recognizes SICT as a key factor in driving sustainability as a competitive differentiator.

This maturity curve serves two important purposes. First, it is the basis of an assessment process that helps to determine the current maturity level. Second, it provides a view of the growth path by identifying the next set of capabilities an organization should develop to drive greater business value from SICT. A contrast of low and high levels of sustainable ICT is offered in Figure 50.1.

50.4.4.3 SICT Capability Building Blocks

While it is useful to understand the broad path to increasing maturity, it is more important to assess an organization's specific capabilities related to SICT. The SICT framework consists of nine capability building blocks (see Table 50.3) across the following four categories:

1. *Strategy and planning,* which includes the specific objectives of SICT and its alignment with the organization's overall sustainability strategy, objectives, and goals
2. *Process management,* which includes the sourcing, operation, and disposal of ICT systems, as well as the provision of systems based on sustainability objectives and the reporting of performance
3. *People and culture,* which defines a common language to improve communication throughout the enterprise and establishes activities to help embed sustainability principles across IT and the wider enterprise
4. *Governance,* which develops common and consistent policies and requires accountability and compliance with relevant regulation and legislation

The first step to systematically develop and manage the nine capabilities within this framework is to assess the organization's status in relation to each one.

The assessment begins with the survey of IT and business leaders to understand their individual assessments of the maturity and importance of these capabilities. A series of interviews with key stakeholders augments the survey to understand key business priorities and SICT drivers, successes

TABLE 50.3   Capability Building Blocks of SICT

Category	Capability Building Block	Description
Strategy and planning	Alignment	Define and execute the ICT sustainability strategy to influence and align to business sustainability objectives
	Objectives	Define and agree on sustainability objectives for ICT
Process management	Operations and life cycle	Source (purchase), operate, and dispose of ICT systems to deliver sustainability objectives
	ICT-enabled business processes	Create provisions for ICT systems that enable improved sustainability outcomes across the extended enterprise
	Performance and reporting	Report and demonstrate progress against ICT-specific and ICT-enabled sustainability objectives, within the ICT business and across the extended enterprise
People and culture	Adoption	Embed sustainability principles across ICT and the extended enterprise
	Language	Define, communicate, and use common sustainability language and vocabulary across ICT and other business units, including the extended enterprise, to leverage a common understanding
Governance	External compliance	Evangelize sustainability successes and contribute to industry best practices
	Corporate policies	Enable and demonstrate compliance with ICT and business sustainability legislation and regulation. Require accountability for sustainability roles and decision-making across ICT and the enterprise

achieved, and initiatives taken or planned. In addition to helping organizations understand their current maturity level, the initial assessment provides insight into the value placed on each capability, which will undoubtedly vary according to each organization's strategy and objectives. The assessment also provides valuable insight into the similarities and differences in how key stakeholders view both the importance and maturity of individual capabilities and the overall vision for success.

Figure 50.8 shows an example of consolidated survey results, resulting in an overall maturity level for each capability building block. This organization is close to level-three maturity overall but is less mature in some individual capabilities. It views alignment and objectives under the strategy and planning category as the most important capability building blocks, but it has not achieved level-three maturity in these areas. It also views operations and life cycle as important capabilities, but its maturity level for that building block is even lower (level 2.4).

50.4.4.4 Assessing and Managing SICT Progress

With the initial assessment complete, organizations will have a clear view of current capability and key areas for action and improvement. However, to further develop SICT capability, the organization should assess and manage SICT progress over time by using the assessment results to:

- Develop a roadmap and action plan
- Add a yearly follow-up assessment to the overall IT management process to measure over time both progress and the value delivered from adopting SICT

Agreeing on stakeholder ownership for each priority area is critical to developing both short-term and long-term action plans for improvement. The assessment results can be used to prioritize the opportunities for quick wins—that is, those capabilities that have smaller gaps between current and desired maturity and those that are recognized as more important but might have a bigger gap to bridge.

The assessment of sustainable IT was carried out in a number of global firms over the last 2 years. The assessment methodology included interviews with stakeholder from both the IT organizations and the business organization, including individuals involved with IT and corporate sustainability programs. The average results for the SICT maturity of the examined organizations are presented in Table 50.4.

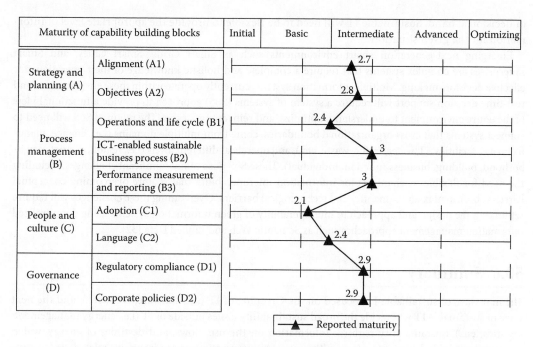

FIGURE 50.8 Aggregated result for the current maturity level from the assessment.

TABLE 50.4 Average SICT Maturity

Category	Capability	AVR CBB	Low	High	Diff	AVR Cat
Strategy and planning	Alignment	2.61	2.38	3.2	0.82	2.51
	Objectives	2.41	2.08	2.8	0.72	
Process management	Operations and life cycle	2.46	2.32	2.8	0.48	2.52
	ICT-enabled business processes	2.70	2.5	3	0.5	
	Performance and reporting	2.40	1.2	3	1.8	
People and culture	Adoption	2.03	1.89	2.3	0.41	2.18
	Language	2.33	2	2.9	0.9	
Governance	External compliance	2.19	1.8	2.9	1.1	2.24
	Corporate policies	2.28	1.4	2.9	1.5	

50.5 Research Directions

Many organizations think that sustainability requires a significant transformational change, yet the ultimate goal is to embed sustainability into business-as-usual activities. Improving sustainability performance [26], especially through changing the way an organization operates, requires a number of practical steps that will include the need for a systematic approach for information-gathering and analysis [30].

While organizations are fighting a data deluge within their information systems [31], there is a significant lack of data on sustainability concerns. A 2010 survey of more than 600 chief information officers and senior IT managers highlighted that few organizations are performing well at measuring the effectiveness of their sustainability efforts [29]. The paucity of sustainable information within organizations is a significant challenge and one that needs to be addressed if sustainable IT efforts are to deliver on their potential. Determining the granularity for effective sustainable data is not well understood, and research is needed to define the appropriate level of usefulness [22]. The appropriateness of information will also be highly dependent on the stakeholders and the task or decision at hand. Sustainable IT will need to be flexible to provide the appropriate level of information for the given situation.

Emerging next-generation smart environments such as Smart Grids, Smart Cities, and Smart Enterprises are complex systems that require a complete and holistic knowledge of their operations for effective decision-making. Multiple information systems currently operate within these environments and real-time decision support will require a system of systems (SoS) approach to provide a functional view of the entire environment to understand, optimize, and reinvent processes. The required SoS will need to connect systems that cross organizational boundaries, come from multiple domains (i.e., finance, manufacturing, facilities, IT, water, traffic, waste, etc.), and operate at different levels (i.e., region, district, neighborhood, building, business function, individual). These SoS pose many significant challenges, including the need for flexible mechanisms for information interoperability that require overcoming conceptual barriers (both syntax and semantic) and technological barriers. Overcoming these challenges will require rethinking the design and approach to interoperability of green information systems using decentralized information management approaches such as Semantic Web and Linked Data [32].

50.6 Summary

The first wave of sustainable IT focused on the Greening of IT (i.e., software, DCs, etc.), and the next wave of sustainable IT will tackle the broad sustainability issues outside of IT (i.e., energy management, logistics, etc.). Sustainability requires information on the use, flows, and destinies of energy, water, and materials including waste, along with monetary information on environment-related costs, earnings, and savings. There is substantial potential for sustainable IT to bring together business processes, resource planning, direct and indirect activities, and extended supply chains to effect positive changes across the entire activities of governments, organizations, and individuals.

Glossary

Energy: A physical quantity that describes the amount of work that can be performed by a force. A device that is energy efficient requires less energy for its "work" or task than its energy-inefficient counterpart. Energy can be expressed in units such as Joules.

Heat: Resistive heat is a natural by-product of running current through a conductor. Heat is a form of energy, and engineers strive to minimize heat release in computer design to minimize the need for cooling (typically by a fan).

Joule: Standard unit of energy measurement or work in the International System of Units.

Power: The rate at which energy is transferred, used, or transformed per unit of time. Within IT, power is typically measured in Watts, where a Watt equals 1 J/s. For example, a light bulb rated at 60 W consumes 60 J in 1 s.

Acknowledgment

This work has been funded in part by Science Foundation Ireland under Grant No. SFI/08/CE/I1380 (Lion-2).

Further Readings

- *IEEE IT Professional*, Special issue on Green IT 13(1): The January/February 2011 issue of IT Professional focused on Green IT and its transition toward sustainable IT. Articles focus on topics including sustainable IT, assessment of Green IT practices, power consumption of end-user PCs browsing the web, and corporate strategies for Green IT.

- *Harnessing Green IT: Principles and Practices*, (2012) S. Murugesan and G. R. Gangadharan (eds.), Wiley—IEEE, ISBN-10: 1119970059: This book presents and discusses the principles and practices of making computing and information systems greener—environmentally sustainable—as well as various ways of using information technology (IT) as a tool and an enabler to improve the environmental sustainability. The book comprehensively covers several key aspects of Green IT—green technologies, design, standards, maturity models, strategies, and adoption—and presents holistic approaches to greening IT encompassing green use, green disposal, green design, and green manufacturing.
- *Green Business Process Management–Towards the Sustainable Enterprise*, (2012), J. vom Brocke, S. Seidel, and J. Recker (eds.), Springer, ISBN-10: 3642274870. This volume consolidates the state-of-the-art knowledge about how business processes can be managed and improved in light of sustainability objectives. This book presents tools and methods that organizations can use in order to design and implement environmentally sustainable processes and provide insights from cases where organizations successfully engaged in more sustainable business practices.
- IT@Intel: IT Best Practices: (http://www.intel.com/content/www/us/en/it-management/intel-it/intel-it-best-practices.html) The IT @ Intel best practices portal gives insights into how an IT department can be run like a business, enabling the organization to take advantage of new technologies. In addition to general IT topics, white paper covers topics on data center energy efficiency, sustainable IT roadmaps, smart buildings, and Green Software.
- *Green IT/Sustainable IT columns in IEEE Computer and IEEE Intelligent Systems*: At the time of writing, both IEEE Computer and IEEE Intelligent Systems had regular columns dedicated to Green IT and sustainable IT. These columns covered many topics including, energy management, server power management, energy awareness, Green Software, and consumption-based metrics.

References

1. S. Elliot, Environmentally sustainable ICT: A critical topic for IS research? in *Pacific Asia Conference on Information Systems (PACIS 2007)*, Auckland, New Zealand, pp. 100–114, 2007.
2. M. Webb, SMART 2020: Enabling the low carbon economy in the information age, *The Climate Group London*, 2008. http://www.smart2020.org/assets/files/02_smart2020Report.pdf (accessed on April 11, 2013).
3. S. Murugesan and P. A. Laplante, IT for a greener planet, *IT Professional*, 13(1), 16–18, 2011.
4. P.-A. Enkvist and J. Rosander, A cost curve for greenhouse gas reduction, *The McKinsey Quarterly*, 1(November 2005), 34, 2007.
5. S. Murugesan, Harnessing green IT: Principles and practices, *It Professional*, 10(1), 24–33, 2008.
6. E. Saxe, Power-efficient software, *Queue*, 8(1), 10, 2010.
7. E. Capra, C. Francalanci, P. Milano, and S. A. Slaughter, Measuring application software energy efficiency, *IEEE IT Professional*, 14(2), 54–61, 2012.
8. B. Steigerwald, C. D. Lucero, C. Akella, and A. R. Agrawal, *Energy Aware Computing: Powerful Approaches for Green System Design*, Intel Press, Hillsboro, OR, 2012.
9. Carbon Disclosure Project & Verdantix Cloud Computing—The IT Solution for the 21st Century, London, U.K.: pp. 1–26, 2011.
10. R. Brown, E. Masanet, B. Nordman, B. Tschudi, A. Shehabi, J. Stanley, J. Koomey et al. Report to Congress on Server and Data Center Energy Efficiency Public Law 109–431.
11. N. Martinez and K. Bahloul, Green IT Barometer European Organisations and the Business Imperatives of Deploying a Green and Sustainable IT Strategy. International Data Corporation, 2008. http://www.dell.com/downloads/global/corporate/environ/comply/IDCWP28Q.pdf (accessed on April 11, 2013).
12. E. Curry, S. Hasan, M. White, and H. Melvin, An environmental chargeback for data center and cloud computing consumers, in *First International Workshop on Energy-Efficient Data Centers*, Madrid, Spain, 2012.

13. T. Lu, X. Lü, M. Remes, and M. Viljanen, Investigation of air management and energy performance in a data center in Finland: Case study, *Energy and Buildings*, 43(12), 3360–3372, 2011.

14. E. Curry, G. Conway, B. Donnellan, C. Sheridan, and K. Ellis, Measuring energy efficiency practices in mature data center: A maturity model approach, in *27th International Symposium on Computer and Information Sciences*, Paris, France, 2012.

15. C. Belady, A. Rawson, J. Pfleuger, and T. Cader, Green grid data center power efficiency metrics: PUE and DCiE. The Green Grid, 2008. http://www.thegreengrid.org (accessed on April 11, 2013).

16. J. Koomey, *Growth in Data Center Electricity Use 2005 to 2010*, Analytics Press, Oakland, CA, 2011.

17. E. Curry and B. Donnellan, Sustainable information systems and green metrics, in *Harnessing Green IT: Principles and Practices*, S. Murugesan and G. R. Gangadharan, Eds. John Wiley & Sons, Inc., Chichester, West Sussex, U.K., 2012.

18. R. Horne, T. Grant, and K. Verghese, *Life Cycle Assessment: Principles, Practice and Prospects*, Csiro Publishing, Collingwood, Victoria, Australia, p. 175, 2009.

19. M. Finkbeiner, A. Inaba, R. Tan, K. Christiansen, and H. J. Klüppel, The new international standards for life cycle assessment: ISO 14040 and ISO 14044, *The International Journal of Life Cycle Assessment*, 11(2), 80–85, 2006.

20. M. Socolof, J. Overly, and J. Geibig, Environmental life-cycle impacts of CRT and LCD desktop computer displays, *Journal of Cleaner Production*, 13(13–14), 1281–1294, 2005.

21. R. T. Watson, M.-C. Boudreau, S. Li, and J. Levis, Telematics at UPS: En route to energy informatics, *MIS Quarterly Executive*, 9(1), 1–11, 2010.

22. R. T. Watson, M.-C. Boudreau, and A. J. Chen, Information systems and environmentally sustainable development: Energy informatics and new directions for the IS community, *MIS Quarterly*, 34(1), 23–38, 2010.

23. N. P. Melville, Information systems innovation for environmental sustainability, *MIS Quarterly*, 34(1), 1–21, 2010.

24. E. Curry, B. Guyon, C. Sheridan, and B. Donnellan, Developing an sustainable IT capability: Lessons from Intel's journey, *MIS Quarterly Executive*, 11(2), 61–74, 2012.

25. E. Curry and B. Donnellan, Understanding the maturity of sustainable ICT, in *Green Business Process Management—Towards the Sustainable Enterprise*, J. vom Brocke, S. Seidel, and J. Recker, Eds. Springer, New York, pp. 203–216, 2012.

26. B. Donnellan, C. Sheridan, and E. Curry, A capability maturity framework for sustainable information and communication technology, *IEEE IT Professional*, 13(1), 33–40, January 2011.

27. M. Curley, *Managing Information Technology for Business Value: Practical Strategies for IT and Business Managers*, Intel Press, Hillsboro, OR, 2004, p. 350.

28. A. Molla et al., E-readiness to G-readiness: Developing a Green Information Technology readiness framework, in *19th Australasian Conference on Information Systems*, Christchurch, New Zealand, pp. 669–678, 2008.

29. A. O'Flynn, *Green IT: The Global Benchmark*, Fujitsu, Tokyo, Japan, 2010.

30. E. Curry, S. Hasan, U. ul Hassan, M. Herstand, and S. O'Riain, An entity-centric approach to green information systems, in *Proceedings of the 19th European Conference on Information Systems (ECIS 2011)*, Helsinki, Finland, 2011.

31. K. Cukier, A special report on managing information: Data, data everywhere, *The Economist*, 12(8), 1–14, Eye on Education, 2010.

32. E. Curry, System of systems information interoperability using a linked dataspace, in *IEEE 7th International Conference on System of Systems Engineering (SOSE 2012)*, Genoa, Italy, pp. 113–118, 2012.

33. E. Curry, S. Hasan, and S. O'Riáin, Enterprise energy management using a linked database for energy intelligence. In 2nd IFIP Conference on Sustainable Internet and ICT for Sustainability (Sustain IT 2012), Pisa, Italy: IEEE.

51

Business Continuity

Nijaz Bajgoric
University of Sarajevo

51.1 Introduction

As more and more businesses are operating in an e-business environment, they are under pressure to "keep their business in business" continuously. It is particularly important to ensure this for data/information availability, because continuous data access and availability are sought by employees, current and prospective customers, suppliers, and other stakeholders. Therefore, most businesses today seek for an information system infrastructure that is running 24/7/365, ensuring that data and applications are continuously available.

Information technologies have brought new opportunities for businesses in their efforts to increase their profits, reduce the costs, enhance the quality of products and services, improve relations with customers, and ease data access. However, at the same time, IT-related problems have produced new business risks as well. Unavailable data, stolen or lost computers, hardware glitches, server operating system crashes, application software errors, damaged backup tapes, broken data communication connections, a destroyed computer center due to several types of natural disasters, hackers' activities, and power outages are just some situations that may cause interruptions in business computing and affect the whole business.

The nature and characteristics of IT-associated risks have dramatically changed over two recent decades. While some 20–30 years ago an IT-related problem on a computer center's mainframe was viewed only as a "technical" problem causing a delay in, for instance, payroll system schedule, today similar problem may affect financial results and a corporation's competitive position or even bring the whole business into a halt. Unlike those times when most businesses operated on a "five-days-a-week" and "9 am–5 pm" basis, currently, it is common for most businesses to run on a 24/7/365 schedule. Hence, the role of business computing is becoming more important than ever with a special emphasis on becoming "continuous computing." It is no longer just about faster data processing, management of data, support for decision making, but rather about "keeping business in business." Business technology—a buzzword coined by Forrester (2007)—represents one of the major shifts in the way people think about computers in IT history. According to Forrester, business technology is based on two key ideas: (1) IT risks are business risks and (2) IT opportunities are now business opportunities.

"Business continuity" (BC) or "business continuance" is a general term that emphasizes the ability of a business to continue with its operations and services even if some sort of failure or disaster on its computing platform occurs. Therefore, BC is rather to be defined as an ultimate objective of modern business with regard to capability of its information system to provide both continuous computing and business resilience. Modern business seeks for an information infrastructure that is going to be always on with zero downtime or highly available with near-zero-downtime. Symantec's Small and Medium Business (SMB) Disaster Preparedness Survey (2011) revealed that disasters can have a significant financial impact on SMBs. At the time of this survey, the median cost of downtime for an SMB was $12,500 per day. Aberdeen Group found recently (Csaplar, 2012) that between June 2010 and February 2012, the cost per hour of downtime increased on average by 65%. In February 2010, the average cost per hour of downtime was reported to be about $100,000.

51.2 Business Continuity Defined and Impact on Practice

51.2.1 Business Continuity Defined

The concept of BC reflects the dependence of modern business on its information system infrastructure. If an enterprise's information system is up with mission-critical applications running and providing services to end users, business is said to be "in business." When an application server running a critical e-business application goes down for any reason, business simply becomes "out of business." Therefore, the process of addressing, mitigating, and managing the IT risks has become one of the major issues in organizational IT management.

Today's IT-related threats/risks that may cause system unavailability and hence affect the business can be classified into the following major categories:

- Technical glitches and/or hardware component failures within the IT infrastructure: processors, memory chips, mainboards, hard disks, disk controllers, tapes and tape drives, network cards, fans, power supply unit.
- Physical damage to/thefts of IT centers, servers, client computers, mobile devices, or communication/networking devices.
- Operating system crashes: crashes of server operating systems that run enterprise servers and make server applications available to end users.
- Internal threats such as intentional malicious or retaliation acts of disgruntled IT staff and other insiders including former employees, operational errors, careless behavior of users and system administrators, human errors such as accidental or deliberate deletion of system and configuration files, unskilled operations, mistakes by well-intentioned users, and hazardous activities including sabotage, strikes, epidemic situations, and vandalism.
- External threats that range from individuals hacking the Internet sites on curiosity basis to organized criminals and professionals working for industrial espionage purposes. These include Internet-based security threats such as viruses, malware, botnets, malicious code, Denial-of-Service or Distributed Denial-of-Service (DoS or DDOS) attacks, spam, intrusion attacks, phishing, and moles.
- File or process-based problems such as corrupted configuration files, broken processes or programs, and file system corruption.
- Application software defects: bugs in programs, badly integrated applications, user interventions on desktop computers, corrupted files, etc.
- LAN/WAN/Internet disconnections, hardware failures on data communication devices, software problems on network protocols, Domain Controllers, gateways, Domain Name Server (DNS) servers, Transmission Control Protocol/Internet Protocol (TCP/IP) settings, etc.

- Loss or leaving of key IT personnel: bad decisions on IT staffing and retention, particularly with regard to system administrators.
- Natural threats such as fire, lightning strike, flood, earthquake, storm, hurricane, tornado, snow, and wind.

These threats are shown in Figure 51.1.

In addition to several types of hardware failures, physical damages on computer systems, and natural threats, a number of threats coming from inside and outside of computer centers may cause system unavailability. These threats are in most cases addressed to enterprise servers given their role in modern business computing. Particularly, external threats require more attention as the servers are, by default, as follows:

- Accessible, available, tend to be always on, and running on continuous basis. Therefore, they are subject to continuous attacks from external attackers. Server-oriented attacks are much more expected than attacks to personal systems.
- Having several configuration files that can be misused (e.g., /etc/passwd file, /etc/hosts file, allow/deny access files on Unix/Linux).

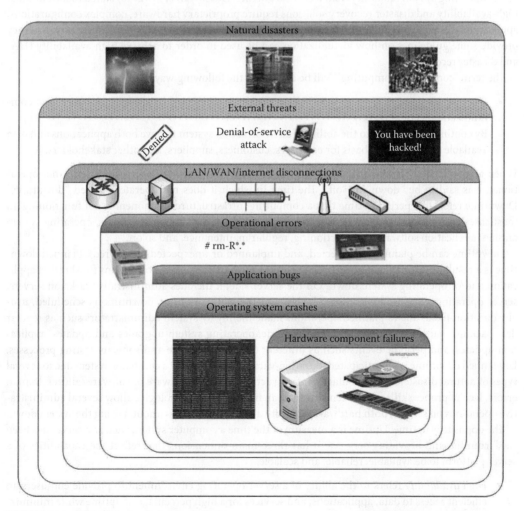

FIGURE 51.1 Major IT-related threats and risks.

- Having a number of services set on the "on" mode (open, active) after installation that opens the door for possible attacks.
- Storing business-critical applications and end user data to be accessed on remote basis, instead of keeping data on desktop/portable computers.

Addressing these threats leads to the process of mitigating (reducing) downtime situations and achieving continuous computing. Operating environment that facilitates continuous computing is a hardware–software platform that encompasses several technologies that are necessary for efficient and effective business-critical applications. Its characteristics will be discussed at a more detailed level in later sections.

Sumner (2009) unveiled the results of a study of top security threats perceived by organizations with over 500 employees as (1) deliberate software attacks, (2) technical software failures or errors, (3) acts of human error or failure, (4) deliberate acts of espionage or trespass, and (5) deliberate acts of sabotage or vandalism. ITIC Report (2009) unveiled the fact that server hardware and server operating system reliability have improved vastly since the 1980s, 1990s, and even in just the last 2–3 years. While technical bugs still exist, the number, frequency, and severity have declined significantly. This report underscores the fact that with few exceptions, common human error poses a bigger threat to server hardware and server operating system reliability than technical glitches. Adeshiyan et al. (2010) stated that traditional high-availability and disaster recovery solutions require proprietary hardware, complex configurations, application-specific logic, highly skilled personnel, and a rigorous and lengthy testing process. They provide some guidelines on how virtualization can be used in order to achieve high availability (HA) and disaster recovery.

The term "continuous computing" will be defined in the following way:

- By computing, we refer to an information system implemented by using information and communication technologies.
- By continuous, we refer to the ability of an information system to have both applications and data available on continuous basis for end users, customers, suppliers, and other stakeholders.

When an application server running business-critical or any other business application is not operational, it is said to be "down" or "off." The time in which it does not operate is called "downtime." Downtime refers to a period of time that a computing infrastructure component is not functioning, in most cases as a result of some system failure such as hardware component failures, operating system crashes, application software malfunctioning, regular maintenance, and so on.

Downtime can be planned or expected, and unplanned or unexpected (undesired). Planned downtime is a result of planned operations due to several administrative tasks or reasons for shutting application and/or operating system down. On the server side, it includes administrative tasks on servers, server operating systems, applications, LANs, and the WAN. Planned downtime is scheduled, usually duration-fixed, due to planned operations made mainly by system administrators such as regular data backups, hardware maintenance and upgrade, operating system upgrades and updates, application updates, and periodic events such as different types of hardware and software testing processes. Unplanned downtime is unanticipated, duration-variable downtime of a computer system due to several types of natural disasters, power outages, infrastructure failures, hardware or software failures, human errors, lack of proper skills, etc. Recent advances in information technologies allow several administrative operations on servers, both hardware and software interventions, without halting the system down.

The opposite is uptime. Uptime is a measure of the time a computer system such as a server has been "up," running and providing services. It has three main dimensions that reflect the capabilities of a server platform to be available, reliable, and scalable.

- *High availability* refers to the ability of a server operating environment to provide end users an efficient access to data, applications, and services for a high percentage of uptime while minimizing both planned and unplanned downtime. It is the ability of a system (server, operating system, application, network) to continue with its operation even in cases of hardware/software failures and/or disasters. IDC (2010) pointed out that HA is critical to building a strong BC strategy. HA is the foundation of all BC strategies.

- *High reliability* refers to the ability of a server operating environment to minimize (reduce) the occurrence of system failures (both hardware and software related) including some fault-tolerance capabilities. Reliability goals are achieved by using standard redundant components and advanced fault-tolerant solutions (hardware, systems software, application software).
- *High scalability* refers to the ability of a server operating environment of scaling up and out. Servers can be scaled up by adding more processors, RAM, etc., while scaling out means additional computers and forming cluster or grid configurations.

The term "high availability" is associated with high system/application uptime, which is measured in terms of "nines" (e.g., 99.999%). The more nines that represent availability ratio of a specific server, the higher level of availability is provided by that operating platform. According to IDC, HA systems are defined as having 99% or more uptime. IDC (2006, 2009) defined a set of different availability scenarios according to availability ratio, annual downtime, and user tolerance to downtime with regard to several continuous computing technologies that are available today (Table 51.1).

Table 51.2 correlates the number of nines to calendar time equivalents.

All these uptime dimensions can be identified and investigated for several components independently such as server hardware, operating system, application servers, network infrastructure, the whole server environment, or whole information system. Ideas International (IDEAS, www.ideasinternational.com) provides detailed analyses of most widely used server operating platforms with regard to several uptime dimensions.

51.2.2 Impact on Practice

In August 1995, Datamation quoted the results of a survey of 400 large companies that unveiled that "… downtime costs a company $1400 per minute, on average. Based on these figures, 43 hours of downtime per year would cost $3.6 million. One hour of downtime per year amounts to $84,000 per year" (Datamation, 1995). Recounting a situation in which eBay's auction site experienced a serious problem,

TABLE 51.1 IDC Classification of Availability Scenarios

	Availability (%)	Average Annual Downtime	User Tolerance to Downtime
Fault-tolerant continuous availability	99.999	5 min	None
Cluster high availability	99.9	8 h 45 min	Business interruption lost transaction
Stand-alone GP or blade server w/RAID	99.5	43 h 23 min	Tomorrow is okay

Source: IDC White Paper, Reducing downtime and business loss: Addressing business risk with effective technology, IDC, August 2009.

TABLE 51.2 Acceptable Uptime Percentage and Downtime Relationship

Acceptable Uptime Percentage	Downtime per Day	Downtime per Month	Downtime per Year
95	72.00 min	36 h	18.26 days
99	14.40 min	7 h	3.65 days
99.9	86.40 s	43 min	8.77 h
99.99	8.64 s	4 min	52.60 min
99.999	0.86 s	26 s	5.26 min

Source: Adapted from http://technet2.microsoft.com/windowsserver/WSS/en/library/965b2f19-4c88-4e85-af16-32531223aec71033.mspx?mfr=true, November 23, 2011.

a study by Nielsen Media Research and NetRatings Inc. (Dembeck, 1999) describes how the average time spent per person at Yahoo's auction site increased to 18 min on Friday from 7 min on Thursday— the night eBay's site crashed. The number of people using Yahoo's site also skyrocketed to 135,000 on Saturday from 62,000 on Thursday. But the number of users fell to 90,000 on Monday. Financial losses to eBay were estimated at $3–$5 million. Starting from the mid-1990s, there have been several stories and studies on the costs of downtime.

Hiles (2004) estimated that IS downtime put direct losses on brokerage operations at $4.5 million per hour, banking industry $2.1 million per hour, and e-commerce operations $113,000. Fortune 500 companies would have average losses of about $96,000 per hour due to the IS downtime. Nolan and McFarlan (2005) quoted that an attack on Amazon.com would cost the company $600,000 an hour in revenue; they also claimed that if Cisco's systems were down for a day, the company would lose $70 million in revenues.

Another event, primarily based on the lack of scalability, occurred in October 2007 when the Beijing 2008 Olympic Committee started the process of online ticket reservation. The server infrastructure went down immediately after the online reservations and sales were activated. The website for online ticket sales received a couple of million page views with an average of 200,000 ticket requests filed every minute. The system was designed to handle a maximum of one million visits per hour and a maximum of 150,000 ticket requests per minute (Computerworld, 2007).

Martin (2011) cited the results of the study by Emerson Network Power and the Ponemon Institute, which revealed that the average data center downtime event costs $505,500 with the average incident lasting 90 min. Every minute the data center remains down, a company is effectively losing $5600. The study took statistics from 41 U.S. data centers in a range of industries, including financial institutions, health-care companies, and collocation providers.

CA Report (2011) provides the results of an independent research study to explore companies' experiences of IT downtime in which 2000 organizations across North America and Europe were surveyed on how they are affected by IT downtime. The key findings include the following:

- Organizations are collectively losing more than 127 million person hours each year through IT downtime and data recovery with each company losing 545 person hours a year.
- Each business suffers an average of 14 h of downtime per year.
- Organizations lose an average of nine additional hours per year to data recovery time.
- 87% of businesses indicated that failure to recover data would be damaging to the business, and 23% said this would be disastrous.

Fryer (2011) explored some characteristics of several downtime incidents that happened in 2011 such as Verizon's LTE mobile data network outage that lasted several days and Alaska Airlines's server outage in March 2011, which resulted in the cancellation of 15% of the scheduled flights. Neverfail Report (2011) revealed results from a survey of 1473 SME and Enterprise IT professionals in the United States regarding their disaster recovery plans and practices. According to this report,

- Twenty-three percent of respondents have had an IT outage for more than one full business day.
- Hardware and software problems cause most of the IT outages (43% or respondents), while others reported power and datacenter outages (35%), natural disasters (8%), human error (6%), and others (8%).

Forrester conducted a joint research study with Disaster Recovery Journal on most common causes of downtime. In their report (Forrester, 2011, p. 2), they pointed out "Most disasters are still caused by mundane events. That headlining disaster that you're watching out for most likely won't be what causes your downtime—instead, it'll be a backhoe operator at the construction site next door who accidently severs your power or network lines."

All these studies and reports demonstrated the importance of downtime costs and emphasized the need for implementing more advanced availability, scalability, and reliability solutions that may help in enhancing BC. In addition to technology solutions, several methodologies, methods, and research approaches have been developed over the last decade in order to address these topics in an academic way, not only from technology perspective. In the section that follows, some of these approaches are briefly elaborated.

51.3 Information Architectures: The Downtime Points

Businesses operating in an e-business environment are in most cases run by different forms of business-critical applications. These applications are installed on application servers and accessed by end users over network, often designed or based on the client–server architecture (c/s). Such an architecture consists of one or more servers hosting business-critical applications and a number of clients accessing them. Depending on the server infrastructure's availability, reliability, and scalability, the whole business operates with more or less downtime that, in turn, has more or less impact on its financial performance.

Several models of information architectures have been identified over the last four decades depending mainly on the evolution of hardware and software.

Turban et al. (2005) defined the notion of "information technology architecture" as a high-level map or plan of the information assets in an organization, which guides current operations and is a blueprint for future directions. Jonkers et al. (2006) defined architecture as "structure with a vision," which provides an integrated view of the system being designed and studied. They build their definition on top of a commonly used definition by the IEEE Standard 1471–2000. Versteeg and Bouwman (2006) defined the main elements of a business architecture as business domains within the new paradigm of relations between business strategy and information technologies. Goethals et al. (2006) argued that enterprises are living things, and therefore, they constantly need to be (re)architected in order to achieve the necessary agility, alignment, and integration. Balabko and Wegmann (2006) applied the concepts of system inquiry (systems philosophy, systems theory, systems methodology, and systems application) in the context of an enterprise architecture. Lump et al. (2008) provided an overview of the state-of-the-art architectures for continuous availability concepts as HA clustering on distributed platforms and on the mainframe. Bhatt et al. (2010) considered IT infrastructure as "enabler of organizational responsiveness and competitive advantage."

Today's information systems are mainly based on the three major types of information architecture: legacy–mainframe environment, several models of client–server architectures, and cloud computing-based platforms.

While mostly considered obsolete, the mainframe-based information architecture has been still in use with the redefined role of today's mainframe configurations as servers or hosts within the c/s architecture. Moreover, today's mainframes are used as servers in a cloud-based computing environment. They are known for their HA ratios, e.g., IBM's z systems (www.ibm.com) with the latest versions bringing the redundant array of independent memory (RAIM) technology as a fault-tolerant RAM.

Client–server architectures consist of servers and clients (desktop computers and portable/mobile devices) with applications being installed and running on server computers. Server's side of such architecture is called "business server" or "enterprise server" or, in a broader sense, "server operating environment." It consists of server hardware, server operating system, application servers, and standard server-based and advanced server technologies (also known as "serverware") that are used to enhance key server platform features such as reliability, availability, and scalability. A hybrid architecture is used as well: a combination of mainframe (legacy) platform with newly implemented "thick" or "thin" client–server applications installed on servers.

Enterprise servers running server operating systems are expected to provide an operating environment that must meet much more rigorous requirements than a desktop operating system such as Windows XP or Windows 7 can provide. System uptime that is used as a measure of the availability is one of the most critical requirements.

No matter which of the c/s models is implemented, some most commonly used layers of the client–server architecture can be identified as follows:

- *Client layer*—a layer that is represented by a client application ("thick" or "thin," depending on the type of client–server architecture that is implemented). Each user, from any business unit, department, or business process, uses appropriate client application or web browser over this client layer.
- *Networking layer*—a layer that consists of a number of different data communication devices, data communication media, hardware and software communication protocols, and applications implemented for data communications.
- *Server operating platform layer*—a layer consisting of one or more servers (data, application, web, messaging, firewall, etc.) accompanied by server operating systems and integrated serverware solutions.
- *Data storage layer*—a layer that contains several data storage, data backup, and recovery solutions intended to be used for data storage.

Several modifications of standard c/s architecture are in use today as well in the forms of newly created computing paradigms or models such as web-enabled legacy systems, utility/on-demand computing, software-as-a-service, web-based software agents and services, subscription computing, grid computing, clustering, ubiquitous or pervasive computing, and cloud computing. They are implemented primarily in order to reduce the costs and enhance uptime.

A number of problems that cause downtime may occur within all types of information architectures such as hardware glitches on server components, operating system crashes, network interface card malfunctions, WAN/Internet disconnections, application bugs, system or network administrator's errors, or natural disasters that hit main computer center or cloud computing provider's site (Figures 51.2 and 51.3).

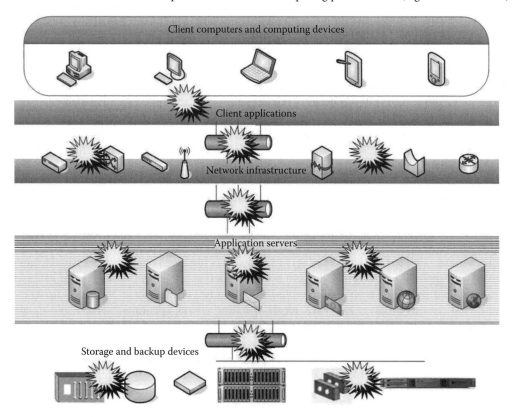

FIGURE 51.2 Downtime points in a client–server architecture.

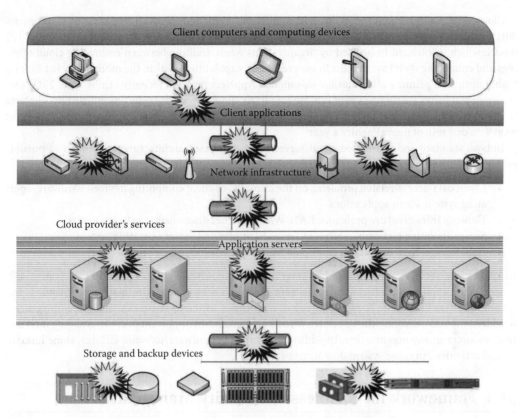

FIGURE 51.3 Downtime points within a cloud-based information architecture.

All these situations may cause unavailability of applications and hence result in a business going "out of business." Therefore, information technologies today play a crucial role by providing several so-called continuous computing technologies that may reduce downtime or enhance the levels of uptime. In all cases, no matter which type of information architecture is used, the server side of the client–server model consisting of one or more application servers plays a crucial role in ensuring the availability of business-critical applications. The downtime points are considered critical in implementing continuous computing solutions for enhancing BC.

Today, cloud computing as a concept tends to replace almost all aforementioned concepts and becomes a platform of choice for many businesses today. Forrester predicts that the global cloud computing market will grow from $40.7 billion in 2011 to more than $241 billion in 2020 (Dignan, 2011). However, the role of servers and server operating environment remains almost the same as business-critical applications are again run by server computers, with the only difference being in the fact that these servers reside within the boundaries of cloud computing provider but not in business's computer center.

Cloud outages such as those of Amazon and Google from April 2011 (Maitland, 2011) showed that even within "by default" highly available infrastructures such as the cloud providers' data centers downtime is still possible and requires attention. In the Amazon's case, huge demand of Lady Gaga fans for the discounted album caused Amazon's servers going down for several hours. In addition, more than 10 h of unavailability of Amazon's servers caused unavailability of some popular websites such as Quora and Reddit. It turned out that these couple of cloud outages could have an impact on the attractiveness of the whole concept of cloud computing (Sisario, 2011).

Cloud requires the same care, skills, and sophisticated management as other areas of IT (Fogarty, 2011). Peiris, Sharma, and Balachandran (2010) developed a Cloud Computing Tipping Point (C2TP) model, which is intended to be used by organizations when deciding between embracing cloud offerings and enhancing their investments in on-premise IT capabilities. Within the model, they list BC and higher uptime as primary attributes for this model. AppNeta revealed recently (Thibodean, 2011) that the overall industry yearly average of uptime for all the cloud service providers monitored by AppNeta is 99.9948%, which is equal to 273 min or 4.6 h of unavailability per year. The best providers were at 99.9994% or 3 min of unavailability a year.

In both standard and web-based client/server and cloud-based architectures, a number of possible problems or critical points can be identified with regard to possible downtime problems:

- Client data-access-related problems on the client PC or client computing devices: hardware, operating system, client applications
- Network infrastructure problems: LAN/WAN/Internet disconnections from servers
- Server operating platform: crashes of server hardware and server operating system
- Data storage: hard disk crash, corrupted file system, magnetic tape broken

As said before, servers can be affected by several types of hardware glitches and failures, system software crashes, application software bugs, hackers' malicious acts, system administrators' mistakes, and natural disasters like floods, fires, earthquakes, hurricanes, or terrorist attacks. Therefore, the process of addressing and managing the IT-related threats has become an important part of managing information resources in any organization. In addition to standard hardware/software failures, some human-related activities can cause system downtime as well.

51.4 Framework for Business Continuity Management

By using Churchman's approach in defining the systemic model (Churchman, 1968, 1971) already applied in explaining information systems for e-business continuance (Bajgoric, 2006), the objective of a continuous computing platform in an organization can be defined as achieving the BC or business resilience. Several information technologies can be applied in the form of continuous computing technologies. This approach leads to identifying the "always-on" enterprise information system that can be defined as an information system with a 100% uptime/zero downtime.

Following Churchman's systemic model, the objective of an "always-on" enterprise information system in an organization can be identified as providing the highest availability, reliability, and scalability ratios for achieving the BC or business resilience. In other words, this refers to the following: continuous data processing with continuous data access and delivery, multiplatform data access, on-time IT services, and better decisions through better data access. Such an information system employs numerous continuous computing information technologies and builds and integrates several components that are aimed at enhancing availability, reliability, and scalability ratios.

A conceptual model used here to illustrate the concepts of BC and continuous computing technologies is given in Figure 51.4.

Continuous computing technologies are considered to be major BC drivers. These technologies include several IT-based solutions that are implemented in order to enhance continuous computing (uptime) ratios such as availability, reliability, and scalability and include the following 10 major technologies:

1. Enterprise servers (server hardware)
2. Server operating systems (server operating environments)
3. Serverware solutions in the form of bundled, in most cases, vendor-specific solutions installed as a set of extensions to standard server operating system
4. Application servers

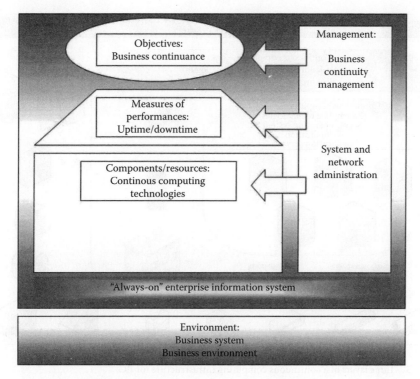

FIGURE 51.4 BC within the systems approach.

5. Server-based fault-tolerance and disaster-tolerance technologies
6. Server clustering
7. Data storage and backup solutions
8. Advanced storage technologies such as mirroring, snapshot, data vaulting
9. Networking and security technologies
10. System administration knowledge and skills

All these technologies are identified as the main building blocks of an always-on enterprise information system. They are represented within the three continuous computing layers as seen in Figure 51.5.

The continuous computing technologies can be implemented within the following three layers:

1. Layer 1—Server and server operating environment
2. Layer 2—Data storage, data backup, and recovery technologies
3. Layer 3—Data communications and security technologies within a networking infrastructure

The first group mainly includes so-called preemptive technologies—standard and specific information technologies and human skills that aim at "preempting" or preventing downtime and keeping system uptime as higher as possible. The second group of continuous computing technologies relate to those solutions that help in recovering from failures and disaster that caused system and application downtime (post-failure). The third group consists of data communications, networking, and security technologies that are used to connect business units and its information system's subsystems and to keep them connected and secure in order to have continuous and secure data access and data exchange. A more detailed depiction of these technologies is given in Figure 51.6 in a revised form of so-called Onion model (Bajgoric, 2010).

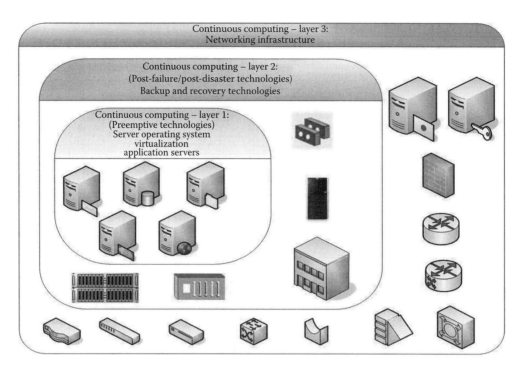

FIGURE 51.5   Three layers of a continuous computing infrastructure for BC.

As can be seen in Figure 51.6, the main technologies that are used to provide HA ratios can be classified into the following categories:

- Server-based technologies
- Server operating systems, enhanced by serverware solutions
- Data storage and backup technologies
- Data communications and networking technologies
- Fault-tolerance and disaster-tolerance technologies
- Redundant units and/or devices such as power supplies and fans, hard disks, network, interface cards, routers, and other communication devices
- Facilities such as UPS units, dual power feeds or generators, and air-conditioning units
- Advanced technologies, such as virtualization, mirroring, snap-shooting, clustering, and data vaulting
- Security and protection technologies
- Predictive system-monitoring capabilities that are aimed at eliminating failures before they occur

Server-based technologies include several types of hardware and software features that support HA technologies such as SMP, clustering, 64-bit computing, storage scalability, RAID and RAIM technologies, fault tolerance, hot spared disk, HA storage systems, online reconfiguration, grid containers, dynamic system domains, virtual machine managers, and hardware and software mirroring. In addition, server platform-based serverware suites include bundled servers, reloadable kernel and online upgrade features, crash-handling technologies, workload management, HA monitors, and Windows/UNIX integration.

Several server operating systems, both commercial proprietary and open source, provide additional HA features. For instance, HP's HP-UX 11.0 server operating platform (http://www.hp.com) provides several HA features such as advanced sys_admin commands (e.g., *fsck, lvcreate, make_recovery, check_recovery*), Veritas's VxFS JFS—Journaled File System, JFS snapshot, dynamic root disk, Bastille, Compartments, fine-grained privileges, role-based access control, and HP Serviceguard. JFS is a recommended file

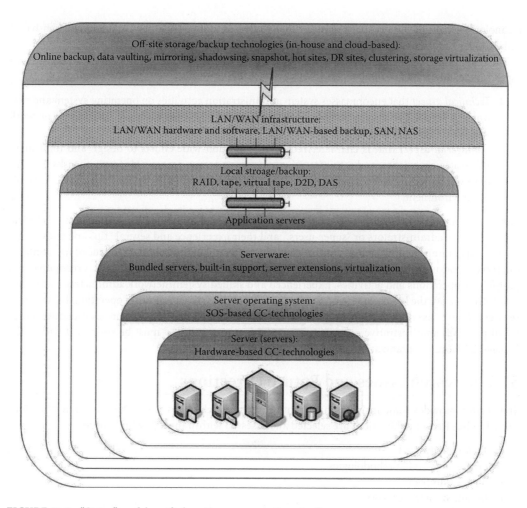

Off-site storage/backup technologies (in-house and cloud-based):
Online backup, data vaulting, mirroring, shadowsing, snapshot, hot sites, DR sites, clustering, storage virtualization

LAN/WAN infrastructure:
LAN/WAN hardware and software, LAN/WAN-based backup, SAN, NAS

Local stroage/backup:
RAID, tape, virtual tape, D2D, DAS

Application servers

Serverware:
Bundled servers, built-in support, server extensions, virtualization

Server operating system:
SOS-based CC-technologies

Server (servers):
Hardware-based CC-technologies

FIGURE 51.6 "Onion" model: unified continuous computing paradigm.

system for high reliability, fast recovery, and online administrative operations. Recently developed new file system by Microsoft for its Windows Server operating system is named the "Resilient File System" (ReFS). It is built on the existing NTFS, enhanced by some additional resiliency features.

Fault-tolerant and disaster-tolerant systems include the following technologies: redundant units, data replication, hot sites, data vaulting, and disaster recovery sites. Some of these technologies overlap with data storage systems, which include standard tape backup technology, Storage Area Networks (SAN), Network Attached Storage (NAS), off-site data protection, Fibre Channel, iSCSI, Serial ATA, etc.

Security and protection technologies that are implemented on computer hardware (servers, desktops, laptops, non-PC computing devices), systems software (restricted use of system privileges, user accounts monitoring, etc.), application software, networking devices such as routers, switches, firewalls, control of network ports, protocols and services, and network applications are also part of unified continuous computing platform shown in Figure 51.6.

The ultimate goal of data communications technologies from BC perspective can be identified as staying connected with employees, customers, and suppliers and ensuring a set of communication technologies to support workforce during disastrous events that interrupt business. These technologies include unified communication technologies that integrate data, e-mail, voice, and video and collaboration technologies supporting personal productivity software on portable devices, scheduling, web-based collaboration, and content sharing.

The management dimension of the systemic model presented in Figure 51.4 consists of two main aspects of BC management:

1. *Technical aspect* that encompasses system administration activities on operating system and networking levels. Two professional roles responsible for these activities are system and network administrators, respectively.
2. *Managerial aspect* that is characterized by organization-wide efforts to design and implement BC strategies, policies, and activities. BC manager is a person responsible for these tasks. He/she, however, has to work closely with the CIO who is responsible for the whole information technology environment.

System and network administration represent core activities of organizational BC efforts. Technical skills of system administrators or "super-users," and their experience and several "tips and tricks" that they use in administering operating system platforms and computer networks are crucial in maintaining servers and networks up and running on always-on basis. This applies for both preemptive and post-failure operations and activities. Experienced and skilled system/network administrators can significantly improve the levels of system/network uptime while system/network is operational and reduce recovery time in case of system/application or network crashes. The fact that system administrator holds an all-powerful administrative account is a potential risk as well and must be addressed within the managerial scope of organizational BC management.

51.5 Research Issues and Future Directions

This section provides some insights on current works, research issues, and future research directions related to BC. Recent publications are selected and assessed within the following aspects:

- Terminology used
- Frameworks and methodologies
- Business impact
- BC and disaster recovery
- BC and related disciplines such as risk management, auditing, and supply chain management
- IT architectures related

51.5.1 Current Works and Research Issues

From historical perspective, business continuity as a concept has been treated extensively over the last 15–20 years. As already stated in Section 51.2, in the very beginning of the e-business era, Datamation (1995) emphasized the importance of minimizing system downtime. Herbane (2010) noted that business continuity management (BCM) has evolved since 1970s in response to the technical and operational risks that threaten an organization's recovery from hazards and interruptions. Regarding the *terminology*, Broder and Tucker (2012) underlined that BC planning is defined in many different ways reflecting its author's particular slant, background, or experience with the process. They make the difference between the meaning of continuity planning and of a continuity plan. Nollau (2009) introduced the term "enterprise business continuity" and argued that each company should have a functional plan that addresses all processes required to restore technology, an individual responsible for that plan, and a disaster response team at the ready. King (2003) introduced a term "Business Continuity Culture" and underscored the fact that "If you fail to plan, you will be planning to fail." Kadam (2010) introduced the concept of personal BC management.

With regard to *BCP frameworks*, Botha and Von Solms (2004) proposed a cyclic approach to BC planning with each cycle concentrated on a specific BCP goal. Gibb and Buchanan (2006) defined a framework for the design, implementation, and monitoring of a BC management program within an information strategy. They assign the CIO a key role in both promoting the philosophy of BCM and ensuring that information management incorporates effective plans, procedures, and policies to protect information assets. Gerber and Solms (2005) emphasized the need for a holistic approach that should include a carefully planned security policy, the use of technology, and well-trained staff. They proposed an integrated approach based on analyzing not only risks to tangible assets but also risks to information or intangible assets. Cerullo and Cerullo (2004, p. 71) stated that "… there is no single recommended plan for business continuity; instead, every organization needs to develop a comprehensive BCP based on its unique situation."

Boehman (2009) presented a model for evaluating the performance of a BCM System according to BS 25999 standard. The model calculates the survivability ex-ante if the key performance indicator for the effectiveness exists. Lindstrom et al. (2010) noted that organizations rely too much on the checklists provided in existing BC standards. These checklists for predefined situations need to be created during the BCP process and kept updated in the following maintenance process. They proposed a multi-usable BC planning methodology based on using a staircase or capability maturity model. Herbane, Elliott, and Swartz (2004) examined the organizational antecedents of BCM and developed a conceptual approach to posit that BCM, in actively ensuring operational continuity, has a role in preserving competitive advantage. Walker (2006) considered outsourcing options for BC. Bartel and Rutkowski (2006) proposed a fuzzy decision support system for IT service continuity (ITSC) management. They argued that ITSC management is typically part of a larger BCM program, which expands beyond IT to include all business services within an organization. Brenes et al. (2011) focused on assessing the impact of BC-related governance structures on family business performances. Avery Gomez (2011) underscored the fact that resilience and the continuity of business are factors that rely on sustainable ICT infrastructures. The author presented a model for alignment of BC and sustainable business processes with regard to ICT usage behavior and response readiness. Sapateiro et al. (2011) presented a model of a mobile collaborative tool that may help teams in managing critical computing infrastructures in organizations with regard to BC management.

As for the *business impact*, in addition to what was said in Section 51.2, several authors still emphasize the impacts of downtime on business. According to Stanton (2007, p. 18), "… it can take less than 60 seconds for a company's reputation to be ruined and its business to be crippled. In just one minute, a server failure or hacker can knock out vital applications." He underscored the saying "Fail to Plan, Plan to Fail." Melton and Trahan (2009) argued that critical risks demand critical attention. Vatanasombut et al. (2008) noted that the proliferation of Internet has not only allowed businesses to offer their products and services through web-based applications but also undermined their ability to retain their customers. Ipe et al. (2010) stated that satisfying decision making in public health often stems from the unavailability of critical information from the surveillance environment. They explored the challenges specific to developing and institutionalizing an IT system for emergency preparedness.

Bertrand (2005) researched the relationships between BC and mission-critical applications. He emphasized the role of Replication Point Objective (RPO) and Replication Time Objective (RTO) in recovering from disasters and stressed the role of replication technologies in providing better RPO and RTO values. IDC (2010) pointed out that HA is critical to building a strong BC strategy meaning that HA is the foundation of all BC strategies. Forrester (2009) unveiled that almost 80% of respondents reported that their firms have had to provide proof of BC readiness.

Zobel (2011) defined an analytic approach in representing the relationship between two primary measures of disaster resilience: the initial impact of a disaster and the subsequent time to recovery. Singhal et al. (2010) presented a solution for optimal BC based on the IP SAN storage architecture for enterprise applications used for data management without burdening application server. In order to achieve zero RPO, they proposed introducing a staging server in addition to DR site with replication performances. Lewis and Pickren (2003) pointed out that disaster recovery is receiving increasing attention as firms grow more dependent on uninterrupted information system functioning. Turetken (2008) developed a decision model for locating redundant facilities (IT-backup facility location) within the activities of BC planning. Bielski (2008) stated that some think of BC as doing what is necessary to set up a "shadow" organization that will take you from incident response through various phases of recovery. Omar et al. (2011) presented a framework for disaster recovery plan based on the Oracle's DataGuard solution that includes remote site, data replication, and standby database.

Pitt and Goyal (2004) saw BCM as a tool for facilities management. They argued that both FM and BCP are important on strategic level and pointed out the need for the involvement of facilities manager in BC planning. Yeh (2005) identified the factors affecting continuity of cooperative electronic supply chain relationships with empirical case of the Taiwanese motor industry. Wu et al. (2004) identified continuity as the most important behavioral factor on business process integration in SCM. Hepenstal and Campbell (2007) provided some insights on transforming Intel's worldwide Materials organization from crisis management and response to a more mature BC approach. According to these authors, the BC-oriented approach improved Intel's ability to quickly recover from a supply chain outage and restoring supply to manufacturing and other operations. Craighead et al. (2007) considered BC issues within the supply chain mitigation capabilities and supply chain disruption severity. They proposed a multiple-source empirical research method and presented six propositions that relate the severity of supply chain disruptions. Tan and Takakuwa (2011) demonstrated how the computer-based simulation technique could be utilized in order to establish the BC plan for a factory. Blackhurst et al. (2011) provided a framework that can be used to assess the level of resiliency and a supply resiliency matrix that can be utilized to classify supply chains according to the level of resiliency realized. Kim and Cha (2012) presented the improved scenario-based security risk analysis method that can create SRA reports using threat scenario templates and manage security risk directly in information systems. Greening and Rutherford (2011) proposed a conceptual framework based on network theories to ensure BC in the context of a supply network disruption. Lavastre et al. (2012) introduced the term supply chain continuity planning framework within the concept of supply chain risk management (SCRM).

Wan (2009) argued that in current business-aligned IT operations, the continuity plan needs to be integrated with ITSC management if an organization is going to be able to manage fault realization and return to normal business operations. He proposed the framework based on using the IT service management portal and network system-monitoring tool. Winkler et al. (2010) focused on the relationships between BCM and business process modeling. They proposed a model-driven framework for BCM that integrates business process modeling and IT management. Kuhn and Sutton (2010) explored the alternative architectures for continuous auditing by focusing on identifying the strengths and weaknesses of each architectural form of ERP implementation.

According to IDC's white paper (2009), by adopting industry best practices (e.g., ITIL, CobIT) and updating their IT infrastructure, companies can lower annual downtime by up to 85%, greatly reducing interruptions to daily data processing and access, supporting BC, and containing operational costs. Umar (2005) explored the role of several business servers in designing an IT infrastructure for "next-generation enterprises." Butler and Gray (2006) underscored the question on how system reliability translates into reliable organizational performance. They identify the paradox of "relying on complex systems composed of unreliable components for reliable outcomes." Adeshiyan et al. (2010) stated that traditional high-availability and disaster recovery solutions require proprietary hardware, complex configurations, application specific logic, highly skilled personnel, and a rigorous and lengthy testing process.

Williamson (2007) found that in BC planning, financial organizations are ahead of other types of businesses. Kadlec (2010) reported on the IT DR planning practices of 154 banks in the United States. Arduini and Morabito (2010) argued that the financial sector sees BC not only as a technical or risk management issue but also as a driver toward any discussion on mergers and acquisitions. Sharif and Irani (2010) emphasized HA, failover, and BC planning in maintaining high levels of performance and service availability of eGovernment services and infrastructure.

51.5.2 Summary of Recent Publications

Table 51.3 provides a summary of recent publications published over the last 10 years about BC, BC management, and related topics. They are listed with regard to research focus and main contribution.

TABLE 51.3 Review of Literature Related to Business Continuity

Article Year	Authors	Research Focus/Main Contribution
2003	Lewis and Pickren	Presented a technique that can be used to assess the relative risk for different causes of IT disruptions
2003	King	Introduced a term "Business Continuity Culture." Underscored the fact "If you fail to plan, you will be planning to fail"
2004	Botha and Von Solms	Proposed a BCP methodology that is scalable for small and medium organizations. Developed a software prototype using this cyclic approach and applied it to a case study
2004	Herbane et al.	Examined the organizational antecedents of BCM and developed a conceptual approach to posit that BCM, in actively ensuring operational continuity, has a role in preserving competitive advantage
2004	Finch	Revealed that large companies' exposure to risk increased by interorganizational networking and that having SMEs as partners in the SCM further increased the risk exposure
2004	Wu et al.	Identified continuity as the most important behavioral factor on business process integration in SCM
2004	Pitt and Goyal	Considered BCM as a tool for facilities management
2004	Cerullo and Cerullo	Provided guidelines for developing and improving a firm's BCP consisting of three components: a business impact analysis, a DCRP, and a training and testing component
2005	Bertrand	Estimated the relationships between BC and mission-critical applications. Emphasized the role of RPO and RTO in recovering from disasters and stressed the role of replication technologies
2005	Yeh	Identified the factors affecting continuity of cooperative electronic supply chain relationships with empirical case of the Taiwanese motor industry
2005	Gerber and Solms	Suggested the need for a holistic approach that should include a carefully planned security policy, the use of technology, and well-trained staff. The article indicated how information security requirements could be established
2005	Umar	Investigated the role of several business servers in designing an IT infrastructure for "next-generation enterprises"
2006	Walker	Considered outsourcing options for BC
2006	Bartel and Rutkowski	Argued that ITSC management is typically part of a larger BCM program, which expands beyond IT to include all business services. Proposed a fuzzy decision support system for ITSC management
2006	Butler and Gray	Underscored the question on how system reliability translates into reliable organizational performance. Identified the paradox of "relying on complex systems composed of unreliable components for reliable outcomes"
2006	Gibb and Buchanan	Defined a framework for the design, implementation, and monitoring of a BC management program within an information strategy
2007	Williamson	Found that in BC planning, financial organizations are ahead of other types of businesses
2007	Hepenstal and Campbell	Provided some insights on transforming Intel's worldwide Materials organization from crisis management and response to a more mature BC approach
2007	Craighead et al.	Proposed a multiple-source empirical research method and presented six propositions that relate the severity of supply chain disruptions
2008	Vatanasombut et al.	Argued that the proliferation of Internet has not only allowed businesses to offer their products and services through web-based applications but also undermined their ability to retain their customers
2008	Bielski	Stated that some think of BC as doing what is necessary to set up a "shadow" organization that will take you from incident response through various phases of recovery

(*continued*)

TABLE 51.3 (continued)   Review of Literature Related to Business Continuity

Article Year	Authors	Research Focus/Main Contribution
2008	Turetken	Developed a decision model for locating redundant facilities (IT-backup facility location) within the activities of BC planning
2008	Lump et al.	Investigated several aspects and relationships among HA, disaster recovery, and BC solutions
2009	Wan	Argued that in current business-aligned IT operations, the continuity plan needs to be integrated with ITSC management
2009	Boehman	Presented a model for evaluating the performance of a BCM System according to BS 25999 standard. The model calculates the survivability ex-ante if the key performance indicator for the effectiveness exists
2009	Nollau	Introduced the term "enterprise business continuity"
2010	Adeshiyan et al.	Stated that traditional high-availability and disaster recovery solutions require proprietary hardware, complex configurations, application-specific logic, highly skilled personnel, and a rigorous and lengthy testing process
2010	Ipe et al.	Stated that satisfying decision making in public health often stems from the unavailability of critical information from the surveillance environment. They explored the challenges specific to developing and institutionalizing an IT system for emergency preparedness
2010	Singhal et al.	Presented a solution for optimal BC based on the IP SAN storage architecture for enterprise applications used for data management without burdening application server
2010	Kuhn and Sutton	Explored the alternative architectures for continuous auditing by focusing on identifying the strengths and weaknesses of each architectural form of ERP implementation
2010	Tammineedi	Categorized the key BCM tasks into three phases of BC: pre-event preparation, event management, and post-event continuity. The Business Continuity Maturity Model of Virtual Corporation presented as a tool
2010	Lindstrom et al.	Proposed a multi-usable BC planning methodology based on using a staircase or capability maturity model
2010	Kadlec	Reported on the IT DR planning practices of 154 banks in the United States
2010	Arduini and Morabito	Argued that the financial sector sees BC not only as a technical or risk management issue but also as a driver toward any discussion on mergers and acquisitions. The BC approach should be a business-wide approach and not an IT-focused one
2010	Winkler et al.	Focused on the relationships between BCM and business process modeling. Proposed a model-driven framework for BCM that integrates business process modeling and IT management
2010	Kadam	Introduced the concept of personal BC management
2011	Tan and Takakuwa	Demonstrated how the computer-based simulation technique could be utilized in order to establish the BC plan for a factory
2011	Zobel	Presented an analytic approach in representing the relationship between two primary measures of disaster resilience: the initial impact of a disaster and the subsequent time to recovery
2011	Omar et al.	Presented a framework for disaster recovery plan based on the Oracle's DataGuard solution that includes remote site, data replication, and standby database
2011	Greening and Rutherford	Proposed a conceptual framework based on network theories to ensure BC in the context of a supply network disruption
2011	Brenes et al.	Focused on assessing the impact of BC-related governance structures on family business performances
2011	Avery Gomez	Presented a model for alignment of BC and sustainable business processes with regard to ICT usage behavior and response readiness

TABLE 51.3 (continued)   Review of Literature Related to Business Continuity

Article Year	Authors	Research Focus/Main Contribution
2011	Sapateiro et al.	Designed a mobile collaborative tool that helps teams managing critical computing infrastructures in organizations. Developed a data model and tool supporting the collaborative update of Situation Matrixes
2011	Blackhurst et al.	Provided a framework that can be used to assess the level of resiliency and a supply resiliency matrix that can be utilized to classify supply chains according to the level of resiliency realized
2012	Broder and Tucker	Made the difference between the meaning of continuity planning and of a continuity plan
2012	Lavastre et al.	Considered BC planning within the "supply chain continuity planning framework." Introduced the concept of Supply Chain Risk Management (SCRM)
2012	Kim and Cha	Presented the improved scenario-based security risk analysis method, which can create SRA reports using threat scenario templates and manage security risk directly in information systems

51.5.3 Future Research Directions

Future research directions with regard to BC can be identified around the following research topics:

- Integrating BC management, IT management, and organizational management
- Positioning BC as part of business risk management
- Exploring relations between BC management and disaster recovery management
- Developing and implementing advanced server technologies for enhancing servers' availability, reliability, and scalability
- Developing frameworks for organization-wide implementation of BC
- Implementing several information technologies as BC drivers
- BC and IT/IS security
- BC and information resources management
- BC and system and network administration
- BC in cloud computing environment
- HRM dimensions of BC management: the roles of system and network administrators, BC managers, IT managers, CIOs, and CSOs
- Implementing BC within the compliance regulations, legislations, and standards on BCM

Given the fact that several terms and buzzwords related to BC have been introduced over the last decade, it would of interest to further explore and develop theoretical aspects and practical implications of those terms such as business continuity, business continuance, business resilience, always-on business, zero-latency business, always-available enterprise, and so on.

51.6 Conclusions

Modern business owes a lot to information technology. It is well known that information technologies provide numerous opportunities for businesses in their efforts to reduce the costs, increase the profits, enhance the quality of products and services, improve relations with customers, and cope with increasing competition. At the same time, modern businesses face several situations in which they may suffer due to some IT-related problems that cause lost data, data unavailability, or data exposed to competition. In addition, some businesses experienced so-called IT-based horror stories due to bad implementations of enterprise information systems.

With advances in e-business, the need for achieving "a near 100%" level of business computing availability was brought up yet again. Consequently, the term "business continuity management" was coined up and became a significant part of organizational information management. BCM has become an integral part of both IT management and organizational management. It involves several measures (activities) that need to be implemented in order to achieve higher levels of the system/application availability, reliability, and scalability ratios. BC relies on several continuous computing technologies that provide an efficient and effective operating environment for continuous computing. Implementation of continuous computing technologies provides a platform for "keeping business in business." Therefore, an enterprise information system should be managed from BC perspective in a way that this process includes both managerial and system administration activities related to managing the integration of BC drivers.

In this paper, an attempt was made to explore the concept of business continuity from both managerial and IT perspective. In addition, several illustrations of downtime costs are provided in order to demonstrate the importance of BC for modern business.

References

Adeshiyan, T. et al. (2010), Using virtualization for high availability and disaster recovery, *IBM Journal of Research and Development*, 53(4), 587–597.

Arduini, F., Morabito, V. (2010), Business continuity and the banking industry, *Communications of the ACM*, 53(3), 121–125.

Avery Gomez, E. (2011), Towards sensor networks: Improved ICT usage behavior for business continuity, *Proceedings of SIGGreen Workshop, Sprouts: Working Papers on Information Systems*, 11(13), http://sprouts.aisnet.org/1099/1/Gomez_ICTUsage.pdf (ISSN 1535-6078).

Bajgoric, N. (2006), Information systems for e-business continuance: A systems approach, cybernetic, *The International Journal of Systems and Cybernetics*, 35(5), 632–652.

Bajgoric, N. (2010), Server operating environment for business continuance: Framework for selection, *International Journal of Business Continuity and Risk Management*, 1(4), 317–338.

Balabko, P., Wegmann, A. (2006), Systemic classification of concern-based design methods in the context of enterprise architecture, *Information Systems Frontiers*, 8, 115–131.

Bartel, V.W., Rutkowski, A.F. (2006), A fuzzy decision support system for IT service continuity threat assessment, *Decision Support Systems*, 42(3), 1931–1943.

Bertrand, C. (2005), Business continuity and mission critical applications, *Network Security*, 20(8), 9–11.

Bhatt, G., Emdad, A., Roberts, N., Grover, V. (2010), Building and leveraging information in dynamic environments: The role of IT infrastructure flexibility as enabler of organizational responsiveness and competitive advantage, *Information and Management*, 47(7–8), 341–349.

Bielski, R. (2008), Extreme Risks, *ABA Banking Journal*, 100(3), 29–44.

Blackhurst, J., Dunn, K.S., Craighead, C.W. (2011), An empirically derived framework of global supply resiliency, *Journal of Business Logistics*, 32, 374–391.

Boehman, W. (2009), Survivability and business continuity management system according to BS 25999, *Proceedings of the 2009 Third International Conference on Emerging Security Information, Systems and Technologies*, Athens, Greece, pp. 142–147.

Botha, J., Von Solms, R. (2004), A cyclic approach to business continuity planning, *Information Management and Computer Security*, 12(4), 328–337.

Brenes, E.R., Madrigal, K., Requena, B. (2011), Corporate governance and family business performance, *Journal of Business Research*, 64, 280–285.

Broder, J.F., Tucker, E. (2012), *Business Continuity Planning, Risk Analysis and the Security Survey*, 4th edn., Elsevier, Oxford, UK.

Butler, B.S., and Gray, P.H. (2006), Reliability, Mindfulness, And Information Systems, *MIS Quarterly*, (30)2, 211–224.

CA Report (2011), The avoidable cost of downtime, CA Technologies, available on: http://www.arcserve.com/us/lpg/~/media/Files/SupportingPieces/ARCserve/avoidable-cost-of-downtime-summary-phase-2.pdf, accessed on January 8, 2012

Cerullo, V., Cerullo, R. (2004), Business continuity planning: A comprehensive approach, *Information Systems Management*, 2004, 70–78.

Churchman, C.W. (1968), *The Systems Approach*, Delacorte Press, New York.

Churchman, C.W. (1971), *The Design of Inquiring Systems: Basic Concepts of Systems and Organizations*, Basic Books, New York.

Computerworld (2007), China abandons plans to sell Olympics tickets online: Crush of buyers overwhelmed system, available on: http://www.computerworld.com/s/article/9045659/China_abandons_plans_to_sell_Olympics_tickets_online, accessed on January 22, 2012.

Craighead, C.W., Blackhurst, J., Rungtusanatham, M.J., Handfield, R.B. (2007), The severity of supply chain disruptions: Design characteristics and mitigation capabilities, *Decision Sciences*, 38(1), 131–151.

Csaplar, D. (2012), The cost of downtime is rising, available at: http://blogs.aberdeen.com/it-infrastructure/the-cost-of-downtime-is-rising/, accessed on March 12, 2012.

Datamation (1995), Simpson, D. Can't Tolerate Server Downtime? Cluster'Em!, *Datamation*, August 15, 45–47.

Dembeck, C. (1999), Yahoo cashes in on Ebay's outage", E-commerce Times, June 18, 1999, available on: http://www.ecommercetimes.com/perl/story/545.html), accessed on September 24, 2012.

Dignan, L. (2011), Cloud computing market: $241 billion in 2020, http://www.zdnet.com/blog/btl/cloud-computing-market-241-billion-in-2020/47702, accessed on January 6, 2012.

Finch, P. (2004), Supply Chain Risk management, *Supply Chain management: An International Journal*, 9(2), 183–196.

Fogarty, K. (2011), Amazon crash reveals 'cloud' computing actually based on data centers, available on: http://www.itworld.com/cloud-computing/158517/amazon-crash-reveals-cloud-computing-actually-based-data-centers?source=ITWNLE_nlt_saas_2011-04-27, accessed on April 27, 2011.

Forrester Report (2007), Business technology defined: Technology management is changing to deliver business results, by Laurie M. Orlov and Bobby Cameron with George F. Colony, Mike Gilpin, Craig Symons, Marc Cecere, and Alex Cullen, available on: http://www.forrester.com/rb/Research/business_technology_defined/q/id/42338/t/2, accessed on January 23, 2012.

Forrester Report (2011), State of enterprise disaster recovery preparedness, Q2 2011, May 2011, available on: http://i.zdnet.com/whitepapers/Forrester_Analyst_White_Paper_The_State_of_Enterprise.pdf, accessed on January 25, 2012.

Fryer, J. (2011), Major downtime incidents illustrate the importance of service availability, available on: http://www.atcanewsletter.com/English/Newsletters/2011/Articles/201104_Article_JohnFryer.html, accessed on January 8, 2012.

Gerber, M., Solms, R. (2005), Management of risk in the information age, *Computers and Security*, 24, 16–30.

Gibb, F., Buchanan, S. (2006), A framework for business continuity management, *International Journal of Information Management*, 26, 128–141.

Goethals, F.G., Snoeck, M., Lemahieu, W., Vandenbulcke, J. (2006), Management and enterprise architecture click: The FAD(E)E framework, *Information Systems Frontiers*, 8, 67–79.

Hepenstal, A., Campbell, B. (2007), Maturation of business continuity practice in the intel supply chain, *Intel Technology Journal*, 11(2), 165–171.

Herbane, B. (2010), The evolution of business continuity management: A historical review of practices and drivers, *Business History*, 52(6), 978–1002.

Herbane, B., Elliott, D., Swartz, E.M. (2004), Business continuity management: Time for a strategic role? *Long Range Planning*, 37, 435–457.

Hiles, A. (2004), *Business Continuity: Best Practices—World-Class Business Continuity Management*, 2nd edn., Kingswell Int., Oxford, UK.

IDC Report (2006), True high availability: Business advantage through continuous user productivity, May 2006.

IDC White Paper (2009), Reducing downtime and business loss: Addressing business risk with effective technology, IDC, August 2009.

Ipe, M., Raghu, T.S., and Vinze, A. (2010), Information intermediaries for emergency preparedness and response: A case study from public health, *Information Systems Frontiers* 12(1), 67–79.

IDC White Paper (2010), Realizing efficiencies: Building business continuity solutions from HP and VMware, April 2010.

ITIC Report (2009), ITIC 2009 global server hardware and server OS reliability survey, Information Technology Intelligence Corp. (ITIC), July 2009

Jonkers, H., Lankhorst, M.M., Doest, H.W.L., Arbab, F., Bosma, H., Wieringa, R.J. (2006), Enterprise architecture: Management tool and blueprint for the organization, *Information Systems Frontiers*, 8, 63–66.

Kadam, A. (2010), Personal business continuity planning, *Information Security Journal: A Global Perspective*, 19(1), 4–10.

Kadlec, C. (2010), Best practices in IT disaster recovery planning among US banks, *Journal of Internet Banking and Commerce*, 15(1), 1–11.

Kim, Y.G., Cha, S. (2012), Threat scenario-based security risk analysis using use case modeling in information systems, *Security Communication Networks*, 5, 293–300.

King, D.L. (2003), *Moving towards a Business Continuity Culture*, Network Security, Elsevier, Amsterdam, The Netherlands, pp. 12–17.

Kuhn, J.R. Jr., and Sutton, S.G. (2010), Continuous Auditing in ERP System Environments: The Current State and Future Directions, *Journal Of Information Systems*, 24(1), 91–112.

Lavastre, O., Gunasekaran, A., Spalanzani, A. (2012), Supply chain risk management in French companies, *Decision Support Systems*, 52, 828–838.

Lewis, W.R., Pickren, A. (2003), An empirical assessment of IT disaster risk, *Communication of ACM*, 2003, 201–206.

Lindstrom, J., Samuelson, S., Hagerfors, A. (2010), Business continuity planning Methodology, *Disaster Prevention and Management*, 19(2), 243–255.

Lump et al. (2008), From high availability and disaster recovery to business continuity solutions, *IBM Systems Journal*, 47(4), 605–619.

Maitland, J. (2011), A really bad week for Google and Amazon, available on: http://searchcloudcomputing.techtarget.com/news/2240035039/A-really-bad-week-for-Google-and-Amazon?asrc=EM_NLN_13718724&track=NL-1324&ad=826828, accessed on April 23, 2011

Martin, N. (2011), The true costs of data center downtime, available on: http://itknowledgeexchange.techtarget.com/data-center/the-true-costs-of-data-center-downtime/, accessed on January 8, 2012, also available on http://www.emersonnetworkpower.com/en-US//Brands/Liebert/Pages/LiebertGatingForm.aspx?gateID=777, accessed on January 08, 2012

Melton, A. and Trahan, J. (2009), Business Continuity Planning, *Risk Management*, 56(10), 46–48.

Neverfail Report (2011), available on: http://www.thefreelibrary.com/Neverfail+Survey+Results%3A+Cloud+Seen+as+Viable+Disaster+Recovery...-a0252114800, accessed on January 14, 2012

Nolan, R. and McFarlan, F.W. (2005), Information technology and board of directors, *Harvard Business Review*, 83(10), 96–106.

Nollau, B. (2009), Disaster recovery and business continuity, *Journal of GXP Compliance*, 13(3), ABI/INFORM Global, 51.

Omar, A., Alijani, D., Mason, R. (2011), Information technology disaster recovery plan: Case study, *Academy of Strategic Management Journal*, 10(2), 2011.

Peiris, C., Sharma, D., Balachandran, B. (2010), C2TP: A service model for cloud, *The Proceedings of CLOUD COMPUTING 2010: The First International Conference on Cloud Computing, GRIDs, and Virtualization*, Lisbon, Portugal, IARIA, pp. 134–144.

Greening, P., Rutherford, C. (2011), Disruptions and supply networks: A multi-level, multi-theoretical relational perspective, *The International Journal of Logistics Management*, 22(1), 104–126.

Pitt, M., Goyal, S. (2004), Business continuity planning as a facilities management tool, *Facilities*, 22(3–4), 87–99.

Sapateiro, C., Baloian, N., Antunes, P., and Zurita, G. (2011), Developing a Mobile Collaborative Tool for Business Continuity Management, *Journal of Universal Computer Science*, 17(2), 164–182.

Sharif, A.M., Irani, Z. (2010), The logistics of information management within an eGovernment context, *Journal of Enterprise Information Management*, 23(6), 694–723.

Singhal, R., Pawar, P., Bokare, S. (2010), Enterprise storage architecture for optimal business continuity, *Proceedings of the 2010 International Conference on Data Storage and Data Engineering*, Bangalore, India, pp. 73–77.

Sisario, B. (2011), Lady Gaga Sale Stalls Amazon Servers, available on: http://www.nytimes.com/2011/05/24/business/media/24gaga.html, accessed on January 14, 2012.

Stanton, R. (2007), Fail to plan, plan to fail, *InfoSecurity*, November/December 2007, Elsevier, Amsterdam, The Netherlands, pp. 24–25.

Sumner, M. (2009), Information Security Threats: A Comparative Analysis of Impact, Probability, and Preparedness, *Information Systems Management*, 26(1), 2–11.

Symantec Disaster Preparedness Survey (2011). Symantec 2011 SMB disaster preparedness survey, available on: http://www.symantec.com/content/en/us/about/media/pdfs/symc_2011_SMB_DP_Survey_Report_Global.pdf?om_ext_cid=biz_socmed_twitter_facebook_marketwire_linkedin_2011Jan_worldwide_dpsurvey, accessed on January 18, 2012.

Tammineedi, R.L. (2010), Business continuity management: A standards-based approach, *Information Security Journal: A Global Perspective*, 19, 36–50.

Tan, Y., Takakuwa, S. (2011), Use of simulation in a factory for business continuity planning, *International Journal of Simulation Modeling (IJSIMM)*, 10, 17–26.

Thibodean, P. (2011), Who gets blame for Amazon outage? available on http://www.itworld.com/cloud-computing/159225/who-gets-blame-amazon-outage?page=0,0&source=ITWNLE_nlt_saas_2011-04-27, accessed on April 27, 2011

Turban, E., Rainer, R.K., Potter, R.E. (2005), *Introduction to Information Technology*, John Wiley & Sons, New York.

Turetken, O. (2008), Is Your Backup IT-Infrastructure in a Safe Location—A Multi-Criteria Approach to Location Analysis for Business Continuity Facilities, *Information Systems Frontiers*, 10(3), 375–383.

Umar, A. (2005), IT Infrastructure to Enable Next Generation Enterprises, *Information Systems Frontiers*, 7(3), 217–256.

Vatanasombut, B., Igbaria, M., Stylianou, A.C., Rodgers, W. (2008), Information systems continuance intention of web-based applications customers: The case of online banking, *Information and Management*, 45, 419–428.

Versteeg, G., Bouwman, H. (2006), Business architecture: A new paradigm to relate business strategy to ICT, *Information Systems Frontiers*, 8, 91–102.

Walker, A. (2006), Business continuity and outsourcing—moves to take out the risk, *Network Security*, May 2006, 15–17.

Wan, S. (2009), Service impact analysis using business continuity planning processes, *Campus-Wide Information Systems*, 26(1), 20–42.

Williamson, B. (2007), Trends in business continuity planning, *Bank Accounting and Finance*, 20(5), 50–53.

Winkler, U., Fritzsche, M., Gilani, W., Marshall, A. (2010), A model-driven framework for process-centric business continuity management, *Proceedings of the 2010 Seventh International Conference on the Quality of Information and Communications Technology*, Porto, Portugal, pp. 248–252.

Yeh Y. P. (2005). Identification of Factors Affecting Continuity of Cooperative Electronic Supply Chain Relationships: Empirical Case of the Taiwanese Motor Industry. *Supply Chain Management: An International Journal*, 10(4), 327–335.

Wu, W.Y., Chiag, C.Y., Wu, Y.J., and Tu, H.J. (2004), The influencing factors of commitment and business integration on supply chain management, *Industrial Management & Data Systems*, 104(4), 322–333.

Zobel, C. (2011), Representing perceived tradeoffs in defining disaster resilience, *Decision Support Systems*, 50(2), 394–403.

52

Technical Foundations of Information Systems Security

Daniela Oliveira
Bowdoin College

Jedidiah Crandall
University of New Mexico

52.1 Introduction

This chapter provides an overview on information systems security. It starts by discussing the key concepts and underlying principles of computer security, common types of threats and attacks, and the origins of software vulnerabilities.

The chapter then reviews UNIX-like security abstractions: the hierarchy of processes in a system, hardware support for separating processes, the UNIX file system hierarchy, and authentication. Then it is shown how these protections extend to the network, how a typical TCP/IP connection is protected on the Internet, and how web servers and browsers use such protections. Security flaws are also covered with a detailed discussion of the most dangerous software vulnerabilities: buffer overflows and memory corruption, SQL injection, cross-site scripting (XSS), and time-to-check-time-to-use race conditions. Following this, the chapter argues that all the vulnerabilities cross multiple layers of abstraction (application, compiler, operating systems (OSes), and architecture) and discusses research directions to mitigate this challenge and better protect information systems.

52.2 What Is Computer Security?

Computer security addresses policies and mechanisms to protect automated computer systems so that the confidentiality, integrity, and availability of their resources are protected. These resources can be software (the OS and application programs), firmware, hardware, information and data (e.g., a file system), and all networked communication between two end systems [34]. The three pillars upon which computer security is based (confidentiality, integrity, and availability) are commonly referred in the literature by the CIA acronym [10,51].

Confidentiality requires that private, confidential, or sensitive information and system resources are not disclosed to unauthorized parties. This can be achieved through access control or encryption. Access control policies manage the access to the system and its resources and have two main functions: authentication and authorization. Authentication is the process that determines the identity of someone attempting to access a system or a set of data [27]. This assessment of the individual or any other entity (e.g., another system) is usually done through credentials, such as a password, fingerprint, or a card. Authorization is the process that determines the set of resources or data a person or a third-party entity (once authenticated) are allowed to access. For example, once authenticated in a certain Linux machine, user bob has access to all files located in his home directory, but he may not have access to the files located in user alice's home directory.

Integrity requires that data or a system itself is altered only in an authorized manner. It is usually described as a two-concept definition: data integrity and system integrity [34]. Data integrity requires that information and programs are changed only in a specified and authorized manner. System integrity requires that a system functions free from any unauthorized change or manipulation.

Availability requires that a system works in a timely manner and its resources and data are promptly available to authorized users.

52.2.1 Vulnerabilities, Threats, Attacks, and Countermeasures

A vulnerability is an error or a weakness in the design or implementation of a system that could be exploited to violate its security policy. Software systems cannot be guaranteed to be free from vulnerabilities because designers and programmers make mistakes, and current verification and testing techniques cannot assure that a significantly complex piece of software meets its specification in the presence of errors or bad inputs. The likelihood of a successful exploitation of a vulnerability depends on its degree and the intelligence of the exploitation. A threat is a potential for violation of security caused by the existence of vulnerabilities. A threat might never be realized, but the fact that it exists requires that measures should be taken to protect the system against this potential danger. An attack is a threat that was realized or a concrete exploitation of a system vulnerability. We call the agent that performed such malicious actions the *attacker*. An attack is active when it attempts to alter system resources and affect its operations, and it is passive when it attempts to learn or leverage system information without affecting its operation or assets [49].

A countermeasure is any procedure taken to reduce a threat, a vulnerability, or an attack by eliminating or preventing it, or minimizing the harm it can cause by discovering, stopping, or reporting it so that some recovery action is taken [49]. Countermeasures can be classified into three types: preventive, detective, and corrective. A preventive countermeasure prevents an attack from happening in the first place. For example, the use of a password prevents non-authorized users to access a system, and network filters prevent known malicious packets from being processed by the network stack at the OS. A detective countermeasure accepts that an attack may occur, but is able to determine when a system is under compromise and possibly stop the attack or at least warn a system administrator. This usually involves the monitoring of several system attributes. After an attack occurs, a corrective countermeasure (usually the hardest to implement) tries to bring the system back to a stable error-free state, for example, by restoring a file system with backup data.

52.2.2 Origins of Current Vulnerabilities

Even though information systems will never be immune to errors, it is important to understand how the interplay of different layers of abstraction exacerbates the problem. Consider the simplest security model: a gatekeeper. An analogy to the gatekeeper model is the bank teller. Suppose a bank customer needs to use a bank service or access their account. They will walk up to the bank teller's window and give them instructions about what they would like to be done (e.g., withdraw money, deposit, transfer, or obtain a cashier's check). The bank teller is a trusted entity that actually carries out these actions on behalf of the customer, while at the same time, ensuring that the bank's policies are being followed. For example, in the case of money withdrawal, the bank policy is to never give out money to a customer before subtracting the same amount from their account.

This abstraction represents the gatekeeper concept from an early OS called MULTICS [17,21]. In MULTICS, processes were separated into several rings, and the lower the ring number some code is executing in, the greater its privileges are. The system must be aware of ring crossing at the architecture layer. When code in ring i attempts to transfer control to code in ring j, for the code in ring j to do something on its behalf, a fault occurs and control is given to the OS. The gatekeeper is the software abstraction that handles this fault. Its name originates from the view of crossing rings as crossing walls separating them. Ideally the gatekeeper should be as simple as possible. Simplicity allows for ease of inspection, testing, and verification. The design is streamlined and the likelihood of an error is reduced. On the other hand, if the gatekeeper is complex, involving a set of abstractions where many high-level, more complex abstractions are built on top of simpler, lower-level ones in multiple tiers, the possibility of introduction of an avenue for deceiving the gatekeeper is greatly increased.

Most modern information systems are based on what can be called UNIX-like systems (including Windows-based systems). These systems combine memory separation of different processes with a hierarchical file system and process hierarchy to achieve security. These systems are designed with complex and cross-layered security abstractions. For example, for security reasons, a process can access memory only inside its address space. However, the implementation of the abstraction of functions uses the process address space (the stack region) at user level. The stack region should be under strict OS control, but processes have access to their address space. Further, sensitive low-level control information (a function return pointer) belonging to the architecture level (value of the program counter register) can be accessed by the OS and processes because it is stored on the stack.

Furthermore, the application of certain security principles can add to the complexity of the gatekeeper and even conflict with one another. In a classic and much-referenced paper from 1975, Saltzer and Schroeder [43] described several design principles, provided by experience, for the design of information systems and security mechanisms.

One of these principles is *fail-safe defaults*. It states that the design of a system or a security mechanism should be very conservative regarding access to objects and functionalities. The default situation should be lack of access, and the security mechanism should identify which subjects should be allowed access and under which circumstances. The intuition behind this principle is that an error in the security mechanism design will fail by restricting access, and this situation would be easily identified by those subjects (a user, program, or another system) who should be granted access to the system. The alternative design, whose default situation is complete access with access control decisions based on exclusion of rights, will fail by allowing access to a subject that should not receive it. Examples of the application of this principle would be to enforce that only well-typed programs are loaded into memory as processes, or to only allow access to memory locations until they are explicitly the right type that the access is meant for. These examples conflict with the principle of *economy of mechanism* that states that the design of a secure system or security mechanism should be as simple and as small as possible. UNIX-like systems give each process its own address space and several ways to communicate with other processes. Each process is responsible for the accesses it makes to its own memory, which errs more on the side of the second principle.

52.2.3 Why Is Securing a System Hard?

There are many reasons that make computer security a challenge. First, as we have mentioned before, systems will never be vulnerability-free because they are becoming more and more complex and diverse and are devised by humans, who are likely to introduce flaws into the system. Another challenge is that the problem of checking if a piece of code contains malicious logic is undecidable [10]. Malware detectors usually target a particular type of attack and may present false positives and false negatives. Also computer security is an arms race between attackers and security researchers and engineers. Unfortunately, the attackers have the edge because their goal is much simpler than those of a security researcher: an attacker needs to find a suitable vulnerability and exploit it; a security engineer must address all types of vulnerabilities a system may have and patch them while avoiding the introduction of new vulnerabilities in the process. Further, we have been witnessing an increase in the complexity of malware and attackers. The current generation of attackers is extremely creative, financially and politically motivated, and structured much like any well-operated criminal organization [1]. Traditional security models and solutions have difficulty keeping up with these attackers' level of innovation and ingenuity. Finally, many systems are not (or were not) designed with security as a requirement. In many cases, security is an afterthought and sometimes is perceived as an inconvenience or a burden, for example, decreasing performance, increasing costs, and making systems harder to operate.

52.3 UNIX-Like Security Abstractions

In this section, we will build a UNIX-like system from the ground up conceptually, to see what the different protection mechanisms are and how they interact with each other through many layers of abstraction.

The protection mechanisms that form the foundation of a UNIX-like system can be divided into two types: *reference-checking* and *reference-omitting*. Reference-checking mechanisms use an access control list to check all accesses to an object to see if the subject trying to access the object has permission to perform that kind of access, where the access could be an operation such as read, write, or execute. Reference-checking mechanisms associate with each object a list of the accesses that various subjects can perform on that object. Reference-omitting mechanisms only give a reference that can be used to make an access to subjects that have permission to make that access, so that every access need not be checked. In other words, every subject is associated with a list of objects that it can access. Note that the references given to a subject to refer to a subject must be protected themselves for reference-omitting mechanisms, however. Reference-omitting mechanisms are often referred to as capabilities and reference-checking as access control lists, but in this section, we will avoid these names to avoid confusion and historical connotations.

An example of a reference-omitting mechanism is a page table. Page table mappings have bits to allow write access, for example, but there is no notion of different processes in these access controls. The mechanism that stops one process from accessing the physical memory of another process is that it has no virtual-to-physical mapping in its page table for the physical page frames of the other process. So it is not a matter of an access control check based on distinguishing one process from another, but rather the process that is not supposed to access the memory has no possible way to calculate or forge a virtual address that refers to the physical memory of the other process. The reference is omitted, rather than checked.

52.3.1 Hardware Support

First, what do we mean by a "UNIX-like" system? We mean a system that combines memory separation of different processes with a hierarchical file system and process hierarchy to enable security. Although we will not discuss Microsoft Windows in this section, under this definition, Windows is a UNIX-like system. Windows has a rich set of security features [42], but the foundations of security in Windows are still based on the abstractions of processes and files that UNIX pioneered. UNIX owes many of its security ideas to MULTICS [44,17]. In MULTICS, processes were separated by segmentations and organized into 64 rings, with inner-more rings having more privileges. A process could only request for higher-privileged processes to do tasks on their behalf by going through a gatekeeper that initiated code defined at a higher privilege, which had the effect of restricting the interaction between the two processes based on this code that resided in an inner-more ring. The code itself would reason about what it should or should not do on behalf of the lower-privileged process with respect to security. UNIX reduced this idea down to two rings, with the kernel in the inner ring and user space processes in the outer ring, and with the processes themselves being kept separate by a different mechanism.

To support the abstractions of UNIX-like security mechanisms, all that is required of the hardware is a mechanism to separate these two rings and then a different mechanism to separate the processes. For modern UNIX-like systems, a supervisor bit that distinguishes kernel mode from user mode is enough for a ring abstraction where there are only two rings. Most modern CPUs that have support for OSes have a bit that indicates when the CPU is in supervisor mode vs. user mode. The exact mechanism varies, but the effect is the same. In the MIPS instruction set architecture, for example, this bit is implied by the virtual address. The x86 architecture provides four rings, but typical UNIX-like system use only two of them (the inner-most and the outer-most) [13].

Not every system is based on UNIX-like abstractions, but UNIX-like abstractions have proven to be very portable. Eros [48] is an OS that is radically different from UNIX in terms of security abstractions and uses all four rings of the x86. Eros is considered a capability-based system. Levy [29] describes historical systems that were based on the notion of capabilities and required much more elaborate hardware for OS security support than is required by UNIX. In fact, for many secure system designs that are not UNIX-like, the hardware and the OS are codesigned for each other, which can limit portability and requires careful thought about the security abstractions whenever two different systems interact.

UNIX-like security mechanisms also need some way for processes to be kept separate. Specifically, one process should not be able to read from or write to other process's memory without such permission being explicitly granted. Originally, UNIX used the PDP-7's segmentation mechanism to separate the memory address spaces of processes. Segmentation divides physical memory into contiguous segments that can be referenced with a base and offset. Because processes could not change their base and the bounds of their offsets are checked, processes cannot access the other segments in memory. Because of external fragmentation, however, paging is now much preferred to segmentation as a way to separate processes' address spaces in physical memory. As discussed earlier, paging is a reference-omitting mechanism, in which processes are given only virtual address mappings in their page tables for those physical memory page frames that they are allowed to access. In this way, there is no way for the process to calculate a virtual address pointer that references the memory of another process unless such a reference is placed in the page table of that process by the OS kernel.

52.3.2 Security Abstractions Implemented in Software

Once the hardware has provided the OS with the two important protection mechanisms that we have now discussed (one to separate the kernel from user space and another to separate processes within user space), kernel data structures can be built to support security at higher levels of abstraction. Individual processes can interact with other processes or with the kernel only through explicit requests to the kernel called system calls (similar to MULTICS's gatekeeper mechanism). Only the kernel has the privileges necessary to directly access system resources such as the network, hard drive, keyboard, and monitor. In this way, the kernel can separate these resources into different files for different users, different network connections, etc., and enforce access controls.

A UNIX-like system uses two key data structures as the basis for security: file system access controls and the process hierarchy. A typical UNIX-like system has different users that must authenticate to the system and then a special user called *root* that can override most access controls. Users own processes, and they also can own file system objects. File system objects can be actual files, network sockets, other interprocess communications mechanisms, special character devices, or anything that processes can interact with through a file abstraction.

The most basic access controls are those on actual files on the file system, for example,

```
-rw-r-----  1  johndoe  faculty  179   2011-11-12  17:35  myfile.txt
drwxr-xr-x  6  johndoe  johndoe  4096  2011-11-12  18:48  Mydirectory
-rwxr-xr-x  1  johndoe  johndoe  253   2011-11-21  10:37  myprogram.pl
```

In this example, the user johndoe can read from or write to `myfile.txt`, and any user in the faculty group can read from this same file, but other users cannot. Anyone is allowed to enter the `mydirectory` directory and list the files there because of the global read and execute permissions (assuming they are already able to reach this point in the directory structure), and anybody can read and execute the executable program `myprogram.pl`. UNIX-style file permissions are a reference-checking mechanism; consequently, all attempts to add a file to a process's file descriptor table must be checked.

It is important to note that these access controls are checked for *processes* owned by users, and they are typically checked only at the time the file is opened. Processes can have any number of files opened for reading or writing at one time, and these are placed in a file descriptor table so that the process can refer to files in system calls to the kernel by the number of the file descriptor entry. The file descriptor table is a reference-omitting mechanism; once a file is placed in the file descriptor table for a certain type of access, the process can perform that access on the file without any additional checks.

In UNIX-like systems, there is a special process called a *shell* that prompts a user for commands and allows the user to cause the system to do a wide range of actions. A typical user shell process in a UNIX-like system has access to tools such as compilers that allow it to basically do arbitrary computations and any sequence of system calls desired. *Thus, the security of a UNIX-like system is based not on what a user process cannot do within its own address space, but on restricting how it interacts with other processes.*

This leads to the hierarchy of processes. *This hierarchy is not necessarily a privilege hierarchy where parents always have more privileges than their children.* This is a very important point, because prevention of privilege escalation and interacting with remote users from around the world over the network in a secure way are both based on limiting the influence that one process can have over another.

UNIX processes can use the `fork()` system call to create new children below them in the hierarchy. These children typically inherit all of the parent's open file descriptors as well as the parent's *real user ID* and *effective user ID*. The parent can also pass arguments to the child. The real user ID is what the ID of the actual user who initiated the process is, while the effective user ID is the one that actually is checked for controlling access to files and other system resources. Typically, these IDs are the same, but they can be different when the child uses the `execve()` system call to mount a binary with the *setuid bit* set, for example,

`-rwsr-xr-x 1 root root 36864 2011-02-14 15:11/bin/su`

The "s" in the permissions indicates that the setuid bit is set for this binary program file. This bit means that any process that mounts this binary (mounting means that the old executable code along with the rest of the address space of the process is discarded in memory and the process is now running the binary code in /bin/su) has their effective user ID set to the owner of the file, in this case root, rather than the process that executed it. For example, *passwd* is a UNIX utility for changing passwords that has the setuid bit set. This means that when user *bob* invoked this utility, the process runs with effective user id set to *root*, instead of *bob*, which is necessary since only root has permissions to write to the /etc/passwd file to update bob's password.

The `execve()` system call is used for mounting new binaries. Most binaries do not have the setuid bit set, so that the child still has the same effective user ID of the parent. Binaries such as /bin/su, which is used to change to another user and possibly the `root` user by entering that user's password, need the effective permissions of root to read the password file and authenticate a new user and then create a shell process for the new user.

Now we have built a conceptual picture that will help to understand the exploits in Section 52.4. The hierarchy of processes comes from parents who fork() children and then those children sometimes execve() to run other binaries. The parents can pass arguments and open file descriptors to the children. The children may sometimes have higher privileges than the parent by mounting setuid binaries. The root of the tree is typically the init process, which forks login children that run as root but can read users' usernames and passwords and fork children that become *shell* processes after dropping their permissions to the appropriate level for that user. So we have a hierarchy of processes, where lower privileged processes can have ancestors, siblings, and descendants with a higher privilege level than themselves and where there are many ways for processes to communicate through files, sockets, pipes, and other mechanisms.

The major thing missing from this conceptual picture is the network. It is possible for a process on one system to open a network socket to a remote system that is treated by both processes the same as any other entry in their file descriptor table. For example, a web browser on one system can request a remote socket with a web server on another system, and if the kernels of these two systems allow this access and successfully do a three-way TCP handshake, the processes can now communicate over the network.

What we will see in Section 52.4 is that when processes can communicate with one another one process can influence the actions of the other through software vulnerabilities and then subvert security mechanisms.

52.4 Vulnerabilities and Countermeasures

In this section, we discuss several common and dangerous vulnerabilities or security flaws that allow attackers to target user-level programs and OSes. The context here is that each of these vulnerabilities allows attackers to violate the assumptions upon which the security abstractions of a UNIX-like system are built.

52.4.1 Buffer Overflows

This vulnerability is one of the most common and dangerous security flaws and was first widely publicized by the Morris Internet Worm in 1988 [2], which compromised thousands of machines and caused millions of dollars in losses.

A buffer is a memory area in which contiguous data (an array or a string) can be stored. A buffer overflow is the condition at an interface under which more data are placed into a particular finite size buffer than the capacity allocated [35]. This location could hold other program or system variables and data structures, function pointers, or important program control information such as a function return address.

To understand buffer overflows, it is important to review how a program is laid out in memory by the OS for execution. The process address space or the memory region that can be accessed by the process is divided into four main regions. The code region is read-only, fixed-sized, and contains the program machine code. The data area is also fixed-sized and contains global and static variables and data structures used by the program. The heap is used for dynamic allocation of program variables and data structures. The stack is used to implement the abstraction of function calls. It grows toward the lower memory addresses of a process and stores function's local variables, parameters, and also control information such as a function return address.

Consider the code snippet illustrated in Figure 52.1. Upon calling *b*, *a* pushes the parameters for *b* on the stack in reverse order of declaration. The return address of *a* (the next instruction that should continue to be executed when *b* returns) is also saved on the stack. Then the address of *a*'s stack frame pointer is saved on the stack. This is necessary to recover it after function *b* returns. Notice that *b*'s local

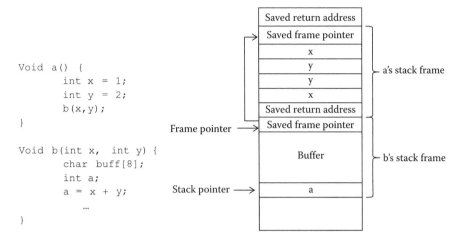

```
Void a() {
        int x = 1;
        int y = 2;
        b(x,y);

}

Void b(int x, int y) {
        char buff[8];
        int a;
        a = x + y;
        ...

}
```

FIGURE 52.1   Function invocation.

variables (including the vulnerable buffer) are stored below the saved value of the frame pointer and the return address. Also when an input is stored in a buffer, the bytes are placed toward higher addresses in the process address space, i.e., toward the saved frame pointer and return address.

A stack-based buffer overflow can cause several types of harm in a system or program. An unintentional buffer overflow may cause a program or an OS to terminate or crash, due to the corruption of data structures, stack frame pointer, or return address. For example, if the return address is overwritten with a random value, this value will be loaded into the PC register (called the *program counter,* which contains the address of the next instruction to be executed by the CPU), and the CPU might identify this value as an illegal memory access (possibly outside the process address space). A malicious buffer overflow will occur when the attacker carefully crafts an input where the return address is overwritten with an address chosen by the attacker. This address could be within the malicious input string and points to malicious instructions that open a *shell* (a command-line interface to the OS), giving the attacker administrator privileges in the system if the compromised program is privileged. The malicious address could also correspond to a library function (e.g., libc) allowing the attacker to perform malicious actions by combining existing functions (also called *return-to-libc* attack [31]).

52.4.1.1 Defenses

There are several different proposed solutions currently in use against buffer overflow attacks, and they can be broadly classified into two types: compile-time and run-time [30].

A compile-time solution involves adding code to programs with the goal of preventing and detecting buffer overflows. Type-safe languages such as Java and Python have the compiler generate code to check array bounds, thus preventing such attacks. The StackGuard gcc extension [15] protects the stack by inserting additional instructions to a function entry and exit code. The goal is to protect the stack by checking if it is corrupted. A function entry code writes a canary value (named after the miner's canary used to detect poisonous air in a mine and warn the miners) below the saved frame pointer on the stack. Upon exiting, the function exit code checks that the canary was not corrupted, as the attacker will need to corrupt the canary to tamper with the return address. This canary value must be randomly generated to avoid the possibility of an attacker guessing the value and overwriting it with the proper value during the attack, thus evading detection.

Run-time solutions protect vulnerable programs by detecting and stopping buffer overflow attacks during execution. One common approach is to protect the address space of a process against improper execution. A buffer overflow attack usually causes a control flow change that leads to the execution of malicious instructions that are part of the malicious input string in the stack. If the stack is marked as nonexecutable, many instances of buffer overflow attacks can be stopped. The assumption is that executable code is supposed to be located in the code portion of a process address space and not on the stack, heap, or data areas. The Solar Designer's StackPatch [18] modifies the address space of a process to make the stack non-executable. The $W \otimes X$ defense [55] ensures that no memory location in a process address space can be marked both as writable (W) and executable (X) at the same time. Most OSes now offer such protection. While this type of defense is considered an effective protection mechanism that is relatively simple to implement, there are situations where a program must place executable code in the stack [51]: just in time compilers (Java), implementation of nested functions in C, and signal handlers.

Another approach is address space randomization. In order to exploit a buffer overflow, an attacker needs to predict the location of a return address or know the exact location of a library function. Randomly arranging the memory positions of key areas such as the heap, stack, and libraries makes this prediction harder for the attacker. This idea is implemented in PAX ASLR [39] and is supported by many OSes.

52.4.2 Vulnerabilities and UNIX-Like Abstractions

Buffer overflows demonstrate an important part of UNIX-like security abstractions, which is that the code that a process runs and the semantics of the program are a key part of the security of a system. In UNIX-like systems, a setuid root binary such as {\tt su} must carefully process all of its inputs to make sure that the terminal it inherited from its parent must provide a valid username and password before a shell for entering commands as the new user is created as its child. Similarly, a web server must check the inputs it receives on a network socket from a remote client to make sure it is a valid request within the HyperText Transfer Protocol (HTTP) protocol and then must only carry out the actions necessary to provide the HTTP service as per that protocol. Since "change the root password on the system" is not part of the HTTP protocol, the remote client cannot directly subvert security within the confines of the protocol. Rather, within the confines of UNIX-like security abstractions, attackers exploit vulnerabilities to have more influence on other processes than the security model for the system allows.

This notion of processes acting on other processes' behalf (perhaps remotely) leads to three kinds of exploits for vulnerabilities: those that directly corrupt a process and take it over through communication, those that confuse other processes at higher levels of abstraction, and those that attack the communication channel between two processes to exploit a trust relationship between them.

52.4.2.1 Directly Corrupting a Process and Taking It Over

Control flow hijacking attacks communicate with other processes in a malicious way that causes that process to effectively be taken over so that its computations and system calls are now controlled by the attacker.

The most famous example of this kind of attack is the buffer overflow, which was explained earlier in this section. These are a special case of a broader problem known as memory corruption attacks and that includes not only buffer overflows [6], but also format string attacks [46] and double `free()`'s [7]. By corrupting the memory of another process through malicious inputs that it cannot handle properly, it is possible to inject the attacker's own instructions to be executed by that process. This can be used to elevate privileges on the current system by corrupting another process with higher privileges, which might be a child with the setuid bit set. This can also be used to take control of a remote machine by corrupting one of its processes. This can be achieved by, for example, connecting to a web server that has a memory corruption vulnerability in its implementation. It can also be achieved by, for example, sending a user a malicious file through e-mail such as a PDF that exploits a software bug in their PDF reader. Mitigations for buffer overflow attacks were discussed earlier in this section.

In these previous examples, a process owned by the attacker sent malicious inputs to another process, either on the same system or on a remote system over the network, where the other process has privileges that the attacker needs in order to subvert security. Another example of directly taking over a process is SQL injection. A typical SQL injection attack involves at least three processes that communicate remotely over the network. The attacker's remote client process sends form inputs to a web server process, which embeds these inputs into an SQL query to interact with a database. The attacker uses SQL delimiters in their form inputs to change the boundaries between the SQL code and their inputs. In this way, they can control the SQL code that the SQL server executes.

For example, web server code might read $name and $password from a web client and then build an SQL query for a database server using the following substitution:

```
SELECT * FROM users WHERE name = '$name' and password = '$password'
```

The single quote (') is a special delimiter in the SQL language. The attacker can leave the password blank and enter this as their username:

```
'; DROP TABLE users--comment…
```

This causes the SQL server to execute the following SQL code that it receives from the web server, with the effect of the entire table of users in the database being deleted:

```
SELECT* FROM users WHERE name =''; DROP TABLE
users-- comment…
```

Fundamentally, SQL injection vulnerabilities are a problem where the inputs from the attacker span across more than a leaf in the parse tree of the resulting query [52]. For example, the name "bob" would be a single leaf in a parse tree of the SQL language for the resulting query, but the name "'; DROP TABLE users -- comment…" will be a tree itself with ";", "DROP", "TABLE", "users", and "--comment" all being separate leaves in the tree. Su and Wassermann [52] describe a dynamic approach to detect SQL injection exploits.

XSS is another type of vulnerability where information being passed between more than two processes leads to the ability for the attacker to control another process via malicious inputs. In the case of XSS, a typical XSS attack allows one web client to pass scripted code to another client through the server. For example, suppose a forum on a web server allows users to add HyperText Markup Language (HTML) to their forum comments so that other users can see formatted text with tags, such as bold text, for example, **This is a bold forum comment**. Suppose also that the forum server does a poor job of filtering out HTML code that could harm other clients, such as Javascript:

```
body onload = " alert('xss')"
```

This is a real example from a college campus course content management system [26] that allows students and faculty to post HTML-formatted messages in forums or as e-mail. An HTML blacklist is applied to not allow HTML tags that lead to the execution of Javascript, but the blacklist only filters out well-formed HTML (in this case well-formed HTML would have a closing angle bracket, ">") while most browsers will execute the Javascript code alert('xss') even without well-formed HTML. In this example, if a student posts this string or sends it as an e-mail to the instructor, the message recipient will receive a pop-up window that simply says, "xss". The ability to execute arbitrary Javascript could, however, give students the ability to cause the instructor's web browser to update the gradebook or do any action that the instructor's browser session has the privileges to perform. For a good explanation of XSS vulnerabilities and why blacklist approaches fail, as well as a dynamic approach that is sound and precise so that XSS vulnerabilities can be removed from source code at development time, see Wassermann and Su [59].

52.4.2.2 Confusing Other Processes at a Higher Level of Abstraction

It is also possible to confuse other processes at a higher level of abstraction, so that the process violates security on the attacker's process behalf of even if the vulnerable process is not corrupted to the point where the attacker has complete control.

An example of this would be a Time of Check to Time of Use (TOCTTOU) race condition vulnerability. This kind of vulnerability occurs when privileged processes are provided with some mechanism to check whether a lower-privileged process should be allowed to access an object before the privileged process does so on the lower-privileged process's behalf. If the object or its attribute can change between this check and the actual access that the privileged process makes, attackers can exploit this fact to cause privileged processes to make accesses on their behalf that subvert security.

The classic example of TOCTTOU is the sequence of system calls of access() followed by open(). The access() system call was introduced to UNIX systems as a way for privileged processes (particularly those with an effective user ID of root) to check if the user who owned the process that invoked them (the real user ID) has permissions on a file before the privileged process accesses the file on the real user ID's behalf. For example,

```
if (access("/home/bob/symlink", R_OK | W_OK) ! = -1)
{
        //Symbolic link can change here
        f = fopen("/home/bob/symlink", "rw");
        ...
}
```

What makes this is a vulnerability is the fact that the invoker of the privileged process can cause a race condition where something about the file system changes in between the call to access() and the call to open(). For example, the file /home/bob/symlink can be a symbolic link that points to a file the attacker is allowed to access during the access() check but is changed to point to a different file that needs elevated privileges for access. The attacker can usually win this race by causing the open() system call to block for a hard drive access to read one of the directory entries (for more details see Borisov et al. [12]). The only effective way to mitigate TOCTTOU vulnerabilities is to identify sequences of system calls that are unsafe with respect to concurrency and not use them in privileged code. For example, secure code should never use access() followed by open(), which means that the access() system call has fallen into disuse since predicating file opens was its purpose.

52.4.2.3 Attacking the Communication Channel between Two Processes

Directly corrupting another process or confusing it at a higher level of abstraction typically entails one process sending malicious inputs to another that the receiving process cannot handle correctly with respect to security due to a bug. It is also possible for malicious agents on the network to corrupt the communication channel between two processes. When two processes communicate over the Internet, there is typically some trust relationship implied, such as that a web browser trusts that it is communicating with the web server a user intended to contact when it passes the user's username and password along on the network. Network attacks seek to exploit this trust relationship.

Network attacks and the security mechanisms to thwart them are too numerous to discuss thoroughly here. For discussion of the most salient attacks on the most common protocols, see Tews et al. [56] regarding Address Resolution Protocol (ARP) injection and Wired Equivalent Privacy (WEP) attacks, Qian and Mao [41] regarding TCP hijacking, Savage et al. [45] regarding a large-scale attack on TCP congestion control, Karlin et al. [25] regarding attacks on Border Gateway Protocol (BGP) routing, and Wright [61] regarding an attack on Domain Name System (DNS).

We will use a typical HTTP connection with a typical network stack to illustrate the different possible points of attack. If a user wants to go to http://www.example.com, the first step is to find an IP address for this URL. The user's system is configured to use a specific DNS server for this lookup, which is typically the DNS server for their organization or Internet Service Provider. But before the DNS server can be contacted by IP address, the user's system must find an ARP address for either the DNS server or a gateway router that can reach the DNS server. ARP is a typical layer 2 protocol of the OSI model [40], and layer 1 is the physical layer.

Above ARP is the IP layer, layer 3. This allows packets to be routed, for example, to the DNS server and back. The user's system makes a DNS request and the DNS server returns the IP address for http://www.example.com. Now the user uses the TCP protocol (layer 4) on top of IP routing to connect to the remote machine where the process for the web server that hosts http://www.example.com is running. The web browser and web server can now communicate through an interprocess communication mechanism called a socket using a layer 7 (application layer) protocol that they have agreed on, in this case HTTP.

Disclosure, deception, disruption, and usurpation [49] are possible at all of these layers. For ARP, any machine on the local network can answer for any IP address, even the gateway. At the IP layer, any of the routers that any packets for the connection go through have arbitrary control over that packet, and attacks on the routing system such as BGP attacks [25] mean that attackers can ensure that certain connections will be routed through their routers. Higher layers can also be attacked, for example, responses to DNS requests can be forged. Because of all of these possible attacks, an important protocol for securing interprocess communication over the Internet in an end-to-end fashion with public key cryptography is Secure Sockets Layer (SSL) at layer 6 (the presentation layer). Using a public key infrastructure (PKI), typically the Certificate Authority (CA) system for the Internet, it is possible for two remote processes to set up an encrypted channel between them. Because SSL is a layer 6 protocol that is implemented in libraries, applications can still communicate with a socket Inter-Process Communication (IPC) mechanism, with the SSL library performing the encryption and decryption relatively transparently. The encryption is end-to-end, meaning that it provides a level of privacy, integrity, and authentication that mitigates many network-based attacks.

In summary, UNIX-like systems start from very simple hardware protections that prevent processes from directly accessing blocks of data on the hard drive or gaining raw access to the network card to interfere with others' network connections. They then build many layers of abstraction on top of this. Process hierarchies, file system hierarchies, network protocols such as ARP, TCP/IP, DNS, and BGP, and many other abstractions build a bridge between the hardware's simple notion of security and the security properties that human users expect for higher-level concepts like URLs (e.g., http://www.example.com). Software bugs called vulnerabilities make it possible for attackers to violate the assumptions that these layers of abstraction are built upon.

52.5 Vulnerabilities and Layers of Abstraction

In this section, we discuss how security vulnerabilities in computer systems cross multiple layers of abstraction (application, compiler, OS, and architecture). This cross-layer existence not only makes the problem of classifying vulnerabilities hard but also hinders the development of strong and flexible solutions to remove or mitigate them.

52.5.1 Challenge of Classifying Vulnerabilities

Bishop and Bailey [11] have analyzed many vulnerability taxonomies and showed that they are imperfect because, depending on the layer of abstraction a vulnerability is being considered, it can be classified into multiple ways.

A taxonomy is a classification system that allows a concept (e.g., a vulnerability) to be uniquely classified. The RISOS [3], the Protection Analysis [24], Landwehr et al. [28], and Aslam [8] taxonomies represented important studies to help systems designers and administrators to understand vulnerabilities and better protect computer systems. The Protection Analysis study [24], for instance, sought to understand OS security vulnerabilities and also identify automatable techniques for detecting them. Bishop and

Bailey [11] also summarized Peter Neumann's presentation of this study [32] with 10 classes of security flaws (which are further summarized here).

1. Improper protection domain initialization and enforcement: vulnerabilities related to the initialization of a system and its programs and the enforcement of the security requirements.
   a. Improper choice of initial protection domain: vulnerabilities related to an initial incorrect assignment of privileges.
   b. Improper isolation of implementation detail: vulnerabilities that allow users to bypass a layer of abstraction (e.g., the OS or the architecture) and write directly into protected data structures or memory areas, for instance, I/O memory and CPU registers.
   c. Improper change: vulnerabilities that allow an unprivileged subject to change the binding of a name or a pointer to a sensitive object so that it can bypass system permissions.
   d. Improper naming: vulnerabilities that allow two objects to have the same name, causing a user to possibly execute or access the wrong object. For example, if two hosts have the same IP address, messages destined to one host might arrive at another one.
2. Improper validation: vulnerabilities related to improper checking of operands or function parameters.
3. Improper synchronization: vulnerabilities arising when a process fails to coordinate concurrent activities that might access a shared resource.
   a. Improper indivisibility: interruption of a sequence of instructions that should execute atomically.
   b. Improper sequencing: failure to properly order concurrent read and write operations on a shared resource.
4. Improper choice of operand and operation: vulnerabilities caused by a wrong choice of the function needed to be called from a process. For example, a cryptographic key generator invoking a weak pseudorandom number generator.

Bishop and Bailey [11] showed how this study and others [3,8,24,28] fail to uniquely identify vulnerabilities using two very common and dangerous security flaws as examples: buffer overflows (described here as an example) and TOCTTOU.

52.5.1.1 Nonunique Classification of Buffer Overflows

At the application layer, this vulnerability can be classified as type 2 (improper validation) because the flaw was caused by the lack of array bounds checking by the programming language compiler and also by the program itself. From the perspective of the subject (in this case the attacker process) who used the vulnerable program, the flaw can be classified as type 4 (improper choice of operand) as the string passed as a parameter was too long (overflowed the buffer at the vulnerable process).

At the OS layer, the vulnerability can be classified as 1b (improper isolation of implementation detail). In this case, the vulnerable process was able to bypass OS controls and overwrite an area of the process address space corresponding to a function return address.

We can also argue that at the architecture layer, this vulnerability can be classified as 1b (improper isolation of implementation detail) as the compromised process was also able to directly write into the PC register. When a function returns, the return address (which was maliciously overwritten) is written into the PC register and the program control flow jumps to that location. It can also be classified as 1c (improper change) as the value of the PC register was maliciously changed.

52.6 Research Issues

Protecting an OS kernel against vulnerabilities is a difficult problem given the complexity and variety of its code. Many proposed solutions employ virtualization. In the traditional virtual machine (VM) usage model [14], it is assumed that the VM is trustworthy and the OS running on top of it can be easily compromised by malware. This traditional usage model comes with a cost: the semantic gap problem. There is a significant difference between the abstractions observed by the guest OS (high-level semantic information) and by the VM (lower-level semantic information). These solutions use introspection to extract meaningful information from the system they monitor/protect [20]. With introspection, the physical memory of the current VM instance is inspected and high-level information is obtained by using detailed knowledge of the OS algorithms and data structures. Another line of research [36] employs active collaboration between a VM and a guest OS to bridge the semantic gap. In this approach, a guest OS running on top of a VM layer is aware of virtualization and exchanges information with the VM through a protected interface to achieve better security. A sizable amount of research is being pursued toward building secure web browsers [9,22,54,58]. We have been witnessing a shift to a web-based paradigm where a user accomplishes most of their computing needs using just a web browser [57]. Original browsers were designed to render static web pages, but as web applications and dynamic content became common, many attackers explored weaknesses in the design of these browsers (e.g., XSS and memory exploitation) to launch attacks. Researchers now view browsers as OSes and are investigating new models where browsers are designed by having their components isolated using the process abstraction, which can contain or prevent web-based vulnerabilities.

Return-oriented programming is a relatively new type of attack [47] inspired by the return-to-lib-c approach of exploiting the stack. The attacker does not inject foreign malicious code but instead combines a large number of short instruction sequences (called gadgets) that return and transfer control to another gadget. Several gadgets crafted from legitimate libraries, OS code, or program instructions can be used to perform malicious computations. This type of attack is dangerous because it does not involve foreign code and can evade important defense approaches such as $W \otimes X$. Current research focuses on randomizing or rewriting code so as to prevent an attacker from finding useful gadgets in them [23,38]. Randomization as a way to thwart attacks is a promising idea that dates back to 1997 [19], but there still is no widely accepted methodology for determining how much more difficult randomization makes it for the attacker to develop exploits for any given vulnerability.

The mitigations discussed for each vulnerability in Section 52.2.2 are specific to one type of vulnerability. Another approach to secure systems is to provide processes with security mechanisms that allow for finer-grained separation. These mechanisms are largely based on Saltzer and Schroeder's Principle of Least Privilege, and the security comes from the fact that when processes are corrupted by an attacker, the attacker is more limited if those processes have very limited privileges. SELinux [4] and Linux capabilities [13] are examples that have already been fully integrated into a modern OS. Capsicum [60] is an example of an effort that is still in the research phase.

Many attempts have been made to secure commodity code without making major changes to it (such as separating it into different privilege domains). One notable effort is dynamic information flow tracking (DIFT) [16,33,53]. DIFT systems have proven effective at detecting existing attacks in full systems without compromising on compatibility, but cannot make any guarantee about attacks designed for DIFT systems. This is because tracking information flow dynamically is difficult to do precisely enough for programs that make heavy use of memory indirections such as pointers and control structures such as for loops [5,50].

52.7 Summary

This chapter discussed the foundations for security in information systems, which are based on the triad of confidentiality, integrity, and availability. We can find the origins of many software security flaws in the intertwinement of different layers of abstraction (application, compiler, OS, and architecture), which hinders the development of simple but strong protection mechanisms for information systems and their resources. Nearly all information systems run on UNIX-like OSes (where Windows-based systems are also included). Knowledge about how such OSes are designed allows us to understand their different protection mechanisms, how the hardware facilitates their security, and also their most common types of vulnerabilities and associated countermeasures. The challenge of countering and even classifying vulnerabilities can also be traced to the fact that they cross multiple layers of abstraction.

Given that security vulnerabilities involve more than one layer of abstraction, defending against them should involve some form of collaboration between layers of abstraction to bridge their semantic gap. Current security solutions operate at one particular level of abstraction. For example, they operate at the application level as a user-level process [37], at the compiler level so that safer binary code is generated [15], at the system level as an OS security extension [4], and also at the architecture level, usually involving a VM layer [20]. Multiple-tiered and collaborative security approaches seem a promising research direction to mitigate these vulnerabilities.

References

1. McAfee Virtual Criminology Report 2009–*Virtually Here: The Age of Cyber Warfare.* Available at http://resources.mcafee.com/content/NACriminologyReport2009NF (accessed on April 12, 2013).
2. C. Schmidt and T. Darby. The What, Why, and How of the 1988 Internet Worm. http://snowplow.org/tom/worm/worm.html (accessed on April 12, 2013).
3. R. P. Abbot, J. S. Chin, J. E. Donnelley, W. L. Konigsford, and D. A. Webb. Security analysis and enhancements of computer operating systems. *NBSIR 76-1041, Institute for Computer Sciences and Technology, National Bureau of Standards*, 1976.
4. National Security Agency. Security-enhanced linux. http://www.nsa.gov/research/selinux/ (accessed on April 12, 2013).
5. M. I. Al-Saleh and J. R. Crandall. On information flow for intrusion detection: What if accurate full-system dynamic information flow tracking was possible? In *New Security Paradigms Workshop*, Bertinoro, Italy, 2010.
6. A. One. *Smashing the stack for fun and profit.* http://www.phrack.com/issues.html?issue=49\&id=14 (accessed on April 12, 2013). 1996.
7. Anonymous. Once upong a free()... http://www.phrack.com/issues.html?issue=57\&id=9 (accessed on April 12, 2013).
8. T. Aslam. A taxonomy of security faults in the UNIX operating system, M.S. Thesis, Purdue University, West Lafayette, IN, 1995.
9. A. Barth, C. Jackson, C. Reis, and The Google Chrome Team. The security architecture of the chrome browser. http://crypto.stanford.edu/websec/chromium/chromium-security-architecture.pdf (accessed on March 29, 2013).
10. M. Bishop. *Computer Security: Art and Science.* Addison Wesley, Boston, MA, 2003.
11. M. Bishop and D. Bailey. A critical analysis of vulnerability taxonomies. Technical Report CSE-96-11, University of California at Davis, Davis, CA, 1996.
12. N. Borisov, R. Johnson, N. Sastry, and D. Wagner. Fixing races for fun and profit: How to abuse a time. In *Proceedings of the 14th Conference on USENIX Security Symposium—Volume 14, SSYM'05*, pp. 20–20, USENIX Association, Berkeley, CA, 2005.
13. D. P. Bovet and M. Cesati. *Understanding the Linux Kernel*, 3rd edn. O'Reilly, Sebastopol, CA, 2005.
14. P. M. Chen and B. D. Noble. When virtual is better than real. *HotOS*, Oberbayern, Germany, May 2001.
15. C. Cowan, C. Pu, D. Maier, J. Walpole, P. Bakke, S. Beattie, A. Grier, P. Wagle, Q. Zhang, and H. Hinton. StackGuard: Automatic adaptive detection and prevention of buffer-overflow attacks. In *USENIX Security*, pp. 63–78, San Antonio, TX, January 1998.
16. J. R. Crandall and F. T. Chong. Minos: Control data attack prevention orthogonal to memory model. In *MICRO*, pp. 221–232, Washington, DC, December 2004.
17. R. C. Daley and J. B. Dennis. Virtual memory, processes, and sharing in MULTICS. In *Proceedings of the First ACM Symposium on Operating System Principles, SOSP '67*, pp. 12.1–12.8, ACM, New York, 1967.

18. Solar Designer. Stackpatch. http://www.openwall.com/linux (accessed on March 29, 2013).

19. S. Forrest, A. Somayaji, and D. Ackley. Building diverse computer systems. In *Proceedings of the Sixth Workshop on Hot Topics in Operating Systems*, pp. 67–72, Los Alamitos, CA, 1997.

20. T. Garfinkel, B. Pfaff, J. Chow, M. Rosenblum, and D. Boneh. Terra: A virtual machine-based platform for trusted computing. In *ACM Symposium on Operating Systems Principles*, pp. 193–206, New York, October 2003.

21. R. M. Graham. Protection in an information processing utility. In *Communications of the ACM*, Vol. 11, ACM, New York, 1968.

22. C. Grier, S. Tang, and S. T. King. Secure web browsing with the OP web browser. In *IEEE Symposium on Security and Privacy*, Oakland, CA, May 2008.

23. J. D. Hiser, A. Nguyen-Tuong, M. Co, M. Hall, and J. W. Davidson. ILR: Where'd my gadgets go? In *IEEE Symposium on Security and Privacy*, San Francisco, CA, May 2012.

24. R. Bisbey II and D. Hollingsworth. Protection analysis project final report. ISI/RR-78-13, DTIC AD A056816, USC/Information Sciences Institute, Marina del Rey, CA, 1978.

25. J. Karlin, S. Forrest, and J. Rexford. Pretty good BGP: Improving BGP by cautiously adopting routes. In *Proceedings of the 2006 IEEE International Conference on Network Protocols, ICNP '06*, pp. 290–299, IEEE Computer Society, Washington, DC, 2006.

26. Personal communication with Jeffrey Knockel., 2011.

27. B. Lampson, M. Abadi, M. Burrows, and E. Wobber. Authentication in distributed systems: Theory and practice. *ACM Transactions on Computer Systems*, 10:265–310, 1992.

28. C. E. Landwehr, A. R. Bull, J. P. McDermott, and W. S. Choi. A taxonomy of computer program security flaws. *ACM Computing Surveys*, 26(3): 211–254, 1994.

29. H. M. Levy. *Capability-Based Computer Systems*. Butterworth-Heinemann, Bedford, MA, 1984.

30. K.-S. Lhee and S. J. Chapin. Buffer overflow and format string overflow vulnerabilities. *Software—Practice and Experience—Special Issue: Security Software*, 33(3):423–460, 2003.

31. Nergal. The advanced return-into-lib(c) exploits: PaX case study, http://www.phrack.org/issues.html?issue=58\%5C\&id=4 (accessed on April 12, 2013).

32. P. Neumann. Computer systems security evaluation. In *National Computer Conference Proceedings (AFIPS Conference Proceedings)*, pp. 1087–1095, Montvale, NJ, 1978.

33. J. Newsome and D. Song. Dynamic taint analysis for automatic detection, analysis, and signature generation of exploits on commodity software. In *NDSS*, San Diego, CA, February 2005.

34. National Institute of Standards and Technology. *A Introduction to Computer Security: The NIST Handbook*, National Institute of Standards and Technology, Washington, DC, 1995.

35. National Institute of Standards and Technology. *Glossary of Key Information Security Terms*, National Institute of Standards and Technology, Gaithersburg, MD, 2011.

36. D. Oliveira and S. F. Wu. Protecting kernel code and data with a virtualization-aware collaborative operating system. In *Annual Computer Security Applications Conference (ACSAC)*, Honolulu, HI, December 2009.

37. S. Ortolani, C. Giuffrida, and B. Crispo. KLIMAX: Profiling memory writes to detect keystroke-harvesting malware. In *RAID*, Menlo Park, CA, 2011.

38. V. Pappas, M. Polychronakis, and A. D. Keromytis. Smashing the gadgets: hindering return-oriented programming using in-place code randomization. In *IEEE Symposium on Security and Privacy*, San Francisco, MA, May 2012.

39. PaX Project. Address space layout randomization, March 2003. http://pageexec.virtualave.net/docs/aslr.txt (accessed on March 29, 2013).

40. L. L. Peterson and B. S. Davie. *Computer Networks: A Systems Approach*, 5th edn. Morgan Kaufmann Publishers Inc., San Francisco, CA, 2011.

41. Z. Qian and Z. M. Mao. Off-path TCP sequence number inference attack. In *IEEE Symposium on S&P*, San Francisco, CA, 2012.

42. M. Russinovich and D. A. Solomon. *Windows Internals: Including Windows Server 2008 and Windows Vista*, 5th edn. Microsoft Press, Redmond, WA, 2009.

43. J. Saltzer and M. Schroeder. The protection of information in computer systems. *Proceedings of the IEEE*, 63(9):338–402, 1975.

44. J. H. Saltzer. Protection and the control of information sharing in MULTICS. *Communications of the ACM*, 17(7):388–402, July 1974.

45. S. Savage, N. Cardwell, D. Wetherall, and T. Anderson. TCP congestion control with a misbehaving receiver. *SIGCOMM Computer Communication Review*, 29(5):71–78, October 1999.

46. scut. Exploiting Format String Vulnerabilities. http://www.cis.syr.edu/~wedu/seed/Labs/Vulnerability/Format_String/files/formatstring-1.2.pdf (accessed on April 12, 2013).

47. H. Shacham. The geometry of innocent flesh on the bone: Return-into-libc without function calls (on the x86). In *ACM CCS*, pp. 552–561, New York, 2007.

48. J. S. Shapiro, J. M. Smith, and D. J. Farber. Eros: A fast capability system. In *Proceedings of the Seventeenth ACM Symposium on Operating Systems Principles, SOSP '99*, pp. 170–185, ACM, New York, 1999.

49. R. Shirey. Internet Security Glossary, RFC 2828, 2000. http://www.ietf.org/rfc/rfc2828.txt (accessed on April 13, 2013).

50. A. Slowinska and H. Bos. Pointless tainting? Evaluating the practicality of pointer tainting. In *Proceedings of the 4th ACM European Conference on Computer systems, EuroSys '09*, pp. 61–74, ACM, New York, 2009.

51. W. Stallings and L. Brown. *Computer Security Principles and Practice*. Pearson, Upper Saddle River, NJ, 2012.

52. Z. Su and G. Wassermann. The essence of command injection attacks in web applications. In *Conference Record of the 33rd ACM SIGPLAN-SIGACT Symposium on Principles of Programming Languages, POPL '06*, pp. 372–382, ACM, New York, 2006.

53. G. E. Suh, J. Lee, and S. Devadas. Secure program execution via dynamic information flow tracking. In *Proceedings of ASPLOS-XI*, Boston, MA, October 2004.

54. S. Tang, H. Mai, and S. T. King. Trust and protection in the Illinois browser operating system. In *Symposium on Operating Systems Design and Implementation (OSDI)*, Vancouver, British Columbia, Canada, October 2010.

55. PAX Team. PaX non-executable pages design & implementation. http://pax.grsecurity.net/docs/noexec.txt (accessed on March 29, 2013).

56. E. Tews, R.-P. Weinmann, and A. Pyshkin. Breaking 104 bit WEP in less than 60 seconds. In *Proceedings of the 8th International Conference on Information Security Applications, WISA '07*, pp. 188–202, Springer-Verlag, Berlin, Germany, 2007.

57. H. J. Wang, A. Moshchuk, and A. Bush. Convergence of desktop and web applications on a multi-service OS. In *USENIX Workshop on Hot Topics in Security*, Montreal, Quebec, Canada, August 2009.

58. H. J. Wang, C. Grier, A. Moshchuk, S. T. King, P. Choudary, and H. Venter. The multi-principal OS construction of the gazelle web browser. In *USENIX Security Symposium*, Montreal, Quebec, Canada, August 2009.

59. G. Wassermann and Z. Su. Static detection of cross-site scripting vulnerabilities. In *Proceedings of the 30th International Conference on Software Engineering, ICSE '08*, pp. 171–180, ACM, New York, 2008.

60. R. N. M. Watson, J. Anderson, B. Laurie, and K. Kennaway. Capsicum: Practical capabilities for UNIX. In *Proceedings of the 19th USENIX Conference on Security, USENIX Security '10*, pp. 3–3, USENIX Association, Berkeley, CA, 2010.

61. C. Wright. Understanding Kaminsky's DNS bug. *Linux Journal*, 25, July 2008. http://www.linuxjournal.com/content/understanding-kaminskys-dns-bug.

53

Database Security and Privacy

Sabrina De Capitani
di Vimercati
*Università degli
Studi di Milano*

Sara Foresti
*Università degli
Studi di Milano*

Sushil Jajodia
George Mason University

Pierangela Samarati
*Università degli
Studi di Milano*

53.1 Introduction

In the last few years, the wide availability of computational and storage resources at low prices has substantially changed the way in which data are managed, stored, and disseminated. As testified by the growing success of data outsourcing, cloud computing, and services for sharing personal information (e.g., Flickr, YouTube, Facebook), both individuals and companies are more and more resorting to external third parties for the management, storage, and (possibly selective) dissemination of their data. This practice has several advantages with respect to the in-house management of the data. First, the data owner needs neither to buy expensive hardware and software licenses nor to hire skilled personnel for managing her data, thus having economic advantages. Second, the external server guarantees high data availability and highly effective disaster protection. Third, even private individuals can take advantage of the avant-garde hardware and software resources made available by providers to store, elaborate, and widely disseminate large data collections (e.g., multimedia files). The main problem of this outsourcing trend is that the data owner loses control over her data, thus increasing security and privacy risks. Indeed, the data stored at an external server may include sensitive information that the external server (or users accessing them) is not allowed to read. The specific security and privacy issues that need to be considered vary depending on the main goal for which the data owner provides her data to a third party. In particular, we identify two scenarios: a *data outsourcing* scenario where the data owner delegates the management and storage of a data collection, possibly including sensitive information that can be selectively accessed by authorized users, to a *honest-but-curious* external server; and a *data publishing* scenario where the data owner delegates the storage of a data collection to an external server for its public dissemination. An honest-but-curious server is typically trusted to properly manage the data and make them available when needed, but it may not be trusted by the data owner to read data content. Both these scenarios are characterized by the interactions among four parties: *data owner*, an organization (or an individual) who outsources her data to an external server; *user*, an individual who can access the data; *client*, the user's front-end in charge of translating access requests formulated by the user in equivalent requests operating on the outsourced data; and *server*, the external third party that stores and manages the data.

The goal of this chapter is to provide an overview of the main data security and privacy issues that characterize the two scenarios mentioned previously along with possible approaches for their solution (Sections 53.2 and 53.3). For each problem considered, we also briefly mention some open issues that still need further consideration and analysis. Clearly, since the data outsourcing and data publishing scenarios have some similarities, we also describe the main issues that are common to the two scenarios (Section 53.4). In the discussion, for simplicity, but without loss of generality, we assume that the outsourced data are stored in a single relation r, defined over relational schema $R(a_1,...,a_n)$, which includes all sensitive information that needs to be protected. The problems as well as the approaches for their solution that we will describe in the following can however be applied to any data model (e.g., XML data or arbitrary set of resources).

53.2 Security Issues in the Data Outsourcing Scenario

The security issues specifically characterizing the data outsourcing scenario are related to three main problems that will be discussed in the following: (1) data confidentiality, (2) efficient evaluation of users' queries at the server side, and (3) access control enforcement.

53.2.1 Data Confidentiality

Collected data often include sensitive information whose protection is mandatory, as also testified by recent regulations forcing organizations to provide privacy guarantees when storing, processing, or sharing their data with others (e.g., [9,59,60]). Since in the data outsourcing scenario these data are not under the direct control of their owners, individuals as well as companies require the protection of their sensitive information not only against external users breaking into the system, but also against malicious insiders, including the storing server. In many cases, the storing server is assumed to be *honest-but-curious*, that is, it is relied upon for ensuring availability of data but it is not allowed to read their content. Ensuring effective and practical data protection in this scenario is complex and requires the design of approaches allowing data owners to specify privacy requirements on data, as well as techniques for enforcing them.

Solutions: The first approach proposed to provide confidentiality of outsourced data consists in wrapping a protective layer of encryption around sensitive data to counteract both outside attacks and the curiosity of the server itself (e.g., [10,36,39,63]).

Encryption represents an effective approach to guarantee proper confidentiality protection. However, since the server is not authorized to decrypt outsourced data, it cannot evaluate users' queries. To minimize the use of encryption and make access to outsourced data more efficient, recent proposals combine *fragmentation* and *encryption* techniques [1,14,15]. These approaches are based on the observation that often data are not sensitive per se but what is sensitive is their association with other data. It is therefore sufficient to protect sensitive associations to preserve data confidentiality. The privacy requirements charactering the outsourced data collection are modeled through *confidentiality constraints* [1]. A confidentiality constraint c over a relational schema $R(a_1,...,a_n)$ is a subset of attributes in R, meaning that for each tuple in r, the (joint) visibility of the values of the attributes in c is considered sensitive and must be protected. As an example, Figure 53.1b illustrates a set of five confidentiality constraints for relation PATIENTS in Figure 53.1a, stating that the list of Social Security

PATIENTS							
SSN	Name	DoB	ZIP	Race	Disease	Doctor
123456789	Alice	1980/02/10	22010	Asian	Flu	I. Smith
234567891	Bob	1980/02/15	22018	Asian	Gastritis	J. Taylor
345678912	Carol	1980/04/15	22043	Asian	Flu	K. Doe
456789123	David	1980/06/18	22040	White	Hypertension	L. Green
567891234	Erik	1980/06/18	22043	White	Asthma	K. Doe
678912345	Frank	1980/10/07	22015	White	HIV	L. Green
789123456	Gary	1980/10/15	22010	White	HIV	L. Green
891234567	Hellen	1980/10/28	22018	Black	Flu	I. Smith

c_1={SSN}
c_2={Name, Disease}
c_3={Name, Doctor}
c_4={DoB, ZIP, Race}
c_5={Disease, Doctor}

(a)   (b)

FIGURE 53.1 An example of relation (a) and of a set of constraints over it (b).

Numbers is considered sensitive (c_1); the associations of patients' names with the diseases they suffer from or their caring doctor are considered sensitive (c_2 and c_3); the association among date of birth, ZIP, and race is considered sensitive (c_4); and the association between the disease of a patient and her caring doctor is considered sensitive (c_5).

Given a relation r and a set C of confidentiality constraints over it, the goal is to outsource the content of r in such a way to obfuscate sensitive associations. The idea is to encrypt the sensitive attributes by making them non-intelligible to non-authorized users, and to break sensitive associations by partitioning the attributes in R in different subsets (fragments) that cannot be joined by non-authorized users. A fragmentation correctly enforces the confidentiality constraints if no fragment stored at the external server represents all the attributes in a constraint in clear form.

The approaches proposed in the literature combining fragmentation and encryption to protect data confidentiality differ in how the original relational schema R is fragmented and in how and whether encryption is used. In particular, existing approaches can be classified as follows.

- *Noncommunicating pair of servers* [1]: R is partitioned into two fragments stored at two non-communicating servers. Encryption is used to protect attributes that cannot be stored at any of the two servers without violating constraints. For instance, consider relation PATIENTS in Figure 53.1a and the confidentiality constraints in Figure 53.1b. A correct fragmentation is represented by F_1 = {tid, Name, DoB, ZIP, SSN^k, Doctor^k}, F_2 = {tid, Race, Disease, SSN^k, Doctor^k}, where F_1 is stored at the first server and F_2 is stored at the second server. Attribute tid is the tuple identifier, introduced in both fragments to guarantee the lossless joint between F_1 and F_2. Attributes SSN^k and Doctor^k contain the encrypted version of attributes SSN and Doctor, respectively.

- *Multiple fragments* [15]: R is partitioned into an arbitrary number of disjoint fragments (i.e., with no common attribute), possibly stored at a same server. Each fragment includes a subset of the original attributes in the clear form and all the other attributes in encrypted form. For instance, consider relation PATIENTS in Figure 53.1a and the confidentiality constraints in Figure 53.1b. A correct fragmentation is represented by F_1 = {salt, Name, DoB, ZIP, enc}; F_2 = {salt, Race, Disease, enc}; and F_3 = {salt, Doctor, enc} in Table 53.1. Here, salt is a randomly chosen value different for each tuple in each fragment, and enc is the result of the encryption of the attributes in PATIENTS not appearing in the clear in the fragment, concatenated with the salt.

- *Departing from encryption* [14]: R is partitioned into two fragments, one stored at the data owner side and one stored at the server side, which can be joined by authorized users only. For instance, consider relation PATIENTS in Figure 53.1a and the confidentiality constraints in Figure 53.1b.

TABLE 53.1   An Example of Correct Fragmentation in the Multiple Fragments Scenario

	F_1				F_2				F_3	
salt	Name	DoB	ZIP	enc	salt	Race	Disease	enc	salt	Doctor	enc
s_{11}	Alice	1980/02/10	22010	bNh67!	s_{21}	Asian	Flu	wEqp8	s_{31}	I. Smith	5tihD]
s_{12}	Bob	1980/02/15	22018	tr354'	s_{22}	Asian	Gastritis	Ap9yt4	s_{32}	J. Taylor	rtF56.
s_{13}	Carol	1980/04/15	22043	7feW[0	s_{23}	Asian	Flu	vl:3rp	s_{33}	K. Doe	Se-D4C
s_{14}	David	1980/06/18	22040	(uhs3C	s_{24}	White	Hypertension	tgz08/	s_{34}	L. Green	eF6hjN
s_{15}	Erik	1980/06/18	22043	@13WcX	s_{25}	White	Asthma	erLK;-	s_{35}	K. Doe	3Ghv8V
s_{16}	Frank	1980/10/07	22015	2Xdc6?	s_{26}	White	HIV	?(iRo4	s_{36}	L. Green	ee%pl;
s_{17}	Gary	1980/10/15	22010	)2okED	s_{27}	White	HIV	+)ie5X	s_{37}	L. Green	Kjh4br
s_{18}	Hellen	1980/10/28	22018	=ieDc2	s_{28}	Black	Flu	Ghi3*;	s_{38}	I. Smith	+ihE67

A correct fragmentation is represented by $F_o = \{\underline{tid}, SSN, Name, Race, Disease\}$, stored at the data owner side; and $F_s = \{\underline{tid}, DoB, ZIP, Doctor\}$, stored at the storing server side. Note that F_o can include in the clear all the attributes composing a constraint since it is stored at a trusted party.

Open issues: Different open issues still remain to be addressed to effectively and efficiently provide confidentiality of outsourced data. For instance, the fragmentation process should take into account dependencies among attributes in the original relation. In fact, dependencies could be exploited by observers to reconstruct the association among attributes appearing in different fragments. As an example, the specialty of a patient's doctor may reveal the disease the patient suffers from (e.g., an oncologist takes care of people suffering from cancer). Furthermore, all the proposals in the literature protect only the confidentiality of static datasets. In real-world scenarios, however, outsourced data collections are subject to frequent changes that should be carefully managed to prevent information leakage. As an example, the insertion of a tuple into relation PATIENTS in Figure 53.1a translates into the insertion of a tuple into each of the fragments in Table 53.1. An observer can therefore easily reconstruct the sensitive associations among the attribute values for the new tuple.

53.2.2 Efficient Query Evaluation

In the data outsourcing scenario, the storing server does not have complete visibility of the outsourced data, as they are fragmented and/or encrypted to preserve confidentiality. Therefore, it cannot evaluate users' queries. Also neither the data owner should be involved in the query evaluation process (this would nullify the advantages of data outsourcing), nor the client should download the complete outsourced data collection to locally evaluate queries. It is therefore necessary to define techniques that permit to partially delegate to the external server the query evaluation process, while not opening the door to inferences.

Solutions: In the last few years, several techniques have been proposed to support the server-side evaluation of a wide set of selection conditions and SQL clauses when the outsourced relation is completely encrypted. These solutions complement the encrypted relation with additional metadata, called *indexes*, on which queries are evaluated at the server side. Relation r, defined over schema $R(a_1,\ldots,a_n)$, is then mapped to an encrypted relation r^k over schema R^k ($\underline{tid}$, enc, $I_{i_1},\ldots,I_{i_j}$), with tid a numerical attribute added to the encrypted relation and acting as a primary key for R^k; enc the encrypted tuple; $I_{i_l}, l = 1,\ldots,j$, the index associated with the i_l-th attribute a_{i_l} in R. For instance, Table 53.2 illustrates the encrypted version of relation PATIENTS in Figure 53.1a, assuming the presence of indexes for DoB (I_{DoB}), ZIP (I_{ZIP}), and Race (I_{Race}). For readability, index values are represented with Greek letters, and we report the tuples in the plaintext and encrypted relations in the same order. Note, however, that the order in which tuples are stored in the encrypted relation is

TABLE 53.2 An Example of
Encrypted Relation with Indexes

PATIENTS^k

id	enc	I_{DoB}	I_{ZIP}	I_{Race}
1	zKZlJxV	α	δ	η
2	AJvaAy1	α	ε	η
3	AwLBAa1	α	ζ	η
4	mHF/hd8	β	δ	θ
5	HTGhoAq	β	ζ	θ
6	u292mdo	γ	ζ	θ
7	ytOJ;8r	γ	δ	θ
8	eWo09uH	γ	ε	ι

independent from the order in which they appear in the plaintext relation. Different indexing techniques support different types of queries. In the following, we illustrate some techniques, partitioning them according to the queries that each of them can manage.

- *Equality conditions* (e.g., [19,39]): The first approach proposed to support equality conditions is represented by *encryption-based indexes* [19]. Given a tuple t, the value of the index for attribute a is computed as $E_k(t[a])$, where E_k is a symmetric encryption function and k the encryption key. As a consequence, any condition of the form $a = v$ is translated as $I_a = E_k(v)$. For instance, consider index I_{Race} in Table 53.2 obtained by adopting this method. Then, condition Race = 'Asian' on relation PATIENTS is translated as $I_{\text{Race}} = E_k$ (Asian), that is, $I_{\text{Race}} = $ 'η' on relation PATIENTS^k. An alternative solution is represented by *bucket-based indexes* [39]. The domain of attribute a is partitioned into nonoverlapping subsets of contiguous values, and each partition is associated with a label. Given a tuple t in the outsourced relation r, the value of the index associated with attribute a is the label of the unique partition containing value $t[a]$. As a consequence, equality condition $a = v$ is translated as $I_a = l$, where l is the label of the partition including v. For instance, index I_{DoB} in Table 53.2 has been obtained by partitioning the domain [1980/01/01, 1980/12/31] of attribute DoB in intervals of 4 months, and assigning, in the order, labels α, β, and γ to the three partitions. Condition DoB = 1980/04/15 on relation PATIENTS is translated as $I_{\text{DoB}} = $ 'β' on PATIENTS^k, since β is the label for the partition [1980/04/01, 1980/07/31]. A third technique efficiently supporting equality conditions is represented by *hash-based indexes* [19]. Given a tuple t in r, the value of the index associated with attribute a is computed as $h(t[a])$, where h is a deterministic hash function that generates collisions. Therefore, condition $a = v$ is translated as $I_a = h(v)$. For instance, consider index I_{ZIP} in Table 53.2 obtained by adopting this method. Then, condition ZIP = 22010 on relation PATIENTS is translated as $I_{\text{ZIP}} = h(22010)$, that is, $I_{\text{ZIP}} = $ 'δ' on relation PATIENTS^k. Both bucket-based and hash-based indexes map different plaintext values to the same index value. Therefore, the result computed by the server may include *spurious tuples* that the client will filter out by executing a query that evaluates the original condition on the tuples received from the server (after their decryption). For instance, only two of the three tuples in the encrypted relation with value 'δ' for attribute I_{ZIP} satisfy condition ZIP = 22010.
- *Range conditions* (e.g., [2,19,64]): To overcome the limitations of indexing techniques that support only equality conditions, in [19], the authors present a $B+$-tree index that allows the evaluation of both equality and range queries at the server side. The $B+$-tree index is built by the data owner over the original plaintext values of an attribute. It is then represented as a relational table, encrypted, and stored at the server. This relation is iteratively queried by the client for retrieving the tuples satisfying the query condition. For instance, Figure 53.2 illustrates the $B+$-tree index built for attribute Name of relation PATIENTS in Figure 53.1a, the relation representing it, and

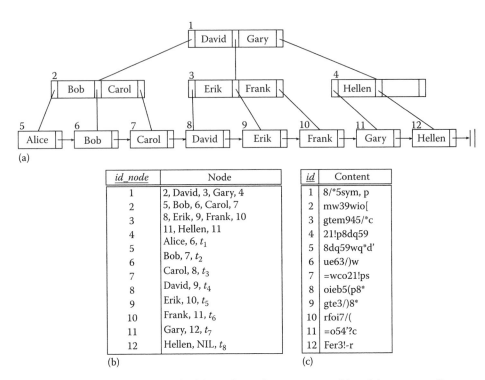

(a)

id_node	Node
1	2, David, 3, Gary, 4
2	5, Bob, 6, Carol, 7
3	8, Erik, 9, Frank, 10
4	11, Hellen, 11
5	Alice, 6, t_1
6	Bob, 7, t_2
7	Carol, 8, t_3
8	David, 9, t_4
9	Erik, 10, t_5
10	Frank, 11, t_6
11	Gary, 12, t_7
12	Hellen, NIL, t_8

(b)

id	Content
1	8/*5sym, p
2	mw39wio[
3	gtem945/*c
4	21!p8dq59
5	8dq59wq*d'
6	ue63/)w
7	=wco21!ps
8	oieb5(p8*
9	gte3/)8*
10	rfoi7/(
11	=o54'?c
12	Fer3!-r

(c)

FIGURE 53.2   An example of *B*+-tree index (a), its relational representation (b), and the corresponding encrypted relation (c).

the encrypted relation stored at the external server. Condition Name LIKE '[E–Z]%' on relation PATIENTS (retrieving the names following 'E' in lexicographic order) is evaluated by executing a set of queries for traversing the *B*+-tree along the path of nodes 1, 3, and 9, and to follow the chain of leaves starting at node 9. For each visited leaf, the client retrieves the tuple associated with it (i.e., tuple t_5 for leaf 9, tuple t_6 for leaf 10, tuple t_7 for leaf 11, and tuple t_8 for leaf 12). To avoid storing additional relations on the server, *order-preserving encryption indexes* have recently been proposed [2,64]. These techniques are based on order-preserving encryption schemas, which take as input a target distribution of index values and apply an order-preserving transformation in a way that the transformed values (i.e., the index values) follow the target distribution.

- *Aggregate operators* (e.g., [33,38]): Privacy homomorphic encryption [61] allows the execution of basic arithmetic operations (i.e., +, −, ×) directly over encrypted data. Their adoption in the definition of indexes allows the server to evaluate aggregate functions and to execute equality and range queries. The main drawback of these approaches is their computational complexity, which makes them not suitable for many real-world applications.

The main challenge that must be addressed in the definition of indexing techniques is balancing precision and privacy: more precise indexes provide more efficient query execution, at the price of a greater exposure to possible privacy violations. As an example, encryption-based indexes permit to completely delegate the evaluation of equality conditions to the server. However, index values have exactly the same frequency distribution as plaintext values, thus opening the door to frequency-based attacks. For instance, it is easy to see that value ι for index I_{Race} represents value *black* for attribute Race, since *black* (ι, respectively) is the only value with one occurrence for attribute Race (index I_{Race}, respectively). Analogously, a higher number of indexes on the same relation improve query evaluation efficiency, while causing a higher risk of inference and linking attacks. As shown in [10], even a limited number of indexes can greatly facilitate the task for an adversary who wants to violate the confidentiality provided by encryption.

The evaluation of queries over outsourced data requires the definition of proper techniques defining how queries on the original table are translated into queries on the encrypted data and indexes over them, or on fragmented data. For instance, consider relation PATIENTS in Figure 53.1a and its fragmentation in Table 53.1. Query "SELECT Name FROM PATIENTS WHERE Disease = 'flu'" is translated into the following query operating on fragment F_2, where attribute Disease is represented in clear from: "SELECT enc FROM F_2 WHERE Disease = 'flu.'" When the client receives the query result, it decrypts attribute enc and applies a projection on patient's names before showing the result to the requesting user. Query plans are designed to minimize the client's overhead in query evaluation. In fact, most query evaluations will require the client's intervention since the server cannot decrypt encrypted attributes. The techniques designed to define efficient query plans depend on the fragmentation approach adopted (e.g., [13,15,31]).

Open issues: Besides the definition of alternative indexing techniques for encrypted data and for efficiently evaluating queries on fragmented data, it still remains to study the possibility of combining fragmentation and indexing approaches. In fact, fragmentation does not permit to delegate the evaluation of conditions involving attributes that do not appear plaintext in a fragment. The association of indexes to fragments could nicely fill this gap, but should be carefully designed to prevent information leakage caused by the plaintext representation in a fragment of an attribute indexed in another.

53.2.3 Access Control Enforcement

In most real-world systems, access to the data is selective; that is, different users enjoy different views over the data. When the data are managed by an external third party, the problem of how to enforce the access control policy defined by the owner becomes crucial. In fact, the data owner cannot enforce the access control policy since this would imply that the data owner has to mediate every access request, thus losing advantages of outsourcing data. Also access control enforcement cannot be delegated to the external server as in traditional systems (e.g., [6]) for privacy reasons. Indeed, the access control policy might be sensitive, and the server might collude with malicious users to gain access to sensitive data. Therefore, it is necessary to define techniques that permit the data themselves to enforce access control policies.

Solutions: The solutions proposed to enforce access control restrictions on outsourced data without the data owner's intervention are based on integrating access control and encryption.

The enforcement of read privileges has been addressed first and has been efficiently solved by mapping selective visibility restrictions in the encryption of the data [23]. The authorization policy is translated into an equivalent *encryption policy*, regulating which data are encrypted with which key and which keys are released to which users. This translation process is performed considering the following two desiderata: (1) at most one key is released to each user, and (2) each tuple is encrypted at most once. To this purpose, the approach in [23] adopts a key derivation method, which permits to compute an encryption key k_i starting from the knowledge of another key k_j and a piece of publicly available information [4]. Key derivation methods are based on the definition of a key derivation hierarchy that specifies which keys can be derived from other keys in the system. A key derivation hierarchy that correctly enforces the access control policy must permit each user to derive from her key all and only the keys used to encrypt the tuples she can access. To this aim, the hierarchy has a node (which represents a key) for each user in the system and a node for each access control list (acl), that is, for each set of users who can access a tuple. The edges in the hierarchy (which correspond to key derivation operations) guarantee the existence of a path connecting each node representing a user to each node representing a set to which the user belongs [23]. Each user knows the key of the node representing herself in the hierarchy and each resource is encrypted with the key representing its access control list. For instance, consider a system with four users {L, M, N, O}

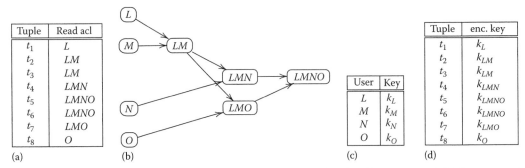

Tuple	Read acl
t_1	L
t_2	LM
t_3	LM
t_4	LMN
t_5	$LMNO$
t_6	$LMNO$
t_7	LMO
t_8	O

(a)

User	Key
L	k_L
M	k_M
N	k_N
O	k_O

(c)

Tuple	enc. key
t_1	k_L
t_2	k_{LM}
t_3	k_{LM}
t_4	k_{LMN}
t_5	k_{LMNO}
t_6	k_{LMNO}
t_7	k_{LMO}
t_8	k_O

(b)   (d)

FIGURE 53.3 An example of access control policy (a), the corresponding key derivation graph (b), keys assigned to users (c), and keys used to encrypt resources (d).

and the tuples composing relation PATIENTS in Figure 53.1a. Figure 53.3b reports the key derivation graph enforcing the access control policy in Figure 53.3a, and Figures 53.3c and d summarize the keys communicated to the users and the keys used to encrypt the tuples. For simplicity and readability, in the key derivation graph, nodes are labeled with the set of users they represent. It is easy to see that each user knows or can derive all and only the keys of the nodes including herself. As a consequence, she can decrypt all and only the tuples in relation PATIENTS that she is authorized to access (i.e., such that the user belongs to the acl of the tuple).

The solution in [25] complements the technique in [23] by associating a write token with each tuple. The write token is encrypted with a key that only users who can modify the tuple and the external server can derive. The server accepts a write operation only if the requesting user proves to be able to correctly decrypt the write token associated with the modified tuple. For instance, Figure 53.4a illustrates the access control policy in Figure 53.3a for relation PATIENTS, extended with write privileges on the tuples. Figure 53.4b illustrates the key derivation graph, which extends the graph in Figure 53.3b introducing the external server S. Figures 53.4c and d summarize the keys communicated to the users and the keys used to encrypt the tuples and the corresponding write tokens.

The enforcement of access control restrictions through encryption raises many other issues that have been recently addressed, as summarized in the following.

- Since the key used to encrypt each tuple depends on the set of users who can access it, updates to the access control policy require data re-encryption, which represents an expensive operation for the data owner. For instance, consider the example in Figure 53.3 and assume that user O is revoked access to tuple t_5. To enforce such a revocation, the data owner should download tuple t_5, decrypt it using key k_{LMNO}, re-encrypt the tuple with key k_{LMN}, and send it back to the server. To prevent re-encryption, while enforcing updates to the access control policy, the proposal

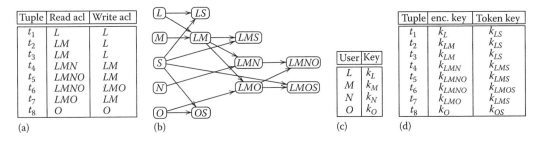

Tuple	Read acl	Write acl
t_1	L	L
t_2	LM	L
t_3	LM	L
t_4	LMN	LM
t_5	$LMNO$	LM
t_6	$LMNO$	LMO
t_7	LMO	LM
t_8	O	O

(a)

User	Key
L	k_L
M	k_M
N	k_N
O	k_O

(c)

Tuple	enc. key	Token key
t_1	k_L	k_{LS}
t_2	k_{LM}	k_{LS}
t_3	k_{LM}	k_{LS}
t_4	k_{LMN}	k_{LMS}
t_5	k_{LMNO}	k_{LMS}
t_6	k_{LMNO}	k_{LMOS}
t_7	k_{LMO}	k_{LMS}
t_8	k_O	k_{OS}

(b)   (d)

FIGURE 53.4 An example of read-and-write access control policy (a), the corresponding key derivation graph (b), keys assigned to users (c), and keys used to encrypt resources (d).

in [23] adopts two layers of encryption: one managed by the data owner and enforcing the initial policy (BEL); and the other managed by the external server and enforcing policy updates (SEL). A user can then access a tuple only if she knows (or can derive) the key used for encrypting the resources at both levels.

- The adoption of key derivation for access control enforcement may reveal to the server the authorization policy. Whenever the policy is considered sensitive, the structure of the key derivation hierarchy should not be revealed to the external server, as proposed in [21].

- Many real-world scenarios are characterized by parties acting both as data producers (i.e., data owners) and as data consumers (i.e., final users). The enforcement of access control should then take into consideration the fact that there are multiple data owners, each regulating access to her data [22].

- The combined use of selective encryption (for access control enforcement) and indexing techniques (for efficient query evaluation) may open the door to inference on attribute values of tuples that users are not authorized to read [24]. In fact, users have also visibility on indexes of tuples they are not allowed to access. Since index values depend on the attribute value they represent, such a visibility could permit users to infer attribute values for tuples they cannot access. For instance, consider plaintext relation PATIENTS in Figure 53.1a, its encrypted version in Table 53.2, and the access control policy in Figure 53.3a. User M, who is authorized to read tuple t_2, can easily infer that value η for index I_{Race} represents plaintext value 'Asian'. As a consequence, M can conclude that t_1 [Race] = 'Asian', thus breaching the confidentiality of tuple t_1 that she is not authorized to access. To limit this risk, the indexing function should be designed to take access control restrictions into consideration [24].

An alternative solution to the combined use of selective encryption and key derivation for access control enforcement has been introduced in [71]. This approach aims at providing systems scalability by adopting *attribute-based encryption*. In this scenario, the traditional access control policy is substituted by the definition of a set of attributes associated with users and tuples regulating which user can access which tuple. More precisely, each tuple is associated with a set of attributes that describe the context in which the tuple should be accessed, and each user is associated with a logical expression over the attributes that represents the properties of the tuples she is authorized to access. To enforce the access control policy, attribute-based encryption techniques define a different public key component for each attribute in the system. Each tuple is then encrypted using a key that reflects all the public key components of the attributes associated with it, while each user knows the secret key that permits to decrypt all the tuples she is authorized to access.

Open issues: Most of the problems studied for the enforcement of read privileges need to be analyzed also for the solutions enforcing write privileges. For instance, it is necessary to define an efficient approach for managing policy updates without the intervention of the data owner and a technique for protecting the confidentiality of the policy when read-and-write operations are restricted to arbitrary subsets of users.

53.3 Security Issues in the Data Publishing Scenario

In the data publishing scenario, information about entities, called *respondents*, are publicly or semi-publicly released. While in the past, released information was mostly in tabular and statistical form (*macrodata*), many situations require today the release of specific data (*microdata*). Microdata provide the convenience of allowing the final recipient to perform on them analysis as needed, at the price of a higher risk of exposure of respondents' private information. In this section, we describe how public data release may cause the unintended disclosure of respondents' identities and/or sensitive attributes, and the solutions proposed to counteract this risk.

53.3.1 Preserving Respondents' Privacy

To protect respondents' identities, data owners remove or encrypt explicit identifiers (e.g., SSN, name, phone numbers) before publishing their datasets. However, data *de-identification* provides no guarantee of anonymity since released information often contains other data (e.g., race, birth date, sex, and ZIP code) that can be linked to publicly available information to reidentify (or restrict the uncertainty about the identity of) data respondents, thus leaking information that was not intended for disclosure. The large amount of information easily accessible today, together with the increased computational power available to data recipients, makes such linking attacks easier [35]. Furthermore, the disclosure of an individual's identity (*identity disclosure*) often implies also the leakage of her sensitive information (*attribute disclosure*). For instance, in 2006, Netflix, an online DVD delivery service, started a competition whose goal was the improvement of its movie recommendation system based on users' previous ratings. To this purpose, Netflix released 100 million records about movie ratings of 500,000 of its subscribers. The released records were anonymized removing the personal identifying information of the subscribers, which was substituted with an anonymous customer id. However, by linking the movie recommendations available on the Internet Movie Database with the anonymized Netflix dataset, it was possible to reidentify individuals, thus revealing potentially sensitive information (e.g., apparent political preferences) [55].

Protecting respondents' identities and sensitive attributes in the today's global interconnected society is a complex task that requires the design of effective techniques that permit data owners to provide privacy guarantees, even without knowing the external sources of information that a possible observer could exploit to reidentify data respondents. Note that the public release of a dataset should also provide utility for data recipients (i.e., the released dataset should include as much information as possible to permit final recipients to obtain representative results from their analysis). Clearly, data utility and privacy are two conflicting requirements that need to be balanced in data release.

Solutions: The solutions proposed to protect respondents' privacy, while providing data recipients with useful data, can be classified in the following two categories:

1. *Approaches based on k-anonymity* (e.g., [46,49,62]) guarantee that the dataset publicly released satisfies properties that provide protection against identity and attribute disclosure (e.g., every combination of values of quasi-identifiers can be indistinctly matched to at least k respondents).
2. *Approaches based on differential privacy* (e.g., [30,41]) guarantee that the released dataset is protected against certain kinds of inference defined before data release (e.g., an observer cannot infer with non-negligible probability whether a subject is represented in the released dataset).

The first approach proposed in the literature to protect respondents' privacy in microdata release is represented by *k-anonymity* [62]. *k*-Anonymity captures the well-known requirement, traditionally applied by statistical agencies, stating that any released data should be indistinguishably related to no less than a certain number of respondents. Since linking attacks are assumed to exploit only released attributes that are also externally available (called *quasi-identifiers*), in [62], this general requirement has been translated into the following *k*-anonymity requirement: *Each release of data must be such that every combination of values of quasi-identifiers can be indistinctly matched to at least k respondents.* A microdata table then satisfies the *k*-anonymity requirement if each tuple in the table cannot be related to less than k respondents in the population and vice versa (i.e., each respondent in the population cannot be related to less than k tuples in the released table). The *k*-anonymity requirement can be checked only if the data owner knows any possible external source of information that may be exploited by a malicious recipient for respondents' reidentification. This assumption is, however, limiting and highly impractical in most scenarios. *k*-Anonymity then takes a safe approach by requiring that each combination of values of the quasi-identifier attributes appears with at least k occurrences in the released table, which is a sufficient (although not necessary) condition to satisfy the *k*-anonymity requirement. Traditional approaches for guaranteeing *k*-anonymity transform the values

TABLE 53.3 An Example of 2-Anonymous Table

				PATIENTS		
<u>SSN</u>	Name	DoB	ZIP	Race	Disease	Doctor
		1980/02	2201*	Asian	Flu	I. Smith
		1980/02	2201*	Asian	Gastritis	J. Taylor
		1980/06	2204*	White	Hypertension	L. Green
		1980/06	2204*	White	Asthma	K. Doe
		1980/10	2201*	White	HIV	L. Green
		1980/10	2201*	White	HIV	L. Green

of the attributes composing the quasi-identifier, while leaving sensitive attribute values unchanged. To guarantee truthfulness of released data, k-anonymity relies on *generalization* and *suppression* microdata protection techniques [17]. Generalization consists in substituting the original values with more general values (e.g., the date of birth can be generalized by removing the day, or the day and the month, of birth). Suppression consists in removing data from the microdata table. The combined use of these techniques guarantees the release of a less precise and less complete, but truthful, data while providing protection of respondents' identities. As an example, Table 53.3 has been obtained from the table in Figure 53.1a by (i) removing explicit identifiers (i.e., SSN and Name); (ii) generalizing attribute DoB removing the day of birth; (iii) generalizing attribute ZIP removing the last digit; and (iv) suppressing the third and eight tuples in the original table. The resulting table is 2-anonymous since each combination of values for attributes DoB, Race, and ZIP appears (at least) twice in the relation. To limit the information loss caused by generalization and suppression (i.e., to improve utility of released data), many k-anonymity algorithms have been proposed (e.g., [5,16,42,43,62]). All these approaches are aimed at minimizing the loss of information caused by generalization and suppression, while providing the privacy guarantees required by respondents. Although k-anonymity represents an effective solution for protecting respondents' identities, it has not been designed to protect the released microdata table against attribute disclosure. Given a k-anonymous table, it may then be possible to infer (or reduce the uncertainty about) the value of the sensitive attribute associated with a specific respondent. This happens, for example, when all the tuples with the same value for the quasi-identifier are associated with the same value for the sensitive attribute. For instance, consider Table 53.3, which is 2-anonymous, and assume that *Susan* knows that her friend *Frank* is a male, born in October 1980, and living in 22015 area. *Susan* can easily infer that *Frank*'s tuple is either the fifth or the sixth tuple in the published table. As a consequence, *Susan* can infer that her friend suffers from HIV. ℓ-Diversity [49] and t-closeness [46] have been proposed to protect released microdata tables against attribute disclosure.

Approaches based on the definition of differential privacy [30] have been recently proposed to guarantee that the released microdata table protects respondents against inference that exploits both published data and external adversarial knowledge. One of the first definitions of privacy in the data publishing scenario states that *Anything that can be learned about a respondent from the statistical database should be learnable without access to the database* [18]. Although this definition has been thought for statistical databases, it is also well suited for the microdata publishing scenario. However, the main problem of this definition of privacy is that only an empty dataset can guarantee absolute disclosure prevention [30]. Differential privacy can be considered as the first attempt of achieving privacy as defined in [18]. It is based on the observation that the release of a dataset may violate the privacy of any individual, independently of whether she is represented in the dataset. For instance, suppose that the released dataset permits to compute the average annual benefits of people living in 22010 area for each ethnic group and suppose that this information is not publicly available. Assume also that *Alice* knows that *Bob*'s annual benefits is 1,000$ more than the average annual benefits of Asian people living in 22010 area. Although this piece of information alone does not permit *Alice* to gain any information about *Bob*'s annual benefits, when it is combined with the released dataset, it allows Alice to infer *Bob*'s annual benefits. Differential privacy aims at preventing an observer from inferring whether a subject is represented or not in the released dataset. A data release is then considered safe if the inclusion in the dataset of tuple t_p, related to respondent p, does not change the probability that a malicious recipient can correctly identify the sensitive attribute value associated with p. Intuitively, differential privacy holds if the removal (insertion, respectively) of one tuple t_p from (into, respectively) the table does not substantially affect the result of the evaluation of a function K on the released table. Most of the techniques proposed to satisfy differential privacy are based on *noise addition*, which does not preserve the truthfulness of released data (e.g., [40,44,51,67]) and may therefore not be suited to different data publishing scenarios. Differential privacy provides protection against inferences on the presence/absence in the published dataset of the record representing a respondent. However, this is not the only cause of privacy breaches. To provide a wider privacy guarantee, a recent line of work has put forward a definition of privacy that permits to protect released data against any kind of inference, provided it is defined (and properly modeled) by the data owner before data release [41].

Open issues: Although many efforts have been made to overcome the assumptions on which the definitions of k-anonymity and differential privacy are based, there are still open issues that deserve further investigation. This is also testified by the fact that there is not a unique definition of respondents' privacy and each definition has some drawbacks that need to be addressed (e.g., external adversarial knowledge is not adequately modeled and addressed by privacy protection techniques). Also all the proposed solutions for protecting respondents' privacy still need to be enhanced, to possibly find a good trade-off between privacy of respondents and utility of released data for final recipients.

53.4 Security Issues in Emerging Scenarios

In this section, we focus on the security and privacy problems that are common to both the data outsourcing scenario and the data publishing scenario. In particular, we consider the (i) confidentiality of users' queries; (ii) integrity of outsourced/published data; and (iii) completeness and correctness of query results.

53.4.1 Private Access

An important issue that arises when data are stored at an external server is preserving the confidentiality of users' queries, independently of whether data are kept confidential or are publicly released. As an example, assume that a user accesses an external medical database looking for the symptoms of a given disease. The server executing the query (as well as any observer) can easily infer that the user (or a person close to her) suffers from the disease in the query condition. Furthermore, queries could be exploited by the server (or by an external observer) to possibly infer sensitive information in the outsourced data collection, thus reducing the effectiveness of the techniques possibly adopted by the data owner to protect data confidentiality. Besides protecting the confidentiality of each query singularly taken, it is also important to protect *patterns* of accesses to the data (i.e., the fact that two or more queries aimed at the same target tuple). Indeed, the frequency of accesses to tuples could be exploited to breach both data and query confidentiality.

Solutions: The solutions proposed to protect the confidentiality of the queries issued by users can be classified depending whether they operate on plaintext data (data publishing) or on encrypted data (data outsourcing).

The line of work first developed to protect query confidentiality is represented by classical studies on *Private Information Retrieval* (PIR) [12], which operates on plaintext datasets stored at an external server. In this model, the database is represented as an N-bit string and a user is interested in retrieving the ith bit of the string without allowing the server to know/infer which is the bit target of her query.

PIR proposals can be classified as follows: *information-theoretic* PIR, which protects query confidentiality against attackers with unlimited computing power; and *computational* PIR, which protects query confidentiality against adversaries with polynomial-time computing power. Unfortunately, information-theoretic PIR protocols have computation and communication costs linear in the size of the dataset (i.e., $\Omega(N)$) as there is no solution to the problem that is better than downloading the whole database [12]. To limit these costs, it is however possible to replicate the dataset on different servers. Intuitively, PIR approaches exploiting data replication require the user to pose a randomized query to each server storing a copy of the dataset, in such a way that neither the servers nor an outside observer can determine which is the bit to which the user is interested. The user will then reconstruct the target bit by properly combining the query results computed by the different servers [3,12]. The main problem of these proposals is that they rely on the unrealistic assumption that the external servers do not communicate with each other. Computational PIR protocols exploit cryptographic properties to reduce both the communication cost and the number of copies of the data with respect to information-theoretic techniques [8,11]. Users then adopt specific functions to encrypt their requests before submitting them to the server. These functions enjoy particular properties that allow the server to compute, without decrypting the request, an encrypted query result that only the requesting user can decrypt. Although more efficient than information-theoretic solutions, also computational PIR schemes suffer from heavy computation costs (i.e., $O(N)$ or $O(\sqrt{N})$) both for the client and for the server, which limit their applicability in real-life scenarios [32]. Recently, traditional PIR protocols have been integrated with relational databases, to the aim of protecting sensitive constant values in SQL query conditions, while providing the client with efficient query evaluation [56]. The original query formulated by the client is properly sanitized before execution, to prevent the server from inferring sensitive data. To extract the tuples of interest from the result of the sanitized query executed by the server, the client then resorts to traditional PIR protocols, operating on the sanitized query result instead of on the whole dataset (thus highly reducing computational costs).

One of the solutions operating on encrypted data, aimed at providing both data and query confidentiality, is based on the definition of a *shuffle index* on the data collection [26,27]. A shuffle index is defined, at the abstract level, as an *unchained B+-tree* (i.e., there are no links connecting the leaves), built on one of the attributes in the outsourced relation and with actual data stored in the leaves of the tree. Figure 53.5a illustrates a graphical representation of the abstract data structure built on attribute Name of relation PATIENTS in Figure 53.1a. At the logical level, the nodes of the tree are allocated to logical addresses that work as logical *identifiers*, which may not follow the same order of the values in the abstract nodes they represent. Figure 53.5b illustrates a possible representation at the logical level of the abstract data structure in Figure 53.5a. In the figure, nodes appear ordered (left to right) according to their identifiers, which are reported on the top of each node. Pointers between nodes of the abstract data structure correspond, at the logical level, to node identifiers, which can then be easily translated at the physical level into physical addresses at the external server. For simplicity and easy reference, in our example, the first digit of the node identifier denotes the level of the node in the tree. Before sending to the server the shuffle index for storing it, the content of each node is encrypted, producing an encrypted block. Since each block is encrypted, the server does not have any information on the content of the node stored in the block or on the parent–child relationship between nodes stored in blocks. Retrieval of the leaf block containing the tuple corresponding to an index value requires an iterative process. Starting from the root of the tree and ending at a leaf, the read block is decrypted retrieving the address of the child block to be read at the next step. To protect access and pattern confidentiality, searches over the B+-tree are extended with the following three protection techniques [26].

- *Cover searches*: fake searches executed together with the target search. The values used as cover searches are chosen to be indistinguishable from actual target values and operate on disjoint paths of the tree. The search process retrieves, for each level in the tree, the same number of blocks (one for the target and the other for the cover searches). Cover searches introduce uncertainty over the leaf block storing the target tuple and do not allow the server to establish the parent–child relationship between blocks retrieved at contiguous levels.

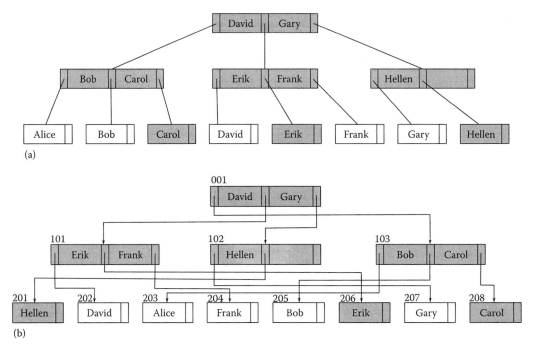

FIGURE 53.5   An example of abstract (a) and corresponding logical shuffle index (b).

- *Cache*: set of blocks/nodes along the path to the target value recently accessed, which are stored at the client side. Cache makes searches repeated within a short time interval not recognizable as being the same search: if the nodes in the target path are already in cache, an additional cover search is executed.
- *Shuffling*: operation performed for changing the block where accessed nodes are stored, thus breaking the correspondence between nodes (contents) and blocks (addresses). Basically, the contents of blocks retrieved at each level and of the blocks at the same level stored in the local cache are mixed. Nodes are then re-encrypted, and the resulting blocks rewritten accordingly on the server.

As an example, consider a search for the name *Carol* over the abstract index in Figure 53.5a that adopts *Hellen* as cover and assume that the local cache contains the path to *Erik* (i.e., (001,101,204)). The nodes involved in the search operation are denoted in gray in the figure. Figure 53.6 illustrates the abstract and logical representation of the shuffle index in Figure 53.5 after the execution of the search operation, which shuffles nodes 101, 102, and 103, and nodes 201, 206, and 208.

Besides the shuffle index technique, other proposals have also been introduced to address data and query confidentiality in data outsourcing scenarios. These approaches are based on the pyramid-shaped database layout of Oblivious RAM [34] and propose an enhanced reordering technique between adjacent levels of the structure [20,29,47,66]. The privacy provided by these approaches is guaranteed however by the presence of a trusted coprocessor at the external server.

Open issues: Among the open issues that still need to be analyzed, there are the needs of providing efficient accesses based on the value of different attributes in the outsourced/published relation (*multiple indexes*) and of supporting possible updates to the data collection.

53.4.2 Data Integrity

Traditionally, the server is assumed to be trusted to correctly store and maintain the data of the owners. There are, however, scenarios where such a trust assumption does not hold. As a consequence, the server

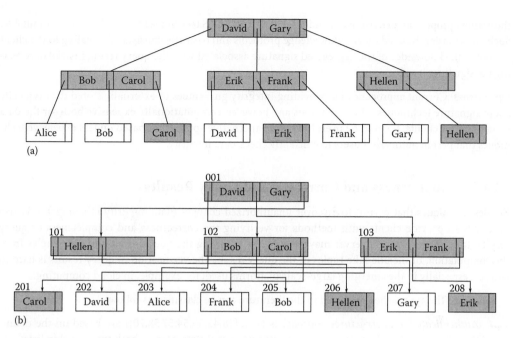

FIGURE 53.6 Abstract (a) and corresponding logical shuffle index (b) in Figure 53.5 after the execution of a search operation.

itself or a user may modify the data collection without being authorized. Since the server directly stores and manages the dataset, it is not possible to physically prevent unauthorized data modifications; there are however techniques that permit the data owner (and any authorized user) to discover non-authorized modifications.

Solutions: Data integrity can be provided at different granularity levels: table, attribute, tuple, or cell level. The integrity verification at the table and attribute level is expensive since it can be performed by the client only downloading the whole table/column. Data integrity at the cell level suffers from a high verification overhead. For these reasons, the majority of the current proposals provide data integrity at the tuple level.

Integrity verification approaches traditionally rely on *digital signatures* (e.g., [37]), that is, the data owner first signs, with her private key, each tuple in the outsourced/published relation, and the signature is concatenated to the tuple. When a client receives a set of tuples from the external server, she can check the signature associated with each tuple to detect possible unauthorized changes. The main drawback of traditional integrity verification techniques relying on digital signature is that the verification cost at the client side grows linearly with the number of tuples in the query result.

To reduce the verification costs of large query results, in [53], the authors propose the adoption of a schema that permits to combine a set of digital signatures. In this way, the client can verify the integrity of the query result by checking one aggregated signature only: the one obtained combining the signatures of the tuples in the query result. In [53], the authors propose three different signature schemes: *condensed RSA*, based on a variation of the RSA encryption schema that allows the aggregation of signatures generated by the same signer; *BGLS* [7], based on bilinear mappings and supporting the aggregation of signatures generated by different signers; batch *DSA signature aggregation*, based on the multiplicative homomorphic property of DSA signature schema. Both condensed RSA and BGLS approaches are *mutable*, meaning that any user who knows multiple aggregated signatures can compose them, obtaining a valid aggregated signature. Although this feature can be of interest in the process of generating aggregated signatures, it also represents a weakness for the integrity of the data. In [52], the authors propose an extension of condensed RSA and BGLS techniques that makes them immutable. Such an extension is based on zero knowledge protocols and basically consists in revealing to the client a proof of the knowledge of the aggregated signature associated with the query result, instead of revealing the signature itself.

Open issues: Current approaches for providing integrity guarantees to externally stored data typically adopt signature techniques. These solutions are, however, computationally expensive both for the data owner and for clients. It could be useful to define alternative approaches based on less expensive techniques that permit authorized users to efficiently check data integrity.

53.4.3 Completeness and Correctness of Query Results

Besides techniques that permit to discover unauthorized changes (data integrity in storage), it is also necessary to provide clients with methods for verifying the correctness and completeness of query results. Indeed, the external server may be lazy in computing the query result and omit tuples from the computation, or include fake tuples in the query result. The verification of query results is hard to enforce, especially in the emerging large-scale platforms used, for example, in cloud computing.

Solutions: The approaches proposed in the literature can be classified as follows.

- *Authenticated data structures* approaches (e.g., [28,45,50,54,57,58,70]) are based on the definition of appropriate data structures (e.g., signature chaining, Merkle hash trees, or skip lists) and provide guarantee of completeness of the result of queries operating on the attribute (set thereof) on which the data structure has been defined. These approaches also guarantee integrity of stored data since unauthorized changes to the data can be detected by the verification process of query results.
- *Probabilistic* approaches (e.g., [48,65,68]) are based on the insertion of sentinels in the outsourced data, which must also belong to the query result. These solutions provide a probabilistic guarantee of completeness of query results.

Most of the authenticated data structure approaches adopt *Merkle hash trees* [50]. A Merkle hash tree is a binary tree, where each leaf contains the hash of one tuple of the outsourced relation, and each internal node contains the result of the application of the same hash function on the concatenation of the children of the node itself. The root of the Merkle hash tree is signed by the data owner and communicated to authorized users. The tuples in the leaves of the tree are ordered according to the value of a given attribute a. Figure 53.7 illustrates an example of a Merkle hash tree built over relation PATIENTS in Figure 53.1a

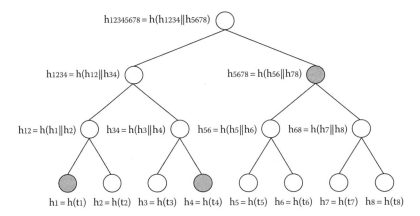

FIGURE 53.7   An example of Merkle hash tree.

for attribute SSN. Whenever the external server evaluates a query with a condition defined on a, it returns to the requesting client the result of the query along with a *verification object* (VO). The VO includes all the information that the client needs to know to verify the completeness of the query result (i.e., to recompute the value of the root node) [28]. If the value of the root node computed by the user is the same as the one received from the owner, the query result is correct and complete. For instance, consider the Merkle hash tree in Figure 53.7: the VO for a query that returns the patients whose SSN starts with either 2 or 3 (i.e., tuples t_2 and t_3) is represented by the gray nodes in the figure. Solutions based on authenticated data structures have the advantage of providing 100% guarantee of completeness of query results. The main problem is that these data structures can be used only for queries with conditions on the specific attribute on which they are built. This implies that the completeness of queries operating on different attributes can be checked only at the price of defining an additional authenticated data structure for each attribute.

Probabilistic approaches are based on the insertion of *fake tuples* or on the *replication* of a subset of the tuples in the outsourced/published relation. The approach based on the insertion of fake tuples [68] checks the completeness of query results by verifying whether all the fake tuples that satisfy the conditions specified in the query belong to the result computed by the server. If at least a fake tuple is missing, the query result is not complete. Clearly, fake tuples must be indistinguishable at the server's eyes from real tuples. As proved in [68], even a limited number of fake tuples ensure a high probabilistic guarantee of completeness. The solution based on the replication of a subset of the outsourced tuples [65] provides guarantee of completeness of the query result by controlling whether tuples in the query result that satisfy the condition for duplication appear twice in the query result. If a tuple that has been duplicated appears only once in the query result, the server omitted at least a tuple. Clearly, the server should not be able to determine pairs of encrypted tuples that represent the same plaintext tuple. The guarantee of completeness provided by probabilistic approaches increases with the number of additional (fake or duplicated) tuples inserted in the dataset.

Recent proposals address the problem of guaranteeing also the *freshness* of query results, meaning that queries are evaluated on the last version of the outsourced relation [45]. To integrate freshness control with solutions based on authenticated data structures, a time stamp is included in the data structure and is periodically updated [69]. If the client knows how frequently the time stamp is updated, it can check whether the VO (and therefore the query result) returned by the server is up-to-date. The solution proposed for probabilistic approaches [68] periodically changes, in a deterministic way, the function that computes fake tuples in the dataset. If the client knows which are the current fake tuples in the outsourced data, it can verify whether the query result includes all and only those valid additional tuples that should be present when the query has been executed.

Open issues: Although the problem of providing guarantees of completeness and freshness of query results is becoming of great interest, there are still many aspects that need to be further investigated. Current solutions consider simple SELECT-FROM-WHERE SQL queries operating on one relation only. However, in many real-world scenarios, it is necessary to assess the correctness and completeness of the result of more complex queries (e.g., queries including GROUP BY and HAVING clauses). Probabilistic approaches provide a good trade-off between completeness guarantee and efficiency in query evaluation. It would be interesting to develop efficient approaches that provide absolute certainty of completeness of query results, while limiting the computational overhead.

53.5 Summary

In this chapter, we illustrated the main security and privacy issues arising in the emerging data outsourcing and data publishing scenarios, where a sensitive data collection is stored and managed by an external third party. For each problem analyzed, the chapter provided an overview of the most important solutions proposed in the literature and described open issues that still need further investigation.

Glossary

Access control list (acl): Set of users who are authorized to access a given resource.

Access control policy: Set of (high-level) rules according to which access control must be regulated.

Confidentiality constraint: Subset of attributes in a relation schema R that should not be jointly visible to unauthorized users.

Data disclosure: Unintended release of (possibly sensitive) information that should not be revealed.

Digital signature: Technique used to prove that a piece of information was created by a known user and that it has not been altered.

Encryption: Process that transforms a piece of information (called plaintext) through an encryption algorithm to make it unintelligible.

Fragmentation: Let R be a relation schema, a fragmentation of R is a set of fragments $\{F_1,\ldots,F_m\}$, where $F_i \subseteq R$, for $i = 1,\ldots,m$.

Index: Piece of additional information stored at the external server together with relation R that can be used to efficiently evaluate queries operating on the attribute on which the index has been defined.

Respondent: Person (or entity) to whom the data undergoing public or semipublic release refer.

Shuffle: Process that randomly changes the assignment of node contents to physical identifiers (i.e., to the block of memory where each node is physically stored).

Verification Object (VO): Piece of information sent to the client by the server executing a query, which can be used by the client to verify whether the query result is correct and complete.

Acknowledgments

This work was partially supported by the Italian Ministry of Research within the PRIN 2008 project "PEPPER" (2008SY2PH4). The work of Sushil Jajodia was partially supported by the National Science Foundation under grant CT-20013A and by the U.S. Air Force Office of Scientific Research under grant FA9550-09-1-0421.

References

1. G. Aggarwal, M. Bawa, P. Ganesan, H. Garcia-Molina, K. Kenthapadi, R. Motwani, U. Srivastava, D. Thomas, and Y. Xu. Two can keep a secret: A distributed architecture for secure database services. In *Proc. CIDR 2005*, Asilomar, CA, January 2005.

2. R. Agrawal, J. Kierman, R. Srikant, and Y. Xu. Order preserving encryption for numeric data. In *Proc. SIGMOD 2004*, Paris, France, June 2004.

3. A. Ambainis. Upper bound on communication complexity of private information retrieval. In *Proc. ICALP 1997*, Bologna, Italy, July 1997.

4. M. Atallah, M. Blanton, N. Fazio, and K. Frikken. Dynamic and efficient key management for access hierarchies. *ACM TISSEC*, 12(3):18:1–18:43, January 2009.

5. R.J. Bayardo and R. Agrawal. Data privacy through optimal k-anonymization. In *Proc. ICDE 2005*, Tokyo, Japan, April 2005.

6. E. Bertino, P. Samarati, and S. Jajodia. Authorizations in relational database management systems. In *Proc. CCS 1993*, Fairfax, VA, November 1993.

7. D. Boneh, C. Gentry, B. Lynn, and H. Shacham. Aggregate and verifiably encrypted signatures from bilinear maps. In *Proc. EUROCRYPT 2003*, Warsaw, Poland, May 2003.

8. C. Cachin, S. Micali, and M. Stadler. Computationally private information retrieval with polylogarithmic communication. In *Proc. EUROCRYPT 1999*, Prague, Czech Republic, May 1999.

9. California senate bill sb 1386, September 2002. http://leginfo.ca.gov/pub/01-02/bill/sen/sb_1351-1400/sb_1386_bill_20020926_chaptered.html (accessed April 12, 2013).

10. A. Ceselli, E. Damiani, S. De Capitani di Vimercati, S. Jajodia, S. Paraboschi, and P. Samarati. Modeling and assessing inference exposure in encrypted databases. *ACM TISSEC*, 8(1):119–152, February 2005.

11. B. Chor and N. Gilboa. Computationally private information retrieval (extended abstract). In *Proc. STOC 1997*, El Paso, TX, May 1997.

12. B. Chor, E. Kushilevitz, O. Goldreich, and M. Sudan. Private information retrieval. *Journal of ACM*, 45(6):965–981, April 1998.

13. V. Ciriani, S. De Capitani di Vimercati, S. Foresti, S. Jajodia, S. Paraboschi, and P. Samarati. Enforcing confidentiality constraints on sensitive databases with lightweight trusted clients. In *Proc. DBSec 2009*, Montreal, Quebec, Canada, July 2009.

14. V. Ciriani, S. De Capitani di Vimercati, S. Foresti, S. Jajodia, S. Paraboschi, and P. Samarati. Keep a few: Outsourcing data while maintaining confidentiality. In *Proc. ESORICS 2009*, Saint Malo, France, September 2009.

15. V. Ciriani, S. De Capitani di Vimercati, S. Foresti, S. Jajodia, S. Paraboschi, and P. Samarati. Combining fragmentation and encryption to protect privacy in data storage. *ACM TISSEC*, 13(3):22:1–22:33, July 2010.

16. V. Ciriani, S. De Capitani di Vimercati, S. Foresti, and P. Samarati. *k*-Anonymity. In T. Yu and S. Jajodia, eds., *Security in Decentralized Data Management*. Springer, Berlin, Germany, 2007.

17. V. Ciriani, S. De Capitani di Vimercati, S. Foresti, and P. Samarati. Microdata protection. In T. Yu and S. Jajodia, eds., *Security in Decentralized Data Management*. Springer, Berlin, Germany, 2007.

18. T. Dalenius. Towards a methodology for statistical disclosure control. *Statistik Tidskrift*, 15:429–444, 1977.

19. E. Damiani, S. De Capitani di Vimercati, S. Jajodia, S. Paraboschi, and P. Samarati. Balancing confidentiality and efficiency in untrusted relational DBMSs. In *Proc. CCS 2003*, Washington, DC, October 2003.

20. T.K. Dang. Oblivious search and updates for outsourced tree-structured data on untrusted servers. *IJCSA*, 2(2):67–84, 2005.

21. S. De Capitani di Vimercati, S. Foresti, S. Jajodia, S. Paraboschi, G. Pelosi, and P. Samarati. Preserving confidentiality of security policies in data outsourcing. In *Proc. WPES 2008*, Alexandria, VA, October 2008.

22. S. De Capitani di Vimercati, S. Foresti, S. Jajodia, S. Paraboschi, G. Pelosi, and P. Samarati. Encryption-based policy enforcement for cloud storage. In *Proc. SPCC 2010*, Genova, Italy, June 2010.

23. S. De Capitani di Vimercati, S. Foresti, S. Jajodia, S. Paraboschi, and P. Samarati. Encryption policies for regulating access to outsourced data. *ACM TODS*, 35(2):12:1–12:46, April 2010.

24. S. De Capitani di Vimercati, S. Foresti, S. Jajodia, S. Paraboschi, and P. Samarati. Private data indexes for selective access to outsourced data. In *Proc. WPES 2011*, Chicago, IL, October 2011.

25. S. De Capitani di Vimercati, S. Foresti, S. Jajodia, S. Paraboschi, and P. Samarati. Support for write privileges on outsourced data. In *Proc. SEC 2012*, Heraklion, Crete, Greece, June 2012.

26. S. De Capitani di Vimercati, S. Foresti, S. Paraboschi, G. Pelosi, and P. Samarati. Efficient and private access to outsourced data. In *Proc. ICDCS 2011*, Minneapolis, MN, June 2011.

27. S. De Capitani di Vimercati, S. Foresti, S. Paraboschi, G. Pelosi, and P. Samarati. Supporting concurrency in private data outsourcing. In *Proc. ESORICS 2011*, Leuven, Belgium, September 2011.

28. P.T. Devanbu, M. Gertz, C.U. Martel, and S.G. Stubblebine. Authentic third-party data publication. In *Proc. DBSec 2000*, Schoorl, the Netherlands, August 2000.

29. X. Ding, Y. Yang, and R.H. Deng. Database access pattern protection without full-shuffles. *IEEE TIFS*, 6(1):189–201, March 2011.

30. C. Dwork. Differential privacy. In *Proc. ICALP 2006*, Venice, Italy, July 2006.

31. V. Ganapathy, D. Thomas, T. Feder, H. Garcia-Molina, and R. Motwani. Distributing data for secure database services. In *Proc. PAIS 2011*, Uppsala, Sweden, 2011.

32. W. Gasarch. A survey on private information retrieval. *Bulletin of the EATCS*, 82:72–107, 2004.

33. C. Gentry. Fully homomorphic encryption using ideal lattices. In *Proc. STOC 2009*, Bethesda, MD, May 2009.

34. O. Goldreich and R. Ostrovsky. Software protection and simulation on oblivious RAMs. *JACM*, 43(3):431–473, May 1996.

35. P. Golle. Revisiting the uniqueness of simple demographics in the US population. In *Proc. WPES 2006*, Alexandria, VA, October 2006.

36. H. Hacigümüs, B. Iyer, and S. Mehrotra. Providing database as a service. In *Proc. ICDE 2002*, San Jose, CA, February 2002.

37. H. Hacigümüs, B. Iyer, and S. Mehrotra. Ensuring integrity of encrypted databases in database as a service model. In *Proc. DBSec 2003*, Estes Park, CO, August 2003.

38. H. Hacigümüs, B. Iyer, and S. Mehrotra. Efficient execution of aggregation queries over encrypted relational databases. In *Proc. DASFAA 2004*, Jeju Island, Korea, March 2004.

39. H. Hacigümüs, B. Iyer, S. Mehrotra, and C. Li. Executing SQL over encrypted data in the database-service-provider model. In *Proc. SIGMOD 2002*, Madison, WI, June 2002.

40. M. Hay, V. Rastogi, G. Miklau, and D. Suciu. Boosting the accuracy of differentially private histograms through consistency. *Proc. VLDB Endowment*, 3(1–2):1021–1032, September 2010.

41. D. Kifer and A. Machanavajjhala. A rigorous and customizable framework for privacy. In *Proc. PODS 2012*, Scottsdale, AZ, 2012.

42. K. LeFevre, D.J. DeWitt, and R. Ramakrishnan. Incognito: Efficient full-domain k-anonymity. In *Proc. SIGMOD 2005*, Baltimore, MD, June 2005.

43. K. LeFevre, D.J. DeWitt, and R. Ramakrishnan. Mondrian multidimensional k-anonymity. In *Proc. ICDE 2006*, Atlanta, GA, April 2006.

44. C. Li, M. Hay, V. Rastogi, G. Miklau, and A. McGregor. Optimizing linear counting queries under differential privacy. In *Proc. PODS 2010*, Indianapolis, IN, June 2010.

45. F. Li, M. Hadjieleftheriou, G. Kollios, and L. Reyzin. Dynamic authenticated index structures for outsourced databases. In *Proc. SIGMOD 2006*, Chicago, IL, June 2006.

46. N. Li, T. Li, and S. Venkatasubramanian. t-Closeness: Privacy beyond k-anonymity and ℓ-diversity. In *Proc. ICDE 2007*, Istanbul, Turkey, April 2007.

47. P. Lin and K.S. Candan. Hiding traversal of tree structured data from untrusted data stores. In *Proc. WOSIS 2004*, Porto, Portugal, April 2004.

48. R. Liu and H. Wang. Integrity verification of outsourced XML databases. In *Proc. CSE 2009*, Vancouver, British Columbia, Canada, August 2009.

49. A. Machanavajjhala, J. Gehrke, and D. Kifer. ℓ-diversity: Privacy beyond k-anonymity. In *Proc. ICDE 2006*, Atlanta, GA, April 2006.

50. R.C. Merkle. A certified digital signature. In *Proc. CRYPTO 1989*, Santa Barbara, CA, August 1989.

51. I. Mironov, O. Pandey, O. Reingold, and S.P. Vadhan. Computational differential privacy. In *Proc. CRYPTO 2009*, Santa Barbara, CA, August 2009.

52. E. Mykletun, M. Narasimha, and G. Tsudik. Signature bouquets: Immutability for aggregated/condensed signatures. In *Proc. ESORICS 2004*, Sophia Antipolis, France, September 2004.

53. E. Mykletun, M. Narasimha, and G. Tsudik. Authentication and integrity in outsourced databases. *ACM TOS*, 2(2):107–138, May 2006.

54. M. Narasimha and G. Tsudik. DSAC: Integrity for outsourced databases with signature aggregation and chaining. In *Proc. CIKM 2005*, Bremen, Germany, October–November 2005.

55. A. Narayanan and V. Shmatikov. Robust de-anonymization of large sparse datasets. In *Proc. IEEE SP 2008*, Berkeley, CA, May 2008.

56. F. Olumofin and I. Goldberg. Privacy-preserving queries over relational databases. In *Proc. PETS 2010*, Berlin, Germany, July 2010.

57. H. Pang, A. Jain, K. Ramamritham, and K.L. Tan. Verifying completeness of relational query results in data publishing. In *Proc. of SIGMOD 2005*, Baltimore, MD, June 2005.

58. H. Pang and K.L. Tan. Authenticating query results in edge computing. In *Proc. ICDE 2004*, Boston, MA, April 2004.

59. Payment card industry (PCI) data security standard, v2.0, October 2010. https://www.pcisecuritystandards. org/documents/pci_dss_v2.pdf (accessed on April 12, 2013).

60. Personal data protection code. Legislative Decree no. 196, June 2003.

61. R.L. Rivest, L. Adleman, and M.L. Dertouzos. On data banks and privacy homomorphisms. In R.A. DeMillo, R.J. Lipton, and A.K. Jones, eds., *Foundation of Secure Computations*. Academic Press, Orlando, FL, 1978.

62. P. Samarati. Protecting respondents' identities in microdata release. *IEEE TKDE*, 13(6):1010–1027, November 2001.

63. P. Samarati and S. De Capitani di Vimercati. Data protection in outsourcing scenarios: Issues and directions. In *Proc. ASIACCS 2010*, Beijing, China, April 2010.

64. H. Wang and L.V.S. Lakshmanan. Efficient secure query evaluation over encrypted XML databases. In *Proc. VLDB 2006*, Seoul, Korea, September 2006.

65. H. Wang, J. Yin, C. Perng, and P.S. Yu. Dual encryption for query integrity assurance. In *Proc. CIKM 2008*, Napa Valley, CA, October 2008.

66. P. Williams, R. Sion, and B. Carbunar. Building castles out of mud: Practical access pattern privacy and correctness on untrusted storage. In *Proc CCS 2008*, Alexandria, VA, October 2008.

67. X. Xiao, G. Wang, and J. Gehrke. Differential privacy via wavelet transforms. *IEEE TKDE*, 23(8):1200–1214, August 2011.

68. M. Xie, H. Wang, J. Yin, and X. Meng. Integrity auditing of outsourced data. In *Proc. VLDB 2007*, Vienna, Austria, September 2007.

69. M. Xie, H. Wang, J. Yin, and X. Meng. Providing freshness guarantees for outsourced databases. In *Proc. EDBT 2008*, Nantes, France, March 2008.

70. Y. Yang, D. Papadias, S. Papadopoulos, and P. Kalnis. Authenticated join processing in outsourced databases. In *Proc. SIGMOD 2009*, Providence, RI, June–July 2009.

71. S. Yu, C. Wang, K. Ren, and W. Lou. Achieving secure, scalable, and fine-grained data access control in cloud computing. In *Proc. INFOCOM 2010*, San Diego, CA, March 2010.

54

Behavioral Information Security Management

Merrill Warkentin
Mississippi State University

Leigh Mutchler
University of Tennessee

54.1 Introduction

Modern organizations' critical dependencies on information assets have resulted in information security being included among the most important academic research issues within the field of information systems (IS) (Dhillon and Backhouse, 2000; Kankanhalli et al., 2003). The massive amounts of information assets being created every day require management and protection. Technical, administrative, and behavioral controls must be implemented to protect information as it is created, captured, stored, and transmitted. Information at each of these four states of information is vulnerable to security threats. Technical controls, such as firewalls, port scanning, automated backups, intrusion detection systems (IDS), and anti-malware software have introduced significant safeguards against external human perpetrators of security threats, but various human behaviors, especially those of organizational insiders, continue to pose the greatest risk to the security of IS and continue to be cited as the greatest challenge for security practitioners (Davis, 2012; Willison and Warkentin, 2013).

The four goals of the management and protection of information assets are to ensure their confidentiality, integrity, availability, and accountability throughout the organization (Siponen et al., 2006; Stanton et al., 2005; Warkentin and Johnston, 2006, 2008). An information asset's confidentiality is maintained by controlling the methods of disclosure within and beyond the organization. The availability goal is maintained by ensuring that only those individuals authorized to access an information asset are afforded such access only when and where it is required. The maintenance of the integrity of an information asset is achieved through use of controls that restrict asset alterations or deletions. Last, the goal of accountability, more formally known as nonrepudiation, is the assurance that an electronic "paper trail" exists and precludes the denial of transaction occurrences or of the actions of any individual.

Achieving information security within an organization requires development of policies and procedures, followed by their successful implementation (Warkentin and Johnston, 2006, 2008). The information security policies (ISPs) specific to the needs of an organization are developed using the information security goals as a framework. Both formal and informal policies should be aligned with the overall organizational policies and strategies. Procedures and performance standards necessary to support the ISP must be developed and will include specification of the mechanisms and methods for implementation. Finally, the ISPs are put into practice through implementation of the procedures. Employees are informed of the ISP and achieve an appropriate level of understanding of the ISP through information security education, training, and awareness (SETA) campaigns, discussed later.

The various sources and types of threats to information security may be categorized, as shown in Figure 54.1, into internal and external threats and further subdivided into human and non-human threats (Loch et al., 1992; Willison and Warkentin, 2013). Information security and therefore information security research typically focuses on either a technical or behavioral viewpoint (Bulgurcu et al., 2010; Tsohou et al., 2008). The non-human threats that exist include potential events that may occur internally such as equipment failure or events due to external forces such as natural disasters. Human threats in the form of computer hackers or corporate spies are commonly seen in news media headlines (Willison and Siponen, 2009) and as favorites of cinematic thrillers, likely making these external threats the most well known. These external human threats can be subdivided into three categories of perpetrators, any of which may cause harm to the organization, but each fueled by a different intent.

The first category of perpetrators within the external human threat is made of individuals believed to be fueled by curiosity, boredom, and the need for notoriety. These individuals are known as script kiddies, a derogatory term for those who seek out system vulnerabilities to exploit by using software tools, or scripts, developed by others and whose antics are typically considered unsophisticated or childish (Vamosi, 2002; Wlasuk, 2012). Although there is no specific age restriction to be considered a script kiddie, they are typically young but are still quite capable of causing great harm. They have been known to have gained unauthorized access to networks and computer systems, built botnets, and committed web tagging, an electronic form of graffiti.

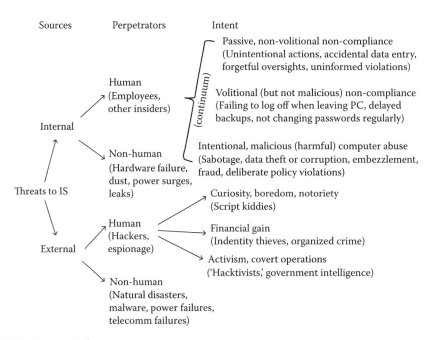

FIGURE 54.1   IS security threat vector taxonomy.

The second category of external human threat perpetrators is made of individuals fueled by the promise of financial gain. These perpetrators have become quite sophisticated and use social engineering methods such as phishing and other forms of deception and psychological techniques to commit fraudulent acts such as theft of intellectual property or of identity theft (Wright and Marett, 2010). These cybercrimes may be committed by individual humans or by members of organized crime syndicates ("Who commits the crime," n.d.). By virtue of the capabilities of the World Wide Web and the Internet, the perpetrators of these crimes can be anywhere in the world, adding to the complexity of their identification and difficulty in punishing them if they are caught.

The last category of perpetrators within the external human threat is made of individuals whose intent is that of activism or covert operations performed by governments. The most well-known perpetrators within this category is the "hactivist" group, Anonymous, who became known to the world in early 2008 when they staged a protest against the Church of Scientology after YouTube was ordered by the church to remove an internal video they claimed was copyrighted (Schwartz, 2012). Since that date, members of Anonymous have been involved in numerous protests and attempts to expose wrongs in the world and have evolved into a culture (Norton, 2011).

Though maintenance of organizational security relies upon employee compliance with ISP, the threats from non-human sources and from humans external to the organization are primarily addressed through maintenance programs and business continuity plans that leverage technological controls such as firewalls and IDS (Cavusoglu et al., 2009), biometrics (Ballard et al., 2007), and encrypted logins (Boncella, 2002). Hardware failures (a prime non-human threat to security) are addressed through both technical and behavioral controls (Malimage and Warkentin, 2010). A great deal of time, effort, and financial resources are spent on protective efforts against these threats, yet research shows that the internal human threat may cause the most harm and be the most difficult to guard against (Doherty and Fulford, 2005; Straub and Welke, 1998; Warkentin and Johnston, 2006; Warkentin and Willison, 2009).

The internal human perpetrators category in Figure 54.1 includes those threats to information security that come from a group of individuals who already have access to the information assets of the organization and are therefore known as the "insider threat." Research in information security (InfoSec) addresses antecedents of and issues related to individual compliance and noncompliance with and criminal violation of IS security policies and procedures. The InfoSec performed to date is the focus of this chapter. We begin with a definition of the term insider threat and a brief discussion of the employee intent continuum shown in Figure 54.1. A review of the theories, methods, and measures used by researchers in this area of study today follows, and the chapter concludes with recently identified major information security research issues and need for future study.

54.2 Information Security Research: The Insider Threat

Who is the insider? Organizational boundaries are dynamic, fluid, and often poorly defined. In the age of the virtual organization and inter-organizational global collaborative teams, many individuals may have access to trusted information resources. In the context of our investigation, there are two essential aspects of an insider: (1) insider access and (2) deep familiarity with organizational processes.

First, the insider must be an individual who has access to information resources not typically provided to the "outsider" (all others in the general public). These resources include systems and their data/information. While this category of individual is typically instantiated as an employee, it should be recognized that he or she may also be a contractor, a virtual partner (supply chain partner or strategic alliance partner with access to resources), or even a constituent who is a customer or other client (e.g., student, medical patient, etc.), provided they have some level of insider privileges. Essentially, the insider is an individual who can easily operate "behind the firewall" who has proper authentication via password or other methods.* This person is thought of as a trusted member of the organization. It should also be recognized that organizations have numerous categories of trusted individuals and use "role-based security" mechanisms to afford greater or lesser control to various individuals based on their respective roles. These may be thought of as concentric circles of access, but with overlap, according to one's need for access.

Second, insiders not only have access to and often control of systems and information, but they also are individuals who have a deep knowledge of their organizations' critical processes because of their dynamic involvement and experiences with these processes. Though this relationship may also exist between companies and their partners, this context is best typified by the employer–employee relationship. In this context, organizational policies and procedures may have an impact on the insider. Organizations seek to hire trustworthy employees. Firms implement SETA programs to train their employees in the use of proper actions and behaviors to ensure compliance with security policies. Managers may also utilize persuasive communications, such as fear appeals (Johnston and Warkentin, 2010), to encourage compliant protective behaviors. Sanctions may also be targeted against noncompliant employees as a deterrent (D'Arcy et al., 2009). Furthermore, the policies and actions of organizational managers can have a strong impact on the psychological perceptions of the employees, as well as their subsequent thought processes that may lead to various actions. Finally, these same principles apply also to former employees, customers, and partners, who often retain insider access and process knowledge and thus can pose continued threats even after the relationship is terminated or expired. The same factors that can influence insiders, presented in our three propositions, can continue to influence former employees. A recent survey of nearly 1900 firms from all major industries and from all over the globe found that 75% of respondents revealed their concern with the possible reprisal from employees recently separated from their organization (Ernst and Young, 2009).

Though extensive evidence suggests that insiders pose the greatest risk to information security (Brenner, 2009; CSO Magazine, 2007; Ernst and Young, 2009; Loveland and Lobel, 2010; Ponemon Institute, 2010; Richardson, 2011; Smith, 2009), many firms continue to spend the lion's share of their time, effort, and budget on perimeter security measures, such as firewalls, IDS, encryption of transmissions, anti-malware software, and content filtering, all of which aim to keep outsiders (e.g., hackers) or other external entities (e.g., malware) from penetrating the perimeter and threatening the organization's information resources. Though a certain degree of this focus is mandated by various regulations (Warkentin et al., 2011a), this allocation of resources is not entirely justifiable.

Behaviors that may result in harm to the organization are often unintentional. Prevention of these behaviors requires that employees be provided with knowledge regarding the behavior expected of them. This requirement is typically addressed through the development and implementation of ISP (Bulgurcu et al., 2010; Karjalainen and Siponen, 2011; Siponen and Vance, 2010; Thomson and von Solms, 1998). To further ensure employee compliance, information security instruction is often performed to explain the policies and achieve an understanding by employees regarding secure behavior (Puhakainen and Siponen, 2010; Siponen, 2000; Siponen and Iivari, 2006; Warkentin and Johnston, 2006; Warkentin et al., 2011a). However, employee compliance with ISP continues to be among the top concerns of organizations (Bulgurcu et al., 2010; D'Arcy and Hovav, 2009; Dodge et al., 2007; Kaplan, 2010; Loveland and Lobel, 2010; Prince, 2009). Therefore, information security research explores the typical controls used by organizations to encourage secure behaviors, such as SETA programs (D'Arcy et al., 2009; Furnell et al., 2002; Peltier, 2005; Wilson and Hash, 2003), development of acceptable use policies (Doherty et al., 2011; Ruighaver et al., 2010), and use of deterrence programs (D'Arcy et al., 2009; Straub and Welke, 1998).

---

* Security practitioners utilize access methods that require "something you know" (passwords, PINs), "something you have" (badges, cards, fobs), and/or "something you are" (biometric signatures such as fingerprints). So-called two-factor authentication is deemed better than methods that employ only one of these three.

Research of this nature historically looked at employee behaviors with the assumptions being that any unsecure behaviors were the result of non-malicious actions on the part of the employees. This assumption has changed over the past few years to include intentional and malicious actions. A continuum of employee intent as illustrated in Figure 54.1 has been identified and broadened the area of information security study.

On the upper end of the continuum, employee intent is passive and, on the whole, behavior is considered secure. The majority of the original insider threat research conducted would reside on this end as it focused on errors or misunderstandings as the underlying cause of unsecure actions or breaches in security. These types of unsecure behaviors may be corrected through improved information security instruction. In the middle of the continuum are intentional (but not malicious) acts and volitional behaviors such as choosing to not log off when leaving a PC unattended or to write down a password. These types of unsecure behaviors may be addressed through persuasion, deterrence, or other forms of behavioral controls. The lower end of the continuum includes intentional behaviors such as theft, embezzlement, and other deliberate actions. These behaviors are no longer benign in their intent and are committed for malicious reasons and or personal gain and are included in the category of crimes known as cybercrime.

The remainder of this manuscript identifies some theoretical tools for exploring InfoSec, along with a very brief assessment of the methods that have been used.

54.3 Theories Applied to Behavioral Information Security Research

The field of IS frequently adapts theories from other fields as is also the case with information security research. The majority of the research to date draws upon a small set of behavioral theories that originated primarily in the fields of psychology, sociology, communications, and criminology. These theories are listed in Table 54.1 along with a brief description and the primary literary sources for each.

Figure 54.2, generated by a word cloud software assessment of the key articles published in the area of IS security behaviors, provides a visual representation of the theories used to investigate such behaviors.

54.3.1 Deterrence Theory

Deterrence theory, commonly called general deterrence theory (GDT) though there are distinctions between general and specific deterrence that we will not address here, is identified as the theory most commonly applied in IS research (D'Arcy and Herath, 2011; Siponen et al., 2008). GDT provides the theoretical foundation necessary to explore and explain ISP compliance issues and IT abuse or misuse issues. Its foundation is the "rational choice model" of human behavior that posits that individual decisions are influenced by the fundamental motivation to apply a rational calculus of maximizing pleasure and minimizing pain. The assumption is that individuals will make behavioral choices based on the potential risks or costs and the potential gains or benefits, with the choice to act hinging on the balance between them. If the benefits are sufficiently large, a person may be willing to assume the risks or costs; but if the costs are sufficiently large, it will deter the behavior in question. Societies impose deterrence measures (costs of actions) in the form of punishments (fines or imprisonment) for illegal acts, whereas organizations impose deterrence measures against policy violations in the form of formal workplace sanctions, such as reprimands, demotions, withholding raises, or termination. These imposed costs are intended to discourage unwanted behavior. The theory states that if sanctions are in place, individuals will make a rational choice through the assessments of the likelihood of being caught (perceived sanction certainty), the harshness of the sanction or punishment (perceived sanction severity), and the speed with which the sanction may be carried out (perceived sanction celerity).

TABLE 54.1 Theories Used in Behavioral Research in Information Systems Security

Theory	Description	Primary Sources
General deterrence theory (GDT)	GDT states that individuals will weigh and balance the risks and gains associated with their actions and will consider the certainty, severity, and celerity of sanctions that exist to discourage the behavior	Gibbs (1975) Straub and Welke (1998)
Rational choice theory (RCT)	RCT states that individuals will make behavioral choices based on balancing the risks and gains of the choices available, and the level of magnitude of the risks and the gains are calculated through subjective individual assessment	Paternoster and Simpson (1996) McCarthy (2002)
Neutralization theory (NT)	NT states that individuals will use neutralization techniques to justify choices to perform behaviors even when the behaviors are not secure or are against policy	Sykes and Matza (1957) Klockars (1974) Minor (1981) Benson (1985) Cressey (1989) Siponen and Vance (2010)
Protection motivation theory (PMT)	PMT states that individuals will process fear appeals (persuasive messages) by performing a threat appraisal (severity and susceptibility of a threat) and a coping appraisal (response efficacy and self-efficacy), which will result in protection motivation behaviors	Rogers (1975) Maddux and Rogers (1983) Floyd et al. (2000)
Extended parallel process model (EPPM)	Related to PMT, the EPPM states that individuals, when presented with a high threat message, may become overwhelmed and will not develop protection motivation behaviors to respond to the threat but instead will develop avoidance behaviors and ignore or reject the threat	Witte (1992)
Elaboration likelihood model (ELM)	ELM states that individuals will process messages via a direct or peripheral route. A message processed via the direct route is more likely to result in stable and predictable outcomes whereas messages processed via the peripheral route are more likely to result in unstable and short-lived outcomes. A receiver's lack of personal involvement in the message topic and the availability of message cues may encourage peripheral route processing	Petty and Cacioppo (1986b)
Information manipulation theory (IMT)	IMT states that when individuals participate in a conversation, a message sender may deceive the other party, the receiver, through the subtle alteration(s) of one or more of the four main components of a message, namely the quantity, quality, clarity, or relevance of the message. The receiver of the message may be deceived if he or she does not detect the alteration(s)	McCornack (1992) McCornack et al. (1992)
Interpersonal deception theory (IDT)	IDT suggests that a sender knowingly wishing to deceive a receiver may strategically insert false information in the course of a conversation. The receiver will judge the credibility of the message and as long as no deception is detected the conversation may continue with the deceiver interactively adjusting the message as needed to remain undetected.	Buller and Burgoon (1996)
Equity theory (ET)	ET addresses human perceptions regarding inequality in social exchanges. The application of ET within organizational settings is known as distributed justice	Adams (1965)
Social cognitive theory (SCT)	Best known as the theory behind the self-efficacy construct, SCT provides a framework by which human behaviors may be predicted and explained through relationships between individuals and the environment	Bandura (1977)

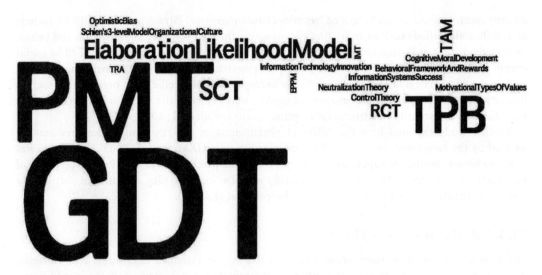

FIGURE 54.2 Visual representation of frequency of theory usage in IS security research.

The original form of GDT included only *formal* sanctions that carried legal consequences such as criminal penalties or civil lawsuits (analogous to workplace punishments). Theories about *informal* sanctions that were introduced by later scholars included those sanctions attributed to societal norms, such as guilt, shame, embarrassment, and social influence (Vance and Siponen, 2012). When moral individuals consider a deviant act, the mere thought of disapproval by one's peer group may deter the commission of the act, and thus informal sanctions likely play a very significant deterrence role, but they remain a largely unexplored antecedent of behavior in the InfoSec context.

The seminal work by Straub (1990) applied GDT within the field of IS and argued that the investments in information security must be balanced with the level of potential information asset risks. The focus was on the intentional human abuses and the common controls to discourage such abuses that were available at the time. The evolution of GDT and its use within IS saw the theory combined with Simon's managerial decision-making model to create the security action cycle (Straub and Welke, 1998), which became what is now the foundation for numerous information security research studies.

The security action cycle combines the generally accepted four stages of system risk reduction and identifies strategies for each where managers may implement security controls to reduce insider computer abuse (Straub and Welke, 1998). Security controls should first focus on *deterrence*, discouraging employees from performing unapproved, undesirable, and noncompliant behaviors. The goal is to minimize abuse; but if deterrence fails, the organization should implement *prevention* measures that inhibit risky employee behaviors that may bring harm to the organization's systems. Third, *detection* should be implemented in cases where employees are not deterred or prevented from behaving in an unsecure manner. Last, *remedies* should be implemented to enable recovery from any harm that may result from an employee's unsecure behavior, as well as punishment of the employee him/herself. GDT tells us that when the risk of being caught and punished outweighs the benefit of the action, individuals will refrain from performing the risky behavior. With this in mind, managers are encouraged to stress the deterrence and prevention security control implementation stages (Shropshire et al., 2010). The security action cycle has been recently expanded by Willison and Warkentin (2013) to look further back in time at the earliest motivations for workplace computer abuse, such as perceptions of organizational injustice, neutralization factors, and expressive motivations for committing abuse.

Information security studies focusing on the deterrence and prevention stages of the security action cycle often apply GDT toward assessing the effectiveness of various methods of information security controls. These controls, termed "security countermeasures" (Straub and Welke, 1998), include both technical methods such as restricting system access through software authorization and behavioral methods such as implementation of SETA programs. While these studies find GDT to be useful overall, inconsistencies across study findings have led researchers to explore additional issues to better explain the effectiveness of security controls. For example, factors specific to an organization such as industry type, company size, and management support were explored and found to be important considerations toward security effectiveness perceptions (Kankanhalli et al., 2003).

Another study explored how the deterrent effectiveness of security countermeasures may be affected by the employees' awareness of the countermeasures (D'Arcy et al., 2009). Awareness was found to have a significant impact particularly through the perceptions of severity and certainty of the sanctions. The level of the employee's morality was also found to play a significant part in how sanctions are perceived (Dinev and Hu, 2007; Shropshire et al., 2012).

54.3.2 Rational Choice Theory

The most serious threats to information security assets may be classified as criminal acts. However, crimes committed within the workplace differ significantly from those upon which theories of criminology were based. The rational choice Theory (RCT) extends GDT and deterrence theory to be applied more generally toward the explanation of individual behaviors by noncriminals within the workplace (Paternoster and Simpson, 1996) and as such has been applied by IS researchers in the context of information security.

The assumption of RCT is that the choice to commit an offense requires that an individual perform an assessment of benefits and costs. When an individual determines that his or her perception of the benefits outweighs his or her perception of the costs, the choice to offend is more likely to be made. When applied to the context of employees and potential offenses committed within an organization, other factors such as business practices and culture are likely to influence the actions taken by individuals (Sabherwal and Robey, 1995). Therefore, perceived costs and benefits assessed may include those for the individual as well as for the employer. Furthermore, costs and benefits may be tangible, intangible, formal, and informal. The moral beliefs of an individual are also believed to be a strong contributing factor within RCT, with individual behaviors more strongly influenced by morality than by any fear of perceived sanctions (Paternoster and Simpson, 1996).

Although RCT's origins suggest an appropriate fit in the area of information security, IS researchers have only recently applied it within an information security context. For example, the costs and benefits of ISP compliance and costs of ISP noncompliance were the focus of a recent study that found RCT significantly explained the effects of an employee's beliefs regarding ISP compliance on the employee's secure behaviors (Bulgurcu et al., 2010). The assumption that moral beliefs are a significant predictor of an individual's behavior was supported within the context of information security (Hu et al., 2011). Along with moral beliefs, the effect of employee self-control was also explored and found that those with low self-control were more likely to commit noncompliant acts when the benefits were perceived to be higher than the perceived costs. In yet another study, once again moral beliefs were found to be significantly related to secure behavior choices, along with perceived benefits (Vance and Siponen, 2012) with higher moral beliefs positively related to ISP compliance intention and higher perceived benefits negatively related to ISP compliance intention.

54.3.3 Neutralization Theory

While GDT has been widely applied across information security studies, inconsistent results have been found and the explanation may lie with the neutralization theory (NT) (Benson, 1985; Cressey, 1989; Sykes and Matza, 1957). NT states that regardless of the values and norms to which an individual prescribes, a process to neutralize, or justify, breaking a rule may be performed when the individual desires

TABLE 54.2 Neutralization Techniques—Neutralization Theory

Technique	Description	Source
Appeal to higher loyalties	An act may be justified when the individual perceives that the act is required for those to whom he or she is loyal or must protect (family, friends)	Sykes and Matza (1957)
Condemnation of the condemners	An act may be justified when the individual perceives that those who condemn the act should themselves be condemned	
Denial of injury	An act may be justified when the individual perceives that no one was injured by the act	
Denial of responsibility	An act may be justified when the individual perceives that committing the act is beyond his or her control	
Denial of the victim	An act may be justified when the individual perceives that the act is a form of retaliation or that the party being injured deserves to be injured because he or she harmed others first	
Metaphor of the ledger	Individuals rationalize their misdeed by pointing to the many good deeds (or compliance) they have done in the past	Klockars (1974)
Defense of the necessity	Required actions, have no choice	Minor (1981)
Avoidance of greater harm	Various justifications	Garrett et al. (1989)
Comparative standards		
Legal rights		
Malicious intentions		
Claim of entitlement	Other rationalizations, some similar to earlier forms	Coleman (1994)
Defense of the necessity of the law		
Claim of individuality	Various justifications	Henry and Eaton (1994)
Claim of relative acceptability		
Justification by comparison	Individuals justify committing the act because it is less harmful than other acts that could be just as easily committed	Cromwell and Thurman (2003)
Postponement	In order to commit the act, individuals put off any thoughts of wrong or right or of guilt feelings about committing the act	

Source: Adapted from Willison, R. and Warkentin, M., *MIS Quarterly,* 37, 1, 2013.

the outcome believed to be attainable by performing the action (Siponen and Vance, 2010). The use of neutralization techniques may explain why individuals behave in an unsecure manner and risk harm to organizational assets. The theory originally proposed that five neutralization techniques exist (Sykes and Matza, 1957) and included denial of responsibility, denial of injury, denial of the victim, condemnation of the condemners, and appeal to higher loyalties. A sixth technique, the metaphor of the ledger, was added about 20 years later (Klockars, 1974) and a seventh, the defense of necessity, was added soon after that (Minor, 1981). Several other neutralization techniques have been identified and are listed in Table 54.2 with the original five techniques with a brief description of selected entries.

54.3.4 Protection Motivation Theory

The study of persuasion and its effects on individual behavior has recently gained popularity with information security researchers in the field of IS. The use of persuasive messages known as fear appeals and the resulting effects of the messages on the attitudes and behaviors of individuals first gained the interest of researchers in the field of psychology during the 1950s (Janis and Feshbach, 1953, 1954). The numerous research projects produced findings that indicated further research was worthwhile, but the explanations of the findings lacked a cohesive theory until 1975 when the protection motivation theory (PMT) was developed (Rogers, 1975). In the years following, researchers in other fields such as communications (Witte, 1992), health care (Kline and Mattson, 2000), marketing (Dillard and Anderson, 2004), and IS security (Anderson and Agarwal, 2010; Herath and Rao, 2009; Johnston and Warkentin, 2010) continued the exploration of the ability of fear appeals to stimulate the appropriate level of fear or concern about an event and change individual attitudes or behaviors. Fear appeals have been shown to be effective in encouraging individual behaviors in the areas of health promotion, societal defense, and information security.

The PMT states that when individuals are exposed to a fear appeal, two cognitive mediating processes will ensue and protection motivation behavior will result. The cognitive mediating processes are the appraisal of the threat and the appraisal of the individual's capacity to cope with the threat. Threat appraisal includes the constructs of perceived threat severity and perceived threat susceptibility. Coping appraisal includes the constructs of perceived response efficacy and perceived self-efficacy. For the fear appeal message to be effective, it must include appropriate details regarding the threat and the recommended response to a threat, specifically that the threat is severe and is likely to occur, and a response that works is available and can be successfully performed (Maddux and Rogers, 1983; Rogers, 1975). When exposed to a fear appeal message, individuals perform threat and coping appraisals, and the individual's perception of the level of threat severity, threat susceptibility, response efficacy, and self-efficacy will affect the level of protection motivation behavior developed. If the threat is perceived to be serious and likely to occur, the recommended response is perceived to be effective against the threat, and the receiver believes he or she is capable of successfully performing the response, motivation toward protective behavior will result.

PMT has been applied in several contexts within the area of information security. Findings in most studies indicate use of PMT is effective toward explaining information security phenomena, but the findings have not been consistent across all contexts. For example, numerous information security threats exist with online activities such as sharing information on social networking sites, participating in blogging, and frequenting any password-secured website (Banks et al., 2010; LaRose et al., 2005, 2008; Marett et al., 2011; Zhang and McDowell, 2009). In the context of social networking, individuals' decisions whether to share personal information were influenced the most by the threat appraisal variables (Marett et al., 2011) along with social influence (Banks et al., 2010). Alternately, when considering overall online safety, the greater influence was found to be self-efficacy and response efficacy (LaRose et al., 2008). Similarly, self-efficacy and response efficacy were found to be strong influences toward online account password management intent along with a strong negative relationship between response cost and intent to implement strong passwords (Zhang and McDowell, 2009).

Other areas of information security research with PMT include explorations of general secure computing practices (Anderson and Agarwal, 2010; Ng et al., 2009; Woon et al., 2005; Workman et al., 2008) and anti-malware software use (Garung et al., 2009; Johnston and Warkentin, 2010; Lee and Larsen, 2009; Liang and Xue, 2010; Stafford and Poston, 2010). Application of PMT within these contexts has also produced varied results. For example, social influence, self-efficacy, and response efficacy were found to be the strongest influences of intent to use anti-malware software in one study (Johnston and Warkentin, 2010). Another study found that threat severity, self-efficacy, and response efficacy were significant indicators of software use, and also found that threat vulnerability and response costs produced no significant contributions (Garung et al., 2009). Particularly interesting results have been found as well, such as relationships between threat severity and self-efficacy and threat severity and response efficacy (Johnston and Warkentin, 2010), two relationships that are generally not tested in PMT-based information security research.

Another common information security threat is related to the great amounts of data being generated each day and the threat of losing the data at any time. A simple response to this threat is to perform data backups. Self-efficacy and response efficacy were found to be positively related, and threat susceptibility and severity were found to be negatively related to performing data backups (Crossler, 2010). Another study, however, found that only threat severity, response efficacy, and self-efficacy were strong indicators of behavioral intent (Malimage and Warkentin, 2010).

PMT research in the context of information security has produced different and varied findings as compared to the prior PMT research in contexts such as health care and marketing. Constructs that are not included in prior PMT research have shown to be strong indicators of secure behavior intent within an information security context, suggesting that the PMT model may require expanding to better fit this context. For example, social influence has been found to be highly significant toward influencing security behaviors (Johnston and Warkentin, 2010; Pahnila et al., 2007a) and descriptive norms have been found to be particularly important in influencing protective behaviors against a collective threat, with subjective norms found to be influential against individual threats (Anderson and Agarwal, 2010).

54.3.5 Extended Parallel Process Model

At times, research has found that certain fear appeals may not produce adaptive (protection motivation) behaviors, but instead produce maladaptive behaviors. A theory developed to explain such results is the extended parallel process model (EPPM) (Vroom and von Solms, 2004; Witte, 1992). EPPM states that fear appeal theories assume message acceptance and therefore only explain behaviors that result from acceptance of the persuasive message. For example, when the threat appraisal is high and not properly balanced with a high coping appraisal, an individual may become overwhelmed by the threat and choose to reject the persuasive message and develop maladaptive avoidance behaviors rather than adaptive protection motivation behaviors. In these unbalanced cases, rather than achieving the intended outcome of the message with acceptance of the persuasion and ensuing adoption of protection motivation behaviors, the outcome may be *defensive avoidance* where the individual responds by avoiding the appeal by choosing to ignore the message or avoid thinking about the threat. Or it may be *reactance* where the individual negatively reacts to the appeal, choosing instead to behave oppositely in a phenomenon called the "boomerang effect." In either case, individuals with insufficient coping levels, either response efficacy ("it won't work") or self-efficacy ("I can't do it"), will exhibit undesirable maladaptive responses that manage the fear itself, rather than the threat. Message rejection is a common phenomenon in which a user suggests that the message is not valid and that the threat is not severe. In the context of information security threats, a user may suggest that reports about viruses are overblown and he or she should not worry about them ("head in the sand" reaction). Or the message recipient may reject the messenger, discrediting the validity of the appeal. EPPM, like PMT, also states that when presented with a fear appeal, an individual will perform two message appraisals. However, rather than the appraisals being those of threat and coping as in PMT, EPPM describes a cognitive process to determine the danger control (which includes threat and coping appraisals) and an emotional process to determine the fear control (Vroom and von Solms, 2004; Witte, 1992).

54.3.6 Elaboration Likelihood Model

A theory that addresses the cognitive processes that individuals perform when presented with information is the elaboration likelihood model (ELM) (Petty and Cacioppo, 1986a,b). The theory states that there are two possible routes from which individuals will choose to process a message: the central route and the peripheral route. The general goal of a persuasive message is to achieve a long-term change in the attitude or behavior of an individual. When a message is processed by way of the central route, a more predictable and stable change of attitude or behavior typically results. When processed by way of the peripheral route, however, the results tend to be unstable and short-lived. Therefore, persuasive messages will be more successful when processed via the central route.

When an individual is confronted with a message that he or she is not motivated to directly process, peripheral route processing may offer a more comfortable "path of least resistance" for them.

For example, an individual may choose the peripheral route to process a message that is too difficult to process, that is not interesting, or that is a subject with which the individual is not comfortable. A message that is processed through the peripheral route relies upon message cues that may be found within or around the message. Cues may be perceptions about the sender of the message such as his or her likability, trustworthiness, or level of authority, or the presentation of the message such as its attractiveness or form of communication media. Rather than spending the cognitive effort toward fully addressing the message, individuals may instead rely on message cues to influence the decisions regarding the message.

Messages processed via the central route will be predictably and more fully processed for maximum understanding. ELM states that to discourage peripheral route processing, a message must be developed with no unnecessary cues that may be used to encourage the peripheral route and must include features that will motivate individuals to process it through the central route. Therefore, knowledge and understanding of these two potential processing routes can aid in the development of a persuasive message to ensure that individuals process the message by way of the direct route such that there is a greater likelihood that an intended and longer-lasting outcome is achieved.

ISP instructional programs may benefit from ELM by concentrating on modeling instructional messages to encourage central route processing and avoiding potential message cues that may encourage peripheral route processing (Puhakainen and Siponen, 2010). This includes developing the programs to ensure that individuals are motivated to expend the cognitive effort to fully process the messages such that longer-lasting behavior change may occur. Concentrating on providing ISP instruction that is relevant and meaningful to individuals has been found to be successful in encouraging employee ISP compliance.

ISP compliance involves numerous information security issues. An increasingly growing threat involves techniques of social engineering used by persons wishing to gain access to systems to which they are not authorized. The social engineering techniques use deceptive communications and encourage individuals to process messages via the peripheral route (Workman, 2007, 2008). When a perpetrator uses social engineering, employees are fooled into performing an act or revealing sensitive information to the perpetrator. Most ISP instructional programs teach employees to be ethical and responsible and to protect information assets but do not prepare them to fend off techniques such as social engineering. ISP instructional programs may benefit from the inclusion of instructional methods for employees to recognize and prevent social engineering attacks.

54.3.7 Information Manipulation Theory

Originating in the field of communications, the information manipulation theory (IMT) (McCornack, 1992; McCornack et al., 1992) has recently been applied to information security research. The theory builds upon conversational theory that states individuals have expectations regarding the quantity, quality, clarity, and relevance of the information transmitted by a conversation. IMT proposes that any or all of these four expectations may be manipulated by the sender in such a way that the receiver of the message is unaware of the manipulation and the sender is then able to deceive the receiver. This can also be stated to mean that the level of deceptiveness of a message may be identified through quantifiable means through measurement of the quantity, quality, manner, and relevance of the information contained in the message.

54.3.8 Interpersonal Deception Theory

Another theory from the field of communications and one gaining interest within information security literature is the interpersonal deception theory (IDT) (Buller and Burgoon, 1996). As with IMT, IDT is built on conversational theory and involves a sender knowingly attempting to deceive receivers, but is expanded to include the receivers' attempts to process the messages and to verify the truthfulness of the messages. IDT states that an interpersonal communication involves two or more individuals dynamically exchanging information. The assumptions of IDT are that communicators enact both strategic and nonstrategic behaviors while encoding and decoding messages. Throughout the conversation, the credibility of the sender and of the message is judged by the receiver. The progression of the conversation is interactive and continuing. A deceiver may strategically insert false information into a conversation and then observe the reaction of the receivers to determine whether or not the deception was recognized. If the deception is not discovered, the conversation continues with the deceiver continuing to insert false information, adjusting as needed to remain undetected. As the conversation continues, familiarity increases and truth biases become stronger, resulting in a situation ripe for abuse. Such exchanges occur in face-to-face so-called social engineering security threats in which an individual tricks another party into revealing sensitive information, such as a password. These exchanges also frequently occur in the virtual space, where emails and websites (such as so-called fake AV sites) deceive individuals into disclosing protected information or into downloading malware in the guise of anti-virus software. Deception plays a large role in the behavioral threat environment, and is receiving fresh attention from scholars.

54.3.9 Equity Theory

The seminal work to advance the equity theory (ET) was conducted by Adams (1965) in which individuals' perceptions of fairness within social exchanges were explored. The application of ET to equity perceptions by employees within organization has become known as distributive justice (Nowakowski and Conlon, 2005). For example, employee pay scales, raises, benefits, and hiring practices are all issues where employees may compare themselves to their coworkers and the potential for perceptions of inequality exists. In those instances when there are such perceptions, employees may become disgruntled and in turn behave in ways that are not beneficial to the organization. Employees are known to be the insider threat to organizations and information security, and the worst cases of employee disgruntlement may result in attempts by employees to harm the organization's information assets. A large percentage of the reported organizational computer abuses have been attributed to employee grievances, suggesting that this is a worthy area of study within information security, yet to date only a few works have addressed information security from this organizational aspect (Willison and Warkentin, 2009, 2013).

A larger domain known as organizational justice includes four constructs in total to describe the various forms of inequality perceptions that employees may develop within a workplace. With the first being distributive justice, second is procedural justice, related to inequalities an employee may perceive to exist within the organizational procedures (Leventhal, 1980; Leventhal et al., 1980). The last two are interactional justice, also known as interpersonal justice, and information justice (Greenberg, 1990, 1993). Interactional justice is related to the interactions between authority figures and their subordinates. Informational justice is related to the completeness and accuracy of the information as it is disseminated from senior members of the organization down through those more junior employees.

54.3.10 Social Cognitive Theory

The social cognitive theory (SCT) provides a framework to explain and predict individual behavior and changes to behaviors through the interaction of an individual with his or her environment (Bandura, 1977). SCT is primarily known as the theory behind the construct of self-efficacy and has been included in numerous studies in the field of IS with the most prominent being the development of the computer self-efficacy construct (Compeau and Higgins, 1995; Marakas et al., 1998). A self-efficacy construct is frequently included in studies within the information security context, most often addressing questions related to ISP compliance (Bulgurcu et al., 2010; Pahnila et al., 2007b). Studies based on the theoretical foundation of PMT (Anderson and Agarwal, 2010; Herath and Rao, 2009; Johnston and Warkentin, 2010), a theory that includes the self-efficacy construct, are perhaps the most common studies found within the context of information security to also rely on SCT.

54.4 Methods Applied to Behavioral Information Systems Security Research

The previous section described the theories used to investigate the research questions explored in the context of individual security-related behaviors, as well as some of the findings from research today. This section very briefly describes some of the methods commonly used in this domain. As expected, the methods are similar to other behavioral research about individual users within the IS context. InfoSec researchers survey individuals about their perceptions, attitudes, and beliefs, as well as their behaviors or intentions to exhibit various behaviors. Scholars have also devised creative lab and field experiments to create environments in which to measure salient constructs that relate various factors to security behavior. However, several unique characteristics about these behaviors and their motivation present unique challenges to valid data collection.

Measuring actual user behavior is the "holy grail" of IS behavioral research, and we often rely on measuring self-reported behavioral intention to act in a specific manner. In InfoSec, we often ask individuals to report their intention to act in a protective manner, to comply with organizationally mandated policies and procedures, or to violate such policies. Beyond the traditional concerns about validity threats from common method bias or social desirability bias, there are unique challenges encountered when asking an individual about committing a crime, which have been addressed by the criminology scientific community, from which we draw our inspiration. (Similar resistance is encountered when asking employees about violation of organizational policies.) One common solution is the implementation of a scenario-based factorial survey method approach, in which subjects are asked to read one or more versions of a scenario in which statements are embedded that represent various levels or values for the independent variables. The scenario character, acting in response to these antecedents, chooses to act in a deviant manner by violating a social norm, law, or organizational policy. The subject is then asked about the likelihood of doing as the scenario character did. In this way, the subject disassociates himself or herself from his or her own moral obligations to a certain extent and will provide a more valid indication of his or her behavioral intent. This method is now used widely within IS security behavior research to investigate the cognitive and affective elements of interest. In other cases, researchers have found creative ways to collect actual behaviors from server logs, from managers, and from other sources (Warkentin et al., 2011b, 2012).

54.5 Conclusions

Recent research in behavioral InfoSec has surfaced extensive knowledge about human behavior within this domain; however, many questions are left unanswered and much work remains. Scholars are beginning to explore new factors and new interpretations of the motivations and other antecedents of human behavior in this context, and more insights are gained every year. Future work in this area is informed by scholarly assessments of this domain (c.f. Willison and Warkentin, 2013 and Crossler et al., 2013) and by the real-world evidence of new threats and new psychological factors at play.

References

Adams, J. S. (1965). Inequity in social exchange. In L. Berkowitz (Ed.), *Advances in Experimental Social Psychology* (Vol. 2, pp. 267–299). New York: Academic Press.

Anderson, C. L. and Agarwal, R. (2010). Practicing safe computing: A multimethod empirical examination of home computer user security behavior intentions. *MIS Quarterly, 34*(3), 613–643.

Ballard, L., Lopresti, D., and Monrose, F. (2007). Forgery quality and its implications for behavioral biometric security. *IEEE Transactions on Systems, Man, and Cybernetics—Part B: Cybernetics, 37*(5), 1107–1118.

Bandura, A. (1977). Self-efficacy: Toward a unifying theory of behavioral change. *Psychological Review*, *84*(2), 191–215.

Banks, M. S., Onita, C. G., and Meservy, T. O. (2010). Risky behavior in online social media: Protection motivation and social influence. *Americas Conference on Information Systems Proceedings*, Lima, Peru, Paper 372.

Benson, M. L. (1985). Denying the guilty mind: Accounting for involvement in white-collar crime. *Criminology*, *23*(4), 583–607.

Boncella, R. J. (2002). Wireless security: An overview. *Communications of AIS*, *9*, 269–282.

Brenner, B. (2009). Why security matters again. The Global State of Information Security: Joint Annual Report, PricewaterhouseCoopers (with CSO Magazine). http://www.pwc.com/en_GX/gx/information-security-survey/pdf/pwcsurvey2010_cio_reprint.pdf (accessed on March 21, 2013).

Bulgurcu, B., Cavusoglu, H., and Benbasat, I. (2010). Information security policy compliance: An empirical study of rationality-based beliefs and information security awareness. [Article]. *MIS Quarterly*, *34*(3), A523–A527.

Buller, D. B. and Burgoon, J. K. (1996). Interpersonal deception theory. *Communication Theory*, *6*(3), 203–242.

Cavusoglu, H., Raghunathan, S., and Cavusoglu, H. (2009). Configuration of and interaction between information security technologies: The case of firewalls and intrusion detection systems. *Information Systems Research*, *20*(2), 198–217.

Coleman, J. W. (1994). *The Criminal Elite: The Sociology of White Collar Crime*, 3rd edn. New York: St. Martin's Press.

Compeau, D. R. and Higgins, C. A. (1995). Computer self-efficacy: Development of a measure and initial test. *MIS Quarterly*, *19*(2), 189–211.

Cressey, D. R. (1989). The poverty of theory in corporate crime research. In W. S. Laufer and F. Adler (Eds.), *Advances in Criminological Theory*. New Brunswick, NJ: Transaction Books.

Cromwell, P. and Thurman, Q. (2003). The devil made me do it: Use of neutralizations by shoplifters. *Deviant Behavior*, *24*, 535–550.

Crossler, R. E. (2010). Protection motivation theory: Understanding determinants to backing up personal data. Paper presented at the *43rd Hawaii International Conference on System Sciences*, Kauai, HI.

Crossler, R. E., Johnston, A. C., Lowry, P. B., Hu, Q., Warkentin, M., and Baskerville, R. (2013). Future directions for behavioral information security research, *Computers and Security*, *32*(1), 90–101.

CSO Magazine. (2007). E-crime watch survey. Retrieved May 20, 2012, from http://www.cert.org/archive/pdf/ecrimesummary07.pdf

D'Arcy, J. and Herath, T. (2011). A review and analysis of deterrence theory in the IS security literature: Making sense of the disparate findings. *European Journal of Information Systems*, *20*, 643–658.

D'Arcy, J. and Hovav, A. (2009). Does one size fit all? Examining the differential effects of IS security countermeasures. *Journal of Business Ethics*, *89*, 59–71.

D'Arcy, J., Hovav, A., and Galletta, D. (2009). User awareness of security countermeasures and its impact on information systems misuse: A deterrence approach. *Information Systems Research*, *20*(1), 79–98.

Davis, M. A. (2012). 2012 Strategic Security Survey Information Week Reports: InformationWeek. http://reports.informationweek.com/abstract/21/8807/Security/2012-Strategic-Security-Survey.html (accessed on March 21, 2013).

Dhillon, G. and Backhouse, J. (2000). Information system security management in the new millennium. *Communications of the ACM*, *43*(7), 125–128.

Dillard, J. P. and Anderson, J. W. (2004). The role of fear in persuasion. *Psychology and Marketing*, *21*(11), 909–926.

Dinev, T. and Hu, Q. (2007). The centrality of awareness in the formation of user behavioral intention toward protective information technologies. *Journal of the Association for Information Systems*, *8*(7), 386–408.

Dodge, R. C., Carver, C., and Ferguson, A. J. (2007). Phishing for user security awareness. *Computers and Security, 26*(1), 73–80.

Doherty, N. F., Anastasakis, L., and Fulford, H. (2011). Reinforcing the security of corporate information resources: A critical review of the role of the acceptable use policy. *International Journal of Information Management, 31*(3), 201–209.

Doherty, N. F. and Fulford, H. (2005). Do information security policies reduce the incidence of security breaches: An exploratory analysis. *Information Resources Management Journal, 18*(4), 21–39.

Ernst and Young. (2009). Outpacing change: Ernst and Young's 12th annual global information security survey. Retrieved May 20, 2012, from http://www.cert.org/archive/pdf/ecrimesummary07.pdf

Floyd, D. L., Prentice-Dunn, S., and Rogers, R. W. (2000). A meta-analysis of research on protection motivation theory. *Journal of Applied Social Psychology, 30*(2), 408–420.

Furnell, S. M., Gennatou, M., and Dowland, P. S. (2002). A prototype tool for information security awareness and training. *Logistics Information Management, 15,* 352–357.

Garrett, D. E., Bradford, J. L., Meyers, R. A., and Becker, J. (1989). Issues management and organizational accounts: An analysis of corporate responses to accusations of unethical business practices. *Journal of Business Ethics, 8*(7), 507–520.

Garung, A., Luo, X., and Liao, Q. (2009). Consumer motivations in taking action against spyware: An empirical investigation. *Information Management and Computer Security, 17*(3), 276–289.

Gibbs, J. P. (1975). *Crime, Punishment, and Deterrence.* New York: Elsevier.

Greenberg, J. (1990). Employee theft as a reaction to underpayment inequity: The hidden cost of pay cuts. *Journal of Applied Psychology, 75*(5), 561–568.

Greenberg, J. (1993). The social side of fairness: Interpersonal and informational classes of organizational justice. In R. Cropanzano (Ed.), *Justice in the Workplace: Approaching Fairness in Human Resource Management* (pp. 79–103). Hilldale, NJ: Earlbaum.

Henry, S. and Eaton, R. (1994). *Degrees of Deviance: Student Accounts of Their Deviant Behavior.* Salem, WI: Sheffield Publishing Company.

Herath, T. and Rao, H. R. (2009). Protection motivation and deterrence: A framework for security policy compliance in organisations. *European Journal of Information Systems, 18*(2), 106–125.

Hu, Q., Xu, Z., Dinev, T., and Ling, H. (2011). Does deterrence work in reducing information security policy abuse by employees? *Communications of the ACM, 54*(6), 54–60.

Janis, I. L. and Feshbach, S. (1953). Effects of fear-arousing communications. *The Journal of Abnormal and Social Psychology, 48*(1), 78–92.

Janis, I. L. and Feshbach, S. (1954). Personality differences associated with responsiveness to fear-arousing communications. *Journal of Personality, 23,* 154–166.

Johnston, A. C. and Warkentin, M. (2010). Fear appeals and information security behaviors: An empirical study. [Article]. *MIS Quarterly, 34*(3), A549–A544.

Kankanhalli, A., Teo, H.-H., Tan, B. C. Y., and Wei, K.-K. (2003). An integrative study of information systems security effectiveness. *International Journal of Information Management, 23*(2), 139–154.

Kaplan, D. (2010). Weakest link: End-user education. *SC Magazine.* Retrieved from SC Magazine for IT Security Professionals website: http://www.scmagazineus.com/weakest-link-end-user-education/article/161685/

Karjalainen, M. and Siponen, M. (2011). Toward a new meta-theory for designing information systems (IS) security training approaches. *Journal of the Association for Information Systems, 12*(8), 518–555.

Kline, K. N. and Mattson, M. (2000). Breast self-examination pamphlets: A content analysis grounded in fear appeal research. *Health Communication, 12*(1), 1–21.

Klockars, C. B. (1974). *The Professional Fence.* New York: Free Press.

LaRose, R., Rifon, N. J., and Enbody, R. (2008). Promoting personal responsibility for internet safety. *Communications of the ACM, 51*(3), 71–76.

LaRose, R., Rifon, N. J., Liu, S., and Lee, D. (2005). Understanding online safety behavior: A multivariate model. Paper presented at the *International Communication Association, Comunication and Technology Division*, New York.

Lee, Y. and Larsen, K. R. (2009). Threat or coping appraisal: Determinants of SMB executives' decision to adopt ant-malware software. *European Journal of Information Systems, 18*(2), 177–187.

Leventhal, G. S. (1980). What should be done with equity theory. In K. J. Gergen, M. S. Greenberg, and R. H. WIllis (Eds.), *Social Exchange: Advances in Theory and Research* (pp. 27–55). New York: Plenum.

Leventhal, G. S., Karuza, J. J., and Fry, W. R. (1980). Beyond fairness: A theory of allocation preferences. In G. Mikula (Ed.), *Justice and Social Interaction: Experimental and Theoretical Contributions from Psychological Research* (pp. 167–218). New York: Springer-Verlag.

Liang, H. and Xue, Y. (2010). Understanding security behaviors in personal computer usage: A threat avoidance perspective. *Journal of the Association for Information Systems, 11*(7), 394–413.

Loch, K. D., Carr, H. H., and Warkentin, M. E. (1992). Threats to information systems: Today's reality, yesterday's understanding. *MIS Quarterly, 16*(2), 173–186.

Loveland, G. and Lobel, M. (2010). Trial By Fire*Connected Thinking: What global executives expect of information security in the middle of the world's worst economic downturn in thirty years Advisory Services: PriceWaterhouseCoopers. http://www.pwc.com/en_US/us/it-risk-security/assets/trial-by-fire.pdf (accessed on March 21, 2013).

Maddux, J. E. and Rogers, R. W. (1983). Protection motivation and self-efficacy: A revised theory of fear appeals and attitude change. *Journal of Experimental Social Psychology, 19*, 469–479.

Malimage, K. and Warkentin, M. (2010). Data loss from storage device failure: An empirical study of protection motivation. Paper presented at the *2010 Workshop on Information Security and Privacy (WISP)*, St. Louis, MO.

Marakas, G. M., Yi, M. Y., and Johnson, R. D. (1998). The multilevel and multifaceted character of computer self-efficacy: Toward clarification of the construct and an integrative framework for research. *Information Systems Reseaerch, 1998*(9), 2.

Marett, K., McNab, A. L., and Harris, R. B. (2011). Social networking websites and posting personal information: An evaluation of protection motivation theory. *AIS Transactions on Human-Computer Interaction, 3*(3), 170–188.

McCarthy, B. (2002). New economics of sociological criminology. *Annual Review of Sociology, 28*, 417–442.

McCornack, S. A. (1992). Information manipulation theory. *Communication Monographs, 59*, 1–16.

McCornack, S. A., Levine, T. R., Solowczuk, K. A., Torres, H. I., and Campbell, D. M. (1992). When the alteration of information is viewed as deception: An empirical test of information manipulation theory. *Communication Monographs, 59*, 17–29.

Minor, W. W. (1981). Techniques of neutralization: A reconceptualization and empirical examination. *Journal of Research in Crime and Delinquency, 18*(2), 295–318.

Ng, B.-Y., Kankanhalli, A., and Xu, Y. C. (2009). Studying users' computer security behavior: A health belief perspective. *Decision Support Systems, 46*(4), 815–825.

Norton, Q. (2011). Anonymous 101: Introduction to the Lulz. *Wired Magazine*. Retrieved from http://www.wired.com/threatlevel/2011/11/anonymous-101/all/1

Nowakowski, J. M. and Conlon, D. E. (2005). Organizational justice: Looking back, looking forward. *International Journal of Conflict Management, 16*(1), 4–29.

Pahnila, S., Siponen, M., and Mahmood, A. (2007a). Employees' behavior towards IS security policy compliance. Paper presented at the *40th Hawaii International Conference on System Sciences*, Big Island, HI.

Pahnila, S., Siponen, M., and Mahmood, A. (2007b). Which factors explain employees' adherence to information security policies? An empirical study. Paper Presented at the *Pacific Asia Conference on Information Systems*, Auckland, New Zealand.

Paternoster, R. and Simpson, S. (1996). Sanction threats and appeals to morality: Testing a rational choice model of corporate crime. *Law and Society Review, 30*(3), 549–584.

Peltier, T. R. (2005). Implementing an information security awareness program. *Information Systems Security*, *14*(2), 37–48.

Petty, R. E. and Cacioppo, J. T. (1986a). *Communication and Persuasion: Central and Peripheral Routes to Attitude Change.* New York: Springer-Verlag.

Petty, R. E. and Cacioppo, J. T. (1986b). The elaboration likelihood model of persuasion. *Advances in Experimental Social psychology*, *19*, 123–162.

Ponemon Institute. (2010). First annual cost of cyber crime study. Retrieved May 20, 2012, from http://www.riskandinsurancechalkboard.com/uploads/file/PonemonStudy(1).pdf

Prince, B. (2009). Survey lists top enterprise endpoint security and compliance holes *eWeek*. Retrieved from eWeek Security Watch website: http://securitywatch.eweek.com/enterprise_security_strategy/survey_lists_top_endpoint_security_and_compliance_holes.html

Puhakainen, P. and Siponen, M. (2010). Improving employees' compliance through information systems security training: An action research study. [Article]. *MIS Quarterly*, *34*(4), A767–A764.

Richardson, R. (2011). 15th Annual 2010/2011 Computer Crime and Security Survey: Computer Security Institute. http://reports.informationweek.com/abstract/21/7377/Security/research-2010-2011-csi-survey.html (accessed on March 21, 2013).

Rogers, R. W. (1975). A protection motivation theory of fear appeals and attitude change. *The Journal of Psychology*, *91*, 93–114.

Ruighaver, A. B., Maynard, S. B., and Warren, M. (2010). Ethical decision making: Improving the quality of acceptable use policies. *Computers and Security*, *29*, 731–736.

Sabherwal, R. and Robey, D. (1995). Reconciling variance and process strategies for studying information system development. *Information Systems Research*, *6*(4), 303–327.

Schwartz, M. J. (2012). Who is anonymous: 10 Key facts. *InformationWeek*. Retrieved from http://www.informationweek.com/news/galleries/security/attacks/232600322?pgno = 1

Shropshire, J. D., Warkentin, M., and Johnston, A. C. (2010). Impact of negative message framing on security adoption. *Journal of Computer Information Systems*, *51*(1), 41–51.

Shropshire, J., Warkentin, M., and Straub, D. (2012). Protect their information: Fostering employee compliance with information security policies. Working Paper, Georgia Southern University, Statesboro, GA.

Siponen, M. (2000). A conceptual foundation for organizational information security awareness. *Information Management and Computer Security*, *8*(1), 31–41.

Siponen, M., Baskerville, R., and Heikka, J. (2006). A design theory for security information systems design methods. *Journal of the Association for Information Systems*, *7*(11), 725–770.

Siponen, M. and Iivari, J. (2006). Six design theories for IS security policies and guidelines. *Journal of the Association for Information Systems*, *7*(7), 445–472.

Siponen, M. and Vance, A. (2010). Neutralization: New insights into the problem of employee information systems security policy violations. *MIS Quarterly*, *34*(3), 487–502.

Siponen, M., Willison, R., and Baskerville, R. (2008). Power and practice in information systems security research. *Proceedings of the 2008 International Conference on Information Systems*, Paris, France.

Smith, R. F. (2009). When good admins go bad: The critical need for log management as a deterrent/detective control, from http://tinyurl.com/cffgx67 (accessed on March 20, 2013).

Stafford, T. F. and Poston, R. (2010). Online security threats and computer user intentions. *Computer*, *43*, 58–64.

Stanton, J. M., Stam, K. R., Mastrangelo, P., and Jolton, J. (2005). Analysis of end user security behaviors. *Computers and Security*, *24*(2), 124–133.

Straub, D. W. (1990). Effective IS security: An empirical study. *Information Systems Research*, *1*(3), 255–276.

Straub, D. W. and Welke, R. J. (1998). Coping with systems risk: Security planning models for management decision-making. *MIS Quarterly*, *22*(4), 441–469.

Sykes, G. and Matza, D. (1957). Techniques of neutralization: A theory of delinquency. *American Sociological Review, 22*(6), 664–670.

Thomson, M. E. and von Solms, R. (1998). Information security awareness: Educating your users effectively. *Information Management and Computer Security, 6*(4), 167.

Tsohou, A., Kokolakis, S., Karyda, M., and Kiountouzis, E. (2008). Investigating information security awareness: Research and practice gaps. *Information Security Journal: A Global Perspective, 17*(5/6), 207–227.

Vamosi, R. (2002). Turning script kiddies into real programmers. *ZDNet.* Retrieved from http://www.zdnet.com/news/turning-script-kiddies-into-real-programmers/120122

Vance, A. and Siponen, M. (2012). IS security policy violations: A rational choice perspective. *Journal of Organizational and End User Computing, 24*(1), 21–41.

Vroom, C. and von Solms, R. (2004). Towards information security behavioural compliance. *Computers and Security, 23*, 191–198.

Warkentin, M. and Johnston, A. C. (2006). IT security governance and centralized security controls. In M. Warkentin and R. Vaughn (Eds.), *Enterprise Information Assurance and System Security: Managerial and Technical Issues* (pp. 16–24). Hershey, PA: Idea Group Publishing.

Warkentin, M. and Johnston, A. C. (2008). IT governance and organizational development for security management. In D. Straub, S. Goodman, and R. L. Baskerville (Eds.), *Information Security Policies and Practices* (pp. 46–68). Armonk, NY: M.E. Sharpe.

Warkentin, M., Johnston, A. C., and Shropshire, J. (2011a). The influence of the informal social learning environment on information privacy policy compliance efficacy and intention. *European Journal of Information Systems, 20*(3), 267–284.

Warkentin, M., Straub, D., and Malimage, M. (2011b). Measuring the dependent variable for research into secure behaviors. *Decision Sciences Institute Annual National Conference*, Boston, MA, November 19.

Warkentin, M., Straub, D., and Malimage, M. (2012). Measuring secure behavior: A research commentary. *Proceedings of the Annual Symposium on Information Assurance*, Albany, NY, June 5–6, pp. 1–8.

Warkentin, M. and Willison, R. (2009). Behavioral and policy issues in information systems security: the insider threat. *European Journal of Information Systems, 18*(2), 101–105.

Who commits the crime. (n.d.). Retrieved May 20, 2012, from http://www.theiacp.org/investigateid/introduction/who-commits-the-crime/

Willison, R. and Siponen, M. (2009). Overcoming the insider: Reducing employee computer crime through situational crime prevention. *Communications of the ACM, 52*(9), 133–137.

Willison, R. and Warkentin, M. (2009). Motivations for employee computer crime: Understanding and addressing workplace disgruntlement through application of organizational justice. In A. Vance (Ed.), *Proceedings of the IFIP TC8 International Workshop on Information Systems Security Research* (pp. 127–144). International Federation for Information Processing, Cape Town, South Africa.

Willison, R. and Warkentin, M. (2013). Beyond deterrence: An expanded view of employee computer abuse. *MIS Quarterly, 37*(1), 1–20.

Wilson, M. and Hash, J. (2003). *Building an Information Technology Security Awareness and Training Program (SP800-50).* Washington, DC: U.S. Government Printing Office.

Witte, K. (1992). Putting the fear back into fear appeals: The extended parallel process model. *Communication Monographs, 59*, 329–349.

Wlasuk, A. (2012). Help! I think my kid is a Script Kiddie. *SecurityWeek.* Retrieved from http://www.securityweek.com/help-i-think-my-kid-script-kiddie

Woon, I., Tan, G.-W., and Low, R. (2005). A protection motivation theory approach to home wireless security. Paper Presented at the *International Conference on Information Systems*, Las Vegas, NV.

Workman, M. (2007). Gaining access with social engineering: An empirical study of the threat. [Article]. *Information Systems Security, 16*(6), 315–331. doi: 10.1080/10658980701788165

Workman, M. (2008). Wisecrackers: A theory-grounded investigation of phishing and pretext social engineering threats to information security. *Journal of the American Society for Information Science and Technology, 59*(4), 662–674.

Workman, M., Bommer, W. H., and Straub, D. W. (2008). Security lapses and the omission of information security measures: A threat control model and empirical test. *Computer in Human Behavior, 24,* 2799–2816.

Wright, R. T. and Marett, K. (2010). The influence of experiential and dispositional factors in phishing: An empirical investigation of the deceived. *Journal of Management Information Systems, 27*(1), 273–303.

Zhang, L. and McDowell, W. C. (2009). Am I really at risk? Determinants of online users' intentions to use strong passwords. *Journal of Internet Commerce, 8,* 180–197.

55

Privacy, Accuracy, and Accessibility of Digital Business

Ryan T. Wright
University of Massachusetts—Amherst

David W. Wilson
University of Arizona

55.1 Introduction

Today's society has been transformed by our ability to do business anywhere using just about any device with Internet connectivity. Digital business refers to the capabilities that make it possible to access information and transact using ubiquitous technologies enabled by the digitization of processes, documents, and services. Technologies facilitating digital business such as cloud computing and mobile access have changed how organizations offer access to information by allowing employees and customers to increase efficiency and task satisfaction by using online tools. As the technology behind digital business has become more and more complex and access become more and more available, severe problems with the security of these systems have emerged.

Many CEOs have ranked security and privacy above financial concerns for their organizations (Boltz 2011). This is not surprising given the publicity surrounding the alleged illicit access to private and sensitive information by several large companies with a strong presence in digital business. Information privacy has also been a major societal/government issue as politicians struggle to protect utility infrastructures that are run using Internet technologies (Bartz 2012). Examining the impact of IT security breaches on privacy, it is evident that this problem is pervasive for both consumers and organizations. For example, Langevin reported that in 2010 cyber-attacks on private information cost nearly $8 billion in the United States (Langevin 2011). Further, over 8 million U.S. residents have been victims of identity theft due to digital attacks (Langevin 2011). It is clear that information privacy is a serious issue for individuals, organizations, and society in general.

There are typically two pillars to combating digital privacy threats. First, we use security technology (e.g., specialized software) to combat attacking technology (e.g., viruses). The Internet environment has been defended for many years by security systems such as firewalls, virtual private network (VPNs), intrusion detection systems, and so on. On the other hand, the human component of protecting organizations' and individuals' privacy looms as an even larger problem. Even the most sophisticated IT security systems can be easily thwarted by human error, incorrect usage, or malicious actions. For example, it was discovered that the Texas government had the private information of over 3 million state workers exposed on the Internet for over a year due to human error (Shannon 2011). Further, in October of 2010 Microsoft servers were commandeered and misused by spammers to promote more than 1000 fraudulent pharmaceutical websites (McMillan 2010). Employee misuse of the systems was again blamed for this security breach.

It is clear that regardless of the technology, the human aspect of information privacy protection remains central in understanding and controlling against breaches, attacks, and misuse. The information systems (IS) discipline should be the right place to address the human–technology–business concerns. IS typically is the interface between business and IT. By providing an understanding of how humans interface with the technology, we can provide a foundation that may help curb information privacy issues. For this reason, the goal of this chapter is to provide a framework to understand the contemporary literature surrounding IS privacy, which is closely tied to the issues of security, and provide a roadmap for future efforts in understanding this important problem of information privacy that threatens digital business. The chapter will focus on the behavioral (as opposed to technical) issues being addressed by IS researchers in the privacy domain.

The chapter will unfold as follows. First, we will define and examine the interplay between information privacy and security. Next, we will provide a foundation for understanding privacy at the individual level. This is then followed by outlining the organizational-level privacy issues. Finally, we will provide a roadmap for future research on privacy and security using the IS research lens.

55.2 Defining IS Privacy and Security

It is difficult to explore issues regarding information privacy without discussing security. Many scholars have argued that privacy and security are integrated and are difficult to tease apart (Chan et al. 2005). To this point, Fair Information Practices have been developed that include the intersection of privacy and security. These global principles are guidelines to balance privacy concerns with organizations' interests (see Table 55.1). This section focuses on the important and significant relationship between privacy and security.

TABLE 55.1   Fair Information Practices

Fair Information Practice	Purpose	Example
Notice	To alert customers to the gathering of their personal information	Your information is collected so we can offer you appropriate goods and services
Choice	To extend to customers the ability to choose whether their information is tracked, used, and reused	You can choose not to receive e-mails that provide information about sales promotions
Access	To offer customers access to their personal files to allow them to request that inaccurate information is corrected	You can request the ability to view your file and make any necessary corrections
Security	To assure customers that their information cannot be accessed by any persons other than those authorized by the organization	We encrypt all information and store it on our secure servers

Source: Chan, Y.E. et al., *Comm. Assoc. Inform. Syst.*, 16(12), 270, 2005.

Companies and individuals have private digital information that they wish to protect from unauthorized access, tampering, corruption, or unplanned disasters. The provision of this protection is referred to as security or, more specifically, information security. The concept of information privacy is rooted in the view that privacy consists of an individual's control of access to the self (Altman 1975). Accordingly, most modern definitions of information privacy equate privacy with the ability to control one's private information and others' access thereto (Smith et al. 2011). A fairly comprehensive definition of this form of privacy is provided by Margulis (1977, p. 10): "Privacy... represents the control of transactions between person(s) and other(s), the ultimate aim of which is to enhance autonomy and/or to minimize vulnerability."

In the early 2000s, marketing organizations started categorizing end users based on their inherent privacy beliefs (Harris Interactive and Westin 2002). Overall, there are three basic categories:

1. Privacy fundamentalists are the users who do not see any benefit in disclosing any personal or sensitive information. These users are extremely careful of their digital information. According to Harris Interactive, this makes up about 25% of the U.S. population.
2. Privacy pragmatists are those users who undertake a cost–benefit analysis to the collect of their personal information. These users evaluate how organizations are going to use their data and weigh this use with the value of the service. The pragmatists often use cognitive economics risk–reward models to help with their decisions. Harris Interactive predicts that about 55% of the U.S. population consists of privacy pragmatists.
3. Privacy unworried are users who are unconcerned in general with how organizations use their personal information. This group generally rejects others' view that there are concerns with personal information. Harris Interaction classifies about 20% of the U.S. population as unworried.

Companies' and individuals' need for security is driven by a desire for one of several privacy concerns, though the relationship between privacy and security is debated in the literature (Belanger et al. 2002). Culnan and Williams (2009) argue that even when organizations take sufficient precautions in protecting personal information (thus providing information security), they might still engage in or permit unauthorized use of that information, which violates users' information privacy. Thus, as Ackerman (2004) states, "security is necessary for privacy, but security is not sufficient to safeguard against subsequent use, to minimize the risk of... disclosure, or to reassure users" (p. 432, cf. Smith et al. 2011). The next section will outline the mechanisms used to understand a user's expectations for security and privacy.

55.2.1 Protecting Users

Protecting users' privacy with security measures is the nexus of understanding both security and privacy issues. This protection is accomplished through providing three security assurances to the end users of an IS: (1) integrity, (2) confidentiality, and (3) authentication (Smith et al. 2011). Taken together, the security assurances provide a framework for end users and organizations to understand the mechanics involved in information security. Since our focus is on the human factors involved in protecting users' privacy, we will break down each assurance in terms of possible threats and opportunities for knowledge advancement (See Table 55.2). In order to do so, we will integrate Smith et al.'s conceptualization of security assurances with Smith et al.'s (1996) research on an individual's information privacy concerns or concern for information privacy (CFIP).

Smith et al.'s (2011) conceptualization of security can be likened to an onion where each assurance is based on the previous order of assurance. Core to the assurance model is integrity. The next layer is confidentiality and finally authentication. With this layered approach, in order to create problems with one of the assurances, the previous assurance must be compromised. For example, if the confidentiality of the private information is inappropriately leaked, first authentication must have been compromised. This is a general rule of thumb when conceptualizing how security threats affect the assurances (see Figure 55.1).

TABLE 55.2   Understanding Privacy Concerns and Security Assurances

Privacy Concerns	Definition	Human Threats	Security Assurance
Collection errors	Provides assurance that information is not altered during transit or storage	Information accuracy Information reliability	Integrity
Secondary use	Ensures that information is disclosed only to authorized individuals and for authorized purposes	Information property	Confidentiality
Unauthorized access	Provides verification of one's identity and permission to access a given piece of information	Information accessibility	Authentication

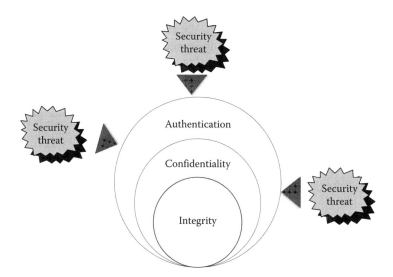

FIGURE 55.1   Relationship between security assurances.

The following section discusses the ways in which the three security assurances provide protection of users' private information. We discuss each of the aspects of Stewart and Segars' (2002) information privacy concerns research. These concerns are summarized in the first column of Table 55.1, which also includes definitions and human threats for each of the three categories of privacy concerns. Table 55.1 also maps these privacy concerns to each of the three security assurances illustrated in Figure 55.1. We now proceed to discuss each grouping of privacy concerns and security assurances in turn, addressing first the privacy concerns, then explaining the related security assurance.

55.2.2 Privacy: Collection and Errors

The collection and accuracy of private information are one potential source of users' concerns regarding information privacy (Stewart and Segars 2002). The act of collecting private information inherently reduces an individual's control over that information, since the individual no longer has direct oversight regarding the information collected. Collection concerns focus on both the amount of personal information being collected and how the information is being stored (Culnan and Bies 2003; Malhotra et al. 2004a). Error concerns are based on the protection against deliberate and accidental error in personal data (Smith et al. 1996). Though the Smith et al. (1996) model separates collection and information errors, these two are often discussed together since most errors in IS take place during a collection process (Clouse et al. 2010). And, in the context of our security assurance framework, both collecting private information and ensuring that it is free from error are related to integrity, as discussed next.

55.2.3 Security: Integrity

Integrity provides assurance that collected private information is not altered during transit or storage. There are two significant human threats associated with integrity. The first threat associated with human interactions is information accuracy. Information accuracy entails ensuring that the fidelity of information stays intact. Security issues often occur when users (or hackers) change data. Hospitals are often faced with accuracy problems as small changes by medical staff in electronic medical records can cause major patient problems and even result in death (Cliff 2012). The second factor that can be manipulated by humans is information reliability. Information reliability is a product of how the users view the information provided by a system. In many instances, we have seen a user question the system's reliability and therefore ignore system requests or system information when making important decisions. One such example is seen in research done on Microsoft's anti-phishing toolbar (Wu et al. 2006). In this study, it was shown that the information provided by the system—in this case the anti-phishing toolbar—was unreliable less than 10% of the time. However, this unreliable information caused users to disregard the information provided by the tool altogether, causing a security threat.

55.2.4 Privacy: Secondary Use

The application of confidentiality in privacy is mainly concerned with the secondary use of data. This may include selling of personal information, marketing using personal information, using personal information to make business decisions, and so on. Although explicit research on secondary use of information is scant, there has been a great deal of popular press about this issue. For example, in late 2011 Google changed their data privacy statement (Garside 2012). This change in policy gained much attention, as Google owns a lot of personal information about their users. It is clear that Google benefits when aggregating personal information from their different businesses. For example, Google can aggregate private information about which videos a user watches from YouTube with geo-location information from their Android phone while scanning the user's Gmail account for certain key words. Using this aggregated data Google can clearly distinguish market segments at a very fine level of detail.

The concerns of users regarding tracking of their online behavior have caused government bodies to start regulating the use of tracking technology such as "cookies," which are text files used to track Internet browsing behavior. In one such case, the European Union's 2011 online privacy law now makes it very difficult for marketers to capture users' information by directing online companies such as Microsoft and Google to add a "do not track" button to their web browsing software (Enright 2012).

There are two main points of emphasis when addressing privacy concerns surrounding secondary use. First is the issue of control by the user of information access by other parties. Control in this instance means that users desire control over who sees what particular piece of personal information. For example, the "do not track" button on browsers allows users to control which information is captured. The second issue is awareness (what Chan et al. 2005 classify as "Notice"). Awareness is the individual's desire to be accurately informed regarding the extent to which the organization shares his or her personal information with third parties. One of the biggest issues for consumers is whether the organization will be allowed to sell this personal information to interested third parties. Being in control and aware in order to reduce users' concerns regarding secondary use requires that organizations manage confidentiality.

55.2.5 Security: Confidentiality

Confidentiality ensures that information is disclosed only to authorized individuals and for authorized purposes. The research around confidentiality in general is difficult to execute due to its sensitive nature. The core threat issue has to do with information property. Information property focuses on who owns the personal information and how this information can be exchanged (Litman 2000).

55.2.6 Privacy: Unauthorized Access

It is clear that access to information is critical to knowledge workers and decision makers in an organization. However, unlimited access to information is a concern as it increases the likelihood of breaches, leaks, and misuse of private information (Malhotra et al. 2004a). As with most security decisions, and as previously mentioned, there is commonly a risk–reward decision pertaining to privacy. With authentication and control these questions are as follows:

1. Who should access the information?
2. For what purpose do they need access?
3. How is this information being used?
4. When do they need access to this information?

These questions most definitively are associated with privacy concerns. Privacy concerns about improper access include trepidations that personal data are readily available to users that are not authorized to access, use, or disclose the personal information (Smith et al. 1996). Understanding users' desire for protection against unauthorized access to private information is a critical security issue that needs to be addressed by the IS authentication process.

55.2.7 Security: Authentication

Authentication provides verification of one's identity and permission to access a given piece of information. Authentication is sometimes seen as a mechanical process whereby a system either provides access or not. Authentication processes are typically set by policies and business rules and enforced by automated systems to provide access to personal or sensitive information. The concept of users' accessibility is the human threat in this assurance category. The concept of information accessibility has been around since the early 1980s. Information accessibility is defined as what information a person or organization has the right to obtain about others and how this information can be accessed and used (Culnan 1984). Accessibility, which is enforced by authentication, is a critical issue for IS managers. Seminal research has posited a positive relationship between perceived accessibility of information and the information use (Culnan and Armstrong 1999).

The next two sections will build on the privacy concerns and security assurances explored earlier to give some guidance on how individual behaviors can influence security behavior. This is followed by scrutinizing how these individual behaviors influence organizational policies and procedures.

55.3 Privacy Behaviors

As the pivotal factor in considering information privacy, humans have been the object of a plethora of privacy and security research and theories in the IS discipline. Organizational policy and outcomes are undoubtedly linked to individual behaviors. Thus, we first synthesize and summarize the literature concerning individual behavior related to information privacy and security. These topics and sample studies are summarized in Table 55.3.

TABLE 55.3 Privacy, Security, and Individual Behavior Theories

Theories	Description	Sample Studies
Trust and risk literature	Common constructs included in theories modeling individual security and privacy behavior. These serve to facilitate or inhibit secure behavior and/or information disclosure	*Trust and distrust* (Gefen et al. 2008; McKnight and Choudhury 2006; McKnight et al. 2002) *Risks* (Dinev and Hart 2006; Featherman and Pavlou 2003; Gabrieal and Nyshadham 2008; Pavlou and Gefen 2004) *Privacy concerns* (Malhotra et al. 2004a; Smith et al. 1996; Xu et al. 2011)
Protection-motivation theory	Self-protection theory adapted from the health behavior domain, which argues that secure behavior is motivated by perceptions of vulnerability, severity of threat, and the effectiveness or efficacy of secure behavior	(Herath and Rao 2009b; Johnston and Warkentin 2010; Lee and Larsen 2009; Ng et al. 2009)
Deterrence theory	Theory adapted from criminology that predicts the effectiveness of sanctions on users' secure behavior. The theory generally concerns a sanction's severity and certainty	(D'Arcy et al. 2009; Harrington 1996; Herath and Rao 2009a; Kankanhalli et al. 2003; Pahnila et al. 2007; Straub 1990)
Neutralization/rational choice	Theoretical lens that accounts for users who justify or neutralize their insecure behavior and/or rationally decide to behave insecurely after weighing the costs and benefits of secure behavior	(Li et al. 2010; Siponen and Vance 2010)
Theory of planned behavior	Common cross-disciplinary theory that predicts a person's behavior according to behavioral intentions, which are influenced by attitudes, social norms, and perceptions of self-efficacy or behavioral control. This has been used as a framework to explain different aspects of security compliance and other individual behaviors	(Ajzen 1991; Bulgurcu et al. 2010; Dinev and Hu 2007)
Privacy disclosure	Rich stream of literature that investigates privacy disclosure behavior, both from the perspective of the organization trying to encourage disclosure for marketing and other purposes and from the perspective of the user who would rather not disclose much information in order to protect privacy	(Acquisti and Grossklags 2005a; Chellappa and Sin 2005; Li et al. 2011; Lu et al. 2004; Malhotra et al. 2004a; Smith et al. 1996; Tsai et al. 2011; Xu et al. 2011)

55.3.1 Theories about Perceptual Beliefs

Much of the IS literature that surrounds privacy and security has been based on individuals' perceptions of certain behaviors that have been developed in conjunction with privacy and security theories. Some examples of research questions about individuals' perceptions are as follows: (1) Do you trust people online (McKnight et al. 2011)? (2) How private do you need to keep your sensitive information (Dinev and Hart 2004, 2006)? (3) How concerned are you about disclosing personal information (Malhotra et al. 2004b)? All of this research uses an individual lens of analysis to understand how users' beliefs impact their interactions with technology systems such as e-commerce, enterprise resource planning (ERP), and so on. The following will outline the three main theoretical perspectives and their respective individual-level behavioral perceptions related to privacy.

55.3.1.1 Risks

In the context of privacy, risk has been defined as "the degree to which an individual believes that a high potential for loss is associated with the release of personal information to a firm" (Smith et al. 2011, p. 13). Perceived risks are often characterized as one's sense of control over their personal information (Featherman and Pavlou 2003), or the general beliefs one has about the likelihood of organizations sharing that information (whether intentionally or through insufficient security precautions) with unauthorized third parties (Dinev and Hart 2006). Risks can either be associated with a specific transaction or transaction partner (Li et al. 2011; Slyke et al. 2006) or can refer to a general measure of risks associated with transacting via the Internet (Dinev and Hart 2006; Gabrieal and Nyshadham 2008). When the likelihood of unauthorized sharing or security breaches is perceived to be high, individuals tend to report higher privacy-related concerns (Dinev and Hart 2004, 2006), are less trusting of the organization (Pavlou and Gefen 2004), and are accordingly less willing to conduct transactions with that organization (Pavlou and Gefen 2004).

55.3.1.2 Concerns

Closely related to privacy and security risks, privacy concerns reflect users' worries about the potential loss of information privacy (Xu et al. 2011). Just as with privacy risks, privacy concerns have been framed both at the general level reflecting broad concerns about electronic transactions (Dinev and Hart 2006; Li et al. 2011; Slyke et al. 2006) and at the situation level representing specific concerns about a given transaction or transaction partner (Xu et al. 2011). Privacy concerns have become the default measurement proxy for measuring users' beliefs or attitudes about privacy in general (Smith et al. 2011). Not surprisingly, individuals who report higher privacy concerns, either at the general or at the specific level, perceive privacy risks to be higher (Li et al. 2011; Malhotra et al. 2004a) and are less willing to engage in transactions via the Internet (Dinev and Hart 2006).

55.3.1.3 Trust

Underlying several of the concepts in this section, trust has been validated as key in facilitating online interactions, both in e-commerce (McKnight et al. 2002) and other (e.g., Posey et al. 2010) contexts. Trust is a many-faceted topic, around which a vast literature has developed. As such, it is beyond the scope of this chapter to provide detailed treatment of the topic. We will thus briefly summarize the trust literature, particularly with respect to the interrelationships between trust and the privacy and security concepts.

McKnight et al. (2002), building on earlier work in off-line contexts (Mayer et al. 1995; McKnight et al. 1998), proposed the most popular framework for online trust formation. This model proposes that trusting beliefs about an entity's benevolence, integrity, and competence will predict the extent to which a user will be willing to transact with the entity. The recipient of trust in the majority of prior research is an e-commerce company (e.g., Pavlou and Gefen 2004), but a growing body of research investigates trust from other perspectives such as social media (Lankton and McKnight 2012), productivity software (Lankton and McKnight 2011; McKnight et al. 2011), and knowledge management systems (Thatcher et al. 2011).

Trust has been shown to be an important part of our understanding of privacy and security in online settings. Trust has been shown to predict feelings of privacy (Belanger et al. 2002), but has more often been modeled as an outcome of privacy (Chellappa and Sin 2005; Malhotra et al. 2004a), wherein a reduction of privacy concerns has been associated with higher levels of trust. Schoenbachler and Gordon (2002) showed that when consumers trust companies, they show lower concern for privacy and are more willing to provide personal information (cf. Smith et al. 2011).

55.3.1.4 Distrust

A concept that is related to but distinct from trust is distrust. Rather than assuming trust and distrust to be two extremes on a singular continuum (Rotter 1980), a mounting body of research is finding that the thought patterns underlying trust and distrust are distinct (Dimoka 2010; McKnight and Choudhury 2006). Trust tends to build slowly, with organizations and online services expending great effort to engender trust among their users (Kim and Benbasat 2003). When trust is violated, however, it is "shattered, and it is replaced with a totally different mind-set, what [we call] distrust" (Gefen et al. 2008, p. 278). Research on distrust in the context of privacy and security remains sparse, with the literature firmly focused on the relationships with trust. As with the e-commerce literature, however, differentiating between trust and distrust in the context of privacy and security can yield more valuable and unique understanding than a focus on trust alone can afford (Gefen et al. 2008).

55.3.2 Theories about Compliance Behaviors

Security compliance behavior has been a popular topic in the IS literature, particularly in the last decade. Compliance research generally deals with factors that predict end-user compliance with organizational privacy and security policies (e.g., Bulgurcu et al. 2010), but has also been examined in home security settings, which impact one's exposure to privacy issues (Anderson and Agarwal 2010). A number of theoretical perspectives have been leveraged to explain compliance behavior. We will address each in turn.

55.3.2.1 Protection-Motivation Theory/Health Belief Model

Researchers have adapted the popular protection-motivation theory (PMT; Rogers 1983) to explain why users adopt secure behavior. Closely related are the health belief model (Rosenstock 1966) and the technology threat avoidance theory (Liang and Xue 2010), the latter of which was recently developed in the IS literature as a partial extension of PMT. These theories treat security threats as hazards about which individuals make judgments regarding such factors as threat severity, perceived vulnerability, the efficacy of suggested protections, or the ability of the individual to carry out those protections. Secure behavior is thus treated similar to healthy eating or exercise habits, with individuals varying in their respective judgments of how dangerous security threats are or how vulnerable they are to being compromised in some way. These theories have been used extensively in the IS security compliance literature, in some cases directly (e.g., Herath and Rao 2009b; Johnston and Warkentin 2010; Lee and Larsen 2009; Ng et al. 2009) and in others less as a framework and more as support for individual hypotheses (e.g., Bulgurcu et al. 2010; Pahnila et al. 2007).

55.3.2.2 Deterrence Theory

A large body of IS literature has investigated the effects of punishments or sanctions in motivating individuals to adhere to organizational security policies and behave securely. The facets of sanctions that have generally been explored include severity (how harsh the punishment will be) and certainty (the perceived probability that punishment will be given) (e.g., Straub 1990). Deterrents have been found to be effective in minimizing security violations in organizational (D'Arcy et al. 2009; Harrington 1996; Herath and Rao 2009a; Kankanhalli et al. 2003; Pahnila et al. 2007) and other (Straub and Welke 1998) settings.

55.3.2.3 Neutralization/Rational Choice

Recent expansion beyond the deterrence literature has introduced new perspectives to the issues associated with end-user behavior. One such perspective acknowledges that deterrence mechanisms are effective in many contexts, but that users will occasionally rationalize or neutralize their insecure behavior even in the presence of imposed sanctions (Siponen and Vance 2010), or that users deliberately choose to act insecurely after weighing the costs of following security policies against the benefits of disregarding them (Li et al. 2010). These perspectives have been validated empirically, and they offer a fresh understanding of the compliance issues that complement the findings regarding the effectiveness of deterrence mechanisms.

55.3.2.4 Theory of Planned Behavior

Several frameworks have been proposed in the IS literature that attempt to frame users' intentions according to the theory of planned behavior (TPB; Ajzen 1991). Dinev and Hu (2007) use the TPB to explore users' intentions to adopt protective technologies. Bulgurcu et al. (2010) frame the idea of information security awareness within the TPB and theoretically justify a comprehensive framework that explains the outcomes of information security awareness and how they relate to a user's behavioral intentions regarding secure behavior.

55.3.3 Privacy Disclosure Behaviors

Research regarding privacy disclosure behaviors generally falls into one of two approaches. The first takes the side of the e-commerce company that would like as much information from its customers and potential customers as it can legally acquire, since such personalized information is highly valuable to for marketing and other business applications (Chellappa and Sin 2005; Lu et al. 2004; Xu et al. 2011). Furthermore, for e-commerce companies to gain revenues, their customers must generally supply at least a minimal amount of personal information (e.g., credit card or shipping information) in order to complete a transaction. Thus, this body of research investigates ways to mitigate privacy concerns so as to facilitate as much sharing of private information by the consumer as possible.

For example, Chellappa and Sin (2005) studied the likelihood of users to supply personal information in exchange for personalized browsing experiences, and trust building factors were proposed to mitigate users' privacy concerns. Lu et al. (2004) proposed using what they call social adjustment benefits (a type of social psychological motivation) to motivate users to disclose information to enable companies to pursue a strategy of focused marketing. In a more recent study (Xu et al. 2011), monetary compensation and multiple forms of industry regulation were proposed as factors that would both increase perceived benefits and reduce perceived risks of disclosing information to location-aware mobile services. In these examples, the objective is to maximize information disclosure for the benefit of the company.

The other dominant approach takes the side of the consumer in defending them against too-frequent or too-generous privacy disclosure behavior. This literature is focused on the so-called privacy paradox (Acquisti 2004; Acquisti and Grossklags 2003, 2005a,b; Norberg et al. 2007), wherein individuals state salient concerns about privacy and sharing private information, but proceed to behave in ways contrary to their stated beliefs (Norberg et al. 2007). Acquisti and colleagues (Acquisti 2004; Acquisti and Grossklags 2003, 2005a,b; Tsai et al. 2011) have investigated probable causes of this contradictory behavior. Their work serves as an example of research that seeks to benefit the users in protecting them from overgenerous disclosure behavior. They address the issue from a behavioral economics perspective and suggest that the privacy paradox is the result of such limitations as psychological distortions (e.g., hyperbolic discounting of risks), limited or asymmetrical information (because of which users do not know the full implications of disclosure), and bounded rationality (i.e., the inability to accurately process all of the probabilities and amounts of benefits and costs related to privacy disclosure) (Acquisti and Grossklags 2003). It is possibly because of these limitations that users have the tendency to give up private information in return for relatively small conveniences or rewards, even when their stated privacy concerns do not support this behavior. This irrational behavior is generally framed negatively. For example, Tsai et al. (2011) show that when privacy policies are made more obvious and accessible (thus reducing information asymmetries), users tend to act more securely and reduce their disclosure behavior.

55.3.4 Privacy Issues with Deception and Fraud in the Twenty-First Century

When privacy is violated or security breaks down, the results can be far-reaching. Businesses lose billions of dollars each year due to cyber-attacks (Langevin 2011), and our increasingly socially networked world has introduced a new generation of privacy issues (Vaidhyanathan 2011). For example, there are more and more attacks via social networks that directly target individuals (Lewis and Kaufman 2008). This section addresses several topics related to security or privacy threats, as well as some countermeasures that are being developed to reduce those threats.

Though deception and fraud are age-old traditions of the human race, these have become increasingly nuanced since the advent of the Internet and electronic communications, with direct impacts on information security and privacy. Deception, which has been defined (Buller and Burgoon 1996) as "a message knowingly transmitted by a sender to foster a false belief or conclusion by the receiver," can take many forms. These range from impression management activities and half-truths to outright lies and financial fraud. Deception and fraud mediated through electronic channels have major implications for both traditional and digital businesses, as well as for individual consumers.

One common way in which deception and fraud can occur is in relation to products sold by online retailers. This type of deception has been called product-related deception, which is when online merchants deliberately manipulate product-related information with the intent to mislead customers (Xiao and Benbasat 2011). Though product-related deception is not unique to online channels, customers shopping online have fewer cues with which to recognize deception. This makes Internet users more vulnerable to deceptive practices such as product misrepresentation, nondelivery of products, obfuscation of warranty or refund policies, or misuse of personal or financial information (Pavlou and Gefen 2005; Xiao and Benbasat 2011).

Fraudsters seeking to gain financially from their deception have found that the Internet provides many opportunities for financial fraud perpetrated against individual users. One of the more popular schemes involves what is known as *phishing*. In this type of scam, the "phishers" create an e-mail that mimics the look and feel of a legitimate business and asks the user to visit a website to correct some purported technical error. A link is provided that takes them to a fraudulent website constructed to match the legitimate website in every way possible. If an inattentive user attempts to sign in to the website, the phishers are able to obtain the person's username and password, which can then be used to defraud him or her with full online account access.

With phishing, we see a direct relationship between individual-level privacy beliefs and security practices. Phishing is indeed a direct attack on one's online privacy. Unfortunately, research on phishing and related online deception is somewhat sparse. Dhamija et al. (2006) investigated various individual characteristics, including gender, age, computer experience, or education. They found no significant differences in the ability of users to detect deception in fraudulent websites. In a series of studies (Wright et al. 2010; Wright and Marett 2010), students were experimentally phished for private information, and the researchers then accounted for both experiential factors (e.g., computer self-efficacy, web experience) and dispositional factors (e.g., trust, suspicion) in determining whether the deceiver was successful in phishing the sensitive information from the student. The results show that while both sets of factors play a role in how effective a phishing scam will be, experiential factors had a stronger influence on the outcome of the deception. In terms of privacy, the behavioral aspect of information disclosure adds to the complexity when trying to guard against attacks on private or personal information.

55.3.4.1 Countermeasures

To counteract the exponential rise of online fraud and deception (e.g., Gyongyi and Garcia-Molina 2005), systems have been developed with the specific purpose of alerting users to fake or otherwise fraudulent websites. In other words, privacy alerts are now a part of many automated systems.

Currently, these systems vary in sophistication and effectiveness, and users tend to discount their usefulness and avoid adoption (Wu et al. 2006).

Most fraud detection systems use a lookup or classification strategy (Abbasi et al. 2010). A lookup strategy consists of a small client (usually a browser toolbar installed on the user's machine) that connects to a repository of known fraudulent websites (i.e., a blacklist) and then warns the user before a black-listed site is accessed (Wu et al. 2006; Zhang et al. 2007). A classifier system uses characteristics of the website itself to classify the site as either safe or fraudulent, according to rule-based heuristics (Zhang et al. 2007). These characteristics include such cues as domain name, host country, or content similarity (compared to frequently spoofed sites such as eBay or PayPal) (Abbasi et al. 2010). These lookup- or classification-based systems are generally ineffective, with reported accuracy rates of 70% or lower (Abbasi et al. 2010; Zhang et al. 2007).

Recent advances in automated fraud detection systems (e.g., privacy attack systems) have been embodied in a new fraud detection system proposed by Abassi et al. (2010). This system utilizes a statistical learning theory approach to analyze a wider set of cues and domain-specific knowledge. The authors demonstrate that this more advanced detection method outperforms other, less sophisticated systems. Although this and other fraud detection systems have not been heavily adopted by the general public, advancements in this area are encouraging and should help to combat the pervasive fraudulent activity on the Internet. Reducing the incidence of successful phishing and other fraudulent activity should be a primary focus in further protecting users' information privacy.

55.3.4.2 Organizational Issues

Organizations play a pivotal role in the information privacy and security discussion. This is because organizations are primarily charged with protecting individuals' information privacy, either in protecting employees' financial or other personal information, or in protecting the private information they collect about their customers or clients in the course of doing business. A transaction in which an organization acquires any form of private information from a customer must, as a precondition, engender enough trust or goodwill such that the customer is comfortable with the risk of releasing that information to the organization. Culnan and Bies (2003) argue that *any* disclosure of personal information constitutes an assumption of risk on the part of the consumer. Recall that security is a necessary but insufficient condition for protection of privacy (Ackerman 2004), since privacy entails the added assurance that an individual's private information will not be used in a way other than originally intended by the sharer of that information, nor shared with unauthorized third parties (Culnan and Williams 2009).

Given the organizations' responsibility to protect their stakeholders' privacy, information security has become a critical topic for organizations. Researchers have noted vital financial and legal implications for failing to protect information privacy (Goel and Shawky 2009; Gordon and Loeb 2002). It has also been proposed that a majority of organizational security issues are related to human error or noncompliance with security policies (Stanton et al. 2005). Accordingly, organizations have reported employees' security-related behavior as a top concern for years (Boltz 2011). The issues that organizations cope with, and the countermeasures that they implement, thus result from tendencies of their employees' individual behaviors.

55.3.5 Security Policies/Compliance

As concerns regarding privacy and security have increased, organizations have implemented privacy and security policies to help inform employees of standards and to enforce security precautions where necessary to ensure the privacy of the organization's data and employees. Hone and Eloff (2002) identify the information privacy and security policy, the most singularly important security control that an organization can implement. This view is also shared by others (Knapp et al. 2009; Whitman 2008).

An information privacy and security policy is established to provide management direction and support for information privacy and security as derived from business requirements and relevant laws and regulations (Knapp et al. 2009). Policies generally consist of a general objective or purpose, a stated scope indicating to whom the policy applies, assigned responsibilities where various organizational roles are given ownership over various portions of the policy, and guidelines for compliance, which provide for the policy's enforcement (Olzak 2006). These are general guidelines that vary according to an organization's needs.

That an entire body of literature (noted earlier) has developed around employee noncompliance with security policies is an indication that many policies or portions of policies are not regularly followed by employees. This represents a serious problem for organizations, since noncompliance by even a few employees can largely negate the benefits of even the most stringent of policies. Organizations must strike a balance between instituting policies that are overly restrictive, which can lower productivity or instill contempt among annoyed employees, and policies that are not restrictive enough, which largely defeats the purpose of a security policy. This balance is elusive, and privacy and security policy compliance continues to fuel discussion and investigation in the literature.

55.3.6 Legal Requirements (SOX, GAAP, Others)

Following the several cases of corporate accounting fraud shortly after the turn of the century, the most notable of which were the Enron and WorldCom scandals, the U.S. government enacted legislation designed to close accounting loopholes and generally tighten controls on accounting practices. This legislation, termed the Sarbanes-Oxley (SOX) Act, has had far-reaching effects, many of which have directly impacted the IS function in many organizations (Volonino et al. 2004). Among the requirements detailed in the SOX bill, companies must ensure transparency, accuracy, timeliness, and reliability of their IT-enabled financial reporting and operations systems. These requirements necessitate security policies and managerial oversight to ensure employee compliance with those policies. Indeed, Volonino et al. (2004) state that "to be in compliance with regulatory boards, companies need to develop and deploy effective information security response and investigation policies. Those policies will require collaboration between corporate IT security teams and IT auditors" (p. 226).

In summary, organizations have the responsibility to ensure the security and privacy of their employees and customers. To accomplish this, most organizations employ official policies requiring certain levels of private and secure behavior. Though noncompliance with these policies is the object of considerable research in the literature, organizations continue to strive to provide for their stakeholders' privacy and security.

55.4 Future Research Agendas

This chapter has identified the major theoretical perspectives and some current privacy- and security-related problems and solutions. Clearly, there are many associated research opportunities that IS scholars can address. The following will outline three areas within privacy and security that we believe have the potential to impact end users, organizations, and society in general. We will first explore interesting questions in the design for privacy and security.

55.4.1 Designing for Privacy and Security

Consistent with the theme of individual behaviors that persists in this chapter, we will look toward design as an avenue that affects privacy and security for individuals. Specifically, we will address the balance between allowing end users to act and automating privacy and security tasks. Within application design, security tools have either focused on being invisible (e.g., filters, automated messages, and so on) or attaching attention (e.g., call outs, security alerts, and so on) (Hong 2012). Although there are mixed results, what is clear is that no single method can achieve the results needed for 100% secure practices that ensure one's privacy. Further, since privacy is individually constructed, one system run by homogenous business rules would likely meet with user resistance or nonadoption. For this reason, the first step toward understanding the design elements for secure IT systems is to develop a framework that includes the invisible (e.g., automatic systems) and the visible (e.g., warnings and human decision points). The need for this framework is echoed in Sheng et al.'s (2010) phishing research, which identifies clear gaps between the heuristics (e.g., the decision process for end users) and machine-learning algorithms (e.g., anti-phishing toolbars). Other general privacy and security research has followed this pattern. In order to solve complex phenomena such as privacy and security issues, it will take a multidisciplinary effort from psychologists, computer scientists, and IS scholars. With multidisciplinary efforts, we will be able to address significant privacy and security questions that are as follows:

1. What privacy decisions are best made using machine learning and which are best left to humans?
2. How should machine-learning algorithms interact with end users?
3. Is there a theoretical perspective that could integrate the need for visibility and invisibility of certain processes?

By attempting to address these questions, the privacy researchers should coalesce around common themes within this important domain.

55.4.2 Understanding Training

Training has been posited as the main defense against directed attacks (Hong 2012). In fact, there has been a variety of training experiments by the anti-phishing group at Carnegie Mellon University (CMU) (Kumaraguru et al. 2009; Sheng et al. 2010). Hong states that "it doesn't matter how many firewalls, encryption software, certificates, or two-factor authentication mechanisms an organization has if the person behind the keyboard falls for a phish" (Hong 2012, p. 76). Training against phishing has taken two lines. First, the CMU group and others have designed microgames that teach employees how to spot phishing messages (Sheng et al. 2007). The second is embedded training in which participants are sent phishing messages in a classroom setting and are asked to evaluate these. Kumaraguru et al. (2009) found this approach to be effective even a month after training. These phishing training mechanisms may offer other areas a clear path for implementation. Some questions around privacy and security include the following:

1. Are some training techniques better than others (e.g., mindfulness training, cognitive trainings)?
2. What are the best delivery mechanisms for training (e.g., online, classroom, games)?
3. How do you combine privacy and security training?
4. What are the biggest gaps in training for employees? For consumers?

Obviously, there are many other opportunities for training research that can be considered. It is important first to address the big IS questions using a multidisciplinary point of view and expertise as stated earlier.

55.4.3 Looking at Information Systems' Policy Differently

A lot of attention is currently paid to employee compliance with privacy and security policies (Hone and Eloff 2002; Knapp et al. 2009; Whitman 2008). In fact, most of the theoretical work is geared toward understanding why an organization's personnel comply with the policy or not (e.g., PMT, deterrence theory, and so on). What is not clear is how the policy itself may enact certain behaviors from employees and consumers. The following research questions are appropriate starting points to address how policy affects behavior:

1. Is there a better way to classify the interplay between IS security policies and privacy beliefs?
2. What is the impact of less-restrictive IS polices on the work force?
3. How do privacy and security policies impact productivity?
4. Do privacy policies influence e-commerce interactions?

In sum, there are multiple interesting angles from which to study privacy and security policy issues that are yet to be explored. These and others represent excellent opportunities for future knowledge creation.

55.5 Conclusion

This chapter outlines the current issues with information privacy and security. In doing so, we have provided a framework that summarizes the interplay between users' concerns regarding information privacy and the provision of that privacy through the three basic components of security assurance: (1) authentication, (2) confidentiality, and (3) integrity. Further, we provide the current state of research at the individual and organizational levels of analysis. Finally, we identify three gaps in privacy and security research that IS scholars should address (i.e., design issues, training, and policies).

References

Abbasi, A., Zhang, Z., Zimbra, D., Chen, H., and Nunamaker, J. F. J. Detecting fake websites: The contribution of statistical learning theory, *MIS Quarterly* (34:3) 2010, 435–461.

Ackerman, M. Privacy in pervasive environments: Next generation labeling protocols, *Personal and Ubiquitous Computing* (8:6) 2004, 430–439.

Acquisti, A. Privacy in electronic commerce and the economics of immediate gratification, in: *Proceedings of the 5th ACM Conference on Electronic Commerce*, ACM, New York, 2004, pp. 21–29.

Acquisti, A. and Grossklags, J. Losses, gains, and hyperbolic discounting: An experimental approach to information security attitudes and behavior, in: *2nd Annual Workshop on Economics and Information Security*, College Park, MD, 2003.

Acquisti, A. and Grossklags, J. Privacy and rationality in individual decision making, *IEEE Security and Privacy* (3:1) 2005a, 26–33.

Acquisti, A. and Grossklags, J. Uncertainty, ambiguity, and privacy, in: *4th Workshop Economics and Information Security*, Cambridge, MA, 2005b.

Ajzen, I. The theory of planned behavior, *Organizational Behavior and Human Decision Processes* (50:2) 1991, 179–211.

Altman, I. *The Environment and Social Behavior: Privacy, Personal Space, Territory, and Crowding*, Brooks/Cole Publishing, Monterey, CA, 1975.

Anderson, C. L. and Agarwal, R. Practicing safe computing: A multimethod empirical examination of home computer user security behavioral intentions, *MIS Quarterly* (34:3) 2010, 613–643.

Bartz, D. 2012. Senators to hear pitch for tougher cyber security, (available at http://www.reuters.com/article/2012/03/07/us-usa-cybersecurity-congress-idUSTRE82621W20120307 [accessed on March 24, 2013]).

Belanger, F., Hiller, J. S., and Smith, W. J. Trustworthiness in electronic commerce: The role of privacy, security, and site attributes, *The Journal of Strategic Information Systems* (11:3–4) 2002, 245–270.

Boltz, M. 2011. Top three security concerns every CEO should know, (available at http://chiefexecutive.net/top-three-security-concerns-every-ceo-should-know [accessed on March 24, 2013]).

Bulgurcu, B., Cavusoglu, H., and Benbasat, I. Information security policy compliance: An empirical study of rationality-based beliefs and information security awareness, *MIS Quarterly* (34:3) 2010, 523–548.

Buller, D. B. and Burgoon, J. Interpersonal deception theory, *Communication Theory* (6) 1996, 203–242.

Chan, Y. E., Culnan, M. J., Greenaway, K., Laden, G., Levin, T., and Smith, H. J. Information privacy: Management, marketplace, and legal challenges, *Communications of the Association for Information Systems* (16:12) 2005, 270–298.

Chellappa, R. K. and Sin, R. G. Personalization versus privacy: An empirical examination of the online consumer's dilemma, *Information Technology and Management* (6:2) 2005, 181–202.

Cliff, B. Q. 2012. Patient safety in Oregon hospitals, The Bulletin (available at http://www.bendbulletin.com/article/20120309/NEWS0107/203090374/ [accessed on March 24, 2013]).

Clouse, S. F., Wright, R. T., and Pike, R. Employee information privacy concerns with employer held data: A comparison of two prevalent privacy models, *Journal of Information Security and Privacy* (6:3) 2010, 47–71.

Culnan, M. J. The dimensions of accessibility to online information: Implications for implementing office information systems, *ACM Transactions on Information Systems* (2:2) 1984, 141–150.

Culnan, M. J. and Armstrong, P. K. Information privacy concerns, procedural fairness, and impersonal trust: An empirical investigation, *Organization Science* (10:1) 1999, 104–115.

Culnan, M. J. and Bies, R. J. Consumer privacy: Balancing economic and justice considerations, *Journal of Social Issues* (59:2) 2003, 323–342.

Culnan, M. J. and Williams, C. C. How ethics can enhance organizational privacy: Lessons from the choicepoint and TJX data breaches, *MIS Quarterly* (33:4) 2009, 673–687.

D'Arcy, J., Hovav, A., and Galletta, D. F. User awareness of security countermeasures and its impact on information systems misuse: A deterrence approach, *Information Systems Research* (20:1) 2009, 79–98.

Dhamija, R., Tygar, J. D., and Hearst, M. Why phishing works, in: *Computer Human Interaction Conference*, Montreal, Quebec, Canada, 2006, pp. 581–590.

Dimoka, A. What does the brain tell us about trust and distrust? Evidence from a functional neuroimaging study, *MIS Quarterly* (34:2) 2010, 373–396.

Dinev, T. and Hart, P. Internet privacy concerns and their antecedents: Measurement validity and a regression model, *Behavior and Information Technology* (23:6) 2004, 413–423.

Dinev, T. and Hart, P. An extended privacy calculus model for E-commerce transactions, *Information Systems Research* (17:1) 2006, 61–80.

Dinev, T. and Hu, Q. The centrality of awareness in the formation of user behavioral intention toward protective information technologies, *Journal of the Association for Information Systems* (8:7) July 2007, 386–408.

Enright, A. 2012. Do-not-track web privacy efforts gain momentum, (available at http://www.internetretailer.com/2012/02/23/do-not-track-web-privacy-efforts-gain-momentum [accessed on March 24, 2013]).

Featherman, M. and Pavlou, P. Predicting E-services adoption: A perceived risk facets perspective, *International Journal of Human-Computer Studies* (59:4) 2003, 451–474.

Gabrieal, I. J. and Nyshadham, E. A cognitive map of people's online risk perceptions and attitudes: An empirical study, in: *41st Annual Hawaii International Conference on System Sciences*, IEEE Computer Society Press, Los Alamitos, CA, 2008.

Garside, J. 2012. Google's privacy policy 'too vague', (available at http://www.guardian.co.uk/technology/2012/mar/08/google-privacy-policy-too-vague?newsfeed=true [accessed on March 24, 2013]).

Gefen, D., Benbasat, I., and Pavlou, P. A research agenda for trust in online environments, *Journal of Management Information Systems* (24:4) 2008, 275–286.

Goel, S. and Shawky, H. Estimating the market impact of security breach announcements on firm values, *Information and Management* (46:7) 2009, 404–410.

Gordon, L. A. and Loeb, M. P. The economics of information security investment, *ACM Transactions on Information and System Security* (5:4) 2002, 438–457.

Gyongyi, Z. and Garcia-Molina, H. Spam: It's not just for inboxes anymore, *IEEE Computer* (38:10) 2005, 28–34.

Harrington, S. The effects of ethics and personal denial of responsibility on computer abuse judgements and intentions, *MIS Quarterly* (20:3) 1996, 257–277.

Harris Interactive and Westin, A. F. *The Haris Poll*, Harris Interactive, Rochester, NY, 2002.

Herath, T. and Rao, H. R. Encouraging information security behaviors in organizations: Role of penalties, pressures and perceived effectiveness, *Decision Support Systems* (47:2) 2009a, 154–165.

Herath, T. and Rao, H. R. Protection motivation and deterrence: A framework for security policy compliance in organisations, *European Journal of Information Systems* (18:2) 2009b, 106–125.

Hone, K. and Eloff, J. H. P. Information security policy—what do international standards say? *Computers and Security* (21:5) 2002, 402–409.

Hong, J. The state of phishing attacks, *Communications of the ACM* (55:1) 2012, 74–81.

Johnston, A. C. and Warkentin, M. Fear appeals and information security behaviors: An empirical study, *MIS Quarterly* (34:3) September 2010, 549–566.

Kankanhalli, A., Teo, H. H., and Tan, B. C. Y. An integrative study of information systems security effectiveness, *International Journal of Information Management* (23:2) 2003, 139–154.

Kim, D. and Benbasat, I. Trust-related arguments in internet stores: A framework for evaluation, *Journal of Electronic Commerce Research* (4:3) 2003, 49–64.

Knapp, K. J., Franklin Morris, R. Jr., Marshall, T. E., and Byrd, T. A. Information security policy: An organizational-level process model, *Computers and Security* (28:7) 2009, 493–508.

Kumaraguru, P., Cranshaw, J., Acquisti, A., Cranor, L., Hong, J., Blair, M. A., and Pham, T. School of phish: A real-world evaluation of anti-phishing training, in: *Proceedings of the 5th Symposium on Usable Privacy and Security*, ACM, Mountain View, CA, 2009, pp. 1–12.

Langevin, J. *Cybersecurity Forum*, University of Rhode Island, Kingston, RI, 2011.

Lankton, N. K. and McKnight, H. D. Examining two expectation disconfirmation theory models: Assimilation and asymmetry effects, *Journal of the Association for Information Systems* (13:2) 2012, 88–115.

Lankton, N. M. and McKnight, D. H. What does it mean to trust Facebook? Examining technology and interpersonal trust beliefs, *The DATABASE for Advances in Information Systems* (42:2) 2011, 32–54.

Lee, Y. and Larsen, K. R. Threat or coping appraisal: Determinants of SMB executives' decision to adopt anti-malware software, *European Journal of Information Systems* (18:2) April 2009, 177–187.

Lewis, K. and Kaufman, J. The taste for privacy: An analysis of college student privacy settings in an online social network, *Journal of Computer-Mediated Communication* (14:1) 2008, 79–100.

Li, H., Sarathy, R., and Xu, H. The role of affect and cognition on online consumers' decision to disclose personal information to unfamiliar online vendors, *Decision Support Systems* (51:3) 2011, 434.

Li, H., Zhang, J., and Sarathy, R. Understanding compliance with internet use policy from the perspective of rational choice theory, *Decision Support Systems* (48:4) 2010, 635–645.

Liang, H. G. and Xue, Y. J. Understanding security behaviors in personal computer u: A threat avoidance perspective, *Journal of the Association for Information Systems* (11:7) 2010, 394–413.

Litman, J. Information privacy/Information property, *Stanford Law Review* (52:1) 2000, 1283–1284.

Lu, Y., Tan, B., and Hui, K.-L. Inducing customers to disclose personal information to internet businesses with social adjustment benefits, in: *International Conference on Information Systems*, Washington, DC, 2004, pp. 272–281.

Malhotra, N. K., Kim, S. S., and Agarwal, J. Internet users' information privacy concerns (IUIPC): The construct, the scale, and a causal model, *Information Systems Research* (15:4) 2004a, 336–355.

Malhotra, N. K., Sung, S. K., and Agarwal, J. Internet users' information privacy concerns (IUIPC): The construct, the scale, and a causal model, *Information Systems Research* (15:4) 2004b, s336–s355.

Margulis, S. T. Conceptions of privacy: Current status and next steps, *Journal of Social Issues* (33:3) 1977, 5–21.

Mayer, R. C., Davis, J. H., and Schoorman, F. D. An integrative model of organizational trust, *Academy of Management Review* (20:3) 1995, 709–734.

McKnight, D. H., Carter, M., Thatcher, J. B., and Clay, P. F. Trust in a specific technology: An investigation of its components and measures, *ACM Transactions on Management Information Systems Quarterly* (2:2) 2011, 1–15.

McKnight, D. H. and Choudhury, V. Distrust and trust in B2C E-commerce: Do they differ? in: *International Conference on Electronic Commerce*, Fredericton, New Brunswick, Canada, 2006.

McKnight, D. H., Choudhury, V., and Kacmar, C. Developing and validating trust measures for e-commerce: An integrative typology, *Information Systems Research* (13:3) 2002, 334–359.

McKnight, D. H., Cummings, L. L., and Chervany, N. L. Initial trust formation in new organizational relationships, *Academy of Management Review* (23:3) 1998, 473–490.

McMillan, R. 2010. Human error gave spammers keys to microsoft systems, (available at http://www.computerworld.com/s/article/9191059/Human_error_gave_spammers_keys_to_Microsoft_systems [accessed on March 24, 2013]).

Ng, B., Kankanhalli, A., and Xu, Y. Studying users' computer security behavior: A health belief perspective, *Decision Support Systems* (46:4) 2009, 815–825.

Norberg, P. A., Horne, D. R., and Horne, D. A. The privacy paradox: Personal information disclosure intentions versus behaviors, *Journal of Consumer Affairs* (41:1) Summer 2007, 100–126.

Olzak, T. *Just Enough Security: Information Security for Business Managers*, Lulu.com, Raleigh, NC, 2006.

Pahnila, S., Siponen, M., and Mahmood, A. Employees' behavior towards IS security policy compliance, in: *40th Annual Hawaii International Conference on System Sciences*, Big Island, 2007, pp. 156b–156b.

Pavlou, P. and Gefen, D. Building effective online marketplaces with institution-based trust, *Information Systems Research* (15:1) 2004, 37–59.

Pavlou, P. and Gefen, D. Psychological contract violation in online marketplaces: Antecedents, consequences, and moderating role, *Information Systems Research* (16:4) 2005, 372–399.

Posey, C., Lowry, P., Roberts, T., and Ellis, T. Proposing the online community self-disclosure model: The case of working professionals in France and the U.K. who use online communities, *European Journal of Information Systems* (19:2) 2010, 181.

Rogers, R. W. Cognitive and physiological processes in fear appeals and attitude change: A revised theory of protection motivation, in: *Social Psychophysiology*, J. Cacioppo and R. Petty (eds.), Guilford Press, New York, 1983.

Rosenstock, I. M. Why people use health services, *Milbank Memorial Fund Quarterly* (44:3) 1966, 94–127.

Rotter, J. B. Interpersonal trust, trustworthiness, and gullibility, *American Psychologist* (35) 1980, 1–7.

Schoenbachler, D. D. and Gordon, G. L. Trust and customer willingness to provide information in database-driven relationship marketing, *Journal of Interactive Marketing* (16:3) 2002, 2–16.

Shannon, K. 2011. Breach in Texas comptroller's office exposes 3.5 million social security numbers, birth dates, (available at http://www.dallasnews.com/news/state/headlines/20110411-breach-in-texas-comptrollers-office-exposes-3.5-million-social-security-numbers-birth-dates.ece [accessed on March 24, 2013]).

Sheng, S., Holbrook, M., Kumaraguru, P., Cranor, L. F., and Downs, J. Who falls for phish? A demographic analysis of phishing susceptibility and effectiveness of interventions, in: *Proceedings of the 28th International Conference on Human Factors in Computing Systems*, ACM, Atlanta, GA, 2010, pp. 373–382.

Sheng, S., Magnien, B., Kumaraguru, P., Acquisti, A., Cranor, L. F., Hong, J., and Nunge, E. Anti-phishing phil: The design and evaluation of a game that teaches people not to fall for phish, in: *Symposium on Usable Privacy and Security (SOUPS)*, ACM Press, Pittsburgh, PA, 2007.

Siponen, M. and Vance, A. Neutralization: New insights into the problem of employee information systems security policy violations, *MIS Quarterly* (34:3) 2010, 487–502.

Slyke, C. V., Shim, J. T., Johnson, R., and Jiang, J. Concern for information privacy and online consumer purchasing1, *Journal of the Association for Information Systems* (7:6) 2006, 415.

Smith, H. J., Dinev, T., and Xu, H. Information privacy research: An interdisciplinary review, *MIS Quarterly* (35:4) 2011, 989–1015.

Smith, H. J., Milberg, S. J., and Burke, S. J. Information privacy: Measuring individuals' concerns about organizational practices, *MIS Quarterly* (20:2) 1996, 167–196.

Stanton, J. M., Stam, K. R., Mastrangelo, P., and Jolton, J. Analysis of end user security behaviors, *Computers and Security* (24:2) March 2005, 124–133.

Stewart, K. A. and Segars, A. H. An empirical examination of the concern for information privacy instrument, *Information Systems Research* (13:1) 2002, 36–49.

Straub, D. W. Effective IS security, *Information Systems Research* (1:3) 1990, 255–276.

Straub, D. W. and Welke, R. J. Coping with systems risk: Security planning models for management decision making, *MIS Quarterly* (22:4) 1998, 441–469.

Thatcher, J. B., McKnight, D. H., Baker, E., and Arsal, R. E. The role of trust in post adoption IT exploration: An empirical examination of knowledge management systems, *IEEETransactions on Engineering Management* (58:1) 2011, 56–70.

Tsai, J., Egelman, S., Cranor, L., and Acquisti, A. The effect of online privacy information on purchasing behavior: An experimental study, *Information Systems Research* (22:2) 2011, 254.

Vaidhyanathan, S. Welcome to the surveillance society, *IEEE Spectrum* (48:6) 2011, 48–51.

Volonino, L., Gessner, G. H., and Kermis, G. F. Holistic compliance with Sarbanes-Oxley, *Communications of AIS* (14:11) 2004, 219–233.

Whitman, M. E. Security policy: From design to maintenance, in: *Advances in Management Information Systems*, D.W. Straub, S. Goodman, and R.L. Baskerville (eds.), M.E. Sharpe, Armonk, NY, 2008, pp. 123–151.

Wright, R., Chakraborty, S., Basoglu, A., and Marett, K. Where did they go right? Investigating deception cues in a phishing context, *Group Decision and Negotiation* (Online:July 2009) 2010, 391–416.

Wright, R. and Marett, K. The influence of experiential and dispositional factors in phishing: An empirical investigation of the deceived, *Journal of Management Information Systems* (27:1) 2010, 273–303.

Wu, M., Miller, R. C., and Garfunkel, S. L. Do security toolbars actually prevent phishing attacks? in: *SIGCHI Conference on Human Factors in Computing Systems*, Montreal, Quebec, Canada, 2006, pp. 601–610.

Xiao, B. and Benbasat, I. Product-related deception in E-commerce: A theoretical perspective, *MIS Quarterly* (35:1) 2011, 169–195.

Xu, H., Dinev, T., Smith, J., and Hart, P. Information privacy concerns: Linking individual perceptions with institutional privacy assurances, *Journal of the Association for Information Systems* (12:12) 2011, 798–824.

Zhang, Y., Egelman, S., Cranor, L., and Hong, J. Phinding phish: Evaluating anti-phishing tools, in: *14th Annual Network and Distributed System Security Symposium*, San Diego, CA, 2007.

56

Digital Forensics

Vassil Roussev
University of New Orleans

56.1 Introduction

Forensic science (or *forensics*) is dedicated to the systematic application of scientific methods to gather and analyze evidence in order to establish facts and conclusions that can be presented in a legal proceeding. Digital forensics (a.k.a. computer or cyber forensics) is a subfield within forensics, which deals specifically with digital artifacts, such as files, and computer systems and networks used to create, transform, transmit, and store them. This is a very broad definition, and there are several closely related fields and sub-fields that share methods and tools with forensics. Indeed, the forensic methods used were originally developed for a different purpose and continue to be shared. Thus, it is primarily the *purpose*, and not their functionality, that classify particular tools and techniques as forensic.

Malware analysis [Mali08] (or *malware forensics*) specializes in understanding the workings and origins of malicious software, such as viruses, worms, trojans, backdoors, rootkits, and bots. Such analysis is often necessary after a security breach to understand the exploited vulnerability vector and the scope of the breach. The analysis depends on well-prepared and executed *incident response* phase, which identifies and collects the source data for the analysis, while restoring the affected IT systems to a known good state.

Reverse engineering, the process of inferring the structure, or run time behavior of a digital artifact, is almost always a critical step in any kind of forensic analysis. Apart from the obvious example of malware, most closed-source software produces artifacts that are rarely publicly documented by the vendor (unless they want to establish them as a standard). For example, the on-disk structure of Microsoft's NTFS file system has never been officially published; the format of the files used by Microsoft Office was not officially released until 2008.*

Data recovery, the process of salvaging useable digital artifacts from partial or corrupted media, file systems, or individual files is another technique frequently employed in forensics during the initial, data acquisition, stage. More broadly, the definition of data recovery often includes the retrieval of encrypted, or otherwise hidden, data.

The intelligence community uses *document and media exploitation* (or *DOMEX*) [Garf07b] to extract actionable intelligence from relevant data sources. In practice, the DOMEX relies heavily on available forensic techniques, and higher-level analysis, to make sense out of large amounts of raw data.

The goal of this chapter is to present an overview of the state of the art in digital forensics from a technical perspective, with an emphasis on research problems. This allows us to treat the topic in a more holistic manner by presenting issues, tools, and methodologies that are common to a broader field in computer security and IT management. The legal aspects of computer forensics, which is clearly of concern to practitioners in law enforcement, can largely be modeled as placing various restrictions on the ways in which the data can be gathered and the way conclusion can be presented in a courtroom. These do present some additional research and development problems; however, these are relatively minor.

56.2 Definitions

The overall objective of digital forensic analysis is to reconstruct a chain of events that have resulted in the currently observable state of a computer system or digital artifact. The purpose of the analysis is to help answer six basic questions that arise in most inquiries: *what* happened? *where* did it happen? *when* did it happen? *how* did it happen? *who* did it? *why* did they do it? Any notion of completeness of the analysis, which can be of legal significance, is tied to the target and purposes of the original investigation and is very difficult to define in technical terms.

Another approach to understanding cyber forensics is to view it from a procedural standpoint—what does a typical investigation entail? This is an important question as one of the cornerstones of scientific processes is that the observed results must be independently reproducible. There are four generally accepted [Kent06] steps in the forensics process: *collection*, *extraction*, *analysis*, and *reporting*. *Examiner preparation*, in which the investigator is trained to perform the specific type of investigation, is sometimes referenced as an added, preliminary step.

During the collection phase, the investigation must identify potential sources of evidence (computers, mobile phones, flash/optical media, etc.) and choose, acquire, and preserve the appropriate ones based on the objective of the inquiry and legal restrictions. We will largely ignore this phase as it is an issue of procedure and training, not science. Similarly, the reporting phase during which investigators prepares a report and testifies will also be outside the scope of this discussion.

The main focus will be on the extraction and analysis phases, which constitute the most technically challenging parts. The investigator goes through the process of forming an initial hypothesis, searching for evidence that supports/disproves it, refining/redefining the hypothesis and going back to the data for proof. This process goes through multiple iterations, until the analyst has reconstructed the sequence of relevant events with the necessary level of certainty.

Throughout, examiners must carefully preserve the evidence and document the tools used and actions taken to demonstrate the reproducibility of their analysis.

For the remainder of the chapter, we look at the investigative process from four different perspectives in an effort to provide a comprehensive view of the field. First, we look at general models and best

---

* http://msdn.microsoft.com//en-us//library//cc313118.aspx

practices of digital forensic inquiries that have slowly emerged over the years and can serve as a useful guide as to how a particular type of investigation should be performed.

Next, we consider forensic methods as they relate to main system components of a general purpose computing infrastructure—secondary storage, main memory, and networks—as well as the additional challenges posed by various integrated devices. Next, we consider the major types of artifacts that are typically of interest to an investigation—text, multimedia data (audio/video), executable code, and system data. Most of these have standardized (file) container formats, which allows their analysis to be decoupled from that of the source. Analytic methods in this domain can readily borrow from other fields like information retrieval and signal processing.

Finally, we consider the specific requirements that digital forensic tool developers face. Some of these, such as the efficient use of commodity parallel computing, are shared with other application domains, while others, like the potential need to comply with legal restrictions, are very specific to forensics.

56.3 Inquiry Models and Tool Uses

Over the past decade, both researchers and practitioners have been working to summarize, classify, and formalize the understanding of digital forensic tools and processes. In broad terms, models coming from practitioners, such as investigators in forensic labs, tend to describe the process as a collection of detailed procedures for routine operations (e.g., disk cloning) and high-level "best practices". Work coming from researchers tends to focus on technical aspects of how to deal with specific problems—data recovery, search and query, etc. These models are difficult to compare with each other as they effectively speak to different parts of the problem and in different languages.

To present a unified picture of the digital forensic process, we adapt the sense-making process originally developed by Pirolli and Card [Piro05] to describe intelligence analysis. This is a cognitive model derived from an in-depth *cognitive task analysis* (CTA), which provides an overall view of an intelligence analyst's work. Although some of the tools used may vary, forensic analysis is very similar in nature to intelligence analysis—in both cases analysts have to go through a mounting of raw data to identify (relatively few) relevant facts and put them together in a coherent story.

The benefit of using the Pirolli and Card model is at least threefold: (a) it provides a fairly accurate description of the investigative process in its own right and allows us to map the various tools to the different phases of the investigation; (b) it provides a suitable framework for explaining the relationships of the various models developed within the area of digital forensics; and (c) it can seamlessly incorporate information from other lines of the investigation.

56.3.1 Pirolli and Card Cognitive Model

The overall process is shown in Figure 56.1. The rectangular boxes represent different stages in the information processing pipeline, starting with raw data and ending with presentable results. Arrows indicate transformational processes that move information from one box to another. The *x* axis approximates the overall level of effort to move information from raw to the specific processing stage. The *y* axis shows the amount of structure (with respect to the investigative process) in the processed information for every stage. Thus, the overall trend is to move the relevant information from the lower left to the upper right corner of the diagram. In reality, the processing can both meander through multiple iterations of local loops, and to jump over phases (for routine cases handled by an experienced investigator).

External data sources include all sources of potential evidence for the specific investigation, such as disk images, memory snapshots, network captures, as well as reference databases, such as hashes of known files. The *shoebox* is a subset of all the data that has been identified as potentially relevant, such as all the email communication between two persons of interest. At any given time, the contents of the shoebox can be viewed as the analyst's approximation of the information content potentially relevant to the case. The *evidence file* contains only the parts that directly speak to the case, such as specific email

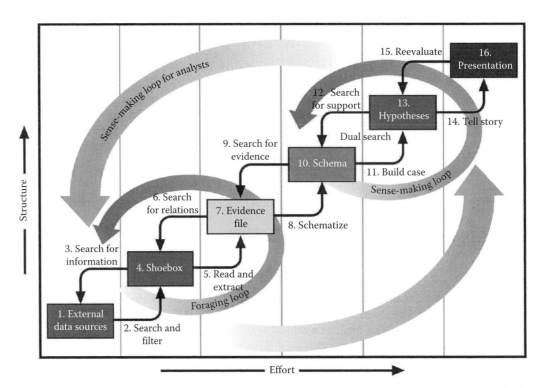

FIGURE 56.1 Notional model of sense-making loop for analysts derived from CTA. (Thomas, J. and Cook, K., (eds.), *Illuminating the Path: The Research and Development Agenda for Visual Analytics*, IEEE CS Press, Piscataway, NJ, 2005.)

exchanges on the topic of interest. The *schema* contains a more organized version of the evidence, such as a timeline of events, or a graph of relationships, which allows higher-level reasoning over the evidence. A *hypothesis* is a tentative conclusion that explains the observed evidence in the schema and, by extension, could form the final conclusion. Once the analyst is satisfied that the hypothesis is supported by the evidence, the hypothesis turns into a *presentation*, which is the final product of the process. The presentation usually takes on the form of an investigator's report that both speaks to the high-level conclusions relevant to the legal case, and also documents those low-level technical steps based on which the conclusion has been formed.

The overall analytical process is split into two main activities loops: (1) a *foraging loop* that involves actions taken to find potential sources of information, query them, and filter them for relevance; (2) a *sense-making* loop in which the analyst develops—in an iterative fashion—a conceptual model that is supported by the evidence. The information transformation processes in the two loops can be classified into bottom-up (organizing data to build a theory) or top-down (finding data based on a theory) ones. In practice, analysts apply these in an opportunistic fashion with many iterations.

56.3.1.1 Bottom-Up Processes

Search and filter: External data sources, hard disks, network traffic, etc. are searched for relevant data based on keywords, time and other constraints, in an effort to eliminate the vast majority of the data that are irrelevant.

Read and extract: Collections in the shoebox are analyzed to extract individual facts and relationships that can support or disprove a theory. The resulting pieces of artifacts (e.g., individual email messages) are usually annotated with their relevance to the case.

Schematize: At this step, individual facts and simple implications are organized into a schema that can help organize and help identify the significance and relationship among a growing number of facts and events. Timeline analysis is one of the basic tools of the trade; however, any method of organizing and visualizing the facts—graphs, charts, and diagrams—can greatly speed up the analysis. This is not an easy process to formalize, and most forensic tools do not directly support it. Therefore, the resulting schemas may exist on a piece of paper, or on a whiteboard, or only the mind of the investigator. Since the overall case could be quite complicated, individual schemas may cover specific aspects of it, such as the sequence of events discovered.

Build case: Out of the analysis of the schemas, the analyst eventually comes up with testable theories that can explain the evidence. A theory is a tentative conclusion and often requires more supporting evidence, as well as testing against alternative explanations.

Tell story: The typical result of a forensic investigation is a final report and, perhaps, an oral presentation in court. The actual presentation may only contain the part of the story that is strongly supported by the digital evidence; weaker points may be established by drawing on evidence from other sources.

56.3.1.2 Top-Down Processes

Re-evaluate: Feedback from clients may necessitate re-evaluations, such as the collection of stronger evidence, or the pursuit of alternative theories.

Search for support: A hypothesis may need more facts to be of interest and, ideally, would be tested against alternative explanations.

Search for evidence: Analysis of theories may require the re-evaluation of evidence to ascertain its significance/provenance, or may trigger the search for better evidence.

Search for relations: Pieces of evidence in the file can suggest new searches for facts and relations on the data.

Search for information: The feedback loop from any of the higher levels can ultimately cascade into a search for additional information—that may include new sources, or the re-examination of information that was filtered out during previous passes.

56.3.1.3 Foraging Loop

It has been observed [Patt01] that analysts tend to start with a high-recall/low-selectivity query, which encompassed a fairly large set of documents—many more than the analyst can afford to read. The original set is then successively modified and narrowed down before the documents are read and analyzed.

The foraging loop is a balancing act between three kinds of processing that an analyst can perform—explore, enrich, and exploit. Exploration effectively expands the shoebox by including larger amounts of data; enrichment shrinks it by providing more specific queries that include fewer objects for consideration; exploitation is the careful reading and analysis of an artifact to extract facts and inferences. Each of these options has varying cost and potential rewards and, according to information foraging theory [Piro09], analysts seek to optimize their cost/benefit trade-off.

56.3.1.4 Sense-Making Loop

Sense-making is a cognitive term and, according to Klein's [Klei06] widely quoted definition, is the ability to make sense of an ambiguous situation. It is the process of creating situational awareness and understanding to support decision making under uncertainty—an effort to understand connections among people, places and events in order to anticipate their trajectories and act effectively.

There are three main processes that are involved in the sense-making loop: *problem structuring*—creation and exploration of hypotheses, *evidentiary reasoning*—the employment of evidence to support/disprove hypothesis, and *decision making*—selecting a course of action from a set of available alternatives.

56.3.2 Model Relationships

Using the previously mentioned cognitive model, we can map different conceptual approaches to modeling the forensic process. *Mathematical* models are suitable for relatively small components of investigative process. They can formalize individual steps in the analysis, such as inferring specific events based on the observed state of the system, or can be used to formalize hypotheses and test them using statistical means. Thus, mathematical models can support specific steps—either bottom-up or top-down—in the analytical process, but it is infeasible to model the entire investigation.

Procedural models come from practitioners and tend to be very linear, essentially, prescribing an overall march from the lower left to the upper right corner (Figure 56.1) while applying the available tools. These are not true scientific models but compilations of best practices based on experience.

Legal models of forensics come from the consumers of the final product of the analysis—legal experts—who have little, if any, technical expertise. They can be viewed as imposing boundary conditions on the process: they restrict what part of the available data may be investigated (e.g., no privileged communication), and formulate the required end result requirements (e.g., prove that the suspect was *knowingly* in possession of contraband material, and had the intent to distribute it). Despite some early efforts, such by Xu et al. [Xu09], these requirements and restrictions are too arbitrary and case-specific to be efficiently translated into technical requirements.

56.3.3 Mathematical Models

An alternative to the Pirolli and Card (P&C) cognitive model is to present the forensic process as reconstructing the history of a computation using mathematical formalisms. Carrier proposed one such model [Carr06a] which makes the assumption that the computer system being investigated can be reduced to a finite state machine (FSM) with set of states Q. The transition function is the mapping between states in Q for each event in the event set, and the machine state changes only as a result of an event.

The point of Carrier's model is not to reduce a system to an enormous FSM but to find basic building blocks from which to build higher-level constructs, while establishing properties using formal methods. In this model, one could formulate a formal hypothesis and test it against the captured state and the known capabilities (transitions) of the system. The model is lower-level than procedural frameworks and analytic taxonomies; therefore, it offers the promise of providing formal proofs to show the completeness of the analytic techniques employed.

Statistical approaches have the potential to be more practical by enabling formal hypothesis testing and quantifying the level of confidence in the conclusions. For example, Kwan et al. [Kwan08] and [Over10] et al. show how Bayesian networks can be used to evaluate the strength of the evidence for some simple cases and study the sensitivity of the model to inputs. The primary challenge, however, is providing reliable estimates of the various probabilities upon which the model depends.

56.4 System Analysis

In this section we outline the various approaches to acquiring and analyzing the content of a computer system; specifically, we consider the examination of persistent storage, main memory, and network communications.

56.4.1 Storage Forensics

Persistent storage in the form of hard drives, optical disks, flash device, etc. is the primary source of evidence for most digital forensic investigations, in terms of both volume and importance. For example, during FY 2010, the average investigation at FBI's Regional Computer Forensics Labs consisted of 470 GB of data [FBI10], and the vast majority of the data comes from hard disks.

Storage analysis, detailed by Carrier [Carr05], can be performed at several levels of abstraction:

- *Physical media*: At the lowest level, every storage device encodes a sequence of bits and it is, in principle, possible to use a custom mechanism to extract the data bit-by-bit. In practice, this is rarely done as it is an expensive and time-consuming process. The only notable exception is second-generation mobile phones for which it is feasible to physically remove (desolder) the memory chips and perform acquisition of content; Willasen [Will05] provides a detailed account of the process. Thus, the lowest level at which most practical examinations are performed is the host bus adapter (HBA) interface. Adapters implement a standard protocol (SATA, SCSI, etc.) through which they can be made to perform low-level operations. For damaged hard drives, it is often possible to perform at least partial forensic repair and data recovery [Moul07]. In all cases, the goal of the process is to obtain a copy of the data in the storage device for further analysis.
- *Block device*: The typical HBA presents a block device abstraction—the medium is presented as a sequence of fixed-size blocks, commonly of 512 or 4096 bytes, and the contents of each block can be read or written using block read/write commands. The media can be divided into partitions, or multiple media may be presented as a single logical entity (e.g., RAIDs). The typical data acquisition process works at the block device level to obtain a working copy of the forensic target—a process known as *imaging*—on which all further processing is performed.
- *File system*: The block device has no notion of files, directories, or even which blocks are considered "used" and which ones are "free"; it is the file system's task to organize the block storage into a file-based storage in which applications can create files and directories with all of their relevant attributes—name, size, owner, timestamps, access permissions, and others. For that purpose, the file system maintains metadata in addition to the contents of user files.
- *Application*: User applications use the file system to store various artifacts that are of value to the end-user—documents, images, messages, etc. The operating system itself also uses the file system to store its own image—executable binaries, libraries, configuration and log files, registry entries—and to install applications.

Overall, analysis at the application level yields the most immediately useful results as it is most directly related to actions and communications initiated by humans. Indeed, most integrated forensics tools present a file system browser as their basic user interface abstraction. Yet, they also provide additional information that is not readily accessible through the normal operating system user interface. Such information is derived from additional knowledge of how a particular operating system works and what the side effects of its storage management techniques are.

One important source of additional data is the recovery of artifacts that have been deleted, or are otherwise not completely recoverable due to hardware errors. An extreme example of that is a used hard disk drive that has been freshly formatted—to the regular file browser interface, it would appear that the device is completely empty. In reality, only a small fraction of the drive has been overwritten—just enough to create blank file system metadata structures. The rest of the drive is marked as available but the previous content is still there, and that includes the complete content of almost all the files. In fact, the content would survive multiple formatting operations and, depending on usage patterns, data could be found years after its intended removal.

56.4.1.1 Target Acquisition

In line with best practices [Kent06], file system analysis is not carried out on a live system. Instead, the target machine is powered down, an exact bit-wise copy of the target disk is created, the original is stored in an evidence locker, and all forensic work is performed on the copy.

There are exceptions to this workflow in cases where it is not practical to shut down the target system and a live disk image from a running target is obtained; evidently, the consistency of data in use cannot be guaranteed. In virtual environments, a standard snapshot of the virtual disk can be trivially obtained.

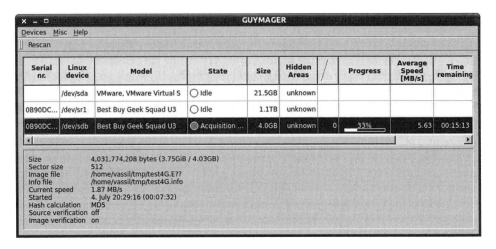

FIGURE 56.2   Screenshot of *guymage* during target acquisition.

The copy process of a physical disk can be accomplished in software, hardware,* or combination of both. The original workhorse of forensic imaging has been the *dd* Unix/Linux general purpose command-line utility, which can produce a binary copy of any file, including special files representing entire storage devices. For example, the simple command dd if=/dev/hda1 of=target.dd, would produce a binary copy of the first partition of the first disk drive and will place it in the *target.dd* file. A hardware *write blocker*† is often installed on the target device to eliminate the possibility of operator error that leads to the accidental modification of the target. Further, cryptographic hashes are computed for the entire image and (optionally) for every block. The latter can be used to demonstrate the integrity of the evidence if the original device suffers a partial failure, which makes it impossible to read its entire contents. To simplify the acquisition process, more specialized versions of *dd*, such as *dcfldd*,‡ provide additional functionality, such as hashing and integrity verification on-the-fly, multiple output streams, and logging. A more advanced tool geared specifically toward forensics is the *guymager*§ target acquisition tool (Figure 56.2), which provides additional output options beside the raw image produced by *dd*.

Ddrescue¶ is another variation on *dd*, which automatically tries to recover from read errors over multiple runs in an effort to extract as much data as possible from a failing hard drive. Virtually all commercial tools, as well as a number of other open source projects, provide the basic disk imaging capability, and it is beyond the scope of this discussion to provide an exhaustive enumeration. The National Institute of Justice sponsors the Computer Forensic Tool Testing (CFTT) project,** which independently tests various basic tools, such as write blockers and image acquisition tools and regularly publishes reports on its findings.††

56.4.1.2 File Carving and Reconstruction

Generally, data stores with dynamic allocation, such as file systems, main memory, and databases, contain left over data that are not immediately accessible via the regular programming interface but are recoverable by direct examination of the raw data. For example, a file deleted by the user is simply marked as deleted and its storage allocation is reclaimed only when the need arises, which could happen anywhere from

---

* http://www.cftt.nist.gov/disk_imaging.htm

† http://www.cftt.nist.gov/hardware_write_block.htm

‡ http://dcfldd.sourceforge.net/

§ http://guymager.sourceforge.net/

¶ https://savannah.gnu.org/projects/ddrescue/

** http://www.ojp.usdoj.gov/nij/topics/technology/electronic-crime/cftt.htm

†† http://www.cftt.nist.gov/disk_imaging.htm

milliseconds to never. Main memory content can also persist for a long time as many computer systems are not reset for long periods of time, and even reboots may not clear memory [Hald08].

This behavior may be surprising to users, but it is primarily the result of performance considerations in the operation of software that manages the data. Sanitizing the de-allocated storage can be very expensive not only when applied to a hard disk but also to main memory management. This is a known security problem and solutions have been proposed [Chow05], but they are not routinely employed. Another reason to find data not actively in use in main memory is caching—perhaps the oldest performance optimization trick. Modern software systems use multiple levels of caching with the goal of minimizing secondary storage access, so data tends to persist where it was not intended to—main memory.

Forensic computing, unlike most other types of computations, is most keenly interested in *all* extractable artifacts, including—and sometimes *especially*—de-allocated ones. Unless a user has taken special measures to securely wipe a drive, it is reasonable to expect that, at any given time, the media contains recoverable applications artifacts—primarily files—that are ostensibly deleted.

File carving is the process of recovering application files directly from block storage without the benefit of file system metadata. The essential observation that makes this possible is twofold: (a) most file formats have specific beginning and end tags (a.k.a. *header* and *footer*); and (b) file systems heavily favor sequential file layout for better performance. We briefly survey the research and tool development in file carving; a more detailed look at the evolution of file carving techniques can be found in [Pal09].

In the simplest case, a file carving tool sequentially scans through a device image and looks for known headers at block boundaries. If a header is found, then the tool looks for a corresponding footer and, upon success, copies (carves out) all the data in between the two tags. For example, a JPEG image always starts with the header (in hexadecimal) FF D8 FF and ends with FF D9. In practice, things are a bit more complicated since file formats were never designed for this type of processing and can have weak distinguishing features or may embed other file types (e.g., JPEGs can have an embedded thumbnail, also in JPEG). Further, actual data layout often has gaps, mixed-in blocks from other files, or outright data losses.

One of the first tools to implement the previously mentioned idea and gain popularity among forensic professionals was *foremost*.* It is a command-line utility which uses a configuration file for header/footer definitions and has some built-in parsing functions to handle Microsoft Office (OLE) files. *Scalpel*[†] [Rich05] is a file carver originally conceived as an improved implementation of *foremost* and utilizes a two-pass strategy to optimize work schedules and reduce the amount of redundant data being carved. Subsequent optimizations [Marz07] allowed the tool to utilize parallel CPU and GPU processing.

The development of new file carving techniques has been pushed forward by the 2006 [Carr06b] and 2007 [Carr07], DFRWS Challenges, which provided interesting scenarios for tool developers to consider. Garfinkel [Garf07a] performed a statistical study of actual used hard drives purchased on the secondary market and found that most fragmented files are split in two pieces and are, therefore, still recoverable with reasonable effort. He also presented a proof-of-concept approach based on using application-level verification.

Sencar and Memon [Senc09] developed a robust algorithm and a specialized tool that can carve out JPEG fragments and successfully reassembles them using image analysis techniques. The tool has been subsequently commercialized and can be useful to regular users to recover images from flash memory used in digital cameras.

Looking ahead, file carving is likely to be a useful tool for the foreseeable future as hard disk sizes continue to grow and it becomes ever more performance-critical for file systems to lay out files sequentially. At the same time, King and Vidas [King11] have established experimentally that file carving would only work in a narrow set of circumstances on modern solid state drives (SSD). The reason lies in the fact that

---

* http://foremost.sourceforge.net
† http://www.digitalforensicssolutions.com/Scalpel/

SSD blocks need to be written twice in order to be reused—the first write resets the state of the block, thereby enabling its reuse. To improve performance, the *trim* command was added to the SATA protocol specification to allow the operating system to garbage collect blocks not in use and make them ready for reuse. Virtualized cloud storage represents another limitation to file carving as de-allocated data are automatically sanitized to prevent data leaks across tenants.

56.4.1.3 File System Analysis

Once a working copy of the investigative target has been obtained, the actual analysis can commence in earnest. The first goal is to identify and gain access to all the artifacts stored on the media—files, directories, email messages and other personal communications, system registry entries, etc. The complete recovery of all useable artifacts requires deep understanding of the physical layout of the on-disk structure of the specific version of the specific operating system. This is a particularly thorny issue with proprietary file systems, such as the MS Windows family, as their structures are not publicly documented and need to be reverse-engineered. Fortunately, the open-source community at large has tackled many of these problems for interoperability reasons, and has developed some robust implementations.

It is useful to place the forensically significant file system data into two categories—system artifacts and application artifacts—which we discuss for the rest of this section.

System artifacts: The operating system continuously generates traces of its users' activities. Some pieces of this information are quite obvious—automatic updates to timestamp file attributes—others are more obscure, such as system log entries. More subtle information can be obtained by observing the physical layout of the files on the block devices. In Unix/Linux, the structure containing the metadata for a file is called an *i*-node and each file is associated with a particular *i*-node. Normally, files created in a quick succession (during installation) would have the same creation timestamp and *i*-node numbers that are close to each other. However, if we find a file which blends in based on timestamp but has an uncorrelated *i*-node number (suggesting creation at a later time), this would be an outlier event worthy of further investigation.

We use *The Sleuthkit**(*TSK*), the most popular open source tool for file system forensics, to illustrate the system artifacts that are commonly used in a forensic investigation. TSK is C/C++ library, as well as a collection of command-line tools. TSK's code is based on The Coroner's Toolkit (TCT), which was originally created by Dan Farmer and Wietse Venema and featured in their classic book on forensics [Farm05]. The scope of TCT was expanded from the original target of Unix files systems to include MS Windows and MacOS formats.

TSK provides over 20 individual tools that can provide file system information at different levels of abstraction. Here, we provide a sampling of the more important ones.

Metadata layer tools process the on-disk metadata structures created by the file system to store standards attributes of a file—name, size, timestamps, etc. Under normal operating system operation, these structures are only available to the kernel. This toolset, however, allows the direct access to the following structures:

- *ifind* finds the metadata structure that has a given file name pointing to it or the metadata structure that points to a given data unit;
- *icat* extracts the data units of a file, which is specified by its metadata address;
- *ils* lists the metadata structures and their contents;
- *istat* displays the statistics and details about a given metadata structure.

Many modern file systems, such as ext3 and NTFS, employ a journaling (log) function, which keeps track of attempted file operation. Should the system crash, the journal allows for the replay of these operations and thus minimizes the loss of data to the user. TSK contains two tools—*jls* and *jcat*—that list the entries in the file system journal and can display the contents of a specific journal block, respectively. This enables an investigator to recover additional information that is not otherwise accessible.

---

* http://sleuthkit.org

The purpose of the *block device* tools is not just to provide access to individual data blocks but also to assign meaning to the content of the block. Specifically, we could target individual blocks that are allocated, unallocated, or we could identify and gain access to the slack space at the end of a block:

- *blkcat* extracts the contents of a given data unit (block);
- *blkls* lists the details about data units and can extract the unallocated space of the file system;
- *blkstat* displays the statistics about a given data unit in an easy-to-read format;
- *blkcalc* calculates where data in the unallocated space image (from *blkls*) exist in the original image. This is used when evidence is found in unallocated space.

Volume system tools take an image as input and analyze its partition structures. These can be used to find hidden data between partitions and to identify the file system offset for TSK:

- *mmls* displays the layout of a disk, including the unallocated spaces;
- *mmstat* displays details about a volume system (the type);
- *mmcat* extracts the contents of a specific volume.

As the previous descriptions suggest, in every group of tools, we can find a "stat" tool, which provides basic information (statistics) for the relevant data structures; an "ls" tool, which enumerates data structures of a particular kind; and "cat" tools, which interpret and extract the data from the structures.

The *Autopsy Forensic Browser* provides a simple Java-based graphical interface to TSK, as well as a framework that allows third-party components to be added to the processing pipeline. The goal of the project, still in its early stages, is to put together an open integrated environment that improves the usability of open source tools and enables further extension of functionality.

The *registry* is a central hierarchical database of key/value pairs used in all versions of Microsoft Windows to store information that is necessary to configure the system for one or more users, applications, and hardware devices. As such, the registry is a treasure trove for an investigator as it records a large number of values that can indicate user and application activity from recently used files, to traces of uninstalled applications, to records of external storage devices that have been connected to the system.

Tools like RegRipper* and RegistryDecoder† offer the option of extracting any available information from the registry. Since a registry contains tens/hundreds of thousands of entries, the investigator must have a fairly specific idea of what to look for. Figure 56.3 shows RegistryDecoder in action, where it has recovered the list of USB devices seen by the system and allows the investigator to tie the computer system to specific devices, including in this case a mobile phone.

Modern operating systems automatically create copies of user and system data in order to both improve reliability, by being able to reset to a known good state, and to support user data backup needs. In MS Windows, the service is called *Volume Shadow Copy Service‡* and captures stable, consistent images for backup on running systems under a variety of circumstances with regular users largely unaware of its operation. As a result, a forensic analyst can retrieve the prior state of the system, including artifacts that the user might have securely deleted [Harg08].

The file systems used by Unix derivatives—HFS+ on MacOS, ext3/ext4 on Linux, and ZFS on Solaris—use journaling as a basic reliability mechanism. By understanding the structure of the journal entries, it is possible to recover the recent history of file operations [Burg08, Fair12, Beeb09].

Application artifacts: An analyst can gain access to all the files created explicitly and implicitly by the user. Most users are well aware of the files they create explicitly (documents, images, emails), and a guilty user may try to cover their tracks by removing incriminating ones. Yet, computer systems tend to create multiple copies of the same information and can be very difficult to dispose them all. Modern applications contain numerous features to help the user by remembering preferences and histories, as well as

* http://regripper.wordpress.com/regripper/
† http://www.digitalforensicssolutions.com/registrydecoder/
‡ http://msdn.microsoft.com/en-us/library/windows/desktop/aa381497.aspx

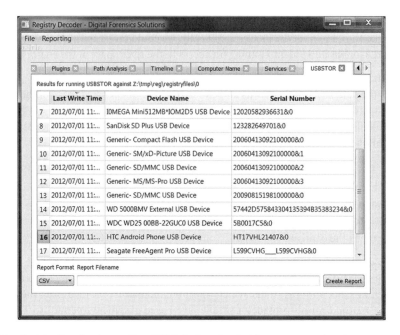

FIGURE 56.3 *RegistryDecoder* shows a list of USB devices that had been connected to the system.

automation features such as recovery of previous document versions and automatic filling of author, date, title, and other properties. These could all be exposed and, together with the documented behavior of the application, could bear witness to what users did. It is fair to say that many applications are also sloppy during uninstall procedures and can leave (sometimes deliberately) traces behind.

Of particular interest are artifacts created as a side effect of the user's web browsing activities. Browsers routinely retain a large amount of cached information about its users' online life—history, searches, bookmarks, cookies, screenshots, open tabs, etc. As Altheide describes in detail [Alth11], most of these data are now stored in common data formats, which makes it trivial to recover and interpret.

Another rich source of information is the metadata that are routinely embedded in various artifacts, such as author information and version information in office documents, and camera and GPS information in image data.

Beverly et al. [Beve11] demonstrated that the average disk contains a significant number of IP and MAC addresses that can be extracted by means of carving and validation. In a sample of 723 hard drives, they found an average of 2279 IP addresses. In addition to the obvious use of connecting a drive to a network, the addresses could be used via geolocation services to map the physical location of a target machine.

Cohen et al. [Cohe11] provide a different perspective on forensics—one in which security monitoring, incident response, and forensic tool are put together in a large distributed system to keep Google's expansive IT infrastructure secure. In building a distributed system to perform these tasks, a number of conflicting requirements, such as performance, privacy, and network connectivity, lead to technical solutions that would be considered incomplete in a classical forensic investigation.

56.4.2 Memory Forensics

The original view of best forensic practices (from a decade ago) was to literally pull the plug on a machine that is to be impounded. The rationale was that this removes any possibility to alert processes running on the host and would preempt any attempts to hide information. Over time, however, experience showed that such concerns were overblown and that the substantial and irreversible loss of important forensic information is simply not justified.

Memory analysis can be performed either in real-time on a live (running) system, or it could be performed on a snapshot (memory dump) of the state of the system. In live forensics, a trusted agent (process) designed to allow remote access over a secure channel is pre-installed on the system. The remote operator has full control over the monitored system and can take snapshots of specific processes, or the entire system. By and large, live investigations are an extension of regular security preventive mechanisms, which allows for maximum control and data acquisition. However, such mechanisms need to be pre-installed, which is only practical in commercial and government entities that can afford the expense of maintaining such a system.

In the general case, memory forensics reduces to forensics of a snapshot (a.k.a. *dead* forensics). It should be evident that analyzing a snapshot is considerably more difficult because we lack the run-time facilities and context provided by the live operating system. We need fundamentally new tools that can rebuild as much as possible of the state of the running system in a forensically sound manner, and, therefore, such tools are the main point of discussion for the rest of this section.

56.4.2.1 Memory Acquisition

Studies have illustrated that data persist for a long time in volatile memory ([Chow04], [Solo07]). Unlike tools that analyze running machines, off-line memory analysis tools extract digital evidence directly from physical memory dumps. These memory dumps may be acquired using a number of different mechanisms (dependent on operating system type and version), from hardware-based approaches, such as Tribble [Carr04] and via the Firewire interface [Bock], to software-only approaches [Vida06], such as using *dd* to access the physical memory device or via insertion of custom kernel modules. Vomel [Vomel1] provides a comprehensive survey of modern acquisition techniques.

These memory dumping mechanisms are not infallible and some high-tech approaches to subverting memory acquisition have been proposed [Rutk07]. Fortunately, unless the subversion mechanism is very deeply embedded in the operating system, a substantial amount of overhead may be incurred to prevent acquisition, potentially revealing the presence of a malicious agent [Korn06a, Korn07]. Ruff and Suiche [Ruff07] developed a tool that provides another alternative for memory acquisition, by converting Windows hibernation files to usable memory dumps. Schatz proposed a novel approach to memory acquisition technique called BodySnatcher, which injects a small, forensic operating system that subverts the running operating system [Scha07]. Sylve's work [Sylv12] on Android memory acquisition* developed ostensibly for mobile devices works equally well on traditional Linux deployments.

Once obtained, a RAM capture contains a wealth of information about the run-time state of the system, with the most important listed in the following:

- *Process information*: We can enumerate and identify all running processes and threads and their parent/child relationships; we can enumerate all loaded systems modules, and can obtain a copy of the processes' code, stack, heap, code, and data segments. All of this information is particularly useful in analyzing compromised machines as it allows us identify suspicious services, abnormal parent/child relationships, and to scan for known malicious code.
- *File information*: We can identify all open files, shared libraries, shared memory, and anonymously mapped memory. This is useful for identifying correlated user actions and file system activities and to prove more difficult implications, such as user intent.
- *Network connections*: We can identify open and recently closed network connections, protocol information, as well as send and receive queues of data not yet sent or delivered, respectively. This information could readily be used to identify related parties and communication patterns among them.
- *Artifacts and fragments*: Just like the file system, memory management systems tend to be reactive and leave a lot of artifact traces behind. This is primarily an effort to avoid any processing that is not absolutely necessary for the functioning of the system. As well, caching of disk and network data can leave traces in memory for a long time.

---

* http://www.digitalforensicssolutions.com/lime/

56.4.2.2 Memory Analysis

The 2005 DFRWS memory analysis challenge[*] has served as an early catalyst to push the development of tools for Windows memory forensics and resulted in several early projects.

KnTTools[†] extracts information about processes, threads, access tokens, the handle table, and other operating system structures from a Windows memory dump. The *memparser*[‡] tool has the added ability to detect hidden objects by cross-referencing different kernel data structures. Schuster's *ptfinder* tools ([Schu06a], [Schu06b]) take a different approach: instead of walking operating system structures, they attempt to carve objects that represent threads and processes directly. Noting the relative dearth of usable tools for deep parsing of Linux memory captures, DFRWS created its 2008 challenge [Geig08] with a major emphasis on Linux memory forensics. Prior to that, the most notable work is Burdach's [Burd] proof-of-concept tool called *idetect*, which parses kernel 2.4-series memory dumps and enumerates page frames, discovers user mode processes, and provides detailed information about process descriptors. Case et al. [Case10] demonstrated the most advance memory parsing techniques, which allows their tool, *ramparser*, to find kernel structures even in custom Linux distributions, which are very common on various devices.

The most influential project in memory forensic has been the development of Volatility Framework[§] led by Aaron Walters [Petr06, Walt07]. It provides core memory processing capabilities for Windows-based systems and a component architecture that allows third-party expansion. Volatility supports multiple input formats (raw, hibernation file, crash dump), and can extract running processes, loaded module, open network connections, open registry items, reconstruct executable binaries, and automatically map virtual to physical addresses. The project has become the focal point for most recent research and development efforts in memory forensics and has the potential to bring all practical techniques into a single package.

56.4.3 Network Forensics

Network forensics uses specialized methods to collect, record, and analyze transmitted data for the purposes of detecting or reconstructing a sequence of relevant events. Obviously, the main source of information is network traffic, which may be monitored in real time, or recorded and analyzed later. From the point of view of forensic reconstruction, there is no fundamental difference between the two cases and the same methods could be applied (subject to performance constraints). Most network traffic analysis is preventive monitoring and aims to identify and neutralize security threats in real time—firewalls and intrusion detection systems observe and actively manipulate network flows to achieve the desired security posture.

It can be useful to distinguish two types of network inquiries—incident response and forensic reconstruction—although they often blend together. Incident response teams deal with identified security breaches, and they function much like an emergency room team—quickly assess the immediate damage, stop the bleeding, and commence a recovery operation. Forensic reconstruction is a deeper analytical process to put together the sequence of events that lead to the incident, assess the wider scope of damages—which systems were compromised and what data were compromised—and try to attribute blame to the ultimate source of the attack.

56.4.3.1 Capture and Reassembly

Among general-purpose network capture and analysis tools, Wireshark[¶] (a.k.a. Ethereal before 2006) stands out as the most popular and complete one in the public domain. It can analyze both live

---

[*] http://dfrws.org/2005/challenge/

[†] http://www.gmgsystemsinc.com/knttools/

[‡] http://sourceforge.net/projects/memparser

[§] https://www.volatilesystems.com/default/volatility

[¶] http://wireshark.org

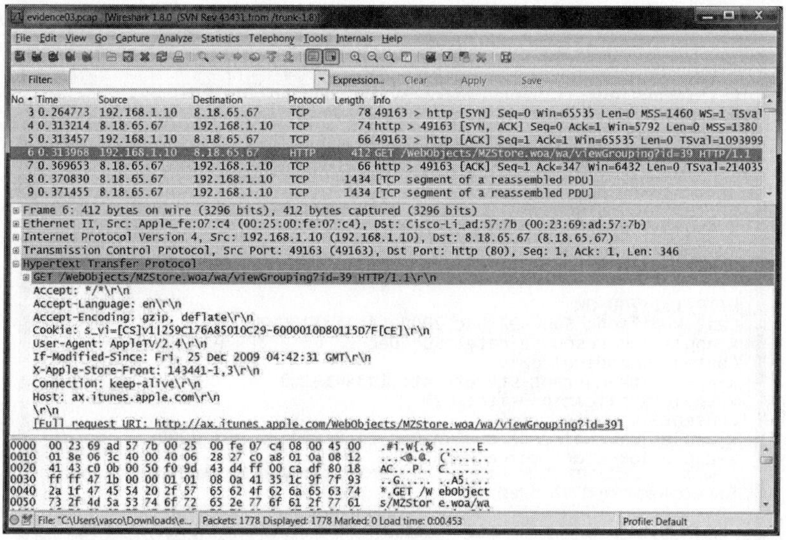

FIGURE 56.4 *Wireshark*'s main window displaying an HTTP request.

traffic and network captures. Its main interface window is shown in Figure 56.4. It consists of three components—a table of captured network packets, a detail panel showing the interpretation of the selected packet, and a raw hexadecimal/text display at the very bottom. Since the number of packets quickly becomes overwhelming, the tool provides a filtering facility, which allows for selection based on any of the fields shown—source, destination, protocol, content, etc.

Recall that the purpose of the Internet protocol stack is to provide end-to-end communication channels for distributed processes that execute on connected hosts. The data sent by processes are transformed by four different layers—application, transport, network, and data link—each of which leaves its footprint on the observed traffic. In particular, the flow of data sent is split into protocol data units and each layer adds protocol-specific header to the original message. During transmission, pieces from different connections are interleaved, which leads to the need to reconstruct the entire conversations to make sense of them; Figure 56.5 shows an example of a reconstructed conversation between a web client and server.

56.4.3.2 Analysis

Network data analysis follows the earlier mentioned process and can be performed at the four different layers of abstraction, depending on the focus of the investigation. At the transport layer, the most important pieces of information added in the header are the source and destination port numbers, which locally identify the communicating processes at the sender and the receiver, respectively. The network layer header contains, among other things, the source and destination IP addresses, which identify the sending and receiving hosts. Finally, at the data link layer, the header contains the physical (MAC) addresses of the source and destination for the current hop on the LAN. For every hop, that information will be different as different hosts, using potentially different link types (Ethernet, 802.11, FDDI) will be forwarding the data.

Thus, if a case revolves around a wireless network breach, the most relevant data are likely to be found at the data link layer. If the issue is a botnet attack coming from the Internet, then most of the relevant data will probably be in the network and transport layers. If the problem is data ex-filtration, the best starting point is likely to be the application layer, as a user is likely to use a ready network application, like email, to transmit the data.

Automating the process of network forensics is an ongoing challenge; the sheer volume of data means that, in order to find the artifacts of interest, one needs to know rather specifically what to look for.

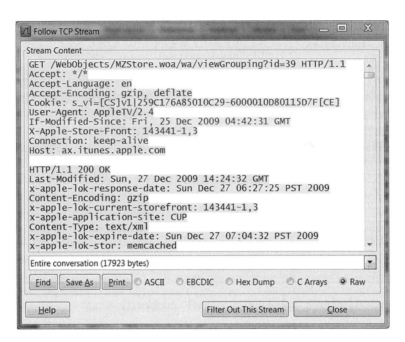

FIGURE 56.5   Reconstructed HTTP client/server conversation in *Wireshark*.

Most network monitoring tools, such as NetWitness,* provide rule-based filtering and aggregation and can alert an analyst that something is amiss. For example, an increase in malformed packets, or abnormal levels of traffic could be an indication of an attack attempt, as would be connections on unusual for the advertised protocol port, connection attempts to known bad hosts, obfuscated content, etc.

56.5 Artifact Analysis

Another approach to understand forensic processing is to consider different methods for processing digital artifacts. We can take advantage of the fact that the representation of many artifacts—text, images, audio, video—remains consistent across different media and devices and we could develop generic methods that could be applied across the board.

56.5.1 Crypto Hashing

The lowest common denominator for all digital artifacts is to consider them a sequence of bits/bytes without trying to parse them, or assign any meaning. Despite this low level of abstraction, there are some very important problems that can be addressed and the most important one of them is to find known content.

Hashing is the first tool of choice in investigating any case, as it provides the reliable means to validate data integrity and to identify known content. At a basic level, hash-based methods are attractive due to their high throughput and memory efficiency. Recall that a hash function takes an arbitrary string of binary data and produces a number, often referred to as digest, in a predefined range. Ideally, given a set of different inputs, the hash function will map them to different outputs.

Hash functions are collision-resistant if it is computationally infeasible to find two different inputs for which the output is the same. Cryptographic hash functions, such as MD5, RIPEMD-160, SHA-1, SHA-256, and SHA-512, are explicitly designed to be collision-resistant and to produce large, 128- to 512-bit results.

---

* http://www.netwitness.com/

Since the probability that two different data objects will produce the same digest by chance is astronomically small, we can assume that, if two objects have the same digest, then the objects themselves are identical.

The current state-of-the-practice is to apply a cryptographic hash function, typically MD5 or SHA-1, either to the entire target (drive, partition, etc.) or to individual files. The former approach is used to validate the integrity of the forensic target by comparing before-and-after results at important points in the investigation. The latter method is used to eliminate known files, such as operating system and application installations, or to identify known files of interest, such as illegal ones. The National Institute of Standards and Technology (NIST) maintains the National Software Reference Library,* which covers the most common operating system installation and application packages. Similarly, commercial vendors of digital forensic tools provide additional hash sets of other known data.

From a performance perspective, hash-based file filtering is very attractive—using a 20-byte SHA-1 hash, we could represent 50 million files in 1 GB. Thus, we could easily load a reference set of that size in main memory and filter out, on the fly, any known files in the set as we read the data from a forensic target. In addition to whole files, we are often interested in discovering file remnants, such as the ones produced when a file is marked as deleted and subsequently partially overwritten.

One routinely used method to address this problem is to increase the granularity of the hashes—we can split up the files into fixed-size blocks and remember the hashes for each individual block. Once we have a block-based reference set, we could view a forensic target as merely a sequence of blocks that can be read sequentially, hashed, and compared to the reference set. Typically, the block size is 4 KB to match the minimum allocation unit used by most operating systems installations. There are two main advantages to this scheme—we can easily identify pieces of known files and avoid reading the target hard drive on a file-by-file basis [Garf10].

56.5.2 Approximate Matching

So far we looked at ways to find artifacts that are identical; however, we could consider the more general problem of finding objects that have similar binary representation. Inspired by earlier work in spam filtering, Kornblum [Korn06b] proposed a fuzzy hash construction. It uses shingling to split up the file into chunks; generates small, 6-bit hashes for every chunk; and concatenates them to produce the final hash, which is base64-encoded. This scheme is a derivative of Rabin's groundbreaking work on data fingerprinting by random polynomials [Rabi81]. For the similarity comparison, the two hashes are treated as text strings and are compared using an edit distance measure, which produces a number between 0 and 100.

An interesting design choice is to limit the overall size of the hash to 80 symbols—this is mainly driven by the aim to give investigators a fixed hash value per file similar to the ones produced by crypto hashes (MD5, SHA-1, etc.). To achieve this, the algorithm uses the file length to estimate the number of piece it needs to break the file into. After the hash calculation is completed, if the result is longer than the target 80 symbols, the estimate is doubled and the calculation is redone from scratch. Due to its sensitivity to object size, the actual file signature produces two hashes for two different resolutions—c and $2c$. This takes the edge off the problem; but if the difference in size between the data objects exceeds a factor of four, the two are not comparable. In practice, the hash does seem to work well in identifying objects that are versions of each other and are not too big and not too dissimilar; however, the hash quickly loses resolution for larger files and cannot be applied to stream data where size is unknown.

Roussev et al. [Rous06] proposed a similarity hash measure, primarily targeted at detecting versions of executable files and libraries. The essential idea is to break up the object into known constituent parts (coded functions and resources), hash each one individually, and combine them into a Bloom filter to produce the similarity hash. Hashes are then compared by counting the number of corresponding

---

* http://www.nsrl.nist.gov

bits in common between the two filters and comparing them with the theoretical expectations—any statistically significant deviation is an indication of similarity. Performance results for system libraries demonstrate that versions are readily detectable by this method, even when filters with very high compression rates are used.

The follow-up work on multi-resolution similarity (MRS) hashing [Rous07] addressed the problem of comparing objects with vastly different size, such as a file and an entire drive. To balance the requirements of accuracy and performance, it was proposed to generate similarity hashes at multiple levels of resolution and to choose an appropriate level, depending on the specific objects being compared.

In [Rous10] Roussev proposed a new approach to generating similarity data fingerprints based on the idea of statistically improbable features. In essence, from every neighborhood the scheme selects a few 64-byte features (byte sequences) that, based on a statistical survey, are least likely to appear by chance. The method relies on the calculation of the Shannon entropy of each sequence and has the added advantage that low-entropy features likely to trigger false positive results are eliminated. This is a different approach from the random polynomials, and the user study in [Rous11] clearly demonstrates that: (a) the approach performs better than random polynomial as low-entropy data are eliminated from consideration; (b) the bit stream similarities found almost always correspond to user observable similarity. Overall precision and recall rates were in the 95% range. Subsequent work further developed the concept to allow parallel computation of the digests [Rous12a] and applied similarity digests as a primary triage technique to a 1.5 TB case study [Rous12b] with over 120 different data sources.

56.5.3 Multimedia

The ability to attribute a multimedia artifact to its source, as well as the ability to identify forged artifacts, is of critical importance to a forensic investigation. Our discussion is focused on images, but similar concerns and methods apply to audio and video artifacts. The problem matching an image and its camera source has been approached from two perspectives: (1) identify the type of camera that has taken this photograph; or (2) prove that the image has been produced by a specific camera.

In [Fang09], the researchers introduce a forensic scheme to distinguish between DSLR and compact images. Since DSLR and compact cameras use different type of sensors and lenses, their camera output quality in terms of sharpness and ISO sensitivity differs significantly. Such differences also affect the sensor noise levels and can be detected through wavelet decomposition and noise analysis. The experimental results show that the proposed forensic scheme has a potential to identify DSLR and compact images even when they are recompressed or down-sampled by 50%.

Lukas et al. [Luka06] proposed a method for identifying the source camera of a digital image based on the imaging sensor's noise pattern. The pattern is extracted by considering a large sample of the images from the camera, and the attribution of images is performed through statistical correlation. To improve the applicability of the method as a forensic tool, Sutcu et al. [Sutc07] proposed an enhancement based on artifacts produced by the demosaicing operation (color interpolation performed by the color filter array of the camera).

Sophisticated image manipulation has become a commodity technology, and it is becoming near impossible to spot forgeries simply by looking at the image. Farid has an excellent survey [Fari09] on the various techniques that could be applied to detect forgery. Here, we offer a brief summary of the different classes of techniques.

Pixel-based techniques work by analyzing the image to detect the most common types of image tampering like cloning, resampling, and splicing. In all cases, the basic observation is that, although the image is perceptually sound, the underlying transformation creates detectable artifacts that can be exposed and detected via statistical classification.

Format-based technique works by considering the effects of lossy image compression, such as the popular JPEG, on the image. Namely, the image is encoded based on a lattice of 8×8 blocks, and the DCT coefficients of the block are quantized and encoded. This process is the main source of loss and produces (not always visible) artifacts along the boundaries. These can be used to detect various forms of tampering.

Camera-based techniques can identify the fact that different pieces of the image do not come from the same camera and this could be established using the attribution techniques outlined earlier. Another technique to spot manipulated images is to consider lighting—it is quite difficult to have two photographs taken under the exactly same conditions. By analyzing the direction from which the light shines on different objects [John 05], it is practical to determine that the image is a composite. For human subjects, the eyes provide a perfect point of reference as their high reflectivity often provides a clear indication of the positions of light sources.

Geometry provides another approach for detecting image tampering. In authentic images, the projection of the camera center onto the image plane is near the center of the image. When a person or object is translated in the image, the principal point is moved proportionally, so any differences in the estimated principal point across the image can be used to detect tampering. In [John07], Johnson and Farid apply this idea of using a pair of eyes, or other suitable objects, to estimate the camera's principal point from the image.

56.5.4 Event/Timeline Reconstruction

Timelines are one of the most frequently used tools in forensic analysis. Typical data sources, like the file system, provide generous amounts of timing information as every creation, access, or modification to a file leads to update of their MAC (modification/access/creation) timestamp attributes. The three main problems related to timelines are provenance, synchronization, and visualization.

The problem of establishing the provenance and reliability of timing information found on a target stems from the fact that timestamps carry no special protection and could be manipulated by users. Users with root access can modify the system clock or manipulate any timestamp directly. Even if no tampering is done, establishing the order of events taken place on the Internet is not as easy as it may appear—no two hardware clocks advance at the same rate, which means that they tend to drift away from real-world time and from each other.

Most hosts use NTP to periodically synchronize with time servers and, thereby, limit the amount of drift. Yet, the error is always present and is complicated by misconfigured clients and servers. Time zones, differences in time-keeping methods on different platforms, and "clock" events like daylight savings time, which are applied very unevenly around the world, further complicate efforts to establish the order and timing of events. In 2007, Buchholz [Buch07] performed a sizeable survey of computer clocks on the Internet and documented these pathologies using a large amount of empirical data.

Forensic timeline visualization tools have considered different approaches to aggregating the timing information—from lower to higher level event and displaying it to the examiner. On the left, Figure 56.6 shows the approach taken by Zeitline [Buch05]—one of the early systems. The essential idea is that investigators can go over the ordered list of events and create different timelines related to different aspects of the investigation. They can also group low-level events and annotate them as higher-level events, ultimately building a hierarchy out of the flat timing information.

TimeLab [Olss09] takes a more automated approach—it uses a number of specialized scanners to process different sources of timing information both from artifact metadata and from system logs. It displays the different timelines next to each other and has basic facilities to zoom in and out to refine granularity of the display.

The *log2timeline* tool [Gu10] is the most comprehensive timeline extraction tool used by practitioners—it parses everything from file time stamps and event logs to registry entries and browser artifacts. It can quickly produce a large amount of data, even for a modest data source. The challenge for the investigator

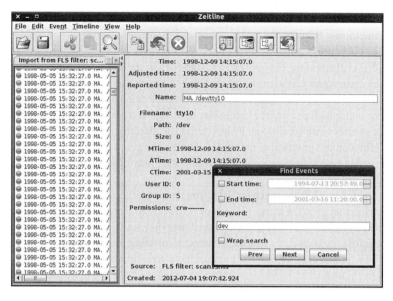

FIGURE 56.6   Forensic timeline visualization with *Zeitline*.

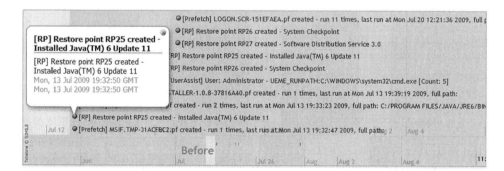

FIGURE 56.7   Visualization of a timeline produced by *log2timeline* using *SIMILE Timeline*.

is soft through, which is typically performed by custom scripts. The visualization is treated as a separate concern and can be performed with general timeline tools like SIMILE Timeline widget* to provide a more intuitive navigation (Figure 56.7) and exploration.

56.5.5 Application Forensics

Application forensics refers to the process of reliably establishing the behavior of applications as the result of user actions, and the identification of associated artifacts. Once such causal links are established, we can look for the artifacts to infer user behavior. As the name suggests, application forensics is inherently tied to the piece of software being used, and it is difficult to generalize this work across applications.

One prominent example of frequently used application forensic analysis is peer-to-peer (P2P) networks. Despite its numerous legitimate uses, P2P has also gained notoriety as one of the major means of distributed contraband material, as well as a major source of copyright infringement complaints. At first glance, one could be tempted to classify this as a network forensics problem. However, a closer look at the methods used

---

* http://www.simile-widgets.org/timeline/

reveals that no actual analysis of network traffic is performed. Since the purpose of P2P is file sharing, it is much easier to handle the problem on the end system by examining the artifacts from the P2P interactions.

P2P Marshal [Adel09] is a tool which automatically detects and analyzes P2P client use found on a particular target machine. The tool aims to automate end-to-end the process of investigating a target for all known P2P activity. It determines what clients are currently installed (or have been installed) on a machine, and then extracts per-user usage information for each client, including lists of shared or downloaded files and peer-servers contacted. Throughout the process, the tool performs its actions in a forensically sound way and maintains a detailed log of all actions performed.

Liberatore et al. [Libe10] provide a more general technical and legal context for investigating P2P networks. Specifically, they study the problem of identifying and tracking individual users in Gnutella and BitTorent and show the problems arising from hearsay information distributed through the networks.

56.6 Forensic Tool Design and Implementation

From a historical perspective, computer forensics tools are relatively young in their development cycle. At this stage, several commercial vendors have established strong presence in forensic labs by providing integrated investigative environments that attempt to support the complete inquiry cycle from acquisition to reporting. Not surprisingly, these vendors would like to retain their competitive advantage. To achieve this, they have created a "walled garden" environment, trying to lock in clients and to establish their standards as the de facto standards for the community.

Much of the advanced research in forensics is produced in academia, which simply lacks the incentives and the resources to develop fully functional integrated environments. Instead, researchers tend to focus their effort on a specific problem and to produce a narrowly targeted tool, which often requires climbing a learning curve before it can be put into practical use. Based on the experience of more mature software domains, the above situation is typical—commercial tools in the field are ready for practical use, but rarely employ the most advanced techniques.

The grand challenge for forensic tool design is to build tools that closely match the cognitive task of forensic analysis. Current generation tools are squarely focused on the information foraging loop (Figure 56.1). Specifically, they emphasize the gathering part of the loop, with most of triage and filtering functions primarily performed by the analyst.

56.6.1 Scalability and Automation

The need for scalable distributed forensic processing was first identified by Roussev and Richard [Rous04, Rich06] as a critical issue that forensic tools will face. Indeed, increasing data volume is already taxing existing systems to the point where slow performance is becoming one of the top concerns for investigators [Hibs11], along with the need for smarter and more automated tools. Coincidentally, more automation and smarter tools also imply more processing, specifically more *compute-intensive* processing. Thus, building *scalable* tools that can handle large cases is a problem of crucial necessity.

The efforts so far have been relatively few despite the early work in [Rous04], which showed that even a modest research prototype can speed processing considerably; in some cases, super-linear speed up has been observed due to cumulative caching benefits. In [Rous09], another research prototype showed the suitability of Map-Reduce style for forensic processing by quantifying its performance on a small cluster using a custom map-reduce system.

56.6.2 Standards and Inter-Operability

Given the wide range of possible forensic analysis tasks, it is difficult to conceive that any particular integrated product would cover everything. Therefore, it is in the interest of the community at large to develop standards that allow for interoperability among tools.

The push for standards has been around for a number of years and has had some high-profile backers, such as the National Institute of Justice, which supported the development of the Digital Evidence Markup Language (DEML). It appears, however, that this effort to define a comprehensive standardized description standard has gained little traction and has all but disappeared into obscurity.

In the real world of forensics, the most pressing concern is the storage of real evidence sources. For some time, the de facto standard for digital evidence containers has been the Expert Witness Format (EWF) developed and maintained by Guidance Software—the maker of the EnCase Forensic product. This is a proprietary format, which has been thoroughly researched and reverse-engineered by the makers of libewf*—an open source library for manipulating EWF containers.

A strong, open alternative has been the afflib project,† which produced the Advanced Forensic Format (AFF) [Garf, Garf06] that, from a purely technical design perspective, is more advanced and forward looking than EWF [Cohe09]. A number of efforts have centered on the idea of using XML containers for digital evidence, most notably XIRAF [Alin06] developed the Netherlands Forensic Institute, and DEX [Levi09] from the University of Massachusetts. The main drawback of using XML is that it is quite verbose and inefficient for the large amounts of data that need to be stored for every case. To become a practical alternative, an XML-based scheme must incorporate compressed storage without compromising the inherent query advantages of XML containers—a non-trivial design challenge.

With the recent push toward the use of memory forensics methods, the need has emerged to define a suitable container format for imaged memory targets. Schuster [Schu08] has commenced an effort to adopt an early standard for such evidence.

The most comprehensive and mature research and development effort is the courtesy of Simson Garfinkel [Garf12]. The Digital Forensics XML (DFXML) is an XML language that enables the exchange of structured forensic information. DFXML provides the means to comprehensively describe any piece of data relevant to an investigation—volumes, files, registry entries, etc.—and maintains provenance information.

56.6.3 Tool Testing and Calibration

One fundamental difference between forensic tools and most other tools is that the legal system and the society at large must be able to trust the results of a forensic inquiry. For that purpose, each tool should be tested and its performance and error rate thoroughly documented. In practice, only simple tools, such as data cloning ones, have been independently tested and this remains a major challenge for researchers and developers.

One of the most important research efforts is the establishment of representative public data sets that can be used for research and testing. This requires a significant effort in any discipline, but forensics presents at least three added challenges: (1) the actual data on which investigations are performed cannot be published; (2) each case is different, so establishing a representative set requires a huge amount of data gathering; (3) we have no credible methods of generating simulated data, especially for complex scenarios.

The DFRWS conference (dfrws.org), through its annual challenges, has contributed several data sets that are often used as benchmarks in evaluating file carving and memory analysis tools. Garfinkel et al. [Garf09] have collected and published several important data sets, such as the *real data corpus* (U.S. and non-U.S. sets), which consists of used drives obtained on the secondary market (6.7 TB), the *govdocs* corpus (1 TB), which consists of a million copyright-free U.S. government documents, and the M57 corpora (1.5 TB), which consists of the data collected during a scripted exercise on a private network.

---

* http://www.sleuthkit.org/informer/sleuthkit-informer-23.txt
† http:// afflib.org/

56.7 Summary

In this chapter we presented a brief overview of the state of the art and state of the practice in digital forensics. Within the constraints of a single chapter, the primary goal has been to provide broad conceptual coverage of this growing research field as opposed to in-depth discussion of specifics.

As a computer science discipline, digital forensics is a smaller and younger field relative to the broader area of security and privacy. Its *initial* development was somewhat ad hoc as it was driven by the daily needs of forensic examiners who did not necessarily have in-depth technical knowledge and had to learn on the job. Over the past decade, however, a growing number of computer science researchers have entered the field and have made progress in establishing more rigorous scientific approach to the development and testing of digital forensic methods.

Researchers and practitioners have developed a large number of specialized methods to identify, extract, and forensically analyze digital artifacts that are resident in main memory, stored on secondary storage, or transmitted over the network.

The foreseeable future promises to continue to bring exponential growth in forensic data, and increasingly complex processing due to the growing number of applications and devices, and their interactions. Increasingly, the type of deep analysis where a human investigator sifts manually through large amounts of data will become infeasible.

56.8 Open Issues and Challenges

Looking forward, digital forensics faces new challenges as new technologies continue to enter our lives. Every new technology presents new challenges and increases the complexity of the forensic analysis, and an exhaustive list of upcoming challenges could be quite long. Therefore, we chose the ones we consider the most fundamental.

56.8.1 Scalability and Automation

Looking forward, the biggest challenge looming over forensics is the ever growing mountain of data that our society produces. FBI statistics show that between 2003 and 2010, the average amount of data per case has grown from 80 to 470 GB—an average annual growth rate of 28%. Such growth, along with technology trends in computing, storage, and communication, renders current single-machine approaches to investigation unsustainable. Forensic computing needs to be able to employ as many resources as necessary to keep pace with the needs of society. To answer the challenge, researchers need to develop new methods that allow for effective automated data reduction, massive parallel processing, and efficient collaboration among investigators.

56.8.2 Sensor and Device Diversity

Another important direction in technology development is the proliferation of various sensors and devices, which drive us toward a pervasive computing environment. Even defining the scope of an examination will become increasingly more difficult as the interconnections among the components open up a large number of potential lines of inquiry. Our current abilities to quickly investigate new devices are quite modest and predominantly focused on mobile phones.

56.8.3 Virtualization and Cloud Computing

Existing methods and procedures are closely tied to the hardware of the investigated system. It is clear that virtualization technology, which decouples a computation from its hardware platform, has become a commodity and will become pervasive in the next few years. Cloud computing is a natural progression of the virtualization idea, and it allows the computation to move the place in the network where it is most efficient to execute.

This is a fundamental change as it negates one of the fundamental assumptions in forensics—that a process is tied to its hardware. Other important consequences include the fact that virtualization means sharing of the same of same physical resources—memory, storage—by numerous, unrelated computations. This forces the infrastructure to sanitize all resources after use and eliminates a vast amount of evidence currently in use.

56.8.4 Pervasive Encryption

Expanding on the previous point, for a long time we have relied on fact that data on most media are readily accessible—encryption was always an option, but the cost was generally too high for most applications. The default state of affairs is about to change—encrypted data is becoming the norm with the introduction of hardware encryption on hard drives. Due to privacy concerns on cloud computing, it is not difficult to envision encryption becoming the norm for in-memory data.

So far, forensic investigators have relied on implementation flaws to get around the encryption problem; however, sooner rather than later, encryption implementations will be cleaned up and the data would be truly inaccessible. Indeed, proper use of encryption should make it inaccessible and there are powerful economic and societal interests that will push for getting it right.

56.8.5 Legal Issues and Proactive Forensics

As better security and privacy mechanisms are deployed ubiquitously, it is increasingly likely that a number of staple forensic techniques would drop in significance. In particular, it is reasonable to expect that operating systems and applications will be more careful about "shredding" de-allocated objects, thereby eliminating traces upon which we currently rely. The trend toward more network-centric cloud computing means that a number of legal issues, such as location of the data, can further reduce the available forensic data.

This leads us to the conclusion that more and more forensically relevant data would have to be collected *pro-actively*, as already suggested by Shields [Shie11] and as already practiced by many large organizations as part of routine monitoring. Further expansion of this data collection will likely require a proper legal framework. What shape this may take is beyond the scope of this chapter but it is fair to say that, currently, the legal system lags considerably behind the pace of technology development and has been in no rush to catch up.

Glossary

Cryptographic hash: A deterministic algorithm that takes an arbitrary block of data and returns a fixed-size bit string, such that *any* change to the data will change the hash value.

Digital/electronic evidence: Probative information stored or transmitted in digital form that can be used in legal proceedings.

File/data carving: The process of extracting and reconstructing digital artifacts from constituent pieces.

Forensic timeline: A sequence of user- and system-initiated events deduced from available time information from sources like logs, file metadata, and file system metadata.

Live forensics: The application of forensic methods to examine the state of a running system.

Memory forensics: The process of reconstructing the live state of the system from a snapshot of RAM contents.

Network forensics: The process of reconstructing the history and content of communication between devices of interest.

Provenance of digital evidence: An explicit description of the set of tools and transformations that led from acquired raw data to the resulting conclusion.

Secure deletion: The process of overwriting de-allocated artifacts such that subsequent reconstruction is not possible.

Similarity/fuzzy hash: A hash function that maps syntactically similar artifact to numerical values that are close to each other.

References

[Adel09] Adelstein, F., Joyce, R., and Powers, J. P2P Marshal: Automatic extraction of peer-to-peer data, ATC-NY, White Paper, July 2009. http://p2pmarshal.com/images/Documents/p2pmar-shalwp_07_16_2009.pdf (accessed on March 25, 2013).

[Alin06] Alinka, W., Bhoedjanga, R., Bonczb, P., and Vriesb, A. XIRAF—XML-based indexing and que-rying for digital forensics, *Proceedings of the 2006 Digital Forensic Research Workshop (DFRWS)*, Pittsburgh, PA, August 2006.

[Alth11] Altheide, C. Making it rain: Examining cloud artifacts, *The Sleuth Kit and Open Source Digital Forensics Conference*, McLean, VA, June 2011. http://www.basistech.com/about-us/events/open-source-forensics-conference/2011/presentations/#cloud-remnants

[Beeb09] Beebe, N., Stacy, S., and Stuckley, D. Digital forensic implications of ZFS, *Proceedings of the Ninth Annual DFRWS Conference*, Montreal, Quebec, Canada, August 2009.

[Beve11] Beverly, R., Garfinkel, S., and Cardwell, G. Forensic carving of network packets and associated data structures, *Proceedings of the Eleventh Annual DFRWS Conference*, pp. S78–S89, New Orleans, LA, August 2011.

[Bock] Bock, B. Firewire-based Physical Security Attacks on Windows 7, EFS and BitLocker. http://www.securityresearch.at/publications/windows7_firewire_physical_attacks.pdf (accessed on March 25, 2013).

[Buch05] Buchholz, F. and Falk, C. Design and implementation of Zeitline: A forensic timeline editor, *Proceedings of the 2007 Digital Forensic Research Workshop (DFRWS)*, Pittsburgh, PA, August 2005.

[Buch07] Buchholz, F. and Tjaden, B. A brief study of time, *Proceedings of the 2007 Digital Forensic Research Workshop (DFRWS)*, Pittsburgh, PA, August 2007.

[Burd] Burdach, M. idetect, http://forensic.seccure.net/tools/idetect.tar.gz

[Burg08] Bughardt, A. and Feldman, A. Using the HFS+ journal for deleted file recovery, *Proceedings of the Eighth Annual DFRWS Conference*, pp. S76–S82, Baltimore, MD, August 2008.

[Carr04] Carrier, B. and Grand, J. Hardware-based memory acquisition procedure for digital investiga-tions, *Journal of Digital Investigation*, 1(1): 50–60, 2004.

[Carr05] Carrier, B. *File System Forensic Analysis*, Addison-Wesley Professional, Boston, MA, 2005, ISBN: 978–0321268174.

[Carr06a] Carrier, B. and Spafford, E. Categories of digital investigation analysis techniques based on the computer history model, *Proceedings of the 2006 DFRWS Conference*, pp. S121–S130, West Lafayette, IN, August 2006.

[Carr06b] Carrier, B., Casey, E., and Venema, W. *DFRWS 2006 Forensics Challenge*, http://dfrws.org/2006/challenge/

[Carr07] Carrier, B., Casey, E., and Venema, W. *DFRWS 2007 Forensics Challenge*, http://dfrws.org/2007/challenge/

[Case10] Case, A., Marziale, L., and Gichard, G. Dynamic recreation of kernel data structures for live forensics, *Proceedings of the Tenth Annual DFRWS Conference*, pp. S32–S40, Portland, OR, August 2010.

[Chow04] Chow, J., Pfaff, B., Garfinkel, T., Christopher, K., and Rosenblum, M. Understanding data life-time via whole system simulation, *Proceedings of the 13th USENIX Security Symposium*, San Diego, CA, August 2004.

[Chow05] Chow, J., Pfaff, B., Garfinkel, T., Christopher, K., and Rosenblum, M. Shredding your garbage: Reducing data lifetime through secure deallocation, *Proceedings of the 14th Conference on USENIX Security Symposium—Volume 14 (SSYM'05)*, Vol. 14. USENIX Association, Berkeley, CA, August 2005.

[Cohe09] Cohen, M., Garfinkel, S., and Schatz, B. Extending the advanced forensic format to accommodate multiple data sources, logical evidence, arbitrary information and forensic workflow, *Proceedings of the Ninth Annual DFRWS Conference*, Montreal, Quebec, Canada, August 2009.

[Cohe11] Cohen, M., Bilby, D., and Caronni, G. Distributed forensics and incident response in the enterprise, *Proceedings of the Eleventh Annual DFRWS Conference*, pp. S101–S110, New Orleans, LA, August 2011.

[Fair12] Fairbanks, K. An analysis of ext4 for digital forensics, *Proceedings of the Twelfth Annual DFRWS Conference*, Washington, DC, August 2012.

[Fang09] Fang, Y., Dirik, A.E., Sun, X., and Memon, N. Source class identification for DSLR and compact cameras, *IEEE International Workshop on Multimedia Signal Processing–MMSP 09*, Rio de Janeiro, Brazil, October 2009.

[Fari09] Farid, H. A survey of image forgery detection, *IEEE Signal Processing Magazine*, 26(2):16–25, 2009.

[Farm05] Farmer, D. and Venema, W. *Forensic Discovery*, Addison-Wesley Professional, Upper Saddle River, NJ, 2005, ISBN: 978–0201634976.

[FBI10] FBI, Regional Computer Forensics Laboratory (RCFL) Program for Fiscal Year 2010, http://www.rcfl.gov/downloads/documents/RCFL_Nat_Annual10.pdf (accessed on March 25, 2013).

[Garf] Garfinkel, S. Providing cryptographic security and evidentiary chain-of-custody with the Advanced Forensic Format, Library, and Tools, http://www.afflib.org/publications.html (accessed on March 25, 2013).

[Garf06] Garfinkel, S., Malan, D., Dubec, K., Stevens, C., and Pham, C. Disk imaging with the advanced forensics format, library and tools, *The Second Annual IFIP WG 11.9 International Conference on Digital Forensics*, National Center for Forensic Science, Orlando, FL, January 29–February 1, 2006.

[Garf07a] Garfinkel, S. Carving contiguous and fragmented files with object validation, *Proceedings of the 7th Annual DFRWS Conference (DFRWS'07)*, Pittsburgh, PA, August 2007.

[Garf07b] Garfinkel, S. Document and media exploitation, *ACM Queue*, 22–30, November/December 2007.

[Garf09] Garfinkel, S., Farrell, P., Roussev, V., and Dinolt, G. Bringing science to digital forensics with standardized forensic corpora, *Proceedings of the Ninth Annual DFRWS Conference*, Montreal, Quebec, Canada, August 2009.

[Garf10] Garfinkel, S., Nelson, A., White, D., and Roussev, V. Using purpose-built functions and block hashes to enable small block and sub-file forensics, *Proceedings of the 10th Annual DFRWS Conference*, pp. S13–S23, Portland, OR, August 2010.

[Garf12] Garfinkel, G. Digital forensics XML and the DFXML toolset. *Journal of Digital Investigation*, 8(3–4): 161–174, February 2012, 10.1016/j.diin.2011.11.002.

[Geig08] Geiger, M., Venema, W., and Casey, E. *DFRWS 2008 Forensics Challenge*, http://dfrws.org/2008/challenge/

[Gu10] Guðjónsson, K. Mastering the super timeline with log2timeline, *SANS Institute InfoSec Reading Room*, http://www.sans.org/reading_room/whitepapers/logging/mastering-super-timeline-log2timeline_33438.

[Hald08] Halderman, J. et al. Lest we remember: Cold boot attacks on encryption keys, *Proceedings of the 17th USENIX Security Symposium*, San Jose, CA, August 2008.

[Harg08] Hargreaves, C., Chiversa, H., and Titheridge, D. Windows vista and digital investigations, *Journal of Digital Investigation*, 5: 34–48, 2008, doi:10.1016/j.diin.2008.08.001

[Hibs11] Hibshi, H., Vidas, T., and Cranor, L. Usability of forensic tools: A user study, *6th International Conference on IT Security Incident Management & IT Forensics*, Stuttgart, Germany, May 10–12, 2011.

[John05] Johnson, M. and Farid, H. Exposing digital forgeries by detecting inconsistencies in lighting, *Proceedings of the ACM Multimedia and Security Workshop*, pp. 1–10, New York, 2005.

[John07] Johnson, M. and Farid, H. Detecting photographic composites of people, *Proceedings of 6th International Workshop on Digital Watermarking*, Guangzhou, China, 2007.

[Kent06] Kent, K., Chevalier, S., Grance, T., and Dang, H. Guide to integrating forensic techniques into incident response, NIST, 2006, http://csrc.nist.gov/publications/nistpubs/800–86/SP800–86.pdf (accessed on March 25, 2013).

[King11] King, C. and Vidas, T. Empirical analysis of solid state disk data retention when used with contemporary operating systems, *Proceedings of the Eleventh Annual DFRWS Conference*, pp. S111–S117, New Orleans, LA, August 2011.

[Klei06] Klein, G., Moon, B., and Hoffman, R. Making sense of sensemaking I: Alternative perspectives, *IEEE Intelligent Systems*, 21(4): 70–73, 2006.

[Korn06a] Kornblum, J. Exploiting the rootkit paradox with windows memory analysis, *International Journal of Digital Evidence*, 5(1), Fall 2006.

[Korn06b] Kornblum, J. Identifying almost identical files using context triggered piecewise hashing, *6th Annual DFRWS Conference*, Lafayette, IN, August 2006.

[Korn07] Kornblum, J. Using every part of the buffalo in windows memory analysis, *Digital Investigation*, 4(1): 24–29, January 2007.

[Kwan08] Kwan, M., Chow, K.-P., Law, F., and Lai, P. Reasoning about digital evidence using Bayesian networks, *Advances in Digital Forensics IV*, Chapter 12, pp. 141–155, Springer, Berlin, 2008.

[Levi09] Levine, B. and Liberatore, M. DEX: Digital evidence provenance supporting reproducibility and comparison, *Proceedings of the Ninth Annual DFRWS Conference*, Montreal, Quebec, Canada, August 2009.

[Libe10] Liberatore, M., Erdely, R., Kerle, T., Levine, B., and Shields, C. Forensic investigation of peer-to-peer file sharing networks, *Digital Investigation*, 7(Suppl): S95–S103, Augut 2010, 10.1016/j.diin.2010.05.012.

[Luka06] Lukas, J., Fridrich, J., and Goljan, M., Digital camera identification from sensor noise, *IEEE Transactions on Information Security and Forensics*, 1(2): 205–214, June 2006.

[Mali08] Malin, C., Casey, E., and Aquilina, J. *Malware Forensics: Investigating and Analyzing Malicious Code*, 1st edn., Syngress, Burlington, MA, 2008, ISBN: 978–1597492683.

[Marz07] Marziale, L., Richard, G., and Roussev, G. Massive threading: Using GPU to increase the performance of digital forensic tools, *Proceedings of the 2007 DFRWS Conference*, pp. 73–81, Elsevier, Pittsburgh, PA, August 2007.

[Moul07] Moulton, S. Re-Animating drives & advanced data recovery, *Def Con 15*, Las Vegas, NV, August 2007, https://www.defcon.org/images/defcon-15/dc15-presentations/Moulton/Whitepaper/dc-15-moulton-WP.pdf

[Olss09] Olsson, J. and Boldt, M. Computer forensic timeline visualization tool, *Proceedings of the Ninth Annual DFRWS Conference*, Montreal, Quebec, Canada, August 2009.

[Over10] Overill, R.E., Silomon, J.A. M., Kwan, Y.K., Chow, K.P., Law, Y.W., and Lai, K.Y. Sensitivity analysis of a Bayesian network for reasoning about digital forensic evidence, *Proceedings of 4th International Workshop on Forensics for Future Generation Communication Environments (F2GC-2010)*, Cebu, Philippines, August 11–13, 2010.

[Pal09] Pal, A. and Memon, N. The evolution of file carving, *IEEE Signal Processing Magazine*, 26(2), March 2009.

[Patt01] Patterson, E., Roth, E., and Woods, D. Predicting vulnerabilities in computer-supported inferential analysis under data overload, *Cognition Technology and Work*, 3: 224–237, 2001.

[Petr06] Petroni, N.L., Jr., Walters, A., Fraser, T., and Arbaugh, W.A. FATKit: A framework for the extraction and analysis of digital forensic data from volatile system memory, *Journal of Digital Investigation*, 3(4): 197–210, December 2006.

[Piro05] Pirolli, P. and Card, S. Sensemaking processes of intelligence analysts and possible leverage points as identified through cognitive task analysis (6 p.), *Proceedings of the 2005 International Conference on Intelligence Analysis*, McLean, VA, 2005.

[Piro09] Pirolli, P. *Information Foraging Theory: Adaptive Interaction with Information*, Oxford University Press, New York, 2009, ISBN: 978–0195387797.

[Rabi81] Rabin, M. *Fingerprinting by Random Polynomials*, Technical Report TR1581, Center for Research in Computing Technology, Harvard University, Cambridge, Massachusetts, 1981.

[Rich05] Richard, G. and Roussev, V. Scalpel: A frugal, high-performance file carver, *Proceedings of the 2005 Digital Forensics Research Workshop (DFRWS)*, New Orleans, LA, August 2005.

[Rich06] Richard, G. and Roussev, V. Next generation digital forensics: Strategies for rapid turnaround of large forensic targets. *Communications of the ACM*, 49(2): 76–80, February 2006.

[Rous04] Roussev, V. and Richard, G. Breaking the performance wall: The case for distributed digital forensics, *Proceedings of the 2004 Digital Forensics Research Workshop (DFRWS)*, Baltimore, MD, August 2004.

[Rous06] Roussev, V., Chen, Y., Bourg, T., and Richard, G. md5bloom: Forensic filesystem hashing revisited. *6th Annual DFRWS Conference*, West Lafayette, IN, August 2006.

[Rous07] Roussev, V., Richard, G., and Marziale, L. Multi-resolution similarity hashing, *Proceedings of the 7th Annual DFRWS Conference (DFRWS'07)*, Pittsburgh, PA, August 2007.

[Rous09] Roussev, V., Wang, L., Richard, G., and Marziale, L. A cloud computing platform for large-scale forensic computing. In Peterson, G., Shenoi, S. (eds.), *Research Advances in Digital Forensics V*, pp. 201–214, Springer, Boston, MA, 2009, ISBN: 978–3–642–04154–9.

[Rous10] Roussevm, V. Data fingerprinting with similarity digests, *Sixth Annual IFIP WG 11.9 International Conference on Digital Forensics*, Hong Kong, China, January 2010.

[Rous11] Roussev, V. An Evaluation of Forensics Similarity Hashes, *Proceedings of the Eleventh Annual DFRWS Conference*, pp. 34–41, New Orleans, LA, August 2011.

[Rous12a] Roussev, V. Managing Terabyte Scale Investigations with Similarity Digests, In Peterson, G., Shenoi, S. (Eds.), *Research Advances in Digital Forensics VIII*, pp. 19–34, Springer, 2012.

[Rous12b] Roussev, V., Quates, C. Content triage with similarity digests: The M57 case study, *Proceedings of the Twelfth Annual DFRWS Conference*, pp. S60–S68, Washington, DC, August 2012.

[Ruff07] Ruff, N. and Suiche, M. Enter Sandman (why you should never go to sleep), *PacSec Applied Security Conference*, Tokyo, Japan, 2007.

[Rutk07] Rutkowska, J. Beyond the CPU: Defeating hardware based RAM acquisition tools (part I: AMD case), *Black Hat* DC 2007 presentation.

[Scha07] Schatz, B. BodySnatcher: Towards reliable volatile memory acquisition by software, *Proceedings of the 2007 Digital Forensic Research Workshop (DFRWS)*, Pittsburgh, PA, 2007.

[Schu06a] Schuster, A. Searching for processes and threads in Microsoft Windows memory dumps, *Proceedings of the 2006 Digital Forensic Research Workshop (DFRWS)*, Lafayette, IN, 2006. http://www.dfrws.org/2006/proceedings/2-Schuster.pdf (accessed on April 13, 2013).

[Schu06b] Schuster, A. Pool allocations as an information source in windows memory forensics, *International Conference on IT-Incident Management and IT-Forensics*, Stuttgart, Germany, October 2006. http://subs.emis.de/LNI/Proceedings/Proceedings97/GI-Proceedings-97-9.pdf (accessed on April 13, 2013).

[Schu08] Schuster, A. Common memory analysis data exchange format, *Open Memory Forensics Workshop (OMFW)*, Baltimore, MD, August 2008.

[Senc09] Sencar, H. and Memon, N. Identification and recovery of JPEG files with missing fragments, *Digital Forensics Research Workshop (DFRWS)*, Montreal, Quebec, Canada, August 2009.

[Shie11] Shields, C., Frieder, O., and Maloof, M. A system for the proactive, continuous, and efficient collection of digital forensic evidence, *Proceedings of the Eleventh Annual DFRWS Conference*, pp. S3–S13, New Orleans, LA, August 2011.

[Solo07] Solomon, J., Huebner, E., Bem, D., and Szezynska, M. User data persistence in physical memory, *Journal of Digital Investigation*, 4(2): 68–72, June 2007.

[Sutc07] Sutcu, Y., Bayram, S., Sencar, H.T., and Memon, N. Improvements on sensor noise based source camera identification, *Proceedings of IEEE ICME*, Beijing, China, 2007.

[Sylv12] Sylve, J., Case, A., Marziale, L., and Richard, G. Acquisition and analysis of volatile memory from android devices, *Journal of Digital Investigation*, 8(3–4): 175–184, 2012.

[Thom05] Thomas, J. and Cook, K., (eds.), *Illuminating the Path: The Research and Development Agenda for Visual Analytics*, IEEE CS Press, Piscataway, NJ, 2005.

[Vida06] Vidas, T. The acquisition and analysis of random access memory, *Journal of Digital Forensic Practice*, 1(4): 315–323, December 2006.

[Vome11] Vomel, S. and Feiling, F. A survey of main memory acquisition and analysis techniques for the windows operating system, *Journal of Digital Investigation*, 8(1): 3–22, July 2011.

[Walt07] Walters, A. and Petroni, N. Volatools: Integrating volatile memory forensics into the digital investigation process, *Black Hat* DC 2007 (February 2007)

[Will05] Willasen, S. Forensic analysis of mobile phone internal memory, *Advances in Digital Forensics*, Springer, Berlin, 2005, pp. 191–204, ISBN 0387300120.

[Xu09] Xu, F., Chow, K.P., He, J., and Wu, X. Privacy reference monitor—A computer model for law compliant privacy protection, *Proceedings of the 15th International Conference on Parallel and Distributed Systems*, Shenzhen, China, December 2009.

VIII

Managing Organizational Information Systems and Technology Capabilities

57

Organizing and Configuring the IT Function

Till J. Winkler
*Copenhagen
Business School*

Carol V. Brown
*Stevens Institute
of Technology*

57.1 Introduction

How to organize and configure the internal information technology (IT) function* has been a critical issue since the beginning of enterprise computing. One of the most important challenges in IT organization design is selecting the extent to which IT decision-making and IT resources (including the IT workforce) are centralized (Brown and Magill 1994). The key rationale for centralization is to leverage economies of scale; the underlying rationale for decentralization is to ensure local responsiveness to internal and external customers (Sambamurthy and Zmud 1999; Agarwal and Sambamurthy 2002; Weill and Ross 2004).

Over the past decades, IT organizations have oscillated between centralized and decentralized forms (Peak and Azadmanesh 1997; Evaristo et al. 2005). In the beginning of enterprise data processing, mainframe computers and magnetic tape devices were commonly organized in central data centers. After the late 1980s and the vast growth of distributed computing (Von Simson 1990), client-server and firm-wide enterprise resource planning applications led to IT recentralizations (Brown 2003; McAdam and Galloway 2005). Many firms further consolidated large parts of their IT infrastructure and application operations into independent shared services organizations (Evaristo et al. 2005). These serve several lines of business to gain further economies of scale advantages as well as to improve the quality of overall IT service delivery through introducing standard IT practices (Schulz et al. 2009). While recent IT reference frameworks—such as ITIL, ISO/IEC20000, CMMI, and COBIT—provide some guidance for designing the IT function and internal processes (Pardo 2010; Marrone and Kolbe 2011), this chapter takes an enterprise-level perspective.

---

* The terms information systems (IS) and information technology (IT) are both used in the literature to describe the IS/IT organization and IS/IT function. In this chapter we will use the term "IT" when referring to an organizational unit performing all or some of the IT functions within an enterprise.

In this chapter we present four IT organization archetypes that differ based on the centralization versus decentralization of both (1) IT decision rights and (2) allocated IT resources. We describe these archetypes based on four additional design dimensions: (1) coordination mechanisms, (2) financial autonomy, (3) sourcing arrangements, and (4) IT-related capabilities and skills. Being mindful that in the past the form of organizing the IT function has been heavily dependent on technological development. We predict that recent technology trends, such as cloud computing and the consumerization of IT, are likely to affect IT organization designs of the near future.

57.2 Six Dimensions of IT Organization Configurations

Organizations (profit as well as non-profit) typically consist of multiple units that may represent different functions or departments, lines of business, markets, or geographies (Daft 2009). We use the term "IT organization" to refer to the collectivity of human resources that perform IT-related tasks, such as planning, building, and operating IT applications and their underlying computer and communications infrastructures, as well as their relationships, practices, norms, and capabilities. This definition does not restrict the notion of an IT organization to the existence of a single organizational unit (i.e., "the IT department"). Rather, it offers the possibility to assume different design options for different IT units, depending on the needs and capabilities of the business unit(s) being supported. We also propose that six important dimensions distinguish an IT organizational design, as described later.*

57.2.1 Allocation of IT Decision Rights

According to IT governance theory, decisions on IT can be made in a more centralized or decentralized fashion (Brown and Grant 2005). In a corporate setting, centralization typically refers to allocating decision-making at the corporate level, while decentralization refers to decision authority at the divisional level or even lower organizational levels (Brown and Magill 1994). A simple scheme includes two primary decision areas: IT applications and IT infrastructure operations. A widely adopted pattern in which infrastructure decisions are centralized, but business application decisions are primarily made by the divisions, has been commonly termed a federated or *federal* model (Sambamurthy and Zmud 1999). More recently, Weill and Ross (2004, p. 6) proposed a five-part classification scheme that distinguishes decisions about business application needs, IT investment and prioritization, IT architecture, IT infrastructure strategies, as well as overall IT principles, with different patterns associated with different business priorities. Defining accountability and the sharing of decision rights between the two extreme poles of centralization and decentralization is commonly seen as a key challenge. However, some studies have demonstrated that companies with well-balanced IT decision rights exhibit better business–IT alignment and thus ultimately achieve superior firm performance (Weill and Ross 2004, p. 202). An IT reference framework such as COBIT can be used to apply overarching accountability schemes to the design of decision rights on the activity and role level.

57.2.2 Allocation of IT Resources

The second dimension captures the structural aspects of the IT organization, i.e., the position and location of the IT human and technology resources within the wider enterprise. Although some prior literature has implied that IT decision rights and IT resources reside together in an organization—we argue that these two dimensions should be considered separately (Boynton et al. 1992; Brown and Grant 2005). For example, IT decisions may be made in a decentralized manner by business units, while

---

* As the focus of this chapter is on organization structure, we refrain from an in-depth discussion of IT processes. However, we will make reference to process-based IT reference frameworks where suitable (see also Chapter 59 in this Volume).

IT resources operate under either divisional or corporate IT authority. Similarly, IT staff may be allocated to a line organization, but these IT resources implement services under centralized authority.

IT resource allocations have also been categorized as either IT *demand* or *IT supply* resources (Thiadens 2005, Mark and Rau 2006). That is, divisional IT units may plan for and formulate the IT resource demand for IT services at a division or business unit level, although a central IT unit (or an external supplier) may have responsibility for actually "supplying" the IT services to meet the specific business demand. Demand activities for IT operations, for example, include monitoring the delivery of IT services and issuing requests for minor changes to the infrastructure. Demand activities for IT application development include business process analysis, requirements definition, and user acceptance testing, as well as overall IT project management and steering. Although the focus of reference frameworks such as ITIL and COBIT is standardizable IT processes for IT supply units, they can provide some guidance also for designing demand-sided IT activities. For example, ITIL defines a dedicated demand management process as a responsibility of a demand manager (reporting to an IT unit).

In practice, the degree of centralization of IT resources differs widely under different IT organization archetypes (Hitt and Brynjolfsson 1997). In highly decentralized IT organizations divisional IT units also accomplish IT supply tasks, while in very centralized IT organizations corporate IT groups also manage much of the IT demand. The distribution of resources has overall been found to reflect the extent to which companies pursue economies of scale, versus enabling local responsiveness through the allocation of resources (Brown and Magill 1994).

The first two dimensions of our framework—allocation of IT decision rights and allocation of IT resources—form the axes for the 2 × 2 matrix in Figure 57.1. In addition to the centralized and

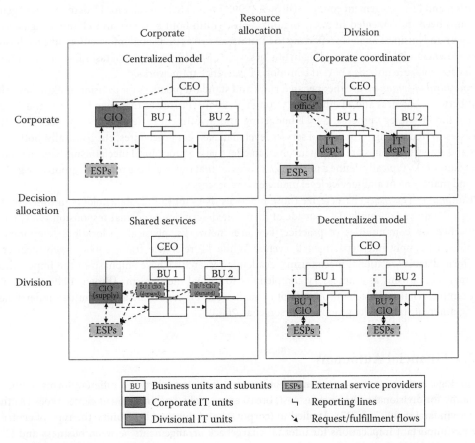

FIGURE 57.1 IT organization archetypes.

decentralized polar extremes, two other IT organization archetypes are defined. In the shared services model, IT decision rights are highly decentralized, but the IT resources that perform IT tasks are highly centralized. In the corporate coordinator model, the IT resources are highly decentralized or outsourced, but a central office holds a higher degree of IT decision rights. Four additional design dimensions for characterizing these archetypes are described later.

57.2.3 Coordination Mechanisms

The mechanisms for coordinating IT tasks across multiple organizational units—e.g., corporate functions, business units or divisions, and/or corporate and divisional IT groups—are an important complementary design dimension to the formal allocation of decision rights and resources (Brown 1999). They can be viewed as an *overlay* of the structural organization, which enables horizontal, not just vertical, information sharing (Daft 2009, p. 95). In general, the more complex and dispersed allocations of decisions and resources are, the more sophisticated coordination mechanisms need to be to effectively coordinate and integrate across the different parties involved in decision-making and execution (Peterson et al. 2000). Three categories of coordination mechanisms have been emphasized in the literature: structural mechanisms, procedural mechanisms, and relational mechanisms (Van Grembergen 2004, p. 20).

Structural mechanisms include "standing" groups or committees (in contrast to temporary teams or task forces), and formal roles that link across different organizational units. Widely used standing groups for IT governance decisions are, for example, IT steering committees with key business representation and IT management councils (Brown 1999). Formal liaison roles for IT demand management have also been implemented in many organizations within both business and IT units, e.g., account managers and business analysts reporting to IT units, as well as divisional information officers, business process owners, and key users residing in business units. Specific examples of tasks for such committees and liaison roles are now also part of common IT reference frameworks.

Procedural mechanisms are the specified rules and standard practices for decision-making and alignment between business and IT units (Peterson et al. 2000). Processes that span business and IT units include the IT strategy process, the IT budgeting and investment review process, project controlling processes, system change request and service level management procedures, etc. Naturally, both formal roles and standing groups are highly involved in effectuating procedural mechanisms. Common reference frameworks typically define a number of processes that involve these roles and groups—e.g., ITIL's demand management and service level management processes.

Relational mechanisms characterize those practices that aim to link stakeholders in different organizational entities *informally* (i.e., outside of their role description or formal responsibility). Common approaches are communities of practice, key user networks, physical co-location, temporary job rotations, or simply interdepartmental events. While IT reference frameworks largely neglect the less "formalizable" relational mechanisms, academic researchers have emphasized the importance of informal mechanisms as a necessary complement to formal mechanisms (Brown 1999; Chan 2002). For example, relational mechanisms are apt to facilitate knowledge sharing and mutual understanding among different stakeholder groups (Peterson et al. 2000).

57.2.4 Financial Autonomy

The strategic management and accounting literature differentiates between different forms of financial autonomy for divisional units, such as cost, break-even, profit, and investment center types (Anthony and Govindarajan 2007, p. 247). Applied to (corporate and divisional) IT units, the type of center not only has important implications for internal chargeback arrangements between business and IT but also determines the degree of financial and managerial autonomy of an IT unit (Venkatraman 1997).

Reference frameworks such as ITIL and COBIT generally acknowledge the importance of this organizational design dimension, but provide minimal design guidance.

In a *cost center* type, the IT unit is led by budget goals and is thus exclusively accounting for the costs of delivering internal IT services. Chargeback mechanisms are typically not in place (thus creating a possible incentive for business managers to underfund their units).

The *break-even center* defines service-based chargebacks based on the actual costs for delivering IT services. Thus, being a mixture between cost and profit centers, the goal of this center type is to close break-even. Since IT costs (e.g., for personnel, hardware, software) are often not directly accrued to an IT service, more complex cost and activity accounting schemes need to be established than in a cost center type. Such cost models often approximate the actually incurred IT cost, combining direct and indirect costs (Ryan and Raducha-Grace 2009).

Profit centers have greater financial autonomy inasmuch as their management carries responsibility for costs and (internal or external) revenues for IT services. Costs are charged to the customers on a more competitive basis, often oriented toward market-based transfer prices. However, in practice, business units are often obligated to contract with an internal IT profit center, so the degree of market competitiveness with external IT service providers is limited.

Investment centers extend profit center responsibilities to include accountability for the investment of accrued capital, so that this type of IT unit can be viewed as an independent company within the company. In large corporations, both profit and investment centers are commonly constituted in separate legal entities, subsidiary to the parent company.

57.2.5 Sourcing Arrangements

IT decision-makers continuously face the question about which tasks can be better and more efficiently performed by an external party. The IS literature provides a large body of literature with relevant considerations related to IT outsourcing (see Lacity et al. 2009 and Chapter 60 of this volume for an overview). Outsourcing arrangements can be differentiated regarding the coordination mode with an external provider—e.g., selected contractual obligations ("arms-length" relationships) for cost efficiency, versus long-term relational partnerships ("embedded") for strengthening IT resources and technological flexibility (Lee et al. 2004). Notably, in recent years, the focus has shifted from long-term, comprehensive IT outsourcing arrangements and purely economic considerations to contracts that also target quality, flexibility, and innovation goals (e.g., Whitley and Willcocks 2011). Recent literature also emphasizes the need for in-house capabilities for governing the different kinds of outsourcing relationships effectively (e.g., Willcocks and Griffiths 2010). One model of nine IS capabilities for modern IT organizations, for example, includes four capabilities that are directly related to managing outsourcing providers: informed buying, contract facilitation, monitoring, and vendor development (Feeny and Willcocks 1998).

While both ITIL and COBIT describe some processes and activities related to managing third-party services, from an enterprise design standpoint, the crucial concern is the *locus of outsourcing governance*, i.e., whether sourcing capabilities are allocated at the business level, the central, or the divisional IT side (Agarwal and Sambamurthy 2002). IT outsourcing decisions can also result in a change in decision rights for that particular IT function, including decentralizing more such rights to business units (Brown 2003). For example, in situations where resources for IT demand already reside in business units or divisional IT groups, this organizational configuration increases the outsourcing readiness of these units and thus the likelihood that an outsourcing relationship will be governed directly by the division. This may as well create more pressure on central IT units to compete with external providers—especially when business units are not obliged to contract internally. Financially autonomous IT supply units that are organized as subsidiary to their parent corporations (i.e., captive IT centers) can therefore also be viewed as transitional structural arrangements prior to outsourcing IT supply to an external party (Kreutter and Stadtmann 2009). In such situations, building appropriate demand-side IT capabilities may become a strategic priority (Feeny and Willcocks 1998).

57.2.6 Capabilities and Skills

We define a capability as *the application of knowledge, competencies, and skills residing in human resources, to accomplish given organizational goals* (Peppard and Ward 2004). Our second dimension, allocation of IT resources, refers to the structural assignment of human resources within the organization, whereas this dimension focuses on the aggregate proficiencies that IT human resources within an enterprise need to have. The IS literature has proposed different categories of capabilities in IT organizations. In addition to the nine-capability framework of Feeny and Willcocks (1998), a common typology derived from marketing research distinguishes between inside-out, outside-in, and spanning capabilities (Wade and Hulland 2004).[*]

Inside-out refers to capabilities that are internally focused, such as IT infrastructure, IT development, and (more generally) cost-effective IT operations—here referred to earlier as *IT supply* capabilities. Outside-in and spanning capabilities are externally oriented, placing an emphasis on requirements and customer relationships, including IT planning and change management, IT/business partnerships, market responsiveness, and external relationship management. These capabilities are likely to be aligned closely with business units and here we characterize them as *IT demand capabilities*.

Some more fine grained competency and skill categories can be found in both the academic and the practitioner literature, including a framework of 36 skills in 5 categories (Zwieg et al. 2006), skills related to roles in ITIL and CMM capabilities, as well as in frameworks such as the Skills Framework for the Information Age promoted by industry groups within the United Kingdom (SFIA 2012).[†] With the increasing pressure of IT organizations to compete on the product and labor markets, the development of appropriate IT demand and IT supply competencies becomes a more important imperative. A wide range of IT human resource practices, such as recruitment, training, and retention, and proactive career development can guide IT organizations to achieve this goal (Luftman 2011).

Table 57.1 summarizes the seminal literature that has motivated our inclusion of each of the six design dimensions.

57.3 IT Organization Archetypes

In Figure 57.1 we presented the four basic archetypes of IT organization configurations that are based on the first two dimensions described earlier: the distribution of IT decision rights and IT resources. In the following, we describe these archetypes in more detail, including their characteristics on the other four dimensions, their occurrence in practice, their strengths, as well as some common challenges.

57.3.1 Centralized Model

In a centralized model, most IT decision rights are allocated to the corporate level and IT resources are reporting to a central IT unit subordinate to corporate control while serving multiple business units. An IT steering or advisory committee has been recognized as an important coordination mechanism for ensuring business leader input into IT decision-making (Brown 2003; Huang et al. 2009). Under this model, the IT function is typically operated as a cost- or break-even center with simple chargeback arrangements. For example, in a corporate setting, a combination of global and business unit–related

---

[*] Although Wade and Hulland (2004) refer to these as categories for *resources*, their definition of resources as "assets and capabilities that are available and useful in detecting and responding to market opportunities or threats" is congruent with the notion of *capabilities* used in this paper.

[†] The SFIA Foundation is a not-for-profit organization that exists to own, promote, develop and maintain the Skills Framework for the Information Age. The members of The Foundation are UK Industry bodies in the field of IT: BCS (The British Computer Society), e-skills UK (e-skills UK Sector Skills Council Ltd), The IET (The Institution of Engineering and Technology), IMIS (The Institute for the Management of Information Systems), and itSMF (IT Service Management Forum).

TABLE 57.1   Design Dimensions and Primary Literature Sources

Dimension	Key Design Questions	Selected Literature
1. Allocation of decision rights	Which decision rights are allocated to business units, corporate, and IT stakeholders?	Brown and Magill (1994); Sambamurthy and Zmud (1999); Weill and Ross (2004); Brown and Grant (2005)
2. Allocation of resources	Which degree of centralization is appropriate? What is the split between IT demand and supply resources?	Boynton et al. (1992); Hitt and Brynjolfsson (1997); Thiadens (2005); Mark and Rau (2006); Daft (2009)
3. Coordination mechanisms	Which integration mechanisms (structural, procedural, relational) are implemented?	Brown (1999); Peterson et al. (2000); Chan (2002); Van Grembergen (2004)
4. Financial autonomy	Which degree of autonomy is appropriate for IT units? Which center type is implemented (cost, break-even, profit, investment center)?	Venkatraman (1997); Anthony and Govindarajan (2007); Ryan and Raducha-Grace (2009)
5. Sourcing arrangements	Which degree of external sourcing is appropriate? Which organizational units govern external relationships?	Agarwal and Sambamurthy (2002); Lee et al. (2004); Lacity et al. (2009); Willcocks and Griffiths (2010); Whitley and Willcocks (2011)
6. Capabilities and skills	Which capabilities are needed for IT demand and IT supply? How are these developed within the organization?	Feeny and Willcocks (1998); Peppard and Ward (2004); Wade and Hulland (2004); Zwieg et al. (2006); Luftman (2011)

IT budgets may be managed together with project-level and person-day based internal pricing. External contractors are typically governed by the corporate IT unit. Therefore, central IT resources not only need to be equipped with IT supply capabilities but also with sufficient IT demand capabilities to identify business needs and translate these into successful delivery by internal resources and external partners (as applicable).

Centralized models were the primary type of IT organization during the early era of mainframe computing and into the late 1980s when relational databases had arisen, however, networking was still limited (Peak and Azadmanesh 1997). A second wave of centralization also occurred in the mid-1990s as large firms initially implemented complex enterprise system packages (Brown 2003; McAdam and Galloway 2005). Today, centralized IT functions are also still the predominant model for small- and medium-sized businesses (Huang et al. 2009). Strengths of this model relate to an inherently high degree of standardization and corresponding efficiency through the sharing of IT resources and an underlying IT architecture across all divisions. Common challenges are business responsiveness and often a (perceived) lack of business contribution, that is, the IT organization may appear to act as a "black box" from a divisional perspective. Many centralized models have experienced improvements in IT responsiveness by enhancing both formal and informal coordination mechanisms, e.g., by introducing dedicated liaison roles and cross-functional IT meetings (Brown 1999; Huang et al. 2009).

57.3.2 Decentralized Model

In a decentralized model, business units make IT decisions (divisional or lower level) and are also responsible for managing IT resources. In the pure decentralized model, a central IT unit does not exist, which means that today it can be viewed as an almost "anarchic" configuration, with no or little coordination on a corporate level (Weill and Ross 2004, p. 58). In small divisions, coordination can even be achieved via informal, relational mechanisms; costs may not be accrued as a separate IT budget; and chargeback arrangements may not be implemented. If decentralized models make use of external suppliers/contractors, potentially for selected IT sourcing or project resources, these are typically governed outside of corporate control.

The decentralized model became more common after the expansion of mini-computers in the late 1970s, when most of the information processing took place in closed (proprietary) systems managed by local IT experts (Peak and Azadmanesh 1997). The rapid growth of desktop computing and more modern distributed computing architectures also facilitated more decentralization (Von Simson 1990). The disadvantages of this model as a "pure" model lie in the cultivation of silo structures and a lack of IT cost transparency. Similar downsides relate to the commonly undesirable phenomenon of "shadow IT," i.e., the existence of ad hoc IT solutions built, used, and managed by the business without central involvement or approval (Raden 2005). However, decentralized configurations can still be appropriate in cases where a strategic independence of a certain business division is desired, which may even include divestment-readiness (Leimeister et al. 2012). This model can also be appropriate for business functions where high innovation through IT and autonomous IT use are a strategic imperative, for example, in the research and development departments in a technology-intensive industry.

57.3.3 Shared Services Model

In the shared services model depicted in Figure 57.1, the IT resources are highly centralized, while the IT decision rights are primarily located at the division level. That is, divisions share the usage of centralized resources to capture advantages associated with a centralized model—including economies of scale and scope and IT architecture planning—without giving up major decision rights to a corporate IT unit. The business divisions typically also participate in steering committees and other decision-making bodies—such as cross-divisional IT boards responsible for IT architecture, IT application prioritizations, and infrastructure management—to set priorities for all of the divisions using the centralized IT resources. Shared services units are financially more autonomous than a purely centralized model and are responsible for their own results (Schulz et al. 2009). IT organizations that transition to this model therefore often need to devote significant efforts to productize their IT services on a competitive cost basis, so that they can retrieve their costs with chargebacks to their business customers (Ryan and Raducha-Grace 2009). External service providers may also be contracted for, and governed by, the shared services unit, especially for infrastructure services. However, depending on their size and maturity—and policies of the overall organization—business divisions may also have sufficient IT demand capabilities (and authorization) to independently contract out to external parties and thus circumvent the shared services unit.

Some of the early roots for this model can be seen in the writings by Von Simson (1990) and others in the 1990s, when organizations sought to better balance the advantages of a centralized model with those of a decentralized model with hybrid approaches. One hybrid approach was to create a federal model with IT application rights *and* resources residing within the divisions or business units, but IT operations (rights and resources) in a corporate IT unit. In contrast, in a "pure" shared services model, IT decision rights are at the division level, but IT (both application and infrastructure operations) resources are centralized. The global implementation of enterprise systems beginning in the late 1990s, which required both centralized application maintenance and process-based customizations, has been one of the catalysts for a wider acceptance of the pure shared services model (Brown 2003).

In many corporations today, shared services have therefore become a dominant model to organize and deliver IT as well as other enterprise support functions (e.g., accounting, physical facilities management), which are therefore sometimes co-located with IT (Schulz et al. 2009). Companies thereby aim to combine the benefits of centralization (economies of scale and scope) for IT applications and operations, with the benefits of outsourcing (e.g., customer focus, quality orientation, and increased variable versus fixed costs at the division level)—without sharing the potential drawbacks of outsourcing to an external supplier (e.g., supplier sustainability, loss of internal know-how, regulatory compliance, and data security concerns, etc.).

Sometimes this model is also seen as an opportunity to generate additional business, or as a strategic step before entirely outsourcing IT operations. Until the mid 2000s, many major corporations set up such IT subsidiaries with the primary goal of generating external revenues during a time of tremendous IT expansion in developed countries—a strategic trend that from today's perspective, with few exceptions, can be counted as a failure (Kreutter and Stadtmann 2009). Reasons for why many of these "captive" players could not sustainably hold ground in an external market include the changing capability requirements for internally versus externally competing service providers, and the rise of mature IT outsourcing firms that utilize cheaper labor.

Some of the inherent challenges of this model also relate to the lengthier channels of communication from business demand to IT supply units (delivery), which may need to be coordinated across multiple division (and country) boundaries. For this reason, sophisticated governance mechanisms, including service level agreements by business units, as well as strong demand-side IT capabilities, are required in order to implement this model successfully (Peterson et al. 2000; Van Grembergen 2004).

57.3.4 Corporate Coordinator Model

In the corporate coordinator model, IT-related tasks are performed externally or by divisional resources (i.e., by divisional IT units or non-IT business users themselves), while a central IT authority (office of the CIO, or in some cases a CTO*) governs through IT decision rights and aligns the IT resource investments with an overall IT architecture strategy. In the "pure" form, the office of the CIO is empowered to develop and enforce standards and monitor adherence via the CIO's direct report to corporate management, but does not possess dedicated resources to provide IT supply. Corporate IT standards differ in extent and range, from the usage of certain technology platform and application standards to guidelines for risk management and security controls. The reliance on committees and other coordination mechanisms to balance corporate and cross-functional priorities is similar to the shared services model. However, in a corporate coordinator model, these governance bodies are under the CIO, who has greater decision-making rights. For example, large IT development projects and sourcing arrangements to be managed at the division level may require pre-approval from the CIO.

External providers are contracted centrally by the office of the CIO or by divisional IT groups, depending if the service being sourced has firm-wide impacts (e.g., infrastructures and communication) or only divisional impacts (e.g., consultants and IT specialists in a project context). The CIO office acts as the mediator of external IT services, which are charged back to the divisions based on the costs of provision. Financial autonomy of the internal, divisional IT units is generally low, costs are accrued to divisional IT cost centers that are consolidated in divisional budgets, and no chargebacks take place at the division level. However, global cost transparency is warranted through oversight by the CIO and a global portfolio of divisional and corporate IT projects. For IT supplier steering and internal as well as external coordination, the central CIO office needs to develop strong demand capabilities (e.g., IT planning and change management, market responsiveness, and external relationship management). IT supply typically takes place through external suppliers or through divisional IT resources (as applicable).

The corporate coordinator model in its pure form is appropriate for several particular contexts, of which we highlight three. First, establishing a CIO office is often used strategically as a first step to advance from very decentralized configurations to more centralized governance and transparency, before actually centralizing resources, consolidating infrastructures, and achieving global scale. Second, for some business models that are based on replication (i.e., different entities with low data integration needs, but similar business processes), a coordinator model is the appropriate choice, due to its ability to leverage standardization potentials and economies of scale in IT sourcing, without integrating IT architectures (Ross et al. 2006, p. 35). Examples for such business models are diversified conglomerates as well as franchise companies.

Finally, the CIO office as a mediator of external IT services enables the ongoing IT outsourcing and industrialization trend. That is, the more (diverse) services are procured from the external market, the

---

* The Chief Technology Officer (CTO) role has evolved from research and development (R&D) management positions in technology-based industries and has recently also attracted more attention as a point of strategic responsibility for long-term goals and guidelines for the use of information technology within organizations (Hunter 2011).

TABLE 57.2 Key Characteristics of Four IT Organization Archetypes

	Centralized	Decentralized	Shared Services	Corporate Coordinator
IT decision rights allocation	CIO with senior management support	Business unit leaders (separately)	Business unit leaders (federally) and central CIO	Central CIO office enforcing standards, local implementation
IT resource allocation	IT resources in corporate IT	IT resources in local divisions	IT resources in shared IT unit, few IT demand resources	IT resources in divisions or external, few strategic IT resources in CIO office
Coordination mechanisms (structural)	Business relationship managers, IT steering committee	Divisional IT heads, divisional management boards	Divisional IT managers, central account managers, cross-divisional IT boards, e.g., IT architecture board	Executive board, divisional IT heads, architecture board
Financial autonomy	Cost or break-even center, simple chargebacks	Cost center or accrued to other budgets, no chargebacks (for small divisions)	Break-even, profit or investment center, productized chargebacks	Chargebacks for external IT services, cost centers for divisional IT, global monitoring
Sourcing arrangements	ESPs[a] governed by corporate IT	ESPs[a] governed by divisional IT	ESPs[a] governed by corporate IT or divisional IT	Firm-wide ESP[a] contracts governed by CIO office, specialist ESPs by divisional IT
Capabilities and skills	Good demand capabilities in corporate IT needed	Demand from business, supply capabilities in division IT	Ideal split of IT demand and IT supply capabilities realized	Demand capabilities in CIO office, supply capabilities in division IT (or externally)
Strengths	Standardization, resource pooling, efficiency	High responsiveness and local innovation, strategic independence	Economies of scale and responsiveness, customer-orientation, IT cost transparency	Expert sourcing by CIO office, standardization, global IT cost transparency, strategic independence
Common challenges	Lack of business value contribution, low flexibility	Lack of efficiency, low cost transparency, silo structures	More complex governance and communication structures, IT supply competes externally	Difficult to empower CIO office, lack of strategic IT competence in business divisions

[a] ESP, external service provider.

higher is the need for expert buyers to steer and manage these providers in order to achieve the desired benefits (e.g., costs, flexibility, and innovation goals). Thus, establishing a corporate coordinator model can be a viable alternative to building the distributed and costly demand capabilities in the business divisions—as required for the shared services model.

The key challenge of the corporate coordinator model is its difficulty in effectively implementing centralized IT governance to leverage economies of scale and standardization via negotiations across division heads. This may explain why this archetype—as a model for the entire IT organization—is still uncommon today in practice.

The four IT organization archetypes and their key characteristics are summarized in Table 57.2.

57.4 Assessing the Business and Technology Contingencies

Past research has proposed traditional business drivers such as a firm's competitive strategy and structure as influencing the "choice" of the archetype of an IT organization (Agarwal and Sambamurthy 2002; Brown and Grant 2005). For example, more globalized firms seeking responsiveness to local markets are likely to decentralize some IT rights and responsibilities, while smaller firms striving for economies of scale are likely to centralize their IT decision rights and resources (Sambamurthy and Zmud 1999; Weill and Ross 2004; Huang et al. 2009). However, more recent literature also emphasizes the complementarities between organizational and technological architectures (Tiwana and Konsynski 2010).

We conjecture that recent technology trends such as cloud computing and IT consumerization are likely to affect the IT organization models for both the IT demand and the supply sides. More specifically, cloud computing and the Internet-based delivery of applications and components as a service will further push the border of what is "core" and what is "commodity" across enterprise application landscapes (Bento and Bento 2011). Thus, on the application level, business units are more likely to manage their own cloud applications in a more decentralized fashion and thus circumvent centralized investment procedures (Winkler et al. 2011). At the backend, fewer IT resources will be needed for the operating infrastructure. However, managing the technological architecture and integrating cloud-based services with internal and external infrastructures will pose increasingly important challenges and the need for new capabilities.

Consumerization of IT superimposes the cloud wave. Employees with increasing IT skills and access to sophisticated client devices for personal use expect to find IT tools in their workplace that they already use in their home environments (Bernnat et al. 2010). As an answer to these new expectations, some companies have created policies for allowing employees to bring their own devices, such as smartphones and tablet PCs, into the work environment and integrate them. This represents a paradigm shift inasmuch as employees are subsidized for using their own hardware and applications. Data security and other related risks need to be diligently addressed by enforcing appropriate firm-wide guidelines.

These and other technology trends suggest greater decentralization of IT responsibilities and more hybrid IT governance designs in the future. More application as well as infrastructure decision rights (e.g., on mobile device use) will shift to tech savvy business users, while IT operations responsibilities are increasingly shared between internal and external suppliers. Managing the diverging ecosystem of user IT demand and entire supply chains of IT service provision will be one of the key IT governance challenges in the future (McDonald 2007). Enterprise-level organizational models that enable a better integration and coordination across users, IT units, and multiple suppliers will need to be developed, which we expect to be reflected in future versions of standard IT reference models (Pardo 2010; Marrone and Kolbe 2011).*

57.5 Further Research Opportunities

Beyond the technology contingencies, other perspectives also appear particularly fruitful for investigating the changing shape of contemporary IT organizational configurations. First, industry-specific approaches have largely been neglected in the past. Organizations in the public sector, for example, national and local governments as well as non-profits in health care and other industries, hold different principles for creating public versus private value, which may also call for different principles of IT governance (Weill and Ross 2004, pp. 185–214; Sethibe et al. 2007). Second, given the increasing dispersion of IT value creation across organizational ecosystems, the understanding of "organizational configurations" needs to be broadened to span entire IT value networks (Leimeister et al. 2010; Iyer and Henderson 2012). This also implies that the extensive, yet separate, literature strands on governance of (internal) IT functions and governance of (external) outsourcing relationships need to be united under a common frame. Third, such governance arrangements may significantly vary depending on the kind of IT subfunctions considered. Various authors have begun to investigate IT organization and governance phenomena regarding certain subdomains, such as governance in system development projects (Tiwana 2009), application governance (Winkler et al. 2011), data governance (Khatri and Brown 2010), and infrastructure sourcing governance (Xue et al. 2011). Taking such modular views and aligning these with overall (networked) governance schemes appear a promising

---

* For example, in its 2011 version ITIL has introduced additional strategic processes and liaison roles to address increasing coordination needs, such as a service strategy manager, a business relationship manager, and a demand manager.

field for future researchers. Finally, having argued that organizational configuration is a dynamic phenomenon influenced by business and technology developments, we conclude that more longitudinal research is needed to study IT organization design phenomena.

In this chapter we conjecture that implementing more characteristics of a corporate coordinator model—i.e., moving operational IT delivery to external specialists while focusing on the alignment of internal standards and decentralized provisioning of specialist resources—may become a viable path for many CIOs to address the design issues related to the IT organization dimensions discussed here. This also implies that the CIO role will continue to evolve from an "inside-out" IT supply manager to an empowered coordinator, who leverages capabilities of the entire IT value network from the outside-in—a role that has also been termed as a *chief business technology strategist* (Carter et al. 2011). Future researchers may build on our enterprise-level view of four archetypes of IT organizations and their dimensions, as well as assess in-depth the impact of emerging IT trends on IT organizations in different, dynamic business and industry contexts.

References

Agarwal, R. and V. Sambamurthy (2002). Principles and models for organizing the IT function. *MIS Quarterly Executive 1*(1), 1–16.

Anthony, R. N. and V. Govindarajan (2007). *Management Control Systems*, 12th edn. McGraw-Hill, Irwin, Columbus, OH.

Bento, A. L. and R. Bento (2011). Cloud computing: A new phase in information technology management. *Journal of Information Technology 22*(1), 39–46.

Bernnat, R., O. Acker, N. Bieber, and M. Johnson (2010). Friendly Takeover—The consumerization of corporate IT. Booz & Co. http://www.booz.com/media/uploads/FriendlyTakeoverVPFINAL.pdf, accessed March 8, 2012.

Boynton, A. C., G. C. Jacobs, and R. W. Zmud (1992). Whose responsibility is IT management. *Sloan Management Review 33*(4), 32–38.

Brown, C. V. (1999). Horizontal mechanisms under differing IS organization contexts. *MIS Quarterly 23*(3), 421–454.

Brown, C. V. (2003). The IT organization of the future. In J. N. Luftman (Ed.), *Competing in the Information Age: Align in the Sand*, 2nd edn. Chapter 8, pp. 191–207. Oxford University Press, New York.

Brown, A. E. and G. G. Grant (2005). Framing the frameworks: A review of IT governance research. *Communications of the AIS 15*, 696–712.

Brown, C. V. and S. L. Magill (1994). Alignment of the IS function with the enterprise: Toward a model of antecedents. *MIS Quarterly 18*(4), 371–403.

Carter, M., V. Grover, and J. B. Thatcher (2011). The emerging CIO role of business technology strategist. *MIS Quarterly Executive 10*(1), 20–29.

Chan, Y. E. (2002). Why haven't we mastered alignment? The importance of the informal organization structure. *MIS Quarterly Executive 1*(2), 97–112.

Daft, R. L. (2009). *Organization Theory and Design*, 10th edn. South-Western Cengage Learning, Mason, OH.

Evaristo, J. R., K. C. Desouza, and K. Hollister (2005). Centralization momentum: The pendulum swings back again. *Communications of the ACM 48*(2), 66–71.

Feeny, D. F. and L. P. Willcocks (1998). Core IS capabilities for exploiting information technology. *Sloan Management Review 39*(3), 9–21.

Hitt, L. M. and E. Brynjolfsson (1997). Information technology and internal firm organization: An exploratory analysis. *Journal of Management Information Systems 14*(2), 81–101.

Huang, R., R. Zmud, and R. Price (2009). IT governance practices in small and medium-sized enterprises: Recommendations from an empirical study. In G. Dhillon, B. C. Stahl, and R. Baskerville (Eds.), *Information Systems—Creativity and Innovation in Small and Medium-Sized Enterprises*, Vol. 301, Chapter 12, pp. 158–179. Springer, Berlin, Germany.

Hunter, M. G. (2011). The duality of information technology roles. *International Journal of Strategic Information Technology and Applications (IJSITA) 2*(1), 37–48.

Iyer, B. and J. C. Henderson (2012). Business value from clouds: Learning from users. *MIS Quarterly Executive 11*(1), 51–60.

Khatri, V. and C. V. Brown (2010). Designing data governance. *Communications of the ACM 53*(1), 148–152.

Kreutter, P. and G. Stadtmann (2009). The captives' end: Lebenszyklusmuster in der entwicklung der deutschen IT-Outsourcing-Industrie. *Research notes working 30, Deutsche Bank Research.* http://ideas.repec.org/p/zbw/dbrrns/30.html (accessed on April 13, 2013).

Lacity, M. C., S. A. Khan, and L. P. Willcocks (2009). A review of the IT outsourcing literature: Insights for practice. *The Journal of Strategic Information Systems 18*(3), 130–146.

Lee, J. N., S. M. Miranda, and Y. M. Kim (2004). IT outsourcing strategies: Universalistic, contingency, and configurational explanations of success. *Information Systems Research 15*(2), 110–131.

Leimeister, J. M., M. Böhm, and P. Yetton (2012). Managing IT in a business unit divestiture. *MIS Quarterly Executive 11*(1), 37–48.

Leimeister, S., C. Riedl, M. Böhm, and H. Krcmar (2010). The business perspective of cloud computing: Actors, roles, and value networks. In *Proceedings of 18th European Conference on Information Systems ECIS 2010*, Pretoria, South Africa.

Luftman, J. N. (2011). *Managing IT Human Resources: Considerations for Organizations and Personnel.* IGI Global, Hershey, PA.

Mark, D. and D. P. Rau (2006). Splitting demand from supply in IT. *The McKinsey Quarterly* (Fall), 22–29.

Marrone, M. and L. M. Kolbe (2011). Impact of IT service management frameworks on the IT organization. *Business & Information Systems Engineering 3*(1), 5–18.

McAdam, R. and A. Galloway (2005). Enterprise resource planning and organisational innovation: A management perspective. *Industrial Management. & Data Systems 105*(3), 280–290.

McDonald, M. P. (2007). The enterprise capability organization: A future for IT. *MIS Quarterly Executive 6*(3), 179–192.

Pardo, C. (2010). A systematic review on the harmonization of reference models. *Engineering* (4), 40–47.

Peak, D. and M. H. Azadmanesh (1997). Centralization/decentralization cycles in computing: Market evidence. *Information and Management 31*(6), 303–317.

Peppard, J. and J. Ward (2004). Beyond strategic information systems: Towards an IS capability. *The Journal of Strategic Information Systems 13*(2), 167–194.

Peterson, R. R., R. O'Callaghan, and P. M. A. Ribbers (2000). Information technology governance by design: Investigating hybrid configurations and integration mechanisms. In *ICIS 2000 Proceedings*, Brisbane, Queensland, Australia, pp. 435–452.

Raden, N. (2005). Shedding light on shadow IT: Is Excel running your business? Hired Brains, Inc. http://www.cioindex.com/nm/articlefiles/69862-ShadowIT.pdf, accessed March 8, 2012.

Ross, J. W., P. Weill, and D. Robertson (2006). *Enterprise Architecture as Strategy: Creating a Foundation for Business Execution.* Harvard Business Press, Boston, MA.

Ryan, R. and T. Raducha-Grace (2009). IT financial management: The business of IT. In *The Business of IT: How to Improve Service and Lower Costs.* IBM Press, Upper Saddle River, NJ.

Sambamurthy, V. and R. W. Zmud (1999). Arrangements for information technology governance: A theory of multiple contingencies. *MIS Quarterly 23*(2), 261–290.

Schulz, V., A. Hochstein, F. Uebernickel, and W. Brenner (2009). Definition and classification of IT-shared-service-center. In *AMCIS 2009 Proceedings*, San Francisco, CA.

Sethibe, T., J. Campbell, and C. McDonald (2007). IT governance in public and private sector organisations: Examining the differences and defining future research directions. In *University of Canberra, 18th Australasian Conference on Information Systems*, Toowoomba, Queensland, Australia.

SFIA (2012). Skills framework for the information age. SFIA Foundation. http://www.sfia.org.uk/, accessed March 8, 2012.

Thiadens, T. (2005). *Manage IT! Organizing IT Demand and IT Supply.* Springer, Dordrecht, the Netherlands.

Tiwana, A. (2009). Governance-knowledge fit in systems development projects. *Information Systems Research 20*(2), 180–197.

Tiwana, A. and B. Konsynski (2010). Complementarities between organizational IT architecture and governance structure. *Information Systems Research 21*(2), 288–304.

Van Grembergen, W. (2004). *Strategies for Information Technology Governance.* IGI Global, Hershey, PA.

Venkatraman, N. (1997). Beyond outsourcing: Managing IT resources as a value center. *Sloan Management Review 38*(3), 51–64.

Von Simson, E. M. (1990). The "centrally decentralized" IS organization. *Harvard Business Review 68*(4), 158–162.

Wade, M. and J. Hulland (2004). The resource-based view and information systems research: Review, extension, and suggestions for future research. *MIS Quarterly 28*(1), 107–142.

Weill, P. and J. Ross (2004). *IT Governance: How Top Performers Manage IT Decision Rights for Superior Results.* Harvard Business Press, Boston, MA.

Whitley, E. A. and L. P. Willcocks (2011). Achieving step-change in outsourcing maturity: Toward collaborative innovation. *MIS Quarterly Executive 10*(3), 95–109.

Willcocks, L. P. and C. Griffiths (2010). The crucial role of middle management in outsourcing. *MIS Quarterly Executive 9*(3), 177–193.

Winkler, T. J., C. Goebel, A. Benlian, F. Bidault, and O. Günther (2011). The impact of software as a service on IS authority—A contingency perspective. In *International Conference on Information Systems (ICIS) 2011 Proceedings*, Shanghai, China, paper 22.

Xue, L., G. Ray, and B. Gu (2011). Environmental uncertainty and IT infrastructure governance: A curvilinear relationship. *Information Systems Research 22*(2), 389–399.

Zwieg, P., K. Kaiser, C. Beath, C. Bullen, K. Gallagher, T. Goles, J. Howland et al. (2006). The information technology workforce: Trends and implications 2005–2008. *MIS Quarterly Executive 5*(2), 47–54.

58

Topics of Conversation: The New Agenda for the CIO

Joe Peppard
*European School of
Management and Technology*

Today, most organizations are so fundamentally dependent on their information systems and technology that they could not operate for very long if their core systems were not available. Organizations such as Tesco, Fed Ex, ING Bank Direct, and Commonwealth Bank of Australia derive competitive advantage from their IT investments while others, such as Amazon, Lastminute, threadless, and eBay, have business models that are totally defined by IT. Many organizations are currently looking at ways to potentially harness technologies such as "the cloud," Web 2.0, social media, and collaborative tools. We also see traditional manufactures like Rolls Royce, MAN Trucks, BAE Systems, and Hilti using IT to enable the delivery of new service propositions to their customers based around availability and capability rather than selling products. Yet despite this crucial role of IT for business today and the opportunities that it provides, research findings are unequivocal: leadership of IT in most organizations is generally weak.*

Many of the CEOs that we speak to continue to lament that they are disappointed with their IT investments and the contribution they feel IT is making to the operational and strategic development of their businesses. They are unanimous in expecting more of their CIOs than merely "keeping the

---

* See, for example, *Leadership, Capability and Culture: The CIO's Silent Revolution*, IBM Corporation, London, 2009; *Strategic Insight Survey: A US IT Leadership Perspective*, Harvey Nash, 2009; *The CIO Profession: Leaders of Change, Drivers of Innovation*, Center for CIO Leadership, 2008; *The CIO Profession: Driving Innovation and Competitive Advantage*, Center for CIO Leadership, October 2007; *The Evolving Role of the CIO*, IBM Global Services, 2008; *The New CIO: Change Partner and Business Leader*, IBM Corporation, 2007; *Global CIO Survey 2007: IT Agility—Enabling Business Freedom*, Cap Gemini, 2007; *Global CIO Survey: The Role of the IT function in Business Innovation*, Cap Gemini, 2008; *CIO 2.0: The Changing Role of the Chief Information Officer*, Deloitte, 2004; *Circa 2015: The CIO of the Future*, Software AG, 2008; *Information Technology II: How To Close the CIO Leadership Credibility Gap*, Accenture, September 2006.

lights on." Increasingly, many are now demanding their CIO to drive the transformation agenda and are looking for IT to seek out opportunities for integration and innovation. However, our research indicates that a considerable number of CIOs are struggling with this new expectation. Whether a consequence or a contributor, many CIOs are reporting to us that strategic conversations are taking place within their organizations without their involvement.

What we have found is that a lot of the CIOs that we have spoken to say they are strategic, but in reality they behave tactically [1]. From our interviews with their business colleagues, many CIOs would seem to lack credibility as strategic decision-makers and consequently are not invited to be part of the "inner sanctum." Even where they do have a "seat" at the top table, they often do not have a "voice," particularly one that is readily listened to. Furthermore, we also observe that many CIOs lack basic influencing skills [2] and have not developed relationships with appropriate parts of the organization or key stakeholders [3].* Consequently, they are seen by their business colleagues as being politically naive [4]. And, perhaps more crucially, many lack the leadership capabilities to drive their organizations forward in the use of technology.

This situation is inhibiting CIOs not only in the performance of their role but more critically hindering the business in optimizing value from information systems and technology investments. In our research, we have talked to dozens of CEOs, CIOs, and other C-level executives in order to better understand why it is that some CIOs do seem to be having significant impact on the performance of their organizations and others are not (see Appendix for overview of this research).

The attitude that the CEO and executive team have toward IT does have an obvious impact on what the CIO can and cannot do. Many portray IT *"as a pain,"* see *"everybody moaning about it,"* and adopt a *"don't bother me attitude"* toward IT. The fact is the level of IT savvy of many CxOs is weak [1,5]. This situation strongly influences their expectations of what CIOs can and cannot be expected to achieve, the extent of their engagement with the incumbent, their contribution to IT decision-making, and their involvement in IT projects and programs. It also shapes their view on the opportunities that technology could potentially provide their organization. Yet, research findings are emphatic that companies reporting a higher level of business-IT alignment have a shared vision between the CIO and CxO team and a common set of core beliefs regarding IT [6].

We have also detected a gap between the rhetoric of CxOs and their actions. That is, they typically say that they recognize the crucial importance of IT, yet they behave in ways that do not back up this position. One newly appointed CIO had a conversation with one of the members of the senior leadership team (he reported into the Chief Financial Officer) in which he congratulated the CIO on his work since joining the company *"as we [the leadership team] have not had to talk about IT since you came on board!"*.

While all CIOs grapple with similar issues, what we have found is that some are more adept at meeting many of the contemporary challenges that they all face. These individuals have become extremely effective in helping their organizations deliver maximum value from their IT investments. These are what we have called *Hi-Impact CIOs*.

Hi-Impact CIOs are those CIOs who are seen by their business colleagues as driving the business forward through the use of IT. These CIOs typically have the trust of their business colleagues; they are seen as a respective member of the executive team; and continuously interact with their business colleagues. Perhaps not unexpectedly, they are also seen as listening to their business colleagues, being empathic to their plight and demonstrating a passion for enabling the business with IT. Their credibility is such is that it enables them to influence business colleagues in IT matters. Not only do they have a seat at the top table, they also have a voice that is listened to.

From our data, there is strong evidence that Hi-Impact CIOs have retooled not just how they interact and communicate with their business colleagues but also have shifted the nature of that interaction. In short, there is strong evidence that they have reframed the dialogue that they have with CxO colleagues. They report that it is not just about having conversation with their colleagues, but having the

---

* See J. Peppard and J. Thorp, 'What every CEO needs to know and do about IT', manuscript currently under review

right conversations. Research shows that high levels of communication between IT and business executives are a direct predictor of alignment of business and IT strategies [7]. And CxOs rank communication as the personal skill most pivotal to success as a CIO.*

To provide context for the research presented in this chapter, we first briefly review the evolution of the CIO role in organizations. This evolution reflects the changing role for IT in organizations and the strategic implications that it now has. This analysis also highlights the ambiguous nature of the role. We then present the changing topics of conversations that Hi-Impact CIOs are having with their C-level colleagues.

58.1 Evolution of the CIO Role†

The genesis of the CIO role and indeed the coining of the "chief information officer" label can be traced to a shift from IT having a supporting role in organizations, automating previously manual tasks, to being a driver of competitive advantage, strategic change, and innovation. This latter objective demands considerably more than just a focus on technology deployment and its management—which the CIO's predecessors did—requiring business-driven approaches for the exploitation of information and IT. Increasingly, information was recognized as a critical resource that required active management, stewardship, and oversight from a senior management perspective [8].‡ The newly created position of CIO [9]§ emphasized information over technology, enterprise over function, and strategy over operations. Indeed, it was not that incumbents were no longer responsible for technology, the role just expanded. In addition to the operational dimensions, the position had a key requirement to provoke executive level discussions across the organization relating to how information and technology could be leveraged, particularly in the pursuit of competitive advantage.

It was in this quest for executive engagement that organizations began to make the distinction between *demand* and *supply* [10]. "Demand management" was concerned with identifying the information required to both shape and deliver the business strategy as well as prioritizing IT spend. This was promoted as a business-led activity. The supply side was concerned with satisfying these requirements through the sourcing and deployment of IT [10,12].¶ While demand would be determined by business executives working closely with their CIOs, or achieved by establishing a demand organization containing relationship and "hybrid" managers [11], efficient and effective supply was the responsibility of the CIO.

Despite this opportunity for CIOs to increase their impact in the C-suite, many have struggled. While clearly some may not be up to the job, this can be seen as a consequence of the polarization of views held by CxOs regarding the role of technology in organizations [13]: at one end is the belief that IT is an administrative expense and a cost to be minimized, while at the other is the stance that it offers significant strategic opportunities. This fundamental dichotomy perhaps manifests itself in

---

*   State of the CIO Survey 2007', *CIO*, January 2007.
†   This section draws heavily from my paper with colleagues "Clarifying the ambiguous role of the CIO," *MIS Quarterly Executive*, Vol. 10, No. 1, 2011, pp. 31–44.
‡   Applegate and Elam [8(a)] suggest that the rise of the CIO marked the transition from the DP era to the information era.
§   Credit for coining the label "Chief Information Officer" is generally attributed to William Synott at the 1980 Information Management Exposition and Conference. Synott predicted that "The manager of information systems in the 1980s has to be Superhuman—retaining his technology cape, but doffing the technical suit for a business suit and becoming one of the chief executives of the firm. The job of chief information officer (CIO)—equal in rank to chief executive and chief financial officers—does not exist today, but the CIO will identify, collect and manage information as a resource, set corporate information policy and affect all office and distributed systems." *Business Week* ran a story in 1986 announcing the arrival of the CIO.
¶   The concept of "the hybrid manager" was also promoted at this time. First coined by Peter Keen in 1988, it was defined by Michael Earl as "people with strong technical skills and adequate business knowledge, or vice versa … hybrids are people with technical skills able to work in user areas doing a line or functional job, but adept at developing and supplementing IT application ideas."

the different perceptions of the role of the CIO, job tasks, reporting relationships and what a CIO can be expected to achieve. It also determines the level of engagement and involvement of CxOs in what might traditionally have been considered as IT issues and decisions. One study has highlighted the "IT decisions the IT people shouldn't make." [14] But even in situations where CxOs subscribe to a more strategic view for IT, they often do not understand what *their* role and the wide responsibilities of "the business" in optimizing value from IT spend. This has been attributed to weak levels of digital literacy [1].

An examination of the prior research regarding the CIO role identifies a number of distinct themes. The early research explored the nature of the CIO position, in particular the mix of business and technical knowledge required of incumbents. Key prescriptions were that the CIO must operate as a corporate focused executive rather than a functional manager and that success in the role required bringing a broad business perspective to the position [8a]. Early research also revealed that those who thrive as CIOs recognize the importance of executive relationships particularly with the CEO [3].

Most of the contemporary research falls into two streams: either advocating that CIOs take a stronger and more proactive leadership role in their organization or exploring the personal competencies of successful CIOs. The former stream of research promotes the need for CIOs to lead change (sometimes using the word "transformation") and drive innovation. It recommends that they must expand their remit beyond merely "keeping the lights on" to having a more strategic role. As some recent papers have expressed: "to find ways for IT to change the company, not just run it," [15] and "take a much broader role in the business, driving business transformation, innovating for competitive advantage and acting as key strategic partners to the CEO and wider organization." [16] Studies in the latter stream explore the competencies, personal attributes, and characteristics that are seemingly required for success. Leadership, influencing, relationship architect, diplomat, etc. are promoted as crucial competencies [2a,17]. Personal attributes identified for success in the role includes communication skills, business acumen, people skills, openness, and conscientiousness [18].* As with the prescriptions from earlier studies, the CIO is envisioned to be more a business expert than a technical expert. Research in this genre has also reported on CIO leadership profiles (mapping "CIO authority" against "leadership capability"), although the construct of CIO leadership capability has not been well developed [19].

While interesting, much of this latter research has tended to add little to help organization as they seek to capitalize on IT. For example, prior research seeks to discover the different competencies (although the term "roles" is used in some studies) of CIOs. Competencies identified include: strategist, relationship architect, integrator, information steward, and (IT) educator. While useful, they are not grounded in a precise understanding of the *job* of the CIO and what incumbents are expected to achieve for their organization. In addition, the assumption of these studies is that if incumbents possess these competencies, they will be successful as CIOs. However, are these competencies not generic and required by all in the C-suite? Should not all members of the C-suite be business leaders, strategic thinkers, driving change and innovation? We would strongly argue that they should.

58.2 Shifting Conversations of the Hi-Impact CIO

What we have found is that Hi-impact CIOs have moved the focus of their attention away from the traditional areas of conversation: Hi-impact CIOs place their emphasis elsewhere. What we see are subtle, but nonetheless profound, shifts away from what have traditionally been seen as areas of focus for CIOs. Our research has identified eight shifts that together contribute to these CIOs having a major business impact (see Appendix for an overview of this research).

---

* Research has reported, for example, that CIO characteristics can explain a large proportion of variance in organizational innovative usage of IT see [18].

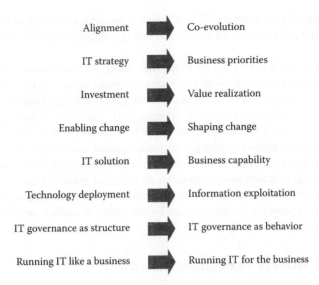

Alignment	Co-evolution
IT strategy	Business priorities
Investment	Value realization
Enabling change	Shaping change
IT solution	Business capability
Technology deployment	Information exploitation
IT governance as structure	IT governance as behavior
Running IT like a business	Running IT for the business

FIGURE 58.1 Shifts in CIO:CxO conversations: the new agenda for the CIO.

These shifts are captured in Figure 58.1. The left hand column lists the traditional areas of CIO attention, illustrating where the focus of their conversations have tended to lie. On the right are the corresponding areas of emphasis placed by Hi-Impact CIOs. These swings point to a new agenda for the CIO. In the remainder of this paper we will have a closer look at these shifts, identifying how each is manifesting itself.

58.2.1 From Alignment to Co-Evolution

For over three decades alignment has been the watchword of the CIO. Indeed, surveys of IT issues over the past 30 years have consistently ranked the challenge of achieving alignment among the top three [20]. CIOs themselves have always seen as one of their primary tasks to aligning their organization's portfolio of investments in information technology (IT) with the strategy of the business. While this is a laudable objective, it, of course, assumes that organizations do actually have an explicit business strategy which can be aligned against. Many CIOs have told us that this unfortunately is often not the case; and, even if a business strategy exists, it might not be shared across the executive management team.

Hi-impact CIOs, however, do not see the issue as one of alignment but one of co-evolution: that is, business and IT strategies co-evolving together [21]. Importantly, and this is the critical issue, the focus on alignment gives the impression that it is a task that is undertaken by the CIO and his team, and that IT requirements are somehow subservient to the demands of the business. One CIO spoke about seeking alignment equal to looking to align the human "heart" with the "body," which he saw as an illogical demand. Both are mutually dependent. As one CIO expressed it: "*just like we don't have IT projects any more, only business projects with an IT component, similarly, we don't have an IT strategy but a business strategy with an IT component.*"

In a similar way with biological organisms [22,23]*, with co-evolution both business and IT co-evolve together. An Australian CIO defined co-evolution as a "psychological state" with "business and

---

* The term "co-evolution" was coined in the 1960s by the American population biologist Paul Ralph Ehrlich and the botanist Peter Hamilton Raven to refer to evolutionary changes that occur in genetically unrelated species as they interact with each other in their environment.

IT working as one toward a common goal." This, however, demands high levels of trust, cognitive alignment, a shared understanding across all parties and strong collaboration [6a], something that is all too often absent.

Co-evolution recognizes that IT can also shape the strategy of the organization through providing innovative opportunities, particularly with new business models defined by IT. For example, many traditional manufacturing companies are today implementing service-based strategies and these new business models are heavily dependent on IT. For example, in the aerospace sector, engine manufacturers such as Rolls-Royce, General Electric, and Pratt & Whitney all offer some form of performance-based contracts with commercial airlines in which their compensation is tied to product availability and the capability it delivers (e.g., hours flown) rather than selling engines. Such contracts provide the airline operator with fixed engine maintenance costs over an extended period of time (e.g., 10 years). Yet to deliver such a service both effectively and profitably demands IT: data on engine performance must be collected in real-time, transmitted, and analyzed by the engine manufacturer at a health monitoring centre using prognostic and modeling tools.

58.2.2 From IT Strategy to Business Priorities

A key prescription for the CIO has always been to develop the organization's IT strategy. Indeed, this task usually appears in their job specification and is an expectation that their C-level colleagues hold of them. The reality is that many struggle to engage their business colleagues in the IT strategy formulation process and often end up "second guessing" the strategy and future business direction, particularly when one is absent. Our data indicate that Hi-Impact CIOs are shifting their dialogue with C-level colleagues from one focused around developing this IT strategy to having a conversation based on eliciting their business needs and priorities. These CIOs tell us that this approach is more likely to get the attention of their colleagues. One CIO we spoke to prefers to engage with his peers about their "initiatives," and to use these to validate their priorities and construct the IT investment portfolio. He determines what their initiatives are for the following 6 months, the next year, and the next 3 years, and then presents these back to them in a coherent fashion and one that makes sense to them before highlighting how IT can support their achievement. This, he also felt, enhanced his relationship and reputation with the leadership team as they see him as listening to them.

Such is the peril of focusing on the IT strategy that the recently retired Global CIO of a global pharmaceutical company warns that CIOs should "beware the IT strategy," as it can become an end in itself. This does not to suggest that a strategy for IT is not required; on the contrary, one is required now more than ever. However, developing this strategy should not be seen as a central focus of interactions with business colleagues. Focusing on the business priorities of colleagues is more likely to gain their interest and attention. From this dialogue, the basis of an IT strategy can be ascertained. It also gives guidance as to the priorities for IT investment as well as setting expectations for IT. The Head of Technology at one U.K. retailer commented that *"[t]here's no point in us wondering later why IT hasn't delivered what was expected if there hasn't been a clear upfront conversation between IT and the rest of the business about business priorities for the next 5/10 years; in particular what the problems are that the business will need IT to help solve."*

Indeed, CIOs often lament the lack of an explicit business strategy and this absence is often used by CIOs to excuse poor performance. One Non-Executive Director of an Australian insurance company recounted to us the visit he had from a newly appointed CIO complaining about the lack of a business strategy and how this was impacting her in her job and her frustration with this situation. This is, unfortunately, the platform from which many CIOs are working from. Her colleagues do have their priorities and it is hard to imagine an insurance company without IT having a significant impact.

We have also encountered situations where prescribed practices can often dictate that a separate IT strategy be developed. The legislation of the State of Queensland in Australia, for example, mandates that an annual IT strategy plan be developed for all public agencies, one that is aligned to the strategy of the agency. It even prescribes the process to ensure that this occurs, with the IT strategy trickling down from the business strategy. This can be problematic and reinforces the old orthodoxy of the IT strategy following the business strategy.

58.2.3 From Investment to Value Realization

Hi-Impact CIOs are telling us that conversations with C-level executives must shift from one centered on determining investments to a dialogue focused on value realization. They explain that value realization has a more dynamic and ongoing connotation than merely determining investments in IT. One CIO wryly lamented the fact that IT investments tend to be characterized based on their cost—"the 50 m dollar ERP project"—rather than described based on the value they will deliver to the business.

This might seem like a subtle shift in emphasis, but its implications are profound. A focus on identifying the investments to make suggests that the decision to spend is made by business executives with delivery managed by the CIO and his/her organization. Unfortunately, unlike other investments that a firm makes, the value of IT is not in its possession. If it was, just building and deploying IT-based systems successfully would result in benefits automatically flowing to the business, and we know that this is not the case. Few IT projects fail because the technology does not work. The reality is that generating value through IT is an organization wide endeavor with executives and employees having a role to play.

Organizations do not buy IT *per se*; what they do buy is value, or at least potential value. Unlike physical assets, IT has no inherent value and the possession of IT itself does not confer any value. This value has to be unlocked. This value emerges through change that the technology both enables and shapes: process change, new work practices, and improved decision-making [24]. Our research suggests that more business managers are getting involved in building business cases for IT investment; but once the approval for the spend has been received, it is then passed on to "IT" for delivery. They tell us that they feel comfortable in so doing as IT organizations tend to be driven by methodologies. Yet, they fail to recognize that such methodologies focus on technology deployment and not on generating value. By emphasizing the process of value realization, the focus is not just on getting approval for the spend, but on managing the life cycle of the investment and the process of value realization. Hi-Impact CIOs also suggest that to promote this behavior the business case must become a living document and accompany the investment through its lifecycle [25].

58.2.4 From Enabler of Change to Shaping Change

To have business relevance, Hi-Impact CIOs look to shape change in their organization, rather than merely responding to business requests. This is in recognition of the power and capability of IT to enable the organization to do things differently, both operationally and strategically. Some CIOs have even taken up leadership of the transformation agenda in their organizations. These CIOs acknowledge the power of innovation though IT and relish the challenge of seeking such opportunities out. However, to do this requires credibility [26]. Hi-Impact CIO tells us that this has nothing to do with one's CV; it is all about the relationship with CxO colleagues.

Hi-Impact CIOs shape change through innovation, seeking opportunities to marry emerging technical capability with business opportunity. They have typically dedicated budget and resources for innovation, constantly searching out new technologies and assessing their business relevance. They also work in close partnership with business colleagues to stimulate discussions and generate ideas for potential IT application.

We have also seen Hi-Impact CIOs recognizing the tension between "business pull" and "IT push." On one hand, they readily acknowledge the challenges of IT push, but their credibility gives them a strong voice. They realize that most "game-changer" innovations come from technology that their business colleagues can be slow to spot. But by having the trust of their business colleagues and credibility, they have the opportunity to "advocate" the idea to them. Of course, as a member of the leadership team, they also seek to "pull" ideas and opportunities from their business colleagues as well as working with the wider ecosystem of customer, suppliers, and business partners.

58.2.5 From IT Solution to Business Capability

In the past, the IT industry was criticized for selling the features and functions of products. In response, CIOs have generally emphasized the provision of solutions to the "problems" of business colleagues. However, what Hi-Impact CIOs are telling us is that the concept of the solution can typically result in business managers abdicating responsibility to the IT organization to provide *the* solution to *their* problems. Too often we hear the comment "*SAP will solve our problems.*" Even today, project management methodologies like PRINCE 2* are designed around building "the product"—the product being a combination of hardware and software. Hi-Impact CIOs prefer to emphasize the capabilities they are providing the business. The implication is that the business has to leverage these capabilities if the expected benefits of the investment are to be achieved.

What CIOs are telling us is that in promoting IT solutions, the focus is on "the IT" not "the problem" that requires solving. Consequently, IT is seen as a solver of problems: this is the magic bullet thesis. They report that it can also cause CxOs and other executives to disengage once the program for execution has been developed as addressing the problem is seen as an IT endeavor. Building the IT capability is just one element of larger program of work. Focusing on providing a business capability is more likely to get the attraction of business managers as it emphasizes business and IT working together to create a capability for the business to exploit.

58.2.6 From Technology Deployment to Information Exploitation

Hi-Impact CIOs recognize that while successfully implementing technology is necessary, on its own it is not a sufficient condition to generate value. Ultimately, value emerges from the usage of information and IT [27]. Yet, if we examine how organizations run a typical IT project, we observe a strong emphasis on the deployment of technology. Project metrics such as on-time delivery, within budget, and meeting the specification are reinforcing this emphasis [28]. While benefits from technology can be achieved through automation of organization processes, taking out time and cost, and other inefficiencies, the real value is through working with information. This is a neglected aspect of IT investment. It is more than just training, but a conversation around how the new information will be capitalized on.

One CIO expressed this as a shift from "delivering IT" to "delivering business value," but also acknowledged that it demands a shift in mindset that should not be underestimated. Not cultivating the ability to work with information is one of the key reasons why many IT projects fail to deliver expected business benefits. Research on CRM, for example, has found that organizations which did not develop competencies among users to work with information did not see much in return for their investment [29].

58.2.7 From IT Governance as Structure to IT Governance as Behavior

Over the years, the pendulum has tended to swing between centralization and decentralization of IT resources and decision-making. This swing is usually driven by a desire to regain control (centralization) or to ensure that IT and resources are more responsive to local business unit requirements (decentralization). The pendulum swings toward increased centralization when unmanaged devolution results in IT costs going out of control, a lack of integration across corporate systems, and, ultimately, suboptimal investments. On the other hand, excessive centralization usually sees

---

* PRINCE2 (PRojects IN Controlled Environments) is a process-based method for effective project management. It is a de facto standard used extensively by the UK Government and is widely recognized and used in the private sector, both in the United Kingdom and internationally. The method is in the public domain, offering non-proprietorial best practice guidance on project management.

dissatisfaction across business units with decisions too often subject to political interference and negative impact on customers. When organizations seek the best of both worlds, a federal model [30] is generally implemented.

Conversations around IT governance typically focus on the allocation of accountabilities for decisions related to IT [31]. Governance is an inevitable necessity in a devolved environment, particularly with global organizations. The IT governance structure is the framework that governs decision-making and ensures that all lead to maximizing the return on all IT investments to the organization as a whole. Mechanisms of governance include establishing steering committees and other cross-organizational forums, chargeback, IT awareness events, and co-location of IT and business staff.

What we find, however, is that many CIOs fail to recognize that governance is ultimately about behavior and *not* structure. In their conversations with management colleagues, they talk about establishing committees and other cross-departmental and divisional forums. Hi-Impact CIOs, however, have reframed the agenda to one focused squarely on behaviors. Just as corporate governance seeks to ensure that executive management act in the best interests of shareholders and owners of the business, IT governance should seek that behaviors about information and IT—decisions, protection, usage—are made in the best interests of the organization as a whole. The challenge that many CIOs face is that business executives often do not feel comfortable making what they perceive as IT decisions; and IT itself making decisions that it should not [32].

In emphasizing behavior, for example, that divisional general managers should prioritize IT spend, the Hi-impact CIOs see this as the quest that guides the choice of governance mechanisms to implement. Additionally, they talk to their business colleagues about information as a core resource that requires stewardship, protection, and exploitation. They report that this can often necessitate educating them and agreeing with them the kind of behaviors that should ideally be exhibited. This includes everything from involvement in IT decision-making to information usage.

Most governance frameworks that are promoted, such as COBIT,* focus on "IT governance of IT." They emphasize the *how*, rather than providing a deep understanding of the *why* of governance. What is missing is "enterprise governance of IT." Indeed, one CIO said to us recently that if the IT governance structure that he had recently instituted works, he will be out of a job as business colleagues would take up responsibilities for IT. He also wryly commented that he saw himself in his role for a very long time, reflecting the difficulty he expected to encounter in getting business managers to take on responsibilities.

58.2.8 From Running IT Like a Business to Running IT for the Business

Over the past decade, a popular prescription that has been promoted is for CIOs to be told that to be effective they must run IT like a business [33]. One CIO of a large bank, plagued by problems with IT, recently boasted at a conference that she saw herself as a CEO of her own organization! Hi-Impact CIOs tell us that this not only sends the wrong message but also strongly influences how business colleagues view the IS organization and their behavior toward it. Running IT like a business instills a certain mindset that can be at odds with the real requirements of IT to generate value to the business. The IS organization can be perceived as wishing to be treated like an other supplier; getting involvement and engagement, particularly for project execution, can be difficult. Hi-Impact CIOs also report that there can be issues with funding, particularly for infrastructure investments.

---

* COBIT (Control Objectives for Information and related Technology) is a set of best practices (framework) for IT. Its mission is "to research, develop, publicize and promote an authoritative, up-to-date, international set of generally accepted information technology control objectives for day-to-day use by business managers and auditors." Managers, Auditors, and users benefit from the development of COBIT because it helps them understand their IT systems and decide the level of security and control that is necessary to protect their companies' assets through the development of an IT governance model. See http://www.isaca.org.

The challenge is not to run IT *like a business* but rather *for the business*. While there is a subtle shift in words, it does illustrate firmly where the focus of the CIOs attention should lie. IT exists to support the business in the achievement of its goals and objectives and to deliver information services. This entails identifying opportunities for innovation, better business integration, and ways to increase customer value. The Hi-Impact CIO acknowledges that when he spends money it is on behalf of the business and for its benefit.

58.3 Conclusion

Few can argue against the importance of IT today. Most businesses could not survive for very long without their IT systems. Yet the history of IT in many organizations is disappointing [34].* We also know that generating business value from IT, both operationally and strategically, is a shared responsibility but that non-IT staff do not always feel comfortable or confident making what they perceive to be IT decisions and getting involved in IT initiatives. Hi-Impact CIOs recognize this and to improve involvement and engagement have engineered subtle shifts in their conversations with CxO colleagues. These shifts establish a new agenda for the CIO, one built around conversations focused on: co-evolution, business priorities, value realization, shaping change, business capability, information exploitation, IT governance as behavior and running IT for the business.

While each shift might seem insignificant, together they represent a fundamental transformation in their dialogue with business colleagues. CIOs report that the outcome is a more rewarding discourse with business colleagues, but more importantly increases the value their organizations generate from IT investments. Language is important. Some CIOs have already embraced this new agenda. The clear message is that CIOs should look at the language that they use when interacting with their CxO colleagues. This might also change the outcomes they are seeking to achieve from these conversations. What my evidence suggests is that in so doing they are likely to get a better response from their business colleagues.

Appendix 58.A: About This Research

Over the past 3 years we have spoken to over 100 CIOs about various aspects of the role and the challenges that they face as CIO. We have also spoken to CxOs about IT and their expectations for CIOs. Many of these interviews have been recorded and transcribed. We have also engaged in in-depth discussions with participants on the Cranfield IT Leadership Program, an executive development program for CIOs and senior IT executives, where findings as reported in this chapter have been presented to test their validity. Publications from this research base include: T. Gerth and J. Peppard, "How newly appointed CIOs take charge," *44th Annual SIMposium*, Dallas, TX, October 2012; J. Peppard, C. Edwards, and R. Lambert "Clarifying the ambiguous role of the chief information officer," *MIS Quarterly Executive*, Vol. 10, 2011, pp. 197–201.

References

1. J. Peppard, Unlocking the performance of the Chief Information Officer (CIO), *California Management Review*, 52(4), 2010, 73–99.
2. (a) H.G. Enns, S.L Huff, and B.R. Golden, How CIOs obtain peer commitment to strategic is proposals: Barriers and facilitators. *Journal of Strategic Information Systems*, 10, 2001, 3–14; (b) H. Enns, S.L. Huff, and C.A. Higgins, CIO lateral influence behaviors: Gaining peers' commitment to strategic

---

* Surveys and reports continue to confirm that the majority of organizations do not realize significant business value from IT-enabled business projects. One recent paper reported that 74% of IT projects from 1994–2002 failed to deliver expected value. See also see *The Challenge of Complex IT Projects*, The Royal Academy of Engineering: London, 2004; and National Audit Office *Delivering Successful IT-enabled Business Change*, Report by the Comptroller and Auditor General, HC 33-1, Session 2006–2007, London, November, 2006.

information systems, *MIS Quarterly*, 27, 2003, 155–176; (c) H.G. Enns, D.B. McFarlin, and S.L. Huff, How CIOs can effectively use influencing behaviours, *MIS Quarterly Executive*, 6(1), 2007, 29–38.

3. D.F. Feeny, B.R. Edwards, and K.M. Simpson, Understanding the CEO/CIO relationship, *MIS Quarterly*, 16, 1992, 435–448.

4. K. Patching and R. Chatham, *Corporate Politics for IT Managers: How to Get Streetwise*, Butterworth-Heinemann, Boston, MA, 2000.

5. P. Weill and J.W. Ross, *IT Savvy: What Top Executives Must Know to Go from Pain to Gain*, Harvard Business Press, Boston, MA, 2009.

6. (a) D. Preston and E. Karahanna, How to develop a shared vision: The key to strategic alignment, *MIS Quarterly Executive*, 8(1), 2009, 1–8; (b) F. Tan and R.B. Gallupe, Aligning business and information systems thinking: A cognitive approach, *IEEE Transactions on Engineering Management*, 53(2), 2006, 223–237.

7. B.H. Reich and I. Benbasat, Factors that influence the social dimension of alignment between business and information technology objectives, *MIS Quarterly*, 24(1), 2000, 81–113.

8. (a) L. Applegate and J. Elam, New information systems leaders: A changing role in a changing world, *MIS Quarterly*, 16(4), 1992, 469–490; (b) E.W. Martin, Critical success factors of chief MIS/DP executives, *MIS Quarterly*, 6(2), 1982, 1–9; (c) M.E. Porter and V. Miller, How information gives you a competitive advantage, *Harvard Business Review*, July–August, 1985, pp. 149–160.

9. G. Bock, K. Carpenter, and J.E. Davis, Management's newest star: Meet the chief information officer, *Business Week*, 2968, October, 1986, pp. 160–172

10. M.J. Earl, *Management Strategies for Information Technology*, Hemel Hempstead: Prentice Hall, New York, 1989.

11. D. Mark and D. Rau, Splitting demand from supply in IT, *McKinsey on IT*, Fall, 2006, 22–19.

12. P.G. Keen, Rebuilding the human resources of IS, in M.J. Earl, *Information Management: The Strategic Dimension*, Oxford University Press, Oxford, NY, 1988.

13. M.L. Kaarst-Brown, Understanding an organization's view of the CIO: The role of assumptions about IT, *MIS Quarterly Executive*, 4(2), 2005, 287–301.

14. J. Ross and P. Weill, Six IT decisions your IT people should make, *Harvard Business Review*, 80(11), 2002, 84–92.

15. D. Mark and E. Monnoyer, Next-generation CIOs, *McKinsey on IT*, Spring, 2004, 2–8. Available at http://www.mckinseyquarterly.com/Next-generation_CIOs_1451 (accessed on April 13, 2013).

16. Today's challenges, Tomorrow's CIO, IBM Global Business Services, 2008.

17. (a) R. Agarwal and C. Beath, Grooming the 2010 CIO, A Report for the Society for Information Management Advanced Practices Council, 2007; (b) M.J. Earl, The chief information officer: Past, present, and future, in M.J. Earl, ed., *Information Management: The Organizational Dimensions*, Oxford University Press, Oxford, NY, 1996, pp. 456–484.

18. Y. Li, C.-H. Tan, H.H. Teo, and B.C. Tan, Innovative usage of information technology in Singapore organizations: Do CIO characteristics make a difference? *IEEE Transactions on Engineering Management*, 53(2), 2006, 177–190.

19. D.S. Preston, D. Leidner, and D. Chen, CIO leadership profiles: Implications of matching CIO authority and leadership capability on IT impact, *MIS Quarterly Executive*, 7(2), 2008, 57–69.

20. (a) J. Luftman and T. Ben-Zvi, Key issues for IT executives 2009: Difficult economy's impact on IT, *MIS Quarterly Executive*, 9(1), 2010, 49–59; (b) J. Luftman, R. Kempaiah, and E.H. Rigoni, Key issues for IT executives, *MIS Quarterly Executive*, 8(3), 2009, 151–159.

21. H. Benbya and B. McKelvey, Using coevolutionary and complexity theories to improve IS alignment: A multi-level approach, *Journal of Information Technology*, 21, 2006, 284–298.

22. D.J. Futuyma, *Coevolution*, Sinauer, Sunderland, MA, 1983.

23. J. Peppard and K. Breu, Beyond alignment—A co-evolutionary view of the IS strategy process, in *Proceedings of the 24th International Conference on Information Systems (ICIS)*, Seattle, WA, December 2003, pp. 743–750.

24. (a) J. Peppard, J. Ward, and E. Daniel. Managing the realization of business benefits from IT Investments, *MIS Quarterly Executive*, 6(1), 2007, 1–11; (b) A. Hughes and M.S. Scott Morton, The transforming power of complementary assets, *MIT Sloan Management Review*, 47(4), 2006, 50–58.

25. J. Ward, E. Daniel, and J. Peppard, Building a better business case for IT investments, *MIS Quarterly Executive*, 7(1), 2008, 1–14.

26. D. Marchand, The role of the chief information officer: Achieving credibility, relevance and business impact, in P. Bottger, ed., *Leading in the Top Team: The CXO Challenge*, Cambridge University Press, Cambridge, NY, 2008, pp. 204–222.

27. (a) D.A. Marchand, W.J. Kettinger, and J.D. Rollins, *Information Orientation: The Link to Business Performance*, Oxford University Press, Oxford, NY, 2001; (b) S. Devaraj and R. Kohli. Performance impacts of information technology: Is actual usage the missing link? *Management Science*, 49(3), 2003, 273–289.

28. R. Ryan Nelson, Project retrospectives: Evaluating project success, failure, and everything in between, *MIS Quarterly Executive*, 4(3), 2005, 361–372.

29. S. Maklan, S. Knox, and J. Peppard, Why CRM fails—And how to fix it, *MIT Sloan Management Review*, 52(4), Summer, 2011, 77–85.

30. S.L. Hodgkinson, The role of the corporate IT function in the federal IT organization, in M.J. Earl, ed., *Information Management: The Organizational Dimension*, Oxford University Press, New York, 1996, pp. 247–269.

31. (a) P. Weill, Don't just lead, govern: How top-performing firms govern IT, *MIS Quarterly Executive*, 3(1), 2004, 1–17; (b) P. Weill and J. Ross, *IT Governance: How Top Performers Manage IT Decision Rights for Superior Results*, Harvard Business School Press, Boston, MA, 2004.

32. J. Ross and P. Weill, Six decisions your IT people shouldn't make, *Harvard Business Review*, November, 2002, 85–91.

33. M.D. Lutchen, *Managing IT as a Business*, John Wiley & Sons, Hoboken, NJ, 2004.

34. (a) D. Shpilberg, S. Berez, R. Puryear, and S. Shah, Avoiding the alignment trap in information technology, *MIT Sloan Management Review*, 49(1), 2007, 52; (b) R. Ryan Nelson, IT project management: Infamous failures, classic mistakes and best practices, *MIS Quarterly Executive*, 6(2), 2007, 67–78.

59

Information Technology Management Frameworks: An Overview and Research Directions

Hillol Bala
Indiana University

V. Ramesh
Indiana University

59.1 Introduction

Information technology (IT) plays a key transformational role at various levels of our lives and society. There are probably very few organizations on the planet that do not have some form of IT and/or IT-enabled organizational processes or routines. Research has repeatedly shown the positive impacts of IT on outcomes at various levels of analysis: country, society, organization, community, and individual (Banker et al. 2011; Brynjolfsson and Hitt 1996; McAfee and Brynjolfsson 2008; Venkatesh et al. 2010). For instance, Dewan and Kramerer (2000) found significant positive impacts of IT on the GDP (Gross Domestic Product) of developed countries. Recent research has found positive influence of IT on organizational outcomes (e.g., Mithas et al. 2011, 2012). In the practitioner literature, although challenges related to the management of IT function have been highlighted (e.g., Heller 2009; Luftman and Ben-Zvi 2011; Wailgum 2009), it is generally well acknowledged that IT has a positive influence on an organization and its members (McAfee and Brynjolfsson 2008; Peppard and Ward 2005). Notwithstanding such general agreements about the favorable role of IT, organizations constantly face challenges related to the effective and efficient management of IT functions for long-term, sustainable organizational benefits.

Challenges associated with managing IT functions were first recognized by organizations in the 1980s when they were concerned about the quality and consistency of services they were getting from their IT units (Arraj 2010; Whittleston 2012). In the 1990s and 2000s, organizations continued to face challenges related to IT management as they attempted to justify and recoup the significant investments they had made for IT. Senior executives had started to question the business value of IT and impacts of IT on organizational productivity and other key performance indicators (KPIs; Brynjolfsson 1993; IT Governance Institute 2003). During this time, academic research reported a phenomenon widely known as the *IT productivity paradox* that captured the notion that despite high investments in IT, organizations failed to realize any measurable benefits from these investments (Brynjolfsson 1993; Brynjolfsson and Hitt 1996). Although later research, particularly at the organizational level, has shown positive impacts of IT on organizational outcomes, it has underscored the importance of effective and efficient management of IT function in organizations (Aral and Weill 2007; Dewan et al. 2007; Dos Santos et al. 2012).

As IT becomes an integral part of today's organizations and the *IT landscape*—the assemblage of IT-related infrastructure, systems, processes, and service components—becomes increasingly more complex, organizations face continuous challenges related to how their IT landscape should be managed and governed so that they are able to leverage their IT capabilities to achieve organizational goals. *IT management frameworks*—a set of best practice-based industry frameworks to manage various facets of organizational IT—have received widespread acceptance by organizations that attempt to manage the complexity of IT landscape and mitigate associated challenges. The diffusion and the widespread popularity of these frameworks have also been catalyzed by the changing operating environment of organizations (Ross et al. 2006). For instance, organizations have started facing intense local and global competition in recent years, and they have wanted to find ways to gain operational efficiency and strategic advantage. Further, organizations are affected by the changes in society and the world caused by the emergence and widespread diffusion of transformational technologies, such as social media and mobile technologies, and significant shifts in global political and economic landscapes. Many organizations are now able to expand their operations in countries that were not accessible even in 1980s. These organizations now face challenges associated with global distribution of resources and distributed execution of processes. Consequently, it has become imperative for such organizations to find ways to achieve and sustain operational efficiency and other organizational goals. Effective management of IT has been touted as one of the key ways to achieve such efficiency and organizational goals.

This chapter provides an overview of IT management frameworks with a particular focus on frameworks that support three specific IT management areas: (a) *IT governance*, (b) *IT service*, and (c) *IT security*. We briefly discuss the underlying principles and processes of these frameworks. In particular, we highlight the richness that these frameworks offer with respect to managing IT provisions, processes, activities, resources, and metrics to help organizations achieve their organizational goals. In addition to providing an overview of these frameworks, we also discuss the impacts of these frameworks on IT management practices using examples of organizations that gain by implementing these frameworks and associated practices. Finally, we offer a set of research directions that can leverage the richness of these frameworks. We focus on future research that, we believe, will have significant implications for theory and practice related to IT management. To our knowledge, this chapter is a first attempt to reduce the gap between IT management practices and research by offering a prognosis of current practice and calling for actionable future research.

59.2 Emergence of IT Management Frameworks

Although IT management frameworks have been around since the 1980s, the popularity of these frameworks among senior IT and business executives has increased in the late 1990s and early 2000s as many organizations started receiving benefits from implementing these frameworks. In recent years, organizations that develop and promote these frameworks have released updated versions of these frameworks in order to keep up with the complexity of today's organizational contexts and operating environments. Further, some of these frameworks have received recognition from international standards bodies such as the International Organization for Standardization (ISO) and the United Nations (Zhang and Chulkov 2011). Although IT management frameworks can be

TABLE 59.1 Overview of IT Management Frameworks

IT Management Frameworks	Definition	Example
IT governance frameworks	Frameworks that help organizations define and manage leadership and organizational structures and processes to ensure that the organization's IT sustains and extends the organization's strategies and objectives (IT Governance Institute 2003)	• COBIT (Control Objectives for Information and Related Technology)[a]
IT service management (ITSM) frameworks	Frameworks that help organizations ensure that the IT services are aligned to the business needs and actively support them (Cartlidge et al. 2007)	• ITIL (Information Technology Infrastructure Library)[b] • ISO 20000
Information security management (ISM) framework	Frameworks that offer best practice recommendations on information security management, risks, and control within the context of an overall information security management system (ISMS) that defines a set of policies related to IT security and risks (IT Governance Institute 2008a)	• ISO/IEC 27000[c] series
IT project management frameworks	Frameworks that provide standardized methodology, guidelines, rules, and characteristics for IT project management (Project Management Institute 2012)	• PMBOK (Project Management Body of Knowledge) • PRINCE2 (Projects in Control Environments 2)

[a] COBIT is a registered trademark of ISACA and ITGI (IT Governance Institute).

[b] ITIL is a registered trademark of the Office of the Government Commerce in the United Kingdom and other countries.

[c] ISO stands for International Organization for Standardization and IEC stands for International Electrotechnical Commission.

classified and grouped in many different ways, we broadly classify them into four categories in keeping with the core objective(s) of these frameworks: (a) *IT governance frameworks*, (b) *IT service management (ITSM) frameworks*, (c) *information security management (ISM) frameworks*, and (d) *IT project management frameworks*. It is important to note that some of these frameworks have provisions and processes that conceptually and operationally overlap with that of the other frameworks. For example, it is possible that frameworks that are primarily developed and implemented for IT governance purposes can have provisions and processes related to ITSM and vice versa. Similarly, ITSM frameworks may have provisions and processes related to ISM. Nonetheless, we use this broad classification to categorize these frameworks because it offers a useful organizing principle for the task. Table 59.1 provides a brief overview of these four types of frameworks and examples.

IT governance is a process of specifying the decision rights (i.e., who holds the rights to make IT-related decisions) and accountability (i.e., who is held responsible for an organization's decision-making about IT assets) to encourage desirable behavior in using IT (Weill and Ross 2004). An IT governance framework helps an organization direct and control the current and future IT endeavors. Such a framework also helps an organization meet high-level IT goals such as alignment of IT with the business, realization of promised benefits, use of IT to exploit opportunities and maximize benefits, responsible use of IT resources, and appropriate management of IT-related risks (IT Governance Institute 2003). Consistent with the practitioner literature associated with the widely used IT governance frameworks and prior academic research on IT governance, we suggest that IT governance should be an inseparable component of corporate governance and a core responsibility of the board of directors and senior executives (IT Governance Institute 2003; Ross and Weill 2002; Sambamurthy and Zmud 1999; Weill and Ross 2004; Zhang and Chulkov 2011). It should not be an isolated discipline or activity performed by the IT leadership. Instead, it should be an integral part of corporate governance and should be considered a core competence of an organization (Agarwal and Sambamurthy 2002; Jewer and McKay 2012; Willcocks et al. 2006). In a nutshell, an IT governance framework offers a set of principles, processes, activities, and measures that help organizations develop and sustain IT governance-related core competencies and outcomes.

The underlying driver of ITSM frameworks is the recognition that IT services are crucial, strategic organizational assets. Consequently, organizations must invest appropriate levels of resources into the support, delivery, and management of these critical assets and the IT systems that underpin these assets (Cartlidge et al. 2007; Gacenga et al. 2010; Kneller 2010; Whittleston 2012). The primary objective of ITSM is to ensure that IT services are aligned to the organizational needs and actively support them. ITSM can be defined in various ways—(a) as a *practice* related to planning, designing, developing, delivering, and optimizing IT services, (b) as a *discipline* that offers processes, methods, activities, measures, functions, and roles that are needed to deliver IT services, and (c) as a *profession* that defines a set of skills needed to deliver high-quality IT services (Kneller 2010; Rai and Sambamurthy 2006). An ITSM framework ensures that effective implementation, management, and support of IT services, that is, IT processes, systems, and resources, will lead to organizational effectiveness in at least two ways: (a) by reducing business disruptions, loss of productive hours, and costs, and (b) by increasing revenue, improving public relations, and achieving its core objectives (Cartlidge et al. 2007). An ITSM framework offers an extensive body of knowledge, capabilities, skills, and best practices for providing value to customers in the form of IT services (Cartlidge et al. 2007; Kneller 2010; Whittleston 2012). It is intended to provide a wide range of benefits to an organization, such as increased satisfaction with IT services, improved service availability, financial savings, strategic advantage, and improved decision-making (Cartlidge et al. 2007; Gacenga et al. 2010; Kneller 2010; Whittleston 2012).

Benefits from effective IT governance practices and service management do not come without risks (Chen et al. 2011; Kouns and Minoli 2010; Spears and Barki 2010). An IT risk is a potential exposure facing an organization as a result of any aspect of the IT environment—that is, IT assets, resources, organization, or processes (Westerman and Hunter 2007). There are numerous examples of IT risks and associated adversarial outcomes faced by today's organizations. In fact, there is hardly an organization that has not faced some type of challenges associated with IT risks such as system failure, network outage, and security breaches. Organizational IT risks can be broadly grouped into four dimensions (Westerman and Hunter 2007): *availability* (i.e., keeping existing processes running and recovering from interruptions), *access* (i.e., ensuring that authorized people have access to information and facilities they need), *accuracy* (i.e., providing accurate, timely, and complete information), and *agility* (i.e., implementing new strategic initiatives). Although IT risks are primarily related to issues connected to IT systems and people (e.g., employees), prior research has found that these risks do not necessarily arise from technical and/or people issues. These risks typically arise from high-level organizational factors, such as ineffective IT governance, uncontrolled complexity in the IT landscape, and inattention to IT-related risks (Westerman and Hunter 2007). These factors clearly indicate a need for a framework and/or standard practices to manage and control IT risks. An ISM framework is intended to provide an organization a mechanism for effective management and control of IT risks. An ISM framework is also expected to help an organization manage all aspects of IT risks through a set of processes, skills, and best practices. In conjunction with managing and controlling IT risks, an ISM framework ensures the confidentiality, integrity, and availability of an organization's assets, information, data, and IT services (Clinch 2009; Kouns and Minoli 2010).

In the remainder of this section, we discuss three widely used IT management frameworks: (a) *COBIT*, an IT governance framework, (b) *ITIL*, an ITSM framework, and (c) *ISO/IEC 27000*, a family of information security standards. In keeping with the core objective of this chapter to focus on frameworks to manage IT functions in organizations, we do not discuss IT project management frameworks because IT projects do not necessarily represent ongoing activities or functions of an IT organization.

59.2.1 IT Governance Framework: COBIT

COBIT is one of the most widely implemented IT governance frameworks. It was developed in 1996 by the Information Systems Audit and Control Association (ISACA) and the IT Governance Institute

(ITGI) as a standard for IT security and control practices (Violino 2005). ISACA that currently goes by its acronym only is an international professional association that deals with IT governance and serves more than 95,000 constituents (members and professionals holding ISACA certifications) in more than 160 countries. While COBIT was first developed as a framework for IT audit and control, it evolved over time and has become the *de facto* guide for defining, implementing, and managing various aspects of IT governance in organizations. Since 1996 COBIT has had five major releases, with COBIT 5 being the most recent version that was released in 2012. COBIT 5 is considered the most comprehensive release of COBIT frameworks because it consolidates and integrates the earlier version of COBIT (4.1) and two other frameworks developed by ISACA—Val IT 2.0 and Risk IT—and draws from ISACA's IT Assurance Framework (ITAF) and the Business Model for Information Security (BMIS). It also aligns with other IT management frameworks such as ITIL, PMBOK, ISO, and PRINCE2 (ISACA 2012).*

59.2.1.1 Overview of COBIT 5

COBIT 5 is based on five core principles (Figure 59.1). These principles dictate the processes and metrics that define the core structure of the framework. COBIT 5 is a value-driven framework that offers a comprehensive goals cascade linking the framework with a set of organizational and IT-related goals. This goals cascade essentially offers a mechanism to translate stakeholder needs (e.g., benefits realization, risk optimization, and resource optimization) into specific, actionable, and customized organizational goals, IT-related goals, and enabler goals. This translation provides the ability to set specific goals at every level and area of an organization in support of the overall goals and stakeholder requirements. Although COBIT 5 allows organizations to set their own goals at every level, it offers a comprehensive set of goals for various levels that organizations can adopt as baseline outcomes and customize them as needed to fit with their unique requirements and contexts.

Building on the core principles shown in Figure 59.1, COBIT 5 prescribes a comprehensive set of 37 processes related to organizational IT activities that are under the purview of IT governance. Figure 59.2 illustrates these core processes of COBIT 5 that are divided into two broad categories: (a) *governance* of enterprise IT and (b) *management* of enterprise IT. As shown in Figure 59.2, the governance processes

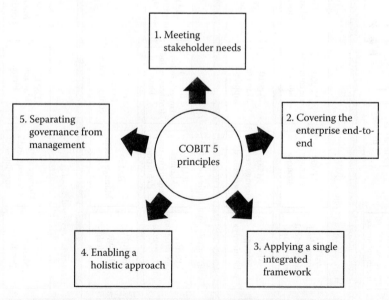

FIGURE 59.1   COBIT 5 principles. (Adapted from http://www.isaca.org/COBIT/Documents/COBIT5-Laminate.pdf)

---

* We discuss the relationship among IT management frameworks later in this chapter.

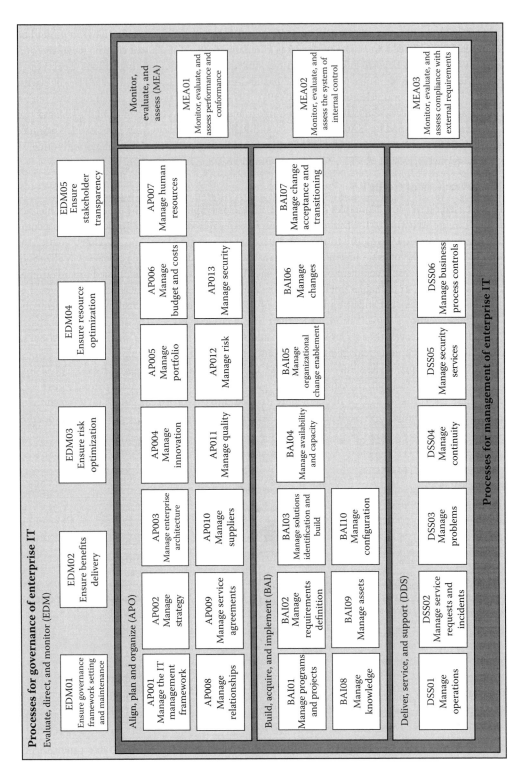

FIGURE 59.2   COBIT 5 process reference model. (Adapted from http://www.isaca.org/COBIT/Documents/COBIT5-Laminate.pdf)

deal with the stakeholder governance objectives such as value delivery, risk optimization, and resource optimization, and include practices and activities aimed at evaluating strategic options, providing direction to IT, and monitoring the outcome. These activities are broadly categorized as *Evaluate, Direct, and Monitor (EDM)* processes or practices. Management processes deal with practices and activities that cover the responsibility areas related to PBRM (plan, build, run, and monitor) of enterprise IT. These processes are grouped into four broad categories: (a) *Align, Plan, and Organize (APO)*, (b) *Build, Acquire, and Implement (BAI)*, (c) *Deliver, Service, and Support (DSS)*, and (d) *Monitor, Evaluate, and Assess (MEA)*. COBIT 5 provides a comprehensive guideline related to each of these governance and management processes with respect to how it should be implemented (i.e., activities, inputs, and outputs), its goals and metrics, its relationship with IT-related goals, and key governance practices (who holds the decision rights and is held accountable for decision-making).

COBIT 5 is expected to offer a set of strategic and operational benefits to organizations that implement these processes (see Figure 59.2). These benefits include, but are not limited to: (a) maintaining high-quality information to support business decisions, (b) achieving strategic goals and realizing business benefits through the effective and innovative use of IT, (c) achieving operational excellence through reliable, efficient application of technology, (d) maintaining IT-related risk at an acceptable level, (e) optimizing the cost of IT services and technology, and (f) supporting compliance with relevant laws, regulations, contractual agreements, and policies (ISACA 2012). However, these intended benefits cannot be achieved if organizations fail to define and manage the enablers that help organizations achieve enterprise and IT-related goals. COBIT 5 provides guidelines related to how each of these enablers—that is: (a) principles, policies, and frameworks, (b) processes, (c) organizational structure, (d) culture, ethics, and behaviors, (e) information, (f) services, infrastructure, and applications, and (g) people, skills, and competencies—should be implemented, monitored, and measured to ensure effective implementation of COBIT and a high degree of alignment between IT and business.

59.2.2 IT Service Management Framework: ITIL

ITIL is considered the *de facto* industry framework for ITSM processes, practices, and delivery. ITIL has been around for more than 20 years. It was originally developed in the early 1980s by a government agency in the United Kingdom, Central Computer and Telecommunications Agency (CCTA), that later merged into the Office of Government Commerce (OGC), a division under the U.K. Treasury. The CCTA and later the OGC recognized the need for developing and implementing standardized processes for IT service delivery. As noted earlier, such standardization was needed because of the increasing complexity in the IT landscape characterized by obsolescence of mainframe-centric infrastructure, distributed computing, and geographically dispersed resources (Arraj 2010). Although such a landscape afforded organizations the potential for increasing flexibility and processing capabilities, it inevitably created a scope for inconsistent IT service delivery across the organization. In an attempt to reduce this inconsistency, organizations started looking for a framework that would help standardize IT service design and delivery.

ITIL was originally published as a collection of 31 associated books covering various aspects of ITSM practices and provisions. The principle of the ITIL framework was based on a process-based view of controlling and managing operations pioneered by W. Edwards Deming using a plan-do-check-act (PDCA) cycle. The early version of ITIL was adopted by many large organizations including government agencies in Europe in the 1990s. ITIL quickly became a cornerstone for ITSM by allowing organizations to take a disciplined approach to service support and delivery. As ITIL grew in popularity in Europe and beyond, and the IT landscape started to become complex, there was a need to revise the early version of ITIL to develop a more closely connected and consistent set of books. ITIL V2 was introduced in the 2000–2001 time period by consolidating the initial library of 31 books to a manageable set of seven books that were more accessible and affordable (Cartlidge et al. 2007; Whittleston 2012). ITIL V2 became the *de facto* standard for ITSM practices around the world. By the mid-2000s, it had been adopted and implemented by thousands of organizations as a basis of effective ITSM practices (Cartlidge et al. 2007; Kneller 2010). In 2007, ITIL V3 was introduced by enhancing and consolidating ITIL V2. ITIL V3 consists of five core books covering the lifecycle of IT services. In addition to these five core books, there is a sixth book, the *Official Introduction* that offers an overview of the five books and an introduction to ITSM as a whole. The core books and the additional complementary publications by the U.K. OGC offer detailed guidance on the implementation of IT service lifecycle and associated processes, principles, practices, and methods (Cartlidge et al. 2007).

59.2.2.1 Overview of ITIL V3

As noted earlier, ITIL V3 adopted a lifecycle approach to IT services. Figure 59.3 illustrates the core stages of this lifecycle and associated processes. The lifecycle starts with the stage of *service strategy* in which IT and business strategists collaborate to develop IT service strategies that support and align with business strategy practices (Cartlidge et al. 2007; Kneller 2010). Some of the key processes related to service strategy are strategy generation, service portfolio management, and demand management. Once the IT service strategies are defined, an IT organization is in a position to enter the second stage, *service design*, in which the IT organization designs the overarching IT architecture and each IT service to meet business objectives practices (Cartlidge et al. 2007; Kneller 2010). Some key processes related to service design are service catalog management, service level management, capacity management, information security management, and IT service continuity management. The third stage,

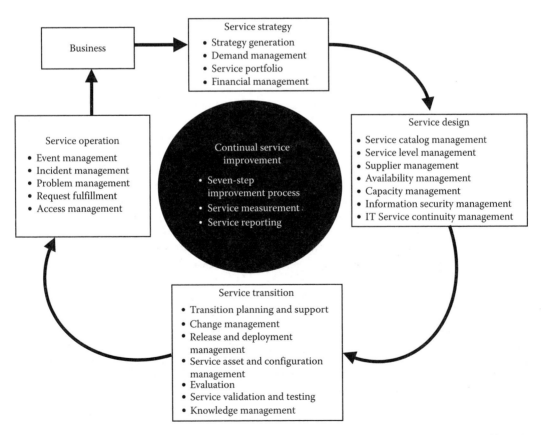

FIGURE 59.3　ITIL V3 process framework. (Adapted from Cartlidge, A. et al., *An Introductory Overview of ITIL® V3*, The UK Chapter of the itSMF, United Kingdom: itSMF Ltd., 2007.)

service transition, helps an IT organization migrate its designed IT services into live environments. The key processes in this stage help an IT organization manage and control new services or changes in IT services. Some of these processes are change management, release and deployment management, transition planning and support, and service assets and configuration management. Once the transition is done, the fourth stage, *service operation*, helps an IT organization deliver and support operational IT services in such a way that they meet organizational needs and expectations (Cartlidge et al. 2007; Kneller 2010). The key processes related to service operation are event management, incident management, request fulfillment, problem management, and access management. Finally, the fifth stage, *continual service improvement*, helps an IT organization learn from experience and adopt an approach, which ensures continual evaluation and improvement of the quality of IT services and the overall maturity of ITSM service lifecycle and underlying processes (Cartlidge et al. 2007; Kneller 2010). The key processes related to continual service improvement are seven-step improvement process, service measurement, and service reporting.

The adoption and effective implementation of ITIL V3 are expected to offer several operational and strategic benefits to organizations. These benefits include, but are not limited to, alignment of IT services with organizational priorities and objectives; manageable IT costs; increased business productivity, efficiency, and effectiveness by making IT services more reliable and better; financial savings from improved resource management and reduced rework; effective change management; improved user and customer satisfaction with IT; and improved end-customer perception and brand image (Cartlidge et al. 2007; IT Governance Institute 2008a; Kneller 2010; Whittleston 2012). However, organizations may not be able to garner these benefits if they fail to implement ITIL effectively. We discuss challenges related to ITIL implementation later in this chapter.

59.2.3 Information Security Framework: ISO/IEC 27000 Series

Although the need for an information security framework and associated policies was felt by organizations in the 1980s and 1990s, it was not until 1995 when the first well-received best practice standard for ISM was published. Known as BS 7799, it was developed by the U.K. Government's Department of Trade and Industry (DTI). BS 7799 had two parts—the first part, BS 7799-1, provided the best practices for ISM and the second part, BS 7799-2, provided specifications and guidance about using information security management systems (ISMS) to manage and control information security in an organization. BS 7799-1 was adopted by ISO as ISO/IEC 17799 in 2000 and was known as the Code of Practice for Information Security Management (Clinch 2009). ISO/IEC 17799 was later revised in 2005. Around this time, organizations across the world started placing significant importance on information security and IT risks because of the operational and strategic value of information and technologies as organizational assets. Recognizing the need for a broad set of standards to support a wide variety of information security-related issues, ISO chose to create a new numbering scheme to accommodate a whole of family of information security standards. Consequently, the ISO/IEC 27000 family of information security standards was created and ISO 17799 was renamed ISO/IEC 27002 and BS 7799-2 became ISO/IEC 27001 (Clinch 2009).

ISO and IEC developed ISO/IEC 27000 as a family of information security standards to specify the fundamental principles, concepts, and vocabulary for the ISO/IEC 27000 series of documents. It is important to note that the scope of ISO/IEC 27000 framework is not limited to IT-related risks and security. These standards also provide best practices and guidelines to manage any aspects of information (e.g., data, documents, messages) that an organization and its members may produce and consume. The key concept of the ISO/IEC 27000 standard is the planning, development, implementation, and operation of a certifiable ISMS. According to these standards, an ISMS provides a model for establishing, implementing, operating, monitoring, reviewing, maintaining, and improving the protection of information assets to achieve business objectives based upon a risk assessment and the organization's risk acceptance levels designed to effectively treat and manage risks (ISO/IEC 2009). Individual standards within the ISO/IEC 27000 family

TABLE 59.2   ISO/IEC 27000 Standards

Standards	Description
ISO/IEC 27000	Provides an overview of and/or introduction to the ISO/IEC 27000 standards as a whole and the specialist vocabulary used in this family of standards
ISO/IEC 27001	Provides the details of the Information Security Management System (ISMS) requirements standard; ISMS provides the underlying foundation for the ISO/IEC 27000 standards and is used as the baseline to certify organizations that are ISO/IEC 27000 compliant
ISO/IEC 27002	Provides the code of practice for information security management describing a comprehensive set of information security control objectives and a set of generally accepted good practice security controls
ISO/IEC 27003	Provides guidance on implementing ISO/IEC 27001 (i.e., ISMS)
ISO/IEC 27004	Provides guidance on information security management measurement
ISO/IEC 27005	Provides details of an information security risk management standard
ISO/IEC 27006	Provides a guide to the certification or registration process for accredited ISMS certification or registration bodies
ISO/IEC 27007	Provides a guide to auditing ISMS
ISO/IEC 27008	Provides guidelines for the auditing of technical security controls
ISO/IEC 27010	Provides guidance on information security management for intersector and interorganizational communications
ISO/IEC 27011	Provides a guideline for information security management for telecommunications organizations (also known as ITU X.1051)
ISO/IEC 27013	Provides guidance on the integrated/joint implementation of both ISO/IEC 20000-1 (derived from ITIL) and ISO/IEC 27001 (ISMS)
ISO/IEC 27014	Provides guidelines for the governance of information security
ISO/IEC 27015	Provides information security management guidelines for organizations in the financial services industry
ISO/IEC 27016	Provides guidelines for the economics of information security management
ISO/IEC 27017	Provides guidelines for the information security aspects of cloud computing
ISO/IEC 27018	Provides guidelines for the privacy aspects of cloud computing
ISO/IEC 27019	Provides guidelines related to information security for process control in the energy industry
ISO/IEC 27031	Provides an ICT (information and communication technology) focused standard on business continuity
ISO/IEC 27032	Provides guidelines for cybersecurity
ISO/IEC 27033	This standard will replace the multipart ISO/IEC 18028 standard on IT network security
ISO/IEC 27034	Provides guidelines on information security for IT applications
ISO/IEC 27035	Provides guidance on information security incident management
ISO/IEC 27036	Provides security guidelines for supplier relationships
ISO/IEC 27037	Provides guidelines for digital evidence
ISO/IEC 27038	Provides specifications for digital redaction
ISO/IEC 27039	Provides guidelines for concerns intrusion detection and prevention systems
ISO/IEC 27040	Provides guidelines on storage security
ISO/IEC 27041	Provides guidelines on assurance for digital evidence investigation methods
ISO/IEC 27042	Provides guidelines on analysis and interpretation of digital evidence
ISO/IEC 27043	Provides guidelines on digital evidence investigation principles and processes
ISO 27799	Provides health sector–specific ISMS implementation guidance based on ISO/IEC 27002

Source: Adapted from http://www.iso27001security.com/html/iso27000.html

provide guidelines and best practices for planning, development, implementation, and operation of an ISMS and associated ISM components (e.g., IT network security or cybersecurity). Table 59.2 provides a list of ISO/IEC 27000 standards that are either published or under development.

59.2.3.1 Overview of ISO/IEC 27001 and 27002

Although individual standards within the ISO/IEC 27000 family provide valuable guidelines for managing and controlling different aspects of information security in an organization, there are two standards—ISO/IEC 27001 and ISO/IEC 27002—that offer the key documents related to requirements and practices associated with ISMS. ISO/IEC 27001 specifies the requirements for establishing, implementing, operating, monitoring, reviewing, maintaining, and improving a documented ISMS, using a continual improvement approach (DiMaria 2012). This standard provides a detailed discussion of the following areas related to ISMS: general requirements, implementation process, management and operation, maintenance and improvement, documentation requirements, management responsibility, resource provision, training awareness, internal audits, management, and continual improvement (Clinch 2009). It also offers mapping with other standards, such as OECD (Organization for Economic Cooperation and Development) principles, ISO 9001 (i.e., a *de facto* quality management standard), and ISO 14001 (an environmental management standard) for organizations that have implemented these other standards (Clinch 2009; DiMaria 2012).

ISO/IEC 27001 is intended to be implemented and used along with ISO 27002, also known as Code of Practice for Information Security Management (DiMaria 2012). ISO/IEC 27002 offers guidance on interpretation and implementation of the list of specific security controls within ISO 27001 (DiMaria 2012). The standard provides a set of security controls in different areas of information security management and offers guidance on how organizations can achieve these control objectives. Figure 59.4 illustrates these security controls grouped into 15 core areas of information security management (e.g., risk management, asset management, access control, software development, incident management, business

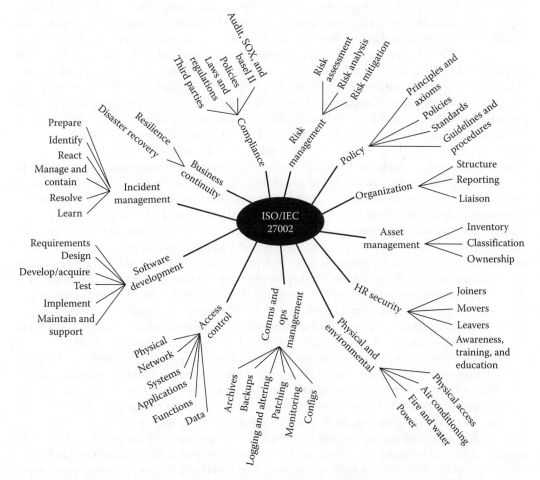

FIGURE 59.4 ISO/IEC 27002 control objectives. (Adapted from http://www.iso27001security.com/html/27002.html)

continuity, and compliance). Together, ISO/IEC 27001 and ISO/IEC 27002 offer several key business benefits to organizations when implemented: increased credibility and trust; improved partner, customer and stakeholder confidence; organizational and trading partner assurance; demonstration to competent authorities that the organization observes all applicable laws and regulations; competitive advantage and market differentiation; and reduced regulation costs (Clinch 2009; DiMaria 2012).

59.2.4 Relationships among the IT Management Frameworks

It is important to note that although IT management frameworks were developed to meet different organizational needs with respect to managing IT assets and capabilities, these frameworks have noticeable overlaps and striking similarities. Indeed, these frameworks have processes and metrics that are conceptually similar. For instance, COBIT 5 has specific processes (and associated subprocesses, activities, inputs, outcomes, and metrics) related to managing service agreements (APO09), availability and capacity (BAI04), changes (BAI06), operations (DSS01), and service requests and incidents (DSS02) that are conceptually similar to some key processes of ITIL V3, such as service level management, availability management, capacity management, change management, event management, and incident management (see Figures 59.2 and 59.3). Similarly, both COBIT 5 and ITIL V3 have processes related to security management, the core focus of ISO/IEC 27000 standards. Although these conceptually similar processes have common goals, these processes are operationally different and offer differential levels of granularity with respect to subprocesses, activities, inputs, and outputs. For example, COBIT offers high-level processes and metrics that are relevant to an entire IT organization and provides a mechanism to align IT and organizational goals. While there are processes, metrics, and IT goals related to IT service and security in COBIT, these are typically not as comprehensive as they are in ITIL or in ISO/IEC 27000 because COBIT focuses on many other IT processes in addition to processes related to IT service and security (IT Governance Institute 2008a,b). Another key difference is that COBIT focuses on the maturity and control of high-level IT processes, whereas ITIL focuses on executional details of service-related processes. Unlike COBIT and ITIL, ISO/IEC 27000 frameworks do not offer detailed processes related to information security management. Instead, it offers a set of control objectives that organizations can incorporate in their security policy to protect their information assets (see Figure 59.4).

Although there are significant differences among IT management frameworks, it is important to note that these frameworks typically do not contradict each other and it is possible for them to peacefully coexist in an organizational setting. In fact, there have been significant efforts in the practitioners' community regarding how these frameworks can be aligned and mapped with each other so that organizations can maximize implementation benefits (e.g., Clinch 2009; IT Governance Institute 2008a,b). These efforts have led into multiple publications by industry associations that provided comprehensive mapping of these frameworks with each other to help organizations implement an IT management framework without sacrificing the benefits they receive from other frameworks that they might have already implemented. For instance, the IT Governance Institute (2008b) offered a detailed mapping of how these frameworks complement each other's different IT governance domains and provided guidelines on how organizations can align these frameworks for organizational benefits. Clinch (2009) offered a detailed overview of how ITIL and ISO/IEC 27000 standards are aligned with each other.

Consistent with the efforts in practitioner literature about the peaceful coexistence of IT management frameworks, we suggest that IT organizations are in a unique position to leverage the IT management frameworks for organizational benefits. Adoption, effective implementation, and mapping of these frameworks will help IT organizations manage and control IT as a routine part of organizational activities. As shown in Figure 59.5, an IT governance framework will help an organization establish a clear standard, consistent procedures, and unambiguous practices regarding the overall IT function of an organization that is aligned with organizational-level activities, goals, and strategies. Once an IT governance framework is established and in effect, other frameworks will be able to leverage its processes, metrics, and control objectives, and deliver additional values to specific areas—e.g., service management,

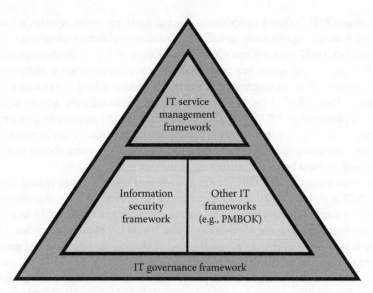

FIGURE 59.5 Relationship among IT management frameworks.

security management, and project management, related to the IT functions of an organization. We view an IT governance framework as an underlying platform that enables an organization to embrace other more specific frameworks that provide much deeper consideration of different elements, such as policies, processes, measurements, and improvements related to specific areas of IT. Together these IT management frameworks will help organizations manage their IT functions effectively and efficiently, and ensure a high degree of IT-business alignment and convergence (King 2010).

59.3 Impacts of IT Management Frameworks

IT management frameworks have profound impacts on how organizations manage and control their IT functions. As noted earlier, these frameworks have gained widespread popularity in the late 1990s and early 2000s. During this time, organizations had started making significant investments in IT and sought ways to ensure positive returns from these investments. There was a need to develop and implement frameworks that could help organizations better manage IT functions, justify investments in IT, and garner benefits from IT. Although there has been limited scientific evidence regarding the impacts of IT management frameworks on organizational outcomes (with a few exceptions that we highlight in this section), the practitioner literature is replete with anecdotal evidence and cases about the positive impacts of IT management frameworks on organizations and its members. In this section, we discuss impacts of IT management frameworks on practice with a particular focus on the measurable benefits that organizations are able to gain from implementing these frameworks. We also discuss the current status of adoption and implementation of these standards and associated challenges.

59.3.1 IT Governance Framework: COBIT

59.3.1.1 Adoption and Implementation of COBIT

Given that COBIT 5 was released recently (2012), our discussion of COBIT adoption and implementation is primarily based on COBIT 4.1 and earlier releases. There are numerous organizations around the world that have implemented different aspects of COBIT and received significant benefits. We will provide a list of some major organizations that have implemented COBIT in the following section. It is important to note that although COBIT offers a comprehensive set of goals, processes, metrics, actions, and implementation details, it allows organizations to fully customize these different components to meet unique organizational needs. COBIT does not enforce implementation of any of these components in any specific sequence. In order to help senior executives make effective decisions about components of COBIT, it offers different tools such as management guidelines and benchmarking tools using a maturity model framework. Further, ISACA and IT Governance Institute offer training programs for organizations that are interested in implementing COBIT. Overall, COBIT offers flexible implementation provisions allowing senior executives to select the specific goals, processes, and metrics that are applicable in their organizational context, determine the cost-benefit ratio of adopting these components, and implement the components that will be most beneficial to their organizations.

We note that implementation of COBIT is not an easy journey. It requires radical changes in senior executives'—both IT and non-IT—mindset regarding the value of IT in their organizations. Given that COBIT is a process-oriented framework that defines management and control of IT assets and decisions, it requires changes in existing processes in order to derive benefits from implementation. Further, COBIT implementation is a resource-intensive process because of changes in organizational processes, and the need for advisory services and extensive training during implementation. In addition to challenges related to change and resource management during implementation, there are other reasons for COBIT implementation failure. Ramanathan (2007) highlighted some of these reasons: (a) absence of a formal, documented strategy during implementation, (b) weak communication of strategy (i.e., there might be major communication gaps between IT and business, and the value of implementation may not be properly communicated to the stakeholders), (c) technology-driven projects (i.e., COBIT is essentially an organizational framework and should not be driven by technology), (d) lack of project ownership by business and senior executives, and (e) informal IT performance assessment (i.e., IT organization is not willing to embrace a formal performance assessment).

59.3.1.2 Impacts of COBIT on Practice

COBIT has been implemented by major organizations such as Oracle, MetLife, Blue Cross Blue Shield, ScotiaBank, National Stock Exchange of India, U.S. Department of Veterans Affairs, U.S. Department of Defense, European Parliament, Harley-Davidson, UNISYS Corporation, Fidelity Investments, Allstate, and Charles Schwab. These organizations have received numerous benefits from COBIT (ISACA 2012). For instance, Oracle was able to use COBIT matrices and mapping documents to make various IT frameworks fit together. It was also able to leverage the concepts in the COBIT-related materials to create discussion of health and maturity self-assessments, provide a line of sight between its activities and its business goals, bring predictability and reliability to how the IT group plans and manages the work across the enterprise, and complement its corporate planning cycle with an IT management cycle (ISACA 2012). Similarly, COBIT helped Unisys standardize its IT strategy to support global operations, align the IT infrastructure with the company's overall business strategy, and help with Sarbanes-Oxley compliance (ISACA 2012). The U.S. Department of Veterans Affairs (VA), a government agency, was able to create a new organizational structure for centralized IT management based on COBIT that provided a framework for IT governance plans, structures, and investments at the VA (ISACA 2012). Recently, Rouyet and Spauwen (2011) suggested that COBIT can help align green business strategy with IT strategy. For example, they suggested that COBIT processes and metrics are flexible enough to align with business goals related to efficient energy use, public image aligned to environmental concerns, and environmental regulatory compliance.

59.3.2 IT Service Management Framework: ITIL

59.3.2.1 Adoption and Implementation of ITIL

There has been an increasing global trajectory of adoption and implementation of ITIL since 2004. Although ITIL was widely used in Europe in the 1990s and early 2000s, awareness and proliferation related to ITIL within North American organizations started to accelerate between 2004 and 2006 (O'Donnell 2011). The adoption and implementation of ITIL are now considered mainstream; a majority of the Fortune 500 corporations have adopted and implemented some form of ITIL (O'Donnell 2011). A study conducted by the Forrester Group in 2011 revealed that about 24% of U.S. organizations have formally embarked on ITIL-based programs more than 5 years ago. These programs were less than 5 years old for the remaining 76% of organizations (Mann 2012). Although effective implementation of ITIL hinges on different technological, organizational, and environmental factors, industry reports have found that senior executives in the organization are the most influential drivers of successful ITIL programs, followed by the core ITSM team that manages the IT service lifecycle in the organization (O'Donnell 2011).

There are several challenges associated with ITIL implementation. Ho (2008) noted that resistance to ITIL is often rooted in fear, and suggested that the key to alleviating this resistance is education. As implementation of ITIL requires commitment to a multi-year project, it is often met with resistance by IT employees (directly) as well as non-IT employees (indirectly). Ho (2008) listed a set of *ITIL fears,* a major source of resistance to ITIL implementation, such as fear of change (i.e., unknown consequences of ITIL implementation), fear of measurement (i.e., a perceived loss of control), fear of perceived process rigidity (i.e., ITIL processes can be perceived to be inflexible), fear of cost (i.e., a large investment is required to implement ITIL), and fear of complexity (i.e., ITIL typically introduces new roles, processes, and complexity in organizations).

Even if an organization is able to overcome these fears through training and other interventions, there are possible implementation mistakes that can still prevent organizations from garnering benefits from ITIL implementation (Guglielmo 2008; Mann 2012; Kneller 2010). For example, organizations may design and develop IT services that do not fit with the organizational processes or do not meet organizational needs. In many cases, organizations fail to understand and take account of the ongoing operational cost that ITIL processes require. Many organizations take into consideration the one-off project cost for ITIL implementation and do not recognize the continual service-improvement stage of the IT service lifecycle. Consequently, these organizations fail to recognize the annual cost of ownership for each IT service (Kneller 2010). In many cases, organizations take a "shortcut" approach to implementing ITIL. These organizations fail to recognize that ITIL implementation requires a strategic vision, top management support, a strong business case, an effective communication strategy, and effective control mechanism (Guglielmo 2008; Kneller 2010). ITIL implementation will change the way IT delivers value to the rest of the organization and the way organizations manage IT. There is a need, therefore, to develop effective strategies and interventions to create a service culture and climate throughout the organization (Guglielmo 2008; Kneller 2010).

59.3.2.2 Impacts of ITIL on Practice

There is plenty of anecdotal evidence and numerous case studies that indicate positive impacts of ITIL implementation on organizations and employees. Major organizations from across the world, such as Microsoft, HP, Fujitsu, IBM, Target, Walmart, Staples, Citi, Bank of America, Barclay's Bank, Sony, Disney, Boeing, Toyota, Bombardier, Eli Lilly, and Pfizer, have adopted and implemented ITIL and received significant benefits (Arraj 2010). In addition to these major organizations, there are literally thousands of other organizations of all industries and sizes that have embraced ITIL practices. A recent industry report revealed that ITIL implementation had significant positive impacts on various aspects of IT service delivery and IT employees who deliver IT services (O'Donnell 2011). A survey of 491 ITSM professionals found that ITIL improved quality of IT service (83% of respondents felt so), productivity (85% of respondents felt ITIL improved IT productivity), reputation with the business (65% of respondents felt ITIL was either beneficial or significantly beneficial to improving reputation with the business), and reduce operational cost (41% of respondents felt ITIL helped reduce operational costs by improving the quality of IT services). At the individual level, this report found that ITSM professionals who are ITIL certified received better salaries than those in the average IT position. In fact, about 70% of ITSM professionals received an increase in their salary over the past year. In the academic literature, Gacenga et al. (2010) conducted a broad-based survey of ITIL implementation in three different countries—Australia, the United Kingdom, and the United States. The results indicated that while ITIL implementation had a positive influence at process-level outcomes, organizations failed to gain broad organizational-level benefits from ITIL because most organizations chose to implement only a subset of ITIL processes.

59.3.3 Information Security Framework: ISO/IEC 27000 Series

59.3.3.1 Adoption and Implementation of ISO/IEC 27000

ISO/IEC 27000 standards are still evolving with many individual standards that are still under development (see Table 59.2). While it is difficult to assess the adoption and implementation drivers and challenges related to these standards, a wide range of organizations have adopted ISO/IEC 27001 and ISO/IEC 27002. These organizations are from different industries and economic sectors, such as IT, software, manufacturing, construction, financial, staffing, shipping, pharmaceuticals, academia, telecom, lottery, security, consulting, insurance, health care, energy, and navigation (IsecT, Inc. 2012; DiMaria 2011). Many government agencies and international organizations, such as the United Nations, have adopted ISO/IEC 27001 for ISM. Given that organizations can receive certification if they comply with ISO/IEC 27001 standards, the number of certificates related to this standard is an acceptable measure of the current adoption level of this standard. There are over 7,200 ISO/IEC 27001 certified organizations in the world as of May 2011 (ISMS International User Group 2012). It is important to note that this total number has potentially been underestimated because reporting of certification status is completely voluntary and many organizations may never report their certification status. Further, ISO/IEC 27001 certificates are issued by multiple certification bodies, so there is no central database of organizations that have adopted these standards and are certified.

59.3.3.2 Impacts of ISO/IEC 27000 on Practice

Although we do not have accurate and complete data regarding the adoption of the ISO/IEC 27001 standard, we found that certifications have been increasing steadily at a rate of about 1000 per year (ISMS International User Group 2012). Again, reporting of certification is voluntary and it is thus possible the rate of increase is underestimated. Such an increasing trajectory of certifications suggests that many organizations that adopted this standard have received benefits from it. Nonetheless, there has been a dearth of academic and practitioner literature related to the impacts of ISO/IEC 27000 standards on organizations. We found a few cases that illustrated how ISO/IEC 27000 standards helped organizations achieve favorable organizational goals (British Standards Institution 2012). For instance, Cleardata, a document scanning and archiving company in the United Kingdom, needed to develop and implement an ISM framework to reassure its clients that it would manage their information and data security to the highest of standards. Cleardata needed to develop a structured framework to ensure that security objectives were met, compliance with laws and regulations was guaranteed, and new business and customer trust was gained. When they adopted ISO/IEC 27001 and got certified, Cleardata was able to win multiple contracts as a direct result of certification. In a different context, a government agency, the NHS Purchasing and Supply Agency of the Department of Health in the United Kingdom, was able to receive official recognition that its security measures were effective; it had a systematic framework for managing sensitive data, and was able to establish a formal standard for the operations of the IT organization. Following the adoption and certification of ISO/IEC 27001, the Agency never had a security incident caused by a staff member, was able to keep the network virus-free and hacker-free, and was able to encourage stakeholders to share information without any fear of security breach and data loss (British Standards Institution 2012).

59.4 Research Opportunities

We suggest that IT management frameworks provide many fruitful avenues for future research. These frameworks offer rich content with respect to principles, provisions, processes, and metrics that can be leveraged in research. Further, there are a multitude of outcomes (i.e., measures of performance indicators) and potential antecedents to these outcomes offered in these frameworks. These relationships also deliver many interesting and fruitful future research opportunities. Although there is a wide range of such opportunities, we provide a set of research directions that we believe will make significant impacts on both theory and practice related to IT management. In addition to leveraging the frameworks that we discuss in previous sections, we also review prior research in the areas of IT governance, ITSM, and information security to unearth scientific gaps in these research streams.

59.4.1 Review of IT Management Research

The literature on IT governance is rich and mature, and it has been around for almost 30 years. This popular stream of research has offered deep insights on issues related to how IT organizations and IT assets should be structured, managed, and controlled for organizational benefits. There have been several recent reviews of this stream that provide a comprehensive description of what we currently know in this literature (e.g., Guillemette and Pare 2012; Jewer and McKay 2012). These reviews have indicated that much prior research has focused on three broad areas: (a) design, development, and implementation of IT governance framework and/or decision-making structures that IT organizations can use to organize their activities (e.g., Kaarst-Brown 2005; Kaarst-Brown and Robey 1999); (b) antecedents and impacts of IT governance practices on organizational outcomes (e.g., Brown 1999; Sambamurthy and Zmud 1999; Weill and Ross 2004); and (c) the role and competencies of board and/or senior executives (both IT and business) in managing and controlling IT governance processes and activities (e.g., Leidner and Mackay 2007; Nolan and McFarlan 2005).

Notwithstanding the richness of research in these three broad areas, we suggest that the IT governance literature in general has overlooked issues related to various components of IT governance frameworks that are widely used in practice. IT governance frameworks such as COBIT offer rich descriptions of processes and their relationships with IT and organizational goals. Although prior research offered insights on governance structure and form, there has been limited research that incorporates various governance processes from COBIT, categorizes them into theoretically meaningful categories, and examines the impacts of these categories on various aspects of organizational outcomes. Such work will be beneficial to practitioners to justify the investments and efforts in IT governance processes and practices. Further, there has been limited research that focused on the measurement of IT governance effectiveness. Although IT governance frameworks such as COBIT offer a comprehensive set of performance indicators for various processes and practices, it is possible that many of these indicators are not scientifically valid and/or practically useful. Future research may organize these measures into theoretically meaningful categories and scientifically validate them for future research and practice.

Like the IT governance literature, the ITSM literature is mature and has offered insights on various aspects of ITSM practices in organizations. ITSM has been a focal topic in two distinct yet related streams of research. The first stream focuses on issues related to designing, developing, implementing, and managing IT services in organizations (e.g., Bardhan et al. 2010; Montoya et al. 2010). In this stream, researchers have concentrated on how organizations design, implement, and access IT services based on emerging IT services platform and architecture, such as application service providers (ASP), web services, service-oriented architecture (SOA), and cloud computing (e.g., Choi et al. 2010; Mueller et al. 2010; Susarla et al. 2003). This stream also deals with issues such as the impact of IT service sourcing and contracts (e.g., Benorach et al. 2010; Kauffman and Sougstad 2008).

Drawing on the service quality literature from marketing (Parasuraman et al. 1994; Zeithaml et al. 1996), the second stream deals with topics related to conceptualizing and measuring IT service quality—in particular, the quality of IT functions in organizations (e.g., Kettinger and Lee 2005; Pitt et al. 1995). When IS researchers conceptualized and operationalized IT service quality, similar to the notion of service quality in the marketing literature (Parasuraman et al. 2005), they looked at people-delivered services. The key difference was that IS researchers focused not only on IT services (e.g., hardware and software installation) delivered by an entire IT department of an organization but also on the quality of service related to an individual IT system itself. In fact, Pitt et al. (1995) noted that there are two possible units of analysis for conceptualizing and measuring IS service quality (SERVQUAL)—the entire IT department and a particular IT application. While people-delivered services were the key focus initially in the IS literature, researchers have also examined technology-delivered services such as e-commerce websites (e.g., Parasuraman et al. 2005; Ziethaml et al. 2002).

Research on ITSM has either focused on IT infrastructure issues or the overall IT function of an organization while overlooking the specific ITSM processes that may offer differential value propositions for organizations. More importantly, prior ITSM research has focused primarily on IT-enabled service delivery. We are not aware of any research that highlights other stages of the ITSM life cycle, such as service strategy, service transition, and continual service improvement. These stages offer a set of processes, practices, and metrics that have been mostly overlooked in the extant literature. Further, there has been limited research that examined how ITSM processes and practices can lead to favorable outcomes for an IT organization in particular and the overall organization in general. Finally, prior research has offered a limited understanding of measuring the quality of ITSM practices. While SERVQUAL has provided a broad-based measure of ITSM quality, it is not an objective measure and does not capture the richness that an ITSM framework, such as ITIL, may offer.

The IT security literature gives rich insights on various aspects of ISM in organization, such as compliance, security behaviors, security policy decisions, and risk management. Research in this stream can be grouped into two broad categories. The first category is related to the behavioral issues of ISM such as individual compliance, security behaviors, and user involvement in ISM in organizations (e.g., Bulgurcu et al. 2010; Johnston and Warkentin 2010; Siponen and Vance 2010). This stream provides a comprehensive understanding of individual-level issues related to ISM processes and practices in organizations. For example, Bulgurcu et al. (2010) examined the factors that employees consider when they decide to comply with organizational security policies. The second category of IT security research is related to the development and implementation of organizational policies related to ISM and IT risk management (e.g., August and Tunca 2011; Chen et al. 2011). This stream of research employed both empirical and analytical approaches to unearth factors that have implications for organizational information security policies.

Although IT security research offers insights on both individual behaviors and organizational policies, we suggest that this research does not offer insights on the characteristics of a comprehensive ISMS, and antecedents and outcomes of establishing an ISMS. With the advances of technology and increasing complexity in the IT landscape, we suggest that ISM should be an important stream of IT research that focuses on both theoretical underpinnings related to ISM and practical considerations when implementing ISMS. Further, there has been little or no research that examined the value proposition of ISMS for organizations and scientifically validated the impacts of ISM practices on organizational outcomes.

59.4.2 Suggested Research Directions

Our goal is to offer a set of future research directions that address the gap in the current literature and leverage the richness that IT management frameworks offer. One of the rich components of each of these frameworks is performance measures—key performance indicators (KPIs)—for various aspects of IT management in organizations. For example, COBIT and ITIL frameworks and associated documents offer a comprehensive selection of measures to assess the performance of various processes and activities. These measures are now widely available and generally accepted among senior IT executives as indicators to measure performance of IT functions and services. While prior research has provided various types of IT function assessment (e.g., Brown and Magill 1994; Chang and King 2005; Saunders and Jones 1992), we suggest that future research could leverage the rich performance indicators that IT management frameworks offer to measure performance of IT function in organizations. Thus, we offer the following opportunities for future research.

Research Opportunity 1: Development of Performance Measures
Development and validation of a new scale of (a) IT service quality measures using the ITIL framework; (b) effectiveness of IT governance practices and processes using the COBIT framework; and (c) effectiveness of ISM practices using the ISO/IEC 27001 and ISO/IEC 27002 standards. Is it possible to develop an overall performance scale for each of these areas, or should there be different scales for practices and processes within each of these areas?

In order to leverage the richness of IT management frameworks, it is important for researchers to dig deeper into the different processes and activities that these frameworks provide. These processes and activities indeed offer a detailed representation of what an IT organization does at strategic, tactical, and operational levels to manage IT functions in an organization and how it attempts to align with organizational activities and strategies. As noted earlier, there has been little or no research that has examined the impacts of these processes on outcomes related to IT and organizations. Given the comprehensive nature of these processes, it is probably not possible to incorporate the processes from an IT management framework in a single research inquiry. Therefore, we suggest that researchers try to categorize these processes into theoretical constructs and operationalize them following the guidelines from respective IT management frameworks. For example, researchers can develop three different categories of ITSM processes—strategic, tactical, and operational—and examine the impacts of these processes on outcomes related to IT and organizations. It is important to note that this categorization should be guided by a theoretical framework. We expect that researchers will be able to develop interesting theoretical constructs based on the IT management frameworks and offer theoretical insights on how these frameworks will influence IT and organizational outcomes. Based on this discussion, we propose the following research direction related to the impacts of different components of IT management frameworks.

Research Opportunity 2: Impacts on Organizational Outcomes
What are the impacts of IT governance/ITSM/ISM processes and activities on IT and organizational outcomes? What processes and activities have the most impacts and why? Is it possible to group these processes into theoretical constructs that will inform current organizational theories?

We suggest that IT management frameworks will have a proximal effect on IT outcomes and a distal effect on organizational outcomes. This distal effect could be mediated by the IT-related outcomes. We believe that these relationships have interesting theoretical implications that researchers should explore in future research. For example, it is important to understand what components of IT management frameworks will have a direct influence on organizational outcomes and what components will have a mediated influence through IT-related outcomes. Such an understanding will help IT organizations focus on certain components to achieve desired outcomes. Further, if researchers could show that certain IT outcomes need to be achieved first before garnering organizational benefits from an IT management framework implementation, IT organizations could focus on specific processes and activities to achieve those IT-related outcomes. Overall, we suggest that the dynamics of the relationship between different components of IT management framework and outcomes (both IT-related and organizational) is a fruitful avenue of future research that will have an immediate impact on practices related to IT management in organizations. Based on this discussion, we propose the following opportunity for future research.

Research Opportunity 3: Impacts of Framework Components on Organizational Outcomes
Do different components of IT management frameworks (e.g., processes and activities) have a direct impact on organizational outcomes such as productivity and profitability (as claimed in practitioner reports) or do they have a mediated impact through IT outcomes? How and why? What are the theoretical mechanisms that explain these relationships?

Gordon and Gordon (2002) discussed the tension between IT function and non-IT functions in organizations during IT service delivery. Both academic and practitioner literatures offer numerous examples of tension and the lack of trust between IT and non-IT functions in organizations. We suggest that IT management frameworks will offer new insights into the understanding of the relationship between IT and non-IT functions. These frameworks have specific processes, activities, and metrics related to how an IT organization should manage change and other IT assets that are exposed to end users and customers who are mostly dealing with non-IT functions. Further, these frameworks have specific goals and processes related to improving user and customer satisfaction. A fruitful area of future research will thus be to look into the issues related to the tension and the lack of trust between IT and non-IT functions and offer mechanisms (e.g., processes, practices, and activities) from IT management frameworks about how to resolve this tension and the lack of trust. Therefore, we offer the following research questions for future inquiries.

Research Opportunity 4: Relationship between IT and Non-IT Functions/Users
What are the most effective components of IT management frameworks to improve user and customer satisfaction and trust between IT and non-IT functions? What are the components that may increase tension between IT and non-IT functions? What are the metrics to assess the relationship between IT and non-IT functions?

We suggest that the effective implementation and practice of IT management frameworks will be a source of competitive advantage. While these frameworks are publicly available, we do not expect that all organizations will be able to equally implement and use them effectively. In fact, prior research has shown that publicly available best practices could be implemented and used differently by different organizations, leading to different outcomes (Bala and Venkatesh 2007; Venkatesh and Bala 2012). Based on this discussion, we suggest that organizations will be able to develop certain capabilities around IT management frameworks. For instance, some organizations will be able to develop capabilities related to effective IT governance (e.g., quick decision-making, formal performance evaluation, and high degree of knowledge sharing between IT and business), whereas other organizations may develop capabilities around ITSM practices. We expect that these capabilities will help organizations achieve favorable outcomes at both IT and organizational levels. Therefore, we offer the following research opportunity related to capabilities stemmed from implementing IT governance frameworks.

Research Opportunity 5: Development of IT Management Capabilities
What are some of the capabilities organizations can develop around IT management frameworks? What are the dimensions of these capabilities? What provisions, processes, and activities will determine and define these capabilities?

In sum, the research opportunities that we discuss here are broad and meant to serve as a guideline for researchers to develop more specific research questions for their research. We believe that these opportunities barely scratch the surface of the potential research space that IT management frameworks may offer. We expect that these opportunities will spur many fruitful research programs in the context of IT management frameworks.

59.5 Concluding Remarks

Our review of both practitioner and academic literatures indicates that although IT management frameworks, such as COBIT, ITIL, and ISO/IEC 27000, have been able to generate a wide range of interests and acceptance among senior executives who are interested in understanding how IT functions in an organization should be governed, managed, operated, and controlled, these frameworks have been conspicuously overlooked in mainstream academic research. While we acknowledge that academic research needs to be theory-driven (as opposed to framework-driven) and needs to offer theoretical insights about a phenomenon of interest, we suggest that IT researchers have a unique opportunity to conduct theory-driven research leveraging the richness of IT management frameworks. Recently, Mithas et al. (2011) employed the Baldridge criteria, an industry framework to assess performance excellence developed by the National Institute of Standards and Technology (NIST), to conduct theory-driven research, and found strong support for the relationship between information management capability and firm performance. We suggest that IT researchers should undertake similar research in the context of IT management frameworks for at least two reasons.

First, we believe that such research will help the IT research community reduce the theory–practice (and the rigor-relevance) divide that is prominent in IT research. The details about this divide and associated debates have been outlined elsewhere (see for example, Benbasat and Zmud 1999; Constantinides et al. 2012; Davenport and Markus 1999). Recently, Rosemann and Vessey (2008) suggested that relevance for practice can be achieved by producing research that is important to practice, is accessible by practitioners, and can be assessed for relevance through applicability checks (Constantinides et al. 2012; see Rosemann and Vessey 2008 for details related to applicability check). Others have called for conducting high-visibility and high-impact research that has impacts on both theory and practice (e.g., Agarwal and Lucas 2005). In light of these views, we suggest that given the widespread acceptance of IT management frameworks, researchers may have a unique opportunity to help practitioners improve these frameworks and implement them effectively by offering insights from rigorous scientific research. We are not aware of any major scientific studies and associated publications that examined the implementation challenges, appropriate configuration, and impacts of these frameworks. Therefore, we suggest that such research will be able to help the IT research community reduce the gap between theory and practice in the context of IT management by helping practitioners improve the quality of these frameworks and the implementation process.

Second, we believe that the richness of IT management frameworks will help researchers develop valuable theoretical insights related to IT management practices in organizations. There has been much debate about the role of *IT artifact* in IT research. Scholars have called for placing an explicit emphasis on IT artifact in order to enhance the legitimacy of the discipline and to alleviate the *identity crisis* that the discipline currently faces (Benbasat and Zmud 2003; Constantinides et al. 2012; Orlikowski and Iacono 2001). We believe that research based on IT management frameworks will help researchers address this call by offering deep insights on how IT should be managed for organizational benefits. Further, prior research has called for incorporating the role of *contexts* in theory development (Johns 2006; Rousseau and Fried 2001). The components of IT management frameworks, such as processes, activities, inputs, outputs, and responsibility assignments, offer rich contexts to study various aspects of IT management practices in organizations. Researchers will be able to discover boundary conditions of existing organizational theories and offer novel theoretical perspectives leveraging the rich contexts that IT management frameworks may offer. For instance, using the resource-based view (RBV) of the firm theory, prior research has developed a comprehensive set of IT capabilities. While these capabilities are important, IT management frameworks can help researchers extend these capabilities by including other dimensions, such as IT service management capabilities, IT governance capabilities, and information security management capabilities. Overall, we suggest that IT management frameworks may offer opportunities for developing novel theoretical perspectives on IT management activities in organizations.

In conclusion, we reiterate our view that IT management frameworks are important for both theory and practice. We believe that these frameworks are here to stay and will improve over time to help organizations address the key concerns related to IT management (Luftman and Ben-Zvi 2011). Notwithstanding the implementation challenges we discuss in this chapter, there is an increasing trend of adoption and implementation of these frameworks. Therefore, we believe that it is high time to develop fruitful research programs around these frameworks. In this chapter we offer an overview of these frameworks and highlight that these frameworks can coexist peacefully in organizations. We also discuss the impacts of these frameworks on practices based on our extensive review of the practitioner literature. Our major contributions are the discussion of prior research and potential research opportunities based on these frameworks. Our hope is that this chapter will fuel future research and intellectual discourse on popular IT management frameworks.

Acknowledgments

We thank Professor Heikki Topi, Volume Editor, for his thoughtful comments and suggestions to improve the chapter. We thank Dr. Suzanne Van Hove, CEO of SED-IT, and Julie Montgomery, Director of Marketing of Plexent, for their support and encouragement for our project on IT management frameworks. We also thank Professor V. Sambamurthy (Michigan State University) for his guidance in helping us think about future research directions on IT management. Finally, we appreciate Akshay Bhagwatwar's help in the literature review and background research on IT management frameworks.

References

Agarwal, R. and H. C. Lucas Jr. 2005. The information systems identity crisis: Focusing on high-visibility and high-impact research. *MIS Quarterly* 29:381–398.

Agarwal, R. and V. Sambamurthy. 2002. Principles and models for organizing the IT function. *MIS Quarterly Executive* 1:1–16.

Aral, S. and P. Weill. 2007. IT assets, organizational capabilities, and firm performance: How resource allocations and organizational differences explain performance variation. *Organization Science* 18:763–780.

Arraj, V. 2010. *ITIL®: The Basics*. Best Practice Management White Paper. Norwich, U.K.: The Stationary Office.

August, T. and T. Tunca. 2011. Who should be responsible for software security? A comparative analysis of liability policies in network environments. *Management Science* 57:934–959.

Bala, H. and V. Venkatesh. 2007. Assimilation of interorganizational business process standards. *Information Systems Research* 18:340–362.

Banker, R., S. Mitra, and V. Sambamurthy. 2011. The effects of digital trading platforms on commodity prices in agricultural supply chains. *MIS Quarterly* 35:599–612.

Bardhan, I. R., H. Demirkan, P. K. Kannan, R. J. Kauffman, and R. Sougstad. 2010. An interdisciplinary perspective on services management and service science. *Journal of Management Information Systems* 26(4):13–64.

Benaroch, M., Q. Dai, and R. J. Kauffman. 2010. Should we go our own way? Backsourcing flexibility in IT services contracts. *Journal of Management Information Systems* 26:317–358.

Benbasat, I. and R. Zmud. 1999. Empirical research in information systems: The practice of relevance. *MIS Quarterly* 23:3–16.

Benbasat, I. and R. Zmud. 2003. The identity crisis within the IS discipline: Defining and communicating the discipline's core properties. *MIS Quarterly* 27:183–194.

British Standards Institution. 2012. ISO/IEC 27001 information security. http://www.bsigroup.com/en/Assessment-and-certification-services/management-systems/Standards-and-Schemes/ISO-IEC-27001/ (accessed on September 15, 2012).

Brown, C. V. 1999. Horizontal mechanisms under differing IS organizational contexts. *MIS Quarterly* 23:421–454.

Brown, C. V. and S. L. Magill. 1994. Alignment of the IS functions with the enterprise: Toward a model of antecedents. *MIS Quarterly* 18:371–403.

Brynjolfsson, E. K. 1993. The productivity paradox of information technology. *Communications of the ACM* 36:66–77.

Brynjolfsson, E. K. and L. M. Hitt. 1996. Paradox lost? Firm-level evidence on the returns to information systems spending. *Management Science* 42:541–558.

Bulgurcu, B., H. Cavusoglu, and I. Benbasat. 2010. Information security policy compliance: An empirical study of rationality-based beliefs and information security awareness. *MIS Quarterly* 34:523–548.

Cartlidge, A., A. Hanna, C. Rudd, I. Macfarlane, J. Windebank, and S. Rance. 2007. *An Introductory Overview of ITIL® V3*, The UK Chapter of the itSMF, United Kingdom: itSMF Ltd.

Chang, J. C.-J. and W. R. King. 2005. Measuring the performance of information systems: A functional scorecard. *Journal of Management Information Systems* 22:85–115.

Choi, J., D. L. Nazareth, and H. K. Jain. 2010. Implementing service-oriented architecture in organizations. *Journal of Management Information Systems* 26:253–286.

Constantinides, P., M. W. Chiasson, and L. D. Introna. 2012. The ends of information systems research: A pragmatic framework. *MIS Quarterly* 36:1–20.

Chen, P.-Y., G. Kataria, and R. Krishnan. 2011. Correlated failures, diversification, and information security risk management. *MIS Quarterly* 35:397–422.

Clinch, J. 2009. ITIL V3 and information security. Best Practice Management White Paper. Norwich, U.K.: The Stationary Office.

Davenport, T. and M. L. Markus. 1999. Rigor vs. relevance revisited: Response to Benbasat and Zmud. *MIS Quarterly* 23:19–23.

Dewan, S. and K. L. Kraemer. 2000. Information technology and productivity: Evidence from country-level data. *Management Science* 46:548–562.

Dewan, S., C. Shi, and V. Gurbaxani. 2007. Investigating the risk-return relationship of information technology investments: Firm-level empirical analysis. *Management Science* 53:1829–1842.

DiMaria, J. 2011. ISO 27001: Information security. Stanorg Limited. http://www.standards.org/standards/listing/iso_27001 (accessed on September 20, 2012).

Dos Santos, B. L., Z. Zheng, V. S. Mookerjee, and Chen, H. 2012. Are new IT-enabled investment opportunities diminishing for firms? *Information Systems Research* 23:287–305.

Gacenga, F., A. Cater-Steel, and M. Toleman. 2010. An international analysis of IT service management benefits and performance measurement. *Journal of Global Information Technology Management* 13:28–63.

Gordon, S. R. and J. R. Gordon. 2002. Organizational options for resolving the tension between IT department and business units in the delivery of IT services. *Information Technology and People* 15:286–305.

Guglielmo, K. 2008. ITIL: 10 deployment mistakes (and fixes). In *Enterprise Innovation eGuide to ITIL*, ed. C. S. Chan, pp. 8–9, Hong Kong, China: Enterprise Innovation.

Guillemette, M. G. and P. Guy. 2012. Toward a new theory of the contribution of the IT function in organizations. *MIS Quarterly* 36:529–551.

Heller, M. 2009. How IT is set up to fail. *CIO Magazine*. December 23, 2009. http://www.cio.com/article/print/511568 (accessed on September 20, 2012).

Ho, L. C. 2008. 10 Reasons why ITIL spooks IT managers. In *Enterprise Innovation eGuide to ITIL*, ed. C. S. Chan, pp. 6–8, Hong Kong, China: Enterprise Innovation.

IsecT, Inc. 2012. *About ISO27k standards*. http://www.iso27001security.com/html/iso27000.html (accessed on September 20, 2012).

ISACA. 2012. ISACA, Rolling Meadows, IL. https://www.isaca.org/Pages/default.aspx (accessed on September 20, 2012).

ISMS International User Group. 2012. *International Register of ISMS Certificates*. http://iso27001certificates.com/ (accessed on September 20, 2012).

ISO/IEC. 2009. *Information Technology—Security Techniques—Information Security Management Systems—Overview and Vocabulary*. Geneva, Switzerland: ISO/IEC.

IT Governance Institute. 2003. *Board Briefing on IT Governance*. Rolling Meadows, IL: IT Governance Institute.

IT Governance Institute. 2008a. *Aligning CobiT® 4.1, ITIL® V3 and ISO/IEC 27002 for Business Benefit*. Rolling Meadows, IL: IT Governance Institute.

IT Governance Institute. 2008b. *COBIT Mapping: Mapping of ITIL v3 with COBIT® 4.1*. Rolling Meadows, IL: IT Governance Institute.

Jewer, J. and K. N. McKay. 2012. Antecedents and consequences of board IT governance: Institutional and strategic choice perspectives. *Journal of the Association for Information Systems* 13:581–617.

Johns, G. 2006. The essential impact of context on organizational behavior. *Academy of Management Review* 31: 386–408.

Johnston, A. C. and M. Warkentin. 2010. Fear appeals and information security behaviors: An empirical study. *MIS Quarterly* 34:549–566.

Kaarst-Brown, M. L. 2005. Understanding an organization's view of the CIO: The role of assumptions about IT. *MIS Quarterly Executive* 4:287–301.

Kaarst-Brown, M. L. and D. Robey. 1999. More on myth, magic and metaphor: Cultural insights into the management of information technology in organizations. *Information Technology and People* 12:192–217.

Kauffman, R. J. and R. Sougstad. 2008. Risk management of contract portfolios in IT services: The profit-at-risk approach. *Journal of Management Information Systems* 25:17–48.

Kettinger, W. J. and C. C. Lee. 2005. Zones of tolerance: Alternative scales for measuring information systems service quality. *MIS Quarterly* 29:607–623.

King, J. 2010. These CIOs go way beyond IT-business alignment. *ComputerWorld*. May 24, 2010. http://www.computerworld.com/s/article/348634/Beyond_Alignment (accessed on September 20, 2012).

Kneller, M. 2010. *Executive Briefing: The Benefits of ITIL®*. Best Practice Management White Paper. Norwich, U.K.: The Stationary Office.

Kouns, J. and D. Minoli. 2010. *Information Technology Risk Management in Enterprise Environments: A Review of Industry Practices and a Practical Guide to Risk Management Teams*. Hoboken, NJ: John Wiley & Sons, Inc.

Leidner, D. E. and J. M. Mackay. 2007. How incoming CIOs transition into their new jobs. *MIS Quarterly Executive* 6:17–28.

Luftman, J. and T. Ben-Zvi. 2011. Key issues for IT executives 2011: Cautious optimism in uncertain economic times. *MIS Quarterly Executive* 10:203–212.

Mann, S. 2012. *Planning Road Map: Adopting ITIL*. Cambridge, MA: Forrester Research, Inc.

McAfee, A. and E. Brynjolfsson. 2008. Investing in the IT that Makes a Competitive Difference. *Harvard Business Review* 86:98–108.

Mithas, S., N. Ramasubbu, and V. Sambamurthy. 2011. How information management capability influences firm performance. *MIS Quarterly* 35:237–256.

Mithas, S., A. R. Tafti, I. R. Bardhan, and J. M. Goh. 2012. Information technology and firm profitability: Mechanisms and empirical evidence. *MIS Quarterly* 36:205–224.

Montoya, M. M., A. P. Massey, and V. Khatri. 2010. Connecting IT services operations to services marketing practices. *Journal of Management Information Systems* 26:65–85.

Mueller, B., G. Viering, C. Legner, and G. Riempp. 2010. Understanding the economic potential of service-oriented architecture. *Journal of Management Information Systems* 26:145–180.

Nolan, R. and F. W. McFarlan. 2005. Information technology and the board of directors. *Harvard Business Review* 83:96–106.

O'Donnell, G. 2011. *The State of IT Service Management in 2011*. Cambridge, MA: Forrester Research, Inc.

Orlikowski, W. J. and C. S. Iacono. 2001. Research commentary: Desperately seeking the "IT" in IT Research—A call to theorizing the IT artifact. *Information Systems Research* 12:121–134.

Parasuraman, A., V. A. Zeithaml, and L. L. Berry. 1994. Reassessment of expectations as a comparison standard in measuring service quality: Implications for further research. *Journal of Marketing* 58:111–124.

Parasuraman, A., V. A. Zeithaml, and A. Malhotra. 2005. E-S-qual: A multiple item scale for measuring electronic service quality. *Journal of Services Research* 7:213–233.

Peppard, J. and J. Ward. 2005. Unlocking sustained business value from IT investments. *California Management Review* 48:52–69.

Pitt, L. F., R. T. Watson, and C. B. Kavan. 1995. Service quality: A measure of information systems effectiveness. *MIS Quarterly* 19:173–187.

Project Management Institute. 2012. *PMBOK® Guide and Standards*. http://www.pmi.org/PMBOK-Guide-and-Standards.aspx (accessed on September 15, 2012).

Rai, A. and V. Sambamurthy. 2006. Editorial notes: The growth of interest in services management: Opportunities for information systems scholars. *Information Systems Research* 17: 327–331.

Ramanathan, S. 2007. IT Governance—Challenges in implementation from an Asian perspective. *Information Systems Control Journal* 5:1–2.

Rosemann, M. and I. Vessey. 2008. Toward improving the relevance of information systems research to practice: The role of applicability checks. *MIS Quarterly* 32:1–22.

Ross, J. and P. Weill. 2002. Six IT decisions your IT people shouldn't make. *Harvard Business Review* 80:84–91.

Ross, J. W., P. Weill, and D. C. Robertson. 2006. *Enterprise Architecture as Strategy: Creating a Foundation for Business Execution*, Boston, MA: Harvard Business School Press.

Rouyet, J. I. and W. J. Spauwen. 2011. *Using COBIT to Achieve Green Business-IT Alignment* ComputerworldUK.com. January 22, 2011. http://www.computerworlduk.com/how-to/infrastructure/3257646/using-cobit-to-achieve-green-business-it-alignment/ (accessed on September 15, 2012).

Rousseau, D. M. and Y. Fried. 2001. Location, location, location: Contextualizing organizational research. *Journal of Organizational Behavior* 22:1–13.

Sambamurthy V. and R. W. Zmud. 1999. Arrangements for information technology governance: A theory of multiple contingencies. *MIS Quarterly* 23:261–291.

Saunders, C. S. and J. W. Jones. 1992. Measuring performance of the information systems function. *Journal of Management Information Systems* 8:63–82.

Siponen, M. and A. Vance. 2010. Neutralization: New insights into the problem of employee information systems security policy violations. *MIS Quarterly* 34:487–502.

Spears, J. L. and H. Barki. 2010. User participation in information systems security risk management. *MIS Quarterly* 34:503–522.

Susarla, A., A. Barua, and A. B. Whinston. 2003. Understanding the service component of application service provision: An empirical analysis of satisfaction with ASP services. *MIS Quarterly* 27:91–123.

Venkatesh, V. and H. Bala. 2012. Adoption and impacts of interorganizational business process standards: Role of partnering synergy. *Information Systems Research* 23:1131–1157.

Venkatesh, V., H. Bala, and T. A. Sykes. 2010. Impacts of information and communication technology implementations on employees' jobs in service organizations in India: A multi-method longitudinal field study. *Production and Operations Management* 19:591–613.

Violino, B. 2005. IT frameworks demystified. *Network World* 22:18–19.

Wailgum, T. 2009. IT management in 2010: Complex, costly and still confounding. *CIO Magazine.* November 30, 2009. http://www.cio.com/article/print/509349 (accessed on September 20, 2012).

Weill, P. and J. W. Ross. 2004. *IT Governance: How Top Performers Manage IT Decision Rights for Superior Results.* Boston, MA: Harvard Business School Press.

Westerman, G. and G. Hunter. 2007. *IT Risk: Turning Business Threats into Competitive Advantage.* Boston, MA: Harvard Business School Press.

Whittleston, S. 2012. *ITIL® is ITIL.* Best Practice Management White Paper. Norwich, U.K.: The Stationary Office.

Willcocks, L., D. Feeny, and N. Olson. 2006. Implementing core IS capabilities: Feeny-Willcocks IT governance and management framework Revisited. *European Management Journal* 24:28–37.

Zeithaml, V. A., L. L. Berry, and A. Parasuraman. 1996. The behavioral consequences of service quality. *Journal of Marketing* 60:31–46.

Zeithaml, V. A., A. Parasuraman, and A. Malhotra. 2002. Service quality delivery through web sites: A critical review of extant knowledge. *Journal of the Academy of Marketing Science* 30:362–375.

Zhang, Y. and N. Chulkov. 2011. *Information and Communication Technology (ICT) Governance in the United Nations System Organizations.* Geneva, Switzerland: Joint Inspection Unit, the United Nations.

60

Sourcing Information Technology Services

Mary Lacity
University of Missouri–St. Louis

Leslie Willcocks
London School of Economics and Political Science

60.1 Introduction

Information technology (IT) sourcing is the acquisition of resources, including human capital resources, to deliver IT services such as application development, application support, systems integration, data management, data center management, telecommunications and network management, and distributed computing (e.g., desktops, laptops, mobile devices, and resources on the cloud). In this chapter, we review the practice of and research on the sourcing of IT services. This chapter is organized around four topics: (1) IT sourcing decisions, (2) the determinants of IT sourcing decisions, (3) IT sourcing outcomes, and (4) the determinants of IT sourcing outcomes (see Figure 60.1).

IT executives make decisions about IT sourcing options and IT sourcing locations, and when outsourcing, decisions about multisourcing versus bundled services, and next-generation sourcing decisions when initial contracts approach expiration. Academic researchers have thoroughly studied the motivations, influence sources, and transaction attributes that determine sourcing decisions. Researchers have also thoroughly studied the outcomes of outsourcing. Overall, the academic evidence finds that outsourcing can deliver value to client organizations, but that it takes a tremendous amount of detailed management by clients and providers to realize expected benefits (Lacity et al. 2010a). Researchers have identified the transaction attributes, contractual governance, relational governance, client-retained capabilities, and provider capabilities that determine IT outsourcing (ITO) outcomes. Despite all that researchers do know about sourcing IT services, there are

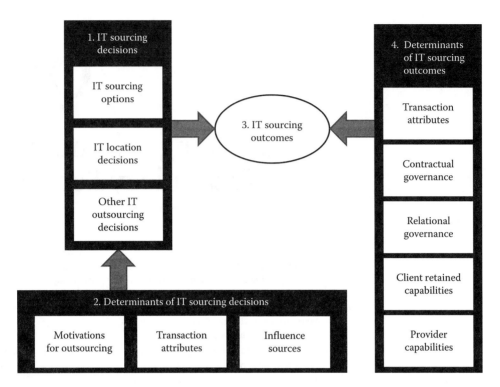

FIGURE 60.1   IT sourcing decisions and outcomes.

still gaps in knowledge and enduring practitioner challenges. Thus, the area of sourcing IT services remains an important topic for research.

60.2 IT Sourcing Decisions

The most fundamental sourcing decision is the "make-or-buy" decision (Williamson 1975). Should organizations insource or outsource IT services? In reality, the make-or-buy decision is more complex because there are many sourcing options, sourcing locations, and providers to be considered. For over a decade, we have argued that a better question is "How can organizations leverage the external services market for business advantage?" (Lacity and Willcocks 2001). Many IT academics, informed closely by both theory and practice, have good answers to this question. In this section, we discuss sourcing options, sourcing locations, and other decisions IT executives consider when sourcing IT services. These choices have different benefits and risks and may require unique practices to ensure positive outcomes.

60.2.1 Sourcing Options

IT executives have numerous sourcing options to consider, including insourcing, staff augmentation, management consulting, shared services, traditional outsourcing, cloud computing, and joint ventures/partnerships (see Table 60.1). The IT sourcing options are not mutually exclusive; many IT organizations use many or even all of these options. Many internal shared services organizations, for example, rely heavily on outsourcing providers. Many client/provider relationships include both staff augmentation (price per full-time equivalent) and fee-for-service pricing components (Fersht et al. 2011). Many joint ventures also have fee-for-service components for the services the venture delivers to the client investor. IT executives provide a portfolio of IT services that, in turn, necessitate a portfolio of sourcing options.

TABLE 60.1 Sourcing Options

Sourcing Options	Description
In-house provision; insourcing	A sourcing option in which the organization owns the IT assets and employs its own IT staff (Lacity and Hirschheim 1995; Hirschheim and Lacity 2000)
Staff augmentation; contract labor	A sourcing option in which an organization buys in low- to mid-level labor to supplement in-house capabilities; the client manages the people, usually at the client site (Ang and Slaughter 2001)
Management consulting	A sourcing option in which an organization buys in high-level expertise to supplement in-house capabilities (Lacity et al. 2003)
Shared services	A sourcing option in which an organization centralizes and standardizes IT delivery (Lacity and Fox 2008; Lacity and Willcocks 2012a)
Traditional outsourcing; fee-for-service outsourcing; exchange-based outsourcing	A sourcing option in which a client pays a fee to a provider in exchange for the management and delivery of specified IT services. The client is in charge of specifying needs and the provider is in charge of managing the resources to deliver those needs (Lacity and Willcocks 1998)
Cloud computing; netsourcing; application service provision	With this utility model, clients pay a usage-based fee to providers in exchange for services being delivered over a network, typically the Internet (Iyer and Henderson 2010; Willcocks et al. 2010b, 2011b)
Joint ventures; strategic partnerships	A specific type of contract entered into by two or more parties in which each agrees to furnish a part of the capital and labor for a business enterprise (DiRomualdo and Gurbaxani 1998; Lacity et al. 2003, 2004)

Insourcing is still the most commonly pursued sourcing option. The annual Society of Information Management survey of Chief Information Officers (CIOs) conducted by Jerry Luftman typically finds that about 33% of an annual IT budget is for internal staff, compared to 9% for consulting services, 15% for domestic contractors/providers, and 3%–4% for offshore providers (Luftman and Kempaiah 2008; Luftman and Zadeh 2011). Increased use of external staff—both domestic and offshore—is growing and obviously gains more public attention than the quieter, yet larger, insourcing practice. Two scholarly works examined insourcing closely (Lacity and Hirschheim 1995; Hirschheim and Lacity 2000). We studied firms that examined the outsourcing market and decided to insource. In some of these firms, the shear threat of outsourcing empowered IT executives to make sweeping—yet unpopular—changes to reduce IT costs on their own through consolidation, standardization, and rationalization of IT resources. The lesson for practitioners is that the external services market serves as a viable competitor that keeps in-house staff motivated.

With *staff augmentation*, an organization buys in low- to mid-level labor to supplement in-house capabilities. In any Fortune 500 company IT headquarters, up to 40% of the staff may be domestic contract laborers who work alongside IT employees. Organizations use a staff augmentation model to meet fluctuations in demand for IT work, to access scarce technical skills, and to avoid the HR headaches associated with employment. Staff augmentation is one of the most expensive options if IT executives buy labor from urban cities in developed countries. Staff augmentation is one of the least expensive options if firms buy labor offshore or from nonurban (rural) areas (Lacity et al. 2010b). Ang and Slaughter (2001) have one of the best studies comparing IT contractors to permanent employees. In a survey, they found that contract workers have lower levels of loyalty, trustworthiness, obedience, and performance than permanent workers. In interviews, they found that contract workers had lower task variety, identity, significance, autonomy, and feedback relative to permanent workers.

Management consulting is a short-term sourcing option in which an organization hires an advisory firm or consulting firm typically for help with new strategic initiatives, like creating a global standard enterprise resource planning (ERP) platform or developing a sourcing strategy. This sourcing option brings in external energy, signals clear commitment to the strategic initiative, and reduces political resistance. Although management consultancy has several major risks, the two most significant ones are potential cost escalation and lack of sustainability because the provider has no long-term commitment.

The result can be a lessened sense of accountability and a lack of alignment between the parties. Furthermore, expertise and knowledge may leave when the consultant leaves (Lacity et al. 2003).

Shared service is defined by Accenture as "the consolidation of support functions (such as human resources, finance, IT, and procurement) from several departments into a standalone organizational entity whose only mission is to provide services as efficiently and effectively as possible" (Accenture 2005). Shared service is one of the most important IT management trends and was listed as one of the seven habits of effective CIOs (Andriole 2007). The recent downturn in the economy has intensified the pressures for organizations in both the public and private sectors to reduce costs, shed headcount, and to do more and more with fewer resources. Shared services are seen as a powerful practice for relieving these pressures. Shared services offer the promises of lower costs, tighter controls, improved service levels, and scalability. Studies have shown, however, that not all organizations achieve the benefits they expect from shared services; many shared service initiatives take years to implement and result in meager cost savings (Lacity and Willcocks 2012a). Among all the advanced practices for successfully implementing shared services, change management may be the most important and the most lacking practice. Based on our case studies, we found that creating shared services requires a coordinated integration of four change programs: business process redesign (BPR), organizational redesign, sourcing redesign, and technology enablement (Lacity and Fox 2008; Lacity and Willcocks 2012a).

Traditional outsourcing is the most common form of outsourcing and has grown each year it has been tracked, with the exception of a slight dip in 2009. Gartner estimates that the global ITO market in 2012 is worth $309 billion and Booz Allen and Hamilton estimate that ITO is growing at 10% per year.* With this option, a client pays a fee to a provider in exchange for the management and delivery of specified IT services. The client is in charge of specifying needs, and the provider is in charge of managing the resources to deliver those needs. Much of the academic literature on ITO falls under this category, and there have been hundreds of published studies, which we reviewed in Lacity et al. (2010a). The fee-for-service model works best when (1) clients have a good understanding of their baseline service levels, costs, and volumes, (2) requirements are stable, (3) a capable provider is selected, and (4) the contract is designed to benefit both parties. However, fee-for-service contracts are notoriously inflexible, and changes are difficult to handle because client/provider incentives are not aligned. Change will usually harm one of the parties. In successful relationships, both parties resolve disputes caused by change by focusing on what is fair, rather than by what is designated in the contract (Lacity and Willcocks 2012b).

Cloud computing is the idea of buying IT services like a utility. The concept of buying IT as a utility service is not new—time sharing was indeed the first outsourcing model back in the 1960s. During the dot com boom, application service provision (ASP) was a business model in which providers hosted and rented standard applications to clients over the Internet. ASP was one way small organizations could access expensive software—like enterprise resource planning software by SAP or Oracle—while avoiding high infrastructure costs, support costs, or hefty software license fees. ASP burst when the dot com bubble burst. Clients liked the idea of renting rather than owning IT resources, but providers had difficulty generating revenues because the value of client contracts was too small, the duration of client contracts too short, the marketing costs to educate clients about ASP too high, the margins from reselling propriety software too thin, and the transaction costs of serving so many needy clients too high (Kern et al. 2002). Today, cloud computing is the reincarnated, supercharged version of ASP. Cloud computing is one of the fastest growing markets in the ITO/business process outsourcing (BPO) space. According to Gartner, cloud computing was a $68.3 billion industry in 2010. IDC and Harris and Nunn predict that cloud-based service revenues will grow globally between $44 billion and $60 billion in 2013. Gartner predicts it will reach $148.8 billion by 2014.

---

* Size of market figures by various consulting firms is in the public domain. Gartner data is also reported in October 2009 issue of *Accenture Outlook: The Journal of High Performance Business*; Booz, Allen, and Hamilton found in Outsourcing for Virtuosos, Webinar with Vinay Couto and Ashok Divaran; Booz, Allen, and Hamilton, http://www.strategy-business.com/media/file/Outsourcing_for_Virtuosos-webinar.pdf

Despite the tremendous public attention to cloud computing, it cannot achieve yet the plug-and-play simplicity of electricity (Willcocks et al. 2011b).

Joint ventures/strategic partnerships are not very common, but the scale and scope of these deals garnered significant media attention, particularly during the 1990s (DiRomualdo and Gurbaxani 1998). In joint ventures, the provider investor intends to sell the client's assets or excess capacity to third parties and share the revenues with the client investor. Examples of these deals included Swiss Bank and Perot Systems, Commonwealth Bank and EDS, Xerox and EDS, and Delta Airlines and AT&T. These deals did not work as planned. The providers had their hands full just servicing the client investors' operational needs. In addition, clients frequently oversold the value and portability of their assets. The IT deals we studied all reverted to fee-for-service relationships or were completely terminated. We found more success with joint ventures/strategic partnerships in other back office services including procurement, human resource management, claims, and policy administration (Lacity et al. 2003, 2004).

60.2.2 Sourcing Locations

IT executives have to make location decisions—where will IT staff be located? Options include domestic, offshore, nearshore, rural, or global (see Table 60.2). Location strategies may or may not involve outsourcing. For example, offshore may involve offshore outsourcing, joint ventures, strategic partnerships, or captive centers (insourcing). Location decisions are often based on a country's business, financial, and human resource attractiveness. A *country's business attractiveness* is the degree to which a country is attractive to organizations because of favorable business environmental factors such as economic stability, political stability, cultural compatibility, infrastructure quality, and security of intellectual property (IP) (e.g., Doh et al. 2009; Malos 2009). A *country's financial attractiveness* is the degree to which a country is attractive because of favorable financial factors such as labor costs, taxes, regulatory, and other costs (e.g., Doh et al. 2009; Malos 2009). A *country's human resource attractiveness* is the degree to which a country is attractive because of favorable people skills and availability factors such as size of labor pool, education, language skills, experience, and attrition rates (e.g., Mehta et al. 2006; Malos 2009).

A country's attractiveness can change rapidly, as happened with the political upheavals in Egypt in 2011, which halted international investment in Egypt's ITO services export market. For this reason, we have advised that clients base location decisions on the organization's strategic objectives and overall commitment to certain destinations (Lacity and Rottman 2008). For example, one aerospace company selected Malaysia as their IT offshore destination because they hoped to sell planes in that country. The Malaysian government requires that some of the manufacturing be done in Malaysia, and the IT presence would certainly help to meet that requirement. Another hardware company selected China because they hoped to sell computers there. Other participants selected offshore locations where they have existing manufacturing or R&D facilities. The existing facilities serve as a launch pad, with current employees serving as guides to the country, providers, and culture (Rottman and Lacity 2006).

TABLE 60.2 Sourcing Location Decisions

Sourcing Locations	Description
Domestic	IT service delivery center located in the same country as the client's business users
Offshore	IT service delivery center located on a different continent than the client's business users (Lacity and Rottman 2008; Oshri 2011)
Nearshore	IT service delivery center located in a nearby country (such as a U.S. client being serviced from a Canadian delivery center) (Carmel and Abbott 2007)
Rural	IT service delivery center located in a rural community (Lacity et al. 2011a)
Global	IT service delivery centers located in several countries (Willcocks and Lacity 2006)

Domestic location of IT staff is pursued when organizations value close proximity between business users and IT staff. Close physical proximity is associated with better service quality, faster response time, better domain understanding, easier communications, and lower transaction costs (Lacity and Rottman 2008). However, domestic staff—whether employees, contract labor, or domestic outsourcing—is the most expensive location option in high-cost locations like urban U.S. or U.K. cities.

Offshore location of IT staff is pursued when organizations seek lower costs, sunrise-to-sunrise services, access to IT talent, and/or geographical risk mitigation. Western-based organizations frequently select India, the Philippines, Eastern Europe, or China as offshore destinations, particularly for the lower cost and talent availability. The offshore outsourcing market has been estimated to be about $80 billion to $100 billion market, with India representing the largest share of IT and business service exports. Despite noise suggesting that India will lose its edge due to rising wages, the Everest Group (2011) predicts that India will maintain its lead for 12 years under a pessimistic scenario and for 23 years under an optimistic scenario. Preferred venues change based on a mix of location attractiveness factors, and a recent development has seen offshore venues such as India also re-outsourcing work to countries like Egypt and China. Research has found that locating staff offshore can deliver on many of its promised benefits (Lacity and Rottman 2008), but researchers have also found that offshore locations pose additional challenges when compared to domestic locations. For example, captive centers or offshore outsourcing are more challenging because of time zone differences (Carmel 2006), increased efforts in knowledge coordination (Kanawattanachai and Yoo 2007; Kotlarsky et al. 2008) and boundary spanning (Levina and Vaast 2008; Mahnke et al. 2008), the need for more controls (Choudhury and Sabherwal 2003; Lacity and Rottman 2008), cultural differences (Carmel and Agarwal 2001; Krishna et al. 2004; Carmel and Tjia 2005), defining requirements more rigorously (Gopal et al. 2002), and difficulties in managing dispersed teams (O'Leary and Cummings 2007; Oshri et al. 2007). Some of these issues are so difficult to manage that practitioners are turning to nearshore alternatives (Carmel and Abbott 2007).

Nearshore location of IT staff is pursued when the organization expects to benefit from one or more of the following constructs of proximity: economic, geographic, temporal (time zone), cultural, linguistic, political, and historical linkages (Carmel and Abbott 2007). Nearshore locations are selected primarily because of lower costs, but may be preferred to "offshore" because the proximity requires less time and money to travel and provides time zone overlap between the business and IT services organizations. For U.S. organizations, the North American Free Trade Agreement (NAFTA) facilitates nearshoring because it is easier to obtain visas from NAFTA partners than it is to obtain visas for India-based staff (Carmel and Abbott 2007).

Rural location of IT staff is an emerging niche trend in several countries, including the United States, India, China, and Israel. We estimated the US ITO "pure-play" rural outsourcing market to be about $200 million in 2011 (Lacity et al. 2011a). The Rockefeller Foundation sizes the global impact sourcing market—which is primarily a rural market—at $4.6 billion in 2010 (Lacity and Willcocks 2012a). The main appeals of rural locations are lower wages and higher retention rates because few or no competitors exist to poach talent. Within the United States, rural locations can reduce IT costs by up to half compared to urban locations like New York City (Lacity et al. 2010b). In India, despite the global economic recession, global demand for Indian ITO and BPO services is still very strong and consequently Indian providers are still experiencing 14%–22% turnover in urban areas (The Everest Group 2011). By building delivery centers in Tier 3 cities, Indian organizations achieve lower costs and attrition rates. Chinese providers also cited lower costs, but not necessarily lower attrition rates, by locating in Tier 3 cities. Specifically, they reported that labor costs are up to 50% lower and real estate costs are 70%–90% lower in Tier 3 cities compared to Tier 1 cities (Lacity et al. 2011a). The major downsides of rural locations are scalability and workforce availability.

Global location, having IT service delivery centers located in several countries, is the norm for large, international companies. Large global companies locate IT services globally using a combination of sourcing options discussed earlier, including captive centers, fee-for-service outsourcing, and joint partnerships. One U.S. global financial services company we studied has various captive centers, joint ventures, and

TABLE 60.3 Other Sourcing Decisions

Sourcing Decisions	Description
Multi-sourcing vs. sole sourcing	The decision pertaining to the number of providers to engage (Lacity and Willcocks 2001, 2012a; Willcocks et al. 2010a)
Next Generation	Before ITO contracts expire, IT executives decide whether to keep the current provider, switch providers, or backsource IT services (Whitten and Leidner 2006; Willcocks et al. 2011a)

fee-for-service relationships with 14 Indian providers. This network of providers enabled the company to quickly adapt to the immense surge in mortgage applications during the refinancing boom. As the refinancing boom burst, the company was able to immediately scale back resources—all without affecting their domestic IT headcount (Lacity and Rottman 2008). Global outsourcing providers like Accenture and IBM can provide vast geographic coverage for their clients. For example, our current research on high-performance BPO is studying how Accenture provides Microsoft with coverage in 94 countries and 37 languages.

60.2.3 Other Outsourcing Decisions

If IT executives decide to outsource, there are two additional decisions that follow: How many providers? What do we do when the contract expires? As an existing outsourcing contract approaches maturity, organizations must decide what to do next: renegotiate with the incumbent provider, switch providers, bring the service back in-house, or terminate the service altogether (see Table 60.3).

Multisourcing or bundled services. When outsourcing, organizations also have to decide about the number of providers to engage. Multisourcing has the advantages of choosing best-of-breed providers, mitigating the risks of relying too much on one provider, and helping clients adapt in changing environments. Multisourcing has several disadvantages, including increased transaction costs as organizations manage more providers, interdependencies, and interfaces (Lacity and Willcocks 2001). The major advantages of bundled services from a single provider include simplified procurement, simplified governance, fewer transaction costs, and economies of scale and scope. But bundled services increase switching costs and the risks of relying on one provider (Willcocks et al. 2010a; Lacity and Willcocks 2012a).

Next generation. Most ITO clients are in their second, third, and even fourth generation of ITO relationships. As contracts begin to mature, organizations need to decide what to do next. Should they renegotiate a contract with the incumbent provider? Switch providers? Backsource by bringing the IT function back in-house? In one of our early surveys of U.S. and U.K. CIOs, we found that 32% of client organizations had canceled one or more outsourcing contracts. Of these half-switched providers, one-third brought the IT service back in-house and 11% ended up renegotiating the contract with the incumbent provider due to prohibitive switching costs (Lacity and Willcocks 2001). More recently, Willcocks et al. (2011a) surveyed Australian CIOs and found that 65% renegotiated with the incumbent, 30% switched providers, and 5% backsourced. In Willcocks et al. (2011a), the authors prescribe how CIOs prepare for the next generation based on hundreds of case studies.

60.3 Determinants of IT Sourcing Decisions

In addition to studying the types of sourcing decisions organizations make, academics have thoroughly studied the determinants of sourcing decisions, including the motives, transaction attributes, and influence sources (see Figure 60.1). These determinants are introduced later (see Lacity et al. 2010a for deeper coverage).

60.3.1 Motives for Sourcing Decisions

Researchers have studied at least 27 motives or reasons driving sourcing decisions. The most frequently found motives are listed in Table 60.4. Research has consistently found that sourcing decisions are

TABLE 60.4   Top Motives for Sourcing Decisions

Sourcing Motives	Description
Cost Reduction	An organization's need or desire to reduce or control Information Systems (IS) costs (e.g., Barthélemy and Geyer 2004)
Focus on Core Capabilities	An organization's desire or need to focus on its core capabilities (e.g., Lacity et al. 1994; Linder 2004)
Access to Skills	An organization's desire or need to access provider skills/expertise (e.g., Lacity et al. 1994; Clark et al. 1995)
Business/Process Improvements	An organization's desire or need to help improve an organization's business, processes, or capabilities (DiRomualdo and Gurbaxani 1998)
Technical Improvements	An organization's desire or need to gain access to leading edge technology that is available through the providers but may not be available in-house (e.g., Sobol and Apte 1995)
Political Reasons	A client stakeholder's desire or need to promote personal agendas such as eliminating a burdensome function, enhancing their career, or maximizing personal financial benefits (e.g., Lacity et al. 1994; Hall and Liedtka 2005; Chakrabarty and Whitten 2011)
Concern for Security/IP	An organization's concerns about security of information, transborder data flow issues, and protection of intellectual property (e.g., Khalfan 2004; Walden 2005; Rao et al. 2006)
Fear of Losing Control	An organization's concerns about losing control over IT (e.g., Patane and Jurison 1994; Collins and Millen 1995)

primarily driven by the need to reduce overall IT costs, usually by at least 10%–15% (IAOP 2010; Lacity et al. 2010a). Research has also found that the greater the fear of losing control of the service, concern for security, or fear of losing IP, the more likely an organization chose insourcing. Besides cost savings, research has found that organizations expect ITO to deliver a number of additional business benefits, including one or more of the following: ability to redirect in-house staff on more strategic activities, access to scarce skill sets, service quality improvements, business process improvements, and technology improvements (Lacity et al. 2010a).

60.3.2 Transaction Attributes

Researchers have examined at least 14 transaction attributes as potential determinants of IT sourcing decisions. Four attributes were commonly found to drive IT sourcing decisions: uncertainty, critical role of IT, transaction costs, and business risks (see Table 60.5). Higher values for any of these transaction attributes were associated with higher frequencies of insourcing (Lacity et al. 2010a).

60.3.3 Influence Sources

Researchers have studied the extent to which mimetic, normative, and coercive influences determine sourcing decisions within organizations. Among these, only mimetic influences were repeatedly examined and found to significantly influence sourcing decisions. Mimetic influences are influences that

TABLE 60.5   Transaction Attributes Affecting Sourcing Decisions

Transaction Attributes	Description
Uncertainty	The degree of unpredictability or volatility of future states as it relates to the definition of IT requirements, emerging technologies, and/or environmental factors (Williamson 1991a,b; e.g., Poppo and Zenger 2002; Aubert et al. 2004)
Critical role of IT	The degree to which an organization views IT as a critical enabler of business success (e.g., Teng et al. 1995; Saunders et al. 1997; Straub et al. 2008)
Transaction costs	The effort, time, and costs incurred in searching, creating, negotiating, monitoring, and enforcing a service contract between buyers and providers (Williamson 1991b; Ang and Straub 1998)
Business risk	The probability that an action will adversely affect an organization (Lacity et al. 2010a)

arise from the perception that peer organizations are more successful and by modeling themselves based on peer organizations, the mimicking organization aims to achieve similar results (e.g., DiMaggio and Powell 1991; Ang and Cummings 1997). In the ITO context, researchers found that client organizations were influenced to outsource IT based on peer institutions that successfully outsourced IT (Lacity et al. 2010a).

Researchers have also examined the determinants of next-generation decisions. Whitten and Leidner (2006) found that both economic and relationship constructs are important determinants of second-generation outsourcing decisions. Clients renewing contracts report high levels of product quality, service quality, relationship quality, and switching costs. Clients that switched providers report high product and service quality but low relationship quality and switching costs. Clients that brought back IT in-house report low levels on all four variables.

60.4 IT Sourcing Outcomes

Many researchers have examined the outcomes of ITO and the determinants of ITO outcomes. In contrast, we are aware of only two studies that examined the effects of insourcing decisions—the decision to source IT services in-house after examining outsourcing options (Lacity and Hirschheim 1995; Hirschheim and Lacity 2000). We conducted 14 in-depth U.S. client case studies on organizations that examined but rejected outsourcing. We wanted to know: Did the internal IT organization change as a consequence of the insourcing decision? Among the 14 cases, two companies backsourced IT services after terminating outsourcing contracts. In both backsourcing cases, IT service levels improved, but IT costs rose in one backsourcing case and IT costs remained the same in the other backsourcing case. In three client organizations, no changes in IT performance resulted from the insourcing decision. However, in nine client companies, IT performance significantly improved. IT costs were reduced between 20% and 54% by replicating provider practices such as centralization, standardization, and rationalization. Business users accepted the drastic changes implemented to achieve such cost savings because they realized that if the internal IT organization did not implement these changes, an outsourcing provider would. The message was: the devil you know is better than the devil you do not know.

In contrast to the small academic literature on insourcing outcomes, hundreds of academic studies have examined outsourcing outcomes. Most of this research is based on large-sample surveys of outsourcing clients or in-depth case studies at client sites. Across these studies, researchers have used many different types of measures to examine the outcomes of outsourcing. The most frequently used measures include outcomes that capture a client's general perceptions of the success or level of satisfaction with outsourcing in general or with offshore outsourcing in particular, perceptions of the quality of relationships, and the effects of outsourcing on a client organization's business performance, such as improvements in stock price performance, return on assets, expenses, or profits after outsourcing (Lacity and Willcocks 2012b). By aggregating findings from both qualitative and quantitative studies in Lacity et al. (2010a), we reported the following: *IT outsourcing decisions resulted in positive outcomes in 63% of the findings, negative outcomes in 22% of the findings, and no changes in performance as a consequence of outsourcing were reported in 15% of the findings.* Many people might consider this statistic quite disappointing because only 63% of ITO engagements were considered positive by clients. The good news is that scholars have studied the determinants of ITO outcomes. We thus have a strong understanding of the practices that differentiate positive from negative outcomes.

60.5 Determinants of ITO Outcomes

Based on our review of the empirical ITO research, we extracted what we call robust practices, practices that have been academically tested and proven to be effective. These practices are grouped into five categories: transaction attributes, contractual governance, relational governance, client-retained capabilities, and provider capabilities (see Figure 60.1).

TABLE 60.6   Transaction Attributes Affecting Outsourcing Outcomes

Transaction Attributes	Description
Uncertainty	The degree of unpredictability or volatility of future states as it relates to the definition of IT requirements, emerging technologies, and/or environmental factors (e.g., Williamson 1991b; Poppo and Zenger 2002; Aubert et al. 2004)
Measurement difficulty	The degree of difficulty in measuring performance of exchange partners in circumstances of joint effort, soft outcomes, and/or ambiguous links between effort and performance (e.g., Eisenhardt 1989)

60.5.1 Transaction Attributes

ITO researchers have studied 17 transaction attributes to determine whether outsourcing certain types of transactions was more or less likely to result in positive outcomes (Lacity et al. 2011a). For example, ITO researchers have asked, "Are large or small transactions likely to result in positive outcomes?" and "How does task complexity affect outcomes?" Overall, only two transaction attributes were repeatedly tested and found to affect ITO outcomes: uncertainty and measurement difficulty. Both of these transaction attributes are negatively and significantly related to ITO outcomes. These two transaction attributes are derived from Transaction Cost Economics (Coase 1937; Williamson 1975, 1991a,b). Of the 15 times uncertainty was studied, 12 times (80%) researchers found that uncertainty adversely affected ITO outcomes (e.g., Barthélemy 2001). Measurement difficulty was examined seven times and was found to adversely affect ITO outcomes six times (86%) (e.g., Poppo and Zenger 2002; Table 60.6).

60.5.2 Contractual Governance

Contractual governance is the formal, written rules that govern client–provider relationships. In scholarly works we reviewed, contractual governance was operationalized most frequently as degree of contract detail (e.g., the types of clauses, number of service-level agreements), contract duration, contract value, and contract type (e.g., fixed price, time, and materials). *Substantial evidence finds that* (1) *more detailed contracts,* (2) *shorter-term contracts, and* (3) *higher-dollar-valued contracts are significantly associated with positive outsourcing outcomes* (Lacity et al. 2010a,b). Detailed contracts that defined the scope of services, prices, service levels, and responsibilities of both parties and prescribed how parties would adapt to changes in character, volume, or market best practices had better outsourcing outcomes than contracts with fewer details. Shorter-term ITO contracts in the 3–5 year range experienced successful ITO outcomes more frequently than contracts with greater than 5 year duration. Higher-valued contracts perform better than lesser-valued contracts because the transaction costs associated with outsourcing are spread over a greater volume of work.

Contract type is a term denoting different forms of contracts used in outsourcing. Examples from the ITO literature include customized contracts, fixed-price contracts, time and materials contracts, fee-for-service contracts, and partnership-based contracts. ITO researchers used different categories of contracts across studies. For example, Lacity and Willcocks (1998) used three categories and found that clients who signed fee-for-service contracts had higher rates of success than clients who signed strategic alliances or loose contracts. In Lacity and Rottman (2008), we found that different contracts incent providers differently. In the offshore context, clients found that time and materials contracts produced better work, but at considerable expense to the client. Providers were not pressured to take shortcuts in order to protect their profit margins as they were with fixed-price deals. Also, some providers placed new employees on client accounts because the client subsidizes the employee's learning curve with time and materials contracts. Ironically, provider employees who are unproductive take more hours to complete tasks, which generates greater provider revenues. In Gopal et al. (2003), contract type was a categorical variable (either fixed price or time and materials) and the authors found that requirements uncertainty

TABLE 60.7 Contractual Governance

Contractual Governance	Description
Contract detail	The number or degree of detailed clauses in the outsourcing contract, such as clauses that specify prices, service levels, benchmarking, warranties, and penalties for nonperformance (e.g., Pinnington and Woolcock 1995; Poppo and Zenger 2002)
Contract duration	The duration of the contract in terms of time (e.g., Lacity and Willcocks 1998)
Contract value	The size of the outsourcing contract usually measured as the total value of the contract in monetary terms (e.g., Oh et al. 2006; Rottman and Lacity 2008)
Contract type	A term denoting different forms of contracts used in outsourcing. Examples include customized, fixed-price, time and materials, fee-for-service, and partnership-based contracts (e.g., McFarlan and Nolan 1995; Poppo and Zenger 2002; Ross and Beath 2006)

was associated with time and materials contracts. They also found that supplier profits were higher for time and materials contracts than for fixed-price contracts (Table 60.7).

60.5.3 Relational Governance

Relational governance comprises the informal rules that manage client–provider relationships. In scholarly works we reviewed, relational governance was operationalized most frequently as effective knowledge sharing, communication, trust, and viewing the provider as a partner (see Table 60.8). In 94% of the findings, the research showed that higher levels of relational governance were associated with higher levels of outsourcing success (Lacity et al. 2010a,b). In some ways, the findings are trivial. Few people would argue that withholding knowledge, closed communications, or distrusting providers would lead to better outsourcing relationships. A more interesting research finding is that contractual governance and relational governance serve as complements, in that both need to be strong to produce positive outsourcing outcomes (Saunders et al. 1997; Sabherwal 1999; Poppo and Zenger 2002; Wűllenweber et al. 2008; Goo et al. 2009). In general, contractual governance and relational governance are not substitutes in that a poorly crafted contract cannot be overcome with friendly, communicative, and trusting account managers. Poor contracts, we have found, can make for poor relationships (Lacity et al. 2010a,b; Lacity and Willcocks 2012a).

60.5.4 Client-Retained Capabilities

Research has found that clients must learn to manage differently after outsourcing in order to achieve expected benefits (see Table 60.9). Clients must become good at managing providers by shifting their capabilities from managing resources and processes to managing inputs and outputs. This is not an easy transition for many clients. The supplier management capability was often found to be lacking in client organizations and seen as a major reason to explain negative outsourcing outcomes. Clients also need a strong contract negotiation capability, which is frequently supplemented with the aid of advisory firms.

TABLE 60.8 Relational Governance

Contractual Governance	Description
Effective knowledge sharing	The degree to which clients and suppliers are successful in sharing and transferring knowledge (e.g., Lee 2001; Murray et al. 2009)
Communication	The degree to which parties are willing to openly discuss their expectations, directions for the future, their capabilities, and/or their strengths and weaknesses (e.g., Klepper 1995)
Trust	The confidence in the other party's benevolence (e.g., Dibbern et al. 2008)
Partnership view	A client organization's consideration of suppliers as trusted partners rather than as opportunistic vendors (e.g., Saunders et al. 1997; Kishore et al. 2003)

TABLE 60.9   Client-Retained Capabilities

Capability	Description
Supplier management capability	The extent to which a client organization is able to effectively manage outsourcing providers (e.g., Feeny and Willcocks 1998; Willcocks et al. 2007)
Contract negotiation capability	The extent to which a client organization is able to effectively bid, select, and negotiate effective contracts with providers (e.g., Feeny and Willcocks 1998)
IS technical/methodological capability	A client organization's level of maturity in terms of technical- or process-related standards including the Capability Maturity Model (CMM), Capability Maturity Model Integrated (CMMI), and the Information Technology Infrastructure Library (ITIL) and best practices such as component reuse (e.g., Davenport 2005; Rottman and Lacity 2006; Kotlarsky et al. 2007)
Cultural distance management capability	The extent to which client and provider organizations understand, accept, and adapt to cultural differences (e.g., Winkler et al. 2008)
Risk management capability	A client organization's practice of identifying, rating, and mitigating potential risks associated with outsourcing (e.g., Smith and McKeen 2004)

Technical and methodological capability is another important client-retained capability. Technical and methodological capability is an operational capability needed by both parties in order to coordinate work effectively. Clients must learn to understand, accept, and adapt to cultural differences between themselves and their providers (cultural distance management capability). A client also needs to be able to identify, rate, and mitigate potential risks associated with outsourcing (risk management capability). Other client capabilities have also been identified as affecting outsourcing decisions and outcomes: absorptive capacity, client outsourcing readiness, change management capability, human resource management capability, and transition management capability (Lacity et al. 2010a). Clearly outsourcing is not about abdicating management responsibility, but about managing differently.

60.5.5 Provider Capabilities

Which provider capabilities contribute to positive outsourcing outcomes? The three most frequently studied and most important provider firm capabilities were human resource management capability, technical and methodological capabilities, and domain understanding (see Table 60.10). A provider's ability to identify, acquire, develop, and deploy human resources to achieve both provider's and client's organizational objectives was found to positively and significantly affect client outcomes 95% of the time it was examined. Clients often engage providers because of their superior human resources in terms of both number and quality of staff. The provider's technical and methodological capability was

TABLE 60.10   Provider Capabilities

Capability	Description
IS human resource management capability	A provider organization's ability to identify, acquire, develop, and deploy human resources to achieve both provider's and client's organizational objectives (e.g., Levina and Ross 2003)
IS technical/methodological capability	A provider organization's level of maturity in terms of technical- or process-related standards including the CMM, CMMI, and ITIL and best practices such as component reuse (e.g., Levina and Ross 2003)
Domain understanding	The extent to which a provider has prior experience and/or understanding of the client organization's business and technical contexts, processes, practices, and requirements (e.g., Clark et al. 1995; Gopal et al. 2002)

the second most frequently studied capability, and it was found to affect outcomes positively. Domain understanding is the extent to which a provider has prior experience and/or understanding of the client organization's business and technical contexts, processes, practices, and requirements. Other provider capabilities were also found to be important: client management capability, managing client expectations, supplier employee performance, risk management capability, security, privacy and confidentiality capability, supplier's core competencies, absorptive capacity, environmental capability, and corporate social responsibility capability (Lacity et al. 2010a). Providers are unlikely to excel in all of these areas, but better capabilities lead to better outcomes.

60.6 Research Opportunities

Despite the maturity of IT sourcing research, academics still have opportunities to advance the theory and practice of IT sourcing. In Lacity et al. (2010a), we identified nine gaps in IT sourcing knowledge. We called for more studies on (1) strategic sourcing decisions, (2) strategic sourcing outcomes, (3) dynamic interactions between client and provider firm capabilities, (4) environmental influences on sourcing decisions and outcomes, (5) configurational and portfolio approaches, (6) alternative locations, (7) emerging trends and models like cloud computing, (8) informing reference discipline theories, and (9) developing endogenous ITO theories. Pertaining to the last two gaps, ITO studies are highly informed by reference theory disciplines, particularly theories from economics (e.g., transaction cost economics, agency theory), management (e.g., resource base view, institutionalism), and sociology (e.g., social exchange theory, social capital theory). Many of these theories fail to adequately explain ITO phenomena—a debate that has been thoroughly discussed in a number of sources, including Lacity and Willcocks (2009), Lacity et al. (2011b), and Karimi Alaghehband et al. (2011).

ITO researchers have found so many anomalies that are counter to TCE logic in the ITO context, that we believe it is more fruitful to develop a theory specific to ITO. We analyzed the TCE anomalies in the ITO context and categorized them by research methods explanations, boundary conditions explanations, TCE assumption violations, and alternative theory explanations. With *research method explanations*, ITO researchers do not assume that their data provide evidence counter to TCE logic but instead attribute lack of empirical support of TCE to measurement problems, or to some TCE effects overpowering other TCE effects. *Boundary condition explanations* attribute lack of empirical support of TCE to the distinctive context of ITO, such as the distinctive nature of IT, the distinctive research setting (e.g., such as public sector IT), or the distinctive attributes of the data collected. *TCE assumption violation explanations* argue that TCE's explicit or implicit assumptions are unsupported. TCE is based on two explicit behavioral assumptions: bounded rationality and opportunism (Williamson 1991a) and ITO authors found evidence that TCE behavioral assumptions were violated in ITO context (Lacity and Willcocks 1995; Tiwana and Bush 2007; Dibbern et al. 2008). Finally, *alternate theory explanations* suggest that other theories are more powerful in explaining ITO. Based on so many anomalies in the ITO empirical tests of TCE, we argued that we are asking too much of TCE—the ITO phenomenon is more complex than can be accommodated by one decision-making theory.

In Lacity et al. (2011b), we further developed the idea of an endogenous ITO theory by arguing that such a theory should be based on three assumptions. (1) We assume that stakeholder alignment must be actively and aggressively designed anew in the face of ITO. (2) We assume that managerial practices such as standardization, centralization, and tight controls contribute more to IT costs than economies of scale. (3) We assume that history matters; merely using the transaction as the unit of analysis overlooks the broader historical context that sheds significant understanding on ITO decisions and outcomes. We found that an organization's prior history with delivering IT, with prior ITO experiences, and client capabilities are all vital to understanding the ITO decisions an organization makes today. Thus, there are many opportunities to advance the theory of IT sourcing.

Researchers also have opportunities to contribute to practice. We describe three enduring practitioner ITO challenges: (1) adaptability, (2) measurement, and (3) innovation. Each of these areas needs rigorous research.

60.6.1 Challenge 1: Adaptability

Adapting to change is a pervasive outsourcing challenge. Many participants from our recent research continue to comment that their contracts were out of date even before the ink was dry (Lacity and Willcocks 2012b). Change comes from many sources:

- The client's business process and IT requirements change.
- The client's volume of services fluctuates unpredictably.
- The market price changes radically as labor costs rise, technology costs decline, or market prices alter with changes in demand, currency value, or political stability.
- The market's best-in-breed service performance improves.
- The political support for outsourcing within client organizations wane with senior executive turnover.

Partners try to design flexible ITO contracts to adapt to such changes, particularly during a long-term relationship. Standard in contracts are clauses for volume fluctuations for when additional resource charges (ARCs) or reduced resource charges (ROOKs) apply, force majeure clauses, change of character clauses, and external benchmarking to reset prices or service levels. To one extent, these contract clauses are effective. As noted earlier, greater contract detail was associated with better outsourcing outcomes in the empirical ITO review. In our case study research, we continue to hear laments that big changes cause severe consequences to the economic viability of the disadvantaged partner. In practice, significant change harms either the client or provider, which can result in a dispute (Lacity and Willcocks 2012b).

60.6.2 Challenge 2: Measurement

The maxim "you can't manage what you can't measure" has been long embraced by ITO practitioners. In contracts, there is no shortage of measures, but the enduring challenge is to measure the right things that matter to clients and to incent providers to deliver what matters most to the client. In our research, we find that providers—quite understandably—will only agree to measure processes in which they have complete control. These measures, however, are typically a small piece of the end-to-end process. Consider, for example, the process of IT user support. Clients want providers to identify root causes and propose fixes, yet the measures may incent the providers to replicate the same easy fix for each user that calls rather than fix it once for the entire user community. In our recent BPO research, high performing relationships measure end-to-end processes. Parties first diagnose the root causes for service issues, propose solutions to improve the end-to-end process, and only then determine what it means for the commercial relationship.

60.6.3 Challenge 3: Innovation

Even satisfied ITO clients invariably ask, "Yes, the suppliers are delivering on the contract, but where is the innovation?" At the beginning of the deal, the usual sticking point is who will pay for innovation. Sometimes, clients establish an innovation fund against which approved client/provider proposals can draw. However, if incarcerated inside a traditional cost-service-focused contract, such an initiative rarely has the size or priority to make inroads. Providers are reluctant to spend time and expert resources on an ancillary part of the contract, especially when clients themselves do not take the positive action required from them to work together with the provider to achieve the more business impactful innovations beyond minor IT operational changes (Lacity and Willcocks 2011).

60.7 Conclusion

We have consistent evidence as to what motivates IT sourcing decisions. IT executives want to reduce costs, to focus on core capabilities, and to inject IT organizations with provider resources such as skills, expertise, and superior technology to improve client IT performance. IT executives are more likely to insource IT activities that have high levels of uncertainty, criticality, business risks, and transaction costs. On the determinants of ITO outcomes, overall we know that both contractual and relational governance are important, that both clients and providers need strong complementary capabilities to make relationships successful, and that certain types of transactions and decisions affect ITO outcomes.

Outsourcing has become an almost routine part of management, but 20 years of research establishes the common denominator that, for management and operational staff, outsourcing requires detailed oversight. Outsourcing itself is not a panacea to a client's IT challenges, but represents a different way of managing IT services. Much depends on experiential learning and sheer hard work by clients and providers alike on a daily basis. Our own work on management practice suggests that back office executives must climb a significant learning curve and build key in-house capabilities in order to successfully exploit outsourcing opportunities. They need to accept that outsourcing is not about giving up management but managing in a different way. We also find providers continually having to readdress their capabilities, their market offerings, and competitive forces. In the face of these difficulties, outsourcing will remain a fascinating and growing area for research for many years to come. It also provides a notable area where academic researchers and the distinctive qualities they bring to bear can continue to provide rich insight and guidelines in an emerging, expanding, but still much muddied field of organizational operation (Lacity and Willcocks 2012b).

References

Accenture (2005), Driving high performance in Government: Maximizing the value of public-sector shared services. http://ebookbrowse.com/accenture-driving-high-performance-in-government-maximizing-the-value-of-public-sector-shared-services-pdf-d365359212 (accessed on April 13, 2013).

Andriole, S. (2007), The 7 habits of highly effective technology leaders, *Communications of the ACM*, 50, 67–72.

Ang, S. and Cummings, L. (1997), Strategic response to institutional influences on information systems outsourcing, *Organization Science*, 8(3), 235–256.

Ang, S. and Slaughter, S. (2001), Work outcomes and job design for contract versus permanent information systems professionals on software development teams, *MIS Quarterly*, 25(3), 321–350.

Ang, S. and Straub, D. (1998), Production and transaction economies and IS outsourcing: A study of the U.S. banking industry, *MIS Quarterly*, 22(4), 535–552.

Aubert, B., Rivard, S., and Patry, M. (2004), A transaction cost approach to outsourcing behavior: Some empirical evidence, *Information & Management*, 41, 921–932.

Barthélemy, J. (2001), The hidden costs of IT outsourcing, *Sloan Management Review*, 42(3), 60–69.

Barthélemy, J. and Geyer, D. (2004), The determinants of total IT outsourcing: An empirical investigation of French and German firms, *Journal of Computer Information Systems*, 44(3), 91–98.

Carmel, E. (2006), Building your information systems from the other side of the world: How Infosys manages time zone differences, *MIS Quarterly Executive*, 5(1), 43–53.

Carmel, E. and Abbot, P. (2007), Why nearshore means that distance matters, *Communications of the ACM*, 50(10), 40–46.

Carmel, E. and Agarwal, R. (2001), Tactical approaches for alleviating distance in global software development, *IEEE Software*, March/April, 22–29.

Carmel, E. and Tjia, P. (2005), *Offshoring Information Technology: Sourcing and Outsourcing to a Global Workforce*, Cambridge University Press, Cambridge, U.K.

Chakrabarty, S. and Whitten, D. (2011), The sidelining of top IT executives in the governance of outsourcing: Antecedents, power struggles, and consequences, *IEEE Transactions on Engineering Management*, 58(4), 799–814.

Choudhury, V. and Sabherwal, R. (2003), Portfolios of control in outsourced software development projects, *Information Systems Research*, 14(3), 291–314.

Clark, T. D., Zmud, R., and McCray, G. (1995), The outsourcing of information services: Transforming the nature of business in the information industry, *Journal of Information Technology*, 10(4), 221–237.

Coase, R. H. (1937), The nature of the firm, *Economica*, 4, 386–405.

Collins, J. S. and Millen, R. A. (1995), Information systems outsourcing by large American industrial firms: Choices and impacts, *Information Resources Management Journal*, 8(1), 5–13.

Davenport, T. (2005), The coming commoditization of processes, *Harvard Business Review*, 83(6), 101–108.

Dibbern, J., Winkler, J., and Heinzl, A. (2008), Explaining variations in client extra costs between software projects offshored to India, *MIS Quarterly*, 32(2), 333–366.

DiMaggio, P. and Powell, W. (eds.). (1991), The iron cage revisited: Institutional isomorphism and collective rationality in organizational fields, in *The New Institutionalism in Organizational Analysis*, The University of Chicago Press, Chicago, IL, pp. 63–82.

DiRomualdo, A. and Gurbaxani, V. (1998), Strategic intent for IT outsourcing, *Sloan Management Review*, 39(4), 67–80.

Doh, J., Bunyaratavej, K., and Hahn, E. (2009), Separable but not equal: The location determinants of discrete services offshoring activities, *Journal of International Business Studies*, 40, 926–943.

Eisenhardt, K. (1989), Agency theory: An assessment and review, *The Academy of Management Review*, 14(1), 57–76.

Feeny, D. and Willcocks, L. (1998), Core IS capabilities for exploiting information technology, *Sloan Management Review*, 39(3), 9–21.

Fersht, P., Herrera, E., Robinson, B., Filippone, T., and Willcocks, L. (2011), *The State of Outsourcing in 2011, Horses for Sources and LSE Outsourcing Unit*, London, U.K., May–July, see http://www.hfsresearch.com

Goo, J., Kishore, R., Rao, H. R., and Nam, K. (2009), The role of service level agreements in relational management of information technology outsourcing: An empirical study, *MIS Quarterly*, 33(1), 1–28.

Gopal, A., Mukhopadhyay, T., and Krishnan, M. (2002), The role of software processes and communication in offshore software development, *Communications of the ACM*, 45(4), 193–200.

Gopal, A., Sivaramakrishnan, K., Krishnan, M., and Mukhopadhyay, T. (2003), Contracts in offshore software development: An empirical analysis, *Management Science*, 49(12), 1671–1683.

Hall, J. and Liedtka, S. (2005), Financial performance, CEO compensation, and large-scale information technology outsourcing decisions, *Journal of Management Information Systems*, 22(1), 193–222.

Hirschheim, R. and Lacity, M. (2000), Information technology insourcing: Myths and realities, *Communications of the ACM*, 43(2), 99–108.

IAOP (2010), *Outsourcing Professional Body of Knowledge*, Van Haren Publishing, Zaltbommel, the Netherlands.

Iyer, B. and Henderson, J. (2010), Preparing for the future: Understanding the seven capabilities of cloud computing, *MIS Quarterly Executive*, 9(2), 117–131.

Kanawattanachai, P. and Yoo, Y. (2007), The impact of knowledge coordination on virtual team performance over time, *MIS Quarterly*, 31(4), 783–808.

Karimi-Alaghehband, F., Rivard, S., Wu, S., and Goyette, S. (2011), An assessment of the use of transaction cost theory in information technology outsourcing, *Journal of Strategic Information Systems*, 20(2), 125–138.

Kern, T., Lacity, M., and Willcocks, L. (2002), *Netsourcing: Renting Business Applications and Services over a Network*, Prentice Hall, New York.

Khalfan, A. M. (2004), Information security considerations in IS/IT outsourcing projects: A descriptive case study of two sectors, *International Journal of Information Management*, 24(1), 29–42.

Kishore, R., Rao, H. R., Nam, K., Rajagopalan, S., and Chaudhury, A. (2003), A relationship perspective on IT outsourcing, *Communications of the ACM*, 46(12), 87–92.

Klepper, R. (1995), The management of partnering development in I/S outsourcing, *Journal of Information Technology*, 10, 249–258.

Kotlarsky, J., Oshri, I., van Hillegersberg, J. and Kumar, K. (2007), Globally Distributed Component-based Software Development: An exploratory study of knowledge management and work division, *Journal of Information Technology*, 22(2), 161–173.

Kotlarsky, J., van Fenema, P., and Willcocks. L. (2008), Developing a knowledge-based perspective on coordination: The case of global software projects, *Information and Management*, 45(2), 96–108.

Krishna, S., Sahay, S., and Walsham, G. (2004), Managing cross-cultural issues in global software outsourcing, *Communications of the ACM*, 47(4), 62–66.

Lacity, M., Carmel, E., and Rottman, J. (2011a), Rural outsourcing: Delivering ITO and BPO services from remote domestic locations, *IEEE Computer*, 44, 44–51.

Lacity, M., Feeny, D., and Willcocks, L. (2004), Commercializing the back office at Lloyds of London: Outsourcing and strategic partnerships revisited, *European Management Journal*, 22(2), 127–140.

Lacity, M., Feeny, D., and Willcocks, L. (2003), Transforming a back-office function: Lessons from BAE systems' experience with an enterprise partnership, *MIS Quarterly Executive*, 2(2), 86–103.

Lacity, M. and Fox, J. (2008), Creating global shared services: Lessons from reuters, *MIS Quarterly Executive*, 7(1), 17–32.

Lacity, M. and Hirschheim, R. (1995), *Beyond the Information Systems Outsourcing Bandwagon: The Insourcing Response*, Wiley, Chichester, U.K.

Lacity, M., Hirschheim, R., and Willcocks, L. (1994), Realizing outsourcing expectations: Incredible promise, credible outcomes, *Journal of Information Systems Management*, 11(4), 7–18.

Lacity, M., Khan, S., Yan, A., and Willcocks, L. (2010a), A review of the IT outsourcing empirical literature and future research directions, *Journal of Information Technology*, 25(4), 395–433.

Lacity, M., Rottman, J., and Khan, S. (2010b), Field of dreams: Building IT capabilities in rural America, *Strategic Outsourcing: An International Journal*, 3(3), 169–191.

Lacity, M. and Rottman, J. (2008), *Offshore Outsourcing of IT Work*, Palgrave, London, U.K.

Lacity, M. and Willcocks, L. (1995), Interpreting information technology sourcing decisions from a transaction cost perspective: Findings and critique, *Accounting, Management, and Information Technologies*, 5(3/4), 203–244.

Lacity, M. and Willcocks, L. (1998), An empirical investigation of information technology sourcing practices: Lessons from experience, *MIS Quarterly*, 22(3), 363–408.

Lacity, M. and Willcocks, L. (2001) *Global Information Technology Outsourcing: In Search of Business Advantage*, Palgrave, London, U.K.

Lacity, M. and Willcocks, L. (2009), *Information Systems and Outsourcing: Studies in Theory and Practice*, Palgrave, London, U.K.

Lacity, M. and Willcocks, L. (2011), PART 1: What suppliers say about clients: Establishing the outsourcing arrangement, *Cutter Consortium*, 12(2).

Lacity, M. and Willcocks, L. (2012a), *Advanced Outsourcing Practice: Rethinking ITO, BPO and Cloud Services*, Palgrave, London, U.K.

Lacity, M. and Willcocks, L. (2012b), Outsourcing business and IT services: The evidence on success, robust practices, and contractual challenges, *Legal Information Management*, 12, 2–8.

Lacity, M., Willcocks, L., and Khan, S. (2011b), Beyond transaction cost economics: Towards an endogenous theory of information technology outsourcing, *Journal of Strategic Information Systems*, 20(2), 139–157.

Lee, J. (2001), The impact of knowledge sharing, organizational capability and partnership quality on IS outsourcing success, *Information & Management*, 38, 323–335.

Levina, N. and Ross, J. (2003), From the vendor's perspective: Exploring the value proposition in IT outsourcing, *MIS Quarterly*, 27(3), 331–364.

Levina, N. and Vaast, E. (2008), Innovating or doing as told? Status differences and overlapping boundaries in offshore collaboration, *MIS Quarterly*, 32(2), 307–332.

Linder, J. (2004), Transformational outsourcing, *Sloan Management Review*, 45(2), 52–58.

Luftman, J. and Kempaiah, R. (2008), Key issues for IS executives, *MIS Quarterly Executive*, 7(2), 99–112.

Luftman, J. and Zadeh, H. (2011), Key information technology and management issues 2010–11: An international study, *Journal of Information Technology*, 26, 193–204.

Mahnke, V., Wareham, J., and Bjorn-Andersen, N. (2008), Offshore middlemen: Transnational intermediation in technology sourcing, *Journal of Information Technology*, 23(1), 18–30.

Malos (2009), Regulatory effects and strategic global staffing profiles: Beyond cost concerns in evaluating offshore location attractiveness, *Employee Responsibilities and Rights Journal*, 22, 113–131.

McFarlan, F. W. and Nolan, R. (1995), How to manage an IT outsourcing alliance, *Sloan Management Review*, 36(2), 9–24.

Mehta, A., Armenakis, A., Mehta, N., and Irani, F. (2006), Challenges and opportunities of business process outsourcing, *Journal of Labor Research*, 27(3), 323–337.

Murray, J., Kotabe, M., and Westjohn, S. (2009), Global sourcing strategy and performance of knowledge-intensive business services: A two stage strategic fit model, *Journal of International Marketing*, 17(4), 90–105.

O'Leary, M. and Cummings, J. (2007), The spatial, temporal and configurational characteristics of geographical dispersion in teams, *MIS Quarterly*, 31(3), 433–452.

Oh, W., Gallivan, M., and Kim, J. (2006), The market's perception of the transactional risks of information technology outsourcing announcements, *Journal of Management Information Systems*, 22(4), 271–303.

Oshri, I. (2011), *Offshoring Strategies: Evolving Captive Center Models*, MIT Press, Boston, MA.

Oshri, I., Kotlarsky, J., and Willcocks, L. (2007), Managing dispersed expertise in IT offshore outsourcing: Lessons from Tata Consultancy Services, *MIS Quarterly Executive*, 6(2), 53–65.

Patane, J. R. and Jurison, J. (1994), Is global outsourcing diminishing the prospects for American programmers? *Journal of Systems Management*, 45(6), 6–10.

Pinnington, A. and Woolcock, P. (1995), How far is IS/IT outsourcing enabling new organizational structure and competences? *International Journal of Information Management*, 15(5), 353–365.

Poppo, L. and Zenger, T. (2002), Do formal contracts and relational governance function as substitutes or complements? *Strategic Management Journal*, 23, 707–725.

Rao, M. T., Poole, W., Raven, P. V., and Lockwood, D. L. (2006), Trends, implications, and responses to global IT sourcing: A field study, *Journal of Global Information Technology Management*, 9(3), 5–23.

Ross, J. and Beath, C. (2006), Sustainable IT outsourcing: Let enterprise architecture be your guide, *MIS Quarterly Executive*, 5(4), 181–192.

Rottman, J. and Lacity, M. (2006), Proven practices for effectively offshoring IT work, *Sloan Management Review*, 47(3), 56–63.

Rottman, J. and Lacity, M. (2008), A US Client's Learning from Outsourcing IT Work Offshore, *Information Systems Frontiers*, Special Issue on Outsourcing of IT Services, 10(2), 259–275.

Sabherwal, R. (1999), The role of trust in outsourced IS development projects, *Communications of the ACM*, 42(2), 80–86.

Saunders, C., Gebelt, M., and Hu, Q. (1997), Achieving success in information systems outsourcing, *California Management Review*, 39(2), 63–80.

Smith, H. A. and McKeen, J. D. (2004), Developments in practice XIV: IT outsourcing—How far can you go? *Communications of the AIS*, 14(1), 508–520.

Sobol, M. and Apte, U. (1995), Domestic and global outsourcing practices of America's most effective IS users, *Journal of Information Technology*, 10, 269–280.

Straub, D., Weill, P., and Schwaig, K. (2008), Strategic dependence on the IT resource and outsourcing: A test of the strategic control model, *Information Systems Frontiers*, 10(2), 195–211.

Teng, J., Cheon, M., and Grover, V. (1995), Decisions to outsource information systems functions: Testing a strategy–theoretic discrepancy model, *Decision Sciences*, 26(1), 75–103.

The Everest Group (2011), http://www.everestgrp.com/2011-10-is-the-arbitrage-of-your-offshore-locations-sustainable-webinar-7855.html (accessed on April 13, 2013).

Tiwana, A. and Bush, A. (2007), A comparison of transaction cost, agency, and knowledge-based predictors of ITO outsourcing decisions: A US–Japan cross cultural field study, *Journal of Management Information Systems*, 24(1), 259–300.

Walden, E. (2005), Intellectual property rights and cannibalization in information technology outsourcing contracts, *MIS Quarterly*, 29(4), 699–721.

Whitten, D. and Leidner, D. (2006), Bringing IT back: An analysis of the decision to backsource or switch vendors, *Decision Sciences*, 37(4), 605–621.

Willcocks, L., Cullen, S., and Craig, A. (2011b), *The Outsourcing Enterprise: From Cost Management to Collaborative Innovation*, Palgrave, London, U.K.

Willcocks, L. and Lacity, M. (2006), *Global Sourcing of Business and IT Services*, Palgrave, London, U.K.

Willcocks, L., Oshri, I., and Hindle, J. (2010a), *To Bundle or Not To Bundle? Effective Decision-Making for Business and IT Services*, OU/Accenture, London, U.K.

Willcocks, L., Reynolds, P., and Feeny, D. (2007), Evolving IS capabilities to leverage the external IT services market, *MIS Quarterly Executive*, 6(3), 127–145.

Willcocks, L., Venters, W., and Whitley, E. (2010b), The coming of the cloud corporation, *Accenture Outlook Point of View*, Accenture, London, U.K.

Willcocks, L., Venters, W., and Whitley, E. (2011a) *Cloud and the Future of Business 1—The Promise*, Accenture/LSE Outsourcing Unit, London, U.K.

Williamson, O. (1975), *Markets and Hierarchies: Analysis and Antitrust Implications*, Free Press, New York.

Williamson, O. (1991a), Strategizing, economizing, and economic organization, *Strategic Management Journal*, 12, 75–94.

Williamson, O. (1991b), Comparative economic organization: The analysis of discrete structural alternatives, *Administrative Science Quarterly*, 36(2), 269–296.

Winkler, J. K., Dibbern, J., and Heinzl, A. (2008), The impact of cultural differences in offshore outsourcing—Case study results from German–Indian application development projects, *Information Systems Frontiers*, 10, 243–258.

Wüllenweber, K., Beimborn, D., Weitzel, T., and Kőnig, W. (2008), The impact of process standardization on business process outsourcing success, *Information Systems Frontiers*, 10(2), 211–224.

61

IS/IT Project Management: The Quest for Flexibility and Agility

Laurie J. Kirsch
University of Pittsburgh

Sandra A. Slaughter
Georgia Institute of Technology

61.1 Introduction

Providing technological capabilities to an organization in support of ongoing or new business initiatives is a core responsibility of the information systems (IS) function. IS/ information technology (IT) project management is therefore a key capability since the delivery of technological capabilities—from understanding user requirements to acquiring a technological solution to implementing the system—often falls to project teams led by an IS/IT project manager. IS/IT project management involves understanding the requirements of a project, developing a plan to deliver a product or service, and implementing the plan. IS/IT project managers are tasked with delivering a particular technological capability (e.g., an information system) to satisfy scope and quality requirements within specific time, budget, and resource constraints.

The context of managing IS/IT projects has changed dramatically over time. When computers were first introduced into organizations, it was common to develop custom information systems in-house. For example, in the United States, the first administrative information system was a payroll system developed by General Electric. In 1954, General Electric purchased a Univac I computer to use in its new Major Appliance Division plant in Louisville, Kentucky. GE Corporate Accounting Services took primary responsibility for managing this project, which involved designing and programming the first payroll information system to be used initially by the Washer and Dryer Department.* As part of its plan to automate its production facilities and make Louisville a showcase location, GE decided to use the computer not only to process payrolls and other accounting applications but also for manufacturing control and planning.

---

* See http://www.softwarehistory.org for more information about the history of the software industry and Burton Grad's first person description of his experiences in programming the first commercial information system at GE (http://www.softwarehistory.org/pdf/x-1stCommCompAppGE.pdf).

From the 1950s well into the 1970s and beyond, sourcing options for information systems were few, and most systems were built by internal IS staff. As the GE example illustrates, information systems were typically designed to meet specific needs, usually for one particular functional area and typically focused on an operational aspect of the business. If a development methodology was used, it was most likely the waterfall (Systems Development Life Cycle) approach. The waterfall process, based on functional decomposition, consumed considerable time and effort and produced a wealth of documentation, including design documents, sign-offs, and responsibility charts, all intended to hold IS professionals accountable for producing a well-designed, well-constructed, and maintainable system. Consistent with the waterfall approach to development, most project managers and developers commonly assumed that information requirements were known, could be elicited, and would be stable. Thus, the waterfall process provided IS/IT project managers with a template for planning and execution, with its series of steps performed in a sequential manner.

The context of computing started to change rapidly, however, beginning in the 1980s. The prevalence of the personal computer and end-user computing, followed by the rise of the Internet and web-based computing, provided firms with many more technical capabilities and options. The business environment was also changing. Firms found themselves in an increasingly global, dynamic, and, in some industries, a highly regulated environment. Global competition, customers, and suppliers became the norm rather than the exception for most firms. Successful firms learned to adapt and evolve in order to respond to ever-changing consumer demands and competitor actions in the marketplace, as well as shifting resource availability and costs along the global supply chain.

The changes in the business environment converged with the rapidly changing technology such that more and more firms today expect information systems to allow for rapid response to both anticipated and unanticipated change (Lee and Xia 2005, 2010; Lyytinen and Rose 2006; Maruping et al. 2009). The complexity, volatility, and uncertainty inherent in the business environment demand *flexibility* and *agility* in the design and delivery of information systems and, correspondingly, the management of IS/IT projects. Flexibility is defined as the ability to adapt and change (Elfatatry 2007, p. 36). Agility, in this context, refers to "the ability to sense and respond swiftly to technical changes and new business opportunities" (Lyytinen and Rose 2006, p. 183).

Satisfying a firm's need for flexibility and agility in its computing environment encompasses various aspects related to information systems acquisition and delivery. Not only must information systems themselves deliver flexibility, but the development methodologies and project management approaches must privilege adaptability and nimble responses to change. Further, firms desire flexibility and agility when they are looking to acquire a new system, and no longer view internal custom development as their only option for sourcing an information system. The waterfall approach as a template for project management becomes less relevant in an environment characterized by change and uncertainty.

This chapter explores IS/IT project management in contemporary firms. We start by considering the underlying principles of IS/IT project management. This includes an overview of modern project management as well as a detailed examination of the context of IS/IT projects, with a particular focus on flexibility and agility. Next, we discuss approaches for managing highly uncertain and volatile projects as IS project managers and teams strive to deliver flexibility and agility. We then explore research opportunities in IS/IT project management. We end this chapter by offering a few concluding remarks.

61.2 Underlying Principles of IS/IT Project Management

Though individuals have been managing projects for centuries, project management, as a profession, is relatively young; it was in 1969 that the Project Management Institute (PMI), a nonprofit association, was established with a mission to foster the profession of project management. By 1990, there were 7500 members of the PMI; today, there are over 600,000 members in more than 185 countries.* The PMI

---

* See http://www.pmi.org for more details about the PMI.

offers various certifications to practicing project managers, including the Certified Associate in Project Management (CAPM) and the Professional Project Manager (PMP) certifications.

As part of its mission, the PMI has identified and codified the project management body of knowledge (PMBOK), the ten knowledge areas that all project managers must master to achieve CAPM or PMP certification. The knowledge areas are scope, time, cost, quality, human resources, communication, risk, procurement, stakeholder, and integration management. Project managers make decisions and take actions related to each knowledge area in an overlapping fashion across project management "process groups": project initiation, planning, execution, monitoring and controlling, and project closing. The PMBOK maps activities associated with each knowledge area into the process groups.

The efforts of the PMI have had a profound impact on professionalizing project management in IS/IT as well as in other disciplines. However, the practice of IS/IT project management has also been greatly influenced by the methodologies used to develop information systems. Methodologies provide templates for managing projects since they articulate activities and processes associated with developing and delivering information systems. As methodologies have evolved, so has IS/IT project management. Therefore, we next examine the information systems development (ISD) process and highlight the relationship between ISD approaches and IS/IT project management.

ISD is the process by which computer software is specified, written, tested, and implemented. The waterfall process, a conventional approach to ISD, involves planning, analysis, design, coding, testing, implementation, and maintenance (George 1999). However, for today's firms, the types of systems being developed, as well as the way in which they are developed, have evolved considerably since the 1950s. Initially, firms generally built single-purpose, stand-alone systems for a particular department or functional unit. Today, we are more likely to find one of two scenarios. The first is the development of smaller systems, often tightly integrated with others. One example is adding a web-based front end to an existing legacy system to improve its usability. The second scenario is the development of very large, complex systems that are often global or inter-organizational in scope. In contemporary firms, the choice of methodology is also changing. The conventional waterfall approach to development is just one of many that may be used; alternatives include the spiral model and agile methods. Over time, project management has also evolved so that today it tends to be more formalized and less ad hoc. Table 61.1 compares the characteristics of traditional systems development and project management with emerging trends found in contemporary firms.

TABLE 61.1 Comparing Traditional and Contemporary Systems Development and Project Management

	Traditional Practices	Contemporary and Emerging Trends
Requirements	Stable and predictable	Emergent and uncertain
System characteristics	Single purpose and stand alone	Smaller and tightly integrated or larger and spanning units
Design philosophy	Design a system for long-term use	Design a system for adaptability
Common development method	SDLC or waterfall	Spiral or agile methods
Methodology approach	Functional decomposition	Component-based design
Methodology assumptions	Predictability and longevity	Uncertainty and flexibility
Methodology principles	Heavyweight	Lightweight
Project management approaches	Ad hoc, individual processes	More formalized approaches
Role of user	Minimal or token user participation	Partner in deployment of IS
Project teams	Colocated	Geographically distributed
Managing project teams	Simplistic understanding and use of control and coordination mechanisms	More nuanced understanding and use of control and coordination mechanisms

In today's quest for flexible and agile systems, three aspects of managing IS/IT projects deserve particular attention: requirements determination, design considerations, and sourcing options. Successfully determining requirements is a crucial step in an IS/IT project, but it is notoriously difficult, particularly when requirements are ambiguous or fluctuating. In addition, developing systems that are adaptable and flexible can heighten the challenge of determining requirements. Similarly, the principles underlying the design of a technical solution have implications for the nature of the solution itself. Whether a system is adaptable over time is, in part, a result of how the system was initially designed. Finally, the way in which a system is sourced—whether it is custom developed or acquired—has implications in terms of flexibility and agility. The method chosen to develop a technical solution varies in its ability to accommodate change: some methods are constraining while others are better suited to handle changes and produce systems that are flexible and agile. Moreover, increasingly organizations are turning away from custom development and considering alternate sourcing options. Accordingly, IS/IT project managers need to understand the advantages and disadvantages of a variety of sourcing options in order to choose the approach that best fits the needs of the system and the organization. In the following sections, we explore requirements determination, design considerations, and sourcing options in some depth, with a focus on flexibility and agility.

61.2.1 Requirements Definition

Research over the last 25 years has repeatedly highlighted the many challenges associated with requirements determination, including ambiguous requirements, communication obstacles, difficulty in reaching consensus on global needs, and the thin spread of domain knowledge across stakeholders (Alvarez 2002; Curtis et al. 1988; Urquhart 2001; Walz et al. 1993). The challenges are compounded in contemporary firms where the business needs are highly volatile and requirements change as systems are developed in response to emerging business needs or increased understanding of a complex business environment (Kirsch and Haney 2006; Maruping et al. 2009). Moreover, needs and requirements may be largely unknown at the start of a project, and may instead emerge as the project unfolds. Dynamic and emergent requirements represent a considerable departure from prior years, when requirements were regarded as specifiable and stable.

IS/IT project managers may increasingly be called upon to develop systems with emergent requirements. For example, there is growing interest and investment in large-scale, multiparty information systems that span disciplines and institutions and are geared toward supporting collaboration among many stakeholders (Berman 2008; Levina 2005). An example of this type of project is a cyberinfrastructure project (or "cyber project"). *Cyberinfrastructure* refers to integrated information technologies (i.e., hardware, software, digital sensors, middleware, networks, and data components) that support scientific research activities (Berman 2008; Bietz et al. 2010; Edwards et al. 2009). A specific objective of cyberinfrastructure is to foster innovative research and discovery through the use of technologies that support distributed collaborative work among geographically dispersed researchers (de la Flor et al. 2010; Ribes and Lee 2010). An example of cyberinfrastructure is the George E. Brown, Jr. Network for Earthquake Engineering Simulation (or NEES), which was created to better understand earthquakes, their causes, and effects.* NEES provides researchers, scholars, and industry access to state-of-the-art technologies and resources, augmenting their ability to study complex questions about earthquakes, to predict when earthquakes will occur, and to provide guidance to governments and emergency responders.

A "cyber project" refers to the information technology (IT) development activities that design, build, integrate, test, and implement a specific cyberinfrastructure. Cyber projects tend to be large and complex: they involve stakeholders from industry, academia, and government with varying goals and requirements, and they often cost millions of dollars and require years to develop (Finholt and Birnholz 2006). Because cyberinfrastructure is meant to promote experimentation and discovery, the requirements are emergent and unpredictable. In fact, the project outcomes are often unknowable and

---

* For more information, see http://www.nees.org

unpredictable: as a cyber project unfolds over time, the vision and goals associated with the project will themselves evolve in response to the ongoing experimentation and discovery.

Though cyber projects may be an extreme example in terms of fluctuating and emergent requirements, they nevertheless are illustrative of the challenges associated with managing volatile requirements. IS/IT project managers must recognize volatility and learn to adapt the way in which these projects are managed (Maruping et al. 2009). With fluctuating requirements, there is increased need for IS/IT project managers to promote ongoing mutual learning between providers of solutions and customers. Some methods are designed to accommodate changing requirements. Prototyping, spiral development methodologies, and rapid application development (RAD) assume that requirements are difficult to elicit and encourage iteration between developer and customer. For example, consider a cyber project called Global Environment for Network Innovation (GENI). The goals of GENI are to develop a virtual laboratory at the frontiers of network science and engineering for exploring future internets at scale (see http://www.geni.net). Given these goals, the project faces a considerable amount of requirements uncertainty. To address this uncertainty, the project is managed using spiral development. Each spiral lasts 1 year and consists of projects that develop different networking technologies. After each spiral, the project manager evaluates the success and failure of the technologies and decides on the technology goals for the next spiral. Spiral development affords the opportunity to explore cutting-edge technologies and provides flexibility to discontinue less promising technologies while affording the ability to explore new technology opportunities. From a project management perspective, however, the spiral approach—particularly as it is implemented in the GENI cyber project—introduces considerable complexity into the management process as there are multiple moving parts to coordinate and oversee.

Newer agile methods also generally recognize the need for intensive user–developer interaction. Using eXtreme programming (XP), for example, customers write brief "user stories" explaining what a system needs to do, and developers estimate the time needed to implement that functionality. Later in the process, the developer receives more details from the customer in order to implement the story.*

In the context of emergent requirements, it is also important for IS/IT project managers to recognize that customers will not be able to fully specify requirements at the start of a project (Curtis et al. 1988; Maruping et al. 2009; Urquhart 2001). A number of empirical studies have highlighted the role of learning and intense analyst–user interactions especially in the context of poorly understood requirements and ill-structured problems (Alvarez 2002; Boland 1978; Kirsch and Haney 2006; Lyytinen and Robey 1999).

61.2.2 Design Approaches

In the past, systems were built for long-term use. The need for flexible, adaptable systems was not paramount as business environments were relatively stable. Contemporary firms, though, face a completely different environment, one in which competitors, products, and regulations are changing quickly and often in unpredictable ways. IS/IT project managers must understand this changing environment and adapt accordingly.

One response to the need for flexibility and agility is a modular approach to designing and implementing systems, based on a specific plan or architecture. An architecture incorporates a high-level understanding of the requirements, the solution, the modules, subsystems, and interface, as well as a broad plan for deployment (Goodhue et al. 1992; McBride 2007). When systems are designed without an architecture, they often evolve haphazardly, are difficult to change, and ultimately fail to meet the needs of the business (McBride 2007). In contrast, when an architecture is in place and followed, systems are designed for change because they are built in modules using standardized parts, which facilitates future adaptations. Thus, an architecture and modular design provide firms with flexibility and agility to deal with highly dynamic requirements.

---

* See http://www.extremeprogramming.org for additional information.

Recent studies have provided insights into the role of modular architectures in system adaptability. In a qualitative study, Collins and Kirsch (1999) observed the use of modular, incremental design and deployment in response to a dynamic and turbulent business environment. For example, in one firm, a complex global system was designed in relatively small increments (typically 6 months or less project duration) for geographically dispersed business units to implement and adapt to local requirements as needed. Moreover, researchers have begun to document the positive impacts of modular design. For example, in a field study of system enhancement activity over a 20 year time period, Barry, Kemerer, and Slaughter (2006) found that greater use of a standard architecture in a system localizes changes to specific components. The findings demonstrate that a standard architecture increases the design integrity and stability of the system while at the same time accommodating change.

Model-driven architecture (MDA) is a relatively recent software design approach that can be adopted for the development of information systems. MDA provides guidelines for specifications, expressed as models, and was launched by the Object Management Group (OMG, http://www.omg.org/mda). It involves producing software code from abstract modeling diagrams (Frankel 2003). MDA can introduce flexibility and agility in the software development process by separating design from architecture. This separation allows the designer to address the functional (use case) requirements while the architecture provides the infrastructure through which nonfunctional requirements like scalability, reliability, and performance are addressed. Separating design from architecture allows developers to take advantage of the latest techniques and technologies for each, while providing an architecture that is robust to changes.

Another design approach that affords considerable flexibility and agility in software projects is component-based development (CBD). An individual software component is a package, service, or module that encapsulates a set of related functions (or data). CBD is a reuse-based approach to defining, implementing, and composing loosely coupled independent components into systems (Heineman and Councill 2001). This approach allows the ability to "plug and play" components into software architectures, and components play a central role in service-oriented architectures (SOA). A recent field study of CBD by Subramanyam, Ramasubbu, and Krishnan (2012) found that CBD approaches are associated with what the authors call "efficient flexibility" in the development of business information systems.

61.2.3 Sourcing Options

Achieving flexibility and agility goes beyond requirements determination and design considerations. It is also important that IS/IT project managers consider sourcing options. We first consider in-house development methodologies and then turn our attention to a variety of external sourcing options.

Custom in-house development is the traditional approach to acquiring an information system in which an organization's IT staff develop and install a system designed to meet the unique needs of the organization. Custom in-house ISD allows for meeting highly specialized requirements, facilitates changes to firms' business processes, and helps to build up personnel skills in the organization. It also provides firms with a certain kind of flexibility—that is, the flexibility to creatively solve business problems, and to develop a solution that fits the particular nuances of a firm's business environment. Custom in-house development enables firms to continually adapt and enhance their information systems to meet unique needs.

However, custom in-house development can also be constraining in that it generally takes a long time to develop a software solution internally. Moreover, custom in-house ISD can tax an organization's resources and may incur significant risk. It is not uncommon for ISD projects to be over budget, run over schedule, and not deliver the required functionality (Barki et al. 2001; Keil 1995). Some projects are abandoned, without developing any useful parts of a system. One recent example is the Virtual Case File (VCF). The VCF project was supposed to automate the FBI's paper-based work environment and allow agents and intelligence analysts to share investigative information. After years of effort, the VCF's contractor, Science Applications International Corp. (SAIC), in San Diego, delivered 700,000 lines of code; however, the software was largely unusable, and the FBI had to scrap the entire $170 million project and start over from scratch (Goldstein 2005).

Because the conventional waterfall approach is often seen as too cumbersome and time-consuming to be effective when flexibility, speed, and agility are paramount (Lyytinen and Rose 2006; Slaughter et al. 2006), firms are increasingly turning to agile methods, a set of practices intended to minimize unnecessary bureaucracy and maximize adaptability and responsiveness. The Agile Manifesto, composed by a group of "independent thinkers about software development" who advocate lightweight, agile methods for development, highlights four distinct value propositions*:

1. Individuals and interactions over processes and tools
2. Working software over comprehensive documentation
3. Customer collaboration over contract negotiation
4. Responding to change over following a plan

While XP and Scrum are the most widely known and used agile methods, there are a number of others, including the Rational Unified Process (RUP) and Dynamic Systems Development Method (DSDM) (Boehm and Turner 2003).†

Despite the advances in development methodologies, custom design and development often are risky, lengthy, and prone to failure (Barki et al. 2001). Thus, organizations often turn to other sourcing strategies. Today, there are many viable alternatives for acquiring IS that a firm and its IS/IT management can consider. These include packaged software, outsourcing, and offshoring. Each is explored as follows.

Packaged software is code written by a vendor for purchase by organizations and use "off the shelf." The package provides common functionality and may allow for customization of certain features. Packaged software is available in a wide variety of systems, sizes, and prices. For example, Enterprise Resource Planning (ERP) systems are very large software packages that enable organizations to integrate business processes across functional areas. ERP packages can be contrasted with smaller, more special purpose software packages such as for payroll and office productivity. Compared with custom development, software packages offer numerous advantages. First and foremost, the software is already written and functioning, and this may save time and reduce the risk of unusable code. The software may be more thoroughly tested and proven as it has been used in many organizations. On the other hand, the organization must pay for and accept the functionality provided, even if not all of the functionality is used. It can be very costly to "customize" a package, even if the vendor allows it (Davis 1988). A larger issue is that the organization may have to significantly change how it does business in order to use the software (Elbanna 2010). Research on implementing software packages, such as ERP systems, suggests that the greater the "misfits" between the package and the organization's needs, the greater the cost and effort of implementation and the higher the risk of failure (Soh et al. 2000). In terms of flexibility and agility, packaged software provides agility in acquisition, as the time to acquire the software is short (compared to in-house custom development), although implementation can still be complex and lengthy. However, the flexibility to change the software is less than for in-house developed systems as the organization must negotiate with the vendor to make the changes.

Outsourcing involves hiring an external vendor, contractor, or service provider to create or support an information system. Information systems outsourcing has been a viable sourcing strategy since the early days of computing. Organizations can outsource their entire IT function (total outsourcing) or can outsource some systems or processes (selective outsourcing). An increasing percentage of organizations are currently outsourcing or considering it.

---

* See http://www.agilemanifesto.org for additional information.

† The degree of agility embodied in any method can vary. See, for example, http://www.ibm.com/developerworks/rational/library/edge/08/feb08/lines_barnes_holmes_ambler for a discussion of how RUP can support an agile approach to development.

Outsourcing offers numerous advantages as a sourcing strategy. By outsourcing, organizations can tap into the potentially broader and deeper IT experience and skill base of a vendor (Lacity and Willcocks 1998). The vendor may be more cost-efficient in providing IT services due to economies of scale in servicing multiple firms (Levina and Ross 2003). Outsourcing also can afford an organization greater agility (Lacity et al. 1995); for example, in the early 1990s Xerox outsourced software maintenance so that Xerox's IT staff could focus on new systems development. However, there are some important disadvantages of outsourcing (for a review, see Lacity et al. 2009). Generally, by hiring an external party to develop the information system, the client can lose control over the development process as well as expertise and can actually have less flexibility in making adaptations to the system. There is also a risk that confidential or strategic information can be compromised. In addition, it is important for the client organization to have capable staff in-house to actively manage the outsourcing relationship with the vendor.

Offshoring is a recent sourcing strategy that can be used by organizations for in-house software development or, more commonly, in outsourcing arrangements. It is a methodology by which an organization exports ISD to another country via offshore development centers (e.g., Target has established an offshore IT development center in Bangalore, India, to complement its IT unit in Minneapolis, Minnesota), captive units (i.e., a business unit of a firm that functions as an independent entity offshore but still maintains close ties with the parent company), outsourcing to locally based vendor (such as Infosys, Tata, or Wipro in India), or outsourcing to a global IT service provider (such as Accenture) that has delivery centers overseas. There are both "near" shore destinations (Canada and Mexico would be near shore offshore destinations for U.S. organizations) and "far" shore destinations (India, Russia, China, and the Philippines would be far shore offshore destinations for U.S. organizations).

Offshore systems development has distinct advantages. By conducting systems development in multiple locations around the globe, organizations can achieve faster development time since they can leverage a 24 × 7 development cycle (Carmel 1999). This can increase an organization's flexibility (because of increased productivity in developing and adapting information systems) and an organization's agility (e.g., by offshoring some projects to one location, IT staff at another location may become available for other projects). Offshoring also allows an organization to diversify its IT talent pool and reduce the risks involved in relying on a single market. Offshoring can facilitate localization of product software and provide proximity to target markets. Finally, offshoring may result in significant cost savings on IT development staff as IT professionals in some parts of the world receive significantly lower salaries. For example, an Indian IT professional costs 1/6 to 1/10 as much as an IT professional in the United States).* On the other hand, the costs of coordinating and communicating can significantly increase when systems development staff are distributed globally (Espinosa, Slaughter, Kraut, and Herbsleb 2007). Other potential issues include language problems, legal, accounting and cultural barriers, loss of control, confidentiality, security, liability, intellectual property, and increased vulnerability to political events (Carmel 1999).

61.3 Implications for Practice: Managing Projects for Flexibility and Agility

In general, strong project management and risk analysis, combined with effective communication and information gathering, help deal with complexity and uncertainty in systems (McBride 2007; Rai et al. 2009). In this section, we offer a number of specific recommendations for managing IS projects with an eye toward flexibility and agility.

---

* For more information about international salaries, see http://www.sourcingline.com/country-data/salaries-software-engineer-web-developer and http://www.payscale.com/research/IN/Job=Computer_Programmer/Salary#by_Years_Experience

TABLE 61.2 Selecting the Best IT Sourcing Strategy

	Custom Develop with Waterfall Approach If:	Custom Develop with Agile Approache If:	Buy Package If:	Outsource If:
Business need	Unique or core/complex, and relatively stable requirements	Unique or core/complex, and volatile requirements	Common	Not core or complex
In-house experience	Have both functional and technical experience		Have functional experience but limited technical experience	No functional or technical experience
Project development skills	Want to build development skills in-house		Development skills are not strategic. Have support and integration skills in-house	Development and support skills are not strategic
Project management skills	Skilled manager of IS/IT development projects. Proven methods available		Manager skilled at coordinating with vendor	Skilled manager can negotiate with and manage outsourcer
Time frame to implement	Flexible	Short	Short	Short or flexible

Source: Adapted from Dennis, A. et al.: *Systems Analysis and Design with UML Version 2.0: An Object-Oriented Approach*, 2nd edn. 2005. Copyright Wiley-VCH Verlag GmbH & Co. KGaA.

It is important for project managers to recognize and consider the various sourcing options. If custom development is called for, managers must decide whether to use the traditional systems development life cycle, or an agile method. A well-understood business need and relatively stable requirements call for using the systems development life cycle. So too do mission-critical systems, or systems being designed for the long term. On the other hand, volatile, emerging, or uncertain requirements suggest the use of an agile method. If time is of the essence, an agile approach is more likely to deliver a system more quickly.

There are many viable alternatives to custom in-house ISD. An important and challenging decision facing managers is which sourcing strategy to select, given a particular project. Although sometimes the sourcing strategy is determined by political or other organizational factors, it may be helpful to compare the strategies along several dimensions in making the best choice. Table 61.2 shows the major IT sourcing alternatives arranged along dimensions including the nature of the business need, the extent of in-house experience and capabilities, and the time frame of the project (Dennis et al. 2005).

As can be seen in Table 61.2, the more unique, complex, or core the business need, or the more volatile the requirements, the more custom in-house development should be favored. In contrast, the more generic or common the business need, the more attractive the strategies to buy the system. Generally, the more skilled, capable, and experienced the in-house IT staff, the more feasible is custom in-house development. Finally, the shorter the time frame, the more attractive it is to buy the system. In terms of software process, the more volatile the requirements, the more advantageous are flexible approaches such as agile methods.

Beyond the sourcing options, IS/IT project managers must learn to live with uncertainty in order to facilitate and encourage flexibility and agility in their teams and the software solutions. One approach is to realize the importance of a common, shared vision for organizing a project (rather than a fixed, inviolate plan) (Kirsch 2004; Kirsch and Haney 2006; Madsen et al. 2006). The vision needs to articulate the business value of the project (Madsen et al. 2006) so that the focus of the project, and ultimately its success, is whether it achieves business value, and not on how well it is implemented. A clear vision gives a project team direction, but does not enslave it to specific behaviors. Thus, it affords flexibility and agility.

Project managers must proactively plan for ways to deal with volatility found in an evolving problem domain, customer expectations, and perhaps the technology itself. In the presence of volatile requirements, it is important to promote ongoing mutual learning between developers and customers. Some methods, such as prototyping or agile approaches, encourage learning and intense customer–developer interaction (Lee and Xia 2010; Maruping et al. 2009). Similarly, Beyer and Holtzblatt (1995) propose an "apprenticeship model" of development in which the analyst and client engage in intense communication, discovery, and learning. Another option is to use participatory design, which recognizes the key role of the user in systems development, and in which both stakeholders together develop an understanding of the problem and solution possibilities (Kyng 1991; Simonsen 2007). Intense and frequent interaction promotes effective requirements gathering and interface definition; this in turn helps stakeholders understand and learn to deal with fluctuating requirements (McBride 2007).

Another means of dealing with volatility is to plan to deliver software solutions iteratively, relying on frequent releases and intense customer interaction and feedback (Harris et al. 2007; McBride 2007). Frequent releases ensure customers have access to the new features as they are developed. As the customers use features, project managers and developers can note their reactions and receive their feedback, learning how to adapt the system as customer needs emerge, evolve, and crystallize. Customers too can develop a better understanding of what is needed in a system.

As noted earlier, a high-level architecture can also help ensure that systems are designed and implemented for flexibility and agility. An architecture provides a high-level understanding of the requirements, solution, modules, subsystems, interfaces, and deployment plan. This helps the project manager and development team deal with system complexity and volatility, and it provides the basis for future adaptability, as it encourages modularization and layering of components based on standardized parts (Barry et al. 2006; McBride 2007). It can be a challenge, however, to ensure that systems are deployed in accordance with an architecture. One tool project managers have is the set of control mechanisms they can put in place to enforce compliance with an architecture (Kirsch 1996, 1997). Such mechanisms include the articulation of specific policies and procedures, and monitoring adherence to them, as well as fostering consensus about the importance of an architecture. Well-designed control mechanisms can help align interests of relevant stakeholders and ensure compliance with policies and standard practices. Another approach to ensure compliance is to add a software architect to projects (McBride 2007). The software architect, as the solution expert, works collaboratively with the business manager and project manager to ensure an initial understanding of the problem domain, elicit high-level requirements, identify and validate unstated expectations (McBride 2007).

Managing projects for flexibility and agility suggests an increased reliance on informal approaches to control and coordination to supplement and augment formal mechanisms (Maruping et al. 2009). Liberal use of ad hoc or casual meetings, hallway conversations, and social events are important means for controlling project teams and coordinating project tasks (Kirsch et al. 2010). Though it is important to rely more on informal and emerging mechanisms, it is equally important to recognize that existing structures, relationships, and methods cannot be ignored (Kirsch 1997; Madsen et al. 2006) as they provide the foundation for an emergent process that is in part predictable. Managers must have or develop an ability to adapt when technological, business, or environmental conditions change (Madsen et al. 2006). For example, systems development methodologies should be adapted for the situation at hand, rather than followed in the same manner for different situations (Fitzgerald 1997; Kirsch 1997; Madsen et al. 2006; Slaughter et al. 2006).

Finally, managers can take advantage of distributed geographic location and expertise by utilizing global distributed teams. However, it is important to foster trusting relationships and take steps to overcome cultural, language, or time barriers (Collins and Kirsch 1999; Rai et al. 2009). Providing access to dispersed team members through the use of tools such as collaborative technologies and videoconferencing will encourage knowledge sharing and joint problem solving (Majchrzak et al. 2005).

61.4 IS/IT Project Management: Research Opportunities

The contemporary environment for IS/IT project management is complex, dynamic, and uncertain (Elbanna 2010; Lee and Xia 2010; Maruping et al. 2009). This calls for an approach for developing and managing information systems projects that is flexible and agile, but our understanding of how to provide flexibility and agility is limited. In this section, we highlight three areas of future research. Each area represents a key role for IS/IT project managers. The first concerns development and design issues. The second concerns managing a distributed project team. The third concerns controlling projects.

61.4.1 Design and Development Issues

Requirements determination is a lynchpin in the development process. Project managers must ensure that a process is in place that not only elicits requirements but is sensitive to changing requirements. Given the increased volatility and uncertainty of the requirements determination task, it is clear that developers need the ability "to effectively and efficiently respond to business and technology changes" (Lee and Xia 2005, p. 77). Research suggests that such flexibility may be hard to find, however. In their study of over 500 ISD projects, Lee and Xia (2005) found that the project teams responded more extensively to business changes than technology changes, but they were more efficient addressing the technology changes. Additional research is needed to better understand how to achieve client–analyst learning in an effective and efficient manner, how to balance the need for customer responsiveness against the need for a disciplined development process, and how to train developers and analysts to thrive in an environment that demands agility and flexibility.

Another means of providing flexibility and agility is an architecture that promotes a modular design via a component-based or service-based approach to design. These approaches are similar in that they can be classified as "requirements adaptation" approaches as opposed to "requirements anticipation" approaches (Elfatatry 2007). However, they differ in that the component-based approach is to assemble existing prefabricated modules into a system during design (before system run-time). A SOA allows system designers to offer services through published or known interfaces. This kind of approach relies on run-time binding of components: when services are invoked, providers are chosen. Thus, a service-based approach provides potentially even more flexibility than a component-based approach. However, a service-based approach is not without its limitations, notably, implementation inefficiency and execution overhead (Elfatatry 2007). Additional research is needed to understand the capabilities and limitations of service and component approaches for achieving flexibility and agility. For example, under what circumstances does it make sense for an organization to adopt either approach?

Finally, there is a need for much more research on the use and effectiveness of agile methods to guide IS/IT project managers in choice of methodologies, and in better understanding the role of IS/IT project managers in agile contexts (Ågerfalk, Fitzgerald and Slaughter 2009). Initial research is promising. For example, in their case study at Intel, Shannon et al. (2006) found that the use of agile practices resulted in reductions in code defect density by a factor of 7, and that longer-term projects (i.e., 6 month and 1 year duration) were delivered ahead of schedule. Unlike some scholars, Fitzgerald et al. (2006) demonstrate that agile methods "are not anti-method, and require an equally disciplined approach, and as much tailoring as any traditional method" (p. 212). A special issue of the *Journal of Database Management* on Agile Information Systems Development (Erickson et al. 2005) contains a number of conceptual and review articles and case studies of agile systems development approaches. For project management, agile methods provide an interesting challenge: it is not clear whether and how formal project management practices (such as earned value analysis, work breakdown structures, and critical path analysis) that are predicated on waterfall development approaches can be applied or adapted to an agile development project. Additional field-based or experimental studies on the adoption and efficacy of agile methods for developing information systems and for managing agile projects would complement this research and provide valuable insights.

61.4.2 Managing Distributed Teams

Historically, project teams were generally collocated, and the management of the effort was typically vested in a single IS stakeholder. However, globalization is the new reality of contemporary firms, which suggests that project stakeholders—managers, users, project team members—are likely to be globally distributed (Herbsleb and Mockus 2003). The use of global teams can provide organizations with flexibility and agility. As previously noted, firms can adopt a 24 × 7 development cycle, leverage distributed knowledge, and respond quickly to local needs (Carmel 1999; Collins and Kirsch 1999).

Managing dispersed teams presents additional challenges for IS/IT project managers, including challenges related to differences in language, time zone, customs, and government regulations (Carmel 1999; Kotlarsky and Oshri 2005; Tractinsky and Jarvenpaa 1995). Moreover, distributed teams can incur higher costs of coordinating, controlling, and communicating: numerous researchers have documented the difficulties and costs associated with geographically dispersed teams (Espinosa et al. 2007; Kirsch 2004; Rai et al. 2009). Rapport and knowledge sharing are essential for successful collaborations but can be difficult to achieve in globally distributed teams (Carmel 1999; Kirsch and Haney 2006). Though technology can help address some of these challenges—e.g., high-speed connections and collaborative tools such as e-mail and instant messaging—they do not necessarily address social needs (Kotlarsky and Oshri 2005), nor do they necessarily compensate for geographic distance (Collins and Kirsch 1999).

Researchers have identified a number of ways to mitigate problems associated with distributed teams. For example, trust among distributed team members can improve personal relationships and productivity (Rai et al. 2009). When team members have electronic access to each other, knowledge sharing and collaboration are encouraged (Majchrzak et al. 2005). Transactive memory, the knowledge possessed by group members along with individual awareness of who knows what in the group (Wegner 1987), has been found to improve project effectiveness because knowledge seekers can quickly find expertise in the group (Faraj and Sproull 2000). Further, social ties and knowledge sharing have been found to improve collaboration among distributed team members (Kotlarsky and Oshri 2005). Additional research is needed to better understand the challenges of managing dispersed project teams, and the effectiveness of various tools and techniques to overcome those challenges.

61.4.3 Exercising Control and Coordination

Successfully controlling and coordinating projects is essential for IS/IT project managers to deliver the technological capabilities sought by their organizations. Control refers to all attempts to motivate individuals to work in accordance with specific objectives (Kirsch 1996). Coordination integrates and links different parts or activities to accomplish a task (Van de Ven et al. 1976; Zmud 1980). Control and coordination, though distinct concepts, share a focus on motivating and integrating people and processes to accomplish particular goals. Mechanisms to do so include formal approaches—rules, policies, methodologies, and standards—as well as informal approaches—ad hoc communication, peer pressure, and common values and norms (Kirsch 1997; Zmud 1980).

In the past, it was not uncommon for IS/IT project managers to rely on a simplistic understanding of control and coordination processes and therefore to use one-size-fits-all mechanisms. That is, control was often viewed as a static and generic cybernetic process in which goals or standards could be precisely specified, progress could be accurately assessed against those objectives, and corrective action could be identified and prescribed (Kirsch 1997). Consistent with this view, managers also may have assumed that a development methodology, precisely specified, could be faithfully followed and produce the desired system. However, it is now recognized that development methodologies are templates or guides for action, rather than a prescription for precise behaviors (Fitzgerald 1998; Fitzgerald et al. 2006; Kirsch 1997; Madsen et al. 2006). Thus, a methodology is best viewed as a "method in action" (Fitzgerald 1997).

In contemporary settings, there is a need to adapt control strategies to particular types or phases of projects or to adopt control approaches appropriate for nonroutine work (Kirsch 2004). Adaptation is particularly important in an environment of uncertainty and volatility. Cyber projects such as NEES and GENI, in which needs and goals are evolving and precise IS-related activities or behaviors can be difficult to articulate, exemplify this environment and present a rich context for studying the complexity of IS/IT project management.

Dealing with increased complexity, uncertainty, and ambiguity also calls for increased reliance on informal communication, coordination, and control (Herbsleb and Moitra 2001; Kirsch et al. 2010; Kotlarsky and Oshri 2005; Kraut and Streeter 1995). For example, the use of informal social or clan control mechanisms early in the project can promote learning and consensus among stakeholders (Kirsch 2004) by focusing attention on common values among project team members and negotiating a consensus on project goals. Research suggests a need for flexibility in control and coordination strategies, adapting them as a project progresses in response to emerging understanding, performance problems, or changes in the project environment (Choudhury and Sabherwal 2003; Kirsch 2004).

Moreover, it is apparent that multiple stakeholders control a project, thus suggesting that no one person completely controls what happens (Kirsch et al. 2010; Madsen et al. 2006). As control strategies evolves over the life of the project, the controller, i.e., the person exercising control, also changes (Kirsch 2004). At any one point in time, any particular manager may be both "in control" and "out of control" in the sense that he or she must grapple both with certainty (e.g., working in a stable environment with regular work, meetings, and patterns) and with uncertainty (e.g., an unpredictable environment with emergent action, conflict, and diversity), what has been termed the "paradox of control" (Madsen et al. 2006). However, it is not clear how managers grapple with a paradox of control. Additional research that examines the ebb and flow of control across project stakeholders over time would be extremely useful. In addition, there is a need for a more nuanced understanding of the types of control and coordination needed to achieve agility and flexibility, specifically the trade-off between too much and too little formal and informal control and coordination as a project unfolds over time.

61.5 Conclusions

Information systems are indispensable to firms, enabling their operational and strategic business processes. In recent years, dramatic changes in the business landscape, coupled with the expanding range and capability of available IT, brought the need for flexibility and agility in corporate computing to the fore. Today's IS/IT project management must facilitate a firm's ability to rapidly respond to competitive strikes and to quickly deploy their own strategic initiatives.

This chapter explored current and emerging trends in IS/IT project management. We discussed the meaning of flexibility and agility in the context of systems development and implementation projects, and highlighted the implications for managing IS projects. Flexibility and agility start with a plan, an architecture, or a blueprint to guide the development and deployment of systems. The architecture or blueprint provides the basis for adaptation by facilitating incremental, modularized development with standardized parts. Flexibility and agility can also be enhanced by using principles from agile development approaches that favor a lightweight, adaptable approach. The numerous sourcing options available to firms provide a great deal of flexibility and agility in system acquisition, as they allow firms the freedom to select an approach that best fits its needs. Firms are no longer constrained to the often high-risk, time-consuming method of developing systems internally. On the other hand, when the situation calls for it, internal development is feasible and can provide firms with flexibility in solving business problems or enabling strategic initiatives.

This chapter discussed the need for flexibility and agility in project management, with a particular focus on design decisions, the management of distributed teams, and on control and coordination of projects teams. Achieving flexibility and agility do not require eliminating the use of standards, rules, and procedures, but it does require judicious use of formal structures so that action is not unnecessarily constrained. Using informal control and coordination mechanisms to supplement formal mechanisms provides a project manager with flexibility. The use of geographically distributed teams can also provide flexibility and agility, but it is not without its challenges.

It is clear that achieving flexibility and agility requires a supple and nimble mindset, as well as a number of specific principles and approaches. Additional research is needed to develop a better understanding of ways to achieve flexibility and agility, as well as ways to overcome barriers preventing developers, managers, and others from embracing flexible and agile approaches.

Acknowledgments

This chapter is based upon work supported by the National Science Foundation under Grant No. 0909611 and 0909833. Any opinions, findings, and conclusions or recommendations expressed in this material are those of the author(s) and do not necessarily reflect the views of the National Science Foundation.

References

Ågerfalk, P., Fitzgerald, B., and Slaughter, S. Flexible and distributed information systems development: State of the art and research challenges, *Information Systems Research* 20(3), 2009, 317–328.

Alvarez, R. Confessions of an information worker: A critical analysis of information requirements discourse, *Information & Organization* (12), 2002, 85–107.

Barki, H., Rivard, S., and Talbot, J. An integrative contingency model of software project risk management, *Journal of Management Information Systems* 17(4), Spring 2001, 37–69.

Barry, E., Kemerer, C., and Slaughter, S. Environmental volatility, development decisions and software volatility, *Management Science* 52(3), March 2006, 448–464.

Berman, F. Got data? A guide to data preservation in the information age, *Communications of the ACM* 51(12), 2008, 50–56.

Beyer, H.R. and Holtzblatt, K. Apprenticing with the customer, *Communications of the ACM* 38(5), 1995, 45–52.

Bietz, M.J., Baumer, E.P.S., and Lee, C.P. Synergizing in cyberinfrastructure development, *Computer Supported Cooperative Work* 19(3–4), 2010, 245–281.

Boehm, B. and Turner, R. *Balancing Agility and Discipline: A Guide for the Perplexed*. Boston, MA: Addison-Wesley, 2003.

Boland, R.J., Jr. The process and product of system design, *Management Science* 24(9), 1978, 887–898.

Carmel, E. *Global Software Teams: Collaborating across Borders and Time Zones*. Upper Saddle River, NJ: Prentice-Hall, 1999.

Choudhury, V. and Saberhwal, R. Portfolios of control in outsourced software development projects, *Information Systems Research* 14(3), 2003, 291–314.

Collins, R.W. and Kirsch, L.J. *Crossing Boundaries*. Cincinnati, OH: Pinnaflex Publishing, 1999.

Curtis, B., Krasner, H., and Iscoe, N. A field study of the software design process for large systems, *Communications of the ACM* 31(11), 1988, 1268–1287.

Davis, G. To buy, build, or customize? *Accounting Horizons*, March 1988, 101–103.

Dennis, A., Wixom, B., and Tegarden, D. *Systems Analysis and Design with UML Version 2.0: An Object-Oriented Approach*, 2nd edn. New York: Wiley, 2005.

Edwards, P., Bowker, G., Jackson, S., and Williams, R. Introduction: An agenda for infrastructure studies, *Journal of the Association for Information Systems* 10(5), 2009, Article 6, http://aisel.aisnet.org/jais/vol10/iss5/6

Elbanna, A. Rethinking is project boundaries in practice: A multiple-projects perspective, *Journal of Strategic Information Systems* 19(1), 2010, 39–51.

Elfatatry, A. Dealing with change: Components versus services, *Communications of the ACM* 50(8), August 2007, 35–39.

Erickson, J., Lyytinen, K., and Siau, K. Agile modeling, agile software development, and extreme programming: The state of research, *Journal of Database Management* 16(4), 2005, 88–102.

Espinosa, A., Slaughter, S., Kraut, R., and Herbsleb, J. Familiarity, complexity and team performance in geographically distributed software development, *Organization Science* 18(4), July–August 2007, 595–612.

Faraj, S. and Sproull, L. Coordinating expertise in software development teams, *Management Science* 46(12), 2000, 1554–1568.

Finholt, T. and Birnholz, J. If we build it, will they come? The cultural challenges of cyberinfrastructure development, in *Managing Nano-Bio-Info-Cogno Innovations*, W.S. Bainbridge and M.C. Roco, eds. Amsterdam, the Netherlands: Springer, 2006, pp. 89–101.

Fitzgerald, B. The use of systems development methodologies in practice, *Information Systems Journal* 7(3), 1997, 201–212.

Fitzgerald, B. An empirical investigation into the adoption of systems development methodologies, *Information & Management* (34), 1998, 317–328.

Fitzgerald, B., Hartnett, G., and Conboy, G. Customising agile methods to software practices at Intel Shannon, *European Journal of Information Systems* 15(2), April 2006, 200–213.

de la Flor, G., Jirotka, M., Luff, P., Rybus, J., and Kirkham, R. Transforming scholarly practice: Embedding technological interventions to support the collaborative analysis of ancient texts, *Computer Supported Cooperative Work* 19(3–4), 2010, 309–334.

Frankel, D.S. *Model Driven Architecture: Applying MDA to Enterprise Computing.* New York: John Wiley & Sons, 2003.

George, J.F. The origins of software: Acquiring systems at the end of the century, in *Framing the Domains of IT Management*, R.W. Zmud (ed.). Cincinnati, OH: Pinnaflex Publishing, 1999, pp. 263–284.

Goldstein, H. Who killed the virtual case file? *IEEE Spectrum*, September 2005, http://www.spectrum.ieee.org/sep05/1455, accessed March 2012.

Goodhue, D.L., Kirsch, L.J., Quillard, J.A., and Wybo, M.D. Strategic data planning: Lessons from the field, *MIS Quarterly* 16(1), 1992, 11–34.

Harris, M., Aebischer, K., and Klaus, T. The Whitewater Process: Software product development in small IT businesses, *Communications of the ACM* 50(5), May 2007, 89–93.

Heineman, G.T. and Councill, W.T. *Component-based Software Engineering: Putting the Pieces Together.* Reading, MA: Addison-Wesley Professional, 2001.

Herbsleb, J.D. and Mockus, A. An empirical study of speed and communication in globally distributed software development, *IEEE Transactions on Software Engineering* 29(6), 2003, 1–14.

Herbsleb, J.D. and Moitra, D. Global software development, *IEEE Software* 18(2), March–April 2001, 16–20.

Keil, M. Pulling the plug: Software project management and the problem of project escalation, *MIS Quarterly* 19(4), December 1995, 421–447.

Kirsch, L.J. The management of complex tasks in organizations: Controlling the systems development process, *Organization Science* 7(1), 1996, 1–21.

Kirsch, L.J. Portfolios of control modes and IS project management, *Information Systems Research* 8(3), 1997, 215–239.

Kirsch, L.J. Deploying common systems globally: The dynamics of control, *Information Systems Research* 15(4), December 2004, 374–395.

Kirsch, L.J. and Haney, M.H. Requirements determination for common systems: Turning a global vision into a local reality, *Journal of Strategic Information Systems* (15), 2006, 79–104.

Kirsch, L.J., Ko, D.-G., and Haney, M.H. Investigating the antecedents of team-based clan control: Adding social capital as a predictor, *Organization Science* 21(3), March–April 2010, 469–489.

Kotlarsky, J. and Oshri, I. Social ties, knowledge sharing and successful collaboration in globally distributed system development projects, *European Journal of Information Systems* 14(1), March 2005, 37–48.

Kraut, R. and Streeter, L. Coordination in software development, *Communications of the ACM* 38(3), March 1995, 69–81.

Kyng, M. Designing for cooperation: Cooperating in design, *Communications of the ACM* 34(12), 1991, 64–73.

Lacity, M.C., Khan, S.A., and Willcocks, L.P. A review of the IT outsourcing literature: Insights for practice, *Journal of Strategic Information Systems* 18(3), 2009, 130–146.

Lacity, M. and Willcocks, L., An empirical investigation of information technology sourcing practices: Lessons from experience, *MIS Quarterly* 22(3), 1998, 363–405.

Lacity, M., Willcocks, L., and Feeny, D. IT outsourcing: Maximize flexibility and control, *Harvard Business Review* May–June, 1995, 84–93.

Lee, G. and Xia, W. The ability of information systems development project teams to respond to business and technology changes: A study of flexibility measures, *European Journal of Information Systems* 14(1), March 2005, 75–92.

Lee, G. and Xia, W. Toward agile: An integrated analysis of quantitative and qualitative field data on software development agility, *MIS Quarterly* 34(1), March 2010, 87–114.

Levina, N. Collaborating on multiparty information systems development projects: A collective reflection-in-action view, *Information Systems Research* 16(2), 2005, 109–130.

Levina, N. and Ross, J. From the vendor's perspective: Exploring the value proposition in IT outsourcing, *MIS Quarterly* 27(3), 2003, 331–364.

Lyytinen, K. and Robey, D. Learning failure in information systems development, *Information Systems Journal* 9, 1999, 85–101.

Lyytinen, K. and Rose, G.M. Information systems development agility as organizational learning, *European Journal of Information Systems* 15(2), April 2006, 183–199.

Madsen, S., Kautz, K., and Vidgen, R. A framework for understanding how a unique and local IS development method emerges in practice, *European Journal of Information Systems* 15(2), April 2006, 225–238.

Majchrzak, A., Malhotra, A., and John, R. Perceived individual collaboration know-how development through information technology-enabled contextualization: Evidence from distributed teams, *Information Systems Research* 16(1), 2005, 9–27.

Maruping, L.M., Venkatesh, V., and Agarwal, R. A control theory perspective on agile methodology use and changing user requirements, *Information Systems Research* 20(3), September 2009, 377–399.

McBride, M.R. The software architect, *Communications of the ACM* 50(5), May 2007, 75–81.

Rai, A., Maruping, L.M. and Venkatesh, V. Offshore information systems project success: The role of social embeddedness and cultural characteristics, *MIS Quarterly* 33(3), September 2009, 617–741.

Ribes, D. and Lee, C.P. Sociotechnical Studies of cyberinfrastructure and e-research: Current themes and future trajectories, *Computer Supported Cooperative Work* 19, 2010, 231–244.

Simonsen, J. Involving top management in IT projects, *Communications of the ACM* 50(8), August 2007, 53–58.

Slaughter, S., Levine, L., Ramesh, B., Baskerville, R., and Pries-Heje, J. Aligning software processes and strategy, *MIS Quarterly* 30(4), December 2006, 891–918.

Soh, C., Kien, S., and Tay-Yap, J. Cultural fits and misfits: Is ERP a universal solution? *Communications of the ACM* 43(4), April 2000, 47–51.

Subramanyam, R., Ramasubbu, N., and Krishnan, M.S., In search of efficient flexibility: Effects of software component granularity on development effort, defects, and customization effort, *Information Systems Research* 23(3), 2012, 787–803.

Tractinsky, N. and Jarvenpaa, S.L. Information systems design decisions in a global versus domestic context, *MIS Quarterly* 19(4), 1995, 507–534.

Urquhart, C. Analysts and clients in organisational contexts: A conversational perspective, *Journal of Strategic Information Systems* 10, 2001, 243–262.

Van de Ven, A.H., A.L. Delbecq, and R. Koenig. Determinants of coordination modes in organizational design, *American Sociological Review* 41(2), 1976, 322–338.

Walz, D.B., Elam, J.J., and Curtis, B. Inside a software design team: Knowledge acquisition, sharing, and integration, *Communications of the ACM* 36(10), 1993, 63–77.

Wegner, D.M. Transactive memory: A contemporary analysis of the group mind, in *Theories of Group Behavior*, G. Mullen and G. Goethals, eds., Volume 2. Greenwich, CT: JAI Press, 1987, pp. 81–123.

Zmud, R.W. Management of large software development efforts, *MIS Quarterly* 4(2), 1980, 45–55.

62

IS/IT Human Resource Development and Retention

Thomas W. Ferratt
University of Dayton

Eileen M. Trauth
*The Pennsylvania
State University*

62.1 Introduction

Over time, information system (IS)/information technology (IT) management has come to be understood as referring to more than the technology alone, or even the information systems based upon it. One of the topics in this broadly understood domain is IS/IT personnel. In fact, in 2012, the Association for Computing Machinery (ACM) Computer Personnel Research Conference became 50 years old. The publications from this conference along with papers published in the ACM publication, *The Database for Advances in Information Systems,* and other journals that publish IS/IT human resource (HR) research, give testament to the sustained interest in the human resource dimension of the computing field. Finally, evidence of the core importance of IS/IT HR is that there is an ACM indexing category for it. Topic K is the computing milieu; within it, K.7 is the computing profession with subcategories including occupations (K.7.1); organizations (K.7.2); testing, certification, and licensing (K.7.3); and professional ethics (K.7.4). Hence, a chapter on IS/IT human resource development and retention clearly has its place also in this handbook.

This chapter is organized as follows. We begin by considering the research to date on the underlying principles for IS/IT human resource development and retention drawn from the extant research. This section is followed by a discussion of IS/IT HR development and retention practices. The final section of this chapter considers emergent themes about IS/IT human resource development and retention that will motivate new research and practice in the future.

62.2 Underlying Principles for IS/IT Human Resource Development and Retention

When considering IS/IT human resource development and retention, an organization's chief information officer (CIO) faces these major questions:

1. What knowledge, skills, and abilities (KSAs) are needed by the personnel for whom I am responsible?
2. For what length of time would I prefer to retain the personnel for whom I am responsible, and what human resource management practices will support that?

We first consider how answers may differ based on the IS/IT organization's alignment with the overall organizational human resource strategy. We then review prior research that addresses these questions.

An IS/IT organization consists of the personnel and supporting resources that provide information and communication services to the larger organization. Examples of supporting resources include IT infrastructure, data, and application systems. Under the assumption that answers to the development and retention questions earlier vary systematically based on the IS/IT organization's alignment with the overall organizational human resource strategy, we first consider research that provides a strategic context for addressing the development and retention questions.

A major decision in an organizational human resource strategy is whether to develop human resources with the desired KSAs from within the organization or to obtain them from the external labor market. The former strategy has been referred to as a "make" strategy and the latter as a "buy" strategy (Miles and Snow 1984; Toh et al. 2008). Similarly, a craft internal labor market (ILM) strategy is similar to the "buy" strategy, whereas an industrial ILM strategy is similar to the "make" human resource management (HRM) strategy (e.g., see Ang and Slaughter 2004). CIOs have long been interested in aligning the IS/IT organization with overall organizational strategy (Luftman and Ben-Zvi 2011; Luftmann and Kempaiah 2008). A body of literature on IS/IT strategy, including alignment of IS/IT with business strategy, has emerged (e.g., see Chen et al. 2010). A natural extension is to the IS/IT organization's alignment with the overall organization's human resource strategy.

Discussing factors that a CIO should consider when determining whether to internally develop or externally acquire IS/IT personnel and deciding how long to keep them is beyond the scope of this chapter. We assume that the CIO will consider a variety of factors to determine the IS/IT human resource strategy, including the organization's overall business and human resource strategy. The IS/IT human resource strategy will provide a context for answering the two questions presented earlier that CIOs face with respect to KSAs and human resource management practices needed to support a specific length of employment relationship for IS/IT personnel. We next examine research on KSAs needed in IS/IT organizations. Following that, we review research on human resource management practices designed to retain IS/IT personnel for different lengths of time.

62.2.1 IS/IT Human Resource Development

62.2.1.1 Studies of Knowledge, Skills, and Abilities

Organizations choosing to build (or "make") their IS/IT personnel through training and development programs would specify the content of their programs based on the answer to the major question facing the organization's CIO: "What KSAs are needed by the personnel reporting to me?" As would be expected in an environment where the technology changes rapidly, the specific technical skills needed by IS/IT personnel also change rapidly. The required skills vary by specific job title. As Huang et al. (2009) report, Java developers need knowledge in Java, JavaScript, or J2EE and sometimes they need HTML, .NET, or XML knowledge. These specific technical skills were not needed by IS/IT personnel in the study reported by Lee et al. (1995); Java had not even been released as a programming language at that time. Besides keeping up with changes in technologies and the associated skills needed to apply those technologies, a challenge for researchers and practitioners has been to specify a set of KSAs that may be used for various IS/IT roles and job titles. Ideally, this set would make it more feasible to identify the mix of KSAs needed by IS/IT personnel in various jobs now and in the future, facilitating planning by CIOs and educators.

Researchers have worked with industry to specify KSAs in various categories, including technical, business, interpersonal, managerial, and project management (e.g., Kaiser et al. 2010; Lee et al. 1995; Trauth et al. 1993). Table 62.1 illustrates KSAs in four such categories reported by Lee et al. (1995). Categories beyond the technical specialties knowledge were defined broadly enough to apply not only in the 1990s, when the study was conducted, but also beyond. Even some of the technical specialties categories are broad enough that they still apply.

A number of researchers have investigated what KSAs IS/IT personnel need (Cheney 1988; Huang et al. 2009; Kaiser et al. 2010; Lee et al. 1995; Nakayama and Sutcliffe 2007). Roles studied include entry-level IS/IT professionals, programmers, analysts, and IS/IT managers. Table 62.2 illustrates results for entry-level IS/IT personnel in high-wage regions of the world (Kaiser et al. 2010). Methods used to identify IS/IT personnel skills have included asking knowledgeable sources (e.g., IS/IT executives/managers, business/user managers, IS/IT consultants), reviewing job advertisements in newspapers or online job postings, and examining practitioner publications.

TABLE 62.1   Categories of Knowledge, Skills, and Abilities

A. Technical specialties knowledge[a]	C. Business functional knowledge
Network	Ability to understand the business environment
Telecommunications	
Relational databases	Ability to learn about business functions
Fourth-generation languages	Knowledge of business functions
Systems integration	Ability to interpret business problems and develop appropriate technical solution
Distributed processing	
Data management (e.g., data modeling)	**D. Interpersonal and management skills[a]**
Structured programming/CASE (computer assisted software engineering) methods or tools	Ability to work closely with customers and maintain productive user or client relationship
Decision support systems	Ability to accomplish assignments
Systems analysis/structured analysis	Ability to plan and execute work in a collaborative environment
B. Technology management knowledge	Ability to be self-directed and proactive
Ability to focus on technology as a means, not an end	Ability to work cooperatively in a one-on-one and project team environment
Ability to learn new technologies	Ability to deal with ambiguity
Ability to understand technological trends	Ability to plan, organize, and lead projects
	Ability to be sensitive to organizational culture/politics
	Ability to plan, organize, and write clear, concise, effective memos, reports, and documentations
	Ability to develop and deliver effective, informative, and persuasive presentations

Source: Lee, D.M.S. et al., *MIS Q.*, 19(3), 313, 1995.
[a] Top 10, based on projecting KSAs needed 3 years beyond the survey.

TABLE 62.2   Top 10 Entry-Level Skills in High-Wage Regions

Skill Category	Skill
Business domain	Communication
Technical	Programming
Technical	Systems analysis
Technical	Desktop support/help desk
Project management	Project planning, budgeting, and scheduling
Business domain	Industry knowledge
Technical	System testing
Business domain	Process knowledge
Project management	User relationship management
Project management	Working with virtual teams

Source: Kaiser, K.M. et al., *Commun. Assoc. Inf. Syst.*, 29(1), 605, 2010.

As Tables 62.1 and 62.2 illustrate, studies of job skills report that IS/IT personnel need a mix of KSAs, not just specific technical KSAs. To illustrate further, Huang et al. (2009) report that the skills needed for programming jobs include not only technical skills but also KSAs in communication and teamwork. IS/IT professionals who move from specific, technical roles to broader roles need a different mix of KSAs for effective performance in their new roles.

62.2.1.2 Capabilities Based on View of IS/IT Human Resources as Strategic Assets

Another perspective on KSAs needed in an IS/IT organization may be derived from research that takes a view of IS/IT human resources as strategic assets. Much of this work is based on the resource-based view of the firm (Bharadwaj 2000; Chen et al. 2010; Mata et al. 1995; Ross et al. 1996). Bharadwaj (2000) identifies two critical skills of human resources in an IS/IT organization. The first is technical skills, such as programming, systems analysis and design, and competencies in emerging technologies. The second is managerial skills, such as effective management of IS/IT functions, coordination and interaction with users, and project management and leadership. Although these skills are similar to those found in the studies of KSAs reviewed earlier, the view of IS/IT human resources as strategic assets incorporates a competitive perspective not necessarily present in the studies discussed earlier.

The implication is that more successful firms will have IS/IT technical and managerial KSAs that are better than competitors. As a result, they will have capabilities to integrate the IS/IT organization and business planning processes more effectively, conceive of and develop reliable and cost-effective applications that support the business needs of the firm faster, communicate and work with business units more efficiently, and anticipate future business needs of the firm and innovate valuable new product features sooner (Bharadwaj 2000). This competitive advantage is based on these capabilities being difficult to acquire and imitate, implying that the needed IS/IT human resource strategy would be to develop IS/IT personnel with these capabilities internally. A related implication is that from the CIO's perspective the desired turnover of IS/IT personnel with these capabilities would be low.

62.2.2 IS/IT Human Resource Retention

A contextual model of turnover for IS/IT personnel presented by Joseph et al. (2007) contains environmental-level, firm-level, and individual-level factors affecting retention. CIOs should understand factors at all levels, but firm-level factors, particularly human resource strategy and human resource practices for IS/IT personnel, are factors that are most likely to have the ability to influence directly. Human resource practices, in turn, affect individual-level factors. For example, the extent to which IS/IT personnel have opportunities for internal promotion is a practice perceived at the individual level as an organizational advancement factor, one of the individual-level factors influencing the turnover process in the Joseph et al.'s (2007) model.

Many human resource management practices that CIOs may consider as part of their human resource strategy have been studied. In the broad domain of management research, Combs et al. (2006) report on the following practices in their meta-analysis of the effects of high performance work practices on organizational performance: flexible work, participation, teams, training, internal promotion, information sharing, compensation level, incentive compensation, employment security, selectivity, HR planning, performance appraisal, and grievance procedures. They reported on these practices because researchers had identified them as high performance work practices and they appeared in at least five studies that could be used in their analysis. Their analysis shows a positive correlation of these practices with retention. For CIOs seeking to develop (or "make") and retain IS/IT personnel, providing more of these practices would be consistent with the intent of establishing a long-term relationship. For CIOs interested in establishing a short-term relationship with IS/IT personnel, providing less of many of these practices would be consistent with that strategy.

Research in the more focused domain of managing IS/IT personnel has also studied human resource management practices that affect retention. Agarwal and Ferratt (1999) asked CIOs and HR executives to identify effective or innovative practices they used to retain IS/IT personnel. The numerous practices identified were grouped into several categories, which Table 62.3 presents.

Based on their survey work and subsequent case studies of several organizations, Agarwal and Ferratt (2001) reported that organizations seeking to establish long-term relationships with IS/IT personnel

TABLE 62.3 Categories of Retention Practices

Retention Practice Category	Explanation
Work arrangements	Nature of the work individuals have the opportunity to experience and how they relate to one another
Employability training and development	Training and development activities provided to IT employees to enhance current skills required by the IT organization or to develop additional skills in current technologies
Quality of leadership	Training for managers and empowerment or participation of employees
Sense of community	Activities undertaken to provide IT professionals with the sense of belonging to and being connected with a larger community
Compensation and benefits systems	Policies that compensate IT employees for work. Includes setting basic compensation levels, incentive pay systems, and bonus systems
Organizational stability and employment security	The extent to which the business enterprise is financially stable and can offer employment security
Longer-term career development	Training and development activities made available to IT employees that focus on developing business and leadership skills
Opportunities for advancement	Career management systems and career paths for IT employees
Opportunities for recognition	Systems that allow supervisors to recognize and reward outstanding or exemplary performance
Lifestyle accommodations	Opportunities available to IT employees to adjust work schedules or otherwise help balance competing demands on their time
Performance measurement	Processes by which IT employee performance is appraised to allocate rewards and identify developmental needs

Source: Agarwal, R. and Ferratt, T.W., *Coping with Labor Scarcity in Information Technology: Strategies and Practices for Effective Recruitment and Retention*, Pinnaflex Press, Cincinnati, OH, 1999.

use different bundles of human resource management practices than organizations desiring short-term relationships. Organizations desiring a long-term relationship follow a long-term investment strategy. This strategy is consistent with the "make" HR strategy. Organizations following this strategy invest more in career development and security than IS/IT organizations interested in a short-term relationship. Organizations following a long-term investment strategy tailor human resource practices, such as compensation and benefits, to make IS/IT personnel reluctant to leave, for example, by providing longer vacation periods than competitors or profit sharing that is available after staying with the organization a specific period of time. In general, these organizations use human resource management practices that reflect concern not only for productivity but also for employees as individuals. For example, they may have greater employee participation, much more community building, and more lifestyle accommodations compared with IS/IT organizations at the short-term extreme.

Using the categorization of human resource management practices of Agarwal and Ferratt (1999), Ferratt et al. (2005) examined the effect of those practices on turnover. A large sample of CIOs described practices used in their organizations and reported turnover. Factor analysis identified five factors representing the human resource management practices used: (1) work environment and career development, (2) community building, including information sharing and social activities, (3) incentives, (4) employment security, and (5) nontechnical skill recruitment (i.e., the extent to which nontechnical skills are used to recruit IS/IT professionals). The first dimension, work environment and career development, includes practices that provide an attractive work environment for IS/IT professionals (e.g., interesting work, employee empowerment, and promotion from within); in addition, it includes training and development opportunities for technical, managerial, and business KSAs.

Using cluster analysis, Ferratt et al. (2005) found different bundles, or configurations, of practices. One of those, the human capital focused (HCF) configuration, was high on all five human resource management practices. The HCF configuration also had the lowest turnover. Findings for other configurations are discussed in conjunction with the study by Ferratt et al. (2012).

Ferratt et al. (2012) extended Agarwal and Ferratt (1999, 2001) and Ferratt et al. (2005). Instead of asking CIOs, they asked IS/IT personnel about the human resource management practices used by their IS/IT organizations. Using factor analysis, they found four factors representing human resource practices—work environment and career development, social support, compensation, and security— similar to four of the five used in Ferratt et al. (2005). They used those as clustering variables to identify configurations of human resource management practices. Instead of investigating the relationship of practices to turnover, they investigated their relationship with job search behavior, a precursor to turnover.

What they found was a configuration that was high on all four human resource management factors and low on job search behavior, which is consistent with the HCF finding from Ferratt et al. (2005). They also found a task-focused (TF) configuration that was low on all four factors and high on job search behavior, which is consistent with a TF configuration found by Ferratt et al. (2005). The other configurations they found were consistent with configurations in Agarwal and Ferratt (2001) or Ferratt et al. (2005) with the exception that they did not find a secure configuration (high on security and low on other practices), which was identified in the 2005 study and had turnover between the HCF and TF configurations. What they also found is that the human resource practices in the two extreme configurations—HCF and TF—have synergistic effects. For the HCF configuration that means that job search is less frequent than would be expected from the independently additive effects of the practices, and for the TF configuration job search is more frequent than would be expected. Furthermore, they found that the human resource practices in all other configurations did not have synergistic effects.

The conclusion we derive from these studies is that retention is related to a variety of human resource management practices, including work environment and career development, social support, compensation, and security. These practices independently affect retention, and in extreme combinations (i.e., all high or all low) they synergistically affect retention. Thus, CIOs and the human resource managers supporting them may consider a number of practices independently and in combination to influence retention.

62.3 IS/IT Human Resource Development and Retention Practices

Practice is reflected in the research presented earlier. For example, KSAs that CIOs build or buy may be found in the results of research (e.g., see Tables 62.1 and 62.2). Similarly, human resource practices that CIOs use to retain IS/IT personnel may be found in the results of research (e.g., see Table 62.3). The research reported in the prior section has implications for practice that we discuss later.

We noted earlier that the IS/IT human resource strategy provides a context for understanding development and retention practices for IS/IT personnel. What may not be so obvious is that changes in how organizations use computer technology also provide such a context. In this section, we suggest that our understanding of who belongs in the category of "IS/IT personnel" and what they do has expanded as technology has changed and organizational use of computers has evolved. Discussion of this coevolution later provides this broader context for understanding IS/IT human resource development and retention practices and serves as a bridge to the final section on research issues for IS/IT personnel.

62.3.1 Implications of Research for Development and Retention Practices

An organization pursuing a strategy to buy its human resources from the external labor market would not expend significant resources to train and develop IS/IT personnel. Instead, it would recruit and hire individuals who possess the desired KSAs. Furthermore, an organization pursuing a buy strategy would not expend significant resources to retain these personnel. Instead, it would simply recruit and hire new IS/IT personnel to replace those who leave. On the other hand, an organization pursuing a strategy to build (or "make") its human resources with the desired KSAs would want to invest significantly in training and development and in other human resource management practices designed to keep those employees from being hired away before the organization has the opportunity to recoup its significant investment in them.

Research on configurations of human resource management practices with respect to IS/IT personnel suggests that the buy and build strategies would have quite different configurations. An implication of findings from that research is that CIOs following a buy strategy and desiring a short-term relationship with IS/IT personnel would implement a TF configuration, which is low on all practices. In contrast, CIOs following a build strategy and desiring a long-term relationship with IS/IT personnel would implement an HCF configuration, which is high on all practices.

Build vs. buy strategies may affect the mix of roles for IS/IT personnel. The mix of roles determines needed KSAs, since KSAs differ by role. Beyond the overall strategy for obtaining human resources, some IS/IT organizations may seek to buy specific KSAs that are not core to their business, for example, via outsourcing, temporary external personnel, or short-term internal hires (Wu and Zmud 2010). They may choose to follow a different approach when obtaining KSAs that are core or necessary for their business strategy. A business strategy that considers the IS/IT organization a strategic asset would be consistent with the resource-based view of IS/IT personnel as strategic assets. An implication would be that IS/IT personnel KSAs that are core or necessary for the business strategy would be developed and retained through using an HCF configuration.

Although not a significant focus in prior studies, change management emerged in Kaiser et al.'s (2010) top 10 skills that are critical to keep within the IS/IT organization, rather than outsource, in high-wage regions of the world. Starkweather and Stevenson (2011) suggest that the ability to deal with ambiguity and change is one of the more important core competencies for project management success. Given the prevalence of changing technologies and organizational change associated with projects involving IS/IT professionals, Markus and Benjamin (1996) have argued for educating IS/IT professionals to become more effective agents of organizational change. An implication for educators is to develop curricula for developing KSAs focused on managing change. Although the IS 2010 model curriculum has some learning objectives and topics on change management (Topi et al. 2010), the focus on change management could be further developed. The implication for CIOs is that they should work with educators to develop the needed educational changes.

KSAs related to change management, including project management, are important for success not only for IS/IT personnel involved in change efforts but also for end users, particularly user managers responsible for leading change efforts. Besides assuring that IS/IT personnel have needed change management KSAs, an implication for CIOs is that they should extend their focus beyond IS/IT personnel. Since the success of projects significantly involving IS/IT personnel is dependent on users, including user managers, CIOs should extend their role to influence the making or buying of non-IS/IT personnel with KSAs needed for leading change efforts.

The KSAs discussed in prior research have included knowledge of technological trends and technology as a means rather than an end. In the context of an organization's business strategy, understanding how IT may be used to maintain and improve the organization's competitive position could be critical to organizational success. An implication for CIOs is that their responsibility for internally developing or externally acquiring KSAs regarding information technologies may go beyond IS/IT personnel. They may need to work to influence managers throughout their organizations to identify and adopt IT to support business processes, with the goal of improving the organization's competitive position.

62.3.2 Coevolution of Organizational Computer Use and IS/IT Personnel

Human resource practice related to IS/IT worker development and retention has coevolved along with the evolution of computing. During the postwar 1950s and 1960s computer technology migrated from exclusively military applications to also include business applications, which became widespread with such computers as the IBM 360. This first expansion of computer use was accompanied by terminology changes as *business data processing* entered the lexicon to characterize the use of the computer to solve finance, accounting, and operations research problems. And accompanying this new use of computers came an expansion in the personnel responsible for operating them. These first IS/IT personnel were required to have the specialized skills of credentialed computer scientists and engineers to create the programs that solved business problems and retrieved business information for others. By the mid-1970s, the pent-up demand for better and more timely information to support management decision making led to the reorientation of IS/IT in organizations as *management information systems* (MIS) and the emergence of IS/IT personnel with MIS degrees from business schools. It also coincided with the emergence of a new generation of smaller and more accessible computers. Two technological innovations that made information more accessible and timely for management decision making were Fourth Generation Languages and minicomputers.

The 1980s and 1990s witnessed increased miniaturization of technology and the merger of computing with communicating. Along with these technological innovations, came new personnel who were part of the computing labor force: *end-user computing personnel* and *data communications personnel*. Thanks to the portable "personal" computer that ran on off-the-shelf software, business people with limited knowledge about computer technology were now able to use them. This represented one of two "revolutions" of this time period: the end-user revolution. The second "revolution" was the telecommunications revolution that resulted from a combination of technological and policy changes. The technological innovations were packet switching, network interoperability, and open protocols. Policy change came in the form of global deregulation of telecommunications that served to increase competition and innovation. Whereas the end-user revolution of the 1980s was enacted primarily in the computing domain, the end-user revolution expanded in the 1990s into the telecommunications domain. By the 1990s, the World Wide Web, commercialization of the Internet, and conceptualization of national information infrastructures led to a further expansion of computer use—and associated computer personnel—signaled by the term *electronic commerce*.

Key technological trends of the 2000s and 2010s have centered around the convergence of telecommunications and computing, mobility, cloud computing, social media, and the ubiquitous IT capabilities that have resulted. Hence, the evolution of IT has resulted in an ever-expanding definition of IS/IT personnel. In the twenty-first century, personnel in this expanded role definition increasingly have come to be labeled as the *IT worker*. These individuals are distributed throughout the organization. Some are IS/IT specialists reporting to other business functions, some work in the centralized IS/IT department of a company, and some work for both new and established IT companies such as IBM, Microsoft, and Google. Finally, given the widespread accessibility of user-friendly computing thanks to PCs, the web, and mobile devices, a new community of IS/IT consumers outside of businesses has emerged.

62.4 Research Issues Related to IS/IT Professionals

Future research will be motivated by drivers of technological evolution, environmental forces, and questions that CIOs face. The pace of technological change shows no sign of slowing down. Mobility, ubiquity, and convergence appear to be the dominant technological trends for the foreseeable future. Further, the implications of social media are just beginning to be realized. At the same time, environmental forces such as increased globalization and an unstable economy will increasingly affect workforce development and retention. Unanswered and newly emerging questions related to needed KSAs and human resource practices for retaining IS/IT professionals for an appropriate length of time will be affected by these technological and environmental forces and will drive future research.

Technological trends and environmental factors will exacerbate the increasingly fuzzy line between users and "IT professionals" as they are increasingly labeled.* Hence, even as the two major IS/IT human resource development and retention questions facing CIOs will remain, new interpretations of these topics are entering the research space and new challenges will emerge. New areas for IS/IT human resources research relate to IS/IT workforce identity in the face of ubiquitous computing, end-user technical sophistication and empowerment, workforce diversity, globalization of the IS/IT workforce, and global economic instability. Hence, the research challenges that this part of the information systems field will face in the future will expand accordingly.

62.4.1 Expanding Boundaries of the IS/IT Profession

New forms of IS/IT work are resulting from new technologies and new uses that are discovered for them. This, in turn, leads to a broader definition of the IS/IT professional and new roles for these individuals. Subsequently, while existing topics related to development and retention will remain (e.g., Joshi and Kuhn 2011), new IS/IT human resource research topics will compete for center stage.

62.4.1.1 Multiple IS/IT Disciplines and Paths to an IS/IT Career

While in the 1950s and 1960s the path to an IS/IT career had a single development path, by the 1970s this was changing. A bifurcation of computing work was in evidence, which further broadened the scope of who was considered to be a "computer person." There was a growing recognition of different categories of "computer personnel": some worked at companies such as IBM while others worked within the information systems departments of companies. This division, in turn, led to the evolution of an additional development path.

The earliest educational path into a career in computing—in the 1950s and 1960—was through computer science. In the 1970s, with the broadening of the focus of computing beyond computation into business applications, programs of study in business schools—MIS—began to appear. Throughout the 1980s, the dominant educational paths were computer science and information systems (the latter was also called MIS or computer information systems). Then, as the century was coming to a close, a third educational path emerged. The information school—or iSchool—movement grew out of computer

---

* The term "IT professional" here is meant in the broadest sense of subsuming "information systems professional" and, hence, is meant to be equivalent to "information professional."

science programs that were adding human issues—such as human computer interaction—to their curricula, as well as library science programs that were increasingly adding computer science and information systems perspectives to their programs. Finally, educational programs—called informatics—emerged that enabled individuals to study computer applications for settings other than business data processing (e.g., healthcare and government).*

As a consequence of this educational splintering, there is an IS/IT human resource development research challenge to better understand the relationship among these different educational paths and the similarities and differences among the KSAs that graduates from these programs will possess. Additional research issues relate to workplace development and retention issues. How will individuals from these varied educational backgrounds relate in the workplace? What additional corporate training will be needed to smooth out differences? With respect to retention, the more in-demand the area, the harder it will be to retain individuals with that expertise, assuming supply does not keep up with demand. Hence, there is a need for research to better understand how to retain these individuals.

62.4.1.2 IT Industry versus IT Occupation

The educational divide that was introduced in the 1970s signaled a larger IT career divide that is coming to considerably greater recognition in the twenty-first century, as explained later. In the late 1970s, the first major clarification about the IT labor force was published. This clarification emerged from a nationwide study conducted in 1977 that sharpened our emerging understanding of the IT labor force. A report to the U.S. Department of Commerce (Porat 1977) introduced the concept of an "information economy" and defined it as that portion of the labor force that is engaged in the production of information and information tools. This analysis divided the IT labor force into the *primary information sector* and the *secondary information sector*. The primary information sector includes IT workers engaged in the production of information and information tools—such as those who work at computer manufacturers, telecommunications companies, or software development firms. It includes those workers engaged in the production of information processing and communication hardware, software, information systems and services, and information content. In contrast, the secondary information sector includes that portion of the IT labor force that is engaged in information processing work in some other industry—such as banking or healthcare. Whereas the output of the primary information sector is the information or information tools such as computer software or communications technology, the output of the secondary information sector is a noninformation good/service such as healthcare, education, or government services (Trauth 2000, p. 5). While similar work such as programming occurs in both sectors, those in the secondary information sector are not considered to be part of the IT industry. Rather, they can be considered to be part of the *IT occupation*.

The development of IS/IT professionals and the enactment of their career paths increasingly occur along the lines of this division. Those individuals interested in careers in the primary information sector would most likely study computer engineering, computer science, and sometimes information science/informatics. Those inclined toward employment in the secondary information sector would study (management/computer) information systems, information science, and informatics. As a result, a number of IS/IT human resource development and retention issues are emerging that require further research.

While the distinction between primary and secondary information labor forces was articulated 35 years ago, it has not been fully incorporated into the development of IS/IT professionals. Further, the emergence of multiple educational paths (as outlined in Section 62.4.1.1) exacerbates the situation. Hence, there is a need for research to provide a better understanding of the human resource dimensions of the primary and secondary IT labor forces. This research will accompany the growing distinction between careers in the IT industry and careers in the IT occupations, and the number of different IT "disciplines" in which one could prepare for either career path. For example, Trauth et al. (2007) examined the implications of this labor force distinction for the notion of an industry cluster. Their conclusion

---

* The School of Informatics and Computing at Indiana University is one such example.

is that in regions without a high density of firms in the primary information sector, Porter's (2000) notion of "industry cluster" would not be helpful. In response, they introduce the concept of an "IT occupational cluster." In addition, although Gallagher et al. (2010) suggest that the same KSAs are needed by all IT professionals, we suggest that more research is needed in better understanding the difference between the KSAs required of students intending upon careers in Google, for example, and students intending upon a career in health informatics.

62.4.1.3 Lateral Entry from Other Disciplines

The growing ubiquity of technology and a generation of increasingly sophisticated users is leading to an increasingly porous boundary between IS/IT professionals and lay people. The emergence of highly computer literate "power users" of IT is resulting in the growth of lateral entry into the IS/IT profession from other fields. And it is leading to yet another widening of the boundary that marks the IS/IT professional. The resulting IS/IT human resource research issues center around the development and career paths for these individuals. Although recent research has provided new insights into the career paths of individuals in the IT workforce (Joseph et al., 2012), many unanswered questions remain. Is this lateral entry occurring through formal education during career preparation (e.g., through students taking a double major or a minor in IS/IT) or is this lateral entry occurring in the workplace? Research is needed to learn the dimensions of this phenomenon and the implications for IS/IT workforce development both in universities and in companies. Further, additional research is needed about the integration of these "lateral entrants" into the IS/IT departments of companies. Research is needed that focuses on both management of the IS/IT workers and worker development to smooth out differences in educational backgrounds and fill in gaps in knowledge and skills. Finally, research is needed about the "make or buy" decisions relative to such training. Examples of questions of interest to IS/IT managers would be whether to fund employees to pursue an existing professional masters degree, contract with universities to provide a specialized degree tailored for a firm's employees, or hire experienced workers who have already completed specific educational programs.

62.4.1.3.1 Globalization and Offshoring

Another expansion of the boundaries of the IS/IT worker has occurred geographically. The movement toward offshore outsourcing began in the 1970s with the first tentative steps by computer hardware manufacturers in high-wage countries such as the United States to low-wage countries such as Ireland (Trauth 2000). Gradually, the offshore work grew to software development and then to information systems processing so that by the 2000s this was a major part of a company's IS/IT labor force strategy. The original "buy" decision with respect to the IS/IT labor force was to hire people with the needed KSAs or to outsource to a firm which could provide them. But in the past 30 years the outsource decision has become more complex as it quite often means *offshore* outsourcing. This globalization of outsourcing has produced a host of IS/IT human resource issues and accompanying themes requiring further research (Kaiser et al. 2010).

Research themes in this domain are related to differences in national IT infrastructure, the role of IT in a nation's economy, public policies related to IT use, and cultural differences. These themes, in turn, need to be connected to identification and enactment of the new KSAs needed to successfully navigate in a global IS/IT workplace. For example, Huang and Trauth (2008, 2010) have focused on issues related to negotiating differences among people from different nationalities in offshoring work groups as well as how different ethnic or national groups react to varied human resource policies. They have done so through examination of cultural influences on the coordination of IS/IT work in geographically and temporally distributed workplaces, and cross-cultural management issues. Other research focuses on differences in the IS/IT profession at the country level.* Many of these studies focus on international samples of IS/IT workers in emerging economies such as China or India, with sporadic reports from

---

* *Journal of Global Information Management* is replete with research studies in this vein.

Southeast Asia, Africa, or South America. This leaves much room for conducting further research into similarities and differences in the preparation of IS/IT workers and their work practices.

62.4.1.4 Diversified Workforce Composition

The movement toward offshore outsourcing that was firmly in place by the turn of the century was an early indicator of another force shaping IS/IT human resource development and retention in the twenty-first century: recognition of the need for a more diversified IS/IT labor force. Recognition of this need is accompanied by a considerable new research agenda for the field.

62.4.1.4.1 Diversity and Workforce Development

The early stereotype of an IS/IT professional (and indeed the image presented) was a middle-class white man with short hair, white shirt, and dark tie. This button-down image eventually gave way to more relaxed images of a computer professional, but they were still men and still white. The mandate for the twenty-first century IS/IT workforce coming from government, underrepresented groups, and private sector firms is for a radical makeover of the IS/IT labor force. In the future, the task of ensuring a supply of qualified IS/IT personnel is increasingly bound up with issues of diversity. As the use of IS/IT spreads and deepens in societies, the IS/IT profession is challenged with meeting the demand to enlarge the IS/IT workforce by recruiting and retaining personnel from historically underrepresented groups (Panko 2008).

Trauth (2011, pp. 561–562) explains that the motivation for greater human diversity in the IS/IT profession comes not only from labor force needs but also from environmental forces. First is the *consumer argument*. In an information society, in which all citizens are engaged in the consumption of information products, it is crucial that the varying needs of the entire consumer base be represented. The impending retirement of the baby boom presents the second, the *demographic argument*. The departure of a significant component of the workforce is compounded by the shift in countries such as the United States from a white majority. This trend coupled with projected growth in the IT sector over the next 10 years will produce a labor force demand that cannot be satisfied by white men alone. The third motivation for greater diversity is the *innovation argument*. As commodity production increasingly shifts to low-wage countries, developed countries are increasingly turning to innovation for economic sustainability. Greater divergent thinking resulting from more diverse groups of people in a supportive environment should lead to greater innovation. Indeed, Florida (2002) documented a connection between tolerance of diversity and the recruitment and retention of individuals with technology talent. Finally, there is the *equity argument*: all individuals, regardless of gender or ethnicity or other identity characteristics, should have the same opportunities to pursue a career in the IS/IT field.

Accompanying the desire to diversify the IS/IT workforce is the need for research to better understand what is keeping segments of society from entering the IS/IT workforce. An understanding of the barriers is the prelude to intervention efforts that must then follow on and be assessed. With a few exceptions, it was not until the late 1990s that IS/IT personnel research began to consistently include diversity. This is when the question of gender and ethnic diversity in the American IS/IT workforce was raised in research presented at IS/IT conferences. For example, as of 2011 the ACM SIGMIS Computer Personnel Research conference has included 43 papers on the topic of gender and the IS/IT workforce. But with the exception of a paper in 1971 and two in 1995, the remainder of the papers were presented since 1997. In recognition of the growing importance of this theme in IS/IT workforce research, the 2003 conference theme was diversity. While the need for research on gender issues and interventions continues (Trauth 2012; von Hellens et al. 2012), the new frontier for diversity research is race, ethnicity, socioeconomic class, and their intersection with gender (Trauth et al., 2012).

62.4.1.4.2 Diversity and Workforce Retention

Closely aligned with the need for research on the barriers to preparation for and entry into the IS/IT profession is the need for research into workforce retention issues as they relate to a more diversified population. Bias and resistance to women and underrepresented minorities is one fruitful avenue of workforce retention research. Another is the influence of work life balance on retention. This topic has historically only been applied to women in the workplace. But the Millennials, including those in the IS/IT workforce, represent a different kind of worker from the baby boomers (Trauth et al. 2010). This generation of workers—men and women alike—is interested in work life balance. Another dimension of this research area is the cohort of seniors, who may have come out of retirement to reenter the labor force or who may want to work for the pleasure of it, and on their own terms. What motivates such workers and what it takes to retain them are not yet fully understood.

62.4.2 Environmental Influences on IS/IT Development and Retention

The evolution of IS/IT human resource research and practice does not exist in a societal vacuum. It is shaped not only by technological innovation but also by environmental forces. In the early part of the twenty-first century, environmental influences such as global terrorism and economic instability have shaped IS/IT human resource behavior.

The emergence of global terrorism as the century dawned and increasing awareness of the value and risks associated with organizational investments in IS/IT have spawned the growth of a new strain of IS/IT professional: the information security expert. Such individuals are employed by both government and the private sector to protect precious information and technology resources from attack. Hence, there has been the concomitant growth of educational programs in security and risk assessment to prepare a cadre of IS/IT professionals to fill these positions.

In similar fashion, the global economic crisis during the first and second decades of this century has resulted in changes in IS/IT organizations as firms have closed, merged, and depended more heavily on IS/IT to produce internal efficiencies. The call to do more with less is familiar to CIOs and IS/IT personnel alike (Watson 2009). Depending on how it is managed, this call could lead to work exhaustion and turnover of IS/IT personnel (Moore 2000). In a weak economy IS/IT workers may find it more difficult to change jobs, however. At the same time, IS/IT organizations may find they can less readily implement human resource practices—even if they wanted to do so—that would make employment with them more attractive. Research is needed that examines human resource management practices of IS/IT organizations and their effects within the context of each organization's overall human resource strategy over various cycles of economic conditions. It would be informative to include in that research examination of at least the correlates of changing labor market conditions throughout the economic cycles. Of particular interest in this part of the research would be the effect of IS/IT worker undersupply and oversupply on turnover and the ability of IS/IT organizations to implement human resource practices that lead to a desired level of turnover. Given the expanding view of who IS/IT workers are, determining labor market supply and demand for IS/IT workers could be challenging. Further, there is a need for research into creative responses to these economic pressures. For example, Beekhuyzen and Bernhardt (2006) describe an innovative IS/IT human resource practice that shifts the focus from the traditional "one-firm–one-role-centric" employment to "multiple-firm–one-role-centric" shared employment across micro and small enterprises in Australia. Employment sharing is the converse of job sharing. Whereas the latter refers to a labor arrangement in which two or more individuals share one full-time job, the former refers to one person working in multiple organizations.

62.5 Summary

This chapter examined IS/IT human resource development and retention from the perspective of a CIO's questions concerning the appropriate (1) KSAs required of the organization's IS/IT personnel and (2) human resource practices to support the employment relationship with those personnel. We expect CIOs to answer these questions in the context of their organization's human resource strategy.

That could lead them to "build" or "buy" KSAs and, correspondingly, implement human resource practices designed to maintain a long- or short-term employment relationship with IS/IT personnel.

We expect CIOs seeking a long-term relationship with productive IS/IT personnel to follow a "build" strategy. Human resource practices supporting a "build" strategy are those providing greater security, more career development, and, in general, greater outcomes for IS/IT personnel who remain with the organization for a longer period. In contrast, we expect CIOs whose organizational human resource strategy is to "buy" KSAs to use human resource practices that provide less security, less career development, and, in general, fewer outcomes that would lead IS/IT personnel to remain with the organization for an extended period of time.

As technology changes, technical KSAs needed by IS/IT personnel will change. Needed business, interpersonal, managerial, and project management KSAs will depend on the mix of IS/IT personnel roles. That mix is influenced by the make or buy decision.

Besides the organization's human resource strategy, other factors influence the context within which CIOs decide on the KSAs needed by IS/IT personnel and the human resource management practices to support the employment relationship with IS/IT personnel. We discussed three such factors: the evolution of IT, the changing understanding of IS/IT personnel, and environmental forces. Given the trajectory of these forces, we expect them to continue as influential factors in the future.

In this chapter, we have also made several assumptions that might be examined and potentially revised in the future. First, we have assumed that the IS/IT organization seeks to align its human resource practices with the organization's overall human resource strategy. In the future, this basic assumption could be challenged as other determinants of IS/IT human resource strategy clearly emerge from research or practice. Second, we have assumed that two divergent approaches to acquiring KSAs—make or buy—are determined by this alignment and will drive the implementation of human resource practices. Under a resource-based view of the IS/IT organization, IS/IT capabilities are considered strategic assets that provide sustainable competitive advantage. Such resources take time to develop and are considered hard to imitate. Buying these KSAs would not be consistent with a resource-based view of the IS/IT organization. Thus, we assume that when the IS/IT organization is viewed as a strategic asset KSAs are developed rather than bought. Future research could challenge this assumption as well. Finally, we have assumed that an IS/IT organization follows a single human resource strategy. But future conditions could force reexamination of this assumption. As intimated in the discussion of scarce KSAs, it continues to be necessary to understand the challenges that are associated with having more than a singularly focused strategy that implies a single build or buy approach to acquiring KSAs and a single set of human resource management practices for IS/IT personnel.

Radical changes in IS/IT human resource development and retention best practice and research may arise not only from rethinking underlying assumptions guiding research and practice but also from significant environmental influences. An example is the increased concern about information security, resulting from global terrorism and global hacking, that has inspired the rapid development of a specialty—information security—within the IS/IT profession. Finally, the spread and duration of global economic cycles may result in a significant rethinking of the employer–employee relationship for IS/IT personnel. Hence, the need for IS/IT personnel research to inform practice will be greater than ever in the future.

Further Information

Arnold, D. and Niederman, F. (Eds.). (2001). *Communications of the ACM*, Special Issue on *The Global IT Workforce*, 44, 7.

Galliers, R. and Currie, W. (Eds.). (2011). *The Oxford Handbook on MIS*. Oxford, U.K.: Oxford University Press.

Howcroft, D. and Trauth, E.M. (Eds.). (2005). *Handbook of Critical Information Systems Research: Theory and Application*. Cheltenham, U.K.: Edward Elgar Publishing.

Igbaria, M. and Shayo, C. (Eds.). (2004). *Strategies for Managing IS/IT Personnel*. Hershey, PA: Idea Group Publishing.

Lowry, G. and Turner, R. (Eds.). (2007). *Information Systems and Technology Education: From the University to the Workplace*. G. Hershey, PA: Idea Group, Inc.

Niederman, F. and Ferratt, T.W. (Eds.). (2006). *IT Workers: Human Capital Issues in a Knowledge-based Environment*. Greenwich, CT: Information Age Publishing.

Trauth, E.M. (Ed.). (2006). *Encyclopedia of Gender and IT*. Hershey, PA: Idea Group Publishing.

Trauth, E.M., Howcroft, D., Butler, T., Fitzgerald, B., and DeGross, J. (Eds.). (2006). *Social Inclusion: Societal and Organizational Implications for Information Systems*. New York: Springer.

Trauth, E.M. and Niederman, F. (2006). *The Data Base for Advances in Information Systems*, Special Issue on *Achieving Diversity in the IT Workforce*, 37, 4.

Trauth, E.M. and Quesenberry, J.L. (2007). Gender and the information technology workforce: Issues of theory and practice, in Yoong, P. and Huff, S. (Eds.), *Managing IT Professionals in the Internet Age*. Hershey, PA: Idea Group Publishing, pp. 18–36.

References

Agarwal, R. and Ferratt, T.W. (1999). *Coping with Labor Scarcity in Information Technology: Strategies and Practices for Effective Recruitment and Retention*. Cincinnati, OH: Pinnaflex Press.

Agarwal, R. and Ferratt, T.W. (2001). Crafting an HR strategy to meet the need for IT workers. *Communications of the ACM* 44(7), 58–64.

Ang, S. and Slaughter, S. (2004). Turnover of information technology professionals: The effects of internal labor market strategies. *The DATA BASE for Advances in Information Systems* 35(3), 11–27.

Beekhuyzen, J. and Bernhardt, S. (2006). Employment sharing for IT micro and small business, in Niederman, F. and Ferratt, T.W. (Eds.), *IT Workers: Human Capital Issues in a Knowledge-based Environment*. Greenwich, CT: Information Age Publishing, pp. 441–460.

Bharadwaj, A.S. (2000). A resource-based perspective on information technology capability and firm performance: An empirical investigation. *MIS Quarterly* 24(1), 169–196.

Chen, D.Q., Mocker, M., Preston, D.S., and Teubner, A. (2010). Information systems strategy: Reconceptualization, measurement, and implications. *MIS Quarterly* 34(2), 233–259, A1–A8.

Cheney, P.H. (1988). Information systems skills requirements: 1980 & 1988. *SIGCPR '88: Proceedings of the ACM SIGCPR Conference on Management of Information Systems Personnel*, College Park, MD, pp. 1–7.

Combs, J., Liu, Y., Hall, A., and Ketchen, D. (2006). How much do high-performance work practices matter? A meta-analysis of their effects on organizational performance. *Personnel Psychology* 59(3), 501–528.

Ferratt, T.W., Agarwal, R., Brown, C.V., and Moore, J.E. (2005). IT human resource management configurations and IT turnover: Theoretical synthesis and empirical analysis. *Information Systems Research* 16(3), 237–255.

Ferratt, T.W., Prasad, J., and Enns, H. (2012). Synergy and its limits in managing information technology professionals. *Information Systems Research*. 23(4), 1175–1194.

Florida, R. (2002). *The Rise of the Creative Class*. New York: Basic Books.

Gallagher, K.P., Kaiser, K.M., Simon, J.C., Beath, C.M., and Goles, T. (2010). The requisite variety of skills for IT professionals. *Communications of the ACM* 53(6): 144–148.

Huang, H., Kvasny, L., Joshi, K.D., Trauth, E.M., and Mahar, J. (2009). Synthesizing IT job skills identified in academic studies. Practitioner publications and job ads. *SIGMIS CPR '09: Proceedings of the Special Interest Group on Management Information System's 47th Annual Conference on Computer Personnel Research*, Limerick, Ireland, pp. 121–127.

Huang, H. and Trauth, E.M. (December 2008). Cultural influences on temporal separation and coordination in globally distributed software development. *Proceedings of the International Conference on Information Systems*, Paper 134, Paris, France.

Huang, H. and Trauth, E.M. (December 2010). Identity and cross-cultural management in globally distributed information technology work. *Proceedings of the International Conference on Information Systems*, Paper 148, St. Louis, MO.

Joseph, D., Boh, W.F., Ang, S., and Slaughter, S.A. (2012). The career paths less (or more) traveled: A sequence analysis of IT career histories, mobility patterns, and career success. *MIS Quarterly* 36(2): 427–452, A1–A4.

Joseph, D., Ng, K., Koh, C., and Ang, S. (2007). Turnover of Information technology professionals: A narrative review, meta-analytic structural equation modeling, and model development, *MIS Quarterly* 31(3) 547–577.

Joshi, K.D. and Kuhn, K. (2011). What determines interest in an IS career? An application of the theory of reasoned action. *Communications of the Association for Information Systems* 29(1), Article 8.

Kaiser, K.M., Goles, T., Hawk, S., Simon, J.C., and Frampton, K. (2010). Information systems skills differences between high-wage and low-wage regions: Implications for global sourcing. *Communications of Association for Information Systems* 29(1): 605–626.

Lee, D.M.S., Trauth, E.M., and Farwell, D. (1995). Critical skills and knowledge requirements of IS professionals: A joint academic/industry investigation. *MIS Quarterly* 19(3): 313–340.

Luftman, J. and Ben-Zvi, T. (2011). Key issues for IT executives 2011: Cautious optimism in uncertain economic times. *MIS Quarterly Executive* 10(4): 203–212.

Luftmann, J. and Kempaiah, R. (2008). Key issues for IT executives 2007. *MIS Quarterly Executive* 7(2): 99–112.

Markus, M.L. and Benjamin, R.I. (1996). Change Agentry—The next IS frontier. *MIS Quarterly* 20(4): 385–407.

Mata, F.J., Fuerst, W.L., and Barney, J.B. (1995). Information technology and sustained competitive advantage: A resource-based analysis. *MIS Quarterly* 19(4): 487–505.

Miles, R.E. and Snow, C.C. (1984). Designing strategic human resource systems. *Organizational Dynamics* 13(1): 36–52.

Moore, J.E. (2000). One road to turnover: An examination of work exhaustion in technology professionals. *MIS Quarterly* 24(1): 141–168.

Nakayama, M. and Sutcliffe, N.G. (2007). Perspective-driven IT talent acquisition. *SIGMIS CPR '07: Proceedings of the 2007 ACM SIGMIS CPR Conference on Computer Personnel Research*, St. Louis, MO, pp. 171–178.

Panko, R.R. (2008). IT employment prospects: Beyond the dotcom bubble. *European Journal of Information Systems* 17: 182–197.

Porat, M. (1977). *Information Economy: Definition and Measurement*. Washington, D.C.: Office of Telecommunications.

Porter, M. (2000). Location, competition, and economic development: Local clusters in a global economy. *Economic Development Quarterly* 14(1): 15–34.

Ross, J.W., Beath, C.M., and Goodhue, D.L. (1996). Developing long-term competitiveness through IT assets. *Sloan Management Review* 38(1): 31–45.

Starkweather, J.A. and Stevenson, D.H. (2011). PMP® certification as a core competency: Necessary but not sufficient. *Project Management Journal* 42(1): 31–41.

Toh, S.M., Morgeson, F.P., and Campion, M.A. (2008). Human resource configurations: Investigating fit with the organizational context. *Journal of Applied Psychology* 93(4): 864–882.

Topi, H., Valacich, J.S., Wright, R.T., Kaiser, K., Nunamaker Jr., J.F., Sipior, J.C., and De Vreede, G.J. (2010). IS 2010: Curriculum guidelines for undergraduate degree programs in information systems. *Communications of the Association for Information Systems* 26(18): 359–428.

Trauth, E.M. (2000). *The Culture of an Information Economy: Influences and Impacts in the Republic of Ireland*. Dordrecht, The Netherlands: Kluwer Academic Publishers.

Trauth, E.M. (2012). Are there enough seats for women at the IT table? *ACM Inroads* 3(4): 49–54.

Trauth, E.M. (2011). Rethinking gender and MIS for the twenty-first century, in R. Galliers and W. Currie (Eds.), *The Oxford Handbook on MIS*. Oxford, U.K.: Oxford University Press, pp. 560–585.

Trauth, E.M., Joshi, K.D., Kvasny, L., Chong, J., Kulturel, S., and Mahar, J. (2010). Millennials and masculinity: A shifting tide of gender typing of ICT? *Proceedings of the 16th Americas Conference on Information Systems*, Paper 73. Lima, Peru.

Trauth, E.M., Cain, C., Joshi, K.D., Kvasny, L., and Booth, K. (May–June 2012). The future of gender and IT research: Embracing intersectionality. *Proceedings of the ACM SIGMIS Computers and People Research Conference*. Milwaukee, WI, pp. 199–212.

Trauth, E.M., Farwell, D., and Lee, D. (September 1993). The IS expectation gap: Industry expectations versus academic preparation. *MIS Quarterly* 17(3): 293–307.

Trauth, E.M., Reinert, M., and Zigner, M. (April 2007). A regional IT occupational partnership for economic development. *Proceedings of the ACM SIGMIS Computer Personnel Research Conference*. St. Louis, MO, pp. 112–120.

Von Hellens, L., Trauth, E.M., and Fisher, J. (2012). Increasing the representation of women in the information technology professions: Research on interventions. *Information Systems Journal* 22(5): 343–353.

Watson, B.P. (2009). Doing MORE with LESS. *CIO Insight* 106: 18–19.

Wu, W.W. and Zmud, R.W. (2010). Facing the challenges of temporary external IS project personnel. *MIS Quarterly Executive* 9(1): 13–21.

63

Performance Evaluation/ Assessment for IS Professionals

Fred Niederman
Saint Louis University

63.1 Overview

In the world of Moore's law, the rate of growth of computing equipment, software, and digital content insures an ever-changing and expanding set of business and consumer activities in cyberspace. Though many are not necessarily conscious of it when experiencing flow and interacting directly with technology, it is clear that generations of laborers have created the building blocks and applications that have become pervasive. In many cases, these workers have dedicated themselves without a profit motive contributing to open source projects and communities, have built technologies "on speculation" hoping with the help of venture capital to make it the next "killer app", or have been employees creating new features and capabilities as part of an IS workforce. For employers of these IS workers, it is important to provide the materials and environment under which the best computing products will emerge—"best" being a convergence of trade-offs between quality, costs, and scope. From that perspective, it is often viewed to be important in managing to assess their performance for purposes of workers in general and IS in particular to provide feedback, to form a basis for compensation decisions, bonuses, and retention, and to aid in matching assignments and promotions to individual workers. For these reasons, many firms have programs for performance evaluation of all workers including IS professionals. From the inception of IS as a part of the business landscape and, therefore, with the hiring and development of the initial crop of IS workers, concern for applying human resource management practices to IS workers has been expressed. For example, in 1962, the first conference on computer personnel research was undertaken in Los Angeles sponsored by the U.S. Navy and organized by the Rand Corporation. From the very beginning of this group, issues about how to define the tasks of computing professionals, measure the performance of these tasks, and organize human resource activities pertaining to these workers were investigated. It is clearly of significant interest to be able to manage IS employees and, as a major part of that, to be able to evaluate and assess performance. However, this is not necessarily as clear and straight-forward a task as it might seem on the surface to be.

Performance evaluation* of the productivity of information systems (IS) professionals is a complex and multifaceted topic. It is complex because (1) evaluation of the performance of any employee has multiple aspects and effects; (2) IS is a broad field suggesting many varied tasks, the active and passive application of many skills, and job titles that tend to be vague and nonstandard from firm to firm and even department to department; (3) in many cases, the tasks of IS professionals require the creation of technologies or applications that have never existed before in that particular or even in any form; (4) the work of the IS individual often cannot be treated as completely separate from other individuals; (5) both the inputs and outputs of the work are frequently not well specified—as we know requirements are difficult to ascertain and users are often unclear about what a preferred outcome is as they will "know it when they see it"; and (6) the benefits to high-quality performance evaluation may be largely intangible (e.g., higher moral and motivation, more targeted future behaviors, and retention of desirable staff members) but the costs tend to be tangible (e.g., the cost of a consultant sponsored program) or at least quantifiable (e.g., the amount of time spent in various related activities translated into estimates of the value of that time).

Addressing this topic is also difficult due to the hierarchical nature of examining IS professionals as a group of workers with the larger class of all knowledge workers or even all workers in general. Some human resource interventions may affect all workers in similar ways. For example, increases in benefits, direct compensation, or advancement opportunities may be viewed as precursors to more satisfaction or organizational commitment for the bulk of workers of all sorts. However, some interventions may affect knowledge workers or more specifically IS professionals to a qualitative or quantitatively different degree. Prior research (e.g., Couger, 1988) has tended to show IS professionals more strongly motivated by challenge and less motivated by working with other people than workers in other job categories. In this chapter, research specific to IS professionals will be emphasized with additional background information regarding findings pertaining to all workers briefly presented as a thorough examination of all knowledge about performance evaluation is outside the scope of this chapter.

63.2 Configuration of IS Personnel Literature on Performance Evaluation

A scan of the ACM Digital Library shows fewer than a dozen papers on performance evaluation with only one of these, Jiang et al. (2001), published since 1990. This is in contrast to literally hundreds of papers on various measures of technical performance of hardware, networks, and software. It is, of course, entirely appropriate to have much attention on performance of technology (note that some of these papers address tasks that involve the interaction of humans and technology, but the emphasis is on measures of the systems, not on the worker as a member of the staff per se—in other words system variance is measured, not the quality of human input). But the vast disproportion of interest is a bit surprising.

The volume of papers on this topic is not much more exhaustive when examining the "basket of 8" leading IS journals (see Table 63.1). Although the search term "performance evaluation" retrieved more than 100 papers, the majority of these did not address personnel evaluation but rather the evaluation of the IS departmental function and its value, particular systems especially virtual teams and decision support, and a wide range of other IS-related artifacts and programs that can be assessed. One paper

---

* In the IS literature, the phenomenon of interest is generally termed "performance evaluation" but the same general concept in the management literature is called "performance appraisal." The two terms are both intended to refer to the same phenomenon of evaluating and communicating individual worker performance.

TABLE 63.1 Literature Search/ABI Informs by Basket of 8 on Performance Evaluation (JSIS Search through Science Direct)[a]

Journal	Raw Returns	Number Pertaining to IS Personnel	Topics
JMIS	49	5	Webmasters, as independent variable antecedent on turnover, job satisfaction, and intention to telecommute; passing reference in career inventory instrument development
MISQ	43	4	Performance measures for differentiating contract and full-time employees; computer-based monitoring, using computers for general HR assessment; effect of performance evaluation on promotion, by gender; measure job performance as a dependent variable (DV)
ISR	29	0	
JAIS	11	0	
EJIS	12	0	
JIT	19	0	
JSIS	20	1	Career outcomes
ISJ	5	0	
Total	188	10	

[a] A note on method. For each journal, I searched in either ABI Informs or Science Direct on personnel evaluation and the journal title. The number of resulting papers is listed in the second column. I then manually went through each paper reading abstracts to ascertain whether the paper pertained to personnel or the evaluation of some other IS artifact or program.

illustrates an effort to develop fundamental data about the development task opening the door to a basis for performance evaluation. Rasch and Tosi (1992) used a survey method to examine a host of factors that might influence the performance of software developers. These factors included effort, ability, personality factors, and the like. Performance was self-reported on a 1–9 scale based on a single question. The two most influential factors on performance were ability and need for achievement. The goal of the paper is very laudable in that valid precursors to performance outcomes would form a strong basis for the development of performance evaluation criteria. However, it does not resolve questions such as how do you motivate people you need but have lesser ability (or keep people with more ability motivated if those with lesser ability or outcomes are also rewarded). The use of a single self-reported question regarding performance may have been necessary in the research context, but it also does not shed much light on the nuances of what constitutes the differences in performance that may have been reported. The responses to the questionnaire may have been as much about confidence or relative performance (e.g., an average developer may look great in a poor group but poor in a great group) as about performance per se.

Of the 10 papers identified, several used performance evaluation as an independent variable. Several papers by Igbaria and colleagues (Igbaria and Baroudi, 1995; Igbaria and Guimaraes, 1999; Igbaria et al., 1995; Igbaria and Wormley, 1992), for example, used the presence or absence of performance evaluation as a precursor to several dependent variables including career satisfaction and intention to turnover (or leave current employment). At most, this sort of study can provide evidence for the importance of performance evaluation in a nomological net of constructs. Unfortunately it does not help much to explain variance (why it works better in some cases than others) nor to provide help for managers considering ways to implement this sort of program. One notable paper (George, 1996) considered the effects of computer-collected data for the evaluation of people doing other jobs, for example, telephone help line staff.

What accounts for the relatively small number of publications on performance evaluation of IS personnel? Logically, we can speculate about a number of reasons. These include (1) the problem is so simple that the solution is obvious; (2) the problem is so difficult that no solution is possible; (3) the problem does not require a precise answer so relatively simple solutions that address part of the problem are sufficient; and (4) the problem is viewed not as a computer-oriented problem but rather as a management problem and, thus, outside the scope of ACM research or concern.

Without discounting the first three explanations, the last one is strongly supported by a quick search of ABI Informs. Such a search shows more than 3000 retrievals on "performance appraisal" when limited to scholarly journals. Clearly, performance appraisal remains a strong managerial topic. It is possible that many IS managers view such evaluation as a purely managerial problem amenable to management solutions without, necessarily, much adjustment for the particulars of IS work per se. Human resource management scholars take up personnel appraisal generally in search of universal precepts that could be applicable across job types and titles. This makes sense from a pragmatic perspective. Imagine the cost of designing different evaluation systems for each distinguishable type of job from doctors to firefighters to actuaries to scuba instructors.

Following such a view, a firm might be inclined to set up a standardized general program across the firm, but vary some details for each division, department, or perhaps even by individual job title. In such a system, a questionnaire might include items pertaining to attitude and skills (e.g., communication) that are common for all employees, but additionally include other items that relate to measures specific to particular jobs (e.g., ability to manipulate HTML, XML, and a UNIX environment by webmasters) (Wade and Parent, 2001/2002).

This approach, however, leaves open questions. How should IS managers adjust such programs for application to employees in the IS function? Is the search for a standard way to fill in the details for IS functions reasonable? Because there can be so much variance between individual jobs and organizational IS tasks, some may consider there to be no substitute for individuals or workgroups assessing their own work environments.

Is the problem of devising a performance appraisal system for IS employees simple or difficult? It is definitely simple if a firm is happy with the default—not doing any performance evaluation. Ultimately, we know of no legal or regulatory requirement that firms provide performance evaluation to their IS personnel so the default value of simply not doing it is pretty simple and straightforward. It also has the advantage of minimizing cost and avoiding some of the conflicts and difficulties that come with performance evaluation programs trying to balance standards for fairness with recognition of individual preferences for another kind of fairness. As pointed out in 1976, however, this strategy has some risks:

> Every time he [the IS manager] is faced with a choice of assignments, promotions, raises, or offices, he ranks his eligible employees. Unfortunately if his subordinates do not know the evaluation criteria, they may not be able to improve. More than likely, they will assume the criteria by observing who gets the rewards, in such cases, the manager runs two immediate risks—employees may misinterpret both the rewards and/or the required behavior (Ledet, 1976).

In some cases, these risks are insufficient to motivate performance assessment that is calibrated to provide the feedback that does move a workforce into directions of mutual benefit to the organization and employee. A poorly designed or executed performance evaluation program can be run with relatively low cost and generates minimal paperwork, but fails to provide much insight or to generate positive employee behaviors and learning. At an extreme, it may push the most productive to experience frustration, diminished job satisfaction, and, eventually, an intention to leave employment. A poorly designed and administered program can cause more problems than it solves. On the other hand, informal signals that no one will take it seriously anyway can sometimes mitigate some of the problems. There is, of course, the concern in such circumstances that some employees, in particular those who take things literally and/or who care a lot about achievement, will think it is serious and become disaffected upon learning that it is not.

When viewed as a task where the firm wants to achieve benefits above costs and is willing to make significant investment to that end, the "simple" problem becomes significantly more difficult. As mentioned, managers may have to choose between treating everyone identically (to the benefit of some and detriment of others) or differentiating treatment based on individual's preferences (and risking some being treated better rather all being equivalent). Other sources of difficulty stem from the need for a number of factors to be present in the organizational setting such as a trustworthy management and workforce willing to trust in the fairness and thoroughness of the process to spur integrity in the process and participation. Where there are difficulties with design of the program, there are also difficulties with its implementation. If the implementation is done in spotty, intermittent, ambiguous, or coercive manner, it may generate a particularly large number of severe unintended consequences as well as, or perhaps instead of, those that are desired. Whatever the purpose and execution of the program, its existence, its content details, and the quality of its implementation will affect each employee sometimes in unanticipated and possibly negative ways; this may offset the benefits of the intended effects.

Somewhat surprisingly, a significant number of studies in human resource management pertaining to performance evaluation take a relatively contrarian perspective expressing concerns regarding performance evaluation as a management tool. Ikramullah et al. (2012), for example, provide a detailed literature review listing many of the issues and difficulties that make performance assessment unpopular in many quarters. "Poorly managed [performance assessments] can cause various problems for organization, like, disputes among employees and management, anger in staff" (p. 144).

One CIO in an informal and personal e-mail put the same idea this way:

> The performance evaluation for IS/IT personnel is probably the single worst thing that can be done in an organization of knowledge workers. There is no upside and the results are generally disastrous. Calculating scores is insanity disguised as rationality—the weighting factors are just BS.

In spite of this sentiment, it is likely that there are situations where performance assessment of IS personnel can be helpful when programmatically applied in organizational settings. Potential benefits would include providing some relief for individuals from arbitrary and incompetent managers and, when done well, with some legitimate guidance regarding organizational priorities and preferences. It is also important to note that in litigious societies performance appraisals can be used to produce a record supporting (or refuting) claims of unjustified employment termination (Malos et al., 2003). It is likely that some organizations undertake performance evaluation primarily, if not solely, for purposes of record keeping and creation of documents for potential dispute resolution. In light of this attitude, a reasonable research question would be: given the necessity of such record keeping (and the costs entailed), how does one get the most possible benefit from it while reducing its risk of harm to the greatest extent? It is also of note that the tendency toward formal performance appraisal and record keeping expands rapidly as firms grow in size and particularly in number of employees (Kotey and Slade, 2005).

From a research perspective, the problem is difficult to shape into a relatively simple formula or theory for a variety of reasons: (1) the purpose of performance evaluation can vary between and within firms and over time; (2) regardless of stated purpose, the actual activities will generate perceived purposes among the personnel involved; (3) standardized programs have trouble-adjusting techniques given that activities that are motivating to some may be equally demotivating to others; (4) implementing programs that are tailored to different groups have the potential to be viewed as unequal and showing favoritism (whether or not such is intended); (5) rating systems have much room for varied interpretation based on either cognitive or interest conflicts; (6) supervisors and workers are in dynamic relationships with some mutual dependencies (that vary with particular situations) and, as game theory may predict, will derive greater benefits from mutual admiration than from frank constructive feedback; and (7) attempting to gather enough detail to fit all standards and differentiate among all variations may create massive overhead that drains away potential benefits.

63.3 Approaches to Performance Assessment for IS Personnel

63.3.1 Rating Based on Outcomes, Characteristics, and/or Abilities

Much of the performance assessment approach revolves around an analysis of jobs and tasks with subsequent rating of performance for each individual for each atomic identified task or characteristic. In the IS literature, early research suggested that an order of magnitude discrepancy existed between high- and low-performing IS workers (programmers in particular, Dickmann, 1964, but with some nod to debugging and other tasks, Woodruff, 1980a,b). Dickmann (1964) led and discussed a detailed study conducted under the auspices of ACM's Special Interest Group on Computer Personnel Research (see Table 63.2 for criteria and sample questions). Four criteria clusters were proposed (professional

TABLE 63.2   Dickmann (1964) Measures and Description for Original Computer Personnel Research Group Performance Appraisal Instrument (Full Instrument Available through ACM Digital Library)

	Sample Questions
Professional preparation and activity	The number of broad areas of application (e.g., programming systems, numerical calculation, management application, Monte Carlo, statistical, etc.) in which he has demonstrated competence is (anchors 1, 2, 3, 4, or more)
Programmer competence	His program output rate is (unsatisfactory to exceptional)
Dealing with people	He helps others with programming difficulties (almost never to "a lot"/often)
Adapting to job	His ability to work independently with limited supervision is (unsatisfactory to exceptional)

preparation and activity, programmer competence, dealing with people, and adapting to job). Additionally, a 42-item questionnaire was produced and validated.

Woodruff (1980a,b) points out that some prior research suggested an order of magnitude difference between higher- and lower-performing IS workers. It was proposed that if the lower performers could be identified "scientifically" and dismissed, then overall performance would be better served. Of course, this can be a self-fulfilling prophesy as dropping the low performers would immediately shift the statistics, but might constrain the total amount of work accomplished. Of course when Woodruff conducted these studies, the range of IS tasks was much narrower. These days the range of needed abilities may be large enough to accommodate clever managers assigning those best suited to different tasks—super programmers may be writing code to integrate systems where better communicators are writing up application requirements or interviewing users to assign security codes. It is noteworthy that Woodruff's findings did not show such dramatic results but in fact showed a rather small standard deviation implying a tight clustering around the average with few high and low performers among his sample.

Woodruff (1980a,b) also proposed a set of criteria for evaluation of IS worker performance (see Table 63.3 for full descriptions). These criteria are quantity of work, quality of work, job knowledge and skills, judgmental ability, job initiative, adaptability, cooperation, and innovativeness. It is interesting to note in a close reading of the descriptions that two of the eight criteria refer to outputs, three refer to characteristics, and another three to abilities of the employee. Presumably one could directly measure outputs, particularly quantity, by counting them rather than necessarily resorting to subjective rating.

TABLE 63.3   Woodruff (1980a,b) Measures and Descriptions

Measure	Description
Quantity of work	Refers to the volume of useful output associated with the ratee's job assignment; output is of sufficient quality as to satisfy requirements of the ratee's job assignment over which he has direct control.
Quality of work	Refers to those characteristics of the output that enhance its usefulness to the recipient; the ratee has direct control over the output quality, and there is a minimum of subsequent rework of the output.
Job knowledge and skills	Refers to those characteristics of the ratee that enable the ratee to sufficiently solve problems, technical, conceptual, structurally oriented, etc. that are normally encountered in the job assignment.
Judgmental ability	Refers to the ability of the ratee to exercise discretionary behavior to arrive at a wise decision when confronted with problem situations, often unstructured and at short notice.
Job initiative	Refers to those displayed characteristics of the ratee to undertake on his own, without specific instructions, actions, and activities deemed to be desirable in the performance of the job assignment.
Adaptability	Refers to the ability of the ratee to adjust properly and expeditiously to changing and unstructured situations and problems encountered in the job environment.
Cooperation	Refers to those displayed characteristics of the ratee to act or operate jointly with facility users and fellow workers.
Innovativeness	Refers to the demonstrated ability of the ratee to introduce something new or novel to effect a desired change to alleviate or solve problems that are generally characterized by their uniqueness or complexity.

In 1980, when the range of tasks performed by an IS department were much narrower than they are now in the 2010s, a narrower set of abilities might be reasonable. It would seem now, however, that a wide range of abilities could be matched to different jobs, some emphasizing technical and abstraction, others planning, decision-making support, auditing, and the like. Characteristics form an interesting set. If someone has a "bad attitude" why were they hired? If such developed in the course of work, might it not be worth investigating why this happened? Cooperation, as a characteristic of the employee, is also interesting. Doesn't this depend in large measure on the others who might be involved? Is this likely to change from one reporting period to the next?

In another study of performance appraisal of IS personnel, Meyer and Stalnaker (1968) (see Tables 63.4 and Table 63.5) make the stunning observation that managers based performance criteria on the top 10 choices in a wide survey, but based hiring and selection decisions on the bottom 10 choices. Even today, more than a few IS faculty scratch their heads when told by IS managers what they look for in IS workers compared to what they see in advertisements for openings and results of applications of their various students.

Jiang et al. (2001) present a detailed alternative schema for performance evaluation. This paper sketches a relatively formal approach to providing a broad but detailed assessment of IS personnel presumably at the level of the entire IS function rather than separately for each individual. Based on

TABLE 63.4 Meyer and Stalnaker (1968, p. 666)
10 Highest Rated Items by Supervisors

Item	Index Rating
Checking Out Programs K/C	74
Planning Programs K/C	73
Defining Problems K/C	72
Understanding Assignments K/C	70
Works Independently W/S	69
Finds Appropriate Programming Methods W/S	68
Diligent TT	67
Can Handle Complexity K/C	66
Able To Work Under Pressure TT	65
Masters Assignments Speedily W/S	65

K/C, programming knowledge/capability; W/S, working style; TT, temperament traits.

TABLE 63.5 Meyer and Stalnaker (1968, p. 666) 10 Items
Rated Least Important in Appraisals

Item	Index Rating
Teaching Formal Classes P/P	10
Age P/P	16
Number of Professional Societies Belongs To P/P	17
Number of Publications, Talks Given P/P	17
Uses Mathematical Analysis Methods K/C	29
Gives On-The-Job Training W/S	30
Number of Machines Can Handle K/C	32
Personal Appearance P/P	33
Initiates Investigations In Math Analysis K/C	34
Evaluates New Hardware K/C	34

K/C, programming knowledge/capability; W/S, working style; P/P, personal/professional item.

personal and local experience with firms, this approach is at the top end of formality, but somewhat approximates what local employers do with a couple of exceptions that will be noted.

This chapter is based on a rather extensive survey of IS employees and other stakeholders at a particular organization. It outlines a procedure for formal performance assessment with the ultimate goal of aligning corporate and IS goals to assure IS behaviors that reflect these goals and priorities. This chapter notes a number of communication links assessed by different stakeholders. When the stakeholders were in agreement based on similarity of observation, this indicated concurrence. When the stakeholders differed in their level of agreement, these defined gaps of varying size based on the deviation in agreement. What to do about such gaps is not addressed. One assumes that the knowledge of such gaps represents an important step allowing managers the opportunity to intervene (or not to intervene) and to consider the sort of intervention they might want to take. This chapter tends to suggest that such intervention might focus on IS employee training but equally might focus on informing users about the actualities of work in an IS department.

Careful reading of this chapter shows some precursors are necessary for effective implementation of this system including detailed knowledge, agreement, and communication on corporate goals and a management structure in place to implement the described program. Such may have been the case for this particular studied organization, but the creation of these two precursors, realistically, may present significant challenge for the IS department managers.

The approach described by Jiang et al. (2001) is based on the measurement of seven criteria (see Table 63.6 for full description of each). These are quality, project work, general tasks, personal qualities, dependability, teamwork and leadership, and career-related training. At a more detailed level, these criteria pertain to following standards, using procedures and tools, implementing new systems, pursuing effectiveness, participating in project planning, control and communications, addressing user concerns, negotiating, being politically astute, solving problems, having a strong work ethic, completing tasks, securing cooperation, handling multiple tasks, interruptions, and diverse assignments.

The implementation of the actual evaluation process is not presented in much detail in this chapter, but seems to involve the comparison on these various specific points of ratings by the employees in contrast to the rating by managers and other stakeholders. It is in this implementation of data collection and analysis that the process differs most from my observation of local companies. Local companies tend to measure against similar criteria but are more inclined to produce goals as a negotiation between supervisor and worker and periodically measure progress against those goals. We will discuss goal setting and performance assessment in the next section.

Regarding Jiang et al. (2001), it is not difficult to raise additional questions and concerns based on the brief summary of this study that really raises as many questions as it answers. For example, it is not completely clear if this survey process is intended as a one-time analysis to identify gaps and provide a basis for selecting among potential ameliorating activities, or if it is intended to be repeated periodically as a way of assessing the effectiveness of interventions as well as getting snapshot pictures at a time. It is also not fully clear if each stakeholder is assessing each individual or the IS function as a whole. Should, for example, the "having a strong work ethic" criterion show a great difference between perceptions of the user community and the IS employee community, it is not clear how this would reflect on any given individual. It is equally unclear how users not knowing how long it takes to optimize an SQL query be able to assess whether an individual IS employee spent weeks of overtime or whipped the problem out in an hour or two. Assuming, however, that assessment of work ethic is a major differentiator in user and employee attitudes, the path for designing the appropriate intervention is not clear.

A number of other issues are skirted somewhat by the methodology of gathering data through surveys and analysis through gap comparisons. For example, the IS department is measured based on its use of standards, but may be frequently asked to bend standards for purposes of getting work done more quickly or at lower cost. Not only might such decisions be outside the IS department's control, but their realization may involve a natural trade-off such that it would be difficult or impossible to optimize both.

TABLE 63.6 Jiang et al. (2001) Measures and Descriptions

Measure	Description
Quality	Many quality problems undermining system performance stem from IS personnel glossing over design and delivery details. Therefore, IS employees should pay careful attention to these details and feel morally obligated to meet their responsibilities. The evaluation items associated with this measure are related to the overall quality of the work performed. Areas in the spotlight include the following of standards, employing procedures and tools, implementing the new system, and pursuing effectiveness.
Project work	Project management by IS personnel is critical to the success of any project. Adhering to users' schedules, budgets, and system specifications is an important aspect of project success and user satisfaction. The measures that help evaluate performance cover planning, control, and communications.
General tasks	Adherence to task is related to dealing with user concerns. IS professionals are often limited in the resources they need to deliver system specifications. A customer-focused IS professional should be prepared to inform users about information technologies, understand problems related to users' jobs, and anticipate users' needs. User relations drive the details that affect communications, persistence, and understanding of user functions.
Personal qualities	The way IS staff members relate to users is an indicator of user satisfaction. IS professionals need a customer-oriented attitude and be adept at such "soft" interpersonal communication skills as negotiating, managing change, being politically astute, and understanding user desires.
Dependability	Users should be able to rely on the IS staff. Successful IS professionals tackle assignments without having to be prodded by project leaders or users. IS professionals need to maintain a high standard of work performance and dedication to quality. IS professionals are expected to have a strong work ethic in order to meet commitments, seek appropriate solutions to problems, and complete tasks.
Teamwork and leadership	System development is often a team activity; team members should take pride in their work and enjoy working cooperatively. A team player is concerned more about achieving team goals than about individual accomplishment. How well an IS staff member secures cooperation and progresses toward user and organizational goals defines this measure.
Career-related training	The kind of IS worker needed to achieve system success should be able to handle multiple tasks, interruptions, and diverse assignments. Interest and willingness are not sufficient skills; knowledge and ability are also needed.

In a scenario illustrating a difficult problem, an IS organization is excellent on enforcing standards, but viewed as poor on providing immediate service to clients. As an intervention, service delivery times are strictly measured, bonuses, perhaps even pay or tenure are adjusted based on this. As a result, IS personnel simply stop bothering with standards, respond immediately to customer demands, and end up with a sloppy, poorly performing, and high maintenance cost platform or portfolio. On the next round of surveys, the rating of customer service rises to excellent but rating of standards falls to poor. On a 5-point scale, the IS department will never gather more than 6 or 7 points (out of 10 possible when the two scores are combined), thus building into the process the assurance that it will *never* be excellent across the board! Even if not drawn out explicitly, it is hard to imagine in such a scenario performance evaluation will ever be highly motivating. On the other hand, one can imagine a courageous management will be unequivocal regarding objectives. For example, they could specify that user response time is paramount and that extra maintenance costs will be tolerated if they will bring down the response time. Such could easily be visualized for a customer-oriented service such as providing financial or legal advice; however, actual brokers will likely also need flawless transaction processing. In such a system, perhaps standard keeping is not measured or is measured for reference but not included in a performance index.

It is also notable how few of the criteria for this approach are unique to IS employees, though their weighting or specification may vary. Surely, many other jobs require attention to detail, finishing tasks, communicating well, and other attributes. I am actually hard pressed to see anything that is unique to the IS function in the listing specified by Jiang et al. (2001), which may be a good thing if it is to be applied across a firm. This raises the question of whether the IS staff will also be rating the users. What is the quality of user work on stating their information requirements and providing quick and accurate feedback on prototypes? How reliable is their commitment on minimizing funding versus maximizing technical quality? It seems if one is going to look at the performance of the IS department, it should be in context of the other departments and their contribution to the management of information for the benefit of the overall firm. We fear some firms assume that the user is always right and any difficulties, by definition, the responsibility of the IS department. Of course, top management may assume all divisions are wrong or culpable, but that is outside the scope of our discussion.

All of this said, in some circumstances this procedure as an exercise in identifying areas of need or potential growth could be helpful to a firm. Such a list of criteria, while perhaps not distinctly customized for IS work, could serve as a basis for discussions about desired results or outcome of IS work and, in turn, the appropriate criteria for the IS department in a particular setting.

Where such programs are compulsory for IS departments, some portions of the HR literature pertain to general principles for various details of the performance appraisal process. In a recent example, Aguinis, et al. (2012) detail the value of addressing employee strengths rather than weaknesses in the communication of feedback to the employee. This chapter illustrates this technique with a variety of scenarios and dialogues. It is interesting to frame papers such as this in the context of research per se. That there is potential value for a manager to read and be aware of these many techniques (all of which on their face value have potentially positive consequences under some conditions) seems self-evident.

63.4 Critical Incident Technique (A Variation of Rating/Ranking)

The idea of the critical incident technique is to add an element of concreteness to the generalization and impressions of rating systems by noting particular incidents or events as part of the rating. As Dickmann (1964) points out, however, even this system has its complexities:

> By way of illustration, there is a professor at American University who states that he can reliably give you the performance rating of any Civil Service worker by merely having him pass by his desk; 99.8 times out of 100 he can tell you the exact rating for the federal worker. As you may know, the rating system for most federal employees has only three levels. They are outstanding, satisfactory, and unsatisfactory. The professor's mystical ability is diluted when you realize that only 0.2% of federal workers are ever rated in the outstanding or unsatisfactory category. It is therefore apparent that a major drawback to a critical incidents type approach is that if it turns out to be too much work, then all the rates become average.

63.5 Goal Setting

Another general approach to individual performance appraisal that can be applied to IS workers is based on goal setting. At time period one, goals are set that specify outcomes at a future time period. When that future time period arrives, the actual outcomes are compared to those specified in the goals. Some or all of the goals may have been exceeded, others reached precisely, and still others may remain unfulfilled. Many variations on implementation can exist in terms of who sets the goals and measures the results, whether or not movement toward the goals is observable, and what the consequences are of meeting or failing to meet the goals. Communication about goals can also vary from individuals setting their own personal goals without revealing these to anyone through public listing and display of goals (e.g., fund raising goals for organizations like United Way that display movement toward a goal as actual funds are pledged). For example, goals may be the result solely of the employee, of the supervisor, or the result of discussion among them. In practice, one would expect the creation of goals to themselves be complex and the quality of their construction to have much influence on their ultimate effect on both performance and net organizational outcomes.

It is reported (Latham and Locke, 2006) that hundreds of studies consistently show that performance on cognitive and physical tasks is enhanced when goals are difficult and clear. It is further reported that there are some explanations for these findings—goals may prompt the exertion of greater effort, help focus on activities toward these goals to the exclusion of other activities, direct the individual to skill formation toward goal achievement or prompt the use of existing knowledge, and encourage persistence as nearness to the goal achievement is reached. When properly administered, each of these can be positive for the mutual benefit of worker and organization. We would caution that too narrow a focus on activities leading toward particular goals can logically cause a tendency to miss opportunities not specific to those goals. For example, measurement of an extraordinary amount of technical outcomes may discourage an individual worker from taking useful time to help a user client or mentor a colleague. Similarly, focus on efficient production of a number of reports may obscure the disappearance of the need for reports at all (or the inefficient restructuring of the report making it provide greater usability). The goal setter has to be careful about creating goals only for those tasks with concrete outcomes that can be counted and checked off when what is important may involve cultivating future transactions or enriching relationships. As noted by Latham (2004, p. 127), one cannot extrapolate directly on the effect of goal setting on "simple" relative to complex tasks:

> When working smarter rather than harder, when one's knowledge rather than one's effort (motivation) is required, participation in decision-making leads to higher performance if it increases the probability of finding an appropriate strategy for performing the task, and if it increases the confidence of people that the strategy can be implemented effectively.

Other cautions and limitations of goal setting are hinted at by Latham and Locke (2006), "So long as a person is committed to the goal, has the requisite ability to attain it, and does not have conflicting goals, there is a positive, linear relationship between goal difficulty and task performance." In this context setting statement, we can infer three requirements or precursors for observation of the positive effects of goal setting. Latham and Locke (2002) report that some research studies indicate that goals set by supervisors or workers did not seem to affect the growth of productivity. This seems reasonable when the overall process of goal setting is used with motivated staff who commit to the goals in either case. A strong test would be contrasting these types of goal setting in a workplace that was reluctant, suspicious, and untrustworthy. In education, one sees students committed to finishing a program, but not equally committed to studying for each examination or excelling at each paper or project as it comes along. The skill with which goals are designed so that individuals may apply both overall and component-based commitment would also seem likely to affect productivity. Along similar lines, it is clear that the ability to perform the requisite tasks to fulfill the goal is critical. In some circumstances, self-knowledge of abilities would also seem important for calibrating goals that are difficult but still obtainable.

On the other hand, it seems reasonable that in the process of goal seeking, individuals might become aware of lack of abilities that can be affected by training, practice, or absorption of explicit knowledge, thus providing something of a positive feedback loop that would have positive effects whether or not it was sufficient for full goal achievement. Locke and Latham (2002) address conflicting goals in terms of hierarchies of groups where individuals may have goals pertaining to departments that are not fully in accord with the larger goals of divisions, for example. Following the logic of agency theory, their own personal goals may not be in full accord with organizational goals. In moving from the individual level concept of difficult but clear goals stimulating higher levels of performance, addressing these potential conflicts would be critical.

Some additional questions pertaining to applicability of these theories in the actual workplace include (1) how the goal program is managed over repeated time periods; (2) how the program addresses those who have the opportunity to surpass the set goals; and (3) how variance in results (in contrast to the comparison of average performances) is understood. It is not clear how the achievement of a difficult goal changes the abilities and expectations over time. It is not difficult to recall scenarios where IS personnel respond to a dramatic challenge with incredible productivity but end up exhausted, depleted, and burned out. This is a classic situation in the IS literature pertaining to the "forced march." After a few of these, retention may be difficult and productivity may return to lower levels. One might argue, following that productivity is likely to regress to the mean after an exceptional instance anyway. This would suggest the goal setting as a "jolt" strategy applied periodically during crisis but used sparingly at other times. It would seem to depend on the size of the gap between performance before and after goal setting and whether there was slack that needed tightening, or whether such productivity is the result of "selling off inventory."

Similarly, how does the firm address those who have the opportunity to perform above their stated goal? Setting higher goals for the next year, to the extent this is not successfully kept as a secret, can well have a chilling effect on anyone working an iota beyond the stated goal. It could also stimulate "game playing" such as hording successful actions above the goal for application to goals in subsequent periods. Interpretation of results in the setting of rewards for individuals based on different scenarios can also be tricky. How do you reward two people each with a productivity of "75 units" where one's goal was 50 and the other 100? Do you provide the same reward or differentiate based on goal achievement? In one case, having a high goal may have inspired greater effort and higher levels of achievement than without the goal; in the other may have had an inhibiting effect once the goal was met. It would be ironic if setting a higher goal (100 units) pushed that individual to create an extra 15 or so units, but the person were "punished" for having failed to achieve the goal. How likely is that person to set a new goal at all?*

What this boils down to is the lack of knowledge of what accounts for variance among both those setting and those not setting goals. Rather than try to add more precursor variables to the nomological net, perhaps it would be equally worthwhile to examine whether the population is heterogeneous and that different people react differently to the same opportunity? One of the reported streams of goal setting relates to "individual" differences, but I am not convinced that another generic characteristic accounts for this difference. Rather, there may be an interaction between individual work preferences and the details of the goal setting program. For example, in a trustworthy environment one individual might profit greatly from such a program but in an untrustworthy one there will be no cooperation; in contrast, some individuals might always or never respond in either trustworthy or untrustworthy environments.

Schweitzer et al. (2004) warn of another potential danger in the use of goal setting. Their experiment involving a production task with self-reporting of results showed a higher level of exaggerated reporting in difficult goal setting than in "do your best" conditions. Interestingly, their results did not show significantly better task performance for those enjoined to difficult goals rather than to the "do your best" condition. They concluded

> … people with unmet goals were more likely to engage in unethical behavior than people attempting to do their best. This relationship held for goals both with and without economic incentives. We also found that the relationship between goal setting and unethical behavior was particularly strong when people fell just short of reaching their goals.

Under a variety of circumstances, goal setting at an individual, group, or organizational level can result in higher levels of productivity. This approach does, however, require a significant understanding of the necessary and helpful prerequisites for its success. For example, is there a threshold of seriousness and follow-through by management that is required to make it work? We recall the "quality circles" of the 1980s and 1990s that were reputed to achieve such great things in a Japanese context but were hard to

---

* Academics, of course, have this dilemma in grading students—do you reward learning or change from initial to final knowledge and ability level (assuming we know where anyone starts) or do you measure results and reward someone taking courses that are below their entry level?

replicate in an American context because a variety of tangential elements were important to its success. At a minimum, we would expect some level of trust and trustworthiness, some ability to define and set well-constructed goals, some flexibility for moving among goals, and for opportunistically responding to new circumstances as critical for a balanced program.

63.6 Ranking and Forced Ranking

Another approach to performance appraisal consists of requiring supervisors to rank all of their employees from best to worst. Interestingly, this method is discussed in a rather humorous, though these days perhaps politically incorrect, manner by Dickmann (1964, p. 46). It is said that in some organizations the bottom set, say the bottom 10%, was automatically fired, or given a year to get out of the bottom 10%, which is interesting because they would only do so if either new people were not as good, someone else declined, or others improved at a significantly slower rate (or, to be cynical, others simply became less popular). In a time of general reduction of staff, this is one way to decide where to slim down. If there is an expectation that untrained newcomers will do better than existing staff, in a Machiavellian way this might also work.*

The obvious drawbacks to such a system are (1) if the hiring department did anything like a good job, all of the employees may be functioning at an acceptable level; (2) one supervisor's worst 10% may be better than half of the workers for another supervisor; (3) the tasks and assignments for each group are not the same; (4) it creates incentives to prosper oneself at the expense of colleagues whether or not this achieves best overall results; and (5) depending on how the ranking is achieved (which may not be standardized supervisor to supervisor), individuals may have incentives to work only on easy small tasks (if the key is number of completed tasks) or only on difficult long-term ones (if the key is having important pending tasks such that the supervisor cannot afford to lose one).

On the other hand, it can be an immense drain on a work group to have an employee dragging along the bottom of the performance totem pole barely achieving enough to satisfy the letter of the minimum job requirements. A ranking system with forced removal of bottom performers may be a very blunt instrument, but may have some benefits if it is the only way managers have the courage to remove those who really cannot keep up.

63.7 Weekly Conferences

Ledet (1976) describes a relatively straightforward approach to personnel evaluation. In essence, this technique posits a weekly meeting between supervisor and each individual staff member where the staff members present a list of the goals they have prepared for the next week. These goals should mostly come from the staff member modified by the supervisor to account for longer-term priorities. Naturally, this is a difficult technique to administer if the supervisor has a dozen or more people reporting or if staff members are working on projects where their simultaneous presence is needed. As a result, variations of biweekly or monthly meetings and/or meeting with multiple staff members at the same time might be employed. This should be particularly effective where the supervisor has technical expertise or skills and where the staff member has not yet developed sufficient skills for independent planning and execution of tasks on a short-term basis.

---

* In the IS arena, this is not an unknown tactic where the expectation is that new college graduates will bring innovation and the latest in technical capabilities to the workforce. An alternative is to hire a relatively small core of permanent employees in whom one invests heavily and supplement this core with numerous contract workers acquired to fulfill specific and well-defined functions. It is then relatively easy to shift training costs to the contract workers who, in principle, would shift between work and skill retooling. This strategy puts a significant responsibility on IS management to divide work between the segments of the workforce, maintain the core, and integrate the results.

I personally experienced this type of weekly conference as a trainer during an intense short-term process for preparing teachers of English as a Second Language in West Africa. Rather than discuss goals, however, each staff member presented a walkthrough of all lessons and activities planned for the following week and received feedback from all other staff members. Each individual was free to use or ignore the feedback, but, in general, the trainers clarified, added, took out, or otherwise improved presentations, and the exercise was viewed as enabling greater performance based on existing high levels of motivation. I cannot attest to whether this would work elsewhere if the same preconditions were not in place, but when reward is not an issue, only learning and improvement, it worked like a charm.

63.8 Conclusion

Performance evaluation is a tricky business in practice, particularly when it comes to knowledge workers and IS workers in particular. Clearly there are cases where such performance evaluation is helpful, but these may rely more heavily on a set of organizational characteristics being solidly in place rather than the details of the program itself.

The IS personnel literature has not focused on the mechanics of performance evaluation since the 1980s. Perhaps this is because the problem is so simple it is solved, so difficult we have given up, so generic we have accepted whatever management literature comes up with, or reconciled to firms either doing nothing (to customize to IS) or bringing in consultants to help them and, thus, do not perceive much need for knowledge creation in this area. In contrast, the human resource management body of scholarship is filled with many studies of precursors of success in performance evaluation programs, specific tactics such as better ways to focus or deliver feedback, and the like. While such knowledge may be helpful for crafting the framework of general programs, there is little help for IS managers in customizing these specifically for IS workers.

Returning to the CIO referenced at the beginning of this chapter, rather than suggesting an entirely skeptical outlook, his eight prescriptions for managing his IS employees sound pretty good. They are as follows:

1. Make sure all employees have the hardware, software, tools, training, and other resources they need to do the job.
2. Ensure a cultural environment that values them.
3. Be clear on the job and insure proper direction and quick management decisions.
4. Deal with any personnel problems contemporaneously. Thank everyone contemporaneously. Sponsor celebrations.
5. Make sure you have the right people on the bus and the wrong people off the bus.
6. If you have to do reviews have them do their own, make sure they are not too different from your mental model, but do not deal with any variance at the time.
7. Give everyone not under a formal HR remediation plan the same raise.
8. Tell your HR department to take their heads out of their books and get real.

To be fair, we can argue or debate about each of these, what they mean, how they might be implemented, and whether they work in all circumstances. Additionally, the CIO as a rule may not be involved in prevention of litigation on a day-to-day basis. In addition, this particular individual does not work for a gigantic corporation where the dampening effect of performance evaluation on strong managers may be compensated by shielding other employees from relatively too weak managers.

But notice that these admonitions are largely about the responsibilities of the manager creating the setting for work rather than the details of productivity of the individual worker. It is perhaps more about accepting the responsibilities of management rather than looking for formulaic procedures and quick fixes that look scientific but can introduce overhead, unfairness, and a distancing rather than integration of a workforce.

References

Aguinis, H., Gottfredson, R.K., and Joo, H. (2012). Delivering effective performance feedback: The strengths-based approach, *Business Horizons*, 55(2): 105.

Couger, J.D. (1988). Motivators vs. demotivators in the IS environment, *Journal of Systems Management*, 39(6): 36.

Dickmann, R.A. (1964). A programmer appraisal instrument, *SIGCPR '64: Proceedings of the Second SIGCPR Conference on Computer Personnel Research*, New York, pp. 45–64.

George, J.F. (1996). Computer-based monitoring: Common perceptions and empirical results, *MIS Quarterly*, 20(4): 459.

Igbaria, M. and Baroudi, J.J. (1995). The impact of job performance evaluations on career advancement, *MIS Quarterly*, 19(1): 107.

Igbaria, M. and Guimaraes, T. (1999). Exploring differences in employee turnover intentions and its determinants among telecommuters, *Journal of Management Information Systems*, 16(1): 147.

Igbaria, M., Meredith, G., and Smith, D.C. (1995). Career orientations of information systems employees in South Africa, *Journal of Strategic Information Systems*, 4(4): 319–340.

Igbaria, M. and Wormley, W.M. (1992). Organizational experiences and career success of MIS professionals and managers: An examination of race differences, *MIS Quarterly*, 16(4): 507.

Ikramullah, M., Shah, B., Khan, S., Hassan, F.S., and Zaman, T. (2012). Purposes of performance appraisal system: A perceptual study of civil servants in District Dera Ismail Khan Pakistan, *International Journal of Business and Management*, 7(3): 142–151.

Jiang, J.J., Sobol, M.G., and Klein, G. (2001). A new view of IS personnel performance evaluation, *Communications of the ACM*, 44(6): 95–102.

Kotey, B. and Slade, P. (2005). Formal human resource management practices in small growing firms, *Journal of Small Business Management*, 43(1): 16–40.

Latham, G.P. (2004). The motivational benefits of goal-setting, *The Academy of Management Executive*, 18(4), 126–129.

Ledet, M. (1976). Personnel performance evaluation techniques, *SIGUCCS Newsletter*, 6(2): 10–12.

Locke, E.A. and Latham, G.P. (2002). Building a practically useful theory of goal setting and task motivation, *American Psychologist*, 57: 705–717.

Latham, G.P. and Locke, E.A. (2006). Enhancing the benefits and overcoming the pitfalls of goal setting, *Organizational Dynamics*, 36(4): 332–340.

Malos, S., Haynes, P., and Bowal, P. (2003). A contingency approach to the employment relationship: Form, function, and effectiveness implications, *Employee Responsibilities and Rights Journal, Supplement: The Evolving Nature of the Employment Relationship*, 15(3): 149–167.

Meyer, D.B. and Stalnaker, A.W. (1968). Selection and evaluation of computer personnel—The research history of SIG/CPR, *ACM '68: Proceedings of the 1968 23rd ACM National Conference*, New York: ACM Press, pp. 657–670.

Rasch, R. and Tosi, H.L. (1992). Factors affecting software developers' performance: An integrated approach, *MIS Quarterly*, 16(3), 395.

Schweitzer, M.E., Ordonez, L., and Douma, B. (2004). Goal setting as a motivator of unethical behavior, *Academy of Management Journal*, 47(3), 422–432.

Wade, M.R. and Parent, M. (Winter 2001/2002). Relationships between job skills and performance: A study of webmasters, *Journal of Management Information Systems*, 18(3): 71–96.

Woodruff, C.K. (1980a). Job performance evaluation of data processing personnel: An empirical study, *SIGCPR Computer Personnel*, 8(4): 7–10.

Woodruff, C.K. (1980b). Job performance evaluation of data processing personnel: An empirical study, *ACM-SE 18: Proceedings of the 18th Annual Southeast Regional Conference*, Talahassee, FL, pp. 185–188.

64

Financial Information Systems Audit Practice: Implications of International Auditing and Accounting Standards

Micheal Axelsen
*The University of
Queensland*

Peter Green
*The University of
Queensland*

Gail Ridley
University of Tasmania

64.1 Introduction

Commentators frequently identify the harmonization of the financial reporting framework across international borders as essential to consistency and transparency in the financial reporting of global corporate entities (Roussey, 1992). Long considered desirable, the alignment of the major international financial reporting standards remained an unachievable dream in the eyes of many observers (Goeltz, 1991). The Mexican currency crisis of 1994 and then, particularly, the Asian crisis of 1997 (Humphrey et al., 2009) increased the prominence of the issue of the international harmonization of the financial reporting framework. In response, the international regulatory bodies set out International Standards on Auditing (ISA) and International Financial Reporting Standards (IFRS) as template standards to encourage international convergence of the financial reporting framework (Smith et al., 2008). Major developed economies have progressively and increasingly adopted these template standards (IFRS Foundation, 2011; International Federation of Accountants, 2011).

Adoption of these financial reporting standards is difficult. Substantial reforms demand measured consideration (Simpkins, 2010), and the adoption of these standards has been debated in the accounting profession (Herbohn et al., 2008). Such changes in the financial reporting framework have implications for financial audit practice (Kinney, 2005). Financial information systems (IS) audit work undertaken in support of the financial audit (Bagranoff and Vendrzyk, 2000; Muthukrishnan, 2008) is an increasingly important component of the financial audit as reliance upon more sophisticated IS increases (Cerullo and Cerullo, 2005). The implications of the adoption of the international financial reporting framework for financial IS audit practice have not been investigated and reported in the literature previously.

Establishing the requirements of international auditing and accounting standards for financial IS audit work is a necessary first step in a research program developing an understanding of the extent of alignment of financial IS audit practice with the international financial reporting framework. This chapter identifies those elements of the international auditing and accounting standards with the highest implications for the practice of IS audit through a direct review of the international auditing and accounting standards. The validity of the review is evaluated with data from interviews with 27 senior public sector auditors in Australia and Canada. These two countries have each adopted country-specific equivalents of the international auditing and accounting standards. This chapter then derives and presents a benchmark model of financial IS audit from this review of the international standards and relevant best practice IS audit frameworks. The benchmark model prescribes that, in a financial IS audit (using COBIT 4.1 terminology), auditors should be giving the majority of their attention to "deliver and support" processes (40%) while approximately a quarter of their attention should be devoted each to processes in the "planning and organizing" and "acquire and implement" areas. Finally, "monitor and evaluate" processes should be reviewed but to a smaller extent (12%).

The remainder of this chapter is set out as follows. Section 64.2 sets out a brief background to the reforms made to the auditing and accounting standards of Australia and Canada in the name of international harmonization. Section 64.2 also discusses the role of IS audit in the context of the financial audit as well as the development of best practice IS audit frameworks. Section 64.3 then develops a benchmark model of financial IS audit through review of the international auditing and accounting standards and professional best practice IS audit frameworks. Section 64.4 discusses the implications of these results for practice and theory. Finally, Section 64.5 identifies the limitations of this research and the implications of these findings for future research activities and concludes this chapter.

64.2 Auditing and Accounting Standard Reforms

This section discusses the motivation for international harmonization of auditing and accounting standards, and the adoption of the template auditing and accounting standards in Australia and Canada. This section also discusses financial IS audit and developments in the establishment and adoption of best practice IS audit frameworks. This discussion provides the basis for the development of a benchmark model for financial IS audit in Section 64.3.

64.2.1 International Harmonization of Auditing and Accounting Standards

Since the end of World War II and the adoption of the General Agreement on Tariffs and Trade, regulators have sought the harmonization of auditing and accounting standards across developed economies to ensure efficient capital markets and consistent "rules of the game" in international business (Hoarau, 1995, p. 218). A key development was the formation of the Independent Accounting Standards Committee (IASC) in 1973 (Camfferman and Zeff, 2007) and the International Auditing Practices Committee (IAPC) that was formed soon after the creation of the International Federation of Accountants (IFAC) in 1977 (Roussey, 1992).

International harmonization became more prominent after the Mexican currency crisis of 1994 and the Asian crisis of 1997 (Humphrey et al., 2009). The Financial Stability Forum was set up in 1998 by the

G7 to set up international standards, including the auditing and accounting standards (Humphrey et al., 2009). Subsequently, the International Auditing and Assurance Standards Board (IAASB) replaced the IAPC in 2002, and the International Accounting Standards Board (IASB) replaced the IASC in 2002 (Camfferman and Zeff, 2007). These two bodies issued revised ISA and IFRS, respectively.

In 2009, the IAASB finalized the Clarity Project. This project substantially changed the drafting convention of the ISA (Dennis, 2010) and provided new template auditing standards. The drafting and formatting of these template auditing standards were specifically chosen to communicate the requirements clearly and concisely. The initial template ISA identified "mandatory requirements" in bold type, with further explanatory material provided in plain type lettering (Dennis, 2010). The Clarity Project reformatted these requirements and explanatory material to remove the distinction between bold type and plain type lettering, and instead presented "requirements" separately to "application and other explanatory material" in the auditing standard (International Auditing and Assurance Standards Board, 2012).

The International Accounting Standards (IAS) released by the IASC prior to 2002 have remained in effect, although some IAS have been withdrawn and replaced by equivalent IFRS. Each IAS and IFRS sets out the standard's main principles in bold type. Although each paragraph has equal authority, paragraphs set out in bold type provide a summary guide of the standard's requirements (Howieson and Langfield-Smith, 2003).

The requirements of the auditing standards, and the main principles of the IFRS accounting standards, summarize the core principles of the international auditing and accounting standards. Many countries, including Australia and Canada, adopted these template auditing and accounting standards or country-specific equivalents.

64.2.2 Auditing and Accounting Standard Reforms in Australia and Canada

The following discussion provides a high-level overview of the current regulatory framework and the status of the adoption of the international auditing and accounting standards in Australia and Canada.

64.2.2.1 Audit and Accounting Reforms in Australia

The Commonwealth Government of Australia reformed the *Corporations Act* (2001) in response to domestic corporate collapses such as One.Tel and HIH (Jubb and Houghton, 2007) through the Corporate Law Economic Reform Program (CLERP). Prior to the adoption of the ninth installment of the CLERP (CLERP 9), the development and application of auditing and accounting standards had been considered the province of the professional accounting bodies. The CLERP 9 reforms instead moved to provide legislative backing to standards set by the government (Jubb and Houghton, 2007).

The ISA issued by the IAASB were used as a basis for the revised auditing standards of the Australian Auditing and Assurance Standards Board (AUASB) (Mifsud and James, 2006). The ISA were amended for the Australian environment as Australian Auditing Standards (AAS) with legal enforceability (Mifsud and James, 2006). Mandatory AAS requirements were initially emphasized in bold type (Mifsud and James, 2006). As the ISA provided a basis for the reformed auditing standards, the content of the auditing standards aligned highly with their international counterparts. The ISA and the AAS diverged however upon the completion of the IAASB Clarity Project (Dennis, 2010) and the AUASB thus revised AAS for consistency in 2010 (Financial Reporting Council, 2009). The revised AAS thus no longer emphasize requirements in bold type.

Similarly, the Australian Accounting Standards Board adopted Australian equivalents to the IFRS from January 1, 2005 (Boymal, 2007). The reformed accounting standards are referred to as the Australian International Financial Reporting Standards (A-IFRS).

64.2.2.2 Audit and Accounting Reforms in Canada

As with most developed nations after the Mexican and Asian currency crises, significant change occurred in the Canadian regulatory framework after 2000 (Carnaghan and Gunz, 2007). Canadian regulatory powers for corporations are shared between the provinces, territories, and the federal government by Section 91 of the *Constitution Act 1867*. The resultant jurisdictional overlap results in differences between the corporate regulatory framework of each jurisdiction (Carnaghan and Gunz, 2007). However, the relevant corporations and securities acts and regulations require compliance with "generally accepted accounting principles," which are defined to be consistent with the Canadian Institute of Chartered Accountants (CICA) Handbook (Carnaghan and Gunz, 2007). The three Canadian regulatory authorities responsible for setting Canadian auditing and accounting standards in the CICA Handbook are the Canadian Auditing and Assurance Standards Board (AASB, Canada), Accounting Standards Board (AcSB), and the Public Sector Accounting Standards Board (PSAB) (Richardson, 2008).

The AASB (Canada), AcSB, and the PSAB have a central role in setting out the generally accepted accounting principles and the financial reporting framework. CICA provides oversight to these three boards. Thus, in Canada, the auditing and accounting standards have "force of law" under the Canadian regulatory framework (Richardson, 2009).

The AASB (Canada) adopted the Clarity audit standards developed by the IAASB for accounting periods ending on or after December 14, 2010 (Auditing and Assurance Standards Board (Canada), 2009). The AcSB and PSAB similarly sought to harmonize accounting standards with the IFRS standards set out by the IASB (Kanagaretnam et al., 2009), and required the adoption of IFRS-based accounting standards over the period January 1, 2011, to January 1, 2013, for publicly accountable enterprises and government business enterprises (Chartered Accountants of Canada, 2012).

64.2.3 Financial IS Audit

Financial IS audit is IS audit work undertaken in support of the financial audit (Muthukrishnan, 2008). Financial IS audit is an increasingly important component of the financial audit as the entity's reliance upon more sophisticated IS increases—for example, where a company's financial accounts are maintained through a large complex, integrated enterprise resource planning (ERP) system like SAP (Cerullo and Cerullo, 2005). The research presented in this chapter examines the relationship between the international auditing and accounting standards and financial IS audit practice. The profession of IS audit has matured greatly, with the development of the best practice IS audit frameworks of Control Objectives for Information Technology (COBIT) and the IT Assurance Framework (ITAF) (Hardy and Guldentops, 2005; Muthukrishnan, 2008). This research requires an understanding of the relationship between the financial audit and IS audit, and the development of best practice IS audit frameworks.

64.2.3.1 Relationship between the Financial Audit and IS Audit

In response to the challenge of delivering value in the financial audit, financial auditors increasingly adopt a business risk audit (BRA) approach to the audit task (Knechel, 2007). BRA requires a holistic approach to the audit (Knechel, 2007), and the audit profession particularly considers that this approach requires an understanding of the strategic systems dynamics and the audit's wider systems context (Bell et al., 1997). The financial audit explicitly considers and addresses risk in audit planning, and, therefore, the dependence of business processes on information technology (IT) increases the need to understand the controls around IS (ISACA, 2007) and the extent to which these controls can be relied upon in audit planning as part of risk. IS audit's role in risk assessment with the entire audit team is increasingly important (ISACA, 2007) as the adoption of BRA methodologies requires an increasing contribution from IS audit in the areas of planning and risk assessment (Vilsanoiu and Serban, 2010).

IS auditing is "the process of collecting and evaluating evidence to determine whether a computer system safeguards assets, maintains data integrity, achieves organizational goals effectively, and consumes resources efficiently" (Weber, 1999, p. 10). Muthukrishnan (2008) suggests that the specialist IS auditor may carry out IS audit work in support of IT-focused, financial, operational, or regulatory audits. IS audit work undertaken in support of the financial audit is the focus of this chapter and is referred to as the "financial IS audit." In the financial IS audit, the IS auditor supports the financial auditor through feedback, assurances, and suggestions to address the traditional financial auditor's attest objectives that IS be available, maintain confidentiality, and ensure integrity (Sayana, 2002; Weber, 1999).

The IS auditor also increasingly undertakes application-level control reviews as BRA methodologies are adopted for financial audit (Vilsanoiu and Serban, 2010). This increased IS audit reliance is due to the higher IS risk of more complex systems (Bagranoff and Vendrzyk, 2000).

The task of financial audit is increasingly reliant upon the support of the IS audit function due to increased system complexity and the adoption of risk-based audit methodologies. The role of IS audit in this support function is to support the financial auditor in the development of an overall audit judgment.

64.2.3.2 Development of "Best Practice" IS Audit Frameworks

The term "best practice" generates considerable debate. In this chapter, the term "best practice" is aligned with the meaning given the phrase in popular parlance—that is, although different organizations have dynamic capabilities that reflect the context of the organization, there remain commonalities across "effective" firms (Eisenhardt and Martin, 2000, p. 1106). The commonalities contrast with the unique requirements of the organization (Eisenhardt and Martin, 2000), but the commonalities between effective firms encapsulate "best practice."

Several best practice IT management frameworks exist. However, the COBIT framework developed by the Information Technology Governance Institute (2007) and the related ITAF from ISACA (2008b) have the highest level of acceptance within the IS audit profession as a base framework (Hardy and Guldentops, 2005; Muthukrishnan, 2008). COBIT seeks to identify "what" needs to be done in managing IT (De Haes and Van Grembergen, 2004), whereas ITAF is a "comprehensive 'one-stop' framework for IT audit and assurance professionals" (Muthukrishnan, 2008, p. 1). ITAF is the preeminent IT management framework specific to the assurance area.

COBIT 4.1 identifies 4 IT process domains and 34 IT processes (Information Technology Governance Institute, 2007). As a more flexible framework, the ITAF identifies three categories of standards (general, performance, and reporting) and also identifies guidelines and tools and techniques for use by the auditor (ISACA, 2008b). The ITAF guidelines are identified as IT assurance processes, and these IT assurance processes are linked to COBIT IT processes.

64.3 Toward a Benchmark Model of Financial IS Audit

This section develops a benchmark model of financial IS audit through consideration of the international auditing and accounting standards and professional best practice IS audit frameworks. A review of the international financial reporting framework is undertaken to identify those auditing and accounting standards with medium or high impact upon IS audit practice. The identified auditing and accounting standards are then evaluated with senior IS audit practitioners in Australia and Canada. These auditing and accounting standards are subsequently mapped to the ITAF IT assurance processes to develop a benchmark model of financial IS audit in terms of the COBIT IT process domains. The implications of the benchmark model for the theory and practice of financial IS audit are then provided in Section 64.4.

64.3.1 Identified Auditing and Accounting Standards with Medium or High Impact

Two expert members of the research team reviewed the auditing and accounting standards to identify those standards with medium or high impact upon IS audit practice.

64.3.1.1 International Auditing Standards with Medium or High Impact upon IS Audit

The requirements of the auditing standards represent the core components of the auditing standards (Dennis, 2010). The requirements were examined to identify the components of the ISA with important implications for IS audit practice.

TABLE 64.1   International Standards on Auditing Identified as Having Medium or High Impact upon IS Audit Practice

Reference	International Standard on Auditing	Impact
ISA 240	The auditor's responsibilities relating to fraud in an audit of financial statements	Medium
ISA 315	Identifying and assessing the risks of material misstatement through understanding the entity and its environment	High
ISA 320	Materiality in planning and performing an audit	High
ISA 330	The auditor's procedures in response to assessed risks	Medium

An auditing standard was considered to directly affect IS audit practice where specific IT-related requirements to be addressed by auditors were stated. The identification of an auditing standard with important implications for IS audit practice required a direct reference to IS audit practice or the strong implication of a required IS audit procedure. The examples provided in ISACA (2008a) provided a guideline in this assessment.

The requirements of each auditing standard were coded by an expert member of the research team to identify the requirement's implications for IS audit practice. Each requirement was coded as having no, medium, or high impact upon IS audit practice. A second expert member of the research team reviewed this coding. A revised coding of the requirements of the auditing standards was then established and agreed by the research team. Table 64.1 identifies the ISA with medium and high impact upon IS audit practice as a result of this coding approach.

This expert assessment highlights the lack of specific requirements for IS audit practice arising from the international auditing standards. The next section provides the results of a similar assessment in the context of accounting standards.

64.3.1.2 International Accounting Standards with Medium or High Impact upon IS Audit

As with the requirements identified in the international auditing standards, the main principles of the international accounting standards represent the core components of the standards (Howieson and Langfield-Smith, 2003). The main principles of the accounting standards were examined to identify the components with important implications for IS audit practice.

The main principles for each accounting standard were coded by an expert member of the research team to identify each principle's implications for common IS audit practice. Each main principle was coded as having no, medium, or high impact upon IS audit practice. This coding was informed by the assessment of De George et al. (2010) that identified A-IFRS that significantly increased financial audit effort and complexity, while having regard for the indirect nature between accounting standards and IS audit practice.

A second expert member of the research team reviewed this coding. The revised coding of the main principles of the auditing standards was then established and agreed by the research team. Table 64.2 identifies the IAS and IFRS with medium and high impact upon IS audit practice as a result of this coding approach.

The assessment, as expected given the indirect relationship between accounting standards and IS audit practice, does not identify any accounting standards with "high" impact.

64.3.2 Evaluation of Auditing and Accounting Standards with High Impact upon IS Audit

A high-level summary of the international auditing and accounting standards identified as having medium or high impact upon IS audit was evaluated through semistructured interviews undertaken

TABLE 64.2 International Accounting Standards and International Financial Reporting Standards Identified as Having Medium or High Impact upon IS Audit Practice

Reference	International Accounting Standard/ International Financial Reporting Standard	Impact
IAS 19	Employee benefits	Medium
IAS 32	Financial instruments: presentation	Medium
IAS 36	Impairment of assets	Medium
IAS 38	Intangible assets	Medium
IAS 39	Financial instruments: recognition and measurement	Medium
IFRS 2	Share-based payment	Medium

with a purposive sample of public sector auditors. This approach evaluated the expert coding of the auditing and accounting standards from the audit practitioner perspective and provided a robust assessment of the impact of auditing and accounting standards upon IS audit practice.

The sample was obtained through the sponsorship of the Australian Research Council and members of the Australian Council of Auditors General. The sample included senior auditors from Australia and Canada at five different public sector audit offices. Altogether, 27 IS and financial auditors were interviewed to evaluate the expert coding of the auditing and accounting standards as outlined in Section 64.3.1. This approach ensured a diversity of opinion, industry experience, and regulatory frameworks in the sample. Table 64.3 provides basic demographic information for the auditors interviewed.

Each auditor was interviewed separately. A semistructured interview protocol was developed to guide the interviews to ensure the purpose of this research activity was addressed. The questions were designed to elicit clear and unambiguous responses from interviewees in the manner of Lillis (1999). The protocol requested each interviewee to agree or disagree with the expert coding of each auditing and accounting standard's impact. The interview protocol mapped the identified international auditing and accounting standards to their local Australian or Canadian equivalent. Each participant identified additional auditing and accounting standards with important implications for IS audit practice where possible.*

Interviews were recorded and transcribed verbatim for analysis. Interview transcripts were coded by individual sentence blocks into themes and concepts by a member of the research team to create a cross-case analytical matrix as set out by Lillis (1999).

Table 64.4 provides the results of this evaluation for auditing standards. Interviewees indicated common support for all auditing standards identified by the expert coding process as having medium or

TABLE 64.3 Demographic Attributes of Auditors Interviewed to Evaluate the Expert Coding of the Impact of Auditing and Accounting Standards upon Financial IS Audit Practice

Office	Interviewees	Average Experience (years)	Minimum (years)	Maximum (years)	Standard Deviation (years)
Australia	18 financial auditors 6 IS auditors	14.8	3	33	11.2
Canada	2 financial auditors 1 IS auditor	19	12	30	9.6
All audit offices	20 financial auditors 7 IS auditors	15.3	3	33	10.9

---

* The full interview protocol is available from the authors upon request.

TABLE 64.4 Results of the Evaluation of the Expert Coding of the International Standards on Auditing as Having "Medium" or "High" Impact as Set Out in Table 64.1

Reference	International Standard on Auditing	Agree (*n* = 27)	Disagree (*n* = 27)	Common Support
ISA 240	The auditor's responsibilities relating to fraud in an audit of financial statements	24	5	Yes
ISA 315	Identifying and assessing the risks of material misstatement through understanding the entity and its environment	24	3	Yes
ISA 320	Materiality in planning and performing an audit	21	7	Yes
ISA 330	The auditor's procedures in response to assessed risks	22	3	Yes

Note: Interviewees may make ambiguous or conflicting statements and so a single interviewee may indicate both agreement and disagreement with the same expert assessment.

TABLE 64.5 Results of the Evaluation of the Expert Coding of the International Accounting Standards and International Financial Reporting Standards as Having "Medium" or "High" Impact as Set Out in Table 64.2

Reference	International Accounting Standard/International Financial Reporting Standard	Agree (*n* = 27)	Disagree (*n* = 27)	Common Support
IAS 19	Employee benefits	23	19	Yes
IAS 32	Financial instruments: presentation	23	17	Yes
IAS 36	Impairment of assets	24	18	Yes
IAS 38	Intangible assets	20	21	No
IAS 39	Financial instruments: recognition and measurement	22	17	Yes
IFRS 2	Share-based payment	18	23	No

Note: Interviewees may make ambiguous or conflicting statements and so a single interviewee may indicate both agreement and disagreement with the same expert assessment.

high impact upon IS audit practice. The interviewees did not identify auditing standards in addition to those identified by the expert coding process.

Table 64.5 similarly provides the results of this evaluation for accounting standards.

Interviewees consistently agreed that the auditing standards identified during the expert coding process had important implications for IS audit practice. In evaluating the assessment of the expert coding of the international accounting standards, the interviewees were more equivocal regarding the relationship between accounting standards and financial IS audit practice. The results of this research indicate the relationship between accounting standards and financial IS audit practice is more limited and indirect than the relationship between auditing standards and financial IS audit practice.

64.3.3 Financial IS Audit Benchmark Model

As outlined previously, the ITAF (ISACA, 2008b) provides a framework for IT assurance activities in support of diverse purposes, including but not limited to financial IS audit (Muthukrishnan, 2008). Singleton (2010) mapped the ITAF to identify the minimum IT controls to assess in a financial audit. The identification of international auditing and accounting standards with medium or high impact

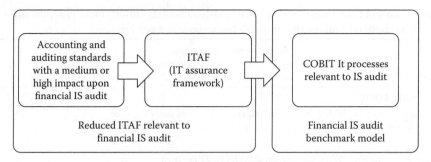

FIGURE 64.1 Process of mapping from the identified accounting and auditing standards with a medium or high impact upon financial IS audit to develop a financial IS audit benchmark model.

upon IS audit practice allows the refinement of this approach to develop a benchmark model of financial IS audit practice. This benchmark model is developed by relating the auditing and accounting standards having an impact upon IS audit with individual ITAF guidelines.

A research assistant experienced in financial IS audit mapped the identified IT-related components of the auditing and accounting standards to the ITAF guidelines. This mapping process was informed by the analysis presented in Singleton (2010). The research team reviewed the map developed by this process. As ITAF identifies the COBIT 4.1 processes related to each guideline, the map provides a basis for the development of the benchmark model for financial IS audit. Figure 64.1 illustrates the process for the development of the benchmark model.

The reduced ITAF financial IS audit benchmark model is presented in Table 64.6. By way of note, the mapping process did not identify any relationship between ITAF guidelines and accounting standards.

The steps of these IS audit methods directly relate to COBIT processes. The reduced ITAF as set out in Table 64.6 identifies the COBIT processes relating to each ITAF guideline. The development of a graphical representation of an IS audit benchmark model using ITAF's mapping to COBIT as a basis is therefore possible. This graphical representation of the financial IS audit benchmark model is provided in Figure 64.2.

As multiple COBIT IT processes support each ITAF guideline, the reduced ITAF implies 213 IT assurance activities in the benchmark model. In this context, an IT assurance activity is the consideration of an IT process in relation to an ITAF guideline. Figure 64.2 thus identifies the relative proportion of COBIT IT process domains on the basis of this map to indicate the relative emphasis of IS audit activity implied by the ITAF. Figure 64.2 presents the financial IS audit benchmark model at a high level by summarizing and aggregating the detail presented in Table 64.6.

In overall terms, the benchmark model provides an indication of the relative proportion of the assurance activities IS auditors "should" be undertaking to support the financial audit in the context of the international financial reporting framework and COBIT 4.1 process domains. The benchmark model prescribes that, in a financial IS audit, auditors should be giving the majority of their attention to "deliver and support" processes (40%) while approximately a quarter of their attention should be devoted to processes in the "planning and organizing" and "acquire and implement" areas each. Finally, "monitor and evaluate" processes should be reviewed but to a smaller extent (12%).

Table 64.7 indicates the specific relative emphasis of the assurance activities of the financial IS audit benchmark model in relation to COBIT 4.1 processes.

The detailed information presented in Table 64.7 allows IS auditors to gauge the relative emphasis of the assurance activities implied by the financial IS audit benchmark model. Taken together, Figure 64.2 and Table 64.7 provide a financial IS audit benchmark model that allows assessment of the alignment of financial IS audit methodologies in use with the international financial reporting framework. In assessing this alignment, it is important to understand that the procedures of audit methodologies do not directly equate to the processes and guidelines of the benchmark model. Each audit methodology procedure is likely to address multiple COBIT 4.1 processes.

TABLE 64.6   Reduced ITAF Model Showing ITAF Guidelines Relating Directly to the Requirements of Those International Standards on Auditing Having a High Impact upon IS Audit Practice

ITAF Guideline	Requirements of International Standard on Auditing (Application and Other Explanatory Material)	Relevant COBIT 4.1 IT Processes
3200 ENTERPRISE TOPICS		
3210 Implication of Enterprise-Wide Policies, Practices, and Standards on the IT Function		
Administrative; Management; Governance	IAS 315 (12) A42–A65	ME2
3230 Implication of Enterprise-Wide Assurance Initiatives on the IT Function		
Impact of General Audit Findings and Recommendation	IAS 240 (17) A12–A13	ME2
3400 IT MANAGEMENT PROCESS		
3410 IT Governance (Mission, Goals, Strategy, Corporate Alignment, Reporting)	IAS 315 (12) A42–A65	PO1, ME1
3600 IT AUDIT AND ASSURANCE PROCESS		
3630 Auditing IT General Controls (ITGCs)		
3630.2 Information Resource Planning	IAS 315 (21) A95–A96	PO1, PO2, PO3, PO4, PO5, PO6, PO7, PO8, PO9, PO10, AI1, AI2, AI3, AI4, AI6, DS1, DS2, DS3, DS4, DS5, DS6, DS7, DS8, DS9, DS10, DS11, DS12, DS13, ME1, ME2, ME3, ME4
3630.3 IT Service Delivery	IAS 315 (21) A95–A96	DS1, DS3, DS4, DS8
3630.4 Information Systems Operations	IAS 315 (21) A95–A96	AI2, DS3, DS4, DS7, DS8, DS9, DS10, DS11, DS12, DS13
3630.5 IT Human Resources	IAS 315 (21) A95–A96	None
3630.6 Outsourced and Third-Party IT Activities	IAS 315 (21) A95–A96	PO4 PO7, PO8, PO9, AI2, AI5, AI6, DS1, DS2, DS4, DS5, DS9, DS10, DS11, DS12, DS13, ME1
3630.7 Information Security Management	IAS 315 (21) A95–A96	PO1, PO4, PO6, PO8, PO9, PO10, AI5, DS5, ME1
3630.8 Systems Development Life Cycle	IAS 315 (21) A95–A96	PO1, PO8, PO10, AI1, AI2, AI3, AI4, AI5, AI6, DS1, DS2, DS3, DS4, DS5, DS7, ME3
3630.9 Business Continuity Plan (BCP) and Disaster Recovery Plan (DRP)	IAS 315 (21) A95–A96	PO4, PO7, PO8, PO9, AI2, AI5, AI6, DS1, DS2, DS4, DS5, DS9, DS10, DS11, DS12, DS13, ME1
3630.10 Database Management and Controls	IAS 315 (21) A95–A96	None
3630.11 Network Management and Controls	IAS 315 (21) A95–A96	PO1, PO2, PO3, PO4, PO6, PO7, PO8, PO9, PO10, AI1, AI2, AI3, AI4, AI5, AI6, DS1, DS2, DS3, DS5, DS8, DS9, DS10, ME1, ME2, ME4
3630.12 Systems Software Support— Other Than Operating Systems	IAS 315 (21) A95–A96	None
3630.13 Hardware Support	IAS 315 (21) A95–A96	None
3630.14 Operating Systems (OSs) Management and Controls	IAS 315 (21) A95–A96	None
3630.15 Physical and Environmental Control	IAS 315 (21) A95–A96	None

TABLE 64.6 (continued)   Reduced ITAF Model Showing ITAF Guidelines Relating Directly to the Requirements of Those International Standards on Auditing Having a High Impact upon IS Audit Practice

ITAF Guideline	Requirements of International Standard on Auditing (Application and Other Explanatory Material)	Relevant COBIT 4.1 IT Processes
3630.17 Identification and Authentication	IAS 315 (21) A95–A96	PO1, PO3, PO5, PO6, PO8, PO9, PO10, AI1, AI2, AI3, AI5, AI6, DS1, DS3, DS4, DS5, DS7, DS9, DS10, DS11, ME1, ME2, ME3
3653 Auditing Traditional Application Controls		
3653.1 Auditing Traditional Application Controls	IAS 315 (21) A97	PO8, PO10, AI1, AI2, AI3, AI4, AI5, AI6, DS1, DS2, DS3, DS4, DS5, DS7
3653.2 Input Authorization	IAS 315 (21) A97	PO6, PO7, PO8, PO9, AI2, AI3, AI4, AI5, AI6, DS1, DS2, DS5, DS8, DS9, DS10, ME1, ME2, ME3, ME4
3653.3 Batch Controls and Balancing	IAS 315 (21) A97	PO8, PO10, AI1, AI2 AI3, AI4, AI5, AI6, DS1, DS2, DS3, DS4, DS5, DS7, ME3
3653.4 Input and Process Editing	IAS 315 (21) A97	None
3653.5 Rejection/Suspense of Transactions	IAS 315 (21) A97	None
3653.6 Batch Integrity in Online or Database Systems	IAS 315 (21) A97	None
3653.7 Processing Procedures and Controls	IAS 315 (21) A97	None
3653.9 Application Access	IAS 315 (21) A97	ME2, ME3, ME4
3653.10 Log Management	IAS 315 (21) A97	None
3653.11 End-user Computing Applications	IAS 315 (21) A97	ME2, ME3, ME4
3653.12 Business Intelligence	IAS 315 (21) A97	None

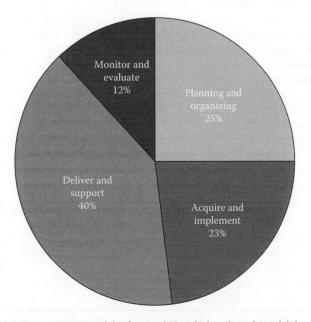

FIGURE 64.2   A graphical representation of the financial IS audit benchmark model developed by mapping the auditing and accounting standards, ITAF guidelines, and COBIT IT processes.

TABLE 64.7   Relative Assurance Activity Emphasis of the Financial IS Audit Benchmark Model in Relation to the 34 IT Processes Identified in COBIT 4.1

COBIT 4.1 Process	Number of ITAF Guidelines Addressed by this COBIT 4.1 Process	Relative Emphasis of Assurance Activities upon this COBIT 4.1 Process
Process domain: Plan and organize		
PO1 Define a strategic IT plan	8	3.8%
PO2 Define the information architecture	2	0.9%
PO3 Determine technological direction	3	1.4%
PO4 Define the IT processes, organization, and relationships	5	2.3%
PO5 Manage the IT investment	2	0.9%
PO6 Communicate management aims and direction	5	2.3%
PO7 Manage IT human resources	5	2.3%
PO8 Manage quality	10	4.7%
PO9 Assess and manage IT risks	7	3.3%
PO10 Manage projects	7	3.3%
Total	54	25.4%
Process domain: Acquire and implement		
AI1 Identify automated solutions	6	2.8%
AI2 Acquire and maintain application software	11	5.2%
AI3 Acquire and maintain technology infrastructure	7	3.3%
AI4 Enable operation and use	6	2.8%
AI5 Procure IT resources	9	4.2%
AI6 Manage changes	9	4.2%
AI7 Install and accredit solutions and changes	0	0.0%
Total	48	22.5%
Process domain: Deliver and support		
DS1 Define and manage service levels	11	5.2%
DS2 Manage third-party services	8	3.8%
DS3 Manage performance and capacity	8	3.8%
DS4 Ensure continuous service	9	4.2%
DS5 Ensure systems security	10	4.7%
DS6 Identify and allocate costs	1	0.5%
DS7 Educate and train users	6	2.8%
DS8 Manage service desk and incidents	5	2.3%
DS9 Manage the configuration	8	3.8%
DS10 Manage problems	7	3.3%
DS11 Manage data	5	2.3%
DS12 Manage the physical environment	4	1.9%
DS13 Manage operations	4	1.9%
Total	86	40.4%
Process domain: Monitor and evaluate		
ME1 Monitor and evaluate IT performance	8	3.8%
ME2 Monitor and evaluate internal control	7	3.3%
ME3 Ensure compliance with external requirements	6	2.8%
ME4 Provide IT governance	4	1.9%
Total	25	11.7%

Source: Information Technology Governance Institute, *COBIT 4.1*, IT Governance Institute, Rolling Meadows, IL, 2007.

64.4 Discussion of Results and Implications for Theory and Practice

This section discusses the results of the research presented in this chapter, and identifies implications of these findings for IS audit theory and practice. These implications inform the discussion presented in Section 64.5.

64.4.1 Discussion of Results

The review of auditing and accounting standards was undertaken with the purpose of supporting the development of a financial IS audit benchmark model. Four ISA (ISA 240, ISA 315, ISA 320, and ISA 330) were identified as having a medium or high impact upon IS audit practice, and the interviewees in this study provided apparent support for the results of this review. These results highlight that the ISA lack specific requirements for IS audit practice.

The relationship between financial IS audit and ISA is stronger than the indirect relationship identified between financial IS audit and the international accounting standards. Although six international accounting standards (IAS 19, IAS 32, IAS 36, IAS 38, IAS 39, and IFRS 2) were identified by expert review as having a medium impact upon financial IS audit practice, the interviewees provided apparent but equivocal support for IAS 19, IAS 32, IAS 36, and IAS 39 only as indicated in Table 64.5. The interviewees participating in this study considered the relationship between accounting standards and financial IS audit practice as limited and indirect.

Relating the international audit and accounting standards to the ITAF in order to identify a reduced ITAF relevant to financial IS audit supports the observation that auditing standards provide limited direct guidance to financial IS audit whereas accounting standards provide limited and indirect guidance. In fact, Table 64.6 indicates that the strongest relationship between the ISA and audit practice exists for ISA 315 paragraph 21, which requires the financial auditor to come to an understanding of the entity's IS.

64.4.2 Implications for IS Audit Theory and Practice

In broad theoretic terms, the financial IS audit benchmark model presented in this chapter considers that the emphasis, or relative audit effort, of financial IS audit methodologies complying with the international financial reporting framework will align with the assurance activity emphasis identified in Figure 64.2 and Table 64.7. The benchmark model presents a relationship between the international reporting framework and IS audit practice. Future research is required to further explore the implications of this theoretical relationship as outlined later.

The results indicate that IS auditors in practice must look to professional guidelines and best practice frameworks such as COBIT 4.1 and the ITAF to develop IS audit methods that address the requirements of the auditing standards. However, both COBIT 4.1 and the ITAF support a wide range of IS audit engagements, not just financial IS audit engagements.

This chapter has developed a benchmark model for financial IS audit by reducing the scope of the ITAF through relating the audit and accounting standards to the ITAF guidelines. The benchmark model provides guidance to auditors seeking to assess the alignment of the financial IS audit methodologies in use with the international financial reporting framework, and additionally guides the development of IS audit methodologies in the financial audit context. Indeed, the benchmark model of Figure 64.2 prescribes that, in a financial IS audit, auditors should be giving the majority of their attention to "deliver and support" processes (40%) while approximately a quarter of their attention should be devoted to processes in the "planning and organizing" and "acquire and implement" areas each. Finally, "monitor and evaluate" processes should be reviewed but to a smaller extent (12%).

64.5 Conclusions, Limitations, and Future Research

This chapter reviewed the international audit and accounting standards increasingly adopted either in whole or as country-specific equivalents by many countries in furtherance of the goal of international harmonization of the financial reporting framework. A detailed analysis of the audit and accounting standards with medium and high impact upon IS audit practice was undertaken and evaluated with public sector auditors, and a financial IS audit benchmark model subsequently developed.

The findings indicate the audit and accounting standards with a medium or high impact upon IS audit practice. The conclusion is that auditing standards have a direct relationship with IS audit practice. The limited number of auditing standards identified in this chapter highlights the absence of auditing standard requirements with important and specific implications for IS audit practice. Further, the relationship between accounting standards and IS audit practice was found to be limited and indirect. This chapter also identifies a benchmark model for financial IS audit that provides for assessment of the alignment of the financial IS audit methodologies in use with the international financial reporting framework, and guides the development of future IS audit methods.

The research has limitations. The benchmark model developed is exploratory, and the evaluation of the assessment of the international audit and accounting standards is limited to senior auditors practicing in Australia and Canada. Both Australia and Canada have similar audit traditions and regulatory frameworks in place. The benchmark model does not consider guidelines and practice requirements other than the international financial reporting framework, and it is apparent that "best practice" resources are widely used in the development of financial IS audit methodologies. Additionally, the evaluation of the expert review sought feedback only from public sector auditors, and thus application of the benchmark model to financial IS audit in private practice may require caution. Nevertheless, the use of a purposive sample with senior auditors from two countries strengthens the representativeness of the sample and the likely applicability to the wider audit profession of the benchmark model developed.

Future research may seek to address the limitations of the analysis presented in this chapter by evaluating the expert review with private sector auditors. Future research assessing financial IS audit practice in comparison with the benchmark model is required to ascertain the extent current practice is in alignment with the international financial reporting framework. Such research also provides an opportunity to consider the robustness of the benchmark model of financial IS audit developed in this chapter. Finally, future refinement of the benchmark model to incorporate "best practice" IS audit resources other than the international financial reporting framework may also yield a more robust benchmark model for financial IS audit as well as for other types of IS audit such as the IT-focused, operational, or regulatory audits identified by Muthukrishnan (2008).

The international harmonization of audit and accounting standards has significant implications for financial IS audit practice. The financial IS audit benchmark model developed in this chapter by reviewing the audit and accounting standards provides an opportunity for financial IS auditors to assess the IS audit approaches in use and to guide their future development. Overall, this chapter presents opportunities for future research in the area of IS audit as well as providing guidance to IS audit professionals looking to align financial IS audit effort with the requirements of the international financial reporting framework set out by the international audit and accounting standards.

Acknowledgments

The material presented in this chapter was supported under Australian Research Council's Linkage Projects funding scheme (Project Number LP0882068). The support of the Partner Organizations CPA Australia, the Institute of Chartered Accountants in Australia, and the Tasmanian Audit Office for this research project is also acknowledged. The authors also gratefully acknowledge the research assistance of Lynne Gerke and Agung Muliawan in the collection and summarization of some of the data for this project.

References

Auditing and Assurance Standards Board (Canada). (2009). Message from the Chair of the Auditing and Assurance Standards Board (AASB) regarding the Adoption of International Standards on Auditing as Canadian Auditing Standards. Retrieved December 06, 2011, from http://www.aasbcanada.ca/item16614.pdf

Bagranoff, N. A. and Vendrzyk, V. P. (2000). The changing role of IS audit among the big five US-based accounting firms. *Information Systems Control Journal*, 5, 33–37.

Bell, T. B., Marrs, F., Solomon, I., and Thomas, H. (1997). *Auditing Organizations Through a Strategic-Systems Lens: The KPMG Business Measurement Process*. Montvale, NJ: KPMG.

Boymal, D. (2007). The work program and priorities of the AASB. *Australian Accounting Review*, 17(42), 3–7.

Camfferman, K. and Zeff, S. A. (2007). *Financial Reporting and Global Capital Markets—A History of the International Accounting Standards Committee 1973–2000*. Oxford, New York: Oxford University Press.

Carnaghan, C. and Gunz, S. P. (2007). Recent changes in the regulation of financial markets and reporting in Canada. *Accounting Perspectives*, 6(1), 55–94.

Cerullo, M. V. and Cerullo, M. J. (2005). How the new standards and regulations affect an auditor's assessment of compliance with internal controls. *JournalOnline* 4, 1–9.

Chartered Accountants of Canada. (2012). IFRSs in Canada. Retrieved August 30, 2012, http://www.cica.ca/applying-the-standards/ifrs/index.aspx/index.aspx

De George, E., Ferguson, C., and Spear, N. (2010). *How much does IFRS Cost? IFRS Adoption and Audit Fees: Evidence from the Australian Experience*. Melbourne, Victoria, Australia: University of Melbourne.

De Haes, S. and Van Grembergen, W. (2004). IT Governance and its mechanisms. *Information Systems Control Journal*, 1, 1–7.

Dennis, I. (2010). 'Clarity' begins at home: An examination of the conceptual underpinnings of the IAASB's clarity project. *International Journal of Auditing*, 14(3), 294–319.

Eisenhardt, K. M. and Martin, J. A. (2000). Dynamic capabilities: What are they? *Strategic Management Journal*, 21(10–11), 1105–1121.

Financial Reporting Council. (2009). *Annual Report 2008–2009*. http://www.frc.gov.au/reports/2008_2009/index.asp (accessed on October 14, 2009).

Goeltz, R. K. (1991). International accounting harmonization: The impossible (and unnecessary?) dream. *Accounting Horizons*, 5(1), 85–88.

Hardy, G. and Guldentops, E. (2005). COBIT 4.0: The new face of COBIT. *Information Systems Control Journal*, 6, 1–4.

Herbohn, K., Ke, Y., and Tutticci, I. (2008). IFRS: Irritating, frustrating and really silly? *InTheBlack*, 78(9), 60–62.

Hoarau, C. (1995). American hegemony or mutual recognition with benchmarks? *European Accounting Review*, 4(2), 217–233.

Howieson, B. and Langfield-Smith, I. (2003). The FRC and accounting standard-setting: Should I still call Australia home? *Australian Accounting Review*, 13(29), 17–26.

Humphrey, C., Loft, A., and Woods, M. (2009). The global audit profession and the international financial architecture: Understanding regulatory relationships at a time of financial crisis. *Accounting Organizations and Society*, 34(6–7), 810–825.

IFRS Foundation. (2011). The move towards global standards. Retrieved March 28, 2012, from http://www.ifrs.org/Use+around+the+world/Use+around+the+world.htm

Information Technology Governance Institute. (2007). *COBIT 4.1*. Rolling Meadows, IL: IT Governance Institute.

International Auditing and Assurance Standards Board. (2012). The clarified standards. Retrieved April 19, 2012, from http://www.ifac.org/auditing-assurance/clarity-center/clarified-standards

International Federation of Accountants. (2011). Basis of ISA adoption. Retrieved December 8, 2011, from http://www.ifac.org/about-ifac/membership/compliance-program/basis-isa-adoption

ISACA. (2007). *CISA Review Manual 2007*. Rolling Meadows, IL: ISACA.

ISACA. (2008a). *G6 Materiality Concepts for Auditing Information Systems*, Vol. 2010. Rolling Meadows, IL: ISACA.

ISACA. (2008b). *Information Technology Assurance Framework: A Professional Practices Framework for IT Assurance*. Rolling Meadows, IL: ISACA.

Jubb, C. and Houghton, K. (2007). The Australian Auditing and Assurance Standards Board after the implementation of CLERP 9. *Australian Accounting Review*, 17(42), 18–27.

Kanagaretnam, K., Mathieu, R., and Shehata, M. (2009). Usefulness of comprehensive income reporting in Canada. *Journal of Accounting and Public Policy*, 28(4), 349–365.

Kinney, W., Jr. (2005). Twenty-five years of audit deregulation and re-regulation: What does it mean for 2005 and beyond? *Auditing: A Journal of Practice & Theory*, 24, 89–109.

Knechel, W. R. (2007). The business risk audit: Origins, obstacles and opportunities. *Accounting, Organizations and Society*, 32(4–5), 383–408.

Lillis, A. M. (1999). A framework for the analysis of interview data from multiple field research sites. *Accounting and Finance*, 39(1), 79–105.

Mifsud, R. and James, B. (2006). Force of law. *InTheBlack*, 76(1), 70.

Muthukrishnan, R. (2008). A prelude to IT assurance framework. *Information Systems Control Journal*, 3, 1–4.

Richardson, A. J. (2008). Due process and standard-setting: An analysis of due process in three Canadian accounting and auditing standard-setting bodies. *Journal of Business Ethics*, 81(3), 679–696.

Richardson, A. J. (2009). Regulatory networks for accounting and auditing standards: A social network analysis of Canadian and international standard-setting. *Accounting Organizations and Society*, 34(5), 571–588.

Roussey, R. S. (1992). Developing international accounting and auditing standards for world markets. *Journal of International Accounting, Auditing and Taxation*, 1(1), 1–11.

Sayana, S. A. (2002). The IS Audit Process. *Information Systems Control Journal*, 1(1), 20-1.

Simpkins, K. (2010). Ringing the changes. *Chartered Accountants Journal of New Zealand*, 89(6), 37.

Singleton, T. W. (2010). The minimum IT controls to assess in a financial audit (Part II). *ISACA Journal*, 2, 1–5.

Smith, L. M., Sagafi-Nejad, T., and Wang, K. (2008). Going International: Accounting and auditing standards. *Internal Auditing*, 23(4), 3–14.

Vilsanoiu, D. and Serban, M. (2010). Changing methodologies in financial audit and their impact on information systems audit. *Informatica Economica*, 14(1), 57–65.

Weber, R. (1999). *Information Systems Control and Audit*. Upper Saddle River, NJ: Prentice Hall.

IX

Information Systems and the Domain of Business Intertwined

65

Strategic Alignment Maturity

Jerry Luftman
Global Institute for IT Management

65.1 Introduction

Business–information technology (IT) alignment refers to applying IT in an appropriate and timely way, in harmony with business strategies, goals, and needs. It has been a fundamental concern of business and IT executives since the 1970s. This definition of alignment addresses:

- How IT is aligned with the business
- How the business should or could be aligned with IT

Mature alignment evolves into a relationship in which IT and other business functions adapt their strategies together. When discussing business–IT alignment, terms such as harmony, linkage, fusion, converged, and integration are frequently used synonymously with the term alignment. It does not matter whether one considers alignment from either a business-driven perspective (IT enabled) or from an IT-driven perspective; the objective is to ensure that the organizational strategies adapt harmoniously. The evidence that IT has the power to transform whole industries and markets is strong.[1-10] In addition to research that will be discussed in this chapter, good examples that have been discussed previously include Amazon.com, Dell, Cisco, and Federal Express. Important questions that need to be addressed include the following:

- How can organizations assess alignment?
- How can organizations improve alignment?
- How can organizations achieve mature alignment?

The purpose of this chapter is to present an approach for assessing the maturity of a firm's business–IT alignment. Until recently, nothing has been available. The alignment maturity assessment described in this chapter provides a comprehensive descriptive and prescriptive vehicle for organizations to evaluate business–IT alignment in terms of where they are and what they can do to improve the alignment. The maturity assessment applies the previous research that identified enablers/inhibitors to achieving alignment[10,11] and the empirical evidence gathered by management consultants who applied the methodology that leverages the most important enablers and inhibitors as building blocks for the evaluation. The foundation for the maturity assessment is also based on the popular work done by the Software Engineering Institute, Keen's "reach and range," and an evolution of the Nolan and Gibson stages of growth.[12–14]

65.2 Why Alignment Is Important

Alignment's importance has been well known and well documented since the late 1970s.[10,15–23] Over the years, it has persisted among the top-ranked concerns of business executives. IT and business alignment was the second highest ranked issue in the recent trends survey of IT leaders from 362 global organizations.[24] Alignment seems more important as companies strive to integrate technology and business in light of dynamic business strategies and the continuously evolving technologies.[2,25] In addition to the importance of alignment, what has not been clear is how to achieve and sustain this harmony between business and IT, how to assess the maturity of alignment, and what the impact of misalignment might be on the firm.[26] To achieve and sustain this synergistic relationship is anything but easy.

There are several reasons why attaining IT–business alignment has been so elusive.

The first reason is that the definition of alignment is frequently focused only on how IT is aligned (e.g., converged, in harmony, integrated, linked, synchronized) with the business. Alignment must also address how the business is aligned with IT. Alignment must focus on how IT and the business are aligned with each other; IT can both enable and drive business change.

The second reason is that organizations (practitioners, consultants, academics) have often looked for a silver bullet. Originally, some thought the right technology (e.g., infrastructure, applications) was the answer. While important, it is not enough. Likewise, improved communications between IT and the business help, but it is not enough. Similarly, establishing a partnership is not enough nor is balanced metrics that combine appropriate business and technical measurements. Clearly, mature alignment cannot be attained without effective and efficient execution and demonstration of value, but this alone is also insufficient. More recently, governance has been touted as the answer—to identify and prioritize projects, resources, and risks. Today, we also recognize the importance of having the appropriate skills to execute and support the environment. Our research has found that all six of these components must be addressed to improve alignment.

The third reason that IT–business alignment has been elusive is that there has not been an effective tool to gauge the maturity of IT–business alignment—a tool that can provide both a descriptive assessment and a prescriptive roadmap on how to improve. As you will see, the insights from the alignment maturity benchmarking provide extensive insights to this longstanding conundrum.

The fourth reason that IT–business alignment has been so difficult to achieve is that there is a tendency in many organizations (even ones where the importance of alignment is recognized) to focus their attention on IT infrastructure considerations. This unbalanced approach can often lead to missed opportunities to identify elements of the business infrastructure that are in need of improvements.

Finally, the fifth reason that the advancement of IT–business alignment has been stalled involves semantic differences in how to refer to it. Disagreements regarding alignment terminology ("linked" vs. "converged"; "integrated" vs. "harmonized") have ironically become a barrier to alignment itself.

Luftman's research suggests that while there is no silver bullet for achieving alignment, progress has been made. In fact, the research demonstrates that "a line" has been drawn. When organizations cross it, they have identified and addressed ways to enhance IT–business alignment. The alignment maturity model is thus both descriptive and prescriptive. CIO's can use it to identify their organization's alignment maturity and identify means to enhance it. Yet, that "line" is dynamic and continually evolving. So alignment can always be improved.

From measuring the six components in organizations in the United States, Latin America, Europe, and India, we found that most organizations today are in Level 3 of a five-level maturity assessment model. Hence, the pronouncement of the "death of alignment" is premature; there is still a long way to go in the journey for aligning IT and business.

Identifying an organization's alignment maturity provides an excellent vehicle for understanding and improving the business–IT alignment. As elaborated on in this chapter, alignment maturity focuses on six important areas. ALL must be simultaneously addressed to improve the harmony among IT and business. Too frequently, consultants and practitioners, looking for the silver bullet, focused their attention on only one or a subset of these important considerations. As companies strive to link technology and business, they must address both

- Doing the right things (effectiveness)
- Doing things right (efficiency)[2,10,25]

In recent years, a great deal of research and analysis focused on the linkages among business and IT,[2,3,10,11,27] the role of partnerships among IT and business management,[13,27] and the need to understand the transformation of business strategies resulting from the competitive use of IT.[28,29] Firms need to change not only their business scope, but also their infrastructure as a result of IT innovation.[30–32] Much of this research, however, was conceptual. Empirical studies of alignment[23,33,34] only examined a single industry and/or firm. Conclusions from such empirical studies are potentially biased and may not be applicable to other industries. These studies lacked the consistent results across industries, across functional positions, and across time. This provided the impetus for defining a vehicle for assessing business, along with providing a roadmap for how best to improve it: IT alignment maturity.

As discussed earlier, alignment maturity evolves into a relationship in which the function of IT and other business functions adapt their strategies together. Achieving alignment is evolutionary and dynamic. IT requires strong support from senior management, good working relationships, strong leadership, appropriate prioritization, trust, and effective communication, as well as a thorough understanding of the business and technical environments. Achieving and sustaining alignment demands focusing on maximizing the enablers and minimizing the inhibitors that cultivate the integration of IT and business.

Alignment of IT strategy and the organization's business strategy is a fundamental principle advocated for several decades.[8,35,36] IT investment has been increasing since its inception, as managers look for ways to manage IT successfully and to integrate it into the organization's strategies. As a result, IT managers need to

- Be knowledgeable about how the new IT technologies can be integrated into the business and with existing/emerging technologies
- Be privy to senior management's tactical and strategic plans
- Be present when corporate strategies are discussed
- Understand the strengths and weaknesses of the technologies in question and the corporate-wide implications[36]

Several proposed frameworks assess the strategic issues of IT as a competitive weapon. They have not, however, yielded empirical evidence, nor have they provided a roadmap to assess and enhance alignment. Numerous studies focus on business process redesign and reengineering as a way to achieve competitive advantage with IT.[37–40] This advantage comes from the appropriate application of IT as a driver and enabler of business strategies.

65.3 Strategic Alignment Maturity

The concept of alignment *maturity* as a necessary precondition for an organization's ability to implement its strategy emerged as a concept in the late 1990s as it became increasingly evident that organizations were, by and large, failing to successfully execute nominally well-defined strategic objectives. Why was this the case? Early research into this issue[41] hypothesized that an organization's ability to successfully implement strategy was related to the "level" of strategic alignment between IT and the business, which reflects both the dynamic nature of alignment and the fact that alignment is, itself, a *process* that reflects key organizational practices that enable (or inhibit, in their absence or misapplication) alignment.[10,41] A model of alignment maturity emerged from this research that reflects these concepts. As Figure 65.1 illustrates, the *Strategic Alignment Maturity* (SAM) model involves the following five conceptual levels of SAM:

1. Initial/ad hoc process—Business and IT are not aligned or harmonized
2. Committed process—The organization has committed to becoming aligned
3. Established focused process—SAM established and focused on business objectives
4. Improved/managed process—Reinforcing the concept of IT as a "value center"
5. Optimized process—Integrated and coadaptive business and IT strategic planning

Each of the five levels of alignment maturity focuses, in turn, on a set of six components based on practices validated in 2001 with an evaluation of 25 "*Fortune 500*" companies. As of the writing of this chapter, 362 Global 1000 organizations from around the world (and several hundred smaller companies) and 2100 business and IT executives have participated in formally assessing their IT business alignment maturity. Some of the insights from these assessments are discussed in the section of this chapter that describes the different maturity components. Assessments continue to be performed.

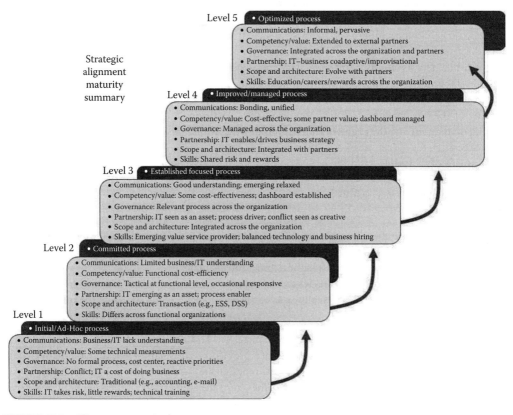

FIGURE 65.1 Alignment maturity summary.

As discussed earlier, organizations have often looked for a silver bullet to improve the alignment of IT–business. Some thought the right technology (e.g., infrastructure, applications) was the answer. While important, it is not enough. Likewise, improved communications between IT and the business help, but are not enough. Similarly, establishing a partnership is not enough, nor is balanced metrics that combine appropriate business and technical measurements. More recently, governance has been touted as the answer—to identify and prioritize projects, resources, and risks. Today, we also recognize the importance of having the appropriate skills to execute and support the environment. Research has found that all six of these components must be addressed to improve alignment.

Additionally, there has not been an effective tool to gauge the maturity of the IT–business alignment— a tool that can provide both a descriptive assessment and a prescriptive roadmap on how to improve. From measuring the six components in organizations in the United States, Latin America, Europe, and India, most organizations today are in a low Level 3 of a five-level maturity assessment model; there are still many opportunities for improvement.

The six IT–business alignment criteria are illustrated in Figure 65.2 and are described at a more detailed level in the following section. All six must be addressed to ensure mature alignment; looking for a single answer will just not do it. These six criteria are

1. *Communications maturity*—Ensuring effective ongoing knowledge sharing across organizations
2. *Competency/value measurement maturity*—Demonstrating the value of IT in terms of contribution to the business
3. *Governance maturity*—Ensuring that the appropriate business and IT participants formally discuss and review the priorities and allocations of IT resources

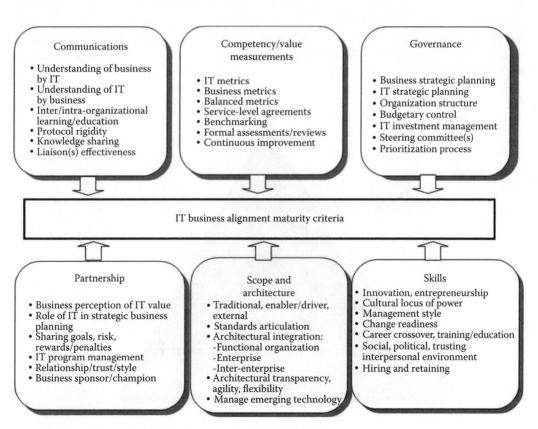

FIGURE 65.2 Alignment maturity criteria.

4. *Partnership maturity*—How each organization perceives the contribution of the other, the trust that develops among the participants, and the sharing of risks and rewards
5. *Scope and architecture maturity*—The extent to which IT is able to
   - Go beyond the back office and into the front office of the organization to directly impact customers/clients and strategic partners
   - Assume a role supporting a flexible infrastructure that is transparent to all business partners and customers
   - Evaluate and apply emerging technologies effectively
   - Enable or drive business processes and strategies as a true standard
   - Provide solutions customizable to customer needs
6. *Skills maturity*—Human resource considerations such as training, salary, performance feedback, and career opportunities are assessed to identify how to enhance the organization's cultural and social environment as a component of organizational effectiveness

Knowing the maturity of its strategic choices and alignment practices makes it possible for a firm to see where it stands with respect to its "alignment gaps" and how it can close these gaps. The pyramid in Figure 65.3 illustrates the alignment gap on each level of alignment maturity vividly. The five levels of alignment maturity are introduced in this section and then will be elaborated in the following section.

Level 1: Initial or ad hoc processes. Organizations at Level 1 generally have poor communications between IT and the business and also a poor understanding of the value or contribution the other provides. Their relationships tend to be formal and rigid, and their metrics are usually technical rather than business oriented. Service-level agreements (SLAs) tend to be sporadic. IT planning or business planning is ad hoc. IT is viewed as a cost center and considered "a cost of doing business." The two parties also have minimal trust and partnership. IT projects rarely have business sponsors or champions. The business and IT also have little to no career crossovers. Applications focus on traditional back-office support, such as e-mail, accounting, and HR, with no integration among them. Finally, Level 1 organizations do not have an aligned IT–business strategy.

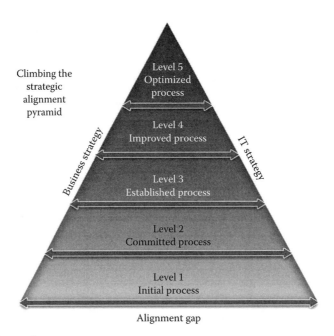

FIGURE 65.3   Alignment gaps.

Level 2: Committed processes. Organizations at Level 2 have begun enhancing their IT–business relationship. Alignment tends to focus on functions or departments (e.g., finance, R&D, manufacturing, marketing) or geographic locations (e.g., United States, Europe, Asia). The business and IT have a limited understanding of each other's responsibilities and roles. IT metrics and service levels are technical and cost-oriented, and they are not linked to business metrics. Few continuous improvement programs exist. Management interactions between IT and the business tend to be transaction-based rather than partnership-based, and IT spending relates to basic operations. Business sponsorship of IT projects is limited. At the function level, there is some career crossover between the business and IT. IT management considers technical skills the most important for IT.

Level 3: Established, focused processes. In Level 3 organizations, IT assets become more integrated enterprise-wide. Senior and mid-level IT management understand the business, and the business's understanding of IT is emerging. SLAs begin to emerge across shared or acted upon. Strategic planning tends to be done at the business unit level, although some inter-organizational planning has begun. IT is increasingly viewed by the business as an asset, but project prioritization still usually responds to "the loudest voice." Formal IT steering committees emerge and meet regularly. IT spending tends to be controlled by budgets, and IT is still seen as a cost center. But awareness of IT's "investment potential" is emerging. The business is more tolerant of risk and is willing to share some risk with IT. At the function level, the business sponsors IT projects and career crossovers between business and IT occur. Both business and technical skills are important to business and IT managers. Technology standards and architecture have emerged at both the enterprise level and with key external partners.

Level 4: Improved, managed processes. Organizations at Level 4 manage the processes they need for strategic alignment within the enterprise. One of the important attributes of this level is that the gap has closed between IT understanding the business and the business understanding IT. As a result, Level 4 organizations have effective decision making and IT provides services that reinforce the concept of IT as a value center. Level 4 organizations leverage their IT assets enterprise-wide, and they focus applications on enhancing business processes for sustainable competitive advantage. SLAs are also enterprise-wide, and benchmarking is a routine practice. Strategic business and IT planning processes are managed across the enterprise. Formal IT steering committees meet regularly and are effective at the strategic, tactical, and operational levels. The business views IT as a valued service provider and as an enabler (or driver) of change. In fact, the business shares risks and rewards with IT by providing effective sponsorship and championing all IT projects. Overall, change management is highly effective. Career crossovers between business and IT occur across functions, with business and technical skills recognized as very important to the business and IT.

Level 5: Optimized processes. Organizations at Level 5 have optimized strategic IT–business alignment through rigorous governance processes that integrate strategic business planning and IT planning. Alignment goes beyond the enterprise by leveraging IT with the company's business partners, customers, and clients, as well. IT has extended its reach to encompass the value chains of external customers and suppliers. Relationships between the business and IT are informal, and knowledge is shared with external partners. Business metrics, IT metrics, and SLAs also extend to external partners, and benchmarking is routinely performed with these partners. Strategic business and IT planning are integrated across the organization, as well as outside the organization.

Figure 65.4 summarizes the results of the 362 Global 1000 companies that have gone through the assessment to date. It illustrates where there is relative agreement regarding which areas are strong and which are weak, and it identifies the gaps between business and IT executive's opinions. The *Y*-axis represents the five levels of maturity; the *X*-axis expands each of the six components of maturity. The maturity elements highlighted in bold tend to be assessed as the strongest, while the italicized elements are those that are assessed as the lowest (hence the areas least aligned). Note that within each of the six components there are diamond lines representing the assessments from IT executives, and circle lines showing the corresponding assessments from business executives. The areas where the circle and

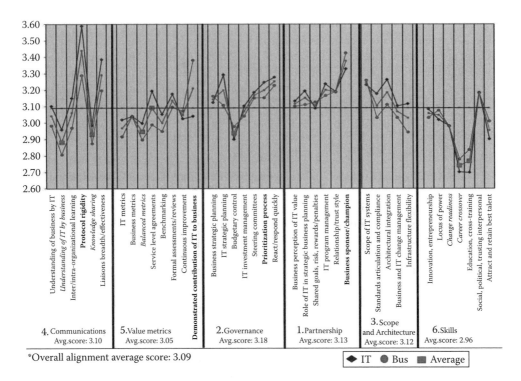

FIGURE 65.4   Overall SAM assessment maturity.

diamond lines converge or overlap depict areas where there is the most agreement (and thus synergy) between business and IT. Conversely, areas with large gaps between the circle and diamond lines are the ones that show disagreement among IT and business executives; these are area that need to be reconciled. For example, Figure 65.4 illustrates a tighter synergy between business and IT in the areas of partnership and skills than for communications. The major elements will be discussed later.

Figure 65.5 summarizes these results by region. A general trend that Figure 65.5 illustrates is that across most components, Asian organizations have higher maturity scores (denoted by circle lines), followed by American and Latin American organizations (denoted by triangle and large diamond lines, respectively), and then European organizations (denoted by small diamond lines). The pattern of maturity scores for Australian organizations (denoted by star lines) reveals that in some dimensions they score as high as or higher than Asian organizations, while for other dimensions they score lower than all other regions. (Since only one African organization is represented in the data, no trends for African organizations are assumed.)

With an overall average maturity score of 3.09, it is clear that there are still opportunities to improve the IT business relationship; alignment is not dead.

A similar graph may be used to plot the responses from an individual organization assessment to identify opportunities for improvement (using the assessment as a prescriptive tool) and to benchmark things such as how a specific organization compares to

- The overall average set of responses
- The responses from exemplar organizations
- Other organizations in their industry (finance, pharmaceutical, utility, retail, healthcare, education)
- Respondents from similar positions (e.g., CIO's, CEO's, CFO's) in other firms

Once the maturity level is understood, the assessment method provides the organization with a prescriptive roadmap that identifies opportunities for enhancing the harmonious relationship of business and IT. This alignment process is expanded in this chapter.

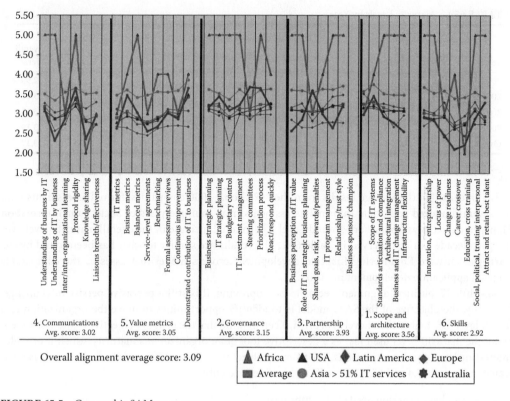

FIGURE 65.5 Geographic SAM summary.

65.4 Six Strategic Alignment Maturity Criteria

This section describes each of the six components (illustrated in Figure 65.2) that are evaluated in deriving the level of SAM. Examples taken from actual assessments illustrate the kinds of insights that can be identified. Most organizations today appear to be around Level 3, as illustrated in Figure 65.6.

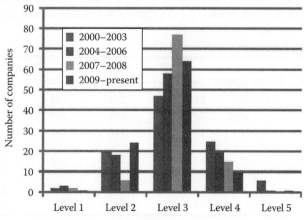

FIGURE 65.6 Distribution of SAM scores.

TABLE 65.1   Maturity Levels by Year

Years	Number of Companies	% of Companies in Level 1	% of Companies in Level 2	% of Companies in Level 3	% of Companies in Level 4	% of Companies in Level 5	Overall Average
2000–2003	83	2	20	47	25	6	2.99
2004–2006	99	3	18	58	20	1	3.06
2007–2008	64	2	6	77	15	0	3.11
2009–Present	116	1	24	64	10	1	3.19
Overall	362	1.25	12.92	56.67	27.92	1.25	3.17

Overall alignment average score: 3.09

That means that the average results from the 362 Global 1000 companies' formal assessments (and the several hundred additional informal assessments from multiple years of Society for Information Management surveys) to date are around Level 3. A gradual increase in the overall maturity level over the past decade can be observed in Table 65.1. The results are similar to what has been found by the Carnegie Software Engineering Institute development process model that assesses the comparable stages of application development maturity.

So, while IT business alignment seems to be improving, it is still a pervasive persistent challenge. Naturally, the objective of the SAM model is to identify opportunities to move the organization to a higher level (i.e., higher than Level 3) of SAM. Keep in mind that the primary objective of the assessment is not the maturity level used just as a descriptive tool of an organizations maturity; albeit it provides interesting benchmark comparisons. The primary objective of the assessment is to understand (as illustrated in Figures 65.4 and 65.5) where IT and business executives

- Agree that a criterion needs to be improved
- Agree that a criterion is good, but can be better
- Disagree with how good/bad a criterion is
- Desire to focus their efforts to improve

As illustrated in Figure 65.5, there were differences in the overall SAM alignment scores by region. On average, Asian organizations had higher scores than their American, Australian, and European counterparts. The SAM scores by criteria and by region are summarized in Table 65.2. It is valuable to benchmark organizations by geography as well as comparing alignment trends across the geographies. This will be discussed later.

When there is agreement among the participants regarding the criteria assessment, the model can be used as a prescriptive roadmap to identify how alignment maturity can be improved. However, when there is disagreement, the key stakeholders (i.e., any groups or individuals who can affect or are affected

TABLE 65.2   Geography Maturity by Component

Geography	Number of Companies	Communication	Competency	Governance	Partnership	Scope of IT Architecture	Skills	Overall Average
Europe	61	2.85	2.63	2.94	2.78	3.01	2.70	2.82
Australia	28	2.88	3.01	3.15	2.96	2.96	2.68	2.94
United States	184	2.93	2.93	3.07	3.09	3.12	2.84	3.00
Latin America	44	3.17	2.94	3.03	3.16	3.27	3.00	3.10
Asia	44	3.52	3.59	3.58	3.64	3.60	3.55	3.58
Africa	1	4.0	3.71	4.13	4.4	4.0	4.0	4.05

Overall alignment average score: 3.09

TABLE 65.3   Industry Maturity by Component

Industry Name	Number of Companies	Communications	Competency	Governance	Partnership	Tech. Scope	Skills	Overall Average
Retail	10	3.54	3.52	3.70	3.75	3.73	3.45	3.62
Hotel/ entertainment	10	3.23	3.39	3.38	3.49	3.77	3.35	3.44
Service	48	3.26	3.23	3.34	3.36	3.34	3.35	3.31
Insurance	11	3.19	3.34	3.51	3.31	3.34	2.89	3.26
Oil/Gas/Mining	13	3.05	3.31	3.36	3.06	3.37	2.98	3.19
Manufacturing	54	3.21	3.03	3.10	3.15	3.24	3.07	3.13
Financial	106	2.93	2.87	3.13	3.17	3.14	2.82	3.01
Utility	7	3.09	3.21	2.64	2.76	3.40	2.75	2.98
Pharmaceutical	15	2.95	2.77	2.98	2.89	3.00	2.97	2.93
HealthCare	18	2.94	2.85	3.02	2.97	3.04	2.73	2.92
Transportation	21	2.77	2.97	2.82	2.93	2.84	2.68	2.84
Government	9	2.91	2.54	3.03	2.99	2.86	2.49	2.80
Chemical	7	2.75	2.64	2.86	2.81	3.08	2.44	2.76
Telecommunication	11	2.68	2.68	2.94	2.69	3.03	2.44	2.74
Agriculture	11	2.54	2.61	3.13	2.63	2.50	2.41	2.64
Educational	11	2.66	2.46	2.83	2.53	2.72	2.56	2.63

Overall alignment average score: 3.09

by IT in the firm) need to understand the points of view of the participants and come to an agreement regarding the criteria and how to enhance it. The organization cannot identify an appropriate road to take if they cannot come to agreement regarding where they want to go. Once the group has identified an agreed to list of areas for improvement, they can proceed to use the model as a prescriptive roadmap. Hence, it is not the maturity "number" that is important. It is what the organization does as a result of identifying how they can work together to improve the alignment maturity.

The next six subsections discuss each of the SAM criteria in more detail and include examples of how they manifest themselves in organizations. These examples have been abstracted from recent research done with a number of major U.S. and global organizations.[11] Table 65.3 summarizes the data from this research across the six SAM components by industry. In terms of their alignment maturity, it is evident that industries can vary considerably in their overall scores. For example, the service sector outperformed the transportation sector by an overall score of 3.31–2.84, while the gap between the retail and educational sectors was almost a full point (3.62 vs. 2.63).

Since this research is still ongoing and the companies that have participated have been assured anonymity, it is not possible to share the specific names of the participating organizations. However, each section illustrates specific issues of SAM that have been uncovered in the research and identifies the industry of the participating organizations.

65.4.1 Communications

Effective exchange of ideas and a clear understanding of the key ideas that ensure successful strategies are high on the list of enablers and inhibitors to alignment. Too often, there is little business awareness on the part of IT or little IT appreciation on the part of the business. The 362 Global 1000 benchmark firm results indicate that 21% of the IT organizations either do not understand or have a limited understanding of business, while 39% of the business executives either do not understand or have a limited understanding of IT. Given the dynamic environment in which most organizations find themselves, ensuring ongoing knowledge sharing across organizations is paramount.

Many firms choose to employ people in formal interunit "liaison" roles or cross-functional teams to facilitate this knowledge sharing. The key word here is "*facilitate.*" Some organizations have facilitators whose role is to serve as the sole conduit of interaction among the different units of the organization. This approach tends to stifle rather than foster effective communications. Rigid protocols that impede discussions and the sharing of ideas should be avoided. The 362 Global 1000 benchmark firm results indicate that 54% of the firms identify liaisons as a major opportunity for improvement.

For example, a large aerospace company assessed its communications alignment maturity at Level 2. Business–IT understanding is sporadic. The relationship between IT and the business function could be improved. Improving communication should focus on how to create the understanding of IT as a strategic business partner by the businesses it supports rather than simply a service provider. The firm's CIO made the comment that there is "no constructive partnership." However, in an interview with the firm's Director of Engineering and Infrastructure, he stated that he views his organization as a "strategic business partner." One way to improve communications and, more importantly, understanding would be to establish effective business function/IT liaisons that facilitate sharing of knowledge and ideas.

In a second case, a large financial services company's communication alignment maturity placed it in Level 2 with some attributes of Level 1. Business awareness within IT is through specialized IT business analysts, who understand and translate the business needs to other IT staff (i.e., there is limited awareness of business by general IT staff). Awareness of IT by the firm's business functions is also limited, although senior and mid-level management are aware of IT's potential. Communications are achieved through biweekly priority meetings of the senior and middle-level managers from both groups, where they discuss requirements, priorities, and IT implementation. But it is still a Level 2 because of the effectiveness of the interaction.

In a third example, a large utility company's communication alignment maturity places it at Level 2. Communications are not open until circumstances force the business to identify specific needs. There is a lack of trust and openness among some business units and their IT team. IT business partners tend to be bottlenecks in meeting commitments. IT's poor performance in previous years left scars that have not healed.

From a geographic perspective (as illustrated in Table 65.2), Asian organizations achieved the highest level of maturity in the communications component with an overall score of 3.52, followed by Latin America with a score of 3.17. The United States, Australia, and European scores were 2.93, 2.88, and 2.85, respectively.

65.4.2 Competency/Value Measurements

Too many IT organizations cannot demonstrate their value to the business in terms that the business understands. Frequently, business and IT metrics of value differ. A balanced "dashboard" that demonstrates the value of IT in terms of contribution to the business is needed. The 362 Global 1000 benchmark firm results indicate that two-thirds of the firms can improve this important area.

Service levels that assess IT's commitments to the business often help. However, the service levels have to be expressed in terms that the business understands and accepts. The service levels should be tied to criteria (see Section 65.4.4) that clearly define the rewards and penalties for surpassing or missing the objectives. The 362 Global 1000 benchmark firm results indicate that 63% of the firms can significantly improve their SLAs.

Frequently, organizations devote significant resources to measuring performance factors. However, they spend much less of their resources on taking actions based on these measurements. For example, an organization that requires analyzing ROI before a project begins, but then does not review how well objectives were met after the project was deployed provides little to the project's success. It is important to assess these criteria to understand (1) the factors that lead to missing the criteria and (2) what can be learned to improve the environment continuously.

For example, a large aerospace company assessed its competency/value measurement maturity to be at Level 2. IT operates as cost center. IT metrics are focused at the functional level, and SLAs are technical in nature. One area that could help to improve maturity would be to add more business-related metrics to SLAs to help form more of a partnership between IT and the business units. Periodic formal assessments and reviews in support of continuous improvement would also be beneficial.

A large software development company assessed its competency/value measurement maturity at Level 3. Established metrics evaluate the extent of service provided to the business functions. These metrics go beyond basic service availability and help desk responsiveness, evaluating such issues as end-user satisfaction and application development effectiveness. The metrics are consolidated on to an overall dashboard. However, because no formal feedback mechanisms are in place to react to a metric, the dashboard cannot be considered to be managed.

At a large financial services company, IT competency/value was assessed at Level 2 because the company uses cost-efficient methods within the business and functional organizations. Balanced metrics are emerging through linked business and IT metrics, and a balanced scorecard is provided to senior management. SLAs are technical at the functional level. Benchmarking is not generally practiced and is informal in the few areas where it is practiced. Formal assessments are done typically for problems and minimum measurements are taken after the assessment of failures.

Table 65.2 shows significantly different IT competency SAM scores across regions. Asian organizations lead the way with an overall score of 3.59, followed by Australian firms with a score of 3.01; Latin American firms (2.94) are followed closely by American firms (2.93). European organizations scored the lowest in this dimension, with a score of 2.63.

65.4.3 Governance

The considerations for IT governance were defined briefly in Figure 65.1. Ensuring that the appropriate business and IT participants formally discuss and review the priorities and allocation of IT resources is among the most important enablers/inhibitors of alignment. This decision-making authority needs to be clearly defined. The 362 Global 1000 benchmark firm results indicate that 57% of the firms should be improving this important component of alignment.

For example, IT governance in a large aerospace company is tactical at the core business level and not consistent across the enterprise. For this reason, they reported a Level 2 maturity assessment. IT can be characterized as reactive to CEO direction. Developing an integrated enterprise-wide strategic business plan for IT would facilitate better partnering within the firm and would lay the groundwork for external partnerships with customers and suppliers.

A large communications manufacturing company assessed its governance maturity at a level falling between 1 and 2. IT does little strategic planning because it operates as a cost center and, therefore, cost reduction is a key objective. In addition, priorities are reactive to business needs as business manager's request services.

A large computing services company assessed their governance maturity at Level 1+. A strategic planning committee meets twice a year. The committee consists of corporate top management with regional representation. Topics or results are neither discussed nor published to all employees. The reporting structure is federated with the CIO reporting to a COO. IT investments are traditionally made to support operations and maintenance. Regional or corporate sponsors are involved with some projects. Prioritization is occasionally responsive.

From a geographic perspective (as illustrated in Table 65.2), Asian organizations achieved the highest level of maturity in the governance component with an overall score of 3.58. Australian organizations came in second with a score of 3.15, followed by American companies with a score of 3.07. Latin American and European organizations earned scores of 3.03 and 2.94, respectively.

65.4.4 Partnership

The relationship that exists between the business and IT organizations is another criterion that ranks high among the enablers and inhibitors. Giving the IT function the opportunity to have an equal role in defining business strategies is obviously important. However, how each organization perceives the contribution of the other, the trust that develops among the participants, ensuring appropriate business sponsors and champions of IT endeavors, and the sharing of risks and rewards are all major contributors to mature alignment. This partnership should evolve to a point where IT both enables AND drives changes to both business processes and strategies. Naturally, this demands having a good business design where the CIO and CEO share a clearly defined vision.

For example, a large software development company assessed their partnership maturity at a level of 2. The IT function is mainly an enabler for the company. But IT does not have a seat at the business table, either with the enterprise or with the business function that is making decisions. In the majority of cases, there are no shared risks because only the business will fail. Indications are that the partnership criterion will rise from Level 2 to Level 3 as top management sees IT as an asset, and because of the very high enforcement of standards at the company.

Partnership for a large communications manufacturing company was assessed at Level 1. IT is perceived as a cost of being in the communications business. Little value is placed on the IT function. IT is perceived only as help desk support and network maintenance.

For a large utility company, partnership maturity was assessed at a level of 1+. IT charges back all expenses to the business. Most business executives see IT as a cost of doing business. There is heightened awareness that IT can be a critical enabler to success, but there is minimal acceptance of IT as a partner.

Partnership for a large computing services company was assessed at Level 2. Since the business executives pursued e-commerce, IT is seen as a business process enabler as demonstrated by the web development. Unfortunately, the business now assigns IT with the risks of the project. Most IT projects have an IT sponsor.

From a geographic perspective (as illustrated in Table 65.2), Asian organizations have a partnership maturity score of 3.64. The next closest region was Latin America, with a partnership score of 3.16. The American, Australian, and European partnership scores were 3.09, 2.96, and 2.78, respectively.

65.4.5 Scope and Architecture

This set of criteria tends to assess IT maturity. The extent to which IT is able to

- Go beyond the back office and into the front office of the organization
- Assume a role supporting a flexible infrastructure that is transparent to all business partners and customers
- Evaluate and apply emerging technologies effectively
- Enable or drive business processes and strategies as a true standard
- Provide solutions customizable to customer needs

Scope and architecture were assessed at a level of 2+ at a large software development company. This is another area where the company is moving from Level 2 to Level 3. ERP systems are installed and all projects are monitored at an enterprise level. Standards are integrated across the organization and enterprise architecture is integrated. It is only in the area of inter-enterprise that there is no formal integration.

A large financial services company assessed their scope and architecture at Level 1. Although standards are defined, there is no formal integration across the enterprise. At best, only functional integration exists.

Once again, Asian companies led in this dimension, scoring 3.6 for the scope and architecture component. Latin America came in second, with a score of 3.27, followed by the United States, which scored 3.12. European and Australian organizations scored 3.01 and 2.96, respectively.

65.4.6 Skills

Skills were defined in Figure 65.1. They include all of the human resource considerations for the organization. Going beyond the traditional considerations such as training, salary, performance feedback, and career opportunities are factors that include the organization's cultural and social environment. Is the organization ready for change in this dynamic environment? Do individuals feel personally responsible for business innovation? Can individuals and organizations learn quickly from their experience? Does the organization leverage innovative ideas and the spirit of entrepreneurship? These are some of the important conditions of mature organizations. The 362 Global 1000 benchmark firm results indicate that 55% of the benchmarked firms do not effectively support career crossover opportunities (IT into the business and the business into IT) and that 55% of the benchmarked firms do not effectively support education cross training.

For example, a large aerospace company assesses its skills maturity at Level 2. A definite command and control management style exists within IT and the businesses. Power resides within certain operating companies. Diverse business cultures abound. Getting to a non-political, trusting environment between the businesses and IT, where risks are shared and innovation and entrepreneurship thrive, is essential to achieve improvements in each of the other maturity tenets. Organizational behavior research has demonstrated that sharing information that is based on expertise is often the most successful approach to influencing others to cooperate and trust one another.[42]

Skills maturity at a large computing services company is assessed at a level of 1. Career crossover is not encouraged outside of top management. Innovation is dependent on the business unit, but in general is not encouraged. Management style is dependent on the business unit, but is usually command and control. Training is encouraged but left up to the individual employee.

Finally, from a geographic perspective, Asian companies earned a maturity score of 3.55. Latin American organizations came in second, earning a score of 3.00. American, European, and Australian organizations received SAM Skill scores of 2.84, 2.70, and 2.68, respectively.

Amazon.com—Alignment Maturity Enables Strategic Transformation

"I buy books from Amazon.com because time is short and they have a big inventory and they're very reliable."

– Bill Gates

"I cannot live without books."

– Thomas Jefferson

By now, everyone knows the brand Amazon.com and its transformation of the staid world of selling books via the Internet. Even the phrase "*being Amazoned*" has entered the business lexicon as synonymous with being blindsided by an unexpected competitor that uses e-business as an enabler of strategic transformation of the business (not to mention the industry). Could Amazon.com have impacted its industry as it did without a high level of SAM? Let us look at some facts about

(continued)

(continued)

Amazon—using the SAM criteria discussed earlier in this chapter—that might support the hypothesis that a high level of SAM contributed to Amazon's ability to transform the book sales industry:

- *Communications*—Jeff Bezos, the founder and CEO of Amazon, understood the power of IT to transform the business of selling books. Understanding of the enabling power of IT by the business (and vice versa) is part of the "warp and woof" of how Amazon operates. Without IT, there is no Amazon.com! Amazon's initial management team continuously reinforces its entrepreneurial culture with messages to its employees from Bezos and the executive team about Amazon's vision and philosophy ("being able to buy anything, anytime, anywhere at the greatest store on Earth …"). Employees understand that being with Amazon is about making history.

- *Competency/value measurements*—Amazon passionately focuses on the metrics of excellent customer service and the effectiveness of its advertising strategy. For example, Amazon produces a weekly report that shows the effectiveness of *each* online advertisement placed in terms of the customer traffic that was generated and the revenue for each customer visit to Amazon's site. Given the intense competition from other web retailers, Amazon continuously monitors its performance against its rivals and adjusts business processes to set the pace rather than react to changes in the environment.

- *Governance*—Amazon has divided its business into three segments: (1) "mature" businesses such as books, music, and DVDs that have to "pay their way," (2) "early-stage" businesses such as games, consumer electronics, and home improvement that need nurturing and feeding, and (3) Amazon's end user–based businesses. This governance model defines how decisions will be made regarding the investment Amazon will make in its IT portfolio and how that investment will be funded (i.e., self-funded or funded from outside investment). It is also clear that Jeff Bezos and a small, tight-knit group of executives make the key strategic business and IT decisions as IT strategy *drives* the business of Amazon.com.

- *Partnership*—It is clear from Amazon's business model—which is enabled through IT—that IT is valued as the engine of industry transformation. The high level of partnership between business and IT is driven by the vision of the CEO—Jeff Bezos—and continuously reinforced by Amazon's ability to coadapt its IT architecture to an evolving business model. Amazon has also heavily invested in other online retailers (members of its "Commerce Network") to extend its retailing capabilities to other products such as jewelry, consumer electronics, etc.

- *Scope and architecture*—The scope of Amazon's business and its IT architecture is driven by the external expectations of its customers. Amazon has implemented standard IT architectures around personalization features and order fulfillment that ensure consistent service to customers around the world.

- *Skills*—Amazon is known in the industry as a firm that aggressively recruits people who are bright, energetic, entrepreneurial, and customer-centric. Amazon's focus on hiring people who are skilled at personalization technologies and other "customer experience" technologies strengthened their brand and their reputation as being "absolutely fanatical about our customers." Amazon also developed deep expertise in fulfillment technologies ("picking, packing, and shipping") that are essential to delivering against high customer expectations of service.

In the discussion in the following section, we look at the different levels of SAM using the earlier criteria. What alignment maturity level would you assign to Amazon.com and why?

Sources: Harvard Business School Case Study 9-897-128, April 9, 1998
Harvard Business School Case Study 9-800-330, September 5, 2000

65.5 Results by Geography and Industry

As noted earlier, results from the assessment from the 362 Global 1000 companies by region reveal higher alignment scores by Asian organizations across all maturity components. As a group, they scored 3.58, as compared to 3.00 for the United States and 2.82 for Europe. A complete illustration of regional SAM scores by component is shown in Table 65.2 and Figure 65.5.

What was it that made Asian organizations score higher in every SAM component than their European, American, and Latin American counterparts? An examination of the factors that have led to the remarkable success of India's service sector offers several lessons. A strong culture that promotes communication between employees, the emphasis of CMM/CMMI-based continuous improvement efforts, and well-planned strategies that promote organizational flexibility are just some of the factors that are illustrated in the Wipro sidebar later in this chapter. These cases offer yet more evidence that achieving IT–business alignment is not a matter of finding a single "silver bullet."

An analysis of SAM data shows that the retail, hotel/entertainment, service, and insurance sectors performed well above the average SAM score of 3.09 in all dimensions. As noted in Table 65.3, these industries scored 3.62, 3.44, 3.31, and 3.26, respectively. (Note: There were relatively few retail and hotel/entertainment companies in the sample, however.) The most well-represented industry in the Global 1000 was the financial industry, which earned an overall SAM rating of 3.01. The manufacturing industry performed closest to the mean, with an overall average of 3.13.

India IT Service Case—Wipro

Wipro is a global IT service company, headquartered in Bangalore, India, that was established in 1945. It entered into IT services in the 1980s. Its revenues have grown at a CAGR of 21% over a six decade period. Today, it is a US$3.47 billion organization with over 66,000 employees with operations in 19 countries.

It is the world's first PCMM Level 5 software company and the first IT service company to use Six Sigma. Among the top 3 offshore business process outsourcing (BPO) service providers in the world, it has almost 600 clients. Wipro is a strategic partner to five of the top 10 most innovative companies in the world. It is also the world's first company outside the United States to receive the IEEE software process award. It is the largest independent R&D service provider in the world. It is the first Indian IT service provider to be awarded Gold-Level status in Microsoft's Windows Embedded Partner Program. It is the first to get the BS15000 certification for its global command center. It has 46 development centers across the globe. It is the pioneer in applying LEAN Manufacturing techniques to IT services.

Communications: Wipro's Foreign Language Initiative enables IT professionals to communicate effectively with international clients. Employees are encouraged to learn one or more foreign languages. The initiative also helps non-English-speaking IT professionals in the use of English for communicating effectively with business executives, since most of Wipro's clients are from English-speaking countries.

Value metrics: Wipro wishes to be the "Toyota of business services" and is on track to becoming the world's most efficient and effective IT service provider. It offers a full portfolio of IT services including systems integration, package implementation, software application development and maintenance, research and development services, and information systems outsourcing across a range of industries delivering benefits for customers with six sigma consistency for global organizations. Using their global delivery model, they have international benchmarks in execution excellence that has translated into measurable results for their global customers, which

(continued)

(continued)

includes 75% faster time to market, 35% cost savings, and 35% productivity enhancements. Wipro is one of the few Indian IT service firms having adopted web services as an independent practice in its business plan. The IT capabilities are being built around web services–oriented applications and services to its customers. In this context, the confidentiality, security, and integrity of organizations' data are paramount, especially as data are exchanged across the Internet. Web service standards have gone a long way to address those concerns.

Skills/HR: Wipro has opened centers in the United States (Atlanta, Georgia and Troy, Michigan) in a continuing trend of "reverse outsourcing." Cultural alignment and closer customer relationships are keys to competing successfully in providing high-end consulting services. The recruits for the Wipro's centers attend 3 months of training in India before starting jobs in the United States in software development and project management. Wipro is also scouting for training sites in the United States. Further, the opening of U.S. centers is also an alternative to getting visas for workers, since getting work visa these days is getting competitive. It has also earmarked as much as $250 million for expansion in Europe through acquisitions (mainly in Germany).

IT Service companies such as Wipro ensure that they are able to have a continuous flow of new engineers and IT people by reaching into India's "second-level" engineering colleges to hire people before their last semester of study and then provide job-related course materials and training for that last semester. International campus hiring has also been initiated across the United States, Europe, and Asia to attract top talent. By doing so, Wipro is able to get commitments from students as they are just becoming ready to take on their first jobs after graduation, and with less in-house training. Only 1 out of 10 candidates gets interviewed, following a 1:50 ratio of resumes scanned, all enabled by IT.

HR counsels every employee on their strengths and weaknesses based on their profile by providing a map of courses to take at Wipro. There is a 40 day "Project Readiness Program" for new IT employees. Also, online study is encouraged. Wipro supplements a continuing education program for those who choose to enroll at leading educational institutes to obtain special skills in areas such as project management. At the leadership institute for senior managers, managers teach other managers business skills. Five percent of billable time is spent on training. The Chairman of Wipro himself spends half a day of his personal time teaching in every leadership program. Wipro offers 100,000 person days of training a year.

Partnership: At Wipro, everyone is encouraged to come up with big or small ideas, which would improve serving the customer/client. The idea is not to break new ground in basic knowledge, but in improving customer service. Wipro ties the rewards to performance.

An entry-level IT person moves onto a higher salary/benefits curve as they progress in completing assigned course/seminars that are geared to transform them from a computer science/engineering/business graduate into a software engineer. The "Wipro Equity Rewards Trust" plan gave Wiproites the benefits of participating in the wealth creation back in the 1980s. The stock option along with Quarterly Performance Linked Compensation (QPLC) provides innovative idea for linking Wipro's performance with employee compensation.

Governance: Wipro's technical competency lies in its ability to apply formal processes to ensure on time delivery, significant investments in accelerators, and partnerships and alliances. For Wipro, the nature of client trust is very important. IT tries to look at the problems the clients (internal and external) are experiencing and invites them to discussions to identify opportunities for the

(continued)

(continued)

future growth of the firm. They push the business verticals to think aggressively about future opportunities to prepare for the challenges the company might face. At Wipro, IT governance practices help clients in the realization of IT and business objectives through process performance optimization and compliance in areas of IT governance. They digitize their leading IT and process initiatives using workflow-based approaches that lead to easy adaptation and adoption of frameworks such as CMMI, Six Sigma, and ITIL.

Technology scope and architecture: Wipro is very cognizant of the fact that they need to remodel their processes and technical foundation to ensure that the IT systems are scalable. They have created autonomous structures combining IT, process, and applications that will allow them to continue the same growth in the future, while making sure that the data from the legacy systems are not lost by incorporating middleware technologies. Mobile applications are also a top priority at Wipro. With more than 50 offices in India and 30 offices abroad, scalability and flexibility are fundamental. Different geographic locations have different IT requirement and Wipro's IT infrastructure conforms to each of the locations in a flexible manner to ensure effectiveness/efficiency. The Internet is the key enabler of their infrastructure. In addition to physical security measures, frequent information audits are carried out to ensure a secure environment. As Wipro keeps hiring more employees, IT enables scalability in the HR process.

In summary, Wipro is one of the largest support service providers worldwide. It has the distinction of being the first PCMM Level 5 and SEI CMM Level 5 certified IT Services organization globally. Wipro provides comprehensive research and development services, IT solutions and services, including systems integration, information systems outsourcing, package implementation, software application development, and maintenance services to corporations globally. IT and business units are aligned well and are able to maximize IT business value, service delivery and are able to reduce the IT cost. They are successful at tracking all requests coming into IT and demonstrate value added to business. There is increased visibility and transparency through dashboards, improved productivity and data accuracy. ROI is high because of the implementation of a rational, sound portfolio management process for selecting more suitable IT projects aligned to business goals.

65.6 Strategic Alignment as a Process

Attaining and sustaining business–IT alignment must first focus on understanding the current level of SAM. This should be followed by steps that concentrate organizational energy on maximizing alignment enablers and minimizing inhibitors. This process embraces the steps[10] illustrated by Figure 65.7 and elaborated in the following text:

1. *Set the goals and establish a team.* Ensure that there is an executive business sponsor and champion for the assessment. Next, assign a team of both business and IT leaders. Obtaining appropriate representatives from the major business functional organizations (e.g., marketing, finance, R&D, and engineering) is critical to the success of the assessment. The purpose of the team is to evaluate the maturity of the business–IT alignment. Once the maturity is understood, the team is expected to define opportunities for enhancing the harmonious relationship of business and IT. Assessments range from 3 to 12 half-day sessions. The time demanded depends on the number of participants, the degree of consensus required, and the detail of the recommendations to carry out.

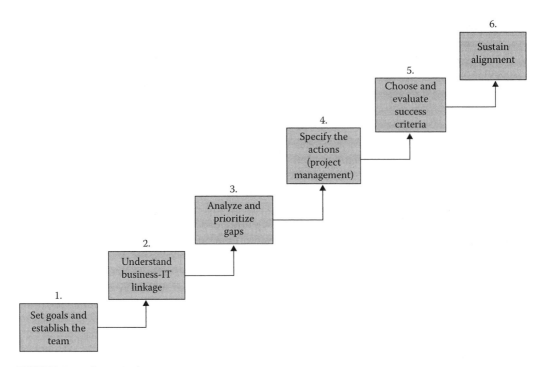

FIGURE 65.7   Strategic alignment as a process.

2. *Understand the business–IT linkage.* The SAM Assessment is an important tool in understanding the business–IT linkage. The team evaluates each of the six criteria. This can be done via executive interviews, group discussion, a questionnaire, or a combination. A trained facilitator can be valuable in guiding the important discussions.

3. *Analyze and prioritize gaps.* Recognize that the different opinions raised by the participants are indicative of the alignment opportunities that exist. Once understood, the group needs to converge on a maturity level. The team must remember that the purpose of this step is to understand the activities necessary to improve the business–IT linkage. The gaps between where the organization is today and where the team believes it needs to be are the gaps that need to be prioritized. Apply the next higher level of maturity as a roadmap to identify what can be done next.

4. *Specify the actions (project management).* Knowing where the organization is with regard to alignment maturity will drive what specific actions are appropriate to enhance IT–business alignment. Assign specific remedial tasks with clearly defined deliverables, ownership, timeframes, resources, risks, and measurements to each of the prioritized gaps.

5. *Choose and evaluate success criteria.* This step necessitates revisiting the goals and regularly discussing the measurement criteria identified to evaluate the implementation of the project plans. The review of the measurements should serve as a learning vehicle to understand how and why the objectives are or are not being met.

6. *Sustain alignment.* Some problems will just not go away. Why are so many of the inhibitors IT related? Obtaining IT–business alignment is a difficult task. This last step in the process is often the most difficult. To sustain the benefit from IT, an "alignment behavior" must be developed and cultivated. The criteria described to assess alignment maturity provide characteristics of organizations that link IT and business strategies. By adopting these behaviors, companies can increase their potential for a more mature alignment assessment and improve their ability to gain business value from investments in IT. Hence, the continued focus on understanding the alignment

maturity for an organization and taking the necessary action to improve the IT–business harmony are keys. Implicit in this is to periodically repeat the process to see how the organization evolves over time.

Fundamental to the effective use of the SAM assessment is not only to measure the maturity level of IT–business alignment but also to identify the problem/opportunity areas; and more important use the model as a roadmap to define specific initiatives for improvement. Repeating the assessment periodically can be insightful.

For example, when the SAM model was first used to assess the level of alignment maturity for a large financial company (fictitiously referred to as Stonehenge), they were assessed at Level 2 (committed processes). At the time, Stonehenge had recently adopted the federated IT organization model, so no one considered that the IT organization structure would be the area to consider in identifying why this financial giant was only at Level 2. After all, the federal (or hybrid) IT organization design has been found to produce higher alignment maturity scores over centralized and decentralized IT organization alternatives, because it captures the benefits of both centralized and decentralized IT organizations. The federated IT organization deployed at Stonehenge essentially centralized IT architecture and common systems, while decentralizing the strategic business unit applications and resources. The centralized IT structure supports the development of strong and efficient IT infrastructures while the decentralized IT group fosters business–IT relationships. Following the aforementioned logic, Stonehenge had decentralized its formally centralized application development staff, expecting that the relationships with the business management would improve. However, the analysis of the Stonehenge SAM assessment data showed the following:

- The indicators that measure the understanding of business by IT and the understanding of IT by business, which are covered in the "communications" area of the SAM model, were very low. Knowledge sharing in the organization was at a minimum to none. IT and business met occasionally (only during major walkthroughs) in a formal setting.
- IT–business relationship and trust measures that are covered under the "partnership" area were also at the minimum. Business viewed IT as a cost of doing business. There was an ongoing conflict between business and IT; they blamed each other for every late or unsuccessful delivery.
- Competency metrics—Measuring value of IT area showed that IT operated as a cost center.
- Social interaction indicator, which is covered under the HR area, was pointing to minimal IT–business interaction.

These and several other criteria used in the assessment suggested that there was conflict in the IT–business relationship in Stonehenge and that trust levels were at a minimum—typical in a centralized IT organization with poor linkages between business and IT. The fact that the company had already adopted the federated model motivated managers to further analyze the data to find out why the relationship with the business management did not improve.

Several other indicators, such as the differences between the IT and the business managers' opinions and the differences between the top and the middle managers' opinions in the SAM model, pointed to the problem in the implementation of the federated model. Looking at the organization charts and the grouping of the departments, they seemed in line with the federated model, meaning that the application development groups were created within the business units and dual reporting relationship for the divisional IT heads was created. Yet, the location of the development teams and the way they were functioning were not different from what they would be like in a typical centralized IT organization. At the end of the study, it was apparent that the management could not diverge from the routine they followed for many years. Indicators such as the tendency of the employees' resistance to change (measured in the HR area) were also in support of this hypothesis.

As illustrated in this example, SAM not only helped identify Stonehenge's maturity score, but it also allowed managers to identify specific problems and opportunities to improve the IT–business alignment.

Once again, organizations should not be in pursuit of a silver bullet. All six components of alignment maturity should be considered to determine the areas that require improvements and the opportunities that exist to help improve the IT–business alignment maturity level of the organization.

The periodical SAM measurement and results at Stonehenge are reviewed by both business and IT managers to ensure appropriate alignment. SAM provides guidance for business changes as well for a better alignment. SAM assessment should be considered as a continuous process of improvement in the organizations facing turbulent changes in business environment to enable organization-led increased SAM in the organization.

65.7 Strategic Alignment Maturity and Business Performance

The concept of performance underlies a lot of the research in strategic management and information science. A broader conceptualization of business performance would include emphasis on indicators of operational performance in addition to indicators of financial performance. Under this conceptualization, it would be logical to treat measurements such as market share, new product introduction, product quality, marketing effectiveness, manufacturing value added, and other measurements of technological efficiency within the domain of business performance.

Research done by Luftman et al. validated the contribution of SAM to company performance based on the data gathered from 362 global organizations across four continents.[24] The research identified that the six SAM components (communications, IT governance, value, partnership, technology scope, and skills) have approximately equal contribution to form the overall SAM score and they are strongly correlated to each other, as illustrated in Figure 65.8 through 65.10. Regarding the relationship of SAM and company performance, the regression weight (0.34) for SAM in the prediction of performance is significant; hence, this proves the contribution of SAM as a major contributor to a company's performance (see Figure 65.8). This relationship was found to be valid across all industry types, cultures, and geographic locations.

In addition, research has shown that the organization's structure—whether it follows a centralized, decentralized, or federated model—also has an impact on SAM maturity (see Figure 65.10). Notably, companies with federated IT structures are able to combine the benefits of centralized structures

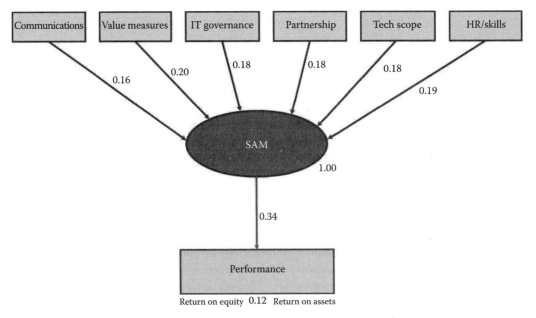

FIGURE 65.8 Structural equation model validation (1).

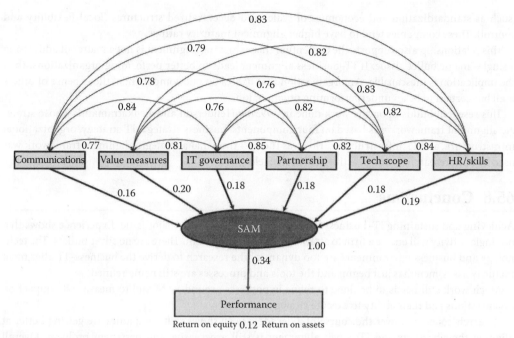

FIGURE 65.9 Structural equation model validation (2).

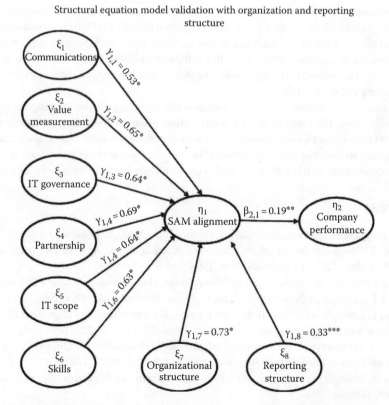

FIGURE 65.10 Structural equation model validation (3). (From Earl, M.J., *Corporate Information Systems Management*, Richard D. Irwin, Inc., Homewood, IL, 1993.)

(such as standardization and economies of scale) and decentralized structures (local flexibility and control). These companies tend to have higher alignment maturity ratings.[24]

This relationship also supports the contention that achieving alignment is not a matter of addressing a single "magic bullet" issue. If IT–business alignment leads to better performing organizations, then the implication is inescapable. An organization that fixates on one component at the expense of others is all but certain to be an underperforming organization.

This research builds upon the work done in 1993 by Henderson and Venkatraman,[23] whose strategic alignment framework was based on four components: business strategy, IT strategy, organizational infrastructure, and IT infrastructure. This was the first time that a strategic alignment framework was used by both researchers and practitioners in the field.

65.8 Conclusions

Achieving and sustaining IT–business alignment continues to be a major issue. Experience shows that no single activity will enable a firm to attain and sustain alignment. There are no silver bullets. The technology and business environments are too dynamic. The research to derive the business–IT alignment maturity assessment has just begun and the tools and processes are still being refined.

Much work still needs to be done to refine hypotheses around SAM and to measure its impact on organizations and their ability to execute strategy.

Research conducted over the course of a decade clearly shows that companies are getting better at aligning their business and IT; albeit alignment is still a pervasive and persistent problem. Overall maturity scores have increased from 2.99 in 2000–2003 to 3.17 in 2009–2010. There is evidence that higher levels of alignment have positive effects on company performance *regardless of industry type or organization structure*. However, results from the assessment of 362 Global 1000 companies demonstrate that some industries clearly do a better job of aligning their IT and business operations than others. Additional studies have linked high alignment maturity levels with better company performance measures, including sales, productivity, ROI, ROA, ROE, and NPM. The research also indicates that there are differences by region. This suggests that the strategic alignment of a company may depend both on industry norms and on local factors.

Achieving significantly higher levels of IT–business alignment across a wider range of organizations is a long-term journey. The journey in each organization begins with a complete assessment of how business views IT, and how IT views business. The journey continues with how business and IT executives work together to close the gaps and improve the performance of the organization. In the quest for continuous improvement within a dynamic global environment, the journey may never end.

References

1. King, J. Re-engineering focus slips, *Computerworld* (March 13), 6, 1995.
2. Luftman, J. *Competing in the Information Age: Practical Applications of the Strategic Alignment Model*, New York: Oxford University Press, 1996.
3. Earl, M.J. *Corporate Information Systems Management*, Homewood, IL: Richard D. Irwin, Inc., 1993.
4. Earl, M.J. Experience in strategic information systems planning, *MIS Quarterly*, 17(1), 1–24, 1996.
5. Luftman, J., Lewis, P., and Oldach, S. Transforming the enterprise: The alignment of business and information technology strategies, *IBM Systems Journal*, 32(1), 198–221, 1993.
6. Goff, L. You say tomayto, I say tomahto, *Computerworld* (November 1), 129, 1993.
7. Liebs, S. We're all in this together, *Information Week* (October 26), 8, 1992.
8 Robson, W. *Strategic Management and Information Systems: An Integrated Approach*, London, U.K. Pitman Publishing, 1994.
9. Luftman, J., Papp, R., and Brier, T. Enablers and inhibitors of business–IT alignment, *Communications of the Association for Information Systems* (1), 11, 1999.

10. Luftman, J. and Brier, T. Achieving and sustaining business–IT alignment, *California Management Review* (1), 109–122, Fall 1999.

11. Luftman, J., Papp, R., and Brier, T. The strategic alignment model: Assessment and validation, in *Proceedings of the Information Technology Management Group of the Association of Management (AoM) 13th Annual International Conference*, Vancouver, BC, Canada, pp. 57–66, August 2–5, 1995.

12. Humphrey, W.S. Characterizing the software process: A maturity framework, *IEEE Software*, 5(2), 73–79, 1988.

13. Keen, P. Do you need an IT strategy? In: J.N. Luftman (ed.), *Competing in the Information Age*, New York: Oxford University Press, 1996.

14. Nolan, R.L. Managing the crises in data processing, *Harvard Business Review*, March 1, 1979.

15. McLean, E. and Soden, J. *Strategic Planning for MIS*, New York: John Wiley & Sons, 1977.

16. IBM. *Business Systems Planning Guide*, GE20-0527, White Plains, NY: IBM Corporation, 1981.

17. Mills, P. *Managing Service Industries*, New York: Ballinger, 1986.

18. Parker, M., and Benson, R. *Information Economics*, Englewood Cliffs, NJ: Prentice Hall, 1988.

19. Brancheau, J. and Wetherbe, J. Issues in information systems management, *MIS Quarterly*, 11(1), 23–45, 1987.

20. Dixon, P. and John, D. Technology issues facing corporate management in the 1990s, *MIS Quarterly*, 13(3), 247–55, 1989.

21. Niederman, F., Brancheau, J., and Wetherbe, J. Information systems management issues for the 1990s, *MIS Quarterly*, 15(4), 475–95, 1991.

22. Chan, Y. and Huff, S. Strategic information systems alignment, *Business Quarterly*, (58), 51–56, 1993.

23. Henderson, J. and Venkatraman, N. Aligning business and IT strategies. In: J.N. Luftman (ed.), *Competing in the Information Age: Practical Applications of the Strategic Alignment Model*, New York: Oxford University Press, 1996.

24 Luftman, J. and Zadeh, H.S. Key Information technology and management issues 2010–11: An international study, *Journal of Information Technology* (April 12), 26, 193–204, 2011.

25. Papp, R. Determinants of strategically aligned organizations: A multi-industry, multi-perspective analysis, PhD Dissertation, Hoboken, NJ: Stevens Institute of Technology, 1995.

26. Papp, R. and Luftman, J. Business and IT strategic alignment: New perspectives and assessments, in *Proceedings of the Association for Information Systems, Inaugural Americas Conference on Information Systems*, Pittsburgh, PA, August 25–27, 1995.

27. Henderson, J., Thomas, J., and Venkatraman, N. *Making Sense of IT: Strategic Alignment and Organizational Context*, Working Paper 3475-92 BPS, Cambridge, MA: Sloan School of Management, Massachusetts Institute of Technology, 1992.

28. Boynton, A., Victor, B., and Pine, B. II. Aligning IT with new competitive strategies. In: J.N. Luftman (ed.), *Competing in the Information Age*, New York: Oxford University Press, 1996.

29. Davidson, W. Managing the business transformation process. In: J.N. Luftman (ed.), *Competing in the Information Age*, New York: Oxford University Press, 1996.

30. Keen, P. *Shaping the Future*, Boston, MA: Harvard Business School Press, 1991.

31. Foster, R. *Innovation: The Attacker's Advantage*, New York: Summit Books, 1986.

32. Weill, P. and Broadbent, M. *Leveraging the New Infrastructure*, Cambridge, MA: Harvard University Press, 1998.

33. Broadbent, M. and Weill, P. Developing business and information strategy alignment: A study in the banking industry, *IBM Systems Journal* (32), 1, 1993.

34. Baets, W. Some empirical evidence on IS strategy alignment in banking, *Information and Management*, 30(4), 155–177, 1996.

35. Rogers, L. Alignment revisited, *CIO Magazine* (May 15), 44–45, 1997.

36. Rockart, J., Earl, M., and Ross, J. Eight imperatives for the new IT organization, *Sloan Management Review*, 38(1), 43–55, 1996.

37. Rockart, J. and Short, J. IT in the 1990's: Managing organizational interdependence, *Sloan Management Review*, 30(2), 7–17, 1989.
38. Davenport, T. and Short, J. The new industrial engineering: Information technology and business process redesign, *Sloan Management Review*, 31(4), 11–27, 1990.
39. Hammer, M. and Champy, J. *Reengineering the Corporation: A Manifesto For Business Revolution*, New York: Harper Business, 1993.
40. Hammer, M. and Stanton, S. *The Reengineering Revolution*, New York: Harper Business, 1995.
41. Luftman, J. Addressing Business–IT Alignment Maturity, *Communications of the Association for Information Systems*, 4(December), 4–15, 2000.
42. Greenberg, J. and Baron, R. *Behavior in Organizations*, New York: Prentice Hall, 1997.

66

Process of Information Systems Strategizing: Review and Synthesis

Anna Karpovsky
Bentley University

Mikko Hallanoro
University of Turku

Robert D. Galliers
Bentley University

66.1 Introduction

This chapter reflects on the developments in research on information systems (IS*) strategy and the process of IS strategizing. In reviewing the literature base, we have developed an integrative framework that captures key concepts and processes that are associated with IS strategizing. By summarizing the research themes and constructs found in the literature, we trust that the framework will serve as a sense-making device for current and future IS strategy researchers; something that sets the scene in terms of the extant research that might form a useful basis for identifying potential future areas of research. By emphasizing the processes relevant for IS strategizing in practice, we also hope the framework will be a practical aid for those who plan and strategize around IS.

---

* As the terms *information technology* (IT) and *information systems* (IS) have been used interchangeably in the literature, the authors will use the terms used by the original authors when referring to the earlier work. Note, however, that we differentiate between the two terms ourselves when analyzing the literature and drawing our conclusions.

First, though, we will present some common definitions associated with the noun—strategy, and the verb—strategize, as a basis for our treatment of the subject matter. The *Webster* and *Oxford* dictionaries define strategy, *inter alia*, as

- A plan of action designed to achieve a long-term or overall aim[*]
- The art of devising or employing plans or stratagems toward a goal
- An adaptation or complex of adaptations (as of behavior, metabolism, or structure) that serves or appears to serve an important function in achieving evolutionary success[†]

Further, *Webster* defines strategize as

- "to devise a strategy or course of action"

From these definitions, we may conclude that these terms relate to a means of achieving a goal or aim but that there is some debate about whether such plans are deliberate and sequential in nature or are more evolutionary and adaptive. This is a debate to which we will return. For now, though, let us assume that the process of strategizing combines both of these elements.

As regards IS, we take the following, inclusive meaning of the term: "the entire infrastructure, organization, personnel, and components that collect, process, store, transmit, display, disseminate, and act on information."[‡] Thus, when we talk of IS strategizing, we see this is a process that will lead to actions being taken that will relate to the sociotechnical concept of infrastructure that is introduced earlier.

We begin this chapter with a historical treatment of the topic in order to provide some background for what follows.

66.2 Historical Background: From Data Processing to Strategic Information Systems and Beyond

66.2.1 Data Processing

At the time of the development of the first commercial applications of information technologies (IT) in the early 1950s,[§] most people regarded computers as massive machines that scientists used to find solutions to computationally intensive equations that would take a human being a number of years to solve. In the late 1960s, companies with highly data-driven tasks operated these large, centralized computer systems that typically ran batch jobs to process the daily transactions for their businesses. This early era is often referred to as the *data processing era*. At that time, the business objective was to automate information-based processes in order to improve operational competence by reducing data processing costs and, consequently, achieving business efficiency (Galliers, 1991). Among the problems of that period were huge maintenance costs, duplication of data, incompatible application systems, and—often—overall user dissatisfaction (Somogyi and Galliers, 1987). In addition, in some cases there was little or no direct link between the business strategy (i.e., objectives and goals) and the IS plan. The IS plan, if one existed, and the IS operations of the organization were predominately concerned with the efficient operation of computer technology for operational purposes and tended to be isolated (Galliers, 2004) or independent (Teo and King, 1997) from the business of the organization.

---

[*] *Source*: http://oxforddictionaries.com/definition/english/strategy

[†] *Source*: http://www.merriam-webster.com/dictionary/strategy

[‡] *Source*: http://www.thefreedictionary.com/information+system

[§] See Caminer et al. (1998) for an account of the world's first business computer, LEO, which was developed in the United Kingdom and introduced in 1951.

66.2.2 MIS

In the 1970s, with the emergence of "minicomputers" that featured increasing power and sophistication, computer systems began to offer solutions to a higher level of decision capability. As more data became stored, managers realized that using the output of IS—information—could increase the effectiveness of their decision-making (Ward and Peppard, 2002). Such developments denoted the beginning of what can be termed the *management information systems* (MIS) *era*. The objective of MIS was to create ways of organizing and delivering information in order to improve management effectiveness. There was growing demand among management to have business-driven IS capable of dealing with business problems and the issues managers faced (Galliers, 1991). MIS reports (printouts) were provided at regular intervals (e.g., monthly); they were intended to provide managers with the information they needed, primarily for control purposes.

The first signals for the need for strategic IS planning appeared in this era since it was considered that strategic planning for the information needs of the organization was both feasible and necessary if MIS were to support its basic purposes and goals (King, 1978). The data processing applications of the earlier era produced "at best fragmented data, at worst a chaotic mess of data with little or no integrity" (Ward and Peppard, 2002, p. 18). As a result, thinking began to focus more on organization-wide information (i.e., what information is used, how it is used in the organization, and what more might be needed).

One of the most influential publications during the MIS era is arguably Nolan's (1979) article that presents the so-called stages of growth model. This had remarkable impact on the development of IS strategizing, being both highly cited in the academic literature and extensively applied in practice, leading to the founding of Nolan, Norton & Co., which was subsequently acquired by KPMG. First published by Nolan (1973) and Gibson and Nolan (1974), the model was not without its critics, however. For example, it was criticized for its lack of empirical substantiation, the overly simplistic assumptions on which it was based, and the limited focus of the original concept (Benbasat et al., 1984; King and Kraemer, 1984). Nonetheless, it had a major impact, and since the original publication of the model was published, it has gone through several refinements and revisions by numerous authors (e.g., Earl, 1988, 1989; Nolan, 1979, 1984), and also later by Galliers and Sutherland (1991). The latter developed a revised model with a broader view concerning strategic, organizational, human resource, and management issues. Thus, notwithstanding the criticisms, the concept—in its various forms—has been applied extensively.

More generally, in an early paper, McFarlan (1971) pointed to the fact that little IS planning was taking place: of the 15 companies he studied, only one had been planning for IS for 4 years while the rest had just considered planning as a potentially necessary activity. During these early days, a handful of articles and books on the topic began to appear, in the United States for the most part (Lincoln, 1975 being an exception), each in their way concerned with the strategic dimensions and potential impacts of computer usage in organizations (e.g., Blumenthal, 1969; King and Cleland, 1975; Kriebel, 1968; McFarlan, 1971; Nolan, 1973; Siegal, 1975; Schwartz, 1970; Young, 1967). The earliest of the journal articles that specifically addressed the process of planning and strategizing for IS are listed in Table 66.1.

Young (1967) listed a number of convincing reasons and benefits of planning for IS. These included the promotion of a better organizational decision-making process through the management of information. Kriebel (1968) argued that an intuitive, ad hoc approach to planning for computers does not work; rather, organizations should develop strategies that define the role of computers in "attaining the strategic objectives of the corporation" and establishing planning objectives on the basis of organizational goals. Assessment of an organization's status in terms of systems development, resource commitment, and management organization for computer systems were all considered to be important decisions in a "company computer strategy" (ibid., 12).

Schwartz (1970) also promoted planning for MIS, stating that the starting point should be a determination of management and user needs. He proposed a specific "systematic and analytic" approach to planning that takes an evolutionary perspective (i.e., a planning–executing–learning–planning

TABLE 66.1   Earliest Articles on IS Planning

		Summary
Author(s)	Young, R.C.	Reasons for planning: long lead time, rapid technological changes, growth, large investments, lack of clear authority and responsibility, need to set a good example, self-protection. Benefits of planning: promote better current decisions, organizational structure, and international communication, teamwork and morale; recognize future needs in time to meet them; and save management's time
Nationality	United States	
Journal	*Computers and Automation*	
Year	1967	
Title	Systems and data processing departments need long-range planning	
Nature	Conceptual	
Focus	Plan/normative	
Author(s)	Kriebel, C.H.	Companies should define the role of computers in "attaining the strategic objectives of the corporation" and establish computer planning objectives on the basis of corporate goals; determine corporate policy for growth, resource commitment, and the management organization for computer systems; and appraise the company's current position in computer systems development
Nationality	United States	
Journal	*Long Range Planning*	
Year	1968	
Title	The strategic dimension of computer systems planning	
Nature	Conceptual	
Focus	Should plan/normative	
Author(s)	Schwartz, M.H.	Planning–executing–learning–planning cycle of evolutionary approach to MIS planning
Nationality	United States	
Journal	*Datamation*	
Year	1970	
Title	MIS planning	
Nature	Conceptual	
Focus	Planning framework	
Author(s)	Zani, W.M.	Proposes a top-down approach to help establishing goals and priorities for MIS development
Nationality	United States	
Journal	*Harvard Business Review*	
Year	1970	
Title	Blueprint for MIS	
Nature	conceptual	
Focus	planning framework	
Author(s)	McFarlan, F.W.	Factors that companies must consider in developing its strategy
Nationality	United States	
Journal	*Harvard Business Review*	
Year	1971	
Title	Problems in planning the information system	
Nature	Conceptual	
Focus	Best practices/descriptive	
Author(s)	Nolan, R.L.	The first publication presenting Nolan's Stages of Growth model, an evolutionary model for the growth of information technology (maturity) in an organization followed with guidelines. The model was further developed to its six-stage form and published by Nolan in 1979
Nationality	United States	
Journal	*Communications of the ACM*	
Year	1973	
Title	Managing the computer resource: a stage hypothesis	
Nature	Conceptual	
Focus	Framework/evolutionary model/guidelines	

TABLE 66.1 (continued)   Earliest Articles on IS Planning

		Summary
Author(s)	Gibson, C. and Nolan, R.L.	Further developing the Nolan 1973 article's idea of depicting
Nationality	United States	advances in IT with four-step stages of growth model, with
Journal	*Harvard Business Review*	the key focus in application portfolio
Year	1974	
Title	Managing the four stages of EDP growth	
Nature	Conceptual	
Focus	Framework/evolutionary model/guidelines	
Author(s)	King, W.R. and Cleland, D.	Systems approach to planning
Nationality	United States	
Journal	*Business Horizons*	
Year	1975	
Title	A new method for strategic systems planning	
Nature	Conceptual	
Focus		
Author(s)	Lincoln, T.	Approach to information system development in the context
Nationality	United Kingdom	of an overall business/IS plan
Journal	*Management Datamatics*	
Year	1975	
Title	A strategy for Information Systems development	
Nature	Conceptual	
Focus	Planning framework	
Author(s)	Mjosund, A.	Two general approaches are suggested: use the information
Nationality	United States	system analysis to guide research, or application of results
Journal	*Computer & Operations Research*	from research, to solve management problems; follow a
Year	1975	strategy in the analysis of information needs such that the
Title	Toward a strategy for information needs analysis	steps in this analysis are closely related to the structure relating the decisions and actions in the organization. A gross classification scheme is proposed to aid in
Nature	Conceptual	determining this strategy
Focus	Planning framework	
Author(s)	Zachman, J.	There are three levels of planning and control functioning in
Nationality	United States	an organization: strategic planning, management control,
Journal	*Journal of Systems Management*	and operational control and it should be clearly identified
Year	1977	to whom the system is to be built
Title	The information systems management system: a framework for planning	
Nature	Conceptual	
Focus	Descriptive/evolution	
Author(s)	Zachman, J.	IS planning must begin with an understanding of the basic
Nationality	United States	components of the IS management system and their
Journal	*DATABASE for Advances in Information Systems*	relationships, which then serves as a framework within which the planning effort can take place. Those basic components are presented
Year	1978	
Title	The information systems management system: a framework for planning	
Nature	Conceptual	
Focus	Planning framework	

cycle framework). Identification of current and potential systems and evaluation of needed resources are present in the Schwartz framework, with implementation and feedback mechanisms being emphasized in this early work. Similarly, King and Cleland (1975) promoted a "systems approach" to planning, where the entire focus was on a feedback (learning) process, where plans depend on the consequences of action just as intrinsically as actions reflect the results of planning.

Other early works on approaches to IS planning included those by Zani (1970), Nolan (1973), and Lincoln (1975), as notable examples. Zani (1970) suggested that the bottom-up approaches of the earlier era produce benefits mostly by chance. Conversely, the top-down approach he advocated helps in establishing goals and priorities for MIS development and therefore focuses IT where it is most needed. While developing the first version of the stages of growth model, Nolan (1973) found a "database/key-task strategy" to be most effective since management would have to consider carefully what the key tasks of the business are, thereby forcing on cross-functional data integration. Lincoln's (1975) approach was built, first and foremost, on an understanding of the business function in terms of objectives and tasks; only after this has been achieved could information requirements to meet defined objectives be examined.

This early research viewed IS plans and strategies as mainly reactive to business strategies, and managers strove to align their IS strategies with their business needs. For example, King and Cleland (1975) framed their discussion around transformation of organizational mission, objectives, and strategies into "IS strategy sets." These ideas were later adopted by IBM as a basis for their Business Systems Planning (BSP) methodology. This methodology was used in many organizations, not just in the United States. It was—frankly—as much a marketing device for IBM mainframes as well as a consulting tool, given the added value of such strategic services IBM was able to offer, compared to other vendors.

66.2.3 SIS

By the late 1980s, there were growing numbers of cases in which IT was shown to have had a strategic impact on organizations. Though the studies undertaken in the United States tended to assume that in almost every instance IS planning is undertaken, and the emphasis was on questions "how" and "why," this was less the case with their British or Australian counterparts, for example, where the emphasis remained on question "whether" any planning was undertaken (Galliers, 1987, p. 51), notwithstanding the early work of the likes of Michael Earl (e.g., Earl, 1988, 1989). Nevertheless, corporate executives became increasingly interested in what were termed strategic information systems (SIS). The objective of the *SIS era* was to change the nature or conduct of business in order to improve the competitiveness of the company—a proactive, rather than reactive approach in other words. The American Airline's SABRE reservation system was one such oft-cited SIS (e.g., McFarlan, 1984; Porter and Millar, 1985). The use of IS began to influence organizations' competitive positions and became a strategic weapon for competitive advantage (Ward and Peppard, 2002).

Research identified a number of organizational and individual factors that promoted the development of SIS. A competitive market was found to be conducive to the development of SIS (King and Sabherwal, 1992). An internal need was found to be the reason for SIS development in 14 well-known SIS (Neo, 1988). Automation to enhance internal efficiency had become crucial for products and processes with high information content (Lindsey et al., 1990). Similarly, businesses with high transaction volumes built systems to improve internal operations (Johnston and Carrico, 1988).

A significant portion of the early SIS literature was devoted to the identification and description of opportunities for SIS. SIS planning (SISP) approaches were developed to identify these strategic opportunities for organizations by applying IT to optimize business performance (Pant and Ravichandran, 2001) and were, as a result, labeled as "impact" mode approaches (Bergeron et al., 1991). For example, Lederer and Sethi (1996, p. 1) define SISP as "the process of identifying a portfolio of computer-based applications that will assist an organization in executing its business plans and realizing its business goals."

During the SIS and SISP era, the purpose of IS strategic planning was both to influence and enable a business strategy and to support it (King and Teo, 1997). IS strategies were viewed as proactive, and attention shifted to how the technology can be employed to increase competitive advantage through analyses of the competitive environment and internal processes (Galliers, 2004).

66.2.4 Summary: Toward Sustainability and Capability

Summarizing the early developments in IT, Galliers (1987, 2004) suggests that the focus of IS strategizing during those earlier periods went through four phases during which attention shifted away from and then back to IT, and from matters of efficiency to matters of effectiveness and competitiveness. The four phases mirror, to some degree, the eras identified earlier: isolation (the data processing era), re-action and prospection (the MIS era), and pro-action (the SIS era). Figure 66.1 describes the development in terms of the degree to which the IS strategy might be viewed as a business-driven, "top-down" process— as opposed to more technology-driven, "bottom-up" concerns—and the extent to which such strategies have been based on short-term problem solving as opposed to more long-term strategic goal setting. Galliers (2004) suggests that, in some respects, current IS strategizing includes characteristics of each of these "phases." Similarly, Peppard and Ward (2004) note that, with the evolution of IT, each "era" displays different characteristics regarding the application of IT and has different objectives. In this sense, every new era or phase in IT evolution and IS strategic planning encompasses the earlier views as well.

More recently, there has been an increased interest in the question of the sustainability of the competitive advantage to be derived from IS (Ward and Peppard, 2002). Even though an organization can gain a competitive advantage over others in the short run, such as from the "first mover advantage" gained through an innovative application (Clemons and Row, 1991; Mata et al., 1995), most of these technologies can be easily copied and, therefore, do not produce a sustainable competitive advantage. Unlike competitive advantage, sustainability is an ongoing state that creates continuous advantage (Barney, 1991).

Valuable, rare, inimitable, and non-substitutable (the so-called VRIN) resources are considered to be a source of sustainable competitive advantage based on the resource-based view (RBV) of the firm (Barney, 1991; Rumelt, 1984; Wernerfelt, 1984). According to Wang and Ahmed (2007, p. 3), "The essence of the RBV lies in the emphasis on resources and capabilities as the genesis of competitive advantage: resources are heterogeneously distributed across competing firms and are imperfectly mobile which, in turn, makes this heterogeneity persist over time." Traditionally IS has been associated with technological systems, which—indeed—are often rather easy to imitate or substitute (cf. recent arguments by

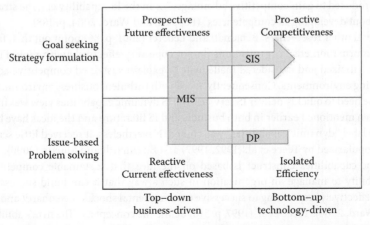

FIGURE 66.1 Developments in IS strategizing. (From Galliers, R.D., in Galliers, R.D. and Leidner, D.E. (Eds.), *Strategic Information Management*, 4th edn., 2009, Oxford University Press, Oxford, U.K.)

the likes of Carr, 2003, for example). However, the RBV and its derivatives such as the capabilities view facilitate the consideration of IS holistically, including, for example, the organization's external environment, internal processes, organizational culture, and employee resources—all of which are considerably harder to imitate or substitute. With respect to strategic planning, the resource-based perspective also places importance on managerial strategies for developing new capabilities (Wernerfelt, 1984).

Ross et al. (1996) and Bharadwaj (2000) introduce the concept of IS capability as something that organizations need to develop to enhance competitiveness. Specifically, Bharadwaj et al. (2002, p. 4) combine earlier works and definitions to define IT capability as a "firm's ability to acquire, deploy, and leverage its IT resources to shape and support its business strategies and value chain activities." Further, Pavlou and El Sawy (2006, p. 198) suggest that "IS researchers should look beyond the direct effects of firm-level IT infrastructures and focus their attention on how business units can leverage IT functionalities to better reconfigure and execute business processes." They list various IT-related resources that, when combined, form IT capability, including many nontechnical resources such as access to capital, proprietary ownership, managerial IT skills, IT–business partnering relationships, IT human resources, complementary IT human and business resources, and business process integration. A somewhat similar perspective is provided by Galliers and Sutherland (1991), based on their amendment of the stages of growth model (Gibson and Nolan, 1974; Nolan, 1973, 1979), which incorporated the so-called 7S framework, used by McKinsey (i.e., shared values, strategy, structure, skills, staff, style, and systems).

Other scholars have provided additional insights on IS capabilities. For example, Kettinger et al. (1994) conclude that gaining IT/IS-based sustained competitive advantage is a matter of building organizational infrastructure. Powell and Dent-Micallef (1997) suggest that some organizations have been successful in using IT to leverage intangibles and human resources such as organizational flexibility, integrating business and strategy planning, and vendor relationships. Mata et al. (1995) argue that IS management skills are the only source of sustained advantage.

The growing complexity of both business processes and IS together with commodity nature of IT has shifted the main focus of attention from technical aspects to the organizational factors enabling effective exploitation of IT. As mentioned earlier, the IS capabilities view emphasizes the development and delivery of all IS-related organizational capabilities rather than the straightforward development of IS systems or simple recognition of promising opportunities. DP-, MIS-, and SIS-era planning methodologies concentrated on introducing more effective IS solutions, recognizing new opportunities, and improving production processes. The IS capability view is, however, concerned mainly with introducing and developing the capabilities to fulfill the DP-, MIS-, and SIS-era objectives. It can, thus, be considered another step forward in the transformation from asking questions of "what" to the questions of "how" and "why." While an SIS-era IS strategy might envision the strategic IS opportunities and systems that will be introduced to gain competitive advantage, "… in the IS capability era, the strategic management of IS is about developing IS competencies" (Peppard and Ward, 2004, p. 188).

Among several other researchers, Benamati and Lederer (2001, p. 86) point out that, facing the challenges of the information era, "organizations do not cope very effectively with rapid IT change." The RBV has been criticized and considered inadequate to explain sustained competitive advantage especially in changing environments. Consequently, faced with turbulent business environments and to further address the need to quickly deliver IS services, the IS dynamic capabilities view was introduced. The concept has been mentioned earlier in both business and IS literature and the ideas have been discussed well before the label "dynamic capabilities" was coined. Nevertheless, it received little serious attention before being popularized by Teece et al. (1992, 1997) and Eisenhardt and Martin (2000).

The dynamic capabilities construct is based on the idea that sustainable competitive advantage requires the ability to manage an organization in such a way that it can build successive temporary advantages by effectively responding to successive environmental shocks (Eisenhardt and Martin, 2000; Peppard and Ward, 2004). Teece et al. (1997, p. 516) define the concept as "The firm's ability to integrate, build, and reconfigure internal and external competences to address rapidly changing environments." Following Barreto's (2010, p. 260) conclusions, dynamic capabilities have been identified as abilities, capacities, and capabilities but also more specifically as processes, routines, learned and stable activities, or as a behavioral orientation that helps organizations to change. The challenge to keep up with the change requirements for IS has led to a growing need for attention for flexibility, agility, and change readiness in all aspects of utilizing IT. Rapid business change places a further challenge to plan for organizational change readiness, agility, and flexibility. Given today's high level of dependence on technology, IS strategizing and planning for IS dynamic capabilities become ever more important.*

Today, few people would disagree that the strategic management of data, information, resources, and knowledge—and the associated IT—represents a major strategic challenge and opportunity for organizations (Galliers, 2004). Of course, the value to an organization is not so much *what* specific technologies it has or develops, but *how* these technologies are employed. As such, strategizing about IS (Galliers, 2007), and the resulting IS strategy, when effectively engaged, may distinguish the performance improvements attributable to IS from one organization relative to others. Without an IS strategy, the contribution of IS to organizational performance is likely to be a result of serendipity (ibid.). Both practice and research have emphasized the need to carefully construct an IS strategy to complement and/or enable organizational objectives—as identified in the business strategy (Reich and Benbasat, 1996). IT strategic planning has remained among IT management's top 10 concerns throughout the last three decades (Luftman and Ben-Zvi, 2010). Understanding the strategic value of IS (i.e., seen holistically) has been a key goal of many IS practitioners and researchers (Galliers, 1993a,b; Luftman et al., 2006). Thus, considerable attention has been paid in the literature to, for example, addressing issues associated with IS strategic planning, alignment of IS and business strategies, and strategy formulation involvement (Chen et al., 2010).

Building on this introduction, the purpose of this chapter is to summarize developments of the field of IS strategy and the process of IS strategizing. First, we address the question of the conceptualization of IS strategy. A variety of terms, such as IT strategy (Gottschalk, 1999), IS/IT strategy (Chan et al., 1997), information management (IM) strategy (Karimi and Konsynski, 1991), and information strategy (Smits et al., 1997), have been used that are related to and sometimes used interchangeably with the term of IS strategy. We aim to clarify the distinctions and the connections among these terms. Second, we describe the evolution of the field and various streams of thought that have been considered in the course of the field's development. Last, in an attempt to synthesize extant research themes, we propose an integrative framework that links the various topics that encompass what we refer to as the research on the *IS strategizing process*.

66.3 Underlying Principles

66.3.1 IS Strategy Components

The foundational work on conceptualization of IS strategy was led—most notably—by Earl (1989) and developed further by Galliers (1991, 1999). Earl's (1989) insightful contribution was to differentiate the "what" from the "how" and the "wherefore" of IS strategy. In so doing, Earl makes clear that IS strategy is very much a business management issue, while IT strategy lies, for the most part, within the domain of the IT function. IS strategy comprises an application development portfolio, a "shopping list" of application and projects—the "demand" side of the IS strategy in other words. IT strategy is regarded as the "supply" side of the IS strategy. IT strategy is concerned with alternative technological solutions (alternative "hows") to support the "what" of business needs, thereby enabling IT to be better aligned with the business strategy. The "wherefore" indicates the kind of developmental path that needs to be taken to achieve this alignment. Other notable authors in this sphere include Lederer, who could

---

* For further information about the construct, recent literature reviews in the organization management literature by Barreto (2010), Ambrosini and Bowman (2009), and Wang and Ahmed (2007) list the main ideas, theories, and criticisms. Further, Schwarz et al. (2010) and Chen et al. (2008) have studied more closely the importance of dynamic capabilities for IS impact and alignment in organizations.

arguably—measured by the number of published and highly cited IS strategy articles in peer-reviewed journals—deserve the title of most influential academic contributor to IS strategy research. His most cited works related to IS strategy cover such topics as SISP methodologies, critical issues, implementation, information resource planning, key prescriptions, and SISP conceptualization (Lederer and Salmela, 1996; Lederer and Sethi, 1988, 1996).

Galliers' (1991) amendment to Earl's earlier conceptualization was to suggest that an *information* strategy might provide a useful distinction from IS or IT strategy in that it focuses on identifying the information that could question taken-for-granted assumptions on which the business strategy is based, as well as providing the information required by the business strategy. Thus, an information strategy is the glue between the business strategy and the IS strategy. It tends to answer questions such as "what is the information required to support primary tasks, or key goals, of the organization strategy?" and "where can this be obtained?" The information strategy also questions the business strategy by attempting to answer the "why" questions—"why might this particular strategy be chosen as against any other?" (Galliers, 2009).

A further distinction provided by Galliers (2009) is the *information services* strategy (also sometimes referred to as the IM strategy). An information services strategy focuses on policy issues regarding the organization of information services (and capabilities), including sourcing decisions—that is, "the organizational arrangements for the provision of IS-related services" (ibid., p. 14). This component of IS strategy mainly answers to the "who?" questions: who is needed to facilitate and enable the strategy to be developed and implemented. Such questions include considerations of the structure, roles, and processes of the IT function. Human resource issues, such as IT personnel requirements and choices between the internal service provision, also need to be considered. In this sense, it is a precursor to Peppard and Ward's (2004) call for a focus on IS capability discussed in the previous section.

Galliers (1999) also stressed the need for an *implementation/change management* strategy as integral to the overall IS strategizing process, and the need for *ongoing evaluation and review*. The former emphasizes the importance of managing the change processes associated with ongoing implementation of the strategy. The latter emphasizes the importance of feedback regarding the impact of past strategic decisions (unintended as well as expected outcomes) and the identification of emergent strategies (i.e., those that might "bubble up" as a result of what Ciborra (1994) calls "tinkering" and "bricolage"). Thus, importantly, the IS strategizing process is viewed as ongoing and iterative, and *not* a one-off initiative, to be undertaken periodically.

In summary, there are distinctions to be made between various commonly used terms in the IS field, and IS strategy is no exception. In Galliers's terms, IS strategy is a broad concept that encompasses the information strategy, the IT strategy, the information services strategy, and the change management and implementation strategy—as well as the assessment and ongoing review of the process outcomes, and the process itself. Collectively, these interlinked strategy components represent fundamental aspects of managing IS in organizations (see Figure 66.2). IS strategy incorporates a range of issues associated with strategy formation and formulation (i.e., the emergent as well as the deliberate), and implementation with respect to IS, and considers social as well as technological aspects. In addition, IS strategy is cross-functional since it encompasses product, process, and human resources (Lefebvre et al., 1997). From this perspective, we would argue that a business strategy is incomplete without its IS strategy.

66.3.2 Review of the Literature

The literature specifically on the processes associated with IS strategy is vast, and it covers 40 years of research, with over 470 articles in peer-reviewed journals and a number of prominent books on the subject (e.g., Blumenthal, 1969; Earl, 1988, 1989; Galliers and Leidner, 2003, 2009; King, 2009; McLean et al., 1977; Siegal, 1975; Ward and Peppard, 2002). More generally, for example, work covering related topics such as the strategic application of IT, or such topics as sourcing, would yield thousands of articles

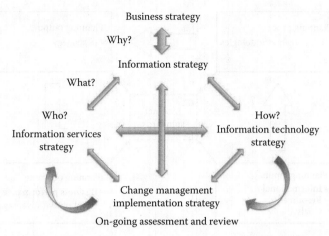

FIGURE 66.2 Components of information systems strategy. (From Galliers, R.D., in Avgerou, C., Ciborra, C., and Land, F. (Eds.), *The Social Study of Information and Communication Technology*, 2004, Oxford University Press, Oxford, U.K.)

(see, e.g., Gable, 2010, for a review of the over 300 articles that have appeared in the *Journal of Strategic Information Systems* alone in the period 1991–2009). Academic articles on IS strategy processes were particularly numerous in the 1990s (with a peak of over 30 articles in 1999), but fewer articles have appeared in recent years. However, we argue that this decline does not necessarily indicate a lack in potential areas of further research in this field, especially—as we have seen—the topic remains a prominent issue for IT executives and their business colleagues (Luftman and Ben-Zvi, 2010). Utilizing the frameworks provided by the extant literature, we derive an integrative framework to help researchers and practitioners alike to further familiarize themselves with key IS strategy themes. Here, as we have noted, we view strategy as a dynamic and ongoing process within an organization and, therefore, we consider the *strategizing process* as our particular focus of study. Through a review of extant frameworks, no single framework has been established that encapsulates all the components relevant to IS strategizing to be found in the literature. To develop such framework, however, we began with a conceptualization that came closest to our view of strategy as a process of relevant and related activities. The input–process–output model (initially introduced by King, 1988) provided a useful building block for our synthesis, given its breadth of coverage.

66.3.3 Comprehensive IS Strategizing Framework

66.3.3.1 Input–Process–Output Models: Early Examples

The traditional IS strategy literature makes the distinction between the process and content of strategy research (Chan and Huff, 1992). While *content* research focuses on linking organizational decisions and structures to performance (or other organizational outcomes), *process* research centers on the actions and activities leading to and supporting strategy (making). The input–process–output model combines these two major components. In the model, strategy is considered to be the *output* of the strategy formulation (including development) process. In other words, it is the deliverable of the process (Ang et al., 1995; Chen et al., 2010; King, 1988; Premkumar and King, 1991; Smits et al., 1997). Furthermore, the *implementation* of IS strategy results in some type of organizational impact: such commonly considered outcomes being competitive advantage, alignment, and firm performance. The *impact* of IS strategy is included as part of the framework in a later version of the input–process–output model (e.g., Lederer and Salmela, 1996). Strategy implementation will potentially change the organizational landscape, i.e., the internal as well as the external environment of the organization. Thus, the cycle starts again with the

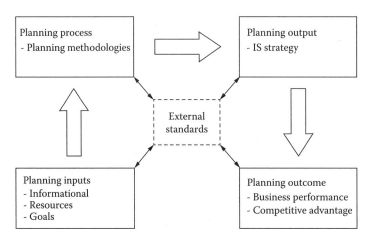

FIGURE 66.3   The input–process–output model.

internal and external environments *influencing* the strategy formulation process and the practices that are used by strategists to assess the needs and objectives of the organization. Consequently, this model incorporates considerations of such contextual factors as the planning process, strategy content, and impact of IS strategy actualization on the organization and its external environment. Figure 66.3 depicts the framework, as adapted from these earlier works.

66.3.3.2 Input–Process–Output Models: Later Extensions

Lederer and Salmela (1996) extended the earlier conceptions of the input–process–output model to provide a theory of SISP. The modified model was intended to provide a parsimonious way of identifying various strands found in the strategic IS planning literature. Specifically, Lederer and Salmela further elaborated on the constructs of the input–process–output model as they apply to SISP. First, they made a distinction among different components of planning input by considering the general context of the strategy development process, which includes elements within the organization's internal environment as well as elements beyond its control. Inputs are also the resources allocated to the planning process such as time, money, and personnel, as well as information and such intangible inputs as motivations toward and expectations of SISP. Newkirk and Lederer (2007) described SISP in terms of three IS resources planning activities: technical resources planning (i.e., planning activities associated with application software, systems software, hardware, and network communications), personnel resources planning activities (i.e., planning activities related to more people-oriented concerns such as technical training, end-user computing, facilities, and the personnel themselves), and data security planning activities (i.e., planning activities associated with protecting the organization from unwanted intrusion and recovering from such intrusions as they occur) (Doherty and Fulford, 2006).

A number of specific planning "input" factors have been mentioned in the literature. Abdul-Gader (1997) and Aladwani (2001) consider factors in the general business environment such as national and regional (e.g., socioeconomic) policies. As attention shifted to planning for inter-organizational systems (and to electronic commerce), suppliers, customers, and partner organizations were recognized as playing a more significant role in organizations' IS planning processes (Finnegan et al., 1999). Teo and King (1997) consider marketplace volatility, industry sector, competitive forces, and environmental uncertainty. Similarly, McFarlan's (1984) strategic grid represents the view that the conditions in the industry in which a firm operates largely set the scene for its IS strategy. The external conditions of a particular industry sector determine the extent of the strategic importance of IT applications (current and future) in the industry.

An explicit emphasis on the industry environment is also considered by Earl (1989), who distinguishes four types of companies in terms of their particular traits and preferences for IT. Earl finds that in some sectors, IT is the means of delivering the goods and services in the sector. In such cases, planning for goods and services and product–market strategy formulation cannot be done without reference to IT, and planning for IT is integral to business planning. In other sectors, business strategies are dependent on IT for their implementation. In these sectors, planning for IT is a derivative of business planning. In the third type of company, IT can yield some strategic advantage and, therefore, planning is likely to include some "IT-push" characteristics, where IT is taken to the users. Last, there are companies for which IT has no strategic impact, and planning is influenced by ad hoc needs.

Additionally, Sullivan (1985) suggests a simple matrix to explain how the IS strategy environment is affected by forces both internal and external to an organization. He describes two dimensions within which an organization can consider the implications of these forces: degree of decentralization of IS control in organization (i.e., diffusion) and degree of dependence on IS of the business (i.e., infusion). Information intensity and the rate of IT change have also been considered by Benamati and Lederer (2001) and Teo and King (1997). Cegielski et al. (2005) emphasize the need for IT managers to scan emerging technologies in order to determine any potential opportunities to which these might give rise. Such studies have identified clarity of corporate strategy, IT planning resources, the available IT budget, present and future impacts of IT (Premkumar, 1992), and uncertainty concerning IS benefits and the availability of IT (Wilson, 1989), as key factors in this regard.

Arguably, business or corporate strategy is also an integral aspect of the internal IS strategy environment. The reconciliation of the IS strategy within the business strategy has found form in research on strategic IS–business *alignment* (Chan and Reich, 2007; Henderson and Venkatraman, 1993) with a clear business strategy being viewed as having a major influence on, and essential to, the IS strategy development process. Others have argued that business strategy has implications not only for the process but also on the IS strategy itself and its impact (e.g., Walsham and Waema, 1994).

Similar to the original input–process–output model, other components of the Lederer and Salmela (1996) framework include the planning process and the "output" of the process—the information plan itself. The major component of the information plan is described as a set of recommendations for new IS. Lederer and Salmela make the point that organizations often fail to develop systems identified in the information plan and, thus, they consider the plan's *implementation* to be key. This is included in their amendment to the input–process model, illustrated in Figure 66.4 and utilized in our analysis.

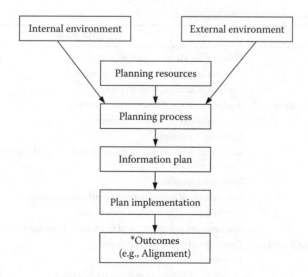

FIGURE 66.4 The Lederer and Salmela (1996, p. 240) model.

66.3.3.3 Elaboration on the Strategic Information Systems Planning Process

Early works viewed SISP as a set of defined activities but considered these activities in broad characteristics and general behaviors; rarely did the literature decompose the planning process into specific activities (McFarlan, 1971; Premkumar and King, 1991; Raghunathan and Raghunathan, 1991; Segars and Grover, 1999). Dealing with this issue, and amending the classification offered by Mentzas (1997), Brown (2004) presents six detailed phases of the planning process: preparation, organizational analysis, external environmental analysis, strategy conception, strategy formulation, and strategy implementation planning. Preparation refers to planning for IS planning process or preplanning. Organizational analysis encompasses the analysis of internal business and IS environments. Similarly, external environment analysis is the analysis of external business and IS environments. Strategy conception includes activities such as scanning the future, identifying alternative scenarios (including information flows and requirements for each), and considering the implications of each scenario (Galliers, 1992, 1993a,b). Strategy formulation is further elaborated into the formulation of agreed organizational recommendations and an associated information architecture (Galliers, 1993a,b), as well as the creation and prioritization of a portfolio of IS application developments. Finally, strategy implementation planning activities include the definition of action plan elements, follow-up, control, and an explanation of the action plan. Brown argues that these activities are necessary to ensure that the new IS are actually placed into production and used. Table 66.2 shows the process phases and the corresponding activities.

66.3.3.4 SISP Methodologies

Early SISP research provided managers with a number of methodologies (tools and techniques, rather) to apply during the planning process. Commonly, these methodologies can be divided into two broad categories: *impact* and *alignment* methodologies (Booth and Philip, 2005; Pant and Hsu, 1995, 1999; see Figure 66.5). The emphasis of impact methodologies is the gaining of competitive advantage through the use of IT—a popular theme throughout the 1980s and 1990s. Impact methodologies focused on creating and justifying innovative, value-added uses of IT. Examples of impact methodologies include critical success factors (Rockart, 1979), customer resource life cycle (Ives and Learmonth, 1984), value chain analysis (Porter, 1985), and strategic thrust analysis (Wiseman, 1985). In retrospect, the value of this kind of argumentation is debatable as much of the evidence presented was anecdotal rather than as a result of systematic studies (see, e.g., Ciborra, 1994).

TABLE 66.2   The Brown (2004, p. 29) Framework

SISP Phases	SISP Stages
Preparation	Planning for planning (pre-planning)
Organization analysis	Analysis of internal business environment
	Analysis of internal IS environment
External environment analysis	Analysis of internal business environment
	Analysis of internal IS environment
Strategy conception	Scanning the future
	Identification of alternative scenarios including information flows and requirements for each
	Scenario elaboration
Strategy formulation	Formulation of agreed organizational recommendations
	Formulation of information architecture
	Synthesis and prioritization of portfolio of information systems to be developed
Strategy implementation planning	Definition of action plan elements
	Elaboration of action plan
	Definition of follow-up and control procedures

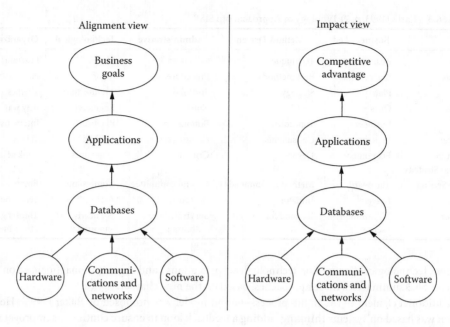

FIGURE 66.5 IS planning methodologies. (From Pant, S. and Hsu, C., Strategic information systems planning: A review, paper presented at the *International Conference on Information Resources Management Association*, Atlanta, GA, 1995, p. 5.)

The main rationale for the alignment school was to establish symbiosis between IS strategy and business objectives (Reich and Benbasat, 2000). The methodologies that fall into this category are extensive. They include Method 1 from Andersen Consulting (Lederer and Gardiner, 1992), BSP from IBM (Zachman, 1982), Information Engineering (Martin, 1989), Robert Holland's Strategic Systems Planning (Holland Systems Corporations, 1986), and Total Information Systems Management (Osterle et al., 1993). These methods take a top-down, business-led approach (Premkumar and King, 1994). Issues related to these methods include the lack of flexibility in responding to a rapidly changing business environments and concerns regarding, for example, "one size fits all" solutions (for a critique, see Galliers, 2007).

Although methodologies for SISP have taken up a large part of academic discussions (Boynton and Zmud, 1987; Lederer and Gardiner, 1992), SISP cannot, in our view, be fully understood by considering formal methods alone. It is the interaction of method applied, process followed, as well as the variety of activities and behaviors (i.e., what is termed "approaches to SISP") that characterize the planning experiences within organizations (Earl, 1993). This literature reveals a number of approaches to strategic IS planning. For example, Pyburn (1983) distinguished between the written (i.e., formal) planning system and the personal (i.e., informal) planning system. The former is a structured, top-down approach and the latter, adaptive and bottom-up. Segars et al. (1998) and Segars and Grover (1999) identified six dimensions to planning: comprehensiveness (extent of solution search), formalization (existence of rules and procedures), focus (extent of innovation versus integration), flow (top-down, bottom-up), participation (number and variety of planners), and consistency (frequency of planning). Earl (1993) identified the characteristics of SISP from other works in what is a fairly comprehensive framework (Doherty et al., 1999; Segars et al., 1998). He differentiated approaches based on nine criteria: emphasis, basis, ends, methods, nature, influencer, relationship with business strategy, priority setting, and the role of IS. Through an examination of 27 companies, Earl identified five SISP approaches: business-led (planning focused on the enterprise), method-driven (focused on the planning technique—often as provided by a vendor or consultancy, as with BSP or Method 1, for example), administrative (focused on the available

TABLE 66.3   Earl's (1993, p. 7) Typology of Approaches to SISP

	Business-Led	Method-Driven	Administrative	Technological	Organizational
Basis	Business	Technique	Resources	Model	Learning
Emphasis	Business plans	Best method	Procedure	Rigor	Partnership
Ends	Plan	Strategy	Portfolio	Architecture	Themes
Methods	Ours	Best	None	Engineering	Any way
Nature	Business	Top-down	Bottom-up	Blueprints	Interactive
Influencer	IS planner	Consultants	Committees	Method	Teams
Relation to Business Strategy	Fix points	Derive	Criteria	Objectives	Look at business
Priority Setting	The board	Method recommends	Central committee	Compromise	Emerge
IS Role	Driver	Initiator	Bureaucrat	Architect	Team member
Metaphor	It's common sense	It's good for you	Survival of the fittest	We nearly aborted it	Thinking IS all the time

resources), technological (focused on technological applications), and organizational (focused on learning). Table 66.3 summarizes these approaches, based on the nine characteristics.

An additional element of the input–process–output model was proposed by Baker (1995). Her contribution was based on systems thinking, adding a feedback loop to ensure continuous improvement in the planning process. Viewing IS strategizing as an iterative, learning process was acknowledged even in the earliest research (e.g., King and Cleland, 1975; Schwartz, 1970). Ongoing assessment and review have also been acknowledged by Galliers (2004) and Earl (1993) and are incorporated in the integrated model we propose.

No matter how detailed the input–process–output model has been extended over the years, however, it does not fully address the complex nature of the process of strategy development in practice. How organizations actually form strategy—the topic of strategy formation—has emerged as an area of intense debate within the strategy field over the years. Mintzberg (1978) points out that formulated or intended strategies may not be necessarily realized, and unplanned and unintended patterns of decisions and actions may emerge (see Figure 66.6). Concepts such as "bricolage" (Ciborra, 1994), gradual enhancement (Senn, 1992), improvisation (Galliers, 1991, 1993a,b), and "muddling through" (Lindblom, 1959) appear alongside the more formal strategy formulation approaches. Our proposed integrative model reflects the emergent nature of the IS strategizing process and distinguishes the notion of intended versus realized strategy. Intended strategy is the output or "result" of a deliberate planning process, whereas the realized strategy is the outcome of strategy implementation process during which some of the intended strategy may be realized while other aspects may not. The realized strategy also reflects some of the emergent strategies that materialize during the formation process.

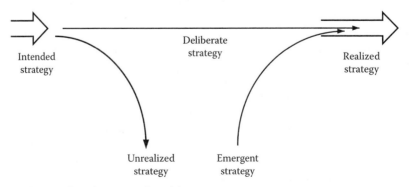

FIGURE 66.6   The Mintzberg (1978, p. 945) model.

The distinction between emergent and deliberate strategies appears in other relevant frameworks. In particular, Whittington (2002) proposes four generic approaches to strategy that differ along two dimensions: the outcomes of strategy and the processes by which it is made. The process dimension reflects whether the strategy is the product of deliberate planning or the result of emergence, accident, muddle, or inertia. The outcome dimension indicates the degree to which the strategy produces unitary (e.g., profit maximization) outcomes, or allows for consideration of alternative, pluralistic outcomes. For example, Segars and Grover (1998) suggest four dimensions of SISP success outcome: alignment (the linkage of the IS strategy with business strategy), analysis (the effort by IS planners to understand the internal process, procedures, technologies), cooperation (the agreement among IS planners concerning priorities, implementation schedules, and managerial responsibilities), and improvement in capabilities (the enhancement in SISP capabilities as a result of organizational learning). Similarly, King (1988) argues that the measurement of SISP impact should be multidimensional, and based on both judgmental and objective assessments. Dimensions proposed by King include the effectiveness of the strategy approach, its relative worth, the role and impact of IS strategy, the performance of IS plans, the relative efficiency of the strategy process, the adequacy of resources made available, and strategic congruence. The outcome of an IS strategizing process thus presents an additional consideration in the literature as it expands the scope of the formation process.

In summary, the proposed framework arising from this literature review includes the contextual dimension of IS strategizing (ISS), which comprises the external as well as the internal environment (see Figure 66.7), and attempts to be more holistic in nature than those that are often presented in existing research. Resources and capabilities are an important component of ISS and are, therefore, explicitly highlighted in the framework. The resource-based view (Barney, 1991; Rumelt, 1984; Wernerfelt, 1984) and the capabilities view (Peppard and Ward, 2004) are gaining increased attention, and they are used to explain the increased complexity of causal relations between IS strategy and firm performance. We have also included ongoing assessment, review, and feedback processes (e.g., Baker, 1995; Earl, 1993; Galliers, 2004) as the link between content and the process of strategizing. Additionally, we incorporate the kind of thinking that Brown (2004) introduces in illustrating specific phases in the SIS process.

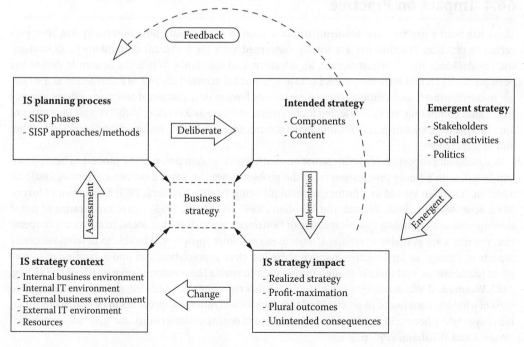

FIGURE 66.7 The proposed framework for IS strategizing.

Galliers (2004) provides components of IS strategy that might be addressed by an organization and are considered as the output of the planning process. These components might guide planners in the areas of strategy they need to plan for in their strategy development process.

The realized strategy is an outcome of the emergent process as well as a deliberate plan (Mintzberg, 1978), so the emergent strategy is incorporated in the framework as a topic that explores the unintended/unplanned social activities of strategy formation. Finally, profit maximization is often seen as the natural outcome of strategy making, but this unitary view is amended by the pluralistic view proposed by Whittington (2002), where strategists might have other interests or objectives than profit.

It is important to note that the IS strategizing "process" does not always occur sequentially in an organization, and there will always be overlap across various components that have been incorporated in the framework. In addition, different businesses and functions within an organization may be involved at different stages of the process. The framework is a conceptual tool—a sense-making device in Weick's (1990) terms—aimed at helping practitioners and researchers in understanding better the various areas encompassing this complex process and particularly those that need attention or further research.

Researchers and practitioners alike can use this framework to assess, in a comprehensive manner, the process of IS strategizing in organizations. For example, the framework points to the importance of resources consumed by the strategizing process. Justification of the use of such resources is critical to organizational survival during downturns because the resources allocated to planning activities are often substantially reduced during such difficult times (King, 1988). With respect to the intended strategy component of the IS strategizing process, a general question that might need to be considered is whether the plan in fact guides the strategic direction for the role of IS in the organization. The impact of IS strategy might be difficult to assess, and the framework suggests evaluation based on multiple outcome parameters. This multidimensionality further suggests the involvement of multiple stakeholders. The notion of "second-order" feedback incorporated in the framework may be applied to the assessment of the IS strategizing process by, for example, determining whether the process has any "self-correcting" characteristics (King, 1988).

66.4 Impact on Practice

There has been a gap between academic discussion on IS strategy and how IS strategy has been perceived in practice. Practitioners are mostly concerned with the technical dimensions of IS strategy, such as decisions on IT infrastructure, architecture, and standards. While the academic debate has been primarily on IT-based competitive advantages and has stressed the role of information as a strategic resource, practitioner conferences and magazines have mainly discussed technology developments (e.g., cloud computing, service-oriented architecture, software as a service, Web 2.0) and new types of applications (e.g., customer relationship management, social software, business intelligence) (Teubner and Mocker, 2008).

A significant proportion of the literature on IS strategizing often presents the process as being problem-based, with a linear progression from the problem toward a desired outcome (Horton, 2003). In addition, it is often viewed as a formal, rational planning activity (Galliers, 1991; Knights and Murray, 1994; Scott-Morton, 1988; Waema and Walsham, 1990; Walsham, 1993). There are a range of noted assumptions underpinning this perspective of IS strategizing such as that social relations are cooperative, resources are available as required, objectives are known, people will be able to develop necessary aspects of strategy in an objective manner, there is a clear appreciation and understanding of cause–effect relationships, and there is enough information to enable the required activities to operate (Kling, 1987; Waema and Walsham, 1990). We argue that this represents an idealized, superficial, and limited view of what has been found in practice and that such formal rational approaches to IS strategy formulation have rightly been criticized as "unrepresentative of organizational reality and generally simplistic" (Waema and Walsham, 1990, p. 30).

During the last decade, researchers have returned to questioning the underlying mechanisms explaining IT/IS-induced competitive advantage. The IS capability view (Peppard and Ward, 2004) and the study of the barriers to the erosion of competitive advantage (Piccoli and Ives, 2005) are good examples of such developments. A number of studies have addressed the *means* and *ends* of IS value creation, but, as with studies concentrating on maintaining IT competitive advantage, there is a paucity of research providing rigorous tests of theoretical propositions, with cause–effect relationships of how IS value is realized being inadequately explained.

The IS capability view concentrates on explaining the "whys?" of IS value creation and suggests, in Peppard and Ward's (2004, p. 188) words, that "the strategic management of IS is about developing IS competencies." The IS capability view guides practitioners to focus on how IS/IT is managed and used and has clear implications for both IS strategic planning theories and practice. The IS strategy impact is explained by more explicit cause–effect relationships, thereby helping organizations to better understand and justify the effects of IS developments. Compared to traditional technological developments, by concentrating on competency development, organizations might be less vulnerable to losing their competitive edge, given that such competencies are considerably harder to imitate or substitute than technological assets. Similarly, IS strategy content will be shifted from technology or alignment to planning for the competencies needed to exploit IS/IT developments. Though the Peppard and Ward (2004) article seems to be one of the most frequently cited works in the IS strategy literature concerning IS capabilities, a decade earlier, several authors contributed to the topic and even adopted the capability view as their "IS strategic platform" to build on (e.g., Ross et al., 1996).

By 2010, the number of articles concerning IS capabilities showed remarkable growth, clearly indicating the shift in IS strategy conceptualization toward more capability-based thinking. The extent to which such thinking is applied in practice remains debatable, but having said that, such centers as MIT's Center for Information Systems Research (CISR) in the United States and Cranfield's IT Leadership Programme in the United Kingdom have a practical orientation, given the financial backing they receive from corporate clients.

66.5 Research Issues

In general, IS strategy implementation has received by far the least attention in the IS strategizing discussion (Teubner and Mocker, 2008). Having said that, a number of scholars have acknowledged the significance of strategy implementation. For example, Ward and Peppard (2002, pp. 125–126) point out that "Despite an understanding of the importance of strategic planning for IS, in the past decade many organizations have developed perfectly sensible IS strategies that have been left to gather dust, or have been implemented in a half-hearted manner …" A survey by Lederer and Sethi (1988) found that after about 2 years into the planning horizon, less than one-fourth of the projects that were defined in the IS strategies had been initiated. A study of Norwegian organizations revealed that, after 5 years, only 42% of the projects that were defined in the strategy had been implemented (Gottschalk, 1999). Stricter implementation measures such as completion on time or realization of intended benefits (Earl, 1993) result in even smaller implementation rates (Gottschalk, 1999).

Smirchich and Stubbart (1985, p. 724) suggest that the problems of strategy implementation stem from the "field's inattention to the fundamentally social nature of the strategy formation and organization processes." Given the vast amount of research that explains different components of IS strategy and the strategizing process, it is perhaps surprising how little literature considers strategizing as a social activity. Only a small portion of the extant research has studied the key figures of the IS strategizing process, such as the chief information officer (CIO), and even fewer publications investigate the activities organizational members engage in during the IS strategy-making process. For example, our literature review of IS strategizing research has identified less than 5% of the articles reviewed considering the role CIOs play in the IS strategizing process.

Strategy is formed by highly skilled workers, and it is important to understand what skills are required for this work, how to acquire these skills, and what the common behaviors and techniques of strategizing actually take place (Whittington, 2003). The limited amount of research that considers the structure, rules, and processes of the IT function, as well as IT personnel requirements, IT skills, and IT personal development, has been mostly conceptual and has provided little more than anecdotal evidence (e.g., Agarwal and Sambamurthy, 2002). More research should focus on how the practitioners of strategy actually act and interact in the process of strategizing (Jarzabkowski, 2004; Whittington, 1996).*

The current economic environment is moving rapidly toward open markets, mobile labor, and information abundance (Johnson et al., 2003). The ease of obtaining resources, falling barrier to market entry, and strategic imitation creates a precarious foundation for a competitive advantage (Barney, 1991). As a result, sustainable advantage must lie in microassets that are hard to discern, imitate, and obtain. There is also a shift to a more "hypercompetitive" environment, in which speed, revelation, and innovation are the bases for competitive advantage. This requires organizational decentralization and more continuous, adaptive, and innovative processes (Brown and Eisenhardt, 1997) in strategy making. More people are involved in the process and research needs to understand their role in the strategizing. As a result, the activities of managers—what they do, how and why they do it—are increasingly becoming central to the strategy debate (Johnson et al., 2003).

66.6 Summary

Considerations of IS strategizing in organizations have developed from the isolated, to the more aligned, and then to the impact modes of planning. The aim of this chapter has been to provide something of a synthesis of the research literature on the concept of IS strategizing process. In doing so, we have taken a broader perspective on IS strategizing and have emphasized the dynamic, continuous, iterative, and interactive nature of the strategizing process (e.g., Auer and Reponen, 1997; Baker, 1995; Ciborra, 1997; Earl, 1993; Reponen, 1993; Salmela et al., 2000). We have described the strategizing process as being quintessentially social—one that includes negotiation for allocation of organizations' resources, requires commitment and support from top management, and involves multiple stakeholders (Earl, 1993). In this regard, we argue that IS strategizing has—or should—become even more inseparable from strategic business planning. Any strategic change in IS requires increasing change in business operations and processes, and thus IS strategizing and strategy implementation require organization-level attention and capabilities well beyond the traditional boundaries of IS. We expect future research in IS strategy to find new ways to connect—and unpack—these already heavily interconnected key aspects of strategic business planning and strategic IS planning—topics that have for too long been considered by their respective academies as being their own "territory." Research in IS capabilities has already taken some steps in this direction, but more is required.

Glossary

Bottom-up planning: An IS planning view in which the driver for planning comes from technological opportunities.
Capabilities: A firm's capacity to acquire, deploy, and leverage resources, usually in combination, using organizational structures, processes, routines, learned and stable activities, and/or behavioral orientation to effect a desired end.
Change management strategy: A component of IS strategy that emphasizes the importance of feedback regarding the impact of past strategic decisions (unintended as well as expected outcomes) and the identification of emergent strategies.

---

* The issue has already recognized by some notable IS journals. For example, at the time of writing, the *Journal of Strategic Information Systems* is planning a special issue on IS strategy and strategizing from a practice perspective.

Data processing: An early (pre) era of IS strategizing in which the business objective was to automate information-based processes to improve operational competence by reducing data processing costs and consequently achieve business efficiency. IS operations and the possible IS plan tended to be isolated from the business of the organization.

Dynamic capabilities: Structures, processes, routines, learned and stable activities, and/or behavioral orientation that help organizations change by creating an ability to integrate, build, and reconfigure internal and external competences to address rapidly changing environments.

Goal seeking: An IS planning view in which the objective and focus are mainly on proactive future effectiveness of IS and delivering new opportunities for business operations.

Implementation strategy: A component of IS strategy that emphasizes the importance of managing the change processes associated with ongoing implementation of the strategy.

Information services strategy/information management strategy: Focuses on policy issues regarding the organization of information services (and capabilities), including sourcing decisions— "the organizational arrangements for the provision of IS-related services." Answers mainly "who?" questions: Who is needed to facilitate and enable the strategy to be developed and implemented? Such questions include considerations of the structure, roles, and processes of the IT function, but also include human resources issues such as IT personnel requirements and choices between the internal service provision.

Information strategy: Focuses on identifying the information that could question taken-for-granted assumptions on which the business strategy is based, as well as providing the information required by the business strategy. Thus, an information strategy is the glue between the business strategy and the IS strategy. It tends to answer questions such as, "What is the information required to support primary tasks, or key goals, of the organization strategy?" and "Where can this be obtained?" The information strategy also questions the business strategy by attempting to answer the "why" questions—"Why might this particular strategy be chosen over any other?"

IS (information systems): IS as a whole is a sociotechnical concept, including not only the technical IT perspective, but also the business processes, social processes, users, and connections to internal and external environment that affect the operation of the systems; "everything that the systems are made of."

IS capabilities: A firm's ability to acquire, deploy, and leverage its IT resources (human as well as technical) to shape and support its business strategies and value chain activities.

IS dynamic capabilities: Dynamic capabilities (see the previous term) related to the IS operations of the organization.

IS strategic planning: An overall term describing the efforts and resources to plan for IS operations of the organization and to create an IS strategy. It should be considered an ongoing process rather than a one-off effort.

IS strategizing/IS strategizing process: The process and details of IS strategic planning, including the objectives, resources, and efforts to keep up a strategically alert cognition for IS and the organizational requirements to implement the IS strategy. An inherent connection to business objectives and strategizing is a crucial part of IS strategizing, and vice versa.

IS strategy: Can be considered the developmental path that needs to be taken to achieve the business objectives related to IS, considering both deliberate and emergent sides of the IS strategy. IS strategy is a cross-functional overall term that encompasses, e.g., product, process, human resources, and several interlinked components such as information strategy, information services strategy, implementation, and change management strategy as well as the strategizing process (see later). Recent developments have taken a more holistic view to what should be taken into account in IS strategy, including IS capabilities (see later).

Issue-based: An IS planning view in which the objective and focus are mainly on reactive effectiveness of IS.

IT (information technology): A term describing the technical/architectural aspects of information systems, mainly the hardware, software, and their physical connections.

IT strategy: The strategic plan for IT (as described earlier), mainly concentrating on technical effectiveness. Concerned with alternative technological solutions (alternative "hows") to support the "what" of business needs.

MIS (management information systems): Another era of IS strategizing in which the objective, in addition to operation efficiency, was to create ways of organizing and delivering information to improve management effectiveness. Also the era when the need for IS strategizing became apparent and focus on IS planning efforts gained increased attention.

RBV (Resource-based view): A theoretical perspective explaining the competitive edge of a company with valuable, rare, inimitable, and non-substitutable (the so-called VRIN) resources and capabilities.

SIS (strategic information systems): The SI strategizing era, with a proactive stance and objective to change the nature or conduct of business to improve the competitiveness of the company.

SISP (strategic information systems planning): The process of identifying a portfolio of computer-based applications that will assist an organization in executing its business plans and realizing its business goals.

Top-down planning: An IS planning view in which the driver for planning comes from holistic business requirements of the organization.

Further Information/Readings

Ciborra, C. U., Braa, K., Cordella, A., Dahlbom, B., Failla, A., Hanseth, O., Hepsø, V., Ljungberg, J., Monteiro, E., and Simon, K. A. (2000). *From Control to Drift: The Dynamics of Corporate Information Infrastructures.* Oxford, U.K.: Oxford University Press.

Galliers, R. D. and Currie, W. (Eds.). (2011). *The Oxford Handbook of Management Information Systems: Critical Perspectives and New Directions.* Oxford, U.K.: Oxford University Press.

Galliers, R. D. and Leidner, D. E. (Eds.). (2009). *Strategic Information Management*, 4th edn. New York: Routledge.

King, W. R. (2009). *Planning for Information Systems.* New York: M.E. Sharpe.

Mansell, R., Avgerou, C., Quah, D., and Silverstone, R. (Eds.). (2007). *The Oxford Handbook of Information and Communication Technologies.* Oxford, U.K.: Oxford University Press.

Ward, J. and Peppard, J. (2002). *Strategic Planning for Information Systems*, 3rd edn. Chichester, U.K.: Wiley.

References

Abdul-Gader, A. (1997). Information systems strategies for multinational companies in Arab Gulf countries. *International Journal of Information Management*, 17(1), 3–12.

Agarwal, R. and Sambamurthy, V. (2002). Principles and models for organizing the IT function. *MIS Quarterly Executive*, 1(1), 1–16.

Aladwani, A. (2001). IT planning effectiveness in a developing country. *Journal of Global Information Technology Management*, 4(3), 51–65.

Ambrosini, V. and Bowman, C. (2009). What are dynamic capabilities and are they a useful construct in strategic management? *International Journal of Management Reviews*, 11(1), 29–49.

Ang, J., Shaw, N., and Pavri, F. (1995). Identifying strategic management information systems planning parameters using case studies. *International Journal of Information Management*, 15(6), 463–474.

Auer, T. and Reponen, T. (1997). Information systems strategy formation embedded into a continuous organizational learning process. *Information Resources Management Journal*, 10(2), 32.

Baker, B. (1995). The role of feedback in assessing information systems planning effectiveness. *Journal of Strategic Information Systems*, 4(1), 61–80.

Barney, J. (1991). Firm resources and sustainable competitive advantage. *Journal of Management*, 17(1), 99–120.

Barreto, I. (2010). Dynamic capabilities: A review of past research and an agenda for the future. *Journal of Management*, 36(1), 256–280.

Benbasat, I., Defier, A., Drury, D., and Goldstein, R. (1984). A critique of the stage hypothesis: Theory and empirical evidence. *Communications of the ACM*, 27(5), 476–485.

Benamati, J. and Lederer, A. L. (2001). Coping with rapid changes in IT. *Communications of the ACM*, 44(8), 83–88.

Bergeron, F., Buteau, C., and Raymond, L. (1991). Identification of strategic information systems opportunities: Applying and comparing two methodologies. *MIS Quarterly*, 15(1), 89–103.

Bharadwaj, A. (2000). A resource-based perspective on information technology capability and firm performance: An empirical investigation. *MIS Quarterly*, 24(1), 169–196.

Bharadwaj, A. S., Sambamurthy, V., and Zmud, R. W. (2002). *Firmwide IT Capability: An Empirical Examination of the Construct and Its Links to Performance*, Emory University, Atlanta, GA.

Blumenthal, S. (1969). *Management Information Systems: A Framework for Planning and Development*. Englewood Cliffs, NJ: Prentice Hall.

Booth, M. and Philip, G. (2005). Information systems management: Role of planning, alignment and leadership. *Behaviour & Information Technology*, 24(5), 391–404.

Boynton, A. C. and Zmud, R. W. (1987). Information technology planning in the 1990s: Directions for practice and research. *MIS Quarterly*, 11(1), 59–71.

Brown, T. (2004). Testing and extending theory in strategic information systems planning through literature analysis. *Information Resources Management Journal*, 17(4), 20–48.

Brown, S. L. and Eisenhardt, K. M. (1997). The art of continuous change: Linking complexity theory and time-paced evolution in relentlessly shifting organizations. *Administrative Science Quarterly*, 42(1), 1–34.

Caminer, D., Aris, J., Hermon, P., and Land, F. (1998). LEO: The incredible story of the world's first business computer. *Journal of Database Management*, 10(1), 44.

Carr, N. (2003). IT doesn't matter. *Harvard Business Review*, 81(5), 41–49.

Cegielski, C. G., Reithel, B. J., and Rebman, C. M. (2005). Emerging information technologies: Developing a timely IT strategy. *Communications of the ACM*, 48(8), 113–117.

Chan, Y. and Huff, S. (1992). Strategy: an information systems research perspective. *The Jounal of Strategic Information Systems*, 1(4), 191–204.

Chan, Y. E., Huff, S. L., Barclay, D. W., and Copeland, D. G. (1997). Business strategic orientation, information systems strategic orientation, and strategic alignment. *Information Systems Research*, 8(2), 125–150.

Chan, Y. and Reich, B. (2007). IT alignment: What have we learned? *Journal of Information Technology*, 22(4), 297–315.

Chen, D., Mocker, M., Preston, D., and Teubner, A. (2010). Information systems strategy: Reconceptualization, measurement, and implications. *MIS Quarterly*, 34(2), 233–259.

Chen, R. S., Sun, C. M., Helms, M. M., and Jih, W. J. (2008). Aligning information technology and business strategy with a dynamic capabilities perspective: A longitudinal study of a Taiwanese Semiconductor Company. *International Journal of Information Management*, 28(5), 366–378.

Ciborra, C. (1994). From thinking to tinkering: The grassroots of IT and strategy. In C. Ciborra and T. Jelessi (Eds.), *Strategic Information Systems: A European Perspective* (pp. 70–83). Chichester, U.K.: Wiley.

Ciborra, C. U. (1997). De profundis? Deconstructing the concept of strategic alignment. *Scandinavian Journal of Information Systems*, 9(1), 2.

Clemons, E. K. and Row, M. C. (1991). Sustaining IT advantage: The role of structural differences. *MIS Quarterly*, 15(3), 275–292.

Doherty, N. F. and Fulford, H. (2006). Aligning the information security policy with the strategic information systems plan. *Computers & Security*, 25(1), 55–63.

Doherty, N., Marples, C., and Suhaimi, A. (1999). The relative success of alternative approaches to strategic information systems planning: An empirical analysis. *Journal of Strategic Information Systems*, 8(3), 263–283.

Earl, M. J. (1988). *Information Management: The Strategic Dimension*. Oxford, U.K.: Oxford University Press.

Earl, M. (1989). *Management Strategies for Information Technology*. Upper Saddle River, NJ: Prentice Hall.

Earl, M. (1993). Experiences in strategic information systems planning. *MIS Quarterly*, 17(1), 1–21.

Eisenhardt, K. and Martin, J. (2000). Dynamic capabilities: What do capabilities come from and how do they matter: A study in the software services industry. *Strategic Management Journal*, 26(1), 25–45.

Finnegan, P., Galliers, R. D., and Powell, P. (1999). Inter-organisational information systems planning: Learning from current practice. *International Journal of Technology Management*, 17(1/2), 129–144.

Gable, G. (2010). Strategic information systems research: An archival analysis. *Journal of Strategic Information Systems*, 19(1), 3–16.

Galliers, R. D. (1987). Information systems planning: A manifesto for Australian-based research. *Australian Computer Journal*, 19(2), 49–55.

Galliers, R. D. (1991). Strategic information systems planning: Myths, reality and guidelines for successful implementation. *European Journal of Information Systems*, 1(1), 55–64.

Galliers, R. D. (1992). Soft systems, scenarios, and the planning and development of information systems. *Systemist*, 14(3), 146–159.

Galliers, R. D. (1993a). IT strategies: Beyond competitive advantage. *Journal of Strategic Information Systems*, 2(4), 283–291.

Galliers, R. D. (1993b). Towards a flexible information architecture: Integrating business strategies, information systems strategies and business process redesign. *Information Systems Journal*, 3(3), 199–213.

Galliers, R. D. (1999). Towards the integration of e-business, knowledge management and policy considerations within an information systems strategy framework. *Journal of Strategic Information Systems*, 8(3), 229–234.

Galliers, R. D. (2004). Reflections on information systems strategizing. In C. Avgerou, C. Ciborra, and F. Land (Eds.), *The Social Study of Information and Communication Technology* (pp. 231–262). Oxford, U.K.: Oxford University Press.

Galliers, R. D. (2007). On confronting some common myths of is strategy discourse. In R. A. Mansell, C. Chrisanthi, D. Quah, and R. Silverstone (Eds.), *The Oxford Handbook of Information and Communication Technologies* (pp. 225–243). Oxford, U.K.: Oxford University Press.

Galliers, R. D. (2009). Conceptual developments in information systems strategy—Reflections on information systems strategizing. In R. D. Galliers and D. E. Leidner (Eds.), *Strategic Information Management*, 4th edn. (pp. 5–33). New York: Routledge.

Galliers, R. D. and Leidner, D. E. (Eds.). (2003). *Strategic Information Management: Challenges and Strategies in Managing Information Systems*, 3rd edn. Oxford, U.K.: Butterworth-Heinemann.

Galliers, R. D. and Leidner, D. E. (Eds.). (2009). *Strategic Information Management: Challenges and Strategies in Managing Information Systems*, 4th edn. New York: Routledge.

Galliers, R. D. and Sutherland, A. (1991). Information systems management and strategy formulation: The 'stages of growth' model revisited. *Information Systems Journal*, 1(2), 89–114.

Gibson, C. and Nolan, R. L. (1974). Managing the four stages of EDP growth. *Harvard Business Review*, 52(1), 76–88.

Gottschalk, P. (1999). Implementation of formal plans: The case of information technology strategy. *Long Range Planning*, 32(3), 362–372.

Henderson, J. and Venkatraman, N. (1993). Strategic alignment: Leveraging information technology for transforming organizations. *IBM Systems Journal*, 32(1), 4–16.

Holland Systems Corporation. (1986). Strategic Systems Planning. Document #M0154–04861986, Ann Arbor, MI.

Horton, K. (2003). Information systems strategy and configural technologies: Cases from the UK Public Sector. *Proceedings: ECIS 2003*. Naples, Italy.

Ives, B. and Learmonth, G. (1984). The information system as a competitive weapon. *Communications of the ACM*, 27(12), 1193–1201.

Jarzabkowski, P. (2004). Strategy as practice: Recursiveness, adaptation, and practices-in-use. *Organization Studies*, 25(4), 529–560.

Johnson, G., Melin, L., and Whittington, R. (2003). Guest editors' introduction. *Journal of Management Studies*, 40(1), 3–22.

Johnston, H. R. and Carrico, S. R. (1988). Developing capabilities to use information strategically. *MIS Quarterly*, 12(1), 37–48.

Karimi, J. and Konsynski, B. R. (1991). Globalization and information management strategies. *Journal of Management Information Systems*, 7(4), 7–26.

Kettinger, W. J., Grover, V., Guha, S., and Segars, A. H. (1994). Strategic information systems revisited: A study in sustainability and performance. *MIS Quarterly*, 18(1), 31–58.

King, W. R. (1978). Strategic planning for management information systems. *MIS Quarterly*, 2(1), 27–37.

King, W. R. (1988). How effective is your information systems planning? *Long Range Planning*, 21(5), 103–112.

King, W. R. (2009). *Planning for Information Systems*. New York: M.E. Sharpe.

King, W. R. and Cleland, D. (1975). A new method for strategic systems planning. *Business Horizons*, 18(4), 55–64.

King, J. and Kraemer, K. (1984). Evolution and organizational information systems: An assessment of Nolan's stage model. *Communications of the ACM*, 27(5), 466–475.

King, W. and Teo, T. (1997). Integration between business planning and information systems planning: Validating a stage hypothesis. *Decision Sciences*, 28, 279–308.

King, W. R. and Sabherwal, R. (1992). The factors affecting strategic information systems applications: An empirical assessment. *Information & Management*, 23(4), 217–235.

Kling, R. (1987). Defining the boundaries of computing across complex organizations. In R. Boland and R. Hirschheim (Eds.), *Critical Issues in Information Systems Research* (pp. 307–361). Chichester, U.K.: Wiley.

Knights, D. and Murray, F. (1994). *Managers Divided: Organisation Politics and Information Technology Management*. Chichester, U.K.: John Wiley & Sons.

Kriebel, C. H. (1968). The strategic dimension of computer systems planning. *Long Range Planning*, 1(1), 7–12.

Lederer, A. L. and Gardiner, V. (1992). Strategic information systems planning—The Method/1-Approach. *Information Systems Management*, 9(3), 13–20.

Lederer, A. and Salmela, H. (1996). Toward a theory of strategic information systems planning. *Journal of Strategic Information Systems*, 5(3), 237–253.

Lederer, A. and Sethi, V. (1988). The implementation of strategic information systems planning methodologies. *MIS Quarterly*, 12(3), 445–461.

Lederer, A. and Sethi, V. (1996). Key prescriptions for strategic information systems planning. *Journal of Management Information Systems*, 13(1), 35–62.

Lefebvre, L. A., Mason, R., and Lefebvre, E. (1997). The influence prism in SMEs: The power of CEOs' perceptions on technology policy and its organizational impacts. *Management Science*, 43(6), 856–878.

Lincoln, T. (1975). A strategy for information systems development. *Management Datamatics*, 4(4), 121–128.

Lindblom, C. (1959). The science of "muddling through". *Public Administration Review*, 19(2), 79–88.

Lindsey, D., Cheney, P. H., Kasper, G. M., and Ives, B. (1990). TELCOT: An application of information technology for competitive advantage in the cotton industry. *MIS Quarterly*, 14(4), 347–357.

Luftman, J. and Ben-Zvi, T. (2010). Key issues for IT executives 2010: Judicious IT investments continue post-recession. *MIS Quarterly Executive*, 9(4), 263–273.

Luftman, J., Kempaiah, R., and Nash, E. (2006). Key issues for IT executives 2005. *MIS Quarterly Executive*, 5(2), 81–99.

Martin, J. (1989). *Information Engineering: Planning and Analysis*, Vol. 2. Upper Saddle River, NJ: Prentice Hall.

Mata, F. J., Fuerst, W. L., and Barney, J. B. (1995). Information technology and sustained competitive advantage: A resource-based analysis. *MIS Quarterly*, 19(4), 487–505.

McFarlan, F. W. (1984). Information technology changes the way you compete. *Harvard Business Review*, 62(3), 98–103.

McFarlan, F. W. (1971). Problems in planning the information system. *Harvard Business Review*, 49(2), 75–89.

McLean, E., Soden, J., and Steiner, G. (1977). *Strategic Planning for MIS*. New York: John Wiley & Sons.

Mentzas, G. (1997). Implementing an IS strategy—A team approach. *Long Range Planning*, 30(1), 84–95.

Mintzberg, H. (1978). Patterns in strategy formation. *Management Science*, 24(9), 934–948.

Mjosund, A. (1975). Toward a strategy for information needs analysis. *Computers & Operations Research*, 2(1), 39–47.

Neo, B. S. (1988). Factors facilitating the use of information technology for competitive advantage: An exploratory study. *Information & Management*, 15(4), 191–201.

Newkirk, H. and Lederer, A. (2007). The effectiveness of strategic information systems planning for technical resources, personnel resources, and data security in environments of heterogeneity and hostility. *Journal of Computer Information Systems*, 47(3), 34.

Nolan, R. L. (1973). Managing the computer resource: A stage hypothesis. *Communications of the ACM*, 16(7), 399–405.

Nolan, R. L. (1979). Managing the crisis in data processing. *Harvard Business Review*, 57(2), 115–126.

Nolan, R. (1984). Managing the advanced stages of computer technology: Key research issues. In F. W. McFarlan (Ed.), *The Information Systems Research Challenge* (pp. 195–214). Boston, MA: Harvard Business School Press.

Osterle, H., Brenner, W., and Hilbers, K. (1993). *Total Information Systems Management: A European Approach*. Chichester, U.K.: John Wiley & Sons.

Pant, S. and Hsu, C. (1995). Strategic information systems planning: A review. Paper presented at the *International Conference on Information Resources Management Association*, Atlanta, GA. http://www.mis.boun.edu.tr/tamers/mis524/strpaper.pdf (accessed on April 16, 2013).

Pant, S. and Hsu, C. (1999). An integrated framework for strategic information systems planning and development. *Information Resources Management Journal*, 12(1), 15–25.

Pant, S. and Ravichandran, T. (2001). A framework for information systems planning for e-business. *Logistics Information Management*, 14(1/2), 85–99.

Pavlou, P. A. and El Sawy, O. A. (2006). From IT leveraging competence to competitive advantage in turbulent environments: The case of new product development. *Information Systems Research*, 17(3), 198–227.

Peppard, J. and Ward, J. (2004). Beyond strategic information systems: Towards an IS capability. *Journal of Strategic Information Systems*, 13(2), 167–194.

Piccoli, G. and Ives, B. (2005). Review: IT-dependent strategic initiatives and sustained competitive advantage: A review and synthesis of the literature. *MIS Quarterly*, 29(4), 747–776.

Porter, M. (1985). *Competitive Advantage: Creating and Sustaining Superior Performance*. New York: Free Press.

Porter, M. and Millar, V. (1985). How information gives you competitive advantage. *Harvard Business Review*, 63(4), 149–160.

Powell, T. C. and Dent-Micallef, A. (1997). Information technology as competitive advantage: The role of human, business, and technology resources. *Strategic Management Journal*, 18(5), 375–405.

Premkumar, G. (1992). An empirical study of IS planning characteristics among industries. *Omega*, 20(5–6), 611–629.

Premkumar, G. and King, W. (1991). Assessing strategic information systems planning. *Long Range Planning*, 24(5), 41–58.

Premkumar, G. and King, W. (1994). Organizational characteristics and information systems planning: An empirical study. *Information Systems Research*, 5(2), 75–109.

Pyburn, P. (1983). Linking the MIS plan with corporate strategy: An exploratory study. *MIS Quarterly*, 7(2), 1–14.

Raghunathan, B. and Raghunathan, T. (1991). Information systems planning and effectiveness: An empirical analysis. *Omega*, 19(2–3), 125–135.

Reich, B. and Benbasat, I. (1996). Measuring the linkage between business and information technology objectives. *MIS Quarterly*, 20(1), 55–81.

Reich, B. and Benbasat, I. (2000). Factors that influence the social dimension of alignment between business and information technology objectives. *MIS Quarterly*, 24(1), 81–113.

Reponen, T. (1993). Information management strategy—An evolutionary process. *Scandinavian Journal of Management*, 9(3), 189–209.

Rockart, J. F. (1979). Critical success factors. *Harvard Business Review*, 57(2), 81–91.

Ross, J., Beath, C., Goodhue, D., and Competitiveness, D. (1996). Develop long-term competitiveness through IT assets. *Sloan Management Review*, 38(1), 31–42.

Rumelt, R. (1984). Towards a strategic theory of the firm. *Resources, Firms, and Strategies: A Reader in the Resource-Based Perspective*, 131–145.

Salmela, H., Lederer, A., and Reponen, T. (2000). Information systems planning in a turbulent environment. *European Journal of Information Systems*, 9(1), 3–15.

Schwartz, M. H. (1970). MIS planning. *Datamation*, 16(17), 28–31.

Schwarz, A., Kalika, M., Kefi, H., and Schwarz, C. (2010). A dynamic capabilities approach to understanding the impact of IT-enabled businesses processes and IT-business alignment on the strategic and operational performance of the firm. *Communications of the Association for Information Systems*, 26, 57–84.

Scott-Morton, M. S. (1988). Strategy formulation methodologies and IT. In M. Earl (Ed.), *Information Management: The Strategic Dimension* (pp. 54–67). Oxford, U.K.: Oxford University Press.

Segars, A. H. and Grover, V. (1998). Strategic information systems planning success: An investigation of the construct and its measurement. *MIS Quarterly*, 22(2), 139–163.

Segars, A. and Grover, V. (1999). Profiles of strategic information systems planning. *Information Systems Research*, 10(3), 199–232.

Segars, A. H., Grover, V., and Teng, J. T. C. (1998). Strategic information systems planning: Planning system dimensions, internal coalignment, and implications for planning effectiveness. *Decision Sciences*, 29(2), 303–341.

Senn, J. A. (1992). The myths of strategic systems. *Information Systems Management*, 9(3), 7–12.

Siegal, P. (1975). *Strategic Planning of Management Information Systems*. New York: Mason and Lips Comp. Publishers.

Smircich, L. and Stubbart, C. (1985). Strategic management in an enacted world. *Academy of Management Review*, 10(4), 724–736.

Smits, M., Van der Poel, K., and Ribbers, P. (1997). Assessment of information strategies in insurance companies in the Netherlands. *Journal of Strategic Information Systems*, 6(2), 129–148.

Somogyi, S. and Galliers, R. D. (1987). From data processing to strategic information systems—A historical perspective. In S. Somogyi and R. D. Galliers (Eds.), *Towards Strategic Information Systems* (pp. 5–25). Cambridge, MA: Abacus Press.

Sullivan, C. H. (1985). Systems planning in the information age. *Sloan Management Review*, 26(2), 3–12.

Teece, D., Pisano, G., and Shuen, A. (1992). Dynamic capabilities and the concept of strategy, University of California at Berkeley Working Paper.

Teece, D., Pisano, G., and Shuen, A. (1997). Dynamic capabilities and strategic management. *Strategic Management Journal*, 18(7), 509–533.

Teo, T. S. H. and King, W. R. (1997). Integration between business planning and information systems planning: An evolutionary-contingency perspective. *Journal of Management Information Systems*, 14, 185–214.

Teubner, R. A. and Mocker, M. (2008). *A Literature Overview on Strategic Information Systems Planning*. Münster, Germany: European Research Center for Information Systems.

Waema, T. and Walsham, G. (1990). Information systems strategy formulation. *Information & Management*, 18(1), 29–39.

Walsham, G. (1993). *Interpreting Information Systems in Organisations*. Chichester, U.K.: Wiley.

Walsham, G. and Waema, T. (1994). Information systems strategy and implementation: A case study of a building society. *ACM Transactions on Information Systems (TOIS)*, 12(2), 150–173.

Wang, C. L. and Ahmed, P. K. (2007). Dynamic capabilities: A review and research agenda. *International Journal of Management Reviews*, 9(1), 31–51.

Ward, J. and Peppard, J. (2002). *Strategic Planning for Information Systems*, 3rd edn. Chichester, U.K.: Wiley.

Weick, K. E. (1990). Technology as equivoque: Sensemaking in new technologies. In P. Goodman and L. Sproull (Eds.), *Technology and Organizations*. San Francisco, CA: Jossey-Bass.

Wernerfelt, B. (1984). A resource-based view of the firm. *Strategic Management Journal*, 5(2), 171–180.

Whittington, R. (1996). Strategy as practice. *Long Range Planning*, 29(5), 731–735.

Whittington, R. (2002). *What is Strategy—and Does it Matter?* 2nd edn. London, U.K.: Thomson Learning.

Whittington, R. (2003). The work of strategizing and organizing: For a practice perspective. *Strategic Organization*, 1(1), 117–125.

Wilson, T. (1989). The implementation of information system strategies in UK companies: Aims and barriers to success. *International Journal of Information Management*, 9(4), 245–258.

Wiseman, C. (1985). *Strategy and Computers*. New York, Dow Jones-Irwin.

Young, R. C. (1967). Systems and data processing departments need long-range planning. *Computers and Automation*, 45, 30–33.

Zachman, J. A. (1977). Control and planning of information systems. *Journal of Systems Management*, 28(7), 34–41.

Zachman, J. (1978). The information systems management system a framework for information systems planning. *DATABASE for Advances in Information Systems*, 9(3), 8–13.

Zachman, J. (1982). Business system planning and business information control study: A comparison. *IBM Systems Journal*, 21, 35–45.

Zani, W. M. (1970). Blueprint for MIS. *Harvard Business Review*, 48(6), 95–100.

67

Information Technology and Organizational Structure

M. Lynne Markus
Bentley University

67.1 Introduction

The first commercial uses of computers—for applications such as accounting, insurance claims processing, and airline reservations—date from the early 1950s. It was not long afterward that futurists began to speculate about the effects of computer use on the form and functioning of organizations. In 1958, Leavitt and Whisler—coiners of the term "information technology"—predicted that IT would have "definite and far-reaching impact on managerial organization" in large- and medium-sized business firms (Leavitt and Whisler, 1958, p. 41).

Leavitt and Whisler made four specific predictions. First, occupational specialization would increase: Business planning work would be taken from mid-level line managers and given to a new class of specialists such as operations researchers, who would develop tools and standardized routines for their work. Second, large organizations would re-centralize, with top managers taking over much of the planning and innovating work that they had formerly delegated to middle managers. Third, while the new specialist functions would increase in status and compensation, many middle management jobs would be downgraded or eliminated; instead of the traditional pyramid shape, the organization of the future might look more like "a football [the top staff organization] balanced on the point of a church bell" (p. 47). Fourth, the invisible line separating the top of the organization and the middle would become more distinct and impenetrable.

For years, these hypotheses about the effects of IT on organizational structure appeared to be as infamously wrong as Thomas Watson's (then IBM's Chairman) 1943 prediction about the world market for computers ("maybe five"). But 30 years later, scholars called Leavitt and Whisler's predictions "downright visionary" (Applegate et al. 1991, p. 128). However, the futurists of the late 1980s noted effects of IT on organizational structure that Leavitt and Whisler had not envisioned; they argued that the adoption of microcomputers would enable increased *de*centralization of decision making at the same time that IT also enabled downsizing, restructuring, and improved centralized control. These paradoxical impacts of IT would make large organizations as flexible, dynamic, and innovative as small firms (Applegate et al., 1991).

This, in a nutshell, is the debate over the impacts of IT on organizational structure, which stimulated much empirical research and theorizing in a number of fields and subfields, including the social issues and impacts of computing, information systems, organization theory, organizational communication, strategic management, international business, sociology, and economics. Although the theories, evidence, and implications are all contested, research about *whether and how* IT affects organizational structure is important for theoretical and practical reasons.

Theoretically, the concept of organizational structure—loosely, the anatomy of an organization, its skeleton and organs, as opposed to its physiology or metabolism—is central to a number of prominent organizational theories such as the information processing theory of organizations and the strategy–structure–performance paradigm. Although some theorists argue that organizational culture and processes are more important for organizational performance than organizational structure is, others counter that organizational structure shapes (but does not determine) information flows and pathways of organizational communication and helps focus the attention of managers, thus influencing organizational behavior. From this point of view, the hypothesis that IT affects organizational structure is important, because it establishes IT as a key factor in organizational performance. In addition, research about the effects of IT on organizational structure is a prominent manifestation of ongoing academic debates about technological determinism versus idealism, that is, about whether and how human behavior is affected by the material aspects of the world versus by people's ideas and beliefs about the world.

Practically, research and theorizing about IT's effects on organizational structure is important for at least two reasons. First, this body of knowledge suggests that, in order to perform well, organizations need information systems that match their organizational structures and strategies. Briefly, organizational structures have been observed to vary not only in degree (e.g., how centralized they are) but also in type (e.g., whether they are organized according to customers, products, or both) even among apparently similar firms (e.g., consumer product manufacturers). A line of argument uniting the strategy–structure–performance paradigm with the IT-affects-organizational-structure hypothesis implies that effective organizations will employ information systems with features that are closely aligned with their structures and strategies. For example, Croteau and Bergeron (2001) found that organizations with information systems matched to their business strategies had higher performance than organizations without such a match. Similarly, Strikwerda and Stoelhorst (2009) observed that organizations with multidimensional structures (i.e., organizations structured to emphasize customers, products, and functions simultaneously) needed to, and typically did, employ centralized information systems that can identify and summarize transactions and forecasts in flexible combinations of customers, products, and functions, whereas organizations with multidivisional structures (i.e., decentralized organizations with business units for particular product–market combinations) did not require or employ such systems. Conversely, the body of knowledge on IT and organizational structure suggests that organizations with information system features inappropriate for their strategies or structures will fail to accomplish their objectives and will perform poorly (Bergeron et al., 2004; Chan and Reich, 2007a,b; Soh et al., 2000; Strong and Volkoff, 2010).

A second practical implication of research on IT and organizational structure concerns the provisions that large and complex organizations make for managing their IT. The body of knowledge on IT and organizational structure suggests that effective strategies *for managing IT in organizations* will vary with organizational structures. Even if an organization relies heavily on external IT service providers, its managers face numerous choices about how to structure IT support. For instance, should they manage IT centrally or allow each major organizational subunit to manage its own IT? Application of research and theorizing on IT and organizational structure suggests that structure should be a major consideration in the design of IT management arrangements.

This chapter aims to survey key concepts, theories, findings, and implications of research on IT and organizational structure. Section 67.2 examines how organizational structure has been conceptualized and discusses findings about the relationship between IT and organizational structure. Section 67.3 elaborates on the practical implications of this body of knowledge. Section 67.4 explores promising directions for future research on IT and organizational structure.

67.2 Underlying Principles

Organizational structure has been studied over many years by scholars in several fields—including sociology, economics, organization theory, business history, strategic management, and organizational communication and information systems—with varied theoretical interests and research questions. Consequently, any attempt to review this vast literature will necessarily oversimplify and risk offending through its categorizations, (mis)characterizations, and omissions. That said, it is important to have at least a cursory appreciation for the ways that organizational structure has been understood before considering its possible relationships with information technology.

67.2.1 Organization Structure: Definitions, Theories, and Questions

Most theories of organizational structure trace their roots to Max Weber's (1864–1920) theory of bureaucracy, which emphasized a hierarchy of authority, written rules, and officials with expert training and career advancement based on technical qualifications. Among other foundational scholars were Nobel laureate Herbert Simon (1916–2001), a major contributor to computer science whose 1947 *Administrative Behavior* became the foundation of American administrative and management sciences, and sociologist Peter M. Blau (1918–2002), whose major empirical studies of organizational structure and technology extended Weber's theory of bureaucracy and its role in society. Research based on the work of these seminal thinkers went in two general directions—focusing on structural characteristics or on configurations of design characteristics—although there was considerable overlap and mutual influence between them.

67.2.1.1 Organizational Characteristics

Shortly after World War II, scholars at the Tavistock Institute in the United Kingdom observed that formal and informal work group structure varied with the nature of coal-mining technology. Others generalized this "sociotechnical systems theory": Woodward (1958) concluded that differences in production technology (e.g., mass production versus process control technology) offered the best explanation for differences in organizational structure. In an influential program of empirical research in the 1960s, known as the Aston studies (e.g., Hickson et al., 1969), researchers focused on dimensions of organizational structure such as role specialization, standardization of procedures, centralization of decision making, and direct supervision versus impersonal control (i.e., written procedures). One finding of the Aston studies was that organizational structure was related to organizational size; larger organizations were more specialized, more formalized, more standardized, but less centralized. A second key conclusion was that of a compensatory relationship between decentralization (delegation of decision making) and structuring in the form of bureaucratic rules and procedures. In other words, organizations can "centralize" decision making either by means of direct supervision or by *de*centralizing decision-making authority within the limits set by "impersonal" rules (Donaldson et al., 1975).

Aston researchers also found that the effects of production technology on organization structure were moderated by organizational size, contradicting Woodward's main conclusion. (Research about the relationship between organizational size and structure is relevant to questions about IT, because one of the major ways in which IT is said to affect organizations is by reducing organizational size; for example, automation replaces workers, enabling downsizing.) Research by Blau and his colleagues (Blau, 1970; Blau et al., 1976) generally confirmed the Aston researchers' findings about the importance of organizational size and the role of technology in organizational structure. From a large-scale empirical study of Employment Security Offices in the United States, Blau (1970) concluded that as organizations increased in size, they increased in differentiation (number of different types of positions), although the rate of increase in differentiation decreased with size. He also found that the size of the administrative component of organizations (the fraction of employees not directly engaged in production) decreased as organizational size increased. In a study of both production technology and computerization of administrative support functions in U.S. manufacturing plants, Blau et al. (1976) found that, whereas mass production technology routinized work and simplified administrative procedures and thereby promoted decentralization, process control technology (e.g., chemical processing) reversed those trends, leading to the development of specialized skills, "a complex administrative apparatus," and a more centralized structure (p. 24).

Some subsequent research has aimed to analyze, model, and classify organizations with detailed attention to their structural characteristics (Burton et al., 2006; McKelvey, 1982). Generally, however, most other organizational scholars have tended to focus less on the fine-grained structural characteristics of organizations and more on a few ideal organizational types, often called configurations or archetypes (Miles and Snow, 1996; Mintzberg, 1979, 1983). Organizational configurations are conceptualized as bundles of interrelated characteristics (Meyer et al., 1993). In addition to structural characteristics, organizational configurations often include nonstructural elements such as organizational strategies, organizational environments, and organizational cultures or informal interaction patterns. Researchers often use the terms "organizational design" or "organizational form" instead of "organizational structure," when they want to refer to configurations of structural and nonstructural characteristics.

67.2.1.2 Organizational Configurations

At about the same time as the Aston studies, Chandler (1962) published a highly influential historical study of structural change in leading American companies such as Standard Oil, General Motors, DuPont, and Sears Roebuck. Whereas Blau was interested in the interplay between the formal structural characteristics and informal patterns of communication and social interaction within organizations, Chandler was interested in differences across organizational configurations or forms. Chandler used the term "organizational structure," but he defined it much more broadly than scholars like Blau. Chandler included in his notion of structure both formal and informal lines of authority and communication, as well as the information and data that flowed through those lines (Whittington, 2002). As a result, Chandler can be considered a configuration theorist.

Chandler documented and offered a historical explanation for the emergence, around 1920, of a new organizational form, called the multidivisional organization. The multidivisional form represented a change in the "basis of organization" in large enterprises. Loosely put, "basis of organization" refers to the boxes on an organizational chart at the level just below that of the chief executive. Prior to the 1920s, most organizations of any size were structured on the basis of "functions," that is, units devoted to activities such as marketing, sales, engineering, and manufacturing (as well as support functions, such as accounting/finance, human resources, and—much more recently—IT). This form of organization is variously called "centralized," "functional," or "unitary" (or "U-form"). The U-form is understood as a "centralized" organizational form, because, for any decision that cuts across functions (such as a decision about a new product to be offered or a new market to be entered), no one functional unit head is able to make a good decision without input from other functional unit heads, and therefore, the final decision must be made at the top. By contrast, M-form (multidivisional) organizations are structured on the basis of "business units," which in the classic definition (Porter, 1980) are organizational units with profit-and-loss responsibility for particular product–market combinations, such as "Asia-Pacific semiconductors" or "International chemicals" or "North American generic drugs." The M-form is understood to be a "decentralized" organizational form. The reason is that, whereas in the U-form chief executives make key decisions about products and markets, in the M-form these decisions are made by business unit managers that report to chief executives. In other words, chief executives delegate some decisions to business unit heads. However, top managers in M-forms generally retain cross-business unit decisions such as which product–markets to enter and how much to invest in each.

Chandler argued that organizations began adopting decentralized forms of organization in the 1920s for two reasons. First, improvements in transportation and communication technology (railroads and telegraphs) had enabled organizations to enter new markets. Second, increased scale of operations and exposure to new markets encouraged organizations to develop and offer new products and services (Chandler, 1962). In the field of strategic management, both developments are understood as strategies of diversification, and these strategies are believed to strain the decision-making capacity of top managers. Spurred by the problems created by diversification, in Chandler's analysis, organizations responded by innovating in their administrative structures and procedures. By decentralizing their structures, large diversified organizations were able to improve their performance.

Chandler's pioneering work stimulated numerous streams of research relevant to the question of IT's relationship with organizational structure. One influence can be seen in the development of the information processing theory of organizations (also powerfully shaped by the work of Herbert Simon and his colleagues), of which Jay Galbraith is a leading proponent (Galbraith, 1977). According to information processing theory, organizational designs (structures plus lateral relationships, including informal ones) are seen as tools for channeling information, communication, and managerial attention. Galbraith is notable for his highly naturalistic, yet holistic, characterizations of organizational design in terms of their primary and secondary "bases of organization." His writings are filled with detailed organizational charts. By looking at the allocation of responsibilities to the most senior executives in these organizational charts, one can easily see each company's strategic priorities.

Galbraith is one of the several prominent authors who studied the internationalization of business and the evolution of multinational enterprises. Other international business scholars such as Bartlett and Ghoshal (1989) and Goold and Campbell (2002) have studied the changing organizational forms of companies with worldwide operations. Today, some observers (Strikwerda and Stoelhorst, 2009) posit the emergence of a new form, the "multidimensional organization," with distinctly different characteristics and management systems than M-form organizations.

Perhaps the most influential program of research stimulated by Chandler's work has been Oliver Williamson's (1985) theory of markets and hierarchies, also known under the shorthand of organizational economics. Williamson viewed the M-form as the most significant organizational innovation of the twentieth century (King, 2011) and developed a theory to explain how, when, and why firms (hierarchical authority systems) differ from, and emerge instead of, arm's length market arrangements. This theory was subsequently expanded to accommodate hybrid or "network" organizational forms involving close partnerships among organizations, such as joint ventures and major "outsourcing" relationships (e.g., contract manufacturing). Theory and empirical research on networked and virtual organizations was, of course, fueled by growing awareness of the pervasiveness of organizational IT.

The many empirical studies in the strategy–structure–performance paradigm of the strategic management field are among the legacy of Chandler's historical research. Although empirical results in this tradition have been mixed (Whittington, 2002), almost all subsequent research on organizational design and new organizational forms owes a debt to Chandler's influence.

67.2.1.3 Summary: Organizational Structures and Configurations

Simplifying greatly, in the years leading up to Leavitt and Whisler's predictions about the effects of IT on organizational structure, some scholars had attempted to explain the detailed formal structural characteristics of organizations and to explore their interactions with informal patterns of behavior and communication. Researchers in this tradition highlighted organizational size and production technology as important determinants of organizational structure. Other scholars tended to focus on organization "designs" or "forms," that is, packages or configurations of elements including structure, strategy, environment, and informal social interactions that commonly occur together. Those researchers emphasized the emergence of new organizational forms and the consequences of organizational forms for performance. Both research streams continue to evolve, influencing research on IT and organizational structure, although the second stream is more active today. The study of organizations includes many other intellectual traditions in addition to the ones reviewed here (including a few discussed later), but those two are the most relevant for understanding research on the links between IT and organizational structure, described more fully next.

67.2.2 Links between IT and Organizational Structure

One of the most interesting aspects of the research on IT and organizational structure is that much theory and research suggest that IT has "dual outcomes" (Pool, 1978), that is, apparently opposite effects nearly simultaneously or coterminously. For example, use of the telephone was said to have promoted the construction of skyscrapers, leading to increased urban density, at the same time that it contributed to urban sprawl (Pool, 1983). Similarly, IT has been argued and found (1) to promote both centralization and decentralization of decision making, (2) to lead both to smaller and larger organizations, and (3) to foster both new horizontal organizational forms and new bureaucratic forms of organization. Each of these three main areas of research is discussed briefly in the following.

67.2.2.1 IT and Centralization/Decentralization

Much of the earliest work on IT and organizational structure concerned structural characteristics. The term "centralization" has multiple meanings (King, 1983), many of which are relevant here. First, the descriptor "centralized" is used by some scholars to refer to a particular organizational design or form—the functional or U-form discussed earlier. Section 67.2.2.3 summarizes literature related to IT and new organizational forms.

A second meaning of the term centralization concerns physical location of organizational activities—concentrated in one geographic location versus dispersed in space and time. Theoretical arguments and empirical examples can be found to link information and communication technology to either increased geographic concentration or greater geographic dispersion. For instance, through the use of telecommunications, some organizations have geographically (and structurally) concentrated organizational support services such as accounting, human resources management, and IT help desks into regional "shared services centers," eliminating the smaller local groups of support personnel dispersed among individual offices or factories (Sia et al., 2010). At the same time, IT is also being used to enable geographically and organizationally dispersed individuals to work together as if they shared geographic location and company affiliation (Ahuja and Carley, 1998; Argyres, 1999; Majchrzak et al., 2000).

Most commonly, the term centralization is used to refer to the level in an organization at which decisions are made. When decisions are made at the top, decision making is labeled centralized; when decision making is delegated to business unit managers or below, the organization is referred to as decentralized. As noted in the introduction to this chapter, Leavitt and Whisler (1958) argued that IT would increase centralization of decision making. Their rationale was that managers prefer centralization, because it gives them greater control over the people and processes of the organization. They decentralize only reluctantly, because in large diversified organizations they cannot be well enough informed about local conditions to make good decisions. Growing computerization increased top managers' ability to make more and better decisions themselves; consequently, use of IT was expected to promote re-centralization (Leavitt and Whisler, 1958).

Computerization was seen as a key tool in the strategy of top-down administrative control, because it gave mangers the ability to formalize tasks and monitor the outcomes of decisions by lower-level staff (Blau and Schoenherr, 1971). Some empirical research (Blau et al., 1976; Whisler, 1970) found support for the centralizing effects of IT (cf. Whisler, 1970). For instance, Blau et al. (1976) found that the use of computers for administrative support generally continued the centralizing trend associated with advanced process control technology (versus mass production).

However, almost immediately, researchers advanced arguments for the "dual" or opposite effect. Klatzky (1970) argued that the analyses leading to routine decisions would be programmed into computer code, giving senior managers enough confidence to delegate responsibility for these areas to lower levels of the organization and allowing senior executives to concentrate on more strategic and less programmable decisions. Blau et al. (1976) found some evidence of IT-related decentralization, as well as centralization, in New Jersey manufacturing plants: "In contrast to the centralizing influence of an advanced production technology [e.g., process control], an in-house computer to automate support functions promotes decentralization, though primarily in the form of granting autonomy to the plant manager [but no lower]" (p. 35).

Of course, one has to ask whether the apparent decentralizing effects of IT constitute true delegation and autonomy for lower-level employees. As noted by Robey (1981) in an early review of IT and organizational structure research, "what appears to be greater decentralization may simply entail the delegation of more routine decisions whose outcomes are more closely controlled" (p. 681). Just as Aston school researchers had concluded that organizations can exert central control both by top-down supervision or by the application of impersonal rules (Donaldson et al., 1975), so it also seems that IT may be used to support greater *centralization* in either way: by enhancing the ability of superiors to make decisions themselves or by enabling subordinates to make decisions within carefully circumscribed limits encoded in information systems.

This reasoning suggests that it is not IT per se that determines whether organizations will move toward centralization or decentralization. Rather, it is *managerial choice* about how to use IT that appears to make the difference (George and King, 1991; Robey, 1981). This observation has been formulated theoretically in terms of statistical moderation (Markus and Robey, 1988; Robey, 1981). In other words, whether executives choose to run an organization in a centralized or decentralized fashion is likely to influence whether the use of IT subsequently leads to centralization or decentralization. But here again, theory and empirical results suggest the possibility of "dual outcomes."

For example, one highly regarded stream of research suggests that IT reinforces the existing regime of power and control in organizations (Kraemer and Dutton, 1979; Pinsonneault and Kraemer, 1997). According to this theory, managers in organizations that are highly centralized to begin with are likely to use IT in ways that will increase central control, whereas decentralized organizations will become more decentralized as a result of using IT. This "reinforcement politics" perspective has received empirical support (Pinsonneault and Kraemer, 1997), especially in the context of the organizational size question, examined later.

On the other hand, some theorists have extended the earlier theory and interpreted empirical evidence to the opposite effect. For instance, in an award-winning paper, Huber (1990) theorized that, because the effects of computerization depend on the choices of decision makers at various organizational levels, the most likely outcome is a more uniform distribution of decision-making authority within and across organizations. In general, Huber predicted that IT would lead to more *de*centralized decision making in *centralized* organizations and more *centralized* decision making in *de*centralized organizations.

For Huber (1990), the prevailing distribution of authority in an organization was the most important contingency in the IT-leads-to-structural-change equation. However, the previously cited study by Blau et al. (1976) suggested that "the influences of automation may be more complex" (p. 37) than Huber proposed. Blau and colleagues compared manufacturing plants that had their own computers on-site and those that used off-site computer systems (probably in that era run by parent company headquarters), and found that the former had greater decision autonomy (were less centralized) than the latter. "[T]he data imply that the location of computer facilities governs the locus of decision-making authority [of manufacturing plants]. If a plant has its own computer, its management is likely to have much autonomy, but if a plant uses an off-site computer, presumably in most cases at corporate head-quarters, chances are the authority is centralized there" (p. 37). In other words, the most important contingency related to IT's effects on organizational structure may be managerial choice about *how to manage IT*, not overall organizational structure/decision-making authority as Huber (1990). (Of course, these two aspects of organizational design may be closely related.) The implications of this possibility are discussed in a later section of this chapter.

Further complicating the picture about the type, nature, and direction of IT's effects is the fact that the power of IT has increased substantially over time, allowing organizations to apply IT to a greater number of decisions and to decisions of ever greater complexity and strategic relevance. For example, organizations began using IT to make decisions about automobile and credit card loans in the early 1980s, but until the second half of the 1990s almost all U.S. residential mortgage loans were decided by human underwriters using judgment and rules of thumb. By 2000, however, nearly 100% of all mortgage lending decisions were made by automated underwriting systems (Markus et al., 2008). Changes in the application of IT over time means that the *ability* of senior organizational executives in large organizations to make more decisions (or to delegate more decision making to subordinates while retaining tight control over the decisions made) has increased substantially in more than 50 years since Leavitt and Whisler first made their predictions. This condition naturally raises questions about the consequences of IT for the employment of middle managers, discussed next.

In short, both theoretical arguments and empirical evidence suggest that the use of IT in organizations has dual outcomes regarding the centralization of decision making.

67.2.2.2   IT and Organizational Size

As discussed earlier, the size of organizations is believed to be strongly related to organizational structure. Larger organizations are generally more decentralized than smaller organizations and more formalized or bureaucratic. This relationship led some scholars to speculate that increased use of IT would lead to smaller firms. Two primary reasons were offered. First, IT can be used to automate some activities performed by middle managers. This labor substitution effect is expected to result in middle management downsizing (Brynjolfsson et al., 1994; Leavitt and Whisler, 1958; Pinsonneault and Kraemer, 1997). Second, following the theory of Oliver Williamson (Williamson, 1985), IT is seen as reducing costs that might make it profitable for an organization to outsource an activity to another organization (Afuah, 2003; Brynjolfsson et al., 1994). In this "make versus buy" decision, three kinds of costs are relevant: the cost of performing the activity, the cost of internal coordination, and the cost of external coordination. How IT changes the relative magnitudes of these costs is theoretically expected to shift the balance toward or away from outsourcing. If the organization outsources, it will be able to eliminate the jobs previously devoted to performing the activity and thus will get smaller.

In this area of research, as with research on centralization of decision making, results are mixed. Brynjolfsson et al. (1994) started with evidence that the number of employees in business establishments had decreased substantially in the prior 15 years, whereas the extent to which IT was used in business had increased enormously. Through analysis of a large dataset of U.S. businesses, they concluded that use of IT was associated with a decline in firm size. Although they took pains not to overgeneralize their find-ings, the authors concluded that IT was contributing to a restructuring of the U.S. economy.

Others, however, took issue with both the premises and the conclusions of Brynjolfsson and colleagues' research. Pryor (2001), for instance, argued that firm size is increasing, not decreasing, despite the waves of downsizing and outsourcing that occurred in the 1990s. Starting with arguments similar to those of Brynjolfsson and colleagues, Afuah (2003) concluded on theoretical grounds that the effects of Internet technology on firm boundaries (organizational size) should be moderated by the nature of organizational technology (such as the interdependence among activities) and by the characteristics of information used. Therefore, in some industries, IT is likely to lead to *increased* organizational size.

Pinsonneault and Kraemer (1997) similarly found theoretical support both for the prediction that organizational size would decrease because of IT and for the prediction that IT would lead to larger organizational size. Therefore, they designed an empirical study to explore factors likely to moderate the relationship between IT and organizational size. They concluded that use of IT led to an increase in the number of middle managers in companies where decision authority (both for organizational and for computer-related decisions) was decentralized and a decrease in the number of middle managers where decision authority was centralized. This finding echoes that of Blau et al. (1976) about the association of computer location with plant managers' decision autonomy— computer location being a key IT-related decision. Pinsonneault and Kraemer further found that use of different types of IT resulted in different outcomes. Organizations that used IT more for coordination purposes had fewer middle managers, whereas those that used IT more for control had more middle managers. Finally, in a configurational argument, they reported that the impacts of IT on organizational size depended not just on individual structural arrangements but also on the congruence among structural arrangements, that is, the compatibility between overall patterns of organizational decision making and the distribution of IT-related decision making.

In short, both theoretical arguments and empirical evidence suggest that the use of IT in organizations has dual outcomes regarding organizational size.

67.2.2.3 IT and New Organizational Forms

A large and still active body of literature has developed around the hypothesis that the use of IT promotes the emergence of new organizational forms. This hypothesis is analogous to Chandler's observation that the railroads and telegraph enabled organizations to grow large, to diversify, and to innovate with the multidimensional form (see also Beniger, 1986; Yates, 1989). Some of the arguments and evidence in this literature overlap with research previously discussed, but generally scholars working in the new organizational forms tradition take a more holistic view of organizational design, focusing on configurations of attributes rather than isolated structural characteristics. However, scholars differ in their explicit or implicit characterizations of new organizational forms, and it is not always clear whether authors are discussing the same thing.

For example, Lambert and Peppard (1993) identified six different organizational arrangements as instances of new organizational forms associated with the use of information technology: (1) "network organizations": collections of legally independent organizations that partner by means of strategic alliances or outsourcing relationships to coproduce products and services that used to be produced by an integrated firm; (2) temporary task-focused teams within individual organizations, in which members may be widely dispersed geographically; (3) relatively stable networked groups inside organizations that do not have a particular task focus; (4) horizontal organizations, understood as involving the creation of customer- and/or supplier-facing coordinating units; (5) learning organizations, in which knowledge-management processes and tools may play a prominent role; and (6) matrix management, in which some organizational members report simultaneously to more than one superior. Lucas and Baroudi (1994) discussed four, somewhat different, "prototypical organizational designs or strategies that have emerged within certain industries or companies during the past two decades" (p. 10). These designs were labeled: (1) "virtual organizations" in which organizational members work together through IT with few constraints of time or space; (2) "negotiated organizations" in which a formerly integrated supply chain is crafted from contractual relationships among independent companies; (3) "traditional companies" in which IT may be used to reduce layers of management and/or restructure business processes; and (4) "vertically integrated conglomerates" in which IT is used as an essential tool for coordination and task forces. (This latter type may be equivalent to the "multidimensional organizational form.")

More recently, Palmer et al. (2007) found five differences in language use and assumptions in the literature about new organizational forms: (1) whether the new forms involve revolutionary or evolutionary change; (2) whether the new forms are structurally simpler (smaller, less differentiated, less bureaucratic) or more complex (as for instance the "multidimensional" form is said to be); (3) whether the drivers of change are believed to be managerial choice or environmental influences (such as the widespread availability of low-cost IT); (4) whether the appropriate level of analysis is intraorganizational or interorganizational; and (5) and whether the "newness" of the organizational form refers to "new in time" or "new in a particular context" such as an industry or even a specific organization.

Palmer and colleagues' point (2) hints at the possibility of "dual effects" from use of IT. Their point (5) encompasses the observation that some of Lambert and Peppard's and Lucas and Baroudi's new organizational forms are arrangements *inside* organizations, whereas others refer to arrangements *between or among* organizations. These differences make it challenging to integrate the theories and findings of the new organizational forms literature; the following review is more illustrative than comprehensive.

Some studies focus on IT's effects on new *intra*organizational arrangements, such as the reduction of layers of hierarchy and the use of more lateral coordination mechanisms. These arrangements have labels like "virtual teams," "distributed organizing," or "internal markets" (DeSanctis and Monge, 1998; Halal, 1994; Orlikowski, 2002). For example, Hitt and Brynjolfsson (1997) theorized that organizations making heavy use of information technology would adopt a package of complementary organizational design arrangements including decentralization of decision making, subjective incentives, and greater reliance on skills and human capital. They found support for this hypothesis in the study of the detailed work practices and IT expenditures in large organizations. Orlikowski (2002) showed how the use of IT augmented direct interpersonal forms of coordination in a distributed new product development team in a global high-tech organization. Sambharya et al. (2005) concluded that the internal structures of multinational enterprises were becoming more "organic" in nature.

However, other studies suggest that the use of IT, particularly in global enterprises, has spurred the evolution of new *bureaucratic* organizational arrangements, characterized by much less decision autonomy for business unit heads and by much more impersonal coordination along with interpersonal coordination across business units. In a study of Irish subsidiaries of global multinationals, Finnegan and Longaith (2002) found large reductions in the autonomy of subsidiaries linked to the ways in which parent company headquarters used IT:

> For many subsidiaries studied, pricing and production decisions used to be within their range of activities as long as they stayed within corporate policies. This approach facilitated a level of local responsiveness. However, six subsidiaries now used enterprise resource planning (ERP) systems located at headquarters. This had resulted in the elimination of subsidiary management's input to pricing decisions. Corporate management also used the ERP system for drawing up a rolling overall production plan for the subsidiary .... IT was also found to play a similar role in selecting suppliers, engaging in research and development and devising the group's strategy. ... IT therefore facilitates the ability of corporate management for excluding subsidiary managers from these types of decisions (p. 156).

Finnegan and Longaigh (2002) also found significant changes in the nature of cross-subsidiary coordination and again argued that the use of IT was responsible for these changes. In the traditional form of coordination, people in different units work directly together and make joint decisions through personal interaction. Finnegan and Longaigh found that the use of e-mail and groupware made traditional coordination more personal than in the past, as Orlikowski (2002) also did. However, they also identified a new form of impersonal coordination that they called "programmed coordination," which they argued to be impossible without the use of IT:

> [Programmed coordination] is coordination without direct communication and it leads to the synchronization of activities as opposed to joint decision making. It is apparent more at an operational rather than a strategic level and, based on the findings of this study, it would be impossible to operate … without the aid of IT. An example … was where subsidiaries were obliged to consult a real time database in order to ensure that subsidiaries were not quoting for the same business (Finnegan and Longaigh, 2002, p. 157).

Finnegan and Longaigh's (2002) study provides a graphic illustration of the operation of management systems in what some experts claim to be a new organizational form, often called the multidimensional enterprise (Ackoff, 1999; Galbraith, 2000, 2008; Goold and Campbell, 2002; Palmisano, 2006; Strikwerda and Stoelhorst, 2009). Many environmental contingencies and strategic considerations have converged to promote the emergence of this form, and some observers suggest that IT is largely irrelevant. Galbraith (2000), for instance, claimed that the information processing demands of multidimensional organizations can be handled with relatively simple information technologies like spreadsheets. Others, however, insisted that integrated systems and corporate databases (e.g., ownership of subsidiary's customer and transaction data by corporate headquarters and a shared general ledger system in use across the firm) are required (Strikwerda and Stoelhorst, 2009), and Finnegan and Longaigh's study provides supporting evidence.

Many writers identify new organizational forms more with *inter*organizational than with the *intra*organizational arrangements just discussed. Here many observers agree that the use of information technology has made it easier for people to coordinate activities across organizational lines and has therefore been a factor in the increased incidence of outsourcing, joint ventures, and other kinds of cooperative interfirm arrangements (Fulk and DeSanctis, 1995; Lucas, 1995; Malone et al., 1987; Zammuto et al., 2007), including the development of interorganizational coordination hubs, in particular business sectors (Markus and Bui, 2012).

Ahuja and Carley (1998), Majchrzak et al. (2000), and Argyres (1999) provide particularly rich case examples of the use of IT to facilitate these new (or at least more widespread) forms of interorganizational networking. For example, the B-2 "Stealth" bomber was an aircraft designed by four independent firms almost entirely by computer. In analyzing this case, Argyres (1999) found that the use of information systems: (1) made coordination across the firms less costly; (2) made the governance of the project more efficient by establishing social conventions that rendered resort to hierarchical authority less necessary; (3) reduced risks that result from poor distribution of information across firms; and (4) decentralized decision making. A comparative study (using data aggregated at the industry level) found evidence that multinational enterprises in information-intensive industries that use IT heavily are more likely to contract with other organizations to perform activities that were traditionally performed internally (Rangan and Sengul, 2009).

The general consensus in the new interorganizational forms literature is that IT promotes horizontal, decentralized, or networked organizational forms. But even here, the possibility of dual effects has been raised. Sahayam et al. (2007) theorized that, because IT can enhance coordination both within and between organizations, IT can promote both tightly coupled organizations (such as vertically integrated companies or ones with strong central control) as well as loosely coupled organizations (such as organizations that rely on alliances with other organizations or those that engage workers on contract instead of through employment). In an industry-level study, they found that the relationship between the use of IT and loosely coupled forms of organization was more frequent in industries with industrywide product or process standards and low technological change.

By extension, Sahayam and colleagues' findings suggest that other organizations may be using IT to support *more tightly coupled* interorganizational coordination. Steinfield et al. (2012) reviewed the arguments and evidence in support of this proposition. Through a case study of an international coordination hub serving the automotive industry, Steinfield et al. illustrated how the use of IT can promote tighter interorganizational coupling.

In short, the literature on IT and new organizational forms also provides some evidence of dual outcomes.

67.2.2.4 Summary

For a half century, scholars of varied theoretical orientations have explored the relationship between IT and organizational structure. The three related issues of centralization or decentralization of decision making, organizational size, and new organizational forms have been the primary foci of research. Trying to make sense of the literature is hampered by differences in theory, concepts, and context, not to mention the fact that the capabilities and applications of IT have grown enormously over the time period.

Nevertheless, regardless of research focus and research time frame, one remarkable consistency runs through the literature. Both theory and empirical evidence strongly suggest that the relationship of IT with organizational structure is one of "dual outcomes," as Pool observed long ago in the context of telephone technology (1978). Robey's conclusion of 30 years ago, "several different organizational structures [are] compatible with computer information systems" (1981, p. 686), seems equally valid today.

This conclusion may be discouraging to some readers, but it should have the opposite effect. When one digs deeper into the findings, one can see that they offer practically important recommendations to practitioners, and they suggest exciting theoretical and methodological challenges for future scholarship.

67.3 Impact on Practice

The research on IT and organizational structure has two major implications for practitioners. First, the literature tells us a great deal about how IT can *constrain* organizations' ability to adopt new organizational designs successfully; conversely, adopting new designs often requires major changes in systems support. Put differently, good organizational design involves making careful decisions about the nature of IT infrastructures and platforms, which is a critical insight of the "enterprise architecture" discipline. Second, the literature tells us that *organizational designs for IT management* are also critical for good organizational performance. IT management design does not have a necessary one-to-one relationship with overall organizational design. However, certain ways of organizing IT management are not capable of producing the systems needed by some organizational designs. Therefore, practitioners should pay close attention to the implications of the literature both for system design or selection and for organizing and managing IT services.

67.3.1 Fit between Organizational Structure and Information Systems

The literature on IT and organizational structure does not support the idea that using IT changes organizational structure in any one particular way (e.g., toward centralization versus toward decentralization). The literature does not even support the idea that using IT changes organizational structure in different ways under different conditions. (Recall that it is not *IT* per se that changes organizational structures; *managerial decisions* about whether, when, how, and why to use IT and *managerial actions* are certainly the most critical factors.)

But the literature does suggest practical implications. First, an organization's legacy of IT may make it difficult, if not impossible, for an organization to adopt new organizational structures or designs. Some years ago, I studied organizations that had redesigned their business processes (Bashein et al., 1997). As a result of process reengineering, an insurance company decided to change its processes for handling claims. The company had been functionally organized; the new design was organized by market segment. That is, the company decentralized into customer-oriented groups, each of which would be free to adopt different policies for claims processing. Fortunately, the company's claims processing system was flexible enough to adapt to this change, but the company's management information systems were not. The information systems produced reports that were useful for functional unit heads, but they did not provide information broken down by customer segment. Every single management report had to be reprogrammed, before the organization was able to move to the new organizational design.

Another example of how IT can hinder organizations' ability to function effectively given their organizational designs is Microsoft, which adopted SAP's financial systems in the mid-1990s (Bashein et al., 1997). Microsoft has a multidimensional organizational design (Strikwerda and Stoelhorst, 2009), meaning that it manages products, markets, and projects—three organizational dimensions that overlap in multiple ways. At the time Microsoft implemented SAP financials, the software product did not provide good support for a three-dimensional company like Microsoft, because it allowed companies to "tag" transactions in only two ways (e.g., by product and market, but not also by project) (Bashein et al., 1997). Unwilling to modify SAP software, Microsoft circumvented the software limitation, painfully, by making people work with a complicated chart of accounts. This constraint was so vexing that Microsoft upgraded to a new version of SAP, which did support three organizational dimensions, as soon as it was available.

A corollary of the first point is that successfully moving to new organizational designs often requires major information systems modifications or entirely new system implementations. The process can be seen in detailed case descriptions of companies like Procter and Gamble (P&G) (Galbraith, 2009; Piskorski and Spandini, 2007) and Nestlé (Steinert-Threlkeld, 2006). In the late 1990s, both of these large multinational companies were structured in a fairly decentralized manner: Managers in different geographic regions had considerable decision-making autonomy. This was believed necessary because preferences for consumer products vary greatly around the globe. But many support functions such as accounting, human resources management, and IT were also decentralized, which encouraged variations in business practices and increased costs.

Several factors motivated a change in organizational design, including recession, declining organizational performance, and perceived needs to standardize global brand images and to be more responsive to the needs of multinational retailers. Consequently, both organizations among many others adopted new organizational designs characterized by much greater centralization and standardization in certain areas.

P&G, for instance, created a new Global Business Services (GBS) unit, with facilities in three geographic locations (San Jose, Costa Rica, and Newcastle) (Delong et al., 2005; Sia et al., 2010). The mission was to provide a wide array of standardized support services to all of P&G's facilities in each region. (Previously, each facility had performed those activities for themselves, and there was less standardization across locations.) The support services provided by GBS included a number of core business activities, such as market research and merchandising, in addition to traditional "support" activities, such as human resources management and IT.

P&G's legacy information systems were not up to the task of supporting this new organizational design, and P&G embarked on a multiyear systems initiative to implement the SAP-integrated enterprise information system. It is clear that IT did not cause this organizational change: The intent to reorganize preceded the system project. However, the new organizational design could not have worked without change in the company's legacy information technology. (Imagine a geographically and organizationally centralized "shared services center" trying to provide standardized services while using a patchwork of nonstandardized information systems running in many dispersed organizational locations. Impossible!)

A third practical implication of the literature on IT organizational structure is that once systems are in place to support a new organizational design, people will discover new ways to use the systems, and the organizational design will evolve. But the organizational design will only evolve up to the point where it is again constrained, this time by the new systems.

In short, the literature on IT and organizational structure offers practical advice that can be summarized as the "Principle of Enterprise Architecture": Design an organization's IT infrastructures and digital platforms to support the organization's "TO BE" (versus "AS IS") organization design. If you design organizational IT to support the existing organizational design, IT may prevent movement toward a preferred organizational design.

A concrete statement of this principle can be found in Ross, Weill, and Robertson (2006). According to Ross and colleagues, organizations with multiple business units can have different "operating models" (organizational design configurations). The "Unification" model is the classic U-form organization like an airline—a single business company with globally standardized business processes. If Unification is the preferred ("TO BE") operating model, Ross and colleagues prescribe enterprise systems that support globally standardized processes and global access to data. The "Replication" operating model is one with multiple, nearly identical, business units: A hotel chain is a good example. For this type of enterprise, a standard IT infrastructure and set of applications is needed that can then be "replicated" in locations around the world. The "Coordination" operating model describes a multi-business company in which the partially independent business units need to coordinate frequently, for example, an insurance company with multiple product lines serving the same customers. Here, the IT platform needs at minimum to be able to provide access to shared data through standard technology interfaces. Organizations with the "Diversification" operating model have independent business units with different technologies and customers (e.g., the Sears holding company, which includes financial services as well as retailing). Here, the need for sharing information across units is low, but IT can provide economies of scale through a light "shared services" platform that does not restrict business units' strategic flexibility.

In short, the literature on organizational design and structure has much to offer practitioners in the way of advice about appropriate systems and technologies.

67.3.2 Managing Information Systems

A second practical implication of the literature on IT and organizational structure concerns, not just the systems and technologies organizations use, but also *how they choose to manage IT* (Earl, 1996; Galliers and Currie, 2011; King, 1983). Managing IT involves many decisions such as (1) whether to provide IT services internally or to contract with external service providers (like IBM or Accenture) for those services; (2) how to organize (structure) people in the organization who perform or manage IT services; (3) *how to make decisions about* which information technologies, systems, and services to adopt, and so forth. In other words, organizations have *designs for IT management* in addition to their overall organizational designs.

Therefore, an important practical question for every organization is: "What should be our organizational design for managing IT?" As you might expect, there is likely to be some relationship between overall organizational design and IT organizational design. For example, if an organization is highly decentralized, perhaps because it has a strategy of unrelated diversification (e.g., the "Diversified" operating model), chances are good that the organization would not only have different IT and systems in different units but would also decentralize the management of IT and systems to the diversified business units. Conversely, if an organization is centralized (e.g., the "Unified" operating model) at the enterprise level, it seems likely that the organization would also choose to have centralized IT management.

And, in fact, a fair number of studies have addressed the relationship between organizational design at the enterprise level and IT management design, particularly in global companies (Alavi and Young, 1992; Jarvenpaa and Ives, 1993; Karimi and Konsynski, 1991; King and Sethi, 2001; Ramarapu and Lado, 1995). This research generally reports a correspondence between organizational strategy and structure, IT management design, and the types of information technologies and systems used.

But the correspondence between organizational design and IT management design is not as close as you might think, because additional considerations enter in. One consideration is history. Computers have not always been as powerful as they are today, and their smaller capacity in the past encouraged decentralized IT management. For example, Southwest Airlines, a "Unified" organization, used to allow each functional unit head to manage its own IT unit and develop its own systems, partly for historical reasons. Today, Southwest has a centralized IT management design (Ross and Beath, 2007), in part because of concerns about IT management costs arising from duplication of effort, lack of standardization, etc., and in part because that way of organizing IT is more "congruent" (Pinsonneault and Kraemer, 1997) with Southwest's overall organizational design. A second reason for the correspondence between organization design and IT management design being less than complete concerns the flexibility of today's centralized IT management design. Organizations can set up centralized IT services and operate them in a way that will produce either (1) standardized and integrated IT, systems, and services for use across an organization or (2) IT, systems, and services that are tailored to the needs of individual business units.

Although a centralized IT organizational design can support either a centralized or decentralized organizational operating model, the reverse is not true. If an organization has a very decentralized design for managing IT (i.e., each business unit makes its own decisions about IT), there is almost no way this IT management design would be able to produce IT systems and services that are integrated, standardized, or commonly used *across the entire organization*. The practical principle here is: IT management design for an organization should be *no less centralized* than the overall organizational design or operating model.

67.3.3 Summary

Findings of the literature on IT and organizational structure do not guarantee practitioners particular organizational outcomes like greater top-down control or improved lateral coordination. But they nevertheless have important practical implications.

For example, Blau et al.'s (1976) finding that managers of plants with their own computers on-site had more decision autonomy than managers of plants with computers at parent company headquarters provides useful insights. This finding tells us that the parent companies of the plants with their own computers (1) had decentralized organizational designs for managing IT, (2) had technologies, systems, and services that were tailored to the needs and preferences of individual plant managers, and (3) therefore, parent company executives would have had great difficulty in exerting stronger top-down control without changing the systems used by plant managers, (4) which in turn would have also required them to change the IT management design in the direction of greater centralization.

As another example, Pinsonneault and Kraemer's (1997) finding that use of IT led to an increase in the number of middle managers in organizations that were decentralized both at the enterprise level and in their IT management also provides practical insights. Business units in these organizations had to add middle managers *to manage IT*. The technologies, systems, and services they produced were not shared with other business units, compounding the increase in middle management staff. A different organizational design for IT management in the same decentralized companies would have led to a much lower increase in middle management staff. In short, research on IT and organizational structure has much to offer to IS and IT practitioners in the way of advice about how to organize and manage IT.

67.4 Research Issues

The consistently mixed findings in research on IT and organizational structure certainly call for more research and theorizing, but not necessarily more of the same. The research record suggests the need for new theoretical perspectives and raises some promising questions for future research.

67.4.1 Need for New Theoretical Perspectives

Organizational structure is a foundational concept in a number of disciplines, including strategic management, organizational theory and economics, and information systems. However, possibly because research results about the role of IT in organizational structure change seem so mixed, scholarly attention has drifted toward other issues, particularly toward IT governance and toward informal, lateral relationships within and across organizations (Brown and Grant, 2005; Martinez and Jarillo, 1989; Whittington and Mayer, 2001). This move is a step backward from the work of scholars like Blau, who explicitly recognized the interplay between organizations' structural characteristics and informal interactions, or Chandler, whose broad conception of organizational structure included both formal/vertical and informal/lateral dimensions.

Without looking at both structure and informal relations, scholars are not able to address how these two dimensions of organizations are (or are not) *aligned* and how they *interact* in different organizational design configurations. The alignment or nonalignment among various dimensions of organizations is a central theme in many literatures. For example, in the IS literature, scholars have proposed the theory of strategic alignment (Bergeron et al., 2004; Chan and Reich, 2007b; Henderson and Venkatraman, 1993) to explain variations in organizational performance. The core hypothesis of this literature, supported by systematic research as well as casual observation, is that organizational elements (e.g., IS strategy and business strategy, organizational structure and systems, or IT policies and IT practices) are not always aligned, and that lack of alignment contributes to intraorganizational conflict and ineffectiveness (Bergeron et al., 2004; Chan and Reich, 2007a).

Despite the centrality of alignment theory, the types, sources, and consequences of alignment and misalignment among aspects of organizational design, including IT and IT management arrangements, have not been fully explored. There are many possible "alignment" relationships among organizational elements: One element may dominate another (Sambamurthy and Zmud, 2000). One element may reinforce another (Argyres and Silverman, 2004). One element may suppress or dampen the effects of another (Markus et al., 2013). Understanding these possibilities is necessary, if research on IT and organizational structure will be able to produce reliable recommendations for practice.

Further, as noted both by IT alignment scholars such as Chan and Reich (2007b) and by organization structure scholars such as Whittington (2002), we need better theories about the dynamics of structural change and alignment. Whittington (2002), for example, recommended further development of the practice perspective on organizational structure. Such a theoretical perspective would be particularly valuable for the study of IT and organizational structure, because it would emphasize "the role of particular functional areas or skills [such as accounting or IT] to the working of structure" (Whittington, 2002, p. 130).

Another promising theoretical perspective for the study of IT and organizational structure is evolutionary (or coevolutionary) theory (Dosi and Marengo, 2007; Lewin and Volberda, 2011; Romanelli, 1991; Williams, 2008), which has also seen some application in the IS literature (Lewis, 2008; Peppard and Breu, 2003). Evolutionary theories posit reciprocal causal relations between concepts such as management intentions and aspects of organizations and their environments such as organizational structures and information technologies. A particularly valuable application of evolutionary theory would be to examine the role of IT in the diffusion of new organizational forms.

For example, I earlier described the difficulties Microsoft encountered in the mid-1990s when it tried to adapt SAP software (an enterprise software package widely used by large organizations) to its multidimensional structure. A few years later, the SAP package did support multidimensionality, and, around that same time, a number of other large companies (e.g., Nestlé and P&G) both restructured and implemented the SAP package enterprise-wide. Twenty years earlier, Golden Triangle Corporation had painfully modified an accounting package to make use of modern database management technology in order to realize its organizational design intentions (Markus, 1983, 2010). Later, accounting packages employing database management technology became widely available commercially. This is precisely the era when the multidivisional organizational form (pioneered in the 1920s) diffused rapidly (Fligstein, 1985).

I believe there is a story here. It is not a story of simple causality (IT causes change in organizational structure), but rather one of reciprocal causation (Wagner et al., 2005) in which organizational needs for new ways of organizing intersected with the software development trajectory, fueling further evolution of both.

In short, much remains to be learned about the role of IT in organizational structure change. New and better theories are one way to get there. In addition, the survey of literature in this chapter suggests a few specific research issues for scholars in the information management tradition.

67.4.2 Unanswered Research Questions

It is now well established that *IT* contributes to organizational effectiveness under certain circumstances (Melville et al., 2004). And scholars have examined how patterns of alignment among multiple organizational design dimensions including *IT management* have contributed to organizational performance (Bergeron et al., 2004). However, there is little comparative research linking alternative IT management designs to organizational performance outcomes.

For example, I noted earlier that a centralized IT-shared services model could be used to provide standardized IT capabilities shared across an entire complex organization or to provide IT capabilities customized to individual business units. Both approaches are likely to be more cost effective than entirely decentralizing IT management to business units. The standardized approach is likely to be less costly than the customized one, but it may involve reduced effectiveness for business units. How much difference do these organizational models make for overall organizational performance? The literature does not, to my knowledge, clearly answer this question.

Second, we know relatively little about how and why *lack of* alignment between various aspects of organization design occurs and what difference it makes. For instance, the World Bank, a highly decentralized organization, adopted an IT-shared services model similar to that of P&G and other more centralized organizations (McFarlan and Delacey, 2003). Does this apparent lack of alignment represent an indicator of, or contributor to, poor organizational performance? Or could it mean that there really is "one best way" to organize IT today in large global organizations (Markus et al., 2012)? If that were the case, we should certainly find out why, because it shakes the foundations of the contingency theory of alignment.

All told, IS scholars have not yet exhausted the topic of IT and organizational structure. Among other possibilities, organizational structure provides a valuable lens for exploring the relationship between IT management and organizational performance.

67.5 Summary

The role of IT and organizational structure has been a topic of speculation, theorization, and empirical research for over 50 years. In particular, research has examined the effects of IT on organizational centralization, organizational size, and new organizational forms. In all three areas, scholars have produced theoretical arguments and empirical results consistent with the "dual effects" hypothesis— the idea that advances in information technology can result in opposite structural consequences simultaneously and coterminously.

Some scholars interpret evidence about dual effects to mean that IT is irrelevant to changes in organizational design. These scholars argue that only managerial intentions and actions ("human agency") are a factor in the outcomes. However, there is considerable evidence that IT does indeed make a difference, even if IT alone does not cause the outcomes observed.

This conclusion has important practical implications. Specifically, organizations are advised to align (or co-design) their information systems and technologies with their strategies and structures. In addition, organizational structure should be an important consideration among several in the design of effective IT management arrangements.

At the same time, much remains to be learned about the processes by which organizational structures intersect with technologies both within individual organizations and across organizations and over time. In addition, much more needs to be learned about organization designs for IT management and how they contribute to, or hinder, organizational effectiveness.

Acknowledgments

This research was supported in part by the National Science Foundation under award number SES-0964909. I wish to thank colleagues at Bentley University, Nanyang Business School (Singapore), and MIT's Center for Information Systems Research for their many contributions to this project.

References

Ackoff, R. L. (1999). *Re-Creating the Corporation: A Design of Organizations for the 21st Century*. New York: Oxford University Press.

Afuah, A. (2003). Redefining firm boundaries in the face of the Internet: Are firms really shrinking? *Academy of Management Review*, 28(1), 34–53.

Ahuja, M. K. and Carley, K. M. (1998). Network structure in virtual organizations. *Journal of Computer-Mediated Communication*, 3(4), http://jcmc.indiana.edu/vol3/issue4/ahuja.html

Alavi, M. and Young, G. (1992). Information technology in an international enterprise: An organizational framework. In S. Palvia, P. Palvia, and R. Zigli (Eds.), *The Global Issues of Information Technology Management* (pp. 495–516). Hershey, PA: IGI Publishing.

Applegate, L. M., Cash, J. I., and Mills, D. Q. (1991). Information technology and tomorrow's manager. *Harvard Business Review*, 66(6), 128–136.

Argyres, N. S. (1999). The impact of information technology of coordination: Evidence from the B-2 "Stealth" bomber. *Organization Science*, 10(2), 162–180.

Argyres, N. S. and Silverman, B. S. (2004). R&D, organization structure, and the development of corporate technological knowledge. *Strategic Management Journal*, 25, 929–958.

Bartlett, C. A. and Ghoshal, S. (1989). *Managing across Borders: The Transnational Solution*. Cambridge, MA: Harvard Business School Press.

Bashein, B. J., Markus, M. L., and Finley, J. B. (1997). *Safety Nets: Secrets of Effective Information Technology Controls*. Morristown, NJ: Financial Executives Research Foundation, Inc.

Beniger, J. (1986). *The Control Revolution: Technological and Economic Origins of the Information Society*. Cambridge, MA: Harvard University Press.

Bergeron, F., Raymond, L., and Rivard, S. (2004). Ideal patterns of strategic alignment and business performance. *Information & Management*, 41, 1003–1020.

Blau, P. M. (1970). A formal theory of differentiation in organizations. *American Sociological Review*, 35, 201–218.

Blau, P. M., Falbe, C. M., McKinley, W., and Tracy, P. K. (1976). Technology and organization in manufacturing. *Administrative Science Quarterly*, 12(1), 20–40.

Blau, P. M. and Schoenherr, R. A. (1971). *The Structure of Organizations*. New York: Basic Books.

Brown, A. and Grant, G. (2005). Framing the frameworks: A review of IT governance research. *Communications of the Association for Information Systems*, 15, 696–712.

Brynjolfsson, E., Malone, T. W., Gurbaxani, V., and Kambil, A. (1994). Does information technology lead to smaller firms? *Management Science*, 40(12), 1628–1644.

Burton, R. M., DeSanctis, G., and Obel, B. (2006). *Organizational Design: A Step-by-Step Approach*. Cambridge, U.K.: Cambridge University Press.

Chan, Y. E. and Reich, B. H. (2007a). IT alignment: An annotated bibliography. *Journal of Information Technology*, 22, 316–396.

Chan, Y. E. and Reich, B. H. (2007b). IT alignment: What have we learned? *Journal of Information Technology*, 22, 297–315.

Chandler, A. D. (1962). *Strategy and Structure: Chapters in the History of the American Industrial Enterprise*. Washington, DC: Beard Books.

Croteau, A.-M. and Bergeron, F. (2001). An information technology trilogy: Business strategy, technological deployment and organizational performance. *Journal of Strategic Information Systems*, 10, 77–99.

Delong, T. J., Brackin, W., Cabans, A., Shellhammer, P., and Ager, D. L. (2005). *Procter & Gamble: Global Business Services*. Boston, MA: Harvard Business School Case #9-404-124.

DeSanctis, G. and Monge, P. (1998). Communication processes for virtual organizations. *Journal of Computer-Mediated Communication*, 3(4), http://jcmc.indiana.edu/vol3/issue4.desanctis.html

Donaldson, L., Child, J., and Aldrich, H. (1975). The Aston findings on centralization: Further discussion. *Administrative Science Quarterly*, 20(3), 453–460.

Dosi, G. and Marengo, L. (2007). On the evolutionary and behavioral theories of organizations: A tentative roadmap. *Organization Science*, 18(3), 491–502.

Earl, M. J. (Ed.). (1996). *Information Management: The Organizational Dimension*. Oxford, U.K.: Oxford University Press.

Finnegan, P. and Longaigh, S. N. (2002). Examining the effects of information technology on control and coordination relationships: An exploratory study in subsidiaries of pan-national corporations. *Journal of Information Technology*, 17, 149–163.

Fligstein, N. (1985). The spread of the multidivisional form among large firms, 1919–1979. *American Sociological Review*, 50(3), 377–391.

Fulk, J. and DeSanctis, G. (1995). Electronic communication and changing organizational forms. *Organization Science*, 6(4), 337–349.

Galbraith, J. R. (1977). *Organization Design*. Reading, MA: Addison-Wesley.

Galbraith, J. R. (2000). *Designing the Global Corporation*. San Francisco, CA: Jossey-Bass.

Galbraith, J. R. (2008). Organization design. In T. G. Cummings (Ed.), *Handbook of Organization Development* (pp. 325–352). Thousand Oaks, CA: Sage Publications.

Galbraith, J. R. (2009). *Designing Matrix Organizations that Actually Work: How IBM, Procter & Gamble, and Others Design for Success*. San Francisco, CA: Jossey-Bass.

Galliers, R. W. and Currie, W. (Eds.). (2011). *The Oxford Handbook of Management Information Systems*. Oxford, U.K.: Oxford University Press.

George, J. F. and King, J. L. (1991). Examing the computing and centralization debate. *Communications of the ACM*, 34(7), 63–72.

Goold, M. and Campbell, A. (2002). *Designing Effective Organizations*. San Francisco, CA: Jossey-Bass.

Halal, W. E. (1994). From hierarchy to enterprise: Internal markets are the new foundation of management. *Academy of Management Executive*, 8(4), 69–83.

Henderson, J. C. and Venkatraman, N. (1993). Strategic alignment: Leveraging information technology for transforming organizations. *IBM Systems Journal*, 32(1), 4–16.

Hickson, D. J., Pugh, D. S., and Pheysey, D. C. (1969). Operations technology and organization structure: An empirical reappraisal. *Administrative Science Quarterly*, 14(3), 378–397.

Hitt, L. M. and Brynjolfsson, E. (1997). Information technology and internal firm organization: An exploratory analysis. *Journal of Management Information Systems*, 14(2), 81–101.

Huber, G. P. (1990). A theory of the effects of advanced information technologies on organizational design, intelligence, and decision making. *Academy of Management Review*, 15(1), 47–71.

Jarvenpaa, S. and Ives, B. (1993). Organizing for global competition: The fit of information technology. *Decision Sciences*, 24(3), 547–580.

Karimi, J. and Konsynski, B. (1991). Globalization and information management strategies. *Journal of Management Information Systems*, 7(4), 7–26.

King, J. L. (1983). Centralized versus decentralized computing: Organizational considerations and managerial options. *Computing Surveys*, 15(4), 319–349.

King, J. L. (2011). CIO: Concept is over. *Journal of Information Technology*, 26, 129–138.

King, W. R. and Sethi, V. (2001). Patterns in the organization of transnational information systems. *Information & Management*, 38, 201–215.

Klatzky, S. R. (1970). Automation, size, and the locus of decision making: The cascade effect. *Journal of Business*, 43(2), 141–151.

Kraemer, K. L. and Dutton, W. H. (1979). The interests served by technological reform: Theoretical perspectives in recent empirical research. *Administration and Society*, 11(1), 80–106.

Lambert, R. and Peppard, J. (1993). Information technology and new organizational forms: Destination but no road map? *Journal of Strategic Information Systems*, 2(3), 180–205.

Leavitt, H. J. and Whisler, T. L. (1958). Management in the 1980's. *Harvard Business Review*, 36(6), 41–48.

Lewin, A. Y. and Volberda, H. W. (2011). Co-evolution of global sourcing: The need to understanding the underlying mechanisms of firm-decisions to offshore. *International Business Review*, 20, 241–251.

Lewis, G. (2008). Concepts from coevolution in information systems. Paper presented at the *Fourteenth Americas Conference on Information Systems*, Toronto, CA. Paper 395. Available at http://aisel.aisnet.org/amcis2008/395 (accessed on April 16, 2013).

Lucas, H. C. (Jr). (1995). *The T-Form Organization: Using Technology to Design Organizations for the 21st Century*. San Francisco, CA: Jossey-Bass.

Lucas, H. C. (Jr). and Baroudi, J. (1994). The role of information technology in organizational design. *Journal of Management Information Systems*, 10(4), 9–23.

Majchrzak, A., Rice, R. E., Malhotra, A., King, N., and Ba, S. (2000). Technology adoption: The case of a computer-supported inter-organizational virtual team. *MIS Quarterly*, 24(4), 569–600.

Malone, T. W., Yates, J., and Benjamin, R. I. (1987). Electronic markets and electronic hierarchies. *Communications of the ACM*, 30(6), 484–497.

Markus, M. L. (1983). Power, politics, and MIS implementation. *Communications of the ACM*, 26(6), 430–444.

Markus, M. L. (2010). On the usage of information technology: The history of IT and organization design in large US enterprises. *Enterprises et Histoire 60 (Numéro Spécial pour le 40 ans du CIGREF)*, pp. 17–28.

Markus, M. L. and Bui, Q. N. (2012). Going concerns: Governance of interorganizational coordination hubs. *Journal of Management Information Systems*, 28(4), 165–199.

Markus, M. L., Dutta, A., Steinfield, C. W., and Wigand, R. T. (2008). The computerization movement in the US Home Mortgage Industry: Automated underwriting from 1980 to 2004. In K. L. Kraemer and M. S. Elliott (Eds.), *Computerization Movements and Technology Diffusion: From Mainframes to Ubiquitous Computing* (pp. 115–144). Medford, NY: Information Today.

Markus, M. L., Jacobson, D. D., Bui, Q. N., Mentzer, K., and Lisein, O. (2013). Organizational and institutional arrangements for e-Government: A preliminary report on contemporary IT management approaches in US State Governments. Paper presented at the *Hawaii International Conference on Systems Sciences* 46, Maui, HI, pp. 2088–2100.

Markus, M. L. and Robey, D. (1988). Information technology and organizational change: Causal structure in theory and research. *Management Science*, 34(5), 583–598.

Markus, M. L., Sia, S. K., and Soh, C. (2012). MNEs and information management: Structuring and governing IT resources in the global enterprise. *Journal of Global Information Management*, 20(1), 1–17.

Martinez, J. I. and Jarillo, J. C. (1989). The evolution of research on coordination mechanisms in multinational corporations. *Journal of International Business Studies*, 20(3), 489–513.

McFarlan, F. W. and Delacey, B. (2003). *Enabling Business Strategy with IT at the World Bank*. Boston, MA: Harvard Business School. Case #9-304-055.

McKelvey, B. (1982). *Organizational Systematics—Taxonomy, Evolution, Classification*. Berkeley, CA: University of California Press.

Melville, N. P., Kraemer, K. L., and Gurbaxani, V. (2004). Review: Information technology and organizational performance: An integrative model of IT business value. *MIS Quarterly*, 28(2), 283–322.

Meyer, A. D., Tsui, A. S., and Hinings, C. R. (1993). Configurational approaches to organizational analysis. *Academy of Management Journal*, 36(6), 1175–1195.

Miles, R. E. and Snow, C. C. (1996). Organizations: New concepts for new forms. In P. J. Buckley and J. Michie (Eds.), *Firms, Organizations and Contracts: A Reader in Industrial Organization* (pp. 429–441). Oxford, U.K.: Oxford University Press.

Mintzberg, H. (1983). *Structure in Fives: Designing Effective Organizations*. Englewood Cliffs, NJ: Prentice Hall.

Mintzberg, H. (1979). *The Structuring of Organizations: A Synthesis of the Research*. Englewood Cliffs, NJ: Prentice Hall.

Orlikowski, W. J. (2002). Knowing in practice: Enacting a collective capability in distributed organizing. *Organization Science*, 13(3), 249–273.

Palmer, I., Benveniste, J., and Dunford, R. (2007). New organizational forms: Toward a generative dialogue. *Organization Studies*, 28(12), 1829–1847.

Palmisano, S. J. (2006). The globally integrated enterprise. *Foreign Affairs*, 85(3), 127–136.

Peppard, J. and Breu, K. (2003). Beyond alignment: A coevolutionary view of the information systems strategy process. Paper presented at the *Twenty-fourth International Conference on Information Systems*, Seattle, WA. Paper 61. Available at http://aisel.aisnet.org/icis2003/61 (accessed on April 16, 2013).

Pinsonneault, A. and Kraemer, K. L. (1997). Middle management downsizing: An empirical investigation of the impact of information technology. *Management Science*, 43(5), 659–679.

Piskorski, M. J. and Spandini, A. L. (2007). *Procter & Gamble: Organization 2005 (A)*. Boston, MA: Harvard Business School. Case #9-707-519.

Pool, I. D. S. (1983). *Forecasting the Telephone: A Retrospective Technology Assessment of the Telephone*. Norwood, NJ: Ablex.

Pool, I. D. S. (1978). *The Social Impact of the Telephone*. Cambridge, MA: MIT Press.

Porter, M. E. (1980). *Competitive Strategy: Techniques for Analyzing Industries and Competitors*. New York: Free Press.

Pryor, F. L. (2001). Will most of us be working for giant enterprises by 2028? *Journal of Economic Behavior and Organization*, 44, 363–382.

Ramarapu, N. and Lado, A. (1995). Linking information technology to global business strategy to gain competitive advantage: An integrative model. *Journal of Information Technology*, 10, 115–124.

Rangan, S. and Sengul, M. (2009). Information technology and transnational integration: Theory and evidence on the evolution of the modern multinational enterprise. *Journal of International Business Studies*, 40, 496–514.

Robey, D. (1981). Computer information systems and organizational structure. *Communications of the ACM*, 24(10), 679–687.

Romanelli, E. (1991). The evolution of new organizational forms. *Annual Review of Sociology*, 17, 79–103.

Ross, J. W. and Beath, C. M. (2007). *Building Business Agility at Southwest Airlines*. Cambridge, MA: MIT Center for Information Systems Research.

Ross, J. W., Weill, P., and Robertson, D. (2006). *Enterprise Architecture as Strategy: Creating a Foundation for Business Execution*. Boston, MA: Harvard Business Press.

Sahayam, A., Steensma, H. K., and Schilling, M. A. (2007). The influence of information technology on the use of loosely coupled organizational forms: An industry-level analysis. *Organization Science*, 18(5), 865–880.

Sambamurthy, V. and Zmud, R. W. (2000). The organizing logic for an enterprise's IT activities in the digital era—A prognosis of practice and a call for research. *Information Systems Research*, 11(2), 105–114.

Sambharya, R. B., Kumaraswamy, A., and Banerjee, S. (2005). Information technologies and the future of the multinational enterprise. *Journal of International Management*, 11, 143–161.

Sia, S. K., Soh, C., and Weill, P. (2010). Global IT management: Structuring for scale, responsiveness, and innovation. *Communications of the ACM*, 53(3), 59–64.

Soh, C., Sia, S. K., and Tay-Yap, J. (2000). Enterprise resource planning: Cultural fits and misfits: Is ERP a universal solution? *Communications of the ACM*, 43(4), 47–51.

Steinert-Threlkeld, T. (2006). Nestle pieces IT together. *Baseline* (January), 36–52.

Steinfield, C. W., Markus, M. L., and Wigand, R. T. (2012). Through a glass clearly: Standards, architecture, and process transparency in global supply chains. *Journal of Management Information Systems*, 28(2), 75–107.

Strikwerda, H. and Stoelhorst, J.-W. (2009). The emergence and evolution of the multidimensional organization. *California Management Review*, 51(4), 11–31.

Strong, D. M. and Volkoff, O. (2010). Understanding organization-enterprise system fit: A path to theorizing the information technology artifact. *MIS Quarterly*, 34(4), 731–756.

Wagner, E., Howcroft, D., and Newell, S. (2005). Special issue part II: Understanding the contextual influences on enterprise system design, implementation, use and evaluation. *Journal of Strategic Information Systems*, 14, 91–95.

Whisler, T. L. (1970). *The Impact of Computers on Organizations*. New York: Praeger.

Whittington, R. (2002). Corporate structure: From policy to practice. In A. M. Pettigrew, H. Thomas and R. Whittington (Eds.), *Handbook of Strategy and Management* (pp. 113–138). London, U.K.: Sage Publications.

Whittington, R. and Mayer, M. (2001). *The European Corporation: Strategy, Structure, and Social Science*. New York: Oxford University Press.

Williams, C. (2008). Comparing evolutionary and contingency theory approaches to organizational structure. In R. M. Burton, B. H. Eriksen, D. D. Hakonsson, T. Knudsen, and C. C. Snow (Eds.), *Designing Organizations: 21st Century Approaches* (pp. 41–56). New York: Springer.

Williamson, O. (1985). *The Economic Institutions of Capitalism: Firms, Markets, Relational Contracting*. New York: Free Press.

Woodward, J. (1958). *Management and Technology*. London, U.K.: Her Majesty's Stationary Office.

Yates, J. (1989). *Control through Communication: The Rise of System in American Management*. Baltimore, MD: The Johns Hopkins University Press.

Zammuto, R. F., Griffith, T. L., Majchrzak, A., Dougherty, D. J., and Faraj, S. (2007). Information technology and the changing fabric of organization. *Organization Science*, 18(5), 749–762.

68

Open Innovation: A New Paradigm in Innovation Management

Sirkka L. Jarvenpaa
*University of
Texas at Austin
Aalto University*

68.1 Paradigm Shift in Innovation Management

Kuhn (1962, p. 103) writes, "[Paradigms] are the source of the methods, problem-field, and standards of solution accepted by any mature scientific community at any given time."

Organizational innovation management is experiencing a paradigm shift. In this chapter, the technological drivers for the open paradigm are discussed followed by key concepts of innovation management including innovation funnel, radical and incremental innovation, exploration and exploitation, and organizational change. We then contrast closed and open innovation and highlight some selected forms, approaches, and practices of open innovation. We review major and promising theoretical models and present questions for further research. Open source will be reviewed as part of open community models.

Innovation means creating and implementing new ideas, services, technologies, business models, applications, or processes (Gupta et al. 2007). Only when an idea or invention is successfully applied to practice does innovation take place (Schumpeter 1942). Innovation can take place in just about any human activity. Innovation is particularly critical when an organization faces one or more of the following: technological discontinuities, global competition, new markets, new competitors, collapsing product and service life cycles, or new customer value systems.

Traditionally, a firm's research and development (R&D) and product development departments generated and managed innovations. Sometimes, this centralized approach would then delegate responsibility to business units organized by products and markets or both. Innovation in services was haphazard at best. Now, there is a greater recognition that innovation affects everything the organization or company does including but not limited to logistics and human resource management, marketing and customer management, production management, information systems, financing and capital management, and public relations. Nearly any job opening highlights innovation management as a core skill set. Never has innovation been as valuable as it is now.

Information and communication technology (ICT) plays a major role in the innovation revolution because it converges multiple technologies enabling access to vast amount of data. ICT complements business and work processes and innovative management techniques that make innovations possible. ICT has fundamentally transformed innovation from an internal activity into an open and network-based business process (Brynjolfsson and Saunders 2010). ICT makes it possible to innovate across geographic and organizational boundaries and reach out to millions of people for their ideas and their assessments. ICT has led to large-scale, fast-cycle experimentation. Real-time data capture, storage, mining, visualization, simulation, and rapid prototyping allow companies to dramatically reduce development cycles and test novel combinations faster than ever before (Dodgson et al. 2006).

Technological changes have opened the innovation activities from inside the firm toward external networks. Shell's GameChanger process brings together ideas from inside and outside the company, including academic researchers, to collaborate on new technologies and business ideas. A number of initiatives have emerged that use virtual worlds and digital games (Schultze et al. 2008). For example, the Institute for the Future uses massively multiparty online war game as a way to collectively address major problems together with the U.S. Office of Naval Research. Much innovation now takes place under an open paradigm: open innovation (Chesbrough 2003), open source (O'Mahony 2003), user innovation (von Hippel 2010), open communities (Faraj et al. 2011), social production (Benkler 2006), and sponsored co-creation (West and O'Mahony 2008). Large-scale unifying communications networks create a global community for producing, distributing, and consuming innovative ideas, products, and processes.

The open paradigm to innovation lowers the barriers to entry and enables small and large companies to participate in the innovation revolution. Global companies such as Procter & Gamble traditionally relied on their internal brick and mortar R&D functions to produce innovations. The company was very insular in terms of the ideas it developed and launched. However, by early 2000, the company adopted a new radical strategy of open innovation. Their "Connect and Develop" model focuses on identifying promising ideas anywhere in the world and bringing them to the company to enhance and capitalize on internal capabilities. The company also launched an Internet-based innovation intermediary (NineSigma). This new model of innovation allows the firm to innovate more with less and shorten its development cycles (Sakkab 2002). In some cases, R&D costs were cut 60% (Huston and Sakkab 2006).

Another company leveraging the paradigm shift in innovation is Threadless, a small Chicago-based company with less than 20 employees. Threadless operates t-shirt design competitions through its website. Globally, over one million people, many of whom were professional designers, submit designs (Howe 2006; Brabham 2010). The designs are then posted and voted on each week. Using all this information, the firm selects the final prints for the t-shirts that it produces.

Open innovation is best thought of as a paradigm shift in innovation management (Huizingh 2011). Openness has been long valued as bringing important diversity in the early phases of innovation, but more recently in other phases of the innovation as well (Granstrand et al. 1997). Openness takes varied forms in different phases (Dahlander and Gann 2010). Open innovation can involve integrating external ideas with internal organizational capabilities and resources, as well as effectively using intellectual property (IP) for appropriating value from innovations (Chesbrough 2003; Lichtenthaler 2011).

Open innovation can also mean exploiting open communities without legal constraints (von Hippel 2010). As problems grow more complex, Internet-based innovation communities make it possible for an almost infinite number of firms and individuals (who may not even know of each other's existence) to work independently yet contribute to the overall innovation (Faraj et al. 2011).

Open innovation in its varied forms and approaches has gained popularity globally and across industries. Early research and writing began in North America and Europe (e.g., Chesbrough and Teece 1996; Granstrand et al. 1997). There is a growing body of open innovation literature in Asia (Asakawa et al. 2010; Lee et al. 2010) and Latin America (Gomes and Kruglianskas 2009). Governmental bodies, eager to improve their economic competitiveness, are very interested in open innovation (de Jong et al. 2008; Almirall and Wareham 2011; Jarvenpaa and Wernick 2011). Leveraging innovation in open networks is not limited to high-technology sectors (e.g., Intel, Apple, and Google) but is increasingly exploited in mature industries and service businesses (Chesbrough 2010; Chiaroni et al. 2010) and in small- and medium-sized firms (van de Vrande et al. 2009). Open is also leveraged in nonprofits (Holmes and Smart 2009; Lakhani et al. 2012). As research continues and is disseminated and as technology continues to converge, more individuals, companies, and governments embrace the open innovation paradigm and reap its benefits.

68.2 Key Concepts in Innovation Management

The field of innovation management is rich with many concepts, frameworks, and models. It bridges many disciplines including organization theory, economics, information systems, communications, sociology, psychology, geography, and history. Successfully connecting these various disciplines requires a holistic management approach. Interested readers are referred to many handbooks that take a holistic view on innovation (e.g., Nooteboom 2000; Burgelman et al. 2004; Fagerberg et al. 2005). With an understanding of the interconnectedness inherent in innovation management, we next review some key concepts that are particularly valuable in understanding open innovation.

68.2.1 Innovation Funnel

The "innovation funnel" (see Figure 68.1), also called stage-gate model, was developed in the 1980s as a way to understand how a firm decides which innovations to pursue and how to pursue them (Cooper 1988). Innovations go through four to five stages such as idea generation, concept development, testing, and commercialization. Each stage typically involves more resources than the previous one, resulting in increased commitments. At the beginning, in the idea generation stage, the firm engages in technology and market assessments that generate a sizable number of ideas and possible opportunities.

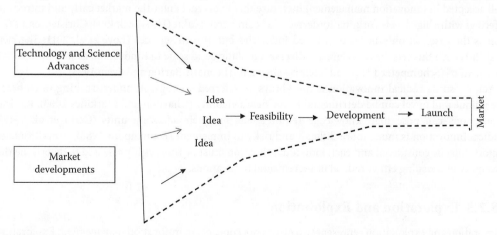

FIGURE 68.1 Closed innovation system.

These may include customer or market need, new product concept, scientific breakthrough, competitive advantage, and so on. At the first gate, the ideas and opportunities are evaluated, and a determination is made whether to proceed into the next stage. A broad set of factors may be considered including the current technology strategy, product/market strategy, competitor analysis, political risks, etc. If the idea passes the first gate then it progresses to the second gate: concept development, if successful there then to the third gate, testing, and if passes that gate on to commercialization (Gronlund et al. 2010).

Many have criticized the funnel as anti-innovative because it creates a linear and rigid process (Nichols 2007). If the innovation fails any of the gates, the innovation is put on the shelf. The gates can involve heavy evaluation processes to the point that more resources are allocated to evaluating and eliminating innovations rather than improving the innovations. The innovation funnel engenders "false negatives." These are ideas that are weeded out by internal groups and remain unused and underutilized by the firm. However, many such ideas might be valued by others outside the firm. The funnel only considers internal paths to market. The funnel does not encourage ideas from outside except in the first stage. There are often strong norms against innovations that are not internally developed, so-called not-invented-here syndrome. The funnel emphasizes alignment between the innovation and the firm's strategy (Gronlund et al. 2010).

68.2.2 Radical versus Incremental Innovation

Different types of innovations (product, process, business models, architectural), as well as innovation processes, are commonly discussed in terms of a continuum from incremental to radical. This continuum reflects the relative newness and uniqueness of the innovation from preexisting alternatives and knowledge bases (Dewar and Dutton 1986) and has a major impact on future developments (Dahlin and Behrens 2005). "An innovation is *radical* when innovators need to acquire extensively *unique* and *novel* technological and process-related *know-what*, *know-why*, and *know-how*" (Carlo et al. 2012). Radical innovations are considered as frame breaking and transformative yet highly risky and uncertain. Radical innovations or radical innovation processes are based on different and unconventional technologies or scientific principles, fundamentally change the cost/benefit ratios, require new cognitive frames, open new use contexts and applications, and lead to new markets (Henderson and Clark 1990).

By contrast, incremental innovation "introduces relatively minor changes to the existing product and exploits the potential of the established design" (Henderson and Clark 1990, p. 9). Incremental innovations are competence enhancing, whereas radical innovations are competence destroying (Henderson and Clark 1990). Radical innovations are new-to-the-world, whereas incremental innovations are new-to-the-firm (Henderson and Clark 1990).

Radical and incremental innovations have different economic and organizational consequences. It is well accepted in innovation management literature that firms that enter the market early and introduce offerings with a high level of originality derive more commercial value (Utterback 1994; Christensen 1997). This is the case not only in product-based industries but also in services (Love et al. 2011; Therrien et al. 2011). All this requires ever-greater diversity, mobility, and scale in innovation, as well as "creative destruction" (Schumpeter 1942) and "creative abrasion" (Leonard-Barton and Swap 1999).

Yet, pursuing radical innovation is not always a preferred strategy. At times, rushing to embrace the unique and novel can be detrimental to the firm's viability (Charitou and Markides 2003). Radical innovations can pose high knowledge requirements for their adopting units (Carlo et al. 2012). Radical innovation is much more difficult and risky to implement (Damanpour 1988). A multitude of factors must be considered and each individual situation must be thoroughly assessed to determine the feasibility of pursuing either radical or incremental innovations.

68.2.3 Exploration and Exploitation

Exploration and exploitation represent two other core concepts in innovation management. Exploration involves the generation of an idea or invention; exploitation involves the conversion of the idea or invention into a solution, business, or application that can be used to generate a profit. All innovative activity involves exploration and exploitation as their main activities (March 1991). Exploration involves "search, variation, risk taking, experimentation, play, flexibility, discovery, and innovation" (March 1991, p. 71). Exploration is about value creation: new learning and knowledge, tasks, functions, and activities that maximize opportunities. Exploration is associated with uncertainty over value appropriation as the future markets, customers, application areas are still unknown. Traditionally, to manage this uncertainty, firms limit their exploration activities to a limited number of parties to maintain control yet leverage informal relationships for the transfer of tacit knowledge.

Exploitation refers to "refinement, choice, production, efficiency, selection, implementation, and execution" (March 1991, p. 71). Exploitation is about efficient use of existing knowledge and assets, routines, standard operating procedures, and formal relationships. Exploitation is about maximizing income from innovation or gaining the maximal utility or use value from innovation. In exploitation, markets are known and focus is on maximizing transaction efficiency.

The organization's ability to achieve a balance between exploration and exploitation is referred to as organizational ambidexterity (Tushman and O'Reilly 1996; Gibson and Birkinshaw 2004). Organizational ambidexterity is challenging as the requirements for leadership, structures, and processes regarding exploration and exploitation are very different even paradoxical (Dougherty 2006; Lavie et al. 2010). Yet, successful innovation requires management of exploration and exploitation at a high level in a complementary fashion (Katila and Ahuja 2002). Organizational ambidexterity requires a holistic view of innovation (e.g., Linder et al. 2003). The innovation value chain (Hansen and Birkinshaw 2007) provides such a view through (1) idea generation, (2) conversion, and (3) diffusion. The innovation value chain interlinks knowledge generation and sourcing by transforming this knowledge into new services and commercialization of these new services to grow business.

68.2.4 Organizational Change

The more radical the innovation, the more challenging the organizational change. Radical innovation means undergoing discontinuous change processes including a cultural change not just in the firm but often also in the broader networks of customers and suppliers. In the vocabulary of radical change theorists, discontinuous change requires *deep structural changes* where basic values, business practices, culture, and organizations change (Tushman et al. 1986). Without a deep structural change, behaviors migrate back toward the old ways of operating, reinforced by the existing processes, structures, incentives, and cognitive frames (see Figure 68.2). The weakest link in the network can undermine accomplishments elsewhere. Some level of identity crisis, disorder, and ambiguity usually precedes a deep structural change that sticks. Therefore, a long-time horizon and high level of organizational, architectural, managerial, and legal competencies are required for successfully implementing a radical innovation. Done in this manner, new incentives and mind frames are adopted and accepted and successfully replace the old behaviors.

For example, Procter & Gamble found that it took 5 years of investments, sourcing ideas externally, and developing online knowledge exchange sites before the firm began to see returns from the transformation of its innovation process from an internal model to an open innovation model (Hansen and Birkinshaw 2007). The transition from closed to open innovation meant radical change for Procter & Gamble. Open innovation required many technological and organizational investments such as partner networks, open innovation business units, platforms to transfer knowledge externally and internally, and IP protection systems (Gassman and Enkel 2004; Huston and Sakkab 2006; Chiaroni et al. 2010; du Chatenier et al. 2010). Procter & Gamble's long-time horizon and high-level organizational, technological, managerial, and legal competencies enabled a successful radical change to open innovation.

Without the correct structures in place, failures are common. A global firm with a large customer base was unable to transfer and leverage even a single external unsolicited R&D even though "it participated in many forms of external innovation such as open source software development, patent donations, and open

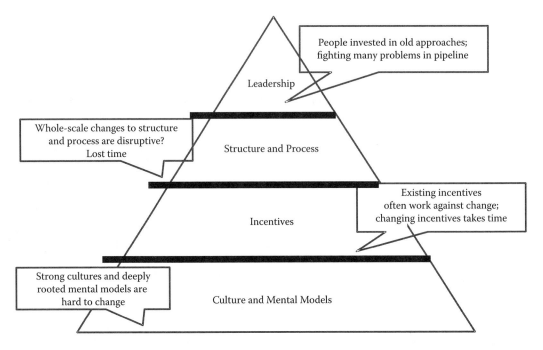

FIGURE 68.2　Organizational challenges to innovation. (Adapted from Cummings, J. Personal Communications, 2011.)

research calls" and was proud of its capabilities at engaging in open innovation (Alexy et al. 2012, p. 122). Such failures abound although they are not currently well captured in the literature.

68.3　Open Innovation

In the 1990s, increased investments in R&D in the automotive, energy, pharmaceutical, and other industries produced declining rates of innovation (e.g., Barrell et al. 2000). Open innovation addressed this issue and became a way to boost a firm's innovative capacity, grow, and increase market share (van de Vrande et al. 2009; Drechsler and Natter 2012). Open innovation augments internal capabilities with complementary capabilities, network externalities, and access to new relationships (Chesbrough and Crowther 2006). If implemented correctly, open innovation can enable growth and increase market share.

68.3.1　Henry Chesbrough on Open Innovation

In the book entitled, *Open Innovation: The New Imperative for Creating and Profiting from Technology*, Henry Chesbrough (2003) popularized the notion of "open innovation." This book has proven to be highly influential. It had more than 5740 citations as of August 2012. There were 1800 citations in July 2010 (Huizingh 2011), just 7 years after the book was published. Chesbrough (2012, p. 20) wrote

> When I wrote *Open Innovation* in 2003, I did a Google search on the term "open innovation," and I got about 200 links that said "company X opened its innovation office at location Y." The two words together really had no meaning. When I conducted a search on that same term last week, I found 483 million links, most of which were about this new model of innovation.

Chesbrough (2003) contrasted open innovation to "closed innovation." Closed innovation refers to a vertically integrated innovation model. The company largely relies on its own innovation investments, processes, and structures for exploration and exploitation (March 1991). In terms of the innovation

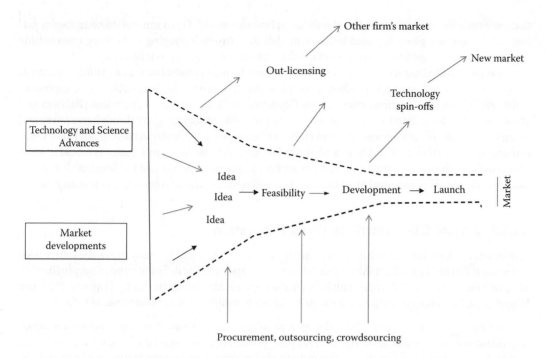

Other firm's market

Out-licensing

New market

Technology
spin-offs

Technology and Science
Advances

Idea

Idea → Feasibility → Development → Launch

Idea

Market

Market
developments

Procurement, outsourcing, crowdsourcing

FIGURE 68.3 Open innovation funnel. (Adapted from Chesbrough, H., *Res. Technol. Manage.*, 55(4), 20, 2012.)

funnel, the closed innovation process is a linear path from ideas to launch. Once fully developed, innovations are channeled to markets or used internally. The firm that creates the innovation is assumed to be in the best position to create value out of it (Kogut and Zander 1992; Wyld 2010).

For Chesbrough (2006), open innovation is about both decoupling and integrating the different phases of the innovation value chain and in so doing seeking radical innovation through business models. Open innovation is a paradigm shift "that assumes that firms can use external ideas as well as internal ideas, and internal and external paths to market" (Chesbrough 2003).

Figure 68.3 illustrates the innovation funnel adapted to open innovation (Gronlund et al. 2010). Ideas, opportunities, and technologies can come from internal or external sources and enter and leave the open innovation funnel at any stage. If the innovation fails one of the gates, it can take a different path and a different business model. Innovations do not just make their way through the firm's sales and marketing channels but as a spin-off, as out-licensing, as a design freedom, as an industry standard. There is a continuous assessment of external technologies, markets, and other developments. There is a continuous assessment of exportation and importation of know-how and technology. The gates are not just cost and risk evaluations but opportunity assessments (Gronlund et al. 2010).

Chesbrough (2012) identified three conditions that must be met to take advantage of the open innovation: workforce mobility, internal R&D, and basic rules on IP. The conditions are important because no significant innovation progresses without people moving; creative abrasion requires the integration of external and internal innovation; and IP protection rules are needed for investors to be willing to invest beyond the nascent stage of an innovation.

A key contribution of Chesbrough's (2006) work has been to bring greater focus on the role of business models in innovation management. A business model is "a structural template of how a focal firm transacts with customers, partners, and vendors" (Zott and Amit 2008). A business model focuses on how the firm commercializes the innovations of others and commercializes innovations through others (Chesbrough 2006). Business models are a way to generate different pathways through the open innovation funnel (Chesbrough 2012). Rather than turning an innovation into a product or service that the firm sells, the firm can think of a different business model. Open innovation often means getting more aggressive about IP-based business models that involve leveraging technology from outside (e.g., in-licensing) or getting others to leverage the firm's technology (e.g., out-licensing).

Chesbrough (2006) argues that the most valuable open business model involves a platform. Platform refers to a system and architecture where a complex system is partitioned into stable core components ("platform"), variable peripheral components ("complement"), and standard interfaces (Baldwin and Woodard 2009; Gawer 2009). Complements provide variability and dynamicity through morphing and mutation. An example of such a platform in mobile technologies is the Android based platform by Open Handset Alliance (Han et al. 2012). In a platform-based model, the company's business model is interconnected with the business models of its key suppliers and customers. Not just the firm but all partners can innovate and share technical and financial risks and rewards. The IP is managed as a strategic asset.

68.3.2 Beyond Chesbrough on Open Innovation

Chesbrough's views have been highly influential but do not represent the only views regarding open innovation. During a paradigm shift, heated debates are expected. The definition and conceptualization of open innovation are still under much discussion (see recent commentaries by Linstone 2010; von Hippel 2010; Lichtenthaler 2011; Chesbrough 2012). We highlight some of these views as follows:

User innovation: To von Hippel (2005), the open paradigm shift is about changing functional relationship between innovator and innovation. In the past, producers were assumed to be the innovators who sold their innovations to users. Now, users innovate by improving and reinterpreting products and services through "sticky information" from repeated use. For example, users renegotiated the *Wii* gaming platform as an in-house exercise platform (Verganti 2009). In some areas as in sports equipment and software design, user innovation does not just complement but competes with producer innovation.

Private-collective innovation: von Hippel and von Krogh (2003) advanced the private-collective innovation model based on their research on open source. The model relies on a group of individuals and organizations collaborating and sharing their private ideas and knowledge and openly revealing innovations: yet gaining higher profits than free riders (von Hippel and von Krogh 2003). The private-collective model of open innovation relies on private investments, open revealing, and tacit learning to iteratively co-create and disseminate the innovation. As some tacit learning remains private, the innovators profit economically from their innovations.

Process transformation: Gassmann and Enkel (2004) promote transformation of three core processes in open innovation: (1) the outside-in process (wide and deep external search and integration), (2) the inside-out process (multiple channels for externalizing innovation including third party commercialization of innovation), and (3) the coupled process (combining outside-in and inside-out by collaborating in strategic networks with other companies). The coupled process requires strategic choices to participate in networks and share IP with other organizations and capabilities and resources to engage in intensive interactions to transfer learning across organizational boundaries.

Organizational permeability: For Dahlander and Gann (2010), openness is about "the permeability of firms' boundaries where ideas, resources, and individuals flow in and out of an organization" (p. 699). Open innovation is about openness in the inbound and outbound processes. The outside-in process of acquiring and inside-out process of selling involve monetary transactions. The outside-in process of sourcing and inside-out process of revealing are non-pecuniary.

Modularity: Baldwin and Clark (2006) and Henkel et al. (2012) emphasize modularity and architectural design decisions to enhance open innovation generation and appropriability (i.e., profit making). Modularity in design refers to the designers being able to design parts independently but working together to support the whole (Baldwin and Clark 2006). Modularity facilitates collaborating with diverse partners with different and even conflicting motives. Modularity can help a group of companies such as a consortia or alliance to jointly develop the concepts for the innovation but then compete to privately appropriate the value (Ritala and Hurmelinna-Laukkonen 2009; Almirall and Casadesus-Masanell 2010; Han et al. 2012).

Disruptive change: A disruptive innovation creates new markets and business models and disrupts and eventually displaces existing technologies, processes, prevailing power and control structures, and institutionalized routines (Christensen 1997). Christensen et al. (2005) view open innovation as a disruptive change. Many unknown and unknowable factors face organizations embracing open innovation. They quote Paivitt (1999) to illustrate the type of uncertainties and ambiguities present, "In many areas, it is not clear before the event who is in the innovation race, where the starting and finishing lines are, and what the race is about."

IP management practices: Pisano and Teece (2007) argue that open innovation makes it much more challenging for firms to capture value (profit) from innovation. Open innovation requires much greater attention and skillful use of IP mechanisms such as patents, copyright, and secrecy in innovation management (Pisano and Teece 2007). At times, IP management is overly emphasized to the point that IP becomes a bottleneck for open innovation (Alexy et al. 2009).

Networks: Open innovation is about network-based innovation. Organizations increase their capacity for open innovation by participating in networks and improving their ability to learn from their partners (Fey and Birkinshaw 2005). However, without continual change in such networks, networks can become as much an obstacle as an enabler of innovation—"the ties that bind may become the ties that blind" (Birkinshaw et al. 2007, p. 68). Building new networks requires right attitudes and resources. It requires engaging in finding, forming, and performing in a wide range of networks. Such networks may include idea networks, funding networks, user innovation networks, cross-industry consortia, etc. (Birkinshaw et al. 2007). Companies have to be highly agile in building networks as there is much uncertainty and ambiguity in new ties.

To recap, we take an inclusive view into the open paradigm in innovation management. The views of Chesbrough and the aforementioned authors are presented and discussed to show that open innovation is multifaceted and complex.

68.3.3 Openness in the Open Innovation Paradigm

Openness is a common theme in the views of open innovation. However, what this openness is and represents is still much under debate. Many have pointed out that the antithesis of open innovation and closed innovation is largely a myth. Throughout history, firms have tapped external sources for innovation including Edison's laboratory in nineteenth century (Mowery 2009). Allen (1983) described the history of the nineteenth century iron and steel industry as collective. Process industries such as chemicals and oil and gas have a long history in relying on external sources of knowledge and technology (Lamoreaux and Sokoloff 1998).

In studying platforms, West (2003) was one of the first to raise the question of "How Open is Open Enough?" He found that in many industries, the industry pioneers initially developed proprietary and vertically integrated platforms to capture profits and protect against imitation (e.g., IBM's 360). Major technical and economic reasons moved the companies to offer more partly open platforms to seek interoperability and attract complementary business. In the partly open platforms, the commodity layers were open and those parts that offered differentiation were retained as proprietary. In a later work, West and O'Mahony (2008) defined "openness" via three indicators: (1) transparency (external parties able to contribute to the innovation), (2) accessibility (external parties able to influence the direction of the innovation), and (3) joint appropriation (external parties have use/development rights). They found that company-sponsored open source communities were more likely to offer transparency than accessibility. Transparency allowed external partners, or complementors, to participate on the platform but limited accessibility and maintained company control over key decisions. In a study of mobile computing platforms, Boudreau (2010) found that granting platform access to independent complementors was preferred over giving up formal company control of the platform itself.

Examining the outside-in process and specifically external search in innovative activities, Laursen and Salter (2006) defined openness in terms of breath and depth of sources. Breadth of external search was defined as the number of different external sources or search channels. The depth was defined in terms of the extensiveness to which firms deeply engage the external sources or search channels. Laursen and Salter (2006) found that those firms that searched both wide and deep were more innovative but only to a point where openness became subject to decreasing returns.

Again focusing on the outside-in process, Drechsler and Natter (2012) built on the Laursen and Salter's (2006) study by examining the underlying drivers of breadth and depth of external search in innovative activities. They found that different factors impact the initial opening of the outside-in process from those that lead the firm to further increasing the openness in the external search. Closed firms that did not engage in external search lacked internal capabilities particularly in the area of market and technology know-how and IP protection. They also faced much rivalry and imitation. The firms that increased openness in the external search were exposed to financial funding opportunities. They also developed effective IP protection mechanisms. This study underscores the internal company competence development both in terms of moving from closed to open and increasing openness.

Dahlander and Gann (2010) conducted a broad review of the open innovation literature to identify a typology of types of openness. They introduced a framework of openness that distinguished activities by their non-pecuniary and pecuniary character and the phase of innovation (outside-in process and inside-out process). Sourcing is the use of external information but without direct monetary payment. Acquisition (i.e., outsourcing) refers to purchasing resources from the market. Revealing is disclosing information without immediate financial rewards whereas selling (e.g., out-licensing) involves monetary exchange. Dahlander and Gann's (2010) review suggests that most studies so far have only examined one or at most two types of openness. Dahlander and Gann called for more research.

Openness also varies by innovation life cycle. In a study of open alliances, Han et al. (2012) speculated that openness is likely to be highest at the early stages and once ideas develop into products and services, with more certainty of markets and customers, openness will be adjusted to minimize spillovers, particularly to rivals. They also propose the possibility that as the products and services become mature, openness might be increased to gather improvements that further extend the life of the offering.

In summary, the discussion of openness has moved beyond a dichotomy to discussing openness by degree, type, and life cycle of the innovation. While no models exist in terms of optimal openness, research is beginning to define some boundaries for openness.

68.4 Open Innovation Forms and Approaches

Openness is impacted by the forms and approaches taken in a particular initiative. According to Huizingh (2011, p. 3), "open innovation [has] become the umbrella that encompasses, connects, and integrates a range of already existing activities." Baldwin and von Hippel (2011) also highlight how open innovation is based on combinations of existing techniques. Next, we highlight some popular forms and approaches. These are mapped to the innovation value chain (Figure 68.4).

68.4.1 Outsourcing of Innovation

Outsourcing of innovation represents limited openness. Companies procure, acquire, and in-license innovation from outside using contractual transactions. Outsourcing of innovation often represents something that the firm may have innovated internally in the past but now is left to another company or entity to innovate and the firm buys or acquires the innovation for a fee. Outsourcing of innovation includes sponsored research, strategic procurement, and development partnerships. While sponsored research addresses the idea generation phase of the innovation value chain, strategic procurement represents the diffusion phase of the innovation. The innovation is well developed with known characteristics.

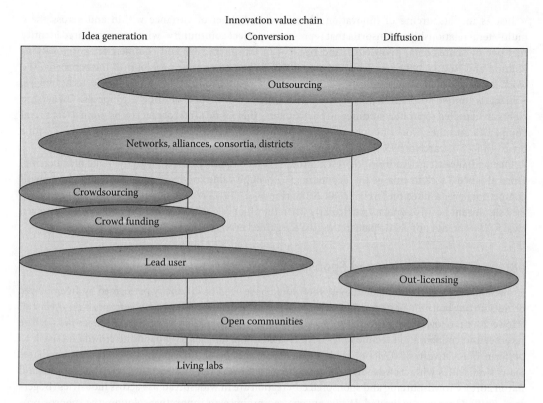

FIGURE 68.4 Illustrating open innovation forms on innovation value chain.

Many industries including the pharmaceutical industry have long turned to various forms of outsourcing of innovation to support their activities.

In 1996, Chesbrough and Teece (1996) asked "When is virtual virtuous?" that is, should firms outsource, namely procure and acquire innovation from outside partners just as they have outsourced production? At this point, the authors cautioned against outsourcing innovation when innovation relates to the firm's core competencies and involves architectural innovation such as platforms.

Linder et al. (2003) conducted a study in 2002 and examined the increasing portion of innovation that is externally sourced. They found that the share of innovative ideas from external sources was accounting for an average of 45% of ideas and the executives expected this percentage to grow. Retail companies reported as high as 90% of innovation from external sources. Shared risk, reduction in costs, and increased growth drove companies toward outside sources of innovation. Linder et al. (2003) found that many firms failed to realize the benefits from outsourcing activities because of a fragmented and project-based approach that lacked a coherent and holistic organizational strategy.

68.4.2 Consortia, Alliances, Networks, Clusters, Districts

Compared to outsourcing of innovation that involves bilateral contractual relationships, firms band together to save costs, share risks, and form larger multilateral relationships such as consortia (Pisano 2006). These are still contract based, but the membership can vary from stable (e.g., significant penalties for premature termination of contracts) to highly dynamic (e.g., open innovation alliances). This is a popular approach to innovation in the oil and petroleum industries, as well as automotive and chemical firms. The life science industry is well known for its clusters (Powell et al. 1996; Bunker-Wittington et al. 2009). Regional and national policy makers incentivize consortia as a way to revitalize or grow new, emerging industry sectors such as clean energy (Jaegersberg and Ure 2011).

Just as in outsourcing of innovation, there is a high level of variance within and across these multilateral relationships. Consortia that represent a "perfect community" with a strong shared identity that openly shares data, knowledge, and results can fail to transfer and commercialize innovations (Gibson and Rogers 1994; Cassier and Foray 2002). When competitors are part of the consortia, the work is divided and structured with clear roles and responsibilities by domains and technologies. Sharing is limited to predetermined areas with little accumulation of knowledge pools. Ownership rights are handled separately by different participants. There is minimal contact by personnel from rival companies. At times, work does not go beyond defining the future research areas as was the case with a breast cancer research consortium (Cassier and Foray 2002).

Open alliances (e.g., Android) represent groups with competing entities with dynamic membership. Open alliances "seek to enlarge the 'economic pie' through value co-creation, rather than fighting with competitors over a 'fixed pie'" (Han et al. 2012). Han et al. (2012) reported that participating firms realized significant positive returns particularly when the alliance engaged in radical innovation. However, rival firms who did not participate in the alliance gained even greater benefits through free riding.

68.4.3 Crowdsourcing and Sponsored Co-Creation

Crowdsourcing involves a job or activity that would have been traditionally performed by an employee or outsourced but now performed by a large group of people, or crowd, in the form of an open call (Howe 2006). Crowds are expected to provide unbiased information ("wisdom of the crowds"). The Internet and collaboration technologies have created opportunities to expand the crowd and use it to perform innovative tasks (Albors et al. 2008; Afuah and Tucci 2012). Sponsored co-creation shares many similarities with crowdsourcing but is often less time paced and less structured (Nambisan and Baron 2010). Crowdsourcing and sponsored co-creation are in widespread use across industries (Bogers et al. 2010). Openness is limited. While anyone can propose an innovative solution, the sponsor proposes a problem and claims ownership rights to the solution.

The Threadless example mentioned at the beginning of this chapter is an example of crowdsourcing. Threadless' design contests allow customers to submit new t-shirt ideas for a chance at prize money. Goldcorp of Canada used the crowd to locate the position of gold on its properties. Dell's IdeaStorm is a firm-sponsored crowdsourcing community. In 2007, the firm launched the site as a way to counter negative feedback toward its business model. The site was introduced with the statement that IdeaStorm was a place "where your ideas reign" (di Gangi et al. 2010, p. 215). Users can post improvement ideas for Dell's products and services and other users can promote or demote the ideas. Users are not given a financial incentive but top users are honored on the Top Idea Makers List.

Tournaments are a common form of crowdsourcing: "each agent from the crowd self-selects to work on its own solution to the problem, and the best solution is chosen as the winning solution" (Afuah and Tucci 2012, p. 355). Netflix used a tournament-based form of crowdsourcing to improve its movie recommendation system. Darpa Grand Challenge and Google Lunar X Prize are other well-known examples. Procter & Gamble, as part of their "Connect and Develop" model, launched NineSigma, a crowdsourcing intermediary that runs tournament-based crowdsourcing (Jeppesen and Lakhani 2010). Procter & Gamble initially set up Innocentive to address internal R&D problems and then turned it into a separate business. Innocentive and other intermediaries attract companies as they can reduce costs of finding solvers, reduce costs of specifying what is needed, and facilitate enforcing rules and agreements. Jeppesen and Lakhani (2010) found that people in fields outside of the problem domain often become successful solvers in tournaments; the distance between the solver's field of technical expertise and the field of the problem owner increased the likelihood of winning; female solvers performed significantly better than men in developing successful science solutions. There are also platforms that facilitate contests inside large and distributed global firms, for example, Innocentive@work.

Crowdsourcing is also used for big social problems. Founded in 2010, the OpenIDEO platform was launched by a global design firm, IDEO. The platform invites anyone in the world to enjoy solving problems associated with social issues. The platform has hosted 16 challenges and 35,000 participants had participated by August 2012. A key incentive is to do social good while staying on your sofa. The OpenIDEO platform does not award prizes. The platform uses IDEO's unique approach to problem solving (Lakhani et al. 2012).

Motivational represents the biggest challenge. The heterogeneous motives and backgrounds make it challenging to motivate and support the participants (Antikainen et al. 2010). Füller (2010) listed a variety of motives including extrinsic reward-oriented, need-driven, curiosity-driven, and intrinsically interested. It is this heterogeneity that makes it challenging to activate and sustain an active community over time (Leimeister et al. 2009; di Gangi et al. 2010; Zheng et al. 2011). Many companies find that they face a major battle trying to sustain co-creation (Jarvenpaa and Tuunainen 2012). The use of awards might crowd out contributors. On sponsored co-creation sites, the frequent contributors may be explicitly identified from less frequent contributors. Even this type of recognition can crowd out participants (Gu and Jarvenpaa 2003). Besides explicit recognition, there are often no immediate benefits except the crowdsourcing or co-creation experience (Nambisan 2002; Nambisan and Baron 2010).

Another challenge is to find the problems that actually can be solved with crowdsourcing. Afuah and Tucci (2012) identify several requirements: Problems need to be easily delineated and modular. Solutions need to require knowledge that is far from what the problem owner currently has available. The crowdsourcing needs to reach a motivated crowd that has problem-solving skills. If the crowd is expected to vote, then the solution has to be appraisable by multiple people.

68.4.4 Crowd Funding

An extension of crowdsourcing is crowd funding. Crowd funding is a collective effort of users or consumers to pool their money together and invest and support an initiative (Ordanini et al. 2011). New funding platforms such as GrowVC collectively fund innovation by opening the opportunity to invest small amounts of capital or effort ("sweat equity") to start-ups around the world. Besides providing funds, they can signal the initiative's market potential and expect their feedback to be heard as the ideas are refined and improved. Voting with money is expected to lead to more thoughtful and credible input and an improvement over traditional crowdsourcing. Crowd funding initiatives can reduce costs of initial development, create visibility, and encourage word of mouth marketing, and even drive initial sales for a new initiative or project idea (Burtch et al. 2012). However, here again, heterogeneous motives make management of the process difficult. Sometimes a crowd can be uninformed, driven by herd behavior, or not understand the initial idea. This then leads to a noisy signal problem in the ideation marketplace. The use of crowd funding is in its infancy and failures are still very common (Burtch et al. 2012).

68.4.5 Lead User

In the lead user approach, it is the user that identifies the problem to be solved and creates the solution (von Hippel 1977). The lead user approach is premised on the idea that users have the most accurate assessment of their needs and hence are ideally suited to be the innovators (von Hippel 1986). Compared to traditional ways, the lead user method has been identified as a much faster and cheaper way of identifying new product concepts (Herstatt and von Hippel 1992). Lüthje et al. (2005) found that users with membership in multiple communities were able to combine previously disparate information; the lead users of mountain bikes were medical doctors who wanted to make bikes safer (Lüthje et al. 2005). IP issues are deemphasized with the lead user approach. The lead users are not usually interested or concerned about protecting their innovations (Shah 2000).

The lead user innovator approach is most applicable in the context of complex products and services involving complex systems that require high levels of customization, iterative design, and adjustment after purchase to meet heterogeneous customer demands. The challenge for the firm is to locate the lead users that operate at the world-class level and identify innovations that have commercial value in the marketplace (Bogers et al. 2010). The firm may provide tool kits to encourage this type of innovation (Franke and von Hippel 2003; Jeppesen and Frederiksen 2006). Toolkits for user innovation allow customers to discover their needs in a structured fashion that enable users to develop new innovations (von Hippel and Katz 2002). The toolkits are specific to a certain design area such as software design (Franke and von Hippel 2003).

68.4.6 Open Community Innovation

Just like the lead user model relies on a non-pecuniary relationship, so do open communities for innovation. There are no legal transactions involved. Whereas the lead user model is driven by the user's private benefit, the open community innovation is driven by generalized reciprocity and community norms (Wasko and Faraj 2000; Roberts et al. 2006). The model relies on open reveling: instead of claiming proprietary rights over the proprietary knowledge that they have, community participants allow others to freely use their knowledge in the community (von Hippel and von Krogh 2006). Open revealing is associated with superior learning opportunities that still allow private benefits such as learning, enjoyment, and reputational gains (Henkel 2006).

There are a multitude of open source communities. Technology open source communities include Linux, GNOME, Apache, and Mozilla. Wikipedia (O'Mahony 2003; Niederman et al. 2006) is another open source community. There are various sporting communities such as kayaking, snowboarding, windsurfing, skateboarding (Franke and Shah 2003), and there are even music remix communities (Jarvenpaa and Lang 2011). The communities are usually characterized as low in rivalry although Franke and Shah (2003) found that even when there is rivalry, assistance is still given and innovations are revealed although not "just not as often as in the less competitive communities" (p. 172). Although the open community can represent radical innovation in its own right such as Wikipedia, particular innovative outputs are often a result of independent entities (firms or individuals) providing incremental contributions.

Challenges relate to finding new contributors and maintaining dynamicity (Faraj et al. 2011). Continuous changing membership, ambiguous identities, rotated leadership, and various technology affordances among other things can maintain necessary dynamics in innovative activities (Faraj et al. 2011). Challenges also relate to building sufficient actionable transparency (Baldwin and von Hippel 2011). Modular architectures and collaborative spaces facilitate such transparency and allow thousands of independent contributors to work independently and in parallel. Others' activities can be transparent without any direct communication taking place between the contributors.

68.4.7 Living Labs

Living labs are hybrid communities of lead user and open community innovation that have become popular in Europe and are particularly encouraged by public sector funders (Almirall and Wareham 2011). Living labs involve an intermediary that brings together the users to engage in exploration (i.e., research), as well as others that are part of the technology implementation, to test out various exploitation options. Almirall and Wareham (2011) analyzed six living labs as part of the European Network of Living Labs. These cases involved media and health, as well as an industrial consortium in the automobile sector. Living labs were found to be valuable in arbitrating incremental innovation by bringing together disparate actors, creating user demand, and testing different business models. The challenges focused on connecting all the dots and coordinating the activities of heterogeneous actors in the innovation value chain.

68.4.8 Out-Licensing

An approach that encourages other companies' use of the firm's innovations includes out-licensing. Out-licensing generates new sources of revenue. Out-licensing can also promote learning of customer and other firms' environments as licensing agreements do not just require legal transactions but also tacit knowledge exchange. Large companies such as IBM and Texas Instruments have had active out-licensing programs since the 1980s. Small companies also have adopted licensing as a key strategy. For example, Rovio has licensed its trademark Angry Birds to over 15,000 different products around the world. But for many companies, out-licensing strategies and activities are still in their infancy (Lichtenthaler 2011). Successful out-licensing programs require a strong involvement from senior management, a dedicated structure to house specialized knowledge and skills (e.g., licensing office, dedicated teams), as well as support and participation of the firm's employees (Lichtenthaler and Ernst 2007; Bianchi et al. 2011a,b). Firms often rely on outside intermediary services (e.g., law firms); however, they are best considered as complementing not substituting internal capabilities (Licthtenthaler and Ernst 2007).

68.5 Theoretical Perspectives to Open Innovation

Open innovation is fundamentally about the boundaries inside and outside the firm, management of knowledge and learning, social relationships, power, identity, and legitimacy. The existing research follows theoretical perspectives commonly used to study inter-organizational relationships (see Parmigiani and Rivera-Santos 2011).

The most common theoretical perspectives relates to the boundaries of the firm from an organizational economics perspective. Open innovation can reduce production costs, bring benefits in the form of complementary assets, and achieve greater incentive alignment through modular platform architectures. Organizational economics perspectives include theories of transaction cost economics (TCE), agency theory, network externalities, and the resource-based view (RBV).

TCE is used to understand when open innovation is a more efficient alternative than governing activities and transactions internally. According to Williamson (1991), the optimal organization of transactions should minimize the production, communication, and opportunity costs, and strive for scale and scope economies. Transactions should be handled internally when probability of opportunistic behaviors is high and coordination costs are substantial.

TCE is commonly used to frame theoretical arguments regarding why certain open innovation forms or approaches are becoming more common or viable (e.g., Cassiman and Valentini 2009; Lazzarotti and Manzini 2009; Baldwin and von Hippel 2011). Baldwin and von Hippel (2011) and Demil and Lecocq (2006) develop governance models for open community innovation (Baldwin and von Hippel 2011). Demil and Lecocq (2006) consider the open community model based on open license to be superior when there is extreme uncertainty, weak incentives, and low levels of control. Baldwin and von Hippel (2011) argue that the open community model is viable in areas where rivalry is low as there are no costs to maintaining secrecy.

Agency theory is defined as aligning incentives across different actors when ownership and control are separated (Eisenhardt 1989). When organizations employ agents to represent them, there are costs from differences in the goals of the organization and agents. Transparent architectures are theorized to reduce agency costs (Baldwin and von Hippel 2011). However, the existing literature on open innovation suggests that agency issues are difficult, unresolved, and poorly understood (e.g., West and O'Mahony 2008). There is also an attempt to develop new notions of agency (Ulhøi 2005).

Open innovation research regarding TCE and agency theory is conceptual or relies on small samples. Compelling empirical evidence proving that open innovation leads to effective governance models is limited or the results are mixed at best. Niederman et al. (2006) note that there is little evidence that open source is a superior form of governance from the efficiency perspectives. Baldwin and von Hippel (2011, p. 1404) limit their developments to "bounds on the viability." Some find that open innovation happens despite very high transaction costs (Christensen et al. 2005). Others (Remneland-Wikhamn and Knights 2012) question the use of transaction costs in shedding insight to open innovation. They argue that TCE has a limited descriptive power and potentially does normative damage to open innovation as the perspective does not consider the broader policy and cultural issues. TCE research in open innovation also suffers from other weaknesses that are commonly associated in applying the theory to information technology–enabled forms. Information technology is rapidly reducing costs in all governance forms including vertically integrated forms. Moreover, information technology is increasing variety in all forms and so variations within forms can be as great as across forms.

68.5.1 Resource-Based View

RBV theorizes that firms are a bundle of resources and capabilities (Barney 1991). Whereas TCE assumes that resources and capabilities are homogeneously distributed across firms, RBV assumes heterogeneity and resource mobility across firms. Firms can develop firm-specific valuable resources, capabilities, competences, and dynamic capabilities, which can be leveraged for competitive advantages and favorably impact firm profitability.

Open innovation provides a way to source resources possessed by the partners, as well as acquire new resources via outsourcing. This, along with internal development, combines to produce uniquely configured capabilities that are valued in the marketplace. Open innovation requires that capabilities coevolve with the speed of market opportunities. The RBV helps explain why open innovation can boost growth via alliances and in-licensing of technology without acquiring the whole business. RBV also helps illuminate risks of open innovation. Open innovation presents risks of unwanted mobility of these capabilities or potential leakage of valuable knowledge to rivals (e.g., open alliances) (e.g., Han et al. 2012).

Case studies highlight the major challenges of employee training and a deep organizational change to deploy open innovation successfully. Chiaroni et al. (2010) studied the journeys of four different companies as they created an infrastructure for open innovation. The companies went through a deep organizational change. The companies developed inter-organizational networks (relationships with universities and outside research centers) and established new roles and skills (liaisons and gatekeepers). New organizational structures (e.g., IP office) were created and new evaluation processes, ICT systems (video conference), and knowledge and project management systems (develop patent filing systems) were implemented. To recap, open innovation presents the ability to tap to critical resources and capabilities outside, but the firm has to make major investments to develop the internal capabilities before the new combinations render imperfectly tradable resources and capabilities.

68.5.2 Absorptive Capacity

RBV suffers from focus on possession rather than the ability to deploy these capabilities. Deployment is the focus of absorptive capacity theory—a behavioral theory of the firm (March and Simon 1958). Absorptive capacity refers to the "ability of a firm to recognize the value of new, external information, assimilate it, and apply it to commercial ends" (Cohen and Levinthal 1990, p. 128). Absorptive capacity depends on the existing stockpiles of knowledge and networks of contacts. The stockpile is a product of past and present managerial choices.

Christensen et al. (2005) found that to successfully leverage open innovation, firms have to have "the abilities or luck to exploit more or less coincidental opportunities emerging outside the boundaries of the firm" (p. 1547). Lichtenthaler and Lichtenthaler (2009) developed a capability-based framework for open innovation. They recognized different knowledge management processes are needed to identify, select, retain, transform, and exploit external and internal knowledge resources for innovation. One significant contribution of their work is the concept of "desorptive" capacity, opposite of absorptive capacity. Desorptive capacity refers to "identifying external knowledge exploitation opportunities and subsequently transferring the knowledge to the recipient" (Lichtenthaler and Lichtenthaler 2009, p. 1322). Desorptive capacity allows the firm to engage in external exploitation. Without desorptive capacity, the firm licensing the innovation will not transfer the necessary tacit knowledge for the licensee to exploit the innovation. The lack of exploitation can lead to reduced licensing revenues and reduced demands for the innovation by other firms. Robertson et al. (2012) developed another capability-based model for process industries that emphasizes deployment of external and internal resources and capabilities. They identified three capacities: accessive capacity, adaptive capacity, and integrative capacity.

Just as with TCE and agency theories, RBV of the firm and absorptive capacities are mainly used to develop conceptual frames that explain why firms succeed in open innovation. Failure stories are sorely lacking—either as standalone or as contrasted with successful cases. Large-scale empirical examinations are hard to find. Dynamic perspectives are still rarely examined although with some exceptions (Han et al. 2012).

68.5.3 Network Externalities

The network externalities theory posits that a product or service becomes more valuable as its user base expands (Katz and Shapiro 1986). This implies that innovations that are associated with large networks will tend to attract more users due to a positive feedback effect (Katz and Shapiro 1994). Adding variety to the platform attracts more users to the platform. More users in turn attract more complementors such as application providers. However, research such as online communities and peer-to-peer file sharing networks have shown positive and negative effects of increased network size due to the additional resources and to increased free riding, congestion, and cognitive load (Asvanund et al. 2004; Butler 2001), suggesting that larger networks may not always provide more superior value.

The network externalities theory (Shapiro and Varian 1999) has been used to study how the owner of the technology should open up the development and commercialization of technology. Opening creates momentum and can attract larger user bases but the opening can also leave the original owner with little control and ability to appropriate value. Using a measure of innovation as the rate at which new handheld devices were introduced, Boudreau (2010) found that granting access to independent complementors to a platform was preferred over giving up formal control over the platform itself. Diversity appeared to be more important than incremental shifts in formal control.

The network externalities theory has been used to study how many complementors (e.g., software solution providers) and contestants should be invited to participate in the open innovation platform. Boudreau (2012) examined a leading handheld computer platform as a two-sided market and studied the number of application producers and the selection of software solutions. Consistent with the network externalities theory, he found that adding producers using different software systems tended to increase innovation incentives, consistent with network effects. However, adding producers created crowding of similar applications that in turn negatively impacted novelty. In another study involving a crowdsourcing tournament, Boudreau et al. (2012) found that as competitors to the contest increased, greater rivalry reduced the incentives of all competitors to exert effort and make investments.

68.5.4 Social Capital

Social capital theories take into consideration embeddedness (Granovetter 1993), connecting with other individuals and organizations, and gaining access to greater and more diverse sources of knowledge. An important question is what type of social capital, or embeddedness, promotes innovation. Relational embeddedness relates to strong ties, repeated interaction, and solidarity with the same partners (Coleman 1990). Relational embeddedness facilitates innovation through the transfer of tacit knowledge but suffers from not-invented-here because of blindness to group norms (Katz and Allen 1982). On the other hand, structural embeddedness, or the presence of structural holes and nonredundant ties, promotes access to a wide range of information sources (Burt 1992). Nonredundant ties increase unique information, the ability to be more informed of opportunities, and have more options for creating unique combinations that result in innovation. Rost (2011) found that both structural and relational embeddedness contributed to the creation of innovation. Individuals who were part of strong social circles but with ties to broader networks came up with more innovations. Relational and structural embeddedness were complements rather than substitutes.

Others have found that emerging leaders in open source initiatives exhibit both structural and relational embeddedness. Fleming and Waguespack (2007) found that emerging leaders were brokers (connected and disconnected actors), boundary spanners (served as technical leaders, knowledge links, respected guardians), and engaged in bonding (enabled community building and community benefit). du Chatenier et al. (2010) found it important for professionals working in open innovation teams to have similar skill sets.

68.5.5 Organizational Paradox Perspective

Researchers widely acknowledge that open innovation paradigm represents a complex balancing act of paradoxes, tensions, competing demands, conflicts, contradictions, and dilemmas (e.g., West and Gallagher 2006; Jarvenpaa and Wernick 2011). There are many tensions in open innovation including balancing how internal innovation units compete for resources with external sources (Chesbrough and Garman 2009). There is the tension between losing control over proprietary knowledge and technology yet the importance of sharing for the benefits of collectivity (West and Gallagher 2006). Almirall and Casadesus-Masanell (2010) examine the trade-off of open and closed approaches to innovation: They find that when complexity is not too high, open innovation is preferred to closed innovation. Yet, the literature has rarely gone beyond recognizing the tension that exists regarding open and closed innovation and how to transcend it in a mutually reinforcing way.

The organizational paradox perspective (Smith and Lewis 2011) is a promising theoretical perspective that has yet to be fully exploited in open innovation research. It is based on dialectics and duality (Farjoun 2010). Paradoxical perspective differs from both TCE and RBV. In simple terms, where TCE focuses to minimize distrust to reduce coordination and opportunity costs and RBV focuses to maximize trust as an organizational capability, the paradoxical perspective argues that value arises from simultaneously maintaining both trust and distrust. This is because trust alone conceals complex mixed motive inter-organizational relationships that require vigilance. Distrust without trust leads to withdrawal from collaboration (Jarvenpaa and Majchrzak 2010).

The paradox perspective treats what is traditionally considered as trade-offs, as something that has to be managed in an integrative fashion (Jelinek and Schoonhoven 1990; Dougherty 2006; Bledow et al. 2009). For example, exploration versus exploitation and radical versus incremental are often viewed as alternatives rather than as something that is embraced simultaneously. The organizational paradox perspective argues more effective outcomes can be achieved if the seemingly contradictory principles or practices are juxtaposed so they remain independent but mutually enabling (Lewis 2000; Smith and Lewis 2011). Success is increased if the firm simultaneously attains exploration and exploitation rather than one over the other or in alternating sequence.

Several theoretical perspectives were reviewed that are present in the open innovation literature. The organizational paradox perspective was presented as a new promising theoretical basis.

68.6 Research Questions

Throughout this chapter, gaps in the current literature are identified. The gaps are also well highlighted in the existing literature. For example, West and Gallagher (2006) identified fundamental challenges in open innovation: (1) What internal resources and capabilities should be shared with the networks? (2) What is the best way to search, discover, and integrate external sources and innovations internally and then best exploit these new innovations in various external markets? and (3) How to motivate the various parties in the network, internal and external, to share their resources and capabilities when the outcomes of these efforts may be available to rivals? How do firms make choices in accessing external knowledge and protecting their own knowledge? Enkel et al. (2009, p. 312) summarized the situation: "we still lack a clear understanding of the mechanisms, inside and outside of the organization, when and who to fully profit from the concept."

Traditionally, there has been a need to gain a better understanding of the costs (transaction costs and production costs) of open innovation and how information technology affects costs. What are the theoretical underpinnings of different open innovation forms and approaches? Following Baldwin and von Hippel (2011), what are the conditions under which these models become viable? We conclude with few questions.

68.6.1 What Are the Bounds of Effective Open Innovation?

Open innovation has been heralded as decreasing the risk of missing market opportunity, not reinventing the wheel, increasing learning, lowering costs, turning customers and users into producers, building broader and deeper networks, and engaging community building. Open innovation is not inherently superior to traditional innovation and is associated with many hazards. However, the boundaries of effective open innovation are still uncertain. Therefore, more research is called for to explore the new frontier of open innovation thereby enabling forward looking firms to put the research into practice.

The viability of open innovation can be examined from the view of transaction efficiency, capability development, social tie formation, and others. Integration of multiple theoretical perspectives should be considered to continue to gain insight into open innovation.

68.6.2 How to Design an Organization for Open Innovation?

A relevant question to ask is: What are the organizational design principles for open innovation? Acha (2008) makes the compelling case that "design shapes open innovation practice." Design can also shape the viable forms and approaches. A key concept in design is task partitioning that reduces the cognitive burden of complexity and allows work to be conducted in parallel (von Hippel 1990). Design is also important for horizontal and self-managing control (O'Mahony and Bechky 2008; Dahlander and O'Mahony 2011). However, design has been given little attention apart from the platform-based studies (Gawer 2009). Design science approaches and methods can and should be leveraged to develop and compare alternative designs (Hevner et al. 2004).

68.6.3 Is Open Innovation about Incremental or Radical Innovation?

A pertinent issue in the open innovation literature is whether the open paradigm promotes radical innovation or continuous improvement. To Chesbrough (2006), a move to open innovation is seen as a way to create discontinuity with the past as new business models create fundamental change. However to others, open innovation is about business experimentation and replication (Davenport 2009; Brynjolsson and Saunders 2010). Experimental approaches reduce the length of the development cycle, making changes continuous and more easily adopted and diffused. The continuous experimental innovation process leverages the capture, measurement, analysis, sharing, and replication capabilities of information technology (Brynjolfsson and Saunders 2010). Davenport (2009) coins it as a "Test and Learn Wheel." Amazon and Google run thousands of experiments daily. Many software companies are using open developer communities to modify, test, and release software in fast cycles. While these innovation outcomes are small, the innovation process itself is revolutionized by information technology (Brynjolfsson and Saunders 2010).

68.6.4 What Is the Role of Information and Communication Technology?

Technological changes have been key drivers as well as enablers of open innovation forms and approaches. ICT has provided ubiquitous global infrastructures that enable the firm to tap into distributed sources of information wherever it exists. Dell used a crowdsourcing platform, Salesforce CRM for its IdeaStorm (di Gangi et al. 2010). Brynjolfsson and Saunders (2010) identify a number of ways that

ICT facilitates innovation. ICT increases productivity and efficiency of transactions. It spurs innovation including new management techniques, business models, work processes, and human resource practices. ICT is changing the way innovation is performed making possible real-time and cost-effective business experimentation and replication where observations are shared and ideas are gathered widely. Dodgson et al. (2006) examined the technological development underlying Procter & Gamble's connect and develop strategy. In this case, ICT was particularly important in enabling data searching, mining, simulation and modeling, and virtual and rapid prototyping. Yet, there is again of paucity of research examining how IT affects failure and success of open innovation, different degrees of openness, and how it affects radical versus incremental innovation in a wide range of contexts. In most studies on open innovation, ICT has taken a nominal role (Orlikowski and Iacono 2001, p. 121): "The IT artifact itself tends to disappear from view, be taken for granted, or presumed to be unproblematic."

68.7 Concluding Note

This chapter covered the emerging open paradigm in organizational innovation management. Several key concepts from innovation management were introduced that are particularly relevant in terms of open innovation. Multiple views regarding open innovation were presented along with emerging notions of openness. Some forms and approaches of open innovation were reviewed. Various theoretical perspectives were briefly highlighted. This chapter concluded with research questions.

References

Acha, V. 2008. Open by design: The role of design in open innovation. *Academy of Management Annual Meeting Proceedings*, Anaheim, CA, pp. 1–6.

Afuah, A. and Tucci, C. L. 2012. Crowdsourcing as a solution to distant search. *Academy of Management Review*, 37(3): 355–375.

Albors, J., Ramos, J. C., and Hervas, J. L. 2008. New learning network paradigms: Communities of objectives, crowdsourcing, wikis and open source. *International Journal of Information Management*, 28(3): 194–202.

Alexy, O., Criscuolo, P., and Ammon, S. 2012. Managing unsolicited ideas for R&D. *California Management Review*, 54(3): 116–139.

Alexy, O., Criscuolo, P., and Salter, A. 2009. Does IP strategy have to cripple open innovation? *MIT Sloan Management Review*, 51: 71–77.

Allen, R. C. 1983. Collective invention. *Journal of Economic Behavior and Organization*, 4(1): 1–24.

Almirall, E. and Casadesus-Masanell, R. 2010. Open versus closed innovation: A model of discovery and divergence. *Academy of Management Review*, 35(1): 27–47.

Almirall, E. and Wareham, J. 2011. Living labs: Arbiters of mid- and ground-level innovation. *Technology Analysis and Strategic Management*, 23(1): 87–102.

Antikainen, M., Makipaa, M., and Ahonen, M. 2010. Motivating and supporting collaboration in open innovation. *European Journal of Innovation Management*, 13(1): 100–119.

Asakawa, K., Nakamura, H., and Sawada, N. 2010. Firms' open innovation policies, laboratories' external collaborations, and laboratories' R&D performance. *R&D Management*, 40(2): 109–123.

Asvanund, A., Clay, K., Krishnan, R., and Smith, M. D. 2004. An empirical analysis of network externalities in peer-to-peer music-sharing networks. *Information Systems Research*, 15(2): 155–174.

Baldwin, C. Y. and Clark, K. B. 2006. The architecture of participation: Does code architecture mitigate free riding in the open source development model. *Management Science*, 52(7): 1116–1127.

Baldwin, C. and von Hippel, E. 2011. Modeling a paradigm shift: From producer innovation to user and open collaborative innovation. *Organization Science*, 22(6): 1399–1417.

Baldwin, C. Y. and Woodard, C. J. 2009. The architecture of platforms: A unified view. In A. Gawer (Ed.), *Platforms, Markets, and Innovation*. Cheltenham, U.K.: Edward Elgar.

Barney, J. 1991. Firm resources and sustained competitive advantage. *Journal of Management*, 17(1): 99.

Barrell, R., Mason, G., and O'Mahony, M. 2000. *Productivity, Innovation, and Economic Performance.* Cambridge, U.K.: Cambridge University Press.

Benkler, Y. 2006. *The Wealth of Networks: How Social Production Transforms Markets and Freedom.* New Haven, CT: Yale University Press.

Bianchi, M., Cavaliere, A., Chiaroni, D., Frattini, F., and Chiesa, V. 2011a. Organisational modes for open innovation in the bio-pharmaceutical industry: An exploratory analysis. *Technovation*, 31(1): 22–33.

Bianchi, M., Chiaroni, D., Chiesa, V., and Frattini, F. 2011b. Organizing for external technology commercialization: Evidence from a multiple case study in the pharmaceutical industry. *R&D Management*, 41(2): 120–137.

Birkinshaw, J., Bessant, J., and Delbridge, R. 2007. Finding, forming, and performing: Creating networks for discontinuous innovation. *California Management Review*, 49(3): 67–84.

Bledow, R., Frese, M., Anderson, N., Erez, M., and Farr, J. 2009. A dialectic perspective on innovation: Conflicting demands, multiple pathways, and ambidexterity. *Industrial and Organizational Psychology*, 2(3): 305–337.

Bogers, M., Afuah, A., and Bastian, B. 2010. Users as innovators: A review, critique, and future research directions. *Journal of Management*, 36(4): 857–875.

Boudreau, K. 2010. Open platform strategies and innovation: Granting access vs. devolving control. *Management Science*, 56(10): 1849–1872.

Boudreau, K. J. 2012. Let a thousand flowers bloom? An early look at large numbers of software App developers and patterns of innovation. *Organization Science*, 23(5): 1409–1427.

Boudreau, K. J., Helfat, C. E., Lakhani, K. R., and Menietti, M. 2012. Field evidence on individual behavior & performance in rank-order tournaments. Harvard Business School Technology & Operations Management Unit Working Paper No. 13-016.

Brabham, D. C. 2010. Moving the crowd at Threadless. *Information, Communication & Society*, 13(8): 1122–1145.

Brynjolfsson, E. and Saunders, A. 2010. *Wired for Innovation: How Information Technology is Reshaping the Economy.* Cambridge, MA: MIT Press.

Bunker-Whittington, K., Owen-Smith, J., and Powell, W. 2009. Networks, propinquity, and innovation in knowledge-intensive industries. *Administrative Science Quarterly*, 54(1): 90–122.

Burgelman, R. A., Christensen, C. M., and Wheelwright, S. C. 2004. *Strategic Management of Technology and Innovation*, 4th edn. Boston, MA: McGraw-Hill Irwin.

Burt, R. S. 1992. *Structural Holes: The Social Structure of Competition.* Cambridge, MA: Harvard University Press.

Burtch, G., Ghose, A., and Wattal, S. 2012. An empirical examination of the antecedents and consequences of contribution patterns in crowd-funded markets. Available at SSRN: http://ssrn.com/abstract=1928168.

Butler, B. S. 2001. Membership size, communication activity, and sustainability: A resource-based model of online social structures. *Information Systems Research*, 12(4): 346–362.

Carlo, J. L., Lyytinen, K., and Rose, G. M. 2012. A knowledge-based model of radical innovation in small software firms. *MIS Quarterly*, 36(3): A865–A810.

Cassier, M. and Foray, D. 2002. Public knowledge, private property and the economics of high-tech consortia. *Economics of Innovation and New Technology*, 11(2): 123–132.

Cassiman, B. and Valentini, G. 2009. Strategic organization of R&D: The choice of basicness and openness. *Strategic Organization*, 7(1): 43–73.

Charitou, C. D. and Markides, C. C. 2003. Responses to disruptive strategic innovation. *MIT Sloan Management Review*, 44(2): 55–63.

du Chatenier, E., Verstegen, J. A. A. M., Biemans, H. J. A., Mulder, M., and Omta, O. S. W. F. 2010. Identification of competencies for professionals in open innovation teams. *R&D Management*, 40(3): 271–280.

Chesbrough, H. 2010. Business model innovation: Opportunities and barriers. *Long Range Planning*, 43(2–3): 354–363.

Chesbrough, H. 2012. Open innovation: Where we've been and where we're going: The father of open innovation offers his assessment of the history and future of the model. *Research Technology Management*, 55(4): 20–27.

Chesbrough, H. and Crowther, A. K. 2006. Beyond high tech: Early adopters of open innovation in other industries. *R&D Management*, 36(3): 229–236.

Chesbrough, H. W. 2006. *Open Business Models: How to Thrive in the New Innovation Landscape*. Boston, MA: Harvard Business School Press.

Chesbrough, H. W. 2003. *Open Innovation: The New Imperative for Creating and Profiting from Technology*. Boston, MA: Harvard Business School Press.

Chesbrough, H. W. and Garman, A. R. 2009. How open innovation can help you cope in lean times. *Harvard Business Review*, 87(12): 68–76.

Chesbrough, H. W. and Teece, D. J. 1996. Organizing for innovation. *Harvard Business Review*, 74(1): 65–73.

Chiaroni, D., Chiesa, V., and Frattini, F. 2010. Unravelling the process from closed to open innovation: Evidence from mature, asset-intensive industries. *R&D Management*, 40(3): 222–245.

Christensen, C. M. 1997. *The Innovator's Dilemma: When New Technologies Cause Great Firms to Fail*. Boston, MA: Harvard Business School Press.

Christensen, J. F., Olesen, M. H., and Kjaer, J. S. 2005. The industrial dynamics of open innovation—Evidence from the transformation of consumer electronics. *Research Policy*, 34(10): 1533–1549.

Cohen, W. M. and Levinthal, D. A. 1990. Absorptive capacity: A new perspective on learning and innovation. *Administrative Science Quarterly*, 35(1): 128–152.

Coleman, J. S. 1990. *Foundations of Social Theory*. Cambridge, MA: Belknap Press of Harvard University Press.

Cooper, R. G. 1988. The new product process: A decision guide for management. *Journal of Marketing Management*, 3(3): 238–255.

Dahlander, L. and Gann, D. M. 2010. How open is innovation? *Research Policy*, 39(6): 699–709.

Dahlander, L. and O'Mahony, S. 2011. Progressing to the center: Coordinating project work. *Organization Science*, 22(4): 961–979.

Dahlin, K. B. and Behrens, D. M. 2005. When is an invention really radical? Defining and measuring technological radicalness. *Research Policy*, 34(5): 717–737.

Damanpour, F. 1988. Innovation type, radicalness, and the adoption process. *Communication Research*, 15(5): 545–567.

Davenport, T. 2009. How to design smart business experiments. *Harvard Business Review*, 87(2): 68–76.

Demil, B. and Lecocq, X. 2006. Neither market nor hierarchy nor network: The emergence of bazaar governance. *Organization Studies*, 27(10): 1447–1466.

Dewar, R. D. and Dutton, J. E. 1986. The adoption of radical and incremental innovations: An empirical analysis. *Management Science*, 32(11): 1422–1433.

Dodgson, M., Gann, D., and Salter, A. 2006. The role of technology in the shift towards open innovation: The case of Procter & Gamble. *R&D Management*, 36(3): 333–346.

Dougherty, D. 2006. Organizing for innovation in the 21st century. In S. Clegg, C. Hardy, and W. Nord (Eds.), *Handbook of Organization Studies*, 2nd edn., pp. 598–617. London, U.K.: Sage Publications.

Drechsler, W. and Natter, M. 2012. Understanding a firm's openness decisions in innovation. *Journal of Business Research*, 65(3): 438–445.

Eisenhardt, K. M. 1989. Agency theory: An assessment and review. *The Academy of Management Review*, 14(1): 57–74.

Enkel, E., Gassmann, O., and Chesbrough, H. 2009. Open R&D and open innovation: Exploring the phenomenon. *R&D Management*, 39(4): 311–316.

Fagerberg, J., Mowery, D. C., and Nelson, R. R. 2005. *The Oxford Handbook of Innovation*. Oxford, NY: Oxford University Press.

Faraj, S., Jarvenpaa, S. L., and Majchrzak, A. 2011. Knowledge collaboration in online communities. *Organization Science*, 22(5): 1224–1239.

Farjoun, M. 2010. Beyond dualism: Stability and change as a duality. *Academy of Management Review*, 35(2): 202–225.

Fey, C. F. and Birkinshaw, J. 2005. External sources of knowledge, governance mode, and R&D performance. *Journal of Management*, 31: 597–621.

Fleming, L. and Waguespack, D. M. 2007. Brokerage, boundary spanning, and leadership in open innovation communities. *Organization Science*, 18(2): 165–180.

Franke, N. and von Hippel, E. 2003. Satisfying heterogeneous user needs via innovation toolkits: The case of apache security software. *Research Policy*, 32(7): 1199–1215.

Franke, N. and Shah, S. 2003. How communities support innovative activities: An exploration of assistance and sharing among end-users. *Research Policy*, 32(1): 157–178.

Füller, J. 2010. Refining virtual co-creation from a consumer perspective. *California Management Review*, 52(2): 98–122.

di Gangi, P. M., Wasko, M. M., and Hooker, R. E. 2010. Getting customers' ideas to work for you: Learning from Dell how to succeed with online user innovation communities. *MIS Quarterly Executive*, 9(4): 213–228.

Gassmann, O. and Enkel, E. 2004. Towards a theory of open innovation: Three core process archetypes. *Proceedings of the R&D Management Conference*, Lisbon, Portugal. Available at https://www.alexandria.unisg.ch/Publikationen/274 (accessed on April 22, 2013).

Gawer, A. 2009. *Platforms, Markets, and Innovation*. Cheltenham, U.K.: Edward Elgar.

Gibson, C. B. and Birkinshaw, J. 2004. The antecedents, consequences, and mediating role of organizational ambidexterity. *Academy of Management Journal*, 47(2): 209–226.

Gibson, D. V. and Rogers, E. M. 1994. *R&D Collaboration on Trial: The Microelectronics and Computer Technology Corporation*. Boston, MA: Harvard Business School Press.

Gomes, C. M. and Kruglianskas, I. 2009. Management of external sources of technological information and innovation performance. *International Journal of Innovation and Technology Management*, 6(2): 207–226.

Granovetter, M. 1993. The nature of economic relationships. In R. Swedberg (Ed.), *Explorations in Economic Sociology*. New York: Russel Sage Foundation.

Granstrand, O., Patel, P., and Pavitt, K. 1997. Multi-technology corporations: Why they have "distributed" rather than "distinctive core" competencies. *California Management Review*, 39(4): 8–25.

Gronlund, J., Sjodin, D. R., and Frishammar, J. 2010. Open innovation and the stage-gate process: A revised model for new product development. *California Management Review*, 52(3): 106–131.

Gu, B. and Jarvenpaa, S. L. 2003. Online discussion boards for technical support: The effect of token recognition on customer contributions. *The Proceedings of the 24th International Conference on Information Systems (ICIS)*, Seattle, WA, pp. 110–120.

Gupta, A. K., Tesluk, P. E., and Taylor, M. S. 2007. Innovation at and across multiple levels of analysis. *Organization Science*, 18(6): 885–897.

Han, K., Oh, W., Im, K. S., Chang, R. M., Oh, H., and Pinsonneault, A. 2012. Value cocreation and wealth spillover in open innovation alliances. *MIS Quarterly*, 36(1): 291–316.

Hansen, M. T. and Birkinshaw, J. 2007. The innovation value chain. *Harvard Business Review*, 85(6): 121–130.

Henderson, R. M. and Clark, K. B. 1990. Architectural innovation: The reconfiguration of existing product technologies and the failure of established firms. *Administrative Science Quarterly*, 35(1): 9–30.

Henkel, J. 2006. Selective revealing in open innovation processes: The case of embedded linux. *Research Policy*, 35(7): 953–969.

Henkel, J., Baldwin, C. Y., and Shin, W. 2012. IP modularity: Profiting from innovation by aligning product architectures with intellectual property. Harvard Business School Working Paper. Available at http://hbswk.hbs.edu/item/7074.html (accessed on April 22, 2013).

Herstatt, C. and von Hippel, E. 1992. From experience: Developing new product concepts via the lead user method: A case study in a "low-tech" field. *Journal of Product Innovation Management*, 9(3): 213–221.

Hevner, A. R., March, S. T., Park, J., and Ram, S. 2004. Design science in information systems research. *MIS Quarterly*, 28(1): 75–105.

von Hippel, E. A. 1977. Has a customer already developed your next product? *Sloan Management Review*, 18(2): 63–74.

von Hippel, E. 1986. Lead users: A source of novel product concepts. *Management Science*, 32(7): 791–805.

von Hippel, E. 1990. Task partitioning: An innovation process variable. *Research Policy*, 19(5): 407–418.

von Hippel, E. 2005. *Democratizing Innovation*. Cambridge, MA: MIT Press.

von Hippel, E. 2010. Comment on "Is open innovation a field of study or a communication barrier to theory development?" *Technovation*, 30(11/12): 555.

von Hippel, E. and Katz, R. 2002. Shifting innovation to users via toolkits. *Management Science*, 48: 821–833.

von Hippel, E. and von Krogh, G. 2003. Open source software and the 'private-collective' innovation model: Issues for organization science. *Organization Science*, 14(2): 209–223.

von Hippel, E. and von Krogh, G. 2006. Free revealing and the private-collective model for innovation incentives. *R&D Management*, 36(3): 295–306.

Holmes, S. and Smart, P. 2009. Exploring open innovation practice in firm-nonprofit engagements: A corporate social responsibility perspective. *R&D Management*, 39(4): 394–409.

Howe, J. 2006. The rise of crowdsourcing, *Wired Magazine*, Vol. 14.06.

Huizingh, E. K. R. E. 2011. Open innovation: State of the art and future perspectives. *Technovation*, 31(1): 2–9.

Huston, L. and Sakkab, N. 2006. Connect and develop. *Harvard Business Review*, 84(3): 58–66.

Jaegersberg, G. and Ure, J. 2011. Barriers to knowledge sharing and stakeholder alignment in solar energy clusters: Learning from other sectors and regions. *Journal of Strategic Information Systems*, 20(4): 343–354.

Jarvenpaa, S. L. and Lang, K. R. 2011. Boundary management in online communities: Case studies of the Nine Inch Nails and ccMixter Music Remix Sites. *Long Range Planning*, 44(5–6): 440–457.

Jarvenpaa, S. L. and Majchrzak, A. 2010. Vigilant interaction in knowledge collaboration: Challenges of Online user participation under ambivalence. *Information Systems Research*, 21(4): 773–784.

Jarvenpaa, S. L. and Tuunainen, V. K. 2012. Company tactics for customer socialization with social media technologies: Finnair's rethink quality and quality hunters initiatives. *The 45th Hawaii International Conference on System Sciences*, Maui, HI, pp. 713–722.

Jarvenpaa, S. L. and Wernick, A. 2011. Paradoxical tensions in open innovation networks. *European Journal of Innovation Management*, 14(4): 521–548.

Jelinek, M. and Schoonhoven, C. B. 1990. *The Innovation Marathon: Lessons from High Technology Firms*. Cambridge, MA: Blackwell.

Jeppesen, L. B. and Frederiksen, L. 2006. Why do users contribute to firm-hosted user communities? The case of computer-controlled music instruments. *Organization Science*, 17(1): 45–63.

Jeppesen, L. B. and Lakhani, K. R. 2010. Marginality and problem-solving effectiveness in broadcast search. *Organization Science*, 21(5): 1016–1033.

de Jong, J. P. J., Vanhaverbeke, W., Kalvet, T., and Chesbrough, H. 2008. *Policies for Open Innovation: Theory, Framework and Cases*. Research project funded by VISION Era-Net, Helsinki: Finland.

Katila, R. and Ahuja, G. 2002. Something old, something new: A longitudinal study of search behavior and new product introduction. *Academy of Management Journal*, 45(6): 1183–1194.

Katz, R. and Allen, T. J. 1982. Investigating the not invented here (nih) syndrome: A look at the performance, tenure, and communication patterns of 50 R&D Project Groups. *R&D Management*, 12(1): 7–20.

Katz, M. L. and Shapiro, C. 1986. Technology adoption in the presence of network externalities. *Journal of Political Economy*, 94(4): 822–841.

Katz, M. L. and Shapiro, C. 1994. Systems competition and network effects. *Journal of Economic Perspectives*, 8(2): 93–115.

Kogut, B. and Zander, U. 1992. Knowledge of the firm, combinative capabilities, and the replication of technology. *Organization Science*, 3(3): 383–397.

Kuhn, T. S. 1962. *The Structure of Scientific Revolutions*. Chicago, IL: University of Chicago Press.

Lakhani, K. R., Lifshitz-Assaf, H., and Tushman, M. L. 2012. Open innovation and organizational boundaries: The impact of task decomposition and knowledge distribution on the locus of innovation. HBS Working Paper No. 12-057.

Lamoreaux, N. R. and Sokoloff, K. L. 1998. Inventors, firms, and the market for technology: U.S. manufacturing in the late nineteenth and early twentieth centuries. In N. R. Lamoreaux, D. Raff, and P. Temins (Eds.), *Learning by Firms, Organizations, and Nations*: pp. 19–57. Chicago, IL: The University of Chicago Press.

Laursen, K. and Salter, A. 2006. Open for innovation: The role of openness in explaining innovation performance among U.K. manufacturing firms. *Strategic Management Journal*, 27(2): 131–150.

Lavie, D., Stettner, U., and Tushman, M. L. 2010. Exploration and exploitation within and across organizations. *The Academy of Management Annals*, 4(1): 109–155.

Lazzarotti, V. and Manzini, R. 2009. Different modes of open innovation: A theoretical framework and an empirical study. *International Journal of Innovation Management*, 13(4): 615–636.

Lee, S., Park, G., Yoon, B., and Park, J. 2010. Open innovation in SMEs—An intermediated network model. *Research Policy*, 39(2): 290–300.

Leimeister, J. M., Huber, M., Bretschneider, U., and Krcmar, H. 2009. Leveraging crowdsourcing: Activation-supporting components for IT-based ideas competition. *Journal of Management Information Systems*, 26(1): 197–224.

Leonard-Barton, D. and Swap, W. C. 1999. *When Sparks Fly: Igniting Creativity in Groups*. Boston, MA: Harvard Business School Press.

Lewis, M. W. 2000. Exploring paradox: Toward a more comprehensive guide. *The Academy of Management Review*, 25(4): 760–776.

Lichtenthaler, U. 2011. Open innovation: Past research, current debates, and future directions. *The Academy of Management Perspectives*, 25(1): 75–93.

Lichtenthaler, U. and Ernst, H. 2007. External technology commercialization in large firms: Results of a quantitative benchmarking study. *R&D Management*, 37(5): 383–397.

Lichtenthaler, U. and Lichtenthaler, E. 2009. A capability-based framework for open innovation: Complementing absorptive capacity. *Journal of Management Studies*, 46(8): 1315–1338.

Linder, J. C., Jarvenpaa, S., and Davenport, T. H. 2003. Toward an innovation sourcing strategy. *MIT Sloan Management Review*, 44(4): 43–49.

Linstone, H. A. 2010. Comment on 'is open innovation a field of study or a communication barrier to theory development?' *Technovation*, 30: 556–556.

Love, J. H., Roper, S., and Bryson, J. R. 2011. Openness, knowledge, innovation and growth in UK business services. *Research Policy*, 40(10): 1438–1452.

Lüthje, C., Herstatt, C., and von Hippel, E. 2005. User-innovators and "local" information: The case of mountain biking. *Research Policy*, 34(6): 951–965.

March, J. G. 1991. Exploration and exploitation in organizational learning. *Organization Science*, 2(1): 71–87.

March, J. G. and Simon, H. A. 1958. *Organizations*. Oxford, U.K.: Wiley.

Mowery, D. C. 2009. Plus ca change: Industrial R&D in the "Third Industrial Revolution." *Industrial & Corporate Change*, 18(1): 1–50.

Nambisan, S. 2002. Designing virtual customer environments for new product development: Toward a theory. *Academy of Management Review,* 27(3): 392–413.

Nambisan, S. and Baron, R. A. 2010. Different roles, different strokes: Organizing virtual customer environments to promote two types of customer contributions. *Organization Science,* 21(2): 554–572.

Nichols, D. 2007. Why innovation funnels don't work and why rockets do. *Market Leader,* (38): 26–31.

Niederman, F., Davis, A., Greiner, M., Wynn, D., and York, P. 2006. A research agenda for studying open source I: A multi-level framework. *Communications of the Association for Information Systems (CAIS),* 18: 129–149.

Nooteboom, B. 2000. *Learning and Innovation in Organizations and Economies.* Oxford; NY: Oxford University Press.

O'Mahony, S. 2003. Guarding the commons: How community managed software projects protect their work. *Research Policy,* 32(7): 1179.

O'Mahony, S. and Bechky, B. A. 2008. Boundary organizations: Enabling collaboration among unexpected allies. *Administrative Science Quarterly,* 53(3): 422–459.

Ordanini, A., Miceli, L., Pizzetti, M., and Parasuraman, A. 2011. Crowd-funding: Transforming customers into investors through innovative service platforms. *Journal of Service Management,* 22(4): 443–470.

Orlikowski, W. J. and Iacono, C. S. 2001. Desperately seeking the "IT" in IT research—A call to theorizing the IT artifact. *Information Systems Research,* 12(2): 121–134.

Parmigiani, A. and Rivera-Santos, M. 2011. Clearing a path through the forest: A meta-review of interorganizational relationships. *Journal of Management,* 37(4): 1108–1136.

Pavitt, K. 1999. *Technology, Management and Systems of Innovation.* Northampton, MA: Edward Elgar Publishing.

Pisano, G. 2006. Profiting from innovation and the intellectual property revolution. *Research Policy,* 35(8): 1122–1130.

Pisano, G. P. and Teece, D. J. 2007. How to capture value from innovation: Shaping intellectual property and industry architecture. *California Management Review,* 50(1): 278–296.

Powell, W., Koput, K., and Doerr, L. 1996. Interorganizational collaboration and the locus of innovation: Networks of learning in biotechnology. *Administrative Science Quarterly,* 41(1): 116–145.

Remneland-Wikhamn, B. and Knights, D. 2012. Transaction cost economics and open innovation: Implications for theory and practice. *Creativity & Innovation Management,* 21(3): 277–289.

Ritala, P. and Hurmelinna-Laukkanen, P. 2009. What's in it for me? Creating and appropriating value in innovation-related coopetition. *Technovation,* 29(12): 819–828.

Roberts, J. A., Il-Horn, H., and Slaughter, S. A. 2006. Understanding the motivations, participation, and performance of open source software developers: A longitudinal study of the Apache projects. *Management Science,* 52(7): 984–999.

Robertson, P. L., Casali, G. L., and Jacobson, D. 2012. Managing open incremental process innovation: Absorptive capacity and distributed learning. *Research Policy,* 41(5): 822–832.

Rost, K. 2011. The strength of strong ties in the creation of innovation. *Research Policy,* 40(4): 588–604.

Sakkab, N. Y. 2002. Connect & develop complements research & develop at P&G. *Research Technology Management,* 45(2): 38–45.

Schultze, U., Hiltz, R., Nardi, B., Rennecker, J., and Stucky, S. 2008. Synthetic worlds in work and learning. *Communications of AIS,* 22(19): 351–370.

Schumpeter, J. A. 1942. *Capitalism, Socialism, and Democracy.* New York: Harper & Brothers.

Shah, S. 2000. Sources and patterns of innovation in a consumer products field: Innovations in sporting equipment. MIT Sloan School of Management Working Paper No. 4105.

Shapiro, C. and Varian, H. R. 1999. *Information Rules: A Strategic Guide to the Network Economy.* Boston, MA: Harvard Business School Press.

Smith, W. K. and Lewis, M. W. 2011. Toward a theory of paradox: A dynamic equilibrium model of organizing. *Academy of Management Review,* 36(2): 381–403.

Therrien, P., Doloreux, D., and Chamberlin, T. 2011. Innovation novelty and (commercial) performance in the service sector: A Canadian firm-level analysis. *Technovation*, 31(12): 655–665.

Tushman, M. L., Newman, W. H., and Romanelli, E. 1986. Convergence and upheaval: Managing the unsteady pace of organizational evolution. *California Management Review*, 29(1): 22–39.

Tushman, M. and O'Reilly, C. 1996. Ambidextrous organizations: Managing evolutionary and revolutionary change. *California Management Review*, 38(4): 8–30.

Ulhøi, J. P. 2005. The social dimensions of entrepreneurship. *Technovation*, 25(8): 939–946.

Utterback, J. M. 1994. *Mastering the Dynamics of Innovation: How Companies can Seize Opportunities in the Face of Technological Change*. Boston, MA: Harvard Business School Press.

Verganti, R. 2009. *Design-Driven Innovation: Changing the Rules of Competition by Radically Innovating What Things Mean*. Boston, MA: Harvard Business Press.

van de Vrande, V., de Jong, J. P. J., Vanhaverbeke, W., and de Rochemont, M. 2009. Open innovation in SMEs: Trends, motives and management challenges. *Technovation*, 29(6–7): 423–437.

Wasko, M. and Faraj, S. 2000. "It is what one does": Why people participate and help others in electronic communities of practice. *The Journal of Strategic Information Systems*, 9(2–3): 155–173.

West, J. 2003. How open is open enough? Melding proprietary and open source platform strategies. *Research Policy*, 32(7): 1259–1285.

West, J. and Gallagher, S. 2006. Challenges of open innovation: The paradox of firm investment in open-source software. *R&D Management*, 36(3): 319–331.

West, J. and O'Mahony, S. 2008. The role of participation architecture in growing sponsored open source communities. *Industry and Innovation*, 15(2): 145–168.

Williamson, O. E. 1991. Comparative economic organization: The analysis of discrete structural alternatives. *Administrative Science Quarterly*, 36(2): 269–296.

Wyld, D. C. 2010. Speaking up for customers: Can sales professionals spark product innovation? *The Academy of Management Perspectives*, 24(2): 80–82.

Zheng, H., Li, D., and Hou, W. 2011. Task design, motivation, and participation in crowdsourcing contests. *International Journal of Electronic Commerce*, 15(4): 57–88.

Zott, C. and Amit, R. 2008. The fit between product market strategy and business model: Implications for firm performance. *Strategic Management Journal*, 29(1): 1–26.

Inter-Organizational Information Systems

Charles Steinfield
Michigan State University

69.1 Introduction

Nearly all modern organizations rely extensively on computer-based information systems to process the data needed to operate their businesses for such tasks as managing employee data, keeping track of sales and inventory, engaging in product development, forecasting future demand, and maintaining customer information. When such information systems extend beyond the borders of one organization and provide for automated information exchange to support linked business processes between two or more organizations, they are considered to be *inter-organizational information systems* (IOS) (Robey et al. 2008). IOS have become increasingly prevalent in the networked world in which businesses operate today. Often these systems are used to connect suppliers and manufacturers to support *just-in-time* (JIT) inventory practices in *supply chains* based on *electronic document interchange* (EDI), a global standard for structured exchange of business data and documents. In the retail sector, *collaborative planning, forecasting, and replenishment* (CPFR) systems are used to improve coordination up and down the value chain of retailers' supply partners, so that suppliers are better aware of consumer demand and retailers can better manage replenishment practices and inventory costs. Moreover, electronic procurement has become pervasive in business-to-business (B2B) exchanges, promoting widespread use of IOS in nearly all types of industries.

The study of IOS has emerged as a distinct area within the broader field of information systems in large part due to the many challenges organizations face when they attempt to implement and use IOS (Markus 2006; Steinfield et al. 2011). Since IOS by definition includes two or more organizations, decisions to adopt, implement, and use IOS are not independent decisions made by one firm and can be influenced by a wide range of factors such as power differences between companies, the degree to which trading partners trust each other, or participants' long-term strategic goals (Robey et al. 2008). In contrast, the management of an internal information system is usually under the control of one organization, simplifying the processes of adoption, implementation, and use. IOS also face inter-operability challenges, since the internal systems of the participating organizations may be based on incompatible and/or proprietary formats (Markus et al. 2006). This chapter provides an overview of the study of IOS, examining the basic types of IOS, implications for practice, and key research questions. An overarching theme of this chapter is that IOS are socio-technical systems, where the combination of technical choices, organizational practices, relational factors, and industry context all influence outcomes of IOS use.

69.1.1 Types of IOS

IOS as a distinct management concern and subject of study began to appear in the early 1980s, with researchers describing and categorizing inter-organizational information sharing systems and their related management concerns (Barrett and Konsynski 1982; Cash and Konsynski 1985). Early IOS were often simple remote ordering systems between trading partners in adjacent steps in a value chain (a manufacturer ordering parts from a supplier, a retailer ordering inventory from a manufacturer, or a travel agent reserving a seat for a client on an airline). More complex and higher-level systems involved more partners, additional business functions, and greater integration of business processes across separate trading partners (Barrett and Konsynski 1982).

69.1.1.1 Electronic Hierarchies and Electronic Markets

One early classification of IOS distinguished between systems connecting a buyer with a specific seller in a hierarchical relationship—termed an *electronic hierarchy*, and systems that connected buyers with a larger set of potential sellers—termed an *electronic market* (Malone et al. 1987). This distinction emphasizes two types of control that govern the behavior of the participants in the IOS. In the case of electronic hierarchies, there is an implication of longer-term, stable business relations, where managerial and/or contractual relations will ensure that the potential for opportunistic behavior, such as providing substandard goods or services, is controlled. In electronic markets, however, market forces and exchange partners' ease of switching to new trading partners provide the necessary *governance* over the behavior of participants in order to reduce the potential for opportunistic behavior (Malone et al. 1987).

Other types of IOS that do not fit so neatly in the electronic hierarchy–electronic market continuum are evident. For example, an ATM network like Cirrus, where the IOS is controlled by a network facilitator that is mainly in the IOS business, rather than set up by a company that is selling the core products and services carried by the network is another form of IOS arrangement (Grover 1993). Choudhury (1997) further extends the electronic hierarchy–electronic market classification to account for the strategic relationships embodied in an IOS, describing three types of linkage patterns: (1) electronic monopolies—where buyers can only obtain a particular good through one seller's IOS, (2) electronic dyads—where buyers establish separate electronic links with a number of different sellers, and (3) multilateral IOS, where buyers and sellers use a shared system for transactions that more closely correspond to spot market purchases.

69.1.1.2 Horizontal, Vertical, and Cross-Linkage IOS

IOS can further be classified according to the industry structure aspects of relationships among the participants—horizontal, vertical, and cross-linkages (Hong and Kim 1998). Those that link businesses that are in adjacent steps in a value chain—where each business progressively adds value to a product or service—can be thought of as vertical IOS. Vertical IOS are best exemplified in supply chain information systems linking suppliers to manufacturers and perhaps to distributors and retailers where the participants are in a dependent relationship with each other. Horizontal IOS link companies that operate at common stages of the value chain, where each company performs similar value adding activities. These are often found in more cooperative forms of business relationships such as alliances or partnerships. Cross IOS span vertical and horizontal dimensions linking both buyers with each other, as well as to sellers at different stages in the value chain (Hong and Kim 1998).

69.1.1.3 Standards-Based versus Proprietary IOS

Recent work extends these earlier typologies by making three additional types of distinctions: (1) proprietary versus standards-based IOS (Markus et al. 2006; Nelson et al. 2005; Zhu et al. 2006), (2) point-to-point configurations versus hub-based architectures linking participants in a supply chain, and (3) IOS systems that are closed and private versus those that are shared and open to new participants (Steinfield et al. 2011). The push toward standards-based IOS has been driven by the rapid expansion of the Internet and associated protocols such as XML into business data networks, making it easier for businesses to interconnect electronically with an ever-larger proportion of trading partners without the need for investment in new proprietary systems for each individual partner (Markus et al. 2003). Over the past decade, data-exchange standards have been developed in a number of different industries, often driven by trade associations and other voluntary industry consortia. Examples include RosettaNet in the electronics industry (Boh et al. 2007), MISMO in the home mortgage industry (Markus et al. 2006), and ACORD in the insurance industry (Jain and Zhao 2003).

69.1.1.4 Point-to-Point versus Hub-Based IOS

The distinction between point-to-point versus hub-based systems has been used to shed light on the fundamental question of *visibility* in supply chains (Steinfield et al. 2011). Lack of visibility in extended supply chains is associated with many information distortions faced by participants leading to such classic coordination problems as the *bullwhip* effect (Lee et al. 1997). Bullwhip effects occur when variations in orders propagate up the supply chain as each company builds in its own buffer inventory under conditions of uncertainty about actual demand, yielding larger and larger variations from actual sales data the further one goes up the chain. Since point-to-point systems only connect one supply participant to another, information about each transaction is not visible to other supply participants, even though they may benefit from having such information. Imagine the situation, for example, when a lower tier supplier experiences a delay in shipping crucial parts to a higher tier assembler. This will in turn impact the availability of completed assemblies to the manufacturer, but if information only flows through separate point-to-point systems, the manufacturer may not be made aware of the delay until later, when it may be too late to adjust production schedules. If the supply chain participants were involved in a hub-based system with appropriate business rules for sharing information about supply chain events, then the manufacturer would know about the delay immediately and can plan accordingly. Such an approach is under development in the automotive industry. This approach, known as the Materials Off-Shore Sourcing standard, is designed as a hub-based system where shipment information is simultaneously made available to the relevant participants as soon as events occur (Steinfield et al. 2011). More broadly, hub-based approaches can support greater collaboration among supply chain participants, as opposed to primarily providing transaction support (Markus and Christiaanse 2003).

69.1.1.5 Open and Shared versus Private IOS Hubs

A final distinction is whether IOS systems are shared and open to new participants or are private and limited to a specific set of invited participants (Christiaanse 2005; Markus 2006; Steinfield et al. 2011). Shared hubs generally are structured as B2B electronic marketplaces, providing the ability to support vertical exchanges among supply chain participants, but incorporating competing companies and their partners as well. Such an arrangement is different from a private hub, which may or may not be standards-based, but is driven by a dominant player that only includes its own trading partners. This arrangement is often associated with *dedicated* supply chains, where a dominant manufacturer or retailer works with a specific set of suppliers who do not do business with other buyers. In many industries, however, supply chains are *interconnected*, such that a supplier does business with multiple buyers who compete with each other. A private hub managed by only one buyer would therefore only accommodate a portion of such a supplier's transactions, requiring additional investment in duplicate systems and processes to support exchanges with other trading partners (Steinfield et al. 2011).

The various combinations of IOS business functions, ownership and governance structures, and technological choices have implications for the extent to which IOS are adopted and implemented, as well as for the kinds of outcomes participants experience. For example, EDI-based IOS often lack participation by small- and medium-sized enterprises (SMEs) that face challenges implementing EDI due to a lack of requisite knowledge, limited resources, and fewer anticipated benefits due to their generally lower volume of transactions (Iacovou et al. 1995; Markus 2006; Segev et al. 1997). When forced to use EDI by larger and more powerful trading partners, SMEs may implement the system only superficially to satisfy partner requirements, which limits the overall integration and coordination benefits that such systems offer (Hart and Saunders 1997). Even the supposedly lower-cost web-based EDI approaches have not led to widespread EDI adoption by SMEs (Beck and Weitzel 2005). On the other hand, developing standards-based IOS and implementing these in shared coordination hubs can also be difficult because competing companies view the systems they use to exchange data with partners as strategic tools that convey competitive advantage, and are therefore prone to pursuing proprietary approaches (Clemons and Row 1988; Mukhopadhyay and Kekre 2002; Young et al. 1999). Additionally, vendors of IT products and services are also reluctant to develop standards-based systems for their clients, since this would presumably make it easier for their clients to switch to new vendors' products (Markus et al. 2006).

69.2 Impact on Practice

There are many well-established cases where companies initiating IOS connections with their trading partners have experienced such benefits as reduced costs of transactions, fewer errors with orders and delivery, greater efficiency due to the reduced need to rekey information into separate systems, and enhanced competitiveness. Examples include McKesson in the wholesale pharmaceutical supply industry (Clemons and Row 1988), Baxter in the medical supply industry (Short and Venkatraman 1992), Chrysler (Mukhopadhyay et al. 1995), Federal Express (Williams and Frolick 2001), Intel (Cartwright et al. 2005; Markus 2006), Japan Airlines (Chatfield and Bjorn-Andersen 1997), and Sabre in the airline reservation industry (Christiaanse and Venkatraman 2002).

Many of the cases cited earlier represent proprietary efforts implemented by a dominant trading partner for the purposes of improving its competitive position. One of the earliest and most heavily cited examples is the well-known case of McKesson's Economost system, introduced in the late 1970s and early 1980s. McKesson, a drug wholesaler, used its system to provide advanced electronic ordering and inventory management services to independent pharmacies that otherwise had little in the way of automated systems. Handheld terminals, communications lines, and a series of business applications were provided to client pharmacies. The system reduced transaction costs, improved inventory practices, solidified McKesson's role as the distributor for these drug retailers, and improved the ability of independent pharmacies to compete against national drugstore chains (Clemons and Row 1988).

The retail sector has also experimented with many IOS approaches, primarily aimed at reducing the proportion of inventory holdings that are *safety stock*—i.e., extra inventory held to protect a company from the negative effects of delayed shipments. Large retailers such as WalMart, as well as consumer products manufacturers like Proctor and Gamble, have experienced significant savings due to such IOS efforts as *vendor managed inventory* (VMI), *efficient consumer response* (ECR), and more recently, CPFR systems (Fliedner 2003). Essentially, these systems strive to support the sharing of retailers' *point-of-sale* (POS) data with partners further upstream in the retail channel, including wholesalers, manufacturers, and suppliers. POS data provides information about consumer demand, enabling each partner in the channel to adjust inventory stock based on real-time information, and speed up ordering and replenishment processes. Despite the efforts of the Voluntary Interindustry Commerce Solutions Association (VICS), an association of retailers and manufacturers, CPFR is not a fully standardized solution. Rather, it is largely limited to specific manufacturer–retailer chains where a strong participant drives implementation among its partners in the channel (Markus and Gelinas 2006).

Cartwright et al. (2005) detail the benefits that Intel has received due to its implementation of an IOS based on RosettaNet standards, enabling electronic connections with its many suppliers around the world. They report that by re-architecting its IT infrastructure around RosettaNet standards, Intel saved as much as $40 million in 2004 alone due to increased process automation and a greater ability to create value through outsourcing to third-party logistics providers. This case study also revealed an important benefit that follows from such an aggressive implementation of standards-based IOS—it supported greater integration of business processes across many disparate information systems *inside* the company. Nonetheless, many of the smaller business partners were unable to bear the costs of converting to RosettaNet standards, precluding Intel from mandating its use. This necessitated the maintenance of a range of information exchange options within Intel's supply chain, including RosettaNet, EDI, Internet-based file transfer protocol (FTP), and web-based transactions (Steinfield et al. 2011).

Industry-wide information system standards can help address the issue of the need to reinvest in duplicative IOS to do business with multiple trading partners, and can enable SME participation particularly when standards are designed around lower-cost Internet protocols such as XML and *service-oriented architectures* (SOA) (Boh et al. 2007; Löhe and Legner 2010; Steinfield et al. 2005; Venkatesh and Bala 2012; Wigand et al. 2005). SOA refers to a new approach to IOS that capitalizes on the Internet's wide reach by using standardized web-based services that work across applications (Löhe and Legner 2010). This allows organizations to have more easily reconfigurable IOS so that new connections and disconnections from trading partners can be more easily accommodated (Löhe and Legner 2010).

For example, XML-based standards for various transactions were developed in the home mortgage industry led by the Mortgage Bankers Association and its subsidiary, the Mortgage Industry Standards Maintenance Organization (MISMO). The availability of such industry-wide or *vertical information system* (VIS) standards has contributed to performance improvements over the industry as a whole, providing significant reductions in the total costs across the value chain for processing mortgage applications as well as the time from application to closing (Steinfield et al. 2005). Moreover, there is further evidence that MISMO standards have contributed to vertical disintegration in the mortgage industry, allowing companies to outsource various mortgage processing functions to SMEs as a result of the reduced coordination costs made possible by VIS (Wigand et al. 2005).

XML-based standards also figured heavily in RosettaNet, which not only helped to lower the costs and time for developing standards, but also enabled participation by the growing number of Asian suppliers to the electronics industry (Boh et al. 2007). More recently, Internet-based standards such as XML implemented into an SOA are forming the basis for inter-organizational business process standards (IBPS) (Venkatesh and Bala 2012). IBPS specify shared and agreed upon "interrelated, sequential tasks and business documents" to enable greater integration among business partners (Venkatesh and Bala 2012).

Not all IOS efforts have been successful. For example, Steinfield et al. (2011) review the history of problematic efforts by one automotive manufacturer to implement proprietary IOS throughout its supply chain. Automotive industry supply chains are characterized by multiple tiers of suppliers, and while there are extensive EDI linkages between top tier suppliers and manufacturers, IOS connections to lower tier suppliers are less common due to their lack of technical expertise and other resources. Moreover, many lower tier suppliers in the automotive industry do business with multiple manufacturers, making the supply chains interconnected rather than dedicated. Hence, it has been difficult to extend proprietary IOS throughout the supply chain. The lack of inter-operable IOS throughout the automotive supply chain imposes significant costs to the industry as a whole. A 2004 study by the National Institute of Standards and Technology (NIST), for example, found that inadequate use of IOS in the automotive supply chain costs the industry more than $5 billion due to errors, delayed shipments, and inadequate or excess inventory holdings (White et al. 2004). Despite the considerable cost savings that have been found to result from EDI use between automobile manufacturers and their suppliers (Mukhopadhyay et al. 1995), the lack of full implementation across the entire supply chain remains a consistent problem.

In summary, there are many well-known cases where use of IOS has contributed to lower costs, greater efficiency, increased business, competitive advantage, and broader performance benefits at the industry level. However, IOS adoption is far from universal, and examples of failed implementations are also evident. Hence, research has investigated the factors associated with successful IOS development, adoption, and implementation, including the identification of both anticipated and unanticipated outcomes of use. An overview of research issues is provided in the following section.

69.3 Research Issues

IOS research can be broadly grouped into studies of IOS adoption, diffusion and implementation, IOS outcomes, and IOS development and governance. A common theme across much of the IOS research is that while the technical aspects of IOS—often called the information system artifact (Benbasat and Zmud 2003; Orlikowski and Iacono 2001)—are important, IOS should be viewed as *socio-technical systems* in order to understand how such systems are developed, adopted, implemented, and used and impact organizations and industries. That is, the social, organizational, and industry contexts in which IOS are embedded must be examined in concert with IOS technical features in order to understand the varying outcomes surrounding systems that contain seemingly equivalent technical features. Indeed, a recent meta-analysis of the EDI literature highlighted the inconsistent findings regarding factors influencing EDI adoption and benefits, positing such factors as industry influence and anticipated benefits as mediating forces that affect the relationship between hypothesized predictors of both adoption and benefit (Narayanan et al. 2009).

69.3.1 IOS Adoption and Implementation

Why do some organizations adopt IOS such as EDI, while others do not, despite the potential benefits they can obtain? Much of the past research has focused on EDI adoption, highlighting a number of antecedents that can be broadly grouped into internal and external factors. Internal factors found to influence adoption decisions include a variety of characteristics of the adopting organization that broadly reflect its *readiness* for IOS. These include having the requisite human, financial, and technical resources needed for implementation, compatible legacy systems, and business processes that have been improved to take advantage of IOS capabilities (Chau and Hui 2001; Iacovou et al. 1995; Markus and Christiaanse 2003; Narayanan et al. 2009; Premkumar and Ramamurthy 1995). External factors generally include aspects of the relationships between an adopting organization and its trading partners. Research has examined the extent of trust and the nature of dependencies among participants, whether there are power differences that might lead to coercion or other pressures to adopt, and the effect of incentives (e.g., promises of increased business) by a dominant partner as predictors of adoption (Chwelos et al. 2001; Hart 1998; Hart and Saunders 1997; Kumar and van Dissel 1996; Kumar et al. 1998; Riggins and Mukhopadhyay 1999; Teo et al. 2003).

As noted earlier, often EDI adoption by SMEs has historically been limited, which is not surprising as SMEs tend to have more limited financial resources, and can lack sufficient managerial and technical expertise needed to implement EDI. Hence, one research issue for IOS adoption researchers is how to increase adoption rates among SMEs. One approach is for the larger, more powerful trading partner to simply mandate adoption in order to do business with it; however, research has demonstrated that such attempts at coercion often do not succeed and generate resistance among smaller trading partners (Hart and Saunders 1997; Steinfield et al. 2011). SMEs may not refuse adoption altogether, but may adopt IOS technologies in a superficial manner, for example, for a limited set of transactions, depriving the partners of potential efficiencies or strategic benefits as a result (Massetti and Zmud 1996). To facilitate adoption, the IOS may be subsidized, or offered at no cost to smaller trading partners (Riggins and Mukhopadhyay 1999). Yet, even when entirely provided by the dominant partner, such systems still impose costs related to learning and the need to maintain redundant systems if the dependent partner does business with other organizations that use a different IOS. To insure that partners do engage in such *relationship-specific* investments, there must be sufficient expectation of future business (Mukhopadhyay and Kekre 2002). Indeed, in one study of IOS use in four industries, electronic transactions with smaller suppliers of key inputs were more common in situations where there were quite strong ties—in some cases extending to ownership ties—between trading partners (Kraut et al. 1999). More broadly, the imposition of costs for adopting and implementing IOS has to be perceived as equitable, and the benefits have to be shared, in order to encourage greater adoption (Jap 1999).

69.3.2 IOS Outcomes

The outcomes of IOS adoption and use have also received a great deal of attention in the literature (Narayanan et al. 2009; Robey et al. 2008). In general, research has explored both tactical and strategic impacts of IOS use (Chatfield and Yetton 2000). Successful IOS use can result in many types of tactical benefits, ranging from reduced lead time for orders, faster deliveries, fewer errors in procurement transactions, fewer out-of-stock situations, reductions in the size of inventory holdings, increased staff efficiency, better monitoring of shipments, lower costs, and other transactional and operational benefits (Narayanan et al. 2009). Research on strategic benefits emphasizes competitive advantage outcomes such as growth in the volume of business with existing or new trading partners, improved relationships with trading partners leading to better communication and information sharing, improved market share, faster product design cycle times, and higher quality of products and services (Chatfield and Yetton 2000; Narayanan et al. 2009; Riggins and Mukhopadhyay 1994).

Not all implementations of IOS realize the benefits anticipated by their initiators, however. Research suggests that any benefits of IOS use, especially strategic ones, are mediated by the breadth and depth of adoption and the extent to which IOS use is integrated into an organization's business processes (Massetti and Zmud 1996; Narayanan et al. 2009). Massetti and Zmud (1996) identify four facets of EDI adoption, for example, that signify the intensity with which it has been integrated into business processes: (1) *volume* is overall extent to which a company's transactions are carried out over EDI, (2) *diversity* measures the range of different types of transactions handled by EDI, (3) *breadth* refers to the extent to which EDI is used with each of a company's trading partners, and (4) *depth* refers to the extent to which EDI has been integrated into the business processes of a company and a particular trading partner. Furthermore, business processes may need to be reengineered to take advantage of IOS capabilities, rather than simply automating existing processes in order to realize benefits (Clark and Stoddard 1996). JIT inventory practices can be enabled with IOS, but only if new business processes by suppliers and manufacturers are developed that support new channel distribution practices such as continuous replenishment. *Business process reengineering* can also help companies integrate across disparate legacy systems, such that enhanced internal efficiencies result from IOS use (Markus 2006).

In order to achieve high-value strategic benefits with more tightly coupled business processes over IOS, companies have to be willing to share proprietary information with each other. Because of the sensitive nature of the data that is shared, including design data, forecast data, operational data and more, there can be understandable reluctance to participate in such highly integrated IOS without a strong degree of trust between partners (Hart 1998; Hart and Saunders 1997). Hence, an ongoing thrust in the IOS research literature explores the nature of the relationships between IOS participants, and the influence of these relationships on outcomes of IOS use. Essentially, this research takes as a starting point that economic exchange can be facilitated by personal relationships among participants—trust and the social obligations that follow from being *embedded* in a network of relationships mitigate opportunistic behavior of participants and encourage reciprocation and cooperative exchange (Granovetter 1985; Kumar et al. 1998; Uzzi 1997). EDI researchers have further shown that when systems exhibit such embeddedness—fostering tight integration among trading partners characterized by high trust—the participants are more likely to experience the kinds of strategic benefits that yield competitive advantage (Chatfield and Yetton 2000).

These findings expose an interesting theoretical tension in IOS research, and reveal how IOS use can lead to impacts at higher levels of analysis such as on market structure, supply chains, and whole industries. On the one hand, there is strong evidence inspired by transaction cost economic theory that increased use of IOS facilitates a "move to the market" encouraging companies to use networks to acquire needed inputs from the lowest cost suppliers that satisfy requirements (Alt and Klein 2011). This is occurring because IOS networks reduce the traditional transaction costs associated with market exchanges; they lower search costs and make it easier to monitor transactions (Malone et al. 1987). On the other hand, achieving strategic benefits seems to require tighter integration that is not consistent with arms-length, market transactions (Chatfield and Yetton 2000; Kraut et al. 1999). More research is needed to resolve this paradox, but one explanation may be that companies are more willing to use electronic market transactions for commodity inputs that are not highly *asset-specific* (i.e., they are not tailored to a particular company's needs). Conversely, they rely on tighter connections to a smaller set of suppliers for core inputs with greater asset specificity and a higher degree of proprietary information sharing. Such an arrangement with core suppliers reflects what has been called a "move to the middle" approach, representing a structure in between hierarchies and markets (Clemons et al. 1993).

Research on the impacts of IOS use at higher levels of analysis beyond the dyad is less common, despite several calls for more attention to the industry structure implications of IT-enabled inter-organizational exchanges (Gregor and Johnston 2001; Johnston and Gregor 2000; Segars and Grover 1995; Steinfield et al. 2005; Wigand et al. 2005). Segars and Grover (1995) studied the effects of IOS on three industries: (1) airlines and computer-based reservation systems, (2) the chemical industry's use of EDI, and (3) the pharmaceutical industry and EDI, finding structural influences such as greater consolidation in all three industries. The degree of structural change was a function of the ease with which competitors could imitate an IOS initiator's innovation; where easily imitated, IOS use proliferated rapidly throughout the industry. Johnston and Gregor (2000) develop a theory of industry-level impacts of IOS, emphasizing the need to look at higher-level factors such as the role of industry groups, government policy makers, and competitive dynamics in order to understand how IOS adoption and use occurs in different industries. These higher-level factors are also viewed as necessary for understanding how IOS use affects both routine business practices and the relationships among actors in the industry.

The availability and widespread acceptance of VIS standards is another important factor in understanding industry-level impacts of IOS. Due to the rise of lower-cost data-exchange protocols made possible by the growing use of Internet-based XML, many industries formed user-led associations with the purpose of establishing VIS standards (Markus et al. 2006). If successful, such efforts have the potential to include SMEs as participants in IOS for a number of reasons. First, if Internet- and XML-based, they can be less costly and less complex. Second, participating in systems built on open standards should presumably make it easier to connect with other trading partners, generating *network externalities* (benefits derived from the greater numbers of other users that are accessible in a network) that increase the attractiveness of using a standardized IOS (Zhu et al. 2006). In the home mortgage industry, for example, widespread usage of VIS standards resulted in significant industry performance benefits due to the aggregation of gains in efficiency among participating organizations (Steinfield et al. 2005). Structural changes in the mortgage industry were also observed, including evidence of consolidation among the larger players as well as growth in the number of smaller players who used VIS-based systems to fill specialty niches in the industry (Steinfield et al. 2005; Wigand et al. 2005).

69.3.3 IOS Development and Governance

Research on IOS development and governance can be vital to understanding the prospects for IOS adoption and use. IOS governance broadly refers to how decisions regarding IOS systems development and usage are made, and how individual and organizational action is coordinated vis-à-vis the IOS (Markus and Bui 2012). This can include decisions regarding who participates in the IOS, how any technical changes to the IOS or its configuration are determined, for what transactions the IOS will be used, and who owns any intellectual property rights associated with the IOS.

A recent focus of IOS development research has investigated the process through which industry consortia develop VIS standards (Boh et al. 2007; Markus et al. 2006). Markus et al. (2006) conceptualize this process as a *collective action* problem. Such problems refer to situations where a group effort is required to achieve a collective good, but such goods have the characteristics of *public goods*, where *free riding* can occur. That is, participants who have not necessarily contributed to the production of the good can still enjoy its benefits, and therefore may have a disincentive to help shoulder the associated costs of production. VIS standards are public goods; organizations do not have to contribute to their development, but nonetheless benefit from their existence if they achieve widespread adoption. Markus et al. (2006) highlight the types of collective action dilemmas that need to be resolved in order for industry consortia to be able to effectively develop standards that have a greater probability of achieving widespread adoption. They point out that standards development and standards diffusion are two linked collective action problems, and it is possible to solve the former in such a way that the latter is less likely. For example, if the consortia privilege only certain types of participants (e.g., large companies, or those from only one segment of an industry) and not others, then there is a greater chance that whatever standard is developed will not adequately address the needs of those who did not participate in the standards development process. This can lead to an industry standard that is not taken up by the underrepresented companies, and hence fails to achieve widespread adoption, depriving all industry participants of potential benefits. Their case study of the home mortgage industry identifies the diverse interests of different types of industry stakeholders—and particularly emphasizes the different needs of IT vendors versus IOS users. Bringing IT vendors into the process, even in user-driven consortia, is essential so that the software solutions available to industry participants will be standards-based. This is especially important for SMEs that are not likely to develop their own IOS. Hence, involvement in the development of the standard by all segments of the industry was a crucial strategy undertaken by MISMO in order to ensure that the outcome was palatable across the key stakeholder groups.

A variety of governance issues come to the fore when IOS standards are examined in this light. For example, standards consortia must deal a priori with intellectual property issues, so that when a standard is developed, adopting companies are not obligated to pay unexpected licensing fees or royalties to participants who had not revealed ownership or granted royalty free use of patents that were used (Steinfield et al. 2007). Processes for ensuring compliance with standards must be in place to avoid the problem of *fragmentation* of the standard, which can occur when there is some leeway in implementing the standard and individual vendors and other organizations make small changes that cause one version of the IOS to be incompatible with another (Damsgaard and Truex 2000). Other direct governance issues related to the function of standards consortia include rules for who can participate in the work of the committees developing the standards, as well as voting rules for selection of officers and for making other decisions (Markus et al. 2003). All of these factors can shape the development process in ways that not only influence IOS standards development, but also the prospects for diffusion throughout the industry. Standards that are largely driven by dominant companies and which fail to take into account the needs of other industry participants are less likely to experience successful diffusion.

Not all governance issues are about standards development. As noted earlier, some IOS are hub-based, and interconnect many different trading partners in a common system. A recent study by Markus and Bui (2012) identifies the governance challenges faced by what they call *inter-organizational coordination hubs* (ICHs). ICHs face such challenges as attracting adequate investment, ensuring participation, and determining rules related to ownership and access to the data they generate (Markus and Bui 2012). In one study of an ICH developed for an automotive supply chain, the participants opted for a third-party owned IOS to avoid perceptions that the system would only cater to the needs of the dominant manufacturer who functioned as the buyer in the supply chain (Steinfield et al. 2011). Moreover, they encouraged participation by allocating costs for the system according to the extent to which participants benefited from its use; in this case since the buyers benefited the most, they absorbed the lion's share of the costs.

69.3.4 Toward a Research Agenda

The previous sections have highlighted a number of significant research questions related to IOS adoption and implementation, IOS outcomes, and IOS development and governance, and offer insights into the key directions for research on IOS in the next several years. In the area of IOS adoption and implementation, there is clearly a need for more research that clarifies how organizations, and particularly SMEs, evaluate the risks and rewards of participating in an IOS. How can industry associations help to diffuse the IOS standards they have developed more effectively, and encourage companies in their industry to adopt them? What approaches can companies take to increase trust and convince smaller trading partners to sign on to use IOS?

In the area of IOS outcomes, more research is needed to better understand why some IOS participants benefit while others do not. What IOS arrangements best contribute to improved outcomes for all participants? How can IOS be structured so that there is not only a reliance on standards to keep costs down, but also ample opportunity for companies to employ IOS strategically in ways that are difficult for competitors to imitate?

IOS development and governance research in the coming years will be critical for understanding the questions posed earlier. For example, research on how to attract more SMEs to participate in VIS standards making will be essential for solving diffusion and adoption problems. With more representative participation, there should be a better match between SME needs and the types of standards that are developed. Moreover, awareness of the availability and utility of standards should be greater and SME involvement in the process may motivate later adoption and use. Research on governance of IOS will be essential as well. With the growth of shared coordination hubs, more research is needed to understand if such shared IOS platforms based on open standards are more likely to find acceptance, or if companies will resist due to perceptions of a loss of control and competitive advantage? Clearly, much work remains to be done over the coming years to provide guidance to IS managers and to IT vendors so that the potential benefits of IOS can be more effectively harnessed.

69.4 Summary

This chapter has provided a broad introduction to the field of IOS. Modern organizations of all types—public and private, profit and nonprofit, large and small—need to rely on IOS to reduce costs, improve quality of products and services, and compete effectively. Implementing IOS, however, is not an independent decision of a single organization, and requires cooperation among the participants in order to achieve desired outcomes.

An overview of the many different types of IOS was provided, ranging from dyadic systems that link a pair of organizations for computer-based transactions to ICHs that can interconnect and support transactions among a supply chain or even an entire industry. Further distinctions between proprietary versus standards-based IOS were described.

The successful use of IOS has had significant impacts in many industries, and is associated with such far-reaching innovations as JIT inventory management practices and the advent of computerized reservation systems. However, success is not guaranteed, and there are many instances where organizations have attempted to implement IOS only to find that partners resist adoption or adopt in ways that limit overall benefits to the various stakeholders. SME adoption of IOS technologies such as EDI in particular has been problematic.

As detailed in this chapter, research has investigated the factors that influence IOS adoption and use, as well as the outcomes that organizations experience from their use of IOS. Such factors as having requisite technical competence, financial resources, and experience with existing internal systems reflect a readiness for IOS that can enhance prospects for adoption. In addition, the review has revealed the importance of trust among IOS participants, anticipated benefits, and the role of incentives in spurring IOS adoption. The research on IOS outcomes detailed in this chapter identifies two broad types of benefits that participants can experience with successful use of IOS: operational benefits that largely result from the ability to engage in more efficient transactions and strategic benefits related to IOS users' relationships with their partners and users' competitive positions in their industry. IOS impacts can be far-reaching, and research has also identified ways in which IOS use can influence entire industries by improving performance and altering the industry structure.

Finally, the study of IOS development and governance was introduced as an important area of research. Questions such as how and why some industries have been able to successfully develop industry-wide IOS standards while other others have failed represent new areas of research that have significant societal and economic implications. As reliance on IOS becomes ever more pervasive, research is needed to more fully understand and solve the governance challenges raised by emerging IOS configurations such as shared coordination hubs. Advances in technology are creating many new opportunities, but the benefits can only be realized when the business and organizational context of IOS use is taken into account.

Glossary

Asset specificity: When goods or services in a transaction are of value only to a particular recipient and cannot be sold as is to other potential buyers, they are considered highly specific assets. In general, when needed inputs exhibit strong asset specificity, transaction cost theory suggests that organizations will seek to produce in-house (i.e., use hierarchy as governance) to protect themselves from threats of opportunism.

Bullwhip effect: Bullwhip effects occur when variations in orders propagate up the supply chain as each company builds in its own buffer inventory under conditions of uncertainty about actual demand, yielding larger and larger variations from actual sales data the further one goes up the chain.

Business process reengineering: The process of redesigning work flows to improve business performance, often motivated by the desire to take advantage of new capabilities made possible by new information systems.

Collaborative Planning, Forecasting, and Replenishment (CPFR): A set of guidelines promoted by the Voluntary Interindustry Commerce Standards (VICS) Association to facilitate the cooperative management of inventory by sharing data on customer demand as well as production and distribution across a supply chain.

Collective action: The set of behaviors by a group of individuals or organizations as they work to achieve a common goal or objective.

Efficient consumer response (ECR): An approach to achieving more timely inventory practices by linking replenishment to actual consumer demand rather than by having retailers make orders based on periodic forecasts.

Electronic data (or document) interchange (EDI): The structured transmission of computerized messages between organizations to support the electronic exchange of standard business documents such as purchase orders or bills of lading. EDI message standards have been developed globally by the International Standards Organization (ISO), as well as by national standards bodies and many different industry associations.

Electronic hierarchy: A type of relationship between two organizations that are connected by an IOS where the two company's business processes are tightly integrated and governed by management control as opposed to market forces.

Electronic market: A form of electronic interconnection among groups of organizations where the actions of the participants are controlled by market forces, enabling companies to engage in transactions with a number of different partners due to the reduced costs of coordination afforded by computer-based networks.

Free riding: Allowing others to bear the costs of producing a collective good while enjoying the benefits of the good.

Governance: In the context of an IOS, governance refers to how decisions regarding IOS systems development and usage are made, and how individual and organizational action is coordinated.

Interconnected versus dedicated supply chains: Interconnected supply chains are those in which suppliers provide goods and services to multiple buyers, while in dedicated supply chains, a supplier typically only sells its output to a single buyer.

Just-in-time (JIT) inventory practices: An inventory management approach that minimizes inventory holding costs making more frequent deliveries of supplies as close as possible to the time when they are needed based on real-time information about stock levels.

Network externality: A benefit from using a good or service that is a function of the number of others that are also using the same good or service, and which is not necessarily captured in the price. An example is the benefit of having more potential trading partners in an IOS that has a larger number of participants, even though each user does not pay more when a new participant joins.

Public good: Such goods are types of goods that have the attributes of being *non-rivalrous* and *non-excludable*. Non-rivalrous means that one person's (or organization's) use of the good does not diminish its supply or prevent others from using it. Non-excludable means that it is not possible to prevent others from using the good once it is produced. A public/open IOS standard is often thought of as a public good, since those who did not contribute to its development can still use it.

Service-oriented architecture (SOA): An emerging approach to information system design utilizing web services accessible through the Internet that makes it easier to reuse software components across applications. Such architectures make IOS more dynamic and reconfigurable.

Supply chain visibility: Supply chain visibility refers to the ability to track goods and related supply chain events from the point of production to the delivery at the end-user's premises.

Vertical information system (VIS) standards: These are a set of data definitions and electronic transaction standards that are focused on the unique needs, products, and services relevant to the value chain of a specific industry. VIS standards have been developed in many industries such as the mortgage industry, insurance industry, automotive industry, and so forth.

Further Information

Eom, S. B. 2005. *Interorganizational Systems in the Internet Age.* Hershey, PA: Idea Group Publishing, Inc.

References

Alt, R. and S. Klein. 2011. Twenty years of electronic markets research—Looking backwards towards the future. *Electronic Markets* 21(1): 41–51.

Barrett, S. and B. Konsynski. 1982. Inter-organization information sharing systems. *MIS Quarterly* 6 (Special Issue on the 1982 *Research Program of the Society for Management Information Systems*): 93–105.

Beck, R. and T. Weitzel. 2005. Some economics of vertical standards: Integrating SMEs in EDI supply chains. *Electronic Markets* 15(4): 313–322.

Benbasat, I. and R. W. Zmud. 2003. The identity crisis within the IS discipline: Defining and communicating the discipline's core properties. *MIS Quarterly* 27(2): 183–194.

Boh, W. F., C. Soh, and S. Yeo. 2007. Standards development and diffusion: A case study of RosettaNet. *Communications of the ACM* 50(12): 57–62.

Cartwright, J., J. Hahn-Steichen, J. He, and T. Millier. 2005. RosettaNet for Intel's trading entity automation. *Intel Technology Journal* 9(3): 239–246.

Cash, J. I. and B. Konsynski. 1985. IS redraws competitive boundaries. *Harvard Business Review* 63(2): 134–142.

Chatfield, A. T. and N. Bjorn-Andersen. 1997. The impact of IOS-enabled business process change on business outcomes: Transformation of the value chain of Japan Airlines. *Journal of Management Information Systems* 14(1): 13–40.

Chatfield, A. T. and P. Yetton. 2000. Strategic payoff from EDI as a function of EDI embeddedness. *Journal of Management Information Systems* 16(4): 195–224.

Chau, P. Y. K. and K. L. Hui. 2001. Determinants of small business EDI adoption: An empirical investigation. *Journal of Organizational Computing and Electronic Commerce* 11(4): 229–252.

Choudhury, V. 1997. Strategic choices in the development of interorganizational information systems. *Information Systems Research* 8(1): 1–24.

Christiaanse, E. 2005. Performance benefits through integration hubs. *Communications of the ACM* 48(4): 95–100.

Christiaanse, E. and N. Venkatraman. 2002. Beyond Sabre: An empirical test of expertise exploitation in electronic channels. *MIS Quarterly* 26(1): 15–38.

Chwelos, P., I. Benbasat, and A. S. Dexter. 2001. Research report: Empirical test of an EDI adoption model. *Information Systems Research* 12(3): 304–321.

Clark, T. and D. B Stoddard. 1996. Interorganizational business process redesign: Merging technological and process innovation. *Journal of Management Information Systems* 13(2): 9–28.

Clemons, E., S. Reddi, and M. Row. 1993. The impact of information technology on the organization of economic activity: The move to the middle hypothesis. *Journal of Management Information Systems* 10(2): 9–35.

Clemons, E. K. and M. C. Row. 1988. McKesson Drug Company: A case study of Economost—A strategic information systems. *Journal of Management Information Systems* 5(1): 36–50.

Damsgaard, J. and D. Truex. 2000. Binary trading relations and the limits of EDI standards: The Procrustean Bed of Standards. *European Journal of Information Systems* 9(3): 142–158.

Fliedner, G. 2003. CPFR: An emerging supply chain tool. *Industrial Management and Data Systems* 103(1): 14–21.

Granovetter, M. 1985. Economic action and social structure: The problem of embeddedness. *American Journal of Sociology* 91(3): 481–510.

Gregor, S. and R. B. Johnston. 2001. Theory of interorganizational systems: Industry structure and processes of change. In *Proceedings of the 34th Annual Hawaii Conference on Systems Sciences,* Vol. 7, p. 7005. Maui, HI: IEEE Computer Society Press.

Grover, V. 1993. An empirically derived model for the adoption of customer-based interorganizational systems. *Decision Sciences* 24(3): 603–640.

Hart, P. M. 1998. Emerging electronic partnerships: Antecedents and dimensions of EDI use from the suppliers' perspective. *Journal of Management Information Systems* 14(4): 87–111.

Hart, P. M. and C. S. Saunders. 1997. Power and trust: Critical factors in the adoption and use of electronic data interchange. *Organization Science* 8(1): 23–42.

Hong, I. B. and C. Kim. 1998. Toward a new framework for interorganizational systems: A network configuration perspective. In *Proceedings of the Thirty-First Hawaii International Conference on Systems Sciences,* Vol. 4, pp. 92–101. Maui, HI: IEEE.

Iacovou, C. L., I. Benbasat, and A. S. Dexter. 1995. Electronic data interchange and small organizations: Adoption and impact of technology. *MIS Quarterly* 19(4): 465–485.

Jain, H. and H. Zhao. 2003. A conceptual model for comparative analysis of standardization of vertical industry languages. In *Proceedings of the Workshop on Standard Making: A Critical Research Frontier for Information Systems,* J. L. King and K. Lyytinen, eds., pp. 210–221. Seattle, WA.

Jap, S. D. 1999. Pie-expansion efforts: Collaboration processes in buyer–supplier relationships. *Journal of Marketing Research* 36(4): 461–475.

Johnston, R. B. and S. Gregor. 2000. A theory of industry-level activity for understanding the adoption of interorganizational systems. *European Journal of Information Systems* 9(4): 243–251.

Kraut, R., C. Steinfield, A. Chan, B. Butler, and A. Hoag. 1999. Coordination and virtualization: The role of electronic networks and personal relationships. *Organization Science* 10(6): 722–740.

Kumar, K. and H. van Dissel. 1996. Sustainable collaboration: Managing conflict and cooperation in interorganizational systems. *MIS Quarterly* 20(3): 279–300.

Kumar, K., H. van Dissel, and P. Bielli. 1998. The merchant of Prato—Revisited: Toward a third rationality of information systems. *MIS Quarterly* 22(2): 199–226.

Lee, H. L., V. Padmanabhan, and S. Whang. 1997. The bullwhip effect in supply chains. *Sloan Management Review* 38(3): 93–102.

Löhe, J. and Legner, C. 2010. SOA adoption in business networks: Do service-oriented architectures really advance inter-organizational integration. *Electronic Markets* 20(3–4): 181–196.

Malone, T., J. Yates, and R. Benjamin. 1987. Electronic markets and electronic hierarchies: Effects of information technology on market structure and corporate strategies. *Communications of the ACM* 30(6): 484–497.

Markus, M. L. 2006. Building successful interorganizational systems: IT and change management. In *Enterprise Information Systems VII*, C. S. Chen, J. Filipe, I. Seruca, and J. Cordeiro, eds., pp. 31–41. Dordrecht, the Netherlands: Springer.

Markus, M. L. and Q. N. Bui. 2012. Going concerns: The governance of interorganizational coordination hubs. *Journal of Management Information Systems* 28(4): 163–198.

Markus, M. L. and E. Christiaanse. 2003. Adoption and impact of collaboration electronic marketplaces. *Information Systems and e-Business Management* 1(April): 1–17.

Markus, M. L. and U. Gelinas. 2006. Comparing the standards lens with other perspectives on IS Innovations: The case of CPFR. *International Journal of IT Standards and Standardisation Research* 4(1): 26–42.

Markus, M. L., C. Steinfield, and R. T. Wigand. 2003. The evolution of vertical IS standards: Electronic interchange standards in the US Home Mortgage Industry. In *Proceedings of the Workshop on Standard Making: A Critical Research Frontier for Information Systems*, J. L. King and K. Lyytinen, eds., pp. 80–91. Seattle, WA.

Markus, M. L., C. Steinfield, R. T. Wigand, and G. Minton. 2006. Industry-wide information systems standardization as collective action: The case of the U.S. Home Mortgage Industry. *MIS Quarterly* 30(August): 439–465.

Massetti, B. and R. W. Zmud. 1996. Measuring the extent of EDI usage in complex organizations: Strategies and illustrative examples. *MIS Quarterly* 20(3): 331–346.

Mukhopadhyay, T. and S. Kekre. 2002. Strategic and operational benefits of electronic integration in B2B procurement processes. *Management Science* 48(10): 1301–1313.

Mukhopadhyay, T., S. Kekre, and S. S. Kalathur. 1995. Business value of information technology: A study of electronic data interchange. *MIS Quarterly* 19(2): 137–156.

Narayanan, S., A. S. Marucheck, and R. B. Handfield. 2009. Electronic data interchange: Research review and future directions. *Decision Sciences* 40(1): 121–163.

Nelson, M. L., M. J. Shaw, and W. Qualls. 2005. Interorganizational systems standards development in vertical industries. *Electronic Markets* 15(4): 378–392.

Orlikowski, W. J. and C. S. Iacono. 2001. Research commentary: Desperately seeking the 'IT' in IT research—A call to theorizing the IT artifact. *Information Systems Management* 12(2): 121–134.

Premkumar, G. and K. Ramamurthy. 1995. The role of interorganizational and organizational factors on the decision mode for adoption of interorganizational systems. *Decision Sciences* 26(3): 303–336.

Riggins, F. J. and T. Mukhopadhyay. 1994. Interdependent benefits from interorganizational systems: Opportunities for business partner reengineering. *Journal of Management Information Systems* 11(2): 37–57.

Riggins, F. J. and T. Mukhopadhyay. 1999. Overcoming EDI adoption and implementation risks. *International Journal of Electronic Commerce* 3(4): 103–123.

Robey, D., G. Im, and J. D. Wareham. 2008. Theoretical foundations of empirical research on interorganizational systems: Assessing past contributions and guiding future directions. *Journal of the Association for Information Systems* 9(9): 497–518.

Segars, A. and V. Grover. 1995. The industry-level impact of information technology: An empirical analysis of three industries. *Decision Sciences* 26(3): 337–369.

Segev, A., J. Porra, and M. Roldan. 1997. Internet-based EDI strategy. *Decision Support Systems* 21(3): 157–170.

Short, J. E. and N. Venkatraman. 1992. Beyond business process redesign: Redefining Baxter's business network. *Sloan Management Review* 34(1): 7–21.

Steinfield, C., M. L. Markus, and R. T. Wigand. 2005. Exploring interorganizational systems at the industry level of analysis: Evidence from the U.S. Home Mortgage Industry. *Journal of Information Technology* 20(4): 224–233.

Steinfield, C., M. L. Markus, and R. T. Wigand. 2011. Through a glass clearly: Standards, architecture, and process transparency in global supply chains. *Journal of Management Information Systems* 28(2): 75–108.

Steinfield, C., M. L. Markus, R. T. Wigand, and G. Minton. 2007. Promoting e-business through vertical IS standards: Lessons from the US home mortgage industry. In *Standards and Public Policy*, S. Greenstein and V. Stango, eds., pp. 160–207. Cambridge, U.K.: Cambridge University Press.

Teo, H.-H., K.-K. Wei, and I. Benbasat. 2003. Predicting intention to adopt interorganizational linkages: An institutional perspective. *MIS Quarterly* 27(1): 19–49.

Uzzi, B. 1997. Social structure and competition in interfirm networks: The paradox of embeddedness. *Administrative Science Quarterly* 42(March): 35–67.

Venkatesh, V. and H. Bala. 2012. Adoption and impacts of interorganizational business process standards: Role of partnering strategy. *Information Systems Research*, 23(4): 1131–1157.

White, W. J., A. C. O'Connor, B. R. Rowe, and R. T. I. International. 2004. *Economic Impact of Inadequate Infrastructure for Supply Chain Integration*. Research Triangle Park, NC: National Institute of Standards and Technology.

Wigand, R. T., C. Steinfield, and M. L. Markus. 2005. Information technology standards choices and industry structure outcomes: The case of the U.S. Home Mortgage Industry. *Journal of Management Information Systems* 22(2): 165–191.

Williams, M. L. and M. N. Frolick. 2001. The evolution of EDI for competitive advantage: The FedEx case. *Information Systems Management* 18(2) (Spring): 47–53.

Young, D., H. H. Carr, and R. K. Kelly. 1999. Strategic implications of electronic linkages. *Information Systems Management* 16(1): 32–39.

Zhu, K., K. L. Kraemer, V. Gurbaxani, and S. X. Xu. 2006. Migration to open-standard interorganizational systems: Network effects, switching costs, and path dependency. *MIS Quarterly* 30(August): 515–539.

70

Future of Information Systems Success: Opportunities and Challenges

American University

Ephraim McLean
Georgia State University

Darshana Sedera
The Queensland University of Technology

70.1 Introduction

Organizations continue to make substantial investments in information systems (IS), expecting positive impact on the organization and its employees. Such investments in contemporary information systems are under increasing scrutiny, and there is strong pressure to justify their value and contribution to productivity, quality, and competitiveness of organizations (Markus et al. 2003), regardless of the state of the economy (Kanaracus 2008). Even though it is difficult, research has also emphasized the importance of systematically measuring information system success. As stated by Peter Drucker (1987), "If you cannot measure it—you cannot manage it" (p. 47). With contemporary organization-wide IS, measuring success takes on special importance since the costs and risks of these large technology investments rival their potential payoffs.

In practice, however, IS investments, though often carefully rationalized in advance, are seldom systematically evaluated after their implementation (Thatcher and Oliver 2001). When post-implementation reviews do occur, their process and measures are often idiosyncratic and lacking credibility or comparability. Moreover, the impacts of information technology (IT) are often indirect and influenced by factors related to the user, organization, and environment; therefore, measurement of information systems success is both complex and illusive. In academic research, however, there is a long-standing tradition of research on system evaluation, dating to 1970s (King and Rodriguez 1978; Rolefson 1978; Matlin 1979). In 1992, DeLone and McLean published their IS success model, which is one of the most widely cited papers of IS success. Moreover, approaches like the Balanced Scorecard also have received great attention over the years. Yet, as Sabherwal et al. (2006, p. 1849) observe, "Despite considerable empirical research, results on the relationships among constructs related to information systems success, as well as the determinants of IS Success, are often inconsistent." In order for the IS success research to be relevant to practitioners and to continue as a vibrant research stream for academia, we must continue to identify opportunities and challenges for future researchers.

The objective of this chapter is to revisit the foundations of system evaluation studies and to identify some opportunities for future studies. In order to identify the opportunities, we make specific observations on five fundamental questions of system evaluations. Taking the DeLone and McLean IS success model as a framework, this chapter will make several recommendations regarding the challenges and opportunities that we must pursue. Our recommendations herein, though based on the DeLone and McLean model, can be generalized to most IS performance measurement approaches.

The opportunities and challenges of IS success research are derived using six simple considerations. As simple as they might sound, these six considerations provide an organized way to focus on the measurement properties of information system success. Our aim in this chapter is not to criticize the DeLone and McLean model. On the contrary, our discussion herein is motivated by the opportunities that have always existed with their framework. We explore what it would take to extend their model into new directions.

70.2 Information Systems Success

Keen (1987, p. 3) described the mission of IS as "the effective design, delivery, use and impact of information technologies in organizations and society." Based on this view of information systems, we believe the evaluation of the "effectiveness" or "success" of information systems is an important aspect of the information systems field, both in research and in practice. However, with the evolution of systems, users, and user requirements, the manner in which we evaluate the success of an information system has changed over time as the context, purpose, and impact of information systems have evolved. It is, therefore, essential to understand what these changes have been and what they mean for the future.

In general, information systems success research evaluates the effective creation, distribution, and use of information via technology. As information systems have developed since the mid-1950s, information has become more voluminous and systems have become ubiquitous and accessible by all. If we believe that information is power, this progress in information availability has changed the power relationships between corporations and consumers, between buyers and suppliers, between small business and large business, and between citizens and their governments.

However, unlike single, one-off investments, information systems are long-term investments, whose performance is subjected to a range of contextual factors. Moreover, IS, being long-term investments, are expected to yield a continuing flow of benefits into the future (Gable et al. 2008). Gable et al. (2008) defined IS success as a measure at a point in time of a stream of net benefits from the IS, to date and anticipated, as perceived by all key user groups. Further complexity arises due to changes in the user base and the access mechanisms, making infinite possibilities in terms of the purpose of an IS and the definition of its stakeholders.

Historically, researchers have employed objective, financial indicators to assess the impact of an information system (e.g., Brynjolfsson and Hitt 1996). However, as many have argued (Davenport 2000; Kaplan and Norton 2000), contemporary information systems provide financially quantifiable benefits as well as substantial nonfinancial benefits. Considering this, to measure the success of these various information systems, organizations are moving beyond traditional financial measures, such as return on investment and return on assets (Rubin 2004). For example, research suggests that large enterprise systems provide tangible as well as substantial intangible benefits to organizations, and highlight the importance of capturing these intangible benefits. In an effort to better understand such benefits of their IS, organizations have turned to methods such as the Balanced Scorecard of Kaplan and Norton (1996). Researchers, too, have developed several methods for assessing IS success using intangible measures (DeLone and McLean 1992, 2003; Gable et al. 2008), most focusing on the subjective assessment of the system by the users of the system. Of these subjective, user-centric system evaluations, the DeLone and McLean (1992) IS success model, in particular, has been widely adopted in a number of research contexts. The DeLone and McLean IS success model has been widely adapted and tested in a number of different system types (e.g., e-commerce, knowledge management systems, ERP) and has been validated across range of geographical settings.

70.3 DeLone and McLean IS Success Model

This chapter will later briefly introduce the DeLone and McLean model since it is the most widely cited and used IS evaluation model in academia. As of September 2012, we found over 8000 citations in Google Scholar for their 1992 (~5000 citations) and 2003 (~3000 citations) papers. We will summarize here its evolution, strengths, potential weakness, and limitations. The DeLone and McLean model is a good example of how the IS success construct has evolved over the years.

Early attempts to define information system success were ill-defined due to the complex, interdependent, and multidimensional nature of IS success. To address this problem, DeLone and McLean (1992) performed a review of the research published during the period 1981–1987, and created a taxonomy of IS success based upon this review. In their 1992 paper, they identified six variables or components of IS success: system quality, information quality, use, user satisfaction, individual impact, and organizational impact. However, these six variables are not independent success measures, but are interdependent variables. Figure 70.1 shows this original IS success model (DeLone and McLean 1992). Figure 70.2 shows the revised model in 2003, where the authors added service quality and consolidated organization and individual impacts to net benefits.

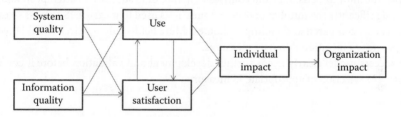

FIGURE 70.1 DeLone and McLean (1992) IS success model.

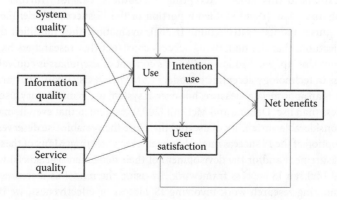

FIGURE 70.2 Revised DeLone and McLean (2003) IS success model.

As the authors explain (DeLone and McLean 2003, pp. 10–11):

> The primary purpose of the 1992 article in Information Systems Research was to synthesize previous research involving MIS success into a more coherent body of knowledge and to provide guidance to future researchers. … The semantic level is the success of the information in conveying the intended meaning. The effectiveness level is the effect of the information on the receiver. In the IS Success Model, SYSTEMS QUALITY measures technical success; INFORMATION QUALITY measures semantic success; and USE, USER SATISFACTION, INDIVIDUAL IMPACTS, and ORGANIZATIONAL IMPACTS measure effectiveness success.
>
> Based on both process and causal considerations, these six dimensions of success are proposed to be interrelated rather than independent. … A temporal, process model suggests that an information system is first created, containing various features, which can be characterized as exhibiting various degrees of system and information quality. Next, users and managers experience these features by using the system and are either satisfied or dissatisfied with the system and/or its information products. The use of the system and its information products then impacts or influences the individual user in the conduct of his or her work, and these individual impacts collectively result in organizational impacts.

The primary conclusions of DeLone and McLean's 1992 paper (1992, pp. 87–88) were quoting the original:

- "The multidimensional and interdependent nature of IS success requires careful attention to the definition and measurement of each aspect of this dependent variable. It is important to measure the possible interactions among the success dimensions in order to isolate the effect of various independent variables with one or more of these dependent success dimensions. Selection of success dimensions and measures should be contingent on the objectives and context of the empirical investigation; but, where possible, tested and proven measures should be used."
- "Despite the multidimensional and contingent nature of IS success, an attempt should be made to reduce significantly the number of different measures used to measure IS success so that research results can be compared and findings validated. More field-study research should investigate and incorporate organizational impact measures."
- "This success model clearly needs further development and validation before it could serve as a basis for the selection of appropriate IS measures."

70.3.1 Model Adoption

Shortly after the publication of the original DeLone and McLean success model, IS researchers began proposing modifications to this model. Accepting the authors' call for "further development and validation," Seddon and Kiew (1996) studied a portion of the IS success model (i.e., system quality, information quality, use, and user satisfaction). In their evaluation, they modified the construct, use, because they "conjectured that the underlying success construct that researchers have been trying to tap is Usefulness, not Use" (p. 93). Seddon and Kiew's concept of *usefulness* is equivalent to the idea of *perceived usefulness* in technology acceptance model (TAM) by Davis (1989). They argued that for voluntary systems use is an appropriate measure; however, if system use is mandatory, usefulness is a better measure of IS success than use. DeLone and McLean (2003) responded that even in mandatory systems, there can still be considerable variability of use and therefore the variable use deserves to be retained.

Researchers' adoption of the IS success model has been overwhelming. Many of these authors' articles positioned the measurement and/or the development of their dependent variable(s) within the context of the DeLone and McLean IS success framework. By using the model as a common framework for reporting and comparing research work involving IS success or effectiveness, we believe one of the primary purposes of the original article has been achieved.

Although many of the cited articles tend to justify their empirical measurement of IS success by citing the DeLone and McLean IS success model, some of them fail to heed DeLone and McLean's cautions. They state "… they [researchers] used the model like a drunkard uses a lamppost—for support rather than for illumination. They overlooked the main conclusion of the article—that IS success is a multidimensional and interdependent construct—and that it is therefore necessary to study the interrelationships among, or to control for, those dimensions, … Researchers should systematically combine individual measures from the IS success categories to create a comprehensive measurement instrument" (1992, pp. 87–88). Although many authors did not choose to measure (or control for) the various dimensions of IS success, a number of other researchers have used multidimensional measures of IS success in their empirical studies and have analyzed the interrelationships among them.

The DeLone and McLean (1992) IS success model has weathered criticism, as well as received widespread acknowledgement as one of the watershed studies in the IS discipline. Their model has been tested in full, and in some cases partially, in over 200 studies since 1992. In 2008, Petter et al. (2008) identified 180 empirical studies that had employed the DeLone and McLean success model. The model has been applied in system evaluation studies, data processing systems, e-business applications, and Enterprise Resource Planning systems. As Petter et al. (2008, p. 237) noted "… As a field, we have made substantial strides towards understanding the nature of IS success. For example, the widely cited DeLone and McLean model of IS success (1992) was updated a decade later based on a review of the empirical and conceptual literature on IS success that was published during this period (DeLone and McLean, 2003) …. [S]ome researchers have synthesized the literature by examining one or more of the relationships in the DeLone and McLean IS success model using the quantitative technique of meta-analysis (Mahmood et al. 2001; Bokhari 2005; Sabherwal et al. 2006) to develop a better understanding of success. Others have started to develop standardized measures that can be used to evaluate the various dimensions of IS success as specified by DeLone and McLean" (e.g., Sedera and Gable 2004).

70.4 Identifying Opportunities in IS Success

Having discussed IS success, the DeLone and McLean IS success model, and some criticisms of it, we will now address some potential opportunities and challenges for future IS success research. Our discussion here is guided by Cameron and Whetten (1983), who suggested five fundamental questions that must be considered in any evaluation. Next, we focus on the opportunities specific to the DeLone and McLean model.

As Cameron and Whetten (1983) suggest, five fundamental questions must be answered before attempting any evaluation of success:

1. WHAT SYSTEM?—On what domain of activity (what system) is the assessment focused?
2. WHO?—From whose perspective is effectiveness being assessed?
3. WHY?—What is the purpose for assessing effectiveness?
4. WHEN?—What time frame is being employed?
5. HOW?—What measures and constructs are being used?

We believe that the evaluator must address the aforementioned questions before embarking on an evaluation. The following section discusses each of those five questions in relation to information system success.

70.4.1 WHAT? The "System" of Evaluation

In defining a "system," there are two points of view to be considered: (i) the type of the system and (ii) the scope of the system. First, in every IS success study, one must specify the *type* of the system or application that they are evaluating and develop approaches and measures appropriate to them. Second, and in most cases concurrently, you must select the *scope* of the system/application.

First, studies on system typologies identify the salient differences between *different types of systems*. For example, McAfee (2006) identifies three types of systems: functional IT, network IT, and enterprise IT. He notes that there are substantial differences among these three in terms of their core purpose, the types of users (i.e., the potential study participants), and the system outcomes. Thus, an IS success study must derive it objectives, approach, and measures according to the types of the IS.

Second, one must establish a clear scope of the system to be evaluated. For example, a system can be defined as narrowly as a particular process, module, or function (e.g., a procurement process) or as broadly as an application portfolio. In each case, noting that neither approach is superior or inferior, the researcher must consider the implications of the scope of the system under investigation. Both system types and scope of the system provide useful boundaries for a system evaluation. Thus, akin to arguments by Burton-Jones and Straub (2006), one must select appropriate measures for the circumstances, without employing "omnibus" measures. The following section provides a summary of McAfee's classification of systems and its implications for IS success studies. Next, we focus on the scope of the system and how it influences IS success studies.

70.4.1.1 Types of Information Systems

At a high level, information systems can be classified as (i) hedonic—developed for pleasure and enjoyment or (ii) utilitarian—developed to improve individual and organizational performance (van der Heijden 2000). However, hedonic applications in organizations are still not common. Thus, in this discussion, we adopt the utilitarian subclassifications of McAfee (2006). Table 70.1 provides a summary of system types, examples, and considerations for system success.

TABLE 70.1 System Classifications and Implications for IS Success

Category	Function IT	Network IT	Enterprise IT
Definition	Assists with the execution of discrete tasks	Facilitates interactions without specifying their parameters	IT that specifies business processes
Characteristics	Can be adopted without complements[a] Impact increases when complements are in place	Does not impose complements,[a] but lets them merge over time Does not specify tasks or sequences Accepts data in many formats Use is optional	Imposes complements[b] throughout the organization. Defines tasks and sequences Mandates data formats Use is mandatory
Examples	Spreadsheets, computer-aided design, statistical software	E-mails, instant messaging, wikis, blogs, and mash-ups	ERP, CRM, and SCM
Automation	Some degree of automation (e.g., spell check)	Very low level of automation	High level of automation
Key user groups	More likely to have a single key user group	More likely to have a single key user group	Multiple key user group using the same system very differently
Considerations for system success	Most users would remain proficient with the basic system features Potential to improve performance through deeper and exploratory use	Limited work-oriented functionality Access to system features is equal across all key user groups Depth of use would not result in substantial improvements	High automation of business processes Many key user groups have different types of uses Must consider mandatory and nonmandatory uses For processes with high automation, frequency of use will only provide observations of efficiency

[a] Complements are defined by McAfee (2006, p. 142) as "*organizational innovations, or changes in the way companies get work done.*"

[b] Examples of complements that allow working performing technologies, according to McAfee (2006, p. 143) are "*better-skilled workers,*" "*higher levels of teamwork,*" "*redesigned processes,*" and "*new decision rights.*"

The characteristics of each system type are explained in Table 70.1. We acknowledge there are other classifications of system types, but we chose McAfee's classification for its simplicity, yet for its discussion of key user groups and system complexity for success measure specifications.

McAfee's (2006) classification of systems highlights the danger in using all-purpose, omnibus constructs and measures for IS success evaluations. The characteristics of the three types of systems show that each system is designed to provide specific, almost nonoverlapping service to a specified group of key users.

70.4.1.2 Scope of the System under Investigation

The second consideration relates to scope of the system. This is more complex than system classifications and can yield multiple overlapping interpretations. With the ever-expanding boundaries of systems, information systems are no longer restricted to "back-office" applications. More and more organizations now use organizational-wide systems (like ERPs), some consumed through web portals, hosted in the "cloud," and managed by third-party service providers.

Thus, defining the "scope" of a system is an important consideration in IS evaluation. The focus will depend on the areas that are generally defined as "the system." For example, Seddon et al. (1999, p. 6) identified six levels of scope that should be evaluated:

1. An aspect of IT use (e.g., a single algorithm or form of user interface)
2. A single IT application (e.g., a spreadsheet, a PC, or a library cataloging system)
3. A type of IT or IT application (e.g., TCP/IP, a GDSS, a TPS, a data warehouse, etc.)
4. All IT applications used by an organization or sub-organization
5. An aspect of a system development methodology
6. The IT function of an organization or sub-organization

As suggested in Seddon et al. (1999), isolating "the system" for evaluation is difficult, but must be done for evaluations to be meaningful. Even after scoping the boundaries of the system being considered for evaluation, certain aspects would still need to be resolved in the minds of both the evaluators and the researchers. They include other systems or portfolio of systems, the infrastructure, the IT function or the IT support service quality, and the administrative area with which the system is most closely associated.

A key challenge for the researcher here is to gain an understanding of the system of interest, without being influenced by aspects that may be a part of the system, but not be a part of the system of interest. As noted in several recent studies (DeLone and McLean 2003; Gable et al. 2008; Petter et al. 2008), most past studies reuse measures and constructs without much considerations as to the type and scope of the system(s). Therefore, this presents a clear opportunity for IS researchers to identify ways to determine the scope of a system. For example, a generalizable taxonomy to identify the system of interest may provide much value for the cumulative tradition of IS success research. The relationship between scope of IS and the type of IS is depicted in Figure 70.3.

Some considerations for developing such a taxonomy are stated here. First, the perceptions of the quality of the system may be influenced by perceptions of the infrastructure; a system may be perceived as slow, because the associated infrastructure is inadequate or underpowered. However, in such circumstances evaluators may not be knowledgeable or aware of the circumstances beyond their immediate work systems, and would not recognize the issues attributable to the IT infrastructure rather than to the system of interest. Second, the scope of the system being evaluated can be defined more narrowly or more broadly, depending on the nature of the evaluation being undertaken. This would make it impossible to compare and benchmark across systems. Third, increasingly when systems services are delivered using web portals, researchers would find it difficult to identify "the system." Similarly, end users will find it even more difficult to identify which system they are accessing and how they will receive information from it. If the researcher evaluates the goodness of the web portal at a high level, such information would not be adequate to address the management and performance issues of the underlying system.

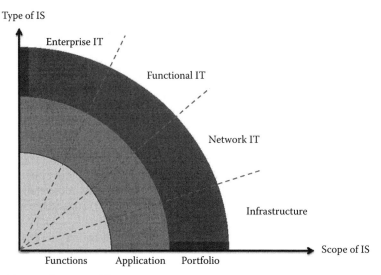

FIGURE 70.3   Selecting the scope of IS success.

70.4.2 WHO? The Stakeholders

The stakeholder is the one who interacts with the system and consumes the information generated by the system. Stockdale and Standing (2006) argue that "... the evaluators must decide which groups are relevant to the project being evaluated. The power associated with stakeholder groups and its implications for effective evaluation is a complex issue that the evaluators should be aware of since there is a danger that outcomes may be skewed to meet the objectives of those holding power" (p. 1093). Stockdale and Standing (2006) identify four types of stakeholder groups relevant for a system evaluation:

1. Initiators of the evaluation
2. The evaluators who conduct the evaluation
3. The users of the systems being evaluated
4. A range of other parties such as trade unions and government agencies

In relation to the Stockdale and Standing (2006) stakeholders, most of the IS success studies have included the users of a system. Seddon et al. (1999, p. 167) also introduced a classification of stakeholders and found that the evaluation of IS effectiveness was generally based on the perceptions of one or more of the following five types of stakeholders:

1. An independent observer, who has no personal stake in the measure
2. An individual user, who evaluates a system from his or her own point of view
3. A group of users, e.g., a group decision support system (GDSS)
4. The management or owners of the organization
5. A country or all of mankind

Here too, many of the IS success studies appear to have concentrated on the post-implementation evaluation of system success, employing the perceptions of the end users. With the transition from in-house, custom-made, stand-alone legacy IS applications to integrated, customizable, and packaged applications, organizations are looking to engage multiple user groups with a single application. For example, Grover et al. (1996) state that there are four key user groups that could take part in a contemporary IS success data collection. The four groups are operational, management, strategic, and technical staff. Similar views have been reported by other researchers as well (Kang and Santhanam 2003;

Grabski et al. 2011). For example, in enterprise systems (e.g., SAP), the operational staff would use an ES for routine business transactions on a daily basis, while middle management would periodically engage the same system for management decision making. This ability to serve multiple key user groups with a single system has provided organizations with great potential in the standardization of information, process automation, and improvements of transparency (Morris and Venkatesh 2010; Seddon et al. 2010; Strong and Volkoff 2010).

How do the multiple stakeholders of a system impact information system success? First, different stakeholders may have different opinions regarding the performance of the same system. As Gable et al. (2008) have demonstrated, different stakeholder groups will have different views of the DeLone and McLean success dimensions. Gable et al. (2008) observed that operational staff tend to place a higher emphasis on system quality when they evaluate information system success, while senior executives (e.g., strategic-level staff) place a higher emphasis on organizational benefits. Second, Gable et al. (2008) stressed the importance of canvassing all stakeholders of a system. They argue that unless all users are canvassed for their views of a system the evaluation will only be partial.

The differences in perspectives across information system stakeholders have been long established in the information system success literature (Cameron and Whetten 1983; DeLone and McLean 1992; Seddon et al. 1999). However, information system research is yet to fully appreciate the impacts of using a single system by multiple user groups. For instance, these different user groups often have multiple and often conflicting objectives and priorities (Gable et al. 2008; Strong and Volkoff 2010). For instance, one user group could have one experience in their interactions with a system, as compared to the interactions of another user group using the same system (Park et al. 2007). Although an enterprise system may be viewed as a success from the standpoint of one user group, it may be interpreted as unsuccessful by another (Urbach et al. 2009).

Robert Anthony (1965) provides a useful classification scheme of management-level activities: strategic, managerial, and operational. The different levels presented by Anthony are important in IS evaluation studies. According to Anthony, the strategic level focuses on organization-wide objectives and allocates the necessary resources to achieve these objectives. The strategic level is involved in complex, irregular decision making, focusing on providing policies to govern the entire organization. At the strategic level, information requirements are ad hoc in nature, with information relevant for long-term organizational planning. At the managerial level, the information requirements are focused on ensuring that the resources, both human and financial, are used effectively and efficiently to accomplish the goals stated at the strategic level. The characteristics of the information required at the managerial level are different from those that are required at the strategic level. The managerial level deals with integrated and procedural information that is necessary for a precise task. Therefore, managers at this level tend to prefer "goal congruent" information systems. At the operational level, employees are involved in highly structured and specific tasks that are structured, routine, and transactional. Tasks carried out at the operational level are precise and are governed by organizational rules and procedures. Operational employees deal with real-time data focused on individual events with little or no emphasis on key organizational performance indicators. The three levels of management described by Anthony tend to be hierarchical on several dimensions: (1) the time horizon of decisions (i.e., long, medium, and short term), (2) the importance of a single action (i.e., critical, important, and routine), and (3) the level of judgment (i.e., strong, moderate, and modest). Using Anthony's framework, Table 70.2 demonstrates possible differences in opinions of the three management levels in relation to enterprise system and information requirements.

Alloway and Quillard (1983) emphasize the importance of operational personnel using a system appropriately in order for the senior managers to be able to make effective use of the system, since the aggregated information at the managerial level depends on the transactions at the operational level. Furthermore, it is argued that enterprise systems, in particular, often fail at the operational and transactional levels because operational users take shortcuts to circumvent the restrictive standards of the enterprise system (Mabert et al. 2001; Umble et al. 2002; Sedera et al. 2003).

TABLE 70.2 Key-User-Group Characteristics

		End-User Cohorts		
	Perspective	Strategic	Managerial	Operational
System	System use	Less direct access	Sporadic, ad hoc	Daily, routine access
	Use of system features	Rare use of high-level reports	Aggregated, summative	Transactions
	Level of interaction	Generally minimal	Moderate	Frequent
	Knowledge of the system	Generally minimal	Moderate	High
	Flexibility of the system	Requires much flexibility	Moderately flexible	Structured, less flexibility
Information	Information focus	Futuristic, one aspect at a time	Whole organization	Single task/transaction
	Information complexity	Many variables	Less complex	Simple, rule-based
	Information structure	Unstructured, irregular	Rhythmic, procedural	Highly structured
	Nature of information	Tailor made, more external, and predictive	Integrated, internal but holistic	Task specific, real time
	Information scope	Long term	Long, medium, and short	Short term

We identify several challenges and opportunities for future studies on the perspective of evaluation. These challenges are ever more important, given the paradigm shift of developing systems for a number of user groups, instead of developing specialized systems for single user groups. First, researchers should identify the interested user groups prior to conducting their empirical assessments. As shown in Table 70.2 and the related discussion, it is evident that each user group has distinct needs that are quite different from one another. Thus, your data collection instruments (e.g., questions or survey items) must be developed with attention to the characteristics of the different user groups. Gable et al. (2008) suggest that certain user groups will be closely aligned with certain success dimensions. They argued, for example, that operational personnel will place a greater emphasis on certain success dimensions to the exclusion of others when they evaluated a system.

Second, there is a real challenge for researchers to identify the relevant respondent characteristics, other than those that relate to their general employment classifications. For example, factors like respondents' exposure with similar systems, their knowledge of the system in question, and their experience with similar industry sectors will all have an impact on how they evaluate a system.

Alternative classifications of user groups are another opportunity for the future research. For example, Sedera and Dey (2008) suggest an alternative classification of respondents based on their skills and degrees of proficiency. They devised a survey instrument—based on knowledge, socio-behavioral factors, and experience—to classify respondents based on degree of proficiency. Their study also suggested that groups with different degrees of proficiency will have different opinions of system success.

Another opportunity comes with the weighted average score of the respondent's views of the system. All past subjective system evaluation studies have treated scores of all respondents equally. This has meant that, regardless of the characteristics of the respondent, all respondents have an equal say of the system evaluation. From a management viewpoint, this is problematic and less valuable. A weighted-average scale would assign weights based on respondent characteristics and the dimensions of success on which they evaluate. When a group of respondents indicate a greater association, and familiarity, with a particular dimension of success, their scores on that dimension could receive a high weighting than the others (Gable et al. 2008).

Finally, we observe that there is potential for "collective" and "group-level" success studies. It has been widely accepted that system success in organization-wide systems depends heavily on the cooperative work of stakeholders, within and across the stakeholder groups. Thus, in addition to addressing success as an individual construct, there is potential to address it as a group-level, collective construct. As such, the questions will focus on "Has the IS helped improve the collective performance," rather than its success at the individual level.

70.4.3 WHY? The Purpose of the Evaluation

Gable et al. (2008) observed that organizations evaluate their information system (IS) for various reasons. A frequently asked question is "Has the IS benefited the organization?" or "Has the IS had a positive impact?" Positive impacts are the ultimate outcome measure of an IS; the "acid test" of the IS. These questions seek a measure of the net benefits or impacts to date; they look backward. The IS, being a long-term investment, is expected, other things being equal, to yield a continuing flow of benefits into the future. Thus other questions of interest include "Is the IS worth keeping?" "Does the IS need changing?" or "What future impacts will the IS deliver?" These questions are forward looking.

Serafeimidis and Smithson (2000, p. 97) reviewed the literature on IS evaluation and found that evaluations are done for a variety of reasons, including (1) "Establishing by quantitative and/or qualitative means the worth of IT to the organization and its growth," (2) "Ranking alternatives," (3) "Forming a central part of an incremental planning and control (diagnosis) process," (4) "Acting as an input to business and IS strategy formulation," (5) "Acting as a feedback function which assists organizational learning," (6) "Acting as a mechanism for gaining commitment and, in highly politically influenced environments, for legitimization," and (7) "Providing a deeper understanding of the interaction between the technology and the underlying organizational processes, culture and politics."

Davern and Kauffman (2000) emphasized the importance of understanding where the future potential value for an IT investment lies, in both project selection evaluation and post-investment evaluation. While evaluation usually has a positive role to play, evaluation can also be for political or social reasons (e.g., to reinforce an existing organizational structure) and be a ritualistic rather than effective process (Stockdale and Standing, 2006, p. 1093). Table 70.3 lists some of the reasons why organizations evaluate IS.

Similar to the scope of the system and stakeholders' opinion sought in an evaluation, the motivation of the assessment (the "why?") is an important aspect of a system evaluation. Yet, there is a severe dearth of studies where the purpose *of* the evaluation is linked to the measures employed *for* the evaluation. For example, the balanced scorecard approach supports this concept; in it, the measures employed in the assessment are derived from the strategy, vision, and mission of the organization.

TABLE 70.3   "Why" of Evaluation

Value Reasons	Leading To	Sources
Appraisal of value	• Improvement in business goals	Farbey et al. (1999)
Measure of success	• Organizational effectiveness	Mirani and Lederer (1998)
Recognition of benefits	• Investment management	Remenyi and Sherwood-Smith (1999)
	• Problem diagnosis	
	• Consensus achievement	Serafeimidis and Smithson (1999)
	• Decision making	
	• Understanding risk	Serafeimidis and Smithson (1998)
	• Gains in organizational and personal learning	Hirschheim and Smithson (1998)

Source: Adapted from Stockdale, R. and Standing, C., *Eur. J. Oper. Res.*, 173(3), 1094, 2006.

As noted earlier, most IS success studies focus mainly on academic contributions, focusing heavily on construct validity and reliability. Thus, such studies rely heavily on constructs and measures that had been validated in prior studies in order to minimize the risk of introducing new measures that do not have a validated "lineage." This myopic focus on construct validation often means that the context of the study is ignored. A challenge—and an opportunity—for future IS success studies is to connect the purpose of the evaluation with the measures and constructs. This means that researchers could introduce new measures and constructs based on the purpose of the evaluation. Furthermore, future IS success studies have the potential to focus more on specific, diagnostic evaluations of an IS. This also means that researchers should be much closer to the study context, management, and user issues.

70.4.4 WHEN? The Timing of an Evaluation

The timing of a system evaluation could easily influence the evaluation outcomes. There seems to be a lag in the effects of IS investments; thus, it is important to consider when the evaluation is being done. For example, Hitt et al. (2002, p. 72) report that for enterprise systems there is "a slowdown in business performance and productivity shortly after the implementation." Markus et al. (2003) suggest there are two phases in an enterprise-system life cycle beyond the go-live date: (1) shakedown and (2) onward/upward. The period immediately after the go-live phase is the *shakedown* phase, during which users learn new system features and functions and adjust their work practices. The *onward/upward* phase follows the shakedown phase and reflects the finished, stable system. Organizations in this phase should see high user acceptance of the system and fewer ongoing technical issues. Also, from here on in, the organizations are also looking to derive additional value from their enterprise system investment through enhancements and value-adding functions. Markus and Tanis (2000) argue that the user experience and system success will be substantially different across these two life cycle phases. Ross and Vitale's (2000) study of 15 organizations found that all organizations underwent a "productivity dip" (p. 237) at the shakedown phase, while gaining productivity improvements in the onward/upward phase.

Researchers attribute the decline in productivity with a system to factors arising from users' lack of familiarity with the new system (Sumner 2000; Strong and Volkoff 2010). Several researchers have demonstrated that users, especially operational personnel, face a steep learning curve in the shakedown phase and it will take several years to master complex software features and functions (Mandal and Gunasekaran 2003; Botta-Genoulaz and Millet 2005). Therefore, during the shakedown phase, most installations restrict access to the system and operations by enabling only mandatory functions to be accessible (Kumar et al. 2003; Amoako-Gyampah and Salam 2004) until users develop a sufficient level of proficiency. On the other hand, users in the onward/upward phase receive more liberal access to value-adding system features and functions (Botta-Genoulaz and Millet 2005) through experience and practice. Social science researchers have identified years of experience, structure of the task, and "deliberate practice" as having a substantial influence on individual performance (Chi et al. 1988; Ericsson and Smith 1991). For example, Ericsson et al. (1993) demonstrate that an individual's performance is a monotonic function of the "deliberate practice" with the system.

Segars and Grover (1993) examined IT impact at an industry level. They found that "in each of the three industries examined no immediate impact was detected. Structural shifts were observed four to five years after the introduction of the technology" (p. 362). In relation to the timing and evaluation, we identify two main streams of opportunities for future researchers: researchers must (1) focus on the longitudinal evaluation of system success and (2) identify benefits in relation to the phases of the life cycle.

First, in relation to conducting longitudinal studies, we note that most IS success studies take a snapshot of the system performance at a point in time, typically using a survey instrument. Given the long-term view of IS in organizations, longitudinal studies will allow organizations to track IS performance over time. For example, despite research suggesting a dip in organizational performance following the go-live date, current research has yet to empirically test and demonstrate the changes to performance across the life cycle. Organizations often postpone an upgrade of aging applications, seeking

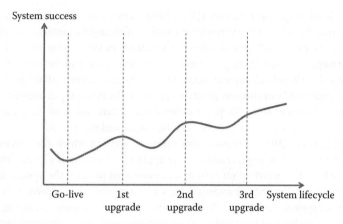

FIGURE 70.4 Systems life cycle.

to delay the relatively more tangible short-term costs. Yet, research by Gable et al. (2003) suggests that it is often optimal to upgrade an aged software earlier rather than later. A validated normative model (e.g., Figure 70.4) would aid organizations to better manage expectations of contemporary IS and to better plan mid- and longer term for evolution of the IS portfolio. This analysis essentially combines the measure(s) of IS success, positioning each application studied on the horizontal axis (the life cycle phase). Sensitivity analysis can then be conducted, through segmenting the overall study data into increasingly homogeneous segments.

Second, studies must continue to observe how benefits occur throughout the systems life cycle. Wang and Sedera (2011) conducted one of few studies that developed a benefits realization life cycle to demonstrate how IS success is achieved throughout the life cycle. Their study employed the DeLone and McLean constructs to illustrate how supply chain management system benefits appeared in a longitudinal study, proposing a benefits life cycle. For example, many studies have shown that the organizational benefits of an information systems will take time to appear, while the benefits from improved system and information qualities will appear earlier in the life cycle. However, it is unreasonable to assume that *all* of the measures of systems quality are realized at the early phase of the life cycle. Similarly, it cannot be assumed that all of the organizational-level benefits are only realized at the latter phase of the life cycle. Therefore, future IS success studies have a real opportunity and a responsibility to contribute to a better understanding of the timing of benefits. Empirical studies depicting a specific timeline for benefits realization using the phases of the life cycle are rare.

70.4.5 HOW? The Constructs and Measures of Success

The value of IT has been measured at various levels of analysis including the economy, industry, firm, business unit, business process, and individual (Barua and Kriebel 1991; Brynjolfsson and Yang 1996; Davern and Kauffman 2000). In the early days of information system, system performance was measured in terms of technical quality using observed speed and accuracy. When information systems evolved to be strategic, researchers looked to see the impact of a system to the organization. Dedrick et al. (2003) note: "For many years, there has been considerable debate about whether the IT revolution was paying off in higher productivity. Several studies in the 1980s found no connection between IT investment and productivity in the U.S. economy, a situation referred to as the productivity paradox. Since then, a decade of studies at the firm and country levels has consistently shown that the impact of IT investment on labor productivity and economic growth is significant and positive" (p. 1). Several researchers have studied productivity improvements through productivity, profitability, and customer surplus (Hitt and Brynjolfsson 1996).

At the industry level, Segars and Grover (1995) have shown how IT has evolved from part of the organizational overhead into a strategic resource capable of changing patterns of competition within industries. They examined the nature and change of structure in three industries during, and after, the introduction of strategic IT. Their findings suggest that in each of these industries, structural characteristics were dramatically altered subsequent to the introduction of competitive-based IT. Brown (2005, p. 169) found that "reported IS evaluation practice appears to be relatively unsophisticated or absent in many organizations." Barua et al. (1995, p. 21) demonstrated that "many of the significant IT impacts occur at low levels in the organization and they can be traced and measured."

Hamilton and Chervany (1981) discussed two different ways in which the success of information systems could be determined by organizations. One approach to evaluating IS success was a goal-centered approach in which the information system was examined to see if the system met the objectives specified by management. The second approach was a system-resource approach that considered the users and whether the system was serving their needs to facilitate communication, improve job satisfaction, or fulfill other needs beyond the primary organizational objectives. This approach highlighted the importance of evaluating information systems from a variety of perspectives including management, users, IS staff, and internal auditors (Hamilton and Chervany 1981).

Subscribing to the second approach, to measure the success of these various IS, organizations are moving beyond traditional financial measures, such as return on investment (Rubin 2004). In an effort to better understand the tangible and intangible benefits of their IS, organizations have turned to methods such as the balanced scorecard (Kaplan and Norton 1996) and benchmarking (Shang and Seddon 2002). Researchers have created models for success (DeLone and McLean 1992; Ballantine et al. 1996; Seddon 1997), emphasizing the need for better and more consistent success metrics.

In this stream of research, researchers have derived a number of models to explain what makes some IS "successful." Davis's (1989) TAM used the Theory of Reasoned Action and Theory of Planned Behavior (Ajzen and Fishbein 1975) to explain why some IS are more readily accepted by users than others. Acceptance, however, is not equivalent to success, although acceptance of an information system is a necessary precondition to success. Early attempts to define information system success were ill-defined due to the complex, interdependent, and multidimensional nature of IS success. The DeLone and McLean IS success model provides researchers and practitioners a model and an approach that focuses on such fundamentals. The remaining part of this chapter outlines future research opportunities based on the DeLone and McLean IS success approach and its applications.

70.4.6 Future Research Opportunities

There exist several opportunities for research in relation to the conception of the DeLone and Mclean IS success model. They include (1) providing a clearer view of the nature of the paths between the key constructs; (2) extending the validity and generalizability of the model constructs and measures; (3) providing a theoretical underpinning of the study model; and (4) improving model constructs.

70.4.6.1 Nature of the Paths between the Key Variables

There has been some criticism of the DeLone and McLean (1992) model for a lack of clarity about the relationships of the constructs—are the relationships among the constructs causal or process? In their 10 year update, DeLone and McLean (2003) responded that their model is both causal/variance and process. They argued that "A temporal, process model suggests that an IS is first created, containing various features, which can be characterized as exhibiting various degrees of system and information quality" (p. 11). Arguing that their model also shows variance/causal relationships, they state that "A causal or variance model studies the covariance of the success dimensions to determine if there exists a casual relationship among them. For example, higher system quality is expected to lead to higher user satisfaction and use" (p. 11). Yet, many users of the DeLone and McLean IS success model employ these constructs blindly, without any regard to the nature of the relationships among the variables.

This suggests that there are several opportunities for future research to extend our understanding of information system success as well as of the model itself. To begin, as demonstrated by Petter et al. (2008) in their meta-analysis of DeLone and McLean studies, no single study has ever addressed the complete DeLone and McLean model. Granted, this would be a rather large undertaking, and perhaps could be addressed in a series of related studies. Second, the nature of the relationship among the constructs provides an opportunity for future research on system success. For example, the necessary and sufficient causality of the model constructs is unclear in the DeLone and McLean model. According to Gable et al. (2008), the directions of the arrows between the constructs are unclear; therefore, one cannot be certain about the positive and negative relationships between the constructs.

70.4.6.2 Extending the Validity

Despite years of research on information systems success, including several hundred publications, there is still no single research study that has tested *all* the constructs and paths in the DeLone and McLean IS success model. Petter et al. (2008) state that "Most past studies focused on a single dimension of success, such as system quality, benefits or user satisfaction. Few studies measure and account for the multiple dimensions of success and the interrelationships among these dimensions." They further state "Until IS empirical studies consistently apply a validated, multidimensional success measure, the IS field will be plagued with inconsistent results and inability to generalize its findings." Gable et al. (2008) make this same observation, encouraging future research to engage in a more holistic approach in employing model constructs.

70.4.6.3 Theoretical Basis

Further opportunities exist in clarifying the relationships between the model constructs. For example, the causal relationship suggested in the model between constructs like system quality and benefits (or individual and organization benefits) requires a theoretical explanation. Moreover, the model does not provide a theoretical explanation for the time lags between constructs. These provide researchers with an opportunity to contribute to a further understanding of IS success.

70.4.6.4 Improving Construct Measurement

In addition to opportunities for improvements in the measurement of IS success variables, several conceptual improvements can also be made. First, researchers should minimize the issues surrounding the use of surrogate, self-reported measures and seek for objective measures to supplement or replace perceptual measures. There is a clear dearth of research in this area, with no studies found that employ objective measures to triangulate subjective measures. Furthermore, paying attention to model constructs that have not been well explored in past studies is another area for future research. Petter et al. (2008) identify use as one of the most under-studied constructs of IS success. Their meta-analysis identified that the relationship between each dimension and use is lower than associations with any other construct. They attribute this inconsistency of attention to the measurement and understanding of the use construct and argue that system use is an important indicator of system success.

70.5 Conclusion

In this chapter, we proposed several opportunities and challenges for IS success researchers. We began by introducing the DeLone and McLean IS success model as a point of reference. The DeLone and McLean model is one of the most widely cited academic studies in the IS domain. However, in spite of the success of the DeLone and McLean model and over three decades of active research on IS success, there are still many opportunities and challenges to the better understanding of information systems success. The body of knowledge for measuring IS success or performance in empirical studies has seen little progress during the past decade.

We presented our arguments for improvements using six considerations, describing each consideration, and followed by ways in which to move forward. We demonstrated that most IS success studies, with few exceptions, do not clarify the positioning of the study, thereby making it difficult to interpret them and use these evaluations for management purposes. First, we highlighted the implications of the evolution of the system. This is a major implication that must be taken into account for IS success studies. More and more corporate-wide systems are accessed through web portals or mobile devices, blurring the boundaries of traditional systems. We then presented several opportunities and challenges for IS researchers within this context. Next, we discussed the stakeholder views, the purpose of the evaluation, timing of the evaluation, and the measurement of success. Finally, we suggested research opportunities in relation to the reconceptualization and expansion of the DeLone and McLean success model.

References

Ajzen, I. and M. Fishbein (1975). *Belief, Attitude, Intention and Behaviour: An Introduction to Theory and Research*. Reading, MA: Addison-Wesley.

Alloway, R. M. and J. A. Quillard (1983). User managers' systems needs. *MIS Quarterly* 7: 27–41.

Amoako-Gyampah, K. and A. F. Salam (2004). An extension of the technology acceptance model in an ERP implementation environment. *Information and Management* 41(6): 731–745.

Anthony, R. N. (1965). *Planning and Control Systems: A Framework for Analysis*. Boston, MA: Harvard University.

Ballantine, J., M. Bonner, M. Levy, and A. Martin (1996). The 3-D model of information systems success: The search for the dependent variable continues. *Information Resources Management Journal* 9(4): 5–14.

Barua, A. and C. H. Kriebel (1991). An economic analysis of strategic information technology investments. *MIS Quarterly* 15(3): 313–331.

Barua, A., C. H. Kriebel, and T. Mukhopadyay (1995). Information technologies and business value: An analytic and empirical investigation. *Information Systems Research* 6(1): 3–23.

Bokhari, R. H. (2005). The relationship between system usage and user satisfaction: A meta-analysis. *The Journal of Enterprise Information Management* 18(2): 211–234.

Botta-Genoulaz, V. and P.-A. Millet (2005). A classification for better use of ERP systems. *Computers in Industry* 56(6): 573–587.

Brown, A. (2005). IS Evaluation in Practice. *The Electronic Journal Information Systems Evaluation*. 8(3): 169–178.

Brynjolfsson, E. and L. M. Hitt (1996). Paradox lost? Firm level evidence on the return of information systems spending. *Management Science* 42(4): 541–558.

Brynjolfsson, E. and S. Yang (1996). Information technology and productivity: A review of literature. *Advances in Computers* 43: 179–214.

Burton-Jones, A. and D. W. Straub (2006). Reconceptualizing system usage: An approach and empirical test. *Information Systems Research* 17(3): 228–246.

Cameron, K. S. and D. A. Whetten (1983). Some conclusions about organizational effectiveness. *Organizational Effectiveness: A Comparison of Multiple Models*. New York: Academic Press, pp. 261–277.

Chi, M. T. H., R. Glaser, and M. J. Farr (1988). *The Nature of Expertise*. Hillsdale, NJ: Erlbaum.

Davenport, T. H. (2000). The future of enterprise system-enabled organizations. *Information Systems Frontiers* 2(2): 163–180.

Davern, M. J. and R. J. Kauffman (2000). Impacts of information technology investment on organizational performance. *Journal of Management Information Systems* 16(4): 121–143.

Davis, F. D. (1989). Perceived usefulness, perceived ease of use, and user acceptance of information technology. *MIS Quarterly* 13(3): 319–340.

Dedrick, J., V. Gurbaxani, and K. L. Kraemer (2003). Information technology and economic performance: A critical review of the empirical evidence. *Journal ACM Computing Surveys* 35(1): 1–28.

DeLone, W. H. and E. R. McLean (1992). Information systems success: The quest for the dependent variable. *Information Systems Research* 3(1): 60–95.

DeLone, W. H. and E. R. McLean (2003). The DeLone and McLean model of information systems success: A ten years update. *Journal of Management Information Systems* 19(4): 9–30.

Drucker, P. F. (1987). The coming of the new organization, *Harvard Business Review on Knowledge Management*. Boston, MA: Harvard Business School Publication, p. 47.

Ericsson, K. A. and J. E. Smith (1991). *Toward a General Theory of Expertise: Prospects and Limits.* Cambridge, England: Cambridge University Press.

Ericsson, K. A., R. T. Krampe, and C. Tesch-Romer (1993). The role of deliberate practice in the acquisition of expert performance. *Psychological Review* 100(3): 363–406.

Farbey, B., F. Land, and D. Targett (1999). Moving IS evaluation forward: Learning themes and research issues. *Journal of Strategic Information Systems* 8(2): 189–207.

Gable, G., D. Sedera, and T. Chan (2008). Re-conceptualizing information system success: The IS-impact measurement model. *Journal of Association for Information Systems* 9(7): 377–408.

Gable, G. G., T. Chan, and W.-G. Tan (2003). Offsetting ERP risk through maintaining standardized application software. *Second-wave Enterprise Resource Planning Systems.* G. Shanks, P. Seddon, and L. Willcocks, eds. Cambridge, U.K.: Cambridge University Press, pp. 220–237.

Grabski, S. V., S. A. Leech, and P. J. Schmidt (2011). A review of ERP research: A future agenda for accounting information systems. *Journal of Information Systems* 25(1): 37–78.

Grover, V., S. R. Jeong, and A. H. Segars (1996). Information systems effectiveness: The construct space and patterns of application. *Information & Management* 31(4): 177–191.

Hamilton, S. and N. L. Chervany (1981). Evaluating information systems effectiveness—Part II: Comparing evaluator viewpoints. *MIS Quarterly* 5(4): 79–86.

Hirschheim, R. and S. Smithson, eds. (1998). *Evaluation of Information Systems: A Critical Assessment. Beyond the IT Productivity Paradox.* Wiley series in information systems. West Sussex: John Wiley & Sons Ltd.

Hitt, L. M. and E. Brynjolfsson (1996). Productivity, business profitability, and consumer surplus: Three different measures of information technology value. *MIS Quarterly* 20(2): 121–142.

Hitt, L. M., D. J. Wu, and X. Zhou (2002). Investment in enterprise resource planning: Business impact and productivity measures. *Journal of Management Information Systems* 19(1): 71.

Kanaracus, C. (2008). Gartner: Global IT spending growth stable. *InfoWorld* April 3, 2008.

Kang, D. and R. Santhanam (2003). A longitudinal field study of training practices in a collaborative application environment. *Journal of Management Information Systems* 20(3): 257–281.

Kaplan, R. S. and D. P. Norton (2000). Having trouble with your strategy? Then map it. *Harvard Business Review* 78(5): 167–176.

Kaplan, R. S. and D. P. Norton (1996). *The Balanced Scorecard: Translating Strategy into Action.* Boston, MA: Harvard Business School Press.

Keen, P. G. W., ed. (1987). MIS research: Current status, trends and needs. *Information Systems Education: Recommendations and Implementation.* Cambridge, New York: Cambridge University Press.

King, W. R. and J. I. Rodriguez (1978). Evaluating management information systems. *MIS Quarterly* 2(3): 43–51.

Kumar, V., B. Maheshwari, and U. Kumar (2003). An investigation of critical management issues in ERP implementation: Empirical evidence from Canadian organizations. *Technovation* 23(10): 793–807.

Mabert, V. A., A. Soni, and M. A. Venkataramanan (2001). Enterprise resource planning: Common myths versus evolving reality. *Business Horizons* 44(3): 69–76.

Mahmood, M. A., L. Hall, and D. L. Swanberg (2001). Factors affecting information technology usage: A meta-analysis of the empirical literature. *Journal of Organizational Computing & Electronic Commerce* 11(2): 107–130.

Mandal, P. and A. Gunasekaran (2003). Issues in implementing ERP: A case study. *European Journal of Operational Research* 146(2): 274–283.

Markus, L., S. Axline, D. Petrie, and C. Tanis (2003). Learning from adopters' experiences with ERP: Problems encountered and success achieved. *Second-Wave Enterprise Resource Planning Systems.* G. Shanks, P. Seddon, and L. Willcocks, eds. Cambridge, U.K.: Cambridge University Press.

Markus, L. M. and C. Tanis, eds. (2000). *The Enterprise Systems Experience—From Adoption to Success. Framing the Domains of IT Research: Glimpsing the Future through the Past.* Cincinnati, OH: Pinnaflex Educational Resources, Inc.

Matlin, G. (1979). What is the value of investment in information systems? *MIS Quarterly* 3(3): 5–34.

McAfee, A. (2006). Mastering the three worlds of information technology. *Harvard Business Review* 84(11): 141–148.

Mirani, R. and A. Lederer (1998). An instrument for assessing the organizational benefits of IS projects. *Decision Sciences* 29(4): 803.

Morris, M. G. and V. Venkatesh (2010). Job characteristics and job satisfaction: Understanding the role of enterprise resource planning system implementation. *MIS Quarterly* 34(1): 143–161.

Park, J.-H., H.-J. Suh, and Yang (2007). Perceived absorptive capacity of individual users in performance of Enterprise Resource Planning (ERP) usage: The case for Korean firms. *Information & Management* 44: 300–312.

Petter, S., W. DeLone, and E. McLeon (2008). Measuring information systems success: Models, dimensions, measures, and interrelationships. *European Journal of Information Systems* 17(3): 236–263.

Remenyi, D. and M. Sherwood-Smith (1999). Maximise information systems value by continuous participative evaluation. *Logistics Information Management* 12(1/2): 14–31.

Rolefson, J. F. (1978). The DP check-up. *Journal of System Management* 29(11): 38–48.

Ross, J. W. and M. R. Vitale (2000). The ERP revolution: Surviving vs. thriving. *Information Systems Frontiers* 2(2): 233–241.

Rubin, H. (2004). Into the light. *CIO Magazine.* http://www.cio.com.au/index.php/id;1718970659 (accessed on July 2004).

Sabherwal, R., A. Jeyaraj, and C. Chowa (2006). Information system success: Individual and organizational determinants. *Management Science* 52(12): 1849–1864.

Seddon, P. B. (1997). A respecification and extension of the DeLone and McLean model of IS success. *Information Systems Research* 8(3): 240–253.

Seddon, P. B., C. Calvert, and S. Yang (2010). A multi-project model of key factors affecting organizational benefits from enterprise systems. *MIS Quarterly* 34(2): A305–A311.

Seddon, P. B. and M.-Y. Kiew (1996). A partial test and development of DeLone and McLean's model of IS success. *Australian Journal of Information Systems* 4(1): 90–109.

Seddon, P. B., S. Staples, R. Patnayakuni, and M. Bowtell (1999). Dimensions of information systems success. *Communications of the AIS* 2, Article 20.

Sedera, D. and S. Dey (2008). Expert performance in information systems. *International Conference on Information Systems (ICIS '08).* Paris, France: AIS.

Sedera, D. and G. Gable (2004). A factor and structural equation analysis of the enterprise systems success measurement model. *Proceedings of the 25th International Conference on Information Systems.* Washington, DC: Association for Information Systems.

Sedera, D., G. Gable, and T. Chan (2003). ERP success: Does organization size matter? *Proceedings of the 7th Pacific Asia Conference on Information Systems.* Association for Information Systems. http://www.pacis-net.org/file/2003/papers/erp/163.pdf (accessed on April 22, 2013).

Segars, A. H. and V. Grover (1993). Re-examining perceived ease of use and usefulness: A confirmatory factor analysis. *MIS Quarterly* 17(4): 517–525.

Segars, A. H. and V. Grover (1995). The industry-level impact of information technology: An Empirical analysis of three industries. *Decision Sciences* 26(3): 337–368.

Serafeimidis, V. and S. Smithson (1998). Analysing information systems evaluation: Another look at an old problem. *European Journal of Information Systems* 3(1): 158–174.

Serafeimidis, V. and S. Smithson (2000). Information systems in evaluation in practice: A case study of organizational change. *Journal of Information Technology* 15(2): 93–105.

Serafeimidis, V. and S. Smithson (1999). Rethinking the approaches to information systems investment evaluation. *Logistics Information Management* 12(1/2): 94–107.

Shang, S. and P. B. Seddon (2002). Assessing and managing the benefits of enterprise systems: The business manager's perspective. *Information Systems Journal* 12(4): 271–299.

Stockdale, R. and C. Standing (2006). An interpretive approach to evaluating information systems: A content, context, process framework. *European Journal of Operational Research* 173(3): 1090–1102.

Strong, D. M. and O. Volkoff (2010). Understanding organization—Enterprise system fit: A path to theorizing the information technology artifact. *MIS Quarterly* 34(4): 731–756.

Sumner, M. (2000). Risk factors in enterprise-wide/ERP projects. *Journal of Information Technology* 15: 317–327.

Thatcher, M. E. and J. R. Oliver (2001). The impact of technology investments on a firm's production efficiency, product quality, and productivity. *Journal of Management Information Systems* 18(2): 17–45.

Umble, E. J., R. R. Haft, and M. M. Umble (2002). Enterprise resource planning: Implementation procedures and critical success factors. *European Journal of Operational Research* 146(2): 241–257.

Urbach, N., S. Smolnik, and G. Riempp (2009). The state of research on information systems success—A review of existing multidimensional approaches. *Business and Information Systems Engineering* 4: 315–325.

van der Heijden, H. (2000). *Measuring IT Core Capabilities for Electronic Commerce: Results from a Confirmatory Factor Analysis.* International Conference of Information Systems. Brisbane, Australia.

Wang, W. and D. Sedera (2011). *A Benefits Expectation Management Framework for Supply Chain Management Systems.* International Conference on Information Systems. Shanghai, China: AIS.

71

Business Value of IS Investments

Ellen D. Hoadley
Loyola University Maryland

Rajiv Kohli
College of William & Mary

71.1 Introduction

The concept and constructs surrounding the business value of information systems (IS) have matured through the 1990s and into the twenty-first century. As data processing grew to information services and the data processing manager evolved to the chief information officer, evaluating the cost of technology has culminated in the valuation of the enterprise IS contributions, architecture, and innovations. The resultant question that the topic addresses is *How do we measure the ultimate impact that our expenditures on IS have on the strategic outcomes of the organization?*

This chapter provides an overview of the underlying principles, research findings, practical applications, and future trends in the efficacy and measurement of the business value of IS investments. Early efforts sought to determine tangible returns as evidenced in the financial statements. Current thinking posits that the investments yield strategic returns when skillfully applied by the chief information officer (CIO) with integrative and innovative vision within an organization that has the knowledge and abilities to use the investments. The understanding of the practice of measuring the business value of IS has moved systematically from the narrow focus of viewing IS as a collection of hardware and software that creates a technology capital investment to a collection of hardware and software that enables organizational processes that may be departmental or global to drive strategy. The former can be calculated succinctly to evaluate efficiency (doing things right); the latter is calculated at the organizational level to measure effectiveness (doing the right things).

71.2 Underlying Principles: What Do We Mean by "Business Value of IS Investments"?

Organizations measure their functional performance using metrics on the financial statements. Marketing is evaluated through revenues and market share. Improvements in operations are demonstrated through decreasing costs, return on assets (Kobelsky et al. 2008), and inventory turns (Dehning et al. 2007). Investments yield returns; profits increase dividends, retained earnings, and owners' equity. When thinking about the business value of IS investments, executives seek comparable metrics to evaluate whether or not the resources put into their information architecture have provided and/or are anticipated to provide tangible value. Additionally, traditional metrics of return on investment (ROI) are often conceptualized as tangible value realized from the actual technology purchase with less consideration of the differential returns resulting from employee development and process improvements (Im and Rai 2008). Generally, executives look at competing investments, the duration of potential payoffs, and the overall economic situation of the organization to justify expenditures in information technology (IT) (Weill 1992; Barua et al. 1995; Brynjolfsson and Hitt 2000; Devaraj and Kohli 2002). Increasingly, the focus of IS investments is on outcomes such as agility (Tanriverdi et al. 2010), flexibility (Saraf et al. 2007), risk management (Dewan and Ren 2011), and interfirm collaboration (Richey and Autry 2009).

Principle 1—The business value of IS investments is best viewed from both a variance and a process approach.

The variance approach to measuring the value of IS investment seeks tangible improvements in organizational metrics such as increased revenues or decreased costs (Weill 1992; Brynjolfsson and Hitt 2000). The process approach recognizes that even when the value of IS investment might be realized in traditional organizational metrics, firms may choose to pass the value on to their consumers as lower prices in order to sustain customer loyalty (Barua et al. 1995). The value of IS may not be identifiable with traditional financial statement metrics. The process approach seeks improvements in the processes affected by the IS investment and views internal process metrics as leading indicators of organizational value. Kohli and Hoadley (2006) present a theoretical framework (see Figure 71.1) that combines the two approaches. The framework recognizes the importance of overall organizational performance as a primary indicator of the value of IS investment while also providing a way to evaluate such an investment in an industry where competitive factors may mask that value or where the value realization may lag the investment by multiple time periods. Whether the business value of an IS investment is passed on to consumers, is mitigated by overall economic conditions, or is lagging the periodic reporting of organizational performance, the measurement of that value is still important to the strategic planning and decision making of the enterprise.

Principle 2—The business value of IS investments is best measured using balanced multiple metrics.

Metrics quantify the value of IS in ways that can be evaluated, compared, and improved. However, there should not be a single set of metrics. Instead, organizations can best measure the value of their IS investments by using and evaluating financial and operational metrics, tangible and intangible metrics, immediate and lagged metrics, and productivity, profitability, and customer value metrics. This may appear to be an overdose of measurement. However, when viewed as a balanced scorecard of metrics, the overall measurement provides the best opportunity to identify the interdependencies of the IS investments and organizational performance (Devaraj and Kohli 2002). Such a balanced scorecard approach would measure the value of the IS investment from the perspective of the customer, the efficiency of the operations, the external positioning of the firm, the innovative capabilities of the employees, and the financial return of the investments. Planning the best metrics to use to view IS investments from multiple perspectives supports the holistic evaluation and improvement of the organization's IS implementation.

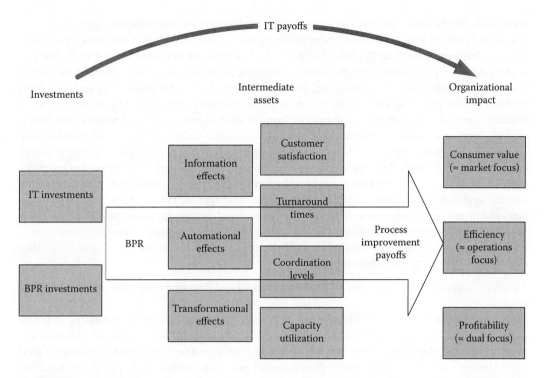

FIGURE 71.1 Process approach to information technology (IT) organizational impact. (From Kohli, R. and Hoadley, E.D., *DATA BASE Adv. Inf. Syst.*, 37(1), 40, 2006.)

Principle 3—The potential for realized business value of IS investments depends on the position of the asset on the technology maturation curve.

Organizations must pay attention to the maturation curves of new technologies as well as assess their own comfort level with new "bleeding edge" technologies and with old tried-and-true technologies. The former technologies offer higher potential for process improvement and return value; the latter technologies offer lower risk of failure. Both may work in the organizational portfolio of interdependent systems, but all will be retired eventually. Organizations must align their technology adoption processes to the risk tolerance of the firm. Successful alignments have been made in organizations that are able to recognize and retire technologies that risk obsolescence, recognize competitors in related fields that may threaten the competitive landscape, and identify usability factors in a new technology that eases acceptance.

Strategic IS implementations may employ disruptive technologies, characterized as emerging technologies that are rapidly developing in ways that are attractive to a small segment of consumers (Bower and Christensen 1995) and are yet invisible to those enterprises that strive to stay close to their current customers. Disruptive technologies exploit novel attributes that are not desirable to current customers but that are quickly adopted by a new customer segment and that lure away a significant portion of the current customer base. Just as Blockbuster was challenged by Netflix and Research in Motion had to deal with Apple's iPhone, so will all organizations be required to be continually vigilant in their evaluation of emerging technologies. According to Gartner, within the next 5–10 years, augmented reality, crowdsourcing, and 3D scanners will be the norm (LeHong and Feng 2012). The values of these technologies are difficult to evaluate within an existing firm and often are viewed as having little or no value to a current customer base. To minimize the risk of disruptive technologies, the enterprise must be vigilant in understanding the strategic significance of its own existing technology portfolio as well as emergent technologies that present creative opportunities for competition.

Principle 4—The business value of IS investments is justifiable using multidimensional approaches.
Principles 1 through 4 lay the foundation for justifying an IS investment. This justification is essentially using the business value of an IS project to lay out the business case to promote investing in that particular project over alternative opportunities for investing the same funds. This is done by converting the value of an IS investment and its project payoff into a logical or mathematical form and comparing the factors of the project against similar factors in the alternative opportunities. Organizations employ differing approaches to justifying resources investment such as intuition-based models, cost-benefit analysis, break-even analysis, net present value, economic value added, and regression-based statistical analysis. For future opportunities, firms use real options–based methodologies to justify IS investment for which NPV or ROI cannot be calculated (Fichman 2004). Each of these approaches quantifies organizational factors of performance and uses these quantifications to compare potential investments. Another approach (Flatto 1996) enhances a net present value approach by incorporating risk management and investment changes into the model. Though each justification approach quantifies like factors of competing projects, the assessment of future value is likely to best predict achieve positive outcomes when approaches are combined to calculate a holistic evaluation with multiple dimensions rather than evaluating each project dimension separately.

Principle 5—Sustained measurement of IS value is accomplished through a comprehensive organizational process.
Measurement of the business value of IS investment is most meaningful when incorporating all factors of the project as implemented. This means that the project must be initially developed with measurement of business value incorporated into the work. One cannot wait until after user acceptance of an IS project before building the measurement structures and gathering evidence of the payoff. Further, when the value of IS investment emerges in other parts of the organization, demonstrating business value is not only the responsibility of the IS managers; rather, t is an organization-wide process (Kohli and Johnson 2011). Therefore, a comprehensive process of value measurement should span the phases from pre-investment to post-investment review.

An exploration phase develops an understanding of what is to be measured, how it will be quantified, and which data techniques are most appropriate for analysis. A discovery phase addresses organizational issues in order to maximize the managerial and political factors of IS business value. The analysis phase collects, analyzes, and reports the data that demonstrates the quantifiable portion of IS business value. The communication phase is the feedback loop that provides the information back to the organization as to the outcome of the analysis and the language that will be used to institutionalize a preference, culture, and bias of measurement of value. Using a comprehensive evaluation process makes the measurement of IS business value a quarter involved with data and three quarters involved with organizational planning, involvement, and communication. Many projects fail, even though they may employ analysis, because they fail in the organizational quadrants. Incorporation of all phases improves the probabilities of implementing demonstrable and sustainable value in one IS investment and supports a culture of linking together like investments over time.

Principle 6—Strategic integration of IS value measurement informs planning and decision making.
Following a process of measuring the business value of an IS investment does not guarantee strategic benefits. It is not until the value measurement is used for planning future strategic initiatives and making decisions for operational improvements that the business value is realized. Organizations must communicate the measurements across multiple levels and through multiple channels. Executive leadership uses measurement to identify immediate and lagging payoffs, to locate impacts that may occur outside the locus of the investment, and to communicate to stakeholders that the vision for the future is achievable through concrete actions that permeate the organization. The integration of business value with senior management motivation, planning, and decision making is an emerging area of research (Salge et al. 2011).

71.3 Trends in the Field

As the measurement of the business value of IS has evolved, different trends have emerged that have generated interest because of their practical application and importance to a fuller understanding. The first trend is a shift from measuring efficiency as a saving of time and materials to measuring effectiveness of the outcomes of systems. Though efficiency remains as an important base for improvement, organizations seek value in systems that achieve the desired goals for themselves, their suppliers, and their consumers. A second trend is the consideration of the sustainability of the business value of IS. It is important to realize first-time, one-time, and current value of the IS investment. It is also becoming more important to sustain that value as long as possible and to recognize the appropriate time to retire or transition an organizational system. A third trend incorporates the global effects of IS and views the business value of IS as the linking mechanism among strategic partners. The goal is to enable synergies that expand the capacities of organizations beyond those that are possible for each one alone. Finally, an emerging trend is to view the business value of IS investments in how they expand the total collection of resources of an enterprise. This view includes the possibility that IS investments enable creative and innovative opportunities above and beyond the streamlined processes of the organization. These opportunities are realized in combinations of organizational assets, strategic vision, and human creativity all synthesized using the firms IS. These trends represent an expansion of the base research presented in the underlying principles of the business value of IS and indicate areas of future investigation that are of interest to practitioners and researchers.

71.3.1 Shift from Efficiency Measures to Effectiveness Measures

The ideas surrounding the importance of the business value of IS have endured since the 1990s. In the 1980s, the Society for Information Management began annually to survey executives in IS to determine the key issues facing them. In 1996, in the midst of the ramp up to Y2K, 70% of the topics reported were issues of IS hardware and software (Brancheau et al. 1996). The issue that eventually was incorporated in the business value of IS—aligning the IS organization within the enterprise—was ranked #9. Since then, the topic label has become "information technology (IT) and business alignment" (Table 71.1). From 2005 to 2010, the topic has remained at or near the top of the list, currently sharing the spot with organizational issues such as "business productivity and cost reduction" and "business agility and speed to market" (see Table 71.1). Additionally, this topic has remained an important focus of study for IS academics as seen in conference proceedings and special issues of journals.

The practitioner press also has evolved in its view of the importance of IS value (Chillingworth 2010). For example, CIO magazine (Overby 2011) stated that those CIOs who want to remain part of a strategic company must shift from using IT simply as a tool to support the strategy of the firm. Likewise, the CIO must become a business leader who can understand business value and communicate the ways that IS can and does create, enhance, and enable that business value (Mitra et al. 2011). Businesses must develop strategy that deploys IT to connect the information assets to their decision makers and customers and then lead the communication of how those assets create value.

The shift from efficiency to effectiveness was presented early on (Soh and Markus 1995), bringing together the constructs of process measurement and IT metrics. The process of the realization of value from investments in IT was articulated as follows: IT expenditures lead to IT assets that have IT impacts that are subsequently realized as organizational performance. The IT expenditures and organizational performance had normally been measured as part of the financial statement. Understanding the intermediate measures provided a way to understand the time delay and mitigating factors between the cause and effect of the variables.

To link efficiency to organizational performance, three categories of metrics were articulated: productivity, business profitability, and consumer surplus (Hitt and Brynjolfsson 1996). The fact of measuring

TABLE 71.1　MISQ Executive Key Issues

MISQ Executive Key Issues for IS Executives (2005–2010)	
2010	• Business productivity and cost reduction
	• Business agility and speed to market
	• IT and business alignment
	• IT reliability and efficiency
	• Business process reengineering
2009	• Business productivity and cost reduction
	• IT and business alignment
	• Business agility and speed to market
	• Business process reengineering
	• IT cost reduction
2008	• IT and business alignment
	• Build business skills in IT
	• IT strategic planning
	• Attracting IT professionals
	• Making better use of information
2007	• Attracting, developing, retaining IT professionals
	• IT and business alignment
	• Build business skills in IT
	• Reduce the cost of doing business
	• Improve IT quality
2005	• IT and business alignment
	• Attracting, developing, and retaining IT professionals
	• Security and privacy (tied with above)
	• IT strategic planning
	• Business process reengineering

at the organization level moved the metrics into the realm of the c-level executives and beyond the limited scope of the CIO. Adding consumer surplus brought the measurement focus on effectiveness. If the IS investments were effective, the organization's customers could realize improvements from those investments. Understanding these organizational performance metrics provided a way to incorporate a broader concept of value to include effectiveness. Additionally, consumer surplus could explain why an organization could realize a large value of effectiveness from IS investments but not see it in profitability because it had passed the value on to its customers.

The interaction between IT implementation, business process reengineering (BPR), and organizational performance was examined across over 300 firms between 1996 and 1999 (Ramirez et al. 2010). The results indicate that synergies between IT and BPR implementations result in improved productivity efficiency and improved market value. These findings support the importance of both efficiency and effectiveness at the organizational level when measuring the business value of IS.

One component of measuring effectiveness is evaluating increases in intangible assets (Tambe et al. 2011). These intangible IT assets include a firm's work practices, information structures, and employee-skill mix that are ancillary to the tangible IT investments. Research in a longitudinal study found that the quantity of intangible IT assets increased from 1987 through 2005 making those assets available for future productivity gains for the organizations. The accompanying low levels of depreciation of the assets indicate a stockpiling. It was concluded that investment in these intangible assets is linked to the growth of significant productivity and should lead to growth. Yet, the importance of these findings is the articulation of specific metrics of intangible IT assets that lend themselves to quantitative analysis similar to tangible assets.

In spite of the years of research into the business value of IS, the concepts, terminologies, and findings remain fuzzy. Some small portion of the perception of long-term vagueness is the reality that "years of research" in the field of IS is not actually as long as "years of research" say, in medieval Scottish poetry. However, it is problematic that theory remains underdeveloped and loosely applied (Schryen 2010). To fully comprehend the phenomenon of the business value of IS for efficiency and effectiveness, programs of research must link studies together and should strive to progressively expand the body of knowledge. Literature reviews must use the theoretical base as a skeletal framework to tie the literature together in a meaningful way that creates knowledge with predictive and explanatory power.

71.3.2 Sustaining the Business Value of IS over Time

An additional focus of research in the area of the business value of IS is that of value sustainability. Organizations may or may not realize immediate or even lagged value from their investments in IS. However, a greater challenge may be to sustain those improvements in effectiveness as seen in profitability, productivity, and consumer value. An examination of firms across multiple industries receiving awards for their realization of value from their IS investments (Masli et al. 2011) looked for improvement of organizational performance over time as well as the duration of improvements. Structural shifts such as Y2K and the dot.com era slowdown caused many of these firms to lose momentum at the turn of the century. However, the study shows that since that time, many of the firms have struggled to sustain the realized value of their IS investments.

Of particular interest are those firms striving to employ sustained IS investment to achieve a low-cost position in a competitive environment (Tallon 2007). The challenge comes from the commoditization of the IT infrastructure that is more easily replicated by others. Those firms that have achieved a level of business value sustainability are those firms with multiple successful IS implementations. This confirms the principle that measuring business value of IS is best accomplished using multiple metrics at the organizational and process levels (see Figure 71.1), and that a culture of using IS to achieve organizational goals contributes to sustaining the business value.

71.3.3 Business Value of Inter-Organizational IS

As businesses seek new ways to achieve their missions and goals, they find that viewing themselves as singular entities competing among others limits their opportunities for success. Firms that seek strategic partners to achieve shared aims realize increasing returns through implementations of interoperable systems. An emerging research question (Grover and Kohli 2012) is asking how separate organizations can exploit their inter-organizational IS to realize values to each that neither could realize on its own. The relational view proposed in the strategy literature (Dyer and Singh 1998; Dyer 2000) presents a model of four factors that relate to value co-creation within multiple organizations:

- *Relationship-specific assets* are those shared assets that enable value co-creation to those firms in the partnership and are protected to ensure their use being limited solely to those organizations.
- *Complementary resources and capabilities* are those within one or more of the organizations in the partnership that enable value co-creation within the partnership but which could not enable value co-creation solely within one of the organizations.
- *Knowledge-sharing routines* enable shared information and expertise to inform value co-creation.
- *Governance* includes contracts and financial protections that streamline costs and boost the incentives for value co-creation.

While numerous examples of these factors and concepts exist, research into value co-creation has a broad horizon of opportunity to add to current theory for expanded understanding and application.

71.3.4 Capabilities of the Firm vs. a Systems Theory View

An alternative framework for understanding and examining the business value of IS is based on the creation of capabilities within a firm (Nevo and Wade 2010). In this framework, existing research takes a resource-based view of the firm explaining that such a view assumes that firms compete based on their own organizational resources. Those organizational resources that have the potential for strategic improvement of the firm must have value, rarity, inimitability, and nonsustainability (Barney 1991). Since IT assets are available and replicable, the strategic potential is minimal. Instead, it is the IS resources that have those factors for creating and sustaining competitive advantage. Therefore, much of the literature that has focused on the business value of IT instead should look at the synergies created between *organizational* characteristics and IT assets for examining the organizational impact.

IT-enabled resources [and IS assets] are developed when IT assets and organizational resources join to achieve effective mission performance (Nevo and Wade 2010). The emergent capabilities may be predicted or not and may have a positive, negative, or neutral impact on the firm. Those positive emergent capabilities are labeled as *synergy* and are sought to achieve organizational goals. The research in the business value of IS should determine the factors in why some potential emergent capabilities are synergistic and others fail to achieve positive outcomes.

Currently, firms face hypercompetition where the industry is marked with products and services that evolve dynamically and stretch the competitive environment of the business. In these organizations, the business value of IS is realized when it supports the ability of the firm to adapt its processes and information flows along with these dynamically evolving products and services (Tallon et al. 2002; Tanriverdi et al. 2010). The adaption includes seeking new areas of competitive advantage while sustaining old products and services only so long as they sustain their business value. The adaptation is mirrored by evolving IS that sustain a series of competitive advantages over time. There is additional evidence that a flexible IT infrastructure that is aligned with the organization's strategic goals yields organizational agility critical to the achievement of those strategic goals (Tallon and Pinsonneault 2011). The overall IS support of business competition plays out like a concerto with slower and faster movements, yet with a sustaining theme of business value.

71.4 Evolving Role of CIO

In addition to measurement of business value of IT, the focus increasingly has been on the organizational structure and the role of the CIO in realizing the value that technology provides to the organization. Recent research has indicated the diverse and occasionally multiple roles of the CIO that have evolved since its inception (Carter et al. 2011). It is the CIO who is charged with recognizing future trends in business and technology; it is the CIO who provides the leadership in realizing the value of the IS investment (Kohli and Johnson 2011) and whose role in the organization is likely to influence business performance (Banker et al. 2011). This leadership role is required within the IS function for system deployment and within the business as a whole for developing technology-enabled strategy (Chen et al. 2010).

The development and focus of the efforts of the CIO were chronicled by Hoadley and McFadden (2010), concluding with the concept that the CIO, to remain relevant, needs to become the chief information innovation officer of the organization. Early data processing managers focused on hardware and software, efficiency of processing, and minimization of downtime of the systems. In 1995 when organizations were beginning to reduce their silo structures by looking at their processes, IS and business executives were called to form partnerships to improve the success of their reengineering efforts (Martinez 1995). A blueprint for building such a partnership was presented that shifted the activities of the CIO from solely managing the IS function internally to aligning the management of IS and business strategy.

Today, the focus has shifted from technical efficiency to strategic leadership. The CIO is to guide the enterprise in its collection and dissemination of vast amounts of internal and external data in information-rich, immediately available, and user-specific formats that lead to timely and strategic deci-sion making. The CIO becomes the executive-level systems thinker who can communicate the value of IS in nontechnical, business-impacting terms that provide direction to the organization. The strength of IS leadership is grounded in a partnership between the IT and business functions with particular atten-tion to strategic planning and after-action-reviews of IS implementations (Tallon 2008).

The strategic leadership role of the CIO is particularly challenging in those firms dominated by specialized mechanization and complex capital investments (Kohli and Johnson 2011). Traditionally, these organizations have used these internal structures to create barriers to entry. The downside of these structures is the limited flexibility, which is antithetical to the nimble processes that are the hallmark of the digital economy. Building operational excellence into redesigned business processes requires both a holistic view of the organization and a technical savvy raising the CIO function to strategic visibility.

The maturity of an enterprise will be modeled around an information orientation in which the effective use of information permeates the decision making at all levels (Kettinger et al. 2011). The CIO serves as a strategic partner with others in the executive suite to create a culture that exploits infor-mation management to increase competitive advantage. The leadership model (Figure 71.2) demon-strates the leadership positions that can drive the organization to an implementation of information orientation (IO) for strategic advantage. With corporate examples, the information orientation guide-lines are presented to the CIO to use to assess the firm's current position, to decide whether to assume leadership, and to select a leadership style appropriate for the firm. Once leadership is assumed, the CIO must determine how to mobilize resources to improve the organization's effectiveness in real-izing the business value of IS.

CIOs should use the specific metrics and language to communicate the value of IS within the organization (Mitra et al. 2011). The metrics are grouped into five domains that address those busi-ness categories that are important to executives. Using the guidelines provided, the CIO can lead the development of a set of metrics that are key performance indicators of IS value and use these indica-tors to improve the impact of IT within that enterprise. The use of these guidelines supports the use of organizational business value measurement with the communication tools to provide the feedback loop for improvement.

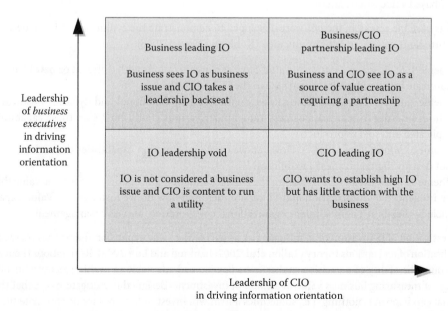

FIGURE 71.2 Leadership positions in driving information orientation. (From Kettinger, W.J. et al., *MIS Q. Exec.*, 10(4), 157, 2011; Reproduced with permission from MISRC—Indiana University.)

71.5 Future Impact on Practice

Practitioners will experience profound effects of measuring the business value of IS. What once was perceived as the *productivity paradox* has become a language and understanding that deploying a supportive and flexible information architecture informs users about the content and context of real-time decision making (Weiss 2011). Some organizations seek to reduce the fragmentation of their systems through the use of an integrated enterprise architecture (Bradley et al. 2011). The architecture is logically designed to link technologies and business processes for current and planned IS development thereby "informating" those processes for strategic purposes. Whether executives choose to implement such a technology/process coupling or choose to implement processes that integrate disparate systems, the interdependencies of technologies and processes require adroit management to yield high value. It is the integrative executive who will lead the most successful organizations to realize the high business value of IS investment.

71.6 Future Research

A review and synthesis of the research literature articulated a theoretical framework in which the business value of IS takes into account the context of the enterprise (Melville et al. 2004). The framework recognized that a firm, as well as its competitive and macro environments, encompasses multiple factors that contribute to and are affected by the organization's performance. To address these factors, the authors posed multiple research questions that have shaped the framework of the business value of IS within those environments.

A consolidation of research synthesized the collection of findings that remain pertinent (Kohli and Grover 2008):

- IT does create value.
- IT creates value under certain conditions; numerous factors mediate IT and value creation.
- IT-based value manifests itself in many ways.
- IT-based value is not the same as IT-based competitive advantage.
- IT-based value can be latent.

From this baseline research, themes were developed to demonstrate the evolution of IT's value and how it is captured, measured, and communicated (Kohli and Grover 2008):

- Theme 1: *IT-based co-creation of value*. How can organizations with different or new IT resources co-create value and equitably partake in the IT-based value?
- Theme 2: *IT pervasiveness*. How can we digitize various functional and dynamic business capabilities in order to increase business value under various conditions? Can these capabilities be replicated and sold as products?
- Theme 3: *Information mindset*. How can we create information capabilities that enhance and do not destroy digital business capabilities?
- Theme 4: *Value expansion*. What are the indirect and intangible paths to economic value that can be influenced by information and IT capabilities, and how do we foster them? Value expansion includes flexibility, agility, inter-organizational collaboration, and risk management.

The investigation of IS investment to mitigate risk as part of the value expansion theme has emerged from the application of real options theory (Tallon et al. 2002; Lankton and Luft 2008). Real options theory takes into account the ability of a decision maker to be flexible and adaptable in making IS investments. The challenge of measuring business value in a chain of investments designed to mitigate risk is that the best thing that can happen is nothing; i.e., sometimes a firm will invest in IS to reduce or eliminate the probabilities of adverse events. The hard metric becomes the estimated cost of potential failure compared to the investment to achieve no failure, reduced exposure to risk, and/or reduced cost of insurance coverage.

Though quantifiably challenging, the application of this type of IS investment is evidenced in practice, specifically in the healthcare industry to improve estimates of complication rates and to reduce malpractice insurance premiums (Menon and Kohli 2012). On the downside, there is additional evidence that the investment in IS may create legal evidence of malpractice, which could increase exposure to risk and malpractice insurance premiums (Ransbotham and Overby 2010). These alternative findings indicate that the IT itself is value neutral, but the IS investment may yield quantifiable risk and mitigation options.

As research has expanded into the areas mentioned earlier, the IT co-creation of value has warranted particular attention. A special issue of *MIS Quarterly* in 2012 presented multiple studies that focus on the particular factors that enhance the development of business value when intangible and tangible IT assets work together. A continuing question of interest is whether or not investments in IS creates measurable improvements in profitability, productivity, and consumer value (Mithas et al. 2012). Yet increasingly, studies explore how inter-organizational systems create and sustain business value for participating partners (Han et al. 2012; Rai et al. 2012; Sarker et al. 2012). The issue includes an essay (Constantinides et al. 2012) proposing that future research consider the pragmatic ends that studies can and should address as well as the decisions that are made regarding data collection, experimental design, and ultimate impact of investigations. The broader question of relevance vs. rigor remains as valid now as it was when the IS field emerged.

Future research also will expand the IS field into interdisciplinarity. Just as organizationally implemented systems expanded into inter-organizational systems and creation of business value expanded into co-creation of business value, so does the field of IS continue to expand into its close business disciplines and broader behavioral disciplines. Ethical questions remain about co-creating value for those who are separated from access to that value as the digital divide separates "have" from "have-not" and digital natives from digital immigrants. What particular value do specific emerging technologies provide? How does consumerization of IS, where the boundaries of personal and professional processes are blurred, influence our value perception of IS and IS investment? How do we create value within virtual business processes such as online diagnosis and planning tools? Is IS value creation different or the same in virtual organizations? These broader questions will require broader thinking to yield greener pastures of study. Yet this is the promise of the future that IT and IS are no longer systems for their own sakes. Instead, they are value-development tools of the business and organizational future. As the speed of technology increases, and the reach of global processes expands, so must our understanding of technical systems combined with human innovation synergistically increase the power organizations to achieve their goals.

71.7 Summary

The study of the business value of IS emerged as a topic of interest as IT infrastructure and personnel have become deeply embedded in business processes. Once it became clear that organizations rely on their IS investments to conduct transactions, make decisions, and monitor their environments, justification of expenditures led to the application of financial analytics to support the investments. Yet large amounts of incoming data (e.g., from social media), the pervasiveness of the technology, and IT's central role in process enablement have exposed limitations of currently available analytical tools. Organizations continue to seek explanations and knowledge from the IS research community to identify how to assess the value of investing in IS vis-à-vis other demands and opportunities for investment. Early studies viewed narrow departmental applications as capital expenses to be amortized and depreciated. Organizational systems enabled interdepartmental processes measuring effectiveness at the firm level. Organizations have come to realize that combining the technology infrastructure with the human capabilities expanded the performance of the assets synergistically. This realization has broadened the investigation of how IS investment creates value for the business.

Currently, organizations continue to justify their expenditures in IS. However, recent innovative uses of IS indicate that the focus of the justification must be on how the organization can use systems to enable and expand business strategy and forge strategic partnerships as a basis for competitive advantage for all the partners and for the consumers. The IS value is embedded within business processes and in inter-organizational, inter-stakeholder processes with the lines between personal and professional applications increasingly blurred. As the technology becomes inseparable from the processes, the enterprise will continue to measure and evaluate the business value of IS through novel ways that may include intangible measures for agility, collaboration, and risk mitigation.

References

Banker, R. D., Hu, N., Pavlou, P. A., and Luftman, J. 2011. CIO reporting structure, strategic positioning, and firm performance. *MIS Quarterly* 35(2): 487–504.

Barney, J. B. 1991. Firm resources and sustained competitive advantage. *Journal of Management* 17(1): 99–120.

Barua, A., Kriebel, C. H., and Mukhopadhyay, T. 1995. Information technologies and business value: An analytic and empirical investigation. *Information Systems Research* 6(1): 3–23.

Bower, J. L. and Christensen, C. M. 1995. Disruptive technologies: Catching the wave. *Harvard Business Review* 73(1): 43–53.

Bradley, R. V., Pratt, R. M. E., Byrd, T. A., and Simmons, L. 2011. The role of enterprise architecture in the quest for IT value. *MIS Quarterly Executive* 10(2): 73–80.

Brancheau, J. C., Janz, B. D., and Wetherbe, J. C. 1996. Key issues in information systems development: 1994–95 SIM Delphi results. *MIS Quarterly* 20(2): 225–242.

Brynjolfsson, E. and Hitt, L. 2000. Beyond computation: Information technology, organizational transformation, and business performance. *Journal of Economic Perspectives* 14(4): 23–48.

Carter, M., Grover, V., and Thatcher, J. B. 2011. The emerging CIO role of business technology strategist. *MIS Quarterly Executive* 10(1): 19–29.

Chen, D. Q., Preston, D. S., and Zia, W. D. 2010. Antecedents and effects of CIO supply-side and demand-side leadership: A staged maturity model. *Journal of Management Information Systems* 27(1): 231–272.

Chillingworth, M. 2010. CIOs looking to increase business value from technology. *CIO Magazine.* http://www.cio.com/article/pring/597360 (accessed January 9, 2012).

Constantinides, P., Chiasson, M. W., and Introna, L. D. 2012. The ends of information systems research: A pragmatic framework. *MIS Quarterly* 36(1): 1–19.

Dehning, B., Richardson, V. J., and Zmud, R. W. 2007. The financial performance effects of IT-based supply chain management systems in manufacturing firms. *Journal of Operations Management* 25(4): 806–824.

Devaraj, S. and Kohli, R. 2002. *The IT Payoff: Measuring the Business Value of Information Technology Investments.* New York: Prentice Hall Publishing.

Dewan, S. and Ren, F. 2011. Information technology and firm boundaries: Impact on firm risk and return performance. *Information Systems Research* 22(2): 369–388.

Dyer, J. H. 2000. *Collaborative Advantage: Winning through Extended Enterprise Supplier Networks.* New York: Oxford University Press.

Dyer, J. H. and Singh, H. 1998. The relational view: Cooperative strategy and sources of interorganizational competitive advantage. *Academy of Management Review* 23(4): 660–679.

Fichman, R. G. 2004. Real options and IT platform adoption: Implications for theory and practice. *Information Systems Research* 15(2): 132–154.

Flatto, J. 1996. The role of real options in valuing information technology projects. *Proceedings of the AIS Conference.* Phoenix, AZ, August.

Grover, V. and Kohli, R. 2012. Cocreating IT value: New capabilities and metrics for multifirm environments. *MIS Quarterly* 36(1): 225–232.

Han, K., Oh, W., Im, K. S., Chang, R. M., Oh, H., and Pinsonneault, A. 2012. Value cocreation and wealth spillover in open innovation alliances. *MIS Quarterly* 36(1): 291–315.

Hitt, L. M. and Brynjolfsson, E. 1996. Productivity, business profitability, and consumer surplus: Three different measures of information technology value. *MIS Quarterly* 20(2): 121–142.

Hoadley, E. D. and McFadden, J. C. 2010. The evolution of the chief information officer (CIO) through the eras of the information age: A case study. *Proceedings of the Conference on Information Systems, Technology, and Management.* Tampere, Finland, August.

Im, G. Y. and Rai, A. 2008. Knowledge sharing ambidexterity in long-term interorganizational relationships. *Management Science* 15(2): 132–154.

Kettinger, W. J., Zhang, C., and Marchand, D. A. 2011. CIO and business executive leadership approaches to establishing company-wide information orientation. *MIS Quarterly Executive* 10(4): 157–174.

Kobelsky, K. W., Richardson, V. J., Smith, R. E., and Zmud, R. W. 2008. Determinants and consequences of firms information technology budgets. *Accounting Review* 83(4): 957–995.

Kohli, R. and Grover, V. 2008. Business value of IT: An essay on expanding research directions to keep up with the times. *Journal of the Association for Information Systems* 9(1): 23–39.

Kohli, R. and Hoadley, E. D. 2006. Toward developing a framework for measuring organizational impact of IT-enabled BPR: Case studies of three firms. *The DATA BASE for Advances in Information Systems* 37(1): 40–58.

Kohli, R. and Johnson, S. 2011. Digital transformation in latecomer industries: CIO and CEO leadership lessons from Encana Oil & Gas (USA), Inc. *MIS Quarterly Executive* 10(4): 141–156.

Lankton, N. and Luft, J. 2008. Uncertainty and industry structure effects on managerial intuition about information technology real options. *Journal of Management Information Systems* 25(2): 203–240.

LeHong, H. and Feng, J. 2012. Key trends to watch in Gartner 2012 emerging technologies Hype Cycle. Available at http://www.forbes.com/sites/gartnergroup/2012/09/18/key-trends-to-watch-in-gartner-2012-emerging-technologies-hype-cycle-2/ (accessed on March 25, 2013).

Martinez, E. V. 1995. Successful reengineering demands IS/business partnerships. *Sloan Management Review* 36(4): 51–60.

Masli, A., Richardson, V. J., Sanchez, J. M., and Smith, R. E. 2011. Returns to IT excellence: Evidence from financial performance around information technology awards. *International Journal of Accounting Information Systems* 12: 189–205.

Melville, N., Kraemer, K., and Gurbaxani, V. 2004. Review: Information technology and organizational performance: An integrative model of IT business value. *MIS Quarterly* 28(2): 283–322.

Menon, N. and Kohli, R. 2012. A risk-management perspective on the role of information systems: Evidence from the healthcare industry. Working Paper, Winter, pp. 1–37.

Mithas, S., Tafti, A., Bardhan, I., and Goh, J. M. 2012. Information technology and firm profitability: Mechanisms and empirical evidence. *MIS Quarterly* 36(1): 205–224.

Mitra, S., Sambamurthy, V., and Westerman, G. 2011. Measuring IT performance and communicating value. *MIS Quarterly Executive* 10(1): 47–59.

Nevo, S. and Wade, M. R. 2010. The formation and value of IT-enabled resources: Antecedents and consequences of synergistic relationships. *MIS Quarterly* 34(1): 163–183.

Overby, S. 2011. IT value is dead. Long live business value. *CIO Magazine.* May 12. http://www.cio.com/article/682226 (accessed September 1, 2012).

Rai, A., Pavlou, P. A., Im, G., and Du, S. 2012. Interfirm IT capability profiles and communications for cocreating relational value: Evidence from the logistics industry. *MIS Quarterly* 36(1): 233–262.

Ramirez, R., Melville, N., and Lawler, E. 2010. Information technology infrastructure, organizational process redesign, and business value: An empirical analysis. *Decision Support Systems* 49(4): 417–429.

Ransbotham, S. and Overby, E. M. 2010. Does information technology increase or decrease hospitals' risk? An empirical examination of computerized physician order entry and malpractice claims. *Proceedings of the Thirty First International Conference on Information Systems.* St. Louis, MO, pp. 1–18.

Richey, R. G. and Autry, C. W. 2009. Assessing interfirm collaboration/technology investments tradeoffs: The effects of technological readiness and organizational learning. *International Journal of Logistics Management* 20(1): 30–56.

Salge, O., Kohli, R., and Barrett, M. 2011. Patterns of technological search: Institutional and behavioral triggers of IS investment. *Proceedings of the Academy of Management Annual Meeting*. San Antonio, TX, August 12–16.

Saraf, N., Langdon, C. S., and Gosain, S. 2007. IS application capabilities and relational value in interfirm partnerships. *Information Systems Research* 18(3): 320–339.

Sarker, S., Sarker, S., Sahaym, A., and Bjorn-Andersen, N. 2012. Exploring value cocreation in relationships between an ERP vendor and its partners: A revelatory case study. *MIS Quarterly* 36(1): 317–338.

Schryen, G. 2010. An analysis of literature reviews on IS business value: How deficiencies in methodology and theory use resulted in limited effectiveness. *Scandinavian Information Systems Research*, K. Kautz and P. A. Nielsen (eds.), Dordrecht, the Netherlands: Springer, pp. 139–155.

Soh, C. and Markus, M. L. 1995. How IT creates business value: A process theory synthesis. *Proceedings of the Sixteenth International Conference on Information Systems*. Amsterdam, the Netherlands, pp. 29–41.

Tallon, P. P. 2007. Does IT pay to focus? An analysis of IT business value under single and multi-focused business strategies. *Strategic Information Systems* 16: 278–300.

Tallon, P. P. 2008. Inside the adaptive enterprise: An information technologies perspective on business process agility. *Information Technology Management* 9: 21–36.

Tallon, P. P., Kauffman, R. J., Lucas, H. C., Whinston, A. B., and Zhu, K. 2002. Using real options analysis for evaluating uncertain investments in information technology. *Communications of the Association for Information Systems* 9(1): 136–167.

Tallon, P. P. and Pinsonneault, A. 2011. Competing perspectives on the link between strategic information technology alignment and organizational agility: Insights from a mediation model. *MIS Quarterly* 35(2): 463–486.

Tambe, P., Hitt, L., and Brynjolfsson, E. 2011. The price and quantity of IT-related intangible capital. *Proceedings of the Thirty Second International Conference on Information Systems*. Shanghai, China, December, pp. 1–14.

Tanriverdi, H., Rai, A., and Venkatraman, N. 2010. Reframing the dominant quests of information systems strategy research for complex adaptive business systems. *Information Systems Research* 21(4): 822–834.

Weill, P. 1992. The relationship between investment in information technology and firm performance: A study of the valve manufacturing sector. *Information Systems Research* 3(4): 307–333.

Weiss, M. 2011. APC forum: Realizing business value from tablets. *MIS Quarterly Executive* 10(2): 93–94.

72

Information Technology and Firm Value: Productivity Paradox, Profitability Paradox, and New Frontiers

Sunil Mithas
University of Maryland

Henry C. Lucas, Jr.
University of Maryland

72.1 Introduction

Interest in the business value of IT has been a dominant theme in the information systems literature since its early days. The literature on IT value has expanded considerably from its origins in the 1970s (Lucas 1975a,b). IT has become far more pervasive throughout the economy since 1970s, changing the way firms do business, enabling new business models that are creating value, and altering everyday life through social networking and Web 2.0 applications (Lucas 2008; Mithas 2012). Our goal in this chapter is to review what the field has learned in the last 40 years and discuss future opportunities for research (see Table 72.1 for several prior reviews of this literature).

Our conclusion is that there is no question any longer that IT, if properly deployed, creates value for organizations and society. We now have a much better understanding of how IT contributes to value, but the changing nature of IT raises some new questions. How do organizations incorporate rapidly developing technologies into their business models? What new kinds of services and products does the technology make possible? How should incumbent firms respond to new technologies that may destroy their value; for example, how does management avoid the outcomes that have befallen Borders, Blockbuster, and Kodak? These and many other related questions make the business value of IT research an enduring, interesting, and exciting area for researchers and practitioners.

TABLE 72.1   Selective Meta Review of Review Articles on "Business Value of IT" Literature

Study	Time Period and Focus of Review	Key Findings	New Opportunities for Research Identified in the Article
Lucas (1993)	1960–1991, Historical perspective on the business value of IT	• Technology should be suitable to the situation in which it is used • To demonstrate business value, the organization and individuals must use technology in an appropriate manner	• Studies of the deployment of one type of technology and of its users during system implementation offer the best chance to demonstrate that an investment in IT has business value
Brynjolfsson (1993)	1983–1993, Empirical studies of IT and productivity	• IT productivity paradox might be due to four reasons: mismeasurement, lags, redistribution, and mismanagement	• Researchers must be prepared to look beyond conventional productivity measurement techniques
Brynjolfsson and Yang (1996)	1982–1995, Empirical studies of IT and productivity	• Several studies document IT's positive effect on productivity performance	• There is little evidence of a positive contribution from IT to other performance measures such as profit and market value
Brynjolfsson and Hitt (2000)	1987–2000, Empirical studies of interaction of IT and organizational complements on productivity	• Organizational capital complements IT investments to increase performance	• Measuring the intangible components of complementary organizational assets
Barua and Mukhopadhyay (2000)	1963–1999, Studies on "process-oriented" and production function approach	• A positive contribution does not tell us whether we over- or underestimate IT impacts • Complementarity between IT and other factors	• Can we empirically distinguish between firms that make identical investments in IT but that may obtain very different returns from such investments? • What are some key factors that might have been ignored in production function and business value studies?
Dehning and Richardson (2002)	1990–2002, Archival studies that use accounting or market measures of firm performance	• Proposes a framework for IS research by suggesting that IT has a direct or indirect effect on business processes, which together determine the overall performance	• Examining the relation between IT and business processes, and business processes and overall firm performance • Understanding the effect of contextual factors on the IT-performance relation • Examining the IT-performance relation in an international context • Examining the interactive effects of IT spending and IT management on firm performance
Dedrick et al. (2003)	1985–2002, Empirical studies based on economic analysis on IT productivity	• At both the firm and the country level, greater investment in IT is associated with greater productivity growth • IT-performance relationship at the firm level can be further explained by complementary investments in organizational capital	• Understanding of the mechanisms by which IT pays off in higher productivity • IT spillover effects • Further examine the IT and profitability relationship • Excess returns at the firm level • Timing of payoff from IT investments • Industry differences on IT returns

TABLE 72.1 (continued)   Selective Meta Review of Review Articles on "Business Value of IT" Literature

Study	Time Period and Focus of Review	Key Findings	New Opportunities for Research Identified in the Article
Kohli and Devaraj (2003)	1990–2000, 66 firm-level empirical studies on IT value	• Meta-analysis • The sample size, data source, and industry in which the study is conducted influence the likelihood of finding positive IT effects on firm performance	• Measuring the payoff accruing to partners using a process measurement approach • Gathering larger samples comprising of longitudinal or panel data to assess the lag effects of IT payoff
Kohli and Grover (2008)	IT business value literature at the level of the firm or network of firms	• IT creates value under certain conditions • IT-based value manifests itself at many levels • IT-based value is not the same as IT-based competitive advantage • Causality for IT value is elusive	• How can companies with different or new IT resources co-create value and equitably partake in IT-based value? • How can we digitize various functional and dynamic business capabilities to increase business value? • How can we create information capabilities that enhance and do not destroy digital business capabilities? • What are the indirect and intangible paths to economic value that can be influenced by IT capabilities?

72.2 Phases and Foci in Business Value of IT Research over Time

Because the onset of modern computing started in 1950s, information systems is a relatively young discipline, compared to other disciplines found in business schools. The field started to create an identity in the 1960s. Gordon Davis wrote one of the first books to introduce business students to computers in 1965. Subsequently, Gordon Davis, Gary Dickson, and Tom Hoffmann started a program at the University of Minnesota in 1967 to offer a master's degree and a PhD in Management Information Systems (Davis 2009). The Sloan School at M.I.T. had a large group of IS faculty in the late 1960s and offered a PhD in information systems. The University of Arizona, NYU and Purdue University also started PhD programs in 1970s. The first premier scholarly journal of the discipline *MIS Quarterly* was launched in 1977 with Gary Dickson as the editor and Hank Lucas as associate editor. The first conference in information systems, ICIS, started in 1980. INFORMS launched the other premier journal of the field, *Information Systems Research*, in 1990.

With this backdrop of the evolution of the discipline, we trace the development of business value of IT research in three overlapping phases over time.

72.2.1 First Phase from 1970s until the Mid-1990s

The first phase of IT value research from 1970s till late 1990s focused on the so-called productivity paradox (Lucas 1999; Panko 1991). During this period, academic studies and the business press periodically revisited this so-called paradox of computers. The productivity paradox referred to the phenomenon that, on the one hand, computing power in the U.S. economy increased significantly over this period, but on the other hand, economy-wide productivity, especially in the service sector, seemed to have stalled. For example, the average labor productivity growth in the U.S. economy averaged only about 1.4% from 1973 to 1995 (Brynjolfsson and Saunders 2010).

Many economy-level studies found broadly negative correlations of IT with productivity. Econometric estimates also suggested low IT capital productivity in different industries (Brynjolfsson and Yang 1996; Loveman 1994; Roach 1991). Multifactor productivity growth, which takes into account changes in capital, had also declined significantly during the period (Baily 1986). The overall negative correlation between economy-wide productivity and the advent of computers was used by some to argue that information technology did not help U.S. productivity or even that IT investments were counterproductive (Baily 1986). This link was further supported by Stephen Roach's (1987, 1991) research focusing specifically on information workers across industries. Roach cites statistics indicating that output per production worker grew by 16.9% between the 1970s and 1986, while output per information worker decreased by 6.6%.

Industry-level studies also suggested that most of the productivity slowdown was concentrated in the service sector (Roach 1987, 1991). Because services use a significant part of computer capital, this was taken as indirect evidence of poor IT productivity. Roach's widely cited research on white collar productivity, discussed earlier, focused principally on IT's performance in the service sector (Roach 1987, 1991). Studies using data from manufacturing also found evidence of a productivity paradox. Berndt and Morrison used a dataset from the U.S. Bureau of Economic Analysis (BEA) that covers the whole U.S. manufacturing sector. Morrison and Berndt (1991) found evidence that every dollar spent on IT produced, on average, less than one dollar of value on the margin, indicating a general overinvestment in IT. They also examined broad correlations of IT with labor productivity and multifactor productivity (Berndt and Morrison 1995) and did not find a significant difference between the productivity of IT capital and other types of capital for a majority of the 20 industry categories.

Some firm-level studies also examined the relationship between IT investment and productivity. Among studies of IT's impact on the performance of financial services firms, a study by Parsons et al. (1993) estimated a production function for banking services in Canada and found that the impact of IT on multifactor productivity was quite low between 1974 and 1987. Another study looked at data on the insurance industry and found a positive, but quite weak, relationship between IT expenses and various performance ratios (Harris and Katz 1991). Among firm-level studies of IT productivity in the manufacturing sector, Loveman (1994) estimated that the impact of IT capital on output was approximately zero over the 5 year period studied in almost every subsample he examined. Weill (1992) was able to disaggregate IT by use. He found that significant productivity could be attributed to transactional types of information technology, but he was unable to identify gains associated with strategic systems or informational investments.

Researchers proposed several explanations for the so-called IT productivity paradox: (1) mismeasurement of outputs and inputs, (2) lags, (3) redistribution and dissipation of profits, and (4) mismanagement of information and technology (Brynjolfsson and Yang 1996). Some even argued that given measurement errors and the nonautomatic nature of payoffs, it was time to stop conducting general IT impact studies for the entire economy and, instead, consider studies of the management of IT impacts for departments, individual firms, and individual users (Panko 1991). Subsequent research addressed some of these issues and led to resolution of the original "productivity paradox" while observing some new paradoxes.

72.2.2 Second Phase from Mid-1990s until the Early 2000s

The second phase of value research from mid-1990s till mid-2000s (and continuing) used newer datasets and more sophisticated methodologies. Some of these studies documented IT's positive effect on productivity (see Table 72.2). Lichtenberg (1995) reported positive effects of IT based on output. Brynjolfsson and Hitt (1996) confirmed the results of Lichtenberg (1995) and their own prior study (Brynjolfsson and Hitt 1995) using the same data source and similar methods based on output and consumer surplus measures.

This phase also began to focus on intangibles, and some of this literature observed a "new productivity paradox" (Anderson et al. 2003). The use of the word "productivity" may not have been entirely accurate here because researchers were characterizing high valuation multiples on Y2K spending, which ranged from 31 to 45 for 1999 to 2002 (about 3–4 times that on R&D spending) as the new paradox of too high returns on IT (Anderson et al. 2003). Tables 72.3 and 72.4 provide a summary of these studies.

TABLE 72.2 Selected Empirical Studies Linking IT and IT-Enabled Capabilities with Productivity

Study	Sample	IT Variable	Dependent or Intermediate Variables	Key Findings
Tambe et al. (2012)	Survey of 253 firms in 2001 matched with IT employment data	IT employment	Product development and productivity	Test a three-way complementarities model that indicates that the combination of external focus, decentralization, and IT is associated with significantly higher productivity. In contrast, firms that have only one or two of these organizational practices in place, instead of all three, are not more productive than firms with none of them
Tambe and Hitt (2012)	1800 firms across 20 years (36,000 firm-years from 1987 to 2006)	IT personnel counts	Value added (sales minus nonlabor variable costs)	IT returns are substantially lower in small- and mid-sized firms than in Fortune 500 firms and materialize more slowly in large firms. The measured marginal product of IT spending is higher from 2000 to 2006 than in any previous period, suggesting that firms, and especially large firms, have been continuing to develop new, valuable IT-enabled business process innovations.Furthermore, the productivity of IT investments is higher in manufacturing sectors
Brynjolfsson and Hitt (1996)	1987–1991	IS spending	Sales	Average marginal product of computer capital was 81% compared to 6.26% for noncomputer capital. Attribute to hidden IT capital and organizational capital

Several studies have stressed the need for better theoretical models that trace the path from IT investments to business value (Bharadwaj 2000). These researchers adopted a "process-oriented" view that examines the effects of IT on intermediate business processes (Barua et al. 1995; Melville et al. 2004; Tallon et al. 2000).

Other researchers proposed that the positive effect of IT on intermediate variables can be examined through internal firm processes, inter-organizational relationships, and customer relationships. First, some studies suggested that IT enables greater automation, streamlining, and rationalization of processes (Santhanam and Hartono 2003). Better governance of systems and streamlined processes also enable firms to make better and more accurate use of their data—leading to improved analytical models and inferences (Davenport 2006) and more effective strategic decision making. IT investments can also facilitate innovation in products and services to help firms leverage organizational learning and knowledge (Kleis et al. 2012; Ravichandran et al. 2011; Tippins and Sohi 2003).

Second, IT investments can also enhance business processes critical in inter-organizational partnerships such as supply-chain relationships and corporate alliances. IT infrastructure provides firms with the ability to implement shared transaction processing and supply chain management across organizational boundaries (Mukhopadhyay and Kekre 2002; Whitaker et al. 2007). IT also makes it possible to reduce transaction hazards that could arise within inter-organizational relationships (Kim and

TABLE 72.3   Selected Empirical Studies Linking IT and IT-Enabled Capabilities with Firm Risk and Returns

Study	Sample	IT Variable	Dependent or Intermediate Variables	Key Findings
Li et al. (2012)	2004–2008 data with 5230 firm-year observations	IT control quality, types of IT material weaknesses	Management forecast accuracy	Firms with IT material weaknesses in their financial reporting system are associated with less accurate management forecasts; systems with IT material weaknesses related to data processing integrity have the least accurate management earnings forecasts
Dewan and Ren (2011)	1987–1994 data over 500 Fortune 1000 firms	IT capital	Firm return and firm risk	Suitable boundary strategies can moderate the impact of IT on firm performance in a way that increases return and decreases risk; this interaction effect is strongest in service firms, in firms with high levels of IT investment, and in more recent time periods
Kim et al. (2012)	1995–2002 data	IT investments	Bond rating and yield spread	Bondholders have different perspective toward risk than equity investors; lower bond ratings and higher yield spread in transform industries than in automate or informate industries
Henderson et al. (2010)	1991–2005 data from *InformationWeek*	IT investments	Market value, accounting performance, firm risk, abnormal stock returns	Information about a firm's IT expenditures help explain its future performance in terms of accounting measures and market measures; mispricing is attributable, at least in part, to the lack of adequate and verified information about firms' investments in IT
Kobelsky et al. (2008)	1992–1997 data from *InformationWeek*	IT investments	Earnings volatility	IT investments increase the volatility of future earnings; this impact is highly contingent upon sales growth, unrelated diversification, and size
Dewan et al. (2007)	1987–1994 data on over 500 Fortune 1000 firms	IT capital	SD (daily stock returns) and SD (realized annual earnings for 5 years after investment), market value of firm	Attribute high returns on IT to higher IT risk. About 30% of gross return on IT investment is due to risk premium associated with IT
Dewan and Ren (2007)	1996–2002 event data	Electronic commerce announcements	Risk-adjusted abnormal return	Wealth effects are not significant after controlling for contemporaneous risk changes

TABLE 72.3 (continued)   Selected Empirical Studies Linking IT and IT-Enabled Capabilities with Firm Risk and Returns

Study	Sample	IT Variable	Dependent or Intermediate Variables	Key Findings
Oh et al. (2006)	1985–1999 event data	IT investment announcements	Cumulative abnormal return	A firm's growth prospects, uncertainty, the strategic role of IT, and discloser information are significantly related to CARs; asset specificity and uncertainty interacts negatively
Anderson et al. (2006)	Y2K investments of Fortune 1000 firms	Y2K spending	Shareholder value	Y2K spending increases firm value; the positive effect of Y2K spending is stronger for firms in transform industries
Anderson et al. (2003)	Y2K investments of Fortune 1000 firms	Y2K spending	Shareholder value	Valuation multiples on Y2K spending range from 31 to 45 for 1999 to 2002, about 3–4 times that on R&D spending; they attribute high multiples to intangible value
Brynjolfsson et al. (2002)	1987–1997 panel data	IT capital	Shareholder value	Valuation multiples on IT spending range 10–15, attribute to organizational capital
Bharadwaj et al. (1999)	1988–1993 data	IT spending	Tobin's q	IT investments have a significantly positive association with Tobin's q value

Mahoney 2006) as well as reduce coordination and procurements costs (Gurbaxani and Whang 1991; Mithas and Jones 2007; Mithas et al. 2008).

Third, IT improves existing customer relationships and helps reach new customers, sometimes involving new channels or new offerings (Mithas et al. 2005). IT systems such as customer relationship management (CRM) facilitate personalization of offerings and services through improved knowledge of customers' wants and needs, leading to better customer response, improved one-to-one marketing effectiveness (Mithas et al. 2006), and higher customer satisfaction (Ansari and Mela 2003; Babakus et al. 2004). In turn, customer satisfaction is associated with higher stock prices and shareholder values and reduced risk (Aksoy et al. 2008, 2012; Fornell et al. 2006, 2009a,b; Tuli and Bharadwaj 2009).

Overall, studies focusing on business processes are examining the activities residing in the black box of microeconomic production theory that transforms a set of inputs into outputs (Melville et al. 2004). These studies suggest that IT can be employed not only to improve individual processes but also to enable process integration across physical and organizational boundaries, thereby improving organizational performance.

72.2.3 Third Phase from Early 2000s Onward

The third phase of value research starting early 2000s began to focus on profitability and shareholder value metrics that also take into account "appropriation of value" aspect of firm performance (see Table 72.5). Previous research, using firm-level IT investment data in early 1990s (Hitt and Brynjolfsson 1996; Rai et al. 1997), failed to find a statistically significant effect of IT investments on firm profitability,

TABLE 72.4 Selected Empirical Studies Linking IT and IT-Enabled Capabilities with Intangibles

Study	Sample	IT Variable	Dependent or Intermediate Variables	Key Findings
Tafti et al. (2013)	2000–2006	IT investments	Alliance formation and alliance value	Adoption of open communication standards is associated with the formation of arms-length alliances, and modularity of IT architecture is associated with the formation of joint ventures. IT architecture flexibility enhances the value of arms-length, collaborative, and joint venture alliances. The contribution of IT flexibility to value is greater in the case of collaborative alliances than in arms-length alliances
Ravichandran et al. (2011)	1990–2005 panel data on manufacturing and service firms	IT investments	Patent counts and patent citations	IT investments have a positive and statistically significant association with patents and patent citations and the effect is stronger for the 1998–2005 period. Also, stronger effect in services than in manufacturing and higher impact on patent citations than on patent counts. Study also finds evidence that IT helps to mitigate diminishing returns to R&D
Gao and Hitt (2012)	1987–1997; 116 manufacturing firms	IT capital stock	Trademarks	More IT capital is associated with more new trademarks and retirement of existing trademarks (these signify differences among similar products) leading to shorter trademark life cycle and greater product variety
Mithas et al. (2011)	An archival data from a conglomerate	Information management capability	Customer, financial, human resources, and organizational effectiveness measures of performance	Information management capability contributes to developing other firm capabilities for customer management, process management, and performance management, which, in turn, favorably influence firm performance measures
Kleis et al. (2012)	1987–1997 data for manufacturing firms	IT investment	Innovation output (patents and citations)	A 10% increase in IT input is associated with a 1.7% increase in innovation output after accounting for R&D
Pavlou and El Sawy (2010)	Surveys of new product development managers in 2002 and 2003	IT-enabled improvisational and dynamic capabilities	Competitive advantage in new product development	While dynamic capabilities are associated with competitive advantage in moderately turbulent environments, improvisational capabilities dominate in highly turbulent environments

TABLE 72.4 (continued)   Selected Empirical Studies Linking IT and IT-Enabled Capabilities with Intangibles

Study	Sample	IT Variable	Dependent or Intermediate Variables	Key Findings
Saraf et al. (2007)	Survey of business units in the high-tech and financial services industries	IS integration and IS flexibility capabilities	Knowledge sharing, process coupling, and BU performance	IS integration with channel partners and customers contributes to both knowledge sharing and process coupling with both types of enterprise partners, whereas IS flexibility is a foundational capability that indirectly contributes to value creation in interfirm relationships by enabling greater IS integration with partner firms
Ray et al. (2005)	Survey of managers in the life and health insurance industry during 2000	Shared knowledge between IT and customer service units	Customer service process performance	Shared knowledge between IT and customer service units affects customer service process performance and moderates the impacts of explicit IT resources such as the generic IT used in the process and IT spending
Mithas et al. (2005)	2001–2002 data from *Information Week*	Use of CRM applications	Customer satisfaction	The use of customer relationship management applications is positively associated with improved customer knowledge and improved customer satisfaction. Gains in customer knowledge are enhanced when firms share their customer-related information with their supply chain partners
Brynjolfsson et al. (2003)		Increased product variety made available through electronic markets	Consumer surplus	Increased product variety of online bookstores enhanced customer welfare by around $1 billion in the year 2000

which Dedrick et al. (2003) called the "profitability paradox" of IT. Hitt and Brynjolfsson (1996) looked for associations between IT spending and various business performance measures. Although they document IT's positive impact on output and consumer surplus, they do not find a significant positive correlation between IT spending and profitability. They proposed that the productivity benefits associated with IT use may be passed on to consumers through lower prices and not lead to greater profitability because of intense competition. However, Dedrick et al. (2003) argued that it was more likely that IT investments do actually affect profitability, but the modeling techniques and datasets used in previous studies were unable to measure the impacts. Mithas et al. (2012b,c), using archival data for the 1998–2003 period, show that IT has a positive impact on profitability and its effect is higher than that of other discretionary investments such as advertising and R&D.

Another stream of research using capital market-based measures examines whether and how firms appropriate value from their IT investments. This research generally shows that IT investments are positively associated with shareholder value and stock returns (Mithas et al. 2012a; Saunders 2010). Some of these studies indicate the need to consider risks involved with IT investments, which might explain excess returns (Dedrick et al. 2003; Dewan et al. 2007; Kim and Mithas 2011; Tanriverdi and Ruefli 2004).

TABLE 72.5   Selected Empirical Studies Linking IT and IT-Enabled Capabilities with Financial Measures and Profitability

Study	Sample	IT Variable	Dependent or Intermediate Variables	Key Findings
Mithas et al. (2012c)	1998–2003 data	IT spending	Profitability	IT has a positive impact on firm profitability; a significant portion of IT impact is accounted by IT-enabled revenue growth
Aral and Weill (2007)	1999–2002	IT investment allocations and IT capabilities	Market valuation, profitability, cost and innovation	Investments in specific IT assets explain performance differences along dimensions consistent with their strategic purpose; a system of organizational IT capabilities drive differences in firm performance
Santhanam and Hartono (2003)	1991–1994 data	IT capability	Profitability and cost	Firms with superior IT capability exhibit superior current and sustained firm performance
Bharadwaj (2000)	1991–1994 data	IT capability	Profitability and cost	Firms with high IT capability tend to outperform a control sample of firms on a variety of profit and cost-based performance measures
Rai et al. (1997)	1994 data	IT investments	Sales, return on assets (ROA), return on equity (ROE), and labor and administrative productivity	Measures of IT investments have differential effects on the various measures of business performance
Hitt and Brynjolfsson (1996)	1988–1992	IT spending	Productivity, profitability, and consumer surplus	IT increases productivity and consumer surplus, but does not increase profitability

72.3 Underlying Theories and Methodologies

While there are a number of theories from which to explain the business value of IT (see Table 72.6 for a summary of some review articles), we focus here on the create–capture value approach. Much of the productivity research focuses on the issue of value creation, while the studies relating to profitability and market value focus on appropriability of value.

Researchers use several theoretical paradigms in explaining the value creation of IT, including production economics and process-oriented models. The theory of production has been particularly useful in providing empirical specifications helping to estimate the economic impact of IT (Brynjolfsson and Hitt 1996; Lichtenberg 1995). Researchers have also employed consumer theory (Hitt and Brynjolfsson 1996), data envelopment theory (Lee and Barua 1999), and Tobin's q (Bharadwaj et al. 1999). Other researchers have developed process-oriented models linking IT to organizational performance in which the impact of IT on firm performance is mediated by intermediate processes (Barua et al. 1995). This approach argues that the enterprise-level impacts of IT can be measured only through a web of intermediate-level contributions.

This body of research suggests several general findings. First, IT creates value. Also, following a complementarity argument, IT infrastructure does not create value in isolation but must be a part of a business value creating process with other organizational factors including IT human capital. The effects of IT occur at many levels (Kohli and Grover 2008), and IT can be treated as an "option" that is valuable because it provides an opportunity to reap benefits if or when the need arises (Benaroch 2002). This argument endows management with flexibility to embrace and manage uncertainty using IT.

TABLE 72.6 Selected Conceptual Reviews or Theories Linking IT and IT-Enabled Capabilities with Firm Performance

Study	Theoretical Lens	IT Variable	Dependent or Intermediate Variables	Proposed Moderating Variables	Key Findings
Melville et al. (2004)	Integrative model	IT resources (technology and human)	Business processes, business process performance, organizational performance	Complementary organizational resources, industry characteristics, and trading partner resources	The high degree of complexity leads to a context-contingent set of synergistic combinations of IT and other organizational resources
Wade and Hulland (2004)	Resource-based view	IS resources (manage external relationships, market responsiveness, etc.)	Competitive advantage	Organizational factors (e.g., top management commitment to IS) and environmental factors (e.g., environmental munificence)	Only IS resources that are inimitable, nonsubstitutable, and imperfectly mobile will have a positive effect on competitive position in the longer term
Piccoli and Ives (2005)	Integrative framework	IT-dependent strategic initiatives	Sustainable competitive advantage		Identify four barriers to erosion for sustained competitive advantage: IT resources barrier, complementary resources barrier, IT project barrier, and preemption barrier. They identify the response lag drivers underpinning these barriers, as well as the process of organizational learning and asset stock accumulation by which they may be strengthened over time
Nevo and Wade (2010)	Systems theory and resource-based view	IT asset	Sustained competitive advantage, synergy, strategic potential		Greater compatibility between an IT asset and an organizational resource positively affects the extent of realized synergy

Although the previous literature focusing on IT value creation perspective provides valuable insights of IT impacts, the external environment of trading partners, industry characteristics, and sociopolitical conditions is also important, but rarely incorporated in analyses. Moreover, prior work has paid less attention to how the IT benefits are appropriated. Some researchers use game theory to examine the role of strategic interaction among competitors in IT business value capture (Belleflamme 2001). Other researchers draw from agency theory and the incomplete contracts literature (Bakos and Nault 1997).

The resource-based view (RBV) of the firm has been a popular framework in the business value of IT literature. This approach is useful in providing a robust framework for analyzing whether and how IT may be associated with competitive advantage (Bharadwaj 2000; Mithas et al. 2012c; Piccoli and Ives 2005; Powell and Dent-Micallef 1997). This body of research suggests that the IT benefits generated can be dissipated from competition. Differential firm value from IT is elusive since it can be copied and competed away, even though value can be created at the industry and economy level. RBV suggests that firms can create differential value by leveraging IT and complementarities to create resources and capabilities that are heterogeneous and imperfectly mobile.

Mithas et al. (2012c) draw on the RBV of the firm as an overarching framework and propose three reasons to explain why overall IT investments are likely to have a positive association with accounting profits. First, an explanation based on virtuous cycle argument suggests that firms that invest in IT in period 1 reap benefits and then invest more in IT in period 2. Over time, these effects become magnified, leading some firms to continue investing more in IT compared with their historical investment and that of their competitors; these firms maintain a more proactive digital strategic posture. Because of their higher investments in IT and greater opportunities to learn from occasional failures in their overall IT portfolio, the firms undergoing the virtuous cycle are also likely to become better at managing IT.

A second, learning-based explanation suggests that years of continued investments in IT and experience in managing these systems may have improved the capability of firms to leverage information and strengthen other organizational capabilities (Grover and Ramanlal 1999, 2004; Mithas et al. 2011). In support of this explanation, several empirical studies show that firms have learned how to make use of IT to improve customer satisfaction, at the same time boosting profitability through the positive effects of customer loyalty, cross-selling, and reduced marketing and selling costs (Fornell et al. 2006, 2009a; Grover and Ramanlal 1999; Mithas and Jones 2007; Mithas et al. 2005).

A third explanation, based on Kohli's (2007) work, suggests that because of a long history of firms viewing IT mainly as an automation-related investment, with a focus on cost reduction rather than revenue generation, firms may have "just about exhausted efficiency gains from IT" (p. 210). To the extent, RBV focuses on differential firm performance; if revenue growth has become a primary driver for differentiation because of exhaustion of cost-based differentiation, tracing the effect of IT on profitability through revenue growth may be more promising than through cost reduction. These three explanations (virtuous cycle, learning, and strategic posture of differentiating through revenue growth rather than through cost reduction) relate to the key tenets of RBV, which uses the notions of social complexity, erosion barriers, path dependence, and organizational learning to explain why resources create and sustain a competitive advantage (see Piccoli and Ives 2005).

Among methodologies, researchers have used quantitative and qualitative approaches to study the business value of IT. The use of longitudinal archival data and panel models has helped to strengthen claims regarding causality. At the same time, use of newer methodologies such as propensity scores has been particularly useful to view causality from a counterfactual perspective (Chang and Gurbaxani 2012; Mithas and Krishnan 2009), complementing other approaches based on explained variance (as in regression), explained covariance (as in structural equation models, such as LISREL), prediction (as in partial least squares), or a comparison of performance at time t with some prior time ($t-1$) (as in "before-and-after" study designs). While large sample studies will continue to be important and availability of even more granular data will provide new opportunities to explore causal mechanisms more closely, we also call for detailed longitudinal case studies of firms (e.g., Mandviwalla and Palmer 2008; Marchand 2002). Sometimes these detailed case studies can provide very useful insights to guide further quantitative research, beyond their value in teaching.

72.4 Impact on Practice

Questions about IT value have forced CIOs to look for more opportunities to enhance revenue as opposed to applications that cut costs. The impact of $1 million additional profits, whether through revenue or through savings, is exactly the same. But contributing to revenue may lead to more sustainable competitive advantage (Mithas et al. 2012b) and may allow CIOs and other executives to leverage newer opportunities enabled by IT, which may not be obvious if they were to narrowly focus on reductions in IT costs alone. Indeed, if marginal increase in IT investments is outweighed by reductions in non-IT costs, then firms are better off increasing their IT investments as some firms have realized (Glazer 2012; Han and Mithas 2013; Worthen 2012). It may also be that cost savings are often not tracked, and when they are, they frequently prove to be less than promised when a system was approved.

Managers with a focus on revenue generation will look for way to support customers and a firm's sales efforts. For example, a firm might implement Salesforce.com or some other CRM system to leverage the efforts of its sales force. Getting a new drug through trials and the approval process is critical for revenue in the pharmaceuticals industry so IT can contribute to earlier revenue recognition with a system to manage and expedite clinical trials. The Internet and the ability to reach customers easily provides a platform for efforts to improve revenues in a variety of ways, from CRM systems to order entry applications that make it easier to do business with the firm.

Since the Internet became available for profit making use in 1995, IT has become an integral part of many business models. Firms like Google, eBay, Amazon, Facebook, Twitter, and others rely on technology to function. Without IT, they could not exist. What is the value of information technology to these firms? Consider the case of Amazon in which the technology enabled it to create and execute a business model for retail sales over the Internet. As a result of its expertise in building and operating infrastructure, Amazon has a booming business of selling cloud computing services to companies like Netflix that do not want to develop their own technology infrastructure. Is there any way to value technology in these cases? Is technology equal to the value of the firm or its annual sales?

Table 72.7 lists the sales and market capitalization of a number of firms with business models intertwined with the Internet. All of these firms have created value through technology. Google, Amazon, and eBay are publicly traded companies and are profitable. Can we attribute their profitability to technology? The answer

TABLE 72.7 When IT and Business Morph into Each Other

Firm	2011 Sales (All Values Are in Millions US$)	2011 Market Cap (All Values Are in Millions US$)	Remarks
1. Amazon	48,077	82,467	Sales—Period ended December 31, 2011
2. Netflix	3,204	5,789	Sales—Period ended December 31, 2011
3. Facebook[a]	3,700[b]	65,000–100,000[b]	*Source*: *USA Today* (Swartz, Martin, and Krantz 2012)
			Date: Sales—Period ended December 31, 2011
4. Twitter[a]	139[c]	7,700[d]	*Source*: *eMarketer* (Fredricksen 2011) and *Reuters* (Reuters 2011)
			Date: Sales—2011
5. Google	37,905	199,077	Sales—Period ended December 31, 2011
6. eBay	11,651	48,246	Sales—Period ended December 31, 2011
7. Groupon	1,624	10,835	Sales—Period ended December 31, 2011
8. Apple	108,249	544,140	Sales—Period ended September 24, 2011

Notes: Annual sales information was collected from Bloomberg. Market capitalization information is as of March 14, 2012, and this data was collected from Bloomberg.

[a] Facebook and Twitter Sales and Market Cap are based on news stories because these firms are yet to trade at the time of writing of this article.

[b] From http://www.usatoday.com/tech/news/story/2012-02-01/facebook-ipo/52921528/1

[c] From http://www.emarketer.com/PressRelease.aspx?R=1008617

[d] From http://www.reuters.com/article/2011/03/04/us-twitter-idUSTRE7221JL20110304

is no because it is the application of technology merged with other business processes that account for their success. Amazon is extremely good at logistics and fulfilling orders. Google has captured the market for search advertising. For these firms, technology is such an important part of their existence that managers simply assume that it will be there like their office or phone. Under these conditions, does it make any sense to try and put a value on IT, or should the focus be on how to manage the technology effectively?

72.5 Research Issues and New Frontiers

We identify several directions for further research. First, the rising use of social networks by businesses raises questions to assess how these networks create value for firms by enabling knowledge exchanges within the firm and with external customers, suppliers, and business partners. Social network data can be very rich and complex with several layers of nesting. For example, one can view individual messages nested in threads, which, in turn, are often nested in forums. Making sense of such nesting can be very complex because messages can also be analyzed at the level of individual contributors, firms, and countries. We expect new methodological innovations and significant extensions of hierarchical linear models to analyze such datasets to generate useful insights.

Second, although we have made significant progress by learning from large sample archival datasets, there is a need for detailed studies of information and technology use inside large and evolving born-digital global firms to understand how firms scale up their information management capabilities to cope with more turbulent and uncertain industry environment.

Third, researchers have paid significant attention to understanding the impact of IT investments, and there have been studies and conceptual frameworks that study IT strategies and IT governance. However, there remains a need to study the joint effects of IT strategies and IT investments on firm performance. In particular, we need to study the interaction between IT investments and ambidextrous strategies (such as the ones that focus on multiple objectives such as revenue growth and cost reduction at the same time) and implications for firm performance (Mithas and Rust 2009). To the extent IT strategies of firms are also shaped by their competitive environment, understanding how competitive environment affects the influence of strategic posture of a firm on its digital business strategy is an important area of research (Mithas et al. 2013).

Fourth, as society digitizes and economies become more globally interconnected, firms are increasingly making use of globally dispersed resources to manage their IT assets. In this context, the role of IT professionals and firm's choice of how to provide IT services using internal IT labor and outsourced IT labor can be investigated by looking at how such choices influence firms' ability to innovate and stay entrepreneurial. We also need a better understanding of how outsourcing service providers create value for firms by helping to reduce non-IT costs and providing a service differentiation advantage (Han and Mithas 2013; Ramasubbu et al. 2008). The role of foreign-born immigrant workers and that of IT professionals located abroad in complementing native workers in firms' profit functions is a fruitful area of research because of significant policy debates and implications that surround movement of people across country borders (Lucas and Mithas 2011; Mithas and Han 2012; Mithas and Lucas 2010).

Fifth, while IT-enabled transformations are spreading across industries, we still do not have adequate theories and methods to investigate such innovations systematically. Although case examples and anecdotal accounts have enriched our understanding of sustaining versus disruptive innovations, more rigorous conceptualizations of core concepts and empirical testing of assertions based on case examples are necessary to draw valid conclusions.

Sixth, because of rising concerns related to climate change there is a need to study how IT can contribute to reducing carbon emissions through more sustainable business operations. This will require broadening the dependent variables for business value of IT research beyond customer satisfaction, profitability, shareholder value, and bondholder value.

Finally, increasing digitization of the economy is accompanied by a rise in services. However, as noted by the prominent economist William Baumol, the unit cost of services increases over time because productivity growth in services generally lags well behind productivity growth in manufacturing, but real wages grow at about the same rate in the two sectors. This phenomenon of growth in costs of services relative to the rate of inflation, which has come to be known as "Baumol's cost disease" (see Baumol 1993, 1996; Baumol and Blackman 1983), afflicts many services such as healthcare, education, auto repair, auto insurance, legal services, restaurant services, and public services. For example, the cost of higher education at private colleges in the United States in 2010 was ~2.5 times than in 1982 in inflation-adjusted dollars (Hacker and Dreifus 2010). Similar trends have also been noted in healthcare, which has witnessed significant increase in costs in excess of inflation (Bohmer and Knoop 2006; Reid 2009). A key question in this context is whether and what types of IT resources can help firms, universities, and governments to bend the cost curve in services. At the same time, the role of IT and e-government services to improve access, transparency, quality of services, and citizen satisfaction needs further investigation. This is particularly the case in emerging economies where governments are often viewed with suspicion and corruption in government services is taken for granted (Gupta et al. 2012; Morgeson and Mithas 2009). There is much that organizations can learn from experience of others in different country settings. Therefore, comparative studies across countries can be particularly helpful.

72.6 Conclusion

Our goal in this chapter was to review the business value of information technology literature in last 50 years since the origin of the academic discipline of information systems in business schools in the 1960s. We identified key milestones, reviewed what we know, and suggested some promising directions for future research. The last 40 years have brought many new insights, theories, and methods to answer the question of how IT creates value and how some of that value can be appropriated by those who invest in IT. However, the changing nature of IT and concomitant changes in the economy due to globalization, the rise of emerging economies such as China and India, concerns about sustainability of our planet, and rising costs of essential services such as healthcare and education suggest that organizations continue to be challenged to define and create new metrics and ways of measuring IT value in the next 40 years. The possibilities for new research opportunities and finding creative solutions to societal and organizational problems using IT are immense and that should be exciting for researchers and practitioners alike.

Acknowledgment

We thank Keongtae Kim for assisting with this research.

References

Aksoy, L., Cooil, B., Groening, C., Keiningham, T.L., and Yalcin, A. The long term stock market valuation of customer satisfaction, *Journal of Marketing* 72(July), 2008, 105–122.

Aksoy, L., Keiningham, T.L., Lariviere, B., Mithas, S., Morgeson, F.V., and Yalcin, A. The satisfaction, repurchase intentions and shareholder value linkage: A longitudinal examination of fixed and firm-specific effects, *Frontiers in Service Conference*, College Park, MD, 2012.

Anderson, M.C., Banker, R.D., and Ravindran, S. The new productivity paradox, *Communications of the ACM* 46(3), March 2003, 91–94.

Anderson, M.C., Banker, R.D., and Ravindran, S. Value implications of investments in information technology, *Management Science* 52(9), 2006, 1359–1376.

Ansari, A. and Mela, C.F. e-Customization, *Journal of Marketing Research (JMR)* 40(2), 2003, 131–145.

Aral, S. and Weill, P. IT assets, organizational capabilities, and firm performance: How resource allocations and organizational differences explain performance variation, *Organization Science* 18(5), 2007, 763–780.

Babakus, E., Beinstock, C.C., and Van Scotter, J.R. Linking perceived quality and customer satisfaction to store traffic and revenue growth, *Decision Sciences* 35(4), Fall 2004, 713–737.

Baily, M.N. What has happened to productivity growth? *Science* 234(4775), 1986, 443–451.

Bakos, J.Y. and Nault, B.R. Ownership and investment in electronic networks, *Information Systems Research* 8(4), 1997, 321.

Barua, A., Kriebel, C.H., and Mukhopadhyay, T. Information technologies and business value—An analytic and empirical investigation, *Information Systems Research* 6(1), 1995, 3–23.

Barua, A. and Mukhopadhyay, T. Information technology and business performance: Past, present, and future, *Framing the Domains of Information Technology Management: Projecting the Future … through the Past*, R.W. Zmud (ed.), Pinnaflex Press, Cincinnati, OH, 2000, pp. 65–84.

Baumol, W.J. Children of performing arts, the economic dilemma: The climbing costs of healthcare and education, *Journal of Cultural Economics* (20), 1996, 183–206.

Baumol, W.J. Social wants and dismal science: The curious case of the climbing costs of health and teaching, *Proceedings of the American Philosophical Society* (250th Anniversary Issue, December 1993), American Philosophical Society, Philadelphia, PA, 1993, pp. 612–637.

Baumol, W.J. and Blackman, S.A.B. Electronics, the cost disease, and the operation of libraries, *Journal of the American Society for Information Science* 34(3), 1983, 181–191.

Belleflamme, P. Oligopolistic competition, IT use for product differentiation and the productivity paradox, *International Journal of Industrial Organization* 19(1–2), 2001, 227–248.

Benaroch, M. Managing information technology investment risk: A real options perspective, *Journal of Management Information Systems* 19(2), Fall 2002, 43–84.

Berndt, E.R. and Morrison, C.J. High-tech capital formation and economic performance in U.S. manufacturing industries: An exploratory analysis, *Journal of Econometrics* 65(1), 1995, 9–43.

Bharadwaj, A. A resource-based perspective on information technology capability and firm performance: An empirical investigation, *MIS Quarterly* 24(1), 2000, 169–196.

Bharadwaj, A.S., Bharadwaj, S.G., and Konsynski, B.R. Information technology effects on firm performance as measured by Tobin's *q*, *Management Science* 45(7), 1999, 1008–1024.

Bohmer, R. and Knoop, C.-I. The Challenge Facing the U.S. Healthcare System, Harvard Business Publishing (9-606-096), Boston, MA, 2006.

Brynjolfsson, E. The productivity paradox of information technology, *Communications of the ACM* 36(12), 1993, 67–77.

Brynjolfsson, E. and Hitt, L. Information technology as a factor of production: The role of differences among firms, *Economics of Innovation and New Technology* (3), 1995, 183–200.

Brynjolfsson, E. and Hitt, L.M. Beyond computation: Information technology, organizational transformation and business performance, *Journal of Economic Perspectives* 14(4), 2000, 23–48.

Brynjolfsson, E. and Hitt, L.M. Paradox lost? Firm-level evidence on the returns to information systems spending, *Management Science* 42(4), 1996, 541–558.

Brynjolfsson, E., Hitt, L.M., and Yang, S. Intangible assets: Computers and organizational capital, *Brookings Papers on Economic Activity: Macroeconomics* (1), 2002, 137–199.

Brynjolfsson, E., Hu, Y., and Smith, M. Consumer surplus in the digital economy: Estimating the value of increased product variety, *Management Science* 49(11), 2003, 1580–1596.

Brynjolfsson, E. and Saunders, A. *Wired for Innovation: How Information Technology is Reshaping the Economy*. The MIT Press, Cambridge, MA, 2010.

Brynjolfsson, E. and Yang, S. Information technology and productivity: A review of literature, *Advances in Computers* (43), 1996, 179–214.

Chang, Y.B. and Gurbaxani, V. IT Outsourcing and Firm Productivity: An Empirical Analysis, *MIS Quarterly* 36(4), 2012, 1043–1063.

Davenport, T.H. Competing on analytics, *Harvard Business Review* 84(1), 2006, 98–107.

Davis, G.B. *History of MIS* (message to ISWorld on September 22, 2009), ISWorld (ed.), 2009.

Dedrick, J., Gurbaxani, V., and Kraemer, K.L. Information technology and economic performance: A critical review of empirical evidence, *ACM Computing Surveys* 35(1), 2003, 1–28.

Dehning, B. and Richardson, V.J. Returns on investments in information technology: A research synthesis, *Journal of Information Systems Management* 16(1), 2002, 7–30.

Dewan, S. and Ren, F. Information technology and firm boundaries: Impact on firm risk and return performance, *Information Systems Research* 22(2), 2011, 369–388.

Dewan, S. and Ren, F. Risk and return of information technology initiatives: Evidence from electronic commerce announcements, *Information Systems Research* 18(4), 2007, 370–394.

Dewan, S., Shi, C., and Gurbaxani, V. Investigating the risk–return relationship of information technology investment, *Management Science* 53(12), 2007, 1829–1842.

Fornell, C., Mithas, S., and Morgeson, F.V. The economic and statistical significance of stock returns on customer satisfaction, *Marketing Science* 28(5), 2009a, 820–825.

Fornell, C., Mithas, S., and Morgeson, F.V. The statistical significance of portfolio returns, *International Journal of Research in Marketing* 26(2), 2009b, 162–163.

Fornell, C., Mithas, S., Morgeson, F.V., and Krishnan, M.S. Customer satisfaction and stock prices: High returns, low risk, *Journal of Marketing* 70(1), 2006, 3–14.

Fredricksen, C. Twitter ad revenues to grow 210% to $139.5 million in 2011, available at http://www.emarketer.com/PressRelease.aspx?R=1008617, *eMarketer*, September 28, 2011.

Gao, G. and Hitt, L.M. Information technology and trademarks: Implications for product variety, *Management Science* 58(6), 2012, 1211–1226.

Glazer, E. P&G's marketing chief looks to go digital, *Wall Street Journal*, Washington, DC, March 14, 2012, p. B7.

Grover, V. and Ramanlal, P. Digital economics and the e-business dilemma, *Business Horizons* 47(4), 2004, 71–80.

Grover, V. and Ramanlal, P. Six myths of information and markets: Information technology networks, electronic commerce, and the battle for consumer surplus, *MIS Quarterly* 23(4), 1999, 465–495.

Gupta, R., Mani, D., Mithas, S., and Shmueli, G. Challenges in impact studies of eGov services: Self-selection and uncertain matching, Working paper, Indian School of Business, Hyderabad, India, 2012.

Gurbaxani, V. and Whang, S. The impact of information systems on organizations and markets, *Communications of the ACM* 34(1), 1991, 59–73.

Hacker, A. and Dreifus, C. *Higher Education?* Times Books, New York, 2010.

Han, K. and Mithas, S. Information technology outsourcing and non-IT operating costs: An empirical investigation, *MIS Quarterly* 37(1), 2013, 315–331.

Harris, S.E. and Katz, J.L. Organizational performance and information technology investment in the insurance industry, *Organization Science* 2(3), 1991, 263–295.

Henderson, B.C., Kobelsky, K., Richardson, V.J., and Smith, R.E. The relevance of information technology expenditures, *Journal of Information Systems* 24(2), 2010, 39–77.

Hitt, L.M. and Brynjolfsson, E. Productivity, business profitability, and consumer surplus: Three different measures of information technology value, *MIS Quarterly* 20(2), June 1996, 121–142.

Kim, K. and Mithas, S. How does bond market view IT investments of firms? An empirical evidence of bond ratings and yield spreads, *Proceedings of the 31st International Conference on Information Systems*, Association for Information Systems, Shanghai, China, December 4–7, 2011.

Kim, K., Mithas, S., and Kimbrough, M. Assessing risks associated with IT investments across industries: Evidence from bond markets, Working paper, Smith School of Business, College Park, MD, 2012.

Kim, S.M. and Mahoney, J.T. Mutual commitment to support exchange: Relation-specific IT system as a substitute for managerial hierarchy, *Strategic Management Journal* 27(5), 2006, 401–423.

Kleis, L., Chwelos, P., Ramirez, R.V., and Cockburn, I. Information technology and intangible output: The impact of IT investment on innovation productivity, *Information Systems Research* 23(1), 2012, 42–59.

Kobelsky, K., Hunter, S., and Richardson, V.J. Information technology, contextual factors and the volatility of firm performance, *International Journal of Accounting Information Systems* (9), 2008, 154–174.

Kohli, R. Innovating to create IT-based new business opportunities at United Parcel Service, *MIS Quarterly Executive* 6(4), 2007, 199–210.

Kohli, R. and Devaraj, S. Measuring Information technology payoff: A meta analysis of structural variables in firm-level empirical research, *Information Systems Research* 14(2), 2003, 127–145.

Kohli, R. and Grover, V. Business value of IT: An essay on expanding research directions to keep up with the times, *Journal of the Association for Information Systems* 9(1), 2008, 23–39.

Lee, B. and Barua, A. An integrated assessment of productivity and efficiency impacts of information technology investments: Old data, new analysis and evidence, *Journal of Productivity Analysis* 12(1), 1999, 21–43.

Li, C., Peters, G., Richardson, V.J., and Watson, M.W. The consequences of information technology control weaknesses on management information systems: The case of Sarbanes-Oxley internal control reports, *MIS Quarterly* 36(1), 2012, 179–204.

Lichtenberg, F.R. The output contributions of computer equipment and personnel: A firm-level analysis, *Economics of Innovation and New Technology* (3), 1995, 201–217.

Loveman, G.W. An assessment of the productivity impact of information technologies, *Information Technology and the Corporation of the 1990s: Research Studies*, T.J. Allen and M.S. Scott-Morton (eds.), MIT Press, Cambridge, MA, 1994, pp. 84–110.

Lucas, H.C. *Information Technology and the Productivity Paradox*, Oxford University Press, New York, 1999.

Lucas, H.C. *Inside the Future: Surviving the Technology Revolution*, Praeger, Westport, CT, 2008.

Lucas, H.C. Performance and the use of an information system, *Management Science* 21(8), 1975a, 908–919.

Lucas, H.C. The use of an accounting information system, action and organizational performance, *The Accounting Review* 50(4), 1975b, 735–746.

Lucas, H.C. The business value of information technology: A historical perspective and thoughts for future research, *Strategic Information Technology Management: Perspectives on Organizational Growth and Competitive Advantage*, R.D. Banker, R.J. Kauffman, and M.A. Mahmood (eds.), Idea Group Publishing, Harrisburg, PA, 1993, pp. 359–374.

Lucas, H.C. and Mithas, S. Foreign-born IT workers in the US: Complements, not substitutes, *IEEE IT Professional* 13(4), 2011, 36–40.

Mandviwalla, M. and Palmer, J. The globalization of Wyeth (Ivey Case 908M17), *Ivey Case Collection*, Ontario, Canada, 2008, pp. 1–19.

Marchand, D.A. Cemex: Global growth through superior information capabilities, IMD International Business School Case, Lausanne, Switzerland, 2002.

Melville, N., Kraemer, K.L., and Gurbaxani, V. Information technology and organizational performance: An integrative model of IT business value, *MIS Quarterly* 28(2), 2004, 283–322.

Mithas, S. *Digital Intelligence: What Every Smart Manager must have for Success in an Information Age* (Amazon.com url link http://amzn.com/0984989617), Finerplanet, North Potomac, MD, 2012.

Mithas, S., Almirall, D., and Krishnan, M.S. Do CRM systems cause one-to-one marketing effectiveness? *Statistical Science* 21(2), 2006, 223–233.

Mithas, S. and Han, K. How do foreign workers affect US workers and firm profits? *Academy of Management Annual Meeting*, Boston, MA, August 3–7, 2012.

Mithas, S. and Jones, J.L. Do auction parameters affect buyer surplus in e-auctions for procurement? *Production and Operations Management* 16(4), 2007, 455–470.

Mithas, S., Jones, J.L., and Mitchell, W. Buyer intention to use Internet-enabled reverse auctions? The role of asset specificity, product specialization, and non-contractibility, *MIS Quarterly* 32(4), 2008, 705–724.

Mithas, S. and Krishnan, M.S. From association to causation via a potential outcomes approach, *Information Systems Research* 20(2), 2009, 295–313.

Mithas, S., Krishnan, M.S., and Fornell, C. Why do customer relationship management applications affect customer satisfaction? *Journal of Marketing* 69(4), 2005, 201–209.

Mithas, S. and Lucas, H.C. Are foreign IT workers cheaper? U.S. visa policies and compensation of information technology professionals, *Management Science* 56(5), 2010, 745–765.

Mithas, S., Ramasubbu, N., and Sambamurthy, V. How information management capability influences firm performance, *MIS Quarterly* 35(1), 2011, 237–256.

Mithas, S. and Rust, R.T. How revenue and cost focus in information technology strategy affect firm performance, *Marketing Strategy Meets Wall Street*, Goizueta Business School, Emory University, Atlanta, GE, January 23–24, 2009.

Mithas, S., Tafti, A., and Kimbrough, M. Information technology capability and stock returns: Theory and evidence, Working paper, Smith school of Business, College Park, MD, 2012a.

Mithas, S., Tafti, A.R., Bardhan, I.R., and Goh, J.M. The impact of IT investments on profits, available at http://sloanreview.mit.edu/x/53302, *MIT Sloan Management Review* 53(3), 2012b, 15.

Mithas, S., Tafti, A.R., Bardhan, I.R., and Goh, J.M. Information technology and firm profitability: Mechanisms and empirical evidence, *MIS Quarterly* 36(1), 2012c, 205–224.

Mithas, S., Tafti, A.R., and Mitchell, W. How a firm's competitive environment and digital strategic posture influence digital business strategy, *MIS Quarterly* 37(2), 2013.

Morgeson, F.V. and Mithas, S. Does e-government measure up to e-business? Comparing end-user perceptions of U.S. Federal Government and e-business websites, *Public Administration Review* 69(4), 2009, 740–752.

Morrison, C.J. and Berndt, E.R. Assessing the productivity of information technology equipment in U.S. manufacturing industries, National Bureau of Economic Research, Working Paper Series (No. 3582), 1991.

Mukhopadhyay, T. and Kekre, S. Strategic and operational benefits of electronic integration in B2B procurement processes, *Management Science* 48(10), 2002, 1301–1313.

Nevo, S. and Wade, M.R. The formation and value of IT-enabled resources: Antecedents and consequences of synergistic relationships, *MIS Quarterly* 34(1), 2010, 163–183.

Oh, W., Kim, J.W., and Richardson, V.J. The moderating effect of context on the market reaction to IT investments, *Journal of Information Systems* 20(1), 2006, 19–44.

Panko, R. Is office productivity stagnant? *MIS Quarterly* (15), 1991, 191–203.

Parsons, D., Gotlieb, C.C., and Denny, M. Productivity and computers in Canadian banking, *Journal of Productivity Analysis* 4(1), 1993, 95–113.

Pavlou, P.A. and El Sawy, O. The "Third Hand": IT-enabled competitive advantage in turbulence through improvisational capabilities, *Information Systems Research* 21(3), 2010, 443–471.

Piccoli, G. and Ives, B. Review: IT-dependent strategic initiatives and sustained competitive advantage: A review and synthesis of the literature, *MIS Quarterly* 29(4), 2005, 747–776.

Powell, T.C. and Dent-Micallef, A. Information technology as competitive advantage: The role of human, business, and technology resources, *Strategic Management Journal* 18(5), 1997, 375–405.

Rai, A., Patnayakuni, R., and Patnayakuni, N. Technology investment and business performance, *Communications of the ACM* 40(7), 1997, 89–97.

Ramasubbu, N., Mithas, S., Krishnan, M.S., and Kemerer, C.F. Work dispersion, process-based learning and offshore software development performance, *MIS Quarterly* 32(2), 2008, 437–458.

Ravichandran, T., Han, S., and Mithas, S. How information technology influences innovation output of a firm: Theory and evidence, IFIP WG 8.2 Organizations and Society in Information Systems (OASIS) Workshop, Shanghai, China, 2011.

Ray, G., Muhanna, W.A., and Barney, J.B. Information technology and the performance of the customer service process: A resource-based analysis, *MIS Quarterly* 29(4), 2005, 625–652.

Reid, T.R. *The Healing of America*, The Penguin Press, New York, 2009.

Reuters. Twitter share auction suggests $7.7 billion valuation, available at http://www.reuters.com/article/2011/03/04/us-twitter-idUSTRE7221JL20110304, *Reuters*, March 4, 2011.

Roach, S.S. America's technology dilemma: A profile of the information economy, Morgan Stanley Special Economic Study, April 1987.

Roach, S.S. Services under siege—The restructuring imperative, *Harvard Business Review* 69(5), 1991, 82–91.

Santhanam, R. and Hartono, E. Issues in linking information technology capability to firm performance, *MIS Quarterly* 27(1), 2003, 125–153.

Saraf, N., Langdon, C., and Gosain, S. IS application capabilities and relational value in interfirm partnerships, *Information Systems Research* 18(3), 2007, 320–339.

Saunders, A. Valuing IT-related intangible capital, *Proceedings of the 30th International Conference on Information Systems*, Association for Information Systems, St. Louis, MO, December 12–15, 2010.

Swartz, J., Martin, S., and Krantz, M. Facebook IPO filing puts high value on social network, available at http://www.usatoday.com/tech/news/story/2012-02-01/facebook-ipo/52921528/1, *USA Today*, February 2, 2012.

Tafti, A., Mithas, S., and Krishnan, M.S. The effect of Information Technology-Enabled flexibility on formation and market value of alliances, *Management Science* 59(1), 2013, 207–225.

Tallon, P.P., Kraemer, K.L., and Gurbaxani, V. Executives' perceptions of the business value of information technology: A process oriented approach, *Journal of Management Information Systems* 16(4), 2000, 145–173.

Tambe, P.B. and Hitt, L.M. The productivity of information technology investments: New evidence from IT labor data, *Information Systems Research* 23(3)(Part 1), 2012, 599–617.

Tambe, P.B., Hitt, L.M., and Brynjolfsson, E. The Extroverted Firm: How external information practices affect innovation and productivity, *Management Science* 58(5), 2012, 843–859.

Tanriverdi, H. and Ruefli, T.W. The role of information technology in risk/return relations of firms, *Journal of the Association for Information Systems* 5(11–12), 2004, 421–447.

Tippins, M.J. and Sohi, R.S. IT competency and firm performance: Is organizational learning a missing link? *Strategic Management Journal* 24(8), 2003, 745–761.

Tuli, K. and Bharadwaj, S. Customer satisfaction and stock returns risk, *Journal of Marketing* 73, November 2009, 184–197.

Wade, M. and Hulland, J. Review—The resource based view and information systems research: Review, extension, and suggestions for future research, *MIS Quarterly* 28(1), 2004, 107–142.

Weill, P. The relationship between investment in information technology and firm performance: A study of the valve manufacturing sector, *Information Systems Research* 3(4), 1992, 307–333.

Whitaker, J., Mithas, S., and Krishnan, M.S. A field study of RFID deployment and return expectations, *Production and Operations Management* 16(5), 2007, 599–612.

Worthen, B. Cut those costs! (But not tech.), available at http://online.wsj.com/article/SB10001424052970204573704577187212484940318.html?KEYWORDS=cut+those+costs, *Wall Street Journal*, Washington, DC, February 27, 2012, p. R5.

Index

W